T5-AFW-950

McGraw-Hill Encyclopedia of Engineering

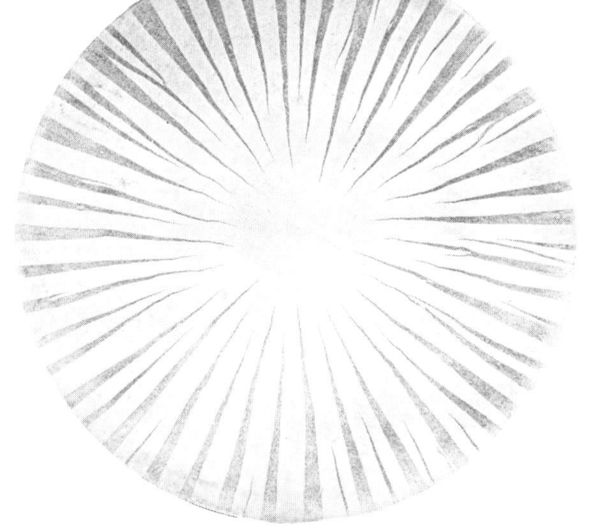

McGraw-Hill Encyclopedia of

Engineering

Sybil P. Parker
EDITOR IN CHIEF

McGraw-Hill Book Company

New York St. Louis San Francisco
Auckland Bogotá Guatemala Hamburg
Johannesburg Lisbon London Madrid Mexico
Montreal New Delhi Panama Paris San Juan
São Paulo Singapore Sydney Tokyo Toronto

1234567890 KPKP 89876543

ISBN-0-07-045486-8

Library of Congress Cataloging in Publication Data

Main entry under title:

McGraw-Hill encyclopedia of engineering.

"All of the material in this volume has been
published previously in the McGraw-Hill encyclopedia
of science & technology, fifth edition" – T.p. verso.
Bibliography: p.
Includes index.
1. Engineering – Dictionaries. I. Parker, Sybil P.
II. McGraw-Hill Book Company. III. McGraw-Hill
encyclopedia of science & technology (5th ed.)
TA9.M36 1983 620'.003'21 82-18663
ISBN 0-07-045486-8

Editorial Staff

Sybil P. Parker, Editor in Chief

Jonathan Weil, Editor
Betty Richman, Editor

Edward J. Fox, Art director

Ann D. Bonardi, Art production supervisor
Cynthia M. Kelly, Art/traffic

Joe Faulk, Editing manager

Thomas M. Siracusa, Editing supervisor
Ruth L. Williams, Editing supervisor

Patricia W. Albers, Senior editing assistant
Judith Alberts, Editing assistant
Barbara Begg, Editing assistant

Project Consultant

Prof. Eugene A. Avallone
Department of Mechanical Engineering
City University of New York

Field Consultants

Vincent M. Altamuro
President, Management Research Consultants
Yonkers, NY

Prof. P. M. Anderson
Department of Electrical and Computer Engineering
Arizona State University

Waldo G. Bowman
Formerly, Editor
"Engineering News-Record"

Prof. Todd M. Doscher
Department of Petroleum Engineering
University of Southern California

Dr. Gary Judd
Vice Provost, Academic Programs and Budget
Dean of the Graduate School
Rensselaer Polytechnic Institute

Prof. Vjekoslav Pavelic
Systems-Design Department
University of Wisconsin

Preface

The profession of engineering integrates a spectrum of knowledge extending from the pure physical sciences to the most advanced and specialized technologies. As a widely pervasive pursuit, both as art and science, engineering touches almost every facet of human life and can be said to have created the physical structure of civilization.

Historically, engineering evolved from ancient practices concerned primarily with military endeavors into a multidisciplinary profession with many subdivisions and specialized branches. Civil engineering, for example, gained distinction as a specialty in the mid-18th century, when it was broadly described as being composed of individuals who dealt with the construction of roads, bridges, waterways, harbors, and drainage systems. Today, there are highway engineers, coastal engineers, sanitary engineers, structural engineers, and drainage engineers, to name but a few specializations of civil engineering. Other major branches of engineering have followed a similar pattern of specialization in modern practice.

These changes reflect the increased interrelationship between science and technology, particularly during the 20th century as the time between scientific discoveries and their application has become progressively shorter. Discoveries in solid-state physics which brought about the huge development of electronics —from transistors to microchips—placed new demands on electrical engineers in the area of computers. Zero-gravity technology, made accessible by the space shuttle, will have a growing impact on metallurgical and materials engineers concerned with the manufacture of new alloys and other advanced materials. Turbines and jet engines created new problems for mechanical engineers in terms of thermodynamics and fluid mechanics. Increasing demands for energy, concerns about environmental protection, and the growing awareness of technological risks have, in general, considerably expanded the range of engineering responsibilities. The discovery of superconductivity, a low-temperature phenomenon occurring in many electrical conductors, challenged the imaginations of electrical and mechanical engineers in developing devices with applications in computers, power transmission, and high-speed levitated trains.

It is obvious from these few examples that the scope of engineering is vast and is based on broad scientific understanding. Cooperation between specialists in all fields of science and engineering is therefore the essence of the unified approach to modern engineering education as well as practice. This concept is treated for the first time in a single, comprehensive reference—the *McGraw-Hill Encyclopedia of Engineering*—an interdisciplinary work designed to provide information on ten major branches of engineering. These include civil engineering, design engineering, electrical engineering, industrial engineering, mechanical engineering, metallurgical engineering, mining engineering, nuclear engineering, petroleum engineering, and production engineering. Covered also are the mechanical, electrical, and thermodynamic principles that are basic to all fields of engineering.

The subjects are discussed in more than 690 alphabetically arranged articles selected from the *McGraw-Hill Encyclopedia of Science and Technology* (5th ed., 1982). Hundreds of drawings, tables, charts, graphs, and photographs serve to amplify the text. All information is readily accessible through a detailed analytical index of 6000 entries and by the use of cross-references. Bibliographies provide information on selected further readings for most entries.

In choosing the articles, we are indebted to the Project Consultant, Prof. E. A. Avallone. Moreover, he and Messrs. Altamuro, Anderson, Bowman, Doscher, Judd, and Pavelic, as Field Consultants on the *Encyclopedia of Science and Technology*, have impressed their high standards on this volume as well.

This Encyclopedia will serve the needs of engineers, students, librarians, technical writers, interested laypersons, and all others concerned with the theory and practice of engineering.

Sybil P. Parker
EDITOR IN CHIEF

McGraw-Hill Encyclopedia of Engineering

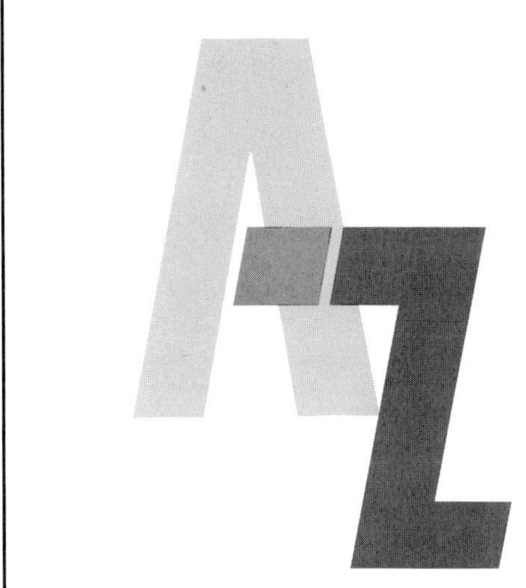

Acceleration-Zone refining

Acceleration

The time rate of change of velocity. Since velocity is a directed or vector quantity involving both magnitude and direction, a velocity may change by a change of magnitude (speed) or by a change of direction or both. It follows that acceleration is also a directed, or vector, quantity. If the magnitude of the velocity of a body changes from v_1 ft/sec to v_2 ft/sec in t sec, then the average acceleration a has a magnitude given by Eq. (1). To designate it fully the direction should be given, as well as the magnitude. *See* VELOCITY.

$$a = \frac{\text{velocity change}}{\text{elapsed time}} = \frac{v_2 - v_1}{t_2 - t_1} = \frac{\Delta v}{\Delta t} \qquad (1)$$

Instantaneous acceleration is defined as the limit of the ratio of the velocity change to the elapsed time as the time interval approaches zero. When the acceleration is constant, the average acceleration and the instantaneous acceleration are equal.

If a body, moving along a straight line, is accelerated from a speed of 10 to 90 ft/sec in 4 sec, then the average change in speed per second is $(90-10)/4 = 20$ ft/sec in each second. This is written 20 ft per second per second or 20 ft/sec^2. Accelerations are also commonly expressed in meters per second per second (m/sec^2), or in any similar units.

Whenever a body is acted upon by an unbalanced force, it will undergo acceleration. If it is moving in a constant direction, the acting force will produce a continuous change in speed. If it is moving with a constant speed, the acting force will produce an acceleration consisting of a continuous change of direction. In the general case, the acting force may produce both a change of speed and a change of direction. [R. D. RUSK]

Angular acceleration. This is a vector quantity representing the rate of change of angular velocity of a body experiencing rotational motion. If, for example, at an instant t_1, a rigid body is rotating about an axis with an angular velocity $\boldsymbol{\omega}_1$, and at a later time t_2, it has an angular velocity $\boldsymbol{\omega}_2$, the average angular acceleration $\bar{\boldsymbol{\alpha}}$ is given by Eq. (2),

$$\bar{\boldsymbol{\alpha}} = \frac{\boldsymbol{\omega}_2 - \boldsymbol{\omega}_1}{t_2 - t_1} = \frac{\Delta \boldsymbol{\omega}}{\Delta t} \qquad (2)$$

expressed in radians per second per second. The instantaneous angular acceleration is given by $\boldsymbol{\alpha} = d\boldsymbol{\omega}/dt$.

Consequently, if a rigid body is rotating about a fixed axis with an angular acceleration of magnitude α and an angular speed of ω_0 at a given time, then at a later time t the angular speed is given by Eq. (3).

$$\omega = \omega_0 + \alpha t \qquad (3)$$

A simple calculation shows that the angular distance θ traversed in this time is expressed by Eq. (4).

$$\theta = \bar{\omega} t = \left[\frac{\omega_0 + (\omega_0 + \alpha t)}{2}\right] t = \omega_0 t + \tfrac{1}{2}\alpha t^2 \qquad (4)$$

In the figure a particle is shown moving in a circular path of radius R about a fixed axis through O with an angular velocity of $\boldsymbol{\omega}$ radians/sec and an angular acceleration of $\boldsymbol{\alpha}$ radians/sec^2. This particle is subject to a linear acceleration which, at any instant, may be considered to be composed of two components: a radial component $\mathbf{a}_r$ and a tangential component $\mathbf{a}_t$.

Radial acceleration. When a body moves in a circular path with constant linear speed at each point in its path, it is also being constantly accelerated toward the center of the circle under the action of the force required to constrain it to move in its circular path. This acceleration toward the center of path is called radial acceleration. In the figure, the radial acceleration, sometimes called centripetal acceleration, is shown by the vector $\mathbf{a}_r$. The magnitude of its value is v^2/R, or ω^2/R, where v is the instantaneous linear velocity. This centrally directed acceleration is necessary to keep the particle moving in a circular path.

Tangential acceleration. The component of linear acceleration tangent to the path of a particle

ACCELERATION

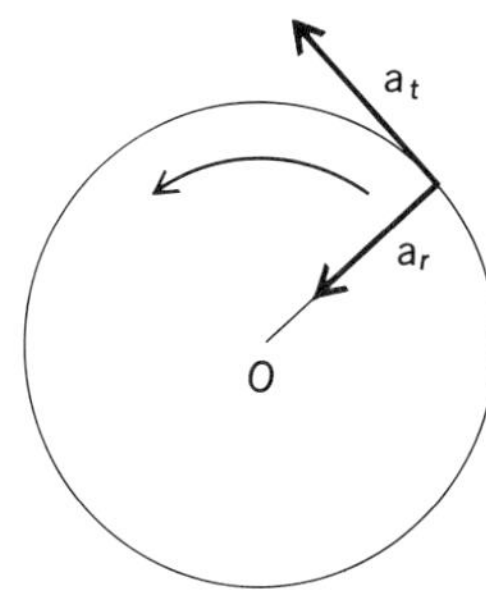

Radial and tangential accelerations in circular motion.

subject to an angular acceleration about the axis of rotation is called tangential acceleration. In the figure, the tangential acceleration is shown by the vector $\mathbf{a}_t$. The magnitude of its value is αR. *See* ACCELERATION ANALYSIS: ROTATIONAL MOTION.

[C. E. HOWE/R. J. STEPHENSON]

Acceleration analysis

A mathematical technique, often done graphically, by which accelerations of parts of a mechanism are determined. In high-speed machines, particularly those that include cam mechanisms, inertial forces may be the controlling factor in the design of members. An acceleration analysis, based upon velocity analysis, must therefore precede a force analysis. Maximum accelerations are of particular interest to the designer. Although an analytical solution might be preferable if exact maxima were required, graphical solutions formerly tended to be simpler and to facilitate visualization of relationships. Today, for advanced problems certainly, a computer solution, possibly one based on graphical techniques, is often more effective and can also produce very accurate graphical output. *See* FORCE; FORCE ANALYSIS; MECHANISM.

Accelerations on a rigid link. On link OB (Fig. 1*a*) acceleration of point B with respect to point O is the vector sum of two accelerations: (1) normal acceleration A^n_{BO} of B with respect to O because of displacement of B along a path whose instantaneous center of curvature is at O, and (2) tangential acceleration A^t_{BO} of B with respect to O because of angular acceleration α.

For conditions of Fig. 1*a* with link OB rotating about O at angular velocity ω and angular acceleration α, the accelerations can be written as Eqs. (1) and (2). The vector sum or resultant is A_{BO} (Fig. 1*b*).

$$A^n_{BO} = (OB)\omega^2 = V_B^2/(OB) \quad (1)$$

$$A^t_{BO} = (OB)\alpha \quad (2)$$

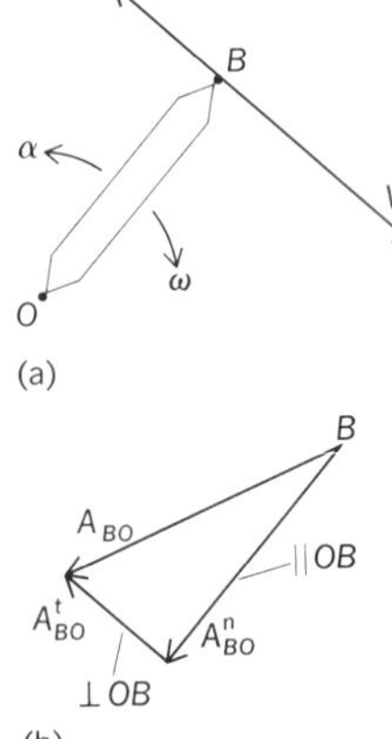

Fig. 1. Elementary condition. (*a*) Point B on rigid member rotates about center O. (*b*) Vector diagram shows normal, tangential, and resultant accelerations.

Accelerations in a linkage. Consider the acceleration of point P on a four-bar linkage (Fig. 2*a*) with $\alpha_2 = 0$ and hence $\omega_2 = k$, the angular velocity of input link 2. First, the velocity problem is solved yielding V_B and by Fig. 2*b*, V_{PB}. Two equations can be written for A_P; they are solved simultaneously by graphical means in Fig. 2*c* by using Fig. 2*b*; that is, normal accelerations of P with respect to B and D are computed first. Directions of tangential acceleration vectors A^t_{PB} and A^t_P are also known from the initial geometry. The tip of vector A_P must lie at their intersection, as shown by the construction of Fig. 2*c*. *See* VELOCITY ANALYSIS.

Explicitly, acceleration A_P of point P is found on the one hand by beginning with acceleration A_B of point B, represented by Eq. (3). To this acceleration is added vectorially normal acceleration A^n_{PB} and tangential acceleration A^t_{PB}, which can be written as Eq. (4). Also for link 2, $\alpha_2 = 0$ and $A^t_B = 0$; and A_B need not be split up.

$$A_P = A_B \leftrightarrow A_{PB} \quad (3)$$

$$A_P = A_B \leftrightarrow A^n_{PB} \leftrightarrow A^t_{PB} \quad (4)$$

On the other hand, A_P can also be expressed as Eq. (5), or in Fig. 2*c*. The meet of the two tangential

$$A_P = A^n_P + \rightarrow A^t_P \quad (5)$$

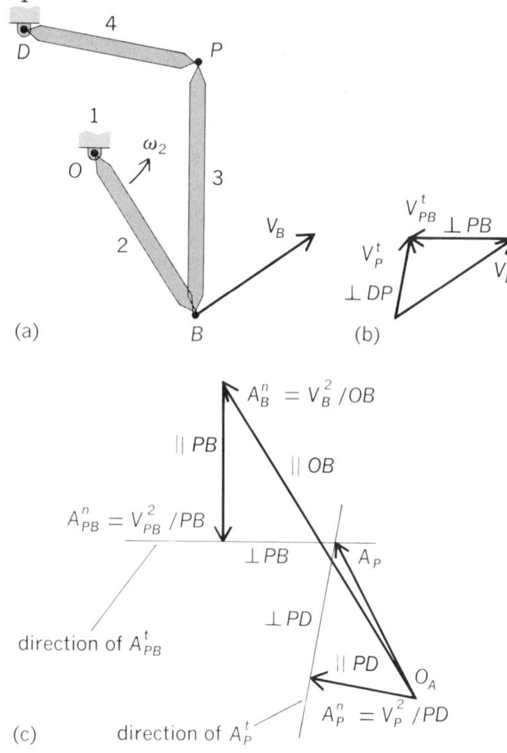

Fig. 2. Four-bar linkage. (*a*) Given are the linkages 2, 3, and 4 between fixed supports 1-1 and the velocity of point B, $\omega_2 = 0$. (*b*) Vector polygon is used to determine the velocity at point P. (*c*) Graphic solution then finds acceleration at P.

components defines the tip of A_P. This problem illustrates the generalization that any basic acceleration analysis can be thoroughly performed only after completion of the underlying velocity analysis.

Acceleration field. Acceleration is a vector and hence has magnitude and direction but not a unique position. A plane rigid body, such as the cam of Fig. 3*a*, in motion parallel to its plane will possess an acceleration field in that plane differing from, but resembling, a rigid-body velocity vector field.

Every acceleration vector in this field, absolute or relative, makes with its instantaneous radius vector the same angle γ (tangent $\gamma = \alpha/\omega^2$) and is proportional in magnitude to the length of this radius vector (Fig. 3*b*). The acceleration field at any instant is thus defined by four parameters: magnitude and direction of accelerations at any two points on the body.

From an acceleration field defined by a_A and a_B (Fig. 3*c*), one can quickly determine the instantaneous center of zero acceleration. From point A', construct vector $A'A''$ parallel and equal to acceleration a_B, represented by vector $B'B$. The relative acceleration a_{AB} is given by the vector of Eq. (6).

$$a_{AB} = a_A - a_B \quad (6)$$

Angle γ between resultant a_{AB} and line AB is common to all radii vectors; therefore, construct radii vectors through A and through B both at γ (properly signed) to their respective acceleration vectors. These lines intersect at Γ, the center of

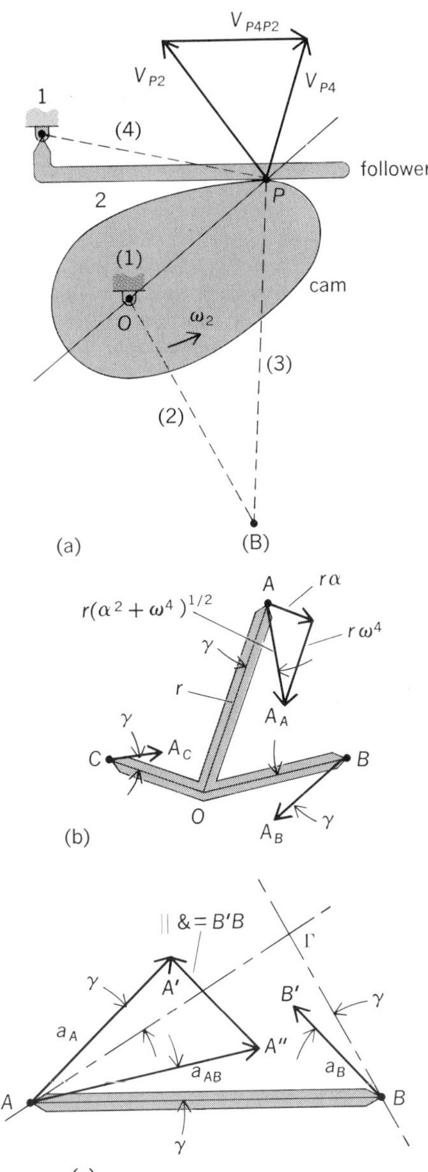

Fig. 3. Accelerations on a cam and follower. (*a*) The cam mechanism. (*b*) The acceleration field about a rigid rotating structure. (*c*) An acceleration field determining an instantaneous center of rotation.

zero acceleration. The geometry contains two similar triangles; $\Delta AB\Gamma$ is similar to $\Delta AA''A'$ because corresponding sides are equally inclined by the angle γ. From these similar triangles comes the generalization that acceleration magnitudes are proportional to radii vectors, just as for the velocity field.

Angle γ ($\tan\gamma = \alpha/\omega^2$) is also used in orientation of acceleration image polygons, especially if the preceding velocity analysis used the image polygon method.

Points not on the same link. Acceleration of follower link 4 of the cam mechanism (Fig. 3*a*) is determined by considering relative accelerations of coincident points not on the same link. Thus point P on link 4, designated P_4, and coincident point P on link 2, designated P_2, have instantaneous velocities, as shown by vectors V_{P_4} and V_{P_2} (Fig. 3*a*). Relative velocity of P_4 with respect to P_2 is $V_{P_4P_2}$ as indicated. Acceleration of P_4 can now be determined by Eq. (7), where the last term is

$$A_{P_4} = A_{P_2} \leftrightarrow A^n{}_{P_4P_2} \leftrightarrow A^t{}_{P_4P_2} \leftrightarrow 2\omega_2 V_{P_4P_2} \qquad (7)$$

the Coriolis component. This component results from referring motion of P_4 on body 4 to P_2 on body 2. Serious errors of analysis can result from omission of the Coriolis component.

Occasionally, mechanisms that appear to require the analysis described in the preceding paragraph may also be analyzed by constructing an instantaneously equivalent linkage shown by dashed lines (2), (3), and (4) in Fig. 3*a*. The instantaneous equivalent linkage is then analyzed by the method of Fig. 2.

Alternative procedures. Other methods may also be used. It is not always necessary to resolve vectors into normal and tangential components; they can be solved by conventional orthogonal component techniques. For points A and B in part m, Eq. (8) holds.

$$a_A^{AB} = a_B^{AB} + a_{AB}^{AB} = a_B^{AB} + \omega^2 \cdot AB \qquad (8)$$

A convenient approximate solution is obtained by considering the difference between velocities due to a small angular displacement. This difference in velocities divided by the elapsed short time interval approximates the acceleration. Typically, the angular displacement is taken to be 0.1 radian (0.05 radian either side of the position under study). [DOUGLAS P. ADAMS]

Bibliography: J. Denavit and R. Hartenberg, *Kinematic Synthesis of Linkages*, 1964; C. W. Ham et al., *Mechanics of Machinery*, 4th ed., 1958; D. Lent, *Analysis and Design of Mechanisms*, 2d ed., 1970; A. Sloane, *Engineering Kinematics*, 1966; C. H. Such and C. W. Radcliffe, *Kinematics and Mechanisms Design*, 1978.

Ackerman steering

Differential gear or linkage that turns the two steered road wheels of a self-propelled vehicle so that all wheels roll on circles with a common center. If a vehicle is to turn without lateral skid of any wheel, the center lines of all wheel axles must intersect, when extended, at every instant in a common center about which the vehicle turns (Fig. 1). This requirement is the Ackerman principle of toe out on turns. It is used universally on wheeled vehicles. For straight, forward motion, the front

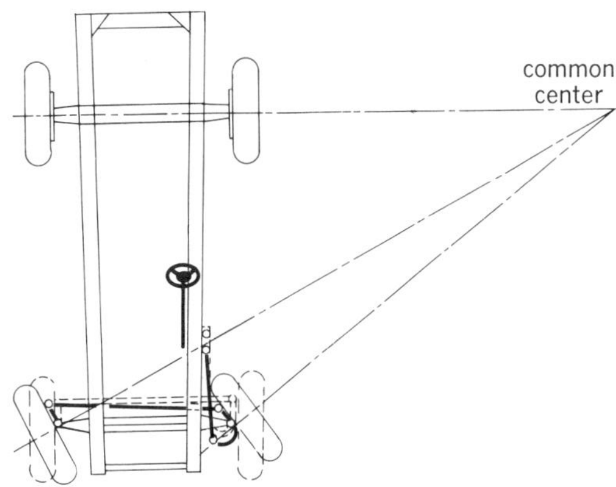

Fig. 1. All wheels turn about a common center.

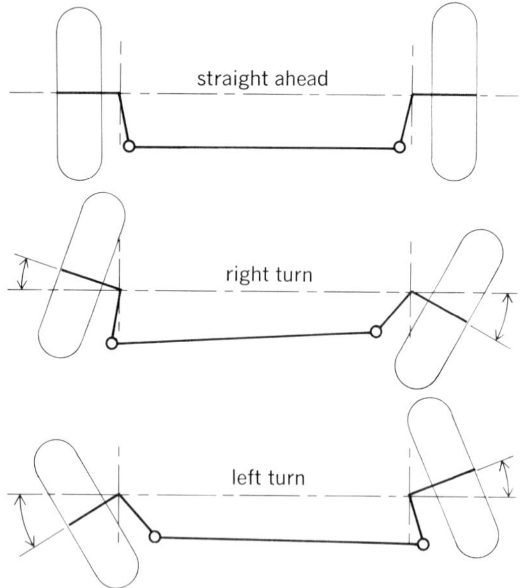

Fig. 2. Inwardly inclined steering knuckles cause the wheels to toe-out on the turns.

wheels are substantially parallel, but as the vehicle enters a curve the inner wheel turns more sharply than the outer wheel. The extreme condition occurs when the vehicle is on a curve of its turning radius. (Turning radius is the arc described by the center of the track made by the outside front wheel of the vehicle when the vehicle makes its shortest turn.)

A common configuration that produces Ackerman steering inclines the knuckle arms inwardly and rearwardly (Fig. 2). The angle of inclination depends on wheelbase and tread of the vehicle. Wheelbase is the distance from front to rear wheels, measured between centers of ground contact. On a vehicle with two rear axles, the rear measuring point is on the ground midway between rear axles. Tread of a vehicle is the distance between front wheels, or between rear wheels, measured from centers of ground contact. *See* AUTOMOTIVE STEERING; FOUR-BAR LINKAGE.

[WILLIAM K. CRESON]

Air and gas compressor

Machines that increase the pressure of a gas or vapor, typically air, by increasing the gas density and delivering the fluid against the connected system resistance. The resistance may be on the suction side, as with vacuum pumps, or on the discharge side, as with air compressors.

Applications. Compressors are typically applied to the operation of pneumatic tools and rock drills, conveying systems, furnace blast systems, ventilation systems, and the inflation of automobile tires. Gas and vapor compressors are represented in the refrigeration, air conditioning, and heat pump fields, where they handle a wide variety of refrigerants near the condensation point, such as Freon, ammonia, carbon dioxide, and sulfur dioxide. Compressors are used in industrial process operations such as nitrogen fixation and gas liquefaction. They are used for repressuring and pumping on natural-gas transmission lines.

The service applications may variously call for suction gas pressures as low as 10^{-8} mm of mercury (with ultrahigh-vacuum pumps) or delivery pressures of 1000 atm (with some chemical process compressors). Volumes handled may be as small as 1 cubic foot per minute (cfm) or as large as 1,000,000 cfm. A wide assortment of machine types is available to meet these diverse conditions and includes positive-displacement units, rotary or reciprocating; free-compression fans and turboblowers, centrifugal and axial; free-compression, hydraulic-, gas-, or vapor-jet pumps; and diffusion and getter pumps. For details and illustrations on many of these types of machines *see* AIR CONDITIONING; COMPRESSOR; FAN; GAS TURBINE; REFRIGERATION; STEAM JET EJECTOR; VACUUM PUMP.

Basic types. The reciprocating compressor is suited to the highest-pressure services, up to about 1000 psi when constructed in three stages with intercoolers. Displacement seldom exceeds 5000 cfm and a 3-ft stroke. Air compressors for the common industrial service of 100 psi are single- or two-stage, selection being determined by economy. Reciprocating compressors are equipped with automatic valves which give accurate timing for the admission and release of gas, and the maintenance of good volumetric efficiency (typically about 75%). Water jackets limit metal temperatures and maintain running clearances on machine parts. On small, portable, and garage-type compressors with free air capacities of about 100 cfm, air jackets may be used.

Rotary compressors are built in capacities as high as 50,000 cfm and compression ratios are usually moderate (less than 3:1). Some designs, such as those with helical lobes or sliding vanes, are good for ratios of 6:1. Rotary compressors are suitable for direct connection to high-speed drivers such as automotive engines and electric motors. Rotary compressors use no valves, and a port construction, with or without liquid seals, controls the cycle kinematics. The rotating lobes are driven through, and maintained in alignment by, gears.

Free-compression devices include fans and blowers, of the centrifugal or axial-flow type, limited to pressure ratios so small that the change in density of the fluid on passage through the unit is negligible. The head gain is customarily measured in inches of water on a manometer, but capacities may be 100–1,000,000 cfm. When higher ratios of compression are required, the multistage centrifugal or axial compressor can be used with pressure ratios as high as 10:1 and frequently operating at speeds of 5000–10,000 rpm.

Jet compressors are free of moving parts; they may be built in sizes as large as 10,000–20,000 cfm, and compression ratios of 5–6:1 in a single stage. They are especially suitable for vacuum service when steam is used as the actuating jet for entrainment of the noncondensible gas. For high vacuum, they are made multistage and equipped with intercoolers and aftercondensers for improved efficiency. [THEODORE BAUMEISTER]

Bibliography: D. W. Anderson, *The Analysis and Design of Pneumatic Systems*, 1976; T. Baumeister (ed.), *Standard Handbook for Mechanical Engineers*, 8th ed., 1978; Trade and Technical Press, *Pneumatic Handbook*, 1978.

Air brake

An energy-conversion mechanism used to retard, stop, or hold a vehicle or, generally, any moving element, the activating force being applied by a difference in air pressure. With an air brake only slight effort by the operator quickly applies full braking force.

An air brake performs the energy conversion like other brakes by friction. The feature that distinguishes an air brake is that the friction-producing device, such as disk or drum, is applied by air in contrast to mechanical, hydraulic, or electrical means. In a particular use the choice between an air brake and any other type of brake depends in part on the availability of an air supply and on the method of brake control. For example, in a motor bus in which compressed air actuates doors, air may also actuate the brakes. On railroad cars compressed air actuates the brakes so that, in the event of a disconnect between cars, air valves can automatically apply brakes on all cars. Air brakes can also be applied mechanically. This allows them to be used while the air pump is not operating; the brakes can be held on even when a vehicle is not in use. Safety regulations require alternate methods of applying brakes. *See* AUTOMOTIVE BRAKE; AUTOMOTIVE VEHICLE; BRAKE; FRICTION.

Because air is compressible, air brakes inherently accommodate to wear, such as in brake shoes. Also, because the force with which brakes are applied depends on the area of an air-driven diaphragm as well as on air pressure, large forces can be developed from moderate pressures. Air pressure can be accurately regulated and can also be controlled by other forces besides the operator's hand or foot; consequently air brakes can be made to respond to static and dynamic forces that influence the efficiency of brake operation.

A totally different kind of air brake is used on aircraft. Energy of momentum is transferred to the air as heat by an air brake that consists of a flap or other device which produces drag. When it is needed, the braking device is extended from an aircraft wing or fuselage into the airstream.

Typical system operation. In a typical air brake system an air compressor takes in and compresses air for use by the brakes and usually for other air-operated components of the vehicle. The compressor may be fitted with a governor to control the air compression within a preselected range, or an air protection valve may allow air to escape if the preselected input pressure is exceeded. Another alternative is for an air pressure reducing valve to deliver a preselected output pressure. The compressed air is stored in an air brake reservoir until needed. A gage indicates to the operator the pressure within the system.

By means of a foot- or hand-operated brake valve the operator controls application of the brake (see figure). A brake rod from the cylinder diaphragm drives the friction element against the moving surface to provide the braking action. When the operator releases the brake valve, an auxiliary reservoir recharges to full pressure through the triple valve. Also, after the braking action is no longer required, a quick release valve assures rapid discharge of air from the brake cylinder. A relay valve may provide automatic application of a trailer brake in case of breakaway or loss of pressure in the trailer line. A relay valve may also serve as a secondary control unit to accelerate the application and release of air pressure in a part of the system. A safety valve or pressure release unit protects the system against excessive pressure. *See* SAFETY VALVE.

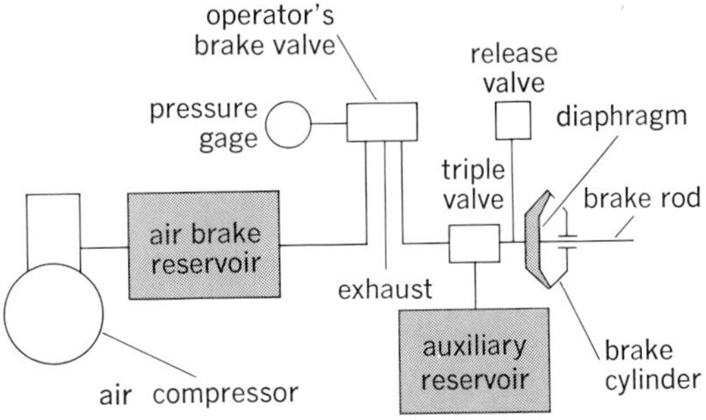

Diagram of air brake. When operator actuates brake valve, pressure to triple valve is reduced. Triple valve then admits compressed air from auxiliary reservoir to brake cylinder, actuating brake rod.

Air brakes may be combined with other systems. In the straight air brake system described above, a mechanical-type brake is actuated by air pressure in a brake cylinder. In a booster system an air-powered master cylinder, controlled by the brake cylinder or chamber, may actuate the brake shoe with added force. In combination with a hydraulic brake system the air system may actuate or assist in control of the hydraulic power unit.

Special features. Air brakes may combine special features. Proportioning valves, by adjusting braking pressures to each actuator cylinder in proportion to axle static and dynamic loads during deceleration, permit shorter stopping distances. The braking force may be balanced with deceleration forces to avoid locking of wheels in an electropneumatic high-speed railroad braking system.

Safety features. To meet international operating requirements for road vehicles, an air brake system may be split. For example, the brake pedal may operate a dual valve with two air lines to the actuator; the actuator may then have two diaphragms, each fully capable of applying the brake. In such a system a warning device alerts the driver if either the service or secondary portion fails. Operation is always arranged to be fail-safe. For example, on railroads, as in the illustrated system, release of pressure in a control line either by the brakeman or by a break in the line actuates the triple valve to connect the auxiliary reservoir to the brake cylinder, with each car carrying its separate auxiliary reservoir.

Critical requirements of air brakes are (1) that the entire system reach pressure stability within the specified minimum differential pressures allowable between separate parts of the system, and (2) that when a valve is transferred or other input change is made to the system, pressure builds up (or reduces) to a new level within a specified minimum time.

Air brakes are required by law on railroads. On passenger trains, where air brakes operate typically at 110 psi, stopping distances are 10% of those produced by hand brakes. Air brakes are widely used on buses and trucks, typically at 80 psi.

Vacuum brake. An alternate form of air brake, the vacuum brake, operates by maintaining low pressure in the actuating cylinder. Braking action is produced by opening one side of the cylinder to the atmosphere so that atmospheric pressure, aided in some designs by gravity, applies the brake. The brake is fail-safe in that it is applied if the vacuum pump stops or if the control line opens. The pump can be the intake of an internal combustion engine.

Vacuum systems are not widely used because of the limited maximum pressure differential. However, one form of automotive power brake provides assistance from a booster cylinder evacuated by the engine air intake.

[FRANK H. ROCKETT]

Bibliography: H. E. Ellinger and R. B. Hathaway, *Automotive Suspension, Steering and Brakes*, 1980.

Air conditioning

The maintenance of certain aspects of the environment within a defined space to facilitate the intended function of that space. Environmental conditions generally encompassed by the term air conditioning include air temperature and motion, radiant heat energy level, moisture level, and concentration of various pollutants, including dust, germs, and gases. Because these environmental factors are associated with air itself, and because air temperature and motion are the factors most readily sensed, simultaneous control of all these factors is called air conditioning, although space conditioning is a better description of the activity.

Comfort air conditioning refers to control of spaces inhabited by people to promote their comfort, health, or productivity. Spaces in which air is conditioned for comfort include residences, offices, institutions, sports arenas, hotels, and factory work areas. Process air conditioning systems are designed to facilitate the functioning of a production, manufacturing, or operational activity. For example, heat-producing electronic equipment in an airplane cockpit must be kept cool to function properly, while the occupants of the cockpit are maintained at comfortable conditions. The environment around a multicolor printing press must have constant relative humidity to avoid paper expansion or shrinkage for accurate registration, while press heat and ink mists must be conducted away for the health of pressmen. Maintenance of conditions within surgical suites of hospitals and in "clean" or "white" rooms of manufacturing plants, where an almost germ- or dust-free atmosphere must be maintained, has become a specialized subdivision of process air conditioning.

Physiological principles. A comfort air conditioning system is designed to help man maintain his body temperature at its normal level without undue stress and to provide him with an atmosphere which is healthy to breathe.

Man's body produces heat at various rates, depending basically upon his weight and degree of activity (see table). Normally more heat is generated than is required to maintain body temperature at a healthful level. Hence, air conditioning always is required to help cool people, never to heat them. Control of a man's body temperature is accomplished by control of the emission of energy from his body by radiation to the space around him, by convection to air currents that impinge on his skin or clothing, by conduction of clothing and objects he touches, and by evaporation of moisture in his lungs and of sweat from his skin. Radiant emission is a function of the amount of clothing (or blankets) worn and the temperature of the surrounding air and objects. Evaporation and convection heat loss are functions of air temperature and velocity. Evaporation is a function, in addition, of relative humidity.

When the amount and type of clothing and the temperature, velocity, and humidity of the air are such that the heat produced by the body is not dissipated at an equal rate, blood temperature begins to rise or fall and discomfort is experienced in the form of fever or chill, in proportion to the departure of body temperature from the normal 98.6°F. Hence, space conditions to promote comfort depend upon the degree of human activity in the space, the amount and type of clothing worn and, to a certain extent, the physical condition of the occupants, because old age or sickness can impair the body's heat-producing and heat-regulating mechanisms.

The heat-dissipating factors of temperature, humidity, and air motion must be considered simultaneously. Within limits, the same amount of comfort (or, more objectively, of heat-dissipating ability) is the result of a combination of these factors in a three-dimensional continuum. Conditions

Estimates of energy metabolism (M) of various types of activity*

Kind of work	Activity	M, Btu/hr
	Sleeping	250
	Sitting quietly	400
Light	Sitting, moderate arm and trunk movements, such as desk work or typing	450–550
	Sitting, moderate arm and leg movements, such as playing organ or driving car in traffic	550–650
	Standing, light work at machine or bench, mostly arms	550–650
Moderate	Sitting, heavy arm and leg movements	650–800
	Standing, light work at machine or bench, some walking about	650–750
	Standing, moderate work at machine or bench some walking about	750–1000
	Walking about, with moderate lifting or pushing	1000–1400
Heavy	Intermittent heavy lifting, pushing, or pulling, such as pick and shovel work	1500–2000
	Hardest sustained work	2000–2400

*Values apply for a 154-lb man and do not include rest pauses.

SOURCE: American Society of Heating, Refrigerating and Air-Conditioning Engineers, Inc., *Handbook of Fundamentals*, 1967.

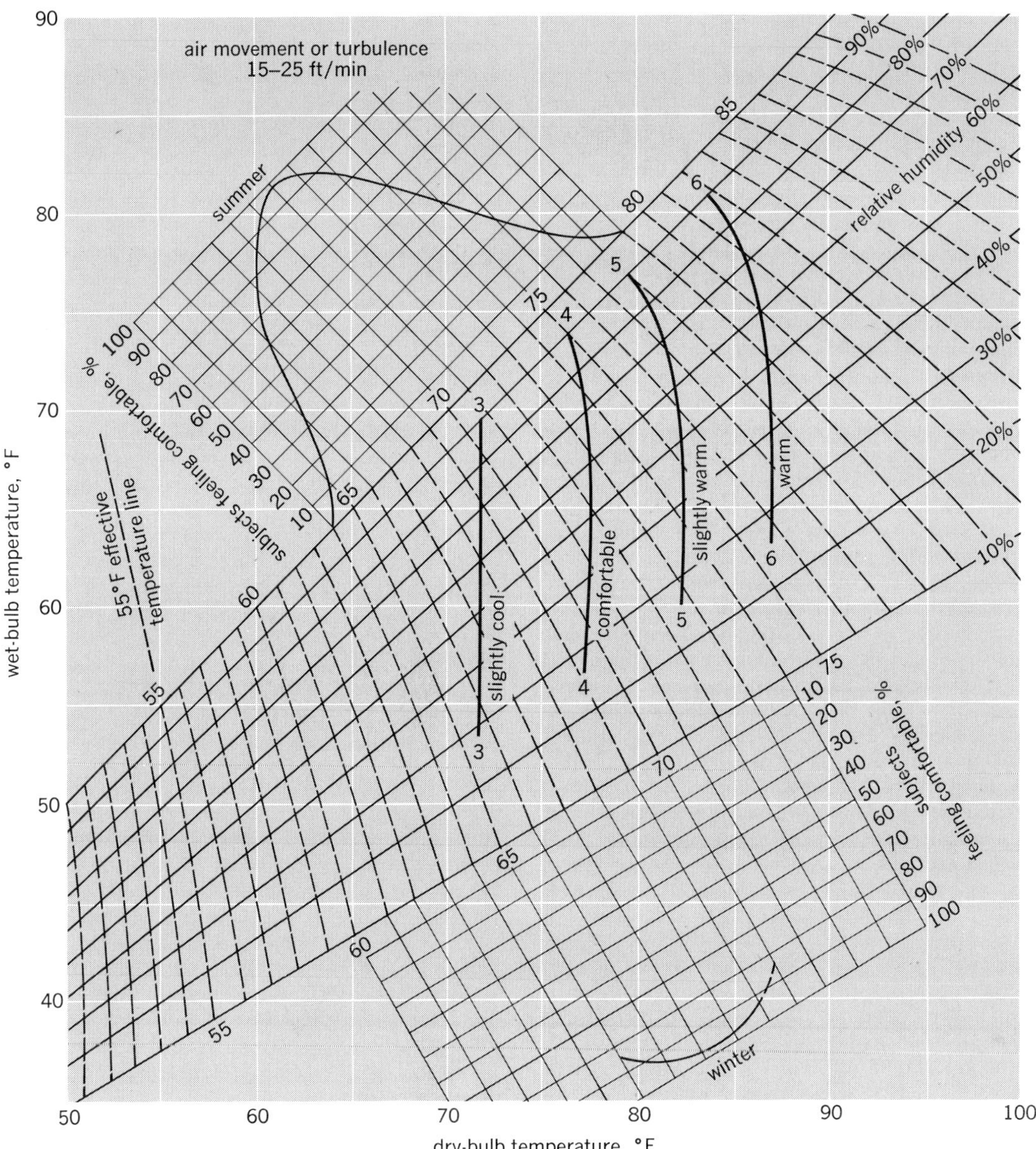

Fig. 1. Revised comfort chart for sedentary adults in the United States. (*American Society of Heating, Refrigerating and Air-Conditioning Engineers, Inc., Handbook of Fundamentals, 1967*)

for constant comfort plot as lines of effective temperature, that is, as combinations that produce equal sensations of comfort (or discomfort). Data on effective temperature have been the subject of intensive research sponsored by American Society of Heating, Refrigerating and Air-Conditioning Engineers, Inc., and its predecessors since 1923. Figure 1 presents the latest data for sedentary adults in the United States.

In practice most air conditioning systems for offices and institutions, schools, and other light-work spaces are designed to maintain a temperature at 75±°F year-round with relative humidity maintained at 50±% by dehumidification in summer. In winter relative humidities within the conditioned spaces are usually much lower, on the order of 20–30%, as a result of heating relatively dry outside air, and the impracticability of maintaining higher humidities which would cause excessive condensation on cold surfaces, such as window panes. Both conditions correspond closely to comfort requirements; hence, a single year-round thermostat setting is feasible.

The comfort chart (Fig. 1) describes average responses to given environmental conditions. Preferences vary considerably from person to person. For instance, in office buildings most women complain of chill or drafts under conditions where men, clad in business suits, are comfortable. Metabolism rates vary with each individual. Whenever a building budget permits, an engineer designs an air conditioning system that is flexible enough to allow individual adjustment by occupants of both temperature and air motion in the air-conditioned space. Conditions that would satisfy all persons within the same space have yet to be determined.

Calculation of loads. Engineering of an air-conditioning system starts with selection of design conditions; air temperature and relative humidity are principal factors. Next, loads on the system are calculated. Finally, equipment is selected and sized to perform the indicated functions and to carry the estimated loads.

Design conditions are selected on the bases discussed above. Each space is analyzed separately. A cooling load will exist when the sum of heat released within the space and transmitted to the space is greater than the loss of heat from the space. A heating load occurs when the heat generated within the space is less than the loss of heat from it. Similar considerations apply to moisture.

Heat generated within the space consists of body heat, approximately 250 Btu/hr/person, heat from all electrical appliances and lights, 3.41 Btu/hr/watt, and heat from other sources such as gas cooking stoves and industrial ovens. Heat is transmitted through all parts of the space envelope, which includes walls, floor, ceiling, and windows. Whether heat enters or leaves the space depends upon whether the outside surfaces are warmer or cooler than the inside surfaces. The rate at which heat is conducted through the space envelope is a function of the temperature difference across the envelope and the thermal conductance of the envelope. Conductances, which depend on materials of construction and their thicknesses along the path of heat transmission, are a large factor in walls and ceilings exposed to the outdoors in cold winters and hot summers. In these cases insulation is added to decrease the overall conductance of the envelope.

Solar heat loads are an especially important part of load calculation because they represent a large percentage of heat gain through walls and roofs, but are very difficult to estimate because solar irradiation is constantly changing. Intensity of radiation varies with the seasons (it rises to 457 Btu/hr/ft^2 in midwinter and drops to 428 in midsummer). Intensity of solar irradiation also varies with surface orientation. For example, the half-day total for a horizontal surface at 40 degrees north latitude on January 21 is 353 Btu/hr/ft^2 and on June 21 it is 1121 Btu, whereas for a south wall on the same dates comparable data are 815 and 311 Btu, a sharp decrease in summer. Intensity also varies with time of day and cloud cover and other atmospheric phenomena.

The way in which solar radiation affects the space load depends also upon whether the rays are transmitted instantly through glass or impinge on opaque walls. If through glass, the effect begins immediately but does not reach maximum intensity until the interior irradiated surfaces have warmed sufficiently to reradiate into the space, warming the air. In the case of irradiated walls and roofs, the effect is as if the outside air temperature were higher than it is. This apparent temperature is called the sol-air temperature, of which tables are available.

In calculating all these heating effects, the object is proper sizing and intelligent selection of equipment; hence, a design value is sought which will accommodate maximums. However, when dealing with climatic data, which are statistical, historical summaries, record maximums are rarely used. For instance, if in a particular locality the recorded maximum outside temperature was 100°, but 95°F was exceeded only four times in the past 20 years, 95°F may be chosen as the design summer outdoor temperature for calculation of heat transfer through walls. In practice, engineers use tables of design winter and summer outdoor temperatures which list winter temperatures exceeded more than 99% and 97.5% of the time during the coldest winter months, and summer temperatures not exceeded 1%, 2.5%, and 5% of the warmest months. The designer will select that value which represents the conservatism required for the particular type of occupancy. If the space contains vital functions where impairment by virtue of occasional departures from design space conditions cannot be tolerated, the more severe design outdoor conditions will be selected.

In the case of solar load through glass, but even more so in the case of heat transfer through walls and roof, because outside climate conditions are so variable, there may be a considerable thermal lag. It may take hours before the effect of extreme high or low temperatures on the outside of a thick masonry wall is felt on the interior surfaces and space. In some cases the effect is never felt on the inside, but in all cases the lag exists, exerting a leveling effect on the peaks and valleys of heating and cooling demand; hence, it tends to reduce maximums and can be taken advantage of in reducing design loads.

Humidity as a load on an air conditioning system is treated by the engineer in terms of its latent heat, that is, the heat required to condense or evaporate the moisture, approximately 1000 Btu/lb of moisture. People at rest or at light work generate about 200 Btu/hr. Steaming from kitchen activities and moisture generated as a product of combustion of gas flames, or from all drying processes, must be calculated. As with heat, moisture travels through the space envelope, and its rate of transfer is calculated as a function of the difference in vapor pressure across the space envelope and the permeability of the envelope construction. To decrease permeability where vapor pressure differential is large, vapor barriers (relatively impermeable membranes) are incorporated in the envelope construction.

Another load-reducing factor to be calculated is the diversity among the various spaces within a building or building complex served by a single system. Spaces with east-facing walls experience maximum solar loads when west-facing walls have no solar load. In cold weather, rooms facing south may experience a net heat gain due to a preponderant solar load while north-facing rooms require heat. An interior space, separated from adjoining spaces by partitions, floor, and ceiling across which there is no temperature gradient, experiences only a net heat gain, typically from people and lights. Given a system that can transfer this heat to other spaces requiring heat, the net heating load may be zero, even on cold winter days.

Air conditioning systems. A complete air conditioning system is capable of adding and removing heat and moisture and of filtering dust and odorants from the space or spaces it serves. Systems that heat, humidify, and filter only, for control of comfort in winter, are called winter air conditioning systems; those that cool, dehumidify, and filter only are called summer air conditioning systems, provided they are fitted with proper controls to maintain design levels of temperature, relative humidity, and air purity.

Design conditions may be maintained by multi-

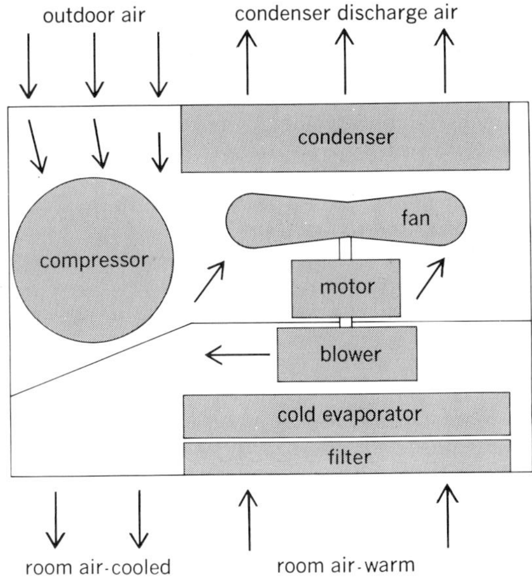

Fig. 2. Schematic of room air conditioner. (*American Society of Heating, Refrigerating and Air-Conditioning Engineers, Inc., Guide and Data Book, 1967*)

ple independent subsystems tied together by a single control system. Such arrangements, called split systems, might consist, for example, of hot-water baseboard heating convectors around a perimeter wall to offset window and wall heat losses when required, plus a central cold-air distribution system to pick up heat and moisture gains as required and to provide filtration for dust and odor.

Air conditioning systems are either unitary or built-up. The window or through-the-wall air conditioner (Fig. 2) is an example of a unitary summer air conditioning system; the entire system is housed in a single package which contains heat removal, dehumidification, and filtration capabilities. When an electric heater is built into it with suitable controls, it functions as a year-round air conditioning system. Unitary air conditioners are manufactured in capacities as high as 100 tons (1 ton of air conditioning equals 12,000 Btu/hr) and are designed to be mounted conveniently on roofs, on the ground, or other convenient location, where they can be connected by ductwork to the conditioned space.

Built-up or field-erected systems are composed of factory-built subassemblies interconnected by means such as piping, wiring, and ducting during final assembly on the building site. Their capacities range up to thousands of tons of refrigeration and millions of Btu per hr of heating. Most large buildings are so conditioned.

Another important and somewhat parallel distinction can be made between incremental and central systems. An incremental system serves a single space; each space to be conditioned has its own, self-contained heating-cooling-dehumidifying-filtering unit. Central systems serve many or all of the conditioned spaces in a building. They range from small, unitary packaged systems to serve single-family residences to large, built-up or field-erected systems serving large buildings.

When many buildings, each with its own air conditioning system which is complete except for a refrigeration and a heating source, are tied to a central plant that distributes chilled water and hot water or steam, the interconnection is referred to as a district heating and cooling system. This system is especially useful for campuses, medical complexes, and office complexes under a single management.

Conditioning of spaces. Air temperature in a space can be controlled by radiant panels in floor, walls, or ceiling to emit or absorb energy, depending on panel temperature. Such is the radiant panel system. However, to control humidity and air purity, and in most systems for controlling air temperature, a portion of the air in the space is withdrawn, processed, and returned to the space to mix with the remaining air. In the language of the engineer, a portion of the room air is returned (to an air-handling unit) and, after being conditioned, is supplied to the space. A portion of the return air is spilled (exhausted to the outdoors) while an equal quantity (of outdoor air) is brought into the system and mixed with the remaining return air before entering the air handler.

Typically, the air-handling unit contains a filter, a cooling coil, a heating coil, and a fan in a suitable casing (Fig. 3). The filter removes dust from both return and outside air. The cooling coil, either containing recirculating chilled water or boiling refrigerant, lowers air temperature sufficiently to dehumidify it to the required degree. The heating coil, in winter, serves a straightforward heating function, but when the cooling coil is functioning, it serves to raise the temperature of the dehumidified air (to reheat it) to the exact temperature required to perform its cooling function. The air handler may perform its function, in microcosm, in room units in each space, as part of a self-contained, unitary air conditioner, or it may be a huge unit handling return air from an entire building. *See* AIR COOLING; AIR FILTER; HUMIDITY CONTROL.

There are three principal types of central air-conditioning systems: all-air, all-water, and air-water. In the all-air system all return air is processed in a central air-handling apparatus. In one type of all-air system, called dual-duct, warm air and chilled air are supplied to a blending or mixing unit in each space. In a single-duct all-air

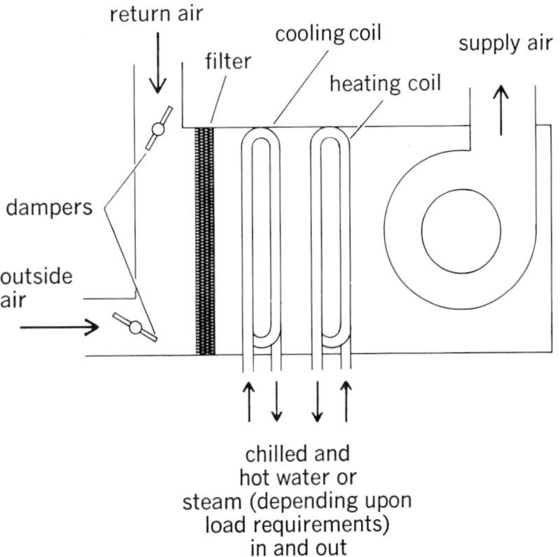

Fig. 3. Schematic of central air-handling unit.

system air is supplied at a temperature for the space requiring the coldest air, then reheated by steam or electric or hot-water coils in each space.

In the all-water system the principal thermal load is carried by chilled and hot water generated in a central facility and piped to coils in each space; room air then passes over the coils. A small, central air system supplements the all-water system to provide dehumidification and air filtration. The radiant panel system, previously described, may also be in the form of an all-water system.

In an air-water system both treated air and hot or chilled water are supplied to units in each space. In winter hot water is supplied, accompanied by cooled, dehumidified air. In summer chilled water is supplied with warmer (but dehumidified) air. One supply reheats the other.

All-air systems preceded the others. Primary motivation for all-water and air-water systems is their capacity for carrying large quantities of heat energy in small pipes, rather than in larger air ducts. To accomplish the same purpose, big-building all-air systems are designed for high velocities and pressures, requiring much smaller ducts.

[RICHARD L. KORAL]

Air cooling

Lowering of air temperature for comfort, process control, or food preservation. Air and water vapor occur together in the atmosphere. The mixture is commonly cooled by direct convective heat transfer of its internal energy (sensible heat) to a surface or medium at lower temperature. In the most compact arrangement, transfer is through a finned (extended surface) coil, metallic and thin, inside of which is circulating either chilled water, antifreeze solution, brine, or boiling refrigerant. The fluid acts as the heat receiver. Heat transfer can also be directly to a wetted surface, such as water droplets in an air washer or a wet pad in an evaporative cooler. *See* AIR CONDITIONING.

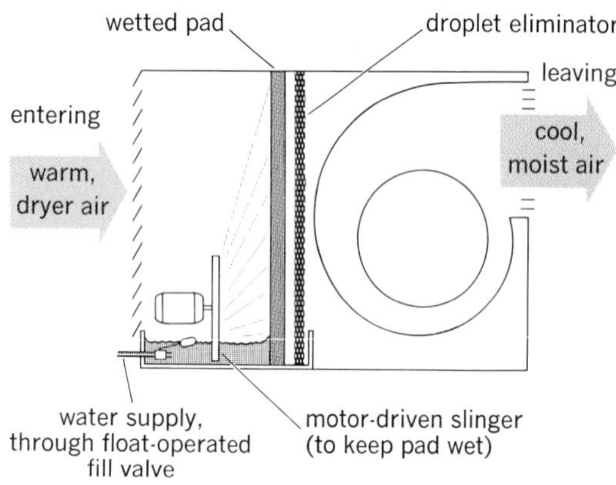

Fig. 1. Schematic view of simple evaporative air cooler.

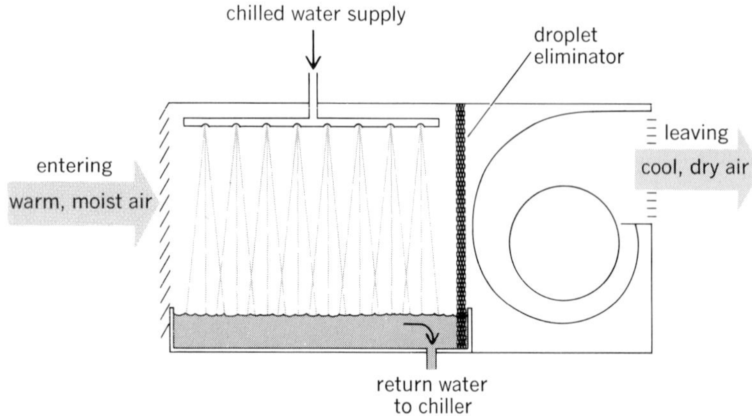

Fig. 2. Schematic of air washer.

Evaporative cooling. For evaporative cooling, nonsaturated air is mixed with water. Some of the sensible heat transfers from the air to the evaporating water. The heat then returns to the airstream as latent heat of water vapor. The exchange is thermally isolated (adiabatic) and continues until the air is saturated and air and water temperatures are equal. With suitable apparatus, air temperature approaches within a few degrees of the theoretical limit, the wet-bulb temperature. Evaporative cooling is frequently carried out by blowing relatively dry air through a wet mat (Fig. 1). The technique is employed for air cooling of machines where higher humidities can be tolerated; for cooling of industrial areas where high humidities are required, as in textile mills; and for comfort cooling in hot, dry climates, where partial saturation results in cool air at relatively low humidity.

Air washer. In the evaporative cooler the air is constantly changed and the water is recirculated, except for that portion which has evaporated and which must be made up. Water temperature remains at the adiabatic saturation (wet-bulb) temperature. If water temperature is controlled, as by refrigeration, the leaving air temperature can be controlled within wide limits. Entering warm, moist air can be cooled below its dew point so that, although it leaves close to saturation, it leaves with less moisture per unit volume of air than when it entered. An apparatus to accomplish this is called an air washer (Fig. 2). It is used in many industrial and comfort air conditioning systems, and performs the added functions of cleansing the airstream of dust and of gases that dissolve in water, and in winter, through the addition of heat to the water, of warming and humidifying the air.

Air-cooling coils. The most important form of air cooling is by finned coils, inside of which circulates a cold fluid or cold, boiling refrigerant (Fig. 3). The latter is called a direct-expansion (DX) coil. In most applications the finned surfaces become wet as condensation occurs simultaneously with sensible cooling. Usually, the required amount of dehumidification determines the temperature at which the surface is maintained and, where this results in air that is colder than required, the air is reheated to the proper temperature. Droplets of condensate are entrained in the airstream, removed by a suitable filter (eliminator), collected in a drain pan, and wasted.

In the majority of cases, where chilled water or boiling halocarbon refrigerants are used, aluminum fins on copper coils are employed. Chief advantages of finned coils for air cooling are (1) complete separation of cooling fluid from airstream, (2) high velocity of airstream limited only by the need to separate condensate that is entrained in the airstream, (3) adaptability of coil configuration to requirements of different apparatus, and (4) compact heat-exchange surface.

Defrosting. Wherever air cooling and dehumidification occur simultaneously through finned coils, the coil surface must be maintained above 32°F to prevent accumulation of ice on the coil. For this reason, about 35°F is the lowest-temperature air that can be provided by coils (or air washers) without ice accumulation. In cold rooms, where air temperature is maintained below 32°F, provision is made to deice the cooling coils. Ice buildup is sensed automatically; the flow of cold refrigerant to the coil is stopped and replaced, briefly, by a hot fluid which melts the accumulated frost. In direct-expansion coils, defrosting is easily accomplished by bypassing hot refrigerant gas from the compressor directly to the coil until defrosting is complete.

Cooling coil sizing. Transfer of heat from warm air to cold fluid through coils encounters three resistances: air film, metal tube wall, and inside fluid film. Overall conductance of the coil, U, is shown in the equation below, where K_o is film conductance

$$\frac{1}{U}=\frac{1}{K_o}+r_m+\frac{R}{K_i}$$

of the outside (air-side) surface in Btu per (hr)(sq ft)(F); r_m is metal resistance in (hr)(sq ft)(F) per Btu, where area is that of the outside surface; K_i is film conductance of the inside surface (water, steam, brine, or refrigerant side) in Btu per (hr)(sq ft)(F); U is overall conductance of transfer surface in Btu per (hr)(sq ft)(F), where area again refers to the outside surface; and R is the ratio of outside surface to inside surface.

Values of K_o are a function of air velocity and typically range from about 4 Btu per (hr)(sq ft)(F) at 100 feet per minute (fpm) to 12 at 600 fpm. If condensation takes place, latent heat released by the condensate is in addition to the sensible heat transfer. Then total (sensible plus latent) K_o increases by the ratio of total to sensible heat to be transferred, provided the coil is well drained.

Values of r_m range from 0.005 to 0.030 (hr)(sq ft)(F) per Btu, depending somewhat on type of metal but primarily on metal thickness.

Typical values for K_i range from 250 to 500 Btu per (hr)(sq ft)(F) for boiling refrigerant. In 40°F chilled water, values range from about 230 Btu per (hr)(sq ft)(F) when water velocity is 1 foot per second (fps) to 1250 when water velocity is 8 fps.

Use of well water. Well water is available for air cooling in much of the world. Temperature of water from wells 30 to 60 ft deep is approximately the average year-round air temperature in the locality of the well, although in some regions overuse of available supplies for cooling purposes and recharge of ground aquifers with higher-temperature water has raised well water temperature several degrees above the local normal. When well water is not cold enough to dehumidify air to the required extent, an economical procedure is to use it for sensible cooling only, and to pass the cool, moist air through an auxiliary process to dehumidify it. Usually, well water below 50°F will dehumidify air sufficiently for comfort cooling. Well water at these temperatures is generally available in the northern third of the United States, except the Pacific Coast areas.

Ice as heat sink. For installations that operate only occasionally, such as some churches and meeting halls, water recirculated and cooled over

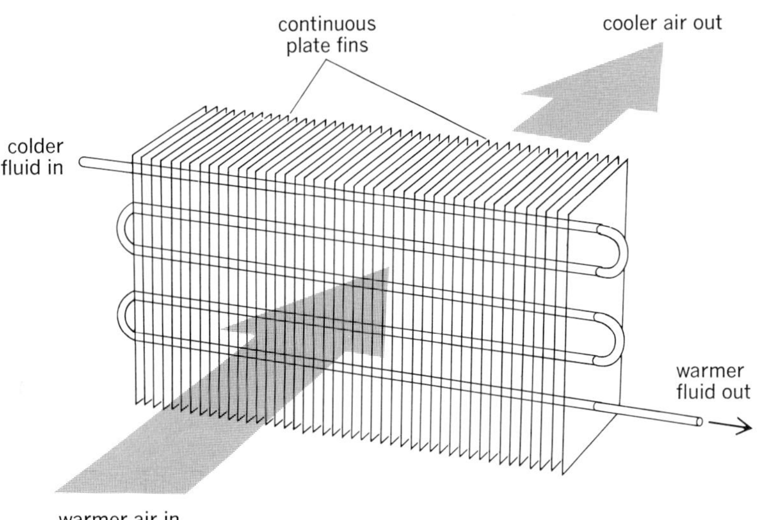

Fig. 3. Typical extended-surface air-cooling coil.

ice offers an economical means for space cooling (Fig. 4). Cold water is pumped from an ice bunker through an extended-surface coil. In the coil the water absorbs heat from the air, which is blown across the coil. The warmed water then returns to the bunker, where its temperature is again reduced by the latent heat of fusion (144 Btu/lb) to 32°F. Although initial cost of such an installation is low, operating costs are usually high.

Refrigeration heat sink. Where electric power is readily available, the cooling function of the ice, as described above, is performed by a mechanical refrigerator. If the building complex includes a steam plant, a steam-jet vacuum pump can be used to cause the water to evaporate, thereby lowering its temperature by the latent heat of evaporation (about 1060 Btu/lb, depending on temperature and pressure). High-pressure steam, in passing through a primary ejector, aspirates water vapor from the evaporator, thereby maintaining the required low pressure that causes the water to evaporate and thus to cool itself (Fig. 5). *See* REFRIGERATION.

Where electric power is costly compared to low-temperature heat, such as by gas, absorption refrigeration may be used. Two fluids are used: an

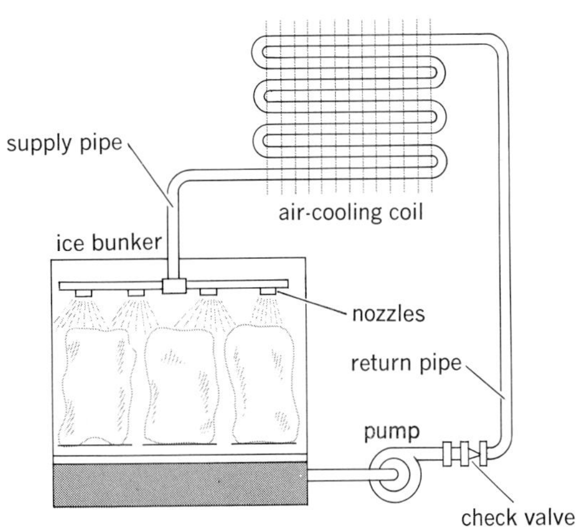

Fig. 4. Air cooling by circulating ice-cooled water.

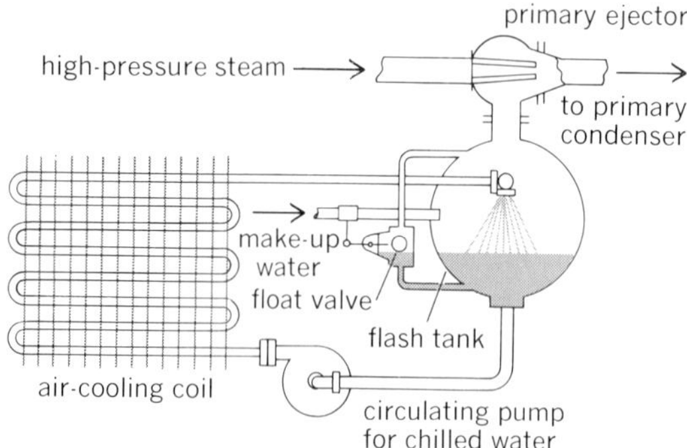

Fig. 5. Air cooling by circulating water that is cooled, in turn, by evaporation in flash tank.

absorbent and a refrigerant. The absorbent is chosen for its affinity for the refrigerant when in vapor form, for example, water is used as the absorber with ammonia as the refrigerant. Concentrated ammonia water is pumped to a high pressure and then heated to release the ammonia. The high-pressure ammonia then passes through a condenser, an expansion valve, and an evaporator, as in a mechanical system, and is reabsorbed by the water. The cycle cools air circulated over the evaporator. [RICHARD L. KORAL]

Bibliography: Air Conditioning and Refrigeration Institute, *Refrigeration and Air Conditioning*, 1979; American Society of Heating and Air Conditioning Engineers, *Guide and Data Book*, annual; S. Elonka and Q. W. Minich, *Standard Refrigeration and Air Conditioning Questions and Answers*, 2d ed., 1973.

Air filter

A component of most systems in which air is used for industrial processes, for ventilation, or for comfort air conditioning. The function of an air filter is to reduce the concentration of solid particles in the airstream to a level that can be tolerated by the process or space occupancy purpose. Degrees of cleanliness required and economics of the situation (life cycle costs) influence the selection of equipment. *See* AIR CONDITIONING; VENTILATION.

Solid particles in the airstream range in size from 0.01 μm (the smallest particle visible to the naked eye is estimated to be 20 μm in diameter) up to things that can be caught by ordinary fly screens, such as lint, feathers, and insects. The particles generally include soot, ash, soil, lint, and smoke, but may include almost any organic or inorganic material, even bacteria and mold spores. This wide variety of airborne contaminants, added to the diversity of systems in which air filters are used, makes it impossible to have one type that is best for all applications.

Three basic types of air filters are in common use today: viscous impingement, dry, and electronic. The principles employed by these filters in removing airborne solids are viscous impingement, interception, impaction, diffusion, and electrostatic precipitation. Some filters utilize only one of these principles; others employ combinations. A fourth method, inertial separation, is finding increasing use as a result of the construction boom throughout most of the Middle East.

Viscous impingement filters. The viscous impingement filter is made up of a relatively loosely arranged medium, usually consisting of spun glass fibers, metal screens, or layers of crimped expanded metal. The surfaces of the medium are coated with a tacky oil, generally referred to as an adhesive. The arrangement of the filter medium is such that the airstream is forced to change direction frequently as it passes through the filter. Solid particles, because of their momentum, are thrown against, and adhere to, the viscous coated surfaces. Larger airborne particles, having greater mass, are filtered in this manner, whereas small particles tend to follow the path of the airstream and escape entrapment.

Operating characteristics of viscous impingement filters include media velocities ranging from 300 to 500 ft/min (1.5–3 m/s), with resistance to airflow about 0.10 in. (2.5 mm) of water gage, or w.g., when clean and 0.50 in. (13 mm) w.g. when dirty. A glass-fiber filter is thrown away when the final resistance is reached, whereas the metal type is washed, dried, recoated with adhesive, and reused. Automatic types, employing rolls of glass fiber material in blanket form, or an endless belt of metal plates which pass through an oil-filled reservoir to remove the accumulated dirt, are often chosen to minimize maintenance labor.

Dry filters. Dry-air filters are the broadest category of air filters in terms of the variety of designs, sizes, and shapes in which they are manufactured. The most common filter medium is glass fiber. Other materials used are cellulose paper, cotton, and polyurethane and other synthetics. Glass fiber is used extensively because of its relatively low cost and the unique ability to control the diameter of the fiber in manufacture (in general, the finer the fiber diameter, the higher will be the air-cleaning efficiency of the medium).

Dry filters employ the principles of interception, in which particles too large to pass through the filter openings are literally strained from the airstream; impaction, in which particles strike and stick to the surfaces of the glass fibers because of natural adhesive forces, even though the fibers are not coated with a filter adhesive; and diffusion, in which molecules of air moving in a random pattern collide with very fine particles of airborne solids, causing the particles to have random movement as they enter the filter media. It is this random movement which enhances the likelihood of the particles coming in contact with the fibers of filter media as the air passes through the filter. Through the process of diffusion a filter is able to separate from the airstream particles much smaller than the openings in the medium itself.

Most dry filters are of the extended surface type, with the ratio of filtered area to face area (normal to the direction of air travel) varing from 7-1/2:1 to as much as 50:1. The higher the filter efficiency, the greater the ratio of areas, in order to hold down the air friction loss. Clean resistance through dry filters can be as low as 0.10 in. (2.5 mm) w.g. or, for very-high-efficiency filters, as much as 1.0 in. (2.5 mm) w.g. These filters are generally allowed to increase 0.5–1.0 in. (13–25 mm) in pressure drop before being changed. Face velocities from 400 ft/min (2 m/s) to 600 ft/min (3 m/s) are common for this class of filter, although newer energy-conscious design favors the lower portion of this range.

Electronic air cleaners. Limited primarily to applications requiring high air-cleaning efficiency, these devices operate on the principle of passing the airstream through an ionization field where a 12,000-volt potential imposes a positive charge on all airborne particles. The ionized particles are then passed between aluminum plates, alternately grounded and connected to a 6000-volt source, and are precipitated onto the grounded plates.

The original design of the electronic air cleaner utilizes a water-soluble adhesive coating on the plates, which holds the dirt deposits until the plates require cleaning. The filter is then deenergized, and the dirt and adhesive film are washed off the plates. Fresh adhesive is applied before the power is turned on again. Other versions of electronic air cleaners are designed so that the plates serve as agglomerators; the agglomerates of smaller particles are allowed to slough off the plates and to be trapped by viscous impingement or dry-type filters downstream of the electronic unit.

Designs of most electronic air cleaners are based on 500 ft/min (2.5 m/s) face velocity, with pressure losses at 0.20 in. (5 mm) w.g. for the washable type and up to 1.0 in. (25 mm) w.g. for the agglomerator type using dry-type after-filters. Power consumption is low, despite the 12,000-volt ionizer potential, because current flow is measured in milliamperes.

Inertial separators. These are devices which utilize the principle of momentum to separate larger particulate micrometers in diameter, primarily that above 10 m altitude, from a moving air stream. The larger particles tend to keep going in the same direction in which they entered the device while the light-density air changes direction. Commonly offered in manifolded multiples of V-shaped pockets with louvered sides or in small-diameter cyclones, the inertial separators are used most frequently in hot, arid, desertlike regions where winds generate significant airborne dust and sand particles. A secondary bleed-air fan is used to discharge the separated particulate to the outside of the building.

Testing and rating. ASHRAE Test Standard 52 is becoming universally accepted as the optimum procedure for the testing and rating of all types of air filters and for comparing the performance of the products of competing manufacturers. An arrestance test, which measures the ability of a filter to remove the weight component of airborne particulate, is used for measuring the effectiveness of inertial separators, all viscous impingement filters, and some of the less effective dry filters. An efficiency test (dust spot efficiency using atmospheric air) provides ratings for dry filters and for electronic air cleaners. Enforcement of the certification procedures, under which manufacturers may claim certified performance, is the responsibility of the Air Conditioning and Refrigeration Institute (ARI) under their Standard 680.

The ASHRAE arrestance test measures the ability of a filter to remove a specified dust sample from the airstream which passes through the test filter. Constituents of the dust sample are carefully controlled to ensure reproducibility of test results. In the dust spot efficiency test, samples of unfiltered atmospheric air and filtered air (downstream of the test filter) are drawn off through chemical filter paper. A photoelectric cell scans the blackness of the dust spots that appear on the two pieces of filter paper; the times required to achieve equal discoloration of both pieces of filter paper, one in the unfiltered air and the other in the filtered air, are translated into a mathematical ratio, which becomes the efficiency of the test filter.

The development of ultra-high-efficiency (HEPA) filters for special purposes (such as protection from radioactive and bacterial contaminants) led to the development of the dioctylphthalate (DOP) test. Of all the methods used to evaluate air filters, the DOP test procedure is the most sophisticated; it is used only for testing filters to be employed in critical air-cleaning applications. In a homogeneous smoke of controlled particle size (0.3-μm particles are used in this test), the DOP vapor passes through the filter. A light-scattering technique counts the number of particles entering the test filter, and a similar device counts those that emerge from the filter. Certain applications call for filters of such high efficiency that only 0.001% of incident 0.3-μm particles is permitted to emerge.

[MORTON A. BELL]

Air heater

A component of a steam-generating unit that absorbs heat from the products of combustion after they have passed through the steam-generating and superheating sections. Heat recovered from the gas is recycled to the furnace by the combustion air and is absorbed in the steam-generating unit, with a resultant gain in overall thermal efficiency. Use of preheated combustion air also accelerates ignition and promotes rapid burning of the fuel.

Air heaters frequently are used in conjunction with economizers, because the temperature of the inlet air is less than that of the feedwater to the economizer, and in this way it is possible to reduce further the temperature of flue gas before it is discharged to the stack. *See* BOILER ECONOMIZER.

Air heaters are usually classed as recuperative or regenerative types. Both types depend upon convection transfer of heat from the gas stream to a metal or other solid surface and upon convection transfer of heat from this surface to the air. In the recuperative type, exemplified by tubular- or plate-type heaters, the metal parts are stationary and form a separating boundary between the heating and cooling fluids, and heat passes by conduction through the metal wall (Fig. 1). In rotary regenerative air heaters (Fig. 2) heat-transferring

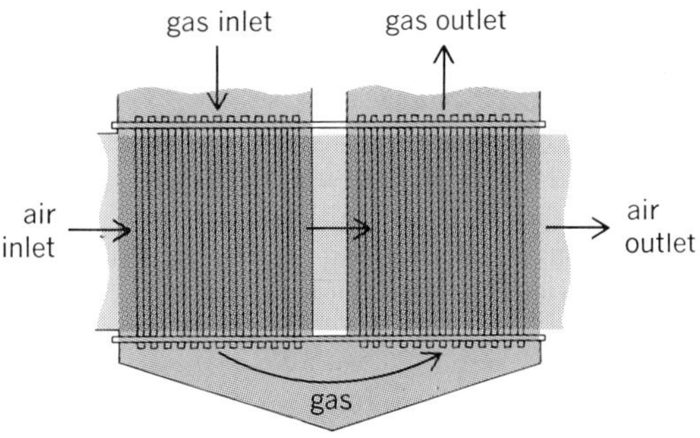

Fig. 1. Tubular air heater, two-gas single-air pass.

members are moved alternately through the gas and air streams, thus undergoing repetitive heating and cooling cycles; heat is transferred to or from the thermal storage capacity of the members. Other forms of regenerative-type air heaters, which seldom are used with steam-generating units, have stationary elements, and the alternate flow of gas and air is controlled by dampers, as in the refractory stoves of blast furnaces; or they may employ, as in the pebble heaters used in the petroleum industry for high-temperature heat exchange, a flow of solid particles which are alternately heated and cooled.

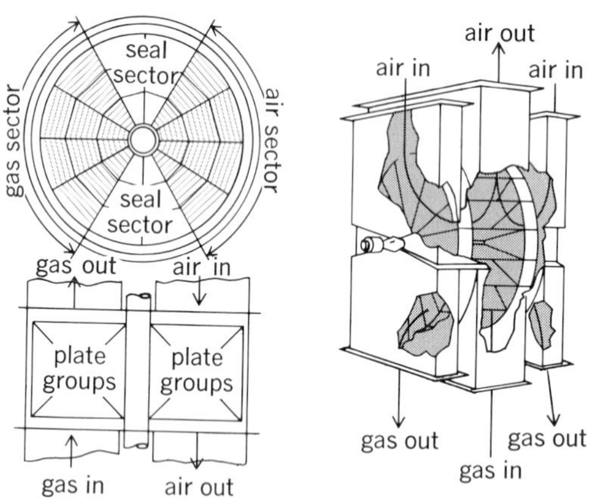

Fig. 2. Two types of rotary regenerative air heaters.

In convection heat-transfer equipment, higher heat-transfer rates and better utilization of the heat-absorbing surface are obtained with a counterflow of gases through small flow channels. The rotary regenerative air heater readily lends itself to the application of these two principles and offers high performance in small space. However, leakage of air into the gas stream necessitates frequent maintenance of seals between the moving and stationary members, and fly ash often is transported into the combustion air system. These problems are not experienced with recuperative air heaters of the tubular type. See STEAM-GENERATING UNIT. [GEORGE W. KESSLER]

Air register

A device attached to an air-distributing duct for the purpose of discharging air into the space to be heated or cooled. These openings are referred to as registers, diffusers, supply outlets, or grills. By common acceptance, a register is an opening provided with means for discharging the air in a confined jet, whereas a diffuser is an outlet which discharges the air in a spreading jet. Both registers and diffusers may be placed at a number of locations in a room, including the floor, baseboard, low or high on the sidewall, window sill, or ceiling.

For heating, the preferred location is in the floor, at the baseboard, or at the low sidewall of the outside wall, preferably under a window. For cooling, the preferred location is high on the inside wall or the ceiling. For year-round air conditioning in homes, a compromise location is the floor, baseboard, or low sidewall at the exposed wall, especially if adequate air velocity in an upward direction is provided at the supply outlet.

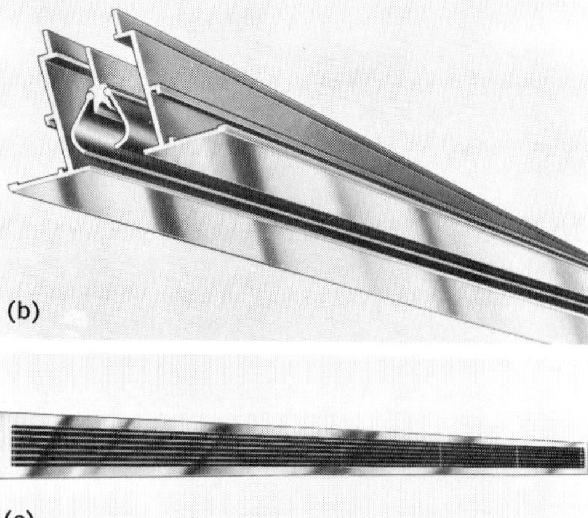

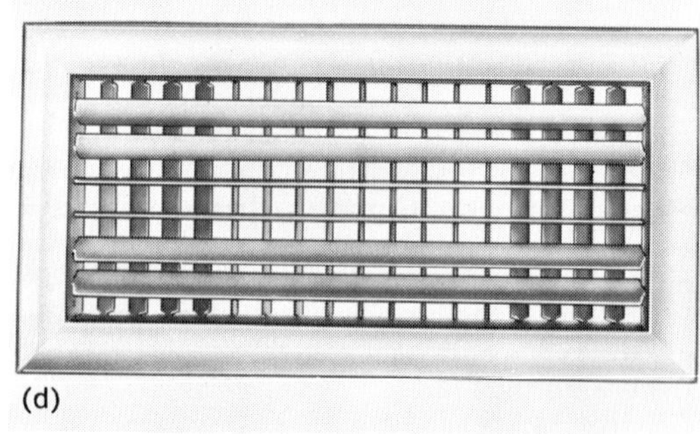

Some of the more common diffusers. (*a*) Round ceiling diffuser. (*b*) Single-slot ceiling diffuser. (*c*) Extruded aluminum grill. (*d*) Aluminum grill with adjustable blades. (*Titus Manufacturing Corp.*)

A well-designed register effectively conceals the hole at the end of the duct, throws or projects the air in the direction and at the distance desired, limits the velocity usually to 1000 ft/min or slower, and deflects the air away from walls and obstructions. The register also adjusts the direction of airflow to provide on-the-spot manipulation of the airstream, and adjusts the airflow rate to lesser amounts. It should accomplish these functions without producing dust streaks on nearby walls and ceilings, excessive air noise, or large pressure losses. Many registers, diffusers, slots, and air panels are commercially available and satisfy a majority of these qualifications. Some of the more common forms are shown in the illustration. *See* OIL BURNER; WARM-AIR HEATING SYSTEM.

[SEICHI KONZO]

Bibliography: American Society of Heating,

Refrigerating, and Air Conditioning Engineers, *Handbook and Product Directory: Equipment*, 1979; N. C. Harris and D. F. Conde, *Modern Air Conditioning Practice*, 2d ed., 1974; B. H. Jennings, *The Thermal Environment: Conditioning and Control*, 1978.

Alidade

An instrument for topographic surveying and mapping by the plane-table method; or the part of the transit which rotates with the telescope about the horizontal circle; or any sighting device or pointer employed for angular measurement. The surveying alidade has a telescope, with attached graduated vertical circle, mounted on a flat base that can be moved about the plane table. The base embodies a straightedge which itself may be a graduated scale for readily plotting the point being sighted (see illustration).

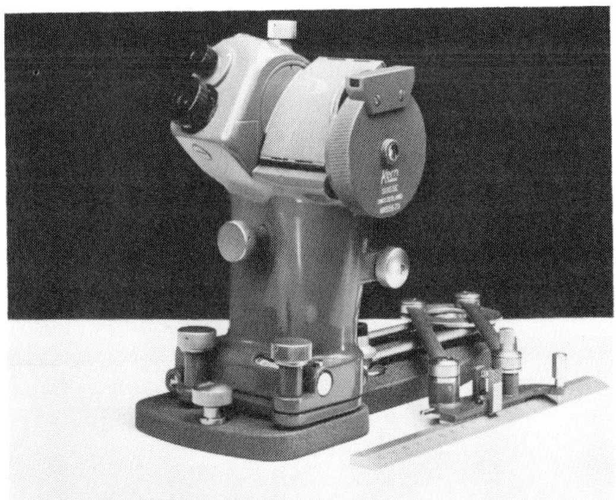

An alidade. (*Kern Instruments, Inc.*)

The stadia technique is usually applied in measuring distances and elevation differences between the plane-table point and observed points. The direction to any point is established by sighting the point through the telescope. A direction line is then drawn along the straightedge, and its distance is laid off with the scale. *See* STADIA; TRANSIT. [B. AUSTIN BARRY]

Allowance

An intentional difference in sizes of two mating parts. With running or sliding fits, in which mating parts move relative to each other, allowance is a clearance, usually for a film of oil. In this sense, allowance is the space "allowed" for motion between parts. To avoid binding between parts, a minimum clearance is critical for a running fit.

With force or shrink fits, in which mating parts, once assembled, are fixed in position relative to each other, allowance is an interference of metal; that is, a portion of metal in one part tends to occupy the same space as the adjacent portion of metal in the mating part. In this sense, allowance is the interference "allowed" to produce pressure between parts. To avoid breaking the external part, a maximum interference is critical for a drive, force, or shrink fit.

Fits are classed as running or sliding fits for parts that move freely against each other. Location fits provide accurate orientation of mating parts. Force fits require appreciable assembly pressure and produce more or less permanent assembly of parts. *See* FORCE FIT; LOCATION FIT; PRESS FIT; RUNNING FIT; SHRINK FIT; TOLERANCE. [PAUL H. BLACK]

Alloy

A metal product containing two or more elements (1) as a solid solution, (2) as an intermetallic compound, or (3) as a mixture of metallic phases. This article will describe alloys on the basis of their technical applications. Alloys may also be categorized and described on the basis of compositional groups. For example, *see* COPPER ALLOYS; IRON ALLOYS.

Except for native copper and gold, the first metals of technological importance were alloys. Bronze, an alloy of copper and tin, is appreciably harder than copper. This quality made bronze so important an alloy that it left a permanent imprint on the civilization of several millennia ago now known as the Bronze Age. Today the tens of thousands of alloys involve almost every metallic element of the periodic table.

Alloys are used because they have specific properties or production characteristics that are more attractive than those of the pure, elemental metals. For example, some alloys possess high strength, others have low melting points, others are refractory with high melting temperatures, some are especially resistant to corrosion, and others have desirable magnetic, thermal, or electrical properties. These characteristics arise from both the internal and the electronic structure of the alloy.

Bearing alloys. These alloys are used for metals that encounter sliding contact under pressure with another surface; the steel of a rotating shaft is a common example. Most bearing alloys contain particles of a hard intermetallic compound that resist wear. These particles, however, are embedded in a matrix of softer material which adjusts to the hard particles so that the shaft is uniformly loaded over the total surface. The most familiar bearing alloy is babbitt, which contains 83–91% tin; the remainder is made up of equal parts of Sb and Cu, which form hard particles of the compounds SbSn and CuSn in a soft tin matrix (Fig. 1). Other bearing alloys are based on cadmium, copper, or silver. For example, a 70% Cu–30% Pb alloy is used extensively for heavily loaded bearings. Bearings made by powder metallurgy techniques are widely used in current technology. These techniques are valuable because they permit the combination of materials which are incompatible as liquids, for example, bronze and graphite. Powdered techniques also permit controlled

ALLOY

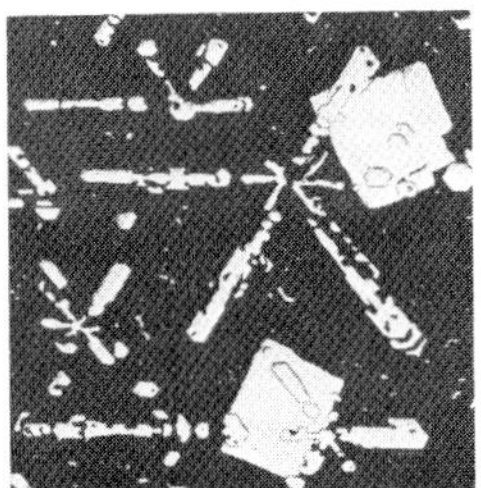

Fig. 1. Babbitt (84% Sn–7% Cu–9% Sb). The white areas are crystals of CuSn (star dendrites) and SbSn (nearly square); these resist wear. The dark matrix is a soft, tin-rich alloy, which adjusts to the hard particles to give uniform loading. (*From R. M. Brick et al., Structure and Properties of Alloys, 3d ed., McGraw-Hill, 1965*)

porosity within the bearings so that they can be saturated with oil before being used, the so-called oilless bearings. *See* ANTIFRICTION BEARING; WEAR.

Corrosion-resisting alloys. Certain alloys resist corrosion because they are noble metals. Among these alloys are the precious-metal alloys, which will be discussed separately. Other alloys resist corrosion because a protective film develops on the metal surface. This passive film is an oxide which separates the metal from the corrosive environment. Stainless steels and aluminum alloys exemplify metals with this type of protection. Stainless steels are iron alloys containing more than 12% chromium. Steels with 18% Cr and 8% Ni are the best known and possess a high degree of resistance to many corrosive environments. Aluminum alloys gain their corrosion-deterring characteristics by the formation of a very thin surface layer of Al_2O_3, which is inert to many environmental liquids. This layer is intentionally thickened in commercial anodizing processes to give a more permanent Al_2O_3 coating. Monel, an alloy of approximately 70% nickel and 30% copper, is a well-known corrosion-resisting alloy which also has high strength. Another nickel-base alloy is Inconel, which contains 14% chromium and 6% iron. The bronzes, alloys of copper and tin, also may be considered to be corrosion-resisting. *See* CORROSION; STAINLESS STEEL.

Dental alloys. Amalgams are predominantly silver-mercury alloys, but they may contain minor amounts of tin, copper, and zinc for hardening purposes, for example, 33% Ag, 52% Hg, 12% Sn, 2% Cu, and less than 1% Zn. Liquid mercury is added to a powder of a precursor alloy of the other metals. After being compacted, the mercury diffuses into the silver-base metal to give a completely solid alloy. Gold-base dental alloys are preferred over pure gold because gold is relatively soft. The most common dental gold alloy contains gold (80–90%), silver (3–12%), and copper (2–4%). For higher strengths and hardnesses, palladium and platinum (up to 3%) are added, and the copper and silver are increased so that the gold content is decreased to 60–70%. Vitallium (Co–27% Cr–5.5% Mo–3% Ni) and other corrosion-resistant alloys are used for bridgework and various special applications.

Die-casting alloys. These alloys have melting temperatures low enough so that in the liquid form they can be injected under pressure into steel dies. Such castings are used for automotive parts and for office and household appliances which have moderately complex shapes. This processing procedure eliminates the need for expensive machining and forming operations. Most die castings are made from zinc-base or aluminum-base alloys. Magnesium-base alloys also find some application when weight reduction is paramount. Low-melting alloys of lead and tin are not common because they lack the necessary strength for the above applications. A common zinc-base alloy contains approximately 4% aluminum and up to 1% copper. These additions provide a second phase in the metal to give added strength. The alloy must be free of even minor amounts (less than 100 ppm) of impurities such as lead, cadmium, or tin, because impurities increase the rate of corrosion. Common aluminum-base alloys contain 5–12% silicon, which introduces hard-silicon particles into the tough aluminum matrix. Unlike zinc-base alloys, aluminum-base alloys cannot be electroplated; however, they may be burnished or coated with enamel or lacquer. Advances in high-temperature die-mold materials have focused attention on the die-casting of copper-base and iron-base alloys. However, the high casting temperatures introduce costly production requirements, which must be justified on the basis of reduced machining costs. *See* METAL CASTING.

Eutectic alloys. In certain alloy systems a liquid of a fixed composition freezes to form a mixture of two basically different solids or phases. An alloy that undergoes this type of solidification process is called a eutectic alloy. A typical eutectic alloy is formed by combining 28.1% of copper with 71.9% of silver. A homogeneous liquid of this composition on slow cooling freezes to form a mixture of particles of nearly pure copper embedded in a matrix (background) of nearly pure silver.

The advantageous mechanical properties inherent in composite materials such as plywood composed of sheets or lamellae of wood bonded together and fiber glass in which glass fibers are used to reinforce a plastic matrix have been known for many years. Attention is being given to eutectic alloys because they are basically natural composite materials. This is particularly true when they are directionally solidified so as to yield structures with parallel plates of the two phases (lamellar structure) or long fibers of one phase embedded in the other phase (fibrous structure). Directionally solidified eutectic alloys are being given serious consideration for use in fabricating jet engine turbine blades. For this purpose eutectic alloys that freeze to form tantalum carbide (TaC) fibers in a matrix of a cobalt-rich alloy are being heavily studied. *See* EUTECTICS; METAL MATRIX COMPOSITE.

Fusible alloys. These alloys generally have melting temperatures below that of tin (232°C), and in some cases as low as 50°C. Using eutectic compositions of metals such as lead, cadmium, bismuth, tin, antimony, and indium achieves these low melting temperatures. These alloys are used for many purposes, for example, in fusible elements in automatic sprinklers, forming and stretching dies, filler for thin-walled tubing that is being bent, and anchoring dies, punches, and parts being machined. Bismuth-rich alloys were formerly used for type metal because these low-melting metals exhibited a slight expansion on solidification, thus replicating the font perfectly for printing and publication.

High-temperature alloys. Energy conversion is more efficient at high temperatures than at low; thus the need in power-generating plants, jet engines, and gas turbines for metals which have high strengths at high temperatures is obvious. In addition to having strength, these alloys must resist oxidation by fuel-air mixtures and by steam vapor. At temperatures up to about 750°C, the austenitic stainless steels (18% Cr–8% Ni) serve well. An additional 100°C may be realized if the steels also contain 3% molybdenum. Both nickel-base and cobalt-base alloys, commonly categorized as superalloys, may serve useful functions up to 1100°C. Nichrome, a nickel-base alloy containing 12–15% chromium and 25% iron, is a fairly simple

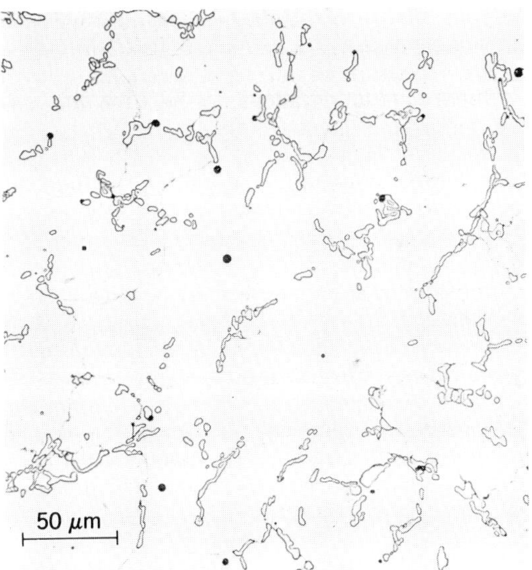

Fig. 2. Cobalt-base superalloy HS31. The dispersed phase is a carbide, $(CoCrW)_6C$, which strengthens the metal. *(From R. M. Brick et al., Structure and Properties of Alloys, 3d ed., McGraw-Hill, 1965; courtesy of Haynes Stellite Co.)*

superalloy. More sophisticated alloys invariably contain five, six, or more components; for example, an alloy called René-41 contains approximately 19% Cr, 1.5% Al, 3% Ti, 11% Co, 10% Mo, 3% Fe, 0.1% C, 0.005% B, and the balance Ni. Other alloys are equally complex. The major contributor to strength in these alloys is the solution-precipitate phase of Ni_3 (Ti,Al), γ'. It provides strength because it is coherent with the nickel-rich γ phase. Cobalt-base superalloys may be even more complex and generally contain carbon which combines with the tungsten and chromium to produce carbides that serve as the strengthening agent (Fig. 2). In general, the cobalt-base superalloys are more resistant to oxidation than the nickel-base alloys are, but they are not as strong. Molybdenum-base alloys have exceptionally high strength at high temperatures, but their brittleness at lower temperatures and their poor oxidation resistance at high temperatures have limited their use. However, coatings permit the use of such alloys in an oxidizing atmosphere, and they are finding increased application. A group of materials called cermets, which are mixtures of metals and compounds such as oxides and carbides, have high strength at high temperatures, and although their ductility is low, they have been found to be usable. One of the better-known cermets consists of a mixture of titanium carbide and nickel, the nickel acting as a binder or cement for the carbide. *See* CERMET.

Joining alloys. Metals are bonded by three principal procedures: welding, brazing, and soldering. Welded joints melt the contact region of the adjacent metal; thus the filler material is chosen to approximate the composition of the parts being joined. Brazing and soldering alloys are chosen to provide filler metal with an appreciably lower melting point than that of the joined parts. Typically, brazing alloys melt above 400°C, whereas solders melt at lower temperatures. A 57% Cu–42% Zn–1% Sn brass is a general-purpose alloy for brazing steel and many nonferrous metals. A silicon-aluminum eutectic alloy is used for brazing aluminum, and an aluminum-containing magnesium eutectic alloy brazes magnesium parts. The most common solders are based on lead-tin alloys. The prevalent 60% Sn–40% Pb solder is eutectic in composition and is used extensively for electrical circuit production, in which temperature limitations are critical. A 35% Sn–65% Pb alloy has a range of solidification and is thus preferred as a wiping solder by plumbers. *See* BRAZING; SOLDERING; WELDING AND CUTTING OF METALS.

Light-metal alloys. Aluminum and magnesium, with densities of 2.7 and 1.75 g/cm³, respectively, are the bases for most of the light-metal alloys. Titanium (4.5 g/cm³) may also be regarded as a light-metal alloy if comparisons are made with metals such as steel and copper. Aluminum and magnesium must be hardened to receive extensive application. Age-hardening processes are used for this purpose. Typical alloys are 90% Al–10% Mg, 95% Al–5% Cu, and 90% Mg–10% Al. Ternary (three-element) and more complex alloys are very important light-metal alloys because of their better properties. The Al-Zn-Mg system of alloys, used extensively in aircraft applications, is a prime example of one such alloy system.

Low-expansion alloys. This group of alloys includes Invar (64% Fe–36% Ni), the dimensions of which do not vary over the atmospheric temperature range. It has special applications in watches and other temperature-sensitive devices. Glass-to-metal seals for electronic and related devices require a matching of the thermal-expansion characteristics of the two materials. Kovar (54% Fe–29% Ni–17% Co) is widely used because its expansion is low enough to match that of glass.

Magnetic alloys. Soft and hard magnetic materials involve two distinct categories of alloys. The former consists of materials used for magnetic cores of transformers and motors, and must be magnetized and demagnetized easily. For alternating-current applications, silicon-ferrite is commonly used. This is an alloy of iron containing as much as 5% silicon. The silicon has little influence on the magnetic properties of the iron, but it increases the electric resistance appreciably and thereby decreases the core loss by induced currents. A higher magnetic permeability, and therefore greater transformer efficiency, is achieved if these silicon steels are grain-oriented so that the crystal axes are closely aligned with the magnetic field. Permalloy (78.5% Ni–21.5% Fe) and some comparable cobalt-base alloys have very high permeabilities at low field strengths, and thus are used in the communications industry. Ceramic ferrites, although not strictly alloys, are widely used in high-frequency applications because of their low electrical conductivity and negligible induced energy losses in the magnetic field. Permanent or hard magnets may be made from steels which are mechanically hardened, either by deformation or by quenching. Some precipitation-hardening, iron-base alloys are widely used for magnets. Typical of these are the Alnicos, for example, Alnico-4 (55% Fe–28% Ni–12% Al–5% Co). Since these alloys cannot be forged, they must be produced in the form of castings. The newest hard magnets are being produced from alloys of cobalt and the rare-earth type of metals.

The compound RCo_5, where R is Sm, La, Ce, and so on, has extremely high coercivity.

Precious-metal alloys. In addition to their use in coins and jewelry, precious metals such as silver, gold, and the heavier platinum metals are used extensively in electrical devices in which contact resistances must remain low, in catalytic applications to aid chemical reactions, and in temperature-measuring devices such as resistance thermometers and thermocouples. The unit of alloy impurity is commonly expressed in karats, where each karat is a 1/24 part. The most common precious-metal alloy is sterling silver (92.5% Ag, with the remainder being unspecified, but usually copper). The copper is very beneficial in that it makes the alloy harder and stronger than pure silver. Yellow gold is an Au-Ag-Cu alloy with approximately a 2:1:1 ratio. White gold is an alloy which ranges from 10 to 18 karats, the remainder being additions of nickel, silver, or zinc, which change the color from yellow to white. The alloy 87% Pt – 13% Rh, when joined with pure platinum, provides a widely used thermocouple for temperature measurements in the 1000 – 1650°C temperature range.

Prosthetic alloys. Metallic implants demand extreme corrosion resistance because body fluids contain nearly 1% NaCl, along with minor amounts of other salts, with which the metal will be in contact for indefinitely long periods of time. Type 316 stainless steels (Fe – 18% Cr – 8% Ni plus Mo) resist pitting corrosion but are subject to crevice corrosion. Vitallium (Co – 27% Cr – 5.5% Mo – 3% Ni) and other cobalt-base alloys have orthopedic applications. Titanium alloys, for example, Ti – 6% Al – 4% V, gained wide usage in Europe during the early 1970s for pacemakers and for retaining devices in artificial heart valves. A protective coating of titanium dioxide, which retards further reaction, forms within a few days. While excellent for corrosion resistance, this alloy is subject to mechanical wear; therefore, it is not satisfactory in hip-joint prostheses and applications with similar frictional contacts.

Shape memory alloys. These alloys have a very interesting and desirable property. In a typical case, a metallic object of a given shape is cooled from a given temperature T_1 to a lower temperature T_2 where it is deformed so as to change its shape. Upon reheating from T_2 to T_1, the shape change accomplished at T_2 is recovered so that the object returns to its original configuration. This thermoelastic property of the shape memory alloys is associated with the fact that they undergo a martensitic phase transformation (that is, a reversible change in crystal structure that does not involve diffusion) when they are cooled or heated between T_1 and T_2.

For a number of years the shape memory materials were essentially scientific curiosities. Among the first alloys shown to possess these properties was one of gold alloyed with 47.5% Cd. Considerable attention has been given to an alloy of nickel and titanium known as nitinal. The interest in shape memory alloys has increased because it has been realized that these alloys are capable of being employed in a number of useful applications. One example is for thermostats; another is for couplings on hydraulic lines or electrical circuits. The thermoelastic properties can also be used, at least in principle, to construct heat engines operating over a small temperature differential and thus of interest in the area of energy conversion.

Superconducting alloys. Early in the 20th century, H. Kamerlingh Onnes discovered the phenomenon of superconductivity when he observed that the resistivity of mercury dropped effectively to zero when cooled below a critical temperature T_c, which was 4.15 K. This effect has been found to occur in a number of other elements. However, the greatest promise for practical application of superconductivity appears to lie in certain alloys of the refractory metals niobium and vanadium. Many of these alloys approach the stoichiometric ratio A_3B, where A corresponds to niobium or vanadium and B to one of the elements Ga, Ge, Al, Si, and Sn. Superconducting alloys are of great interest in the design of certain fusion reactors which require very large magnetic fields to contain the plasma in a closed system. The advantage of the use of a material with a resistivity approaching zero is obvious. However, two significant problems are involved in the use of superconducting alloys in large electromagnetics. First is the critical temperature; the higher the critical temperature, the easier it will be to operate the magnets. Second is the fact that above a certain critical current density the superconducting materials tend to become normal conductors with a finite resistance. The alloys based on composition such as Nb_3Sn are superior in this regard, and in some cases critical temperatures as high as 23 K have been obtained. Serious materials problems still have to be solved before these materials can be used successfully.

Thermocouple alloys. These include Chromel, containing 90 % Ni and 10% Cr, and Alumel, containing 94% Ni, 2% Al, 3% Cr, and 1% Si. These two alloys together form the widely used Chromel-Alumel thermocouple, which can measure temperatures up to 2200°F (1204°C). Another common thermocouple alloy is constantan, consisting of 45% Ni and 55% Cu. It is used to form iron-constantan and copper-constantan couples, used at lower temperatures. For precise temperature measurements and for measuring temperatures up to 1650°C, thermocouples are used in which one metal is platinum and the other metal is platinum plus either 10 or 13% Rh. *See* STEEL.

[LAWRENCE H. VAN VLACK; ROBERT E. REED-HILL]

Bibliography: American Society for Metals, *Metals Handbook*, 9th ed., vol. 1, 1978, vol. 2, 1979; American Society for Testing and Materials, *Materials Testing Standards*, updated triennially; G. S. Brady, *Materials Handbook*, 10th ed., 1971; R. M. Brick et al., *Structure and Properties of Alloys*, 3d ed., 1965; C. T. Lynch (ed.), *Handbook of Materials Sciences*, 1st ed., vol. 2, 1975; J. Perkins (ed.), *Shape Memory Effects in Alloys*, 1975; R. H. Perry and C. M. Chilton (eds.), *Chemical Engineers' Handbook*, 5th ed., 1973; R. E. Reed-Hill, *Physical Metallurgy Principles*, 1973; L. H. Van Vlack, *Elements of Materials Science and Engineering*, 3d ed., 1975; N. E. Woldman, *Engineering Alloys*, 5th ed., 1973.

Alternating current

Electric current that reverses direction periodically, usually many times per second. Electrical energy is ordinarily generated by a public or a private utility organization and provided to a customer, whether industrial or domestic, as alternating current.

One complete period, with current flow first in one direction and then in the other, is called a cycle, and 60 cycles per second (60 hertz, or Hz) is the customary frequency of alternation in the United States and in all of North America. In Europe and in many other parts of the world, 50 Hz is the standard frequency. On aircraft a higher frequency, often 400 Hz, is used to make possible lighter electrical machines.

When the term alternating current is used as an adjective, it is commonly abbreviated to ac, as in ac motor. Similarly, direct current as an adjective is abbreviated dc.

Advantages. The voltage of an alternating current can be changed by a transformer. This simple, inexpensive, static device permits generation of electric power at moderate voltage, efficient transmission for many miles at high voltage, and distribution and consumption at a conveniently low voltage. With direct (unidirectional) current it is not possible to use a transformer to change voltage. On a few power lines, electric energy is transmitted for great distances as direct current, but the electric energy is generated as alternating current, transformed to a high voltage, then rectified to direct current and transmitted, then changed back to alternating current by an inverter, to be transformed down to a lower voltage for distribution and use.

In addition to permitting efficient transmission of energy, alternating current provides advantages in the design of generators and motors, and for some purposes gives better operating characteristics. Certain devices involving chokes and transformers could be operated only with difficulty, if at all, on direct current. Also, the operation of large switches (called circuit breakers) is facilitated because the instantaneous value of alternating current automatically becomes zero twice in each cycle and an opening circuit breaker need not interrupt the current but only prevent current from starting again after its instant of zero value.

Sinusoidal form. Alternating current is shown diagrammatically in Fig. 1. Time is measured horizontally (beginning at any arbitrary moment) and the current at each instant is measured vertically. In this diagram it is assumed that the current is alternating sinusoidally; that is, the current i is described by Eq. (1), where I_m is the maximum instantaneous current, f is the frequency in cycles per second (hertz), and t is the time in seconds.

$$i = I_m \sin 2\pi ft \qquad (1)$$

A sinusoidal form of current, or voltage, is usually approximated on practical power systems because the sinusoidal form results in less expensive construction and greater efficiency of operation of electric generators, transformers, motors, and other machines.

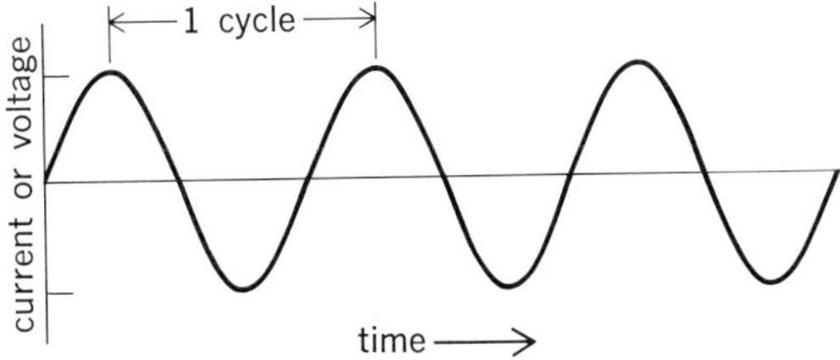

Fig. 1. Diagram of sinusoidal alternating current.

Measurement. Quantities commonly measured by ac meters and instruments are energy, power, voltage, and current. Other quantities less commonly measured are reactive volt-amperes, power factor, frequency, and demand (of energy during a given interval such as 15 min).

Energy is measured on a watt-hour meter. There is usually such a meter where an electric line enters a customer's premises. The meter may be single-phase (usual in residences) or three-phase (customary in industrial installations), and it displays on a register of dials the energy that has passed, to date, to the system beyond the meter. The customer frequently pays for energy consumed according to the reading of such a meter.

Power is measured on a wattmeter. Since power is the rate of consumption of energy, the reading of the wattmeter is proportional to the rate of increase of the reading of a watt-hour meter. The same relation is expressed by saying that the reading of the watt-hour meter, which measures energy, is the integral (through time) of the reading of the wattmeter, which measured power. A wattmeter usually measures power in a single-phase circuit, although three-phase wattmeters are sometimes used.

Current is measured by an ammeter. Current is one component of power, the others being voltage and power factor, as in Eq. (5). With unidirectional (direct) current, the amount of current is the rate of flow of electricity; it is proportional to the number of electrons passing a specified cross section of a wire per second. This is likewise the definition of current at each instant of an alternating-current cycle, as current varies from a maximum in one direction to zero and then to a maximum in the other direction (Fig. 1.) An oscilloscope will indicate instantaneous current, but instantaneous current is not often useful. A dc (d'Arsonval-type) ammeter will measure average current, but this is useless in an ac circuit, for the average of sinusoidal current is zero. A useful measure of alternating current is found in the ability of the current to do work, and the amount of current is correspondingly defined as the square root of the average of the square of instantaneous current, the average being taken over an integer number of cycles. This value is known as the root-mean-square (rms) or effective current. It is measured in amperes. It is a useful measure for current of any frequency. The rms value of direct current is identical with its dc value. The rms value of sinusoidally alternating current is $I_m/\sqrt{2}$, where I_m is the maximum instantaneous current. See Fig. 1 and Eq. (1).

Voltage is measured by a voltmeter. Voltage is the electrical pressure. It is measured between one point and another in an electric circuit, often between the two wires of the circuit. As with current, instantaneous voltage in an ac circuit reverses each half cycle and the average of sinusoidal voltage is zero. Therefore the root-mean-square (rms) or effective value of voltage is used in ac systems. The rms value of sinusoidally alternating voltage is $V_m/\sqrt{2}$, where V_m is the maximum instaneous voltage. This rms voltage, together with rms current and the circuit power factor, is used to compute electrical power, as in Eqs. (4) and (5).

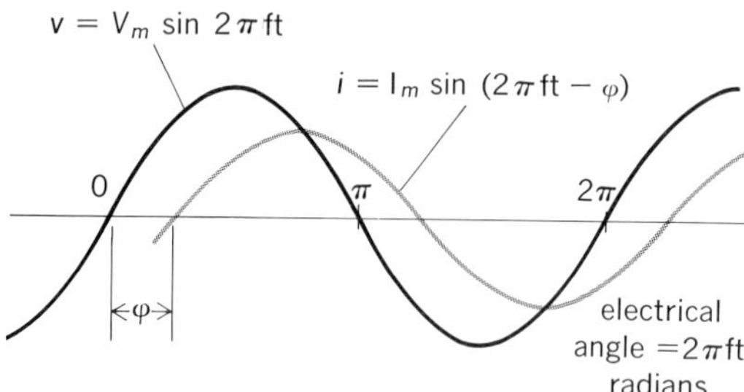

Fig. 2. The phase angle φ.

The ordinary voltmeter is connected by wires to the two points between which voltage is to be measured, and voltage is proportional to the current that results through a very high electrical resistance within the voltmeter itself. The voltmeter, actuated by this current, is calibrated in volts.

Phase difference. Phase difference is a measure of the fraction of a cycle by which one sinusoidally alternating quantity leads or lags another. Figure 2 shows a voltage v which is described in Eq. (2) and a current i which is described in Eq. (3).

$$v = V_m \sin 2\pi ft \quad (2)$$

$$i = I_m \sin (2\pi ft - \varphi) \quad (3)$$

The angle φ is called the phase difference between the voltage and the current; this current is said to lag (behind this voltage) by the angle φ. It would be equally correct to say that the voltage leads the current by the phase angle φ. Phase difference can be expressed as a fraction of a cycle or in degrees of angle, or as in Eq. (3), in radians of angle, with corresponding minor changes in the equations.

If there is no phase difference, and $\varphi = 0$, voltage and current are in phase. If the phase difference is a quarter cycle, and $\varphi = \pm 90$ degrees, the quantities are in quadrature.

Power factor. Power factor is defined in terms of the phase angle. If the rms value of sinusoidal current from a power source to a load is I and the rms value of sinusoidal voltage between the two wires connecting the power source to the load is V, the average power P passing from the source to the load is shown as Eq. (4). The cosine of the phase

$$P = VI \cos \varphi \quad (4)$$

angle, cos φ, is called the power factor. Thus the rms voltage, the rms current, and the power factor are the components of power.

The foregoing definition of power factor has meaning only if voltage and current are sinusoidal. Whether they are sinusoidal or not, average power, rms voltage, and rms current can be measured, and a value for power factor is implicit in Eq. (5). This gives a definition of power factor when V and I are not sinusoidal, but such a value for power factor has limited use.

$$P = VI \text{(power factor)} \quad (5)$$

If voltage and current are in phase (and of the same waveform), power factor equals 1. If voltage and current are out of phase, power factor is less than 1. If voltage and current are sinusoidal and in quadrature, power factor equals zero.

The phase angle and power factor of voltage and current in a circuit that supplies a load are determined by the load. Thus a load of pure resistance, as an electric heater, has unity power factor. An inductive load, such as an induction motor, has a power factor less than 1 and the current lags behind the applied voltage. A capacitive load, such as a bank of capacitors, also has a power factor less than 1, but the current leads the voltage, and the phase angle φ is a negative angle.

If a load that draws lagging current (such as an induction motor) and a load that draws leading current (such as a bank of capacitors) are both connected to a source of electric power, the power factor of the two loads together can be higher than that of either one alone, and the current to the combined loads may have a smaller phase angle from the applied voltage than would currents to either of the two loads individually. Although power to the combined loads is equal to the arithmetic sum of power to the two individual loads, the total current will be less than the arithmetic sum of the two individual currents (and may, indeed, actually be less than either of the two individual currents alone). It is often practical to reduce the total incoming current by installing a bank of capacitors near an inductive load, and thus to reduce power lost in the incoming distribution lines and transformers, thereby improving efficiency.

Three-phase system. Three-phase systems are commonly used for generation, transmission, and distribution of electric power. A customer may be supplied with three-phase power, particularly if he uses a large amount of power or if he wishes to use three-phase loads. Small domestic customers are usually supplied with single-phase power.

A three-phase system is essentially the same as three ordinary single-phase systems (as in Fig. 2, for instance) with the three voltages of the three single-phase systems out of phase with each other by one-third of a cycle (120 degrees) as shown in Fig. 3. The three voltages may be written as Eqs. (6), (7), and (8), where $V_{an(\max)}$ is the maximum

$$v_{an} = V_{an(\max)} \sin 2\pi ft \quad (6)$$

$$v_{bn} = V_{bn(\max)} \sin 2\pi (ft - 1/3) \quad (7)$$

$$v_{cn} = V_{cn(\max)} \sin 2\pi (ft - 2/3) \quad (8)$$

value of voltage in phase an, and so on. The three-phase system is balanced if relation (9) holds,

$$V_{an(\max)} = V_{bn(\max)} = V_{cn(\max)} \quad (9)$$

and if the three phase angles are equal, 1/3 cycle each as shown.

If a three-phase system were actually three separate single-phase systems, there would be two wires between the generator and the load of each system, requiring a total of six wires. In fact, however, a single wire can be common to all three sys-

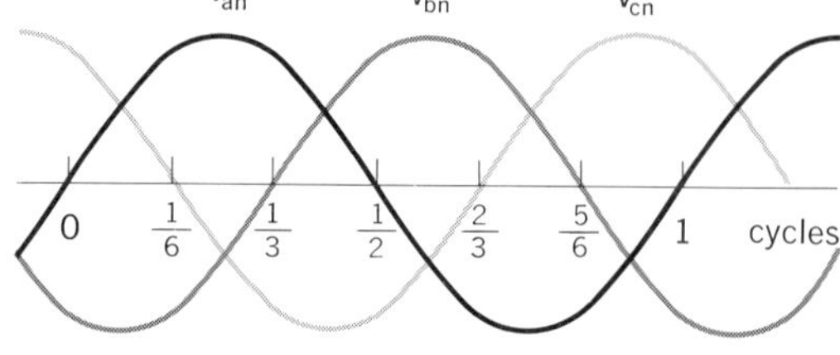

Fig. 3. Voltages of a balanced three-phase system.

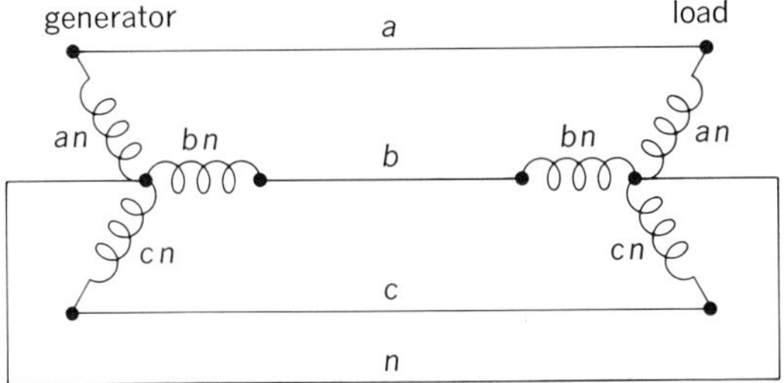

Fig. 4. Connections of a simple three-phase system.

tems, so that it is only necessary to have three wires for a three-phase system (a, b, and c of Fig. 4) plus a fourth wire n to serve as a common return or neutral conductor. On some systems the earth is used as the common or neutral conductor.

Each phase of a three-phase system carries current and conveys power and energy. If the three loads on the three phases of the three-phase system are equal and the voltages are balanced, then the currents are balanced also. Figure 2 can then apply to any one of the three phases. It will be recognized that the three currents in a balanced system are equal in rms (or maximum) value and that they are separated one from the other by phase angles of 1/3 cycle and 2/3 cycle. Thus the currents (in a balanced system) are themselves symmetrical, and Fig. 3 could be applied to line currents i_a, i_b, and i_c as well as to the three voltages indicated in the figure. Note, however, that the three currents will not necessarily be in phase with their respective voltages; the corresponding voltages and currents will be in phase with each other only if the load is pure resistance and the phase angle between voltage and current is zero; otherwise some such relation as that of Fig. 2 will apply to each phase.

It is significant that, if the three currents of a three-phase system are balanced, the sum of the three currents is zero at every instant. Thus if the three curves of Fig. 3 are taken to be the currents of a balanced system, it may be seen that the sum of the three curves at every instant is zero. This means that if the three currents are accurately balanced, current in the common conductor (n of Fig. 4) is always zero, and that conductor could theoretically be omitted entirely. In practice, the three currents are not usually exactly balanced, and either of two situations obtains. Either the common neutral wire n is used, in which case it carries little current (and may be of high resistance compared to the other three line wires), or else the common neutral wire n is not used, only three line wires being installed, and the three phase currents are thereby forced to add to zero even though this requirement results in some inbalance of phase voltages at the load.

It is also significant that the total instantaneous power from generator to load is constant (does not vary with time) in a balanced, sinusoidal, three-phase system. Power in a single-phase system that has current in phase with voltage is maximum when voltage and current are maximum and it is instantaneously zero when voltage and current are zero; if the current of the single-phase system is not in phase with the voltage, the power will reverse its direction of flow during part of each half cycle. But in a balanced three-phase system, regardless of phase angle, the flow of power is unvarying from instant to instant. This results in smoother operation and less vibration of motors and other ac devices.

Three-phase systems are almost universally used for large amounts of power. In addition to providing smooth flow of power, three-phase motors and generators are more economical than single-phase machines. Polyphase systems with two, four, or other numbers of phases are possible, but they are little used except when a large number of phases, such as 12, is desired for economical operation of a rectifier.

Power and information. Although this article has emphasized electric power, ac circuits are also used to convey information. An information circuit, such as telephone, radio, or control, employs varying voltage, current, waveform, frequency, and phase. Efficiency is often low, the chief requirement being to convey accurate information even though little of the transmitted power reaches the receiving end.

An ideal power circuit should provide the customer with electric energy always available at unchanging voltage of constant waveform and frequency, the amount of current being determined by the customer's load. High efficiency is greatly desired. *See* ALTERNATING-CURRENT CIRCUIT THEORY; CIRCUIT (ELECTRICITY); ELECTRICAL IMPEDANCE; ELECTRICAL RESISTANCE; INDUCTANCE; OHM'S LAW; RESONANCE (ALTERNATING-CURRENT CIRCUITS).

[H. H. SKILLING]

Alternating-current circuit theory

The mathematical description of conditions in an electric circuit when the circuit is driven by an alternating source or sources. *See* ALTERNATING CURRENT.

The alternating quantity is often assumed to be sinusoidal. With this assumption, an alternating voltage v may be described by Eq. (1), where t is

$$v = V_m \cos 2\pi ft \qquad (1)$$

time (seconds), f is frequency (in cycles per second, or hertz), and V_m is the maximum instantaneous value of the alternating voltage (volts). In some treatments of circuit theory the sine function is used, rather than the cosine function, as above, but this difference is immaterial because it does not affect the voltage, except as to an arbitrary time reference. Although a sine function of voltage is perhaps easier to visualize, the cosine function provides a readier graphical interpretation.

Phasors. It is usual to represent a sinusoidally varying quantity by a rotating line. In Fig. 1a a line of length V_m rotates at an angular speed ω where $\omega = 2\pi f$; the line therefore makes one revolution for each electrical cycle represented. The projection of the rotating line on the fixed horizontal axis (Fig. 1a) is $V_m \cos \omega t$, and because $\omega = 2\pi f$, this projection is identically equal to voltage v of Eq. (1). Only at one particular instant does Fig. 1a represent Eq. (1); the line in the figure must be visualized as rotating at synchronous speed.

At the instant when t was zero, the rotating line was horizontal and its projection was V_m, for at this

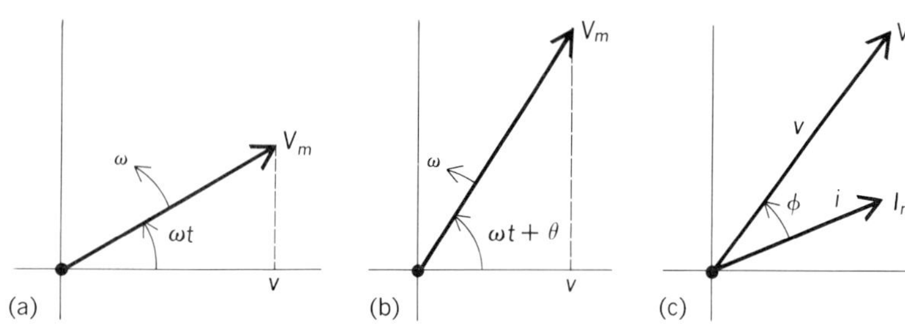

Fig. 1. Phasor representations. (*a*) Of voltage $v = V_m \cos \omega t$. (*b*) Of voltage $v = V_m \cos(\omega t + \theta)$. (*c*) Of voltage V_m and current I_m separated by phase angle ϕ.

instant $\cos \omega t = \cos 0 = 1$. To represent a voltage that is not zero at zero time, a phase angle θ is introduced. Such a voltage is shown by Eq. (2),

$$v = V_m \cos(2\pi f t + \theta) = V_m \cos(\omega t + \theta) \quad (2)$$

where θ is in radians of angle; Fig. 1*b* shows a graphical representation of Eq. (2).

Such rotating lines are called phasors. (The term vector was formerly used, and sinor is employed by some authors.) Phasors can represent any sinusoidally varying quantities, such as alternating voltage or current. Thus Fig. 1*c* shows voltage v with maximum instantaneous value V_m, and current i of the same frequency with maximum instantaneous value I_m; the phase angle between v and i is φ. This angle φ remains unchanging, as the two phasors rotate with the same velocity ω.

In a phasor diagram such as Fig. 1*c*, it is understood that the phasors shown are all rotating together at the same speed. Hence, their lengths and their angles relative to each other are important, but their angles relative to a fixed horizontal axis are not significant; consequently, a fixed axis is usually not shown in the diagram. However, in drawing such a diagram, it is always necessary to assume an arbitrary angular position for one of the phasors, called the reference phasor, and to draw the others with correct relative angles.

A phasor diagram can be drawn with the measured length of a line proportional to the maximum instantaneous value of the voltage or current represented, as in Fig. 1, or with the measured length proportional to the root-mean-square (rms) value of that voltage or current. Because for sinusoidal quantities the rms value equals the maximum value divided by $\sqrt{2}$, either method can be used.

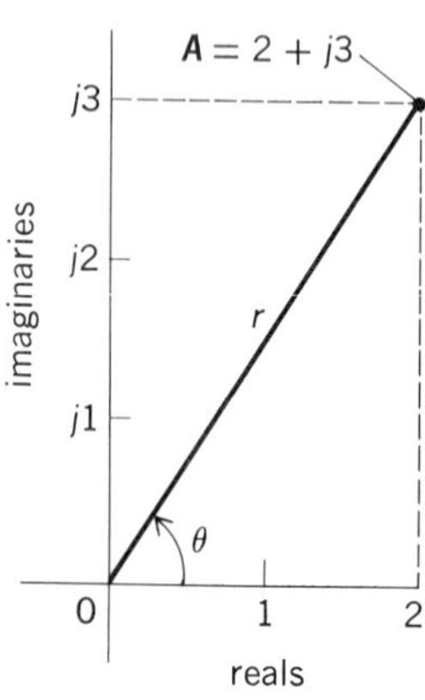

Fig. 2. Number $2+j3$ represented on the complex plane.

Complex notation. Common practice in all modern writing on alternating currents is to use a convention (introduced by C. P. Steinmetz in 1893) that assumes the phasors to lie in the mathematical complex plane. The actual instantaneous value of voltage or current is then the real component of the phasor. A brief discussion of the algebra of complex quantities is given below.

Algebra deals with numbers and with operations, such as addition and multiplication, performed on numbers. A number can be real (positive or negative), imaginary (positive or negative), or complex (the sum of a real number and an imaginary number). A complex number can be written in the form of Eq. (3), where symbol j indicates that the associated number is imaginary.

$$Z = x + jy \quad (3)$$

Thus $3 + j4$ is a complex number of which the real component is 3 and the imaginary component is 4. (The symbol i is often used in place of j, but electrical engineers commonly use j, reserving i as a symbol for current.)

By definition, the product of two positive imaginary numbers is real and negative. Thus $(j5)(j7) = -35$. Most notably, $(j1)(j1) = -1$, or the relation is expressible as Eq. (4). By this definition the square

$$\sqrt{-1} = j1 \quad (4)$$

root of a negative real number is defined and this is the basis of complex algebra.

In general, operations of complex algebra are those of the algebra of real numbers, with the additional rules shown below.

Equality. Complex numbers are equal if their real components are equal and their imaginary components are also equal.

Addition and subtraction. The sum of two complex numbers has a real component equal to the sum of the real components, and an imaginary component equal to the sum of the imaginary components. Thus, $(3 + j4) + (5 + j6) = (8 + j10)$. To subtract, change the signs of both components of the subtrahend and add.

Multiplication. The product of a real number and an imaginary number is imaginary; thus $3(j4) = j12$. The product of two positive imaginary numbers is real and negative (as stated above); thus $(j3)(j4) = -12$; but $(-j3)(j4) = +12$.

Complex numbers are multiplied by the rules of real algebra supplemented by the foregoing rules. These rules give rise to Eq. (5), and Eq. (6) is a numerical example.

$$(a + jb)(c + jd) = (ac - bd) + j(ad + bc) \quad (5)$$

$$(3 + j4)(5 + j6) = -9 + j38 \quad (6)$$

Conjugates. Complex numbers are conjugate if their real components are equal and their imaginary components are equal in magnitude but opposite in sign. Thus $(2 + j7)$ and $(2 - j7)$ are conjugate. An asterisk is commonly used to indicate the conjugate; thus $\mathbf{A}$ and $\mathbf{A}^*$ are conjugates. Equations (7) and (8) indicate that the sum of conjugates

$$(a + jb) + (a - jb) = 2a \quad (7)$$

$$(a + jb)(a - jb) = a^2 + b^2 \quad (8)$$

is real, and that the product of conjugates is real.

Division. Division is defined as the inverse of multiplication. It is possible to divide by the following technique. To divide $(5 + j10)$ by $(2 + j1)$, multiply each by the conjugate of the latter, Eq. (9).

$$\frac{5+j10}{2+j1} = \frac{(5+j10)(2-j1)}{(2+j1)(2-j1)}$$
$$= \frac{(10+10)+j(20-5)}{4+1} = 4 + j3 \quad (9)$$

Thus a ratio can be evaluated by rationalization of the denominator. However, both multiplication and division of complex numbers are easier in the exponential or polar form, as given below.

Complex plane. Two axes are drawn at right angles to each other (Fig. 2), calibrated in equal divisions with the point of intersection (origin) being zero on each axis. All real numbers are represented by points on the horizontal axis, and all imaginary numbers by points on the vertical axis. Each point in the plane represents a complex number, its projection on the axis of reals being its real component and its projection on the axis of imaginaries being its imaginary component. Thus any complex number $\mathbf{A}=a+jb$ is represented by a point in the plane; the point represented in Fig. 2 is $\mathbf{A}=2+j3$.

Addition (and subtraction) are easily visualized in the complex plane. Following the foregoing rule for addition, real components are added and imaginary components are added. Figure 3 shows $\mathbf{A}+\mathbf{B}$ where $\mathbf{A}=2+j3$ and $\mathbf{B}=4+j1$; the sum, as shown by the parallelogram, is $6+j4$.

A line marked r can be drawn from the origin to the point $\mathbf{A}$, as in Fig. 2. Its length, called A, is $(a^2+b^2)^{1/2}$, and it makes an angle θ with the real axis. By using trigonometry, Eq. (10) is obtained. In

$$\mathbf{A}=a+jb=A(\cos\theta+j\sin\theta) \qquad (10)$$

the example of Fig. 2, $a=2$, $b=3$, $A=\sqrt{13}$, and θ is the angle whose tangent is 3/2, written mathematically as $\theta=\tan^{-1} 3/2$.

With this interpretation, another notation is used for complex numbers. This notation is shown in Eq. (11), where $a=A\cos\theta$, $b=A\sin\theta$, $A=$

$$\mathbf{A}=a+jb=A\underline{/\theta} \qquad (11)$$

$(a^2+b^2)^{1/2}$, and $\theta=\tan^{-1} b/a$. The notation using a and b is called rectangular, and that using A and θ is called trigonometric or polar.

Euler's theorem. Equations (12*a*) and (12*b*) show

$$\cos\theta=\tfrac{1}{2}(e^{j\theta}+e^{-j\theta}) \qquad (12a)$$

$$j\sin\theta=\tfrac{1}{2}(e^{j\theta}-e^{-j\theta}) \qquad (12b)$$

the relation of the sine and cosine functions to the exponential e, which is the base of natural logarithms ($e\approx 2.718$). By adding Eq. (12*a*) and (12*b*), Eq. (13) is obtained, which is Euler's theorem. By

$$e^{j\theta}=\cos\theta+j\sin\theta \qquad (13)$$

using Euler's theorem in Eq. (10), Eq. (14) is ob-

$$\mathbf{A}=Ae^{j\theta} \qquad (14)$$

tained, where $\mathbf{A}$ is any complex number and A and θ are real with values found as in Eq. (11. The expression for $\mathbf{A}$ in Eq. (14) is the exponential form of a complex quantity.

Multiplication in exponential form. Multiplication and its inverse, division, are easier when complex numbers are expressed in exponential form. To multiply two complex numbers $\mathbf{A}=Ae^{j\alpha}$ and $\mathbf{B}=Be^{j\beta}$, the familiar rules of algebra give the relationships of Eq. (15). Hence the product of com-

$$\mathbf{A}\cdot\mathbf{B}=(Ae^{j\alpha})(Be^{j\beta})=ABe^{j(\alpha+\beta)} \qquad (15)$$

plex numbers is the product of the magnitudes at the sum of the angles. In polar form the relationship is written as indicated in Eq. (16). Division fol-

$$\mathbf{A}\cdot\mathbf{B}=(AB)\,\underline{/\alpha+\beta} \qquad (16)$$

lows the same rule. That is, as shown in Eq. (17),

$$\mathbf{A}/\mathbf{B}=(A/B)\,\underline{/\alpha-\beta} \qquad (17)$$

magnitudes are divided and angles subtracted.

Multiplication (and division) are easily visualized in the complex plane. Equation (16) or, employing exponential notation, Eq. (15) can be illustrated, as shown in Fig. 4; the multiplication $(0.9\underline{/30°})(1.2\underline{/70°})=1.08\underline{/100°}$. Figure 4 could also illustrate the division $(1.08\underline{/100°})/(0.9\underline{/30°})=1.2\underline{/70°}$.

In accordance with custom, arrowheads have been added to radial lines in Fig. 4. Actually, only a point in the plane is meaningful, but the radial line and arrowhead are perhaps helpful graphically to indicate the location of the point relative to the origin. A point at unit distance from the origin, at the tip of a radial line of unit length, as in Fig. 5*a*, that makes an angle θ with the horizontal axis of reals, represents a complex quantity of magnitude 1 and angle θ.

Similarly, a complex quantity of magnitude A and angle ωt is indicated by a radial line in Fig. 5*b*. Because t is time, the angle of the radial line increases with time. Thus the complex quantity $A\underline{/\omega t}$, which can also be written as $Ae^{j\omega t}$, is represented by a line of length A in the complex plane that revolves about the origin with angular velocity ω. Such a line represents an electrical phasor.

Powers and roots. The raising of complex quantities to powers follows the same rules as multiplication. Thus powers of $\mathbf{A}$, where $\mathbf{A}=A\underline{/\theta}$, are shown by Eq. (18). For instance, the square of a complex

$$\mathbf{A}^n=A^n\underline{/n\theta} \qquad (18)$$

number is the square of the magnitude at twice the angle.

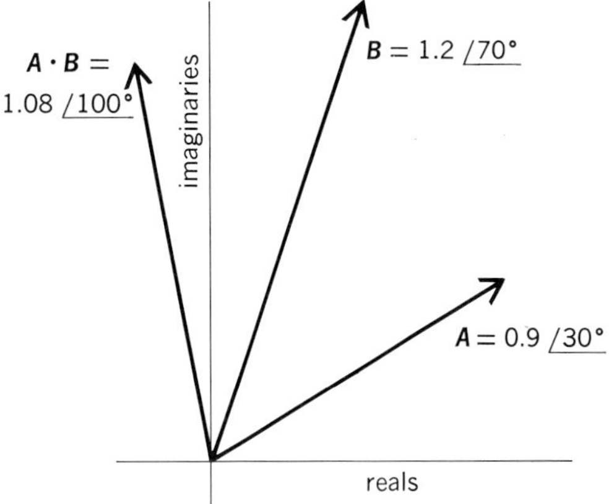

Fig. 4. Multiplication of complex numbers.

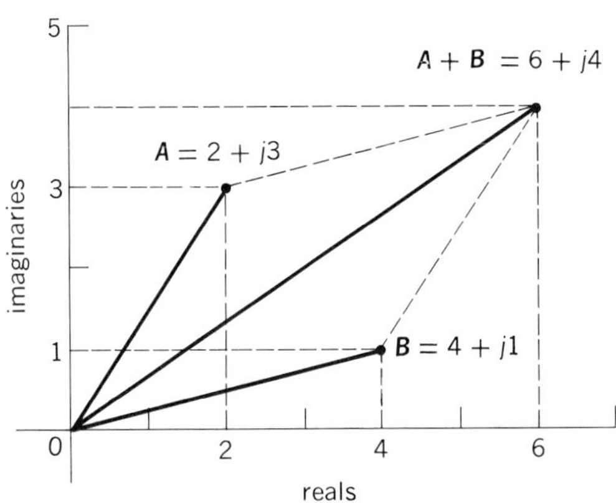

Fig. 3. Addition of complex numbers.

ALTERNATING-CURRENT CIRCUIT THEORY

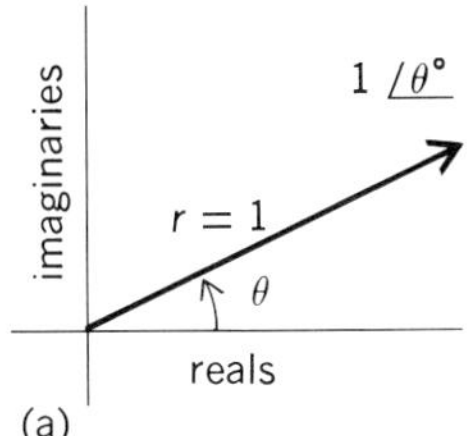

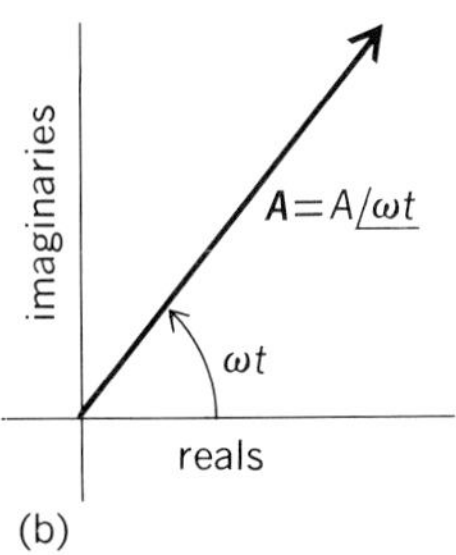

Fig. 5. Representations on the complex plane. (*a*) A line of unit length at angle θ. (*b*) A line revolving at rate ω that has reached at time t an angle ωt.

The most important powers are those of $j1$. The first four are tabulated for reference by Eqs. (19).

$$\begin{array}{ll} (j1)^2=-1 & (jx)^2=-x^2 \\ (j1)^3=-j1 & (jx)^3=-jx^3 \\ (j1)^4=+1 & (jx)^4=x^4 \\ (j1)^5=j1 & (jx)^5=jx^5 \end{array} \quad (19)$$

A complex number has multiple roots; it has two distinct square roots, three distinct cube roots, and so on. Thus if $\mathbf{A}=8\underline{/60^\circ}$, its cube roots are shown by Eq. (20). This is true because

$$(\mathbf{A})^{1/3}=2\underline{/20^\circ} \text{ or } 2\underline{/140^\circ} \text{ or } 2\underline{/260^\circ} \quad (20)$$

$(2\underline{/140^\circ})^3=8\underline{/420^\circ}$, which is indistinguishable from $8\underline{/60^\circ}$ if each is considered a point in the complex plane, because $420^\circ-360^\circ=60^\circ$. In this way three distinct cube roots of a complex number can be computed.

The algebra of complex quantities is applied below to alternating electrical quantities as represented by phasors in the complex plane.

Alternating voltage and current. By using Euler's theorem, Eq. (13), one has the equivalency shown by Eq. (21), and Eq. (1) can be written as

$$e^{j\omega t}=\cos \omega t+j \sin \omega t \quad (21)$$

Eq. (22), where $\mathcal{Re}$ means "the real component of."

$$v=V_m \mathcal{Re}\, e^{j\omega t} \quad (22)$$

In Eq. (21) $\cos \omega t$ is identically the real component of $e^{j\omega t}$. With this notation, Eq. (2) can be written in the form of Eq. (23).

$$v=V_m \mathcal{Re}\, e^{j(\omega t+\theta)}=V_m \mathcal{Re}\, e^{j\omega t} e^{j\theta} = \mathcal{Re}\sqrt{2}\,(Ve^{j\theta})\, e^{j\omega t} \quad (23)$$

The quantity $Ve^{j\theta}$ is called the transform of voltage v, and is commonly symbolized by $\mathbf{V}$, as shown in Eq. (24), where V is the rms value of

$$\mathbf{V}=Ve^{j\theta} \quad (24)$$

voltage, and θ is the phase angle of voltage (relative to an assumed reference), as in Eq. (2). By using this symbol, Eq. (23) becomes Eq. (25).

$$v=\mathcal{Re}\sqrt{2}\,\mathbf{V}e^{j\omega t} \quad (25)$$

In this expression $\sqrt{2}\,\mathbf{V}e^{j\omega t}$ is a line revolving about the origin of the complex plane: It is the phasor of voltage v, as discussed in the first section of this article. (The phasor may be defined without $\sqrt{2}$ if preferred, but the definition given here is more usual.) Transform $\mathbf{V}$ is a complex quantity. It is not voltage v, although it is often loosely called a voltage in ac circuit theory.

As an alternative notation, with the same meaning, Eq. (24) can be written as Eq. (26).

$$\mathbf{V}=V\underline{/\theta} \quad (26)$$

Parameters. There is a relation between the electric current in an element or network of elements and the voltage between the terminals of that element or network (Fig. 6). In the figure the arrow beside a conductor indicates the nominal positive direction of current i; the choice of direction is arbitrary, but all equations must be consistent with this arbitrary choice. The nominal positive direction of voltage is indicated in the figure in two alternative ways. (1) An arrow from the upper terminal to the lower terminal marks the direction in which positive voltage produces electric field in the space between the terminals and, hence, the direction in which there would be positive current in a resistive load, such as a voltmeter, from terminal to terminal. (2) Alternatively, the + beside the upper terminal indicates the same conditions. Both these conventions are used to indicate the nominal positive direction of voltage. (A third convention, sometimes seen, uses an arrow with the head pointing toward the + terminal, opposite to the arrow shown in Fig. 6.)

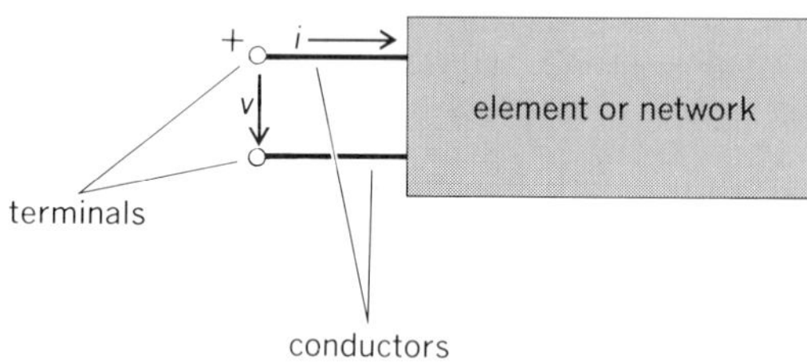

Fig. 6. Current i entering a network with voltage v across the terminals.

It is found by experiment and measurement that most actual networks, as in Fig. 6, can be treated theoretically as if they were made up of various combinations of resistance, capacitance, and inductance, including mutual inductance, and sources that are either independent of all currents and voltages (independent sources) or that are proportional to specified currents or voltages (controlled sources). For analysis an actual electrical device or network is idealized, being replaced by a model that behaves in a similar manner but that is composed of idealized parameters.

To define the parameters, the voltage-current relations, shown by Eqs. (27)–(29) and which are taken from the general theory of electricity, are used for ac circuit theory. Voltage across resistance R is shown by Eq. (27). Voltage across con-

$$v=Ri \quad (27)$$

stant inductance L is shown by Eq. (28). Current

$$v=L(di/dt) \quad (28)$$

entering constant capacitance C is shown by Eq. (29). In these equations R is resistance (ohms); L is

$$i=C(dv/dt) \quad (29)$$

inductance (henrys); C is capacitance (farads); v is voltage (volts); i is current (amperes) with nominal directions, as in Fig. 6; and t is time (seconds).

Theorem. An important theorem relating to steady-state ac operation of networks of linear elements is based on the mathematical fact that the sum of sinusoidal functions is itself sinusoidal. The theorem is: Because a sinusoidal current through an element requires a sinusoidal voltage across that element, it follows that if voltage or current at any part of any linear network is sinusoidal, voltages and currents at every part of the network are sinusoidal with the same frequency. This theorem applies only to steady-state operation, after all transient components of current and voltage have died away.

Voltage and current phasors. Equation (30) is

the same as Eq. (25) and is an expression for alternating voltage. Similarly, alternating current is expressed by Eq. (31). In these equations **V** and **I**

$$v = \mathcal{R}e\sqrt{2}\,\mathbf{V}e^{j\omega t} \tag{30}$$

$$i = \mathcal{R}e\sqrt{2}\,\mathbf{I}e^{j\omega t} \tag{31}$$

are complex quantities, the transforms of voltage and current. When these expressions are inserted into Eq. (27), it is found that Eq. (30) expresses the relation between them across pure resistance, Eq. (32). This identity (true at all values of time and

$$\mathcal{R}e\,\sqrt{2}\,\mathbf{V}e^{j\omega t} = R\,\mathcal{R}e\sqrt{2}\,\mathbf{I}e^{j\omega t} \tag{32}$$

with R constant) can be valid only if Eq. (33) is

$$\mathbf{V} = R\mathbf{I} \tag{33}$$

true. Hence the transform or phasor for voltage across a purely resistive element is equal to the resistance of the element times the current transform or phasor (Fig. 7*a*).

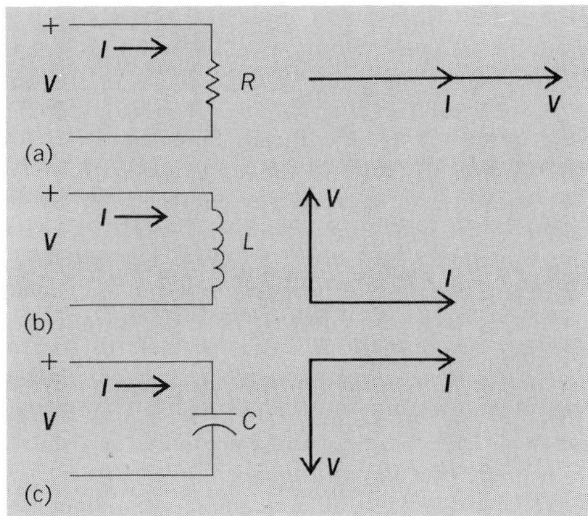

Fig. 7. Phasors. (*a*) For a resistive element. (*b*) For an inductive element. (*c*) For a capacitive element.

Inductance. In the same way Eqs. (30) and (31) are inserted into Eq. (28), which relates to inductance, giving Eq. (34). The derivative of the real

$$\mathcal{R}e\,\sqrt{2}\,\mathbf{V}e^{j\omega t} = L\frac{d}{dt}\,(\mathcal{R}e\,\sqrt{2}\,\mathbf{I}e^{j\omega t}) \tag{34}$$

component is equal to the real component of the derivative; differentiating and dividing each side by $\sqrt{2}$ gives Eq. (35). This identity (to be true for all

$$\mathcal{R}e\,\mathbf{V}e^{j\omega t} = \mathcal{R}e\,j\omega L\mathbf{I}e^{j\omega t} \tag{35}$$

t) requires that Eq. (36) hold true. Hence the pha-

$$\mathbf{V} = j\omega L\mathbf{I} \tag{36}$$

sor for voltage across a purely inductive element is proportional to the current phasor, to the inductance of the element, and to the frequency of the current; furthermore, the voltage phasor leads the current phasor by 90° because of the right-angle relationship in the complex plane that is indicated by the symbol j (Fig. 7*b*).

Capacitance. Likewise, Eqs. (30) and (31) are inserted into Eq. (29), which relates to capacitance, giving Eqs. (37) and (38). This identity (to be true for all t) requires that Eq. (39) hold true.

$$\mathcal{R}e\sqrt{2}\,\mathbf{I}e^{j\omega t} = C\frac{d}{dt}\,(\mathcal{R}e\sqrt{2}\,\mathbf{V}e^{j\omega t}) \tag{37}$$

$$\mathcal{R}e\;\mathbf{I}e^{j\omega t} = \mathcal{R}e\,j\omega C\,\mathbf{V}e^{j\omega t} \tag{38}$$

$$\mathbf{I} = j\omega C\mathbf{V} \tag{39}$$

Hence the phasor for current to a purely capacitive element is proportional to the voltage phasor, to the capacitance, and to the frequency; furthermore, the current phasor leads the voltage phasor by 90° (Fig. 7*c*).

Impedance and admittance. Impedance of an element or network is the ratio of the phasor of applied voltage to the phasor of entering current (Fig. 6). (Impedance is not the ratio of instantaneous voltage to instantaneous current.) Admittance is the ratio of the current phasor to the voltage phasor. Thus impedance **Z** and admittance **Y** are defined by Eq. (40). Because **V** and **I** are com-

$$\mathbf{Z} = \frac{\mathbf{V}}{\mathbf{I}} \qquad \mathbf{Y} = \frac{\mathbf{I}}{\mathbf{V}} = \frac{1}{\mathbf{Z}} \tag{40}$$

plex quantities, Z and Y are complex quantities also, but they are not phasors; that is, they do not represent sinusoidally varying quantities.

The foregoing paragraph defines self-impedance and self-admittance, because **V** and **I** are taken at the same pair of entering terminals. If **V** represents a voltage between any two points in a network and **I** represents current at some other point, Eq. (40) gives transfer impedance and transfer admittance between the respective locations. More generally, a transfer function can be a ratio of phasors of voltage to voltage, voltage to current, current to voltage, or current to current at different points in a network.

Real and imaginary parts of impedance are called resistance R and reactance X. Real and imaginary parts of admittance are called conductance G and susceptance B. Thus are obtained Eqs. (41) and (42).

$$\frac{\mathbf{V}}{\mathbf{I}} = \mathbf{Z} = R + jX = |\mathbf{Z}|\;(\cos\varphi + j\sin\varphi) \tag{41}$$

$$\varphi = \tan^{-1}\frac{X}{R}$$

$$\frac{\mathbf{I}}{\mathbf{V}} = \mathbf{Y} = G + jB = |\mathbf{Y}|\;(\cos\theta + j\sin\theta) \tag{42}$$

$$\theta = \tan^{-1}\frac{B}{G}$$

Equation (43) follows, and the components of **Y** are therefore expressed by Eqs. (44).

$$\mathbf{Y} = \frac{1}{\mathbf{Z}} = \frac{1}{R + jX} = \frac{R - jX}{R^2 + X^2} \tag{43}$$

$$G = \frac{R}{R^2 + X^2} \qquad B = \frac{-X}{R^2 + X^2} \tag{44}$$

Similarly, Eq. (45) follows, and its components are expressed by Eqs. (46).

$$\mathbf{Z} = \frac{1}{\mathbf{Y}} = \frac{1}{G + jB} = \frac{G - jB}{G^2 + B^2} \tag{45}$$

$$R = \frac{G}{G^2 + B^2} \qquad X = \frac{-B}{G^2 + B^2} \tag{46}$$

It should be observed that G is not the reciprocal of R unless X is zero, nor is R the reciprocal of G unless B is zero.

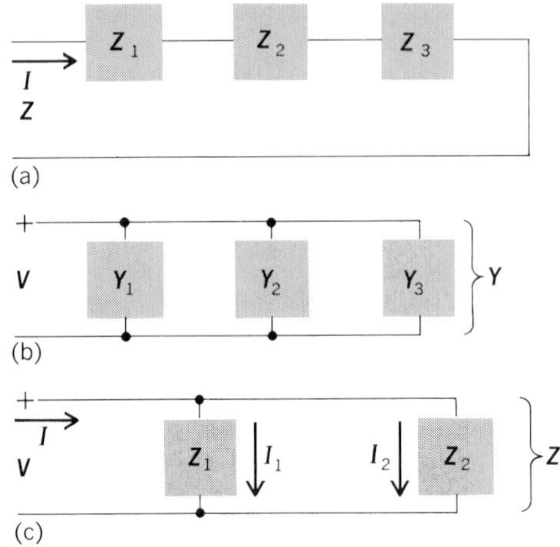

Fig. 8. Circuits of elements. (*a*) Elements in series. (*b*) Elements in parallel. (*c*) Two impedances in parallel.

The unit of resistance, reactance, or impedance is the ohm. The unit of conductance, susceptance, or admittance is called the mho.

Series and parallel connections. Elements are said to be in series if the same current passes through them (Fig. 8*a*). Elements are said to be in parallel if the same voltage is across them (Fig. 8*b*).

The total impedance of elements or networks in series is the sum of the impedances of the several elements or networks. Thus Eq. (47) expresses the total impedance shown in Fig. 8*a*.

$$\mathbf{Z} = \mathbf{Z}_1 + \mathbf{Z}_2 + \mathbf{Z}_3 \quad (47)$$

The total admittance of elements or networks in parallel is the sum of the admittances of the several elements or networks. Thus Eq. (48) expresses

$$\mathbf{Y} = \mathbf{Y}_1 + \mathbf{Y}_2 + \mathbf{Y}_3 \quad (48)$$

the total admittance shown in Fig. 8*b*.

The impedance of several elements or networks in parallel (Fig. 8*b*) is found as shown by Eq. (49).

$$\mathbf{Z} = \frac{1}{\mathbf{Y}} = \frac{1}{\mathbf{Y}_1 + \mathbf{Y}_2 + \mathbf{Y}_3} = \frac{1}{1/\mathbf{Z}_1 + 1/\mathbf{Z}_2 + 1/\mathbf{Z}_3} = \frac{\mathbf{Z}_1\mathbf{Z}_2\mathbf{Z}_3}{\mathbf{Z}_1\mathbf{Z}_2 + \mathbf{Z}_2\mathbf{Z}_3 + \mathbf{Z}_3\mathbf{Z}_1} \quad (49)$$

An important special case is that of two elements or networks in parallel (Fig. 8*c*). If the impedances are $\mathbf{Z}_1$ and $\mathbf{Z}_2$, the impedance of the two in parallel is shown by Eq. (50). Total current $\boldsymbol{I}$ divides (Fig.

$$\mathbf{Z} = \frac{\mathbf{Z}_1\mathbf{Z}_2}{\mathbf{Z}_1 + \mathbf{Z}_2} \quad (50)$$

8*c*) as indicated by Eqs. (51) and (52). Equations

$$\boldsymbol{I}_1 = \frac{\mathbf{V}}{\mathbf{Z}_1} = \boldsymbol{I}\frac{\mathbf{Z}}{\mathbf{Z}_1} = \frac{\mathbf{Z}_2}{\mathbf{Z}_1 + \mathbf{Z}_2}\boldsymbol{I} \quad (51)$$

$$\boldsymbol{I}_2 = \frac{\mathbf{Z}_1}{\mathbf{Z}_1 + \mathbf{Z}_2}\boldsymbol{I} \quad (52)$$

(50), (51), and (52) are frequently used in ac circuit theory.

Power. The power entering an ac load at its two terminals is computed from the rms current into the load, the rms voltage between the terminals, and the power factor. If current and voltage vary sinusoidally with time, the power factor is cos φ, where φ is the angle of phase difference between current and voltage. Thus the power, averaged through an integer number of cycles, is shown by Eq. (53).

$$P = VI \cos \varphi \quad (53)$$

Power is expressed usefully in several ways. For instance, the power to impedance $\mathbf{Z} = R + jX$ when the impedance carries current I is shown by Eq. (54). The final form of this equation is obtained by

$$P = (|\mathbf{Z}|I)I \cos \varphi = I^2|\mathbf{Z}| \cos \varphi = I^2R \quad (54)$$

noting that φ, the angle between voltage and current in Eq. (53), is also the angle of the impedance in Eq. (41), and therefore $|\mathbf{Z}| \cos \varphi = R$.

Another useful formula for power to a resistor, found from Eq. (54), is shown by Eq. (55), where V is the voltage across resistance alone.

$$P = I^2R = \left|\frac{V}{R}\right|^2 R = \frac{V^2}{R} \quad (55)$$

Devices and models. Frequently the best model of a practical device is obtained from parameters in series. Thus a coil of wire is commonly represented by resistance and inductance in series (Fig. 9*a*), although in reality both resistance and inductance reside in the whole coil. This model of a coil using lumped parameters (Fig. 9*a*) is quite satisfactory if, in the actual wire, current is essentially the same from end to end. The model fails if appreciable current is carried by distributed capacitance between one part of the wire and another, or between part of the wire and ground, as may well happen at radio frequency.

A battery is usually represented as a source of constant, unidirectional voltage in series with resitance. An ac generator is represented as a source of voltage that varies sinusoidally with time, connected in series with resistance and inductance (Fig. 9*b*). (This model can be made more exact by using different values of L under different circumstances.) A medium-length transmission line (up to 30 mi long at 60 hertz) can be represented satisfactorily by a model using lumped resistance, inductance, and capacitance (Fig. 9*c*).

Impedance of a coil. The impedance $\mathbf{Z}$ of the model of a coil (Fig. 9*a*) is shown by Eq. (56), where

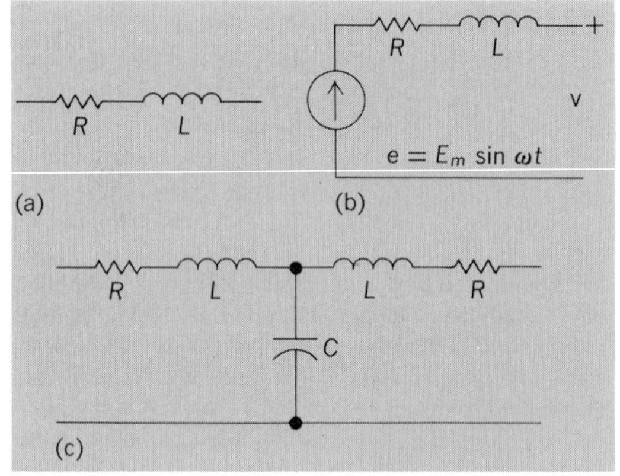

Fig. 9. Models of circuit elements. (*a*) Model of a coil. (*b*) Model of an ac generator. (*c*) Model of medium-length single-phase transmission line.

$$\mathbf{Z} = R + j\omega L \quad (56)$$

R is the resistance of the coil and ωL is its reactance. Resistance R is basically independent of frequency, although for accuracy it may be best to distinguish between the dc resistance and the slightly higher ac resistance, which takes also into account other losses, such as skin effect in the wire and core losses due to eddy currents or hysteresis if the coil has a ferromagnetic core. *See* CORE LOSS.

Inductance L is nearly independent of frequency, except as L may be slightly reduced by skin effect. As a first approximation, R and L are usually considered to be independent of frequency.

Impedance of a generator. Figure 9*b* shows a simple model of an ac generator. There is an independent source of sinusoidal electromotive force e in series with R and L, and the output voltage is v *See* GENERATOR.

By using transforms, the transform of the electromotive force is $\boldsymbol{E}$, that of the output voltage is $\mathbf{V}$, that of the current is $\boldsymbol{I}$, impedance is $\mathbf{Z} = R + j\omega L$, and the transform equivalent is shown by Eq. (57).

$$\mathbf{V} = \boldsymbol{E} - \mathbf{Z}\boldsymbol{I} \quad (57)$$

Electromotive force appears in a circuit where energy of some other nature is transformed into electrical energy. In the generator, mechanical energy is changed into electrical energy. In a battery, chemical energy is changed into electrical energy. Voltage results from the presence of an electromotive force. Both voltage and electromotive force are measured in volts.

Symbols. Instantaneous values of electromotive force and voltage are represented by e and v, respectively, as in Eq. (1). For current (originally called intensity), i is used. The maximum instantaneous value of an alternating quantity is often distinguished as V_{max} or V_m, as in Eq. (1). An italic letter without subscript indicates the rms value, as V or I; Eq. (24) illustrates the use. A phasor or transform is boldface italic, as $\mathbf{V}$ or $\boldsymbol{I}$ in Eq. (57). The nominal positive direction of every current, voltage, and electromotive force, either direct or alternating, must be indicated on circuit diagrams.

Instantaneous power is usually p, and power averaged through one or more complete cycles or periods is P. Time is t. Numbers represented by symbols in this article are the ratio of circumference to diameter $\pi \approx 3.1416$, and the base of natural logarithms $e \approx 2.718$. (The numerical value of e is not required in this article; its importance comes from Euler's law, Eq. (13). Also, the unit imaginary quantity $j1 = \sqrt{-1}$ is used in ac circuit theory.

The complex number representing impedance is also bold-face italic, as $\mathbf{Z}$ in Eq. (57). When only the real component is referred to, an italic capital letter is used.

Graphical symbols for resistance R, inductance L, and capacitance C are used in Figs. 7 and 9. Figure 10 shows forms of the common symbols.

It is convenient to follow the convention adopted by some authors of using circles and narrow rectangles to indicate independent sources. A circle represents an independent voltage source, meaning that the source has an electromotive force within it, and hence a voltage between its terminals, that has always a specified value. In particular, the voltage across the source is independent of current through the source. The value of the independent electromotive force or voltage must be given. For example, the model of a storage battery may contain an independent source of 12 volts in series with some amount of resistance; a model of an ac generator (Fig. 9*b*) may contain an independent source of $e = 162 \sin \omega t$ volts in series with appropriate amounts of R and L.

An independent current source, represented by a narrow rectangle, has within it such an electromotive force that the current through the source always has the value specified for that current source. Thus an independent ac current source may have a current that is always $5 \sin \omega t$ amperes.

The specified value of an independent source is usually indicated on a diagram beside the symbol for the source.

It must be understood that an independent voltage source is not without current. Its voltage is specified and its current depends on the circuit in which the source is connected. Similarly, an independent current source is not without voltage. Its current is given and its voltage depends on the circuit to which the source is connected.

Ideal sources cannot exist alone in actuality, but they are useful in linear network models. An independent voltage source is a physical impossibility; if short-circuited, it would be required to develop an unlimited (infinite) current. Similarly, an independent current source would have to supply unlimited (infinite) voltage if its terminals were open-circuited.

A type of source used in models of electronic circuits (but not needed in this article) is a dependent or controlled source, the voltage or current of which depends on some other voltage or current elsewhere in the system. For example, the voltage in the collector circuit of a transistor model might be proportional to the emitter current.

Units. For electrical work the meter-kilogram-second (mks) system of mechanical units is extended by adopting a fundamental electrical unit. This unit is commonly either the coulomb or the ampere. The resulting mksc or mksa system is compatible (not requiring dimensional constants) and includes volts of potential or electromotive force, ohms of impedance, mhos of admittance, henrys of inductance, and farads of capacitance. The compatible unit of angle is the radian, and that of frequency (symbol ω) is the radian per second. (Frequency measured in cycles per second, or hertz, is incompatible; the symbol is f and $\omega = 2\pi f$.)

Prefixes. It is common practice to use prefixes to

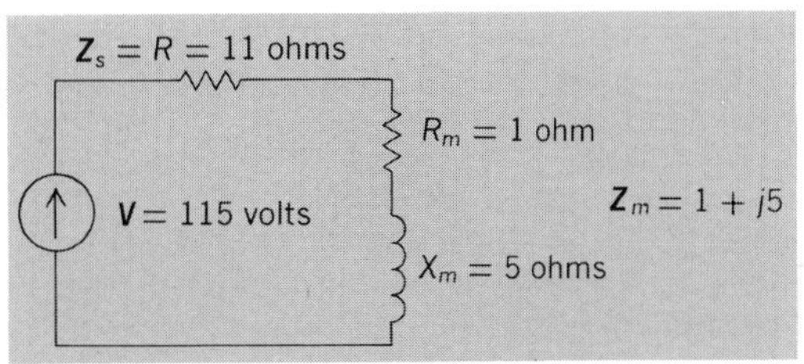

Fig. 11. Impedances in series: $\mathbf{Z}_s$ is starter resistance; R_m and X_m are model of motor with stalled rotor.

ALTERNATING-CURRENT CIRCUIT THEORY

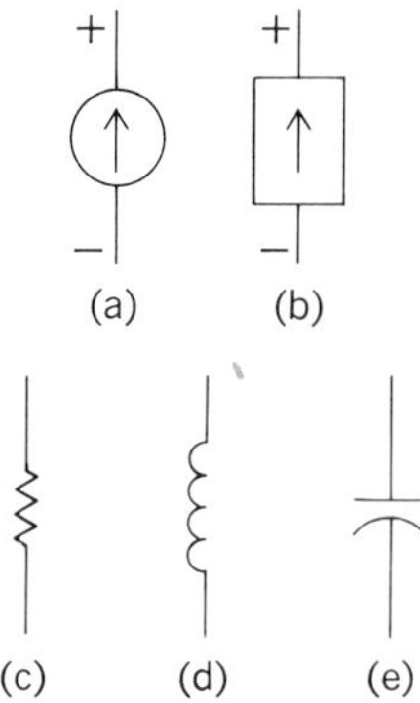

Fig. 10. Commonly used graphic circuit symbols. (*a*) Independent voltage source. (*b*) Independent current source. (*c*) Resistance. (*d*) Inductance. (*e*) Capacitance.

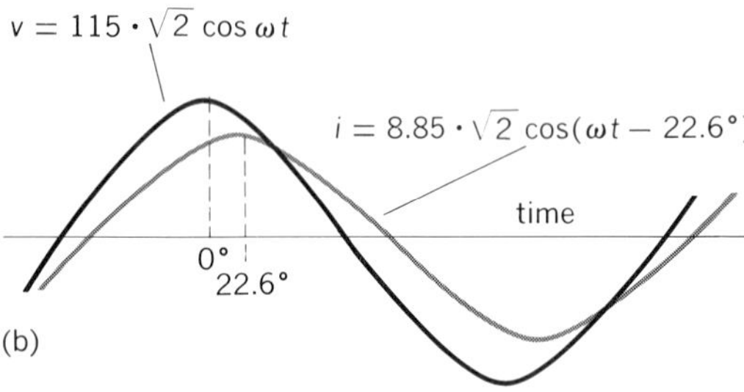

Fig. 12. For the circuit of Fig. 11: (*a*) phasors and (*b*) corresponding waves.

indicate powers of 10. Thus 1 kilowatt means 10^3 watts or 1000 watts. Similarly 1 microampere means 10^{-6} ampere or one-millionth of an ampere. The prefixes are:

tera-	$= 10^{12}$	milli-	$= 10^{-3}$
giga-	$= 10^{9}$	micro-	$= 10^{-6}$
mega-	$= 10^{6}$	nano-	$= 10^{-9}$
kilo-	$= 10^{3}$	pico-	$= 10^{-12}$

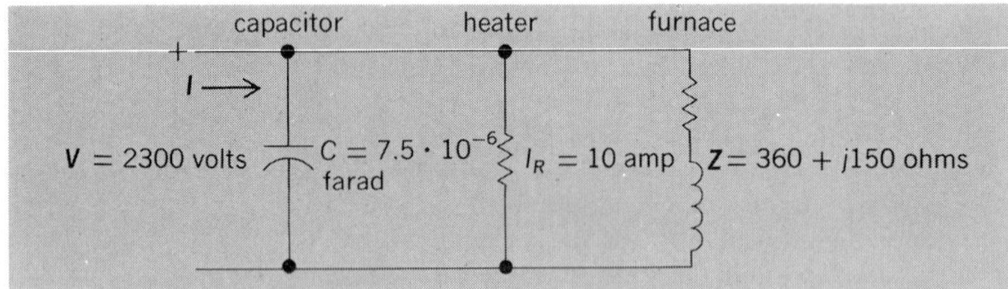

Fig. 13. Capacitor, heater, and furnace connected in parallel.

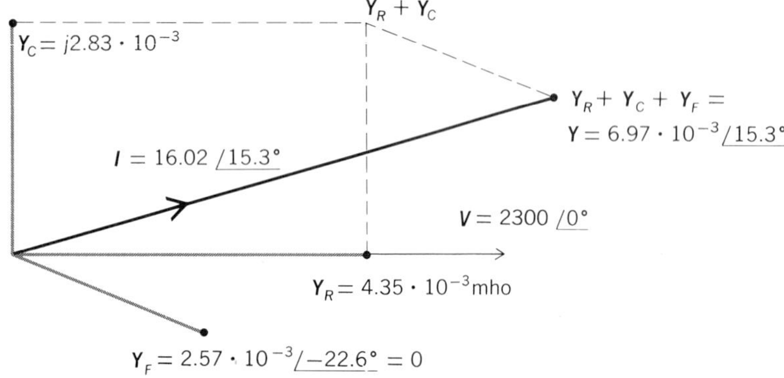

Fig. 14. Phasors for circuit of Fig. 13.

Examples. The following examples illustrate alternating-current circuit theory. The first example concerns elements in series and division of voltage; the second example concerns elements in series and parallel, division of current, and computation of power.

Example 1. An incoming electric line provides power at 115 volts, 60 Hz. This is the amount of ac voltage as read on an ordinary voltmeter, and is therefore the rms value. Both voltage and frequency are essentially independent of the amount of current received from this line by the load that is about to be considered. The voltage, known to be alternating, is assumed to be sinusoidal.

Onto this line is connected a motor with 60-Hz impedance (at standstill) of $\mathbf{Z}_m = 1 + j5$ ohms in series, with a starting resistance of $\mathbf{Z}_s = 11$ ohms. The motor does not turn, being too heavily loaded. Compute the current that will flow steadily. Also, find voltage across the motor alone with this starting resistance in the circuit.

Solution. Draw Fig. 11 and start Fig. 12. Take voltage to be the reference phasor with zero angle, $\mathbf{V} = 115/\underline{0°}$. Obtain the total impedance $\mathbf{Z}$ by adding, as shown in Eq. (58*a*). The complex number $12 + j5$ can also be written in polar form, as shown by Eq. (58*b*). Using this polar form and dividing, it is found, as shown in Eq. (58*c*), that the steady

$$\mathbf{Z} = \mathbf{Z}_s + \mathbf{Z}_m = 11 + 1 + j5 = 12 + j5 \qquad (58a)$$

$$\mathbf{Z} = 12 + j5 = 13/\underline{22.6°} \qquad (58b)$$

$$\mathbf{I} = \frac{\mathbf{V}}{\mathbf{Z}} = \frac{115/\underline{0°}}{13/\underline{22.6°}} = 8.85/\underline{-22.6°} \qquad (58c)$$

current has an rms value of 8.85 amp, and that the sinusoidal wave of current lags (is delayed behind) the sinusoidal wave of voltage by 22.6°. Voltage and current phasors are shown in Fig. 12*a*, and the corresponding waves are shown in Fig. 12*b*. Usually, phasors are shown but not waves (time functions of voltage and current); this is consistent with the usual practice of calling the phasor a current and leaving the reader to understand that the magnitude is the rms value of current and the angle is phase relative to an assumed reference.

In the present example, one is also required to find voltage across the motor with the starting resistance in the circuit. This voltage $\mathbf{V}_m$ is shown by Eq. (59). The phasor representing motor voltage is included in Fig. 12*a*.

$$\begin{aligned}\mathbf{V}_m &= \mathbf{Z}_m\mathbf{I}\\ &= (1 + j5)(8.85/\underline{-22.6°})\\ &= (5.10/\underline{78.7°})(8.85/\underline{-22.6°}) = 45.1/\underline{56.1°}\end{aligned} \qquad (59)$$

Check. As a check on the accuracy of the solution, if one computes voltage across the starting resistor and then adds this to the motor voltage, the resultant sum should be the line voltage. This computation is shown in Eqs. (60). The sum of the

$$\begin{aligned}\mathbf{V}_s &= \mathbf{Z}_s\mathbf{I} = 11(8.85/\underline{-22.6°}) = 97.4/\underline{-22.6°}\\ &= 89.9 - j37.4\\ \mathbf{V}_m &= 25.1 + j37.4\\ \mathbf{V} &= 115.0 + j0\end{aligned} \qquad (60)$$

voltage phasors is $115.0 + j0$, which is the correct line voltage. Graphical addition is illustrated in Fig. 12*a*.

Example 2. An electric furnace of the induction type has impedance $\mathbf{Z} = 360 + j150$ ohms at its rated frequency of 60 Hz. It is connected to an incoming line that provides 2300 volts at this frequency. A purely resistive electric heater is also connected to the line (Fig. 13), and draws 10 amp at this voltage. To improve the power factor, management has connected to the incoming line a capacitor with 7.5 microfarads (μf) capacitance. Find the current supplied from the incoming line; also find total power supplied and power to each load.

Solution. Draw Fig. 13 and start Fig. 14. Take the voltage of the incoming line as the reference phasor, $2300\underline{/0°}$. Compute the admittance of each element of the network and add. Computations are shown in Eqs. (61).

For the capacitor,

$$\begin{aligned} Y_C &= j\omega C = j2\pi(60)(7.5)10^{-6} \\ &= j(2.83)10^{-3} \text{ mho} \end{aligned} \quad (61a)$$

For the resistor,

$$Y_R = \frac{10}{2300} = (4.35)10^{-3} \text{ mho} \quad (61b)$$

For the furnace,

$$\begin{aligned} Y_F &= \frac{1}{360+j150} = \frac{1}{390\underline{/22.6°}} \\ &= (2.57)10^{-3}\underline{/-22.6°} \end{aligned} \quad (61c)$$

$$\text{or } Y_F = \frac{360-j150}{(360)^2+(150)^2} = (2.37-j0.986)10^{-3} \text{ mho}$$

Total line current is then

$$\begin{aligned} I &= VY = 2300(j2.83+4.35+2.37-j0.986)10^{-3} \\ &= 2.30(6.72+j1.84) = 16.02\underline{/15.3°} \end{aligned} \quad (61d)$$

Power computations shown in Eqs. (62)–(64) are derived from earlier equations as indicated. Power from line by Eq. (53) is

$$\begin{aligned} P &= VI\cos\varphi = (2300)(16.02)\cos 15.3° \\ &= 35{,}520 \text{ watts} \end{aligned} \quad (62)$$

Power to resistor by Eq. (53) is

$$P_R = (2300)(10) = 23{,}000 \text{ watts} \quad (63)$$

Power to impedance of furnace by Eq. (54) is

$$P_F = I^2R = \left(\frac{2300}{390}\right)^2 360 = 12{,}520 \text{ watts} \quad (64)$$

As a check, no power is supplied to the capacitor, and the sum of power to the resistor, 23.0 kw, and power to the furnace, 12.52 kw, equals the incoming line power, 35.52 kw, which is correct. *See* RESONANCE (ALTERNATING-CURRENT CIRCUITS). [H. H. SKILLING]

Bibliography: H. H. Skilling, *Electric Networks*, 1974; H. H. Skilling, *Electro-Mechanics*, 1962; H. H. Skilling, *Fundamentals of Electric Waves*, 2d ed., 1947.

Alternating-current generator

A machine which converts mechanical power into alternating-current electric power. The most common type, sometimes called an alternator, is the synchronous generator, so named because its operating speed is proportional to system frequency. Another type is the induction generator, the speed of which varies somewhat with load for constant output frequency.

Synchronous generators. These generators usually have the field winding mounted on the rotor and a stationary armature winding mounted on the stator. In small ratings where high reliability is imperative, the magnetic field may be created by permanent magnets. Small alternators are also being built with stationary field windings and rotating armatures with their leads brought out through collector rings or fed directly to rotating rectifiers, which supply direct current to the rotating field of a much larger alternator, as in the brushless excitation system. Still another type, the inductor alternator described below, has both its field and armature windings in the stator. Although synchronous generators may be single-phase, they are usually two- or three-phase; most are three-phase. Single-phase generators are rare because they are larger than polyphase machines of the same kilovolt-ampere ratings, because they have a pulsating torque and are noisy, and because single-phase power is not well suited to self-starting ac motors other than those of fractional horsepower sizes. For theory of synchronous machines *see* SYNCHRONOUS MOTOR.

In the usual type with rotating field windings, a pair of field poles must pass a given point on the armature in 1 cycle. Hence the number of poles required are determined from the frequency f in hertz (Hz) and speed by Eq. (1). The speed se-

$$\text{Number of poles} = 120f/\text{rpm} \quad (1)$$

lected is that best suited to the prime mover. Steam turbines operate most economically at high speed, hence two- or four-pole generators are used for this service, running at 3600 or 1800 rpm, respectively, for an output frequency of 60 Hz. For hydraulic turbine or engine drive, slow-speed machines having many poles are customary.

High-speed synchronous generators. Generators of this type have cylindrical rotors of solid alloy-steel forgings, with radial slots machined along their length to receive the field windings, as shown in Fig. 1. The field coils are of bare, hard-drawn, strip copper installed turn by turn, within an insulating channel in each slot. Mica plate, epoxy-bonded woven-glass laminate, or similar insulation is commonly used between turns.

The slot portion of the windings is supported against centrifugal force by nonmagnetic wedges, while the coil ends are retained by nonmagnetic metal rings lined with insulation. Since most high-speed generators over 15,000 kVA are now either hydrogen-cooled or water-cooled, the insulated field leads are brought out to the collector rings through an axial hole in the shaft, with the aid of gastight radial studs. Fans or blowers are mounted on the rotor in most cases, employing hydrogen as a cooling medium.

The stator of a large, high-speed, synchronous generator uses a steel yoke, which also serves as an enclosure for the ventilating medium. A cylindrical core of laminated electrical sheet steel is stacked on dovetail bars within the yoke and tightly clamped to minimize magnetic noise. The insulated armature coils of large machines are usually of one turn, lap wound in rectangular slots around the inner periphery of the core. Coil ends are securely lashed to supporting brackets, and both ends of each phase are brought out through terminal bushings. Hydrogen seals are mounted in the end

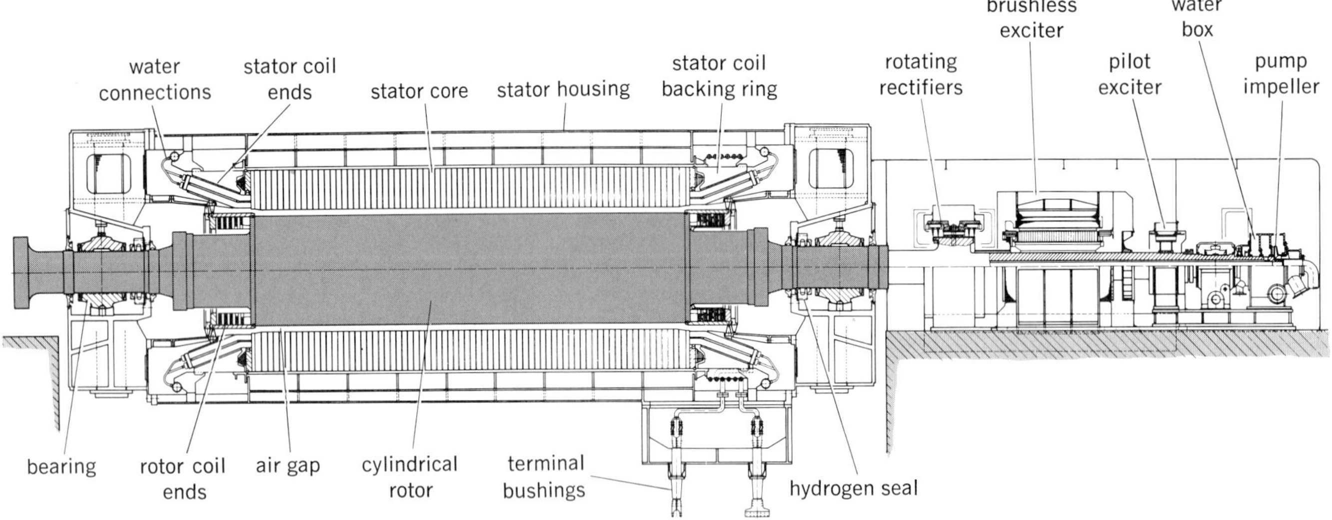

Fig. 1. An 1800-MVA four-pole (1800 rpm) synchronous generator; water-cooled stator and rotor windings. (*Allis-Chalmers Power Systems*)

covers, and the entire stator is designed to resist explosion pressures. *See* ARMATURE; WINDINGS IN ELECTRIC MACHINERY.

Cooling methods. Air-cooled and hydrogen-cooled generators under 100 MVA are most commonly indirectly cooled, that is, the cooling medium contacts metal surfaces and exterior surfaces of the coil insulation. Heat generated in the windings must flow through the major insulation, which is a poor heat conductor. In 1948 S. Beckwith designed a 60-MW 3600-rpm rotor in which hydrogen was forced by a powerful blower to flow at high velocity through ducts within the conductors. A tremendous increase in output was made possible by this highly effective cooling method, which is now called direct cooling. When applied also to the stator windings and augmented with higher hydrogen pressures, much greater gains in generator output were achieved.

In parallel with this development, direct water cooling of stator windings was also adopted in the 1950s, and later applied to rotors as well. Figure 1 illustrates an 1800-MVA four-pole generator with water-cooled stator and rotor windings, together with its brushless excitation equipment.

Slow-speed synchronous generators. The field poles of these generators are usually the salient-pole type. When driven by reciprocating engines, they sometimes require flywheels to minimize the pulsations transmitted to the power system. In hydroelectric generators the rotors are sometimes required to withstand high overspeed because of the time delay in closing the gates.

Generated voltage. The voltage generated per phase in a synchronous generator can be derived from Faraday's law. If it is assumed the flux distribution over each pole is sinusoidal, a sinusoidal voltage is induced in each coil side. However, the coil voltages must be added vectorially because of their time-phase displacement. The effective root-mean-square voltage per phase can be shown to be that in Eq. (2), where Φ_m is peak flux per

$$E = 2.22(a/c)\Phi_m f k_d k_p \qquad (2)$$

pole in webers, a is total conductors per phase, f is frequency in Hz, c is number of parallel circuits, k_d is distribution factor, and k_p is pitch factor.

If the flux distribution is not a perfect sine wave, any irregularites appear as odd harmonics in the generated voltage. Although seldom harmful to the generator, harmonics sometimes interfere with telephone communication. Perhaps the most troublesome harmonics result from pulsation in airgap reluctance as the poles move across the slots and teeth. These are known as slot harmonics. They occur in pairs at frequencies equal to (the slots per pair of poles $\pm$ 1) times the fundamental. Slot harmonics can be minimized by careful design of pole contour, fractional-slot windings, and skewed stator slots. Interference effects are often reduced with external resonant shunts or wave traps.

Characteristics. The characteristic curves of a synchronous generator are shown in Fig. 2. The open-circuit and short-circuit curves are readily found from tests at no-load. The saturation curve at rated amperes and 0% power factor is obtained by electrical connection to another synchronous machine. The field currents of both machines are adjusted at each successive test voltage point to circulate rated amperes between them, overexcited on the unit being tested and underexcited on the other machine. The curves show how field current varies with load and voltage, and indicate some of the constants. The Potier reactance drop, used in determining saturation, is the height MH of the Potier triangle OHD.

The triangle is found from the 0% power factor saturation curve DE as follows: (*a*) Lay off length OD to left of point E to find L. (*b*) Draw line LG parallel to the airgap line to find G, its intersection with the open circuit saturation curve. (*c*) Complete the triangle by drawing line GE. (*d*) Make OH equal to LG, thus establishing the Potier reactance drop MH.

An approximate value of the Potier reactance may be calculated from Eq. (3), where x_l = leakage reactance and x'_d = transient reactance.

$$x_p = x_l + 0.63(x'_d - x_l) \qquad (3)$$

Unsaturated per unit synchronous reactance is

OD/JK, and the short circuit ratio, whose reciprocal is often used in steady-state stability studies, is *JA/OD*. The voltage regulation is given by *FN/OJ* for 85% power-factor lagging load.

Inductor alternator. A synchronous generator in which the field winding is fixed in magnetic position relative to the armature conductors is known as an inductor alternator. There are two types: the homopolar, in which the dc field coil is concentric with the shaft, and the heteropolar, in which the dc windings are distributed. In both types the ac windings are distributed, and generate their induced voltage from the pulsation in the flux caused by the change in position of the salient poles on the rotor. Inductor alternators are used for high-frequency power and, in conjunction with static rectifiers, as a maintenance-free power source for ac excitation systems.

Induction generators. These nonsynchronous ac generators are driven above synchronous speed by external sources of mechanical power. The construction of these machines is identical to that of induction motors. They can operate either as motors or generators, depending on whether the speed is below or above synchronous speed. In some frequency-changer sets, induction generators may operate as a motor part of the time. This is accomplished by coupling them to a two-speed synchronous machine capable of operating at system frequency as a motor at the higher speed and as a generator at the lower speed. Induction generators are not common in large sizes because of their poor power factor. They can deliver only leading currents. Moreover, they require a power supply from which to obtain their magnetizing current. This normally requires that they be operated in parallel with a synchronous source. Capacitors may be used to minimize the current taken from the source. Under such conditions the frequency of the induction generator is determined by the frequency of the synchronous source and the output of the generator is determined by the mechanical input to its shaft. Induction generators do not require any dc excitation and their control can be very simple. For this reason they are well suited to small, unattended hydroelectric units, where they are easily operated by remote control. See INDUCTION MOTOR.

[LEON T. ROSENBERG]

Bibliography: D. G. Fink and H. W. Beaty (eds.), *Standard Handbook for Electrical Engineers*, 11th ed., 1978; A. E. Fitzgerald et al., *Electric Machinery*, 3d ed., 1971; S. B. Hammond and D. W. Gehmlich, *Electrical Engineering*, 2d ed., 1971; I. L. Kosow, *Electric Machinery and Transformers*, 1972; E. Levi and M. Panzer, *Electromechanical Power Conversion*, 1974; J. Rosenblatt and M. H. Friedman, *Direct and Alternating Current Machinery*, 1963.

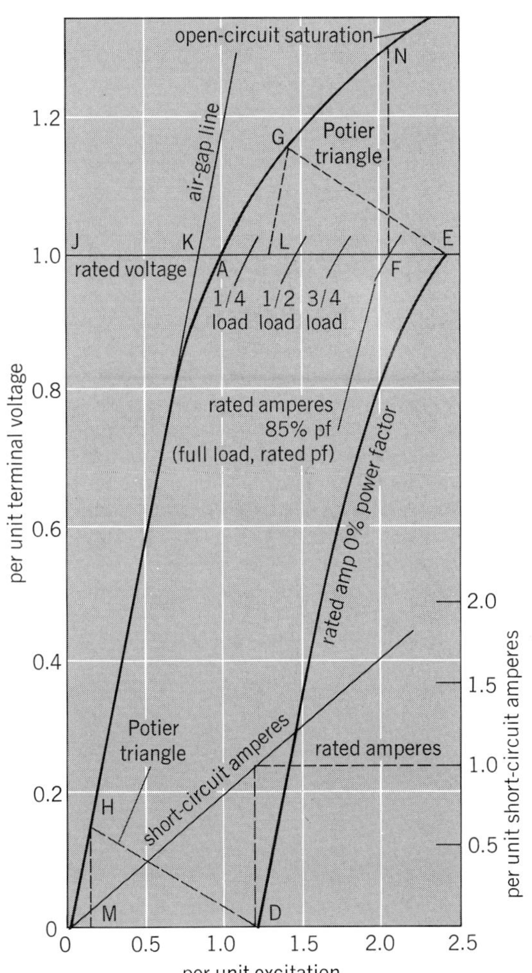

Fig. 2. Characteristic curves of a synchronous generator.

Alternating-current motor

An electric rotating machine which converts alternating-current (ac) electric energy to mechanical energy; one of two general classifications of electric motor. Because ac power is widely available, ac motors are commonly used. They are made in sizes from a few watts to thousands of horsepower (hp) (Fig. 1). *See* DIRECT-CURRENT MOTOR.

CLASSIFICATIONS OF AC MOTORS

Each type of ac motor has special properties. These motors are generally classified by application, construction, principle of operation, or operating characteristics.

Induction motor. The most common type of ac motor is the induction motor. Current is induced in a rotor as its conductors cut lines of magnetic flux created by currents in a stator. Three-phase induction motors are simple, reliable motors with a fairly constant speed over the rated load range. They are self-starting and are widely used in industry. Single-phase induction motors require special means for starting, but are widely used in fractional and small integral horsepower sizes, especially in homes. *See* INDUCTION MOTOR.

Synchronous motor. Where constant speed is essential, a synchronous motor is used. It runs at a fixed speed in synchronism with the frequency of the power supply. Large synchronous motors used in industry employ dc fields on their rotors and three-phase armature (or stator) windings. Efficiency and power factor of these motors are high. The reluctance motor and the Permasyn motor, either single- or three-phase, come in fractional and lower integral hp sizes. The reluctance motor has low efficiency and power factor, but is simple and inexpensive. The Permasyn has permanent magnets embedded in the squirrel-cage rotor

ALTERNATING-CURRENT MOTOR

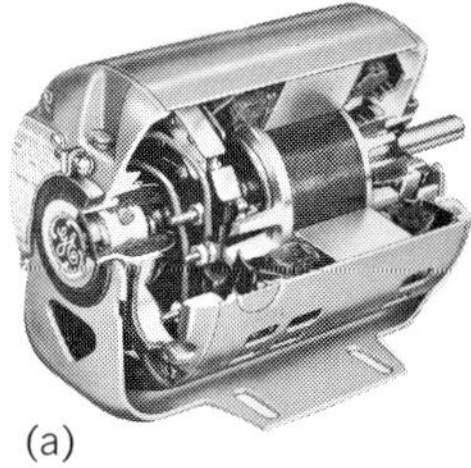

(a)

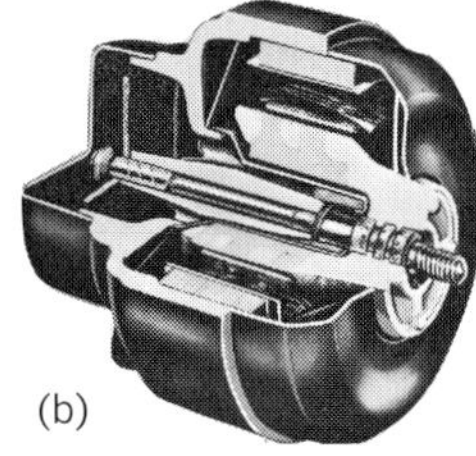

(b)

Fig. 1. Typical ac motors. (*a*) One-half-hp split-phase induction motor, about 6 in. in diameter. (*b*) One-hundredth-hp motor, about 1 in. in diameter.

to provide the equivalent of a dc field. The hysteresis motor is used only in small sizes where its quiet operation is especially desired, such as in phonographs. *See* SYNCHRONOUS MOTOR.

AC series motor. For operation from either ac or dc power, the series motor has its field winding connected in series with the armature winding through a commutator and brush arrangement, as in a dc series motor. Field and armature iron are laminated. This universal motor has high starting torque. Speed can be controlled by adjusting the applied voltage. They are used in sizes up to 2000 hp in electric railways, and in small sizes for domestic appliances. *See* UNIVERSAL MOTOR.

Repulsion motor. This is also a commutator motor, but the brushes and commutator are short-circuited and not connected in series with the stator. Armature current is set up by induction from the stator rather than by conduction, as in the series motor. The rotor of a repulsion motor differs from the squirrel-cage rotor of the single-phase induction motor; it is similar to a dc armature with commutator and brushes. Torque is developed by action of induced armature current on stator flux. The repulsion motor has high starting torque. Its speed can be controlled by changing the applied voltage or by shifting the brushes or both. Its no-load speed is above synchronism, but is lower than the no-load speeds of ac series and universal motors. In normal operation the brushes are located 15–25 electrical degrees off the stator position. Rotation is reversed by shifting the brushes to the opposite side of the stator axis. Although repulsion motors have been used on single-phase electric railways, by far the widest application in the past has been as a starting arrangement for single-phase induction motors. When used for this purpose, a centrifugal device short-circuits the commutator bars at about 70% of full speed. Combinations of series and repulsion connections have been used for controlling speed, commutation, and character-

Standard ranges of hp, speeds, and slips at 60 Hz*

No. of phases	Type of motor	Power output, hp	Synchronous speed, rpm	Slip, %
Single	Capacitor	1–10	1800–1200	0–5
	Split-phase	0–0.5	1800–1200	0–5
	Shaded-pole	0–0.25	1800–900	11–14
	Repulsion-start	0–25	1800–1200	0–5
Three	Induction	0–100,000	3600–450	0–5
Three	Synchronous	20–30,000	3600–80	0

*National Electrical Manufacturers Association.

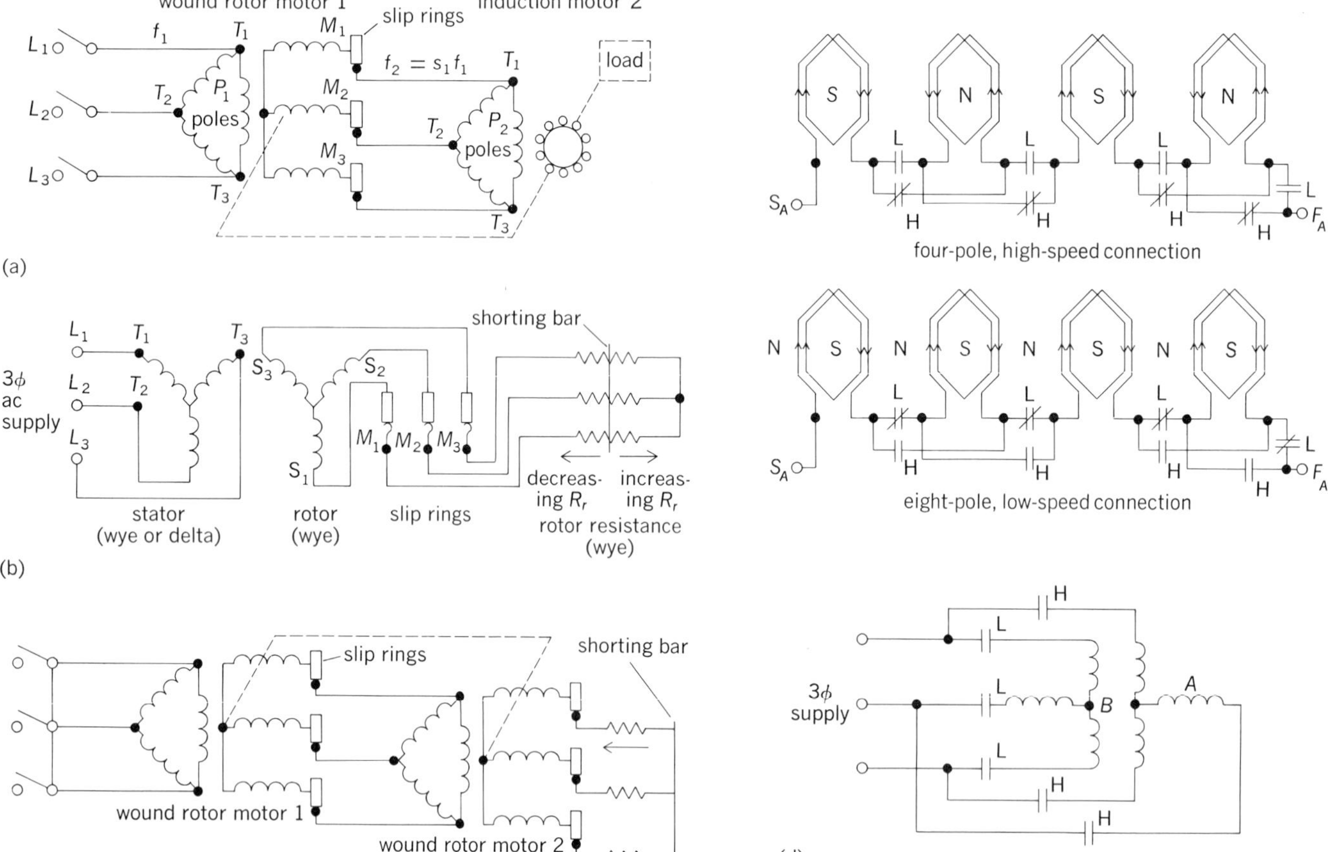

Fig. 2. Schematics of methods of speed control for polyphase ac motors. (*a*) Frequency control using cascade-wound rotor and squirrel-cage motors. (*b*) Slip control using wound-rotor induction motor external rotor resistance. (*c*) Foreign voltage slip control using secondary concatenation. (*d*) Phase *A* of a three-phase winding showing consequent pole connections for multispeed induction motors.

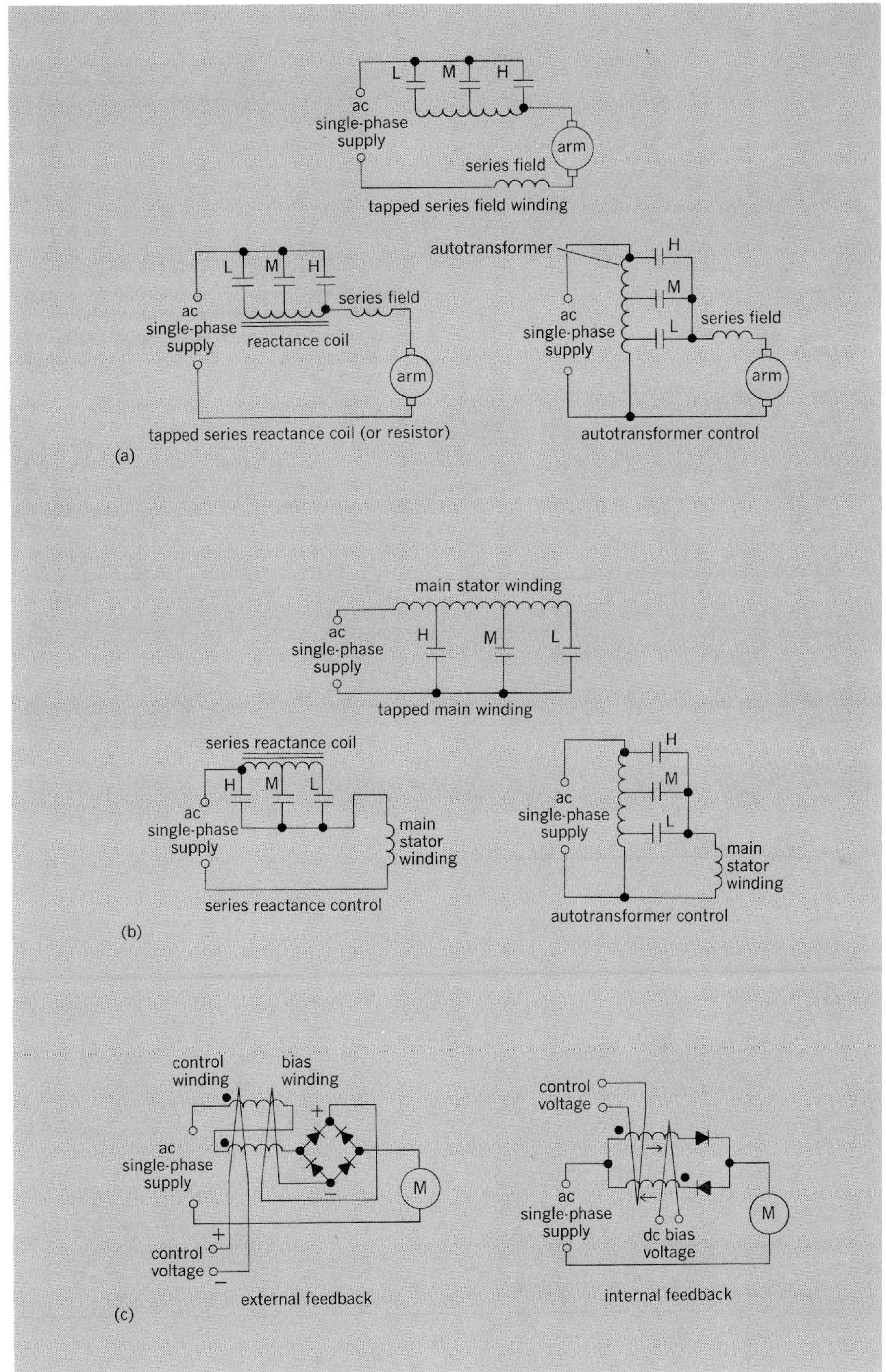

Fig. 3. Schematics for various methods of speed control for single-phase ac motors operating from fixed supplies. (*a*) Controls for ac series motors or universal motors. (*b*) Slip control for shaded-pole, reluctance, and split-phase motors. (*c*) Magnetic amplifier for voltage control of single-phase motors.

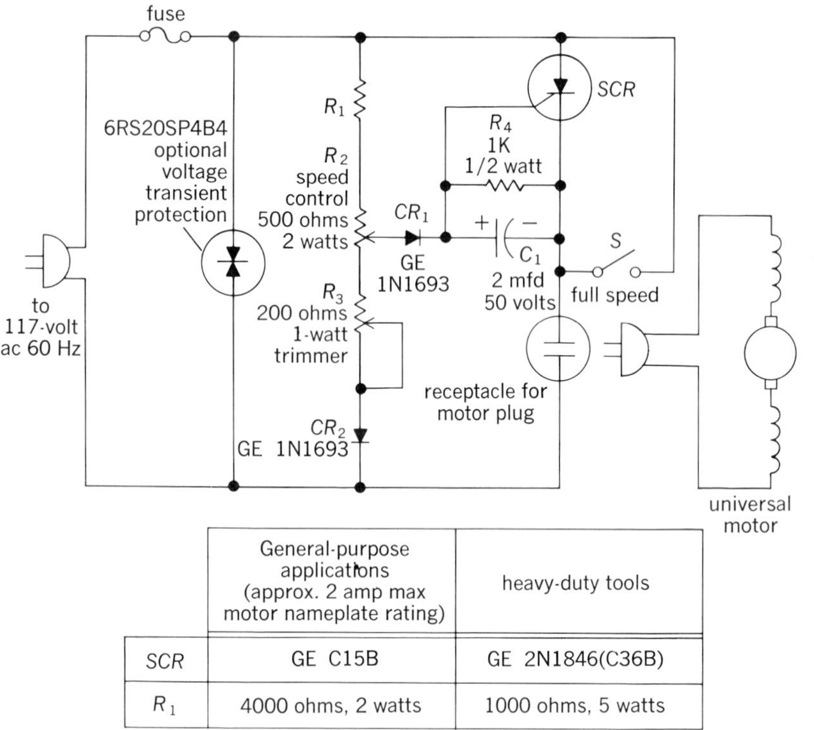

	General-purpose applications (approx. 2 amp max motor nameplate rating)	heavy-duty tools
SCR	GE C15B	GE 2N1846(C36B)
R_1	4000 ohms, 2 watts	1000 ohms, 5 watts

Fig. 4. Plug-in speed control with feedback using a silicon-controlled rectifier (SCR) for voltage control of single-phase motors.

istics; occasionally a squirrel-cage winding is added on the armature. For starting duty, the repulsion motor draws low starting current, but despite this advantage the repulsion motor has been superseded by the simpler, cheaper capacitor motor. *See* REPULSION MOTOR.

The table gives a comparison of available sizes and characteristics of some common ac motors.

For general principles *see* ELECTRIC ROTATING MACHINERY; MOTOR; WINDINGS IN ELECTRIC MACHINERY. [ALBERT F. PUCHSTEIN]

SPEED CONTROL OF AC MOTORS

The polyphase synchronous motor is a constant-speed (synchronous), variable-torque, doubly excited machine. The stator armature is excited with polyphase ac of a given frequency, and the rotor field is excited with dc. The rotor speed of the synchronous motor is a direct function of the number of stator and rotor field poles and the frequency of the ac applied to the stator. Since the number of rotor poles of the polyphase synchronous motor is not easily modified, the change of frequency method is the only way to control synchronous speed of the motor.

The polyphase induction motor is also a doubly excited machine, whose stator armature is excited with polyphase ac of line frequency and whose rotor is excited by induced ac of variable frequency, depending on rotor slip. The speed of the poly-

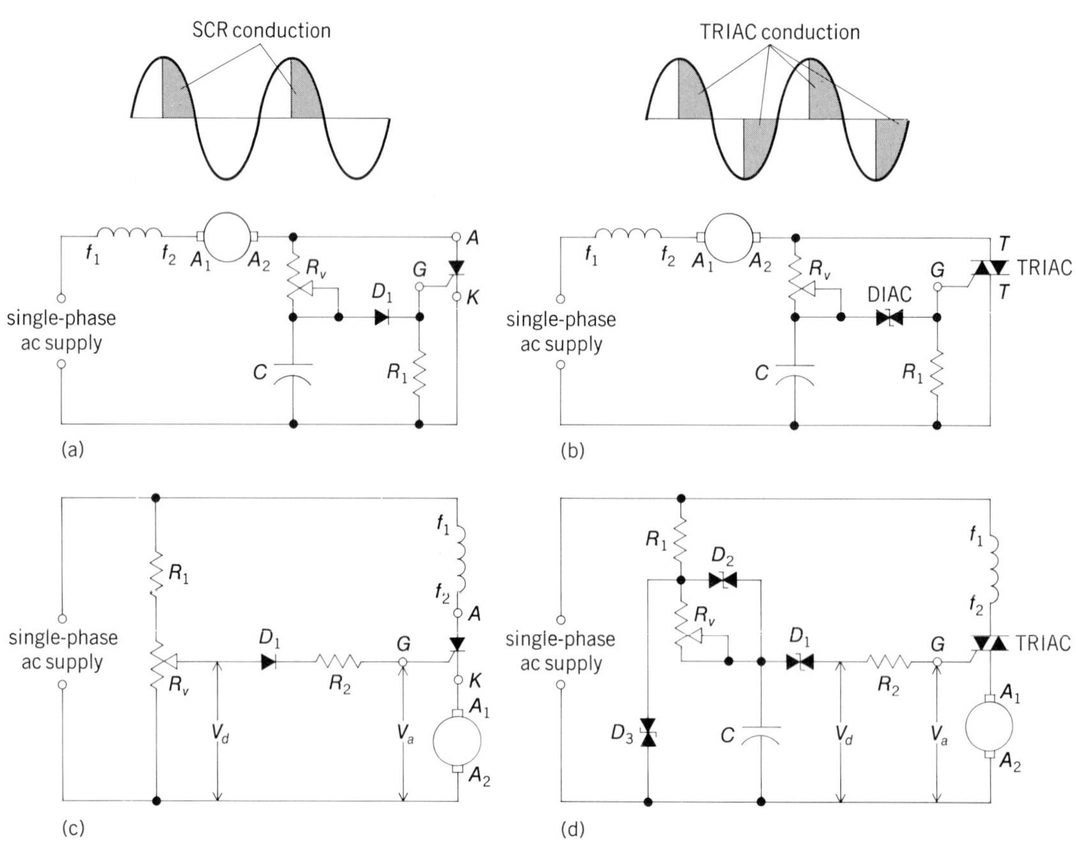

Fig. 5. Single-phase, universal or dc motor control using half- and full-wave (ac) controls, with and without voltage feedback. (*a*) Half-wave circuit. (*b*) Full-wave circuit. (*c*) Half-wave circuit with voltage feedback. (*d*) Full-wave circuit with voltage feedback. (*From I. L. Kosow, Control of Electric Machines, 1973*)

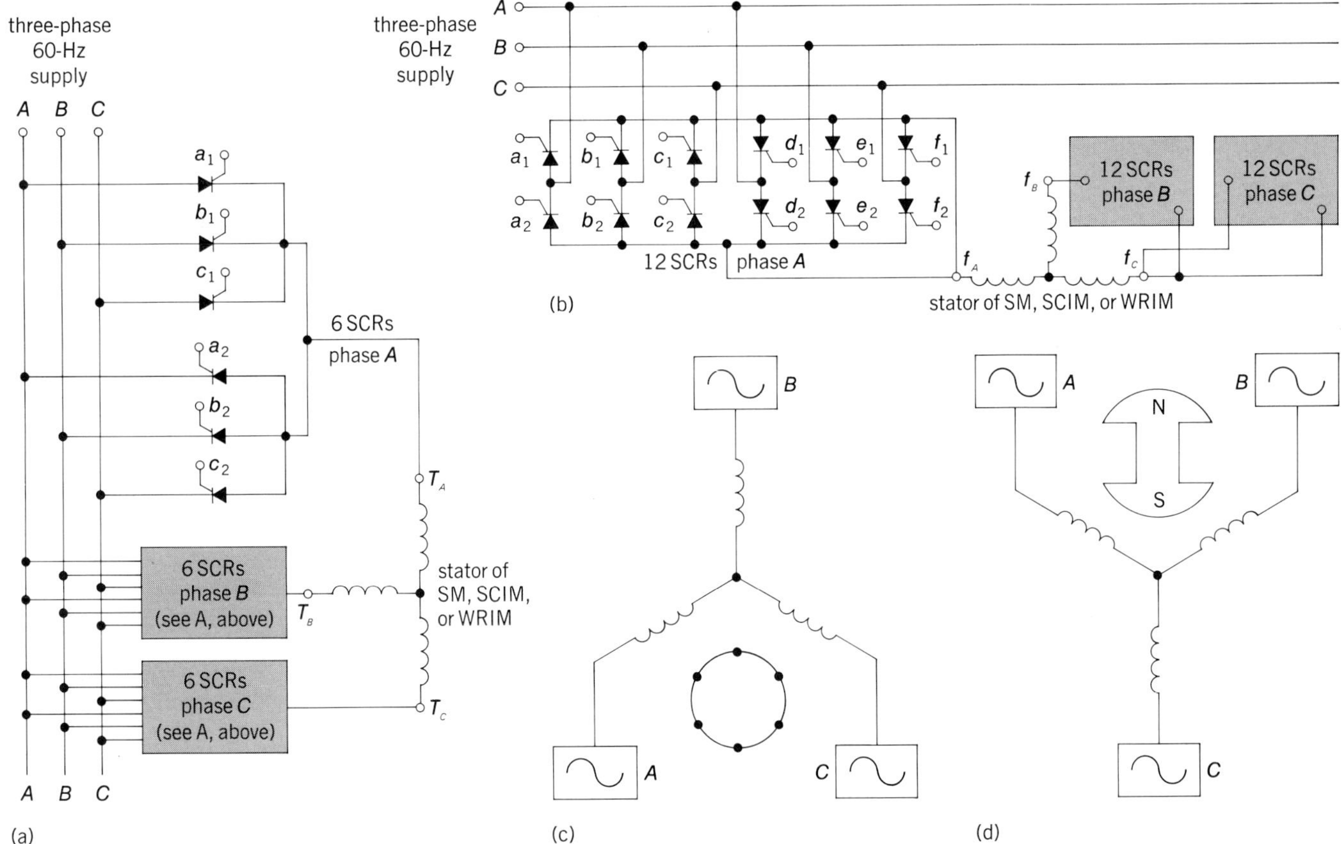

Fig. 6. Half-wave and full-wave cycloconverters with graphical symbols for simplified representation. (*a*) Half-wave cycloconverter (exclusive of triggering of SCR gates). (*b*) Full-wave cycloconverter (exclusive of SCR gating). (*c*) Symbol for SCIM driven by a cycloconverter. (*d*) Symbol for SM driven by a cycloconverter. (*From I. L. Kosow, Control of Electric Machines, 1973*)

phase induction motor is an asynchronous speed varied by one or more of the following methods: (1) by changing the applied frequency to the stator (same method as for a synchronous motor, already discussed); (2) by controlling the rotor slip by means of rheostatic rotor resistance control (used for wound-rotor induction motors); (3) by changing the number of poles of both stator and rotor; and (4) by "foreign voltage" control, obtained by conductively or inductively introducing applied voltages of the proper frequency to the rotor. Figure 2 illustrates with schematics these methods of speed control.

It should be noted that a number of electromechanical or purely mechanical methods also serve to provide speed control of ac motors. One is the Rossman drive, in which an induction motor stator is mounted on trunnion bearings and driven with an auxiliary motor, providing the desired change in slip between stator and rotor. Polyphase induction and synchronous motors having essentially constant speed characteristics at rated voltage are also assembled in packaged drive units employing gears, cylindrical and conical pulleys, and even hydraulic pumps to produce a variable speed output. In addition to reversal of rotation, some of these units employ magnetic slip clutches and solenoids to control the various mechanical and hydraulic arrangements through which a relatively smooth control of speed is achieved. A discussion of such electromechanical and mechanical speed-control methods is beyond the scope of this article.

The principal method of speed control used for fractional hp single-phase induction-type, shaded-pole, reluctance, series, and universal motors is the method of primary line-voltage control. It involves a reduction in line voltage applied to the stator winding (of the induction type) or to the armature of series and universal motors. In the former this produces a reduction of torque and increase in rotor slip. In the latter it is simply a means of controlling speed by armature voltage control or field flux control or both. The reduction in line voltage is usually accomplished by any one of five methods shown in Figs. 3 and 4; these are autotransformer control, series reactance control, tapped main-winding control, saturable reactor (or magnetic amplifier) control, and silicon-controlled rectifier feedback control.

Electronic speed control techniques. The development of the thyristor, or silicon-controlled rectifier (SCR), has created unlimited possibilities for control of virtually all types of motors (single-phase, dc, and polyphase). SCRs in sizes up to 1600 amperes (rms) with voltage ratings up to 1600 volts are available.

SCR control of series motors. Fractional horsepower universal, ac and dc series motors may be speed-controlled from a single-phase 110- or 220-volt supply using the half-wave circuit involving a diode D_1 and SCR (Fig. 5*a*). The trigger point of the SCR is adjusted via potentiometer R_v, which phase-shifts the gate turn-on voltage of the SCR whenever its anode *A* is positive with respect to cathode *K*. During the negative half-cycle of input voltage, SCR conduction is off and the gating pulse is blocked by diode D_1. When the positive half-cycle is initiated once again, capacitor *C* charges to provide the required gating pulse at the time preset by the time constant R_vC.

Replacing diode D_1 by a DIAC and the SCR by a TRIAC converts the circuit from half-wave to full-wave operation and control (Fig. 5*b*). Compared with Fig. 5*a*, this circuit provides almost twice the torque and improved speed regulation. Neither circuit, however, is capable of automatic maintenance of desired speed because of the inherently poor speed regulation of series-type ac, universal, or dc motors, with application or variation of motor loading.

Automatic speed regulation is achieved by using the half-wave feedback circuit shown in Fig. 5*c*. This circuit requires, however, that both field leads (f_1, f_2) and both armature leads (A_1, A_2) are separately brought out for connection as shown in Fig. 5*c*. The desired speed is set by potentiometer R_v, in terms of reference voltage V_d across diode D_1. The actual motor speed is sensed in terms of the armature voltage V_a. Diode D_1 conducts only when the reference voltage V_d (desired speed) exceeds the voltage V_a to restore the motor speed to its desired setting. Thus, when the motor load is increased and the speed drops, the SCR is gated earlier in the cycle because diode D_1 conducts whenever V_d exceeds V_a.

As with the previous half-wave circuit, the feedback circuit may be converted to full-wave operation by replacing diode D_1 with a DIAC and the SCR with a TRIAC (Fig. 5*d*). This last circuit also is shown with additional minor modifications to improve regulation of speed at lower speeds, namely addition of DIACs D_2, D_3 and capacitor *C*. The advent of high-power TRIACs and SCRs has also extended use of the circuit shown in Fig. 5*d* to larger, integral horsepower ac series motors.

When dc motors are operated by using electronic techniques shown in Fig. 5, it is customary to derate motors proportionately because of heating effect created by ac components in both half- and full-wave waveforms. Note that there is no reversal of direction in either the ac or dc series motor under full-wave operation because both the armature and field currents have been reversed.

SCR control of polyphase ac motors. There are fundamentally only three types of polyphase ac motors, namely synchronous motors (SMs), squirrel-cage induction motors (SCIMs), and wound-rotor induction motors (WRIMs). All three employ identical stator constructions. As a result, larger polyphase motors (up to 10,000 hp) are presently being controlled by SCR packages which employ some so-called universal method of speed control. In these methods, both the frequency and stator voltages are varied (in the same proportion)

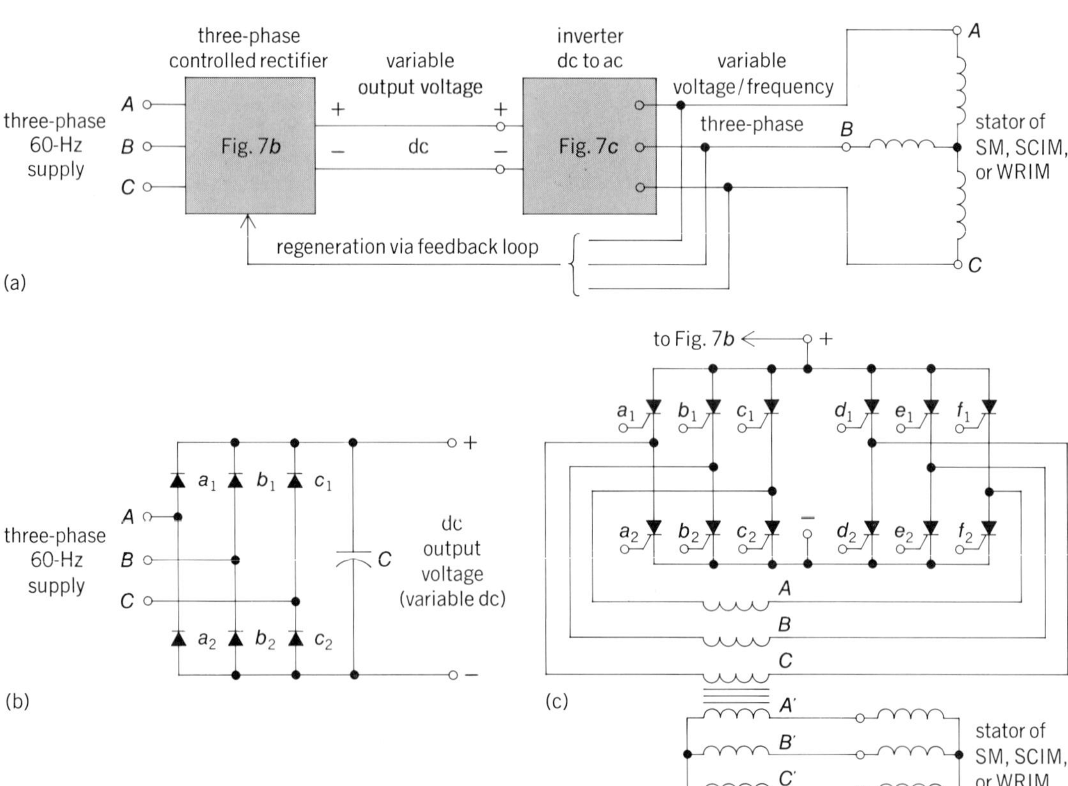

Fig. 7. Rectifier-inverter circuitry for variable voltage/frequency control of three-phase motors. (*a*) Block diagram of rectifier-inverter. (*b*) Three-phase full-wave controlled rectifier. (*c*) Basic inverter dc to three-phase ac for phase-shift voltage control. (*From I. L. Kosow, Control of Electric Machines, 1973*)

to maintain constant polyphase stator flux densities and saturation, thus eliminating the possibility of overheating.

Two major classes of solid-state adjustable voltage/frequency drives have emerged, namely the cycloconverter (an ac/ac package) and the rectifier-inverter (an ac-dc-ac package). Both packages convert three-phase fixed-frequency ac to three-phase variable voltage/variable frequency.

The half-wave cycloconverter shown in Fig. 6*a* is capable of supplying from zero to 20 Hz to an ac polyphase motor. It uses 6 SCRs per phase but is incapable of either phase-sequence reversal or frequencies above 20 Hz. The full-wave converter, shown in Fig. 6*b*, uses twice the number of SCRs (12 per phase) and possesses advantages of wider frequency variation (from +30 Hz down to 0 and up to −30 Hz), potentialities for dynamic braking, and capability of power regeneration.

The symbol for a polyphase SCIM driven by a cycloconverter is shown in Fig. 6*c*, and an SM in Fig. 6*d*. Note that the symbol implies frequency/voltage control of identical stators, and the nature of the motor is determined solely by its rotor.

The second class of solid-state drive package is the rectifier-inverter, which converts fixed-frequency polyphase ac to dc (variable voltage). The dc is then inverted by a three-phase inverter (using a minimum of 12 SCRs, commercially) to produce variable-voltage, variable-frequency three-phase ac for application to the motor stator. The block diagram of a solid-state rectifier-inverter package is shown in Fig. 7*a*. The three-phase 60-Hz supply is first rectified to produce variable dc (Fig. 7*b*) by appropriate phase shift of SCR gates. The variable dc is then applied to the dc bus of Fig. 7*c*. Inversion is accomplished by appropriate phase shift of gates of the 12 SCRs to produce phase and line voltages displaced, respectively, by 120°. This three-phase output voltage is applied to a transformer whose secondary is applied to the stator of the motor (Fig. 7*c*). The rectifier-inverter is capable of power regeneration as noted by the feedback loop shown in Fig. 7*a*. There is no clear-cut choice between cycloconverters and rectifier-inverter packages, as of this writing.

Brush-shifting motors. By use of a regulating winding, a commutator, and a brush-shifting device, speed of a polyphase induction motor can be controlled similarly to that of a dc shunt motor. Such motors are used for knitting and spinning machines, paper mills, and other industrial services that require controlled variable-speed drive. The primary winding on the rotor is supplied from the line through slip rings. The stator windings are the secondary windings (S_1, S_2, S_3), and the third winding, also in the rotor, is an adjusting winding provided with a commutator (Fig. 8). Voltages collected from the commutator are fed into the secondary circuit. Brushes 1, 2, and 3 are mounted 120 electrical degrees apart on a movable yoke. Brushes 4, 5, and 6 are similarly mounted on a separate movable yoke. Each set of brushes can be moved as a group. Thus both the spacing between sets of brushes and the angular position of the brushes are adjustable. Brush spacing determines the magnitude of the voltage applied to the secondary. When brush sets are so adjusted that pairs of brushes are in contact with the same commutator

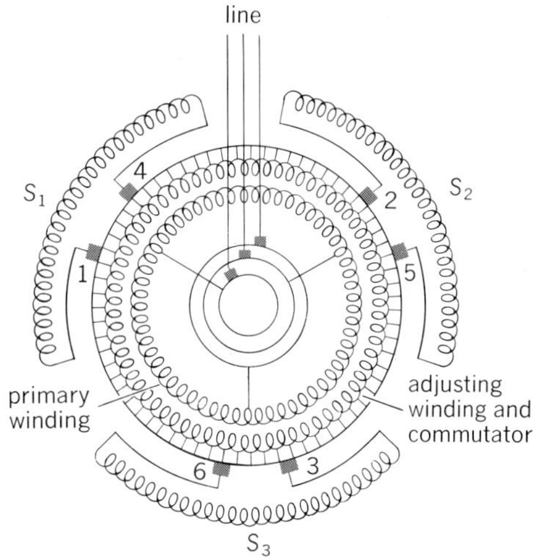

Fig. 8. Schematic of adjustable-speed brush-shifting motor. *(From A. E. Fitzgerald and C. Kingsley, Jr., Electric Machinery, McGraw-Hill, 1952)*

segment, the secondary is short-circuited and no voltage is supplied. Under these conditions the motor behaves as an ordinary induction motor. The speed can be reduced by separating the brushes so that secondary current produces a negative torque. The machine can be operated above synchronism by interchanging the position of the brushes, so the voltage collected is in a direction to produce a positive torque. The motor can be reversed by reversing two of the leads supplying the primary. [IRVING L. KOSOW]

Bibliography: D. G. Fink and H. W. Beaty (eds.), *Standard Handbook for Electrical Engineers*, 11th ed., 1978; A. E. Fitzgerald et al., *Electric Machinery*, 3d ed., 1971; I. L. Kosow, *Electric Machinery and Transformers*, 1972; I. L. Kosow, *Control of Electric Machines*, 1973.

Alternator

A machine that generates an alternating voltage when its rotating portion is driven by a motor, engine, or other means. It is thus a means for converting mechanical power into electric power. The frequency of the generated alternating voltage is exactly proportional to the speed at which the alternator is driven. An alternator generates one cycle of alternating current each time one of its coils passes a pair of magnetic poles in the field structure. *See* ALTERNATING-CURRENT GENERATOR.

[JOHN MARKUS]

Aluminum alloys

Solid solutions of one or more metals in aluminum. The principal alloying elements in aluminum-base alloys are magnesium, silicon, copper, zinc, and manganese. In wrought products, which constitute the greatest use of aluminum, the alloys are identified by four-digit numbers of the form NXXX, where the value of N denotes the alloy type and the principal alloying element(s) as follows: 1 (Al; at least 99% aluminum by weight), 2 (Cu), 3 (Mn), 4 (Si), 5 (Mg), 6 (Mg + Si), 7 (Zn), 8 (other).

Iron and silicon are commonly present as impurities in aluminum alloys, although the amounts may be controlled to achieve specific mechanical or physical properties. Minor amounts of other elements, such as Cr, Zr, V, Pb, and Bi, are added to specific alloys for special purposes. Titanium additions are frequently employed to produce a refined cast structure.

Aluminum-base alloys are generally prepared by making the alloying additions to molten aluminum, forming a liquid solution. As the alloy freezes, phase separation occurs to satisfy phase equilibria requirements and the decrease in solubility as the temperature is lowered. The resultant solidified structure consists of grains of aluminum-rich solid solution and crystals of intermetallic compounds. Elements which lower the freezing point of aluminum, such as Cu, Mg, and Si, tend to segregate to the portions of the grains which freeze last, such as the cell boundaries. Elements which raise the freezing point, such as Cr or Ti, segregate in the opposite manner. *See* EUTECTICS.

A decrease in solubility with falling temperature also provides the basis for heat treatment of solid aluminum alloys. In this operation, the alloy is held for some time at a high temperature to promote dissolution of soluble phases and homogenization of the alloy by diffusion processes. The limiting temperature is the melting point of the lowest melting phase present. The time required depends both on temperature and on the distances over which diffusion must occur to achieve the desired degree of homogenization. Times of several hours can be necessary with coarse structures such as sand castings. Only a few minutes may be adequate, however, for rapidly heated thin sheet. *See* HEAT TREATMENT (METALLURGY).

The solution heat treatment is followed by a quenching operation in which the article is rapidly cooled, for example, by plunging it into cold or hot water or by the use of an air blast. This produces a supersaturated metallic solid solution that is thermodynamically unstable at room temperature. In several important alloy classes, such as 2XXX, 6XXX, and 7XXX, the supersaturated solution decomposes at room temperature to form fine, submicroscopic segregates or precipitates that are precursors to the equilibrium phases predicted by phase diagrams. The precipitation phenomenon, occurring over periods of days to years, produces substantial increases in strength. Additional precipitation strengthening can be obtained by heating the alloy at temperatures in the range 250–450°F (121–232°C), the time and temperature varying with alloy composition and the objectives with respect to mechanical properties and other characteristics such as corrosion resistance.

Table 1 lists the nominal compositions of a number of commercially important wrought alloys and the type of products for which they are used. The alloys are generally classified in two broad categories depending upon their response to heat treatment as described in the preceding paragraph. Those having no or minor response are identified as non-heat-treatable alloys and include the 1XXX, 3XXX, and 5XXX compositions. Those that do respond are known as heat-treatable alloys and include the 2XXX, 6XXX, and 7XXX compositions.

Included in the non-heat-treatable group is 1350, a special grade used for electrical conductor products. Alloy 1100 is a grade of 99.0% minimum aluminum content with particular controls on Fe, Si, and Cu contents, available in a variety of product forms such as sheet, foil, wire, rod, and tube, used for packaging, fin stock, and a variety of sheet metal applications. The Mn-containing alloy 3003 is a moderate-strength, very workable alloy for cooking utensils, tube, packaging, and lithographic sheet applications. The stronger Al-Mn-Mg alloy 3004 is used for architectural applications, for storage tanks, and especially for drawn and ironed beer and beverage containers.

Alloy 5052 is a workable, corrosion-resistant Al-Mg alloy for many metalworking and metal-forming purposes and marine applications. Where higher strength is required, the higher-Mg-content, weldable 5086 or 5083 alloys may be used. The latter is employed in construction of welded tanks for liquefied gas (cryogenic) transport and storage and

Table 1. Nominal composition and forming processes for common wrought aluminum alloys

Alloy	Form†	Si	Cu	Mg	Mn	Zn	Other
1350	b–d	—	—	—	—	—	99.5 Al (min.)
1100	b–d	—	0.1	—	—	—	99.00 Al (min.)
2011	b–d	—	5.5	—	—	—	0.5 Pb, 0.5 Bi
2014	b–e	0.8	4.4	0.4	0.8	—	—
2219	b, e	—	6.3	—	0.3	—	0.1 V, 0.1 Zr
2024	b–e	—	4.4	1.5	0.6	—	—
2036	b	—	2.6	0.45	0.25	—	—
3003	b–d	—	—	—	1.2	—	—
3004	b–d	—	—	1.0	1.2	—	—
5052	b–d	—	—	2.5	—	—	0.25 Cr
5657	b, c	—	—	0.8	—	—	—
5083	b–e	—	—	4.5	0.7	—	0.15 Cr
5086	b–d	—	—	4.0	0.5	—	0.15 Cr
5182	b	—	—	4.5	0.35	—	—
6061	b–e	0.6	0.25	1.0	—	—	0.25 Cr
6063	b–e	0.4	—	0.7	—	—	—
7005	b–e	—	—	1.5	0.5	4.5	0.15 Cr, 0.14 Zr
7050	b, e	—	2.4	2.3	—	6.2	0.12 Zr
7075	b–e	—	1.6	2.5	—	5.6	0.25 Cr

†b = rolled; c = drawn; d = extruded; and e = forged.

Table 2. Nominal composition and casting procedure for common aluminum casting alloys

Alloy	Form*	Si	Cu	Mg
413.0	D	12	–	–
B†443.0	B, C	5.3	.15 max.	–
F†332.0	C	9.5	3.0	1.0
355.0	B, C	5.0	1.3	0.5
356.0	B, C	7.0	–	0.3
380.0	D	8.5	3.5	–
390.0	C, D	17	4.5	0.55

*B = sand casting; C = permanent mold casting; and D = die casting.

†The letter indicates modifications of alloys of the same general composition or differences in impurity limits, from alloys having the same four-digit numerical designations.

for armor plate in military vehicles. Alloy 5182 is also a high-strength Al-Mg alloy that is employed primarily in a highly strain-hardened condition for beverage can ends. Alloy 5657 is a lower-strength material produced with a bright anodized finish for automobile trim and other decorative applications.

The heat-treatable alloys 2014, 2024, and 7075 have high strengths and are employed in aircraft and other transportation applications. Modifications of these basic alloys, such as 2124 and 7475, have been developed to provide increased fracture toughness. High toughness at high strength levels is achieved in thick-section products with 7050. Where elevated temperatures are involved, the 2XXX alloys are preferred. One such alloy, 2219, also has good toughness at cryogenic temperatures and is weldable. This alloy was prominently employed in the fuel and oxidizer tanks serving as the primary structure of the Saturn space vehicle boosters.

Alloy 6061 is widely used for structural applications where somewhat lower strengths are acceptable. For example, 6061 may be used for trailers, trucks, and other transportation applications. Alloy 6063 is a still-lower-strength, extrudable, heat-treatable alloy for furniture and architectural applications. The use of Pb and Bi in 2011 produces a heat-treatable, free-machining alloy for screw-machine products. Alloy 2036 is a moderate-strength alloy with good workability and formability, employed as body sheet in automotive applications.

Table 2 shows the nominal compositions of several important casting alloys. The major alloying addition is silicon, which improves the castability of aluminum and provides moderate strength. Other elements are added primarily to increase the tensile strength. Most die castings are made of alloy 413.0 or 380.0. Alloy 443.0 has been very popular in architectural work, while 355.0 and 356.0 are the principal alloys for sand castings. Number 390.0 is employed for die-cast automotive engine cylinder blocks, while alloy F332.0 is used for pistons for internal combustion engines.

Casting alloys are significant users of secondary metal (recovered from scrap for reuse). Thus, casting alloys usually contain minor amounts of a variety of elements; these do no harm as long as they are kept within certain limits. The use of secondary metal is also of increasing importance in wrought alloy manufacturing as producers take steps to reduce the energy required in producing fabricated aluminum products. *See* ALLOY.

[ALLEN S. RUSSELL]

Antifriction bearing

A machine element that permits free motion between moving and fixed parts. Antifrictional bearings are essential to mechanized equipment; they hold or guide moving machine parts and minimize friction and wear. Friction wastefully consumes energy, and wear changes dimensions until a machine becomes useless.

Simple bearings. In its simplest form a bearing consists of a cylindrical shaft, called a journal, and a mating hole, serving as the bearing proper. Ancient bearings were made of such materials as wood, stone, leather, or bone, and later of metal. It soon became apparent for this type of bearing that a lubricant would reduce both friction and wear and prolong the useful life of the bearing. Convenient lubricants were those of animal, vegetable, or marine origin such as mutton tallow, lard, goose grease, fish oils, castor oils, and cottonseed oil. Egyptian chariot wheels show evidence of the use of mutton tallow for bearing lubrication.

The use of mineral oils dates back principally to the discovery of Drake's well at Titusville, Pa., in 1859. Petroleum oils and greases are now generally used for lubricants, sometimes containing soap and solid lubricants such as graphite or molybdenum disulfide, talc, and similar substances.

Materials. The greatest single advance in the development of improved bearing materials took place in 1839, when I. Babbitt obtained a United States patent for a bearing metal with a special alloy. This alloy, largely tin, contained small amounts of antimony, copper, and lead. This and similar materials have made excellent bearings. They have a silvery appearance and are generally described as white metals or as Babbitt metals. For many decades they have served as the measure of excellence against which other bearing materials have been compared.

Wooden bearings are still used, however, for limited applications in light-duty machinery and are now frequently made of hard maple which has been impregnated with a neutral oil. Wooden bearings made of lignum vitae, the hardest and densest of all woods, are still used. Lignum vitae is native only to the Caribbean area. This wood weighs approximately 80 lb/ft^3 and has a resin content of some 30% by volume; thus it is remarkably self-lubricating. The grain is closely interwoven, giving the material high resistance to wear and compression and making it difficult to split. Applications are currently found in chemical processing and food industries where lignum vitae wood can successfully resist the destructive action of mild acids, alkalies, oils, bleaching compounds, liquid phosphorus, and many food, drug, and cosmetic compounds.

About 1930 a number of significant developments began to occur in the field of bearing metals. Some of the most successful heavy-duty bearing metals are now made of several distinct compositions combined in one bearing. This approach is based on the widely accepted theory of friction, which is that the best possible bearing material would be one which is fairly hard and resistant but which has an overlay of a soft metal that is

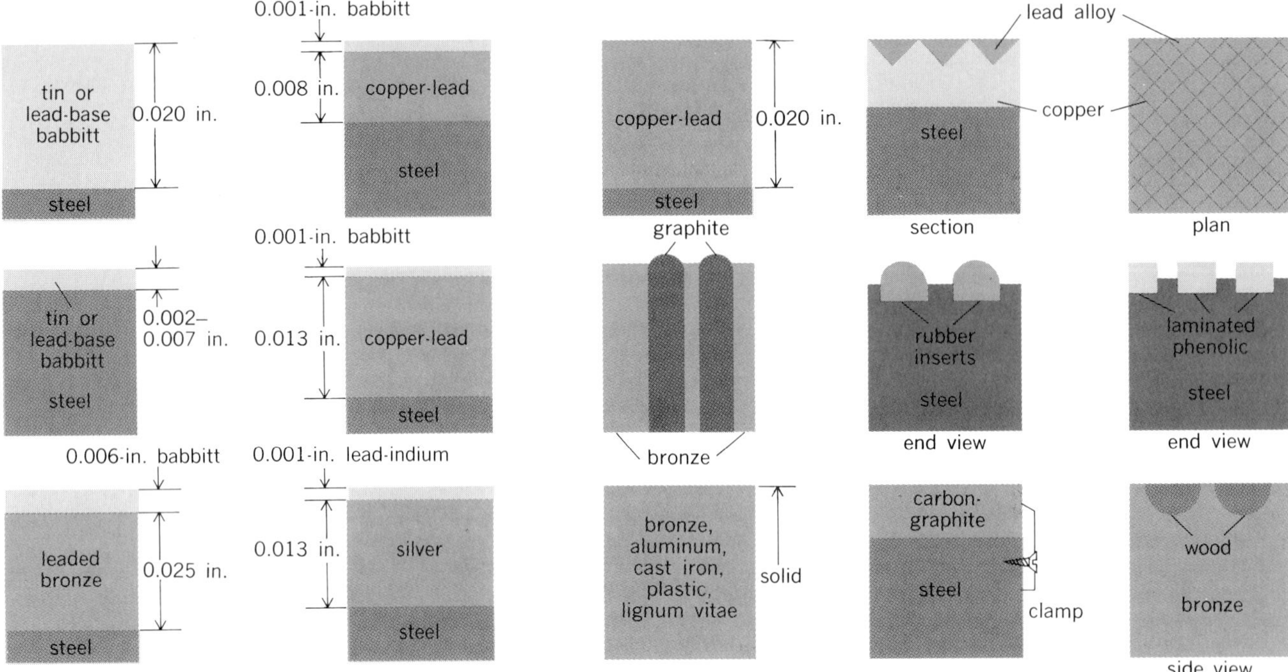

Fig. 1. Schematic of a number of combinations of bearing materials made up of hard and soft materials. *(From J. J. O'Connor, ed., Power's Handbook on Bearings and Lubrication, McGraw-Hill, 1951)*

easily deformed. Figure 1 shows a number of combinations of bearing materials made up as indicated of combinations of hard and soft materials. Figure 2 shows actual bearings in which graphite, carbon, plastic, and rubber have been incorporated into a number of designs illustrating some of the material combinations that are presently available.

Rubber has proved to be a surprisingly good bearing material, especially under circumstances in which abrasives may be present in the lubricant. Rubber bearings have found wide application in the stern-tube bearings of ships, on dredge cutter heads, on a number of centrifugal pumps, and for shafting bearings on deep-well pumps. The rubber used is a tough resilient compound similar in texture to that in an automobile tire. These bearings are especially effective with water lubrication, which serves as both coolant and lubricant.

Cast iron is one of the oldest bearing materials. Iron bearings were used in ancient India and China. With the advent of more complicated machinery during the industrial revolution, cast iron became a popular bearing material. It is still used where the duty is relatively light.

Porous metal bearings are frequently used when plain metal bearings are impractical because of lack of space or inaccessibility for lubrication. These bearings have voids of 16–36% of the volume of the bearing. These voids are filled with a lubricant by a vacuum technique. During operation they supply a limited amount of lubricant to the sliding surface between the journal and the bearing. In general, these bearings are satisfactory for light loads and moderate speeds.

In some areas recent research has shown that, surprisingly enough, very hard materials when rubbed together provide satisfactory bearing characteristics for unusual applications. Materials such as Stellite, Carboloy, Colmonoy, Hastelloy, and Alundum are used. Because of their hardness, these bearings must be extremely smooth and the geometry must be precise for there is little possibility that these materials will conform to misalignment through wear.

Lubricants. Petroleum oils and greases have been fortified by chemical additives so that they are effective in reducing wear of highly stressed machine elements such as gears and cams. The additives include lead naphthenate, chlorine, sulfur, phosphorus, or similar materials. In general, compounds containing these elements are used as additives to form—through reaction with the metal surfaces—chlorides, sulfides, and phosphides which have relatively low shear strength and protect the surface from wear and abrasion.

The method of supplying the lubricant and the quantity of lubricant which is fed to the bearing by

ANTIFRICTION BEARING

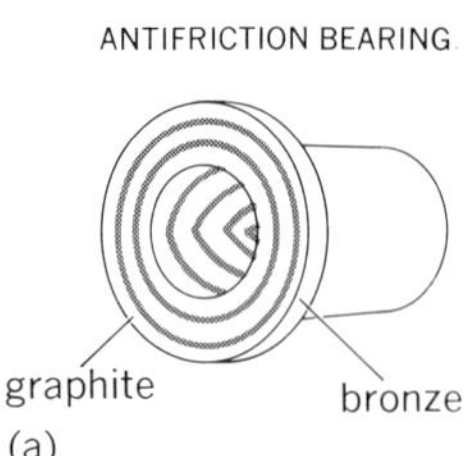

(a)

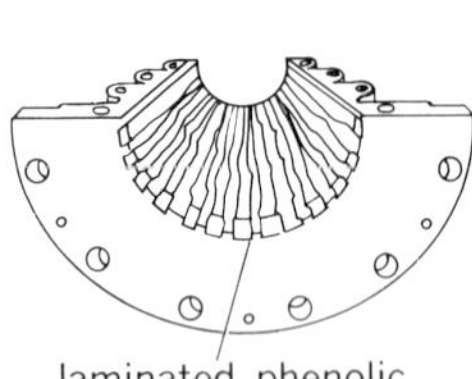

(b)

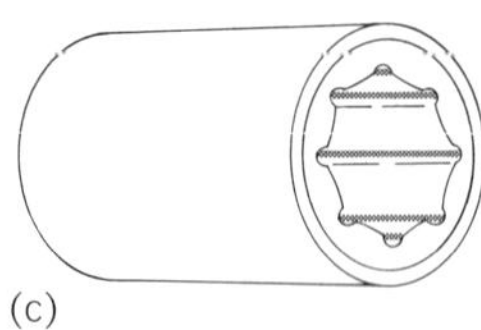

(c)

Fig. 2. Bearings with (*a*) graphite, (*b*) wood, plastic, and nylon *(from J. J. O'Connor, ed., Power's Handbook on Bearings and Lubrication, McGraw-Hill, 1951)*; (*c*) rubber *(Lucian Q. Moffitt, Inc.)*.

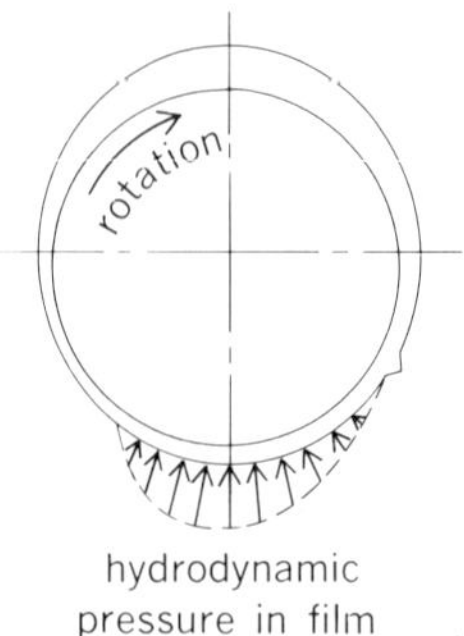

Fig. 3. Hydrodynamic fluid-film pressures in a journal bearing. *(From W. Staniar, ed., Plant Engineering Handbook, 2d ed., McGraw-Hill, 1959)*

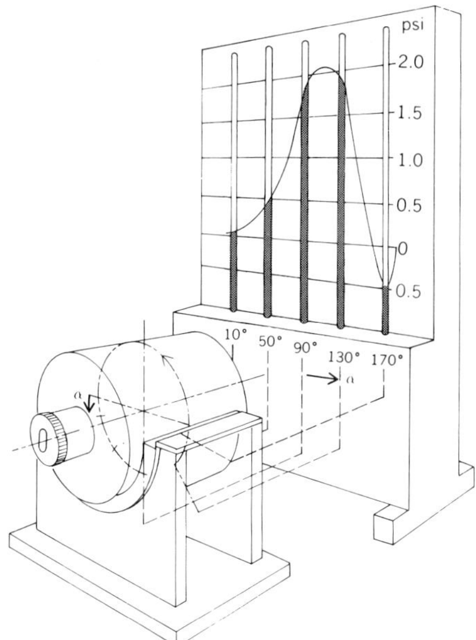

Fig. 4. Test device for determining pressure in a journal bearing. *(From H. Drescher, Air lubrication bearings, Eng. Dig., 15:103–107, 1954)*

the supplying device will often be the greatest factor in establishing performance characteristics of the bearing. For example, if no lubricant is present, the journal and bearing will rub against each other in the dry state. Both friction and wear will be relatively high. The coefficient of friction of a steel shaft rubbing in a bronze bearing, for example, may be about 0.3 for the dry state. If lubricant is present even in small quantities, the surfaces become contaminated by this material whether it be an oil or a fat, and depending upon its chemical composition the coefficient of friction may be reduced to about 0.1. Now if an abundance of lubricant is fed to the bearing so that there is an excess flowing out of the bearing, it is possible to develop a self-generating pressure film in the clearance space as indicated in Fig. 3. These pressures can be sufficient to sustain a considerable load and to keep the rubbing surfaces of the bearing separated. This is the type of bearing that is found on the crank shaft of a typical automobile engine. Unit pressures on these bearings reach and exceed 2500 psi and with little attention will last almost indefinitely, provided that the oil is kept clean and free from abrasive particles and that the bearing materials themselves do not deteriorate from fatigue, erosion, or corrosion. Figure 4 shows a schematic diagram of a simple test device which indicates the pressure developed in the converging clearance space of the journal bearing. The position shown where the center of the journal is eccentric to the center of the bearing is that which the journal naturally assumes when loaded. If the supply of lubricant is insufficient to fill the clearance space completely or if the load and speed of operation are not favorable to the generation of a complete fluid film, the film will be incomplete and there will be areas within the bearing which do not have the benefit of a fluid film to keep the rubbing surfaces apart. These areas will be only lightly contaminated by lubricant.

The types of oiling devices that usually result in insufficient feed to generate a complete fluid film are, for example, oil cans, drop-feed oilers, waste-packed bearings, and wick and felt feeders.

Oiling schemes that provide an abundance of lubrication are oil rings, bath lubrication, and forced-feed circulating supply systems. The coefficient of friction for a bearing with a complete fluid film may be as low as 0.001. Figure 5 shows typical devices which do not usually supply a lubricant in sufficient quantity to permit the generation of a complete fluid film. Figure 6 shows devices in which the flow rate is generally sufficient to permit a fluid film to form. Figure 6*a*, *b*, and *d* show some typical forms of fluid-film journal bearings. The table shows current design practice for a number of bearings in terms of mean bearing pressure. This is the pressure applied to the bearing by the external load and is based on the projected area of the bearing. *See* ENGINE LUBRICATION; JEWEL BEARING; LUBRICANT.

Fluid-film hydrodynamic types. If the bearing surfaces can be kept separated, the lubricant no longer needs an oiliness agent, such as the fats, oils, and greases described above. As a consequence, many extreme applications are presently

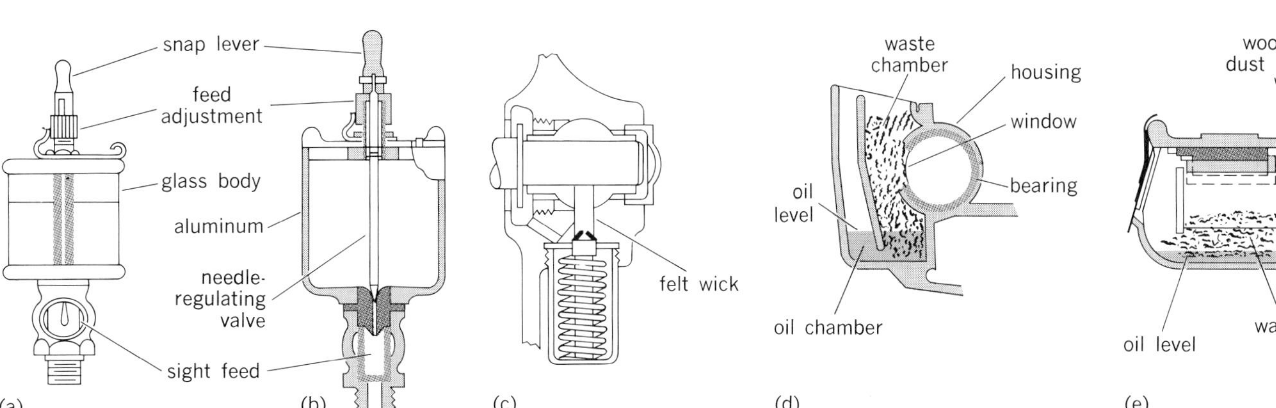

Fig. 5. Schematic showing some of the typical lubrication devices. (*a–b*) Drop-feed oilers. (*c*) Felt-wick oiler for small electric motors. (*d*) Waste-packed armature bearing. (*e*) Waste-packed railroad bearing. *(From W. Staniar, ed., Plant Engineering Handbook, 2d ed., McGraw-Hill, 1959)*

Current practice in mean bearing pressures

Type of bearing	Permissible pressure, psi of projected area
Diesel engines, main bearings	800–1500
Crankpin	1000–2000
Wrist pin	1800–2000
Electric motor bearings	100–200
Marine diesel engines, main bearings	400–600
Crankpin	1000–1400
Marine line-shaft bearings	25–35
Steam engines, main bearings	150–500
Crankpin	800–1500
Crosshead pin	1000–1800
Flywheel bearings	200–250
Marine steam engine, main bearings	275–500
Crankpin	400–600
Steam turbines and reduction gears	100–220
Automotive gasoline engines, main bearings	500–1000
Crankpin	1500–2500
Air compressors, main bearings	120–240
Crankpin	240–400
Crosshead pin	400–800
Aircraft engine crankpin	700–2000
Centrifugal pumps	80–100
Generators, low or medium speed	90–140
Roll-neck bearings	1500–2500
Locomotive crankpins	1500–1900
Railway-car axle bearings	300–350
Miscellaneous ordinary bearings	80–150
Light line shaft	15–25
Heavy line shaft	100–150

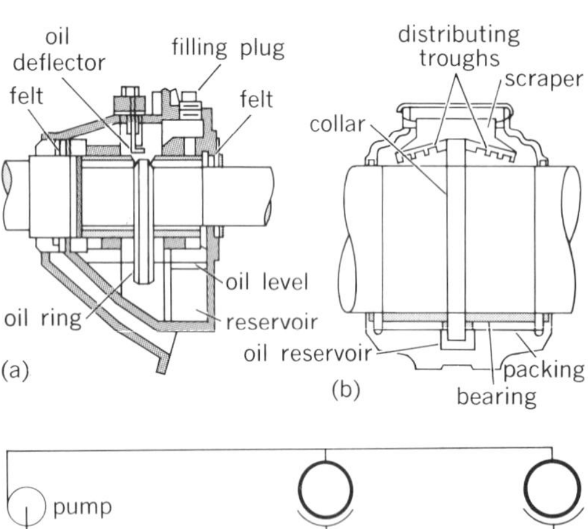

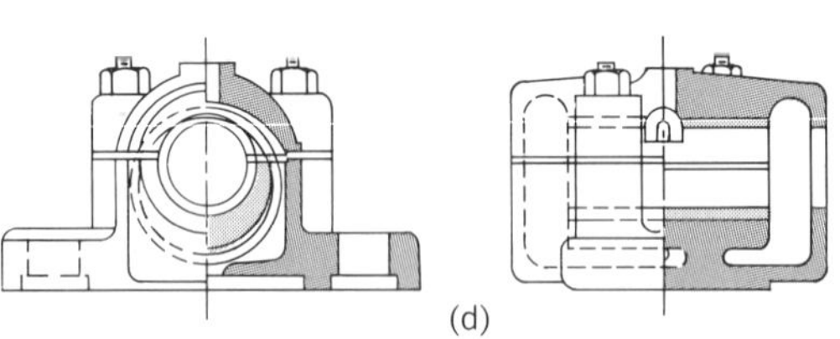

Fig. 6. Lubrication devices. (*a*) Ring-oiled motor bearings. (*b*) Collar-oiled bearing (*from W. Staniar, ed., Plant Engineering Handbook, 2d ed., McGraw-Hill, 1959*). (*c*) Circulating system for oiling bearings (*from J. J. O'Connor, ed., Power's Handbook on Bearings and Lubrication, McGraw-Hill, 1951*). (*d*) Rigid ring-oiling pillow block (*from T. Baumeister, ed., Standard Handbook for Mechanical Engineers, 7th ed., McGraw-Hill, 1967*)

found in which fluid-film bearings operate with lubricants consisting of water, highly corrosive acids, molten metals, gasoline, steam, liquid refrigerants, mercury, gases, and so on. The self-generation of pressure in such a bearing takes place no matter what lubricant is used, but the maximum pressure that is generated depends upon the

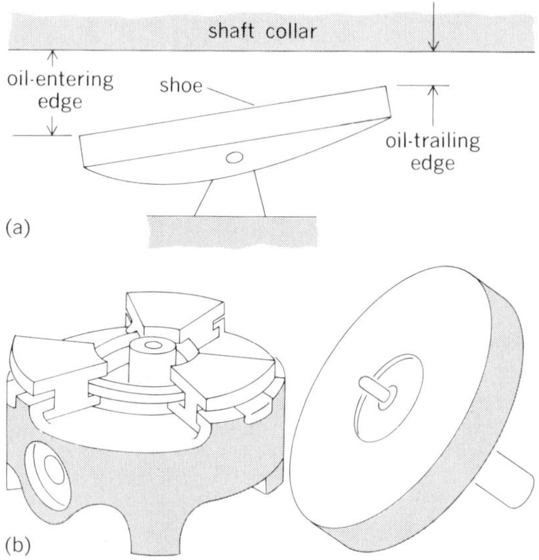

Fig. 7. Tilting-shoe-type bearing. (*a*) Schematic (*from W. Staniar, ed., Plant Engineering Handbook, 2d ed., McGraw-Hill, 1959*). (*b*) Thrust bearing (*from D. D. Fuller, Theory and Practice of Lubrication for Engineers, Wiley, 1956*).

viscosity of the lubricant. Thus, for example, the maximum load-carrying capacity of a gas-lubricated bearing is much lower than that of a liquid-lubricated bearing. The ratio of capacities is in direct proportion to the viscosity.

Considerable research has been directed toward the operation of machinery at extremes of temperature. On the low end of the scale this may mean −400°F. On the upper end of the scale expectation is that some devices may be called upon to function at 2000–3000°F. Gas is the only presently known lubricant that could possibly be used for such extremes of temperature. Because the viscosity of gas is so low, the friction generated in the bearing is correspondingly of a very low order. Thus gas-lubricated machines can be operated at extremely high speeds because there is no serious problem in keeping the bearings cool. A rotor system has been operated on gas-lubricated bearings up to 433,000 rpm.

The self-generating pressure principle is applied equally as well to thrust bearings as it is to journal bearings. One of the first commercial applications of the self-generating type of bearing was in the tilting-pad type of thrust bearing. One tilting pad is shown schematically in Fig. 7*a*. There is certainly no question regarding the great value of the tilting-pad thrust bearing. It excels in low friction and in reliability. A model of a typical commercial thrust bearing is shown in Fig. 7*b*. This thrust bearing is made up of many tilting pads located in a circular position. One of the largest is on a hydraulic turbine at the Grand Coulee Dam. There, a bearing 96 in. in diameter carries a load of 2,150,000 lb with a coefficient of friction of about 0.0009. Large marine thrust bearings are of this type and transfer the entire thrust of the propeller to the hull of the ship.

Fluid-film hydrostatic types. Sleeve bearings of the self-generating pressure type, after being brought up to speed, operate with a high degree of efficiency and reliability. However, when the rotational speed of the journal is too low to maintain a complete fluid film, or when starting, stopping, or reversing, the oil film is ruptured, friction increases and wear of the bearing accelerates. This condition can be eliminated by introducing high-pressure oil to the area between the bottom of the journal and the bearing itself, as shown schematically in Fig. 8. If the pressure and quantity of flow are in the correct proportions, the shaft will be raised and supported by an oil film whether it is rotating or not. Friction drag may drop to one-tenth of its original value or even less, and in certain kinds of heavy rotational equipment in which available torque is low, this may mean the difference between starting and not starting. This type of lubrication is called hydrostatic lubrication and, as applied to a journal bearing in the manner indicated, it is called an oil lift. Synchronous condensers need the oil lift when the unit is of large size. Rolling-mill and foil-mill bearings may be equipped with an oil lift to reduce starting friction when the mills are under load. Occasionally, hydrostatic lifts are used continuously on bearings that are too severely overloaded to maintain a normal hydrodynamic or self-pressurizing oil film.

Hydrostatic lubrication in the form of a step bearing has been used on various machines to carry thrust. Such lubrication can carry thrust whether the shaft is rotating or not and can maintain complete separation of the bearing surfaces. Figure 9 shows a schematic representation of such a step-thrust bearing. High-speed machines such as ultracentrifuges have used this principle by employing air as the lubricant at speeds upward of 100,000 rpm.

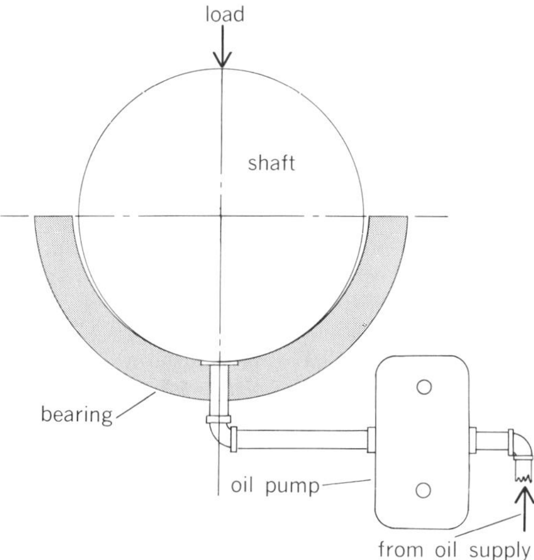

Fig. 8. Fluid-film hydrostatic bearing. Hydrostatic oil lift can reduce starting friction drag to less than one-tenth of usual starting drag. (*From W. Staniar, ed., Plant Engineering Handbook, 2d ed., McGraw-Hill, 1959*)

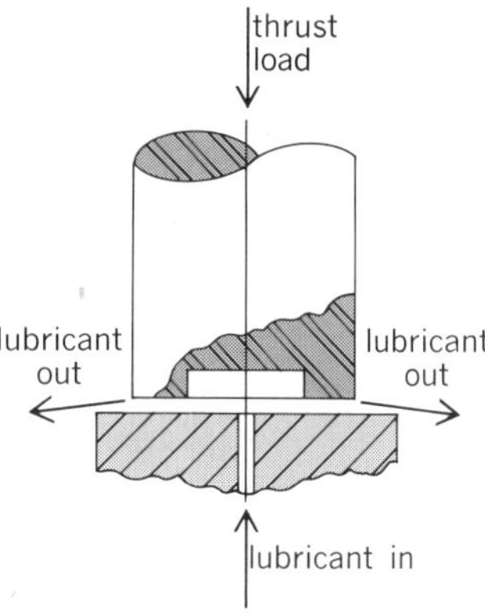

Fig. 9. Step bearing. (*From W. Staniar, ed., Plant Engineering Handbook, 2d ed., McGraw-Hill, 1959*)

Large structures have been floated successfully on hydrostatic-type bearings. For example, the Hale 200-in. telescope on Mount Palomar weighs about 1,000,000 lb; yet the coefficient of friction for the entire supporting system, because of the hydrostatic-type bearing, is less than 0.000004. The power required is extremely small and a 1/12-hp clock motor rotates the telescope while observations are being made. Hydrostatic bearings are currently being applied to large radio telescopes and radar antennas, some of which must sustain forces of 5,000,000 lb or more, considering wind loads as well as dead weight. One such unit constructed at Green Bank, W.Va., by the Associated Universities has a parabolic disk or antenna 140 ft in diameter.

Rolling-element types. Everyday experiences demonstrate that rolling resistance is much less than sliding resistance. The wheelbarrow, the two-wheeled baggage truck, and similar devices are striking examples of the reduction in friction by the use of the wheel. Heavy crates and similar objects are easily moved by introducing rollers under the leading edge of the load while the crate or object is pushed along. Egyptian engineers building the pyramids transported huge blocks of stone from the quarry to the building site by means of rollers. This principle is used in the rolling-element bearing which has found wide use.

The first major application of these bearings was to the bicycle, the use of which reached its peak just before the year 1900. In the development of the automobile, ball and roller bearings were found to be ideal for many applications, and today they are widely used in almost every kind of machinery.

Structure. These bearings are characterized by balls or cylinders confined between outer and inner rings. The balls or rollers are usually spaced uniformly by a cage or separator. The rolling elements are the most important because they transmit the loads from the moving parts of the machine to the stationary supports. Balls are uniformly spherical, but the rollers may be straight cylinders, or they may be barrel- or cone-shaped or of other forms, depending upon the purpose of the design. The rings, called the races, supply smooth, hard, accurate surfaces for the balls or rollers to roll on. Some types of ball and roller bearings are made

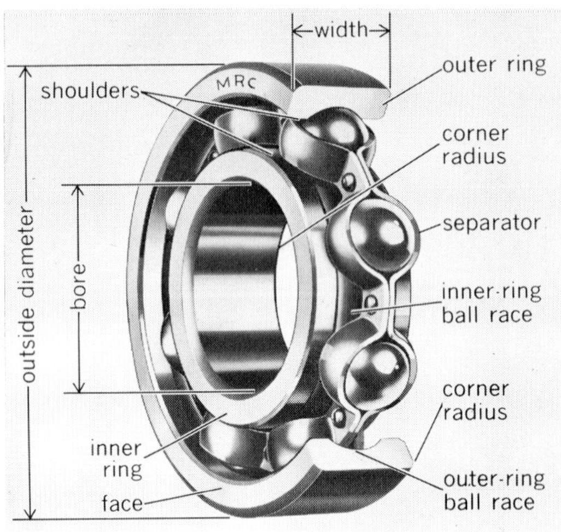

Fig. 10. Deep-groove ball bearing. (*Marlin-Rockwell*)

without separators. In other types there is only the inner or the outer ring, and the rollers operate directly upon a suitably hardened and ground shaft or housing. Figure 10 shows a typical deep-grooved ball bearing, with the parts that are generally used.

These bearings may be classified by function into three groups: radial, thrust, and angular-contact bearings. Radial bearings are designed principally to carry a load in a direction perpendicular to the axis of rotation. However, some radial bearings, such as the deep-grooved bearings shown in Fig. 10, are also capable of carrying a thrust load, that is, a load parallel to the axis of rotation and tending to push the shaft in the axial direction. Some bearings, however, are designed to carry only thrust loads. Angular-contact bearings are especially designed and manufactured to carry heavy thrust loads and also radial loads.

Life. A unique feature of rolling-element bearings is that their useful life is not determined by wear but by fatigue of the operating surfaces under the repeated stresses of normal use. Fatigue failure, which occurs as a progressive flaking or pitting of the surfaces of the races and rolling elements, is accepted as the basic reason for the termination of the useful life of such a bearing.

Because the load on a bearing determines the severity of the stress in the surfaces of the races and the rolling elements, it follows that if the load is increased, the life of the bearing will be decreased, and conversely if the load is decreased, the life of the bearing will be increased. This is usually expressed by the relationship that the life of the bearing is inversely proportional to the load cubed. Thus, doubling the load will reduce the life of the bearing by a factor of eight.

The life of a bearing under a given load will therefore be a certain number of revolutions. If this number of revolutions is used up at a relatively high rate, the life of the bearing will be correspondingly short. If the total number of revolutions is used up at a low rate, the life of the bearing is correspondingly longer; that is, the life is inversely proportional to the speed. Life expectancy is a prediction based on results obtained from tests of a large number of identical bearings under identical loads and speeds. Individual bearings may deviate from this figure on a statistical basis, but manufacturers have followed the law of averages in establishing their ratings. For example, some manufacturers specify load and speed ratings for their bearings based on 3000 hr of operation. The bearing manufacturer who uses 3000 hr as a life-expectancy figure assumes that at least 90% of all bearings will last 3000 hr under the specified conditions of load and speed. Based on statistical averages, however, this means that 10% of the bearings will fail before reaching the 3000-hr life expectancy, 50% of the bearings will attain five times the design life, and a few may reach 20–30 times the design life of 3000 hr.

Characteristics. The various types of rolling-contact bearings can be identified in terms of their broad general characteristics (Fig. 11). A separable or a magneto-type bearing is useful where disassembly is frequent (Fig. 11*a*). The outer race is pressed firmly in the removable housing; the inner race may be pressed against a shoulder on the shaft. A deep-grooved bearing with a filling slot (Fig. 11*b*) allows more balls to be used in the bearing than are shown in Fig. 10 and will thus carry heavier radial loads. Because of the filling slot, however, it should be used only for light thrust loads. If the thrust is in one direction, the bearing should be mounted with the slot away from the thrust direction. The double-row radial bearing with deep grooves handles heavier radial and thrust loads than a single-roll bearing of the same dimensions. Internal self-aligning double-roll bearings (Fig. 11*d*) may be used for heavy radial loads where self-alignment is required. The self-aligning feature should not be used to correct poor design or assembly because excessive misalignment will harm the bearing. Thrust loads should be light because a thrust load will be sustained by only one row of balls. Figure 11*e* shows an external self-aligning bearing which requires a larger outside diameter, but has the advantage of being able to carry thrust in either direction, as well as providing a self-aligning feature. Angular contact bearings (Fig. 11*f*) provide for a maximum thrust and modest radial loads. They may be mounted back to back as duplex bearings and carry thrust in either direction. To minimize axial movement of such a bearing and the shaft that it restrains, these bearings may be preloaded to take up any possible slack or clearance. A ball bushing is shown in Fig. 11*g*. This is used for translational motion or motion of a shaft in its axial direction. Any such motion should be at a very low speed and with light radial loads. Roller bearings with short straight rollers are shown in Fig. 11*h*. They permit free movement in the axial direction when the guide lips are on either the outer or inner race. Needle bearings (Fig. 11*i*) have rollers whose length is at least four times the diameter. They may be furnished with a retainer but have their maximum load capacity when there is no retainer and the clearance space is completely filled with rollers. They are with or without an inner race and are most useful where space must be saved. If the shaft is used as an inner race, it should be hardened.

The tapered roller bearings shown in Fig. 11*j* permit heavy radial and thrust loads. These may be procured as double tapered rollers with two inner races and one outer or two outer races and

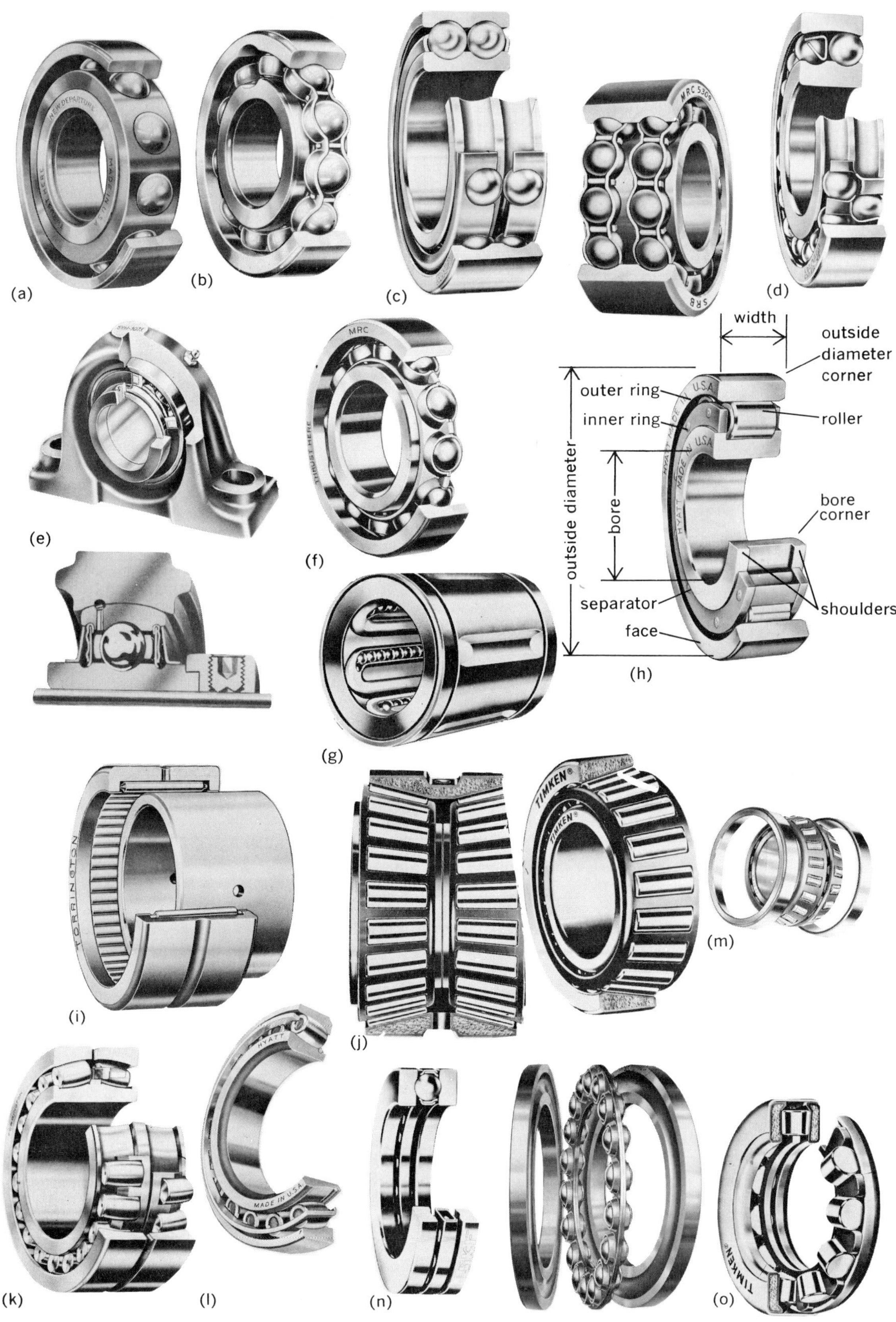

Fig. 11. Views of the various ball and roller bearings. (*a*) Separable angle-contact ball bearing (*New Departure*). (*b*) Deep-groove ball bearing with filling slot (*New Departure*). (*c*) Double-row deep-groove ball bearings (*SKF; Marlin-Rockwell*). (*d*) Internal self-aligning bearing (*SKF*). (*e*) Deep-groove ball bearing with spherical outer ring. (*f*) Standard angular-contact ball bearing (*Marlin-Rockwell*). (*g*) Ball bushing that permits unlimited travel, linear motion (*Thomson Industries*). (*h*) Roller bearing showing nomenclature (*Hyatt*). (*i*) Needle roller bearing (*Torrington*). (*j*) Tapered roller bearings, two-row and single-row (*Timken*). (*k*) Radial spherical roller bearings (*Torrington*). (*l*) Barrel roller bearing (*Hyatt*). (*m*) Concavex roller bearing (*Shafer*). (*n*) Single-direction ball-bearing thrust (*SKF; Marlin-Rockwell*). (*o*) Cylindrical roller thrust bearing (*Timken*).

one inner. The race cones intersect at the axis of the bearing. Figure 11*k* shows a self-aligning bearing with two rows of short barrel rollers. The spherical outer race is one piece, or it may be in two parts for preloading. The thrust is taken on the spherical surfaces of the center flange of an inner race. Figure 11*l* shows a barrel roller with spherical outer race. Figure 11*m* is a self-aligning bearing with hourglass rollers. This bearing is also built as a two-row bearing with a one-piece inner race.

The ball thrust bearing shown in Fig. 11*n* is used only for carrying thrust loads acting in the direction of the axis of the shaft. It is used for low-speed applications, while other bearings must support the radial load. Straight roller thrust bearings are made of a series of short rollers to minimize twisting or skewing. The tapered roller thrust bearing shown in Fig. 11*o* eliminates the twisting that may take place with the straight rollers but cause a thrust load between the ends of the rollers and the shoulder on the race.

Mountings. Many types of mountings are available for rolling-element bearings, and their selection will depend upon the variety of service conditions encountered. In the preferred method of mounting, the inner and outer races are held securely in place to prevent creeping or spinning of the races in the housing or on the shaft. In rigid mountings, provision must be made for expansion either by building slip clearances into one mount or by using a straight roller bearing on the free end of the shaft. A bearing can be mounted on a shaft with a snap ring on the outer race which locates the bearing against a shoulder. Bearings may also be mounted on shafts by means of an eccentric ring cut on the side of the extended inner race. A set screw can be used to hold the bearings in place.

[DUDLEY D. FULLER]

Bibliography: T. Baumeister (ed.), *Standard Handbook for Mechanical Engineers*, 8th ed., 1978.

Arc heating

The heating of a material by the heat energy from an electric arc. The electric arc is a component of an electric circuit, like a resistor, but with its own peculiar characteristics. Its resistance decreases as its current increases and vice versa. The electric arc is characterized also by its high temperature and high concentration of heat energy. Millions of watts are controllably dissipated in rather confined space, producing temperatures measured in thousands of degrees Celsius. At these temperatures most materials in the Earth's crust melt quickly. The molten materials can be processed under an oxidizing, neutral, or reducing slag or combination of slags to produce a variety of finished materials which are used for many purposes.

The electric arc is the heating element in a number of heating, melting, and smelting appliances classified as direct-arc furnaces, submerged-arc furnaces, indirect-arc furnaces, air-arc furnaces, arc welders, arc-machining tools, and arc-boring machines. *See* FURNACE CONSTRUCTION.

Direct-arc furnaces. In iron and steel foundries, in melt shops of mini-steel and large steel mills, and in some nonferrous melt shops, direct-arc furnaces are used for their ability to bring the burden quickly to pour temperature. Essentially, a direct-arc furnace consists of (1) a refractory lined shell to hold the burden, or material to be melted; (2) means to charge the burden into the shell; (3) a set of electrodes (three for usual three-phase operation) arranged to move vertically and make electrical contact with the burden; (4) means to energize the electrodes; (5) means to regulate the position of the electrodes to draw arcs under them off the burden (the burden acts as a common electrode in the circuit); and (6) means to tilt the shell rearward to pour off slag, and forward to pour out the burden when it is molten, at the proper temperature, and metallurgically finished. The direct arcing from electrodes to burden gives the direct-arc furnace its name.

Iron foundry uses. In the iron foundry, the direct-arc furnace is used as a batch melter, a duplexer, or continuous melter for making gray iron, malleable iron, nodular iron, and alloy iron castings of closely controlled composition within a wide range of any metallurgical formulation.

The batch melter usually has equipment which raises and swings the roof aside. The burden (iron and steel scrap) is placed in a drop-bottom bucket, lifted over the open chamber, and tripped, and the furnace is charged in one drop. The roof is swung back and lowered to reclose the furnace. The energized electrodes automatically lower to strike arcs and bore down through the burden until a molten pool is formed on the bottom. Continued arcing on the pool superheats it. Also, the rest of the burden that still envelopes the electrodes soaks up radiant heat from the long melt-down arcs formed under the electrodes and melts down.

When the burden is fully melted down to a flat bath, continued radiation from the long melt-down arcs would overheat the sidewall and roof refractory lining before the molten bath reached metallurgical processing temperature. Therefore, the arcs are shortened by reducing their voltage and increasing their current. More heat is then delivered down into the molten burden and less up into the exposed refractory lining.

Arc voltage is further reduced during metallurgical processing, except perhaps for short duration when the slag temperature is too low for the desired slag-metal chemical reactions to develop.

On melt-down, a preliminary analysis (called the prelim) is made of the burden. From the prelim, necessary metallurgical and temperature adjustments are made, and if the second analysis shows the desired result, the burden is poured out for distribution into molds. The furnace is recharged for the next batch or heat.

Little, if any, metallurgical adjustment is done, or required, in iron foundry practice, which is essentially a melt-down and pour-out operation. In other melt shops, the burden, or heat as it is called after melt-down, may be refined from 15 min to several hours before it is poured.

Where molds require continuous pouring, batches from the arc furnace are transferred to a heated or unheated vessel, called a duplexer, with which continous pouring is accomplished while the arc furnace is working on the next batch.

For example, in the illustration a direct-arc furnace is shown pouring a heat of iron having a carbon equivalent of 4.29%. The burden for this heat consisted entirely of steel scrap and alloy additions, the carbon being added first with the cold

A typical direct-arc furnace as used in an iron foundry, steel foundry, or mini-steel plant.

charge and the silicon last in the ladle. The metal in this case is poured into a channel-type induction holding furnace, from which it is poured continuously into molds on an adjacent conveyor. Arc furnaces of this size (11-ft inside shell diameter) pour 15–20-ton heats regularly and often up to 30-ton heats. With a transformer capacity up to 13,000 kva, the arc furnace can produce up to 15 tons of iron or steel per hour around the clock.

The direct-arc furnace used as a duplexer in the iron foundry is practically the same type of furnace as the batch melter, except that the top charge is not essential. The burden is already molten, and only a means of pouring it into the furnace is required. The arc-furnace substation energizing the electrodes is smaller, because the burden usually needs only nominal superheating and metallurgical adjustments. Cold additions may be made through the door to the molten bath for further metallurgical adjustment. Metal is poured from the furnace continuously or in batches as required.

Prior to melt-down use of the direct-arc furnace, practically all tonnage in iron foundries had been melted in the cupola and poured directly into molds or duplexed in the direct-arc furnace duplexer. With the need to reduce smoke and dust contamination of the atmosphere, the procedure has changed. The cupola emits four times the volume of gases as the direct-arc furnace for the same tonnage output. Cost of a collector capable of reducing flue discharge to a level comparable to that from an arc furnace is appreciable. The arc furnace also uses lower-cost scrap which cannot be melted economically in the cupola. As a result, much of the new tonnage is melted in the arc furnace and duplexed in another arc furnace or a channel induction furnace.

Some new tonnage is melted also in the crucible-type induction furnace, also known as the coreless type. (The latter term is a misnomer, cores being essential to both channel and crucible types of induction furnaces.) For safety and economy of operation, the crucible furnace is provided, almost universally, with a gas-fired scrap preheater for safely drying and more economically preheating the burden before it is charged into the furnace. A scrap preheater is not necessary to dry the burden charged into the arc furnace, nor is it justified economically for preheating the arc-furnace burden, particularly when the arc furnace is sufficiently energized with a large transformer and when it is properly controlled with a responsive electrode positioner to maintain the arcs at the highest efficiency.

The arc furnace used as a continuous melter in the iron foundry also maintains a hot metal bath in the furnace. Scrap is charged into the bath and hot metal is poured out as required.

Other uses of direct-arc furnaces. In the steel foundry, the direct-arc furnace is used almost exclusively as the batch melter for making steel castings of nearly any metallurgical requirement.

In steel mills, the direct-arc furnace is also used almost exclusively as a batch melter, although the furnace may be partially or entirely charged with molten iron from a cupola or blast furnace, or with molten steel from a converter. The finished metal is poured into ingots for rolling into steel shapes. *See* STEEL MANUFACTURE.

In nonferrous melt shops, the direct-arc furnace is commonly used for either a batch melter or continuous melter for metals, such as nickel, copper, and cobalt for making billets, cakes, ingots, and castings.

The direct-arc furnace is used also in vacuum melting of various metals to improve purity and physical properties.

Submerged-arc furnaces. Arcs may be completely submerged under the charge or in the molten bath under the charge. Furnaces operated in this manner are generally smelting furnaces with fixed shells, into which ore granules are fed. The bath is tapped at one, two, or three levels as required for the product, slag, and impurities. Many ores are smelted in the submerged-arc furnace and the products are numerous.

Some submerged-arc furnaces have open top shells; others are enclosed and exhausted to collect the product or by-product. Some have tilting shells and some have shells which rotate one revolution in 24 hr or more. Most are cylindrical in shape.

Indirect-arc furnaces. To use only the radiant heat from an arc, the furnace has a refractory-lined cylindrical shell with its axis horizontal. The shell is mounted on rollers so that it may roll or rock through something less than one revolution. Two electrodes, one from each end, enter the cylinder along the axis and touch and draw an arc inside the furnace. The heat of the arc is radiated to the refractories. The burden in the furnace, as it rocks back and forth, "washes" the heat off the refractories. This indirect heating of the burden gives the indirect-arc furnace its name. It is particularly applicable to melting low-temperature metals and alloys, such as brass, where the difference in the melting points of the burden and the refractory is enough to maintain a high rate of heat transfer, and where the heat of the arc is spread before reaching the burden to prevent large losses of volatile elements, such as zinc, in the burden. The furnace is a single-phase unit requiring multiple units, a special unit generator, or special circuitry and equipment to spread the load among the three phases of a power line.

Air-arc furnace. An entirely different application of the arc furnace is to power wind tunnels. The furnace employs a cylindrical shell and a pair of electrodes arranged similarly to those in the indirect-arc furnace; however, the shell is fixed and mounted usually with its axis vertical. There is no refractory; instead, a whorl of water surrounds the arc. Air (or other gas) is blown or pumped through the space between water and arc. The air is superheated to 20,000K and expanded to emerge at supersonic speeds.

Other applications of the arc-furnace technique include arc welding, boring and machining. The electric arc is employed in arc welding to heat and melt a welding rod and the work to be joined. *See* ARC WELDING.

The electric arc is used also in boring machines and in other machining operations. The boring bar is one electrode and the work is the other. A nonconducting liquid carries away the material melted by the arc. The boring bar is fed into the work as the melted material is flushed away. The boring bar may be of any cross section. The bore has the same shape as the boring bar and is slightly larger, but it is held to close tolerance. [C. W. VOKAC]

Bibliography: AIME, *Proceedings of Electric Furnace Steel Conferences*, annual; AIME, Iron and Steel Division, *Electric Furnace Steelmaking*, vol. 1: *Design, Operation, and Practice*, vol 2: *Theory and Fundamentals*, 1962–1963.

Arc lamp

A type of electric-discharge lamp in which an electric current flows between electrodes through a gas or vapor. In most arc lamps the light results from the luminescence of the gas; however, in the carbon arc lamp the light is produced by the incandescence of one or both electrodes. The color of the arc depends upon the electrode material and the surrounding atmosphere. Most lamps have a negative resistance characteristic so that the resistance decreases after the arc has been struck. Therefore some form of current-limiting device is required in the electric circuit. For other electric-discharge lamps *see* VAPOR LAMP.

The carbon arc lamp was the first practical commercial electric lighting device, but the use of arc lamps at present is limited. In many of its previous functions, the carbon arc lamp has been superseded by the high-intensity mercury vapor lamp. Arc lamps are now used to obtain high brightness from a concentrated light source, where large amounts of radiant energy are needed, and where spectral distribution is an advantage. Typical uses of arc lamps are in projectors, searchlights, blue-printing, photography, therapeutics, and microscope lighting, and for special lighting in research.

Carbon arc lamp. The electrodes of this lamp are pure carbon. The lamp is either open, enclosed, or an intensified arc with incandescence at the electrodes and some light from the luminescence of the arc. The open-arc form of the carbon arc is obsolete. It is unstable on a constant voltage supply, although on a constant current system the operating characteristics are good. In the enclosed type of restriced supply of air slows the electrode consumption and secures a life approximately 100 hr. The intensified type used a small electrode, giving a higher intensity output that is more white in color and more steady than either the open or enclosed types. The electrode life is approximately 70 hr. In the high-intensity arc the current may reach values of 125–150 amp with positive voltampere characteristics, and these may operate directly from the line without a ballast. The illustration shows two basic differences in brightness characteristics of the low- and high-intensity arc lamps. The brightness of the high-intensity arc is higher and depends markedly on current.

Although carbon arc lamps are built for alternating-current operation, direct current produces a steadier and more satisfactory operation. To start the arc, the carbons are brought together for a short period, or a third, starting electrode is used. As the carbon burns away, the arc gap must

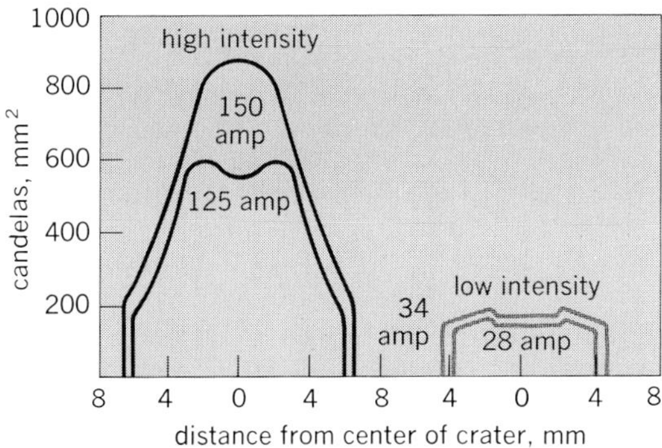

Crater brightness distribution in forward direction for typical low- and high-intensity carbons.

be adjusted for efficient operation. This may be done by manual adjustment; however, an automatic feed mechanism is more efficient and satisfactory.

Flame arc lamp. The flame arc lamp radiates light from the arc instead of from the electrode. The carbon is impregnated with chemicals, which are more volatile than the carbon and, when driven into the arc, become luminous. The chemicals commonly used are calcium, barium, titanium, and strontium, which produce their characteristic color of light as chemical flames and radiate efficiently in their specific wavelengths in the visible spectrum. Some flames use several chemicals, some for light and others for high arc temperature and steady arc operation. Other chemicals may be introduced to produce radiation outside the visible spectrum for specific commercial or therapeutic purposes.

The flame arc is available in short-burning, medium-electrode-life type and long-burning type. With enclosure and automatic adjusting mechanism, and with consideration of the proper chemical composition, the electrode life of the long-burning type may be more than 100 hr.

Metallic electrode arc lamp. In this type of lamp, light is produced by luminescent vapor introduced into the arc by conduction from the cathode. The positive electrode is solid copper, and the negative electrode is formed of magnetic iron oxide with titanium as the light-producing element and other chemicals to control steadiness and vaporization. These lamps are limited to direct-current use. A regulating mechanism adjusts the arc to a constant length.

[JOHN O. KRAEHENBUEHL]

Bibliography: D. G. Fink and M. W. Beaty, *Standard Handbook for Electrical Engineers*, 11th ed., 1978; J. E. Kaufman and J. F. Christensen, *Lighting Handbook*, 5th ed., 1972.

Arc welding

A welding process utilizing the concentrated heat of an electric arc to join metal by fusion of the parent metal and the addition of metal to the joint usually provided by a consumable electrode. This discussion of the arc-welding process is divided into five general subjects: kinds of welding current, methods of welding, arc-welding processes, types of arc welders, and classification of electrodes. Figure 1 presents a fundamental arc-welding circuit, and Fig. 2 shows the elements of the weld at the arc. For the metallurgical aspects of welding *see* WELDING AND CUTTING OF METALS.

Welding current. Electric current for the welding arc may be either direct or alternating, depending upon the material to be welded and the characteristics of the electrode used. The current source may be a rotating generator, rectifier, or transformer and must have transient and static volt-ampere characteristics designed for arc stability and weld performance.

On direct current the distribution of heat in the arc generally produces a concentration of heat at the positive terminal. The work generally has the larger mass and requires more heat for proper fusion; therefore the work is made positive and the electrode negative. This condition is known as straight polarity. When certain covered electrodes are used, the electrode is made positive and this condition is referred to as reverse polarity. With alternating current there is no polarity because of the current reversal each half cycle.

Methods. There are three basic welding methods; manual, semiautomatic, and automatic. Each has its benefits and limitations.

Manual welding. This is the oldest method, and though its proportion of the total welding market diminishes yearly, it is still the most common. Here an operator takes an electrode, clamped in a hand-held electrode holder, and manually guides the electrode along the joint as the weld is made. Usually the electrode is consumable; as the tip is consumed, the operator manually adjusts the position of the electrode to maintain a constant arc length.

Semiautomatic welding. This method is fast becoming the most popular welding method. The electrode is usually a long length of small-diameter bare wire, usually in coil form, which the welding operator manually positions and advances along the weld joint. The consumable electrode is normally motor-driven at a preselected speed through the nozzle of a hand-held welding gun or torch.

Automatic welding. This method is very similar to semiautomatic welding, except that the electrode is automatically positioned and advanced along the prescribed weld joint. Either the work may advance below the welding head or the mechanized head may move along the weld joint.

Welding processes. There are, in addition to the three basic welding methods, many welding processes which may be common to one or more of

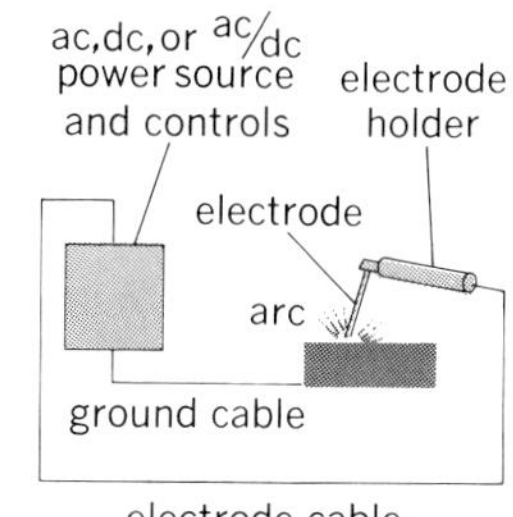

Fig. 1. Fundamental connections of electric equipment and workpiece for arc welding.

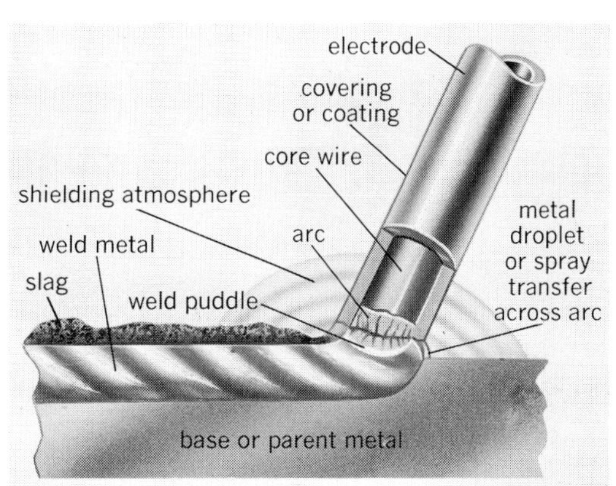

Fig. 2. Metallic welding arc.

these methods. A few of the more common are described below.

Carbon-electrode arc welding. This process has been superseded, to a great extent, by other welding processes, but is still in limited use for welding ferrous and nonferrous metals. Normally, the arc is held between the carbon electrode and the work. The carbon arc serves as a source of intense heat and simply fuses the base materials together, or filler metal may be added from a separate source. A modification in which the arc is formed between two carbon electrodes is known as twin carbon arc welding, where the work does not form a part of the electrical circuit and only the heat of the arc is played against the work. The carbon electrodes are slowly consumed, and adjustments must be made while welding to maintain a constant arc length.

Shielded metal arc welding. This is the most widely used arc-welding process. A coated stick electrode, usually 14 in. (35 cm) long, is consumed during the welding operation, and therefore provides its own filler metal. The electrode coating burns in the intense heat of the arc and forms a blanket of gas and slag that completely shields the arc and weld puddle from the atmosphere. Its use is generally confined to the manual welding method.

Submerged-melt arc welding. In this process a consumable bare metal wire is used as the electrode, and a granular fusible flux over the work completely submerges the arc. This process is particularly adapted to welding heavy work in the flat position. High-quality welds are produced at greater speed with this method because as much as five times greater current density is used. Automatic or semiautomatic wire feed and control equipment is normally used for this process (Fig. 3). A modification of this process uses twin arcs (lead and trail) supplied from separate power sources but phased to reduce magnetic interaction of the arc stream.

Tungsten–inert gas welding. This process, often referred to as TIG welding, utilizes a virtually nonconsumable electrode made of tungsten (Fig. 4). Impurities, such as thorium, are often purposely added to the tungsten electrode to improve its emissivity for direct-current welding. The necessary arc shielding is provided by a continuous stream of chemically inert gas, such as argon, helium, or argon-helium mixtures, which flows axially along the tungsten electrode that is mounted in a special welding torch. This process is used most often when welding aluminum and some of the more exotic space-age materials. When filler metal is desired, a separate filler rod is fed into the arc stream either manually or mechanically. Since no flux is required, the weld joint is clean and free of voids.

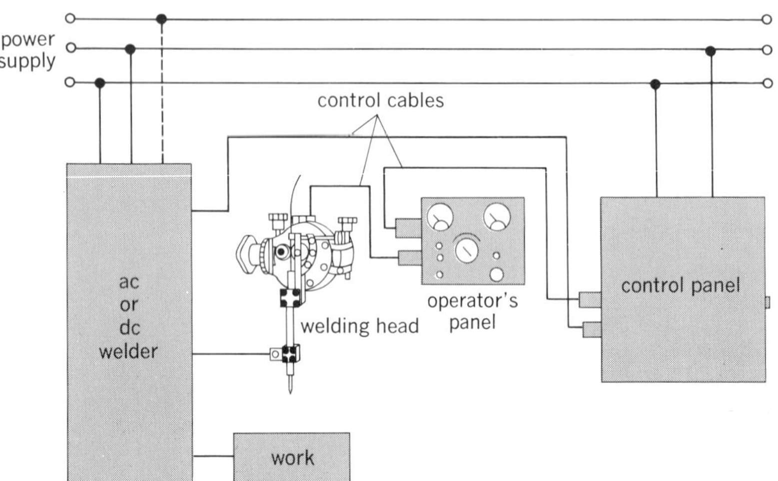

Fig. 3. Schematic diagram of automatic wire feed and control for arc welding.

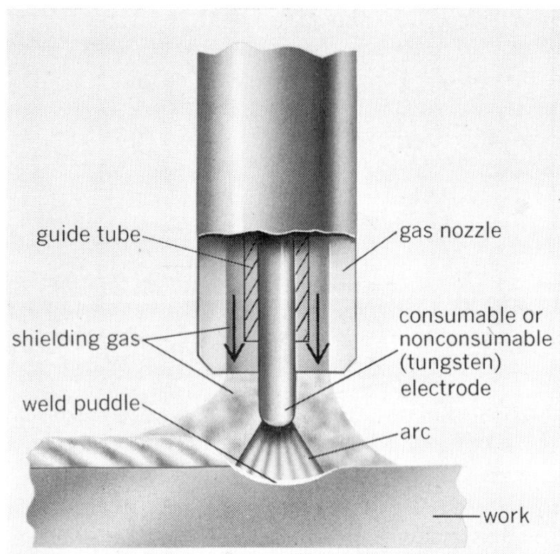

Fig. 4. Inert-gas-shielded metal welding arc.

Metal–inert gas welding. This welding process, often referred to as MIG welding, saw its greatest growth in the 1960s. It is similar to the TIG welding process, except that a consumable metal electrode, usually wire in spool form, replaces the nonconsumable tungsten electrode. This process is adaptable to either the semiautomatic or the automatic method. In addition to the inert gases, carbon dioxide has become increasingly common as a shielding means. In MIG welding, the bare, but sometimes coated, wire is fed into the arc at current densities as high as 10 times those used for conventional shielded metal arc welding. Automatic wire feed and controls are used to drive the wire at speeds as high as 1000 in./min. This extremely versatile process produces high-quality welds at high speeds and with little or no spatter, thus eliminating cleaning.

Space does not allow describing all of the other welding processes, but a few of those most commonly referred to are electroslag welding, flux-cored welding, vapor-shielding welding, and plasma-arc welding.

Arc welders. An arc welder is an electric power generator, or conversion unit, for supplying electric current and voltage of the proper characteristics to a welding arc. Welders may be classified by the type of current supplied to the arc, alternating or direct. Most arc welders are designed for single-operator use. Multiple-operator dc arc welding has limited application and is used primarily where operators are concentrated in a welding area.

Alternating-current welders. These are generally of the static type employing a two-winding trans-

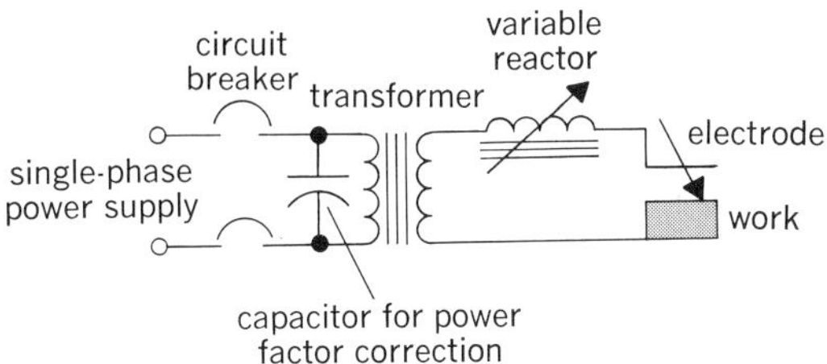

Fig. 5. Schematic diagram of a typical alternating-current arc welder.

former and a variable inductive reactor, which is connected in series with the welding circuit. The reactor may be combined with the transformer to provide adjustable leakage reactance between windings (Fig. 5). The transformer isolates the welding circuit from the primary line, and steps down the line voltage to an open-circuit voltage of 80 volts or less for manual and 100 volts or less for automatic welding.

The ac arc is extinguished each half cycle as the current wave passes through zero. The reactor, by reason of its stored inductive energy, provides the voltage required for reignition of the arc at each current reversal.

Several types of reactor construction are employed, including a movable core with variable air gap, a movable coil to vary leakage between windings, and a saturable reactor with dc control in combination with a tap changer.

Direct-current welders. These may be of the rotating dc generator or static rectifier type, with either constant-current or constant-voltage characteristics.

The dc generator type is usually driven by a directly coupled induction motor or internal combustion engine. The conventional design employs a generator with a combination shunt-field rheostat control and a differentially compounded series field tapped to give an adjustable welding current with drooping volt-ampere characteristic.

Other schemes of current adjustment are used, most of which employ the principle of field control. These include excitation from a third brush utilizing the effect of armature reaction or a movable core to vary the leakage between poles of the generator. Because of the extreme and rapid fluctuation of the arc voltage during welding, the welding generator must be designed for rapid response to load changes.

The rectifier-type arc welder is a combination of transformer, variable reactor, and rectifier. The transformer and reactor designs are similar to those employed in ac welders, except that dc welders are generally used on a three-phase supply (Fig. 6). Some single-phase welders of this type are used, particularly where a dual ac/dc output is required. The rectifiers are the selenium dry plate or silicon diode full-wave bridge type.

Both the rotating generator and rectifier types are also available with constant-voltage characteristics. On these units the output voltage is adjustable, and the welders are designed for either flat or adjustable droop volt-ampere characteristics. In addition, a direct-current inductor is often included in the welder output to tailor the welder's response to the process requirements. The constant-voltage type of welder is most often used with the MIG welding process where the welding current is adjusted by varying the speed of the electrode wire feed. An addition to the continually expanding line of constant-potential type of welders is a direct-current arc welder with a pulsing output superimposed on a relatively smooth background characteristic. This welder provides the power source requirements imposed by a new MIG process, the applications of which have not yet been fully explored.

Multiple-operator dc welding. This method employs one or more power units with approximately 80 load volts and relatively flat voltage regulation. The units operate in parallel to supply a number of resistor outlets, one for each operator. The resistor units provide ballast impedance and a means for adjusting the welding current. This system effects apparatus economy by utilizing the load diveristy of multiple welding arcs.

Electrodes. An arc welding electrode is a wire or rod of current conducting material, with or without a covering or coating, forming one terminal of the electric arc used in arc welding. Electrodes may be classified as nonconsumable and consumable. Nonconsumable electrodes (carbon, tungsten) are consumed at a very low rate and add no filler metal to the weld. Consumable electrodes melt with the heat of the arc, and the metal is transferred across the arc to form part of the weld metal.

Consumable electrodes may be in the form of rods 9–18 in. long or continuous wire. Practically all welding electrodes for manual welding are coated or covered. This covering (usually extruded on the core wire) serves a number of important functions. The gas and slag shielding produced by the covering protects the molten metal from the atmos-

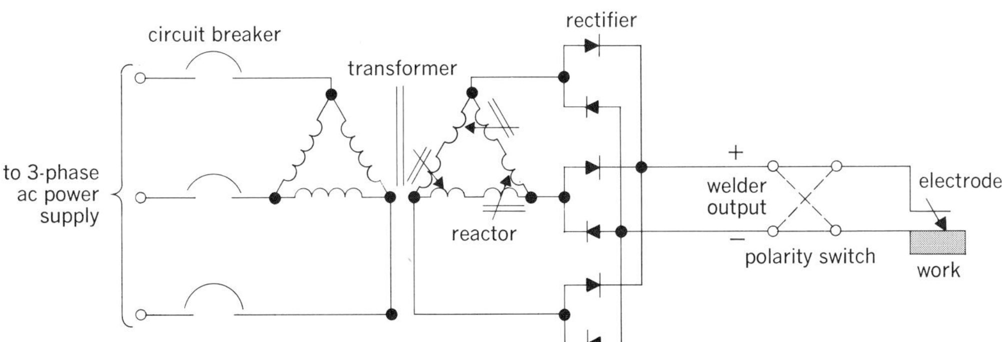

Fig. 6. Schematic diagram of a typical direct-current arc welder.

phere during the transfer across the arc and after deposit.

The materials in the electrode covering are intended to perform one or more of the following: (1) purify the weld metal; (2) control melting rate, penetration, bead shape, and cooling rate; (3) add alloying elements to the weld metal; (4) stabilize the arc; and (5) provide insulating coating to reduce shock hazard.

The covering of some electrodes includes powdered iron to increase the speed of welding by stepping up the rate of metal deposit.

Welding electrodes have been classified according to tensile strength of the deposited weld metal, type of covering, welding position, and type of current. The electrodes are identifiable by electrode classification numbers under specifications prepared jointly by the American Welding Society and the American Society for Testing and Materials. *See* RESISTANCE WELDING.

[EMIL F. STEINERT]

Bibliography: R. Bakish and S. S. White, *Handbook of Electron Beam Welding*, 1964; O. Blodgett and J. Scalzi, *Design of Welded Structural Connections*, 1961; A. L. Phillips (ed.), *Welding Handbook*, 4th ed., 1957.

Arch

A structure, usually curved, that when subjected to vertical loads causes its two end supports to develop reactions with inwardly directed horizontal components. The commonest uses for an arch are as a bridge, supporting a roadway, railroad track, or footpath, and as part of a building, where it provides a large open space unobstructed by columns. Although arches are usually built of steel, reinforced concrete, or timber, the Saguenay River Bridge is an all-aluminum arch spanning 290 ft (88 m) from center to center of skewbacks.

The designations of the various parts of an arch are given in Fig. 1. The skewback is the abutment or pier surface upon which the arch rests. Because the arch springs from its skewback, the intersection of the arch and the skewback is called the springing line. The upper surface of the arch is the extrados, the inner surface the intrados. The line passing through the center of gravity of each section is the arch axis. The crown is the highest section.

The advantage of the arch over a simple truss or beam of the same span is the reduction in the positive moment by the negative moment resulting from the horizontal thrust at the supports. The arch rib is subjected to large axial forces and small bending moments. To minimize the moments, the center line of the rib should closely approximate the funicular polygon for dead load, plus perhaps some portion of the live load. For a uniformly distributed load the funicular polygon is a parabola.

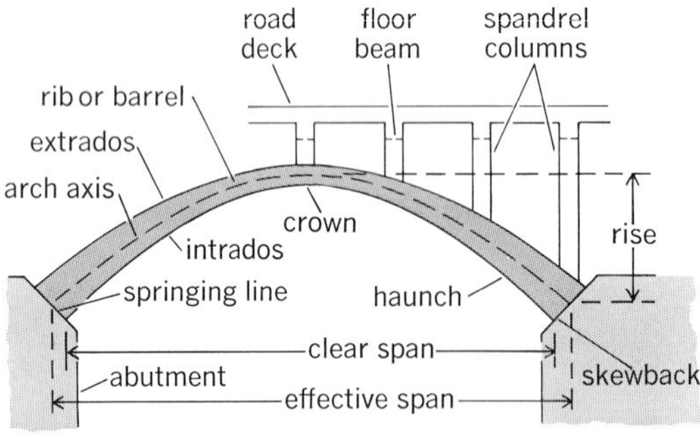

Fig. 1. An open-spandrel, concrete, fixed-arch bridge.

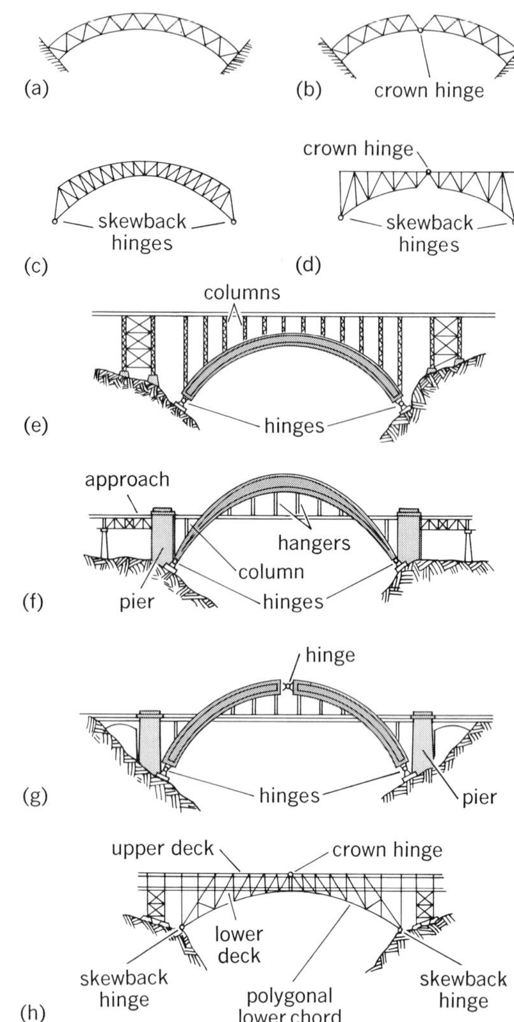

Fig. 2. Various types of bridge arches (*a*) Fixed. (*b*) Single-hinged. (*c*) Two-hinged. (*d*) Three-hinged. (*e*) Parallel curve rib arch. (*f*) Half-through, two-hinged, crescent-rib arch. (*g*) Parallel curve, three-hinged rib arch. (*h*) Double-deck spandrel-braced arch. (*From G. A. Hool and W. S. Kinne, Movable and Long-span Steel Bridges, 2d ed., McGraw-Hill, 1943*)

The principal dimensions of the center line of the arch are span and rise. These may be dictated by conditions at the site, if the structure is a bridge, or by architectural requirements, if the arch is to form part of a building. A rise of from one-fifth to one-third of the span may prove economical.

On the basis of structural behavior, arches are classified as fixed (hingeless), single-hinged, two-hinged, or three-hinged (Fig. 2). An arch is considered to be fixed when rotation is prevented at its supports. Reinforced concrete ribs are almost always fixed. For long-span steel structures only fixed solid-rib arches are used. The Rainbow Bridge at Niagara Falls, with a span of 950 ft (290 m) and a rise of 150 ft (46 m) is the longest fixed steel arch in the world. The Henry Hudson Bridge in New York City is a fixed arch of 800-ft (244-m) span. Because of its greater stiffness the fixed arch is better suited for long spans than hinged arches.

A hinge introduced into the arch rib produces an axis of zero moment forcing the thrust line (funicular polygon) to pass through the hinge. An arch hinged at the crown is a single-hinged arch; it is rarely used because it has no distinct advantages. The two-hinged arch has hinges at each skewback. Because foundations for two-hinged arches receive thrust forces only, the abutments are more easily planned. A hinge introduced at the crown of a two-hinged arch forms a three-hinged arch. The three-hinged arch is statically determinate, and no stresses result from temperature, shrinkage, or rib-shortening effects. *See* STRUCTURAL ANALYSIS.

Because the horizontal component of an arch reaction is large, excellent foundation conditions must exist at the site. Abutments constructed on a steeply sloping gorge of strong, sound rock transmit the thrust directly to the bedrock. Abutments on rock can usually be designed to resist the moment developed at the skewback of a fixed-end arch. Foundations on earth or piles may rotate and relieve part of the assumed restraint. A slight yielding of foundations usually causes no serious harm to three-hinged arches. If suitable foundation conditions are not available, the horizontal reaction component may be provided by a tie between skewbacks. An arch so constructed is classified as a tied arch (Fig. 3). The tied arch is usually more expensive than the two-hinged type.

Concrete arches. Concrete is relatively weak in tension and shear but strong in compression and is therefore ideal for arch construction. Figure 1 is an sketch of an open-spandrel concrete fixed-arch bridge. The arch proper consists of two or more solid ribs or a single solid barrel whose width is equal to that of the bridge. The solid-rib arch supports its roadway by means of a system of beams and spandrel columns. The rib type is particularly desirable for skew crossings because each rib may be arranged to act independently and is designed as a right arch. Ribs are usually interconnected by cross struts. Hollow box sections have been used in place of solid ribs to form a hollow-rib arch.

The roadway in the barrel-type arch is supported on an earth fill, which is confined by spandrel walls, extending from the barrel to the road deck. The barrel type, frequently called a spandrel-filled arch, is generally less economical of material than the rib type. Where for architectural reasons a solid wall effect is desired for a low, flat arch, the spandrel-filled design is often adopted. Placing solid curtain walls over the spandrel openings of an open-spandrel arch achieves the same effect.

Because the rib or barrel is subject to compression and some bending, it is designed as an eccentrically loaded column, with reinforcement placed near both the intrados and extrados. Steel percentages should be kept small.

Precast reinforced concrete arches of the three-hinged type have been used in buildings for spans up to 160 ft (49 m).

Steel arches. Steel arches are solid-rib or braced-rib arches. Solid-rib arches usually have two hinges but may be hingeless. A parallel curved rib (with a constant depth throughout its length) is the most commonly used form of plate girder or solid webbed rib.

Fig. 3. Fort Pitt highway bridge, a double-deck tied-arch bridge across Monongahela River at Pittsburgh, Pa. Unusual are a truss instead of girders as ties, and box sections as arch ribs. (*Lead Industries Association*)

The braced-rib arch has a system of diagonal bracing replacing the solid web of the solid-rib arch. The world's longest arch spans are both two-hinged arches of the braced-rib type, the Sidney Harbor Bridge in Australia and the Bayonne Bridge at Bayonne, N.J. (Fig. 4), which have spans of 1650 and 1652 ft (502.9 and 503.5 m), respectively. Both are of the half-through type. Other classifications according to the method by which the roadway is supported by the arch proper are through arches and deck arches. Through and half-through arches are usually of the rib type.

The spandrel-braced arch is essentially a deck truss with a curved lower chord, the truss being capable of developing horizontal thrust at each

Fig. 4. The Bayonne Bridge across the Kill Van Kull in New Jersey. (*Port of New York Authority*)

support. This type of arch is generally constructed with two or three hinges because of the difficulty of adequately anchoring the skewbacks.

Wood arches. Wood arches may be of the solid-rib or braced-rib type. Solid-rib arches are of laminated construction and can be shaped to almost any required form. Arches are usually built up of nominal 1- or 2-in. (2.5- or 5-cm) material because bending of individual laminations is more readily accomplished. For wide members a lamination may consist of two or more pieces placed side by side to form the required width. All end-to-end joints are scarfed and staggered. Although the cross section is usually rectangular, I-beam sections have been used. Because of ease in fabrication and erection, most solid-rib arches are of the three-hinged type. This type has been used for spans of more than 200 ft (60 m).

The lamella arch has been widely used to provide wide clear spans for gymnasiums and auditoriums. The wood lamella arch is more widely used than its counterpart in steel. The steel lamella roof for the civic auditorium in Corpus Christi, Tex. has a clear span of 224 ft (68 m). The characteristic diamond pattern of lamella construction provides a unique and pleasing appearance. Basically, lamella construction consists of a series of intersecting skewed arches made up of relatively short straight members. Two members are bolted, riveted, or welded to a third piece at its center. These structures are erected from scaffolds because no supporting action develops until a large section is in place. Construction starts from each sill and moves up until both sides meet in the center. *See* BUILDINGS.

The horizontal thrust of the lamella roof must be resisted by steel tie rods or wood ties at wall height or by buttresses. The thrust component developing in the longitudinal direction of the building may be resisted by ties extending the length of the building or by the roof decking. *See* BRIDGE; TRUSS.

[CHARLES N. GAYLORD]

Bibliography: V. Leontovich, *Frames and Arches*, 1959.

Architectural engineering

That branch of engineering dealing primarily with the properties and characteristics of building materials and components, and with the design of structural systems for buildings rather than the structural design of heavy construction, such as bridges, dams, and highways. The term is also used to indicate a field embracing all of the engineering aspects of building design, including mechanical and electrical equipment, architectural acoustics, illumination, air conditioning, and layout for usage, in addition to the historic primary interest in design of the building structure. When used in this broader sense, architectural engineering connotes building structures as its specialty, since the growing complexity of the other fields generally requires their engineering design to be accomplished by specialists trained in the pertinent branches of engineering. However, the coordination of the structural, mechanical, and electrical systems with each other and with the architectural scheme is often the responsibility of the architectural engineer. *See* BUILDINGS.

Because of the complexity of modern building, the need for engineers trained to appreciate the architectural problems involved in building design is greater today than it was in the early 1890s, when the first architectural engineering curriculum was established at the University of Illinois. The development of such curricula has varied widely among colleges and universities offering degrees in this field and has been influenced strongly by the attitudes of the various departments of architecture and engineering. The undergraduate architectural engineering curriculum usually offers a thorough grounding in mathematics, structural theory, and building materials; sufficient work in the engineering sciences, such as thermodynamics and electrical theory, to lay a foundation for intelligent collaboration with mechanical and electrical engineers; and enough study of architectural design or building-type analysis to understand the importance of technological factors in meeting the occupancy requirements of a building.

Criticism has been leveled that such curricula (especially where they are only 4 years in length) do not have sufficient depth to warrant accreditation as an engineering curriculum. One school of thought contends that graduates in civil, mechanical, and electrical engineering might better supplement their undergraduate work by graduate study, perhaps in a school of architecture, in order to focus their previous technical training on the special problems of building design. A more comprehensive background in the engineering sciences would thus be achieved than is possible in a 4-year undergraduate architectural engineering curriculum. This criticism has been met in some schools by expanding the architectural engineering curriculum to 5 years.

Development of education in this field is fostered by the architectural engineering division of the American Society for Engineering Education, and curricula in architectural engineering are accredited through the Engineers Council for Professional Development. Graduates of such programs usually become registered as professional engineers. Some state registration laws permit the designation architectural engineer as a professional specialty while others, such as New York, prohibit the use of this title unless the individual qualifies for licensure as both an architect and an engineer. To dispel the professional and public confusion, the American Institute of Architects has suggested the term building engineering be substituted for architectural engineering to designate engineering curricula devoted primarily to the engineering aspects of building design. Whether this proposal will eventually receive widespread acceptance among the design professions of engineering and architecture is not yet clear.

[HAROLD D. HAUF]

Bibliography: W. C. Huntington and R. E. Mickadeit, *Building and Construction: Materials and Types of Construction*, 4th ed., 1975.

Armature

That part of an electric rotating machine which includes the main current-carrying winding. The armature winding is the winding in which the electromotive force (emf) produced by magnetic flux rotation is induced. In electric motors this emf is known as the counterelectromotive force.

On machines with commutators, the armature is

Fig. 1. A rotor armature of a direct-current generator or motor. (*General Electric*)

Fig. 2. A stator armature of an ac induction motor. (*Allis-Chalmers*)

normally the rotating member, as shown in Fig. 1. On most ac machines, the armature (Fig. 2) is the stationary member and is called the stator. The core of the armature is generally constructed of steel or soft iron to provide a good magnetic path, and is usually laminated to reduce eddy currents. The armature windings are placed in slots on the surface of the core. On machines with commutators, the armature winding is connected to the commutator bars. On ac machines with stationary armatures, the armature winding is connected directly to the line. *See* COMMUTATOR; CORE LOSS; WINDINGS IN ELECTRIC MACHINERY.

[ARTHUR R. ECKELS]

Armature reaction

The reaction of the magnetic field produced by the armature current on the magnetic field produced by the field current. This causes a distorted flux-density distribution in the air gap. Figure 1 shows the magnetic field produced in the air gap of a two-pole dc motor or generator by (*a*) the mainfield magnetomotive force (mmf), (*b*) armature mmf, and (*c*) both armature and mainfield mmf, respectively. Figure 2 shows the distorted flux distribution in the air gap, curves *a*, *b*, and *c* correspond to the above three cases. Due to the saturation of the armature teeth, the flux density is decreased by a greater amount under pole tip *e* than it is increased under pole tip *f*, and therefore the cross-magnetizing armature reaction produces a demagnetizing effect, and the generated voltage or counter-emf will be reduced when the armature is loaded. In a generator this degrades the voltage regulation. In a motor it tends to increase the speed and may cause instability.

For machines subject to heavy overloads, rapidly changing loads, or operation with a weak mainfield, the resultant flux-distribution distortion by excessive cross-magnetizing armature reaction will cause nonuniform distribution of voltage between commutator segments, and may result in flashover between commutator segments. A poleface (or compensating) winding, embedded in slots in the pole face and excited by armature current, is provided to neutralize the cross-magnetizing effect under the pole faces.

In polyphase synchronous machines, balanced polyphase load current flowing in the armature winding produces a revolving mmf. This mmf rotates at a speed (revolutions per minute) equal to 120 times the line frequency (hertz) divided by the number of poles.

This is exactly the speed of the rotating field poles. Hence the mmf of armature reaction reacts with the mmf of the field winding to produce a displaced resultant magnetic field. The angle of the armature reaction mmf with respect to the pole axis depends upon the power factor and kilovolt-amperes of the load. That component of armature reaction mmf which acts along the pole axis is known as direct-axis armature reaction and has a direct magnetizing or demagnetizing effect. That component located at right angles to the pole axis

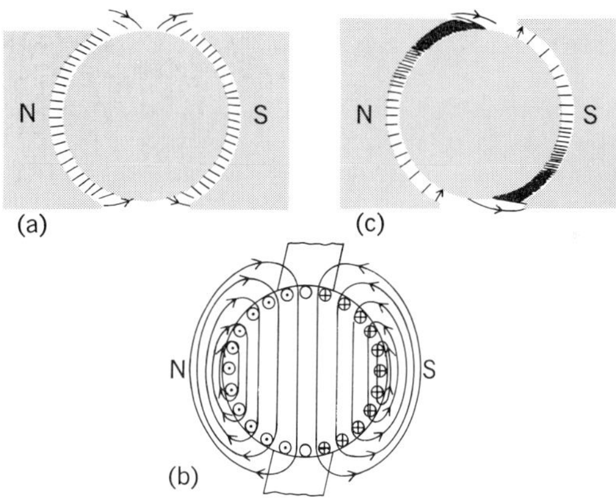

Fig. 1. Magnetic fields in air gap of two-pole machine. (*a*) Main field. (*b*) Armature field. (*c*) Load conditions.

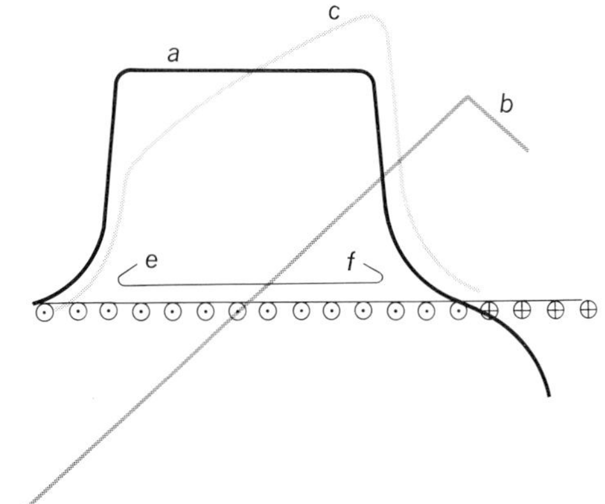

Fig. 2. Distorted flux distribution in air gap.

is known as quadrature-axis armature reaction. It is analogous to the armature reaction effect in modern dc machines. *See* ELECTRIC ROTATING MACHINERY. [ARTHUR R. ECKELS]

Assembly methods

Techniques used to assemble a manufactured product. The development of optimum assembly methods has a great influence on cost and end use of a product. In ancient times skilled craftsmen produced products singly and of varying quality. With the development of modern assembly techniques, products can be produced with better quality at lower cost. Modern assembly methods and the extent to which they are applied are determined by production volume rather than operator skill. The categories of assembly methods are hand assembly, fixture or work-station assembly, progressive line assembly, conveyorized assembly, and automatic assembly.

Hand assembly. In hand assembly, products are assembled by hand without the use of fixtures. It is usually reserved for low-volume assembly if tolerance requirements do not warrant special fixtures. Assembly of large products and of custom-built apparatus is usually by hand. The assembly aids used are of a general nature such as power cranes, screwdrivers, nut runners, and riveters. Hand assembly will always have a definite place in assembly methods, but as volume increases, an effort is made to upgrade hand assembly to either work-station or conveyorized assembly.

Fixture or work-station assembly. In fixture or work-station assembly, jigs and fixtures are provided as operator aids. Convenient parts bins and power tools are also provided to reduce lost motion and increase production. Products may vary in size from large aircraft, where jigs play an important part in assuring the close tolerances required, to small-instrument assembly, the criterion for selection being the most effective use of labor and capital equipment. Usually no attempt is made to transfer the product mechanically between assembly stations, and the range of required operator skill varies with the amount and type of assembly done at each station. The majority of assembly work in manufacturing is by this method. It is basically, like hand assembly, a batch process with little reduction of in-process inventory. *See* JIG, FIXTURE, AND DIE DESIGN.

Progressive line. The next upward step in method implies an effort to integrate the various assembly operations to provide a flow of product assembly and reduce in-process inventory. The product is transferred manually between operators as it is being assembled. This type is called a progressive line. The line may provide only transfer between assembly operations, or it may include fixturing to allow assembly on the line. Properly paced progressive lines increase production and reduce the range of assembly skills required by each operator. One requirement of this method is that each operator's assembly task be balanced with the line. In progressive assembly, machines and work stations are arranged approximately in the order of the assembly sequence even though the product does not warrant a conveyor. Progressive assembly can accomplish paced production and can lower inventory costs without a large capital investment.

Conveyorized line. The next step in assembly methods is the conveyorized line. A conveyor transfers the product between assembly operations. This is a specialized version of the progressive line; it must be designed and installed for a particular product. It usually does not have the flexibility of the more general types of assembly methods discussed above. A limitation of this method is that a breakdown at any station along the line due to a shortage of parts or mechanical difficulty stops the entire line. When volume justifies it, this type of assembly enables operators to be most productive and uses inventory most efficiently.

Mechanized or automatic assembly. Where product characteristics and high volume dictate, the assembly operation may be transferred to a machine in whole or in part. This is referred to as mechanized or automatic assembly. In mechanized assembly, the operator feeds parts individually to the machine. In automatic assembly, the materials handling is also mechanized with the assembly parts bulk fed to the machine where they are automatically oriented and assembled. Mechanized or automatic assembly is justified only with high-volume production. Requirements on quality of parts and on machine maintenance are greater than for hand assembly. The benefits are product uniformity and high production rates. An example of automatic assembly is the electric lamp industry. The general types of assembly methods do not rule out combinations of different types in producing the same product. Standard subassemblies may be assembled mechanically or automatically to be assembled later in a range of different products at a work station or on a progressive line.

Other improvements in assembly include changes in product design or in assembly method to reduce assembly costs. Common assembly processes such as fastener insertion and assembly may be mechanized to cover a range of products. But, in any case, the assembly methods used and the improvements contemplated must be weighed against the savings and other benefits that they may provide. *See* PRODUCTION ENGINEERING.

[DONALD C. CUMMING]

Automation

A coined word having no precise, generally accepted technical meaning but widely used to imply the concept, development, or use of highly automatic machinery or control systems.

Concepts. Common concepts of automation are:

1. Production systems so integrated that materials move through the required operations with little or no human assistance. The degree or character of control is not a criterion. Hence, automation implies highly automatic manufacturing, however achieved.

2. Control systems that automatically maintain machinery performance within desired limits. Regulation is usually accomplished by control devices that sense the condition of significant variables of the system and then translate this information into an appropriate command (feedback) to the machinery control. Automation, therefore, implies control systems that adjust machinery, without human attention, to yield the desired performance despite internal or external disturbances.

3. Machinery that mathematically manipulates information (usually in the form of numbers), storing, selecting, presenting, and recording input data or internally generated data as required. Therefore, automation implies the application of computing machinery to the storage and processing of business, engineering, and scientific data. The term thus used is a synonym for data processing.

4. Any machinery or equipment (or changes in product design, materials, or processing technique) that reduces the labor content in a product or service. Automation thus is sometimes used as a synonym for technological advance that alters the former degree or character of mechanization and labor content.

Historical development. D. S. Harder of the Ford Motor Co. originated the word automation in late 1946 to describe machinery then being developed by Ford to move engine blocks automatically into and from transfer machines and to remove large stampings from presses. Ford formed the first automation department in April, 1947, and appointed to it manufacturing engineering specialists called automation engineers. Automation first appeared in print in *American Machinist*, Oct. 21, 1948, in a description of the efforts of Ford's automation group. It was defined as "the art of applying mechanical devices to manipulate work pieces into and out of equipment, turn parts between operations, remove scrap, and to perform these tasks in timed sequence with the production equipment so that the line can be put wholly or partially under pushbutton control at strategic stations."

Automation began to take on a new meaning with the publication in 1950 of *The Human Use of Human Beings* by Norbert Wiener of the Massachusetts Institute of Technology. Wiener credited the vacuum tube as being the major instrument of a new era of technological advance by making self-regulating controls feasible. He predicted that automatic controls would make the automatic factory a common reality within 25 years and conveyed to many readers that automatic manufactering would be substantially the result of feedback control and computers applied to production machinery. He added that this would bring about a depression. These statements began to give automation an ominous tone in many quarters.

Automation was defined by John Diebold in *Automation: The Advent of the Automatic Factory* as denoting both automatic operation and the process of making things automatic.

Since 1956 the claims for automation as a unique and devastating innovation have declined. The dire predictions of automatic factories, of increased skill requirements for common labor, and of unemployment have not materialized. Progress in mechanization of any activity – mental or physical – continues to be a matter of evolution. Automation has become commonly accepted to describe significant accomplishments in this area.

Contrary to many claims, progress in automatic production has not been solely or even principally due to the feedback control of machinery. Gains in automatic manufacturing come from five sources:

1. Rearrangements of facilities and equipment to minimize distances to be moved and the manual handling required and to enable one operation to feed the next.

2. Changes to materials using manufacturing processes that are simpler or easier to mechanize (for example, the use of sheet-metal stampings for television cabinets rather than wood).

3. Changes to processes that are easier to mechanize or that lend themselves to mechanization in multiple, simultaneous actions (for example, dip soldering of television chassis connections).

4. Changes in the product design that eliminate production tasks or that make them easier to mechanize (for example, elimination of parts by redesign or the development of transistors to replace complicated vacuum tubes.).

5. More highly automatic machinery, whether based on self-regulating control or not.

Automatic machinery in itself is a vague and imprecise term. Mechanization of a system has three qualities: level, or degree of mechanical accomplishment of all required actions; span, or extent to which mechanization continues through a sequence of necessary actions without interruption; and penetration, or extent to which secondary or tertiary supporting tasks such as lubrication, setup, and repair are themselves mechanized.

Levels of mechanization. Seventeen levels of mechanization have been identified:

1. Hand. Body members are used to perform the required action without tools; for example, manual packaging.

2. Hand tool. A nonpowered tool is added to supplement the force, strength, forming action, or manipulative ability of body members; for example, a screwdriver.

3. Powered hand tool. Mechanical power is added to supplement body strength; for example, a portable drill.

4. Power tool, hand control. A framework is added to guide and limit processing action to specific spatial limits. Control of tool application remains in the operator's hands; for example, a drill press.

5. Power tool, fixed cycle, single function. The machine repeats a single action without human attention; for example a belt conveyor.

6. Power tool, program control, multifunction. The machine performs a fixed sequence of actions without human attention; for example, an automatic home laundry.

7. Power tool, remote control. The point of control is separated from the point of application, enabling consolidation or integration of many controls in one station; for example, a power-plant control board.

8. Power tool, activated by workpiece. The machine is triggered into its cycle by the work, rather than the operator; for example, automotive transfer machine tools.

9. Measurement. The machine identifies quantitatively some characteristic of operation; for example, a boiler temperature gage.

10. Signaling selected values of performance. The machine measures and then signals significant performance actions by making a comparison to established standards; for example, a grinder may be equipped to flash a red light when an off-dimension part is produced.

11. Performance recording. The machine measures and creates permanent records of its action; for example, a strip or dial chart recording of chemical-process characteristics.

12. Machine action altered through measure-

ment. The measurement of performance is translated into a signal that adjusts some simple machine action such as speed, volume, or direction; for example, a boiler flow altered by thermostat control.

13. Segregating or rejecting according to measurement. The machine makes physical disposition of work produced according to some measured characteristics; for example, bottling machines that reject improperly filled units.

14. Selection of appropriate action cycle. The machine selects one of several sets of production actions (of the level 6 order) through identification of some environment or product characteristic; for example, multipurpose transfer machines that automatically choose a proper processing sequence for a succession of intermixed parts.

INITIATING CONTROL SOURCE	TYPE OF MACHINE RESPONSE		POWER SOURCE	LEVEL NUMBER	LEVEL OF MECHANIZATION
From a variable in the environment	Responds with action	Modifies own action over a wide range of variation	Mechanical (nonmanual)	17	Anticipates action required and adjusts to provide it
				16	Corrects performance while operating
				15	Corrects performance after operating
		Selects from a limited range of possible prefixed actions		14	Identifies and selects appropriate set of actions
				13	Segregates or rejects according to measurement
				12	Changes speed, position, direction according to measurement signal
	Responds with signal			11	Records performance
				10	Signals preselected values of measurement (includes error detection)
				9	Measures characteristic of work
From a control mechanism that directs a predetermined pattern of action	Fixed within the machine			8	Actuated by introduction of work piece or material
				7	Power tool system, remote controlled
				6	Power tool, program control (sequence of fixed functions)
				5	Power tool, fixed cycle (single function)
From man	Variable			4	Power tool, hand control
				3	Powered hand tool
			Manual	2	Hand tool
				1	Hand

Fig. 1. Levels of mechanization and their relationship to power and control sources. (*Harvard Bus. Rev.*)

15. Correcting performance after operating. The machine examines its output and adjusts itself to correct it; for example, centerless grinders equipped with diameter inspection and self-setting control adjustment.

16. Correcting performance while operating. The machine measures and adjusts continually during the operation to maintain output within standards; for example, annealing over with an automatic time and temperature control system.

17. Anticipatory performance control. The machine analyzes environmental requirements and adjusts to achieve desired performance prior to operation; for example, a guided-missile system or elements of oil refinery equipment.

These levels can be arranged in a chart (Fig. 1) that reveals a distinct evolution. First, there is the substitution of mechanical power for manual effort, which takes some burden from the worker (after level 2). As increasing degrees of fixed control yield the desired machine action, the worker guides the tool less and less (levels 5–8). As the ability to measure is added to the machine, a portion of the control decision information is mechanically obtained for the operator (after level 8). As the machine is given still higher degrees of automaticity, more and more of the decision making and appropriate follow-up actions are performed mechanically. For instance, as the selection of proper machine speeds, feeds, temperature control, safety controls, and so on is mechanized, further decision-making, judgment, experience, responsibility, and even alertness demands are lifted from the worker (levels 12–14). Finally, the machine is given the power of self-correction to a minor, and then to a greater, degree (levels 15–17), until the need to operate the machine has been completely removed from the worker by full automatic action.

This evolution explains why the effect of increasing automaticity generally is to lower the skill requirements of the operating work force rather than to raise them, as proclaimed in much automation literature. It is true, of course, that the increased complexity of automatic machinery and control may increase the amount of maintenance personnel and the degree of maintenance skill required to the point that the increase of skilled personnel in the total work force becomes significant. New types of jobs requiring higher skill on the part of operators also are occasionally created. However, few studies of changes in work force skill requirements caused by automation have been made, and fewer still definitively have related skill to the character of the mechanization employed. Present studies seem to indicate that (1) the effect of higher levels of mechanization generally is to reduce skill and training requirements for operators and to increase these requirements for some maintenance trades (electricians who must master electronic equipment are particularly affected), and (2) the size of the maintenance task is more nearly proportional to the amount, condition, and degree of perfection of the machinery used, rather to the absolute degree, or level, of automatic performance.

Mechanization profile. By using levels of mechanization as a vertical scale, a mechanization profile (Fig. 2) can be constructed for any system by plotting the level of each required activity in

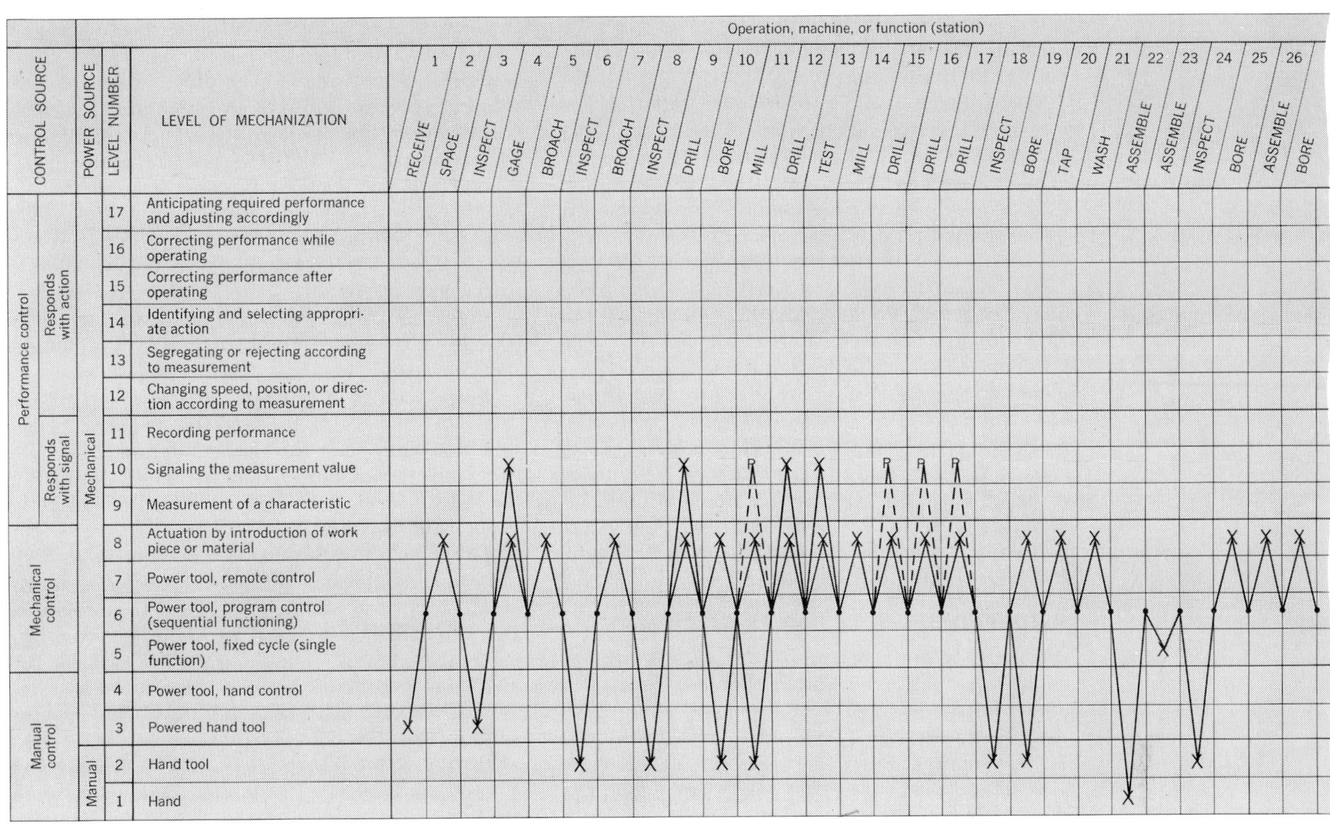

Fig. 2. Mechanization profile of a cylinder-block line. (*Harvard Bus. Rev.*)

sequence. Mechanization profiles of so-called automated plants reveals that the spans of mechanization are quite short, that the levels of mechanization fluctuate widely, that very little self-correcting action (levels 12–17) is found in most systems, and that automatic operations are most frequently achieved by mechanization on levels 5–8. Often growth in the span is more significant than an increase in the level of mechanization in achieving automatic operations.

Trends in automatic maufacturing. Highly automatic production systems exhibited these trends in the 1950–1960 decade:

1. Mechanization of work feeding and removal techniques.
2. Mechanization of material handling between machines, departments, and buildings.
3. Mechanization of more direct labor tasks, especially assembly operations.
4. Mechanization of indirect labor tasks, expecially inspection and testing.
5. Compounding of the production equipment (compression of several or many production functions on one machine base and the performance of multiple operations while the workpiece is clamped in one location.).
6. Centralization of the control of the production system through remote-control panels.
7. Use of feedback control to maintain automatically optimum performance or to acheive extremely precise process control.
8. Use of program control devices, such as magnetic and punched tapes and cards, to direct intricate action sequences without the need for human attention.
9. Mechanization of trouble detection, prevention, and correction by safety controls, indicating lights, built-in control tests circuits, and so on.
10. Mechanization of information collection, transmission, and interpretation through computers and other data-processing machines.
11. Integration of production equipment with information-handling systems.

Some or all of these developments, in many combinations, are interwoven to create automatic systems for manufacturing or data processing. Understandably, therefore, any definition of automation is difficult to use accurately if it is intended to credit all the advances in automatic systems to any one phenomenon such as feedback control. It must be recognized, however, that the refinement and spread of automatic control (not necessarily feedback control) lies at the heart of many current and potential advances in mechanization.

In 1959–1960, the first three chemical plants under digital-computer control were built. Computer process control subsequently has extended to oil refineries, many types of chemical processes, steel and brass rolling mills, blast furnaces, paper mills, and at least one parts-producing line (semiconductors). It is also controlling one automatic bakery warehousing system (Sara Lee Kitchens). More than 500 computer (or computer-monitored) control systems were installed as of 1966, and this was recognized as an important new trend. Numerical control represents an important expansion of this trend. *See* TOOLING.

Another development has inspired new concern about automation: the tape-controlled transfer arm, which can be readily programmed to perform a complex set of actions such as feeding a furnace or performing an assembly operation. This machine is expensive, but as the annual cost of labor increases, it will continue to become more economically attractive to industry.

The growing demand for product variation has forced industry to adopt more flexible manufacturing systems. Thus, automation of some assembly lines has decreased, although automation of many individual parts-manufacturing systems has increased.

[JAMES R. BRIGHT]

Bibliography: American Assembly, *Automation and Technological Change*, 1962; *Automation and Technological Change*, Report of the National Commission on Technology, Automation, and Economic Progress, 1966; J. R. Bright, *Automation and Management*, 1958; E. M. Grabbe, S. Ramo, and D. E. Wooldridge (eds.), *Handbook of Automation, Computation and Control*, vols. 1–3, 1958; A. M. Hilton, *Logic, Computing Machines and Automation*, 1963; V. L. Lossiyevskii (ed.), *The Automation of Production Processes*, 1963; J. F. Schuh, *Principles of Automation*, 1965.

Automobile

A generic term for a self-propelled, trackless personal or public carrier which encompasses passenger cars, recreational vehicles, taxis, and buses used to transport people in cites, on highways, or across country. Passenger cars are available in several body styles, with optional components, and they are made in various sizes.

Major components. Ladder, swept-perimeter, x-type, and unitized frames of various designs are the foundation upon which the chassis is assembled. The chassis consists of all components essential to vehicle operation, except the body. These components include the engine, or power plant, the power train (clutch, transmission, and drive shaft), and the rear axle. In addition, there are the various systems: steering, suspension, brake, fuel-exhaust-emission control, electrical (ignition and lighting), cooling, and lubrication. *See* AUTOMOTIVE CHASSIS; AUTOMOTIVE FRAME; AUTOMOTIVE VEHICLE.

Power plant. American-made passenger cars are equipped with four-stroke internal combustion engines as the source of motive power. Many vehicles are equipped with diesel engines, including both American-made and imported passenger cars.

Four- and six-cylinder engines are available in in-line, slant, or V types; eight-cylinder power plants are V type only. In-line eight-cylinder engines and V-12 and V-16 engines were abandoned for various reasons; aluminum blocks also were tried and discontinued, because of unsatisfactory performance. The vertical profile of the engine is reduced as much as possible to achieve a low hood line and thus an unobstructed view for the driver. Engine, transmission, and axle are matched for performance requirements, and engines are rated for output by the number of cylinders, cubic-inch displacement (cid), horsepower, and miles per gallon. *See* AUTOMOTIVE ENGINE; INTERNAL COMBUSTION ENGINE.

Power train. Single-plate or multiple-disk clutches transfer the engine output to the transmission, drive shaft, rear axle, and rear wheels. Conventional three- or four-speed transmissions may be manual, semiautomatic, or automatic types, with overdrive available as an optional added speed. American-made and imported automobiles have either front- or rear-wheel drive. When engine, clutch, and transmission are integrated for either front-wheel or rear-wheel drive, the unit is known as a transaxle. *See* AUTOMOTIVE TRANSMISSION.

Steering. Among the numerous types of steering systems, with variations in design, are reciprocating-ball, worm-and-sector, or worm-and-roller units. These systems provide vehicle stability in turns and directional control. Power options reduce steering effort and are a distinct advantage when the driver is maneuvering into tight turns or limited parking spaces. *See* AUTOMOTIVE STEERING.

Suspension. Coil springs, leaf springs, air-suspension systems, or torsion bars are used in conjunction with shock absorbers to improve ride comfort and automobile roadability. The introduction of independent front suspension eliminated the front axle; mounting the differential, rather than the rear axle, on the frame has made it possible for each rear wheel to move vertically, independently of the other rear wheel. The automobile body and the passengers are protected from the bouncing effect of chuckholes and road irregularities by the snubbing action of the shock absorbers in both shock and rebound. *See* AUTOMOTIVE SUSPENSION.

Brakes. Service brakes may be drum-type on all four wheels, fixed- or floating-caliper types on front wheels, or a combination of both for mechanical manual operation or optional power assist. Higher hydraulic operating pressures for disk brakes make power assist necessary; front-disk brakes, as an additional safety feature, became mandatory in 1976 by Federal order. Parking brakes are usually integrated mechanically with rear-wheel drum service brakes, or they can be a separate drive shaft–type unit. *See* AUTOMOTIVE BRAKE.

Body types. With emphasis on passenger car size, rather than model, basic body styles are available in most luxury, standard, intermediate, compact, and subcompact groups. Some models are restricted to one size only; others are available in several sizes, as detailed briefly below. A former popular model known as the "rag-top" or convertible has been phased out, as have the running boards, rumble seats, roadsters, and touring cars of the past. *See* AUTOMOTIVE BODY.

Sedan. This body style comes in two- or four-door models for five or six passengers. Optional bucket or bench seats in front accommodate two or three occupants; rear bench seats accommodate three more. Access to rear seats in two-door models is gained by tilting the front seats forward. Door windows may be raised or lowered manually or by optional power assists.

Hardtop. A metal-roof model, usually with a vinyl top, the hardtop replaced the convertible, offering some of its features as well as the all-weather advantages of the sedan. Partial C pillars, in place of the conventional sedan top- and bottom-door hinging, permit an unobstructed open area when the door glass is lowered. The seating capacity of two- and four-door models is the same as that of sedans.

Station wagon. This is a utility vehicle for both passengers and commodities. One or two seats behind the front seat carry five to nine passengers comfortably, with two or four doors for occupants and a swing or lift door for rear loading of packages and luggage. When rear seats are not needed, they can be folded down flush with the floor to provide a level deck for transporting a variety of goods.

Hatchback. Advantages of sedans and station wagons prevail in these small three-door models for passenger- and cargo-carrying capability, particularly in the subcompact group. The back door flips up for convenient loading of packages.

Sports car. A car with a metal or soft top and bucket seats for the driver and passenger qualifies for this category.

Van. Californians first adopted this originally commercial vehicle for personal use as recreational units, rolling sales offices, sound studios, and so on. Elaborate interior and exterior embellishment and customizing are frequently added.

[PAUL A. OTT, JR.]

Automotive body

An enclosure mounted on and attached to the frame of an automotive vehicle. The passenger car body is fabricated from sheet metal, glass, interior trim, and upholstery materials for the convenience and protection of the occupants from the weather and from annoyances such as dirt, dust, noise, and smoke. The body is the showcase which sells the automobile.

All-steel body shells are assembled from numerous stampings in automated welding bucks with a minimum of mechanical fasteners in order to reduce noise accountable to moving parts. Roof, quarter panels, doors, deck lid, and hood undergo similar welding operations in fabrication and are then cleaned and undercoated prior to internal and external painting. Body interior, doors, deck lid, and hood are sprayed with sound-deadening materials before door glass, interior trim, upholstery, and floor and deck carpeting are installed; backup lights, windshield, and similar items are added during the final assembly process.

Unitized bodies are a departure from conventional body-frame construction in that shorter side rails are used for a frame that is integrated with the body and floor pan; in conventional construction, full-length standard side rails with fore-and-aft extensions beyond the body shell proper are used. Fore-and-aft stub frames and their assembled components are attached to the side rails of unitized bodies. The front stub frame includes the steering and suspension systems, power plant, and power train; the rear stub frame contains the suspension system and rear axle (see illustration).

Commercial bodies and cabs have comparable accommodations for the operator and passengers but a much broader range than passenger cars in both size and weight for commodities to be carried. Consequently, the design of commercial bodies incorporates a comprehensive use of steel for strength and aluminum for lightness to assure economically high capacity-to-weight factors wherever possible. This requires studies of the type and volume of cargo to be carried, transit distances, climate, and characteristics of commodities in terms of perishability and protection. Obviously, commercial bodies are customized to their job to a much greater degree than passenger cars are, with the type of vehicle ranging from light pickup trucks, panel trucks, and vans to huge bus and tractor units for tankers and trailers, many having refrigeration, temperature-control equipment, and so on for critical commodities. *See* AUTOMOTIVE VEHICLE.

[PAUL A. OTT, JR.]

Automotive brake

A system with mechanical and hydraulic components to slow or stop a vehicle by means of the action of friction upon wheel drums, disks, or drive shafts. Drum-type brakes are the most popular, and front-disk brakes have been optional; but effective Jan. 1, 1976, a Federal order made front-disk brakes mandatory as standard equipment on all new cars.

Power brakes. In an automotive power brake assembly, movement of the brake pedal operates a valve which allows one side of a piston to be exposed to atmospheric pressure, while engine-intake-manifold vacuum is applied to the other side. The difference in pressure causes the piston to move. This movement is transmitted to the piston in the brake master cylinder, causing the brakes to be applied. One such assembly is shown

Unitized construction of a body for a four-door sedan. Body sections and box girders are welded to form an integral unit, eliminating the need for a separate frame. Engine and power train are carried by this unit. (*American Motors Corp.*)

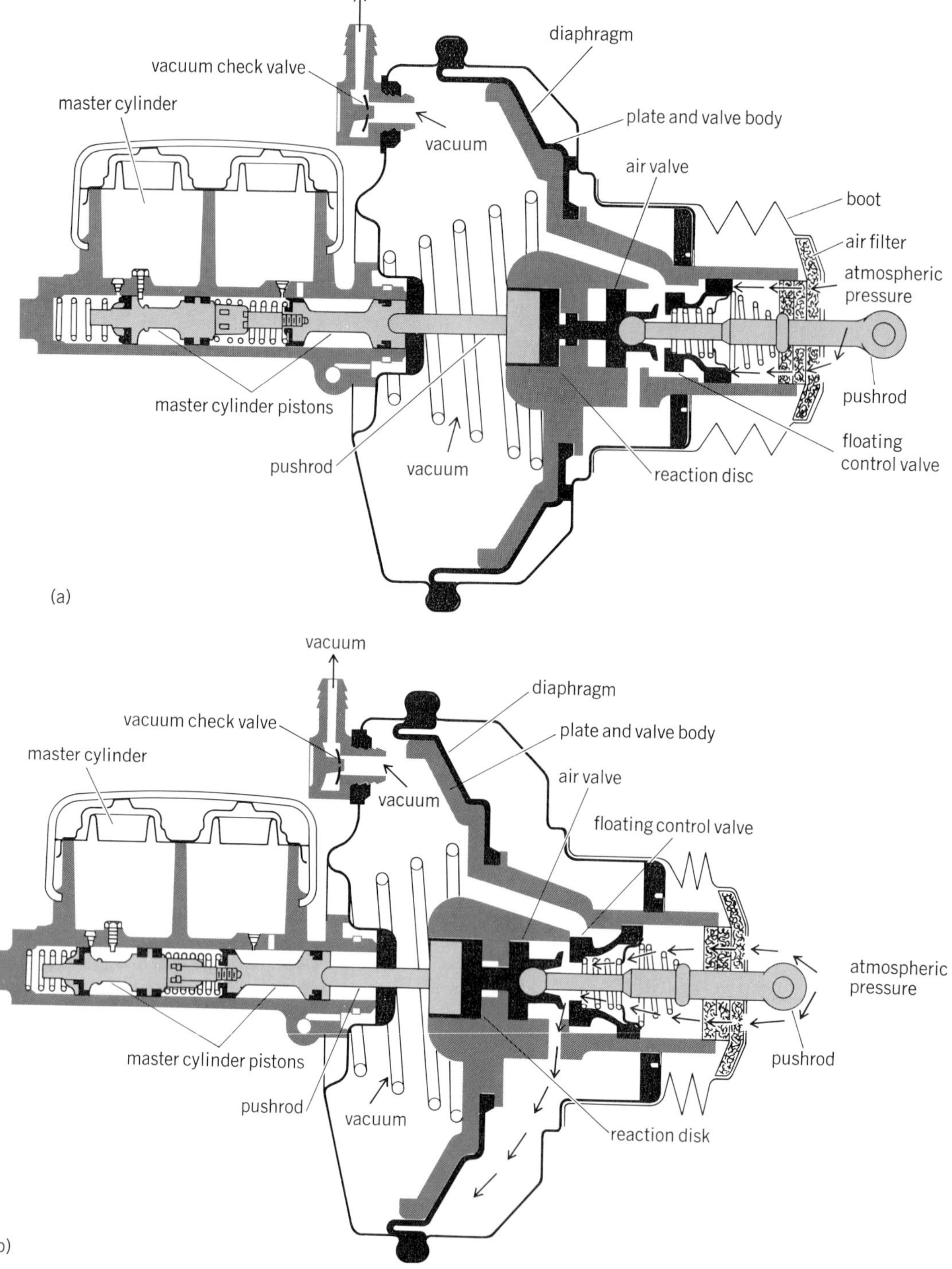

Fig. 1. Schematic views of a power-brake assembly in (*a*) unapplied position and (*b*) applied position. (*From W. H. Crouse and J. D. Helsel, Automotive Transparencies: Automotive Brakes, McGraw-Hill, 1971*)

in Fig. 1. In the unapplied position, the vacuum valve is open and engine-intake-manifold vacuum is applied to both sides of the piston. In the applied position, movement of the brake pedal has forced the push rod into the cylinder, causing the vacuum valve to close. Air pressure on the piston causes it to move, and this movement forces the hydraulic plunger into the hydraulic brake cylinder. Hydraulic pressure is transmitted through the brake lines to the cylinders at the wheels.

Air brakes. In an air-brake system, an engine-driven air compressor provides the pressure to the braking system. Operation of the brake pedal admits high-pressure air to brake chambers at each wheel. A diaphragm or piston in the brake chambers is displaced, forcing the brake shoe against the brake drum. Figure 2 shows a typical air-brake system schematically.

Electric brakes. Brakes can also be applied by electromagnets located at each wheel. The driver

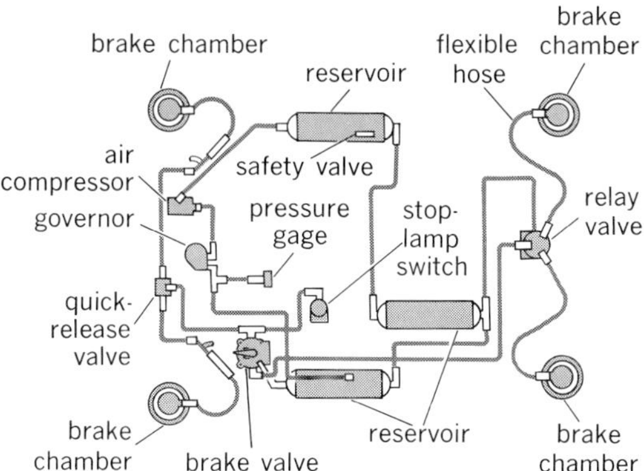

Fig. 2. Schematic diagram of typical air-brake system. (*International Harvester Co.*)

operates a controller that connects the electromagnet to the vehicle battery. The electromagnet is attracted to disks on the rotating wheels, causing the electromagnet to shift through a limited arc. This movement actuates the brake shoes. Further movement of the controller allows more current to flow in the electromagnets, producing stronger magnetic fields and greater braking action. [WILLIAM H. CROUSE]

Drum brakes. Conventional drum-type service brakes have two steel shoes, faced with heat- and wear-resistant lining, for each wheel, which are forced against the inner brake drum surfaces when pedal pressure is applied by the driver (Fig. 3). A dual master cylinder operates the hydraulic systems for front and rear brakes separately to assure braking action in the event that one fails. Pedal pressure forces fluid from the master cylinder to flow through the brake lines to hydraulic cylinders in each wheel, where plungers press the brake shoes against the drums; release springs return the shoes to the inactive position when pedal pressure is released. Some drums are cast aluminum and finned for rapid dissipation of generated heat. Power brakes, which are standard in some cars and optional in others, reduce braking effort by means of intake manifold vacuum or atmospheric pressure.

Parking brakes are independent of service brakes, but they may be integral with wheel brakes, or used on the drive shaft. Parking brakes are set on rear wheels or the drive shaft by the foot pedal and are released manually.

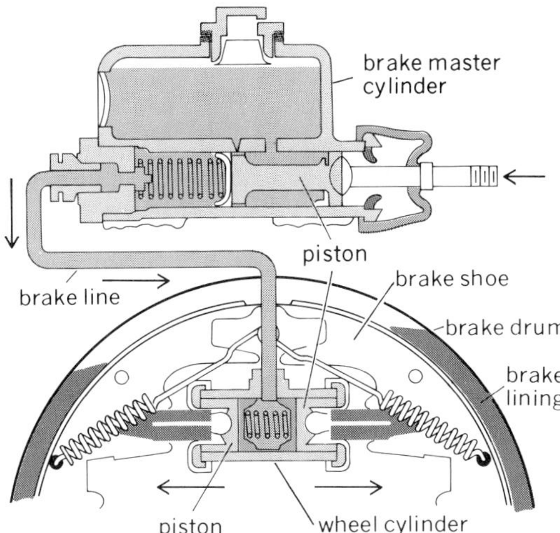

Fig. 3. Schematic view of hydraulic braking system for one wheel on an automobile. Brakes are shown applied. Large arrows indicate motions of pistons, small arrows the direction of motion of hydraulic fluid. (*Pontiac Motor Division, General Motors Corp.*)

Disk brakes. Disk-type brakes are less subject to water fading and more resistant to heat fading due to high speed or repeated stops than drum brakes are, and they are capable of straight-line stops; that is, hard braking effort does not cause them to pull to the right or left. They are available in pressure-plate and fixed- or floating-caliper types. Pressure plates faced with brake lining material are expanded against inner and outer housings (Fig. 4). Fixed-caliper brake assemblies consist of pressure pads faced with brake lining material and hydraulic cylinders anchored in position; pedal pressure forces the pistons to clamp pads against the rotor which is attached to the wheel and which turns with it. As the name implies, floating-caliper brakes are not in a fixed position; they tilt to exert clamping pressure on the rotor, returning to inactive position when released. Because high hydraulic operating pressure is necessary for disk brakes, power-assist units must be added.

[PAUL A. OTT, JR.]

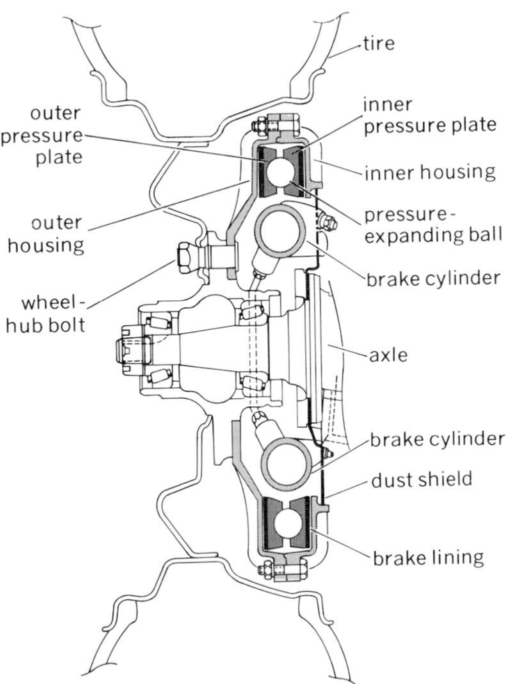

Fig. 4. Schematic view of disk-brake assembly. (*Chrysler Sales Division, Chrysler Corp.*)

Automotive chassis

A bare vehicle (minus body) with frame, power plant, power train, brakes, and wheels, and cooling, fuel, exhaust, electrical, lubrication, steering, and suspension systems assembled and operational. Factory chassis assembly begins with an upside-down frame to which running gear components are attached progressively and which is then rolled over for engine and transmission mounting. Body drop and final assembly operations produce the finished vehicle.

Current trends are toward unitized bodies (without separate frames). The chassis components are attached directly to frame members which are integrated with the body shell and which are adapted to serve the functions of conventional frames. In heavy load-bearing commercial vehicles the frames are retained as the foundation for chassis assembly and body attachment. *See* AUTOMOTIVE BODY; AUTOMOTIVE FRAME; AUTOMOTIVE VEHICLE.

[PAUL A. OTT, JR.]

Automotive engine

An integral major component of and source of power for automotive vehicles. Several types of engines are available for passenger and commercial vehicles, but most vehicles are driven by reciprocating internal combustion gasoline engines operating upon the four-cycle principle worked out by Nikolaus Otto in 1876. Many large commercial vehicles use diesel engines, with the fuel ignited by the heat caused by high air compression, for more economical operation and longer life, rather than gasoline engines with ignition by spark plugs. Diesel engines are available in some American and imported passenger cars. *See* DIESEL ENGINE; OTTO CYCLE.

Domestic passenger car engines evolved from an assortment of electrics and steamers, designed with two or four cycles. Present four-, six-, and eight-cylinder engines are four-cycle types of in-line, V-block, slant, or horizontally opposed "pancake" configurations. In 1900, electrics and steamers far outnumbered gasoline engines, but within 2 decades they faded from the scene. Formerly popular in-line eights and V-12 and V-16 powerplants were discontinued when higher compression ratios were developed. Redesign of six- and eight-cylinder engines provided equivalent horsepower, with incidental reduction in weight.

As vehicles grew in size and type, and greater road speed was demanded, engines tended to increase in size and weight and had to improve in performance. In an attempt to reduce engine weight, aluminum cylinder blocks were tried, but found faulty. Present cylinder blocks and heads are gray iron castings, as are numerous other engine parts. Weight reduction has been achieved by decreasing cylinder wall sections and substituting stampings for castings, and plastics for metal parts.

Higher engine compression ratios produced greater combustion temperatures, and ignition knocks which could be eliminated only with higher-octane fuel, made possible by introduction of gasoline additives. Such developments created a new family of high-performance engines. This developmental trend was modified by the Clean Air Act of 1970, which mandated specific emission controls and use of nonleaded fuel, and encouraged manufacturers to get more miles per gallon.

Engine compartments have limited space available and this has been encroached upon by increasing "hang-on" hardware and by front suspension units and similar space-robbing components. The current tendency toward compact and subcompact cars accentuates this problem, which will undoubtedly require revolutionary changes. Rotary and other types of engines have been available for more than 2 decades, with approximately 40% fewer parts and 50% less bulk. *See* INTERNAL COMBUSTION ENGINE.

[PAUL A. OTT, JR.]

Automotive frame

The basic structure of an automotive vehicle. Frames are composites of heavy steel stampings which include fabricated side rails, cross members, and mounting brackets assembled into a foundation for the major components and bodies of passenger and commercial vehicles.

Side rails have either channel or box sections, with a swept perimeter conforming to body contours. Heavy cross members support the engine in the front and the loads carried in the rear; additional cross members tie the side rails together through their length; bumpers are attached at both ends. Butyl rubber body mounts insulate occupants from noises or vibrations coming from the power plant and power train.

Severe contortions at various points in the frame permit steering and axle clearances at both ends; most of the side rail length is depressed in order to achieve a low center of gravity. The frame width at the front of the vehicle narrows and bends upward to make turning possible and to provide extra height for suspension mounting. Also, at the rear of the vehicle a "camel hump" provides room for axle bounce and suspension mounting; the end section of the frame is higher than the center to avoid bumper scraping when the vehicle is driven up a ramp or steep driveway. Side rail sections and cross members have greater depth and width in the front than in the rear to accommodate the deadweight of the engine and to compensate for the loss of rigidity that occurred when independent suspension units replaced axles (suspension units did not, however, provide the support to the frame given by axles).

For the frame to withstand bending, twisting, lateral, and longitudinal stresses encountered on irregular road surfaces or open terrain, side rails and cross members are designed with these factors and vehicle weight distribution in mind. Side rail channel sections are much deeper in the center for weight support; they taper toward both ends. Larger and heavier side rails are used in commercial vehicles than in passenger cars.

There are several departures from standard ladder-type box frames in both passenger and commercial vehicles. One design incorporates an X for the two major frame elements normally used as side rails; the other uses an X with the box frame, particularly in heavy commercial vehicles, for additional strength and stiffness (see illustration). Another variation has a heavy central tube with a fork at both ends and central cross members to support the body.

Unitized body construction requires a somewhat different frame arrangement, composed of side rails integrated with the body floor pan, to which stub frames are attached at front and rear. Chassis assembly of power plant, power train, and

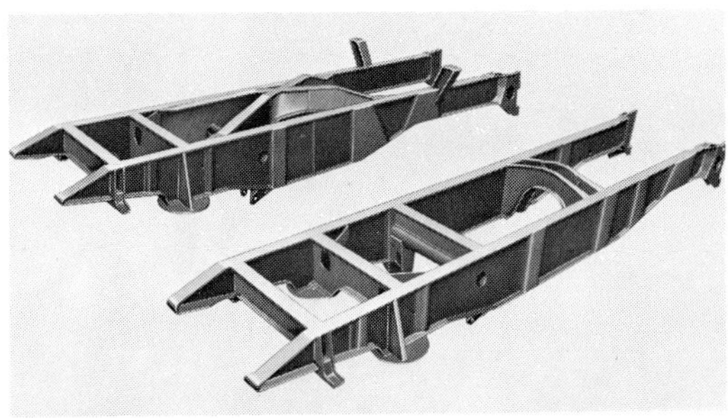

X frame (*above*) and ladder-type frame (*below*) for truck tractors. Both types are made from three-piece fabricated side members with top and bottom flanges continuously machine-welded to form an I beam. (*GMC Truck and Coach Division, General Motors Corp.*)

suspension units follows a procedure similar to that for conventional frames, except that bogeys are used to support the subframe during assembly, after which they are attached to the body. The chassis is attached to the body in this case, whereas the reverse is true with conventional frames. This type of body is used for many passenger cars and some light commercial vehicles. *See* AUTOMOTIVE BODY.

In the past, conventional frame components were fastened with rivets and bolts which suffered from fatigue after considerable rough mileage and handling of vehicles. As a result, these made-to-order noisemakers were a constant source of annoyance. Like modern body shells, the frame is now an almost one-piece creation with extensive welding that eliminates former shortcomings in frame fabrication. Butyl body mounts reduce sound and vibration formerly accepted as a part of driving that had to be endured. Further reduction of sound telegraphed from the frame to the floor pan, deck lid, and hood is accomplished with a variety of sound-deadening materials. *See* AUTOMOTIVE CHASSIS; AUTOMOTIVE VEHICLE; TRUCK.

[PAUL A. OTT, JR.]

Automotive steering

Mechanical means by which a driver controls the course of an automobile or similar vehicle such as a bus, truck, or tractor. A typical system (Fig. 1) consists of a manually operated steering wheel in the driver's compartment connected by a steering column to the steering gear from which linkages attach to steerable road wheels mounted at the front of the vehicle. Clockwise rotation of the steering wheel actuates the steering system to turn the vehicle to the right; rotation counterclockwise turns the vehicle to the left. In manual steering the driver provides all the force necessary to turn the steered road wheels; in power steering an auxiliary mechanism assists the driver by supplying part of the steering force. *See* POWER STEERING.

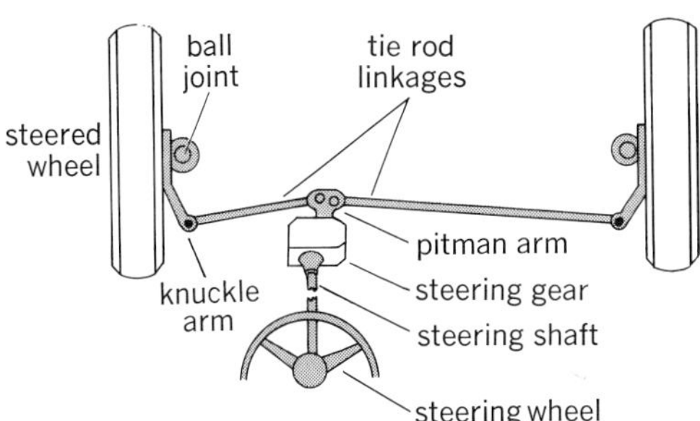

Fig. 1. Components of a typical steering system.

A steering system provides four essential features: (1) sufficient mechanical advantage for driver to steer vehicle without excessive physical effort, (2) steering of road wheels without hindering their free movement on the spring suspensions, (3) differential turning of steered wheels so that both roll on circles with a common center, and (4) tendency for steered wheels to return to the straight forward position when steering wheel is released. *See* ACKERMAN STEERING.

Steering gear. This gear is normally fixed to the vehicular chassis. Rotary motion of the hand wheel is changed by the steering gear and pitman arm to translational movement to position the steering linkage. Together, the gear and linkage provide a mechanical advantage so that steering effort is small; a small tangential pull on the steering wheel rim develops a large force to turn the road wheels. The mechanical advantage also gives such fineness of control that substantial motion of the hand wheel turns the road wheels only slightly.

The reduction mechanism of the steering gear is semireversible. Some reversibility is essential to road sense and recovery. Excessive reversibility allows road shocks to be transmitted to the steering wheel.

At the end of the steering shaft is a housing which encloses the steering gear immersed in lubricant. The output shaft is fitted with the pitman arm to actuate the steering linkage.

Three common types of manual steering gears (Fig. 2) are worm and roller, screw and ball nut, and cam and lever. A rack and pinion, used for steering on some European and American subcompact cars, is fully reversible.

Principal features of the gear are illustrated by the worm and roller. The worm is at the end of the steering shaft, and tapered roller bearings hold it fixed on its axis of rotation. The roller, which is clamped to the output shaft, has parallel grooves that mesh with the spiral gear of the worm. So that the worm and roller will continue to mesh as the output shaft rotates, the contour of the underside of the worm gear is an arc with the axis of the output shaft as its center. Pitch of the worm controls reversibility. The ability of the roller to rotate about its own axis reduces friction below that of a simple worm and gear.

In the screw and ball nut, a segment of a gear on the output shaft maintains meshing contact with a rack milled on one side of a nut. This nut rides on a screw at the end of the steering column. Ball bearings interposed between the screw threads on the screw and in the nut drive the nut axially along the threaded steering shaft. The balls reduce friction between the screw and nut and thus lessen steering effort.

In the cam and lever steering gear, a screw-type cam is fixed to the lower end of the steering shaft. The cam is milled with a continuous spiral groove in which the studs of the lever shaft are engaged. The spiral groove may, by design, provide either constant or variable ratio of movement between input and output shafts. The degree of reduction provided depends on the size of the vehicle to be

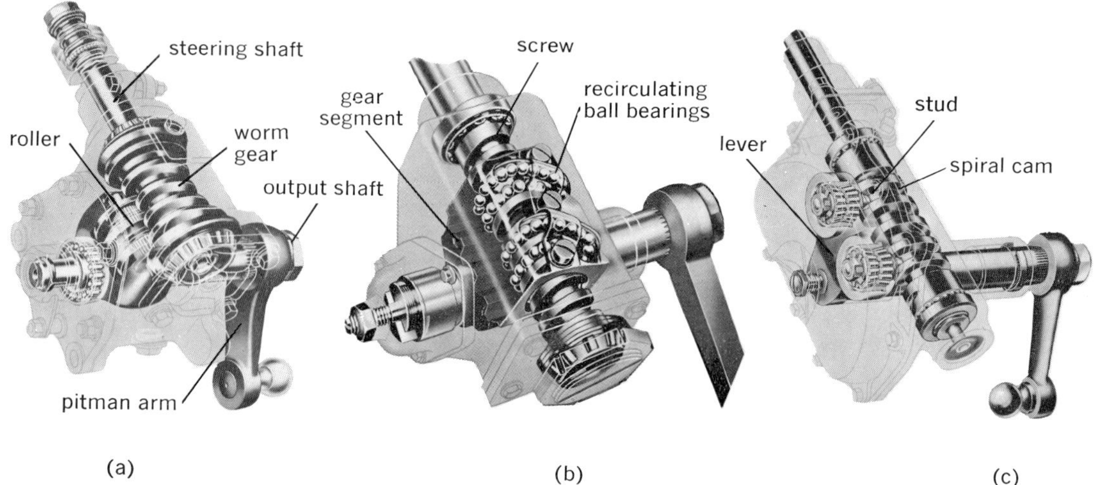

Fig. 2. Drawings of three common manual steering gears. (*a*) Ross-Gemmer worm and roller gear. (*b*) Saginaw screw and nut gear. (*c*) Ross cam and lever gear. (*TRW Inc.*)

steered. At midposition, both studs of the lever shaft are engaged.

Linkage. The steering gear pitman arm connects to the tie rods or drag links that actuate the steered road wheels. The links are rods or tubes and normally have ball joints at each end to permit free movement in any plane with minimum friction and lost motion. These articulated linkages couple from the steering gear that is fixed to the chassis to the road wheels that are suspended to move with road irregularities. The steering geometry of these linkages allows the wheels freedom of motion on their suspensions, yet holds them firmly in the direction determined by the steering wheel. This arrangement permits the vehicle to jounce or roll on rough terrain without causing involuntary self-steering movements, wandering, or abrupt reactions at the hand wheel. Similarly, it permits severe brake application with consequent brake dive without causing self-steering (as distinguished from swerving from faulty brake adjustment).

Steerable road wheels. To roll smoothly when the vehicle is directed into the desired course, the steerable wheels are mounted at slight angles to the normal, as shown in Fig. 3. This drawing shows the front left wheel; all angles are greatly exaggerated for clarity. As viewed from the driver's position in Fig. 3*a*, the wheel tilts outward from the vertical. This tilt is called camber and is positive if outward as shown. The wheels are given slight positive camber (usually less than a degree) initially so that, when the vehicle is loaded, the wheels become almost vertical.

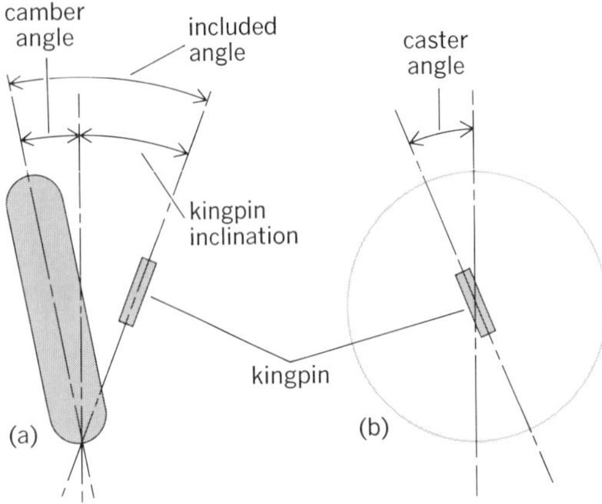

Fig. 3. Front left wheel. (*a*) Viewed from driver's position. (*b*) Viewed from inside hood. Angles exaggerated.

The kingpin, or the axis of the balls on ball suspension, is tilted inward. This kingpin inclination causes the chassis to rise when the wheel turns. Thus the weight of the car tends to keep the wheel turned straight ahead or to return it to that position after a turn. This self-return of a steering system to straight-line travel is called recovery.

Camber and kingpin inclination together are called the included angle (Fig. 3*a*). If the apex of this angle lies on the road surface, road resistance on the tire and forward push at the kingpin are along the line of roll. If the apex is not on the road, the tire is forced to toe in or out. To take up play in the front wheel supports, the front wheels are purposely toed in slightly so that the planes of the wheels intersect ahead of the vehicle. Road resistance then forces them to roll parallel and takes up any play in the steering system.

The inclination of the kingpin or its equivalent forward or backward is called caster. Caster is positive if the kingpin inclines backward, negative if it inclines forward, and zero if the kingpin is vertical as viewed along the axis of the front wheels. Positive caster aids in directional stability and recovery. *See* AUTOMOTIVE VEHICLE.

[WILLIAM K. CRESON]

Bibliography: J. W. Durstine, *The Truck Steering System from Hard Wheel to Road Wheel*, SAE Publ. no. 730039, January 1973.

Automotive suspension

The springs and related parts intermediate between the wheels and frame of an automotive vehicle that support the car body and frame on the wheels and absorb road shock caused by passage of the wheels over irregularities. The four types of spring have been used: coil, leaf, torsion bar, and air.

Coil- and leaf-spring suspension. Figure 1 illustrates a typical automotive suspension arrange-

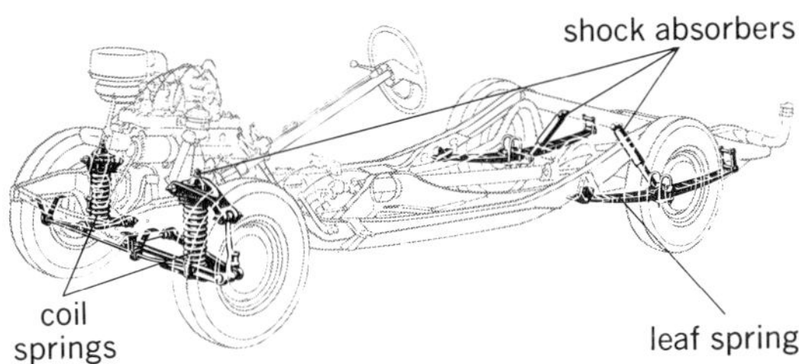

Fig. 1. Phantom view of an automotive chassis, showing location of coil springs (at left, or front of car) and of leaf springs (at right, or rear of car).

ment using coil springs at the front wheels and leaf springs at the rear. Weight of the vehicle applies an initial compression to the springs. When the wheels encounter irregularities in the road, the springs further compress or expand to absorb most of the shock. The suspension at the rear wheels is relatively simple in most cars. It is considerably more complex at the front because the front wheels must swing from side to side in steering. This calls for multiple-point attachments that permit the wheel to move up and down and still be swung away from straight ahead for steering. Figure 1 shows such a system of front suspension.

Shock absorbers are used at each wheel to restrain spring movement and prevent prolonged spring oscillations. The shock absorber contains a piston that moves in a cylinder as the car wheel moves up and down with respect to the car frame. As the piston moves, it forces a fluid through an orifice, imposing a restraint on the springs. Spring-loaded valves open to permit more rapid flow of the fluid if fluid pressure rises high enough, as it may when rapid wheel movements take place. *See* SHOCK ABSORBER. [WILLIAM H. CROUSE]

Air suspension. In this system, suspension units on each wheel contain a cushion of air which supports the sprung weight of the vehicle while maintaining road clearance, alignment of wheels with chassis, and protection of car and passengers from road shock. As in other types of suspension systems, shock absorbers are attached to the frame and suspension unit for additional passenger comfort and prevention of roll on curves and severe

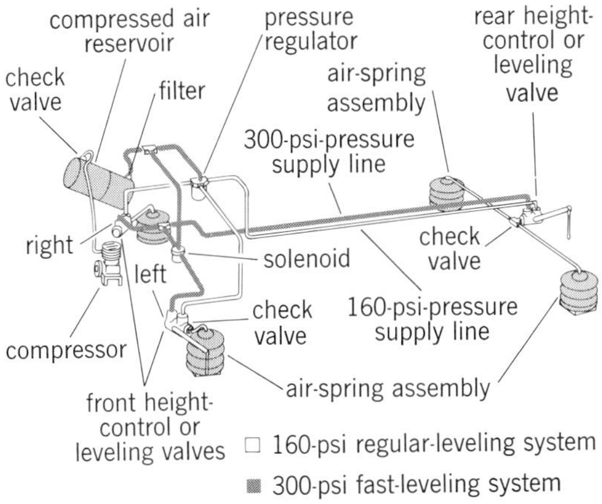

Fig. 2. Basic components of an air-suspension system. Vehicle height is independent of weight.

shock at all speeds for better steering control and safety. This system is diagrammed schematically in Fig. 2; a typical front-wheel air suspension unit is shown in Fig. 3.

Output from the compressor to the reservoir, and from that point to the system, is controlled by the pressure regulator and the check and leveling valves. A filter prevents oil and water from entering and damaging the system. A solenoid actuates the leveling valves rapidly when weight is shifted or when doors are opened.

The air pressure within the bellows of each suspension unit performs the function of mechanical springs, with similar flexing action inside the air chamber. Check valves compensate for loss of air by maintaining constant pressure from the reservoir. An inflow of air from the reservoir prevents the body level from being lowered by increased loading. Conversely, as the body rises with decreasing load, leveling valves release air from the bellows. Thus, air pressure within the system automatically maintains both body level and riding comfort, regardless of load. [PAUL A. OTT, JR.]

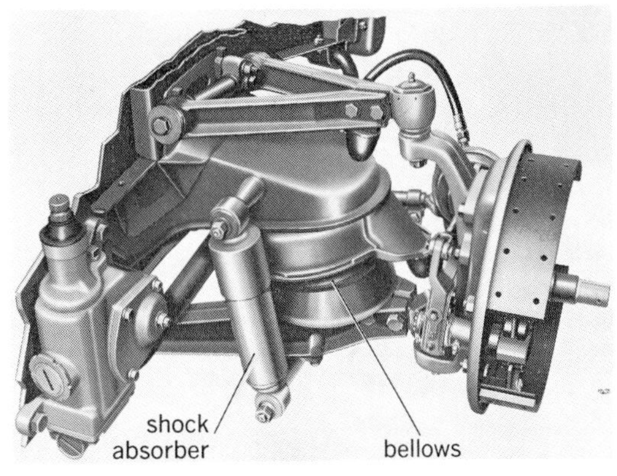

Fig. 3. Left front piston-type air-bellows assembly. (*GMC Truck and Coach Division, General Motors Corp.*)

Torsion-bar suspension. In a torsion-bar system the spring is a bar or rod stressed in torsion about its long axis. *See* TORSION BAR.

A typical torsion-bar system is shown in Fig. 4. The bar is rigidly fixed to the car frame at the rear end and is free to rotate at its front end, near which the lower wheel-control arm is splined to the bar. The longitudinal stiffness of the bar aids steering

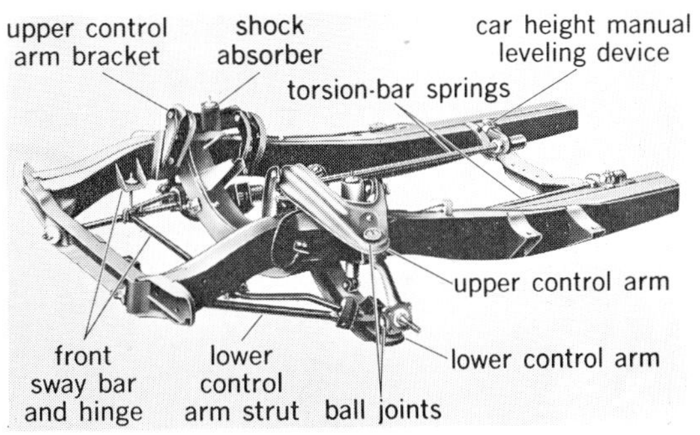

Fig. 4. Torsion-bar suspension. (*Chrysler Corp.*)

and directional stability. The neutral, or normal rest, position of an automobile with torsion-bar suspension is easily changed by adjusting the torque on the bar. [NEIL MAC COULL]

Bibliography: H. E. Barnacle, *Mechanics of Automobiles*, 1964; W. H. Crouse, *Automotive Chassis and Body*, 3d ed., 1967; W. H. Crouse, *Automotive Mechanics*, 5th ed., 1965.

Automotive transmission

The device for providing different gear or drive ratios between the engine and drive wheels of an automotive vehicle. A principal function of the transmission is to enable the vehicle to accelerate from rest up to a maximum speed through a wide speed range while the engine operates within its most effective range. The transmission in most applications is placed in the vehicle power train between the engine and the propeller shaft. Power passes through the transmission and propeller shaft and is delivered to the differential and drive axles. *See* AXLE; DIFFERENTIAL.

There are two general types of transmission in passenger cars: manual and automatic. Because transmissions in trucks and off-the-road vehicles come in such a variety of designs and number of speeds, only passenger car units will be described here. *See* TRUCK TRANSMISSION.

Manual transmission. Transmissions employing layshaft or countershaft gearing are used in conjunction with a clutch. This mechanism disengages the engine from the transmission while the sliding of gears in the transmission is controlled manually. *See* CLUTCH.

Manual transmissions are mainly of the three-speed type; once they were made with synchronizers only in second and third speed or direct drive, to avoid gear-clashing when shifting. Transmissions are now built with synchronizers in all forward speeds (Fig. 1). This type carries a clutch slideably splined on the main shaft (at the left of center in each part of Fig. 1) to obtain third-speed engagement by shifting to the left, and second speed by shifting to the right. The clutch at the right, also slideably splined on the main shaft, obtains first speed when shifted to the left and reverse when shifted to the right. The two clutches are actuated by forks mechanically connected to the hand shift lever, which may be located on the steering column or on the floor directly over the transmission.

Because synchronizers for each of the three speeds operate exactly the same, only the shift into direct drive will be described. After depressing the foot clutch pedal, the operator starts to move the left clutch to the left (Fig. 1). The first movement causes an internal cone, ordinarily bronze, which has slots accommodating three inserts (located in slots in the clutch hub) on which the clutch sleeve slides, to contact an external cone integral with the drive shaft at the left. The frictional contact of the cones causes the bronze cone on the clutch member to rotate slightly, its slots being wider than the projecting inserts. This rotation causes chamfers, ordinarily 45° on the end of the splines of the clutch sleeve, to come into contact with like-angle chamfers on the end of the teeth on the outside of the bronze cone. Pressure now applied on the

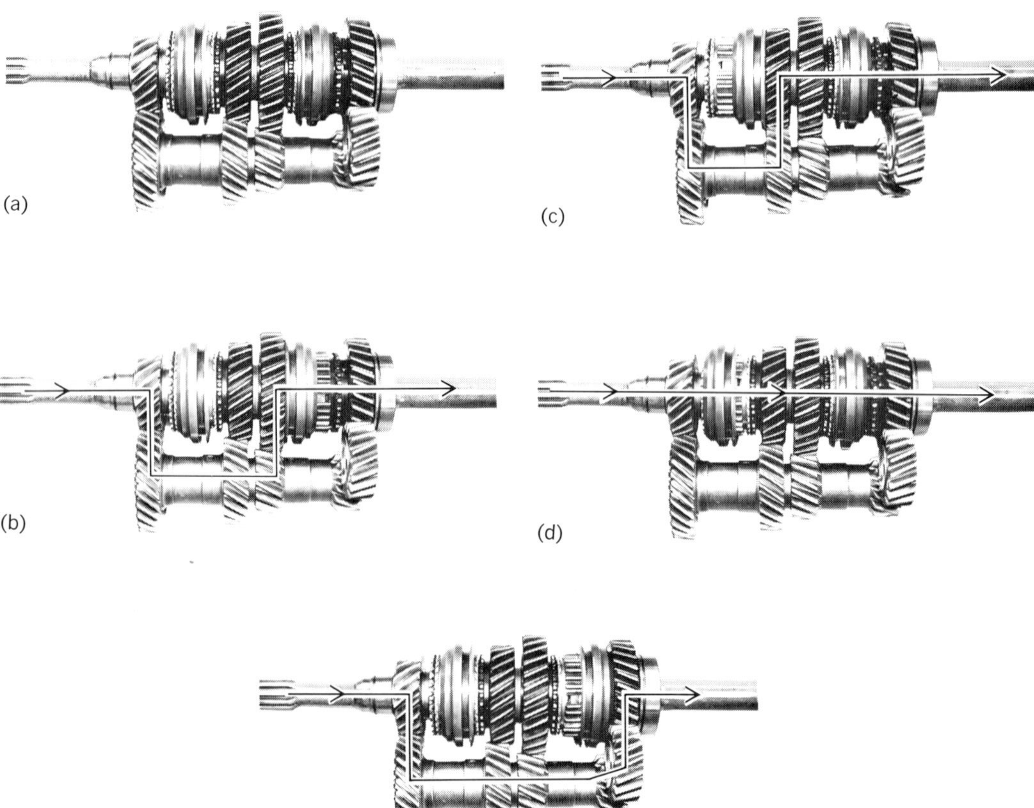

Fig. 1. Three-speed manual transmission gear set, with synchronizers in all forward speeds. The arrows show the direction of the power flow. (*a*) Neutral. (*b*) First. (*c*) Second. (*d*) Third. (*e*) Reverse. (*General Motors*)

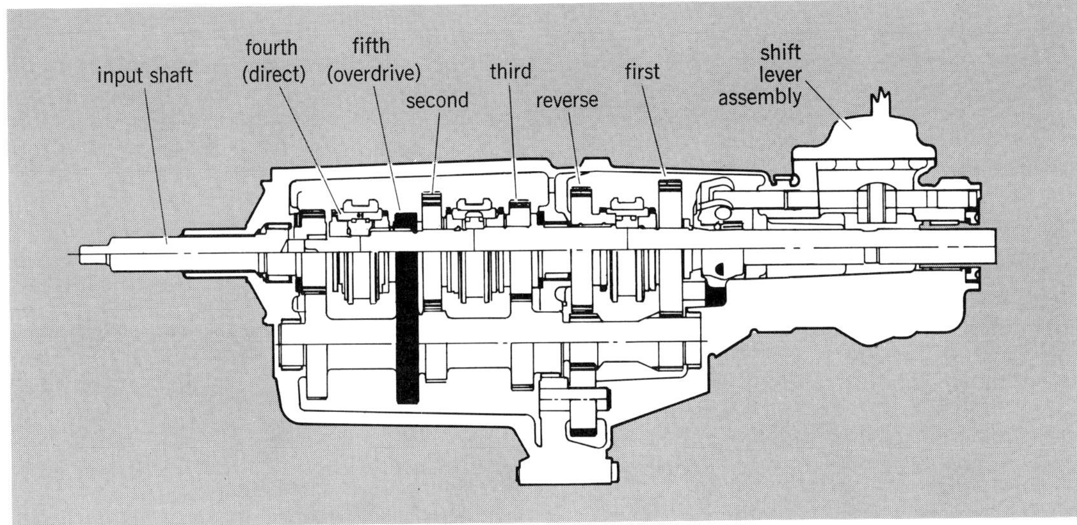

Fig. 2. Borg-Warner T50 five-speed manual transmission.

cones through the chamfers causes the drive shaft to come to the speed of the clutch sleeve. With the speeds thus synchronized, the splines of the clutch sleeve can pass through the spaces of the splines on the exterior of the bronze cone and then proceed to engagement with the splines on the drive shaft. Gear ratios are in the order of 1.5:1 for second speed, 2.5:1 for first speed, and 2.7:1 for reverse. Speed ratios may be varied to better accommodate engine powers.

Four-speed manual-shift transmissions have enjoyed popularity, particularly in sports cars. Generally they are synchronized in all forward speeds as described for three-speed manual-shift transmissions. Reverse generally involves sliding a gear on the main shaft into mesh with the reverse idle by a separate lever. Gear ratios are in the order of 1.28:1 for third speed, 1.64:1 for second speed, 2.2:1 for first speed, and 2.25:1 for reverse.

In 1974, the first five-speed manual transmission for automobiles (Fig. 2) to be mass-produced in the United States was introduced. It has overdrive in fifth speed with its ratio either 0.8:1 or 0.84:1. One gear set, for cars with engines under 175 lb/ft torque, has ratios of 3.41:1 in first speed, 2.08 in second, 1.40 in third, 1:1 in fourth (direct drive), and 0.8:1 in fifth or overdrive. The ratio for reverse is 3.36:1.

The other gear set, covering torque up to about 205 lb/ft, has ratios of 3.10 in first, 1.89 in second,

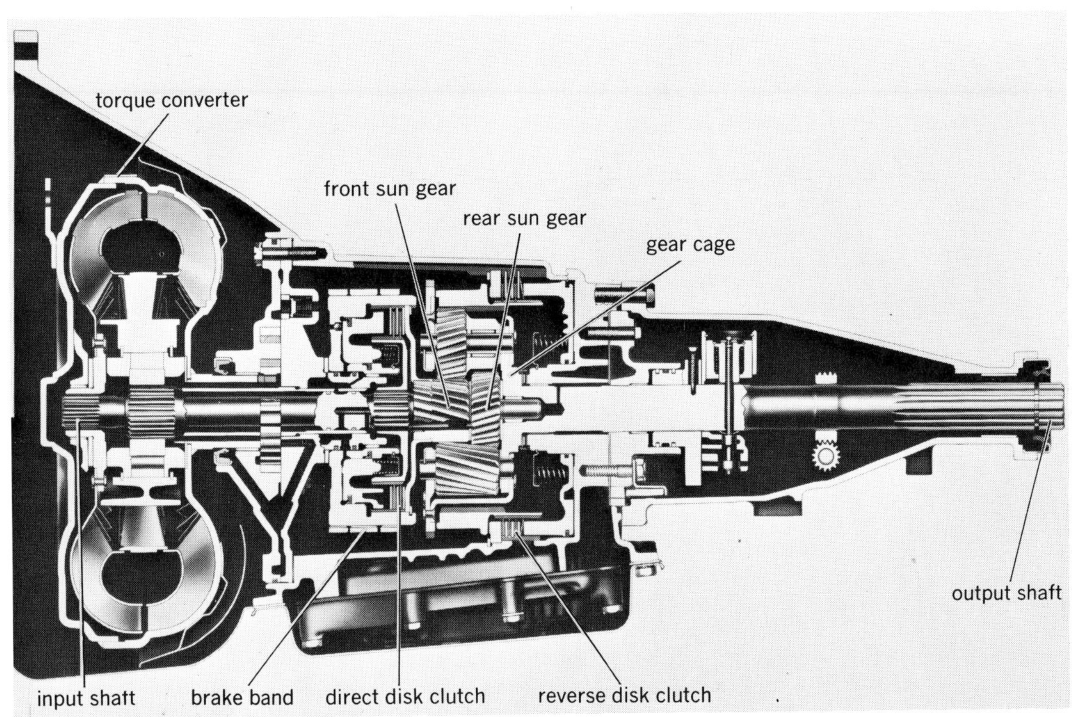

Fig. 3. Typical two-speed automatic transmission. (*Chevrolet*)

1.27 in third, 1:1 in fourth, and 0.84:1 in fifth. Reverse is 3.06:1. A representative five-speed manual transmission is shown in Fig. 2.

Principal manufacturers of passenger-car manual transmissions are the automobile manufacturers and the Borg Warner Corp.

Automatic transmission. In American passenger cars, automatic transmissions generally employ a three-element hydrodynamic torque converter with planetary gear sets of two or three speeds. The torque converter provides a smooth start with a torque (turning effort) multiplication within itself that is continuously variable, depending upon the degree of engine input. The torque may rise as high as 2.0:1 or 2.4:1 at stall speed (full engine torque, the driven member of converter, or turbine, stationary). As car speed increases, the maximum converter torque ratio available decreases until it becomes 1:1 with little slippage. Converters provide an additional advantage in isolating the transmission gear set from torsional vibrations of the engine, which generally make special dampers necessary when engine clutches are used.

Two-speed transmission. The last two-speed automatic transmission provided by one car manufacturer, in addition to its three-speed automatic transmission line, employed a three-element torque converter and planetary gearing (Fig. 3). The torque converter drove the rear sun gear while the front sun gear was held stationary by the band brake. The gear cage was the output member. A torque multiplication in the gear set of approximately 1.8:1 was obtained.

Shift into direct, controlled by car speed and engine throttle opening, was accomplished by simultaneous release of the band holding the front sun gear and engagement of a multiple disk clutch connecting the front sun gear to the input shaft. This transfer locked up the gear set, giving a 1:1 drive through it. Reverse with a gear ratio of approximately 1.8:1 was obtained by holding the ring gear stationary by a multiple disk clutch, thus causing the gear cage which is fastened to the output shaft to rotate rearwardly. *See* PLANETARY GEAR TRAIN.

Three-speed transmission. For three-speed automatic transmissions a three-element torque converter is employed. The converter has two elementary planetary gear sets, of identical size, each consisting of sun gear, carrier with planet gears, and ring gear (Fig. 4). The torque converter drives the ring gear of one planetary set, this set being described as the first set regardless of location, upon application of a forward clutch. The planet carrier of the first set is connected to the output shaft, giving a torque multiplication due to this set only of approximately 1.5:1. The sun gear of the first gear set takes the reaction of the planet pinions and delivers it to the sun gear of the second planetary set. The reaction causes the sun gears of both sets to rotate rearwardly, and in so doing causes the ring gear of the second set to move forwardly; the second carrier is restrained from mov-

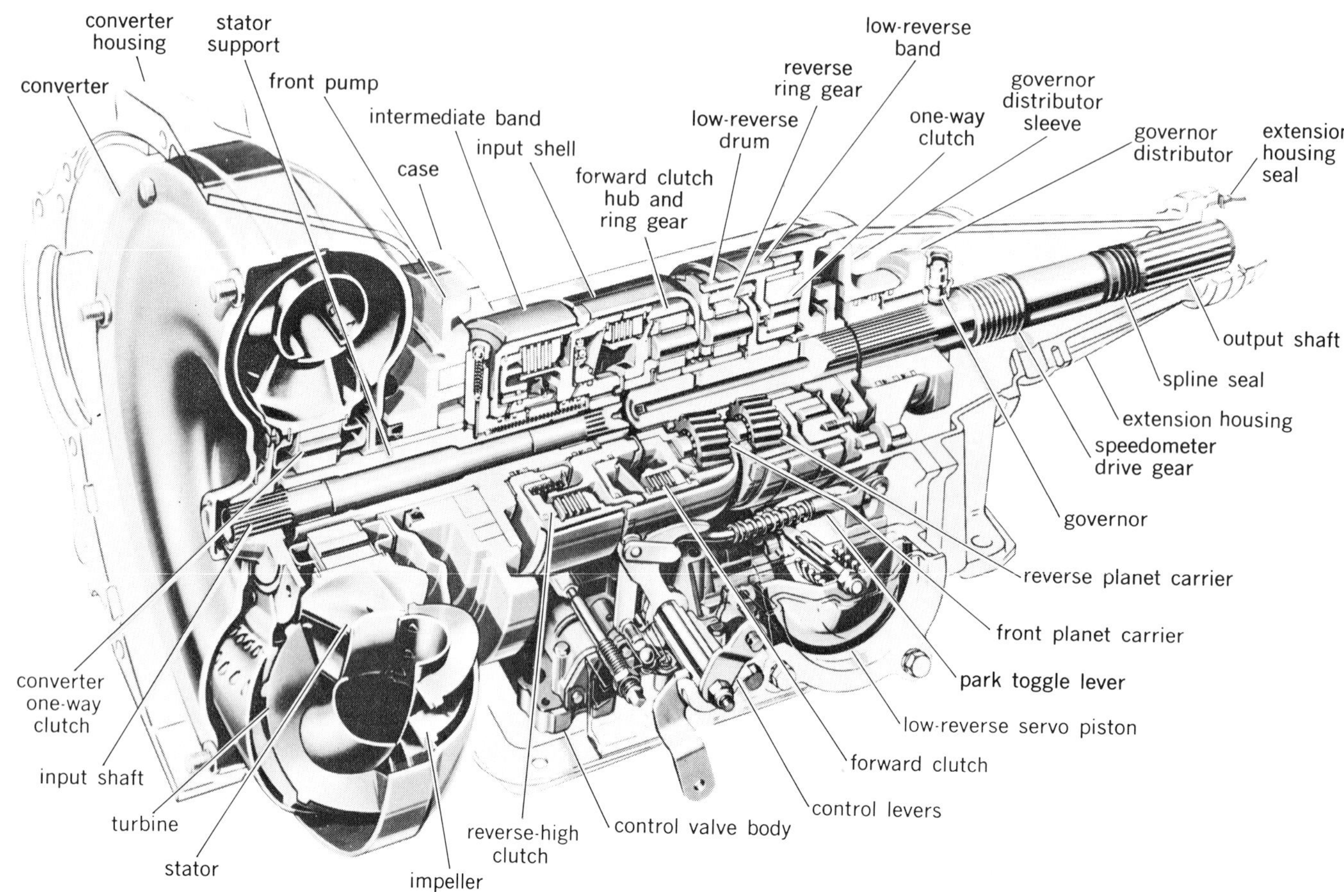

Fig. 4. Three-speed automatic transmission. (*Ford*)

ing rearward by a one-way clutch that fixes it to the transmission case. The ring gear of the second set is attached to the output shaft. Because the diameter of the second ring gear is the same as that of the first, the output torque of the second ring gear is the same as the output torque to the first ring gear. An overall output torque ratio of 2.5:1 is thus obtained as the output torque from the two gear sets for first speed ratio.

Second speed is obtained by fixing the sun gears, previously rotating rearwardly in first speed, to the transmission case by brake band or multiple disk brake. This causes only the first planetary gear set, its ring gear driven by the converter, to operate and provides a speed ratio of 1.5:1 for second speed.

Direct drive is obtained by clutching the sun gears to the input shaft locking up the gear set, so that a 1:1 ratio is obtained. In the design in Fig. 4 the release of the brake band previously fixing the sun gear to the transmission case is simultaneous with the clutch engagement. In a different three-speed automatic transmission design the sun gear is picked up from a one-way clutch, requiring no brake release.

Reverse of approximately 2.1:1 ratio is obtained by engaging the sun gears to the converter output member and locking the carrier of the second gear set to the transmission case by a brake band, the ring gear being fastened to the output shaft. As the sun gear now rotates forwardly, the ring gear now rotates rearwardly.

The transmissions with the engine provide for hill braking in first speed or second speed by use of appropriate band brakes to bridge over the one-way clutches and give solid drives. In some sports cars that have automatic transmissions, shifts may be made manually to simulate manual transmissions.

Automatic control. Hydraulic controls for obtaining second and third speed are designed so that the greater the accelerator depression the higher will be the shift speed. Shifting back from direct drive or third speed into second speed, up to approximately 65 miles per hour, for passing is accomplished by depressing the accelerator pedal past the full throttle position. On fairly steep hills, first speed can likewise be obtained up to 25–30 mph. The controls provide for locking the transmission into second or first speed for hill braking by selecting the proper letter on the shift quadrant.

Transmission controls have become rather sophisticated in securing smooth shifts while maintaining adequate life of the friction members. Measurement of engine torque by intake manifold vacuum has come into general use. One refinement of such vacuum modulators is an additional sensing of atmospheric pressure, which decreases on increase in altitude, resulting in approximately 3% decrease in engine power per 1000 ft. This variation, originally introduced by General Motors, has had acceptance by others in the industry. The unit decreases the pressure of the fluid actuating the friction elements to maintain the smoothness of engagement at altitude which is normally present at sea level, while maintaining the necessary pressures at the latter level.

[FOREST R. MC FARLAND]

Bibliography: N. A. Carter (ed.), *Proceedings of a Symposium on Automatic Transmissions*, 1964; W. H. Crouse, *Automotive Transmissions and Power Trains*, 3d ed., 1967; I. G. Giles, *Automatic and Fluid Transmissions*, 1961; J. S. McGrath, *The Automobile Transmission and Drive Line*, 1962; Society of Automotive Engineers, Passenger Car Automatic Transmissions, *Design Practices*, vols. 1 and 2.

Automotive vehicle

A trackless, self-propelled vehicle for land transportation of people or commodities or for moving materials. Passenger and commercial vehicles have comparable components but separate identities, primarily because of their different functions. Passenger vehicles are built for the comfort and protection of occupants, whereas commercial vehicles are designed more ruggedly to withstand daily abuse and are available in an infinite variety of forms. *See* AUTOMOBILE; EARTHMOVER; TRUCK.

Among the types of passenger cars are sedans, hardtops, hardtop or soft-top sports cars, and station wagons. Additional types are recreation vehicles (RVs) and all-terrain vehicles (ATVs) such as campers and motor homes with sleeping accommodations, motor bikes, motorcycles, scooters, trail bikes, dune buggies, and snowmobiles. After 1975 convertibles became obsolete.

Included in the broad range of body types and configurations of commercial vehicles are ambulances, airport limousines, taxicabs, buses, vans, tankers, light- and heavy-duty highway and industrial trucks, tractors, and self-propelled farm equipment, off-highway construction and excavation units, and various military vehicles.

A typical motor vehicle begins to operate when the ignition switch is turned on. Battery current is released so that the starter motor can crank the engine, as the carburetor delivers an explosive fuel mixture to the engine cylinders. Simultaneously, the ignition system sends high-tension current to the spark plugs to ignite the fuel. Connecting rods transmit energy from the pistons to the crankshaft, while the camshaft (geared to the crankshaft) times intake or exhaust valve operation and firing intervals for spark plugs. Power impulses from each piston rotate the crankshaft, which is a source of motion for the vehicle. *See* VALVE TRAIN.

Power plant energy is transmitted to the power train, consisting of the clutch, transmission, drive shaft, and rear axle. In front- or rear-drive vehicles the power train is known as a transaxle when its elements are located in the front or rear. Engine output to the wheels, controlled by the driver, is governed by the power train and the accelerator for required speed and direction.

Major components. Vehicles with conventional body frames are first built up from frame to chassis, prior to the attachment of the body, which includes all assemblies and systems to make the vehicles operational. Similar procedures are used in building unitized body vehicles, but in this case the frame elements are integral with the body. *See* AUTOMOTIVE BODY; AUTOMOTIVE CHASSIS; AUTOMOTIVE FRAME.

Power plant. Since automotive vehicles first appeared, at the beginning of the 20th century, engines with 1 to 16 cylinders have been tried; most have been abandoned for various reasons.

Among the types of power plants presently available are diesel, free-piston, gas-turbine, and reciprocating or rotary internal combustion engines of various designs. Redesigned four-, six-, and eight-cylinder engines with in-line, V-block, and horizontally opposed "pancake" conformations provide much-needed additional engine compartment space to compensate for "hang-on" hardware which increases annually with new emission controls and options. *See* AUTOMOTIVE ENGINE; DIESEL CYCLE; INTERNAL COMBUSTION ENGINE.

Power train. This unit provides the means of transferring engine output to the wheels and of selectively changing speed. More improvements have been made on these components than on engines; in the past, cone clutches grabbed the flywheel and sheared mounting bolts, unsynchromeshed transmission gears grated and lost their teeth, and universal joints and rear axles failed without warning.

Single-, double-, and multiple-disk dry-plate clutches or wet-disk clutches in an oil bath are used to transfer engine output to the transmission. Separate clutches are not required in automatic transmissions; their function is performed by fluid couplings and torque converters. *See* CLUTCH; TORQUE CONVERTER.

Manual, semiautomatic, and automatic transmissions are available in basic sliding-gear or planetary types for speed changes involving movement of heavy loads and for additional power for climbing hills or passing other vehicles. Manual units are three- or four-speed, or four-on-the-floor for high-performance engines; transmissions and engines are usually matched with selective options for best performance. An overdrive option gives three-speed units a fourth speed for use in higher speed ranges to reduce engine speed and improve mileage. Commercial vehicles have manual and automatic transmissions with or without power assists. *See* AUTOMOTIVE TRANSMISSION; OVERDRIVE.

The power transfer link between the transmission and the rear axle (or front drive) is a rotating drive line with universal joints at both ends. Changes in drive line angularity due to power torque or irregular road surfaces, involving either vertical or twisting action, are compensated for by U-joint flexibility when the vehicle is in motion.

Various types of full-floating, semifloating, two-speed, and double-reduction axles convert drive line motion through 90° to the wheels. Two gear trains perform this action: ring gear and pinion, and differential gear cluster. The ring gear transmits power to the axles, and the differential compensates for resistance to rotation by one wheel while the other is traveling at a greater speed in negotiating turns by applying equal power to both wheels when they are rotating at unequal speeds. The limited-slip differential is available as an option to transmit greater power to the wheel with the greatest traction and less power to the slipping wheel for better power distribution. *See* DIFFERENTIAL.

Brakes, wheels, suspension units, and shock absorbers are mounted to flanges and brackets at the outer ends of axle housings.

Systems. In addition to the major components, automotive vehicles have several operating systems vital to proper functioning which are described briefly below.

Fuel system. The fuel system includes the fuel tank, fuel lines leading to the fuel pump, filter, and carburetor, with an air cleaner mounted on the carburetor to supply adequate clean air. The fuel and air mixture from the carburetor is drawn through the intake manifold and into all cylinders by the engine. Gasoline from the fuel tank is pumped to the filter and carburetor by the fuel pump. Camshaft rotation causes a pushrod to reciprocate against the spring-loaded linkage; the vibration of the fuel pump diaphragm creates this pumping action. The proper fuel and air mixture is prepared automatically by the carburetor for cold weather starting and idle and acceleration conditions; this mixture ranges from rich to lean.

Conventional carburetors and intake manifolds have been replaced by fuel-injection systems in some automobiles that have a metering computer which controls fuel flow to each cylinder. This system produces better combustion and less fuel waste than conventional systems, which distribute fuel to all cylinders. The distribution is based on air and engine temperatures and manifold pressure and speed required, with automatic controls for all operating conditions. *See* CARBURETOR; FUEL INJECTION.

Conventional exhaust systems were ordered changed by California, with emission-control legislation in 1966, and by the Federal Clean Air Act of 1970. There are three basic sources of pollution: fuel tank vapor, crankcase fumes, and exhaust fumes. Although conventional intake and exhaust systems remain in use, emission-control equipment and no-lead gasoline are now required in new vehicles. Positive crankcase ventilation (PCV) uses engine suction to draw crankcase and fuel tank fumes, as well as additional quantities of air, from the air cleaner into the system. Normal fuel and air mixtures for these added combustibles are controlled by the PCV valve; a vapor separator returns liquid to the fuel tank, or vapor to the PCV unit. The exhaust manifold has an air-injection system driven by an air pump belted to the crankshaft, to create afterburn of unexploded hot engine exhaust gas for more complete combustion. A catalytic converter inserted in the exhaust pipe line ahead of the muffler breaks down noxious nitrogen oxides, hydrocarbons, and carbon monoxide into harmless or inert matter. Satisfactory operation of catalytic converters is still a controversial matter subject to future correction. Sulfur in petroleum was not given due consideration by either manufacturers or Federal agencies. Neither emission systems nor catalytic converters remove this element, and as a result sulfur oxides combine with water to form sulfuric acid and other pollutants that had not been anticipated. Moreover, fouled spark plugs or inefficient engines increase catalytic converter surface temperatures from their normal operating temperature of 500°F (260°C) to more than 1200°F (650°C), causing the converter to be declared a fire hazard in many areas by several Federal agencies.

Electrical system. At the heart of the whole electrical system are the ignition switch, battery, alternator, voltage regulator, and battery discharge sensor; without them none of the ancillary circuits would function. The ignition switch controls a half-dozen circuits which use these components: the starting, charging, and ignition circuits, lighting system, accessories, and warning sensors that keep the driver informed of changes in electrical, mechanical, fuel, oil, and other operating conditions that require correction.

The starting circuit includes the ignition switch, battery, starter relay, solenoid, and starting motor, which cranks the engine to provide power.

The charging circuit is composed of the battery, alternator, voltage regulator, and battery-discharge sensor. The alternator is belt-driven by the engine to produce alternating current that is converted by an integral diode rectifier unit into direct current; earlier cars had dc generators to perform this function. Prolonged use of the starter, lighting, or accessory circuits with the engine not running will discharge the battery. Therefore, the alternator keeps the battery charged and available for other circuits in the electrical system.

Conventional ignition circuits comprise the battery, ignition coil, resistor, distributor with breaker points and condenser, spark plugs, and high- and low-tension wiring. In the early 1970s electronic ignition systems appeared and not only eliminated the resistor, breaker points, and condenser but produced higher voltage for better performance than that provided by conventional systems. With the electronic system, spark plugs are cleaner and have longer lives, and power plant operation is considerably more trouble-free. *See* IGNITION SYSTEM.

The lighting system is a complex arrangement of many individual circuits which use the vehicle frame for a common return, or ground; thus it is termed a single-wire system. Passenger vehicles use more than enough bulbs to light a house (usually about six dozen); they vary in size from miniatures to powerful sealed-beam headlights. The possibility of overload and short circuits is always present under such conditions; thus fuse boxes or circuit breakers are necessary in order to protect the entire system until corrections are made. Whereas the ignition system is controlled by the ignition switch, the lighting system is controlled by the main lighting switch, which directly governs headlights, parking lights, license lights, taillights, interior lights, and brilliance of instrument panel lights with a built-in rheostat. Other switches operate backup lights, directional lights, courtesy lights, sidelights, and warning flashers; warning sensors operate the instrument panel pilot lights to indicate the location of operational difficulties.

Cooling. Power plants are cooled by water or a solution of water and permanent antifreeze for year-round protection against boiling over in summer and freezing in winter. The system includes a water pump and fan driven by the engine, water jackets around cylinders in the engine head and block, a radiator to dissipate heat from the coolant, and a thermostat to control operating temperatures. Also included in cooling systems are vehicle heaters for passenger comfort, as well as a transmission oil cooling element, mounted at the bottom of the radiator, which has finned tubes through which transmission oil flows for cooling. The thermostat restricts coolant circulation to engine water jackets during cold weather until the proper operating heat is reached; thereafter, the coolant circulates throughout the whole system. The water pump draws coolant from the radiator outlet (bottom), driving it through the system and radiator inlet (top). The operation of the radiator is similar to that of a condenser; fins permit heat dissipation by air pulled through the radiator and the engine by the fan, as coolant flows into tubes toward the radiator outlet and water pump. *See* ENGINE COOLING.

Steering. Several types of steering systems are used for manual or hydraulic power-assist control when the vehicle is in motion, with provisions for manual operation in the event of failure of the power unit. The introduction of power steering reduced the manual effort needed, and subsequently, variable-ratio steering has produced faster vehicle response when full right or left turns are negotiated. *See* AUTOMOTIVE STEERING; POWER STEERING.

Suspension system. Suspension systems are the means by which passenger and vehicle are protected from road shock due to irregular street and highway surfaces. Current systems include coil- and leaf-spring, air suspension, and torsion-bar types for passenger comfort, ride control, and driving safety. Shock absorbers are designed for air or hydraulic operation, or for either type with auxiliary spring-loaded units for heavy-duty usage in conjunction with suspension systems. Mounting methods may be axle to frame or suspension unit to frame, or within the suspension unit itself. Shock absorbers are mounted in vertical or angular positions, inboard or outboard, for better stability in turns; some vehicles have two shock absorbers in front and four at the rear for better ride control. *See* AUTOMOTIVE SUSPENSION.

Brakes. Conventional four-wheel drum brake systems have progressed to disk brakes; some imports are equipped with four-wheel disk brakes. A Federal braking regulation, effective Jan. 1, 1976, required front-disk brakes as standard equipment on all new cars. *See* AUTOMOTIVE BRAKE; BRAKE.

Lubrication. Two lubricating systems are used in passenger and commercial vehicles to reduce friction and wear of moving engine and chassis parts.

The engine lubrication system includes an oil pan and sump for oil supply and return, oil pump, and filter; oil level is measured by a dipstick and checked periodically for replenishment. Lubricant circulates through galleries, or drilled passages, in the camshaft, crankshaft, rocker arm shaft, rocker arm, connecting rod and camshaft bearings, valve lifters, and piston pins. Cylinder block and head channels permit oil to drip back into the crankcase for recirculation through the system. The oil pump drives oil through the filter and system, where it also acts as a cleansing agent to remove dirt and contaminants in the filter; it also carries engine operating heat away from exposed piston and cylinder surfaces.

Lubricant is applied to the fittings of strategic working parts of the chassis at specified points and at prescribed service intervals for satisfactory vehicle operation. The rear-axle lubricant level is also checked occasionally, and lubricant is replaced as needed.

[PAUL A. OTT, JR.]

Automotive voltage regulator

A device in the automotive electrical system to prevent generator or alternator overvoltage. There are two general types: one for direct-current and another for alternating-current generators.

Direct-current voltage regulator. The device contains a set of spring-loaded contacts and a winding shunted across the generator (Fig. 1). When the generated voltage exceeds the preset voltage of the regulator, the magnetic flux is sufficient to open the contacts, thereby inserting resistance into the generator field. The generator voltage is reduced, the contacts close, and the cycle is repeated. The contacts vibrate up to 200 times a second.

Included with the voltage regulator are a current regulator and a cutout relay (Fig. 1). The current regulator prevents generator overload. The cutout relay opens the circuit when the generator is inoperative to prevent battery discharge through the generator.

Alternating-current voltage regulator. The wiring diagram of one regulator of this type is shown in Fig. 2. The regulator has two sets of contacts, lower and upper. At intermediate speeds, the lower contacts open and close, inserting and removing the resistance in the alternator field to provide voltage-limiting action. At higher speeds, the resistance is not sufficient to hold the voltage down. It increases slightly and the upper contacts begin to make and break. When they close, the alternator field is grounded by the contacts. With both ends of the field grounded, additional regulation is achieved.

The alternator shown in Fig. 2 has a field relay which connects the alternator field to the battery when the ignition switch is closed and disconnects it when the ignition switch is opened to shut off the engine. This keeps the battery from discharging through the alternator field. The indicator lamp comes on when the ignition switch is first turned on to indicate that the alternator is not charging. However, as soon as the alternator begins to charge the battery, the alternator voltage closes the field relay contacts. With both sides of the indicator lamp connected to the insulated terminal of the battery, the light goes out, indicating that the alternator is charging the battery. Many automotive alternators have integral transistorized voltage regulators which do not use vibrating contacts and do not require adjustment. They prevent excessive alternator voltage by inserting and removing field resistance; this is accomplished by transistor action. *See* GENERATOR.

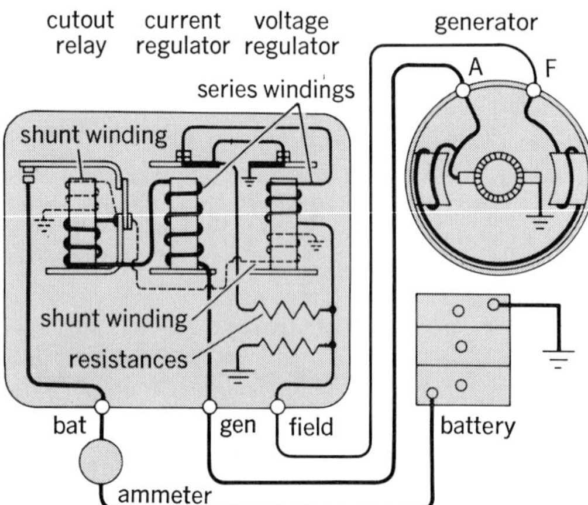

Fig. 1. A direct-current wiring circuit of voltage regulator which includes a current regulator and a cutout relay. (*Delco-Remy Div., General Motors Corp.*)

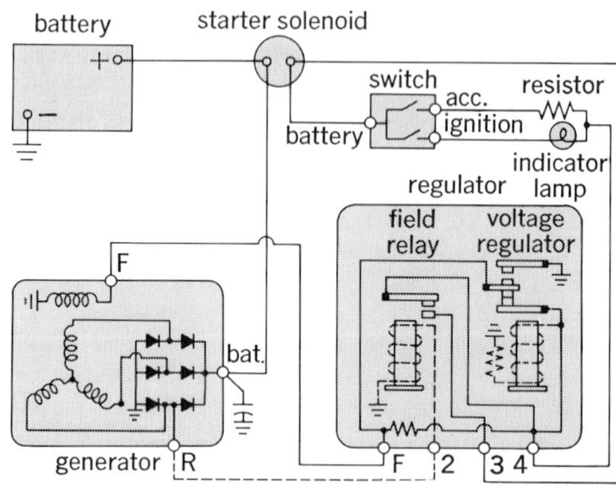

Fig. 2. Wiring circuit of an alternating-current voltage regulator with field relay and indicator lamp. (*Delco-Remy Div., General Motors Corp.*)

[WILLIAM H. CROUSE]

Bibliography: W. H. Crouse, *Automotive Electrical Equipment*, 6th ed., 1966; W. H. Crouse, *Automotive Mechanics*, 5th ed., 1965.

Autotransformer

A special form of transformer having one winding, a portion of which is common to both the primary and the secondary circuits. The current in the high-voltage circuit flows through the series and common windings (see figure).

The current in the low-voltage circuit flows through the common winding and adds vectorially to the current in the high-voltage circuit to give the common winding current. Thus, an electrical connection exists between high-voltage and low-voltage windings. Because of this sharing of parts of the winding, an autotransformer having the same kilovolt-ampere (kVa) output rating is generally smaller in weight and dimensions than a two-winding transformer. The equivalent size of a two-winding autotransformer without taps is given by the coratio times the output kva, where the coratio equals $(HV - LV)/HV$. When the coratio is small, that is, when the high-voltage and low-voltage magnitudes are close together, the cost advantage in favor of an autotransformer is large. As the coratio becomes large, the equivalent size and the cost of an autotransformer approach that of a normal transformer.

One possible disadvantage of autotransformers is that the windings are not insulated from each other and that the autotransformer provides no isolation of the primary and secondary circuits.

Types. Autotransformers of large sizes are used for interconnecting high-voltage power systems. They are used in small sizes for intermittent-duty starting of motors. For this use the motor is connected for a short time to the common winding voltage, and then connected to the full line voltage. Small, variable-ratio autotransformers are used in testing and as components of other apparatus.

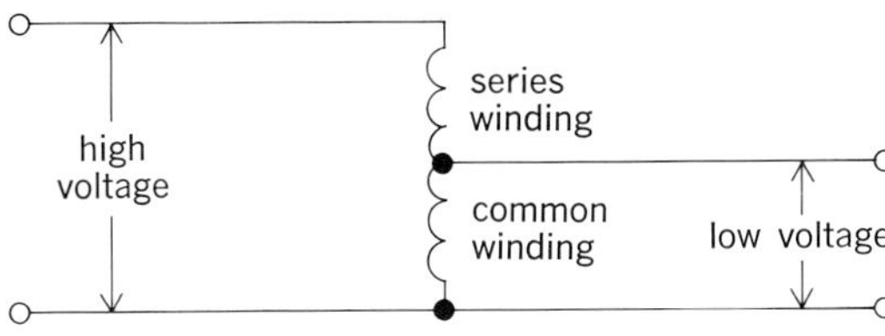

Typical autotransformer circuit.

Characteristics. Because of their smaller equivalent size, autotransformers generally have lower no-load loss, exciting current, and load loss than the corresponding transformers. In addition, the impedance and regulation normally are lower because of the connection between the two circuits.

Taps. Taps can be provided in the autotransformer to adjust the turns ratio. This provides control of the output voltage over the operating range of the transformer. These taps may be placed in the series winding or in the common winding. The problems of switching from tap to tap under load are similar to those encountered in tap changing on a two-winding transformer. The switching from tap to tap can be done without interruption of service.

Small variable-ratio autotransformers used for testing have a brush contact which serves as the common line. This contact may be moved across the burns to give a common line voltage from approximately 100 to 0% of the high voltage. *See* TRANSFORMER.

[J. R. SUTHERLAND]

Axle

A supporting member that carries a wheel. An axle may rotate with the wheel to transmit power to or from it, as does the rear axle of an automobile, or it may allow the wheel to rotate freely, as does the axle of a trailer.

Types. Rear axles in American cars are usually of the unsprung type in which the rear-axle housing carries the right and left rear-axle shafts and the wheels are mounted at the outer end of each shaft.

Rear axles in many European and some English cars are of the sprung type, in which the rear-axle gearing housing at the center is mounted to the frame or body, while the wheels and axle shaft outer ends are attached by linkage to the frame, body, or central gear housing. One popular sprung axle has been the swinging type, in which the structure carrying each wheel swings from a center point: one example of this type is the Mercedes-Benz car.

Each type of rear axle has its own characteristics and is selected for suitability to the road conditions found in the countries where the car is to be sold.

The unsprung axle is usually simpler and cheaper, permits easier alignment of the rear wheels, and provides an easier design to reduce lowering of the rear end of the car when accelerating or to reduce raising of the rear end when braking. The sprung rear axle permits reducing the unsprung weight at the rear wheels, generally considered an advantage on rough roads. It also permits lowering the tunnel in the floor of the car. Unless special insulating precautions are taken, cars with sprung

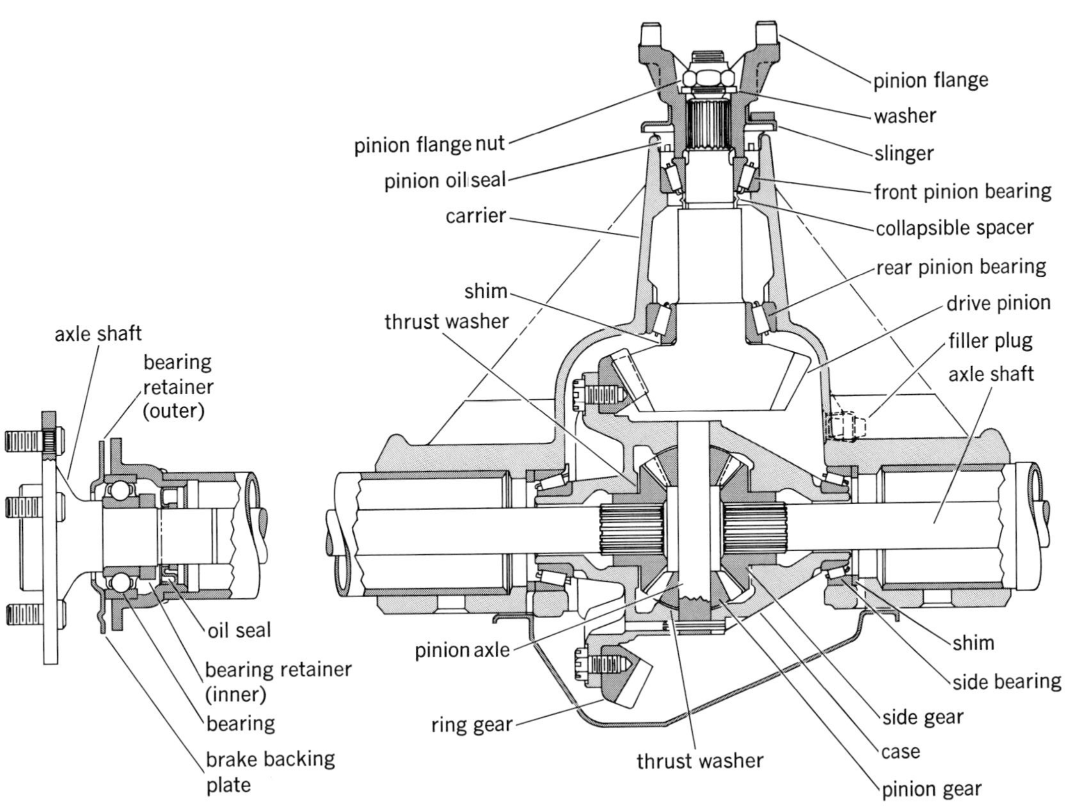

Semifloating rear axle typical of American design. (*Buick Division, General Motors*)

rear axles transfer more noise into the body than those with unsprung axles.

General construction. A typical rear axle is shown in the illustration. This Salisbury-type axle housing comprises a carrier casting with tubes pressed and welded into it to form a complete carrier and tube assembly. (The ball bearings installed in the tubes carry the axle shafts at the outer ends.) A removable steel cover bolted to the rear of the carrier permits service of the differential assembly without removing the entire assembly from the car.

The rear wheels, hubs, and tires are carried at the extremities of the axle shafts of the live type, so called because they rotate. The axle shafts are carried on bearings at the outer ends of the housing by ball bearings (as in the illustration), each of which takes radial and thrust loading or by tapered roller bearings, each of which takes radial and outward thrust loads, the inward thrust loads being taken through the axle shaft, across a center thrust block, and through the other axle shaft to its outer bearing.

Semifloating axles are supported at their inner ends by the splined coupling to the differential side gears which are, in turn, supported by bearings in the carrier. A semifloating axle carries torque and the wheel loads at its outer end. Other types not now in general use include the full-floating, in which the axle carries only torque, the wheel vertical and horizontal skidding loads being carried by the rear-axle housing; and the three-quarters floating, in which the axle carries torque and the wheel skidding loads, while the housing carries the wheel vertical and horizontal driving or coasting loads. *See* DIFFERENTIAL.

The rear-axle drive gear set (the ring gear, mounted on the differential case, held in position by the lateral bearings in the rear-axle carrier, and the driving pinion, held in position by the forward bearings in the carrier) of all American and many European axles is of the hypoid type. This type of gearing generally carries the pinion 1.5–2 in. or more below the centerline of the gear. It results in sliding along the face of the teeth not present in the earlier spiral bevel gears and necessitates use of an extreme-pressure lubricant. The design permits lowering the tunnel in the rear floor compartment and provides additional rear-seat cushion at the center of the car. *See* AUTOMOTIVE TRANSMISSION; GEAR; TRUCK AXLE.

[FOREST MC FARLAND]

Bibliography: W. H. Crouse, *Automotive Transmissions and Power Trains: Construction, Operation and Maintenance*, 3d ed., 1967.

Babbitt metal

A white alloy of tin, lead, antimony, and copper used as an antifriction metal in sleeves and bearings. Bearings of babbitt metal are used where high pressures need not be sustained and where high temperatures are not likely. The alloy is a moderately soft matrix carrying cubic crystals of a hard compound of either copper or lead with the antimony. Wear soon relieves the softer matrix, exposing the hard crystals which carry the load. Their small exposed area decreases friction, and the undercut matrix permits lubricant to circulate between crystals. *See* ANTIFRICTION BEARING; TIN ALLOYS.

[FRANK H. ROCKETT]

Ball-and-race-type pulverizer

A grinding machine in which balls rotate under pressure to crush materials, such as coal, to a fine consistency. The material is usually fed through a chute to the inside of a ring of closely spaced balls. In most designs the upper spring-loaded race applies pressure to the balls, and the lower race rotates and grinds the coarse material between it and the balls (see illustration). The finely ground material discharges along the outer periphery of the ball races.

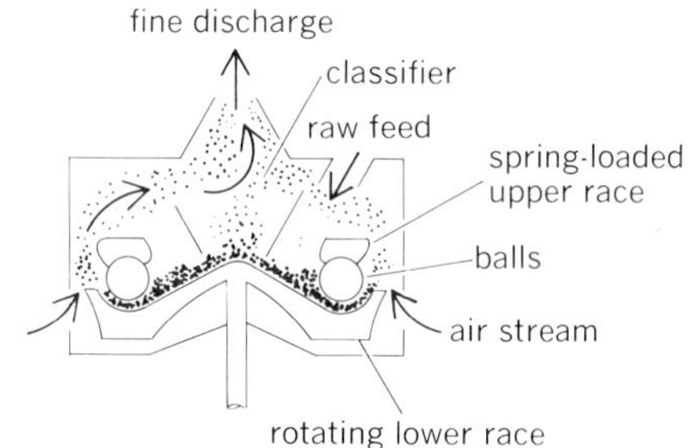

Coarse raw material is ground by crushing and attrition between balls and races and is then withdrawn from the pulverizer by an airstream.

For the pulverization of coal, hot air, introduced between the lower race and the pulverizer housing, lifts or carries the fines to a cyclone classifier at the center of the pulverizer. There the finer particles discharge from the pulverizer while the larger particles return to the grinding zone for further reduction in size. Two or more rings of balls can be cascaded in one machine to obtain greater capacity or output. Counterrotating top and bottom rings also are used to increase pulverizer capacity. Such pulverizers are compact and the power required per ton of material ground is relatively low. CRUSHING AND PULVERIZING.

[GEORGE W. KESSLER]

Ballast resistor

A resistor that has the property of increasing in resistance as current through it increases and decreasing in resistance as current decreases. Thus it tends to maintain a constant current through it, despite variations in applied voltage or changes in the rest of the circuit. A ballast resistor acts as a variable load on the system and thus differs from load resistors, which maintain a constant resistance.

The ballast action is obtained by using resistive material that increases in resistance as temperature increases. Any increase in current then causes an increase in temperature, which results in an increase in resistance and reduces the current. Ballast resistors are usually mounted in an evacuated envelope to reduce heat radiation; they are therefore also called ballast tubes.

Ballast resistors have been used to compensate for variations in line voltage or to compensate for negative volt-ampere characteristics of other devices, such as fluorescent lamps and other vapor lamps. *See* VAPOR LAMP.

[ARTHUR A. WELCH]

Beam

A structural member extending between two reaction points and supporting loads such as floors, walls, and other types of surfacing is a simple beam. The loads are generally placed to produce a bending action about a principal axis of the beam cross section. When the structural member extends over several reaction points, it is referred to as a continuous beam. *See* LOADS, TRANSVERSE.

[JOHN B. SCALZI]

Beam column

A structural member subjected to the combined action of axial and transverse loads which produce compression and bending moments in the member. Both compression and bending moments may also be produced by eccentrically applied axial loads. Beam columns exist in frame structures.

Secant formula. The secant formula defines the critical load for an eccentrically loaded column as the load which first produces an external fiber stress equal to the yield strength of the material calculated on the basis of the linear elastic theory. This load is conservatively considered to be the buckling load or collapse load for the member. However, to evaluate properly the buckling load for an eccentrically loaded member, it is necessary to consider the inelastic behavior which occurs at stresses beyond the proportional limit.

The secant formula for the case of an eccentrically loaded member in which, due to equal end eccentricities, the plane of the bending moment coincides with the plane of buckling is stated in Eq. (1) in terms of the maximum stress at the extreme fiber. The formula was developed for the stress-strain characteristics of steel.

$$f_m = \frac{P}{A}\left(1 + \frac{ec}{r^2}\sec\frac{L}{2r}\sqrt{\frac{P}{AE}}\right) \qquad (1)$$

where f_m is the maximum stress, psi
P is the axial load, lb
A is the area of the cross section, in.^{2}
c is the distance from the centroidal axis to the extreme fiber, in.
L is the length of the member, in.
r is the radius of gyration in the plane of bending, in.
e is the eccentricity of the load, in.
E is the modulus of elasticity, psi.

Interaction formulas. Another method of calculating the strength of the combined action of axial loads and bending moments is expressed by a functional relationship in terms of ratios of actual load to the strength of the member under pure axial and pure bending load. One form of interaction formula is given by Eq. (2), where P and M are the

$$\frac{P}{P_u} + \frac{M}{M_u} = 1 \qquad (2)$$

actual load and bending moment, P_u is the axial limit or collapse load, and M_u is the ultimate or limit bending moment when the bending moment is acting separately.

When the design is based on the allowable stress method, the formula may be expressed in terms of stresses and expanded to include the secondary moment effect produced by the axial load and the lateral deflection due to the combined bending moments. The formula for the combined effect of axial load and bending about two axes is given by Eq. (3),

$$\frac{f_a}{F_a} + \frac{C_{mx} f_{bx}}{\left(1 - \frac{f_a}{F'_{ex}}\right)F_{bx}} + \frac{C_{my} f_{by}}{\left(1 - \frac{f_a}{F'_{ey}}\right)F_{by}} \leq 1.0 \qquad (3)$$

where f_a is the axial stress, kips per square inch (ksi)
F_a is the axial compressive stress permitted in the absence of a bending moment, ksi
C_{mx}, C_{my} are coefficients applied to the bending terms about the x and y axes, respectively
f_{bx}, f_{by} are the bending stresses about axes x and y, ksi
F_{bx}, F_{by} are the bending stresses permitted in the absence of an axial force, ksi
F'_{ex}, F'_{ey} are the Euler stresses divided by a factor of safety, ksi.

The application of the interaction formulas requires a consideration of the conditions and limitations on the axial and bending stresses caused by lateral-torsional buckling. *See* COLUMN.

[JOHN B. SCALZI]

Bibliography: B. Bresler, T. Y. Lin, and J. B. Scalzi, *Design of Steel Structures*, 2d ed., 1968; E. H. Gaylord and C. N. Gaylord, *Design of Steel Structures*, 2d ed., 1972; B. G. Johnston (ed.), *Guide to Stability Design Criteria for Metal Structures*, 3d ed., 1976; W. McGuire, *Steel Structures*, 1968.

Belt drive

The lowest-cost means for transmitting power between shafts that are not necessarily parallel. Belts run smoothly and quietly, and they cushion motor and bearings against load fluctuations. Belts typically are not as strong or durable as gears or chains. However, improvements in belt construction and materials are making it possible to use belts where formerly only chains or gears would do.

Advantages of belt drive are: They are simple. They are economical. Parallel shafts are not required. Overload and jam protection are provided. Noise and vibration are damped out. Machinery life is prolonged because load fluctuations are cushioned (shock-absorbed). They are lubrication-free. They require only low maintenance. They are highly efficient (90–98%, usually 95%). Some misalignment is tolerable. They are very economical when shafts are separated by large distances. Clutch action may be obtained by relieving belt tension). Variable speeds may be economically obtained by step or tapered pulleys.

Disadvantages include: The angular-velocity ratio is not necessarily constant or equal to the ratio of pulley diameters, because of belt slip and stretch. Heat buildup occurs. Speed is limited to usually 35 meters per second (7000 feet per minute). Power transmission is limited to 370 kilowatts (500 horsepower). Operating temperatures are usually restricted to −35 to 85°C. Adjustment of center distance or use of an idler pulley is necessary for wear and stretch compensation. A means of disassembly is provided to install endless belts.

Types and general use. There are four general types of belts, each with its own special characteristics, limitations, advantages, and special-purpose

BELT DRIVE

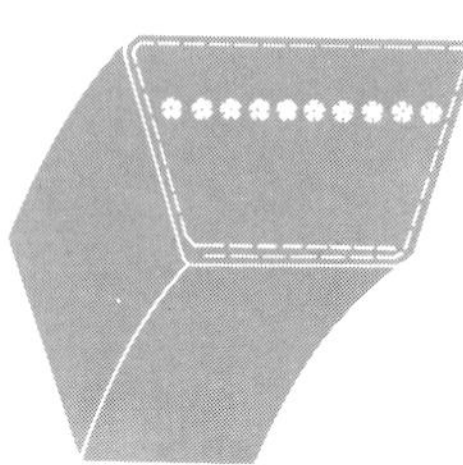

Fig. 1. V-belt in cross section.

variations for different applications.

Flat belts. Flat belts, in the form of leather belting, served as the basic belt drive from the beginning of the Industrial Revolution. They can transmit large amounts of power at high speeds—up to 370 kw (500 hp) and 50 mps (10,000 fpm)—if wide belts are used over large pulleys. Such drives are cumbersome; they require high belt tensions, which in turn produce large bearing loads. For these reasons V-belts have generally replaced flat belts in power delivery applications. Flat belts find their widest application where high-speed motion, rather than power, is the main concern. Flat belts are very useful where large center distances and small pulleys are involved. They can engage pulleys on both inside and outside surfaces, and both endless and jointed construction is available.

V-belt. V-belts are the basic power-transmission belt. They provide the best combination of traction, operating speed, bearing load, and service life. The belts are typically endless, with a trapezoidal cross section which runs in a pulley with a V-shaped groove (Fig. 1). The wedging action of the belt in the pulley groove allows V-belts to transmit higher torque at less width and tension than flat belts. V-belts are far superior to flat belts at small center distances and high reduction ratios. Center distances greater than the largest pulley diameter and less than three times the sum of the two pulley diameters are preferred. Optimum speed range is 5–35 mps (1000–7000 fpm). V-belts require larger pulleys than flat belts because of their greater thickness. Several individual belts running on the same pulley in separate grooves are often used when the power to be transmitted exceeds that of a single belt. These are called multiple-belt drives.

Fig. 2. Timing belt and pulleys. *(Morse Chain Division of Borg-Warner Corp.)*

Jointed and link V-belts are available when the use of endless belts is impractical. Jointed belts are not as strong and durable as endless and are limited to speeds of 20 mps (4000 fpm). A link V-belt is composed of a large number of rubberized-fabric links jointed by metal fasteners. The belt may be disassembled at any point and adjusted to length by removing links.

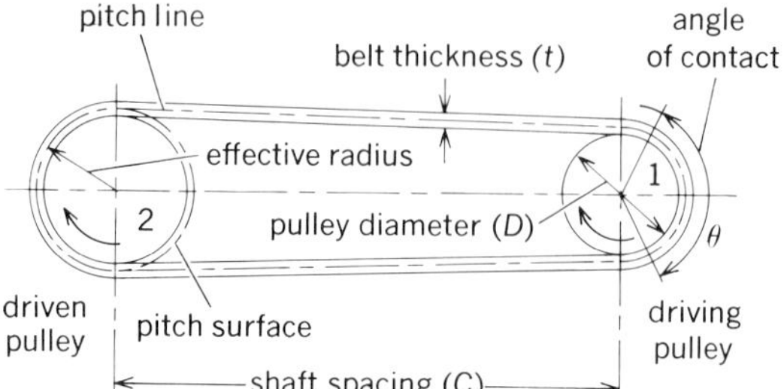

Fig. 3. Open belt drive.

Film belts. Film belts are often classified as a variety of flat belt, but actually they are a separate type. Film belts consist of a very thin (100–4000 micrometers or 0.5–15 mils) strip of material, usually plastic but sometimes rubber. They are well suited to low-power (up to 7 kW or 10 hp), high-speed application where they provide high efficiency (up to 98%) and long life. One of the great advantages is very low heat buildup. Business machines, tape recorders, and other light-duty service provide the widest use of film belts.

Timing belts. Timing belts have evenly spaced teeth on their bottom side which mesh with grooves cut on the periphery of the pulleys to produce a positive, no-slip, constant-speed drive, similar to chain drives (Fig. 2). They are often used to replace chains or gears, reducing noise and avoiding the lubrication bath or oiling system requirement. They have also found widespread application in miniature timing applications. Timing belts, known also as synchronous or cogged belts, require the least tension of all belt drives and are among the most efficient. They can transmit up to 150 kW (200 hp) at speed up to 800 mps (16,000 fpm). There is no lower limit on speed. Disadvantages are high first cost, grooving the pulleys, less overload and jam protection, no clutch action, and backlash.

Belt–pulley arrangements. The most common arrangement, by far, is the open belt drive (Fig. 3). Here both shafts are parallel and rotate in the same direction. The cross-belt drive of Fig. 4 shows parallel shafts rotating in opposite directions. Timing and standard V-belts are not suitable for cross-belt drives because the pulleys contact both the inside and outside belt surfaces. Nonparallel shafts can be connected if the belt's center line, while approaching the pulley, is in the pulley's central plane (Fig. 5). *See* ROLLING CONTACT.

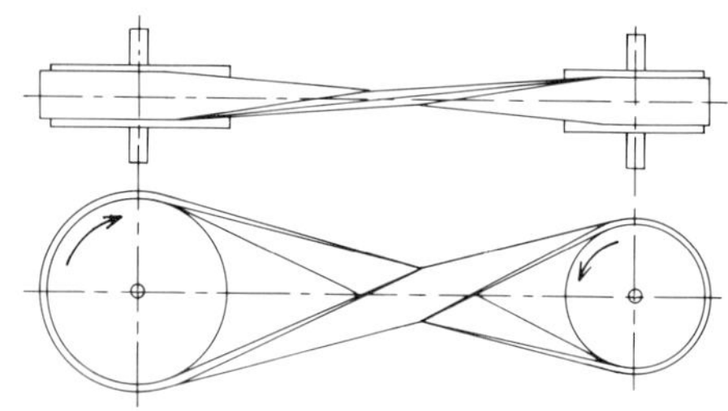

Fig. 4. Crossed belt drive.

Belt materials. Industrial belts are usually reinforced rubber or leather, the rubber type being predominant. Nonreinforced types, other than leather, are limited to light-duty applications. General belt material properties are given in the table.

Speed relationship between pulleys. Accurate calculation of the speed relationship between pulleys is based on pitch diameter, the effective pulley diameter. As a belt bends around a pulley, the outer fibers are in tension, the inner in compression. In between is the pitch line, which is subject to neither stress nor change in length (Fig. 3). For film and flat belts, the pitch line is midway through the belt surface. For timing and V-belts, the pitch line depends on cross-section shape and size. The factors to convert pulley diameters to pitch diameters are typically listed in engineering handbooks and manufacturer's literature. The speed relationship between pulleys is given by Eq. (1), where N_1 is the angular speed of pulley 1, N_2 is the angular speed of pulley 2, PD_1 is the pitch diameter of pulley 1, and PD_2 is the pitch diameter of pulley 2.

The actual pulley speeds are often 0.5 – 1% less than predicted by Eq. (1), because of belt slip and

$$\frac{N_1}{N_2}=\frac{PD_2}{PD_1} \quad (1)$$

stretch. The exact speed relationship for timing belts is also the inverse ratio of the number of teeth on the pulleys.

The speed in mps of the belt itself is given by Eq. (2), where PD is the pitch diameter of the given

$$V=\pi\,(PD)(N) \quad (2)$$

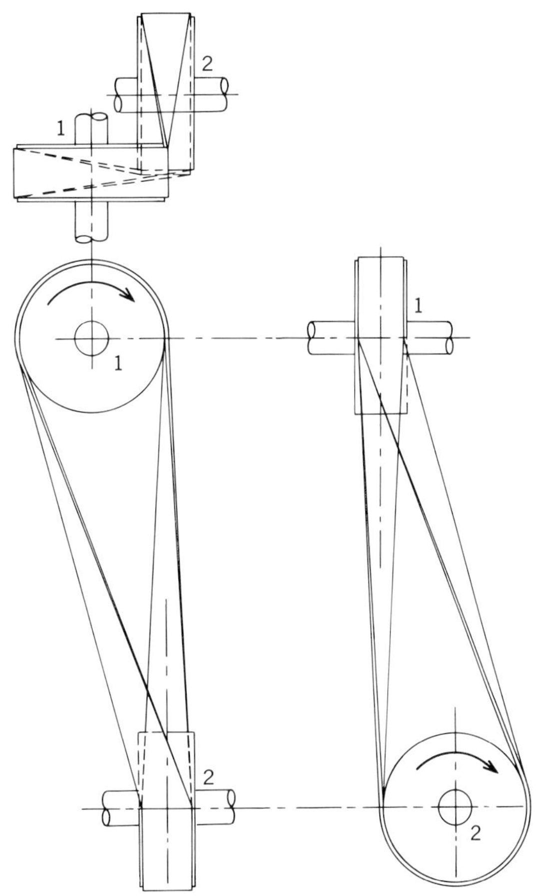

Fig. 5. Quarter-turn belt.

Belt material properties

Material	Properties	Uses
Nonreinforced rubber	Low-power and low-speed applications; very useful for fixed-center drives	Flat and film belts, vulcanized joints
Nonreinforced plastics	Low-power and low-speed application	Flat, film, and miniature timing belts, molded endless
Nonreinforced leather	High power; long service life; high cost, must be cleaned and dressed; prone to stretching and shrinking; limited to low and moderate speeds	Flat belts only; often laminated, with glued, riveted, or wire laced joints
Fabric	Ability to track uniformly; low friction is a disadvantage	Endless flat belts constructed from cotton or synthetic fabric, with or without chemical or rubber coatings
Steel	Very-high-speed, low-power applications; high tension required	Thin steel bands with riveted or silver-soldered joints
Reinforced rubber, leather, and plastics	Nylon, polyamide, glass fiber, or steel wire tensile members covered with rubber, leather, or plastic; speed and power capabilities are greatly improved by the reinforcement	All but film belts
Rubberized fabric	Consists of plies of cotton or synthetic fabric impregnated with rubber; low cost; good abrasion resistance, but has a lower power-transmitting ability and service life than leather; limited to speeds below 300 mps (6,000 fpm)	All but film belts
Rubberized cord	Constructed with rubber-impregnated cords running lengthwise; carries 50% more power than rubberized fabric belts; fair shock-absorption capacity	Timing, flat, and V-belts, endless only
Rubberized cord and fabric	Combines strength of cord with abrasion resistance of fabric; well suited to minimum-stretch, high-speed operations.	All but film belts, endless only

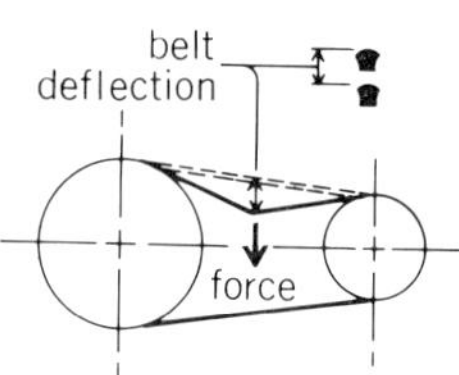

Fig. 6. Tension measurement by deflection.

pulley in meters and N is the angular speed of that pulley in rpm.

Selection criteria. To design a belt drive, the following information is required: speeds of drive and driven unit; power transferred between drive and driven unit: power (kW) = torque (newton-meters) × rpm × 1.05 × 10^{-4}; desired distance between shafts; and operating conditions.

The drive horsepower is corrected for various conditions such as: speed ratio; distance between shafts (long or short); type of drive unit (electric motor, internal combustion engine); service environment (oily, wet, dusty); driven unit loads (jerky, shock, reversed); and pulley-belt arrangement (open, crossed, turned). The correction factors are listed in engineering handbooks and manufacturer's literature. The corrected horsepower is compared to rated horsepowers of standard belt cross sections at specific belt speeds to find a number of combinations that will do the job. At this point, the pulley diameters are chosen. A trade-off must be made between larger pulley diameters or larger belt cross sections, since larger belts transmit this same power at low belt speeds as smaller belts do at high speeds. To keep the drive unit as small as possible, minimum-diameter pulleys are desired. Minimum pulley diameters are limited by the elongation of the belt's outer fibers as the belt bends around the pulleys. Small pulleys increase this elongation, greatly reducing belt life. Minimum pulley diameters are often listed with each cross section and speed, or listed separately by belt cross section. Once the most economical pulley diameters and belt section are chosen, the belt length is calculated. If endless belts are used, the desired shaft spacing may need adjusting to accommodate standard length belts. It is often more economical to use two or more V-belts side by side, rather than one large belt.

If a large speed ratio or a small distance between shafts is used, the angle of contact between the belt and pulley may be less than 180°. If this is the case, the drive power must be further increased, according to manufacturer's tables, and the selection process repeated. The drive power must be increased because belt power capacities are specified assuming 180° contact angles. Smaller contact angles mean less area for the belt to obtain traction, and thus the belt carries less power.

Belt specifications. To fully specify a belt, the material, length, and cross-section size and shape are required. Timing belts, in addition, require that the size of the teeth be given.

For all belt types, and for open and crossed belt drives, an accurate approximation to belt length is Eq. (3). (Symbol definitions are given in Fig. 3.)

$$L = 2C + 1.57(D_2 + D_1) + \frac{(D_2 \pm D_1)^2}{4C} \qquad (3)$$

The sign of the second D_1 is plus for crossed belt drives and minus for open belt drives.

The angle of contact in radians between belt and pulley (Fig. 3) is given by Eq. (4); the plus sign is for

$$\theta = \pi \pm 2 \text{ arc sin}\left(\frac{D_2 - D_1}{2C}\right) \qquad (4)$$

the larger pulley, the minus sign for the smaller pulley.

For the crossed belt drive, both angles of contact are calculated by Eq. (5).

$$\theta = \pi + 2 \text{ arc sin}\left(\frac{D_2 + D_1}{2C}\right) \qquad (5)$$

When correcting for angle of contact, the smallest angle calculated should always be used.

Belt tension. Belts need tension to transmit power. The tighter the belt, the more power transmitted. However, the loads (stress) imposed on the belt and supporting bearings and shafts are increased. The best belt tension is the lowest under which the belt will not slip with the highest load condition. Belt manufacturers recommend operating belt tensions according to belt type, size, speed, and pulley diameters. Belt tension is determined by measuring the force required to deflect the belt a specific distance per inch of span length (Fig. 6). Timing belts require minimum tension, just enough to keep the belt in contact with the pulleys.

Belt wear. Belts probably fail by fatigue more often than by abrasion. The fatigue is caused by the cyclic stress applied to the belt as it bends around the pulleys. Belt failure is accelerated when the following conditions are present: high belt tension; excessive slippage; adverse environmental conditions; and momentary overloads caused by shock, vibration, or belt slapping. *See* CHAIN DRIVE; GEAR DRIVE; PULLEY.

[ARTHUR ERDMAN; RAYMOND GIESE]

Bibliography: T. Baumeister (ed.), *Standard Handbook for Mechanical Engineers*, 8th ed., 1978; W. S. Miller (ed.), Drive systems, *Mach. Des.*, 50(15):26–33, June 29, 1978; P. B. Schubert (ed.), *Machinery's Handbook*, 21st ed., 1979; J. E. Shigley, *Mechanical Engineering Design*, 3d ed., 1977.

Bending moment

The algebraic sum of all moments that are on one side of a cross section of a structural member. For example, in the diagram of a floor joist, a load on the joist tends to bend it at any particular cross section of the joist. At section A-A the bending moment is the reactive upward force of the left wall on the joist times the distance from wall to section A-A. At section B-B the bending moment is the upward reactive force of the left wall times the distance from wall to B-B plus (in the algebraic sense) the moment produced by the downward load acting at its distance from section B-B. The lower portion of

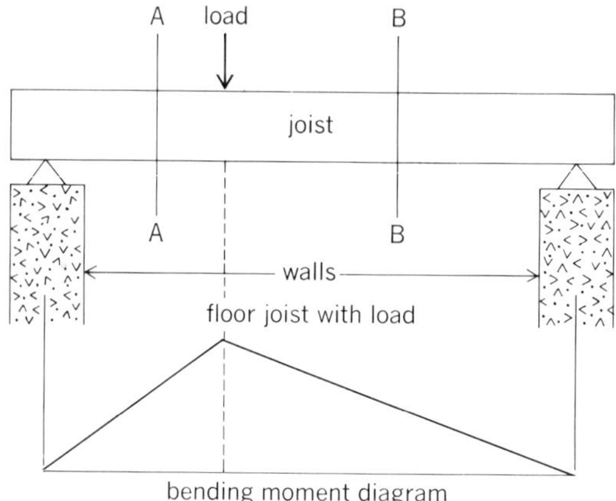

Schematic of bending moment on an end-supported joist with a concentrated load.

the diagram shows the resultant bending moment at each cross section of the joist. A bending moment that bends a beam convex downward is positive, and that bends a beam convex upward is negative. *See* LOADS, TRANSVERSE; SHEAR; STRENGTH OF MATERIALS.

[FRANK H. ROCKETT]

Blasting

The process of loosening or fragmenting solid materials such as rock, masonry, or hard clays through the use of explosives to permit their excavation. Blasting consists of drilling and cleaning blast holes, loading the holes with a suitable explosive and igniter or detonator, stemming the hole (tamping sand or clay over the charge to help prevent dissipation of its explosive force), and firing or setting off the charge.

The most common explosives are ANFO (ammonium nitrate–fuel oil), water gels, and dynamite. Black powder, the original commercial explosive, is almost never used today. About 80% of all blasting is done with ANFO because of its low cost. Dynamite and water gels are used when conditions require small-diameter blast holes or water resistance, and for other special situations. Both dynamite and water gels are supplied in cartridges; water gels also are available in bulk for pumping into blast holes. ANFO, a granular material, is used in bags or poured into holes.

Because water gels are less sensitive than dynamite, they are safer to handle. Water gels do not produce the headache-causing nitroglycerine fumes associated with dynamite, and therefore often are preferred for use where ventilation is poor.

Dynamite and some water gels are detonated with blasting caps or detonating cord, while other water gels and ANFO require a high-explosive primer detonated with a cap or detonating cord. Detonating cord is flexible and made of a tough, abrasion-resistant cover surrounding an extruded core of a high explosive. It is detonated with a blasting cap and is made in several explosive strengths for various uses.

Today most blasting caps are electrical, replacing the less safe and less versatile oil-style powder-core fuse and fuse caps. Electrical blasting caps are detonated with blasting machines, either hand-driven or battery-powered, or from power lines through special switches. New nonelectrical systems reduce the possibility of premature detonation because they are not affected by stray electric currents.

Explosives in most blasts involving more than a few holes are detonated in sequence to provide better rock breakage and reduce vibration or shock in the earth that could cause damage. The sequential delays between holes are provided by delay blasting caps, delay devices in detonating cord lines, sequential blasting machines with electronic delay circuits, or a combination of these methods. Delays range from a few milliseconds to 2 seconds, with the longer delays being used in tunneling. Delay blasts are designed so that individual holes will not be "cut off" by earlier detonations which can disturb adjacent holes or break cap wires or detonating cord lines.

The blast holes near an exposed face in construction or quarrying are detonated first to move rock out of the way of material which will be moved by explosives further from the face. In a tunnel, holes near the center of the face are detonated first to provide space for the rock nearer the edges to move into. For a discussion of drilling *see* CONSTRUCTION METHODS.

[WILLIAM DIGEL]

Block and tackle

Combination of a rope or other flexible material and independently rotating frictionless pulleys; the pulleys are grooved or flat wheels used to change the direction of motion or application of force of the flexible member (rope or chain) that runs on the pulleys. As in a lever, the summation of torques about the axis of rotation of the pulley equals zero for static equilibrium. Tension T in the rope is the same on both sides of the pulley.

For example, in the particular block and tackle illustrated here, at static equilibrium, the summa-

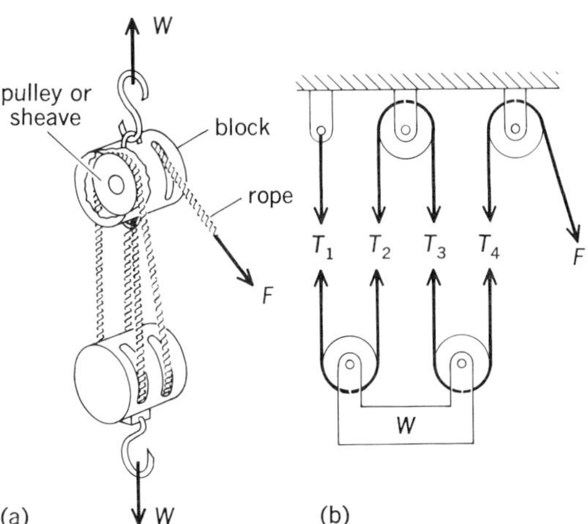

Block and tackle. (*a*) Actual view. (*b*) Schematic.

tion of forces in any direction equals zero. Each vertical rope carries one-fourth the weight; algebraically $T_1 = T_2 = T_3 = T_4 = W/4$, and therefore the applied force F also equals one-fourth the weight W. For this case mechanical advantage MA is shown by the equation below. The block and tack-

$$MA = \frac{W}{F} = \frac{W}{W/4} = 4$$

le is used where a large multiplication of the applied forces is desirable. Examples are: lifting weights, sliding heavy machinery into position, and tightening fences. *See* SIMPLE MACHINE.

[RICHARD M. PHELAN]

Blowout coil

A coil that produces a magnetic field in an electrical switching device for the purpose of lengthening and extinguishing an electric arc formed as the contacts of the switching device part to interrupt the current. The magnetic field produced by the coil is approximately perpendicular to the arc. The interaction between the arc current and the magnetic field produces a force driving the arc in

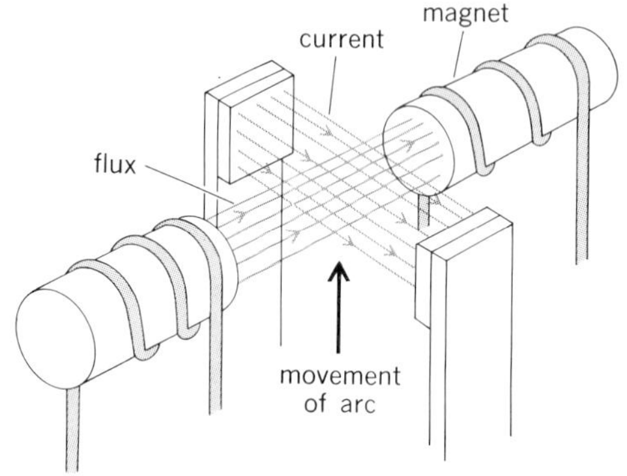

Relation of directions of current, magnetic flux, and movement of arc.

the direction perpendicular to both the magnetic flux and the arc current (see illustration).

In alternating-current circuits the arc usually extinguishes at a natural current zero (when the current passes through zero before reversing its direction). If, within a short time around current zero, sufficient energy can be removed from the arc, conduction will cease. The function of the blowout coil is to move the arc into a region inside a circuit interrupter, such as arc chutes in circuit breakers where the energy removal process takes place. *See* CIRCUIT BREAKER.

In direct-current circuits there are no natural current zeros. When a switching device opens, an arc strikes. As long as the voltage across the open contacts is sufficient to sustain the arc, current will flow. To interrupt this dc arc, the arc must be converted to a form that, to continue, requires more than the available voltage. Thus, in a dc switching device, the function of the blowout coil is to lengthen the arc, which increases the arc voltage, and also to move the arc into regions, inside the interrupter, where the arc voltage can be further increased.

The blowout coil is either connected in series with the contacts or is inserted as the arc moves along an arc runner.

[THOMAS H. LEE]

Boiler

A pressurized system in which water is vaporized to steam, the desired end product, by heat transferred from a source of higher temperature, usually the products of combustion from burning fuels. Steam thus generated may be used directly as a heating medium, or as the working fluid in a prime mover to convert thermal energy to mechanical work, which in turn may be converted to electrical energy. Although other fluids are sometimes used for these purposes, water is by far the most common because of its economy and suitable thermodynamic characteristics.

The physical sizes of boilers range from small portable or shop-assembled units to installations comparable to a multistory 200-ft-high (60-m) building equipped, typically, with a furnace which can burn coal at a rate of 6 tons/min (90 kg/s). In terms of steam production capacities, commercial boilers range from a few hundred pounds (1 lb = 0.45 kg) of steam per hour to more than 6,000,000 lb/hr (750 kg/s). Pressures range from 0.5 psi (3.4 kPa) for domestic space heating units to 5000 psi (34 MPa) for some power boilers. The latter type will deliver steam superheated to 1100±°F (593±°C) and reheated to similar values at intermediate pressures. Large units are field-assembled at the installation site but small units (frequently referred to as package boilers) are shop-assembled to minimize the overall boiler price.

Boilers operate at positive pressures and offer the hazardous potential of explosions. Pressure parts must be strong enough to withstand the generated steam pressure and must be maintained at acceptable temperatures, by transfer of heat to the fluid, to prevent loss of strength from overheating or destructive oxidation of the construction materials. The question of safety for design, construction, operation, and maintenance comes under the police power of the state and is supplemented by the requirements of the insurance underwriters. The ASME Boiler Construction Code is the prevalent document setting basic standards in most jurisdictions.

Being in the class of durable goods, boilers that receive proper care in operation and maintenance function satisfactorily for several decades. Thus the types of boilers found in service at any time represent a wide span in the stages of development in boiler technology.

The earliest boilers, used at the beginning of the industrial era, were simple vats or cylindrical vessels made of iron or copper plates riveted together and supported over a furnace fired by wood or coal. Connections were made for offtake of steam and for the replenishment of water. Evolution in design for higher pressures and capacities led to the use of steel and to the employment of tubular members in the construction to increase economically the amount of heat-transferring surface per ton of metal. The earliest improvement was the passage of hot gases through tubes submerged in the water space of the vessel, and later, arrangements of multiple water-containing tubes which were exposed on their outer surface to contact with hot gases. *See* FIRE-TUBE BOILER; WATER-TUBE BOILER.

The overall functioning of steam-generating equipment is governed by thermodynamic properties of the working fluid. By the simple addition of heat to water in a closed vessel, vapor is formed which has greater specific volume than the liquid, and can develop increase of pressure to the critical value of 3208 psia (22.1 MPa absolute pressure). If the generated steam is discharged at a controlled rate, commensurate with the rate of heat addition, the pressure in the vessel can be maintained at any desired value, and thus be held within the limits of safety of the construction. *See* STEAM.

Addition of heat to steam, after its generation, is accompanied by increase of temperature above the saturation value. The higher heat content, or enthalpy, of superheated steam permits it to develop a higher percentage of useful work by expansion through the prime mover, with a resultant gain in efficiency of the power-generating cycle. *See* SUPERHEATER.

If the steam-generating system is maintained at pressures above the critical, by means of a high-pressure feedwater pump, water is converted to a vapor phase of high density equal to that of the water, without the formation of bubbles. Further heat addition causes superheating, with corresponding increase in temperature and enthalpy. The most advanced developments in steam-generating equipment have led to units operating above critical pressure, for example, 3600–5000 psi (25–34 MPa).

Superheated steam temperature has advanced from 500±°F (260±°C) to the present practical limits of 1050–1100°F (566–593°C). Progress in boiler design and performance has been governed by the continuing development of improved materials for superheater construction having adequate strength and resistance to oxidation for service at elevated temperatures. For the high temperature ranges, complex alloy steels are used in some parts of the assembly.

Steam boilers are built in a wide variety of types and sizes utilizing the widest assortment of heat sources and fuels. *See* NUCLEAR POWER; STEAM-GENERATING UNIT.

[THEODORE BAUMEISTER]

Bibliography: American Society of Mechanical Engineers, *Boiler Code*, 1980; Babcock and Wilcox Co., *Steam: Its Generation and Use*, 39th ed., 1978; T. Baumeister, *Standard Handbook for Mechanical Engineers*, 8th ed., 1978.

Boiler economizer

A component of a steam-generating unit that absorbs heat from the products of combustion after they have passed through the steam-generating and superheating sections. The name, accepted through common usage, is indicative of savings in the fuel required to generate steam.

An economizer is a forced-flow, once-through, convection heat-transfer device to which feedwater is supplied at a pressure above that in the steam-generating section and at a rate corresponding to the steam output of the unit. The economizer is in effect a feedwater heater, receiving water from the boiler feed pump and delivering it at a higher temperature to the steam generator or boiler. Economizers are used instead of additional steam-generating surface because the feedwater, and consequently the heat-receiving surface, is at a temperature below that corresponding to the saturated steam temperature; thus, the economizer further lowers the flue gas temperature for additional heat recovery. *See* FEEDWATER; THERMODYNAMIC CYCLE.

Generally, steel tubes, or steel tubes fitted with externally extended surface, are used for the heat-absorbing section of the economizer; usually, the economizer is coordinated with the steam-generating section and placed within the setting of the unit.

The size of an economizer is governed by economic considerations involving the cost of fuel, the comparative cost and thermal performance of alternate steam-generating or air-heater surface, the feedwater temperature, and the desired exit gas temperature. In many cases it is economical to use both an economizer and an air heater. *See* AIR HEATER; STEAM-GENERATING UNIT.

[GEORGE W. KESSLER]

Boiler feedwater regulation

Addition of water to a steam-generating unit at a rate commensurate with the removal of steam from the unit. The addition of water to a boiler requires a feedwater pump or some other device that will develop a pressure higher than that of the steam generated. Means also are required to control the rate at which water is added.

Pressurization. Reciprocating plunger pumps, driven by steam from the boiler itself, often are used for small-capacity boilers. They have the advantage of simplicity and are not dependent upon other sources of power. Such pumps may have single or multiple cylinders, with single- or double-acting plungers, and they can be operated at variable speeds to deliver water at a steady and controlled rate to maintain a specified water level in the steam drum of the boiler.

Pumps for high-pressure high-capacity boilers are usually of the centrifugal type, with multiple stages connected in series to develop the required feedwater pressure. They may be driven by electric motors, by auxiliary steam turbines, or directly from the main turbine shaft. When a pump is operated at constant speed, the flow of water to the economizer or boiler is controlled by regulating valves installed in the discharge piping, with a bypass around the pump to handle the excess flow from the pump.

Injectors are sometimes used to supply water to portable fire-tube boilers. The injectors utilize the energy of a jet of steam, supplied from the boiler itself, to develop a high velocity by expanding the steam through a nozzle. The kinetic energy imparted by the high-velocity steam to the entrained supply water is converted to pressure in a discharge tube which is connected to the boiler shell.

Regulation. The water level in the steam drum must be clearly indicated by an illuminated gage glass, which is required by law for reasons of safety. It is attached to a water column or vertical chamber which is connected, in turn, without obstruction, between upper and lower nozzle openings in the steam drum. In high-pressure boilers two water gages are directly connected to the steam drum, one at each end, thus eliminating the use of water columns.

Intermittent manual regulation of the feedwater flow satisfactorily maintains the steam drum water level within acceptable limits in boilers having relatively large water-storage capacity, such as shell-type or multiple drum water-tube units; but automatic control provides better regulation.

For water-tube boilers having a single steam drum of relatively small diameter, a continuous and exact regulation of the feedwater is required; and automatic single-element control devices can be used to operate the supply valve or to change the pump speed in response to a change in water level.

Variations of water level in the steam drum usually are due to changes in the rate of steam generation because such changes affect the steam output, the volume of the steam below the water level, and the consequent displacement of water into or out of the drum. Changes in water level can be compensated by use of an automatic three-element control which, primarily, regulates the rate of feedwater flow to be equal to the rate of steam output, as determined by metering equipment; the control then readjusts the rate of feedwater flow to maintain the water level within the prescribed normal range. *See* FEEDWATER; STEAM-GENERATING UNIT.

[GEORGE W. KESSLER]

Boiler water

Water in the steam-generating section of a boiler unit. Boiler water is heated to produce steam; it also cools the heat-absorbing surfaces as it circulates through the boiler unit. The steam withdrawn from the steam drum is of high purity and, therefore, contaminants that enter with the boiler feedwater, even in small concentrations, accumulate in the boiler water. These contaminants, if not removed, can interfere with boiler operation. They can cause corrosion, adversely affect boiler circulation or steam-water separation, or form deposits on the internal surfaces of the heat-absorbing components. Such deposits increase metal temperatures and result in eventual failure of the pressure parts due to overheating. Most soluble contaminants are readily removed by the blowdown of boiler water. However, insoluble contaminants must be treated by the addition of chemicals to the boiler water to change their characteristics. *See* STEAM-GENERATING UNIT.

Insoluble contaminants. In most applications it is necessary to remove all but trace quantities of the troublesome insoluble impurities from the water before it is pumped into the boiler. Hardness (calcium and magnesium salts) is removed from the raw water prior to its introduction to the boiler cycle as makeup. Metal oxides (iron and copper) are frequently removed from the condensate returns by a filter process. However, residual amounts of these deposit-forming contaminants are inevitably present. Further, additional contamination may occur periodically from the leakage of raw cooling water into heat exchangers, such as the turbine condenser. *See* FEEDWATER.

Calcium hardness. At the lower operating pressures (below 1500 psi), residual calcium hardness is treated by introducing sodium phosphate to the boiler. This treatment precipitates the calcium as a phosphate sludge, which can be removed with the boiler water blowdown before it deposits on the heat-transfer surfaces. At high operating pressures (above 1500 psi), the rate of deposition is so rapid that the use of blowdown is ineffective. Even at low pressures some deposition occurs, and the most effective method of minimizing boiler deposits is to reduce the deposit-forming materials in the feedwater.

Silica contamination. Silica, which may form an adherent and highly insulating scale in combination with other contaminants, has the additional property of vaporizing at significant and increasing rates as the temperature of the boiler water exceeds 500°F. The concentration of silica must be held within limits by blowdown, the allowable concentration decreasing as pressure increases. Silica in vapor solution with the output steam is not arrested by mechanical means. Consequently, it can cause troublesome deposits on turbine blades, leading to lower turbine capacity and costly outages for deposit removal. *See* STEAM SEPARATOR.

Corrosion prevention. Under normal conditions internal corrosion of boiler steel is prevented by maintaining the boiler water in an alkaline condition. This usually is accomplished at the lower operating pressures by the addition of sodium hydroxide to the boiler water to produce a pH within the range 10.5–11.5. At higher operating pressures the presence of strong alkalies in the boiler water can cause boiler corrosion in those zones where a local concentrating mechanism exists. In addition, sodium hydroxide volatilizes at high pressure to the extent that it produces turbine deposits and loss of turbine capability. Thus, at the higher operating pressures the modern practice is to maintain only a few parts per million (ppm) of sodium phosphate or a volatile amine (ammonia, morpholine, or cyclohexalamine) in the boiler water to produce a pH in the range 9.0–10.0. *See* WATER TREATMENT.

Trace quantities of oxygen that are not removed by deaeration can cause corrosion and are usually scavenged chemically in the feedwater or boiler water with hydrazine or sodium sulfite. Sulfite is seldom used at pressures above 1500 psi because the thermal decomposition of sulfite produces undesirable acidic gases.

In forced-flow, once-through boilers, boiler water is not recirculated; consequently, impurities entering with the feedwater must either leave with the steam or be deposited within the unit. Thus, such units require high purity makeup water, which should be evaporated or demineralized water. [GEORGE W. KESSLER]

Bibliography: T. Baumeister, *Marks' Standard Handbook for Mechanical Engineers*, 8th ed., 1978; L. I. Pincus, *Practical Boiler Water Treatment, Including Air Conditioning Systems*, 1962; S. T. Powell, *Water Conditioning for Industry*, 1954.

Bolt

A rod, usually of metal, with a head at one end and a screw thread on the other. A bolt is used to fasten objects together. A bolt is passed through clearance holes in two or more parts, a nut is engaged on the threaded end, and the parts are drawn together. Bolts can be obtained in a variety of forms, each suited for a specific application (see illustration). Standard hexagon wrench-head bolts and nuts are available in three degrees of finish: Unfinished bolts are finished only on the thread, semifinished bolts are machined under the head, and finished bolts are fully machined on all surfaces. The two standard weights for bolts are regular, for general use, and heavy, for applications where a greater bearing surface is needed. For the same nominal size, a heavy bolt or nut has larger head and nut dimensions than one belonging to the regular series. Wrench-head bolts are specified by giving the diameter, number of threads per inch, series, class of thread, length, finish, and type of head; for example, $\frac{1}{2}$-13 UNC-2A × $1\frac{3}{4}$ SEMI-FIN. HEX. HD. BOLT.

BOLT

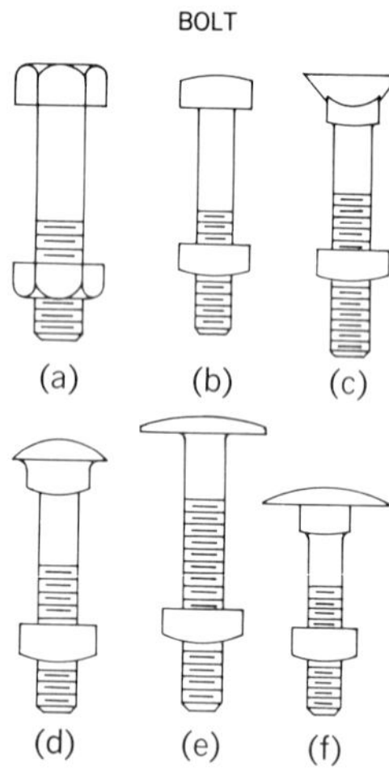

Examples of bolts. (*a*) Heavy-weight standard bolt. (*b*) Regular-weight standard bolt. (*c*) Counter sunk carriage bolt. (*d*) Round-head carriage bolt. (*e*) Elevator bolt. (*f*) Step bolt.

Machine bolts are used in the automotive, aircraft, and machinery fields, where a snug full-bodied fit is needed. They are available with such heads as square, hexagon, and flat, and with either square or hexagon nuts.

Stove bolts are used on electrical equipment and household appliances, either in tapped holes or with nuts. They are available with round, oval, truss, or flat heads. Normally, they come with square nuts.

The carriage bolt is a round-head type with nut; it is used as a through bolt. The head is essentially of the truss type. For normal application, the square-neck style is used, although carriage bolts can be obtained with oval or ribbed necks.

Special bolts are elevator, plow, step, strut, tire, T-head, and track bolts. Plow bolts are employed on agricultural equipment. *See* NUT; SCREW FASTENER. [WARREN J. LUZADDER]

Bolted joint

The assembly of two or more parts by a threaded bolt and nut or by a screw that passes through one member and threads into another (see illustration). A bolted joint can be disassembled more readily than welded or riveted joints.

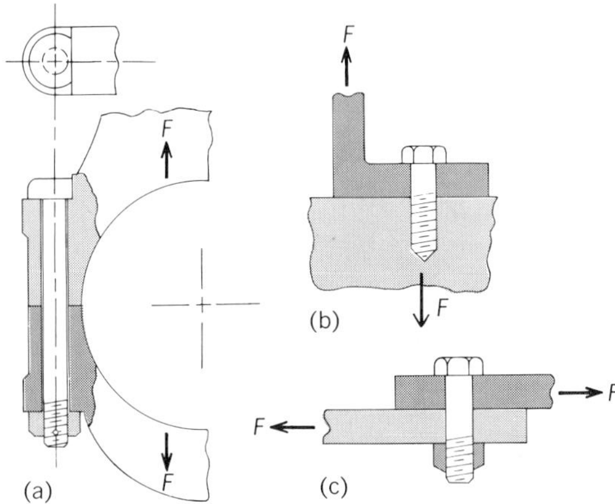

Forms of bolted joint (*F* represents force). (*a*) Direct tension. (*b*) Eccentric loading. (*c*) Single shear.

The stress in a bolted joint is similar to that of a riveted joint, and hole spacings and bolt sizes can be calculated by the same rules and equations. The bolt may be a single or double shear, or it may be in tension. The surface of the cylindrical section of a bolt is also in compression on one side when the bolt is in shear. Bolts or screws may be so loaded that they are in axial tension only. The bolt holding the cap on a connecting rod is an excellent example of a bolt substantially in pure tension. *See* RIVETED JOINT.

Structural members, such as I beams, H beams, angles, and plates, may be joined by bolting. When properly tightened, such joints are as strong and reliable as riveted ones. For joints of this type, high-strength alloy-steel bolts and nuts tightened with impact wrenches are generally used. No locking feature is necessary for properly selected and tightened bolts.

The strength of a bolted assembly depends to a great extent on the initial loading and stress placed on the bolt by the assembly torque. Devices such as torque wrenches are used to determine tension as a function of the torque applied to the assembly of nut and bolt or cap screw and machine part despite the uncertain coefficient of friction between nut, bolt, and assembled members. A better measure is to gage the elongation of the bolt as the tightening torque is applied. Using Young's modulus of elasticity for the bolt material, it is easy to calculate the tensile stress. Bolt and nut assemblies often include special lock washers to prevent accidental loosening of the fastening by vibration. Where critical loads are to be carried and safety is paramount, bolts or screws smaller than 1/2 in. in diameter are dangerous to use because it is easy to overload the bolt or screw in assembly and actually twist it off with standard wrench sizes. *See* JOINT; SCREW FASTENER; STRUCTURAL CONNECTIONS; WELDED JOINT.

[L. SIGFRED LINDEROTH, JR.]

Boring

A machining operation that increases the size of an existing hole in a workpiece. The usual purpose of boring is to machine a hole to a desired diameter while obtaining required accuracy and finish. In metalworking, boring is performed after making a circular hole in the piece by some method such as coring, drilling, burning, punching, or trepanning. *See* DRILLING MACHINE.

Boring machine of the horizontal type. (*Giddings and Lewis Machine Tool Co.*)

Boring machines may be listed in two categories, horizontal and vertical. The former type holds the workpiece stationary on a movable table. A rotating spindle, holding the cutting tool, can be fed into the work from a vertically adjustable head (see illustration). On vertical machines the work is revolved on a circular table while tools can be fed in for boring as well as facing operations. *See* MACHINING OPERATIONS.

[ALAN H. TUTTLE]

Boring and drilling (mineral)

The drilling of holes for exploration, extractive development, and extraction of mineral deposits. Exploration holes are drilled to locate mineral deposits, define their extent, determine their ore dressing characteristics, determine mining or quarrying conditions to be met in their development or extraction, and otherwise to establish the commercial value of the deposit. Holes for extractive development are drilled for blasting, mine dewatering, access for utility lines, grouting for water control, pilot holes for shaft sinking, and as small shafts for backfilling, ventilation, portals, and escapeways.

Elements of drilling. The two essentials are a process to detach particles of soil or rock from the floor of the hole, and a way to remove the loosened particles from the hole. The principles that can be applied to these processes are limited by the confines of the drill hole, but they can be combined into an innumerable variety of drilling methods, applicable to mineral work.

Drills and equipment. Energy and action required for the two processes of drilling are furnished in various ways. Drills are designed to create motion, rotary, linear, or combined rotary and linear motion. Auxiliary drill-rig equipment (pumps, compressors, generators) may introduce additional energy into the hole in the form of hydraulic, pneumatic, or electric power. Another source of energy may be created at the bit by conversion of gases into heat.

The energy of motion created by the drill may be transmitted to the drill bit by a shaft (drill rods, drill pipe, drill stem, drill steel) or by a flexible cable (wire-line, rope). Drill rods, drill pipe, and drill steel are tubular so that they may be used for the injection of fluids, compressed air, or gases.

Principal drilling methods. Five general categories may be classified: rotary drilling, penetration by the abrasive action of a drill bit in rotary motion; core drilling, rotary drilling an annular groove leaving a central core; percussion drilling, penetration by the chipping or crushing action produced by a drill bit in linear motion; rotary-percussion drilling, combination of rotary and linear motions to produce penetration; fusion piercing, penetration by flaking or melting caused by the application of heat. Many drills may be adapted to a variety of drilling methods, particularly those drills designed to produce penetration by rotary motion and which also incorporate dual linear motion mechanisms, one for control and one for hoisting. Some drills are specifically designed for a combination of drilling methods (Fig. 1).

Fig. 1. Combined auger-churn-core drill. (*Penndrill*)

Drilling methods have terminology generally indicative of the outstanding feature of the method, but the same term may be used for a method that is also used for a type of drilling, as in the use of the term rotary drilling.

Rotary drilling. This is vertical drilling through a rotary table, and the typical drill is that used for drilling exploratory and producing wells for the petroleum minerals (oil and gas). The rotary table turns a square kelly, with mud swivel at the top and connection to drill pipe below. Penetration is by rotation of drill bits of two types: roller bits, which have rolling cutters with projecting hard teeth; and drag bits, with fixed chisel-type hard cutting edges. Cuttings removal is by circulation of drilling mud, which is also used to protect the wall of the hole and eliminate the need for casing. Controlled vertical motion of the tools is provided by wire-line drum-type hoist with multisheave blocks in the derrick for multiplying hoisting capacity, controlling weight on the bit by partial support of the weight of drill pipe and heavy drill collars.

Smaller rotary drills, with a thrust mechanism added to react against the weight of the drill and increase bit pressure, are used for seismograph prospecting, core drilling, water wells, and blast holes. Circulation of a large volume of air, rather than drilling mud, is commonly used for cuttings removal in shallow rotary drill holes. Water is almost always used as the circulating medium with core drills, to dissipate heat as well as to remove cuttings.

Auger drilling. Rotary-type auger methods penetrate by the cutting or gouging action of chisel-type cutting edges forced into the formation by rotation of the bit. Cuttings are removed by mechanical action. Auger rods are made with a continuous helical projecting surface so that they act as screw conveyors to remove the cuttings. Auger drills are of various sizes, and usually incorporate a hydraulic cylinder to increase bit pressure and to withdraw the tools. The method is used for drilling blast holes and for reconnaissance prospecting. A special type is used as a mining tool.

Diamond core drilling. Another rotary type, this drilling utilizes the extreme hardness of the diamond to penetrate rock by abrasive action. Core drills rotate an annular bit to cut a narrow kerf around a central core and thus obtain unaltered samples of the formation drilled. Diamond drills are the principal tools for exploration of mineral deposits.

The diamond drill furnishes rotary motion for penetration, and generally also has dual linear motion mechanisms, a hydraulic cylinder or screw mechanism to control the feed of the bit against the rock, and a wire-line hoist or pneumatic cylinder for tool withdrawal. The drill head usually swivels for drilling at any angle. Tubular drill rods transmit rotary motion to the core barrel, which has a core-retaining device just above the diamond bit. The ring-shaped diamond bit is set with industrial-grade diamonds, those with imperfections that prohibit their use as gems. Water or mud circulation cools the bit and removes the cuttings. Double-tube core barrels may be used to keep water circulation from washing the core (see table).

A few variations are developed. Surface-set bits contain a single layer of diamonds. Impregnated bits have diamond fragments distributed through the crown so that fresh cutting points are exposed as the bit wears. Noncoring diamond bits may be used where a core is not desired, as in diamond blast-hole drilling.

Shot (abrasive) drilling. This method of rotary core drilling uses hard, chilled steel shot or other loose abrasive as a cutting medium. Penetration is accomplished by the grinding action of the abrasive when dragged over the rock by a rotating core barrel with a blank steel bit, much as any grinding process utilizing a loose abrasive.

The method is employed to obtain cores in large-diameter holes where abrasive bit cost is sufficiently lower than diamond bit cost to offset the increased operating costs of slower progress, and to drill shafts and other large-diameter holes where it is more economical or desirable to drill a groove and hoist out the core than to expend drilling energy over the entire cross-sectional area (Fig. 2).

Cuttings are removed from under the bit by water circulation but without sufficient velocity to carry them to the surface. Shot core barrels are therefore built with a chip box (calyx) on top so that the cuttings are retained when they settle out under the velocity drop at the top of the core barrel.

Air percussion drilling. Penetration of this drilling is by crushing action under pneumatically powered impact. The drill bit is chisel type, commonly with four cutting edges in the form of a cross. The drill produces linear impact motion by means of a reciprocating pneumatically powered piston to strike a rapid series of blows against a tubular drill steel, which transmits the energy to the bit. The drills also provide pneumatically powered rotary motion constantly to change the position of the bit against the floor of the hole. Cuttings are usually blown out of the hole by compressed air injected through the hollow drill steel and bit, although water may be substituted with drills designed for hazardous dust conditions. Dust collectors may be used to reduce hazard and also to collect samples.

Standard diamond-core bit sizes in inches*

Hole diameter	Design and core diameter		
	Lifter in bit	Lifter in case	Thin wall
$1\frac{5}{32}$			$\frac{3}{4}$
$1\frac{15}{32}$	$\frac{27}{32}$	$\frac{27}{32}$	$\frac{29}{32}$
$1\frac{7}{8}$	$1\frac{3}{16}$	$1\frac{3}{16}$	$1\frac{9}{32}$
$2\frac{11}{32}$	$1\frac{21}{32}$	$1\frac{21}{32}$	$1\frac{3}{4}$
$2\frac{31}{32}$	$2\frac{5}{32}$	$2\frac{5}{32}$	$2\frac{5}{16}$
$3\frac{57}{64}$	3	3	$3\frac{3}{16}$
$3\frac{7}{8}$		$2\frac{11}{16}$	
$5\frac{1}{2}$		$3\frac{15}{16}$	
$7\frac{3}{4}$		$5\frac{15}{16}$	

*Data from Diamond Core Drill Manufacturers Association. 1 in = 25.4 mm.

Fig. 2. Shot drill core. (*Pennsylvania Drilling Co.*)

Air percussion drills are primarily used for blast-hole drilling, in various sizes and mountings, for both surface and underground work. The method is limited in hole size and depth by two considerations — constant loss of bit gage and absorption of dynamic energy by the drill steel. Tungsten carbide cutting inserts help retain gage, and special down-the-hole drills are of aid in the latter problem. Detachable drill bits are commonly used in pneumatic drilling operations.

Cable tool (churn) drilling. A different percussion-type method, this penetrates by crushing impact of a falling heavy chisel bit. The drill creates an oscillating vertical linear motion, which is transmitted to the drilling tools by wire-line cable, so that they are alternately lifted and dropped. The cuttings are suspended in water by the churning action of the tools, and then periodically removed by bailing. Churn drills are sometimes used for prospecting where analysis of the cuttings gives sufficient information. In mineral exploitation work they are used for drilling blast holes in quarrying, for drilling water wells, and for drilling access holes for utility lines into mines.

Fusion piercing methods. For these, an oxygen-acetylene blow pipe applies intense heat to the floor of the hole and penetrates by flaking or melting. Drills must produce both linear and rotary motion to handle the blow pipe and break up slag formed in the process. The cuttings and slag are blown from the hole. Fusion piercing methods have been used successfully for drilling blast holes in hard rocks, such as taconite and granite.

Other drilling methods. Innumerable combinations of the basic drilling methods described above provide a wide range of variations in drilling methods.

Air or natural gas is employed as the circulating medium in rotary-type drilling. Faster rates of penetration have been obtained with these methods, but groundwater conditions may prohibit their use by creating too much fluid head for the available air pressure to work against. Techniques are under development to introduce foaming agents into the fluid column to reduce its weight.

Auger mining utilizes the conveyor action of continuous-flight augers for actual mining operations in large-diameter horizontal holes, such as for mining coal in the high wall beyond the economical limits of strip mining. The science of electronics has been applied to guide such an auger and keep it within the bed. *See* MINING EXCAVATION.

Auger stem drilling uses a short helix, run on a solid or telescoping stem, as a bit for gouging out and collecting cuttings. With a derrick higher than the hole is deep, or high enough to withdraw the entire string of tools in telescoped position, the auger can be withdrawn quickly and spun rapidly to throw off the cuttings by centrifugal force. The method is suitable for holes 12–72 in. or larger diameter. By substituting a bit incorporating a steel cylinder to confine the cuttings, the terminology is changed to bucket drilling.

Reverse-circulation drilling usually refers to a variation of the rotary drilling method in which the cuttings are pumped up and out of the drill pipe. This is advantageous in certain large-diameter holes. The term is also applied to diamond core drilling when the water is injected through a stuffing box into the annular space around the drill rods and thus forced up special large drill rods. Here the water forces the core up through the drill rods as a piston through a cylinder.

Rotary blast-hole drilling is a term commonly applied to two types of drilling. In quarrying and open-pit mining it implies rotary drilling with roller-type bits, using compressed air for cuttings removal, either conventional rotary table drive or hydraulic motor to produce rotation, and with hydraulic or wire-line mechanisms to add part of the weight of the drill to the weight of the tools and thereby increase bit pressure. In underground mining, and sometimes above ground, it implies the drilling of small-diameter blast holes with a diamond drill, using either coring or noncoring diamond bits.

Rotary-percussion drilling increases the penetration speed of a roller bit by adding pneumatic impact in linear motion to the normal abrading action of the bit in rotary motion.

Shaft drilling is shaft sinking by drilling a hole the size of the shaft, as contrasted with conventional shaft-sinking methods of drilling small holes and blasting. The perfect arch action of the smooth-cut circular wall of a drilled shaft makes shaft lining unnecessary in hard massive rocks, and much safer and more economical in others, because a blasted wall is shattered and requires more concrete for lining due to overbreak. Both shot core drilling and rotary drilling methods are used in shaft drilling, with variations such as the use of carbide or rolling cutters on the core barrel, rotary reaming a pilot hole in successively larger stages, use of reverse circulation, and down-the-hole-type rotary drills. A core barrel has been developed for shaft drilling that brings the core to the surface in the core barrel (Fig. 2).

Down-the-hole drilling is air percussion type (for large-diameter blast holes) with the reciprocating pneumatic piston placed in the drill tools close to the bit for minimizing energy losses.

Sonic drilling methods are percussion or rotary-percussion type, utilizing for impact the energy of a drill stem vibrating at sonic frequency. One method of producing sonic vibration is by use of eccentric weights driven by mud turbine.

Turbodrilling is rotary drilling with rotary motion created in the hole close to the bit by a turbine driven by the circulating mud.

Wash boring or jet drilling utilizes a chopping bit with a water jet run on a string of hollow drill rods to chop through soils and wash the cuttings to the surface. It is percussion-type drilling for which churn drills are ideally suited, but is often used on other drills because of the need for rotary coring methods in another part of the same hole. When a combination churn and diamond drill is not available, the method may be used with a diamond drill with a cathead by snubbing a manila rope to create the churning action. By introducing a check valve at the bit, the churning action may also be used to pump the cuttings up the drill rods, using natural groundwater or water added to the hole, a prospecting method called hollow-rod drilling.

Wire-line coring is a method of removing core by pulling the inner tube of the core barrel, with the bit, core barrel, and drill rods remaining in the hole. The inner tube is dropped or pumped down through the drill rods, and recovered by running a retriever on a wire line, so the periodic removal of core can be done in deep drill holes in less time than required for removal and replacement of drill pipe or drill rods. In rotary drilling, the inner tube was designed to pass through the drill pipe and seat above a ring-type roller cutter core bit. In adapting the method to diamond core drilling, special drill rods were developed with maximum possible inside diameter so that the inner tube and the core could be kept as large as possible. Most of the deep diamond drill holes are now being drilled with wire-line diamond core drilling equipment. *See* CORE DRILLING; OIL AND GAS WELL DRILLING.

[FRANK C. STURGES]

Brake

A machine element for applying a force to a moving surface to slow it down or bring it to rest in a controlled manner. Brakes are used in cars, trains, airplanes, elevators, and other machines. Most brakes are of a frictional type in which a fixed surface is brought into contact with a moving part that is to be slowed or stopped. Brakes in general connect a moving and a stationary body, whereas clutches and couplings usually connect two moving bodies. *See* CLUTCH; COUPLING.

The limitations on the applications of brakes are similar to those of clutches, except that the service conditions are more severe because the entire energy is absorbed by slippage which is converted to heat that must be dissipated. The important thing is the rate at which energy is absorbed and heat dissipated. With frictional brakes, if the temperature of the brake becomes too high, the result is a lowering of the friction force, called fading.

There are also electrical and hydrodynamic brakes. The electrical type may be electromagnetic, eddy-current, hysteresis, or magnetic-particle. The hydrodynamic type works somewhat like a fluid coupling with one element stationary. Another type of brake, used on electric trains, is the regenerative brake. The dynamo on this machine can be used as a motor or a generator. As a generator it brakes the train and stores the generated electricity in an accumulator. Another type of brake is the air brakes (flaps) on an airplane.

Friction types. Friction brakes are classified according to the kind of friction element employed and the means of applying the friction forces. *See* FRICTION.

Single-block. The simplest form of brake consists of a short block fitted to the contour of a wheel or drum and pressed up against the surface by means of a lever on a fulcrum, as widely used on railroad cars. The block may have the contour lined with friction-brake material, which gives long wear and a high coefficient of friction. The fulcrum may be located with respect to the lever in a manner to aid or retard the braking torque of the block. The lever may be operated manually or by a remotely controlled force (Fig. 1*a*).

Double-block. Two single-block brakes in symmetrical opposition, where the operating force on the end of one lever is the reaction of the other, make up a double-block brake (Fig. 1*b*). External thrust loads are balanced on the rim of the rotating wheel.

External-shoe. An external-shoe brake operates in the same manner as the block brake, and the designation indicates the application of externally contracting elements. In this brake the shoes are appreciably longer, extending over a greater portion of the drum (Fig. 1*c*). This construction allows more combinations for special applications than the simple shoe, although assumptions of uniform pressure and concentrated forces are no longer possible. In particular, it is used on elevator installations for locking the hoisting sheave by means of a heavy spring when the electric current is off and the elevator is at rest. Self-energization is possible, as with block brakes, depending upon the arrangement of the supporting mechanism.

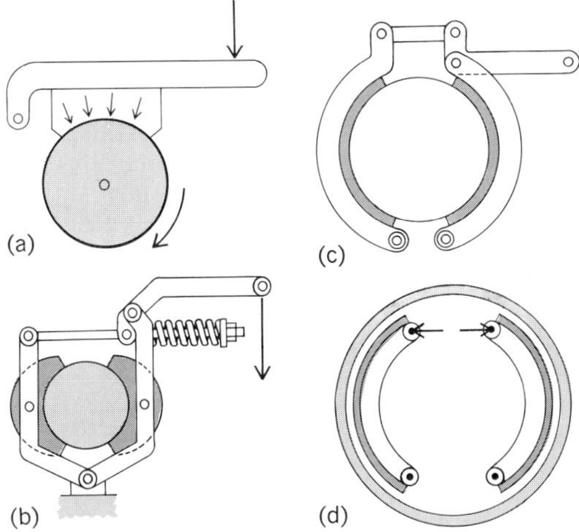

Fig. 1. Brakes. (*a*) Single-block brake. Block fixed to operating lever; force in direction of arrow applies brake. (*b*) Double-block brake. Blocks pivoted on their levers; force in direction of arrow releases brake. (*c*) External shoe brake. Shoes are lined with frictional material. (*d*) Internal shoe brake with lining.

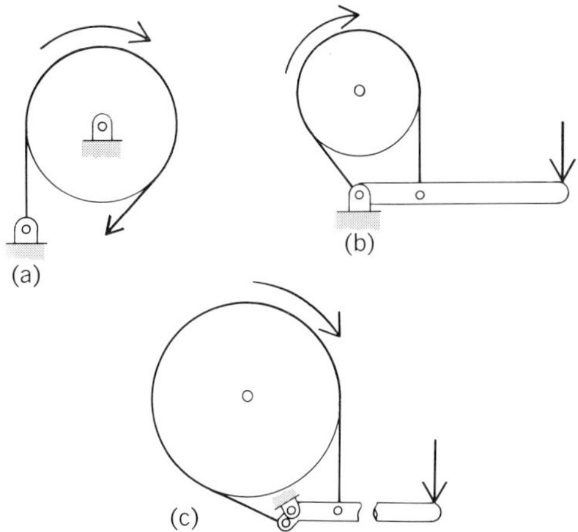

Fig. 2. Band brakes. (*a*) Basic structure. (*b*) With direct-action lever. (c) Differential self-assisting feature.

Internal-shoe. An internal shoe has several advantages over an external shoe. Because it works on the inner surface of the drum, it is protected from water and grit (Fig. 1*d*). It may be designed in a more compact package, is easily activated, and is effective for drives with rotations in both directions. The internal shoe meets the requirement of the automobile. Early mechanical brakes had an elliptical cam; modern brakes have hydraulic piston-actuated shoes.

Band. Hoists, excavating machinery, and hydraulic clutch-controlled transmissions have band brakes. They operate on the same principle as flat belts on pulleys. In the simplest band brake, one end of the belt is fastened near the drum surface and the other end is then pulled over the drum in the direction of rotation so that a lever on a fulcrum may apply a tension to the belt.

The belt may be a thin metallic strip with a friction lining. The method of applying the lever on the fulcrum and attaching the belt determines the structural operation of the brake. These variations make possible a sensitive differential brake, self-energizing brakes, and brakes operating with equal effectiveness in both directions (Fig. 2). The radial pressure of the brake is proportional to the tension in the band, and decreases in the direction of rotation. The band brakes on automatic transmissions are completely circular and are loaded by springs which are cam-operated to lock and unlock.

Disk. Structurally similar to disk clutches, disk brakes have long been used on hoisting and similar apparatus (Fig. 3). Because more energy is absorbed in prolonged braking than in a clutch start-up, additional heat dissipation must be provided in equivalent disk brakes. A significant new application of disk brakes is on the wheels of aircraft, where segmented rotary elements are pressed against stationary plates by hydraulic pistons. Flexibility, self-alignment, and rapid cooling are inherent in this design. A familiar application is the bicycle coaster brake.

The caliper disk brake consists of a revolving disk which can be gripped between two brake pads. A brake on a bicycle is an example. The automotive caliper disk brake is hydraulically operated, and the pads cover between one-sixth and one-ninth of the swept area of the disk. It is used on the front axle of a car. *See* AUTOMOTIVE BRAKE.

Actuators. A brake can be arranged so that it is normally released or applied, a spring usually forcing it into the normal condition. A controllable external force then applies or releases the brake. This force can be, for example, mechanical through a linkage or cable to a lever, pneumatic or hydraulic through a cylinder to a piston, or electric through a solenoid to a plunger.

Air brake. Railway brakes are normally applied air brakes; if the air coupling to a car is broken, the brakes are applied automatically. To apply the brakes, the brakeman releases the compressed air that is restraining the brakes by means of a diaphragm and linkage. Over-the-road trucks and buses use air brakes. Another form of air brake consists of an annular air tube surrounding a jointed brake lining that extends completely around the outside of a brake drum. Air pressure expands the tube, pressing the lining against the drum. *See* AIR BRAKE.

Electromagnetic brake. In commonly encountered electromagnetic brakes, the actuating force is applied by an electric current through a solenoid (Fig. 4). Direct current gives greater braking force than alternating current, and is therefore used almost exclusively; ac is rectified if used. Usually the electromagnetic force releases the brake against a compression spring, thus providing braking action if the power fails and overspeed protection when the brake is used with dc series motors.

The plunger and its stop are cone-shaped to provide a higher force than without a stop, but without an abrupt increase in force as the gap closes. Because the solenoid force decreases somewhat when the brake is applied and the gap is thereby increased, the retraction current is slightly higher than the current at which the spring overcomes the solenoid and applies the brake. This operation is stable.

The solenoid is wound with high impedance for shunt operation across a power line. Shunt operation is common where loads and motor speeds vary widely or where a compound motor is used. Where a series motor is used for high starting torque, the solenoid is wound with low impedance for series operation in the circuit of the motor it protects. In this way the brake will be applied if the motor starts to run away under light load.

[HARRY P. HALE]

Bibliography: R. H. Creamer, *Machine Design*, 2d ed., 1976; J. E. Shigley, *Mechanical Engineering Design*, 3d ed., 1977.

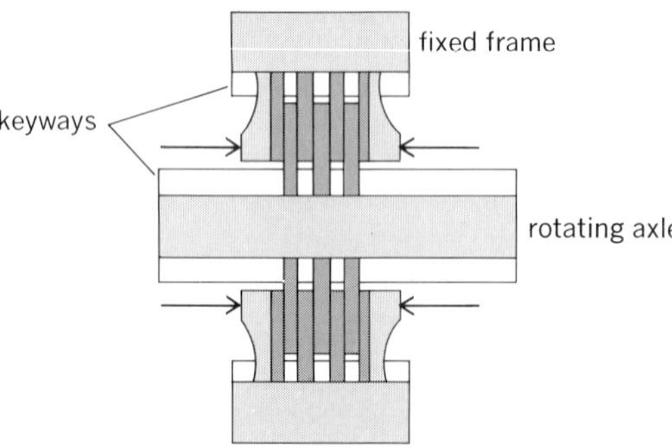

Fig. 3. Forces in direction of arrows press disks keyed to axle against disks keyed to frame.

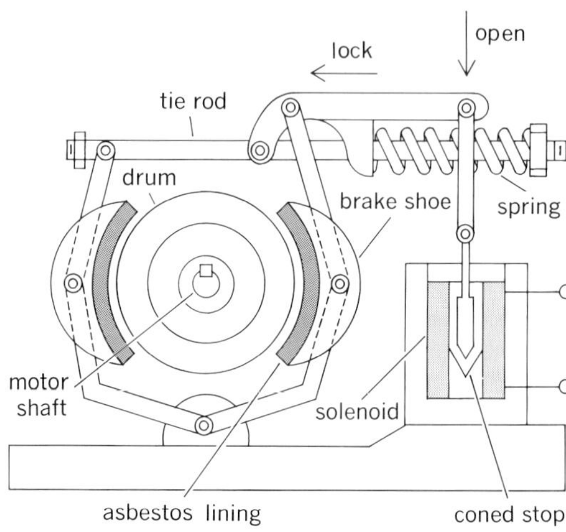

Fig. 4. Electromagnetic brake. Actuating force applied to double-block brake through solenoid.

Branch circuit

The portion of an electric wiring system that follows the final circuit overcurrent protector and that terminates at the utilization device (lighting fixture, motor, or heater) or a plug receptacle for the connection of portable lamps and appliances.

Motor branch circuits must have conductors with current-carrying capacity at least 125% of the motor full-load current rating. Overcurrent protection must be capable of carrying the starting current of the motor. Maximum values are tabulated for various currents in the National Electrical Code. *See* ELECTRICAL CODES.

Branch circuits, other than motor branch circuits, are classified by the maximum rating of the overcurrent device. Thus a branch circuit with a 20-amp circuit breaker is designated a 20-amp branch circuit, even though the conductors may be capable of carrying higher current.

Individual branch circuits serving a single fixed appliance, such as a clothes dryer or a water heater, may be of any size if the overcurrent protection (fuse or circuit breaker) does not exceed 150% of the rated current of the appliance (where 10 amp or more). For continuous loads of 3-hr or longer duration the branch circuit rating must be at least 125% of the continuous load current.

Branch circuits serving more than one outlet or load are limited by the National Electrical Code to three types:

1. Circuits of 15 or 20 amp may serve lights and appliances; the rating of one portable appliance may not exceed 80% of the circuit capacity; the total rating of fixed appliances may not exceed 50% of circuit capacity if lights or portable appliances are also supplied.

2. Circuits of 30 amp may serve fixed lighting units with heavy-duty lampholders in other than dwellings or appliances in any occupancy.

3. Circuits of 40 or 50 amp may serve fixed lighting with heavy-duty lampholders in other than dwellings, fixed cooking appliances, or infrared heating units.

Multiwire branch circuits consist of two or more ungrounded conductors having a potential difference (voltage) between them and a (neutral) grounded conductor having equal potential difference between it and each ungrounded conductor.

Conventionally the current-carrying capacity of conductors may be the same as the branch-circuit rating (for example, no. 12 conductors in a 20-amp branch circuit); however, larger conductors may be required to avoid excessive voltage drop.

[WILLIAM T. STUART; J. F. MC PARTLAND]

Bibliography: H. P. Richter, *Practical Electrical Wiring: Residential, Farm, and Industrial*, 11th ed., 1978.

Brass

An alloy of copper and zinc. In manufacture, lump zinc is added to molten copper, and the mixture is poured into either castings ready for use or into billets for further working by rolling, extruding, forging, or similar process. Brasses containing 75–85% copper are red-gold and malleable; those containing 60–70% are yellow and also malleable; and those containing 50% or less copper are white, brittle, and not malleable. Alpha brass contains up to 36% zinc; beta brass contains nearly equal proportions of copper and zinc. Specific brasses are designated as follows: gilding (95% copper: 5% zinc), red (85:15), low (80:20), and admiralty (70:29, with balance of tin). Naval brass is 59–62% copper with about 1% tin, less than that of lead and iron, and the remainder zinc. The nickel silvers contain 55–70% copper and the balance nickel. With small amounts of other metals, other names are used. Leaded brass is used for castings.

Addition of zinc to copper produces a material that is harder and stronger than copper, yet retains the malleability, ductility, and corrosion resistance of copper. Mechanical and heat treatment greatly vary the properties of the finished product. Because brass with 64% or more copper (α-brass) forms a single solid solution, it combines high strength, ductility, and corrosion resistance. The β solid solution is harder and more brittle. With 61% zinc the compound Cu_2Zn_3 forms; it is brittle, having the lowest tensile strength of the brasses. Higher percents of zinc produce other brittle constituents.

Brass stains in moist air; however, the oxide so formed is sufficiently continuous and adherent to retard further oxidation. When brass is required to remain bright, it is either washed in nitric acid and then coated with clear lacquer, or it is regularly polished and waxed. For brightwork or other corrosion-resistant applications, α-brass is usually used. In atmospheres containing traces of ammonia, cold-worked brass or annealed brass highly stressed in tension may fail by cracking. This is called season cracking or stress corrosion cracking. Susceptibility of cold-worked brass to failure of this kind is reduced by proper annealing.

Brass is widely used in cartridge cases, plumbing fixtures, valves and pipes, screws, clocks, and musical instruments. *See* ALLOY; BRONZE; COPPER ALLOYS; CORROSION.

[FRANK H. ROCKETT]

Bibliography: R. M. Brick et al., *Structure and Properties of Engineering Materials*, 4th ed., 1977; K. G. Budinski, *Engineering Materials: Properties and Selection*, 1979.

Brayton cycle

A thermodynamic cycle (also variously called the Joule or complete expansion diesel cycle) consisting of two constant-pressure (isobaric) processes interspersed with two reversible adiabatic (isentropic) processes (Fig. 1). The ideal cycle performance, predicated on the use of perfect gases, is given by relationships (1) and (2). Thermal efficiency η_T, the work done per unit of heat added, is given by Eq. (3). In these relation-

$$V_3/V_2 = V_4/V_1 = T_3/T_2 = T_4/T_1 \qquad (1)$$

$$\frac{T_2}{T_1} = \frac{T_3}{T_4} = \left(\frac{V_1}{V_2}\right)^{k-1} = \left(\frac{V_4}{V_3}\right)^{k-1} = \left(\frac{p_2}{p_1}\right)^{\frac{k-1}{k}} \qquad (2)$$

$$\eta_T = [1 - (T_1/T_2)] = \left[1 - \left(\frac{1}{r^{k-1}}\right)\right] \qquad (3)$$

ships V is the volume in cubic feet, p is the pressure in pounds per square foot, T is the absolute

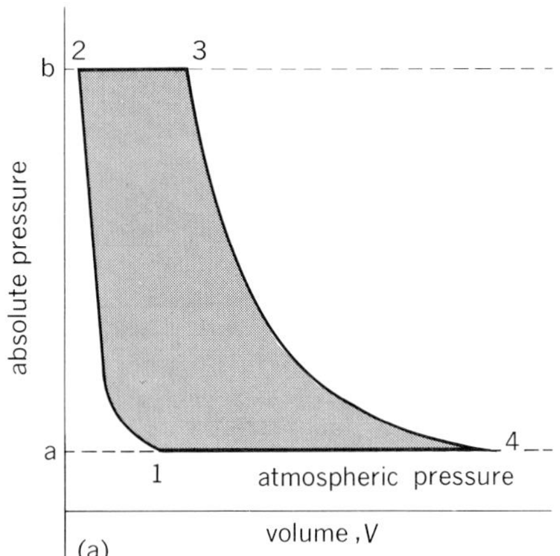

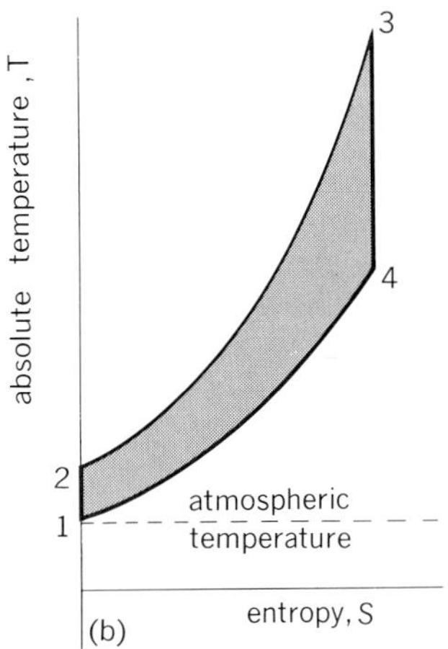

Fig. 1. Brayton cycle, air-card standard. (*a*) *p-V* cycle diagram. (*b*) *p-S* diagram. Phases: 1–2, compression; 2–3, heat addition; 3–4, expansion; and 4–1; heat abstraction.

temperature in degrees Rankine, k is the c_p/c_v, or ratio of specific heats at constant pressure and constant volume, and r is the compression ratio, V_1/V_2.

The thermal efficiency for a given gas, air, is solely a function of the ratio of compression. This is also the case with the Otto cycle. For the diesel cycle with incomplete expansion, the thermal efficiency is lower. The overriding importance of high compression ratio for intrinsic high efficiency is clearly demonstrated by these data.

A reciprocating engine, operating on the cycle of Fig. 1, was patented in 1872 by G. B. Brayton and was the first successful gas engine built in the United States. The Brayton cycle, with its high inherent thermal efficiency, requires the maximum volume of gas flow for a given power output. The Otto and diesel cycles require much lower gas flow rates, but have the disadvantage of higher peak pressures and temperatures. These conflicting elements led to many designs, all attempting to achieve practical compromises. With a piston and cylinder mechanism the Brayton cycle, calling for the maximum displacement per horsepower, led to proposals such as compound engines and variable-stroke mechanisms. They suffered overall disadvantages because of the low mean effective pressures. The positive displacement engine consequently preempted the field for the Otto and diesel cycles.

With the subsequent development of fluid acceleration devices for the compression and expansion of gases, the Brayton cycle found mechanisms which could economically handle the large volumes of working fluid. This is perfected today in the gas turbine power plant. The mechanism (Fig. 2) basically is a steady-flow device with a centrifugal or axial compressor, a combustion chamber where heat is added, and an expander-turbine element. Each of the phases of the cycle is accomplished with steady flow in its own mechanism rather than intermittently, as with the piston and cylinder mechanism of the usual Otto and diesel cycle engines. Practical gas-turbine engines have various recognized advantages and disadvantages which are evaluated by comparison with alternative engines available in the competitive market place. *See* AUTOMOBILE; GAS TURBINE; INTERNAL COMBUSTION ENGINE.

The net power output P_{net}, or salable power, of the gas-turbine plant (Fig. 2) can be expressed as shown by Eq. (4), where W_e is the ideal power out-

$$P_{\text{net}} = W_e \times \text{eff}_e - \frac{W_c}{\text{eff}_c} \tag{4}$$

put of the expander (area b34a, Fig. 1), W_c the ideal power input to the compressor (area a12b, Fig. 1), eff_e the efficiency of expander, and eff_c the efficiency of compressor. This net power output for the ideal case, where both efficiencies are 1.0, is represented by net area (shaded) of the p-V cycle diagram of Fig. 1*a*. The larger the volume increase from point 2 to point 3, the greater will be the net power output for a given size compressor. This volume increase is accomplished by utilizing the maximum possible temperature at point 3 of the cycle.

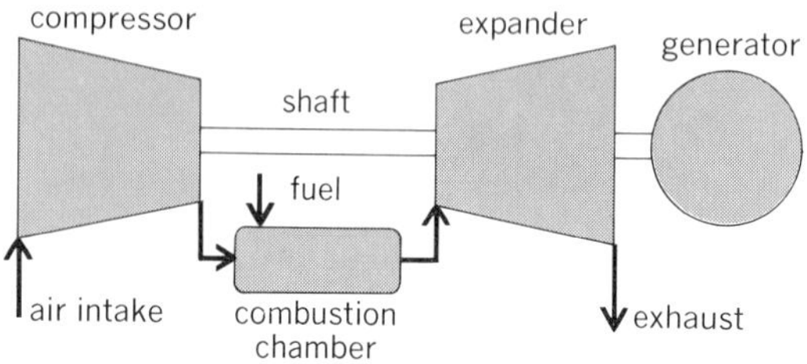

Fig. 2. Simple, open-cycle gas-turbine plant.

The difference in the two terms on the right-hand side of Eq. (4) is thus basically increased by the use of maximum temperatures at the inlet to the expander. These high temperatures introduce metallurgical and heat-transfer problems which must be properly solved.

The efficiency terms of Eq. (4) are of vital practical significance. If the efficiencies of the real compressor and of the real expander are low, it is entirely possible to vitiate the difference in the ideal powers W_e and W_c, so that there will be no useful output of the plant. In present practice this means that for adaptations of the Brayton cycle to acceptable and reliable gas-turbine plants, the engineering design must provide for high temperatures at the expander inlet and utilize high built-in efficiencies of the compressor and expander elements. No amount of cycle alteration, regeneration, or reheat can offset this intrinsic requirement for mechanisms which will safely operate at temperatures of 1500°±F. *See* CARNOT CYCLE; DIESEL CYCLE; OTTO CYCLE; THERMODYNAMIC CYCLE. [THEODORE BAUMEISTER]

Bibliography: T. Baumeister (ed.), *Standard Handbook for Mechanical Engineers*, 8th ed., 1978; J. B. Jones and G. A. Hawkins, *Engineering Thermodynamics*, 1960; H. Cohen et al., *Gas Turbine Theory*, 2d ed., 1975; G. N. Hatsopoulos and J. H. Keenan, *Principles of General Thermodynamics*, 1965; M. J. Zucrow and J. D. Hoffman, *Gas Dynamics*, vol. 1, 1976, vol. 2, 1977.

Brazing

A method of joining metals, and other materials, by applying heat and a brazing filler metal. The filler metals used have melting temperatures above 840°F (450°C), but below the melting temperature of the metals or materials being joined. They flow by capillary action into the gap between the base metals or materials and join them by creating a metallurgical bond between them, at the molecular level. The process is similar to soldering, but differs in that the filler metal is of greater strength and has a higher melting temperature.

Design. When properly designed, a brazed joint will yield a very high degree of serviceability under concentrated stress, vibration, and temperature loads. It can be said that in a properly designed brazement, any failure will occur in the base metal, not in the joint.

There are many design variables to be considered. First among them is the mechanical configuration of the parts to be joined, and the joint area itself. All brazements can be categorized as having one of two basic joint designs: the lap joint or the butt joint. Others are adaptations of these two. Stress concentration points should be carefully considered. The lap joint, or a modification of it, provides the best joint strength because the area of the joint interface is easily increased. The overlap area can be adjusted to equal $4T$, where T is the thickness of the thinner member (Fig. 1). Figure 2 illustrates the American Welding Society's (AWS) "Standard Method of Evaluating the Strength of Brazed Joints." It shows the load-carrying properties of the lap shear joint with increasing joint overlap, as indicated by the curve rising from the lower left to the upper right. This curve is used by the designer. The load on the filler metal in the joint is indicated by the curve starting at the upper left and dropping to the lower right.

The butt joint is successfully used when high-strength filler metals are employed that will produce strong joints, or with lower-strengh filler metals where less strength is required in the brazement. The joint clearance is also an important design consideration and is a function of the base and filler metals being used. The clearances are important because they influence the effectiveness of the capillary action, acting to distribute the molten filler metal between the surfaces of the base metals. As the clearance increases above the optimum, the joint strength is reduced (Fig. 3). In general, a press fit up to 0.002 in. (0.051 mm) clearance is used for brazing steel with copper filler metal. For other filler metals, a 0.002 to 0.005 in. (0.51 to 0.127 mm) clearance is desirable.

Selection of materials. Design considerations should include the informed selection of the base and filler metals. In addition to the basic mechanical requirements, the base metals used in the brazement must retain the integrity of their physical properties throughout the heat of the brazing cycle. Temperatures used in brazing copper with silver filler metals generally are in the range of 1100 to 1600°F (590 to 870°C). Wherever the copper is annealed in this manner, its strength will be lowered. Where service conditions require that brazements withstand high temperatures, as in aerospace components, nickel-based filler metals are used, typically brazed at temperatures ranging from 1800 to 2200°F (980 to 1205°C).

No universal filler metal that will satisfy all design requirements is possible, but there are many types available, ranging from pure metals such as copper, gold, or silver to complex alloys of aluminum, gold, nickel, magnesium, cobalt, silver, and palladium. The designer is able to choose one for almost any condition found in service.

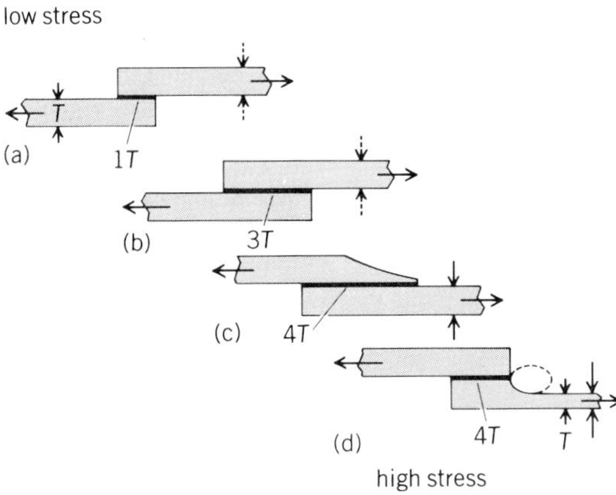

Fig. 1. Lap joint designs a–d to be used at low and high stress. Stress caused by flexure of right member in c and d will be distributed through the base metal by judicious shaping of the members at critical points.

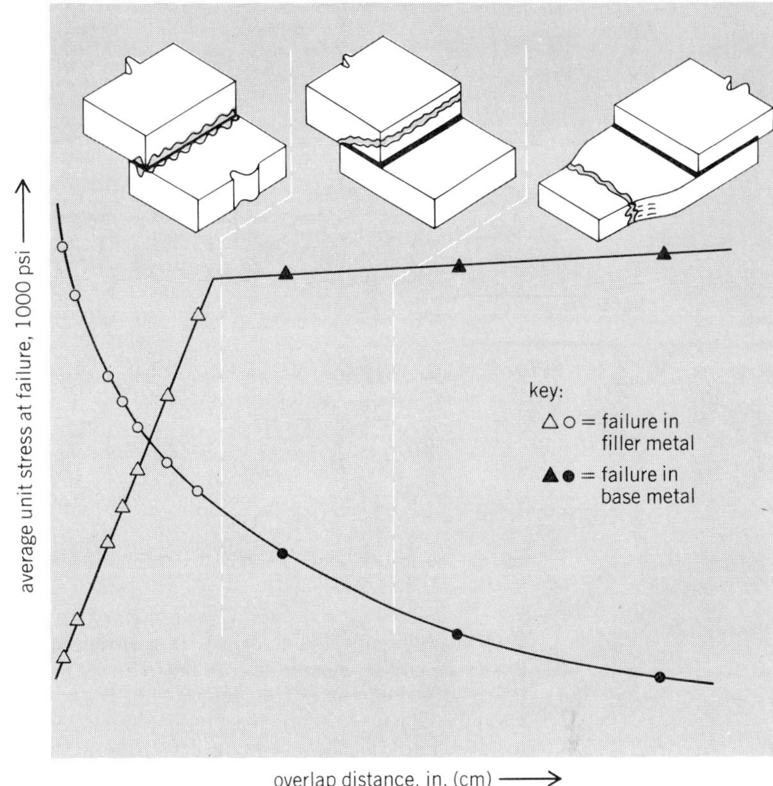

Fig. 2. Average unit shear stress (circles) and average unit tensile strength (triangles) as function of overlap distance.

Information on coefficients of expansion is important in the selection of materials, particularly when the base metals to be joined are widely dissimilar. If the expansion rates are at great variance, there can be a marked effect on the strength of the brazement and its resistance to fracture as the base materials cool. The fluxes for brazing are given in Table 1.

Brazing processes. There are 11 basic brazing processes.

Torch brazing. Heat is applied by flame, from some type of torch, directly to the base metal. A mineral flux is normally used in this process, which serves to prevent the formation of oxides and to absorb those that are formed as the metal is heated. The brazing filler metal may be preplaced in the joint, or face-fed into the joint when the parts have reached a suitable temperature.

Induction brazing. Brazing temperatures are developed in the parts to be brazed by placing them in or near a source of high-frequency ac electricity. Flux and preplaced filler metals are normally employed. However, a controlled atmosphere can replace the flux.

Resistance brazing. Electrodes are arranged so that the joint forms a part of an electric circuit. Heat is developed by the resistance of the parts to the flow of the electric current.

Dip brazing. The brazing filler metal is preplaced in or at the joint, and the assembly is immersed in a bath of molten salt or flux until the

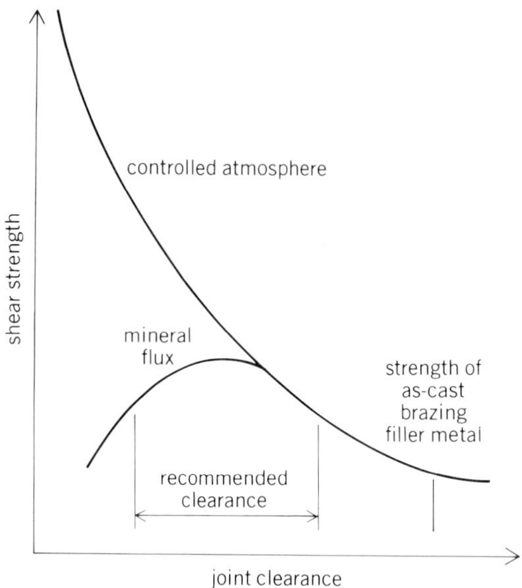

Fig. 3. The effect of joint clearance on shear strength.

brazing temperature is achieved. In a variation of this process, the assembly is prefluxed and dipped into a bath of molten brazing filler metal. When the assembly has heated sufficiently, the filler material will flow into the joint, whereupon the assembly is removed and the filler metal allowed to solidify.

Infrared brazing. High-intensity quartz lamps, capable of delivering up to 5000 W of radiant energy, are directed on the metals to be joined. This method of brazing has been developed for use in the fabrication of honeycomb-section aircraft components.

Furnace brazing. This is a widely used technique, especially useful where the parts to be brazed are machined or formed to their final dimensions, or constitute a complex assembly that has already been lightly joined or fixtured. The atmosphere within a brazing furnace is usually controlled, which permits a great deal of flexibility. The need to apply a mineral flux to the joint is eliminated in a properly controlled atmosphere of combusted city gas, vacuum, (dry) hydrogen, nitrogen, gas mixtures of nitrogen and hydrogen, or argon, because these gases (when properly controlled) prevent oxidation of the base metals and filler metal. A furnace with vacuum atmosphere is used for the more difficult base metals which may contain aluminum or titanium or base metals such as titanium, zirconium, or aluminum.

An important advantage is that potential distortion of metal, created by heating and cooling, can be predicted and controlled and thereby minimized or eliminated. Subsequent stress relieving or cleaning of the parts is not required since this is accomplished during the atmosphere furnace brazing operation. The capacity for automation is facilitated in the furnace brazing process. The atmospheres for brazing are given in Table 2.

Diffusion brazing. Unlike the foregoing, this process is defined not by the method of heating but rather by the degree of mutual filler-metal solution and diffusion with the base metal resulting from the temperature used, and the time interval at heat. In diffusion brazing, temperature, time, in some cases pressure, and selection of base and filler materials are so controlled that the filler metal is partially or totally diffused into the base metal. The joint properties then closely approach those of the base metal. For example, a nickel filler with a solidus temperature of 1760°F (960°C), when partially diffused at a temperature of 2000–2100°F (1095–1150°C), will have a remelt temperature of the filler metal in excess of 2500°F (1370°C).

Table 1. Fluxes for brazing

AWS brazing flux type number	Recommended base metals	Recommended useful temperature range	Ingredients
1	All brazeable aluminum alloys	700–1190°F 371–643°C	Chlorides, fluorides
2	All brazeable magnesium alloys	900–1200°F 482–649°C	Chlorides, fluorides
3A	All except those listed for 1, 2, and 4*	1050–1600°F 566–871°C	Boric acid, borates, fluorides, fluoborates, wetting agent
3B	All except those listed for 1, 2, and 4	1350–2100°F 732–1149°C	Boric acid, borates, fluorides, fluoborates, wetting agent
4	Aluminum bronze, aluminum brass and iron or nickel-base alloys containing minor amounts of Al or Ti†	1050–1600°F 566–871°C	Chlorides, fluorides, borates, wetting agent
5	All except those listed for 1, 2, and 4	1400–2200°F 760–1204°C	Borax, boric acid, borates, wetting agent

*Some type 3A fluxes specifically recommended for base metals are listed under type 4.
†In some cases, type 1 flux may be used on base metals listed for type 4.
SOURCE: American Welding Society.

Table 2. Atmospheres for brazing

AWS brazing atmosphere type number	Base metals	Source	Maximum dew point of incoming gas
1	Copper, brass[a]	Combusted fuel gas (low hydrogen)	Room temp.
2	Copper[b], brass[a], low-carbon steel, nickel, monel, medium carbon steel[c]	Combusted fuel gas (decarburizing)	Room temp.
3	Same as 2 plus medium- and high-carbon steel, monel, nickel alloys	Combusted fuel gas, dried	−40°F (−40°C)
4	Same as 2 plus medium- and high-carbon steels	Combusted fuel gas, dried (carburizing)	−40°F (−40°C)
5	Same as 1, 2, 3, 4, plus alloys containing chromium[d]	Dissociated ammonia	−65°F (−54°C)
6	Same as 2	Cylinder hydrogen	Room temp.
7	Same as 5 plus cobalt, chromium, tungsten alloys, and carbides[d]	Deoxygenated and dried hydrogen	−75°F (−59°C)
8	Brasses	Heated volatile materials	Inorganic vapors (such as zinc, cadmium, lithium, volatile fluorides)
9	Same as 5 plus titanium, zirconium, hafnium	Purified inert gas	Inert gas (such as helium, argon)
			Pressure[e]
10	Copper	Vacuum	Vacuum above 2 torr
10A	Low carbon steel, copper	Vacuum	0.5 – 2 torr
10B	Carbon and low-alloy steels, copper	Vacuum	0.001 – 0.5 torr
10C	Heat-and corrosion-resistant steels, aluminum, titanium, zirconium, refractory metals	Vacuum	1×10^3 torr and lower

[a]Flux required in addition to atmosphere when filler metals containing volatile components are used.
[b]Copper should be fully deoxidized or oxygen-free.
[c]Heating time should be kept at a minimum to avoid objectionable decarburization.
[d]Flux must be used in addition to the atmosphere if appreciable quantities of aluminum, titanium, silicon, or beryllium are present.
[e]1 torr = 133 pascals.
SOURCE: American Welding Society.

Other methods. Other, less used processes include arc brazing, block brazing, flow brazing, and twin carbon arc brazing. Brazing processes have many variations and refinements. Selecting an appropriate method involves the consideration of the properties of the base and filler metals or materials, the joint design, the number of the parts to be joined, and the size and configuration of the parts or assemblies.

Inspection. Inspection of brazed joints is facilitated if consideration is given to this subject in the design stage. The most widely used (and probably the most effective) technique is visual inspection, where filler metal that has been applied to one side of a joint can be seen to have penetrated to the opposite side of the joint. It is sometimes convenient to punch or drill holes in one member to facilitate visual inspection. The x-ray technique is effective when silver-class filler metals have been used, but is more difficult to use on joints brazed with nickel and copper fillers, since they show low density to x-rays. Sonic inspection, pressure-proof testing, and heat pattern inspections are sometimes employed also.

Applications. The use of brazing as a means of joining metals of similar and dissimilar properties will continue to expand, partly because brazing processes are easily automated. Skill is required for manual torch brazing. However, for automated machine torch brazing, furnace brazing, and similar methods, most of the real skill involved lies in the design and engineering of the joint, not in precise positioning of the filler metal and so forth. If designed correctly, the joint tends to "make itself" through capillary action, and little operator skill is required. Through carefully controlled furnace brazing techniques, complex assemblies of aerospace components (which include stampings, weldments, and machine parts of exotic metals) are successfully joined.

Many familiar products, such as bicycles, faucets, automobile radiators, and air-conditioning equipment, are manufactured through using the brazing process extensively. *See* SOLDERING; WELDING AND CUTTING OF METALS.

[R. L. PEASLEE]

Bibliography: American Welding Society, *Brazing Manual*, 3d ed., 1976; *Welding Res. Council Bull.*, no. 182, April 1973; *Welding Res. Council Bull.*, no. 187, August 1973.

Breakwater

A structure to protect against wave action of the seas and to create comparatively calm water behind for (1) harbors where adjacent land projections and nearby islands are inadequate to give protection, (2) coastal lands subject to damage by the sea, and (3) offshore structures. Openings, left to provide for the passage of ships between the open seas and the harbor, must give consideration to the propagation of waves through the openings into calm water, currents produced by tidal action, and ship maneuverability.

The usual breakwater is a mound of stone rubble dumped on the ocean or harbor bed and rising to a height above mean sea level. Sometimes it is capped on the top and sea face with variously shaped concrete blocks, regularly or irregularly placed. In some instances, breakwaters do not rise above the water and provide only partial protection.

Other common forms of breakwaters are vertical-faced structures built of solid blocks of masonry or concrete, precast-concrete cellular walls floated into place and filled with earth or stone or steel-sheet-piling cells filled with stone (common on the Great Lakes). Masonry and concrete walls are placed on low rubble mounds or on the sea bed if firm. Scouring of the sea bed in front may require a stone protection bed. *See* COASTAL ENGINEERING.

[EDWARD J. QUIRIN]

Bridge

A structure built to provide ready passage over natural or artificial obstacles, or under another passageway. Bridges serve highways, railways, canals, aqueducts, utility pipelines, and pedestrian walkways. In many jurisdictions, bridges are defined as those structures spanning an arbitrary minimum distance, generally about 10–20 ft (3–6 m); shorter structures are classified as culverts or tunnels. In addition, natural formations eroded into bridgelike form are often called bridges. This article covers only bridges providing conventional transportation passageways.

At present, the longest single span provided by a bridge—the Verrazano-Narrows Bridge in New York City—is 4260 ft (1298 m); the longest multiple-span bridge—the Lake Pontchartrain Causeway at New Orleans, LA—is 126,055 ft (38,422 m). A suspension bridge in England,—the Humber Bridge—will have a main span of 4626 ft (1410 m).

History. Bridges undoubtedly have been built since the origin of humankind, perhaps first as trees felled over waterways and later as structures of timber or stone. The art of constructing stone bridges reached a high degree of development during the height of the Roman era, and for a thousand years or so thereafter, it continued as an empirical art rather than a science. During the 19th century the theories of physics and mathematics were first applied to bridges in efforts to produce structures which would be rationally and economically proportioned to take the intended loads. During the mid-19th century, with application of wrought iron as a material for construction (in the Brittania railway bridge in England), model testing and materials testing were initiated in a scientific manner. At this time too, and continuing nearly until the end of the century, many firms in the United States developed proprietary bridges which were competitively peddled to railroads and governmental divisions. Eventually the failures of bridges, particularly railroad bridges, either because of faulty design or skimpy construction intended to lower cost, led in the 1880s to the establishment of consulting bridge engineering as a specialized discipline of civil engineering.

Parts. Bridges generally are considered to be composed of three separate parts: substructure, superstructure, and deck. The substructure or foundation of a bridge consists of the piers and abutments which carry the superimposed load of the superstructure to the underlying soil or rock. The superstructure is that portion of a bridge or trestle lying above the piers and abutments. The deck or flooring is supported on the bridge superstructure; it carries and is in direct contact with the traffic for which passage is provided.

Types. Bridges are classified in several ways. Thus, according to the use they serve, they may be termed railway, highway, canal, aqueduct, utility pipeline, or pedestrian bridges. If they are classified by the materials of which they are constructed (principally the superstructure), they are called steel, concrete, timber, stone, or aluminum bridges. Deck bridges carry the deck on the very top of the superstructure. Through bridges carry the deck within the superstructure. The type of structural action is denoted by the application of terms such as truss, arch, suspension, stringer or girder, stayed-girder, composite construction, hybrid girder, continuous, cantilever, or orthotropic (steel deck plate).

The main load-carrying member or members of a bridge are almost invariably parallel to the alignment of the bridge. When the alignment of the bridge and the obstacle being bridged are not square with one another, the main structural members may not be opposite one another, and the deck may be a parallelogram in plan; in this case the bridge is said to be a skewed bridge. Otherwise, it is known as square.

Many bridges are also designed on horizontally curved alignments to conform with curved approach roadways.

The two most general classifications are the fixed and the movable. In the former, the horizontal and vertical alignment of the bridge are permanent; in the latter, either the horizontal or vertical alignment is such that it can be readily changed to permit the passage beneath the bridge of traffic, generally waterbound, which otherwise could not pass because of restricted vertical clearance. Movable bridges are sometimes called drawbridges in reference to an obsolete type of movable bridge spanning the moats of castles.

A singular type of bridge is the floating or pontoon bridge, which can be a movable bridge if it is designed so that a portion of it can be moved to permit the passage of water traffic.

The term trestle is used to describe a series of short spans supported by braced towers, and the term viaduct is used to describe a high structure of short spans, often of arch construction.

FIXED BRIDGES

Fixed-bridge construction is selected when the vertical clearance provided beneath the bridge exceeds the clearance required by the traffic it spans. For very short spans, construction may be a solid slab or a number of beams; for longer spans, the choice may be girders or trusses. Still longer spans may dictate the use of arch construction, and if the spans are even longer, stayed-girder bridges are used. Suspension bridges are used for the longest spans.

Each of the above types of bridge is generally designed so that the substructure, superstructure, and deck are each considered to carry only the loads directly imposed upon them. In certain types of construction, the deck and the main load-carrying superstructure members are made to participate in carrying the load in order to make the bridge more economical than ordinary stringer or girder bridges with a concrete deck. This is known as composite or hybrid construction, depending upon the type of stringers. If there is girder-type construction, with a deck partly of steel topped by an asphalt surfacing material, the bridge is known as an orthotropic bridge. When the substructure and superstructure act together, the bridge is described as being of rigid frame construction.

The longer the span of a bridge, the greater is the relative cost per unit of deck area.

The choice of type of bridge superstructure may depend not only on the obstacle to be spanned but also on the substructure. Thus, if an expensive substructure is required because of water depth or unsatisfactory foundation conditions, the selection of a longer-span superstructure may be indicated even though it may not in itself be the economical choice.

Beam bridge. Beam bridges consist of a series of beams, usually of rolled steel, supporting the roadway directly on their top flanges. The beams are placed parallel to traffic and extend from abutment to abutment. When foundation conditions permit the economical construction of piers or intermediate bents, a low-cost multiple-span structure can be built. Spans of 50 ft (15 m) for railroad beam bridges and 100 ft (30 m) for highway beam bridges may be economical.

Composite I-beam bridges are beam bridges in which the concrete roadway is mechanically bonded to the I beams by means of shear connectors, which develop horizontal shear between the concrete slab and the beam. Such a connection forces a portion of the slab to act with the beam, resulting in a composite T beam. This construction yields a saving in the weight of the beams. Rolled shapes such as the channel, angle, and I, bars in serpentine form or in the form of a longitudinal helix, and steel studs are used as shear connectors. These connectors are usually welded to the flange of the steel beam and should extend at least halfway into the slab. *See* COMPOSITE BEAM.

Plate-girder bridge. Plate-girder bridges are used for longer spans than can be practically traversed with a beam bridge. In its simplest form, the plate girder consists of two flange plates welded to a web plate, the whole having the shape of an I. The railroad deck plate girder consists of two girders which support the floor system for a single track directly on their top flanges. A double-track bridge consists of two single-track bridges placed side by side on common abutments or piers. Through plate-girder bridges are used when clearance below the structure is limited. For railroad traffic, the floor system consists of two floor beams which are supported by the girders just above their lower flanges.

Box-girder bridge. Steel girders fabricated by welding four plates into a box section have been used for spans from 100 to more than 850 ft (30 to 259 m). The Rhine River crossing at Cologne, Germany, is an example of an 850-ft span. In the United States, a 750-ft (229 m) box-girder span is used in the San Mateo–Hayward Bridge in California. A conventional floor beam and stringer can be used on box-girder bridges, but the more economical arrangement is to widen the top flange plate of the box so that it serves as the deck. When this is done, the plate is stiffened to desired rigidity by closely spaced bar stiffeners or by corrugated or honeycomb-type plates. These stiffened decks, which double as the top flange of the box girders, are termed orthotropic. The wearing surface on such bridges is usually a relatively thin layer of asphalt. Single lines of box girders with orthotropic decks can be used for two-lane bridges, but when wider decks are required, two or more box-girder bridges can be placed parallel to each other.

Truss bridge. Truss bridges, consisting of members vertically arranged in a triangular pattern,

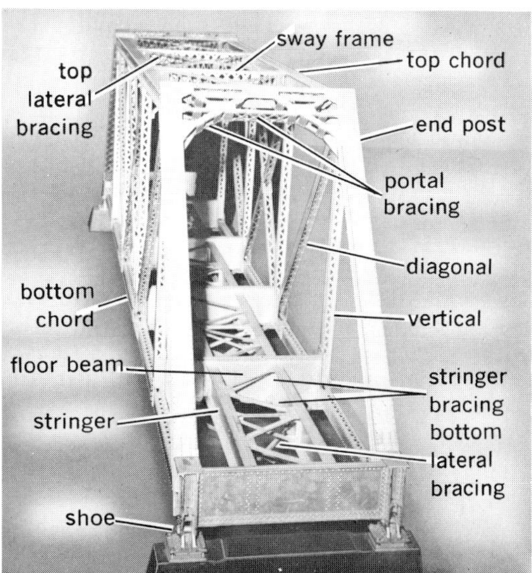

Fig. 1. Model of a through truss railroad bridge.

can be used when the crossing is too long to be spanned economically by simple plate girders. Where there is sufficient clearance underneath the bridge, the deck bridge is more economical than the through bridge because the trusses can be placed closer together, reducing the span of the floor beams. For multiple spans, a saving is also effected in the height of piers.

Through-truss. A through-truss bridge is illustrated in Fig. 1. The top and bottom series of truss members parallel to the roadway are called top chords and bottom chords, respectively. The diagonals and verticals form the web system and connect the top and bottom chords. The point at which web members and a chord intersect is called a panel point. Gusset plates connect the members intersecting at a panel point.

Lateral bracing of a bridge ties the two trusses together and assures a stable and rigid structure. The top lateral bracing consists of cross struts connecting the top chords at opposite panel points and the diagonals joining the ends of adjacent cross struts. It decreases the unsupported length of the top chord members, reducing the cross-sectional area they require. The floor beams and the diagonals connecting the ends of adjacent floor beams constitute the bottom lateral system. Although the floor system of a highway bridge can take over the function of a lateral truss, the lateral bracing must be provided to stiffen the structure during erection and to furnish wind resistance until the steel or concrete floor slab is in place.

The stringers of an open-floor railroad bridge must be braced to relieve them of bending due to lateral forces from the train. In addition, the floor beams should be provided with bracing to relieve the bending due to tractional forces.

Portal and sway bracing are systems of bracing in transverse vertical planes of a bridge. A sway frame is formed in the transverse planes at each end of the deck truss bridge by two diagonals, the verticals, the floor beam, and the bottom strut. This frame transmits the end reaction of the top lateral system to the abutment. Intermediate sway frames in the plane of opposite verticals give added rigidity.

The end posts of the through-truss bridge are tied together to form a rigid frame or portal capable of transferring the end reaction of the top lateral system to the abutments. To keep the bending stresses in the end posts as small as possible and to provide a rigid portal, portal bracing should be as deep as headroom allows; and the end post should be braced by brackets or diagonal members.

Simple-span trusses. Common types of simple-span bridge trusses are shown in Fig. 2. The Pratt truss, with its various modifications (Fig. 2*a–c*), has tension diagonals and compression verticals. The diagonal in every other panel of a Warren truss (Fig. 2*d* and *e*) is in compression. The depth of short-span trusses is usually determined by the depth necessary for clearance at the portal. For long spans, it is usually economical to make the depth of the truss greater at the center than at the ends. If the depth is increased in proportion to the increase in the forces tending to bend the bridge, the force in the chord members can be more nearly equalized. Figure 2*b* shows a curved-chord Pratt truss.

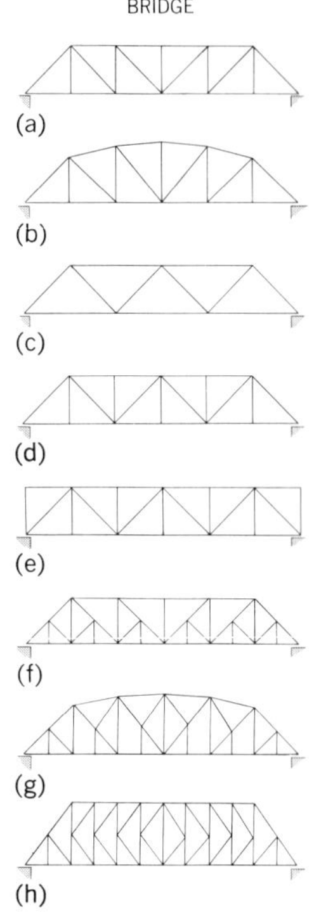

Fig. 2. Simple-span trusses. (*a–c*) Pratt trusses. (*d, e*) Warren trusses. (*f–h*) Subdivided trusses.

Trusses of economical proportions usually result if the ratio of depth of truss to length of span is approximately 1:5 to 1:8 and if the diagonals make angles of 45–60° with the horizontal. The panel length produced in long-span trusses when both of these factors are considered results in an uneconomical floor system. Subdivided trusses (Fig. 2*f–h*) are used to get reasonably long panels.

Continuous bridge. The continuous bridge is a structure supported at three or more points and capable of resisting bending and shearing forces at all sections throughout its length. The bending forces in the center of the span are reduced by the bending forces acting oppositely at the piers. Trusses, plate girders, and box girders can be made continuous. The advantages of a continuous bridge over a simple-span bridge (that is, one that does not extend beyond its two supports) are economy of material, convenience of erection (without need for falsework), and increased rigidity under traffic. Its relative economy increases with the length of span. No increase in rigidity is obtained by making more than three spans continuous. The disadvantages are its sensitivity to relative change in the levels of supporting piers, the difficulty of constructing the bridge to make it function as it is supposed to, and the occurrence of large movements at one location due to thermal changes.

Cantilever bridge. The cantilever bridge consists of two spans projecting toward each other and joined at their ends by a suspended simple span. The projecting spans are known as cantilever arms, and these, plus the suspended span, constitute the main span. The cantilever arms also extend back to shore, and the section from shore to the piers offshore is termed the anchor span (Fig. 3*a*). Trusses, plate girders, and box girders can be built as cantilever bridges.

The chief advantages of the cantilever design are the saving in material and ease of erection of the main span, both due to the fact that no falsework is needed. By adding continuity of members after erection, the cantilever bridge is made to act as a continuous structure under live load. *See* CANTILEVER.

Cable-stayed bridge. A modification of the cantilever bridge which has come into modern use resembles a suspension bridge, and it is termed a cable-stayed bridge. It consists of girders or trusses cantilevering both ways from a central tower and supported by inclined cables attached to the tower at the top or sometimes at several levels (Fig. 4). Usually two such assemblies are placed end to end to provide a bridge with a long center span.

Suspension bridge. The suspension bridge is a structure consisting of either a roadway or a truss suspended from two cables which pass over two towers and are anchored by backstays to a firm foundation (Fig. 3*b* and *c*). If the roadway is attached directly to the cables by suspenders, the structure lacks rigidity, with the result that wind loads and moving live loads distort the cables and produce a wave motion on the roadway. When the roadway is supported by a truss which is hung from the cable, the structure is called a stiffened suspension bridge. The stiffening truss distributes the concentrated live loads over a considerable length of the cable.

Cables of the larger sizes, up to 36 in. (91 cm) in diameter for the George Washington Bridge in

Fig. 3. Some major United States bridges. (*a*) Greater New Orleans cantilever bridge, across the Mississippi River (*Bethlehem Steel Co.*). (*b*) Looking north from south tower of Mackinac Bridge during construction (*Mackinac Bridge Authority*). (*c*) Verrazano-Narrows Bridge, New York, N.Y. (*Triborough Bridge and Tunnel Authority*). (*d*) Vertical-lift span and swing bridge across Arthur Kill, between Staten Island and New Jersey. The vertical-lift bridge replaced the swing bridge, in use since 1888 (*Baltimore and Ohio Railroad Co.*).

New York City, are assembled (or spun) in the field of pencil-thick wires laid parallel. For smaller cables, strands of wire wound spirally in the factory are assembled into cables. Factory-made strands of parallel wires were used for the first time, in 1968, for the 15-in.-diameter (38-cm) cables of the Newport Bridge in Rhode Island. The seven longest bridge spans in the world are all of the suspension type, and all are in the United States. Until the Humber Bridge is completed, the longest main span, 4260 ft (1298 m), is that of the Verrazano-Narrows Bridge in New York City (Fig. 3*c*). The second longest, 4200 ft (1280 m), is that of San Francisco's Golden Gate Bridge. The Mackinac Bridge (connecting Michigan's upper and lower pennisulas) is third, with a center-span length of 3800 ft (1158 m; Fig. 3*b*).

Pontoon bridge. The pontoon bridge is a floating bridge supported by pontoons. The structure may be a temporary one, as for military usage, or a

Fig. 4. Dames Point Bridge in Jacksonville, FL. (*Jacksonville Transportation Authority and Howard Needles Tammen & Bergendoff*)

permanent one, if the level of the water can be carefully controlled. A pontoon bridge may be advantageous where deep water and adverse bottom conditions make piers expensive. Seattle's Lake Washington is crossed by three floating concrete bridges. Typically, the pontoons are 360 ft (110 m) long, 60 ft (18 m) wide, and 14.5 ft (4.4 m) deep. A 378-ft (115-m) pontoon is used as a floating draw span to give a clear channel opening 200 ft (61 m) wide. Adjacent pontoons are bolted together, and each is secured by a pair of anchors, one on each side. Reinforced concrete was used for the pontoons of the first bridge, but the later bridges are of prestressed concrete.

Concrete bridge. The bridges that have been discussed are usually made of steel, although they may carry a concrete roadway. Increasingly, however, since the development of the prestressing method, bridges of almost every type are being constructed of concrete. Prior to the advent of prestressing, these bridges were of three types: (1) arches, which were built in either short or long spans, even up to 1000 ft (305 m) for the Gladesville Bridge in Sydney, Australia; (2) slab bridges of quite short spans, which were simply reinforced concrete slabs extending from abutment to abutment; and (3) deck girder bridges, consisting of concrete slab built integrally with a series of concrete girders placed parallel to traffic. The advent of prestressed concrete greatly extended the utility and economy of concrete for bridges, particularly by making the hollow box-girder type practicable. One of the longest such spans, 682 ft (208 m), is in a bridge over the Rhine River near Koblenz, Germany.

The objective in prestressing concrete is to reduce or eliminate tensile stresses in the concrete by applying a force that greatly increases internal compressive stress in the concrete member. The force is applied by stretching the reinforcing, which may be either wires or bars, by means of hydraulic jacks that react against the concrete member. Prestressing may be done to precast members, or it may be done as the concrete is being placed in the field. It may be done by either pretensioning or posttensioning. In pretensioning, the reinforcing is stretched prior to the placing of the concrete. In posttensioning, the reinforcing tendons are installed in tubes so that they are isolated from the concrete while it is being placed. When the concrete hardens, the tendons are stretched by jacks reacting against and compressing the concrete member.

MOVABLE BRIDGES

Modern movable bridges are either bascule, vertical lift, or swing; with few exceptions, they span waterways. They are said to be closed when set for the traffic they carry, and open when set to permit traffic to pass through the waterway they cross.

At the present state of development, both bascule and vertical-lift bridges operate well and reliably. Sometimes the bascule is chosen because its appearance can be pleasing and creates less of a visual impact than do the towers and ropes of vertical-lift bridges.

Swing bridges are now considered almost obsolete because the center pier, on which the span rotates, occupies the most desirable portion of the waterway from the standpoint of mariners. However, one feature of a swing bridge, which dictates its use near airports, is that it does not encroach on flight paths as would the towers of lift bridges or the raised leaves of bascule bridges.

Bascule, swing, and lift bridges may be of either stringer, girder, or truss construction, depending upon the length of their span.

Other types of movable spans, such as the retractable, are generally obsolete.

Bascule and swing bridges provide unlimited vertical clearance in the open position. The vertical clearance of a lift bridge is limited by its design.

Bascule bridge. The bascule bridge consists primarily of a cantilever span, which may be either a truss or a plate girder, extending across the channel (Fig. 5). It is generally chosen for spans up to about 175 ft (53 m) for highway use. Because of the

Fig. 5. Stanley Stroffolino Bridge in Norwalk, CT. (*Connecticut Department of Transportation and Hardesty & Hanover*)

large deflection of the cantilevered leaves of the double-leaf bascule, which is excessive under railway loadings and intolerable to railroad operation, bascule spans for railway usage are exclusively single-leaf bridges with maximum spans of about 250 ft (76 m).

Bascule bridges rotate about a horizontal axis parallel with the waterway. The portion of the bridge on the land side of the axis, carrying a counterweight to ease the mechanical effort of moving the bridge, drops downward, while the forward part of the leaf opens up over the channel much like the action of a playground seesaw. Bascule bridges may be either single-leaf, where rotation of the entire leaf over the waterway is about one axis on one side of the waterway, or double-leaf, where the leaves over the waterway rotate about two axes on opposite sides of the waterway. The two leaves of double-leaf bascule bridges are locked together where they meet when the bridge is closed. If the bridge actually rotates about an axis, it is called a trunnion-type bascule. If it rolls back on a track, it is called a rolling lift span. Proprietary types of bascule bridges, those formerly patented but now all more or less in the public domain, are variations of these. These include the Strauss, Scherzer, and Hopkins, among others less well known.

Vertical-lift bridge. The vertical-lift bridge has a span similar to that of a fixed bridge, and is lifted by steel ropes running over large sheaves at the tops of its towers to the counterweights, which fall as the lift span rises and rise as it falls. If the bridge is operated by machinery on each tower, it is known as a tower drive. If it is driven by machinery located on the lift span, it is known as a span drive. The 585-ft (178-m) span of the lift bridge over the Arthur Kill, an arm of New York harbor, is the longest of this type in the world (Fig. 3*d*). An example of a more recently constructed vertical-lift bridge is shown in Fig. 6.

Fig. 6. Stratford Avenue Bridge in Bridgeport, CT. (*Connecticut Department of Transportation and Hardesty & Hanover*)

Swing bridge. Swing bridges revolve about a vertical axis on a pier, called the pivot pier, in the waterway (Fig. 3*d*). There are three general classes of swing bridges: the rim-bearing, the center-bearing, and the combined rim-bearing and center-bearing. Rim-bearing bridges are supported on circular girder drums on rollers, center-bearing on a single large bearing at the center of rotation. In each of these classes, a bridge may have either one or two supporting points per truss at the pivot pier. Center-bearing draws are generally arranged so as to carry the dead load on the pivot and the live load on either a drum or four carriages formed of groups of rollers. Combined rim and center-bearing draws carry a portion of the dead load on the pivot and the remainder on a drum, with the live load supported as in the last described case. Swing bridges have been classified also as to the character of their main girders—that is, plate girder swings, open-webbed girder or truss swings, and pin-connected-truss swings. Another general division of swing bridges is in relation to their continuity or lack thereof in the travel of live load across the pivot pier. Most structures are more or less continuous in this regard, but a few have been designed in such a manner that, when the ends of the arms are raised to their normal position for the closed span, the two halves of the structure are simple spans and are entirely independent of each other in respect to all kinds of loading and all conditions thereof. Swing spans may be divided into through, half-through, and deck.

Machinery. Almost all movable bridges are driven by electric motors which operate gear trains that convert the high-speed low-torque output of the motor to a low-speed high-torque output of the gear train at a pinion, usually mounted on the stationary portion of the bridge, acting on a rack usually mounted on the movable portion.

Originally, movable bridges were operated by steam engines, and until recently, many still were. At least one former steam-operated bridge is operated by compressed air driving the old steam engine. A very few bridges have been operated hydraulically. Some small bridges are operated by hand power. Provisions are still made on most movable bridges to operate by hand power in the event of power failure or malfunction of the bridge's electrical controls, even though it is an impractical means of operation since it takes some 8 hours or more to open or close the bridge.

Machinery for early movable bridges consisted of simple custom-made components produced by the numerous small foundries, forge shops, and machine shops supplying the numerous fabricators of movable bridges. Such components consisted of cast and rough-cut gears of various tooth profiles, babbitt and bronze bushed bearings, custom-designed mounts for individual bearings, and common frames for the mounting of multiple bearings for open sets of reduction gears and various combination drive assemblies (Fig. 7).

Substantial progress was made just after World War II, when machinery manufactures began to market standard components of various kinds suitable for bridge drives. Commercial speed reducers, offering high-quality helical and herringbone gears turning on antifriction bearings in enclosed gear cases, became available (Fig. 8). Various special features, such as differential gearing, special shaft extensions, and forced lubrication for operating positions other than horizontal,

Fig. 7. Exposed gear reductions on an early bascule bridge.

made these units well suited for bridge use.

While the main drive machinery has been fairly simple and well standardized for years on most types of movable bridges, the auxiliary drives for lock bars, end lifts, rail locks, third-rail disconnects, signal-system disconnects, and other such special devices have taxed the ingenuity of many designers and resulted in all kinds of solutions.

Welded components have replaced cast components. Speed-reducer housings, which formerly were castings, are essentially all welded now. Base frames tend to be welded instead of cast. Limited success has been achieved in using weldments in lieu of castings for the huge sheaves of vertical lift bridges.

Electrical equipment. Except on the simplest small movable bridges, the electrical control of a movable bridge is so interlocked that the bridge cannot be moved until a series of prior operations have been made in correct order. Thus, on a highway bridge, first the traffic lights must be turned to red, then a set of traffic gates lowered, next the barrier gates set, and finally the locks or wedges pulled before the bridge can be moved.

Fig. 8. Enclosed gear reducers on a later bascule bridge. *(Earle Gear and Machine Company)*

The movable span generally can be completely opened or closed in from 1 to 2 min. However, the prior operations, particularly closing the bridge to traffic and making sure that the bridge is cleared of all vehicles between traffic gates, may run the cycle of operation up to 15 min or longer, depending upon the speed of passage through the span of the vessel for which the bridge has opened.

The last 50 years have seen electrical equipment for bridges develop from relatively crude simple devices to modern sophisticated equipment, including solid-state circuitry.

Throughout this period, the characteristic that remains unchanged and is common to all bridge electrical drives is that the span movement should be under complete control, regardless of whether the load is motoring or overhauling, and particularly when it is approaching the fully open and seated positions. The latter requirement is of prime importance, since it requires that the control system provide accurate movement of the span near the extremes of travel. A well-designed control system must bring the span smoothly to the end of its travel and seat it without undue shock, without allowing the span to rebound from the fully seated position. As the size of the span motors is determined by extreme conditions of loading on the span deck, the motors will usually be only lightly loaded during the normal cycle of opening and closing the span.

The drive systems on movable bridges have progressed from the earlier forms of technology to the latest solid-state control devices permitting push-button operation.

A further step into automatic operation, and one which has already been used in other industries, is the programmable controller. This device will control all the sequential functions in the opening and closing of a bridge without the complex system of multiple-contact limit switches and relays currently used.

DESIGN

Bridges are designed according to the laws of physics pertaining to statics. The primary members of bridges act in tension, compression, shear, or bending, or in combinations of these. Secondary members act in the same fashion, sometimes participating in carrying the principal loads, especially if failure of any primary members should occur. Redundancy, so that failure of a single member or part of a member will not cause immediate collapse of the bridge, is now recognized as a desirable feature of design, particularly since collapse of the eyebar suspension bridge across the Ohio River at Point Pleasant, WV, in 1967.

The present design of bridges follows standards established by their principal users, the state highway or transportation departments and the railroads. In the United States, these standards are those of the American Association of State Highway and Transportation Officials and the American Railway Engineering Association, respectively. These specifications are developments of those for bridge design formulated by the early bridge engineers. While they are advisory, they are generally followed, albeit at times with modifications. Other countries have similar specifications.

The traffic load, called live load, is given in these specifications. It depends on the service to which the bridge will be subjected. The highway loadings are in terms of a simulated conventional truck or, for long spans, a uniform load with a roving concentrated load representing a line of average traffic. Railroad loadings are a simulated conventional locomotive load followed by a uniform load representing loaded freight cars. In both cases, in design, the live loads are positioned to give maximum load in the member being designed. In addition, dead load, the structure's own weight, and loads from impact, wind, temperature changes, ice, traction or braking of traffic, earthquake, and, in the case of highway bridges, pedestrians are specified. Allowable stresses and limiting deflections, too, are specified.

In addition to the above specifications, those of the American Society for Testing and Materials, the American Welding Society, the American Institute of Steel Construction, the construction specifications of the individual states, and special specifications, when no other covers the situation, are followed in the design or construction of bridges.

VIBRATION

Bridges are generally considered to be statically loaded structures under their own dead load, with dynamic loadings from the live load and from the wind.

Vibrations of bridges or individual components of a bridge occur when a resonant frequency of the bridge or a component is excited by one or more of the applied dynamic loadings. Excitation from live loads is more apt to happen on bridges carrying rail traffic than on those carrying highway traffic because of the uniform spacing of railroad cars. Excitation from wind is caused by the repeated formation of eddies or vortices as the wind travels past nonstreamlined members. Such aerodynamically induced vibrations caused the failure of the Tacoma Narrows Bridge in the state of Washington in 1941. A similar failure is recorded for a suspension bridge over the Ohio River at Wheeling, WV, in 1854. In both cases, the formation of vortices built up as the torsional movements of the deck increased, ultimately leading to failure. Long, thin, H-shaped truss members have also failed from similar wind-induced torsional vibrations. The wind velocities necessary to excite such vibrations need not be excessively high. The Tacoma Narrows Bridge failed under a 40-mph (18 km/sec) wind, less than half the equivalent static wind load for which it was designed, and less than 20% of the load which would have caused structural distress.

Light lateral bracing on many spans will vibrate under the passage of live load.

Sufficent data exist on aerodynamically induced vibration to predict the possibility of its occurrence and thus modify the design when necessary to prevent serious problems.

SUBSTRUCTURE

Bridge substructure consists of those elements that support the trusses, girders, stringers, floorbeams, and decks of the bridge superstructure. Piers and abutments are the primary bridge substructure elements. Other types of substructure, such as skewbacks for arch bridges, pile bents for trestles, and various forms of support wall, are also commonly used for specific applications.

The type of substructure provided for a bridge is greatly affected by the conditions of the site. Studies must be made on topography, stream currents, floating drift and ice, seismic potential, wind, and soil conditions. Forces and loads encountered in the design of substructure elements include dead load, live load, impact loads, braking forces, earth pressure, buoyancy, wind forces, centrifugal forces, earthquake loads, stream flow, and ice pressure.

Both piers and abutments are generally supported on either spread footings or pile footings. A spread footing is usually a concrete pad large enough in area to transmit all superimposed loads and forces directly to the soil on which it is founded. The size of the spread footing is related to the bearing capacity of the soil on which it rests and the external forces on the substructure which will be transmitted to the spread footing.

A pile footing is usually a large concrete block supported on piles, so that the superimposed substructure loads and forces are transmitted to the support piles through the footing. The footing piles are used primarily to transmit loads through soil formations having poor supporting properties into or onto formations that are capable of supporting the loads. Piles may be point-bearing or friction types or the two in combination, and they may be timber, steel, precast concrete, cast-in-place concrete, or prestressed concrete. They may be driven by the use of a pile driver equipped with a hammer; they may be augered, jetted or prebored, or predrilled and cast-in-place. *See* FOUNDATIONS.

Bridge piers. Bridge piers are the intermediate support systems of bridges and viaducts. They may be located in water or on dry land. When located in water, piers may be subjected to scour by current and collision by vessels. Bridge piers support the superstructure and must carry dead loads and live loads, and withstand braking forces and other induced forces peculiar to the location of the pier, such as wind, ice, earthquake, and stream flow. A major consideration in pier design is stability and the ability to support all loads without appreciable settlement.

The shape, type, and location of piers are based on many factors; the major ones are horizontal and vertical clearance requirements, subsurface conditions, architectural and esthetic considerations, political and urban planning factors, traffic, and cost.

The most common pier shapes are solid shafts, multiple columns and portal, two columns and portal, separate columns, T or hammerhead, and cantilever. There are many variations of these pier shapes, which are constructed using concrete, steel, or wood (Fig. 9). A pier should have sufficient horizontal area at its top to receive the superstructure bearings. Architecturally, it should give the appearance of strength; it should not look weak and flimsy, although calculations may have shown the design to be adequate.

Piers in water are sometimes faced with stone, steel, or other protective devices below the high-water line to protect the pier from scour, ice, and other floating matter.

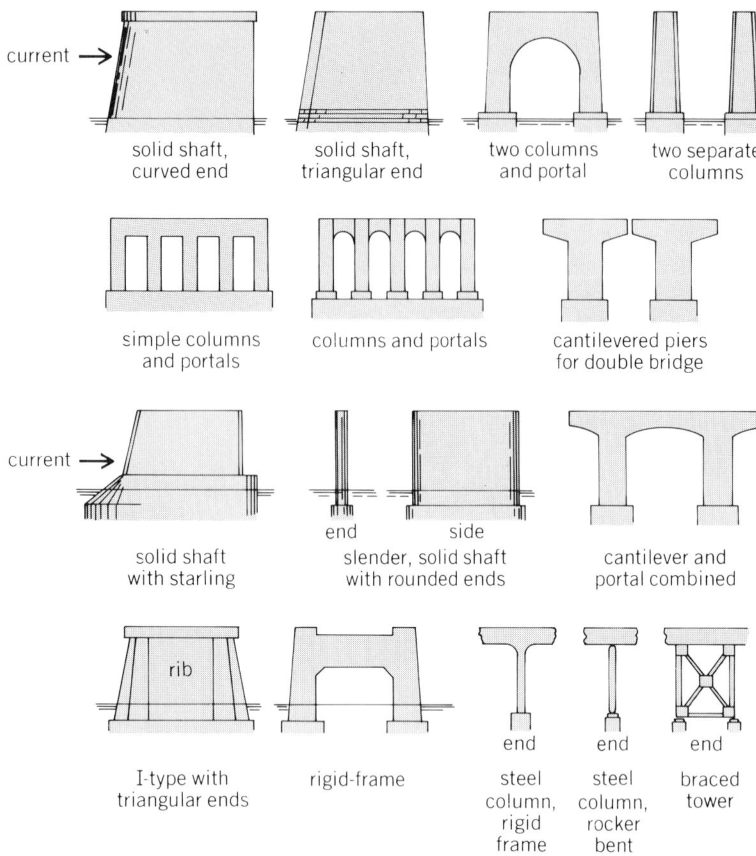

Fig. 9. General shapes of some piers for small bridges. (*From C. W. Dunham, Foundations of Structures, McGraw-Hill, 2d ed., 1962*)

can sustain slight yield of the foundation. Skewbacks for these may be supported on piles or on spread footings on earth.

Caissons. A caisson is a boxlike structure, round or rectangular, which is sunk from the surface of either land or water to the desired depth as excavation proceeds inside or under it. It is an aid in making excavations for bridge piers or abutments and remains in place as part of the permanent structure.

The most common types are open caissons, which, as the name implies, have no top or bottom. Pneumatic caissons have permanent or temporary tops and are so arranged that people can work in the compressed air trapped under the structure.

Generally, the ultimate purpose of caisson construction is to reach a bearing stratum which will carry the load of supporting piers and abutments. *See* CAISSON FOUNDATION.

Cofferdam. A cofferdam, generally, is an enclosed temporary structure used to protect an excavation against lateral earth pressure or water pressure during the process of construction. The material within the confines of the cofferdam is removed to allow the construction of piers and abutments below ground or water level. *See* COFFERDAM.

Tremie seal. In constructing cofferdams to counterbalance the upward pressure which may exist at the bottom of the caissons or cofferdams, a concrete seal, also known as tremie seal, is placed prior to dewatering. Concrete is placed in water using vertical pipes known as tremies. These pipes are continually filled with concrete. The bottom of the pipe is submerged in the plastic concrete while the concrete is poured through a funnellike top.

Abutments. The abutments of a bridge are the substructure elements that support the ends of a bridge (Fig. 10). Bridge abutments are generally constructed of concrete or masonry and are designed to be architecturally, esthetically, and functionally pleasing. Major loads to which abutments are subjected are dead loads, live loads, braking forces, ice, wind, earthquake, stream flow, and earth pressure. The last is applied at the rear of the abutment wall.

An abutment is generally composed of a footing, a wall with a bridge seat supporting the superstructure bearings, and a backwall to retain the earth. Abutments may have wingwalls to retain the earth of the approach fill to the bridge. Proper drainage behind abutment walls is essential to avoid increasing the lateral pressure forces.

Skewbacks. A skewback is a common expression for an abutment for an arch; it is practically nothing more than an inclined footing that receives the thrust from the arch superstructure (Fig. 10).

A skewback differs from common abutments in the type and direction of forces that are applied to it. The horizontal component of an arch reaction is usually very large and exceeds the vertical component.

The large thrusts which are supported by skewbacks require good foundations. The ideal foundations are rock gorges which have strong, sound, and suitably sloping rock.

In general, a skewback for a fixed arch must provide practically no yielding. Some hinged arches, which are statically determinate structures,

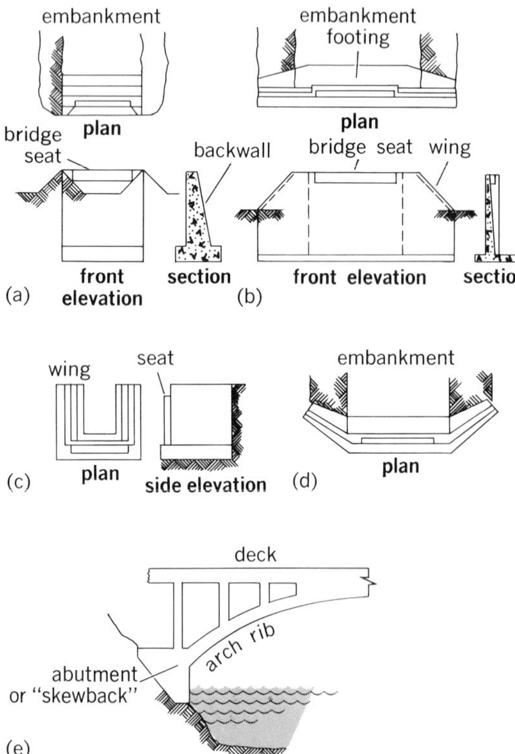

Fig. 10. Abutment designs for bridges. (*a*) Wingless. (*b*) Straight wing. (*c*) U-shaped. (*d*) Beveled wing. (*e*) Abutment for arch bridge.

During the pour, the tremie pipe is pounded, vibrated, and raised and lowered to increase the hydraulic head of the concrete in the tube and to cause the concrete to flow. The depth of the tremie concrete pour is a function of the upward pressure encountered after dewatering.

[E. R. HARDESTY; H. W. FISCHER; R. W. CHRISTIE; B. HABER]

Brittleness

That characteristic of a material that is manifested by sudden or abrupt failure without appreciable prior ductile or plastic deformation. A brittle fracture occurs on a cleavage plane which has a crystalline appearance at failure because each crystal tends to fracture on a single plane. On the other hand, a shear fracture has a fibrous appearance because of the sliding of the fracture surfaces over each other. Brittle failures are caused by high tensile stresses, high carbon content, rapid rate of loading, and the presence of notches. Materials such as glass, cast iron, and concrete are examples of brittle materials.

[JOHN B. SCALZI]

Broaching

The machine shaping of metal by pushing or pulling a multiple-tooth, barlike tool across a surface or through an existing hole in a workpiece. The peripheries or cutting edges of the broach teeth are shaped to give a desired surface or contour, whereas cutting action results from the fact that each tooth is progressively higher or projects further than the preceding one.

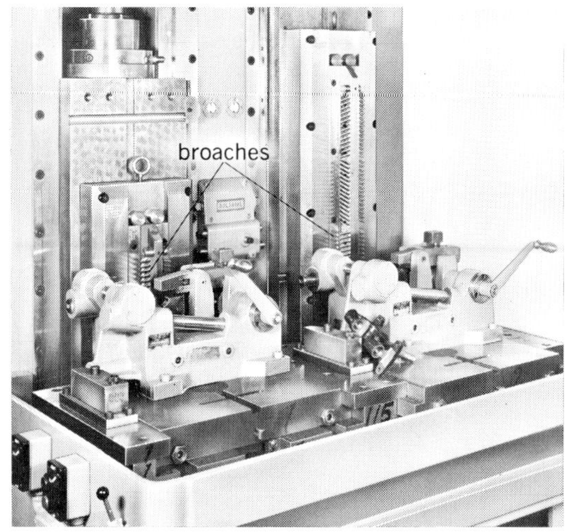

Dual ram-type vertical broaching machine with fixtures. (*Colonial Broach and Machine Co.*)

Broaching machines are classed as either horizontals or verticals, depending on the mounting plane of the broach. Most horizontal machines are of the pull type, while vertical models are made to use pull-up, pull-down, push, or surface broaches. Multiple rams increase output (see illustration). Continuous broaching machines permit large-quantity production by moving the workpieces on rotary tables or chain conveyors past stationary broaches. *See* MACHINING OPERATIONS.

[ALAN H. TUTTLE]

Bronze

Usually an alloy of copper and tin. Bronze is used in bearings, bushings, gears, valves, and other fittings both for water and steam. Lead, zinc, silver, and other metals are added for special-purpose bronzes. Tin bronze, including statuary bronze, contains 2–20% tin; bell metal 15–25%; and speculum metal up to 33%. Gun metal contains 8–10% tin plus 2–4% zinc.

The properties of bronze depend on its composition and working. Phosphor bronze is tin bronze hardened and strengthened with traces of phosphorus; it is used for fine tubing, wire springs, and machine parts. Lead bronze may contain up to 30% lead; it is used for cast parts such as low-pressure valves and fittings. Manganese bronze with 0.5–5% manganese plus other metals, but often no tin, has high strength. Aluminum bronze also contains no tin; its mechanical properties are superior to those of tin bronze, but it is difficult to cast. Silicon bronze, with up to 3% silicon, casts well and can be worked hot or cold by rolling, forging, and similar methods. Beryllium bronze (also called beryllium copper) has about 2% beryllium and no tin. The alloy is hard and strong and can be further hardened and strengthened by precipitation hardening; it is one of the few copper alloys that responds to heat treatment, approaching three times the strength of structural steel.

Tin bronze is harder, stronger in compression, and more resistant to corrosion than the brasses. Bronze that will not be exposed to extremes of weather can be protected from corrosion by warming it to slightly over 100°C in an oxygen atmosphere. A thin layer of oxide or patina forms to prevent further oxidation. A patina may be formed on art objects by exposure first to acid fumes and then drying as above. While still warm, the object can be further protected by a spray of wax in a solvent.

For bearings, sintered bronze is compacted from 10% tin, up to 2% graphite, and the balance by weight of copper. The aggregate is formed under pressure at a temperature below the melting point of its constituents but high enough to reduce their oxides. After forming, a bearing is repressed or sized and impregnated with oil, the pores retaining the lubricant until needed. In place of copper, alpha-bronze powder may be used; zinc may replace some of the tin. At forming temperatures below 700°C, properties depend primarily on compacting pressure. At higher temperatures, properties depend first on temperature, although heat treatment beyond 30 min has minor influence. *See* ALLOY; ANTIFRICTION BEARING; COPPER ALLOYS.

[FRANK H. ROCKETT]

Buffing

The smoothing and brightening of a surface by an abrasive compound pressed against the work by means of a soft wheel or belt. The abrasive is a fine powder or flour mixed with tallow or wax to form a smooth composition or paste. This is applied as required to the buffing wheel or belt, which is made of a pliable material, such as soft leather, linen, muslin, or felt. Buffing is accomplished on the same types of machines as polishing and frequently both types of wheels are included on one machine. *See* POLISHING.

[ALAN H. TUTTLE]

Buhrstone mill

A mill for grinding or pulverizing, in which a flat siliceous rock, generally of cellular quartz, rotates against a stationary stone of the same material. The Buhrstone mill is one of the oldest types of mill and, with either horizontal or vertical stones, has long been used to grind grains and hard materials. Grooves in the stones facilitate the movement of the material. Fineness of the product is controlled by the pressure between the stones and by the grinding speed. A finely ground product is achieved by slowly rotating the stone at a high pressure against the materials and its mate (see illustration). The capacity, or output, of a Buhrstone mill is low and its power requirements are high. The stones require frequent maintenance, even when grinding only slightly abrasive materials. *See* CRUSHING AND PULVERIZING.

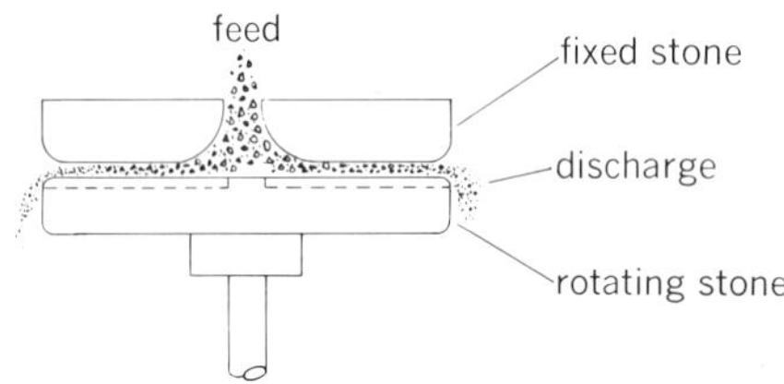

Buhrstone mill. Material is fed at center of fixed stone and moves toward outer edge of the stones where finely ground product is discharged.

[GEORGE W. KESSLER]

Buildings

Fixed structures for human occupancy and use. The space and structure must be planned to produce the environment and facilities required for the purpose of the building. Items such as light, heat, air conditioning, acoustics, and color must be given careful consideration.

Fig. 1. Simple, or conventional, type of steel skeleton-frame construction. (*American Institute of Steel Construction*)

Many governments, states, professional organizations, and municipalities have adopted building codes to establish minimum load capacities and to ensure safety to the occupants under normal use and under common hazards like fire and windstorms. Buildings are usually classified according to occupancy and use as follows: (1) public (auditoriums, churches, theaters), (2) institutional (hospitals, schools, penitentiaries), (3) residential (hotels, apartments, studios), (4) business (stores, factories, office buildings), and (5) storage (garages, warehouses).

Loads. The loads that a building supports are called dead load, which includes the weight of walls, columns, partitions, floors, and roof; and live load, which includes the weight of the physical equipment and occupants of the building. Typical live load for residential buildings is 40 pounds per square foot (1 psf = 47.85 N/m^2); for office buildings, 50 psf; and for auditoriums and such public spaces, 100 psf. Snow load on the roof is considered a live load. Recommended minimum load varies from 10 psf on steep roofs in the Southern states to 40 psf on flat roofs in New England.

Because there is little likelihood of applying the full specified live load all over large areas, codes permit a reduction in the basic design live load when areas exceed a certain minimum. A live load reduction is therefore allowed for supporting members such as girders, columns, piers, and foundations. An allowance for the dynamic effect of all moving loads, such as elevator machinery and cranes, is usually specified as a fixed percentage of the load.

The evaluation of pressures exerted by wind on a building is complex, and code requirements for wind forces are only rough approximations. The recommended design wind pressure on plane surfaces which are normal to the wind increases with the height of the structure above ground level. Wind on the windward roof slope may produce suction or pressure, depending on the slope of the roof. Leeward slopes are always subject to the effects of suction.

In earthquake areas the building code provides for special seismic design requirements. On the basis of predicted earthquake shocks, all components of a building must be designed to take into account the lateral forces that might be induced by such ground motion. There must be assurances that, in case the expected maximum earthquake occurs, the building may be structurally damaged but will not actually collapse.

For the design of blast-resistant structures the Federal Civil Defense Administration recommends two classes of equivalent static loads. Class I loads are designated for buildings housing personnel and critical equipment. Class II loads are designated for buildings in which damage resulting from failure of walls or other effects of blast can be repaired easily.

Design standards have been developed which set forth the minimum requirements consistent with current practice. These standards cover such items as design loads, design stresses, methods of analysis, proportioning of structures and their components, fabrication, and erection.

Fig. 2. Long-span construction using cables to support the roof. (*Bill Engdahl, Hedrich-Blessing, Chicago; photo courtesy of Howard N. Kaplan Architectural Photographs, Chicago*)

Framing systems. The roof and the floors of buildings are generally supported by one of the three common framing systems: (1) bearing wall, (2) skeletal framing composed of beams and columns or slabs and columns, and (3) special long-span roof systems.

In bearing-wall construction, walls support the ends of the floor and roof beams and slabs in addition to enclosing partitioning space. The bearing wall can be made of bricks, concrete blocks, cast-in-place reinforced concrete, or precast concrete panels. Although bearing-wall construction is more common for one-family homes and some industrial and commercial buildings, reinforced concrete bearing walls, commonly known as shear walls, are increasingly being used for tall buildings up to 75 stories.

The skeletal frame construction can be either of steel framing or of reinforced concrete. Skeletal frames of reinforced concrete normally have monolithic rigid connections between beams and columns. However, steel skeletal frame construction can be simple, rigid, or semirigid. Generally, simple connections are used between beams and girders, and rigid or semirigid connections are used between girders and columns (Fig. 1). In simple framing, beams and girders are assumed to be free to rotate at their connections, whereas in rigid frame structures the angles members make at each joint are assumed to remain fixed under all loadings. Frames of intermediate rigidity, in which it is assumed that the connections possess a known and dependable moment capacity, are usually called semirigid frames. *See* WALL CONSTRUCTION.

Long-span construction is used for auditoriums, hangars, gymnasiums, and other structures requiring a wide unobstructed floor area that cannot be economically spanned with standard steel beams or typical reinforced concrete or solid-timber girders. The appropriate forms of construction are plate girders, prestressed girders, trusses, arches, rigid frames, domes, cantilever suspension frames, space frames, and suspended cables, the last two types gaining in popularity in recent years. A space frame is a three-dimensional system of intersecting trusses that forms a series of octahedrons. One of these frames with a depth of 14 ft (1 ft = 0.3 m) spans Denver's Exhibition Hall, an area 685 × 240 ft which contains no columns. Buildings in which interior columns are eliminated by using cables to support the roof are characterized by a heavy compression ring on the top of the wall columns to which one end of the cables is anchored. The other end is attached to a tension ring at the center of the building. A notable structure of this type is the Coliseum in Oakland, CA. Its 420-ft diameter is spanned by a series of 2½-in.-diameter (1 in. = 2.54 cm) radial steel cables on which precast prestressed concrete pie-shaped panels attach to form the ribbed catenary circular roof. A recent example of an umbrella-type cable-supported roof is the Baxter Laboratory Dining Hall in

Fig. 3. Haj Terminal at the New Jeddah International Airport in Saudi Arabia. (*Photo by Jack Horner, New York*)

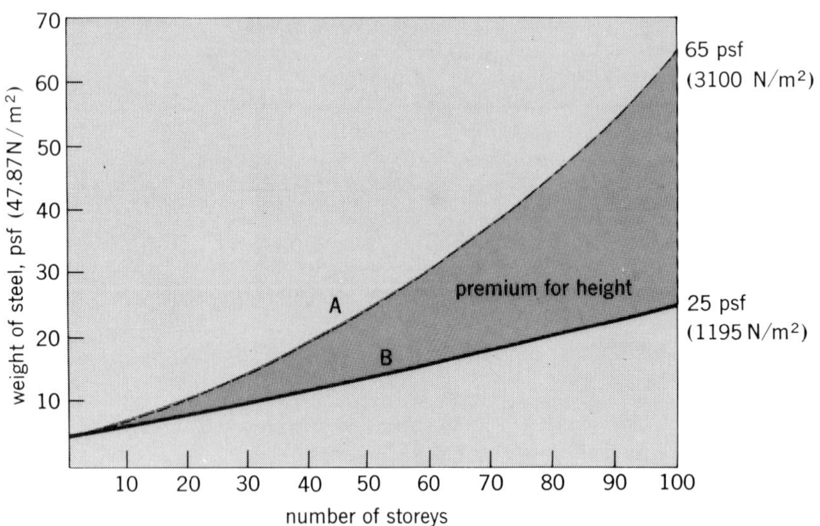

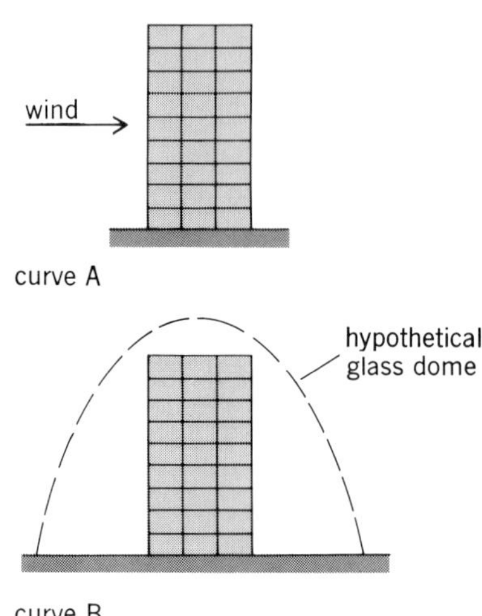

Fig. 4. Graphical relationship between design quantities of steel and building heights for a typical building frame. Curves A and B correspond to the boundary conditions indicated at the right.

Deerfield, IL, of 300 × 150 ft, suspended from only two column shafts (Fig. 2). *See* ARCH; CANTILEVER; ROOF CONSTRUCTION; TRUSS.

High-strength fiber-glass fabric coated with Teflon on both sides is increasingly being used for large permanent roof structures. The inflated roof of the Silver Dome Stadium in Pontiac, MI, is an excellent example. Another example of Teflon-coated fiber glass used as a structural material interacting with steel cables is the permanent tent-like roof structure for the Haj Terminal at the New Jeddah International Airport in Saudi Arabia. This roof covers a total area of 127 acres or 51 ha (Fig. 3).

Fig. 5. Belt trusses increase stiffness of the First Wisconsin Bank building, Milwaukee. (*ESTO*)

Industrial buildings. It is usually economical to support heavy industrial buildings with a steel framework consisting of a series of trusses supported on columns. Such industrial buildings often attain huge dimensions, for example, the Boeing Aircraft plant in Everett, WA, which is framed with 300-ft clear span trusses 25 ft deep and encloses a volume of 200,000,000 ft^3 (5,664,000 m^3).

Natural illumination in an industrial building may be increased considerably by the addition of one or more monitors along the length or width of the building. Such arrangements are usually called high-low bay roofs. A uniform level of illumination is achieved with a sawtooth roof having either vertical or sloping windows. Monitors are of questionalbe value for manufacturing processes requiring more than 20 ft-candles of illumination (215.2 lumen/m^2).

Tall buildings. A new surge of activity in the construction of tall buildings has taken place within the last few years. Although there have been many advancements in building construction technology in general, spectacular achievements have been made in the design and construction of ultra-high-rise buildings.

The early development of high-rise buildings began with structural steel framing. Reinforced concrete has since been economically and competitively used in a number of structures for both residential and commercial purposes. The newer high-rise buildings, ranging from 50 to 110 stories and now being built all over the United States, are the result of very recent innovations and development of new structural systems.

Greater height entails increased column and beam sizes to make buildings more rigid so that under wind load they will not sway beyond an acceptable limit. Excessive lateral sway may cause serious recurring damage to partitions, ceilings, and other architectural details. In addition, excessive sway may cause discomfort to the occupants of the building because of their perception of such motion. New structural systems of reinforced concrete, as well as steel, take full advantage of the inherent potential stiffness of the total building

Fig. 6. Exterior diagonal system is used in the John Hancock Center, Chicago. (*ESTO*)

and therefore do not require additional stiffening to limit the sway.

In a steel structure, for example, the economy can be defined in terms of the total average quantity of steel per square foot of floor area of the building. Curve A in Fig. 4 represents the average unit weight of a conventional frame with increasing numbers of stories. Curve B represents the average steel weight if the frame is protected from all lateral loads. The gap between the upper boundary and the lower boundary represents the premium for height for the traditional column-and-beam frame. Within the last few years structural engineers have developed new structural systems with a view to eliminating this premium.

Systems in steel. Tall buildings in steel have developed as a result of several types of structural innovations. The innovations have been applied to the construction of both office and apartment buildings.

Frames with rigid belt trusses. In order to tie the exterior columns of a frame structure to the interior vertical trusses, a system of rigid belt trusses at mid-height and at the top of the building has been developed. A good example of this system is the First Wisconsin Bank Building (1974) in Milwaukee (Fig. 5).

Framed tube. The maximum efficiency of the total structure of a tall building, for both strength and stiffness, to resist wind load can be achieved only if all column elements can be connected to each other in such a way that the entire building acts as a hollow tube or rigid box in projecting out of the ground. This particular structural system was probably used for the first time in the 43-story reinforced concrete DeWitt Chestnut Apartment Building in Chicago. The most significant use of this system is in the twin structural steel towers of the 110-story World Trade Center building in New York.

Column–diagonal truss tube. The exterior columns of a building can be spaced reasonably far apart and yet be made to work together as a tube by connecting them with diagonal members intersecting at the center line of the columns and beams. This simple yet extremely efficient system was used for the first time on the John Hancock Center in Chicago (Fig. 6), using as much steel as is normally needed for a traditional 40-story building.

Bundled tube. With the future need for larger and taller buildings, the framed tube or the column–diagonal truss tube may be used in a bundled form to create larger tube envelopes while maintaining high efficiency. The 110-story Sears Roebuck Headquarters Building in Chicago has nine tubes, bundled at the base of the building in three rows. Some of these individual tubes terminate at different heights of the building, demonstrating the unlimited architectural possibilities of this latest structural concept. The Sears tower, at a height of 1450 ft, is the world's tallest building.

Systems in concrete. While tall buildings of steel had an early start, present development of tall buildings of reinforced concrete has progressed at a fast enough rate to provide a competitive challenge to structural steel systems for both office and apartment buildings.

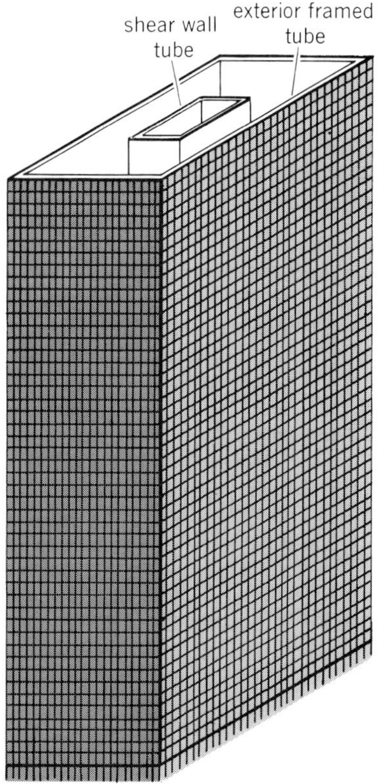

Fig. 7. Schematic sketch of tube-in-tube system.

Framed tube. As discussed earlier, the first framed-tube concept for tall buildings was used for the 43-story DeWitt Chestnut Apartment Building. In this building, exterior columns were spaced at 5-ft 6-in. centers, and interior columns were used as needed to support the 8-in.-thick flat-plate concrete slabs.

Tube in tube. The most recent system in reinforced concrete for office buildings combines the traditional shear-wall construction with an exterior framed tube. The system consists of an outer framed tube of very closely spaced columns and an interior rigid shear-wall tube enclosing the central service area. This system (Fig. 7), known as the tube-in-tube system, made it possible to design the world's present tallest (714 ft) lightweight concrete building (the 52-story One Shell Plaza Building in Houston) for the unit price of a traditional shear-wall structure of only 35 stories.

New systems combining both concrete and steel have also been developed, the latest of which is the composite system developed by Skidmore, Owings & Merrill in which an exterior closely spaced framed tube in concrete envelops an interior steel framing, thereby combining the advantages of both reinforced concrete and structural steel systems. The 52-story One Shell Square Building in New Orleans is based on this system. *See* STRUCTURAL ANALYSIS. [FAZLUR R. KHAN]

Bibliography: Architectural Record Editors, *Building for Commerce and Industry*, 1978; W. C. Huntington, *Building Construction: Materials and Types of Construction*, 4th ed., 1975; F. R. Khan, *Civil Eng.*, October 1967; F. R. Khan, *Inland Arch.*, July 1967; F. R. Khan, in *Proceedings of the ACI Annual Convention*, April 1970; F. R. Khan, in *Proceedings of the Tall Building Conference*, Vanderbilt University, 1974; W. J. McGuiness et al., *Mechanical and Electrical Equipment for Buildings*, 6th ed., 1980; Modular Buildings Standards Association, *Modular Practice*, 1962.

Bulk-handling machines

A diversified group of materials-handling machines specialized in design and construction for handling unpackaged, divided materials.

Bulk material. Solid, free-flowing materials are said to be in bulk. The handling of these unpackaged, divided materials requires that the machinery both supports their weight and confines them either to a desired path of travel for continuous conveyance or within a container for handling in discrete loads. Wet or sticky materials may also be handled successfully by some of the same machines used for bulk materials. Characteristics of materials that affect the selection of equipment for bulk handling include (1) the size of component particles, (2) flowability, (3) abrasiveness, (4) corrosiveness, (5) sensitivity to contamination, and (6) general conditions such as dampness, structure, or the presence of dust or noxious fumes.

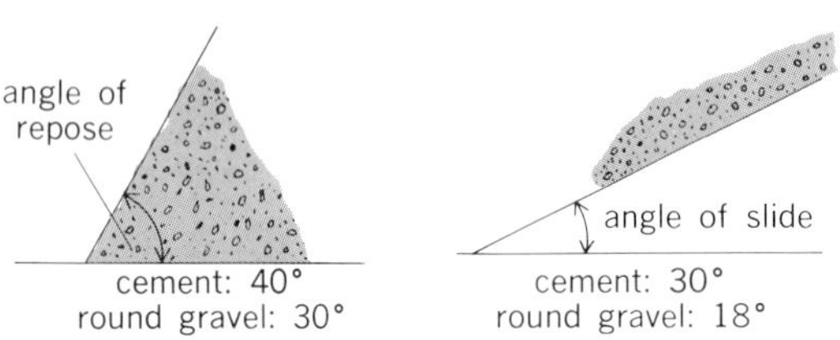

Fig. 1. Critical angles for bulk materials.

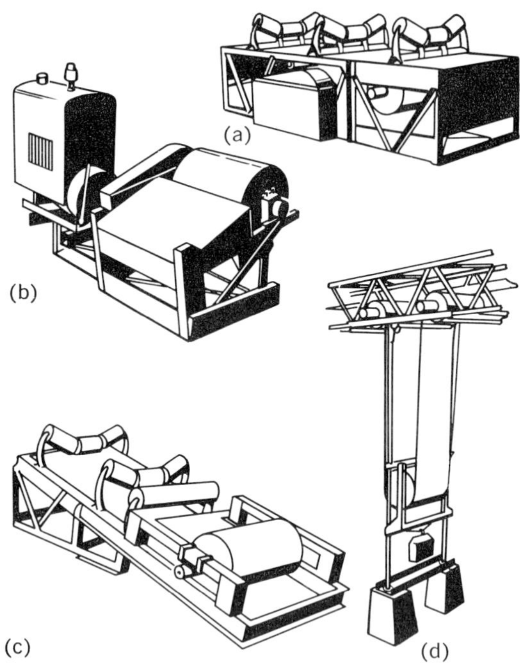

Fig. 2. Typical components of belt conveyor. (*a*) Electric motor wrap drive. (*b*) Gasoline engine head-end drive. (*c*) Screw type take-up. (*d*) Vertical gravity take-up.

Particles range in size from those that would pass through a fine mesh screen to those that would be encountered in earth-moving and mining processes. Fine granular materials are usually designated by their mesh size, which is an indication of the smallest mesh screen through which all or a specified percent of the particles will pass. As an example, a 100-mesh screen is one in which there are 100 openings per linear inch. Where the size of the bulk material varies, it is customary to indicate the percentage of each size in each mixture.

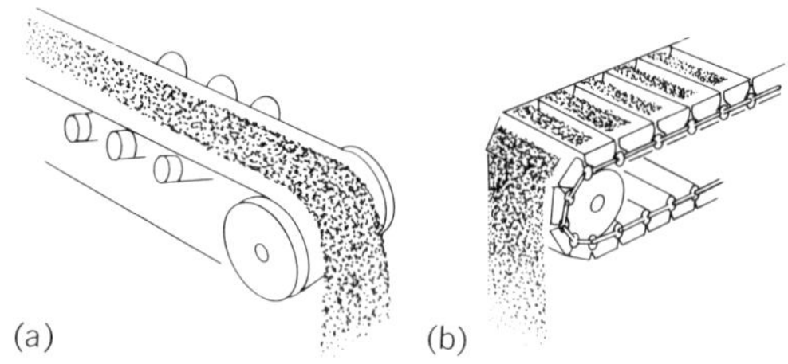

Fig. 3. Conveyors for abrasive materials, with material and conveying surface moving together. (*a*) Belt conveyor. (*b*) Apron conveyor.

Flowability, corrosiveness, abrasiveness, and similar terms are relative and are usually modified by adjectives to indicate the degree of the characteristic, such as "mildly" corrosive or "highly" abrasive. Many products are sensitive to contamination; this characteristic may be a determining factor in selecting bulk-handling equipment and its material of construction. For example, an enclosed

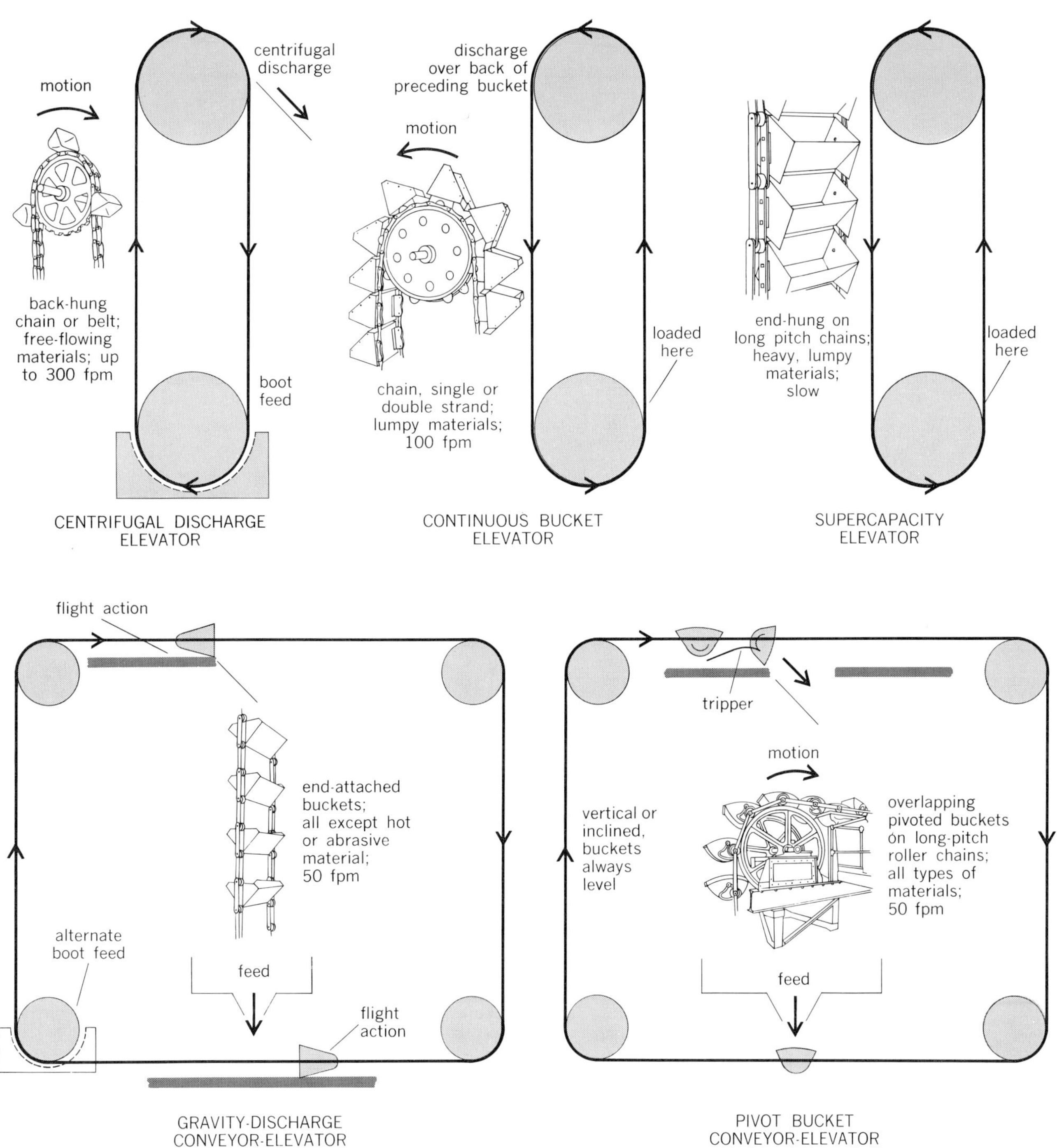

Fig. 4. Components of bucket conveyor-elevators and typical paths of travel for which they are used.

conveyor is used to protect the material from exterior contamination; and construction from noncorrosive material, such as stainless steel, may be required to protect the handled material from interior contamination.

Consideration also must be given to the angle of repose and the angle of slide of the material (Fig. 1). The angle of repose is the maximum slope, expressed in degrees, which piled material will stand without sliding on itself. The angle of slide, also expressed in degrees, is the angle at which material flows freely on an inclined surface such as a chute.

Corrosiveness and other handling characteristics of materials are compiled and updated in literature provided by manufacturers of bulk-handling machines. Trade associations also publish technical bulletins on the relations of machines to jobs, applications, and operating costs. Among these organizations are the Power Crane and Shovel Association and the Conveyor Equipment Manufacturers Association. Related data are published by various governmental departments, especially the U.S. Department of Commerce.

Continuous bulk conveyor. Equipment that transports material continuously in a horizontal, inclined, or vertical direction in a predetermined path is a form of conveyor. The many different

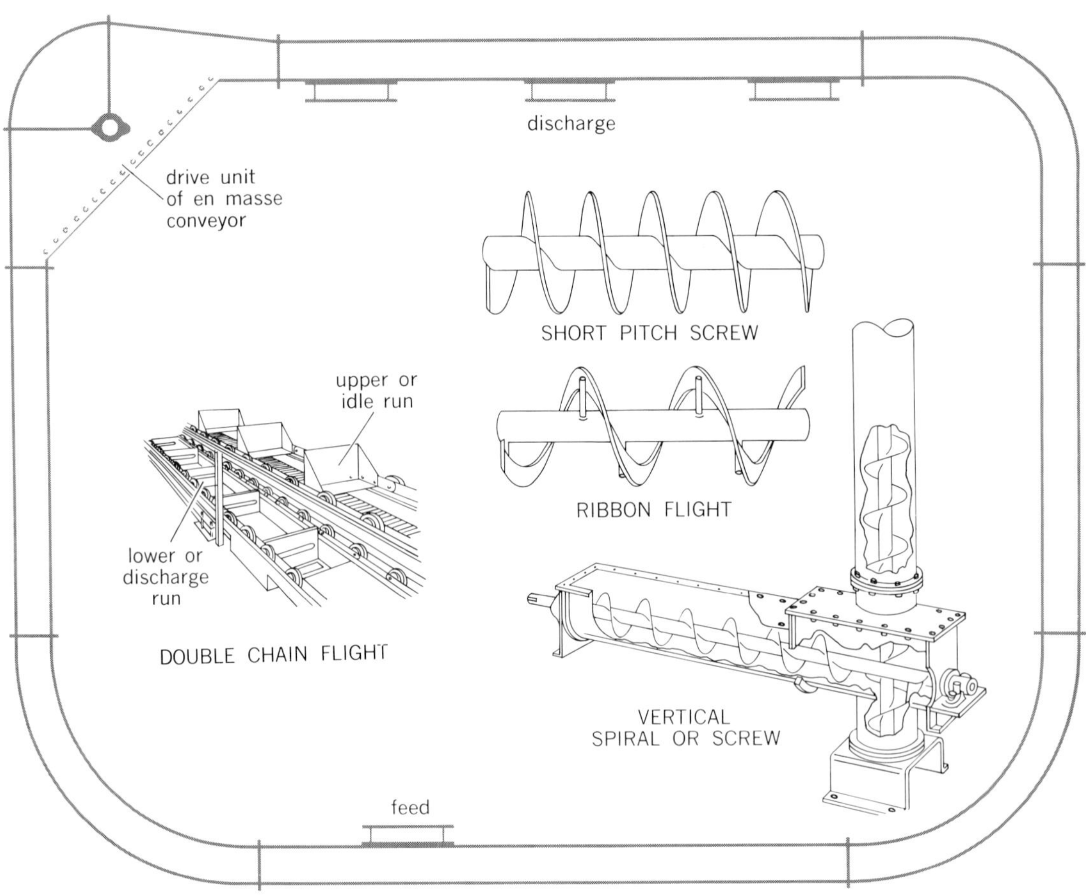

Fig. 5. Diagram of an en masse conveyor trough. The insets show some of the flight conveyors used to move bulk materials: double chain flight, and horizontal and vertical screw or spiral conveyors.

means used to convey bulk materials include gravity, belt, apron, bucket, skip hoist, flight or screw, dragline, vibrating or oscillating, and pneumatic conveyors. Wheel or roller conveyors cannot handle bulk materials.

Gravity chutes are the only unpowered conveyors used for bulk material. They permit only a downward movement of material. Variations such as chutes with steps or cleats are employed as a means of slowing product movement.

Belt conveyors of many varieties move bulk materials. Fabric belt conveyors have essentially the same operating components as those used for package service; however, these components are constructed more ruggedly to stand up under the more rigorous conditions imposed by carrying coal, gravel, chemicals, and other similar heavy bulk materials (Fig. 2). In the latter type, the belt runs on a bed of closely spaced rollers positioned to form a flat or troughed conveyor bed. Belts may also be made of such materials as rubber, metal, or open wire. Belt conveyors are only used within angles of 28° from horizontal. Materials feed onto belt conveyors from hoppers or from storage facilities overhead and may discharge over the end (Fig. 3*a*) or, if the belt is flat, by being diverted off one side. Belt conveyors can handle most materials over long distances and up and down slopes. Their advantages include low power requirements, high capacities, simplicity, and dependable operation.

An apron conveyor is a form of belt conveyor, but differs in that the carrying surface is constructed of a series of metal aprons or pans pivotally linked together to make a continuous loop. The pans, which may overlap one another, are usually attached at each end to two strands of roller chain. The chain runs on steel tracks, movement being provided by suitable chain sprockets. Turned-up edges or side wings provide a troughed carrying surface (Fig. 3*b*). As in the case of the belt conveyor, the top strand of the apron is the carrying surface. This type of conveyor is suitable for handling large quantities of bulk material under severe service conditions and can operate at speeds up to approximately 100 ft per minute, handling up to approximately 300 tons per hour; with the addition of cleats, an apron can convey up inclines to as steep as 60°. Apron conveyors are most suitable for heavy, abrasive, or lumpy materials.

Bucket conveyors, as the name implies, are constructed of a series of buckets attached to one or two strands of chain or in some instances to a belt. These conveyors are most suitable for operating on a steep incline or vertical path, sometimes being referred to as elevating conveyors (Fig. 4). Bucket construction makes this type of conveyor most ideal for bulk materials such as sand or coal. Buckets are provided in a variety of shapes and are usually constructed of steel.

Flight conveyors employ the use of flights, or bars attached to single or double strands of chain. The bars drag or push the material within an

BULK-HANDLING MACHINES

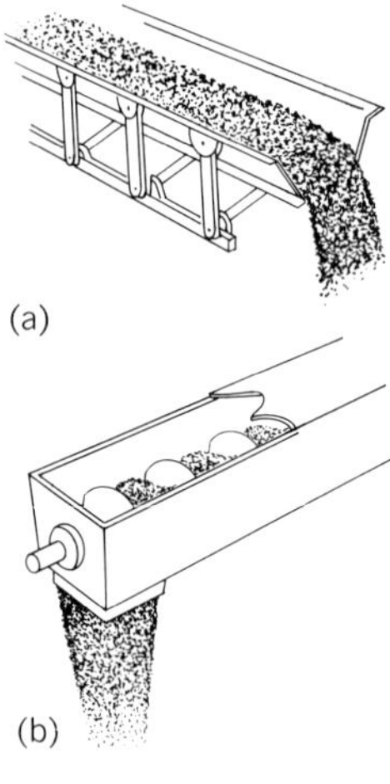

Fig. 6. Open and closed troughs. (*a*) Oscillating conveyor, with whole trough oscillating. (*b*) Screw conveyor, with helix driving material along trough.

enclosed duct or trough. These are frequently referred to as drag conveyors. Although modifications of this type of conveyor are employed in a number of ways, its commonest usage is in a horizontal trough in which the lower strand of the flights actually move the material. This type of conveyor is commonly used for moving bulk material such as coal or metal chips from machine tools (Fig. 5). Constant dragging action along the trough makes this type of conveyor unsuitable for materials which are extremely abrasive. Materials can be fed to the conveyor from any desired intermediate point and, by the use of gating, can be discharged at any desired point.

Dragline conveyors operate on basically the same principle as flight conveyors, as previously described. The chain, ruggedly made for this service, drags the material, such as clinkers and slag, along the bottom of a concrete trough. The dragging is done entirely by the links of the chain.

Spiral or screw conveyors rotate upon a single shaft to which are attached flights in the form of a helical screw. When the screw turns within a stationary trough or casing, the material advances (Fig. 6*b*). These conveyors are used primarily for bulk materials of fine and moderate sizes, and can move material on horizontal, inclined, or vertical planes (Fig. 5). The addition of bars or paddles to the flight conveyor shaft also makes it ideal for mixing or blending the materials while they are being handled. In addition, enclosed troughs may be water- or steam-jacketed for cooling, heating, drying, and so forth.

Vibrating or oscillating conveyors employ the use of a pan or trough bed, attached to a vibrator or oscillating mechanism, designed to move forward slowly and draw back quickly (Fig. 6*a*). The inertia of the material keeps the load from being carried back so that it is automatically placed in a more advanced position on the carrying surface. Adaptations of this principle are also used for moving material up spiral paths vertically. Mechanical (spring), pneumatic (vibrator), and electrical (vibrator) devices provide the oscillating motion. These conveyors can handle hot, abrasive, stringy, or irregularly shaped materials. The trough can be made leakproof or, by enclosing it, dustproof.

Pneumatic, or air, conveyors employ air as the propelling media to move materials. One implementation of this principle is the movement within an air duct of cylindrical carriers, into which are placed currency, mail, and small parts for movement from one point to discharge at one of several points by use of diverters. Pneumatic pipe conveyors are widely used in industry, where they move granular materials, fine to moderate size, in origi-

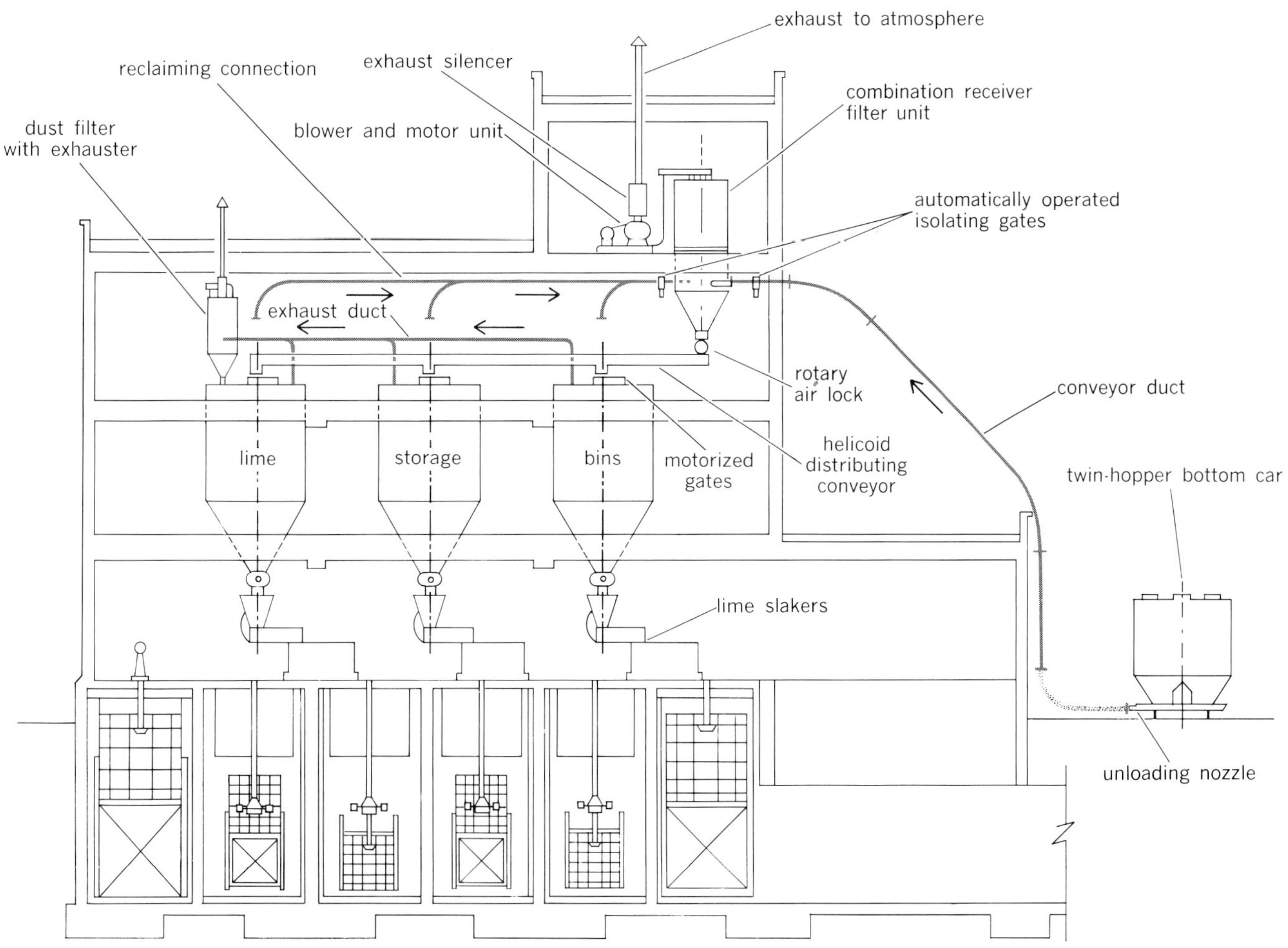

Fig. 7. Typical conveyor system, with hopper car delivering material, pneumatic lines unloading and lifting material, screw flight transferring it to bins, and gravity chutes discharging it. (*National Conveyors Co.*)

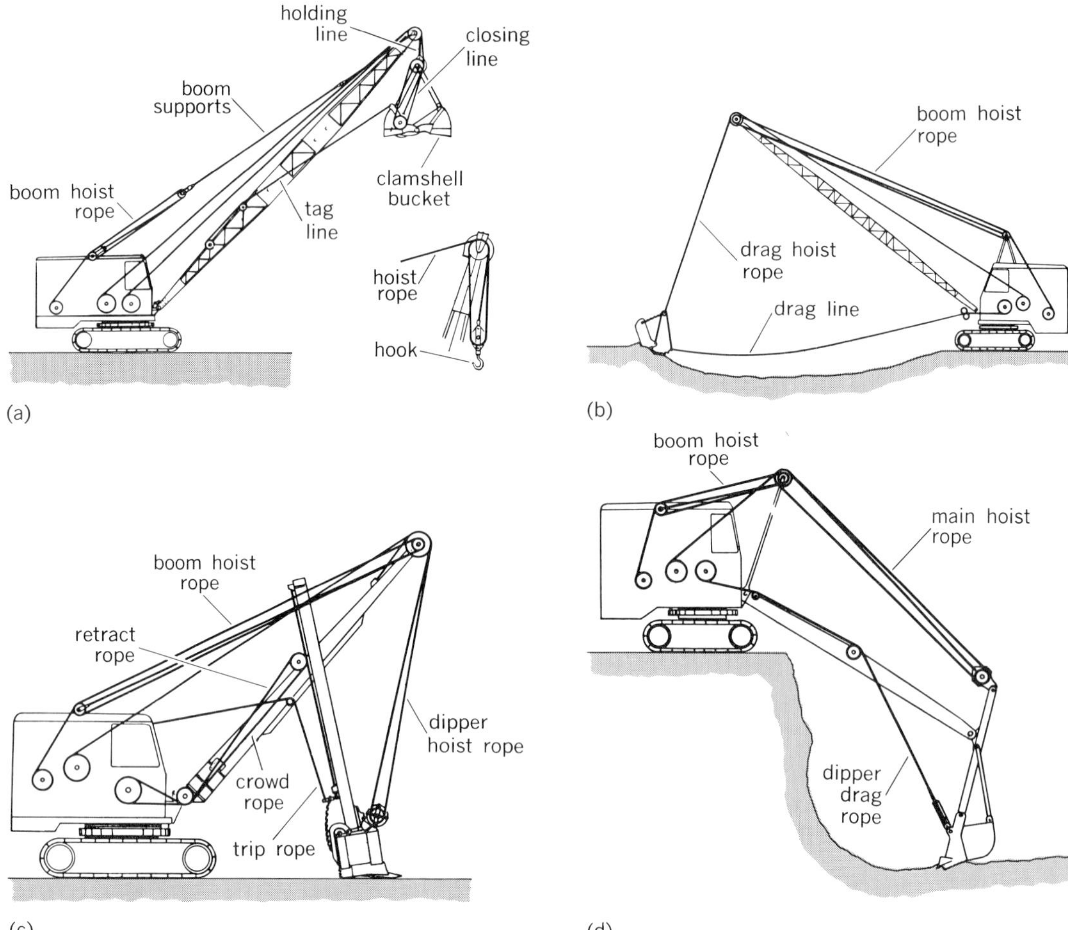

Fig. 8. Reeving diagrams for power crane fitted (*a*) with clamshell or (*b*) with scoop and for shovel fitted (*c*) with forward dipper or (*d*) with back hoe (backdigger or dragshovel). (*United States Steel Corp.*)

nal bulk form without need of internal carriers. An air compressor provides the air to move the material either by pressure or vacuum. Materials are introduced to the system by means of air locks, which are of the rotary or slide type. These locks are designed to permit the entry of the material with negligible loss of air. By means of diverted valves in the pneumatic pipe, a discharge can be effected at any one of a number of predetermined points. The small number of moving parts, the ability to move material in any direction, and the need of minimal prime plant floor space are among the advantages of this type of bulk-handling machine.

A pneumatic conveyor may form part of a conveyor system. The system comprises a network of machines, each handling the same product through its various processing stages; a single system may well employ numerous varieties of conveyors, such as pneumatic, screw, and vibrating conveyors. Figure 7 shows a system for handling pebble lime in a water-treatment plant. The lime is received in railroad cars or trucks, is unloaded and conveyed by a pneumatic vacuum system to a receiver filter, and is discharged to a screw conveyor, from which it is selectively delivered to any desired storage bin. From the bins, the lime can be delivered on demand to the slakers below.

Another adaptation of the pneumatic conveyor is to activate a gravity conveyor. Such a conveyor handles dry pulverized materials through slightly inclined chutes. Air flows through the bottom of the chute, which is usually constructed of a porous medium, fluidizing the material and causing it to flow in the manner of a liquid. An advantage of this conveyor is that there are no moving parts.

Aerial tramways and cableways employ the use of a cab or carrier suspended by a grooved wheel on an overhead cable to transport materials over long distances, particularly where the terrain is such that truck or rail transportation is impractical. They are used primarily to meet the needs of such activities as dam construction, loading ships, bringing coal to and from power houses, and stock piling in open country.

Discontinuous bulk handlers. Power cranes and shovels perform many operations moving bulk materials in discrete loads. When functioning as cranes and fitted with the many below-the-hook devices available, they are used on construction jobs and in and around industrial plants. Such fittings as magnets, buckets, grabs, skullcrackers, and pile drivers enable cranes to handle many products.

The machines of the convertible, full-revolving type are mounted on crawlers, trucks, or wheels. Specialized front-end operating equipment is required for clamshell, dragline, lifting-crane, pile-driver, shovel, and hoe operations. Commercial sizes of these machines are nominally from $\frac{1}{4}$

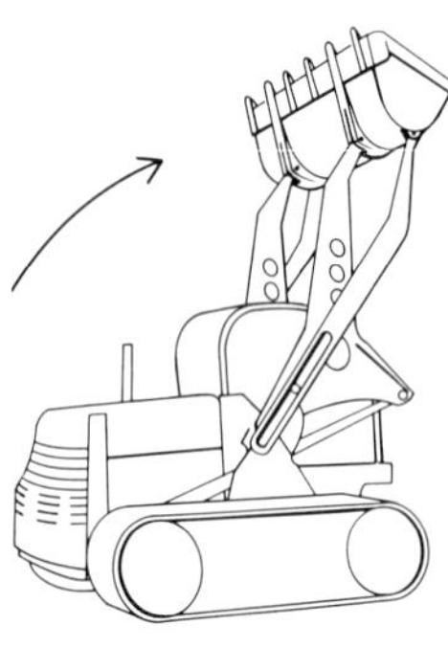

Fig. 9. Tractor fitted with shovel for up-and-over operation. (*Service Supply Corp.*)

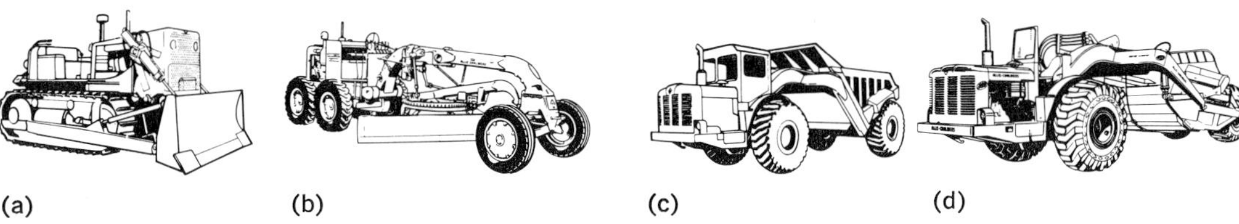

Fig. 10. Road-making machines. (*a*) Front-end shovel. (*b*) Motor grader. (*c*) Motor wagon. (*d*) Motor scraper.

to $2\frac{1}{2}$ yd^3 as shovels and from $2\frac{1}{2}$ to 60 tons as cranes.

The revolving superstructure consists of the rotating frame and the operating machinery thereon. It may be carried on a crawler mount, consisting of two continuous parallel crawler belts. A truck mount is a heavy-framed, rubber-tired carrier supported by two or more axles and having the general characteristics of a heavy-duty truck. The carrier may be controlled for road travel from a cab mounted on the carrier or located on the revolving superstructure. Machines of this type can also be secured on railroad mountings.

Six types of front-end operating equipment are standard: crane, clamshell, dragline, pile driver, shovel, and hoe. Common crane-boom equipment is used with crane, clamshell, dragline, and pile driver. The boom usually consists of two sections, between which additional sections may be inserted. Shovel and hoe equipment have their distinctive mechanisms for operation (Fig. 8).

Scoops and shovels are used for handling bulk materials in plants and yards. Lighter models are usually wheel-mounted, while those for heavier duty are apt to be carried on tractors. Two basic types are recognized: those that load and dump only at the front or at the rear, and those with an up-and-over action which permits loading at the front end and discharging at the rear (Fig. 9). This last arrangement is frequently a time-saver in that it eliminates the need for maneuvering when loading wagons.

Specialized equipment for mechanized pit mining has been developed. Power cranes, shovels, and scoops are actively engaged in strip mines, quarries, and other earth-moving operations.

The rapid expansion of highway systems has led to the widespread use of road-making machines (Fig. 10). Some of these are modifications of equipment standard to this kind of work; others are highly specialized. Among the former are machines that have evolved by the addition of attachments to wheel and crawler tractors. Machines such as bulldozers and graders do not function as true handling equipment in that they do not pick up and transport materials but push them. On the other hand, self-loading scrapers do transport materials and are usually constructed so that they are either side- or end-dumping. Other general-purpose machines used in highway construction are trench diggers, hole diggers for utility poles, and cable-laying machines. Highly specialized equipment is used for surfacing the road with concrete or other material. *See* CONVEYING MACHINES; CONVEYOR; ELEVATING MACHINES; HOISTING MACHINES; INDUSTRIAL TRUCKS; MATERIALS-HANDLING MACHINES; MONORAIL.

[ARTHUR M. PERRIN]

Bibliography: H. A. Bolz and G. E. Hageman, *Materials Handling Handbook*, 1958; K. E. Booth and C. G. Chantrill, *Materials Handling with Industrial Trucks*, 1962; M. N. Kraus, *Pneumatic Conveying of Bulk Materials*, 1968.

Bulkhead

A structure primarily used to retain earth fill or protect embankments along the shore. Bulkheads sometimes also serve as wharves. They are usually used in comparatively sheltered areas and therefore require less strength against sea action than sea walls. Sheet piling of timber, steel, or concrete is the commonest form of construction, with tiebacks to "deadmen" buried in the earth. Pile-supported platforms behind sheet piling help to relieve the earth pressure against the sheeting, and platforms on the water side help to reduce the unsupported height of the sheeting. Masonry or concrete gravity wall construction and stone-filled crib construction are also used. *See* COASTAL ENGINEERING; SEA WALL.

[EDWARD J. QUIRIN]

Bulldozer

An assembly consisting of a reinforced, curved steel plate or blade and a frame which is attached to the front of a wheeled or crawler tractor for the purpose of pushing dirt, rock, or other materials as illustrated. Blade widths vary from roughly 6 ft (1.8 m) to about 25 ft (8 m), and blade heights range from 2 ft (0.6 m) to about 7 ft (2.1 m). The blade may be straight, angling, or U-shaped.

The U blades and straight blades mount perpendicular to the longitudinal axis of the tractor and push material straight ahead; the U blade, because of its shape, can contain and move more material than the straight blade. The angle blade

Hydraulic-actuated blade which is mounted on crawler tractor. (*Caterpillar Tractor Co.*)

can be set so that either of its ends is ahead of the other; thus it is useful in windrowing material or backfilling trenches. Blades are controlled either by cables or hydraulic units, although the latter are far commoner. The straight and U blades can be raised and lowered, worked with one side lower than the other, and tipped forward or backward. The angle blade cannot be tipped. In common parlance, the term bulldozer erroneously includes the tractor. *See* EARTHMOVER. [EDWARD M. YOUNG]

Bus-bar

An electric conductor that serves as a common connection between load circuits and the source of electric power of one polarity in direct-current systems or of one phase in alternating-current systems. Two or more bus-bars which serve all polarities or phases are collectively called a bus.

Bus-bars must be designed to carry the continuous current without overheating. The highest continuous current can be in the order of hundreds of thousands of amperes. Bus-bars must also be designed to withstand the mechanical forces caused by short-circuit currents. Special metallic shields are designed to enclose the bus, thus minimizing the effect of short-circuit forces. Such a bus is known as an isolated phase bus.

[THOMAS H. LEE]

Bushing

A removable metal lining, usually in the form of a bearing to carry a shaft. Generally a bushing is a small bearing in the form of a cylinder and is made of soft metal or graphite-filled sintered material. Bushings are also used as cylindrical liners for holes to preserve the dimensional requirements, such as in the guide bearings in jigs and fixtures for drilling holes in machine parts.

[JAMES J. RYAN]

Caisson foundation

A permanent substructure that, while being sunk into position, permits excavation to proceed inside and also provides protection for the workers against water pressure and collapse of soil. The term caisson covers a wide range of foundation structures. Caissons may be open, pneumatic, or floating type; deep or shallow; large or small; and of circular, square, or rectangular cross section. The walls may consist of timber, temporary or permanent steel shells, or thin or massive concrete. Large caissons are used as foundations for bridge piers, deep-water wharves, and other structures. Small caissons are used singly or in groups to carry such loads as building columns. Caissons are used where they provide the most feasible method of passing obstructions, where soil cannot otherwise be kept out of the bottom, or where cofferdams cannot be used. *See* BRIDGE; PILE FOUNDATION.

Fig. 1. Underside of open caisson for Greater New Orleans bridge over Mississippi River. (*Dravo Corp.*)

The bottom rim of the caisson is called the cutting edge (Fig. 1). The edge is sharp or narrow and is made of, or faced with, structural steel. The narrowness of the edge facilitates removal of ground under the shell and reduces the resistance of the soil to descent of the caisson.

Open caisson. An open caisson is a shaft open at both ends. It is used in dry ground or in moderate amounts of water. A bottom section, having a cutting edge, is set on the ground and soil is removed from inside while the caisson sinks by its own weight. Sections are added as excavation and sinking proceed. After excavation is completed, the hollow interior of the shaft is filled with concrete.

Open caissons are usually constructed of reinforced concrete, but if steel shells are used, concrete may also be required to provide weight. When an open caisson is to be towed into place, hollow walls or false bottoms are provided to give it buoyancy; they are removed after the caisson is in position. Open caissons can be sunk in water to practically any depth.

In deep water an open caisson is sometimes sunk by the sand-island method. A cofferdam is constructed and filled with sand to form an artificial island. The island serves as a working platform and guide for sinking the caisson. This method also avoids the necessity of transporting a fabricated shell and holding it in position.

Figure 2 shows the major types of caissons in use. Chicago caissons (open well) are formed by excavating a few feet of soil, installing short vertical pieces of wood or concrete lagging beveled to form a circle, bracing with segmental hoops, and repeating. Such caissons are used in clay that is not strong enough to carry the permanent load alone but will permit excavation for the lagging without flowing in at the bottom. The clay is hoisted out in buckets or elevators. The bottom may be belled out. Lagging and bracing are generally left in place.

Sheeted caissons are similar to Chicago caissons, except that vertical sheeting is continuous and is driven by hand or pile-driving hammers either in advance of excavation or as it proceeds. Bracing rings are placed every few feet. Sheeting may be wood, concrete plank, or steel, and is left in place. Sheeted caissons may be round or square.

Small cylindrical caissons, 2–12 ft (0.6–3.6 m) in diameter, consist of a concrete shell or steel pipe sunk into the ground by its own or by a temporarily superimposed weight. Excavation is done

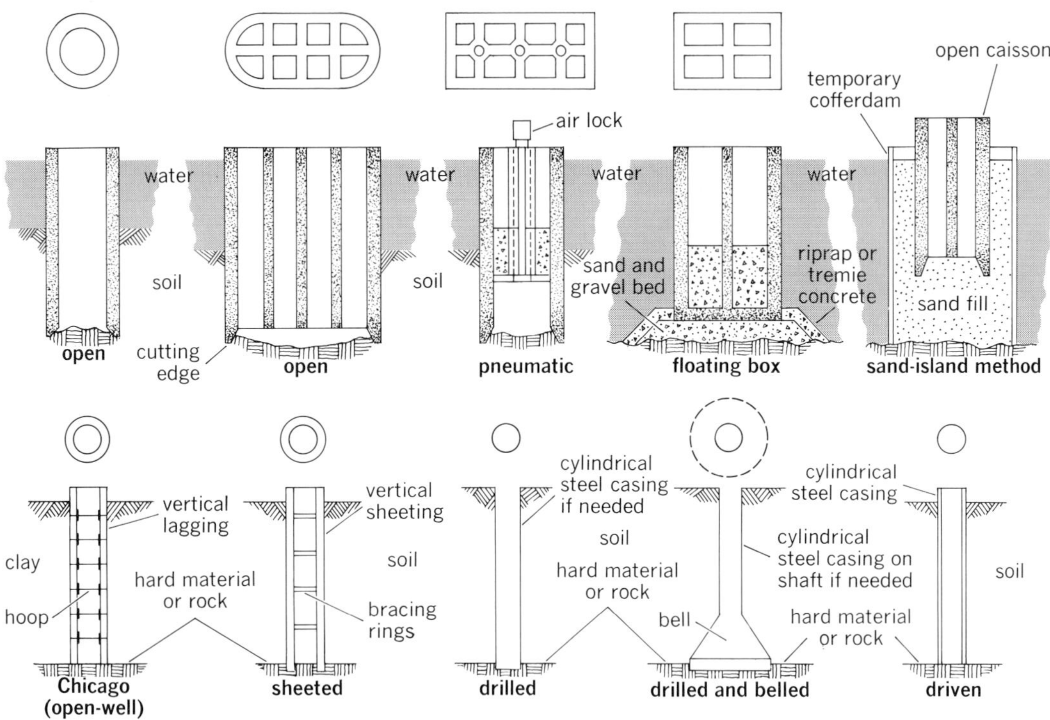

Fig. 2. Major types of caissons as they appear from top and in cross section from side.

by hand or mechanically, and the caisson is finally filled with concrete.

Drilled caissons are drilled holes that are filled with concrete. Rock socketing may be used at the bottom, or the bottom may be belled up to 30 ft (9 m) in diameter. The caisson may be as long as 150 ft (46 m). Cohesive soils are drilled with augers or bucket-type drills. Granular soils are drilled in the same manner with the aid of a binding agent or by rotary drills using drilling mud or water to keep the hole from caving in. Steel-shaft casings are used where needed to shut off water or keep soil out of the hole. Casings are removed before concreting, or during concreting in unstable ground.

Driven caissons are formed by driving a cylindrical steel shell with one or more pile-driving hammers. After excavation by grabs, buckets, or jetting, concrete is placed inside. The shells usually are withdrawn as the concrete rises, unless the soil is so soft or water conditions are such that they must be left in place.

Pneumatic caisson. A pneumatic caisson is like a box or cylinder in shape; but the top is closed and thus compressed air can be forced inside to keep water and soil from entering the bottom of the shaft. A pneumatic caisson is used where the soil cannot be excavated through open shafts (for instance, where there are concrete, timbers, boulders, or masonry lying underwater) or where soil conditions are such that the upward pressure must be balanced. The air pressure must balance or slightly exceed the hydrostatic head and is increased as the caisson descends. The maximum depth is about 120 ft (37 m), corresponding to an air pressure of 52 psi or about 3 1/2 atm, which is about the limit of human endurance. Workers and materials must pass through air locks. Too rapid decompression may result in caisson disease or "bends," resulting from the expansion of bubbles of air trapped in joints, muscles, or blood. The length of time that workers can perform under pressure and the speed of decompression are regulated by law. Pneumatic caissons may be started as open caissons, then closed and air applied. They are made of reinforced concrete, which may be faced on one or both sides with steel plates.

Floating caisson. A floating or box caisson consists of an open box with sides and closed bottom, but no top. It is usually built on shore and floated to the site where it is weighted and lowered onto a bed previously prepared by divers. The caisson is then filled with sand, gravel, or concrete. This type is most suitable where there is no danger of scour. Floating caissons may be built of reinforced concrete, steel, or wood.

Small box caissons usually consist of a single cell. Large caissons are usually divided into compartments; this braces the side walls and permits more accurate control during loading and sinking. *See* FOUNDATIONS. [ROBERT D. CHELLIS]

Bibliography: *See* FOUNDATIONS.

Caliper

An instrument with two legs used for measuring linear dimensions. Calipers may be fixed, adjustable, or movable. Fixed calipers are used in routine inspection of standard products; adjustable calipers are used similarly but can be reset to slightly different dimensions if necessary. Movable calipers can be set to match the distance being measured. The legs may pivot about a rivet or screw in a firm-joint pair of calipers; they may pivot about a pin, being held against the pin by a spring and set in position by a knurled nut on a threaded rod; or the legs may slide either directly (caliper rule) or along a screw (micrometer caliper) relative to each other. *See* MICROMETER.

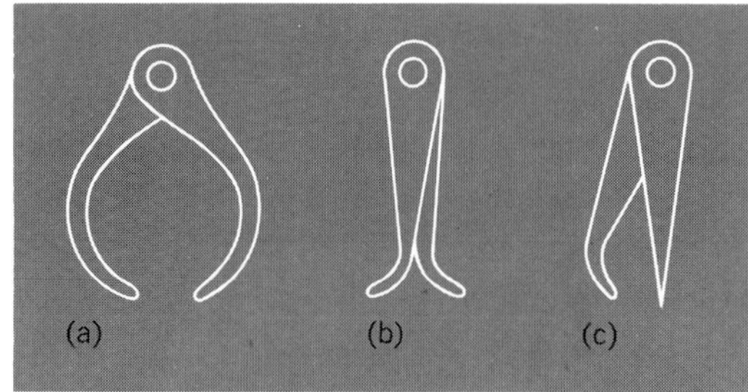

Some typical machinist's calipers. (*a*) Outside. (*b*) Inside. (*c*) Hermaphrodite. (*R. J. Sweeney, Measurement Techniques in Mechanical Engineering, Wiley, 1953*)

The legs may be shaped to facilitate measuring outside dimensions, inside dimensions, surface dimensions as between points on a plate, or from a surface into a hole as in a keyway (see illustration). Other forms are adapted to special needs. *See* GAGES.

[FRANK H. ROCKETT]

Cam mechanism

A mechanical linkage whose purpose is to produce, by means of a contoured cam surface, a prescribed motion of the output link of the linkage, called the follower. Cam and follower are a higher pair. *See* SLIDING PAIR.

A familiar application of a cam mechanism is in the opening and closing of valves in an automotive engine (Fig. 1). The cam rotates with the cam shaft, usually at constant angular velocity, while the follower moves up and down as controlled by the cam surface. A cam is sometimes made in the form of a translating cam (Fig. 2*a*). Other cam mechanisms, employed in elementary mechanical analog computers, are simple memory devices, in which the position of the cam (input) determines the position of the follower (output or readout).

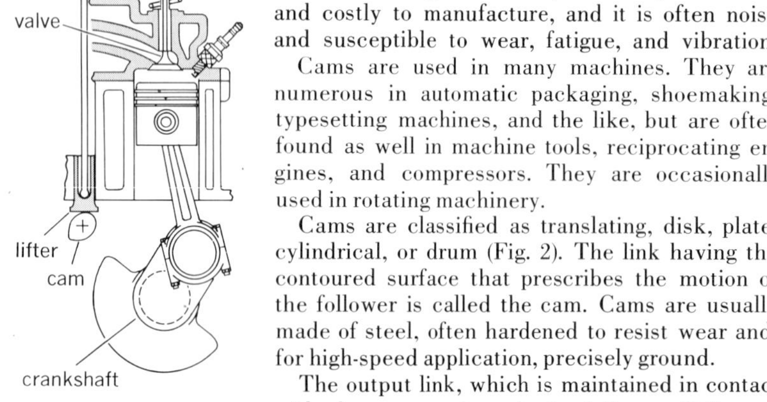

Fig. 1. Cam mechanism for opening and closing valves in automotive engine. (*Texaco, Inc.*)

Although many requisite motions in machinery are accomplished by use of pin-jointed mechanisms, such as four-bar linkages, a cam mechanism frequently is the only practical solution to the problem of converting the available input, usually rotating or reciprocating, to a desired output, which may be an exceedingly complex motion. No other mechanism is as versatile and as straightforward in design. However, a cam may be difficult and costly to manufacture, and it is often noisy and susceptible to wear, fatigue, and vibration.

Cams are used in many machines. They are numerous in automatic packaging, shoemaking, typesetting machines, and the like, but are often found as well in machine tools, reciprocating engines, and compressors. They are occasionally used in rotating machinery.

Cams are classified as translating, disk, plate, cylindrical, or drum (Fig. 2). The link having the contoured surface that prescribes the motion of the follower is called the cam. Cams are usually made of steel, often hardened to resist wear and, for high-speed application, precisely ground.

The output link, which is maintained in contact with the cam surface, is the follower. Followers are classified by their shape as roller, flat face, and spherical face (Fig. 3). The point or knife-edge follower is of academic interest in developing cam profile relationships. Followers are also described by the nature of their constraints, for example, radial, in which motion is reciprocating along a radius from the cam's axis of rotation (Fig. 1); offset, in which motion is reciprocating along a line that does not intersect the axis of rotation (Fig. 2*b*); and oscillating, or pivoted (Fig. 2*a*). Three-dimensional cam-and-follower systems are coming into more frequent use, where the follower may travel over a lumpy surface.

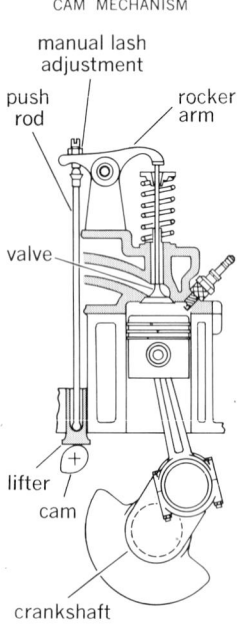

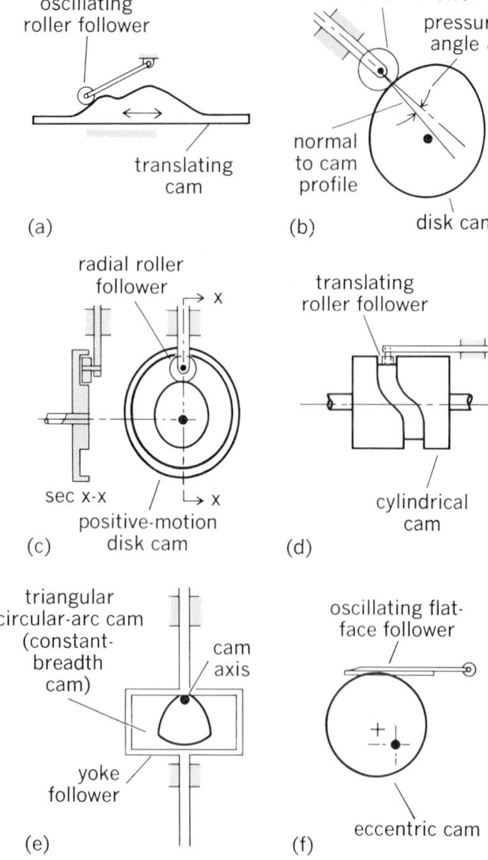

Fig. 2. Classification of cams. (*a*) Translating. (*b*) Disk. (*c*) Positive motion. (*d*) Cylindrical. (*e*) With yoke follower. (*f*) With flat-face follower.

Motion of cam follower. The first step in the design of a cam mechanism is the determination of the motion of the cam follower. In a packaging machine, for example, the ends of a carton may be folded by cam-operated fingers that advance at the proper times, retract as soon as the fold has been made, and then rest or dwell until the next carton is in position. The motion of the cam follower, which in turn moves the folder fingers, can be represented by a displacement-time diagram (Fig. 4). The time axis is usually laid off in degrees of cam

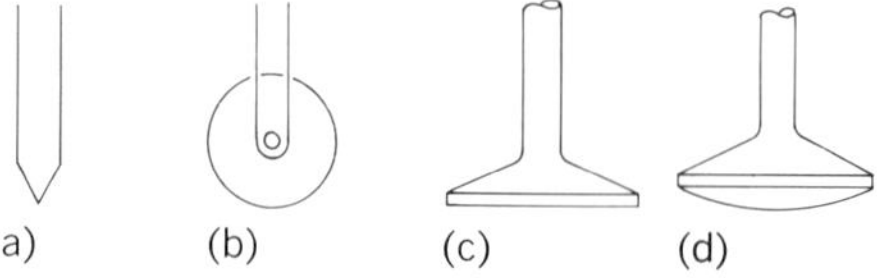

Fig. 3. Cam followers. (*a*) Knife edge. (*b*) Roller. (*c*) Flat face. (*d*) Spherical face.

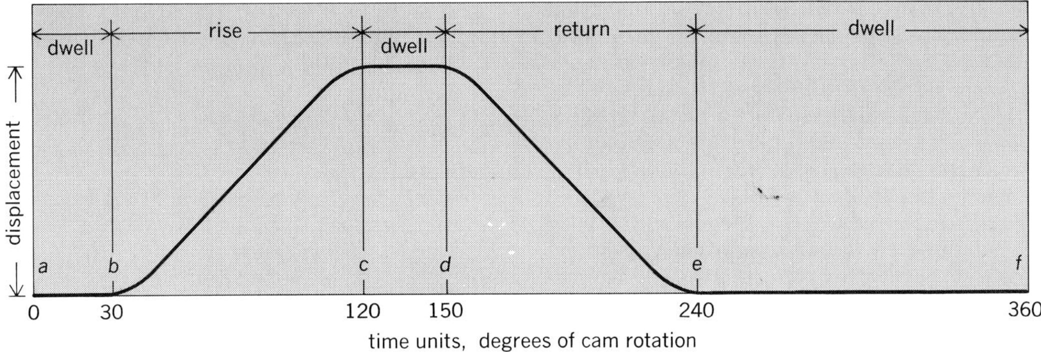

Fig. 4. Displacement-time diagram for a cam, with motion of cam follower indicated.

rotation. The conventional meanings of follower dwell, rise, and return are indicated in the figure.

The maximum displacement of the follower and the periods of dwell are determined, more or less arbitrarily, by the designer. He has the choice of any curve to connect the dwell portions of the complete displacement-time diagram. The practical form for this curve is determined largely by the maximum acceleration that can be tolerated by the follower linkage. In addition, the pressure angle (α in Fig. 2*b*) must be kept fairly small, usually less than 30°, to avoid undue friction and possible jamming of the reciprocating follower in its guides. The space that is available for the cam will affect the maximum pressure angle. Usually, a small cam is preferred; yet the larger the cam can be made (which in effect physically increases the length of the time axis for the same time interval), the smaller the maximum pressure angle will be. The final form may further represent a compromise to make possible economical manufacture of the cam. *See* ACCELERATION ANALYSIS.

If the diagram of Fig. 4 were laid out on and cut out of steel and a knife-edge follower were constrained to move vertically, the translating cam mechanism of Fig. 5*a* would result. The process of wrapping this translating cam around a disk (Fig. 5*b*), thus producing a disk cam whose follower action would be similar to that of Fig. 5*a*, can be visualized readily. The introduction of a roller or flat-face follower complicates the determination of the actual cam contour that will produce a desired follower displacement-time relationship; but recognition of the similarity between the displacement-time diagram and the final cam contour makes it easier to visualize the conditions that must be met to design a cam that will operate satisfactorily.

Consider the dwell-rise-dwell portion of the curve of Fig. 4. Displacement of the follower might be plotted arbitrarily as a straight line (Fig. 6*a*).

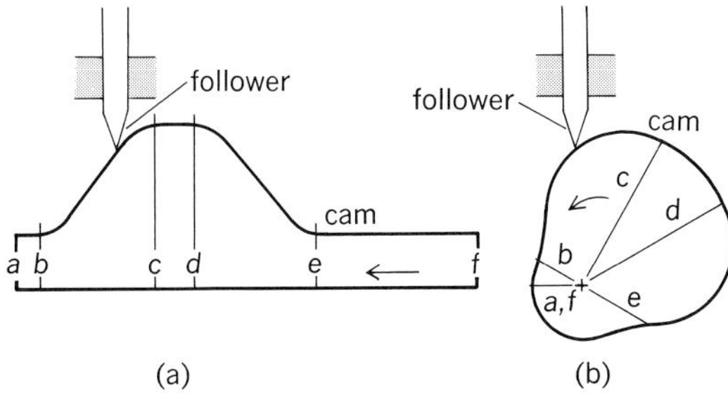

Fig. 5. Converting (*a*) translating to (*b*) disk cam.

The slope $\Delta s/\Delta t$ of a displacement-time (s-t) curve is equal to velocity ($s/t = v$) so that velocity of the follower from A to B (dwell) will be zero, from B to C it will be constant and finite, and from C to D velocity will be zero again (Fig. 6*b*).

The slope $\Delta v/\Delta t = a$ of the velocity-time curve is correspondingly equal to acceleration. Thus, the acceleration of the follower necessary to increase velocity from zero to a finite value in zero time (B to B in Fig. 6*b*) is infinite. Likewise, the deceleration that occurs at C must also be infinite. The acceleration along the constant velocity line, from B to C, would be zero. Thus the curve chosen in Fig. 6*a* for displacement is unrealistic because of the high inertial forces that would result from abrupt changes of velocity.

Choice of acceleration curve. Therefore, a curve having a gradual transition from dwell to maximum velocity is necessary. Three such curves are plotted in Fig. 7 and are superimposed in Fig. 8 for comparison. The derived curves for velocity and acceleration are also plotted so comparisons may be made.

The constant acceleration – constant deceleration curve, in which displacement s is proportional to t^2, is desirable except for the instantaneous reversal of acceleration at the point of maximum velocity; such a reversal would cause high stresses in the mechanism. If the follower were spring-loaded, a heavy spring would be required to prevent the follower's leaving the cam face momentarily, with resulting shock to the linkage as it returned.

The simple harmonic displacement curve is plotted by projecting onto the diameter (equal to follower displacement) of a circle a point moving with constant velocity around the circle's circumference. Although maximum acceleration is higher than in the preceding curve, the abrupt changes of acceleration occur only at the beginning and end of the rise. Both of these curves have been used in cam design; both are satisfactory if speeds are low to moderate and follower mass is not large. However, serious difficulties are encountered when high speeds or heavy followers accentuate the stresses resulting from acceleration.

The cycloidal curve is plotted by projecting points from a cycloid whose generating circle has a diameter equal to follower displacement divided by π, as in Fig. 7. This curve has desirable acceleration characteristics, but requires that the cam

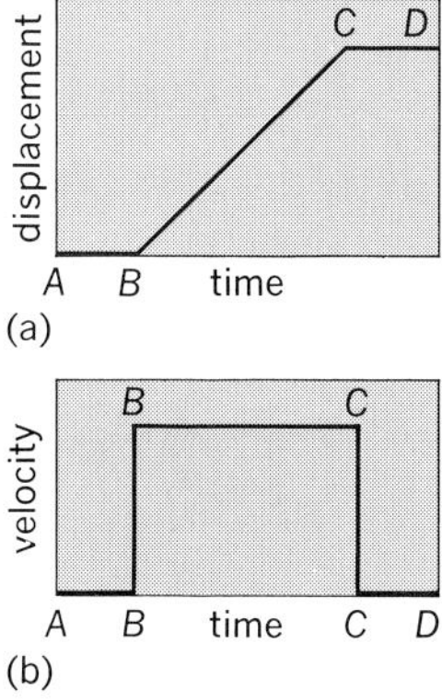

Fig. 6. Effect of (*a*) displacement on (*b*) velocity.

face be accurately machined at the beginning and end of rise to accomplish in fact the theoretical performance.

Comparison of the three curves is shown in Fig. 8. The cycloidal curve exhibits a higher pressure angle α for a radial translating follower than the other curves, but its acceleration characteristics are much superior.

Today's extensive use of cams and the extreme demands upon them under increasing speeds of modern manufacturing have yielded still another form of cam profile, which might be called catenoidal because of its association with the catenary (the curve in which a rope or chain (catena) hangs freely). This curve has an equation expressed primarily in terms of exponential functions ϵ^x, whose slope variation is a curve with the remarkable property that it partakes of the same general form as the original function. Hence the velocity and acceleration curves show patterns similar to that of displacement rather than being so different, as was markedly so with the constant acceleration-deceleration profile, and less so with others. This curve (catenoidal) also is shown in the figures; it is alleged to give even smoother performance and less vibration than the cycloidal type. Its slope (jerk) will be zero four times in each cycle.

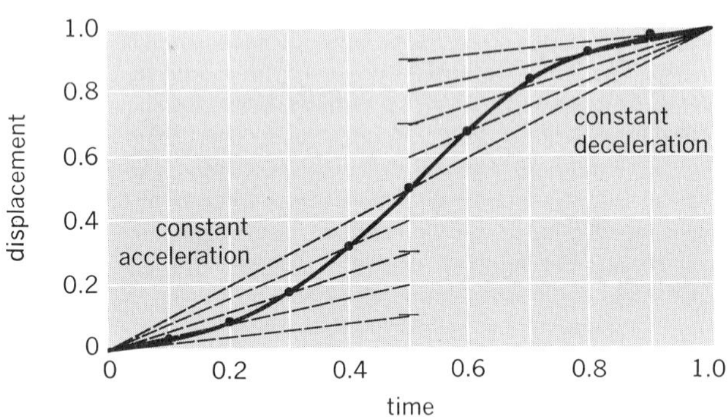

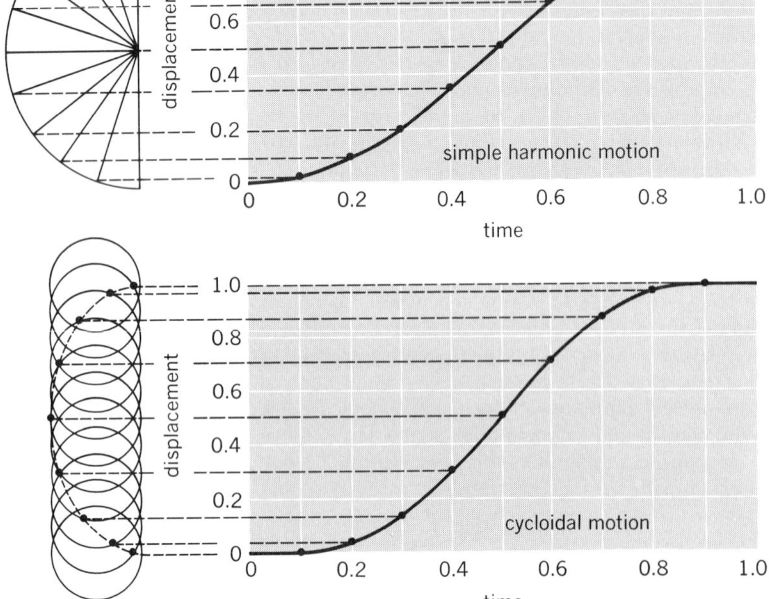

Fig. 7. Displacement-time diagrams for cam contours.

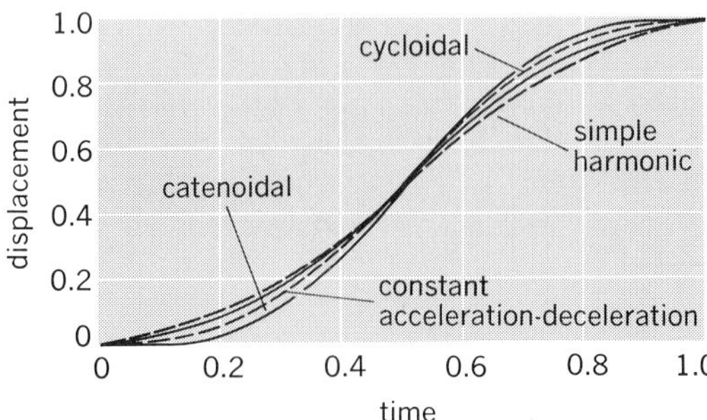

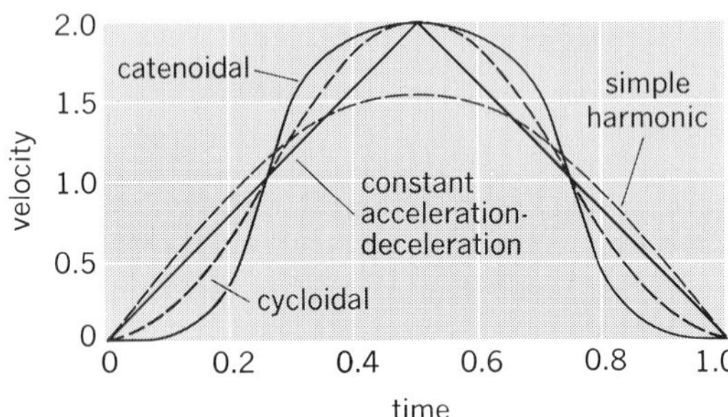

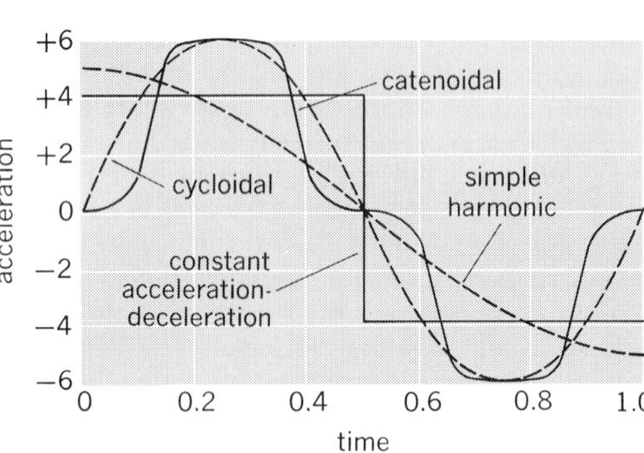

Fig. 8. Comparison of motions for three shapes of rise-return regions. Fourth motion, for catenoidal shape, also superposed. (*From G. L. Guillet and A. H. Church. Kinematics of Machines, 5th ed., copyright © 1950 by John Wiley and Sons. Inc.; used with permission*)

The manufacture of cam and follower from a master model is dwindling in favor of manufacture using a computer tape, which accurately directs the cutting of the surface. This procedure avoids the expensive manufacture of master cams.

In certain high-speed cam mechanisms, for example, an automotive engine valve gear, the elasticity and vibration characteristics of the follower linkage must be taken into account if faulty operation is to be avoided. Presented by W. M. Dudley in 1948 and elaborated by D. A. Stoddart in 1953, the polydyne method derives its name from employment of a polynomial displacement curve that suits the dynamic characteristics of the follower linkage.

Construction of cam profile. Empirically, a cam profile can be plotted to as large a scale as desired if the displacement curve and the configuration of the follower linkage are known. The method consists essentially in inverting the mechanism by fixing the cam and rotating the follower linkage about it, plotting only enough of the follower linkage to establish the successive positions of the follower face that will bear on the cam. Figure 9 shows the method of constructing the profile for a disk cam having a radial translating roller follower whose displacement curve is given. The cam profile is faired in, being at every point tangent to the follower roller. (Accuracy is improved by using a larger scale and by plotting additional positions.)

The method of constructing a cam profile for a flat face, offset, or oscillating follower is similar to that shown in the figure. It is important, however, that the location of the point in the follower linkage whose displacement is described by the displacement-time curve be kept constantly in view while the follower linkage is plotted in various successive positions.

Analytically, a profile can be calculated to any desired accuracy and the cam profile may be shown in tabular form, giving, for example, displacement of the follower for each degree of cam rotation. If a milling cutter or grinding wheel of the same size and shape as the follower is then used to cut the cam contour, the resulting contour will be true except for the small ridges that remain between given positions. These ridges can be removed by hand, using a file or a stone.

[DOUGLAS P. ADAMS]

Bibliography: R. G. Fenton, Reducing noise in cams, *Mach. Des.*, Apr. 14, 1966; R. G. Fenton, Determination of the minimum base radius for disk cams with reciprocating flat faced followers, *Auto. Eng.*, May 1967; F. J. Ogozalek, *Theory of Catenoidal-Pulse Motion and Its Application to High-Speed Cams*, ASME Publ. 66-Mech-45, 1966.

CAM MECHANISM

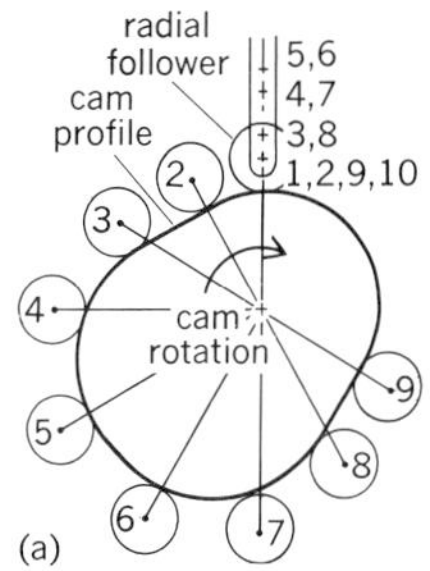

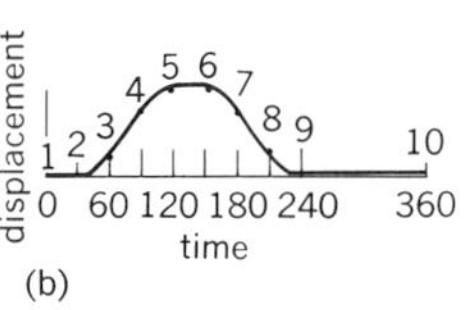

Fig. 9. Method of construction of (*a*) cam profile for a disk cam having a radial translating roller follower from (*b*) displacement curve.

Canal

An artificial open channel usually used to convey water or vessels from one point to another. Canals are generally classified according to use as irrigation, power, flood-control, drainage, or navigation canals or channels. All but the last type are regarded as water conveyance canals.

Canals may be lined or unlined. Linings may consist of plain or reinforced concrete, cement mortar, asphalt, brick, stone, buried synthetic membranes, or compacted earth materials. Linings serve to reduce water losses by seepage or percolation through pervious foundations or embankments and to lessen the cost of weed control. Concrete and other hard-surface linings also permit higher water velocities and, therefore, steeper gradients and smaller cross sections, which may reduce costs and the amount of right-of-way required.

Water conveyance canals. The character of material along the bottom and sides of a canal must be considered in determining whether lining is required. The velocity in an irrigation canal must be sufficiently low to avoid erosion of the canal banks and bottom, and it must be high enough to prevent deposition of silt: Velocities of 1.5–3.5 ft/sec are normally used for earth canals, and about 7 ft/sec is the usual maximum for concrete-lined irrigation canals. The velocity depends on canal slope, boundary roughness, and proportions of the cross-sectional area. Canal capacity Q in ft³/sec may be expressed by the formula $Q=AV$,

Fig. 1. A segment of the San Luis irrigation canal in California. A major waterway in the western United States, it is 103 mi long, 200 ft wide, and 36 ft deep. (*U.S. Bureau of Reclamation and State of California*)

where A is the cross-sectional area in square feet, and V is the average velocity in ft/sec.

Most irrigation canals are trapezoidal in cross section. Their side slopes are generally either 1.5:1 or 2:1 (horizontal to vertical) depending on the type of soil, but flatter slopes are used in unstable materials. Side slopes of concrete-lined irrigation canals are generally 1.25:1 or 1.5:1. They occasionally may be as high as 2:1, as for example in the San Luis Canal, shown in Fig. 1. Freeboard, or the vertical distance between maximum water surface and top of canal bank, generally ranges from 1 ft for small earth laterals to 4 ft for earth canals. Freeboard for concrete-lined canals ranges from 0.5 to 3.5 ft to the top of the lining, depending on capacity.

Irrigation canal intakes usually consist of a headworks and a river control structure with some form of intake gates to regulate the quantity of water admitted or stop the flow into the canal. Frequently, when the canal diverts from a stream that transports considerable sediment, the intake may be designed to keep out as much sediment as possible. This is accomplished by (1) locating the canal intake on the stream where the bed load is the smallest; (2) using weirs that divert water from the less silty upper layers of the flowing stream called skimming weirs; or (3) interposing basins in which much of the sand and silt are deposited before the water enters the canal. The deposited material is removed from these settling or desilting basins by occasional sluicing or dredging.

Irrigation canals usually require check structures in them to regulate the elevation of the water on the upstream side and wasteways as a safety device to carry off surplus water. The checks hold back the water in case of a break in the canal banks and retard the flow when the canal is being emptied to prevent a sudden evacuation and resulting uplift in the lining or sloughing of saturated banks. Hard-surfaced linings are frequently protected from uplift by automatic relief valves placed in the bottom and sides of the canal.

Although generally similar to irrigation canals in design, power, flood-control, and drainage canals or channels have special requirements. Power canals are placed on minimum grade to conserve head for power development; hence velocities are usually low. Rectangular wooden, concrete, or steel flumes with high ratio of depth to width are often used for this purpose. Because power canals are subject to sudden changes in flow, overflow wasteways are usually provided as well as ample freeboard.

Banks of flood-control channels constructed in earth are lined with grass or rock riprap for protection against erosion by flood flows. Grass protection is used for velocities from about 3 to 8 ft/sec, and riprap for velocities of 8 to 18 ft/sec. Either rectangular or trapezoidal concrete-lined channels are used for velocities exceeding 18 ft/sec, although in some cases concrete lining is used for lower velocities to reduce the size of channel.

Drainage canals are deeply excavated to facilitate the drainage of surrounding land. They usually have a minimum grade and small depths of water flow.

Fig. 2. The Walter F. George Lock, Dam and Powerhouse on the Chattahoochee River, Alabama-Georgia boundary. The lock chamber can be seen at right center of the photograph. (*U.S. Army Corps of Engineers*)

Navigation canals and canalized rivers. Navigation canals are artificial inland waterways for boats, barges, or ships. A canalized river is one that has been made navigable by construction of one or more weirs or overflow dams (Fig. 2) to impound river flow, thereby providing navigable depths. Locks may be built in navigation canals and canalized rivers to enable vessels to move to higher or lower water levels.

Navigation canals are often built along portions of canalized rivers or located so as to connect two such rivers. They are adapted to the topography by a series of level reaches connected by locks. Sea-level navigation canals, connecting two tidal bodies of water, are excavated sufficiently deep to preclude the need for locks, if the tidal flow permits.

The dimensions of a navigation canal are determined primarily by the size, and to some extent by the speed, of the vessels that are to use it. Depth must be sufficient to assure bottom clearance under all operating conditions, and ordinarily the width allows vessels to pass each other safely. The canal cross section is usually trapezoidal, with side slopes ranging from 1.5:1 to 3:1 or flatter, depending on the stability of the bank material. Sections cut in rock may have vertical or near-vertical sides. Earth banks of some navigation canals are protected against erosion from wave action by placing rock riprap or similar protection near the water surface.

Locks. A lock (Fig. 2) is a chamber equipped with gates at both upstream and downstream ends. Water impounded in the chamber is used to raise or lower a vessel from one elevation to another. The lock chamber is filled and emptied by means of filling and emptying valves and a culvert system usually located in the walls and bottom of the lock.

After a vessel enters the lock chamber, the afterward gates are closed and the water level is lowered (if the vessel is headed downstream) by operation of the emptying valve, or raised (if the vessel is headed upstream) by operation of the filling valve. When the water level in the chamber reaches the water level forward of the vessel, the forward set of gates is opened and the vessel leaves the lock.

Maximum lift of a lock is the vertical distance from the normal pool upstream of the lock to the low-water surface downstream of the lock. Low lifts simplify design problems, but generally in developing a major waterway it will be more economical to use fewer higher-lift locks. Lock lifts vary from a few feet in tidal canals to over 100 ft in major rivers, such as the Columbia and Snake. Lock widths vary from 56 to 110 ft and usable lengths vary from 400 to 1200 ft, except that smaller locks are used when all the traffic consists of small craft.

On some canalized rivers having low-lift locks, dams contain movable sections which are lowered during periods of moderate or high flow to permit the unobstructed passage of vessels and barges over the dam. The locks are used in these rivers only during periods of low flow when the movable sections of the dams are closed to create sufficient depth of water for navigation. *See* TRANSPORTATION ENGINEERING; WATER SUPPLY ENGINEERING.

[CORPS OF ENGINEERS; BUREAU OF RECLAMATION]

Bibliography: V. T. Chow, *Open Channel Hydraulics*, 1959; H. W. King and E. F. Brater, *Handbook of Hydraulics*, 1963; J. K. Vennard, *Elementary Fluid Mechanics*, 1961; S. M. Woodward and C. J. Posey, *Hydraulics of Steady Flow in Open Channels*, 1941.

Cantilever

A beam supported at one end and supporting a load along its length or at its free end, the upper portion of the beam, if horizontal, being everywhere in tension and the lower portion being everywhere in compression. Familiar examples are the symmetrical paired cantilevers of a seesaw or teeterboard, the beam in a chemical balance, and the unsymmetrical cantilever in the overhang of a roof. The longest cantilever structure is the railroad bridge at Quebec; opened in 1918 after redesign, it spans 1800 ft. Many drawbridges are basically cantilevers. The balcony in a theater may be designed as a cantilever surface. Hammerhead and similar type cranes are cantilevers (see illustration). For any of these structures to be stable, the beam must sustain at any cross section both the tensions and compressions and the couple between them.

Cantilever construction permits spanning wide distances without obstructing supports. Airport hangars of cantilever design provide wide unobstructed entry for planes through open sides. The overhead cantilever roof may support sliding hanging panels or doors by which the building can be closed. In office buildings floor beams may cantilever outward beyond supporting columns to support a lightweight weather wall. Horizontal mementire building may be designed in similar fashion as a network of cantilevers with elevators and service ducts in the central supporting column.

A flagpole is also a cantilever, being supported at its base and being deflected by wind loading along its length. From the same viewpoint, a tall building is often analyzed as a vertical cantilever supported at its foundation against wind loading on any side. An airplane wing is an example of a horizontal cantilever which bears compression loads on its upper surface and tension loads on its lower surface. *See* BRIDGE; HOISTING MACHINES; ROOF CONSTRUCTION.

[FRANK H. ROCKETT]

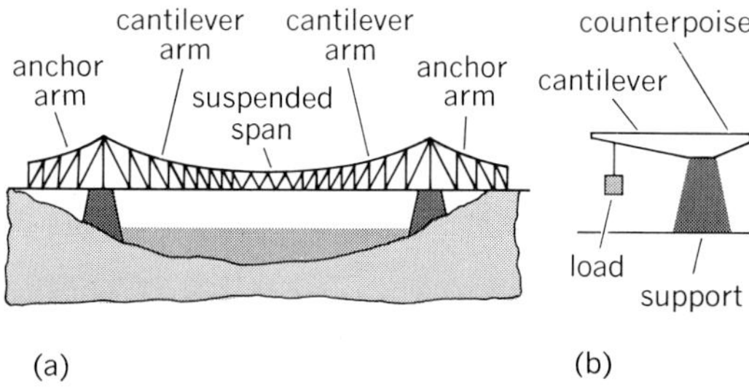

Two different applications of the cantilever. (*a*) Bridge. (*b*) Hammerhead crane.

Carburetor

The device that controls the power output and fuel feed of internal combustion spark-ignition engines generally used for automotive, aircraft, and auxiliary services. Its duties include control of the engine power by the air throttle; metering, delivery, and mixing of fuel in the airstream; and graduating the fuel-air ratio according to engine requirements in starting, idling, and load and altitude changes.

Engine air charge. A simple updraft carburetor illustrates basic carburetor action (Fig. 1). Intake air charge, at full or reduced atmospheric pressure

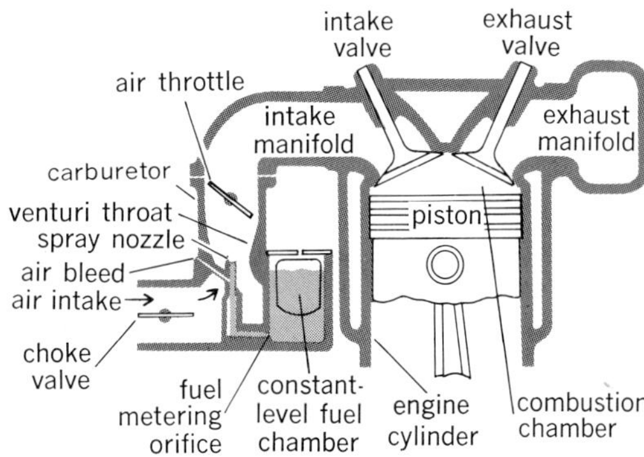

Fig. 1. Elements which basically determine air and fuel charges received by engine through carburetor.

as controlled by the throttle, is drawn into the cylinder by the downward motion of the piston to mix with the unscavenged exhaust remaining in the combustion chamber from the previous explosion. The total air-charge weight per unit time is approximately proportional to the square root of its pressure drop in the carburetor venturi throat (with a square root correction, direct for change in air pressure, and inverse for change in air absolute temperature). Also, each individual cylinder air-charge volume follows the intake pressure minus about one-sixth the exhaust pressure, so that the intake manifold pressure and temperature are often taken as an approximate indication of the engine torque; this factor times the engine speed is similarly taken as a measure of the engine power output. Any given fixed part-throttle opening gives higher intake manifold pressure, increased individual-cylinder air charge, and greater engine

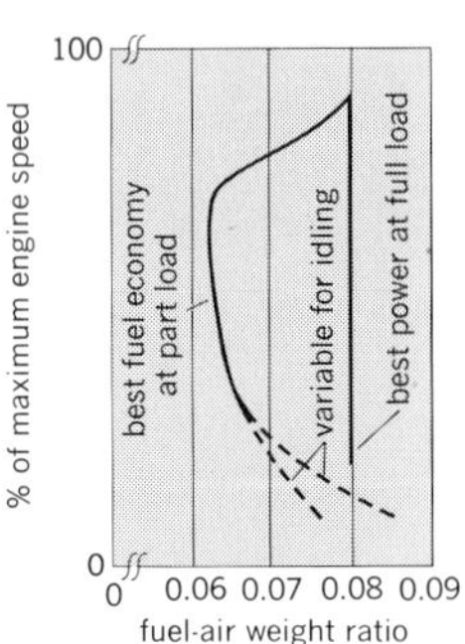

Fig. 2. Fuel-air ratios required for engine operation.

torque, as the engine speed is decreased.

A cylinder is most completely filled with the fuel-air mixture when no other cylinder is drawing in through the same intake at the same time, so that, for high speed, best power is usually obtained with multiple carburetor and intake-duct units. However, in practice it is more difficult to throttle all the cylinders equally when the multiple units are used.

Fuel charge. The fuel is usually metered through a calibrated metering orifice at a differential pressure derived from the pressure drop in a venturi throat in the intake air passage. Since this fuel flow also follows the square root of its differential pressure, the fuel and air rates tend to be accurately parallel, particularly if a small bleed of air is fed into the fuel delivery or spray nozzle to overcome the retarding effect of surface tension.

Fuel-air requirements. For normal steady speed, the limits of consistent ignition lie between 0.060 and 0.083 fuel-air by weight (Fig. 2). Richer mixtures (more fuel) give higher power, apparently cooler combustion, and less detonation. The weaker mixtures give better fuel economy.

A narrowly controlled rich mixture is required for idling because low charge densities and relatively high contamination with residual exhaust make combustion conditions unfavorable. Individual adjustments are required both for the idling fuel feed and for the throttle closure stop to keep the engine firing at minimum speed (Fig. 3). Slight intake or exhaust valve leaks can make smooth idling impossible.

Cold-engine operation. Motor fuels are a mixture of components varying in volatility, usually selected so that they evaporate completely in the intake manifold at ordinary operating temperatures at the reduced pressures of part throttle, but not at full throttle. There is a tendency toward vapor saturation at full power and low engine speed so that richer fuel-air ratios are required, especially with a cold engine. For starting at low temperatures and for the first few ignitions with cylinder walls cold, a great increase in fuel feed is momentarily required since only the most volatile part of the fuel can vaporize and burn.

The higher metering suction necessary to give the excess fuel feed required for cold starting is usually supplied by a choke valve, which may be manually or automatically operated. Partial closure of this valve also yields the moderate enrichment needed during the warm-up period. While it is usually impossible to overchoke a cold engine, an additional connection is generally supplied on passenger car carburetors so that pressing the accelerator pedal all the way down will open the choke valve and permit return to normal fuel feed.

Automobile carburetion. An automobile engine performs through a wider range of loads and speeds than most prime movers. Furthermore, its transient response to changes of load is highly important. For these reasons, the functions of the automobile carburetor are complex. Figure 3 is a section through one barrel of a typical downdraft carburetor with the throttle closed to the idling position. The multiple main and boost venturi structure is used to increase the fuel metering force for a given air delivery; also, it yields better metering regulation at the higher air velocities.

Idling system. To obtain adequate metering forces at low air speeds and small throttle openings, the fuel, after passing through the main metering orifice, is bypassed to an idle metering orifice, then to an idling delivery orifice located in the high suction at the edge of the throttle valve. Some graduation of fuel feed from low to high idling speed is obtained as the throttle valve edge passes across the idling delivery holes.

Because of increased piston and other friction with a cold engine, a greater throttle opening as well as more fuel is required for idle at that time, and it is common to interconnect the choke valve so that, in its partly closed or cold-engine position, the minimum-throttle area is increased.

Power enrichment. The richer fuel feeds necessary for best power are customarily obtained by varying the area of the main metering orifice, preferably by a valve responsive to intake manifold pressure beyond the throttle, as shown on the right side of Fig. 4. At light loads the lowered manifold pressure draws the enrichment valve down, but a rise in manifold pressure resulting either from a drop in engine speed or added throttle opening will allow the plunger spring to open the metering orifice further. An approximation of this function is obtained in some designs by a direct mechanical connection between the enrichment valve and the throttle.

Transitional requirements. Proper regulation under change of throttle is a major problem in obtaining what is commonly called good carburetor action. Any lag in evaporation of the fuel when the throttle is opened tends to give a momentary delay

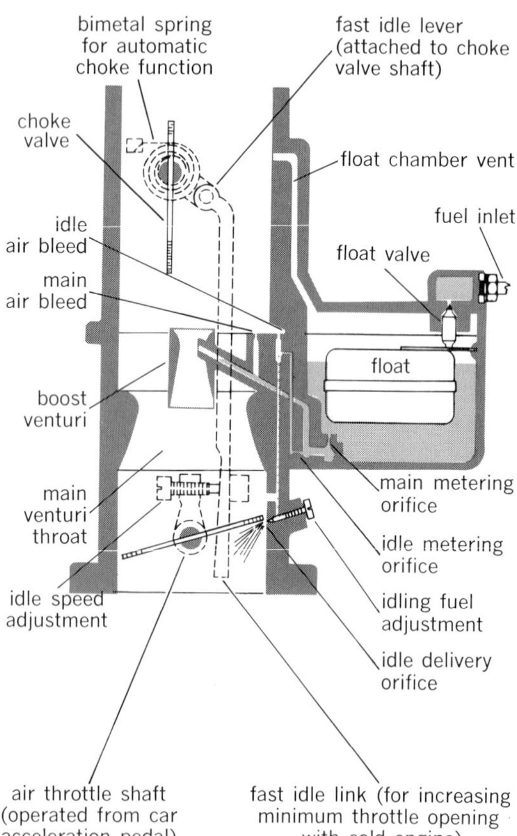

Fig. 3. Idling fuel circuit and choke device for starting and warm-up in typical downdraft carburetor.

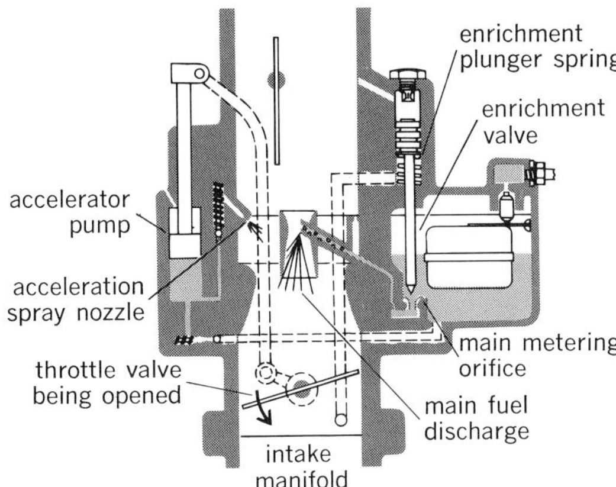

Fig. 4. Special acceleration and power enrichment devices used on the automobile carburetor.

in power response, which is usually compensated by an accelerating pump, shown on the left side of Fig. 4. The pump gives a quick squirt of fuel as the throttle is opened. The apparent willingness of the engine to respond to the accelerator pedal is largely determined by how accurately the accelerating pump discharge is proportioned to the existing engine temperature and to the volatility of the fuel.

A converse transitional problem is associated with sudden closing of the throttle, when wet fuel is present on the intake manifold walls. The sudden drop in pressure causes the liquid to flash into vapor, resulting in a temporarily overrich condition in the cylinder, with subsequent misfiring and incompletely burned combustion products in the exhaust. Better manifold heating helps, as does injection of fuel at cylinder intake ports.

Aircraft carburetion. Early aviation carburetors followed automobile carburetor practice, but in the 1940s the pressure or injection type came into use. These incorporated the following advantages: (1) The whole metering system is kept under pressure, to prevent formation of bubbles and disturbances to the metering under the reduced pressures at altitude. (2) The fuel spray can be delivered into a heated part of the intake system so that ice formation resulting from fuel evaporation becomes impossible. (3) Correction of the metering function under changes of altitude and temperature is automatically provided. (4) The closed system permits a design that avoids disturbance of the fuel metering, or leakage from vents, during vigorous maneuvers of the airplane.

In the most commonly used form, the metering function of the automobile carburetor is retained in that the fuel metering differential, between chambers D and C of Fig. 5, is derived from the multiple venturi air differential A−B by the system of opposed synthetic-rubber diaphragms as illustrated. The idling spring adds a small positive increment to the A−D force to provide the required rich idle fuel flow.

The fuel pressure in chamber C is held to a desired value above the surrounding atmosphere by the fuel-pressure valve and its regulating diaphragm and spring.

Beyond the pressure valve the fuel may be led to a spinner ring which has multiple discharge orifices. The ring is mounted on, and rotates with, the supercharger shaft. On other common installations the fuel, after metering as here illustrated, is taken to an injection pump which divides the charge equally and delivers a portion at much higher pressure to the interior of each engine cylinder on its intake stroke.

Pressure from the engine-driven fuel pump, as regulated by its bypass valve, must be adequate to maintain flow through the system at the highest powers, but beyond this it does not affect the fuel metering.

Fuel-air ratio, as desired for light or full load or as selected by the pilot, is controlled by one or more metering valves which collectively determine the flow area between pressures D and C. A variety of combinations have been used, but all are directed toward the regulation illustrated in Fig. 2.

Correction of the fuel-air ratio at increased altitude or temperature is accomplished by bringing the pressure in chamber A down nearer to that in chamber B, thus decreasing both the A−B and the D−C differentials. A small fixed-size depression vent V is provided between chamber A and the venturi communication channel, while a large pressure vent between the main air intake and chamber A is varied by the aneroid valve. This valve is connected to the aneroid capsule, a metallic bellows filled with inert gas and extended by a spring. As the airplane gains altitude, or as the temperature of the entering air rises, the capsule extends to reduce the area of the entering air vent to chamber A, and thereby reduces the A−B differential at a rate determined by the profile and adjustment of the aneroid valve.

Deviations from the form of Fig. 4 have been used on smaller American aircraft engines. Occasionally in both England and the United States, similar construction has been used to regulate fuel

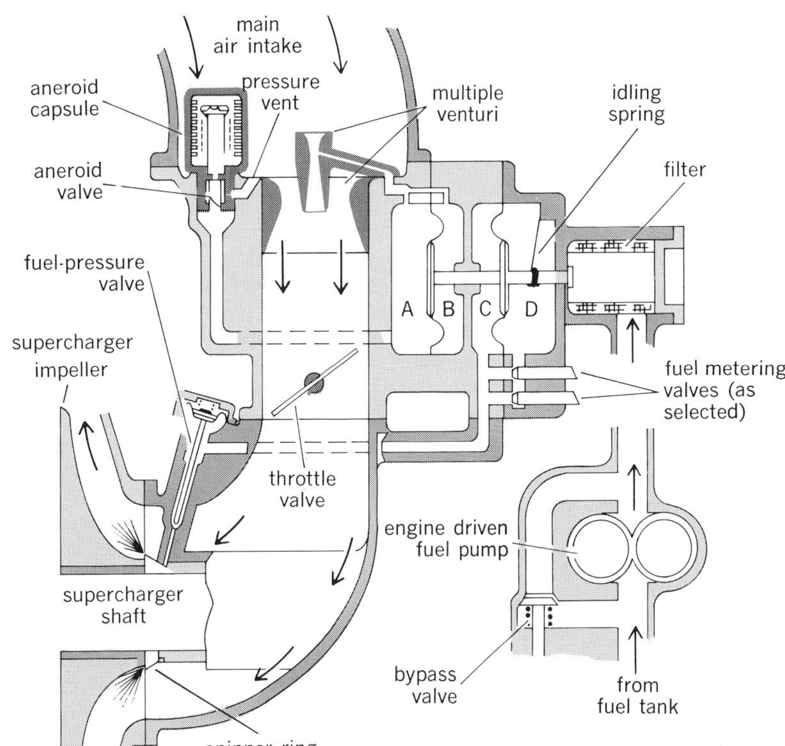

Fig. 5. Schematic view of an aircraft injection-type or pressure-type carburetor.

feed by the parameters of engine speed and intake manifold pressure; this is commonly called speed-density metering.

Carburetor icing. There are two reasons for condensation of atmospheric moisture and formation of ice in the intake system. First, at partly closed throttle, the adiabatic pressure drop across the throttle orifice generates a temperature drop. When atmospheric humidity is high, ice sometimes partly clogs the throttle orifice, requiring further throttle opening to keep the engine running for a short period after starting and before the engine heat has had time to build up. With carburetors having a fuel discharge at the throttle edge (Fig. 3), the addition of water-soluble antifreeze to the fuel may reduce such icing tendency.

The other source of temperature drop in the intake system is the evaporation of fuel. Ice tends to form wherever fuel spray impinges upon an unheated surface. Such icing was formerly encountered frequently in aircraft operation and was dealt with by heating the intake air, which reduced the engine power and sometimes caused detonation. Later the problem was largely eliminated either by use of the injection carburetor, which delivers fuel only in the warm parts of the intake system, or by direct injection into the engine cylinders.

[F. C. MOCK/J. A. BOLT]

Bibliography: A. W. Judge, *Motor Series*, vol. 2: *Carburetors and Fuel Systems*, 1963; W. B. Larew, *Carburetors and Carburetion*, 1967.

Carnot cycle

A hypothetical thermodynamic cycle originated by Sadi Carnot and used as a standard of comparison for actual cycles. The Carnot cycle shows that, even under ideal conditions, a heat engine cannot convert all the heat energy supplied to it into mechanical energy; some of the heat energy must be rejected. In a Carnot cycle, an engine accepts heat energy from a high-temperature source, or hot body, converts part of the received energy into mechanical (or electrical) work, and rejects the remainder to a low-temperature sink, or cold body. The greater the temperature difference between the source and sink, the greater the efficiency of the heat engine.

The Carnot cycle (Fig. 1) consists first of an isentropic compression, then an isothermal heat addition, followed by an isothermal expansion, and concludes with an isentropic heat rejection process. In short, the processes are compression, addition of heat, expansion, and rejection of heat, all in a qualified and definite manner.

Processes. The air-standard engine, in which air alone constitutes the working medium, illustrates the Carnot cycle. A cylinder of air has perfectly insulated walls and a frictionless piston. The top of the cylinder, called the cylinder head, can either be covered with a thermal insulator, or, if the insulation is removed, can serve as a heat transfer surface for heating or cooling the cylinder contents.

Initially, the piston is somewhere between the top and the bottom of the engine's stroke, and the air is at some corresponding intermediate pressure but at low temperature. Insulation covers the cylinder head. By employing mechanical work from the surroundings, the system undergoes a reversible adiabatic, or an isentropic, compression. With no heat transfer, this compression process raises both the pressure and the temperature of the air, and is shown as the path *a-b* on Fig. 1.

After the isentropic compression carries the piston to the top of its stroke, the piston is ready to reverse its direction and start down. The second process is one of constant-temperature heat addition. The insulation is removed from the cylinder head, and a heat source, or hot body, applied that is so large that any heat flow from it will not affect its temperature. The hot body is at a temperature just barely higher than that of the gas it is to heat. The temperature gradient is so small it is considered reversible; that is, if the temperature changed slightly the heat might flow in the other direction, from the gas into the hot body. In the heat addition process, enough heat flows from the hot body into the gas to maintain the temperature of the gas while it slowly expands and does useful work on the surroundings. All the heat is added to the working substance at this constant top temperature of the cycle. This second process is shown as *b-c* on Fig. 1.

Part way down the cylinder, the piston is stopped; the hot body is removed from the cylinder head, and an insulating cover is put in its place. Then the third process of the cycle begins; it is a frictionless expansion, devoid of heat transfer, and carries the piston to the bottom of its travel. This isentropic expansion reduces both the pressure

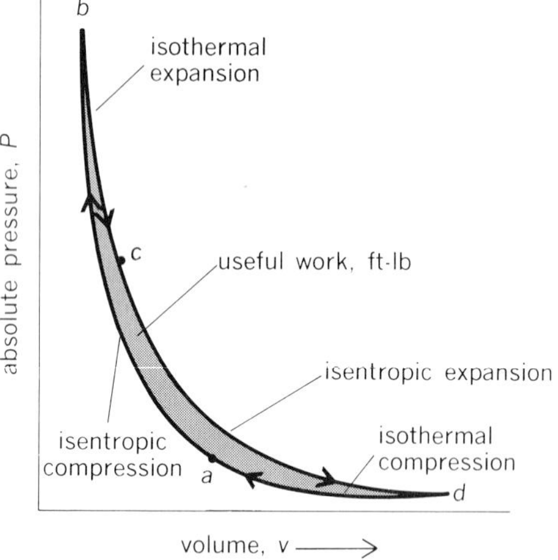

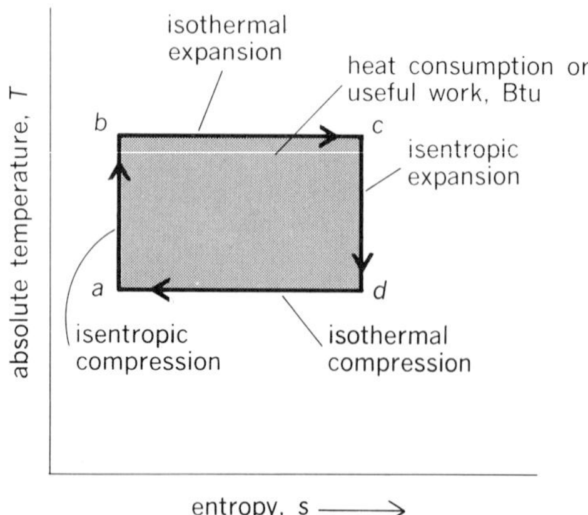

Fig. 1. Carnot cycle for air. (*a*) Pressure-volume diagram. (*b*) Temperature-entropy diagram.

and the temperature to the bottom values of the cycle. For comparable piston motion, this isentropic expansion drops the pressure to a greater extent than the isothermal process would do. The path *c-d* on Fig. 1 represents this third process, and is steeper on the *P-v* plane than process *b-c*.

The last process is the return of the piston to the same position in the cylinder as at the start. This last process is an isothermal compression and simultaneous rejection of heat to a cold body which has replaced the insulation on the cylinder head. Again, the cold body is so large that heat flow to it does not change its temperature, and its temperature is only infinitesimally lower than that of the gas in the system. Thus, the heat rejected during the cycle flows from the system at a constant low-temperature level. The path *d-a* on Fig. 1 shows this last process.

The net effect of the cycle is that heat is added at a constant high temperature, somewhat less heat is rejected at a constant low temperature, and the algebraic sum of these heat quantities is equal to the work done by the cycle.

Figure 1 shows a Carnot cycle when air is used as the working substance. The *P-v* diagram for this cycle changes somewhat when a vapor or a liquid is used, or when a phase change occurs during the cycle, but the *T-s* diagram always remains a rectangle regardless of phase changes or of working substances employed.

It is significant that this cycle is always a rectangle on the *T-s* plane, independent of substances used, for Carnot was thus able to show that neither pressure, volume, nor any other factor except temperature could affect the thermal efficiency of his cycle. Raising the hot-body temperature raises the upper boundary of the rectangular figure, increases the area, and thereby increases the work done and the efficiency, because this area represents the net work output of the cycle. Similarly, lowering the cold-body temperature increases the area, the work done, and the efficiency. In practice, nature establishes the temperature of the coldest body available, such as the temperature of ambient air or river water, and the bottom line of the rectangle cannot circumvent this natural limit.

The thermal efficiency of the Carnot cycle is solely a function of the temperature at which heat is added (phase *b-c*) and the temperature at which heat is rejected (phase *d-a*) (Fig. 1). The rectangular area of the *T-s* diagram represents the work done in the cycle so that thermal efficiency, which is the ratio of work done to the heat added, equals $(T_{hot} - T_{cold})/T_{hot}$. For the case of atmospheric temperature for the heat sink ($T_{cold} = 500°R$), the thermal efficiency, as a function of the temperature of the heat source, T_{hot}, is shown in Fig. 2.

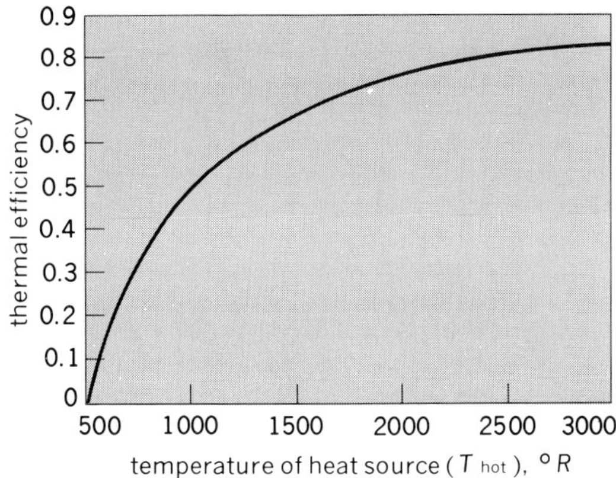

Fig. 2. Thermal efficiency of the Carnot cycle with heat-rejection temperature T_{cold} equal to 500°R.

Carnot cycle with steam. If steam is used in a Carnot cycle, it can be handled by the following flow arrangement. Let saturated dry steam at 500°F flow to the throttle of a perfect turbine where it expands isentropically down to a pressure corresponding to a saturation temperature of the cold body. The exhausted steam from the turbine, which is no longer dry, but contains several percent moisture, is led to a heat exchanger called a condenser. In this device there is a constant-pressure, constant-temperature, heat-rejection process during which more of the steam with a particular predetermined amount of condensed liquid is then handled by an ideal compression device. The isentropically compressed mixture may emerge from the compressor as completely saturated liquid at the saturation pressure corresponding to the hot-body temperature. The cycle is closed by the hot body's evaporating the liquid to dry saturated vapor ready to flow to the turbine.

The cycle is depicted in Fig. 3 by *c-d* as the isentropic expansion in the turbine; *d-a* is the constant-temperature condensation and heat rejection to the cold body; *a-b* is the isentropic compression; and *b-c* is the constant-temperature boiling by heat transferred from the hot body. *See* STEAM ENGINE.

Carnot cycle with radiant energy. Because a Carnot cycle can be carried out with any arbitrary system, it has been analyzed when the working substance was considered to be a batch of radiant energy in an evacuated cylinder. If the system boundaries are perfectly reflecting thermal insulators, the enclosed radiant energy will be reflected and re-reflected with no loss of radiant energy and no change of wavelength.

The electromagnetic theory of radiation asserts that the radiant energy applies pressure P to the cylinder walls. This radiation pressure is equal to $u/3$, where u represents the radiant energy density, or the amount of radiant energy per unit volume.

The piston moves so that the cylinder boundaries expand, and additional radiant energy is supplied to the system so that the temperature remains constant. Then, cutting off any further supply of radiant energy, the system undergoes a further infinitesimal expansion with its associated pressure drop and temperature drop. The third process is an isothermal compression at the low-temperature level, accompanied by some rejection of radiant energy. One last process closes the cycle with a reversible compression that raises the temperature to the original level. The assumption is made throughout this analysis that energy density is a function of temperature alone. Thus, because this last compression process increases the energy density, it raises the temperature.

A record of all energy quantities in this radiant-energy Carnot cycle indicates that energy density u is proportional to the fourth power of the absolute temperature. Consequently the total rate of emission of radiant energy from the surface of a blackbody is also proportional to the fourth power of its absolute temperature, thereby using a Carnot

CARNOT CYCLE

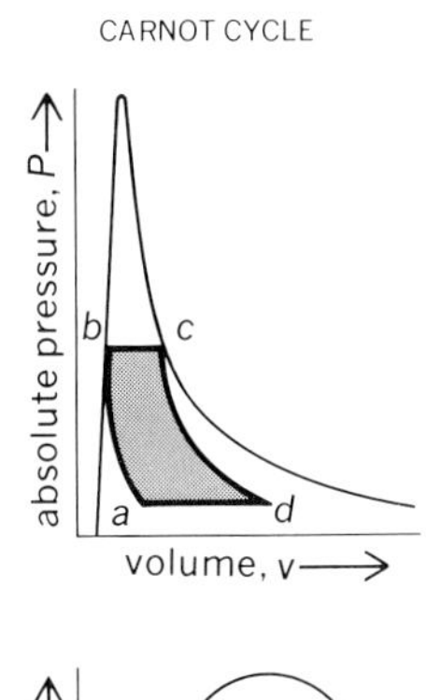

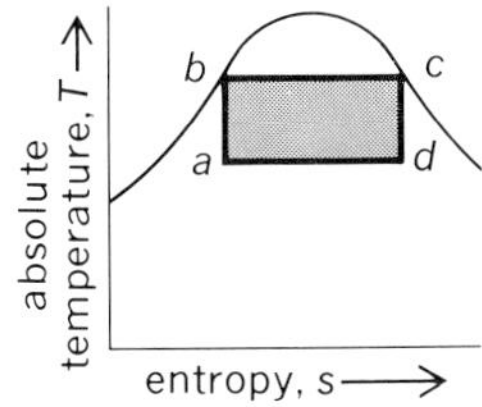

Fig. 3. Carnot cycle with steam. (*a*) Pressure-volume diagram. (*b*) Temperature-entropy diagram.

cycle to provide a relationship by theoretical analysis, the same relationship which had previously been determined experimentally and labeled as Stefan's law.

Conversion of heat to electricity. Several physical phenomena convert heat energy directly into electrical energy. The extent of this direct conversion of heat to electricity is limited by the temperature levels between which the process operates. The ideal efficiency of such direct-conversion thermoelectric cycles equals the efficiency of a Carnot cycle that operates between the same temperature limits of heat source and heat sink. However, the conversion efficiency obtained in practice is only a small fraction of the ideal efficiency at the present stage of development.

The most widely known physical arrangement for direct thermoelectric generation is the thermocouple, which produces an electromotive force, or voltage, when one junction of two dissimilar conductors is heated while the other junction is kept cool. Thermocouples made of metals are inefficient converters of heat energy to electric energy, because metals that have good electrical conductivity unfortunately have equally good thermal conductivity, which permits heat loss by conduction from the hot to the cold junction. In contrast, thermocouples made of semiconductors offer the prospect of operation at high temperatures and with high temperature gradients, because semiconductors are good electrical conductors but poor heat conductors. Semiconductor thermocouples may have as large a junction potential as the metal couples do.

Thermionic emission is another phenomenon that permits the partial conversion of heat energy directly into electrical energy. Externally applied heat imparts kinetic energy to electrons, liberating them from a surface. The density of the emission current is a function of the absolute temperature and work function of the emitter material.

For many years thermionic emission received little attention as a source of power because of its very low conversion efficiency. However, interest has been stimulated by development of a contact potential thermionic emission cell. In this device, current flows between the surfaces of two materials which have different work functions. These materials are held at different temperatures, and the gap between the electrode surfaces is filled with gas at low pressure.

Such direct-conversion techniques show promise of becoming small-scale, if unconventional power sources. Although these techniques ideally can convert heat to electrical energy with the efficiency of a Carnot cycle operating between the same temperature levels, laboratory devices do not surpass about 8% efficiency, which is just a fraction of the ideal performance.

Reversed Carnot cycle. A Carnot cycle consists entirely of reversible processes; thus it can theoretically operate to withdraw heat from a cold body and to discharge that heat to a hot body. To do so, the cycle requires work input from its surroundings. The heat equivalent of this work input is also discharged to the hot body.

Just as the Carnot cycle provides the highest efficiency for a power cycle operating between two fixed temperatures, so does the reversed Carnot cycle provide the best coefficient of performance for a device pumping heat from a low temperature to a higher one. This coefficient of performance is defined as the ratio of the quantity of heat pumped to the amount of work required, or it equals $T_{hot}/(T_{hot}-T_{cold})$ for a warming machine, and $T_{cold}/(T_{hot}-T_{cold})$ for a cooling machine, where all temperatures are in degrees absolute. This is one of the few engineering indices with numerical values greater than unity. *See* HEAT PUMP; REFRIGERATION CYCLE.

Practical limitations. Good as the ideal Carnot cycle may be, there are serious difficulties that emerge when one wishes to make an actual Carnot engine. The method of heat transfer through the cylinder head either limits the operation of the engine to very low speeds, or requires an engine with a huge bore and small stroke. Moreover, the material of the cylinder head is subjected to the full top temperature of the cycle, imposing a metallurgical upper limit on the cycle's temperature.

A practical solution to the heat transfer difficulties which beset the Carnot cycle is to burn a fuel in the air inside the engine cylinder. The result is an internal combustion engine that consumes and replaces its working substance while undergoing a periodic sequence of processes.

The working substance in such an internal combustion engine can attain very high temperatures, far above the melting point of the metal of the cylinder walls, because succeeding lower-temperature processes will keep the metal parts adequately cool. Thus, as the contents change temperature rapidly between wide extremes, the metal walls hover near a median temperature. The fuel-air mixture can be ignited by a spark, or by the rise in temperature produced by the compression stroke. *See* DIESEL CYCLE; DIESEL ENGINE; INTERNAL COMBUSTION ENGINE; OTTO CYCLE.

Even so, the necessarily high peak pressures and temperatures limit the practical thermal efficiency that an actual engine can achieve. The same theoretical efficiency can be obtained from a cycle consisting of two isobaric processes interspersed by two isentropic processes. The isobaric process requires that the cycle handle large volumes of low-pressure gas, which can best be done in a rotating turbine. *See* BRAYTON CYCLE; GAS TURBINE.

Although the Carnot cycle is independent of the working substance, and hence is applicable to a vapor cycle, the difficulty of efficiently compressing a vapor-liquid mixture renders the cycle impractical. In a steam power plant the sequence of states assumed by the working substance is (1) condensate and feedwater are compressed and pumped into the boiler, (2) heat is added to the water first at constantly increasing temperature and then at constant pressure and temperature, (3) the steam expands in the engine, and (4) the cycle is completed by a constant-temperature heat-rejection process which condenses the exhausted steam. *See* VAPOR CYCLE.

By comparison to the Carnot cycle, in which heat is added only at the highest temperature, this steam cycle with heat added over a range of temperatures is necessarily less efficient than is theoretically possible. An analysis of engine operations in terms of thermodynamic cycles indicates what efficiencies can be expected and how the operations should be modified to increase engine performance, such as high compression ratios for reciprocating internal combustion engines and high steam temperatures for steam engines. *See* POWER PLANT; THERMODYNAMIC CYCLE.

[THEODORE BAUMEISTER]

Bibliography: T. Baumeister, *Standard Handbook for Mechanical Engineers*, 1967; J. B. Jones and G. A. Hawkins, *Engineering Thermodynamics*, 1960; H. C. Weber and H. P. Meissner, *Thermodynamics for Chemical Engineers*, 2d ed., 1957; M. W. Zemansky, *Heat and Thermodynamics*, 4th ed., 1957.

Cast iron

A generic term describing a family of iron alloys containing 1.8–4.5% carbon. Cast iron usually is made into specified shapes, called castings, for direct use or for processing by machining, heat treating, or assembly. In special cases it may be forged or rolled moderately. Generally, it is unsuitable for drawing into rods or wire, although to a limited extent it has been continuously cast into rods and shapes from a liquid bath or swaged from bars into smaller-dimensional units. Silicon usually is present in amounts up to 3%, but special compositions are made containing up to 6% (Silal) and up to 12% (Duriron).

Cast iron of the above composition range is often made into blocks or rough shapes and called pig iron. It is an intermediate form of cast iron used for remelting into iron castings. Pig iron usually is produced from iron ore and coke smelted into liquid iron in the blast furnace. In some cases this liquid iron is poured into sand molds of specific design which, when cleaned and trimmed, become ingot molds for steel manufacture. In these instances the cast iron may be referred to as direct metal, indicating its production in the blast furnace without the intermediate remelting in the cupola furnace of the foundry, which is common practice in the manufacture of iron castings.

Classification. Cast iron may be purchased in several commercial grades called gray iron, chilled iron, mottled iron, white iron, malleable iron, ductile iron, spheroidal graphite iron, nodular iron, and austenitic cast iron.

Originally, cast iron was graded by examination of fracture characteristics of the surfaces produced when a sample was broken; its true composition was unknown. This method of grading first was applied to pig iron, which was classified similarly as white, mottled, or gray. The color of the fracture governed the classification of the product until about 1888, when extensive work by W. J. Keep revealed that the silicon content in a large measure determined the character of the fracture. Further work clarified the relationships between composition, microstructure, and fracture characteristics, and gradually the importance of the examination of fractures receded until the present time, when it is used only to classify the products generally. The chemistry of cast iron today is quite well known, and gray, white, and mottled iron can be selected on the basis of composition charts which mark the areas of their occurrence.

Further developments in cast irons have adapted heat-treatment processes to produce special properties that permit classification of special grades of gray, malleable, and white cast irons.

More recently metals such as cerium and magnesium have been used to produce the spheroidal graphite structures which are characteristic of ductile irons.

Properties and uses. Gray iron is an iron-base material within the broad composition limits mentioned above. Of its total carbon content, more than one-half is in the form of graphite flakes. It is the presence of the graphite flakes which produce a gray color in a fresh fracture. The remaining carbon is dissolved in the metal and is called combined carbon. It may be present in amounts ranging from 0.05 to 1.20%. Gray iron is the mainstay of iron foundry production. Its application covers practically all engineering fields involving construction, machinery, engines, power, transport, mining, and many daily uses, such as stove parts, pots and pans, and hardware.

Mottled iron solidifies as a mixture of white and gray irons. The fracture is a mottled mixture of white and gray areas. The fracture may contain as few as 10% white areas in a gray background or the reverse, namely, 10% gray areas in a white background. Mottled irons are useful for abrasion- or erosion-resisting services, where their composite structure provides a hard surface with some ability to deform. The high frictional resistance of its surface is occasionally utilized.

Chilled iron is gradually disappearing as a descriptive term. It was applied to cast iron which was poured into a metal mold or chill. Frequently, a gray iron would be processed in this manner, and if the cooling rate was rapid enough, a white-iron casting, or a surface zone of white iron on the casting would develop, backed by the less rapidly cooled gray iron inside. Since glass molds, for example, are often made by casting one of their surfaces against a chill, such castings would be called chilled iron. Unfortunately, the term spread to the part-white, part-gray types of cast iron and more inaccurately to a solid white iron which had been cast in a chill.

White iron is an iron-base material within the broad composition limits mentioned above whose total carbon content is in the combined form. It therefore contains little or no graphite. A fresh fracture is white (ground-glass white). Its characteristic structure usually is obtained by lowering the carbon content or the silicon content or both, so that the graphitizing power of the composition is subdued.

Malleable cast iron is made as a white-iron casting within the narrower composition range that the term implies. Annealing produces a tough, bendable, and machinable cast iron. During the anneal, all or most of the carbon is precipitated in ragged nodules of graphite. The end product is available in a number of grades that vary in strength, hardness, and toughness. Malleable iron is used widely in the transportation, valve, fitting, and hardware industries for innumerable small parts.

Ductile iron (also called spheroidal graphite iron or nodular iron) is a product of high carbon and high silicon content in which most of the carbon has been coagulated into spheres during processing in the liquid state. During processing, an ingredient such as magnesium or cerium is added, followed by another substance such as silicon, calcium, or combinations of these ingredients. The resulting product acquires the structure of steel peppered with spheres of graphite. Its strength closely approaches that of steel, and it may be heat-treated to develop a ductility exceeding that of malleable cast iron. Its properties make it useful for a great many engineering applications requiring strength and toughness levels which previously were beyond the reach of iron foundry products.

Austenitic cast irons are available in the gray-iron and ductile-iron classifications. When gray or ductile iron is alloyed with substantial amounts of nickel, manganese, silicon, or other elements to alter the basic crystalline structure from magnetic alpha-iron to nonmagnetic gamma-iron, the resulting product becomes austenitic cast iron. These irons generally offer better corrosion- and heat-resistance than the low-alloy irons. *See* ALLOY; HEAT TREATMENT (METALLURGY); IRON ALLOYS; METAL CASTING; STEEL.

[JAMES S. VANICK]

Bibliography: American Foundrymen's Association, *Alloy Cast Irons*, 2d ed., 1944; Metallurgical Society of the American Institute of Mining, Metallurgical and Petroleum Engineers, *History of Iron and Steelmaking in the United States*, 1961; L. H. Van Vlack, *Elements of Materials Science*, 2d ed., 1964.

Catalytic converter

An aftertreatment device used for pollutant removal from automotive exhaust. Since the 1975 model year, increasingly stringent government regulations for the allowable emission levels of carbon monoxide (CO), hydrocarbons (HC), and oxides of nitrogen (NO_x) have resulted in the use of catalytic converters on most passenger vehicles sold in the United States. The task of the catalytic converter is to promote chemical reactions for the conversion of these pollutants to carbon dioxide, water, and nitrogen.

By definition a catalyst is an agent which promotes the rates at which chemical reactions occur, without affecting the final equilibrium as dictated by thermodynamics, but which itself remains unchanged. For automotive exhaust applications, the pollutant removal reactions are the oxidation of carbon monoxide and hydrocarbons and the reduction of nitrogen oxides. Metals, base and noble, are the catalytic agents most often employed for this task. Small quantities of these metals, when present in a highly dispersed form (often as individual atoms), provide sites upon which the reactant molecules may interact and the reaction proceed.

In addition to the active metal, the converter contains a support component whose functions include yielding structural integrity to the device, providing a large surface area for metal dispersion, and promoting intimate contact between the exhaust gas and the catalyst. Two types of supports are used: pellets and monoliths. The pelleted converter consists of a packed bed of small, porous, ceramic spheres whose outer shell is impregnated with the active metal. The monolith is a honeycomb structure consisting of a large number of channels parallel to the direction of exhaust gas flow. The active metals reside in a thin layer of high-surface-area ceramic (usually γ-alumina) placed on the walls of the honeycomb. In either system the support is contained in a stainless steel can installed in the exhaust system upstream of the muffler.

Two types of catalyst systems, oxidation and three-way, are found in automotive applications. Oxidation catalysts remove only CO and HC, leaving NO_x unchanged. An air pump is often used to add air to the engine exhaust upstream of the catalyst, thus ensuring an oxidizing atmosphere. Platinum and palladium are generally used as the active metals in oxidation catalysts. Three-way catalysts are capable of removing all three pollutants simultaneously, provided that the catalyst is maintained in a "chemically correct" environment that is neither overly oxidizing or reducing. To achieve this requires that the engine air-fuel ratio always be at, or very near, stoichiometry under all vehicle operating conditions. Feedback air-fuel ratio control systems are often used to satisfy this requirement. Platinum, palladium, and rhodium are the metals most often used in three-way catalysts. In addition, base metals are frequently added to improve the ability of the catalyst to withstand small, transient perturbations in air-fuel ratio. In both oxidation and three-way catalyst systems, the production of undesirable reaction products, such as sulfates and ammonia, must be avoided.

Maintaining effective catalytic function over long periods of vehicle operation is often a major problem. Catalytic activity will deteriorate due to two causes, poisoning of the active sites by contaminants, such as lead and phosphorus, and exposure to excessively high temperatures. Catalyst overtemperature is often associated with engine malfunctions such as excessively rich operation or a large amount of cylinder misfire. To achieve efficient emission control, it is thus paramount that catalyst-equipped vehicles be operated only with lead-free fuel and that proper engine maintenance procedures be followed. In such cases catalytic converters have proved to be a very effective means for reducing emissions without sacrificing fuel economy. *See* AUTOMOTIVE ENGINE.

[NORMAN OTTO]

Central heating and cooling

The use of a single heating or cooling plant to serve a group of buildings, facilities, or even a complete community through a system of distribution pipework that feeds each structure or facility. Central heating plants are basically of two types: steam or hot-water. The latter type uses high-temperature hot water under pressure and has become the more usual because of its considerable advantages. Steam systems are only used today where there is a specific requirement for high-pressure steam. Central cooling plants utilize a central refrigeration plant with a chilled water distribution system serving the air-conditioning systems in each building or facility.

Benefits. Advantages of a central heating or cooling plant over individual ones for each building or facility in a group include reduced labor cost, lower energy cost, less space requirement, and simpler maintenance. Though a central plant may require a 24-hr shift of operators, the total number of employees can be substantially less than that required to operate and maintain a number of individual plants.

Firing efficiencies of 85–93%, dependent upon such factors as fuel, boiler, and plant design, are usual with large central heating plants. Corresponding efficiencies for small individual heating boiler plants average 50–70%. Fuel in bulk quantity has a lower unit cost, and single handling for one large plant as distinct from multiple handling for many small plants saves appreciably in labor and transportation. Maintenance costs on a single central plant are considerably lower than for the aggregate of small plants of equal total capacity.

The disadvantages of a central heating plant concern mainly the maintenance of the distribution system where steam is used. Corrosion of the condensate water return lines shortens their life, and the steam drainage traps need particular attention. These disadvantages do not occur with high-temperature hot-water installations.

Central cooling plants, using conventional, electrically driven refrigeration compressors, have the advantage of utilizing bulk electric supply, at voltages as high as 13.5 kV, at wholesale rates. Additionally, their flexible load factor, resulting from load divergency in the various buildings served, results in major operating economies.

Design. Winter heat-load requirements are calculated by the addition of the following for each individual building or facility: (1) winter heat losses, (2) domestic water-heating requirements, and (3) industrial or other special heat requirements. To the sum of these for all buildings or facilities must be added the system distribution heat losses.

The summer load on the central cooling plant is calculated by the addition for each individual building or facility of the summer air-conditioning requirements. To the sum of these for all buildings or facilities must be added the system summer distribution heat losses.

To the individual winter and summer totals a diversity factor of 60–80% is applied because not all loads peak simultaneously. Winter heat loss due to weather conditions must be taken at its maximum. Water-heating and industrial requirements vary throughout the daily cycle. The summer air conditioning load must also be taken at its maximum. System distribution losses must be calculated both for the winter and the summer loads. The individual characteristics of the system must be considered in the diversity factor used.

Individual boiler sizing should allow the best arrangement to meet load variations as between individual 24-hr peaks for winter loads. Standby capacity is essential. Fuel selection, usually oil, coal, or gas, depends on local conditions and costs, taking into account labor and firing efficiencies to be expected with each fuel. Individual refrigeration machine sizing must allow for cooling load variation over the 24-hr cycle and the turn-down range of the machine itself. Considerable flexibility in machine choice thus results.

Distribution pipework sizing follows normal practice using suitable pressure drops with allowance for load variations and diversity factors as indicated.

Economics. The economics of each system must be individually computed because of the many factors involved. There is no average cost unitary basis applicable. The following major factors affect each plant's economics study: (1) system type, that is, steam, low pressure or high pressure; medium- or high-temperature hot water; chilled water temperature used; and so forth; (2) cost and type of energy used, that is, coal, gas, oil, and electricity; (3) type and occupancy of facilities served; (4) labor costs and conditions; (5) terrain; (6) climatic conditions; (7) plant first cost; (8) system life and maintenance costs. *See* DISTRICT HEATING.

Boiler plant. Both high-pressure (125 psi or 860 kPa saturated) and low-pressure (15 psi or 103 kPa saturated) steam plants are used, although the former is the more common. Both types follow conventional design. Feedwater and firing auxiliaries are of conventional type. Chemical treatment of feedwater is necessary.

Either conventional hot-water heating boilers or high-temperature hot-water boilers, depending on size, are used. Design and components for high-temperature hot-water plants are more complex and include nitrogen pressurization and gland-cooled circulating pumps. However, standard manufactured equipment is available for both conventional and high-temperature plants. Hot-water circulation, due to losses in the distribution pipework, should be at maximum temperature-pressure limitation. Circulation through the distribution system is by centrifugal pump with standby equipment being furnished. High-temperature hot-water installations may operate in 400–500°F (200–260°C) range.

Fuel handling, firing, and control arrangements also follow conventional design. Capacity of fuel storage should be on a minimal 3–4 week basis.

Central refrigeration plant. The central refrigeration plant may be electrically driven, or it may be of the steam-turbine-driven centrifugal type or steam absorption type. Steam as the prime energy source can be used where the plant is installed adjacent to a central boiler plant. Otherwise, medium-tension electricity, bought at bulk rates, forms an economical energy source. Gas turbines may also power centrifugal refrigeration compressors.

All equipment, including cooling towers and pumps, is of conventional design.

Distribution systems. Both overhead and underground pipework are used for distribution, although the latter is more usual except for industrial plants. Overhead mains must be strongly supported, insulated, and weatherproofed. Underground heating mains must be insulated and carefully waterproofed, particularly in damp areas. They must be structurally adequate. Steam distribution requires proper drainage of mains and often pumped return of condensate when gravity flow is not practical. In municipal distribution systems in larger cities where steam is used, it is frequently considered uneconomical to return the condensate to the central boiler plant due to pumping costs.

Hot-water distribution systems have the advantage that they are not affected by grade variations, that is, they can be run both uphill and downhill. The circulating pump pressures must be calculated accordingly.

A number of prefabricated insulated piping systems are commercially available. Each requires special handling for its installation and jointing. Hydrocarbon fill may also be used for heating mains insulation. Proper depth of burial must be arranged to give adequate protection against surface loads. Arrangements for expansion by loops or sliding fittings in access manholes are essential, and frequently cathodic protection is necessary. Where high-temperature hot water is used, it may be necessary to provide heat exchangers at each building or facility to furnish secondary heat at more moderate temperatures for uses such as heating systems and hot water. Where low-pressure steam is required, such exchangers may furnish this on the secondary side, if the primary hot water is at sufficiently high temperature.

For chilled water distribution systems, asbestos-cement plastic-lined piping can be used. The plastic lining has a low flow friction coefficient. If the piping is buried 4 ft (1.2 m) or deeper, no insulation is required, provided soil is not waterlogged.

Heat sales. Various methods of heat sale are in use where central plants service public communities or facilities. With steam distribution steam meters to each individual building or facility served are usual, and a utility type of sliding scale rate per pound of steam sold is charged.

For high-temperature and chilled water distribution a combination meter measuring both water flow and temperature differential between supply and return mains may be used. This measures directly Btu per hour furnished. If a constant temperature differential is maintained between supply and return water mains, metering may be by flow only, although this is not very accurate. *See* AIR CONDITIONING; BOILER; COMFORT HEATING; REFRIGERATION; STEAM; STEAM HEATING; WARM-AIR HEATING SYSTEM.

[JOHN K. M. PRYKE]

Bibliography: *ASHRAE Systems Handbook*, 1976; P. L. Geiringer, *High Temperature Water Heating*, 1963.

Centrifugal pump

A machine for moving fluid by accelerating it radially outward. More fluid is moved by centrifugal pumps than by all other types combined. The smooth, essentially pulsationless flow from centrifugal pumps, their adaptability to large capacities, easy control, and low cost make them preferable for most purposes. Exceptions are those in which a relatively high pressure is required at a small capacity, or in which the viscosity of the fluid is too great for reasonable efficiency.

Centrifugal pumps consist basically of one or more rotating impellers in a stationary casing which guides the fluid to and from the impeller or from one impeller to the next in the case of multistage pumps. Impellers may be single suction or double suction (Fig. 1). Additional essential parts of all centrifugal pumps are (1) wearing surfaces or rings, which make a close-clearance running joint between the impeller and the casing to minimize the backflow of fluid from the discharge to the suction; (2) the shaft, which supports and drives the impeller; and (3) the stuffing box or seal, which prevents leakage between shaft and casing.

Characteristics. The rotating impeller imparts pressure and kinetic energy to the fluid pumped. A collection chamber in the casing converts much of the kinetic energy into head or pressure energy before the fluid leaves the pump. A free passage exists at all times through the impeller between the discharge and inlet side of the pump. Rotation of the impeller is required to prevent backflow or draining of fluid from the pump. Because of this, only special forms of centrifugal pumps are self-priming. Most types must be filled with liquid, or primed, before they are started.

Every centrifugal pump has its characteristic curve, which is the relation between capacity or rate of flow and pressure or head against which it will pump (Fig. 2). At zero pressure-difference, maximum capacity is obtained, but without useful work. As resistance to flow external to the pump increases, capacity decreases until, at a high pressure, flow ceases entirely. This is called shut-off head and again no useful work is done. Between these extremes, capacity and head vary in a fixed relationship at constant rpm. Input power generally increases from a minimum at shut-off to a maximum at a capacity considerably greater than that at which best efficiency is realized. The operating design point is set as close as practical to the point of best efficiency. Operation at higher or lower speed results in a change in the characteristic curves, with capacity varying directly with the

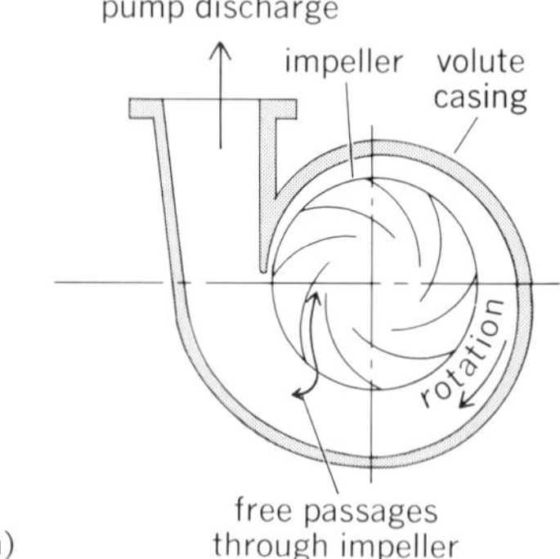

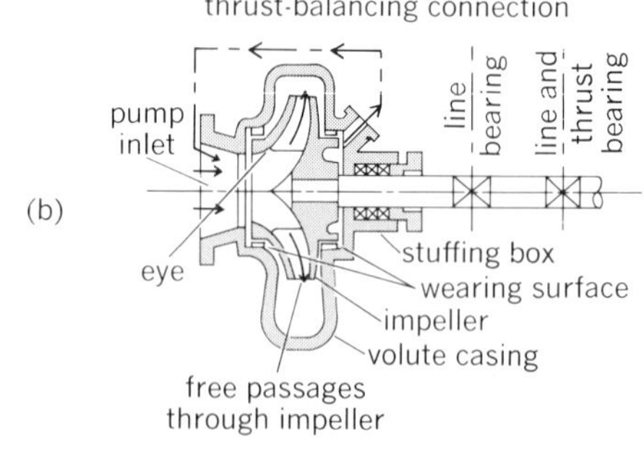

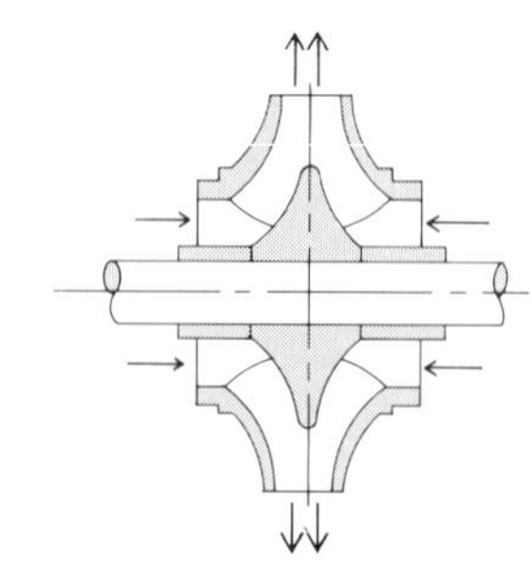

Fig. 1. Views of centrifugal pumps. (*a*) Section across the axis of a single-suction volute pump. (*b*) Section along the axis of a single-suction volute pump. (*c*) Double-suction impeller.

speed, and head varying as the square of speed. Since the power required is proportional to the product of the head and capacity, it varies as the cube of the speed. These relations are essentially constant as long as viscosity of the fluid is low enough to be negligible.

Centrifugal impeller (and casing) forms vary with the relation of desired head and capacity at a practical rotating speed. Impellers and pumps are commonly classified as centrifugal, mixed-flow, and axial-flow or propeller (Fig. 3). There is a continuous change from the centrifugal impeller to the axial-flow impeller. For maximum practical head at small capacity, the impeller has a large diameter and a narrow waterway with vanes curved only in the plane of rotation. As the desired capacity is increased relative to the head, the diameter of the impeller is reduced, the width of the waterway is increased, and the vanes are given a compound curvature. For higher capacities and less head, the mixed-flow impeller is used. It discharges at an angle approximately midway between radial and axial. For maximum capacity and minimum head, the axial-flow or propeller-type impeller is used.

Capacity. Specific speed is used to identify the place occupied by an impeller in this range of form or type. The narrow, purely radial discharge impeller is at the low end of the specific speed scale while the axial-flow impeller is at the high end. The highest pump efficiency is obtainable at a specific speed of approximately 2000 when the capacity is expressed in gallons per minute (gpm), or 94 when the capacity is expressed in cubic feet per second (cfs). Specific speed $N_s = NQ^{1/2}H^{-3/4}$ where N is rpm, H is heat in feet, and Q is capacity in gpm or cfs.

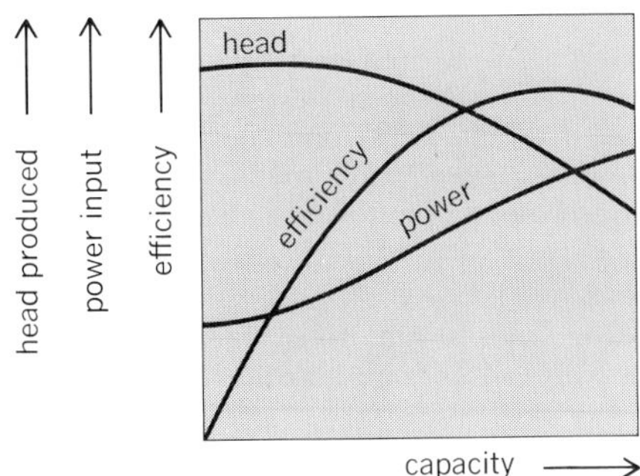

Fig. 2. Some of the typical characteristics of centrifugal pump at constant speed of rotation.

Since the highest efficiency is obtained at a medium specific speed, theoretically an optimum speed of rotation may be calculated for any head and capacity. Such a selection must be adjusted for practical considerations. Because most pumps are driven directly by electric motors, available motor speeds govern operating speeds to a large extent. For small capacities, a lower specific speed is usually necessary. The low limit is set by casting limitations for long vanes and extremely narrow impellers, by the hydraulic losses from flow through the small passages, and by the friction of a large disk rotating in the fluid. At the other end of the scale, for large capacities and low heads, high specific speed is necessary to keep the driver speed as high as practical. In large pumps the head per stage may be limited by mechanical considerations. At common motor speeds, single-stage pumps are limited to heads of about 600 ft per stage. However, at higher speeds heads up to about 2000 ft per stage have been attained in multistage boiler feed pumps.

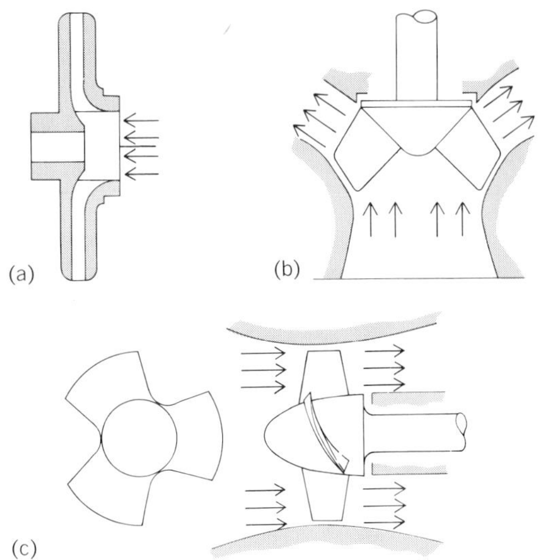

Fig. 3. Common classification of impellers. (*a*) Centrifugal for low speed. (*b*) Mixed-flow for intermediate speed. (*c*) Axial-flow or propeller for high speed.

Cavitation limits also play a major part in limiting the operating speed and head per stage for any specified condition. High speeds of rotation require high velocities at the entrance to the impeller. Since the static pressure in a closed stream of fluid drops as velocity is increased, there is a limiting velocity at which the absolute pressure approaches the vapor pressure of the fluid. Beyond this, partial vaporization occurs and the subsequent collapse of these pockets of vapor causes noise, vibration, and ultimately destruction of the surrounding walls. This form of vaporization in a moving stream of fluid is called cavitation. Higher operating speeds require higher suction pressure

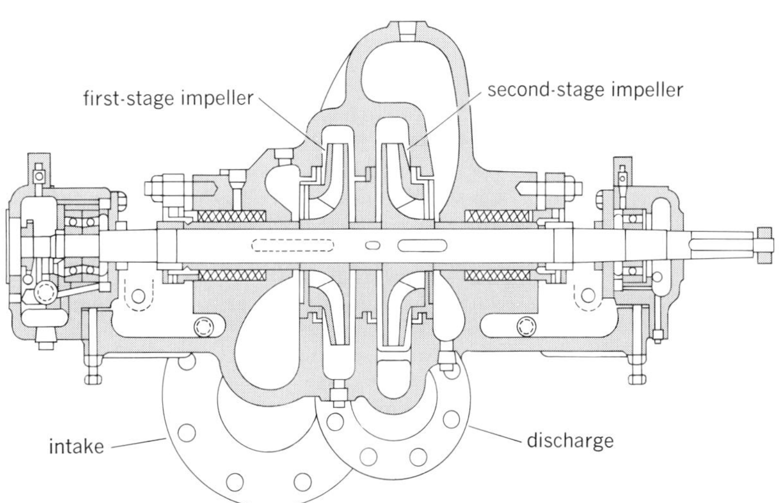

Fig. 4. Two-stage horizontally split casing pump.

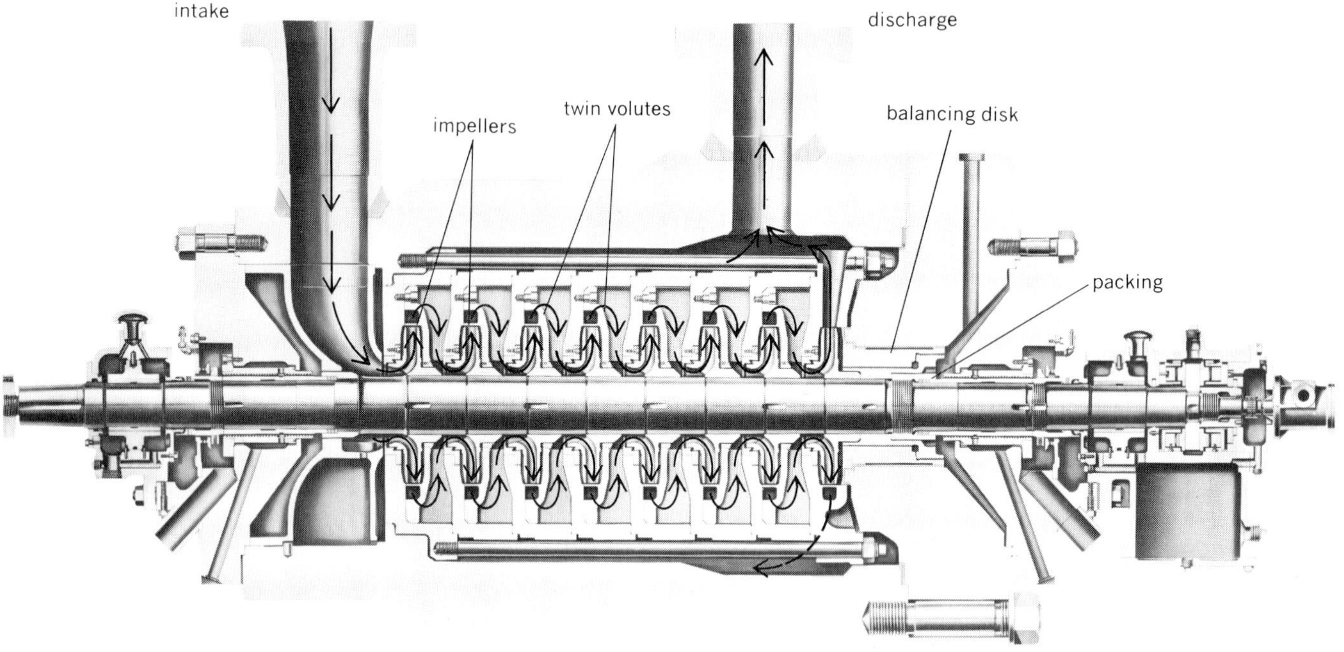

Fig. 5. High-pressure multistage centrifugal boiler-feed pump.

over the vapor pressure, spoken of as a net positive suction head.

Multistage pumps. When the required head exceeds that practical for a single-stage pump, several stages are employed. Figure 4 shows a typical two-stage pump with single-suction impellers mounted back to back to obtain an approximate balance of the hydraulic thrust developed by pressure acting on the greater area of the back side of the impeller. A modern high-pressure multistage centrifugal pump enclosed in a forged steel barrel with all the single-suction impellers facing in the same direction is shown in Fig. 5. In this design axial thrust is balanced by a rotating disk mounted at the high-pressure end of the pump. Leak-off from this balancing device, which is piped back to suction pressure, serves also to reduce, practically to suction pressure, the load on the stuffing box packing or seal at this end of the pump. Pumps of this type are built with 4 to 10 or more stages, depending on the speed, pressure, and capacity desired.

Another widely used type of multistage pump is the vertical turbine or deep-well pump. The details of design and the choice of an impeller of relatively high specific speed are the result of a drastic limitation of the outside diameter required to fit inside drilled-well casings. Pumps of this type are built with as many as 20 or 30 stages for high lifts from relatively small-diameter wells. Pumps with this same arrangement are built with a closed-bottom cylindrical housing with the inlet connected near the top of the housing, for applications at moderate head and limited net positive suction head. The cylindrical housing, or can, is then located in a pit or dry well with suction and discharge connections near ground or floor level. *See* PUMP; PUMPING MACHINERY.

[ELLIOTT F. WRIGHT]

Bibliography: A. J. Stepanoff, *Centrifugal and Axial Flow Pumps: Theory, Design and Application*, 2d ed., 1957; P. C. Tramm and R. C. Dean, Jr. (eds.), *Centrifugal Compressor and Pump Stability, Stall and Surge*, 1976.

Cermet

A group of composite materials consisting of an intimate mixture of ceramic and metallic components.

Fabrication. Cermets can be fabricated by mixing the finely divided components in the form of powders or fibers, compacting the components under pressure, and sintering the compact to produce physical properties not found solely in either of the components. Cermets can also be fabricated by internal oxidation of dilute solutions of a base metal and a more noble metal. When heated under oxidizing conditions, the oxygen diffuses into the alloy to form a base metal oxide in a matrix of the more noble metal. *See* CORROSION; POWDER METALLURGY.

Representative cermets

Class	Ceramic	Metal addition
Oxides	Al_2O_3	Al, Be, Co, Co-Cr, Fe, stainless steel
	Cr_2O_3	Cr
	MgO	Al, Be, Co, Fe, Mg
	SiO_2	Cr, Si
	ZrO_2	Zr
	UO_2	Zr, Al, stainless steel
Carbides	SiC	Ag, Si, Co, Cr
	TiC	Mo, W, Fe, Ni, Co, Inconel, Hastelloy, stainless steel, Vitallium
	WC	Co
	Cr_3C_2	Ni, Si
Borides	Cr_3B_2	Ni
	TiB_2	Fe, Ni, Co
	ZrB_2	Ni
Silicides	$MoSi_2$	Ni, Co, Pt, Fe, Cr
Nitrides	TiN	Ni

Components. Ceramic components may be metallic oxides, carbides, borides, silicides, nitrides, or mixtures of these compounds; the metallic components include a wide variety of metals whose selection depends on the application of the respective cermet. Some of the component materials under investigation are shown in the table.

Interactions. The reactions taking place between the metallic and ceramic components during fabrication of cermets may be briefly classified and described as follows:

1. Heterogeneous mixtures with no chemical reaction between the components, characterized by a mechanical interlocking of the components without formation of a new phase, no penetration of the metallic component into the ceramic component, and vice versa, and no alteration of either component (example, MgO-Ni).

2. Surface reaction resulting in the formation of a new phase as an interfacial layer that is not soluble in the component materials. The thickness of this layer depends on the diffusion rate, temperature, and time of the reaction (example, Al_2O_3-Be).

3. Complete reaction between the components, resulting in the formation of a solid solution characterized by a polyatomic structure of the ceramic and the metallic component (example, TiC-Ni).

4. Penetration along grain boundaries without the formation of interfacial layers (example, Al_2O_3-Mo).

Bonding behavior. One important factor in the selection of metallic and ceramic components in cermets is their bonding behavior. Bonding may be by surface interaction or by bulk interaction. In cermets of the oxide-metal type, for example, investigators differentiate among three forms of surface interaction: macrowetting, solid wetting, and wetting assisted by direct lattice fit.

Characteristics. The combination of metallic and ceramic components can result in cermets characterized by increased strength and hardness, higher temperature resistance, improved wear resistance, and better resistance to corrosion, each characteristic depending on the variables involved in composition and processing.

The yield strength σ of a cermet depends on the component materials, the volume fraction f of the dispersed phase, the particle diameter d of the dispersed material, and the mean spacing λ between particles. These variables are related by Eqs. (1) and (2), where A and B are material constants.

$$\lambda = \frac{2d}{3f}(1-f) \qquad (1)$$

$$\sigma = -A \log \lambda + B \qquad (2)$$

Applications. Friction parts as well as cutting and drilling tools have been successfully made from cermets for many years. Certain nuclear reactor fuel elements, such as dispersion-type elements, are also made as cermets. Gas turbine parts and jet engine components in cermet composition, however, have not yet been developed satisfactorily. These materials should be corrosion resistant, have high-temperature strength, temperature (thermal) shock resistance, and a certain ductility. Although it has been possible to develop cermets to satisfy three of these requirements, it is not yet possible to satisfy all four of them.

Metal ceramic combinations. One distinguishes basically between four different combinations of metal and ceramic components: (1) the formation of continuous interlocking phases of the metallic and ceramic components, (2) the dispersion of the metallic component in the ceramic matrix, (3) the dispersion of the ceramic component in the metallic matrix, and (4) the interaction between the metallic and ceramic components.

Fiber reinforcement. Fiber-reinforced cermets consist either of a metallic matrix reinforced by ceramic fibers (for high-temperature strength) or a ceramic matrix with metallic fibers inserted (for better heat conductivity). The term fiber refers to a multicrystalline material approximately 0.5–2.5 micrometer (0.02–0.1 mil) diameter. Whiskers which can be used instead of fibers are short single crystals approximately 1–10 μm in diameter. Combinations of powders and fibers can be made with short or continuous fibers, randomly dispersed or aligned (oriented). Special materials with directional properties can be made by insertion of aligned fibers.

Bulk interactions. Two different forms of bulk interaction between metals and ceramics can be distinguished; solid solution and formation of chemical compounds. Usually solid-solution bonding involves the addition, or formation, within the cermet of a small amount of the appropriate ceramic form of the metal constituent; examples of the type of phase involved are provided by the systems Al_2O_3-Cr_2O_3 and NiO-MgO for oxides, and TaC-TiC and NbC-ZrC for carbides. The systems form continuous series of solid solutions, but there are also many suitable systems in which solid solution occurs over a limited range only.

Formation of a compound in a bonding phase can best be shown by the examples of spinels, having the generalized formula $RO \cdot R'_2O_3$, where R could, for example, stand for Ni^{++}, Mg^{++}, Fe^{++}, or Co^{++}, and R′ for Al^{+++}, Cr^{+++}, or Fe^{+++}. Much study has been made of system Al_2O_3-Fe, without achieving the combination of properties required for the high-temperature engineering applications in view.

[HENRY H. HAUSNER]

Bibliography: J. J. Burke (ed.), *Strengthening Mechanisms, Metals and Ceramics*, 1966; D. Peckner (ed.), *The Strengthening of Metals*, 1967; G. V. Samsonov (ed.), *High Temperature Cermets: Refractory Transition Metal Compounds*, 1964; J. R. Tinklepaugh and W. B. Crandall (eds.), *Cermets*, 1960.

Chain drive

A flexible device of connected links used to transmit power. A drive consists of an endless chain which meshes with sprockets located on the shaft of a driving source, such as an electric motor/reducer, and a driven source, such as the head shaft of a belt conveyor.

Chains have been used for more than 2000 years. However, the modern contribution of the chain to industrialization began in 1873 with the development of a cast detachable chain. It was a simple cast-metal chain composed of identical links which could be coupled together by hand. This chain so greatly improved the performance of power takeoff drives for farm implements that mechanization became a practical reality. The first development after the success of the cast detachable chain was the cast pintle chain with a closed barrel design and steel pins. Other cast chain designs and variations evolved until an all-steel chain was developed.

Chain types. The new development, now called the roller chain (Fig. 1), found uses in the early 1900s on bicycles and other forms of conveyances. Constantly refined and improved, today's roller chain meets the demands of heavy-duty oil well drilling equipment, high-production agricultural machinery (Fig. 2), construction machinery, and similar equipment. It also meets the precise timing requirements of lighter-duty equipment such as printing, packaging, and vending equipment.

Another type of chain also evolved, called the engineering steel chain (Fig. 3). In a broad sense, the early designs of the engineering steel chain were a blend of the other two, the cast and the roller chain. The drive chains in the engineering steel chain category are usually identified by their offset/cranked link sidebar design. Generally, larger pitch sizes as compared to the roller chain and higher-strength chains characterize this group. A third group used for chain drives is the inverted tooth/silent chains (Fig. 4). A familiar application is their use as automotive timing chains in automobile engines.

Advantage. The use of chains for power transmission rather than another device, such as V-belts or a direct coupling to the power source, is usually based on the cost effectiveness and economy of chains and sprockets. Chains and sprockets offer the following advantages: large speed ratios; sufficient elasticity to absorb reasonable shocks; a constant speed ratio between the driving and driven shaft; long life without excessive maintenance; mechanical understandability regarding installation and functionality; coupling and uncoupling with simple tools; and a simple means to get power from its source to the location where needed. *See* BELT DRIVE.

Fig 2. Roller chain drive. (*FMC Corp.*)

Drive design. The design of a chain drive consists primarily of the selection of the chain and sprocket sizes. It also includes the determination of chain length, center distance, method of lubrication, and in some cases the arrangement of chain casings and idlers. Chain and sprocket selection is based on: the horsepower and type of drive; the speeds and sizes of the shafting; and the surrounding conditions. A properly selected chain, following prescribed chain manufacturers' and the American Chain Association's techniques, is usually based on 15,000 hours of operation without breakage of components, considering chain wear not to exceed that which can be accommodated by the sprockets. Generally, chains are considered worn out when the roller chain exceeds approximately 3% elongation, the engineered steel chain exceeds approximately 5% elongation, and inverted tooth/silent chain exhibits malfunction characteristics. To achieve the rated 15,000-hour life, the environment must be clean and the chain lubricated as recommended by the manufacturer for the speed, the horsepower capacity, and the number of teeth on the smaller sprocket.

Chain and sprocket design. The particular design of the sideplate/sidebar/cranked link, pin, bushing, and roller of chains used in chain drives has been standardized to a substantial degree. The dimensional parameters regarding pitch, pin, bushing, roller, and side-plate sizes have been standardized (see table). Interference between parts is usually controlled by the manufacturer, as is hardness of the chain parts, which determine the life of the chain. The selection of materials and

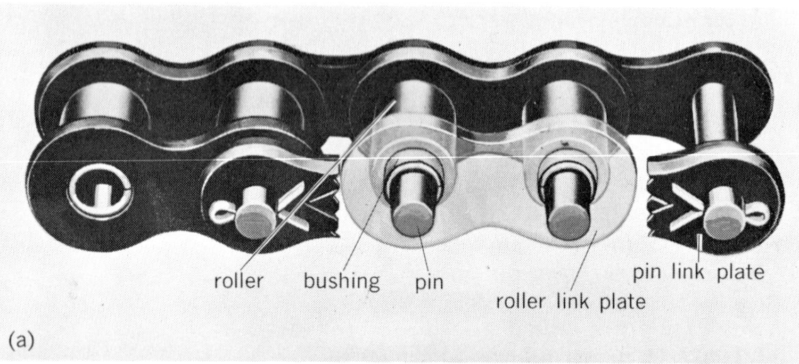

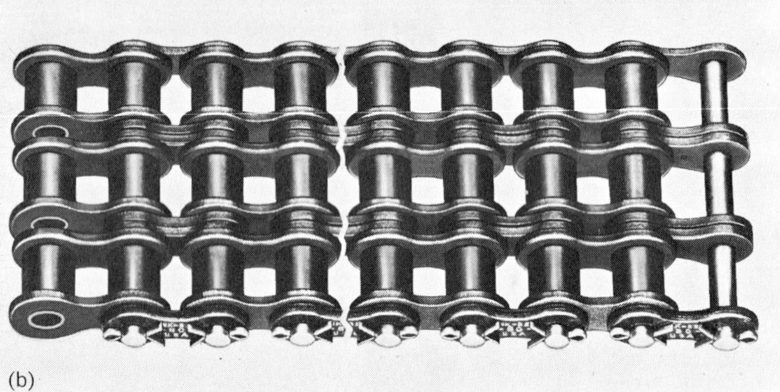

Fig. 1. Roller chains. (*a*) Single-strand. (*b*) Triple-strand.

ANSI and ISO standards for drive chains

Chain description	ANSI standard number	ISO standard number	Title
Roller chain	B29.1	R606	(ANSI) Transmission roller chains and sprocket teeth (ISO) Short-pitch transmission roller chain
Inverted-tooth (silent) chain	B29.2	–	(ANSI) Inverted-tooth (silent) chains and sprocket teeth (ISO) No standard available
Engineering steel chain	B29.10	ISO3512	(ANSI) Heavy-duty offset sidebar power transmission roller chains and sprocket teeth (ISO) Heavy-duty crank-link transmission chains

heat treatments to obtain desired hardnesses are selected by manufacturers, but generally certain minimum hardnesses are required to meet minimum ultimate strength or breaking load criteria.

A sprocket is a wheel with teeth shaped to mesh with the chain. The sprocket tooth form, when properly selected, assures the success of the chain drive. Space limitations often determine the chain length and the number of teeth on the sprocket. Usually more than one combination of chain type, chain size, and number of sprocket teeth will satisfy a requirement. The final determination is based on economics and availability.

Lubrication. Lubrication of chains reduces joint wear as the links flex onto and off the sprockets. Lubricated bearing surfaces, that is, the chain joints, can carry high loads without galling. Chain capacity or horsepower ratings are determined for operation in a clean environment with proper lubrication. Lubrication extends the wear life of chains and sprockets operating in any environment, no matter how dirty or abrasive. Chains should not be greased. A nondetergent petroleum-base oil is recommended. For operation at high speeds, oil stream–force feed lubrication is required for cooling. An oil bath is effective at intermediate speeds, and manual lubrication is acceptable at slow speeds.

ANSI and ISO standards. The American National Standards Institute has established a committee for the standardization of transmission chains and sprockets. Also, the International Standards Organization has established a committee to promulgate international standards for chains and chain wheels for power transmission and conveyors. The table identifies those standards which apply to drive chains. Typically, the standards identify those characteristics of a chain which ensure that one manufacturer's chain will couple with another and that the minimum ultimate strength or breaking load characteristics have been established for each chain. The ANSI standard includes a supplemental section with horsepower rating tables and selection information. *See* CONVEYOR; GEAR DRIVE.

[VICTOR D. PETERSHACK]

Bibliography: American Chain Association, *Design Manual Roller and Silent Chain Drives*, 1974; American Chain Association, *Engineering Steel Chains: Application Handbook*, 1973.

Fig. 3. Engineering steel chain drive. *(Rexnord Inc.)*

CHAIN DRIVE

Fig. 4. Inverted tooth/silent chain drive.

Circuit (electricity)

A general term referring to a system or part of a system of conducting parts and their interconnections through which an electric current is intended to flow. A circuit is made up of active and passive elements or parts and their interconnecting conducting paths. The active elements are the sources of electric energy for the circuit; they may be batteries, direct-current generators, or alternating-current generators. The passive elements are resistors, inductors, and capacitors. The electric circuit is described by a circuit diagram or map showing the active and passive elements and their connecting conducting paths.

Devices with an individual physical identity such as amplifiers, transistors, loudspeakers, and generators, are often represented by equivalent circuits for purposes of analysis. These equivalent circuits are made up of the basic passive and active elements listed above.

Electric circuits are used to transmit power as in high-voltage power lines and transformers or in low-voltage distribution circuits in factories and homes; to convert energy from or to its electrical form as in motors, generators, microphones, loudspeakers, and lamps; to communicate information as in telephone, telegraph, radio, and television systems; to process and store data and make logical decisions as in computers; and to form systems for automatic control of equipment.

Electric circuit theory. This includes the study of all aspects of electric circuits, including analysis, design, and application. In electric circuit theory the fundamental quantities are the potential differences (voltages) in volts between various points, the electric currents in amperes flowing in

CIRCUIT (ELECTRICITY)

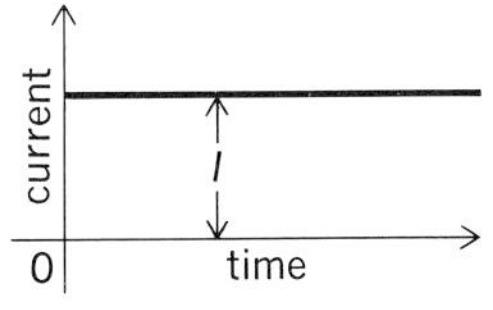

Fig. 1. Direct current.

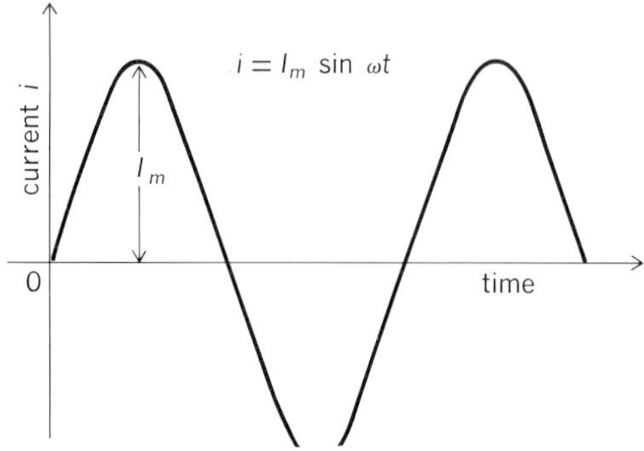

Fig. 2. Alternating current.

CIRCUIT (ELECTRICITY)

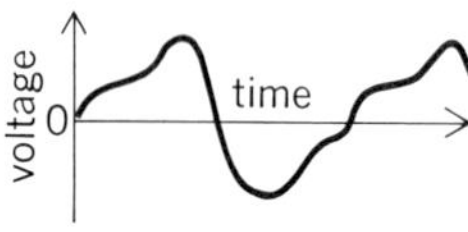

Fig. 3. Nonsinusoidal voltage wave.

the several paths, and the parameters in ohms or mhos which describe the passive elements. Other important circuit quantities such as power, energy, and time constants may be calculated from the fundamental variables. For a discussion of these parameters *see* ELECTRICAL IMPEDANCE; ELECTRICAL RESISTANCE; REACTANCE.

Electric circuit theory is an extensive subject and is often divided into special topics. Division into topics may be made on the basis of how the voltages and currents in the circuit vary with time; examples are direct-current, alternating-current, nonsinusoidal, digital, and transient circuit theory. Another method of classifying circuits is by the arrangement or configuration of the electric current paths; examples are series circuits, parallel circuits, series-parallel circuits, networks, coupled circuits, open circuits, and short circuits. Circuit theory can also be divided into special topics according to the physical devices forming the circuit, or the application and use of the circuit. Examples are power, communication, electronic, solid-state, integrated, computer, and control circuits.

CIRCUIT (ELECTRICITY)

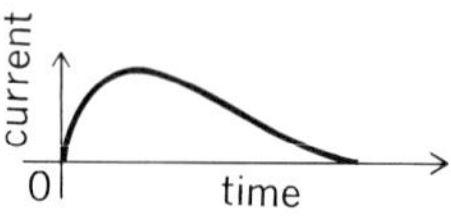

Fig. 4. Transient electric current.

Direct-current circuits. In dc circuits the voltages and currents are constant in magnitude and do not vary with time (Fig. 1). Sources of direct current are batteries, dc generators, and rectifiers. Resistors are the principal passive element. For a discussion of direct-current circuits *see* DIRECT-CURRENT CIRCUIT THEORY.

Magnetic circuits. Magnetic circuits are similar to electric circuits in their analysis and are often included in the general topic of circuit theory. Magnetic circuits are used in electromagnets, relays, magnetic brakes and clutches, computer memory devices, and many other devices.

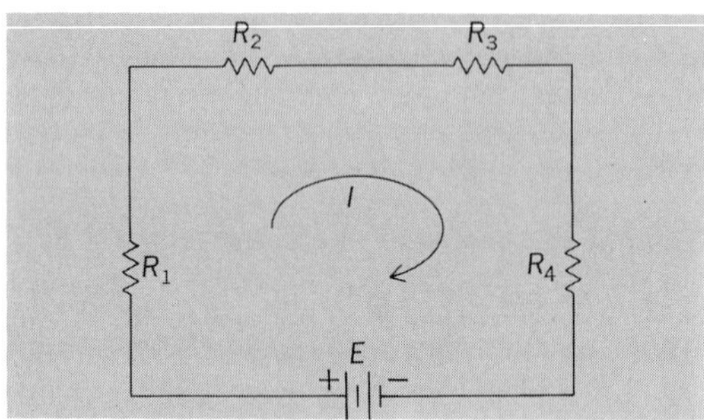

Fig. 5. Series circuit.

Alternating-current circuits. In ac circuits the voltage and current periodically reverse direction with time. The time for one complete variation is known as the period. The number of periods in 1 sec is the frequency in cycles per second. A cycle per second has recently been named a hertz (in honor of Heinrich Rudolf Hertz's work on electromagnetic waves).

Most often the term ac circuit refers to sinusoidal variations. For example, the alternating current in Fig. 2 may be expressed by $i = I_m \sin \omega t$. Sinusoidal sources are ac generators and various types of electronic and solid-state oscillators; passive circuit elements include inductors and capacitors as well as resistors. The analysis of ac circuits requires a study of the phase relations between voltages and currents as well as their magnitudes. Complex numbers are often used for this purpose. For a detailed discussion *see* ALTERNATING-CURRENT CIRCUIT THEORY.

Nonsinusoidal waveforms. These voltage and current variations vary with time but not sinusoidally (Fig. 3). Such nonsinusoidal variations are usually caused by nonlinear devices, such as saturated

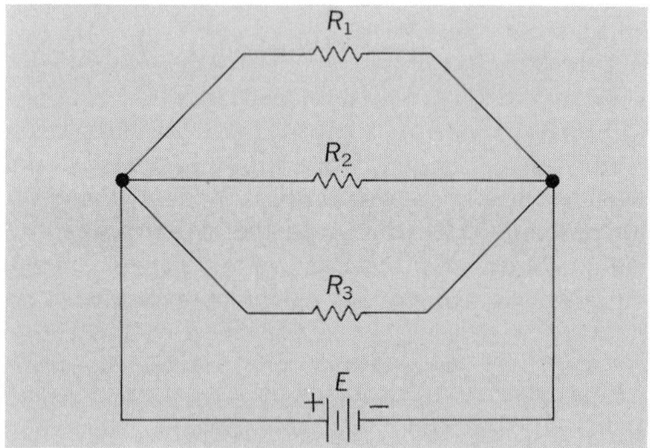

Fig. 6. Parallel circuit.

magnetic circuits, electron tubes, and transistors. Circuits with nonsinusoidal waveforms are analyzed by breaking the waveform into a series of sinusoidal waves of different frequencies known as a Fourier series. Each frequency component is analyzed by ac circuit techniques. Results are combined by the principle of superposition to give the total response.

Electric transients. Transient voltage and current variations last for a short length of time and do not repeat continuously (Fig. 4). Transients occur when a change is made in the circuit, such as opening or closing a switch, or when a change is made in one of the sources or elements.

Series circuits. In a series circuit all the components or elements are connected end to end and carry the same current, as shown in Fig. 5. *See* SERIES CIRCUIT.

Parallel circuits. Parallel circuits are connected so that each component of the circuit has the same potential difference (voltage) across its terminals, as shown in Fig. 6. *See* PARALLEL CIRCUIT.

Series-parallel circuits. In a series-parallel circuit some of the components or elements are

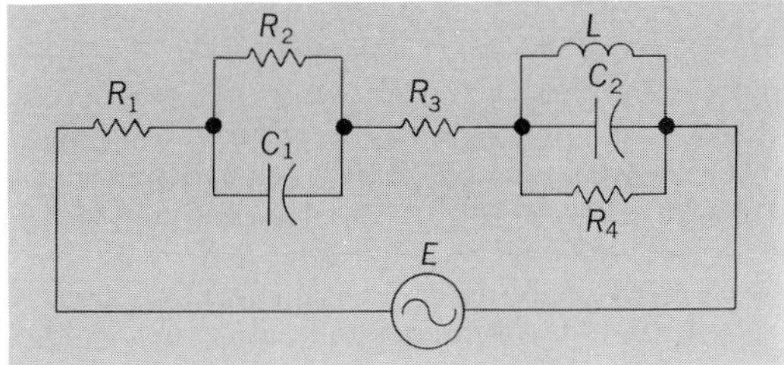

Fig. 7. Series-parallel circuit.

connected in parallel, and one or more of these parallel combinations are in series with other components of the circuit, as shown in Fig. 7.

Electric network. This is another term for electric circuit, but it is often reserved for the electric circuit that is more complicated than a simple series or parallel combination. A three-mesh electric network is shown in Fig. 8.

Coupled circuits. A circuit is said to be coupled if two or more parts are related to each other through some common element. The coupling may be by means of a conducting path of resistors or capacitors or by a common magnetic linkage (inductive coupling), as shown in Fig. 9.

Open circuit. An open circuit is a condition in an electric circuit in which there is no path for current flow between two points that are normally connected. *See* OPEN CIRCUIT.

Short circuit. This term applies to the existence of a zero-impedance path between two points of an electric circuit. *See* SHORT CIRCUIT.

Integrated circuit. The integrated circuit is a recent development in which the entire circuit is contained in a single piece of semiconductor material.

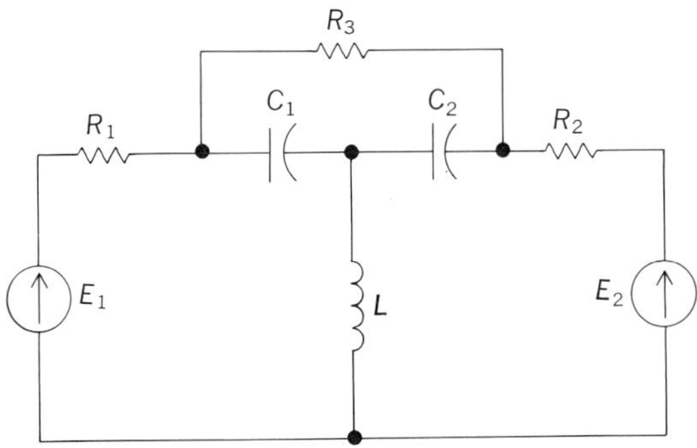

Fig. 8. A three-mesh electric network.

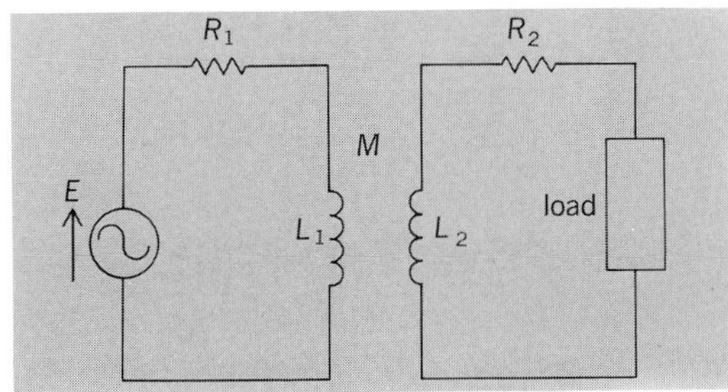

Fig. 9. Inductively coupled circuit.

Sometimes the term is also applied to circuits made up of deposited thin films on an insulating substrate.

[CLARENCE F. GOODHEART]

Bibliography: R. Boylestad, *Introductory Circuit Analysis*, 3d ed., 1977; E. Brenner and M. Javid, *Analysis of Electric Circuits*, 2d ed., 1967; P. Chirlian, *Electronic Circuits: Physical Principles, Analysis, and Design*, 1971; A. E. Fitzgerald et al., *Basic Electrical Engineering*, 4th ed., 1975; W. Hayt, Jr., and J. E. Kemmerly, *Engineering Circuit Analysis*, 3d ed., 1978; W. W. Lewis and C. F. Goodheart, *Basic Electric Circuit Theory*, 1957; R. E. Scott, *Elements of Linear Circuits*, 1965; R. Smith, *Circuits, Devices, and Systems*, 1966.

Circuit breaker

A device to open or close an electric power circuit either during normal power system operation or during abnormal conditions. A circuit breaker serves in the course of normal system operation to energize or deenergize loads. During abnormal

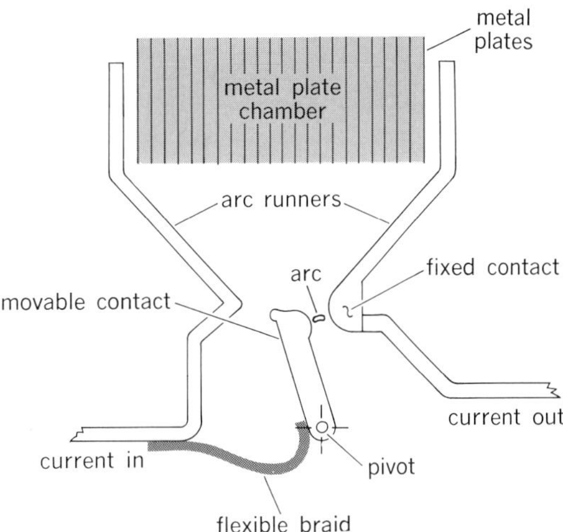

Fig. 1. Cross section of interrupter for a typical medium-voltage circuit breaker.

conditions, when excessive current develops, a circuit breaker opens to protect equipment and surroundings from possible damage due to excess current. These currents are usually the result of short circuits created by lightning, accidents, deterioration of equipment, or sustained overloads.

Formerly, all circuit breakers were electromechanical devices. In these breakers a mechanism operates one or more pairs of contacts to make or break the circuit. The mechanism is powered either electromagnetically, pneumatically, or hydraulically. The contacts are located in a part termed the interrupter. When the contacts are parted, opening the metallic conductive circuit, an electric arc is created between the contacts. This arc is a high-temperature ionized gas with an electrical conductivity comparable to graphite. Thus the current continues to flow through the arc. The function of the interrupter is to extinguish the arc, completing circuit-breaking action.

Fig. 2. Bulk oil circuit breaker for 138-kV application.

Current interruption. In alternating-current circuits, arcs are usually extinguished at a natural current zero, when the ac voltage applied across the arcing contacts reverses polarity. Within a short period around a natural current zero, the power input to the arc, equal to the product of the instantaneous current and voltage, is quite low. There is an opportunity to remove more energy from the arc than is applied to it, thus allowing the gas to cool and change from a conductor into an insulator.

In direct-current circuits, absence of natural current zero necessitates the interrupter to convert the initial arc into one that could only be maintained by an arc voltage higher than the system voltage, thus forcing the current to zero. To accomplish this, the interrupter must also be able to remove energy from the arc at a rapid rate.

Different interrupting mediums are used for the purpose of extinguishing an arc. In low- and medium-voltage circuits (110–15,000 volts) the arc is driven by the magnetic field produced by the arc current into an arc chute (Fig. 1). In the arc chute the arc is either split into several small arcs between metal plates or driven tightly against a solid insulating material. In the former case the splitting of the arc increases the total arc voltage, thus increasing the rate of energy dissipation. In the latter case the insulating material is heated to boiling temperatures, and the evaporated material flows through the arc, carrying a great deal of energy with it.

Fig. 3. Air blast circuit breaker rated for 500 kV.

Fig. 4. SF_6 circuit breakers rated for 500 kV 3 kA.

For outdoor applications at distribution and subtransmission voltage (10 kV and above), oil breakers are widely used. In the United States, bulk oil breakers are used (Fig. 2), and in Europe and many other countries, "low-oil-content" breakers are quite popular. The principles of operation of both kinds of oil breakers are basically the same. Only the amount of oil used and the detailed engineering design differ. In oil breakers, the arc is drawn in oil. The intense heat of the arc decomposes the oil, generating high pressure that produces a fluid flow through the arc to carry energy away. At transmission voltages below 345 kV, oil breakers used to be popular. They are increasingly losing ground to gas-blast circuit breakers such as air-blast breakers (Fig. 3) and SF_6 circuit breakers (Fig. 4). Even though a 765-kV low-oil-content breaker design has been announced, it has seen little practical application.

In air-blast circuit breakers, air is compressed to high pressures (50 atm approximately; 1 atm = 1.01×10^5 Pa). When the contacts part, a blast valve is opened to discharge the high-pressure air to ambient, thus creating a very-high-velocity flow near the arc to dissipate the energy. In SF_6 circuit breakers, the same principle is employed, with SF_6 as the medium instead of air. In the "puffer" SF_6 breaker, the motion of the contacts compresses the gas and forces it to flow through an orifice into the neighborhood of the arc. Both types of SF_6 breakers have been developed for ehv (extra high voltage) transmission systems.

For ehv systems, it was discovered that closing of a circuit breaker may cause a switching surge which may be excessive for the insulation of the system. The basic principle is easy to understand. If a breaker is closed at the peak of the voltage wave to energize a single-phase transmission line which is open-circuited at the far end, reflection can cause the transient voltage on the line to reach twice the peak of the system voltage. If there are trapped charges on the line, and if it happens that at the moment of breaker-closing the system voltage is at its peak, equal in magnitude to but opposite to the polarity of the voltage due to the trapped charges left on the line, the switching surge on the line can reach a theoretical maximum of three times the system voltage peak.

One way to reduce the switching surge is to insert a resistor in series with the line (Fig. 5) for a short time. When switch A in Fig. 5 is closed, the voltage is divided between the resistance and the surge impedance Z of the line by the simple relationship $V_L = V[Z/(R+Z)]$, where V is voltage impressed, V_L is voltage across the transmission line, R is resistance of the resistor, and Z is surge impedance of the line. Only V_L will travel down the line and be reflected at the other end. The magnitude of the switching surge is thus considerably reduced.

Vacuum and solid-state breakers. Two other types of circuit breakers have been developed. The vacuum breaker, another electromechanical device, uses the rapid dielectric recovery and high dielectric strength of vacuum. A pair of contacts is hermetically sealed in a vacuum envelope

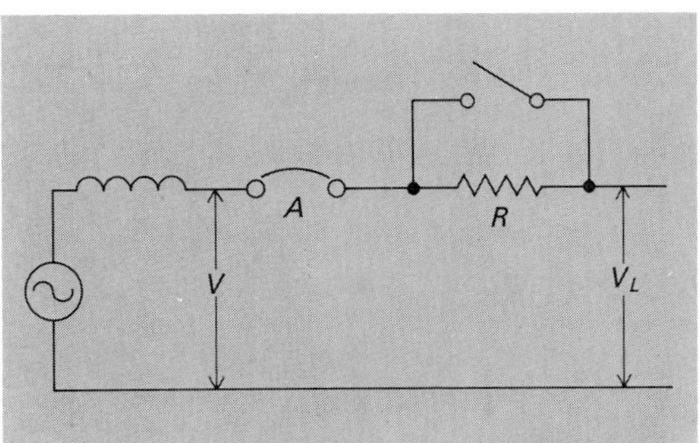

Fig. 5. Resistor insertion to reduce switching surge.

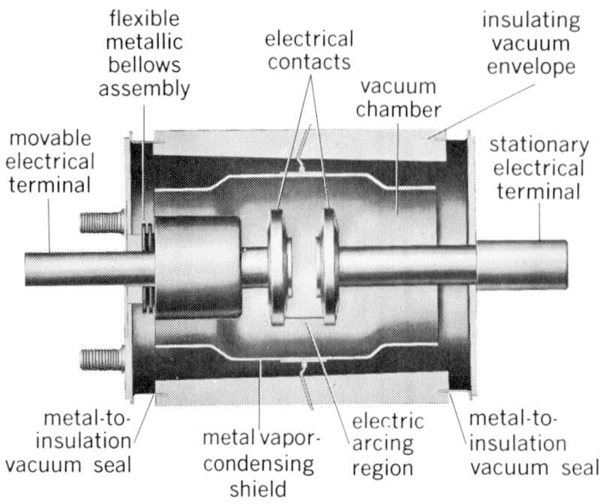

Fig. 6. Cutaway view of vacuum interrupter.

(Fig. 6). Actuating motion is transmitted through bellows to the movable contact. When the contacts are parted, an arc is produced and supported by metallic vapor boiled from the electrodes. Vapor particles expand into the vacuum and condense on solid surfaces. At a natural current zero the vapor particles disappear, and the arc is extinguished. Vacuum breakers of up to 242 kV have been built (Fig. 7).

The other type of breaker uses a thyristor, a semiconductor device which in the off state prevents current from flowing but which can be turned on with a small electric current through a third electrode, the gate. At the natural current zero, conduction ceases, as it does in arc interrupters. This type of breaker does not require a mechanism. Semiconductor breakers have been built to carry continuous currents up to 10,000 A.

Semiconductor circuit breakers can be made to operate in microseconds if the commutation principle is applied. Figure 8 illustrates the commutation principle for an hvdc (high-voltage direct-current) circuit, but it can easily be extended to ac circuits. During normal operation, the circuit breaker (CB in the diagram) would be closed and the load would be supplied with the current from the hvdc source. In this diagram, inductances L_1 and L_2 represent the circuit inductance on either side of the breaker. Suppose that a fault occurs which applies a short circuit between points A and B. The current will commence to increase, its rate being determined by L_1 and L_2. When the increased current is detected, the contacts of the circuit breakers are opened, drawing an arc, and the switch (S in the diagram) is closed, causing the precharged capacitors C to discharge through the circuit breaker. The current I_2 so produced is traveling in such a direction as to oppose I_1 and drive it to zero, thereby giving the circuit breaker an opportunity to interrupt.

Fig. 7. Vacuum circuit breaker, 242 kV 40 kA.

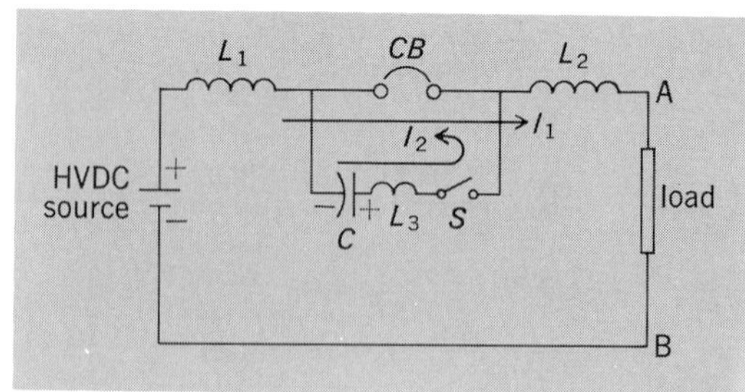

Fig. 8. Commutation principles of circuit interruption.

Thyristors can be used for both circuit breakers and switches. The closing and opening operations are, of course, not mechanical but are controlled by the gates. Such a circuit breaker has been built for very special switching applications, such as thermonuclear fusion research. [THOMAS H. LEE]

Bibliography: Thomas H. Lee, *Physics and Engineering of High Power Switching Devices*, 1975.

Circuit testing (electricity)

The testing of electric circuits to determine and locate any of the following circuit conditions: (1) an open circuit, (2) a short circuit with another conductor in the same circuit, (3) a ground, which is a short circuit between a conductor and ground, (4) leakage (a high-resistance path across a portion of the circuit, to another circuit, or to ground), and (5) a cross (a short circuit or leakage between conductors of different circuits). Circuit testing for complex systems often requires extensive automatic checkout gear to determine the faults defined above as well as many quantities other than resistance.

In cable testing, the first step in fault location is to identify the faulty conductor and type of fault. This is done with a continuity tester, such as a battery and flashlight bulb or buzzer (Fig. 1), or an ohmmeter.

Murray loop test. Useful for locating faults in relatively low-resistance circuits, the Murray loop is shown in Fig. 2 with a ground fault in the circuit under test. A known "good" conductor is joined to the faulty conductor at a convenient point beyond the fault but at a known distance from the test connection. One terminal of the test battery is grounded. The resulting Wheatstone bridge is then balanced by adjusting R_B until a null is obtained, as indicated by the detector in Fig. 2. Ratio R_A/R_B is

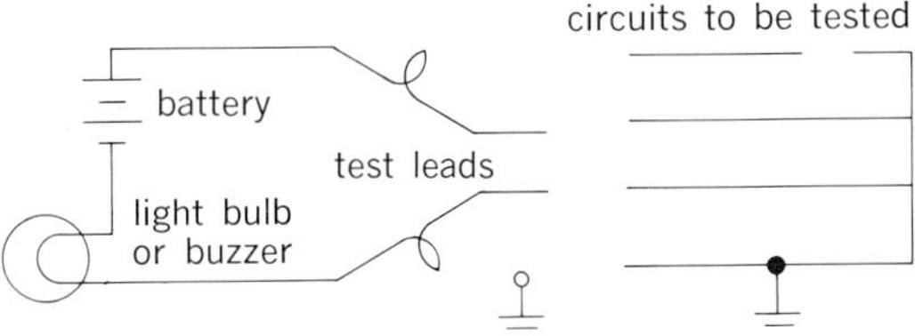

Fig. 1. Simple continuity test setup.

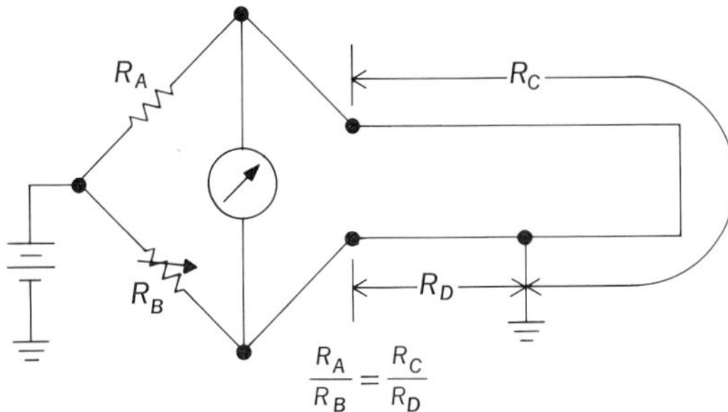

Fig. 2. Murray loop for location of ground fault.

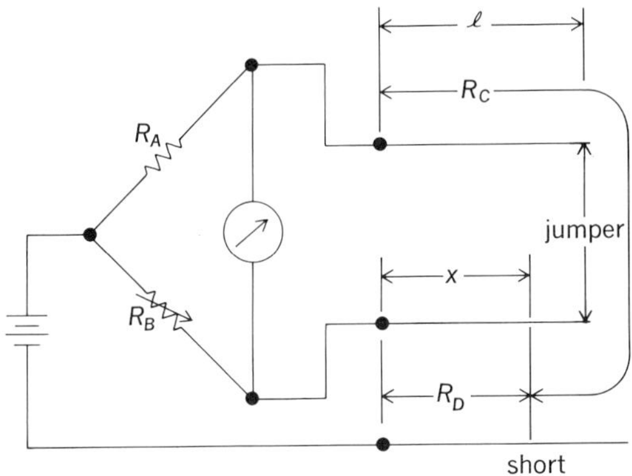

Fig. 3. Murray loop for location of short or cross fault.

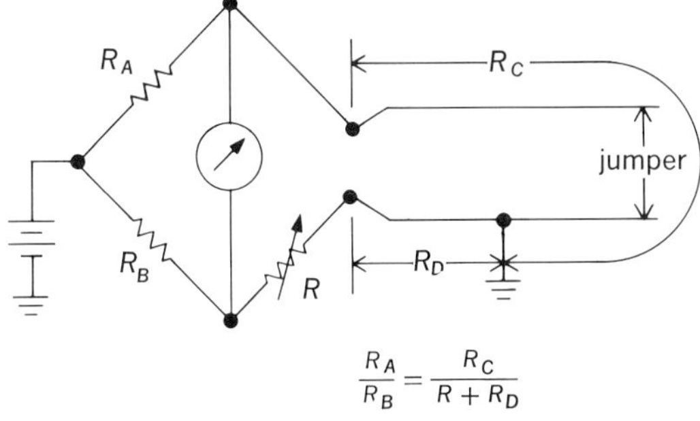

Fig. 4. Varley loop for location of leakage to ground.

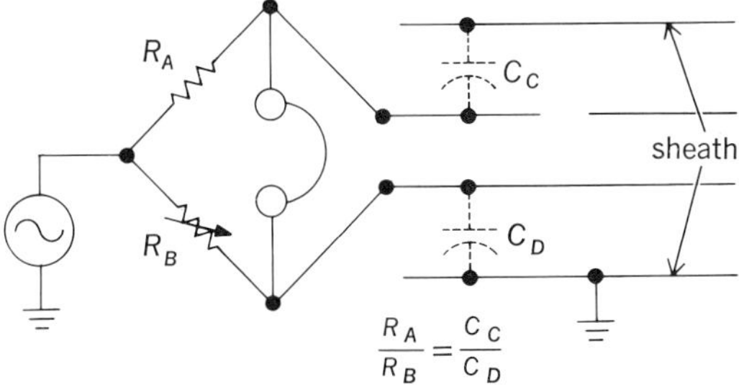

Fig. 5. Alternating-current capacitance bridge used in location of an open circuit in one conductor.

then known. For a circuit having a uniform ratio of resistance with length, circuit resistance is directly proportional to circuit length. Therefore, the distance to the fault is determined from the procedure given by Eq. (1)–(3). From Eq. (3) and a

$$R_C \propto l + (l - x) \qquad R_D \propto x \tag{1}$$

$$R_A/R_B = r = R_C/R_D = (2l - x)/x \tag{2}$$

$$x = 2l/(r - 1) \tag{3}$$

knowledge of total length l of the circuit, once ratio r has been measured, the location of the fault x is determined. If circuit resistances are not uniform with distances, as when the known faultless conductor is different in size from the faulty conductor, additional calculation taking into consideration the resistance per unit length of the conductors is necessary.

If the fault is a short circuit or a cross instead of a ground, the battery is connected to the conductor to which the short or cross has taken place (Fig. 3). The test circuit is then equivalent to the one used to locate a ground (Fig. 2). The bridge is balanced and the calculations carried out as before.

Varley loop test. This is similar to the Murray loop test except for the inclusion of the adjustable resistance R. The Varley loop (Fig. 4) is used for fault location in high-resistance circuits.

Open circuit. An alternating-current capacitance bridge can be used for locating an open circuit as shown in Fig. 5. One test terminal is connected to the open conductor and the other terminal to a conductor of known continuity in the cable. All conductors associated with the test are opened at a convenient point beyond the fault but at a known distance from the test connection. An audio oscillator supplies the voltage to the bridge, which is balanced by adjusting R_B for a null as detected by the earphones. Measured ratio R_A/R_B equals the ratio of capacitances between the lines and the grounded sheath. Because each capacitance is proportional to the length of line connected to the bridge, the location of the open circuit can be determined from Eq. (4).

$$R_A/R_B = C_C/C_D \tag{4}$$

For convenience in carrying out the tests for fault location, the fault location bridge has switches for setting up the various loop test circuits. Basically it is a bridge like other bridges used in circuit testing. *See* CIRCUIT (ELECTRICITY). [CHARLES E. APPLEGATE]

Bibliography: M. Braccio, *Basic Electrical and Electronic Tests and Measurement*, 1979; W. D. Cooper, *Electronic Instrumentation and Measurement Techniques*, 2d ed., 1978; F. K. Harris, *Electrical Measurements*, 1975.

Civil engineering

The planning, design, construction, and management of all types of works and facilities, including buildings and structures, transportation facilities, water resource development projects, power generation plants, and other facilities for enhancement of the environment. These functions are generally categorized as public works, although the civil engineer also serves industry and private clients. The Englishman John Smeaton, builder of the Eddystone lighthouse, was the first to identify himself as a "civil engineer," in 1782.

Buildings. The civil engineer plans, designs, and constructs residential, industrial, and commercial buildings ranging from warehouses to skyscrapers. Planning involves site selection, allocation of space to building functions, and provision for circulation of people and materials, all in compliance with zoning laws and building codes. Using steel and concrete as the main materials, the civil engineer designs a foundation suited to the geological conditions and loads to be accommodated, and then conceives a frame capable of resisting fire, wind forces, and seismic shocks to support the floors, walls, and roof. The first metal-frame high-rise structure was the 10-story Home Insurance Building in Chicago, erected in 1893.

The largest building in the world is the Vehicle Assembly Building at the NASA Space Center in Cape Canaveral, FL. A steel box 710 ft × 518 ft × 552 ft high (216 m × 158 m × 168 m), its 128,-000,000 ft^3 (11,900,000 m^3) would be able to enclose the Great Pyramid. *See* BUILDINGS; STRUCTURES.

Transportation. The urgent need in colonial America for movement of passengers and goods was an early responsibility of the civil engineer. The need was first met in the canal era which lasted until about 1830, when the railroad age began. The world today demands a broad spectrum of efficient and economical transportation facilities and terminals, including streets, highways, expressways, and parking facilities; rapid transit systems and subways; transmission systems and pipelines; waterways and harbors; airports; and the launch facilities for vehicles to outer space. *See* TRANSPORTATION ENGINEERING.

The design of highways and expressways is dictated by terrain features, the nature of the area to be traversed and served, availability of right-of-way, and existing and potential traffic volumes. The interstate highway system exemplifies the best available engineering as applied to alignment, slopes, drainage, slab design, appurtenant structures, and safety features. *See* HIGHWAY ENGINEERING; TRAFFIC ENGINEERING.

Bridges, tunnels, and other special structures appurtenant to highways and railroads are among the most challenging and spectacular projects of the civil engineer. The most creative skills are needed to meet all the parameters of load, topography, geology, social constraints, esthetics, and economy. *See* BRIDGE.

Water transportation, the most economical method of all, calls upon the civil engineer for channel improvements, canals, locks, dams, ports, harbors, and navigational facilities. Works for controlling erosion by rivers and by ocean tides and currents are also important. *See* COASTAL ENGINEERING; RIVER ENGINEERING.

Pipelines for carrying water, oil, gas, and slurries of solids have drawn increasingly upon the skills of the civil engineer in recent years. Transport of pulverized coal as a pipeline slurry is especially promising. *See* PIPELINE.

Air travel involves the civil engineer in airport planning, design, and construction. Site election, runway design, and terminal traffic control for planes, passengers, and baggage must meet demanding constraints of convenience, time-saving, safety, noise control, and esthetics.

The vehicle assembly buildings, roadways, pads, and special structures making up the rocket launching complex are demanding the most sophisticated planning, design, and construction techniques of the civil engineer.

Water resources. The harnessing of rivers to control floods, produce hydroelectric power, and provide domestic, industrial, and agricultural water supplies is a major area of civil engineering. Groundwater resource development is also important. Efforts toward exploitation of ocean resources for minerals, aquatic life, and power will demand a wide spectrum of civil engineering technology.

Flood control is usually accomplished by channelization to increase stream capacity and by dams to impound and store peak flows. Natural storage areas in the floodplain may often be utilized.

Water released from impounded or elevated natural reservoirs may be passed through turbines to generate electric power. About 5% of the electrical energy produced in the United States comes from this source.

As recently as 1930, waterborne typhoid fever was still a significant public health problem in the United States. The modern water treatment plant, using either a surface water or groundwater source, produces clean safe water free from suspended matter, pathogenic bacteria, and objectionable minerals, tastes, and odors. By means of elevated storage facilities and pumping stations, the potable water is carried through a pipe distribution network to the homes, industries, and fire hydrants of the city.

The civil engineer not only provides the potable water, but also plans, designs, and constructs the sewer system and treatment facilities for disposal of the used wastewater so as to avoid hazard to the public health and pollution of natural waterways. *See* SANITARY ENGINEERING.

Where it is necessary to transport water long distances from its source to the point of utilization, the civil engineer develops the necessary pipelines, aqueducts, canals, and pumping stations. The system for distribution and control of irrigation water in agriculture is also the civil engineer's responsibility.

The Grand Coulee Dam and Columbia Basin project exemplifies the utilization of the civil engineering art in multipurpose development of the Columbia River for flood control, hydropower, and large-scale irrigation. *See* DAM; HYDRAULICS.

Other fields. All large construction projects are planned, directed, and managed by the civil engineer, even though they may require major input by other engineering specialists and architects. Fossil fuel and nuclear power plants, sports complexes and stadiums, and chemical plants and other industrial installations are examples.

Some of the most sophisticated civil engineering works are out of sight beneath the ground, such as complicated foundations and tunnels. Knowledge of geology and soil science is required to solve these problems. *See* ENGINEERING GEOLOGY; FOUNDATIONS.

Civil engineers have a major role in the urban and regional planning process. They are involved in the solution of air, water, and land pollution problems and in the control of some insect-borne diseases. Management of solid wastes, including collection, treatment, and disposal—with attendant by-product recovery—is a civil engineering function. In summary, the overall quality of the environment is largely dependent upon the facilities and services that are provided through civil engineering.

[WILLIAM H. WISELY]

Use of computers. Each civil engineering activity requires the collection of large quantities of data and the determination therefrom of the most effective (optimum) design. In this respect, civil engineering is a decision-making operation. The civil engineer is called upon to determine the feasibility of a project, alternative locations, competitive materials, performance versus cost, safety and reliability, means for preserving such natural resources as air and water, and probable course of future changes. In making these decisions, he must utilize imagination, judgment, intuition, and experience.

With the large amount of information involved, the civil engineer often uses the computer to help evaluate alternatives. The large-scale automatic data processer is ideally suited for this role, because it is basically a logical device and an information organizer. Much of the information handling and processing formerly performed within engineering firms by large numbers of technicians is now accomplished efficiently by the computer. Operations include computation, secondary decision making (the kind made by technicians), drafting and data display, searches for alternative designs, storage of complete files of previous information, exhaustive checking and testing of solutions, and communications.

Almost every facet of modern-day civil engineering practice is influenced by the availability of high-speed computers. Large consulting firms purchase or rent their own computer equipment. Smaller firms and individuals purchase computing time on large computers operated by data-processing organizations.

Many civil engineering calculations can be reduced to the solution of equations that characterize mathematically the engineering process involved. These equations are frequently solved on the computer by use of mathematical approximations. Generally, the more precise the computations, the more computer time required, with consequent larger expenditure of funds for computer usage. In programming the problem for the computer, the civil engineer, therefore, considers use of the computer on the basis of the accuracy of the solution compared to the cost of acquiring this increased accuracy. After reaching a decision as to the level of precision to be sought, the engineer must then exercise careful judgment in the interpretation of the results from the computer.

Among the areas of civil engineering practice affected by the computer are bridge design and analysis, building design, earthquake- and blast-resistant design of structures, area computations in highway earthwork design, analysis of traffic surveys, soil analysis, slope stability of earth embankments, complex surveying computations, critical path scheduling of a construction project, analysis of fluid flow in sewer and drainage systems, and the computer-assisted drawing of blueprints. *See* DIGITAL COMPUTER.

[RICHARD L. JENNINGS]

Bibliography: American Society of Civil Engineers, *The Civil Engineer: His Origins*, 1970; A. R. Golze, *Your Future in Civil Engineering*, 1965; F. S. Merritt, *Standard Handbook for Civil Engineers*, 1978; W. H. Wisely, *The American Civil Engineer—1852–1974*, 1975.

Cladding

An old jewelry art, now employed on an industrial scale to add the desirable surface properties of an expensive metal to a low-cost or strong base metal. In the process a clad metal sheet is made by bonding or welding a thick facing to a slab of base metal; the composite plate is then rolled to the desired thickness. The relative thickness of the layers does not change during rolling. Cladding thickness is usually specified as a percentage of the total thickness, commonly 10% (Fig. 1).

Uses. Gold-filled jewelry has long been made by this process; the surface is gold, the base metal bronze or brass with the cladding thickness usually 5%. The process is used to add corrosion resistance to steel and to add electrical or thermal conductivity, or good bearing properties, to strong metals. One of the first industrial applications, about 1930, was the use of a nickel-clad steel plate for a railroad tank car to transport caustic soda; by 1947 stainless-clad steels were being used for food and pharmaceutical equipment. Corrosion-resistant pure aluminum is clad to a strong duralumin base, and many other combinations of metals are widely used in cladding; a development includes a technique for cladding titanium to steel for jet-engine parts.

Today's coinage uses clad metals as a replacement for rare silver. Since 1965 dimes and quarters have been minted from composite sheet consisting of a copper core with copper-nickel facing. The proportion of core and facing used duplicates the weight and electrical conductivity of silver so the composite coins are acceptable in vending machines (Fig. 2).

A three-metal composite sole plate for domestic steam irons provides a thin layer of stainless steel on the outside to resist wear and corrosion. A thick core of aluminum contributes thermal conductivity and reduces weight, and a thin zinc layer on the inside aids in bonding the sole plate to the body of the iron during casting.

Cladding supplies a combination of desired

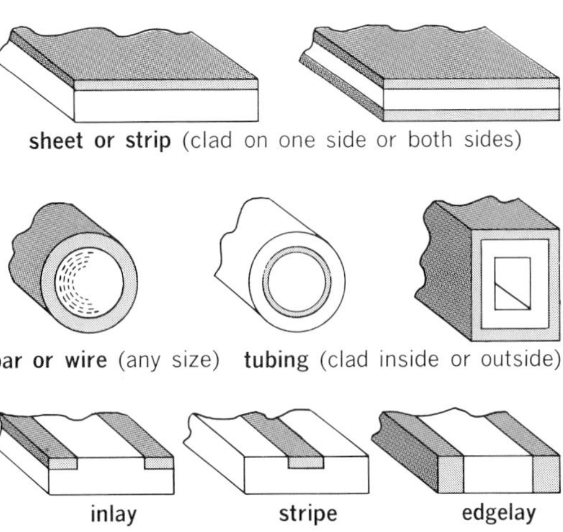

Fig. 1. Types of cladding.

Fig. 2. Rolls of copper clad with nickel silver for minting composite coins.

properties not found in any one metal. A base metal can be selected for cost or structural properties, and another metal added for surface protection or some special property such as electrical conductivity. Thickness of the cladding can be made much heavier and more durable than obtainable by electroplating.

Combinations. The following clad materials are in common use.

Stainless steel on steel. Provides corrosion resistance and attractive surface at low cost for food display cases, chemical processing equipment, sterilizers, and decorative trim.

Stainless steel on copper. Combines surface protection and high thermal conductivity for pots and pans, and for heat exchangers for chemical processes.

Copper on aluminum. Reduces cost of electrical conductors and saves copper on appliance wiring.

Copper on steel. Adds electrical conductivity and corrosion resistance needed in immersion heaters and electrical switch parts; facilitates soldering.

Nickel or monel on steel. Provides resistance to corrosion and erosion for furnace parts, blowers, chemical equipment, toys, brush ferrules, and many mechanical parts in industrial and business machines; more durable than electroplating.

Titanium on steel. Supplies high-temperature corrosion resistance. Bonding requires a thin sheet of vanadium between titanium and steel.

Bronze on copper. Usually clad on both sides, for current-carrying springs and switch blades; combines good electrical conductivity and good spring properties.

Silver on copper. Provides oxidation resistance to surface of conductors, for high-frequency electrical coils, conductors, and braiding.

Silver on bronze or nickel. Adds current-carrying capacity to low-conductivity spring material; cladding sometimes is in form of stripes or inlays with silver areas serving as built-in electrical contacts.

Gold on copper. Supplies chemical resistance to a low-cost base metal for chemical processing equipment.

Gold on nickel or brass. Adds chemical resistance to a stronger base metal than copper; also used for jewelry, wristbands, and watchcases.

Applications. Cladding can be added to both sides of a sheet or strip of base metal. Tubing can be supplied with a clad surface on inside or outside; round and rectangular wire can be clad similarly.

For some forms of electrical contacts, the composite materials are bonded side by side, or silver is inset as a stripe on one side or along the edges. This construction can place solid silver just where it is needed to form an electrical contact with no waste of costly metal.

Another related form of cladding is found in thermostatic bimetals in which equal thicknesses of low- and high-expansion metals are bonded together. With a change in temperature, differing expansion rates of the two metals cause the composite material to bend and thus operate valves in automobile cooling systems, or electrical contacts in room thermostats.

Clad wires with properly chosen proportions of materials of different thermal-expansion rates can match the thermal expansion of types of glass used for vacuum-tight seals for conductors in lamp bulbs and hermetically sealed enclosures.

In making parts from clad metal, the composite material can be bent, drawn, spun, or otherwise formed just the same as the base metal without breaking the bond. The maximum service temperature is limited by the melting point of the material at the juncture of the two metals; annealing temperatures during manufacturing are similarly limited.

Cut edges of clad sheets expose the base metal and thus may require special protection. Welding of thick clad sheets sometimes offers special problems in maintaining the integrity of the protective surface, but joining methods are generally the same as for the base metal. *See* ELECTROPLATING OF METALS; METAL COATINGS; NUCLEAR FUELS.

[ROBERT W. CARSON]

Clevis pin

A fastener with a head at one end and a hole at the other used to join a clevis to a rod. A clevis is a yoke with a hole formed or attached at one end of a rod (see illustration). When an eye or hole of a second rod is aligned with the hole in the yoke, a clev-

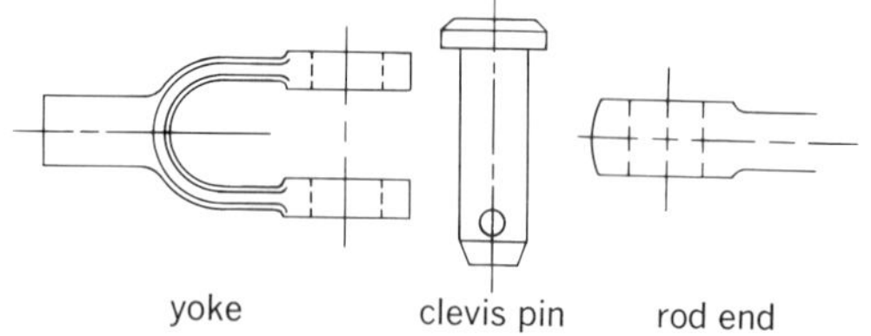

Clevis pin which joins yoke to rod end.

is pin can be inserted to join the two. A cotter pin can then be inserted in the hole of the clevis pin to hold it in, yet the fastening is readily detachable. This joint is used for rods in tension where some flexibility is required. [PAUL H. BLACK]

Bibliography: P. H. Black and O. E. Adams, Jr., *Machine Design*, 3d ed., 1969; V. H. Laughner and A. D. Hargan, *Handbook of Fastening and Joining of Metal Parts*, 1956; Society of Automotive Engineers, *SAE Handbook*, revised annually.

Clutch

A machine element for the connection and disconnection of shafts in equipment drives. If both shafts to be connected can be stopped or made to move relatively slowly, a positive-type mechanical clutch may be used. If an initially stationary shaft is to be driven by a moving shaft, friction surfaces must be interposed to absorb the relative slippage until the speeds are the same. Likewise, friction slippage allows one shaft to stop after the clutch is released.

There are many limitations on the applications of clutches for transmitting power through shafts. These limitations are created by the rotational speed of the shafts, the quantity of torque delivered, the space required for the mechanical parts, the availability of materials having the proper characteristics for absorbing frictional energy, and the heat dissipation qualities of assemblies for heavy-duty applications. Clutches may be created by the individual designer, or they may be purchased from manufacturers who have developed many specialized units.

Positive type. When positive connection of one shaft with another in a given position is needed, a positive clutch is used. This clutch is the simplest of all shaft connectors, sliding on a keyed shaft section or a splined portion and operating with a shift lever on a collar element (Fig. 1). Because it does not slip, no heat is generated in this clutch. Interference of the interlocking portions prevents engagement at high speeds; at low speeds, if connection occurs, shock loads are transmitted to the shafting.

Square jaw. Positive clutches may consist of two or more jaws of square section which mesh together in the opposing clutches. Because the faces, sides, and bottoms are rectangular, meshing does not result on contact unless the jaws are aligned. Thus, although square jaw clutches are the strongest and most elementary to construct, the difficulty of engagement limits their use.

Spiral jaw. A modification of the square-jaw clutch to permit more convenient engagement and to provide a more gradual movement of the mating faces toward each other produces the spiral jaw clutch. Although the most common positive clutch is the spiral-jaw type, ratchet and gear-teeth elements may be used in the clutch faces. Most power presses use these positive clutches to drive the punch-press parts for exact positioning and for automatically releasing the drive after completion of the working stroke. The starting impact on such equipment is not considered harmful, although wear and fatigue stresses from shock loads must be considered in the design.

Friction type. When the axial pressure of the clutch faces on each other serves to transmit torque instead of the mating shape of their parts, the clutch operates by friction. This friction clutch is usually placed between an engine and a load to be driven; when the friction surfaces of the clutch are engaged, the speed of the driven load gradually approaches that of the engine until the two speeds are the same. The rate at which the load can be applied to the engine depends upon the coefficient of friction of the materials in contact, the normal pressure on these materials, the mean radius of the tangential force, and the ability of the clutch elements and the housing to dissipate the heat generated during the slipping period. If this rate of application is too rapid, the rotational inertia resistance becomes high.

A friction clutch is necessary for connecting a rotating shaft of a machine to a stationary shaft so that it may be brought up to speed without shock and transmit torque for the development of useful work. The three common designs for friction clutches, combining axial and radial types, are cone clutches, disk clutches, and rim clutches. All these types require a common activation mechanism, one that has a stationary lever engaging a collar which, by its displacement, loads and unloads the friction surfaces to operate the clutch.

Cone clutch. In one form of friction clutch, the surfaces are sections of a pair of cones (Fig. 2). This shape uses the wedging action of the mating surfaces under relatively small axial forces to transmit the friction torque. These forces may be established by the compression of axial springs or by the outward displacement of bell-crank levers to apply axial thrust to the conical surfaces. To operate the clutch, a cylindrical spool mounted on the main shaft activates thrust levers and self-locking toggles when displaced by the shifting lever and yokes. The cone clutch is an elementary device and is limited in its versatility and power range.

Industrial applications of the cone clutch include uses as a power takeoff on couplings for driving pulleys, sheaves, and hoisting drums which are run intermittently from a continuously rotating shaft.

Machine-tool applications of the cone clutch arose from the requirements for reversing the spindle on high-speed machines. With double clutches in opposition and a plastic impregnated woven material for the inner cone, the lower inertia of the cone clutch elements permits rapid reversal with low energy loss.

CLUTCH

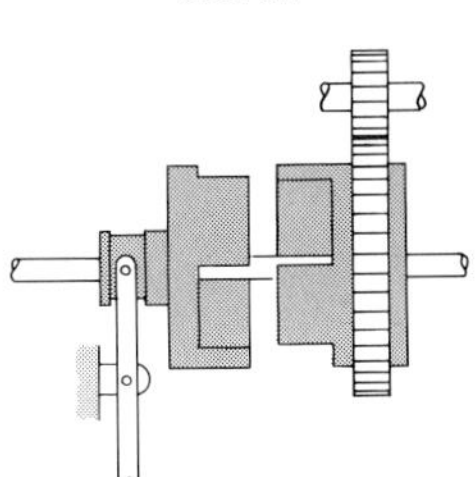

Fig. 1. Square-jaw-type positive clutch.

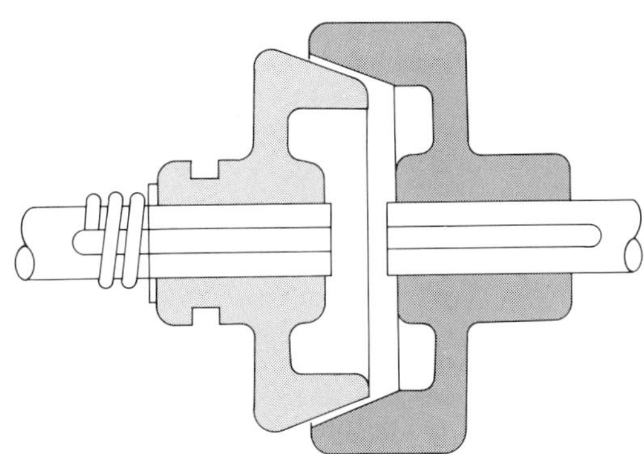

Fig. 2. Cone-type friction clutch.

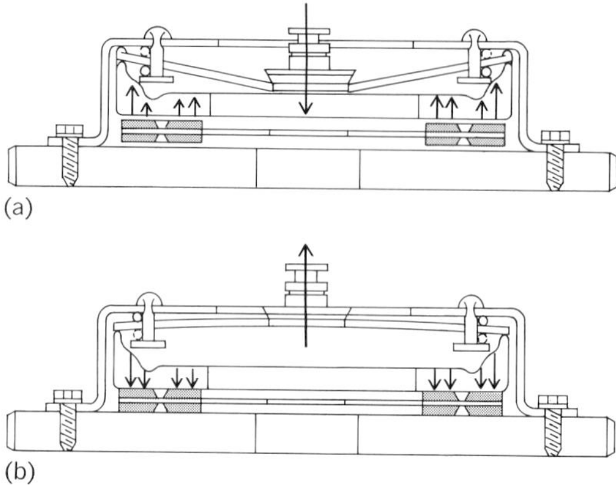

Fig. 3. Disk-type friction clutch. (*a*) In disengaged position. (*b*) In engaged position.

Disk clutch. With the advent of improved friction materials, disk clutches have become more common than cone clutches. They are capable of wide application in the industrial and automotive fields (Fig. 3). Disk clutches are not subject to centrifugal effects, present a large friction area in a small space, establish uniform pressure distributions for effective torque transmission, and are capable of effectively dissipating the heat generated to the external housing.

The disk clutch consists essentially of one or more friction disks connected to a driven shaft by splines. These friction disks have specially prepared friction material riveted to both sides. Contact plates keyed to the inner surface of an external hub separate the friction disks. The axial force, necessary to establish tangential friction between the friction disks and the contact plates for the transmission of the torque, is applied through an end pressure plate by springs or toggles in the assembly.

The disk clutch may be operated dry, as in most automobile drives, or wet by flooding it with a liquid, as in heavier automotive power equipment and in industrial engines. The advantage of wet operation is the ability to remove heat by circulating the liquid enclosed in the clutch housing.

Of particular importance are the friction materials used on the clutch disks; many special mechanical designs have been developed for large power drives and machine tools.

Rim clutch. Another form of frictional contact clutch has surface elements that apply pressure to the rim either externally or internally. Clutches of this nature may have hinged shoes connected together by an expanding or contracting kinematic mechanism on a hub fixed to one shaft and riding on a drum or rim attached to a hub on a second shaft. They have clutch expansion rings or external bands which when displaced are capable of transmitting torque to the clutch rim. The rims may be grooved to increase the surface; the clutch may have a double grip, internal and external; or pneumatically expanding flexible tubes may set the friction surfaces against the rim. Because the clutch rotates with the shaft when it is engaged, its balancing must be considered. *See* BRAKE.

Overrunning type. The driven shaft can run faster than the driving shaft with an overrunning clutch. This action permits freewheeling as the driving shaft slows down or another source of power is applied. Effectively this is a friction pawl-and-ratchet drive, wherein balls or rollers become wedged between the sleeve and recessed pockets machined in the hub. The clutch does not slip when the second shaft is driven, and is released automatically when the second shaft runs faster than the driver. Specially shaped struts or sprags may replace the balls; springs may be employed to hold the pawl elements in position (Fig. 4). Mechanisms have been devised for reversing the direction of operation. In the design, the effective wedging angle must be slightly greater then the coefficient of friction of the wedging elements. *See* RATCHET.

CLUTCH

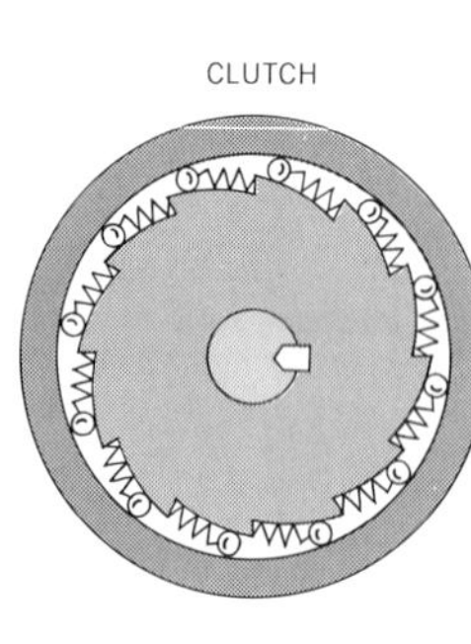

Fig. 4. Overrunning clutch with spring-constrained rollers or balls.

Centrifugal type. Centrifugal force from the speed of rotation can operate the clutching mechanism. Heavy expanding friction shoes may act by centrifugal force on the internal surface of a rim clutch. Flyball-type mechanisms are used to activate clutching surfaces on cones and disks. This type of clutch is not normally used because it becomes unwieldy and unsafe with increasing size.

Hydraulic type. Clutch action is also produced by hydraulic couplings, with a smoothness not possible with a mechanical clutch. Automatic transmissions in automobiles represent a fundamental use of hydraulic clutches. *See* FLUID COUPLING; TORQUE CONVERTER.

Electromagnetic type. Magnetic coupling between conductors provides a basis for several types of clutches. The magnetic attraction between a current-carrying coil and a ferromagnetic clutch plate serves to actuate a disk-type clutch. Slippage in such a clutch produces heat that must be dissipated and wear that reduces the life of the clutch plate. Thus the electromagnetically controlled disk clutch is used to engage a load to its driving source. A typical unit 24 in. in diameter and weighing about 600 lb develops 2400 lb-ft of torque when excited at 2 amp and 115 dc volts; at a maximum safe speed of 1200 rpm it transmits 540 hp. Multiple interleaved disks alternately splined to the driving and driven shafts provide a compact

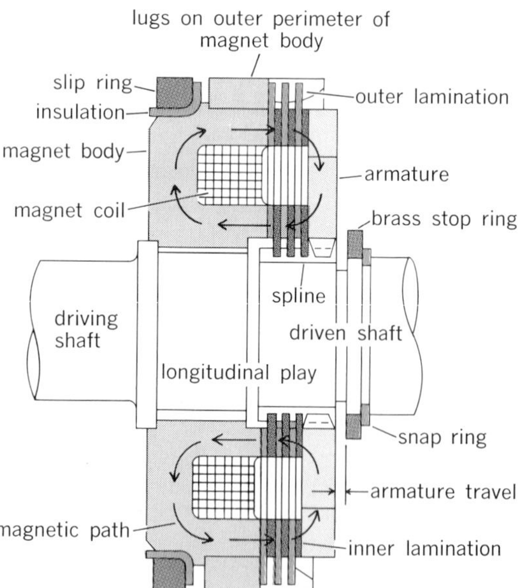

Fig. 5. Electromagnetically activated multiple disk clutch. (*I.T.E. Circuit Breaker Co.*)

structure (Fig. 5). *See* ELECTROMAGNET.

Magnetic fluid and powder clutches. Finely divided magnetic particles in an oil carrier or dry can fill the space between what, in effect, is a disk clutch. Upon the application of a magnetic field, the particles tend to coalesce, thereby creating viscous or friction forces between the clutch members. The wet type provides the smoother operation; the dry type is freer from loss of particles. Torque at a fixed moderate excitation is independent of slip.

Eddy-current clutch. The current induced by a magnetic field can serve as the torque-transmitting means. In the eddy-current clutch, a coil carrying direct current in one rotating member establishes poles that generate currents in the adjacent smooth conductive ring of the mating member. Torque increases as the slip and the excitation increase. Efficiency is high, better than 96% at full load. Excitation depends on size, and is typically 2% of the total power handled. The coupling acts as an untuned damper and therefore does not transmit torsional vibration. This characteristic and the rising torque-slip characteristic are desirable in such applications as power shovels, drag lines, industrial trucks, punch presses, and wind tunnels. Such a clutch consists of a separately excited motor and generator combined into a single machine. Clutch action can be obtained by electrically connecting a generator on the driving shaft to a motor on the driver or by various intermediate designs. *See* DYNAMOMETER; ELECTRIC ROTATING MACHINERY.

Hysteresis clutch. For low-power applications where continuous control without slip is needed, as in instrument servos, a hysteresis clutch may be used. Direct-current excitation of a coil in one part generates a steady flux field that magnetizes an iron ring on the other part, the ring having high magnetic retentivity. The action induces poles in the iron ring. Attraction between the induced poles in the ring and the control field produces a torque that opposes rotation between the two parts of the clutch. Within its load limit the hysteresis clutch transmits rotation without slip. Like the eddy-current clutch, there is no contact surface to wear so that the characteristics are stable and the life is long. *See* COUPLING.

[JOHN E. GIBSON; JAMES J. RYAN]

Bibliography: V. L. Doughtie and A. Valance, *Design of Machine Members*, 4th ed., 1964; I. Levinson, *Machine Design*, 1978.

Coal mining

The technical and mechanical job or removing coal from the earth and preparing it for market. Coal, the most abundant and traditionally the most economical source of power in the world, is found in varying amounts throughout the globe. It lies in veins of various thickness and richness beneath the crust of the Earth. The product of fossilized plant material mixed with various mineral matter, coal rests in giant subterranean sandwiches, shallow or deep, flat or pitched. At the present time, there are coal mines in the United States operating at the 1500-ft (450-m) level in Virginia and shafts at the 1800-ft (540-m) level in Alabama. Obtaining coal in sufficient quantities at a competitive expense and making it the proper grade for the market demand is, in its simplest terms, the science of coal mining. Mammoth mining machines require technically qualified manpower. The pick-and-shovel miner has all but disappeared from the American scene, along with the stereotyped characterization of him as an oppressed laborer living in a company town. Today's miner is a skilled, well-paid technician who handles complex, costly, and highly efficient machines.

Prospecting and planning. Of major importance in coal property development is the accumulation of seam information from borehole drilling. Chemical analysis of the cores will provide details concerning moisture, volatile material, fixed carbon, ash, sulfur, Btu per pound, and fusion temperature of the ash. Washability data will give the percentage of float (coal) and sink (foreign material) for each size coal at each specific gravity used (generally from 1.35 to 1.60), together with the amounts of ash and sulfur and the Btu value on a dry basis for each increment of specific gravity. *See* BORING AND DRILLING (MINERAL): PROSPECTING.

After preliminary analyses and correlations, the engineering details are planned. Production requirements and the extent of coal reserves generally determine the method of mining and the type and capacity of equipment to be used. Amortization of the investment and an adequate return on the investment are also influential factors. The cost of opening, developing, and operating the mine is estimated. A reasonable evaluation can then be based on the projected costs and markets available.

Development is commonly planned for the life of the property. This involves projecting and estimating the working sequence of the various parts of the area; the type of mining equipment to be used; transportation; water drainage; ventilation and roof control (underground); overburden analysis

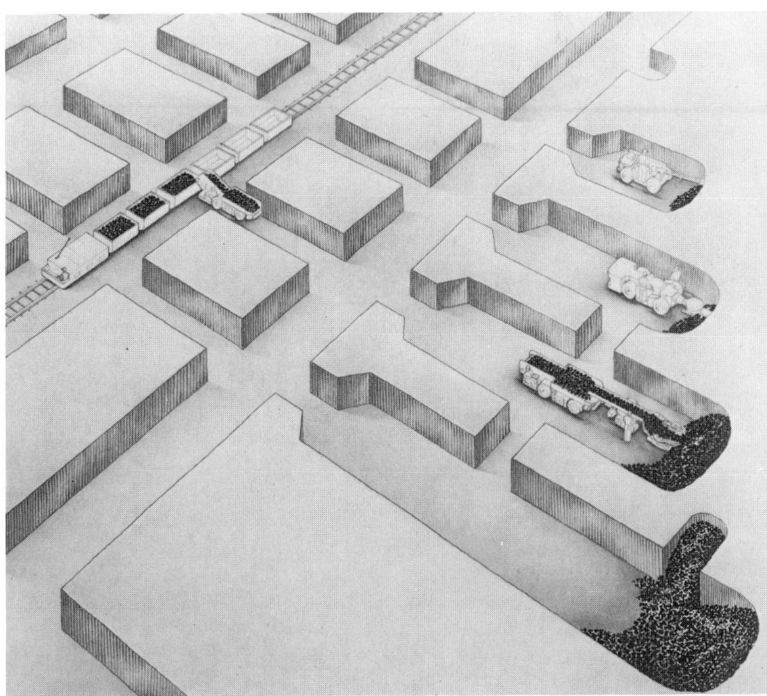

Fig. 1. Conventional coal mining system, for underground mining, shuttle car and train haulage. (*Joy Manufacturing Co.*)

Fig. 2. Coal mining system with continuous miner using shuttle car and belt haulage. (*Joy Manufacturing Co.*)

(surface); and all the interrelations of these factors with costs and quality. *See* MINING OPERATING FACILITIES.

Coal is extracted from the earth by three basic methods: (1) underground, (2) strip, and (3) auger. Approximately two-thirds of America's coal comes from underground mining. The remaining one-third is mined from the surface, either by strip or auger mining.

Underground mining. The systems of underground coal mining generally in use are room-and-pillar, longwall, and, in a few places in Europe, hydraulic. In the room-and-pillar system, tunnels are carved into the seam, leaving pillars of coal for support. In some mines these pillars are removed in subsequent mining, allowing the overlying strata to collapse; in others the pillars are not recovered. In longwall mining, widely spaced tunnels are driven, leaving large blocks of coal. Later, these blocks are completely extracted, allowing the roof material to collapse behind the coal face as it is removed. In hydraulic mining, as now practiced, a stream of water is directed against the coal face with sufficient pressure to dislodge the coal. The water also acts as a transporting medium.

The room-and-pillar system of mining has been the most widely practiced method in the United States, whereas the longwall system has been used for many years in Europe. However, the demand for metallurgical coal has necessitated mining deeper into the earth. With the advent of mechanical plows and shearers, it has become economically feasible to use the longwall system in the United States.

Fig. 3. Continuous miner. (*Lee-Norse Co.*)

Underground mining equipment. The type of equipment used at the mining face is governed by a complex of factors. Outstanding are the relative difficulties of supporting the immediate roof, the height of the seam, grades of coal, maintenance required on machinery, and productivity expected of manpower using different types of machines. Today, mechanical equipment falls into two classes, the so-called conventional and the continuous miner machine (Figs. 1 and 2).

In the conventional method several machines are used in a cycle of operation: undercutting or top-cutting the seam with cutting machines; blasting down the face; loading coal with a mechanical loader; and transporting it from the face by shuttle car or conveyor. The continuous miner (Fig. 3) has been used in many seams. It bores or rips to dislodge the coal from the face without blasting, then loads the coal and puts it into the transportation system.

Fig. 4. Longwall mining equipment. Plow works back and forth across working face. (*Mining Progress, Inc.*)

The longwall mining machine (Fig. 4) employs a plow or shearer which is pulled back and forth across a working face several hundred feet long. The loosened coal is dropped into a conveyor. Self-advancing hydraulic jacks support the roof and follow the machine as it slices into the coal on a wide front. With longwall mining in overburdens of 1000 ft or more, it is possible to mine extremely gassy seams. Because of the large tonnage being extracted from a single face, large volumes of air are employed to dilute the liberated gases. The massive jacks used in longwalling control the weight of the roof and also create the falls necessary to remove the pressure at the coal face, thereby providing protection from roof falls.

The power source for operating mining machinery can be liquid fuels, seldom used in coal mining, or electricity, either alternating or direct current. Power distribution for a mining property has to be planned for a whole mine in the same manner as the actual mining operation.

Prospecting information and actual experience determine the amount and type of roof control necessary for the protection of workmen. The old type of roof support is so-called timber; this may be simply wooden posts, or may consist of beams of steel or wood supported by posts or fastened to the walls. Currently, the use of roof bolts has replaced a great amount of timber support. Holes are drilled into the roof, and steel rods, held by expansion devices, are screwed in until the rod head is tight against the surface. The size, spacing, and length of the bolts are governed by the roof conditions to be controlled. Roof bolting combines many roof materials into one large beam instead of letting the various strata act separately. For permanent support at some mines the roof is gunited, particularly when the roof materials are subject to deterioration by weathering if exposed to the air currents.

Mined coal is transported from the working face to the main transportation system by shuttle cars (underground trucks) (Fig. 5) or by conveyor. Shuttle cars are used predominantly, although many properties with steep grades or thin coal find it more economical to employ conveyors. In hydraulic mining in Europe, the water used to break down the coal also conveys the coal away from the face.

Fig. 5. Shuttle car hauling from working face area in a coal mine to the main transportation system. (*National Mine Service Co.*)

The main transportation to the surface or to shafts or slopes is generally provided by mine cars which are pulled along tracks by locomotives. Although not as common, there are many instances where coal is transported on conveyor belts. Mainline belt transportation eliminates a great deal of grading required for track haulage but requires essentially the same amount of maintenance and roof protection.

The largest cars that seam height and width of working areas will permit are used in track systems. Modern mine cars hold 6–20 tons of material. Most have solid bodies that are emptied by tipping or rotary dumping. Motive power is generally provided by electric locomotives of 20-tons weight.

When the coal does not have a surface outcrop, it must be elevated by a slope or shaft. Almost all slopes are on inclinations that permit the use of belts to transport the coal, which is dumped at the bottom of the slope, to the surface. These belts are installed with sufficient capacity to carry the maximum production of the mine on a daily or shift basis. When the coal is brought to a shaft, it is dumped into skip buckets and hoisted to a dump at the surface. Modern installation in these shafts are completely automatic as to loading and dumping.

An underground coal mine consists of a great number of spaces and openings in which people must work safely. In many cases, however, explosive gas is emitted. Hence, artificial ventilation over the entire mine is required to maintain a normal atmosphere and to dilute and carry away such gases. Numerous shafts and fans provide the necessary volume of air.

To prevent water from entering and to eliminate it from the mine are the aims of mine drainage. Removing excess water from mine properties may be difficult and expensive; in extreme cases over 30 tons of water have to be removed for each ton of coal mined. *See* MINING MACHINERY; UNDERGROUND MINING.

Strip mining. Where the coal seam lies close to the surface, it is more economical to remove the overburden of earth and rock that covers the coal seam. For this job, power shovels, draglines, or wheel excavators are used. As a farmer plows his field in furrows, a shovel excavates a "furrow" and casts the overburden parallel to the cut. Draglines sit on the bank above the coal seam and remove the overburden from the seam. Bucket-wheel machines excavate the material from above the coal seam. It then flows in a continuous stream via a transfer to the conveyor system, which in turn transports it to the discharge point. When the seam is exposed, the coal is loaded by smaller power shovels into trucks and hauled to preparation plants or loading bins. *See* STRIP MINING.

Strip mining is an efficient way to mine coal. Seventy-foot-thick seams of coal in Wyoming are being strip-mined; seams such as these will furnish the United States with a substitute for natural gas. Gigantic machines have been developed to efficiently mine coal by the strip-mining method (Fig. 6). Off-the-highway haulage trucks in the 100-ton category are common; units capable of carrying 240 tons are available. Auxiliary equipment used to complement the excavating machines has correspondingly increased in capacity.

Augering. For coal seams which continue under rising land too thick for economical strip mining and where underground mining cannot burrow further to the surface because of the shallow and more treacherous roof conditions, a relatively new procedure of mining, the auger method, was developed in 1951. The auger miner (Fig. 7) twists huge drills like carpenter's bits into a hillside coal seam, drawing out the coal to a conveyor which loads it into trucks. Section by section, the drills

Fig. 6. Sketch of 220-yd^3 dragline. (*Bucyrus-Erie Co.*)

Fig. 7. Single-head auger. (*Salem Tool Co.*)

bore into the hillside. Augering in general will recover 40–60% of the coal seam. The development of the dual-headed auger and the multiheaded auger, with progressive increase in size until the giants of today have been manufactured with head diameters up to 96 in. (2.4 m), has made it possible to penetrate to a depth of 300 ft (90 m) in a level seam. Auger recovery gives additional tonnage at minimal cost for equipment and labor and permits recovery of coal that might not be recovered by other methods.

Cleaning, grading, and shipping. After the coal is mined, it may be loaded directly into transportation facilities for the market if the foreign material in it is not excessive. However, coal from most seams requires preparation to provide a desired and uniform quality.

A number of washing devices are used for cleaning coal, all of which operate on the basic principle of the difference of specific gravity between coal and foreign material. The coal is floated in a vessel and the foreign material, being heavier, drops to the bottom. Washed coal in sizes less than 3/8 in. (9.3 mm) carries sufficient water from the washing circuit so that it must be dewatered to be marketable. Where a market can accept a moisture content of 7–8%, mechanical devices can meet the requirements. If a lower moisture is required, the product is thermally dried of surface moisture to the required extent after being mechanically dewatered. Once the coal is ready for market, it is loaded for shipment to the customer by railroad, water transportation, or in some cases belt conveyor systems.

The 108-mi (174-km) coal slurry pipe line between Cadiz and Cleveland, both in Ohio, pioneered a new concept in coal transportation. A 273-mi (439-km) coal slurry pipe line is also in operation between Arizona and Nevada. This innovation in coal transportation has proved to be a more economical method than building railroads for coal shipment.

One innovation in the industry has been the adoption of the unit train concept for coal shipments. A unit train, in its true sense, is a complete train of cars (usually privately owned) with assigned locomotives. It operates only on a regularly scheduled cycle movement between a single origin and a single destination each trip.

Future trends. In view of improvements in the past, the following future developments can be expected: (1) Conversion of coal into oil. Present coal reserves in the United States are known to be 3.2×10^{12} tons (2.9×10^{12} metric tons). With 1 ton of coal converting into approximately 4.5 bbl of oil, present reserves will yield more than 14×10^{12} bbl (2.2×10^{12} m^3), which is more oil than presently known to exist in the world. (2) Conversion of coal into gas to be placed in the underground storage fields and delivered into the major pipelines which intersect the coal fields. (3) Further research into the possibilities of using coal tars and shales as a source for synthetic oils.

[JAMES D. REILLY]

Bibliography: National Coal Association, *International Coal*, annually; J. R. Rowe (ed.), *Coal Surface Mining: The Impact of Reclamation*, 1979; Society of Mining Engineers, *Surface Mining*, 1968.

Coastal engineering

A branch of civil engineering concerned with the planning, design, construction, and maintenance of works in the coastal zone. The purposes of these works include control of shoreline erosion; development of navigation channels and harbors; defense against flooding caused by storms, tides, and seismically generated waves (tsunamis); development of coastal recreation; and control of pollution in nearshore waters. Coastal engineering usually involves the construction of structures or the transport and possible stabilization of sand and other costal sediments.

The successful coastal engineer must have a working knowledge of oceanography and meteorology, hydrodynamics, geomorphology and soil mechanics, statistics, and structural mechanics. Tools that support coastal engineering design include analytical theories of wave motion, wave-structure interaction, diffusion in a turbulent flow field, and so on; numerical and physical hydraulic models; basic experiments in wave and current flumes; and field measurements of basic processes such as beach profile response to wave attack, and the construction of works. Postconstruction monitoring efforts at coastal projects have also contributed greatly to improved design practice.

Environmental forces. The most dominant agent controlling coastal processes and the design of coastal works is usually the waves generated by the wind. Wind waves produce large forces on coastal structures, they generate nearshore currents and the alongshore transport of sediment, and they mold beach profiles. Thus, a primary concern of coastal engineers is to determine the wave climate (statistical distribution of heights, periods, and directions) to be expected at a particular site. This includes the annual average distribution as well as long-term extreme characteristics. In addition, the nearshore effects of wave refraction, diffraction, reflection, breaking, and runup on structures and beaches must be predicted for adequate design.

Other classes of waves that are of practical importance include the astronomical tide, tsunamis, and waves generated by moving ships. The tide raises and lowers the nearshore water level and thus establishes the range of shoreline over which coastal processes act. It also generates reversing currents in inlets, harbor entrances, and other locations where water motion is constricted. Tidal currents which often achieve a velocity of 1 – 2 m/s can strongly affect navigation, assist with the maintenance of channels by scouring sediments, and dilute polluted waters.

Tsunamis are quite localized in time and space but can produce devastating effects. Often the only solution is to evacuate tsunami-prone areas or suffer the consequences of a surge that can reach elevations in excess of 10 m above sea level. Some attempts have been made to design structures to withstand tsunami surge or to plant trees and construct offshore works to reduce surge velocities and runup elevations.

The waves generated by ships can be of greater importance at some locations than are wind-generated waves. Ship waves can cause extensive bank erosion in navigation channels and undesirable disturbance of moored vessels in unprotected marinas.

On coasts having a relatively broad shallow offshore region (such as the Atlantic and Gulf coasts of the United States), the wind and lower pressures in a storm will cause the water level to rise at the shoreline. Hurricanes have been known to cause storm surge elevations of as much as 5 m for periods of several hours to a day or more. Damage is caused primarily by flooding, wave attack at the raised water levels, and high wind speeds. Defense against storm surge usually involves raising the crest elevation of natural dune systems or the construction of a barrier-dike system.

Other environmental forces that impact on coastal works include earthquake disturbances of the sea floor and static and dynamic ice forces. Direct shaking of the ground will cause major structural excitations over a region that can be tens of kilometers wide surrounding the epicenter of a major earthquake. Net dislocation of the ground will modify the effect of active coastal processes and environmental forces on structures.

Ice that is moved by flowing water and wind or raised and lowered by the tide can cause large and often controlling forces on coastal structures. However, shore ice can prevent coastal erosion by keeping wave action from reaching the shore.

Coastal processes. Wind-generated waves are the dominant factor that causes the movement of sand parallel and normal to the shoreline as well as the resulting changes in beach morphology. Thus, structures that modify coastal zone wave activity can strongly influence beach processes and geometry.

Active beach profile zone. Figure 1 shows typical beach profiles found at a sandy shoreline (which may be backed by a cliff or a dune field). The backshore often has one or two berms with a crest elevation equal to the height of wave runup during high tide. When low swell, common during calm conditions, acts on the beach profile, the beach face is built up by the onshore transport of sand. This accretion of sand adds to the seaward berm. On the other hand, storm waves will attack the beach face, cut back the berm, and carry sand offshore. This active zone of shifting beach profiles occurs primarily landward of the −10-m depth contour. If the storm tide and waves are sufficiently high, the berms may be eroded away to expose the dunes or cliff to erosion. The beach profile changes shown in Fig. 1 are superimposed on any longer-term advance or retreat of the shoreline caused by a net gain or loss of sand at that location.

Any structure constructed along the shore in the active beach profile zone may retain the sand behind it and thus reduce or prevent erosion. However, wave attack on the seaward face of the structure causes increased turbulence at the base of the structure, and usually increased scour which must be allowed for or prevented, if possible, by the placement of stone or some other protective material.

It is desirable to keep all construction of dwellings, recreational facilities, and such landward of the active beach profile zone, which usually means landward of the frontal dunes or a good distance back from retreating cliffs. It is also desirable to maintain and encourage the growth of the frontal dune system by planting grass or installing sand fencing.

In addition to constructing protective structures and stabilizing the dune system, it is common practice to nourish a beach by placing sand on the beach face and nearshore area. This involves an

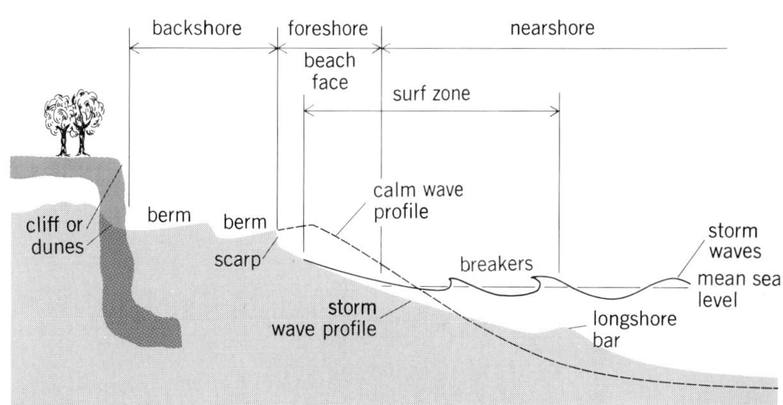

Fig. 1. Typical beach profiles (vertical scale is exaggerated).

initial placement of sand to develop the desired profile and periodic replenishment to make up for losses to the profile. A common source for sand, which should be clean and at least as coarse as the native sand, is the offshore area near the nourishment site.

Alongshore current and transport. Waves arriving with their crest oriented at an angle to the shoreline will generate a shore-parallel alongshore current in the nearshore zone. The current flows in the direction of the alongshore component of wave advance and has the highest velocity just inside the breaker line. It may be assisted or hindred by the wind and by tidal action, particularly along sections of the shore adjacent to tidal inlets and harbor entrances. There is a continuous accumulation of flow in the downcoast direction which may be relieved by seaward-flowing jets of water known as rip currents.

The alongshore current transports sand in suspension and as bed load, and is assisted by the breaking waves, which place additional sand in suspension. Also, wave runup and the return flow transports sand particles in a zigzag fashion along the beach face. Coastal structures that obstruct these alongshore transport processes can cause the deposition of sand. They do this by blocking wave action from a section of the shore and thus removing the wave energy required to maintain the transport system; by interfering with the transport process itself; or by directly shutting off a source of sand that feeds the transport system (such as a structure that protects an eroding shoreline).

The design of most coastal works requires a determination of the volumetric rate of alongshore sand transport at the site—both the gross rate (upcoast plus downcoast transport) and the net rate (upcoast minus downcoast transport). The most reliable method of estimating transport rates is by measuring the rate of erosion or deposition at an artificial or geomorphic structure that interrupts the transport. Also, field studies have developed an approximate relationship between the alongshore transport rate and the alongshore component of incident wave energy per unit time. With this, net and gross transport rates can be estimated if sufficient information is available on the annual wave climate. Typical gross transport rates on exposed ocean shorelines often exceed 500,000 yd^3 (382,000 m^3) per year.

Primary sources of beach sediment include rivers discharging directly to the coast, beach and cliff erosion, and artificial beach nourishment. Sediment transported alongshore from its sources will eventually be deposited at some semipermanent location or sink. Common sinks include harbors and tidal inlets; dune fields; offshore deposition; spits, tombolos, and other geomorphic formations; artificial structures that trap sand; and areas where beach sand is mined.

By evaluating the volumetric transports into and out of a segment of the coast, one can develop a sediment budget for the coastal segment. If the supply exceeds the loss, shoreline accretion will occur, and vice versa. When a coastal project modifies the supply or loss to the segment, geomorphic changes can be expected. For example, when a structure that traps sediment is constructed upcoast of a point of interest, the shoreline at the point of interest can be expected to erode as it resupplies the longshore transport capacity of the waves.

Harbor entrance and tidal inlet control structures built to improve navigation conditions, stabilize navigation channel geometry, and assist with the relief of flood waters will often trap a large portion of the alongshore transport. This can result in undesirable deposition at the harbor or inlet entrance and subsequent downcoast erosion. The solution usually involves designing the entrance structures to trap the sediment at a fixed and acceptable location and to provide protection from wave attack at this location so a dredge can pump the sand to the downcoast beach.

Coastal structures. Coastal structures can be classified by the function they serve and by their structural features. Primary functional classes include seawalls, revetments, and bulkheads; groins; jetties; breakwaters; and a group of miscellaneous structures including piers, submerged pipelines, and various harbor and marina structures.

Seawalls, revetments, and bulkheads. These structures are constructed parallel or nearly parallel to the shoreline at the land-sea interface for the purpose of maintaining the shoreline in an advanced position and preventing further shoreline recession. Seawalls are usually massive and rigid, while a revetment is an armoring of the beach face with stone rip-rap or artificial units. A bulkhead acts primarily as a land-retaining structure and is found in a more protected environment such as a navigation channel or marina.

A key factor in the design of these structures is that erosion can continue on adjacent shores and flank the structure if it is not tied in at the ends. Erosion on adjacent shores also increases the exposure of the main structure to wave attack. Structures of this class are prone to damage and possible failure caused by wave-induced scour at the toe. In order to prevent this, the toe must be stabilized by driving vertical sheet piling into the beach, laying stone on the beach seaward of the toe, or maintaining a protective beach by artificial nourishment. Revetments that are sufficiently porous will allow leaching of sand from below the structure. This can lead to structure slumping and failure. A proper stone or cloth filter system must be developed to prevent damage to the revetment. *See* BULKHEAD; REVETMENT.

Groins. A groin is a structure built perpendicular to the shore and usually extending out through the surf zone under normal wave and surge-level conditions. It functions by trapping sand from the

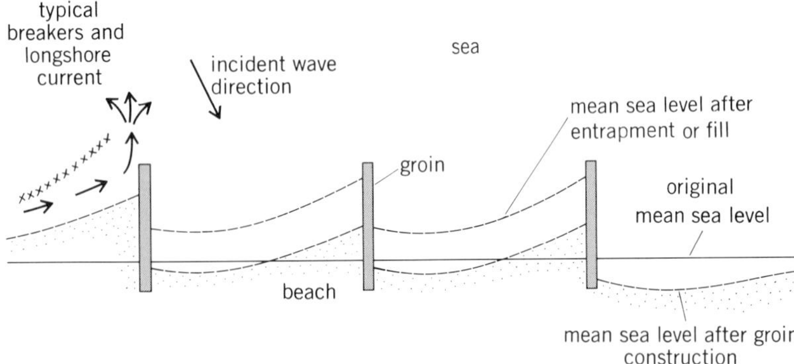

Fig. 2. Groin system and beach response.

alongshore transport system to widen and protect a beach or by retaining artificially placed sand. The resulting shoreline positions before and after construction of a series of groins and after nourishment is placed between the groins are shown schematically in Fig. 2. Typical groin alongshore spacing-to-length ratios vary from 1.5:1 up to 4:1.

There will be erosion downcoast of the groin field, the volume of erosion being approximately equal to the volume of sand removed by the groins from the alongshore transport system. Groins must be sufficiently tied into the beach so that downcoast erosion superimposed on seasonal beach profile fluctuations does not flank the landward end of a groin. Even the best-designed groin system will not prevent the loss of sand offshore in time of storms. *See* GROIN.

Jetties. Jetties are structures built at the entrance to a river or tidal inlet to stabilize the entrance as well as to protect vessels navigating the entrance channel. Stabilization is achieved by eliminating or reducing the deposition of sediment coming from adjacent shores and by confining the river or tidal flow to develop a more uniform and hydraulically efficient channel. Jetties improve navigation conditions by eliminating bothersome crosscurrents and by reducing wave action in the entrance.

At many entrances there are two parallel (or nearly parallel) jetties that extend approximately to the seaward end of the dredged portion of the channel. However, at some locations a single updrift or downdrift jetty has been used, as have other arrangements such as arrowhead jetties (a pair of straight or curved jetties that converge in the seaward direction). Jetty layouts may also be modified to assist sediment-bypassing operations. A unique arrangement is the weir-jetty system in which the updrift jetty has a low section or weir (crest elevation about mean sea level) across the surf zone. This allows sand to move over the weir section and into a deposition basin for subsequent transport to the downcoast shore by dredge and pipeline. *See* JETTY.

Breakwaters. The primary purpose of a breakwater is to protect a shoreline or harbor anchorage area from wave attack. Breakwaters may be located completely offshore and oriented approximately parallel to shore, or they may be oblique and connected to the shore where they often take on some of the functions of a jetty. At locations where a natural inland site is not available, harbors have been developed by the construction of shore-connected breakwaters that cover two or three sides of the harbor.

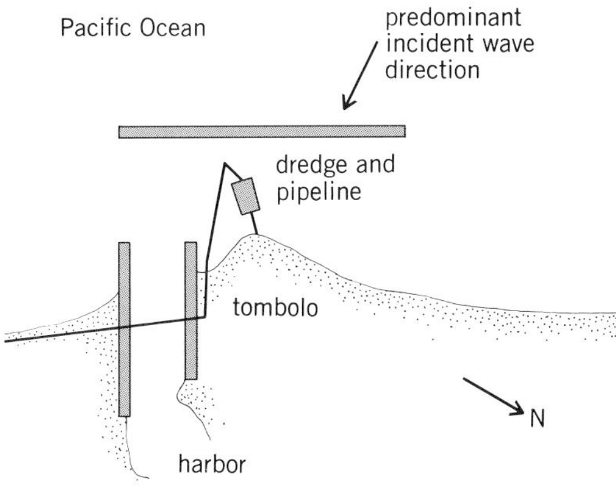

Fig. 3. Overhead view of breakwater and jetty system of Channel Islands Harbor, in California.

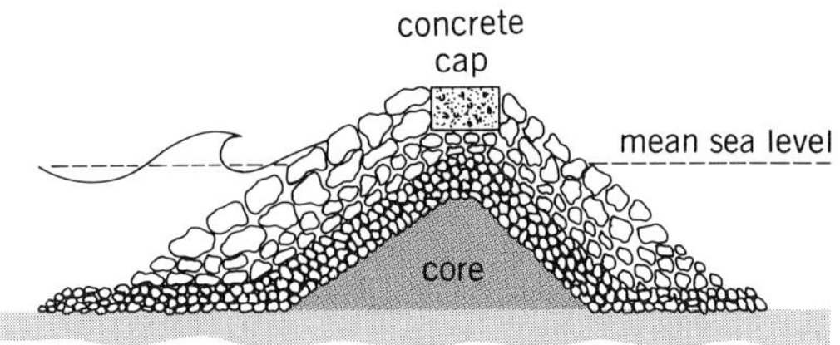

Fig. 4. Rubble mound breakwater cross section.

Figure 3 shows the breakwater and jetty system at the entrance to Channel Islands Harbor in southern California. The offshore breakwater intercepts incident waves, thus trapping the predominantly southeastern longshore sand transport; it provides a protected area where a dredge can operate to bypass sediment; and it provides protection to the harbor entrance. A series of shore-parallel offshore breakwaters (with or without artificial nourishment) has been used for shore protection at a number of locations. If there is sufficient fill or trapped material, the tombolo formed in the lee of each breakwater may grow until it reaches the breakwater.

Breakwaters are designed to intercept waves, and often extend into relatively deep water, so they tend to be more massive structures than are jetties or groins. Breakwaters constructed to provide a calm anchorage area for ships usually have a high crown elevation to prevent overtopping by incident waves and subsequent regeneration of waves in the lee of the breakwater.

Groins, jetties, and breakwaters are most commonly constructed as rubble mound structures. Figure 4 shows the cross section of a typical rubble mound breakwater placed on a sand foundation. The breakwater has an outer armor layer consisting of the largest stones or, if sufficiently large stones are not available, the armor units may be molded of concrete with a special shape. Stone sizes decrease toward the core and base in order to develop a filter system so that the fine core stone and base sand are not removed by wave and current action. The core made of fine stone sizes is provided to diminish wave transmission through the structure. Jetties and groins have a simpler cross section, consisting typically of only armor and core layers. Breakwaters, groins, and jetties have also been constructed of steel or concrete caissons with sand and gravel fill; wood, steel, and concrete sheet piles; and sand-filled bags.

A different type of breakwater that can be effective where incident wave periods are short and large water-level fluctuations occur (as in a reservoir marina) is the moored floating breakwater. This type has been constructed of hollow concrete prisms, scrap tires, logs, and a variety of other materials.

In an attempt to develop low-cost shore protection, a number of novel materials have been used for shoreline revetments, including cinder blocks, tires, sand-filled rubber tubes, woven-fiber mattresses, and soil-cement paving. *See* BREAKWATER.

[ROBERT M. SORENSEN]

Bibliography: American Society of Civil Engineers, *Journal of the Waterway, Port, Coastal and Ocean Engineering Division* and *Proceedings of the International Conferences on Coastal Engineering*; P. D. Komar, *Beach Processes and Sedimentation*, 1976; A. D. Quinn, *Design and Construction of Ports and Marine Structures*, 1972; R. Silvester, *Coastal Engineering*, 2 pts., 1974; R. M. Sorensen, *Basic Coastal Engineering*, 1978; U.S. Army Coastal Engineering Research Center, *Shore Protection Manual*, 3 vols., 1977.

Coaxial cable

A two-conductor transmission line with an outer metal tube or braided shield concentric with and enclosing the center conductor. The inner conductor is supported by some form of dielectric insulation, solid, expanded plastic, or semisolid. Semisolid supports are polyethylene disks (Fig. 1), helical tapes, or helically wrapped plastic strings. Beads, supporting pins, or periodically crimped plastic tubes are used in some designs.

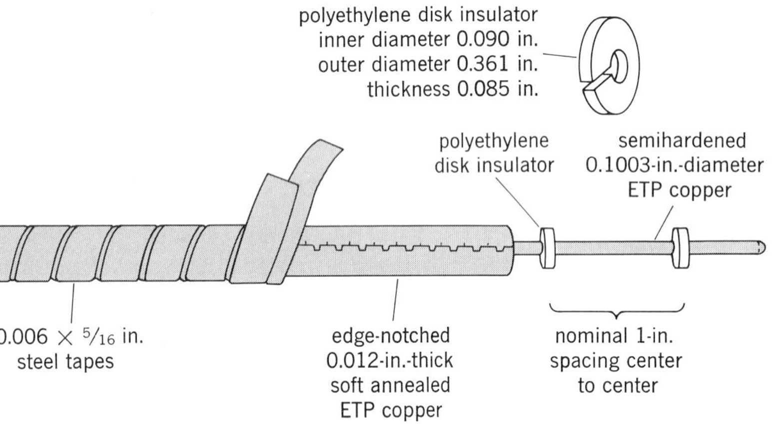

Fig. 1. An air dielectric disk-insulated coaxial.

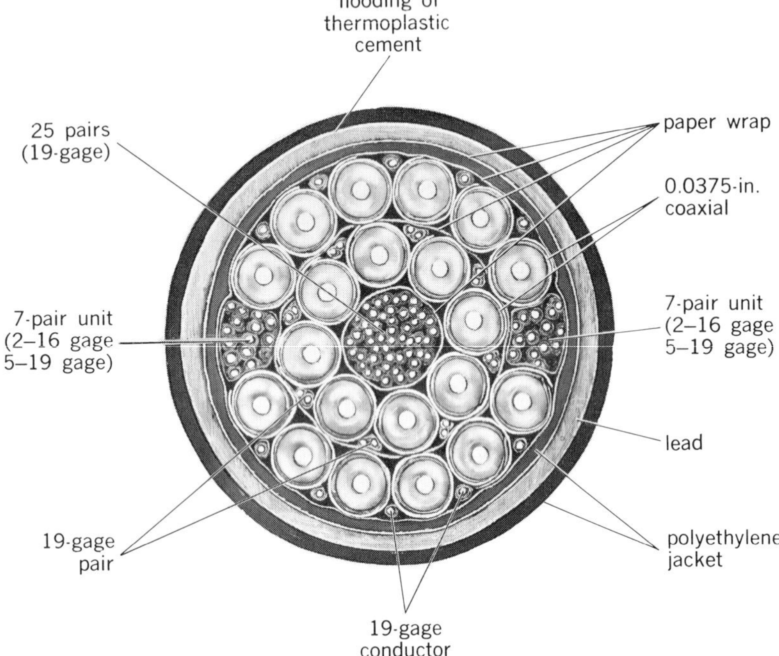

Fig. 2. Construction of coaxial transmission line.

The significant feature of the coaxial cable is that it is a shielded structure. The electromagnetic field associated with each coaxial unit is nominally confined to the space between the inner and outer conductors. Since alternating current concentrates on the inside of the outer conductor as the frequency of the current increases (skin effect), a coaxial unit is a self-shielded transmission line whose shielding improves at higher frequencies. Unshielded lines, such as the pairs of multipair cable, share the space for the electromagnetic fields. Thus, for equivalent transmission loss, pairs occupy less space than coaxials. The major use of the coaxial cable is for transmitting high-frequency broadband signals. Coaxial cables are rarely used at or near voice frequency since the shielding properties are poor and they are more expensive than twisted pairs having the same transmission loss.

Types and uses. Coaxial units are made in three general types for different applications: flexible, semirigid, or rigid. In general, the more rigid the unit, the more predictable and stable its electrical properties. Since loss on a transmission line increases approximately as the square of the frequency, low loss in a coaxial cable is important. Also, physical and electrical irregularities, especially if they are periodic, must be kept to a minimum, or the transmission loss of the coaxial will be very high at certain frequencies.

Flexible coaxials. Flexible coaxials have a braided outer conductor, solid or expanded plastic dielectric, and a stranded or solid inner conductor. The electrical properties of flexible coaxials, particularly those due to physical irregularities, vary widely. Flexible coaxial cable is used mostly for short runs, as in interconnecting high-frequency circuits within electrical equipment.

Semirigid coaxials. This class of coaxials has thin tubular outer conductors which may be corrugated to improve bending. The dielectric can be either insulating disks or an expanded plastic insulation to lower the dielectric constant and hence the loss of the coaxial. Expanded insulation can be as much as 80% air in the form of very fine bubbles. This design finds wide usage in closed-circuit television applications. It is commonly used as the distribution cable from the transmitter to the subscriber's home.

Another variation of the semirigid coaxial, insulated with polyethylene disks approximately 1 in. (1 in. = 2.54 cm) apart and enclosed inside a 10- or 12-mil-thick (1 mil = 0.0000254 m) copper tube, is widely used for transcontinental carrier transmission. One version used in Canada has a corrugated tubular outer conductor. The disk-insulated coaxial unit has very low loss and is now used in carrier systems in the United States, Japan, and Europe to transmit up to 10,800 two-way voice-frequency channels per pair of coaxial units. Figure 2 shows a 20-unit coaxial cable with 9 working coaxial pairs and 2 standby coaxials, which automatically switch in if the electronics of the regular circuits fail.

At these broad bandwidths the slightest physical irregularities introduced during manufacture of the unit or stranding of the cable can produce catastrophic variation in loss. Much worldwide development effort has been expended on developing manufacturing processes, quality control procedures, and installation techniques to make these cables suitable for signals as high as 60 MHz. These coaxial carrier systems, together with microwave radio, provide virtually all long-distance communications facilities in the continental United States.

An interesting example of a semirigid coaxial cable, with a bending radius of 4 ft (1 ft = 0.3 m), is the solid dielectric coaxial units 1.3 to 1.7 in. in diameter used for ocean cable systems. Instead of heavy external armoring for strength, these coaxials have center conductors composed of a copper tube welded around a stranded core of high-strength steel wires (Fig. 3). The center conductor provides the strength to lay and recover the cable at depths up to 3000 fathoms (1 fathom = 1.83 m).

Special cable-laying machinery which grips the outer polyethylene jacket was developed and installed on a specially designed cable-laying ship to enable this unique undersea coaxial design to be used.

Rigid cables. When utmost electrical performance is required and no bending is needed to install the cable, a heavy, rigid, precise outer conductor is used. The inner conductor is either solid wire or a hollow tube which can be liquid-cooled in high-power applications. As little dielectric as possible is used to minimize losses, generally a semisolid such as pins or disks. The rigid construction allows the dielectric supports to be spaced far apart. This coaxial can have propagation velocities up to 99.7% of the speed of light. For high-voltage applications, the unit can be gas-filled to reduce corona and to increase dielectric strength.

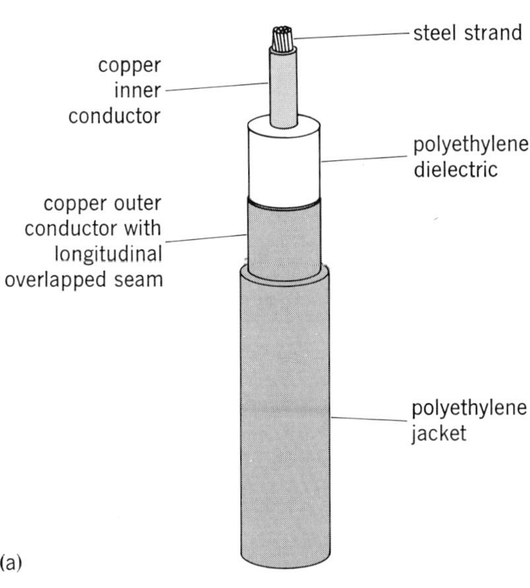

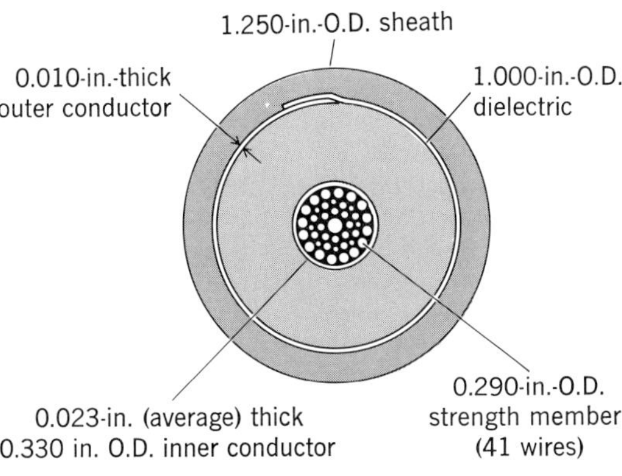

Fig. 3. Armorless ocean cable coaxial construction. (*a*) Vertical view; (*b*) cross section.

Special coaxials. Coaxial units with periodic slots in the outer conductor have been developed since the mid-1960s to provide mobile communications with moving trains in tunnels and subways. These "leaky" coaxials transmit signals along their length while radiating a controlled amount of energy to antennas on passing trains. The Japanese have developed 3-in.-diameter slotted coaxials that are used in the United States for paging systems, vehicular tunnels, and subway communications, and for control of mine locomotives.

[JOHN R. APEN]

Cofferdam

A temporary, wall-like structure to permit dewatering an area and constructing foundations, bridge piers, dams, dry docks, and like structures in the open air. A dewatered area can be completely surrounded by a cofferdam structure or by a combination of natural earth slopes and cofferdam structure. The type of construction is dependent upon the depth, soil conditions, fluctuations in the water level, availability of materials, working conditions desired inside the cofferdam, and whether located on land or in water. An important consideration in the design of cofferdams is the hydraulic analysis of seepage conditions, and erosion of the bottom in streams or rivers.

Where the cofferdam structure can be built on a layer of impervious soil (which prevents the passage of water), the area within the cofferdam can be completely sealed off. Where the soils are pervious, the flow of water into the cofferdam cannot be completely stopped economically, and the water must be pumped out periodically and sometimes continuously.

Types. The illustration shows the types of cofferdam construction: *a*, *b*, and *e* are used in rectangular form to enclose small areas, such as individual bridge piers where cross bracing, consisting of steel, wood, or concrete beams and struts, to the opposite wall is practical; the other types are for large areas required for dams, river locks, or large buildings, where cross bracing is impractical or undesirable. Cross bracing can be replaced with steel or concrete rings acting in compression when a circular form is used for *a* and *e*. *See* CANAL; DAM.

Sheeted cofferdams are made of sheets of steel, timber, or concrete, driven vertically or placed horizontally. The sheeting may be made reasonably watertight by using interlocking, overlapping, or grouted joints. In gravel or boulders, where sheeting cannot be driven to the required depth, steel H piles are driven vertically into the ground and timber sheeting is placed horizontally between or against the vertical piles (illustration *d*) as the excavation progresses. Watertight cofferdams are designed for a full hydrostatic head of water, plus the pressure of earth where on land. Nonwatertight cofferdams have a reduced hydrostatic pressure because of seepage.

Another type of cofferdam suitable for use in

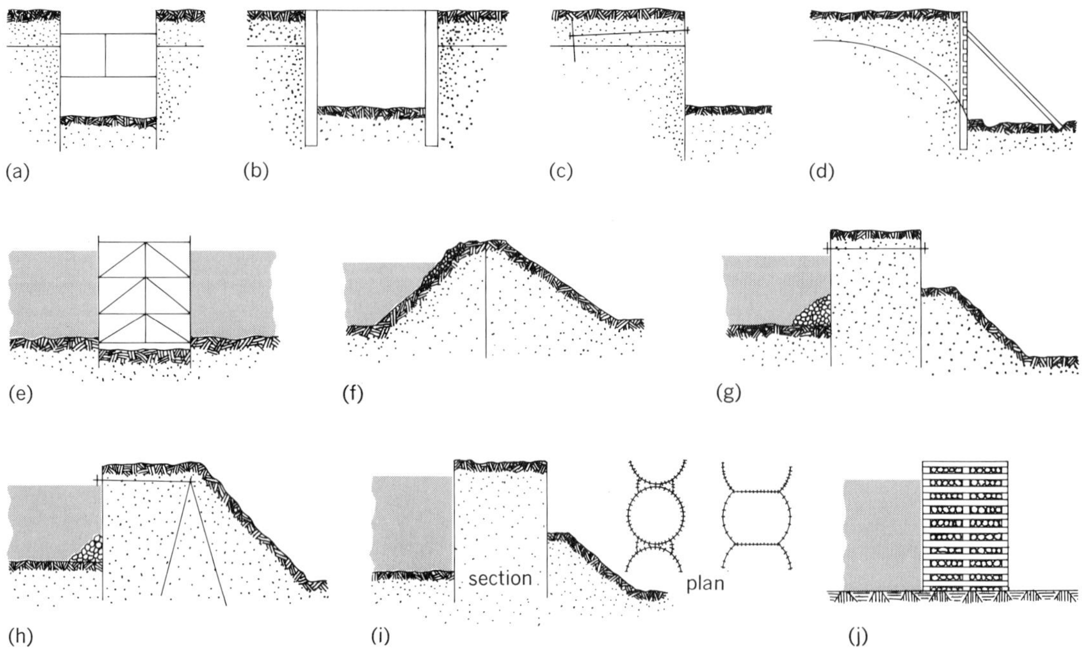

Types of cofferdams. For use on land: (*a*) cross-braced sheet piles; (*b*) cast-in-place concrete cylinder; (*c*) anchored sheet piles; (*d*) braced vertical piles with horizontal sheeting. For use in water: (*e*) cross-braced sheet piles; (*f*) earth dam; (*g*) tied sheet piles; (*h*) anchored sheet piles with earth berm; (*i*) steel sheet-pile cellular cofferdam; (*j*) rock-filled crib.

gravel or boulders is constructed of cast-in-place concrete. Illustration *b* shows this type in a cylindrical form.

Most sheet-pile cofferdams, with the exception of the cellular type (*i*), require a soil into which the sheets may be imbedded to give the walls stability. The cellular cofferdam, consisting of interlocking steel sheet piling driven as a series of interconnecting cells, is independently stable when filled and may be constructed on either rock or soil. The cells of the circular type may be filled individually. The cells of the straight-wall diaphragm type must be filled systematically to avoid a differential height of fill that would deform the diaphragms.

The filled crib (*j*) is an open, boxlike structure of timber, precast concrete, or fabricated steel sections, filled with a suitable material. A rock-filled crib is suitable for construction on a rock bottom in swiftly flowing waters. Sheeting or an impervious core provides the necessary watertightness. Earth dams (*f*) can be constructed of soil or rock, but if pervious, a cutoff is required to control seepage.

A recent type for a large open building area has been a slurry wall anchored back to the earth (illustration *c*), constructed by pouring a concrete mix into a trench dug along the proposed wall before excavating the cofferdam area.

Uses. When cofferdams are required for year-round construction in waters subject to flooding, they are made sufficiently high to exclude the flood waters. If overflow is to be permitted, then the cofferdam fill and the bottom of the dewatered area must be protected against scour (eroding action of the water). Scour can also be prevented along the stream side of cofferdams in swiftly flowing waters by protecting the stream bed (*g*, *h*). The berm (embankment) shown in the illustration *g* and *h* on the dewatered side, furnishes stability and reduces the flow of water under the cofferdam.

The cofferdam shown in illustration *b* has been used to form a mine shaft 220 ft (66 m) deep. Type *d* was used extensively in subway construction in New York City.

A nautical application of the term cofferdam is a watertight structure used for making repairs below the waterline of a vessel. The name also is applied to void tanks which protect the buoyancy of a vessel. [EDWARD J. QUIRIN]

Bibliography: A. Brinton Carson, *Foundation Construction*, 1965; W. C. Huntington, *Earth Pressures and Retaining Walls*, 1957; D. H. Lee, *Sheet Piling, Cofferdams, and Caissons*, 1946; E. A. Prentis and L. White, *Cofferdams*, 1950.

Column

A slender compression member which fails by instability when the applied axial load reaches a critical value. Instability or buckling is said to occur when the system cannot attain an equilibrium condition and the load causes a deformation of an indeterminate amount. The collapse load is a function of the end restraints, the ratio of length to the least radius of gyration (L/r, the slenderness ratio), and the combined action of axial and lateral loads applied to the member. Columns are present in all types of structures, such as bridges, buildings machinery, cranes, and airplanes. Different materials may be used in the various structures, such as steel, concrete, wood, aluminum, and alloy steels of all types.

Column design. A column is designed to support a required load in compression. Columns of different slenderness ratio behave differently under load and thus the criteria for maximum capacity differ. Long columns (those with high slenderness ratio) become unstable at a critical load, usually the maximum resistance, with compressive stresses less than the elastic limit, so that failure is a phenomenon of elastic instability. The column buckles with a lateral deflection or with a twist. Columns of intermediate length, with slenderness ratios between long columns and short blocks, develop inelastic stresses with ultimate failure by inelastic buckling. Short blocks have such dimensions that lateral deflections can be neglected, failure being determined by the yield strength of the material.

The load capacity of columns is further influenced by the properties of the material, end conditions, initial crookedness, defects, and residual stresses, particularly in the intermediate range of slenderness.

Elastic buckling. An abrupt increase of lateral deflection of a compression member at a critical load while the stresses are wholly elastic constitutes elastic buckling. An initially straight column under axial load remains straight until a condition of neutral equilibrium is reached, when the column will continue to support the full load with considerable lateral deflection. A small increase of load above this condition causes a large increase in deflection, leading to collapse.

Leonhard Euler's expression for the critical buckling load P_{cr} of an ideally straight, axially loaded column with no rotational restraint at the ends is $P_{cr} = \pi^2 EI/L^2$, from which the critical average stress f_{cr} is defined by Eq. (1), where E is

$$f_{cr} = \frac{\pi^2 E}{(L/r)^2} \qquad (1)$$

modulus of elasticity of the perfectly elastic material, I is the moment of inertia, L/r is the slenderness ratio, and r is the least radius of gyration of the section.

Euler's formula is directly applicable to the design of long columns whose safe load is Euler's predicted critical load divided by a factor of safety. Elastic buckling loads for columns with other restraining conditions modifying the elastic curve are found by using the length between points of contraflexure of the deflected member as the effective length KL, where $K = 1$ for rotationally free ends, $K = 0.5$ for both ends fixed, $K = 2$ for one end fixed with the other free to rotate and translate, and $K = .70$ for one end rotationally fixed with the other only fixed in position. The buckling load depends only on the stiffness E and the slenderness ratio KL/r, which is determined by end conditions. In design, end conditions are designated restraint coefficients, $C = 1/k^2$, with $P_{cr} = C\pi^2 EI/L^2$. The coefficient is taken to provide partial restraint for various types of end connections.

Eccentrically loaded columns. Off-axis loads subject a column to combined bending because of end moments and axial load. Deflections are produced immediately on application of the load and increase rapidly until the member yields under compression because of the combined loading, with inelastic buckling ultimately producing the collapse. Buckling may be accompanied by twisting, depending upon the shape of the section, lateral bracing, and slenderness ratio.

An estimate of the load capacity can be made according to either initial yield, or relations involving ratios of actual to ultimate thrust and moment, called interaction formulas. When initial yield under the combined stress is assumed to be the critical condition leading to collapse, the corresponding load is taken as a minimum value of the collapse load. The limiting average stress for design is defined by the secant formula, derived for free-end columns of material having a distinct yield point, loaded with equal end moments. As used, actual load eccentricity is increased to account for crookedness, defects, accidental eccentricity, and the effect of residual stresses. This formula is prescribed in bridge specifications for steel, with modifications to provide for effective length and double eccentricity. *See* BEAM COLUMN.

Interaction formulas are more convenient for general application. They depend upon the formulas adopted for predicted column and bending resistance.

Inelastic buckling. When the compressive stress in the column reaches the elastic limit before elastic buckling develops, any sudden increase of deflection or twist leads to collapse. The behavior of columns under inelastic stress depends on the shape of the stress-strain curve as determined by testing a short length of the actual column section. Because of residual stresses caused by unequal cooling of rolled sections, structural rolled steel has a proportional limit approximately half the yield point for annealed material, and the stress-strain curve has a gradual knee similar to alloy steels, aluminum, or magnesium alloys. Euler's formula is not applicable to these materials because the modulus of elasticity E is not constant. An analytical approach, called the tangent modulus theory, leads to prediction of the buckling load. *See* STRESS AND STRAIN.

The tangent modulus formula (Engesser formula), is a modification of Euler's formula and is shown in Eq. (2), where E_T is the slope of the stress-strain curve at the critical stress (Fig. 1).

$$f_{cr} = \frac{\pi^2 E_T}{(KL/r)^2} \qquad (2)$$

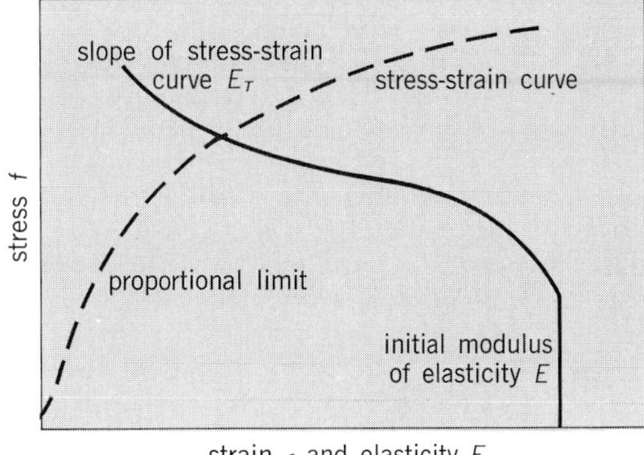

Fig. 1. Stress-strain curve for an aluminum alloy.

Values of E_T and stress f are interdependent. Solution is accomplished by trial or graphically for a given value of KL/r. Also, values of $\pi\sqrt{E_T/f}$ are consistent with particular values of KL/r, and a curve can be constructed to represent column resistance (Fig. 2). The resistance curve in the inelastic range can be represented by a parabola. For columns of wide-flange steel sections, in which residual stress is an important factor, a recommend-

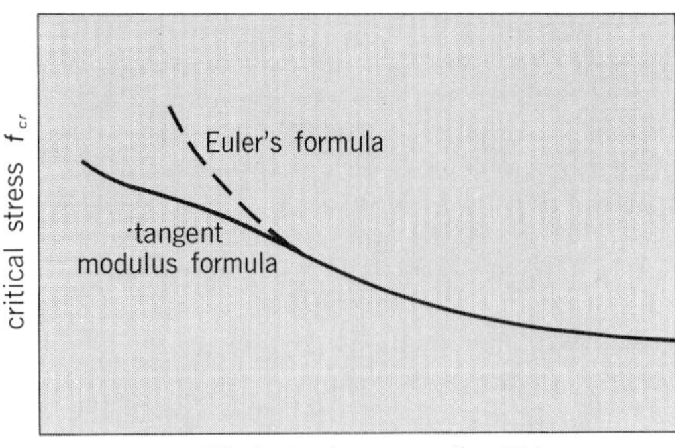

Fig. 2. Column resistance to inelastic buckling.

ed column strength equation is Eq. (3), where F_y is the stress of the material at its yield point.

$$f_{cr} = F_y - \frac{F_y^2}{4\pi^2 E}\left(\frac{KL}{r}\right)^2 \qquad (3)$$

Local buckling. Relatively thin elements of a column section may buckle locally in a series of waves or wrinkles before the member as a whole buckles elastically. The outstanding legs of thin angles, thin flanges of channels, or thin plates connecting parts of the section are susceptible to local buckling. Thin-wall tubes buckle in a similar manner. Usually, local buckling reduces the load capacity of the member, but redistribution of stresses may produce conditions favorable to increased column loads. The likelihood of local buckling can be predicted by the theory of flat plates, which involves the thickness-to-width ratio and edge supporting conditions. See STRUCTURAL PLATE.

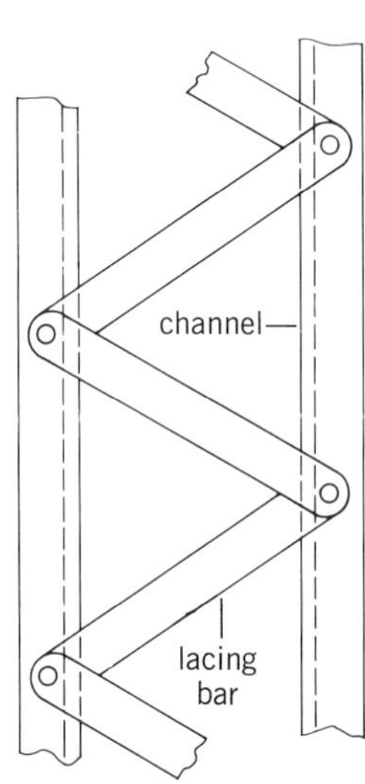

Fig. 3. Laced channel section.

Shear is produced in columns by variation of bending moment associated with initial and buckling curvatures causing a shear force, lateral external forces, and lateral forces transmitted by distortion of the connecting members of the frame of which the column forms a part. An allowance for shear is taken as 2–2½% of the axial load for columns composed of aluminum alloys or structural steel. In rolled sections the web provides shear resistance, whereas in open-web members resistance is provided by lacing or batten plates (Fig. 3).

Lacing. A system of diagonal bars connected to the elements of an open-web column forms a truss capable of resisting transverse shear (Fig. 3). These bars, tying the separated elements together, are usually inclined at 45° or 60° and are designed to act in either tension or compression. The slenderness ratio of the flange elements between lacing-bar connections is limited by specification to avoid local buckling.

Batten plates. Alternatively, flat plates can be attached transversely to the flange of an open-web column, at intervals along the length (Fig. 4). These batten plates serve as ties holding the elements in position, and, together with the column flanges, form a continuous frame resisting transverse shear. The flanges and plates are subject to bending and shear. The unsupported length between batten plates is limited to provide smaller slenderness in the flange elements than in the overall column, to avoid local buckling.

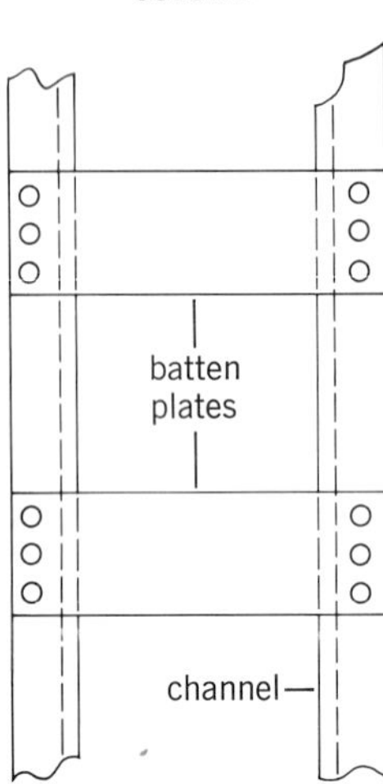

Fig. 4. Batten plates.

Torsional buckling. The dimensions and shape of a column section should provide a large radius of gyration to reduce KL/r and increase the allowable stress. A large radius of gyration requires that the area be spread far from the centroid, and for a given area the elements of the section become thin. Columns with open sections composed of thin elements, particularly angle and T sections, are susceptible to failure either by local buckling or by twisting.

A short compression member may fail by twisting of the section while the axis remains straight. This is called primary torsional buckling, in which the member becomes torsionally unstable at a condition of neutral equilibrium similar to a long column in bending. The critical load may be determined by energy methods. Columns which bend upon approaching maximum load are subjected to transverse shear, which may also produce twist buckling if the shear center does not coincide with the centroid. Local buckling strength determined by plate theory approximates the twist buckling load.

Column baseplates. Column loads and moments are transferred by baseplates to footings and foundations. The plate must be large enough to distribute the load without excessive bearing stresses. The column is connected to the plate by welded or bolted connections. Where no moment is involved, the connection only maintains alignment. The pressure of the plate on the foundation is nonuniformly distributed and depends on relative stiffnesses.

An approximate design procedure assumes uniform pressure and cantilever action of the plate projecting beyond the column boundaries. Anchorage is not required for axially loaded columns or when the resultant pressure due to moment and load falls within the mid-third of the base width. However, bolts are used to position the column. Where lateral forces, such as wind pressures, are involved, the resultant moment requires anchor bolts. [JOHN B. SCALZI]

Bibliography: B. Bresler, T. Y. Lin, and J. B. Scalzi, *Design of Steel Structures*, 2d ed., 1968; E. H. Gaylord and C. N. Gaylord, *Design of Steel Structures*, 2d ed., 1972; Bruce G. Johnston (ed.), *Guide to Stability Design Criteria for Metal Structures*, 3d ed., 1976; W. McGuire, *Steel Structures*, 1968.

Combustion chamber

The space at the head end of an internal combustion engine cylinder where most of the combustion takes place.

Spark-ignition engine. In the spark-ignition engine, combustion is initiated in the mixture of fuel and air by an electrical discharge. The resulting reaction moves radially across the combustion space as a zone of active burning, known as the flame front. The velocity of the flame increases nearly in proportion to engine speed so that the distance the engine shaft turns during the burning process is not seriously affected by changes in speed. See SPARK PLUG.

For high efficiency it is desirable to burn the mixture as quickly as possible while the piston is near top center position (constant volume combustion). Short burning time is achieved in modern combustion chambers by a combination of several factors. The spark plug is located in a central position to minimize the distance the flame front must travel. The chamber itself is made as compact as

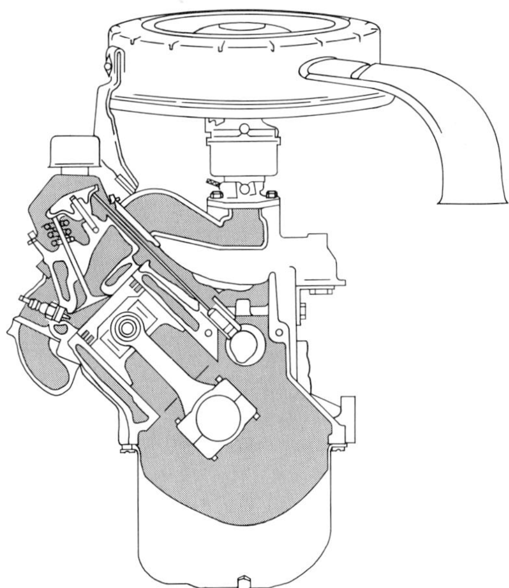

Fig. 1. Combustion chamber as found in a spark-ignition engine. *(From L. C. Lichty, Internal-Combustion Engines, 6th ed., McGraw-Hill, 1951)*

possible for the required volume, to keep the flame paths short. A compact chamber presents less wall area to the enclosed volume, thereby reducing the heat loss from the flame front. Because motion of the flame depends upon transferring heat from the flame to the adjacent layers of unburned mixture, the reduced heat loss increases the flame velocity. One important method for increasing flame velocity is to provide small-scale turbulence in the cylinder charge, often by designing the chamber so that part of the piston head comes close to the cylinder head at top center position (Fig. 1). The squish that results forces the mixture in this region into the rest of the combustion chamber at high velocity. The turbulence so produced increases the rate of heat transfer from the flame to the unburned mixture, greatly increasing the flame velocity.

Occasionally a high burning rate, or too rapid change in burning rate, gives rise to unusual noise and vibration called engine roughness. Roughness may be reduced by using less squish or by shaping the combustion chamber to control the area of the flame front. A short burning time is helpful in eliminating knock because the last part of the charge is burned by the flame before it has time to ignite spontaneously. *See* COMBUSTION KNOCK.

For high-power output the shape of the combustion chamber must permit the use of large valves with space around them for unrestricted gas flow. *See* VALVE TRAIN.

Studies have shown that pockets or narrowed sections in the combustion chamber which trap thin layers of combustible mixture between adjacent surfaces interfere with combustion and increase the small but important quantities of unburned fuel hydrocarbons normally present in the exhaust gases. Because unburned hydrocarbons are one of the constituents of smog-forming pollution, the expectation is that future combustion chambers will be built with less squish and with less wall area per unit of volume. With regard to squish, some compromise will be necessary between requirements for efficiency and quietness and a clean exhaust.

Compression-ignition engine. In compression-ignition (diesel) engines the fuel is injected late in the cycle into highly compressed air; mixing must take place quickly if the fuel is to find oxygen and burn while the piston remains near top center. Rapid mixing is particularly important in the smaller-sized engines, which are usually operated at higher rotative speeds. In simple combustion chambers, mixing can be improved by producing a general swirl in the cylinder so that the air is moving relative to the fuel spray during injection. Swirl is often produced by masking part of the periphery of the inlet valves or by giving an angle to the inlet ports so that the air enters with a tangential component of velocity. This circular motion persists until injection takes place; swirl velocity increases with the higher inlet velocities present at high engine speeds. Thus mixing becomes more rapid as the time available for mixing becomes less.

In some compression-ignition engines a considerable fraction of the cylinder air is forced into a small auxiliary chamber (prechamber) during the compression stroke. The entrance to the auxiliary chamber may be a short tangential passage that imparts a rapid swirl to the air in the chamber (Fig. 2). The fuel is injected into the swirl chamber where rapid mixing and combustion take place, and the burning mixture issues into the main chamber where any remaining fuel is burned.

In compression-ignition engines the injected fuel ignites from contact with the hot cylinder air after a short delay. There is no flame front travel to limit the combustion rate, and if mixing of fuel and air is too thorough by the end of the delay period, high rates of pressure rise result, and the operation of the engine is rough and noisy. To avoid this condition, the prechambers of most compression-ignition engines operate at high temperature so that the fuel ignites soon after injection begins. This reduces the amount of fuel present and the degree of mixing at the time ignition takes place.

High rates of pressure rise can also be reduced by keeping most of the fuel separated from the chamber air until the end of the delay period.

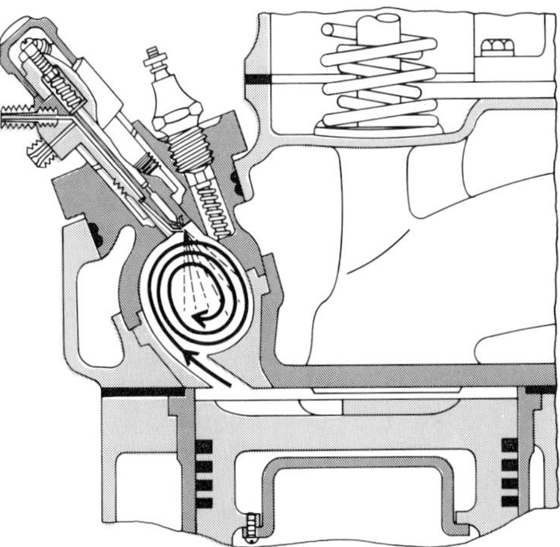

Fig. 2. Combustion chamber for diesel engine with high-swirl prechamber. *(From J. B. Fisher, Development of a combustion chamber for medium- and high-speed diesel engines, SAE J., 57(5):59–60, 1949)*

Rapid mixing must then take place to ensure efficient burning of the fuel near top center. A development which appears to result in smooth combustion without loss in efficiency due to late or slow burning operates as follows. The piston head contains a fairly large cup-shaped cavity. The fuel is injected directly against the sides of this cavity instead of into the chamber air. Mixing of the fuel after the end of the delay is accomplished by a combination of chamber air motion and the rapid evaporation of the fuel from the hot piston cavity walls. [AUGUSTUS R. ROGOWSKI]

Bibliography: T. Pipe, *Gas Engine Manual*, 2d ed., 1977.

Combustion knock

In spark-ignition engines, the sound and other effects associated with ignition and rapid combustion of the last part of the charge to burn, before the flame front reaches it.

As the flame travels from the spark plug toward the far end of the combustion space, the gas behind the flame front, heated by combustion, expands and rapidly compresses the unburned mixture ahead of the flame. In this way the temperature of the unburned mixture is raised above its self-ignition temperature. Time-consuming chemical reactions must take place in this gas, however, before actual ignition takes place. If the flame front travels rapidly enough to consume the unburned mixture before these reactions are completed, normal combustion without knock takes place. If the reactions proceed too rapidly or if the flame front travels too slowly, the entire last part of the charge may burn almost instantaneously, producing a strong pressure wave, which is reflected back and forth across the combustion space at the speed of sound. The motion of the cylinder head and walls, under the action of this shock wave, causes the high-pitched sound known as combustion knock (see illustration). (Some authorities refer to the above phenomenon and its effects as detonation, although the process is not the same as detonation in explosives.)

The results of combusion knock or engine detonation include undesirable noise and overheating or mechanical damage to engine parts such as pistons and spark plugs. Most of the power loss commonly attributed to combustion knock is probably an indirect effect, the result of overheated spark-plug electrodes, which ignite the mixture far in advance of the normal ignition spark.

Effect of engine. Because of these undesirable effects, much effort is made to avoid combustion knock. Any change in engine design or operating conditions that lengthens the delay period tends to allow the flame to burn the last part of the charge normally. The delay period may be lengthened by reducing the temperature or pressure to which the last part of the charge is subjected during the time of flame travel. Closing the throttle, retarding the ignition timing, or reducing the compression ratio reduces the pressures and temperatures in the last part of the charge and lengthens the delay, thereby diminishing or eliminating combustion knock. The power or efficiency of the engine is reduced by these control measures. The delay period may also be lengthened and knock eliminated by using a very rich or a very lean fuel-air ratio, but again power or efficiency is adversely affected.

Another method for reducing combustion knock is to reduce the time required for the flame front to cross the combustion space. If the burning time is short, the last part of the charge will be consumed by the flame before the delay reactions are completed. Burning time may be reduced by using a compact combustion chamber with centrally located or multiple spark plugs. One factor in combustion chamber design which is most effective in eliminating knock by increasing flame velocity is the production of small-scale turbulence. The usual method of producing turbulence is to design one part of the piston head to come close to the cylinder head at top center, forcing part of the charge into the rest of the chamber. Some authorities believe that the close approach of the parts increases the cooling of the gas left in the space between them and slows the delay reactions if the last part of the charge is in this quench area.

Eliminating knock by a change in one engine variable will often permit a more nearly optimum adjustment of another. For example, a retarded

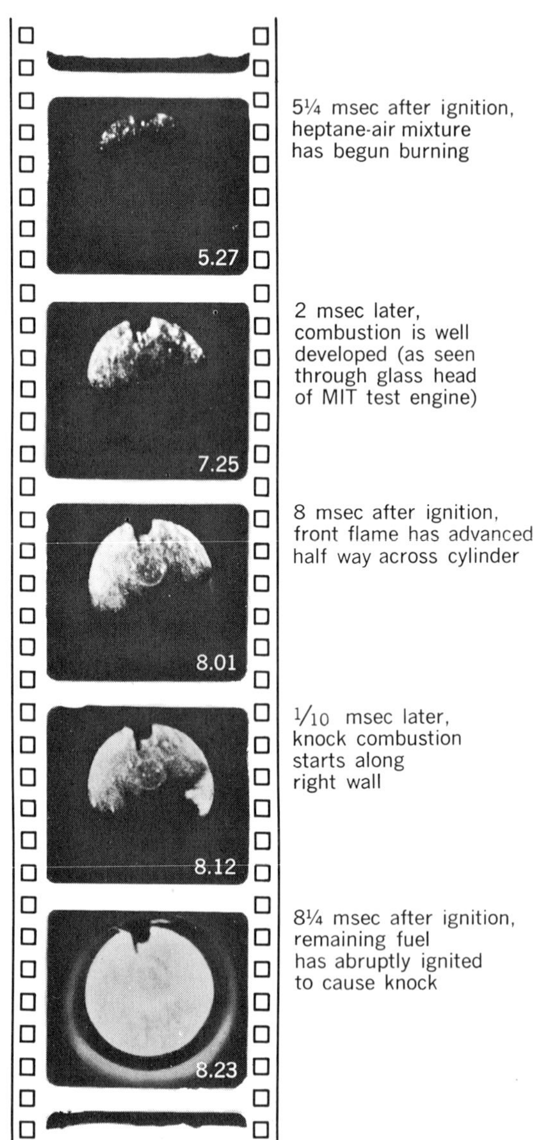

5.27
5¼ msec after ignition, heptane-air mixture has begun burning
7.25
2 msec later, combustion is well developed (as seen through glass head of MIT test engine)
8.01
8 msec after ignition, front flame has advanced half way across cylinder
8.12
1⁄10 msec later, knock combustion starts along right wall
8.23
8¼ msec after ignition, remaining fuel has abruptly ignited to cause knock

Photographic sequence showing (top to bottom) slow normal fuel burning ending in rapid knock combustion.

spark or a rich mixture will permit a wider-open throttle to be used, which will give more power; a better combustion chamber will permit a higher compression ratio with increased engine efficiency.

Effect of fuel. A fuel of high octane number resists combustion knock principally because it has a longer self-ignition delay than other fuels under a given set of operating conditions. High-octane gasoline burns in the same manner and with the same flame velocity as low-octane gasoline. Combustion knock can thus be eliminated by using high-octane fuel, without loss of power or efficiency.

The long delay of high-octane fuel results either from its natural chemical makeup or from the addition of an antiknock material. Hydrocarbon fuels with compact molecular structures are less likely to knock. One of the most effective antiknock additives is tetraethyllead, $Pb(C_2H_5)_4$, which is added to almost all gasoline sold for automobile or aircraft use. In general, fuels of lower natural octane number show a somewhat greater response to the addition of a given amount of tetraethyllead.

High rates of pressure rise or sudden increases in the rate of pressure rise sometimes produce noises, which are confused with combustion knock. In spark-ignition engines these effects may be caused by abnormally high burning rates. In compression-ignition engines they may be caused by the rapid combustion of the first part of the fuel to be injected.

[AUGUSTUS R. ROGOWSKI]

Comfort heating

The maintenance of the temperature in a closed volume, such as a home, office, or factory, at a comfortable level during periods of low outside temperature. Two principal factors determine the amount of heat required to maintain a comfortable inside temperature: the difference between inside and outside temperatures and the ease with which heat can flow out through the enclosure.

Heating load. The first step in planning a heating system is to estimate the heating requirements. This involves calculating heat loss from the space, which in turn depends upon the difference between outside and inside space temperatures and upon the heat transfer coefficients of the surrounding structural members.

Outside and inside design temperatures are first selected. Ideally, a heating system should maintain the desired inside temperature under the most severe weather conditions. Economically, however, the lowest outside temperature on record for a locality is seldom used. The design temperature selected depends upon the heat capacity of the structure, amount of insulation, wind exposure, proportion of heat loss due to infiltration or ventilation, nature and time of occupancy or use of the space, difference between daily maximum and minimum temperatures, and other factors. Usually the outside design temperature used is the median of extreme temperatures.

The selected inside design temperature depends upon the use and occupancy of the space. Generally it is between 66 and 75°F (19 and 24°C).

The total heat loss from a space consists of losses through windows and doors, walls or partitions, ceiling or roof, and floor, plus air leakage or ventilation. All items but the last are calculated from $H_l = UA\,(t_i - t_o)$, where heat loss H_l is in British thermal units per hour (or in watts), U is overall coefficient of heat transmission from inside to outside air in Btu/(hr)(ft²)(°F) (or J/s · m² · °C), A is inside surface area in square feet (or square meters), t_i is inside design temperature, and t_o is outside design temperature in °F (or°C).

Values for U can be calculated from heat transfer coefficients of air films and heat conductivities for building materials or obtained directly for various materials and types of construction from heating guides and handbooks.

The heating engineer should work with the architect and building engineer on the economics of the completed structure. Consideration should be given to the use of double glass or storm sash in areas where outside design temperature is 10°F (−12°C) or lower. Heat loss through windows and doors can be more than halved and comfort considerably improved with double glazing. Insulation in exposed walls, ceilings, and around the edges of the ground slab can usually reduce local heat loss by 50–75%. Table 1 compares two typical dwellings. The 43% reduction in heat loss of the insulat-

Table 1. Effectiveness of double glass and insulation*

Heat-loss members	Area, ft²†	Heat loss, Btu/hr‡: With single-glass weather-stripped windows and doors	Heat loss, Btu/hr‡: With double-glass windows, storm doors, and 2-in. (5.1-cm) wall insulation
Windows and doors	439	39,600	15,800
Walls	1,952	32,800	14,100
Ceiling	900	5,800	5,800
Infiltration		20,800	20,800
Total heat loss		99,000	56,500
Duct loss in basement and walls (20% of total loss)		19,800	11,300
Total required furnace output		118,800	67,800

*Data are for two-story house with basement in St. Louis, Mo. Walls are frame with brick veneer and 25/32-in. (2.0-cm) insulation plus gypsum lath and plaster. Attic floor has 3-in. (7.6-cm) fibrous insulation or its equivalent. Infiltration of outside air is taken as a 1-hr air change in the 14,400 ft³ (408 m³) of heating space. Outside design temperature is −5°F (−21°C); inside temperature is selected as 75°F (24°C). †1 ft² = 0.0929 m². ‡1 Btu/hr = 0.293 W.

ed house produces a worthwhile decrease in the cost of the heating plant and its operation. Building the house tight reduces the large heat loss due to infiltration of outside air. High heating-energy costs may now warrant 4 in. (10 cm) of insulation in the walls and 8 in. (20 cm) or more in the ceiling.

Humidification. In localities where outdoor temperatures are often below 36°F (2°C), it is advisable to provide means for adding moisture in heated spaces to improve comfort. The colder the outside air is, the less moisture it can hold. When it is heated to room temperature, the relative humidity in the space becomes low enough to dry out nasal membranes, furniture, and other hygroscopic materials. This results in discomfort as well as deterioration of physical products.

Various types of humidifiers are available. The most satisfactory type provides for the evaporation of the water to take place on a mold-resistant treated material which can be easily washed to get rid of the resultant deposits. When a higher relative humidity is maintained in a room, a lower dry-bulb temperature or thermostat setting will provide an equal sensation of warmth. This does not mean, however, that there is a saving in heating fuel, because heat from some source is required to evaporate the moisture.

Some humidifiers operate whenever the furnace fan runs, and usually are fed water through a float-controlled valve. With radiation heating, a unitary humidifier located in the room and controlled by a humidistat can be used.

Insulation and vapor barrier. Good insulating material has air cells or several reflective surfaces. A good vapor barrier should be used with or in addition to insulation, or serious trouble may result. Outdoor air or any air at subfreezing temperatures is comparatively dry, and the colder it is the drier it can be. Air inside a space in which moisture has been added from cooking, washing, drying, or humidifying has a much higher vapor pressure than cold outdoor air. Therefore, moisture in vapor form passes from the high vapor pressure space to the lower pressure space and will readily pass through most building materials. When this moisture reaches a subfreezing temperature in the structure, it may condense and freeze. When the structure is later warmed, this moisture will thaw and soak the building material, which may be harmful. For example, in a house that has 4 in. (10 cm) or more of mineral wool insulation in the attic floor, moisture can penetrate up through the second floor ceiling and freeze in the attic when the temperature there is below freezing. When a warm day comes, the ice will melt and can ruin the second floor ceiling. Ventilating the attic helps because the dry outdoor air readily absorbs the moisture before it condenses on the surfaces. Installing a vapor barrier in insulated outside walls is recommended, preferably on the room side of the insulation. Good vapor barriers include asphalt-impregnated paper, metal foil, and some plastic-coated papers. The joints should be sealed to be most effective.

Thermograms of building structures: *(a–c)* masonry buildings; *(d)* glass-faced building. Black indicates negligible heat loss; gray, partial loss; and white, excessive loss. *(Courtesy of A. P. Pontello)*

Thermography. Remote heat-sensing techniques evolved from space technology developments related to weather satellites can be used to detect comparative heat energy losses from roofs, walls, windows, and so on. A method called thermography is defined as the conversion of a temperature pattern detected on a surface by contrast into an image called a thermogram (see illustration). Thermovision is defined as the technique of utilizing the infrared radiation from a surface, which varies with the surface temperatures, to produce a thermal picture or thermogram. A camera can scan the area in question and focus the radiation on a sensitive detector which in turn converts it to an electronic signal. The signal can be amplified and displayed on a cathode-ray tube as a thermogram.

Normally the relative temperature gradients will vary from white through gray to black. Temperatures from −22° to 3540°F (−30° to 2000°C) can be measured. Color cathode-ray tubes may be used to display color-coded thermograms showing as many as 10 different isotherms. Permanent records are possible by using photos or magnetic tape.

Infrared thermography is used to point out where energy can be saved, and comparative insulation installations and practices can be evaluated. Thermograms of roofs are also used to indicate areas of wet insulation caused by leaks in the roof.

Infiltration. In Table 2, the loss due to infiltration is large. It is the most difficult item to estimate accurately and depends upon how well the house is built. If a masonry or brick-veneer house is not well caulked or if the windows are not tightly fitted and weather-stripped, this loss can be quite large. Sometimes, infiltration is estimated more accurately by measuring the length of crack around windows and doors. Illustrative quantities of air leakage for various types of window construction are shown in Table 2. The figures given are in cubic feet of air per foot of crack per hour.

Table 2. Infiltration loss with 15-mph outside wind

Building item	Infiltration, ft³/(ft)(hr)
Double-hung unlocked wood sash windows of average tightness, non-weather-stripped including wood frame leakage	39
Same window, weather-stripped	24
Same window poorly fitted, non-weather-stripped	111
Same window poorly fitted, weather-stripped	34
Double-hung metal windows unlocked, non-weather-stripped	74
Same window, weather-stripped	32
Residential metal casement, 1/64-in. crack	33
Residential metal casement, 1/32-in. crack	52

Design. Before a heating system can be designed, it is necessary to estimate the heating load for each room so that the proper amount of radiation or the proper size of supply air outlets can be selected and the connecting pipe or duct work designed. *See* AIR REGISTER; CENTRAL HEATING AND COOLING; HOT WATER HEATING SYSTEM; OIL BURNER; RADIATOR; STEAM HEATING; WARM AIR HEATING SYSTEM.

Heat is released into the space by electric lights and equipment, by machines, and by people. Credit to these in reducing the size of the heating system can be given only to the extent that the equipment is in use continuously or if forced ventilation, which may be a big heat load factor, is not used when these items are not giving off heat, as in a factory. When these internal heat gain items are large, it may be advisable to estimate the heat requirements at different times during a design day under different load conditions to maintain inside temperatures at the desired level.

Cost of operation. Design and selection of a heating system should include operating costs. The quantity of fuel required for an average heating season may be calculated from

$$F = \frac{Q \times 24 \times \mathrm{DD}}{(t_i - t_o) \times \mathrm{Eff} \times H}$$

where F = annual fuel quantity, same units as H
Q = total heat loss, Btu/hr (or J/s)
t_i = inside design temperature, °F (or °C)
t_o = outside design temperature, °F (or °C)
Eff = efficiency of total heating system (not just the furnace) as a decimal
H = heating value of fuel
DD = degree-days for the locality for 65°F (19.3°C) base, which is the sum of 65 (19.3) minus each day's mean temperature in °F (or °C) for all the days of the year.

If a gas furnace is used for the insulated house of Table 1, the annual fuel consumption would be

$$F = \frac{56{,}500 \times 24 \times 4699}{[75 - (-5)] \times 0.80 \times 1050} = 94{,}800 \text{ ft}^3 \ (2684 \text{ m}^3)$$

For a 5°F (3°C), 6- to 8-hr night setback, this consumption would be reduced by about 5%. *See* THERMOSTAT.

[GAYLE B. PRIESTER]

Bibliography: American Society of Heating, Refrigerating, and Air Conditioning Engineers, *Handbook of Fundamentals*, 1977; A. P. Pontello, Thermography: Bringing energy waste to light, *Heat./Piping/Air Condit.*, 50(3):55–61, 1978.

Commutation

The process of current reversal in the armature windings of a direct-current (dc) electric rotating machine to provide dc at the brushes. *See* COMMUTATOR.

The basic process of commutation may be explained with the aid of the elementary motor diagram illustrated; N and S indicate the north and south magnetic poles; the force on conductor A is f_A, and on conductor B is f_B. With the rotor position shown at a, the battery will force current into conductor B and out of conductor A to produce a clockwise torque and rotation. The generated electromotive force (emf) in the conductors is opposite to the direction of current. With the rotor position of b, the conductors are not cutting lines of flux; no emf is induced in the coil; no tangential forces are produced in the coil; and the torque is zero. With the rotor position of c, the battery will force current into conductor A and out of conductor B. The torque is again clockwise; clockwise rotation will continue, and the generated emf will have a polarity opposite to the conductor current. Note that the coil current and the coil emf are reversed from a to c. The switching is accomplished by the commutator. Practical machines have many coils and many commutator segments on the armature. *See* DIRECT-CURRENT GENERATOR; DIRECT-CURRENT MOTOR; WINDINGS IN ELECTRIC MACHINERY.

COMMUTATION

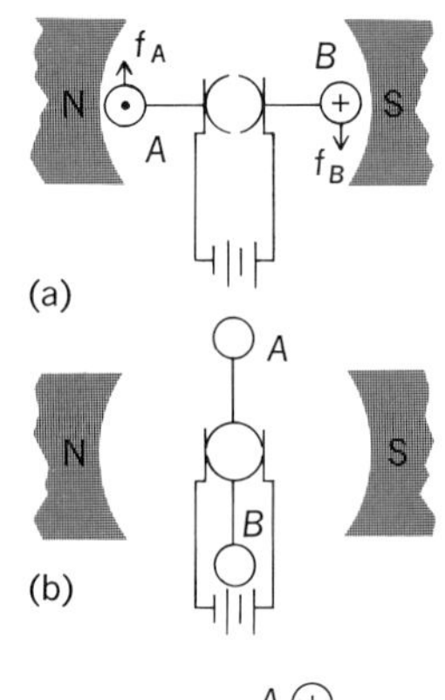

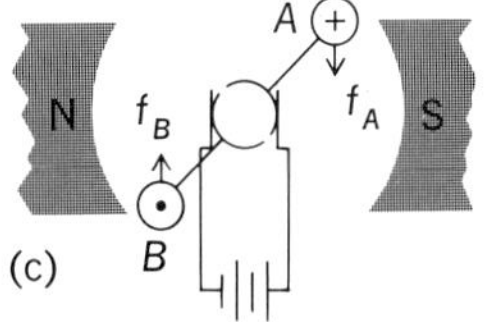

Views of an elementary motor with two-segment commutator; *a*, *b*, and *c* explained in text. The symbol f_A indicates the force on conductor *A*, and f_B the force on conductor *B*.

For commutation to be sparkless, however, this apparently simple process presents severe problems. First, the coil undergoing commutation is short-circuited by the brush, so the brush must be able to tolerate the circulating current. The net voltage induced in the coil while short-circuited should be negligible, to minimize circulating current. Second, because of the self-inductance of the coil, there will be induced in the coil an inductive voltage proportional to the rate of change of current but opposing the change of current. This inductive voltage can be minimized by increasing the time allowed for current reversal and by providing a linear variation of current with time.

A linear transfer of current is achieved by means of brushes with relatively high resistance which span at least one commutator segment. As the commutator segment area covered by a brush changes with armature rotation, the accompanying change in contact resistance effects the desired current change. This is known as resistance commutation. A film forms on the surface of the commutator to aid in the process.

In all but small machines, narrow interpoles are provided between the main field poles to overcome the inductive voltage induced in the coil under commutation. The conductors undergoing commutation cut the interpole flux, and a voltage equal and opposite to the inductive voltage is generated in these conductors. [ARTHUR R. ECKELS]

Bibliography: A. E. Fitzgerald et al., *Electric Machinery*, 3d ed., 1971; S. A. Nasar and L. E. Unnewehr, *Electric Machines*, 1979; Jack Rosenblatt and M. H. Friedman, *Direct and Alternating Current Machinery*, 1963.

Commutator

That part of a dc motor or generator which serves the dual function, in combination with brushes, of providing an electrical connection between the rotating armature winding and the stationary terminals, and of permitting the reversal of the current in the armature windings. For explanation of the necessity of this function *see* COMMUTATION.

A commutator (Fig. 1) is composed of copper bars assembled to form a drumlike cylinder which is concentric with the axis of rotation. Insulation, commonly mica, to provide exceptional mechanical and electrical stability, is placed between commutator bars and between the bars and the shaft.

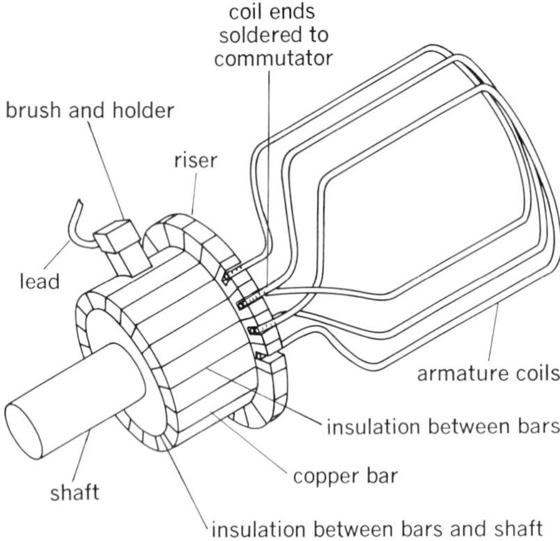

Fig. 1. Commutator and brush assembly with coil connections for lap winding.

Conducting brushes (Fig. 2), commonly carbon, sufficient in size and number to carry the current, are spaced at intervals of 180 electrical degrees about the surface of the commutator and held in contact with the surface of the commutator by spring tension. In special cases, as when only the top or bottom half of a dc motor is accessible, only one positive brush set and one negative brush set are used when the armature is wave-wound. This is permissible, as wave windings characteristically have two parallel paths between any two commutator bars separated 180 electrical degrees. Machines employing commutators are usually restricted to voltages below the kilovolt range because flashover between the closely spaced commutator segments may result from deposition of carbon dust from the brushes and from uneven voltage distribution between commutator bars under heavy loading. For machines using commutators *see* DIRECT-CURRENT GENERATOR; DIRECT-CURRENT MOTOR; REPULSION MOTOR; UNIVERSAL MOTOR. *See also* WINDINGS IN ELECTRIC MACHINERY. [ARTHUR R. ECKELS]

Fig. 2. Direct-current generator showing the typical brush assembly. (*Allis-Chalmers*)

Comparator

A device used to inspect a gaged part for deviation from a specific dimension. The term comparator is also used to identify control-system and analog-computer devices which compare two information signals for such characteristics as simultaneity, size, direction, or rate of change.

To check a gaged part, a comparator is usually preset to the basic critical dimension to be inspected, for example, to 1.505 in the dimension 1.505 ± 0.005 or 2.000 in 2.000 ± 0.002. A master having this dimension is used for presetting the comparator. A comparator may also be set by two masters, one for the maximum limit and one for minimum limit. Two classes of comparator are in common use: one in which the comparison is made visually or optically, the other in which comparison is made by contact. The three common types of contact comparators are mechanical, electrical, and pneumatic. *See* INSPECTION AND TESTING.

Contact types. Comparators that depend on contact are similar in physical arrangement, differing only in the method of amplifying and indicating the deviation from the master setting. The limits of tolerance may be indicated on a dial for visual observation or by a signal such as a light or buzzer or, in the case of automated equipment, by actual physical rejection from the lot of those pieces outside of tolerance.

A general-purpose comparator of the contact type usually consists of (1) a fixed contact point or surface; (2) an indicating contact or measuring head; and (3) a means for physically maintaining the distance between. This last means takes the form of a frame in a gage making external measurements, or the form of a plug in a gage making internal measurements. The indicator or meter, which records the results, may be an integral part of the measuring head or a separate unit. A general-purpose comparator can usually be set with gage blocks, as well as with a master. *See* GAGES.

The indicator or measuring head amplifies the lineal movement of the contact point enough so that minute movements are clearly visible. Comparator heads are frequently rated by the amount of the magnification; thus a 1000:1 comparator would have a 1-in. (25-mm) travel of the indicating hand or point or marker for each 0.001-in. (25-μm) movement of the contact point.

Contact comparators are used to check individual single measurements or groups of single measurements. Accuracy depends on the measuring head. Several contact comparators can be assembled as a single gaging device to perform multiple measurements. Frequently all dimensions on a part are inspected on one machine simultaneously. When the amplifying heads in such a machine are arranged to actuate sorting equipment and to separate within-tolerance and out-of-tolerance parts, inspection is automated.

Mechanical comparators. Movement in mechanical comparators is amplified usually by a rack, pinion, and pointer, or by a parallelogram arrangement. Usually, mechanical comparators are applied to tolerances of 0.001 in. (25 μm), although heads are available which are accurate to 0.0001 in. (2.5 μm).

Electrical comparators. Movement in electrical comparators is amplified by several methods. The most popular method consists of a floating core in a solenoid attached to the contact point. Lineal movement of the contact changes the penetration of the core into the field of the solenoid coil. The resulting change in coil reactance is amplified electrically. Electrical comparators are applied to tolerances in the range of 0.00005–0.001 in. The accuracy of the instrument itself may be as fine as 0.000001 in.

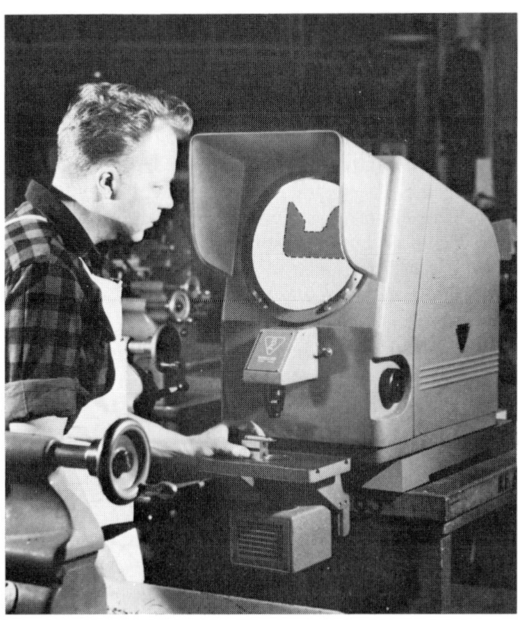

Fig. 2. Optical comparator. Part is checked against master form on screen. (*Bausch and Lomb Optical Co.*)

Pneumatic types. Intermediate in operation between contact and optical comparators are pneumatic units. They sense the surface of the part by means of an airstream. A precision orifice is brought so close to the surface to be measured that the flow of air from the orifice is restricted (Fig. 1). The closer the orifice comes to the surface, the greater the restriction; a measurement of this restriction in flow is, in effect, a measurement of this distance. Two methods of measurement are employed: measurement of the variation in flow and measurement of variation in back pressure. Measurement of back pressure is by a conventional, though highly accurate, pressure gage. Measurement of flow is by a tube, accurately tapered on the inside from top to bottom, in which a round cylindrical bobber is free to move. The top or larger end is attached to the gaging orifice. The bottom or smaller end is attached to the air source. Air is introduced into the bottom, flows by the edge of the bobber, and goes out the top through the gage orifice. The amount of air leaving the orifice is the exact amount flowing by the bobber and, because the area of the tube varies lineally, a variation in flow is made visible by a variation in the height of the bobber in the tube.

The range of accuracy of pneumatic gages is approximately that of electric gages.

Optical types. Visual or optical comparators of the general-purpose type consist of (1) a light source, (2) a stage on which to place the part being inspected, (3) a series of lenses and mirrors for magnifying and projecting the shadow of the object, and (4) a screen on which the image is projected (Fig. 2). The screen is usually a frosted glass. Sometimes the master outline is drawn on the screen. Some master outlines may be purchased already constructed on the screen; circle diagrams and standard thread forms are examples. Sometimes a master form is drawn on vellum or tracing paper to a scale corresponding to the magnification of the optical system and is taped on the screen.

Optical or visual comparators are usually used to inspect complex forms or the relationship of one or more individual surfaces. For most purposes they are limited to a part tolerance of about 0.001 in. (25 μm). To some extent optical comparators are self-limiting. Most measurements are across, parallel, or tangential to the surface. In general the more accurate surfaces are more highly finished and consequently more reflective. This reflection creates an aberration or fuzziness on the screen which makes accurate identification of the actual surface image difficult.

Automated inspection. Automated production equipment and automated inspection equipment are interdependent and have developed together. The contact comparator head has been the basic unit of equipment around which automated inspec-

Fig. 3. Automated inspection machine for ball-bearing races. (*Sheffield Corp.*)

COMPARATOR

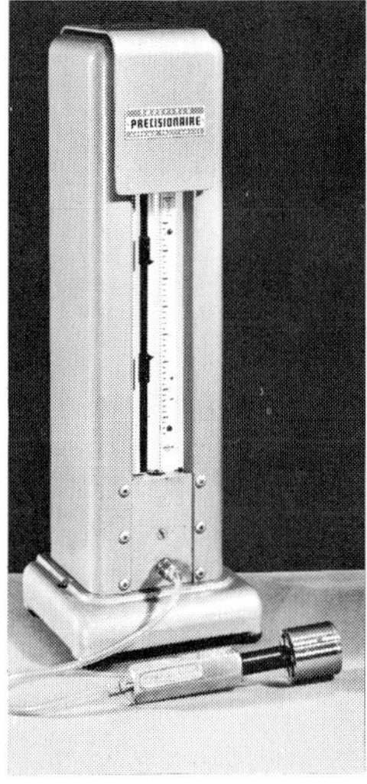

Fig. 1. Pneumatic comparator in a plug gage for inspection of an inside diameter. (*Sheffield Corp.*)

tion has been built. In some cases individual elements, checking individual dimensions, are incorporated in the production equipment. In others they are assembled in a single machine, and this machine is introduced into an automated production line.

A machine of this latter type is shown in Fig. 3. This machine inspects and either physically accepts or rejects several different sizes of inner ball races. The races are loaded on the shelf at the top. They pass through the machine in which their important dimensions are checked. They leave the machine by running down one of the four troughs in the foreground, either accepted or rejected. *See* AUTOMATION. [RUSH A. BOWMAN]

Composite beam

Composite beams are so called because they provide beam action of two materials joined together in such a way as to act as a unit. In civil engineering the term is most commonly given to the beam action developed by a concrete slab resting on a steel beam, usually an I or a wide flange shape, the slab and the beam being made to act together by shear connectors. *See* CONCRETE SLAB.

In noncomposite design the structural function of the concrete slab is only to span transversely across a series of parallel steel beams. In composite design the concrete slab performs its normal function and in addition increases to a considerable extent the capacity of the structure to resist bending in the direction of the steel beams. The unit stresses in the concrete slab are not added algebraically from the dual function the slab performs, because the flexural stresses are in mutually perpendicular directions. Thus the concrete slab thickness and area of reinforcing steel are not necessarily greater for this type of design than for the noncomposite. *See* STRESS AND STRAIN.

Shear connectors are devices that provide unified action between the concrete slab and the steel beam by resisting the horizontal shear developed in beam bending (Fig. 1). They take the form of steel projections welded to the top flange of the steel beam and around which the concrete slab is poured. The connectors perform a necessary function in the development of composite beam action, for without them the two elements, the concrete slab and the steel beam, would act separately as two beams and the beam action would be analogous to two unconnected wooden planks placed one on top of the other and slipping on each other when bent under a transverse load. There are many types of shear connectors, but three

Fig. 1. Steel part of composite beam; spiral shear connectors welded to I beams. *(Porete Manufacturing Co.)*

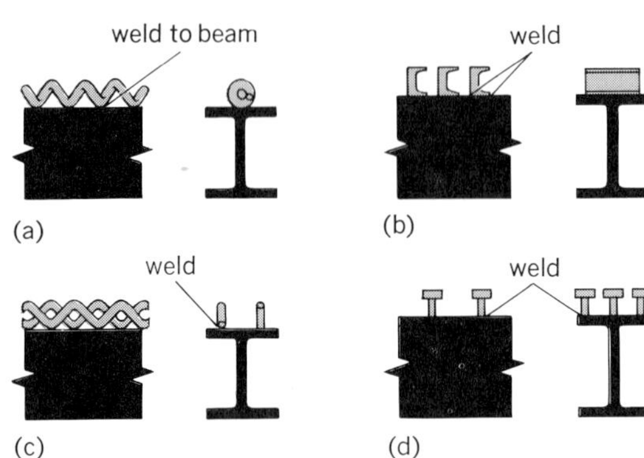

Fig. 2. Typical shear connectors. (*a*) Spiral shear connector. (*b*) Channel shear connector. (*c*) Serpentine shear connector. (*d*) Stud bolt connector.

commonly used types are stud bolts, a helical coil of steel rod, and short lengths of a standard channel shape. Figure 2 shows several types of shear connectors. A typical stud bolt might be 1/2 in. in diameter and 5 in. in length embedded in a 6-in.-thick concrete slab. The bolt would have a cylindrical head 3/4 in. in diameter and perhaps 1/2 in. high. All shear connectors must have some provision to prevent the concrete slab from lifting vertically from the steel beam when bent under load. In the case of the stud bolt the head provides this restraint. The bolts are welded to the beam flange, several in a grouping, and spaced longitudinally along the beam in accordance with the design based on the intensity of the horizontal shearing stress at each position along the beam. A short channel length welded across the beam flange provides the same shear connection function as the stud bolts, or a helical coil of rod with a pitch of 6 or 8 in. and a diameter of 4 or 5 in. can connect the two elements. A typical rod diameter is 1/2 in. *See* SHEAR.

Composite beams are used both in building and highway bridge construction and can be used in any type of construction where it is appropriate to use a concrete slab resting on steel beams or stringers. Definite and considerable economy is achieved in composite construction when the spans are of sufficient length to offset the increased cost of fabricating the shear connectors with the saving from the lesser beam size required. Composite construction is generally not considered economically feasible for spans under 35 or 40 ft but is used on simply supported spans up to 100–120 ft or longer with specially fabricated steel beams. In the longer spans the size of the steel stringers can be reduced at least to the next lower standard rolled steel section. In most size ranges the beams are listed in groupings of beam depths in 3-in. multiples. For example, a beam design for noncomposite action which required a depth of 36 in. could be expected to be reduced to a depth of 30–33 in. if composite action is provided.

[HENRY L. KINNIER]

Bibliography: B. Bresler, T. Y. Lin, and J. B. Scalzi, *Design of Steel Structures*, 2d ed., 1967; I. M. Viest, R. S. Fountain, and R. C. Singleton, *Composite Construction in Steel and Concrete*, 1958; C. K. Wang and C. G. Salmon, *Reinforced Concrete Design*, 1965; G. Winter et al., *Design of Concrete Structures*, 7th ed., 1964.

Compression ratio

In a cylinder, the piston displacement plus clearance volume, divided by the clearance volume. This is the nominal compression ratio determined by cylinder geometry alone. In practice, the actual compression ratio is appreciably less than the nominal value because the volumetric efficiency of an unsupercharged engine is less than 100%, partly because of late intake valve closing. In spark ignition engines the allowable compression ratio is limited by incipient knock at wide-open throttle. Knock in turn depends on the molecular structure of the fuel and on such engine features as the temperature of the combustible mixture prior to ignition, the geometry and size of the combustion space, and the ignition timing. For example, isooctane, benzene, and alcohol can be burned at much higher compression ratios than *n*-heptane. In compression ignition engines critical compression ratio is that necessary to ignite the fuel and depends on fuel and cylinder geometry. *See* COMBUSTION CHAMBER; DIESEL ENGINE; INTERNAL COMBUSTION ENGINE; VOLUMETRIC EFFICIENCY.

[NEIL MAC COULL]

Compressor

A machine that increases the pressure of a gas, vapor, or mixture of gases and vapors. The pressure of the fluid is increased by reducing the fluid specific volume during passage of the fluid through the compressor. When compared with centrifugal or axial-flow fans on the basis of discharge pressure, compressors are generally classed as high-pressure and fans as low-pressure machines.

Compressors are used to increase the pressure of a wide variety of gases and vapors for a multitude of purposes (Fig. 1). A common application is the air compressor used to supply high-pressure air for conveying, paint spraying, tire inflating, cleaning, pneumatic tools, and rock drills. The refrigeration compressor is used to compress the gas formed in the evaporator. Other applications of compressors include chemical processing, gas transmission, gas turbines, and construction. *See* GAS TURBINE; REFRIGERATION.

Characteristics. Compressor displacement is the volume displaced by the compressing element per unit of time and is usually expressed in cubic feet per minute (cfm). Where the fluid being compressed flows in series through more than one separate compressing element (as a cylinder), the displacement of the compressor equals that of the first element. Compressor capacity is the actual quantity of fluid compressed and delivered, expressed in cubic feet per minute at the conditions of total temperature, total pressure, and composition prevailing at the compressor inlet. The capacity is always expressed in terms of air or gas at intake (ambient) conditions rather than in terms of arbitrarily selected standard conditions.

Fig. 1. Mobile air compressor unit which is driven by engine in the forward compartment. (*Ingersoll-Rand*)

Air compressors often have their displacement and capacity expressed in terms of free air. Free air is air at atmospheric conditions at any specific location. Since the altitude, barometer, and temperature may vary from one location to another, this term does not mean air under uniform or standard conditions. Standard air is at 68°F (20°C), 14.7 psia (101.3 kPa absolute pressure), and a relative humidity of 36%. Gas industries usually consider 60°F (15.6°C) air as standard.

Types. Compressors can be classified as reciprocating, rotary, jet, centrifugal, or axial-flow, depending on the mechanical means used to produce compression of the fluid, or as positive-displacement or dynamic-type, depending on how the mechanical elements act on the fluid to be compressed. Positive-displacement compressors confine successive volumes of fluid within a closed space in which the pressure of the fluid is increased as the volume of the closed space is decreased. Dynamic-type compressors use rotating vanes or impellers to impart velocity and pressure to the fluid.

Reciprocating compressors are positive-displacement types having one or more cylinders, each fitted with a piston driven by a crankshaft through a connecting rod. Each cylinder also has

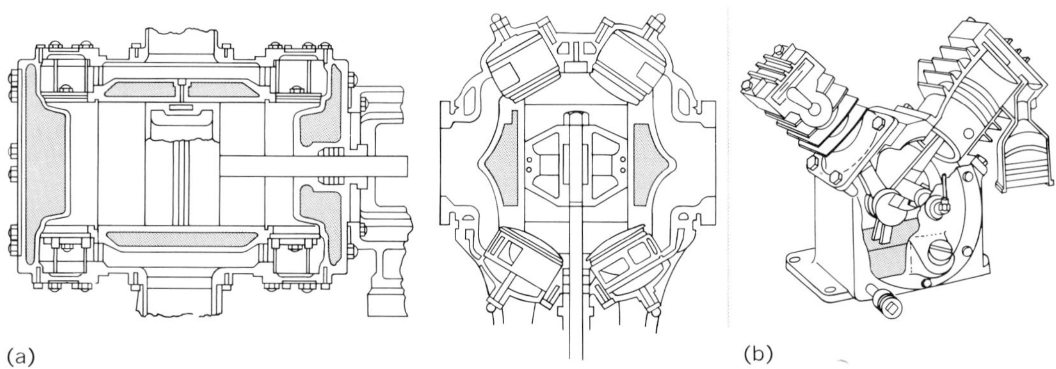

Fig. 2. Compressor cylinders, (*a*) water-cooled and (*b*) air-cooled.

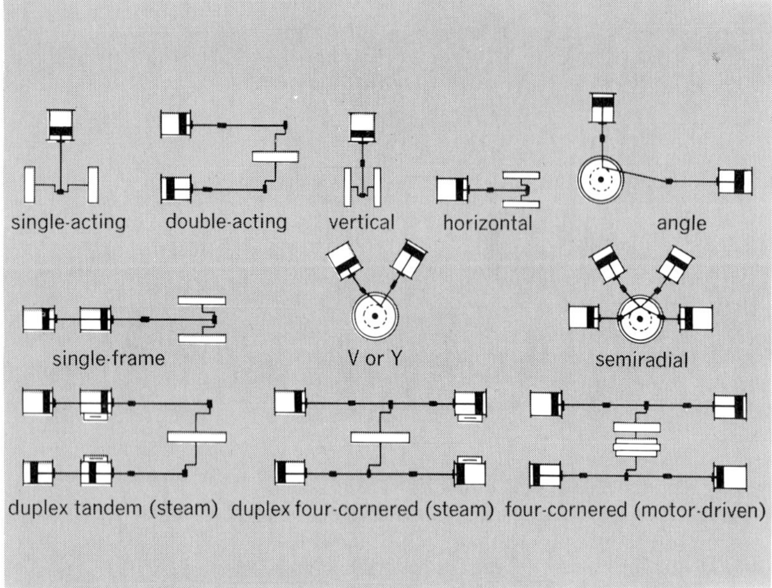

Fig. 3. Frame arrangements of positive-displacement piston compressors.

intake and discharge valves, and a means for cooling the mechanical parts (Fig. 2). Fluid is drawn into the cylinder during the suction stroke. At the end of the suction stroke, motion of the piston reverses and the fluid is compressed and expelled during the discharge stroke. When only one end of the piston acts on the fluid, the compressor is termed a single-acting unit. When both ends of the piston act on the fluid, the compressor is double-acting. The double-acting compressor discharges about twice as much fluid per cylinder per cycle as the single-acting (Fig. 3).

Single-stage compressors raise the fluid pressure from inlet to discharge on each working stroke of the piston in each cylinder. Two-stage compressors use one cylinder to compress the fluid to an intermediate pressure and another cylinder to raise it to final discharge pressure. When more than two stages are used, the compressor is called a multistage unit.

Vertical and horizontal compressors may be single-cylinder or multicylinder units. The angle type is multicylinder with one or more horizontal and vertical compressing elements (Fig. 4). Single-frame (straight-line) units are horizontal or vertical, double-acting, with one or more cylinders in line with a single frame having one crankthrow, connecting rod, and crosshead. The V or Y type is a two-cylinder, vertical, double-acting machine with cylinders usually at a 45° angle with the vertical. A single crank is used. Semiradial compressors are similar to V or Y type, but have horizontal double-acting cylinders on each side. Duplex compressors have cylinders on two parallel frames connected through a common crankshaft. In duplex-tandem steam-driven units, steam cylinders are in line with air cylinders. Duplex four-cornered steam-driven compressors have one or more compressing cylinders on each end of the frame and one or more steam cylinders on the opposite end. In four-cornered motor-driven units, the motor is on a shaft between compressor frames.

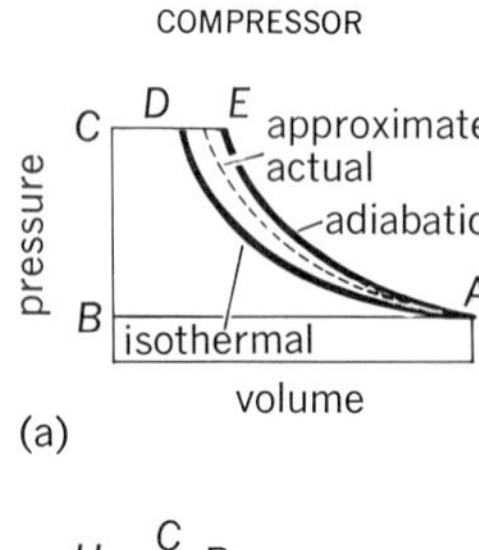

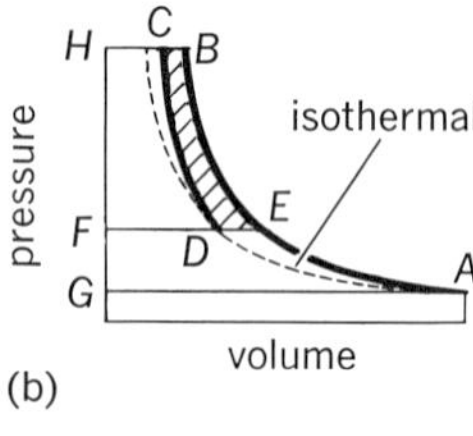

Fig. 5. Compression curves for *(a)* theoretical and actual units and *(b)* effect of intercooling.

Reciprocating compressors are built to handle fluid capacities ranging to 100,000 cfm; pressures range to over 35,000 psi. Special units can be built for larger capacities or higher pressures. Water is the usual coolant for cylinders, intercoolers, and aftercoolers, but other liquids, including refrigerants, may also be used.

Compressor thermodynamics. Compression efficiency in any compressor is compared against two theoretical standards—isothermal and adiabatic. Neither occurs in an actual compressor because of unavoidable losses. A plot of compression process on a pressure-volume diagram shows that an actual unit works between these two standards (Fig. 5). Isothermal compression has perfect cooling—air remains at inlet temperature while being compressed. Work input to compressor, measured by area *ABCD*, is least possible. Adiabatic compression has no cooling; temperature rises steadily during compression. Discharge pressure is reached sooner than with isothermal compression. Since air pressure is higher during every part of the piston's stroke, more work input is needed, as shown by area *ABCE*.

If compression is divided into two or more steps or stages, air can be cooled between stages. This intercooling brings the actual compression line closer to the isothermal line. Area *BCDE* shows the power saved.

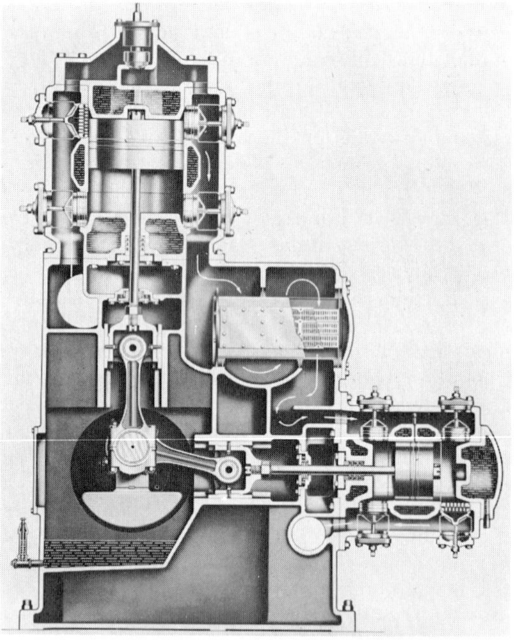

Fig. 4. Cross section of stationary angle-type compressor with first stage vertical and second stage horizontal.

Water vapor in air entering the compressor leaves as superheated vapor because its temperature is in excess of that corresponding to its pressure. It can be converted to water only by cooling to a temperature below the saturation temperature corresponding to its pressure. Immediate cooling of air after it leaves the compressor proves best, because this prevents vapor from reaching the distribution system. Either air- or water-cooled heat exchangers, called aftercoolers, are used for this.

The usual single-stage reciprocating compressor is built for pressures to about 150 psi; two-stage types for up to about 500 psi; three-stage for up to about 2500 psi; four- and five-stage for up to 15,000 psi, and higher.

Rotary compressors. Other forms of positive-displacement compressor are the rotary types.

Sliding-vane type. In the sliding-vane rotary compressor, fluid is trapped between vanes as the rotor passes the inlet opening (Fig. 6). Further rotation of the rotor reduces the volume of the space in which the fluid is trapped. Fluid pressure rises until the discharge port is reached, when discharge occurs.

Depending on design, the compressor may be cooled with atmospheric air, oil, or water. For low pressures one stage of compression is used, two stages for higher pressures. Capacities range to about 5000 cfm. Some compressors of this type have rings around the vanes to keep them from bearing on cylinder walls. Others have vanes bearing against the cylinder.

Lobe type. In the lobed rotary compressor, fluid is trapped between two or more rotors held in fixed relationship to each other (Fig. 7). Rotation of the impellers reduces the volume in which the fluid is trapped, producing a pressure rise. Fluid is discharged when the rotors pass the outlet port. Either two or three rotors are used. Capacities range from about 5 to 50,000 cfm. Pressures above about 15 psi are obtained by operating two or more lobe-type compressors in series. Rotors are straight, as shown, or slightly twisted.

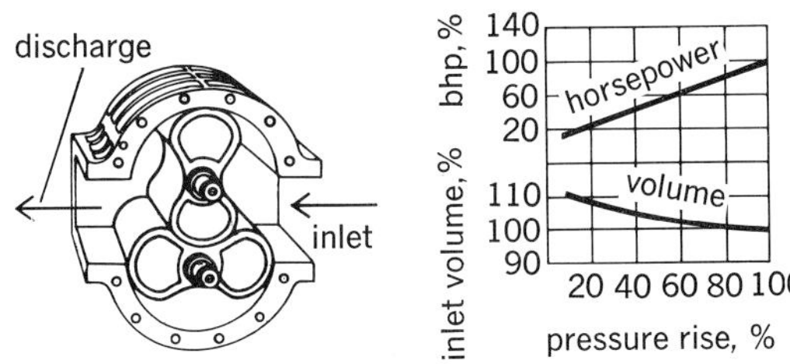

Fig. 7. Two-lobe compressor and its performance curves at constant speed.

Liquid-piston type. In the liquid-piston rotary compressor, a round multiblade rotor revolves in an elliptical casing partly filled with liquid, usually water. When the rotor turns, it carries liquid around with it, the blades forming a series of buckets (Fig. 8). Because the liquid follows the casing contours, it alternately leaves and returns to the space between blades (twice each revolution). As the liquid leaves the bucket, the fluid to be compressed is drawn in. When the liquid returns, it compresses the fluid to discharge pressure.

Liquid-piston compressors handle up to about 5000 cfm. Single-stage units can develop pressures to about 75 psi; multistage designs are used for higher pressures. Water or almost any other low-viscosity liquid serves as the compressant. For exacting services, the compressor may be sealed with chilled water to prevent condensation in lines.

Radial-flow units. Dynamic-type centrifugal compressors use rotating elements to accelerate the fluid radially. By diffusing action, velocity is converted to static pressure. Thus, the static pressure is higher in the enlarged section. Centrifugal compressors usually take in fluid at the impeller eye (or central inlet of the circular impeller) and accelerate it radially outward (Fig. 9). Some static-pressure rise occurs in the impeller, but most of the pressure rise occurs in the diffuser section of the casing, where velocity is converted to static pressure. Each impeller-diffuser set is a stage of the compressor. Centrifugal compressors are built with from 1 to 12 or more stages, depending on the final pressure desired and the volume of fluid to be handled. The pressure ratio, or compression ratio, of a compressor is the ratio of the absolute discharge pressure to the absolute inlet pressure.

In the typical compressor (Fig. 9), centrifugal action of the impeller A_1 produces pressure rise BC and a large increase in air velocity EF. In the diffuser B_1, velocity energy is converted to static pressure. Velocity falls from F to G, pressure rises from C to D.

Pressure-volume curves for a centrifugal compressor operating at different speeds show that, for example, at speed M the unit delivers volume V_M at pressure P_M (point A on the pressure-volume diagram, Fig. 9). Increasing speed to H raises volume to V_H at P_M, or old volume V_M can be delivered at higher pressure P_H.

Pressure delivered by a centrifugal compressor is practically constant over a relatively wide range of capacities.

Pumping limit, also called surge point or pulsation point, is the lower limit of stable operation. Percentage stability is 100 minus the pumping limit in percent of design capacity.

Multistage centrifugal compressors handle 500 to more than 150,000 cfm, at pressures as high as 5000 psi, but are limited to compression ratios in the order of 10.

Axial-flow units. Compressors that accelerate the fluid in a direction generally parallel to the rotating shaft consist of pairs of moving and stationary blade-rows, each pair forming a stage. The pressure rise per stage is small, compared with a radial-flow unit. Hence the usual axial-flow compressor has more stages than a centrifugal compressor working through the same pressure range. Single-stage axial compressors have capacities from a few to more than 100,000 cfm at pressures from less than 1 to several psi. Multistage axials compress air to 150 psi or more. Some special machines handle over 2,000,000 cfm. Pressure rise per stage is generally relatively small, so units for higher pressures frequently have a considerable number of stages—20 or more (Fig. 10).

While centrifugal machines deliver practically constant pressure over a considerable range of capacities, axials have a substantially constant delivery at variable pressures. In general, centrifugals have a wider stable operating range than axials. Because of their more or less straight-through flow, axials tend to be smaller in diameter than centrifugals and are apt to be longer. Efficiency of axials usually runs slightly higher.

To prevent surging at extremely low loads, large-capacity axials are sometimes fitted with a blowoff system that discharges excess air to the atmos-

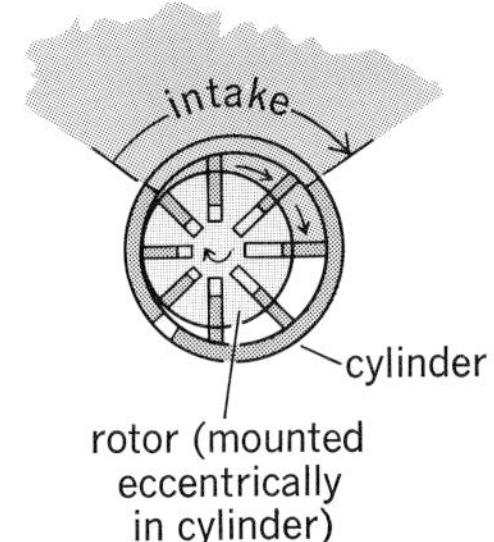

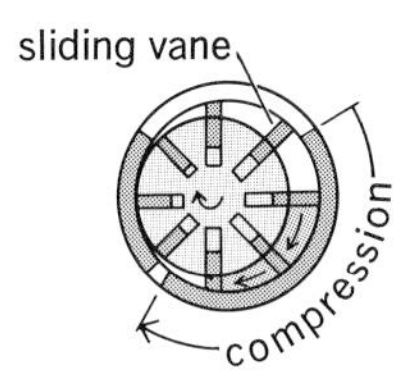

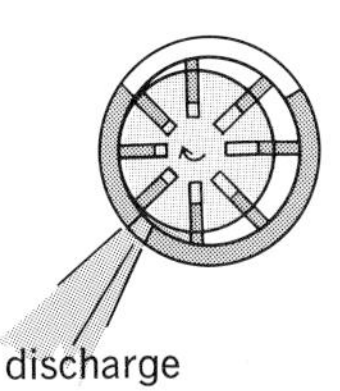

Fig. 6. Operation of sliding-vane rotary compressor.

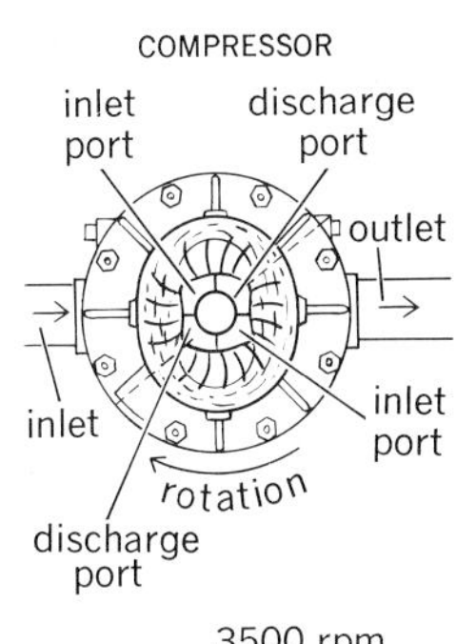

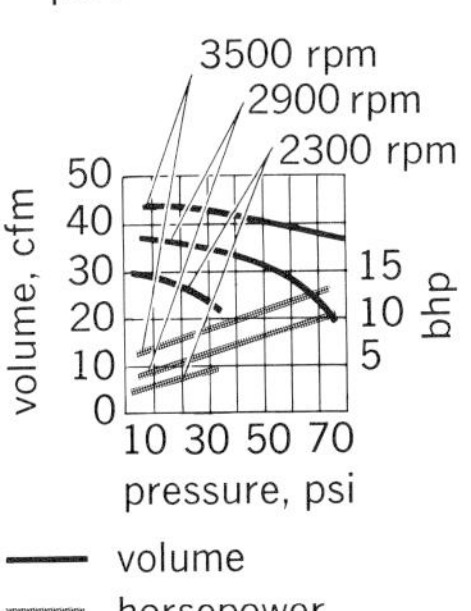

Fig. 8. Liquid-piston rotary compressor and its constant-speed performance curves.

phere. Then there is always enough air passing through the machine to keep it in its stable range.

Because of difficulty in accurately predicting the performance curves of centrifugal and axial compressors, only one capacity and one discharge-pressure rating, together with corresponding power input, are normally guaranteed. Shape of the curve may be indicated but is never guaranteed. *See* FAN; PUMPING MACHINERY.

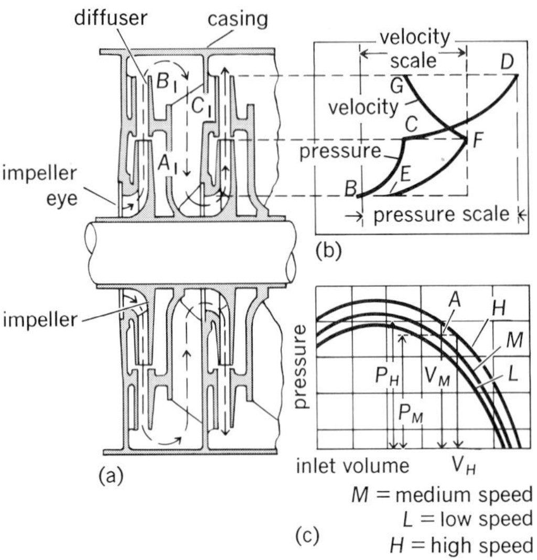

Fig. 9. Performance curves for radial-flow or centrifugal compressor. (*a*) Air flow through centrifugal compressor following paths shown by arrows. (*b*) Pressure and velocity relationships of air in a typical centrifugal compressor. (*c*) Volume and pressure relationships of a centrifugal compressor at various speeds.

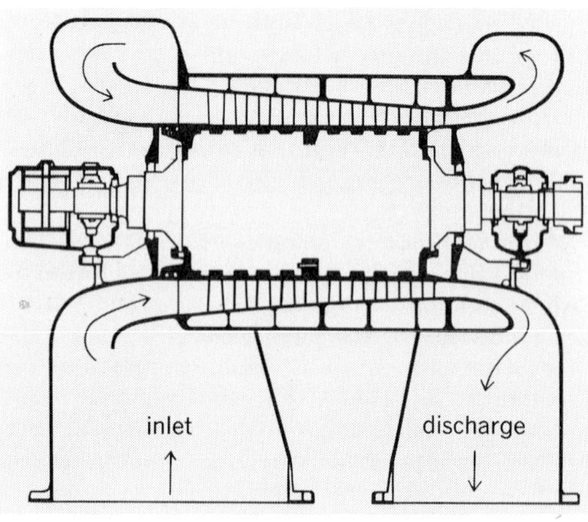

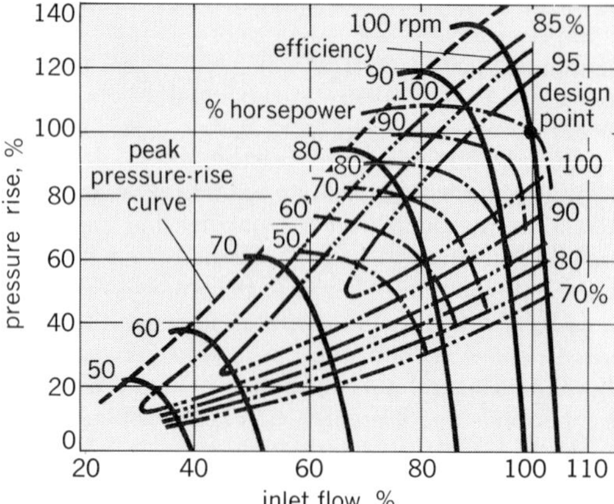

Fig. 10. Axial-flow compressor and performance curves of its constant-speed operation.

[TYLER G. HICKS]

Bibliography: Compressed Air and Gas Institute, *Compressed Air and Gas Handbook*, 4th ed., 1973; *Compressor Handbook for the Hydrocarbon Processing Industries*, 1979; L. Sheel, *Gas Machinery*, 1972.

Computer-aided design and manufacturing

Application of the computer in combining design and manufacturing functions continues to gain momentum in diminishing the time between concept and finished product. Specifically, computer-aided design (CAD) refers to the use of computers to perform design calculations for determining an optimum shape and size for a variety of applications ranging from mechanical structures and tiny integrated circuits to maps of huge areas. Computer-aided manufacturing (CAM) employs computers to communicate the work instructions for automatic machinery in the handling and processing technology used to produce a workpiece. Computer-based automation is drastically changing the way things are made and profoundly changing the jobs of people who make them. Among the benefits of an integrated CAD/CAM system are increased productivity, a significant reduction in nonproductive time, improved product quality, and a payback potential obtained by lowering the cost per piece. This modern concept of manufacturing management has led to important advances in the design and production of components used in aerospace, automotive, electronics, and other industries throughout the world

CAD. This first component of the totally integrated CAD/CAM system yields an impressive number of benefits. One benefit is that the designer is equipped with vastly more efficient computer equivalents of the many drafting tools needed in the performance of the job. Computer programs can be written to generate hard copy of drawings at speeds and accuracies far beyond human skills. The computer has liberated the designer from countless, tedious drafting details. Designers are free from the repetitious task of drawing various lines and, often, even from tasks such as calculating workpiece sizes. In fact, a CAD system not only eliminates the need for drawing dimensions on the views of a part, but also saves the monotonous labor involved in making the arrowheads.

Another time-saving feature of considerable importance is that sectional views as well as auxiliary views can be automatically conceived and drawn by computer methods. The unique system accepts commands like "erase line," "move circle," or "insert dotted or crosshatch lines." CAD systems can be programmed to generate and display a variety of symbols, characters, and points, lines, arcs, and circles—in virtually any form required for the construction of a geometric image. Once developed, part drawings can be stored in computer memory as dynamic, three-dimensional forms or as conventional, shop-type multiview projections.

Another significant benefit of CAD is that a

computer-created design can be instantly recalled, either partially or totally. The graphic image can be displayed on a CRT (cathod-ray tube) screen to permit an analysis of the workability of the designer's ideas. So versatile is a computer-graphics system that mirror images of mating parts may be quickly produced and displayed. Computer-refreshed views on a CRT may be manipulated, reduced, enlarged, or viewed from different angles for possible design modifications. With this technique, a great number of design alternatives may be examined within an astonishingly short period of time. Also, when required, a CAD system can automatically generate a control tape that in some manufacturing systems can be employed to drive a numerical control metal cutting or forming machine.

When all of the elements of the design concept are in final and acceptable form, an assembly drawing can be readily generated by recalling each part from computer memory and placing the parts in their appropriate assembled graphical position. A hard copy bill of material may be automatically produced with lines ruled off, correctly spaced, and with all of the components of the design accurately recorded in the proper position.

CAM. The second component of CAD/CAM enables the utilization of processes that allow the machines to perform productive chip removal operations over a much larger percentage of time than heretofore. Using a concept which some machine tool firms call total processing, the manufacturing engineer identifies each processing requirement that interfaces with the computer data base corresponding to the original design. As an example, functional tool design data (that is, proper tool path generation ensured by selected datum plane locations on the workpiece) is integrated with program data relating to the final part design. CAM programs are written that automatically command an optimum machining sequence of processing operations, control the cutter type and size, turn coolant on or off, select appropriate feeds and speeds, and regulate a number of other machining parameters. Correlation of the design phase with the part-processing phase has significantly affected the production cycle. CAM permits a part to progress at a more rapid rate from raw material to finished product.

Programming languages. Computer programs, called software, are the principal form of communication between the programmer and the computer. Compatability of a computer-integrated design with the machine and control begins at the design stage with each aspect taking full advantage of the relative capabilities of the other. While there are many languages available for this purpose, most CAD/CAM software programs use either FORTRAN, COBOL, BASIC, PASCAL PL/1, or NUFORM.

Of all the programs, FORTRAN is currently considered by most to be the easiest programming language to work with in engineering applications. It is the program ordinarily used for solving complex numerical calculations and is a particularly useful language for solving engineering analysis problems. COBOL is mainly a business language. Its principal application is commercial data processing. Despite a somewhat limited arithmetic capability, COBOL is an effective programming language for certain kinds of engineering applications. BASIC, an acronym for Beginners All-purpose Symbolic Instruction Code, was originally developed as an educational tool to be used in teaching the use of remote-control time-sharing systems. The original language has recently undergone significant improvements and standardization. PASCAL was also originally introduced as a teaching tool for computer programming. The principal advantage of PASCAL is in its effectiveness in structured programming. It is currently being evaluated by several agencies as a possible standard. PASCAL is used extensively in universities and colleges and is gaining wide acceptance in industry and government installations. The PL/1 language in some ways resembles FORTRAN. There are also similarities in block structures to ALGOL, while the data types suggest the influence of COBOL. PL/1 is a large general-purpose language well adapted to engineering and scientific applications. NUFORM features a fixed format with numeric input. This versatile system significantly reduces the syntax and vocabulary requirements associated with an alphanumeric input. NUFORM was designed to simplify the programmer's task, thus saving time and promoting accuracy.

Vendor computer programs. Most CAD/CAM programs that have been developed by industry are proprietary and thus are not available for general use. Often the applications of such programs are limited because of their specialized and restricted nature. Commercially available preprogrammed software in the area of mechanism design has been found to be helpful in certain specific problem-solving applications. There are a number of software firms which can supply a full range of computer services. These range from a surprisingly large assortment of comparatively simple, straightforward software programs to a complete data-base manufacturing management system that, in addition to an engineering and manufacturing system, includes software for inventory control bill of material processing, material requirements planning, work in progress, and production costing modules.

Examples of software that can be purchased from software vendors are Automatic Dynamic Analysis of Mechanical Systems (ADAMS), Dynamic Response of Articulated Machinery (DRAM), Integrated Mechanisms Program (IMP), and Kinematic Synthesis (KINSYN). Examples of other software programs currently available are NASTRAN and CSMP. These programs deal with aspects of engineering analysis. There are also two computer programs in extrusion die engineering currently available at no cost. Developed by Battelle-Columbus under the sponsorship of the U.S. Army, they are ALEXTR and EXTCAM. Both programs have been written in FORTRAN language and operate on a minicomputer. Also available are programs such as STRUDL, ANSYS, and SAP which relate to a newly developed analytical technique called finite element analysis.

Numerical control. Before the advent of numerical control (NC) the regulation of product accuracy and repeatability was directly related to the skill of the operator. The maintenance of precision is no longer considered a serious problem. The limits of accuracy on a workpiece are now controlled entirely by NC machines. The result is that the operator is no longer responsible for positioning and repeating the operation of the tool. Instead of manual controls, electric signals on NC machines precisely guide the movements of the tools and control the position of the workpiece.

NC machines may be operated by manually dialing the machine setting at a console and letting the electronic signals execute the operation. As might be expected, manual operation not only is a very slow production method but is very inefficient for large-volume production of parts. Improved versatility for NC machines is possible by using a punched paper or Mylar tape. In this method the tool and machine instructions are punched onto the tape according to an alphanumeric code. In operation, signals from the punched tape are sent to a data storage unit. As the tape unwinds past the tape reader, electric impulses are sent to the drive mechanisms which automatically control the machine functions. Information on tape can be used to establish the feed rate, spindle speed, machine table positions, traverse rates, tool selection, depth of cut, stops, and dwell intervals for selected periods of time. Unfortunately, there are some inherent problems associated with punched tape. Some of these include difficulties in tape preparation, tape breakage, limitations in adapting programs for a sufficiently wide range of workpiece conditions, and problems associated with editing and with making program changes.

The NC machine was first introduced in the 1950s. It was found that NC machines could be depended upon to operate more hours a day than was possible with traditional machine tools. Also, NC machines were more accurate. Finally, of considerable advantage was the fact that tools and workpieces could be installed and positioned automatically. As programming requirements increased in volume and complexity, however, the tape control machine proved impractical—if not impossible—for many potential jobs. Computer-assisted programming for machine tools appeared to offer a solution. In 1964 graphic computer display terminals were introduced, thus giving the system engineering staff a unique opportunity to visually verify each step in the production sequence of a workpiece.

Direct numerical control. The first industrial direct numerical control (DNC) system became operational in 1968. A DNC system, the lifeline of the CAM concept, includes both the hardware and the software required to drive one or more NC machines simultaneously while connected to a common memory in a computer. The computer may be a minicomputer, several minicomputers linked together, a minicomputer connected to a large computer, or a single large computer. Unlike an NC system, a punched tape is not used. An important advantage of the DNC system is that more than one machine can be operated at a time because the computer can be time-shared. The early DNC systems were produced out of the need to ensure optimum utilization of NC machines.

Some DNC systems consist of combinations with capabilities that encompass CRT display, part program storage, part program edit, and maintenance diagnostics. On some machines, as the program runs, the CRT provides visible part-program information, operating data, current-status messages, error messages, and diagnostic instructions. A typewriterlike keyboard facility located at the NC machine permits the operator to input, delete, or correct data. After the editing and optimizing phases have been completed, the program changes are stored in memory. The sequence of processing operations in CAM systems begins immediately after the program is recalled from storage in the computer memory. DNC programs are commanded entirely by electronic signals sent from the computer memory.

In addition to the primary function, that of controlling the manufacturing sequence of parts, DNC computers can perform a wide range of other useful data transmission functions: parts program development and verification and job scheduling. Thus, DNC systems may be applied to an almost unlimited range of product management activities.

Computer numerical control. The newest CAM innovation is called computer numerical control (CNC). This sophisticated manufacturing concept first appeared in 1970. Unlike DNC, each machine tool has its own computer. The principal advantage of CNC over DNC is that the computer software systems are less expensive. While one standard control can be used for a range of different types of machine tools, the software can be written in such a way as to adapt the control to each particular machine requirement.

Today, extremely large and versatile, totally integrated CAM machining centers are available from a number of well-established machine tool firms. Advantages cited for this trend toward complete manufacturing centers include shorter design lead time and shorter manufacturing make-ready time. In combination, these aspects result in a dramatic reduction of production throughput time. The economic pressures for computer-managed parts technology have never been more apparent. The proof is currently reflected in substantial investments in CAD/CAM research and development activities by machine tool builders both in the United States and in many other countries. The field is changing at an ever-increasing rate, and most forward-thinking parts processers are changing with it. *See* TOOLING.

[HERBERT W. YANKEE]

Bibliography: Happy marriage of CAD and CAM, *Machine Design*, 51(2):36–42, Jan. 25, 1979; J. Harrington, *Computer Integrated Manufacturing*, 1974, reprinted 1979; G. L. Petoff, The look of modern CAM standards, *Manufact. Eng.*, 82(4): 78–80, April 1979; Society of Manufacturing Engineers, *Numerical Control in Manufacturing*, 1975.

Concrete beam

A structural member of reinforced concrete placed horizontally to carry loads over openings.

Because both bending and shear in such beams induce tensile stresses, steel reinforcing tremendously increases beam strength. Usually, beams are designed under the assumption that tensile stresses have cracked the concrete and the steel reinforcing is carrying all the tension. *See* STRESS AND STRAIN.

Two design theories are currently used, elastic design and ultimate-load design.

Elastic design. The following assumptions are made for elastic design:

1. Plane sections remain plane after bending and are perpendicular to the longitudinal fibers.

2. The stress-strain curve is a straight line.

3. The ratio n of the modulus of elasticity of steel E_s to that of concrete E_c is a constant $n = E_s/E_c$.

4. The concrete does not carry tensile stress.

Transformed section. One approach to elastic design of reinforced concrete beams is to convert the steel to concrete. Because the steel and concrete are assumed to be firmly bonded and thus strained the same amount, the stress in the steel is n times the concrete stress. Hence the steel area may be replaced by an equivalent concrete area which is n times as large (Fig. 1).

If the equivalent area is placed at the same level as the steel and the moment of inertia I computed for the transformed section, the bending stresses can be computed from the simple flexural formula $f = Mc/I$, where M is the bending moment and c the distance from the neutral axis to the level at which stresses are to be computed.

Rectangular beams. The following formulas can be derived from the basic assumptions of elastic theory. Equation (1) locates the neutral axis, given

$$\frac{nf_c}{f_s} = \frac{k}{1-k} \qquad (1)$$

steel and concrete extreme stresses, where n is the ratio of the modulus of elasticity of steel to that of concrete, f_c is the stress in the extreme fiber of the concrete, f_s is the stress in the steel, and k is the ratio of the distance between the top of the beam and the neutral axis to the distance between the top of the beam and the steel (Fig. 2).

The design equation for equal moment resistance of concrete and steel (balanced design) is Eq. (2). Equations (3)–(6) are review equations.

$$k = \frac{1}{1 + f_s/nf_c} \qquad (2)$$

$$k = \sqrt{2np - (np)^2} - np \qquad (3)$$

$$j = 1 - k/3 \qquad (4)$$

$$M_c = \tfrac{1}{2} f_c kjbd^2 \qquad (5)$$

$$M_s = f_s A_s jd \qquad (6)$$

Here p is the ratio of effective area of tension reinforcement to effective area of concrete in beams, j is the ratio of lever arm of resisting couple to depth d, M_c is the moment resistance of the concrete, b is the width of the rectangular beam or width of flange of the T beam, d is the depth from the compressive surface of beam or slab to center of longitudinal tension reinforcement, M_s is the moment resistance of the steel, and A_s is the steel area.

When M_s is less than M_c, the capacity of the steel determines the maximum moment that the beam will carry. The beam is called underreinforced. If the beam is loaded to failure, the steel rather than the concrete determines the maximum load that is sustained.

When M_c is less than M_s, the concrete determines the maximum moment the beam will carry. The beam is said to be overreinforced. Usually, overreinforced beams are avoided because they are not considered economical and failure may occur without warning.

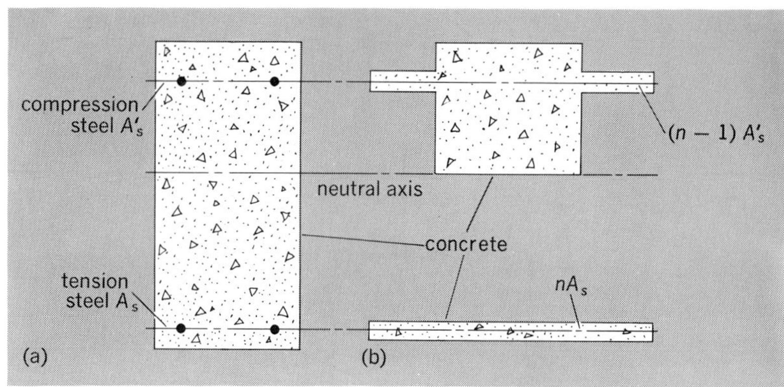

Fig. 1. Sketch of concrete beam. (*a*) Actual beam section. (*b*) Same section transformed. The use of the term $(n-1)A'_s$ prevents compression steel area from being included twice when computing moment of inertia of section; $(n-1)A'_s = nA'_s - A'_s$, because computations require the equivalent steel area to be added to the original concrete area above the neutral axis.

To design a rectangular beam by the elastic theory: (1) Select allowable unit stresses and determine k from the formula for balanced design; (2) compute j; (3) assuming the resisting moment of the concrete equal to the bending moment on the section, calculate bd^2 and select values for b and d; and (4) determine the steel area by equating the resisting moment of the steel to the bending moment on the section. This equation indicates that the amount of steel required can be reduced by increasing the depth. If the depth is fixed and an overreinforced beam results, it may be economical to use compression steel.

Compression-reinforced rectangular beams. Reinforcement may be added in the compression zone of concrete beams when the resisting moment of the concrete is less than the bending moment on the section or to avoid an overreinforced design. The compression steel is assumed to act as in plastic design; the Building Code Requirements of the American Concrete Institute allow the steel to take twice the stress given by elastic-theory formulas, provided the allowable tensile stress of the steel is not exceeded. Formulas for design of beams with both tension and

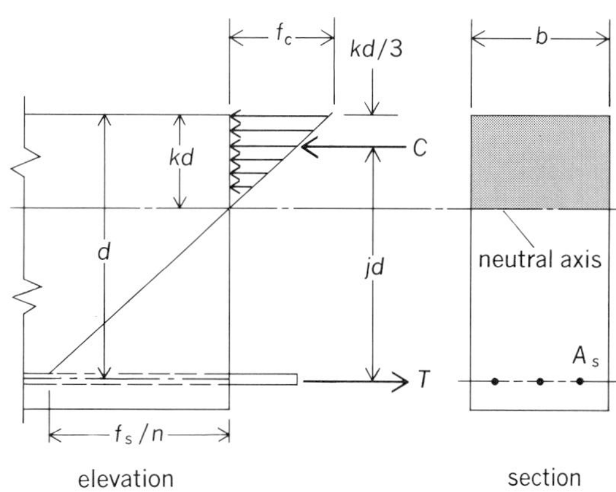

Fig. 2. Stress distribution in a rectangular beam designed according to elastic theory. C is total compressive force in concrete and T is total tensile stress in longitudinal reinforcement.

compression reinforcement can be derived from the elastic theory in the same manner as for beams with tension reinforcement only, but they are too complicated for ordinary use. Equations (7)–(12) are either exact or a close approximation.

$$k=\frac{1}{1+f_s/nf_c} \tag{7}$$

$$f'_s=nf_c\frac{kd-d'}{d-kd} \tag{8}$$

$$A_s=M/f_s jd \tag{9}$$

$$M_c=\tfrac{1}{2}f_c bkd\left(d-\frac{kd}{3}\right) \tag{10}$$

$$M'_s=M-M_c \tag{11}$$

$$A'_s=\frac{M'_s}{2f'_s(d-d')} \tag{12}$$

k = ratio of distance between top of beam and neutral axis to the distance between top of beam and tension steel
f_s = allowable steel stress
f_c = allowable concrete stress
n = ratio of modulus of elasticity of steel to that of concrete
f'_s = stress in compression steel computed from elastic theory formula
d' = distance from top of beam to compression steel
A_s = area of tension steel
M = bending moment on the section
jd = moment arm of the tensile reinforcement (assumed)
M_c = resisting moment of the concrete
b = width of beam
d = distance from top of beam to tension steel
M'_s = resisting moment of the compression steel
A'_s = area of compression steel

Concrete T beams. A slab cast integrally with a rectangular concrete beam usually is assumed to assist the beam in carrying loads. In regions of positive bending moment, the two act together as a T beam. In regions of negative moment, the beam is designed as a rectangular section because the slab is in tension and is assumed not to be able to resist such stresses, which must be taken by the reinforcing. *See* CONCRETE SLAB.

If the neutral axis of a T beam is within the slab, it is designed as a rectangular beam, with width b the same as that of the flange, to resist bending moments. For shear, however, only the width of stem b' can be assumed to be effective.

If the neutral axis falls within the stem, the section can be designed as a T beam. However, the compression in the stem is negligible and can be ignored to simplify computations.

The American Concrete Institute's code recommends the following limits for the part of the slab that can be considered effective as the flange: (1) b shall be less than one-fourth the span length, (2) the overhanging width of flange shall not exceed eight times the slab thickness, and (3) the overhanging width shall not exceed one-half the clear distance between beams.

If t is the flange thickness and the remaining symbols are the same as for elastic-theory design of rectangular beams, Eqs. (13) through (20) can be used for T beams when stem compression is neglected.

$$k=\frac{1}{1+f_s/nf_c} \tag{13}$$

$$\frac{f_s}{f_c}=\frac{bt(2kd-t)}{2A_s kd} \tag{14}$$

$$kd=\frac{2ndA_s+bt^2}{2nA_s+2bt} \tag{15}$$

$$z=\frac{t(3kd-2t)}{3(2kd-t)} \tag{16}$$

$$jd=d-z \tag{17}$$

$$M_s=A_s f_s jd \tag{18}$$

$$M_c=\frac{f_c btjd}{2kd}(2kd-t) \tag{19}$$

$$f_c=\frac{f_s}{n}\left(\frac{kd}{d-kd}\right) \tag{20}$$

The shear unit stress in T beams is computed from Eq. (21), where V is the shear on the section. *See* ELASTICITY; HOOKE'S LAW; YOUNG'S MODULUS.

$$v=V/b'd \tag{21}$$

Ultimate-load design. Other names for ultimate-load design are ultimate-strength design, limit-load design, and the plastic design. Design assumptions differ from those of the elastic theory principally in that the stress-strain curve is not a straight line. Instead of being designed to carry allowable unit stresses, as in the elastic theory, beams are proportioned to carry at ultimate capacity the design load multiplied by a safety factor.

Among the reasons for using ultimate-load design are the following:

1. The elastic theory is not corroborated with sufficiently great accuracy by beam tests.
2. It is logical to use different safety factors for live and dead loads. Different factors can easily be used with ultimate-load design but not with elastic design.
3. Column design is based on a modified ultimate-load theory. To avoid inconsistency, structural members subjected to both bending and compressive stress should also be designed by ultimate-load theory.
4. The ultimate strength of beams carrying both bending and axial compression, as determined in tests, conforms closely with ultimate-load theory.

Rectangular beams. The compressive stress distribution may be assumed to be a rectangle, parabola, trapezoid, or any other shape that conforms to test data. Maximum concrete stress is assumed to be $0.85f'_c$, where f'_c is the compressive strength of a standard-test concrete cylinder at 28 days (Fig. 3).

When the moment resistance of the steel is less than that of the concrete, the bending moment that a rectangular beam with only tension reinforcement can sustain under ultimate load is Eq. (22).

$$\begin{aligned} M_u &= \phi[A_s f_y(d-a/2)] \\ &= \phi[bd^2 f'_c q(1-0.59q)] \end{aligned} \tag{22}$$

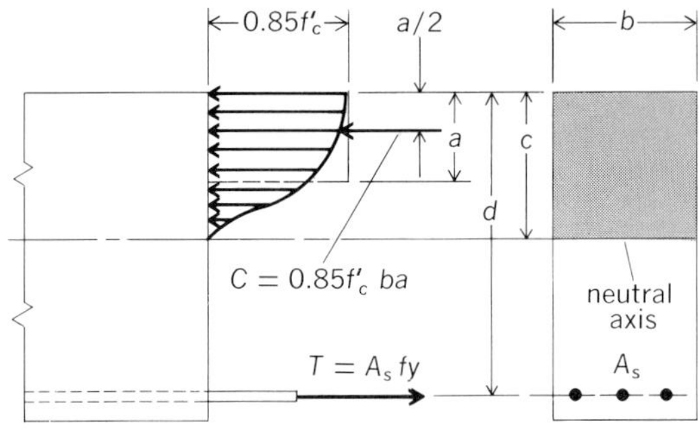

Fig. 3. Stress distribution in a rectangular beam under ultimate load; T and C are as in Fig. 2.

A_s = area of tension reinforcement
f_y = yield point stress of steel
f'_c = compressive strength of standard-test concrete cylinder at 28 days
ϕ = reduction factor = 0.90 for this type of beam in flexure
b = width of beam
d = distance from top of beam to centroid of the steel
a = depth of rectangular stress block
$q = A_s f_y / bdf'_c$

Reinforcement ratio $p = A_s/bd$ should not exceed $0.75p_b$, where p_b is given by Eq. (23), and k_1 =

$$p_b = \frac{0.85k_1 f'_c}{f_y}\left(\frac{87{,}000}{87{,}000+f_y}\right) \qquad (23)$$

0.85 for values of f'_c up to 4000 psi, 0.80 for 5000 psi, and 0.75 for 6000 psi.

Compression-reinforced rectangular beams. Based on a nonlinear stress-strain relation, the bending moment that a rectangular beam with both compression and tension reinforcement can sustain under ultimate loads is shown in Eq. (24).

$$M_u = \phi[(A_s - A'_s)f_y(d - a/2) + A'_s f_y(d - d')] \qquad (24)$$

A_s = area of tension steel
A'_s = area of compression steel
f_y = yield point stress of the steel
d = distance from top of beam to centroid of tension steel
d' = distance from top of beam to centroid of compression steel
$a = (A_s - A'_s) f_y / 0.85f'_c b$
b = width of beam
f'_c = compressive strength of standard-test concrete cylinder at 28 days

Equation (24) holds only if Eq. (25) is true, where

$$(p-p') \geq 0.85k_1 \frac{f'_c d'}{f_y d}\left(\frac{87.000}{87{,}000-f_y}\right) \leq 0.75p_b \qquad (25)$$

$p = A_s/bd$; $p' = A'_s/bd$; and p_b and k_1 are the same as for beams with tension steel only.

Concrete T beams. Two cases should be considered, one with a relatively thick slab and one with a thin slab. If the flange thickness exceeds $1.18qd/k_1$, where $q = pf_y/f'_c$, the bending moment under ultimate load may be taken to be the same as that for a rectangular beam with tension reinforcement only. The value of p used in computing q should be A_s/bd, with b the width of the flange. For thinner flanges, use Eq. (26) to compute bending moment.

$$M_u = \phi[(A_s - A_{sf})f_y(d - a/2) + A_{sf} f_y (d - 0.5t)] \qquad (26)$$

$A_{sf} = 0.85f'_c(b - b')t/f_y$
f_y = yield point stress of the steel
t = flange thickness
b = width of flange
b' = width of stem
d = distance from top of beam to centroid of tension steel
$a = (A_s - A_{sf})f_y / 0.85f'_c b'$

Equation (26) holds only if $p_w - p_f$ does not exceed $0.75p_b$, where $p_w = A_s/bd$, $p_f = A_{sf}/bd$, and p_b is the same as for beams with tension reinforcement only, as in Eq. (23).

Shear and bond. Maximum unit shear stress v acting on a section of a beam subjected to total shear V is given by Eq. (27), where b is the beam width, and d is depth from top of beam to the tensile steel. The bond stress can be computed from Eq. (28), where Σ_o is the sum of the bar perimeters.

$$v = V/bd \qquad (27)$$

$$u = V/\Sigma_o jd \qquad (28)$$

This formula applies only to tension steel. Bond stresses will be at a maximum where shear is a maximum and the steel is in the tension side of the concrete.

To develop a given bar stress through bond, a bar should be embedded a length L at least equal to Eq. (29), where a is the side of a square bar or the diameter of a round bar. This length of embedment is called anchorage. Usually, bars are extended 10 diameters past the section where they are no longer required for bending stress.

$$L = f_s a / 4u \qquad (29)$$

Shear in itself is not as important in the design of a concrete beam as the tensile stresses that accompany it on a diagonal plane. To resist these stresses, concrete beams should be reinforced with bent-up bars or with stirrups. The latter are bars placed vertically or on an incline in a beam. They may be the legs of a single bar bent into a U shape or the sides of a rectangle.

The shear V' taken by stirrups is assumed to be the total shear on the section at which the stirrups are to be placed, less the shear taken by the concrete, vbd. The cross-sectional area of the stirrups needed at a section is given by Eq. (30), where f_v is the allowable tensile stress of the steel, and s is the stirrup spacing.

$$A_v = V's/f_v d \qquad (30)$$

If the stirrups, instead of being vertical, are laid at an angle α with the horizontal, greater steel area is required, as in Eq. (31). Every potential 45° crack should be crossed by at least one line of reinforcement.

$$A_v = V's/d(\sin\alpha + \cos\alpha)f_v \qquad (31)$$

If the area of the stirrups is given, the spacing can be computed from the two formulas above.

The first stirrup is usually placed as close to the support as practical, generally 2 in. Stirrups should be placed throughout a beam, even if theoretically they are not needed. They serve also as supports for the longitudinal steel. When not required for diagonal tension, stirrups should be spaced at most 18 in. apart.

It is common practice to cut off bars or bend them up where they are no longer needed to resist tension at the bottom of the beam. The bent bars serve as diagonal-tension reinforcement and tensile reinforcement at the top of the beam over the support.

[FREDERICK S. MERRITT]

Bibliography: American Concrete Institute, *Building Code Requirements for Reinforced Concrete*, 1977; C. Davies, *Steel-Concrete Composite Beams for Building*, 1975. C. W. Dunham, *Theory and Practice of Reinforced Concrete*, 4th ed., 1966; G. Winter et al., *Design of Concrete Structures*, 9th ed., 1979.

Concrete column

A structural member subjected principally to compressive stresses. Concrete columns may be unreinforced, or they may be reinforced with longitudinal bars and ties (tied columns) or with longitudinal bars and spiral steel (spiral-reinforced columns). Sometimes the columns may be a composite of structural steel or cast iron and concrete (see illustration).

Plain concrete columns. Unreinforced concrete columns are seldom used, because of the existence of transverse tensile stresses and the possibility of longitudinal tensile stresses being induced by buckling or unanticipated bending. Because concrete is weak in tension, such stresses are generally avoided.

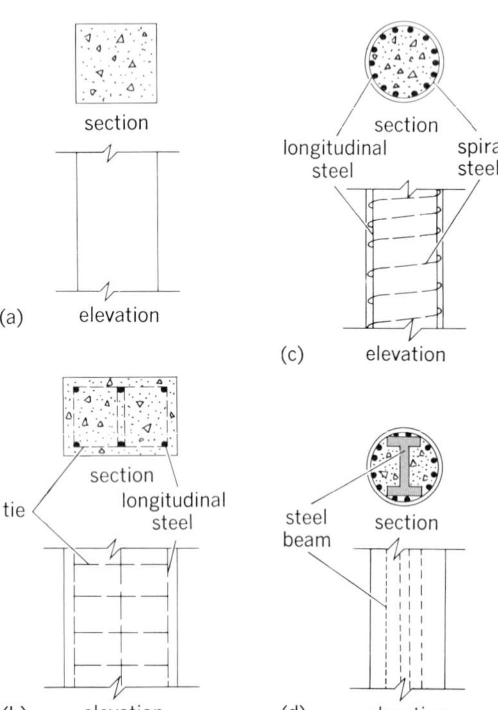

Column types. (*a*) Plain concrete. (*b*) Tied column. (*c*) Spiral-reinforced column. (*d*) Composite column.

When plain concrete columns are used, they usually are limited in height to five or six times the least thickness. Under axial loading, the load divided by the cross-sectional area of the concrete should not exceed the allowable unit compressive stress for the concrete.

Axially loaded reinforced columns. Reinforced concrete columns are designed by ultimate-load theory. Two types of column are considered: short and long. Those whose length is three to ten times their least lateral dimension are called short columns.

For spiral-reinforced short columns, the American Concrete Institute code gives the formula for the allowable load in pounds shown in Eq. (1),

$$P = 0.25 f'_c A_g + f_s A_s \tag{1}$$

where f'_c is the 28-day compressive strength of a standard concrete test cylinder in pounds per square inch (psi), A_g the gross area of section in in.^{2}, f_s the allowable unit stress for steel in psi, and A_s the area of reinforcing steel in in.^{2}

For a tied column, the allowable load is 85% of that for a spiral-reinforced column. Spiral-reinforced columns are stronger because columns tend to fail by a lateral bursting of the concrete as the longitudinal bars bend outward, and spiral reinforcing is more effective than ties in restraining the concrete. If spirals are used, the longitudinal bars should be arranged in a circle.

For long columns, the allowable load is reduced from that permitted for short columns because of the possibility of buckling. When h/d, the ratio of unsupported length of column to least lateral dimension, is equal to or greater than 10, one should use the appropriate reduction factor given in Sec. 916 of the Building Code Requirements for Reinforced Concrete (ACI 318–63).

Spiral reinforcement is determined from Eq. (2),

$$p' \geqq 0.45(R-1) f'_c/f'_s \tag{2}$$

where p' is the ratio of the volume of spiral reinforcement to the volume of the spiral core (out-to-out of the spiral), R is the ratio of gross area to core area, and f'_s is the yield-point stress of spiral steel, with a maximum of 60,000 psi.

Composite columns. A concrete compression member having a structural steel or cast iron core with a cross-sectional area not exceeding 20% of the gross area of the column is called a composite column. Spiral and longitudinal reinforcing also may be incorporated in the concrete.

The allowable load in pounds on a composite column is given by Eq. (3), where A_c is the net area

$$P = 0.225 A_c f'_c + f_r A_r + f_s A_s \tag{3}$$

of the concrete section ($A_g - A_r - A_s$), A_s is the area of longitudinal reinforcement other than the metal core, A_r is the cross-sectional area of the structural steel or cast iron core, f_r is the allowable unit stress for the core, and f_s is the allowable unit stress for the longitudinal reinforcement.

If the core area is 20% or more of the gross area and the concrete is at least 2 in. thick over all metal except rivet heads, the member is called a combination column. If the concrete is to be allowed to share the load with the core, the concrete must be reinforced; usually wire fabric is wrapped completely around the core.

Combined bending and axial load. Because columns are designed by ultimate-load theory and this theory has also been developed for beams, it is logical to design columns subjected to both bending and axial loads by ultimate-load theory. The ultimate load is given by Eqs. (4) and (5), where f'_c

$$P_u = \phi(0.85f'_c ba + A'_s f_y - A_s f_s) \quad (4)$$

$$P_u e = \phi[0.85f'_c ba(d - a/2) + A'_s f_y (d - d')] \quad (5)$$

is the compressive strength of concrete at 28 days; f_y is the yield strength of reinforcement; f_s is the calculated stress in reinforcement when less than yield strength; A'_s is the area of compression reinforcement; A_s is the area of tension reinforcement; e is the eccentricity of axial load at the end of the member measured from centroid of tension reinforcement; a is the depth of equivalent rectangular compression-stress block; b is the width of compression flange; d is the distance from extreme compression fiber to centroid of tension steel; d' is the distance from extreme compression fiber to centroid of compression steel; and ϕ is a capacity reduction factor. It equals 0.75 for spiral-reinforced members and 0.70 for tied members.

For strength reduction factors when length-thickness ratio exceeds 10, one should consult Sec. 916 of the Building Code Requirements for Reinforced Concrete (ACI 318–63). *See* CONCRETE SLAB.

[FREDERICK S. MERRITT]

Concrete slab

A shallow, reinforced-concrete structural member that is very wide compared with depth. Spanning between beams, girders, or columns, slabs are used for floors, roofs, and bridge decks. If they are cast integrally with beams or girders, they may be considered the top flange of those members and act with them as a T beam. *See* CONCRETE BEAM.

One-way slab. A slab supported on four sides but with a much larger span in one direction than in the other may be assumed to be supported only along its long sides. It may be designed as a beam spanning in the short direction. For this purpose a 1-ft width can be chosen and the depth of slab and reinforcing determined for this unit.

Some steel also is placed in the long direction to resist temperature stresses and distribute concentrated loads. The area of the steel generally is at least 0.20% of the concrete area.

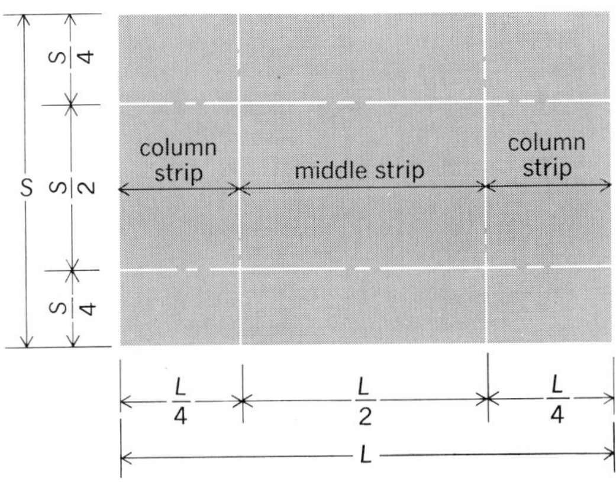

Fig. 1. A two-way slab in strips for design purposes.

Two-way slab. A slab supported on four sides and with reinforcing steel perpendicular to all sides is called a two-way slab. Such slabs generally are designed by empirical methods. A two-way slab is divided into strips for design purposes.

The American Concrete Institute (ACI) code recommends the following design method: Divide the slab in both directions into a column strip and middle strip (Fig. 1). If the ratio S/L of the short span to the long span is equal to or greater than 0.5, the width of the middle strip extending in the short direction equals $L/2$, as shown. If S/L is less than 0.5, the width of the middle strip in the short direction is $L - S$; the remaining width is divided equally between the two column strips. However, when S/L is less than 0.5, most of the load would be carried in the short direction, and it would be desirable to design the slab as a one-way slab.

A table in the ACI code gives coefficients for calculation of the bending moments in the middle strip for different values of S/L and different types of panels. The moment in the column strip is assumed to be two-thirds that in the middle strip. The reinforcing steel area is determined from $A_s = M/f_s jd$, where M is the bending moment, f_s the tensile unit stress in longitudinal reinforcement, and j the ratio of lever arm of resisting couple to d, the depth from compressive surface of beam or slab to center of longitudinal tension reinforcement.

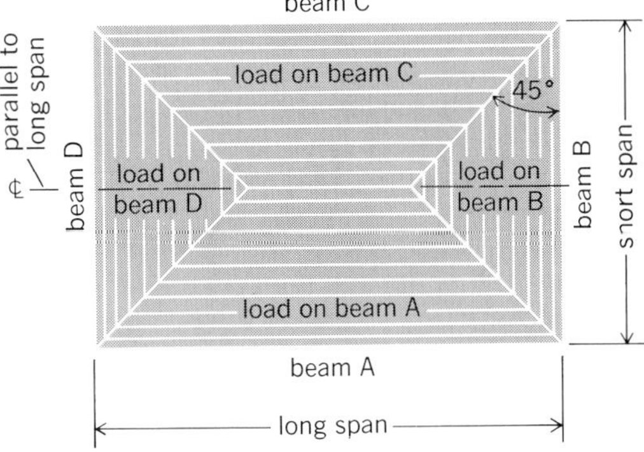

Fig. 2. Load distribution assumed for beams supporting a uniformly loaded two-way slab.

In the design of the beams that support a two-way slab along its sides, the load is assumed to be uniform over the slab and distributed to the beams as shown in Fig. 2. If w is the slab load in psf, moments in the beams may be approximated by assuming the beams loaded with the following equivalent uniform loads: for the short span, $wS/3$; for the long span, $wS(3 - S^2/L^2)/6$.

Flat slabs. When a slab is supported directly on columns, without beams and girders, it is called a flat plate or flat slab (Fig. 3).

A flat plate generally is of uniform thickness throughout. Usually, the columns also have constant dimensions throughout the story height below the slab (excluding the portion forming capitals, which are discussed later). In designing such construction, it is necessary to investigate shear and diagonal tension in the vicinity of the columns; frequently, reinforcements which are known as

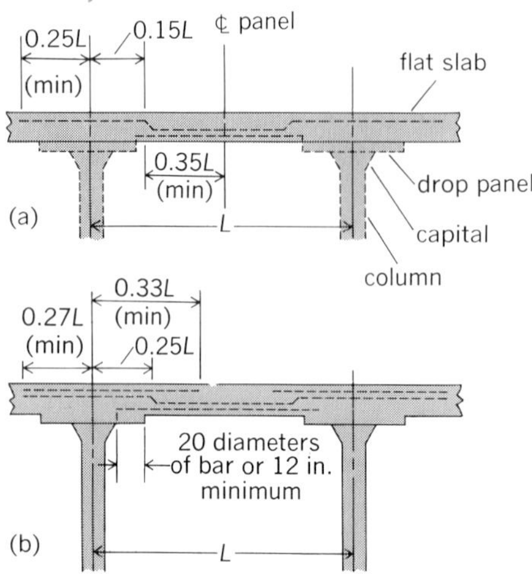

Fig. 3. Details of a flat slab drawn in cross section at two points. (*a*) Middle strip. (*b*) Column strip. *L* is the center-to-center distance between columns.

shear heads are embedded at the column tops.

In flat-slab construction it is customary to flare out the columns at the top to form capitals so as to give the slab-column junction greater rigidity. The capital is usually sloped at 45°. For exterior columns the capital is sometimes only a bracket that projects inward.

Flat slabs generally are thickened in the region around the columns. The thickened portion, called a drop panel, may be extended until it reaches from column to column, forming shallow beams and giving the effect of a paneled ceiling.

Although thicker and more heavily reinforced than slabs in beam-and-girder construction, flat slabs are advantageous because they offer no obstruction to passage of light (as beam construction does); savings in story height and in the simpler formwork involved; less danger of collapse due to overload; and better fire protection with a sprinkler system because the spray is not obstructed by beams.

Flat slabs may be reinforced in several ways:

1. Two-way system. When the columns are arranged to form rectangular bays, the reinforcing steel may be placed in two directions, perpendicular to the column lines. Design is based on column strips and middle strips, similar to that for a two-way slab.

2. Four-way system. Column strip steel is similar to that for a two-way slab. But middle-strip reinforcing consists of diagonal bands of steel extending over the columns.

3. Circumferential system. Bars are placed in the top of the slab in concentric rings around the columns, also radially. Similar radial and circular bars are placed in the bottom of the slab in the central portion.

4. Three-way system. When columns are arranged so that lines joining them would form triangles, the reinforcing steel may be laid parallel to the column lines.

Design of flat slabs and flat plates is based on empirical formulas. The ACI code presents design rules that are widely used. *See* CONCRETE COLUMN.

[FREDERICK S. MERRITT]

Bibliography: *See* CONCRETE BEAM.

CONDUCTOR (ELECTRICITY)

19-strand

7-strand

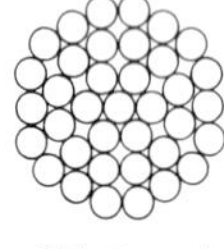

37-strand

Fig. 1. End views of stranded round conductor.

Conductor (electricity)

Metal wires, cables, rods, tubes, and bus-bars used for the purpose of carrying electric current. Although any metal assembly or structure can conduct electricity, the term conductor usually refers to the component parts of the current-carrying circuit or system.

Types of conductor. The most common forms of conductors are wires, cables, and bus-bars.

Wires. Wires employed as electrical conductors are slender rods or filaments of metal, usually soft and flexible. They may be bare or covered by some form of flexible insulating material. They are usually circular in cross section; for special purposes they may be drawn in square, rectangular, ribbon, or other shapes. Conductors may be solid or stranded, that is, built up by a helical lay or assembly of smaller solid conductors (Fig. 1).

Cables. Insulated stranded conductors in the larger sizes are called cables. Small, flexible, insulated cables are called cords. Assemblies of two or more insulated wires or cables within a common jacket or sheath are called multiconductor cables.

Bus-bars. Bus-bars are rigid, solid conductors and are made in various shapes, including rectangular, rods, tubes, and hollow squares. Bus-bars may be applied as single conductors, one bus-bar per phase, or as multiple conductors, two or more bus-bars per phase. The individual conductors of a multiple-conductor installation are identical.

Sizes. Most round conductors less than ½ in. (1 in. = 2.54 cm) in diameter are sized according to the American wire gage (AWG)—also known as the Brown & Sharpe gage. AWG sizes are based on a simple mathematical law in which intermediate wire sizes between no. 36 (0.0050-in. diameter) and no. 0000 (0.4600-in. diameter) are formed in geometrical progression. There are 38 sizes between these two diameters. An increase of three gage sizes (for example, from no. 10 to no. 7) doubles the cross-sectional area, and an increase of six gage sizes doubles the diameter of the wire.

Sizes of conductors greater than no. 0000 are usually measured in terms of cross-sectional area. Circular mil (cmil) is usually used to define cross-sectional area and is a unit of area equal to the area of a circle 1 mil (0.001 in.) in diameter.

Wire lengths are usually expressed in units of feet or miles in the United States. Bus-bar sizes are usually defined by their physical dimensions—height and width in inches or fractions of an inch, and length in feet.

Materials. Most wires, cables, and bus-bars are made from either copper or aluminum. Copper, of all the metals except silver, offers the least resistance to the flow of electric current. Both copper and aluminum may be bent and formed readily and have good flexibility in small sizes and in stranded constructions. Typical conductors are shown in Fig. 1.

Aluminum, because of its higher resistance, has less current-carrying capacity than copper for a given cross-sectional area. However, its low cost and light weight (only 30% that of the same volume of copper) permit wide use of aluminum for bus-bars, transmission lines, and large insulated-cable installations.

Metallic sodium conductors were used in 1965

on a trial basis for underground distribution insulated for both primary and secondary voltages. Sodium cable offered light weight and low cost for equivalent current-carrying rating compared with other conductor metals. Because of marketing problems and a few safety problems—the metal is reactive with water—the use of this cable was abandoned temporarily.

For overhead transmission lines where superior strength is required, special conductor constructions are used. Typical of these are aluminum conductors, steel reinforced (ACSR), a composite construction of electrical-grade aluminum strands surrounding a stranded steel core. Other constructions include stranded, high-strength aluminum alloy and a composite construction of aluminum strands around a stranded high-strength aluminum alloy core (ACAR).

For extra-high-voltage (EHV) transmission lines, conductor size is often established by corona performance rather than current-carrying capacity. Thus special "expanded" constructions are used to provide a large circumference without excessive weight. Typical constructions use helical lays of widely spaced aluminum strands around a stranded steel core. The space between the expanding strands is filled with paper twine, and outer layers of conventional aluminum strands are applied. In another construction the outer conductor stranding is applied directly over lays of widely separated helical expanding strands, without filler, leaving substantial voids between the stranded steel core and the closely spaced outer conductor layers. Diameters of 1.6 to 2.5 in. are typical. For lower reactance, conductors are "bundled," spaced 6–18 in. apart, and paralleled in groups of two, four, or more per phase. Figure 2 shows views of typical aluminum-conductor–steel-reinforced and expanded constructions.

Bare conductors. Bare wires and cables are used almost exclusively in outdoor power transmission and distribution lines. Conductors are supported on or from insulators, usually porcelain, of various designs and constructions, depending upon the voltage of the line and the mechanical considerations involved. Voltages as high as 765 kV are in use, and research has been undertaken into the use of ehv transmission lines, with voltages as high as 1500 kV.

Bare bus-bars are used extensively in outdoor substation construction, in switchboards, and for feeders and connections to electrolytic and electroplating processes. Where dangerously high voltages are carried, the use of bare bus-bars is usually restricted to areas accessible only to authorized personnel. Bare bus-bars are supported on insulators which have a design suitable for the voltage being carried.

Insulated conductors. Insulated electric conductors are provided with a continuous covering of flexible insulating material. A great variety of insulating materials and constructions has been developed to serve particular needs and applications. The selection of an appropriate insulation depends upon the voltage of the circuit, the operating temperature, the handling and abrasion likely to be encountered in installation and operation, environmental considerations such as exposure to moisture, oils, or chemicals, and applicable codes and standards. *See* ELECTRICAL INSULATION.

Magnet wires, used in the windings of motors, solenoids, transformers, and other electromagnetic devices, have relatively thin insulations, usually of enamel or cotton or both. Magnet wire is manufactured for use at temperatures ranging from 105 to 200 °C. *See* MAGNET WIRE.

Conductors in buildings. Building wires and cables are used in electrical systems in buildings to transmit electric power from the point of electric service (where the system is connected to the utility lines) to the various outlets, fixtures, and utilization devices. Building wires are designed for 600-volt operation but are commonly used at utilization voltages substantially below that value, typically 120, 240, or 480 volts. Insulations commonly used include thermoplastic, natural rubber, synthetic rubber, and rubberlike compounds. Rubber insulations are usually covered with an additional jacket, such as fibrous braid or polyvinyl chloride, to resist abrasion. Building wires are grouped by type in several application classifications in the National Electrical Code.

Classification is by a letter which usually designates the kind of insulation and, often, its application characteristics. For example, Type R indicates rubber or rubberlike insulation. TW indicates a thermoplastic, moisture-resistant insulation suitable for use in dry or wet locations; THW indicates a thermoplastic insulation with moisture and heat resistance. Other insulations in commercial use include silicone, fluorinated ethylene, propylene, varnished cambric, asbestos, polyethylene, and combinations of these.

Building wires and cables are also available in duplex and multiple-conductor assemblies; the individual insulated conductors are covered by a common jacket. For installation in wet locations, wires and cables are often provided with a lead sheath (Fig. 3).

For residential wiring, the common constructions used are nonmetallic-sheath cables, twin- and multiconductor assemblies in a tough abrasion-resistant jacket; and armored cable with twin-and multiconductor assemblies encased in a helical, flexible steel armor as in Fig. 4.

Power cables. Power cables are a class of electric conductors used by utility systems for the dis-

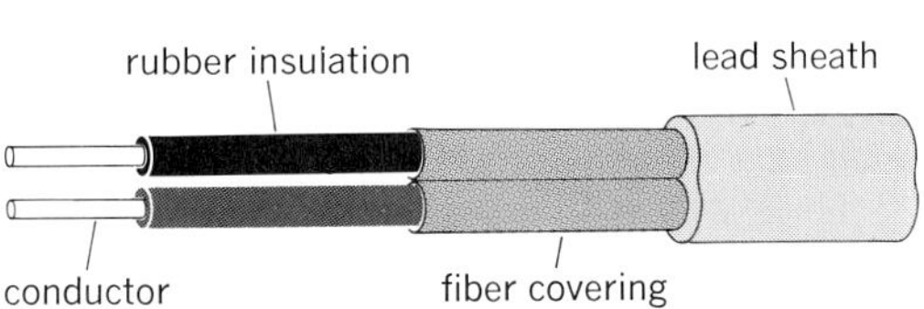

Fig. 3. Rubber-insulated, fiber-covered, lead-sheathed cable, useful for installation in wet places.

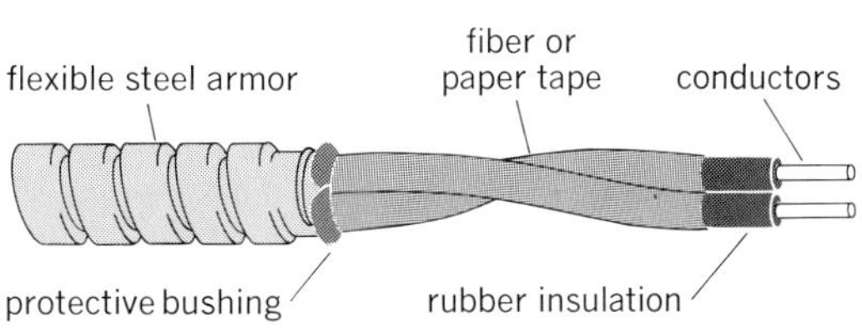

Fig. 4. Two-conductor armored cable.

CONDUCTOR (ELECTRICITY)

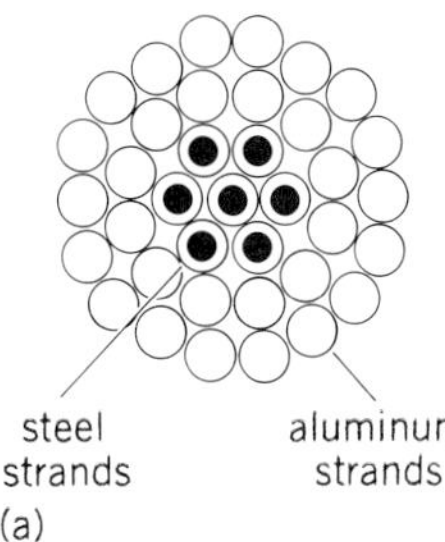

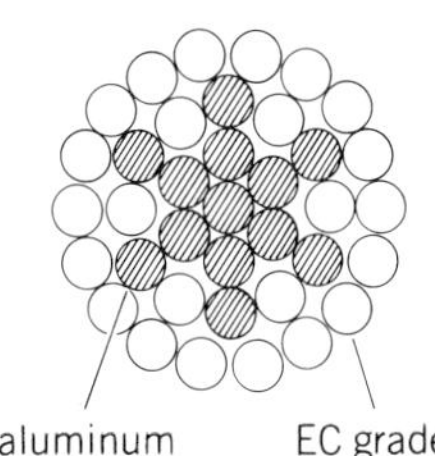

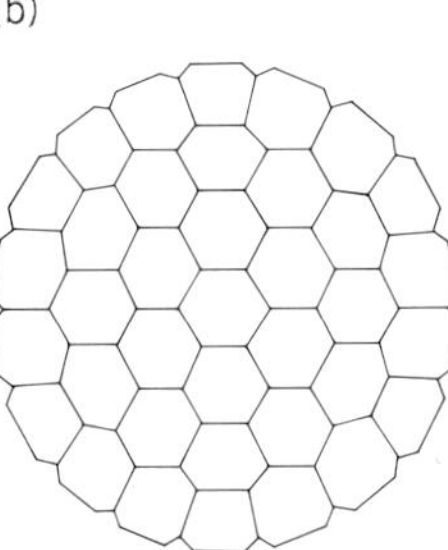

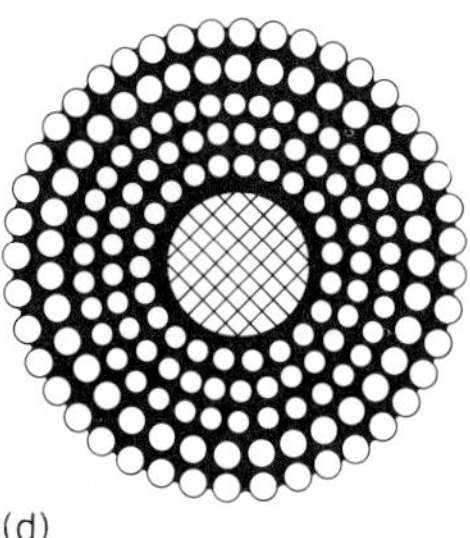

Fig. 2. Aluminum conductors. (*a*) ACSR. (*b*) ACAR. (*c*) Compact concentric stranded conductor. (*d*) Expanded-core concentric stranded conductor.

tribution of electricity. They are usually installed in underground ducts and conduits. Power cables are also used in the electric power systems of industrial plants and large buildings.

Power-cable insulations in common use include rubber, paper, varnished cambric, asbestos, and thermoplastic. Cables insulated with rubber (Fig. 5), polyethylene, and varnished cambric (Fig. 6) are used up through 69 kV, and impregnated paper to 138 kV. The type and thickness of the insulation for various voltages and applications are specified by the Insulated Power Cable Engineer Association (IPCEA).

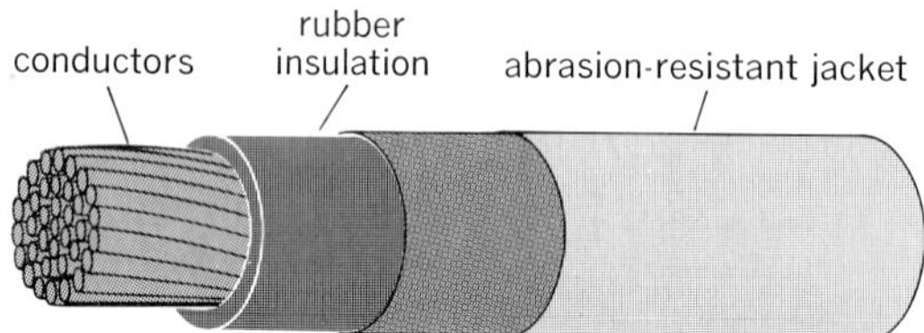

Fig. 5. Rubber-insulated power cable.

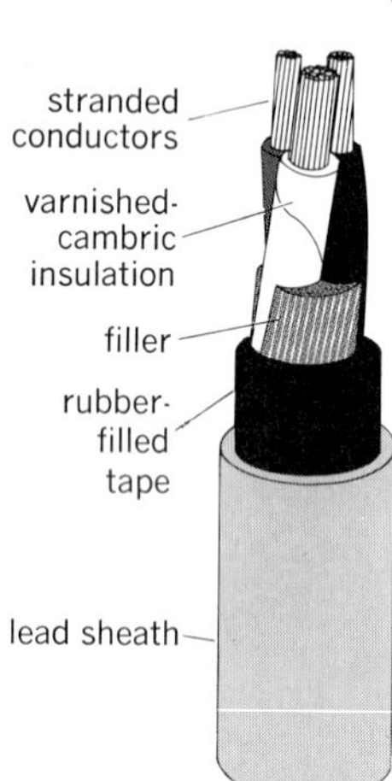

Fig. 6. Varnished cambric-insulated, lead-sheathed power cable, for voltages up to 28 kV.

Spaced aerial cable. Spaced aerial cable systems are employed for pole-line distribution at, typically, 5 to 15 kV, three-phase. Insulated conductors are suspended from a messenger, which may also serve as a neutral conductor, with ceramic or plastic insulating spacers, usually of diamond configuration. For 15 kV, a typical system may have conductors with 10/64-in. polyethylene insulation, 9-in. spacing between conductors, and 20 ft (6 m) between spacers.

High-voltage cable. High-voltage cable constructions and standards for installation and application are described by IPCEA. Insulations include (1) paper, solid type; (2) paper, low-pressure, gas-filled; (3) paper, low-pressure, oil-filled; (4) pipe cable, fully impregnated, oil-pressure; (5) pipe cable, gas-filled, gas-pressure; (6) rubber or plastic with neoprene or plastic jacket; (7) varnished cloth; (8) AVA and AVL (asbestos-varnished cloth).

Underground transmission cables are in service at voltages through 345 kV, and trial installations have been tested at 500 kV. Research has also been undertaken to develop cryoresistive and superconducting cables for transmitting power at high density. Preliminary designs have been prepared for a three-phase, 345-kV, 3660-MVA cable.

Enclosed bus-bar assemblies, or busways, are extensively used for service conductors and feeders in the electrical distribution systems of industrial plants and commercial buildings. They consist of prefabricated assemblies in standard lengths of bus-bars rigidly supported by solid insulation and enclosed in a sheet-metal housing.

Busways are made in two general types, feeder and plug-in. Feeder busways have no provision for taps or connections between the ends of the assembly. Low-reactance feeder busways are so constructed that conductors of different phases are in close proximity to minimize inductive reactance. Plug-in busways have provisions at intervals along the length of the assembly for the insertion of bus plugs.

Voltage drop in conductors. In electric circuits the resistance and (in ac circuits) the reactance of the circuit conductors result in a reduction in the voltage available at the load (except for capacitive loads). Since the line and load resistances are in series, the source voltage is divided proportionally. The difference between the source voltage and the voltage at the load is called voltage drop.

Electrical utilization devices of all kinds are designed to operate at a particular voltage or within a narrow range of voltages around a design center. The performance and efficiency of these devices are adversely affected if they are operated at a significantly lower voltage. Incandescent-lamp light output is lowered; fluorescent-lamp light output is lowered and starting becomes slow and erratic; the starting and pull-out torque of motors is seriously reduced.

Voltage drop in electric circuits caused by line resistance also represents a loss in power which appears as heat in the conductors. In excessive cases, the heat may rapidly age or destroy the insulation. Power loss also appears as a component of total energy use and cost. *See* COPPER LOSS.

Thus, conductors of electric power systems must be large enough to keep voltage drop at an acceptable value, or power-factor corrective devices—such as capacitors or synchronous condensers—must be installed. A typical maximum for a building wiring system for light and power is 3% voltage drop from the utility connection to any outlet under full-load condition.

[H. WAYNE BEATY]

Bibliography: Aluminum Association, *Aluminum Electrical Conductor Handbook*, September 1971; American National Standards Institute, *National Electrical Safety Code*, ANSI C2, 1977; Electric Power Research Institute, *Research Progress Report TD-3*, September 1975; D. Fink and J. Carroll (eds.), *Standard Handbook for Electrical Engineers*, 11th ed., 1978; J. McPartland, *McGraw-Hill's National Electric Code Handbook*, 16th ed., 1979.

Connecting rod

A link in several kinds of mechanisms. Usually one end of a connecting rod is intended to follow a circular path, while the other end follows a path along a straight line or a curve of large radius. The term is sometimes applied, however, to any straight link that transmits motion or power from one linkage to another within a mechanism. Figure 1 shows conventional arrangements of connecting rods in typical mechanisms. In some applications (for example, the connecting rod between the crank and an overhead oscillating member or walking beam in a well-drilling rig, or between the steering column and cross links in an automobile) the con-

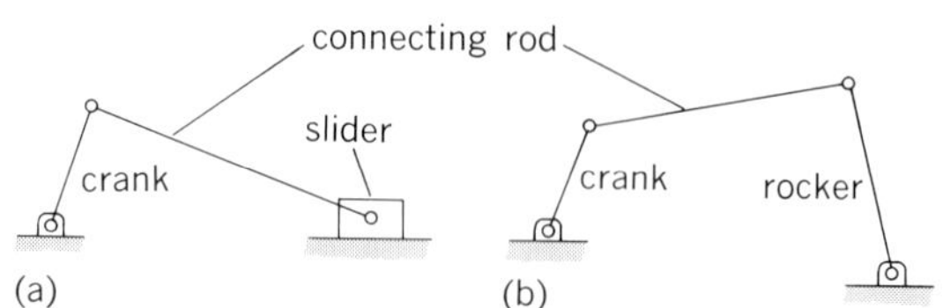

Fig. 1. Connecting rod in (*a*) slider-crank mechanism and (*b*) crank-and-rocker mechanism.

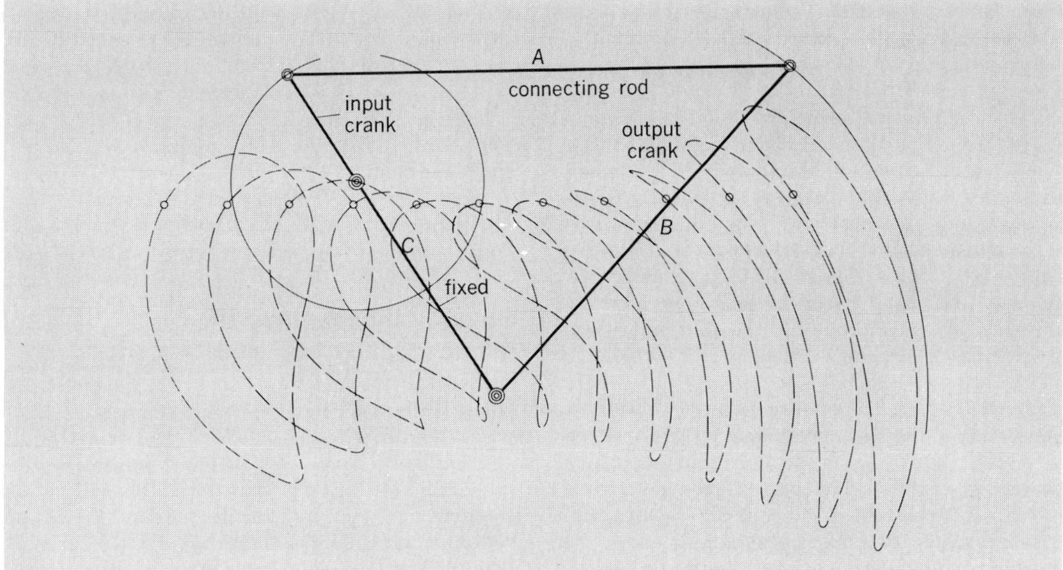

Fig. 2. Paths of points on connecting-rod plane traced by four-bar crank-and-rocker mechanisms for $A = 4.0$, $B = 3.5$, and $C = 2.0$. (*From J. A. Hrones and G. L. Nelson, Analysis of Four-Bar Linkage, copyright © 1951 by John Wiley and Sons, Inc.; used with permission*)

necting rod is called a pitman. *See* AUTOMOTIVE STEERING.

The connecting rod of the four-bar linkage, often called the coupler, has special significance. The motion of its plane can now be synthesized to furnish desired paths for points not necessarily on the straight line AB (coupler-point paths), or desired positions of the entire connecting-rod plane. The connecting rod is then not primarily used for transmission of force or motion from input to output crank, but the entire mechanism is employed to impart to the connecting-rod plane certain displacements, and sometimes velocities and accelerations. Figure 2 illustrates paths traced by various points in the plane of the rod, also revealing some velocities and accelerations there. *See* FOUR-BAR LINKAGE; MECHANISM. [DOUGLAS P. ADAMS]

Bibliography: E. A. Dijksman, *Motion Geometry of Mechanisms*, 1976; R. S. Hartenberg and J. Denavit, *Kinematic Synthesis of Linkages*, 1964; A. Ramous, *Applied Kinematics*, 1972.

Conservation of energy

The principle of conservation of energy states that energy cannot be created or destroyed, although it can be changed from one form to another. Thus in any isolated or closed system, the sum of all forms of energy remains constant. The energy of the system may be interconverted among many different forms—mechanical, electrical, magnetic, thermal, chemical, nuclear, and so on—and as time progresses, it tends to become less and less available; but within the limits of small experimental uncertainty, no change in total amount of energy has been observed in any situation in which it has been possible to ensure that energy has not entered or left the system in the form of work or heat. For a system that is both gaining and losing energy in the form of work and heat, as is true of any machine in operation, the energy principle asserts that the net gain of energy is equal to the total change of the system's internal energy.

Application to life processes. The energy principle as applied to life processes has also been studied. For instance, the quantity of heat obtained by burning food equivalent to the daily food intake of an animal is found to be equal to the daily amount of energy released by the animal in the forms of heat, work done, and energy in the waste products. (It is assumed that the animal is not gaining or losing weight.) Studies with similar results have also been made of photosynthesis, the process upon which the existence of practically all plant and animal life ultimately depends.

Conservation of mechanical energy. There are many other ways in which the principle of conservation of energy may be stated, depending on the intended application. Examples are the various methods of stating the first law of thermodynamics, the work-kinetic energy theorem, and the assertion that perpetual motion of the first kind is impossible. Of particular interest is the special form of the principle known as the principle of conservation of mechanical energy (kinetic E_k plus potential E_p) of any system of bodies connected together in any way is conserved, provided that the system is free of all frictional forces, including internal friction that could arise during collisions of the bodies of the system. Although frictional or other nonconservative forces are always present in any actual situation, their effects in many cases are so small that the principle of conservation of mechanical energy is a very useful approximation. Thus for a missile or satellite traveling high in space, the dissipative effects arising from such sources as the residual air and meteoric dust are so exceedingly small that the loss of mechanical energy $E_k + E_p$ of the body as it proceeds along its trajectory may, for many purposes, be disregarded. *See* ENERGY.

Mechanical equivalent of heat. The mechanical energy principle is very old, being directly derivable as a theorem from Newton's law of motion. Also very old are the notions that the disappearance of mechanical energy in actual situations is always accompanied by the production of heat and that heat itself is to be ascribed to the random motions of the particles of which matter is composed. But a really clear conception of heat as a form of energy came only near the middle of the 19th century, when J. P. Joule and others demonstrated the equivalence of heat and work by showing experimentally that for every definite amount of work done against friction there always appears a definite quantity of heat. The experiments usually were so arranged that the heat generated was absorbed by a given quantity of water, and it was observed that a given expenditure of mechanical energy always produced the same rise of temperature in the water. The resulting numerical relation between quantities of mechanical energy and heat is called the Joule equivalent, or mechanical equivalent of heat. The present accepted value is one 15° calorie $= 4.1855 \pm 0.0004$ joules.

Conservation of mass-energy. In view of the principle of equivalence of mass and energy in the restricted theory of relativity, the classical principle of conservation of energy must be regarded as a special case of the principle of conservation of mass-energy. However, this more general principle need be invoked only when dealing with certain nuclear phenomena or when speeds comparable with the speed of light (3×10^{10} cm/sec) are involved.

If the mass-energy relation, $E = mc^2$, where c is the speed of light, is considered as providing an equivalence between energy E and mass m in the same sense as the Joule equivalent provides an equivalence between mechanical energy and heat, there results the relation, 1 kg $= 9 \times 10^{16}$ joules.

Laws of motion. The law of conservation of energy has been established by thousands of meticulous measurements of gains and losses of all known forms of energy. It is now known that the total energy of a properly isolated system remains constant. Some parts or particles of the system may gain energy but others must lose just as much. The actual behavior of all the particles, and thus of the whole system, obeys certain laws of motion. These laws of motion must therefore be such that the energy of the total system is not changed by collisions or other interactions of its parts. It is a remarkable fact that one can test for this property of the laws of motion by a simple mathematical manipulation that is the same for all known laws: classical, relativistic, and quantum mechanical.

The mathematical test is as follows. Replace the variable t, which stands for time, by $t + a$, where a is a constant. If the equations of motion are not changed by such a substitution, it can be proved that the energy of any system governed by these equations is conserved. For example, if the only expression containing time is $t_2 - t_1$, changing t_2 to $t_2 + a$ and t to $t_1 + a$ leaves the expression unchanged. Such expressions are said to be invariant under time displacement. When daylight-saving time goes into effect, every t is changed to $t + 1$ hr. It is unnecessary to make this substitution in any known laws of nature, because they are all invariant under time displacement.

Without such invariance laws of nature would change with the passage of time, and repeating an experiment would have no clear-cut meaning. In fact, science, as it is known today, would not exist.

[DUANE E. ROLLER/LEO NEDELSKY]

Bibliography: K. R. Atkins, *Physics*, 3d ed., 1976; D. Halliday and R. Resnick, *Physics*, 3d ed., 1978; G. Laundry et al., *Physics: An Energy Introduction*, 1979; F. W. Sears et al. *University Physics*, 5th ed., 1976; E. P. Wigner, Symmetry and conservation laws, *Phys. Today*, 17(3):34–40, March 1964.

Construction engineering

A specialized branch of civil engineering concerned with the planning, execution, and control of construction operations for such projects as highways, buildings, dams, airports, and utility lines.

Planning consists of scheduling the work to be done and selecting the most suitable construction methods and equipment for the project. Execution requires the timely mobilization of all drawings, layouts, and materials on the job to prevent delays to the work. Control consists of analyzing progress and cost to ensure that the project will be done on schedule and within the estimated cost.

Planning. The planning phase starts with a detailed study of construction plans and specifications. From this study a list of all items of work is prepared, and related items are then grouped together for listing on a master schedule. A sequence of construction and the time to be allotted for each item is then indicated. The method of operation and the equipment to be used for the individual work items are selected to satisfy the schedule and the character of the project at the lowest possible cost. For a discussion of methods of operation *see* CONSTRUCTION METHODS. For a discussion of various types of equipment *see* CONSTRUCTION EQUIPMENT.

The amount of time allotted for a certain operation and the selection of methods of operation and equipment are generally determined by the equipment that is readily available to the contractor. After the master or general construction schedule has been drawn up, subsidiary detailed schedules or forecasts are prepared from the master schedule. These include individual schedules for procurement of material, equipment, and labor, as well as forecasts of cost and income.

Execution. The speedy execution of the project requires the ready supply of all materials, equipment, and labor when needed. The construction engineer is generally responsible for initiating the purchase of most construction materials and expediting their delivery to the project. Some materials, such as structural steel and mechanical equipment, require partial or complete fabrication by a supplier. For these fabricated materials the engineer must prepare or check all fabrication drawings for accuracy and ease of assembly and often inspect the supplier's fabrication.

Other construction engineering duties are the layout of the work by surveying methods, the preparation of detail drawings to clarify the design engineer's drawings for the construction crews, and the inspection of the work to ensure that it complies with plans and specifications.

On most large projects it is necessary to design and prepare construction drawings for temporary construction facilities, such as drainage structures, access roads, office and storage buildings, formwork, and cofferdams. Other problems are the selection of electrical and mechanical equipment and the design of structural features for concrete material processing and mixing plants and for compressed air, water, and electrical distribution systems.

Control. Progress control is obtained by comparing actual performance on the work against the desired performance set up on the master or detailed schedules. Since delay on one feature of the project could easily affect the entire job, it is often necessary to add equipment or crews to speed up the work. *See* PERT.

Cost control is obtained by comparing actual unit costs for individual work items against estimated or budgeted unit costs, which are set up at the beginning of the work. A unit cost is obtained by dividing the total cost of an operation by the number of units in that operation.

Typical units are cubic yards for excavation or concrete work and tons for structural steel. The actual unit cost for any item at any time is obtained by dividing the accumulated costs charged to that item by the accumulated units of work performed.

Individual work item costs are obtained by periodically distributing job costs, such as payroll and invoices to the various work item accounts. Payroll and equipment rental charges are distributed with the aid of time cards prepared by crew foremen. The cards indicate the time spent by the job crews and equipment on the different elements of the work. The allocation of material costs is based on the quantity of each type of material used for each specific item.

When the comparison of actual and estimated unit costs indicates an overrun, an analysis is made to pinpoint the cause. If the overrun is in equipment costs, it may be that the equipment has insufficient capacity or that it is not working properly. If the overrun is in labor costs, it may be that the crews have too many men, lack proper supervision, or are being delayed for lack of materials or layout. In such cases time studies are invaluable in analyzing productivity. [WILLIAM HERSHLEDER]

Bibliography: G. E. Deatherage, *Construction Scheduling and Control*, 1965; T. C. Kavanagh, F. Muller, and J. J. O'Brien, *Construction Management: A Professional Approach*, 1978; J. J. O'Brien, *CPM in Construction Mangement*, 2d ed., 1971; R. L. Peurifoy, *Construction Planning, Equipment, and Methods*, 3d ed., 1979.

Construction equipment

A wide variety of relatively heavy machines which perform specific construction (or demolition) functions under power. The power plant (which is treated in later paragraphs) is commonly an integral part of an individual machine, although in some cases it is contained in a separate prime mover, for example, a towed wagon or roller. It is customary to classify construction machines in accordance with their functions such as hoisting, excavating, hauling, grading, paving, drilling, or pile driving. There have been few changes for many years in the basic types of machines available for specific jobs, and few in the basic configurations of those that have long been available. Design emphasis for new machines is on modifications that increase speed, efficiency, and accuracy (particularly through more sophisticated controls); that improve operator comfort and safety; and that protect the public through sound attenuation and emission control. The selection of a machine for a specific job is mainly a question of economics and depends primarily on the ability of the machine to complete the job efficiently, and secondarily on its availability.

Hoisting equipment. This class of equipment is used to raise or lower materials from one elevation to another or to move them from one point to another over an obstruction. The main types of hoisting equipment are derricks, cableways, cranes, elevators, and conveyors. *See* BULK-HANDLING MACHINES; HOISTING MACHINES.

Derricks. The two main types of derricks are the guy and the stiff-leg. The former has a mast that is held in a vertical position by guy wires and a boom that can rotate with the mast 360°. In the stiff-leg the mast is tied to two or more rigid structural members, and the rotation of the boom is limited by the position of these members. The derrick is practical only where little mobility is required as in some types of steel erection, excavation of a shaft, or hoisting of materials through a shaft.

Cableway. This is a combination hoist and tram system comprising a trolley that runs on main load cables stretched between two or more towers which may be stationary or tiltable. Since even with towers that can be tilted slightly from side to side the cableway is limited almost entirely to linear movement of materials, its use in construction is restricted to dams, to some dredging and dragline operations, and in special cases to bridges.

Elevator. The construction elevator, like the passenger elevator, consists of a car or platform that operates within a structural framework and is raised and lowered by cables. Because the elevator can move materials only in a vertical linear direction and because the materials so moved must subsequently be moved horizontally by other means, the elevator is limited to moving relatively light materials on jobs of small areas.

Fig. 1. Hydraulic truck crane.

Fig. 2. Conveyor carrying rock over a railroad yard.

Crane. A crane is basically a fast-moving boom mounted on a frame containing the power supply and mechanisms for moving the boom and for raising and lowering the load-bearing cables that run through sheaves at the top of the boom. Once all cranes in the United States were mobile; that is, they were attached to a chassis mounted on pneumatic tires, crawler treads, or flanged (railroad) wheels. This type of machine is still predominant in all operations where mobility is a must. Figure 1 shows a hydraulic truck crane. However, European and Australian tower cranes and climbing cranes have become popular throughout the United States and the world for high-rise building construction where mobility requirements are negligible. These machines, which carry horizontal or diagonal masts or booms, can "grow" with the construction by climbing with it or by adding sections to their own towers. They have the fast boom and line action of the mobile crane but are not restricted in boom reach.

Conveyor. Occasionally a conveyor is used to raise materials needed for construction of low buildings. This machine, which has a movable endless belt mounted on a frame, can carry only light loads such as bricks and cement sacks and is not practical for buildings of more than one or two floors. A conveyor carrying rock over a railroad yard is shown in Fig. 2. *See* CONVEYING MACHINES.

Excavating equipment. This type of equipment is divided into two main classifications: standard land excavators and marine dredges, each of which has many variations.

Standard land excavator. This comprises machines that merely dig earth and rock and place it in separate hauling units, as well as those that pick up and transport the materials. Among the former are power shovels, draglines, backhoes, cranes with a variety of buckets, front-end loaders, excavating belt loaders, trenchers, and the continuous bucket excavator. The second group includes such machines as bulldozers, scrapers of various types, and sometimes the front-end loader.

Power shovel. This has for years been regarded as the most efficient machine for digging into vertical banks and for handling heavy rock. Generally mounted on crawler treads, the power shovel carries a short boom on which rides a movable dipper stick carrying an open-topped bucket. The bucket digs in an upward direction away from the machine and dumps its load by lowering the front of its hinged bottom. Figure 3 shows power shovels in the process of loading dump trucks.

Front-end loader. This has made heavy inroads into the domain of the power shovel and in time may well replace it in all but the biggest operations. Loaders are built that can handle buckets having 12, 14, and 24 yd^3 capacities, sizes which a

Fig. 3. Power shovels loading over-the-road rear-dump trucks.

Fig. 4. A loader operating in a swamp.

decade ago were considered huge even for power shovels. The loader, basically an articulated bucket mounted on a series of movable arms at the front of a crawler or rubber-tired tractor, has the advantage of mobility, speed, economy of cost and operation, and light weight. A loader of any given capacity can cost as little as one-half or even one-third as much as a power shovel of the same capacity and requires only one person to run it as opposed to the two or more normally required for the power shovel. Figure 4 shows a loader operating in a swamp.

Backhoe. This is fundamentally an upside-down power shovel. Its bucket, mounted on a hinged boom, digs upward toward the machine and unloads by being inverted over the dumping point. Backhoes are manufactured as individual machines but more often are attachments mounted on a crawler crane, a rubber-tired truck crane, or a tractor (which usually has a front-end loader bucket at the other end). It is particularly well adapted to the digging of deep trenches.

Dragline. This is a four-sided bucket that is used mainly on soil too wet to support an excavating machine. It is usually carried on a mobile machine mounting a crane boom, but it can also be worked from a cableway where excavating distances are great. The bucket is carried or cast to a point ahead of its support and dropped to the ground. The dragline is designed to excavate and to fill itself as it is drawn across the ground. It empties when its front end is lowered.

Clamshell. A clamshell is a two-sided bucket that can dig only in a vertical direction. The bucket is dropped while its leaves are open and digs as they close. Formerly, clamshells were suspended from cables on cranes or gantries and worked only under their own weight. Therefore they were imprecise in digging and were practical only in relatively soft soil or loose rock. Presently many clamshell buckets are attached directly to powered booms, are closed hydraulically, and can work with greater precision in dense soils.

Orange peel. This is a multileaved bucket, generally round in configuration. It is normally cable-supported and, like the clamshell, works under its own weight. Like the clamshell, however, it too is being mounted on powered booms and closed hydraulically. One of the commonest uses of the orange peel is the cleaning of small shafts, sewers, and storm drains. Larger sizes are often used for handling broken rock.

Grapple. This group includes a wide variety of special-purpose tined grabs that work on the principle of the clamshell and orange peel. The grapple is used mainly for handling rock, pipe, and logs.

Excavating belt loader. This type of machine is used mainly for loading granular materials from a stockpile, or as a vertical or inclined belt, or as a pair of chains on which are mounted small buckets. Some belt loaders are mobile and can move into a bank under their own power as the material is moved. Others are fixed and are fed by bulldozers which push material onto the belt from above. Their advantage over most excavators lies in their continuous operation.

Trenching machine. This equipment for digging trenches ranges in size from small hand-pushed units used for getting small pipes from streets to private homes to monsters of many tons capable of cutting a 4 ft-wide swath for 2000 mi of transcontinental pipelines. All, however, work on basically the same principle: A series of buckets, mounted on a chair or a wheel, lift dirt from the ground and deposit it alongside the trench being dug.

Continuous bucket excavator. This excavator works like a trenching machine but is designed to remove earth and loose rock from a wider area and at shallower depths. These machines are self-propelled; they can load trucks in a continuous operation and can switch from a filled to an empty truck in less than 2 sec. The continuity of loading and the absence of intermittent activities, such as those necessary in shovel or loader operations, make the continuous bucket excavator a fast, economical machine for large excavating jobs.

Combination excavators and haulers. This classification of construction equipment includes bulldozers, carrying scrapers, self-loading or elevating scrapers, and sometimes front-end loaders.

Bulldozer. This is a curved blade mounted on the front of a crawler or rubber-tired tractor for the purpose of digging and pushing earth and broken rock from one place to another. It is often used in conjunction with a ripper, a heavy tooth or series of teeth mounted on the rear of a tractor to break up rock so that it can be handled by the dozer. Because the amount of dirt the dozer can move at one time is limited by the size of its blade, it is not economical to use the dozer for moving dirt more than a few hundred feet.

Front-end loader. This is sometimes used for moving dirt in the same manner as the dozer, but with the same limitations.

Scraper. This is the most economical of the combination excavating-hauling units for hauls of over a few hundred feet. A wheel-mounted, open-top box or "bowl," the scraper has a hinged bottom that is lowered to scoop up dirt as the machine moves forward. When the bowl is full, the bottom is closed and the unit is moved to the dumping area, where the bottom is lowered and a hydraulic pushing unit in the back of the bowl moves the dirt out through the opening in the bottom, to spread the dirt as needed. Scrapers are pulled by tractors, most of which are an integral part of the unit. A scraper is shown in Fig. 5.

Self-loading or elevating scraper. This group of machines does the same work as the conventional scraper, but in front of the bowl is a series of horizontal plates mounted on moving chains. As the chains move, the plates lift the dirt into the bucket in the same manner as a bucket loader. The advantage of the elevating scraper is that it needs less tractor power during the loading operation.

Marine excavator. Usually called a dredge, the marine excavator is an excavating machine mounted on a barge or boat. Two common types are similar to land excavators, the clamshell and the bucket excavator. The suction dredge is different; it comprises a movable suction pipe which can be lowered to the bottom, usually with a fast-moving cutter head at the bottom end. The cutter churns up the bottom so that the pumps on the barge or boat suck up water and the earth suspended in it. When practical, the pumped material is then piped to land, where the solids settle out, allowing the liquids to run back to the body of water being dredged. When distance to land is too great or where there is no empty land on which to discharge the material, the effluent is pumped into barges in which the settled-out solids are towed to a dump area at a distant point.

Fig. 5. A scraper in operation, as it is pulled by a tractor.

Fig. 6. A 100-ton rear-dump truck.

Hauling equipment. Excavated materials are moved great distances by a wide variety of conveyances. The most common of these are the self-propelled rubber-tired rear-dump trucks, which are classed as over-the-road or off-the-road trucks. The main difference between these two is weight and carrying capacity. Many states restrict the total weight of highway-using trucks (truck and load) to as little as 30 tons. For jobs where trucks remain on site, such as at dams and some highway construction areas, trucks with a carrying capacity of 50 tons are common and some capable of carrying loads in excess of 100 tons are in use. A 100-ton rear-dump truck is shown in Fig. 6. Wagons towed by a rubber-tired prime mover are also used for hauling dirt. These commonly have bottom dumps which permit spreading dirt as the vehicle moves. In special cases side-dump trucks are also used.

Conveyors, while not commonly used on construction jobs for hauling earth and rock great distances, have been used to good advantage on large jobs where obstructions make impractical the passage of trucks. One such conveyor which was utilized on a New Jersey highway job exceeded 2 mi in length.

Graders. Graders are high-bodied, wheeled vehicles that mount a leveling blade between the front and rear wheels. The principal use is for fine-grading relatively loose and level earth. Commonly considered a maintenance unit, the grader is still an important machine in highway construction. While the configuration of graders has not changed in many years, some models have been built with articulated main frames that permit faster turns of shorter radii and safer operation at the edges of steep slopes.

Pavers. These place, smooth, and compact paving materials. They may be mounted on rubber tires, endless tracks, or flanged wheels. Asphalt pavers embody tamping pads that consolidate the material; concrete pavers use vibrators for the same purpose. Concrete pavers had been required to use forms to contain the material until the advent of the slip-form pavers. These drag their own short forms behind them and consolidate the concrete sufficiently to stand without slumping after removal of the forms. Many high-speed conventional pavers have been produced with highly sophisticated automated control devices that provide extremely accurate and consistent surface elevations.

Drilling equipment. Holes are drilled in rock for wells and for blasting, grouting, and exploring. Drills are classified according to the way in which they penetrate rock, namely, percussion, rotary percussion, and rotary. The first two types are the most common for holes up to 6 in. in diameter and under 40 ft in depth. In smaller sizes these drills may be hand-held, but for production work they are mounted on masts which are supported by trucks or, as shown in Fig. 7, special tracks mounted on drill rigs. In tunnel work the drills are commonly mounted on platform-supported movable arms or posts that are hydraulically or pneumatically controlled to permit drilling in horizontal or angled positions, with a minimum of effort by the drill operator. The rotary drill is most common in larger sizes, but drills as small as 3 in. in diameter are used for coring rock to obtain samples.

Specialized equipment. Among the most common specialized construction equipment are augers, compactors, and pile hammers.

Augers. These are used for drilling wells, dewatering purposes, and cutting holes that can be filled with concrete for foundations. Augers up to 6 ft in diameter are common.

Compactors. These machines are designed solely to consolidate earth and paving materials to sustain loads greater than those sustained in the uncompacted state. They range in size from small pneumatic hand-held tampers to multiwheeled machines weighing more than 60 tons. Actual compaction may be achieved by heavily loaded rubber tires or steel rollers. The steel rollers may be solid cylinders, or have separate pads or grids, or contain protrusions (sheep-foot roller). Many machines induce a vibratory action into the compacting units so that compaction is achieved by impact force rather than sheer weight.

Pile hammers. These machines are used to install bearing piles for foundations, or sheet piles for cofferdams and retaining walls. The prehistoric pile hammer consisted of a heavy weight lifted by a rope and allowed to drop on top of a pile. This drop hammer is still in use, basically unchanged. More efficient modern hammers are true machines activated by steam, air, oil, hydraulic fluid, or electricity. Conventional hammers contain pistons that are raised by one of these means and then either allowed to fall freely because of gravity (singe-acting) or to be driven downward by the same means (double-acting) to impart an impact to the pile. Vibratory hammers have been developed which use electrically activated eccentric cams to vibrate piles into place. Relatively rare are machines that use hydraulic action to install piles.

Road planer. This machine looks like a conventional motor grader, but instead of a blade it carries near its center a horizontal drum that has on its periphery many rows of replaceable hardened steel teeth. As the drum revolves, the teeth pulverize cracked asphalt or decayed concrete pavement to depths of up to 4 in. This permits the placement of a smooth new riding surface on a sound base without raising the road surface elevation, having to raise manhole covers, and reducing the clearances under bridges.

Bore tunneling machine. This is a machine capable of placing water, sewer, or utility pipes accurately for great distances underground. Key to the operation is a series of intermediate hydraulic jacking rings that expand or contract as controlled to permit the movement of any section of pipe independent of the sections before or behind. This independent movement of short sections reduces the friction on the line and therefore the amount of pressure that would be required if the entire pipe were moved from a single push point.

Fig. 7. Crawler-mounted drill. (*Gardner-Denver*)

Power plants. Steam is seldom used as a source of power today except in some marine applications and rarely for pile driving. Gasoline and diesel engines of the piston type are the most common source of power for construction machines, with the diesel attaining an increasingly greater prominence for at least two major reasons. First, the size of many construction machines has increased to proportions formerly thought impractical. With this increase in size, the economy of operation and maintenance of the diesel has come to far outweigh its greater initial cost. Second, the increased popularity of the diesel lies in the fact that manufacturers have been able to build lightweight economical diesel engines in very small sizes, thus making practical their application in small compressors, pumps, portable electric power plants, and so on.

Electricity is sometimes, though not often, used as a primary source of power for some construction machines, but usually only at the site of large dams and strip-mining operations where mobility is not a primary requirement. Electricity is making its appearance as a secondary source of power on several types of mobile construction machinery, such as trucks and scrapers, which use diesel or gas engines to turn generators that provide power for electric wheels.

A more common type of secondary power, however, is the hydraulic motor which, activated by a gasoline or diesel-powered hydraulic pump, is being used to provide direct power for virtually every movement, including travel, of construction machines.

Some manufacturers are experimenting with turbine engines for construction applications. While they are being used with success as primary power sources for large stationary electric power plants on job sites, their practical use in mobile equipment is not yet evident. *See* CONSTRUCTION ENGINEERING.

[EDWARD M. YOUNG]

Bibliography: R. D. Chellis, *Pile Foundations*, 1961; H. L. Nichols, Jr., *Moving the Earth*, 1962; F. W. Stubbs, Jr. (ed.), *Handbook of Heavy Construction*, 1971.

Construction methods

The procedures and techniques utilized during construction. Construction operations are generally classified according to specialized fields. These include preparation of the project site, earthmoving, foundation treatment, steel erection, concrete placement, asphalt paving, and electrical and mechanical installations. Procedures for each of these fields are generally the same, even when applied to different projects, such as buildings, dams, or airports. However, the relative importance of each field is not the same in all cases. For a description of tunnel construction, which involves different procedures, *see* TUNNEL.

Preparation of site. This consists of the removal and clearing of all surface structures and growth from the site of the proposed structure. A bulldozer is used for small structures and trees. Larger structures must be dismantled.

Earthmoving. This includes excavation and the placement of earth fill. Excavation follows preparation of the site, and is performed when the existing grade must be brought down to a new elevation. Excavation generally starts with the separate stripping of the organic topsoil, which is later reused for landscaping around the new structure. This also prevents contamination of the nonorganic material which is below the topsoil and which may be required for fill. Excavation may be done by any of several excavators, such as shovels, draglines, clamshells, cranes, and scrapers. For a discussion of their application *see* CONSTRUCTION EQUIPMENT.

Efficient excavation on land requires a dry excavation area, because many soils are unstable when wet and cannot support excavating and hauling equipment. Dewatering becomes a major operation when the excavation lies below the natural water table and intercepts the groundwater flow. When this occurs, dewatering and stabilizing of the soil may be accomplished by trenches, which conduct seepage to a sump from which the water is pumped out. Dewatering and stabilizing of the soil may in other cases be accomplished by wellpoints and electroosmosis. *See* WELLPOINT SYSTEMS.

Some materials, such as rock, cemented gravels, and hard clays, require blasting to loosen or fragment the material. Blast holes are drilled in the material; explosives are then placed in the blast holes and detonated. The quantity of explosives and the blast-hole spacing are dependent upon the type and structure of the rock and the diameter and depth of the blast holes. *See* BLASTING.

After placement of the earth fill, it is almost always compacted to prevent subsequent settlement. Compaction is generally done with sheep's-foot, grid, pneumatic-tired, and vibratory-type rollers, which are towed by tractors over the fill as it is being placed. Hand-held, gasoline-driven rammers are used for compaction close to structures where there is no room for rollers to operate.

Foundation treatment. When subsurface investigation reveals structural defects in the foundation area to be used for a structure, the foundation must be strengthened. Water passages, cavities, fissures, faults, and other defects are filled and strengthened by grouting. Grouting consists of injection of fluid mixtures under pressure. The fluids subsequently solidify in the voids of the strata. Most grouting is done with cement and water mixtures, but other mixture ingredients are asphalt, cement and clay, and precipitating chemicals. *See* FOUNDATIONS.

Steel erection. The construction of a steel structure consists of the assembly at the site of mill-rolled or shop-fabricated steel sections. The steel sections may consist of beams, columns, or small trusses which are joined together by riveting, bolting, or welding. It is more economical to assemble sections of the structure at a fabricating

Steel erection with guy derricks on a high structure. (*Bethlehem Steel Co.*)

shop rather than in the field, but the size of preassembled units is limited by the capacity of transportation and erection equipment. The crane is the most common type of erection equipment, but when a structure is too high or extensive in area to be erected by a crane, it is necessary to place one or more derricks on the structure to handle the steel (see illustration). In high structures the derrick must be dismantled and reerected to successively higher levels to raise the structure. For river bridges the steel may be handled by cranes on barges or, if the bridge is too high, by traveling derricks which ride on the bridge being erected. Cables for long suspension bridges are assembled in place by special equipment that pulls the wire from a reel, set up at one anchorage, across to the opposite anchorage, repeating the operation until the bundle of wires is of the required size.

Concrete construction. Concrete construction consists of several operations: forming, concrete production, placement, and curing. Forming is required to contain and support the fluid concrete within its desired final outline until it solidifies and can support itself. The form is made of timber or steel sections or a combination of both and is held together during the concrete placing by external bracing or internal ties. The forms and ties are designed to withstand the temporary fluid pressure of the concrete.

The usual practice for vertical walls is to leave the forms in position for at least a day after the concrete is placed. They are removed when the concrete has solidified or set. Slip-forming is a method where the form is constantly in motion, just ahead of the level of fresh concrete. The form is lifted upward by means of jacks which are mounted on vertical rods embedded in the concrete and are spaced along the perimeter of the structure. Slip forms are used for high structures such as silos, tanks, or chimneys.

Concrete may be obtained from commercial batch plants which deliver it in mix trucks if the job is close to such a plant, or it may be produced at the job site. Concrete production at the job site requires the erection of a mixing plant, and of cement and aggregate receiving and handling plants. Aggregates are sometimes produced at or near the job site. This requires opening a quarry and erecting processing equipment such as crushers and screens.

Concrete is placed by chuting directly from the mix truck, where possible, or from buckets handled by means of cranes or cableways, or it can be pumped into place by special concrete pumps.

Curing of exposed surfaces is required to prevent evaporation of mix water or to replace moisture that does evaporate. The proper balance of water and cement is required to develop full design strength.

Concrete paving for airports and highways is a fully mechanized operation. Batches of concrete are placed between the road forms from a mix truck or a movable paver, which is a combination mixer and placer. A series of specialized pieces of equipment, which ride on the forms, follow to spread and vibrate the concrete, smooth its surface, cut contraction joints, and apply a curing compound. *See* HIGHWAY ENGINEERING.

Asphalt paving. This is an amalgam of crushed aggregate and a bituminous binder. It may be placed on the roadbed in separate operations or mixed in a mix plant and spread at one time on the roadbed. Then the pavement is compacted by rollers. [WILLIAM HERSHLEDER]

Contact condenser

A device in which a vapor is brought into direct contact with a cooling liquid and condensed by giving up its latent heat to the liquid. In almost all cases the cooling liquid is water, and the condensing vapor is steam. Contact condensers are classified as jet, barometric, and ejector condensers, as illustrated. In all three types the steam and cooling water are mixed in a condensing chamber

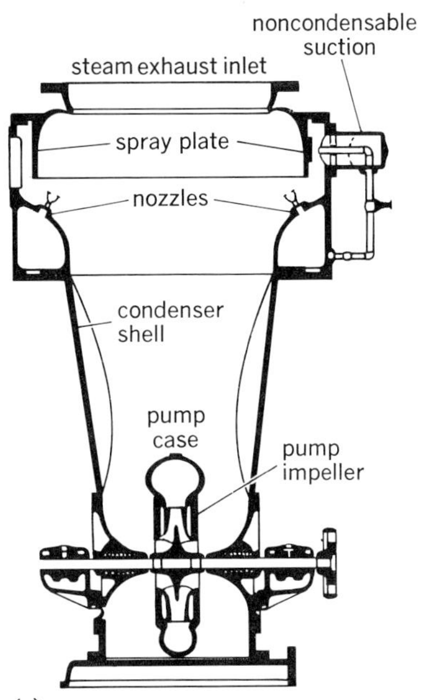

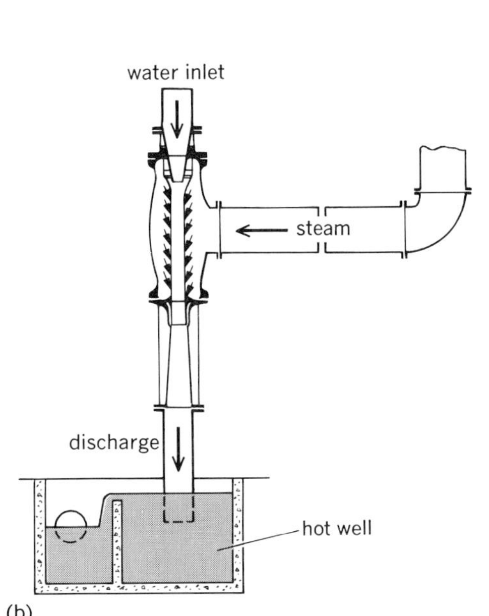

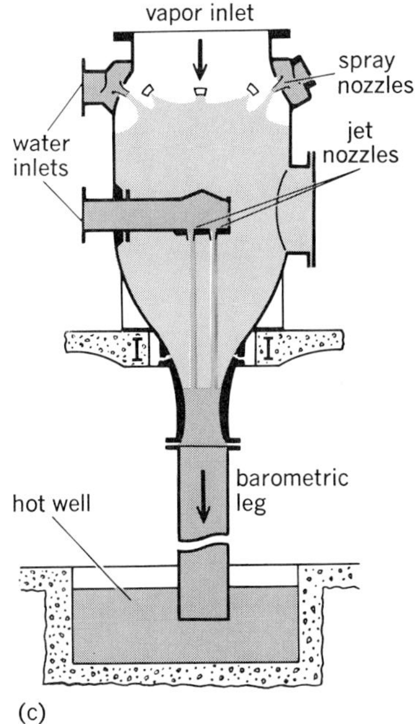

Three basic types of contact condenser. (*a*) Low-level jet condenser (*C. H. Wheeler Manufacturing Co.*). (*b*) Single-jet ejector condenser; (*c*) multijet barometric condenser (*Schutte and Koerting Co.*).

and withdrawn together. Noncondensable gases are removed separately from the jet condenser, entrained in the cooling water of the ejector condenser, and removed either separately or entrained in the barometric condenser. The jet condenser requires a pump to remove the mixture of condensate and cooling water and a vacuum breaker to avoid accidental flooding. The barometric condenser is self-draining. The ejector condenser converts the energy of high-velocity injection water to pressure in order to discharge the water, condensate, and noncondensables at atmospheric pressure. *See* VAPOR CONDENSER.

[JOSEPH F. SEBALD]

Control chart

A chart for the analysis and presentation of data obtained from production processes and research investigations. It is also used to display data in other commercial applications, such as labor turnover, office efficiency, costs, and accident rates. In the simplest charts, results are plotted in sequence over time, so that trends and changes may be identified and acted upon. In a more advanced form, limits calculated by statistical methods are placed on the chart. Provided that results stay within these limits, no action is taken, but if a result or series of results fall outside the limits, some action is indicated. Such statistical control charts are known as Shewhart charts, after W. A. Shewhart who initiated them in the 1930s.

Due to natural variations in raw materials and processes, it is rare to find any two manufactured units which are exactly alike. There is also a tendency for industrial processes to change over time (due to tool wear and changes in external conditions, for example). In order to detect such changes, quality-control personnel take samples from the process output and make measurements on important characteristics of the units in the sample. In isolation the results of individual samples are not very informative, but when plotted on a control chart they provide an effective cumulative picture of process behavior. Together with a history of changes made to the process, they provide an invaluable diagnostic tool to production supervision.

Consider, for example, the chart shown in Fig. 1. The solid horizontal line near the center of the chart represents the design target value for the characteristic in question. The broken horizontal lines represent values which the average of the process must not exceed. The first 9 points plotted

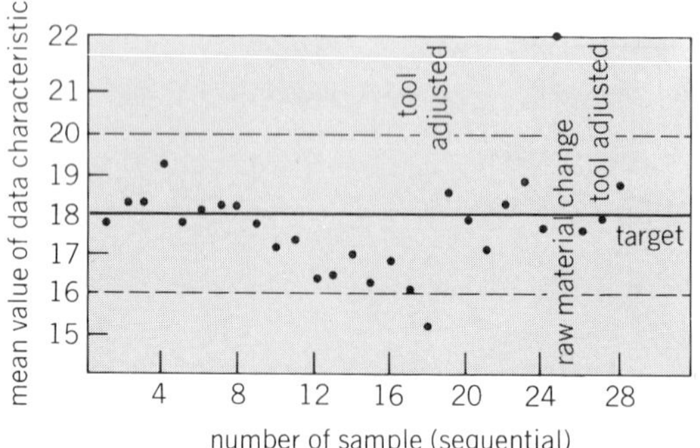

Fig. 1. Control chart showing changes in average of process.

Sample no.	Observations or measurements				Total	Average ($\bar{X}$)	Largest value	Smallest value	Range (R)
1	223	207	229	229	888	222	229	207	22
2	248	248	248	235	979	245	248	235	13
3	248	248	228	228	952	238	248	228	20
4	201	248	197	217	863	216	248	197	51
5	229	207	241	241	918	229	241	207	34
6	235	228	217	217	897	224	235	217	18
7	228	217	192	217	854	213	228	192	36
8	192	212	212	207	823	206	212	192	20
9	207	207	241	223	878	219	241	207	34
10	228	228	229	241	926	231	241	228	13

Section: W-5

Machine: HEAT TREAT #3

Date: MARCH 10-14, 1982

Dimension controlled: BRINELL HARDNESS

Fig. 2. Typical work sheet for $\bar{X}$ and R chart.

on this chart are each the average of 5 units drawn at random from the process at hourly intervals, and they will be seen to be grouped closely around the target value. Points 10 to 17 indicate a downward drift in the process, and point 18 actually falls outside the lower broken line. The process was adjusted at this point and brought back on target. It then ran satifactorily for another 6 hours, and then the average increased suddenly to a point above the upper line. Reference to production records show that the first change occurred due to a tool setting drifting out of specification, and the second occurred when a new batch of raw material was used.

It will be noticed that even when the process was running satisfactorily there was still some variation in results about the target value. Such variations are due to a system of chance causes, and provided they can be tolerated in the finished product, it is unnecessary and usually uneconomical to try and reduce them. Under these conditions the process is said to be in a state of statistical control. However, changes which move the process average away from its target, such as those shown in Fig. 1, must be identified and dealt with. These changes are due to assignable causes, as distinct from chance causes. The object of Shewhart control charts is to provide rules for determining whether changes are more likely due to assignable causes or chance variation.

The statistical theory on which Shewhart charts are based depends on the assumption that the underlying data are normally distributed, and this must be kept in mind when interpreting charts which are constructed by the methods given below. Misleading indications can result from the assumptions being violated, and therefore a thorough analysis of the data by skilled statisticians may be necessary. Custom-designed charts may be required in some cases.

Average and range. The charts most frequently used by the quality control engineer are the $\overline{X}$ (X bar) and R charts. These charts display the average characteristic $\overline{X}$ of production units and the range R of the characteristic. The first step toward $\overline{X}$ and R charts is to collect the data and make the calculations. In Fig. 2 four measurements were made on each lot samples, and $\overline{X}$ and R computed for each sample. The next step is to calculate for the 10 samples grand average X and average range $\overline{R}$ by using Eqs. (1) and (2). From

$$\overline{\overline{X}}=\sum_{i=1}^{i=10}\frac{\overline{X}_i}{10}=\frac{2243}{10}=224 \qquad (1)$$

$$\overline{R}=\sum_{i=1}^{i=10}\frac{\overline{R}_i}{10}=\frac{261}{10}=26 \qquad (2)$$

these data control limits are calculated based on the normal distribution curve. Tabulated chart factors simplify such calculations (see table). The control limits for X are shown by Eqs. (3) and (4), and for R, by Eqs. (5) and (6).

$$\text{Upper limit}=\overline{\overline{X}}+A_2\overline{R} = 224+(0.73\times 26)=243 \qquad (3)$$

$$\text{Lower limit}=\overline{\overline{X}}-A_2\overline{R} = 224-(0.73\times 26)=205 \qquad (4)$$

$$\text{Upper limit}=D_4\overline{R}=2.28\times 26=59 \qquad (5)$$

$$\text{Lower limit}=D_3\overline{R}=0\times 26=0 \qquad (6)$$

Finally an X and an R chart are constructed from the calculations (Fig. 3). With the control limits thus established, subsequent lots are inspected and $\overline{X}$ and R (dotted lines in Fig. 3) are recorded. A point outside the control limits, such as for lot 2 (on $\overline{X}$ chart), indicates need for a change in the manufacturing process.

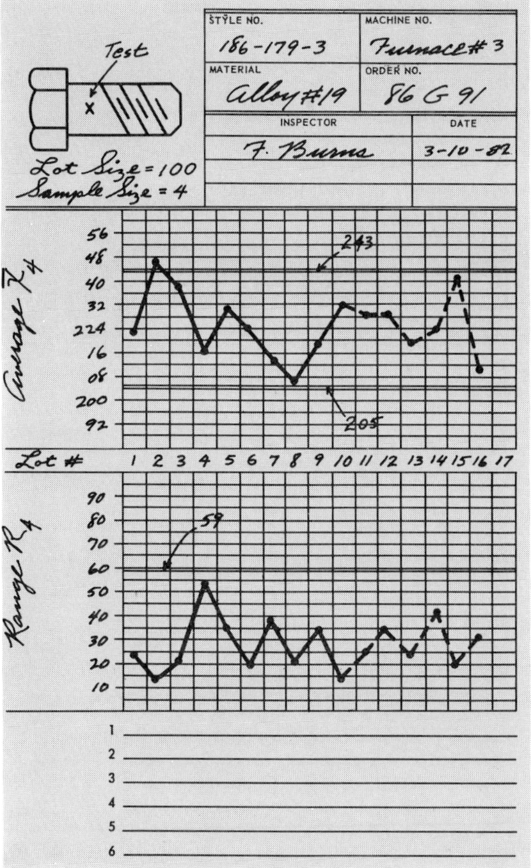

Fig. 3. $\overline{X}$ and R chart constructed from the data and calculations which are given in Fig. 2.

Fraction defective. In some situations the quality characteristic being inspected is judged as accepted or not accepted. This is inspection by attributes. The inspected unit is either passed or rejected without reference to any degree of goodness. The product may be inspected for one characteristic or for several (Fig. 4).

The average fraction defective is represented by $\bar{p}$ and is the ratio of the number of rejected units to the number of units inspected. Control limit lines are established by expression (7), where $\bar{p}$ and $\bar{n}$ are defined by Eqs. (8) and (9), respectively.

$$\bar{p}\pm 3\sqrt{\frac{\bar{p}(1-\bar{p})}{\bar{n}}} \qquad (7)$$

$$\bar{p}=\frac{\text{total units rejected}}{\text{total units inspected}} \qquad (8)$$

$$\bar{n}=\frac{\text{total units inspected}}{\text{number lots inspected}} \qquad (9)$$

The cause of points falling outside control limit lines should be determined, and action should be taken to correct the situation.

Defects per unit. The c chart, generally known as a defects-per-unit chart, is used to show the number of defects found in one unit of product. For this purpose the item is inspected, and the number of defects found is recorded. One or more items in a lot may be inspected, but a separate record is made of the number of defects found on each unit, and this number is plotted on the chart (Fig. 5). Control limits for a c chart are shown by expression (10). Here $\bar{c}$ is the average number of defects per unit, which is calculated from the in-

Factors for computing control chart lines

	Chart for averages	Chart for ranges		
Number of observations in sample, n	Factor for control limits	Factor for central line	Factors for control limits	
	A_2	d_2	D_3	D_4
2	1.88	1.13	0	3.27
3	1.02	1.69	0	2.58
4	0.73	2.06	0	2.28
5	0.58	2.33	0	2.12
6	0.48	2.53	0	2.00
7	0.42	2.70	0.08	1.92
8	0.37	2.85	0.14	1.86
9	0.34	2.97	0.18	1.82
10	0.31	3.08	0.22	1.78

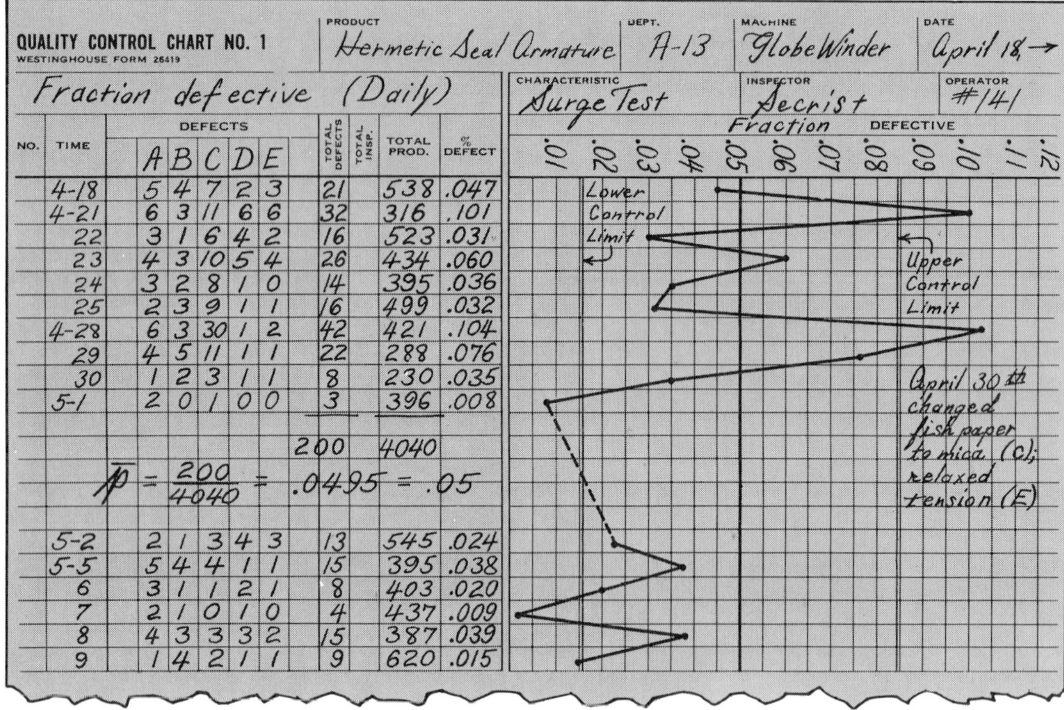

QUALITY CONTROL CHART NO. 1 — WESTINGHOUSE FORM 26419

PRODUCT: Hermetic Seal Armature | DEPT.: A-13 | MACHINE: Globe Winder | DATE: April 18 →

Fraction defective (Daily)

CHARACTERISTIC: Surge Test | INSPECTOR: Secrist | OPERATOR: #141

NO.	TIME	A	B	C	D	E	TOTAL DEFECTS	TOTAL INSP.	TOTAL PROD.	% DEFECT
	4-18	5	4	7	2	3	21		538	.047
	4-21	6	3	11	6	6	32		316	.101
	22	3	1	6	4	2	16		523	.031
	23	4	3	10	5	4	26		434	.060
	24	3	2	8	1	0	14		395	.036
	25	2	3	9	1	1	16		499	.032
	4-28	6	3	30	1	2	42		421	.104
	29	4	5	11	1	1	22		288	.076
	30	1	2	3	1	1	8		230	.035
	5-1	2	0	1	0	0	3		396	.008
							200		4040	
	5-2	2	1	3	4	3	13		545	.024
	5-5	5	4	4	1	1	15		395	.038
	6	3	1	1	2	1	8		403	.020
	7	2	1	0	1	0	4		437	.009
	8	4	3	3	3	2	15		387	.039
	9	1	4	2	1	1	9		620	.015

$\bar{p} = \frac{200}{4040} = .0495 = .05$

Fig. 4. Work sheet and p chart. Defects are classified to assist in determination of corrective action.

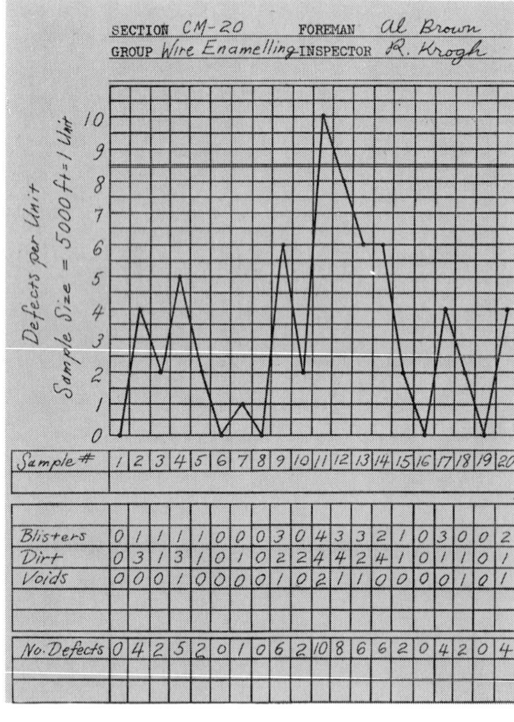

SECTION CM-20 FOREMAN Al Brown
GROUP Wire Enamelling INSPECTOR R. Krogh

Sample #	1	2	3	4	5	6	7	8	9	10	11	12	13	14	15	16	17	18	19	20
Blisters	0	1	1	1	1	0	0	0	3	0	4	3	3	2	1	0	3	0	0	2
Dirt	0	3	1	3	1	0	1	0	2	2	4	4	2	4	1	0	1	1	0	1
Voids	0	0	0	1	0	0	0	0	1	0	2	1	1	0	0	0	0	1	0	1
No. Defects	0	4	2	5	2	0	1	0	6	2	10	8	6	6	2	0	4	2	0	4

Fig. 5. Control chart for defects per unit.

spection of a number of units, as determined by Eq. (11).

$$\bar{c} \pm 3\sqrt{\bar{c}} \qquad (10)$$

$$\bar{c} = \frac{\text{total number of defects}}{\text{total number of units inspected}} \qquad (11)$$

Recording different classes of defects facilitates corrective action when required. In Fig. 5, 64 defects were found in 20 units inspected. Therefore $\bar{c} = 64/20 = 3.2$, giving an upper control limit of 8.6 and a lower control limit of 0, because a negative control is meaningless in this case.

Average number of defects per unit. The $\bar{c}$ chart is used to show the average number of defects per unit in a sample. For this purpose a number of items, generally four or five, are inspected, and a record is made of the number of defects found in all units in the sample. Then the number of defects is divided by the number of units in the sample, as shown in Eq. (12), and this average number is

$$\bar{c} = \frac{\text{number defects found in sample}}{\text{number units in sample}} \qquad (12)$$

plotted on the $\bar{c}$ chart. Control limits are shown by expression (13) where n is the number of units in the sample, and $\bar{\bar{c}}$ is the average number of defects in all units in a number of samples, usually 10 or more, as shown in Eq. (14).

$$\bar{\bar{c}} \pm 3\sqrt{\bar{\bar{c}}/n} \qquad (13)$$

$$\bar{\bar{c}} = \frac{\text{total number of defects}}{\text{total number of units inspected}} \qquad (14)$$

Figure 6 shows a simple form of $\bar{c}$ chart, which provides for recording the size of sample inspected and the number of defects found in each sample. In this case, 294 defects were found in 500 units. Therefore $\bar{c} = 294/500 = 0.6$, giving upper and lower control limits of 1.06 and 0.14, respectively.

The quality-control engineer finds many uses for control charts. The $\bar{X}$ and R charts are widely used. The $\bar{c}$ chart and c chart are as informative

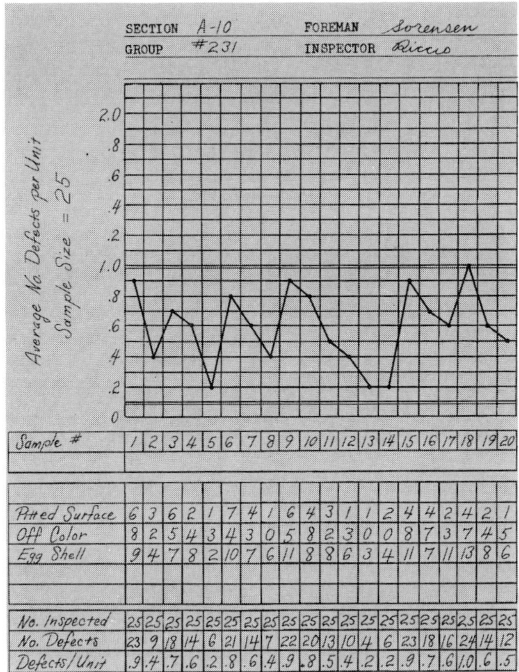

SECTION A-10 FOREMAN Sorensen
GROUP #231 INSPECTOR Riccio

Sample #	1	2	3	4	5	6	7	8	9	10	11	12	13	14	15	16	17	18	19	20
Pitted Surface	6	3	6	2	1	7	4	1	6	4	3	1	1	2	4	4	2	4	2	1
Off Color	8	2	5	4	3	4	3	0	5	8	2	3	0	0	8	7	3	7	4	5
Egg Shell	9	4	7	8	2	10	7	6	11	8	8	6	3	4	11	7	11	13	8	6
No. Inspected	25	25	25	25	25	25	25	25	25	25	25	25	25	25	25	25	25	25	25	25
No. Defects	23	9	18	14	6	21	14	7	22	20	13	10	4	6	23	18	16	24	14	12
Defects/Unit	.9	.4	.7	.6	.2	.8	.6	.4	.9	.8	.5	.4	.2	.2	.9	.7	.6	1.0	.6	.5

Fig. 6. Control chart for recording sample size and average number of defects per unit in each sample.

and could be used more widely. For unusual situations the engineer may have to design special charts. *See* GAGES; INDUSTRIAL COST CONTROL; INSPECTION AND TESTING; QUALITY CONTROL.

[JOHN A. CLEMENTS]

Bibliography: American Society for Quality Control, *J. Qual. Technol.* and *Qual. Progr.*, periodically; A. V. Feigenbaum, *Total Quality Control*, 1961; E. L. Grant and R. Leavenworth, *Statistical Quality Control*, 5th ed., 1979.

Conveying machines

A family of materials-handling machines designed to move individual articles of solid, free-flowing bulk materials over a fixed horizontal, inclined, declined, or vertical path of travel with continuous motion. The actuating medium may be the force of gravity, air, vibration, or power-operated components of the machine such as continuous belts, chains, or cables. Of the many varieties, some are used almost exclusively to convey bulk materials. The kinds described in this article fall largely in the category popularly referred to as package conveyors, and consist of gravity and powered belt, chain, and cable conveyors. *See* BULK-HANDLING MACHINES.

Gravity conveyors. The most economical means for lowering articles and materials is by gravity conveyors. Chutes depend upon sliding friction to control the rate of descent; wheel and roller conveyors use rolling friction for this purpose (Fig. 1).

With a body resting on a declined plane, friction F opposes component P parallel to the surface of the plane of weight W of the body. When the angle of elevation equals the angle of repose, the body is just about to start sliding, or to express it another way, when the inclination of the plane equals the angle of repose, F equals P. In this position the tangent of the angle of repose equals the coefficient of friction. For metal on metal the coefficient of friction is 0.15 to 0.25, and for wood on metal it is 0.2 to 0.6, which is why steel tote boxes slide more readily on metal chutes than do wooden cases.

When the body is a smooth-surfaced container and the inclined plane is replaced by rollers, the rollers are spaced so that at least three rollers support the smallest container to be conveyed; then each roller supports one-third the weight of the load. Component P of the weight W which acts downward and parallel to the inclination of the plane produces a turning moment Pr about the roller's axle, where r is the radius of the roller. As the inclination of the conveyor is increased, P increases until, at the point where rolling is about to start, Pr equals Fr. In this case, however, the force of friction is made up of the friction between the contacting surfaces and the friction in the bearings of the rollers. The mass of a tubular roller is concentrated near its circumference; hence rollers have greater starting inertia than wheels, but after they have started rolling, they have a flywheel effect that tends to keep products in motion, especially when packages follow each other in close succession. Inclination (or declination) is expressed as the number of inches of rise or fall per foot of conveyor or per 5-ft or 10-ft length.

Gravity chutes. Gravity chutes may be made straight or curved and are fabricated from sheet metal or wood, the latter being sometimes covered with canvas to prevent slivering. The bed of the chute can be shaped to accommodate the products to be handled. In spiral chutes centrifugal force is the second controlling factor. When bodies on a

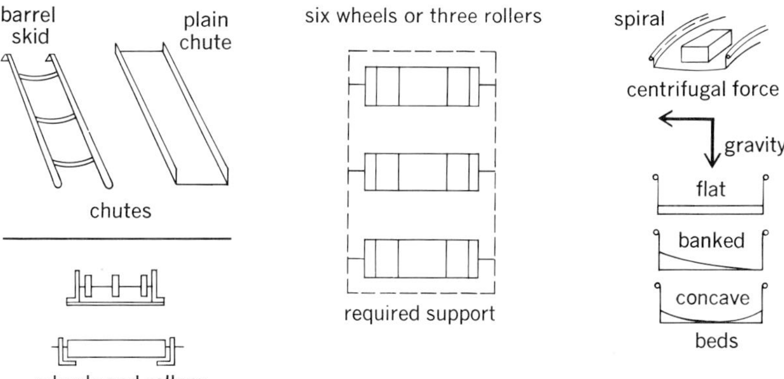

Fig. 1. Elements used in gravity conveyors.

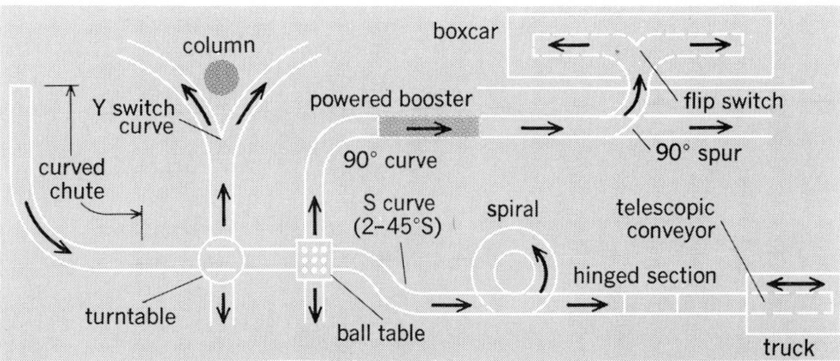

Fig. 2. Gravity components for conveyor lines.

spiral move too fast, they are thrown out toward a guard rail, and contact with the rail increases the friction thereby causing the bodies to slow down somewhat and settle back into the center of the runway and continue their descent at a controlled rate. Spirals with roller beds or wheels provide smooth descent of an article and tend to maintain the position of the article in its original starting position. Rollers may be constructed of metals, wood, or plastic and can be arranged in an optimum position, depending upon the articles to be carried.

Changes in direction can be effected by turntables, ball transfer tables, or gradual S curves (Fig. 2). Generally speaking, wheel-type conveyors are less expensive than the roller varieties; however, the latter can withstand severer service. For successful use of wheeled conveyors, the conveyed article must have a smooth and firm riding surface.

Powered conveyors. Gravity conveyors are limited to use in conditions where the material is to be lowered in elevation. The horizontal conveying distance capabilities of the gravity conveyor depend on the difference in elevation and the angle of sliding or rolling friction of the material to be conveyed.

To move loads on level or inclined paths, or declining paths that exceed the angle of sliding or rolling friction of the particular material to be conveyed, powered conveyors must be employed. The following are various types of powered conveyors.

Belt conveyors. Loads are moved on a level or inclined path by means of power-driven belts. Belt conveyors with rough-top belts make possible inclines up to 28°; cleated belts are limited on degree of incline only by the position of the center of gravity of the conveyed item.

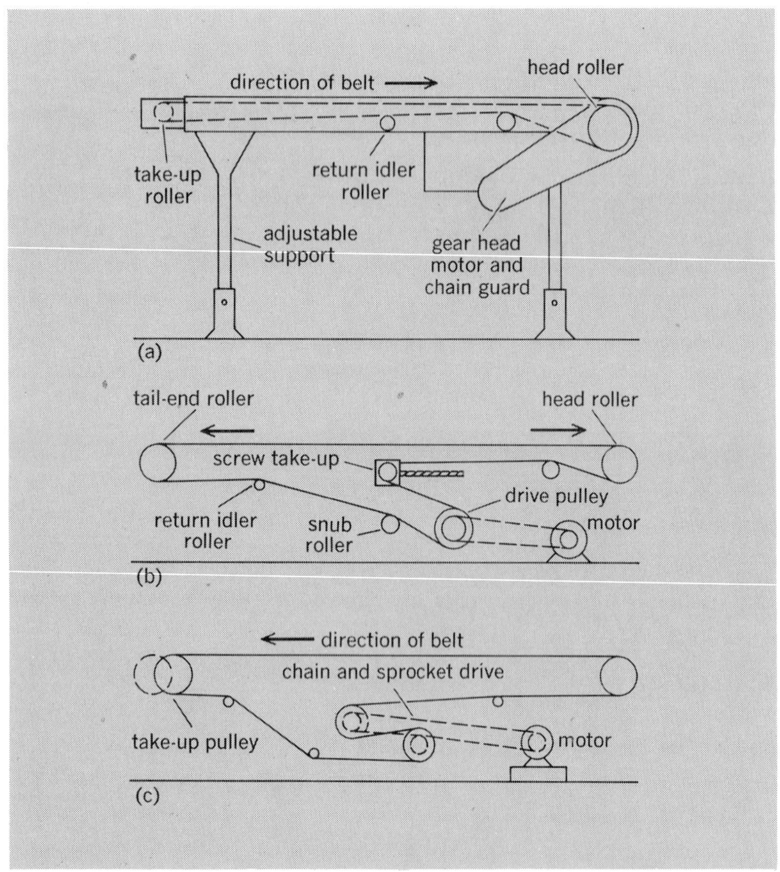

Fig. 3. Diagrams of three driven-belt conveyors. (*a*) Components of a fabric belt conveyor. (*b*) Wrap drive. (*c*) Tandem drive.

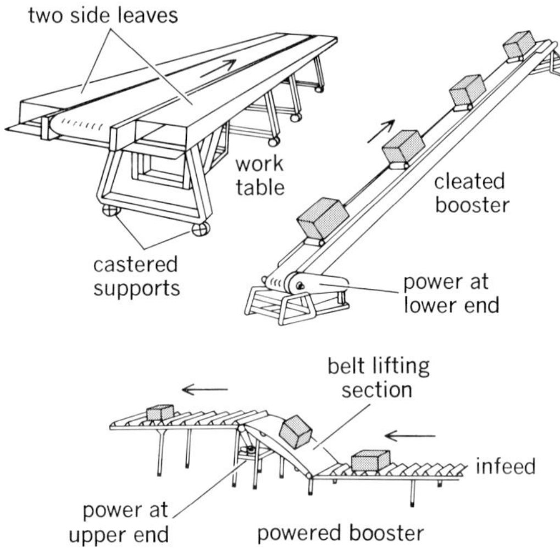

Fig. 4. Fabric belt conveyors.

Essential components of belt conveyors in addition to the belt itself are (1) a bed, which may be a combination bed and frame; (2) end rollers; (3) a take-up to adjust belt tension as it varies with age and atmospheric conditions; and (4) a power unit (Fig. 3). The bed of a fabric belt conveyor may be wood, or more usually flat or concave sheet metal; or it may be made up of rollers. The former are called slide beds, the latter roller beds. Slide beds provide quiet, smooth conveying, but they require more power than roller beds. The two end rollers are mounted in antifriction bearings with provision for lubrication.

Where the tail roller is mounted in adjustable end arms to provide take-up, the length of the conveyor is not constant. Where this feature is objectionable, automatic gravity, screw, or spring take-ups can be used. Another required adjustment is provided by mounting the rollers so that their bearings can be shifted slightly either forward or backward to train the belt to run true.

Belts can be powered in several ways. A gearhead motor can be connected directly to the head roller, or the drive may be through a chain and sprockets. These arrangements necessitate placing the drive outside the frame. When this is undesirable, one of two drives can be located under the frame. Power units usually provide fixed or variable speeds from a few feet per minute (fpm) to over 50 fpm; 50 fpm is usual with package conveyors. Normally 1/4-hp motors power these units, but there are other, more powerful motors which are available for heavier duty.

Powered package conveyors can be combined in various ways to meet specific conditions. Powered curves are used, but wheel and roller sections are more usually employed to change direction. Inclined belt conveyors usually have a hump at the upper end to reduce the slapping by packages on the belt as they move from inclined to horizontal travel (Fig. 4). Cleats can be used, but they complicate the feeding of packages onto the line and require special construction features so that the cleats can pass around pulleys and be properly

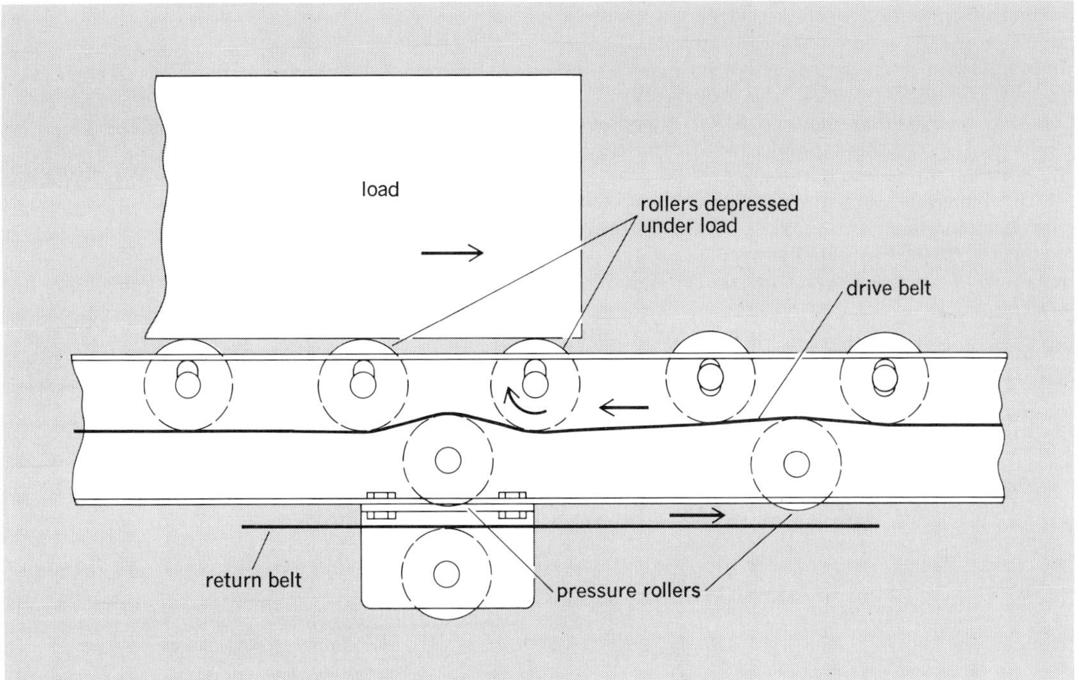

Fig. 5. Live rollers provide power for gravity curves.

supported on the return portion of the conveying cycle.

Metal belts made of woven and flat wire, linked rods, and steel bands are engineered to make up conveyors, which are essentially the same as fabric belt conveyors. For example, woven wire belts can be driven by pulleys and these may be lagged, that is, covered with a material to improve traction. Contact with the drive pulley can be increased by snubbing with small-diameter idler pulleys which guide the belt to increase the angle of contact. Flat wire, linked rod, and woven wire belts with chain edges are driven positively by pulleys with sprocket teeth or by true sprockets. Steel band conveyors are driven by pulleys with unusually large diameters. Other components are similar to those used in fabric belt conveyors but are more rugged in construction to handle the heavier carriers and to meet the more exacting conditions under which metal belts operate. Metal belts are used, for example, in high-temperature operations such as baking, annealing, brazing, and heat-treating.

The choice of a belt for a given situation depends on the effect of the belt's surface on the product, amount of drainage required, and similar factors.

Live-roller conveyors. Objects are moved over series of rollers by the application of power to all or some of the rollers. The power-transmitting medium is usually belting or chain. In one arrangement a belt drive running under and in contact with the rollers turns them to propel the load forward (Fig. 5). In another variation small sprockets are attached to the ends of some or all of the roller shafts, the sprockets being powered by a chain drive.

Chain conveyors. There are three basic types of chain conveyors: (1) those that support the product being conveyed, (2) those that carry actuating elements between the two chains, and (3) those that operate overhead or are set into the floor (Fig. 6).

In the sliding chain conveyor either plain links or links with such special attachments as lugs are made up into the conveyors to handle cases, cans, pipes, and other similar products.

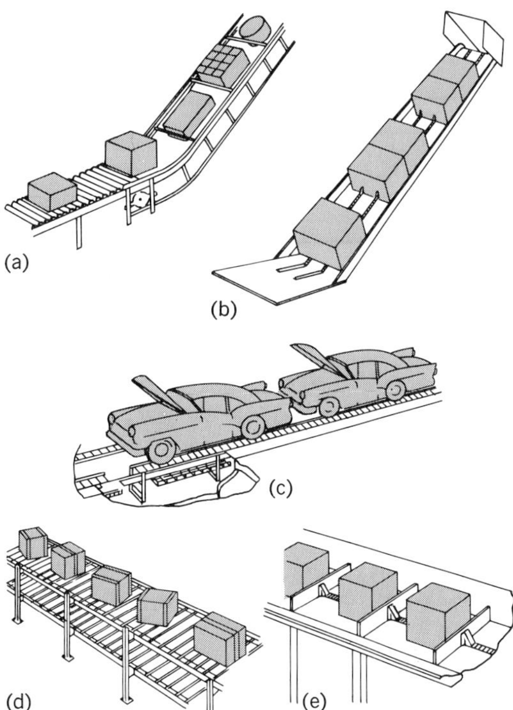

Fig. 6. Five examples of the basic types of chain conveyors. (*a*) Push-bar conveyor. (*b*) Twin chains carrying cartons. (*c*) Heavy-duty plates for assembly conveyor. (*d*) Apron or slat conveyor to carry miscellaneous articles. (*e*) Push-bar or cleat conveyor. (*From Conveyor Equipment Manufacturers Association, Conveyor Terms and Definitions, 3d ed., 1966*)

Slats or apron conveyors are fitted with slats of wood or metal, either flat or in special shapes, between two power chains. The slats handle such freight as barrels, drums, and crates. With the addition of cleats these conveyors carry articles up steep inclines. In this type of conveyor the article rides on the top strand of the slats.

Push-bar conveyors are a variation of the slat conveyor, in which two endless chains are cross-connected at intervals by pusher bars which propel the load along a stationary bed or trough of the conveyor.

[ARTHUR M. PERRIN]

Bibliography: Conveyor Equipment Manufacturers Association, *Conveyor Terms and Definitions*, 3d ed., 1979.

Conveyor

A horizontal, inclined, declined, or vertical machine for moving or transporting bulk materials, packages, or objects in a path predetermined by the design of the device and having points of loading and discharge fixed or selective. Included in this category are skip hoist and vertical reciprocating and inclined reciprocating conveyors; but in the strictest sense this category does not include those devices known as industrial trucks, tractors and trailers, cranes, hoists, monorail cranes, power and hand shovels or scoops, bucket drag lines, platform elevators, or highway or rail vehicles. *See* BULK-HANDLING MACHINES; CONVEYING MACHINES.

Normal operating characteristics of typical conveyors

	Paths						Typical products											Bulk materials							
Conveyor Type	horizontal	declined	inclined	vertical	straight	curved	cases, boxes, etc.	barrels and kegs	drums	textile bags	paper bags	bottles and jars	cans	small parts	food products	lumber, pipe, etc.	towing trucks	free flowing	sluggish	dry	wet	cold	hot	nonabrasive	abrasive
GRAVITY CONVEYORS																									
Sliding friction																									
Skid		X			X			X	X																
Chute		X			X	X	X			X	X				X			X		X		X	X	X	
Spiral chute		X				X	X			X	X				X										
Rolling friction																									
Wheel	X	X			X	X	X				X					X									
Roller	X	X			X	X	X	X	X							X									
Spiral wheel		X				X	X																		
Spiral roller		X				X	X	X	X																
POWERED CONVEYORS																									
Continuous belt																									
Fabric	X	X	X		X		X	X	X	X	X	X	X	X	X	X		X	X	X	X	X	X	X	X
Flexible tube with zipper	X	X	X	X	X	X												X		X		X	X	X	X
Steel band	X	X	X		X		X	X	X	X	X	X	X	X	X	X		X	X	X	X	X	X	X	X
Woven or flat wire	X	X	X		X							X	X	X	X										
Linked rod	X	X	X		X	X						X	X	X	X										
Live roller (drive)																									
Flat belt	X	X	X		X	X	X	X	X	X	X	X	X	X	X										
V belt	X	X	X		X	X	X	X	X	X	X	X	X	X	X										
Sprocket	X	X	X		X	X	X	X	X	X	X	X	X	X	X										
Chain																									
Apron (slat)	X	X	X		X		X	X	X	X	X														
Free roller	X	X	X		X		X	X	X																
Pan	X	X	X		X								X	X	X										
Pusher bar	X	X	X		X		X																		
Hinged plate	X	X	X										X	X	X										
Flat top plate	X				X	X						X	X												
Bucket*	X	X	X	X	X													X		X	X	X	X	X	X
Flight	X	X	X	X	X													X		X		X	X	X	
Drag chain																									
Plain links with lugs	X	X	X		X	X	X		X				X			X									
Overhead trolley†	X	X	X		X	X											X								
Infloor trolley	X				X	X											X								
Cable																									
Overhead trolley†	X	X	X		X	X																			
Table-mounted trolley	X	X	X																						
Cableway‡	X	X	X		X	X																			
Spiral (screw)	X	X	X	X	X	X												X	X	X	X	X	X	X	X
Vertical																									
Rigid arm				X	X		X	X	X	X	X														
Pendant carriage				X	X		X	X	X	X	X														
Pneumatic																									
Tube (with carriers)	X	X	X	X	X	X								X											
Pressure or suction	X	X	X	X	X	X												X		X		X	X	X	X
Vibrating	X	X	X	X	X													X		X		X	X	X	X

*Some bucket conveyors may use endless rubber belts.

†Most products or bulk materials can be conveyed by selection of attachments to the trolleys.

‡Cableways or aerial tramways are used primarily for cross-country conveying of coal, ashes, lumber, concrete, and similar products used in construction.

The more usual basic types of conveyors and their normal, rather than exceptional, operating characteristics are shown in the table. There are, of course, variations of all the basic types shown.

[ARTHUR M. PERRIN]

Bibliography: J. M. Apple, *Plant Layout and Materials Handling*, 3d ed., 1977.

Cooling tower

A tower- or buildinglike device in which atmospheric air (the heat receiver) circulates in direct or indirect contact with warmer water (the heat source) and the water is thereby cooled. A cooling tower may serve as the heat sink in a conventional thermodynamic process, such as refrigeration or steam power generation, or it may be used in any process in which water is used as the vehicle for heat removal, and when it is convenient or desirable to make final heat rejection to atmospheric air. Water, acting as the heat-transfer fluid, gives up heat to atmospheric air, and thus cooled, is recirculated through the system, affording economical operation of the process.

Basic types. Two basic types of cooling towers are commonly used. One transfers the heat from warmer water to cooler air mainly by an evaporation heat-transfer process and is known as the evaporative or wet cooling tower. The other transfers the heat from warmer water to cooler air by a sensible heat-transfer process and is known as the nonevaporative or dry cooling tower. These two basic types are sometimes combined, with the two cooling processes generally used in parallel or separately, and are then known as wet-dry cooling towers.

Cooling process. With the evaporative process, the warmer water is brought into direct contact with the cooler air. When the air enters the cooling tower, its moisture content is generally less than saturation; it emerges at a higher temperature and with a moisture content at or approaching saturation. Evaporative cooling takes place even when the incoming air is saturated, because as the air temperature is increased in the process of absorbing sensible heat from the water, there is also an increase in its capacity for holding water, and evaporation continues. The evaporative process accounts for about 65–75% of the total heat transferred; the remainder is transferred by the sensible heat-transfer process.

The wet-bulb temperature of the incoming air is the theoretical limit of cooling. Cooling the water to within 5–20°F (1°F = 0.56°C) above wet-bulb temperature represents good practice. The amount of water evaporated is relatively small. Approximately 1000 Btu (1 Btu = 1055 J) is required to vaporize 1 lb (0.45 kg) of water at cooling tower operating temperatures. This represents a loss in water of approximately 0.75% of the water circulated for each 10°F cooling, taking into account the normal proportions of cooling by the combined evaporative and sensible heat-transfer processes. Drift losses may be as low as 0.01–0.05% of the water flow to the tower (recent performance of 0.001% has been achieved) and must be added to the loss of water by evaporation and losses from blowdown to account for the water lost from the system. Blowdown quantity is a function of makeup water quality, but it may be determined by regulations concerning its disposal. Its quality is usually expressed in terms of the allowable concentration of dissolved solids in the circulating cooling water and may vary from about two to six concentrations with respect to the dissolved solids content of the cooling water makeup.

With the nonevaporative process, the warmer water is separated from the cooler air by means of thin metal walls, usually tubes of circular cross section, but sometimes of elliptical cross section. Because of the low heat-transfer rates from a surface to air at atmospheric pressure, the air side of the tube is made with an extended surface in the form of fins of various geometries. The heat-transfer surface is usually arranged with two or more passes on the water side and a single pass, cross flow, on the air side. Sensible heat transfer through the tube walls and from the extended surface is responsible for all of the heat given up by the water and absorbed by the cooling air. The water temperature is reduced, and the air temperature increased. The nonevaporative cooling tower may also be used as an air-cooled vapor condenser and is commonly employed as such for condensing steam. The steam is condensed within the tubes at a substantially constant temperature, giving up its latent heat of vaporization to the cooling air, which in turn is increasing in temperature. The theoretical limit of cooling is the temperature of the incoming air. Good practice is to design nonevaporative cooling towers to cool the warm circulating water to within 25 to 35°F of the entering air temperature or to condense steam at a similar temperature difference with respect to the incoming air. Makeup to the system is to compensate for leakage only, and there is no blowdown requirement or drift loss.

With the combined evaporative-nonevaporative process, the heat-absorbing capacity of the system is divided between the two types of cooling towers, which are selected in some predetermined proportion and usually arranged so that adjustments can be made to suit operating conditions within definite limits. The two systems, evaporative and nonevaporative, are combined in a unit with the water flow arranged in a series relationship passing through the dry tower component first and the wet tower second. The airflow through the towers is in a parallel-flow relationship, with the discharge air from the two sections mixing before being expelled from the system. Since the evaporative process is employed as one portion of the cooling system, drift, makeup, and blowdown are characteristics of the combined evaporative-nonevaporative cooling tower system, generally to a lesser degree than in the conventional evaporative cooling towers.

Of the three general types of cooling towers, the evaporative tower as a heat sink has the greatest thermal efficiency but consumes the most water and has the largest visible vapor plume. When mechanical-draft cooling tower modules are arranged in a row, ground fogging can occur. This can be eliminated by using natural-draft towers, and can be significantly reduced with modularized mechanical-draft towers when they are arranged in circular fashion.

The nonevaporative cooling tower is the least efficient type, but it can operate with practically no consumption of water and can be located almost anywhere. It has no vapor plume.

The combined evaporative-nonevaporative cooling tower has a thermal efficiency somewhere be-

Fig. 1. Counterflow natural-draft cooling tower at Trojan Power Plant in Spokane, WA. (*Research–Cottrell*)

tween that of the evaporative and nonevaporative cooling towers. Most are of the mechanical-draft type, and the vapor plume is mitigated by mixing the dry warm air leaving the nonevaporative section of the tower with the warm saturated air leaving the evaporative section of the tower. This retards the cooling of the plume to atmospheric temperature; visible vapor is reduced and may be entirely eliminated. This tower has the advantage of flexibility in operation; it can accommodate variations in available makeup water or be adjusted to atmospheric conditions so that vapor plume formation and ground fogging can be reduced.

Fig. 2. Cross-flow mechanical-draft cooling towers. (*Marley Co.*)

Evaporative cooling towers. Evaporative cooling towers are classified according to the means employed for producing air circulation through them: atmospheric, natural draft, and mechanical draft.

Atmospheric cooling. Some towers depend upon natural wind currents blowing through them in a substantially horizontal direction for their air supply. Louvers on all sides prevent water from being blown out of these atmospheric cooling towers, and allow air to enter and leave independently of wind direction. Generally, these towers are located broadside to prevailing winds for maximum sustained airflow.

Thermal performance varies greatly because it is a function of wind direction and velocity as well as wet- and dry-bulb temperatures. The normal loading of atmospheric towers is 1–2 gal/min (0.06–0.13 liter/s) of cooling water per square foot of cross section. They require considerable unobstructed surrounding ground space in addition to their cross-sectional area to operate properly. Because they need more area per unit of cooling than other types of towers, they are usually limited to small sizes.

Natural draft. Other cooling towers depend for their air supply upon the natural convection of air flowing upward and in contact with the water to be cooled. Essentially, natural-draft cooling towers are chimneylike structures with a heat-transfer section installed in their lower portion, directly above an annular air inlet in a counterflow relationship with the cooling air (Fig. 1), or with the heat-transfer section circumscribing the base of the tower in a cross-flow relationship with the cooling airflow (Fig. 2). Sensible heat absorbed by the air in passing over the water to be cooled increases the air temperature and its vapor content and thereby reduces its density so that the air is forced upward and out of the tower by the surrounding heavier atmosphere. The flow of air through the tower varies according to the difference in specific weights of the ambient air and the air leaving the heat-transfer surfaces. Since the difference in specific weights generally increases in cold weather, the airflow through the cooling tower also increases, and the relative performance improves in reference to equivalent constant-airflow towers.

Normal loading of a natural-draft tower is 2–4 gal/(min)(ft^2) [1.4–2.7 liters/(s)(m^2)] of ground-level cross section. The natural-drafting cooling tower does not require as much unobstructed surrounding space as the atmospheric cooling tower does, and is generally suited for both medium and large installations. The natural-draft cooling tower was first commonly used in Europe. Subsequently, a number of large installations were built in the United States, with single units 385 ft (117 m) in diameter by 492 ft (150 m) high capable of absorbing the heat rejected from an 1100-MWe light-water-reactor steam electric power plant.

Mechanical draft. In cooling towers that depend upon fans for their air supply, the fans may be arranged to produce either a forced or an induced draft. Induced-draft designs are more commonly used than forced-draft designs because of lower initial cost, improved air-water contact, and less air recirculation (Fig. 3). With controlled airflow, the capacity of the mechanical-draft tower can be adjusted for economic operation in relation to heat load and in consideration of ambient conditions.

Normal loading of a mechanical-draft cooling tower is 2–6 gal/(min)(ft^2) [1.4–4.1 liters/(s)(m^2)] of cross section. The mechanical-draft tower requires less unobstructed surrounding space to obtain adequate air supply than the atmospheric cooling tower needs; however, it requires more surrounding space than natural-draft towers do. This type of tower is suitable for both large and small installations.

Nonevaporative cooling towers. Nonevaporative cooling towers are classified as air-cooled condensers and as air-cooled heat exchangers, and are further classified by the means used for producing air circulation through them. *See* HEAT EXCHANGER; VAPOR CONDENSER.

Air-cooled condensers and heat exchangers. Two basic types of nonevaporative cooling towers are in general use for power plant or process cooling. One type uses an air-cooled steam surface condenser as the means for transferring the heat rejected from the cycle to atmospheric cooling air. The other uses an air-cooled heat exchanger for this purpose. Heat is transferred from the air-cooled condenser, or from the air-cooled heat exchanger to the cooling air, by convection as sensible heat.

Nonevaporative cooling towers have been used for cooling small steam electric power plants since the 1930s. They have been used for process cooling since 1940; a complete refinery was cooled by the process in 1958. Until recently, the nonevaporative cooling tower was used almost exclusively with large steam electric power plants in Europe. Interest in this type of tower is increasing in the United States, and a 330-MWe plant in Wyoming

Fig. 3. Mechanical-draft cooling towers. (*a*) Conventional rectangular cross-flow evaporative induced-draft type; (*b*) circular cross-flow evaporative induced-draft type. (*Marley Co.*)

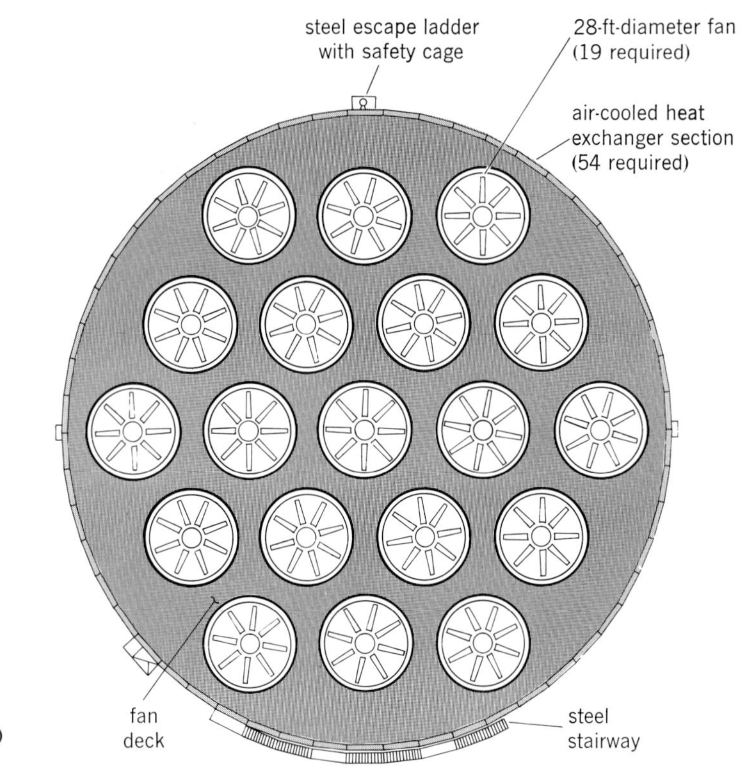

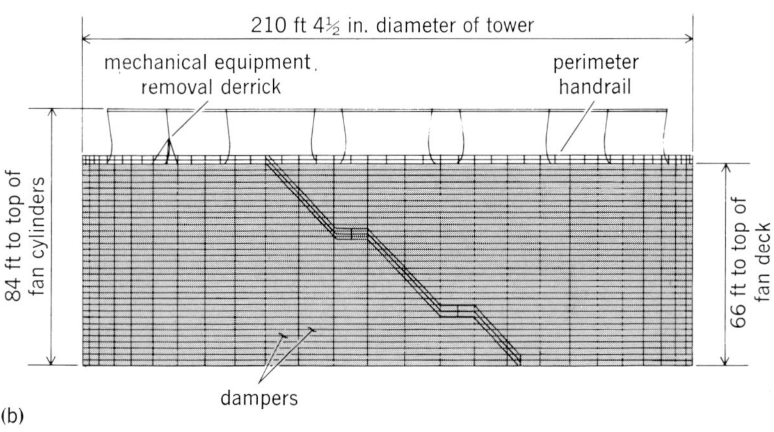

Fig. 4. Multifan circular nonevaporative tower. (*a*) Plan. (*b*) Elevation. (*Marley Co.*)

Fig. 5. Wet-dry cooling tower at Atlantic Richfield Company. (*Marley Co.*)

using nonevaporative cooling towers was completed in the late 1970s.

The primary advantage of nonevaporative cooling towers is that of flexibility of plant siting. There is seldom a direct economic advantage associated with the use of nonevaporative cooling tower systems in the normal context of power plant economics. They are the least efficient of the cooling systems used as heat sinks.

Cooling airflow. Each of the two basic nonevaporative cooling tower systems may be further classified with respect to type of cooling airflow. Both types of towers, the direct-condensing type and the heat-exchanger type, can be built as natural-draft or as mechanical-draft tower systems.

Design. The heat-transfer sections are constructed as tube bundles, with finned tubes arranged in banks two to five rows deep. The tubes are in a parallel relationship with each other and are spaced at a pitch slightly greater than their outside fin diameter, either in an in-line or a staggered pattern. For each section, two headers are used, with the tube ends secured in each. The headers may be of pipe or of a box-shaped cross section and are usually made of steel. The bundles are secured in an open metal frame.

The assembled tube bundles may be arranged in a V shape, with either horizontal or inclined tubes, or in an A shape, with the same tube arrangement. A similar arrangement may be used with vertical tube bundles. Generally, inclined tubes are shorter than horizontal ones. The inclined-tube arrangement is best suited to condensing vapor, the horizontal-tube arrangement best for heat-exchanger design. The A-shaped bundles are usually used with forced-draft airflow, the V-shaped bundles with induced-draft airflow. With natural-draft nonevaporative cooling towers there is no distinction made as to A- or V-shaped tube bundle arrangement; the bundles are arranged in a deck above the open circumference at the bottom of the tower, in a manner similar in principle to that used in the counterflow natural-draft evaporative cooling tower (Fig. 1). A natural-draft cooling tower with a vertical arrangement of tube bundles has been built, but this type of structure is not generally used.

Modularization of the bundle sections has become general practice with larger units; a recent installation of a circular module–type unit is shown in Fig. 4.

Combined evaporative-nonevaporative towers. The combination evaporative-nonevaporative cooling tower is arranged so that the water to be cooled first passes through a nonevaporative cooling section which is much the same as the tube bundle sections used with nonevaporative cooling towers of the heat-exchanger type. The hot water first passes through these heat exchangers, which are mounted directly above the evaporative cooling tower sections; the water leaving the nonevaporative section flows by force of gravity over the evaporative section.

The cooling air is divided into two parallel flow streams, one passing through the nonevaporative section, the other through the evaporative section to a common plenum chamber upstream of the induced-draft fans. There the two airflow streams are combined and discharged upward to the atmosphere by the fans. A typical cooling tower of this type is shown in Fig. 5.

In most applications of wet-dry cooling towers attempts are made to balance vapor plume suppression and the esthetics of a low silhouette in comparison with that of the natural-draft evaporative cooling tower. In some instances, these cooling towers are used in order to take advantage of the lower heat-sink temperature attainable with evaporative cooling when an adequate water supply is available, and to allow the plant to continue to operate, at reduced efficiency, when water for cooling is in short supply.

The more important components of evaporative cooling towers are the supporting structure, casing, cold-water basin, distribution system, drift eliminators, filling and louvers, discharge stack, and fans (mechanical-draft towers only). The coun-

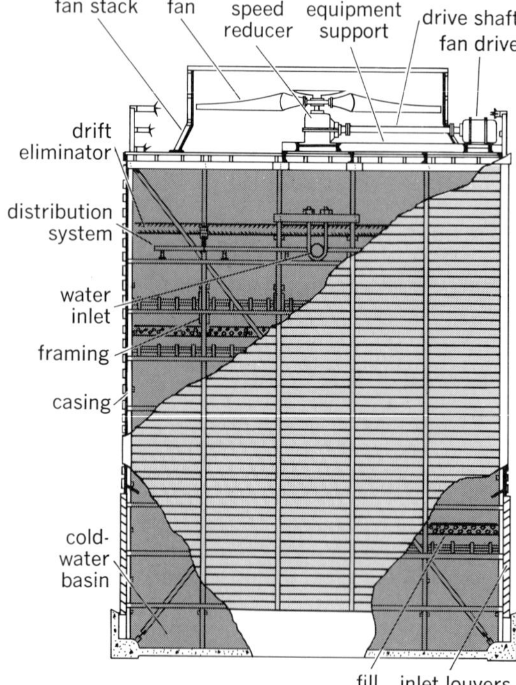

Fig. 6. Induced-draft counterflow cooling tower, showing the component parts.

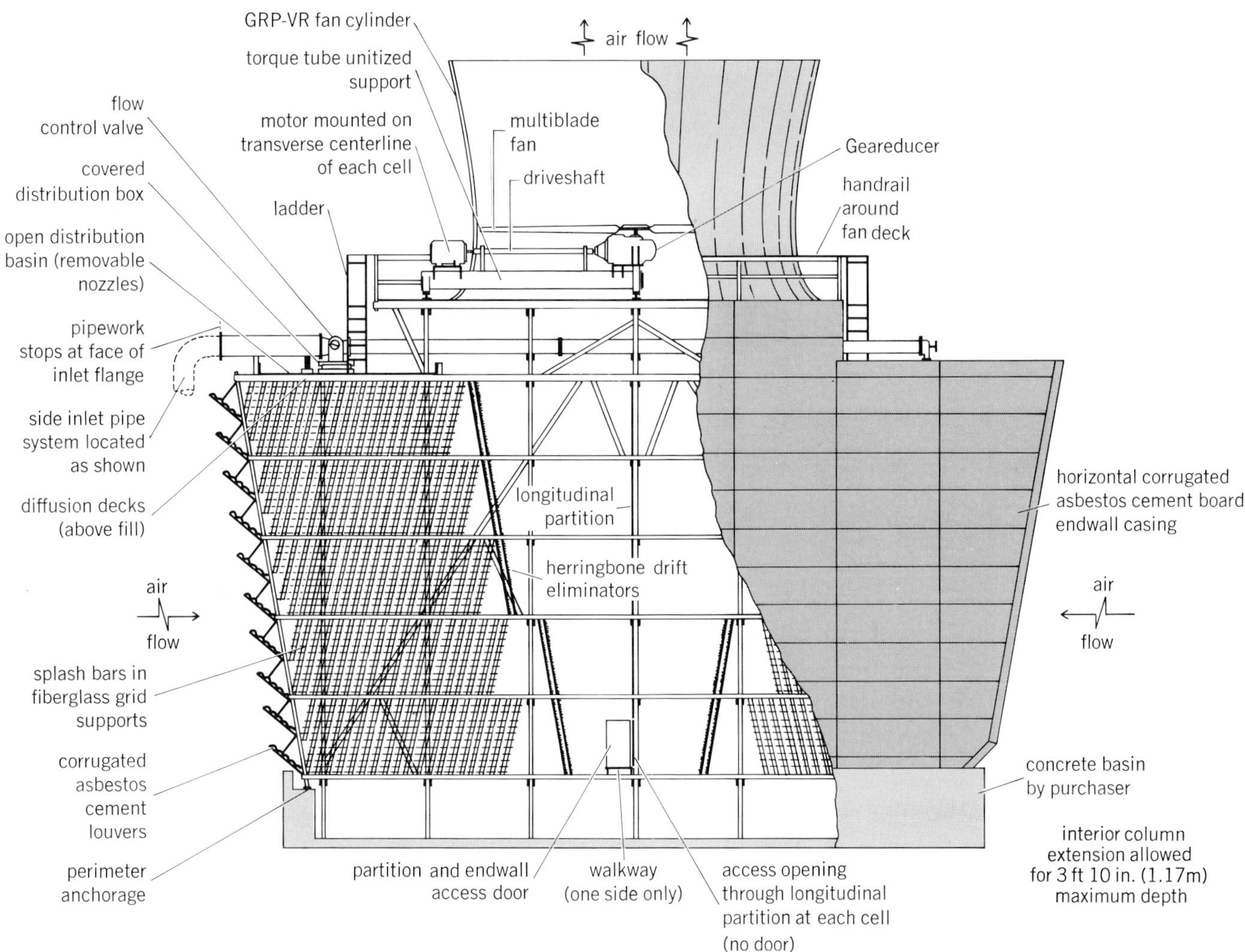

Fig. 7. Transverse cross section of a cross-flow evaporative cooling tower.

terflow type is shown in Fig. 6; the cross-flow type is shown in Fig. 7.

Treated wood, especially Douglas fir, is the common material for atmospheric and small mechanical-draft cooling towers. It is used for structural framing, casing, louvers, and drift eliminators. Wood is commonly used as filler material for small towers. Plastics and asbestos cement are replacing wood to some degree in small towers and almost completely in large towers where fireproof materials are generally required. Framing and casings of reinforced concrete, and louvers of metal, usually aluminum, are generally used for large power plant installations. Fasteners for securing small parts are usually made of bronze, copper-nickel, stainless steel, and galvanized steel. Distribution system may be in the form of piping, in galvanized steel, or fiber glass–reinforced plastics; and they may be equipped with spray nozzles of noncorrosive material, in the form of troughs and weirs, or made of wood, plastic, or reinforced concrete. Structural framing may also be made of galvanized or plastic-coated structural steel shapes. Natural-draft cooling towers, especially in large sizes, are made of reinforced concrete. The cold-water basins for ground-mounted towers are usually made of concrete; wood or steel is usually used for roof-mounted towers. Fan blades are made of corrosion-resistant material such as monel, stainless steel, or aluminum; but most commonly fiber glass–reinforced plastic is used for fan blades.

The heat-transfer tubes used with nonevaporative cooling towers are of an extended-surface type usually with circumferential fins on the air side (outside). The tubes are usually circular in cross section, although elliptical tubes are sometimes used. Commonly used tube materials are galvanized carbon steel, ferritic stainless steel, and various copper alloys. They are usually made with wrapped aluminum fins, but steel fins are commonly used with carbon steel tubes and galvanized. Most designs employ a ratio of outside to inside surface of 20:25. Outside-diameter sizes range from $3/4$ to $1\frac{1}{2}$ in. (19 to 38 mm).

Tube bundles are made with tube banks two to five rows deep. The tubes are in parallel relationship with each other and secured in headers, either of steel pipe or of weld-fabricated steel box headers. The tube bundle assemblies are mounted in steel frames which are supported by structural steel framework. The bundle assemblies may be arranged in a V pattern, requiring fans of the induced-draft type, or in an A pattern, requiring fans

Fig. 8. Nonevaporative cooling tower, Utrillas, Spain. (*GEA*)

of the forced-draft type, Figure 8. Fans and louvers are similar to those described for evaporative cooling towers.

Nonevaporative cooling towers may also be used with natural airflow. In this case, the tube bundles are usually mounted on a deck within the tower and just above the top of the circumferential supporting structure for the tower. In this application, the tower has no cold-water basin, but otherwise it is identical with the natural-draft tower used for evaporative cooling with respect to materials of construction and design.

Performance. The performance of an evaporative cooling tower may be described by the generally accepted equation of F. Merkel, as shown below, where a = water-air contact area, ft²/ft³; h =

$$\frac{KaV}{L} = \int_{T_2}^{T_1} \frac{dT}{h'' - h}$$

enthalpy of entering air, Btu/lb; h'' = enthalpy of leaving air, Btu/lb; K = diffusion coefficient, lb/(ft²)(hr); L = water flow rate, lb/(hr)(ft²); T = water temperature, °F; T_1 = inlet water temperature, °F; T_2 = outlet water temperature, °F; and V = effective volume of tower, ft³/ft² of ground area. The Merkel equation is usually integrated graphically or by Simpson's rule.

The performance of a nonevaporative cooling tower may be described by the generally accepted equation of Fourier for steady-state unidirectional heat transfer, using the classical summation of resistances formula with correction of the logarithmic temperature difference for cross-counterflow design in order to calculate the overall heat transfer. It is usual to reference the overall heat-transfer coefficient to the outside (finned) tube surface.

Evaluation of cooling tower performance is based on cooling of a specified quantity of water through a given range and to a specified temperature approach to the wet-bulb or dry-bulb temperature for which the tower is designed. Because exact design conditions are rarely experienced in operation, estimated performance curves are frequently prepared for a specific installation, and provide a means for comparing the measured performance with design conditions.

[JOSEPH F. SEBALD]

Bibliography: J. D. Guerney and I. A. Cotter, *Cooling Towers in Refrigeration*, International Ideas, Philadelphia, 1966; D. Q. Kern and A. D. Kraus, *Extended Surface Heat Transfer*, 1972; R. D. Landon and J. R. Houx, Jr., Plume abatement and water conservation with the wet-dry cooling towers, *Proc. Amer. Power Conf.*, Chicago, 35:726–742, 1973; Marley Company, *Cooling Tower Fundamentals and Application Principles*, 1969; J. I. Reisman and J. C. Ovard, Cooling towers and the environment: An overview, *Proc. Amer. Power Conf.*, Chicago, 35:713–725, 1973; J. F. Sebald, *Site and Design Temperature Related Economies of Nuclear Power Plants with Evaporative and Nonevaporative Cooling Tower Systems*, Energy Research and Development Administration, Division of Reactor Research and Development, C00-2392-1, January 1976; E. C. Smith and M. W. Larinoff, Power plant siting, performance and economies with dry cooling tower systems, *Proc. Amer. Power Conf.*, Chicago, 32:544–572, 1970; W. Stanford and G. B. Hill, *Cooling Towers: Principles and Practice*, 1970.

Copper alloys

Solid solutions of one or more metals in copper. Many metals, although not all, alloy with copper to form solid solutions. Some insoluble metals and nonmetals are intentionally added to copper alloy to enhance certain characteristics.

Copper alloys form a group of materials of major commercial importance because they are characterized by such useful mechanical properties as high ductility and formability and excellent corrosion resistance. Copper alloys are easily joined by soldering and brazing. Like gold alloys, copper alloys have decorative red, pink, yellow, and white colors. Copper has the second highest electrical and thermal conductivity of any metal. All these factors make copper alloys suitable for a wide variety of products.

Wrought alloys. Copper and zinc melted together in various proportions produce one of the most useful groups of copper alloys, known as the brasses. Six different phases are formed in the complete range of possible compositions. The relationship between composition and phases alpha, beta, gamma, delta, epsilon, and eta are graphically shown in the well established constitution diagram for the copper-zinc system. Brasses containing 5–40% zinc constitute the largest volume of copper alloys. One important alloy, cartridge brass (70% copper, 30% zinc), has innumerable uses, including cartridge cases, automotive radiator cores and tanks, lighting fixtures, eyelets, rivets, screws, springs, and plumbing products. Tensile strength ranges from 45,000 psi as annealed to 130,000 psi for spring temper wire.

Lead is added to both copper and the brasses, forming an insoluble phase which improves machinability of the material. Free cutting brass (61% copper, 3% lead, 36% zinc) is the most important alloy in the group. In rod form it has a strength of 50,000–70,000 psi, depending upon temper and size. It is machined into parts on high-speed (10,000 rpm) automatic screw machines for a multiplicity of uses.

Increased strength and corrosion resistance are obtained by adding up to 2% tin or aluminum to various brasses. Admiralty brass (70% copper, 1% tin, 0.025% antimony, phosphorus, or arsenic, 29% zinc) and aluminum brass (76% copper, 2% aluminum, 0.023% arsenic, 21.97% zinc) are two useful condenser-tube alloys. The presence of phosphorus, antimony, or arsenic effectively inhibits these alloys from dezincification corrosion.

Alloys of copper, nickel, and zinc are called nickel silvers. Typical alloys contain 65% copper, 10–18% nickel, and the remainder zinc. Nickel is added to the copper-zinc alloys primarily because of its influence upon the color of the resulting alloys; color ranges from yellowish-white, to white with a yellowish tinge, to white. Because of their tarnish resistance, these alloys are used for table flatware, zippers, camera parts, costume jewelry, nameplates, and some electrical switch gear.

Copper forms a continuous series of alloys with nickel in all concentrations. The constitution diagram is a simple all alpha-phase system.

Nickel slightly hardens copper, increasing its strength without reducing its ductility. Copper with 10% nickel makes an alloy with a pink cast. More nickel makes the alloy appear white.

Three copper-base alloys containing 10%, 20%, and 30% nickel, with small amounts of manganese and iron added to enhance casting qualities and corrosion resistance, have gained commercial importance. These alloys are known as cupronickels and are well suited for application in industrial and marine installations as condenser and heat-exchanger tubing because of their high corrosion resistance and particular resistance to impingement attack. Heat-exchanger tubes in desalinization plants use the cupronickel 10% alloy.

Copper-tin alloys (3–10% tin), deoxidized with phosphorus, form an important group known as phosphorus bronzes. Tin increases strength, hardness, and corrosion resistance, but at the expense of some workability. These alloys are widely used for springs and screens in papermaking machines.

Silicon (1.5–3.0%), plus smaller amounts of other elements, such as tin, iron, or zinc, increases the strength of copper, making alloys useful for hardware, screws, bolts, and welding rods.

Sulfur (0.35%) and tellurium (0.50%) form insoluble compounds when alloyed with copper, resulting in increased ease of machining.

Precipitation-hardenable alloys. Alloys of copper that can be precipitation-hardened have the common characteristic of a decreasing solid solubility of some phase or constituent with decreasing temperature. Precipitation is a decomposition of a solid solution leading to a new phase of different composition to be found in the matrix. In such alloy systems, cooling at the appropriate rapid rate (quenching) from an equilibrium temperature well within the all-alpha field will preserve the alloy as a single solid solution possessing relatively low hardness, strength, and electrical conductivity.

A second heat treatment (aging) at a lower temperature will cause precipitation of the unstable phase. The process is usually accompanied by an increase in hardness, strength, and electrical conductivity. Some 19 elements form copper-base binary alloys that can be age- or precipitation-hardened.

Two commercially important precipitation-hardenable alloys are beryllium- and chromium-copper. Beryllium-copper (2.0–2.5% beryllium plus cobalt or nickel) can have a strength of 200,000 psi and an electrical conductivity of less than 50% of IACS (International Annealed Copper Standard). Cobalt adds high temperature stability, and nickel acts as a grain refiner. These alloys find use as springs, diaphragms, and bearing plates and in other applications requiring high strength and resistance to shock and fatigue.

Copper chromium (1% chromium) can have a strength of 80,000 psi and an electrical conductivity of 80%. Copper-chromium alloys are used to make resistance-welding electrodes, structural members for electrical switch gear, current-carrying members, and springs.

Copper-nickel, with silicon or phosphorus added, forms another series of precipitation-hardenable alloys. Typical composition is 2% nickel, 0.6% silicon. Strength of 120,000 psi can be obtained with high ductility and electrical conductivity of 32% IACS.

Zirconium-copper is included in this group because it responds to heat treatment, although its strength is primarily developed through application of cold deformation or work. Heat treatment restores high electrical conductivity and ductility and increases surface hardness. Tensile strength of 70,000 psi coupled with an electrical conductivity of 88% can be developed. Uses are resistance welding wheels and tips, stud bases for transistors and rectifiers, commutators, and electrical switch gear.

Cast alloys. Copper alloy castings of irregular and complex external and internal shapes can be produced by various casting methods, making possible the use of shapes for superior corrosion resistance, electrical conductivity, good bearing quality, and other attractive properties. High-copper alloys with varying amounts of tin, lead, and zinc account for a large percentage of all copper alloys used in the cast form. Tensile strength ranges from 36,000 to 48,000 psi, depending upon composition and size. Leaded tin-bronzes with 6–10% tin, about 1% lead, and 4% zinc are used for high-grade pressure castings, valve bodies, gears, and ornamental work. Bronzes high in lead and tin (7–25% lead and 5–10% tin) are mostly used for bearings. High tin content is preferred for heavy pressures ·or shock loading, but lower tin and higher lead for lighter loads, higher speeds, or where lubrication is less certain. A leaded red brass containing 85% copper and 5% each of tin, lead, and zinc is a popular alloy for general use.

High-strength brasses containing 57–63% copper, small percentages of aluminum and iron, and the balance zinc have tensile strengths from 70,000 to 120,000 psi, high hardness, good ductility, and resistance to corrosion. They are used for valve stems and machinery parts requiring high strength and toughness.

Copper alloys containing less than 2% alloying elements are used when relatively high electrical conductivity is needed. Strength of these alloys is usually notably less than that of other cast alloys.

Aluminum bronzes containing 5–10.5% aluminum, small amounts of iron, and the balance copper have high strengths even at elevated temperature, high ductility, and excellent corrosion resistance. The higher-aluminum-content castings can be heat-treated, increasing their strength and hardness. These alloys are used for acid-resisting pump parts, pickling baskets, valve seats and guides, and marine propellers.

Additives impart special characteristics. Manganese is added as an alloying element for high-strength brasses where it forms intermetallic compounds with other elements, such as iron and aluminum. Nickel additions refine cast structures and add toughness, strength, and corrosion resistance. Silicon added to copper forms copper-silicon alloys of high strength and high corrosion resistance in acid media. Beryllium or chromium added to copper forms a series of age- or precipitation-hardenable alloys. Copper-beryllium copper alloys are among the strongest of the copper-base cast materials. *See* ALLOY; METAL CASTING; METAL FORMING.

[RALPH E. RICKSECKER]

Bibliography: C. M. Dozinel, *Modern Methods of Analysis of Copper and Its Alloys*, 2d ed., 1963; J. H. Mendenhall, *Understanding Copper Alloys*, 1980.

Copper loss

The power loss due to the flow of current through copper conductors. When an electric current flows through a copper conductor (or any conductor), some energy is converted to heat. The heat, in turn, causes the operating temperature of the device to rise. This happens in transformers, generators, motors, relays, and transmission lines, and is a principal limitation on the conditions of operation of these devices. Excessive temperature rises lead to equipment failure.

If the resistance R (in ohms) and the current I (in amperes) are both known, the copper loss in watts may be found from the equation $P = I^2R$. An alternative form is to know the voltage across the resistor V (in volts) and to determine the power loss by the equation $P = V^2/R$. V and I are the effective values of the voltage and current, respectively. The resistance R varies with both frequency and temperature.

In electric machines, there are typically constant losses, as well as those that vary with the square of the load, such as the copper loss. Maximum efficiency is obtained when the load is adjusted so as to make the constant and varying losses equal.

Much research has been undertaken to reduce the loss in machines and thus to increase the output capability and efficiency for a given amount of conductor. For large machines and for transmission lines, the concept of superconductivity appears to hold much promise for substantial improvements in operating efficiencies. Superconducting materials are characterized by negligible I^2R losses. Thus, they show the possibility of having very large power transmission, generation, and conversion capabilities for given amounts of material, and it appears reasonable to expect that the energy required to maintain superconducting materials in the superconducting state will be less than the copper loss in conventional machines. *See* SUPERCONDUCTING DEVICES; WIRING.

[EDWIN C. JONES, JR.]

Bibliography: Electric Power Research Institute, *Cost Components of High-Capacity Transmission Options*, Rep. EL-1065, May 1979; J. L. Kirtley, Jr., and M. Furuyama, A design concept for large superconducting alternators, *IEEE Trans.*, PAS-94(4):1264–1269, July/August 1975.

Core drilling

The boring of a hole in the earth by drilling a circular groove and leaving a central core. It is usually done to obtain the core as a sample for visual examination and to test its physical and chemical properties. But core drilling is also done for making the hole itself whenever it is more economical to drill the annular groove and remove the core than to drill the cross-sectional area of the hole.

The greatest use for core drilling is in prospecting for mineral deposits. It is usually done with a diamond core drill, so called because it utilizes the extreme hardness of the diamond to cut the annular groove in rock. The cores carved out by the diamond bit hundreds of feet in the earth are brought to the surface as true and unaltered specimens of the rocks and their mineral contents (see illustration). Diamond core drilling is also widely

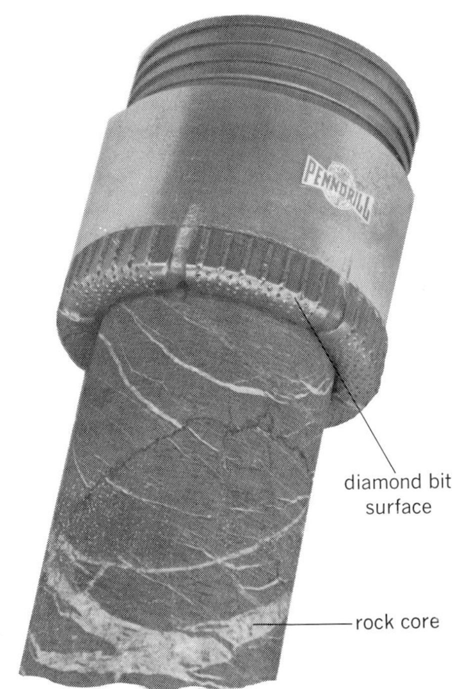

Diamond drill core and diamond bit.

used in civil engineering work to determine the strength of the rocks that must support the tremendous weights of buildings, bridges, and dams, and to determine how expensive it will be to excavate foundations and highway cuts.

In shaft sinking by the core-drill method, and in core drilling other large-diameter holes, chilled steel shot is used as a cutting medium instead of diamonds. Other cutting mediums sometimes used are hard alloy-steel teeth, sintered carbides, and corundum grit. *See* BORING AND DRILLING (MINERAL); PROSPECTING. [FRANK C. STURGES]

Core loss

The rate of energy conversion into heat in a magnetic material due to the presence of an alternating or pulsating magnetic field. It may be subdivided into two principal components, hysteresis loss and eddy-current loss. *See* MAGNETIC HYSTERESIS.

Hysteresis loss. The energy consumed in magnetizing and demagnetizing magnetic material is called the hysteresis loss. It is proportional to the frequency and to the area inside the hysteresis loop for the material used. Hysteresis loss can be approximated empirically by using Eq. (1), where

$$P_h = K_h f B_{\max}^n \qquad (1)$$

K_h is a constant characteristic of the material, f is the frequency, $B_{\max}$ is the maximum flux density, and n, called the Steinmetz coefficient, is often taken as 1.6 although it may vary from 1.5 to 2.5.

Most rotating machines are stacked with silicon steel laminations, which have low hysteresis losses. The cores of large units are sometimes built up with cold-reduced, grain-oriented, silicon iron punchings having exceptionally low hysteresis loss, as well as high permeability when magnetized along the direction of rolling.

Eddy-current loss. Induced currents flow within the magnetic material because of variations in the flux. This eddy-current loss may be expressed to close approximation by using Eq. (2), where K_e is a

$$P_e = K_e (B_{\max} f \tau)^2 \qquad (2)$$

constant depending on the volume and resistivity of the iron, $B_{\max}$ the maximum flux density, f the frequency, and τ the thickness of the laminations in the core. For 60-cycle rotating machines, core laminations of 0.014–0.018 in. are usually used to reduce eddy-current loss.

Measured core loss. The measured core loss in a rotating machine also includes eddy-current losses in solid structural parts, such as the frame, ventilating duct spacers, pole faces, and damper windings, as well as those due to burrs or other contacts between punchings. A number of precautions are taken to minimize these components, which become appreciable in large machines. Structural parts close to the core, including clamping plates, I-beam spacers, dovetail bars, and shields, may be made of nonmagnetic material. Some of these parts may be shielded from the variable flux by low-resistance plates or by flux traps of laminated, high-permeability, low-loss steel. The core punchings are carefully deburred and are coated with a baked-on insulating varnish. The finished cores are often tested with an ac magnetizing coil to locate and correct any hot spots caused by damage during assembly. Pole-face losses may be reduced by avoiding excessive slot-width to air-gap ratios, by surface grooving of the poles, and by low-resistance amortisseur windings. *See* ELECTRIC ROTATING MACHINERY.

[LEON T. ROSENBERG]

Corrosion

The destruction, degradation, or deterioration of material due to the reaction between the material and its environment. Generally, the reaction is chemical or electrochemical, but there are often important physical and mechanical factors in the corrosion process. Corrosion represents an enormous financial loss and wastage of material. With decreasing reserves of metal ores and other raw materials and the energy resources needed to produce new material, the corrosion problem is beginning to receive its deserved attention.

CHEMICAL-MECHANICAL CORROSION

A survey of corrosion problems occurring during a 10-year period was conducted by a large German chemical plant. The results of the survey are given in Table 1. A total of 40% of the corrosion problems are due to stress-corrosion cracking, hydrogen embrittlement, and corrosion fatigue. All of these corrosion phenomena have a mechanical, as well as a chemical, aspect to the failure. They constitute some of the worst corrosion problems because failure of the structure occurs often without warning and with sometimes catastrophic results. Understanding these forms of corrosion is the goal of much current research. Other forms of corrosion attack include fretting corrosion, liquid-metal embrittlement, and corrosion by gases.

Table 1. Survey of corrosion phenomena occurring during the period 1963–1972 in a German chemical industry

Corrosion phenomenon	% of total failures*
General corrosion	33
Stress-corrosion cracking	25
Corrosion fatigue	11
Hydrogen damage	3
Pitting corrosion	5
Intergranular corrosion	4
Wear, erosion, cavitation	6
Nonaqueous high-temperature corrosion	3
Others	10

*Note that in 17% of the failures two or more corrosion phenomena were combined.

Stress-corrosion cracking. The conjoint action of mechanical stress and chemical reaction of the structure with the environment results in the formation of cracks which propagate through the structure, rendering it useless for supporting a load or containing a gas or liquid. Although stress corrosion is usually thought to be a problem in metallic structures, glasses and ionic solids exhibit a formation of cracks which is called crazing.

In most cases involving stress corrosion, the metal is not severely attacked by the environment. The metal usually exhibits passivity under the circumstances; in fact, it is this passive behavior which makes the material useful for the particular application. Stress corrosion thus often involves rather high-quality materials, and therefore relatively costly ones. Stress corrosion can be a problem with stainless steels, nickel-base alloys, alloy steels, aluminum-base alloys, copper-base alloys, and titanium-base alloys, all of which are important construction materials.

Because stress corrosion commonly occurs when the metal is passive, one explanation for its occurrence is that the action of the stress and the environment is to rupture the protective passive films and expose the underlying metal to severe local attack. This explanation is the basis of the film rupture mechanism, shown schematically in Fig. 1*a* and *b*. There are other mechanisms that are fowarded to explain stress corrosion cracking in addition to the film rupture mechanism. However, this mechanism does explain most of the observed phenomena and consequently is presented here.

A stressed metal undergoes deformation. The metal deforms preferentially along certain planes (called slip planes). The protective passive film on the metal is usually an oxide whose mechanical properties are much less ductile than those of the metal. An area of bare metal is thus exposed to the solution. The important question is how fast the environment will cause repassivation of the material.

In case I of Fig. 1*b*, the environment is relatively corrosive to the metal. A significant amount of dissolution occurs which could ultimately result in failure of the structure by general corrosion. However, case I does not lead to stress corrosion because the morphology of the attack does not appreciably increase the intensity of the stress due to the blunt and shallow penetration. In case III, repassivation occurs very rapidly; there is almost no penetration into the metal and no prospect for stress corrosion. In case II, the ideal conditions for stress corrosion are met. The repassivation process is delayed long enough for some significant dissolution to occur. Because the attack is sufficiently restricted, the intensity of the local stress increases (like the effect of a notch on a smooth surface). In addition, the localized and restricted area of dissolution establishes a geometric condition in which the localized chemistry in the corroded area may differ significantly from the chemistry in the bulk solution. In particular, the local chemistry may be more acid and more concentrated in ions such as chloride that further stifle repassivation. Continued deformation of metal along the slip plane repeats the rupture–dissolution–slow repassivation process, initiating a surface defect that becomes the basis of a growing crack.

Thus, it appears from the film-rupture model that the ideal conditions for stress-corrosion initiation occur when the passive film is marginally stable. If the film is stable, case III occurs; if the film is unstable, case I obtains. The stability of passive films, which are usually oxides such as Cr_2O_3, Al_2O_3, NiO, CuO, Fe_2O_3, and TiO_2, depends on the electrochemical potential.

Electrochemical potential. The electrochemical potential depends largely upon the composition of the alloy, the chemical species present in the environment, and the temperature. Stress has relatively little to do with the electrochemical potential as such, but under some conditions the stress can influence the metallurgical microstructure and distribution of constituents and phases in the alloy (particularly minor constituents like carbon, sulfur, and phosphorus which are present in most commercial alloys). Therefore, stress can in-

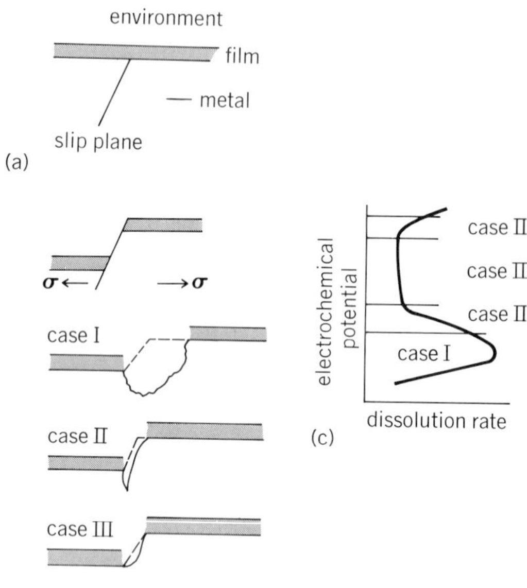

Fig. 1. Principles of the film-rupture mechanism of stress-corrosion cracking. (*a*) Initial condition is metal covered by thin, brittle passive film. (*b*) Application of stress σ breaks the film by slip process. In case I, metal dissolution is rapid and film repassivation is slow, resulting in more or less general corrosion. In case II, metal dissolution occurs under restricted conditions. In case III, repassivation is so rapid relative to dissolution that no crack can be initiated. (*c*) Potential-dissolution rate curve showing the region where cases I, II, and III occur.

directly influence the potential. *See* ALLOY.

It is important to emphasize that the stress on the metal or other solid can come from the service stress (the load that the metal must bear in service), as well as from fabrication stresses that are residual in nature (such as stresses developed by rolling or forging the metal components and by welding or joining the components). In addition to producing stress, these processes can also change the metallurgical microstructure of the alloy, which in turn may influence the susceptibility to stress corrosion. Another source of stress is thermal stresses which occur in installations such as heat exchangers.

The electrochemical potential for a given metal in solution measured relative to a standard electrode determines the dissolution rate of the metal or alloy, as shown in Fig. 1*c*. This generalized curve applies to stainless steels, some copper alloys, titanium and its alloys, some aluminum alloys, and nickel alloys which show an active to passive transition. The engineering usefulness of these alloys occurs when they are in the passive state. Some of these alloys show a passive to transpassive transition where dissolution of the alloy occurs once again. Where the alloys are imperfectly passive, as at these transitions, stress corrosion is possible.

Stress corrosion of a given alloy often occurs only in very specific environments. For instance, many stainless steels fail in the presence of chloride or hydroxide ions; brasses and bronzes fail when both ammonia and dissolved oxygen are present in water; ordinary carbon steels are susceptible to nitrates and hydroxides; some nickel alloys fail when small amounts of oxygen are present; and titanium stress-corrosion cracks when exposed to a combination of chlorides and methanol. The specificity of the alloy-environment combination appears to be related to the electrochemical potential. This approach of predicting bad combinations of alloys and environments is the subject of much corrosion research work being conducted in industrial, governmental, and university laboratories.

The stress-corrosion crack can proceed along the grain boundaries of a metal (intergranular stress corrosion) or through the grains of a metal (transgranular stress corrosion). The route of the stress-corrosion crack suggests possible causes and remedies. For instance, many aluminum alloys fail by intergranular stress corrosion. Pure aluminum is too soft to be of use in construction, but aluminum alloys are useful because the alloying elements harden the metal. Unfortunately, the alloying elements often concentrate along the grain boundaries and create a condition where the grain boundary region is attacked severely and restrictively (as in case II, Fig. 1*b*). This condition favors stress corrosion. Dispersion of the alloying elements results in a less susceptible condition. In stainless steels, carbides precipitate along grain boundaries following certain heat treatment processes (such as stress relieving) and in the material adjacent to welds. Again, this action (termed sensitization) sets up an unfavorable preferential dissolution at the grain boundary.

Stress-corrosion testing. The objective of laboratory testing is to determine under what circumstances stress corrosion is likely to occur. An accelerated test is often used, in which the environment is chosen to present the most extreme conditions that may occur in a notch or surface flaw, and the stress is considerably higher than the nominal service stress. Three different arrangements for applications of the stress have been developed: the constant-load, constant-deformation, and constant-strain-rate techniques. The tensile component of the stress causes stress corrosion; the compressive component tends to reduce or eliminate stress corrosion. In fact, one of the common ways of mitigating against stress corrosion is to apply a surface compressive stress to the metal, such as by shot peening, sand blasting, or hammering.

In the constant-load technique, the stress is applied in simple tension. The nominal stress is the applied load per unit of cross-sectional area of the specimen. The load is often just a simple weight. As the stressed specimen deforms or as a surface crack initiates, the actual stress increases (due to the reduced cross section). Thus, the crack propagates rapidly.

In the constant deformation arrangement, the specimen is flexed in the shape of a U between two constraints. The convex surface is in tension; the concave surface is in compression. Cracks always initiate from the tensile-stressed side. In contrast to the constant-load test, initiation of the crack in the constant-deformation test causes a relaxation in the stress so that the crack propagates relatively slowly. This arrangement is perhaps more closely related to the actual stress in service, because components are often forced to fit together.

The third arrangement is the constant-strain-rate technique. This is a relatively new technique in which the load is applied gradually over a period of time instead of all at once as in the constant-load technique. With regard to stress-corrosion susceptibility, the constant-strain-rate technique is the most severe of the methods, because the alternate film-rupture—repassivation process is repeated continuously. This type of stress application may occur whenever the service stress is transient, such as in the start-up or shut-down of process equipment.

Another way of conducting accelerated tests which may simulate localized conditions in a real system is to notch the surface of the specimen. The stress intensity is considerably higher at the base of a notch than at a smooth surface. The stress intensity becomes more severe as the sharpness of the notch increases. A notched specimen offers the additional advantage that the crack will be generated from the notch instead of at random from a macroscopically smooth specimen. Control of the crack location is advantageous if the crack propagation rate is to be measured (for example, visually, by means of a traveling microscope, or electronically, by means of an attenuator gage).

Much useful information can be obtained by the examination of cracked specimens. The optical microscope and the scanning electron microscope are used. Microscopy reveals the crack mode (intergranular, transgranular), preferential crack paths in alloy phases (many alloys contain more than one metallurgical phase, each with varied chemical compositions), the relationship to precipitated particles and inclusions, and the number of cracks and their tendency to branch.

Also, the basic aspects of the fracture (brittle or ductile) can be ascertained, particularly by the scanning electron microscope.

Hydrogen damage. Closely related to stress-corrosion cracking is damage to the metal or alloy caused by the entry of hydrogen. In many environments, the principal cathodic reaction is the evolution of hydrogen gas from hydrogen ions (protons). It is believed that hydrogen enters the metal in the form of a proton or as a single atom. Because of the small size of these particles, hydrogen is able to diffuse readily in many metals, particularly ferrous alloys. Entry of hydrogen is favored by an acidic pH and the presence of compounds of sulfur, phosphorus, and arsenic. These compounds are called cathodic poisons; hydrogen sulfide (H_2S) is a particularly important and effective one. A tensile stress dilates the crystal lattice and enhances the amount of hydrogen that can enter the metal. Iron, mild steels, alloy steels, and stainless steels with body-centered cubic structure are most prone to hydrogen damage. The damage takes the form of blisters in the low-strength and more ductile materials, and the form of embrittlement in the higher-strength steels. Titanium and zirconium and their alloys are subject to hydrogen embrittlement which is caused by a chemical reaction between the hydrogen and the metal to form a brittle hydride. In ferrous materials, no brittle hydrides have been found, but hydrogen appears to reduce the cohesive strength between iron atoms in the steel. Blisters are formed by condensation of hydrogen absorbed in the steel around structural defects in the steel. These defects may be vacancy sites, dislocations, grain boundaries, and phase boundaries. In the more ductile materials, some accommodation of the condensed hydrogen is allowed, but the blisters increase the local stress intensity.

Corrosion fatigue. Corrosion fatigue is the synergistic action of corrosion and fatigue, and results in crack formation. Corrosion fatigue and stress corrosion share many similarities. In corrosion fatigue, the load is applied cyclically (tension-compression). In fact, stress corrosion can be regarded as a limiting case of corrosion fatigue in which only half a stress cycle is considered. But an important difference is that while stress corrosion occurs only in the presence of rather specific environmental species, corrosion fatigue may occur in a much wider range of environments. Corrosion fatigue occurs in components such as gears, propellers, impellers, turbines, and the shafts, housings, and attachments to these components that are subjected to cyclic stress and a corrosive atmosphere. Another outstanding example of corrosion fatigue is failure of metal implants in the human body. These bone and joint substitutes are exposed to severe load cycling in an environment whose salinity is much like that of sea water.

The corrosion-fatigue resistance of many steels is lower than that of the usual fatigue limit (tests conducted in air). The corrosion-fatigue life is determined as the number of cycles needed to cause failure at a given stress. The data are graphed in the form shown in Fig. 2, which is called an S-N curve. The area below the curve represents a safe-operating region, while the area above the curve represents imminent failure. A limiting condition sometimes occurs where fatigue will not initiate below a certain stress despite the number of stress cycles. This limit is the fatigue limit. A true fatigue limit does not exist for materials fatigued in corrosive environments, as opposed to those fatigued in air.

Another important consideration is the frequency of the stress cycle. A low frequency is usually more severe than a high frequency because the fatigued metal is exposed to the environment for a relatively longer period of time. As for stress corrosion, the relative rates of metal dissolution and film repassivation are an important concern in

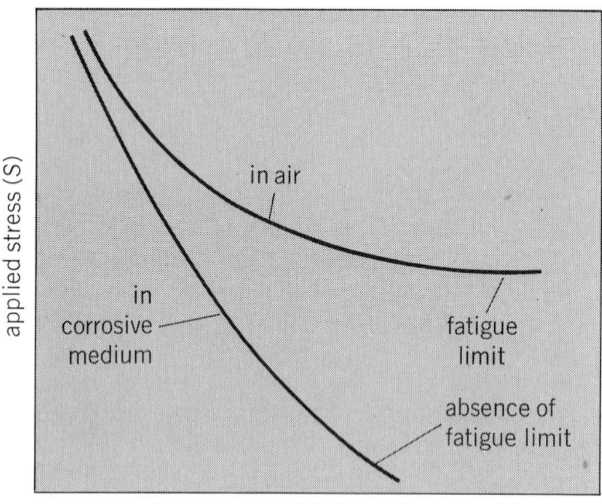

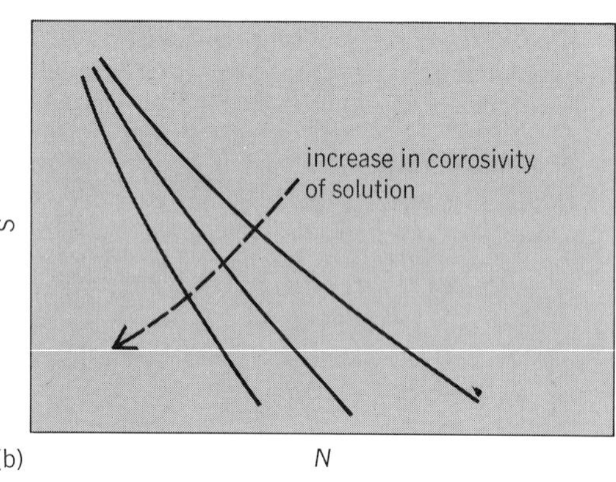

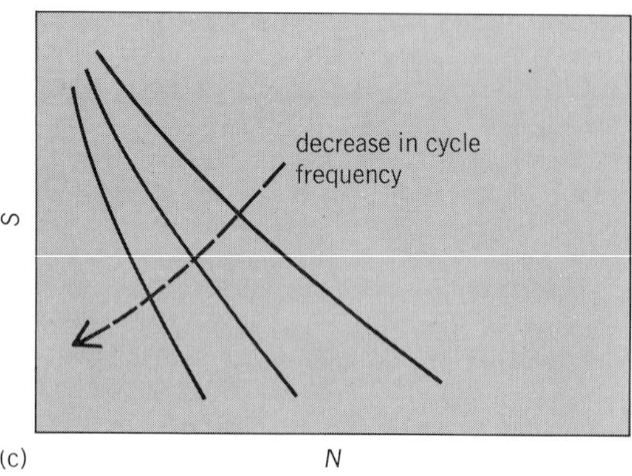

Fig. 2. Corrosion-fatigue curves, typical of data obtained for an 18% chromium–8% nickel stainless steel fatigue-loaded in acidified chloride solutions. (*a*) Fatigue in corrosion environment is more rapid than in air. Fatigue life decreases with (*b*) increase in corrosivity of the environment and (*c*) decrease in the frequency of the stress cycle.

predicting susceptibility. For many steels, hydrogen entry plays an important role in initiating corrosion-fatigue susceptibility.

Corrosion-fatigue cracks may propagate intergranularly or transgranularly. The fractured surface has a brittle fracture appearance. Often, fatigue striations can be observed, these correspond to crack fronts being alternately stopped and started by the stress cycling.

Corrosion fatigue may be eliminated by proper design considerations to reduce the stress or increase the frequency. Elimination of the eccentricity of moving parts is an important practice. With some steels, lower-strength steels may give better fatigue life than the higher-strength ones. Abatement of the corrosivity of the environment is another method. Application of surface compressive stresses reduces the tensile component of the stress and increases fatigue life. Also, coatings are sometimes employed to eliminate the corrosion aspect. *See* METAL, MECHANICAL PROPERTIES OF; METAL COATINGS; PLASTIC DEFORMATION OF METAL.

Fretting corrosion. Fretting corrosion is phenomenon which occurs between mated surfaces which are stressed and in small relative motion or "slip" to one another. Mechanical vibration is often sufficient to cause the effect; fretting is frequently a problem in engine and machine components. The sequence of events is illustrated in Fig. 3. The debris, which is usually an oxide or is quickly converted to an oxide because of the frictional heat, accumulates in the narrow space between surfaces. The ultimate result is galling or seizing of the two surfaces or initiation of a fatigue crack whose propagation is formed by continued buildup of the debris and increased stress.

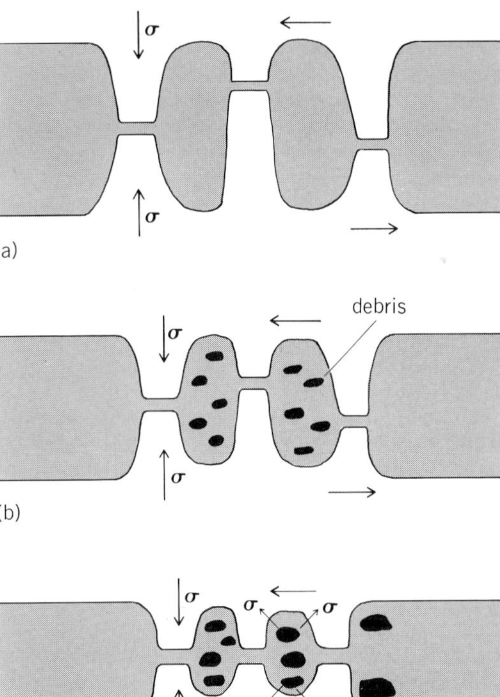

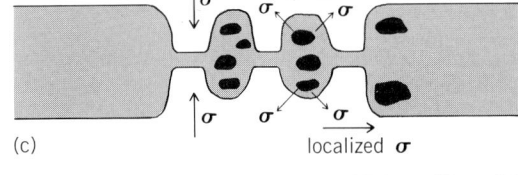

Fig. 3. Fretting-corrosion process. (a) Asperities exist between mated surfaces under stress σ. (*b*) The asperities are broken off during relative motion. Debris is accumulated in spaces between the points of contact. (c) Oxidation of the chafed metal results in voluminous corrosion products which upon further slip between surfaces cause localized stresses that initiate cracks.

A key feature of fretting damage is that the surfaces must be in small relative motion. For example, automobiles which were transported across country by rail were inoperable at their destination because the ball-bearing races had "frozen" during shipment because of the vibrations encountered in transit. Under normal operation where there is large and continuous motion between bearing surfaces (that is, complete revolutions) fretting does not occur.

Fretting damage can be minimized by: (1) eliminating the motion or vibration between parts; (2) lubricating the surfaces to lower the friction and prevent oxygen entry; (3) using a gasket material such as rubber or Teflon; (4) increasing the load at the surface to discourage relative motion between parts; and (5) increasing the hardness of one or both components. Certain combinations of materials have good fretting resistance; other combinations are poor. Generally, harder materials are more resistant.

Liquid-metal embrittlement. Liquid-metal embrittlement (LME) is the rapid loss in mechanical properties (fracture stress and ductility) of a metal or alloy due to the contact with certain liquid metals. As with stress corrosion, the alloy–liquid-metal combination is very specific. Occurrence of liquid-metal embrittlement results from joining operations, such as soldering or brazing, and from coating operations, such as galvanizing and tin dipping. Further, the use of liquid-metal coolants (such as liquid sodium in nuclear reactors) can possibly endanger the construction material of the heat exchanger. Another source of problems is mercury contamination of aluminum pipelines. It appears that certain embrittling metals can embrittle the basis metal even when they are present in solid form. Thus, such metals as zinc, cadmium, tin, and lead can embrittle certain steels after they have been soldered or electroplated onto the steel.

There is no criterion based on the physical properties of the alloy–liquid-metal couple which can be used to predict whether or not a particular liquid metal will embrittle a particular alloy. Embrittlement of an alloy seems to require complete wetting of the surface by the liquid metal. Sometimes, oxide films on metal are helpful in that they prevent complete wetting. In many instances, alloys are more susceptible to liquid-metal embrittlement than pure metals. Many steels are susceptible to embrittlement by liquid sodium, lithium, tin, cadmium, lead, zinc, indium, and antimony, whereas pure iron is not. *See* EMBRITTLEMENT. [R. D. MC CRIGHT]

CORROSION BY GASES

Chemical reactions between metals and gases produce two effects on the metal: Metal is consumed, and the metal's properties are changed. Although corrosion by gases is similar to corrosion by liquids, two important differences are found: First, in gases the products of corrosion usually remain on the surface, forming in some systems a protective film, and second, local electrochemical cell reactions are suppressed.

Types and rate. Four types of reactions are observed. (1) The gas is dissolved in the metal as molecules, atoms, or ions, and solution is followed by diffusion into the metal. (2) A solid reaction product is formed on the surface, in which case reaction continues by diffusion of the gas or metal atoms or ions through the surface film. (3) A volatile reaction product is formed. (4) A liquid reaction product is formed. Most corrosion reactions involve two or more of these types. Rapid corrosion occurs if liquid or volatile corrosion products are formed or if the solid compound is nonadherent (spalling). For these conditions, new metal is continually being exposed to the gaseous environment.

Thermochemical data can be used to calculate (1) the oxygen or chemical potential of the gas mixture, (2) the thermochemical stability of the condensed solid reaction products, and (3) the equilibrium pressures of the volatile species over the condensed phases. At 25°C and at 1 atm O_2 pressure (101,325 N/m²) the following generalizations can be made: All metals react with oxygen except gold. Most metals react with the halogen gases or with sulfur vapor. Many metals react with nitrogen or hydrogen. The tendency of metals to react with water vapor. carbon dioxide, sulfur dioxide, and other gaseous compounds is much less than with the elementary gases mentioned above. For many metals the type of gas corrosion processes and mechanisms can be predicted with the aid of thermochemical analyses.

Large amounts of heat are evolved when some of the metals react with oxygen or the halogens. When 1 mole of zirconium reacts with oxygen at 25°C, 262 kcal/mole (1,096,208 KJ/mole) of heat are evolved. Smaller heats of reaction are associated with the less reactive metals.

The rate of corrosion is not determined by the magnitude of the heat or free energy of the reaction, but by kinetic processes such as diffusion. Diffusion in solids takes place by means of imperfections or defects in the solid. These may be divided into two groups: point or lattice defects, and line and surface defects. The former include vacancies, interstitial atoms, and misplaced atoms. The latter include dislocations, twin planes, twist planes, grain boundaries, and inner and outer surfaces. At low temperatures corrosion occurs largely by diffusion through line and surface defects, while at high temperatures corrosion occurs both through lattice defects and line and surface defects. Impurity atoms and alloying elements of valence differing from components of the oxide can have a major effect on the rate of corrosion through their effects on lattice defect concentration.

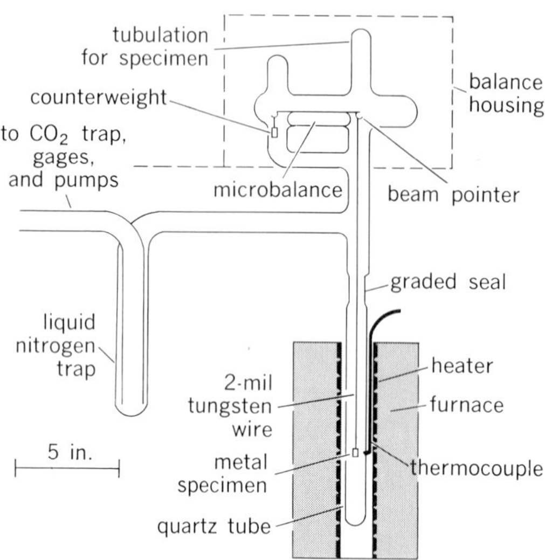

Fig. 4. Microbalance apparatus.

Reaction is often reduced by the formation of a thin oxide film which limits access of oxygen, water vapor, and other gases. In dry atmospheres at 25°C the oxide film on iron is 15–50 A thick and is invisible. Its rate of growth is negligible. In humid atmospheres the oxide film breaks down locally with the formation of hydrated oxides called rust. In dry atmospheres at higher temperature the oxide film grows in thickness, and metal is consumed. Protection of metals depends upon controlling the rate of growth of the oxide film and upon maintaining its adherence to the oxide. To provide good protection at high temperatures, alloys are often used. These alloys are designed to give a low rate of reaction, good adherence of the oxide, low volatility of the metal and oxide, and a high melting point. Chromium, nickel, aluminum, and silicon are some of the alloying elements used to improve the corrosion resistance of iron.

At some temperature, time, or oxide thickness a more rapid reaction may occur for some metals, indicating a change in reaction mechanism. This phenomenon is called breakaway corrosion. This type of corrosion is sometimes related to loss of adhesion of the oxide layer to the metal.

Several conclusions concerning the behavior of metals at elevated temperatures can be given. First, transport of oxygen, metal atoms and ions, lattice vacancies, and electrons through the oxide lattice and at oxide grain boundaries determines the rate of reaction in the initial stage. Second, adhesion of the oxide to the metal and structural changes within the oxide layer determine the oxidation resistance after the initial stage.

Experimental methods. The experimental investigation of the corrosion reactions by gases involves a study of the following problems: (1) rate of corrosion, (2) crystal structure and chemical composition of the corrosion product, (3) crystal habit and metallurgical structure of the corrosion product, reaction product–metal interface, and metal, and (4) mechanism of reaction.

The kinetics of corrosion reactions deals with the effects of time, temperature, gas composition, pressure, and surface conditions on the rate of corrosion. Kinetic studies are important for comparison of metals and alloys, for prediction of the use of metals and alloys, and for determination of the mechanisms of corrosion. Several methods used include (1) change in weight of metal samples, (2) change in volume or pressure of the reacting gas, and (3) change in thickness of the reaction product. Most kinetic data have been obtained by using weight-change methods. Because the weight changes may be small, sensitive microbalances are necessary.

Weight changes. Figure 4 shows a diagram of a vacuum microbalance reaction system. The balance is placed inside the reaction system. One lay-

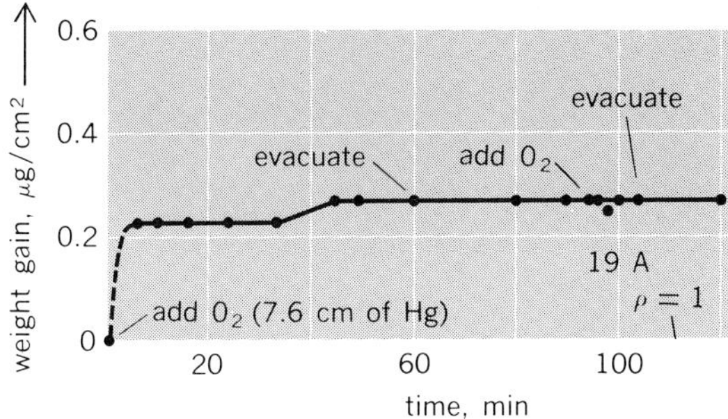

Fig. 5. Course of the reaction of pure iron with oxygen at 25°C and stability to high vacuum.

er of oxygen atoms on a metal surface of 15 cm² can be measured. A thin sheet of metal to be tested is suspended on one end of the balance and a counterweight of the same material from the other. Movements of the balance due to chemical reaction are followed by a micrometer microscope. Figure 5 shows the course of the reaction of oxygen with an oxide-free iron sample at 25°C. An oxide film of 19 A is formed. After the initial reaction, the rate of reaction is so slow that the oxide film can be said to protect the metal. This experiment shows the importance of a thin oxide film in the protection of metals. Figure 6 shows the time course of the reaction of nickel with oxygen at temperatures of 400–750°C. A rapid initial reaction occurs. As the oxide film grows, the rate of reaction decreases. The oxide film limits the rate of reaction of nickel with oxygen. From such data mathematical equations can be derived to express the influence of time, temperature, and pressure on the rate of reaction.

Crystal structure. The crystal structure of the oxide or corrosion film can be determined by x-ray or electron diffraction analyses. For the study of surfaces, electron diffraction has been particularly useful. Low-energy electrons with an energy of the order of 100 volts are scattered largely by the first layer of atoms of a crystal. Their diffraction patterns have been used to study two-dimensional structures on solid surfaces, such as the adsorption of oxygen or nickel. High-voltage electrons having energies of 50,000–5,000,000 volts are useful for the study of the crystal structure of thicker surface films. In the reflection method, in which the beam is directed at grazing angle, the penetration of electrons is limited to a few layers of surface atoms.

Table 2 shows the oxides formed by oxidation of iron, cobalt, nickel, chromium, and copper at

Table 2. Oxide films formed on metals at various temperatures

Metal	200°C	300°C	400°C	500°C	600°C	700°C
Iron	α-Fe_2O_3	α-Fe_2O_3	Fe_3O_4 + α-Fe_2O_3	Fe_3O_4	FeO	FeO
Cobalt		Co_3O_4 + CoO	Co_3O_4 + CoO	CoO	CoO	CoO
Nickel		NiO	NiO	NiO	NiO	NiO
Chromium		Cr_2O_3	Cr_2O_3	Cr_2O_3	Cr_2O_3	Cr_2O_3
Copper		Cu_2O	Cu_2O	Cu_2O	Cu_2O	

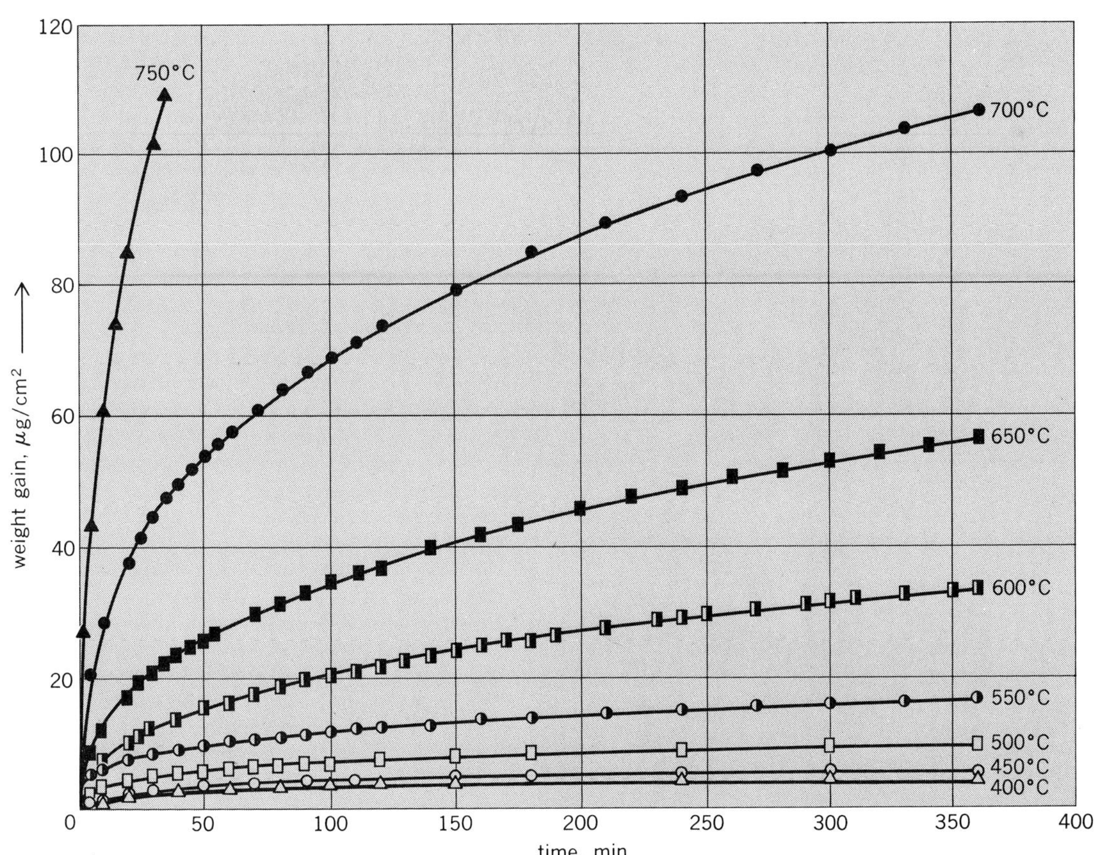

Fig. 6. Effect of temperature on oxidation of nickel at 400–750°C (oxygen pressure is 7.6 cm Hg).

Fig. 7. Crystal habit of oxide formed in the oxidation of Armco iron at 500°C in tank oxygen for 220 hr; no applied stress.

temperatures between 200 and 700°C. Four oxides are formed on iron (three are shown in the table). Transformation in the oxides on iron occurs when the metal is heated or cooled. The poor protective properties of oxide films on iron can be related in part to these transformation reactions. Chromium and nickel form only one oxide over the temperature range. The oxide films on these metals have much superior protective properties.

The chemical composition of the corrosion product film on metals can be inferred from crystal structure studies. For alloys it is necessary to determine the chemical composition by other methods such as microchemical analyses, x-ray fluorescence analysis, electron-beam microprobe analysis, and direct image mass analysis. The latter two methods are especially useful for identification of small quantities of second-phase particles in the oxide film.

Crystal habit. Crystal habit refers to the size and shape of the crystals in the corrosion film and the relationships of these crystals to the metal. From these studies information can be obtained on the mechanism of reaction and the role of dislocations, precipitates, and grain boundaries in the metal on the corrosion reaction. In many systems the localized corrosion product, rather than the uniform corrosion product, is of primary interest. This localized attack occurs in gaseous corrosion as well as in liquid-phase corrosion.

The crystal habit can be studied by the light and electron microscopes in several ways. The reflection electron microscope can be used to study the crystal habit of the surface directly. To study thick corrosion layers by conventional transmission electron microscopes the metal can be encased in plastic. A cross section of the metal with corrosion layer is made and studied. For thin films the corrosion layer can be covered by a thin plastic film and removed from the metal by chemical or electrochemical methods. The film is then studied by the electron microscope. Results show that the thin oxide film is made up of many small crystals on each metal grain. For thick films the presence of pores, cracks, and second-phase particles can be observed. *See* METALLOGRAPHY.

Crystal orientation. Crystal structure and crystal habit studies of the corrosion product have shown that the metal orientation has a major influence on the extent of corrosion. These effects are observed in chemisorption of oxygen on metals, in the initial oxide nucleation and growth of oxide films, and in thick oxide films The degree of oxide orientation is a function of the reaction conditions and the oxide thickness. For copper oxidized at 350°C, the effect of substrate orientation decreases rapidly for oxide thicknesses above 500 A. Oxidation at low oxygen pressures favors the growth of oriented oxide.

Direct observation of the edges of oxidized specimens by the electron microscope has shown new and unsuspected crystal habits in the corrosion film. Figure 7 shows an electron micrograph of the edge of an iron specimen heated in oxygen for 220 hr. Oxide whiskers up to 300,000 A long and 200 A thick grow from the surface. Each whisker contains a screw dislocation. These studies show the importance of dislocations in the mechanisms of oxidation and that chemical reaction occurs at highly localized areas on the surface.

Environment and stress play a major role in determining the crystal habit of the oxide on iron. Long, narrow, blade-shaped oxide crystals form when water vapor is added to the oxidizing atmosphere. Here rapid corrosion occurs at twin interfaces in the blade-shaped oxide platelet. Rounded oxide platelets are formed when stress is applied during oxidation, as shown in Fig. 8. Here rapid reaction occurs through twist planes.

Studies of the crystal habit of oxide films have shown the influence of stress, environment, and substructure of the metal or alloy on the corrosion

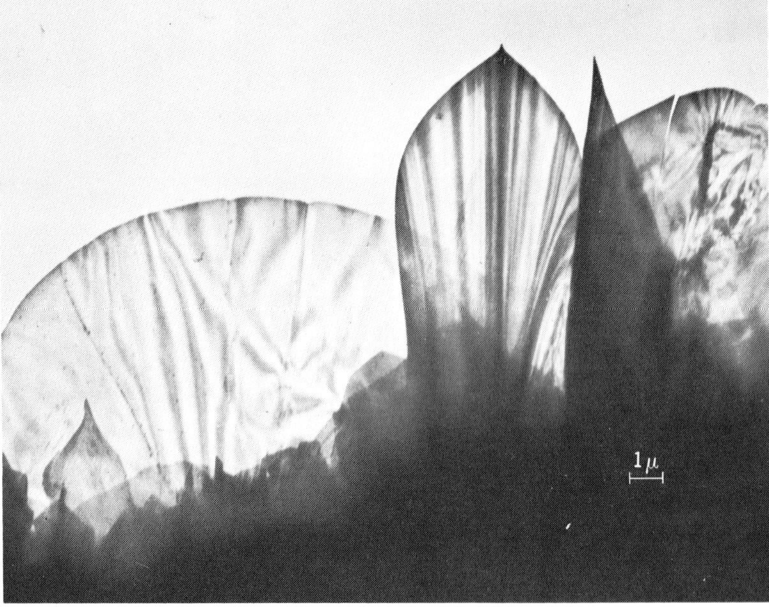

Fig. 8. Crystal habit of oxide formed in oxidation of 18-8 stainless steel at 600°C in wet oxygen plus trace of HCl vapor for 138 hr; prestressed to 59,000 psi.

process and the major role of imperfections in the growth mechanisms.

Sulfur vapor and sulfur-containing gases. The high-temperature sulfurization reactions are similar in many respects to oxidation. In general, higher corrosion rates are found. These higher rates are related to higher defect concentration in the sulfide lattices and to the different thermochemical stabilities of the sulfide and oxide scales. The morphologic structure of the sulfide and oxide scales formed on pure metals is a function of the sulfur and oxygen potential of the gas atmosphere, temperature, and geometrical parameters of the reaction system. The surface scale may consist of several layers, the growth of which may proceed by the outward diffusion of metal or by diffusion of both metal and reactant gas.

Hot corrosion refers to the combined sulfidation-oxidation attack found in the combustion gases of a jet engine or gas turbine operating near the sea. Sodium sulfate is formed by the reaction of sodium chloride in sea air with oxidized sulfur in jet fuel. This film of sodium sulfate deposited at high temperature on the alloy reacts with the oxide scale and speeds up the corrosion reaction.

[EARL A. GULBRANSEN]

Bibliography: American Society for Testing and Materials, *Atmospheric Factors Affecting the Corrosion of Engineering Metals*, 1978; ASTM, *Corrosion Fatigue Technology*, 1978; ASTM, *Intergranular Corrosion of Stainless Alloys*, 1978; G. R. Brubaker and P. B. Phipps (eds.), *Corrosion Chemistry*, 1979; M. G. Fontana and N. D. Greene, *Corrosion Engineering*, 2d ed., 1978; M. G. Fontana and R. W. Staehle (eds.), *Advances in Corrosion Science and Technology*, vols. 1–7, 1970–1979; J. C. Scully, *The Fundamentals of Corrosion*, 2d ed., 1975.

Cotter pin

A cotter pin is a split pin (as illustrated) formed by doubling a piece of wire semicircular in cross section to form a loop at one end. After insertion in a hole or a slot in a nut and through a mating crosswise hole in a bolt, the ends of the pin are separated and bent to hold it in place. This fastener is not adaptable to quick assembly but is widely used in locations where service inspection is difficult and where failure would be disastrous. Cotter pins are similarly used in many other applications to prevent relative rotation or sliding. *See* NUT.

[PAUL H. BLACK]

COTTER PIN

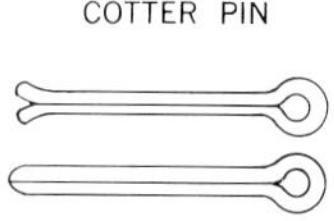

Two of the common forms of cotter pin.

Couple

A system of two parallel forces of equal magnitude and opposite sense (Fig. 1). Forces P at normal distance p constitute a counterclockwise couple C_z (viewed from $+Z$) in the OXY plane; forces Q at arm q, constitute a clockwise couple C_x (from $+X$) in OYZ. Under a couple's action a rigid body tends only to rotate about a line normal to the couple's plane. This tendency reflects the vector properties of a couple.

Vector properties. The total force of a couple is zero. The total moment $\mathbf{C}$ of a couple is identical about any point. Accordingly, $\mathbf{C}$ is the moment of either force about a point on the other and is perpendicular to the couple's plane. *See* STATICS.

In Fig. 1, $\mathbf{C}_z$, at the origin for convenience, is the moment of couple C_z. Its magnitude $|C_z| = +P_p$. Its sense, by the convention of moment, is $+Z$. Also $\mathbf{C}_x$, of sense $-X$, represents couple C_x; $|C_x| = +Q_q$.

Scalar moment. The moment of a couple about a directed line is the component of its total moment in the line's direction. For example, the moment of couple C_z about line L is $M_L = |C_z| \cos \theta_L = +P_p \cos \theta_L$. Also $M_x = 0$, $M_y = 0$, and $M_z = +P_p \cos 0° = +P_p$.

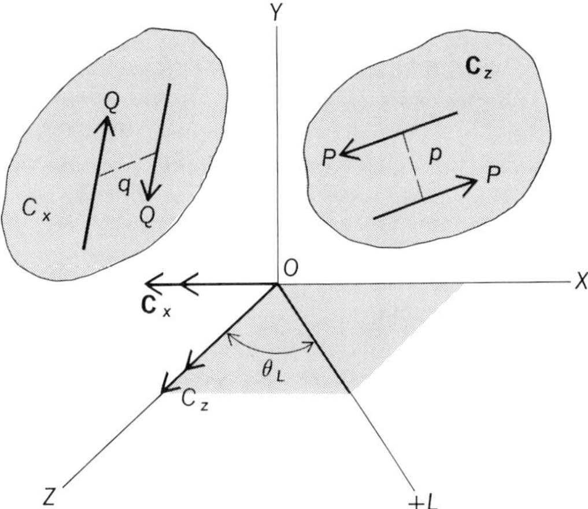

Fig. 1. Vector properties of a couple.

Equivalent couples. Couples are equivalent whose total moments are equal. In Fig. 2, the paired linear forces and the counterclockwise curl represent counterclockwise couples C_1, C_2, and C_3 in parallel planes OXY, AXY, and BXY. When their total moments are directed the same ($+Z$), and their magnitudes $|C_1| = P_1p_1$, $|C_2| = P_2p_2$, and $|C_3|$ are equal, then $\mathbf{C}_1 = \mathbf{C}_2 = \mathbf{C}_3$ and these couples are equivalent. Thus $\mathbf{C}_1$ can represent these or any number of other equivalent couples.

[NELSON S. FISK]

COUPLE

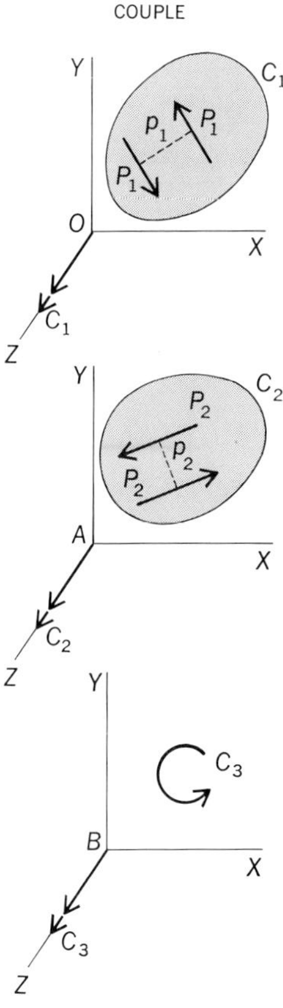

Fig. 2. Equivalent couples.

Coupling

The mechanical fastening that connects shafts together for power transmission, often to form long sections of shafting or to connect the shaft of a machine to an external drive. Couplings are classified with respect to the kind of alignment and the centerline position of the connecting shafts. Rigid couplings are used where the axes of the shafts are directly in line, flexible couplings where the axes may be at a slight angle and slightly displaced, and universal joint couplings for large angularity or large displacement.

Rigid coupling. The mechanical connection can be rigid for shafts in close alignment or held in alignment without inducing destructive forces in shafts or bearings.

For commercial shafting, the coupling may be a sleeve with the shafts pressed into each end, or it may be a clamping sleeve. The sleeve on each shaft end may have an external flange with bolt holes. For large power machines, shafts are usually forged from billets with flanged couplings shaped integrally on their ends. These couplings are bolted together to hold the shafts rigidly; hence the shafts must be accurately aligned before assembly.

Flange coupling consists of a hub, keyed or

shrunk to a shaft. A larger-diameter flange on one end of the hub can be bolted to a mating flange and hub to form a continuous shaft. The bolts in the flange are usually countersunk, or the flange has a shroud for safety. The face of the flange must be checked for trueness on the shaft after assembly; a counterbore fit for a mating projection is necessary to assure alignment of the centers of the two connected shafts. The coupling bolts are often fitted into reamed holes in the flanges to distribute the stress when large torques are to be carried.

Clamping couplings have split cylindrical elements which clamp the shaft ends together by direct compression, through bolts or rings, and by the wedge action of conical sections. These couplings are not considered a permanent part of the shaft because they can be removed and reassembled with hand tools. They are generally used on line shafting rather than on the shafting of machines.

Flexible coupling. The mechanical connection for shafts at a slight angle and displaced somewhat, and particularly where power machines may set up vibratory forces, must be flexible. Such couplings are used when alignment is difficult to obtain, when a nonrigid support makes alignment impossible to maintain, or if shaft position changes because of operation.

Flexible couplings consist of symmetrical hubs on each end of the joined shafts with a flexible connecting member between. This member may take such forms as the floating gear, the embedded spring, the flexible pin, and the flexible disk.

Flexible couplings not only provide for lateral and angular misalignment, but also reduce the transmission of shock loads and change the vibration characteristics and critical speeds of the shafts.

Oldham's coupling is the original flexible coupling, and presents the basic principle from which most flexible couplings are derived. It consists of two hubs on shaft ends having single radial slots into which fits a floating plate having tongues at right angles on each side. This coupling allows kinematic movement to compensate for the lateral displacement of the shafts and also accommodates slight angular misalignment. Because the tongues move about in the slots, the coupling must be well lubricated and can be used only for low speeds. *See* OLDHAM'S COUPLING.

The geared Fast coupling, which is modified from Oldham's, uses two interior hubs on the shafts with circumferential gear teeth surrounded by a casing having internal gear teeth to mesh and connect the two hubs. Considerable misalignment can be tolerated because the teeth inherently have little interference. This completely enclosed coupling provides means for better lubrication and is thus applicable to higher speeds. A similar coupling originated by the Clark Manufacturing Co. and known as the Clark type consists of a roller chain wrapped around the teeth of two sprockets, but it is less versatile.

Spring couplings are flexible couplings with resilient parts. The Falk flexible coupling was developed by slotting axially the periphery of the two hubs on the shaft ends, and threading a continuous steel spring back and forth in hairpin style. The Westinghouse-Nuttall flexible coupling has a series of coil springs nested tangentially around the flanges of the coupling to transmit torque and allow misalignment in angularity and displacement.

Flexible pin couplings use pins as the support for connecting materials between the coupling faces. The Ajax flexible coupling has rigid pins from one flange fitting into holes lined with brass bushings and rubber sleeves in the other. The Francke coupling uses laminated steel pins between the flanges, the pins being relatively flexible with one end fixed and the other end free. Another coupling uses a leather belt which threads in and out through pins in the flanges to transmit the torque and adjust for misalignment. A flexible disk coupling is used between alternate pins on the two flanges to transmit low power; it is commonly employed on automotive drive shafts. The disk may be made of leather or rubber-impregnated canvas. Most pin couplings have pins set alternately in the two flanges for stability and symmetry. They are available in many sizes.

Universal joint. A flexible coupling for connecting shafts that have appreciable permanent angularity is termed a universal joint. The initial universal joint, credited to Robert Hooke, was a swivel arrangement by which two pins at right angles allowed complete angular freedom between two connected shafts. However, modifications are necessary to transmit constant angular velocity. *See* CLUTCH; FLUID COUPLING; SHAFTING; UNIVERSAL JOINT. [JAMES J. RYAN]

Crane hoist

A mobile construction machine built principally for lifting loads by means of cables. A crane hoist consists of three principal elements: (1) an undercarriage on which the unit moves, (2) a cab or house which envelops the main frame and which contains the power units and controls, and (3) a movable boom over which the cables run and to which miscellaneous working attachments are connected. A crane is usually typed according to its undercarriage. *See* EXCAVATOR.

The crawler crane is a self-propelled unit mounted on continuous tracks similar to those on army tanks. This crane is ideally suited for work on soft ground where wide distribution of weight is more important than rapid mobility.

The truck crane is a unit consisting of a crane house and boom mounted on a truck chassis, as illustrated. Originally assembled by contractors from crawler crane and truck parts, the truck crane has for years been manufactured and sold as

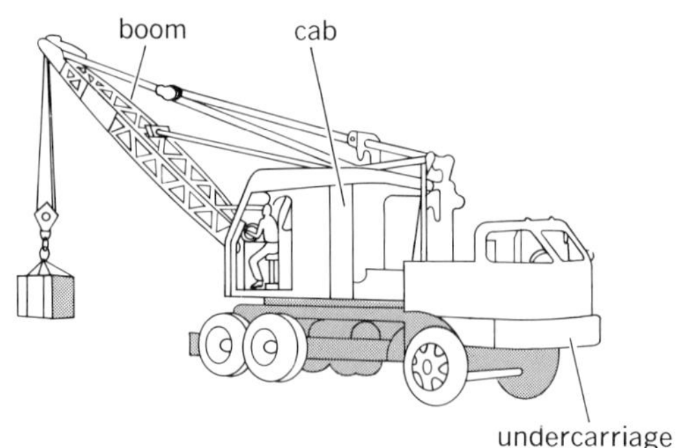

Typical truck crane. (*Link-Belt Co.*)

a unit. Although the truck crane is difficult to move on soft or slippery ground, it is highly mobile on a firm footing and is easily moved over roads and highways.

The wheeled crane is self-propelled and rides on a rubber-tired chassis that is an integral part of the unit. Power for transport is provided by the same engine that is used for hoisting, and the wheels are controlled from the crane cab.

The rail or locomotive crane is mounted on a railroad flatcar or a special chassis with flanged wheels for use on standard- or special-gage tracks. Rail cranes may be of either the towed type or self-propelled. For transport over long distances, both types are pulled by locomotives, often as part of a freight train.

The floating crane has a barge or scow for an undercarriage. It is used for water work and for work on waterfronts and generally lacks its own motive power, being towed from place to place by tugboats. On a job site, it is moved short distances by means of cables attached to anchors set some distance off the four corners of the barge. Most floating cranes are powered by steam. For the land cranes, however, in which size and weight are prime considerations in over-the-road transport, power is usually supplied by internal combustion engines operating on either gasoline or diesel fuel. Cranes used exclusively for pile driving are, however, frequently driven by steam or air. *See* HOISTING MACHINES.

[E. M. YOUNG]

Crank

In a mechanical linkage or mechanism, a link that can turn about a center of rotation. The crank's center of rotation is in the pivot, usually the axis of a crankshaft, that connects the crank to an adjacent link. A crank is arranged for complete rotation (360°) about its center; however, it may only oscillate or have intermittent motion. A bell crank is frequently used to change direction of motion in a linkage (see illustration).

In mechanisms where energy input fluctuates, it is sometimes useful to design a crank stoutly, as a form of flywheel, so that considerable rotational energy is stored in it, tending thereby to make the output smoother inasmuch as the inertia of such a crank absorbs peak input and releases energy more smoothly to output.

Input and output cranks of a four-bar linkage may express in their relative angular placement a relation $\beta = \beta(\alpha)$, which is excellent in control operations where electric current computers would be vulnerable. *See* FOUR-BAR LINKAGE.

[DOUGLAS P. ADAMS]

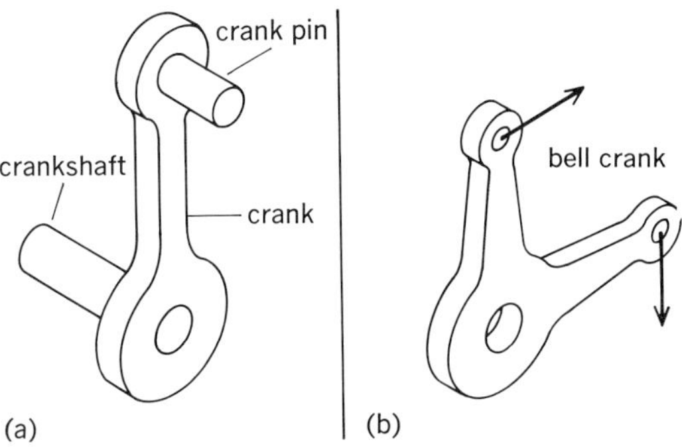

Cranks (*a*) for changing radius of rotation, and (*b*) for changing direction of translation.

Critical mass

That amount of fissile material (U^{233}, U^{235}, or Pu^{239}) which permits a self-sustaining chain reaction. The critical mass ranges from about 950 g of U^{235} or a smaller amount of Pu^{239} for dissolved compounds through 16 kg for a solid metallic sphere of U^{235}, and up to hundreds of tons for some power reactors. It is increased by the presence of such neutron absorptive materials as admixed U^{238}, aluminum pipes for flow of coolant, and boron or cadmium control rods. It is reduced by a moderator, such as graphite or heavy water, which slows down the neutrons, inhibits their escape, and indirectly increases their chance to produce fission. *See* REACTOR PHYSICS.

[JOHN A. WHEELER]

Critical path method (CPM)

A diagrammatic network-based technique, similar to the program evaluation and review technique (PERT), that is used as an aid in the systematic management of complex projects. The technique is useful in: organizing and planning; analyzing and comprehending; problem detecting and defining; alternative action simulating; improving (replanning); time and cost estimating; budgeting and scheduling; and coordinating and controlling. It has its greatest value in complex projects which involve many interrelated events, activities, and resources (time, money, equipment, and personnel) which can be allocated or assigned in a variety of ways to achieve a desired objective. It can be used to complete a multifaceted program faster or with better utilization of resources by reassignments, trade-offs, and judiciously using more or less assets for certain of the individual activities composing the overall project. *See* PERT.

CPM was introduced in 1957–58 by E. I. du Pont de Nemours & Company. Present applications include: research and development programs; new product introductions; facilities planning and designs; plant layouts and relocations; construction projects; equipment installations and start-ups; major maintenance programs; medical and scientific researches; weapons systems developments; and other programs in which cost reduction, progress control, and time management are important. It is best suited to large, complex, one-time or first-time projects rather than to repetitive, routine jobs.

A simplified example will show the features of the critical path method and the steps involved in constructing and using the network-based system:

Project: Design personal home-use computer.

Objective: Complete project as soon as possible.

Procedure: The CPM procedure for this project involves eight steps:

Step 1: List required events; arrange as to best guess of their sequence; assign a code letter or number to each. (See center column of Table 1.) [Events are accomplished portions or phases of the project. Activities are the resources applied to progress from one event to the next. This example

Table 1. CPM events table

Prior event	Event	Following event
–	A = Authorization to start received	B and C
A	B = Computing circuits designed	D and E
A	C = Video circuits designed	D
B and C	D = Keyboard and cabinets designed	G
B	E = Programming completed	F and H
E	F = Operating systems completed	G
D and F	G = Testing and debugging finished	H
E and G	H = Design specifications and programs finalized	–

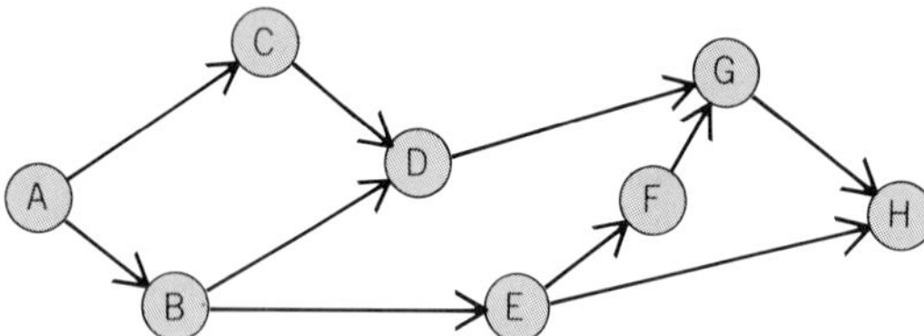

Fig. 1. Network of relationship between events (circles) and their activities (arrows).

has been greatly simplified for illustrative purposes. A real case could have several hundred events and activities and would be a candidate for computer analysis, for which programs are available.]

Step 2: Determine which event(s) must precede each event before it can be done and which event(s) must await completion of that event before it can be done. Construct a table to show these relationships. (See left and right columns of Table 1.)

Step 3: Draw the network, which is a diagram of the relationship of the events and their required activities that satisfies the dictates of Table 1 (see Fig. 1).

Step 4: List the required activities and enter the best estimate available of the time required to complete each activity. In PERT, three time estimates are entered: the most optimistic, the most likely, and the most pessimistic. This is the basic way in which PERT differs from CPM. Some users of CPM enter a second time estimate for each activity, the first being the normal time and the second being the crash (or fast at any price) time. Table 2 shows the list of activities (AB designates the activity required in going from event A to event B), the expected normal time, the maximum number of weeks that each activity can be shortened under a crash program, and the cost per week to shorten, if it can be shortened.

Step 5: Write the normal expected time to complete each activity on the diagram (Fig. 2). Note that the relative lengths of the arrows (activities) have no relationship to their time magnitudes.

Step 6: Find the critical path, that is, the *longest* route from event A to event H. The longest path is the total time that the project can be expected to take unless additional resources are added or resources from some of the other paths in which there may be slack can be reallocated to activities which lie in the critical path. In the example shown in Fig. 2, the initial critical path is A – B – D – G – H, or 5 + 12 + 18 + 5 = 40 weeks. If the critical path

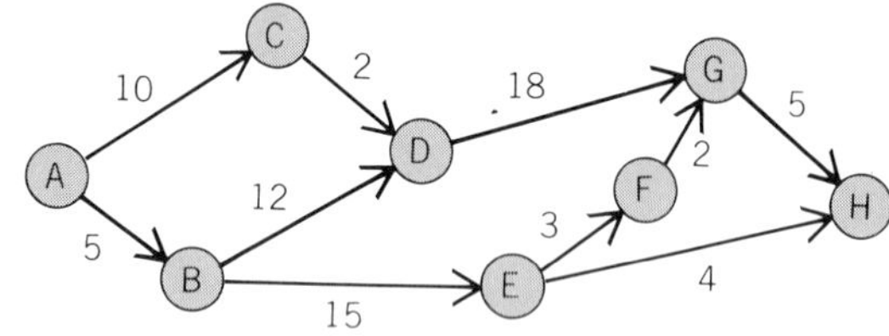

Fig. 2. Network diagram modified to show time to complete activities. The critical path is indicated by the black arrows.

Table 2. Time and cost factors for activities

Activity	Expected time to complete (weeks)	Maximum possible time reduction (weeks)	Cost (per week) to reduce time
AB	5	1	$1000
AC	10	3	500
BD	12	3	1000
BE	15	2	500
CD	2	0	–
DG	18	4	500
EF	3	0	–
EH	4	1	2000
FG	2	0	–
GH	5	2	500

Table 3. Costs of shortening activities

	Reduce		Cost	
AB from	5 weeks to	4 weeks	1 week	× $1000 = $1000
BD	12	9	3	× 1000 = 3000
DG	18	14	4	× 500 = 2000
GH	5	3	2	× 500 = 1000
TOTAL	40	30	10	$7000

time works out to be less or more than the target time to complete the project, there is then a positive or negative float.

Step 7: Shorten the critical path. The stated objective for this example was to complete the project as soon as possible. Other projects might have different objectives. An examination of Table 2 and Fig. 2 suggests that the activities lying along the critical can be shortened for the costs shown in Table 3. The total expected time to complete the project can be shortened from 40 weeks to 30 weeks—a 25% improvement. A judgment can be made by those responsible for the project as to whether the 10-week time savings is worth the additional cost of $7000. In some cases, the critical path is so shortened that it is no longer the longest path, and another path becomes the critical one. Only reductions in the critical path time reduce the total project time, and activities in noncritical paths may run over or late to the limits of their floats without causing the project to be delayed. The floats are the differences between the total expected time required for each possible path in the network and the total time for the critical path, and are measures of spare time available.

Step 8: Construct a schedule of the project showing the earliest and latest permissible start and completion dates of each activity. To do this, the individual floats are calculated. With the slack in the critical path set at zero at the onset of the project's implementation, any slippage in the completion of any event along that path will result in a delay in the completion of the whole project, unless made up later.

The use of CPM adds another step to a project, and it requires continual updating and reanalysis as conditions change, but experience has shown that the effort can be a good investment in completing a project in less time, with less resources, with more control, and with a greater chance of on-time, within-budget completion. *See* INDUSTRIAL ENGINEERING.

[VINCENT M. ALTAMURO]

Crushing and pulverizing

The reduction of materials such as stone, coal, or slag to a suitable size for their intended uses such as road building, concrete aggregate, or furnace firing. Crushing and pulverizing are processes in ore dressing needed to reduce valuable ores to the fine size at which the valueless gangue can be separated from the ore. These processes are also used to reduce cement rock to the fine powder required for burning, to reduce cement clinker to the very fine size of portland cement, to reduce coal to the size suitable for burning in pulverized form, and to prepare bulk materials for handling in many processes. *See* MATERIALS-HANDLING MACHINES.

Equipment suitable for crushing large lumps as they come from the quarry or mine cannot be used to pulverize to fine powder, so the operation is carried on in three or more stages called primary crushing, secondary crushing, and pulverizing (see table). The three stages are characterized by the size of the feed material, the size of the output product, and the resulting reduction ratio of the material. The crushing-stage output may be screened for greater uniformity of product size.

Reduction in size is accomplished by five principal methods: (1) crushing, a slow application of a large force; (2) impact, a rapid hard blow as by a hammer; (3) attrition, a rubbing or abrasion;

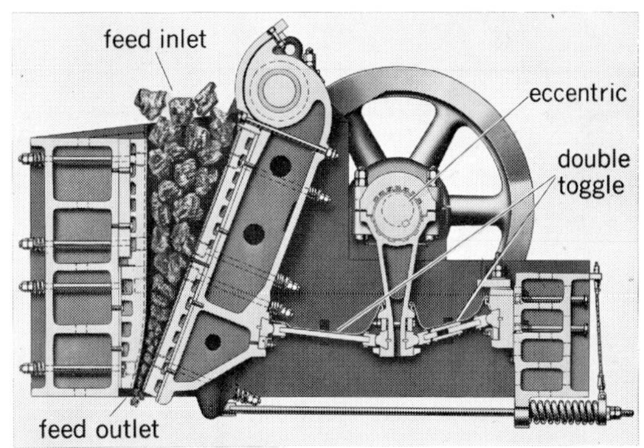

Fig. 1. Blake-type jaw crusher. (*Allis Chalmers Co.*)

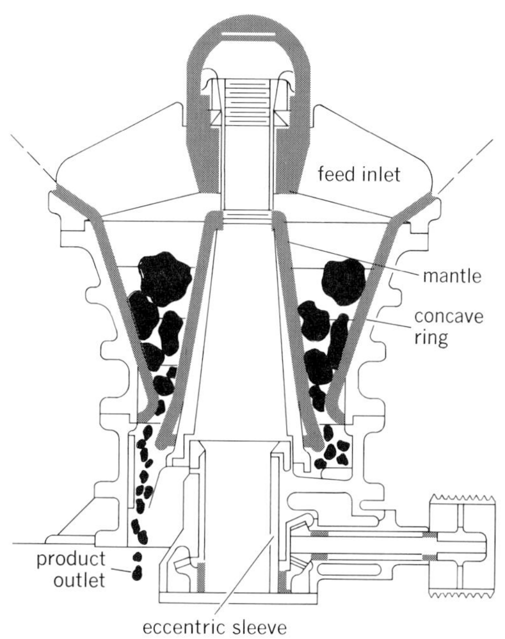

Fig. 2. Gyratory crusher, Gates type. (*Allis Chalmers Co.*)

(4) sudden release of internal pressure; and (5) ultrasonic forces. The last two methods are not in common use.

Crushers. All the crushers used in primary and secondary crushing operate by crushing as defined above except the hammer mill, which is largely impact with some attrition.

Primary crushers. The Blake jaw crusher using a double toggle to move the swinging jaw is built in a variety of sizes from laboratory units to large sizes

Crushing specifications

Category	Feed, in. (cm)	Product	Reduction ratio	Equipment used
Primary crushing	27–12 (69–30)	9–4 in. (23–10 cm)	3:1	Jaw, gyratory, cone
Secondary crushing (one or two stages)	9–4 (23–10)	1–½ in. (25–13 mm)	9:1	Hammer mill, jaw, gyratory, cone, smooth roll, and toothed rolls
Pulverizing	1–½	60–325 mesh	60:1	Ball and tube, rod, hammer, attrition, ball race, and roller mills

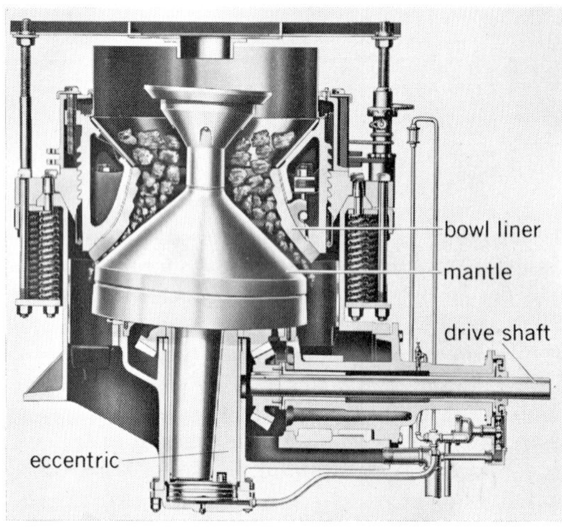

Fig. 3. Symons cone crusher. (*Nordberg Co.*)

Fig. 4. Types of roll crusher. (*a*) Single-roll. (*b*) Double-roll. (*Link Belt Co.*)

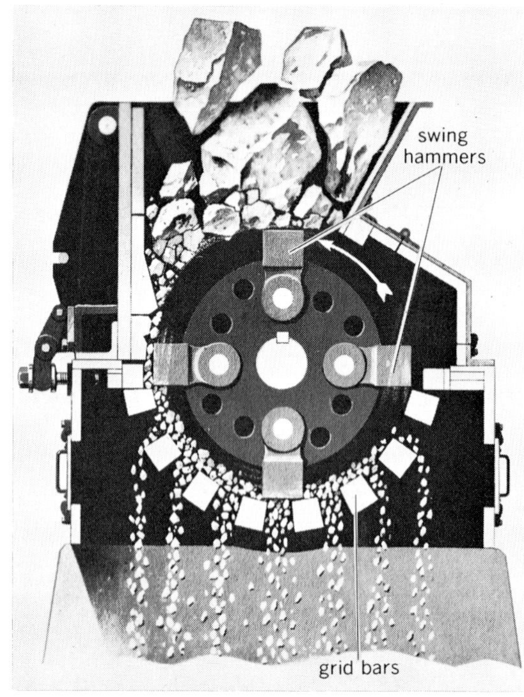

Fig. 5. Hammer crusher, a secondary crusher for ore, rock, and coal. (*Jeffrey Co.*)

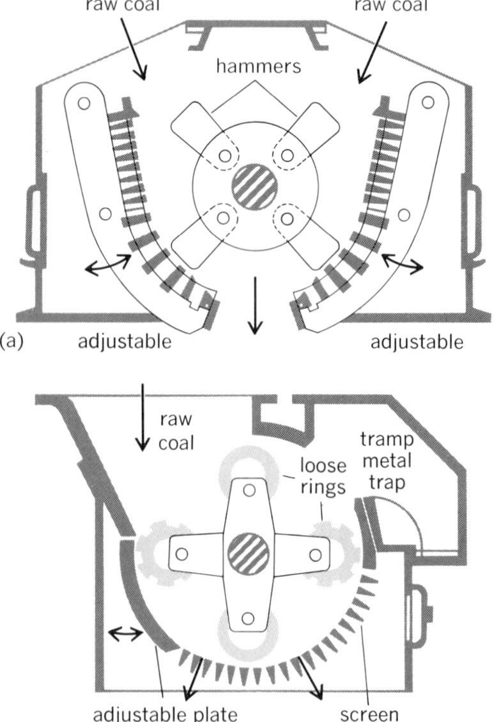

Fig. 6. Two examples of coal crusher. (*a*) Hammer mill type. (*b*) Ring type. (*Penna Crusher*)

having a feed inlet 84 by 120 in. or 213 by 305 cm (Fig. 1). Large units have a capacity of over 1000 tons/hr (900 metric tons/hr) and produce a 9-in. (23-cm) product. The Dodge jaw crusher uses a single toggle or eccentric and is generally built in smaller sizes.

The Gates gyratory crusher has a cone or mantle that does not rotate but is moved eccentrically by the lower bearing sleeve (Fig. 2). A 42- by 134-in. (107- by 340-cm) crusher has a capacity of 850 tons/hr (770 metric tons/hr), crushing rock from 27 to 8 in. (69 to 20 cm).

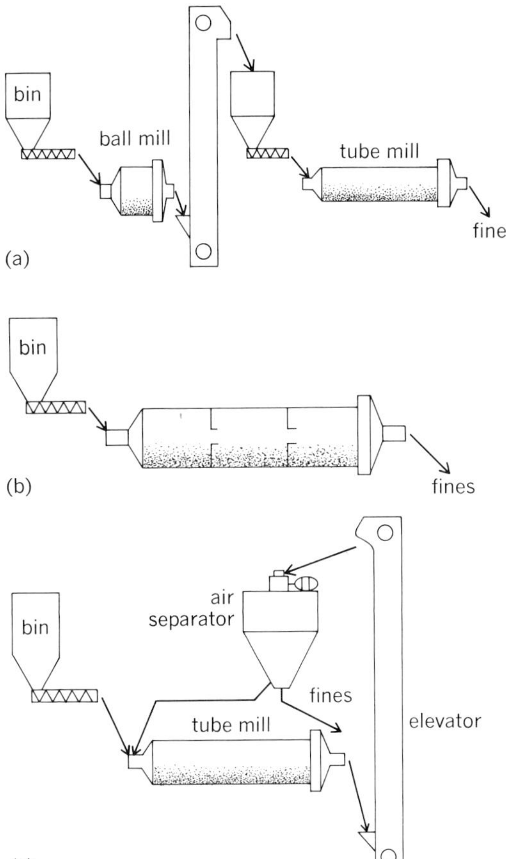

Fig. 7. Pulverizers. (*a*) Open-circuit type with preliminary ball mill stage followed by tube mill. (*b*) Open-circuit three-compartment tube mill. (*c*) Closed-circuit tube mill with an air separator.

The Symons cone crusher also has a gyratory motion, but has a much flatter mantle or cone than does the gyratory crusher (Fig. 3). The top bowl is spring-mounted. It is used as a primary or secondary crusher.

Angle of nip is the largest angle that will just grip a lump between the gyratory mantle and ring, between the smooth jaws of a jaw crusher, or between the pair of smooth rolls. Depending on the material, it is 8–48°, but 32° is commonly used.

Secondary crushers. The single-roll crusher and the double-roll crusher have teeth on the roll surface and are used mainly for coal (Fig. 4). Single-roll crushers 36 in. (91 cm) in diameter by 54 in. (137 cm) long have a capacity of 275 tons/hr (250 metric tons/hr) crushing run-of-mine coal to $1\frac{1}{4}$-in. (32 mm) ring size. Smooth rolls without teeth are sometimes used for crushing ores and rocks.

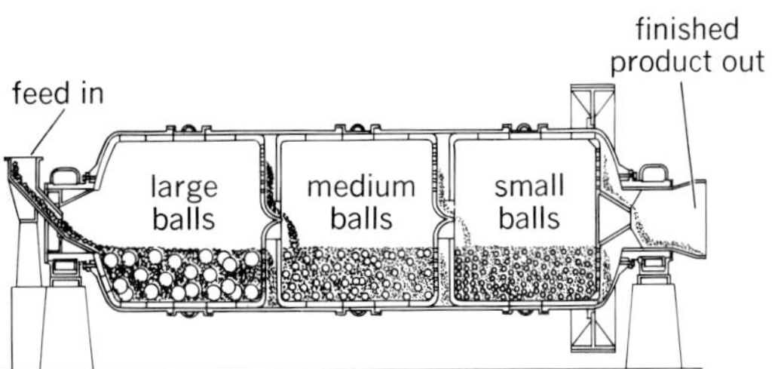

Fig. 8. Three-compartment tube mill pulverizer, containing three sizes of balls. (*Hardinge Co., Inc.*)

The hammer crusher is the type of secondary crusher most generally used for ore, rock, and coal (Fig. 5). The reversible hammer mill can run alternately in either direction, thus wearing both sides of the hammers (Fig. 6*a*). A hammer mill 42 in. (107 cm) in diameter by 82 in. (208 cm) long crushes 300 tons/hr (270 metric tons/hr) of coal to 1/4 in. (6 mm) for coke oven feed. The ring coal crusher is a modification of the hammer mill using rings instead of hammers. (Fig. 6*b*).

Pulverizers. In open-circuit pulverizing, the material passes through the pulverizer once with no removal of fines or recirculation.

In closed-circuit pulverizing, the material discharged from the pulverizer is passed through an external classifier where the finished product is removed and the oversize is returned to the pulverizer for further grinding.

Ball and tube mills, rod mills, hammer mills, and attrition mills are pulverizers operating by impact and attrition. In ball race and roller pulverizers, crushing and attrition are used.

Buhrstone pulverizer. The Buhrstone pulverizer is one of the oldest forms of pulverizers; a few still operate in flour and feed mills. The pulverizing is done between a rotating lower stone and a floating upper stone, which is restrained from rotating. Modern pulverizers use tumbling or rolling action instead of the grinding action of such earlier mills. For example, smooth double rolls are used for pul-

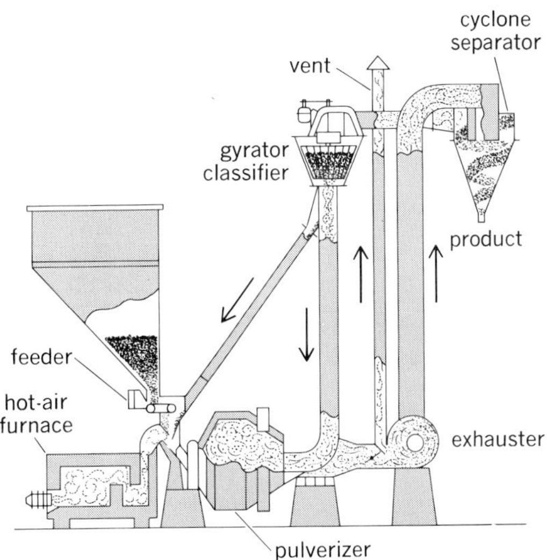

Fig. 9. Conical ball mill pulverizer in closed circuit with classifier. (*Hardinge Co., Inc.*)

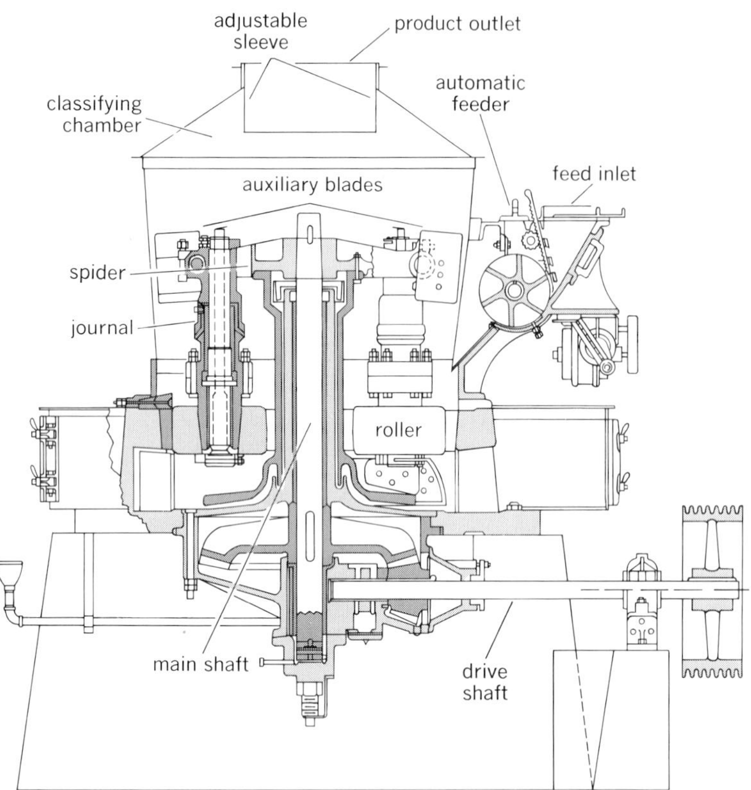

Fig. 10. A pulverizer of the airswept pendulum type. (*Raymond Pulverizer Division, Combustion Engineering Co.*)

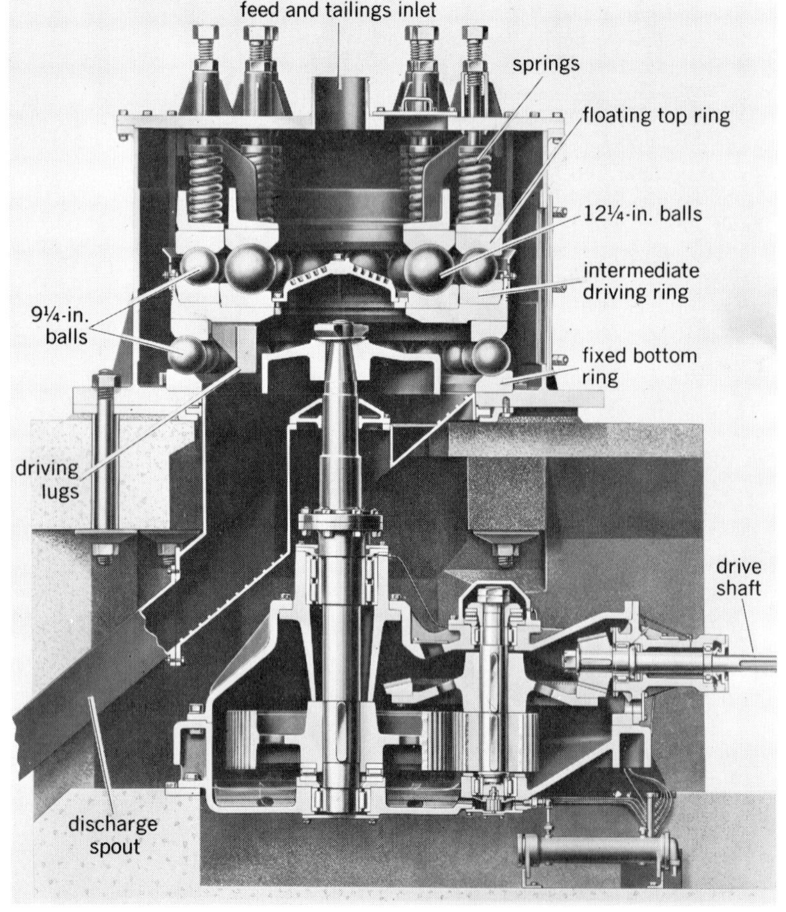

Fig. 11. Ball race rock pulverizer for use in closed circuit. (*Babcock and Wilcox Co.*)

verizing flour and other materials. *See* BUHRSTONE MILL.

Tumbling pulverizers. There are two principal classes of tumbling pulverizer. Both consist of a horizontal rotating cylinder containing a charge of tumbling or cascading steel balls, pebbles, or rods. If the length of the cylinder is less than 2–3 diameters, it is a ball mill. If the length is greater than 3 diameters, it is a tube mill. Both types are extensively used in ore and rock pulverizing where the material is abrasive, as it is easy to add balls during operation. *See* TUMBLING MILL.

Ball mills charged with large steel balls 2–3 in. (5–8 cm) in diameter are often used as preliminary pulverizers in series with a tube mill in open circuit (Fig. 7*a*). The Krupp mill is a preliminary ball mill which discharges the material through a screen which forms the cylindrical surface.

Pebble mills use balls and liners of quartz for application where contamination of the product

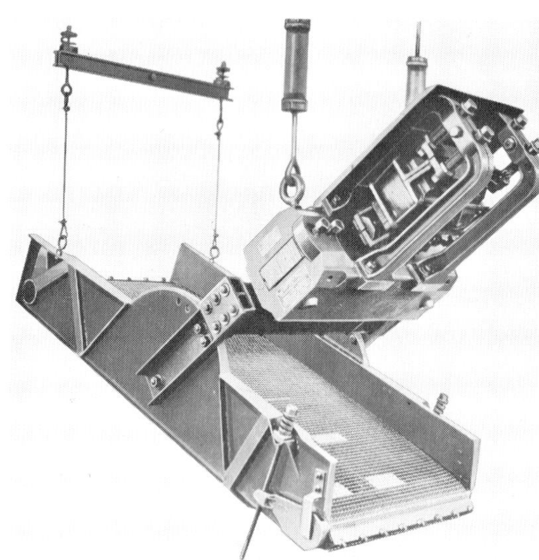

Fig. 12. Vibrating screen classifier. (*Jeffrey Co.*)

with iron would be detrimental. For products coarser than 6 mesh, steel rods are sometimes used instead of balls. *See* PEBBLE MILL.

In some cement plants and many mining operations, ball, tube, rod, and pebble mills operate wet, water being added with the feed and the product being discharged as a slurry. Rake or cone type classifiers are used in closed circuits that employ wet grinding.

Tube mills for fine pulverization are used extensively in open circuit, or as compartment mills, or as closed-circuit mills (Fig. 7).

A three-compartment, grate-discharge tube mill has larger balls in the first, medium-size balls in the second, and smaller balls in the third compartment (Fig. 8).

Cascade pulverizers are another form of tumbling pulverizer; they use the large lumps to do the pulverizing. The Aerfall pulverizer is built in diameters up to 16 ft. (4.9 m). Feed is 12- to 18-in. (30–46-cm) lumps and the unit is airswept with an external classifier. A few steel balls are sometimes used but not over $2\frac{1}{2}\%$ of the volume.

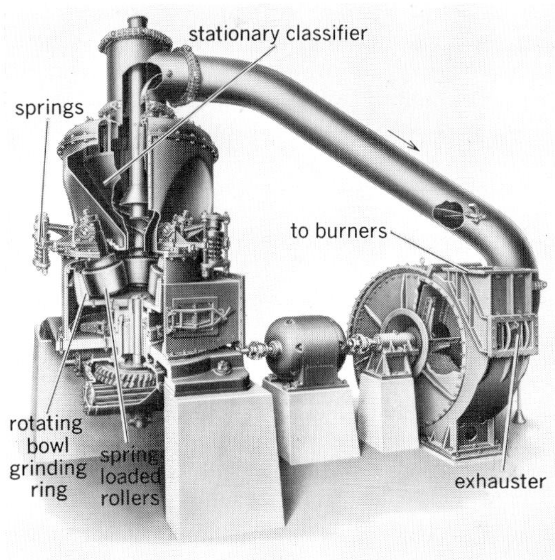

Fig. 13. Medium-speed bowl-mill coal pulverizer, which feeds burners. (*Combustion Engineering Co.*)

The airswept conical ball mill is used in closed circuit with an external air classifier (Fig. 9). The conical shape of the mill classifies the balls, keeping the large balls at the feed end, where they are needed to crush the larger lumps, and the smaller balls at the small discharge end where the material is finer. This conical ball mill is also used extensively in open circuit in rocks and ores.

Roller pulverizers. The airswept pendulum roller pulverizer has rollers pivoted from a rotating spider on a vertical shaft; the centrifugal force of the rollers supplies the pressure between rollers and ring. Hot air enters below the rollers and picks up the fines. The rotating rollers and spider provide a centrifugal classifying action which returns the oversize to the pulverizing zone. For higher fineness a rotating classifier is added in the top housing (Fig. 10).

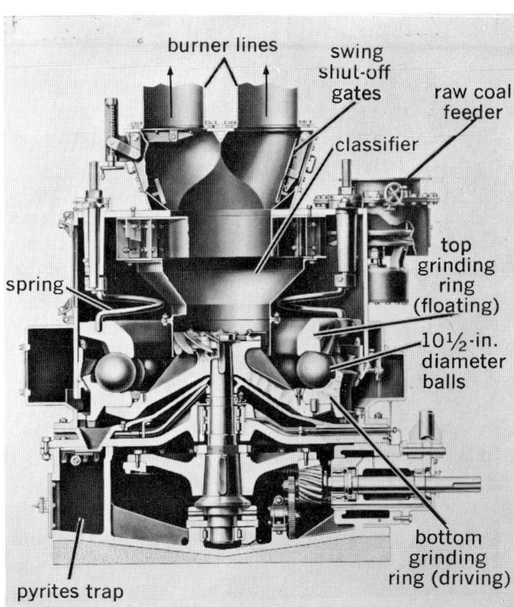

Fig. 14. Medium-speed ball race coal pulverizer, operated under pressure. $10\frac{1}{2}$ in. = 27 cm. (*Babcock and Wilcox Co.*)

The ball race pulverizer uses large steel balls between races (Fig. 11). The bottom race is fixed, the intermediate race is rotated through driving lugs, and the top races are restrained from rotating by stops and are loaded by the springs. The material flows by centrifugal force from the center outward through the $12\frac{1}{4}$-in. (31.1-cm) and $9\frac{1}{2}$-in. (23.5-cm) balls. A ball race unit in closed circuit with an air separator is used for pulverizing cement rock.

Screens. Material is separated into desired size ranges by screening. For products 2 in. (5 cm) or larger, an open-end perforated cylindrical rotary screen is used. It is on a slight inclination, and the undersize goes through the holes and the oversize passes out through the open lower end.

For finer products from 2 in. (5 cm) down to fine powder passing through a 200-mesh sieve, shaking or vibrating screens of woven wire are used.

Shaking screens have either an eccentric drive or an unbalanced rotating weight to produce the shaking. Vibrating screens have electric vibrators (Fig. 12). They are often hung by rods and springs

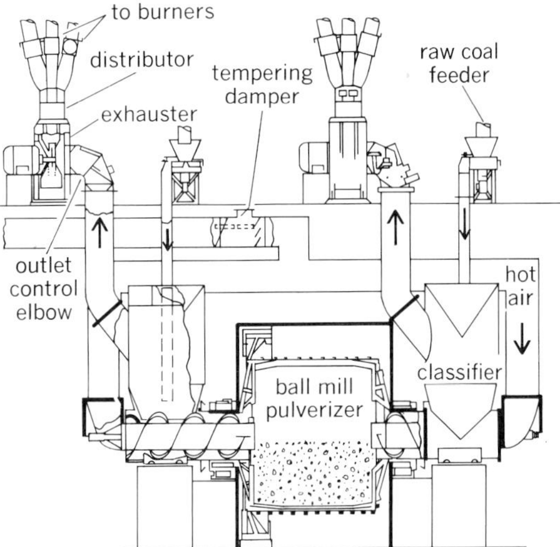

Fig. 15. Slow-speed coal pulverizer of the ball mill type; classifier sorts out oversize. (*Foster Wheeler Corp.*)

from overhead steel, giving complete freedom to vibrate. In some applications the screen is double decked, thus producing three sizes of product.

Direct-fired pulverizers for coal. Most modern large coal-fired furnaces are fed directly by the pulverizer, the coal rate being regulated by the rate of pulverizing. Air at 300–600°F (150–315°C) is supplied to the pulverizer to dry and preheat the coal. In a bowl-type medium-speed pulverizer, the springs press the pivoted stationary rolls against the rotating bowl grinding ring, crushing the coal between them (Fig. 13). The self-contained stationary cone classifier returns the oversize to the pulverizing zone.

The medium-speed pulverizer of the ball race type uses closely spaced 18-in.-diameter (46-cm) balls between a lower rotating race and a floating top race (Fig. 14). The single coil springs restrain the top race and also apply the pressure needed. A stationary cone classifier is used. This pulverizer operates under pressure, using a blower on

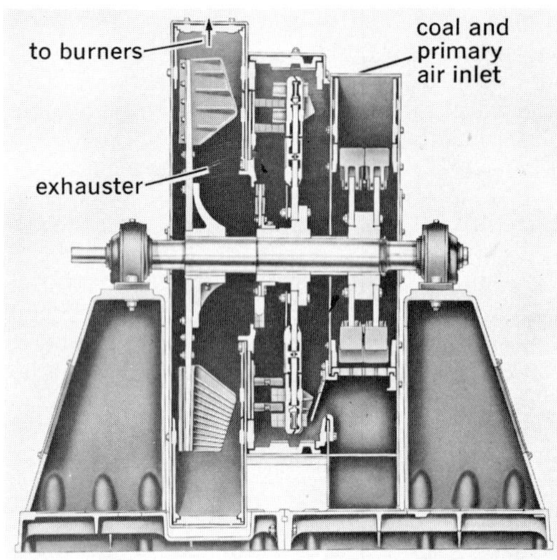

Fig. 16. High-speed coal pulverizer. (*Riley Stoker Corp.*)

the clean inlet air instead of an exhauster on the coal-laden outlet air.

Slow-speed pulverizers of the ball mill type have classifiers and return the oversize for regrinding on both ends (Fig. 15).

High-speed pulverizers usually incorporate multiple stages such as a preliminary hammer mill and a secondary pulverizing stage. A built-in exhaust fan exhausts the fuel and air from the unit (Fig. 16). *See* BALL-AND-RACE-TYPE PULVERIZER.

[RALPH M. HARDGROVE]

Bibliography: J. H. Perry and C. H. Chilton (eds.), *Chemical Engineers' Handbook*, 5th ed., 1973; W. Staniar (ed.), *Plant Engineering Handbook*, 2d ed., 1959.

Cyclone furnace

A water-cooled horizontal cylinder in which fuel is fired and heat is released at extremely high rates. When firing coal, the crushed coal, approximately 95% of which is sized at 1/4 in. or less, is introduced tangentially into the burner at the front end of the cyclone (see illustration). About 15% of the combustion air is used as primary and tertiary air to impart a whirling motion to the particles of coal. The whirling, or centrifugal, action on the fuel is further increased by the tangential admission of high-velocity secondary air into the cyclone.

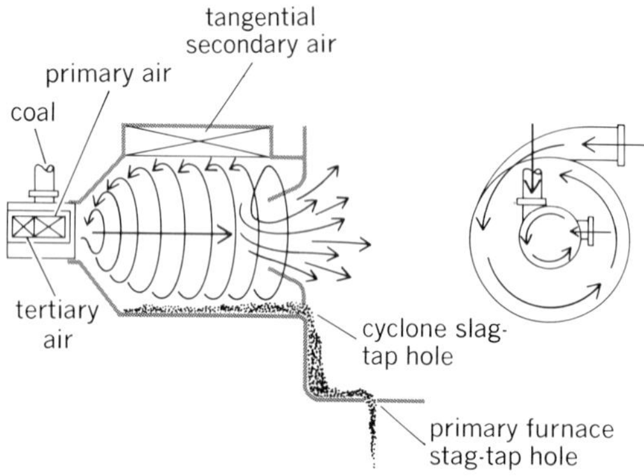

Schematic diagram of cyclone furnace. (*From T. Baumeister, ed., Standard Handbook for Mechanical Engineers, 7th ed., McGraw-Hill, 1967*)

The combustible is burned from the fuel in the cyclone furnace, and the resulting high-temperature gases melt the ash into liquid slag, a thin layer of which adheres to the walls of the cyclone. The incoming fuel particles, except those fines burned in suspension, are thrown to the walls by centrifugal force and caught in the running slag. The secondary air sweeps past the coal particles embedded in the slag surface at high speed. Thus, the air required to burn the coal is quickly supplied and the products of combustion are rapidly removed.

The products of combustion are discharged through a water-cooled reentrant throat at the rear of the cyclone into the boiler furnace. The part of the molten slag that does not adhere to the cyclone walls also flows toward the rear and is discharged through a taphole into the boiler furnace.

Essentially, the fundamental difference between cyclone furnaces and pulverized coal–fired furnaces is the manner in which combustion takes place. In pulverized coal–fired furnaces, particles of coal move along with the gas stream; consequently, relatively large furnaces are required to complete the combustion of the suspended fuel. With cyclonic firing, the coal is held in the cyclone and the air is passed over the fuel. Thus, large quantities of fuel can be fired and combustion completed in a relatively small volume, and the boiler furnace is used to cool the products of combustion.

Gas and oil can also be burned in cyclone furnaces at ratings and with performances equaling those of coal firing. When oil is the fuel, it is injected adjacent to the secondary air ports and directed downward into the secondary airstream. The oil is picked up and sufficiently atomized by the high-velocity air. Gas is fired similarly through flat, open-ended ports located in the secondary air entrance. *See* BOILER; STEAM-GENERATING FURNACE; STEAM-GENERATING UNIT.

[G. W. KESSLER]

Dam

A structure that bars or detains the flow of water in an open channel or watercourse. Dams are constructed for several principal purposes. Diversion dams divert water from a stream; navigation dams raise the level of a stream to increase the depth for navigation purposes (Fig. 1); power dams raise the level of a stream to create or concentrate hydrostatic head for power purposes; and storage dams store water for municipal and industrial use, irrigation, flood control, river regulation, recreation, or power production. A dam serving two or more purposes is called a multiple-purpose dam. Dams are commonly classified by the material from which they are constructed, such as masonry, concrete, earth, rock, timber, and steel. Most dams now are built either of concrete or of earth and rock.

Concrete dams. Concrete dams may be typed as gravity, arch, or buttress type. Gravity dams depend on weight for stability against overturning and for resistance to sliding on their foundations (Figs. 2 and 3). An arch dam may have a near-vertical face or, more usually, one that curves concave downstream (Figs. 4 and 5). The dam acts as an arch to transmit most of the horizontal thrust from the water pressure against the upstream face

Fig. 1. John Day Lock and Dam, looking upstream across the Columbia River at the Washington shore. In the foreground the navigation lock may be seen, then the spillway dam beyond it, and then the powerhouse. The John Day multiple-purpose project boasts the highest single-lift navigation lock in the United States. (*U.S. Army Corps of Engineers*)

of the dam to the abutments of the dam. The buttress type of concrete dam includes the slab-and-buttress, or Ambursen, type; round- or diamond-head buttress type; multiple-arch type; and multiple-dome type. Buttress dams depend on the weight of the structure and of the water on the dam to resist overturning and sliding.

Forces acting on concrete dams. Principal forces acting on a concrete dam are (1) vertical forces from weight of the structure and vertical component of water pressure against the upstream and downstream faces of the dam, (2) uplift pressures under the base of the structure, (3) horizontal forces from the horizontal component of the water pressure against the upstream and downstream faces of the dam, (4) forces from earthquake accelerations in regions subject to earthquakes, (5) temperature stresses, (6) pressures from silt deposits and earth fills against the structure, and (7) ice pressures.

The uplift pressure under the base of a dam varies with the effectiveness of the foundation drainage system and the perviousness of the foundation.

Earthquake loads are usually selected after consideration of the accelerations which may be expected at the site as indicated by the geology, proximity to major faults, and the earthquake history of the region. Conventionally, earthquake forces have been treated as static forces representing the effects of the acceleration of the dam itself and the hydrodynamic force produced against the dam by water in the reservoir. Such horizontal forces often are assumed to equal 0.05–0.10 the force of gravity, with a somewhat smaller vertical force. Dynamic analysis procedures have been developed which determine the structure's response to combined effects of the contemplated ground motion and the structure's dynamic properties.

Stresses resulting from temperature changes

Fig. 2. Green Peter Dam, a concrete gravity type on the Middle Santiam River, Willamette River Basin, OR. Gate-controlled overflow-type spillway is constructed through crest of dam; powerhouse is at downstream toe of dam. (*U.S. Army Corps of Engineers*)

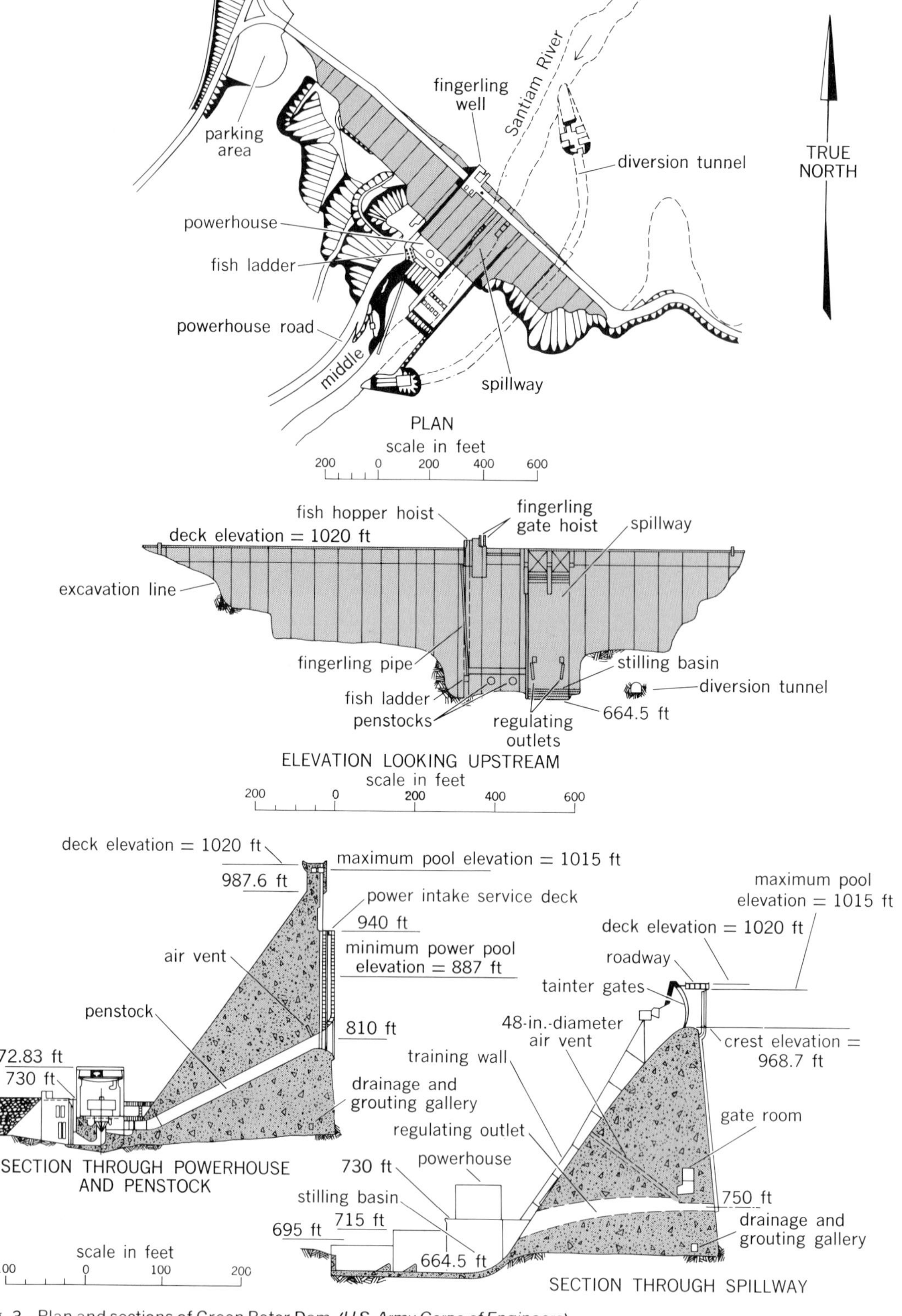

Fig. 3. Plan and sections of Green Peter Dam. *(U.S. Army Corps of Engineers)*

must be considered in analyzing arch dams. These stresses are usually disregarded in the design of concrete gravity dams, but must be controlled to acceptable limits by concreting and curing methods, discussed below.

Pressure from silt deposited in the reservoir against the dam is considered only after sedimentation studies indicate that it may be a significant factor. Backfill pressures are important where a concrete gravity dam ties into an embankment.

Ice pressure, applied at the maximum elevation at which the ice will occur in project operations, is considered when conditions indicate that it would be significant. The pressure, commonly assumed to be 10,000–20,000 lb per linear foot, results from the thermal expansion of the ice sheet and varies with the rate and magnitude of temperature rise and thickness of the ice.

Stability and allowable stresses. Stability of a concrete gravity dam is evaluated by analyzing the available resistance to overturning and sliding. To satisfy the former, the resultant of forces is required to fall within the middle third of the base under normal load conditions. Sliding stability is assured by requiring available shear and friction resistance to be greater by a designated safety factor than the forces tending to produce sliding. The strengths used in computing resistance to sliding are based on investigation and tests of the foundation. Bearing strength of the foundation for a gravity dam is a controlling factor only for weak foundations or for high dams. Because an arch dam depends on the competency of the abutments, the rock bearing strength must be sufficient to provide an adequate safety factor for the compressive stresses, and the resistance to sliding along any weak surface must be great enough to provide an adequate safety factor.

Concrete stresses control the design of arch dams, but ordinarily not gravity dams. Stresses adopted for concrete arch dams are conservative. A safety factor of 4 on concrete compressive strength is commonly used for normal load conditions.

Concrete temperature control. Volume changes accompanying temperature changes in a concrete dam tend to cause the development of tensile stresses. A major factor in development of temperature changes within a concrete mass is the heat developed by chemical changes in the concrete after placement. Uncontrolled temperature changes can cause cracking which may endanger the stability of a dam, cause leakage, and reduce durability. Temperatures are controlled by using cementing materials having low heat of hydration, and by artificial cooling by precooling the concrete mix or circulating cold water through pipes embedded in the concrete or both.

Concrete dams are constructed in blocks, with the joints between the blocks serving as contraction joints (Fig. 6). In arch dams the contraction joints are filled with cement grout after maximum shrinkage has occurred to assure continuous bearing surfaces normal to the compressional forces set up in the arch when the water load is applied to the dam.

Quality control. During construction, continuing testing and inspection are performed to ensure that the concrete will be of required quality. Tests are also made on materials used in manufacture of the concrete, and concrete batching, mixing, transporting, placing, curing, and protection are continuously inspected.

Earth dams. Earth dams have been used for water storage since early civilizations. Improvements in earth-materials techniques, particularly the development of modern earth-handling equipment, have brought about a wider use of this type of dam, and today as in primitive times the earth embankment is the most common dam (Figs. 7 and 8). Earth dams may be built of rock, gravel, sand, silt, or clay in various combinations.

Most earth dams are constructed with an inner impervious core, with upstream and downstream zones of more pervious materials, sometimes including rock zones. Earth dams limit the flow of water through the dam by use of fine-grained soils. Where possible, these soils are formed into a relatively impervious core. When there is a sand or gravel foundation, the core may be connected to bedrock by a cutoff trench backfilled with compacted soil. If such cutoffs are not economically feasible because of the great depth of pervious foundation soils, then the central impervious core is connected to a long horizontal upstream impervious blanket that increases the length of the seepage path. The impervious core is often encased in pervious zones of sand, gravel, or rock fill for stability. When there is a large difference in the particle sizes of the core and pervious zones, transition zones are required to prevent the core material from being transported into the pervious zones by seeping water. In some cases where pervious soils are scarce, the entire dam may be a homogeneous fill of relatively impervious soil. Downstream pervious drainage blankets are provided to collect seepage passing through, under, and around the abutments of the dam.

Fig. 4. East Canyon Dam, a thin-arch concrete structure on the East Canyon River, UT. Note uncontrolled over-flow-type spillway through crest of dam at right center of photograph. (*U.S. Bureau of Reclamation*)

Materials can be obtained from required excavations for the dam and appurtenances or from borrow areas. Rock fill is generally used when large quantities of rock are available from required excavation or when soil borrow is scarce.

Earth-fill embankment is placed in layers and compacted by sheepsfoot rollers or heavy pneumatic-tire rollers. Moisture content of silt and clay soils is carefully controlled to facilitate optimum compaction. Sand and gravel fills are compacted in slightly thicker layers by pneumatic-tire rollers, vibrating steel drum rollers, or placement equipment. The placement moisture content of pervious fills is less critical than for silts and clays. Rock fill usually is placed in layers 1–3 ft deep and is

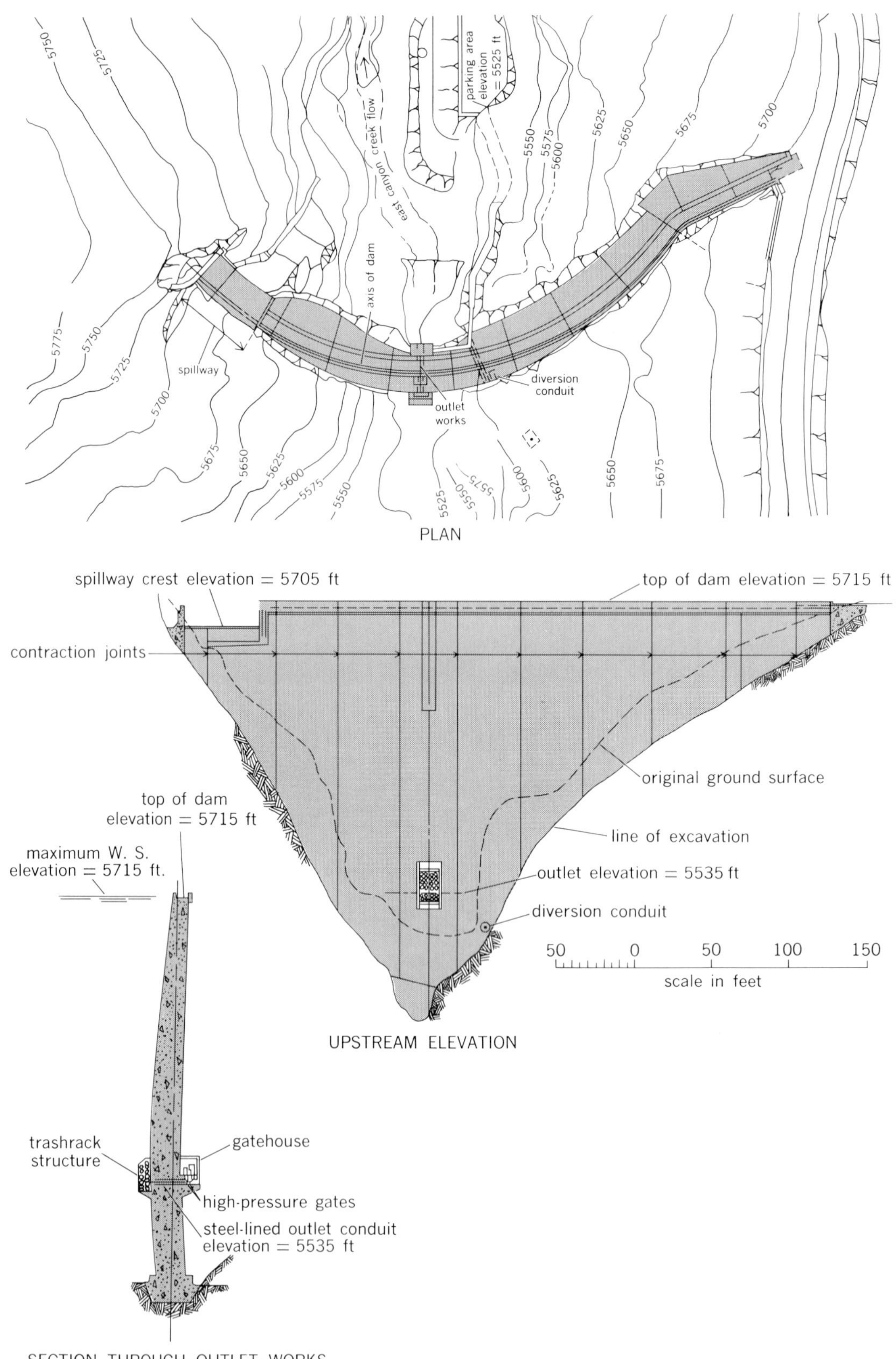

Fig. 5. Plan and sections of East Canyon Dam. *(U.S. Bureau of Reclamation)*

compacted by placement equipment and vibrating steel drum rollers.

Spillways. A spillway releases water in excess of storage capacity so that the dam and its foundation are protected against erosion and possible failure. All dams must have a spillway, except small ones where the runoff can be safely stored in the reservoir without danger of overtopping the dam. Ample spillway capacity is of particular importance for large earth dams, which would be destroyed or

Fig. 6. Block method of construction on a typical concrete gravity dam. *(U. S. Army Corps of Engineers)*

severely damaged by being overtopped. Failure of a large dam could result in severe hazards to life and property downstream.

Types. Spillways are of two general types: the overflow type, constructed as an integral part of the dam; or the channel type, located as an independent structure discharging through an open chute or tunnel. Either type may be equipped with gates to control the discharge. Various control structures have been used for channel spillways, including the simple overflow weir, side-channel overflow weir, and drop or morning-glory inlet where the water flows over a circular weir crest and drops directly into a tunnel.

Unless the discharge end of a spillway is remote from the toe of the dam or erosion-resistant bedrock exists at shallow depths, some form of energy dissipator must be provided to protect the toe of the dam and the foundation from spillway discharges. For an overflow spillway the energy

Fig. 7. Aerial view of North Fork Dam, a combination earth and rock embankment on the North Fork of Pound River, VA. Channel-type spillway (left center) has simple overflow weir. *(U.S. Army Corps of Engineers)*

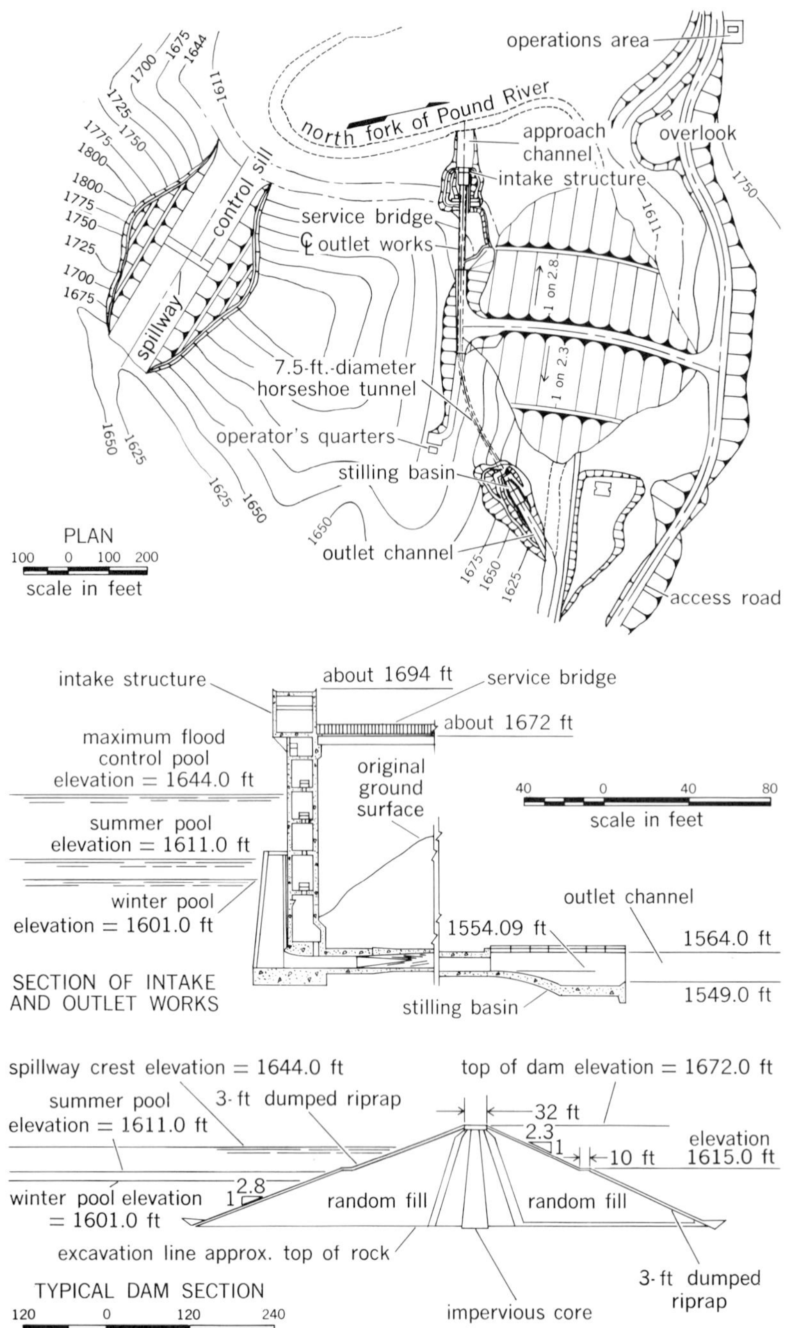

Fig. 8. Plan and sections of North Fork of Pound Dam. *(U.S. Army Corps of Engineers)*

dissipator may be a stilling basin, a sloping apron downstream from the dam, or a submerged bucket. When a channel spillway terminates near the dam, it usually has a stilling basin. A flip bucket is used for both overflow and channel spillways when the flow can be deflected far enough downstream, usually onto rock, to prevent erosion at the toe of the dam or end of the spillway.

Gates. Several types of gates may be used to regulate and control the discharge of spillways (Fig. 9). Tainter gates are comparatively low in cost and require only a small amount of power for operation, being hydraulically balanced and of low friction. Drum gates, which are operated by reservoir pressure, are costly but afford a wide, unobstructed opening for passage of drift and ice over the gates. Vertical-lift gates of the fixed-wheel or roller type are sometimes used for spillway regulation, but are more difficult to operate than the others. Floating ring gates control the discharge of morning-glory spillways. Like the drum gate, this type offers a minimum of interference to the passage of ice or drift over the gate and requires no external power for operation.

Reservoir outlet works. These are used to regulate the release of water from the reservoir; they consist essentially of an intake and an outlet connected by a water passage, and are usually provided with gates. Outlet works usually have trashracks at the intake end to prevent clogging by debris. Bulkheads or stop logs are commonly provided to close the intakes so that the passages may be unwatered for inspection and maintenance. A stilling basin or other type of energy dissipator is usually provided at the outlet end.

Locations. Outlets may be sluices through concrete dams with control valves located in chambers in the dam or on the downstream end of the sluices, tunnels through the abutments of the dam, or cut-and-cover conduits extending along the foundation through an earth-fill dam. In the last case, the control valves are usually located within the dam or at the upstream end of the conduit, and special precautions must be taken to prevent leakage of water along the outside of the conduit.

Outlet control gates. Various gates and valves are used for regulating the release of water from reservoirs, including high-pressure slide gates, tractor gates (roller or wheel), and radial or tainter gates (Fig. 10); also needle valves of various kinds, butterfly valves, fixed cone dispersion valves, and cylinder or sleeve valves. They must be capable of operating, without excessive vibration and cavitation, at any opening and at any head up to the maximum to which they may be subjected. They also must be capable of opening and closing under the maximum operating head. Emergency gates generally are used upstream of the operating gates, where stored water is valuable, so that closure can be made if the service gate should fail to function.

The slide gate, which consists of a movable leaf that slides on a stationary seat, is the most commonly used control gate. The high-pressure slide gate is of rugged design, having corrosion-resisting metal seats on both the movable rectangular leaf and the fixed frame. This gate has been used for regulating discharges under heads of over 600 ft.

Provision of low-level outlet. The usual storage reservoir has low-level outlets near the elevation of the stream bed to enable release of all the stored water. Some power and multiple-purpose dams have relatively high-level dead storage pools and do not require low-level outlets for ordinary operation. In such a dam, provision of a capability for emptying the reservoir in case of an emergency must be weighed against the additional cost.

Penstocks. A penstock is a pipe that conveys water from a forebay, reservoir, or other source to a turbine in a hydroelectric plant. It is usually made of steel, but reinforced concrete and wood-stave pipe have also been used. Pressure rise and speed regulation must be considered in the design of a penstock.

Pressure rise, or water hammer, is the pressure change that occurs when the rate of flow in a pipe or conduit is changed rapidly. The intensity of this pressure change is proportional to the rate at which the velocity of the flow is accelerated or decelerated. Accurate determination of the pressure changes that occur in a penstock involves consideration of all operating conditions. For example, one important consideration is the pressure rise that occurs in a penstock when the turbine wicket gates are closed after the loss of load. *See* CANAL.

Selection of dam site. This depends upon such factors as hydrologic, topographic, and geologic conditions; storage capacity of reservoir; accessibility; cost of lands and necessary relocations of prior occupants or uses; and proximity of sources of suitable construction materials. For a storage dam the objective is to select the site where the desired amount of storage can be most economically developed. Power dams must be located to develop the desired head and storage. For a diversion dam the site must be considered in conjunction with the location and elevation of the outlet canal or conduit. Site selection for navigation dams involves special factors such as desired navigable depth and channel width, slope of river channel, natural river flow, amount of bank protection, amount of channel dredging, approach and exit conditions for tows, and locations of other dams in the system.

Unless topographic and geologic conditions for a proposed storage, power, or diversion dam site are satisfactory, hydrological features may need to be subordinated. Important topographic characteristics include width of the floodplain, shape and height of valley walls, existence of nearby saddles for spillways, and adequacy of reservoir rim to retain impounded water. Controlling geologic conditions include the depth, classification, and engineering properties of soils and bedrock at the dam site, and the occurrence of sinks, faults, and major landslides at the site or in the reservoir area. The elevation of the groundwater table is also significant because it will influence the construction operations and suitability of borrow materials. The beneficial effect of reservoir water on groundwater recharge may become an important consideration, as well as the adverse effects on existing or potential mineral resources and developments that would be destroyed or require relocation at the site or within the reservoir. *See* SURVEYING.

Selection of type of dam. This is made on the basis of the estimated costs of various types. The most important factors are topography, foundation conditions, and the accessibility of construction materials. In general, a hard-rock foundation is suitable for any type of dam, provided the rock has no unfavorable jointing, there is no danger of movement in existing faults, and foundation underseepage can be controlled at reasonable cost. Rock foundations of high quality are essential for arch dams because the abutments receive the full thrust of the water pressure against the face of the dam. Rock foundations are necessary for all medium and high concrete dams. An earth dam may be built on almost any kind of foundation if properly designed and constructed.

The chance of an embankment dam being most economical is improved if large spillway and outlet capacities are required and topography and foundation are favorable. In a wide valley a combination of an earth embankment dam and a concrete dam section containing the spillway and outlets often is economical. Availability of suitable construction materials frequently determines the most economical type of dam. A concrete dam requires adequate quantities of suitable concrete aggregate and reasonable availability of cement, while an earth dam requires sufficient quantities of both pervious and impervious earth materials. If quantities of earth materials are limited and enough rock is available, a rock-fill dam with an impervious earth core may be the most economical.

Determination of dam height. The dam must be high enough to (1) store water to the normal full-pool elevation required to meet intended functions of the project, (2) provide for the temporary storage needed to route the spillway design flood through the dam, and (3) provide sufficient freeboard height above the maximum surcharge elevation to assure an acceptable degree of safety against possible overtopping from waves and runup.

Physical characteristics of the dam and reservoir site or existing developments within the reservoir area may impose upper limits in selecting the normal full-pool level. In other circumstances economic considerations govern.

With the normal full-pool elevation established, flood flows of unusual magnitude may be passed by providing spillways and outlets large enough to discharge the probable maximum flood or other

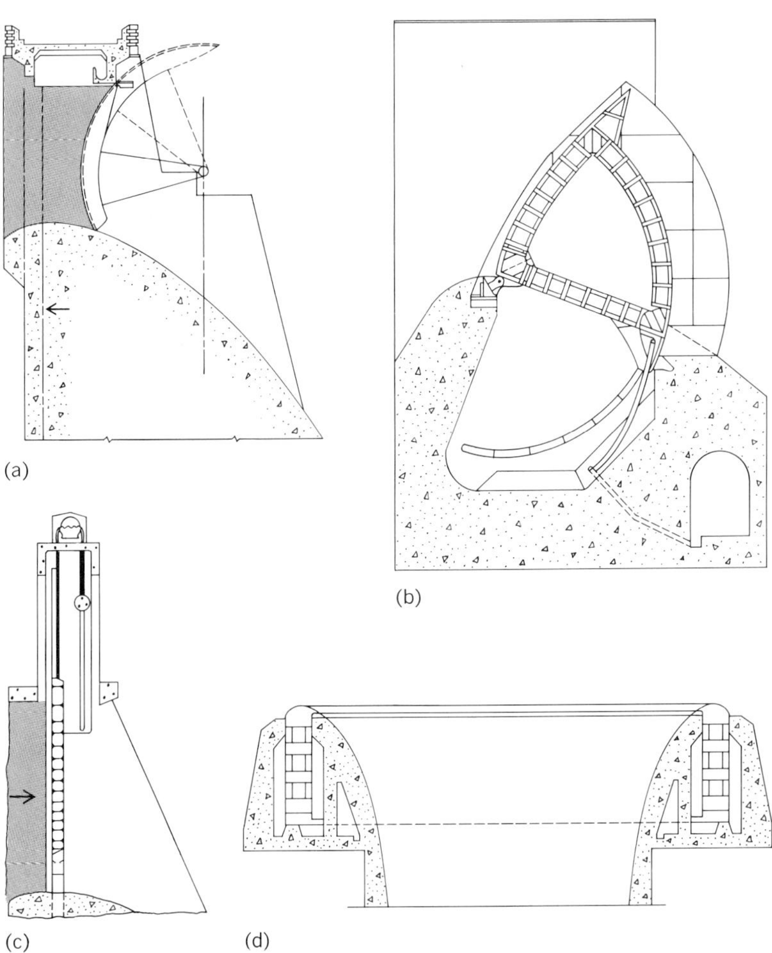

Fig. 9. Spillway gates. (*a*) Tainter gate. (*b*) Drum gate. (*c*) Vertical lift gate. (*d*) Ring gate. (*U.S. Army Corps of Engineers and U.S. Bureau of Reclamation*)

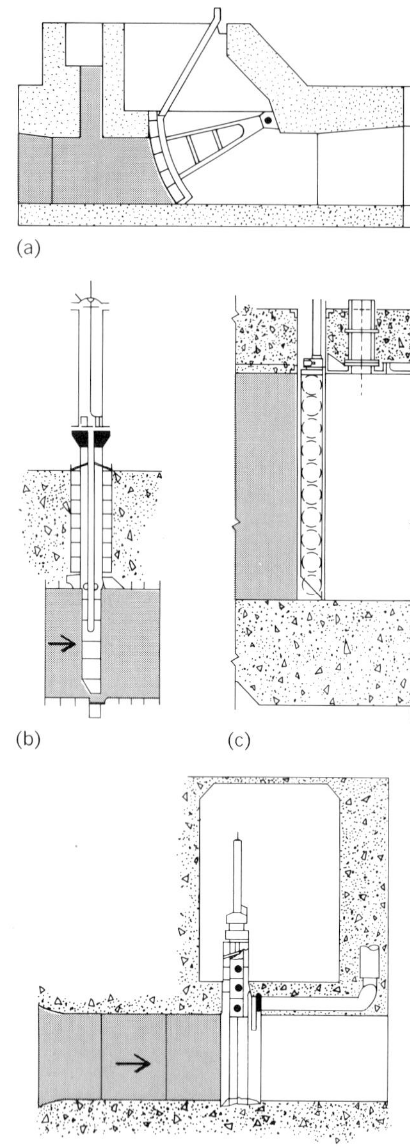

Fig. 10. Outlet gates. *(a)* Tainter gate. *(b)* High-pressure slide gate. *(c)* Tractor gate. *(d)* Jet flow gate. *(U.S. Army Corps of Engineers and U.S. Bureau of Reclamation)*

spillway design flood without raising the reservoir above the normal full elevation or, if it is more economical, by raising the height of the dam and obtaining additional lands to permit the reservoir to temporarily attain surcharge elevations above the normal pool level during extreme floods. Use of temporary surcharge storage capacity also serves to reduce the peak rates of spillway discharge.

Freeboard height is the distance between the maximum reservoir level and the top of the dam. Usually 3 ft or more of freeboard is provided to avoid overtopping the dam by wind-generated waves. Additional freeboard may be provided for possible effects of surges induced by earthquakes, landslides, or other unpredictable events.

Diversion of stream. During construction the dam site must be unwatered so that the foundation may be prepared properly and materials in the structure may be moved easily into position. The stream may be diverted around the site through tunnels, passed through or around the construction area by flumes, passed through openings in the dam, or passed over low sections of a partially completed concrete dam. Diversion may be conducted in one or more stages, with a different method used for each stage. Initial diversion is conducted during a period of low flow to avoid the necessity for passing large flows.

Foundation treatment. The foundation of a dam must support the structure under all operating conditions. For concrete dams, following removal of unsatisfactory materials to a sound foundation surface, imperfections such as adversely oriented rock joints, open bedding planes, localized soft seams, and faults lying on or beneath the foundation surface receive special treatment. Necessary foundation treatment prior to dam construction may include "dental excavation" of surface weaknesses, or shafting and mining to remove deeper localized weaknesses, followed by backfilling with concrete or grout. Such work is sometimes supplemented by pattern grouting of foundation zones after construction of the dam. Foundation features such as rock joints, bedding planes, or faults that do not require preconstruction treatment are made relatively water-tight by curtain grouting from a line of deep grout holes located near the upstream heel and extending the full length of the dam. Although a grout curtain controls seepage at depth, the effectiveness of the grouting or its permanency cannot be relied upon alone to reduce hydrostatic pressures acting on the base of the dam. As a result, drain holes are drilled into the foundation just downstream of the grout curtain to intercept seepage passing through it and to reduce hydrostatic pressure. Occasionally chemical solutions such as acrylamide, sodium silicate, chromelignin, and polyester and epoxy resins are used for consolidating soils or rocks with fine openings.

The foundation of an earth dam must safely support the weight of the dam, limit seepage of stored water, and prevent transportation of dam or foundation material away or into open joints or seams in the rock by seepage. Earth-dam foundation treatment may include removal of excessively weak surface soils to prevent both potential sliding and excessive settlement of the dam, excavation of a cutoff trench to rock, and grouting of joints and seams in the bottom and downstream side of the cutoff trench. The cutoff trenches and grouting extend up the abutments, which are first stripped of weak surface materials.

When weak soils in the foundation of an earth dam cannot be removed economically, the slopes of the embankment must be flattened to reduce shear stress in the foundation to a value less than the soil strength. Relief wells are installed in pervious foundations to control seepage uplift pressures and to reduce the danger of piping when the depth of the pervious material is such as to preclude an economical cutoff.

Instrumentation. Instruments are installed at dams to observe structural behavior and physical conditions during construction and after filling, to check safety, and to provide information for design improvement.

In concrete dams instruments are used to measure stresses either directly or to measure strains from which stresses may be computed. Plumb lines are used to measure bending, and clinometers to measure tilting. Contraction joint openings are measured by joint meters spanning between two adjacent blocks of a dam. Temperatures are measured either by embedded electrical resistance thermometers or by adapting strain, stress, and joint measuring instruments. Water pressure on the base of a concrete dam at the contact with the foundation rock is measured by uplift pressure cells. Interior pressures in a concrete dam are measured by embedded pressure cells. Measurements are also made to determine horizontal and vertical movements; strong-motion accelerometers are being installed on and near dams in earthquake regions to record seismic data.

Instruments installed in earth-dam embankments and foundations are piezometers to determine pore water pressure in the soil or bedrock during construction and seepage after reservoir impoundments; settlement gages to determine settlements of the foundation of the dam under dead load; vertical and horizontal markers to determine movements, especially during construction; and inclinometers to determine horizontal movements along a vertical line.

Inspection of dams. Because failure of a dam may result in loss of life or property damage in the downstream area, it is essential that dams be inspected systematically both during construction and after completion. The design of dams should be reviewed to assure competency of the structure and its site, and inspections should be made during construction to ensure that the requirements of the design and specifications are incorporated in the structure.

After completion and filling, inspections may vary from cursory surveillance during day-to-day operation of the project to regularly scheduled comprehensive inspections. The objective of such inspections is to detect symptoms of possible distress in the dam at the earliest time. These symptoms include significant sloughs or slides in embankments; evidence of piping or boils near embankments; abnormal changes in flow from drains; unusual increases in seepage quantities; unexpected changes in pore water pressures or uplift pressures; unusual movement or cracking of embankments or abutments; significant cracking of concrete structures; appearance of sinkholes or localized subsidence near foundations; excessive deflection, displacement, erosion, or vibration of concrete structures; erratic movement or excessive deflection or vibration of outlet or spillway gates or valves; or any other unusual conditions in the structure or surrounding terrain.

Detection of any such symptoms of distress should be followed by an investigation of the causes, probable effects, and remedial measures required. Inspection of a dam and reservoir is particularly important following significant seismic events in the locality. Systematic monitoring of the instrumentation installed in dams is essential to the inspection program.

[JACK R. THOMPSON]

Bibliography: American Concrete Institute, *Symposium on Mass Concrete*, Spec. Publ. SP-6, 1963; W. P. Creager, J. D. Justin, and J. Hinds, *Engineering for Dams*, 1945; C. V. Davis, *Handbook of Applied Hydraulics*, 3d ed., 1968; A. R. Golze (ed.), *Handbook of Dam Engineering*, 1977; J. L. Sherard et al., *Earth and Earth-Rock Dams*, 1963; G. B. Sowers and G. F. Sowers, *Introductory Soil Mechanics and Foundations*, 3d ed., 1970; U.S. Bureau of Reclamation, *Design of Small Dams*, 2d ed., 1973; U.S. Bureau of Reclamation, *Trial Load Method of Analyzing Arch Dams*, Boulder Canyon Proj. Final Rep., pt. 5, Bull. no. 1, 1938; H. M. Westergaard, Water pressures on dams during earthquakes, *Trans. ASCE*, 98:418–472, 1933.

Dehumidifier

Equipment designed to reduce the amount of water vapor in the atmosphere.

The atmosphere is a mechanical mixture of dry air and water vapor, the amount of water vapor being limited by air temperature. Water vapor is measured in either grains per pound of dry air or pounds per pound of dry air (7000 gr = 1 lb).

There are three methods by which water vapor may be removed: (1) the use of sorbent materials, (2) cooling to the required dew point, and (3) compression with aftercooling.

Sorbent type. Sorbents are materials which are hygroscopic to water vapor; they are available in both solid and liquid forms. Solid sorbents include silica gels, activated alumina, and aluminum bauxite. Liquid sorbents include halogen salts such as lithium chloride, lithium bromide, and calcium chloride, and organic liquids such as ethylene, diethylene, and triethylene glycols and glycol derivatives.

Solid sorbents may be used in static or dynamic dehumidifiers. Bags of solid sorbent materials within packages of machine tools, electronic equipment, and other valuable materials subject to moisture damage constitute static dehumidifiers. An indicator chemical may be included to show by a change in color when the sorbent is saturated. The sorbent then requires reactivation by heating at 300–350°F for 1–2 hr before reuse.

A dynamic dehumidifier for solid sorbent consists of a main circulating fan, one or more beds of sorbent material, reactivation air fan, heater, mechanism to change from dehumidifying to reactivation, and aftercooler.

A single-bed dehumidifier (Fig. 1) operates on an intermittent cycle of dehumidifying for 2–3 hr and then swtiches to the reactivation cycle for 15–45 min. No dehumidification is obtained during the reactivation cycle. A single-bed unit is used

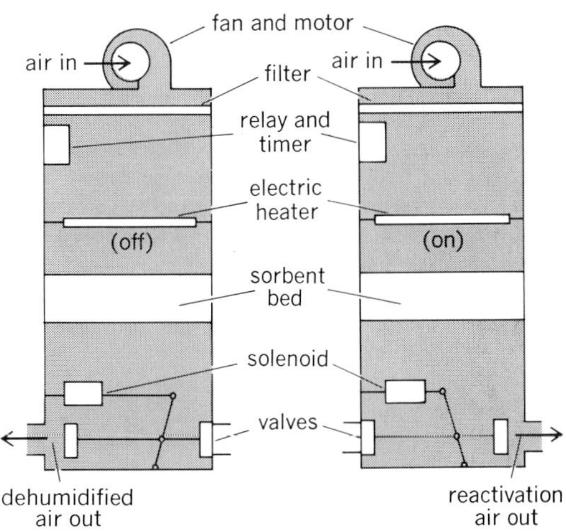

Fig. 1. Single-bed solid-sorbent dehumidifier. Dehumidifying cycle on left and reactivation cycle on right.

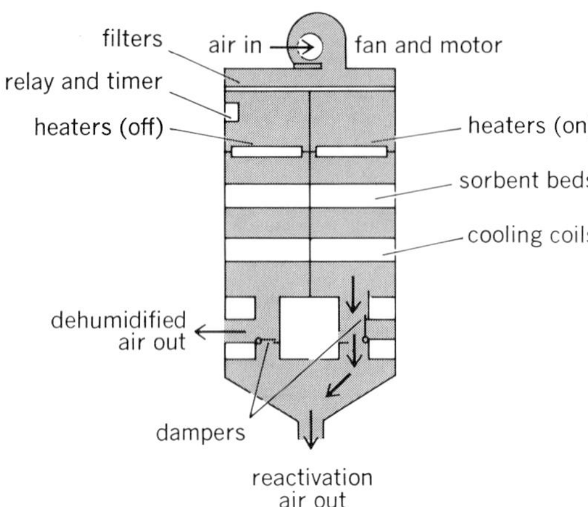

Fig. 2. Dual-bed solid-sorbent dehumidifier. Air is being dehumidified through the left bed at the same time that the right bed is being reactivated.

for small areas where moisture is a problem. The moist reactivation air is discharged to the outside.

The dual-bed machine is larger in capacity than the single-bed unit and has the advantage of providing a continuous supply of dehumidified air (Fig. 2). While one bed is dehumidifying, the other bed is reactivating. After a predetermined time interval, the air cycle is switched to pass the air through the reactivated bed for dehumidification and to reactivate the saturated bed.

The dew point of the effluent air of a fixed-bed machine is lowest at the start of a cycle immediately after the reactivated bed has been placed in service. The dew point gradually rises as the bed absorbs the water vapor and eventually would be the same as the entering dew point when the vapor pressure of the sorbent reached the vapor pressure of the air and could no longer absorb moisture from the air. The cutoff point at which the absorbing bed is changed over to reactivation is fixed by the maximum allowable effluent dew point.

A multibed unit with short operating cycles will reduce the range of effluent dew point to within a few degrees. A multibed unit with rotating cylindrical bed maintains a reasonably constant effluent dew point.

The liquid-sorbent dehumidifier consists of a main circulating fan, sorbent-air contactor, sorbent pump, and reactivator including contactor, fan, heater, and cooler (Fig. 3). This unit will control the effluent dew point at a constant level because dehumidification and reactivation are continuous operations with a small part of the sorbent constantly bled off from the main circulating system and reactivated to the concentration required for the desired effluent dew point.

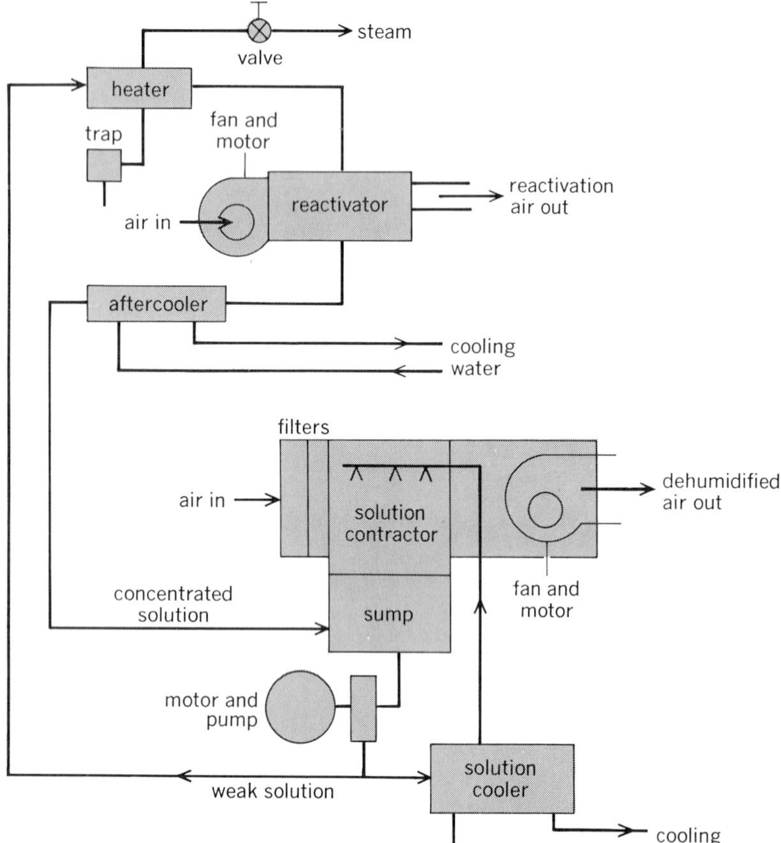

Fig. 3. Liquid-sorbent dehumidifier with continuous dehumidifying and reactivation.

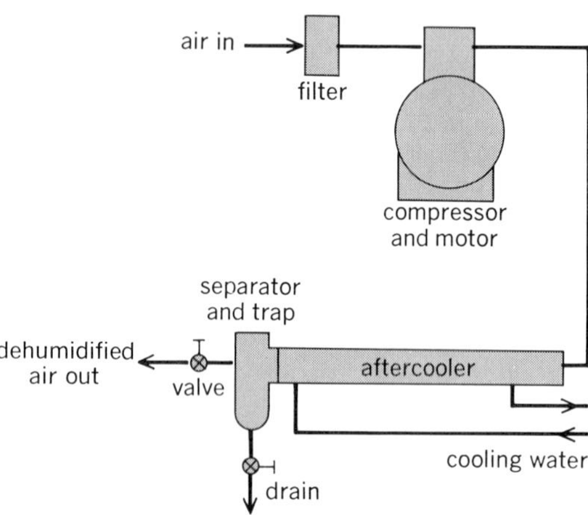

Fig. 4. Dehumidifying by compression and aftercooling.

Cooling type. A system employing the use of cooling for dehumidifying consists of a circulating fan and cooling coil. The cooling coil may use cold water obtained from wells or a refrigeration plant, or may be a direct-expansion refrigeration coil. In place of a coil, a spray washer may be used in which the air passes through two or more banks of sprays of cold water or brine, depending upon the dew-point temperature required.

When coils are used, the leaving dew point is seldom below 35°F because of possible buildup of ice on the coil. When it is necessary to use coils for temperatures below 35°F, as in cold-storage rooms, either two coils are used so one can be defrosted while the other is in operation, or only one coil is used and dehumidifying is stopped during the defrost period.

A brine-spray dehumidifier or brine-sprayed coil can produce dew-point temperatures below 35°F without frosting if properly operated and maintained.

Compression type. Dehumidifying by compression and aftercooling is used when the reduction of water vapor in a compressed-air system is required. This is particularly important, for example, if the air is used for automatic control instruments or cleaning of delicate machined parts.

If air is compressed and the heat of compression removed to bring the temperature of the air back to the temperature entering the compressor, condensation will take place and the remaining water vapor content will be directly proportional to the absolute pressure ratio of the compressed air (Fig. 4).

For example, if saturated air at 70°F (111 gr/lb of dry air) is compressed from atmospheric pressure (14.7 pounds per square inch absolute, psia) to 88 psia (6:1 compression ratio) and cooled to 70°F, the remaining water vapor in the compressed air will be 111/6 = 18.5 gr/lb of dry air. If the air is expanded back to atmospheric pressure and 70°F, the dew point will be 24°F.

The power required for compression systems is so high compared to power requirements for dehumidifying by either the sorbent or refrigeration method that the compression system is not an economical one if dehumidifying is the only end result required.

[JOHN EVERETTS, JR.]

Bibliography: American Society of Heating, Refrigerating, and Air-Conditioning Engineers, *Handbook and Product Directory*: *Equipment*, 1979; ASHRAE, *Heating, Ventilating, Air-Conditioning Guide*, annual; ASHRAE, Symposium on dehumidification: Journal section, *Heat. Piping Air Cond.*, 29(4):152–162, 1957; J. Everetts, Jr., Dehumidification methods and applications, *Heat. Piping Air Cond.*, 18(12):121–124, 1964.

Density

The mass per unit volume of a material. The term is applicable to mixtures and pure substances and to matter in the solid, liquid, gaseous, or plasma state. Density of all matter depends on temperature; the density of a mixture may depend on its composition, and the density of a gas on its pressure. Common units of density are grams per cubic centimeter, and slugs or pounds per cubic foot. The specific gravity of a material is defined as the ratio of its density to the density of some standard material, such as water at a specified temperature, for example, 60°F, or, for gases the basis may be air at standard temperature and pressure. Another related concept is weight density, which is defined as the weight of a unit volume of the material.

[LEO NEDELSKY]

Derrick

A hoisting machine consisting usually of a vertical mast, a slanted boom, and associated tackle (see illustration). Derricks have a wide variety of forms. The mast may be no more than a base for the boom; it may be a tripod, an A-frame, a fixed column, and so on. Fixed stays may guy it in place. The boom may be fixed, it may pivot at the base of the mast, it may swing horizontally from near the top of the mast, or it may be omitted. The derrick may be permanently fixed, temporarily erected, or mobile on a cart or truck.

Derricks are widely used in construction, in cargo handling, and in shops. Their lifting action is intermittent compared to bucket conveyors, and their coverage is limited by the reach of the boom. A manual or powered winch provides the lifting action by coiling in the running tackle, the load swinging free as it rises. *See* BULK-HANDLING MACHINES; CRANE HOIST; HOISTING MACHINES; OIL AND GAS WELL DRILLING.

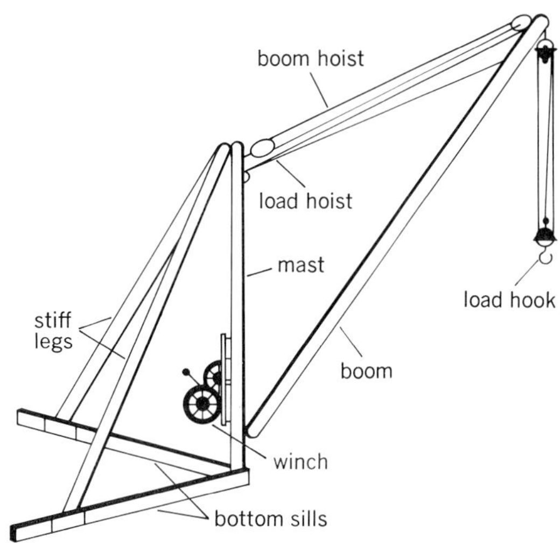

A boom derrick equipped with a swinging mast and anchored legs.

[DAVID O. HAYNES/FRANK H. ROCKETT]

Bibliography: W. E. Rossnagel, *Handbook of Rigging*, 3d ed., 1964; H. I. Shapiro, *Cranes and Derricks*, 1980.

Design standards

Specifications of materials, physical measurements, processes, performance of products, and characteristics of services rendered. Design standards may be established by individual manufacturers, trade associations, and national or international standards organizations. The general purpose is to realize operational and manufacturing economies, to increase the interchangeability of products, and to promote uniformity of definitions of product characteristics.

Individual firms. Individual firms often maintain extensive and detailed standards of parts that are available for use in their product designs. Usually the standards have the effect of restricting the variety of parts to certain sizes and materials. In this way the production lots required for inventory purposes are increased, and production economies may thereby be realized through the wider use of mass production. However, even if the larger quantities needed of the relatively few sizes do not in themselves lead to a cheaper manufacturing process, the costs of carrying inventory and setting up for production runs are reduced. If a given rate of usage is equally allocated among k different sizes rather than one, these costs increase by a factor of $\sqrt{k}$. For many firms, therefore, standardization has also meant simplification, defined as reducing a standard to the minimum of different sizes or other specifications that will cover user requirements. For instance, light bulbs and industrial electric motors have thus been limited to relatively few sizes. The practice has proved highly effective for many firms, especially when they had to reduce the "birthrate" of new-line items that threatened to clog even the most sophisticated inventory control system. A further advantage is that, provided the standards can be made sufficiently versatile, it is often possible to buy generic parts; for example, a certain bolt or screw can be bought as a standard hardware product rather than being a special job based on where it is installed.

A further development of this design approach may lead to the modulization of the entire product line, by reducing it to certain major subassemblies that are common to as many products as possible. Special jobs then typically require only a few added features, and cost savings may be realized. One such effort is that of most of the world's major automobile manufacturers to create a "world car" in which major components such as engines, frames, and transmissions are standardized, leaving certain features of final trim and bodywork to establish distinction between various models and makes.

A possible disadvantage is that the extensive use of general-purpose parts may jeopardize the spare parts business, especially where outside manufacturers can skim off the market for the more commonly used and profitable spare parts once the original patents, if any, have expired, and then leave the more complex and slow-moving spares to the original manufacturer. However, it is precisely this aspect of standardization which is often welcomed by the users of the product. For example, transfer machines, such as are used in automobile and other large-scale manufacture of piece parts, consist of many work heads joined by a conveyor, thus automating the materials-handling part of the process. Under pressure from the major auto makers, the mounting dimensions of the work heads have had to be standardized so that the entire machine can be reassembled out of the modules of several manufacturers instead of starting anew at each model change. Major manufacturers of diesel locomotives have likewise modulized and standardized their products, with great cost reductions relative to the former, highly individualistic steam locomotives, an action which helped displace them after 1945.

Standardization also determines the nature of design practice. Especially when the specifications also give data on strength and performance as well as the usual dimensions, it is only necessary to compute loads approximately and then select the nearest standard sizes. Much design effort is thereby saved, especially on detail drawings, bills of material, and so forth. This approach also simplifies programming when computer-aided design is used. *See* COMPUTER-AIDED DESIGN AND MANUFACTURING.

Trade associations. Trade associations are the principal sources of American industrial standards. These involve standardization over an entire product line. In general, their scope is considerably less than that within firms with extensive standardization programs, but the technical and policy considerations in the two levels of standardization are quite similar. Trade standards are primarily concerned with specifying overall dimensions, so that products of different manufacturers may be used interchangeably; with performance, so that customers know what they are buying; and with certain design features, such as major materials, in order to assure proper function. In some cases, dimensional standards particularly must be related to standards in other industries; for instance, an American butter dish must accommodate the standard 4-oz (113.6-g) sticks in which butter is packed. Like national standards to which they are closely related, trade association standards should be established on the basis of as broad a consensus as possible within the industry. If standards were established such that any required burden of retooling and product change would fall in a discriminatory fashion upon only certain members of the industry, legal remedy would certainly be sought under the American antitrust laws.

In large part for this reason, a major problem in the development of standards by trade associations is the resistance to simplification which often arises. A line of least resistance then is simply to take everybody's prestandardization product lines, put them together, and call the result a standard. Historically, this has resulted in some standards having so many different products as scarcely to merit the name. For instance, there are dozens of steel compositions bearing standard designations but virtually identical to several others; in spite of these redundancies, they remain as standards.

In structural steel, some shapes were eliminated soon after the major standards were set about 1905, but simplification such as that in the equivalent British standards, for instance, never took place. Thus, the "nominal sizes" of steel beams are turned out in several weights per foot. However, economic analysis of weight-bearing capacity indicates that the lightest version of each size is usually best, except in the relatively unusual cases where there is a dimensional restriction such as a need for increased clearance.

National standards organizations. The principal industrial countries have official agencies that approve, consolidate, and in some cases establish standards. Among them are the British Standards Institution, German Institute for Norms, and the American National Standards Institute, which issue the BSS, DIN, and ANSI (formerly ASA) standards, respectively. ANSI is a federation of 900 companies, large and small, and about 200 trade, technical, professional, labor, and consumer organizations. It does not itself develop standards, but coordinates and promotes the voluntary development of national standards by industries provided that these have been established according to detailed rules for achieving a consensus among the producing industries, consumers (through ANSI's Consumer Council), relevant government agencies, and other interested parties.

There are about 10,000 approved ANSI standards, dealing with dimensions, ratings, terminology and symbols, test methods, and performance and safety specifications for equipment, components, and products. Major applications are in construction, electrical, electronic, mechanical, and nuclear products and processes, piping and welding, heating, air conditioning and refrigeration, information systems, medical devices, physical distribution, photography and motion pictures, and textiles. Most activities in the private and public sector use standardized products at some stage and specify them routinely in their purchases. ANSI standards have been embodied in building codes and many other government regulations, notably those of the Occupational Health and Safety Administration (OSHA). Related activities by government agencies also establish standards, as in agricultural grading.

A problem in setting standards in general, but particularly national ones, is the frequent need to anticipate future technologies and needs at an early stage of product development, knowing that later any substantial changes would be very difficult. For instance, domestic electrical supply is standardized at 110 V ac, 60 Hz, in the United States, and 210–250 V ac, 50 Hz, in Europe. These standards were set before 1900 and were, in part, the result of perceived limits on the mechanical capabilities of electrical machinery. Today, it would be better to have 400-Hz current such as used in aircraft systems because it would allow the use of smaller, faster, more efficient, and less expensive motors, as well as better functioning fluorescent lights. But any change would require costly conversions of transmission systems and consuming equipment.

Another example is that of railroad gages, which were set at 4 ft 8½ in. (1435 mm) in most of Europe and North America and narrower in much of the rest of the world. Yet a wider gage of 7 ft (2 m) would allow for more efficient use and higher speeds, albeit at higher construction costs. Such changes have never been made, but other changes in established gages have often proved expensive, requiring both new construction and inconvenient changes for passengers and transshipments of freight.

International standards. The national standardization agencies are members of a wide variety of international groupings and United Nations agencies. The principal ones are the International Organization of Standardization (ISO) and the International Electrotechnical Commission (IEC). These attempt to coordinate national activities and promote cooperation in the area of standardization. Several of the more than 50 organizations deal with weights and measures, which have always been the subject of extensive international coordination and agreement. Others engage in transnational or international activities in the standardization of many products or cover specific regional issues and requirements. Some, like the European Economic Community (EEC), the International Telecommunications Union (ITU), the International Civil Aviation Organization (ICAO), or the administration of the General Agreement on Tariffs and Trade (GATT) mainly have other political, economic, and scientific concerns but must necessarily take note of standardization as part of their work. *See* ENGINEERING; ENGINEERING AND ARCHITECTURAL CONTRACTS; ENGINEERING DESIGN.

[JOHN E. ULLMANN]

Diesel cycle

An internal combustion engine cycle in which the heat of compression ignites the fuel. Compression-ignition engines, or diesel engines, are thermodynamically similar to spark-ignition engines. The sequence of processes for both types is intake, compression, addition of heat, expansion, and exhaust. Ignition and power control in the compression-ignition engine are, however, very different from those in the spark-ignition engine.

Usually, a full unthrottled charge of air is drawn in during the intake stroke of a diesel engine. A compression ratio between 12 and 20 is used, in contrast to a ratio of 4 to 10 for the Otto spark-ignition engine. This high compression ratio of the diesel raises the temperature of the air during the compression stroke. Just before top center on the compression stroke, fuel is sprayed into the combustion chamber. The high temperature of the air ignites the fuel, which burns almost as soon as it is introduced, adding heat. The combustion products expand to produce power, and exhaust to complete the cycle.

Performance of a diesel engine is anticipated by analyzing the action of an air-standard diesel cycle. An insulated cylinder equipped with a frictionless piston contains a unit air mass. The metal cylinder head is alternately insulated and then uncovered for heat transfer.

Air is compressed until the piston reaches the top of the stroke. Then the air receives heat through the cylinder head and expands at constant pressure along path *a-b* as shown in Fig. 1, moving the piston part way down through the cylinder. Then the cylinder head is insulated, and the air completes its expansion along path *b-c* at constant entropy. The cylinder head is uncovered, and with the piston at the bottom of its stroke, a constant-volume heat rejection takes place on path *c-d*. The insulation is replaced, and the cycle is completed with an isentropic compression on path *d-a*.

An increase in compression ratio $r = v_d/v_a$ increases efficiency η, the increase becoming less at higher compression ratios. Another characteristic of the diesel cycle is the ratio of volumes at the end and at the start of the constant-pressure heat-addi-

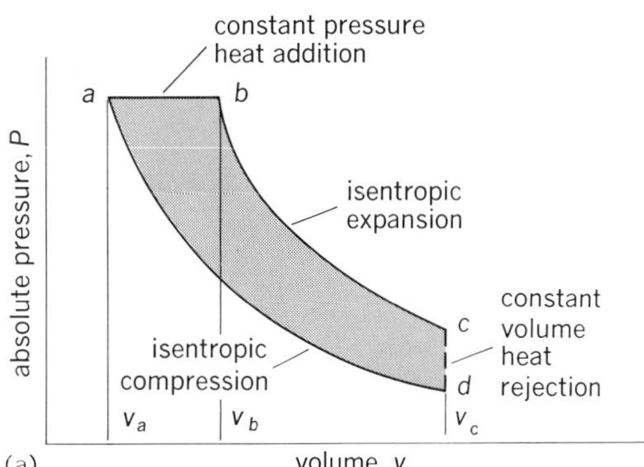

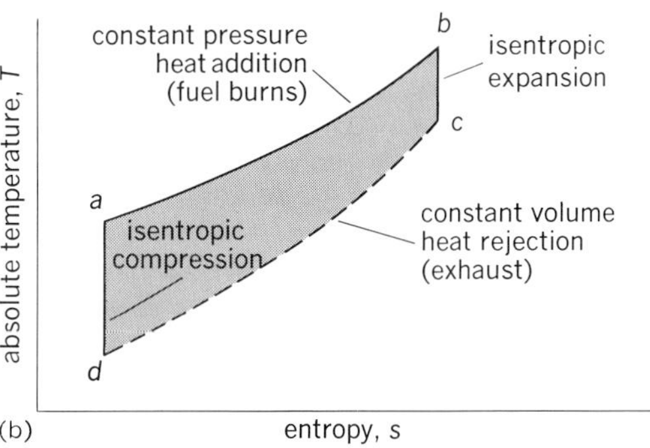

Fig. 1. Ideal diesel cycle, with (*a*) pressure-volume and (*b*) temperature-entropy bases.

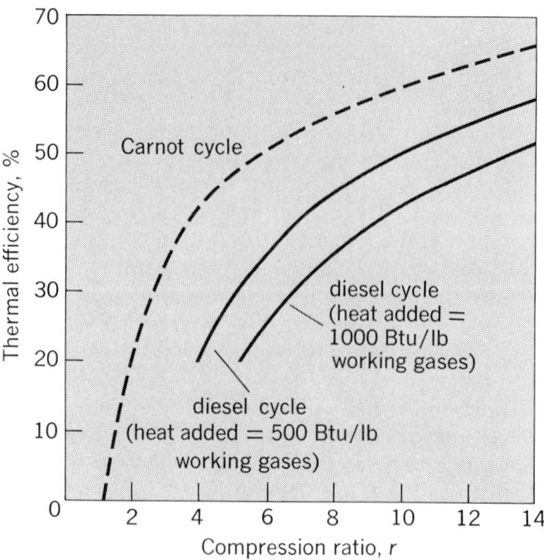

Fig. 2. Thermal efficiency of an ideal diesel cycle.

tion process. This cutoff ratio $r_c = v_b/v_a$ measures the interval during which fuel is injected. For an engine to develop greater power output, the cutoff ratio is increased and heat continues to be added further into the expansion stroke. The air-standard cycle shows that, with less travel remaining during which to expend the additional heat energy as mechanical energy, the efficiency of the engine is reduced. Conversely, efficiency increases as the cutoff ratio decreases, so that a diesel engine is most efficient at light loads. Specifically, the equation below may be written, where $k = c_p/c_v$ the ratio

$$\eta = 1 - \frac{1}{r^{k-1}}\left[\frac{r_c^{\,k} - 1}{k(r_c - 1)}\right]$$

of specific heat of the working substance at constant pressure to its specific heat at constant volume (Fig. 2). In the limiting case when cutoff ratio r_c approaches unity, diesel cycle efficiency approaches Otto cycle efficiency for cycles of the same compression ratio.

In an acutal engine with a given compression ratio, the Otto engine has the higher efficiency. However, fuel requirements limit the Otto engine to a compression ratio of about 10, whereas a diesel engine can operate at a compression ratio of about 15 and consequently at a higher efficiency.

In addition, heat can be added earlier in the cycle by injecting fuel during the latter part of the compression process *d-a*. This mode of operation is the dual-combustion or semidiesel cycle. With most of the heat added near peak compression, semi-diesel efficiency approaches Otto cycle efficiency at a given compression ratio.

[THEODORE BAUMEISTER]

Bibliography: T. Baumeister (ed.), *Standard Handbook for Mechanical Engineers*, 8th ed., 1978; L. Lichty, *Combustion Engine Processes*, 7th ed., 1967; D. S. Williams (ed.), *The Modern Diesel: Development and Design*, 14th ed., 1977.

Diesel engine

An internal combustion engine operating on a thermodynamic cycle in which the ratio of compression ($R_v = 15\pm$) of the air charge is sufficiently high to ignite the fuel subsequently injected into the combustion chamber. The engine differs essentially from the more prevalent mixture engine in which an explosive mixture of air and gas or air and the vapor of a volatile liquid fuel is made externally to the engine cylinder, compressed to a point some 200°F below the ignition temperature, and ignited at will as by an electric spark. The diesel engine utilizes a wider variety of fuels with a higher thermal efficiency and consequent economic advantage under many service applications. The true diesel engine, as projected by R. Diesel and as represented in most low-speed engines, such as about 300 rpm, uses a fuel-injection system where the injection rate is delayed and controlled to maintain constant pressure during combustion. Adaptation of the injection principle to higher engine speeds, such as 1000–2000 rpm, has necessitated departure from the constant pressure specification because the time available for fuel injection is so short (milliseconds). Combustion proceeds with little regard to the constant-pressure specification. High peak pressures may be developed. Yet nonvolative (distillate) fuels are burned to advantage in these engines which cannot be rigorously identified as true diesels but which properly should be called commercial diesels. In ordinary parlance all such engines are classified as diesels. *See* DIESEL CYCLE; OTTO CYCLE.

Identifying alternative features of diesel engine types include: (1) two-cycle or four-cycle operation; (2) horizontal or vertical piston movement; (3) single or multiple cylinder; (4) large (5000 hp) or small (50 hp); (5) cylinders in line, opposed, V, or radial; (6) single acting or double acting; (7) high (1000–2000 rpm), low (100–300 rpm), or medium speed; (8) constant speed or variable speed; (9) reversible or nonreversible; (10) air injection or solid injection; (11) supercharged or unsupercharged; and (12) single or multiple fuel. Section drawings of two representative engines are given in Figs. 1 and 2, and selected performance data are given in Table 1.

Maximum diesel engine sizes (5000 kw) are less than steam turbines (1,000,000 kw) and hydraulic turbines (300,000 kw). They give high instrinsic and actual thermal efficiency (20–40%); a sample comparative heat balance is shown in Table 2; variation in performance with load is shown in Fig. 3. Control of engine output is by regulation of the fuel supplied but without variation of the air supply (100±% excess air at full load). Supercharging (10–15 psi) increases cylinder weight charge and consequently power output for a given cylinder size and engine speed. With two-cycle constructions, scavenging air (approximately 5 psi) is delivered by crankcase compression, front end compression, or separate rotary, reciprocating, or centrifugal blowers. The engine cylinder may be without valves and with complete control of admission of scavenging air and release of spent gases in a two-port construction, the piston covering and uncovering the ports; or the cylinder may have a single port (for admission or release) uncovered by the main piston at the outer end of its stroke and conventional cam-operated valve in the cylinder head. The objective is to replace spent gases with fresh air by guided flow and high turbulence. The four-cycle engine, with its complement of admis-

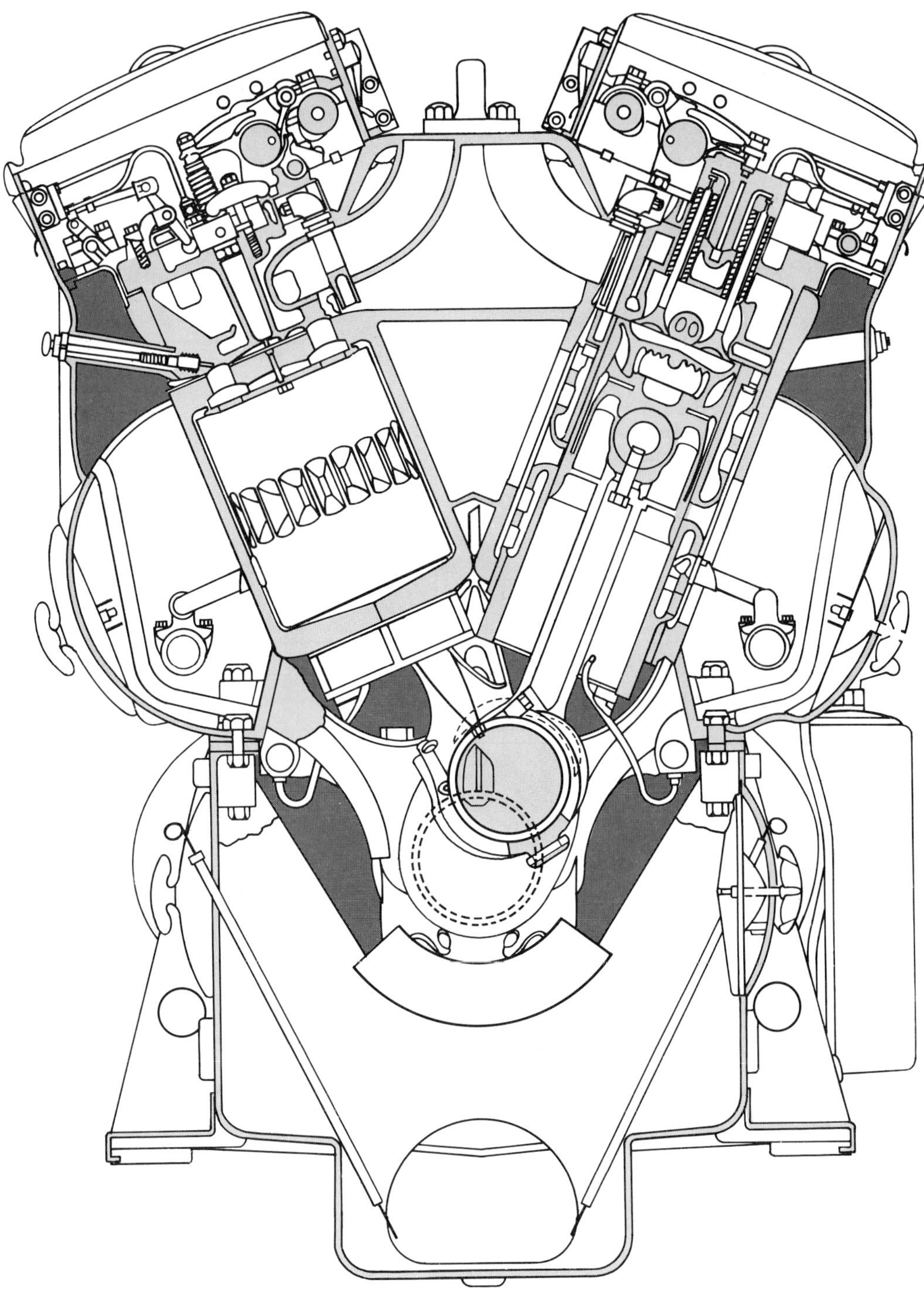

Fig. 1. Section through a locomotive diesel engine. (*General Motors, Electromotive Division*)

sion and exhaust valves on each cylinder, is most effective in scavenging. But the sacrifice of one power stroke out of every two is a frequent deterrent to its selection. Valves are exclusively of the poppet type with the burden of tightness and cooling dominant in the exhaust valve designs. Cylinder heads become complicated structures because of valve porting, jacketing, and spray-valve locations and the accommodation of these to effective combustion, heat transfer, and internal bursting pressures.

Distillate fuel (40° API, 19,000–19,500 Btu/lb, 135,000–140,000 Btu/gal) prevails with locomotive, truck, bus, and automotive applications. Lower-speed engines (stationary and motorship service) burn heavier fuels (for example, 20° API, 18,500–19,000 Btu/lb, 145,000–150,000 Btu/gal). Alternative fuels are burned in dual-fuel and gas diesel engines for stationary service. The main fuel is typically natural gas (90–95%) with oil (5–10%) used to control burning and to stabilize ignition. In the more prevalent liquid-fuel-injection system, the technical problems are numerous and embrace such elements as pumps, spray nozzles, and com-

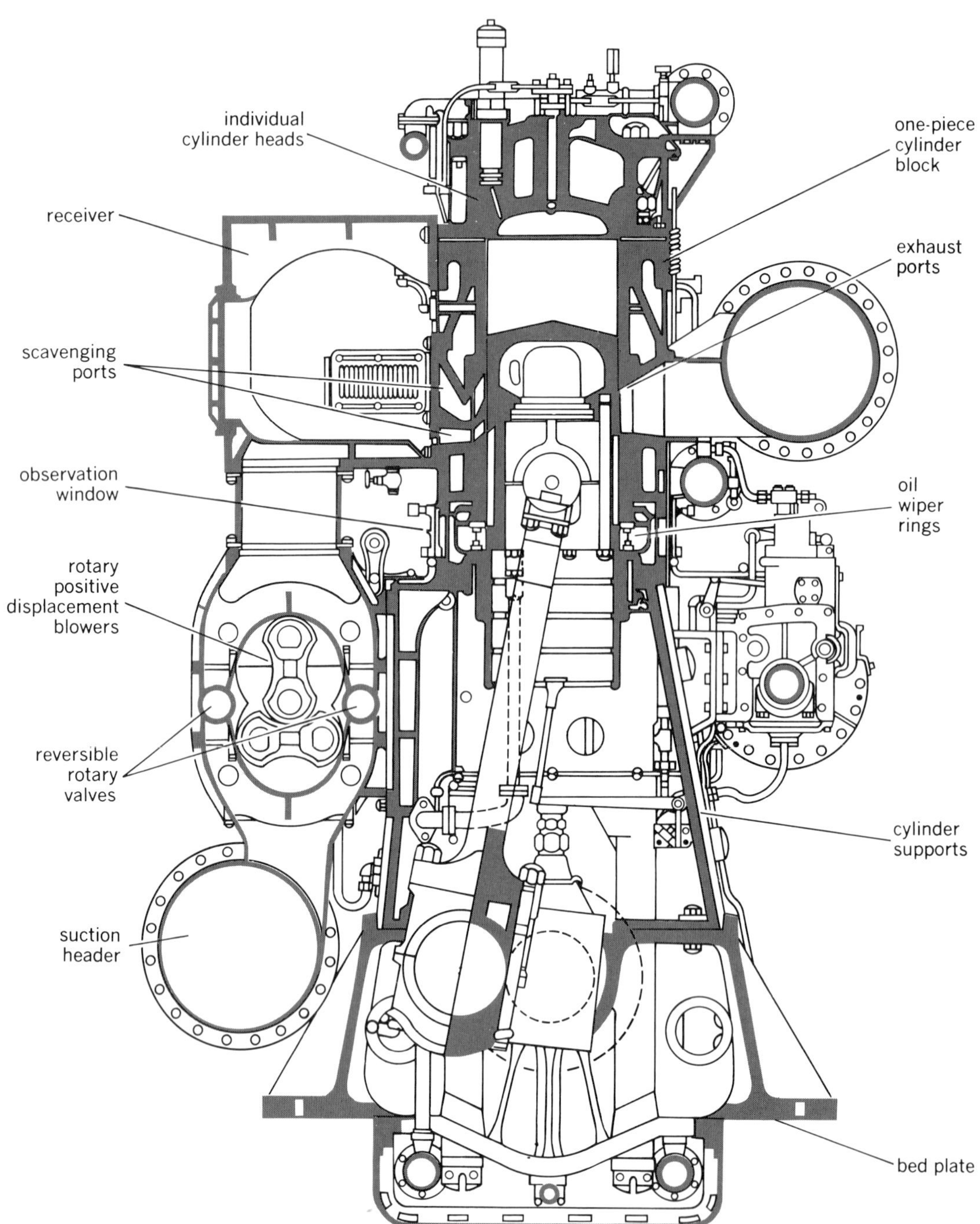

Fig. 2. Section through a Busch-Sulzer two-cycle diesel engine.

Table 1. Performance of selected diesel engine plants

Type of plant	Shaft horse-power (shp)	Ratio of compres-sion, R_v	Brake mean effective pressure, psi	Piston speed, ft/min	Weight, lb/in.³ displace-ment	Weight, lb/shp	Overall thermal efficiency, %
Air injection engine	300–5000	12–15	50–75	600–1000	3–8	25–200	30–35
Solid injection, compression ignition							
Automotive	20–300	12–15	75–100	800–1800	2.5–4	7–25	25–30
Railroad	200–2500	12–15	60–90	800–1800	2.5–4	10–40	30–35
Stationary							
Unsupercharged	50–2500	12–15	70–80	600–1500	2.5–5	10–100	30–35
Supercharged	60–4000	10–13	110–125	600–1500	2.5–5	7.5–75	32–40
Dual fuel, stationary							
Unsupercharged	50–2500	12–15	80–90	600–1500	2.5–5	10–100	30–35
Supercharged	60–4000	10–13	120–135	600–1500	2.5–5	7.5–75	32–40

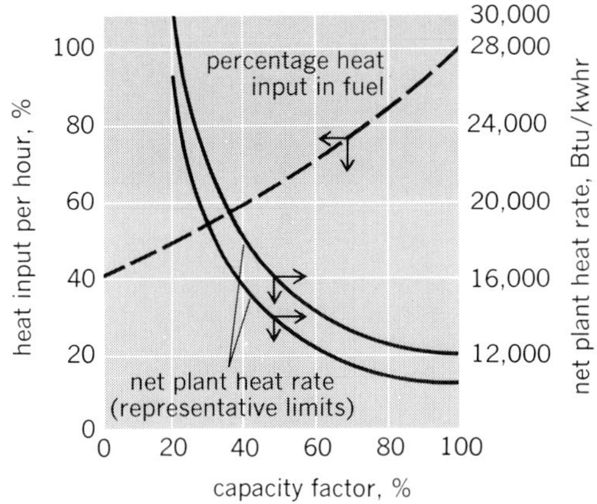

Fig. 3. Heat rate and heat input curves for selected diesel-engine generating plants.

Table 2. Approximate allocation of losses in internal combustion engine plants

Type of loss	Mixture engines, %	Injection (diesel) engines, %
Output	20	33
Exhaust losses	40	33
Cooling system losses	40	33
Other	<1	1
Total (input)	100	100

bustion chambers for the delivery, atomization, and burning of the fuel in the hot compressed air. There must be accurate timing (measured in milliseconds) for the entire process to give clean, complete combustion without undue excess air. Combustion characteristics of fuels are defined by rigorous specifications and include such factors as viscosity, flash point, pour point, ash, sulfur, basic sediment, water, Conradsen carbon number, cetane number, and diesel index.

Small size (<200 hp) engines are conveniently started by an electric motor and storage battery. Larger engines use compressed air (about 200 psi) introduced through valves in the cylinder head. Starting, with engine-driven generator sets, may be accomplished by motoring the generator.

Cooling systems use water at 120–180°F with radiators, cooling towers, and cooling ponds employed for conservation and reclamation. Lubrication costs can become prohibitive with inadequate engine maintenance. Foundations must be designed to handle stress loadings and to reduce vibration. Exhaust systems should be equipped with wavetrap silencers or mufflers. Filters on air and fuel supply are good insurance for engine reliability. *See* INTERNAL COMBUSTION ENGINE.

[THEODORE BAUMEISTER]

Bibliography: T. Baumeister (ed.), *Standard Handbook for Mechanical Engineers*, 8th ed., 1978; Diesel Engine Manufacturers Association, *Standard Practices for Stationary Diesel and Gas Engines*, 1958; L. C. Lichty, *Combustion Engine Processes*, 7th ed., 1967; D. S. Williams (ed.), *The Modern Diesel: Development and Design*, 14th ed., 1977.

Differential

A mechanism which permits a rear axle to turn corners with one wheel rolling faster than the other. An automobile differential is located in the case carrying the rear-axle drive gear (Fig. 1).

The differential gears consist of the two side gears carrying the inner ends of the axle shafts, meshing with two pinions mounted on a common pin located in the differential case. The case carries a ring gear driven by a pinion at the end of the drive shaft. This arrangement permits the drive to be carried to both wheels, but at the same time as the outer wheel on a turn overruns the differential case, the inner wheel lags by a like amount.

Special differentials (Dana, Borg Warner, Eaton) permit one wheel to drive the car by a predetermined amount even though the opposite wheel is on slippery pavement; they have been used on racing cars for years and are now used by a number of car manufacturers (Fig. 2).

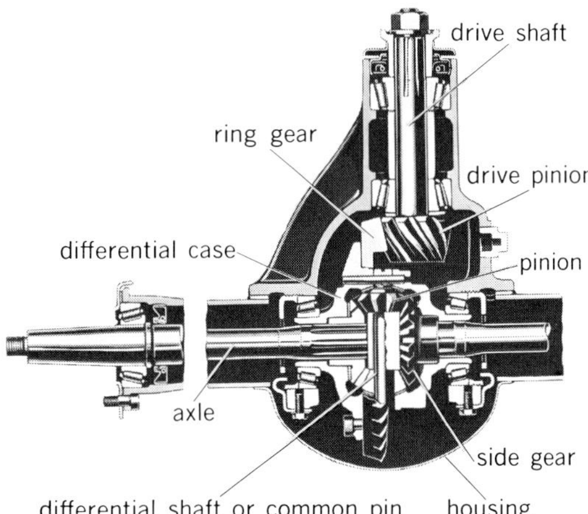

Fig. 1. Commonly used rear-axle differential. (*Chrysler*)

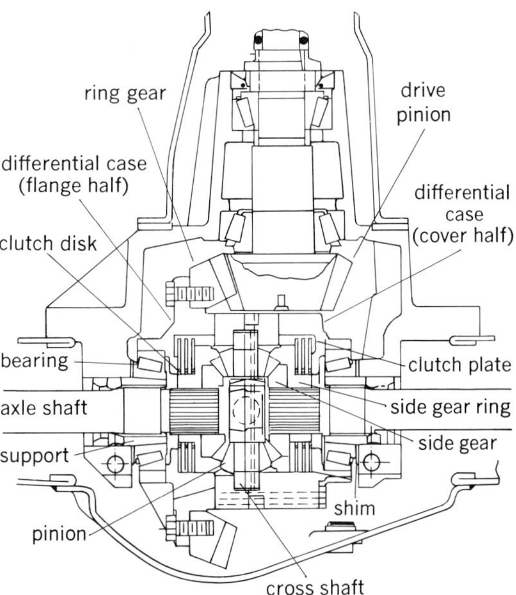

Fig. 2. One type of nonspinning differential. (*Salisbury Axle Division, Dana Corp.*)

In operation, engine torque applied to the differential case causes the angular contacts on the case to bear on corresponding angles on the differential pinion pins. Two contacts 180° apart push one pin to the right; the other two contacts, 180° apart, spaced 90° from the first, push the other pin to the left. On one of the pins are two pinions meshing with the right-side gear; on the other are two pinions meshing with the left-side gear. Diameters on the pinions concentric with the teeth bear against the side gear rings splined to the inner end of the axle shafts and force the rings to apply clutches which connect the differential case directly to the rings. The pinions also load the side gears axially, the latter contributing their thrusts to the clutches. The drive thus passes from the differential case to the rings splined to the respective rear-axle shafts, providing a frictional drive to one wheel, even though the other is on ice or slippery pavement. This nullifies the conventional differential action by the predetermined amount. *See* AUTOMOTIVE TRANSMISSION; TRUCK TRANSMISSION. [HAROLD FISCHER]

Bibliography: D. W. Dudley (ed.), *Gear Handbook: The Design, Manufacture and Application of Gears*, 1962; B. A. Shtipelman, *Design and Manufacture of Hypoid Gears*, 1978.

Differential transformer

An iron-core transformer with movable core. A differential transformer produces an electrical output voltage proportional to the displacement of the core. It is used to measure motion and to sense displacements. It is also used in measuring devices for force, pressure, and acceleration which are based on the conversion of the measured variable to a displacement.

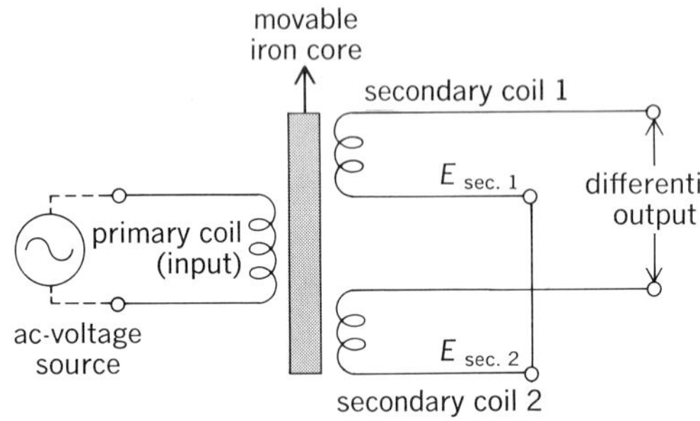

Fig. 1. Basic circuit of differential transformer.

Various available configurations, some translational and others rotational, all employ the basic circuit of Fig. 1: a primary winding, two secondary windings, and a movable core. The primary winding is energized with alternating voltage. The two secondary windings are connected in series opposition, so that the transformer output is the difference of the two secondary voltages. When the core is centered, the two secondary voltages are equal and the transformer output is zero. This is the balance or null position. When the core is displaced from the null point, the two secondary voltages are no longer alike and the transformer produces an output voltage. With proper design, the output voltage varies linearly with core position over a small range. Motion of the core in the opposite direction produces a similar effect with 180° phase reversal of the alternating output voltage. *See* TRANSFORMER.

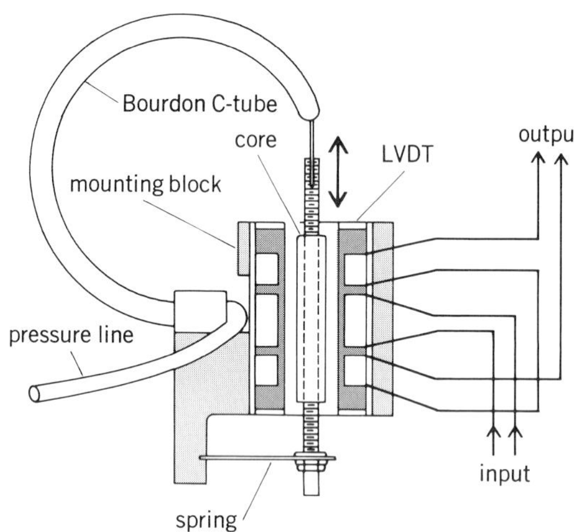

Fig. 2. Bourdon-tube pressure transducer employing a linear variable differential transformer (LVDT). *(From E. E. Herceg, Handbook of Measurement and Control, Schaevitz Engineering, 1976)*

The principal advantages of the differential transformer over other displacement transducers, such as the resistance potentiometer, are absence of contacts and infinite resolution. No friction is introduced by the measurement, and movement smaller than a microinch (25 nm) can be sensed. The separation between coil and core makes the differential transformer useful in difficult and dangerous environments. Stability of the null makes it ideal as a null sensor in self-balancing devices and servomechanisms. Typical applications are machine tool inspection and gaging, pressure measurement (Fig. 2), liquid level control, load cells, and gyroscopic instruments.

The linear variable differential transformer (LVDT) (Fig. 3) is the commercially prevalent form. A rotary version is also manufactured. The E-pickoff (Fig. 4) is an older device. Its linearity is not as good, and its principal use is as a null sensor. Both translational and rotational E-pickoffs are made.

The amplitude of the ac output voltage forms a V-shaped curve when plotted against core position; the phase angle abruptly reverses by 180° at the null point. When the bottom of the V-curve is ex-

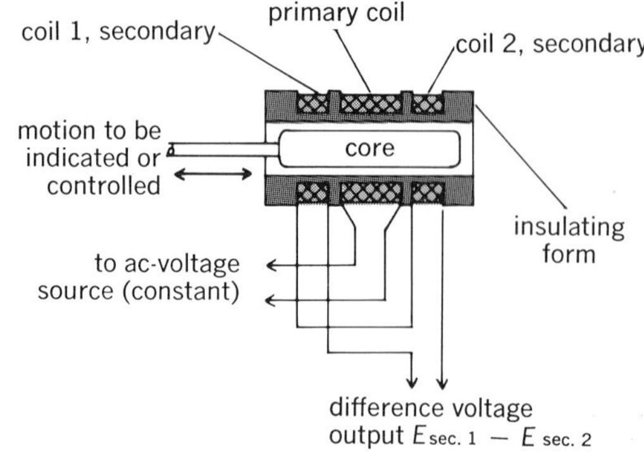

Fig. 3. Linear variable differential transformer (LVDT). *(Schaevitz Engineering)*

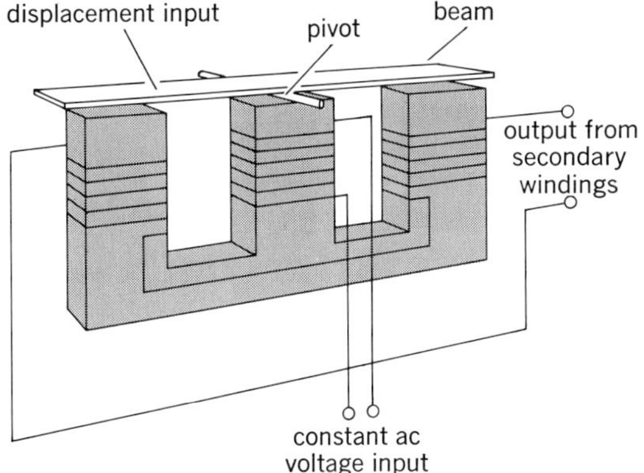

Fig. 4. E-shaped differential transformer. (*From P. J. O'Higgins, Basic Instrumentation, McGraw-Hill, 1966*)

amined in closer detail, it is seen that the output voltage at balance is not exactly zero. The small residual null voltage consists of higher harmonics of the input frequency, as well as a fundamental frequency component 90° out of phase (called the quadrature component).

Electronic signal conditioning is commonly employed to eliminate the residual and to make the transducer usable with standard dc instrumentation. The electronic circuit consists of an ac oscillator (carrier generator) to drive the input windings, plus a demodulator and an amplifier to convert the output into dc. The microelectronics can be built into the transformer housing, and the resulting package is sold as a dc-LVDT.

[GERALD WEISS]

Bibliography: W. R. Ahrendt and C. J. Savant, Jr., *Servomechanism Practice*, 1960; E. O. Doebelin, *Measurement Systems*, 1966; E. E. Herceg, *Handbook of Measurement and Control*, 1976.

Digital computer

Any device for performing mathematical calculations on numbers represented digitally; by extension, any device for manipulating symbols according to a detailed procedure or recipe. The class of digital computers includes microcomputers, conventional adding machines and calculators, digital controllers used for industrial processing and manufacturing operations, store-and-forward data communication equipment, and electronic data-processing systems.

In this article emphasis is on electronic stored-program digital computers. These machines store internally many thousands of numbers or other items of information, and control and execute complicated sequences of numerical calculations and other manipulations on this information in accordance with instructions also stored in the machine. The first section of this article discusses digital system fundamentals, reviewing the components and building blocks from which digital systems are constructed. The following section introduces the stored-program general-purpose computer in more detail and indicates the characteristics by which system performance is measured. The final section traces the history of stored-program digital computer systems and shows how the requirements of new applications and the development of new technologies have influenced system design.

DIGITAL SYSTEM FUNDAMENTALS

A digital system can be considered from many points of view. At the lowest level it is a network of wires and mechanical parts whose voltages and positions convey coded information. At another level it is a collection of logical elements, each of which embodies certain rules, but which in combination can carry out very complex functions. At a still higher level, a digital system is an arrangement of functional units or building blocks which read (input), write (output), store, and manipulate information.

Codes. Numbers are represented within a digital computer by means of circuits that distinguish various discrete electrical signals on wires inside the machine. Theoretically, a signal on a wire could be made to represent any one of several different digits by means of the magnitude of the signal. (For example, a signal from 0 to 1 volt could represent the digit zero, a signal between 1 and 2 volts could represent the digit one, and so on up to a signal between 9 and 10 volts, the digit nine.) In practice, the most reliable and economical circuit elements distinguish between only two signal levels, so that a signal between 0 and 5 volts may represent the digit zero and a signal between 5 and 10 volts, the digit one. These two-valued signals make it necessary to represent numbers and symbols using a corresponding base-two or binary system. Table 1 lists the first 20 binary numbers and their decimal equivalents.

Data are stored and manipulated within a digital computer in units called words. The binary digits (called bits), which make up a word, may represent either a binary number or a collection of binary-coded alphanumeric characters. For example, the two-letter word "it" may be stored in a 16-bit computer word as follows, making use of the code shown in Table 2:

0100100101010100

The computer word merely contains a binary pattern of alternating 1's and 0's, and it is up to the computer user to determine whether that word

Table 1. Counting from 0 to 19 by decimal and binary numbers

Decimal number	Binary number
00	00000
01	00001
02	00010
03	00011
04	00100
05	00101
06	00110
07	00111
08	01000
09	01001
10	01010
11	01011
12	01100
13	01101
14	01110
15	01111
16	10000
17	10001
18	10010
19	10011

Table 2. American standard alphabetic code for binary representation of letters

11000001	A	11001110	N
11000010	B	01001111	O
01000011	C	11010000	P
11000100	D	01010001	Q
01000101	E	01010010	R
01000110	F	11010011	S
11000111	G	01010100	T
11001000	H	11010101	U
01001001	I	11010110	V
01001010	J	01010111	W
11001011	K	01011000	X
01001100	L	11011001	Y
11001101	M	11011010	Z

should be interpreted as the English word "it" or as the decimal number 18,772.

Logical circuit elements. Two kinds of logical circuits are used in the design and construction of digital computers: decision elements and memory elements. A typical decision element provides a binary output as a function of two or more binary inputs. The AND circuit, for example, has two inputs and an output which is 1 only when both inputs are 1. A memory element stores a single bit of information and is set to the 1 state or reset to the 0 state, depending on the signals on its input lines. And because such a circuit can be caused alternately to store 0's and 1's from time to time, a memory element is commonly called a flip-flop.

These two basic logical elements are all that are required to construct the most elaborate and complex digital arithmetic and control circuits. A simple example of such a circuit is shown in Fig. 1. Here the object is to perform a simple binary count, as shown in the table at the bottom of Fig. 1. As long as control signal C is equal to 1, the counting continues. When the control input is 0, the counter is to remain in whatever state it had last counted to. Two flip-flops are used, labeled $Q1$ and $Q2$, and will be made to count through the sequence 0,1,2,3,0,1, ... To understand the design, it is necessary to introduce one more concept, the complementary output of a flip-flop. Each flip-flop generally has two output wires, which are always of opposite polarity. When flip-flop $Q1$ is storing a 1, output $Q1$ is 1 and the complementary output (which is labeled $\overline{Q1}$ and pronounced Q1 bar) is 0. When the flip-flop contains a 0, the $\overline{Q1}$ output is 1 and the $Q1$ output is 0.

To analyze the circuit, note first that, when control input C is 0, the outputs of all AND gates are 0 and, because the reset and set inputs to both flip-flops are 0, the flip-flops will remain in whatever state they last reached. Now suppose that $Q1$ and $Q2$ both contain 0 and that the control input becomes 1. While flip-flop $Q2$ contains a 0, its $Q2$ output is also 0 and AND gate number 1 (labeled AND 1) is effectively turned off so that the reset and set inputs to $Q1$ are both 0. Thus flip-flop $Q1$ will remain in the 0 state. For the same reason AND gate 4 will also be turned off, and the reset input to flip-flop $Q2$ will be 0. However, from flip-flop $Q2$ complementary output $\overline{Q2}$ will be in the 1 state, and (while the control input is 1) AND gate 5 will be turned on and the set input to $Q2$ will be 1. Flip-flop $Q2$ will thus be turned on by the first clock pulse to occur after C is turned on; and from one clock pulse time to the next the two flip-flops will change from the (0,0) state to the (0,1) state. A careful review of the indicated circuits will show that the counter will indeed go through the count sequence as shown, as long as the control input is 1. The logic equations in Fig. 1 represent another way of describing the circuit and may be used in place of the more cumbersome diagram.

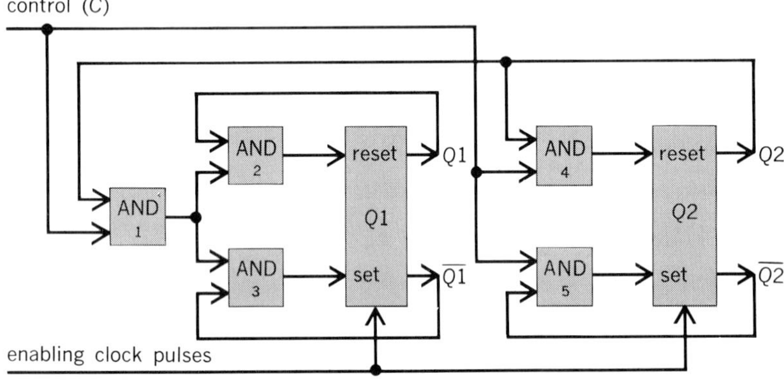

Binary count:

Q1	Q2
0	0
0	1
1	0
1	1
0	0
0	1
. . .	. . .

Logic equations:

$SQ1 = C \cdot Q2 \cdot \overline{Q1}$

$RQ1 = C \cdot Q2 \cdot Q1$

$SQ2 = C \cdot \overline{Q2}$

$RQ2 = C \cdot Q2$

Note: $C \cdot Q2$ means C AND $Q2$

Fig. 1. Simple digital counting circuit.

Physical components. The logical elements described in the paragraphs above are the fundamental conceptual components used in virtually all digital systems. The actual physical components which were used to realize conceptual gates and flip-flops in some specific piece of equipment are dependent on the status of electronic technology at the time the equipment was designed. In the 1950s the earliest commercial computers used vacuum tubes, resistors, and capacitors as components. A flip-flop typically required a dozen or more such components in these first-generation computers. Between the late 1950s and middle 1960s, solid-state transistors and diodes replaced the vacuum tubes, and the resulting second-generation systems were considerably more reliable than their first-generation predecessors; they were also smaller and consumed less power. But the number of electronic components per conceptual logical component remained about the same—a dozen or more for a flip-flop.

Since the mid-1960s the integrated circuit (IC) has been the principal logical building block for digital systems. Early digital integrated circuits contained a single flip-flop or gate, and the use of these components permitted designers of the early third-generation systems to provide much more capability per component than was possible with the first- or second-generation technology. Since the mid-1960s integrated circuit technology has consistently improved, and by 1980 typical large-scale-integration (LSI) integrated circuits contained thousands of flip-flops and gates.

System building blocks. On a completely different conceptual level, a digital computer can be regarded as being composed of functional, system building blocks, containing (among other things) subassemblies of the fundamental logical components. A computer viewed at this level may be described in an oversimplified fashion by the diagram of Fig. 2. The computation and control block (often called the central processing unit, or CPU) is constructed entirely of logical elements of the kind described above. The main memory, which may store from a few thousand to several million binary digits, and the input/output and auxiliary memory devices (the so-called peripheral equipment) are specialized devices available over a range of speeds and operating characteristics.

Main memory is a building-block capable of storing data or instructions in bulk for use by the computation and control portion of the computer. The important characteristics of a memory are capacity, access time, and cost. Capacity is the amount of data that the computer can store. Access time is the maximum interval between a request to the memory for data and the moment when the memory can provide that data. Cost is measured for each bit stored. For first-generation systems, designers used a variety of technologies in realizing main memory: mercury delay lines, electrostatic storage tubes, and magnetic drums all appeared in various products. But second- and third-generation systems were almost exclusively built using magnetic core main memories. Starting in the early 1970s, the integrated circuit memory was introduced, and is now the most widely used technology.

Input/output and auxiliary memory peripherals represent the other major computer building blocks. Equipment is now and has from the beginning been available for feeding information to the computer from paper tape and punched cards, and for receiving data from the computer and printing it, or punching it on tape or cards. But in the intervening years, designers have provided additional output devices which record computer data on microfilm, or plot data on graphs, or use data to control physical devices such as valves or rheostats. They have also designed input equipment which feeds the computer data from laboratory instruments, and from devices which scan documents and "read" printed characters. Data can be transmitted to and from the computer over ordinary telephone lines, and a wide variety of devices generally called terminals, make it possible for people to send data to, or receive requested data from, a computer system located hundreds or thousands of miles away.

The earliest auxiliary memory equipment recorded data on reels of magnetic tape. Magnetic tape units are still very widely used, for although they are slow in comparison to the operating speeds of modern computers—it typically takes 2–30 min to read all the data on a 2400-ft (732-m) reel of tape, depending on the speed of the tape unit—they make it possible to store large volumes of data at low cost by virtue of the low cost of the tape itself. The other widely used auxiliary memory devices are the magnetic disk and drum, both of which provide faster access to data than do the tape units, but at higher cost per bit of data stored.

STORED PROGRAM COMPUTER

Components and building blocks described in the preceding paragraphs could be organized in a multitude of different ways. The first practical electronic computers, constructed during the latter part of World War II, were designed with the specific purpose of computing special mathematical functions. They did their jobs very well, but even while they were under construction, engineers and scientists had come to realize that it was possible to organize a digital computer in such a way that it was not oriented toward some particular computation, and could in fact carry out any calculation desired and defined by the user. The basic machine organization invented and constructed at that time was the stored-program computer, and it continues to be the fundamental basis for each of the hundreds of thousands of computing systems in use today. It has also become a system component, since the microcomputer is simply a stored program computer on a single integrated-circuit chip.

The concept of the stored-program computer is simple and can be described with reference to Fig. 2. Main memory contains, in addition to data and the results of intermediate computations, a set of instructions (or orders, or commands, as they are sometimes called); these specify how the computer is to operate in solving some particular problem. The computation and control section reads these instructions from the memory one by one and performs the indicated operations on the specified data. The instructions can control the reading of data from input or auxiliary memory peripherals, and (when the prescribed computations are completed) can send the result to auxiliary memory, or to output devices where it may be printed, punched, displayed, plotted, and so forth. The feature that gives this form of computer organization its great power is the ease with which instructions can be changed; the particular calculations carried out by the computer are determined entirely by a sequence of instructions stored in the computer's memory; that sequence can be altered completely by simply reading a new set of instructions into the memory through the computer input equipment.

Instructions. To understand better the nature of the stored-program computer, consider in more detail the kinds of of instructions it can carry out and the logic of the computation and control unit which interprets and implements the instructions. Because the instructions, like the data, are stored in computer words, one begins by examining how an instruction is stored in a word. As an example, assume one is looking at a small computer with words 16 bits long, and assume further that an instruction is organized as shown in Fig. 3. In this simple computer an instruction has two parts: the first 5 bits of the word specify which of the computer's repertoire of commands is to be carried out.

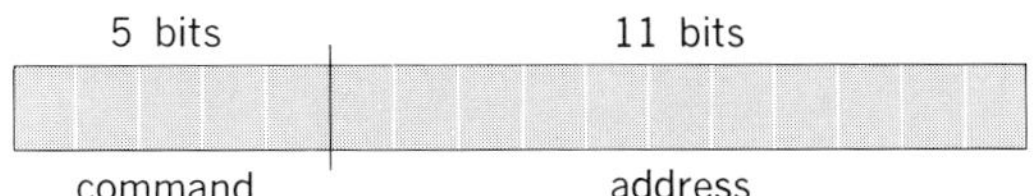

Fig. 3. Sixteen-bit instruction.

DIGITAL COMPUTER

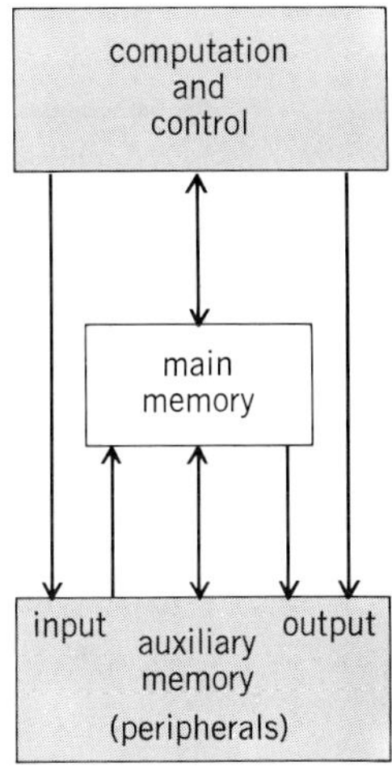

Fig. 2. Block diagram of a digital computer.

and the last 11 bits generally specify the address of the word referred to by the command. A 5-bit command permits up to 32 different kinds of instructions in the computer, and an 11-bit address permits one to address up to 2048 different memory locations directly.

Typical instruction types for a computer of this kind are listed below.

Load. Load the number from the prescribed memory location into the arithmetic unit.

Store. Store the number from the arithmetic unit in the memory at the prescribed memory location.

Add. Add the contents of the addressed memory location to the number in the arithmetic unit, leaving the result in the arithmetic unit.

Subtract. Subtract the contents of the addressed memory location from the number in the arithmetic unit, leaving the result in the arithmetic unit.

Branch. If the number in the arithmetic unit is zero or positive, read the next instruction from the address in the next instruction register as usual. If the number in the arithmetic unit is negative, store the address from the branch instruction itself in the next-instruction register, so that the next instruction carried out will be retrieved from the address given in the branch instruction.

Halt. Stop; carry out no further instructions until the operator presses the RUN switch on the console.

Input. Read the next character from the paper tape reader into the addressed memory location and then move the tape so a new character is ready to be read.

Output. Type out the character whose code is stored in the right-hand half of the addressed memory location.

With the exception of the branch command, the preceding instructions are easy to interpret and to understand. The load and store commands move data to and from the arithmetic unit, respectively. The add and subtract commands perform arithmetic operations, each using the number previously left in the arithmetic unit as one operand, and a number read from a designated memory location as the other. The halt command simply tells the computer to stop and requires intervention by the operator to make the computer initiate computation again. The input and output commands make possible the reading of information into the computer memory from a paper tape input device, and the printing out of the results from previous computations on an output typewriter.

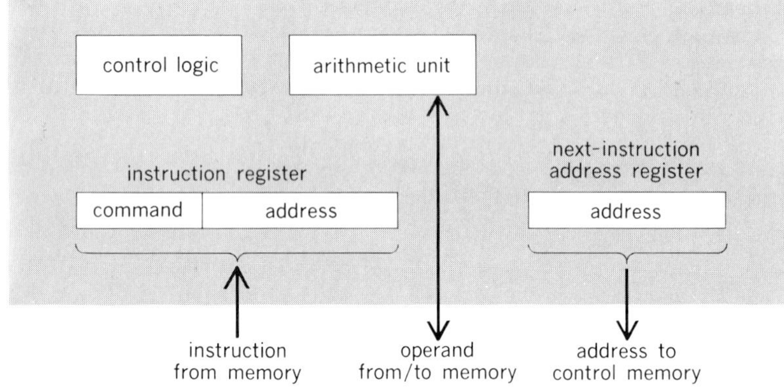

Fig. 4. Computation and control unit.

To understand the branch command, consider how the computation and control unit of Fig. 4 uses the instructions in the memory. To begin with, the instructions which are to be carried out must be stored in consecutive storage locations in memory. Assume that the first of a sequence of commands is in memory location 100. Then the "next-instruction address register" in the computation and control unit (Fig. 4) contains the number 100, and the following sequence of four events takes place: (1) read, (2) readdress, (3) execute, and (4) resume.

Read. The control logic reads the next instruction to be carried out from the memory location whose address is given by the next instruction address register. The instruction coming from memory is stored in another register called the instruction register. (In this example the next-instruction address register started out containing the number 100, and so the instruction in memory location 100 is transferred to the instruction register.)

Readdress. The control logic now adds unity to the number in the next instruction address register. (In the present example this changes the number in the next instruction address register from 100 to 101. The result is that, when the computer has interpreted and carried out the instruction from location 100, following the rules given in the third and fourth steps below, it will return to the read step above and next interpret and carry out the instruction from location 101.)

Execute. The instruction from location 100 is now in the instruction register and must be carried out. The control logic first looks at the command portion of the instruction in the left-hand 5 bits of the register and interprets or decodes it to determine what to do next. If the instruction is add, subtract, load, or output, the control logic first uses the address in the instruction register—the address associated with the command—and reads the word from that addressed location in memory; it then proceeds to load the word into the arithmetic unit, add it to or subtract it from the number in the arithmetic unit, or transfer it to the output typewriter, depending on the command. If the command is store or input, the control logic collects a number from the arithmetic unit or the paper tape reader and then transfers that number to a location in the memory whose address is given in the instruction register. If the command is halt, the control logic simply prevents all further operations, pending a signal from the operator console.

If the command is branch, the control logic begins by looking at the number in the arithmetic unit. If that number is zero or positive, the control logic goes on to the fourth step below. If the number is negative, however, the control logic causes the address in the instruction register to be transferred to the next-instruction address register before going on to the fourth step. The computer will then continue with one sequence of commands if the previous arithmetic result was positive, and with another if the result was negative. This seemingly simple operation is one of the most important features a computer possesses. It gives the computer a decision-making capability that permits it to examine some data, compute a result, and continue with one of two sequences of calculations or operations, depending only on the computed result.

Resume. As the fourth and final step in the sequence, when the command has been interpreted and carried out properly, the control logic returns to the read step and repeats the entire series of steps.

A sequence of instructions intended to carry out some desired function is called a program; collections of such programs are called software (as distinguished from the equipment, or hardware), and the act of preparing such programs is called programming. Because a computer can perform no useful function until someone has written a program embodying that function, the programming activity is an exceedingly important one and provides a basic limitation to the facility with which the computer can be applied to new areas.

Computer characteristics. A computer installation is complex. Consequently it is difficult to describe a system or to compare the characteristics of two systems without listing their instruction types and describing their modes of operation at some length. Nevertheless, certain important descriptors are commonly used for comparison purposes and are shown in Table 3, where salient characteristics of two typical systems are shown. Definitions of these characteristics can be stated as follows.

Table 3. Typical computer characteristics

Characteristics	Typical systems	
	Large	Small
Memory cycle time	0.08 μs	0.90 μs
Add time	0.17 μs	2.84 μs
Main memory storage capacity	750,000 words	128,000 words
Word length	64 bits	16 bits

Memory cycle time is the time required to read a word from main memory. Most modern computers have integrated-circuit memories with cycle times in the ranges shown in Table 3. Add time is the time required to perform an addition, including the time necessary to extract the addition instruction itself and the operand from memory. Main memory storage capacity is the number of words of storage available to the computation and control unit. Typically, a computer manufacturer gives the buyer some choice; the buyer can purchase enough memory to meet the needs of the expected application. This internal capacity refers to the high-speed internal storage only, and does not include disks, drums, or magnetic tape.

Word length is the number of bits in a computer word.

System cost may vary over a range of 5 or even 10 to 1 for a particular computer because of the great variety of options offered the buyer by the manufacturer—options such as memory size, special instructions for efficiency in certain calculations, and number and type of peripheral devices.

There are obviously a number of other measures which may be used to describe a computer. They include such characteristics as multiplication time, transfer rates between input/output equipment and memory, physical size, power consumption, and the availability of a variety of computing options and special features.

EVOLUTION OF CAPABILITIES

The process by which new circuit and peripheral equipment technologies led to the development of a series of generations of computers was discussed above. But simultaneous with the changes in technology, there came changes in the structure or architecture of computers. These changes were introduced to improve the capability and efficiency of systems, as designers came to understand how computers were actually used.

Computer efficiency. One way of looking at system efficiency is indicated in Fig. 5, where the operation of a computer is shown broken down into the following four parts. (1) Operator time includes such activities as inserting cards into a card reader, loading magnetic tapes onto a tape unit, setting up controls on a computer operator's panel, and reviewing printed results. (2) Input comes to the computer from peripheral devices or from auxiliary memory. The inputs include instructions from the operator, inputs of programs to be run, and inputs of data. (3) Computation, being the principal activity, should occupy relatively much of the total time. (4) Output includes storage of intermediate and final results in auxiliary memory, and printing of results along with instructions or warnings to the computer operator.

In the first generation of computer equipment only one of these activities could be carried out at a time. Between jobs the computer was idle while an operator made ready for the next task. When the operator was ready, the program was read into the computer from some input device and the input data were then loaded. The program operated upon the data and performed necessary calculations. When the calculations were complete, the computer printed out answers, and the operator took steps to set up the next problem.

This series of operations was inefficient, and the designers of second-generation equipment removed some of the inefficiency by arranging input and output operations to be performed directly between the input/output peripherals and the computer memory without interfering with computations. As a result, second-generation computers were able to perform computations while reading in data and printing out replies, and efficiency was greatly enhanced. Figure 6 indicates schematically the organizational change between generations of computers.

First-generation equipment was most efficient while performing tedious and lengthy computations. The input/output capabilities of the second generation made them useful in applications where large volumes of data had to be handled with relatively little computation—applications such as billing, payroll, and inventory control. At the same time, the great capability and increased reliability of second-generation systems encouraged engineers to apply them to situations where the computer acts as a control element. In military aircraft, in oil refineries and chemical plants, in research laboratories, and in factories, the computer received data directly from measuring instruments, performed appropriate calculations, and as a result made adjustments in the aircraft engine

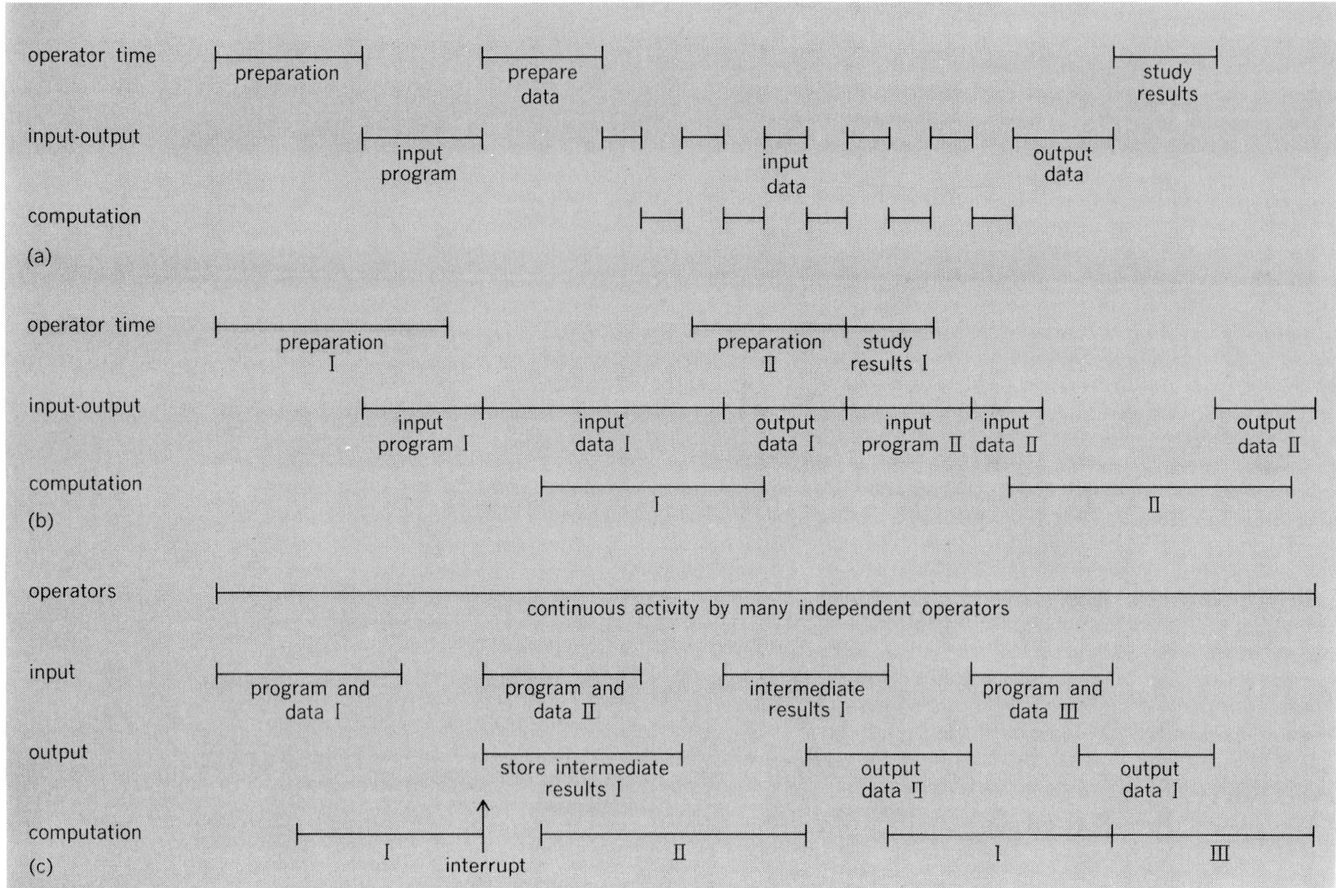

Fig. 5. Comparison of efficiency for three generations of computers: (*a*) first, (*b*) second, and (*c*) third.

thrust, the flow of raw materials in the plant or factory, or the experimental setup in the laboratory. These new application areas led to two important developments in computer design. The first was a new set of input/output equipments that could be connected to process instruments, converting instrument signals into digital quantities and back again.

The second development, which evolved from the use of the computer in second-generation control applications, was the interrupt. The processes or activities under control provided data to, and required action from, the computer at random times. These random requests to the computer required rapid response, either because the process required quick control action on the part of the computer or because the process data supplied by instruments were rapidly changing and had to be stored before the data were lost. The interrupt feature, built into the computation and control unit, solved these problems by providing circuits that could stop the computer after any given instruction, enable it to carry out some special program in response to the external interrupt, and finally permit it to return to precisely the position in the program where it had originally been interrupted.

The logic to achieve these ends is quite simple. When an interrupt signal is received from an external device, the logic circuitry waits until the computer has completed the current instruction and then stores the next instruction address in a specially reserved storage location. The logic substitutes a standard predetermined address in the next instruction address register and continues. The computer of course next executes the instruction at the standard address, and the rest is up to the programmer. The programmer must have inserted a special program at this standard address, and the program must respond to the conditions that caused the interrupt (by inputting data from instrumentation or by taking previously specified control action, for example) and then must return control to the original program at exactly the same place where the program was interrupted. Though the interrupt was originally used largely in control applications, it is now employed in virtually all systems to notify the computer when transfers to and from peripheral equipment are completed.

Third-generation computer. In the mid-1960s a new set of trends in computer applications was becoming apparent, and a third generation of computer systems became available to cope with those trends. The usefulness and flexibility of the stored-program computer, together with its improved cost-performance ratio, made it apparent that the computer had the basic power to perform a great variety of small and large tasks simultaneously. For example, the speed and capability of a computer were such that it could simultaneously collect data from a test run or experiment; maintain a file of inventory records on a disk memory; answer inquiries on status of specific items in inventory, such as inquiries entered at random from a dozen different cathode-ray-tube display devices; and assemble or compile programs for users at numerous terminals, all remote from the computer and all working independently on different problems.

A computer system which serves a number of users in this way is called a time-sharing system. To perform in such complex applications, third-generation computers required elaborations of the features found in second-generation computers (Fig. 6). Main and auxiliary memories became bigger and cheaper, more input/output equipments became available, input/output channels improved so that a larger number of simultaneous operations were possible, and interrupt structures grew more flexible. In addition, many new features appeared. Three of them, memory protection circuitry, rapid context switching, and the operating system, are worth discussing briefly.

To understand the implications of third-generation computers and the usefulness of these new features, consider the last portion of Fig. 5. Here the computer is engaged in a variety of different tasks, and it switches back and forth between them as it finishes one portion of a job or as it is interrupted to perform some higher-priority job. This rapid switching from job to job led to the development of context-switching equipment. At the time of a changeover from one program to another, the programmer is able with a single instruction to interchange the instruction address register, together with the contents of various arithmetic and control registers, between the job he or she had previously been working on and the new job to be performed. Second-generation computers, without this context-switching feature, require a long sequence of commands every time a change is made from one task to another.

The memory protect feature is important for other reasons. Some programs executed in a time-sharing system such as that depicted by the third-generation portion of Fig. 6 would be new programs being run for the first time. Errors are common in such new programs, and it is the nature of the computer that such errors could have serious effects on system operation. For example, a user's program might accidentally store data in memory space reserved for supervisory or monitor programs—the programs that determine job priorities and reconcile conflicting input/output requirements. To keep such critical memory areas from being destroyed or modified by unchecked programs, designers have made it possible for the user to designate certain areas of memory as protected, and have ensured that only the monitor or supervisory programs can access these particular areas.

The operating system is the set of supervisory programs which manages the system, keeps it operating efficiently, and takes over many of the scheduling and monitoring functions which had previously been the function of the computer operator. It is typically supplied by the computer hardware manufacturer, and was first used with some second-generation systems, where it included little more than interrupt processing and error-handling routines, together with programs to control input/output operations for the user. Its size and importance have both grown with time, and it currently includes those early functions, as well as facilities for accessing compilers and utilities, scheduling jobs to maximize system throughput, managing data-communication facilities, controlling user access to protected data, maintaining records on system usage, and so forth.

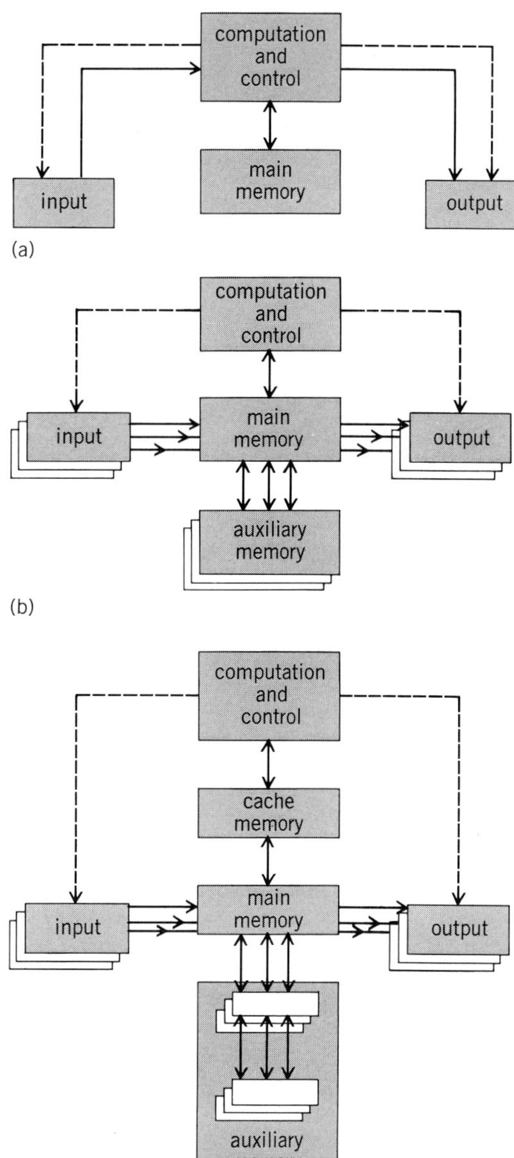

Fig. 6. Evolution of organization for four generations of computers: (*a*) first; (*b*) second and third; (*c*) fourth.

Fourth-generation systems. Computer memory has become increasingly important with the passing generations, as users have found it useful to store more and more business, government, engineering, and scientific data in machine-readable form. Starting with the third generation, but increasingly with computers introduced in the early 1970s, computer-system architects have provided a hierarchy of memory devices to improve system performance and give the user access to very large memory capacities. Figure 6*c* shows how a very large system may make use of such a hierarchy. A so-called cache memory has been interposed between the central processing unit and the main memory, and auxiliary memory has been split into two parts.

The cache is a relatively small but very-high-speed memory which stores the data and programs currently being used by the central processing unit. When the central processing unit needs an instruction or data, it sends a main memory address to the cache memory. If the requested information is stored in the cache, it is immediately delivered to the central processing unit; if it is not in the cache, special cache memory hardware requests a block of information from main memory, and that block, which includes the data requested by the central processing unit, is delivered to the cache and displaces another block stored there. Because programs contain many branches, a typical program may repeatedly access a relatively small portion of the main memory, and so a small fast cache can be very effective in increasing the central processing unit's instruction-processing rate. (The cache's great speed makes it correspondingly expensive, and therefore not economical for use as main memory.)

A typical large fourth-generation system may make use of a similar hierarchical arrangement between main and auxiliary memory. Main memory in a large system may contain 750,000 words. But a system equipped with a virtual storage translator can give the user-programmer the illusion of working with 16,000,000 words. This is done by supplying the 750,000 words of real memory with temporary virtual addresses, and transferring data automatically from auxiliary to main storage as it is needed. For example, suppose the central processing unit needs a program which starts at address 2,500,000. The virtual storage translator may as- the virtual memory address 2,500,000 to the real address 700,000, and then transfer a block of (say) 1000 words from auxiliary memory to the main memory locations starting with 700,000. Now when the central processing unit next requests, for example, information from location 2,500,001, the translator locates it in real memory location 700,001, and sends it off to the cache.

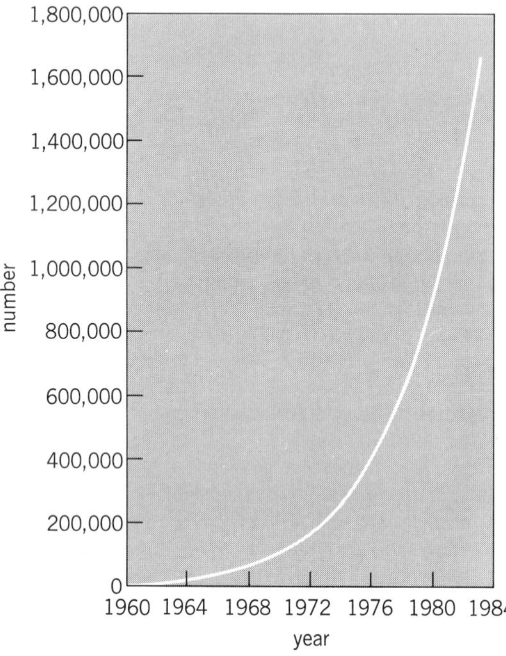

Fig. 7. Computers in use worldwide, made by United States companies.

Finally, in some very large systems there may also be an auxiliary memory hierarchy, in which a very large, cheap, but relatively slow auxiliary memory delivers data or programs as required to the primary auxiliary memory.

In all these hierarchical arrangements, the guiding principle is that the larger, slower, cheaper memory supplies data as needed to the smaller, faster, more expensive memory. The various levels of memory are invisible to the user; the hardware and, where necessary, the operating system make the various hierarchical levels deliver data and programs to the user without the necessity for any special action on his or her part.

Industry growth. The versatility of the digital computer has led to its application in a wide variety of industries and activities. The evolution of the integrated circuit has made it possible for designers to provide ever cheaper systems, and as a result the computer has become economical for use even in small organizations. Since they are so numerous, the number of digital computers in use has increased remarkably. The development of the microcomputer—the stored-program computer on an integrated-circuit chip—has made possible the introduction of the personal computer, which can supply entertainment, educational, record-keeping, and computing capabilities in the home at a relatively low cost. The net result of these developments is indicated in Fig. 7, which shows the growth in the number of computers in use each year, and forecasts how that growth will continue.

The influence of the digital computer is extending far beyond the realm of computing. The microcomputer is being used, and will increasingly be used, as a component in a variety of apparently noncomputing applications. Already it is used in games and toys. Increasingly it will find use in automobiles, appliances, tools, and many other artifacts.

[MONTGOMERY PHISTER, JR.]

Bibliography: C. G. Bell and A. Newell, *Computer Structures: Readings and Examples*, 1971; M. Phister, Jr., *Data Processing Technology and Economics*, 1979; Special issue on computer architecture, *Commun. ACM*, vol. 21, no. 1, January 1978.

Dimensioning

Assigning of dimensions on a mechanical drawing. A dimension on a drawing is a labeled measure in a straight line of the breadth, height, or thickness of a part, the angular position of a line, or the location of a detail such as a hole or boss. Units for machine-part dimensions are inches or, when over 72 in., feet and inches. Dimensions specify the size and shape of the part as required by the designer and aid the workman in constructing the part.

Two plans of dimensioning are used on mechanical drawings, as illustrated. One is called point-to-point or chain dimensioning; each length is dimensioned from the end of the preceding one. These dimensions are usually taken directly from the designer's sketch; however tolerances are cumulative and, rather than averaging out, usually add or subtract so that intermediate clearances may be disturbed. In the other plan, called datum dimensioning or the reference-line method, all dimension

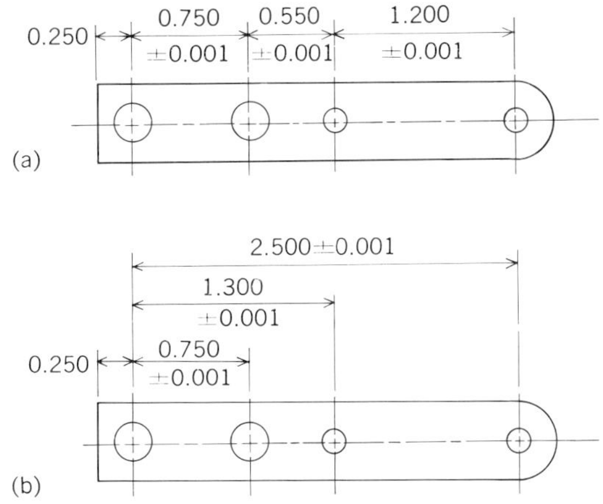

Two standard methods for marking dimensions on mechanical drawings. (*a*) Point-to-point dimensioning. (*b*) Datum dimensioning.

lines in each direction extend from a datum or reference line or plane which is usually the first machined surface. Here, tolerances are not cumulative and a mating part will be within the specified limits. *See* DRAFTING; TOLERANCE.

[PAUL H. BLACK]

Bibliography: T. E. French and C. J. Vierck, *Fundamentals of Engineering Drawing and Graphic Technology*, 3d ed., 1972.

Direct current

Electric current which flows in one direction only through a circuit or equipment. The associated direct voltages, in contrast to alternating voltages, are of unchanging polarity. Direct current corresponds to a drift or displacement of electric charge in one unvarying direction around the closed loop or loops of an electric circuit. Direct currents and voltages may be of constant magnitude or may vary with time.

Batteries and rotating generators produce direct voltages of nominally constant magnitude (illustration *a*). Direct voltages of time-varying magnitude are produced by rectifiers, which convert alternating voltage to pulsating direct voltage (illustration *b* and *c*). *See* GENERATOR.

Direct current is used extensively to power adjustable-speed motor drives in industry and in transportation (illustration *d*). Very large amounts of power are used in electrochemical processes for the refining and plating of metals and for the production of numerous basic chemicals. *See* ELECTROPLATING OF METALS.

Direct current ordinarily is not widely distributed for general use by electric utility customers. Instead, direct-current (dc) power is obtained at the site where it is needed by the rectification of commercially available alternating current (ac) power to dc power. Solid-state rectifiers ordinarily are employed to supply dc equipment from ac supply lines. Rectifier dc supplies range from tiny devices in household electronic equipment to high-voltage dc transmission links of at least hundreds of megawatts capacity. *See* ELECTRIC POWER SYSTEMS.

Many high-voltage dc transmission systems have been constructed throughout the world since 1954. Very large amounts of power, generated as ac and ultimately used as ac, are transmitted as dc power. Rectifiers supply the sending end of the dc link; inverters then supply the receiving-end ac power system from the link. High-voltage dc transmission often is more economical than ac transmission when extremely long distances are involved. *See* ALTERNATING CURRENT; TRANSMISSION LINES.

[D. D. ROBB]

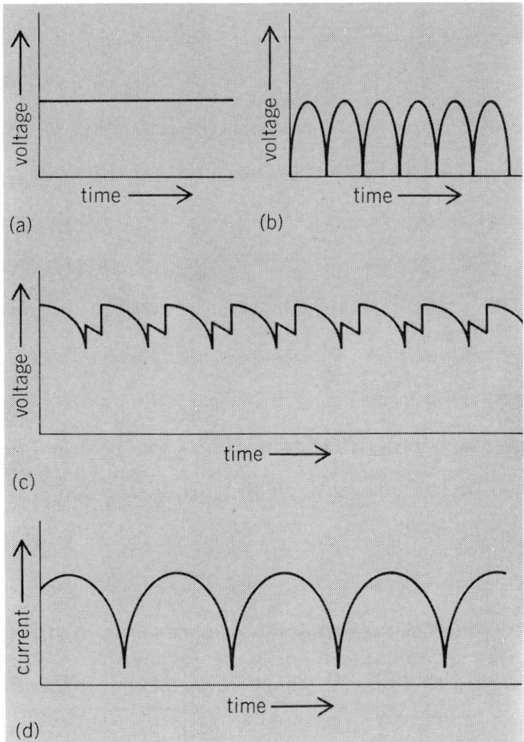

Typical direct currents and voltages. (*a*) Output from a battery; (*b*) full-wave rectified voltage; (*c*) output of the rectifier station of a high-voltage dc transmission link (*from E. W. Kimbark, Direct Current Transmission, vol. 1, Wiley-Interscience, 1971*). (*d*) Current in a rectifier-supplied dc motor (*from A. E. Fitzgerald, C. Kingsley, and A. Kusko, Electric Machinery, 3d ed., McGraw-Hill, 1971*).

Direct-current circuit theory

An analysis of relationships within a dc circuit. Any combination of direct-current (dc) voltage or current sources, such as generators and batteries, in conjunction with transmission lines, resistors, inductors, capacitors, and power converters such

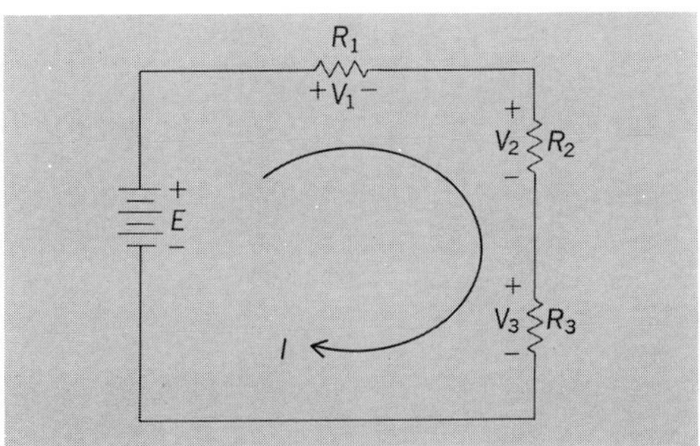

Fig. 1. Simple series circuit.

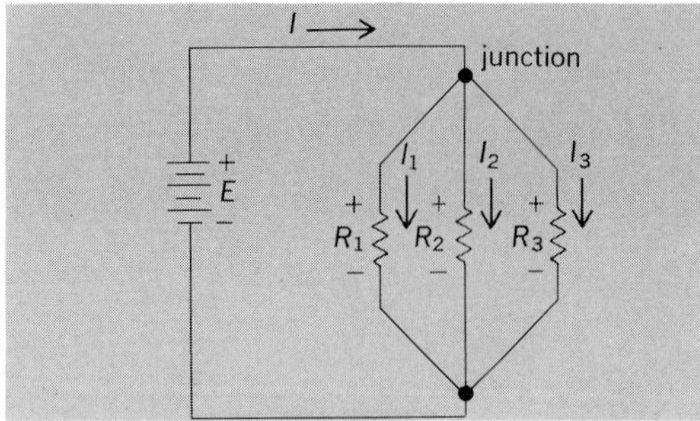

Fig. 2. Simple parallel circuit.

as motors is termed a dc circuit. Historically the dc circuit was the first to be studied and analyzed mathematically. *See* CIRCUIT (ELECTRICITY).

Classification. Circuits may be identified and classified into simple series and parallel circuits. More complicated circuits may be developed as combinations of these basic circuits.

Series circuit. A series circuit is illustrated in Fig. 1. It consists of a battery of E volts and three resistors of resistances R_1, R_2, and R_3, respectively. The conventional current flows from the positive battery terminal through the external circuit and back to the negative battery terminal. It passes through each resistor in turn; therefore the resistors are in series with the battery.

Parallel circuit. The parallel circuit, shown in Fig. 2, consists of a battery paralleled by three resistors. In this case the current leaving the positive terminal of the battery splits into three components, one component flowing through each resistor, then recombining into the original current and returning to the negative terminal of the battery.

Physical laws of circuit analysis. The operation of the basic series and parallel circuits must obey certain fundamental laws of physics. These laws are referred to as Ohm's law and Kirchhoff's laws in honor of their originators.

Voltage drops. When an electric current flows through a resistor, a voltage drop appears across the resistor, the polarity being such that the voltage is positive at the end where the conventional current enters the resistor. This voltage drop is directly proportional to the product of the current in amperes and the resistance in ohms. This is Ohm's law, expressed mathematically in Eq. (1).

$$V = IR \tag{1}$$

Thus in Fig. 1 the drop across R_1 is V_1 and has the polarity shown. *See* OHM'S LAW.

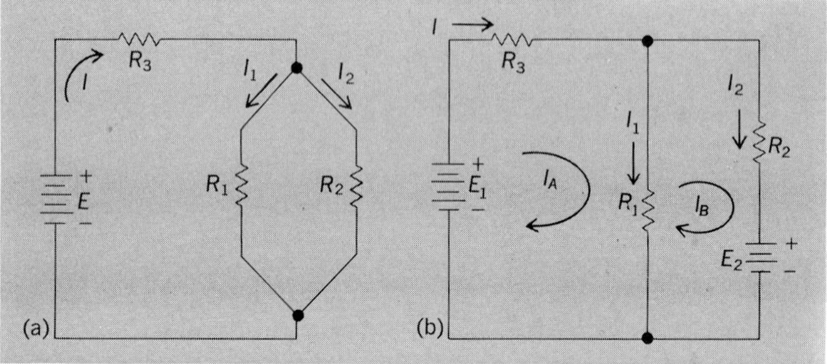

Fig. 3. Series-parallel circuits. (*a*) Single source. (*b*) Multiple sources.

Summation of voltages. The algebraic sum of all voltage sources (rises) and voltage drops must add up to zero around any closed path in any circuit. This is Kirchhoff's first law. In Fig. 1, the sum of the voltages about this closed circuit is as given by Eqs. (2), where the minus signs indicate a voltage

$$E - V_1 - V_2 - V_3 = 0 \tag{2a}$$

$$E = V_1 + V_2 + V_3 \tag{2b}$$

drop. Written in terms of current and resistance, this becomes Eq. (3). From this results the impor-

$$E = I(R_1 + R_2 + R_3) = IR_{\text{total}} \tag{3}$$

tant conclusion that resistors in series may be added to obtain the equivalent total resistance, as shown in Eq. (4). *See* KIRCHHOFF'S LAWS OF ELECTRIC CIRCUITS.

$$R_{\text{total}} = R_1 + R_2 + R_3 + \cdots \tag{4}$$

Summation of currents. The algebraic sum of all currents flowing into a circuit junction must equal the algebraic sum of all currents flowing out of the junction. In the circuit shown in Fig. 2, the current flowing into the junction is I amperes while that flowing out is the sum of I_1 plus I_2 plus I_3. Therefore the currents are related by Eq. (5). This is

$$I = I_1 + I_2 + I_3 \tag{5}$$

Kirchhoff's second law. In this case the same voltage appears across each resistor. If the currents are expressed in terms of this voltage and values of the individual resistors by means of Ohm's law, Eq. (5) becomes Eq. (6). The equivalent resistance

$$I = \frac{E}{R_1} + \frac{E}{R_2} + \frac{E}{R_3} = \frac{E}{R_{\text{eq}}} \tag{6}$$

R_{eq} that can replace the resistances in parallel can be obtained by solving Eq. (6) in the form of Eq. (7).

$$R_{\text{eq}} = \left[\frac{1}{R_1} + \frac{1}{R_2} + \frac{1}{R_3} + \cdots\right]^{-1} \tag{7}$$

Therefore, resistances in parallel are added by computing the corresponding conductances (reciprocal of resistance) and adding to obtain the equivalent conductance. The reciprocal of the equivalent conductance is the equivalent resistance of the parallel combination.

Sources. Sources such as batteries and generators may be connected in series or parallel. Series connections serve to increase the voltage; the net voltage is the algebraic sum of the individual source voltages.

Sources in parallel provide the practical function of increasing the net current rating over the rating of the individual sources; the net current rating is the sum of the individual current ratings.

Series-parallel circuits. More complicated circuits are nothing more than combinations of simple series and parallel circuits as illustrated in Fig. 3.

Single source. Circuits that contain only a single source are readily reduced to a simple series circuit. In the circuit of Fig. 3*a* the parallel combination of R_1 and R_2 is computed and used to replace the parallel combination. The resultant circuit is now a simple series circuit consisting of R_3 and R_{eq} and can readily be solved for the series current if the voltage is known.

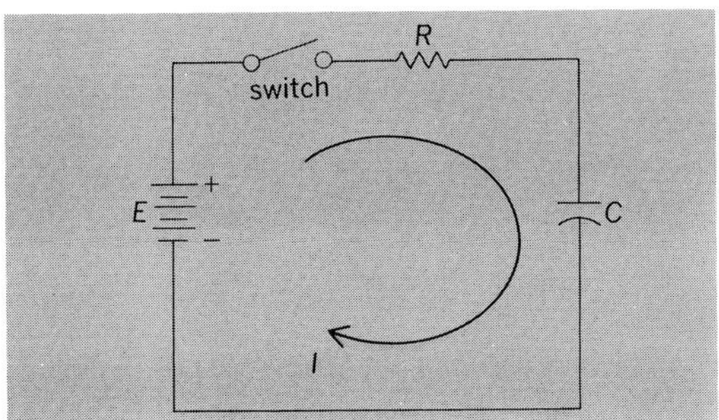

Fig. 4. A series RC circuit.

Multiple sources. Circuits that contain two or more sources located in various branches cannot be reduced to a simple series circuit (Fig. 3*b*). The three basic laws of circuit theory still hold and may be directly applied to provide a simultaneous solution to the loop currents I_A and I_B flowing in each basic series circuit present in the overall network.

In this example the summations of voltages around the individual series circuits or loops are given by Eqs. (8) and (9), which may be solved for

$$E_1 = (R_1 + R_3)I_A - R_1 I_B \quad (8)$$

$$-E_2 = -R_1 I_A + (R_1 + R_2)I_B \quad (9)$$

the mathematical loop currents I_A and I_B. These loop currents can in turn be identified by reference to the circuit where I_A is identical to I, and I_B therefore, as stated by Eq. (10).

$$(I_A - I_B) = I_1 \quad (10)$$

This method may be used to solve any complicated combination of simple circuits. Other methods are also available to the circuit analyst.

Power. The electric power converted to heat in any resistance is equal to the product of the voltage drop across the resistance times the current through the resistance, as stated by Eq. (11*a*). By means of Ohm's law, Eq. (11*a*) may also be written as Eq. (11*b*). The total power dissipated in a circuit

$$P = VI \quad (11a)$$

$$P = VI = V^2/R = I^2R \quad (11b)$$

is the arithmetic sum of the power dissipated in each resistance.

Circuit response. In the circuits mentioned thus far, the circuit responds in an identical manner from the moment the circuit is excited (switches closed) through any extended period of time. This is not true of circuits typified by those of Figs. 4 and 5.

For instance, when the switch of Fig. 4 is first closed, a momentary current limited only by resistance R flows. As time passes, the capacitor, with capacitance C, charges and the voltage across it increases, eventually reaching a value equal to the applied voltage, at which time all flow of current ceases. The circuit current, given by Eq. (12), is in

$$i = (E/R)\epsilon^{-t/RC} \quad (12)$$

amperes. The product RC is known as the time constant of the circuit. The energy W in joules stored in a capacitance at any time is given by Eq. (13).

$$W = CE^2/2 \quad (13)$$

For the circuit of Fig. 5, the initial current upon closing the switch is zero, since any attempt to cause a rate of change of current through the coil whose inductance is L induces a counter emf across the coil or inductor. Eventually this counter emf disappears and a steady-state current E/R flows indefinitely in the circuit.

The current at any time after closing the switch is given by Eq. (14) in amperes. The factor R/L is the

$$i = (E/R)(1 - \epsilon^{-Rt/L}) \quad (14)$$

time constant of the circuit. The energy W stored in an inductance at any time is given by Eq. (15).

$$W = LI^2/2 \quad (15)$$

[ROBERT L. RAMEY]

Bibliography: W. Hayt and J. Kemmerly, *Engineering Circuit Analysis*, 1978; S. L. Oppenheimer and J. P. Borchers, *Direct and Alternating Currents*, 2d ed., 1973; C. S. Siskind, *Electrical Circuits*, 2d ed., 1964.

Direct-current generator

A rotating electric machine which delivers a unidirectional voltage and current. An armature winding mounted on the rotor supplies the electric power output. One or more field windings mounted on the stator establish the magnetic flux in the air gap. A voltage is induced in the armature coils as a result of the relative motion between the coils and the air gap flux. Faraday's law states that the voltage induced is determined by the time rate of change of flux linkages with the winding. Since these induced voltages are alternating, a means of rectification is necessary to deliver direct current at the generator terminals. Rectification is accomplished by a commutator mounted on the rotor shaft. *See* COMMUTATION; ELECTRIC ROTATING MACHINERY; GENERATOR; WINDINGS IN ELECTRIC MACHINERY.

Carbon brushes, insulated from the machine frame and secured in brush holders, transfer the armature current from the rotating commutator to the external circuit. Brushes are held against the commutator under a pressure of 2–2½ psi. Armature current passes from the brush to brush holder through a flexible copper lead. In multipolar machines all positive brush studs are connected together, as are all negative studs, to form the positive and negative generator terminals. In most dc generators the number of brush studs is the same as the number of main poles. In modern machines

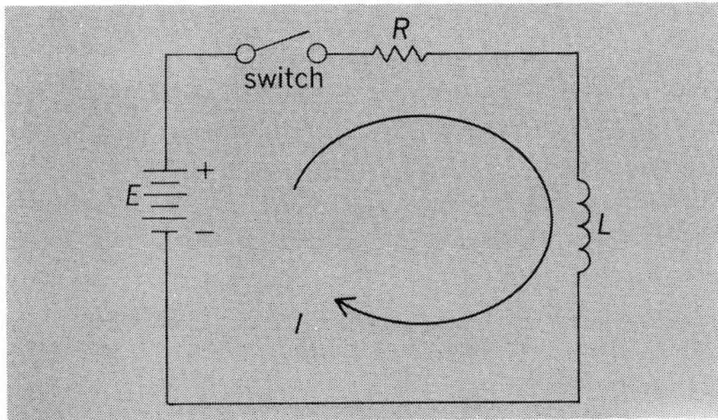

Fig. 5. A series RL circuit.

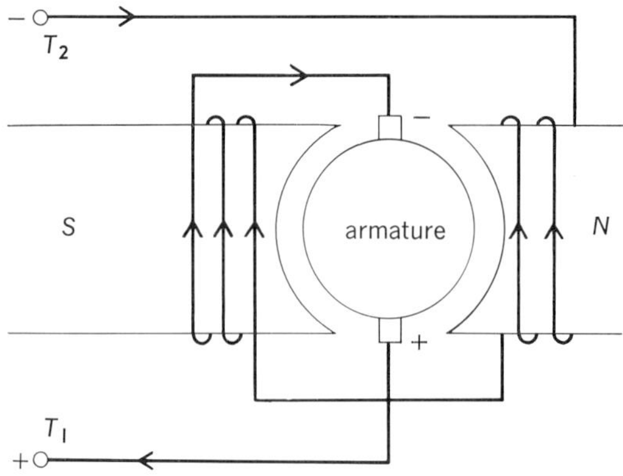

Fig. 1. Series generator.

brushes are located in the neutral position where the voltage induced in a short-circuited coil by the main pole flux is zero. The brushes continuously pick up a fixed, instantaneous value of the voltage generated in the armature winding. *See* COMMUTATOR.

The generated voltage is dependent upon speed n in revolutions per minute, number of poles p, flux per pole Φ in webers, number of armature conductors z, and the number of armature paths a. The equation for the average voltage generated is given below.

$$E_g = \frac{np\Phi z}{60a} \quad \text{volts}$$

The field windings of dc generators require a direct current to produce a magnetomotive force (mmf) and establish a magnetic flux path across the air gap and through the armature. Generators are classified as series, shunt, compound, or separately excited, according to the manner of supplying the field excitation current. In the separately excited generator, the field winding is connected to an independent external source. Using the armature as a source of supply for the field current, dc generators are also capable of self-excitation. Residual magnetism in the field poles is necessary for self-excitation. Series, shunt, and compound-wound generators are self-excited, and each produces different voltage characteristics.

When operated under load, the terminal voltage changes with change of load because of armature resistance drop, change in field current, and armature reaction. Interpoles and compensating, or pole-face, windings are employed in modern generators in order to improve commutation and to compensate for armature reaction. *See* ARMATURE REACTION.

Series generator. The armature winding and field winding of this generator are connected in series, as shown in Fig. 1. Terminals T_1 and T_2 are connected to the external load. The field mmf aids the residual magnetism in the poles, permitting the generator to build up voltage. The field winding is wound on the pole core with a comparatively few turns of wire of large cross section capable of carrying rated load current. The magnetic flux and consequently the generated emf and terminal voltage increase with increasing load current. Figure 3 shows the external characteristic or variation of terminal voltage with load current at constant speed. Series generators are suitable for special purposes only, such as a booster in a constant voltage system, and are therefore seldom employed.

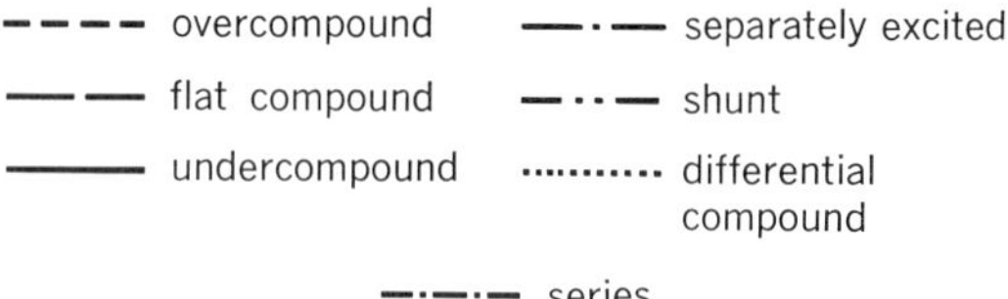

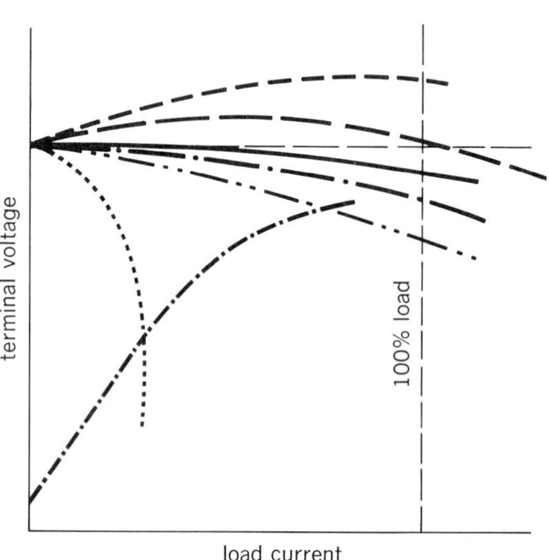

Fig. 3. External characteristics of dc generators.

Shunt generator. The field winding of a shunt generator is connected in parallel with the armature winding, as shown in Fig. 2. The armature supplies both load current I_t and field current I_f. The field current is 1–5% of the rated armature current I_a, the higher value applying to small machines. The field winding resistance is fairly high since the field consists of many turns of small-cross-section wire. For voltage buildup the total field-circuit resistance must be below a critical value; above this value the generator voltage cannot build up. The no-load voltage to which the generator builds up is varied by means of a rheostat in the field circuit. The external voltage characteristic (Fig. 3) shows a reduction of voltage with in-

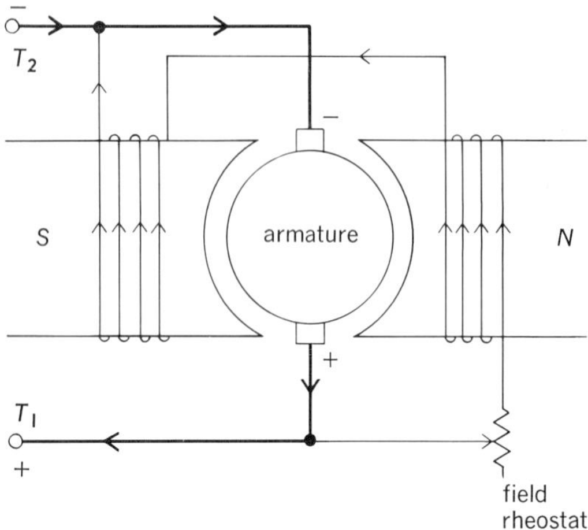

Fig. 2. Shunt generator.

creases in load current, but voltage regulation is fairly good in large generators. The output voltage may be kept constant for varying load current conditions by manual or automatic control of the rheostat in the field circuit. A shunt generator will not maintain a large current in a short circuit in the external circuit, since the field current at short circuit is zero.

The shunt generator is suitable for fairly constant voltage applications, such as an exciter for ac generator fields, battery charging, and for electrolytic work requiring low-voltage and high-current capacity. Prior to the use of the alternating-current generator and solid-stage rectifying devices in automobiles, a shunt generator, in conjunction with automatic regulating devices, was used to charge the battery and supply power to the electrical system. Shunt-wound generators are well adapted to stable operation in parallel.

Compound generator. This generator has both a series field winding and a shunt field winding. Both windings are on the main poles with the series winding on the outside. The shunt winding furnishes the major part of the mmf. The series winding produces a variable mmf, dependent upon the load current, and offers a means of compensating for voltage drop. Figure 4 shows a cumulative-compound connection with series and shunt fields aiding. A diverter resistance across the series field is used to adjust the series field mmf and vary the degree of compounding. By proper adjustment a nearly flat output voltage characteristic is possible. Cumulative-compound generators are overcompounded, flat-compounded, or undercompounded, as shown by the external characteristics in Fig. 3. The shunt winding is connected across the armature (short-shunt connection) or across the output terminals (long-shunt connection). Figure 4 shows the long-shunt connection.

Voltage is controllable over a limited range by a rheostat in the shunt field circuit. Compound generators are used for applications requiring constant voltage, such as lighting and motor loads. Generators used for this service are rated at 125 or 250 volts and are flat or overcompounded to give a regulation of about 2%. An important application is in steel mills which have a large dc motor load. Cumulative-compound generators are capable of stable operation in parallel if the series fields are connected in parallel by an equalizer bus.

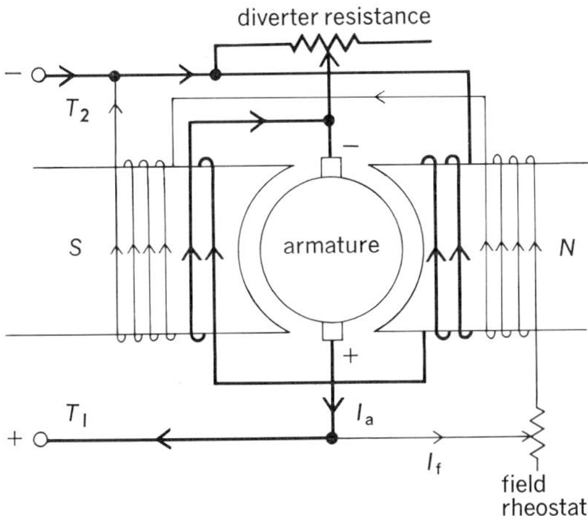

Fig. 4. Cumulative compound generator. The long-shunt connection is seen in this example.

In the differentially compounded generator the series field is connected to oppose the shunt field mmf. Increasing load current causes a large voltage drop due to the demagnetizing effect of the series field. The differentially compounded generator has only a few applications, such as arc-welding generators and special generators for electrically operated shovels.

Separately excited generator. The field winding of this type of generator is connected to an independent dc source. The field winding is similar to that in the shunt generator. Separately excited generators are among the most common of dc generators, for they permit stable operation over a very wide range of output voltages. The slightly drooping voltage characteristic (Fig. 3) may be corrected by rheostatic control in the field circuit. Applications are found in special regulating sets, such as the Ward Leonard system, and in laboratory and commercial test sets.

Special types. Besides the common dc generators discussed in this article, a number of special types may be found in the bibliographical references. These include the homopolar, third-brush, diverter-pole, and Rosenberg generators. For discussion of the Amplidyne, Regulex, and Rototrol *see* DIRECT-CURRENT MOTOR.

Commutator ripple. The voltage at the brushes of dc generators is not absolutely constant. A slight high-frequency variation exists, which is superimposed upon the average voltage output. This is called commutator ripple and is caused by the cyclic change in the number of commutator bars contacting the brushes as the machine rotates. The ripple decreases as the number of commutator bars is increased and is usually ignored. In servomechanisms employing a dc tachometer for velocity feedback, the ripple frequency is kept as high as possible.

[ROBERT T. WEIL, JR.]

Bibliography: D. G. Fink and H. W. Beaty (eds.), *Standard Handbook for Electrical Engineers*, 11th ed., 1978; A. E. Fitzgerald and C. Kingsley, *Electric Machinery*, 3d ed., 1971; S. A. Nasar and L. E. Unnewehr, *Electromechanics and Electric Machines*, 1979.

Direct-current motor

An electric rotating machine energized by direct current and used to convert electric energy to mechanical energy. It is characterized by its relative ease of speed control and, in the case of the series-connected motor, by an ability to produce large torque under load without taking excessive current. Output of this motor is given in horsepower, the unit of mechanical power. Normal full-load values of voltage, current, and speed are generally given.

Direct-current motors are manufactured in several horsepower-rating classifications: (1) subfractional, approximately 1–35 millihorsepower (mhp); (2) fractional, 1/40–1 horsepower (hp); and (3) integral, 1/2 to several hundred horsepower.

The standard line voltages applied to dc motors are 6, 12, 27, 32, 115, 230, and 550 volts. Occasionally they reach higher values.

Normal full-load speeds are 850, 1140, 1725, and 3450 rpm. Variable-speed motors may have limiting rpm values stated.

Protection of the motor is afforded by several types of enclosures, such as splash-proof, drip-proof, dust-explosion-proof, dust-ignition-proof, and immersion-proof enclosures. Some motors are totally enclosed.

The principal parts of a dc motor are the frame, the armature, the field poles and windings, and the commutator and brush assemblies (Fig. 1). The frame consists of a steel yoke of open cylindrical shape mounted on a base. Salient field poles of sheet-steel laminations are fastened to the inside of the yoke. Field windings placed on the field poles are interconnected to form the complete field winding circuit. The armature consists of a cylindrical core of sheet-steel disks punched with peripheral slots, air ducts, and shaft hole. These punchings are aligned on a steel shaft on which is also mounted the commutator. The commutator, made of hard-drawn copper segments, is insulated from the shaft. The segments are insulated from each other by mica. Stationary carbon brushes in brush holders make contact with the segments of the commutator. Copper conductors which are placed in the insulated armature slots are interconnected to form a reentrant lap or wave style of winding. *See* COMMUTATION; WINDINGS IN ELECTRIC MACHINERY.

PRINCIPLES

Rotation of a dc motor is produced by an electromagnetic force exerted upon current-carrying conductors in a magnetic field. For basic principles of motor action *see* MOTOR.

In Fig. 2, forces act on conductors on the left path of the armature to produce clockwise rotation. Those conductors on the right path, whose current direction is reversed, also will have forces to produce clockwise rotation. The action of the commutator allows the current direction to be reversed as a conductor passes a brush.

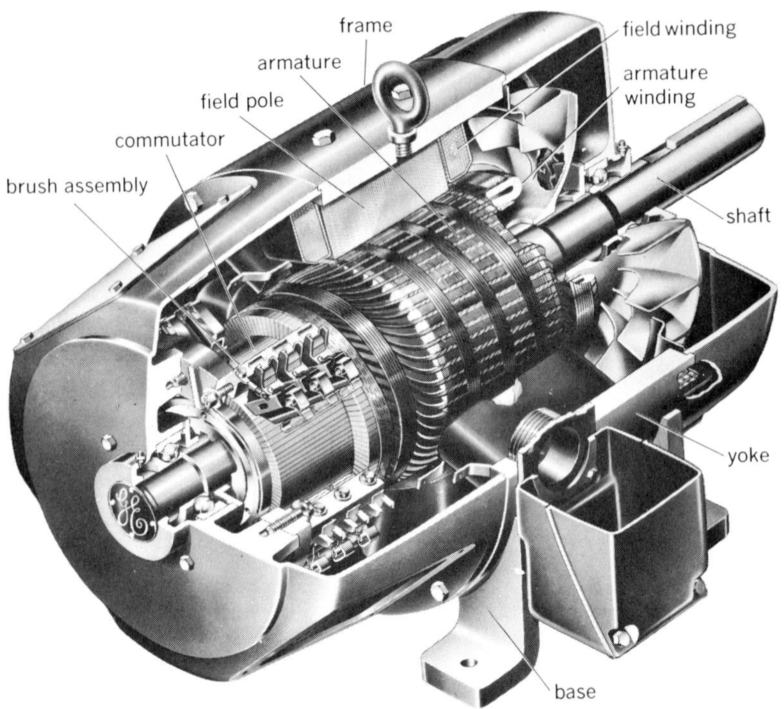

Fig. 1. Cutaway view of typical dc motor. (*General Electric*)

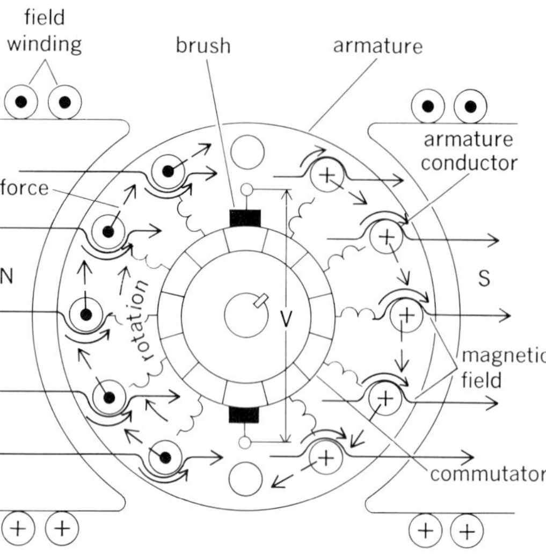

Fig. 2. Rotation in a dc motor.

The net force from all conductors acting over an average radial length to the shaft center produces a torque T given by Eq. (1), where K_t is a conver-

$$T = K_t \Phi I_a \tag{1}$$

sion and machine constant, Φ is net flux per pole, and I_a is the total armature current.

The voltage E, induced as a counter electromotive force (emf) by generator action in the parallel paths of the armature, plus the voltage drop I_aR_a through the armature due to armature current I_a and armature resistance R_a, must be overcome by the total impressed voltage V from the line. Voltage relations can be expressed by Eq. (2).

$$V = E + I_aR_a \tag{2}$$

The counter emf and motor speed n are related by Eq. (3), where K is a conversion and machine constant.

$$n = \frac{E}{K\Phi} \tag{3}$$

Mechanical power output can be expressed by Eq. (4), where n is the motor speed in rpm and T is the torque developed in pound-feet.

$$\text{HP} = \frac{2\pi nT}{33{,}000} \quad \text{horsepower} \tag{4}$$

By use of these four equations, the steady-state operation of the dc motor may be determined.

TYPES

Direct-current motors may be categorized as shunt, series, compound, or separately excited.

Shunt motor. The field circuit and the armature circuit of a dc shunt motor are connected in parallel (Fig. 3*a*). The field windings consist of many turns of fine wire. The entire field resistance, including a series-connected field rheostat, is relatively large. The field current and pole flux are essentially constant and independent of the armature requirements. The torque is therefore essentially proportional to the armature current.

In operation an increased motor torque will be

produced by a nearly equal increase in armature current, Eq. (1), since K_t and Φ are constant. Increased I_a produces an increase in the small voltage I_aR_a, Eq. (2). Since V is constant, E must decrease by the same small amount resulting in a small decrease in speed n, Eq. (3). The speed-load curve is practically flat, resulting in the term "constant speed" for the shunt motor. Typical characteristics are shown in Fig. 3*b*.

Typical applications are for load conditions of fairly constant speed, such as machine tools, blowers, centrifugal pumps, fans, conveyors, wood- and metalworking machines, steel, paper, and cement mills, and coal or coke plant drives.

Series motor. The field circuit and the armature circuit of a dc series motor are connected in series (Fig. 4*a*). The field winding has relatively few turns per pole. The wire must be large enough to carry the armature current. The flux Φ of a series motor is nearly proportional to the armature current I_a which produces it. Therefore, the torque, Eq. (1), of a series motor is proportional to the square of the armature current, neglecting the effects of core saturation and armature reaction. An increase in torque may be produced by a relatively small increase in armature current.

In operation the increased armature current, which produces increased torque, also produces increased flux. Therefore, speed must decrease to produce the required counter emf to satisfy Eqs. (1) and (3). This produces a variable speed characteristic. At light loads the flux is weak because of the small value of armature current, and the speed may be excessive. For this reason series motors are generally connected permanently to their loads through gearing.

The characteristics of the series motor are shown in Fig. 4*b*. Typical applications of this motor are to loads requiring high starting torques and variable speeds, for example, cranes, hoists, gates, bridges, car dumpers, traction drives, and automobile starters.

Compound motor. A compound motor has two separate field windings. One, generally the predominant field, is connected in parallel with the armature circuit; the other is connected in series with the armature circuit (Fig. 5).

The field windings may be connected in long or short shunt without radically changing the operation of the motor. They may also be cumulative or differential in compounding action. With both field windings, this motor combines the effects of the shunt and series types to an extent dependent upon the degree of compounding. Applications of this motor are to loads requiring high starting torques and somewhat variable speeds, such as pulsating loads, shears, bending rolls, plunger pumps, conveyors, elevators, and crushers. *See* DIRECT-CURRENT GENERATOR.

Separately excited motor. The field winding of this motor is energized from a source different from that of the armature winding. The field winding may be of either the shunt or series type, and adjustment of the applied voltage sources produces a wide range of speed and torque characteristics. Small dc motors may have permanent-magnet fields with armature excitation only. Such motors are used with fans, blowers, rapid-transfer switches, electromechanical activators, and programming devices.

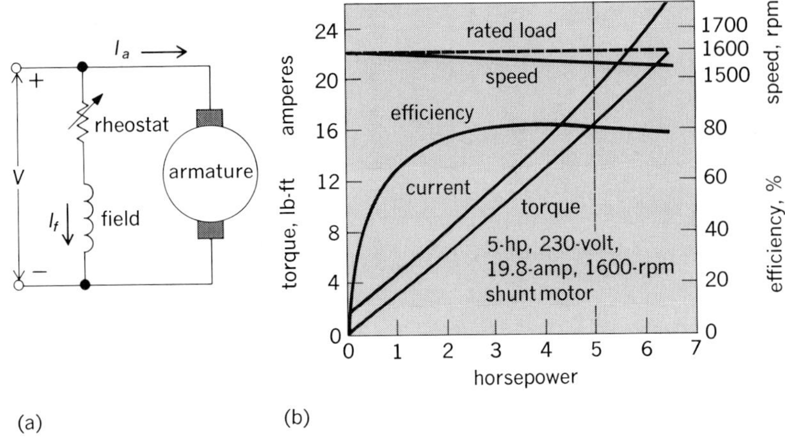

Fig. 3. Shunt motor. (*a*) Connections. (*b*) Typical operating characteristics.

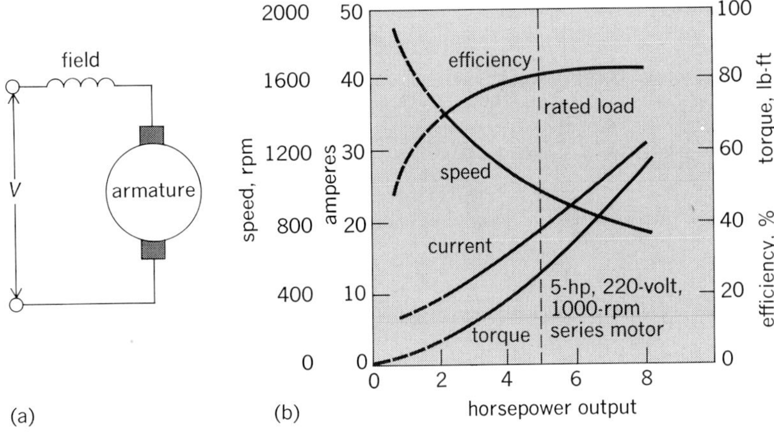

Fig. 4. Series motor. (*a*) Connection. (*b*) Typical operating characteristics.

STARTING AND SPEED CONTROL

Except in small dc motors, it is necessary to limit armature current when the motor is started. Therefore a load resistance must be in the armature circuit until the motor reaches full speed.

Starting. Direct-current motors are usually started with a rheostat in series with the armature circuit. This motor-starting resistor is of the proper rating in watts and ohms to withstand starting currents.

When a dc motor is started, the field winding is fully excited. Since there is no rotation of the armature, no counter emf is generated. Therefore, the armature current would be dangerously high unless an additional starting resistance were placed in the armature circuit, Eq. (2). This rheostat is manually or automatically cut out of the circuit as the motor approaches full speed. Small motors which have low armature inertia reach full speed rapidly and they do not require starting resistors. Separately excited motors may be started by control of the voltages which are applied to the armature.

Speed control. Speed of a dc motor may be controlled by changing the flux or counter emf of the motor, Eq. (3). Adjustment of the armature voltage V will affect the counter emf E, Eq. (2), by approximately the same amount. The speed n is affected by the change in counter emf according to Eq. (3).

DIRECT-CURRENT MOTOR

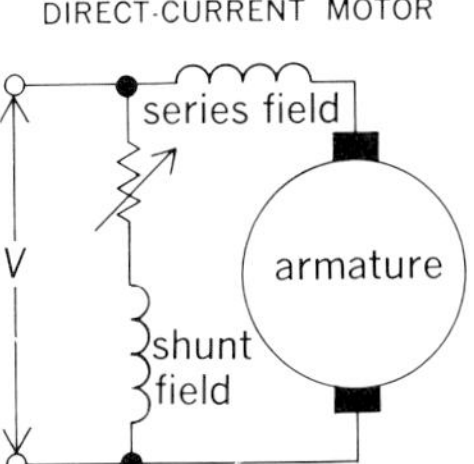

Fig. 5. Connection of a compound motor.

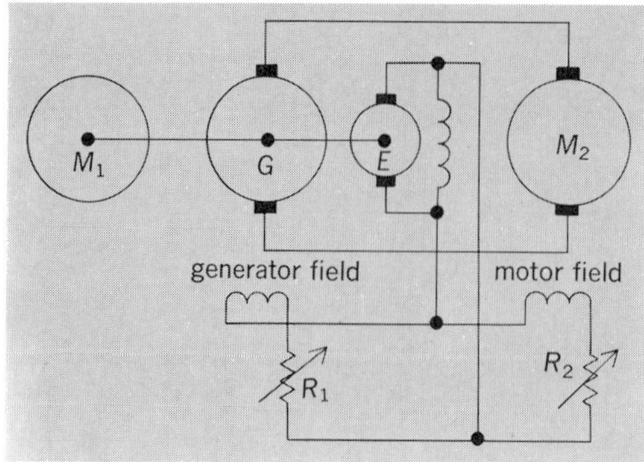

Fig. 6. Ward Leonard speed control system.

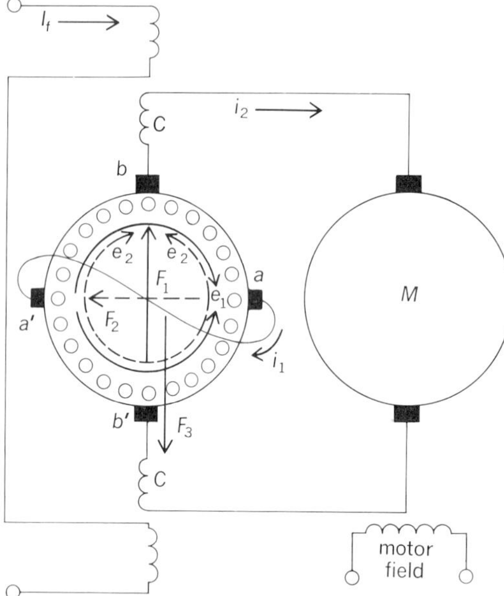

Fig. 7. Operation of the Amplidyne.

Insertion of a resistor in the armature circuit would also affect the speed but is seldom used because of the large power losses in the resistor. Speed control by adjustment of the applied armature voltage is used extensively where separate, adjustable voltage sources are available.

A change in flux Φ will also affect speed n, Eq. (3). Flux may be changed by a variable resistor in series with the shunt field of a shunt or compound motor. This field rheostat should have a total resistance comparable to that of the shunt field and be of sufficient capacity to withstand the relatively small shunt-field current.

Ward Leonard. In this system the armature voltage of a separately excited dc motor is controlled by a motor-generator set. A typical circuit (Fig. 6) shows a prime mover M_1, often a three-phase induction motor, mechanically coupled to a dc generator G and to an exciter generator E. The latter provides field excitation for the dc machines. Control of the generator field rheostat R_1 affects the output voltage of the generator G. This voltage may be smoothly varied from a low value to a value above normal. When this voltage is applied to the armature of the motor M_2, the speed of this motor will be variable over a wide range. Additional speed control of motor M_2 may be gained by adjustment of rheostat R_2.

The disadvantages of this system are the added equipment and maintenance costs it entails. However, the wide range and fineness of control in a low-current circuit make it applicable to high-speed passenger elevators, large hoists, power shovels, steel-mill rolls, drives in paper or textile mills, and the propulsion of small ships.

Amplidyne. The dynamoelectric amplifier (Amplidyne) is a rotating, two-stage, power amplifier in which a small change in field power in a dc generator results in a large change in output armature power. A large motor connected to the output of the generator may be controlled in speed by adjustment of the relatively small field power of the Amplidyne.

In Fig. 7 the control field current I_f produces an mmf F_1. The resultant flux and the short circuit of brushes aa' cause the induced voltage e_1 to force a large current i_1 through armature circuit. Because of the magnetic core design, current i_1 produces mmf F_2 and its resultant flux which induces voltage e_2 between brushes bb'. Motor M connected across brushes bb' will draw a current i_2 which produces an mmf F_3 tending to weaken the original mmf F_1. However, compensating windings C energized by i_2 will produce an mmf to oppose F_3 and restore the value of F_1.

This dynamic amplifier may produce amplifications of 10,000 to 1 or higher. It is applied to a variety of servomotors to control starting, acceleration, and deceleration. Other typical applications include voltage regulation of large ac generators, dc voltage control in cold-strip mills, speed control of paper mills, positioning control of gun turrets, machine-tool drives, and power-factor control of synchronous generators.

Regulex. The regulating exciter (Regulex) is a dc generator acting as a power amplifier. By proper design of the machine magnetic core, an extensive linear portion of the voltage buildup curve is obtained (Fig. 8). A small change in mmf F will produce a large change in induced voltage E resulting in a degree of amplification. Critical-value adjustment of the field rheostat R (Fig. 9) will cause the generator to operate on this linear portion. A reference field F_2 and an opposing field F_3 combine with field F_1 to establish a point of operation, such as point a in Fig. 8.

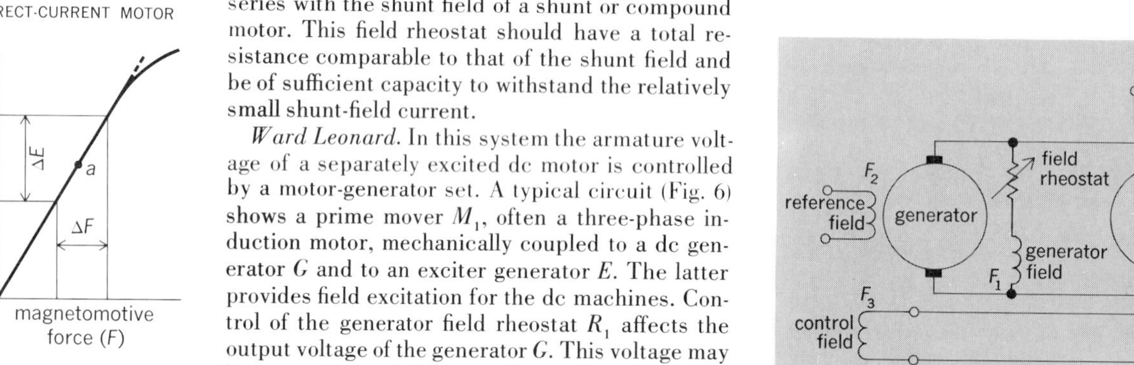

Fig. 9. Typical Regulex circuit.

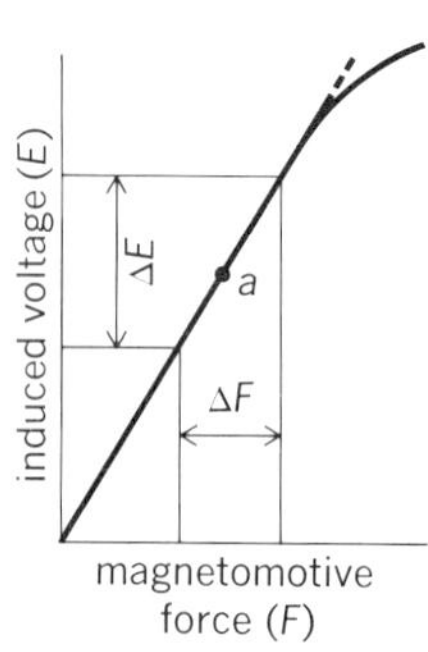

Fig. 8. Regulex magnetization curve.

Departure from this balance because of a variation in the control field F_3 will produce the large change in voltage E. The output of this device may be used to drive a dc motor M, which has its speed translated into voltage by means of a small pilot generator coupled to the motor shaft. By proper feedback of this voltage to control field F_3, the motor speed may be maintained at a constant value.

Rototrol. The rotating control (Rototrol) is a dc generator acting as a power amplifier. It is similar to the Regulex, but the self-excited field is a series type in contrast to the shunt-type field of the Regulex.

Solid-state control for motor speed. Solid-state devices such as diodes and thyristors, including silicon-controlled rectifiers (SCRs), may be used in a number of circuit applications to control the speed of dc motors.

Pulse control. One such circuit supplies a number of unidirectional voltage pulses to the motor whose speed is adjusted by variation of pulse frequency or pulse width. In the circuit of Fig. 10, the thyristor acts as an ON-OFF switch. The relative on-to-off time determines the width of the voltage

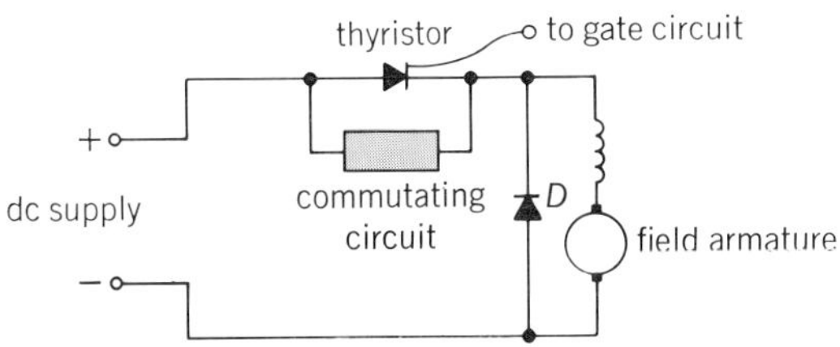

Fig. 10. Elementary pulse-control circuit.

pulses. The start of thyristor conduction is determined by a gating voltage, but a negative voltage must be applied across the thyristor by the commutating circuit in order to stop its conduction. Controlling the width of the voltage pulses or the number of pulses per second (several hundred) determines the average value of the voltage applied to the motor and hence its speed. During the thyristor OFF period, diode D allows the energy stored in the motor field coils to be discharged through the motor armature. Current is continuous, increasing during the ON period and decreasing during the OFF period when the current is the result of the energy coming from the coils, but the average voltage is independent of the current.

Pulse control is applied to several types of electrically operated vehicles and affords an efficient and smooth control of speed.

Full-wave rectifier control. Thyristors (SCRs) are used in full-wave rectifier circuitry to control the motor speed, as indicated in Fig. 11.

Diodes D_1, D_2, D_3, and D_4 form a bridge rectifier to supply a full-wave rectified dc voltage to the field of the motor. A second bridge rectifier circuit also using diodes D_3 and D_4 as well as SCR_1 and SCR_2 provides adjustable current to the armature.

Control of the firing of the SCRs is accomplished by amplitude changes in the SCR gate voltages due to adjustment of potentiometer R. If the gate voltages exceed the back emf generated in the

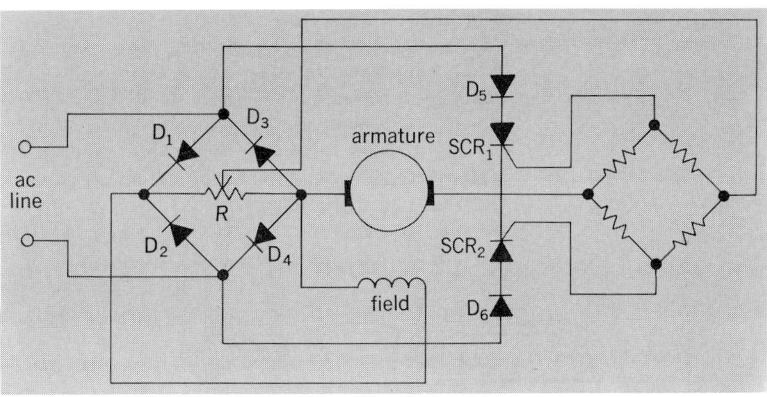

Fig. 11. SCR motor speed control.

armature of the motor for a given load condition, the SCRs will fire. The average value of armature current and hence the speed is thus under control. Damage to the SCRs due to large values of reverse currents is prevented by diodes D_5 and D_6, which block the flow of such reverse currents.

[L. F. CLEVELAND]

Bibliography: A. E. Fitzgerald, D. E. Higginbotham, and A. Grabel, *Basic Electrical Engineering*, 4th ed., 1975; A. E. Fitzgerald, C. Kingsley, and A. Kusko, *Electric Machinery*, 3d ed., 1971; B. C. Kuo and J. Tal (eds.), *Incremental Motion Control: DC Motors and Controls*, vol. 1, 1978; J. Rosenblatt and M. H. Friedman, *Direct and Alternating Current Machinery*, 1963; A. Kusko, *Solid-State DC Motor Drives*, 1969.

Direct-current transmission

The conduction of electric power by means of electric currents which flow in one direction only. *See* DIRECT CURRENT.

Early use. The first commercial applications of electric power used direct current (dc). They included groups of series-connected arc lamps and groups of parallel-connected carbon-filament incandescent lamps energized by electricity from dc generators, usually driven by steam engines. Another early application was street railways.

Sometimes storage batteries were used with the low-voltage dc systems to provide against emergencies such as failures of a generator or of the conductors (mains) from the generator to the load. The chief limitation of these early electric power systems was their low voltage, which limited the economic distance between generator and loads. One improvement was the adoption of the three-wire circuit consisting of both a positive conductor and a negative conductor instead of one or the other. The load could be divided fairly equally between these two conductors, giving a much lower current in the neutral conductor, which was grounded. This system effectively doubled the voltage of the distribution system without creating the hazard of a higher voltage to ground at the incandescent lamps accessible to the customers.

Replacement of dc by ac. Still more improvement in efficiency was required by increased loads and by the demands for service in suburban and rural areas having lower load density than the city. This problem was solved by the invention of the transformer and the use of alternating current (ac). The transformer is both efficient and simple. Its use permits any number of voltage levels. For example, the generators can have one voltage, the transmission lines another, higher voltage, the local distribution system a lower but moderately high voltage, and the loads the same low voltage that they had when dc was used. The ac generators have no commutators, and this fact makes it possible to build more powerful generators operating at higher speed. They are driven by steam turbines instead of reciprocating steam engines. *See* ALTERNATING CURRENT; TRANSFORMER.

Another notable improvement in electric power systems was the use of polyphase systems, especially of three-phase systems. A three-phase generator has a higher rated power than does a single-phase generator of equal size. The polyphase induction motor, especially that with a squirrel-cage rotor, is simple, rugged, and self-starting, and most of these motors can be started safely by applying their full rated voltages. *See* ALTERNATING-CURRENT GENERATOR; ALTERNATING-CURRENT MOTOR.

Polyphase high-voltage transmission lines make it practical to generate electric power from water power at remote sites and to transmit this power efficiently over long distances to the sites (usually in or near large towns) where there is a demand for electric power.

Low-voltage dc networks persisted in the central districts of large cities into the 1970s, but most of these have now been converted to ac networks.

Advantages. After the almost complete replacement of dc by ac, however, interest in dc power transmission began increasing in 1954. One reason is that the cost of a dc line is about two-thirds that of the corresponding ac line. This is because the usual dc bipolar line has two conductors, compared with three conductors of an ac single-circuit three-phase line. In neither case are the overhead ground wires for intercepting lightning strokes counted. In emergencies the bipolar line can be operated for considerable time by use of one pole with return path through the ground.

The dc line requires additional apparatus at its terminals, not required for an ac line, which increases the cost. These are the converters for changing ac to dc at the sending end and dc to ac at the receiving end. Considering the combined costs of the line and of the terminal equipment, the dc link is thus more expensive than ac for a short line but less expensive for a long line. The costs are equal at a "break-even" distance of approximately 500 mi (800 km), but this distance varies because of the variability of the costs. The costs of converters are expected to decrease in the future because of the increasing number of suppliers and the experience that they will gain. This would decrease the break-even distance. However, the choice between ac and dc usually depends on other factors.

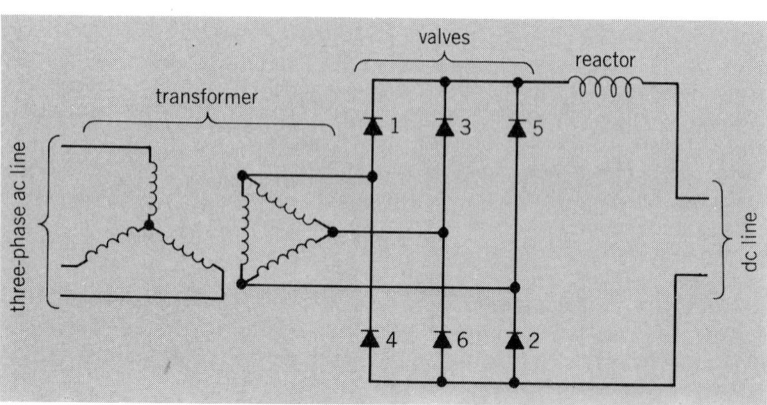

Three-phase two-way or Graetz converter circuit.

Transmission over submarine cables. Direct-current lines are used to perform functions for which ac lines are unsuited, such as transmitting power across a wide body of water, using submarine cables. Cables have a much larger shunt capacitance per unit of length than overhead lines have. When an ac voltage is applied between the two conductors of a cable, an alternating "charging" current flows even if the distant end is not connected to a load. When the length of the cable exceeds some distance—about 30 mi (50 km)—the thermal limit is reached, even though no power is transmitted. This was a significant factor in the choice of dc for connecting the island of Gotland to the mainland of Sweden, Denmark to Sweden, England to France, and the South Island of New Zealand to the North Island.

System interconnections. Direct-current lines are also used to provide interconnections of two ac systems of different nominal frequencies, for example, 50 Hz and 60 Hz, as was done in Japan, or interconnections between two ac systems having the same nominal frequency but different frequency controls. An ac tie of the latter type might require a much higher power rating than the greatest power which was desired to be interchanged. In some cases, the two converters are installed in the same station, so that really there is no dc line. Such stations are called frequency-changer stations if there are two different nominal frequencies, and asynchronous ties if the nominal frequencies are equal, as in New Brunswick, Canada, or Steagle, Nebraska.

Corona. Overhead dc lines differ from ac in the effect of corona in producing radio interference. An important difference between ac and dc corona is the effect of rainfall, which greatly increases the amount of radio noise on an ac line but makes a negligible change on a dc line. In addition, even with equal readings on a noise meter, on an ac line the modulation at the power frequency increases the annoyance to a human observer by an amount equivalent to 3 dB.

Converters. A converter consists of transformers and of valves which conduct current in only one direction, called the forward direction, but not in the opposite direction, called the inverse direction. The valves used in dc transmission are controlled valves; each valve has a control electrode which can prevent conduction from starting even though the voltage across the valve is in the proper direction of conduction. However, once the valve begins to conduct, the control electrode cannot stop conduction or control the value of the instantaneous current.

The early dc transmission schemes (1954 to 1974) used mercury-arc valves in steel tanks with mercury-pool cathodes. Most of these valves are still in use. However, all schemes commissioned since the late 1970s use thyristors, which are solid-state devices. These require less maintenance and less expensive auxiliary devices.

The most advantageous method of interconnecting the several valves in a converter is the three-phase two-way circuit, also known as the Graetz circuit, having six valves (see illustration). They are ignited in the sequence of the valve numbers.

The direction of power flow is usually reversed on a dc line by reversing the polarity of the direct voltage, since the current cannot be reversed except by switching.

Fault clearance. Most faults on dc lines, as well as on ac lines, are temporary and are cleared by deenergizing the line for a short while to extinguish the fault arc. Circuit breakers are not used for this purpose on dc lines. Instead, the valves are controlled so as to tend to reverse the current in the faulted conductor. Since the valves can conduct only in the normal direction, the fault current is brought to zero. After allowance of a short time for deionization of the arc path, the line is reenergized. *See* ELECTRIC POWER SYSTEMS; TRANSMISSION LINES. [EDWARD W. KIMBARK]

Bibliography: C. Adamson and N. G. Hingorani, *High Voltage Direct Current Power Transmission*, 1960; B. J. Cory (ed.), *High Voltage Direct Current Convertors*, 1965; E. W. Kimbark, *Direct Current Transmission*, vol. 1, 1971; E. Uhlmann, *Power Transmission by Direct Current* (transl. by R. Clark), 1975.

Displacement (mechanics)

When an object is moved from one position to another, it is said to be displaced, and the linear distance from the initial to the final position, regardless of the length of path followed, is called the displacement. The displacement is always in a particular direction, and consequently displacement is a vector quantity involving direction as well as magnitude. In rectilinear motion the magnitude of the displacement is also the length of path or the distance traversed, but since length of path does not involve direction, it is a scalar quantity. For instance, the shortest distance or length of path is the same from Washington to Philadelphia as it is from Philadelphia to Washington, but the displacement of an object in one direction is entirely different from its displacement in the opposite direction.

When a body is rotated about any axis, it is said to undergo angular displacement. Angular displacement is commonly measured in radians or degrees, one radian being equal to 57.3°. *See* ROTATIONAL MOTION. [R. D. RUSK]

Displacement pump

A pump that develops its action through the alternate filling and emptying of an enclosed volume.

Reciprocating types. Positive-displacement reciprocating pumps have cylinders and plungers or pistons with an inlet valve, which opens the cylinder to the inlet pipe during the suction stroke, and an outlet valve, which opens to the discharge pipe during the discharge stroke. Reciprocating pumps may be power-driven through a crank and connecting rod or equivalent mechanism, or direct-acting, driven by steam or compressed air or gas.

Figure 1 shows a small high-speed plunger-type power pump for high-pressure service. The three-throw crankshaft is carried in roller bearings at each end. The manifolds below the suction valves and above the discharge valves connect the three pumping cylinders to the suction and discharge piping.

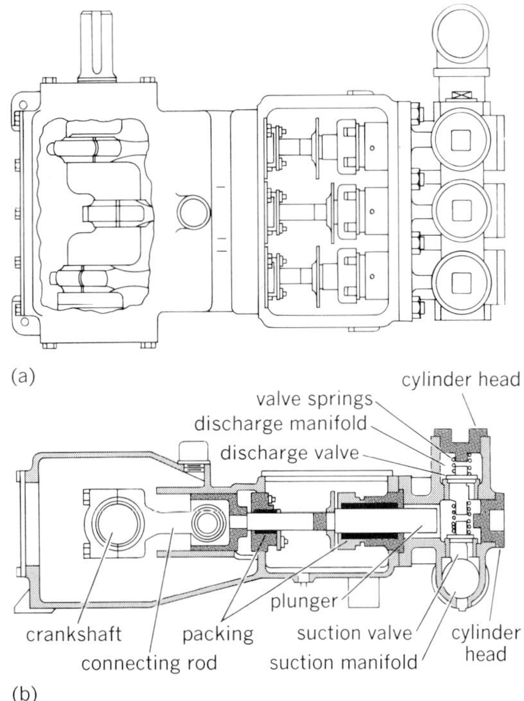

Fig. 1. Horizontal triplex power pump of reciprocating type. (*a*) Plan. (*b*) Elevation.

Power pumps are frequently built with one or two throw cranks and double-acting liquid ends, or with five, seven, or even nine cranks where smoother flow is desirable. Power-driven reciprocating pumps are highly efficient over a wide range of discharge pressures. Except for some special designs with continuously variable stroke, reciprocating power pumps deliver essentially constant capacity over their entire pressure range when driven at constant speed. In some applications this is an advantage, but in others it complicates the controls required.

Direct-acting steam types. A reciprocating pump is readily driven by a reciprocating engine; a steam or power piston at one end connects directly to a fluid piston or plunger at the other end. Direct-acting reciprocating pumps are simple, flexible, low-speed machines which are low in efficiency unless the heat in the exhaust steam can be used for heating. Steam pumps can be built for a wide range of pressure and capacity by varying the relative size of the steam piston and the liquid piston or plunger. The delivery of a steam pump may be varied at will from zero to maximum simply by throttling the motive steam, either manually or by automatic control. Direct-acting pumps are built as (1) simplex, having one steam and one liquid cylinder; and (2) duplex, having two steam and two liquid cylinders side by side. As indicated by Fig. 2, each steam valve of a duplex pump is positively driven by the motion of the opposite piston rod by means of cranks and links. In the case of a simplex pump, to avoid stalling at low speed, the valve

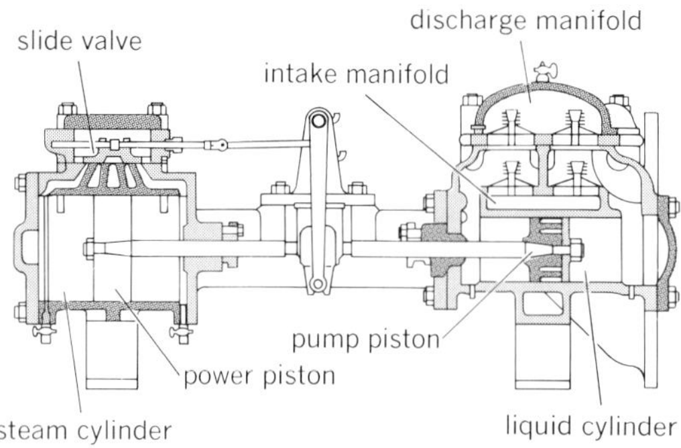

Fig. 2. Duplex type of direct-acting steam-driven feed-water pump.

linkage operates a small pilot valve which in turn controls a piston-thrown main steam valve.

Reciprocating pumps are used for low to medium capacities and medium to highest pressures. They are useful for low- to medium-viscosity fluids, or high-viscosity fluids at materially reduced speeds. Specially fitted reciprocating pumps are used to pump fluids containing the more abrasive solids.

Rotary types. Another form of displacement pump consists of a fixed casing containing gears, cams, screws, vanes, plungers, or similar elements actuated by rotation of the drive shaft. Most forms of rotary pumps are valveless and develop an almost steady flow rather than the pulsating flow of a reciprocating pump. Three of the many types of rotary pumps are shown in Fig. 3.

DISPLACEMENT PUMP

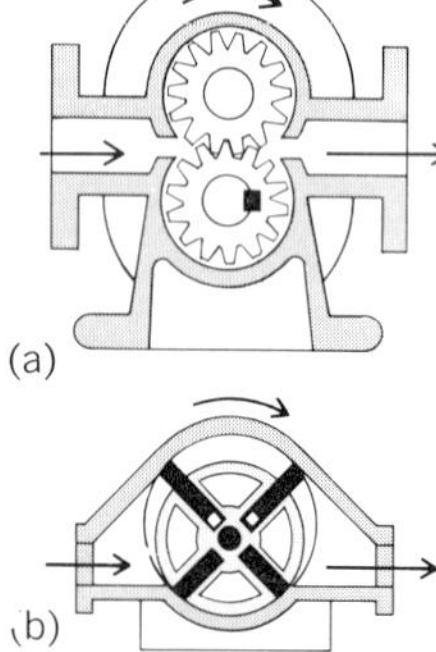

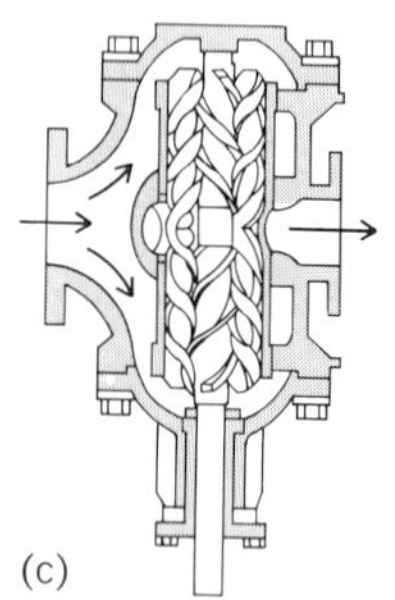

Fig. 3. Rotary pumps. (*a*) Gear. (*b*) Sliding vane. (*c*) Screw. (*From T. Baumeister, ed., Standard Handbook for Mechanical Engineers, 7th ed., McGraw-Hill, 1967*).

Rotary pumps require very close clearances between rubbing surfaces for their continued volumetric efficiency. Consequently they are most useful for pumping clean oils or other fluids having lubricating value and sufficient viscosity to prevent excessive leakage. On petroleum oils of suitable viscosity, rotary pumps are highly efficient at moderate pressure and speed, while at reduced speed they can pump, with lower efficiency, the most viscous materials. The increasing use of hydraulic actuation of machine tools and mechanisms, such as power steering of automobiles, has extended the use of rotary pumps and similar hydraulic motors.

Vacuum types. Although vacuum pumps actually function as compressors, displacement pumps are used for certain vacuum pump applications. Simplex steam pumps with submerged piston pattern fluid ends similar to the fluid end in Fig. 2 are used as wet vacuum pumps in steam heating and condensing systems. Sufficient liquid remains in the cylinder to fill the clearance volume and drive the air or gas out ahead of the liquid. Certain types of rotary pumps are arranged to retain a quantity of sealing oil when operating as vacuum pumps.

Air-lift types. In handling abrasive or corrosive waters or sludges, where low efficiency is of secondary importance, air-lift pumps are used. The pump consists of a drop pipe in a well with its lower end submerged and a second pipe which introduces compressed air near the bottom of the drop pipe. The required submergence varies from about four times the distance from the water level to the surface for a low lift to an equal distance for a relatively high lift. The mixture of air and water in the drop pipe is lighter than the water surrounding the pipe. As a result, the mixture of air and water is forced to the surface by the pressure of submergence. *See* COMPRESSOR; PUMP; VACUUM PUMP.

[ELLIOTT F. WRIGHT]

Bibliography: J. Karassik and W. C. Krutzch, *Pump Handbook*, 1976.

Distributor

A rotary switch that directs the high-voltage ignition current in the proper firing sequence to the various cylinders of an internal combustion engine. In automotive practice the distributor housing usually contains, in addition, apparatus for timing the ignition to occur when each piston is at optimum position in the cycle. This apparatus includes a set of cam-operated contacts called breaker points, the opening of which triggers the ignition pulse. The timing of the breaker point opening is made earlier at high engine speeds, by the centrifugal action of small weights that are driven by the breaker cam shaft (distributor shaft). Timing is also varied with engine load, by the movement of a diaphragm exposed to the pressure in the engine intake manifold. *See* IGNITION SYSTEM.

[AUGUSTUS R. ROGOWSKI]

District heating

The supply of heat, either in the form of steam or hot water, from a central source to a group of buildings. As an industry, district heating began in the early 1900s with distribution of exhaust steam, previously wasted to atmosphere, from power plants located close to business and industrial districts. Advent of condensing-type electrical generating plants and development of long-distance electrical transmission led to concentration of electrical generation in large plants remote from business districts. Most district heating systems in the United States rely on separate steam generation facilities close to load centers. In some cities, notably New York, high-pressure district steam (over 120 psi) is used extensively to feed turbines that drive pumps and refrigerant compressors. Although some district heating plants serve detached residences, the cost of underground piping and the small quantities of heating service required makes this service generally unfeasible.

District heating, apart from utility-supplied systems in downtown areas of cities, is accepted as efficient practice in many prominent colleges, universities, and government complexes. Utility systems charge their customers by metering either the steam supply or the condensed steam returned from the heating surfaces.

District heating has grown faster in Europe than in the United States. A substantial part of the heat distribution is by hot water from plants combining power and heat generation. Heat from turbine exhaust or direct boiler steam is used to generate high-temperature (270–375°F or 132–191°C) water, often in cascade heat convertors, where live steam is mixed with return hot water. Hot water distribution systems have the advantage of high thermal mass (hence, storage for peak periods) and freedom from problems of condensate return.

On the other hand, pumping costs are high and two mains are required, supply and return. (The cost of condensate return piping and pumping is so high that many steam distributors in the United States make no provision for condensate return at all.)

Reykjavik, Iceland, is completely heated from a central hot water system supplied from hot springs. Average water temperature is 170°F (77°C).

The most extensive European developments in district heating have been in the Soviet Union, mainly in the form of combination steam-electric plants serving new towns. The Soviets have developed a single-pipe high-temperature water district heating system. At each building served, superheated water from the main is diluted with return water from the building system. Excess water is used in kitchens, laundries, bathrooms, and swimming pools.

An important and relatively recent extension of the principle of district heating is the development of district heating and cooling plants. Usually owned and operated by the natural gas utilities, central plants distribute chilled water for air conditioning in parallel conduits with hot water or steam for heating. The system serving Hartford, Conn., is the pioneer installation. Because the costs of these four-pipe distribution systems are extremely high, development is slow and restricted to small areas of high density usage within a few blocks of plant. *See* COMFORT HEATING.

[RICHARD L. KORAL]

Bibliography: International District Heating Association, *District Heating Handbook*, 4th ed., 1981.

Drafting

The making of drawings of objects, structures, or systems that have been visualized by engineers, scientists, or others. Such drawings may be executed in the following ways: manually with drawing instruments and other aids such as templates and appliqués, freehand with pencil on paper, or with automated devices, described below.

Drafting is done by persons with varied backgrounds. Engineers often draft their own designs to determine whether they are workable, structurally sound, and economical. However, much routine drafting is done under the supervision of engineers by technicians specifically trained as draftsmen. *See* ENGINEERING DRAWING.

[CHARLES J. BAER]

Templates. As the complexity of designs has increased and the use of standard parts has become widespread, drawings have become increasingly stylized. Graphic symbols have replaced pictorial representations. The next step in the evolution of drafting techniques has been the introduction of templates that carry frequently used symbols, from which the draftsman quickly traces the symbols in the required positions on the drawing. (Fig. 1).

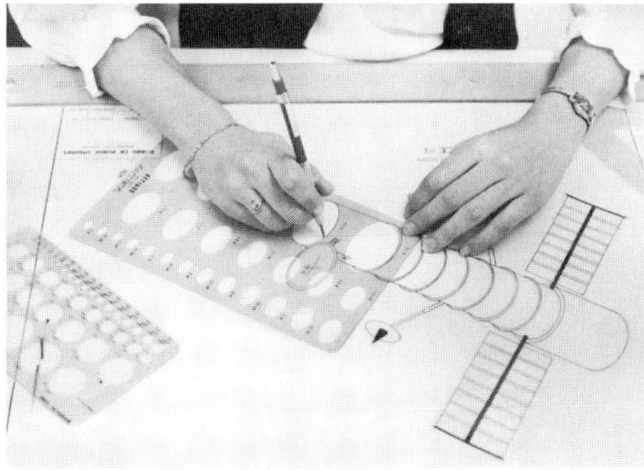

Fig. 1. Use of template in drafting.

Automated drafting. Where the design procedures from which drawings are developed are repetitive, computers can be programmed to perform the design and to produce their outputs as instructions to automatic drafting equipment. Essentially, automated drafting is a method for creating an engineering drawing or similar document consisting of line delineation either in combination with, or expressed entirely by, alphanumeric characters.

The computer receives as input a comparatively simplified definition of the product design in a form that established a mathematical or digital definition of the object to be described graphically. The computer then applies programmed computations, standards, and formatting to direct the graphics-producing device. This method provides for close-tolerance accuracy of delineation and produces at speeds much greater than possible by manual drafting. In addition, the computer can be programmed to check the design information for

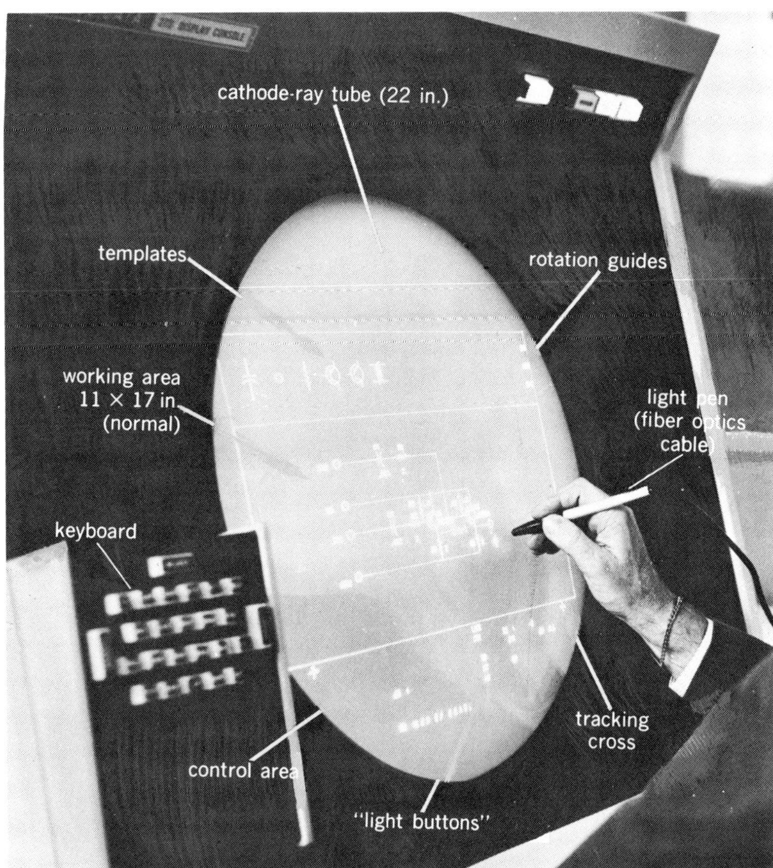

Fig. 2. Electronic light pen is used by designer on a cathode-ray tube to instruct computer in the drafting of a schematic. (*Lockheed*)

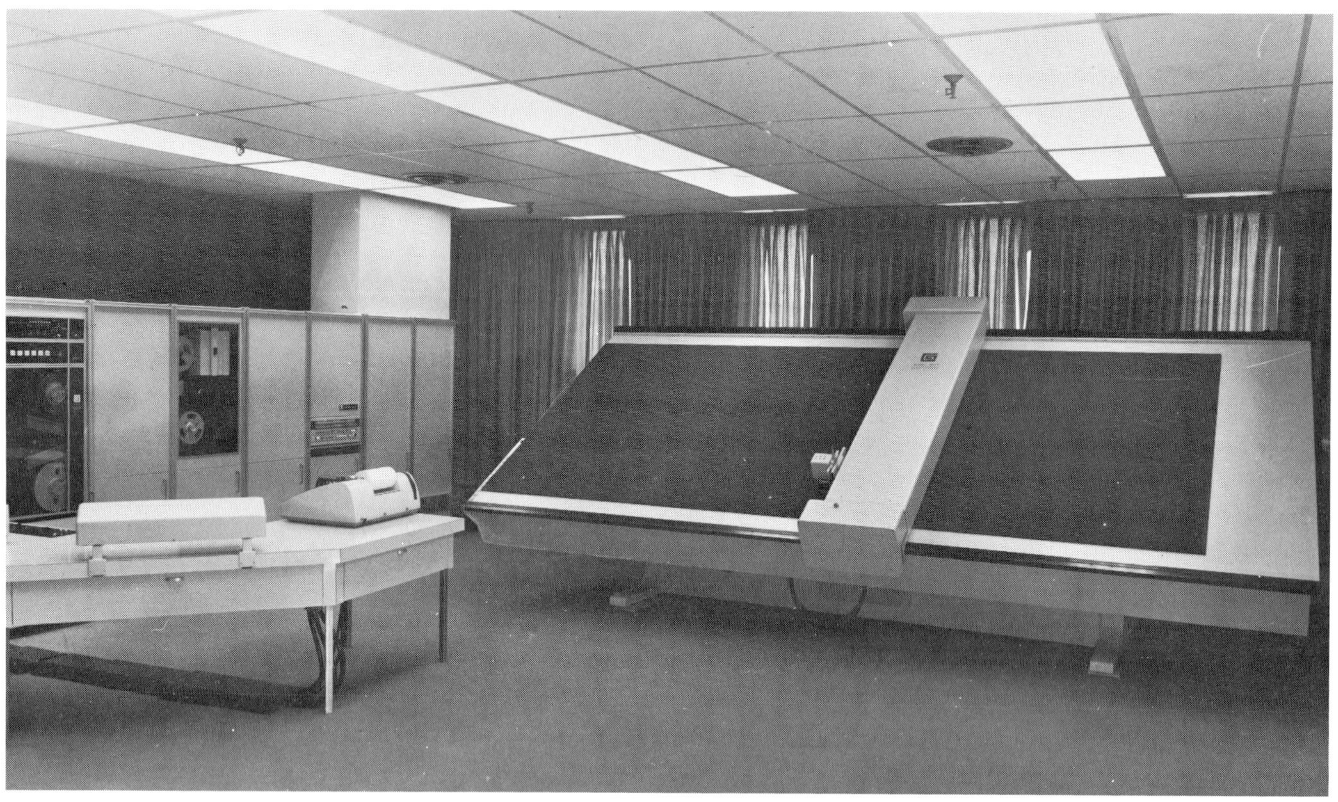

Fig. 3. XY plotting setup produces precise drawings at high speed. (*Gerber Scientific Instruments*)

accuracy, completeness, and continuity during the processing cycle.

Transmittal of the design definition information from the engineer to the computer is in algebraic terms that can be understood and processed by the computer program. This may be accomplished in two ways: (1) by the engineer trained in computer-understandable language, or (2) an intermediary who translates the engineer's notes and sketches into computer language for processing by the computer to drive a graphics output device.

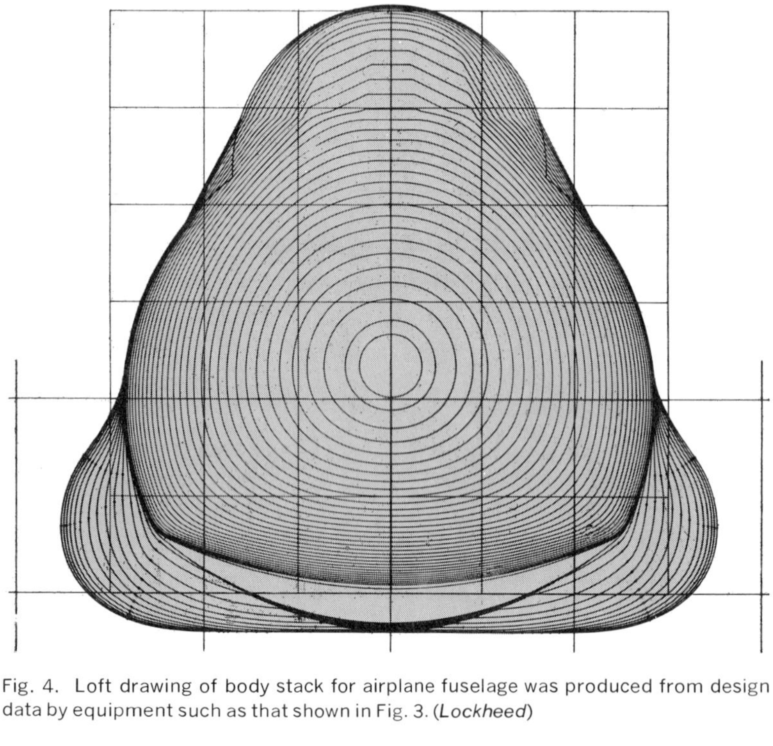

Fig. 4. Loft drawing of body stack for airplane fuselage was produced from design data by equipment such as that shown in Fig. 3. (*Lockheed*)

In increasing use is the cathode-ray tube (CRT) console on-line with the computer to provide direct interaction between a designer and a computer to develop the definition of a design (Fig. 2). The console contains a CRT on which the design is displayed as it is developed by the engineer writing with an electronic light pen to create line delineation and position characters; control keys to initiate the action to be performed on the CRT; function controls to call out program-stored information; and a keyboard, or typewriter, to apply the alphanumeric information required.

The engineering designer uses the light pen in a freehand mode to sketch the drawing and to position lines and program-stored symbols and characters. After a point is located or digitized by the light-sensing pen (which is a digitizer), the designer may type in a code for a symbol on the keyboard or may touch a symbol on the menu with the pen. A menu is a coded chart that shows graphic symbols. The symbols are displayed on the face of the CRT and are moved in sequential position by direction of the light pen. When the required symbols are in place, interconnecting lines are applied by direction with the light pen. During the development of the drawing, graphical images can be repositioned, or erased, instantaneously. The designer can observe the drawing as it progresses and call on the computer to provide computational assistance in developing the design. The information displayed on the CRT is stored in the computer memory, available to drive a graphics output device to produce a conventional engineering drawing when the design is complete.

Currently in use as graphics-producing devices are XY coordinate plotters, CRT plotters, and photocomposing units, each driven either on-line or off-line by a computer direction. XY coordinate plotters are available in models that produce highly precise drawings rapidly. Figure 3 is an illustration of an XY plotter, showing the drawing surface, transverse beam, and drawing head. In the background are the computer and tape readers that drive the transverse beam and drawing head in XY directions to produce drawings. Alphanumerics and symbols are either drawn by movement of the beam and drawing head, or the equipment may include a character-generator unit to imprint characters or symbols in a single stroke at the position directed by the computer.

Figure 4 is a loft drawing of the body stack, or station lines, in end view of an airplane fuselage. The drawing was produced on an XY plotter of the type shown in Fig. 3 to very close tolerances, and can be reproduced by the plotter in any scale required.

CRT plotters produce drawings by exposing an image on sensitized film, usually 35 mm or smaller. By computer direction, the drawing image is displayed on a precision CRT face and recorded on film by a microfilm camera. There is also available on CRT plotter that produces full-size drawings up to 40 × 60 in. by sweeping the CRT head several times across the surface of a large sheet of sensitized film.

Photocomposing units are used primarily to produce drawings consisting of straight lines, symbols, and alphanumeric characters. Through a method of digitizing that identifies characters, lines, and their position in relation to each other, the drawing information is supplied to the computer where it is formatted to drive the photocomposing unit. The photo unit then projects a light beam successively through masks of the required symbols onto sensitized film.

The computer program required for processing design information to drive the graphics devices is unique for each type of equipment to be used. Through the use of these automated drafting methods, corrections and revisions can be made by providing the computer with the change information only for revision to the instructions for driving the output device. [TRAM C. PRITCHARD]

Bibliography: R. W. Mann and S. A. Coons, Computer-aided design, *McGraw-Hill Yearbook of Science and Technology*, 1965; R. A. Siders et al., *Computer Graphics: A Revolution in Design*, 1966.

Drafting machine

A mechanical aid to drafting having two straightedges which, although fixed at right angles to each other, maintain a preset angular relationship to the work when moved from place to place over the drafting surface.

Engineering drawings often require making numerous parallel lines. These lines may be horizontal, vertical, or at some other angle. Electrical schematics, orthographic projections, charts, and many architectural drawings contain numerous horizontal and vertical lines. Isometric and trimetric projections contain numerous parallel lines at specific angles. Drafting machines simplify the making of these and other types of drawings and substantially reduce the man-hours required.

Fig. 1. Drafting machine using vertical and horizontal bars. (*Keuffel and Esser Co.*)

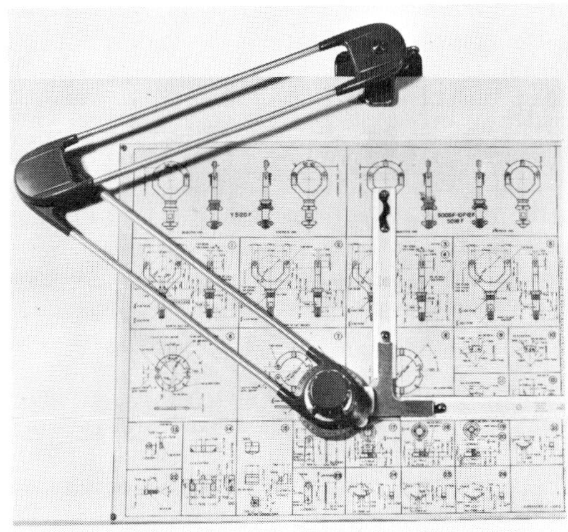

Fig. 2. Drafting machine using mechanical linkages. (*Keuffel and Esser Co.*)

On most drafting machines the straightedges can be pivoted to any desired angle, then locked. Usually a detent enables the draftsman to set the machine to the horizontal-vertical position quickly and accurately. Other angles can be selected by using a built-in protractor scale.

Two types of mechanisms are in use to allow the straightedges to move over the drafting surface. One consists of vertical and horizontal tracks or bars (Fig. 1). The other consists of mechanical linkages which may take the form of a double set of parallel four-bar linkages having a common member at the elbow, or two sets of equal-diameter pulleys tightly coupled by metal bands (Fig. 2). *See* DRAFTING; ENGINEERING DRAWING; FOUR-BAR LINKAGE.

[WILLIAM W. SNOW]

Drawing of metal

An operation wherein the workpiece is pulled through a die, resulting in a reduction in outside dimensions. This article deals only with bar and wire drawing and tube drawing; deep drawing and other processes, as performed on sheet metal, are

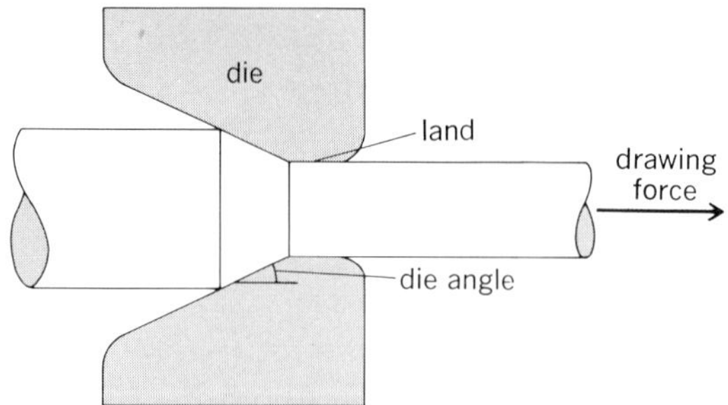

Cross section of drawing die.

described in another article. *See* SHEET-METAL FORMING.

Wire and bar drawing. Among the variables involved in the drawing of wires and bars are properties of the original material, percent reduction of cross-sectional area, die angle and geometry, speed of drawing, and lubrication. The operation usually consists of swaging the end of a round rod to reduce the cross-sectional area so that it can be fed into the die; the material is then pulled through the die at speeds as high as 8000 feet per minute (fpm). Short lengths are drawn on a draw bench, while long lengths (coils) are drawn on bull blocks. Most wire drawing involves several dies in tandem to reduce the diameter to the desired dimension.

Die angles usually range from 6 to 15°, the actual angle depending on die and workpiece materials. Reductions in cross-sectional area vary 10–45% in one pass, although theoretically the maximum reduction per pass for a perfectly plastic material is 63%.

Die materials are usually alloy steels, carbides, and diamond. Diamond dies are used for drawing fine wires. The purpose of the die land is to maintain dimensional accuracy (see illustration).

The force required to pull the workpiece through the die is a function of the strength of the material, die angle, coefficient of friction, and reduction in cross-sectional area. The work applied to the process comprises three components: ideal work of deformation, friction work, and redundant work due to nonuniform deformation within the material. Depending on a number of factors, there is an optimum die angle for which the drawing force is a minimum. In cold-drawing, the strength of the material increases due to work hardening. The terminology for the increasing hardness is shown in the table.

Temperature rise in drawing is important because of its effect on die life, lubrication, and residual stresses. Also, a defect in drawn rods is the rupturing of the core, called cuppy core. The tendency for such internal rupturing increases with increasing die angle, friction, and inclusions in the original material, and with decreasing reduction per pass.

Hardness terminology for drawn wire

Terminology	Reduction in area in drawing, %
1/8 hard	11
1/4 hard	21
1/2 hard	37
Hard	60
Extra hard	75
Spring	84
Extra spring	90
Special spring	94

The magnitude of residual stresses in a drawn material depends on the die geometry and reduction. The surface residual stresses are generally compressive for light reductions and tensile for intermediate or heavy reductions.

Extensive study has been made of lubrication in rod and wire drawing. The most common lubricants are various oils containing fatty or chlorinated additives, chemical compounds, soap solutions, and sulfate and oxalate coatings. The original rod to be drawn is usually surface-treated by pickling to remove scale, which can be abrasive and thus considerably reduce die life. For drawing of steel, chemically deposited copper coatings are also used. If the lubricant is applied to the wire surface, it is called dry drawing; if the dies and blocks are completely immersed in the lubricant, the process is called wet drawing.

Tube drawing. Tubes are also drawn through dies to reduce the outside diameter and to control the wall thickness. The thickness can be reduced and the inside surface finish can be controlled by using an internal mandrel (plug). Various arrangements and techniques have been developed in drawing tubes of many materials and a variety of cross sections. Dies for tube drawing are made of essentially the same materials as those used in rod drawing.

[SEROPE KALPAKJIAN]

Bibliography: J. Neely, *Practical Metallurgy and Materials of Industry*, 1979; V. P. Severdenko and V. V. Klubovich (eds.), *Ultrasonic Rolling and Drawing of Metals*, 1972.

Dredge

A machine used to excavate underwater and discharge into a containment vessel or a transport pipeline. The digging machinery is installed on a floating hull, and the entire assembly constitutes the dredge.

A dredge may be either mechanical or hydraulic. The mechanical dredge picks up material with a scoop, shovel, or bucket and deposits it in a barge for transport to an unloading site. Mechanical dredges include the revolving crane with clamshell or dragline bucket, the backhoe, the front-opening dipper, and the endless-chain bucket dredge.

The hydraulic dredge picks up material with a pump which creates a scouring action at the open end of a suction pipe. The solids are carried up the pipe with the flow of water. The slurry may be discharged into a hopper barge or diked landfill. The suction pipe is equipped with an entrance fitting such as a rugged grate to screen out large objects, or a very wide scoop as on a dustpan dredge. Scarifiers such as high-pressure water jets or a powered cutter with blades or teeth may be installed at the end of the suction pipe to dislodge and facilitate the scouring action of the hydraulic flow. Hydraulic dredges include the trailing suction hopper dredge, the cutterhead or pipeline dredge, and the dustpan dredge. *See* CONSTRUCTION EQUIPMENT.

[KEITH W. LAWRENCE]

Drilling machine

A motor-driven device fitted with an end cutting tool that is rotated with sufficient power either to create a hole or to enlarge an existing hole in solid material. One or more flutes or grooves in the drill tool conduct coolant to the cutting lips and also provide chip relief.

Twist drills with two spiral flutes are commonly used to originate holes, while 3- and 4-flute non-center-cutting drills are used to enlarge holes. Other types are center, core, hognose, and gun-barrel drills. Cylindrical saws or pin drills are used in trepanning to cut large circular holes.

Drilling machines range in size and complexity from small sensitive drill presses through upright and multiple-spindle models designed for mass production (see illustration).

Four-spindle drilling machine, designed for mass production. (*Fosdick Machine Tool Co.*)

Drilling speeds usually decrease with material hardness while feed per revolution increases with drill diameter. Spot-facing to finish the area around a hole, counterboring to enlarge the diameter over part of the depth, and countersinking to chamfer edges of a hole are operations frequently performed during drilling setups. *See* BORING; MACHINING OPERATIONS; REAMER.

[ALAN H. TUTTLE]

Ducted fan

A propeller or multibladed fan inside a coaxial duct or cowling, also called a ducted propeller or a shrouded propeller, although in a shrouded propeller the ring is usually attached to the propeller tips and rotates. The duct serves to protect the fan blades from adjacent objects and to protect objects from the revolving blades, but more importantly, the duct prevents radial flow of the fluid at the blade tips. Fan efficiency remains high over a wider speed range with a properly shaped duct than without. However, fan efficiency is sensitive to duct shape at off-center design conditions. Without a well-rounded inlet lip and a variable-area exit, off-center performance may be worse than without the duct. *See* FAN.

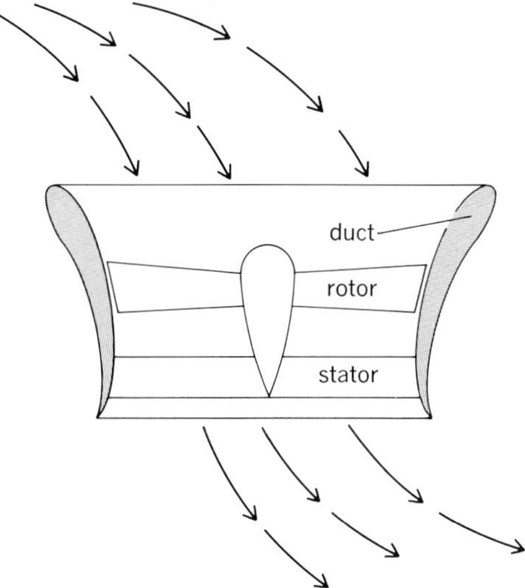

Fig. 1. Static thrust and operating speed range of fan is improved by a duct.

With a duct, static thrust for a given power input is higher than without one (Fig. 1). For this reason, propellers of vertical takeoff and landing aircraft may be ducted. At low speeds, a stator of radial airfoils downstream from the propeller or an oppositely turning coaxial propeller improve efficiency by converting slipstream rotation into axial velocity. The duct may also form a nozzle to further increase exit jet velocity. Airflow past the outer contour of the duct influences overall performance (Fig. 2).

Fig. 2. Two ducted fans provide propulsion for Bell-Navy Hydroskimmer. (*Bell Aerosystems Co.*)

Ducted fans are used in axial-flow blowers or compressors of several stages for turbine engines. In such applications, solidities are higher than for usual propellers, and stators or countervanes are usual in each stage. A ducted fan engine is a gas turbine arranged to move a larger mass of air than passes through the turbine, the additional air leaving at lower exit velocity and hence higher jet propulsion efficiency for moderate-speed aircraft than obtainable with a simple turbojet. *See* GAS TURBINE.

[FRANK H. ROCKETT]

Dynamic braking

A technique of electric braking in which the retarding force is supplied by the same machine that originally was the driving motor. Dynamic braking is effective only in high inertial systems wherein the kinetic energy of the motor rotor with its connected load is converted into electrical energy and dissipated mainly as I^2R losses or returned to the source.

The commonest type of dynamic braking will be explained for a direct-current (dc) motor. To accomplish braking action, the supply voltage is removed from the armature of the motor but not from the field. The armature is then connected across a resistor. The electromotive force generated by the machine, now acting as a generator driven by the kinetic energy of the rotating system, forces current in the reverse direction through the armature. Thus a torque is produced to oppose rotation, and the load decelerates as its kinetic energy is dissipated, mostly in the external resistor but to some extent in core and copper losses of the machine.

Electric braking can also be accomplished by causing the kinetic energy of the rotating system to be converted in the armature to electrical energy and then returned to the supply lines. This mode of operation, called regenerative braking, occurs when the counter electromotive force exceeds the supply voltage.

Interchanging two of the lines supplying a three-phase alternating-current (ac) induction motor also produces braking. In this case, called plugging, the direction of the electromagnetic torque on the rotor is reversed to cause deceleration. Both the kinetic energy of the system and the energy drawn from the supply lines are expended in copper and core losses in the machine.

Sometimes the term dynamic braking is applied only when kinetic energy is dissipated in an external resistor, but its more general interpretation includes regenerative braking and plugging.

The induction motor, when employed for dynamic braking in either the regenerative or plugging mode, represents an interesting departure from the normal operation of an induction motor. Curve *a* of the figure is a typical torque-speed characteristic of an induction motor, with N_s the synchronous speed. Curve *b* is the torque-speed characteristic when the rotation of the stator flux is opposite to the direction of rotation of the motor.

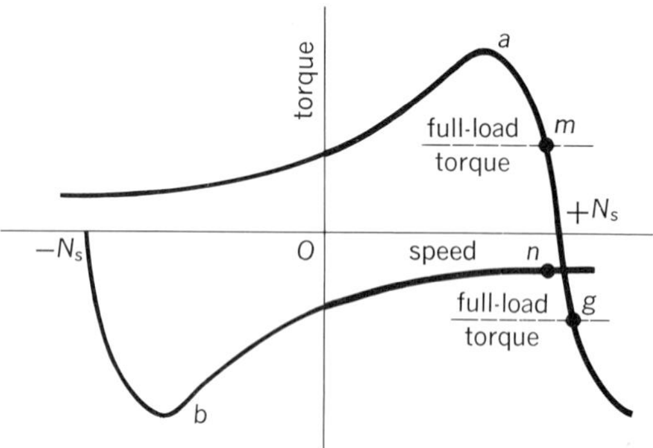

When the induction motor converts electric energy into mechanical torque for positive rotation, it operates along curve *a* in the first quadrant. When the induction motor converts mechanical torque into electrical energy, it operates along curves *a* or *b* in the fourth quadrant; curve *b* for plugging and curve *a* for regenerative braking.

Interchanging two of the ac lines results in plugging due to reversal of the direction of rotation of the magnetic field. If the motor develops full-load torque at *m*, operation will be at point *n* on curve *b* immediately after plugging. The negative torque will decelerate the load, and the kinetic energy plus the energy drawn from the line will be dissipated in the rotor resistance. Unless the supply line is disconnected when the speed reaches zero, operation will be in the third quadrant along curve *b*, and the machine will become a motor operating in the reverse direction.

For regenerative braking to occur, the power transmitted across the air gap of the machine is also negative, and it results when the rotor speed is greater than synchronous speed. Point *g* on curve *a* corresponds to the condition of regenerative braking with rated torque. The occurrence of negative torque and power across the air gap means that the machine is an induction generator. The kinetic energy is converted into electrical energy and, less internal losses, is returned to the supply. When operating as a generator, an induction machine operates with a leading power factor; when operating as a motor, it operates with a lagging power factor. This behavior is a consequence of the fact that, for all conditions of operation, the power lines must supply the magnetizing current. *See* ALTERNATING-CURRENT GENERATOR; DIRECT-CURRENT MOTOR; INDUCTION MOTOR; SYNCHRONOUS MOTOR. [ARTHUR R. ECKELS]

Dynamical analogies

Analogies are useful when one is comparing a familiar system with an unfamiliar one. An electrical circuit can be considered to be a vibrating system, and this immediately suggests analogies between electrical circuits and vibrating systems. The work of an acoustical engineer involves the study of acoustical, electroacoustical, mechanoacoustical, or electromechanoacoustical systems, while a mechanical engineer studies vibrating systems involving masses, springs, and friction. The analogies between systems of these types and electrical circuits are known as dynamical analogies.

Vibration problems may be solved by establishing an equivalent electrical circuit. This is known as an electromechanical analogy method. By this experimental technique it is possible to study the effect of varying certain parts of a mechanical system, such as damping or spring rate. The corresponding electrical components are easily controlled and are inexpensive to provide. The electrical equivalent system will not only save time but may be the only means to solve some complex mechanical problems.

The analogy is based on a similarity of the equations of electrical circuits with those of the mechanical system. The circuit equations are generally established by a method based on Kirchhoff's second law, which states that in any network the algebraic sum of the potential difference around any closed circuit is zero. Such a system is shown in illustration *a*. An examination of this system shows it to be similar to a mechanical system for a forced vibration in a single-degree-of-freedom sys-

Equivalents for mechanical and electrical systems

Mechanical quantity	Electrical quantity
m, mass	L, inductance
k, spring constant	$\frac{1}{C}$, $\frac{1}{\text{capacitance}}$
c, damping factor	R, resistance
x, displacement	q, charge
dx/dt, velocity	dq/dt, current
F, force	v, voltage
ω, frequency	ω, frequency

tem (illustration *b*). The mass m in the mechanical system is equivalent to the inductance L, the damping factor c to the resistance R, and the spring constant k to the capacitance C. The forcing function $F_0 \cos \omega t$ is analogous to the impressed voltage $v_0 \cos \omega t$.

It should be noted in the mechanical system that the forces acting on the mass are in parallel, whereas in the electrical system the components of the circuit are in series. If the forces in the mechanical system were in series, then the equivalent electrical system would be put in parallel. The equivalents for the mechanical system and the electrical system are summarized in the table. *See* ALTERNATING-CURRENT CIRCUIT THEORY; KIRCHHOFF'S LAWS OF ELECTRIC CIRCUITS; VIBRATION DAMPING.

[K. W. JOHNSON]

Bibliography: R. E. D. Bishop, *Vibration*, 2d ed., 1979; A. H. Church, *Mechanical Vibrations*, 2d ed., 1963; H. F. Olson, *Solutions of Engineering Problems by Dynamical Analogies*, 1966.

Dynamo

A machine for the conversion of electrical energy into mechanical energy or, conversely, mechanical energy into electrical energy. It is called a generator if it converts mechanical into electrical energy, and it is called a motor if it converts electrical into mechanical energy. The term "dynamo" is now largely deprecated in favor of "electric machine." *See* ELECTRIC ROTATING MACHINERY; GENERATOR; MOTOR.

[ARTHUR R. ECKELS]

Dynamometer

A special type of electric rotating machine used to measure the output torque or driving torque of rotating machinery. Most dynamometers consist of a direct-current (dc) machine with the stator cradle-mounted in antifriction bearings. The rotor is connected to the rotor of the machine under test. The field current is introduced through flexible leads. The stator is constrained from rotating by a radial arm of known length to which is attached a scale for measuring the force required to prevent rotation.

The torque of the connected machine is found from the product of the lever arm length and the scale reading, after correcting the scale reading by the amount of the zero torque reading. By using a tachometer to measure the rotor speed, the power may be found from the equation $\text{hp} = 2\pi NT/33{,}000$, in which N is the shaft speed in rpm and T is the torque in foot-pounds. *See* TORQUE.

If the machine under test is a motor, the dynamometer will act as a generator. The dynamometer output is absorbed by a loading resistance or by feeding it into a dc line. The amount of the output is easily adjusted by changing the loading resistance or by changing the field excitation. If the machine under test is a generator or mechanical load, the dynamometer will act as a motor. The speed is adjusted by changing the armature voltage or the field excitation. The dynamometer method is direct-reading and is more accurate than measuring the electrical output and correcting for the losses. Except for the inaccuracy caused by friction in the stator mounting bearings and by windage loss not reflected in the stator torque, all the shaft torque is accounted for in the scale reading. *See* DIRECT-CURRENT GENERATOR; DIRECT-CURRENT MOTOR; ELECTRIC ROTATING MACHINERY; GENERATOR; MOTOR.

When the machine under test is a motor, a mechanical device known as a prony brake may be employed to convert the output energy to heat through friction. A drum, which may be water-cooled, is driven by the machine under test. A brake arm, which is constrained from rotating by a scale at the outer end, is tightened around the drum to increase the friction and to produce the desired torque. The torque is the product of the scale reading and the length of the brake arm. This is an inexpensive and accurate method of measuring the torque of small motors. With high-horsepower motors, however, it becomes difficult to dissipate the large amounts of energy. Large-capacity units, which utilize a liquid brake in place of the friction drum, have been constructed. These are smaller and less expensive than electric units of like capacity but lack the flexibility and ease of recovering the energy.

[ARTHUR R. ECKELS]

Earthmover

Any of a variety of construction machines designed to move or transport earth. Earthmovers include heavy-duty trucks with high-sided dump bodies, self-propelled or towed scrapers, wagons, and bulldozers. The bulldozer mounted on a wheeled or crawler tractor is suitable for moving large quantities of earth for distances of several hundred feet. Scrapers and wagons are efficient for moving earth over relatively level terrain for distances up to 1 or 2 mi. For longer distances or for grades in excess of 5%, trucks are the most practical. Scrapers have the advantages of being able to load themselves without help from a crane or power shovel and of discharging their loads in finely controlled layers. Dump trucks can be designed to unload either to the side or to the rear, and the latter type can distribute loads in a swath the width of the dump body. Wagons can have side, bottom, or rear dumping mechanisms. *See* BULK-HANDLING MACHINES; CONSTRUCTION EQUIPMENT.

[EDWARD M. YOUNG]

Efficiency

The ratio, expressed as a percentage, of the output to the input of power (energy or work per unit time). As is common in engineering, this concept is defined precisely and made measurable. Thus, a gear transmission is 97% efficient when the useful

DYNAMICAL ANALOGIES

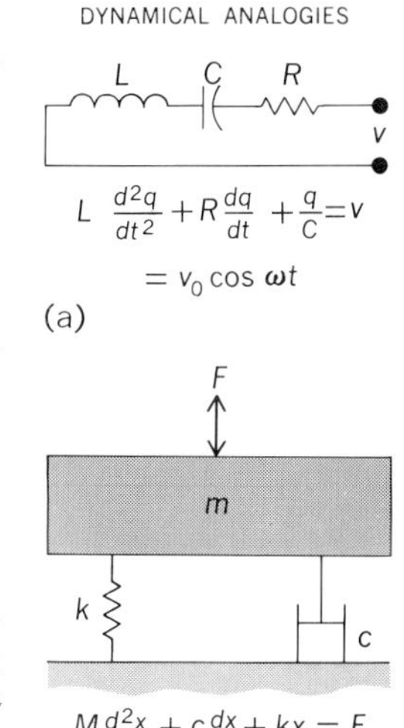

Diagrams of (*a*) electrical circuit and (*b*) mechanical system.

energy output is 97% of the input, the other 3% being lost as heat due to friction. A boiler is 75% efficient when its product (steam) contains 75% of the heat theoretically contained in the fuel consumed. All automobile engines have low efficiency (below 30%) because of the total energy content of fuel converted to heat; only a portion provides motive power, while a substantial amount is lost in radiator and car exhaust.

In such simple cases the value is clear. However, in some others it can be difficult to calculate exactly. For example, the efficiency of the process for converting corn to alcohol for use as automobile fuel can be computed as the ratio of the heat value of the alcohol to the heat value of all the energy used to produce it, including the fuel for the tractor to plow and harvest the cornfield and even the fuel used to create the steel and fabricate the tractor. The question is then, how much to include in the overall efficiency determination. *See* SIMPLE MACHINE. [F. R. E. CROSSLEY]

Elastic limit

The largest stress that a material can sustain without permanent deformation or strain. The proportional limit, it may be noted, is the maximum stress at which the stress is directly proportional to the strain.

Actual materials are not perfectly elastic upon first application of load because they lack homogeneity and have residual stresses produced by the manufacturing process. Upon subsequent loadings, the material will exhibit elastic properties. *See* STRESS AND STRAIN. [JOHN B. SCALZI]

Elasticity

The property whereby a solid material changes its shape and size under the action of opposing forces, but recovers its original configuration when the forces are removed. The theory of elasticity deals with the relations between the forces acting on a body and the resulting changes in configuration, and is important in many branches of science and technology, for instance, in the design of structures, in the theory of vibration and sound, and in the study of the forces between atoms in crystal lattices.

Elastic constants. The forces acting on a body are expressed as stresses and measured as force per unit area. Thus if a bar *ABCD* of square cross section (Fig. 1*a*) is fixed at one end and subjected to a force *F* uniformly distributed over the other end *DC*, the stress is $F/(DC)^2$. This stress causes the bar to become longer and thinner and to assume the shape $A'B'C'D'$. The strain is measured by the ratio (change in length)/(original length), that is, by $(B'C' - BC)/(BC)$. According to Hooke's law, stress is proportional to strain, and the ratio of stress to strain is therefore a constant, in this case the Young's modulus, denoted by *E*, so that $E = F(BC)/(DC)^2(B'C' - BC)$. *See* HOOKE'S LAW; STRESS AND STRAIN; YOUNG'S MODULUS.

Poisson's ratio σ is defined as the ratio of lateral strain to longitudinal strain, so that $\sigma = BC(DC - D'C')/DC(B'C' - BC)$. The bar of Fig 1*a* is in a state of tension, and the stress is tensile; if the force *F* were reversed in direction, the stress would be compressive. Stresses of this type are called direct or normal stresses; a second type of stress, known as tangential or shear stress, is illustrated in Fig. 1*b*. In this case, the configuration *ABCD* becomes $ABC'D'$, with the shear forces *F* acting in the directions *AB* and *CD*. The shear strain is measured by the angle θ, and if the body is originally a cube, the shear stress is $F/(DC)^2$. The ratio of stress to strain, $F/(DC)^2\theta$, is the shear or rigidity modulus *G*, which measures the resistance of the material to change in shape without change in volume.

A further elastic constant, the bulk modulus *k*, measures the resistance to change in volume without changes in shape, and is illustrated in Fig. 1*c*. The original configuration is represented by the circle *AB*, and under a hydrostatic (uniform) pressure *P*, the circle *AB* becomes the circle $A'B'$. The bulk modulus is then $k = Pv/\Delta v$, where $\Delta v/v$ is the volumetric strain. The reciprocal of the bulk modulus is the compressibility.

Determination of values. The elastic constants may be determined directly in the way suggested by their definitions; for instance, Young's modulus can be determined by measuring the relative extension of a rod or wire subjected to a known tensile stress. Less direct methods are, however, usually more convenient and accurate. Prominent among these are the dynamic methods involving frequency of vibration and velocity of sound propagation. The elastic constants can be expressed in terms of frequency of (or velocity in) regularly shaped specimens, together with the dimensions and density, and by measuring these quantities, the elastic constants can be found.

The elastic constants can also be determined from the flexure and torsion of bars. As an illustration, consider a bar *AB* (Fig. 2) of breadth *b* (in the x_1 direction) and depth *d* (in the x_2 direction) supported by forces *F* at the ends, and loaded symmetrically by forces *F* at points *C* and *D*. Over the portion *CD* there is a uniform bending moment $M = Fl_2$, and the theory of bending shows that the portion *CD* is bent into the arc of a circle such that Eq. (1) applies, where *R* is the radius of curvature,

$$R = EI/M \tag{1}$$

E is Young's modulus and *I* is the moment of inertia of cross section, equal to $bd^3/12$ for a rectangular cross section. The longitudinal stress at the lower face of the bar is tensile, and at the upper face, compressive. The middle plane of the bar is free of stress, and is the neutral axis. The stress at a distance x_2 from the neutral axis is shown in Eq. (2).

$$T = Ex_2/R = Mx_2/I \tag{2}$$

It is thus possible to determine *E* from Eq. (1) by

ELASTICITY

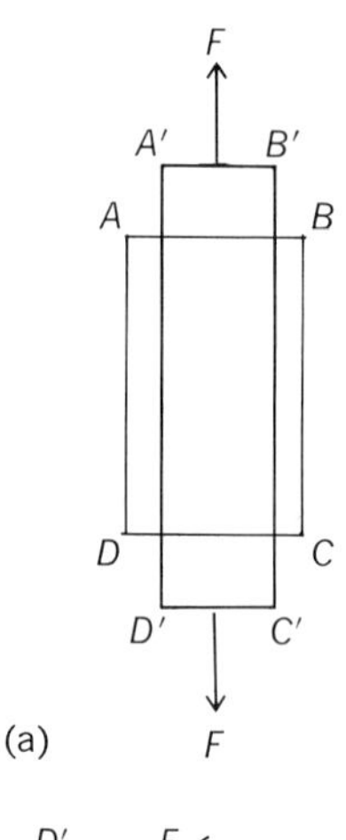

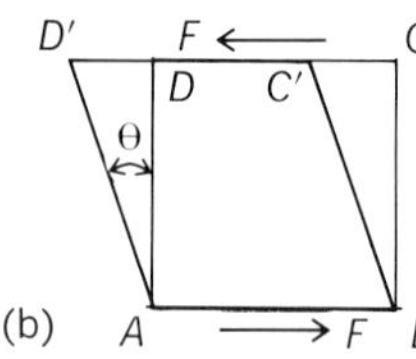

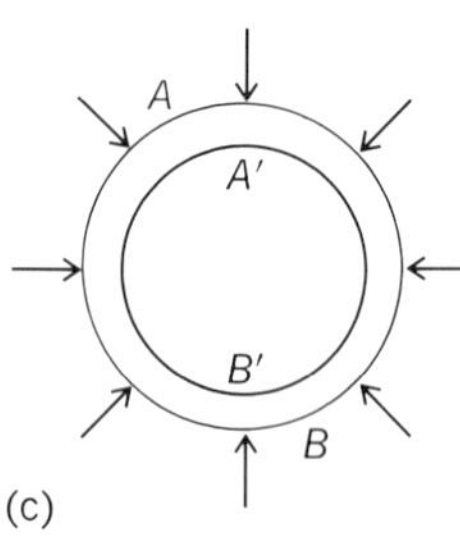

Fig. 1. Stresses on a bar. (*a*) Direct or normal stress. (*b*) Tangential or shear stress. (*c*) Change in volume with no change in shape. (All deformations are exaggerated.)

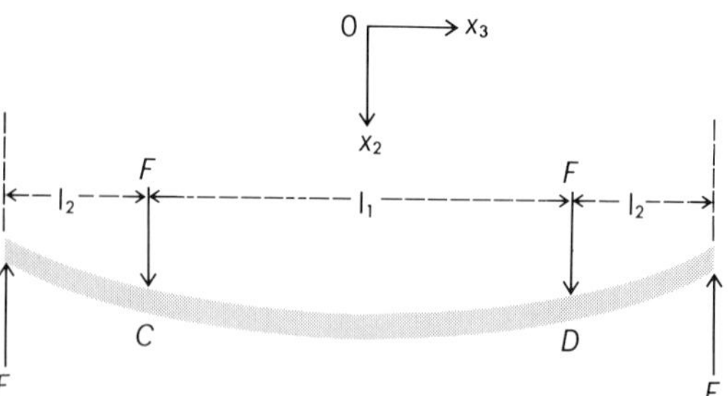

Fig. 2. Flexure and torsion in a bar.

measuring I, R, and M; conversely, if E is known, the stress may be determined from Eq. (2). *See* LOADS, TRANSVERSE.

Practical limitations. In practice, stress is only proportional to strain, and the strain is only completely recoverable within certain limits called the elastic limits of the material. The stress below which the strain is completely recoverable is sometimes called the limit of perfect elasticity, and the stress up to which Hooke's law is obeyed is sometimes called the proportional limit or limit of linear elasticity. Above the elastic limits, the material is subject to time-dependent effects, and as the stress is further increased, the ultimate strength of the material is approached. *See* PLASTICITY; STRENGTH OF MATERIALS.

Theory of elasticity. In classical elasticity theory, it is assumed that the strains are always small; Hooke's law is therefore obeyed; the strains are completely recoverable and, moreover, are superposable, so that the strain produced by the joint action of two or more stresses is the sum of the strains produced by them individually.

In order to develop the theory, it is necessary to specify the stresses and strains more closely. Figure 3 shows the stress components T_{ij} (where i, j may take the values 1, 2, or 3) acting on the faces of a cube, parallel to the coordinate axes x_1, x_2, x_3. The first suffix indicates the direction of the stress component and the second the direction of the normal to the plane under consideration. Stresses of the type T_{11} are normal stresses, and of the type T_{12}, shear stresses. The conditions for zero rotation of the cube are $T_{12}=T_{21}$, $T_{13}=T_{31}$, $T_{23}=T_{32}$, and there are therefore six independent stress components.

In addition to the stresses T_{ij}, body forces proportional to volume (for instance, forces due to the weight of the body) may also be acting. If the stresses T_{ij} vary with position, application of Newton's second law leads to Eq. (3) for the x_1 direction

$$\frac{\partial T_{11}}{\partial x_1}+\frac{\partial T_{12}}{\partial x_2}+\frac{\partial T_{13}}{\partial x_3}+X_1=\rho f_1 \qquad (3)$$

where ρ is the density, f_1 is the acceleration, and X_1 the body force component per unit volume along x_1, together with two similar equations for the x_2 and x_3 directions. If $f_1=f_2=f_3=0$, these equations become the equations of equilibrium, and if, further, $X_1=X_2=X_3=0$, they become the equations of equilibrium in the absence of body forces. The preceding equations are important in many branches of elastic theory and, for example, provide a starting point in the study of vibrating bodies and of the twisting of cylinders and prisms with cross sections of various shapes. *See* TORSION.

The components of strain are specified in a similar way to the stresses. There are six independent strain components: S_{11}, S_{22}, S_{33}, S_{23}, S_{13}, and S_{12}. If, as a result of strain, the coordinates of a point x_1, x_2, x_3 become x_1+u_1, x_2+u_2, x_3+u_3, the quantities u_1, u_2, and u_3 are the components of the displacement vector, and the strain components are as displayed in Eq. (4), so that, for example the relations shown by Eq. (5) would hold true.

$$S_{ij}=\tfrac{1}{2}\left(\frac{\partial u_i}{\partial x_j}+\frac{\partial u_j}{\partial x_i}\right) \qquad (4)$$

$$S_{11}=\frac{\partial u_1}{\partial x_1} \qquad S_{12}=\tfrac{1}{2}\left(\frac{\partial u_1}{\partial x_2}+\frac{\partial u_2}{\partial x_1}\right) \qquad (5)$$

By eliminating the displacements from these equations, the so-called compatibility equations are obtained with three of the type shown in Eq. (6)

$$\frac{\partial^2 S_{22}}{\partial x_3{}^2}+\frac{\partial^2 S_{33}}{\partial x_2{}^2}=2\frac{\partial^2 S_{23}}{\partial x_2\,\partial x_3} \qquad (6)$$

and three of the type shown in Eq. (7).

$$\frac{\partial^2 S_{11}}{\partial x_2\,\partial x_3}=\frac{\partial}{\partial x_1}\left(-\frac{\partial S_{23}}{\partial x_1}+\frac{\partial S_{13}}{\partial x_2}+\frac{\partial S_{12}}{\partial x_3}\right) \qquad (7)$$

The stresses and strains have so far been denoted by two suffixes. This is essential if the methods of tensor analysis are to be applied to elasticity problems, but for many purposes a single suffix notation is adequate. The change from a two- to a one-suffix notation for the stresses is simply $T_{11}=T_1$, $T_{22}=T_2$, $T_{33}=T_3$, $T_{23}=T_4$, $T_{13}=T_5$, $T_{12}=T_6$. The change of notation for the strains is $S_{11}=S_1$, $S_{22}=S_2$, $S_{33}=S_3$, $2S_{23}=S_4$, $2S_{13}=S_5$, $2S_{12}=S_6$; the factor 2 is required to make the strains S_4, S_5, and S_6 conform with the usual definition of shear strain (Fig. 1*b*).

Hooke's law generalized. Hooke's law may be generalized to the statement that each stress component is proportional to each strain component, equivalent to the six equations

$$T_1=c_{11}S_1+c_{12}S_2+c_{13}S_3+c_{14}S_4+c_{15}S_5+c_{16}S_6$$
$$\cdots\cdots\cdots\cdots\cdots\cdots\cdots\cdots$$
$$T_6=c_{61}S_1+c_{62}S_2+c_{63}S_3+c_{64}S_4+c_{65}S_5+c_{66}S_6$$

which may be written more concisely as Eq. (8),

$$T_q=\sum_r c_{qr}S_r \qquad (8)$$

where the summation extends over $r=1, 2, 3, 4, 5$, and 6. The elastic constants c_{qr} are termed the stiffnesses; there are altogether 36 of them but they are subject to the reciprocal relations $c_{qr}=c_{rq}$ imposed by thermodynamic requirements, and the number is thus reduced to 21.

Additional relations can be derived from the three assumptions that the interatomic forces act along the lines joining the centers of atoms in the lattice, that the atoms are situated at centers of symmetry, and that the lattice is initially at zero stress. These relations, called Cauchy relations, are $c_{23}=c_{44}$, $c_{13}=c_{55}$, $c_{12}=c_{66}$, $c_{14}=c_{56}$, $c_{25}=c_{46}$, $c_{45}=c_{36}$ and, if true, would reduce the number of stiffnesses to 15. Experiment shows, however, that they are not true in general; nevertheless, their investigation provides an indication of the extent to which these three assumptions hold in any particular case.

The generalized Hooke's law can also be written to express the strains in terms of the stresses given in Eq. (9), in which the quantities s_{qr} are the elastic compliances.

$$S_q=\sum_r s_{qr}T_r \qquad (r=1, 2, 3, 4, 5, 6) \qquad (9)$$

If the six simultaneous equations of Eq. (8) are solved for the strains, the compliances are obtained in terms of the stiffnesses as in Eq. (10), where Δc is the determinant shown in Eq. (11)

$$s_{qr}=\Delta c_{qr}/\Delta c \qquad (10)$$

ELASTICITY

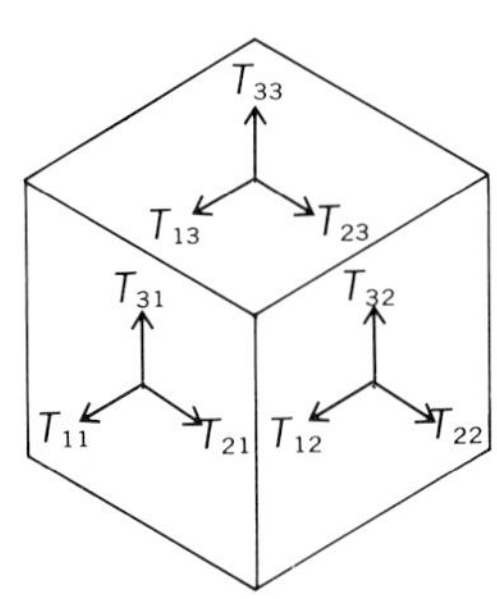

Fig. 3. Stress components.

Stress-strain relations

Stress	Strain: Orthorhombic	Hexagonal	Cubic	Isotropic
$T_1=$	$c_{11}S_1+c_{12}S_2+c_{13}S_3$	$c_{11}S_1+c_{12}S_2+c_{13}S_3$	$c_{11}S_1+c_{12}S_2+c_{12}S_3$	$c_{11}S_1+c_{12}S_2+c_{12}S_3$
$T_2=$	$c_{12}S_1+c_{22}S_2+c_{23}S_3$	$c_{12}S_1+c_{11}S_2+c_{13}S_3$	$c_{12}S_1+c_{11}S_2+c_{12}S_3$	$c_{12}S_1+c_{11}S_2+c_{12}S_3$
$T_3=$	$c_{13}S_1+c_{23}S_2+c_{33}S_3$	$c_{13}S_1+c_{13}S_2+c_{33}S_3$	$c_{12}S_1+c_{12}S_2+c_{11}S_3$	$c_{12}S_1+c_{12}S_2+c_{11}S_3$
$T_4=$	$c_{44}S_4$	$c_{44}S_4$	$c_{44}S_4$	$(c_{11}-c_{12})S_4/2$
$T_5=$	$c_{55}S_5$	$c_{44}S_5$	$c_{44}S_5$	$(c_{11}-c_{12})S_5/2$
$T_6=$	$c_{66}S_6$	$(c_{11}-c_{12})S_6/2$	$c_{44}S_6$	$(c_{11}-c_{12})S_6/2$

$$\Delta c=\begin{vmatrix} c_{11} & c_{12} & c_{13} & c_{14} & c_{15} & c_{16} \\ c_{12} & c_{22} & c_{23} & c_{24} & c_{25} & c_{26} \\ c_{13} & c_{23} & c_{33} & c_{34} & c_{35} & c_{36} \\ c_{14} & c_{24} & c_{34} & c_{44} & c_{45} & c_{46} \\ c_{15} & c_{25} & c_{35} & c_{45} & c_{55} & c_{56} \\ c_{16} & c_{26} & c_{36} & c_{46} & c_{56} & c_{66} \end{vmatrix} \quad (11)$$

and Δc_{qr} is the cofactor obtained by deleting the row and column containing c_{qr} from the determinant Δc.

The 21 stiffnesses (or compliances) of the generalized Hooke's law describe the elastic behavior of a material belonging to the triclinic crystal system. The existence of symmetry elements reduces the number of independent elastic constants in the other crystal systems to the following numbers: monoclinic, 13; orthorhombic, 9; tetragonal, 7 or 6; trigonal, 7 or 6; hexagonal, 5; and cubic, 3. Materials belonging to all of these systems are anisotropic, and the elastic properties depend upon direction within the material. If the properties are independent of direction, the material is isotropic and its elastic behavior is completely described by two independent stiffnesses (or compliances).

The stress-strain relations, referred to the principal axes in the orthorhombic, hexagonal, cubic, and isotropic systems, are given in the table. The equations involving the compliances are completely analogous, with S and T interchanged, and s_{qr} written for c_{qr}, except where $T_q=\frac{1}{2}(c_{11}-c_{12})\,S_q$, in which case $S_q=2(s_{11}-s_{12})\,T_q$.

Rochelle salt is an example of an orthorhombic crystal; materials which, although not crystalline, possess the same symmetry and matrix of elastic constants as orthorhombic crystals are said to be orthotropic. Wood and plywood are materials of this description, and orthotropic elastic theory has also been applied to laminated plastics and reinforced concrete.

Single-crystal zinc, cobalt, magnesium, and ice are hexagonal materials; they are transversely isotropic because the properties are independent of direction in all of the planes normal to the hexagonal axis.

Single-crystal copper, gold, silver, nickel, and the alkali halides (for example, sodium chloride) are important cubic materials. The stress-strain equations are derived from those of the orthorhombic system by superimposing the condition that the three principal directions are all equivalent. This does not mean that the properties are independent of direction; for example, the compliance s'_{11} in an arbitrary direction is given by Eq. (12), where a_1, a_2, a_3 are the cosines of the angles between the arbitrary direction and the cubic axes. This equation shows that s'_{11} depends on orientation unless $s_{44}/2=(s_{11}-s_{12})$.

$$s'_{11}=s_{11}-2(s_{11}-s_{12}-s_{44}/2)\cdot(a_1^2a_2^2+a_2^2a_3^2+a_3^2a_1^2) \quad (12)$$

[R. F. S. HEARMON]

Bibliography: A. P. Boresi and P. P. Lynn, *Elasticity in Engineering Mechanics: With an Introduction to Numerical Stress Analysis*, 1974; S. F. Borg, *Fundamentals of Engineering Elasticity*, 1962, reprint 1973; R. F. S. Hearmon, *An Introduction to Applied Avisotropic Elasticity*, 1961; J. F. Nye, *Physical Properties of Crystals: Their Representation by Tensors and Matrices*, 1957; S. P. Timoshenko and J. M. Gere, *Theory of Elastic Stability*, 2d ed., 1961; S. P. Timoshenko and J. N. Goodier, *Theory of Elasticity*, 3d ed., 1969.

Electric contact

A part, in an electrical switching device, made of conducting material, for the purpose of closing, opening, or changing the conductive path of an electrical circuit. To open or close a circuit, an electric contact is made to come in contact with or separate from its mating part. Devices embodying contacts for these purposes are electric switches, relays, contactors, and circuit breakers. Contacts may be actuated directly or through a linkage that is driven either manually, mechanically, electromagnetically, hydraulically, or pneumatically.

An electric contact is also an essential part of a variable resistor. In such an application the contact or wiper provides a moving connection to the resistive element. For such service the contact material is chosen for low abrasion combined with absence of a high-resistance film such as would be formed by an adherent oxide. *See* POTENTIOMETER (VARIABLE RESISTOR); RHEOSTAT.

Requirements for contacts differ markedly depending on applications. For contacts in relays and low-power applications, reliability in completing a circuit may be of the utmost importance. Surface films and contaminants are not tolerable. Noble metals such as gold and platinum are sometimes sealed hermetically for this purpose. For the purpose of carrying high continuous currents, electrical resistance must be kept to a minimum. Contacts with high silver content are usually used. High contact forces, up to hundreds of pounds, may be applied when the contacts are in the closed position. For contacts that must interrupt high currents, refractory materials such as tungsten and molybdenum are often used. These materials also possess antiwelding characteristics, which are essential for contacts that must close in on short-circuit currents and later open under normal operating force.

On many occasions all the aforementioned properties are desired in one material. Special alloys such as silver cadmium oxide or refractory materials impregnated with silver have been developed for this purpose.

A related electrical part is the brush on rotating machines. Pairs of brushes provide continuity between the external circuit and slip rings or commutator segments on the rotating portion of the machine. Similar arrangements of brushes and slip rings connect to continuously rotating radar antenna mounts. Graphite, because of its inherent lubricating nature, is a usual constituent of such sliding contacts for high-current operation; corrosion-resistant metals or metal alloys are used for low-current applications. [THOMAS H. LEE]

Bibliography: R. Holm and E. Holm, *Electric Contacts*, 4th ed., 1967.

Electric-discharge machining

A metal-removal process in which materials that conduct electricity can be removed by an electric spark. It is used to form holes of varied shape in materials of poor machinability and to form cavities in steel dies.

Originally developed as a method of removing broken taps or drills from holes in castings, it is also called electric-spark machining.

The spark is a transient electric discharge through the space between the tool (cathode) and the workpiece (anode), as shown in Fig. 1. The sparking circuit is controlled by a resistance-capacitance charging circuit or the pulse circuit of Fig. 2.

The 0.001 – 0.004-in. (25 – 100-μm) space between tool and workpiece is filled with a dielectric hydrocarbon oil, which serves as a cooling medium and flushes away the metal particles. Sparking rate may be from about 20,000 to several hundred thousand discharges per second.

It is believed that each spark melts or vaporizes the small point of the workpieces that it strikes and forms a pit mark or crater, as seen under magnification. These craters make up the workpiece surface, the quality of which for any material varies with the spark gap, current, voltage, and material of the cathode tool.

The tool may be brass, copper, steel or other metals, or graphite. The tool wears and loses its shape with use. A 90% tungsten – 10% silver tool has longer usable life and cuts much faster than a brass tool. The rate of feed and proper gap size are maintained by a servo device.

Cavities from a few thousandths of an inch to 2 or 3 in. (50-75 μm) in width may be machined by this process. High surface quality and accuracy can be obtained with a small gap between electrodes, but larger gaps allow maximum metal removal of 0.05 in.^{3}/min (14 mm^3/sec) with correspondingly rougher surfaces. For other metal-removal processes *see* MACHINING OPERATIONS.

[ORLAN W. BOSTON]

Bibliography: J. M. Alexander (ed.), *Machine Tool Design and Research: Proceedings*, 18th ed., 1978; F. J. DeMaine, *Electrical Discharge Machining: Considerations for Profitable Utilization*, Amer. Soc. Tool Mfg. Eng. Pap. MR68-118, May 2, 1968; F. L. King, *EDM Goes Adaptive*, Amer. Soc. Tool Mfg. Eng. Pap. MR68-119, May 2, 1968.

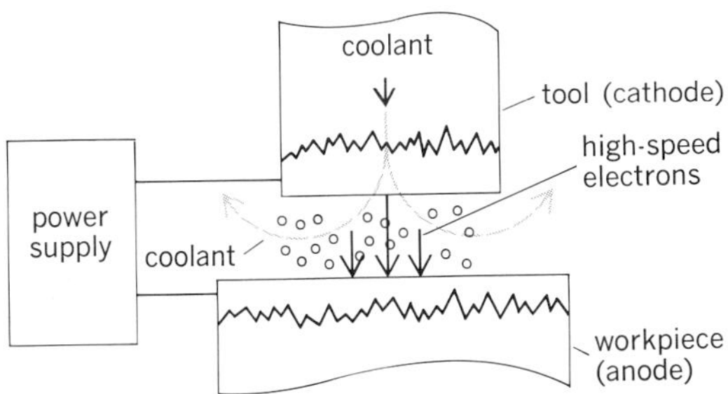

Fig. 1. Electric-discharge, or electric-spark, machining process. (*Elox Corporation of Michigan*)

Electric distribution systems

That part of an electric power system that supplies electric energy to the individual user or consumer. The distribution system includes the primary circuits and the distribution substations that supply them; the distribution transformers; the secondary circuits, including the services to the consumer; and appropriate protective and control devices. The four general classes of individual users are residential, industrial, commercial, and rural.

Systems. The three-phase, alternating-current (ac) system (Fig. 1) is practically universal, although a small amount of two-phase and direct-current systems from early days are still in operation. Three-phase transmission and substransmission lines require three wires, termed phase conductors. Most of the three-phase distribution systems consist of three phase conductors and a common or neutral conductor, making a total of four wires. Single-phase branches (consisting of two wires) supplied from the three-phase mains are used for single-phase utilization in residences, small stores, and farms. Loads are connected in parallel to common supply circuits. *See* ELECTRIC POWER SYSTEMS.

Substation. The distribution substation is an assemblage of equipment for the purpose of switching, changing, and regulating the voltage from subtransmission to primary distribution. More important substations are designed so that the failure of a piece of equipment in the substation or one of the subtransmission lines to the substation will not cause an interruption of power to the load.

Primary voltages. The primary system leaving the substation is most frequently in the 11,000–15,000 volt range. A particular voltage used is 12,470-volt line-to-line and 7200-volt line-to-neutral (conventionally written 12,470Y/7200 volts). Some utilities use a lower voltage, such as 4160Y/2400 volts. The use of voltages above the 15-kv class is increasing. Several percent of primary distribution circuits are in the 25- and 35-kv classes; all are four-wire systems. Single-phase loads are connected line-to-neutral on the four-wire systems.

Secondary voltages. Secondary voltages are derived from distribution transformers connected to the primary system and they usually correspond to utilization voltages. Residential and most rural loads are supplied by 120/240-volt single-phase three-wire systems (Fig. 2). Commercial and small industrial needs are supplied by either 208Y/120-volt or 480Y/277-volt three-phase four-wire sys-

ELECTRIC-DISCHARGE MACHINING

electrode
capacitor
dc source
workpiece
pulsing tube
pulser

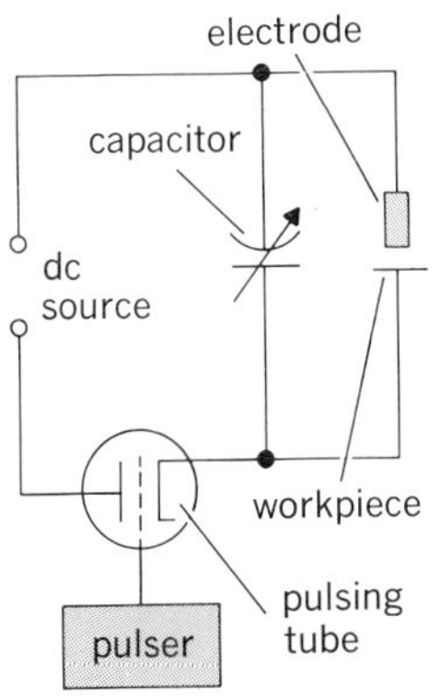

Fig. 2. Pulse-type circuit to control sparking circuit. (*Cincinnati Milling Machine Co.*)

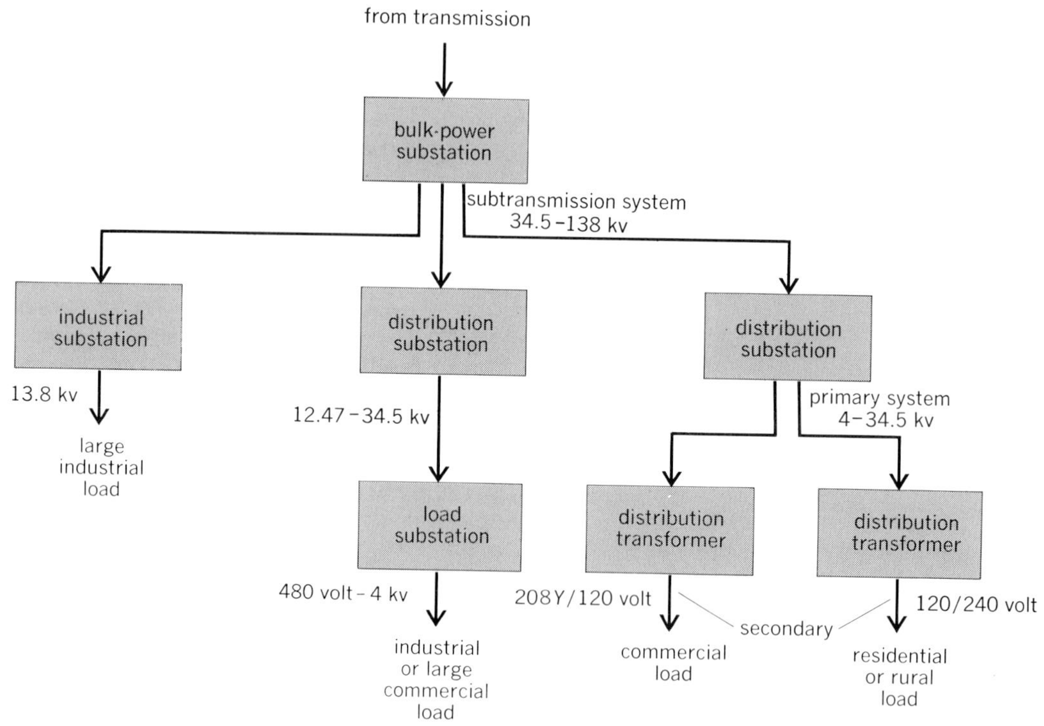

Fig. 1. Typical three-phase power system from bulk-power source to consumer's switch.

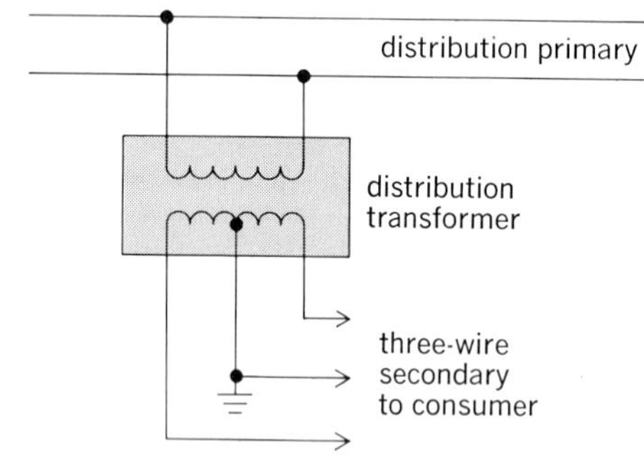

Fig. 2. Single-phase three-wire secondary circuit.

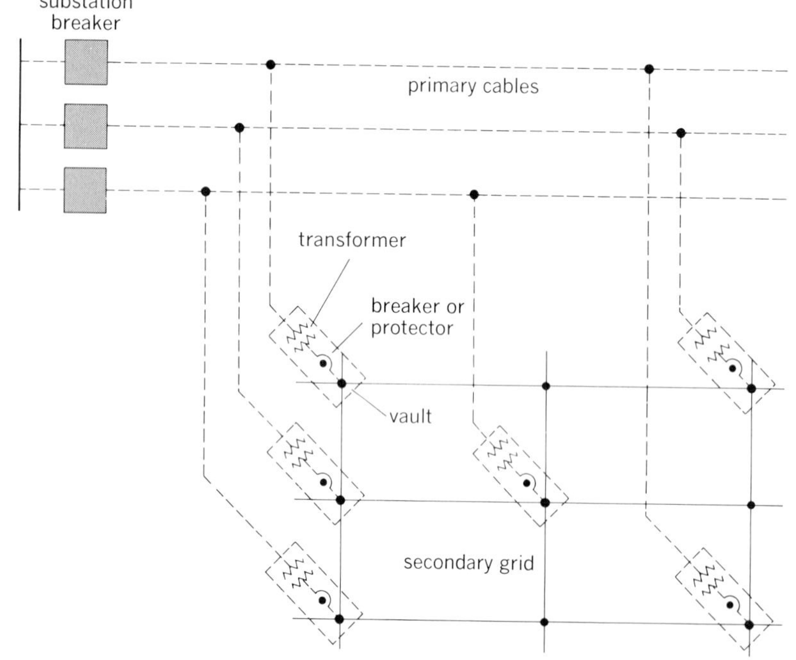

Fig. 3. Typical three-phase 208Y/120-volt secondary grid network system.

tems. The secondary voltage is usually used to supply multiple street lights, in addition to service to consumers. Photoelectric controls are employed to turn the street lamps on and off.

Good voltage. Good voltage means that the average voltage level is correct, that variations do not exceed prescribed limits, and that sudden momentary changes in level do not cause objectionable light flicker. Utilization voltage varies with changing load on the system, but a voltage variation of less than 5% at the consumer's meter is common. To achieve this result, distribution systems are designed for a plus and minus voltage spread from the nominal voltages shown above. This is accomplished by proper wire size for the circuits, application of capacitors, both permanently connected and switched, and use of voltage regulators. Voltage regulators may be used at the substation or at a point along the circuit.

Good continuity. Service continuity is the providing of uninterrupted electric power to the consumer; therefore, good continuity is doing this a high percentage of the time. This is accomplished for large industrial and commercial loads by use of some form of duplicate power supply. Downtown commercial areas are supplied from three-phase 208Y/120-volt grid networks (Fig. 3). These networks are fed from a number of primary feeders, stepping down through network transformers and automatic-reclosing secondary circuit breakers (protectors) to the secondary grid formed by cables under the streets. The system is arranged so that the failure of a primary feeder will not cause a loss of load on the secondary. Commercial buildings and shopping centers are often served by spot networks. All of the transformers and protectors are at the same location. Residential and rural loads are usually supplied by a radial system. Good continuity for them is obtained by sectionalizing the system with fuses, circuit breakers, and manual switches to reduce the extent of an outage due to a failure.

Elements of distribution systems. In distribution systems there are a number of elemental parts or subsystems, which are discussed below.

Primary feeders. Power is carried by primary feeders from distribution substations to the load areas where the consumers are located.

Distribution transformers. The distribution transformer located near the consumers changes the voltage from the primary distribution voltage to the secondary distribution voltage.

Secondary mains. This element of electric distribution is a low-voltage system which connects the secondary winding of distribution transformers to the consumers' services.

Overhead construction. The majority of distribution systems already built in residential, industrial, and rural areas has been overhead construction. This construction utilizes poles treated with pentachlorophenol or creosote, distribution transformers mounted near the top of the pole, and bare primary and secondary conductors strung from pole to pole. Aluminum conductors have been used in place of copper because of their lower cost.

For aesthetic reasons, underground distribution systems are being installed in most new residential developments. *See* TRANSMISSION LINES.

Underground construction. Commercial areas and certain main thoroughfares usually have underground construction. Construction for these areas is usually a system of concrete or fiber ducts and vaults. Impregnated paper protected by lead sheaths or synthetic compounds is generally used as conductor insulation.

In residential areas the cable is usually insulated with polyethylene or chemically crosslinked polyethylene, which is a thermosetting material. Primary cables supplying distribution transformers in the development area are likely to be single-phase, consisting of an insulated phase conductor with a concentric bare neutral conductor taken from a 12,470-volt three-phase primary feeder. The cables are often directly buried in the earth. The distribution transformers used are either the submersible type installed in a hole with a liner or pad-mounted and located on the surface. *See* ELECTRIC POWER SUBSTATION. [HAROLD E. CAMPBELL]

Bibliography: Edison Electric Institute, *Underground Systems Reference Book*, EEI Publ. no. 55-15, 1967; B. G. A. Skortizki (ed.), *Electric Transmission and Distribution*, vol. 3, 1954; U.S. Standards Institute, *Definition of Electrical Terms: Transmission and Distribution (Group 35)*, USASI C42.35, 1957; U.S. Standards Institute, *Preferred Voltage Ratings for A-C Systems and Equipment: Guide for (EEI R-6-1949) (IEC38)*, USASI C84.1, 1954; E. Vennard, *The Electric Power Business*, 2d ed., 1970; B. M. Weedy, *Electric Power Systems*, 3d ed., 1979.

Electric energy measurement

The measurement of the integral, with respect to time, of the power in an electric circuit. The absolute unit of measurement of electric energy is the joule, or the charge in coulombs times the potential difference in volts. The joule, however, is too small (1 watt-second) for use in commercial practice, and the more commonly used unit is the watt-hour (3.6×10^3 joules). The most common measurement application is in the utility field.

Electric energy is one of the most accurately measured commodities sold to the general public. Many methods of measurement, with different degrees of accuracy, are possible. The choice depends on the requirements and complexities of the measurement problems. Basically, measurements of electric energy may be classified into two categories, direct-current power and alternating-current power. The fundamental concepts of measurement are, however, the same for both.

Methods of measurement. There are two types of methods of measuring electric energy: electric instruments and timing means, and electricity meters.

Electric instruments and timing means. These make use of conventional procedures for measuring electric power and time. The required accuracy of measurement dictates the type and quality of the measuring devices used (for example, portable instruments, laboratory instruments, potentiometers, stopwatches, chronographs, and electronic timers). Typical methods are listed below. *See* ELECTRIC POWER MEASUREMENT.

1. Measurement of energy on a direct-current circuit by reading the line voltage and load current at regular intervals over a measured period of time. The frequency of reading selected (such as one per second, one per 10 seconds) depends upon the steadiness of the load being measured, the time duration of the test, and the accuracy of measurement desired. The first dc electric energy meter was an electroplating cell in series with the load. It deposited a mass of metal on an electrode exactly proportional to the total charge transported to the load. The electrode was weighed periodically to determine the energy used. Errors were introduced if line voltage was not constant. It was replaced by more convenient instruments. *See* CURRENT MEASUREMENT.

In electric energy measurements, the losses in the instruments must be considered. Unless negligible from a practical standpoint, they should be deducted from the total energy measured. If the voltmeter losses are included in the total energy measured, then watt-hours $= (VI - V^2R)t/3600$, where V is the average line voltage (volts), I is the average line current (amperes), R is the voltmeter resistance (ohms), and t is the time (seconds).

2. Measurement of energy on a direct-current circuit by controlling the voltage and current at constant predetermined values for a predetermined time interval. This method is common for controlling the energy used for a scientific experiment or for determining the accuracy of a watt-hour meter. For best accuracy, potentiometers and electronic timers are desirable.

3. Measurement of energy on an alternating-current circuit by reading the watts input to the load at regular intervals over a measured period of time. This method is similar to the first, except that the power input is measured by a wattmeter.

4. Measurement of energy on an alternating-current circuit by controlling the voltage, current, and watts input to the load at constant predetermined values. This method is similar to the second, except that the power input is measured by a wattmeter. A common application of this method is to determine the standard of measurement of electric energy, the watt-hour.

5. Measurement of energy by recording the watts input to the load on a linear chart progressed uniformly with time. This method makes use of a conventional power record produced by a recording wattmeter. The area under the load record over a period of time is the energy measurement.

Electricity meters. These are the most common devices for measuring the vast quantities of electric energy used by industry and the general public. The same fundamentals of measurement apply as for electric power measurement, but in addition the electricity meter provides the time-integrating means necessary for electric energy measurement.

A single meter is sometimes used to measure the energy consumed in two or more circuits. However, multistator meters are generally required for this purpose. Totalization is also accomplished with fair accuracy, if the power is from the same voltage source, by paralleling secondaries of instrument current transformers of the same ratio at the meter. Errors can result through unbalanced loading or use of transformers with dissimilar characteristics.

Watt-hour meters are generally connected to measure the losses of their respective current circuits. These losses are extremely small compared to the total energy being measured and are present only under load conditions.

Other errors result from the ratio and phase angle errors in instrument transformers. With modern transformers these errors can generally be neglected for commercial metering. If considered of sufficient importance, they can usually be compensated for in adjusting the calibration of the watt-hour meter. For particularly accurate measurements of energy over short periods of time, portable standard watt-hour meters may be used. Errors may also arise due to integral-cycle control. This is a well-established method of power control in which a silicon-controlled rectifier circuit acts to turn the power on for several cycles and off for a different number of cycles. The main source of error in the induction meter is the difference between the mechanical time-constants for two on-off intervals, making the meter read high.

Watt-hour meters used for the billing of residential, commercial, and industrial loads are highly developed devices. Over the last decade many significant improvements have been made, including improvements in bearings, insulating materials, mechanical construction, and new sealing techniques which exclude dust and other foreign material. As a result of the higher degree of accuracy and dependability achieved by modern meters, the utility industries have adopted statistical sampling methods for testing of in-service accuracy as sanctioned by ANSI C12-1975.

Automatic remote reading. Various aspects of the energy crisis have spurred active development of automatic meter-reading systems with the functional capability of providing meter data for proposed new rate structures, for example, time-of-day pricing, and initiating control of residential loads, such as electric hot-water heaters.

Automatic meter-reading systems under development generally consist of a utility-operated, minicomputer-controlled reading center, which initiates and transmits commands over a communication system to a terminal at each residential meter. The terminal carries out the command, sending the meter reading back to the reading center, or activating the control of a residential load.

Several communication media are being proposed and tested by system developers, including radio, CATV, use of the existing subscriber phone lines, and communication over the electric distribution system itself.

The rapid advances of technology, coupled with the increasing needs for improved management of energy usage, indicate that automatic meter reading and control systems may start replacing conventional manual meter reading within the next few years.

Quantities other than watt-hours. Included in the field of electric energy measurement are demand, var hours, and volt-ampere hours.

Demand. The American National Standards Institute defines the demand for an installation or system as "the load which is drawn from the source of supply at the receiving terminals, averaged over a suitable and specified interval of time. Demand is expressed in kilowatts, kilovolt-amperes, amperes, kilovars and other suitable units" (ANSI C12-1975).

This measurement provides the user with information as to the loading pattern or the maximum loading of equipments rather than the average loading recorded by the watt-hour meter. It is used by the utilities as a rate structure tool.

Var hour. ANSI defines the var hour (reactive volt-ampere hour) as the "unit for expressing the integral, of reactive power in vars over an interval of time expressed in hours" (ANSI C12-1975).

This measurement is generally made by using reactors or phase-shifting transformers to supply to conventional meters a voltage equal to, but in quadrature with, the line voltage.

Volt-ampere hour. This is the unit for expressing the integral of apparent power in volt-amperes over an interval of time expressed in hours. Measurement of this unit is more complicated than for active or reactive energy and requires greater compromises in power-factor range, accuracy, or both. Typical methods include: (1) Conventional watt-hour meters with reactors or phase-shifting transformers tapped to provide an in-phase line voltage and current relationship applied to the meter at the mean of the expected range of power-factor variation. (2) A combination of a watt-hour and a var-hour meter mechanically acting on a rotatable sphere to add vectorially watt-hours and var-hours to obtain volt-ampere hours, volt-ampere demand, or both.

Measurement of volt-ampere hours is sometimes preferred over var-hours because it is a more direct measurement and possibly gives a more accurate picture of the average system power factor. This would not necessarily be true, however, where simultaneous active and reactive demand are measured and recorded.

[WILLIAM H. HARTWIG]

Bibliography: T. S. Banghart and R. E. Riebs, Practical aspects of large-scale automatic meter reading using existing telephone lines, *Proceedings of the American Power Conference*, vol. 36, pp. 945–951, 1974; M. Braccio, *Basic Electrical and Electronic Tests and Measurements*, 1978; *Code for Electrical Metering*, ANSI C12-1975, 6th ed., 1975; W. C. Downing, Watthour meter accuracy on SCR controlled resistance loads, *IEEE Trans. Power*

App. Syst., 93(4):1083–1089, 1974; D. Fink and H. W. Beaty (eds.), *Standard Handbook for Electrical Engineers*, 11th ed., sec. 3: Measurements and instrumentation, 1978; A. E. Emanuel, B. M. Hynds, and F. J. Levitsky, Watthour meter accuracy on integral-cycle-controlled resistance loads, *IEEE Trans. Power App. Syst.*, 98(5):1583–1590, 1979; Instrument Society of America, *ISA Standards and Practices for Instrumentation*, 6th ed., 1980.

Electric furnace

An enclosed space heated by electric power. The furnace may be in such forms as a refractory crucible, a large tiltable refractory basin with a capacity of 100 tons and a removable roof, or a long insulated chamber equipped with a continuous conveyor. Heat is provided by an arc to the charge or melt (direct-arc furnace), by an arc between electrodes (indirect-arc furnace), or by an arc confined for concentrated heating by an electromagnetic field (plasma-arc furnace). Heat may also be produced by current flowing in the melt. This current may be between a pair of electrodes inserted into the charge (resistance- or submerged-arc furnace), or it may be induced into the charge from a surrounding coil (induction furnace). The heat may also be radiated to the charge from an electrical resistance near it. These furnaces perform their smelting or refining functions by the heat that they produce in the material to be treated. In an electrolytic furnace, such as that used in refining metals from fused salts of aluminum, lithium, or sodium, the electric current performs its function by setting up ionic migration in the melt, although high temperature may be necessary to fuse the charge. *See* ELECTROMETALLURGY; PYROMETALLURGY.

Because the source of heat is nonchemical, electric furnaces are especially desirable in melting alloys of controlled composition. Temperature is also readily controlled. The arc furnace may be used to smelt ores or to refine metals or alloys. High rates are obtained; for example, in a 50-ton arc furnace, total melting rates of 20 tons/hr are reached. Induction furnaces are widely used to melt alloys for castings. Because electric furnaces can be enclosed, they are used for operations that require controlled or inert atmospheres, such as growing crystals or annealing. When sealed and evacuated, they are used in degassing metals. Furnaces with hearth resistors are used for operations below melting temperatures, such as annealing, and with infrared heat lamps, for drying paints or setting glues. The electric furnace is not used for central heating of buildings because electricity is usually more efficiently used for such low-grade heat in other ways. *See* ARC HEATING; ELECTRIC HEATING; HEAT PUMP; HEAT TREATMENT (METALLURGY); INDUCTION HEATING.

[FRANK H. ROCKETT]

Electric heating

Methods of converting electric energy to heat energy by resisting the free flow of electric current. Electric heating has several advantages: it can be precisely controlled to allow a uniformity of temperature within very narrow limits; it is cleaner than other methods of heating because it does not involve any combustion; it is considered safe because it is protected from overloading by automatic breakers; it is quick to use and to adjust; and it is relatively quiet. For these reasons, electric heat is widely chosen for industrial, commercial, and residential use.

Types of electric heaters. There are four major types of electric heaters: resistance, dielectric, induction, and electric-arc.

Resistance heaters. Resistance heaters produce heat by passing an electric current through a resistance—a coil, wire, or other obstacle which impedes current and causes it to give off heat. Heaters of this kind have an inherent efficiency of 100% in converting electric energy into heat. Devices such as electric ranges, ovens, hot-water heaters, sterilizers, stills, baths, furnaces, and space heaters are part of the long list of resistance heating equipment. *See* RESISTANCE HEATING.

Dielectric heaters. Dielectric heaters use currents of high frequency which generate heat by dielectric hysteresis (loss) within the body of a nominally nonconducting material. These heaters are used to warm to a moderate temperature certain materials that have low thermal conducting properties; for example, to soften plastics, to dry textiles, and to work with other materials like rubber and wood.

Induction heaters. Induction heaters produce heat by means of a periodically varying electromagnetic field within the body of a nominally conducting material. This method of heating is sometimes called eddy-current heating and is used to achieve temperatures below the melting point of metal. For instance, induction heating is used to temper steel, to heat metals for forging, to heat the metal elements inside glass bulbs, and to make glass-to-metal joints. *See* INDUCTION HEATING.

Electric-arc heaters. Electric-arc heating is really a form of resistance heating in which a bridge of vapor and gas carries an electric current between electrodes. The arc has a property of resistance. Electric-arc heating is used mainly to melt hard metals, alloys, and some ceramic metals. *See* ARC HEATING.

General design features. All electrical parts must be well protected from contact by operators, work materials, and moisture. Terminals must be enclosed within suitable boxes, away from the high heat zone, to protect the power supply cables. Repairs and replacements should be possible without tearing off heat insulations.

Resistance heaters are often enclosed in pipes or tubes suitable for immersion or for exposure to difficult external conditions. Indirect heating is done by circulating a heat transfer medium, such as special oil or Dowtherm (liquid or vapor), through jacketed vessels. This permits closer control of heating-surface temperature than is possible with direct heating.

Some conducting materials can be heated by passing electric current through them, as is done in the reduction of aluminum. Some conducting liquids can be heated by passing an electric current between immersed electrodes. Heat is produced by the electrical resistance of the liquid.

The supply of necessary electric power for large heating installations necessitates consultation with the utility company. The demand, the power factor of the load, and the load factor all affect the power rates. Large direct-current or single-phase alternating-current loads should be avoided. Polyphase power at 440–550 volts permits lower current and reduced costs. *See* FURNACE CONSTRUCTION.

Electric heating for houses. In the past, electricity has not been a popular choice for heating houses. Power plants lose 60–65% efficiency in generating electricity, and another 10% in transmission and distribution. With only a 35% efficiency rate, electricity has long been overlooked in favor of the direct use of gas and oil for heating homes. As these fossil fuels diminish, however, electric heating is becoming a more attractive option.

Electric heating devices have been installed in houses in rural areas, where people can no longer depend on the ready availability of coal, gas, and oil to meet their heating needs. If nuclear power becomes more widespread, so will electric heating; the generation of nuclear power will make electric heating far more economical than the direct use of gas and oil.

A popular electric heater for houses is the heat pump. The heat pump can be used alone or to supplement the output of direct resistance heating (heat strips). The heat pump works on a reversed refrigeration cycle; the pump pulls the heat from the cold outside air and forces it into the house. The efficiency of this process is directly affected by the ambient temperature and the desired interior temperature; the larger the difference in these temperatures, the lower the efficiency of the pump. Generally, however, the efficiency of the heat pump is higher than the efficiency of resistance heating, so that the efficiency of the heat pump is actually higher than 100%.

Since the heat pump performs both heating and cooling and is more efficient than other forms of electric heat, it can become the most economical way to heat houses in the future. *See* HEAT PUMP.

Common electric heating systems in houses are: central heating employing an electric furnace with forced air circulation; central heating employing an electric furnace with forced water circulation; central heating using radiant cables; electrical duct heaters; space (strip) heaters which use radiation and natural convection for heat transfer; and portable space heaters. [MO-SHING CHEN]

Bibliography: D. G. Fink and H. W. Beaty, *Standard Handbook for Electrical Engineers*, 11th ed., 1978; P. Sporn, E. R. Ambrose, and T. Baumeister, *Heat Pumps*, 1947.

Electric power generation

The production of bulk electric power for industrial, residential, and rural use. Although limited amounts of electricity can be generated by many means, including chemical reaction (as in batteries) and engine-driven generators (as in automobiles and airplanes), electric power generation generally implies large-scale production of electric power in stationary plants designed for that purpose. The generating units in these plants convert energy from falling water, coal, natural gas, oil, and nuclear fuels to electric energy. Most electric generators are driven either by hydraulic turbines, for conversion of falling water energy; or by steam or gas turbines, for conversion of fuel energy. Limited use is being made of geothermal energy, and developmental work is progressing in the use of solar energy in its various forms. Electric power generating plants are normally interconnected by a transmission and distribution system to serve the electric loads in a given area or region. *See* GENERATOR; PRIME MOVER.

An electric load is the power requirement of any device or equipment that converts electric energy into light, heat, or mechanical energy, or otherwise consumes electric energy as in aluminum reduction, or the power requirements of electronic and control devices. The total load on any power system is seldom constant; rather, it varies widely with hourly, weekly, monthly, or annual changes in the requirements of the area served. The minimum system load for a given period is termed the base load or the unity load-factor component. Maximum loads, resulting usually from temporary conditions, are called peak loads. Electric energy cannot feasibly be stored in large quantities; therefore the operation of the generating plants must be closely coordinated with fluctuations in the load.

Actual variations in the load with time are recorded, and from these data load graphs are made to forecast the probable variations of load in the future. A study of hourly load graphs (Figs. 1 and 2) indicates the generation that may be required at a given hour of the day, week, or month, or under unusual weather conditions. A study of annual load graphs and forecasts indicates the rate at which new generating stations must be built. Load graphs and forecasts are an inseparable part of utility operation and are the basis for decisions that profoundly affect the financial requirements and overall development of a utility.

Generating plants. Often termed generating stations, these plants contain apparatus that converts some form of energy to electric energy in bulk. Three significant types of generating plants are hydroelectric, fossil-fuel-electric, and nuclear-electric.

Hydroelectric plant. This type of generating plant utilizes the potential energy released by the weight of water falling through a vertical distance called head. Ignoring losses, the power, in horsepower (hp) and kilowatts (kW), obtainable from falling water is shown in the equations below (metric quantities in brackets).

$$\text{hp} = \frac{\begin{pmatrix}\text{quantity of water}\\ \text{in ft}^3\text{/s [m}^3\text{/s]}\end{pmatrix}\begin{pmatrix}\text{vertical head}\\ \text{in ft [m]}\end{pmatrix}}{8.8\ [0.077]}$$

$$\text{kW} = 0.746\ \text{hp}$$

A plant consists basically of a dam to store the water in a forebay and create part or all of the head, a penstock to deliver the falling water to the turbine, a hydraulic turbine to convert the hydraulic energy released to mechanical energy, an alternating-current generator (alternator) to convert the mechanical energy to electric energy, and all accessory equipment necessary to control the power flow, voltage, and frequency, and to afford the protection required (Fig. 3).

Pumped storage hydroelectric plants are being used increasingly. Under suitable geographical and geological conditions, electric energy can, in effect, be stored by pumping water from a low to a higher elevation and subsequently releasing this water to the lower elevation through hydraulic turbines. These turbines and their associated generators are reversible. The generators, operating in reverse direction as motors, drive their turbines as pumps to elevate the water. When this water is released through the turbines, electric power is produced by the generators. A relatively high overall cycle efficiency can be attained, usually of the order of 65–75%.

Since system peak loads are usually of relatively short duration (Figs. 1 and 2), the high output available for a short time from pumped storage can be used to supply this peak. During off-peak hours, that is, 10 P.M. to 7 A.M., the surplus generating capacity of
the most economical system energy resources can be used to return the water by pumping to the elevated storage space for use on the next peak. This type of operation assists in maintaining a high capacity factor on prime generation with resulting best economy.

Pumped storage plants can be brought up to load much faster than large steam plants and, hence, contribute to system reliability by providing an immediately available reserve against the unscheduled loss of other generation. *See* WATERPOWER.

Fossil-fuel-electric plant. This type utilizes the energy of combustion from coal, oil, or natural gas. A typical large plant (Fig. 4) consists of fuel processing and handling facilities, a combustion furnace and boiler to produce and superheat the steam, a steam turbine, an alternator, and the accessory equipment required for plant protection and for control of voltage, frequency, and power flow. A steam plant can frequently be built near a convenient load center, provided an adequate supply of cooling water and fuel is available, and is usually readily adaptable to either base loading or

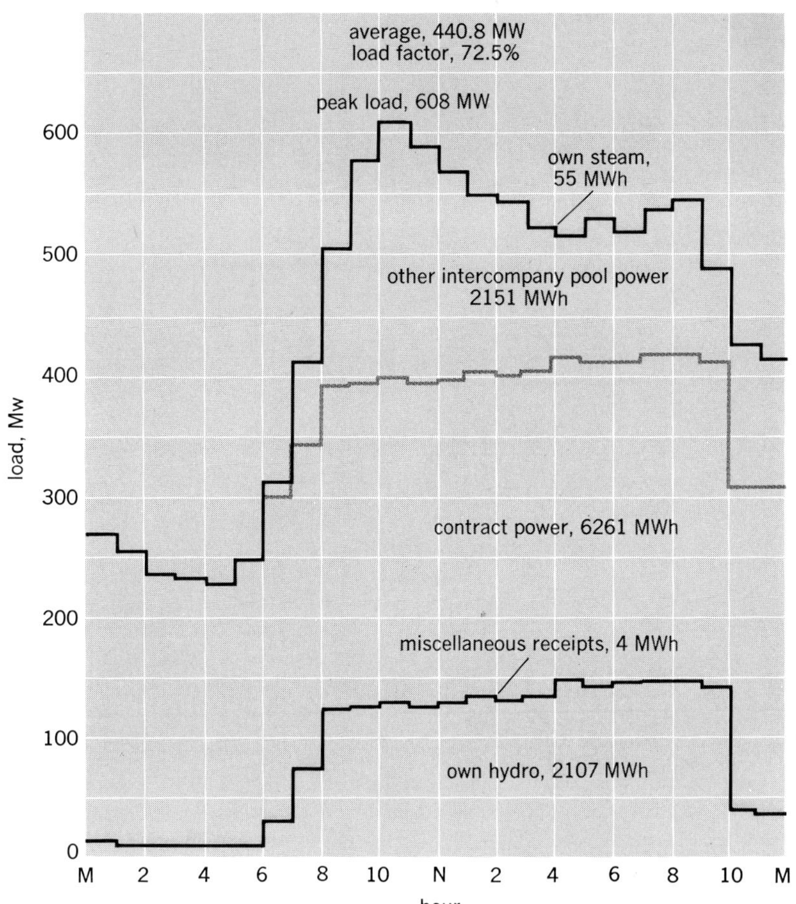

Fig. 1. Load graph indicates net system load of a metropolitan utility for typical 24-hr period (midnight to midnight), totaling 10,578 MWh. Such graphs are made to forecast probable variations in power required.

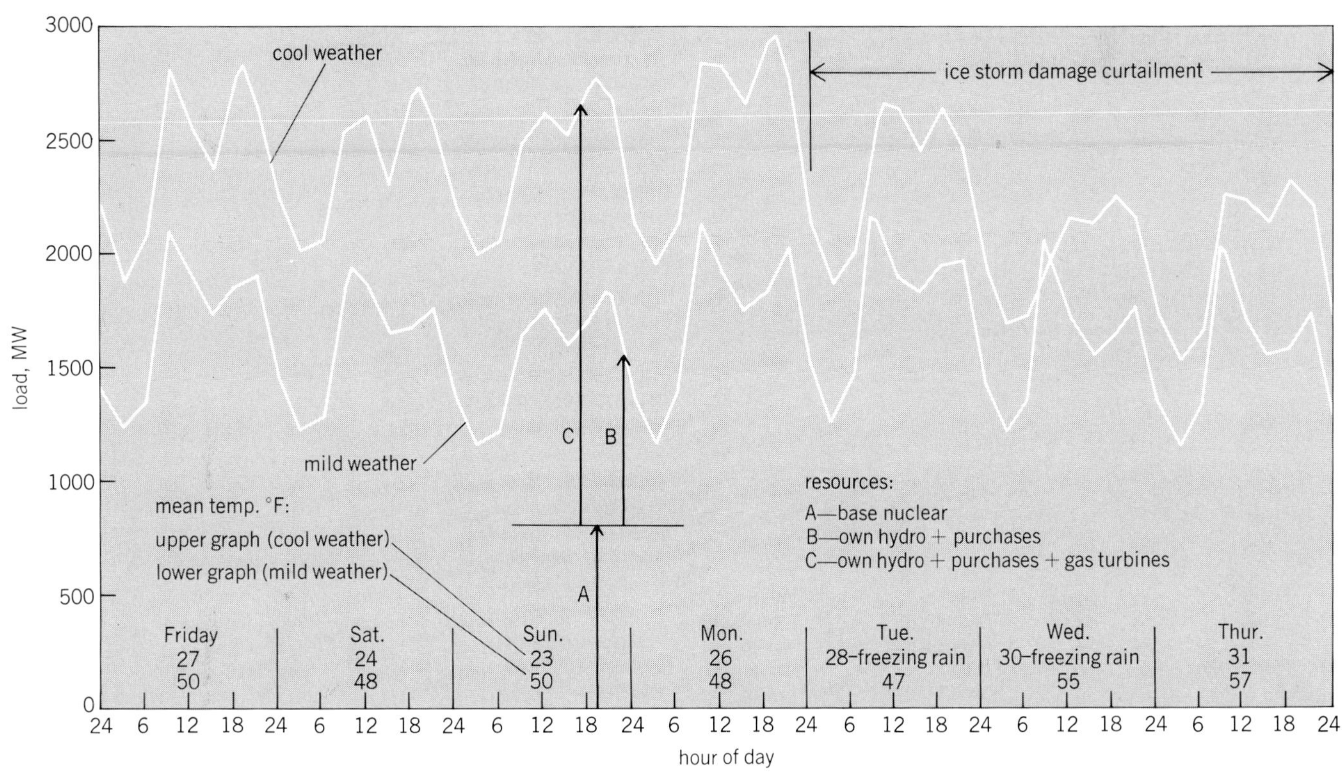

Fig. 2. Examples of northwestern electric utility weekly load curves showing same-year weather influence.

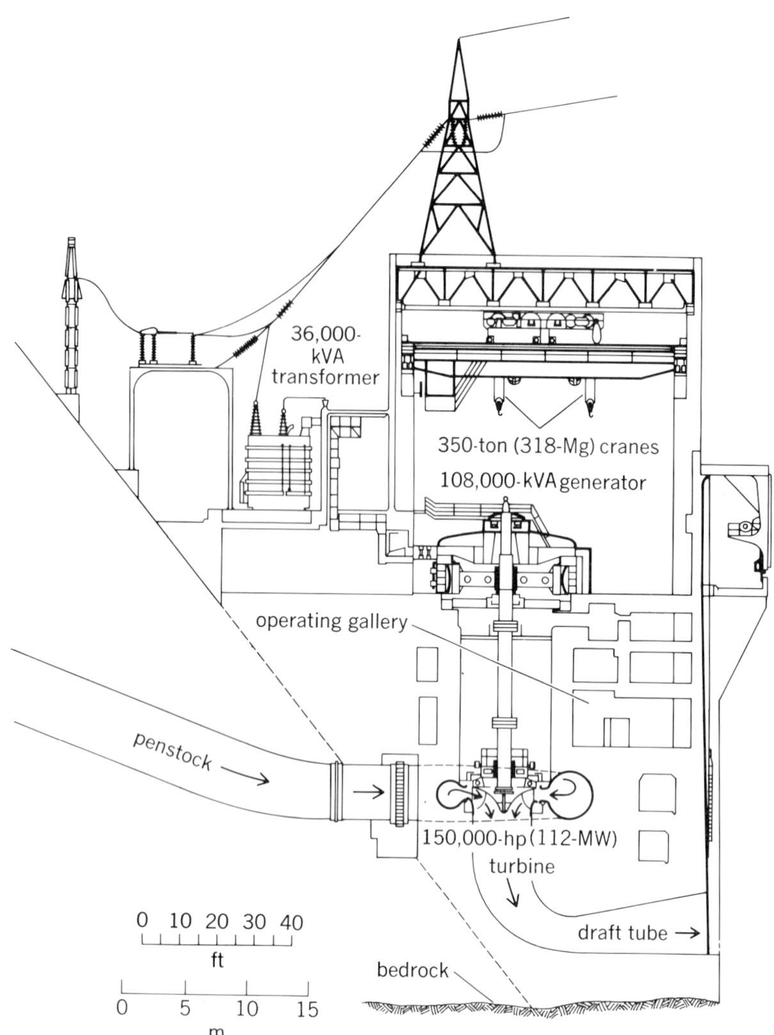

Fig. 3. Typical layout and apparatus arrangement in a hydroelectric generating plant. Cross section is made through the powerhouse and dam at a main generator position in the original Grand Coulee plant.

intermediate or peak loading. Environmental constraints require careful control of stack emissions with respect to sulfur oxides and particulates. Cooling towers or ponds are often required for waste heat dissipation. Gas turbine plants do not require condenser cooling water (unless combined with a steam cycle), have a relatively low unit capital cost and relatively high unit fuel cost, and are widely used for peaking service. Progress is being made in the development of magnetohydrodynamic (MHD) "topping" generators to be used in conjuction with normal steam turbines to improve the overall thermal conversion efficiency.

Nuclear electric plant. In this type of plant one or more of the nuclear fuels are utilized in a suitable type of nuclear reactor, which takes the place of the combustion furnace in the typical steam electric plant. The heat exchangers and boilers (if not combined in the reactor), the turbines, and alternating-current generators, complete with controls, accessories, and auxiliaries, make up the atomic electric plant. Large-scale fission reaction plants have been developed to the point where they are economically competitive in much of the United States, and many millions of kilowatts of capacity are under construction and more on order. The current and projected future growth of the nuclear power industry in the United States is shown graphically in Figs. 5 and 6. Although in 1978 only approximately 9% of the total generating capacity was nuclear, by 1989 it is predicted to become over 21%. However, in 1978 about 14% of the net electric energy production was nuclear and is forecast to become more than 22% by 1988. The coal-fired share in 1978 was approximately 45% and is forecast to become about 50% in 1988. Conventional hydro generation as a source of prime electric energy had reached near-saturation by 1978, and growth in fossil-fuel generation shows a declining trend. However, nuclear generation exhibits a doubling time of about 51/2 years during a considerable period after 1978. Recent operating events may slow this growth somewhat.

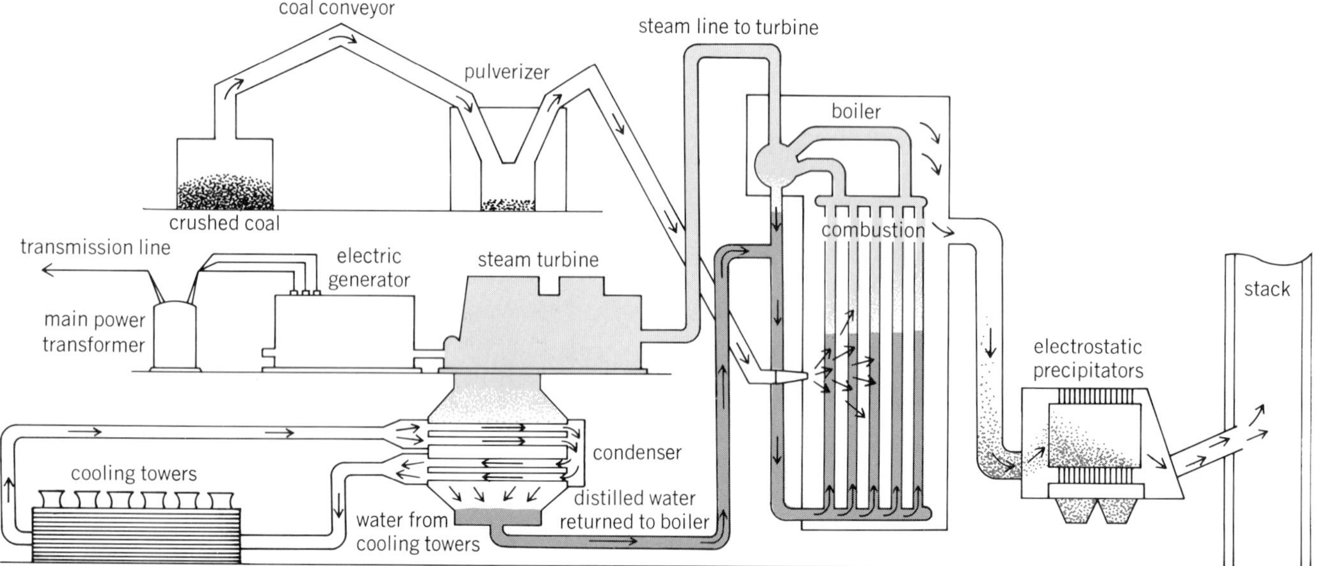

Fig. 4. Schematic of typical coal-fired steam electric power plant. *(Pacific Power & Light Co.)*

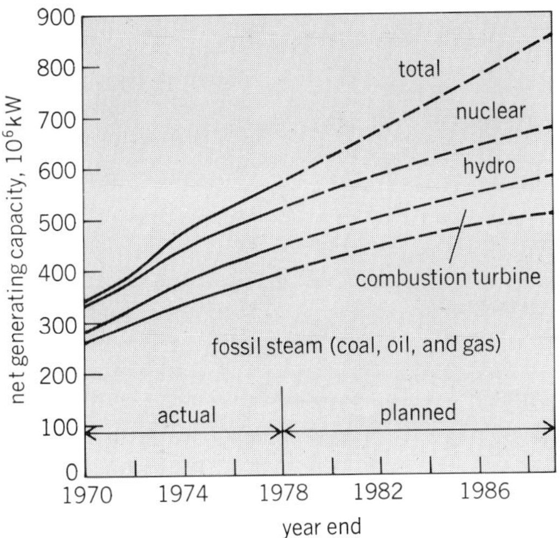

Fig. 5. Net generating capacity of the United States electric power industry. Hydro includes pumped storage. Combustion turbine includes internal combustion.

Three types of nuclear plants are in general use and a fourth is being developed actively. The pressurized and boiling light-water types, shown schematically in Figs. 7 and 8, have been highly developed in the United States. The pressurized heavy-water type has been well developed in Canada. The gas-cooled type, one version of which is shown schematically in Fig. 9, has been highly developed and extensively used in Great Britain. The liquid-metal (sodium) cooled type is under intensive development and will be featured prominently in the fast or breeder reactor field. It is shown schematically in Fig. 10. As in the high-temperature gas-cooled reactor system, steam temperatures and pressures in the liquid-metal systems can be essentially the same as those recorded in the plants powered by fossil fuel and can give similar thermal efficiencies. *See* NUCLEAR REACTOR.

Several other types of nuclear fission reactor systems are receiving attention, and some may be expected to become commercially competitive. Fusion reaction nuclear plants are in the early research and development stage with possible commercialization early in the next century. Direct conversion from nuclear reaction energy to electric energy on a commercial scale for power utility service is a future possibility but is not economically feasible at present.

Generating unit sizes. The size or capacity of electric utility generating units varies widely, depending upon type of unit; duty required, that is base-, intermediate-, or peak-load service; and system size and degree of interconnection with neighboring systems. Base-load nuclear or coal-fired units may be as large as 1200 MW each, or more. Intermediate-duty generators, usually coal-, oil-, or gas-fueled steam units, are typically of 200 to 600 MW capacity each. Peaking units, combustion turbines or hydro, range from several tens of megawatts for the former to hundreds of megawatts for the latter. Hydro units, in both base-load and intermediate service, range in size up to 700 MW.

The total installed generating capacity of a system is typically 20 to 30% greater than the annual predicted peak load in order to provide reserves for maintenance and contingencies.

Power-plant circuits. Both main and accessory circuits in power plants can be classified as follows:

1. Main power circuits to carry the power from the generators to the step-up transformers and on to the station high-voltage terminals.
2. Auxiliary power circuits to provide power to the motors used to drive the necessary auxiliaries.
3. Control circuits for the circuit breakers and other equipment operated from the control room of the plant.
4. Lighting circuits for the illumination of the plant and to provide power for portable equipment required in the upkeep and maintenance of the plant. Sometimes special circuits are installed to supply the portable power equipment.
5. Excitation circuits, which are so installed that they will receive good physical and electrical protection because reliable excitation is necessary for the operation of the plant.
6. Instrument and relay circuits to provide values of voltage, current, kilowatts, reactive kilovolt-amperes, temperatures, and pressures, and to serve the protective relays.
7. Communication circuits for both plant and system communications. Telephone, radio, transmission-line carrier, and microwave radio may be involved.

It is important that reliable power service be provided for the plant itself, and for this reason station service is usually supplied from two or more sources. To ensure adequate reliability, auxiliary power supplies are frequently provided for start-up, shut-down, and communication services.

Generator protection. Necessary devices are installed to prevent or minimize other damage in cases of equipment failure. Differential-current and ground relays detect failure of insulation, which may be due to deterioration or accidental overvoltage. Overcurrent relays detect overload currents that may lead to excessive heating; over-

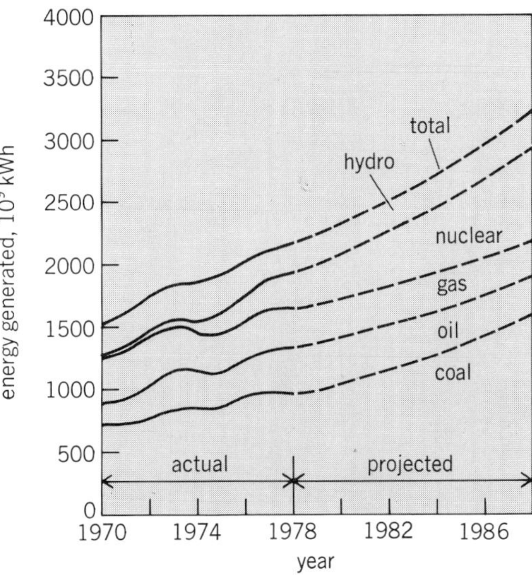

Fig. 6. Net energy production of the United States electric power industry.

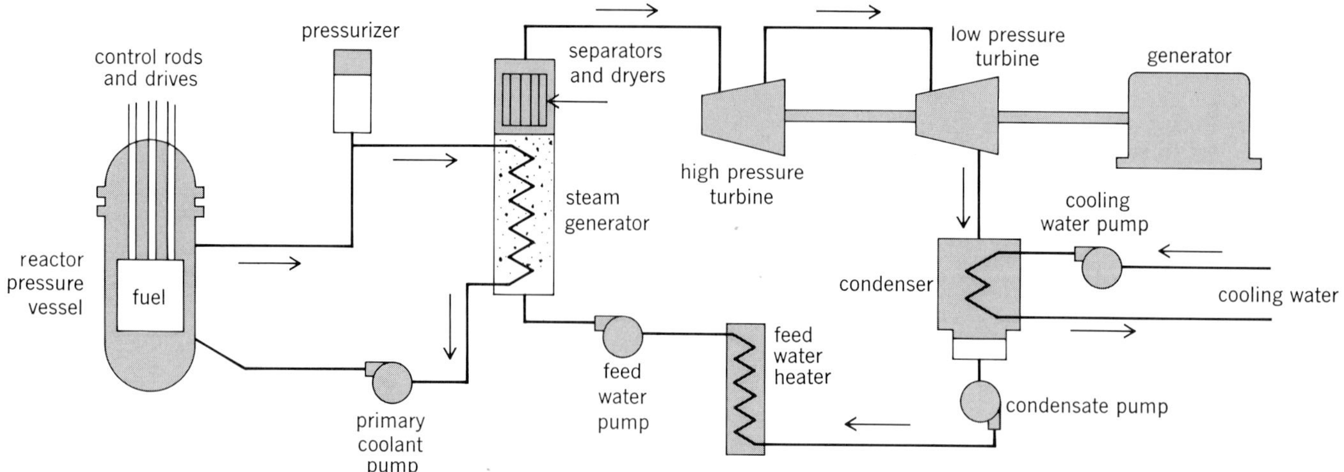

Fig. 7. Schematic of a pressurized-water reactor plant. *(From the Nuclear Industry, USAEC, WASH 1174–73, 1973)*

voltage relays prevent insulation damage. Loss-of-excitation relays may be used to warn operators of low excitation or to prevent pulling out of synchronism. Bearing and winding overheating may be detected by relays actuated by resistance devices or thermocouples. Overspeed and lubrication failure may also be detected.

Not all of these devices are used on small units or in every plant. The generator is immediately deenergized for electrical failure and shut down for any over-limit condition, all usually automatically.

Voltage regulation. This term is defined as change in voltage for specific change in load (usually from full load to no load) expressed as percentage of normal rated voltage. The voltage of an electric generator varies with the load and power factor; consequently, some form of regulating equipment is required to maintain a reasonably constant and predetermined potential at the distribution stations or load centers. Since the inherent regulation of most alternating-current generators is rather poor (that is, high percentagewise), it is necessary to provide automatic voltage control. The rotating or magnetic amplifiers and voltage-sensitive circuits of the automatic regulators, together with the exciters, are all specially designed to respond quickly to changes in the alternator voltage and to make the necessary changes in the main exciter output, thus providing the required adjustments in voltage. A properly designed automatic regulator acts rapidly, so that it is possible to maintain desired voltage with a rapidly fluctuating load without causing more than a momentary change in voltage even when heavy loads are thrown on or off.

Electronic voltage control has been adapted to some generator and synchronous condenser installations. Its main advantages are its speed of operation and its sensitivity to small voltage variations.

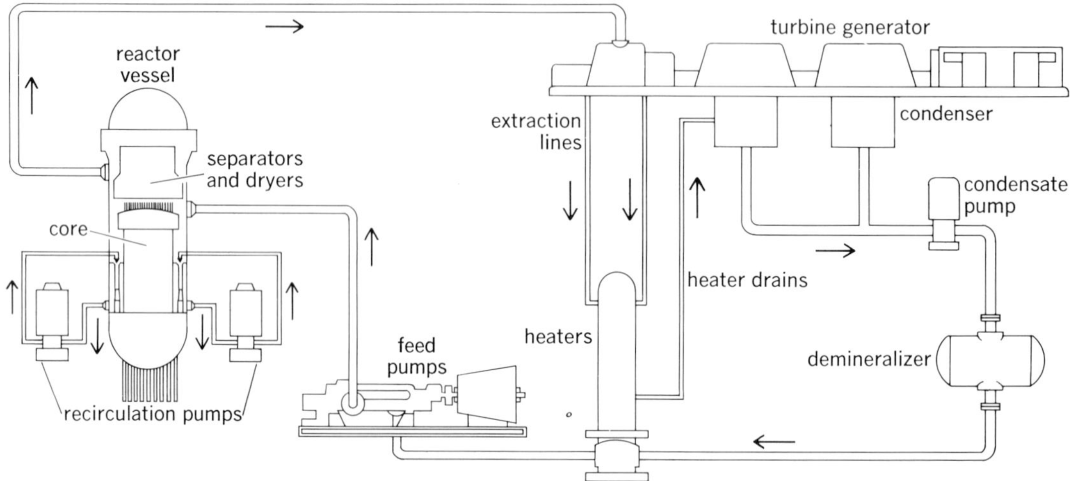

Fig. 8. Single-cycle boiling-water reactor system flow diagram. *(General Electric Co.)*

As the reliability and ruggedness of electronic components are improved, this form of voltage regulator will become more common.

Generation control. Computer-assisted (or on-line controlled) load and frequency control and economic dispatch systems of generation supervision are being widely adopted, particularly for the larger new plants. Strong system interconnections greatly improve bulk power supply reliability but require special automatic controls to ensure adequate generation and transmission stability. Among the refinements found necessary in large, long-distance interconnections are special feedback controls applied to generator high-speed excitation and voltage regulator systems.

Synchronization of generators. Synchronization of a generator to a power system is the act of matching, over an appreciable period of time, the instantaneous voltage of an alternating-current generator (incoming source) to the instantaneous voltage of a power system of one or more other generators (running source), then connecting them together. In order to accomplish this ideally the following conditions must be met:

1. The effective voltage of the incoming generator must be substantially the same as that of the system.
2. In relation to each other the generator voltage and the system voltage should be essentially 180° out of phase; however, in relation to the bus to which they are connected, their voltages should be in phase.
3. The frequency of the incoming machine must be near that of the running system.
4. The voltage wave shapes should be similar.
5. The phase sequence of the incoming polyphase machine must be the same as that of the system.

Synchronizing of ac generators can be done manually or automatically. In manual synchronizing an operator controls the incoming generator while observing synchronizing lamps or meters and a synchroscope, or both. Voltage (potential) transformers may be used to provide voltages at lamp and instrument ratings. Lamps properly connected between the two sources are continuously dark when voltage, phase, and frequency are properly matched. Wave shape and phase sequence are determined by machine design and rotation or terminal sequence. Large units generally are provided with voltmeters and frequency meters for matching these quantities, and a synchroscope connected to both sources to indicate phase relationship. Lamps may also be included. The standard synchroscope needle revolves counterclockwise when the incoming machine is slow and clockwise when fast. The needle points straight up when the two sources are in phase. The operator closes the connecting switch or circuit breaker as the synchroscope needle slowly approaches the in-phase position.

Automatic synchronizing provides for automatically closing the breaker to connect the incoming machine to the system, after the operator has properly adjusted voltage (field current), frequency (speed), and phasing (by lamps or synchroscope).

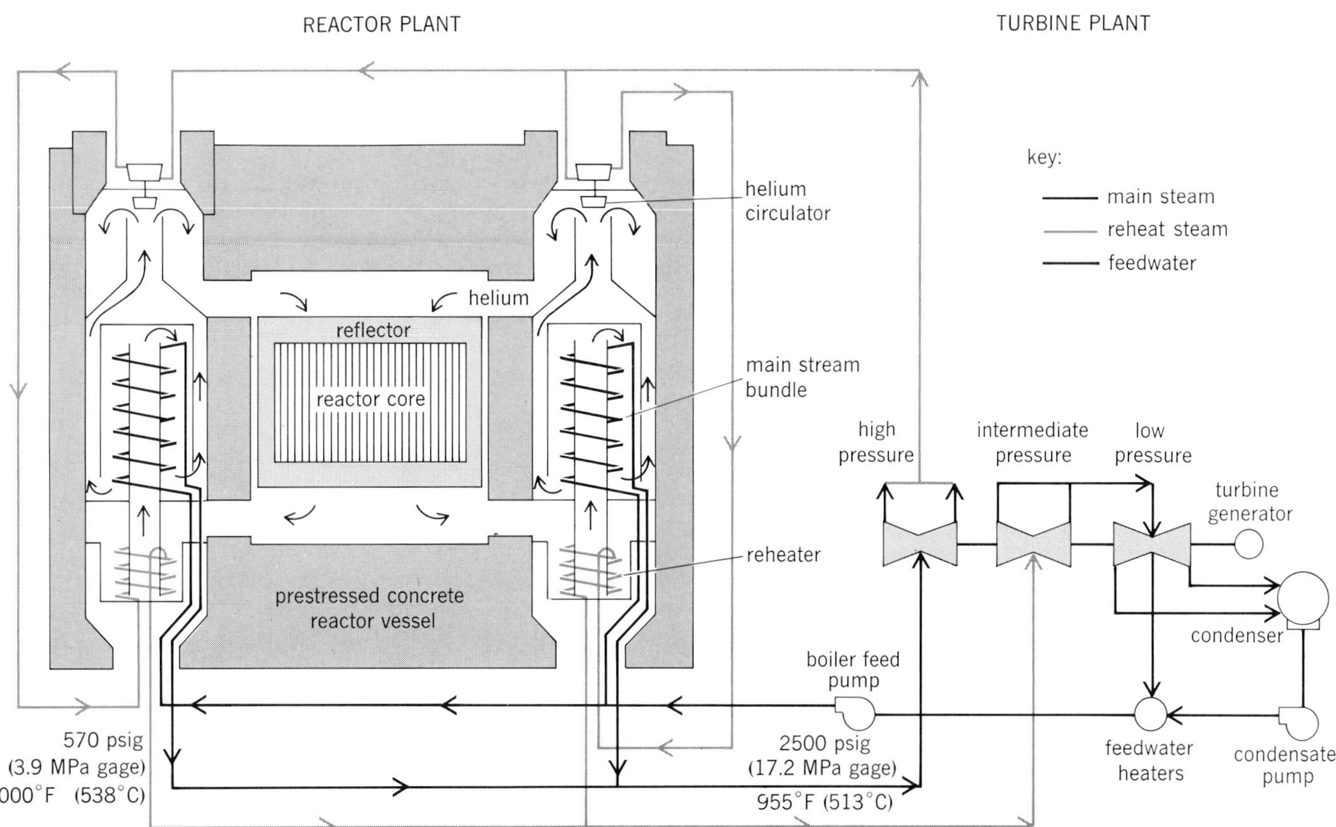

Fig. 9. Schematic diagram for a high-temperature gas-cooled reactor power plant. (*General Atomic Corp.*)

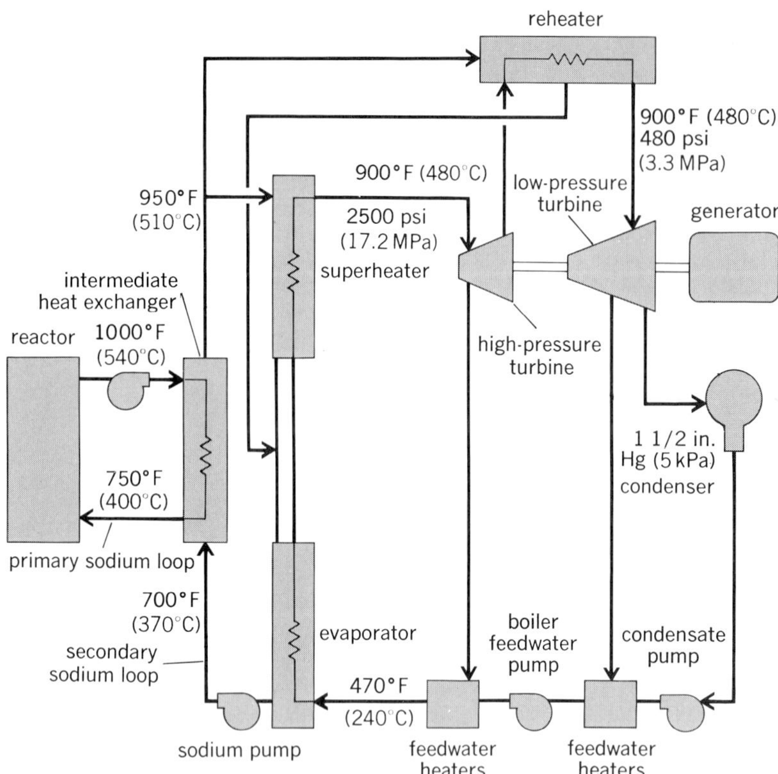

Fig. 10. Diagram for liquid-metal-cooled fast breeder reactor power plant. *(North American Rockwell Corp.)*

A fully automatic synchronizer will initiate speed changes as required and may also balance voltages as required, then close the breaker at the proper time, all without attention of the operator. Automatic synchronizers can be used in unattended stations or in automatic control systems where units may be started, synchronized, and loaded on a single operator command. *See* ALTERNATING-CURRENT GENERATOR; ELECTRIC POWER SYSTEMS.

[EUGENE C. STARR]

Bibliography: R. S. Brown, *Hydro for the Eighties: Bringing Hydroelectric Power to Poor People: The Workbook*, 1980; D. G. Fink and H. W. Beaty (eds.), *Standard Handbook for Electrical Engineers*, 11th ed., 1978; R. F. Grundy (ed.), *Magnetohydrodynamic Energy for Electric Power Generation*, Noyes Data Corporation, 1978; International Atomic Energy Agency, *Reliability of Nuclear Power Plants*, 1976; P. Kruger and C. Otte (eds.), *Geothermal Energy: Resources, Production, Stimulation*, 1973; S. Miller, *The Economics of Nuclear and Coal Power*, 1976; P. H. Nowill, *Productivity and the Technological Change in Electric Power Generating Plants*, 1979; R. Noyes (ed.), *Cogeneration of Steam and Electric Power*, 1978; R. Noyes (ed.), *Small and Micro Hydroelectric Power Plants: Technology and Feasibility*, 1981; E. Pederson, *Nuclear Power: Nuclear Power Plant Design*, vol. 1, 1978; J. Vardi and B. Avi-Itzhak, *Electrical Energy Generation: Economics, Reliability and Rates*, 1980; J. M. Willenbrock and M. R. Thomas (eds.), *Planning, Engineering and Construction of Electric Power Generation Facilities*, 1980; E. B. Woodruff and H. B. Lammers, *Steam-Plant Operation*, 4th ed., 1976.

Electric power measurement

The measurement of the time rate at which work is done or energy is dissipated in an electric system. The work done in moving an electric charge is proportional to the charge and the voltage drop through which it moves. Charge per unit time defines electric current; electric power p is therefore defined as the product of the current i in a curcuit and the voltage e across its terminals at a given instant. Expressed symbolically, $p = ei$.

A second important definition of power follows directly from Ohm's law. $p = i^2R$, where R is the resistance of the circuit.

The practical unit of electric power is the watt. The watt represents a rate of expending energy, and thus it is related to all other units of power: for example, in mechanics 1 watt = 1 joule/s and 746 watts = 1 horsepower. Commonly used small units are the milliwatt (0.001 watt) and the microwatt (0.000001 watt). Large units are the kilowatt (1000 watts) and the megawatt (1,000,000 watts).

Power measurements must cover the frequency spectrum from direct current through the conventional power frequencies, the audio and the lower radio frequencies, to the highest frequencies (up to 25,000 gigahertz). In general, different techniques are required in each frequency range, and this article is divided into sections dealing with these frequency ranges.

In the measurement of power in a dc circuit not subject to rapid fluctuations, there is usually no difficulty in making simultaneous observations of the true values of voltage and current using common types of dc voltmeters and ammeters. The product of these observations then gives a sufficiently accurate measure of power in the given circuit, except that, if great accuracy is required, allowance must be made for the power used by the instruments themselves.

If in a circuit the voltage e, or current i, or both are subject to rapid variations, instantaneous values of power are difficult to measure and are usually of no interest. The important value is the average value, which is expressed mathematically in Eq. (1), where T is the period or time interval

$$P = \frac{1}{T}\int_0^T ei\,dt \qquad (1)$$

and t is time. This relation holds true for any waveform of current and voltage. In circuits with rapidly varying direct currents, pulsating rectified current, or, in general, alternating currents, the continuous averaging over short periods of time and the automatic multiplication of current and voltage values is accomplished by the wattmeter.

In ac circuits with steady effective values of voltage and current, the voltmeter-ammeter method may be used as in the dc case, except that, of course, ac meters are used, and a phase meter is also required to measure phase angle unless current and voltage are in phase. Because ac ammeters and voltmeters actually measure root-mean-square, or effective, values, these lead directly to values of average power.

Sinusoidal ac waves. Figure 1 illustrates the case of a sinusoidal voltage and current in a circuit containing only a resistive load. Here the current wave is entirely symmetrical with the voltage wave, and the power curve formed from the product of the voltage and current at each instant appears as a double-frequency wave on the positive side of the zero axis.

In Fig. 2, an inductance is assumed in the measured circuit. The current wave lags behind the voltage wave in what is called the negative out-of-phase, or quadrature, condition. A pure capacitance, however, would produce a positive out-of-phase, or quadrature, condition in which the current wave leads the voltage wave. In general, circuits contain elements of resistance, inductance, and capacitance in varying amounts and they must, therefore, assume some intermediate condition of phase angle between voltage and current. In this case the power wave, as shown in Fig. 2, dips below the zero line and becomes negative, indicating that during that part of the cycle power feeds back into the circuit. Measurements based on readings of conventional ac ammeters and voltmeters do not account for these negative

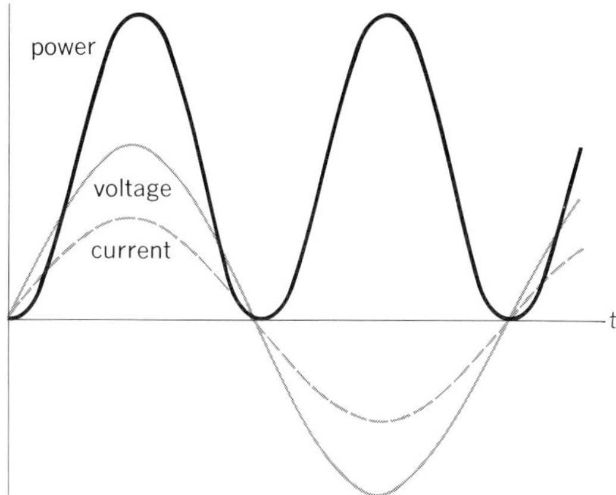

Fig. 2. Curves of instantaneous current, voltage, and power in an ac circuit; current and voltage out of phase.

excursions, and therefore, in general, the product of steady effective voltage and current readings in an ac circuit differs from, and is greater than, the reading of a wattmeter in such a circuit.

These relationships in the general ac circuit may be expressed as Eqs. (2) and (3), where I_m and E_m

$$i = I_m \sin 2\pi ft = \sqrt{2}\, I \sin 2\pi ft \qquad (2)$$

$$e = E_m \sin (2\pi ft + \phi) = \sqrt{2}\, E \sin (2\pi ft + \phi) \qquad (3)$$

are maximum values of current and voltage, f is frequency in hertz, and ϕ is the phase angle by which the current leads (+) or lags behind (−) the voltage in the circuit.

But by definition Eq. (1) holds. Substituting and carrying out the indicated operations, Eq. (4) is

$$P = EI \cos \phi \qquad (4)$$

obtained where E and I are effective values of voltage and current.

The expression for P in Eq. (4) is the real or active power in the circuit and is distinguished from the simple product EI, which is called the appar-

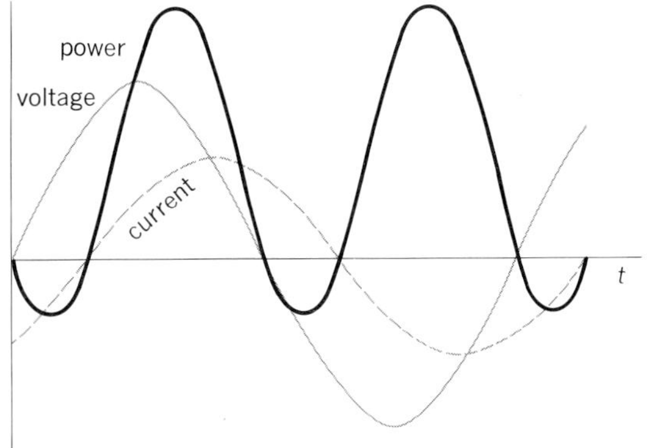

Fig. 2. Curves of instantaneous current, voltage, and power in an ac circuit; current and voltage out of phase.

ent or virtual power, by the factor $\cos \phi$, which is called the power factor. It is obvious from the previous formula that Eqs. (5) hold.

$$\cos \phi = P/EI \qquad (5a)$$

or

$$\text{Power factor} = \frac{\text{real power}}{\text{apparent power}} \qquad (5b)$$

Negative, or reactive, power due to inductance and capacitance in a circuit, is given by the relation $EI \sin \phi$.

The units for these quantities are, for real power, watts; for apparent power, volt-amperes; and, for reactive power, reactive volt-amperes or vars.

Polyphase power measurement. Summation of power in the separate phases of a polyphase circuit is accomplished by combinations of single-phase wattmeters, or wattmeter elements, disposed according to the general rule, called Blondel's theorem, as follows: If energy is supplied to any system of conductors through N wires, the total power in the system is given by the algebraic sum of N wattmeters, so arranged that each of the N wires contains one current coil, the corresponding potential coil being connected between that wire and some point on the system which is common to all the potential circuits. If this common point is on one of the N wires and coincides with the point of attachment of the potential lead to that wire, the measurement may be effected by the use of $N-1$ wattmeters.

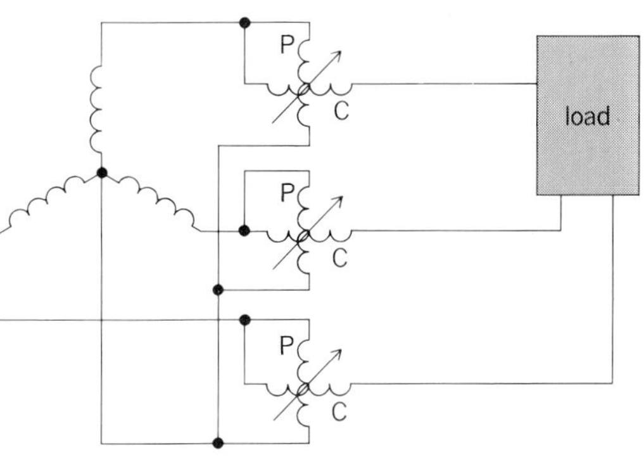

Fig. 3. Three wattmeters in three-phase, three-wire circuit. C and P refer to current and potential coils.

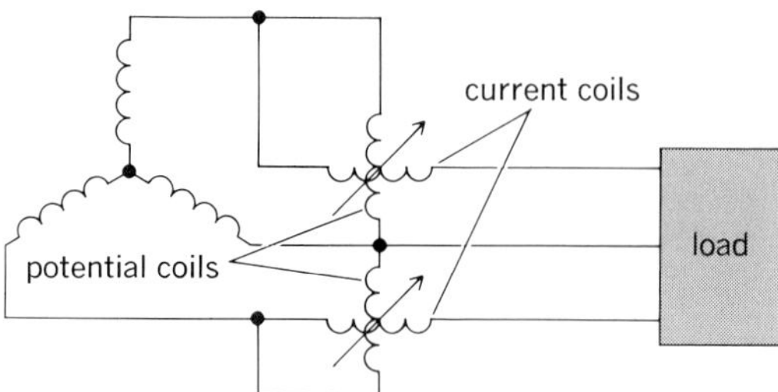

Fig. 4. Wattmeters in three-phase, three-wire circuit.

Considering measurement in three-phase circuits as an example of polyphase practice in wide application in the power industry, the two most common systems are the three-phase, three-wire system in which the source may be Y-connected or delta-connected to three load wires; or the three-phase, four-wire system which also has three load wires and, in addition, a fourth wire, or neutral, which may or may not carry current to the load. If the neutral does not carry current, the circuit may be treated as a three-wire system.

Before applying the rule to commonly used circuits, an exception may be noted in the case of a balanced circuit in which the effective values of the currents and voltages and the phase relationships between them remain constant. In other words, the loads on the separate phases are equal. In this special case, power may be measured by a single wattmeter connected in one phase and the reading multiplied by three.

To measure power in either three-wire or four-wire systems, a wattmeter may be connected in each of the power-receiving circuits, as in Fig. 3. The sum of the three readings gives the total power.

Alternatively, in a three-wire system total power may be measured by two wattmeters, each having its current coil connected in one of the line conductors and its potential circuit connected between the line conductor in which its current coil is connected and the third line conductor (Fig. 4). The algebraic sum of the readings of the two wattmeters indicates the total power in the three power-receiving circuits.

In a four-wire system, three wattmeters may also be effectively used by connecting the current coils in each of two of the line conductors and in the neutral conductor, as in Fig. 5. The potential coils are connected between each of the line conductors and the neutral conductor in which the respective current coils are connected and the third line conductor.

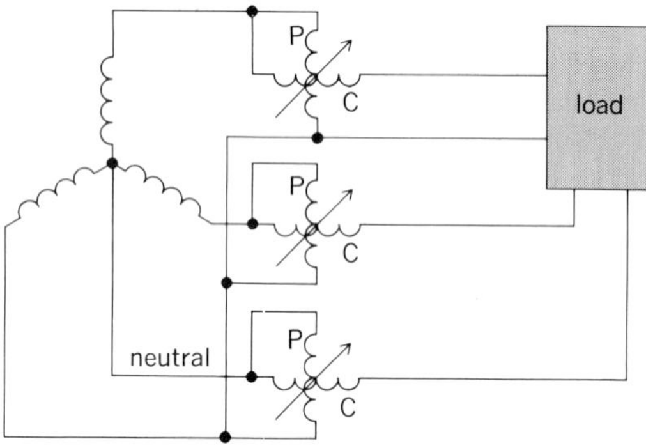

Fig. 5. Wattmeters in three-phase, four-wire circuit. C and P refer to current and potential coils.

In the last three cases the methods are correct for any value of balanced or unbalanced load and for any value of power factor.

A variety of other circuit connections is available for polyphase power measurement for various special conditions of use.

[WILLIAM H. HARTWIG]

Bibliography: M. Braccio, *Basic Electrical and Electronic Tests and Measurements*, 1978; *Code for Electrical Metering*, ANSI C12-1975, 6th ed., 1975; D. Fink and H. W. Beaty (eds.), *Standard Handbook for Electrical Engineers*, 11th ed., sec. 3: Measurements and instrumentation, pp. 1–212, 1978; Instrument Society of America, *ISA Standards and Practices for Instrumentation*, 6th ed., 1980.

Electric power substation

An assembly of equipment in an electric power system through which electrical energy is passed for transmission, distribution, interconnection, transformation, conversion, or switching. *See* ELECTRIC POWER SYSTEMS.

Specifically, substations are used for some or all of the following purposes: connection of generators, transmission or distribution lines, and loads to each other; transformation of power from one voltage level to another; interconnection of alternate sources of power; switching for alternate connections and isolation of failed or overloaded lines and equipment; controlling system voltage and power flow; reactive power compensation; suppression of overvoltage; and detection of faults, monitoring, recording of information, power measurements, and remote communications. Minor distribution or transmission equipment installation is not referred to as a substation.

Classification. Substations are referred to by the main duty they perform. Broadly speaking, they are classified as: transmission substations (Fig. 1), which are associated with high voltage levels; and distribution substations (Fig. 2), associated with low voltage levels. *See* ELECTRIC DISTRIBUTION SYSTEMS.

Substations are also referred to in a variety of other ways, which may be described as follows.

Transformer substations. These are substations whose equipment includes transformers.

Switching substations. These are substations whose equipment is mainly for various connections and interconnections, and does not include transformers.

Customer substations. These are usually distribution substations located on the premises of a larger customer, such as a shopping center, large office or commercial building, or industrial plant.

Converter stations. The main function of converter stations is the conversion of power from ac to dc and vice versa. Converter stations are complex substations required for high-voltage direct-current (HVDC) transmission or interconnection of two ac systems which, for a variety of reasons, cannot be connected by an ac connection. The main equipment includes converter valves usually

Fig. 1. A typical transmission substation (345 kV).

located inside a large hall, transformers, filters, reactors, and capacitors. Figure 3 shows a ±250-kV 500-MW HVDC converter station together with a 245-kV ac substation. The building in the center of the converter station contains all the thyristor valves. On each side of the building against the wall are converter transformers connected to the valves inside. Further away from the building are filters and capacitors. Going away from the building is the dc line, with a typical dc line tower in the uppermost part of the figure. *See* DIRECT-CURRENT TRANSMISSION.

Air-insulated substations. Most substations are installed as air-insulated substations, implying that the bus-bars and equipment terminations are generally open to the air, and utilize insulation properties of ambient air for insulation to ground. Modern substations in urban areas are esthetically designed with low profiles and often within walls, or even indoors.

Metal-clad substations. These are also air-insulated, but for low voltage levels; they are housed in metal cabinets and may be indoors or outdoors (Fig. 2).

Gas-insulated substations. Acquiring a substation site in an urban area is very difficult. Land in many cases is either unavailable or very expensive. Therefore, there has been a definite trend toward increasing use of gas-insulated substations, which occupy only 5–20% of the space occupied by the air-insulated substations. In gas-insulated substations, all live equipment and bus-bars are housed in grounded metal enclosures, which are sealed and filled with sulfur hexafluoride (SF_6) gas, which has excellent insulation properties. These substations have the appearance of a large-scale plumbing (Fig. 4). The disk-shaped enclosures in Fig. 4 are circuit breakers.

Mobile substations. For emergency replacement or maintenance of substation transformers, mobile substations are used by some utilities. They may range in size from 100 to 25,000 kVA, and from 2400 to 220,000 V. Most units are designed for travel on public roads, but larger substations are built for travel on railroad tracks. These substations are generally complete with a transformer, circuit breaker, disconnect switches, lightning arresters, and protective relaying.

Substation switching arrangement. An appropriate switching arrangement for "connections" of generators, transformers, lines, and other major equipment is basic to any substation design. There are seven switching arrangements commonly used (Fig. 5). Each breaker is usually accompanied by two disconnect switches, one on each side, for maintenance purposes. Selecting the switching arrangement involves considerations of cost, reliability, maintenance, and flexibility for expansion.

Fig. 2. View of a residential outdoor distribution substation with double-ended underground supply to transformers which are positioned at either end of the metal-clad switchgear.

Fig. 3. View of ±250-kV 500-MW HVDC converter station (upper part) and 245-kV ac substation (lower part). The large building of the converter station contains all the thyristor valves. (*General Electric*)

Single bus. This involves one common bus for all connections and one breaker per connection. This is the least costly arrangement, but also the least desirable for considerations of reliability and maintenance.

Double bus, single breaker. This involves two common buses, and each connection of the line equipment and so forth can be made to either bus through one or the other breaker. However, there is only one breaker per connection, and when a breaker is out for maintenance, the connected equipment or line is also removed from operation.

Double bus, double breaker. This arrangement has two common buses and two breakers per connection. It offers the most reliability, ease of maintenance, and flexibility, but is also the most expensive.

Main and transfer bus. This is like the single-bus

Fig. 4. A 345-kV ac gas-insulated substation. (*Gould-BBC*).

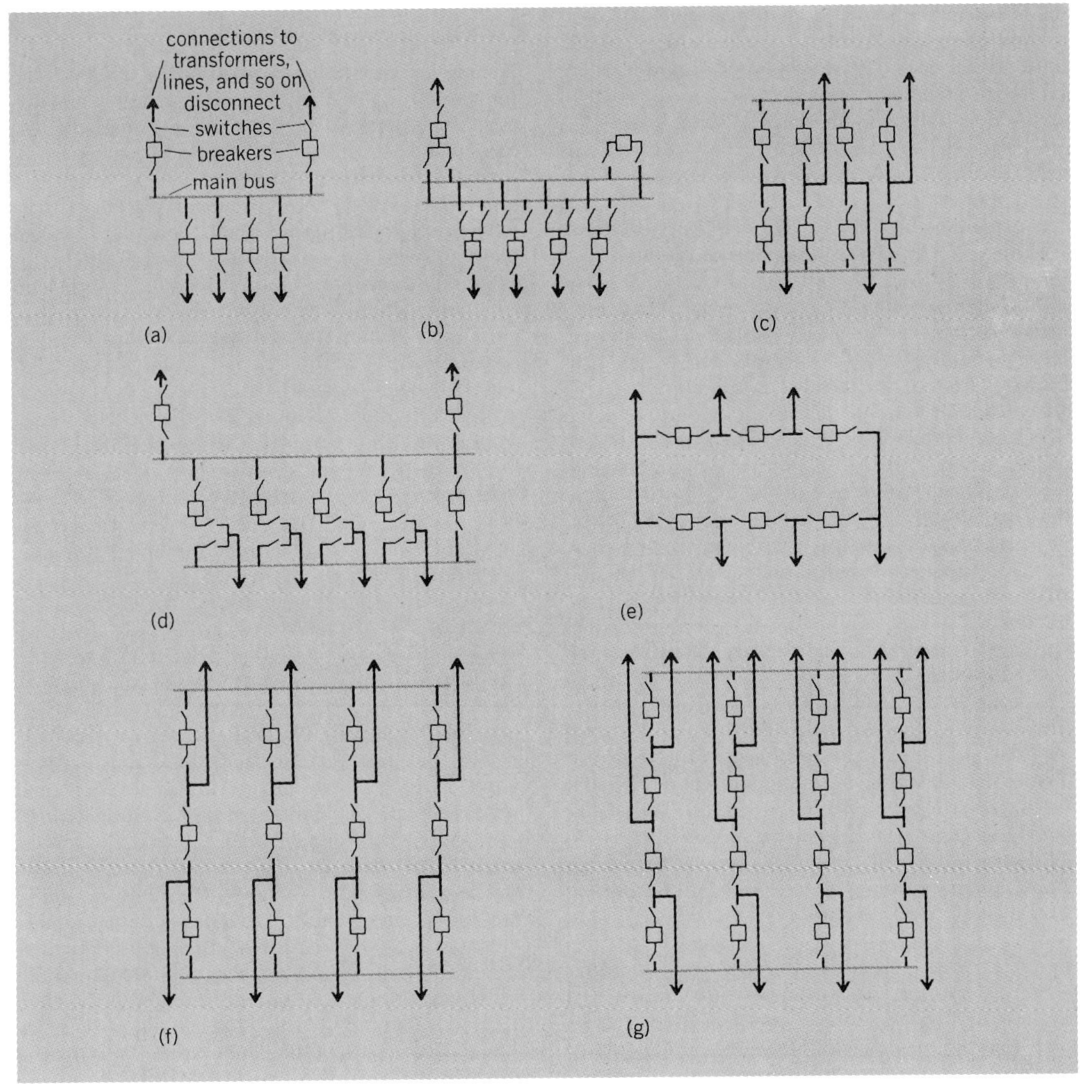

Fig. 5. One-line diagrams of substation switching arrangements. (*a*) Single bus. (*b*) Double bus, single breaker. (*c*) Double bus, double breaker. (*d*) Main and transfer bus. (*e*) Ring bus. (*f*) Breaker-and-a-half. (*g*) Breaker-and-a-third.

arrangement; however, an additional transfer bus is provided so that a breaker can be taken out of service by transferring its connection to the transfer bus which is connected to the main bus through a breaker between the two buses.

Ring bus. This may consist of four, six, or more breakers connected in a closed loop, with the same number of connection points. In this case, a breaker can be taken out of service without also taking out any connection.

Breaker-and-a-half. This arrangement involves two buses, between which three breaker bays are installed. Each three-breaker bay provides two circuit connection points—thus the name breaker-and-a-half.

Breaker-and-a-third. In this arrangement, there are four breakers and three connections per bay.

Substation equipment. A substation includes a variety of equipment. The principal items are listed and briefly described below.

Transformers. These involve magnetic core and windings to transfer power from one side to the other at different voltages. Substation transformers range from small sizes of 1 MVA to large sizes of 2000 MVA. Most of the transformers and all those above a few MVA size are insulated and cooled by oil, and adequate precautions have to be taken for fire hazard. These precautions include adequate distances from other equipment, fire walls, fire extinguishing means, and pits and drains for containing leaked oil. *See* TRANSFORMER.

Circuit breakers. These are required for circuit interruption with the capability of interrupting the highest fault currents, usually 20–50 times the normal current, and withstanding high voltage surges that appear after interruption. Switches with only normal load-interrupting capability are referred to as load break switches. *See* CIRCUIT BREAKER.

Disconnect switches. These have isolation and connection capability without current interruption capability. *See* ELECTRIC SWITCH.

Bus-bars. These are connecting bars or conductors between equipment. Flexible conductor buses are stretched from insulator to insulator, whereas more common solid buses (hollow aluminum alloy tubes) are installed on insulators in air or in gas-enclosed cylindrical pipes. *See* BUS-BAR.

Shunt reactors. These are often required for compensation of the line capacitance where long lines are involved.

Shunt capacitors. These are required for compensation of inductive components of the load current.

Current and potential transformers. These are for measuring currents and voltages and providing proportionately low-level currents and voltages at ground potential for control and protection.

Control and protection. This includes (a) a variety of protective relays which can rapidly detect faults anywhere in the substation equipment and lines, determine what part of the system is faulty, and give appropriate commands for opening of circuit breakers; (b) control equipment for voltage and current control and proper selection of the system configuration; (c) fault-recording equipment; (d) metering equipment; (e) communication equipment; and (f) auxiliary power supplies. *See* ELECTRIC PROTECTIVE DEVICES; RELAY.

Many of the control and protection devices are solid-state electronic types, and there is a trend toward digital techniques using microprocessors. Most of the substations are fully automated locally with a provision for manual override. The minimum manual interface required, along with essential information on status, is transferred via communications channels to the dispatcher in the central office.

Other items which may be installed in a substation include: phase shifters, current-limiting reactors, dynamic brakes, wave traps, series capacitors, controlled reactive compensation, fuses, ac to dc or dc to ac converters, filters, and cooling facilities.

Substation grounding and shielding. Good substation grounding is very important for effective relaying and insulation of equipment; however, the safety of the personnel is the governing criterion in the design of substation grounding. It usually consists of a bare wire grid, laid in the ground; and all equipment grounding points, tanks, support structures, fences, shielding wires and poles, and so forth, are securely connected to it. The grounding resistance is reduced to be low enough that a fault from high voltage to ground does not create such high potential gradients on the ground, and from the structures to ground, to present a safety hazard. Good overhead shielding is also essential for outdoor substations, so as to virtually eliminate the possibility of lightning directly striking the equipment. Shielding is provided by overhead ground wires stretched across the substation or tall grounded poles. *See* GROUNDING; LIGHTNING AND SURGE PROTECTION.

[NARAIN G. HINGORANI]

Bibliography: C. Adamson and N. G. Hingorani, *High Voltage Direct Current Power Transmission*, 1960; C. H. Flurscheim, *Power Circuit Breaker Theory and Design*, 1975; E. W. Kimbark, *Direct Current Transmission*, 1971; S. A. Stigant and A. C. Franklin, *THE J&P Transformer Book*, 10th ed., 1974.

Electric power systems

A complex assemblage of equipment and circuits for generating, transmitting, transforming, and distributing electrical energy. Principal elements of a typical power system are shown in Fig. 1.

Generation. Electricity in the large quantities required to supply electric power systems is produced in generating stations, commonly called power plants. Such generating stations, however, should be considered as conversion facilities in which the heat energy of fuel (coal, oil, gas, or uranium) or the hydraulic energy of falling water is converted to electricity. *See* ELECTRIC POWER GENERATION; POWER PLANT.

Steam stations. About 89% of the electric power used in the United States is obtained from generators driven by steam turbines. 650-, 800-, and 950-MW units are commonplace for new fossil-fuel–fired stations, and 1150–1300-MW units are the most commonly installed units in nuclear stations. *See* STEAM TURBINE.

Coal is the fuel for more than 50% of steam turbine generation, and its share should increase somewhat because of the projected long-term shortage of natural gas and both the sharp rise in the cost of fuel oil and governmental policy of restricting firing of oil. Natural gas, used extensively in the southern part of the United States, fuels about 16% of the steam turbine generation, and heavy fuel oil, 19%, largely in power plants able to take delivery from ocean-going tankers or river barges. The remaining 14% is generated from the radioactive energy of slightly enriched uranium, which, for many power systems, produces electricity at a lower total cost than either coal or fuel oil. As more nuclear units go into commercial operation, the contribution of uranium to the electrical energy supply will rise, probably to more than 25% of the total fuel generated output by the mid-1980s.

Nuclear steam systems used by United States utilities are mostly of the water-cooled-and-moderated type, in which the heat of a controlled nuclear reaction is used to convert water into steam to drive a conventional turbine generator; such units are presently limited to about 1300 MW by the thermal limit placed on nuclear reactors by the Nuclear Regulatory Commission of the U.S. Department of energy. *See* NUCLEAR REACTOR.

Hydroelectric plants. Waterpower supplies about 10.3% of the electric power consumed in the United States. But this share can only decline in the years ahead because very few sites remain undeveloped where sufficient water drops far enough in a reasonable distance to drive reasonably sized hydraulic turbines. Consequently, the generating capability of hydro plants is slated to fall off by the mid-1980s because of its very small share of the planned additions. Much of this additional hydro capability will be used at existing plants to increase their effectiveness in supplying peak power demands, and as a quickly available source of emergency power. *See* HYDRAULIC TURBINE.

Some hydro plants totaling 10,640 MW actually draw power from other generating facilities during light system-load periods to pump water from a river or lake into an artificial reservoir at a higher

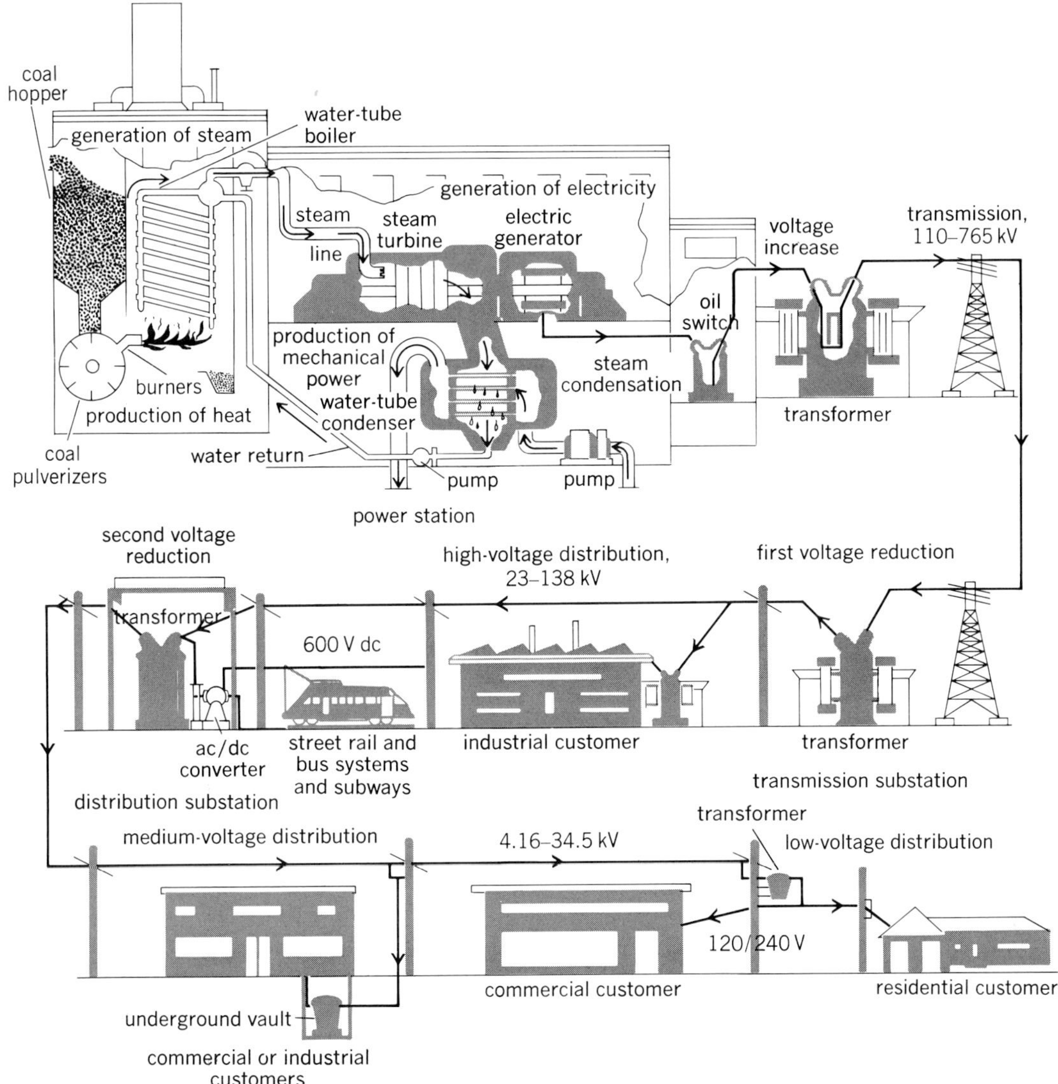

Fig. 1. Major steps in the generation, transmission, and distribution of electricity.

elevation from which it can be drawn through a hydraulic station when the power system needs additional generation. These pumped-storage installations consume about 50% more energy than they return to the power system and, accordingly, cannot be considered energy sources. Their use is justified, however, by their ability to convert surplus power that is available during low-demand periods into prime power to serve system needs during peak-demand intervals—a need that otherwise would require building more generating stations for operation during the relatively few hours of high system demand. Installations now planned should double the existing capacity by the late 1980s to 22,684 MW.

Combustion turbine plants. Gas-turbine-driven generators, now commonly called combustion turbines because of the growing use of light oil as fuel, have gained wide acceptance as an economical source of additional power for heavy-load periods. In addition, they offer the fastest erection time and the lowest investment cost per kilowatt of installed capability. Offsetting these advantages, however, is their relatively less efficient consumption of more costly fuel. Typical unit ratings in the United States have climbed rapidly in recent years until some units in operation are rated at 80 MW. Some turbine installations involve a group of smaller units totaling, in one case, 260 MW. Combustion turbine units, even in the larger rating, offer extremely flexible operation and can be started and run up to full load in as little as 10 min. Thus they are extremely useful as emergency power sources, as well as for operating during the few hours of daily load peaks. Combustion turbines total about 8.5% of the total installed capability of United States utility systems and supply less than 3% of the total energy generated.

In the years ahead, however, combustion turbines are slated for an additional role. Several installations have used their exhaust gases to heat boilers that generate steam to drive steam turbine generators. Such combined-cycle units offer fuel economy comparable to that of modern steam plants and at considerably less cost per kilowatt. In addition, because only part of the plant uses steam, the requirement for cooling water is considerably reduced. A number of additional combined-cycle installations have been planned, but wide acceptance is inhibited by the doubtful availability of light fuel oil for them. This barrier should be resolved, in time, by the successful development of systems for fueling them with gas derived from coal.

Internal combustion plants. Internal combustion engines of the diesel type drive generators in many small power plants. In addition, they offer the ability to start quickly for operation during peak loads or emergencies. However, their small size, commonly about 2 MW per unit although a few approach 10 MW, has limited their use. Such installations account for about 1% of the total power-system generating capability in the United States, and make an even smaller contribution to total electric energy consumed. *See* INTERNAL COMBUSTION ENGINE.

Three-phase output. Because of their simplicity and efficient use of conductors, three-phase 60-Hz alternating-current systems are used almost exclusively in the United States. Consequently, power-system generators are wound for three-phase output at a voltage usually limited by design features to a range from about 11 kV for small units to 30 kV for large ones. The output of modern generating stations is usually stepped up by transformers to the voltage level of transmission circuits used to deliver power to distant load areas.

Transmission. The transmission system carries electric power efficiently and in large amounts from generating stations to consumption areas.

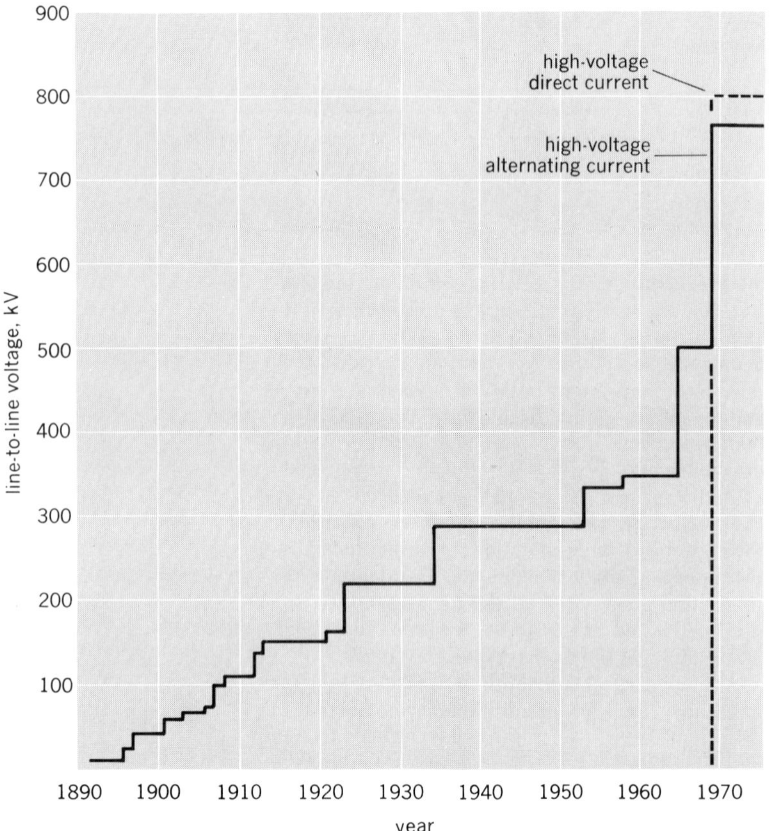

Fig. 2. Growth of ac transmission voltages from 1890.

Power capability of typical three-phase open-wire transmission lines

Line-to-line voltage, kV	Capability, MVA
115 ac	60
138 ac	90
230 ac	250
345 ac	600
500 ac	1200
765 ac	2500
800 dc*	1500

*Bipolar line with grounded neutral.

Such transmission is also used to interconnect adjacent power systems for mutual assistance in case of emergency and to gain for the interconnected power systems the economies possible in regional operation. Interconnections have expanded to the point where most of the generation east of the Rocky Mountains, except for a large part of Texas, regularly operates in parallel, and over 90% of all generation in the United States, exclusive of Alaska and Hawaii, and in Canada can be linked.

Transmission circuits are designed to operate up to 765 kV, depending on the amount of power to be carried and the distance to be traveled. The permissible power loading of a circuit depends on many factors, such as the thermal limit of the conductors and their clearances to ground, the voltage drop between the sending and receiving end and the degree to which system service reliability depends on it, and how much the circuit is needed to hold various generating stations in synchronism. A widely accepted approximation to the voltage appropriate for a transmission circuit is that the permissible load-carrying ability varies as the square of the voltage. Typical ratings are listed in the table.

Transmission as a distinct function began about 1886 with a 17-mi (27-km) 2-kV line in Italy. Transmission began at about the same time in the United States, and by 1891 a 10-kV line was operating (Fig. 2). In 1896 an 11-kV three-phase line brought electrical energy generated at Niagara Falls to Buffalo, 20 mi (32 km) away. Subsequent lines were built at successively higher levels until 1936, when the Los Angeles Department of Water and Power energized two lines at 287 kV to transmit 240 MW the 266 mi (428 km) from Hoover Dam on the Colorado River to Los Angeles. A third line was completed in 1940.

For nearly 2 decades these three 287-kV lines were the only extra-high-voltage (EHV). lines in North America, if not in the entire world. But in 1946 the American Electric Power (AEP) System inaugurated, with participating manufacturers, a test program up to 500 kV. From this came the basic design for a 345-kV system, the first link of which went into commercial operation in 1953 as part of a system overlay that finally extended from Roanoke, VA, to the outskirts of Chicago. By the late 1960s the 345-kV level had been adopted by many utilities interconnected with the AEP System, as well as others in Illinois, Wisconsin, Minnesota, Kansas, Oklahoma, Texas, New Mexico, Arizona and across New York State into New England.

The development of 500-kV circuits began in 1964, even as the 345-kV level was gaining wide acceptance. One reason for this was that many utilities that had already adopted 230 kV could gain only about 140% in capability by switching to 345 kV, but the jump to 500 kV gave them nearly 400% more capability per circuit. The first line energized at this new level was by Virginia Electric & Power Company to carry the output of a new mine mouth station in West Virginia to its service area. A second line completed the same year provided transmission for a 1500-MW seasonal interchange between the Tennessee Valley Authority and a group of utilities in the Arkansas-Louisiana area. Lines at this voltage level now extend from New Jersey to Texas and from New Mexico via California to British Columbia.

The next and latest step-up occurred in 1969 when the AEP System, after another cooperative test program, completed the first line of an extensive 765-kV system to overlay its earlier 345-kV plant to a major switchyard several thousand feet (1 ft = 0.3 m) away. And 765-kV cable, after extensive testing above the 500-kV level at the Waltz Mill Cable Test Center operated by Westinghouse Electric Corporation for the Electric Power Research Institute (EPRI), will be available when required.

In anticipation of the need for transmission circuits of higher load capability, an extensive research program is in progress, spread among several large and elaborately equipped research centers. Among these are: Project UHV for ultra-high-voltage overhead lines operated by the General Electric Company near Pittsfield, MA, for EPRI; the Frank B. Black Research Center built and operated by the Ohio Brass Company near Mansfield; the above-mentioned Waltz Mill Cable Test Center; and an 1100 kV, 1-mi (1.6 km) test line built by the Bonneville Power Authority. All include equipment for testing full-scale or cable components at well over 1000 kV. In addition, many utilities and specialty manufacturers have test facilities related to their fields of operation.

A relatively new approach to high-voltage long-distance transmission is high-voltage direct current (HVDC), which offers the advantages of less costly lines, lower transmission losses, and insensitivity to many system problems that restrict alternating-current systems. Its greatest disadvantage is the need for costly equipment for converting the sending-end power to direct current, and for converting the receiving-end direct-current power to alternating current for distribution to consumers. Starting in the late 1950s with a 65-mi (105-km) 100-kV system in Sweden, HVDC has been applied successfully in a series of special cases around the world, each one for a higher voltage and greater power capability. The first such installation in the United States was put into service in 1970. It operates at 800 kV line to line, and is designed to carry a power interchange of 1440 MW over a 1354-km overhead tie line between the winter-peaking Northwest Pacific coastal region and the summer-peaking southern California area. These HVDC lines perform functions other than just power transfer, however. The Pacific Intertie is used to stabilize the parallel alternating-current transmission and lines, permitting an increase in their capability; and back-to-back converters with no tie line between them are used to tie together two systems in Nebraska that otherwise could not be synchronized. The first urban installation of this technology was energized in 1979 in New York.

In addition to these high-capability circuits, every large utility has many miles of lower-voltage transmission, usually operating at 110 to 345 kV, to carry bulk power to numerous cities, towns, and large industrial plants. These circuits often include extensive lengths of underground cable where they pass through densely populated areas. Their design, construction, and operation are based upon research done some years ago, augmented by extensive experience. *See* TRANSMISSION LINES.

Interconnections. As systems grow and the number and size of generating units increase, and as transmission networks expand, higher levels of bulk-power-system reliability are attained through properly coordinated interconnections among separate systems. This practice began more than 50 system. The first installation in this voltage class, however, was by the Quebec Hydro-Electric Commission to carry the output at 735 kV (the expected international standard) from a vast hydro project to its load center at Montreal, some 375 mi (604 km) away.

Transmission engineers are anticipating even higher voltages of 1100 to 1500 kV, but they are fully aware that this objective may prove too costly in space requirements and funds to gain wide acceptance. Experience already gained at 500 kV and 765 kV verifies that the prime requirement no longer is insulating the lines to withstand lightning discharges, but insulating them to tolerate voltage surges caused by the operation of circuit breakers. Audible noise levels, especially in rain or humid conditions, are high, requiring wide buffer zones. Environmental challenges have been brought on the basis of possible negative biological effects of the electrostatic field produced under EHV lines, though research to date has not shown any such effects.

Experience has indicated that, within about 10 years after the introduction of a new voltage level for overhead lines, it becomes necessary to begin connecting underground cable. This has already occurred for 345 kV; the first overhead line was completed in 1953, and by 1967 about 100 mi (160 km) of pipe-type cable had been installed to take power received at this voltage level into metropolitan areas. The first 500-kV cable in the United States was placed in service in 1976 to take power generated at the enormous Grand Coulee hydro years ago with such voluntary pools as the Connecticut Valley Power Exchange and the Pennsylvania-New Jersey-Maryland Interconnection. Most of the electric utilities in the contiguous United States and a large part of Canada now operate as members of power pools, and these pools (except one in Texas) in turn are interconnected into one gigantic power grid known as the North American Power Systems Interconnection. The operation of this interconnection, in turn, is coordinated by the North American Power Systems Interconnection Committee (NAPSIC). Each individual utility in such pools operates independently, but has contractual arrangements with other members in re-

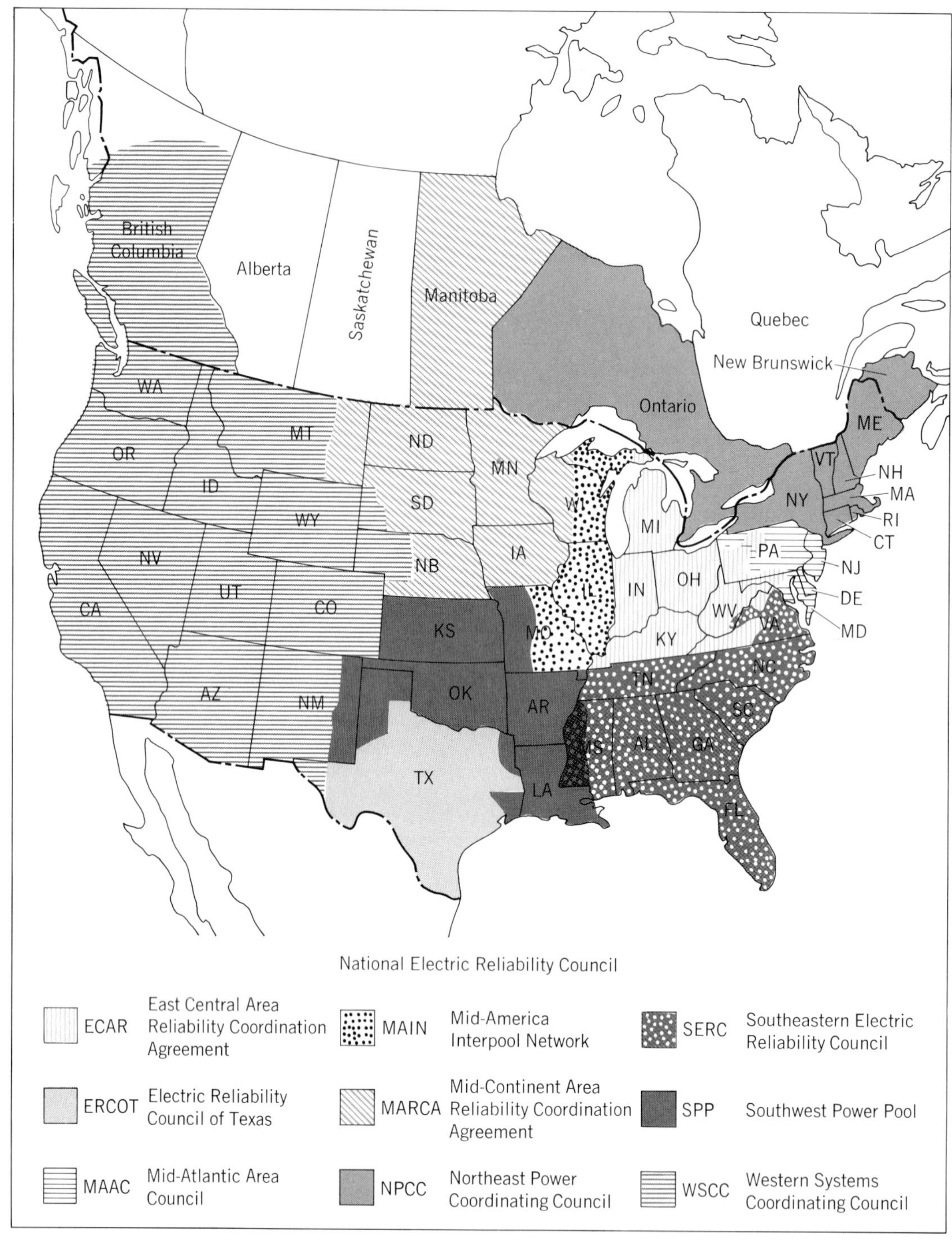

Fig. 3. Areas served by the nine regional reliability councils, coordinated by the National Electric Reliability Council, that guide the coordination and operation of generation and transmission facilities.

spect to generation additions and scheduling of operation. Their participation in a power pool affords a higher level of service reliability and important economic advantages.

Regional and national coordination. The Northeast blackout of Nov. 9, 1965, stemmed from the unexpected trip-out of a key transmission circuit carrying emergency power into Canada and cascaded throughout the northeastern states to cut off electric service to some 30,000,000 people. It spurred the utilities into a chain reaction affecting the planning, construction, operation, and control procedures for their interconnected systems. They soon organized regional coordination councils, eventually nine in number, to cover the entire contiguous United States and four Canadian provinces (Fig. 3). Their stated objective was "to further augment reliability of the parties' planning and operation of their generation and transmission facilities."

Then, in 1968, the National Electric Reliability Council (NERC) was established to serve as an effective body for the collection, unification, and dissemination of various reliability criteria for use by individual utilities in meeting their planning and operating responsibilities. NAPSIC was reorganized shortly afterward to function as an advisory group to NERC.

Increased interconnection capability among power systems reduces the required generation reserve of each of the individual systems. In most utilities the loss-of-load probability (LOLP) is used to measure the reliability of electric service, and it is based on the application of probability theory to unit-outage statistics and load forecasts. A common LOLP criterion is 1 day in 10 years when load may exceed generating capability. The LOLP decreases (that is, reliability increases) with increased interconnection between two areas until a saturation level is reached which depends upon the amount of reserve, unit sizes, and annual load shape in each area. Any increase in interconnection capability beyond that saturation level will not cause a corresponding improvement in the level of system reliability.

Traditionally, systems were planned to withstand all reasonably probable contingencies, and operators seldom had to worry about the possible effect of unscheduled outages. Operators' normal security functions were to maintain adequate generation on-line and to ensure that such system variables as line flows and station voltages remained within the limits specified by planners. However, stronger interconnections, larger generating units, and rapid system growth spread the transient effects of sudden disturbances and increased the responsibilities of operators for system security.

System security is concerned with service continuity at standard frequency and voltage levels. The system is said to be insecure if a contingency would result in overloading some system components, in abnormal voltage levels at some stations, in change of system frequency, or in system instability, even if there is adequate capability as indicated by some reliability index.

Energy control centers. The majority of large electric systems have computerized energy centers whose functions are to control operation of the system so as to optimize economy and security. In some large interconnected networks of individual utility companies, such as those in the Pennsylvania–New Jersey–Maryland power pool, regional control centers have been established to perform the same function for a group of companies. New generation centers now use digital computers, or hybrid analog-digital units, rather than the analog machines previously used, and redundant systems are required to ensure security.

Control can be be broken into two phases: operations and security. In the operational area, a typical center provides a completely updated analysis of power flows in the system at 10-min intervals. In real time, this means calculation of the cost penalty factors for line losses, machine efficiencies, and fuel costs, thereby minimizing cost of power. Modern systems also include predictive programs that produce probabilistic forecasts of hourly system load for several days into the future, using historical data on weather-load correlations. This permits scheduling startup and shutdown of available generating units to optimize operating economy and to establish necessary maintenance schedules.

System security and analysis programs permit the system operators to simulate problems on the

Fig. 4. Matrix mimic board and CRT displays. *(Courtesy of Ferranti-Packard Inc. and Cleveland Electric Illuminating Co.)*

system as it actually exists currently, thereby preparing them for appropriate actions should such emergencies occur. The most advanced centers today use techniques such as state estimation, which processes system data to calculate the probabilities of emergencies in the near-term future, thus permitting the operators to take preventive action before the event occurs.

Interface between the operators and computers are through data loggers, cathode-ray tubes (CRTs), plotters, or active mimic boards (Fig. 4). Dynamic mimic boards, in which the displays are driven by the computer output, have generally displaced the older static representation systems which used supervisory lights to display conditions. The CRTs display the system in up to seven colors, and the operator can interact with them by using a cursor or light pen either to analog the effect of changes to the system or actually to operate the equipment.

Substations. Power delivered by transmission circuits must be stepped down in facilities called substations to voltages more suitable for use in industrial and residential areas. On transmission systems, these facilities are often called bulk-power substations; at or near factories or mines, they are termed industrial substations; and where they supply residential and commercial areas, distribution substations.

Basic equipment in a substation includes circuit breakers, switches, transformers, lightning arresters and other protective devices, instrumentation, control devices, and other apparatus related to specific functions in the power system. *See* ELECTRIC POWER SUBSTATION.

Distribution. That part of the electric power system that takes power from a bulk-power substation to customers' switches, commonly about 35% of the total plant investment, is called distribution. This category includes distribution substations, subtransmission circuits that feed them, primary circuits that extend from distribution substations to every street and alley, distribution transformers, secondary lines, house service drops or loops, metering equipment, street and highway lighting, and a wide variety of associated devices.

Primary distribution circuits usually operate at 4160 to 34,500 V line to line, and supply large commercial institutional and some industrial customers directly. The lines may be overhead open wire on poles, spacer or aerial cable, or underground cable. Legislation in more than a dozen states now requires that all new services to developments of five or more residences be put underground. The bulk of existing lines are overhead, however, and will remain so for the indefinite future.

At conveniently located distribution transformers in residential and commercial areas, the line voltage is stepped down to supply low-voltage secondary lines, from which service drops extend to supply the customers' premises. Most such service is at 120/240 V, but other common voltages are 120/208 V, 125/216, and, for larger commercial and industrial buildings, 240/480, 265/460, or 277/480 V. These are classified as utilization voltages. *See* ELECTRIC DISTRIBUTION SYSTEMS.

Electric utility industry. In the United States, which has the third highest per-capita use of electricity in the world and more electric power capability than any other nation, the electric capability systems as measured by some criteria are the largest industry.

[WILLIAM C. HAYES]

Bibliography: A. S. Brookes et al., *Proceedings of International Conference on Large High Voltage Electric Systems*, Paris, Aug. 21–29, 1974; J. F. Dopazo, State estimator screens incoming data, *Elec. World*, 185(4):56–57, Feb. 15, 1976; *EHV Transmission Line Reference Book*, Edison Electric Institute, 1968; The electric century, 1874–1974, *Elec. World*, 181(11):43–431, June 1, 1974; Electric Research Council, *Electric Transmission Structures*, Edison Electric Institute, 1968; *Electrical World Directory of Electric Utilities, 1977–1978*, 1977; *Electrostatic and Electromagnetic Effects of Ultra-High-Voltage Transmission Lines*, Electric Power Research Institute, 1978; L. W. Eury, Look one step ahead to avoid crises, *Elec. World*, 187(10):50–51, May 15, 1977; G. F. Friedlander, 15th Steam Station Design Survey, *Elec. World*, 190(10):73–88, Nov. 15, 1978; G. F. Friedlander, 20th Steam Station Cost Survey, *Elec. World*, 188(10):43–58, Nov. 15, 1977; 1979 Annual Statistical Report, *Elec. World*, 191(6):51–82, Mar. 15, 1979; W. P. Rades, Convert to digital control system, *Elec. World*, 186(1):50–51, July 1, 1976; N. D. Reppon et al., *Proceedings of the Department of Energy's System Engineering for Power: States and Prospects Conference*, Henniker, NH, Aug. 17–22, 1975; W. D. Stevenson, Jr., *Elements of Power System Analysis*, 1975; R. L. Sullivan, *Power System Planning*, 1977; *Transmission Line Reference Book, HVDC to ±600 kV*, Electric Power Research Institute & Bonneville Power Administration, 1977; 29th Annual Electric Utility Industry Forecast, *Elec. World*, 190(6):62–76, Sept. 15, 1978.

Electric power systems engineering

Electricity plays a key role in future energy strategy because of its versatility with respect to input energy form. Electricity can be produced with coal, nuclear, hydro, geothermal, fission, fusion, biomass, wind, or solar energy as well as oil or gas. Electrical supply also offers the opportunity of total environmental enhancement compared to other energy use patterns.

In meeting the need of an increasing electrical penetration, the electrical utility industry faces critical driving forces of increased capital costs, financial and environmental restraints, increasing fuel costs, and regulatory delays. The totality of these driving forces leads to the need for more comprehensive understanding and analysis of electric utility systems. Developments in systems analysis and synthesis as well as in related digital, analog, and hybrid computer techniques provide important tools aiding the systems engineer in meeting these more exacting challenges.

This article discusses tools that aid long-range electric utility conceptual planning leading to design studies and equipment specification. Also discussed are concepts in systems engineering applied to improving system operation. The specific computer programs used to illustrate the discussion were developed by the General Electric Company; however, the general concepts and techniques covered here are used throughout the electrical utility industry.

SYSTEM PLANNING

The tools to be discussed are directed toward the bulk power system consisting of generating sources, loads which are supplied from substations, interconnection points with neighbors, and a transmission system which interconnects these elements.

In the area of planning, considerations facing the system planner include: the amount and type of generating capacity to be added; the optimum size of generating units; the generation types or combinations thereof to be used (nuclear, gas turbine, conventional steam, pumped hydro, solar, wind, and so forth); the environmental impact of various generation alternatives; the location of new generation; the size of the interconnections with neighboring systems; the transmission lines to be constructed and their most economic voltage levels; the impact of major facility additions upon the financial structure of the utility; and the effect of utility requirements on targets of performance for new technologies.

Simulation programs. The various system simulation programs used in long-range system planning are shown in Fig. 1. The load shape modeling program tabulates the daily peak loads throughout the year for use in the generation expansion program. The load shape program also determines daily load shapes for weekdays and weekends throughout the year for use in the production costing program.

The generation expansion program, based upon generation and load data, undertakes a probabilistic simulation to determine the size and installation dates for future units and interconnections to achieve adequate reliability levels throughout the expansion.

Generation production costs are computed by the production costing program. This program simulates the operation of future systems to provide a knowledge of fuel costs and start-up and maintenance costs. Generation investment costs are treated in the investment cost program, which determines annual costs of investments. Transmission expansion programs have been developed to synthesize the transmission system required to serve future loads and connect future generation into the network. Transmission investment costs are treated in the investment cost program. The programs described determine the annual costs and the present worth of future revenue requirements, typically for a 20-year future period. However, engineering economic analyses of this type are frequently insufficient to fully judge alternative plans. The translation of various technical options into measures meaningful to chief executives is accomplished by a corporate model, which predicts the effects of technological decisions on the utility's financial statements.

Generation expansion program. The generation expansion program determines the timing and sizing of generation additions and interconnections from an analysis of system reserves through the use of probability mathematics to determine the ability of generation to meet daily peak loads. Capacity and load models measure the service reliability expressed as a loss of load probability, which states the probability of insufficient capacity to meet the load. This modeling may be accomplished on either a single-area or multiarea basis.

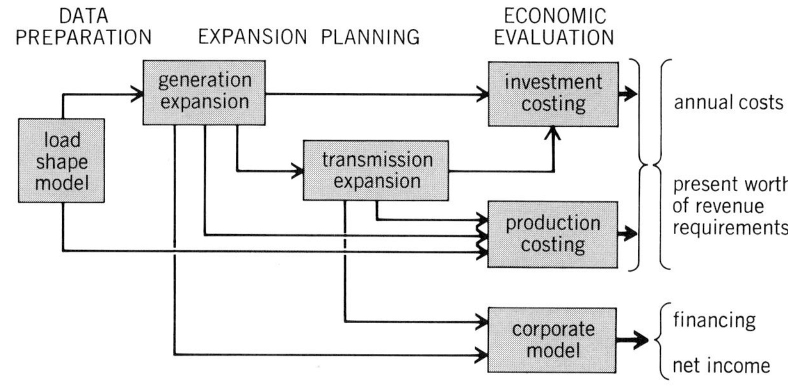

Fig. 1. System planning programs.

In the single-area program, generation is assumed to be located on a single bus. Subsequent programs plan an integrated transmission system to meet this assumption. Principal input data are the unit size, forced outage rate, maintenance requirements, peak load characteristics, and load-forecasting uncertainity.

The output of this single-area program includes the system maintenance schedule, the system loss of load probability, the load-carrying capability of the system for a given risk level, and the effective capability of the added units. This program may be used to automatically add units to meet a given index of reliability. Multiarea modeling is used to examine in detail the interconnection requirements between areas. *See* ELECTRIC POWER GENERATION.

Production costing program. The production costing program simulates the minimum-cost daily economic operation of a power system to determine the expense incurred for fuel, start-up, and operation and maintenance labor. Capacity factors are also determined. As the initial step in this program, a load model is prepared, with the peaks for each month of the year represented as a fraction of the annual peak for each of the 12 months. Next, annual maintenance schedules are developed. Units are taken out for maintenance to levelize and minimize risk throughout the year. The next step involves modeling the daily load cycles. Each month is represented in terms of a Sunday, peak weekday, average weekday, and Saturday shape. The procedure next involves unit commitment. In operating the electric utility system, the total generating capacity placed in service exceeds the load by an amount designated as spinning reserve. This reserve is required to safeguard the operation of the system, considering various unexpected forced outages that may occur. Based upon spinning reserve rules, the minimum-cost thermal commitment is determined. The actual thermal commitment will exceed the minimum because of operating constraints such as minimum downtimes and uptimes. Once the unit commitment is given, the units are dispatched according to equal incremental costs in order to achieve minimum overall costs.

Production costing output may be yearly or monthly, and includes information for a given unit and the overall system as well. Unit information includes unit identity, rating, maintenance period,

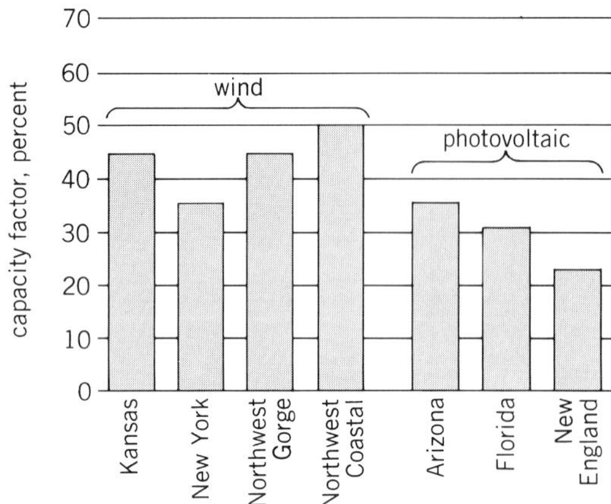

Fig. 2. Wind and photovoltaic capacity factors based on weather data for a typical year.

number of starts, number of hours on the line, capacity factor, average operating heat rate, energy produced, start-up costs, operation and maintenance costs, and total costs. System output includes the energy produced, fuel cost, total cost, and energy sold or purchased.

Multiarea production models have also been developed to study interconnection requirements based upon economy flows between areas.

Application to wind and solar generation. In the use of the generation expansion, production cost, and investment costing programs, various alternatives are formulated by the planner and the effect of various input assumptions determined. Application studies have involved the assessment of wind and solar photovoltaic generating sources. Plant performance models were developed that calculated the hourly output of such plants based upon hourly weather data and the plant characteristics. The generation expansion program measured the expected number of hours per year of capacity deficiency. This calculation was performed hourly to recognize the hourly performance of solar and wind plants. Capacity and energy benefits were evaluated in order to determine power plant break-even values.

Figure 2 presents the wind and photovoltaic capacity factors for various locations. The capacity factor is determined by calculating the ratio of the kilowatt-hours generated to the kilowatt-hours that would be generated if the plant operated at full load for all of the hours of the year. The results shown are for the best candidate wind and photovoltaic plant designs. The plant capacity factors vary from 20 to 50% and are very site-specific.

Effective capability for 5% penetration is shown in Fig. 3 for the same sites as Fig. 2. The percent penetration is the percent of total system generating capacity represented by the designated type of generating unit. The percent effective capability represents the amount of capacity displacement that corresponds to a perfect conventional unit. For example, at Northwest Coastal, with 20% effective capability, the addition of 100 MW of wind generation would be equivalent to adding a perfect conventional unit of 20 MW. Of particular importance is the point that capacity credits are present for these intermittent forms of generation. Dedicated storage for photovoltaic and wind is often advocated. Dedicated storage can increase the effective capability, but dedicated storage is not economically attractive.

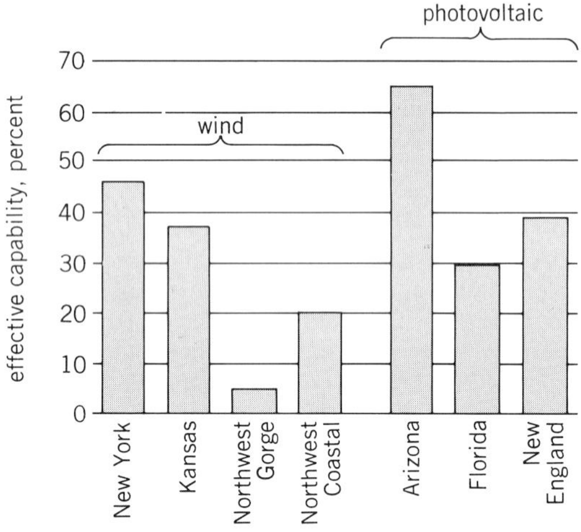

Fig. 3. Wind and photovoltaic effective capability for 5% penetration, based on weather data for a typical year.

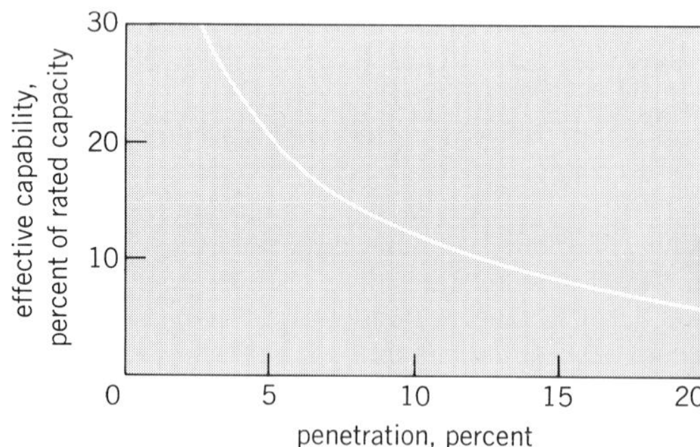

Fig. 4. Effective capability versus penetration for wind and photovoltaic power.

Figure 4 presents illustrative results showing effective capability in percent of rated capacity plotted as a function of penetration. The effective capability and therefore the capacity credit decrease rapidly with penetration. This function is very site-specific and is dependent upon the load, wind, and solar characteristics.

The final results of such studies are power plant break-even values, such as those shown in Fig. 5. The bottom part of each column is the energy or production cost value; the top part, the capacity value. The energy value results from the savings in fuel costs. The capacity value results from the investment savings realized by the reduction in installed generation of the other types of units in the system. These data show that photovoltaic power will require a technological breakthrough to compete as an energy source. Wind generation could compete in some locations. However, long blade life has not been demonstrated. The studies indicate that photovoltaic and wind power will play a modest role before the year 2000. *See* WIND POWER.

Optimized generation planning program. The generation expansion, generation production cost, and generation investment cost programs have

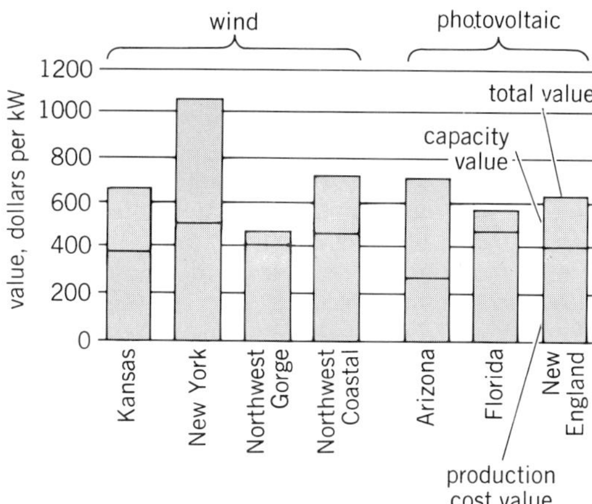

Fig. 5. Wind and photovoltaic power plant breakeven values, based on weather data for a typical year (in 1978 dollars per kilowatt).

been combined into one overall program designated as the optimized generation planning program (OGP; Fig. 6). Based on economics, the OGP automatically seeks an optimum mix of various power generation types for a given reliability, and equipment and fuel costs assumptions. New generation unit data, giving performance and costs, together with data on the existing system, are inputs to the OGP. Output from each run of the program is an optimum expansion pattern listing unit sizes, installation dates, annual cost details, and the total present worth of the expansion. Time and effort expended in finding an optimum is a fraction of that required with earlier, separate step-by-step cut-and-try programs and procedures. Studies can be made quickly and easily, showing cost sensitivity to changes in input parameters such as load shape and growth, unit size, unit types, installed unit costs, fuel costs, reliability index, forced outage rate, and environmental constraints.

The OGP provides environmental outputs on a system-wide basis, including waste heat rejection, emission discharges (sulfur dioxide, nitrogen oxides, carbon monoxide, and particulates), and cooling water consumption. Scrubbers, cooling towers, and precipitators are modeled with allowance for a total system-wide evaluation of these abatement devices. Environmental system dispatching schemes to minimize environmental effects can be modeled and assessed.

The OGP synthesizes a future electric generation system and provides economic results utilizing the present worth of future revenue requirements. To analyze the financial aspects of an electric utility system, a financial simulation program (FSP) has been computer-linked to the OGP, as is illustrated in Fig. 6. The FSP performs a financial analysis of the electric utility and presents the results in terms of balance sheets, income statements, and cash reports. The FSP is a simplified corporate model, developed to produce a realistic utility financial simulation but with less detailed input data requirements than a comprehensive corporate model. When the FSP is used with the OGP, much of the input data is developed by the OGP and transferred to the FSP automatically.

Applications of the combined OGP-FSP include optimum generation mix, parametric sensitivity tests, long-range fuel requirements, load management, evaluation of new technologies, unit slippages, unit size, and the financial impact on nonoptimum additions.

Transmission planning. The next major area of electric utility planning is that of transmission planning. The transmission planning program is a key tool directed toward determining: when new lines will be required; the location of lines in the network; and the voltage level to be used.

The direct-synthesis procedure illustrated in Fig. 7 starts with the initial system, and a projection of loads and generation for the horizon year (say, 20 years hence). A horizon plan is synthesized, with recognition of normal as well as line and unit outage conditions. The program then moves through time, year by year, to specify the annual additions. As appropriate, elements from the horizon-year design are used to meet the yearly requirements. The total plan will require elements in addition to the horizon elements to recognize intermediate-year generation load unbalances.

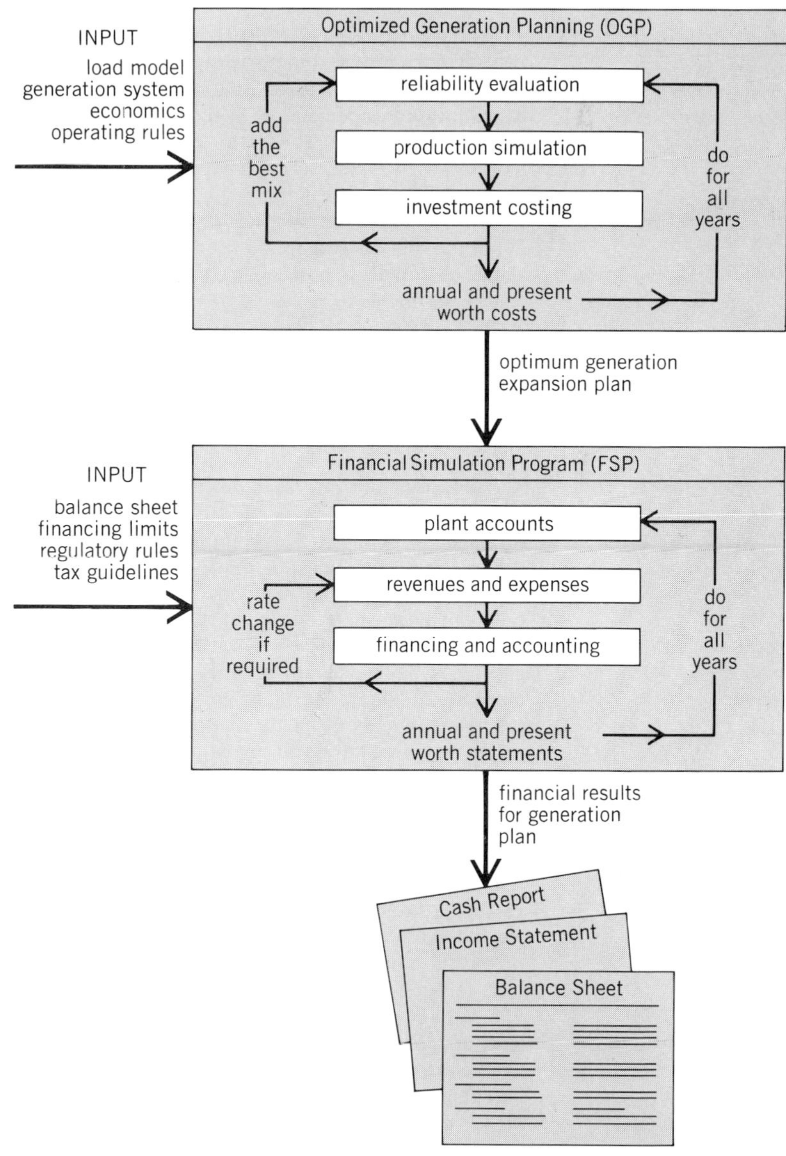

Fig. 6. Overall structure of the linked optimized generation planning program and financial simulation program.

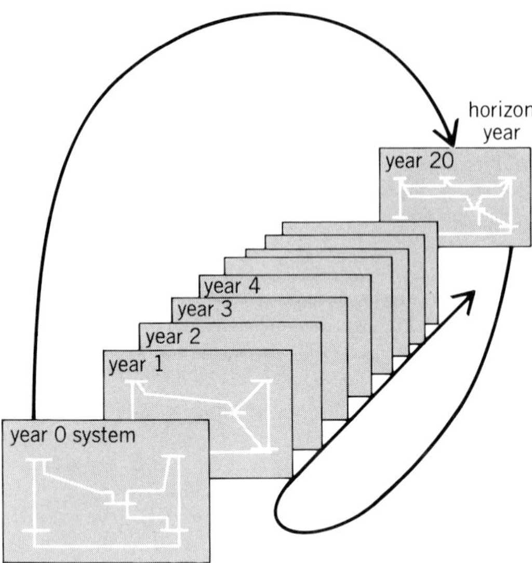

Fig. 7. Transmission expansion program, using direct-synthesis procedure.

Two essential blocks make up the synthesis procedure: a flow estimation program and rules for circuit selection. Flow estimation is a linear approximation to a power flow designed to answer the following questions about the transmission network: (1) Does the network have capacity shortages? (2) If so, what is the most economic route to utilize, thereby eliminating the shortages? (3) What are the shortage magnitudes? Circuit additions are made according to programmed rules that recognize both normal and emergency operating conditions.

The results are presented as a computer-drawn map as well as a list of circuits added each year, including the terminals involved, voltage level, length, and reason for the addition. Potential loading information for new rights-of-way is also given.

Illustrative studies include evaluation of the effect of generation siting and transmission voltage levels on transmission system costs. One study utilized the transmission expansion program to determine the transmission feasibility and requirements of large-scale energy parks as compared to conventional dispersed generation. Also, this program has been used to identify transmission requirements of small solar and wind generation plants dispersed on an electric utility system.

Power flow and stability programs. In general, after a transmission plan is selected based upon the transmission expansion program, a thorough check of system performance is made with a large-capacity digital ac power flow program and a companion stability program. The load flow program predicts the steady-state ac behavior of the network. Systems planners through use of power flow programs confirm the adequacy of the proposed transmission network, considering line loadings, bus voltages, and reactive power supply under normal and outage conditions.

Of course, the adequacy of the system must also be examined upon the occurrence of faults or the loss of major blocks of generation. Digital transient stability programs evaluate system performance after the disturbance to ascertain both transient stability and dynamic stability. Stability studies start with the steady-state solution calculated by the power flow program for the conditions existing immediately before the system disturbance.

Power flow and stability programs define line and transformer loadings, generator excitation requirements, series and shunt compensation needs, fault clearing and reclosing time requirements, and the performance of interconnected ac/dc systems. These programs were used to demonstrate the superior stability characteristics of the Square Butte Project, which is the first solid-state high-voltage direct-current (HVDC) transmission system in the United States. Square Butte generates electric power at the lignite coal fields in Center, ND, and delivers power by dc over a distance of 456 mi (734 km) to Duluth, MN. The project is rated ± 250 kV dc, 1000 A, 500 MW. The improvement in stability is shown in Fig. 8. Generator rotor angle is shown as a function of time after initiation of a disturbance. The generator recovers after the disturbance with the combined HVDC and ac system and is less stable with the all-ac system. *See* DIRECT-CURRENT TRANSMISSION.

Research studies affecting stability calculations include the determination of load characteristics, the development of system equivalents, and the development of the capability of modeling multiterminal dc systems. Future HVDC applications, particularly in the western part of the United States, will utilize multiterminal concepts.

System behavior under severe conditions. Following a major disturbance, a power system may experience large variations in frequency and voltage and heavy loadings on the transmission system. These abnormal conditions impose stresses on the system and its components which may result in cascading, thereby leading to the formation of electrical islands, loss of generation, and underfrequency load shedding.

The next planning step involves analyzing the system behavior under extremely severe conditions where system split-up and possible shutdown are imminent. Research efforts examining a large number of actual occurrences have identified the predominant processes involved in the long-term dynamics of power systems. Quantitative models of these processes have been developed leading to

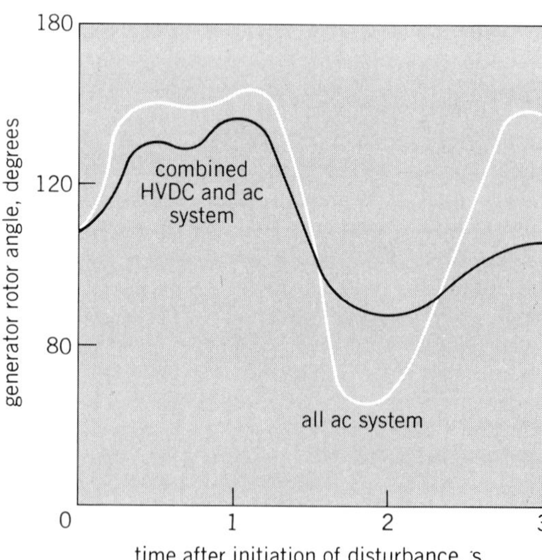

Fig. 8. Stability results for Square Butte Project.

a long-term dynamics computer simulation program. This program has been used to evaluate measures for preventing breakups. Also, simulations of actual disturbances have been used to test and validate the program.

Figure 9 illustrates the elements of the power system represented in the program. The prime mover models include fossil-fired steam, nuclear, hydro, and gas turbine units. The generator and excitation system models determine the voltage and reactive power output and represent the controls in both automatic and manual regulation models. The block that is designated "system connection relations" solves the transmission system power flow with acceleration constraints to determine the system frequency, acceleration, line real and reactive power flows, and bus voltages. Dispatch center activities are modeled, thereby providing signals to change the generation of the controlled prime movers.

This program is a powerful tool for system planners to assess the effect of long-term disturbances.

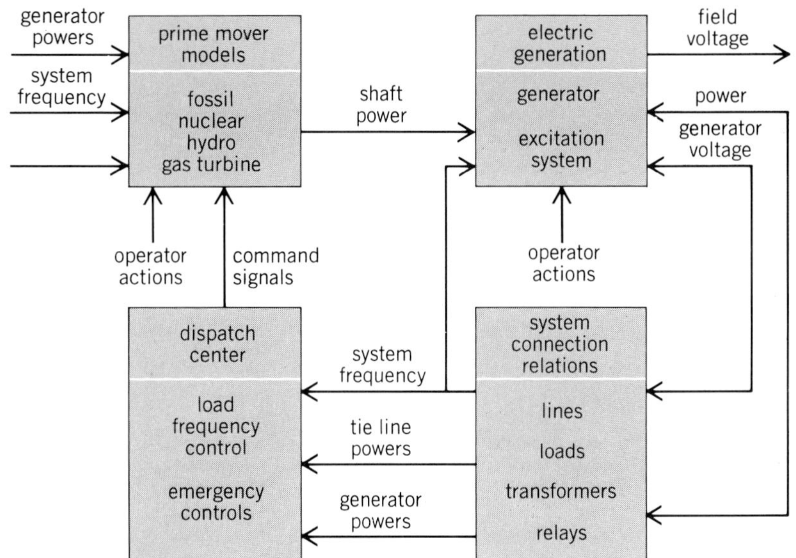

Fig. 9. Elements of long-term dynamics computer simulation program.

SYSTEM DESIGN

Once the transmission system configuration has been selected, the major transmission elements require careful study. The evaluation of switching overvoltages is an important initial step.

Transient network analyzer. The transient network analyzer (TNA) is a special analog computer used to predict the transient and harmonic overvoltages which can occur from switching operations. The TNA has been used to design the vast majority of the extra-high-voltage systems in the United States and Canada. It is a three-phase analog of a transmission system, with each element of the actual system represented by a model counterpart. All components are interconnected at a common connection panel to form the desired system.

TNA studies provide guidance on such design considerations as line insulation levels, lightning arrester duties, apparatus insulation levels, reactor ratings, and locations. They also help define improved system operating procedures.

Parallel to improvements to the analog TNA, developmental work has been devoted to extending the digital simulation of such transients. The digital program has been found particularly helpful in studying the behavior of circuit breakers and other equipment where the analysis of rapidly changing voltages and currents is required. Two outstanding advantages of analysis by digital means are the ease in changing equipment parameters in the study and the ability to model low-loss transmission lines and transformers.

Improved transmission-line models extend the range of the TNA up to 1500-kV systems. Totally electronic shunt reactor models represent linear or nonlinear reactors and are continuously adjustable over a wide range of inductances, saturation levels, saturated inductances, and quality factors. Surge arrester models have been developed to electronically represent present silicon carbide arresters with current-limiting gaps as well as arresters using zinc oxide valve elements. The digital data acquisition system provides a fast means of obtaining statistical distributions of switching surge magnitudes. A statistical presentation of switching surge overvoltages is of primary importance in the probabilistic designs of extra-high-voltage transmission lines, and will be essential when evaluating transmission-line designs at the ultra-high-voltage level. *See* ELECTRIC PROTECTIVE DEVICES; LIGHTNING AND SURGE PROTECTION.

Transmission-line design. With respect to line design, a series of digital programs examine statistical strength/stress criteria of line insulation, including weather influence. Transmission line design programs also examine environmental aspects of line design, including corona phenomena, radio/television noise, audible noise, and ground-level electric fields. These programs enable optimum designs of new lines, or changes needed to uprate existing lines. The analysis of transmission-line performance with such digital programs is considerably enhanced by the use of research results from the Project UHV research facility (Pittsfield, MA). These data cover the area of transient and steady-state insulation, corona performance, and environmental concerns such as ground-level electric fields. Figure 10 illustrates some of the facilities of Project UHV being used to study the behavior of high-voltage dc transmission. *See* TRANSMISSION LINES.

HVDC simulator. The HVDC simulator plays a key role during the design stages of a dc system. It may be interconnected with the TNA. The HVDC simulator is a model HVDC system complete with converter control, dc and ac lines, ac and dc harmonic filters, HVDC circuit breakers, and independent nonsynchronous generation sources. In addition, it has a switching and sequencing system to control disturbances and a measurement system to record system response. This simulator was initially constructed for use in the design stages of the West Coast HVDC Intertie. Since that time, in modified and improved form, it has been extensively used to study a wide range of HVDC system application problems. Improvements have included adding General Electric HVDC system control, additional converter bridges with circuitry facilitating short-pulse firing of silicon controlled rectifiers, low-loss filter and line components, and instrumentation advances. The development of an

Fig. 10. Project UHV dc test set built for the Department of Energy.

electronic machine model representing the direct and quadrature axis machine equations and coupled through an amplifier to the model ac/dc system has further improved system representation capability.

The HVDC simulator is used for such studies as insulation coordination, surge arrester design requirements, ac and dc filter performance, diode/thyristor systems, parallel ac/dc transmission systems, HVDC control techniques, recovery of ac/dc systems from faults, multiterminal dc systems, and HVDC breaker application problems.

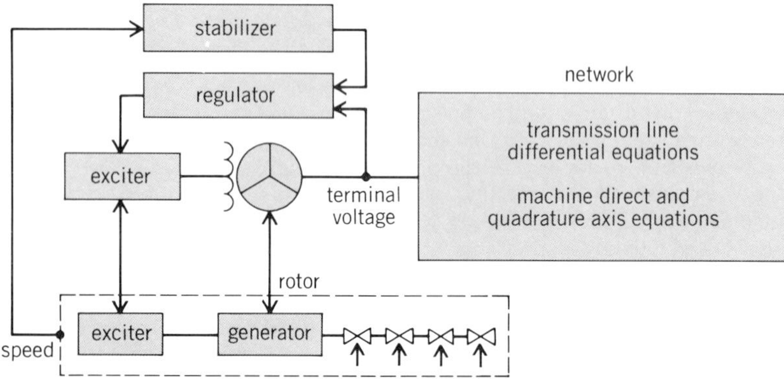

Fig. 11. MANSTAB program.

Hybrid computer. Another useful tool is the hybrid facility. This configuration consists of two general-purpose analog computer consoles and an associated digital computer for monitoring and controlling the analog simulation.

This analytical tool is extensively used for control and dynamical studies, and provides fast turnaround time. The hybrid computer is very cost-effective for the analysis of system dynamics, particularly when high-frequency components are of interest. Thus, the hybrid computer is used extensively for studies of systems involving converters and other thyristor-controlled components in which current and voltage waveshapes are important. Examples include studies of excitation systems, static reactive power systems, and HVDC controls.

Machine and network stability. A program, entitled MANSTAB (machine and network stability), makes it possible to study in one simulation many complex interactions over an extended frequency range which occur between turbine generators and the transmission system. As shown in Fig. 11, the MANSTAB program models balanced transmission networks by differential equations and represents the torsional dynamics of the turbine generator as well. In addition to time domain results, MANSTAB generates root locus and frequency response characteristics. A fundamental aspect of this approach is a technique of linearizing a system about its operating point and constructing the matrix equations of the linearized system directly from the nonlinear time-simulation model. The state matrix equations are then used to obtain eigenvalues and eigenvectors of the system, and frequency response of selected transfer functions. This program has been particularly helpful in the study of subsynchronous resonance.

Subsynchronous resonance. Subsynchronous resonance between the transmission system electrical natural frequency and a shaft torsional natural frequency may occur in systems with series-compensated lines, and has resulted in cracked turbine generator shafts. This phenomenon can be analyzed with the MANSTAB program by representing the turbine generator as a mechanical spring-mass system connected through a transformer to a transmission line which has series compensation. Because of the presence of series compacitors, the electrical system has a natural frequency generally in the range of 15–40 Hz. The turbine generator also has torsional natural frequencies in this range. Electrical transients normally decay away in about half a second, but while they are doing so they produce mechanical torques in the shaft. When these torques are on, or close to, one of the natural shaft torsional frequencies, the stresses produced in the shaft can be enormous. Furthermore, the torsional oscillation of the shaft, in response to these torques, sets up voltages in the generator which act to reinforce the electrical disturbances. As a consequence, torsional instability may result.

Analysis of this phenomenon with the MANSTAB program led to the development of a new concept of protection utilizing a series-blocking power filter between the generator and the power transmission system. This subsynchronous filter blocks the electrical frequencies that correspond to the mechanical torsional resonances. The first subsynchronous filter was installed in 1976 at the

Navajo Project (Page, AZ), and field tests have confirmed the validity of the digital simulation.

MANTRAP. For problems requiring representation of the network as separate phases, the program MANTRAP (machine and network transient program) may be applied to provide a time simulation. Examples of problems to be studied by MANTRAP include unbalanced faults, multiple faults, reclosing, resynchronizing, generator load rejection, and generator out-of-step operation.

SYSTEM OPERATION

Analytical techniques and automation technology are applied to improve the system operation of electric power system facilities. Day-to-day operation and control entails utilizing these installed facilities to meet customer load requirements at maximum reliability and minimum costs within constraints such as equipment limitations, maintenance needs, and interconnection agreements.

Applying the system planning techniques described previously has led to the interconnection of individual power systems into large grids, thereby obtaining economies in capital and operating expenses as well as improved reliability. Fully exploiting these benefits presents ever more complex problems to the power system operator.

It is helpful to think of the system operation problems in terms of operations planning, operations control, and operations accounting and review as shown in Fig. 12. These problems involve economic and security aspects as well as environmental factors. Operations planning involves the problems of looking ahead to the next hour, day, week, or month. Examples are daily load forecasting and unit commitment scheduling. Operations control is concerned with second-by-second real-time actions such as frequency control. Operations accounting and review is concerned with after-the-fact evaluations such as interconnection billing and postdisturbance analysis.

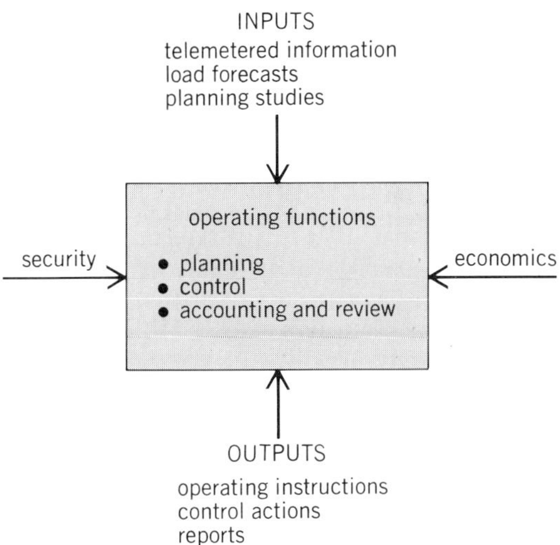

Fig. 12. Operating functions.

Automation technology. There has been a rapid evolution of automation technology in assisting the operator in solving these problems. The operator's first tool was the telephone involving manual instructions. Then came remote supervisory systems.

Following this, analog dispatch computers were applied which allocated and controlled generation to minimize total system fuel costs and held system frequency and interchange at desired values. These functions are called economic dispatch and load frequency control. With the development of process-control digital computers, digital computers were next used to drive analog circuits to provide the operations control functions of load frequency control and economic dispatch. This same computer was also used in a shared-time manner to undertake problems of operations planning, and accounting and review as well. As digital process-control computers increased in capability, the analog regulating functions were taken over by the digital computer. Next the digital computer assumed control of the supervisory systems as well. *See* DIGITAL COMPUTER.

Early justification of computers in the dispatch center was primarily based on improvements in fuel economy, whereas emphasis now primarily involves security of system operation. However, with rapidly escalating costs of fuel, the benefits of economic dispatching are greatly increasing.

System security. In thinking about power system security, it is helpful to consider the system to be in one of the four states shown in Fig. 13. In the normal state, all customer demands are met, no apparatus or lines are overloaded, and there are no impending emergencies. The objective in this state is to continue to meet customer demands, to operate the apparatus and lines within limits, to operate the system at minimum costs within the constraints, and to minimize the effects of possible future contingent events.

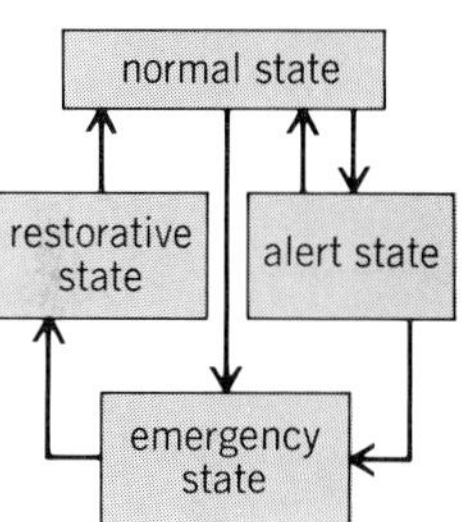

Fig. 13. System states.

The alert state is similar to the normal state in that all customer demands are met and no apparatus or lines are overloaded. However, a potential emergency has been detected by noting that a line loading has reached a limit set below its rating or that an assumed contingency would result in a transmission overload. The objective is to impose constraints and return to the normal state in a minimum time.

There will, however, be situations in which the system will reach the emergency state where customer demands are not met or apparatus is overloaded. In this case, the objective is to prevent the spread of the emergency. In the restorative state, the emergency has been stabilized; yet customer demands may not be served or apparatus may be overloaded. The objective here is to return to normal in minimum time.

The normal state is the only secure state since it implies a low risk of failure. Steps leading to improved system security have the goals of minimizing the probability of leaving the normal state, and minimizing the time required to return to the normal state once the system is in some other state. It is apparent that the first goal requires that steps be taken while the system is in the normal state.

A step-by-step computer application approach has been developed involving: (1) status monitor and display; (2) contingency evaluation; (3) corrective strategy; and (4) automatic control. Significant improvement in system security is obtained through increased monitoring of information and more meaningful display of system information. In the second step, the computer is called upon to predict the effect of contingencies and planned outages and to alert the operator to potential problems in the system. Here the computer is used primarily in a predictive mode rather than in the mode of reacting after a disturbance. As a third step, the computer formulates corrective strategies. At this stage, the operator calls for the execution of the strategy he or she chooses. In the fourth step, the computer uses the communication network to execute automatically the computer-formulated strategies.

Hierarchy of control centers. A hierarchy of action centers as shown in Fig. 14 exists within a typical power company. Overall responsibility for the bulk generation-transmission system resides with a company dispatching center. This center directs the second-by-second control of generation to maintain tie-line schedules and frequency, and is responsible for minimizing the cost of power generation within constraints imposed by security.

Operation of the substation and transmission facilities is accomplished within transmission divisions. Unstaffed substations are monitored and controlled at a division dispatching office by means of supervisory control equipment. Staffed stations may or may not have telemetry equipment installed to enable automatic monitoring by the division.

Generating plants, with the exception of remote hydro or gas-turbine installations, are staffed with operators and maintenance personnel. The operation of large thermal plants involves the local control of many plant variables, and the high degree of automation achieved has significantly contributed to their reliability and economy.

Viewed from the company dispatch center, these plants provide a point at which generation is controlled as well as a source of data on plant status.

As interconnections among utilities become stronger, power pools have been formed in which the facilities of individual member companies are scheduled and operated in a manner which minimizes the total cost of the pool and improves its reliability. The pool is responsible for improving overall economics, and has direct control over total generation for the purpose of maintaining net pool tie flow and frequency on schedule. For this reason, the general responsibilities and facilities of a power pool center are much like those of a company dispatch center, the difference being one of degree.

More recently, the electric utility industry has organized regional coordination centers which have responsibilities over very large geographic areas. At present, these centers are largely concerned with studies of future operating conditions, coordinating emergency procedures, and analyzing disturbances.

The four-step plan previously described is being implemented with data collection, communication devices, and computers organized in a multilevel structure such as the hierarchy of Fig. 14.

Local levels are associated with local control and the collection of primary data. Higher levels receive data from levels below, process it, and act as a data source for levels above. Reliability is enhanced by placing direct control at the lowest feasible level. Such a multilevel computer configuration is designed to supplement automatic control already in place such as normal protective relays, load shedding relays, and excitation controls. Computer technology is becoming of increasing importance in improving system operations ranging from microcomputers at the local primary levels to large-scale scientific computers at the company, pool, and regional levels. *See* ELECTRIC POWER SYSTEMS. [L. K. KIRCHMAYER]

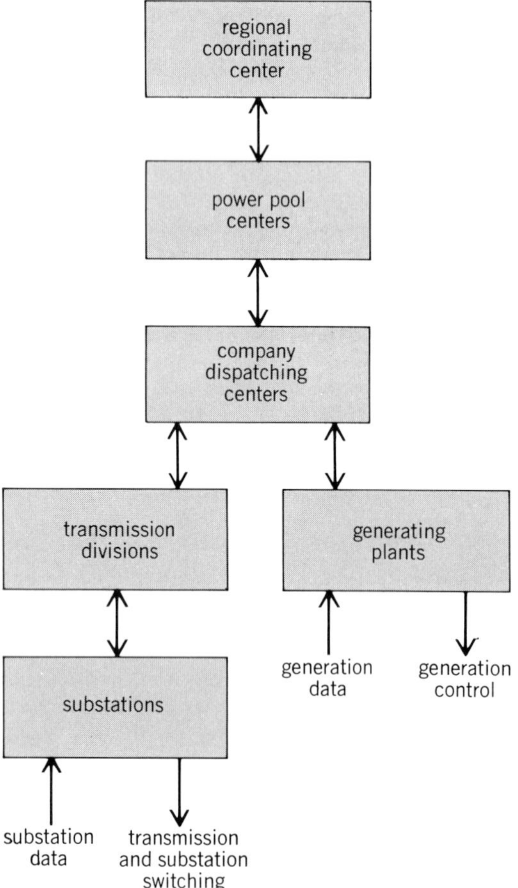

Fig. 14. Hierarchy of action centers.

Electric protective devices

A particular type of equipment applied to electric power systems to detect abnormal and intolerable conditions and to initiate appropriate corrective action. From time to time, disturbances in the normal operation of electric power systems occur. These may be caused by natural phenomena, such as lightning, wind, or snow; by accidental means traceable to reckless drivers, inadvertent acts by plant maintenance personnel, or other acts of human beings; or by conditions produced in the system itself, such as switching surges, load swings, or equipment failure. Protective devices must therefore be installed on a power system to ensure continuity of electrical service, to limit injury to personnel, and to limit damage to equipment when abnormal situations develop.

Protective devices, like any type of insurance, are applied commensurately with the degree of protection desired. For this reason, application of protective devices varies widely.

Zone protection. For the purpose of applying protection, the electric power system is divided into five major protection zones: generators; transformers; buses; transmission and distribution lines; and motors (Fig. 1). Each block represents a set of protective relays and associated equipment selected to initiate correction or isolation of that area for all anticipated intolerable conditions or trouble. The detection is done by protective relays with a circuit breaker used to physically disconnect the equipment. For other areas of protection *see* GROUNDING; LIGHTNING AND SURGE PROTECTION.

Protective relays. These are compact analog or digital networks connected throughout the system to detect intolerable conditions within their assigned area or zone. They operate on voltage, current, current direction, power factor, power, impedance, temperature, and so forth, as well as combinations of these. In all cases there must be a measurable difference between the normal or tolerable operation and the intolerable or unwanted condition. System faults for which the relays respond are generally short circuits between the phase conductors, or between the phases and grounds. Some relays operate on unbalances between the phases, such as an open or reversed phase. A fault in one part of the system affects all other parts. Therefore relays throughout the power system must be coordinated to ensure the best quality of service to the loads and to avoid operation in the affected areas unless the trouble is adequately cleared in a specified time.

The most fundamental and widely used protection technique is the differential principle (Fig. 2). The current flowing into the equipment to be protected is compared with the current flowing out. Two circuits are shown, but there may be many, as in a bus section. For normal and permissible operation, currents all sum to essentially zero (Fig. 2*a*). However, for internal trouble, they add up to flow through the relay (Fig. 2*b*). The equipment to be protected can be a generator, transformer, bus, motor, or line. In the last case, with many miles between the circuit breakers, telephone-type low-energy pilot wires, radio frequency on the power line, or a microwave channel is used. The quantities at the line terminals are thus differentially compared via the channel, in what is known as pilot relaying.

While the application and protection principles are the same, the relay units may be either electromechanical or solid-stage type. Originally, all relays were electromechanical, but in the late 1960s solid-state components began to be used. Both types are now widely used.

Electromechanical relays. Relays of this type can be illustrated by two common basic units: the plunger and the induction disk. Both of these are used in a number of relay designs. The plunger type (Fig. 3) operates on the electromagnetic attraction principle. This relay is composed of a coil, plunger, and set of contacts. When current I flows in the coil, a force is produced that causes the plunger to move and close the relay contacts. These relays are characterized by their fast operating time.

The electromagnetic induction-disk relay responds to alternating current only, whereas the relays discussed above respond to either direct or alternating current. Briefly, an induction relay consists of an electromagnetic circuit, a disk or other form of rotor made of a nonmagnetic current-carrying material, and contacts. A schematic of an induction-type relay is shown in Fig. 4. The main coil is connected to an external source. When current flows in the main coil, transformer action induces current in the secondary circuit connected to the upper poles. Fluxes produced by the currents flowing in the upper pole circuit induce eddy currents in the rotor disk. Interaction between rotor eddy currents and the flux from the lower pole produces torque on the rotor, causing it to move and thus closing the contacts. This will be recognized as the split-phase motor principle, where two out-of-phase fluxes produce torque in a rotor. The

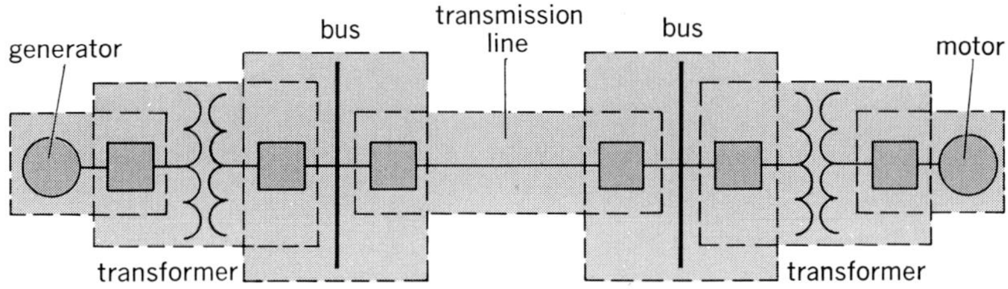

Fig. 1. Zones of protection on simple power system.

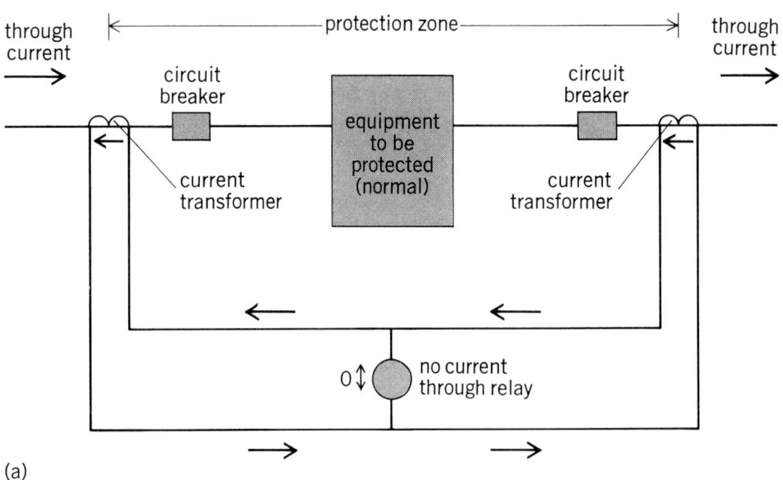

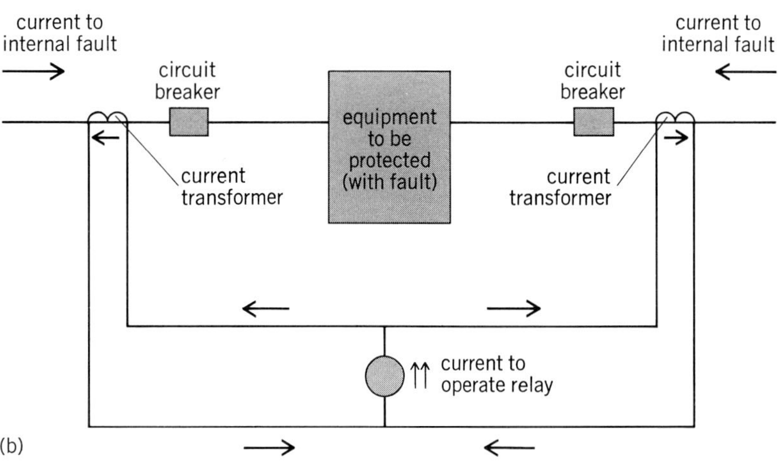

Fig. 2. Simple differential protection scheme. The system compares currents to detect faults in the protection zone. Current transformers reduce the primary current to a small secondary value. (*a*) Normal operation. (*b*) With fault.

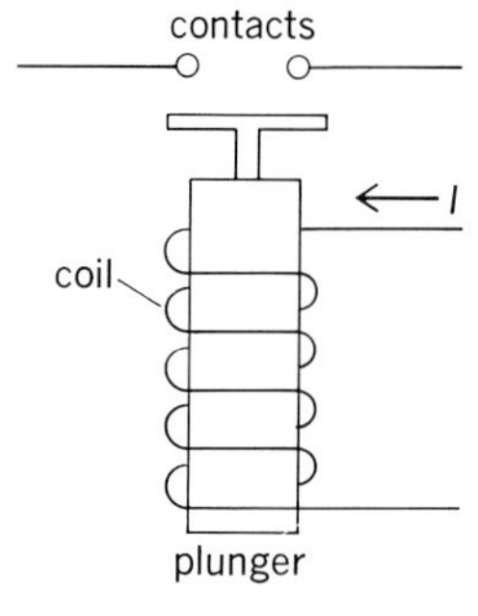

Fig. 3. Plunger relay.

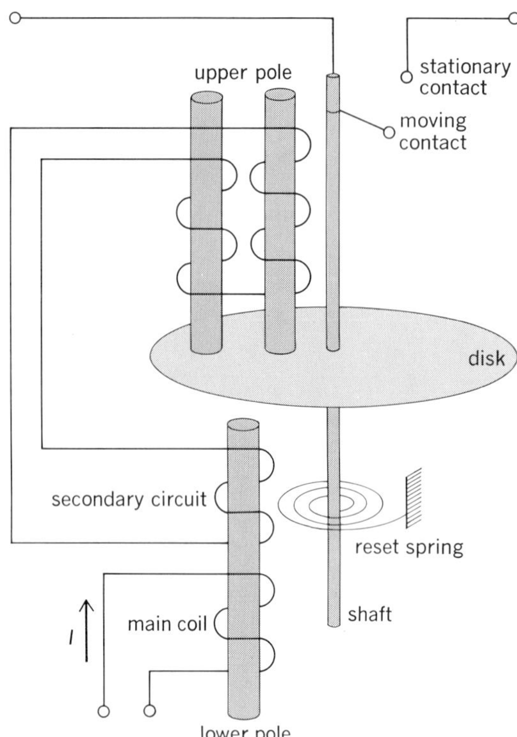

Fig. 4. Induction relay.

upper pole may be supplied from another source to permit comparison of two quantities. A spring automatically resets the disk after the relay has operated. By use of the principles of electromagnetic attraction and electromagnetic induction, protective relays can be built to respond to all abnormal conditions that may occur. *See* RELAY.

Solid-state relays. Considerable care must be taken with solid-state relays, also known as static relays, as the solid-state components must be able to correctly detect the trouble, yet must not be damaged during normal system operation and disturbances. Both conditions can produce severe transient overvoltages. Discrete components are used such as transistors and diodes. An example of this type of relay is shown in Fig. 5. This logic circuit is a dc-level detector used as an instantaneous overcurrent unit. It is functionally equivalent to Fig. 3. The ac input current is transformed into a current-derived voltage by the input transformer. This voltage is limited by Zener clipper Z_1 and R_2. At low input currents, the voltage is proportional to the current and determined by R_1 and R_3. Minimum operating current is adjusted by R_1; a low R_1 diverts more current through R_1 and R_3 and less to the phase splitter. The phase splitter has a resistor-capacitor network, a transformer, and a bridge rectifier. Its output voltage with time is shown as a ripple. When this voltage equals the Zener voltage of Z_2, Z_2 will conduct to provide base current to transistor Q_1. Q_1 turns on transistor Q_2 to provide an output current through D_2 and R_9. Q_2 provides positive feedback through R_7 and D_1 to provide a constant output and snap action. The dropout is adjusted by R_7.

Integrated circuits functioning as operational amplifiers are used in later designs and may replace the discrete circuit designs.

Computer relays. There has been considerable study and experimental work in the use of computers in relaying, but there have been almost no practical economic applications. There is a very narrow time window for a correct decision during a fault or trouble in which a considerable transient may occur. Typical high-speed electromechanical and solid-state relays operate within 4–18 ms. Since trouble in the power system can occur at any moment, relay systems require a very high degree of availability. These factors, combined with economics, make computer relaying difficult.

Overcurrent protection. This must be provided on all systems to prevent abnormally high currents from overheating and causing mechanical stress on equipment. Overcurrent in a power system usually indicates that current is being diverted from its normal path by a short circuit. In low-voltage, distribution-type circuits, such as those found in homes, adequate overcurrent protection can be provided by fuses that melt when current exceeds a predetermined value. *See* FUSE (ELECTRICITY).

Small thermal-type circuit breakers also provide overcurrent protection for this class of circuit. As the size of circuits and systems increases, the problems associated with interruption of

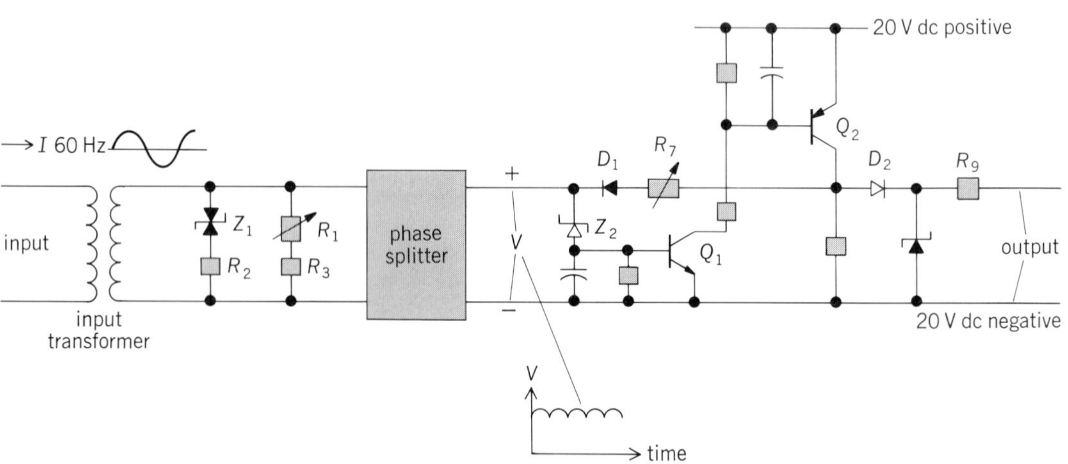

Fig. 5. Solid-state instantaneous overcurrent unit, using magnitude-comparison dc-level detector.

large fault currents dictate the use of power circuit breakers. Normally these breakers are not equipped with elements to sense fault conditions, and therefore overcurrent relays are applied to measure the current continuously. When the current has reached a predetermined value, the relay contacts close. This actuates the trip circuit of a particular breaker, causing it to open and thus isolating the fault. *See* CIRCUIT BREAKER.

Induction-disk or plunger relays can be used to detect overcurrent conditions. As the current in either type of relay increases, the resultant force also increases. When sufficient force is available, the relay contacts close. Induction-disk relays have an inverse time-current characteristic; that is, a longer time is required to close the contacts on a slight overcurrent. A shorter time is required to close the contacts on a heavy overcurrent. Another type of relay used in high-voltage lines operates on the combination of reduced voltage and increased current occasioned by a fault. This is a distance relay.

Overvoltage protection. Lightning in the area near the power lines can cause very-short-time overvoltages in the system and possible breakdown of the insulation. Protection for these surges consists of lightning arresters connected between the lines and ground. Normally the insulation through these arresters prevents current flow, but they momentarily pass current during the high-voltage transient to limit overvoltage. Overvoltage protection is seldom applied elsewhere except at the generators, where it is part of the voltage regulator and control system. In the distribution system, overvoltage relays are used to control taps of tap-changing transformers or to switch shunt capacitors on and off the circuits.

Undervoltage protection. This must be provided on circuits supplying power to motor loads. Low-voltage conditions cause motors to draw excessive currents, which can damage the motors. If a low-voltage condition develops while the motor is running, the relay senses this condition and removes the motor from service.

Undervoltage relays can also be used effectively prior to starting large induction or synchronous motors. These types of motors will not reach their rated speeds if started under a low-voltage condition. Relays can be applied to measure terminal voltage, and if it is below a predetermined value, the relay prevents starting of the motor.

Underfrequency protection. A loss or deficiency in the generation supply, the transmission lines, or other components of the system, resulting primarily from faults, can leave the system with an excess of load. Solid-state and digital-type underfrequency relays are connected at various points in the system to detect this resulting decline in the normal system frequency. They operate to disconnect loads or to separate the system into areas so that the available generation equals the load until a balance is reestablished.

Reverse-current protection. This is provided when a change in the normal direction of current indicates an abnormal condition in the system. In an ac circuit, reverse current implies a phase shift of the current of nearly 180° from normal. This is actually a change in direction of power flow and can be directed by ac directional relays.

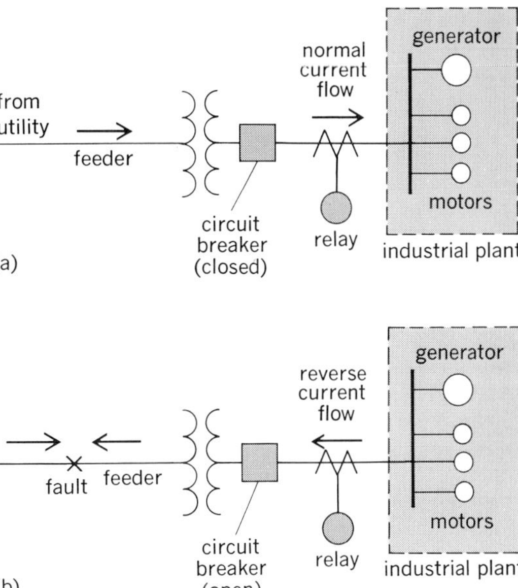

Fig. 6. Reverse-current protection. (*a*) Normal load conditions. (*b*) Internal fault condition; relay trips circuit breaker under reverse-current condition.

A common application of reverse-current protection is shown in Fig. 6. In this example, a utility supplies power to an industrial plant having some generation of its own. Under normal conditions, current flows from the utility to the plant (Fig. 6*a*). In the event of a fault occurring on the utility feeder (Fig. 6*b*) the current reverses direction and flows from the plant to the fault location. The relay operates and trips the circuit breaker, isolating the plant from the utility, thus preventing an excessive burden on the plant generator. Usually in these cases, the plant generator cannot carry the plant load, so that underfrequency relays are used to shed noncritical load. When the utility tie is restored, the shed loads then can be reconnected for full plant service.

Phase unbalance protection. This protection is used on feeders supplying motors where there is a possibility of one phase opening as a result of a fuse failure or a connector failure. One type of relay compares the current in one phase against the currents in the other phases. When the unbalance becomes too great, the relay operates. Another type monitors the three-phase bus voltages for unbalance. Reverse phases will operate this relay.

Reverse-phase-rotation protection. Where direction of rotation is important, electric motors must be protected against phase reversal. A reverse-phase-rotation relay is applied to sense the phase rotation. This relay is a miniature three-phase motor with the same desired direction of rotation as the motor it is protecting. If the direction of rotation is correct, the relay will let the motor start. If incorrect, the sensing relay will prevent the motor starter from operating.

Thermal protection. Motors and generators are particularly subject to overheating due to overloading and mechanical friction. Excessive temperatures lead to deterioration of insulation and increased losses within the machine. Temperature-sensitive elements, located inside the machine, form part of a bridge circuit used to supply current to a relay. When a predetermined temperature is reached, the relay operates, initiating opening of a circuit breaker or sounding of an alarm. [J. LEWIS BLACKBURN]

Bibliography: C. R. Mason, *The Art and Science of Protective Relaying*, 1956; Westinghouse Electric Corp., *Applied Protective Relaying*, 1979; Westinghouse Electric Corp., *Electrical Transmission and Distribution Reference Book*, 4th ed., 1950.

Electric rotating machinery

Any form of apparatus, having a rotating member, which generates, converts, transforms, or modifies electric power. The most common forms are motors, generators, synchronous condensers, synchronous converters, rotating amplifiers, phase modifiers, and combinations of these in one machine. The capacity, or rating, is usually indicated on a nameplate and denotes the maximum continuous duty which can be sustained without overheating or other injury. Motors are rated in horsepower. They are built in sizes from a small fraction of a horsepower to more than 230,000 hp. Generators are rated in kilowatts or kilovolt-amperes (kVA). The maximum output of alternating-current generators exceeds 1,250,000 kVA. Other types of rotating machines fall within these limits.

Construction. Most rotating machines consist of a stationary member, called the stator, and a rotating member, called the rotor. The rotor may be supported in bearings at both ends, or it may be supported at one or both ends by the shaft of another machine. *See* GENERATOR; MOTOR.

The illustration shows a typical rotating machine having a bracket bearing at one end and an arrangement for coupling to a turbine shaft at the other. Although small machines sometimes employ antifriction bearings, larger units are built with sleeve bearings generally lined with babbitt. Vertical shaft machines use thrust bearings to support the rotating member. Lubrication in slow- or medium-speed units is often supplied from an oil reservoir contained within the bearing housing. Where bearing losses are high, water-cooling coils may be immersed in the oil to prevent overheating. High-speed machines are often lubricated from a pressurized oiling system, which also supplies the shaft seals in hydrogen-cooled units.

To function properly, rotating machines must have a magnetic circuit, usually involving both rotor and stator, and one or more insulated electrical circuits which interlink the magnetic circuit. To afford a low-reluctance magnetic path, the rotor and stator are separated only by a small clearance, called the air gap.

The windings are insulated electrically with materials such as enamel, cotton, varnished cambric, mica, asbestos, dacron, and glass fabric. The most common impregnants are shellac, asphaltum base varnish, and epoxy, polyester, or phenolic resins. External partially conducting varnish is sometimes applied to high-voltage coils for corona shielding.

Electrical-mechanical energy conversion. The force F in newtons produced on a conductor located at right angles to a magnetic field is $F = BIL$ newtons, where B is the flux density in webers per square meter in the vicinity of the conductor, I is the conductor current in amperes, and L is the length of the conductor, in meters, exposed to the flux. In a motor the magnetic field created by one member exerts a force on the current-carrying conductors of the other, producing a mechanical torque which drives the load. In a generator the changing magnetic field induces voltage in the armature windings when the rotor is driven by a source of mechanical power. Little power is required at no-load, but as the load current builds up, the prime mover must supply the torque to overcome the forces in the equation between the field and conductors.

Ventilation. Rotating machinery must be ventilated to avoid overheating from internal losses. The principal cooling medium, usually air or hydrogen, is circulated by fans or blowers mounted on the rotor or separately driven. The illustration shows axial-flow fans at each end of the rotor, with arrows indicating the path taken by the gas. With conventional cooling, the cooling medium is blown over

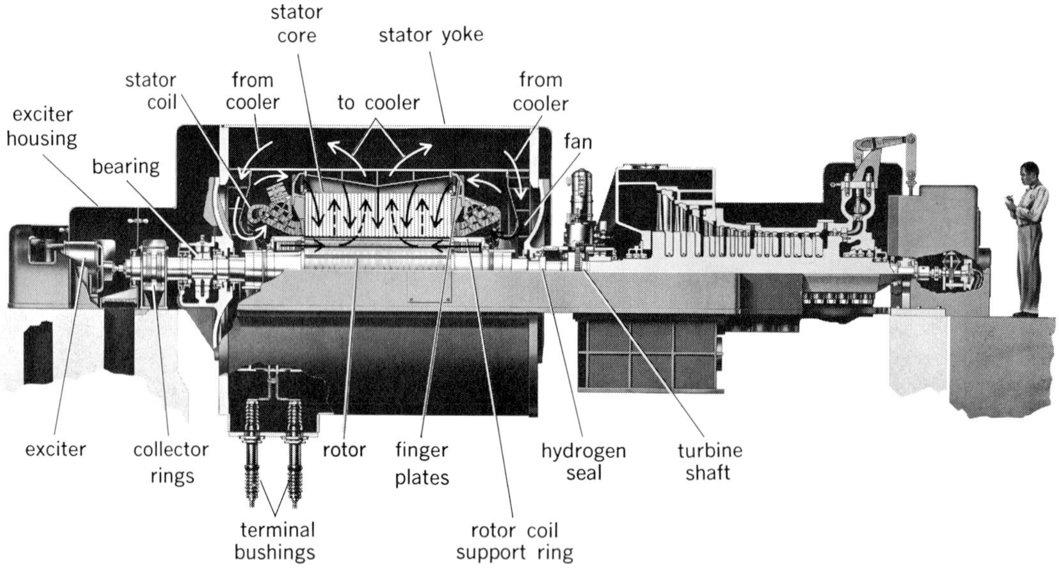

Cross section of a typical electric rotating machine. (*Allis-Chalmers*)

exposed surfaces of the insulated windings and core. In conductor cooling, the cooling medium flows in ducts within the major insulation wall.

In large machines the superior effectiveness of conductor cooling is essential. In addition to hydrogen at pressures up to several atmospheres for cooling the rotor and the stator core iron, hydrogen at far greater pressure or a liquid such as oil or water is circulated through the stator conductors in the largest ratings. Generators having liquid-cooled rotor conductors are also being built. *See* ALTERNATING-CURRENT GENERATOR.

Losses. In all rotating machines, losses occur. Among them are I^2R losses, called copper losses, in the windings, connections, and brushes; stray load losses in windings, solid metal structures, and frame; core loss in the magnetic material and structural parts; windage and friction loss; and exciter and rheostat losses.

I^2R losses (in watts) in each path of the windings are equal to the square of the effective current in amperes times the resistance in ohms. Brush I^2R loss is the product of the potential drop in volts times the current in amperes. Stray load losses are caused mainly by eddy currents, due to variable magnetic fields (produced by the load current) within the conductors, pole surface, structural members, end shields, frame, and so forth.

Windage and friction losses are the result of circulation and turbulence of the cooling medium and friction of bearings, seals, and brushes. Windage loss is relatively large in air-cooled high-speed machines. In hydrogen the loss is only 7–15% of that in air within the operating range of purity. Bearing and seal friction losses are generally absorbed by the lubricating oil. To avoid excessive friction or overheating of bearings, an inlet oil temperature of 100–120°F is often recommended for large machines, with discharge at about 150°F.

[LEON T. ROSENBERG]

Bibliography: D. G. Fink and H. W. Beaty (eds.), *Standard Handbook for Electrical Engineers*, 11th ed., 1978; E. Levi and M. Panzer, *Electromechanical Power Conversion*, 1974; J. Rosenblatt and M. H. Friedman, *Direct and Alternating Current Machinery*, 1963; S. Seely, *Electromechanical Energy Conversion*, 2d ed., 1973.

Electric switch

A device that makes, breaks, or changes the course of an electric circuit. Basically, an electric switch consists of two or more contacts mounted on an insulating structure and arranged so that they can be moved into and out of contact with each other by a suitable operating mechanism.

The term switch is usually used to denote only those devices intended to function when the circuit is energized or deenergized under normal manual operating conditions; as contrasted with circuit breakers, which have as one of their primary functions the interruption of short-circuit currents. Although there are hundreds of types of electric switches, their application can be broadly classified into two major categories: power and signal.

Power switches. In power applications, switches function to energize or deenergize an electric load. On the low end of the power scale, wall switches are used in homes and offices for turning lights on and off; dial and push-button switches control power to electric ranges, washing machines, and dishwashers. On the high end of the scale are load-break switches and disconnecting switches in power systems at the highest voltages (several hundred thousand volts).

For power applications, when closed, switches are required to carry a certain amount of continuous current without overheating, and in the open position they must provide enough insulation to isolate the circuit electrically. The latter function is particularly important in high-voltage circuits because it is the practice in the electrical industry to forbid people from working on electrical equipment unless it is isolated from the electrical supply system by a visible break in air. As an added precaution, a grounding switch is often used to connect the equipment to the ground before permitting any direct contact by a worker.

Load-break switches are required also to have the capability of interrupting the load current. Although this requirement is easily met in low-voltage and low-current applications, for high-voltage and high-current circuits, arc interrupters, similar to those used in circuit breakers are needed. *See* CIRCUIT BREAKER.

In medium-voltage applications the most popular interrupter is the air magnetic type, in which the arc is driven into an arc chute by the magnetic field produced by the load current in a blowout coil. The chute may be made either of ceramic materials or of organic materials such as plexiglass. For voltages higher than 15,000 volts, gaseous interrupters are sometimes used. Special gases such as sulfur hexafluoride (SF_6) have been found to have good interrupting capability for load currents without a high-speed flow. For short-circuit currents, even with SF_6, a high-speed flow is needed. More recently, vacuum interrupters have been produced with high-voltage load-breaking duties. *See* BLOWOUT COIL.

Some load-break switches may also be required to have the capability of holding the contacts in the closed position during short-circuit conditions so that the contacts will not be blown open by electromagnetic forces when the circuit breaker in the system interrupts the short-circuit current. For most contact configurations, the interaction between the short-circuit current and the magnetic field tends to force the contacts to part. Special configurations such as finger contacts are sometimes used to overcome this problem (Fig. 1).

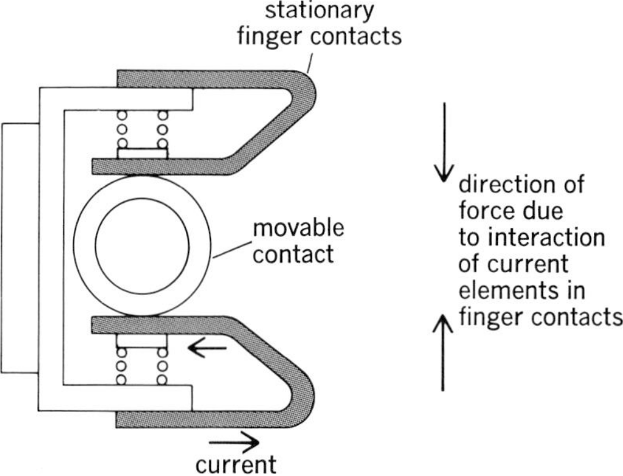

Fig. 1. With paired fingers for fixed contacts, short-circuit forces tend to hold contacts closed.

Signal switches. For signal applications, switches are used to detect a specified situation that calls for some predetermined action in the electrical circuit. For example, thermostats detect temperature; when a certain limit is reached, contacts in the thermostat energize or deenergize another electrical switching device to control power flow. Centrifugal switches prevent overspeeds of motors. Limit switches prevent cranes or elevators from moving beyond preset positions. Many different types of switches perform such signaling purposes in communication systems, computers, and control systems for industrial processes.

Switches for signaling purposes are often required to have long life, high speed, and high reliability. Contaminants and dust must be prevented from interfering with the operation of the switch. For this purpose, switches are usually enclosed and are sometimes hermetically sealed.

Among the many different arrangements for contacts, the knife switch (Fig. 2), because of its early widespread use, is the basis for the elemental switch symbol. In the knife switch, a metal blade is hinged to a stationary jaw at one end and contacts a similar jaw at the other end.

In the leaf-spring switch, parallel strips of spring metal are sandwiched between blocks of insulation at one end and pushed into or out of contact at the other end. The sliding contact switch takes the form of a dial or drum with metal segments engaging contact fingers that wipe over the dial or drum surface. A butt-contact switch (Fig. 3) consists of one movable contact in the form of a disk or bar and one or two stationary contacts of similar form. In the mercury switch, a pool of mercury is made either to bridge the contacts or to separate from them (Fig. 4).

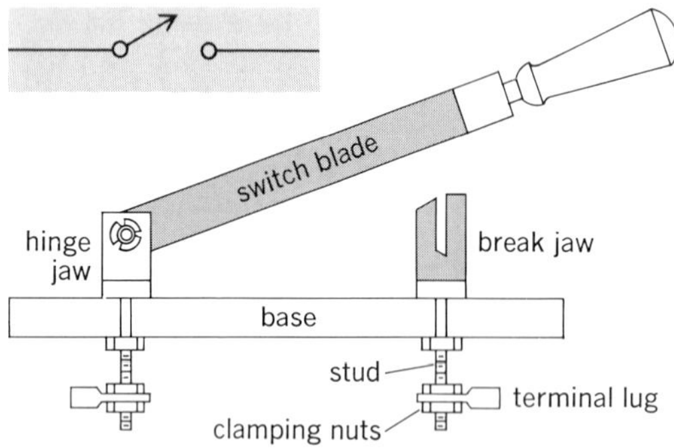

Fig. 2. Early knife switch with schematic symbol of single switch shown at top.

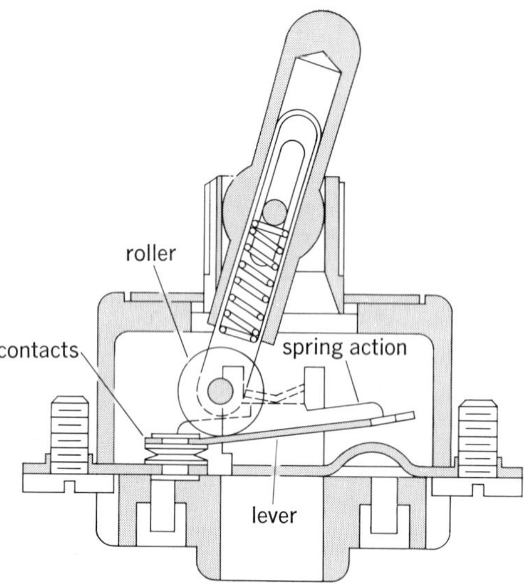

Fig. 3. Toggle mechanism actuates butt-contact switch.

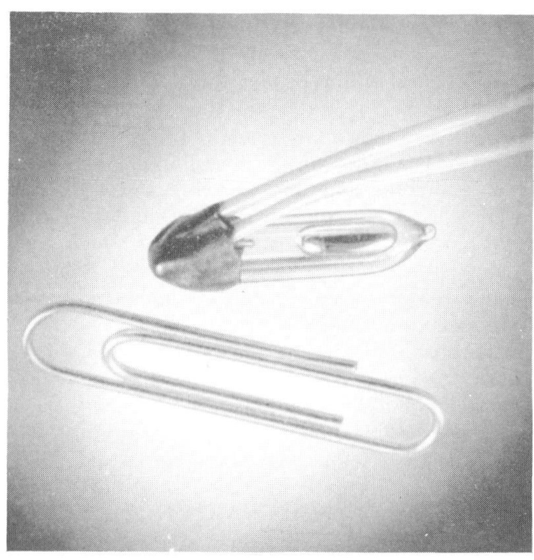

Fig. 4. Miniature mercury element may be used with a variety of tilting mechanisms to cause switch action. (*Minneapolis-Honeywell Regulator Co.*)

Switches frequently are composed of many single circuit elements, known as poles, all operated simultaneously or in a predetermined sequence by the same mechanism. Switches used in complex machines, such as computers, may have a large number of poles. Switches used in power circuits usually have from one to four poles, depending on the kind of circuit. Switches are often typed by the number of poles and referred to as single-pole or double-pole switches, and so on. It is also common to express the number of possible switch positions per pole, such as a single-throw or double-throw switch. [THOMAS H. LEE]

Bibliography: J. M. Carroll and H. W. Beaty (eds.), *Standard Handbook for Electrical Engineers*, 11th ed., 1978.

Electric uninterruptible power system

A system that provides protection against commercial power failure and variations in voltage and frequency. Uninterruptible power systems (UPS) have a wide variety of applications where unpredictable changes in commercial power will adversely affect equipment. This equipment may include computer installations, telephone exchanges, communications networks, motor and sequencing controls, electronic cash registers, hospital intensive care units, and a host of others. The uninterruptible power system may be used on-line between the commercial power and the sensitive load to provide transient free well-regulated power, or off-line and switched in only when commercial power fails.

Types of system. There are three basic types of uninterruptible power system. These are, in order of complexity, the rotary power source, the standby power source, and the solid-state uninterruptible power system.

Rotary power source. The rotary power source consists of a battery-driven dc motor that is mechanically connected to an ac generator. The battery is kept in a charged state by a battery charger that is connected to the commercial power line. In the event of a commercial power failure, the battery powers the dc motor which mechanically drives the ac generator. The sensitive load draws its power from the ac generator and operates through the outage.

Standby power source. The standby power source (Fig. 1) consists of a battery connected to a dc-to-ac static inverter. The inverter provides ac power for the sensitive load through a switch. A battery charger, once again, keeps the battery on full charge. Normally, the load operates directly from the commercial power line. In the event of commercial power failure, the switch transfers the sensitive load to the output of the inverter.

Fig. 1. Standby power source. (*Topaz Electronics*)

Solid-state system. The solid-state uninterruptible power system has a general configuration much like that of the standby power system with one important exception. The sensitive load operates continually from the output of the static inverter. This means that all variations on the commercial power lines are cleaned and regulated through the output of the uninterruptible power system. A commercial power line, known as a bypass, is provided around the uninterruptible power system through a switch. Should the uninterruptible power system fail at some point, the commercial power is automatically transferred to the sensitive load through the switch. This scheme is known as an on-line automatic reverse-transfer uninterruptible power system (Fig. 2).

Subsystems. An uninterruptible power system consists of four major subsystems (Fig. 3): a method to put energy into a storage system, a battery charger; an energy storage system, the battery; a system to convert the stored energy into a usable form, the static inverter; and a circuit that electrically connects the sensitive load to either the output of the uninterruptible power system or to the commercial power line, the transfer switch.

Fig. 2. A 250-kW on-line automatic reverse-transfer uninterruptible power system. (*Exide Electronics*)

Battery charger. The purpose of the battery charger, also called the rectifier charger, is to deliver the power required to drive the static inverter and to charge the battery. When the commercial power line is present, the charger circuit supplies all the power used by the inverter and maintains the battery at full charge. Additionally the charger is sized to allow a safety margin for overload conditions at the output of the static inverter. Uninterruptible power systems that have a very long battery charge time require a charger that is only slightly larger than that of the inverter. In an uninterruptible power system where a short recharge time is required, the charger becomes much larger.

The battery charger is generally of the float type and exhibits very-well-regulated output voltage over a wide output current demand. This tight regulation is performed through the use of silicon con-

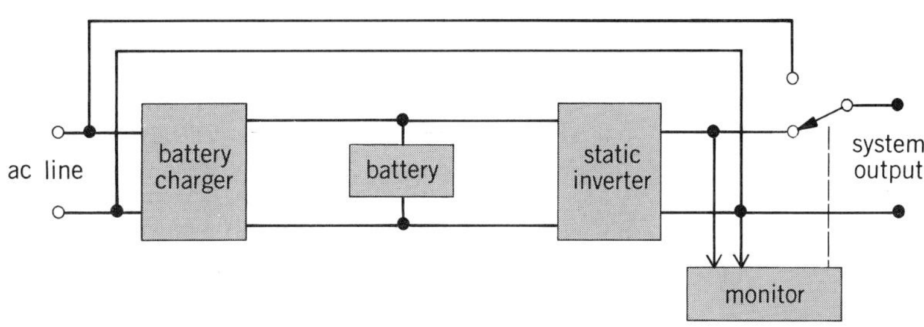

Fig. 3. Block diagram of an on-line automatic reverse-transfer uninterruptible power system, showing subsystems. (*Topaz Electronics*)

trolled rectifiers for voltage regulation and silicon rectifiers to convert ac to dc. Current limiting is built into the circuit to protect the charger. The rating of the charger is determined by the requirements of the static inverter and the current necessary to recharge the battery after an uninterruptible power system has gone through a commercial power outage.

Batteries. The battery used on an uninterruptible power system is sized to provide power to the inverter any time that the commercial power is removed. Power may be provided for anywhere from a few seconds to an hour or more. The battery necessary for power is dependent upon the sensitive load to be powered and the desired backup time.

The batteries are arranged in a series of racks called a battery bank. Each battery is made up of individual cells containing lead alloy plates and an electrolyte. Conductor size between batteries and current to be handled by the static inverter determine the battery bank voltage. Small uninterruptible power systems use low voltages, while large uninterruptible power systems use battery banks of several hundred volts. In the smaller uninterruptible power systems the battery bank may be mounted adjacent to or inside the system.

Batteries may be any one of four different types. They are, in order of their frequency of use: lead antimony, nickle cadmium, gel electrolyte, and lead calcium. The cells within the lead antimony battery contain grids constructed of an alloy of lead and antimony and are surrounded by an electrolyte of sulphuric acid. This type of battery requires specialized charging techniques and a high level of maintenance. The nickle cadmium battery offers a high storage level at high cost. It is best used where the battery bank is subject to extremes of cold and physical space constraints. Once again, specialized charging techniques are required for this type of battery. The gel electrolyte type of battery is usually used in smaller uninterruptible power system installations where voltage and current demands will be at a relatively low level. Although generally more expensive than a liquid electrolyte lead acid battery, it is sealed and maintenance-free. In the lead calcium battery each cell uses calcium alloyed with lead and is filled with an electrolyte of sulfuric acid. This battery type has wider recharge tolerances, longer life, and substantially lower maintenance.

Static inverter. The purpose of the static inverter is to change the dc voltage derived from the battery charger or battery bank to an ac voltage in order to power the sensitive load. The static inverter determines the quality of power used to drive the sensitive load. The ac output voltage must be stable and free from all interruptions. It is the most important subsystem of the uninterruptible power system. If the inverter should fail, the system is out of operation.

The most common static inverter types are, in order of complexity, the ferroresonant type, the quasi-square-wave type, the pulse-width-modulated type, and the step-wave type. The ferroresonant approach starts with a square-wave inverter system and a tuned output transformer. The output transformer performs all filtering, voltage regulation, and current limiting with magnetic regulation, allowing for the design of a very simple inverter.

The technique is generally found in smaller, lower-power uninterruptible power systems. The quasi-square-wave approach uses a true electronic regulation technique. It consists of two square waves superimposed on each other to approximate the form of an ac sine wave. Regulation is achieved with a silicon controlled rectifier bridge and control circuit that change the relationship between the two square waves. This changes the pulse width and amplitude of the square wave, achieving regulation. An output LC filter is employed to filter and wave-shape the output sine wave. The pulse-width-modulated approach is essentially a square-wave inverter operating at high frequency. The pulse width, and not the amplitude, of the square wave is adjusted to approximate the sine wave. Once again, output filtering is employed to shape the output waveform. The stepped-wave approach is an extension of the pulse-width-modulated approach. Several pulses are provided per half cycle of the sine wave and are combined to develop an output voltage resembling a sine wave that needs very little filtering. This approach is complex and is found only in high-powered uninterruptible power system units. *See* TRANSFORMER.

Transfer switch. The switch that connects the sensitive load to either the inverter or the commercial power line is called the transfer switch. The major function of the switch is to provide an alternate source of ac power to the critical load should any of the components in the uninterruptible power system fail. The position of the switch is controlled by a monitor circuit as shown in Fig. 3. Generally the switch in an uninterruptible power system is a high-speed solid-state device that can transfer the load from one ac source to another with little or no break in power. The switch may be designed for uninterrupted make-before-break or interrupted break-before-make transfers. *See* ELECTRIC POWER SYSTEMS; ELECTRIC SWITCH.

[JOHN SULLIVAN]

Bibliography: *AC Power Handbook of Problems and Solutions*, 1979; K. G. Brill, *Mini-Micro Syst.*, 10(7):38–45, July 1977; N. L. Conger, *Instrum. Technol.*, 20(9):57–63, September 1973; J. J. Waterman, *Specifying Eng.*, 43(2):60–64, February 1980.

Electrical codes

Systematic bodies of rules governing the practical application, installation, and interconnection of electrically operated equipment, devices, and electrical wiring systems.

National Electrical Code. The basic code used throughout the United States is the National Electrical Code, prepared under the direction of the National Fire Protection Association (NFPA). It is approved by the American National Standards Institute, and the 1981 edition of the code is known as NFPA 70-1981 (ANSI).

The National Electrical Code was originally drawn in 1897 as a result of the united efforts of various insurance, electrical, architectural, and allied interests. The original code was prepared by the National Conference on Standard Electrical Rules, composed of delegates from various interested national associations.

In 1911 the National Conference of Standard Electrical Rules was disbanded, and since then the NFPA has acted as the code's sponsor. Beginning with the 1920 edition, the National Electrical Code has been under the further auspices of the American National Standards Institute with the NFPA continuing in its role as administrative sponsor. Since that date, the NFPA committee that produces the code has been identified as ANSI Standards Committee C1 (formerly USAS C1 or ASA C1).

The provisions of the National Electrical Code are under constant review by a number of panels whose members are selected to provide broad representation of electrical, industrial, and public interests. The code is amended in its periodic republication every 3 years or by tentative interim amendments which are announced by bulletins and through the technical press.

The National Electrical Code is purely advisory as far as the National Fire Protection Association is concerned, but it is very widely used for legal regulatory purposes. The code is administered by various local inspection agencies, whose decisions govern its application to individual installations. Local inspectors are largely members of the International Association of Electrical Inspectors. This organization, the National Electrical Manufacturers Association, the National Electrical Contractors Association, the Edison Electric Institute, the Underwriters' Laboratories, Inc., the International Brotherhood of Electrical Workers, governmental groups, and independent experts all contribute to the development and application of the National Electrical Code.

Compliance with the provisions of the code can effectively minimize fire and accident hazards in any electrical design. It sets forth requirements that constitute a minimum standard for the framework of electrical design. As stated in its introduction, the code is concerned with the "practical safeguarding of persons and of buildings and their contents from hazards arising from the use of electricity for light, heat, power, radio, signalling, and for other purposes." The National Electrical Code is recognized as a legal criterion of safe electrical design and installation. It is used in court litigation and by insurance companies as a basis for insuring buildings.

Other standards. In addition to the National Electrical Code volume itself, other standards and recommended practices are made available in pamphlet form by the National Fire Protection Association. These cover such special subjects as hospital operating rooms, municipal fire alarm systems, garages, aircraft hangars, and other equipment with great potential hazards due to improper design.

The National Electrical Safety Code (to be distinguished from the National Electrical Code) is published by the Institute of Electrical and Electronic Engineers, Inc. Designated as ANSI C2, this code presents basic provisions for safeguarding persons from hazards arising from the installation, operation, or maintenance of (1) conductors and equipment in electric supply stations, and (2) overhead and underground electric supply and communications lines. Basically, this code applies to the outdoor circuits of electric utility companies and to similar systems or equipment on commercial and industrial premises.

Municipal codes. The National Electrical Code is incorporated bodily or by reference in many municipal building ordinances, often with additional provisions or restrictions applicable in the particular locality. Some large cities have independent electrical codes; however, the actual provisions in most such codes tend to be basically similar to the National Electrical Code.

Testing of electrical products. Standards on the construction and assembly of many types of electrical equipment, materials, and appliances are set forth in literature issued by the Underwriters' Laboratories, Inc. The Underwriters' Laboratories examines, tests, and determines the suitability of materials and equipment to be used according to code regulations. Each year, it publishes three volumes listing commercially available electrical products which have been found acceptable with reference to fire and accident hazards and which conform with the application and installation requirements of the code. The three volumes are titled: *Electrical Construction Materials*, *Electrical Appliance and Utilization Equipment*, and *Hazardous Location Equipment*. The Underwriters' Laboratories publishes other literature such as *Gas and Oil Equipment* and *Fire Protection Equipment*, dealing with special equipment involving hazard to life or property.

Administration. Electrical codes are administered locally by inspectors who review plans and specifications and examine electrical work during installation and after the work is completed to ensure compliance with applicable rules or ordinances.

Electrical inspection bureaus are maintained in many cities by the National Board of Fire Underwriters. In communities where codes are enforced by ordinance, inspections may be performed by municipal electrical inspectors. Utility inspectors examine the service entrance and metering installation for compliance with prevailing utility regulations.

Federal and state buildings are usually inspected by authorized government electrical inspectors. In these instances inspection includes both safety consideration and the requirements of the particular job specifications. Other specification (by underwriters or municipal inspectors) is often waived. *See* WIRING. [J. F. MC PARTLAND]

Bibliography: J. F. McPartland, *National Electrical Code Handbook*, 16th ed., 1979; *National Electrical Code*, NFPA 70–1981 (ANSI), 1981.

Electrical degree

A unit equal to 1/360 of a complete cycle of electric current or voltage. In an electric machine it is 1/360 of the angle subtended at the axis by two consecutive field poles of like polarity, since the voltage wave generated in a conductor completes one cycle when it traverses one pair of poles. The term mechanical degree is used to designate the space angle between two positions about the axis of the machine. The number of electrical degrees between two positions about the axis equals the number of mechanical degrees multiplied by the number of pairs of poles on the machine. *See* ALTERNATING CURRENT. [ARTHUR R. ECKELS]

Electrical engineering

A branch of engineering dealing primarily with electricity and magnetism and devoted to utilization of the forces of nature and materials for the benefit of mankind. Electrical engineering encompasses many phases of other engineering sciences and the physical sciences; it includes research, invention, development design, application, and education. Many phases of electrical engineering are based on applications of higher mathematics. *See* ENGINEERING.

The great advances of the past in electrical engineering are closely associated with certain inventions and discoveries which have made practical uses of electricity and magnetism. Throughout the history of electrical engineering, there have been eras of accelerated engineering activity that are closely identified with important discoveries by a relatively few scientists and engineers. In considering the historical development and present scope of electrical engineering, it is convenient to consider five eras of development.

First era. As early as the latter part of the 16th century, experimenters were exploring the behavior of static electricity. W. Gilbert (1540–1603), personal physician to Queen Elizabeth I, experimented with electric charges and discharges. In 1750 Benjamin Franklin proved that lightning was electrical in nature. Neither investigator discovered anything that was significant from the standpoint of the applications of electricity. Discovery of the presence of magnetism in certain rocks preceded the earliest knowledge of electricity. Such knowledge was common about 600 B.C. Applications of electrical knowledge were completely absent in this era.

Second era. The second era had its beginning in electrochemical developments. Electrochemical deposition was discovered by W. Nicholson and A. Carlisle in 1800, and in the same year A. Volta discovered the principle of the electric battery. The voltaic cell was one of the most important discoveries in the history of the electrical art, because it provided a continuous source of appreciable amounts of electric power at reasonably low voltage. It was an essential component of the early communication systems, such as the telephone and telegraph.

The most significant developments of the second era centered around the field of communications. The first United States patent on the electrical telegraph was obtained by J. Groat in 1800. The invention of a practical electromagnet was announced by Joseph Henry in 1827. These inventions by Groat and Henry opened the way for a still more significant invention, the electromagnetic telegraph. The principle of this forerunner of the communications industry was conceived in 1831, proven practical in 1837, and patented in 1840 by Samuel F. B. Morse.

Few developments have had greater impact on American life than Morse's invention. His idea paved the way for the first system of electrical communication, the telegraph. This in turn led to the telephone and later to the wireless telegraph. The growth of electrical communications resulted in extensive engineering, production of electrical equipment, and the birth of an electrical industry, adding much to the wealth of the United States and at the same time making possible rapid communications throughout the nation.

Third era. The discovery of electromagnetic induction by Michael Faraday in 1831 established many principles upon which modern machines function. Motors, generators, transformers, and many other electrical devices found in heavy electrical industry were made possible by the discoveries of Faraday. The contributions of Faraday in the electrical power industry are comparable to those of Morse in the field of communications.

One of the first important developments based on the disclosures of Faraday was the electric dynamo. English patent no. 1858 describes the principle of operation. In the following years many types of dc generators were developed and used commercially. The Gramme-ring armature was one of the first used in conjunction with a commutator. This machine was somewhat inefficient, but it provided a source of relatively high voltage at a reasonably large power capacity (up to 100 kW). *See* ELECTRIC ROTATING MACHINERY.

With the development of the high-resistance carbon filament lamp by Thomas Edison in 1880, the dc generator became one of the essential components of the constant-potential lighting system. Commercial lighting and residential lighting became practical and the electric light and power industry was born. One of the most common uses for direct current during this period was for street lighting. These lamps were of the carbon arc type. Many lamps were operated in series from a constant-current generator. The generators have long since been replaced by the constant-current ac transformer and the lamps have been replaced by low-voltage incandescent, sodium vapor, or fluorescent types, but the constant-current system of power supply for street lighting still prevails.

The first transformer was announced by L. Gaulard and J. D. Gibbs in 1883. This device probably did more to revolutionize the systems of power transmission than any other. The advantages of high-voltage low-current systems over the low-voltage high-current systems of power transmission were well known. Following the discovery of the transformer, power could be generated at low voltages, transformed to higher voltages for transmission over great distances (several hundred miles), and then reduced by transformers to lower values for utilization. This system of high-voltage transmission (110,000–750,000 volts) made possible the generation of electric energy in one part of the country and the utilization of that energy in another part. This method of power transmission is of great significance in the development of American industry, providing better efficiency and dependability in the utilization of electrical energy. It also permits interconnection of power systems. *See* ELECTRIC POWER SYSTEMS.

The first direct-current central station in the United States (Pearl Street Station, New York) began operation in 1882. In 1886 the first alternating-current station was placed in operation. The output of this station was limited essentially to lighting because no suitable ac motor was available. In 1888, however, N. Tesla was granted a patent on the polyphase ac induction motor, which soon became the most commonly used motor for supplying large amounts of power; in its improved state, it is most extensively used today.

Early alternating-current systems were designed for many different frequencies (25, 33⅓, 40, 50, 60, 90, 130, and 420 cycles, or Hz in today's terminology). In 1891, through the efforts of the American Institute of Electrical Engineers, studies were made to determine the possibilities of standardizing equipment to standard frequency and voltage ratings, with the result that 60 Hz was made the standard frequency in the United States. Similar studies in Europe resulted in the selection of 50 Hz. This standardization resulted in more universally adaptable equipment and great savings on equipment costs. These frequencies are still considered standard and are prevalent today.

The power industry made rapid strides in this era, but the field of communications was not dormant. In 1876 Alexander Graham Bell invented the telephone. This device was soon put into use and, as a result, another huge industry was established.

Throughout this period of development, another outstanding contributor was Thomas A. Edison. His work included research, invention, development, and production. His activities extended into chemistry, electrical dynamos, systems of transmission, sound recording and reproduction, and electrical lamps. Perhaps one of his most important discoveries was one that he did not pursue sufficiently to realize its vast importance, a discovery known in later years as the Edison effect.

Fourth era. The fourth era of electrical engineering began with the announcement in 1883 of the Edison effect. Edison discovered that, when a voltage of proper polarity is applied between two electrodes, one hot and one cold, placed in an evacuated enclosure, current flows from the hot to the cold element in an external circuit joining the two. This phenomenon was the first indication of thermionic emission of electrons. This discovery opened the new field of electronics, which has since grown enormously. The discovery resulted from keen observations by Edison while he was pursuing research on incandescent lamps. Edison was not searching for this effect; it was a discovery made more or less by accident.

Lee DeForest introduced the use of the third element (grid) in the vacuum tube in 1906. This development opened an entirely new field of engineering. It made possible new systems of communication and methods of control and indicated the possibility of the multielement tube. It provided the basis for future developments in electronics.

Fifth era. The fifth era of electrical engineering can be classified as that of engineering research. With production methodology being well established, there was rapid expansion in research engineering in the first half of the 20th century. Industrial research laboratories expanded in size and in number. College faculties became increasingly aware of the importance of research to education. To administer necessary training in research, extensive research laboratories were constructed by American universities. Academic appointments have been made of many faculty members who are trained in the systematic pursuit of scientific and engineering knowledge. Today research is an essential ingredient in the education of the engineering student—the agent by which the student develops originality, inventive genius, and an understanding of the world.

Research today is a big business, no longer carried out by isolated individuals working over long periods. It is conducted by highly organized groups of investigators who have been selected because of their competence in certain areas of investigation. The lapse of 30 years between invention and production which seemed to prevail in the 19th century has been shortened to several years and sometimes to several months, a saving in time which can be attributed largely to better systems of communication between scientists and engineers in the engineering profession.

Since 1945 great advances have been made as the result of the invention of the transistor by J. Bardeen, W. H. Brattain, and W. Shockley. This solid-state device has made possible the miniaturization of many components in computers, integrated circuits, and calculators. During this same period, research in electron optics has preceded the development of lasers and holography.

The rate of growth of research in electrical engineering was enhanced in the 1940s as a result of support of Federal agencies. Many ideas associated with the military effort of that period are now being used commercially and for research purposes. Microwaves have become part of modern communication systems. The development of semiconductors has made possible more rugged, smaller, and cheaper systems. Research in miniaturization has greatly increased the speed of modern computers. The laser has provided communications systems operative over millions of miles. Integrated circuits have reduced size and weights and made practical interplanetary and satellite communications. Planetary radar astronomy and radio astronomy are also the result of adaptations to engineering systems of electrical components developed through research.

The need for better communications between electrical engineers led to the establishment of the American Institute of Electrical Engineers (AIEE) in 1884. In 1913 the Institute of Radio Engineers (IRE) was founded. In later years these two organizations were merged into a single organization which is now known as the Institute of Electrical and Electronics Engineers (IEEE). Its present membership is approximately 170,000.

The references cited in this article will lead the reader to other articles discussing certain branches of electrical engineering in more detail. For other areas not previously mentioned *see* CIRCUIT (ELECTRICITY); ELECTRIC HEATING.

Electrical engineers apply their abilities in other engineering fields that are not strictly electrical. There is hardly a field of technology to which electrical engineering has not made a contribution.

Electrical impedance

The total opposition that a circuit presents to an alternating current. Impedance, measured in ohms, may include resistance R, inductive reactance X_L, and capacitive reactance X_C. *See* REACTANCE.

The impedance of the series RLC circuit is given by Eq. (1).

$$Z = \sqrt{R^2 + (X_L - X_C)^2} \text{ ohms (magnitude)} \quad (1)$$

In terms of complex quantities, this impedance is given by Eq. (2). The two components of Z are

$$Z = R + j(X_L - X_C) \tag{2}$$

at right angles to each other in an impedance diagram. Therefore, impedance also has an associated angle, given by Eq. (3). The angle is called the

$$\theta = \arctan \frac{X_L - X_C}{R} \tag{3}$$

phase, or power-factor, angle of the circuit. The current lags or leads the voltage by angle θ depending on whether X_L is greater or less than X_C.

Impedance may also be defined as the ratio of the rms voltage to the rms current, $Z = E/I$. This is a form of Ohm's law for ac circuits. For further discussion of impedance *see* ALTERNATING-CURRENT CIRCUIT THEORY. [BURTIS L. ROBERTSON]

Electrical insulation

A nonconducting material that provides electric isolation of two parts at different voltages. To accomplish this, an insulator must meet two primary requirements: it must have an electrical resistivity and a dielectric strength sufficiently high for the given application. The secondary requirements relate to thermal and mechanical properties. Occasionally, tertiary requirements relating to dielectric loss and dielectric constant must also be observed. A complementary requirement is that the required properties not deteriorate in a given environment and desired lifetime. *See* CONDUCTOR (ELECTRICITY).

Electric insulation is generally a vital factor in both the technical and economic feasibility of complex power and electronic systems. The generation and transmission of electric power depend critically upon the performance of electric insulation, and now plays an even more crucial role because of the energy shortage.

INSULATION REQUIREMENTS

The important requirements for good insulation will now be discussed.

Dielectric strength. The basic difference between a conductor and a dielectric is that free charge has high mobility on and in a conductor, whereas free charge has little or no mobility on or in a dielectric. Dielectric strength is a measure of the electric stress required to abruptly move substantial charge on or through a dielectric. It deteriorates with the ingress of water and with elevated temperature. For high-voltage (on the order of kilovolts) applications, dielectric strength is the most important single property of the insulation.

Steady-state strength. The voltage at which a sudden high-current discharge punctures an insulator is called the breakdown voltage. In a uniform field this voltage divided by the electrode gap is called the dielectric strength, the breakdown field strength, the electric strength, or the breakdown stress. In a nonuniform field it is important to differentiate between the maximum field and the average field at breakdown.

Any insulator with an operational dielectric strength of 100–300 V/mil (40–120 kV/cm) may be considered good. Many of the insulators listed with a dielectric strength more than an order of magnitude higher than this are ordinarily stressed at these lower values in practical operation. The reason is that imperfections in the form of voids, cracks, filamentary defects, and so forth occur whenever long lengths and thick sections of insulating material are manufactured.

Whenever two dielectrics are in series in an electric field, the stress E is higher in the medium of lower dielectric constant κ by the ratio of dielectric constants, $E_2/E_1 = \kappa_1/\kappa_2$. Therefore the electric field is higher in a defect, which generally has a dielectric constant of 1, by a factor of 2 to 3 than it is in the insulator. Furthermore, as is evident in Table 1 and Fig. 1, the dielectric strength is much less in the defect than in the insulator. It is as if the whole burden is placed on the weak link. Thus, in addition to operating insulators at much lower than intrinsic dielectric strength, schemes such as oil impregnation are used when possible to fill the defects to increase both their dielectric constant and their dielectric strength.

Although solid extruded dielectric cable is now available at higher voltages and electric stress than ever before, the more traditional oil-impregnated-paper wound-tape insulation can be used at even higher values. The probability that defects in

Table 1. Properties of various good insulators

Insulator	Resistivity, 10^{14} ohm-cm	Dielectric strength, 10^3 V/mil*	Power factor $\times 10^{-4}$	Dielectric constant	Tensile strength, 10^3 psi†	Chemical resistance	Flammability
Lexan	10–1000	8–16	10–30	3	8	Good	Self-extinguishing
Kapton H	1000–10,000	3–8	20–50	3	22	High	Self-extinguishing
Kel-F	10–1000	2–6	20–40	2–3	5	High	No
Mylar	0.1–1	4–16	30–120	3	20	Good	Yes
Parylene	10–10,000	6–10	2(N)–200(C)	2–3	10	Very good	Yes
Polyethylene	100–500	1–17	3–30	2	4–6	Good	Yes
Teflon	0.1–1000	1–7	1–4	2	4–5	High	Very low
Air (1 atm, 1 mm gap)	—	0.1	10^{-3}	1	0	Stable	No
Sulfur hexafluoride (1 atm, 1 mm gap)	—	0.2–0.3	10^{-3}	1	0	Stable	No
Vacuum (<10^{-5} torr or 1.3×10^{-3} Pa, 1 mm gap)	—	2–3	0	1	0	Stable	No

*10^3 V/mil = 4×10^5 V/cm.
†10^3 psi = 6.9 MPa.

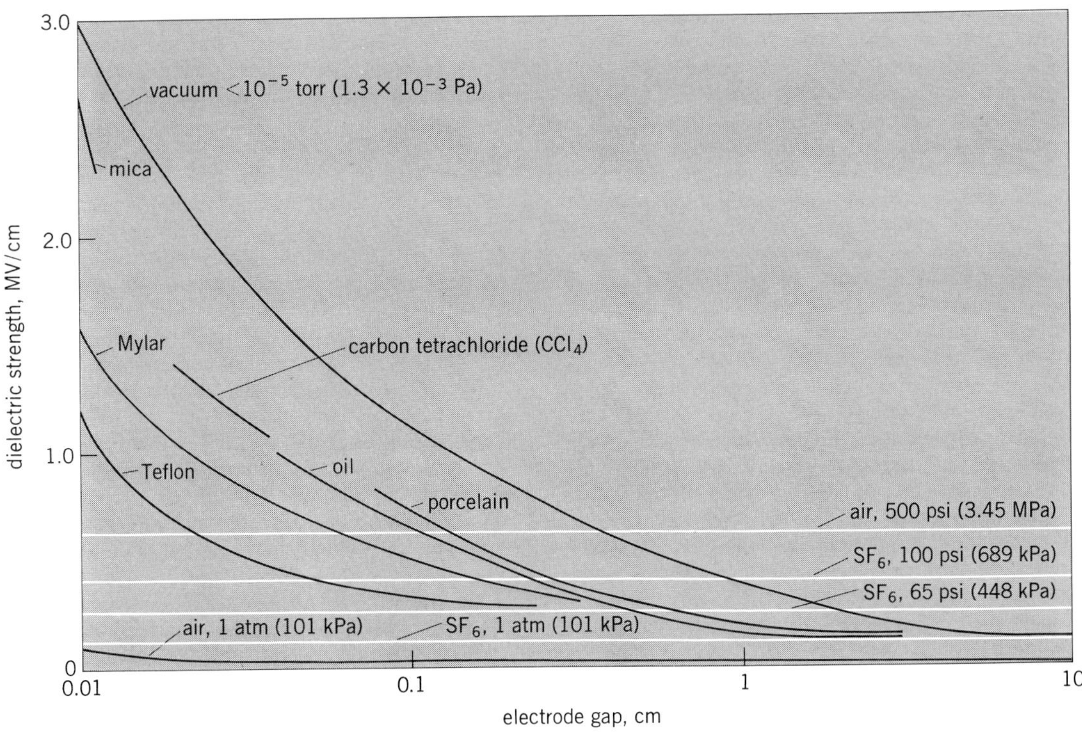

Fig. 1. Uniform field dielectric strength for solid, liquid, gas, and vacuum insulation.

the tape will line up in close proximity is much less than in a solid piece of the same insulation material. The tape thickness also limits the size of the defect in the direction of the electric field.

Laboratory testing of insulation over a wide range of conditions is conducted to determine the effects of aging, high temperature, heat cycling, ultraviolet radiation, moisture, and so forth. A statistical analysis of the data based upon the weak-link Weibull statistics is then used to determine safe operating voltages and electric fields for the required lifetimes and environments. Operational withstand stress is sometimes arbitrarily defined as being at least two standard deviations below the average dielectric strength in the laboratory.

Impulse strength. Impulse strength is a measure of an insulator's ability to withstand a rapid rise and rapid fall (measured in microseconds) of electric stress. In addition to steady-state dielectric strength, an insulator must be able to meet a basic impulse level (BIL) and a basic switching level (BSL) surge (not as high a rate of voltage increase as for BIL) requirement. The impulse strength is highest for solids, and generally decreases in going from solids to liquids to gases to vacuum. For example, a new development in tape insulation consisting of a combination of cellulose paper bound to polypropylene film or fibers and impregnated with oil has an impulse strength of 2.7 MV/cm, whereas the impulse strength of sulfur hexafluoride from 15 to 60 psi (100 to 400 kPa) goes only from 0.2 to 0.45 MV/cm. Thus, in designing transmission cables, sulfur hexafluoride cables are limited by their BIL, whereas most solid insulation cables are limited only by their steady-state dielectric strength. This is why solid insulation cables are so much more compact than sulfur hexafluoride cables for the same ratings, in addition to the fact that they have higher steady-state dielectric strength. However, when a solid dielectric breaks down electrically, the punctured and surrounding regions must be replaced. After sulfur hexafluoride breaks down, the gas is self-healing and recovers dielectric strength. If no damage has been incurred by the conductors, operation may resume.

The impulse strength for sulfur hexafluoride increases with increasing pressure, but tends to level off above 60 psi (400 kPa). The impulse strength of liquids increases less rapidly with increasing pressure, leveling off above 300 psi (2 MPa). For example, carbon tetrachloride's impulse strength goes from 2.2 to 2.6 MV/cm in going from 15 to 300 psi (100 kPa to 2 MPa). This pressure dependence for liquids tends to disappear for rates of field increase greater than 4 MV/cm-μs. This is experimental evidence that a "bubble" mechanism (see below) does not have time to be operative in such short times.

Resistivity. Resistivity is a measure of how much current will be drained away from the conductor through the bulk or along the surface of the dielectric.

An insulator with resistivity equal to or greater than 10^{13} ohm-cm may be considered good. While resistivity measurements of conductors are straightforward, measurements on insulators are much more difficult and are subject to a much wider range of values for a given material. A dc measurement of current is made after application of a uniform electric field for a specified period of time. A time variation from 1 s to 1 min can lead to orders-of-magnitude variation in the measured resistivity.

Since it is nondestructive, a resistivity measurement can be used as an on-line quality and uniformity check during the manufacturing process. During actual operation in the field, a sharp decrease in resistivity usually signals an ensuing service failure. *See* INSULATION RESISTANCE TESTING; NONDESTRUCTIVE TESTING.

Dielectric loss. When a dielectric is subjected to an alternating field, a time-varying polarization of the atoms and molecules in the dielectric is produced. The alternation of both the permanent and induced polarization in the dielectric results in power dissipation within the dielectric of which the power factor is a measure. This dielectric power loss is proportional to the product of the dielectric constant and the square of the electric field in the dielectric. Although it may be a loss that is small relative to other losses in most ambient-temperature applications, and even though it generally decreases at low temperatures, it is a relatively important loss for dielectrics to be used at cryogenic temperatures.

Dielectric loss is a function of both temperature and frequency and generally decreases as these two variables decrease. However, it is not a monotonic function of these parameters, and at various discrete values of temperature and frequency there are quasi-resonances where the loss is locally very high. A sharp rise in dielectric loss, usually accompanied by a rapid temperature increase, is an indication of impending breakdown.

Neglecting these high-loss sites, Table 2 indicates the power factor in the three temperature regions of general interest, 300 K (conventional), 77 K (cryoresistive), and 4K (superconducting). Values of less than 10^{-3} are good at 300 K, values less than 10^{-4} are good at 77 K, and values less than 10^{-5} are good at 4 K. With the increasing need to deliver large blocks of power (greater than 3000 MW) for long distances (greater than 80 km) from remote energy sources, advanced power transmission must have insulation capable of meeting cryogenic requirements. Cryogenic electronic systems such as the high-density, ultrafast superconducting computer are emerging with similar cryogenic dielectric requirements. A power factor of less than 10^{-5} at the operating cryogenic temperature is generally desirable, since the overall power loss is amplified by the refrigeration inefficiency, which may require as much as 500 W per watt of dissipation. *See* SUPERCONDUCTING DEVICES.

Ideally the power factor should be independent of electric field stress. However, owing to processes unrelated to polarization, such as electron emission and ionization, the dielectric loss increases with increasing electric field, and this is reflected in an increasing power factor.

The power factor is the cosine of the phase angle ϕ between the voltage and current, or equivalently

Table 2. Power factor of insulators at three temperatures

Insulator	Power factor $\times 10^{-4}$		
	300 K	77 K	4 K
Kapton H	20–50	5	0.5
Polyethylene (low density)	3–30	0.1	0.02
Teflon	1–4	0.9	0.02
Kraft paper	50	20	7
Vacuum ($<10^{-5}$ torr or 1.3×10^{-3} Pa)	0	0	0

the ratio of resistance to impedance for the medium. Occasionally confusion arises regarding terminology when the dielectric loss is related to the complementary angle δ, which satisfies Eq. (1) for

$$\text{Power factor} = \cos\phi = \sin\delta \doteq \tan\delta \doteq \delta \qquad (1)$$

small δ. Thus the power factor is often used interchangeably with the terms loss tangent and loss angle.

Dielectric constant. The dielectric constant, also known as the relative permittivity or specific inductive capacity (SIC), is a measure of the ability of the dielectric to become polarized, taken as the ratio of the charge required to bring the system to the same voltage level relative to the charge required if the dielectric were vacuum. It is thus a pure number, but is in fact not a constant, and may vary with temperature, frequency, and electric-field intensity. For highly polar materials such as water, the variation with temperature and frequency is dramatic, going from 80 at 300 K to 3 at 77 K, or from 80 at 60 Hz to 20 at 10 kHz.

In addition to the problem of intensification of the electric field in regions of relatively lower dielectric constant, a low-dielectric-constant insulation is desirable for two more compelling reasons. In ac transmission cables, the lower the dielectric constant, the more the current and the voltage will be in phase. This means that more usable power will be delivered, without the need for reactive compensation. Furthermore, in reducing the charging current (which is proportional to the dielectric constant), concomitant power losses (related to the square of the charging current) are also reduced. A high dielectric constant is desirable in capacitors, since the capacitance is proportional to it.

INSULATOR PROPERTIES

Table 1 summarizes the properties of various good insulators at or near 60 Hz and ambient temperature (about 300 K). The range of values indicates not only differences between measurements of different investigators, but differences resulting from small variations in a parameter. For example, the dielectric strength depends on the thickness and area of the sample being tested. It generally decreases with increasing thickness and area. The voltage at which electrical breakdown occurs increases less than linearly with thickness. Thus the corresponding value of electric field decreases.

All insulators may be classified as either solid or fluid. Solid insulation is further divided into flexible and rigid types.

Flexible insulation. Flexible hydrocarbon insulation is generally either thermoplastic or thermosetting. Thermosets are initially soft, and can be extruded by using only pressure. Following heat treatment, when they return to ambient temperature, they are tougher and harder. After thermosetting, nonrubber thermosets are harder, stronger, and have more dimensional stability than the thermoplastics. Thermoplastics are softened by heating, and when cool become hard again. They are heat-extruded.

Thermoplastics. The first seven insulators listed in Table 1 are thermoplastics, as are nylon, polystyrene, and polyvinyl chloride. Those listed in Table 1 are solid polymers that are used at ambient temperature and have good potential for cryogenic applications, and their monomeric structures are shown in Fig. 2. Polymers come in amorphous and ordered states which have different dielectric properties. The ordered state is frequently called "crystalline", but is not the same as the crystalline state of solids.

Lexan is a trade name for a group of polyesters formed from carbonic acid, and generically called polycarbonate (PC). Polycarbonate has good electrical and mechanical properties, good dimensional stability, good resistance to creep, high distortion temperature, and ease of molding and extrusion.

Kapton H is a trade name for polypyromellitimide (PPMI). PPMI has exceptional resistance to thermal degradation, because it is one of the polyimides that incorporates multiple bonds along the backbone of the chain. It has good mechanical strength and good dimensional stability at both high and low temperatures, retaining ductility down to cryogenic temperatures.

Kel-F is a trade name for polychlorotrifluoroethylene (CTFE). It is one of the best insulators in its ability to exclude moisture, in high-temperature strength, in chemical inertness, and in low coefficient of friction.

Mylar is a trade name for polyethylene terephthalate (PET). PET is a high-molecular-weight polyester with a stiff polymer chain and resilient interchain bonds that retain good mechanical properties beyond 150°C. Because of their toughness, PET films are used not only for electrical insulation, but also for magnetic tapes.

Parylene is a trade name for polyparaxylylene (PPX). PPX is linear and highly regular, and was developed to be highly radiation-resistant. It has good dimensional stability down to cryogenic temperatures. It has not had widespread use owing to its cost.

Polyethylene (PE) is a widely used inexpensive electrical insulator with good mechanical and excellent dielectric properties. Polyethylene is cross-linked chemically or by irradiation to increase its mechanical strength and improve its high-temperature properties. Chemical cross linking leaves impurities and produces microscopic voids which may decrease its dielectric strength. Cross linking by irradiation appears to increase its dielectric strength. Although polyethylene has a higher dielectric strength than chemically cross-linked polyethylene (XLP), its low melting point of 118°C makes it less desirable for extruded dielectric cables than XLP. Ethylene propylene rubber (EPR) can also be used in extruded solid dielectric cables, but has a lower dielectric strength than XLP.

Teflon is a trade name for polytetrafluoroethylene (TFE), and its copolymer with hexafluoropropylene. It is a highly ordered and orientable polymer made of chains of CF_2 with its strong C-F bond. This is why its structure involves little or no branching or cross linking. Thus TFE has a low coefficient of friction, is stiff, inert, and insoluble, and has a high melting point of 327°C. Because of its high symmetry and tight bonding, TFE has one of the lowest known power factors for a solid, as well as other excellent dielectric properties. It may be processed by powder and sinter techniques, as well as by extrusion or injection molding.

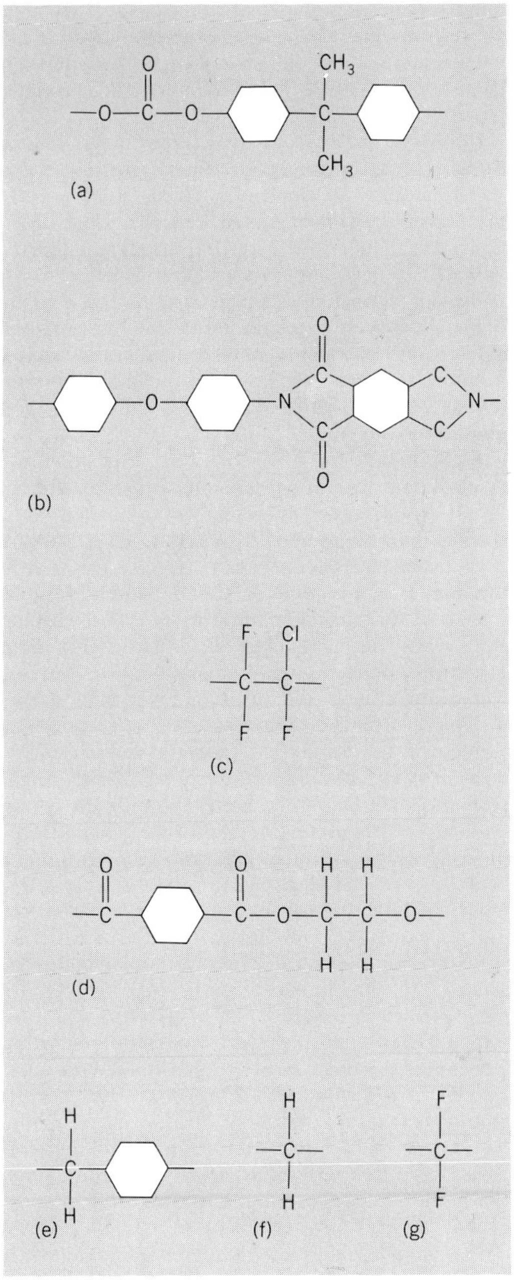

Fig. 2. Monomeric structures of polymers used as electrical insulators. (*a*) Lexan. (*b*) Kapton H. (*c*) Kel-F. (*d*) Mylar. (*e*) Parylene. (*f*) Polyethylene. (*g*) Teflon.

Polystyrene is a good insulator that is used in thin films, or where it will not be stressed or bent, because it is stiff and brittle when it is thick. Fillers in the form of silica, alumina, mica, and so forth reduce the brittleness (probably by inhibiting crack propagation), improve the thermal conductivity, and reduce thermal expansion. Hydrated fillers, such as aluminum trihydrate, inhibit tracking, since the waters of hydration tend to be arc-quenching.

Thermosets. The various rubber compounds, neoprene, and epoxies are thermosets. Natural rubber, butyl rubber, buna S rubber, and silicone rubber are all good insulators which are tough and flexible. Although chemically and mechanically tough, neoprene does not have good dielectric properties. The epoxies are not too flexible, but are used at joints of flexible insulation because of their good bonding, electrical, and mechanical properties.

Cellulose paper. Cellulose paper insulation is neither thermoplastic nor thermosetting. It is widely used in cables and rotating machinery in multilayers and impregnated with oil. It has a relatively high dielectric loss that hardly decreases with decreasing temperature (Table 2), which rules it out for cryogenic applications. Because of its high dielectric strength, the high loss has not been a deterrent to its use in conventional ambient-temperature applications. However, the high dielectric strength deteriorates quickly if moisture permeates the paper.

Rigid insulation. This second major group of solid insulators includes glass, mica, epoxies, ceramoplastics, porcelain, alumina, and other ceramics. Rather than being used to insulate wires and cables, except for mica, these materials are used in equipment terminations (potheads) and as support insulators (in tension or compression) for overhead lines whose primary dielectric is air. These rigid structures must be shock-resistant, be relatively water-impervious, and be able to endure corona discharges over their surfaces.

For use as termination insulators, such as for transformers, underground transmission lines, reactors, and generators, both high bulk dielectric strength and high surface dielectric strength are important. For use in connection with equipment such as overhead transmission lines, the requirement for bulk dielectric strength is less severe, and the emphasis is on the surface. Surface dielectric strength of several kilovolts per inch is adequate, and the creepage paths are usually designed for approximately 1 kV/in. (400 V/cm).

In addition to normal electromagnetic operating forces and windage stress, these insulators must be able to withstand switching surges and shocks from short-circuit currents. These fault currents may be 10 times higher than the normal current, resulting in forces that are 100 times higher, since the force is proportional to the square of the current.

Porcelain is widely used for such applications. It is a hard, brittle material made by firing feldspar, quartz, clay, and other minerals at high temperature. Glass is a completely amorphous material containing 50–90% silica. Glass has a dielectric strength of about 2 MV/cm at 300 K. Mica is a mineral that occurs naturally in a laminated form. For thin sections of 0.002 cm thickness, mica has a dielectric strength as high as 10 MV/cm perpendicular to the direction of lamination.

Fluid insulation. Liquids, gases, and vacuum fall in the category of fluid insulation. For all of these, the electrical structure must be such as to contain the fluid in the regions of high electric stress.

Liquids. The main types of insulating liquids are the mineral oils, silicones, chlorinated hydrocarbons, and the fluorocarbons with dielectric strengths on the order of megavolts per centimeter. Many other liquids also have good dielectric strength, such as carbon tetrachloride, toluene, hexane, benzene, chlorobenzene, alcohol, and even deionized water. However, special problems with liquids, such as chemical activity, high thermal expansion, thermal instability, low boiling point, and tracking (conducting carbon residue) after arcing, rule out many liquids. Even transformer oil in combination with the insulating paper in activated transformers produces water which must be filtered out.

Gases. Most gases have a dielectric constant of about 1, and low dielectric loss. Table 3 compares the relative dielectric strengths of various gases. Those containing fluorine or chlorine are strongly electronegative. Of course, the temperature at which the gas condenses out of the gaseous state is an important consideration, as well as chemical stability, tracking, reactivity, and cost.

Air is used as a dielectric in a wide variety of applications, ranging from electronics to high-voltage (765-kV) and high-power (2000-MW) electric transmission lines. Dry air is a reasonably good insulator (Table 1). However, its dielectric strength decreases with increasing gap (Fig. 1).

Sulfur hexafluoride (SF_6 ; Table 1) has (at 1 atm or 101 kpa) more than twice the dielectric strength of air, carbon dioxide, or nitrogen, with this difference increasing with pressure and gap in uniform electric fields (Fig. 1). Sulfur hexafluoride is an electronegative gas which impedes electrical breakdown by capturing free electrons and forming negative ions. Not only does sulfur hexafluoride possess a high dielectric strength, but it is 100 times as effective as air and other gases in quenching arcs. This latter property makes it a very effective arcing medium in high-power circuit breakers. It also is used frequently in electrical utility substations as an insulating medium.

Vacuum. Vacuum (that is, pressures of less than 10^{-5} torr or 10^{-3} Pa) has one of the highest dielectric strengths in the gap ranging 0.1 to 1 mm (Fig. 1; Table 1). However, as the gap increases, its dielectric strength decreases rapidly. A perfect vacuum might be expected to be a perfect insulator, since there would be no charge carriers present to contribute to electrical conductance. That this is not so in practice arises because of the effects of a high electric field or high voltage at the surface of electrodes in vacuum, rather than because a perfect vacuum is far from being realized in the laboratory.

The dielectric properties of vacuum can de-

Table 3. Comparison of various gaseous dielectrics

Gas	Condensation temperature (at 1 atm or 101 kPa), °C	Dielectric strength relative to nitrogen
N_2	−195.8	1.0
CO_2	−78	0.9
SF_6	−63.8	2–2.5
CCl_4	76	6.3
$CFCl_3$	23.8	3–4.5
CF_2Cl_2	−29.8	2.4–2.5
CF_2ClCF_2Cl	3.6	2.8
CF_3CN	−63	4
C_2F_5CN	−30	5
C_3F_7CN	1	6
C_3F_8	−36.7	2–2.9
C_4F_6	−5	4
C_4F_{10}	−2	2.5
C_5F_8	25	6
C_5F_{10}	22	>4
He	−269	<0.2

generate rapidly because vacuum offers no resistance to the motion of charge carriers, once they are introduced into the vacuum region. The collision mean free path is of the order of meters at pressures less than 10^{-4} torr (10^{-2} Pa), and this is below the pressure range for a gas discharge. However, avalanche processes occur at the electrodes, and vapor from the electrodes becomes the arcing medium.

Vacuum has found increasing applications as an insulator in electron tubes, photocells, high-frequency capacitors, electron microscopes, particle accelerators, circuit breakers, and so forth. It has also found use as an arcing medium with exceptional rate of recovery of dielectric strength. The realization of its full potential is strongly linked to the constitution of the electrodes.

NATURE OF ELECTRICAL BREAKDOWN

Electrical breakdown in gases first began to be understood in 1889, and breakdown in solids in 1935. Breakdown in liquids is less well understood. Breakdown in vacuum (in terms of an encompassing predicative theory) is least well understood, despite experimental investigations going back to 1897.

Gases. The Townsend avalanche criteria satisfactorily accounts for the threshold for electrical breakdown in nonelectronegative gases. This theory may be modified to include the process of electron capture by electronegative gases. When the space-charge field becomes large, other mechanisms enter in, such as streamers. At very high pressures, the electric field strength for breakdown is so high that field emission from the electrodes enters into the process.

The Townsend criterion for avalanche is given by Eq. (2). Here α, the first Townsend coefficient,

$$\gamma(e^{\alpha\delta} - 1) = 1 \qquad (2)$$

is the number of ionizing collisions per unit length in the direction of the field made by an electron; and γ, the second Townsend coefficient, is the number of secondary electrons produced at the cathode per electron produced by impact ionization in the gap, δ. Thus the dielectric strength of a gas is a function of both the gas species and the cathode material.

Although impact ionization in the gas produces equal numbers of positive ions and electrons, a space charge would develop even with equal displacement of the negative and positive charges. Since the electrons have greater mobility, this further separates the charges. In those situations when enough positive-ion space charge is produced at the original avalanche head to cause the space, charge field to be almost as large as the applied field, streamer breakdown occurs.

In highly nonuniform fields, where the electric field is very high at one electrode and low at the other, localized breakdown (called corona) which does not bridge the gap can occur. Corona can reduce the electric field around the pointed electrode, in effect rounding it off. This results in a higher breakdown voltage than otherwise, a phenomenon known as corona stabilization.

Most systems in the field become contaminated with conducting particles which greatly reduce the gaseous dielectric strength. This problem is important enough to warrant incorporating particle traps in such equipment, particularly those using electronegative gases.

Solids. For ambient temperature and pressure, the densities of solids are 10^3-10^4 times those of gases. This results in dielectric strengths $10-10^2$ times higher than gases. At these higher fields, quantum-mechanical processes such as electron tunneling enter in. Additionally, through excitation of lattice vibrations a new mechanism for reducing the energy of the accelerated electrons is provided.

For the purpose of considering electrical breakdown, solids may be put in the categories of polar crystals, nonpolar crystals, quasicrystalline (highly ordered), and amorphous. On the basis of only density, the dielectric strength of solids and liquids may be expected to be comparable, and this is so. The alkali halides such as potassium bromide, KBr, and sodium chloride, NaCl, are examples of polar crystals. Mica is an example of a nonpolar crystal. The polymeric solids such as polyethylene can vary between highly ordered and amorphous. Glass is totally amorphous. In their absence of structure, the amorphous materials more nearly resemble and behave like highly viscous liquids.

Above 0 K some electrons in the high-energy tail of the electron energy distribution have enough energy to find themselves in the conduction band of an insulator. This is why insulators have a slight conductivity. A. R. Von Hippel assumed that above a critical electric field, these thermally excited electrons would gain energy faster in the conduction band than they would lose it. Thus they would gain enough energy to ionize the solid and cause breakdown. The values of critical field, calculated from this theory, agreed reasonably well with the measured values of the breakdown field for the alkali halides.

A simplifying assumption in Von Hippel's treatment is that the electrons in the conduction band all have the same energy. A consequence of this assumption is that since there is no ionization below the critical field, the prebreakdown current can be associated only with these free electrons, implying that (1) the prebreakdown current should be extremely small, (2) the prebreakdown current should be independent of the electric field below the critical field, and (3) the ionization coefficient should be discontinuous—going from zero to a large value at the critical field. Since these predictions are at variance with experiment, new theories were developed. H. Fröhlich assumed an energy distribution for the electrons in the conduction band. C. M. Zener assumed tunneling of electrons into the conduction band because of the high electric field.

Both the Von Hippel and Fröhlich theories neglected interaction of the conduction electrons with the lattice vibrations (phonons). Introducing a further refinement, F. Seitz demonstrated that the phonon interaction is dominant in nonpolar crystals, and is nonnegligible in polar crystals.

Despite the success of the above approaches, there are solids in which other mechanisms also occur, such as thermal, electromechanical, gas-discharge, and electrochemical breakdown. In thermal breakdown, part of the solid reaches a critical temperature at the breakdown field, causing chemical deterioration or melting. Even when thermal breakdown is avoided by using thin specimens, massive electrodes, and pulsed voltages, a gas-discharge mechanism of breakdown may initiate in voids in the specimen, or between the electrodes and the solid. Such discharges may cause treeing (Lichtenberg figures) in the solid, and may chemically and mechanically degrade the solid to the point of breakdown.

Liquids. Strictly speaking, liquids do not have an intrinsic dielectric strength. It is necessary to specify the electrode composition, geometry, and the time duration of the applied field. For very long stress durations, the time is important even for solids. Thus, except for electrode composition, this distinction between liquids and solids may be only a matter of degree. In liquids, microprojections on the cathode and migration of suspended particles (due to the gradient in the field) to the high-field regions contribute to a decrease in dielectric strength. Dissociation of the liquid by field-emitted electrons, electrochemical processes, and thermal bubble formation further complicate electrical breakdown in liquids.

A conditioning effect occurs, also common to vacuum, in which the breakdown strength increases progressively with the first few breakdowns, provided that the breakdown current is limited. Other commonalities with vacuum are curvature and area effects. Contrary to simple expectation, up to a point the dielectric strength is higher for electrodes of smaller radius of curvature, and higher for electrodes of smaller area.

The fact that the breakdown strength increases with pressure is indicative that a bubble is formed during the breakdown process. However, if the voltage is applied in the nanosecond range, the pressure dependence vanishes and another breakdown mechanism becomes dominant. A piece of evidence in support of the bubble mechanism of breakdown for longer stress duration is the direct relationship between the boiling point and the breakdown strength for the *n*-alkanes.

Vacuum. As pointed out above, breakdown in vacuum has a number of features in common with liquid breakdown. L. Cranberg suggested that breakdown is initiated when a charged clump of loosely adhering material is removed from one electrode surface by the electric field, strikes the opposite electrode, and thus causes a high enough temperature to produce local evaporation. A discharge then ensues in the metal vapor. (The clump here is analogous to particle-initiated gas or liquid breakdown.)

Cranberg's model predicts that $VE \geq K$, where V is the breakdown voltage, E is the macroscopic electric field at the electrode where the clump originates, and K is a constant characteristic of a given pair of electrodes. For a uniform field and gap d, this implies that V is proportional to $d^{1/2}$, which agrees reasonably well with experiment. However, it does not predict the area effect, nor the polarity effect where the breakdown voltage is much higher when the pointed electrode is the anode. Moreover, it incorrectly predicts the curvature effect.

L. B. Snoddy was the first to suggest anode vaporization as a result of electron bombardment to account for breakdown. A. J. Ahearn was the first to suggest local heating of the cathode as initiating breakdown. L. C. Van Atta, R. J. Van de Graaff, and H. A. Barton were the first to hypothesize a particle interchange multiplication process. The basic assumption here is that, at a critical voltage, a free charged particle upon striking an electrode produces an avalanche of charged particles by secondary emission, with photoemission also playing a role.

There have since been many variations on these models. However, none has yet correctly predicted (even qualitatively) the area and curvature effects, as well as the gap dependence of approximately $d^{1/2}$. M. Rabinowitz introduced a hypothesis that predicts at least qualitatively the known experimental results, without reference to any particular model in terms of processes that are assumed to occur. He observed that the initiation of vacuum breakdown and gap conduction processes occurs so fast (on the order of 10^{-9} s), compared with the time constants of most breakdown circuits, that only the capacitively stored energy of the electrodes and electrode supports discharges within this time. (Light travels less than 1 ft or 30 cm in 10^{-9} s.) His assumption was simply that breakdown cannot occur until there is sufficient stored energy in the electric field in the gap, because only this energy (or a fraction of it) is available to break down the gap.

The energy in the field is the capacitively stored energy, $\frac{1}{2} CV^2$. The hypothesis leads to Eq. (3),

$$W = \tfrac{1}{2} fCV^2 \quad (3)$$

where W is the critical energy needed to initiate breakdown (characteristic of the electrodes), C is the capacitance of the electrode system, and f is the fraction of the capacitively stored energy available to produce breakdown, $0 < f \leq 1$. For a uniform field, this predicts that the breakdown voltage V is proportional to $d^{1/2}$. The curvature and area effects are also qualitatively predicted.

[MARIO RABINOWITZ]

Bibliography: J. D. Cobine, *Gaseous Conductors*, 1941; M. J. Druyvesteyn and F. M. Penning, Mechanisms of electrical discharges in gases at low pressures, *Rev. Mod. Phys.*, 12:87, 1940; H. Fröhlich, *Theory of Dielectrics: Dielectric Constant and Dielectric Loss*, 1949; H. S. W. Massey and E. H. S. Burhop, *Electronic and Ionic Impact Phenomena*, 1952; N. F. Mott and H. S. W. Massey, *The Theory of Atomic Collisions*, 1950; M. Rabinowitz, Electrical breakdown in vacuum: New experimental and theoretical observations, *Vacuum*, 15(2):59–66, 1965; R. E. Schramm, A. F. Clark, and R. P. Reed, *A Compilation and Evaluation of Mechanical, Thermal, and Electrical Properties of Selected Polymers*, National Bureau of Standards Monogr. no. 132, 1973; A. H. Sharbaugh, J. C. Devins, and S. J. Rzad, Progress in the field of electric breakdown in dielectric liquids, *IEEE Trans. Electr. Insul.*, EI-13(4):249–276, 1978; A. R. Von Hippel, *Dielectrics and Waves*, 1954.

Electrical resistance

That property of an electrically conductive material that causes a portion of the energy of an electric current flowing in a circuit to be converted into heat. In 1774 A. Henley showed that current flowing in a wire produced heat, but it was not until 1840 that J. P. Joule determined that the rate of conversion of electrical energy into heat in a conductor, that is, power dissipation, could be expressed by the relation given in notation (1).

$$H/t \propto I^2R \qquad (1)$$

The day-to-day determination of resistance by measuring the rate of heat dissipation is not practical. However, this rate of energy conversion is also VI, where V is the voltage drop across the element in question and I the current through the element, as in Eq. (2), from which the more conventional

$$H/t \propto I^2R = VI \qquad (2)$$

relationship implied by Ohm's law, Eq. (3), is apparent.

$$R = V/I \qquad (3)$$

[CHARLES E. APPLEGATE]

Electroless plating

A chemical reduction process which, once initiated, is autocatalytic. The process is similar to electroplating except that no outside current is needed. The metal ions are reduced by chemical agents in the plating solutions, and deposit on the substrate. Electroless plating is used for coating nonmetallic parts. Decorative electroless plates are usually further coated with electrodeposited nickel and chromium. There are also applications for electroless deposits on metallic substrates, especially when irregularly shaped objects require a uniform coating. Electroless copper is used extensively for printed circuits, which are produced either by coating the nonmetallic substrate with a very thin layer of electroless copper and electroplating to the desired thickness or by using the electroless process only. Electroless iron and cobalt have limited uses. Electroless gold is used for microcircuits and connections to solid-state components. Deeply recessed areas which are difficult to plate can be coated by the electroless process.

Nonmetallic surfaces and some metallic surfaces must be activated before electroless deposition can be initiated. Activation on nonmetals consists of the application of stannous and palladium chloride solutions. Once electroless plating is begun, it will continue to a desired thickness; that is, it is autocatalytic. The process thus differs from a displacement reaction, in which a more noble metal is deposited while a less noble one goes into solution; this ceases when the more noble deposit, if pore-free, covers the less noble substrate.

Electroless copper and gold deposits consist of very small crystals. Electroless nickel deposits are really highly supersaturated alloys containing phosphorus or boron, depending on the reducing agent used. These deposits, which are so fine-grained that they are almost amorphous, can be precipitation-hardened. Electroless nickel deposits are generally harder and more brittle than the electroplated variety. Corrosion resistance of electroless nickel depends, among other factors, on the uniformity of distribution of the phosphorus or boron. Adhesion to nonmetallic substrates is achieved primarily by mechanical means—by plating into pores which are created by selectively etching the substrate. An advantage of electroless plating over plating with current is the more uniform thickness of the surface coating. *See* ELECTROPLATING OF METALS.

[ROLF WEIL]

Electromagnet

A soft-iron core that is magnetized by passing a current through a coil of wire wound on the core. Electromagnets are used to lift heavy masses of magnetic material and to attract movable magnetic parts of electric devices, such as solenoids, relays, and clutches.

The difference between cores of an electromagnet and a permanent magnet is in the retentivity of the material used. Permanent magnets, initially magnetized by placing them in a coil through which current is passed, are made of retentive (magnetically "hard") materials which maintain the magnetic properties for a long period of time after being removed from the coil. Electromagnets are meant to be devices in which the magnetism in the cores can be turned on or off. Therefore, the core material is nonretentive (magnetically "soft") material which maintains the magnetic properties only while current flows in the coil. All magnetic materials have some retentivity, called residual magnetism; the difference is one of degree.

A magnet, when brought near other susceptible material, induces magnetic poles in the susceptible material and so attracts it. A force is developed in the susceptible material that tends to move it in a direction to minimize the reluctance of the flux path of the magnet. The reluctance force may be expressed quantitatively in terms of the rate of change of reluctance with respect to distance.

In an engineering sense the word electromagnet does not refer to the electromagnetic forces incidentally set up in all devices in which an electric current exists, but only to those devices in which the current is primarily designed to produce this force, as in solenoids, relay coils, electromagnetic brakes and clutches, and in tractive and lifting or holding magnets and magnetic chucks.

Electromagnets may be divided into two classes: traction magnets, in which the pull is to be exerted over a distance and work is done by reducing the air gap; and lifting or holding magnets, in which the material is initially placed in contact with the magnet. For examples of the first type *see* BRAKE; CLUTCH; RELAY; SOLENOID (ELECTRICITY).

Examples of the latter type are magnetic chucks and circular lifting magnets. The illustration shows a cross-sectional view of a typical circular lifting

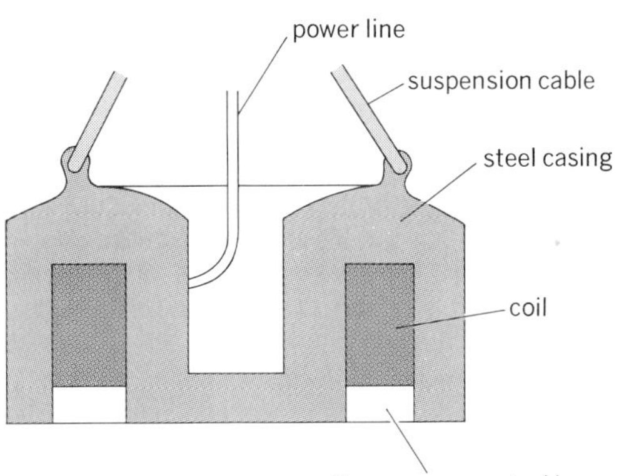

Cross section of circular lifting electromagnet.

magnet. The outer rim makes up one pole and the inner area is the opposing pole. The coil is wound cylindrically around the center pole. Manganese steel, used as a protective cover plate for the coil, is nonmagnetic and thus forces the flux through the magnetic member being lifted.

The mechanical force between two parallel surfaces is given by Maxwell's equation, shown below, where B is the flux density (in Wb/m^2), A is

$$F = B^2A/(2\,\mu_0) \quad \text{(newtons)}$$

the cross-sectional area (in m^2) through which the flux passes, and μ_0 is the permeability of free space ($4\pi \times 10^{-7}$ H/m). When two poles are active, the force produced by each is calculated to find the total force. An interesting result of this relation is that the force is not simply the result of the total flux (BA) but also of the flux density. Thus if the same flux can be forced through one-half the cross-sectional area, the net pull will be doubled. In practice it is difficult, if not impossible, to calculate the actual lifting capacity of the magnet by using Maxwell's equation since the capacity varies with the shape and kind of material lifted, how the material is stacked, and other factors. Therefore, lifting magnets are usually rated on their all-day average lifting capacity.

Since currents are large (10–20 A) and the circuit is highly inductive, control of a lifting magnet is a problem. If the line switch were simply opened, a destructive arc would result. Therefore, the controller employed with a lifting magnet usually does the following things automatically: (1) reduces magnet current after initial high value to reduce heating of the magnet, (2) introduces a shunt discharge resistor across the magnet coil before allowing the line to be opened when the operator turns the magnet off, and (3) causes a reduced current of reverse polarity to flow in the magnet coil for a short time after the operator turns the switch off. Thus the residual magnetism is canceled and scraps and small chunks that might have continued clinging to the magnet are released.

[JEROME MEISEL]

Bibliography: D. G. Fink and H. W. Beaty (eds.), *Standard Handbook for Electrical Engineers*, 11th ed., 1978; B. W. Jones, *Relays and Electromagnets*, 1935; H. C. Roters, *Electromagnetic Devices*, 1941; M. G. Say (ed.), *The Electrical Engineers' Reference Book*, 13th ed., 1973.

Electromagnetic pumps

Pumps that operate on the principle that a force is exerted upon a conductor (the fluid) carrying current in a magnetic field. The high electrical conductivity of liquid metals (used as heat-transfer media in some nuclear reactors) makes it possible to pump them by electromagnetic means. For use in nuclear reactors, where a minimal amount of maintenance is desirable, electromagnetic pumps are often preferable to conventional mechanical pumps because they have no moving parts, bearings, or seals. Various methods are employed to cause current to flow in the liquid metal.

Direct-current conduction pumps. These pumps are a direct application of the right-hand rule, which states that a current passing at right angles to a magnetic field will produce a force at right angles to both. Pump performance depends upon the magnitude of the current, magnetic field intensity, and the geometry of the pump duct. In its simplest form, a pump of this type consists of a rectangular tube with electrodes attached to the short sides of the rectangular section and with the long axis of the section placed between the poles of a magnet. Thus, current flowing through the fluid along the long axis is cut by the magnetic field and produces a longitudinal thrust on the fluid in the tube. Corrections must be made for the magnetic field produced by the flow of current through the duct walls, and provision must be made to minimize end losses (flow of current through the fluid but outside the magnetic field). The disadvantage of this type of pump is the very high current (thousands of amperes) at low voltage (1–2 volts) required.

Alternating-current induction pumps. Large currents can be developed in the liquid metal by electromagnetic induction. An ac induction pump consists of a duct in the form of a flattened tube extending between two core sections containing a three-phase ac winding. The winding is similar to that of an induction motor stator except that the field structure is flat and a sliding rather than a rotating magnetic field is produced. This pump employs conventional power supplies (60-hertz ac), but the field winding must be cooled to protect the electrical insulation.

Other types of electromagnetic pumps have been developed and employed for laboratory use. However, the two types described here have received the most attention, and their development has been carried to the most advanced levels, including large commercial-size units. *See* NUCLEAR REACTOR; PUMP. [LEONARD J. KOCH]

Bibliography: H. Etherington (ed.), *Nuclear Engineering Handbook*, 1958; J. M. Harrer and J. B. Beckerley, *Nuclear Power Reactor Instrumentation Systems Handbook*, National Technical Information Service, TID-25952-P1 and -P2, vol. 1, 1973, vol. 2, 1974; A. Sesonske, *Nuclear Power and Plant Design Analysis*, National Technical Information Service, TID-26241; 1973; J. G. Yevick (ed.), *Fast Reactor Technology: Plant Design*, 1966.

Electrometallurgy

The branch of process metallurgy dealing with the use of electricity for smelting or refining of metals. The electrochemical effect of an electric current brings about the reduction of metallic compounds, and thereby the extraction of metals from their ores (electrowinning) or the purification of the metals (electrorefining).

In other metallurgical processes, electrically produced heat is utilized in smelting, refining, or alloy manufacturing. For a discussion of electrothermics, that is, the theory and applications of electric heating to metallurgy, *see* ELECTRIC FURNACE; ELECTRIC HEATING; STEEL MANUFACTURE.

Electrowinning. This metallurgical process involves the recovery of a metal, usually from its ore, by dissolving a metallic compound in a suitable electrolyte and reducing it electrochemically through passage of a direct electric current.

Following beneficiation and, sometimes, chemical pretreatment, the metal-bearing constituent of the ore is dissolved in an aqueous solution or in a molten salt. The electrolysis of the purified electrolyte with direct current yields the reduced metal at the cathode (negative electrode of an electrolytic cell), and the nonmetal is the oxidation product at the anode (positive electrode). Since the process is very selective, the purity of the electrowon metal is high, and no further refining is needed.

Water is a suitable solvent for electrolysis of metals less active than zinc and manganese, and sulfuric acid is the preferred leaching agent. The high mobility of the solvated proton provides high-conductivity solutions, and the sulfate ion is electrochemically inert; acid is regenerated by anodic oxidation and recycled for leaching.

Zinc. The process of electrowinning of zinc was developed in 1915 for treating complex ores not amenable to thermal processing, and it is presently the process of choice for most new plants. Changes in the chemistry have been minor, but steady progress has been made in the areas of process instrumentation and control, automation, and pollution-free operation.

The zinc sulfide ores are concentrated, roasted, and leached in several stages with sulfuric acid, and the electrolyte is purified. Lead and silver are insoluble through the acid leach; iron precipitates by oxidation in the neutral leach, and it coprecipitates impurities such as arsenic, antimony, and germanium. Other impurities more noble than zinc, such as copper, cadmium, cobalt, and nickel, are removed by galvanic precipitation on zinc dust.

The reduction potential of zinc is far more cathodic (−0.763 V) than that of the hydrogen ion, and its efficient electrodeposition depends upon maintaining a high hydrogen overvoltage. Metals more noble than zinc deposit on the cathode, and most promote hydrogen evolution and zinc corrosion. This reduces the yield of zinc, which is expressed by the current efficiency, namely, the ratio of the amount of electricity (coulombs) theoretically required to yield a given quantity of metal to the amount actually consumed. Current efficiencies equal to 90% are achieved by bringing the concentration of the most harmful impurities such as germanium, arsenic, and antimony down to 0.01 mg/liter.

The zinc electrolyte, containing 120–160 g/liter of zinc as sulfate, is circulated in lead-lined or plastic-lined (for example, polyvinyl chloride) concrete or wooden tanks in which the electrolysis proceeds between vertically suspended electrodes. The insoluble anodes are lead that contains small amounts of silver, and the cathodes are aluminum sheets from which the zinc deposits are stripped every 24–28 hr. The submerged cathode area varies from 1.3 to 2.6 m^2 The current density varies from 300 to 600 A/m^2. From the Gibbs free-energy change for the chemical reaction shown in Eq. (1), a reversible voltage of 2 V can be calcu-

$$ZnSO_4 + H_2O = Zn + \tfrac{1}{2}O_2 + H_2SO_4 \qquad (1)$$

lated. The operating cell voltage is about 3.5 V; the difference is due to the anodic and cathodic overvoltages and the ohmic voltage drop in the electrolyte and at the electrode contacts. The power consumption averages 3.2 kWhr/kg zinc for modern plants. This relatively high energy cost is acceptable since special high-grade zinc is produced (99.995%).

Copper. Copper electrowinning is similar to zinc electrowinning, but the pyrometallurgical reduction is preferred for sulfide ores. Most of the electrowon copper comes from the direct leaching of low-grade sulfide ores, or from oxide ores. The recent development of liquid ion-exchange reagents specifically for copper makes possible an upgrading, through solvent extraction of dilute solutions, from dump or in-place leaching in order to obtain a concentrated solution suitable for electrolysis. The electrowinning of copper is carried out in sulfate solutions containing from 20 to 40 g/liter of copper and about 100 g/liter of acid at current densities from 160 to 200 A/m^2. The anodes are made of lead, alloyed with antimony, and copper is deposited on copper starting sheets.

Aluminum. Electrowinning has been the only process used for the commercial production of aluminum since it was developed independently by C. M. Hall and by P. L. T. Héroult in 1885. Alumina extracted from bauxite ores and purified by digestion in a caustic soda solution is dissolved in a mixture of cryolite ($AlF_3 \cdot 3NaF$) and other fluorides (for example, CaF_2 and LiF). The molten-salt electrolyte floats on the molten aluminum, which is the cathode, and carbon anodes dip into the bath from above.

The carbon reacts with the oxygen produced by anodic oxidation. The overall cell reaction, shown in Eq. (2), corresponds to a theoretical reversible

$$Al_2O_3 + \tfrac{3}{2}C = 2Al + \tfrac{3}{2}CO_2 \qquad (2)$$

voltage of 1.2 V, but the cell operates under a total voltage of about 4.2 V. The heat generated by irreversible electrochemical and ohmic losses maintains the cell temperature at about 960°C.

Current densities are about 1 A/cm^2, with total currents of 100,000 A for a modern cell. Alumina is periodically added to the melt to maintain its concentration between 2 and 5%. The current efficiency averages 89%; the main losses are attributed to the recombination of aluminum by reduction of carbon dioxide to carbon monoxide. The energy consumption is about 14 kWhr/kg aluminum.

Magnesium. The electrowinning of magnesium, another well-established process, accounts for about 80% of the metal output. The electrolyte is a mixture of chlorides of potassium, sodium, calcium, and magnesium. Since the reduced metal has a lower specific gravity than the electrolyte has, it floats to the surface; the cells are designed so that it does not come in contact with air or with the chlorine gas produced at the anode.

Other metals. Cadmium, and in smaller quantities, thallium and indium, are recovered by electrolysis in sulfate solutions. Cobalt is also electrowon by a similar process, and some of the nickel production is achieved by electrolysis in sulfate-chloride solutions. Pure metallic chromium and manganese are produced by electrowinning; these metals are rather difficult to electrodeposit, and the acid produced at the anode must be kept away from the cathode by a diaphragm. Antimony is electrowon from a sodium sulfide electrolyte, and gallium is produced by electrolysis of a caustic solution.

Metals more active than manganese can not be reduced cathodically from aqueous solutions, but most can be electrowon from mixtures of molten salts. Usually, a compound of the metal is dissolved in salts of still more active metals in order to improve the conductivity and lower the melting point of the melt. Solid metallic deposits are difficult to recover, and all the commercially successful processes yield the liquid metal. The solubility and reactivity of the reduced metal in the melt is often important, and the cells must be designed to protect the reduced metal from contact with the anodic reaction product or with air. The cell electrolyte cannot be easily recycled, and the feed must be carefully purified and dehydrated.

All the sodium metal produced today comes from electrolysis of a molten chloride mixture between a steel cathode and a graphite anode separated by a perforated steel diaphragm. Lithium is produced by a similar process. The other alkali and alkaline-earth metals, as well as the rare earths, can be electrowon from molten chlorides. Processes have been developed for electrowinning titanium, tantalum, and niobium in molten salts, and uranium, thorium, and zirconium have been obtained in the laboratory. Beryllium and boron are also produced by electrolysis.

Electrorefining. This is a purification process in which an impure metal anode is dissolved electrochemically in a solution of a salt of the metal being refined; the pure metal is recovered by electrodeposition at the cathode.

Electrorefining is the most economical method for securing the high purity required for many uses of nonferrous metals. Since the same electrochemical reaction proceeds in opposite directions at the anode and the cathode, the overall chemical change is a small change in the activity of the metal at the two electrodes, and the reversible cell voltage is practically zero. The anodic and cathodic overvoltages, required for an economical reaction rate, are small, and the operating cell voltage consists mainly of the ohmic voltage drop through the electrolyte and at the electrode contacts. The power consumption is moderate, and electrorefining is less sensitive to the cost of electric power than other electrometallurgical processes are.

Electrochemical refining is a more efficient purification process than other chemical methods are because it is very selective. In particular, for metals such as copper, silver, gold, and lead, the operating electrode potential is close to the reversible potential, and a sharp separation is accomplished, both at the anode, where more noble metals do not dissolve, and at the cathode, where more active metals do not deposit.

Lifting load of electrorefined copper from refining cell. (*Anaconda Copper*)

The first commercial electrolytic copper refinery was established in 1871, and today most of the copper is electrorefined. Minor quantities of some impurities lower the electrical conductivity of copper markedly, and the refining process ensures that the metal will meet the specifications of the electrical industry. Furthermore, silver, gold, and other precious metals are removed by the electrolytic refining, and their recovery is thus an economic asset of the process.

The impure copper is cast in the shape of anodes approximately 0.9 by 1.0 m and 3.5 to 4.5 cm thick. The starting cathodes are pure copper sheets, prepared by electrodeposition (see illustration). The electrodes are placed vertically in cells which are lead-lined or plastic-lined wooden or concrete tanks. The anodes are replaced at regular intervals of 20–28 days, and two successive cathodes are produced during the same period. The solution contains about 45 g/liter of copper as copper sulfate and approximately 200 g/liter of sulfuric acid. The temperature is maintained at 55–60°C to lower the resistance of the electrolyte which is circulated through the cell. The current density is usually 200–250 A/m^2, and the current efficiency averages 95%. The cell voltage is 0.2–0.3 V, and the power consumption varies from 0.18 to 0.25 kWhr/kg copper.

Metals less noble than copper, such as iron, nickel, cobalt, zinc, and manganese, dissolve from the anode. Lead is oxidized, and it precipitates as lead sulfate. Other impurities such as arsenic, antimony, and bismuth partly remain as insoluble compounds in the slimes and partly dissolve as complexes in the electrolyte. Precious metals do not dissolve at the potential of the anode, and they remain as metals in the slimes, along with insoluble selenium and tellurium compounds and some metallic copper particles that fall from the anode. Metals less noble than copper do not deposit at the cathode. In particular, nickel and arsenic accumulate in the electrolyte, and their concentration is controlled by partial withdrawal and purification of the solution.

Nickel is electrorefined in a sulfate-chloride electrolyte. Its anodic dissolution requires a significant overvoltage, and both copper and nickel dissolve at the anode potential. To prevent plating of the dissolved copper at the cathode, a diaphragm cell is used, and the anodic solution is circulated through a purification circuit before entering the cathode compartment. Cobalt is electrorefined by a similar process.

Very-high-purity lead is produced by electrorefining in which a fluosilicate electrolyte is used, and a sulfate bath is used for purifying tin. Silver is electrorefined in a copper and silver nitrate electrolyte, and gold is refined by electrolysis in a chloride bath.

The electrorefining of many metals can be carried out with molten-salt electrolytes, but the processes are usually expensive, and the only industrial application is the electrorefining of aluminum by the "three-layer" process. The density of the electrolyte is adjusted so that a pure molten aluminum cathode will float on the molten salt, which in turn will float on an anode consisting of a molten mixture of impure aluminum and copper. *See* PYROMETALLURGY.

[PAUL DUBY]

Bibliography: A. Kuhn, *Industrial Electrochemical Processes*, 1971; A. G. Robiette, *Electric Smelting Processes*, 1973.

Electromotive force (emf)

The electromotive force, represented by the symbol $\mathcal{E}$, around a closed path in an electric field is the work per unit charge required to carry a small positive charge around the path. It may also be defined as the line integral of the electric intensity around a closed path in the field. The abbreviation emf is preferred to the full expression since emf, also called electromotance, is not really a force. The term emf is applied to sources of electric energy such as batteries, generators, and inductors in which current is changing.

The magnitude of the emf of a source is defined as the electrical energy converted inside the source to some other form of energy (exclusive of electrical energy converted irreversibly into heat), or the amount of some other form of energy converted in the source into electrical energy, when a unit charge flows around the circuit containing the source. In an electric circuit, except for the case where an electric current is flowing through resistance and thus electrical energy is changed irreversibly into heat energy, electrical energy is converted into another form of energy only when current flows against an emf. On the other hand, some other form of energy is converted into electrical energy only when current flows in the same sense as an emf. *See* DIRECT-CURRENT MOTOR.

[RALPH P. WINCH]

Bibliography: E. M. Purcell, *Electricity and Magnetism*, Berkeley Physics Course, vol. 2, 1965; R. Resnick and D. Halliday, *Physics*, 3d ed., 1978; F. W. Sears et al., *University Physics*, 5th ed., 1976.

Electroplating of metals

The process of electrodepositing metallic coatings to alter the existing surface properties or the dimensions of an object. Electroplated coatings are applied for decorative purposes, to improve resistance to corrosion or abrasion, or to impart desirable electrical or magnetic properties. Plating is also used to increase the dimensions of worn or undersized parts. An example of a decorative coating is that of nickel and chromium on automobile bumpers. However, in this application, corrosion and abrasion resistance are also important. An example of electrodeposition used primarily for corrosion protection is zinc plating on such steel articles as nuts, bolts, and fasteners. Since zinc is more readily attacked by most atmospheric corrosive agents, it provides galvanic or sacrificial protection for steel. An electrolytic cell is formed in which zinc, the less noble metal, is the anode, and steel, the more noble one, the cathode. The anode corrodes, and the cathode is protected. Deposited metals which are more noble than the substrate, the part upon which the coating is applied, protect against corrosion only if they are completely continuous. Should a small area of the substrate be exposed, it corrodes very rapidly.

An example of an electroplated coating applied primarily for wear resistance is hard chromium on a rotating shaft. To impart desirable electrical properties, gold is often electroplated on contacts, for example. The absence of oxide films, which would raise the electrical resistance, is desired on such contacts. In the case of permalloy, an alloy of iron and nickel plated on copper wires which is used for computer information storage, the magnetic properties are of greatest importance.

The electroplating process consists essentially of connecting the parts to be plated to the negative terminal of a direct-current source and another piece of metal to the positive pole, and immersing both in a solution containing ions of the metal to be deposited. The part connected to the negative terminal becomes the cathode, and the other piece the anode. In general, the anode is a piece of the same metal that is to be plated. Metal dissolves at the anode and is plated at the cathode. If the applied current is used only to dissolve and deposit the metal to be plated, the process is 100% efficient. Often, fractions of the applied current are diverted to other reactions such as the evolution of hydrogen at the cathode, which results in lower efficiencies as well as changes in the acidity (pH) of the plating solution. In some processes, such as chromium plating, a piece of metal which is essentially insoluble in the plating solution is the anode. When such insoluble anodes are used, metal ions in the form of soluble compounds must also be added periodically to the plating solution. The anode area is generally about the same as that of the cathode; in some applications it is larger.

Most plating solutions are of the aqueous type. There is a limited use of fused salts or organic liquids as solvents. Nonaqueous solutions are employed for the deposition of metal with lower hydrogen overvoltages; that is, hydrogen rather than the metal is reduced at the cathode in the presence of water.

In addition to metal ions, plating solutions contain relatively large quantities of various substances used to increase the electrical conductivity, to buffer, and, in some instances, to form complexes with the metal ions. Relatively small amounts of other substances, which are called addition agents, are also present in plating solutions to level and brighten the deposit, to reduce internal stress, to improve the mechanical properties, and to reduce the size of the metal crystals or grains or to change their orientation.

The quantity of metal deposited, that is, the thickness, depends on the current density (amp/m^2), the plating time, and the cathode efficiency. The current is determined by the applied voltage, the electrical conductivity of the plating solution, the distance between anode and cathode, and polarization. Polarization potentials develop because of the various reactions and processes which occur at the anode and cathode, and depend on the rates of these reactions, that is, the current density. If the distance between anode and cathode varies because the part to be plated is irregular in shape, the thickness of the deposit may vary. A quantity called the throwing power represents the degree to which a uniform deposit thickness is attained on areas of the cathode at varying distances from the anode. Good throwing power results if the plating efficiency is low due to polarization where the current density is high. A plating solution such as an alkaline cyanide bath, in which the polarization of the cathode increases strongly with increasing current density, has good throwing power. On the other hand, a plating solution such as an acid-sulfate copper bath, which is almost 100% efficient, has poor throwing power.

Two other deposition processes are closely related to electroplating: electroless plating and displacement plating. Both processes require no applied current. Electroless plating is a process which, once initiated, continues; that is, it is autocatalytic. Displacement plating occurs when the metal deposited is more noble than the substrate, and the substrate dissolves. The reaction ceases when the substrate is completely covered by a pore-free deposit. *See* ELECTROLESS PLATING.

PREPARATION AND EQUIPMENT

In order for adherent coatings to be deposited, the surface to be plated must be clean, that is, free from all foreign substances such as oils and greases, as well as oxides or sulfides. The two essential steps are cleaning and pickling.

Cleaning. Three principal methods are employed to remove grease and attached solids. (1) In solvent cleaning, the articles undergo vapor degreasing, in which a solvent such as tri- or tetrachloroethylene is boiled in a closed system, and its vapors are condensed on the metal surfaces. (2) In emulsion cleaning, the metal parts are immersed in a warm mixture of kerosine, a wetting agent, and an alkaline solution. (3) In electrolytic cleaning, the articles are immersed in an alkaline solution, and a direct current is passed between them and the other electrode, which is usually steel. Heavily soiled articles are cleaned in solvent or alkaline spray machines. Cleaning solutions may contain sodium hydroxide, carbonate, phosphate, and metasilicate, plus wetting agents and chelating agents. More highly alkaline solutions are used for steel than for other metals. Most of the cleaning is accomplished by the scrubbing action of the evolved gases and the detergency of the components of the solution. The articles may be connected as anodes, as is usual for steel, or as cathodes, as is usual for other metals. Electrolytic cleaning is usually the last cleaning step.

Ultrasonic cleaning is also used extensively, especially for blind holes or gears packed with soils. Ultrasonic waves introduced into a cleaning solution facilitate and accelerate the detachment of solid particles embedded in crevices and small holes. Frequencies from about 18,000 to 24,000 Hz are usually employed. They produce cavitation, which causes rapid local circulation.

Pickling. In this process, oxides are removed from the surface of the basis metal. For steel, warm, dilute sulfuric acid is used in large-scale operations because it is inexpensive; but room-temperature, dilute hydrochloric acid is also used for pickling because it is fast-acting. In cathodic pickling of steel, attack of the metal is retarded while the oxide is being dissolved. In addition to rough pickling, acid treatments to activate the surface just prior to plating are often used.

Hydrogen embrittlement may be caused by the diffusion of hydrogen in steel during pickling and also in certain plating operations. Especially with high-carbon steels, hydrogen causes cracking (a reduction in the fatigue strength and ductility). Hydrogen is gradually evolved on standing, and more rapidly evolved by heating to about 200°C. *See* EMBRITTLEMENT.

Electropolishing. Electropolishing is used when a thin, bright deposit is to be produced. In this case, the substrate surface contours are essentially copied by the deposit. The substrate surface therefore must have a bright finish which can be attained by electropolishing. *See* ELECTROPOLISHING.

Equipment used in plating. The important parts of typical electroplating equipment are shown in the diagram.

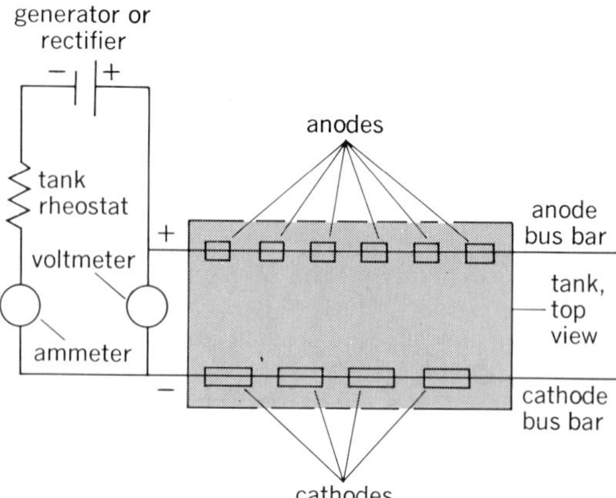

Typical connections for simple electroplating process.

Electrical equipment. Most plating operations are conducted with direct current at potentials of 3–12 V. Since electricity is delivered as alternating current at 220 V, it is necessary to reduce the voltage and to rectify the current. Both motor generators and rectifiers are used. Selenium rectifiers have been used extensively in plating. Since 1955 germanium and silicon rectifiers have also been employed. *See* DIRECT-CURRENT GENERATOR.

Mechanical equipment. The plating tanks are either polyvinyl chloride (PVC) or steel, which requires no lining for alkaline solutions. For neutral and acid solutions, steel tanks are lined with rubber or plastic. For chromic acid baths, lead or plastic linings are used. Most large plating operations are conducted in conveyor tanks. In semiautomatic conveyors, the cathode racks are carried only through the plating tank. In fully automatic conveyors, the cathodes are carried successively through tanks that contain cleaning, pickling, and plating solutions, with intermediate rinses. Plating of continuous strip is another very important operation. It is used extensively for tin plating.

Small objects are plated in "barrels," usually of hexagonal shape, with perforated plastic sides. The barrels rotate on a horizontal axis in a tank containing the plating solution and anodes. The articles contact cathode connections as they tumble during the rotation of the barrels.

PLATING OF SPECIFIC METALS

Most metals can be electroplated from either aqueous or fused-salt solutions. The more important metals plated from aqueous baths are chromium, copper, gold, nickel, silver, tin, and zinc. Alloys can also be electroplated. Electrodeposited copper-zinc and lead-tin alloys are used extensively.

Chromium. Chromium electrodeposits used for decorative purposes are applied as thin coatings over much thicker nickel plate. Chromium deposits are corrosion- and wear-resistant and preserve a bright surface. Chromium deposits are often cracked and thus expose the undercoat (the anode) to corrosive mediums. In order to increase the nickel area which is thus exposed and to reduce thereby corrosion at a particular site intentionally microcracked or microporous chromium is plated. Thick chromium electrodeposits are also used in application in which wear resistance is of primary importance. Chromium plating solutions consist of chromic acid, sulfuric acid, and water. The throwing power and current efficiency are rather poor. Some trivalent chromium baths are also used, but on a smaller scale.

Copper. Copper electrodeposits are frequently used as undercoats for decorative nickel-chromium plates. Copper plating also has wide usage in the electrical and electronics industries.

Several methods involving plating are used to produce printed circuits on an insulating-material substrate. The entire copper circuit may be produced by activating the surface with stannous and palladium-chloride solutions and then applying electroless copper plating. Rather thin electroless deposits can serve as the substrate for thicker electrodeposits. Copper sheet laminated to plastic is also used extensively; in this case, the electroless plating and electroplating of copper are performed only to make through holes. The actual circuit can be produced by selective etching or selective plating involving techniques such as photosensitizing, photoresist, and etch resist, which also use masking. The copper segments of the printed circuit may be coated with an electrodeposit of a lead-tin alloy to facilitate subsequent soldering. Gold deposits can also be used for this purpose. The use of gold is limited to special applications partly because of cost.

Acid sulfate and alkaline cyanide plating solutions are used most widely. The acid sulfate solution has a very high efficiency and is easy to use, but it has poor throwing power and cannot be directly applied to steel because a displacement coating, which adheres poorly, would result. The alkaline cyanide solution has good throwing power and can be applied directly to steel as well as to zinc-base die castings, but it has the disadvantage of being a pollutant. Fluoborate, pyrophosphate, and sulfamate copper plating solutions are also used.

The most widely used solution for electroless copper plating contains formaldehyde as the reducing agent and copper ions from a sulfate salt. The solution is used at a pH of about 11. Compounds called stabilizers are also present to prevent electroless copper plating from forming on small particles floating in the bath or on other areas which are not to be coated. A chelating agent and cyanide, to reduce the inclusion of hydrogen gas bubbles in the deposit, are also present in electroless copper plating solutions. The presence of gas bubbles can markedly reduce the ductility of electroless copper deposits.

Gold. Gold plating is used for jewelry and for electrical contacts which must remain oxide-free. Connections for microcircuits, certain information-storage devices, and solid-state components are also gold-plated.

Acid gold-plating baths operated at pH values from 3 to 6 and containing potassium gold cyanide and acetic or citric acid are popular. As insoluble anodes are generally used, periodic replenishment of metal ions is required. Alkaline cyanide plating baths have a more limited application, but those based on sulfide complexes are becoming more important. Various color shades can be attained by codepositing nickel, silver, copper, cobalt, or indium with gold. An electroless gold-plating solution is also in use.

Nickel. Nickel plating is carried out extensively. Nickel coatings covered with chromium provide corrosion-resistant and decorative finishes for steel, brass, and zinc-base die castings. The most widely used plating solution is the Watts bath, which contains nickel sulfate, nickel chloride, and boric acid. All-chloride, sulfamate, and fluoborate plating solutions are also used. There are a number of compounds, mostly organic, which can be added to nickel-plating baths. When specularly bright nickel is plated, at least one from each of two groups of compounds called class I and class II brighteners must be added to the plating solution. Class I compounds, for example, benzene and naphthalene disulfonic acids, contain sulfur. Class II brighteners, for example, butyne diol, do not contain sulfur. Class II brighteners can also cause leveling and, if added to the plating bath without the sulfur-containing agents, result in semibright nickel deposits. For improved corrosion protection, a duplex nickel deposit, that is, a relatively thick, semibright, sulfur-free plate covered with a thinner bright coating, is frequently used. A thin chromium layer is plated over the bright nickel. If corrosion begins through a pore or crack in the chromium, it will penetrate to the more noble semibright nickel and then spread laterally. Thus unsightly rust spots which result from penetration to the steel are less likely to develop.

Electroless nickel deposits are used for decorative purposes on nonmetallic substrates. Adhesion occurs mainly mechanically by the penetration of the deposit into crevices caused by a prior selective-etch treatment of the substrate. The reducing agents in electroless nickel-plating solutions are either hypophosphites or borohydrides. The deposits are highly supersaturated solid solutions of phosphorus or boron in nickel. The crystal size of these alloys is so small that they can be considered to be essentially amorphous. Electroless nickel plated on metallic substrates produces a uniform thickness over irregularly shaped parts.

Silver. The principal use of silver electrodeposits is for tableware because of their corrosion resistance (except to sulfur-containing foods) and pleasing appearance. Other important uses are for bearings and electrical circuits, waveguides, and hot-gas seals. The plating solutions are of the cyanide type and generally contain additives that produce bright deposits.

Tin. The principal purpose of plating tin is to coat steel sheet to make electrolytic tin plate. Tin is more noble than steel in the atmosphere and therefore does not provide sacrificial corrosion protection; but it can do so in hermetically sealed cans. Tin-plated steel is therefore used extensively for food preservation. Other tin-plating applications are for refrigerator coils, bearing surfaces, and parts which are to be subsequently soldered. A bright pore-free coating can be produced by melting a thin tin plate on the steel and allowing it to "reflow."

Stannous chloride and sulfate are the main components of acid tin-plating solutions. The alkaline bath contains sodium or potassium stannate. Both types are used for making electrolytic tin plate.

Zinc. The sacrificial protection of steel against corrosion is the main reason for zinc plating. Electrodeposited zinc coatings can be applied as thinner coatings, and with higher purity than the hot-dipped type. A conversion film formed by dipping in a chromate solution inhibits the formation of white corrosion products. Cadmium is widely used as a substitute for zinc.

The alkaline cyanide zinc-plating solution is still the most frequently used solution because of its good throwing power, relative ease of operation, and the availability of brighteners. The pollution problem has spurred interest in alkaline cyanide–free baths. Acidic baths are also used, especially for continuous-wire and strip plating and electrowinning.

PROPERTIES OF ELECTRODEPOSITS

In the various applications of electrodeposits, certain properties must be controlled and, therefore, must be measured. Specifications have been supplied by the American Society for Testing and Materials, and by other technical societies and some governmental agencies, establishing the various properties and the methods of determining them. The properties of electroplated metal which should be considered, depending on the use of the deposit, are thickness, adhesion to the substrate, brightness, corrosion resistance, wear resistance, the mechanical properties of yield strength, tensile strength, ductility, and hardness, internal stress, solderability, density, electrical conductivity, and the magnetic characteristics.

Thickness. Several properties depend on the thickness of an electrodeposit and how it varies over the substrate surface. Several types of gages, as well as microscopic, gravimetric, and chemical methods, are used to measure thickness.

Magnetic gages can be used if either the deposit or the substrate is ferromagnetic. The force of attraction between the sample and a magnet is measured. If the substrate is magnetic, the force of attraction decreases with the thickness of the plate; if the deposit is magnetic, the force of attraction increases with the thickness of the plate. Eddy-current gages are based on the principle that a high-frequency alternating current is sensitive to the thickness of a deposit if the electrical conductivity of the deposit differs greatly from that of the substrate, provided that neither is magnetic. Beta backscatter gages employ radioisotopes which emit beta rays. A detector measures the intensity of beta rays which are backscattered. If the atomic numbers of the coating and substrate are sufficiently different, as is the case for gold and nickel, the backscattered intensity is sensitive to the deposit thickness. This is the only method used for measuring the thickness of gold plate. The density of the plated metal must also be known.

The measurement of the thickness of a deposit using a metallographically prepared cross section can be performed microscopically. The cross section must be precisely normal to the deposit surface, or, as in the case of very thin deposits, to a known angle of inclination. Generally a calibrated (filar) eyepiece is used. The scanning electron microscope can also be used to determine the thickness of the cross section.

Chemical thickness-measurement methods involve the determination of the time required to dissolve the deposit, either by allowing a reagent to drip onto the sample at a calibrated rate or by electrolytic means. It is important that a well-recognizable change occur when the deposit is dissolved. X-ray fluorescence can also be used for thickness measurements. Gravimetric methods consist of separating the deposit from the substrate and weighing the deposit, or weighing the substrate before plating and again with the deposit attached to it. In both gravimetric methods, the density and area of the deposit must be accurately known.

Adhesion. Good bonding between substrate and deposit is very important in almost all plating applications. Poor adhesion results in peeling or blistering of the deposit. Generally, the adhesion between two metals is strong. Poor bonding results if the substrate surface was not clean prior to plating, if foreign substances were adsorbed, or if brittle phases form as a result of interdiffusion at elevated temperatures. As already discussed, the bond between a nonmetallic substrate and the electroless plate is primarily mechanical and can therefore vary considerably.

Adhesion tests are not well standardized. In the peel test, which is widely used, part of the deposit is loosened from the substrate, and the force required to pull off the rest is measured. A tensile test, in which the bonded interface is placed perpendicular to the applied force, is also used. Scraping tests and attempts to interpose a knife edge between substrate and deposit are examples of qualitative tests.

Brightness. For an electrodeposit to be specularly bright, the hills and valleys of the surface morphology should not vary in height by more than the wavelength of light. This requirement is generally fulfilled when the crystal size is very small. Addition agents in the plating solutions are also adsorbed on faster-growing sites, permitting the microscopic recessed region to catch up. Quantitative measurements of brightness are rarely performed; comparisons by eye are generally made.

Corrosion. With sacrificial coatings, corrosion protection occurs as long as enough of the coating remains; thus protection depends mainly on the thickness of the plate. In order for a substrate to be protected against corrosion by a deposit which is more noble, the coating must be completely free of pores, cracks, or discontinuities. The substrate must not be exposed at any point to the corrosive medium. When the less noble substrate is completely covered, the corrosion resistance of the coating itself is enhanced if the structure and chemical compositions are uniform.

Corrosion testing is performed under either actual or simulated service conditions or under accelerated exposure. The conditions under which accelerated exposure tests are performed are more severe than those encountered in service so that the results can be seen more quickly. Salt-spray and acetic acid–modified salt-spray accelerated corrosion tests are most frequently used. The CASS test and the Corrodkote test, in which parts are exposed to a slurry of clay in a salt solution, are examples of the second type. These tests are used to duplicate under accelerated conditions the corrosion of plated automotive parts in winter conditions. *See* CORROSION.

Wear resistance. Although a high hardness value is an indication of good wear resistance, there are other factors relating to wear resistance which are not well understood. There are also several types of wear, the two most common types being adhesive and abrasive wear. Corrosion and lubrication can greatly affect wear. Wear resistance is generally determined by moving two contacting surfaces relative to one another and noting the weight loss after some period of time.

Mechanical properties. In many respects, the mechanical properties of electrodeposits are similar to those of metals which have been heavily plastically deformed, that is, cold-worked, in that they exhibit higher strength and hardness and lower ductility than the same materials in the as-cast or annealed state. The number of such crystal defects as dislocations and lattice vacancies in electrodeposits is of the same order of magnitude as that found in severely cold-worked metals. Electrodeposited metals also recrystallize on heating. The small crystal size which electrodeposited metals frequently possess, and the inclusion in the crystal lattice of foreign materials from the plating-bath additives and side reactions, contributes to the strength. Many electrodeposited metals also have a large number of growth twins, which can increase strength. In general, the same factors that increase strength reduce ductility. Annealing electrodeposits can reduce strength and improve ductility. However, both strength and ductility are reduced upon annealing if a brittle grain-boundary phase is formed, as in the case of nickel deposits from plating solutions to which certain sulfur-containing agents have been added. Pores due to included gas also lower both strength and ductility. *See* METAL, MECHANICAL PROPERTIES OF.

The mechanical properties of electrodeposits can be determined by a tensile test in which the same type of specimen used to test sheet is employed. Care must be taken so that the specimen is accurately aligned and the edges are not serrated from cutting. The bulge test is being used more often, especially because it is less sensitive to porosity than the standard tensile test. Bend tests to assess ductility are of little quantitative value. Hardness is usually measured by an indentation test on a metallographically prepared cross section of the deposit. The Knoop indentor is particularly useful for evaluating relatively thin deposits. *See* HARDNESS SCALES.

Internal stress. Most electrodeposits are in a state of internal tensile stress. The stress can be reduced or made compressive by adding certain materials to the plating bath. The addition of saccharin to nickel-plating solutions has this effect. The causes of the internal tensile stress are not well understood, but there are indications that the coalescence of crystals or crystallites is one important cause.

Internal stress can be measured from the bending of a strip plated on only one side or from the change in length of a specimen plated on both sides. A long strip wound into a spiral, with one end fixed and the other free, is the most frequently used stress-measuring device. If this spiral is plated on one side, stress causes the free end to rotate. The rotation is calibrated to yield a stress value.

Solderability. Solderability is an important property, especially for plated segments of printed circuits. Solderability depends on the cleanliness of the surface, especially the degree of oxidation after prolonged storage. Coating copper deposits by plating them with solder, tin, or gold can prevent oxidation and thus improve solderability. Solderability is generally determined by how well molten solder pools spread. The spreading is also affected by the temperature and the flux. Because of the different coefficients of thermal expansion of the printed-circuit substrate and the copper deposit, high stresses can develop on heating for soldering. These stresses can result in the fracture of thin and brittle plated segments, especially at through holes. *See* SOLDERING.

Density, and electrical and magnetic properties. There are no special methods of measuring the density or electrical and magnetic properties of electrodeposits. The density of electrodeposits can be reduced because of the presence of gas or other foreign materials. Lattice vacancies can result in higher electrical resistivities of plated metals. The very fine grain size of electrodeposited or electroless plated ferromagnetic metals results in high coercive forces.

SPECIAL APPLICATIONS

Special processes, such as electroforming and anodizing, are required for certain applications, as discussed below.

Electroforming. Electroforming is a special type of plating in which thick deposits are subsequently removed from the substrate, which acts as a mold. The process is particularly suitable for forming parts which require intricate designs on inside surfaces, for example, waveguides. Intricate machining operations can be performed much more easily on outside surfaces. First, the outside surface of the substrate mold is machined; then, the contours of the design can be transferred to the inside surface of the deposit, and, finally, the deposit is separated from the mold. The mold or matrix can be either metallic or nonmetallic. Nonmetallic molds must be rendered electrically conductive by the application of a powder or by chemical reduction, electroless plating, or vapor deposition. For nonadherent deposits, substrate removal is easy; otherwise, the substrate must be dissolved or melted away. Important applications of electroforming are in the production of phonograph record masters, printing plates, and some musical instruments and fountain pen caps as well as in waveguides.

Anodizing. In anodizing, a process related to plating, an oxide is deposited on a metal which is the anode in a suitable solution. The process is primarily used with aluminum, but it can be applied to beryllium, magnesium, tantalum, and titanium. Relatively thick oxide deposits can be produced, even though they are electrically insulating. The presence of small pores through which the solution can reach the metal surface permits the continuation of the reaction. The electrical current is carried by the electrolyte in the pores.

Solutions for anodizing aluminum generally contain either chromic or sulfuric acid. After anodizing, the pores should be sealed, to improve protection of the substrate, by a hot-water or steam treatment which causes hydration and a resulting volume expansion of the oxide. Colored coatings can be produced by the incorporation of dyes.

[ROLF WEIL]

Bibliography: W. Blum and G. B. Hogaboom, *Principles of Electroplating and Electroforming*, 1949; A. Brenner, *Electrodeposition of Alloys*, 1963; D. G. Foulke and F. D. Crane (eds.), *Electroplaters Process Control Handbook*, 1963; A. K. Graham (ed.), *Electroplating Engineering Handbook*, 1971; F. A. Lowenheim (ed.), *Modern Electroplating*, 1974; R. Sard et al. (eds.), *Properties of Electrodeposits: Their Measurement and Significance*, 1975.

Electropolishing

A method of polishing metal surfaces by applying an electric current through an electrolytic bath in a process that is the reverse of plating. The metal to be polished is made the anode in an electric circuit. Anodic dissolution of protuberant burrs and sharp edges occurs at a faster rate than over the flat surfaces and crevices, possibly because of locally higher current densities. The result produces an exceedingly flat, smooth, brilliant surface.

Electropolishing is used for many purposes. The brilliance of the polished surface makes an attractive finish. Because the polished surface has the same structural properties as the base metal, it serves as an excellent surface for plating. Electropolishing avoids causing differential surface stresses, one of the requirements for the formation of galvanic cells which cause corrosion. Because no mechanical rubbing is involved, work hardening is avoided. Contaminants, which often are associated with the use of abrasives and polishing compounds, are also avoided. The surface is left clean and may require little or no preparation for subsequent treatment or use. Electropolishing also minimizes loss of high-temperature creep-rupture strength.

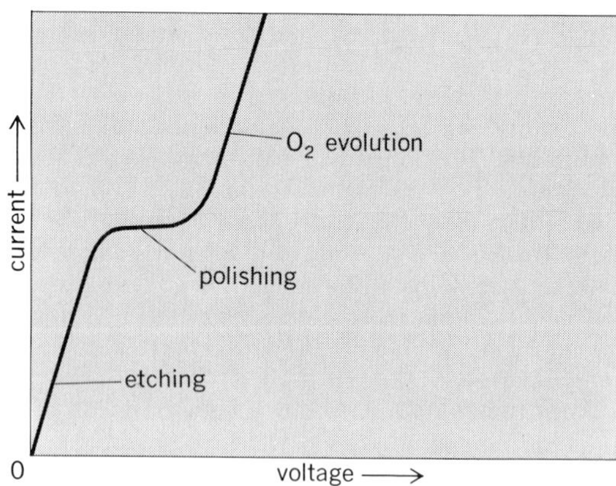

Anode current versus anode voltage in electropolishing.

In electropolishing, the work is submerged in an electrolyte and connected to the positive terminal (anode) of a source of direct-current (dc) power. The negative terminal is connected to an electrode

Electropolishing processes

Metal to be polished	Electrolyte	Temperature, °F	Current density, amp/ft²	Polishing time, min
Aluminum	Phosphoric acid and ethylene glycol	180	140	2–5
Brass and copper	Copper cyanide, zinc cyanide, and potassium cyanide	80	50	0.1–0.5
Molybdenum and niobium	Sulfuric acid and hydrofluoric acid in methyl alcohol	80–130	4500	0.05–0.4
Silver	Silver (powder), potassium cyanide, and potassium carbonate	80–125	15–20	0.1–0.2
Steel	Sulfuric acid and phosphoric acid, chromic or humic acids may be added	160	15–25	2–5
Tantalum and tungsten	Sulfuric acid and hydrofluoric acid in methyl alcohol	80–130	4500	0.05–0.4
Zinc*	Chromic oxide	65–90	1500	1–1.5

*After treatment, zinc surfaces must be successively dipped in dilute solutions of potassium chromate (2 min) and hydrochloric acid (5 sec).

that will resist chemical interaction with the electrolyte. Carbon is often used. The electrolyte is usually a concentrated acid, although alkaline solutions, dilute acids, and salts have been successfully employed.

Current density on the work surface is crucial. The relationship of anode current to anode voltage is shown in the illustration. Below a certain level of voltage, etching occurs. Above that level a constant-current region is reached in which polishing occurs. One mechanism proposed to explain this effect suggests that a film, or a systematic structuring of solution contents and ions, forms over the work surface and that this film, being thinner over burrs and sharp protuberances, allows selective dissolution of the surface to cause flatness. Whether or not this is correct, raising the voltage to a level sufficient to cause oxygen evolution results in solution agitation at the work surface and thereby interferes with polishing.

Acids commonly used as electrolytes, either alone or in combination, include acetic, chromic, citric, hydrofluoric, nitric, phosphoric, and sulfuric, with phosphoric playing a major role in many processes. In most processes the temperature of the electrolyte is usually maintained in the range 90–250° F. Current densities vary widely, but 100–500 amp/ft^2 of surface will remove moderate amounts of metal. Current densities up to 5000 amp/ft^2 may be employed if large amounts of metal are to be removed. Polishing times vary from a few seconds to 15 min or longer (see table).

In general, higher electrolyte temperatures, higher current densities, and longer polishing times all tend to produce brighter finishes. The range of finishes includes all gradations from satin to mirror, control being obtained through selection of electrolyte, solution concentration, operating temperature, and polishing time.

Numerous metals and alloys electropolish well. Aluminum, both wrought and cast, can be electropolished and anodized to yield a finish that rivals chromium plate. Brass, copper, chromium plate, gold, nickel, nickel alloys, carbon steel, the 300 and 400 stainless steels, and zinc can all be electropolished. *See* ELECTROPLATING OF METALS.

[WILLIAM W. SNOW]

Bibliography: R. M. Burns and W. W. Bradley, *Protective Coatings for Metals*, 1967; D. R. Gabe, *Principles of Metal Surface Treatment and Protection*, 1972.

Elevating machines

Materials-handling machines that lift and lower a load along a fixed vertical path of travel with intermittent motion. In contrast to hoisting machines, elevating machines support their loads instead of carrying them suspended, and the path they travel is both fixed and vertical. They differ from vertical conveyors in operating with intermittent rather than continuous motion. Industrial lifts, stackers, and freight elevators are the principal classes of elevating machines.

Industrial lifts. A wide range of mechanically, hydraulically, and electrically powered machines are classified simply as lifts (Fig. 1). They are adapted to such diverse operations as die handling and feeding sheets, bar stock, or lumber. In some locations with differences in floor level between adjacent buildings, lifts take the form of broad platforms to serve as floor levelers to obviate the need for ramps. They are also used to raise and lower loads between the ground and the beds of carriers when no loading platform exists. Lifting tail gates attached to the rear of trucks are similarly used for loading or unloading merchandise on sidewalks or roads and at points where the lack of a raised dock would make loading or unloading difficult. These units are usually driven by battery-operated motors on a power takeoff from the drive transmission of the vehicle.

Adjustable loading ramps are necessary because heights of truck and trailer beds vary. Advances in mechanized loading and unloading of vehicles have made necessary sturdier, more efficient dock boards or bridge plates, mechanically or hydraulically operated.

Stackers. Tiering machines and portable elevators used for stacking merchandise are basically portable vertical frames that support and guide the carriage, to which is attached a platform, pair of forks, or other suitable lifting device (Fig. 2). The operation of these units varies in the sense that the carriage can be raised and lowered by hand, by an electrically driven winch, or by a hydraulic cylinder, which actuates the system of chains or cables, operative by hand lever, pedal, or push button. Early electric motors used on stackers were plugged into adjacent power lines to receive current. This limited the flexibility of the stackers. Since the trend is to make the machines independent of this source, there are now models powered by either storage batteries or by small gasoline engines. Horizontal movement is effected by casters on the bottom of the vertical frame, and can be accomplished manually, or mechanically, by using

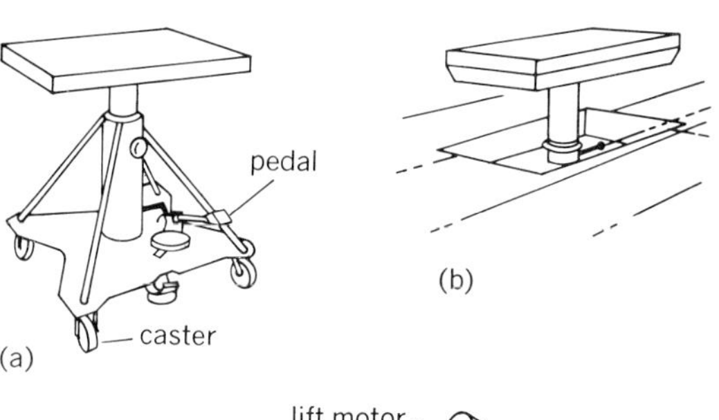

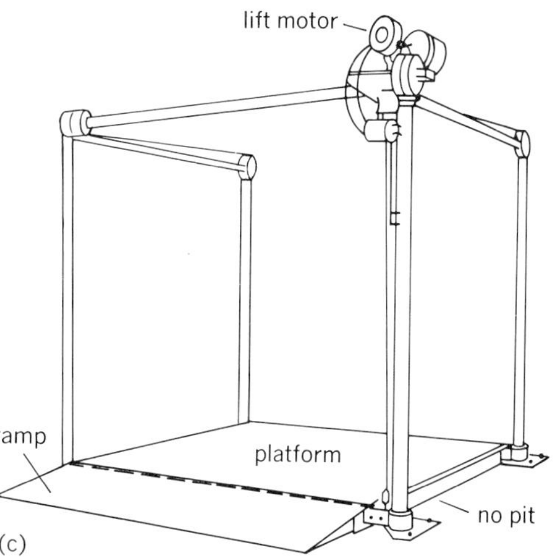

Fig. 1. Examples of industrial lifts. (*a*) Hydraulic elevating work table. (*b*) Hydraulic lift floor leveler. (*c*) Motor-driven floor leveler.

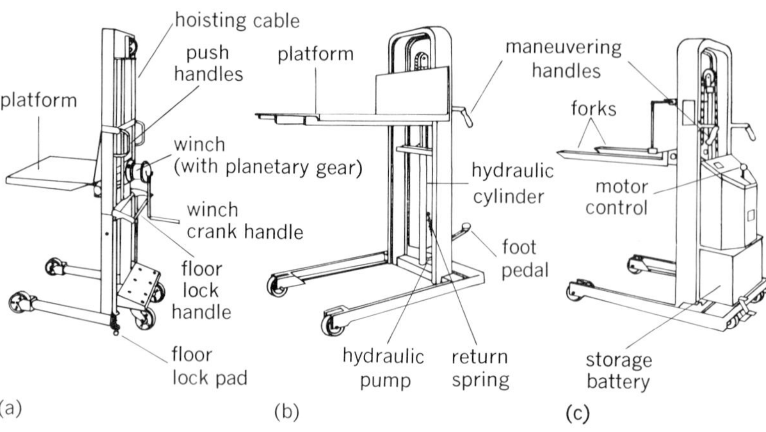

Fig. 2. Three types of electric and hydraulic stackers. (*a*) Hand type. (*b*) Hydraulic foot type. (*c*) Electric lift type.

the same power source as the lifting mechanism. These casters are usually provided with floor locks bolted in position during the elevating or lowering operation.

The basic type of stackers can be varied in several particulars. Masts, which are part of the frame, can be hinged or telescopic, and the platforms can be plain, equipped with rollers, or constructed specially to handle a specific product. Some stackers have devices for tilting barrels and drums or for lifting and dumping free-flowing bulk materials. Used in conjunction with cranes, they are widely applied to the handling of materials on storage racks and die racks. Stackers have a significant place in the development of materials-handling equipment. They are the prototypes of completely powered noncounterbalanced platform and forklift trucks. *See* INDUSTRIAL TRUCKS.

Industrial elevators. Examples of industrial elevators range from those set up temporarily on construction jobs for moving materials and personnel between floors to permanent installations for mechanized handling in factories and warehouses. Dumbwaiters are a type of industrial elevator, having capacities up to approximately 500 lb (227 kg), with a maximum floor space of 9 ft^2 (0.84 m^2); they carry parts, small tools, samples, and similar small objects between buildings, but are not permitted to carry people.

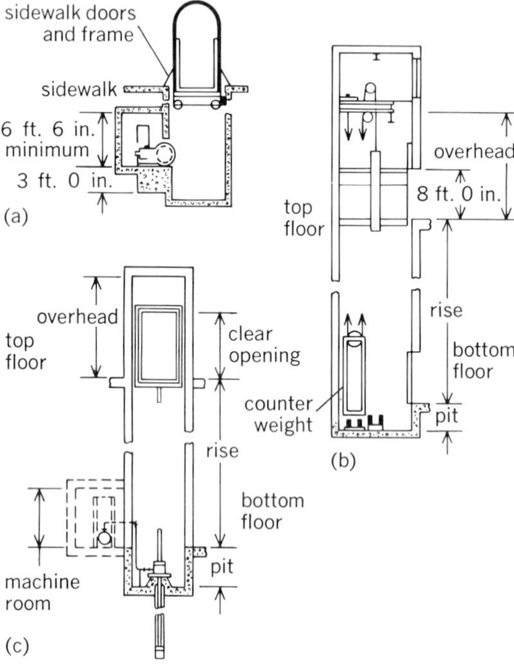

Fig. 3. Three types of industrial elevator. (*a*) Sidewalk elevator type. (*b*) Heavy-duty freight elevator type. (*c*) Hydraulic electric elevator type.

Oil hydraulic plunger electric elevators are designed for low-rise, light- or heavy-duty freight handling. Although they can be installed without special building alterations, they are restricted to buildings with only a few floors because of the limitations of the plunger length and design.

The most common and economical elevator employs electric motors, cables, pulleys, and counterweights (Fig. 3). The use of powered machines imposes a severe operating condition on elevators. Elevator platforms and structures are subjected to impact loading, off-balance loading, and extra static loading. To meet these forces, which are in addition to load forces, freight elevator design and construction provide greatly increased ruggedness over that of passenger units.

Special-purpose freight-handling elevators are equipped with platforms or arms for carrying specific articles such as rolls of paper, barrels, or drums. Some of these elevators load and discharge automatically and are arranged so that they can operate at any selected floor by means of remote control. *See* MATERIALS-HANDLING MACHINES.

[ARTHUR M. PERRIN]

Bibliography: American National Standards Institute, *Safety Code for Elevators, Dumbwaiters and Escalators*, a17.1–1978, ASME, 1978; F. A. Annett, *Elevators*, 3d ed., 1960.

Embrittlement

A general set of phenomena whereby materials suffer a marked decrease in their ability to deform (loss of ductility) or in their ability to absorb energy during fracture (loss of toughness), with little change in other mechanical properties, such as strength and hardness. Embrittlement can be induced by a variety of external or internal factors, for example, (1) a decreasing or an increasing temperature; (2) changes in the internal structure of the material, namely, changes in crystallite (grain) size, or in the presence and distribution of alloying elements and second-phase particles; (3) the introduction of an environment which is often, but not necessarily, corrosive in nature; (4) an increasing rate of application of load or extension; and (5) the presence of surface notches. This article is restricted to metals and alloys, and describes embrittlement in terms of the above factors.

Effect of test temperature. In body-centered cubic metals (for example, iron, tungsten) and hexagonal close-packed metals (for example, zinc, magnesium), a critical temperature exists below which the metal exhibits limited toughness. Fracture is usually brittle in nature, occurring either through the crystal lattice (cleavage) or along the boundaries separating the crystallites or grains (intergranular fracture). In simple terms, low-temperature embrittlement results from a competition between deformation and brittle fracture, with the latter becoming preferred at a critical temperature. For a material to be useful structurally, it is desirable that this critical temperature be below the minimum anticipated service temperature; in most cases, this is room temperature. At high temperatures, internal structural changes that lead to intergranular embrittlement can occur. Embrittlement usually occurs in the creep temperature range, a temperature at which deformation can occur under very low stresses; and the two processes are believed to be connected. *See* PLASTIC DEFORMATION OF METAL.

Effect of heat treatment. In many metals, particularly structural steels, annealing or heat treating in certain temperature ranges sensitizes the grain boundaries in such a way that intergranular embrittlement subsequently occurs during service. In steels, for example, high strength is achieved by quenching from the stable high-temperature face-centered cubic phase (called austenite). This can produce a room-temperature structure called martensite, a distorted, body-centered cubic phase which is extremely hard and strong, but brittle. The transformation of austenite to martensite is controlled by several metallurgical variables, including the alloy composition and the quench rate. To reduce the brittleness, the steel undergoes an annealing treatment called tempering, which, while decreasing the strength, usually increases the toughness. The exception to this trade-off occurs when the steel is tempered at 1000°F (538°C). This can lead to a mode of intergranular fracture called temper embrittlement; such a process has led to catastrophic failures in turbines, rotors, and other high-strength steel parts. It is associated with the preferential segregation of undesirable "tramp" elements such as tin and antimony to the grain boundaries; these elements should be kept to as low a level as possible to help alleviate the problem. Tempering at 550°F (287°C) can also lead to embrittlement in steels. In other metals, there are less specific but similar types of embrittlement resulting from critical heat treatments. *See* HEAT TREATMENT (METALLURGY); STEEL; TEMPERING.

Effect of environment. Metals can fracture catastrophically when exposed to a variety of environments. These environments can range from liquid metals to aqueous and nonaqueous solutions to gases such as hydrogen. The phenomenology of these processes and some of the corrective procedures used or envisaged are described below.

Liquid metal embrittlement. If a thin film of a liquid metal is placed on the oxide-free surface of a solid metal, the tensile properties of the solid metal will not be affected, but the fracture behavior can be markedly different from that observed in air. In some cases, specimens stressed above a critical value will fracture catastrophically. Although the embrittlement is usually intergranular, transgranular cleavage has also been observed. Such fractures can be induced in highly ductile metals such as aluminum and copper. Although many different liquid metals are capable of inducing embrittlement in a variety of solid metals, some of the more common couples, many of which have important engineering and design consequences are mercury embrittlement of brass (an alloy of copper and zinc), lead embrittlement of steel, and gallium embrittlement of aluminum. The mechanism of liquid metal embrittlement is thought to result from the reduction by the liquid metal atoms of the force necessary to break apart two solid metal atoms at a crack tip.

Stress corrosion cracking. If a metal is stressed and simultaneously exposed to an environment which may be, but is not necessarily, corrosive in nature, cracking and fracture can occur. Both stress and environment are required; if only one of these elements is present, the metal usually displays no embrittlement. *See* CORROSION.

The analogy of stress corrosion cracking to liquid metal embrittlement is obvious, but in this case, the environment can be any aqueous or nonaqueous solution. In specific metals cracking has been observed in such diverse solutions as high-purity distilled water, salt water, caustics (for example, lye), ammonia, molten anhydrous salts, and organics. Stress corrosion cracking can occur over a wide temperature range, and can be a very serious problem in many service applications. For example, brass cartridge cases which were internally stressed by a forming operation cracked in storage because of the presence of trace amounts of ammonia in the air; and titanium alloys used in aerospace applications cracked when cleaned with methyl alcohol. There are a host of situations in which moisture or salt water led to cracks or total fracture of structural materials.

A mechanism which has been proposed as an explanation for many of these cases has been associated with the stress-assisted preferential dissolution of metal along specific metallurgical features, leading to intergranular or transgranular cracks. This process can be likened to the action of a battery. In this case, the cracking region is the anode; when the anode is coupled to a cathodic region, the circuit is completed. This electrolytic behavior has been exploited as a protective device. Critical parts are made to serve as cathodes in local cells, allowing other noncritical parts to act as sacrificial anodes. This procedure, termed cathodic protection, is used to protect not only against stress corrosion cracking but also against general corrosion; for example, zinc sacrificial anodes are used to minimize corrosion of ship propellers. Other mechanisms, such as corrosion, have also been put forward to explain the stress-corrosion cracking phenomena.

Hydrogen embrittlement. This form of embrittlement is often considered to be a type of stress corrosion cracking. Hydrogen atoms can enter a metal, causing severe embrittlement, again with little effect on other mechanical properties. This phenomenon was originally observed, and is most critical in, steels, but it is now documented to occur in titanium and nickel alloys, and may lead to cracking in other alloy systems as well.

The source of hydrogen can be extremely varied, making this type of embrittlement difficult to control. For example, hydrogen can be retained internally during the melting and casting of alloys; it can be discharged at cathodic areas in electrolytic cells (it is apparent that, at both the anode and the cathode, events which may lead to embrittlement occur); it can enter the metal during a plating operation; and it can come from an external molecular gas environment, even at very low partial pressures of hydrogen. When it is realized that coal gasification plants produce hydrogen or methane (a hydrogen-containing gas) as a fuel, that long-distance transport of hydrogen is envisaged

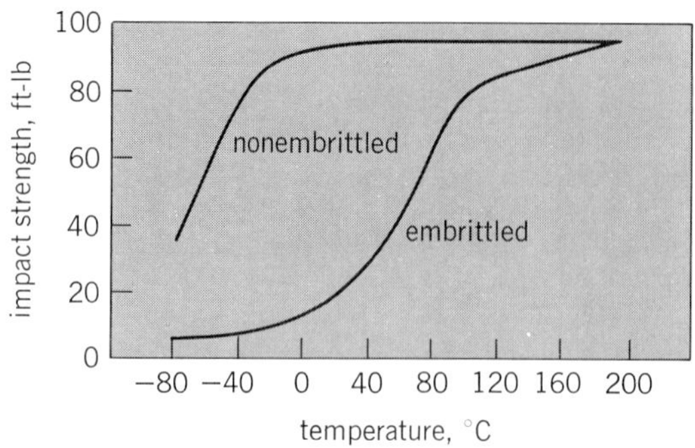

The effect of temper brittleness on the impact resistance of a low-alloy steel. (*From R. E. Reed-Hill, Physical Metallurgy Principles, Van Nostrand, 1973*)

as a means of alleviating the energy crisis, that liquid hydrogen is used as a rocket fuel, and that hydrogen is a by-product of many chemical reactions, it becomes apparent that structural materials that are resistant to hydrogen embrittlement must be developed.

In all forms of environmental embrittlement, the best remedial practice is to physically isolate the metal from the dangerous environment. This is not always practical or possible. Studies have indicated that the susceptibility of a given material to environmental embrittlement can be reduced by controlling metallurgical factors, such as alloy composition and internal structure. This type of alloy design should help control environmental embrittlement, even where there is uncertainty as to the controlling mechanism involved in the process. *See* ALLOY.

Other factors. It has been pointed out that factors such as notches (which both raise and modify the local stress state) and the rate of application of stress can modify the response of a material to a specific type of embrittlement. In general, notches or surface flaws always enhance embrittlement, both by acting as a stress raiser and by providing a preexisting crack. An increasing loading rate (impact) enhances low-temperature embrittlement, and possibly thermal embrittlement, but, interestingly, suppresses environmental embrittlement. For environmental embrittlement, species such as hydrogen must keep up with the moving crack, and they do so by moving through the crystal lattice at a rate largely determined by the temperature. If the crack is moving too fast, as under impact conditions, hydrogen can not keep pace, and the severity of embrittlement decreases. For normal rates of loading, the maximum effect of hydrogen is unfortunately at, or near, ambient temperatures. The combined effects of various embrittlement phenomena are shown in the illustration. Temper embrittlement increases the impact ductile-to-brittle transition temperature from about −50 to 80°C, making this steel unacceptable for room-temperature applications.

Summary. The many forms of embrittlement share the property of a catastrophic loss in ductility or toughness, with little change in other mechanical properties. However, since each type is seemingly controlled by different parameters, it is difficult, at the present level of understanding, to develop general remedial design guidelines. [I. M. BERNSTEIN]

Bibliography: American Society for Testing and Materials, *Atmospheric Factors Affecting the Corrosion of Engineering Materials*, 1978; C. R. Barrett, W. D. Nix, and A. S. Tetelman, *The Principles of Engineering Materials*, 1973; M. G. Fontana and N. D. Greene, *Corrosion Engineering*, 2d ed., 1978; H. H. Uhlig, *Corrosion and Corrosion Control*, 2d ed., 1971.

Energy

The ability of one system to do work on another system. There are many kinds of energy: chemical energy from fossil fuels, electrical energy distributed by a utility company, radiant energy from the Sun, and nuclear energy from a reactor. The units of energy include ergs, joules, kilowatts, foot-pounds, and foot-poundals. Work and heat have the same units as energy, but are entirely different physical concepts.

Work. If a particle is moved from a point in space $\vec{r}_1$ to a $\vec{r}_2$ along a curve C by a force $\vec{F}$, the force does an amount of work $W_{12}(C)$ given by Eq. (1), where $d\vec{R}$ parameterizes the path along C and

$$W_{12}(C) = \int_{\vec{r}_1}^{\vec{r}_2} \vec{F} \cdot d\vec{R} \qquad (1)$$

$\vec{F} \cdot d\vec{R}$ is the dot product of $\vec{F}$ and $d\vec{R}$, that is, the component of $\vec{F}$ along $d\vec{R}$. Since forces cause changes in motion through Newton's laws, the work in Eq. (1) could impart motion to a stationary object, stop a moving object, or change the magnitude or direction of the object's velocity. Thus work is a technical term, and requires both a force and a displacement. A person traveling around the Earth at constant speed would undergo a displacement of approximately 40,000 km, but would do no work since this displacement is perpendicular to the gravitational force of the Earth. Similarly, a person who pushes on a wall without moving it does no work because the wall has no displacement. *See* FORCE; WORK.

A force is called a conservative force if the work which it does in displacing a particle from one point to another is independent of the path of the particle for all paths. Conservative forces can be expressed as the gradient of a scalar function $V_\alpha(\vec{r})$, which is called the potential energy function for the conservative force. Each type of conservative force corresponds to a kind of interaction in nature. Only five potentials, or interactions, are required for all present understanding in fundamental physics: gravitational potential energy; electromagnetic potential energy; weak, or neutron decay, potential energy; superweak potential energy; and nuclear potential energy. The first two potential energy functions are fairly well understood. S. Weinberg and A. Salam have done work toward unifying electromagnetism with weak interactions.

Any particle or system of particles subject to conservative forces has two kinds of energy, potential energy and kinetic energy. Potential energy is the energy due to position or configuration, and kinetic energy is the energy due to motion.

Potential energy. The electric field $\vec{E}$ and the magnetic field $\vec{B}$ in a region of space can be calculated from an electrostatic potential and a magnetic vector potential. Then the force on a charged particle can be calculated, and the work done on the particle along a specific path by the force can be obtained. The electromagnetic potentials in Schrödinger's equation account for the atomic structure of matter. Thus, macroscopic changes in potential energy that result from pushing an object from one place to another with a stick or pulling it with a string have their microscopic origin in the electromagnetic potential energy which holds the atoms in place in the stick or string together. However, there are macroscopic forces which produce different potential energy functions, such as the elastic spring (Fig. 1). The potential energy function for a spring is given by Eq. (2), where k is in

$$V(x) = \tfrac{1}{2} k(x - x_0)^2 \tag{2}$$

units of N/m and x_0 is the equilibrium spacing. The hookean force law $F = -k(x - x_0)$ can be derived by taking the negative derivative, $-d/dx$, of $V(x)$ in Eq. (2). There are also simple functional forms for the five fundamental interactions. Each potential energy function specifies the dependence of the force on position, and each one contains a constant, such as the x_0 in Eq. (2), which is a reference potential energy. This constant part of the potential energy simply specifies a reference point and does not contribute to the force (Fig. 2). *See* HOOKE'S LAW.

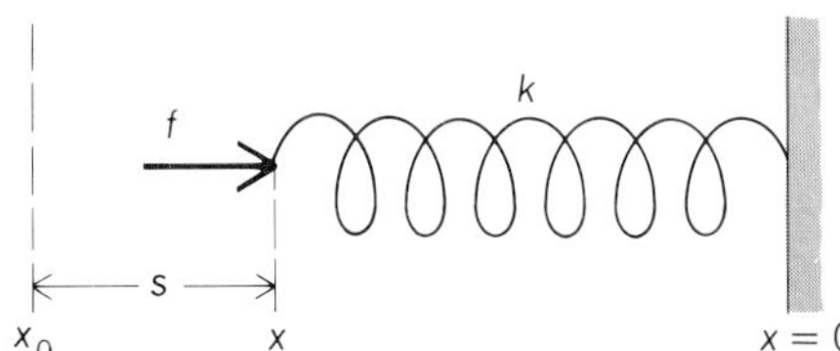

Fig. 1. Potential energy of a compressed spring. Here $s = x - x_0$ is the distance that the spring has been compressed from its equilibrium length x_0. The spring constant is k, and the external force f compresses the spring.

Kinetic energy. The kinetic energy of a body is due to its motion. A point particle with mass m and velocity v has kinetic energy given by Eq. (3). Any

$$E = \tfrac{1}{2} mv^2 \tag{3}$$

body in motion has Eq. (3) as its translational kinetic energy. The velocity must be specified with respect to some inertial reference frame.

A rigid body can move as in Eq. (3), but in addition it can rotate about its center. The rotational kinetic energy E_R, for a body with moment of inertia I and angular frequency ω, in units of radians per second is given by Eq. (4). The most general

$$E_R = \tfrac{1}{2} I\omega^2 \tag{4}$$

kinetic energy for a solid body in motion is given by Eq. (5). Understanding of the significance of

$$E_k = \tfrac{1}{2} mv^2 + \tfrac{1}{2} I\omega^2 \tag{5}$$

Eqs. (3)–(5) is aided by consideration of the order of magnitude of the quantities appearing there. The mass of a large automobile is about 900 kg. A translational speed of 55 mph is about 25 m/s. The translational kinetic energy of this automobile is 275,000 J. The moment of inertia of this auto is $I = mk^2$, where k is the radius of gyration ($k \sim 1.5$ m). If the auto rotates about its center at $\omega = 16.7$ rad/s, which is a little more than 2.5 revolutions per second, its rotational kinetic energy will also equal 275,000 J. *See* ROTATIONAL MOTION.

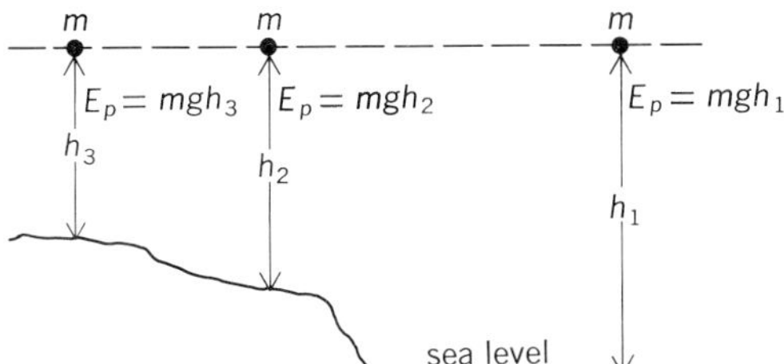

Fig. 2. Gravitational potential energy $E_p = mgh$ for an object of mass m and weight mg shown at the same height h for different reference levels. The different values of E_p correspond to the additive constant discussed in the text.

Reference frames. The galilean principle of relativity is that all of the mechanical properties of systems are the same in all inertial systems which move at constant velocity with respect to each other. Thus the velocity must be measured in some particular inertial frame. A person in an auto traveling at a constant velocity of 55 mph (or 25 m/s) as measured from the ground is at rest in an inertial frame traveling at 25 m/s attached to the car. However, if the car collides with a large tree, the driver gains no advantage from his or her zero velocity because the tree is racing toward the car at 55 mph in this inertial frame.

Conservation of energy. Energy is conserved for all isolated mechanical systems. This is because if a system A is isolated, there is no other system B that it can give any energy to, and its total energy must remain constant. This system A can convert kinetic energy to potential energy, and it can convert one form of potential energy to another, but the total energy must remain the same. The meaning of conserved total energy is that the system has the same value of total energy at all times. The definition of time translational invariance is that the system as a whole does not change in time, that is, the system is isolated. *See* CONSERVATION OF ENERGY.

Internal energy. Macroscopic bodies and processes are not isolated in reality, so that some small energy transfer is present. In the laboratory, the isolation of a gas, solid, or fluid sample can be approximated. Usually, one object is in contact with another one, such as a book on a table or a pan of water on a stove burner. Energy can be transferred in these cases, and dissipations can occur. A new variable, the temperature T, is required to describe equilibrium of a new physical quantity, heat. If two objects have the same temperature, they are in thermal equilibrium and no heat will flow between them. If their temperatures are not equal, heat will flow from the body with a higher temperature to the body with a lower temperature. Heat has the same units as work and energy, but is physically different.

The first law of thermodynamics is a generalized conservation-of-energy law for all macroscopic processes. It may be expressed by Eq. (6), where

$$dU = đW + đQ \tag{6}$$

dU is the internal energy of the system, $đW$ is the

work done on the system, and $đQ$ is the heat added to the system. The symbol $đ$ emphasizes that the work and heat are path-dependent, or are inexact differentials, whereas the internal energy $U(S,V)$ is an exact or path-independent differential, where S is the entropy and V is the volume. The interpretation of the first law of thermodynamics, Eq. (6), is that of generalized conservation of energy, generalized to include dissipations, $đQ$. The microscopic origin of the macroscopic internal energy is that of the average energy $\bar{\epsilon}$ given by Eq. (7), where $P(\epsilon_i)$ is

$$U = \bar{\epsilon} = \sum_i \epsilon_i P(\epsilon_i) \qquad (7)$$

the probability that the N molecules of the system have energy ϵ_i. It is the large number of particles in macroscopic systems, N of order equal to or greater than 10^{28}, which sharpens the statistical distribution of energies and makes the internal energy a single number. The large numbers also allow pressure, volume, temperature, and entropy to be treated as simple variables when they are really statistical distributions. The average statistical energy is path-independent, and is equal to the sum of a path-dependent work term which involves only forces and displacements (or pressure and volume) and a path-dependent heat term which involves temperature and entropy. The work term includes all mechanical effects, and the heat term involves only thermal variables.

Mass-energy. In 1905 A. Einstein showed that at high velocities near the speed of light important modifications must be made in physical concepts. One particularly radical idea which he advanced was that space and time are not independent, but rather are two aspects of the same object, a space-time manifold. This necessitated a reexamination of the concept of energy and led to the conclusion that the inertia, or mass m, depends upon its energy through the mass-energy relation, Eq. (8), where

$$E = mc^2 \qquad (8)$$

c is the speed of light in vacuum. Furthermore, energy and momentum conservation become joined in a single four-momentum conservation law in special relativity. Equation (8) is valid for all known physical and chemical processes.

Nuclear energy, first in the atomic bomb and then in peaceful applications such as energy-generating reactors, was developed in part due to Eq. (8). Since it was experimentally known that large mass changes Δm occur in nuclear reactions, Eq. (8) implies that the energy released is $\Delta E = \Delta mc^2$. In the nuclear reaction $^2H + {}^3H \rightarrow {}^1n + {}^4He + Q$ among the light nuclei, the energy released, Q, is almost 20 MeV. This reaction is prominent in fusion technology because 20 MeV, approximately 20,000,000 times the energy released in one TNT or dynamite process, is especially large. In one fission process of ^{235}U, over 200 MeV are released, but are distributed among two or three neutrons and one or more heavier nuclei, so that "gathering" the energy for use is more difficult. *See* NUCLEAR POWER.

Field energy. Whereas the particles and rigid bodies of mechanics are localized objects which have positions and velocities, there are quantities called fields which are defined throughout a region, such as fluids, the electromagnetic field, and sound and water waves. Fields are recognizable disturbances. Each field satisfies some field equation which determines what sort of disturbance is created under various conditions. A fluid or water wave is a disturbance in the fluid or water medium. A sound wave is a pressure wave in a solid, liquid, or gas. The electromagnetic wave consists of an electric vector $\vec{E}(\vec{r}, t)$ and a magnetic vector $\vec{B}(\vec{r}, t)$, and can propagate through a vacuum; that is, no medium is needed. Maxwell's field equations for the electromagnetic field require that $\vec{E}$ and $\vec{B}$ are mutually perpendicular and are orthogonal to $\vec{k}$, the direction of propagation of the wave. If a field is set up in an isolated region of space Ω, an energy density $e(\vec{r})$ exists such that the total energy E_T is given by Eq. (9), and is conserved (since Ω is iso-

$$E_T = \int_\Omega e(\vec{r})d^3r \qquad (9)$$

lated from external agents once the field is established). For an electromagnetic field, Eq. (9) becomes Eq. (10), where ϵ and μ are the permittivity

$$E_T = \frac{1}{2}\int_\Omega (\epsilon\,\vec{E}^2 + \frac{1}{\mu}\vec{B}^2)\,d^3r \qquad (10)$$

and permeability of the region and $\vec{E}$ and $\vec{B}$ are the electromagnetic field vectors. In the case of fields, the conserved total energy E_T is constant because of time translational invariance of the field equations.

Atomic energy. An atomic or subatomic particle subject to the potential energy $V(\vec{r})$ is described by a function $\psi(\vec{r})$ whose behavior is governed by Schrödinger's equation. For a one-dimensional particle of mass m in a potential $V(x)$, Schrödinger's equation is given by Eq. (11), where $\hbar = 1.054 \times$

$$H\Psi(x,t) = \left\{-\frac{\hbar^2}{2m}\frac{d^2\Psi}{dx^2} + V(x)\Psi(x,t)\right\} = i\hbar\left(\frac{\partial\Psi}{\partial t}\right) \qquad (11)$$

10^{-34} J-s is Planck's constant, x is the space coordinate, t is the time parameter, and H is the hamiltonian operator for the particle. If $\psi(x,t) = \phi(x)T(t)$, it is straightforward to show that $T(t) = e^{-iEt/\hbar}$ and Eq. (11) reduces to Eq. (12), where E_n is the value

$$H\phi_n(x) = E_n\phi_n(x) \qquad (12)$$

of the total energy of the particle. Depending upon which H and V are used, a form of Eq. (12) describes electronic states in atoms, proton and neutron states in a nucleus, or atoms in a molecule. In all of these cases the hamiltonian acts as an energy operator, and the time translational invariance of the system requires that the energy be conserved in Schrödinger's equation too.

Time translational invariance. For each case discussed in this article, there are systems or states of systems which do not change in time; and each of these systems has a conserved total energy. Conversely, when systems or states of systems change in time, the total energy must change. Thus a single concept, time translational invariance, unifies many kinds of energy in many theories.

[BRIAN DE FACIO]

Bibliography: P. C. W. Davies, *The Forces of Nature*, 1979; D. Halliday and R. Resnick, *Physics*, 3d ed., 1978; F. A. Kaempffer, *The Elements of Physics: A New Approach*, 1967; Physical Science Study Committee, *Physics*, 1960.

Energy conversion

The process of changing energy from one form to another. There are many conversion processes that appear as routine phenomena in nature, such as the evaporation of water by solar energy or the storage of solar energy in fossil fuels. In the world of technology the term is more generally applied to man-made operations in which the energy is made more usable, for instance, the burning of coal in power plants to convert chemical energy into electricity, the burning of gasoline in automobile engines to convert chemical energy into propulsive energy of a moving vehicle, or the burning of a propellant for ion rockets and plasma jets to provide thrust.

There are well-established principles in science which define the conditions and limits under which energy conversions can be effected, for example, the law of the conservation of energy, the second law of thermodynamics, the Bernoulli principle, and the Gibbs free-energy relation. Recognizable forms of energy which allow varying degrees of conversion include chemical, atomic, electrical, mechanical, light, potential, pressure, kinetic, and heat energy. In some conversion operations the transformation of energy from one form to another, more desirable form may approach 100% efficiency, whereas with others even a "perfect" device or system may have a theoretical limiting efficiency far below 100%.

The conventional electric generator, where solid metallic conductors are rotated in a magnetic field, actually converts 95–99% of the mechanical energy input to the rotor shaft into electric energy at the generator terminals. On the other hand, an automobile engine might operate at its best point with only 20% efficiency, and even if it could be made perfect, might not exceed 60% for the ideal thermal cycle. Wherever there is a cycle which involves heat phases, the all-pervading limitation of the Carnot criterion precludes 100% conversion efficiency, and for customary temperature conditions the ideal thermal efficiency frequently cannot exceed 50 or 60%. *See* CARNOT CYCLE.

In the prevalent method of producing electric energy in steam power plants, there are many energy-conversion steps between the raw energy of fuel and the electricity delivered from the plant, for example, chemical energy of fuel to heat energy of combustion; heat energy so released to heat energy of steam; heat energy of steam to kinetic energy of steam jets; jet energy to kinetic energy of rotor; and mechanical energy of rotor to electric energy at generator terminals. This is a typical, elaborate, and burdensome series of conversion processes. Many efforts have been made over the years to eliminate some or many of these steps for objectives such as improved efficiency, reduced weight, less bulk, lower maintenance, greater reliability, longer life, and lower costs. For a discussion of major technological energy converters *see* POWER PLANT. *See also* ELECTRIC POWER GENERATION.

Efforts to eliminate some of these steps have been stimulated by needs of astronautics and of satellite and missile technology and need for new and superseding devices for conventional stationary and transportation services. Space and missile systems require more compact, efficient, self-contained power systems which can utilize energy sources such as solar and nuclear. With conventional services the emphasis is on reducing weight, space, and atmospheric contamination, on improving efficiency, and on lowering costs. The predominant objective of energy conversion systems is to take raw energy from sources such as fossil fuels, nuclear fuels, solar energy, wind, waves, tides, and terrestrial heat and convert it into electric energy. The scientific categories which are recognized within this specification are electromagnetism, electrochemistry (fuel cells), thermoelectricity, thermionics, magnetohydrodynamics, electrostatics, piezoelectricity, photoelectricity, magnetostriction, ferroelectricity, atmospheric electricity, terrestrial currents, and contact potential. *See* ENERGY SOURCES.

The electromagnetism principle today dominates the field. Electric batteries are an accepted form of electrochemical device of small capacity, for example, 1 kw in automobile service. Other categories are in various stages of development. Extensive efforts and funds are being given to some fields with attractive prospects of practical adaptation. *See* ELECTRIC ROTATING MACHINERY; MAGNETOHYDRODYNAMIC POWER GENERATOR.

[THEODORE BAUMEISTER]

Bibliography: R. C. Bailey, *Energy Conversion Engineering*, 1978; T. Baumeister (ed.), *Standard Handbook for Mechanical Engineers*, 8th ed., 1978; S. S. L. Chang, *Energy Conversion*, 1963; W. Mitchell, Jr. (ed.), *Fuel Cells*, 1963.

Energy sources

Sources from which energy can be obtained to provide heat, light, and power. Sources of energy have evolved from human and animal power to fossil fuels, uranium, water power, wind, and the Sun.

The principal fossil fuels are coal, lignite, peat, petroleum, and natural gas—all of which were formed in finite amounts millions of years ago. Other potential sources of fossil fuels include oil shale and tar sands. Nonfuel sources of energy include wastes, water, wind, geothermal deposits, biomass, and solar heat. As fossil fuels become depleted, nonfuel sources and fission and fusion sources will become of greater importance since they are renewable. Nuclear power based on the fission of uranium, thorium, and plutonium, and the fusion power based on the forcing together of the nuclei of two light atoms such as deuterium, tritium, or helium-3, could become principal sources of energy in the 21st century.

As the world has become more industrialized, the consumption of fuels to produce power and energy has increased at a rapid rate. World energy demand amounted to 132 quads (132×10^{15} Btu or 139×10^{18} J; 1 quad = 10^{15} Btu = 1.055×10^{18} J) in 1961 and increased to 238 quads (251×10^{18} J) in 1972, for an average annual growth of 5.5%. During the same period of time, population and real gross national product increased at rates of 1.9 and 5.1%, respectively. In 1973 the Arab embargo and the subsequent quadrupling of the price of oil threw the world into economic turmoil, and as a result, industrial growth slowed drastically and in some countries there was no growth for several years.

Petroleum. Crude petroleum and natural gas liquids production in the United States peaked at

4×10^9 bbl in 1970 and had been declining ever since.

The decline in petroleum production in the United States is the result of less and less drilling after 1960 and lower finding rates. This decline in drilling activity was caused by government actions which made it uneconomical to drill for oil and gas in the United States. These actions included wellhead ceiling prices on oil and gas, repeal of depletion allowance for the oil and gas industry, and a generally unfavorable climate for the oil industry in the United States. As a result, the oil companies sent their drilling rigs to foreign countries, particularly in the Middle East and Africa, where the cost of drilling was less and the expectations and probability of finding oil in large quantities were greater.

Estimates of conventionally recoverable oil and gas indicate that an amount (Btu equivalent) approximately equal to present proved and produced reserves remain to be found. These potential resources probably will be found in many of the presently productive areas, and in unexplored or frontier basins which are located primarily in the arctic regions, offshore areas of the continental shelf, possibly the deep sea, and many areas which in the past have been relatively inaccessible. Worldwide geologic distribution and per capita consumption of oil and gas reserves are not uniform geographically or politically. Estimates indicate that future discoveries will include a greater percentage of relatively smaller accumulations and more gas than in the past, and will probably not be uniform in distribution. For the world in general, it appears that the addition of new reserves of oil is less than growth in consumption; that is, oil may be topping out.

Natural gas. The annual production of natural gas in the United States peaked in 1973 and then declined at the rate of over 5% per year. Since 1975, the production of natural gas in the United States has stabilized at about 20×10^{12} ft^3 (5.6×10^{11} m^3) per year, and it appears that the partial decontrol of wellhead prices for natural gas has had a beneficial effect on its production. In other words, a free market price and full decontrol of all new natural gas production should keep the production at this level.

The American Gas Association forecast that a number of supplemental sources of gas could be made available by the year 2000.

Coal. While coal production in the United States declined during the late 1970s, it is expected that production will increase, with forecasts of future United States coal production by 1985 ranging from 850×10^6 to 10^9 short tons (770 to 910×10^6 metric tons) per year. In order to attain this future level of production, it will be necessary to develop hundreds of new deep mines in the east and hundreds of new surface mines in the east and west.

The U.S. Bureau of Mines estimated that as of Jan. 1, 1974, the demonstrated coal reserve base in the United States was 434×10^9 tons (394×10^9 metric tons) of coal, of which 297×10^9 tons (269×10^9 metric tons) is minable by surface methods. The world's total proved and currently recoverable reserves of coal in place amount to 2.376×10^{12} short tons (2.155×10^{12} metric tons) of coal, and on this basis, the United States contains 18% of the world's proved and currently recoverable reserves. *See* COAL MINING.

Nuclear energy. In 1978, nuclear power provided 12.5% of the electric power production and 3.8% of the total United States energy supply. Seventy-one nuclear reactors with a total capacity of 50,721 MWe were in existence in the United States as of June 1979. At one time it was projected that by 1985 at least 200,000 MWe of nuclear plant capacity would be completed and operating and would be supplying 20% of the total United States electric power production. Nuclear power originally was expected to supply over 11% of the total United States energy by 1985. However, due to the accident at Three Mile Island near Harrisburg, PA, in March 1979, there has been a substantial slowdown in nuclear plant construction in the nation and in future plans for nuclear plants by electric utilities.

The present United States nuclear program is based principally on light-water reactors. The cutback in orders for power plants in 1975 and in subsequent years by electric utilities resulted in a number of both light-water reactors and high-temperature gas-cooled reactors being delayed or canceled.

Reserves of uranium are deemed to be inadequate to support light-water reactor (LWR) nuclear power plants in the numbers planned for the next 25 years since these reactors make use of only a small percentage of the energy which is potentially available in nuclear fuels. Maximum use of the energy potential of nuclear fuels would require development of the breeder reactor, which produces more fuel than it consumes. Two different types of breeder reactors are being considered: the thermal breeder, which operates on a thorium-uranium fuel cycle and would employ either water or molten salt as a coolant; and the fast breeder, which operates best with a uranium-plutonium fuel cycle and would use gas or liquid metal as a coolant. *See* NUCLEAR POWER.

Fusion power. Fusion power is expected to be a major source of energy in the future, but commercialization is not expected until the early part of the 21st century. The fusion process is based on the principle that when the nuclei of two light atoms are forced together to form one or two nuclei with a smaller mass, energy is released. The principal fuels in a fusion reactor are the gases deuterium, tritium, or helium-3. Deuterium, a heavy isotope of hydrogen, is found in ordinary sea water. It can be extracted relatively cheaply as deuterium oxide, or heavy water. When it can be used in a fusion reactor, the amount of deuterium in sea water will provide an inexhaustible supply of energy for the world.

Oil shale. Oil shale deposits have been found in many areas of the United States, but the only deposits of sufficient potential oil content considered as near-term potential resources are those of the Green River Formation in Colorado, Wyoming, and Utah. The U.S. Geological Survey estimates that about 80×10^9 bbl of oil are recoverable under present economic conditions. When marginal and submarginal reserves of all shale are included, it is estimated that reserves of 600×10^9 bbl of oil equivalent are present in economically recoverable oil shale deposits in the United States. It is projected that production of synthetic crude oil (syncrude) from United States oil shale will total 148,000 barrels per day by 1990 and 374,000 bpd by 2000.

Tar sands. Tar sands of the world represent the largest known supply of liquid hydrocarbons. Extensive resources are located throughout the world, but primarily in the Western Hemisphere. The total world reserves in 1971 were estimated to contain 2×10^{12} bbl.

The best-known deposit is the Athabasca tar sands in northeastern Alberta, Canada. The oil in place is estimated to amount to 95×10^9 metric tons (over 700×10^9 bbl). Commercial operations have been under way for a number of years at a production level of 50,000 bpd and by 1979 had reached a level of 150,000 bpd. Several other commercial plants were announced in 1979. It is projected that by 1995 production of syncrude from Canadian tar sands will total 570,000 bpd and will reach 850,000 bpd by 2000.

The principal United States reserves are located in Utah. It is estimated that the resources in place amount to 18 to 28 $\times 10^9$ bbl in five major formations.

Solar energy. Solar energy has always been a potential source of limitless, clean energy. However, commercial development of solar energy has been slow because of storage requirements and high capital cost requirements. Past availability of low-cost fossil fuels resulted in solar energy being considered uneconomical. With the sudden massive increase in world oil and gas prices, solar energy economics, particularly for low-level heating and cooling, are reaching the stage where commercial applications are economical. Nevertheless, only a small percentage of the total United States energy supply will come from solar energy even by the year 2000.

The U.S. Department of Energy predicts that by 1985 solar energy could provide between 0.3 and 0.6 quads (0.3 and 0.6×10^8 J) of space heating. This would still amount to only one-fourth to one-half of 1% of the energy required by the United States in 1985. It is well to keep the use of commercial solar energy in perspective and recognize that the conventional fossil fuels—coal, oil, and gas—are still going to be providing the country with most of its energy until 2000.

Geothermal energy. The heat content of the Earth is immense and would appear to many to be an inexhaustible and plentiful supply for much of the energy needs. However, for a variety of reasons, it appears highly unlikely that the amount of geothermal energy will ever supply more than a few percent of the United States energy requirements.

The only area in the nation where geothermal energy has been commercialized is in California, yet there is considerable doubt that geothermal energy will be able to supply more than a few percent of the state's electrical generating requirements.

It is important, nevertheless, to develop all available economic geothermal resources since they represent an important source of renewable energy. *See* GEOTHERMAL POWER.

Synthetic fuels. Fuels which do not exist in nature are known as synthetic fuels. They are synthesized or manufactured from varieties of fossil fuels which cannot be used conveniently in their original forms. Substitute natural gas (SNG) is manufactured from coal, peat, or oil shale. Synthetic liquid fuels can be produced from coal, oil shale, or tar sands. Both gaseous and liquid fuels can be synthesized from renewable resources, collectively called biomass. These carbon sources are trees, grasses, algae, plants, and organic waste. Production of synthetic fuels, particularly from renewable resources, increases the scope of available energy sources.

It should always be kept in mind that the capital investment required for synthetic fuels is many times more than the investment required to drill for oil and gas and obtain production by conventional means. Before embarking on a very capital-intensive program such as synthetic fuels, attention should be paid to the possibility of a shortage of capital which could result from the demands of the program. Full decontrol of prices for all forms of petroleum, including crude oil and refined products, and full decontrol of natural gas prices must be instituted before embarking on the very expensive program of synthetic fuel production.

Other energy sources. Several other sources of energy are worth considering.

Waste. Refuse in the form of residential, commercial, industrial, and agricultural wastes has been underutilized in the past. However, as the cost of conventional fossil fuels increases, the use of refuse as an alternative fuel becomes economically attractive. A number of plants utilizing wastes as fuel have been built to produce steam and electric energy, and more plants are being considered.

Wind. Wind power's energy potential is very large. However, wind power is not expected to add significantly to the United States energy capability by 1990 or 2000.

Tides. Tidal power, while potentially large, has not become a commercial reality except in a few areas of the world such as France, which has developed one site to 240 MWe of capacity.

Biomass. Biomass conversion to energy is technically feasible and is potentially economical. Use of biomass on a significant scale would not require a large research or demonstration effort or excessively large capital and worker-power investments. Production of alcohol from grains and other biomass material became economic in 1979 and sparked the sale of gasohol in the United States. Even though alcohol for this use will increase in the future, usage will still remain relatively small and insignificant.

Energy management. Energy has become a major item of cost in every industrial process, commercial establishment, and home. Energy management now includes not only the procurement of fuels on the most economical basis, but the conservation of energy by every conceivable means. Whether this is done by squeezing out every Btu through heat exchangers, or by room-temperature processes instead of high-temperature processes, or by greater insulation to retain heat which has been generated, each has a role to play in requiring less energy to produce the same amount of goods and materials.

Conservation of energy is as important as the

finding of new sources of energy. The alternative is higher and higher levels of imported oil which will eventually become too expensive to use.

[GERARD C. GAMBS]

Bibliography: Atomic Industrial Forum, *Status Report: Energy Resources and Technology*, March 1975; Council on Environmental Quality, *The Good News about Energy*, 1979; Federal Energy Administration, *Project Independence Report*, November 1974; Institute of Gas Technology, *International Gas Technology*, September 1979; National Electric Reliability Council, *Forecast of Fuel Requirements*, July 1979; National Petroleum Council, *U.S. Energy Outlook*, December 1972; *9th World Energy Conference*, September 1974; Power Magazine, *The 1975 Energy Management Guidebook*, August 1975; U.S. Bureau of Mines, *Mineral Industry Surveys*, 1979; U.S. Bureau of Mines, *Minerals in the U.S. Economy*, July 1975; U.S. Department of Energy, *Energy Data Report: Annual Energy Balance 1978*, April 1979.

Energy storage

The general method and specific techniques for storing energy derived from some primary source in a form convenient for use at a later time when a specific energy demand is to be met, often in a different location.

In the past, energy storage on a large scale had been limited to storage of fuels. For example, large amounts of natural gas are routinely stored under pressure in underground reservoirs during the summer and used to meet increased demands for heating fuel in the colder seasons. Petroleum and its products are stored at several points in the energy system, from the strategic petroleum reserve to the fuel tanks of automobiles. Since gasoline is a highly concentrated and readily portable form of energy, this method of energy storage makes the automobile independent of the supply system for appreciable distances and times. On a smaller scale, electric energy is stored in batteries that power automobile starters and a great variety of portable appliances. *See* OIL AND GAS STORAGE.

In the future, energy storage in many forms is expected to play an increasingly important role in shifting patterns of energy consumption away from scarce to more abundant and renewable primary resources. For example, automobiles are likely to store transportation energy in the form of coal-derived synthetic fuels or as electricity in batteries charged with electric power from coal or nuclear power plants. Solar energy can already be made more usable by accumulating it during the day as warm water that can be stored for later use. An example of growing importance is the storage of electric energy generated at night by coal or nuclear power plants to meet peak electric loads during daytime periods. This is achieved by pumping water from a lower to a higher reservoir at night and reversing this process during the day, with the pump then being used as a turbine and the motor as a generator. As shown in Fig. 1, this example can be used to illustrate the conversion and storage functions of an energy storage system.

Broader application and use of new methods of energy storage could reduce oil consumption in the major energy use sectors collectively by perhaps 2.5×10^6 bbl (4×10^5 m³) per day in the year 2000, as much as $5-8 \times 10^6$ bbl ($8-13 \times 10^5$ m³) per day if all likely applications of storage can be fully developed. Probably the largest practical potential for oil displacement is in the heating and cooling of buildings through storage of heat or coolness generated on site with solar energy or off-peak electricity (singly or in combination). In principle, the potential for oil displacement is largest in transportation, since highway vehicles consume around 9×10^6 bbl (1.43×10^6 m³) of fuel per day, more than 50% of the United States' oil use. In practice, extensive displacement of conventional automobiles will be very difficult because of the very large energy storage capacity of fossil fuels compared to the most attractive alternative, electric storage batteries: a 20-gal (76-liter) tank of gasoline will give the average United States car a range of 200–300 mi (300–500 km), at minimal extra cost for the tank itself. On the other hand, a lead-acid battery providing a range of just 30–60 mi (50–100 km) weighs nearly a ton (0.9 metric ton) and costs between $1200 and $1500. The contrast between the two systems is illustrated in Fig. 2, which compares their specific ranges. Yet, because of their independence of oil and their more efficient as well as cleaner use of energy, electric vehicles powered by batteries are a possible alternative to conventional automobiles, especially for urban driving.

Energy storage in electric power systems has excellent potential for reducing the use of oil and gas for power generation, especially if pumped hydro storage can be supplemented by more broadly applicable techniques such as underground pumped hydro, compressed-air storage, and batteries. These techniques also could provide utilities with special operating advantages not found in conventional generating plants. Storage technologies that could find broad application in one or more of the major energy-use sectors are discussed below.

Pumped hydro storage. Until the late 1970s the only bulk energy storage method used by electric utilities, pumped hydro goes back to 1929 in the

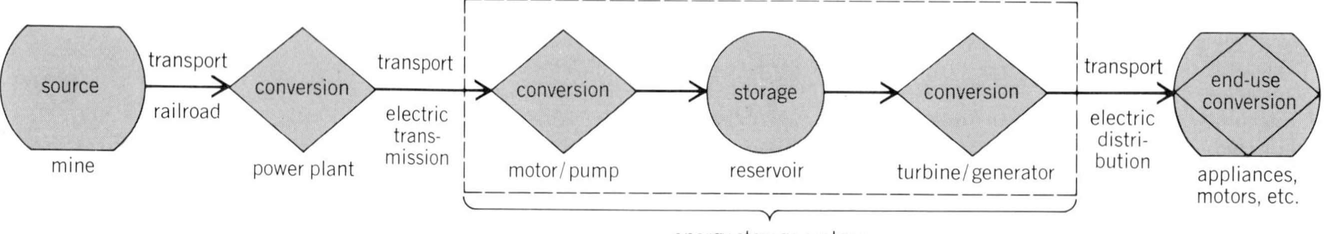

Fig. 1. Principal components of an energy storage system with specific reference to an electric power system.

United States; the largest such facility in the world—with a capacity of 15×10^6 kWh—is the Ludington pumped storage plant on Lake Michigan. The construction of such facilities is increasingly limited by a variety of geographic, geologic, and environmental constraints, all of which may be considerably reduced if the power generation equipment and discharge reservoir are placed deep underground (Fig. 3). The size of the surface reservoir required for such an underground pumped hydroelectric system will be several times smaller than that used in the conventional system because of the greater pressure available to run the generators. The difference in height between reservoirs in the underground system may be several thousand feet (1 ft = 0.3 m), compared to several hundred feet common in the conventional pumped hydroelectric system.

Compressed-air storage. Off-peak electric energy can also be converted into mechanical energy by pumping air into a suitable cavern where it is stored at pressures up to 80 atm (8 MPa). Compared to pumped hydro storage, compressed air offers several advantages, including a wider choice of suitable geologic formations, lower volume requirements, and a smaller minimum capacity to be economically attractive. The world's first commercial compressed-air energy storage system (Fig. 4) is in operation in Huntorf, West Germany. In this installation, expanding air must be heated by the combustion of natural gas before it can be used to drive the generator and turbines. In a more efficient concept, the heat evolved during air compression would be stored in a bed of pebbles and reused to heat the expanding air.

Batteries. The development of advanced batteries with characteristics superior to those of the familiar lead-acid battery could result in use of battery energy storage on a large scale. For example, batteries lasting 2000 or more cycles could be used in installations of several-hundred-thousand-kilowatt-hour capacity in various locations on the electric power grid, as an almost universally applicable method of utility energy storage. Batteries combining these characteristics with energy densities (storage capacity per unit weight and volume) well above those of lead-acid batteries could provide electric vehicles with greater range, thus removing a major barrier to their broader use.

The development of several new battery types has been undertaken to achieve systems with superior characteristics. Candidates include nickel-zinc and nickel-iron systems that promise modest improvements over lead-acid. The zinc-chlorine, sodium-sulfur, and lithium–iron sulfide systems could yield substantial improvements over lead-acid; of these more advanced systems the zinc-chlorine battery (Fig. 5) is farthest along in development. The battery operates near ambient temperature with an aqueous solution of zinc chloride as the electrolyte. On charge, zinc is deposited on a graphite substrate and chlorine is released as a slightly soluble gas at the opposite electrode, composed of porous graphite. Chlorine is stored in a separate vessel by freezing it out of the aqueous solution as a solid compound of water (ice) and chlorine at about 9°C; in this form, chlorine is relatively safe to handle. On discharge, zinc redissolves and chlorine is reduced at the porous graphite cathode, reforming the original zinc chloride aqueous solution. The active materials and other components offer the promise of a low-cost battery; however, the complexity of the system (which uses pumps, valves, and other auxiliaries) suggests that operating reliability could be a possible problem.

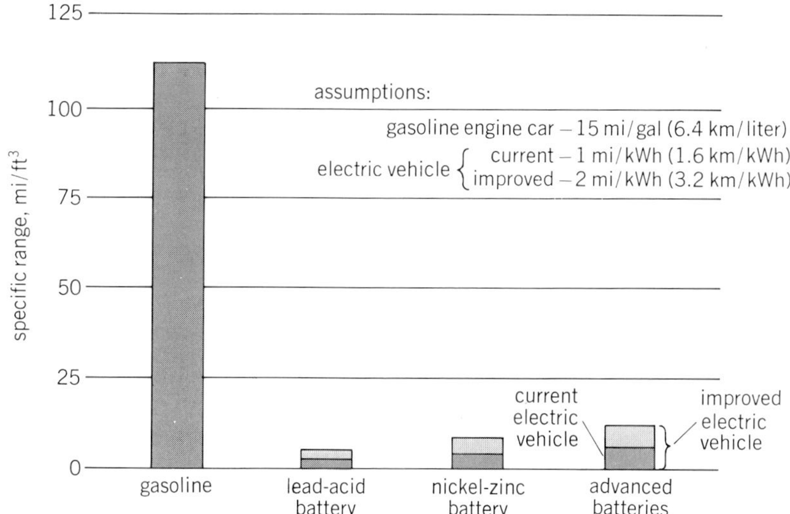

Fig. 2. Specific range of vehicles using energy stored in various forms. 1 mi/ft³ = 57 km/m³.

The sodium-sulfur battery is another battery under intensive development in the United States and Europe. Sodium-sulfur operates at temperatures above 300°C, where its electrodes exist as liquids separated by a unique solid ceramic electrolyte. Small experimental cells have already been tested through 1000 charging cycles while showing no appreciable degradation. A disadvantage of the sodium-sulfur battery, however, is that it must operate at a temperature of 300–350°C. The operating costs and safety characteristics of large batteries consisting of thousands of single cells in electric series and parallel connection are also potential drawbacks.

Two important national programs have been aimed at bringing advanced batteries into commercial use. The Electric and Hybrid Research, Development and Demonstration Act of 1976 provided funds for a 5-year program to demonstrate viable electric vehicles. This act provided major funding for the development and evaluation of batteries such as nickel-zinc and nickel-iron that could become available in the mid-1980s to increase the range of electric vehicles. To assist ongoing efforts to develop and commercially introduce advanced batteries for utility energy storage service on the multimegawatt level, the establishment of a major new facility, the Battery Energy Storage Test (BEST) Facility, has been undertaken to test a zinc-chlorine battery, and other advanced batteries when they become available, in the service area of Public Service Electric and Gas Company of Newark, NJ.

In parallel with this facility is the Storage Battery Electric Energy Demonstration (SBEED) project, designed to demonstrate a 10-MW, 30-MWh lead-acid battery energy storage system connected to the electrical system of the Wolverine Power Cooperative in northern Michigan. This project

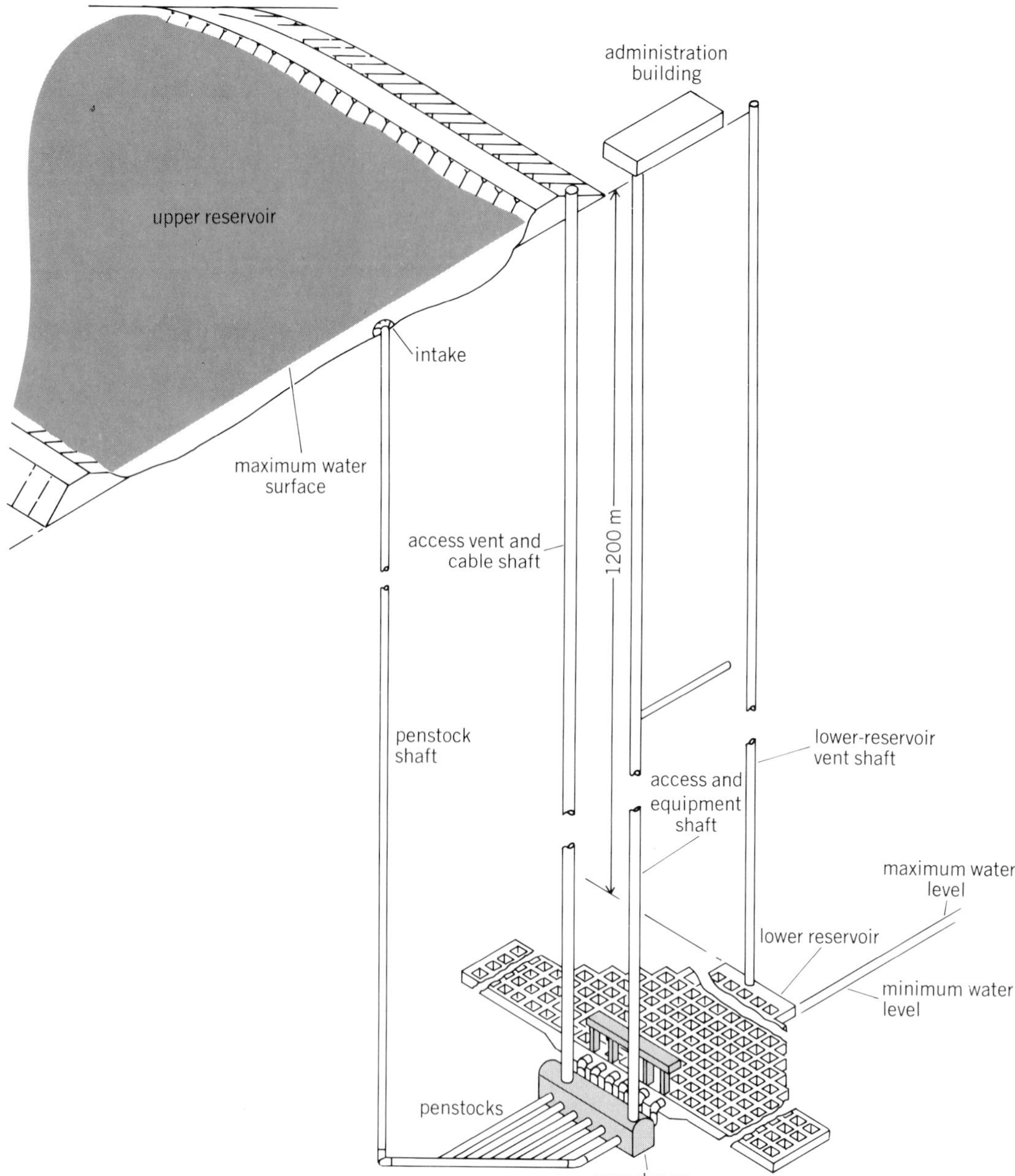

Fig. 3. Concept of an underground pumped hydroelectric storage system.

should go far in establishing the basic characteristics and advantages of battery energy storage for utilities.

Thermal storage. Ceramic brick "storage heaters" that store off-peak electricity in the form of heat have gained wide acceptance for heating buildings in Europe, and the barriers to their increased use in the United States are more institutional and economic than technological.

Testing has been undertaken on prototype "coolness" storage systems, which use electric refrigeration to chill water or produce ice at night. Experimental installations indicate that daytime electric power demand for air conditioning could be reduced up to 75% by using such systems, but these systems are still relatively bulky and expensive.

Solar hot-water storage is technically simple and commercially available. However, the use of solar energy for space heating requires relatively large storage systems, with water or rock beds as storage media, and difficulties can arise in integrating this storage with existing buildings while keeping costs within acceptable limits. Innovative designs that may help overcome these problems include new types of heat transfer equipment, storage of heat in the walls and ceilings of buildings (passive solar heating), and the use of low-cost, roof-mounted water bags. *See* SOLAR HEATING AND COOLING.

Chemical reaction systems. Heat or electricity may be stored by using these energy forms to force certain chemical reactions to occur. Such reactions are chosen so that they can be reversed readily with release of energy; in some cases the products can be transported from the point of generation to that of consumption, adding flexibility to the ways the stored energy can be used. For example, reactions which produce hydrogen could become attractive since hydrogen could be stored

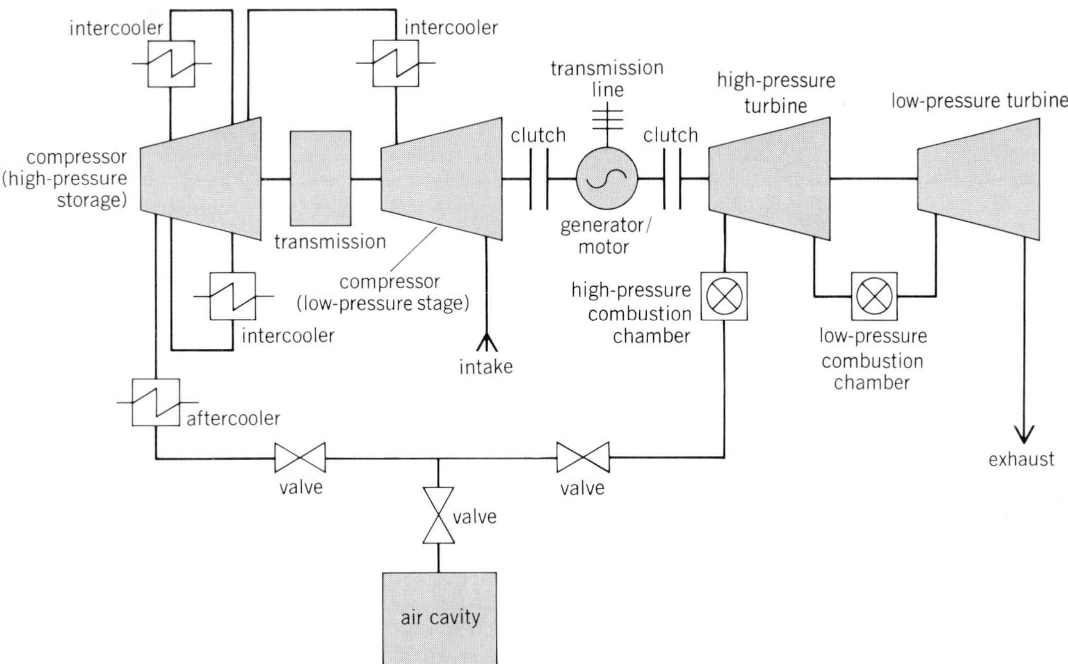

Fig. 4. Schematic layout of compressed-air storage plant at Huntorf, West Germany.

for extensive periods of time and then conveniently used in either combustion devices or in fuel cells.

Another possibility is to apply heat to a mixture of methane and water, converting them to hydrogen and carbon monoxide, which can be stored and transported to the end-use site, where a catalyst permits the reverse reaction to occur spontaneously with the release of heat. While not yet in practical use, chemical reaction systems are under extensive investigation as economic and flexible ways of storing energy.

Superconducting magnets. Electrical energy can be stored directly in the form of large direct

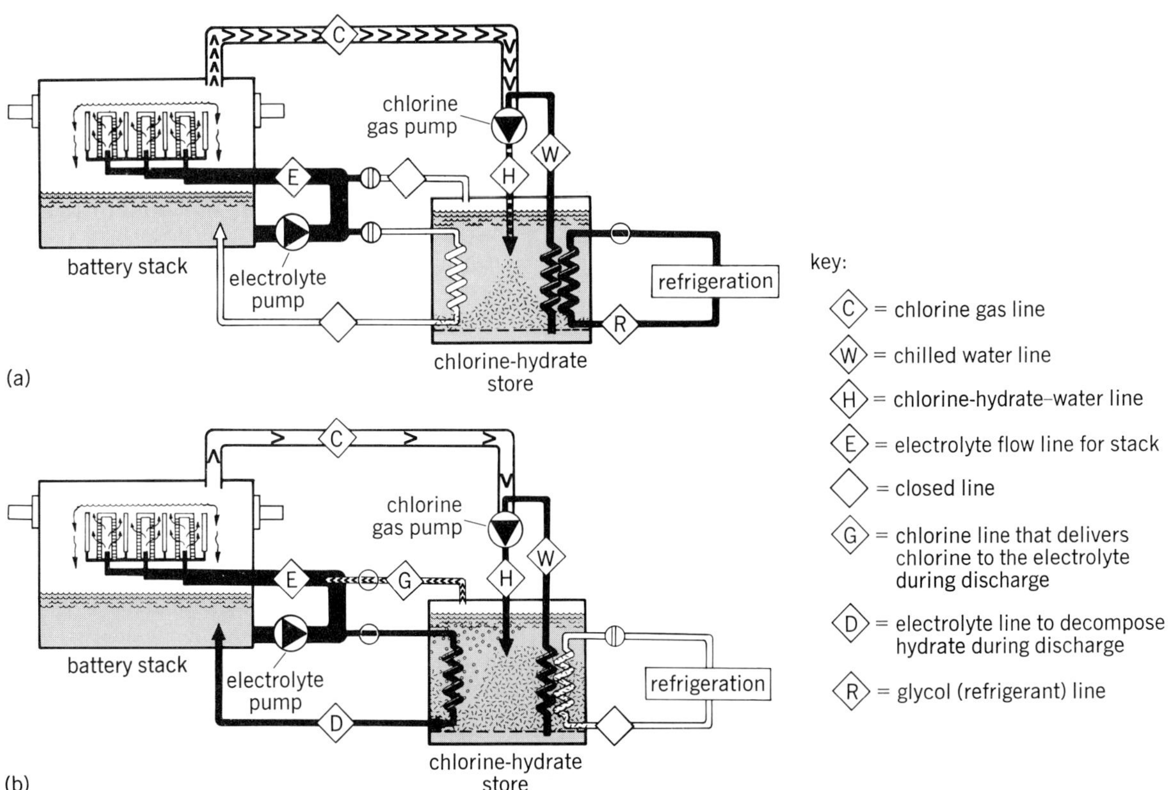

Fig. 5. Schematic layout of zinc-chloride battery, showing two main compartments in the design, the battery stack and the chlorine hydrate store. (*a*) Charge cycle. (*b*) Discharge cycle. (*Gulf and Western Company*)

currents used to create fields surrounding the superconducting windings of electromagnets. In principle such devices appear attractive because their storage efficiency is high, plant life could be long, and utilities would have few difficulties establishing the necessary conversion equipment. However, the need for maintaining the system at temperatures approaching absolute zero and, particularly, the need to physically restrain the coils of the magnet when energized require auxiliary equipment (insulation, vacuum vessels, and structural supports) which will represent a large expenditure.

Development of such systems is still in the early stages, and even the discovery of new superconducting materials with high critical currents and higher cryogenic temperatures may not be sufficient to reduce costs to acceptable levels since this would not significantly affect the structural containment requirements. *See* SUPERCONDUCTING DEVICES.

Flywheels. Storage of kinetic energy in rotating mechanical systems such as flywheels is attractive where very rapid absorption and release of the stored energy is critical. However, research indicates that even advanced designs and materials are likely to be too expensive for utility energy storage on a significant scale, and applications will probably remain limited to systems where high power capacity and short charging cycles are the prime consideration. Such applications do exist in pulse power supplies and in electric transportation for recovery of braking energy.

Combined systems. The rising importance of energy storage comes from its potential for shifting demand from scarce to plentiful primary energy sources. As such, the most successful storage devices are likely to be those that are adopted as components of larger systems designed specifically for resource conservation. In electric power generation, for example, system-wide storage may be combined with solar-electric systems to flatten load curves and reduce oil consumption (Fig. 6). If such complex systems are eventually to be developed and commercially introduced, financial incentives and more coherent national energy policy planning must be adopted. Thus, while advanced energy storage systems could have a major impact on United States energy consumption patterns before the end of the century with continued, successful research, nontechnical factors such as regulatory strategies and pricing policies are likely to become decisive to their large-scale use.

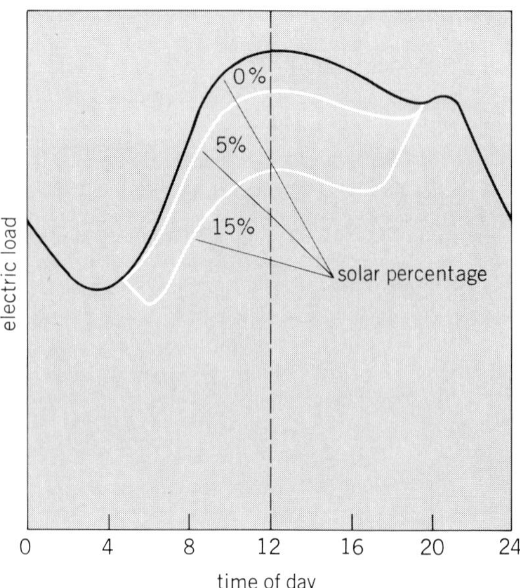

Fig. 6. Impact of solar-electric generation on utility load curve (schematic).

[FRITZ KALHAMMER; THOMAS R. SCHNEIDER]

Bibliography: J. R. Birk, K. Klunder, and J. C. Smith, Superbatteries: A progress report, *IEEE Spectrum*, 16(3):49–55, March 1979; J. R. Bolton, Solar fuels, *Science*, 202(4369):705–711, Nov. 17, 1978; E. J. Cairns and H. Shimotake, High temperature batteries, *Science*, 164(3886):1347–1355, June 20, 1969; R. A. Hein, Superconductivity: Large-scale applications, *Science*, 185(4147):211–222, July 19, 1974; F. R. Kalhammer, Energy-storage systems, *Sci. Amer.*, 241(6):56–65, December 1979; F. R. Kalhammer and T. R. Schneider, Energy storage, *Annu. Rev. Energy*, 1:311–343, 1976; G. D. Whitehouse, M. E. Council, and J. D. Martinez, Peaking power with air, *Power Eng.*, 72(1):50–52, January 1968.

Engine

A machine designed for the conversion of energy into useful mechanical motion. The principal characteristic of an engine is its capacity to deliver appreciable mechanical power, as contrasted to a mechanism such as a clock or analog computer whose significant output is motion. By usage an engine is usually a machine that burns or otherwise consumes a fuel, as differentiated from an electric machine that produces mechanical power without altering the composition of matter. Similarly, a spring-driven mechanism is said to be powered by a spring motor; a flywheel acts as an inertia motor. By this definition a hydraulic turbine is not an engine, although it competes with the engine as a prime source of mechanical power. *See* HYDRAULIC TURBINE; MOTOR; PRIME MOVER; WATERPOWER.

Applications. A fuel-burning engine may be stationary, as a donkey engine used to lift cargo between wharf and ship, or it may be mobile, like the engine in an aircraft or automobile. Such an engine may be used for both fixed service and mobile operation, although accessory modifications that adapt the engine to its particular purpose are preferable. For example, the fan that draws air through the radiator of a water-cooled fixed engine is large and fitted in a baffle, whereas the fan of a similar but mobile engine can be small and unbaffled because considerable air is driven through the radiator by means of ram action as the engine propels itself along. *See* AUTOMOTIVE ENGINE.

Some types of engine can be designed for economic efficiencies in fixed service but not in mobile operation. Thus, the steam engine is widely used in central electric generator stations but is obsolete in mobile service. This is chiefly because, in a large ground installation, the furnace and boiler can be fitted with means for using most of the available heat. The engine proper can be a reciprocating (piston) or a rotating (turbine) type. Because shaft rotation is by far the most used form of mechanical motion, the turbine is the more common form of modern steam engine. For railroad service the steam engine has given place to diesel and gasoline internal combustion engines and to electric motors. *See* BOILER; POWER PLANT; STEAM ENGINE; STEAM-GENERATING UNIT; STEAM TURBINE.

Types. Traditionally, engines are classed as external or internal combustion. External combustion engines consume their fuel or other energy source in a separate furnace or reactor. *See* NUCLEAR REACTOR; STEAM-GENERATING FURNACE.

Strictly, the furnace or reactor releases chemical or nuclear energy into thermal energy, and the engine proper converts the heat into mechanical work. The principal means for the conversion of heat to work is a gas or vapor, termed the working fluid. By extension, an engine which derives its heat energy from the Sun by solar radiation to working fluid in a boiler can be considered an external combustion type. To avoid loss of or contamination from nuclear fuel, the reactor and boiler are separated from (and may also be shielded from) the engine.

The working fluid takes on energy in the form of heat in the boiler and gives up energy in the engine, the engine proper being a thermodynamic device. The device may be a turbine for stationary power generation on a nozzle for long-range vehicular propulsion. *See* NUCLEAR POWER.

In an engine used for propulsion, the rearward velocity with which the working fluid is ejected and, thus, the forward acceleration imparted to the vehicle depend on the temperature of the fluid. For practical purposes, temperature is limited by the engine materials that serve to contain the chemical combustion or nuclear reaction. To achieve higher exhaust velocity, the working fluid may be contained by nonmaterial means such as electric and magnetic fields, in which case the fluid must be electrically conductive.

The engine proper is then a magnetohydrodynamic device receiving electric energy from a separate fuel-consuming source such as a gas turbine and electric generator or a nuclear reactor and electric generator, or possibly from direct electric conversion of nuclear or solar radiation.

A further basis of classification concerns the working fluid. If the working fluid is recirculated, the engine operates on a closed cycle. If the working fluid is discharged after one pass through boiler and engine, the engine operates on an open cycle. Closed-cycle operation assures the purity of the working fluid and avoids the discharge of harmful wastes. The open cycle is simpler. Thus the commonest types of engine use atmospheric air in open cycles both as the principal constituent of their working fluids and as oxidizer for their fuels.

If open-cycle operation is used, the next modification is to heat the working fluid directly by burning fuel in the fluid; the engine becomes its own furnace. Because this internal combustion type engine uses the products of combustion as part of the working fluid, the fuel must be capable of combustion under the operating conditions in the engine and must produce a noncorrosive and nonerosive working fluid. Such engines are the common reciprocating gasoline and diesel units. *See* DIESEL ENGINE; INTERNAL COMBUSTION ENGINE; ROTARY ENGINE; STIRLING ENGINE.

At low speeds the combustion process is carried out intermittently in a cylinder to drive a reciprocating piston. At high speed, however, friction between piston and cylinder walls and between other moving parts dissipates an appreciable portion of the developed power. Thus, where high power is developed at high speed, performance is improved by continuous combustion to drive a turbine wheel. *See* BRAYTON CYCLE; CARNOT CYCLE; DIESEL CYCLE; GAS TURBINE; OTTO CYCLE.

Engine shaft rotation may be used in the same way as in a reciprocating engine. However, for high-velocity vehicular propulsion, the energy of the working fluid may be converted into thrust more directly by expulsion through a nozzle. Once the vehicle is in motion, the turbine can be omitted. Alternatively, instead of drawing atmospheric oxygen into the combustion chamber, the engine may draw both oxidizer and fuel from storage tanks within the vehicle, or the combustion chamber may contain the full supply of fuel and oxidizer.

Despite all the variation in structure, mode of operation, and working fluid—whether of moving parts, moving fields, or only moving working fluid—these machines are basically means for converting heat energy to mechanical energy.

[FRANK H. ROCKETT]

Bibliography: W. H. Crouse and D. L. Anglin, *Automotive Engines*: *Workbook*, 6th ed., 1980; J. Day, *Engines*: *The Search for Power*, 1980.

Engine cooling

Cylinder gas temperatures in internal combustion engines may reach 4500°F (2500°C). This is well above the melting point of the engine parts in contact with the gases, so that it is necessary to control the temperature of the parts or they will become too weak to carry the gas pressure stresses. The lubricating oil film on the cylinder wall can fail because of chemical changes at wall temperatures above about 400°F (200°C). Complete loss of power may take place if some spot in the combustion space becomes sufficiently heated to ignite the charge early on the compression stroke. *See* INTERNAL COMBUSTION ENGINE.

Fortunately a thin protective boundary of relatively stagnant gas of poor conductivity exists on the inner surfaces of the combustion space. If the outer cylinder surface is placed in contact with a cool fluid such as air or water and there is sufficient contact area to cause a rapid heat flow, the resulting temperature drop produced by the heat flow in the inside boundary layer keeps the cylinder wall temperature much closer to the coolant temperature than to the combustion gas temperature. The quantity of heat that crosses the stagnant boundary layer and must be carried away by the coolant is a function of the Reynolds number of the gas existing in the cylinder. In terms of practical engine quantities the heat flow to the coolant varies approximately as (charge density × piston speed)$^{0.8}$. At full throttle and normal piston speed this heat flow amounts to about 15% of the energy of the incoming fuel.

Liquid cooling. If the coolant is water, it is usually circulated by a pump through jackets surrounding the cylinders and cylinder heads. The water is circulated fast enough to remove steam bubbles that may form over local hot spots and to limit the water's temperature rise through the engine to about 15°F (8°C). The warmed coolant is piped to an air-cooled heat exchanger called a radiator (Fig. 1). The airflow required to remove the heat from the radiator is supplied by an engine-driven fan, and also by the forward motion of the vehicle in automotive applications. In liquid cooling, the engine and radiator may be separated and each placed in the optimum location, being connected through piping. To prevent freezing, the water coolant is often mixed with alcohol or ethylene-glycol.

Low water-jacket temperature is conducive to corrosive wear of the engine parts and increases the piston friction losses. High water-jacket temperature increases the coolant loss by evaporation or by actual boiling. Temperature of the water jacket is often automatically maintained near 160°F (70°C) by a thermostat placed in the line from engine to radiator. When the engine outlet water is too cool, it is prevented from entering the radiator, and is usually recirculated in the engine block until warm enough to open the thermostat.

Air cooling. Engines are often cooled directly by a stream of air without the interposition of a liquid medium. The heat-transfer coefficient between the cylinder and airstream is much less than with a liquid coolant, so that the cylinder temperatures must be much greater than the air temperature to transfer to the cooling air the heat flowing from the cylinder gases. To remedy this situation and to reduce the cylinder wall temperature, the outside area of the cylinder, which is in contact with the cooling air, is increased by finning. The heat flows easily from the cylinder metal into the base of the fins, and the great area of finned surface permits heat to be transferred to the cooling air (Fig. 2). The ideal fin shape depends upon the conductivity of the fin material. In general, the fin is thickest at the base to permit heat flow from the cylinder. The fin should taper to a thin edge to give a good temperature gradient along its length. For reasons of mechanical strength, fins are usually made thicker than necessary for heat-transfer considerations. *See* HEAT EXCHANGER.

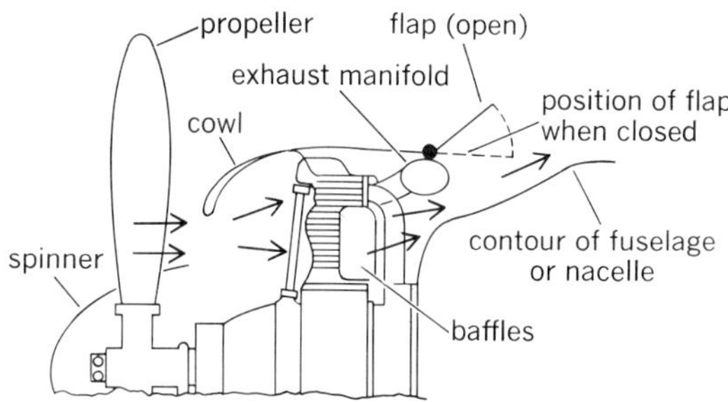

Fig. 2. Airflow in a radial aircraft engine.

High-output cylinders require many closely spaced fins. In these engines the area between adjacent cylinders is blocked off with sheet metal baffles, which are also shaped to follow the fin tips part way around the cylinder. A pressure is built up in front of the baffles by means of a fan or because of the forward motion of the vehicle (ram effect). The pressure differential between front and rear of the engine forces the cooling air through the spaces between the fins.

The power required to cool depends upon the quantity of cooling air used and the velocity at which it passes the fins. For minimum cooling power, the fins should be long and close together so that a large heat-transfer area is served by a small coolant flow area. The temperature difference between fins and cooling air should be kept as high as possible so that less air velocity will be required. Cylinder temperatures of air-cooled engines are sometimes controlled by louvers or flaps, which may be set to restrict the cooling airflow until the engine becomes warm.

[AUGUSTUS R. ROGOWSKI]

Bibliography: T. Pipe, *Audel's Gas Engine Manual*, 2d ed., 1977.

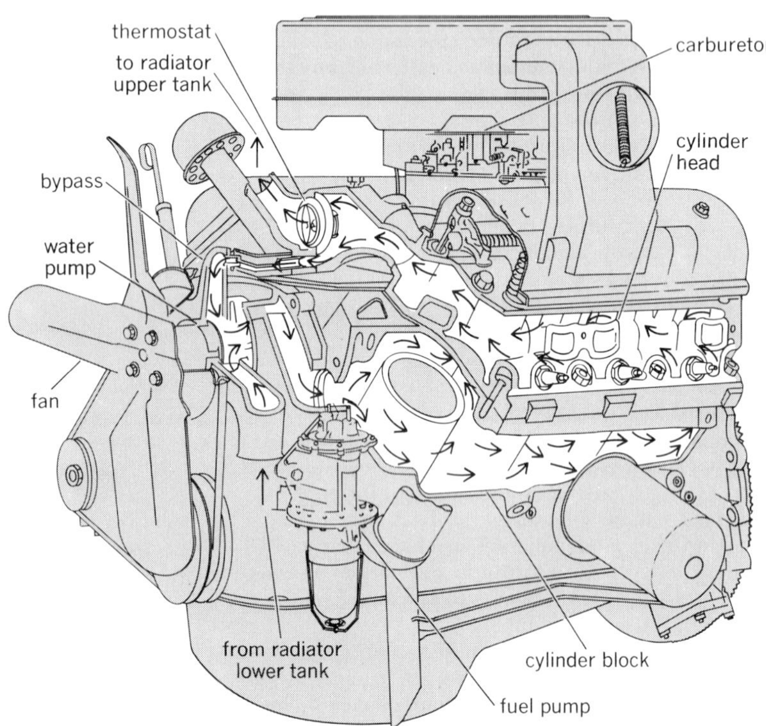

Fig. 1. Cooling system of a V-8 engine. Engine is partly cut away to show jackets and, by arrows, circulation of cooling water. Radiator is not shown.

Engine lubrication

Continuous supply of a viscous film between moving surfaces in an engine during its operation. Relative motion between unlubricated surfaces of engine parts would result in excessive wear, overheating, and seizure, with destruction of the contacting surfaces. The energy loss associated with the wear and overheating would be evident in reduced engine power and efficiency. The purpose of lubrication is to minimize these effects by interposing a viscous fluid called the lubricant between the parts. Hydrodynamic forces are produced in the lubricant to keep the surfaces separated, even though heavy loads are tending to bring them into contact. The thickness of the separating film, for a given part geometry, depends upon the relative velocity of the two surfaces, the absolute viscosity of the lubricant, and the load per unit of surface area.

Lubricant. Engines are generally lubricated with petroleum base oils containing chemical additives to improve their natural properties. Refining methods and composition may be varied to suit special engine design operating conditions. Perhaps the most important oil property is the absolute viscosity, which is a measure of the force required to move one layer of the oil film over another. With low viscosity a protecting oil film between the parts is not formed. With high viscosity, too much power is required to shear the oil film; in addition, the flow of oil through the engine is retarded.

The viscosity of oil tends to decrease as its temperature is raised. It is therefore difficult to maintain the proper viscosity as the engine warms up, or as other operating conditions change. Viscosity index is a number which indicates the resistance of an oil to changes in viscosity with temperature. The smaller the change in viscosity with temperature, the higher is the viscosity index of the oil.

Lubrication system. The function of the lubrication system is to supply clean oil cooled to the proper viscosity to critical points in the engine, where the motion of the parts produces hydrodynamic oil films to separate and support the various rubbing surfaces. In all but the most primitive lubrication systems, the oil, which has been heated by friction and by heat transfer from the hot engine parts, is collected, cooled, and pumped back into the engine. In automobiles the oil returns by gravity to the crankcase. The air motion over the outside of the crankcase is usually sufficient to cool the oil. It is then pumped from the crankcase back to the oil distributing galleries in the engine. In high-output engines the oil is pumped from the crankcase into a storage reservoir and then through an air- or water-cooled heat exchanger on the way back to the engine.

A positive oil distribution method called force-feed or full-pressure lubrication is used in most engines (see illustration). In this system the pump forces the oil at considerable pressure through passages to each main crankshaft bearing. The crankshaft is drilled to convey the oil from the main bearings to the connecting rod bearings. Sometimes the connecting rods are drilled to carry the oil to the wrist pins. Except in the largest engines, the cylinder walls are lubricated by leakage spray from the ends of the connecting rod bearings. In the force-feed system the pump pressure ensures a steady replacement of the oil which has been heated by friction work, with fresh cool oil from the crankcase or reservoir. In this way the temperature, and therefore the working viscosity of the oil in the bearings, is controlled.

Wear. Since the lubricating oil flows to all parts of the engine, it is important that it does not carry abrasive or corrosive material with it. Such material may come from the combustion of the fuel, from dirt in the inducted air, or from parts of the engine itself. It is common practice to filter all or part of the oil as it flows through the system, to reduce wear. In some engines a small fraction of the oil leaving the pump is continually bypassed through a filter and returned to the crankcase. The rest of the oil flows directly to the bearings. Since considerable time is required for all of the oil to be filtered by this method, most automobile engines pump all of the oil through a full-flow filter placed in the line between the pump and the bearings.

A large part of the cylinder wear occurs during starting and before the engine has warmed up. Cold oil, being quite viscous, is slow to reach the bearings, and the end leakage spray onto the cylinder walls is greatly reduced. At low engine temperatures, there is an increased tendency for corrosive products of combustion to condense on the cylinder walls. Chemical additives are used in modern engine oils to increase the viscosity index, reduce corrosive action, wash away accumulations of sludge or other deposits on engine parts or in oil passages, and improve oil performance in various other ways.

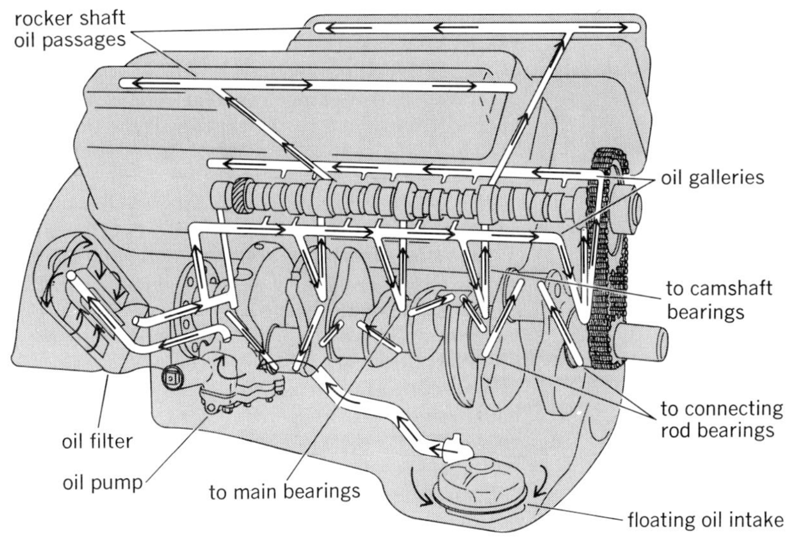

Full-pressure lubrication system in eight-cylinder V-type engine. (*Chrysler Motor Corp.*)

Excessive use of oil may be caused by wear of rings or cylinder bore, oil of improper viscosity, leakage past the shaft seals, or loss of vapor from the crankcase breather. By design and maintenance, oil consumption can be made small. The principal objection to high oil flow into the combustion space is smoky exhaust, fouling of spark plugs, and carbon deposits in the cylinders. *See* INTERNAL COMBUSTION ENGINE; LUBRICANT.

[AUGUSTUS R. ROGOWSKI]

Bibliography: A. Cameron, *Principles of Lubrication*, 2d ed., 1980; W. H. Crouse, *Automotive Fuel, Lubricating, and Cooling Systems*, 5th ed., 1976; D. D. Fuller, *Theory and Practice of Lubrication for Engineers*, 1956; A. Schilling, *Automobile Engine Lubrication*, 1972.

Engine manifold

The branch pipe arrangement which connects the valve ports of multicylinder engines to a single carburetor or to a muffler. The manifold may have considerable effect on engine performance. The intermittent nature of the gas flow through each cylinder may develop resonances (similar to the vibrations in organ pipes) in the airflow at certain speeds. These may increase the volumetric efficiency and thus the power at certain engine speeds at the expense of loss at other speeds, depending on manifold dimensions. This action applies to the exhaust pipes as well as to the intake pipes. *See* INTERNAL COMBUSTION ENGINE.

Another very important influence which intake manifolds may have on engine performance results from the fact that it is here most of the heat is supplied to vaporize the fuel droplets metered to the airstream as it passes through the carburetor. This is usually accomplished by an area of the manifold directly under the carburetor which is heated by exhaust gas. The resulting temperature of the fuel mixture should be held within a comparatively narrow range for best engine performance with a fuel of a given volatility. If the temperature is not high enough, the unvaporized fuel in the manifold makes engine response to the throttle during acceleration sluggish and probably unsatisfactory. If the temperature is unnecessarily high, it causes loss of power, due to the reduced weight of the mixture supplied to the cylinders and the increased tendency of the engine to develop vapor lock from fuel boiling in the metering system; moreover, fuel knock can develop. Provision is therefore generally made to reduce the flow of exhaust gas over this vaporizing area, or "hot spot," by means of a thermostatically controlled valve in the exhaust system.

Even when provision has been made for the desired heating of the intake manifold during normal engine operation, the problem remains of obtaining the best performances between the time an engine is started cold and the time it reaches normal temperatures. This is important for many automobile drivers who want to get under way immediately.

During this warm-up period, liquid fuel flows along the walls of the manifold at a lower velocity than the airstream driving it, and its distribution to the various cylinders is a different and more difficult matter than the distribution of a "dry" mixture containing no liquid fuel. Acceleration would be poor except for the supplemental fuel momentarily supplied by the carburetor accelerator pump, because (1) the liquid stream does not change speed as rapidly as the airstream, and (2) the drop in manifold pressure during acceleration causes some of the fuel previously vaporized to condense on the manifold walls and thus to starve the cylinders of the desired fuel still further.

On some engines it has been found that supplemental fuel required from the accelerator pump is reduced by water-jacketing at least part of the intake manifold with some of the hot cylinder-jacket water. This source of heat does not develop as soon after an engine is started cold as the exhaust-heated "hot spot." However, it retains heat longer from a previous run and stabilizes the performance when the engine is started while there is still some heat in the jacket water.

When the throttle of an engine is closed suddenly, as in a quick stop, with the intake manifold at a temperature which permits some liquid fuel to be present, the events occurring during acceleration are reversed and part or all the liquid on the manifold walls vaporizes almost instantly at the lower manifold pressure. The cylinders then receive a super-rich mixture which discharges undesired and offensive combustion products for a few engine cycles, or which may be too rich to burn and results in a few misfires. *See* CARBURETOR. [NEIL MAC COULL]

Engineering

Most simply, the art of directing the great sources of power in nature for the use and convenience of humans. In its modern form engineering involves people, money, materials, machines, and energy. It is differentiated from science because it is primarily concerned with how to direct to useful, economical ends the natural phenomena which scientists discover and formulate into acceptable theories. Engineering therefore requires above all the creative imagination to innovate useful applications of natural phenomena. It is always dissatisfied with present methods and equipment. It seeks newer, cheaper, better means of using natural sources of energy and materials to improve the standard of living and to diminish toil.

Types of engineering. Traditionally there were two divisions or disciplines, military engineering and civil engineering. As man's knowledge of natural phenomena grew and the potential civil applications became more complex, the civil engineering discipline tended to become more and more specialized. The practicing engineer began to restrict his operations to narrower channels. For instance, civil engineering came to be concerned primarily with static structures, such as dams, bridges, and buildings, whereas mechanical engineering split off to concentrate on dynamic structures, such as machinery and engines. Similarly, mining engineering became concerned with the discovery of, and removal from, geological structures of metalliferous ore bodies, whereas metallurgical engineering involved extraction and refinement of the metals from the ores. From the practical applications of electricity and chemistry, electrical and chemical engineering arose.

This splintering process continued as narrower specialization became more prevalent. Civil engineers had more specialized training as structural engineers, dam engineers, water-power engineers, bridge engineers; mechanical engineers as machine-design engineers, industrial engineers, motive-power engineers; electrical engineers as power and communication engineers (and the latter divided eventually into telegraph, telephone, radio, television, and radar engineers, whereas the power engineers divided into fossil-fuel and nuclear engineers); mining engineers as metallic-ore mining engineers and fossil-fuel mining engineers (the latter divided into coal and petroleum engineers).

As a result of this ever-increasing utilization of technology, humans and their environment have been affected in various ways—some good, some bad. Sanitary engineering has been expanded from treating the waste products of humans to also treating the effluents from technological processes. The increasing complexity of specialized machines and their integrated utilization in automated processes has resulted in physical and mental problems for the operating personnel. This has led to the development of bioengineering, concerned with the physical effects upon man, and management engineering, concerned with the mental effects.

Integrating influences. While the specialization was taking place, there were also integrating influences in the engineering field. The growing complexity of modern technology called for many specialists to cooperate in the design of industrial processes and even in the design of individual machines. Interdisciplinary activity then developed to coordinate the specialists. For instance, the design of a modern structure involves not only the static structural members but a vast complex including moving parts (elevators, for example); electrical machinery and power distribution;

communication systems; heating, ventilating, and air conditioning; and fire protection. Even the structural members must be designed not only for static loading but for dynamic loadings, such as for wind pressures and earthquakes. Because people and money are as much involved in engineering as materials, machines, and energy sources, the management engineer arose as another integrating factor.

Typical modern engineers go through several phases of activity during their career. Their formal education must be broad and deep in the sciences and humanities underlying their field. Then comes an increasing degree of specialization in the intricacies of their discipline, also involving continued postscholastic education. Normal promotion thus brings interdisciplinary activity as they supervise the specialists under their charge. Finally, they enter into the management function as they interweave people, money, materials, machines, and energy sources into completed processes for the use of humans.

For specific articles on various engineering disciplines *see* CIVIL ENGINEERING; ELECTRICAL ENGINEERING; INDUSTRIAL ENGINEERING; MANUFACTURING ENGINEERING; MECHANICAL ENGINEERING; METHODS ENGINEERING; MINING; NUCLEAR ENGINEERING. *See also* TECHNOLOGY.

[JOSEPH W. BARKER]

Engineering, social implications of

The rapid development of human ability to bring about drastic alterations of the environment has added a new element to the responsibilities of the engineer. Traditionally, the ingredients for sound engineering have been sound science and sound economics. Today, sound sociology must be added if engineering is to meet the challenge of continued improvement in the standard of living without degradation of the quality of the environment.

Despite the fact that present and evolving engineering practices must meet the criteria of scientific and economic validity, these same practices generally cause societal problems of new dimensions. Consider, for example, exhaust gases emitted from tens of millions of internal combustion engines, both stationary and moving; stack gases from fossil-fuel-burning plants generating steam or electric power; gaseous and liquid effluents and solid waste from incinerators and waste-treatment systems; strip mining of coal and mineral ores; noise issuing from automotive vehicles, aircraft, and factory and field operations; toxic, nondegradable or long-lived chemical and particulate residues from ore reduction, chemical processing, and a broad spectrum of factory and mill operations; dust storms, soil erosion, and disruption of groundwater quality and quantity accompanying intensified mechanized farming in conjunction with massive irrigation and fresh-water diversion.

Progress often results in the substitution of one set of problems for another. For example, in nuclear electric power generating plants, replacement of fossil fuels by nuclear fuels relieves the burden of atmospheric pollution from stack gas emissions. Lower thermal efficiency of a nuclear plant, however, results in higher heat rejection rates and increased thermal pollution of sources of cooling water or air. The attendant consequences on atmospheric conditions or on the viability of aquatic life in the affected water are of great concern in the short and long terms. Ultimately, the cost and benefit considerations of nuclear power must be all-inclusive; in addition to usual considerations of economic length of plant life and so forth, one must account for all the economic and societal costs of the entire fuel cycle, from mining and refinement through use and ultimate recycling or safe disposal. The long-term effects of very low levels of radiation exposure (as such studies become available) will be an additional factor to consider.

The modern engineer must be increasingly conscious of the societal consequences of technological innovation. *See* ENGINEERING.

[EUGENE A. AVALLONE]

Engineering and architectural contracts

The legal documents pursuant to which most professional services are rendered. The contract should set forth the terms, provisions, and obligations of each party. If a dispute should arise, the contract is the first document to be interpreted by the courts. The importance of a succinct and properly written contract cannot be overstressed. A carefully written contract is the best way to avoid disputes and litigation. Before the contract is executed, it is advisable to seek proper legal advice. It must be realized that most governmental and quasi-governmental bodies have separate statutes, rules, legislation, and legal precedent; legal obligations in one state might be totally different from those in another.

Engineering and architectural licenses. In order to engage in the professional practice of engineering or architecture, an individual must obtain a license. The licensing requirements of each state differ in wording and court interpretation. To render engineering or architectural services without a license could be the basis for criminal penalties. To render services without a license can also result in an illegal contract, and the person rendering the services may be precluded from recovering compensation for the work performed. It must be stressed that even though the architect or engineer (A/E) is licensed in one state, when the services are rendered in a different state, the contract might be deemed illegal, and the A/E might be practicing without a license. It is strongly recommended that the A/E have a license in every state in which services are being rendered, or that the A/E comply, in all other respects, with the specific prevailing laws.

In the California case of *Palmer v. Brown*, 127 Cal. App.2d 44, 273 P.2d 306, an architectural partnership was denied recovery for work done in the name of the firm by an unlicensed member. The court held that, although the partnership could render architectural services by licensed architects, it could not recover for services rendered by an unlicensed partner. In the case of *Wineman v. Blueprint, Inc.*, 75 Misc.2d 665 (1973), a New York trial court was faced with the question of whether to permit recovery for services rendered as an "architectural consultant," although the person had no architectural license. The court noted that the purpose of licensing statutes is to safeguard the health, life, and property of citizens, and that the statute would be strictly enforced. Holding oneself out as an "architectural consultant," when one is not a licensed architect, is prohibited. The court ruled that it was against public policy to permit such as person to recover any compensation.

Another question which is often raised is whether an architectural or engineering license is required of a construction or manufacturing firm which is rendering certain A/E services as part of other services under a larger contract. A New York decision held that the contract was valid when the overall contract was not primarily the rendering of A/E services, and when a licensed person was actually engaged by the firm to perform those tasks for which the law requires licensed skills. *Eisen und Stahlwerke v. Modular Building and Development Corp.*, 64 Misc.2d 1050 (1970).

Formation and existence. Various legal events must be proved before a valid and enforceable contract comes into existence. The initial two events are the offer and acceptance. The offer must be clear and certain; it must not be ambiguous. The acceptance must accept specifically what was offered; otherwise, the acceptance is actually deemed to be a counteroffer. The offer to render professional services, or the offer to retain those services, must be accepted by the other party. In addition, consideration must flow from one party to the other. Consideration is normally monetary compensation, although it may take other forms. For a valid contract to exist, there must be a "meeting of the minds" of all contracting parties. In determining whether a meeting of the minds has occurred, the courts will carefully review the contract to determine whether both parties are legally aware of their respective obligations. An essential ingredient of the A/E contract is the scope of the work, the obligations, and the responsibilities of the A/E.

Scope of A/E professional services. The most important aspect of the A/E contract is the clear definition of the A/E's scope of work. The contract should succinctly state what services are to be rendered. In an A/E contract, services may be rendered over a long period of time and during several phases of design and construction. The A/E's responsibilities during each of the phases should be stated explicitly. These phases have generally been categorized as (1) schematic or preliminary design; (2) design development; (3) construction or bid document; (4) contracting; and (5) construction. A basic statement should be set forth in the A/E contract listing the general scope of the work to be performed by the A/E during each of these phases. As part of the general statement, the contract should indicate whether or not other professionals such as surveyors and environmentalists are to be employed by the A/E and incorporated as part of the A/E's overall responsibility and obligation, or whether the client will furnish any necessary additional services. In addition to general statements, the contract should indicate the specific obligations of the A/E during each phase.

Preliminary design phase. During the schematic or preliminary design phase, the A/E normally prepares conceptual study reports and analyses for the client. The specific depth and specificity of these documents should be stated. In addition, the general outlines and parameters of the client's requirements should be set forth so that the A/E has a guide. A statement of estimated costs of the final project should be set forth very carefully at this point.

Design development. During the design development phase, the contract should specify whether the A/E is obligated to prepare final documents from which the client can obtain construction or manufacturing prices. An all-important consideration is whether or not the A/E is impliedly or expressly guaranteeing the eventual costs of the project, or whether the A/E is representing the "estimated" cost of the project.

Bid and construction documents. After the final design is completed, whether the A/E is to be involved in the actual preparation of the bid and construction documents should be specified. The responsibility for defective bids and contracting documents must be seriously weighed before the A/E undertakes this aspect of the work. Many times, this responsibility is divided between the client and the A/E. It is normal at this junction for the A/E to assist in obtaining necessary governmental approvals of the design and project. During the contracting stage the professional may or may not have a function. It is not unusual for the A/E to conduct the bidding and negotiation on behalf of the client. When this function is assumed by the A/E, the A/E must be aware of numerous applicable bidding laws and regulations. The laws and precedents differ in most states regarding public as opposed to private contracts. The bid documents prepared by an A/E for a private contract might be legally insufficient to obtain prices for a public contract. Numerous cases exist as to the propriety in the acceptance of bids on public contracts. It is recommended that a legal consultant be available during all phases of public bidding.

Construction phase. The final major phase relates to construction. The A/E contract, as well as the contract between the client and contractor, should specify whether or not the A/E is to be involved in the actual contract performance stage. If the A/E is to be the representative of the client, the limits of that responsibility should be delineated. Consideration must be given as to whether the A/E has full or limited responsibility and authority to act.

The responsibilities and liabilities of the A/E during the construction phase raise numerous questions, all of which should be resolved in the contract. Among these questions are: (1) When and how often must the A/E consult with the client? (2) Is the A/E obligated and responsible for the supervision and inspection of the work, and if so, to what limits? (3) How often must the A/E visit the site? (4) Is the A/E obligated to review, check, and approve shop drawings, diagrams, and other data? (5) Does the A/E have the authority to issue modifications of, or revisions to, the final contract on behalf of the client during actual performance of the work, and if so, must the client be consulted? (6) Is the A/E to be the interpreter of the contract documents? (7) Is the A/E responsible for reviewing the work for the purpose of approving or disapproving payments to the contractor? (8) Is the A/E responsible to see that the work is being performed in strict or substantial compliance with the contract documents? (9) Is the A/E obligated to perform field tests? (10) Is the A/E obligated to furnish expert testimony if necessary? (11) Does the A/E have any contractual relationship, expressed or implied, with the contractor performing the work?

When these items are reviewed, consideration must be given as to whether the A/E is to be the judge of the performance of the work, a neutral arbitrator showing preference to neither side, or an advocate and agent of the owner. In order to avoid placing the A/E in the precarious position of a neutral, while at the same time having to pass judgment on his own plans and specifications, many clients and courts acknowledge that the A/E's decision is not final and binding in the interpretation of the contract documents. This is especially true when the A/E's decision is arbitrary or capricious.

The scope of the A/E's professional services is of utmost importance; it has a direct effect on the A/E's professional responsibility and liability.

Professional responsibility. By holding himself out as a licensed A/E, and entering into a contract for the rendering of professional services, the A/E implies that he possesses, within reasonable expectations, all the requisite skills and knowledge necessary to obtain the license, and is able to exercise ordinary and reasonable A/E skill, care, and due diligence in his performance. These implied representations and standards are the basis upon which all of the A/E's actions will be judged. The client is entitled to rely upon and expect that the A/E possesses the necessary skills, and that the finished product will be suitable for its intended use. The A/E is liable for breach of contract and negligence if he fails to exercise the ordinary skill of his profession. The A/E has frequently been held liable for defects in design. Not only is professional responsibility owed to the person with whom the A/E has contracted, but the A/E has the responsibility to provide reasonable professional skills to any person who might reasonably come in contact with the work product. An A/E can be professionally liable to third persons many years after the work is completed and accepted by the client. *Totten v. Gruzen*, 52 N.J. 202, 245 A.2d 1 (1968).

Codes, laws, rules and regulations. One of the implied professional skills of the A/E is knowledge of all applicable building, structural, safety, health, zoning, and other laws, rules, and regulations which have a direct effect upon rendering of services. Designs and plans which are prepared must comply with regulations, and if they do not, the A/E may not be able to recover for services rendered and may be liable for damages incurred. *Bott v. Moser*, 175 Va. 11, 7 S.E.2d 217 (1940).

In the case of *Rozny v. Marnul*, 250 N.E.2d 656 (1969), the Supreme Court of Illnoisw was faced with the situation of a surveyor who had prepared an inaccurate survey of a vacant lot in 1953 for a real estate firm. Several years later a plat survey was issued by a surveyor who guaranteed the accuracy of the survey. Years later the survey was relied upon by the owner of a home when he extended a driveway and built a new garage. In 1962 it was found that the new driveway and garage encroached on adjacent property. The surveyor was sued for the cost to correct these conditions. The surveyor was liable, even though no direct contract existed, since he had failed to exercise reasonable care, and failed to fulfill the reasonable duties and obligations of one of his profession.

Cost estimates and guarantees. An A/E contract often stipulates that the eventual construction costs shall not exceed a maximum figure. Depending upon the specific words, if the eventual construction costs exceed the specified amount, or a reasonable margin thereof, the A/E will be deemed to have breached the contract and is precluded from recovering for services rendered. *Bueche v. Eickenroht*, 220 S.W.2d 911 (Texas Civ. App. 1949).

The question becomes more complex when the A/E furnishes an "estimate" of probable construction costs. Although several court rulings have held that merely exceeding the estimate does not constitute a breach, the trend appears to be that a gross underestimation constitutes a breach. In the case of *Kostohryz v. McGuire*, 212 N.W.2d 850 (1973), the Supreme Court of Minnesota acknowledged that the architect's estimates of probable construction costs did not constitute an expressed guarantee or firm figures, but ruled that the estimates represented the best judgment of a design professional familiar with the construction industry. The court held "that an architect who substantially underestimates through the lack of skill and care the cost of a proposed structure, which representation as to cost is relied upon by the owners in entering the contract and proceeding with the construction, may not only forfeit his right to compensation but he may also become liable to the owner for damages."

Defective contract documents. It is generally recognized by the courts that a contractor is not liable for defects in the work which result from defective plans and specifications prepared by the owner's A/E. Rather, it is the owner who warrants the adequacy to the contractor. *United States v. Spearin*, 248 U.S.132 (1918). It is the A/E who prepared these documents on behalf of the owner, and it is the A/E who will be liable to the client when the defective plans and specifications result from the A/E's failure to exercise reasonable care and skill. *White v. Pallay*, 247 P. 316, 119 Or. 97.

Work supervision and inspection. When an A/E contracts to perform supervisory functions, his liability is greatly enlarged, and now extends substantially beyond the issue of design. The A/E is now responsible to supervise the work of the contractors. As stated by a New York court: "An architect, where supervision is part of the duties assumed by him under the contract of employment, is bound to exercise due care in the performance of such duty and negligence on his part gives the employer cause of action for damages." *Lindberg v. Hodgens*, 152 N.Y.S. 229, 89 Misc. 454 (1915).

Satisfactory work provision. Many construction contracts contain a provision to the effect that the contractor's work must satisfy the A/E. Such provision does not authorize the A/E to arbitrarily reject material or the contractor's work. In the case of *Midgley v. Cambell Building Co.*, 38 Utah 293, 112 P. 820 (1911), the court held that an architect cannot arbitrarily refuse and condemn the use of plumbing fixtures manufactured by a concern other than the one originally intended to be used if the quality of the fixtures is as good as that originally submitted. The arbitrary refusal to permit the substitution of materials might well create liability on the part of the owner. If the owner-client becomes liable, then the responsibility and liability of the A/E comes into question. The general rule of law is that a contract provision to the effect that work must be done to the satisfaction of the owner or architect merely means that the work should be satisfactory to a reasonably prudent person. *Fielding & Shepley, Inc. v. Dow*, 163 P.2d 908 (1945). The A/E must carefully scrutinize his acts so as not to be viewed as having acted arbitrarily or capriciously.

Settlement of disputes. The A/E is often retained to act as the arbiter of disputes which arise between the client and the contractor relating to performance of work and payment. The specific scope of this authority is to be set forth in the A/E's contract, as well as in the contract between the client and contractor.

Under most construction contracts the A/E has the authority to issue certificates of payment. If the certificate is not issued, the contractor normally is not entitled to payment. However, the certificate cannot be arbitrarily withheld. A general rule of law is that the final certificate of the A/E is conclusive unless fraud, bad faith, or obvious mistake can be shown. It is equally well established that the final certificate is not binding upon the contractor when the A/E has erred in his legal interpretation of the contract. *Uvalde Contracting Co. v. City of New York*, 154 N.Y.S. 604 (1914). In the New Jersey case of *Terminal Construction Corp. v. Bergen County Hackensack River Sanitary Sewer District Authority*, 18 N. J. 294, 113 A.2d 787 (1955), the court held that the engineer's acts in the exercise of the authority vested in him under the terms of the contract were valid unless legally fraudulent. This rule applies even when the owner is not a direct participant in the engineer's fraud. It must be noted, however, that the term fraud as applied to a construction contract has a broader meaning than usual, since it encompasses arbitrary action or gross mistake. The court stated "that the architect or engineer occupies the position of trust and confidence and that he should act in absolute and entire good faith throughout, and when he acts under a contract as the official interpreter of its conditions and judge of its performance, he should side neither with the Owner or Contractor, but exercise impartial judgment." If the A/E is found to have acted arbitrarily, capriciously, or in bad faith, the question of the A/E's liability comes to the forefront. The A/E should carefully consider the possible consequences before accepting the responsibilities of arbitrator and judge.

Payment for professional services. The A/E contract also specifies the obligations of the client. The all-important clause is the one dealing with payment for services. Services are rendered under different forms of contracts, and each contract calls for different methods of payment. The four primary contract payment forms are:

1. Set fee or lump sum. This contract provides for a fee in a set dollar amount. In most instances, the fee is not dependent upon the amount of time expended by the A/E, or the final cost of construction. However, as noted previously, if the A/E's contract is based upon a guaranteed or estimated construction cost, and if that obligation is breached by the A/E, then there will be questions as to the right of the A/E to receive any payment.

2. Percentage of construction costs. Compensation under this contract form is directly proportionate to the final cost of the project. This contract leaves certain variables open, and the final fee is often not subject to computation until the entire project is completed.

3. Multiple of direct personnel expense. This type of contract compensation clause requires great detail in the contract form. The A/E's compensation is normally the cost of direct salaries times a mutually agreed-upon factor. To avoid conflict, great care should be used in writing the A/E contract so that all items of expense to be multipled by the factor are known beforehand. This might require a specific listing of those personnel and items of expense which will be considered. The factor used to multiply the expense items should be based upon recognized accounting principles so that the arrangement will be equitable to both contracting parties.

4. Cost plus fixed fee. This form of contract is really a combination of those discussed above. It provides for reimbursement of actual cost to the A/E, in addition to payment of a fixed fee. Again, a list of the itemized costs which will be allowed should be given. The fixed fee is subject to negotiation by the parties and should have a direct relationship to the size and duration of the project.

The determination for the basis of payment is only the first of two important steps. The second is the method for payment. It is very unusual for an A/E contract to provide for full payment after the entire project is completed. Rather, progress payments are made at various stages during the course of the project. Many times, payments coincide with the various phases of the A/E's work. The A/E must be sure to include within the contract the stipulation that he has the right to receive some payment whether or not the client finally proceeds with the project.

Payment for plans. In the case of *Bergstedt, Wahlberg, Berquist Associates, Inc. v. Rothchield*, 225 N.W.2d 261 (1975), the Supreme Court of Minnesota had to decide whether an architect was entitled to any compensation when no formal agreement specifying a fee arrangement had been entered into. However, there did exist a letter from the architect to the owner setting forth a fee arrangement, but there was no response to the letter. A/E services were expended over a period of 2 years. In reaching its final decision that an implied contract existed and that the architect was entitled to payment, the court stated, "Where the evidence fails to disclose an expressed agreement, the law may imply a contract from the circumstances or acts of the parties. Words are not the only medium of commercial expression, and no legal distinction can be made in the effect of a promise whether it is expressed in writing, orally, in the acts of the parties, or in a combination of means. It is the objective thing, the manifestation of mutual assent, which is essential to the making of a contract." Reviewing the facts, the court felt that it was obvious that the two parties had worked very closely for a period in excess of 2 years. The owner was continuously advised of all facts and events. "In short, a normal architect-client relationship was maintained during this period," according to the court, and the architect was entitled to be paid for services rendered.

Ownership of plans and design. A/E designs, plans, and drawings are protected from their inception by a common-law copyright. *Smith v. Paul*, 174 Cal. App.2d 744, 345 P.2d 546 (1959). It has generally been held by the courts that upon publication of the plans, drawings, and designs, this common-law copyright protection is lost. The question arises, however, as to when a general publication has taken place. Each case must rest on its specific facts. Such cases have involved the exhibition of models, the drawings and designs appearing in brochures and descriptive literature, photographs, and the filing of plans with public building departments. As long as the physical plans, drawings, and designs remain in the possession of the A/E, they are the property of the A/E. However, with most standard contract forms, when the A/E has prepared the plans and specifications for a client, and has been paid, they become the property of the owner. If the A/E prefers another arrangement, his wishes must be specifically set forth in the A/E contract. Maintaining the protection of the ownership of plans and design until payment is of utmost importance. Often, payment becomes a problem, and prior to that time the A/E should zealously maintain individual ownership of his work.

Insurance. The responsibilities and liabilities of both the A/E and the client not only flow to each other but also flow to and are for the benefit of third parties. It is extremely important for each party to consider obtaining insurance. The contract should clearly specify the kinds of insurances each party is to obtain and maintain. Potential liabilities are great, and often insurance is the only means to reasonably protect oneself. *See* ARCHITECTURAL ENGINEERING; ENGINEERING

[MICHAEL S. SIMON]

Bibliography: H. D. Hauf, *Building Contracts for Design and Construction*, 1968; W. Jabine, *Case Histories in Construction Law*, 1973; McGraw-Hill, *Legal Briefs for Architects, Engineers and Contractors*, 1976, N. Walker and T. K. Rohdenburg, *Legal Pitfalls in Architecture, Engineering and Building Construction*, 1968.

Engineering design

Engineering is concerned with the creation of systems, devices, and processes useful to, and sought by, society. The process by which these goals are achieved is engineering design.

The process can be characterized as a sequence of events as suggested in Fig. 1, with the recognition that no final, universally accepted description of so complex an intellectual and physical exercise, applicable to an enormously broad spectrum of products and processes, is possible. The process may be said to commence upon the recognition of, or the expression of, the need to satisfy some human want or desire, the "goal," which might range from the detection and destruction of incoming ballistic missiles to a minor kitchen appliance or fastener.

Concept formulation. Since the human aspiration is usually couched in nonspecific terms as a sought-for goal, the first obligation of the engineer is to develop more detailed quantitative information which defines the task to be accomplished in order to satisfy the goal, labeled on Fig. 1 as task specification. At this juncture the scope of the problem is defined, and the need for pertinent information is established. The source of the original request is questioned to establish the correspondence between the developing specifications and the initial definition of the goal.

But to know that a need exists and to have started on the task of qualitatively and quantitatively defining its substance and bounds should not be confused with the generation of ideas for possible solutions to the problem. This creative stage is called the concept formulation. When great strides in engineering are made, this represents ingenious, innovative, inventive activity; but even in more pedestrian situations where rational and orderly approaches are possible, the conceptual stage is always present.

The concept does not represent a solution, but only an idea for a solution. Initially evisceral and ephemeral, it can only be described in broad, qualitative, frequently graphical terms. Concepts for possible solutions to engineering challenges arise initially as mental images which are recorded first as sketches or notes and then successively tested, refined, organized, and ultimately documented by using standardized formats.

At this point it is important to note that the necessarily two-dimensional description of Fig. 1 and this sequential textural description should not be construed as an implication that the goal, task specification, and concept always appear in a simple, sequential temporal order. In fact, a signal characteristic of the design process is the seemingly random unpredictable emergence of, and iterations between, the various steps. This stochastic character can be suggested by drawing two-way connections between the various steps. For example, the definition of a task and emergence of a concept might precipitate restudy and possible alteration of the original goal. Consider the not infrequent experience of the serendipitous emergence of a new and interesting concept, and then subsequent search for a goal (application or market) to which to apply the idea.

Concepts are accompanied and followed by, sometimes preceded by, acts of evaluation, judgment, and decision. It is in fact this testing of ideas against physical, economic, and functional reality that epitomizes engineering's bridge between the

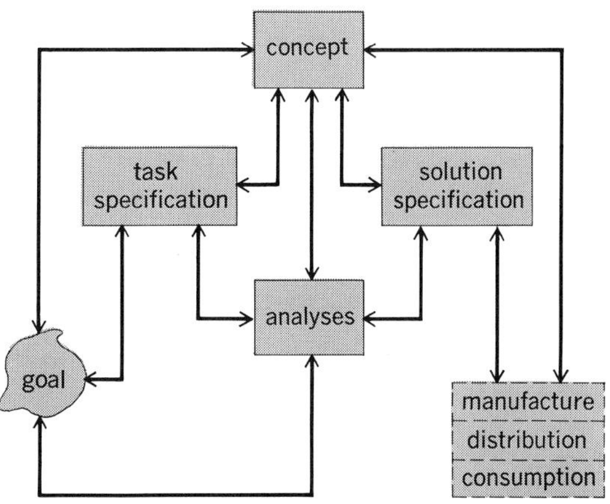

Fig. 1. The engineering design process.

art of innovation and science. The process of analysis is sometimes intuitive and qualitative, but it is often mathematical, quantitative, careful, and precise. The iteration of concept and analysis invariably gives rise to a focusing and sharpening, possibly a complete change, of the concept. Frequently, in light of the analyses conducted, what can or might be accomplished is illuminated and task specifications changed. Even goals may be altered. That which was sought may be beyond accomplishment, or perhaps solutions more sophisticated or more useful than initially undertaken may prove achievable.

Production considerations can have a profound influence on the design process, especially when high-volume manufacture is anticipated. Evolutionary products manufactured in large numbers, such as the automobile, are tailored to conform with existing production equipment and techniques such as assembly procedures, interchangeability, scheduling, and quality control. New techniques such as those associated with space exploration, where volume production is not a central concern, factor into the engineering design process in a very different fashion.

Similarly, the design process must anticipate and integrate provisions for distribution, maintenance, and ultimate replacement of products. Well-conceived and executed engineering design will encompass the entire product cycle from definition and conception through realization and demise and will give due consideration to all aspects.

The iterations of analysis and concept and considerations of manufacture develop information which defines a sequence of progressively specific solutions. In final form the solution specification consists of all drawings, materials and parts lists, manufacturing information, and so on necessary for construction of the device, system, or process.

It is important to note that, while the culmination of a particular engineering design process defines but one solution, inspection of the process during its evolution indicates a complex series of discrete partial, temporary, interim solutions which are compared one with another, and out of this comparison emerges the final compromise solution. The limitations on this scrutiny of alternative approaches are the practical contingencies of time and manpower, or other ways of measuring resources. When these assets reach their budgetary limits, the most satisfactory solution is accepted; however, in as dynamic a field as engineering the final design is not necessarily an ultimate or optimal best.

Figure 1 provides a broad overview of the engineering design process. Some expansion of the interaction between conceptualization and analysis is warranted, as shown in Fig. 2. On the left- and right-hand sides is concept; everything between is characterized as analysis.

Any physical entity, existing or hypothesized, of any degree of complexity cannot be analyzed in its entirety because of inadequate knowledge of the relevant physical laws, or inadequate time or facilities for the required computation, or a combination of these shortcomings. For these reasons, plus its inherent initial vagueness, the concept cannot be analyzed completely. Instead, simplified models are deliberately and precisely defined by applying established physical principles and laws to describe the model via mathematical equations. By using numerical values from the task specifications for parameters, the requisite computational tasks are performed.

In many engineering situations, particularly those where there is no body of experience with similar geometries, materials, and so on, the model from which the analysis is derived cannot be confidently assumed to characterize completely all significant attributes of the ultimate physical system. In some cases one does not have physical knowledge of certain relevant processes; other times adequate resources to perform all pertinent analyses are not available. Thus, one takes recourse to experiment. Since nature is the final arbiter of all physical proposals, however analyzed, experiment or test always precedes final acceptance of any proposal.

As previously suggested for Fig. 1, where all possible interconnections occurred between various stages and the process underwent a dynamic evolution, complex interactions occur also in the concept-analysis loop of Fig. 2. For example, one defines the model based on the concept recognizing existing physical knowledge about included processes and phenomena, and anticipates the effort that will be involved in the reduction of the resulting mathematical equations into quantitative results. For example, an initial model might be a greatly simplified, or perhaps oversimplified, characterization of the concept in order that the computational results might quickly identify the utility of the idea or its ranges of applicability. Subse-

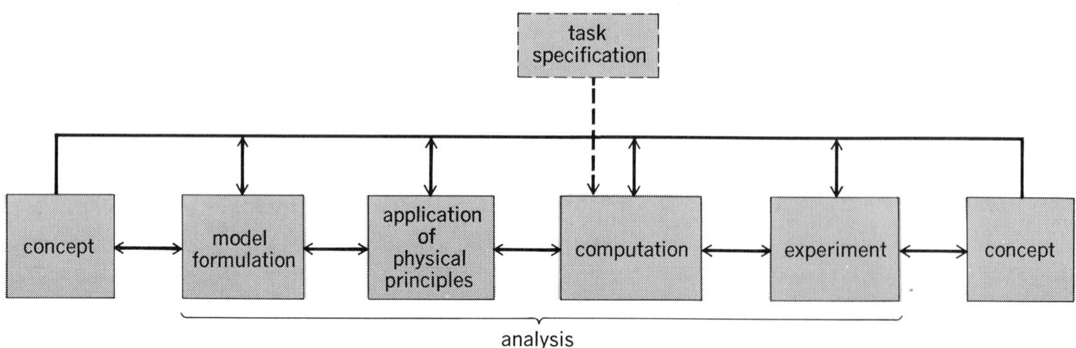

Fig. 2. The role of analysis in concept formulation.

Fig. 3. Flight configuration of *Mariner 2* spacecraft.

quently, more refined models of aspects of the concept of greater subtlety might be subject to scrutiny.

Where the function of the part is critical or the physical knowledge is inadequate, one might, in fact, eliminate computation and move rapidly to experiment, perhaps on an analog or scaled-down version of important attributes of the concept.

Hierarchy of design. An adequate description of the engineering design process must have both general validity and applicability to a wide variety of engineering situations: tasks simple or complex, small- or large-scale, short-range or far-reaching. That is to say, there is a hierarchy of engineering design situations.

Systems engineering occupies one end of the spectrum. The typical goal is very broad, general, and ambitious, and the concepts are concerned with the interrelationships of a variety of subsystems or components which, taken together, make up the system to accomplish the desired goal.

At a subsidiary level of the design problem hierarchy, the same engineering design process applies to creation of a device which might be one component of the overall system. And at the most detailed end of this hierarchy the same process diagrams the engineering design of a single element of a component. Obviously, as one applies the engineering design process to create one of these several elements, components, or systems, one exercises different phases of the process in different ways and to different degrees, depending upon the particular problem.

As an illustration of this hierarchy of design problems, all of which exhibit the schema of the engineering design process, consider a particular goal, the interplanetary inspection of the atmosphere of Venus by the *Mariner 2* space probe. At the system level this study emcompasses considerations of possible Earth-Venus trajectories based on astronomical events, launching times, booster-spacecraft weight-thrust combinations, and so on. Projections of foreseeable booster capabilities balanced against estimates of the spacecraft weight, combined with the astronomical "windows" to Venus which result from Earth-Venus juxtaposition, give rise to detailed analyses of conceivable space paths which pose the need for midcourse maneuver capabilities.

The scientific measurements desired and the assumed Venus atmosphere postulate the functional characteristics of instruments: weight, volume, and power consumption. Control and telemetry requirements to and from the spaceship likewise augment projected weight, volume, and power consumption needs. Combinations of these requirements and others give rise to concepts for the spaceship design, the flight configuration of which is shown in Fig. 3. *Mariner 2* is an interesting example of design *nouveau*, there being no established precedent as to what a Venus probe should look like.

Each major component of the spacecraft—structure, internal power supply, solar collecting vanes, radio telemetry antenna, and so on—undergoes a similar evolution from requirements and concepts through analysis. Each major subsystem is composed of subcomponents, each of which is composed of elements; Fig. 4 shows the detailed configuration of the planetary boom. The same process by which the major spacecraft arrangement decisions were made applies again; individual concepts and analyses are permutated in order to arrive at the design configuration of elements, such as this planetary boom. Thus, the same general overall diagrams of Figs. 1 and 2 characterizing the engineering design process can be applied in an ever broadening and encompassing fashion from the most minor components to the most major system aspects of an overall design. In point of fact, in an actual situation the order is usually reversed, the process being applied first to the most sweeping question, and progressively to the more detailed aspects, although one should never lose sight of the fact that the overall design is the sum of all its parts, that each part makes a necessary contribution to the whole, and that the performance of the whole is dependent on each part.

The Mariner project is interesting as an example, not only because of its new characteristics, adventuresome quest, and dramatic success, but because it also demonstrates the dynamics of the design process. The spaceship weight, of course, depended upon the particular booster rocket des-

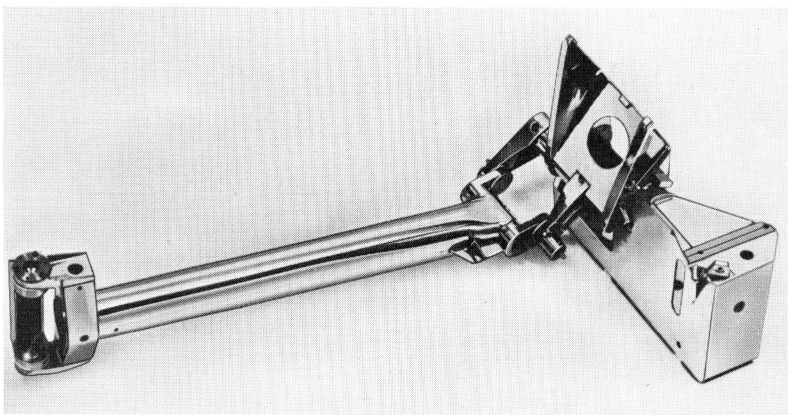

Fig. 4. Planetary boom of the *Mariner 2* spacecraft.

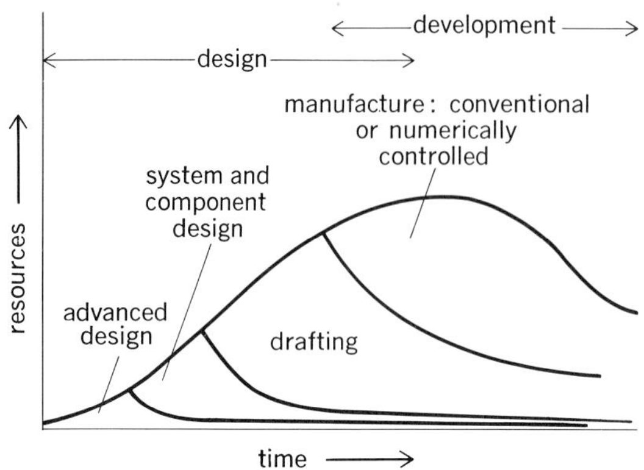

Fig. 5. Elapse of time and resources in an engineering design project, showing various stages in sequence.

ignated to bear it into interplanetary space. As the booster itself was under development during the spaceship development, the permissible weight of the Mariner vehicle varied in accordance with the predicted capability of the booster. Shortly before one of the scheduled departure times, major reductions in the vehicle were made mandatory by a change in the booster configuration. But this impermanence and the variability of the "ground rules" and "boundary conditions" of a design situation are always present, although perhaps not as dramatically as in this case.

Another aspect of the Mariner project particularly appropriate to this discussion is the importance of time and schedule to the success of the mission. Trajectory "windows" to Venus are only available at certain discrete, foreseeable times. Either the vehicle-booster combination would be ready to launch at the designated time, or else there would be no possibility of success. Again, most design situations are characterized by this urgency of time and schedule.

Finally, while the government-funded NASA program is perhaps not the best example of cost as the vital common denominator of all industrial and commercial engineering ventures, even in this case there was a budget, not a matter of profit or return on investment, but rather the limited skilled man-hours available to meet the commitment.

Time–worker-power resource dynamics. Another dimension of the dynamics of the engineering design process is the elapse of time and expenditure of worker-hours in the evolution of an engineering design project. Figure 5 plots time as the abscissa and resources (worker-power or dollars) as the ordinate. The various stages of the engineering design process are set out in time sequence from left to right.

Goal refinement, task specification, and first-order concept and analyses iterations are conducted by one to a few engineers in the early stages to establish the feasibility of the idea and to block out possible approaches. This is usually called the advanced design stage.

As the design concept becomes more specific and substantive, more and more engineers, technicians, and draftsmen become involved in the project. For example, in the case of a modern aircraft probably fewer than a dozen extremely talented engineers carry through the early feasibility and configuration studies, while at a latter stage hundreds of engineers, draftsmen, expediters, and coordinating personnel are involved. This deployment aspect of the design process cannot be overemphasized. In projects of significant size, the problem of coordinating and integrating the efforts of the many participants of different talents and skills becomes itself a major consideration. While some of the coordination involves judgment and decisions, much of this coordination is purely clerical and some involves prosaic application of standard reference material. *See* PERT.

Ultimately, the process culminates crescendo-like in the solution specification—manufacturing drawings and specifications, parts lists, and so on. Where automated manufacturing techniques are warranted, as in the aircraft industry, the efforts of computer programmers are devoted to the transformation of graphic and numeric manufacturing information into a form decipherable by computers and machine-tool directors. Ultimately, the physical parts are realized through the manufacturing process.

The efforts of this large group of people demand

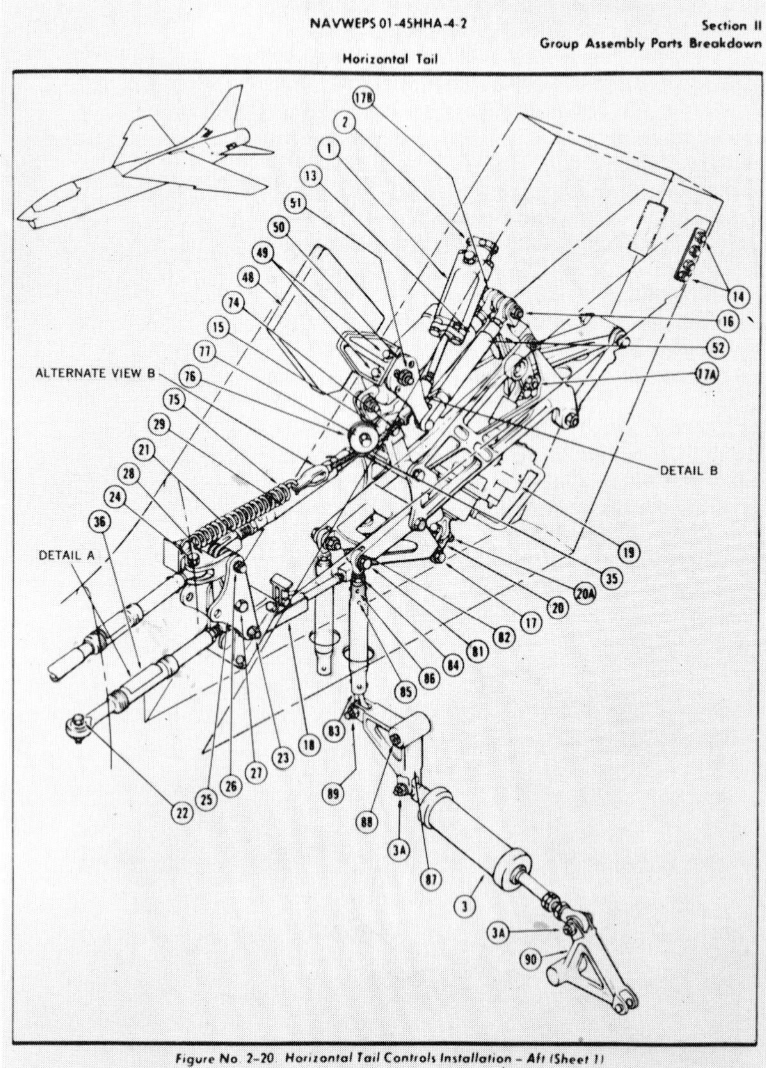

Fig. 6. One of the 10,000 sheets involved in the design engineering of a military aircraft. Numbers refer to the various components and are an indication of the tremendous bookkeeping involved in a project of this type.

a high level of coordination and integration if each of the thousands of separate parts of the aircraft are to satisfy its function, be compatible with one another, and be adequately strong and yet of minimum weight and volume. Figure 6 suggests a few of the individual elements which must undergo design, one sheet of a 2-ft-high pile of similar diagrams for a typical jet aircraft. The parts shown represent only a few of the better than 10,000 parts that make up the typical jet aircraft.

Use of the computer in design. Reference to the role of analysis in Fig. 1 and to computation in Fig. 2 highlights the ever increasing use of the computer, both analog and digital, in the engineering design process. As a high-powered successor to the slide rule and desk calculator, the computer is used routinely for much of the calculation, computation, and data reduction which constitute a major activity in design. Where repetitive series of calculations are carried out, programs are prepared to relegate to the computer more and more of the responsibility for analysis. Where economically justified, the overall engineering design process for a product is mechanized via computer programming. For example, Fig. 7 describes the flow chart of a branching computer program which, given vital data on the requirements of an electrical power transformer, carries out the design automatically. In the process the program, having made assumptions on core size and copper windings, calculates temperature gradient, impedance, and so on; checks these against preprogrammed constraints; and optimizes its choices considering, among other things, the current cost of copper. Figure 8 illustrates typical iterations of the computer design as it "homes" in on the lowest weight and cost design which satisfies the initial data and the internal constraints.

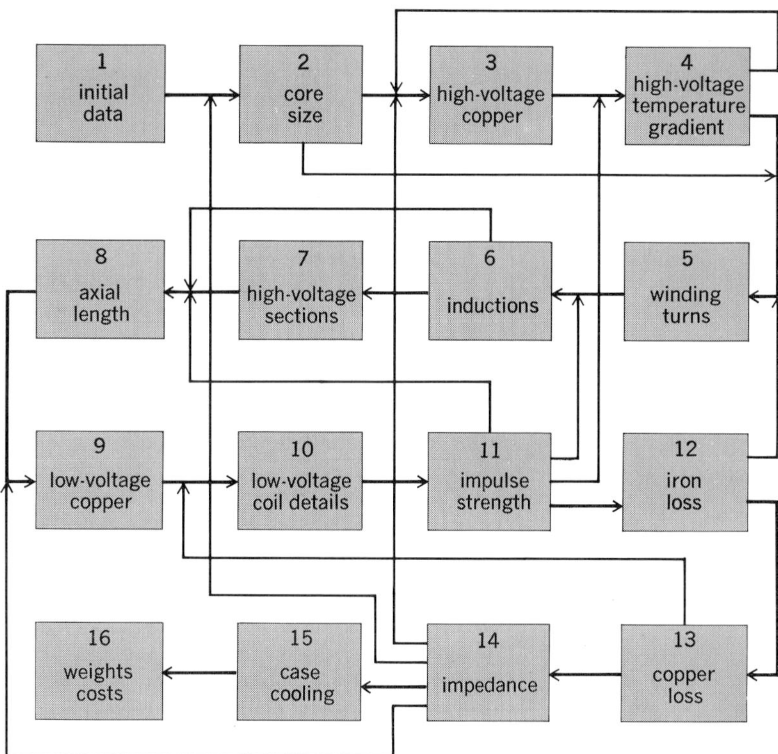

Fig. 7. Flow chart of a computer program.

The speed, memory, and accuracy of the computer to iteratively calculate, store, sort, collate,

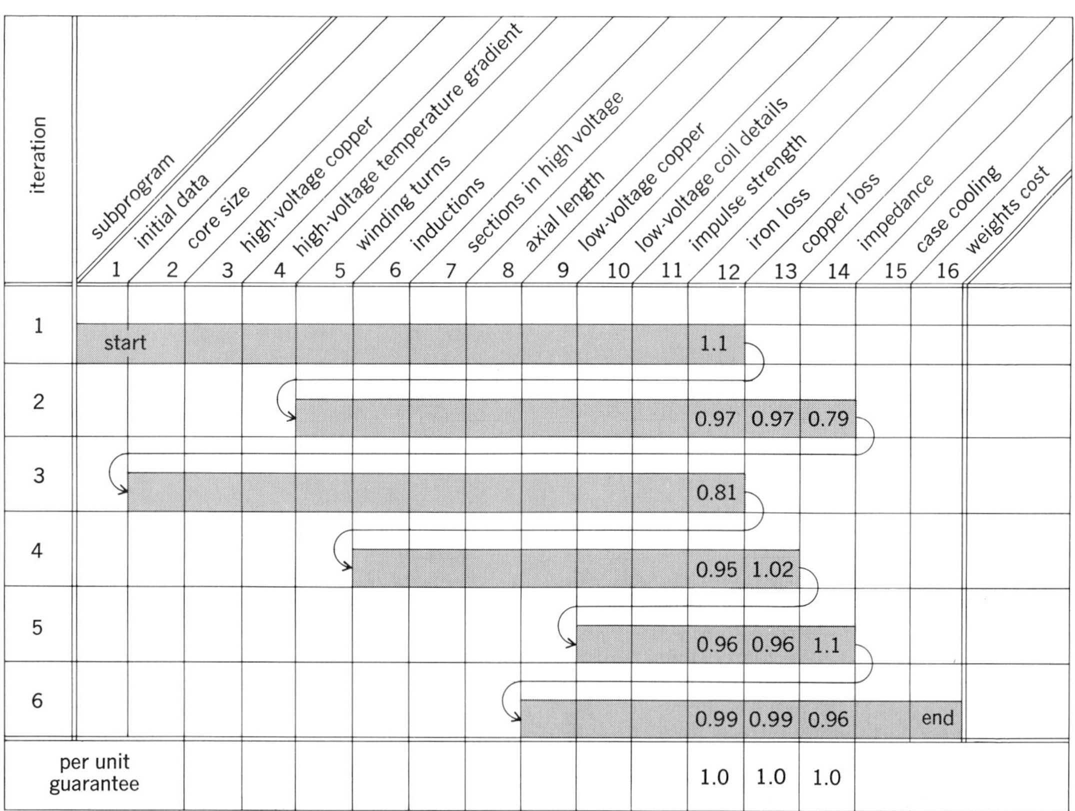

Fig. 8. Iterations of the computer program shown by the flow chart in Fig. 7.

and tabulate have greatly enhanced its use in design and encouraged the study, on their own merits, of the processes and subprocesses exercised in the design process. These include optimization or sensitivity analysis, reliability analysis, and simulation as well as design theory. *See* DIGITAL COMPUTER.

Optimization analyses, given a model of the design and using linear and nonlinear programming, determines the best values of the parameters consistent with stated criteria and further studies the effects of variations in the values of the parameters.

Reliability is a special case of optimization where the emphasis is to choose or evaluate a system so as to maximize its probability of successful operation, for example, the reliability of electronics.

Simulation, as of dynamic systems, is mathematical modeling to study the response of a design to various inputs and disturbances. The analog computer has been widely used for simulation through its physical modeling of the mathematic analytical relationships of the proposed design. The digital computer's use of numerical data adapts it more readily to nonlinear or probabilistic situations by using random or Monte Carlo techniques, as well as to those situations requiring higher accuracy.

Decision theory deals with the general question of how to choose between a great number of alternatives according to established criteria. It proposes models of the decision process as well as defining techniques, that is, programs or algorithms, of calculation by which to make choices.

Although each of these techniques represents either a gross oversimplification or but a part of the overall design process of Fig. 1, their study and application is useful. Even where not strictly applicable, the awareness of a model of a process enhances its qualitative evaluation by providing a guide to action. The utility and comprehensiveness of these techniques are already advancing and will mature rapidly, largely because of the ability of the high-speed digital computer to carry out evaluations which heretofore were impractical. Finally, the record-keeping capacity of the computer makes possible the recording and reutilization of past actions. This integration of prior results into new endeavors represents a "mechanization of experience."

Computer-aided design. Notwithstanding the present contribution and unlimited promise of the computer to the design process, one must, however, recognize certain limitations. In reflecting on the model of the engineering design process and especially on the interaction between concept and analysis, it is useful to consider the nature of the activities involved. The concepts usually do not come about as a result of any ponderous, systematic, organized effort; rather they appear in an undisciplined, creative, spectacular fashion. The short-duration exultation of concept creation is then followed by usually very much longer, more regular, systematic periods of analysis. The typical use of a computer does not lend itself to the kind of man-analysis interaction implicit in the engineering design process and essential to its successful negotiation. Usually, the solution process for a piece of work must first be well thought out and known in the most intricate detail; only then can the user write a program to describe to the computer the sequence of steps necessary. But when the nature of the problem is only vaguely grasped and much learning has to be done, the computer may not be as helpful as pencil and paper. *See* DRAFTING.

The reason is that in the thinking process one needs to advance in steps and to test these steps frequently. While the steps may be large, as in the first gross examination of an idea, or small, as in an exercise of refinement, these tests need not be elaborate or even precise. What is wanted first is only a qualitative result or perhaps a quantitative result of only moderate precision, a confirmation or a denial of tentative guess work.

By conventional computer programming, it is virtually impossible to obtain quick answers to small discrete problems, even though the computer can work very rapidly. To solve a problem, the user must first prepare a program, a detailed ritual of calculations and comparisons of quantities for the computer to carry out. It is hoped that data so processed will yield the solution. Since computers, especially large powerful ones, are very expensive to own or rent, they are usually shared by users. As it costs very little more to operate a computer than to have it lie idle, prospective users usually must wait their turns to try out their programs. Ordinarily the programs are in the form of a deck of punched cards, which customarily is run by the computer operator, and one returns for the results later.

If there is a mistake in the program, the computer either will detect the mistake and refuse to waste time by trying to run the program, or if the mistake is more subtle, perhaps an error in principle, will go ahead and compute a mass of meaningless nonsense. One can easily see how this dependence on programming, waiting, and error finding and correcting frustrates the formulation and testing of concepts essential to a creative effort.

Another point is that the creative process is, virtually by definition, unpredictable. The sequence of the steps is never known at the beginning. If it were, the whole process could be accomplished by the computer since the information prerequisite to the computer program would be available. Indeed, the creative process is the process of learning how to accomplish the desired result.

Clearly what is needed if the computer is to be of greater use in the creative process is a more intimate and continuous interchange between man and machine. This interchange must be of such a nature that all forms of thought congenital to man, whether verbal, symbolic, numerical, or graphical, are also understood and acted upon by the machine in ways appropriate to man's purpose. In reflecting on this man-machine symbiosis with the goal of designing a system with which to design, it is useful to identify the special attributes of each partner.

The speed of the computer is prodigious when measured against any term of reference and especially so when measured against the appropriate one, the human mind. Operating at megahertz or higher rates, a computer performs tens of thousands of arithmetic or logical operations per second. This rate outstrips by many orders of magnitude man's neurological responses to premeditated cerebrations.

The machine's memory is likewise extraordinary, although restricted to an extremely narrow but nonetheless useful class of memory impression. It can store binary numbers of 36 digits and recall them unerringly at microsecond speeds, while the human memory system is very much slower and not nearly so reliable. Of course, in contrast, the human memory encompasses an indescribable universe of types and kinds, far beyond the range of a computer. The reliability routinely exercised by the computer is unattainable by man. Once properly instructed, the computer executes its routines faithfully, untiringly, yielding to no boredom or carelessness, taking no coffee breaks and introducing no humanlike errors.

Counterposed against the computer's assets stands an open-ended list of human attributes. In terms of the engineering design process, these human characteristics would include reflection on, and evaluation of, the social, esthetic, and economic aspects of the original goals; the formulation of previously unforeseen and unanticipated questions at many points in the investigation; an unflagging curiosity about the way things are done and about the way they are proposed to be done; the flexibility of mind to shift from one approach to another, to sort out the significant from a great mass of information and misinformation on the basis of very few discernible criteria; the devising and structuring of new and unusual approaches; the willingness, ability, and intellectual integrity to make decisions, frequently on the basis of inadequate data; the ability to recognize that the job must get on and some decisions must be made rapidly, though the right to reverse decisions whenever facts so indicate is reserved; exercise of judgment based on prior experience in making decisions; and finally the desire and willingness to develop such adjudicative ability through experience. While the computer demonstrates the capability of encroaching upon some of these attributes, but only to the extent that man thoroughly understands them and can program them, clearly others are forever denied it.

Considerable effort is being devoted to creating computer systems which interact with the creative designer in truly effective ways. *See* COMPUTER-AIDED DESIGN AND MANUFACTURING.

Time-sharing a computer. Already it is possible to communicate directly with a computer by means of a kind of typewriter keyboard and to have the computer reply immediately by taking over the keyboard and typing out an answer. Thereby the obstacle to thought caused by waiting hours or days for a reply has been removed, and the user has no compunctions about posing smaller problem fragments to the computer, in a way more consonant with the small steps of the creative process. The results are immediately available, and the course of further action is guided more precisely.

Such intimate "conversations" with a computer would scarcely be economically feasible if only one person could use the computer at a time because, measured against the speed of a computer, human beings are intolerably slow. A man likes to ponder a problem and often needs time to decide what to try next. Even when he reaches a decision, his next instruction to the computer takes several seconds to type on the keyboard, and during even this short time the computer could perform hundreds of thousands of calculations were it free to do so. It is logical to arrange to have the computer in conversation with a large number of people, so that when one user is idle the computer can turn to another and answer his question. Should all users be idle at once, the computer can work on some large problem left as a backlog. This type of effective computer utilization is becoming quite routine.

The computer devotes only a very short time to any one user. Depending upon the length of the program to be run, it spends from 0.2 sec on short programs to as much as 0.5 sec on long ones. The computer spends these short time bursts on each problem in a sort of round-robin sequence, like a chess master playing several games with lesser experts simultaneously. The net result is that each user seems to have the undivided attention of the entire computer system.

This technique is called time-sharing. On a rough average, if a user is intensely busy for 1 hr, the computer can discharge its responsibilities to him in short bursts that total about 3 min. The rest of the time it serves the other users.

Graphical input/output. In many fields of design—notably architecture; design of airplanes, automobiles, and ships; and almost all mechanical design—the designer works largely in visual terms. But a conventional limitation of computers was the impossibility of communicating with them graphically. Originally, the computer was limited to numbers; then it was set up to recognize and correctly interpret words, and to perform certain manipulations of words and numbers. Finally a way has been found to set up the computer to interpret drawings. Moreover, the computer has become an active partner in the act of drawing, so that it can provide a certain superskill in preparing the drawing once the human operator has made clear his intentions.

The pioneering work in providing a graphical link with the computer was done by Ivan E. Sutherland at MIT. His computer program is known as Sketchpad. Far beyond being merely an expensive way to do drafting, the program has brought out new philosophies about the proper use of computers; these thoughts can profitably be applied to nongraphic modes of communication as well.

The effect of using computers in the design and development stages of a manufacturing process is twofold. First, the costs are greatly reduced because fewer people are involved in the stages, and less time is required to accomplish the tasks. For example, the translation of a designer's sketch of the shape of an airplane fuselage into a full-size, precise, geometrical shape description suitable for manufacturing purposes used to take about 6 months. Conventional computer methods have reduced this time to a few weeks. But the emergent computer techniques will cause a further reduction of this time, perhaps to seconds. The results will be far more accurate than have previously been obtained and will be tailored much more closely to manufacturing requirements.

Second, an important hidden benefit is the greatly increased lead time which a computerized manufacturer gains over a noncomputerized (or less computerized) competitor; that is, the former can be ready much sooner to begin production. Both the national military and the commercial aspects of this acceleration are obvious.

[ROBERT W. MANN]

Bibliography: J. H. Faupel and F. E. Fisher, *Engineering Design*, 2d ed., 1980; L. E. Hulpert (ed.), *Interactive Computer Graphics in Engineering*, 1977; E. V. Krick, *An Introduction to Engineering and Engineering Design*, 2d ed., 1969; T. T. Woodson, *Introduction to Engineering Design*, 1966.

Engineering drawing

A graphical language used by engineers and other technical personnel associated with the engineering profession. The purpose of engineering drawing is to convey graphically the ideas and information necessary for the construction or analysis of machines, structures, or systems.

In colleges and universities, engineering drawing is usually treated in courses with titles like Engineering Graphics. Sometimes these courses include other topics, such as computer graphics and nomography.

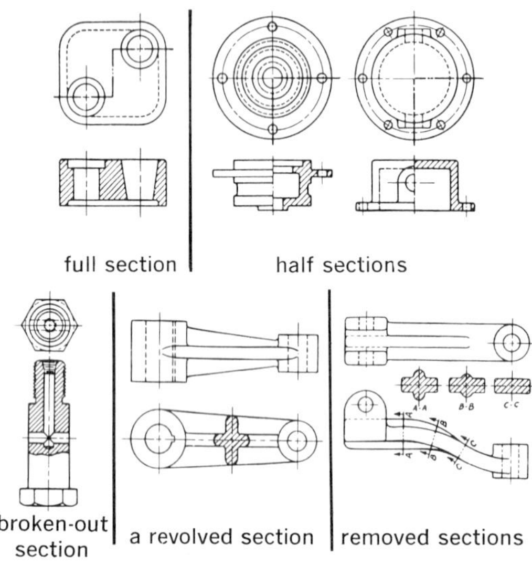

Fig. 1. Section views. (*From T. E. French and C. J. Vierck, Engineering Drawing, McGraw-Hill, 1953*)

The basis for much engineering drawing is orthographic representation (projection). Objects are depicted by front, top, side, auxiliary, or oblique views, or combinations of these. The complexity of an object determines the number of views shown. At times, pictorial views are also shown. *See* PICTORIAL DRAWING.

Engineering drawings often include such features as various types of lines, dimensions, lettered notes, sectional views, and symbols. They may be in the form of carefully planned and checked mechanical drawings, or they may be freehand sketches. Usually a sketch precedes the mechanical drawing. Final drawings are usually made on tracing paper, cloth, or Mylar film, so that many copies can be made quickly and cheaply by such processes as blueprinting, ammonia-developed (diazo) printing, or lithography.

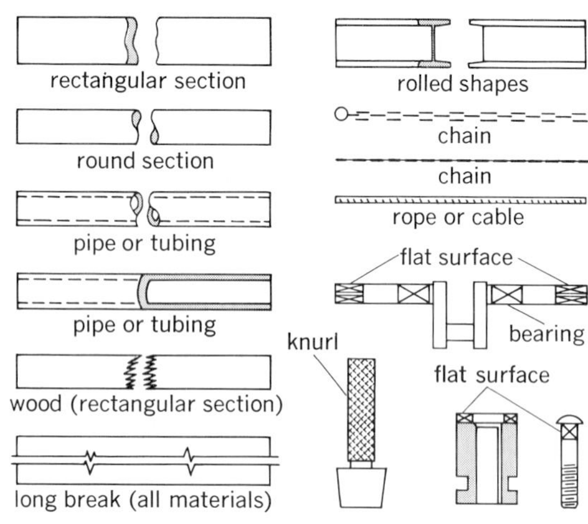

Fig. 2. Conventional breaks and other symbols to indicate details. (*From T. E. French and C. J. Vierck, Engineering Drawing, McGraw-Hill, 1953*)

Section drawings. Many objects have complicated interior detail which cannot be clearly shown by means of front, top, side, or pictorial views. Section views enable the engineer or detailer to show the interior detail in such cases. Features of section drawings are cutting-plane symbols, which show where imaginary cutting planes are passed to produce the sections, and section-lining (sometimes called cross-hatching), which appears in the section view on all portions that have been in contact with the cutting plane. When only a part of the object is to be shown in section, conventional representation such as a revolved, rotated, or broken-out section is used (Fig. 1). Details such as flat surfaces, knurls, and threads are treated conventionally, which facilitates the making and reading of engineering drawings by experienced personnel (Fig. 2). Thus, certain engineering drawings will be combinations of top and front views, section and rotated views, and partial or pictorial views.

Dimensioning. In addition to describing the shape of objects, many drawings must show dimensions, so that workers can build the structure or fabricate parts that will fit together. This is accomplished by placing the required values (measurements) along dimension lines (usually outside the outlines of the object) and by giving additional information in the form of notes which are referenced to the parts in question by angled lines called leaders. In drawings of large structures the major unit is the foot, and in drawings of

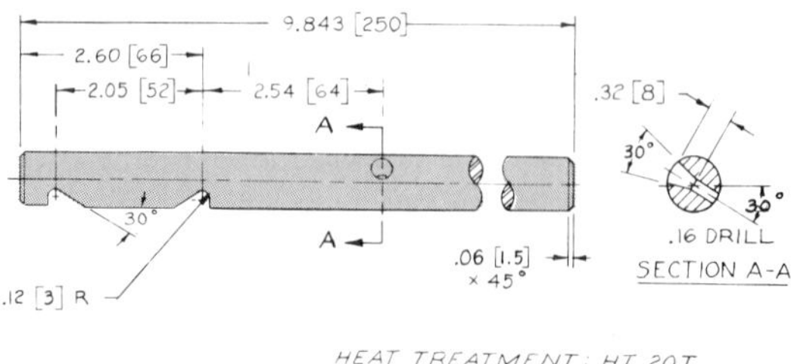

Fig. 3. Machine part which has been dimensioned in inches and millimeters (brackets).

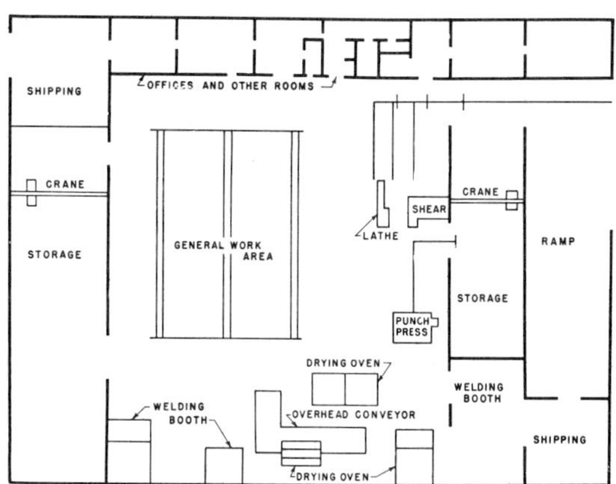

Fig. 4. A plant-layout drawing. (*From F. Zozzora. Engineering Drawing, McGraw-Hill, 2d ed., 1958*).

small objects the unit is the inch. A drawing containing dual dimensioning, inches and millimeters (in brackets), is shown in Fig. 3. In metric dimensioning, the basic unit may be the meter, the centimeter, or the millimeter, depending upon the size of the object or structure.

Working types of drawings may differ in styles of dimensioning, lettering (inclined lowercase, vertical uppercase, and so on), positioning of the numbers (aligned, or unidirectional—a style in which all numbers are lettered horizontally), and in the type of fraction used (common fractions or decimal fractions). If special precision is required, an upper and a lower allowable limit are shown. Such tolerance, or limit, dimensioning is necessary for the manufacture of interchangeable mating parts, but unnecessarily close tolerances are very expensive. *See* DIMENSIONING.

Layout drawing. Layout drawings of different types are used in different manufacturing fields for various purposes. One is the plant layout drawing, in which the outline of the building, work areas, aisles, and individual items of equipment are all drawn to scale (Fig. 4). Another type is the aircraft, or master, layout, which is drawn on glass cloth or on steel or aluminum sheets. The object is drawn to full size with extreme accuracy. The completed

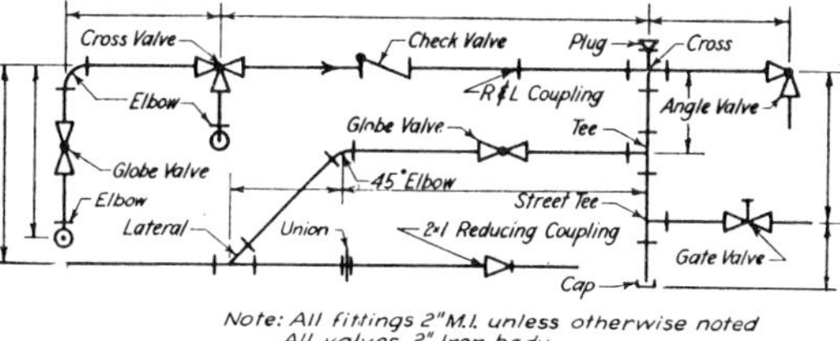

Fig. 6. Piping diagrammatic drawing. (*From T. E. French and C. J. Vierck, Engineering Drawing, McGraw-Hill, 1953*)

drawing is photographed with great precision, and a glass negative made. From this negative, photo templates are made on photosensitized metal in various sizes and for different purposes, thereby eliminating the need for many conventional detail drawings. Another type of layout, or preliminary assembly, drawing is the design layout, which establishes the position and clearance of parts of an assembly.

Assembly drawings. A set of working drawings usually includes detail drawings of all parts and an assembly drawing of the complete unit. Assembly drawings vary somewhat in character according to their use, as: design assemblies or layouts; working drawing assemblies; general assemblies; installation assemblies; and check assemblies. A typical general assembly may include judicious use of sectioning and identification of each part with a numbered balloon (Fig. 5). Accompanying such a drawing is a parts list, in which each part is listed by number and briefly described; the number of pieces required is stated and other pertinent information given. Parts lists are best placed on separate sheets and typewritten to avoid time-consuming and costly hand lettering.

Schematic drawings. Schematic or diagrammatic drawings make use of standard symbols and single lines between symbols which indicate the direction of flow. In piping and electrical schematic diagrams, symbols recommended by the American National Standards Institute (ANSI), other agencies, or the Department of Defense (DOD) are used. In contrast to Fig. 6, the fixtures or compo-

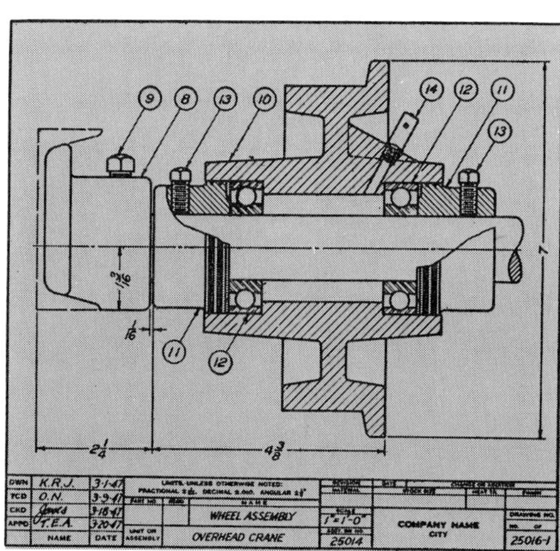

Fig. 5. A unit, or general, assembly. (*From T. E. French and C. J. Vierck, Engineering Drawing, McGraw-Hill, 1953*)

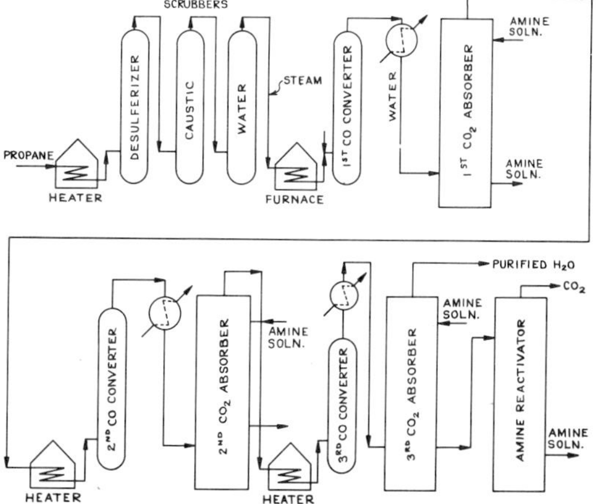

Fig. 7. Chemical engineering flow diagram. (*From P. H. Groggins, ed., Unit Processes in Organic Synthesis, 5th ed., McGraw-Hill, 1958*)

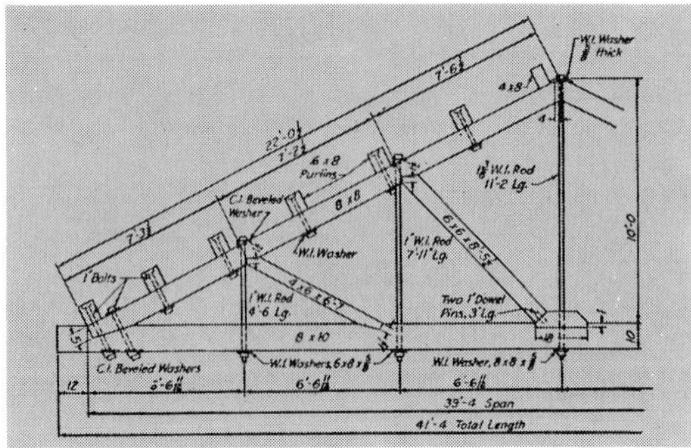

Fig. 8. Structural drawing of wood truss. A drawing of a steel truss would look similar, but would include symbols for such structural shapes as angles and channels. *(From T. E. French and C. J. Vierck, Engineering Drawing, McGraw-Hill, 1953)*

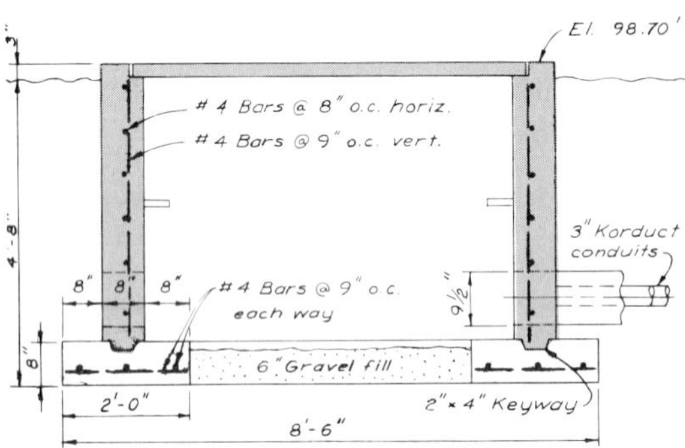

Fig. 9. Drawing of a reinforced concrete structure.

nents are not labeled in most schematics because readers usually know what the symbols represent.

Additional information is often lettered on schematic drawings, for example, the identification of each replaceable electrical component. Etched-circuit drawing has revolutionized the wiring of electronic components. By means of such drawing, the wiring of an electronic circuit is photographed on a copper-clad board, and unwanted areas are etched away. On electrical and other types of flow diagrams, all single lines (often with arrows showing direction of flow) are drawn horizontally or vertically (Fig. 7); there are few exceptions. In some flow diagrams, rectangular enclosures are used for all items. Lettering is usually placed within the enclosures. *See* SCHEMATIC DRAWING; WIRING DIAGRAM.

Structural drawings. Structural drawings include design and working drawings for structures such as buildings, bridges, dams, tanks, and highways (Figs. 8 and 9). Such drawings form the basis of legal contracts. Structural drawings embody the same principles as do other engineering drawings, but use terminology and dimensioning techniques different from those shown in previous illustrations. *See* DRAFTING. [CHARLES J. BAER]

Bibliography: C. J. Baer, *Electrical and Electronic Drawing*, 4th ed., 1980; T. E. French and C. L. Svensen, *Mechanical Drawing*, 6th ed., 1957; T. E. French and C. J. Vierck, *Engineering Drawing and Graphic Technology*, 12th ed., 1979; T. E. French and C. J. Vierck, *Fundamentals of Engineering Drawing*, 2d ed., 1966; F. E. Giesecke, A. Mitchell, and H. C. Spencer, *Technical Drawing*, 5th ed., 1967; R. P. Hoelscher, C. H. Springer, and J. S. Dobrovolny, *Basic Drawing for Engineering Technology*, 1964; F. Zozzora, *Engineering Drawing*, 2d ed., 1958.

Engineering geology

The application of education and experience in geology and other geosciences to solve geological problems posed by civil engineering works. The branches of the geosciences most applicable are surficial geology, petrofabrics, rock and soil mechanics, geohydrology, and geophysics, particularly exploration geophysics and earthquake seismology. This article discusses some of the practical aspects of engineering geology.

The terms engineering geology and environmental geology often seem to be used interchangeably. Specifically, environmental geology is the application of engineering geology in the solution of urban problems; in the prediction and mitigation of natural hazards such as earthquakes, landslides, and subsidence; and in solving problems inherent in disposal of dangerous wastes and in reclaiming mined lands.

Another relevant term is geotechnics, the combination of pertinent geoscience elements with civil engineering elements to formulate the civil engineering system that has the optimal interaction with the natural environment.

Engineering properties of rock. The civil engineer and the engineering geologist consider most hard and compact natural materials of the earth crust as rock, and their derivatives, formed mostly by weathering processes, as soil. A number of useful soil classification systems exist. Because of the lack of a rock classification system suitable for civil engineering purposes, most engineering geology reports use generic classification systems modified by appropriate rock-property adjectives.

Rock sampling. The properties of a rock element can be determined by tests on cores obtained from boreholes. These holes are made by one or a combination of the following basic types of drills: the rotary or core drill (Fig. 1), the cable-tool or churn drill, and the auger. The rotary type generally is used to obtain rock cores. The rotary rig has a motor or engine (gasoline, diesel, electric, or compressed air) that drives a drill head that rotates a

Fig. 1. Rotary or core drill on damsite investigation.

drill rod (a thick-walled hollow pipe) fastened to a core barrel with a bit at its end. Downward pressure on the bit is created by hydraulic pressure in the drill head. Water or air is used to remove the rock that is comminuted (chipped or ground) by the diamonds or hard-metal alloy used to face the bit. The core barrel may be in one piece or have one or two inner metal tubes to facilitate recovery of soft or badly broken rock (double-tube and triple-tube core barrels). The churn-type drill may be used to extend the hole through the soil overlying the rock, to chop through boulders, occasionally to deepen a hole in rock when core is not required or to obtain drive samples of the overburden soils. When the rock is too broken to support itself, casing (steel pipe) is driven or drilled through the broken zone. Drill rigs range in size from those mounted on the rear of large multiwheel trucks to small, portable ones that can be packed to the investigation site on a person's back or parachuted from a small plane. *See* BORING AND DRILLING (MINERAL).

The rock properties most useful to the engineering geologist are compressive and triaxial shear strengths, permeability, Young's modulus of elasticity, erodability under water action, and density (in pounds per cubic foot, or pcf).

Compressive strength. The compressive (crushing) strength of rock generally is measured in pounds per square inch (psi) or kilograms per square centimeter (kg/cm^2). It is the amount of stress required to fracture a sample unconfined on the sides and loaded on the ends (Fig. 2). If the load P of 40,000 lb is applied to a sample with a diameter of 2 in. (3.14 in.^{2}), the compressive stress is $40{,}000 \div 3.14 = 12{,}738$ psi ($177{,}920$ N $\div$ 0.00203 m^2 $= 87{,}645$ kN/m^2). If this load breaks the sample, the ultimate compressive strength equals the compressive stress acting at the moment of failure, in this case 12,738 psi. The test samples generally are cylindrical rock cores that have a length-to-diameter ratio (L/D) of about 2. The wide variety of classification systems used for rock results in a wide variation in compressive strengths for rocks having the same geologic name. The table gives a statistical evaluation of the compressive strengths of several rocks commonly encountered in engineering geology.

Most laboratory tests show that an increase in moisture in rock causes a decrease in its compressive strength and elastic modulus; what is not generally known, however, is that the reverse situation shown in Fig. 3 has been encountered in certain types of volcanic rocks. In sedimentary rock the compressive strength is strongly dependent upon the quality of the cement that bonds the mineral grains together (for example, clay cement gives low strength) and upon the quantity of cement (a rock may have only a small amount of cement, and despite a strong bond between the grains, the strength is directly related to the inherent strength of the grains). Strength test results are adversely affected by microfractures that may be present in the sample prior to testing, particularly if the microfractures are oriented parallel to the potential failure planes.

The value of compressive strength to be used in an engineering design must be related to the direction of the structure's load and the orientation of the bedding, discontinuities, and structural weaknesses in the foundation rock. This relationship is important because the highest compressive strength usually is obtained when the compressive stress is normal to the bedding. Conversely, the highest Young's modulus of elasticity (E) usually results when the compressive stress parallels the bedding. When these strength and elastic properties apparently are not affected by the direction of applied load, the rock is described as isotropic; if load applied parallel to the bedding provides physical property data that are significantly different than those obtained when the load is applied normal (perpendicular) to the bedding, the rock is anisotropic or aeolotropic. If the physical components of the rock element or rock system have equal dimensions and equal fabric relationships, the rock is homogeneous; significant variance in these relationships results in a heterogeneous rock. Most rocks encountered in foundations and

ENGINEERING GEOLOGY

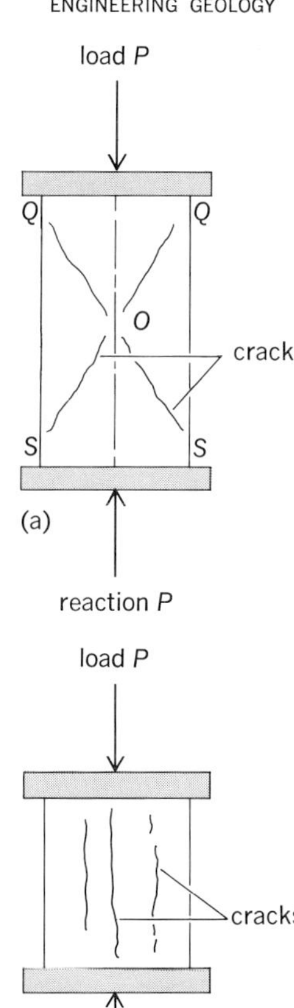

Fig. 2. Unconfined compression test. (*a*) Shear failure showing failure planes *QS*. (*b*) Tension failure. (*From D. P Krynine and W. R. Judd, Principles of Engineering Geology and Geotechnics, McGraw-Hill, 1957*)

Compressive strength of rocks*

Type of rock	No. of tests	Weighted mean, kN/m^2	Minimum and maximum values, kN/m^2
Igneous			
Basalt	195	112,390	4,140 – 383,360
Diabase	6	150,655	4,140 – 275,100
Diorite	26	167,550	78,600 – 310,275
Gabbro	10	245,875	110,000 – 399,220
Granite	140	160,515	17,925 – 332,340
Rhyolite	44	162,100	11,720 – 453,690
Sedimentary			
Dolomite	62	87,840	6,205 – 358,540
Limestone	211	74,190	1,380 – 260,630
Sandstone	257	63,430	2,070 – 328,200
Shale	67	66,605	690 – 230,980
Siltstone	14	108,525	3,445 – 315,790
Metamorphic			
Amphibolite	14	152,240	24,820 – 280,625
Gneiss	103	133,760	35,850 – 292,350
Marble	31	100,600	30,340 – 273,730
Quartzite	25	292,350	25,510 – 628,825
Schist	16	50,265	6,895 – 162,030

*Based on data in W. R. Judd, *Statistical Relationships for Certain Rock Properties*, DDC AD735376, 1971, and *Statistical Methods to Compile and Correlate Rock Properties*, Purdue Res. Tech. Rep. no. 2, 1969.

underground works are anisotropic and heterogeneous.

Shear in rocks. Shearing stresses tend to separate portions of the rock (or soil) mass. Faults and folds are examples of shear failures in nature. In engineering structures, every compression is accompanied by shear stresses. For example, an arch dam compresses the abutment rock and, if the latter is intersected by fissures or weak zones, it may fail in shear with a resulting tensile stress in the dam concrete that may rupture the concrete. The application of loads over long periods of time on most rocks will cause them to creep or even to flow like a dense fluid (plastic flow).

*Ambient stress.*This type of stress in a rock system is actually potential energy, probably created by ancient natural forces, recent seismic activity, or nearby human-caused disturbances. Ambient (residual, stored, or primary) stress may remain in rock long after the disturbance is removed. An excavation, such as a tunnel or quarry, will relieve the ambient stress by providing room for displacement of the rock, and thus the potential energy is converted to kinetic energy. In tunnels and quarries, the release of this energy can cause spalling, the slow outward separation of rock slabs from the rock massif; when this movement is rapid or explosive, a rock burst occurs. The latter is a different phenomenon from a rock bump, which is a rapid upward movement of a large portion of a rock system and, in a tunnel, can have sufficient force to flatten a steel mine car against the roof or break the legs of a person standing on the floor when the bump occurs. *See* ROCK BURST; UNDERGROUND MINING.

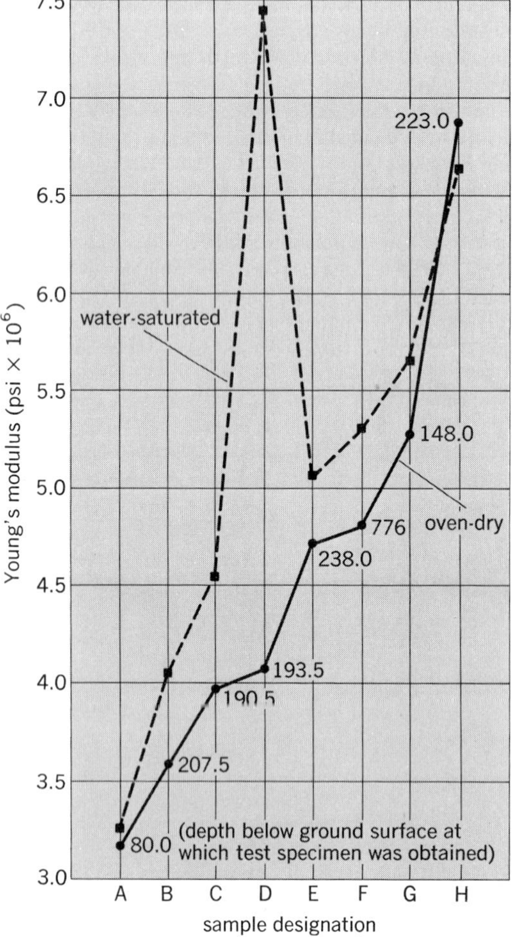

Fig. 3. Increase in Young's modulus caused by saturation of dacite porphyry. (*After J. R. Ege and R. B. Johnson, Consolidated tables of physical properties of rock samples from area 401, Nevada Test Site, U. S. Geol. Surv. Tech. Letter Pluto-21, 1962*)

One of two fundamental principles generally is used to predict the possible rock load on a tunnel roof, steel, or timber supports, or a concrete or steel lining: (1) The weight of the burden (the rock and soil mass between the roof and the ground surface) and its shear strength control the load, and therefore the resultant stresses are depth-dependent, or (2) the shear strength of the rock system and the ambient stresses control the stress distribution, so the resultant loading is only indirectly dependent upon depth. The excavation process can cause rapid redistribution of these stresses to produce high loads upon supports some distance from the newly excavated face in the tunnel. The geometry and span of the opening also influence the stress distribution. Lined tunnels can be designed so that the reinforced concrete or steel lining will have to carry only a portion of the ambient or burden stresses. *See* TUNNEL.

Construction material. Rock as a construction material is used in the form of dimension, crushed, or broken stone. Broken stone is placed as riprap on slopes of earth dams, canals, and river banks to protect them against water action. Also, it is used as the core and armor stone for breakwater structures. For all such uses, the stone should have high density (±165 pcf or 2650 kg/m^3), be insoluble in water, and be relatively nonporous to resist cavitation. Dimension stone (granite, limestone, sandstone, and some basalts) is quarried and sawed into blocks of a shape and size suitable for facing buildings or for interior decorative panels. For exterior use, dimension stone preferably should be isotropic (in physical properties), have a low coefficient of expansion when subjected to temperature changes, and be resistant to deleterious chemicals in the atmosphere (such as sulfuric acid). Crushed stone (primarily limestone but also some basalt, granite, sandstone, and quartzite) is used as aggregate in concrete and in bituminous surfaces for highways, as a base course or embankment material for highways, and for railroad ballast (to support the ties). When used in highway construction, the crushed stone should be resistant to abrasion as fine stone dust reduces the permeability of the stone layer; the roadway then is more susceptible to settling and heaving caused by freezing and thawing of water in the embankment. Concrete aggregate must be free of deleterious material such as opal and chalcedony; volcanic rocks containing glass, devitrified glass, and tridymite; quartz with microfractures; phyllites containing hydromica (illites); and other rocks containing free silica (SiO_2). These materials will react chemically with the cement in concrete and release sodium and potassium oxides (alkalies) or silica gels. Preliminary petrographic analyses of the aggregate and chemical analysis of the cement can indicate the possibility of alkali reactions and thus prevent construction difficulties such as expansion, cracking, or a strength decrease of the concrete. *See* BREAKWATER.

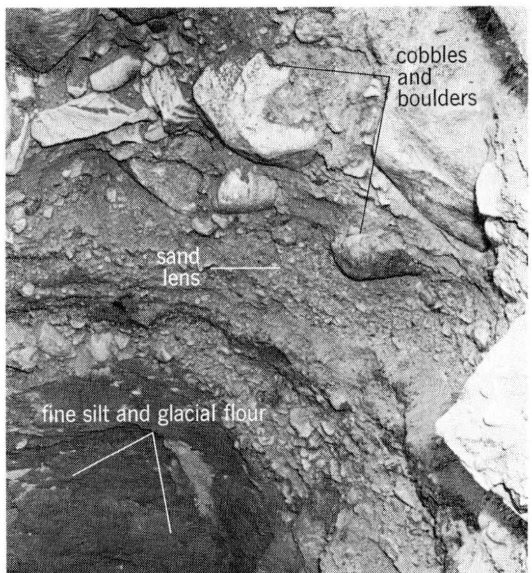

Fig. 4. Glacial deposit in test pit.

Geotechnical significance of soils. Glacial and alluvial deposits contain heterogeneous mixtures of pervious (sand and gravel) and impervious (clay, silt, and rock flour) soil materials (Fig. 4). The pervious materials can be used for highway subgrade, concrete aggregate, and filters and pervious zones in earth embankments. Dam reservoirs may be endangered by the presence of stratified or lenticular bodies of pervious materials or ancient buried river channels filled with pervious material. Deep alluvial deposits in or close to river deltas may contain very soft materials such as organic silt or mud. An unsuitable soil that has been found in dam foundations is open-work gravel. This material may have a good bearing strength because of a natural cement bond between grains, but it is highly pervious because of the almost complete lack of fine soil to fill the voids between the gravel pebbles.

Concrete or earth dams can be built safely on sand foundations if the latter receive special treatment. One requirement is to minimize seepage losses by the construction of cutoff walls (of concrete, compacted clay, or interlocking-steel-sheet piling) or by use of mixed-in-place piles three feet (0.9 m) or more in diameter. The latter are constructed by augering to the required depth but not removing any of the sand. At the desired depth, cement grout is pumped through the hollow stem of the auger, which is slowly withdrawn while still rotating; this mixes the grout and the sand into a relatively impervious concrete pile. The cutoff is created by overlapping these augered holes. Some sand foundations may incur excessive consolidation when loaded and then saturated, particularly if there is a vibratory load from heavy machinery or high-velocity water in a spillway. This problem is minimized prior to loading by using a vibrating probe inserted into the sand or vibratory rollers on the sand surface or by removing the sand and then replacing it under vibratory compaction and water sluicing.

Aeolian (windblown) deposits. Loess is a relatively low-density ($\pm$ 1.4 metric tons/m^3) soil composed primarily of silt grains cemented by clay or calcium carbonate. It has a vertical permeability considerably greater than the horizontal. When a loaded loess deposit is wetted, it rapidly consolidates, and the overlying structure settles. When permanent open excavations ("cuts") are required for highways or canals through loess, the sides of the cut should be as near vertical as possible: Sloping cuts in loesses will rapidly erode and slide because of the high vertical permeability. To avoid undesirable settlement of earth embankments, the loess is "prewetted" prior to construction by building ponds on the foundation surface (Fig. 5). Permanently dry loess is a relatively strong bearing material. Aeolian sand deposits present the problem of stabilization for the continually moving sand. This can be done by planting such vegetation as heather or young pine or by treating it with crude oil. Cuts are traps for moving sand and should be avoided. The failure of Teton Dam in 1976 indicated, among other factors, that when loessial or silty soils are used for core materials in dam embankments, it is important to take special measures to prevent piping of the silts by carefully controlled compaction of the core, by using up- and downstream filters, and by extraordinary treatment of the foundation rock.

Organic deposits. Excessive settlement will occur in structures founded on muskeg terrain. Embankments can be stabilized by good drainage, the avoidance of cuts, and the removal of the organic soil and replacement by sand and gravel or, when removal is uneconomical, displacement of it by the continuous dumping of embankment material upon it. Structures imposing concentrated loads are supported by piling driven through the soft layers into layers with sufficient bearing power.

Residual soils. These soils are derived from the in-place deterioration of the underlying bedrock. The granular material caused by the in-place disintegration of granite generally is sufficiently thin to cause only nominal problems. However, there are regions (such as California, Australia, and Brazil) where the disintegrated granite (locally termed "DG") may be hundreds of feet thick; although it may be competent to support moderate loads, it is unstable in open excavations and is pervious. A thickness of about 200 ft (60 m) of DG and weathered gneiss on the sides of a narrow canyon was a

Fig. 5. Prewetting loess foundation for earth dam. *(U.S. Bureau of Reclamation)*

major cause for construction of the Tumut-1 Power Plant (New South Wales) in hard rock some 1200 ft (365 m) underground. Laterite (a red clayey soil) derived from the in-place disintegration of limestone in tropical to semitropical climates is another critical residual soil. It is unstable in open cuts on moderately steep slopes, is compressible under load, and when wet produces a slick surface that is unsatisfactory for vehicular traffic. This soil frequently is encountered in the southeastern United States, southeastern Asia, and South America.

Clays supporting structures may consolidate slowly over a long period of time and cause structural damage. When clay containing montmorillonite is constantly dried and rewetted by climatic or drainage processes, it alternately contracts and expands. During the drying cycle, extensive networks of fissures are formed that facilitate the rapid introduction of water during a rainfall. This cyclic volume change of the clay can produce uplift forces on structures placed upon the clay or compressive and uplift forces on walls of structures placed within the clay. These forces have been known to rupture concrete walls containing ³⁄₄-in.-diameter (19-mm) steel reinforcement bars. A thixotropic or "quick" clay has a unique lattice structure that causes the clay to become fluid when subjected to vibratory forces. Various techniques are used to improve the foundation characteristics of critical types of clay: (1) electroosmosis that uses electricity to force redistribution of water molecules and subsequent hardening of the clay around the anodes inserted in the foundation; (2) provision of adequate space beneath a foundation slab or beam so the clay can expand upward and not lift the structure; (3) belling, or increasing in size, of the diameter of the lower end of concrete piling so the pile will withstand uplift forces imposed by clay layers around the upper part of the pile; (4) treatment of the pile surface with a frictionless coating (such as Teflon or a loose wrapping of asphalt-inpregnated paper) so the upward-moving clay cannot adhere to the pile; (5) sufficient drainage around the structure to prevent moisture from contacting the clay; and (6) replacement of the clay by a satisfactory foundation material. Where none of these solutions is feasible, the structure then must be relocated to a satisfactory site or designed so it can withstand uplift or compressive forces without expensive damage.

Silt may settle rapidly under a load or offer a "quick" condition when saturated. For supporting some structures (such as residences), the bearing capacity of silts and fine sands can be improved by intermixing them with certain chemicals that will cause the mixture to "set" or harden when exposed to air or moisture; some of the chemicals used are sodium silicate with the later application of calcium chloride, bituminous compounds, phenolic resins, or special cements (to form "soil cement"). The last mixture has been used for surfacing secondary roads, for jungle runways in Vietnam, and as a substitute for riprap of earth dams. Some types of silt foundations can be improved by pumping into them soil-cement or clay mixtures under sufficient pressures to create large bulbs of compacted silt around the pumped area.

Geotechnical investigation. For engineering projects these investigations may include preliminary studies, preconstruction or design investigations, consultation during construction, and the maintenance of the completed structure.

Preliminary studies. These are made to select the best location for a project and to aid in formulating the preliminary designs for the structures. The first step in the study is a search for pertinent published material in libraries, state and Federal agencies, private companies, and university theses. Regional, and occasionally detailed reports on local geology, including geologic maps, are available in publications of the U.S. Geological Survey; topographic maps are available from that agency and from the U.S. Army Map Service. Oil companies occasionally will release the geologic logs of any drilling they may have done in a project area. Sources of geologic information are listed in the *Directory of Geological Material in North America* by J. A. Howell and A. I. Levorsen (NAS-NRC Publ. 556, 1957). Air photos and other remote sensing techniques such as pulsed or side-looking radar or false color can be used to supplement map information (or may be the only surficial information readily available). The U.S. Geological Survey maintains a current index map of the air-photo coverage of the United States. The photos are available from that agency, the U.S. Forest Service, the Soil Conservation Service, and commercial air-photo companies; for some projects the military agencies will provide air-photo coverage. The topographic maps and air photos can be used to study rock outcrop and drainage patterns, landforms, geologic structures, the nature of soil and vegetation, moisture conditions, and land use by humans (cultural features). Airborne geophysical techniques using magnetometers or gravimeters also may be useful to delineate surface and subsurface geologic conditions.

Field reconnaissance may include the collection of rock and soil specimens; inspection of road cuts and other excavations; inspection of the condition of nearby engineering structures such as bridges, pavements, and buildings; and location of sources of construction material. Aerial reconnaissance is essential at this stage and can be performed best in helicopters and second-best in slow-flying small planes.

Preconstruction. Surface and subsurface investigations are required prior to design and construction. Surface studies include the preparation of a detailed map of surficial geology, hydrologic features, and well-defined landforms. For dam proj-

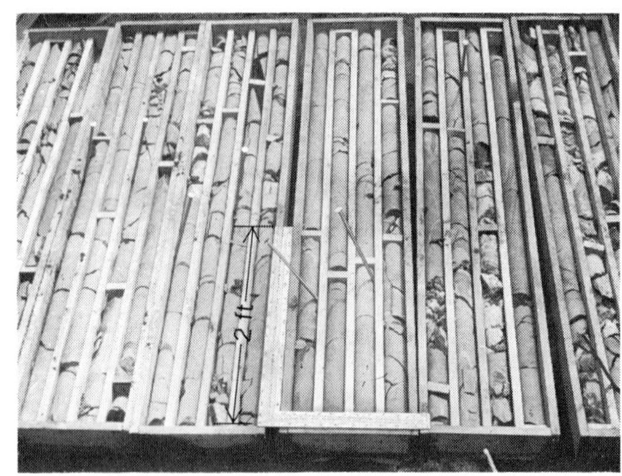

Fig. 6. Drill core obtained in dam foundation.

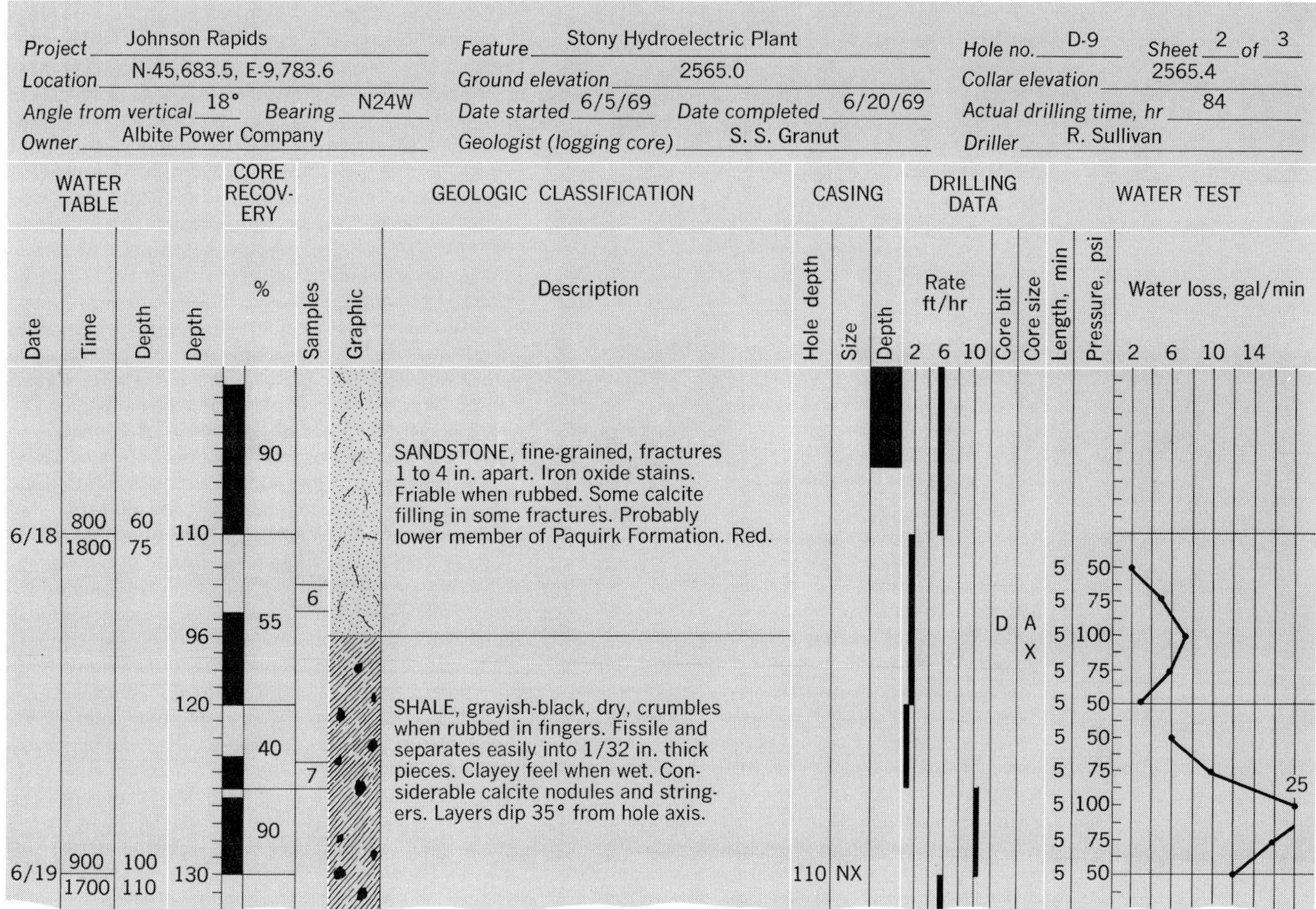

Project Johnson Rapids
Feature Stony Hydroelectric Plant
Hole no. D-9
Sheet 2 of 3
Location N-45,683.5, E-9,783.6
Ground elevation 2565.0
Collar elevation 2565.4
Angle from vertical 18°
Bearing N24W
Date started 6/5/69
Date completed 6/20/69
Actual drilling time, hr 84
Owner Albite Power Company
Geologist (logging core) S. S. Granut
Driller R. Sullivan
WATER TABLE
CORE RECOVERY
GEOLOGIC CLASSIFICATION
CASING
DRILLING DATA
WATER TEST
Date
Time
Depth
Depth
%
Samples
Graphic
Description
Hole depth
Size
Depth
Rate ft/hr
2 6 10
Core bit
Core size
Length, min
Pressure, psi
Water loss, gal/min
2 6 10 14
90
SANDSTONE, fine-grained, fractures 1 to 4 in. apart. Iron oxide stains. Friable when rubbed. Some calcite filling in some fractures. Probably lower member of Paquirk Formation. Red.
6/18
800 60
1800 75
110
6
55
96
120
40
7
90
SHALE, grayish-black, dry, crumbles when rubbed in fingers. Fissile and separates easily into 1/32 in. thick pieces. Clayey feel when wet. Considerable calcite nodules and stringers. Layers dip 35° from hole axis.
6/19
900 100
1700 110
130
110 NX
D
A X
5 50
5 75
5 100
5 75
5 50
5 50
5 75
5 100
5 75
5 50
25

Fig. 7. Log recording of geological information from a borehole.

ects, a small-scale geologic map (for example, 1:5000) is made of the reservoir area and any adjacent areas that may be directly influenced by the project; in addition, a large-scale geologic map (for example, 1:500) is required of the specific sites of the main structures (the dam, spillway, power plant, tunnels, and so on). [This preferred means of designating map scales can be used for either British or metric units. It means 1 unit of measurement on the map is equal to 5000 similar units on the ground; for example, 1 cm measured on the map is equal to 5000 cm measured on the ground.] These maps can be compiled by a combination of field survey methods and aerial mapping procedures. They should have a grid system (coordinates) and show the proposed locations for subsurface investigations.

Subsurface investigations are required to confirm and amplify the surficial geologic data. These may include test pits, trenches, short tunnels (drifts or adits), and the drilling of vertical, horizontal, or oblique (angle) boreholes (Fig. 6). Geologic data obtained by these direct methods can be supplemented by indirect or interpreted data obtained by geophysical methods on the surface or in subsurface holes and by installation of special instruments to measure strain or deformation in a borehole or tunnel.

The geology disclosed by subsurface investigations is "logged" on appropriate forms. Tunnel logs display visual measurements of fractures and joint orientations (strike and dip); rock names and a description of their estimated engineering properties; alteration, layering, and other geologic defects; the location and amount of water or gas inflow; the size and shape of blocks caused by fracturing or jointing and the width of separation or the filling material between blocks; and the irregularities in the shape of the tunnel caused by the displacement of blocks during or after excavation (rock falls, rock bursts, chimneying, and overbreakage). Geophysical seismic methods may be used to define the thickness of loosened rock around the tunnel; geoacoustical techniques that detect increases in microseismic noise during tunneling may be used to determine if the excavation is causing excessive loosening in the tunnel rock. This detection of "subaudible rock noise" occasionally is used to detect the potential movement of rock slopes in open excavations.

The borehole data can be logged on a form such as shown in Fig. 7. These data can be obtained by direct examination of the core, by visual inspection of the interior of the borehole using a borehole camera (a specially made television camera) or a stratoscope (a periscopelike device), or by geophysical techniques. Direct viewing of the interior of the hole is the only positive method of determining the in-place orientation and characteristics of separations and of layering in the rock system. The geophysical techniques include use of gamma-gamma logging that evaluates the density of the

Fig. 8. Kortes Dam, Wyo. Arrows show diverter walls protecting power plant against rock falls.

rock surrounding the borehole or at depths as great as 150 ft (45 m) beneath the gamma probe; neutron logging to determine the moisture content of the rock system by measuring the depth of penetration of the neutrons; traversing the borehole with a sonic logger that, by calibration, measures differences in the velocity of wave propagation in different strata (and thus can determine in place Young's modulus of elasticity and the thickness of each stratum encountered by the borehole); and electric logging that uses differences in the electrical resistivity of different strata to define their porosity, moisture content, and thickness.

Occasionally a hole is drilled through a talus deposit containing the same type of rock as the underlying rock in place (bedrock). Because of the similarity in rock types, the talus-bedrock contact sometimes is best identified by determining the orientation of the remnant magnetism in the core: the magnetic lines in the core will have a regular orientation, but the talus magnetism will have random directions. This method is useful only in rocks that contain appreciable remnant magnetism such as some basalts.

Geophysical seismic or electrical resistivity methods also can be used on the ground surface to define the approximate depth of bedrock or various rock layers. The results require verification by occasional boreholes, but this is an inexpensive and satisfactory technique for planning and design investigations. The seismic methods are not useful when it is necessary to locate soft strata (wherein the seismic waves travel at relatively low velocity) that are overlain by hard strata (that have higher wave velocity); the latter conceal and block the signal from the soft strata. Also, difficulties may occur when the strata to be located are overlain by soil containing numerous large boulders composed of rock having higher velocities than the surrounding soil, or when the soil is very compact (such as glacial till) because its velocity characteristics may resemble those of the underlying bedrock. Another problem is that the seismic method seldom can identify narrow and steep declivities in the underlying hard rock (because of improper reflection of the waves).

Construction. Geotechnical supervision is desirable during construction in or on earth media. The engineering geologist must give advice and keep a record of all geotechnical difficulties encountered during the construction and of all geological features disclosed by excavations. During the operation and maintenance of a completed project the services of the engineering geologist often are required to determine causes and assist in the preparation of corrective measures for cracks in linings of water tunnels, excessive settlement of structures, undesirable seepage in the foundations of dams, slides in canal and other open excavations, overturning of steel transmission-line towers owing to a foundation failure, and rock falls onto hydroelectric power plants at the base of steep canyon walls (Fig. 8). The engineering geologist also is considered an important member of the team assigned to the task of assessing the safety of existing dams as now required by Federal legislation.

Legal aspects. An important consideration for the engineering geologist is the possibility of a contractor making legal claims for damages, purportedly because of unforeseen geologic conditions (generally referred to legally as charged conditions) encountered during construction. Legal support for such claims can be diminished if the engineering geologist supplies accurate and detailed geologic information in the specifications and drawings used for bidding purposes. These documents should not contain assumptions about the geological conditions (for a proposed structure), but they should show all tangible geologic data obtained during the investigation for the project: for example, an accurate log of all boreholes and drifts and a drawing showing the boundaries of the outcrops of all geological formations in the project area. The engineering geologist should have sufficient experience with design and construction procedures to formulate an investigation program that results in a minimum of subsequent uncertainties by a contractor. Numerous uncertainties about geologic conditions not only can result in increased claims but also may cause a contractor to submit a higher bid (to minimize risks) than if detailed geologic information were available.

Special geotechnical problems. In arctic zones, structures built on permafrosted soils may be heaved or may cause thawing and subsequent disastrous settlement (Fig. 9). The growth of permafrost upward into earth dams seriously affects their stability and permeability characteristics. Obtaining natural construction materials in perma-

Fig. 9. Door-frame distortion and floor settlement which occurred as a result of permafrost thaw and heave. (*U.S. Geological Survey*)

frosted areas requires thawing of the borrow area to permit efficient excavation; once excavated, the material must be protected against refreezing prior to placement in the structures. Permafrost in rock seldom will cause foundation difficulties. In planning reservoirs, it is essential to evaluate their watertightness, particularly in areas containing carbonate or sulfate rock formations or lava flows. These formations frequently contain extensive systems of caverns and channels that may or may not be filled with claylike material or water. Where extensive openings occur, grouting with cement slurry or chemicals can be used as a sealant; however, as demonstrated by the 1976 failure of the Teton Dam, such measures are not always successful. Sedimentation studies are required for the design of efficient harbors or reservoirs because soil carried by moving water will settle and block or fill these structures. In areas with known earthquake activity, aseismic (earthquakeproof) design requires knowledge of the intensity and magnitude of earthquake forces. The prevention and rehabilitation of slides (landslides) in steep natural slopes and in excavations are important considerations in many construction projects and are particularly important in planning reservoirs, as was disastrously proved by the Vaiont Dam catastrophe in 1963.

Geohydrologic problems. In the foundation material under a structure, water can occur in the form of pore water locked into the interstices or pores of the soil or rock, as free water that is moving through openings in the earth media, or as included water that is a constituent or chemically bound part of the soil or rock. When the structure load compresses the foundation material, the resulting compressive forces on the pore water can produce undesirable uplift pressures on the base of the structure. Free water is indicative of the permeability of the foundation material and possible excessive water loss (from a reservoir, canal, or tunnel); uplift on the structure because of an increase in hydrostatic head (caused by a reservoir or the like); or piping, which is the removal of particles of the natural material by flowing water with a consequent unfilled opening that weakens the foundation and increases seepage losses.

The possibility of excessive seepage or piping can be learned by appropriate tests during the boring program. For example, water pressure can be placed on each 5-ft (2-m) section of a borehole, after the core is removed, and any resulting water loss can be measured. The water pressure is maintained within the 5-ft section by placing an expandable rubber ring (packer) around the drill pipe at the top of the test section and then sealing off the section by using mechanical or hydraulic pressure on the pipe to force expansion of the packer. When only one packer is used, because it is desired to test only the section of hole beneath it, it is a "single-packer" test. In a double-packer test a segment of hole is isolated for pressure testing by placing packers at the top and the bottom of the test section. The best information on the permeability characteristics of the rock can be obtained by the use of three or more increments of increasing and then decreasing water pressure for each tested length of hole. If the water loss continues to increase when the pressure is decreased, piping of the rock or filling material in fractures may be occurring or fractures are widening or forming. The water-pressure test can be supplemented by a groutability test in the same borehole. This test is performed in the same way as the water test except, instead of water, a mixture of cement, sand, and water (cement grout) or a phenolic resin (chemical grout) is pumped under pressure into the test section. The resulting information is used to design cutoff walls and grout curtains for dams. The pressures used in water-pressure or grouting tests should not exceed the pressure exerted by the weight of the burden between the ground surface and the top of the test section. Excessive test pressure can cause uplift in the rock, and the resulting test data will be misleading.

Included or pore water generally is determined by laboratory tests on cores; these are shipped from the borehole to the laboratory in relatively impervious containers that resist loss of moisture from the core. The cores with their natural moisture content are weighed when received and then dried in a vacuum oven at about 110° F (45°C) until their dry weight stabilizes. The percentage of pore water (by dry weight) is (wet weight − dry weight) × 100 ÷ dry weight. Temperatures up to 200°F (about 90°C) can be used for more rapid drying, provided the dried specimens are not to be used for strength or elastic property determinations. (High temperatures can significantly affect the strength because the heat apparently causes internal stresses that disturb the rock fabric or change the chemical composition of the rock by evaporation of the included moisture.)

Protective construction. Civilian and military structures may be designed to minimize the effects of nuclear explosions. The most effective protection is to place the facility in a hardened underground excavation. A hardened facility, including the excavation and its contents, is able to withstand the effects generated by a specified size of nuclear weapon. These effects include the amount of displacement, acceleration, and particle velocity that occurs in the earth media and the adjacent structure. Desirable depths and configurations for hardened facilities are highly dependent upon the shock-wave characteristics of the surrounding earth media, for example, the type of rock, discontinuities in the rock system, free water, and geologic structure. Therefore, prior to the design and construction of such facilities, extensive geotechnical field and laboratory tests are performed, including an accurate geologic map of the surface and of the underground environment that will be affected by the explosion. The map should show the precise location and orientation of all geologic defects that would influence the wave path, such as joints, fractures, and layers of alternately hard and soft rock.

Application of nuclear energy. The use of nuclear energy for the efficient construction of civil engineering projects has been investigated in the Plowshare Program. Examples include rapid excavation, increasing production of natural gas by opening fractures in the reservoir rock, expediting production of low-grade copper ore by causing extensive fracturing and possible concentration of the ore, and by the underground "cracking" of oil shale. The production of electrical energy by nuclear fission requires engineering geology inputs during the planning and design of the power plant;

for example, a major question to be answered is the presence or absence of faults and an estimate of when the last movement on the fault occurred. This question of "active" faults also is of increasing concern in the siting of dams.

Waste disposal. Another geotechnical problem occurs in the use of nuclear energy for generation of power or radioisotopes: safe disposal of the radioactive waste products. These products can be mixed with concrete and buried in the ground or ocean, but geohydrologic or oceanographic conditions must not be conducive to the deterioration of the concrete. One proposed solution is to excavate large caverns in rock or salt a thousand or more meters deep; however, such a solution must consider possible contamination of groundwater in the event the waste products' containers should leak. The disposal of toxic chemical or biological products in deep wells no longer is considered safe because such a disposal process near Denver, CO, between 1962 and 1965 apparently disturbed the ambient stress regime sufficiently to trigger a succession of small earthquakes. This latter phenomenon instigated ongoing research on whether fluid pressures such as might be induced by a surface reservoir could induce seismic activity. *See* NUCLEAR POWER; NUCLEAR REACTOR.

[WILLIAM R. JUDD]

Bibliography: T. Fluhr and R. F. Legget (eds.), *Reviews in Engineering Geology*, vol. 1, Geological Society of America, 1962; Geological Society of America, *Engineering Geology Case Histories*, vols. 1–11, 1957–1978; M. E. Harr, *Groundwater and Seepage*, 1962; Idaho State University Department of Geology, *Proceedings of the 5th Annual Engineering Geology and Soil Engineering Symposium*, 1967; W. R. Judd, Geological factors in choosing underground sites, in J. J. O'Sullivan (ed.), *Protective Construction in a Nuclear Age*, 2 vols., 1961; D. P. Krynine and W. R. Judd, *Principles of Engineering Geology and Geotechnics*, 1957; U. Langefors and B. Kihlström, *Rock Blasting*, 1963; R. F. Legget, *Geology and Engineering*, 2d ed., 1962; D. S. Parasins, *Mining Geophysics*, 1966; R. L. Schuster and R. J. Krizek (eds.), *Landslides—Analysis and Control*, Transportation Research Board Special Report, 1978 (also HRB Rep. 216, 1959, and 236, 1960); U.S. Coast and Geodetic Survey, *Earthquake Investigations in the U.S.*, Special Publication no. 282, 1965; Q. Zaruba and V. Mencl, *Engineering Geology*, 1976, and *Landslides and Their Control*, 1969.

Environmental engineering

The discipline which evaluates the effects of humans on the environment and develops controls to minimize environmental degradation.

In the 1960s the United States became acutely aware of the deterioration of its air, water, and land. The roots of the problem lie in the rapid growth of the national population and the industrial development of natural resources which has given Americans the highest standard of living in the world. Since 1964 there has been enactment of national, state, and local legislation directed toward the preservation of these resources.

The technology which provided society with all the necessities and luxuries of life is now expected to continue providing these services without degradation to the environment. How well this goal is attained will depend in great measure on the environmental engineers, who must cope with the enormous challenges presented by society.

It is the feeling of industry and governmental agencies that the ultimate goals should be the design of processes and systems which need minimal treatment for pollution control and the ultimate recycling of all wastes for reuse. This philosophy is both logical and necessary in a society that is rapidly depleting its nonrenewable natural resources.

Governmental policy. The Federal government enacts environmental air, water, and land-use laws which in most cases require the individual states to develop programs and enforcement policies to ensure that the goals are fulfilled. Much environmental legislation was enacted in the 1970s by both the state and Federal governments.

The significant Federal legislation has included the Wilderness Act of 1964; the Air Quality Act of 1967, followed by the Clean Air Act of 1970 and the very significant Clean Air Act Amendments of 1977; the Water Quality Act of 1965, replaced by the Water Pollution Control Act of 1972 with amendments to the act in 1977; the National Environmental Policy Act of 1969; Executive Order No. 11574 of December 1970, restating the Refuse Act of the Rivers and Harbors Act of 1899; the Noise Control Act and the Coastal Zone Management Act, both of 1972; the Resources Recovery and Conservation Act of 1976; the Surface Mining Control and Reclamation Act and the Toxic Substances Control Act, both of 1977.

Environmental legislation enacted from 1970 to 1977 has had an unmistakable impact upon the community. Standards and limitations governing the release of pollutants to the environment have placed severe restrictions on industrial operations. The sweeping Federal legislation program for environmental protection has also led to the establishment of complementary legislation, regulations, and requirements at the state level. All states now have extensive programs for environmental control and protection. The confrontation between government and industry as a result of the legislation must be eliminated and replaced by a cooperative effort toward developing the technology necessary for environmental improvement in the best interests of all concerned.

The National Environmental Policy Act is an example of the all-inclusiveness of government regulation. This act makes it mandatory that an in-depth study of environmental impacts be made in connection with any new industrial or government activity that may involve the Federal government, directly or indirectly. Similarly, state environmental policy acts, patterned after this Federal act, often require more detailed studies which must withstand more vigorous scrutiny.

These requirements extend to a wide range of activities such as:

Construction of electric power plants and transmission lines, gas pipelines, railroads, highways, bridges, nuclear facilities, airports, and mine facilities.

Any industrial facility releasing effluents to navigable waters and their tributaries.

Rights-of-way, drilling permits, mining leases, and other uses of Federal lands.

Applications to the Interstate Commerce Commission for the approval of transportation rate schedules.

In December 1970, by presidential order, the Environmental Protection Agency (EPA) was formed. Under Public Law 91-604, which extended the Clean Air and Air Quality acts, this agency now embodies under one administrator the responsibility for setting standards and a compliance timetable for air and water qualities improvement. This agency also administers the Noise Control Act, the Resources Recovery and Conservation Act, and the Toxic Substances Control Act. In addition, the agency administers grants to state and local governments for construction of wastewater treatment facilities and for air and water pollution control.

Industrial policy. Through its interdisciplinary environmental teams, industry is directing large amounts of capital and technological resources both to define and resolve environmental challenges. The solution of the myriad complex environmental problems requires the skills and experience of persons knowledgeable in health, sanitation, physics, chemistry, biology, meteorology, engineering, and many other fields.

Each air and water problem has its own unique approach and solution. Restrictive standards necessitate high retention efficiencies for all control equipment. Off-the-shelf items, which were applicable in the past, no longer suffice. Controls must now be specificially tailored to each installation. Liquid wastes can generally be treated by chemical or physical means, or by a combination of the two, for removal of contaminants with the expectation that the majority of the liquid can be recycled. Air or gaseous contaminants can be removed by scrubbing, filtration, absorption, or adsorption and the clean gas discharged into the atmosphere. The removed contaminants, either dry or in solution, must be handled wisely, or a new water- or air-pollution problem may result.

Industries that extract natural resources from the earth, and in so doing disturb the surface, are being called on to reclaim and restore the land to a condition and contour that is equal to or better than the original state. The Surface Mining Control and Reclamation Act, which now covers only the mining of coal, requires the states to have an EPA-approved program for controlling surface mining operations and the reclaiming of abandoned mined lands. Most states already have reclamation laws which cover all types of mining activities; most require an approved restoration plan and bonding to assure that restoration is accomplished.

Air quality management. The air contaminants which pervade the environment are many and emanate from multiple sources. A sizable portion of these contaminants are produced by nature, as witnessed by dust that is carried by high winds across desert areas, pollens and hydrocarbons from vegetation, and gases such as sulfur dioxide and hydrogen sulfide from volcanic activity and the biological destruction of vegetation and animal matter.

The greatest burden of atmospheric pollutants resulting from human activity comprises carbon monoxides, hydrocarbons, particulates, sulfur oxides, and nitrogen oxides, in that order. Public and private transportation using the internal combustion engine is a major source of these contaminants. It has been conservatively estimated that 50% of the major pollutants in the United States come from the use of the internal combustion engine; restrictive regulations by EPA to control these emissions have been imposed on the automobile industry. Expensive emission control measures such as catalytic converters and the mandatory use of lead-free gasoline are the two most significant requirements of the Clean Air Act of 1970 directed toward control of automobile emissions.

Industrial and fuel combustion sources (primarily utility power plants) together contribute approximately 30% of the major pollutants. Interestingly, it has been reported that sulfur dioxide and suspended particulate levels in New York City have been reduced since 1965 by nearly 90%; however, no demonstrable improvement has been seen in the occurrence of sickness and death from respiratory diseases. This would suggest, possibly, that the annual expenditure of billions of dollars by the public for achieving the goals of the Clean Air Act may benefit esthetics by improving visibility and reducing damage to materials and crops more than providing the planned health benefits.

All states have ambient air and emission standards directed primarily toward the control of industrial and utility power plant pollution sources. The general trend in gaseous and particulate control is to limit the emissions from a process stack to a specified weight per hour based on the total material weight processed to assure compliance with ambient air regulation. Process weights become extremely large in steel and cement plants and in large nonferrous smelters. The degree of control necessary in such plants can approach 100% of all particulate matter in the stack. Retention equipment can become massive both in physical size and in cost. The equipment may include high-energy venturi scrubbers, fabric arresters, and electrostatic precipitators. Each application must be evaluated so that the selected equipment will provide the retention efficiency desired.

Sulfur oxide retention and control present the greatest challenges to industrial environmental engineers. Ambient air standards are extremely low, and the emission standards calculated to meet these ambient standards place an enormous challenge on the affected industries. Many copper smelters and all coal-fired utility power plants have large-volume, weak-sulfur-dioxide effluent gas streams. Scrubbing these weak-sulfur-oxide gas streams with limestone slurries or caustic solutions is extremely expensive, requires prohibitively large equipment, and creates water and solid waste disposal problems of enormous magnitude. Installations employing dry scrubbing have been used on very-low-sulfur-oxide gas streams. A number of utilities had elected to burn low-sulfur coal to meet the existing sulfur dioxide regulations. This may not be an attractive alternative for utility plants constructed or modified after June 11, 1979. EPA promulgated on that date a minimum 70% reduction in potential sulfur oxide emissions for utilities burning low-sulfur coal.

Copper smelters are required to remove 85–95% of the sulfur contained in the feed concentrate. Smelters using old-type reverbatory furnaces produce large volumes of gas containing low concentrations of sulfur dioxide which is not amenable to removal by acid making. However, gas streams from newer-type flash and roaster-electric furnace operations can produce low-volume gas streams containing more than 4% sulfur dioxide which can be treated more economically to obtain elemental sulfur, liquid sulfur dioxide, or sulfuric acid. Smelters generally have not considered the scrubbing of weak-sulfur-dioxide gas streams as a viable means of attaining emission limitations because of the tremendous quantities of solid wastes that would be generated.

The task of upgrading weak smelter gas streams to produce products which have no existing market has led to extensive research into other methods of producing copper. A number of mining companies piloted, and some have constructed, hydrometallurgical plants to produce electrolytic-grade copper from ores by chemical means, thus eliminating the smelting step. These plants have generally experienced higher unit costs than smelters, and a number have been plagued with operational problems. It does not appear likely that hydrometallurgical plants will replace conventional smelting in the foreseeable future. Liquid ion exchange, followed by electrowinning, is also being used more extensively for the heap leaching of low-grade copper. This method produces a very pure grade of copper without the emission of sulfur dioxide to the atmosphere.

Water quality management. The Federal effort in water-pollution control has been assumed by the EPA and its office of water quality under the National Pollutant Discharge Elimination System (NPDES). The essential function of NPDES is compliance with effluent standards adopted for municipal and industrial waste dischargers. Many states already have a permit program in effect and have been granted authority to administer these programs; however, permit conditions and compliance schedules are still subject to Federal approval. Originally, EPA guidelines called for universal standards, but some flexibility has been permitted under sections of the law addressing receiving-water-quality criteria, and at least in some cases permit conditions are being altered to treat more site-specific situations when permits expire and are renewed. In the case of new source permitting, however, flexibility in permit conditions continues to be lacking even in the face of good background survey information. Some mining industry environmental engineers can be expected to take issue with the philosophy of wasting a portion of the assimilative capacity of receiving waters by not utilizing part of this natural phenomenon and thus producing finished products at a lower cost to the consumer.

The state of the art of treatment of wastewaters containing metals has barely advanced beyond neutralization and chemical precipitation. As water quality standards continue to become more stringent, especially concerning dissolved solids, it is apparent that chemical treatment of metals wastes will be unacceptable. Chemical treatment in many instances substitutes a different molecular species for another with no reduction in total dissolved solids.

Some segments of the mining industry are considering physical treatment methods of desalination techniques such as evaporation, reverse osmosis, and electrodialysis. These techniques require vast quantities of energy and are prohibitive in both capital and operational costs. Continued experimentation will undoubtedly improve the benefit-cost ratio.

For the most part the mining industry tends toward a policy of complete recycle of water. Some metallurgical processes are amenable to reuse of process waters with only minimal treatment, such as removal of suspended and settleable solids. Other processes require higher-quality water with lower dissolved-solids content. Even this generally requires removal of only a portion of the solids from a waste stream in order to maintain an acceptable process water quality. Many of these solids can also be reclaimed, thus recovering values now lost. This approach is not only logical but economical. If receiving-water quality criteria are more stringent than process-water treatment requirements, it is impractical to comply with those standards and then waste the water to the nearest surface drainage. In the more arid parts of the world, recycle of water has become a practice by necessity; in water-rich areas it will become practice by governmental decree.

Originally, the Federal Clean Water Act required best practicable treatment technology by 1977, best available technology (BAT) by 1983, and (stated as a goal) no pollutant discharge by 1985. The deadline for achieving BPT was extended under limited conditions to Apr. 1, 1979. The deadline applicable to BAT was extended to July 1, 1984. In addition to better definition of treatment of conventional pollutants, recent changes in regulations have focused EPA attention on toxic and hazardous materials. Most mining and milling wastewaters contain many of these "priority pollutants," and more stringent permit conditions are certain to be a result. Hopefully, in the case of both BAT and ND, definition and application of this requirement and goal can be tempered by demonstrated need and economic practicality. As stated before, a significant portion of the nation's energy could be needlessly wasted, especially in efforts to attain ND. *See* WATER POLLUTION.

Land reclamation. It has been stated that strategic mineral development is the most productive use of land because of the great values that are received by the nation from such small land areas. Open-pit and strip mining have recently been criticized for their impact on the surrounding ecosystems. Some preservationists and conservationists feel that the Earth's surface should not be disturbed by either exploration or mining activities. Mining companies must reverse the spoiler image that has been created in the past, especially in the large-scale stripping of coal in the eastern and mid-central states. Most states now have regulations that require that all stripped land be reclaimed. Strip mining of coal has often produced an attendant acid mine drainage problem. *See* OPEN-PIT MINING; STRIP MINING.

Ecosystem studies are now being conducted by mining companies during exploration and prior to the commencement of mining operations. These studies lead to the effective planning for the most desirable method of mining that will least disturb the environment and yet lend itself to reclamation.

Much controversy erupted with the passage of the Federal Surface Mining Control and Reclamation Act of 1977, pitting states, which had already enacted reclamation legislation, against the Office of Surface Mining, which was charged with the administration of the act. Many states are attempting to retain control of reclamation and are busily engaged in amending legislation and rewriting regulations to ensure compliance with the Federal act.

Even prior to state or Federal legislation, many mining companies had taken the initiative in planning operations so as to limit the adverse impact on the environment. Notable examples of this forward-looking and concerned approach are programs carried on by American Cyanamid Company, American Metal Climax, Inc., Anamax Company (formerly Anaconda), Bethlehem Steel Company, and Peabody Coal Company.

Typical of these industrial programs is that of Anamax which undertook the development of the Twin Buttes Copper Mine in the desert area near Tucson, AZ. In excess of 240,000,000 tons of alluvium were removed from this open-pit site before production could begin. This material and future waste will be used to form large dikes that will impound tailings. The dikes are terraced and planted with vegetation indigenous to the area and, when completed, will be 1000 ft wide (1 ft = 0.3 m) at the base and 250 ft wide at the apex, with a maximum height of 230 ft. These dikes take on the appearance of mesas and blend into the desert landscape.

Moreover, American Cyanamid has found that restoration of Florida's phosphate-mined lands can best be accomplished by reclaiming the major portions of those areas simultaneously during mining operations.

Land reclamation is made an integral part of mine planning by advanced consideration and decisions regarding what the area should look like upon completion of mining activity. By systematically forming eventual lakes and distributing stripped wastes into previously mined cuts, grading the area and restoring it to desirable land become relatively easy. Many of these reclaimed areas have become useful as parks, recreational areas, wildlife sanctuaries, and agricultural and residential development sites.

Again, if a reasonable and empirical approach to mined-land reclamation is allowed, following the dictates of physical, chemical, and biological laws, the very small area of the Earth's surface disturbed by mining can be restored to beneficial use at minimal cost to the consumer and optimum conservation of energy. [LEWIS N. BLAIR; JOHN C. SPINDLER; WALTER H. UNGER]

Bibliography: M. Eisenbud, *Environment, Technology, and Health: Human Ecology in Historical Perspective*, 1978; J. E. McKee and H. W. Wolf (eds.), *1963 State of California Water Quality Criteria*, California Water Quality Board and U.S. Public Health Service, 2d ed., 1963; *Modern Pollution Control Technology*, vols. 1 and 2, Research and Education Association, 1978; F. W. Schaller and P. Sutton (eds.), *Reclamation of Drastically Disturbed Lands*, American Society of Agronomy, Crop Science Society of America, and Soil Science Society of America, 1978.

Eutectics

The microstructures that result when a solution of metal of eutectic composition solidifies. The eutectic reaction must be distinguished from eutectic microstructures.

Eutectic reaction. The eutectic reaction is a reversible transformation of a liquid solution to two or more solids, under constant pressure conditions, at a constant temperature denoted as the eutectic temperature T_E. The eutectic phase diagram is notable for displaying negative slopes of both pairs of solidus and liquidus lines, relative to the pure metal or compound terminal components A and B (Fig. 1).

Eutectic reactions often constitute only a portion of a complex phase diagram, as in the Cu-Si system; other alloy systems, for example, the Cu-Mg system, may exhibit multiple eutectic reactions between intermediate phases. The transformation may occur between metals only, between metals and nonmetals (as in Ni-TaC), or between nonmetals only (KCl-NaCl).

In all cases the central liquid phase L touches the eutectic temperature line from above, as shown in Fig. 1; the eutectic alloy, at composition c_e, has the lowest melting point of any mixture of components A and B. Solidification of the eutectic composition occurs with the simultaneous crystallization of the two solid phases α and β.

In terms of Gibbs' phase rule, at constant pressure, $F = C + 1 - P$, where F = number of degrees of freedom, C = number of components, and P = number of phases. Binary equilibrium involving three phases allows no degrees of freedom: $F = 2 + 1 - 3 = 0$. Consequently, the eutectic reaction is known as an invariant reaction; the three phases must be at the equilibrium temperature and have fixed compositions.

A very similar invariant reaction involving liquid and solid phases is the peritectic reaction, in which two solids and one liquid are at equilibrium; in this case the central solid phase meets the invariant line from below ($S_1 + L \xrightarrow{\text{Cooling}} S_2$). In the solid state the equivalent reactions with decreasing temperature are the eutectoid ($S_1 \rightarrow S_2 + S_3$) and peritectoid ($S_4 + S_5 \xrightarrow{\text{Cooling}} S_6$), where S_i denotes solid phases.

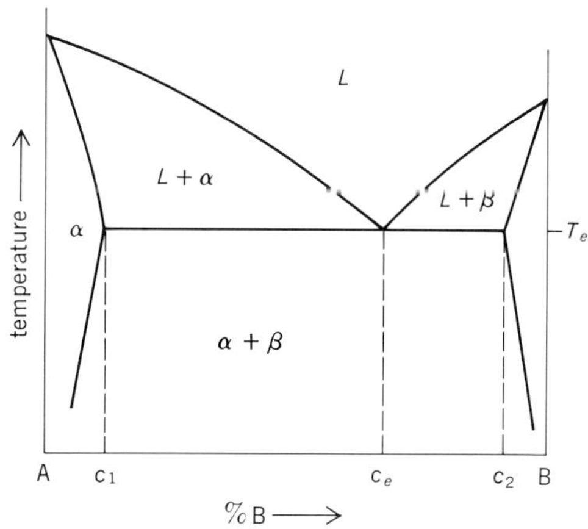

Fig. 1. Schematic eutectic phase diagram between components A and B.

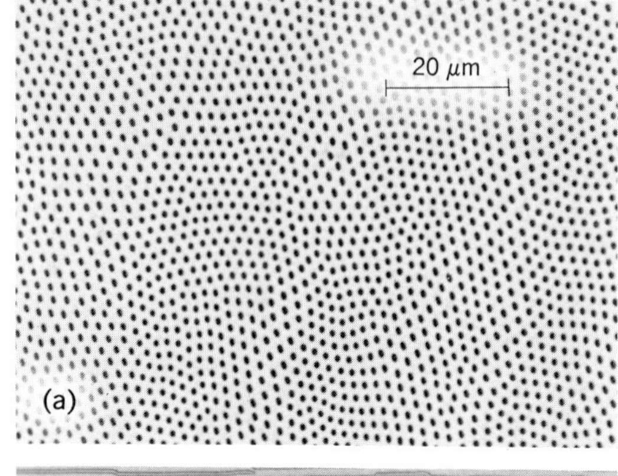

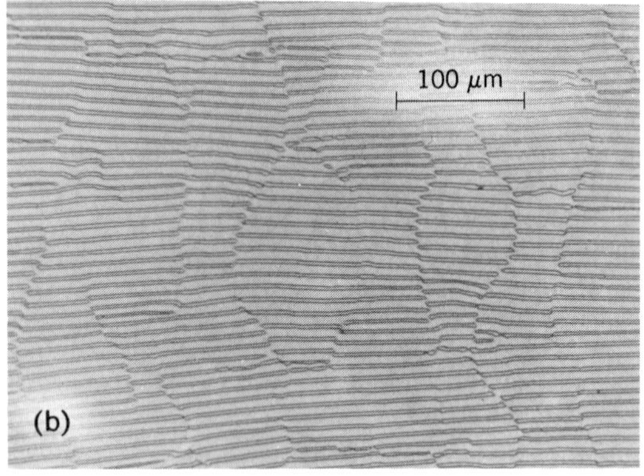

Fig. 2. Cross sections of aligned eutectic composites. (*a*) Fibers of Al_3Ni in Al matrix. (*b*) Alternating lamellae of Mg and $Mg_{17}Al_{12}$. Lighter etching phase is $Mg_{17}Al_{12}$.

Eutectic microstructures. Microstructures which are wholly eutectic in nature can occur only for a single, fixed composition c_e in each alloy system demonstrating the reaction (Fig. 1). However, any alloy whose composition passes through the eutectic invariant line (composition c_1 to c_2) undergoes the eutectic reaction as part of its solidification process. Initially, however, the liquid forms some primary crystals of the nearest end-point phase. Whatever liquid remains in the two-phase mixture at the equilibrium temperature then transforms by the eutectic reaction, and the primary crystals are left unchanged.

Microstructure classification. The great variety of patterns observed in eutectic microstructures, in some cases even in a single alloy system, has led to several attempts to develop classification systems. The most successful of these is due to E. Scheil, who classified eutectics on the basis of their mode of crystallization rather than on morphological characteristics, and this classification will be followed here.

Binary eutectic microstructures can be divided into normal and abnormal classes. Normal microstructures comprise primarily lamellar or fibrous types that are formed by simultaneous growth of the two solid phases in the form of parallel fibers in a continuous matrix (Fig. 2*a*) or as parallel lamellae (Fig. 2*b*). Normal growth occurs when the two solid phases grow locally at the same velocity, resulting in a planar solid-liquid interface. The fiber or lamellar axis tends to be perpendicular to the interface. When freezing begins at more than one center, each center gives rise to a eutectic grain, within which all lamellae or fibers have the same crystallographic orientation. Simple crystallographic relationships between coexisting eutectic phases have been noted; for example, in the lamellar Cd-Zn eutectic the following planes are parallel: $(0001)_{Cd}$ and $(0001)_{Zn}$; and the following directions are parallel: $[01\bar{1}0]_{Cd}$ and $[01\bar{1}0]_{Zn}$. Preferred growth directions have been reported for many systems, but there is no general agreement concerning their existence.

Abnormal eutectics are those in which the two solid phases do not grow at equal velocity. The faster-growing phase (usually that present in smaller volume) grows in a branching or dendritic-like pattern, and the other phase forms from the melt remaining between the branches.

Eutectic fusible alloys. Although technologically important alloy systems, particularly the Fe-C system (all cast irons used commercially pass through a eutectic reaction during solidification), exhibit at least one eutectic reaction, there has been little exploitation of wholly eutectic microstructures for structural purposes. A number of relatively low-melting eutectics are found in binary and higher-order mixtures of bismuth, lead, cadmium, tin, and indium. Table 1 shows the compositions and melting temperatures of eutectic fusible alloys which are used as solders, as heat-transfer media, for punch and die mold and pattern applications, and as safety plugs. A silver-copper eutectic alloy is also used for high-temperature soldering applications. *See* SOLDERING.

Eutectic composites. During the early 1960s it was recognized that directional solidification of eutectic alloys so as to create a microstructure well aligned parallel to the growth direction could produce high-strength, multiphase composite materials. High-temperature gradients at the melt-solid interface, necessary to produce aligned alloys, are achieved by cooling the base of the crucible containing the liquid alloy as the crucible is withdrawn from the furnace. Excellent mechanical properties have been reported for metal-metal as well as nonmetal-reinforced metallic systems; examples of such alloys, all of which melt above

Table 1. Compositions and melting temperatures of eutectic fusible alloys

	Composition, percentage by weight				
T_E,°C	Bi	Pb	Sn	Cd	Other
46.8	44.70	22.60	8.30	5.30	19.10 In
58	49.00	18.00	12.00	—	21.00 In
70	50.00	26.70	13.30	10.00	—
91.5	51.60	40.20	—	8.20	—
95	52.50	32.00	15.50	—	—
102.5	54.00	—	26.00	20.00	—
124	55.50	44.50	—	—	—
138.5	58.00	—	42.00	—	—
142	—	30.60	51.20	18.20	—
144	60.00	—	—	40.00	—
177	—	—	67.75	32.25	—
183	—	38.14	61.86	—	—
199	—	—	91.00	—	9.00 Zn
221.3	—	—	96.50	—	3.50 Ag
236	—	79.7	—	17.7	2.60 Sb
247	—	87.0	—	—	13.00 Sb

Table 2. High-strength—high-temperature eutectic alloys

System	T_E,°C	Form	V_f
Ni-W	1500	Fibrous	0.06
Ni,Cr,Al-Tac	1348	Fibrous	0.05
Co-Tac	1402	Fibrous	0.10
Ni-Ni_3Nb	1270	Lamellar	0.26
Ni,Al,Cr-Ni_3Nb	1250	Lamellar	0.33
Ni_3Al-Ni_3Ta	1360	Fibrous	0.35
Ni_3Al-Ni_3Nb	1280	Lamellar	0.44
Co-Cr_7C_3	1303	Fibrous	0.30
Co-NbC	1365	Fibrous	0.12

1000°C, are given in Table 2. Note that lamellar eutectics are generally characterized by much higher-volume fractions V_f of reinforcing phase than fibrous alloys are. Many of the alloys listed demonstrate elevated temperature creep, stress-rupture, and fatigue properties which are superior to those of the best current conventional alloys.

Several factors have been shown to exert major influences on the mechanical behavior of eutectic composites. These include: (1) Degree of alignment; the more perfect the alignment the stronger the alloy. (2) Growth speed; $\lambda^2 R =$ constant, where $\lambda =$ interphase separation and $R =$ growth rate; strength is inversely proportional to $\lambda^{-1/2}$. (3) Perfection of reinforcement; faults in the lamellae or fibers tend to reduce strength. (4) Ductility of coexisting phases; cracks nucleate readily in brittle phases, thereby limiting ductility of the composite. (5) Presence of intentional alloying additions; eutectics with up to nine components have been successfully aligned and reveal superior mechanical properties.

Among the major advantages of these alloys are extraordinary thermal stability of unstressed microstructures, retention of high strength to very close to the eutectic temperature of the respective alloys, and the ability to optimize strength by appropriate alloying additions to induce either solid-solution strengthening or intraphase precipitation of additional phases.

Future applications. The most likely future applications for aligned eutectics are as gas turbine engine materials (turbine blades or stator vanes) or in nonstructural applications such as superconducting devices in which directionality of physical properties is important. *See* ALLOY; METAL, MECHANICAL PROPERTIES OF.

[NORMAN S. STOLOFF]

Bibliography: P. M. French et al. (eds.), *Microstructural Science*, vol. 3, 1975; R. K. McCrone (ed), *Treatise on Materials Science*, vol. 11: *Properties and Microstructure*, 1977.

Excavator

Any of a variety of bucket-equipped construction machines used for digging earth and rock. The most common excavator is basically a modified crane. When modified for digging, the unit made up of crane and its digging attachment is commonly called by the name of the attachment. Thus a crane carrying a dragline bucket is called a dragline. *See* CRANE HOIST.

Standard excavators. There are four principal modifications of cranes for use as an excavator: dragline, clamshell, power shovel, and backhoe.

The dragline uses a long latticed crane boom, two working cables, and a bucket designed for horizontal digging. The bucket is cast into the area to be dug. The drag cable pulls the bucket toward the machine. Because of its weight and shape, the bucket fills itself. The filled bucket is lifted by the second cable and swung by the machine to the disposal area where release of the drag cable tips the bucket to empty it. The dragline is particularly suited for long-reach digging in wet and muddy areas where scrapers and bulldozers cannot work.

The clamshell uses a crane boom and a clamshell bucket. Digging with a clamshell is not a power operation; the clamshell is filled only by its own weight and the bite of its jaws as they are closed. For this reason, the clamshell is best suited for digging soft or loosely packed materials. It can, however, dig deeper and raise materials higher than any other crane-type excavator. Clamshell buckets with powered closure are available for digging moderately packed materials. *See* GRAB BUCKET.

The power shovel uses a heavy but relatively short boom and a dipper stick to which is attached a hinged-bottom digging bucket or dipper. The bucket is filled by a forward and upward motion. It is emptied by the opening of its hinged bottom. The power shovel is the most efficient excavator for hard or heavy digging within short reaches.

The backhoe is a short-boomed machine with an inverted bucket at the end of its stick (see illustration). The bucket is filled by being pulled in an arc downward and backward toward the machine. The backhoe bucket is emptied by being raised and extended with the open side down. The backhoe is generally a smaller and lighter machine than the dragline or the power shovel; most backhoe buck-

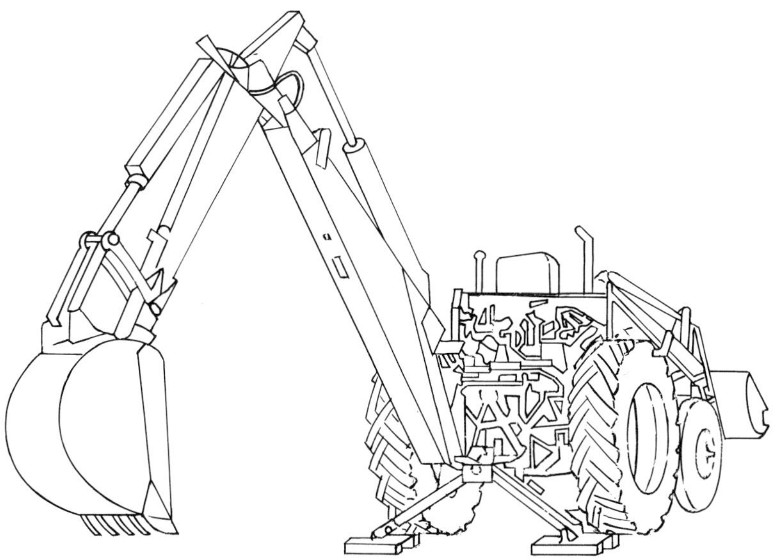

Hydraulic backhoe used in trenching. (*Case*)

ets carry only fractions of 1 yd^3, as illustrated. The machine excels at digging trenches.

Special excavators. The hydraulic backhoe is a variation in which the operations are activated by hydraulic pistons, as illustrated, rather than cables. The basic machine may have either tracks or wheels and may be powered by either a gasoline or diesel engine.

The front-end loader consists of a continuous-track or wheeled tractor with a wide hydraulically operated bucket mounted on jointed arms at its front end. The bucket is filled by the forward motion of the tractor. It is raised and emptied by hydraulic pistons. The front-end loader is ideal for moving earth short distances or for loading trucks when long or high reaches are not involved. Its efficiency, speed, and economy are so great that it has all but replaced power shovels of less than 20 yd^3 capacity.

The trencher is designed especially for excavating trenches. Basically it is a crawler tractor mounting either a movable wheel or a continuous chain, on which there are numerous buckets. The digging assembly is raised or lowered, sometimes by cables, more often by hydraulic pistons or motors, to achieve the desired trench depth. As the assembly turns, the buckets scoop up earth, carry it upward, then spill it out onto a transverse conveyor belt which dumps it beside the trench. *See* BULK-HANDLING MACHINES; CONSTRUCTION EQUIPMENT. [EDWARD M. YOUNG, JR.]

Explosive forming

The shaping or modifying of metals by means of explosions. The explosives may be of either the detonating or deflagrating type. Explosive gas mixtures or stored gas at high pressure may also provide the motive power (Fig. 1).

Most types of explosive forming involve a mold into which a flat sheet of metal is pressed by the explosion. The metal is stretched uniformly by the explosive impulse. Even welds in the original blank survive the deformation without damage. The advantages of explosive forming over conventional forging methods accrue especially in the case of intricate shapes of which only a few items are required. Tooling costs are low.

Cold welds can be made between dissimilar metals by driving the two parts together under explosive impact. In other applications of explosive-forming methods, powders are pressed into solid billets.

Fig. 2. Typical metal shapes produced by explosive forming. Note weld in bowl. (*National Northern Corp.*)

In a quite different application, high explosives are used to cut large blocks of metal and even to split thin sheets into two layers of exactly one-half the original thickness. When an explosive charge is detonated in contact with the metal, it produces a compression wave. On reflection from a free surface, the compression wave turns upside down to produce a tension wave. Along the planes in the metal where the tension in the wave front exceeds the strength of the metal, rupture occurs.

Explosives can also be employed to extrude metal shapes and to punch hard metals with the aid of dies. Shapes produced explosively are very exact and free from the fine cracks that sometimes result when pressure is slowly applied (Fig. 2). Forces far exceeding those of the largest hydraulic presses can be applied by explosives.

Metals can also be hardened under explosive impact with results that compare favorably with those from slow cold-working methods. *See* METAL FORMING.

[WILLIAM E. GORDON]

Extrusion

The forcing of solid metal through a suitably shaped orifice under compressive forces. Extrusion is somewhat analogous to squeezing toothpaste through a tube, although some cold extrusion processes more nearly resemble forging, which also deforms metals by application of compressive

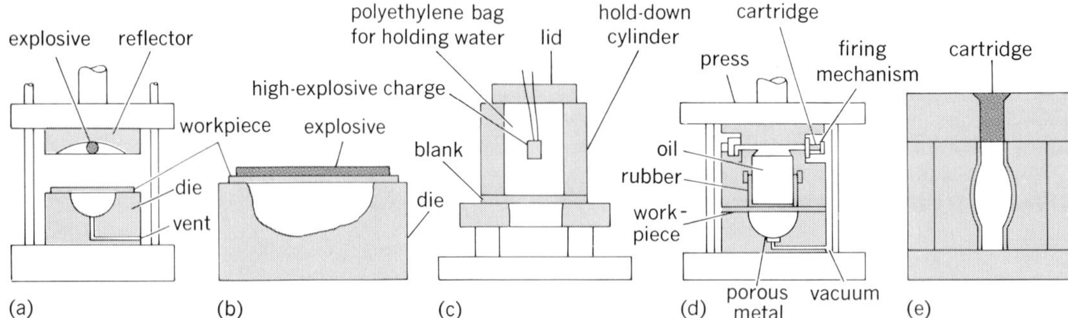

Fig. 1. Five methods of explosive forming. (*a*) Shaped charge and parabolic reflector. Shock and pressure are motive power. Reflector and charge must be shaped to direct explosive force toward workpiece. (*b*) Flat high explosive is simply placed on top of sheet which lies on female die. Forming over a punch produces wrinkles. (*c*) The hold-down cylinder is often a carton of water or a bolted ring on lower die. (*d*) Gunpowder cartridges are usually applied in a press or enclosed die. This one uses oil and rubber to distribute force. Vacuum prevents air pocket which would retard action. (*e*) Bulge forming with cartridge power is comparatively simple and safe. Such devices properly vented are less noisy than many standard forming operations. (*From Explosives form space age shapes, Steel, Aug. 25, 1958*)

forces. Most metals can be extruded, although the process may not be economically feasible for high-strength alloys.

Hot extrusion. The most widely used method for producing extruded shapes is the direct, hot extrusion process. In this process, a heated billet of metal is placed in a cylindrical chamber and then compressed by a hydraulically operated ram (Fig. 1). The opposite end of the cylinder contains a die having an orifice of the desired shape; as this die opening is the path of least resistance for the billet under pressure, the metal, in effect, squirts out of the opening as a continuous bar having the same cross-sectional shape as the die opening. By using two sets of dies, stepped extrusions can be made. The small section is extruded first to the desired length, the small split die is replaced by the large die, and the large section is then extruded.

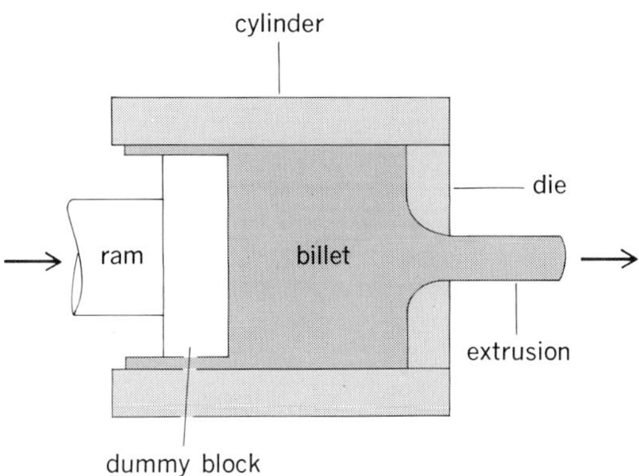

Fig. 1. Schematic representation of the widely used direct-extrusion process (hot).

Extrusion pressures and speed vary considerably, depending upon the size and shape of the section and the mechanical properties of the metal. Some metals, such as magnesium and some aluminum alloys, require slow speeds of a few feet per minute, while others, such as some copper alloys, lead, and steel, are extruded at speeds of over 1000 ft/min. The extrusion speed is also somewhat dependent upon the temperature of the alloy. Considerable heat is generated by the process; if the extrusion speed is high, this heat cannot be dissipated, resulting in a rise in temperature. In some instances the rise in temperature may be sufficient to melt or at least weaken the metal to the point at which the frictional stresses at the surface cause cracking.

The flow of metal is not uniform during extrusion of the billet. Because of the restraining effect of the die face and the frictional effects at the cylinder walls, the outer zones of the billet resist deformation and the flow occurs most rapidly at the center of the billet. Eventually, as the billet shortens, the different rates of flow at the center and surface result in the billet's becoming hollow. If the extrusion ram travels further, defects appear at the center of the extruded section. Therefore, a portion of the billet may be left in the cylinder and discarded. To prevent the oxidized surface layers of the billet from getting into the extruded product, the dummy block (or follower plate) in front of the ram is of slightly smaller diameter than that of the cylinder; thus, a thin sleeve of the billet is extruded out of the ram end of the cylinder and is subsequently discarded.

Lubricants are used to minimize friction and protect the die surfaces. Graphite is a common lubricant for nonferrous alloys, whereas for hot extrusion of steel, glass is an excellent lubricant.

Indirect, or inverted, extrusion was developed to overcome such difficulties as surface friction and entrainment of surface oxide of direct extrusion. In the indirect process the ram is hollow, the die opening being in the dummy block, and the opposite end of the cylinder is closed. As the ram advances, the billet does not move as in the case of direct extrusion, and the metal is extruded backward through the die and the hollow ram. However, the process is not very popular because the hollow ram is weaker, resulting in lower machine capacity; trouble-free operation requires that the extruded product be straight and not hit the inside of the ram.

Tubular shapes are produced by a mandrel of the desired inside shape of the product. The mandrel may be fastened to the dummy block in a direct extrusion press if a pierced billet is used; or, in the case of a solid billet, a mandrel must first pierce the billet, after which the main ram advances.

In the production of lead cable sheathing or cored solder wire, the core material (cable or rosin flux) passes through a core tube (or die block) around which is heated lead under pressure. The core material passes from the core tube into a die cavity and the lead is extruded out with it, forming a casing around the cable or rosin core. The process is semicontinuous, molten lead being added to a vertical cylinder and pressure applied periodically. The lead is solid when it reaches the core material.

Cold extrusion. The extrusion of cold metal is variously termed cold pressing, cold forging, cold extrusion forging, extrusion pressing, and impact extrusion. The term cold extrusion has become popular in the steel fabrication industry, while impact extrusion is more widely used in the nonferrous field.

The original process (identified as impact extrusion) consists of a punch (generally moving at high velocity) striking a blank (or slug) of the metal to be extruded, which has been placed in the cavity of a die. Clearance is left between the punch and die walls; as the punch comes in contact with the blank, the metal has nowhere to go except through the annular opening between punch and die. The punch moves a distance that is controlled by a press setting. This distance determines the base thickness of the finished part. The process is particularly adaptable to the production of thin-walled, tubular-shaped parts having thick bottoms, such as toothpaste tubes.

A process requiring less pressure than backward extrusion is the forward-extrusion process, originally called the Hooker process (Fig. 2). A formed blank (usually a thick-walled cup) is placed in a die cavity and struck by a punch having a shoulder or enlarged section a short distance from the end. Upon contact with the blank, the nose or end of the punch starts to push the center of the blank through the die cavity, in a manner similar to the action occurring in deep drawing of sheet met-

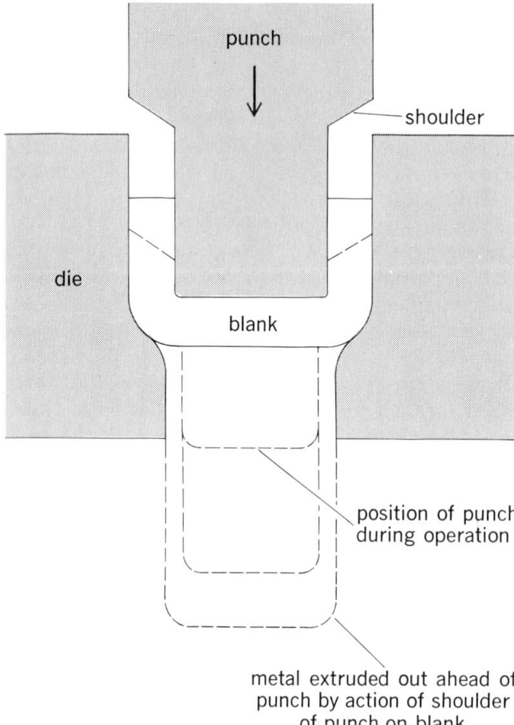

Fig. 2. Principles of forward cold-extrusion process, originally called the Hooker process.

al. After the punch has advanced a short distance, the shoulder comes in contact with the top of the thick wall of the blank. The punch shoulder then extrudes the metal through the annular space between the die and the end of the punch. Thus, in forward extrusion, the metal moves in the same direction as the punch, whereas in backward extrusion the metal moves in the opposite direction. *See* SHEET-METAL FORMING.

The application of the cold-extrusion process to steel was developed in Germany about 1930 but was not released until after World War II. The success of cold extrusion of steel hinged on the discovery of a suitable lubricant and surface treatment. The steel is given a phosphate surface coating, which absorbs and holds the lubricant (soap emulsion, vegetable oils, or dry metal stearates) and prevents seizure between the metal and tools.

Advantages of cold extrusion are higher strength because of severe strain-hardening, good finish and dimensional accuracy, and economy due to fewer operations and minimum of machining required. *See* METAL FORMING.

[RALPH L. FREEMAN]

Fan

A fan moves gases by producing a low compression ratio, as in ventilation and pneumatic conveying of materials. The increase in density of the gas in passing through a fan is generally negligible; the pressure increase or head is usually measured in inches of water.

Blowers are fans that operate where the resistance to gas flow is predominantly downstream of the fan. Exhausters are fans that operate where the flow resistance is mainly upstream of the fan.

Fans are further classified as centrifugal or axial (Fig. 1). The housing provides an inlet and an outlet and confines the flow to the region swept out by the rotating impeller. The impeller imparts velocity to the gas, and this velocity changes to a pressure differential under the influence of the housing and ducts connected to inlet and outlet.

Performance. In selecting a fan for a particular application, requirements of primary interest are the quantity Q of gas to be delivered by the fan and the head H which must be developed to overcome the resistance to flow of the quantity Q in the connected system. These operating conditions establish the fan dimensions of diameter D and rotational speed N. Performance of a fan of diameter D is rigorously described by its characteristic curves (Fig. 2). Fans of different types and sizes are conveniently compared by converting their characteristics to dimensionless form. Figures 3 and 4 show selected fan performance on the percentage basis, where 100% rating is defined as the peak of the efficiency curve.

The load placed on the fan must correspond to a condition on its operating characteristic. Thus if, for example, the system in which the fan operates presents less resistance than is overcome by the head developed by the fan at the required capacity, additional resistance must be introduced by a damper, or the fan speed must be changed, or the excess capacity must be diverted elsewhere.

Performance of a given fan, operating at a given point on its efficiency curve, varies with speed in accordance with the following rules:

Capacity Q proportional to speed N
Head H proportional to speed squared N^2
Horsepower P proportional to speed cubed N^3

For a series of similar fans, operating at a given point on their efficiency curves and at the same speed, performance varies in accordance with the following rules:

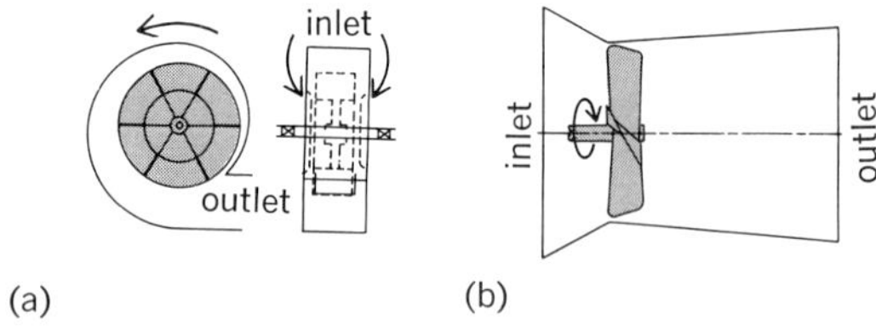

Fig. 1. Fan types. (*a*) Centrifugal. (*b*) Axial.

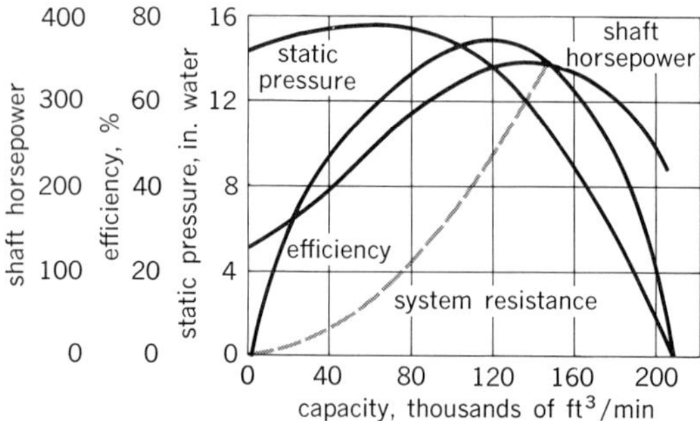

Fig. 2. Static pressure characteristics of backwardly curved blade centrifugal fan at constant speed.

Capacity Q proportional to diameter cubed D^3
Head H proportional to diameter squared D^2
Horsepower P proportional to diameter to fifth power D^5

These six relations can be combined to express the performance of a family of similar fans in terms of dimensionless coefficients, as in Eqs. (1), (2), and (3).

$$\text{Capacity coefficient } C_Q = Q/ND^3 \quad (1)$$
$$\text{Head coefficient } C_H = gH/N^2D^2 \quad (2)$$
$$\text{Power coefficient } C_P = gP/wN^3D^5 \quad (3)$$

Here Q = capacity, ft³/sec
H = head, ft of fluid
P = shaft horsepower, ft-lb/sec
g = gravitational acceleration, ft/sec²
N = revolutions/sec
w = weight density, lb/ft³
D = wheel diameter, ft

Values of these coefficients for a selected group of fans are listed in the table.

Comparison of different types of fans is further facilitated by the elimination of the diameter term. Equations for capacity coefficient and head coefficient are solved simultaneously to give specific speed N_S, as in Eq. (4). Specific speed is an

$$N_S = NQ^{1/2}/(gH)^{3/4} \quad (4)$$

inherent performance criterion; it is usually employed as a dimensionally impure coefficient. In the selection of a fan for a given application, a fan is chosen that has the desired value of N_S in its region of peak efficiency. Low specific speed corresponds to a fan of low rotative speed and large impeller diameter.

Alternatively, speed can be eliminated, giving specific diameter D_S, as in Eq. (5).

$$D_S = (gH)^{0.25}D/Q^{0.5} \quad (5)$$

Types. The shape of a fan characteristic as a function of capacity depends on the fan type. For

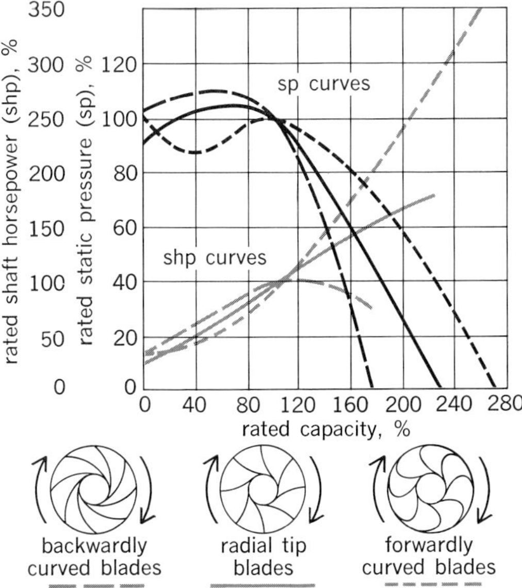

Fig. 3. Percentage characteristics compare performance of three forms of centrifugal fans.

Dimensions and performance of a group of fans

Fan dimensions and performance	Centrifugal fans: Backwardly curved blades	Centrifugal fans: Steel plate	Centrifugal fans: Forwardly curved blades	Axial fan: Pressure blower
Performance				
Specific speed, N_S	0.25	0.11	0.21	0.75
Peak efficiency, %	75	65	65	72
Tip speed, fpm × 10³	18	11	3.5	16
C_H at shutoff	3.8	6.3	11.2	3.2
C_Q at peak eff	0.42	0.18	1.2	0.4
C_P at peak eff	2.1	1.7	2.1	0.55
Proportions as functions of wheel diameter, D				
Inlet area (XD^2)	0.7	0.4	0.8	0.8
Outlet area (XD^2)	0.6	0.3	0.7	
Axial blade length (XD)	0.3	0.4	0.6	
Number of blades	18	8	64	6

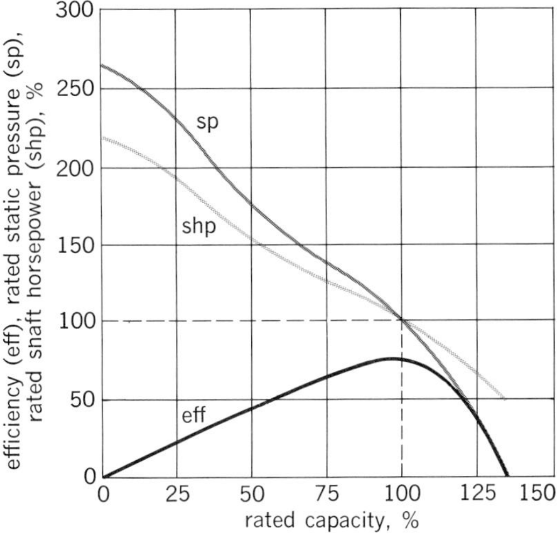

Fig. 4. Percentage characteristics of one axial fan.

example, backwardly curved blade centrifugal fans have a steeply falling head characteristic and a self-limiting horsepower characteristic as shown in Fig. 3.

Fans with forwardly curved blades have rising head and horsepower characteristics. The axial fan has a falling horsepower characteristic so that at shutoff the fan may require more power than at high flow rate (Fig. 4).

[THEODORE BAUMEISTER]

Bibliography: T. Baumeister (ed.), *Standard Handbook for Mechanical Engineers*, 8th ed., 1978; A. H. Church, *Centrifugal Pumps and Blowers*, 1944, reprint 1972; V. L. Streeter and B. Wylie, *Fluid Mechanics*, 1979.

Feedwater

Water supplied to a boiler unit for the generation of steam. Feedwater should be virtually free of impurities that are harmful to the boiler and its associated system. Generally, natural waters are unsuitable for direct use as feedwater because of their contamination by contact with the earth, the atmosphere, or other sources of pollution. These contaminants can be removed or altered by chemical treatment and other means to provide satisfactory feedwater. *See* RAW WATER; WATER TREATMENT.

The need for and the type of feedwater treatment vary considerably in different plants, depending on the nature of the water supply and the percentage of makeup water required. Its use is critical for steam-generating units operating at high pressures with high rates of heat transfer.

Components. Usually, feedwater is a mixture of relatively pure condensate and treated water. The proportions vary with the amount of makeup water required to compensate for losses from the system served by the boiler. In a closed power-generating system, losses may be less than 1% of the cycle flow. But, when steam is used for industrial processes, the return of part or all of the condensate may be impractical and, thus, larger percentages of treated makeup water are needed. *See* STEAM CONDENSER.

Processing. The processing of raw water for use as treated makeup water involves the removal or the conversion of the suspended and dissolved solids that could form scale or insulating deposits on the heat-transfer surfaces, or could be transported with the steam to other parts of the system. Contaminants that might, under certain conditions, be corrosive also must be removed or converted to noncorrosive substances. The effectiveness of the treatment varies with the methods used and with the care taken in operation, but water can be treated to meet the most exacting requirements. *See* BOILER WATER.

Corrosion. Internal corrosion may be initiated by dissolved gases, such as oxygen and carbon dioxide, picked up by the water at any point of contact with air. Therefore, in the treatment of feedwater the dissolved oxygen and other gases are usually removed just before the water is fed to the boiler. Most of these gases can be removed by boiling the water in open heaters and discharging the noncondensable gases through a vent condenser. However, the most effective removal is obtained with spray- or tray-type deaerating heaters arranged for the countercurrent scavenging of the released gases to prevent their going back into solution in the water.

In high-purity water systems, care must be taken to prevent corrosion of the piping, heaters, and other parts of the preboiler system. Thus, an alkaline condensate should be maintained by adding ammonia or volatile amines. *See* BOILER; STEAM-GENERATING UNIT. [GEORGE W. KESSLER]

Ferroalloy

An important group of metallic raw materials required for the steel industry. Ferroalloys are the principal source of such additions as silicon, Si, and manganese, Mn, which are required for even the simplest plain-carbon steels; and chromium, Cr, vanadium, V, tungsten, W, titanium, Ti, and molybdenum, Mo, which are used in both low- and high-alloy steels. Also included are many other more complex alloys. Ferroalloys are unique in that they are brittle and otherwise unsuited for any service application, but they are important as the most economical source of these elements for use in the manufacture of the engineering alloys. These same elements can also be obtained, at much greater cost in most cases, as essentially pure metals. The ferroalloys contain significant amounts of iron and usually have a lower melting range than the pure metals and are therefore dissolved by the molten steel more readily than the pure metal. In other cases, the other elements in the ferroalloy serve to protect the critical element against oxidation during solution and thereby give higher recoveries. Ferroalloys are used both as deoxidizers and as a specified addition to give particular properties to the steel. *See* STEEL MANUFACTURE.

Many ferroalloys contain combinations of two or more desirable alloy additions, and well over 100 commercial grades and combinations are available. Although of less general importance, there are other sources of these elements for steelmaking, such as metallic nickel, Ni, silicon carbide, molybdic oxide, and even mischmetal (a mixture of rare earths). Analyses of a few typical ferroalloys are given in the table.

The three ferroalloys which account for the major tonnage in this class are the various grades of silicon, manganese, and chromium. For example, 13 lb of manganese is used on the average in the United States for every ton of open-hearth steel produced. Elements supplied as ferroalloys are among the most difficult metals to reduce from ore.

The most common grade of ferromanganese is a blast-furnace product (standard ferromanganese), the major variation from pig iron production being the use of manganese-rich ore. Other grades of ferromanganese with low C are made by the processes described in the next sections. A low-Si ferroalloy can be made in the same way, but the tonnage grades are made by other methods.

The most general method of ferroalloy manufacture is the submerged-arc furnace (see illustration). Its use will be described in connection with the production of 50% ferrosilicon. A modern furnace of this type is 26 ft in diameter and 10 ft deep with the three carbon electrodes supplying an average of 5 kwhr per pound of silicon produced. The furnace produces about 2 tons of alloy per hour. The furnace is charged intermittently from the top with

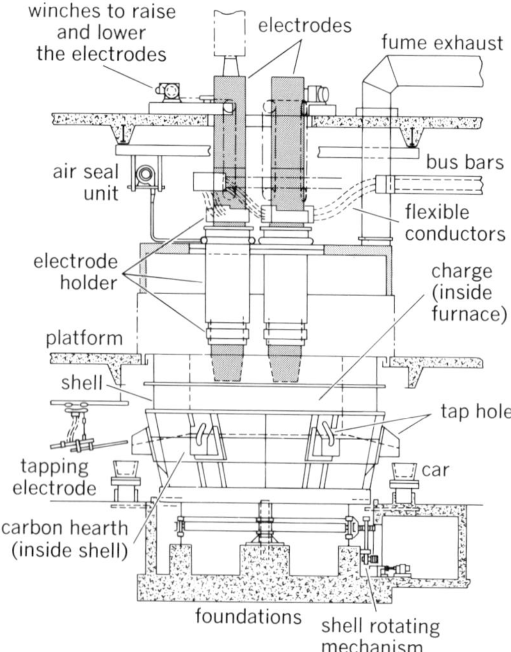

Submerged-arc furnace used in ferroalloy manufacture. *(Lectromelt Furnace Division, McGraw-Edison Co.)*

Analysis of typical ferroalloys, % weight

Type of ferroalloy*	Mn	Si	C	Cr	Mo	Al	Ti	V
Ferromanganese								
Standard	78–82	1.25	7.5					
Medium carbon	80–85	1.25–2.5	1–3†					
Low carbon	80–85	1.25–7.0	0.75					
Ferrosilicon								
50% regular		47–52	0.15					
75% regular		73–78	0.15					
Ferrochromium								
High carbon		1–2	4.5–6.0†	67–70				
Low carbon		0.3–1.0	0.03–2.0†	68–71				
SM low carbon	4–6	4–6	1.25	62–65				
Ferromolybdenum								
High carbon		1.5	2.5		55–70			
Ferrovanadium								
High carbon		13.0	3.5			1.5		30–40
Ferrotitanium								
Low carbon		3–5	0.1			6–10	38–43	

*In all cases the balance is Fe, with the exception of minor impurities. The latter are usually specified, such as 0.10% max P.

†In several specified grades within this range.

a mixture of the required amounts of SiO_2 (quartzite rock), C (coke), and Fe (as turnings), and the ferrosilicon is periodically tapped from the bottom. The overall operation is essentially continuous, like that of a blast furnace, but the heat is supplied by the resistance of the charge and some arcing, and the coke serves primarily as a reducing agent. After solidification in flat molds, the ferroalloy is crushed to specified sizes for delivery. The overall reaction for the process can be written as reaction (1). The Si reduced in this way dissolves

$$SiO_2 + C \rightarrow Si + 2CO \qquad (1)$$

in the molten iron which is also present and simplifies the operation of the process.

Because SiO_2 is present as gangue in many ores, and because silicate slags have sufficiently low melting points to be controlled conveniently, many other ferroalloys contain appreciable amounts of Si when they are manufactured by this same process. The inherent low solubility of C in Si alloys makes this combination desirable when Si is not objectionable. Low-silicon grades can be made, however, by further addition of metal oxide as illustrated by the general reaction (2). Thus

$$2MO + Si \rightarrow SiO_2 + 2M \qquad (2)$$

ferrosilicon can be used as an intermediate to produce other ferroalloys, and this scheme is common.

Likewise, aluminum can be used as the reducing agent by a process called the thermit reaction, as indicated in (3), in which all or part of the

$$2Al + Fe_2O_3 \rightarrow 2Fe + Al_2O_3 + \text{heat} \qquad (3)$$

heat required results from the reduction reaction. The oxide of various desirable alloys may be substituted for Fe_2O_3, and Ca or Mg may be used for Al, a variety of such combinations being possible. The product is low in both Si and C.

Thus, depending upon the reducibility of the ore, and the amounts of C, Si, and Fe that can be tolerated in the product, as well as economic considerations, ferroalloys are produced in the blast furnace, or in the submerged-arc or similar variations of the electric furnace, or by aluminothermic or silicothermic reactions. These ferroalloys are then supplied to the steel industry for use as deoxidizers or alloy-addition agents. *See* ARC HEATING; STEEL. [GERHARD DERGE]

Bibliography: AIME, *Proceedings of the Conference on Electric Furnace Steel*, 1958; A. Hanson and J. Gordon Parr, *The Engineer's Guide to Steel*, 1965; Physical Chemistry of Steelmaking Committee, *Basic Open Hearth Steelmaking*, AIME, 1951; B. D. Saklatwalla, Thermal reactions in ferro-alloy metallurgy, the basis of alloy steel development, *Trans. Electrochem. Soc.*, 84:13–32, 1943; E. E. Thum, Ferrosilicon manufacture at Marietta, *Metal Progr.*, 70(4):65–72, 1956.

Finite element method

A numerical analysis technique for obtaining approximate solutions to many types of engineering problems. The need for numerical methods arises from the fact that for most practical engineering problems analytical solutions do not exist. While the governing equations and boundary conditions can usually be written for these problems, difficulties introduced by either irregular geometry or other discontinuities render the problems intractable analytically. To obtain a solution, the engineer must make simplifying assumptions, reducing the problem to one that can be solved, or a numerical procedure must be used. In an analytic solution, the unknown quantity is given by a mathematical function valid at an infinite number of locations in the body, while numerical methods provide approximate values of the unknown quantity only at discrete points in the body. In the finite element method, the region of interest is divided up into numerous connected subregions or elements within which approximate functions are used to represent the unknown quantity.

The physical concept on which the finite element method is based has its origins in the theory of structures. The idea of building up a structure by fitting together a number of structural elements is quite a natural one (Fig. 1) and was used in the early truss and framework analysis approaches

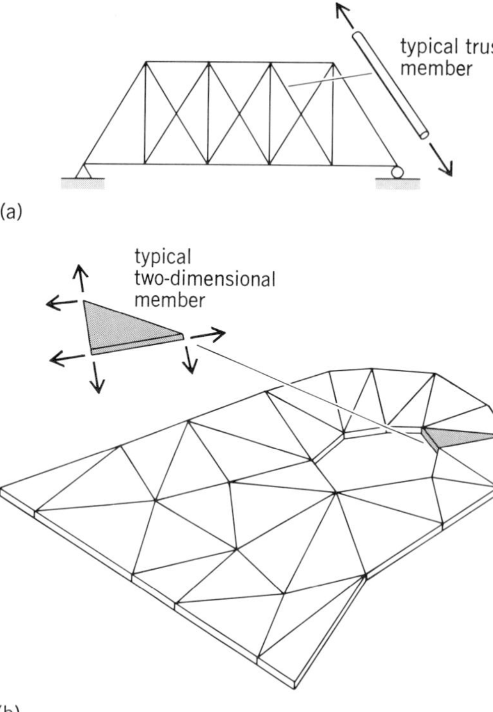

Fig. 1. Structures modeled by fitting together structural elements: (a) truss structure; (b) two-dimensional planar structure.

employed in the design of bridges and buildings in the early 1900s. By knowing the characteristics of individual structural elements and combining them, the governing equations for the entire structure are obtained. The limitation on the number of simultaneous algebraic equations that could be solved posed a severe restriction to the analyses of that period.

The finite element method evolved into a practical analysis tool as a result of development work carried out in the structural analysis of aircraft. This was due to demanding analysis requirements coupled with several technological advances which took place in the early 1950s. The first of these was the formulation of the structural problem in terms of matrix algebra, thus forming the governing linear algebraic equations in a routine manner. Next, the introduction of the digital computer made possible the solution of the large-order systems of equations. Finally, the theoretical concepts of finite element analysis also crystallized during the same period.

Development work since then caused the finite element method to emerge as one of the most powerful approaches for the approximate solution of a wide range of problems in mathematical physics. As such, its application has extended beyond structural analysis to soil mechanics, heat transfer, fluid flow, magnetic field calculations, and other areas.

Due to the broadened scope of the finite element method and the recognition that it is one of the most powerful methods for the approximate solution of boundary value problems, it has attracted the attention of mathematicians who have endeavored to establish the method on a solid mathematical footing. It is, in essence, a variational method of approximation, employing the Rayleigh-Ritz-Galerkin approach. A given region is represented as a collection of a number of geometrically simple regions (finite elements) connected together at nodal points. Simple mathematical functions, generally polynomials, are chosen for each element, and the solution over the entire region is obtained by fitting together the individual elements or local functions. An important aspect of this process is that it embodies a systematic procedure for constructing the solution, which is independent of the geometric complexity and boundary conditions of the problem.

Element formulation and types. The formulation of the individual finite elements is central to the method, since the total solution to a problem is constructed of linear combinations of the individual element behavior. In structural analysis, the unknown field variables of displacements or stresses are defined in terms of values at the node points, which are the unknowns of the problems.

The accuracy of the solution depends not only on the number and size of the finite elements used but also on the interpolation functions selected for the elements. These functions must be carefully chosen and must satisfy certain criteria to ensure convergence to the "correct" solution.

Using these functions, a number of approaches can be used to formulate the element properties (stiffness and stress matrices). The first of these is the "direct approach," because physical reasoning is used to establish the element equations in matrix form, which are then combined to form the governing equations for the entire problem. The other approaches are more advanced and versatile in that application of the method is extended to other fields. Here the calculus of variations is used to derive various energy principles which are then employed in deriving the element formulations.

In solid mechanics problems, the element functions are usually chosen to represent displacements within the element (commonly called the displacement method). They could also be chosen to represent stresses (force or equilibrium method), or a combination of displacements and stresses (the hybrid method). For most problems the displacement method is the simplest to apply; and consequently, it is the most widely used approach.

Some of the more commonly used finite elements are depicted in Fig. 2. The first (Fig. 2*a*) is the simplest type of element and can resist only axial loads. They can be used by themselves to model truss-type structures. The beam element (Fig. 2*b*) is similar to the bar element, but can also take moments and shears necessary for representing frame structures. The next two elements (Fig. 2*c*) are grouped together and are basic ones in finite element analysis because of their usefulness in a wide range of problems and their position in the development of the method. They are thin-membrane elements, triangular and quadrilateral in shape, that are loaded only in their own plane.

Extension of the concept to three dimensions results in a family of solid elements (Fig. 2*d*) of which the tetrahedron and hexahedron are the most common shapes. These elements often present the only practical approach to three-dimensional stress-analysis problems. Another type of solid element of importance is the solid of revolution (Fig. 2*e*).

The remaining elements depicted in Fig. 2 can be classified as bending elements. They can all be loaded with bending moments and normal loads in addition to in-plane loads. The first two (Fig. 2*f*) are used to study the behavior of flat plates and shells or thin-walled members. A more accurate representation of a shell structure can be attained by using curved-shell elements (Fig. 2*g*). Finally, the shell-of-revolution element (Fig. 2*h*) is applicable to a restricted class of thin-shell problems.

Application and implementation procedure. The method in which the elements are employed, as well as the scale and complexity of the problems which can be addressed by the finite element technique, are illustrated in representative applications. Many of the early developments and applications of the method were due to work in aircraft structural analysis. The aerospace industry continues to lead in these areas and in system implementation of the method.

Figure 3 shows a finite element model of a business jet. It is typical of the type of modeling used for aircraft structures. Both the fuselage and wing construction consist of thin metal sheet wrapped around a framework. Bending effects are neglected in the thin skin, and it is modeled with membrane elements (Fig. 2*c*). For the framework structure, bar and beam elements are used (Fig. 2*a* and *b*), as well as membrane elements. The model shown is of medium complexity; larger aircraft, such as the Boeing 747, require more complicated models. The finite element analysis of the 747 wing body region required the solution of over 7000 equations.

While aircraft companies were the first to use the finite element method extensively, today all major shipbuilders, auto manufacturers, and designers of power plants, bridges, and other structures employ the method. In addition to applications in static analysis, it is used to handle thermal stress problems, to compute vibration modes, to perform dynamic analysis, and to predict buckling loads. Moreover, as previously mentioned, the finite element method has found applications in many other fields besides structural analysis.

The finite element procedure is most often implemented with large general-purpose programs capable of handling many types of problems. These programs are very costly to develop and are often key components of computer-based design systems incorporating interactive graphics and data management systems. Every major aerospace company has its own proprietary code, but perhaps the most widely used at this time is the NASTRAN program developed by NASA.

The tasks necessary to perform a finite element analysis are:

1. Model generation. The first step requires a great deal of modeling expertise. The level of model complexity required to meet the intent of the analysis must be established, as well as the types of elements to be employed. Input data describing the coordinates of the grid points and the element topology and properties, as well as boundary conditions and load data, must next be generated. Today these tasks are done automatically with mesh generating programs, digitizing devices, and interactive graphics scopes.

2. Data verification. The large amounts of data must be checked, since the consequences of error are costly. Computer graphics is a must here, with the use of both display devices and plotters.

3. Analysis. In this task the equations describing the finite element model are generated and solved. For static structural analysis problems, a set of

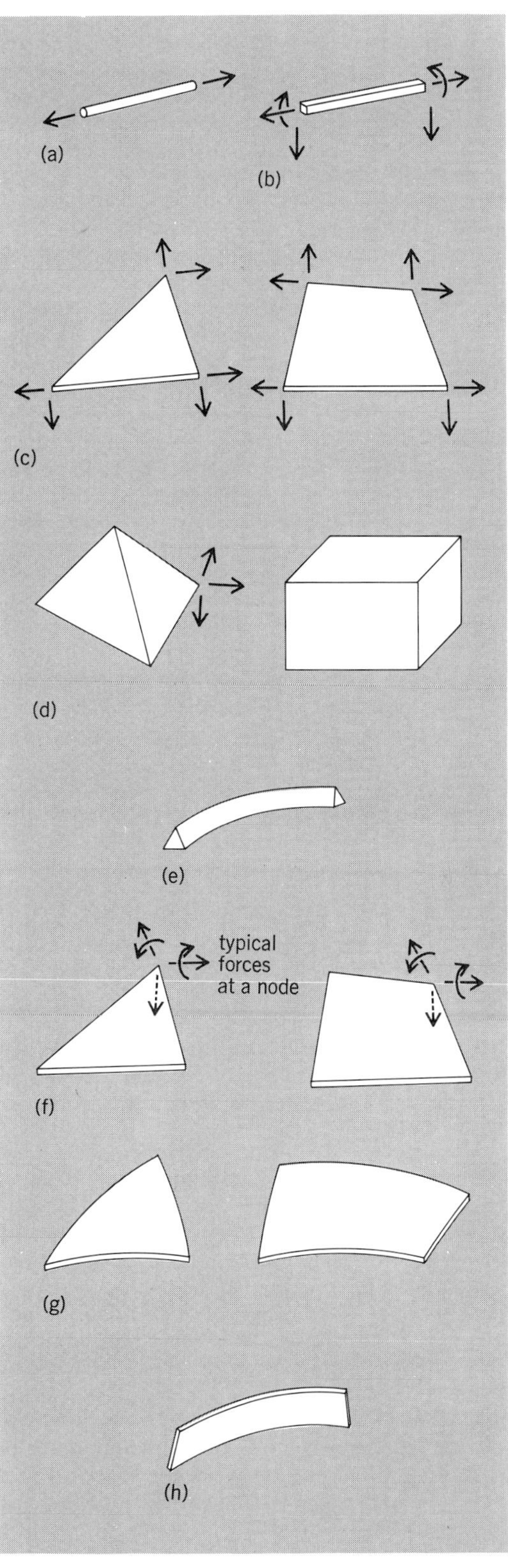

Fig. 2. Element types: (*a*) bar; (*b*) beam; (*c*) membrane elements; (*d*) solid elements; (*e*) axisymmetric solid; (*f*) plate-bending elements; (*g*) curved-shell elements; (*h*) axisymmetric thin-shell element.

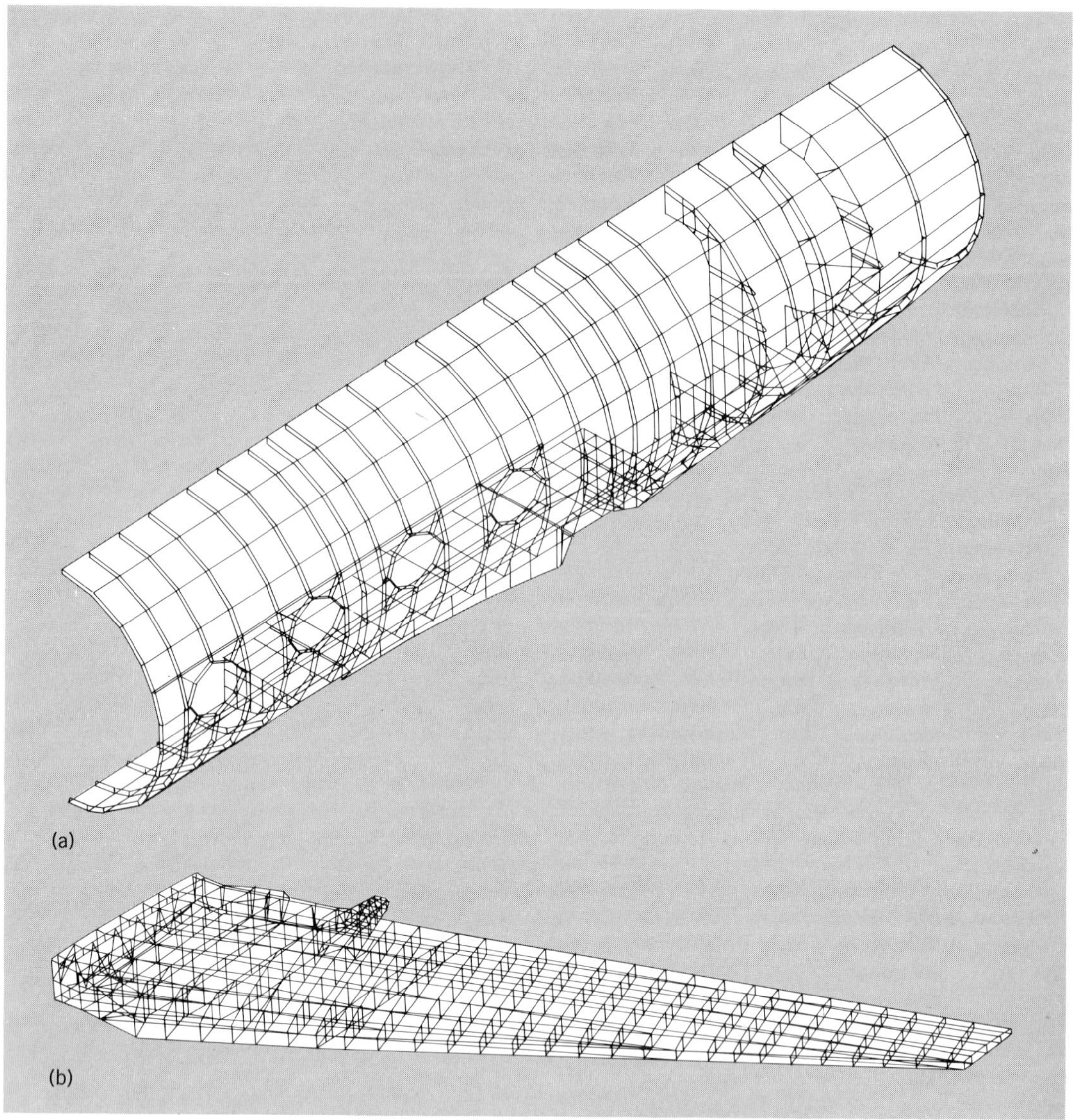

Fig. 3. Finite element model of (*a*) fuselage section and (*b*) wing of a business jet.

linear algebraic equations are solved, yielding the deflections at the grid points. Substitution into an additional set of equations gives the stresses or forces in the individual elements.

4. Postprocessing. The output of the analysis must be manipulated and edited in order that it can be presented in efficient-to-use formats. This includes conversion of the data to report formats, and plots of internal loads and deflections.

The finite element method has found a tremendous number of applications. This is due to its generality in analyzing almost any type of structure without restrictions on material properties or boundary conditions. In addition, it is readily adaptable to the design process, wherein crude models are employed in the conceptual stages, and more detailed models as the design evolves. Moreover, it is used to reduce the cost of test programs by using analytical models to minimize the number of tests required.

[T. BALDERES]

Bibliography: C. S. Desai and J. F. Abel, *Introduction to the Finite Element Method*, 1972; R. H. Gallagher, *Finite Element Analysis Fundamentals*, 1975; J. T. Oden and J. N. Reddy, *An Introduction to the Mathematical Theory of Finite Elements*, 1976; O. C. Zienkiewicz, *The Finite Element Method*, 1977.

Fire-tube boiler

A steam boiler in which hot gaseous products of combustion pass through tubes surrounded by boiler water. The water and steam in fire-tube boilers are contained within a large-diameter drum or shell, and such units often are referred to as shell-type boilers. Heat from the products of combustion is transferred to the boiler water by tubes or flues of relatively small diameter (approximately 3–4 in. or 7.5–10 cm) through which the hot gases flow. The tubes are connected to tube sheets at each end of the cylindrical shell and serve as structural reinforcements to support the flat tube sheets against the force of the internal water and steam pressure. Braces or tension rods also are used in those areas of the tube sheets not penetrated by the tubes.

Fire-tube boilers may be designed for vertical,

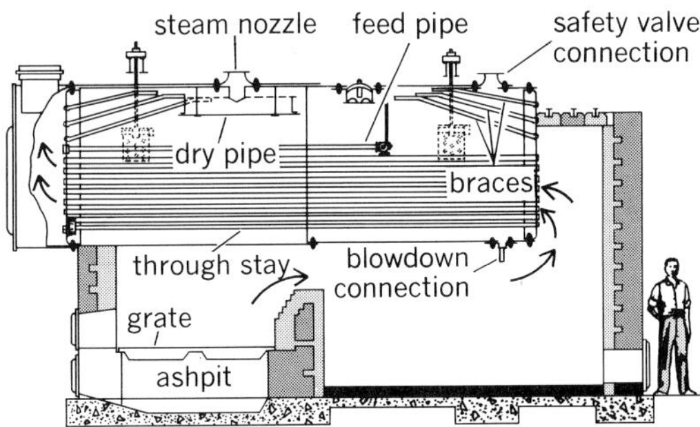

Fig. 1. Horizontal-return-tube boiler.

inclined, or horizontal positions. One of the most generally used types is the horizontal-return-tube (HRT) boiler (Fig. 1). In the HRT boiler, part of the heat from the combustion gases is transferred directly to the lower portion of the shell. The gases then make a return pass through the horizontal tubes or flues before being passed into the stack.

In the Scotch marine boiler, one or more large flues (approximately 18–24 in. or 45–60 cm in diameter) are used for furnaces or combustion chambers within the shell, and the hot gases are returned through groups of small-diameter tubes or flues. The flues that form the combustion chamber are corrugated to prevent collapse when subjected to the water and steam pressure. These boilers may be oil-fired, or solid fuel can be burned on grates set within the furnace chambers. Scotch marine boilers have, with few exceptions, been superseded by water-tube marine boilers.

Gas-tube boilers are sometimes used for the absorption of waste heat from process gases or the exhaust from internal combustion engines, particularly in those cases in which their installation provides a simple and economical means for the recovery of low-grade heat. The boiler shell may be augmented by an external steam-separating drum and downcomer and riser connections to provide for proper circulation (Fig. 2).

Shell-type boilers are restricted in size to 14 ft in diameter and an operating pressure of 300 psi by the stresses in the large-diameter shells and the necessity to design the flat-tube sheets for practicable thicknesses. They are best suited for low-pressure heating service and for portable use because the integrated structure can be transported easily and the generous reserve in water capacity requires minimum attendance. However, there is the risk of catastrophic damage if overheating and rupture of the shell or tube sheet occur as a result of low water levels or the formation of an insulating layer of internal scale produced by water impurities. *See* STEAM-GENERATING UNIT.

[GEORGE W. KESSLER]

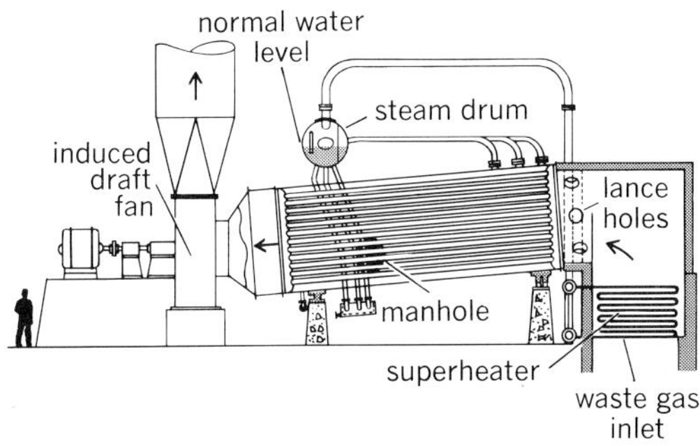

Fig. 2. Fire-tube waste-heat boiler.

Flash welding

A form of resistance welding that is used for mass production. The welding circuit consists of a low-voltage, high-current energy source (usually a welding transformer) and two clamping electrodes, one stationary and one movable.

The two pieces of metal to be welded are clamped tightly in the electrodes, and one is moved toward the other until they meet, making light contact (Fig. 1*a*). Energizing the transformer causes a high-density current to flow through small areas that are in contact with each other (Fig. 1*b*). Flashing starts, and the movable workpiece must accelerate at the proper rate to maintain an increasing flashing action. After a proper heat gradient has been established on the two edges to be welded, an upset force is suddenly applied to complete the weld. This upset force extrudes slag, oxides, molten metal, and some of the softer plastic metal, making a weld in the colder zone of the heated metal (Fig. 2). Flash and upset material are fre-

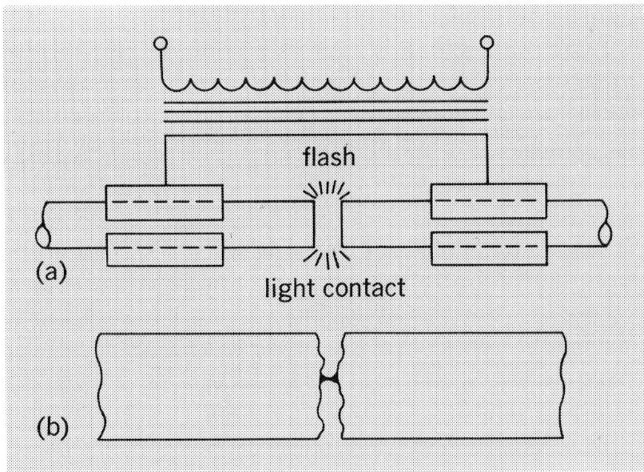

Fig. 1. Elements of flash-welding operation. (*a*) Two pieces of horizontally clamped metal are brought together to make a light contact. At that time the welding circuit is closed to the transformer. (*b*) Current starts flowing when small areas of the workpiece make initial contact. (*From A. L. Phillips, ed., Welding Handbook, 4th ed., American Welding Society, 1957*)

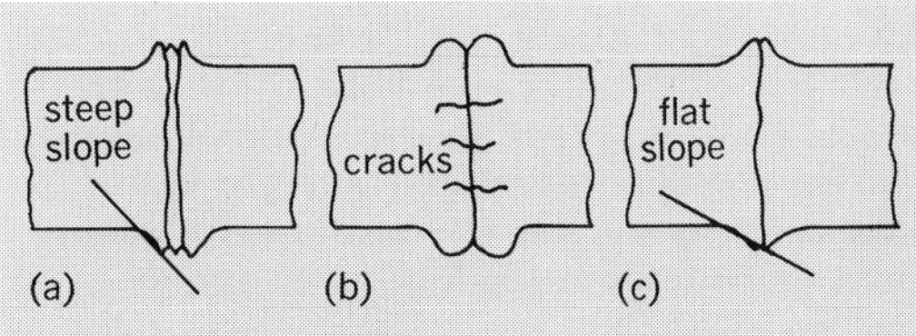

Fig. 2. Flash-weld quality indicated on the surface. (*a*) Satisfactory heat and upset. (*b*) Cracks caused by insufficient heat. (*c*) Insufficient heat or upset force, or both. (*From A. L. Phillips, ed., Welding Handbook, 4th ed., American Welding Society, 1957*)

quently removed, although when the design permits, a portion of the upset material may be retained as a reinforcement. *See* ARC WELDING; RESISTANCE WELDING. [EUGENE J. LIMPEL]

Bibliography: J. W. Giachino, W. R. Weeks, and E. J. Brune, *Welding Skills and Practices*, 2d ed., 1973.

Floor construction

The selection of the type of construction for the floors of a building should be based on the building's architectural and structural requirements and on cost.

A medium-priced home might contain a wooden floor supported by wooden joists. A floor system frequently used in small commercial buildings, apartment buildings, and higher-priced homes consists of a solid concrete slab supported on steel or concrete beams or open-web steel joists. Open-web joists are available for clear spans up to 120 ft (36 m). A thin, lightly reinforced concrete slab cast in one piece with its supporting reinforced concrete joists is particularly economical for short spans. The joists are formed by using metal pans, or clay-tile, gypsum-tile, or concrete-block fillers. A grid system in which the reinforced concrete joists run in two directions may be economical for floor panels that are approximately square. For heavy loads on moderate spans a two-way flat slab supported on columns arranged in square or rectangular patterns is appropriate. Flaring heads on the columns and plinthlike dropped panels are an integral part of the system. Two-way flat plates, very similar to flat slabs except that the drop panels and column capitals are omitted, provide much more acceptable framing in apartments and hotels.

Light-gage steel cellular floor decking is often used for steel-framed buildings. One type of unit consists of a flat sheet of steel spot-welded to a second sheet bent into a series of troughs. Another consists of two sheets of steel bent into a series of troughs and spot-welded together so that the troughs of one sheet lie against the shoulders of the other. [CHARLES N. GAYLORD]

Fluid coupling

A device for transmitting rotation between shafts by means of the acceleration and deceleration of a hydraulic fluid. Structurally, a fluid coupling consists of an impeller on the input or driving shaft and a runner on the output or driven shaft. The two contain the fluid (Fig. 1). Impeller and runner are bladed rotors, the impeller acting as a pump and the runner reacting as a turbine. Basically, the impeller accelerates the fluid from near its axis, at which the tangential component of absolute velocity is low, to near its periphery, at which the tangential component of absolute velocity is high (Fig. 2). This increase in velocity represents an increase in kinetic energy. The fluid mass emerges at high velocity from the impeller, impinges on the runner blades, gives up its energy, and leaves the runner at low velocity (Fig. 3).

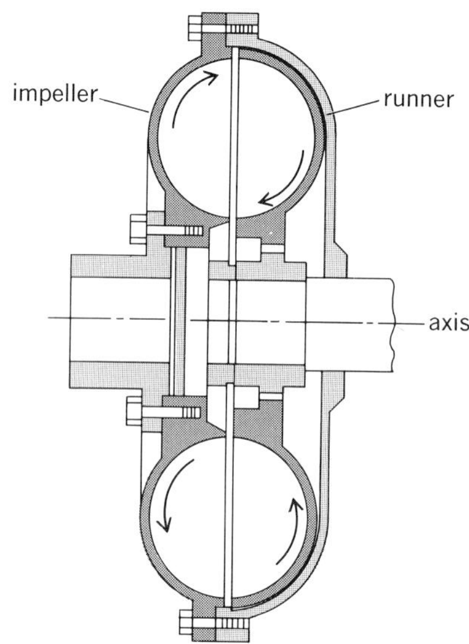

Fig. 1. Basic fluid coupling.

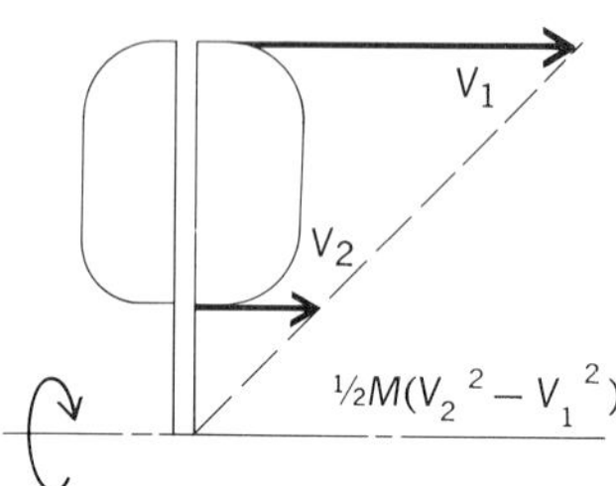

Fig. 2. Hydrokinematic principle.

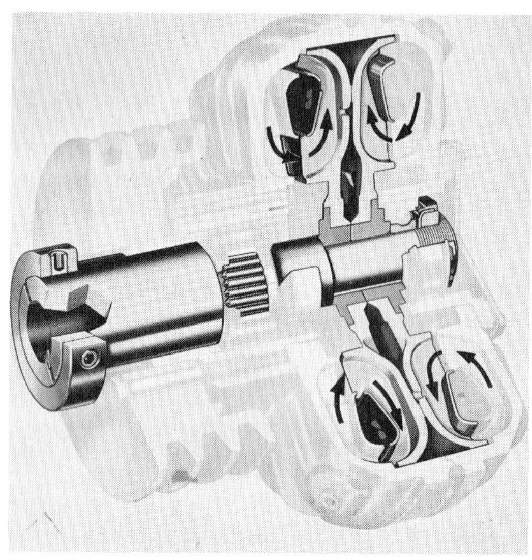

Fig. 3. Sheave with fluid coupling. Dark portion of the figure is driven directly by motor; light portion drives the load. (*Twin Disc Inc.*).

Coupling characteristics. A fluid coupling absorbs power from an engine by means of the centrifugal pump or impeller, which increases the moment of momentum of the mass of fluid. As a consequence, the input power required to drive the coupling increases as the cube of the speed and as the fifth power of the outside diameter of the impeller. Similarly, the capacity of the coupling to absorb torque varies as the square of its input speed.

Because the efficiency of a coupling is a function of slip, couplings normally operate at 95–98% efficiency. The fluid follows a continuous toroidal helix, the pitch of which depends on the slip, or relative difference in rotational velocity, between

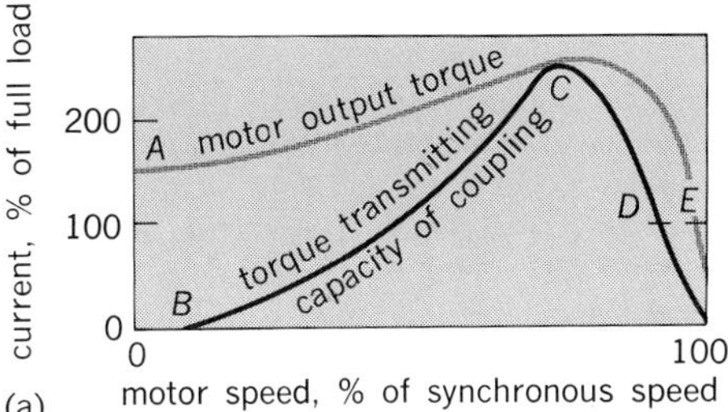

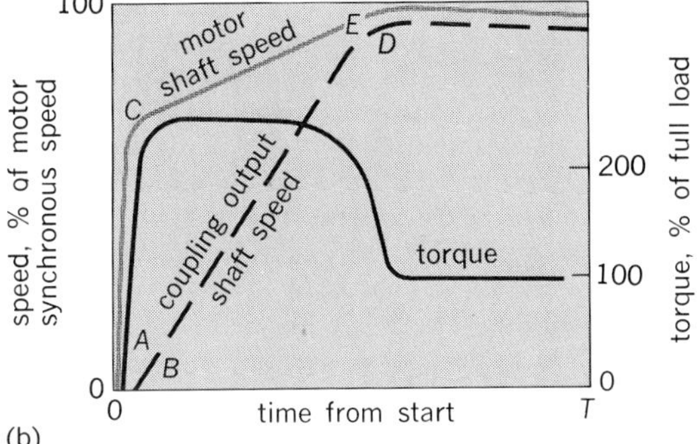

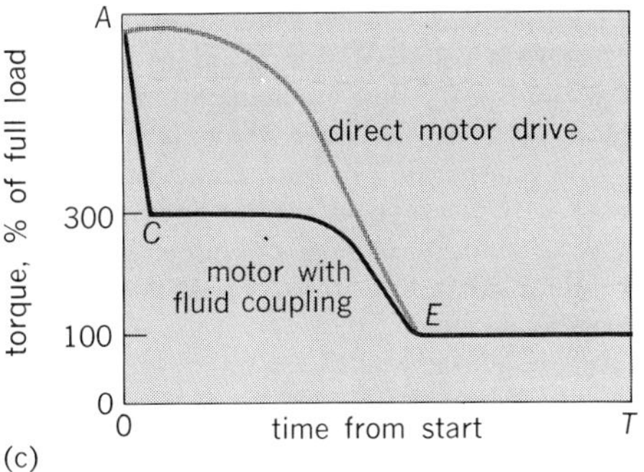

Fig. 4. Starting characteristics of electric motor with fluid coupling. (*a*) Torque speed during acceleration. (*b*) Acceleration. (c) Current.

impeller and runner. Some slip is necessary for energy transfer. As the slip increases at a given input speed, the output shaft turning slower than the input shaft, the quantity of fluid pumped between impeller and runner increases, with consequent increase in torque absorption and transmission. However, because the fluid flow cannot change abruptly, a fluid coupling absorbs rather than transmits shock loads from output to input. Torsional vibrations are similarly absorbed.

A fluid coupling is designed for a normal running load at 2–5% slip. Conservation of momentum requires that, although shaft speeds may differ, input and output torques must be equal. Thus, for example, in starting a machine from standstill, with the output from the fluid coupling locked, the engine or motor is completely unloaded initially. As the engine starts up toward running speed, the drag torque characteristic of the fluid coupling lies below the developed torque characteristic of the engine. The engine can reach running speed faster with consequent saving in fuel or electrical consumption. The slip of the fluid coupling enables the engine to operate near its maximum-torque speed while accelerating the load, although this characteristic differs from that of a variable-torque transmission. *See* TORQUE CONVERTER.

Typical application. The effect of a fluid coupling can be described in terms of a specific application. If an electric motor is starting with a high inertia load that requires nearly full-motor torque, the motor accelerates slowly when connected directly to its load. However, with fluid coupling, the motor accelerates rapidly along *A* to *C* in Fig. 4. Meanwhile, the torque-transmitting capacity of the coupling builds up along *B* to *C* until it equals the developed motor torque. The motor is then able to deliver well over full load torque to accelerate the load; the torque demand falls from *C* to *D* as the load comes up to speed *D*. Meanwhile, the motor further accelerates *C* to *E* until steady state is reached with coupling slip represented by differential speed *D* to *E*. Use of the fluid coupling improves performance by faster load acceleration and results in appreciably less power loss due to motor heating. *See* AUTOMOTIVE TRANSMISSION.

[HENRY J. WIRRY]

Bibliography: W. Bober and R. A. Kenyon, *Fluid Mechanics*, 1980; W. H. Crouse, *Automotive Transmissions and Power Trains*, 5th ed., 1976.

Fluidized-bed combustion

A method of burning fuel in which the fuel is continually fed into a bed of reactive or inert material while a flow of air passes up through the bed, causing it to act like a turbulent fluid. Fluidized beds

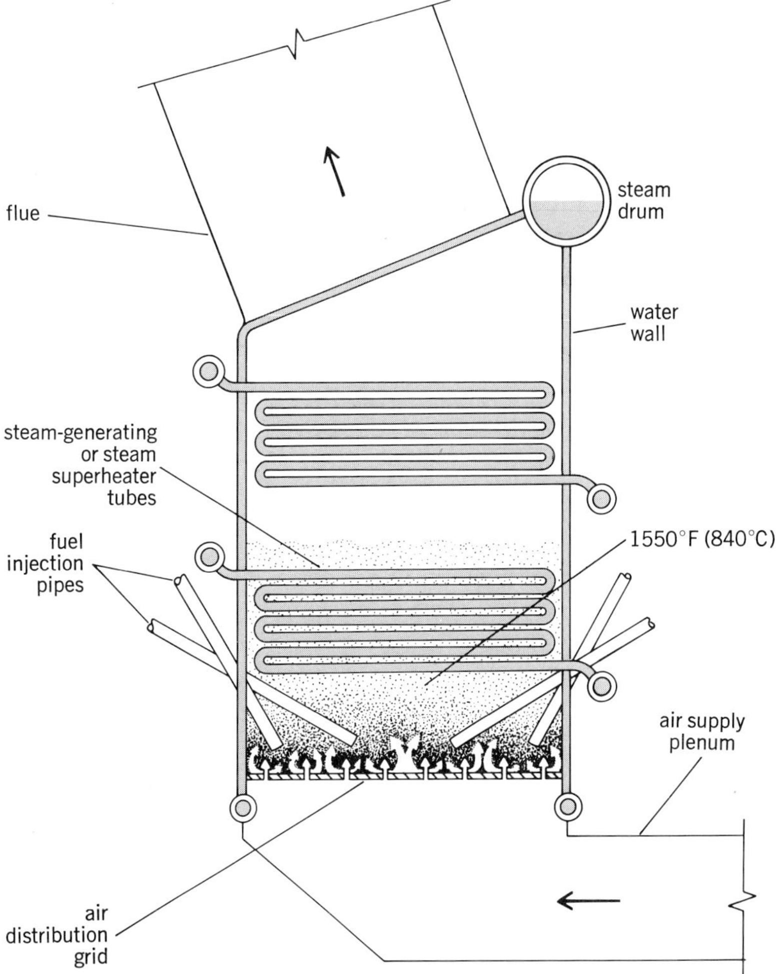

Fig. 1. Fluidized-bed steam generator.

have long been used for the combustion of low-quality, difficult fuels and have become a rapidly developing technology for the clean burning of coal.

Fluidization process. A fluidized-bed combustor is a furnace chamber whose floor is slotted, perforated, or fitted with nozzles. Air is forced through the floor and upward through the chamber. The chamber is partially filled with particles of either reactive or inert material, which will fluidize at an appropriate air flow rate. When fluidization takes place, the bed of material expands (bulk density decreases) and exhibits the properties of a liquid. As air velocity increases, the particles mix more violently, and the surface of the bed takes on the appearance of a boiling liquid. If air velocity were increased further, the bed material would be blown away.

Once the bed is fluidized, its temperature can be increased with ignitors until a combustible material can be injected to burn within the bed. Proper selection of air velocity, operating temperature, and bed material will cause the bed to act as a chemical reactor. In the case of coal combustion the bed is generally limestone, which reacts with and absorbs the sulfur in the coal, reducing sulfur dioxide emissions.

Fluidized-bed combustion can proceed at low temperatures (1400–1500°F, or 760–840°C), so that nitrogen oxide formation is inhibited. In the limestone/sulfur-absorbing designs the waste product is a relatively innocuous solid, although the large volume produced does have environmental implications. Both Federal and privately funded programs are exploring techniques for utilizing the waste product, for example, determining its potential value as an agricultural supplement or bulk aggregate. Low combustion temperatures minimize furnace corrosion problems. At the same time, intimate contact between the hot, turbulent bed particles and the heat-transfer surfaces results in high heat-transfer coefficients. The result is that less surface area is needed, and overall heat-transfer tube requirements and costs are lower. Intensive development of fuel injection systems has been undertaken, particularly for pressurized operation, where the movement of solids through the pressure barrier poses some special problems. These efforts hold the promise of achieving early solutions.

Although frequently identified as a sulfur-reduction process, fluidized-bed combustion has application to low-sulfur fuels as well. In firing a low-sulfur lignite fuel, for example, the bed material would simply be coal ash.

Applications. There are three broad areas of application: incineration, gasification, and steam generation.

Coal gasification. The original application of fluidized combustion was to the gasification of coal. A gasifier burns coal in a fluidized bed with less air than that required for complete combustion. This results in a high concentration of combustible gases in the exhaust. This principle is being applied in a number of pilot plants which are for the most part directed at producing low-sulfur-content gases of medium heating value (150–300 Btu per standard cubic foot, or 5590–11,170 kJ/m³).

Incineration. The use of this technology for effi-

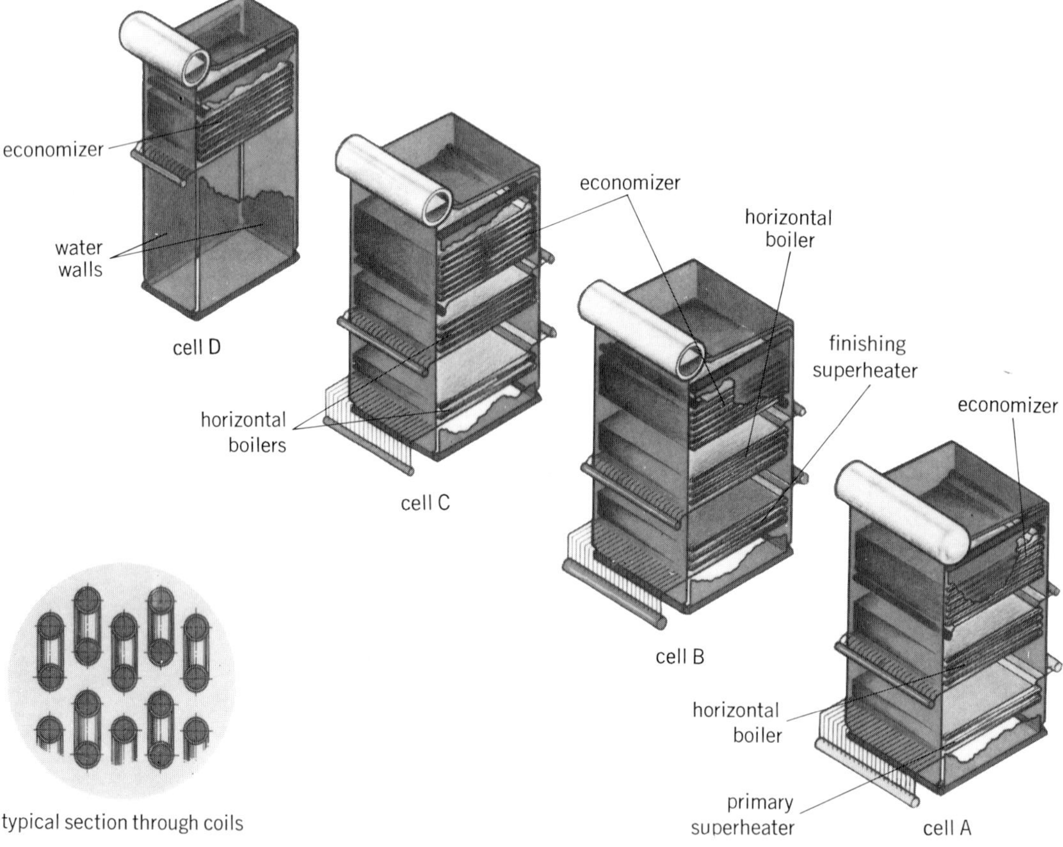

Fig. 2. Multicell fluidized-bed boiler.

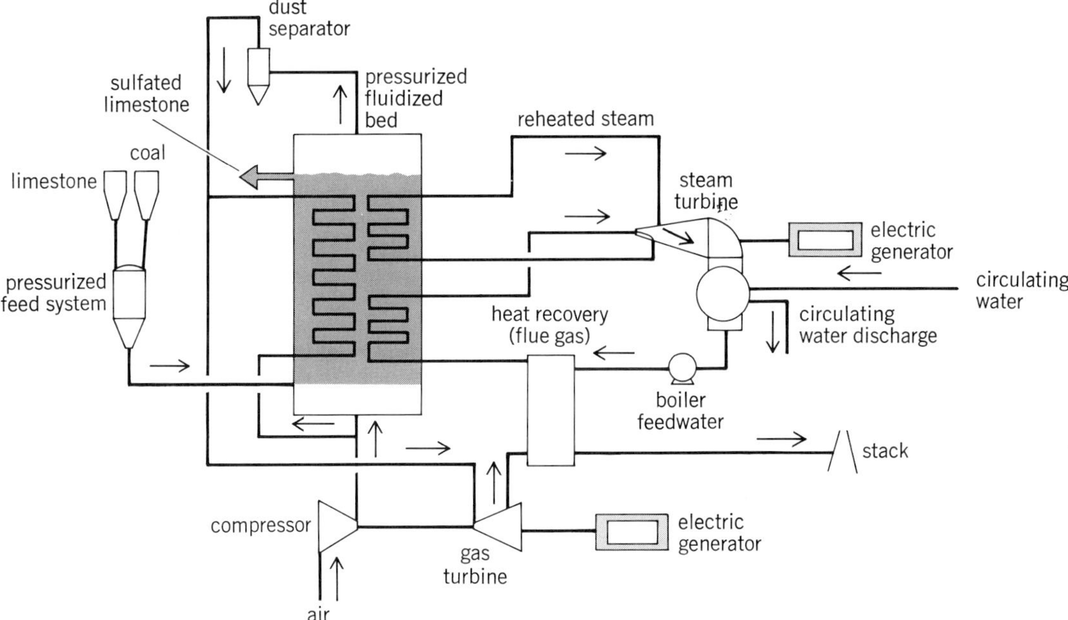

Fig. 3. Flow diagram of 600-MW combined-cycle system.

cient incineration of waste products, such as food processing wastes, kraft process liquors, coke breeze, and lumber wastes, has increased rapidly. The process has the advantages of controllability, lower capital cost, more compact design, and reduced emissions.

Steam generation. Immersing heat-transfer tubes within the hot fluidized bed converts the furnace from a simple chemical reactor to a steam boiler (Fig. 1). The use of a number of relatively small cells or beds permits the construction of large-capacity boilers (Fig. 2). Cells may be arranged horizontally, vertically, or both ways.

The largest fluidized-bed boiler in operation is at Rivesville, WV. This unit generates 300,000 lb/hr (136,200 kg/hr) of high-pressure superheated steam for power generation. The boiler is a horizontally arranged multicell unit. Since initial startup of this boiler in March 1977, tests performed while burning high-sulfur fuels resulted in both NO_x and SO_2 emissions below Environmental Protection Agency (EPA) limits.

Designs of steam generators for larger plants, supplying 200, 600, and 800 MWe (megawatts of electric power), have also been undertaken; the first such unit, planned by the Tennessee Valley Authority, should be operational in 1985. The steam generator furnace pressure designs are generally for operation at or near atmospheric pressure.

Application of the technology to small boilers should permit the burning of coal instead of gas and oil in industrial-sized boilers. A 100,000 lb/hr (45,000 kg/hr) high-sulfur-coal–burning boiler was put into operation at Georgetown University in Washington, DC. In Renfrew, Scotland, a conventional 40,000 lb/hr (18,160 kg/hr) stoker-fired boiler was converted to fluidized combustion by the substitution of an air distribution floor for the stoker.

In the so-called pressurized fluidized-bed combustor, operating the combustion chamber at elevated pressures permits the use of the heated exhaust gases to drive gas turbines. Tests have been undertaken on a 13-MWe pilot plant using this combined cycle. Designs for larger plants have also been undertaken (Fig. 3). *See* GAS TURBINE; STEAM-GENERATING UNIT.

Other applications. Fluidized-bed combustion techniques have had a long history of application in the chemical industries, especially in catalytic petroleum cracking, and in extractive metallurgy, for example, in ore roasters and calciners. Development of a fluidized-bed combustor for the spent graphite fuel rods from high-temperature gas-cooled nuclear reactors has been undertaken.

Competitiveness of process. Since nearly any combustible product, including such difficult fuels as oil shales, anthracite wastes, and residual oils, can be effectively burned, the potential of fluidized-bed combustion as a major energy conversion process is great. Economically and environmentally, its competitiveness with flue gas desulfurization, synthetic fuel preparation, and similar alternatives has been demonstrated.

[MICHAEL POPE]

Fluorescent lamp

A lamp which produces light largely by conversion of ultraviolet energy from a low-pressure mercury arc to visible light. Chemicals that absorb radiant energy of a given wavelength and reradiate at longer wavelengths are called phosphors. The phosphors produce most of the light provided by fluorescent lamps manufactured today.

The fluorescent lamp consists of a glass tube containing two electrodes, a coating of powdered phosphor, and small amounts of mercury. The glass tube seals the inner parts of the lamps from the atmosphere. The electrodes provide a source of free electrons to initiate the arc and are connected to the external circuit through the ends of the lamp. The phosphor is a chemical or mixture of chemicals that converts short-wave ultraviolet energy into light. The mercury, when vaporized in

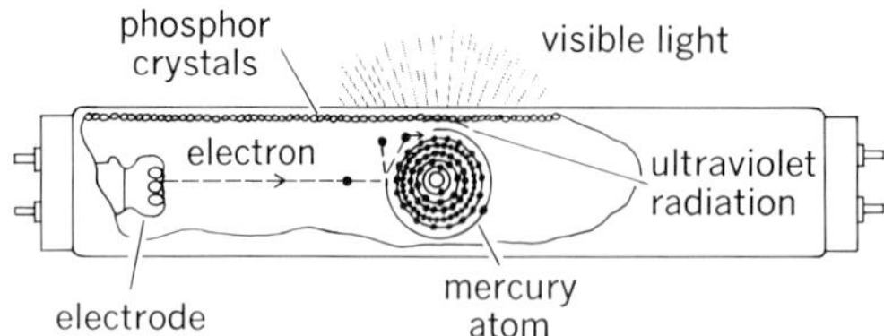

Fig. 1. Parts of a typical fluorescent lamp. (*Westinghouse Lighting Handbook*)

the arc, produces the ultraviolet radiation that causes fluorescence. Argon gas, introduced in small quantities, provides the ions that facilitate starting of the lamp (Fig. 1).

Operation. To start the flow of current that forms the arc in the tube, free electrons must be introduced into the tube. This is accomplished by thermionic emission, field emission, or a combination of these techniques of obtaining electron emission from a cathode.

Current passed through the electrodes of a preheat lamp causes emission of electrons from the electrodes into the tube. When sufficient electrons are released, the resistance of the gap between the electrodes is low enough to permit striking the arc across the gap.

The application of a high potential difference across opposite instant-start cathodes in a lamp draws electrons from the negative electrode and attracts them toward the positive electrode. On alternating-current (ac) circuits, the electrodes are alternately negative and positive in each half-cycle, so both electrodes emit electrons. Instant-start cathodes may be either hot or cold. The hot cathode permits greater lamp current and produces lower overall lighting costs; it consists of a coiled wire coated with a material that yields electrons freely. The cold cathode is well suited to sign tubing and lighting lamps of special lengths or shapes; it consists of a thimble-shaped iron cup.

Electrodes that are specially designed to heat quickly can be used in conjunction with a moderately high voltage to attract electrons into the tube. This technique is used with rapid-start electrodes. Rapid-start electrodes are continuously heated during lamp operation.

Basic electric circuits for operation of the three types of fluorescent lamps include the preheat, instant-start, and rapid-start circuits. Each includes a ballast, which provides starting voltage and limits current. Many circuit variations are used for the operation of one or more lamps from a single ballast, with and without circuit elements

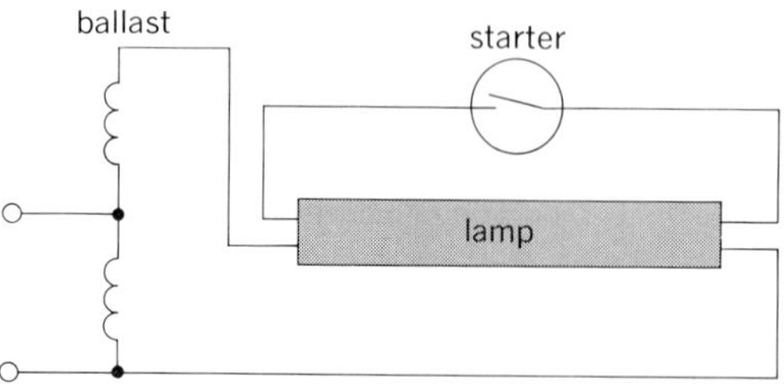

Fig. 2. Preheat lamp circuit.

for the correction of power factor. *See* VAPOR LAMP.

Preheat circuit. In the preheat circuit (Fig. 2), a starting switch, usually an automatic starter, is used to heat the electrodes. The most common starter employs a small argon glow tube with one fixed electrode and one electrode made of a bent bimetallic strip. When the circuit is energized, a low-current glow discharge forms in the starter; the heat from this glow is sufficient to expand the bimetallic electrode until it contacts the fixed electrode, forming a short circuit through the starter. The full output voltage of the ballast then causes current to flow through the lamp electrodes, heating them and causing them to emit electrons. The starter cools because the glow is no longer present, and the switch opens, impressing full ballast voltage between the lamp electrodes. If there are enough electrons in the tube, the arc is formed; if not, the glow-switch process is repeated. This process may require several repetitions, accounting for the delay and flickering present at the starting of preheat lamps. When the arc has formed and full lamp current is flowing, the ballast absorbs about half its initial voltage, and there is not sufficient voltage to cause the glow switch to operate; it then becomes an inactive circuit element.

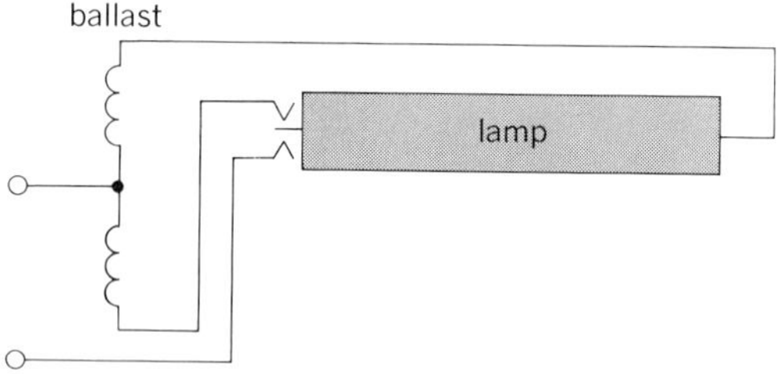

Fig. 3 Instant-start lamp circuit.

Preheat fluorescent lamps are commercially available in lengths from 6 in. to 5 ft (15 cm to 1.5 m), with wattages from 4 to 90. Longer lamps or lamps of higher wattage generally use instant-start or rapid-start circuits, which have more favorable starting characteristics. Each size of fluorescent lamp requires a ballast designed to match the requirements of the lamp in starting voltage and current, the operating current.

Instant-start circuit. In the instant-start circuit of Fig. 3 the ballast voltage is much higher than in the preheat circuit. Immediately upon energizing the circuit this voltage attracts enough electrons into the tube to form the arc. For a given lamp wattage and current, the ballast must absorb a higher voltage than with preheat ballasts; hence the ballast is usually larger and dissipates greater wattage. Lampholders for instant-start circuits are connected so that the ballast is disconnected when the lamp is removed, eliminating the hazard of high voltage during maintenance operations.

Instant-start lamps for general lighting purposes are available in lengths from 24 in. to 8 ft (0.6 to 2.4 m), with wattages from about 15 to 74. Shorter lamps are not usually economical for general lighting service, because they require larger and more costly ballasts than preheat circuits. Higher wattages would require overly large ballasts and cathodes of costly construction.

Rapid-start circuit. The rapid-start ballast contains transformer windings that continuously provide the required voltage and current for electrode heating. When the circuit is energized, these windings quickly heat the electrodes, releasing enough electrons into the tube for the lamp to arc from the voltage of the main windings. This combination of heat and moderately high voltage permits quick lamp starting with smaller ballasts than those for instant-start lamps, eliminates the annoying flicker associated with the starting of preheat lamps, and eliminates the starter and its socket from the lighting system (Fig. 4).

The continuously heated cathode of the rapid-start lamp is better adapted to higher lamp currents and wattages. Therefore, rapid-start lamps are available in wattages up to 215 for 8-ft (2.4-m) lamps. The smallest conventional lighting lamps are 3-ft (0.9-m), 30-W lamps; circular rapid-start lamps are available in 22-, 32-, and 40-W sizes.

Special ballasts and circuits are available that permit the economical dimming and flashing of rapid-start lamps, providing a range of applications and control that was not possible with previous fluorescent lamp types. Rapid-start lamps are commonly used in flashing signs, and in residential and commercial lighting where continuously variable illumination levels are desired.

High-frequency fluorescent lighting. Fluorescent lamps are usually operated on ac circuits with a frequency of 60 hertz (Hz). However, higher frequencies permit higher-efficiency lamp operation with simpler ballasts of lower power dissipation per watt of lamp. Consequently, systems have been developed for the operation of fluorescent lamps at frequencies from 360 to 3000 Hz. These systems employ various types of frequency converters to obtain the higher-frequency power. Motor-generator sets are most common, but static magnetic converters and converters using transistors and controlled rectifiers have also been developed. High-frequency systems can often provide lower overall lighting cost for fluorescent lamp loads of 25 to 50 kW or more.

Lamp colors. Fluorescent lamps provide light at several times the efficiency of incandescent lamps, the exact ratio depending on the fluorescent lamp color. Lamp color is determined by the selection of chemicals used in the phosphors; various chemicals respond to the ultraviolet energy in the arc by producing different colors of light. Several types of essentially white fluorescent lamps are available commercially, as well as a range of tinted and saturated colors.

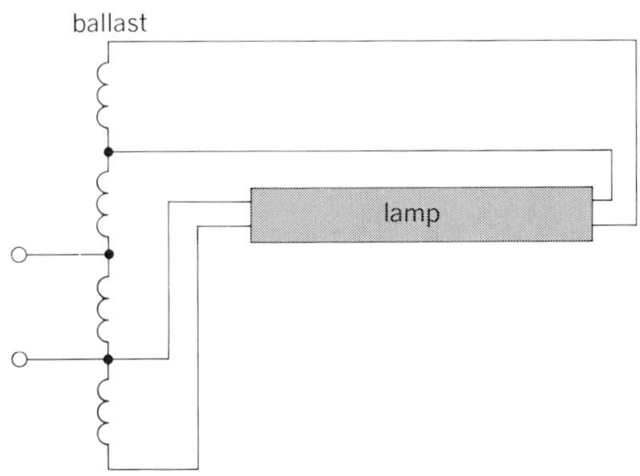

Fig. 4. Rapid-start lamp circuit.

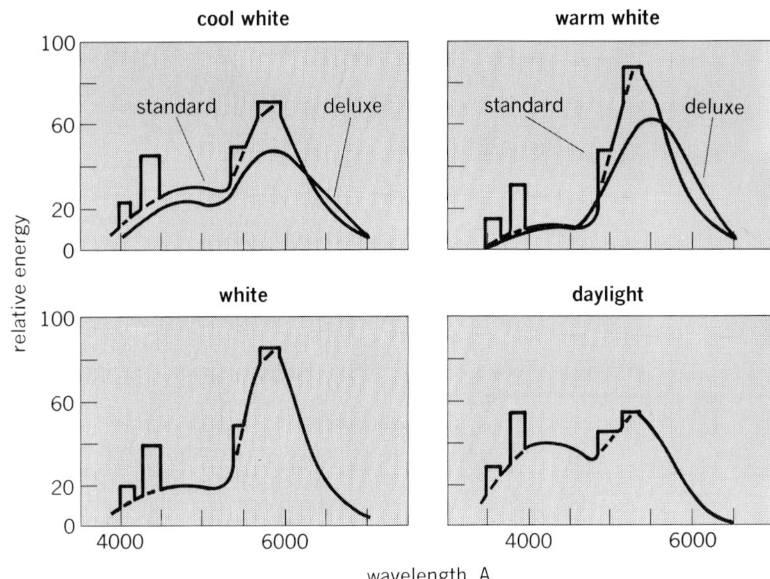

Fig. 5. Graphs indicating spectral energy distribution for fluorescent lamps. The discontinuities in the curves are characteristic lines of the mercury emission spectrum. (*Westinghouse Lighting Handbook*)

One of the earliest fluorescent "whites" was the so-called daylight lamp, which produced a bluish-white light of poor color rendition. The most widely used fluorescent color is called cool white; its lighted appearance is whiter than that of the daylight lamp, it produces more light per watt, and its color rendition is better. In commercial and residential lighting, where the faithful rendition of colors is important, deluxe cool white and deluxe warm white lamps are commonly used, providing vastly superior color rendition at a slight sacrifice in lamp efficiency. Figure 5 shows the relative spectral energy for lamps of equal wattage.

[ALFRED MAKULEC]

Flywheel

A rotating mass used to maintain the speed of a machine between given limits while the machine releases or receives energy at a varying rate. A flywheel is an energy storage device. It stores energy as its speed increases and gives up energy as the speed decreases. The specifications of the machine usually determine the allowable range of speed and the required energy interchange.

Theory. The energy, speed, and size of a flywheel are related by Eq. (1), where W is weight

$$Wk^2 = \frac{182.4gE}{n_1^2 - n_2^2} \qquad (1)$$

in pounds, k is radius of gyration in feet, g is the gravity constant in ft/sec², E is the energy change in foot-pounds, and n_1 and n_2 are the maximum and minimum speeds in revolutions per minute. The term $n_1^2 - n_2^2$ is sometimes replaced by the product $2c_f n^2$, which gives Eq. (2), where c_f, the

$$Wk^2 = \frac{91.2gE}{c_f n^2} \qquad (2)$$

coefficient of fluctuation, has the value given by Eq. (3), and n, the average speed, is given by Eq. (4).

$$c_f = \frac{n_1 - n_2}{n} \tag{3}$$

$$n = \frac{n_1 + n_2}{2} \tag{4}$$

The energy interchange cycle of a flywheel may occupy any amount of angular motion about its center of rotation. Most frequently it is a cycle of one or more revolutions, but in internal combustion engines with multiple cylinders, it may be only a small part of a revolution.

All rotating components in a machine contribute to the flywheel effect and should be considered in a complete analysis of the design. The term Wk^2 in Eqs. (1) and (2) is the flywheel effect and may be made up of the sum of the many rotating weights and their respective radii of gyration squared. Equation (2) can therefore be written in more general terms, as shown by Eqs. (5) and (6).

$$W_1k_1^2 + W_2k_2^2 + \cdots = \Sigma Wk^2 \tag{5}$$

$$\Sigma Wk^2 = \frac{91.2gE}{c_f n^2} \tag{6}$$

Flywheel energy. In the design of a flywheel the first step is to determine the energy interchanged. This can be done best by plotting the torque required by the machine against the angular motion of the flywheel for one cycle of θ_0 radians (Fig. 1).

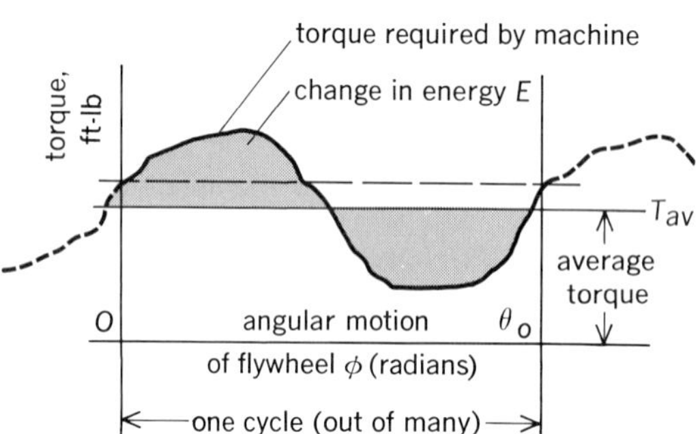

Fig. 1. Determination of energy change for a flywheel.

In this plot, the area under the torque curve represents the total energy requirements of the machine during the cycle. Then, average torque T_{av} defined as the total energy of the cycle E_{tot} divided by the length of the cycle θ_0 in radians is drawn on the same torque scale. The areas between the original torque curve and the average torque curve above and below this average torque (shaded areas on the drawing) should be equal and are the E, or change in energy, in Eq. (1).

The second step is to establish the speed range n_1 and n_2 of the machine or to select the appropriate coefficient of fluctuation from the table.

The third step is to calculate the Wk^2 required by substituting the values of E, c_f, and n previously determined. From this total Wk^2 required is subtracted the values of Wk^2 for the several rotating components in the machine if these are significantly large. The remaining Wk^2 must be supplied by an appropriate flywheel or flywheels.

Coefficients of fluctuation

Machine	c_f
Crushers, hammers, punches, shears	0.100–0.200
Machine tools	0.020–0.030
Electric generators	0.003–0.005
Compressors and pumps	0.040–0.050
Textile machinery	0.010–0.030

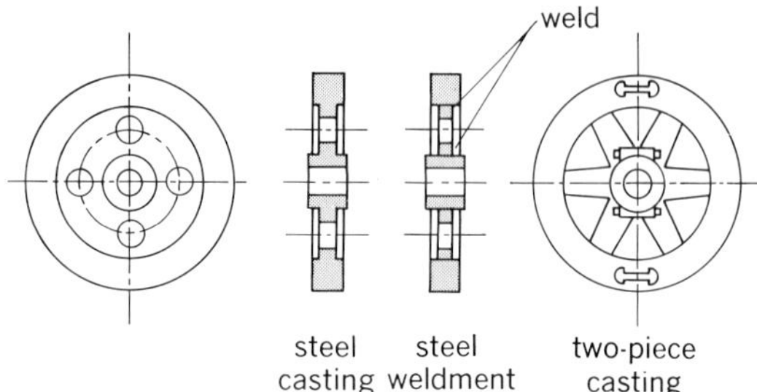

Fig. 2. Typical flywheel structures.

The designer can pick any number of values of W and k whose quotient Wk^2 will be the required amount. The limiting factors are weight and diameter. A suitable decision is based on space availability and weight restrictions. It is sometimes necessary to gear the flywheel to a higher speed than the machine to reduce Wk^2 so that weight and size restrictions can be met.

Design of flywheel. The difficulty of casting stress-free spoked flywheels leads the modern designer to use solid web castings or welded structural steel assemblies. For large, slow-turning flywheels on heavy-duty diesel engines or large mechanical presses, cast spoked flywheels of two-piece design are standard (Fig. 2).

Because it is difficult to calculate the stresses in a spoked flywheel, these are usually designed on the basis of a maximum allowable peripheral speed of 5000–7000 feet per minute (fpm) for cast iron. The higher value is for rims cast by special techniques to eliminate rim porosity or blowholes. Higher rim speeds to 10,000 fpm may be used with cast steel. Structural steel plate welded flywheels of the solid web type can be accurately analyzed for stress by applying the equations of rotating disks to the components of the flywheel and allowing a factor for the stresses caused by welding. *See* ENERGY; MOMENT OF INERTIA; MOMENTUM.

[L. SIGFRED LINDEROTH]

Bibliography: C. Carmichael (ed.), *Kent's Mechanical Engineers Handbook*, pt. 1: *Design and Production*, 12th ed., 1950; R. C. Johnson, *Mechanical Design Synthesis: Creative Design and Optimization*, 2d ed., 1978; J. E. Shigley, *Mechanical Engineering Design*, 12th ed., 1972.

Force

Force may be briefly described as that influence on a body which causes it to accelerate. In this way, force is defined through Newton's second law of motion.

This law states in part that the acceleration of a

body is proportional to the resultant force exerted on the body and is inversely proportional to the mass of the body. An alternative procedure is to try to formulate a definition in terms of a standard force, for example, that necessary to stretch a particular spring a certain amount, or the gravitational attraction which the Earth exerts on a standard object. Even so, Newton's second law inextricably links mass and force. *See* ACCELERATION.

Many elementary books in physics seem to expect the beginning student to bring to his study the same kind of intuitive notion concerning force which Isaac Newton possessed. One readily thinks of an object's weight, or of pushing it or pulling it, and from this one gains a "feeling" for force. Such intuition, while undeniably helpful, is hardly an adequate foundation for the quantitative science of mechanics.

Newton's dilemma in logic, which did not trouble him greatly, was that, in stating his second law as a relation between certain physical quantities, he presumably needed to begin with their definitions. But he did not actually have definitions of both mass and force which were independent of the second law. The procedure which today seems most free of pitfalls in logic is in fact to use Newton's second law as a defining relation.

First, one supposes length to be defined in terms of the distance between marks on a standard object, or perhaps in terms of the wavelength of a particular spectral line. Time can be supposed similarly related to the period of a standard motion (for example, the rotation of the Earth about the Sun, the oscillations of the balance wheel of a clock, or perhaps a particular vibration of a molecule). Although applying these definitions to actual measurements may be a practical matter requiring some effort, a reasonably logical definition of velocity and acceleration, as the first and second time derivatives of vector displacement, follows readily in principle.

Absolute standards. Having chosen a unit for length and a unit for time, one may then select a standard particle or object. At this juncture one may choose either the absolute or the gravitational approach. In the so-called absolute systems of units, it is said that the standard object has a mass of one unit. Then the second law of Newton defines unit force as that force which gives unit acceleration to the unit mass. Any other mass may in principle be compared with the standard mass (m) by subjecting it to unit force and measuring the acceleration (**a**), with which it varies inversely. By suitable appeal to experiment, it is possible to conclude that masses are scalar quantities and that forces are vector quantities which may be superimposed or resolved by the rules of vector addition and resolution.

In the absolute scheme, then, Eq. (1) is written

$$\mathbf{F} = m\mathbf{a} \tag{1}$$

for nonrelativistic mechanics; here boldface type denotes vector quantities. The quantities on the right of Eq. (1) are previously known, and this statement of the second law of Newton is in fact the definition of force. In the absolute system, mass is taken as a fundamental quantity and force is a derived unit of dimensions MLT^{-2} (M= mass, L= length, T= time).

Gravitational standards. The gravitational system of units uses the attraction of the Earth for the standard object as the standard force. Newton's second law still couples force and mass, but since force is here taken as the fundamental quantity, mass becomes the derived factor of proportionality between force and the acceleration it produces. In particular, the standard force (the Earth's attraction for the standard object) produces in free fall what one measures as the gravitational acceleration, a vector quantity proportional to the standard force (weight) for any object. It follows from the use of Newton's second law as a defining relation that the mass of that object is $m = w/g$, with g the magnitude of the gravitational acceleration and w the magnitude of the weight. The derived quantity mass has dimensions FT^2L^{-1}.

Because the gravitational acceleration varies slightly over the surface of the Earth, it may be objected that the force standard will also vary. This may be avoided by specifying a point on the Earth's surface at which the standard object has standard weight. In principle, then, the gravitational system becomes no less absolute than the so-called absolute system.

Composition of forces. By experiment one finds that two forces of, for example, 3 units and 4 units acting at right angles to one another at point 0 produce an acceleration of a particular object which is identical to that produced by a single 5-unit force inclined at arccos 0.6 to the 3-unit force, and arccos 0.8 to the 4-unit force (see figure). The laws of vector addition thus apply to the superposition of forces.

FORCE

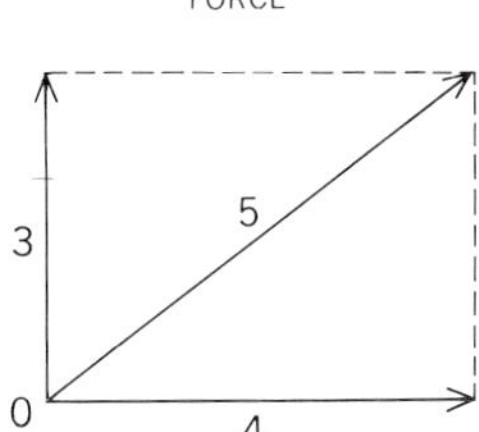

Vector addition of forces.

Conversely, a single force may be considered as equivalent to two or more forces whose vector sum equals the single force. In this way one may select the component of a particular force which may be especially relevant to the physical problem. An example is the component of a railroad car's weight along the direction of the track on a hill.

Statics is the branch of mechanics which treats forces in nonaccelerated systems. Hence, the resultant of all forces is zero, and critical problems are the determination of the component forces on the object or its structural parts in static equilibrium. Practical questions concern the ability of structural members to support the forces or tensions. *See* STATICS.

Specially designated forces. If a force is defined for every point of a region and if this so-called vector field is irrotational, the force is designated conservative. Physically, it is shown in the development of mechanics that this property requires that the work done by this force field on a particle traversing a closed path is zero. Mathematically, such a force field can be shown to be expressible as the (conventionally negative) gradient of a scalar function of position V, Eq. (2).

$$\mathbf{F} = -\nabla V \tag{2}$$

A force which extracts energy irreversibly from a mechanical system is called dissipative, or nonconservative. Familiar examples are frictional forces, including those of air resistance. Dissipative forces are of great practical interest, although they are often very difficult to take into account precisely in phenomena of mechanics.

The force which must be directed toward the

center of curvature to cause a particle to move in a curved path is called centripetal force. For example, if one rotates a stone on the end of a string, the force with which the string pulls radially inward on the stone is centripetal force. The reaction to centripetal force (namely, the force of the stone on the string) is called centrifugal force.

Methods of measuring forces. Direct force measurements in mechanics usually reduce ultimately to a weight comparison. Even when the elastic distortion of a spring or of a torsion fiber is used, the calibration of the elastic property will often be through a balance which compares the pull of the spring with a calibrated weight or the torsion of the fiber with a torque arising from a calibrated weight on a moment arm.

In dynamic systems, any means of measuring acceleration—for example, through photographic methods or radar tracking—allows one to calculate the force acting on an object of known mass.

Units of force. In addition to use of the absolute or the gravitational approach, one must contend with two sets of standard objects and lengths, the British and the metric standards. All systems use the second as the unit of time. In the metric absolute system, the units of force are the newton and the dyne. The newton, the unit of force in the International System (SI), is that force which, when applied to a body having a mass of 1 kilogram, gives it an acceleration of 1 m/s^2. The poundal is the force unit in the British absolute system, whereas the British gravitational system uses the pound. Metric gravitational systems are rarely used. Occasionally one encounters terms such as gram-force or kilogram-force, but no corresponding mass unit has been named.

[GEORGE E. PAKE]

Bibliography: F. Bueche, *Principles of Physics*, 3d ed., 1977; M. L. Bullock, Systems of units in mechanics: A summary, *Amer. J. Phys.*, 22:291–299, 1954; C. Kittel, W. D. Knight, and M. A. Ruderman, *Berkeley Physics Course*, vol. 1: *Mechanics*, 2d ed., 1973; R. Resnick and D. Halliday, *Physics*, 3d ed., 1977.

Force analysis

An analysis yielding the respective forces acting at any point of any member, or part of a member, of a mechanism, obtained by using relationships for dynamic equilibrium in a plane rigid body subject to external forces within this plane and to internal forces due to its motion in this plane. The following treatment is a careful selection and condensation of essential features of the reasoning.

The basic relationships for equilibrium are that the forces acting upon a body, or their resultants, must act through a point (their force polygon is closed); and that the couples acting on the body also reduce to zero (the moment of the existing forces on any member about any point is zero). If there are no more than three unknowns, a solution can be found, utilizing the three conventional equations summing forces in two perpendicular directions to zero and summing moments to zero. If there are more unknowns, further information must be acquired from other members.

The method of approach may, in practice, be purely algebraic, for instance, using the three mentioned equations of force and moment equilibrium; or it may be graphical, using diagrams adapted to yielding quantitative results. It is often advantageous to combine the two methods where some of the quantities used graphically are cumbersome and can be more easily computed.

Free-body diagram. A fundamental tool in solving problems is the free-body diagram. In this, each part of the mechanism is isolated and the forces acting upon each part are established. It is then often possible to piece this knowledge together until the entire solution is known.

In Fig. 1*a*, a four-bar linkage with frictionless pins and weightless members is acted upon by two forces *S* and *P* on members 3 and 4, respectively. The forces on all pins and the couple on member 2 are sought. Figure 1*b* shows the three moving members isolated in their respective free-body diagrams. Using these diagrams, if one starts with 4, he can successfully work backward for the entire solution. In Fig. 2*a*, a moment diagram with fulcrum at O_4 permits solution for F_{34}^{T4}, the normal force component of member 3 on member 4 at pin *B*. In Fig. 2*b*, (F_{43}^{T4}, *S*, and *R* are now known), F_{23} and F_{43}^{N4} can be found. In Fig. 2*c*, knowledge of F_{43}^{N4} yields the direction of *R*. Figure 2*d* thus reveals the direction and magnitude of F_{14} on pin O_4, as well as the pin force (equal and opposite to F_{23}), and hence the couple on 2. *See* FOUR-BAR LINKAGE.

Force diagram. Force diagrams for many friction and other devices common to mechanisms may be helpful. The force between two gears is one through the pitch point in the direction of the common tangent to the two base circles and inclined at the pressure angle to the common tan-

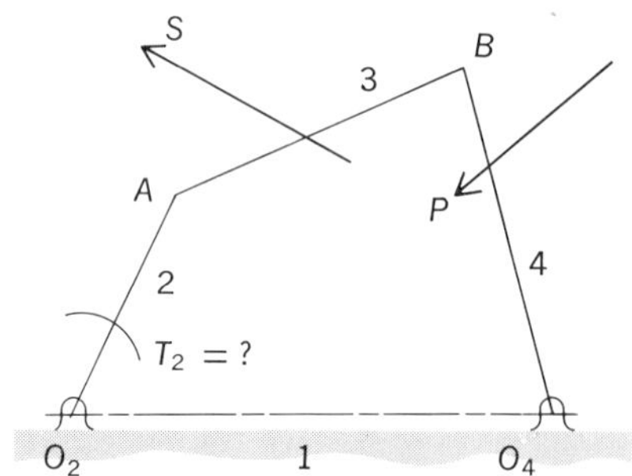

(a)

S
3
F_{43}
F_{32}
F_{34}
F_{23}
2
P
4
F_{12}
F_{14}

(b)

Fig. 1. Construction of free-body diagram. (*a*) Forces *S* and *P* acting on members 3 and 4 of a four-bar linkage. (*b*) Free-body diagrams of the three moving members of *a*.

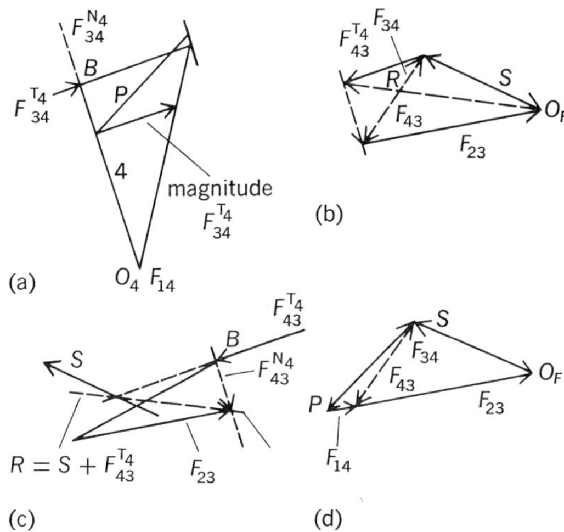

Fig. 2. Successive steps (*a–d*) in solution of forces.

gent to the pitch circles (Fig. 3*a*). The force acting on a pivot or pin of radius R will act along the center line and through the center point unless there is friction in the pin (Fig. 3*b*), in which case the force will lie at the friction angle where the center line meets the moving head, at angle ϕ ($\tan \phi = \mu$). Hence it is tangent to a small circle, the friction circle, $r = R \sin \phi$ (Fig. 3*b*). The force on the upper or lower side of a slider will vary in direction with forward and return strokes, but will lie at the friction angle to the normal and through the point of concurrency of other forces (Fig. 3*c*). *See* GEAR.

In Fig. 4, a press mechanism with massless members but with pin and slider friction is treated. There is a known resistance force of P units at output slider block 7. The solution polygon showing all forces appears in Fig. 4*b*. All joint frictions have been assumed equal, with gear friction not being treated. The general method of solution in Fig. 4*a* observes that polygon triangles can be evaluated backward, starting with the three forces acting on 7. Respective part directions must first be ascertained followed by general force directions under impending motion. Next, in order to fix the placement and direction of friction forces, it is necessary to decide whether the angles are respectively increasing or decreasing. This can be done by inspection, by diagrams, or by computations. With the direction of each of the forces at the pins known, it can be determined upon which side of the pin the force must pass to induce a rotation opposite to that due to joint friction. Thus at point A, the geometry shows that angle α is expanding; hence friction must be acting clockwise (to oppose it), and hence F_{43} must act on the upper side of the friction circle to turn it counterclockwise, as shown. Corresponding patterns apply for angles β, γ, and δ. Thus at D, since δ is diminishing under the impending motion, pin motion is counterclockwise, pin friction is clockwise, and the force F_{67} must yield a counterclockwise effect. At D there is a backward, upward friction force on the left side of the slider, establishing the friction force.

Once the positions of the forces are known, as in Fig. 4*a*, three basic force triangles are successively solved; in each of these, the direction and magnitude of one force and the directions of two others are known. Thus with P known, F_{65} is ascertained for the triangle of forces involved with body 7. Then for the force triangle for body 5, F_{56} has just been found, and hence F_{45} can be found in direction and magnitude. For the force triangle for body 3, F_{45} is known in position and magnitude. Hence, with knowledge of the directions of F_{32} and F_{13}, the complete triangle is known, and F_{12} can be ascertained in direction and magnitude.

When inertia forces and moments are involved, as occurs in moving parts due to their masses, the situation can be met by regarding the forces momentarily as static forces in equilibrium. By using such devices, the existing forces (such as compression forces or shear forces) and couples on mechanisms can frequently be found at points of interest. It is assumed that this mass is concentrated at the known center of gravity and that its linear and angular accelerations have been learned from kinematic studies. Kinematics is the inevitable background ingredient of every preparation for a force study where mass is present.

The force $F = MA_g$ is thus known in direction and magnitude, and Eq. (1) holds. Here I is the

$$T = I\alpha \tag{1}$$

mass moment of inertia (the inertia of rotation) and T is the turning force, torque, or inertia moment.

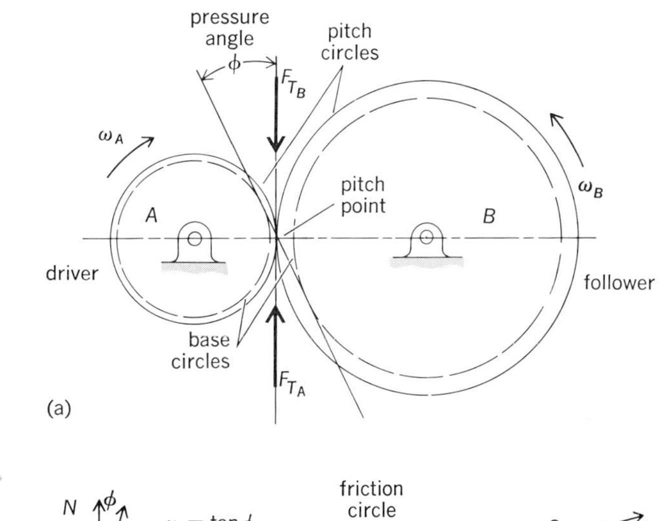

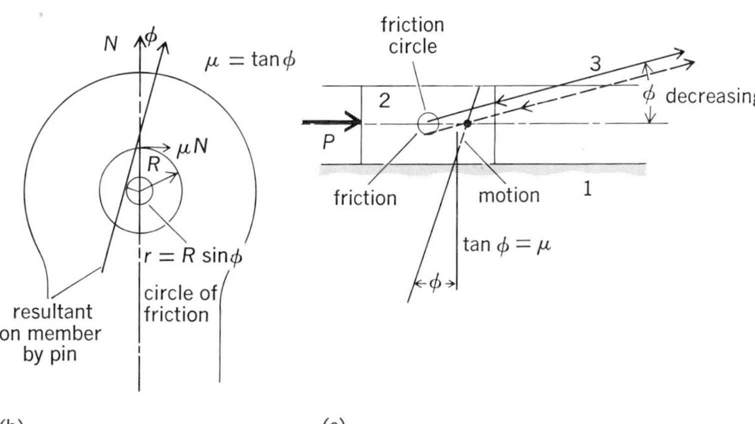

Fig. 3. Force diagrams. (*a*) Force between two gears. (*b*) Force on a pivot or pin of radius *R*. (*c*) Force on the upper or lower side of a slider.

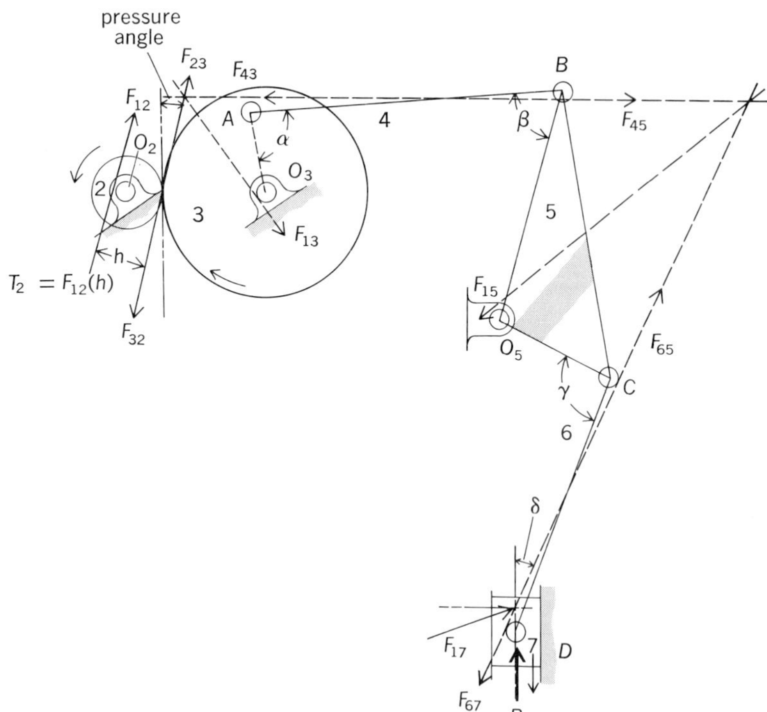

(a)

(b)

Fig. 4. Solution of (*a*) force problem by (*b*) force triangles.

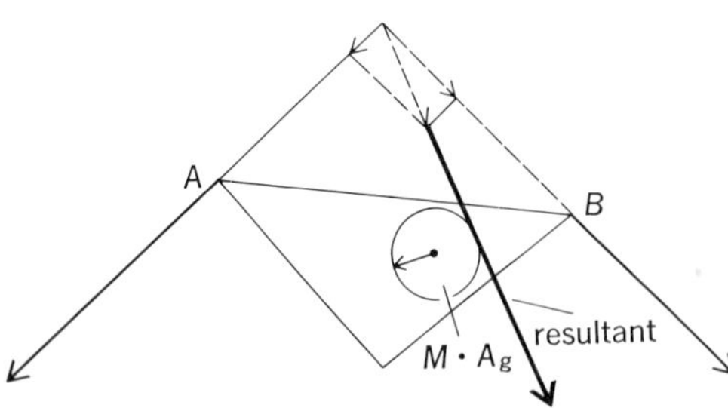

Fig. 5. Representation of the forces acting on a mass to cause the resultant force.

Then Eq. (2), or Eq. (3), can be written. Here k is the radius of gyration.

$$MA_gH = I\alpha \quad (2)$$

$$h = I\alpha/MA_g = k^2\alpha/A_g \quad (3)$$

Equations (1)–(3) can also be shown in circle form (Fig. 4*a*), where the force acting upon the mass (to produce the known acceleration) is represented as the resultant of the other forces upon the piece (Fig. 5). *See* MOMENT OF INERTIA.

[DOUGLAS P. ADAMS]

Bibliography: E. A. Dijksman, *Motion Geometry of Mechanism*, 1976; H. H. Mabie and F. W. Ocvirk, *Mechanisms and Dynamics of Machinery*, 3d ed., 1978; B. Paul, *Kinematics and Dynamics of Planar Machinery*, 1979.

Force fit

A means for holding mating mechanical parts in fixed position relative to each other. In a force fit of cylindrical parts, the inner member has a greater diameter than the hole of the outer member; that is, the metals of the two parts interfere. *See* ALLOWANCE.

Light drive fits require assembly pressure and produce a more or less permanent assembly of thin sections or long fits. They are suitable for cast iron external members. For a tighter fit, medium or heavy drive fits are used, although such fits may require more than direct mechanical pressure for assembly. In a true force fit, the parts are highly stressed, the interference amounting to 0.002 or 0.003 in. for parts with a basic diameter of 1 in. *See* SHRINK FIT.

[PAUL H. BLACK]

Forging

The plastic deformation of metals, usually at elevated temperatures, into desired shapes by compressive forces exerted through a die. Forging processes are usually classified either by the type of equipment used or by the geometry of the end product. The simplest forging operation is upsetting, which is carried out by compressing the metal between two flat parallel platens. From this simple operation, the process can be developed into more complicated geometries with the use of dies. A number of variables are involved in forging; among major ones are properties of the workpiece and die materials, temperature, friction, speed of deformation, die geometry, and dimensions of the workpiece.

The simple upsetting test has been quite useful in identifying the role of variables in forging. This test demonstrates that the force to forge is a function of the strength of the material, coefficient of friction, and ratio of the lateral to thickness dimensions of the workpiece. Pressure exerted on the workpiece increases toward the center of the part, this distribution being known as the friction hill. The overall slope of this hill depends on the coefficient of friction and dimensions of the part.

One basic principle in forging is the fact that the material flows in the direction of least resistance. Thus a part will expand more in its short dimension than it will in its longest dimension; hence a cubic specimen eventually will acquire the shape of a pancake. Also, friction between dies and workpiece (and, in hot forging, the more rapid cooling of the surfaces of the workpiece that are in contact with relatively cooler dies) causes barreling. Barreling can be reduced by lowering friction and also by carrying out the forging process in a more isothermal condition.

Forgeability. In practice, forgeability is related to the material's strength, ductility, and friction. Because of the great number of factors involved, no standard forgeability test has been devised. For steels, which constitute the majority of forgings, torsion tests at elevated temperatures are the most predictable; the greater the number of twists of a round rod before failure, the greater is its ability to be forged. A number of other tests, such as simple upsetting, tension, bending, and impact tests, have also been used. Typical forging temperatures are: aluminum alloys, 750–850°F (400–450°C); copper alloys, 1500°F (820°C); alloy steels, 1700–2300°F (930–1260°C); titanium alloys, 1400–1800°F (760–980°C); and refractory metal alloys, 1800–3000°F (980–1650°C).

In terms of factors such as ductility, strength, temperature, friction, and quality of forging, various engineering materials can be listed as follows in order of decreasing forgeability: aluminum alloys, magnesium alloys, copper alloys, carbon and low-alloy steels, stainless steels, titanium alloys, iron-base superalloys, cobalt-base superalloys, columbium alloys, tantalum alloys, molybdenum alloys, nickel-base superalloys, tungsten alloys, and beryllium. *See* METAL, MECHANICAL PROPERTIES OF.

Forging dies. Design and selection of die materials for forging involve considerable experience. Both static and impact strength are required, in addition to resistance to abrasion and to heat checking. Dies are generally heated to below 1000°F in hot forging.

Some of the terminology in forging is shown in Fig. 1. Draft angles facilitate the removal of the forging from the die cavity; they usually range between 3 and 10°. The purpose of the saddle or land in the flash gap is to offer resistance to the lateral flow of the material so that die filling is encouraged. Die filling increases as the ratio of land width to thickness increases up to about 5; larger ratios do not increase filling substantially and are undesirable due to increased forging loads and excessive die wear. The purpose of the gutter is to store excess metal. The flash is removed either by cold or hot trimming or by machining.

Filling of the die cavities is an important aspect and requires a careful balance of die geometry, temperature, lubrication, and speed of forging. Lubrication is important not only for reducing friction but also in its role as a parting agent, in addition to serving as a heat barrier between the die and the hotter forging. The commonest lubricant is graphite suspended in grease, oil, or water; other lubricating materials are molybdenum disulfide, talc, mica, sawdust, salt water, and glasses. Some metals, such as tungsten and molybdenum, form soft oxides upon heating, and these serve as lubricants. Forging lubricants should be nontoxic and economical to use. The exact composition of many lubricants used in the forging industry is proprietary.

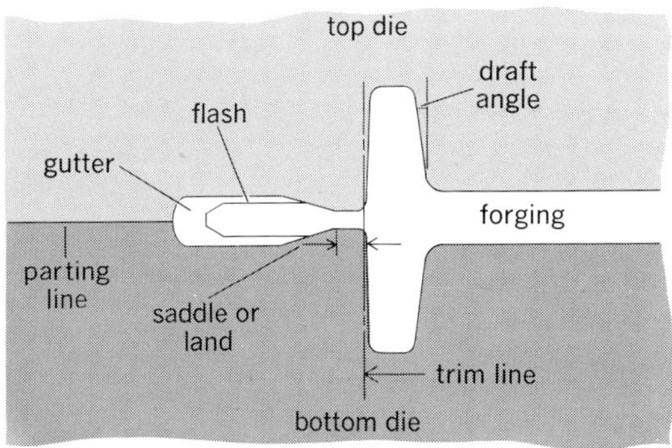

Fig. 1. Closed-die forging terminology.

Because of the many factors involved, great variations in strength and ductility have been observed in the forging of the more complex shapes. Incorrect design of dies or improper operation can result in forged products with internal defects such as laps (Fig. 2). Other flaws are hot tears, cracks, and coarse-grain wrinkles.

Forging design. There are in general four types of forging designs: blocker type, commercial, close tolerance, and precision. The blocker type involves the use of large radii and draft angles, smooth contours, and generous allowances. Commercial designs have more refined dimensions, such as standard draft angles of 5–7°, and smaller radii and allowances. In close-tolerance designs draft angles are in the order of 1–3°, tolerances are less than half of those for commercial designs, and there is little or no allowance for finish. Precision designs usually involve close dimensional tolerances; they require special tooling and equipment and are used mainly for aluminum and magnesium forging.

Forging costs are usually broken down into four basic categories: material, die, forging, and machining costs. Economical designs usually require careful consideration of factors important both to the consumer and to the forging facility. The cost of forgings depends on items such as the alloy to be forged, metallurgical specifications, quality desired, dimensional tolerances, surface finish, and number of parts desired.

Forging equipment. A number of methods produce the necessary force and die movement for forging. Two basic categories are open-die and closed-die forging. Drop hammers supply the energy through the impact of a falling weight to which the upper die is attached. Two common types of drop hammers are the board hammer and the steam hammer. In the former the ram is attached to wooden boards which slide between two rollers; after the ram falls freely on the forging, it is raised by friction between the rotating rollers. In the steam hammer the ram is raised and lowered by a double-acting steam cylinder; the forging force is a combined effect of gravity and steam pressure. The speed of hammers usually ranges up to 20 ft/sec, and the rated capacities are from 500 to 50,000 lb for a steam hammer.

Another type of forging equipment is the mechanical press, varieties of which are crank or eccentric, knuckle, percussion, and toggle. For large forgings the hydraulic press (with existing capacities of 50,000 tons and future capacities in the order of 200,000 tons) is the only equipment with sufficient force. However, the speed for such presses is about one-hundredth that of hammers. Other types of forging equipment are counterblow hammers (vertical) and impacters (horizontal).

Hammers are generally preferred for copper, steels, titanium, and refractory alloys; presses are preferred for aluminum, magnesium, beryllium, bronze, and brass. Before a selection of equipment is made, many considerations need to be taken into account, such as sensitivity of the workpiece material to speed of deformation, amount of reduction, size of forging, and economy. The majority of metals and alloys can be forged interchangeably in either presses or hammers. More recent develop-

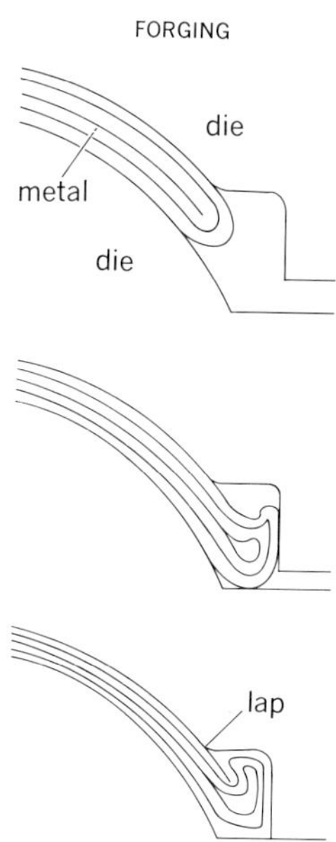

Fig. 2. Formation of a defect in closed-die forging.

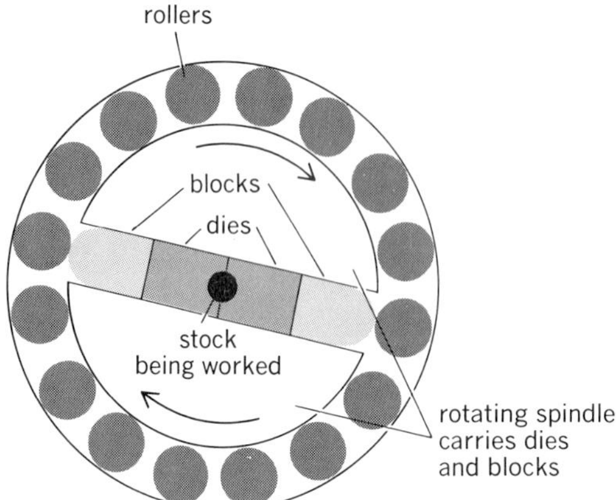

Fig. 3. Operation principle of swaging machine.

ments in forging equipment involve the pneumatic-mechanical system where the ram is made to travel at speeds up to 200 ft/sec by special designs using gas pressure. Capacities of such equipment have been as high as 500,000 ft-lb.

Swaging. Swaging or, as it is also called, rotary swaging consists of two or four dies which are activated radially by blocks in contact with a series of rollers (Fig. 3). The rotation of the block-die assembly causes the curved ends of the blocks (either circular or modified sine curve) to be in contact with the rollers; relative motion between the blocks and the roller housing then gives a reciprocating motion to the dies. The hammering action on the outer surface of the stock reduces its diameter; the stock is generally prevented from rotating.

A variety of geometries is obtained in this process: (1) reduction of wall thickness of tubes; (2) sinking of tubes, whereby both the outside and inside diameters are reduced, the final wall thickness being controlled by a mandrel; (3) tapered or stepped sections; and (4) a variety of internal geometries, such as fins or hexagonal or spiral grooves, each with the use of a specially designed mandrel. Die angles are usually a few degrees, maximum outside diameter is generally limited to 1–2 in., and hot or cold operations may be carried out. Capacities of equipment are as high as 75 tons per die. Cold swaging produces excellent surface finish and dimensional accuracy, together with improved mechanical properties. *See* METAL FORMING.

[SEROPE KALPAKJIAN]

Bibliography: B. Avitzur, *Metal Forming*, 1980; National Safety Council, *Forging Safety Manual*, 1970; A. M. Sabroff et al., *Forging Materials and Practices*, 1968.

Foundations

Structures or other constructed works are supported on the earth by foundations. The word "foundation" may mean the earth itself, something placed in or on the earth to provide support, or a combination of the earth and the elements placed on it. The foundation for a multistory office building could be a combination of concrete footings and the soil or rock on which the footings are supported. The foundation for an earth-fill dam would be the natural soil or rock on which the dam is placed. Concrete footings or piles and pile caps are often referred to as foundations without including the soil or rock on which or in which they are placed. The installed elements and the natural soil or rock of the earth form a foundation system; the soil and rock provide the ultimate support of the system. Foundations that are installed may be either soil-bearing or rock-bearing. The reactions of the soil or rock to the imposed loads generally determine how well the foundation system functions. In designing the installed portions, the designer must determine the safe pressure which can be used on the soil or rock and the amount of total settlement and differential settlement which the structure can withstand.

The installed parts of the foundation system may be footings, mat foundations, slab foundations, and caissons or piles, all of which are used to transfer load from a superstructure into the earth. These parts, which transmit load from the superstructure to the earth, are called the substructure (Fig. 1).

Footings. Footings or spread foundations are used to spread the loads from columns or walls to the underlying soil or rock. Normally, footings are constructed of reinforced concrete; however, under some circumstances they may be constructed of plain concrete or masonry. When each footing supports only one column, it is square. Footings supporting two columns are called combined footings and may be either rectangular or trapezoidal. Cantilever footings are used to carry loads from two columns, with one column and one end of the footing placed against a building line or exterior

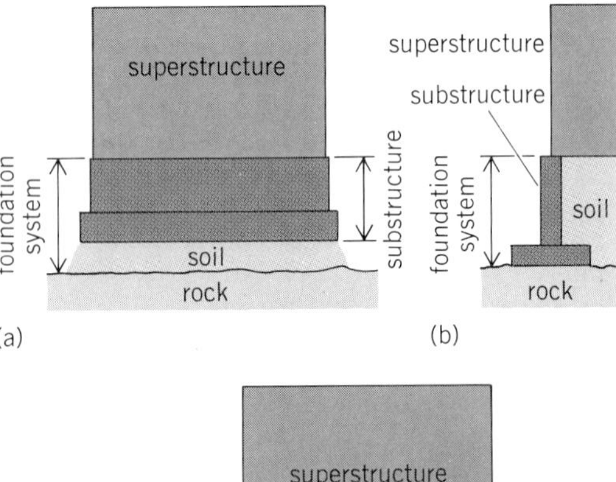

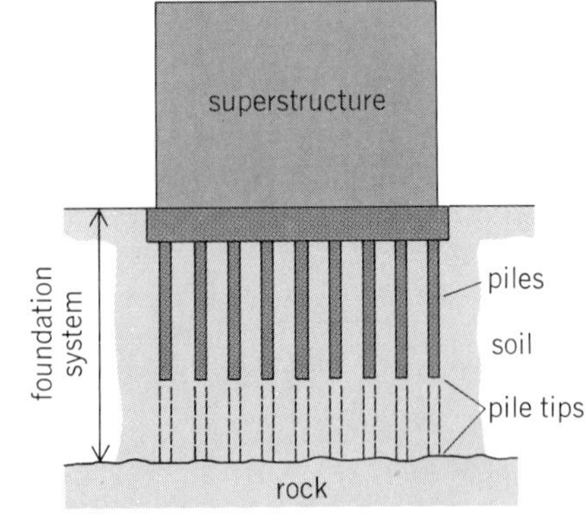

Fig. 1. Examples of foundation systems. *(a)* Structure supported on a foundation bearing on soil. *(b)* Structure supported on a foundation bearing on rock. *(c)* Structure supported by a pile foundation. The piles may be installed so that the pile tips terminate in soil and all support is derived from the soil, or they may be installed so that the tips penetrate to rock.

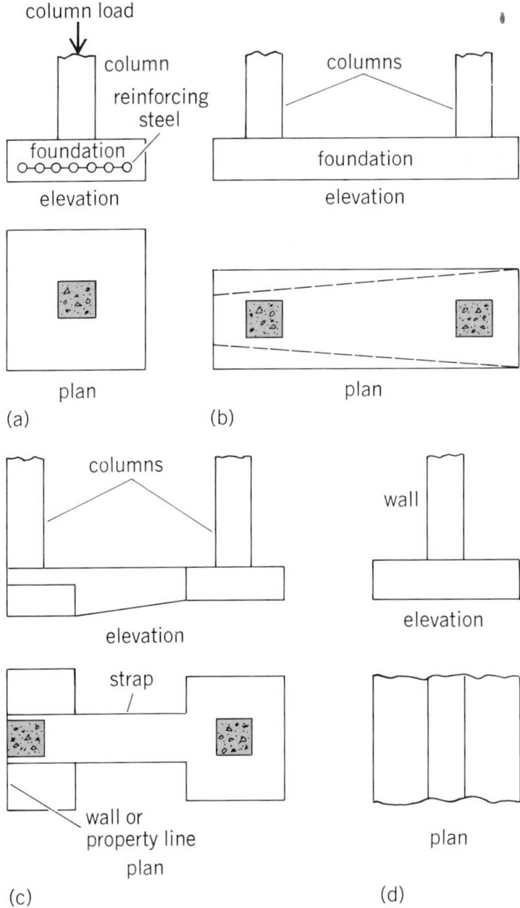

Fig. 2. Types of footings. (*a*) Column footing. (*b*) Combined footing, which can be either rectangular or trapezoidal. (*c*) Cantilevered footing. (*d*) Wall footing.

wall. Footings supporting walls are continuous footings (Fig. 2).

The sizes of footings are determined by dividing the loads to be imposed at the base of the footings by the allowable bearing pressure which can be imposed on the soil or rock of the earth. Most building codes and textbooks on foundations contain tables listing allowable bearing pressures for various types of soil and rock; however, these tables give only general classifications and descriptions of the soil or rock and must be used with caution. More specific information about the soil or rock is normally obtained by drilling test borings, extracting soil or rock samples, performing laboratory tests on the samples, and making engineering analyses to determine suitable bearing pressures. In addition to bearing pressure, consideration must be given to the amount of settlement which may occur and the capability of the structure to withstand such settlement. If settlement is a problem it may be necessary to use an alternate foundation type rather than footings or to enlarge the footings and decrease the bearing pressure.

Grade beams may be used between exterior column footings to support walls, with the beams transferring the weight of the walls to the column footings. Beams are also used between interior column footings to act as braces or ties or to support interior walls. Retaining walls are those walls subjected to horizontal earth pressures due to the retention of earth behind them. The foundations for these walls must have sufficient frictional resistance with the soil or rock on which they rest so that they will not slide when subjected to the horizontal earth pressure. In addition, retaining walls must be designed so they will not overturn (Fig. 3). In frost-susceptible areas, footings must be placed below the frost line.

Mat foundations. Mat or raft foundations are large, thick, and usually heavily reinforced concrete mats which transfer loads from a number of columns or columns and walls to the underlying soil or rock. Mats are also combined footings, but are much larger than a footing supporting two columns. They are continuous footings and are designed to transfer a relatively uniform pressure to the underlying soil or rock. Mats are rigid and will act as a bridge over discontinuities in the soil or rock on which they are founded. Mats founded several meters below the ground surface, when combined with external walls, are termed floating foundations (Fig. 4). The weight of the soil excavated from the ground surface to the bottom of the mat may be equal to or approach the total weight of the structure. In this situation, little or no new load is applied to the underlying supporting soil, and settlements of a structure may be minimal after construction.

Slab foundations. Slab foundations are used for light structures wherein the columns and walls are supported directly on the floor slab. The floor slab is thickened and more heavily reinforced at the places where the column and wall loads are imposed (Fig. 5).

Special problems. Groundwater is a major problem in connection with the design and installation of foundations where a substructure is to be placed below the groundwater level. Well points, pumping from deep wells, or pumping from sumps are methods used to dewater construction sites during foundation installation. Other methods which are less often used are freezing of the water in the soil, removal of water by electroosmosis, and the installation of cutoff walls made of piling or grout around the periphery of an excavation with water removed by pumping from within the excavation. If dewatering operations are performed in an area surrounded by existing structures, precautions must be taken to protect them, as the lowering of the groundwater may cause the soil on which they are supported to subside.

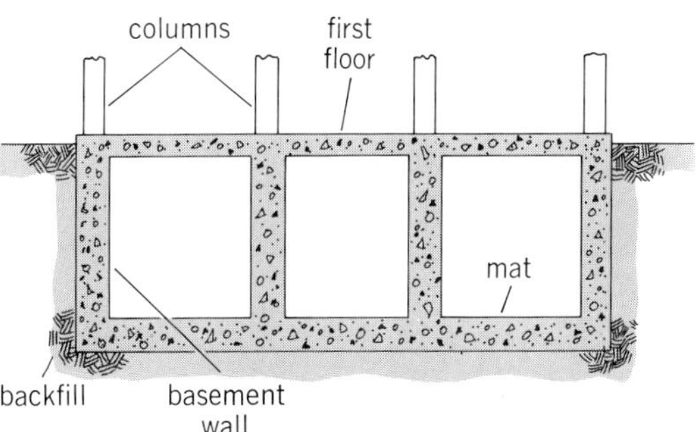

Fig. 4. Mat foundation, sometimes called a floating foundation, generally heavily reinforced in both directions.

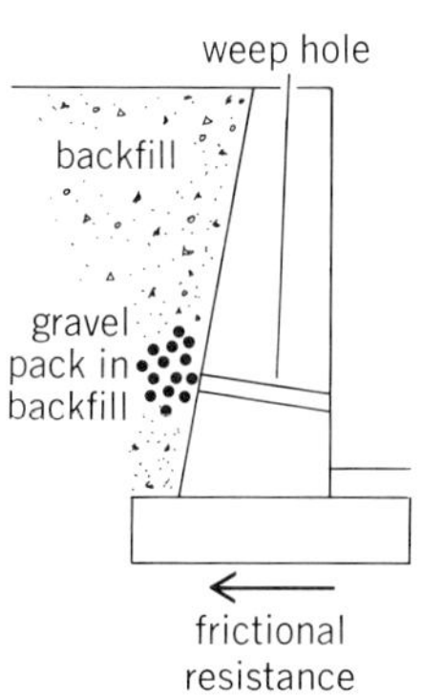

Fig. 3. Retaining wall, Gravel pack behind the wall prevents the weep hole from clogging with soil.

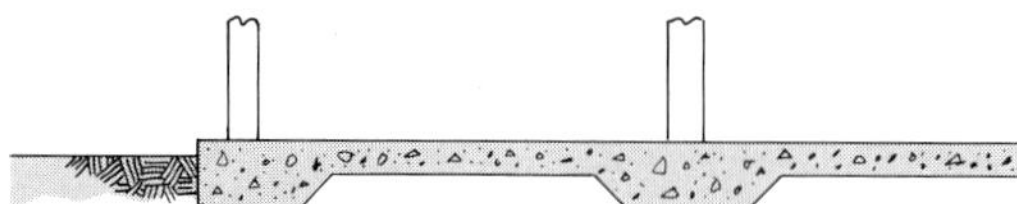

Fig. 5. Slab foundation.

If a basement is partially or totally below the groundwater level its walls must be designed to withstand the hydrostatic pressures of the water on the outside in addition to the pressure from the soil backfill. An alternate procedure is to install a permanent system to remove water outside the walls. Some substructures below groundwater level may at times be subjected to hydrostatic uplift forces which are greater than the downward forces imposed by the structures. In these cases, provisions must be made to anchor the structures to prevent them from floating upward.

Groundwater also causes problems by infiltrating through basement walls, slabs, and joints into the basement itself. This can be prevented or reduced by providing an external permanent drainage system that carries water away from the basement, by encasing the walls and slabs in an impermeable plastic membrane, or by coating the external walls with an asphaltic mastic to lower their permeability (Fig. 6). Combinations of the foregoing are also used. Retaining walls and abutments often have weep holes in the lower sections through which water accumulating behind the walls or abutments can escape. The water pressure behind the walls is relieved as the water flows through the walls into an open external drainage system.

Foundations placed on expansive soils are often subjected to distressing movements unless special precautions are taken. Expansive soils are those which swell and contract excessively with varying amounts of moisture. Problems can be overcome by installing foundations below the zone of significant change in moisture content and backfilling with nonexpansive materials, by altering the soil with an admixture such as lime or cement so that volume changes do not occur, or by providing flexibility in the structure to accommodate movements.

Underpinning of foundations is often necessary, and it may be either remedial or precautionary. Remedial underpinning is used to correct defects in existing foundations which may have settled excessively. If the structure is to be saved or returned to its original state, additional foundation support must be provided. Precautionary underpinning is used when new structures are to be installed adjacent to or beneath existing structures, as in the construction of city subways. Underpinning of foundations is a specialized construction technique. The work is generally performed in a confined space, such as the basement of a building, or in small pits excavated outside a building area. It is necessary to provide support for the loads of the existing structure while new foundations are installed. The new foundations may be footings which are placed deeper in the ground than the original foundations, or they may be piles or caissons.

Underpinning of a wall footing may be performed by excavating pits adjacent to and beneath existing foundations. The pits are small, some 0.9 m wide by 1.2 m long. Horizontal sheeting is placed in the pits as excavation proceeds to prevent caving of the walls and undermining of the structure being underpinned. When the new bearing stratum is reached, forms are placed in the pit, and concrete is poured from the new bearing stratum up to within 3 in. (76 mm) of the bottom of the old footing. After the new concrete has hardened, the 3-in. (76-mm) space is packed by hand with a mixture of sand, cement, and a small amount of water. Called grout, the mixture is packed very tightly into the space between the top of the new footing and the underside of the old footing. The pit underpinning process is repeated throughout the entire length of the wall footing. The resulting new foundation may be a continuous wall or intermittent piers. *See* BRIDGE; BUILDINGS; CAISSON FOUNDATION; DAM; PILE FOUNDATION; RETAINING WALL.

[GARDNER M. REYNOLDS]

Bibliography: J. E. Bowles, *Foundation Analysis and Design*, 2d ed., 1975; G. Fletcher and V. A. Smoots, *Construction Guide for Soils and Foundations*, 1974; R. B. Peck, W. E. Hanson, and T. H. Thornburm, *Foundation Engineering*, 2d ed., 1974; H. F. Winterkorn and H. Fang, *Foundation Engineering Handbook*, 1975.

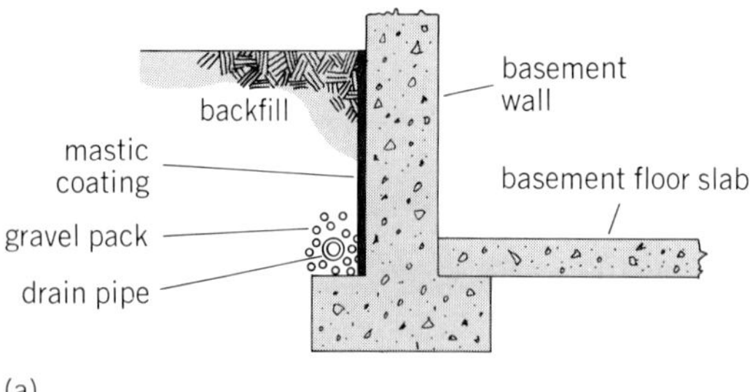

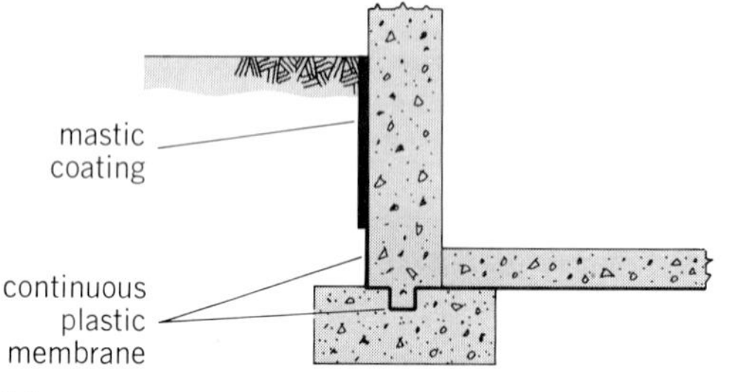

Fig. 6. Methods of reducing water infiltration into basements. *(a)* External drain pipe. *(b)* Walls, floor slab, and joints encased in a plastic membrane.

Four-bar linkage

A basic linkage mechanism used in machinery and mechanical equipment. The term has been applied to three types of linkages: plane, spherical, and skew.

The plane four-bar linkage (Fig. 1) consists of four pin-connected links forming a closed loop, in which all pin axes are parallel. The spherical four-bar linkage (Fig. 2) consists of four pin-connected links forming a closed loop, in which all pin axes intersect at one point. The skew four-bar linkage (Fig. 3) consists of four jointed links forming a closed loop, in which crank 2 and link 4 are pin-connected to ground 1 and the axes of the pins are

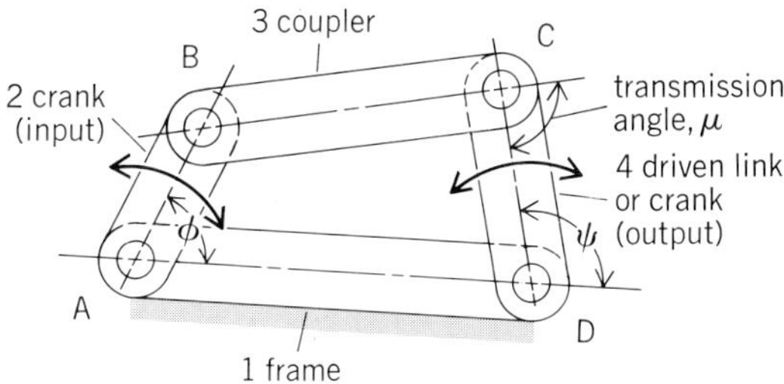

Fig. 1. Plane four-bar linkage.

generally nonparallel and nonintersecting; coupler 3 is connected to crank 2 and link 4 by ball joints.

Mechanical function. Four-bar linkages are most frequently used to convert a uniform, continuous rotation (the motion of crank 2) into a nonuniform rotation or oscillation (the motion of link 4). Depending on whether the axes of the connected shafts (attached to links 2 and 4) are parallel, intersecting, or skew, a plane, spherical, or skew four-bar linkage is used. In instrument applications the primary function of the linkage is the conversion of motion, while in power applications both motion conversion and power transmission are fundamental. Four-bar linkages also are used to impart a prescribed motion to the coupler, as well as for other purposes.

Types of motions. Each of the above linkages can be proportioned for three types of motion, or linkage types: crank-and-rocker, drag, and double-rocker.

Crank-and-rocker linkages. In this motion the crank (link 2) is capable of unlimited rotation, while the output link (link 4) oscillates or rocks through a fraction of one turn (usually less than 90°). This is the most common form of the plane and the skew four-bar linkage, and is used in machinery and appliances of all types, as for example, in an electric razor (Fig. 4). The most common spherical crank-and-rocker motion is known as the wobble plate. In wobble-plate motion the angle between the pin axes on the crank is less than 90°, while for each of the other links the angle between the pin axes is equal to 90°. The wobble motion is that of the coupler. It has been used in axial engines and elsewhere, for example, in dough kneading (Fig. 5), in which the dough is stirred by an agitator integral with the coupler. A skew four-bar linkage of crank-and-rocker proportions is shown

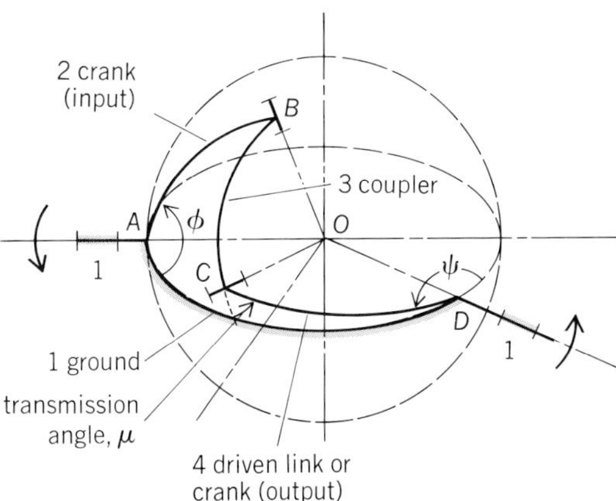

Fig. 2. Spherical four-bar linkage.

in Fig. 6; the angle between the axes of crank 2 and rocker 4 is a right angle. Such motions have been used in textile machinery, tobacco-processing machinery, and generally for "going around corners."

Drag linkages. In drag linkages the motions of cranks 2 and 4 are both capable of unlimited rotations. The plane drag linkage has been used for quick-return motions, but care must be taken to avoid mechanical interference of the cranks with the coupler. The skew drag linkage is rarely used. The most common drag linkage is the spherical drag linkage. One such linkage is the Hooke-type universal joint, or Hooke joint, in which the angle between the pin axes of each moving link is a right angle (Fig. 7). The Hooke joint is used to connect intersecting shafts, for example, in automotive drive lines. The angular-velocity ratio between the connected shafts depends on the shaft angle (which is usually less than or equal to 30°) and is not constant. *See* UNIVERSAL JOINT.

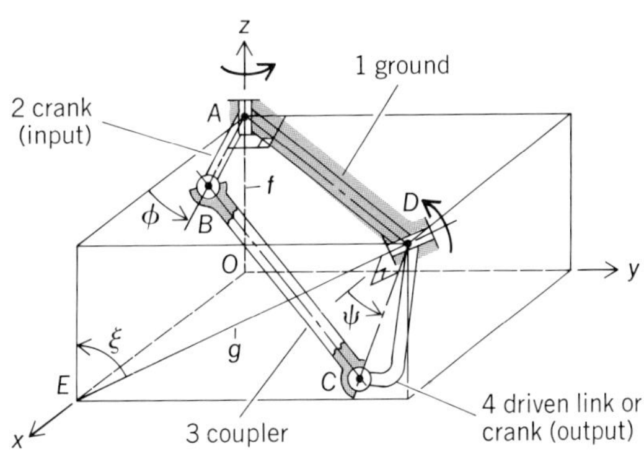

Fig. 3. Skew four-bar linkage. $OA = f$; $ED = g$; OE = common perpendicular between axes of pin joints at A and D.

Double-rocker linkages. In such linkages (Fig. 8) neither crank 2 nor 4 is capable of complete rotations. Such motions occur in hand tools and mechanical equipment in which only limited rotations are required, such as the oscillating motion in certain household fans.

Grashaf's inequality. In the case of plane four-bar linkages the proportions which determine the type of motion are governed by Grashof's inequality: the sum of the lengths of the longest and shortest links is less than the sum of the lengths of the two intermediate links. If the inequality is satisfied and either link 2 or link 4 is the shortest link, the linkage is a crank-and-rocker (the shortest link functioning as the crank). If the equality is satisfied and the frame is the shortest link, the linkage is a drag linkage. In all other cases the linkage is a double-rocker linkage or a folding linkage. A folding linkage occurs when Grashof's inequality becomes an equality. Such linkages include those having parallelogram proportions, such as occur in the parallel and antiparallel equal-crank linkages. The latter can be used only, however, if provision is made for a positive drive while passing through the folded position or positions.

For spherical four-bar linkages a simple extension of Grashof's inequality is available, but for

FOUR-BAR LINKAGE

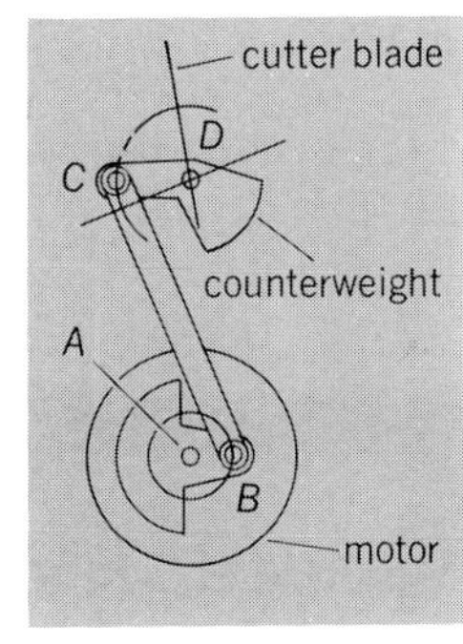

Fig. 4. Plane crank-and-rocker as used in a Sunbeam electric razor.

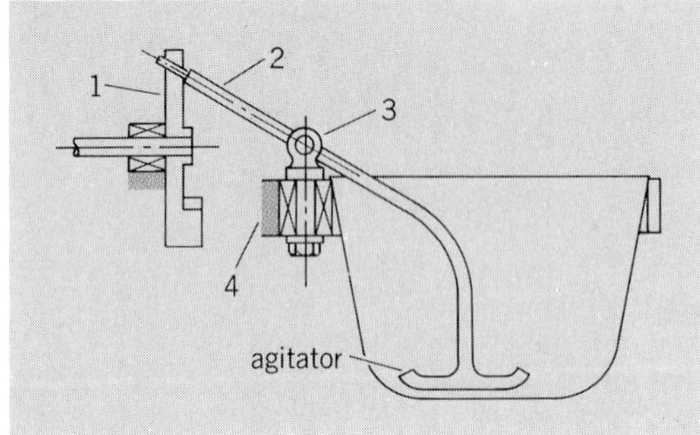

Fig. 5. Dough-kneading mechanism.

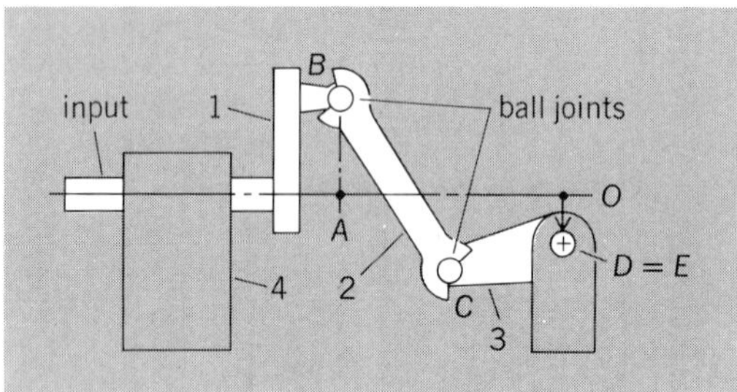

Fig. 6. Skew crank-and-rocker linkage.

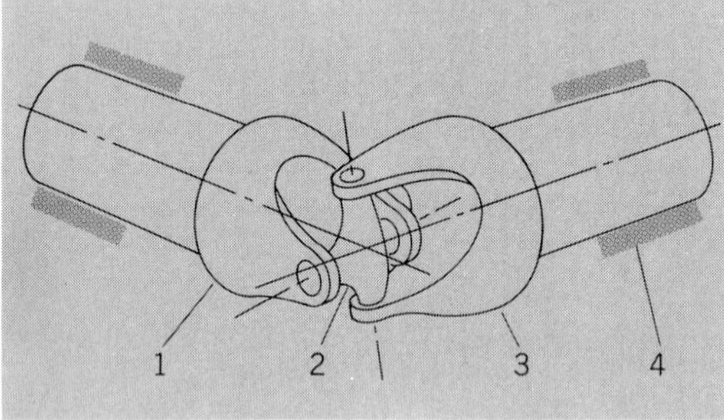

Fig. 7. Hooke joint.

skew four-bar linkages the criteria governing motion types are more involved.

Transmission characteristics. For the transfer of power and motion in crank-driven four-bar linkages, the linkages should be proportioned to avoid lockup positions. In such positions a turning effort applied to the crank will not produce a turning effort applied to the output link. In plane and spherical four-bar linkages, a lockup position occurs when the pin axes on the coupler and output link are coplanar. The transmission angle μ (Figs. 1 and 2) is then equal to 0° or 180°. For efficient force transmission in plane and spherical four-bar linkages the maximum deviation Δ of the transmission angle μ from 90° should be minimized (usually $\Delta \leqslant 30°$, at least in high-speed operation). Analogous but more complicated conditions apply to skew four-bar linkages. The Hooke joint and wobble plate are well proportioned for force transmission, as are the "centric" plane crank-and-rocker linkages for which $(AB)^2+(AD)^2=(BC)^2+(CD)^2$. The determination of crank-and-rocker proportions yielding optimum force transmission as a function of rocker swing angle and corresponding crank rotation has been solved as well.

Angular displacements. Equation (1) gives the angular displacements of crank 2 and link 4 for the plane four-bar linkage and for a special case of the skew four-bar linkage—the "simply skewed" ($f=g=0$; Fig. 3).

$$f(t,u)=\{[(a-b-d)^2-c^2]+[(a+b-d)^2-c^2]t^2\}u^2 -8\lambda bdtu+[(a+b+d)^2-c^2]t^2 +[(a-b+d)^2-c^2]=0 \quad (1)$$

Equation (2) gives the displacements of crank 2 and link 4 for the spherical four-bar linkage.

$$f(t,u)=\{[\cos(a-b-d)-\cos c] +[\cos(a+b-d)-\cos c]t^2\}u^2+4tu\sin b\sin d +[\cos(a+b+d)-\cos c]t^2 +[\cos(a-b+d)-\cos c]=0 \quad (2)$$

These equations are useful not only in the determination of the angular displacements of links 2 and 4, but also in the determination of their angular velocities and accelerations, which follow by differentiation.

Link sizes are expressed in ordinary length units, such as inches (for the plane and skew four-bar linkages) and in great circular arcs (in degrees, say, for the spherical four-bar linkage). Let a, b, c, d denote the lengths (linear of arc) of ground (AD), crank, coupler, and output link, respectively, and let ξ denote the angular offset between pin axes of links 2 and 4 of the simply skewed four-bar linkage. $\lambda=1$ (in a plane four-bar linkage) or $\cos\xi$ (in a simply skewed four-bar linkage). If ϕ and ψ denote the angular positions of links 2 and 4, respectively, with $t=\tan\phi/2$ and $u=\tan\psi/2$, the displacement equations, $f(t,u)=0$, which give the relationship between the angular displacements of links 2 and 4, are as in Eqs. (1) and (2).

Design. This involves proportioning the linkages for given motion and power requirements and considering kinematics, dynamics, strength, lubrication, wear, balancing, tolerances, actuation, control and other factors. At low speeds the links can often be treated as rigid bodies, but at high speeds the elasticity needs to be included. The design procedures can be analytical, graphical, experimental, or a combination of these.

Due to the nonlinearity of the motion of four-bar linkages, their design is often time-consuming and difficult. However, the determination of linkage proportions for certain standard design requirements has been worked out in an efficient manner. These include transmission-angle optimization, the coordination of the rotations of links 2 and 4, and the generation of a particular coupler motion

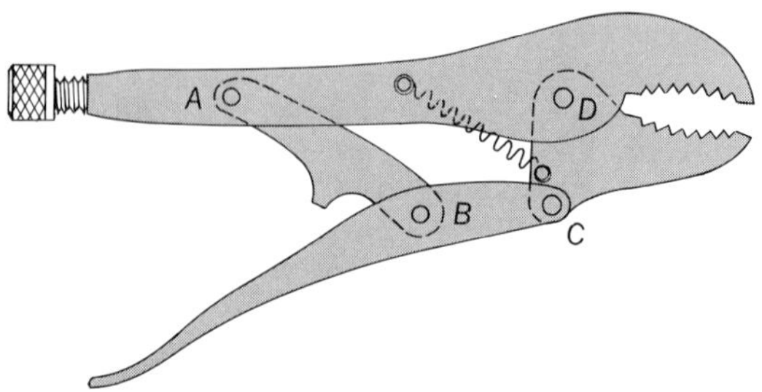

Fig. 8. Linkage-type pliers. (*Vise-Grip, Peterson Manufacturing Co., DeWitt, NE*)

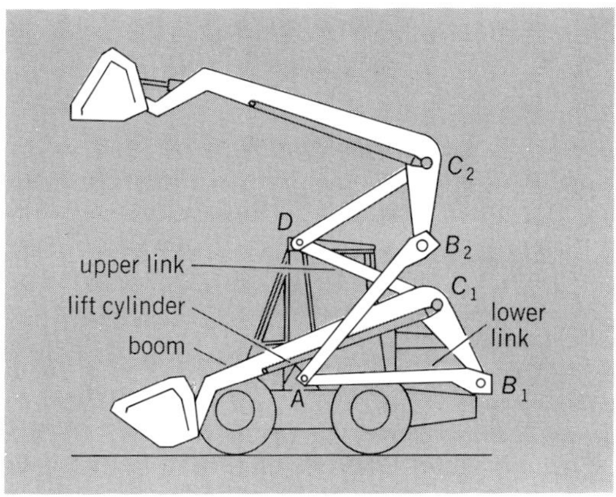

Fig. 9. Sperry New Holland front loader. The bucket is mounted on coupler *BC* of the four-bar linkage *ABCD*.

or point path (Fig. 9). In addition, general-purpose computer codes for the kinematic design and dynamic analysis of linkages, including four-bar linkages, have been developed.

Design trends emphasize high-speed operation and increasingly sophisticated motion and power-transmission requirements. *See* MECHANISM; SLIDER CRANK MECHANISM; STRAIGHT-LINE MECHANISM. [FERDINAND FREUDENSTEIN]

Bibliography: A. Erdman and G. N. Sandor, *Mechanism Design: Analysis and Synthesis*, 1982; B. Paul, *Kinematics and Dynamics of Planar Machinery*, 1979; S. B. Tuttle, *Mechanisms for Engineering Design*, 1967.

Friction

Resistance to sliding, a property of the interface between two solid bodies in contact. Many everyday activities like walking or gripping objects are carried out through friction, and most people have experienced the problems that arise when there is too little friction and conditions are slippery. However, friction is a serious nuisance in devices that move continuously, like electric motors or railroad trains, since it constitutes a dissipation of energy, and a considerable proportion of all the energy generated by humans is wasted in this way. Most of this energy loss appears as heat, while a small proportion induces loss of material from the sliding surfaces, and this eventually leads to further waste, namely, to the wearing out of the whole mechanism. *See* WEAR; WORK.

In stationary systems, friction manifests itself as a force equal and opposite to the shear force applied to the interface. Thus, as in Fig. 1, if a small force S is applied, a friction force P will be generated, equal and opposite to S, so that the surfaces remain at rest. P can take on any magnitude up to a limiting value F, and can therefore prevent sliding whenever S is less than F. If the shear force S exceeds F, slipping occurs. During sliding, the friction force remains approximately equal to F (though often it is smaller by about 20%) and always acts in a direction opposing the relative motion. The friction force is proportional to the normal force L, and the constant of proportionality is defined as the friction coefficient f. This is expressed by the equation $F = fL$.

The "laws" of friction are essentially statements about the friction coefficient, and have been worked out over the past 5 centuries by a number of distinguished engineers and scientists, among them Leonardo da Vinci and Charles de Coulomb. These laws, which are approximately, but never perfectly, obeyed by typical sliding systems are as follows: (1) The friction coefficient is independent of the normal force. This is another way of saying that the friction force is proportional to the normal force. (2) The friction coefficient is independent of the sliding velocity. However, as was stated above, the friction for surfaces at rest is often about 20% greater than for the same system when sliding. Such systems are often spoken of as having larger static friction than kinetic friction. (3) The friction coefficient is independent of the apparent area of contact. Thus, flat surfaces and surfaces with contacting asperites but of the same projected area give the same friction coefficient. (4) The friction coefficient is independent of surface roughness.

Actually, the friction coefficient is primarily a property of the contacting materials and the contaminants or lubricants at the interface.

Closely related to the friction coefficient is the concept of friction angle. The friction angle θ_f is the largest angle relative to the horizontal at which a surface may be tilted, so that an object placed on the surface does not slide down. Leonhard Euler was the first to show that $\tan \theta_f$ was equal to the friction coefficient.

Mechanism. A number of processes occur at the interface between two solids that tend to inhibit sliding or to use up energy during sliding. All these contribute to friction. Over the last 3 centuries, there has been much discussion over which one of these processes is the most important and can thus be considered the main cause of friction. Until about 1940 it was considered that surface roughness was the main cause of friction, and that work had to be done during sliding to lift one sliding surface over the high spots on the other one. Modern work has largely discounted the importance of surface roughness: first, from theoretical considerations, because the work done in sliding up a high spot on a surface is largely recovered on the down side, and second, because experimental testing shows that for most sliding systems friction coefficients are largely independent of the roughness of the surfaces. Very smooth surfaces, like cleaved mica, which is smooth to within one atomic diameter, give friction at least as great as that of ordinary surfaces. For reasons that are not clear, most people find it very hard to accept the fact that smooth surfaces give as much friction as rough ones.

The modern theory of friction attributes friction to adhesion between surface atoms. These surface atoms form bonds, similar in strength to those that hold the solids themselves together, and friction represents the force required to break the bonds.

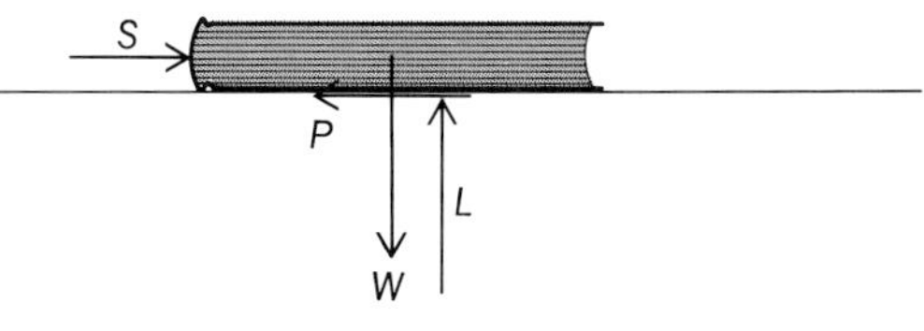

Fig. 1. The forces acting on a book resting on a flat surface when a sheer force *S* is applied. The friction from *P* is equal to *S* (up to a limiting value *F*), while *L*, the normal force, is equal to the weight *W* of the book.

Atomic forces are known to be of very short range (about 10^{-10} m), and accordingly the bonds are appreciable only at those small patches (called junctions) at which the surface atoms come within 10^{-10} m of each other. All the friction originates at these junctions, which make up the real area of contact. Over the rest of the apparent area of contact the separation of the surfaces is generally many hundreds or thousands of atomic diameters, and essentially none of the friction originates there. In a typical sliding system involving relatively large surfaces, the real area of contact is less than 1% of the apparent area.

According to the adhesion theory, lubricants at the surface reduce friction because they form bonds of reduced strength. Since hard solids generally form strong bonds while liquids and soft solids forms weak bonds, it comes as no surprise to discover that good lubricants are generally either liquids or soft solids.

Besides the adhesion mechanism, which generally accounts for more than 90% of the total friction force, there are a few minor mechanisms that act so as to increase the friction. One is a plowing mechanism, when a hard surface slides against a soft one, and plows out a series of grooves. A second one, with very rough surfaces, consists of a component of the motion perpendicular to the interface. (This is the roughness effect, which at one time was thought to be the main component of friction). Third, there is elastic hysteresis, as result of the fact that there is elastic compression and then relaxation of the material near the contacting interface, and not all the energy is recoverable. Fourth, there is an electrostatic attraction between the surfaces, but this is only significant when electrically insulating materials are used.

Almost all the frictional energy appears as heat at the interface between the sliding surfaces. This frictional heat was used by humans in prehistoric times to light fires, and this use survives today in the striking of matches, with chemical combustion initiated by the temperature rise produced by sliding. However, frictional heat usually is a nuisance, and sliding surfaces must often be cooled to prevent heat damage.

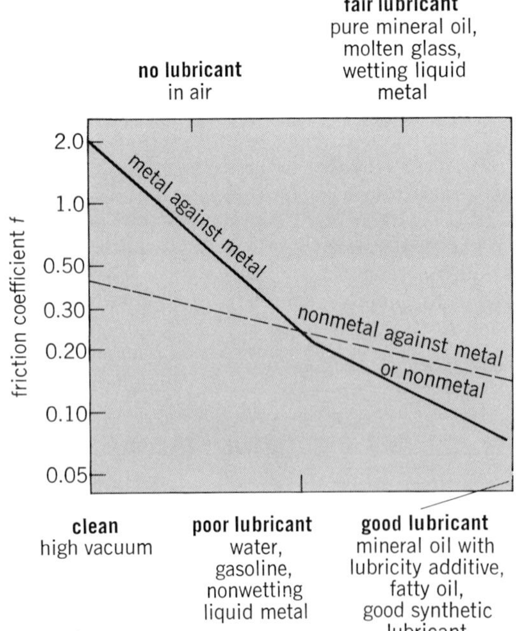

Fig. 2. Typical friction coefficient values.

Friction coefficient values. Values of the friction coefficient are required in analyzing many problems in mechanics, in order to estimate the frictional forces at interfaces and to compute their consequences. Figure 2 shows typical friction coefficient values, plotted as a function of the state of cleanliness or of lubrication of the surfaces. As will be seen, friction coefficients range from about 1.5 to about .07 depending on the sliding conditions. Unlubricated metals give higher friction than nonmetals, but well-lubricated metals give less. In consequence, when sliding mechanisms giving low friction are desired, metals are used when a lubricant is available, but nonmetals are preferred when the surfaces must operate in an unlubricated condition.

It should be emphasised that the friction coefficient values shown in Fig. 2 are typical ones, and that in any specific case friction values differing by as much as 30% are quite likely to be found. Indeed, it is not yet possible, even for an expert in the friction field, to be able to estimate the friction coefficient in any given case to within 10% or better.

A number of sliding systems are exceptional, in that they give friction values that are rather different from those shown in Fig. 2, and much use is made of their exceptional friction properties. Elastomeric materials like rubber give friction values that are about twice as great as those of other nonmetals, and this explains the use of rubber in shoe soles and heels, in automobile tires, and in other cases in which good traction is required. Solids that give exceptionally low friction, about half the values for other nonmetals, include graphite, Teflon, and ice. Teflon and graphite are used in low-friction coatings, while the low friction of ice is made use of in skating and skiing. Perhaps the lowest friction values are found in mammalian joints (like the human knee joint), where it has been found that friction coefficient values are generally as low as .02. It is not clear how such a low value is achieved.

Two special cases of low friction should be mentioned. One involves the use of rolling rather than sliding action. This generally produces much lower friction than is possible during sliding. For example, ball bearings generally give friction coefficients in the range .002 to .005. The other involves fluid-lubricated sliding systems at high sliding speeds, in which a hydrodynamic effect allows separation of the surfaces by a full fluid film. Here friction coefficient values of .001 – .003 are common.

Stopping distances. For a sliding object moving with velocity v, the distance required to bring it to a stop through friction action is governed by the equation $s = v^2/2gf$, where g is the acceleration of gravity. The distance is independent of the weight of the object. The typical stopping distances given in the table assume an initial speed of 60 mph (100 km/hr). The distances are minimal, assuming in the case of rolling vehicles that the brakes are applied severely enough to cause locking of the wheels, or a braking action just short of that.

Many traffic accidents at level railroad crossings occur because drivers of cars are not aware of the

Stopping distances at 60 mph (100 km/hr)

Typical system	Friction coefficient	Stopping distance
Car on dry road	.8	151 ft (46 m)
Car on wet road	.5	242 ft (74 m)
Train on dry rails	.3	403 ft (123 m)
Train on wet rails	.2	605 ft (184 m)
Car on icy road	.10	1210 ft (369 m)

great length of stopping distances for trains. Road accidents frequently take place because drivers used to high friction coefficient values under dry road conditions do not realize that weather conditions have changed and that lower friction values of wet surfaces apply.

Human interactions with friction. For normal walking, humans require friction coefficient values of .20 or above. When surfaces with lower friction coefficient values are encountered, people instinctively slow down, shorten their stride, and are able to walk satisfactorily when the friction is as low as .10. Slipping generally occurs when someone walking normally on a high friction surface encounters a low-friction spot (like a banana peel), and there is no opportunity to adjust stride length.

When lifting an object by hand, a neurophysiological mechanism comes into play so that the object is gripped with a force that is only a little larger than would allow it to slip.

Frictional oscillations. It was noted above that the laws of friction are not perfectly obeyed. This generally means that if one of the variables (for example, load or apparent area) is varied by a factor of 10, the friction coefficient is changed by less than 10%. Only one of these departures from constancy has practical significance, namely, if the friction goes down when the sliding speed goes up (a so-called negative characteristic), frictional vibrations may be produced. At slow sliding speeds the vibrations take the form of relaxation oscillations (creaking, chattering, and groaning noises), while at high speeds they are harmonic oscillations (squeaking). As a general rule, unlubricated and poorly lubricated surfaces give a negative characteristic and show a tendency toward oscillation, while well-lubricated surfaces give a positive characteristic and are immune to oscillations.

Friction oscillations are generally referred to as "stick-slip," but strictly speaking this term applies only to relaxation oscillations, because during part of each oscillation cycle the sliding surfaces actually stick and have zero relative velocity. There is always some interfacial slip during harmonic oscillations.

Frictional oscillations can generally be avoided by changing one or more of the parameters. Helpful changes include reducing the normal load, increasing the sliding speed, introducing damping into the system, stiffening all components acting as springs, changing materials, or introducing good lubricants.

Frictional oscillations can be useful. They are the mechanism of music production in bowed stringed instruments, and they have been used occasionally to detect the entry of intruders into a home. More significantly, in many sliding systems they act as a warning signal that the lubricant has disappeared, that excessive wear damage is imminent, and that relubrication is in order.

Significant friction problems. In prehistoric and early historic times, humans' main interest in friction was to reduce the friction coefficient, to reduce the labor involved in dragging heavy objects. This led to the invention of lubricants, the first of which were animal fats and vegetable oils. A great breakthrough was the use of rolling action, first in the form of rolling logs and then in the form of wheels, to take advantage of the lower friction coefficients of rolling systems. *See* LUBRICANT.

In modern engineering practice available materials and lubricants reduce friction to acceptable values. In special circumstances when energy is critical, determined efforts to minimize friction are undertaken. Friction problems of practical importance are those of getting constant friction in brakes and clutches, so that jerky motion is avoided, and avoiding low friction in special circumstances, such as when driving a car on ice or on a very wet road. Also, there is considerable interest in developing new bearing materials and new lubricants that will produce low friction even at high interfacial temperatures and maintain these properties for long periods of times, thus reducing maintenance expenses. Perhaps the most persistent problem is that of avoiding frictional oscillations, a constant cause of noise pollution of the environment.

[ERNEST RABINOWICZ]

Bibliography: F. P. Bowden and D. Tabor, *Friction: An Introduction to Tribology*, 1973; D. H. Buckley, *Friction, Wear and Lubrication in Vacuum*, National Aeronautics and Space Administration, NASA SP-277, 1971; S. Halling (ed.), *Principles of Tribology*, 1975; D. F. Moore, *The Friction and Lubrication of Elastomers*, 1972; J. J. O'Connor and J. Boyd (eds.), *Standard Handbook of Lubrication Engineering*, 1968

Fuel injection

Implicitly, the delivery of fuel to an internal combustion engine by pressure from a mechanical pump. It must be used in diesel engines because of the high pressures necessary to deliver fuel against the highly compressed air in the engine cylinders at the end of the compression stroke, as required for combustion in the diesel cycle. In addition, considerable pressure is required to break up the fuel oil, which has low volatility and is often viscous, into the fine-spray pattern required for good combustion. *See* DIESEL CYCLE; DIESEL ENGINE; INTERNAL COMBUSTION ENGINE.

The fuel pumps are of the reciprocating piston type, fitted to very close tolerances. One pump is usually provided for each cylinder, driven directly by the engine, with the delivery timed as needed for combustion in the cylinder. The load carried by the engine, that is, the mean effective pressure (mep) developed by combustion, is controlled by adjustment of the volumetric capacity of the pumps. Since the oil pressure may be so high as to cause oscillation waves in the lines to the spray valves, which would interfere with the desired fuel delivery to individual cylinders, all fuel lines between the pumps and spray valves are usually made of equal length. The oscillation waves in long fuel-delivery lines are eliminated in another injection system, which supplies fuel for all cylinders from a high-pressure manifold or "rail"; injector valves at each cylinder control the amount of fuel injected by the length of time they are held open by cams on a suitably located camshaft. Some early diesel engines used an air-injection system, in which a metered quantity of fuel was delivered to the injection valves while they were closed and was atomized into the cylinder at the time and rate desired for combustion by opening the air injector valve from an air supply at very high pressure. This usually required air compressors with as many as four stages with intercoolers and water separators. They were expensive, but they did provide excellent smoke-free combustion. *See* MEAN EFFECTIVE PRESSURE.

Fuel injection has been used also for spark ignition engines with volatile fuels, notably racing automobile engines and reciprocating piston airplane engines. The advantages over the almost universally used aspirated carburetors on these engines are the additional power which may be obtained from an engine of a given size, elimination of the warm-up time, and freedom from stalling due to ice formation on the throttle valves. The additional power results from the elimination of both the throttling effect of the venturis, which are required for the pressure drop to draw fuel through the metering system and atomizing nozzle, and the heating required to convert the spray of liquid fuel into a vapor which may be distributed more evenly to each cylinder from a manifold when more than one cylinder is supplied by one carburetor. The heat not only expands the air and reduces the weight of combustible charge per cycle, but it also decreases the allowable compression ratio which may be used with a fuel of given antiknock value. Fuel-injection systems minimize these power losses characteristic of aspirating carburetors, as well as the warm-up time required for the development of sufficient heat for fuel evaporation. Until sufficient heat is attained, there is always a possibility of ice forming in the carburetor because of the refrigeration due to the latent heat of vaporization of the fuel when the temperature of air entering the carburetor is 35–45°F, if there is considerable moisture in the air.

While fuel injection minimizes these disadvantages characteristic of aspirated carburetors, the additional mechanical parts required (such as metering pumps to control the metering rate of fuel with change in engine speed and load, especially if the change is rapid) are expensive and liable to misadjustment; consequently it is used on few engines for the general public.

Three methods of fuel injection have been developed for spark ignition engines: direct injection, metered port injection, and continuous-flow injection. Direct injection supplies fuel to individual cylinders, either by individual metering pumps as in diesel practice or by a single metering pump with a distributor which directs each discharge stroke to an individual cylinder in sequence. In metered port injection the fuel required for a working cycle is metered to each intake port before or while the intake valve is open. This may be accomplished by a variable displacement device or by control of the length of time the spray valve is opened to a constant-pressure fuel supply. Continuous-flow injection supplies fuel to the intake ports, the quantity being controlled by variation of the fuel supply pressure. *See* CARBURETOR.

[NEIL MAC COULL]

Bibliography: C. F. Taylor, *The Internal Combustion Engine in Theory and Practice*, vol. 2, 1977.

Fuel pump

A pump for drawing fuel from a storage tank and forcing it to an engine or furnace. Choice of pump type for fuel depends to a great extent on the volatility of the liquid to be pumped.

In a gasoline engine the fuel is highly volatile at ambient temperature; therefore a sealed diaphragm positive-displacement pump is used. With such a pump the fuel line is completely sealed from tank to carburetor to prevent escape of fuel and to enable the pump to purge the line of vapor in event of a "vapor lock" caused by fuel vaporizing due to abnormally high ambient temperature. *See* FUEL SYSTEM.

In a diesel engine, where fuel is injected at high pressure through a nozzle into the combustion cylinder, a piston serves as its own inlet valve and as the compression member of the fuel pump. A spring-biased needle controls the outlet when fuel pressure reaches the required high level. In an oil-fired furnace, although nozzle pressures need not be so high as in diesel engines, a piston pump is also used to provide positive shutoff of the fuel line when the pump stops. *See* DIESEL ENGINE.

In a liquid-fueled rocket, with pump feed, both fuel pump and oxidizer pump are typically centrifugal or turbine types for high capacity. The pumps are usually driven by an accessory turbine engine that burns a small quantity of pumped fuel and oxidizer. Among design considerations is the necessary high flow rate, which in turn requires high pump speed to obtain the required pressure rise with a lightweight structure. The pumps must also withstand the corrosive liquids and have close clearances, preferably with no leakage. Flow rates may be limited by cavitation at the pump inlet, especially because the fuel or cyrogenic fluid, such as liquid oxygen, may be near its vapor pressure because of its high velocity.

Fuel pumps must operate before the engine or furnace that they serve can function; consequently an auxiliary or starter motor is usually an essential adjunct to a fuel pump. [FRANK H. ROCKETT]

Fuel system

A system which stores fuel for present use and delivers it as needed to an engine. The commonly used components for automobiles and stationary gasoline engines are fuel tank, remote reading fuel gage, filter, fuel pump (usually engine driven), and carburetor (Fig. 1).

Automobile. In automobiles the fuel pump is usually of the diaphragm type (Fig. 2). The rotation of an eccentric on the engine cam shaft actuates a lever that pulls the pump diaphragm down against the pressure of the diaphragm spring. This lowers the pressure in the pumping chamber, drawing fuel up from the tank through the inlet valve into the chamber. As the eccentric moves away from the lever, the diaphragm is moved upward by the diaphragm spring, and fuel is forced through the outlet valve to the carburetor. When the carburetor fuel chamber is filled, its float valve closes and a back pressure is created in the fuel-pump chamber, holding the diaphragm down against the pressure of the diaphragm spring. The lever still rides on the eccentric but without further pumping effect. *See* FUEL PUMP.

Because the pump is usually located higher than the tank level, there may be some tendency, particularly during short engine stops on hot days, for the volatile fuel to boil in the supply line and pumping chamber to such an extent that the full travel of the diaphragm will not raise enough pressure in the vapor content to expel it onward to the carburetor. The system is then said to be vapor-locked; the usual remedy is to open the engine hood until the fuel system cools off.

One way of avoiding vapor lock is to mount a remotely driven fuel pump close to, and at the bottom level of, the fuel tank (as shown by the broken outline in Fig. 1) so that there is no additional boiling between the tank and pump. Such pumps

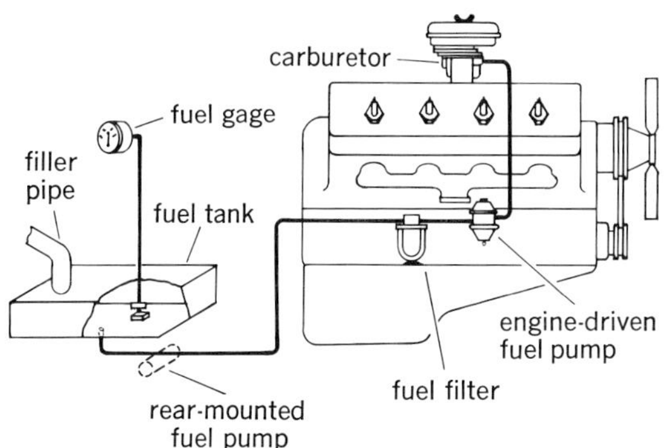

Fig. 1. Elements of an automobile fuel system.

usually have electric drive units which operate plungers, diaphragms, or metallic bellows pumping elements, and are designed so that their speed of operation and current consumption depend upon the rate of fuel use.

Aircraft. The presence of multiple engines and multiple tanks complicates the aircraft fuel system. Also, the reduction of pressure at altitude necessitates the regular use of boost pumps, sub-

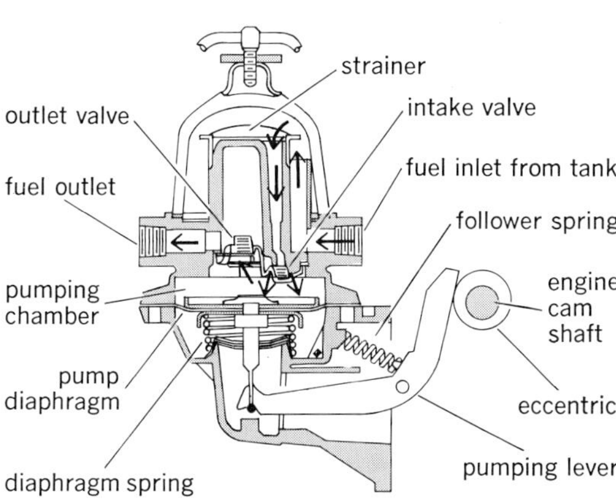

Fig. 2. Typical automobile fuel supply pump.

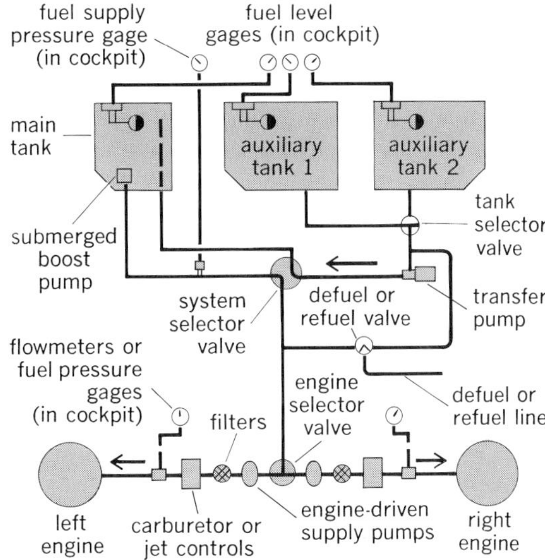

Fig. 3. Diagram of a typical aircraft fuel system.

merged in the fuel tanks, which are usually of the centrifugal type and electrically driven. These supplement the engine-driven fuel supply pumps, which are usually of the gear or eccentric vane type.

Components of a typical aircraft fuel system include one main and two auxiliary tanks with their gages; booster, transfer, and engine-driven pumps; various selector valves; and a fuel jettisoning or defuel valve and connection, which is typical also of what would be needed for either single-point ground or flight refueling (Fig. 3). The arrangement is usually such that all the fuel supply will pass to the engines by way of the main tank, which is refilled as necessary from the auxiliary tanks. In case of emergency, the system selector valve may connect the auxiliary tanks to the engines directly. Tank vents, not shown, are arranged so that overflow will go safely overboard.

Because of the large quantities of fuel used, aircraft fuel systems are often contaminated with dirt, metal chips, and lint; adequate filtration is therefore essential to reliable service. *See* CARBURETOR.

[F. C. MOCK/J. A. BOLT]

Furnace construction

A furnace is an apparatus in which heat is liberated and transferred directly or indirectly to a solid or fluid mass for the purpose of effecting a physical or chemical change. The source of heat is the energy released in the oxidation of fossil fuel (commonly known as combustion) or the flow of electric current through adjacent semiconductors or through the mass to be heated. In recent years, scientific and engineering effort has been made to utilize nuclear and solar energy for heating purposes. Therefore, according to the source of heat and method of its application, there are four categories of furnaces: combustion, electric, nuclear, and solar, in the order of their present commercial or industrial importance.

Furnaces employing combustion vary widely in construction, depending upon the application of the heat released, whether direct or indirect. Direct heat transfer is used, in regenerative refractory-type heaters, in flow systems in which reactants are injected into the combustion gases. Indirect heat transfer is employed in heaters in which the mass to be heated is kept separate from the combustion gases and made to flow in tubes which absorb and transmit the heat to the fluid to be heated.

Furnaces developed for indirect heat transfer can be divided into two classes. One class of heaters is used solely for general utility purposes, such as all types of boilers; and the second class is applied in the petroleum and chemical industries as an essential unit operation in refining or processing plants. The second category of furnaces will be discussed in this article.

Directly fired furnaces are employed in oil refineries and chemical process units whenever the temperature level to which a fluid must be heated is above that attainable with utility steam. Furnaces as a heat-transfer apparatus generally cost more per thermal unit of heat transferred than conventional tubular heat exchangers. According to the kind of service, furnaces in process units should be divided into two classes:

1. Those which perform solely a heating duty, that is, raising the temperature of a fluid and effecting essentially no change of state or of chemical composition. Furnaces which perform this one duty may be termed conventional heaters.

2. Those which handle a fluid undergoing a change during heating. The physical or chemical changes constitute an essential performance requirement. Typical applications of this order are associated with such processes as distillation or preheating of temperature-sensitive materials, pyrolysis of hydrocarbons or organic chemicals, and catalytic steam-gas reforming for the production of synthesis gas.

The conventional tubular heater came into industrial use, especially in oil refineries, about 1925 and was termed a tube still to distinguish it from the shell still, a horizontal cylindrical vessel mounted on top of a firebox. The design of these tube stills or furnaces, which is essentially the same as originally conceived by analogy with developments in steam-generating equipment, consists of pipes connected by 180° return bends forming a continuous coil and arranged in a refractory furnace setting, partly in the combustion chamber. Heat is absorbed mainly by radiation and partly in a confined flue-gas passage from the combustion chamber in which heat is absorbed mainly by convection. The fluid flow through the coil is generally countercurrent to the flow of combustion gases, first through convection tubes and then through the radiant-tube section; thus a reasonable thermal efficiency can readily be attained by providing convection tube surface to an economically justifiable extent. *See* ELECTRIC FURNACE; STEAM-GENERATING FURNACE.

The furnaces provided in modern process units vary considerably in outer shape, arrangement of tubes, and type and location of burners, depending mainly upon the desire of the designer or manufacturer to be identified with a specific model. Some distinctive furnace designs are shown in Fig. 1.

Design. The furnace design for a given performance or thermal efficiency is usually evolved by the following procedure: (1) determination of the composition of the combustion products and the amount of the liberated heat which must be utilized to meet the postulated thermal efficiency; (2) allocation of heat to be absorbed by the heating elements located in the combustion or radiant chamber and in the convection section; (3) determination of the heat-transfer rate and heating surface area in the radiant section; and (4) determination of the heat-transfer rate and tube surface area in the convection section or sections.

Source of heat. The combustion products vary in composition according to the type of fossil fuel burned and the excess air used in the oxidation process. The carbon and hydrocarbon content of the fuel governs the combustion-gas composition, and many formulas relating it to the elemental, oxidizable constituents of the fuel have appeared in the literature. Complete combustion is a prerequisite for high thermal efficiency; to ensure this, air is used in excess of the minimum or stoichiometric requirements, depending upon the type of fuel and the combustion equipment or system (Table 1).

Table 1. Fuel-air mixtures

Type of fuel	Type of burner	Normal range of excess air, %
Natural or refinery gas	Air and fuel gas premixed	5–15
Natural or refinery gas	Air induced by natural draft	10–20
Distillate fuel oils	Steam atomization air by natural draft	15–25
Distillate fuel oils	Mechanical atomization air by natural draft	15–30
Residual fuel oils	Steam atomization air by natural draft	20–35

The type of fuel, its heating value, and the excess air applied in combustion determine the theoretical flame temperature which would prevail if the oxidation were instantaneous. However, combustion is a rate process requiring time, and the actual flame temperature is considerably lower because of the radiation from the combustion zone. The theoretical flame temperature can readily be determined from the heating value of the fuel and the amount of combustion products evolved, including the excess air. Accurate enthalpy data on the combustion gases are required and are presented in the literature.

In order to meet a required thermal efficiency, the heat losses must be appraised. There is a certain loss through the furnace setting which varies with the surface area of the enclosure, the heating capacity, and the atmospheric environment. The wall construction of most types of furnaces, that is, application of insulating material, is such that the specific heat loss, Btu/(ft^2)(hr) of outer wall area, is practically the same for all operating temperature levels; hence the heat loss will vary principally only with the furnace capacity. Table 2 shows the average values that will be encountered in industrial furnace constructions. The setting heat loss ΔH_{sl} can be calculated from the relation $\Delta H_{sl} = \lambda H_{fl}$, where λ is the fraction of heat released, and

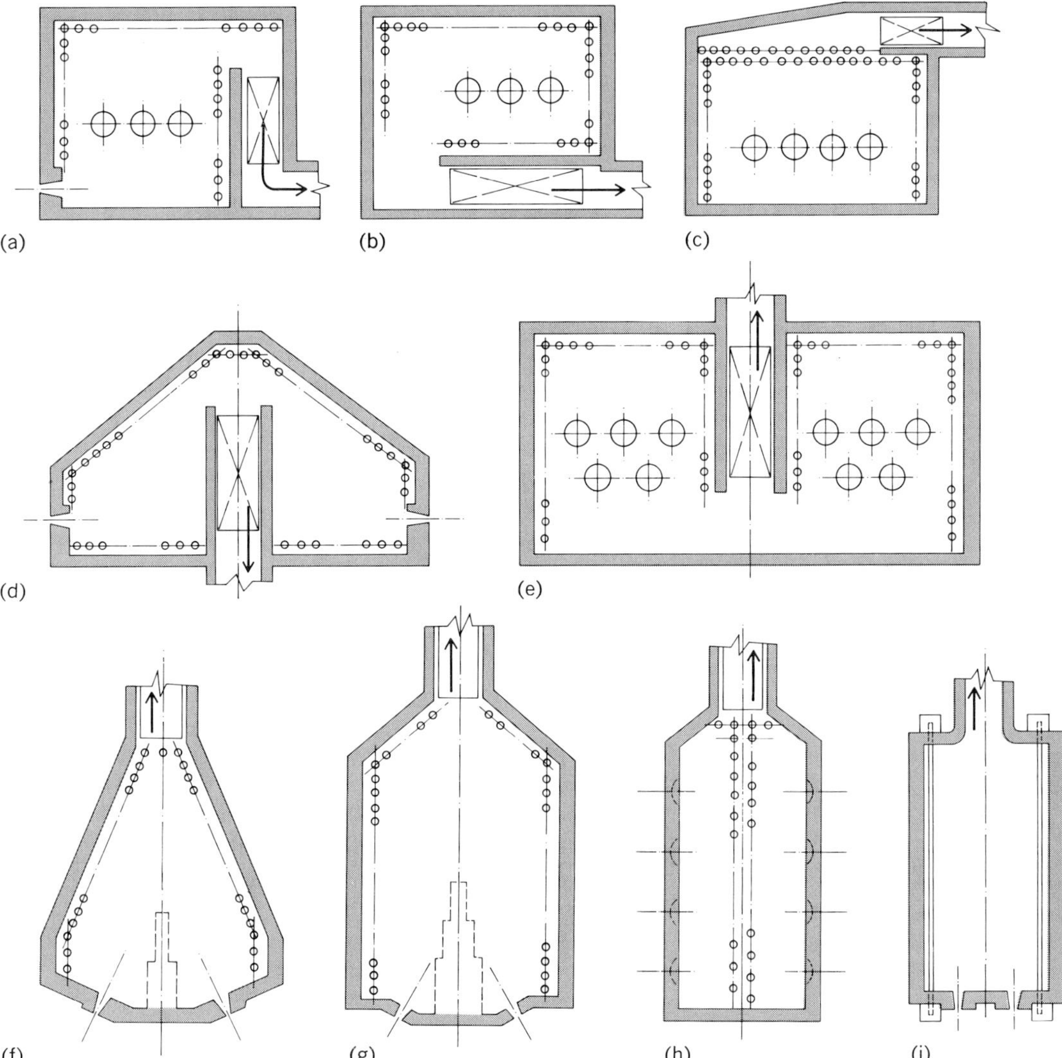

Fig. 1. Typical process furnace designs and characteristic features. (*a*) Radiant, updraft, vertical-convection bank. (*b*) Radiant, downdraft, horizontal-convection bank. (*c*) Radiant, updraft, with high convection effect in radiant roof tubes. (*d*) Large refractory heater, dual radiant sections, sloping roofs. (*e*) Large conventional refinery heater, dual radiant sections. (*f*) Updraft furnace, slanting walls, two parallel coils. (*g*) Updraft furnace, vertical walls, two parallel coils. (*h*) Updraft furnace, tubes in center, burners in wall, and with good heat-intensity distribution. (*i*) Updraft circular, all-radiant furnace, with the burners in the floor.

H_{fl} is the enthalpy of the combustion gas at the theoretical flame temperature.

The flue-gas effluent temperature from the furnace can then be determined in accordance with the postulated thermal efficiency η, as shown in the equation below, where H_{st} is the enthalpy of

$$H_{st} = H_{fl} \times (1 - \eta) - \Delta H_{sl}$$

the combustion gas at the convection-bank exit temperature.

Table 2. Heat loss from setting

Heat liberated, Btu/hr	Loss as a fraction of heat released, λ $\Delta H_{sl} = \lambda H_{fl}$
15,000,000	0.030
30,000,000	0.0275
60,000,000	0.0225
75,000,000+	0.020

Heat load. The distribution of heat load between the radiant and convection sections of conventional heaters is associated with the manufacturer's design approach and service requirements. There are all-radiant heaters, and furnaces with tubes arranged mostly for heat absorption by convection. The amount and disposition of heating surface in the combustion or radiant chamber has a distinct influence upon the radiant-heat absorption and the supplementary heat to be recovered by convection for the required thermal efficiency. Therefore, the allocation of heating duties to the radiant and convection sections is more or less an empirical matter. Conventional process heaters are generally designed for approximately 75% thermal efficiency in the United States, where fuel costs are relatively low. Most box-type heaters are designed for moderate radiant-heat-transfer rates; approximately 65–75% of the heat load is carried by the radiant section and 25–35% by the convection bank. In some cases, very high radiant-heat intensities are employed as a matter of design principle, and in these furnaces, the heat absorbed in the radiant section may be only 50% of the total duty.

Radiant section. The present procedure for determining the radiant-heat-transfer rate and heating surface rests on a long series of scientific investigations and evaluation of operating data.

The combustion gases radiate to the solid-body environment, tubes, and refractories. Carbon dioxide and water vapor are the principal constituents with emissive power. Their emissivity depends upon the partial pressure and the thickness of the gas layer, and at 1900°F has a value of approximately 25% of the black-body radiation intensity. The terms luminous- and nonluminous-flame radiation used in scientific treatises are somewhat misleading; the mere luminosity or light has no significance as far as heat radiation is concerned, because practically all emissivity and certainly carbon dioxide, CO_2, and water, H_2O, radiation is in the infrared wavelength region, even at flame temperatures encountered in process heaters. The maximum gas radiation for an infinite value of the product of the partial pressure and thickness of the gas layer, frequently termed black gas, amounts to only 21–33% of the complete blackbody emissivity at temperatures of 2500 and 1100°F, respectively. The mean radiating temperature in the combustion chamber lies within this temperature range.

The consideration of heat transmission by radiation alone, from the combustion gas and refractory walls by reradiation, has not resulted in the derivation of a universally applicable formula for the heat-transfer rate in the radiant section. This is because heat is also transmitted to the tubes by convection which is in the order of 10% to as high as 35% of the heat absorbed in the radiant section, depending upon (1) the shape of the combustion chamber, (2) the arrangement of tubes, and (3) the type and location of burners. The convection effect is a linear function of the temperature difference (not the difference of temperatures to the fourth power) and can be varied for the same furnace design by the amount of excess air employed.

Convection heating surface. The rate of heat transfer solely by convection, with flow of gas at a right angle to tubes, increases with the gas velocity and the temperature level, and is greater for small than for large tubes.

There are radiation effects in the convection-tube bank augmenting the heat absorption, especially when the flue-gas temperatures or the temperature differences between tube wall and flue gases are large. These include (1) radiation to the front two rows of tubes, which can be appreciable if they are exposed to the combustion chamber; (2) radiation from the hot gases surrounding the tubes; and (3) the radiation from the refractory enclosure to the tube bank as a whole.

Process furnaces. The design of process furnaces, although following generally the procedure outlined earlier for conventional heaters, requires careful consideration of the transitory state of the fluid being heated.

Vacuum distillation of a temperature-sensitive fluid or the stripping of solvent from the raffinate and extract of a selective solvent-separation process are typical cases in which temperature limitations are encountered in the heating operation. Partial vaporization of the fluid must take place in the tubes under relatively mild heat-transfer rates and tube-wall temperatures, not greatly in excess of the bulk-flowing temperature. Aside from the knowledge and postulation of the heat-intensity pattern and distribution on the combustion-gas side, the proper design of the furnace coil involves exacting pressure-drop calculation for mixed vapor-liquid flow in conjunction with phase-equilibrium determination of the amount and composition of vapor and liquid at any given point in the coil. The pressure drop for mixed vapor-liquid flow is reasonably well established, and vapor-liquid equilibrium relationships are known for many binary and ternary systems. The determination of temperature, pressure, and transition state of the fluid throughout the coil involves stepwise trial-and-error calculations which can best be executed by a modern digital computer for various coil dimensions and heat-intensity patterns. The coil design which satisfies all technical criteria and also cost aspects can then be selected from the computer evaluations.

Gas reforming is another industrially important heating process involving a chemical change of the fluid. Methane or natural gas and steam are decomposed in the presence of excess steam in tubes filled with a catalyst. The principal products are hydrogen, H_2, carbon monoxide, CO, and CO_2, and maximum conversion of hydrocarbons is attempted. The chemical transformation proceeds at approximately 1400–1450°F and is highly endothermic, having a heat of reaction of 87,500 Btu/(lb)(mole of methane converted). The tubes are located in the center of the radiant or combustion chamber and are heated from two sides to effect

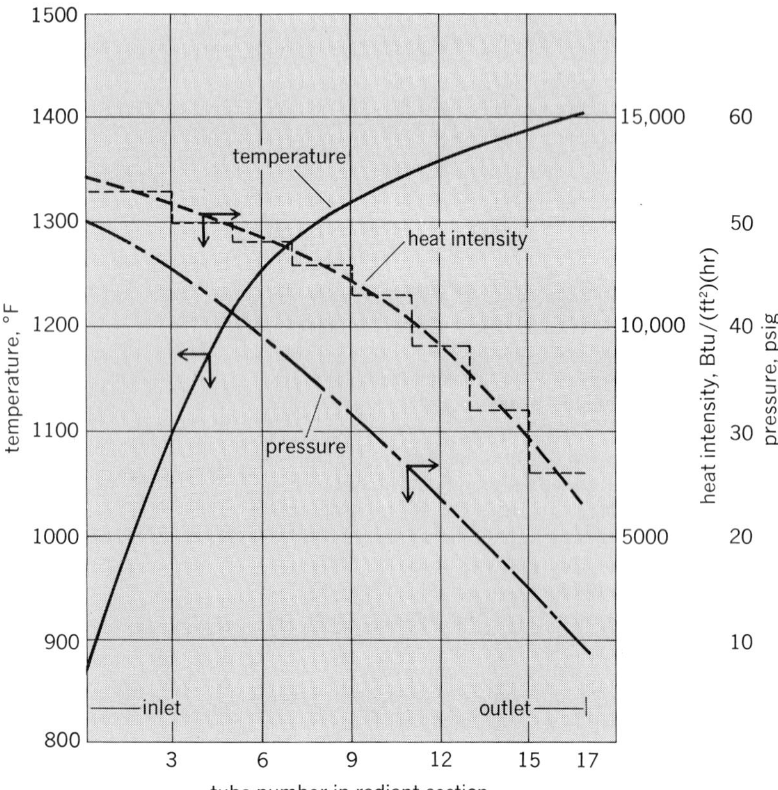

Fig. 2. Typical temperature and heat-flux pattern for naphtha pyrolysis coil.

an even circumferential heat-intensity distribution. The average rate of heat transfer is in the order of 18,000 Btu/(ft^2)(hr).

Pyrolysis of hydrocarbons, the thermal conversion of ethane, propane, butane, and heavier hydrocarbons to produce olefins and diolefins for the petrochemical industry, is another example of a process furnace. The modern pyrolysis furnace is an outgrowth of the oil-cracking heater, the mainstay of motor-fuel production from petroleum before the advent of catalytic cracking. The design of a pyrolysis furnace must take thermochemical equilibria and reaction rates into consideration in order to realize a high conversion and a product distribution most favorable with respect to yield of olefins. The conversion of higher-molecular-weight hydrocarbons which have a high rate of decomposition involves the problem of attaining the coil outlet temperature dictated by equilibrium relationships. A shorter coil-residence time and correspondingly higher heat-transfer rates are means to this effect, but still the basic principle of moderating the heat intensity in the pyrolysis coil zone, in which the reactant is already in a high state of conversion, must be observed. Figure 2 gives a typical temperature-pressure and heat-flux pattern for a pyrolysis coil cracking a naphtha boiling in the range 190–375°F for maximum ethylene yield. The design of such a pyrolysis coil can be executed with greatest precision or refinement by the use of a modern digital computer, permitting the rapid evaluation of all design variables.

Mechanical construction. The design of the furnace setting and application of refractories and insulating materials generally follow standard practice prevailing in boiler-furnace construction. Fully suspended wall construction and the use of light-weight firebrick with bulk densities of 30–65 lb/ft^3 (as compared with 125–150 lb/ft^3 for fireclay bricks) are generally preferred for process heaters to reduce the heat storage in the setting, start-up, and shutdown time.

The burners are carefully selected and must offer great flexibility with respect to firing rate and excess air. The use of a multiplicity of small-capacity burners is rather common for specific process heaters such as gas-reforming and pyrolysis furnaces.

The design of the heating coil for high-temperature service involves the application of alloy steels. Tube materials are usually selected according to the schedule in Table 3. Tubes of carbon and low-chrome steel are usually connected by forged-welding return bends, and those of stainless steel by cast 25/12 chrome-nickel alloy steel bends, also weldable to the tubes.

Table 3. Heating-coil materials

Tube-wall temperature range, °F	Tube material*
875 and under	Carbon steel, A-161
875–1150	Low-chrome steel, A-200 (1.25–2.25% Cr, 0.5% Mo)
1150–1550	18/8 stainless steel, A-271, type 304
1550–1800	25/20 stainless steel, A-271, type 310

*American Society for Testing and Materials specifications.

The flue-gas exhaust is normally carried to the atmosphere by a stack of sufficient height to induce a draft equivalent to the pressure drop in the flue-gas passage. For highest heat economy a process furnace can readily be equipped with an air preheater or waste-heat boiler usually requiring air and flue-gas exhaust blowers. Preheated combustion air results in higher flame and mean effective radiation temperatures which aid in attaining high radiant-heat-transfer rates required in certain pyrolysis operations. These furnace accessories are provided in the same manner as in boiler design practice. *See* GAS FURNACE; HEAT EXCHANGER; OIL FURNACE; PYROMETALLURGY.

[H. C. SCHUTT/H. L. BEGGS]

Bibliography: J. D. Gilchrist, *Fuel, Furnaces, and Refractories*, 1976; D. Q. Kern, *Process Heat Transfer*, 1950; L. Lichty, *Combustion Engineering Processes*, 7th ed., 1967; W. H. McAdams, *Heat Transmission*, 3d ed., 1954; W. L. Nelson, *Petroleum Refinery Engineering*, 4th ed., 1958; R. D. Reed, *Furnace Operations*, 1976.

Fuse (electricity)

An expendable device for opening an electric circuit when the current therein becomes excessive. An electric fuse consists principally of a section of conductor, known as a fusible element, of such properties and proportions that excessive current melts it and thereby severs the circuit. Fuses are used in nearly all types of electric circuits to protect circuit conductors and apparatus from damage that could result from sustained excessive current.

Fuses are rated according to the voltage of the circuit for which they are designed, the current they can carry continuously, and the amount of excessive current they can successfully interrupt. Voltage ratings range from a few volts to more than 100,000 volts; current ratings extend from a fraction of an ampere to several thousand amperes. At lower voltages interrupting ratings may be hundreds of thousands of amperes.

General classification. The most familiar type of fuse is the screw-plug fuse used in domestic electric systems (Fig. 1). It consists of a shell, similar to the base of an incandescent lamp, a fusible element connected between the screw shell and the center terminal, and a transparent protective cover.

Another widely used type is the cartridge fuse, in which the fusible element is connected between metal ferrules at either end of an insulating tube (Fig. 2). In an expendable cartridge fuse, this tube is usually filled with an insulating material, such as chalk, sand, or a suitable liquid, which cools and quenches the arc that forms when the fusible element melts. Some fuses, known as renewable fuses, are so constructed that a new fusible element can be inserted to replace one which has been melted.

In some types the cartridge is unfilled and open at one end to allow the arc gases to escape to the atmosphere. These are known as expulsion fuses. Some expulsion fuses and liquid-filled fuses have a spring arranged to pull the terminals of the fusible element apart quickly when it melts, thus rapidly lengthening the arc and expediting its

FUSE (ELECTRICITY)

Fig. 1. Common plug-type fuse. (*Bussman Manufacturing Division, McGraw Edison Co.*)

FUSE (ELECTRICITY)

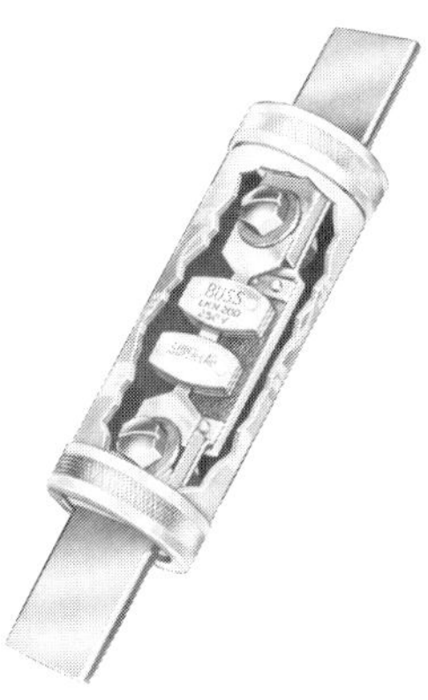

Fig. 2. Renewable cartridge fuse, 200 amp at 250 volts. (*Bussman Manufacturing Division, McGraw Edison Co.*)

extinction. Expulsion fuses are usually used in high-voltage circuits where the lengthening of the arc is necessary for interruption of the circuit.

There are many ratings and types of cartridge fuses. They range in size from about 1 in. long, such as those used in automobile circuits, to high-voltage power fuses, some of which are several feet long and several inches in diameter.

Less commonly used are open-link fuses, which are merely strips of fuse material bolted to open terminal blocks.

Operation. The fundamental relationship between the time required for the fuse to melt and the value of the current, known as the time-current curve, is an important property of the fuse. To protect an apparatus, the fuse should melt and interrupt the circuit before the apparatus is overheated. This action must take place under any combination of current and time. Coordination of current and time is provided by varying the size, shape, and material of the fusible element. The time required to quench the arc after the element melts depends upon the nature of the arc-quenching material and the fusible element. The sum of the melting and arcing times is the interrupting time.

Special types. In large power systems available short-circuit current can be enormous. In such systems, it is often desirable to have a fuse that can melt and interrupt rapidly. With such a fast-acting fuse, the current never reaches the value that the system would be capable of delivering if current were allowed to flow for a longer time. These fuses are known as current-limiting fuses. To accomplish the current-limiting effect, the fuse must be able to melt rapidly at high currents and to form an arc that the system voltage cannot sustain. One way to accomplish rapid melting is to cut a notch in the silver fusible element and to attach a small drop of soft solder to the element at the notch. To generate a quenchable arc (one having an arc voltage higher than the available circuit voltage), several notches of this kind are provided on the element. The element is packed tightly in sand. When the element melts, several arcs form. The close contact of arcs with sand causes a large amount of energy to be removed from the arcs, increasing the arc voltage.

Fuses are frequently used in circuits in which it is required that moderately excessive currents, such as during motor starting, be permitted to flow intermittently without melting the fuse. Such fuses, called time-delay or time-lag fuses, have relatively longer melting time at moderate overcurrent than those used in circuits in which occasional overcurrents do not normally occur.

Many fuses are equipped with a device to give visual evidence that the fusible element has melted. These are called indicating fuses.

Fuses used on utility-line poles are sometimes mounted in such a way that, when the fuse melts, the fuse holder drops to an open circuit position readily distinguishable from the ground. These are called drop-out fuses (Fig. 3). They spring open when the fuse melts, not only to give visual indication but also to provide rapid arc extinction.

FUSE (ELECTRICITY)

Fig. 3. Flip-open cutout fuse rated 50 amp at 7.8 kv, used on utility-line poles. (*General Electric Co.*)

A major limitation to the use of fuses arises in polyphase circuits. A fault in a polyphase circuit equipped with fuses will usually cause only one fuse to interrupt its phase circuit. With the other phase circuits still intact, single-phase power is applied to any load connected to these lines. Such a condition can damage polyphase motors left running single-phase. To avoid such damage, in some applications the polyphase circuit is arranged so that an open-circuited fuse trips a multipole automatic switch or circuit breaker to open all the phase circuits. In this arrangement, the current-limiting-type fuse is often used to interrupt only those currents that exceed the interrupting rating of the circuit breaker and that seldom occur. Interruption of the more frequently occurring lower values is left to the circuit breaker.

For other means of protecting against excessive current *see* CIRCUIT BREAKER; ELECTRIC PROTECTIVE DEVICES.

[THOMAS H. LEE]

Bibliography: D. G. Fink and H. W. Beaty (eds.), *Standard Handbook for Electrical Engineers*, 11th ed., 1978.

Gages

Devices for determining the relative size or shape of objects. The function of gages is to determine whether parts are within or outside of the specified tolerances, which are expressed in a linear unit of measurement. Gages are the most widely used production tools for controlling linear dimensions during manufacture and for assuring interchangeability of finished parts. A gage may be an indicating type that measures the amount of deviation from a mean or basic dimension, or it may be a fixed type that simply accepts parts within tolerance and rejects parts outside tolerance. *See* TOLERANCE.

Gage tolerance. If parts are to be interchangeable, it is necessary that, when one device is used to control the accuracy of another, the controlling device itself ordinarily must be accurate to the next significant figure. Thus, if a dimension of a part being manufactured is to be controlled accurately in thousandths of an inch (1 in. = 2.5 cm), the gages to do this must be accurate in ten-thousandths of an inch. The gages must be set or measured by instruments accurate in hundred-thousandths of an inch. The instruments must be set by gage blocks accurate in millionths of an inch. The size of the gage blocks is verified by an interferometer reading in ten-millionths of an inch.

The tolerance allowed the gage maker and its effect on the size of the part can be illustrated as follows (Fig. 1). If the diameter of a 2.000-in. shaft is to be controlled during manufacture to 0.004-in. tolerance, the "go" ring gage would have a wear allowance of 0.0002 and a gage maker's tolerance of 0.0002 in. Its size would be 1.9998 − 0.0002 in. The "not go" gage would have no wear allowance, but a gage maker's tolerance of 0.0002 in. Its size would be 1.9960 + 0.0002 in. Thus, because of the necessity for a gage tolerance, if the part tolerance must be maintained absolutely, something less than the full part tolerance is available as a working tolerance (in this case 85–95%), depending upon the actual sizes of the "go" and "not go" gages.

Gage blocks. The standard of lineal measurement for most manufacturing processes is the gage block (Fig. 2). Gage blocks are used for setting gages, for setting machines, and for comparative measurements. *See* COMPARATOR.

Gage blocks are made of chrome steel and may be chrome-plated or have carbide wear faces. Individual blocks have two surfaces which are flat and parallel. The parallel distance between the surfaces is the size marked on the block to a guaranteed accuracy of 0.000002, 0.000005, or 0.000008 in. at an ambient temperature of 68°F (20°C). Blocks are sold in sets so that combinations of individual blocks, when wrung firmly together, will produce a new end standard equal to the sum of the sizes of the individual blocks. With a set of 81 blocks, any length expressed in four decimal places from 0.2000 in. to approximately 12.0000 in. can be built up. Between 0.3000 in. and 4.0000 in., three to four gage groups can be built at the same time from a set to an identical length expressed in four decimal places.

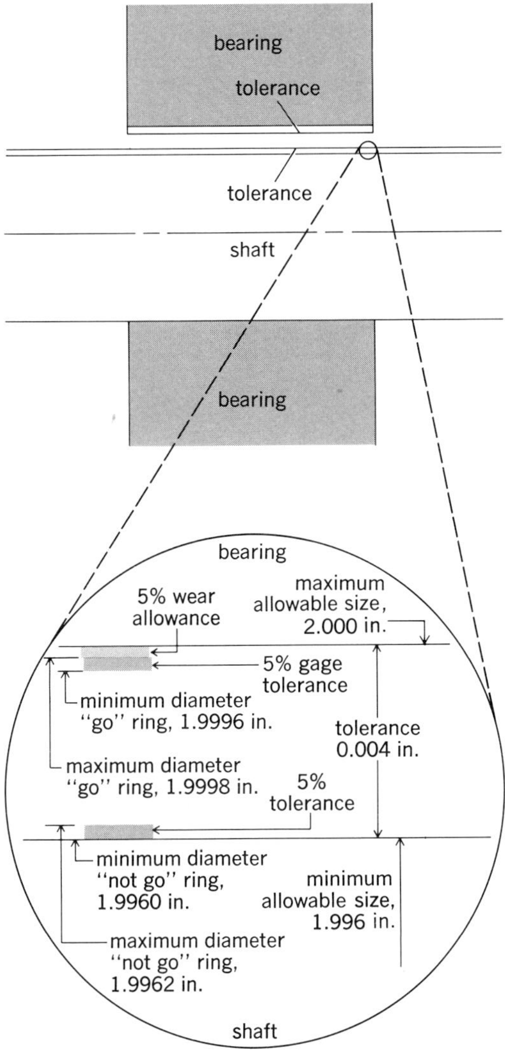

Fig. 1. Makeup of gage tolerances (1 in. = 2.5 cm).

Gage blocks are measured by interferometry. Their actual sizes are determined by comparison with the wavelength of red light in the color spectrum, the ultimate standard of lineal measurement being the wavelength of light. Gage blocks translate this basic lineal standard into a practical form for shop use. *See* OPTICAL FLAT.

Ring gages. Cylindrical rings of steel whose inside diameters are ground and then honed or lapped to gage tolerance are used for checking the diameter of a cylindrical object such as a shaft. The gaging surface is sometimes chrome-plated or made of carbide for wear resistance. A single gage controls one limit of tolerance of the diameter. The "go" gage will pass over the largest acceptable part but will not pass over any larger part. The "not go" gage will not pass over any part within tolerance but will pass over an undersize part. A ring gage is superior for checking the "go" dimension because it checks an infinite number of diameters simultaneously and, to the extent of its length, straightness too. Conversely, it is not preferred as a "not go" gage; an out-of-round or bent part could have many diameters undersize but still be "accepted" because the gage would "not go" if the part were bent or if even one diameter were up to the low limit. Ring gages are made to order.

Plug gage. A cylinder of steel whose diameter is ground, then honed or lapped to gage tolerance, a plug gage is used for checking the diameter of a cylindrical cavity such as a bolt hole or bearing bore. The materials, dimensions, and tolerances of plug gages are comparable to those of ring gages. A single plug gage controls one limit of tolerance of an inside diameter. Plug gages are usually used in pairs, a "go" gage and a "not go" gage. Plug gages are also made to order.

Screw-thread gage. This type of gage may be used to inspect the pitch diameter, major diameter, minor diameter, lead, straightness, and thread angle of a screw thread (Fig. 3). Screw-thread gages may be plug gages, ring gages, snap gages, or indicating gages. A "go" plug or ring gage inspects all these characteristics. A "not go" plug or ring gage checks only pitch diameter and lead. Snap or indicating gages check pitch diameter and one or more of the other characteristics. Ring gages are adjustable and can be set with a "setting" plug.

Taper pipe thread gage. Standards for pitch diameter and taper of a pipe thread were first set to control threads used in petroleum service, where safety depended on accuracy of the thread dimensions. Standards were set jointly by the American Petroleum Institute and the U.S. Bureau of Standards. Size is determined by measuring the distance that a plug gage will screw into a coupling or that a ring gage will screw onto a pipe. A manufacturer of pipe or couplings has a master plug and master ring gage certified by an approved laboratory. These are used to control working gages. As gages wear, the distance of engagement changes and limits are reset as long as thread form remains satisfactory.

Snap gage. A snap gage has two surfaces that are flat and parallel; these are spaced to control one limit of tolerance of an outside diameter or a length. A progressive snap gage has the "not go" surface just behind the "go" surface so that in a single motion both limits of tolerance can be checked. Adjustable snap gages have adjustable gaging surfaces so that the distance between gaging surfaces can be set with gage blocks and used to gage any dimension falling within the limits of the frame size.

Snap gaging is faster than ring gaging, but it checks only one diameter or length at a time. It is frequently used as a "not go" gage only, in conjunc-

GAGES

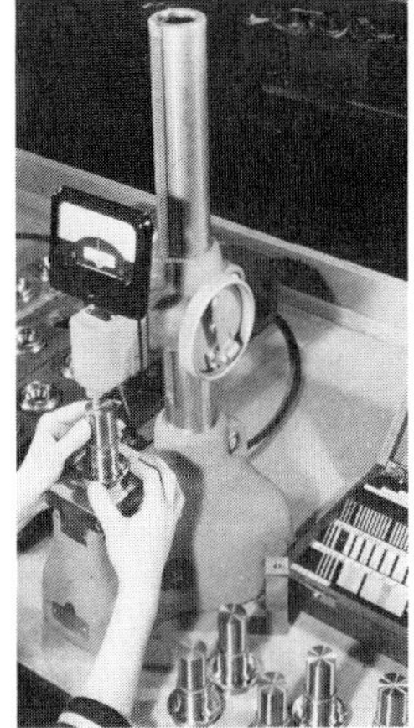

Fig. 2. General-purpose comparator of the electrical type being set with gage blocks (in box to right). (*Pratt and Whitney Co.*)

GAGES

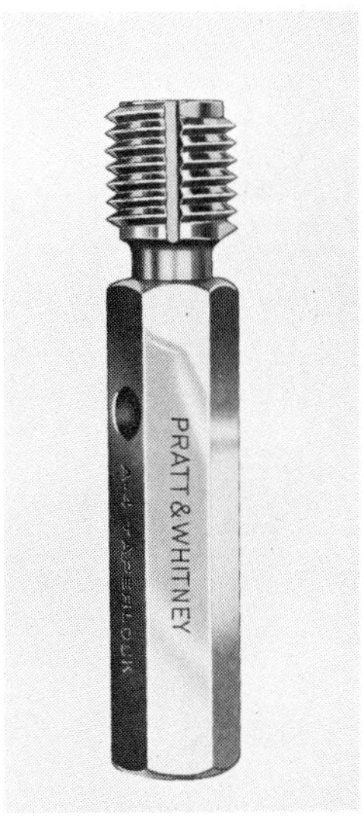

Fig. 3. Thread plug gage. (*Pratt and Whitney Co.*)

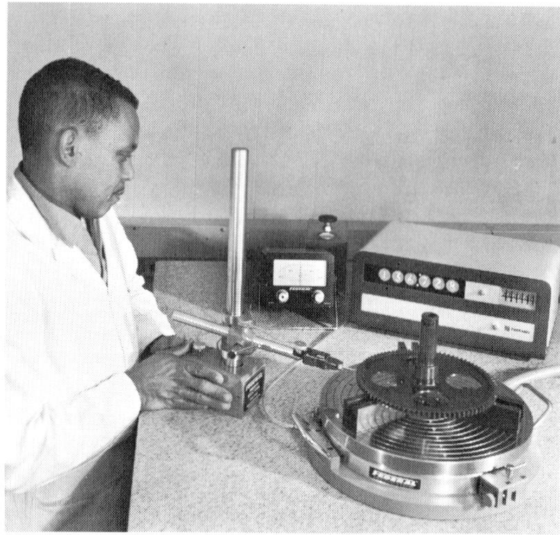

Fig. 4. Indicating gage of the electronic type being used to inspect a part. (*Federal Products Corp.*)

tion with a ring gage for "go" limit to detect out-of-round condition in cylindrical objects. Adjustable gages have an advantage over fixed gages because allowances for wear and gage tolerance may be disregarded in their setting.

Receiving gage. A fixed gage designed to inspect a number of dimensions and also their relation to each other is termed a receiving gage because of its similarity to the cavity which receives the part in actual service. The gage checks on "go" limits. A typical example is a chamber gage which resembles the chamber of a rifle but which is made to the largest limits of size of a cartridge. Thus any cartridge which fits the receiving gage will fit any rifle chamber of that caliber.

Indicating gage. An indicating-type gage has contact points that move as they contact the part being inspected (Fig. 4). The movement is amplified on an indicator whose scale designates the limits of tolerance. Initial setting is done with a master or check gage. An indicator often may be substituted for a fixed or adjustable contact in which a single dimension is being checked. Snap gages are commercially available with indicators instead of fixed contact points. Indicators are also used for inspecting out-of-roundness, centerline runout, or taper, by rotating the part on one axis or diameter and indicating the "run out" of another axis or diameter. Indicating gages do not require allowances for wear or gage tolerance in their setting. *See* INSPECTION AND TESTING.

[RUSH A. BOWMAN]

Bibliography: C. W. Kennedy and D. A. Andrews (eds.), *Inspection and Gaging*, 5th rev. ed., 1977; National Bureau of Standards, *Contact Deformation in Gage Block Comparisons*, Publ. NBS104-TN962, 1978.

Galvanizing

The generic term for any of several techniques for applying thin coatings of zinc to iron or steel stock or finished products to protect the ferrous base metal from corrosion; more specifically, the hot dipping that is widely practiced with mild steel sheet for garbage cans and corrugated sheets for roofing, sheathing, culverts, and iron pipe; and with fencing wire. During dipping, molten zinc reacts with the steel to form a brittle zinc-iron alloy. Control of temperature and the addition of aluminum reduce formation of the alloy, resulting in a more ductile coating. For marine use, magnesium is added.

An electrolytic process (also called cold galvanizing or electrogalvanizing) is also used for wire, as well as for applications requiring deep drawing. An alloy layer does not form, hence the smooth electroplated coating does not flake in the drawing die. *See* METAL COATINGS. [FRANK H. ROCKETT]

Bibliography: R. M. Burns and W. Bradley, *Protective Coatings for Metals*, 3d ed., 1967; R. E. Kirk and D. F. Othmer (eds.), *Encyclopedia of Chemical Technology*, 3d ed., vol. 11, 1980; F. Lowenheim, *Modern Electroplating*, 3d ed., 1974.

Gantt chart

A visual control device developed by Henry L. Gantt and used in production planning and control. In concept, the Gantt chart is a bar chart of several interrelated activities with time on the horizontal axis. The chart specifies the work required by a project (comprising a series of tasks) and the order in which the tasks must be done.

There are two basic types of Gantt charts: the planning chart and the progress chart. Preparation for the chart begins by listing the tasks that must be done to complete the project. A technological sequence of the activities for each task is required. After this step is completed, time estimates for each activity must be obtained. Next the Gantt planning chart (Fig. 1) is prepared, making a definite plan for each task necessary. This is one

Project Number	Time Scale
1	task #1 task #2 task #3 scheduled
2	task #1 task #2 scheduled

Fig. 1. Gantt planning chart.

of the advantages of Gantt charts. It forces a thinking-through of the things that will be encountered and for which provisions must be made. The chart depicted in Fig. 1 is then used as a further management tool. As work progresses, a bar is drawn on the chart with a length proportional to the amount of work done during each time interval. Work done is initially plotted on the time interval during which it is accomplished. Later a cumulative bar may be added to summarize the status of work for an extended period. This second type of Gantt chart is known as the progress chart (Fig. 2).

The result is a dynamic chart depicting graphically what has happened and displaying in

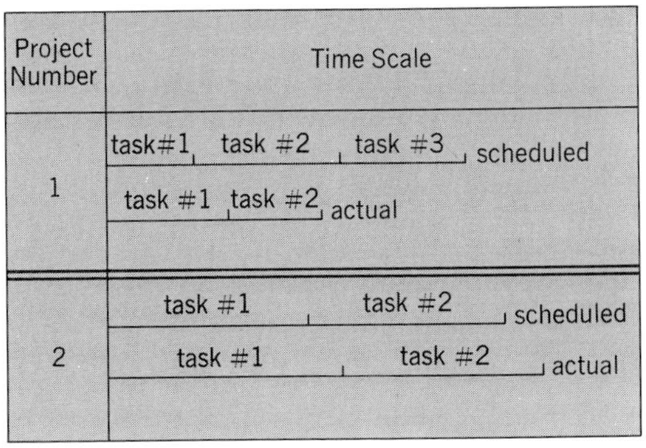

Project Number	Time Scale
1	task #1 task #2 task #3 scheduled
	task #1 task #2 actual
2	task #1 task #2 scheduled
	task #1 task #2 actual

Fig. 2. Gantt progress chart.

easily interpreted form the status of each project. Many firms have found that maintaining Gantt charts can be costly. Also, the time required to make the necessary posting can take so long that the chart may not depict the current status of the project. Nevertheless, Gantt charts are easy to understand and use, and their applicability spans machine loading, personnel assignments, project planning, order scheduling, and research and development programs. The chart can be an effective tool for operations that involve a minimum of interrelationships; but it does not make provision for treatment of uncertainty, nor does it provide ways for treatment and forecasting when interrelationships between the various tasks create constraints on performance. *See* CRITICAL PATH METHOD (CPM); INDUSTRIAL COST CONTROL; PERT. [NELSON M. FRAIMAN]

Bibliography: C. Heyel (ed.). *The Encyclopedia of Management*, 1973; H. B. Maynard (ed.). *Industrial Engineering Handbook*, 3d ed., 1971.

Gas field and gas well

Petroleum gas, one form of naturally occurring hydrocarbons of petroleum, is produced from wells that penetrate subterranean petroleum reservoirs of several kinds. Oil and gas production are commonly intimately related, and about one-third of gross gas production is reported as derived from wells classed as oil wells. If gas is produced without oil, production is generally simplified, in part at least because the gas flows naturally without lifting, and also because of fewer complications in reservoir problems. As for all petroleum hydrocarbons, the term field designates an area underlain with little interruption by one or more reservoirs of commercially valuable gas. *See* OIL AND GAS FIELD EXPLOITATION; OIL AND GAS WELL DRILLING; PETROLEUM RESERVOIR ENGINEERING.

[CHARLES V. CRITTENDEN]

Gas furnace

An enclosure in which a gaseous fuel is burned. Domestic heating systems may have gas furnaces. Some industrial power plants are fired with gases that remain as a by-product of other plant processes. Utility power stations may use gas as an alternate fuel to oil or coal, depending on relative cost and availability. Some heating processes are carried out in gas-fired furnaces. *See* STEAM-GENERATING FURNACE.

Among the gaseous fuels are natural gas, producer gas from coal, blast furnace gas, and liquefied petroleum gases such as propane and butane. Crude industrial heating gases carry impurities that corrode or clog pipes and burners. Solid or liquid suspensoids are removed by cyclones or electrostatic precipitators; gaseous impurities are removed chemically. The cleaned gas may be mixed with air in the furnace, in the burners, or in a blower before going to the burners. The gas and air may be supplied at moderate pressure, or one or both at high pressure. The high-pressure component may serve to induce the other component into the furnace. The burner may be a single center-fire type or a multispud type with numerous small gas parts, depending on how the heat is to be concentrated or distributed in the furnace. Crude uncleaned gases are fired through burners with large ports, the burners being removable for cleaning. [RALPH M. HARDGROVE]

Gas turbine

A heat engine that converts some of the energy of fuel into work by using gas as the working medium and that commonly delivers its mechanical output through a rotating shaft.

Cycle. In the usual gas turbine, the sequence of thermodynamic processes consists basically of compression, addition of heat in a combustor, and expansion through a turbine. The flow of gas during these thermodynamic changes is continuous in the basic, simple, open-cycle arrangement (Fig. 1*a*).

This basic open cycle can be modified through the addition of heat exchangers and multiple components for reasons of efficiency, power output, and operating characteristics (Fig. 1*b*–*d*). A regenerator recovers exhaust heat and returns it to the cycle by heating the air after compression and before it enters the combustor. An intercooler reduces the work of compression by removing some of the heat of compression. A second stage of heating can be added between sections of the turbine, called an afterburner in aircraft turbines. In various combinations, the auxiliary features provide means for meeting a wide range of operating needs.

Types. The various gas-turbine cycle arrangements can be operated as open, closed, or semiclosed types.

Open cycle. In the open-cycle gas turbine, there is no recirculation of working medium within the structural confines of the power plant, the inlet and exhaust being open to the atmosphere (Fig. 1). This cycle offers the advantage of a simple control and sealing system. It also can be designed for high power-to-weight ratios (aircraft units) and for operation without cooling water. Most gas turbine plants are of this type.

Closed cycle. In the closed-cycle gas turbine, essentially all the working medium (except for seal leakage, bleed loss, and any addition or extraction of working medium for control purposes) is continuously recycled (Fig. 2*a*). Heat from a source such as fossil fuel (or, possibly, nuclear reaction) is transferred through the walls of a closed heater to the cycle. The closed cycle can be charged with gases other than air such as helium, carbon dioxide, or nitrogen. This is a particular advantage with a nuclear heat source. Other advantages of the closed cycle are (1) clean working fluid; (2) control of the pressure and composition of the working fluid; (3) high absolute pressure and density of the working fluid; and (4) constant efficiency over wide

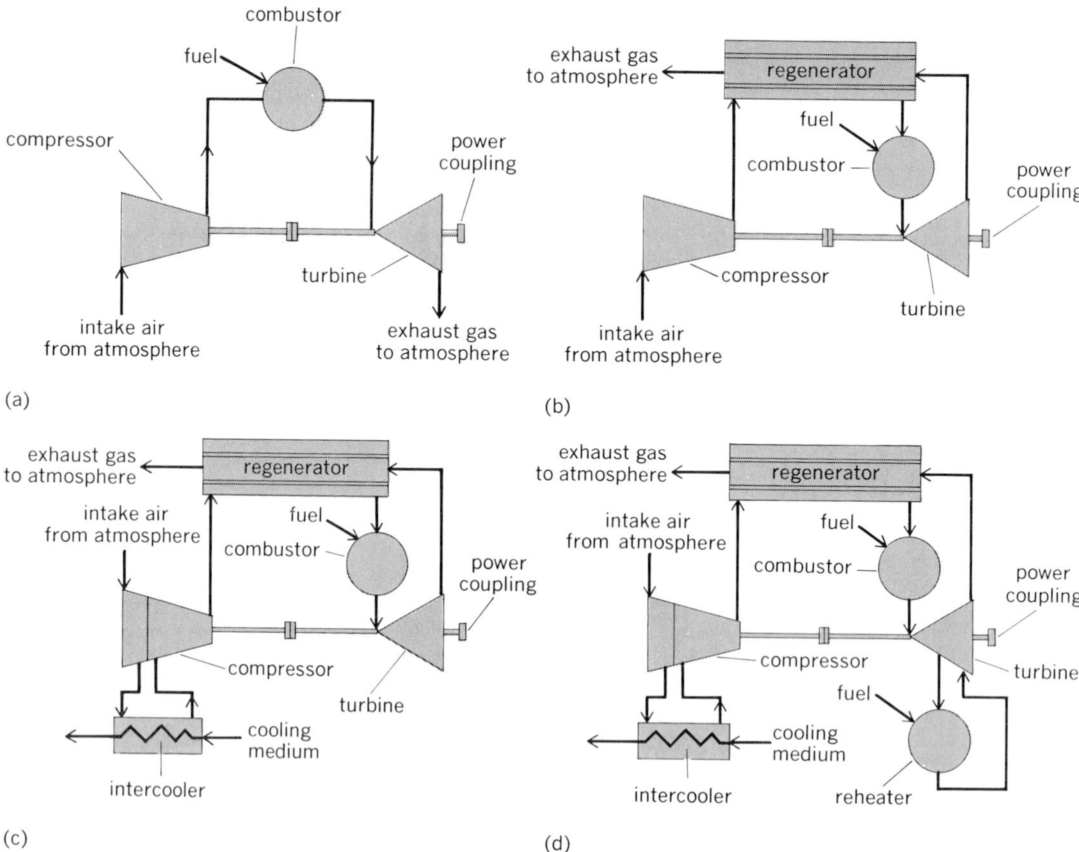

Fig. 1. Series-flow, single-shaft, gas-turbine power plant. (*a*) Simple open cycle. (*b*) Open cycle with regenerator. (*c*) Open cycle with intercooler and regenerator. (*d*) Open cycle with intercooler, regenerator, and reheater.

load range. A precooler is required to reduce the temperature of the working fluid before recompression. The higher densities of the working fluid increase the horsepower capacity of a plant of given volume. Changing the absolute pressure level at the compressor inlet changes the weight of working fluid circulated without changing the compression ratio or the temperatures, which results in relatively constant efficiency over a wide load range. The major disadvantage of the closed-cycle gas-turbine plant is the cost and size of the required high-temperature heater.

Semiclosed cycle. In the semiclosed cycle gas turbine, a portion of the working fluid is recirculated (Fig. 2*b*). This type requires a precooler for the recirculated gas, and a charging compressor to provide the necessary air for combustion. The semiclosed cycle can operate at high densities. The major disadvantages of this cycle are the corrosion and fouling which occur with the recirculation of the products of combustion, particularly when the fuels used have high sulfur or ash content.

Generalized performance. The overall performance of a given gas-turbine power-plant cycle depends basically on component efficiencies, pressure and leakage losses, pressure ratio (of the highest to the lowest pressure), and temperature level.

For the simple cycle, the influence of pressure ratio on performance is illustrated in Fig. 3*a*. The curves are based on ambient inlet conditions of 80°F and 1000 ft altitude, 85% compressor efficiency, 90% turbine efficiency, 95% combustion efficiency, and 5% combustor pressure loss.

The effect of pressure ratio on plant efficiency for the four basic gas-turbine cycle arrangements illustrated in Fig. 1 is shown in Fig. 3*b*; the effect of inlet temperature on cycle efficiency for the four basic cycles is shown in Fig. 3*c*. The curves are drawn for the optimum pressure ratio for each cycle arrangement at the various turbine inlet temperatures.

Components. To achieve such overall performance, each process is carried out in the engine by a specialized component (Fig. 4). Air for the combustion chamber is forced into the engine by a compressor. In an aircraft, the intake may advance into the air fast enough to ram air into the engine. Fuel is mixed with the compressed air and burned in combustors. The heat energy thus released is converted by the turbine proper into rotary energy. Because of the high initial temperature of the combustion products, excess air is used to cool the combustion products to the allowable turbine inlet design temperature. To improve efficiency, heat exchangers can be added on the gas turbine exhaust to recover heat energy and to return it to the working medium after compression and prior to its combustion.

Compressors. Two basic types of compressors are used in gas turbines: axial and centrifugal. In a few special cases a combination type known as a mixed wheel, which is partially centrifugal and partially axial, has been used. The axial-flow compressor is the most widely used because of its ability to handle large volumes of air at high efficiency.

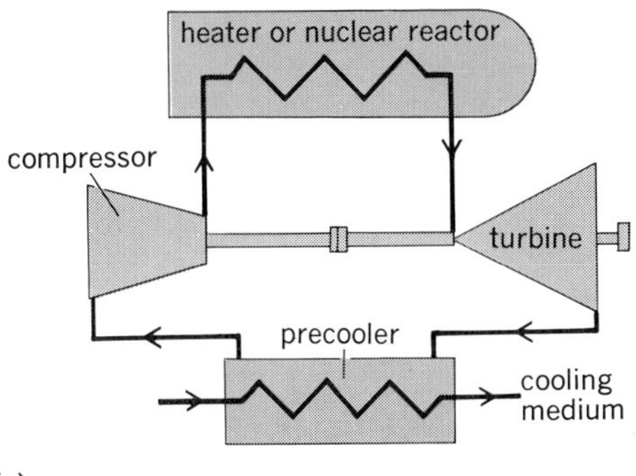

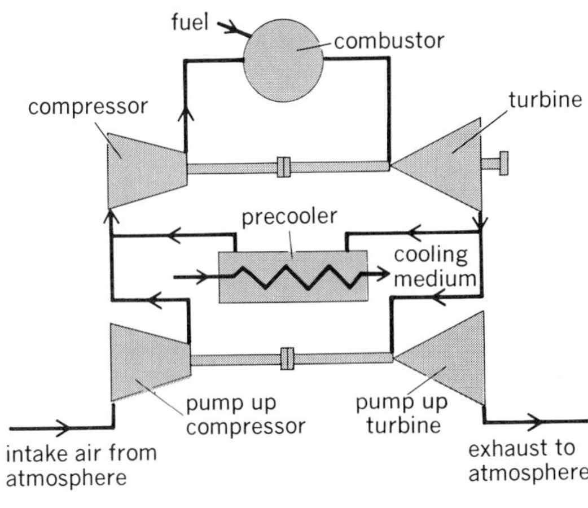

Fig. 2. Gas-turbine power plant. (*a*) Closed cycle with precooler, series flow, single shaft. (*b*) Semiclosed cycle with precooler, series flow, single shaft.

For small gas turbines in the range of 500 hp and less, the centrifugal replaces the axial because it has comparable efficiency when handling reduced volume flow, and is smaller and more compact. *See* COMPRESSOR.

Axial and centrifugal compressors both have a stall or pumping limit where flow reverses. This limit is usually encountered on starting with stationary power units, and at high altitude and high speed with aircraft units. This stall limit is shifted out of the operating range of the gas turbine by the use of compressor bleed, variable compressor vanes, or water injection. All three methods are applied to reduce the aerodynamic loading on the stalled compressor stages.

Ram effect. Aircraft gas turbines moving at high speeds obtain a pressure rise from the ram effect in addition to the compressor pressure rise. The ram effect is the recovery of part of the air velocity, due to the forward motion of the plane, and conversion of this velocity energy to pressure.

Combustors. Combustors, sometimes referred to as combustion chambers, for gas turbines take a wide variety of shapes and forms. All contain fuel nozzles to introduce and meter the fuel to the gas stream and to atomize or break up the fuel stream for efficient combustion (Fig. 5). All are designed for a recirculating flow condition in the region of the nozzle to develop a self-sustaining flame front in which the gas speed is lower than the flame propagation speed. In addition to being designed to burn the fuel efficiently they also uniformly mix excess air with the products of combustion to maintain a uniform turbine-inlet temperature.

The combustor must bring the gas to a controlled, uniform temperature with a minimum of impurities and a minimum loss of pressure. In the open-type gas turbine, the large excess of air must be controlled to avoid chilling the flame before complete combustion has taken place. Extremely high rates of combustion are common, the heat release being 1,000,000–5,000,000 Btu/ft³/hr/atm or 5–20 times that of high-output, steam-boiler furnaces. Refractory linings are not suitable because they cannot withstand the vibrations and velocities. Metal liners and baffles cooled by the incoming air are commonly used.

Turbine wheels. Two types of gas turbine wheels are used: radial-inflow and axial-flow. Small gas turbines use the radial-flow wheel. For large vol-

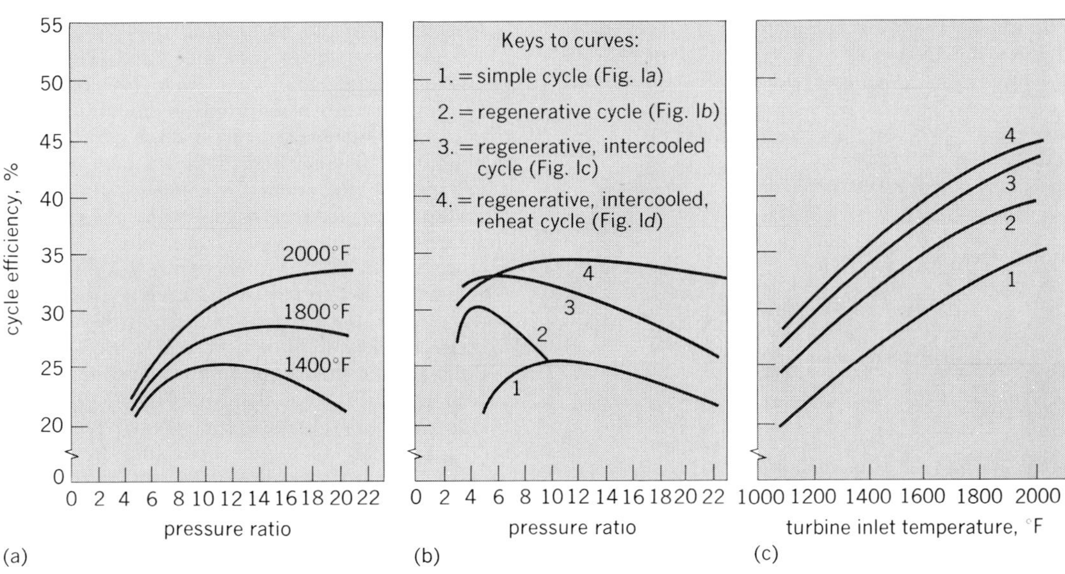

Fig. 3. Effects of pressure ratio on thermal efficiency. (*a*) Simple gas-turbine cycle at various turbine-inlet temperatures. (*b*) Various gas-turbine cycles at constant turbine-inlet temperature. (*c*) Effect of turbine-inlet temperature on thermal efficiency for various gas-turbine cycles at optimum pressure ratios.

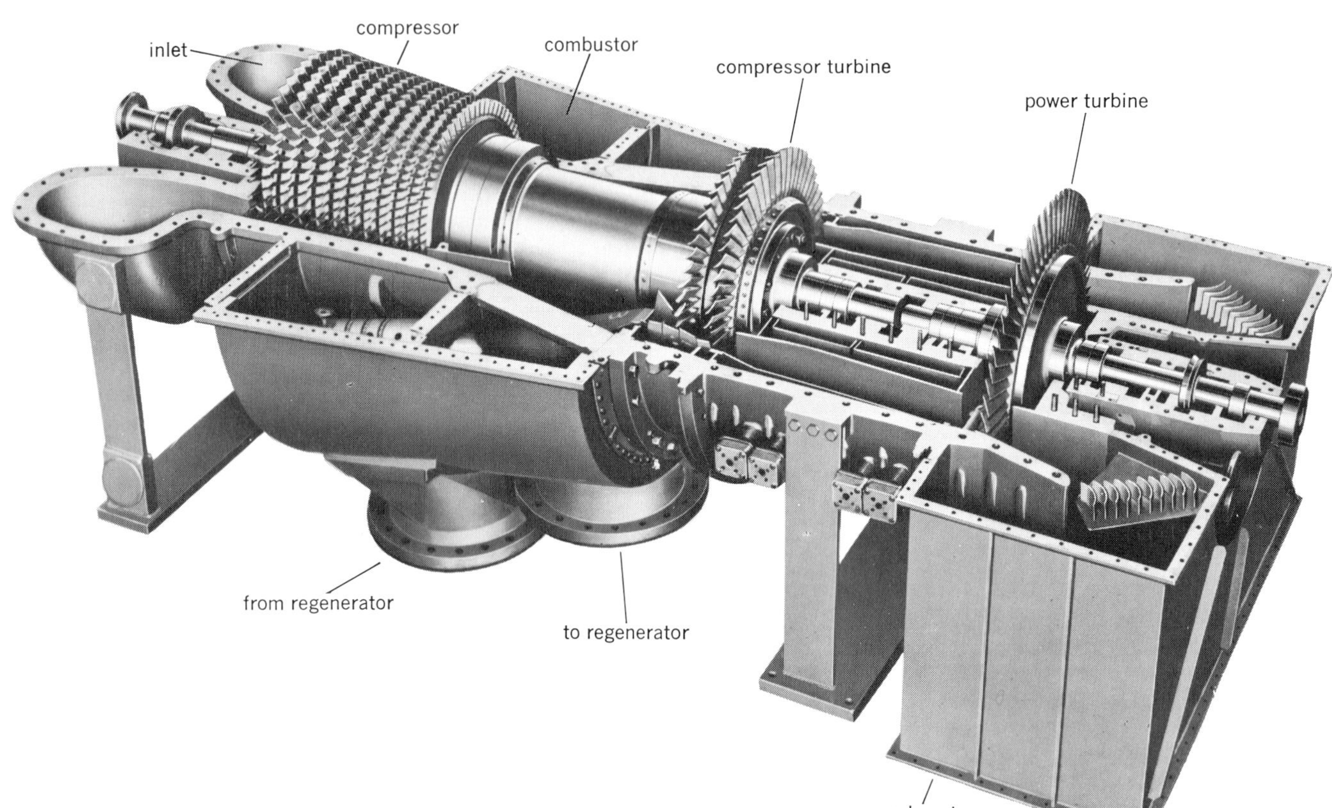

Fig. 4. Dual-shaft open-cycle 5000-hp gas turbine with regenerator.

ume flows, axial turbine wheels are used almost exclusively. Although some of the turbines used in the small gas turbine plants are of the simple impulse type, most high performance turbines are neither pure impulse nor pure reaction. The high-performance turbines are normally designed for varying amounts of reaction and impulse to give optimum performance. This usually results in a blade which is largely impulse at the hub diameter and almost pure reaction at the tip. *See* TURBINE.

Turbine cooling. Gas turbines all employ cooling to various extents and use a liquid or gas coolant to reduce the temperature of the metal parts. The cooling systems vary from the simplest form where only first-stage disk cooling is involved to the more complex systems where the complete turbine (rotor, stator, and blading) is cooled. At high temperatures of operation the turbine material is designed to have a predetermined life, which can vary from a few minutes for missile applications to 100,000 hr or more for industrial units.

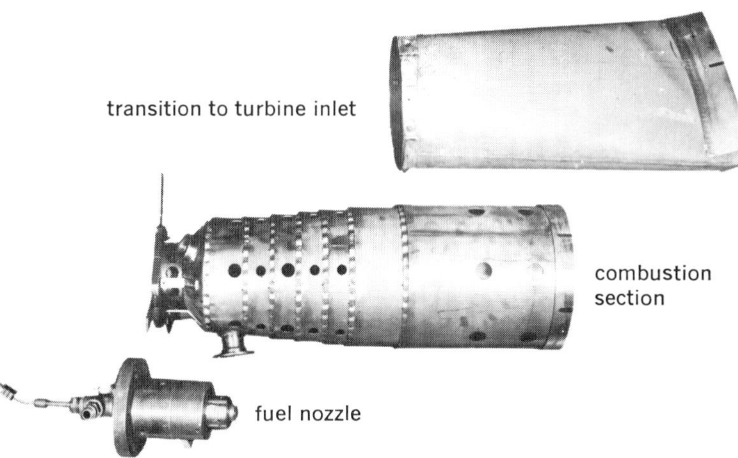

Fig. 5. Gas-turbine combustor components.

Heat exchangers. Two basic types of heat exchangers are used in gas turbines: gas-to-gas and gas-to-liquid. An example of the gas-to-gas type is the regenerator, which transfers heat from the turbine exhaust to the air leaving the compressor. The regenerator must withstand rapid large temperature changes and must have low pressure drop. Regenerators of both the shell-and-tube and the extended-surface type are used (Fig. 6). Rotary regenerators having high performance and reduced weight are under development. Intercoolers, which are used between stages of compression, are air-to-liquid units. They reduce the work of compression and the final compressor discharge temperature. When used with a regenerator, they increase both the capacity and efficiency of a gas turbine power plant of a given size.

Controls and fuel regulation. The primary function of the subsystem that supplies and controls the fuel is to provide clean fuel, free of vapor, at a rate appropriate to engine operating conditions. These conditions may vary rapidly and over a wide range. As a consequence, fuel controls for gas turbines are, in effect, special-purpose computers employing mechanical, hydraulic, or electronic means, frequently all three in combination.

In aircraft the pilot controls the turbine engine through a throttle lever, which usually establishes engine speed for each position of the lever. To achieve this relation between throttle position and engine speed under the design envelope range of flight conditions, the fuel control senses numerous engine conditions in addition to engine speed in revolutions per minute. Among these conditions are compressor inlet pressure and ambient air temperature. From these inputs, together with the position of the pilot's throttle, the fuel control continuously adjusts the rate of fuel flow to the engine When the pilot moves the throttle to change engine power, the control senses the rate of engine acceleration or decleration and provides for the required rate of change in fuel rate.

The fuel control, as well as maintaining the engine at the speed called for by the throttle, should also prevent overspeeding, excessive temperature, or loss of the fire. Under some conditions, the rate at which a turbine engine can accelerate may be limited by the rate at which heat can be released by fuel combustion: Excessive internal engine temperatures could damage such engine parts as the turbine wheels. Under other conditions, the rate at which an engine can decelerate may be limited by the necessity of maintaining sufficient fuel flow to hold the flames; too low a fuel rate for the prevailing gas velocity in the combustors could blow out the flame.

Arrangements. Gas turbines can be constructed for single- or multiple-shaft arrangements. They can be arranged to supply power, high-pressure air, or hot-exhaust gases, either singly or in combination.

A single-shaft unit, consisting of a compressor, combustor, and turbine, is a compact, lightweight power plant. It is capable of rapid starting and loading and has no standby losses. It can be arranged to use little or no cooling water, which makes it particularly attractive for powering transportion equipment and as a mobile standby and emergency power plant. This type of plant can compete efficiently with small steam plants. Its simplicity involves a minimum of station operating personnel; some of these plants are arranged for completely remote operation.

To improve efficiency. the energy in the exhaust gases can be used either in a waste-heat boiler, or in combination with other processes. For example, a unit arranged to supply process air at 35 psig for operation of blast furnaces can use blast-furnace gas for its fuel (Fig. 7*a*).

Gas turbines might also be arranged to use a nuclear heat source. For aircraft application, the cycle is open because of weight and space considerations. The combustor is replaced with a nuclear reactor (Fig. 2*a*). For shipboard and stationary units, the cycle is closed so that fission products can be contained within the power plant in the event of a nuclear accident or fuel element failure. A closed-cycle, gas-turbine power plant for marine service could operate directly with a high-temperature, gas-cooled reactor (Fig. 7*b*).

Fuels. In the open-cycle plant, products of combustion come in direct contact with the turbine blading and heat-exchanger surfaces. This requires a fuel in which the products of combustion are relatively free of corrosive ash and of residual solids that could erode or deposit on the engine surfaces. Natural gas, refinery gas, blast-furnace gas, and distillate oil have proved to be ideal fuels for open-cycle gas turbines. Combustion chamber and fuel nozzle maintenance are negligible with these types of fuel. Residual fuel oil treated to avoid hot ash corrosion and deposition is also satisfactory although it requires frequent cleaning of fuel nozzles and higher combustor maintenance. Vanadium pentoxide and sodium sulfate are the principle ashes that have been found to cause corrosion and deposition in the 1250–1600°F temperature ranges of modern-day gas turbines. The treatment of residual oil to make it suitable consists of washing with water to remove sodium and introducing additives to raise the fusion temperature of the ash.

Fig. 6. Air-to-gas regenerator, partly cut away.

In closed-cycle turbines, gaseous fuel, distillate oils, and coal are satisfactory. Residual fuels also must be treated to avoid corrosion in the heater parts. Ash erosion is not a major problem because of the lower velocity of the combustion gases over the heater surfaces.

Fuels for semiclosed-cycle turbines must be selected to minimize both erosion and corrosion. Because the products of combustion are circulated through the entire cycle, the fuel must be selected to avoid corrosion at all pressures and temperatures of the cycle. These requirements limit the use of the semiclosed plant to fuels of low ash and sulfur content.

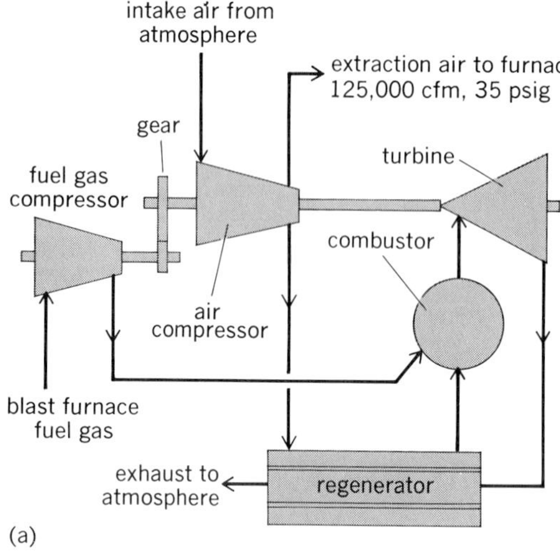

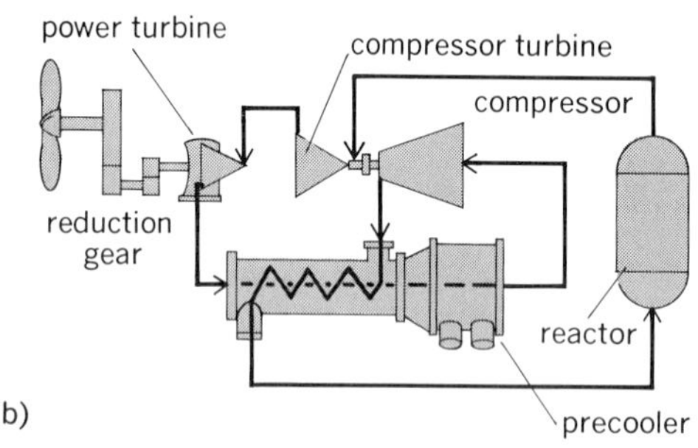

Fig. 7. Cycle layouts (*a*) for a 125,000-ft³/min (59 m³/s) extraction unit and (*b*) for a closed-cycle gas turbine utilizing a gas-cooled reactor. 35 psig = 240 kPa gage pressure.

Applications. Gas-turbine power plants have been successfully applied in the following industries, where their characteristics have proved superior to competitive power plants.

Aviation. The gas turbine finds its most important application in the field of aviation. In the propulsion of aircraft, the gas-turbine power plant as a turboprop or turbojet engine has replaced the reciprocating engine in large, high-speed airplanes. This change is due primarily to its high power-to-weight ratio and its ability to be built in large horsepower sizes with high ratio of thrust per frontal area (Fig. 8). It makes use of the ram effect on the compressor inlet to give almost constant thrust at all aircraft speeds.

Gas pipeline transmission. Gas turbines have been installed along piplines to drive centrifugal compressors. Turbine sizes range from 1800–14,000 hp. Gas fuel is normally used, though units have been arranged for dual fuel firing using either natural gas or distillate oil.

Petroleum. In the petroleum industry the gas turbine generates power, drives air and gas compressors in refineries, supplies extraction air or exhaust gases for process, and in the oil fields drives gas compressors to maintain well pressure.

Steel. In the steel industry gas turbines drive air compressors to furnish extraction air for processes such as blast-furnace operation, as mentioned above, and drive electric generators for various plant services. The primary fuel in these applications is blast-furnace gas with a number of units arranged for dual fuel firing.

Marine. Prototype gas turbines are being tested for marine use in ratings up to 6000 ship horsepower (shp). These units operate with residual oil as fuel processed through fuel treatment equipment. A regenerative-cycle gas turbine with a separate power turbine provides the wide speed range required by such applications as marine service (Fig. 4).

Marine installations of gas turbines range from small units used for propulsion of short-range, high-speed craft, emergency generator drive, minesweeper boat propulsion and generator drive, deicing, smoke generation, fire-pump drive, and pneumatic power applications, to large boost engines used in combination with steam turbines.

Electric utilities. Gas turbines perform a variety of functions in electric-power generation. Peaking service uses simple single-shaft units in sizes up to 25,000 kW. Gas turbines have been installed as hydro-standby. Aircraft propulsion elements can be applied as gas generators, driving electric generator units of up to 300,000 kW. Rail-mounted mobile power plants provide 5000–6000 kW for emergency service. The compactness and simplicity of the gas turbine make it possible to house these units in a single cab. Units from 5000 to 30,000 kW carry base load. These units have been simple cycle machines in low fuel cost areas; units with regenerative, intercooled cycles are used in high fuel cost areas.

Combined steam and gas plants. Efficiency of a power plant is improved by combining a gas turbine with a steam turbine. Various combinations are possible; gas-turbine exhaust can heat feed water for the steam turbine, fuel-fired boilers can use gas turbine exhaust as combustion air (Fig. 9*a*), or in supercharged boiler plants the high-pressure boiler can serve as the pressurized combustor for the gas turbine (Fig. 9*b*). The gain in efficiency by using turbine exhaust (Fig. 9*a*) depends on efficiency of the overall steam plant to which the combined cycle is compared. For the most efficient

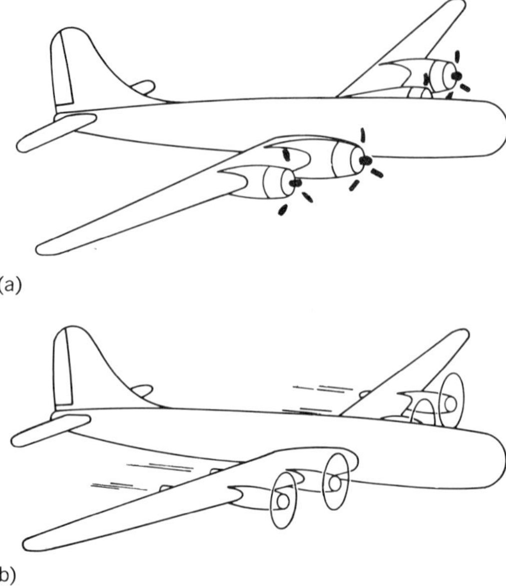

Fig. 8. Types of aircraft engine. (*a*) Reciprocating engine has greater frontal area for a given horsepower than (*b*) internal-combustion gas-turbine engine.

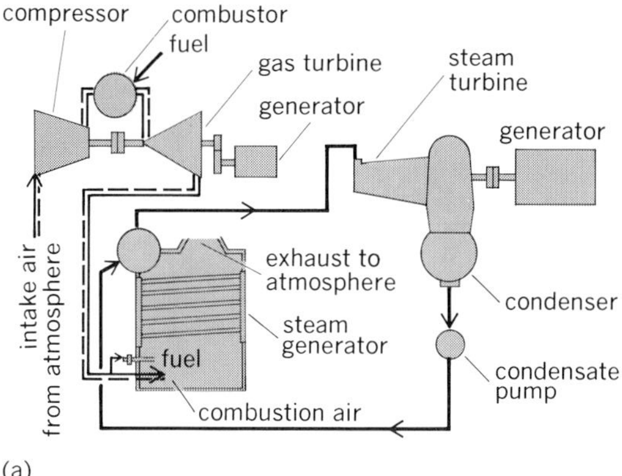

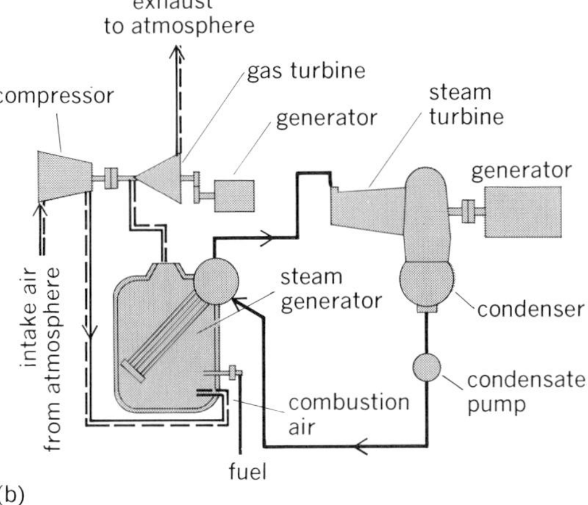

Fig. 9. Cycle arrangements (*a*) for combined steam- and gas-turbine cycle with gas-turbine, exhaust-fired steam generator and (*b*) for combined steam- and gas-turbine cycle utilizing a pressurized-steam generator.

plant using the best steam conditions and largest number of feedwater heaters, a gain of 2% is possible, while for the less efficient plant, a gain of 5% is possible.

Where the steam generator is pressurized with air from the gas turbine compressor and the hot gases from the boiler are then expanded through the gas turbine (Fig. 9*b*), a gain in overall efficiency of 4–8% is realized, depending on the plant used for comparison. These improvements can be made to all steam plant cycles including the most efficient in operation, which employ the superpressure cycle. [THOMAS J. PUTZ/JAY A. BOLT]

Bibliography: T. Baumeister (ed.), *Standard Handbook for Mechanical Engineers*, 8th ed., 1978; H. Cohen et al., *Gas Turbine Theory*, 2d ed., 1979; D. G. Fink and H. W. Beaty (eds.), *Standard Handbook for Electrical Engineers*, 11th ed., 1978; C. W. Grennan (ed.), *Gas Turbine Pumps*, 1972; W. R. Hawthorne and W. T. Olson (ed.), *Designs and Performance of Gas Turbine Power Plants*, 1960; A. W. Judge, *Small Gas Turbines and Free Piston Engines*, 1960; D. G. Shepherd, *Introduction to the Gas Turbine*, 2d ed., 1960; C. W. Smith, *Aircraft Gas Turbines*, 1956; H. A. Sorensen, *Gas Turbines*, 1951.

Gasket

Deformable material used to make a pressure-tight joint between stationary parts, such as cylinder head and cylinder, that may require occasional separation. Gaskets are known as static seals, as compared with packing or dynamic seals. In packings the parts are frequently in motion, as in piston rods and valve stems. *See* PRESSURE SEAL.

Gaskets are made of sheet materials such as natural or synthetic rubber, cork, vegetable fiber such as paper, asbestos and plastic pastes, or of soft metallic materials such as lead and copper. Rubber in the form of O-rings is used for light pressure. [PAUL H. BLACK]

Bibliography: P. H. Black and O. E. Adams, Jr., *Machine Design*, 3d ed., 1968; E. Oberg et al., *Machinery's Handbook*, 20th rev. ed., 1975; J. E. Shigley, *Mechanical Engineering Design*, 3d ed., 1976; Society of Automotive Engineers, *SAE Handbook*, (revised annually).

Gear

A machine element used to transmit motion between rotating shafts when the center distance of the shafts is not too large. Toothed gears provide a positive drive, maintaining exact velocity ratios between driving and driven shafts, a factor that may be lacking in the case of friction gearing which is subject to slippage. While the motion transmitted to mating gears is kinematically equivalent to that of rolling surfaces identical with the gear pitch surfaces, the action of one gear tooth on another is generally a combination of rolling and sliding motion. When the distance between shafts is large, other methods of transmission are used. *See* BELT DRIVE; CHAIN DRIVE; ROLLING CONTACT.

The application of gears for power transmission between shafts falls into three general categories: those with parallel shafts, those for shafts with intersecting axes, and those whose shafts are neither parallel nor intersecting but skew (see table).

Chief types of application of gears for power transmission

Shaft relationship	Type of gearing
Parallel	Spur Helical Herringbone Double helical
Intersecting	Bevel
Skew	Helical Worm Hypoid

Principal features. Figures 1 and 2 illustrate the principal features of toothed gears. Such terms as pitch circle, addendum circle, and root circle, being geometrical, are defined by the diagrams.

Definitions for other commonly used terms for describing gears are addendum, the radial distance between the pitch circle and the addendum circle; dedendum, the radial distance between the pitch circle and the root circle; face, the tooth surface outside the pitch circle; flank, the tooth surface between the pitch circle and the root circle; clearance, the amount by which the dedendum exceeds the addendum of the mating gear; whole depth, addendum plus dedendum; and working depth, whole depth minus clearance; circular pitch, the distance from a point on one tooth to the corresponding point on the next, measured along

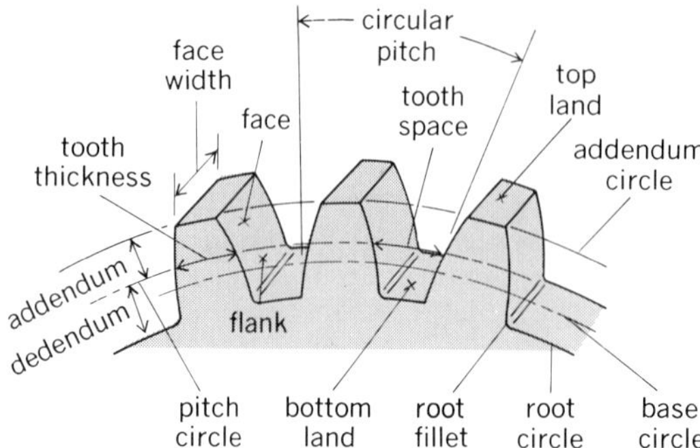

Fig. 1. Principal features of gear teeth.

the pitch circle; and base pitch, the similar measurement along the base circle.

Terms descriptive of a pair of spur gears in mesh are shown in Fig. 2. Where the pitch circles are tangent is pitch point P. Each pair of teeth always has its point of contact on the pressure line. The angle between this pressure line and the common tangent to the pitch circles through P is the pressure angle. As Fig. 2 illustrates, the pressure angle at which two gears actually operate depends on the distance between centers at which they are mounted. Initial contact between each pair of teeth occurs at point C, where the addendum circle of the driven gear crosses the pressure line; and final contact is at D, where the driver's addendum circle crosses the pressure line. Distance CD is the length of the path of contact. It must exceed the value of the base pitch if a pair of teeth is to come into contact before the previous pair has gone out of contact. The ratio of the length of the path of contact to the base pitch is called the contact ratio, which can be viewed as approximately the average number of pairs of teeth in contact. It is usual to design for a contact ratio of 1.4 or above to ensure smooth, continuous tooth action.

Gear tooth sizes are designated by diametral pitch, which is the number of teeth per inch of

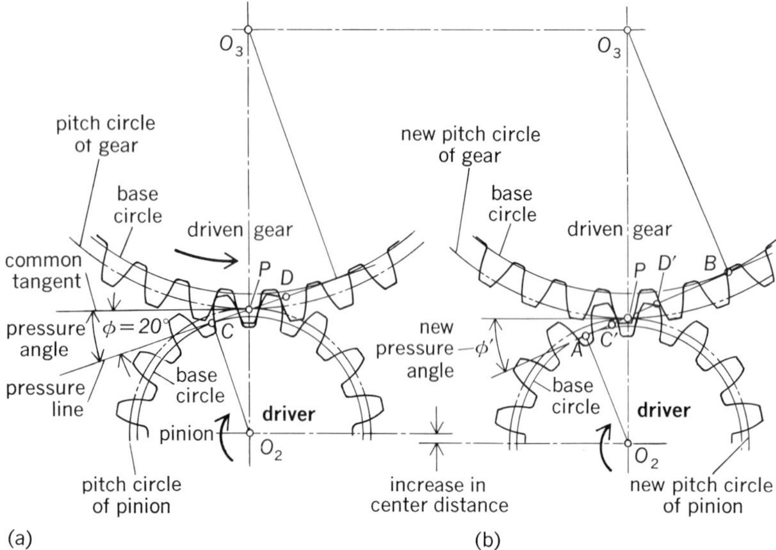

Fig. 2. Action of involute gearing. (*a*) Gear and pinion mounted at normal center distance. (*b*) Gear and pinion mounted at greater than normal center distance. (*From J. E. Shigley, Theory of Machines, McGraw-Hill, 1961*)

diameter of the pitch circle. Pitch circle is, in turn, the circle whose periphery is the pitch surface, or surface of an imaginary cylinder that would transmit by rolling contact the same motion as the toothed gear. A gear with a 20-in. pitch diameter and teeth having a diametral pitch of 2 has 40 teeth. Diametral pitch P times circular pitch p (Fig. 1) equals π. The diametral pitches most widely used are: 1, $1\frac{1}{2}$, 2, $2\frac{1}{2}$, 3, 4, 5, 6, 8, 10, 12, 16, 20, 24, 32, 40, 48, 64, 72, 96, 120, 160.

The smaller of two gears in mesh is the pinion. It has the fewer teeth and is the driving gear in a speed reduction unit. The minimum number of teeth an involute pinion can have and still run without interference between its flanks and the tips of the mating gear teeth is fixed by the tooth system. Smaller pinions are possible only if the pinion's flanks are undercut.

Backlash. The amount by which the tooth space of a gear exceeds the tooth thickness of the mating gear at the pitch circle is the backlash. It can be determined in the plane of rotation or, for helical gears, in the plane normal to the tooth face.

If mating gears have zero backlash, gears and mountings need to be dimensionally perfect. To retain zero backlash with varying operation conditions, all parts need exactly the same thermal expansion characteristic. Because of the difficulty of meeting these requirements and for lubrication, freedom—backlash—is provided between gear teeth. The usual practice is to reduce the tooth thickness on each gear by an amount equal to half the desired backlash. However, in the case of a gear and a small pinion it is customary to reduce the tooth thickness on the gear only, leaving the pinion with standard tooth thickness. Backlash can also be adjusted by slight changes in the center distance between gears. Except for a small change in the pressure angle, the action of involute gear teeth is not affected by backlash or center distance adjustment.

In the case of precision gearing for control systems and similar applications, backlash results in a nonlinear relation between input and output. Several methods of reducing backlash are in use. One method is to place two identical spur gears on the same shaft, one fixed to the shaft, one free. The loose gear is attached to the fixed one by springs, which keep the composite gear in positive contact with the pinion at all times. A second method is to use tapered-tooth gearing (beveloid) with adjustment along the shaft to eliminate excessive backlash.

Gear action. A principal function of gears is to change the speed of rotation. This action is described by the velocity ratio of the gears in mesh. Ratio VR is the number of revolutions N_1 of the driving gear divided by the number of revolutions N_2 of the driven gear in the same time interval. For gears with teeth T_1 and T_2, respectively, VR is expressed as in Eq. (1). When two curved surfaces,

$$VR = N_1/N_2 = T_2/T_1 \qquad (1)$$

such as the mating surfaces of two gear teeth, are in driving contact, there is a definite velocity ratio between the bodies. The angular velocities of the two bodies are inversely proportional to the segments into which their line of centers is divided by a line passing through their point of contact and normal to their surfaces at this point. Thus a con-

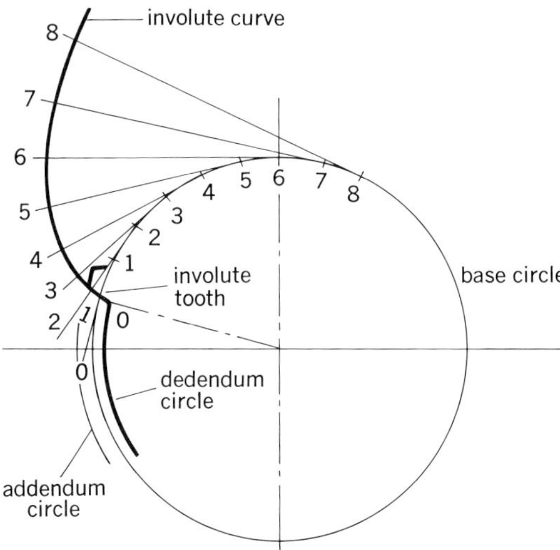

Fig. 3. Method of generation for face of involute tooth.

stant angular velocity ratio between bodies in driving contact demands that the common normal to the profiles at the point of contact cut the center line at a fixed point, the pitch point. This latter statement is frequently referred to as the fundamental law of gear tooth action. Pure rolling contact between gear teeth occurs only when they are in contact at the pitch point. At all other positions the teeth have some sliding with the maximum sliding at the first and last instants of contact. Although there are a number of tooth forms that will satisfy the fundamental law, only two of them have been used to any great extent. The cycloidal tooth predominated until the late 1800s but has been replaced to a great extent by the involute gear tooth. Cycloidal teeth are still found in instruments, watches, clocks, and, occasionally, in cast and cut gears.

Gears are said to be interchangeable when any gear of a set will run with any other gear of the same set. Actually there is little need today for gears that are interchangeable. In the manufacture of the overwhelming proportion of machines pairs or groups of gears are designed to mesh with each other and no others. Standardization in manufacture is common, however. Each of the gears is made from one of a set of standardized cutters. Thus it is the standardization of the production tools rather than of the gears themselves that is of most importance.

Involute gear teeth. An involute tooth is laid out along an involute (Fig. 3), which is the curve generated by a point on a taut wire as it unwinds from a cylinder. The generating circle is called the base circle of the involute. Proportions of the tooth are fixed by the gear tooth system and the diametral pitch. The involute curve establishes the tooth profile outward from the base circle. From the base circle inward, the tooth flank ordinarily follows a radial line and is faired into the bottom land with a fillet. The basic rack form of the involute tooth has straight sides, an important property from the manufacturing standpoint.

Gear teeth may interfere with each other, especially where pinions have a relatively small number of teeth. In Fig. 2, point of initial contact C occurs on the pressure line to the right of (after) the point of tangency of the pressure line and the pinion's base circle. Here there is no interference. But if point C were to the left of the tangency point, this would indicate premature contact between the teeth, a contact occurring on the noninvolute portion of the pinion tooth flank below the base circle. The tip of the gear tooth, in such a case, digs into the flank of the pinion tooth.

There are several ways to eliminate interference: (1) Reduce addendum of the gear. (2) Increase pressure angle. (3) Increase backlash by increasing center distance between gears. (4) Undercut flank of the pinion. (5) Relieve or modify face of the gear tooth.

Spur gears. In the truest sense, spur gears are only those that transmit power between parallel shafts and have straight teeth parallel to the gear axis (Fig. 4*a*). It is common practice, however, to group helical gears that have parallel shafts under the heading of spur gears (Fig. 4*b*). The pitch sur-

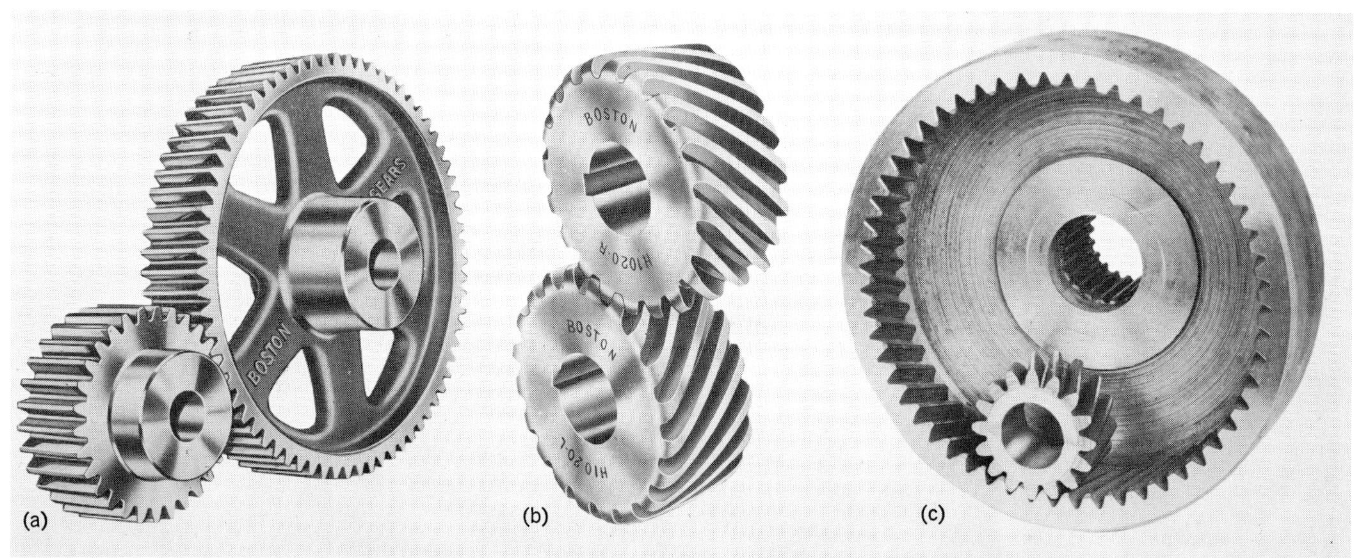

Fig. 4. Spur gears. (*a*) External spur gear and pinion, the commonest type of spur gear. (*b*) Single helical gear-set (*Boston Gear Works*). (*c*) Helical internal gear and pinion (*Fellows Gear Shaper Co.*).

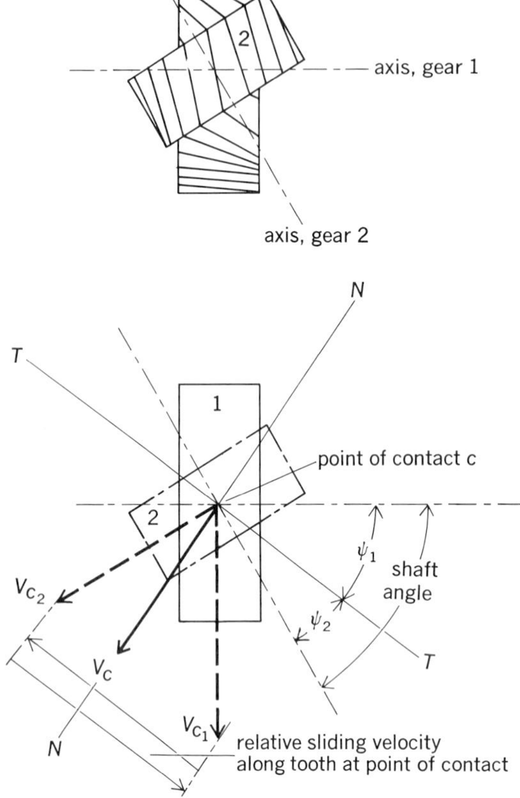

Fig. 5. Action of crossed helical gears.

faces of gears with parallel axes are rolling cylinders, and the motion these gears transmit is kinematically equivalent to that of the rolling pitch cylinders. Spur gears are classified as external, internal, and rack and pinion. External spur gears, the most common, have teeth which point outward from the center of the gear. Internal or annular gears have teeth pointing inward toward the gear axis (Fig. 4*c*). A rack may be considered as a gear having an infinite pitch circle radius. Thus its pitch surface is a plane. A rack and pinion running together transform rectilinear motion into rotary motion, or vice versa.

To standardize manufacture of gears, the American Gear Manufacturers Association has adopted three basic types of spur-gear tooth, with pressure angles of 20° and 25° for coarse-pitch gears (with a diametral pitch under 20) and 20° for fine-pitch gears (diametral pitch of 20 and above).

An internal gear has the positions of the addendum and dedendum reversed from those of an external gear. This results in a different tooth action and less slippage than with an equivalent external spur gear. For a given tooth ratio, the arc of action of an internal gear is slightly greater than that of an external gear of the same size and the tooth is stronger. The nature of an internal gear makes it especially suited to closer center distances than could be used with an external gear of the same size. When it is necessary to maintain the same sense of rotation for two parallel shafts, the internal gear is especially desirable because it eliminates the need for an idler gear. These conditions make the internal gear highly adaptable to epicyclic and planetary gear trains.

Noncircular gears are used to obtain velocity ratios that vary in a precise manner. Elliptical gears are an example of noncircular gears. They provide a convenient method of obtaining a quick return for machines that do most of their work during only a portion of the drive shaft revolution. Noncircular gears are used in computing mechanisms and other devices where a prescribed varying output function is to be obtained using a linear input.

Helical gears. Gears running on parallel axes and with teeth twisted oblique to the gear axis are essentially spur gears. Because of the twist, contact is progressive across the tooth surface, starting at one edge and proceeding across the face of the tooth. The action results in reduced impact and quieter operation, particularly at high speed. Herringbone gears are equivalent to two helical gears of opposite hand placed side by side. They are especially suited for high-speed operation and eliminate the axial thrust produced by single helical gears. Double helical gears have a central groove for tool runout, making it possible to finish the teeth by a shaving operation. They can be run at even higher speeds than herringbone gears. Dimensions of a helical gear are determined on both the plane of rotation and on the plane perpendicular to the helix angle of the tooth. Thus, a helical gear has a circular pitch measured on the plane of rotation and a normal circular pitch measured on the normal plane.

Crossed helical gears. Where shafts cross obliquely, motion is transmitted by crossed helical gears (Fig. 5). The teeth are helical but differ from the teeth of worm gears in that no one tooth (thread) makes a complete turn on the pitch circle.

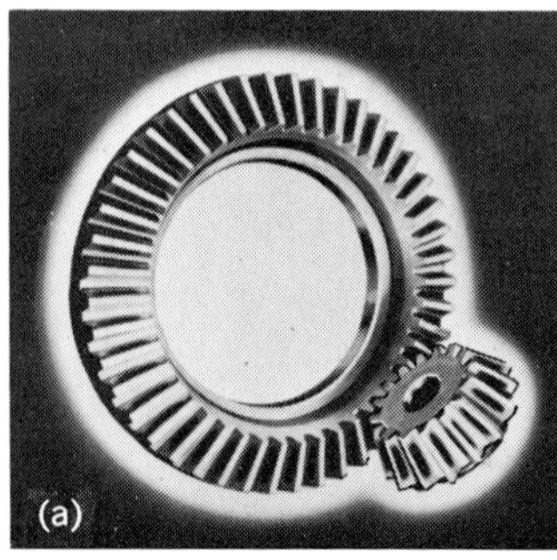

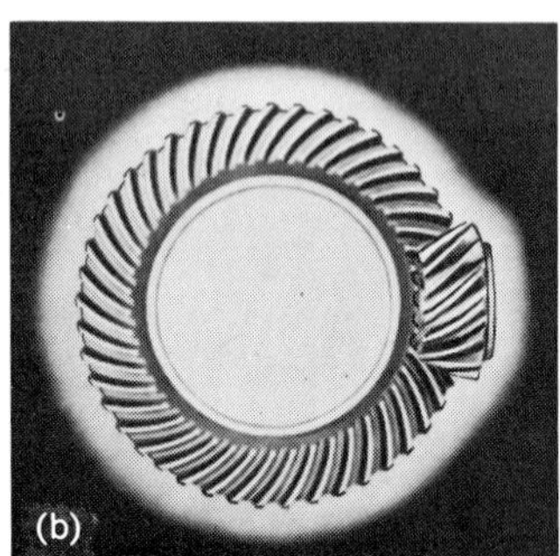

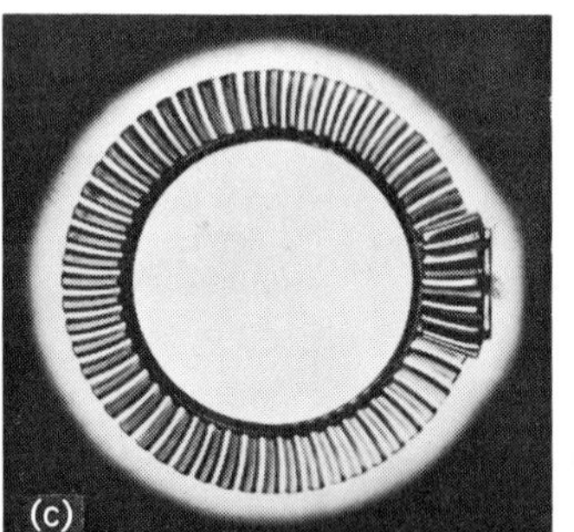

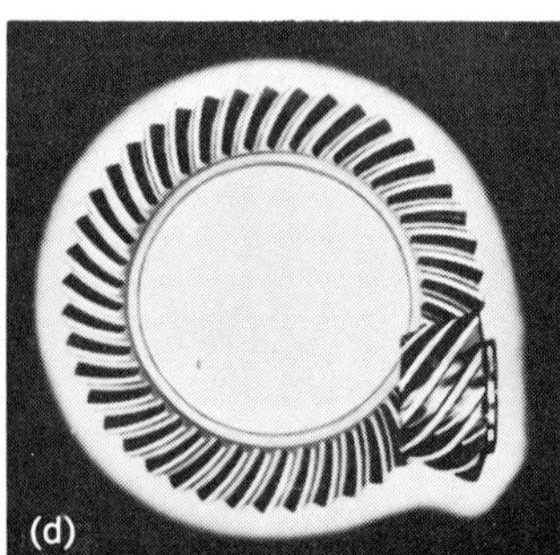

Fig. 6. Bevel gears with pinions. (*a*) Straight. (*b*) Spiral. (*c*) Zero. (*d*) Hypoid. (*Gleason Gear Works*)

Pitch surfaces of crossed helical gears are cylindrical as with spur gears. However, with crossed shafts, the oblique teeth have point contact rather than the line contact that occurs with parallel shafts. Analysis of the gears is based on equal components of the pitchpoint velocity on each mating gear along common normal *N-N*. Sliding occurs in direction *T-T* of the tooth elements. As with spur gears, the revolutions per unit time are inversely proportional to the numbers of teeth. Velocity ratio *VR* may also be expressed as in Eq. (2), where *D* is

$$VR = \frac{N_1}{N_2} = \frac{D_1 \cos \psi_1}{D_2 \cos \psi_2} \qquad (2)$$

pitch diameter and ψ helix angle. Helix angle is between the shaft axis and a line tangent to the tooth through the pitch point.

Helical gears are referred to as right- or left-hand in the same manner as screw threads, a right-hand gear being one on which the teeth twist clockwise as they recede from an observer looking along the axis.

Bevel gears. Where shafts intersect, bevel gears transmit the motion. Such gears may be used only to change the shaft axis direction or to change speed as well as direction. Two bevel gears with equal numbers of teeth and running together with their shaft axes intersecting at 90° are called miter gears. Several forms of bevel gears are in use, including straight-tooth, spiral, and skewed bevel gears (Fig. 6).

External bevel gears have pitch angles less than 90° (Fig. 7*a*). Internal bevel gears have pitch angles greater than 90°, hence their pitch cones are inverted (Fig. 7*b*). A crown gear is one having a pitch angle of 90° (Fig. 7*c*). Thus its pitch surface is a plane, and the crown gear corresponds in this respect to a rack in spur gearing.

Straight bevel gears. The simplest form of bevel gear has straight teeth which, if extended inward, would come together at the intersection of the shaft axes. This point is also the apex of the rolling cone, which forms the pitch surface of the gear. Much of the terminology applied to bevel gears is the same as that used for spur gears. Additional terms used in reference to bevel gears are given in Fig. 8. Diametral pitch of a bevel gear is not constant across the full width of the tooth. The diametral pitch at the pitch diameter is used in fixing tooth proportions. The formative number of teeth in a bevel gear is the number of teeth that would be on a spur gear whose pitch radius equaled the back cone distance of the bevel gear.

Speeds of the shafts of bevel gears (velocity ratio) are inversely proportional to the numbers of teeth on the gears or to the sines of the pitch angles, but not to the formative number of teeth. Use of straight bevel gears is limited to low-speed operations, ordinarily below 1000 surface feet per minute or, in the case of small gears, 1000 rpm.

Because each point on a straight tooth bevel gear remains a fixed distance from the pitch cone apex, there is no sliding along the tooth as it engages. Contact across the full tooth face occurs instantaneously—as with spur gears—as the teeth come into mesh.

Spiral bevel gears. To provide a gradual engagement, as contrasted to the full line engagement of straight bevel gears, the teeth of spiral bevel gears are curved and oblique. Theoretically the curve is a spiral but, to facilitate manufacture, the curve is actually a circular arc which, within the tooth face width, closely approximates a spiral. This tooth inclination brings more teeth in contact at any one time than with an equivalent straight-tooth bevel gear. The result is smoother and quieter operation, particularly at high speeds, and greater load-carrying ability than with straight bevel gears of the same size.

Spiral bevel gears are used in sewing machines, motion picture equipment, machine tools, and other applications where quiet, smooth operation is essential. They should, in general, be mounted on antifriction bearings because of the axial thrust due to the oblique teeth. In the past they have been used extensively in the rear axle drives of automobiles, but are being replaced by hypoid gears.

Zero bevel gears. A special form of bevel gear has curved teeth with a zero-degree spiral angle. Thus the teeth are not oblique as is the case with spiral bevel gears. Rather, the teeth lie in the same general direction as those of an equivalent straight-tooth bevel gear, and so the gears are usu-

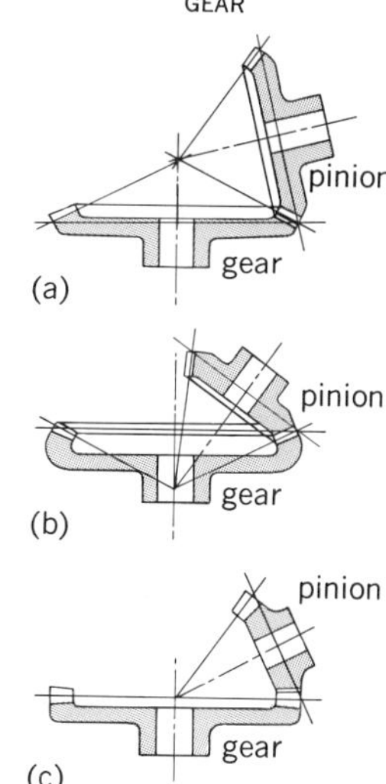

Fig. 7. Bevel gears. (*a*) External. (*b*) Internal. (*c*) Crown. (*From G. L. Guillet and A. H. Church, Kinematics of Machines, 5th ed., Wiley, 1950*)

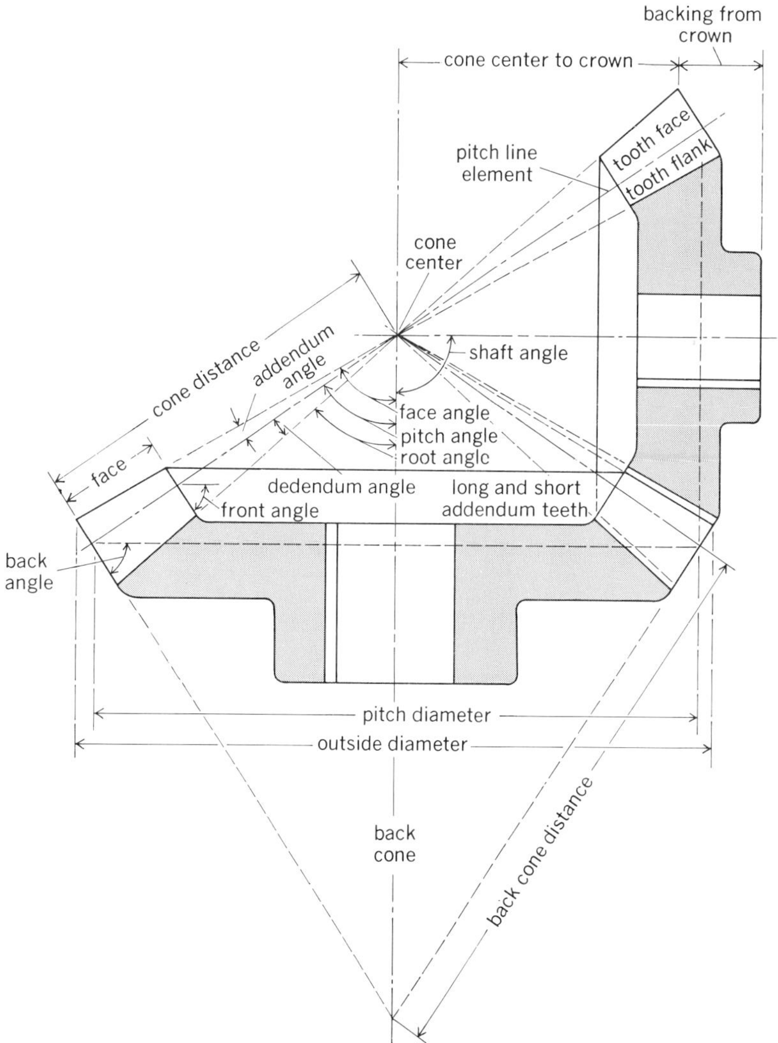

Fig. 8. Characteristics of bevel gears. (*From C. W. Ham, E. J. Crane, and W. L. Rogers, Mechanics of Machinery, 4th ed., McGraw-Hill, 1948*)

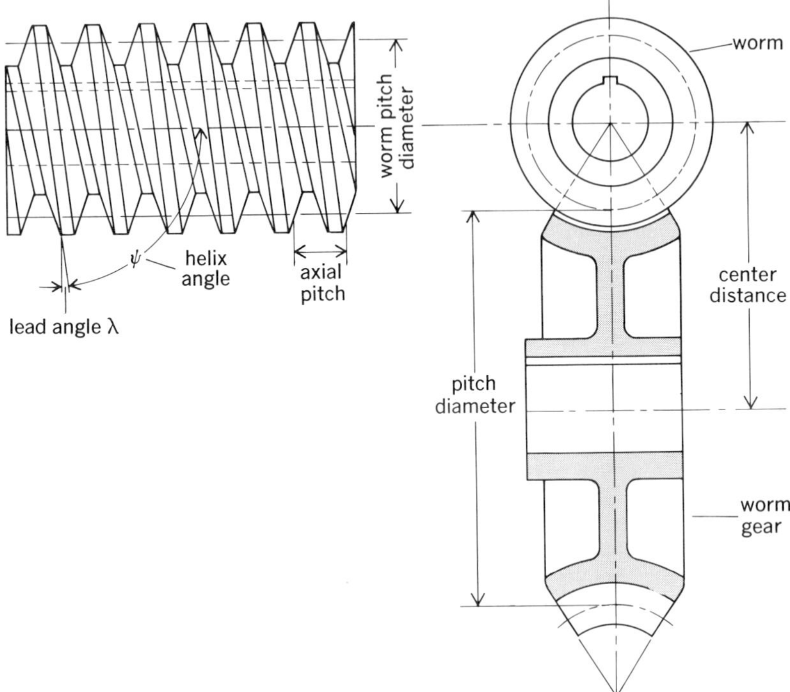

Fig. 9. Nomenclature of a single-enveloping worm gearset. Teeth of worm gear are called threads. (*From J. E. Shigley, Theory of Machines, McGraw-Hill, 1961*)

ally used in the same types of drives as the straight-tooth gear. As with straight bevels, they produce no axial thrust and, therefore, may be used without thrust bearings. The fact that they may be produced on the same equipment as spiral, bevel, and hypoid gears makes them economically desirable.

Tooth proportions for bevel gears follow the standards established by the Gleason Works and adopted as the recommended standard of the American Gear Manufacturers Association. Tooth proportions are a function of the velocity ratio. Thus, bevel gears are not as interchangeable as spur gears.

Hypoid gears. To connect nonparallel, nonintersecting shafts, usually at right angles, hypoid gears are used. They are similar to spiral bevel gears in their general appearance. The axis of the hypoid pinion may be offset above or below the axis of the gear. The shape of the tooth is similar to that of the spiral bevel gear and gives progressive contact across the tooth. In operation these gears run even more smoothly and quietly than spiral bevel gears. To maintain line contact of the teeth, with the offset shaft, the pitch surface of the hypoid gear is a hyperboloid of revolution rather than a cone as in bevel gears.

One of the first uses of hypoid gears was in the rear-axle drive of Packard automobiles. The operating smoothness of hypoid gears, along with the lower body lines made possible by the offset pinion shaft, has made them extremely popular for automotive use. Industrial applications of the hypoid gear also take advantage of the pinion offset, which allows the mounting of any number of pinions on a single continuous shaft, a feat not possible with bevel gears. The shaft arrangement of the hypoid gear and pinion enables bearings to be placed on both sides of the gear and of the pinion. The offset axis results in a larger and, consequently, stronger hypoid pinion tooth than on an equivalent straight-tooth or spiral bevel gear. Expressed in terms of pitch diameter, a hypoid pinion has fewer teeth than a spiral bevel pinion of the same pitch diameter. It is possible to use hypoid pinions having 7, 8, or 9 teeth in contrast to a minimum of 12, 13, or 14 teeth—depending on the velocity ratio—on a spiral bevel pinion. Lubricants must withstand the higher loading and the sliding that occurs along the teeth of hypoid gears.

Hypoid gears are suitable for large velocity reductions; reduction ratios of 60:1 and higher are entirely feasible. In general, shaft offset should not exceed 40% of the equivalent bevel gear back cone distance and, when the loading is heavy, as in truck and tractor drives, the offset should be nearer 20% of this distance. Direction of the offset, above or below center, must be specified for any given installation. The gears should, in general, be mounted on antifriction bearings in an oiltight case. Thrust bearings must be provided. Because of the sliding tooth action, the efficiency of hypoid gears is somewhat less than that of equivalent bevel gears.

Worm gears. A chief way to connect nonparallel, nonintersecting shafts that are at right angles is through a worm gear. The worm, ordinarily the driver, is similar to a crossed helical gear except that it has at least one complete tooth (thread) around the pitch surface. The mating gear is the worm wheel or worm gear. Worm gearing is generally used to obtain large velocity reductions with the worm as the driver and the worm wheel as the driven gear, although occasional applications, for example cream separators, have the worm wheel as the driver. The pitch surfaces of straight worms are cylinders and the involute teeth have point contact. Because the appearance of the worm is similar to that of a screw, the teeth are called threads.

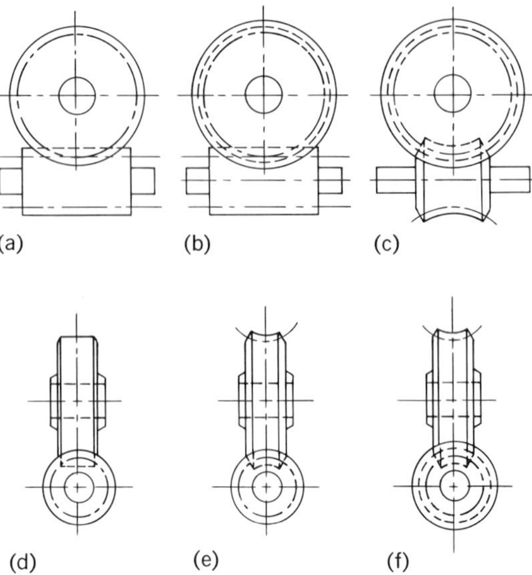

Fig. 10. Nature of contact for worm gears. (*a*) Non-throated. (*b*) Single-throated. (*c*) Double-throated on cone. (*d*) Point contact. (*e*) Line contact. (*f*) Area contact. (*Michigan Tool Co.*)

The pitch of the worm is the axial distance from any point on one tooth to the corresponding point on the next tooth (Fig. 9). This must equal the circular pitch of the mating worm wheel. Lead is the axial distance the worm helix advances in one complete revolution around the pitch surface. A single thread worm has the pitch and lead equal; one revolution of such a worm will, for shafts at right angles, advance the worm wheel $1/N$ revolutions if N is the number of teeth on the worm wheel. A double-threaded worm has the lead equal to twice the pitch and will then advance the worm wheel $2/N$ revolutions per turn of the worm. Thus, worm gears follow the general rule of angular velocity ratio inversely proportional to the ratio of the numbers of teeth. Worms are right- and left-hand in the same sense as helical gears. Changing hand of the worm reverses the relative rotation of the worm wheel.

Improved load-carrying capacity and wear characteristics are made possible by increasing the contact between the worm and wheel (Fig. 10). Line contact is obtained by making the worm wheel surface concave to conform to the tooth profile of the worm. Still greater contact is obtained by using a concave worm as well. Known as a cone-drive or Hindley worm, this design permits greater contact surface and allows more teeth to be in contact at one time. *See* GEAR CUTTING; GEAR TRAIN; MECHANISM; PLANETARY GEAR TRAIN.

[JOHN R. ZIMMERMAN]

Bibliography: N. P. Chironis (ed.), *Gear Design and Application*, 1967; D. W. Dudley (ed.), *Gear Handbook*, 1962; G. W. Michalec, *Precision Gearing: Theory and Practice*, 1966, reprint 1979; J. W. Patton, *Kinematics*, 1979; B. A. Shtipelman, *Design and Manufacture of Hypoid Gears*, 1978.

Gear cutting

The cutting or forming of a uniform series of toothlike projections on a surface of a workpiece, the teeth being designed to mesh with a mating tooth series in order to transmit power or motion.

Gear-cutting methods may be divided into two general categories. The first is gear generating in which the tooth is produced by the conjugate or total cutting action of the tool plus the rotation of the workpiece. The second method is gear forming in which the desired tooth shape is produced by a tool whose cutting profile matches the tooth form. Frequently gear teeth may be rough cut by a generating method and finished to size by a form tool.

Gear-cutting machines are designed to hold rigidly and to position accurately in relation to each other both a gear blank and a cutting tool so that the designed operation of the machine will produce the desired tooth series on the piece.

Hobbing. Gear-hobbing machines revolve the gear blank in a horizontal plane while the rotating hob, or cutter, moves downward across the face of the gear to generate the teeth. The hob, similar to a work gear, is made with a spiral thread on its periphery. Gashes across the thread provide the cutting edges. The power arbor, which holds the hob, is tilted so that the cutting teeth line up with the teeth on the gear blank. A system of index gearing maintains the relative turning ratio between the blank and the hob, and changing gear combinations permits the desired number of teeth to be cut. Other gearing or additional mechanism is provided so that the hob and gear blank are either advanced or retarded in respect to each other to cut a helix angle on the blank. The arbor must be tilted an additional amount so that the hob thread angle coincides with the helix angle of the gear. The hobbing process is used for roughing as well as for finish cutting of gears.

Generating. Gear-generating machines produce straight-bevel gear teeth by reciprocating a cutting tool across the gear face while a generating motion is attained by slowly rotating the work and the tool holder, or cradle, at a synchronized rate. The gear blank is automatically withdrawn from the tools at the end of each roll so that both the cradle and work spindle roll back to starting position to cut the next tooth.

Two rectangular tools are used for cutting. The tools are ground and positioned so that each side of one gear tooth is cut during a generating roll.

Spiral and curved-toothed bevel gears and pinions are generated by the cutting action of a face milling cutter synchronized to rotate with a gear blank (Fig. 1). The cutter is a circular tool holder with a series of cutting blades around the edges of its face (Fig. 2). After each generating roll of the tool and blank, the machine indexes to bring the cutter into the next tooth slot.

Planing-type generating machines, using a single reciprocating tool, are designed to cut large spiral bevel gears.

Shaping. Gear-shaping machines generate gear teeth on a gear blank by means of a pinion-shaped cutting tool. A cutting edge is provided along one side of the teeth. The cutter generates by simultaneously reciprocating across the gear face while it also rolls with the blank. Because the cutter must return to its starting position at the end of each stroke, it is usually relieved on the return stroke by automatically increasing the center distance between work and cutter spindles. Large gears are frequently finish-lapped after shaping. *See* LAPPING.

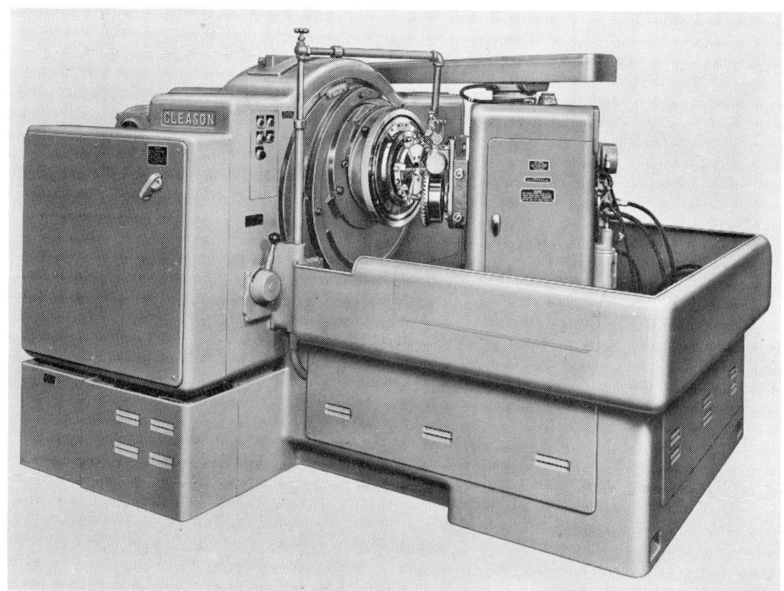

Fig. 1. Hypoid gear generator. (*Gleason Works*)

Fig. 2. Gear blank and cutter as used on a hypoid generator. Cutter has a series of cutting blades. (*Gleason Works*)

Gear shaping requires only a small amount of run out at the end of each cutting stroke permitting shaped teeth to be located close to shoulders.

Milling. Gear-milling machines produce gears by use of a rotating cutter whose cutting profile matches the contours of the gear teeth. During the cutting process the gear blank remains stationary because the cutter forms both sides of the space between two gear teeth. At the completion of the cut the cutter is returned to its starting position, and the blank is indexed or rotated for the next cut. Spur and helical gears, worms, and sometimes bevel gears may be produced by milling. Standard rather than special machines often are used.

Milling does not usually produce gears of great accuracy; often the gears are hardened and finish-ground after milling.

Grinding. Gear-grinding machines shape gears both by formed grinding wheels and by generation. Although some fine-pitch gear teeth are finish-ground from solid stock, gear grinding is primarily a finishing operation. Grinding is the only way to finish gears which have been brought to a high degree of hardness. *See* GRINDING.

A form-grinding disk wheel grinds both sides of the space between two gear teeth as it moves across the face of the gear (Fig. 3). An involute tooth form is dressed into both sides of the grinding wheel.

Fig. 3. An automatic gear-grinding machine. (*Gear Grinding Machine Co.*)

Generating grinding is done with a flat-sided disk wheel. As the disk rotates it both reciprocates across the gear face and moves sideways as the gear rotates. Collectively these movements result in an involute tooth form.

Shaving. Gear shaving is used for finishing and for improving profile accuracy. Shaving is accomplished by using both rotary and rack-type cutters whose teeth are serrated with many small notches or cutting edges. The axis of the cutters is set at a slight angle to the axis of the workpiece so that, as the two are rolled together, the notched cutter teeth shave the gear teeth. *See* BROACHING.

Usually only gears of machinable hardness are shaved, although gears may be shaved before or after they are hardened.

Broaching. Gear-broaching machines are generally used to produce internal gears in addition to racks and gear segments. Small internal gears are frequently made in one pass of the broach. Larger gear require several passes plus indexing. Either spur or helical teeth may be produced by broaching. *See* MACHINING OPERATIONS.

[ALAN TUTTLE]

Gear drive

Transmission of motion or torque from one shaft to another by means of direct contact between toothed wheels. The active parts of the gear teeth are usually made in the form of involutes and transmit the force smoothly from a tooth on one gear to a tooth on another gear (see illustration).

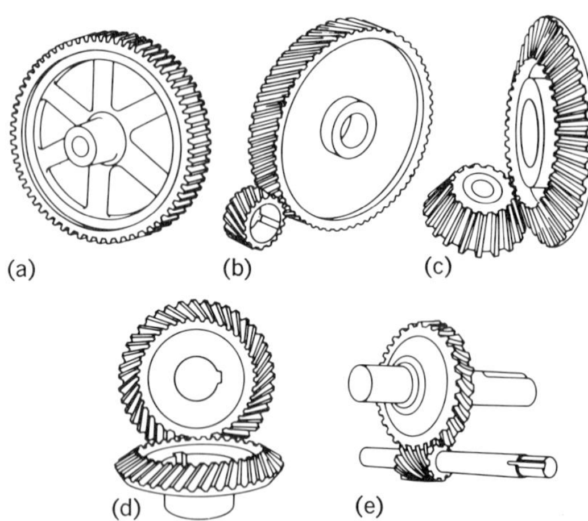

Gear drives. (*a*) Spur gear. (*b*) Helical gears. (*c*) Bevel gears. (*d*) Skew bevels. (*e*) Worm gears. (*W. H. Crouse, Automotive Mechanics, 5th ed., McGraw-Hill, 1965*)

If a 20-tooth pinion (the smaller of a pair of gears) meshes with and drives a 40-tooth gear, the pinion must rotate twice for each revolution of the gear. In general, the ratio of the angular velocity ω of gears A and B is as shown in Eq. (1), where N_B

$$\frac{\omega_A}{\omega_B}=\frac{N_B}{N_A} \tag{1}$$

and N_A are the numbers of teeth on the respective gears. If friction is negligible, as it usually is in gear drives transmitting power between parallel shafts, input power to A equals output power from B. Power is thus expressed as the product of torque and angular velocity, as shown in Eqs. (2),

$$T_A\omega_A = T_B\omega_B \qquad T_B = T_A\frac{\omega_A}{\omega_B} \tag{2}$$

where T_A and T_B are the torques on shafts A and B respectively. *See* GEAR; GEAR TRAIN; SIMPLE MACHINE.

[RICHARD M. PHELAN]

Gear loading

The power transmitted or the contact force per unit length of a gear. If gear speed, size, and tooth contour are fixed, the power-handling capacity may be increased by increasing the axial length of the gear. However, where space is restricted (as in aircraft and many ground vehicles), the gear is usually loaded to its safe limit.

To obtain satisfactory tooth-surface durability from highly loaded gears, experience indicates that several items of the gear set must be properly designed and manufactured, namely, the tooth profile which must be properly modified from a true involute to suit the operating conditions; index of teeth and parallelism of teeth, which must be held within close limits; gearing, which must be mounted so the teeth will not deflect out of line; and gear tooth surfaces, which must be of sufficient hardness and proper finish and which should have good lubrication, particularly on start of initial operation.

Gear design is a compromise between tooth strength and surface durability. Larger teeth give more tooth strength but less surface durability; smaller teeth result in less tooth strength but more surface durability.

Surface distress on the teeth becomes particularly important at loadings in the order of thousands of pounds per inch of face.

In designing the gear tooth surfaces, one mesh of teeth is assumed to carry the load through the action, and the action is balanced so that the product of Hertz maximum compressive pressure P in pounds per square inch and sliding velocity V in feet per second is the same at each end of the assumed single-mesh action. This product, known as the PV value, may be 3,000,000 or higher in gears with high pitch-line speeds for which the designer may use PVT values. The design may produce a smaller contact ratio than is conventional, but this result indicates that contact ratio alone is an unreliable measure of gear capacity.

A highly loaded tooth of adequate rigidity deflects about a point in the middle of the rim, bending as a rigid body under load rather than as a nonuniform beam only. Relief or other modification of tooth profile provides clearance so as to avoid excessive loading at teeth tips due to deflection of the preceding mesh, and ramps at the tooth tips assure that first contact does not extend to the tips.

This design refinement necessitates corresponding care in production, as in minimizing distortion during carburizing. Teeth can be held parallel within 0.0003 in. (7.5 μm) in the width of the tooth, and the index can be maintained within 0.0002 in. (5 μm) between adjacent teeth of a gear.

To assure removal of all possible error in grinding when this method of finishing is used and to achieve a more satisfactory working surface on the gear tooth face, grinding wheels are dressed to produce a finish which is coarser than usual. Slight surface roughness (Profilimeter readings between 15 and 37) produces less surface distress, or scuffing, of highly loaded steel gears than do smoother finishes, although a steel gear running with a bronze one should be smoother.

Quiet operation is an indication of efficient operation free from abrupt tooth engagement and severe scuffing of one tooth face by another. Helical gears provide smooth transition of effort from tooth to tooth. The gear is designed with a modified involute to provide soft contact at the start of engagement and end of engagement to reduce the pressure on the teeth when the sliding of the tooth surfaces is highest. For quietest operation they should be cut and shaved accurately for lead, profile, and index. A helical contact ratio of 2 or more is desirable. This ratio can be obtained by increasing the width or the helix angle, although effort expended to increase helical contact ratio reduces deflection and PV values more than does the same effort expended to increase involute contact ratio. As tooth size and resultant sliding increase, this proportioning of effort becomes more important. Helix angles in the order of 45° on constant mesh gears require more expensive mountings but improve quietness. Rigid and accurate mounting is essential to achieve best operating condition.

With other gears subject to deflection, such as hypoid rear-axle gearing, conventional practice is to crown the tooth and modify the tooth profile to keep contact away from the tooth ends and tips. The gear set is usually designed for stiffness rather than for strength, and the shafts and mountings should also be designed on this criterion so that little crowning is needed. [FOREST R. MC FARLAND]

Bibliography: J. O. Almen and J. C. Straub, Transmission gears, *Automot. Ind.*, vol. 75, 1937; American Gear Manufacturing Association, *PVT Values of Gear Teeth*, AGMA Progr. Rep. no. 101.02, 1951; S. A. Couling, *Industrial and Marine Gearing*, 1962; D. W. Dudley, *Gear Handbook*, 1962; F. R. McFarland, Experiences, highly-loaded gearing, *Proc. Ind. Math. Soc.*, 1959; S. P. Timoshenko, *Theory of Elasticity*, 2d ed., 1951; W. A. Tuplin, *Gear Load Capacity*, 1962.

Gear train

A combination of two or more gears used to transmit motion between two rotating shafts or between a shaft and a slide. In theory two gears can provide any speed ratio in connecting shafts at any center distance, but it is often not practical to use only two gears. If the ratio is large or if the center distance is relatively great, the larger of the two gears may be excessively large. Moreover, an additional gear may be necessary simply to give the proper direction to the output gear. Belt, rope, and chain drives are frequently used in conjunction with gear trains. *See* BELT DRIVE; CHAIN DRIVE; PLANETARY GEAR TRAIN.

Classification. The most important distinction in classifying gear trains is that between ordinary and epicyclic gear trains. In ordinary trains (Fig. 1*a*), all axes remain stationary relative to the frame. But in epicyclic trains (Fig. 1*b*), at least one

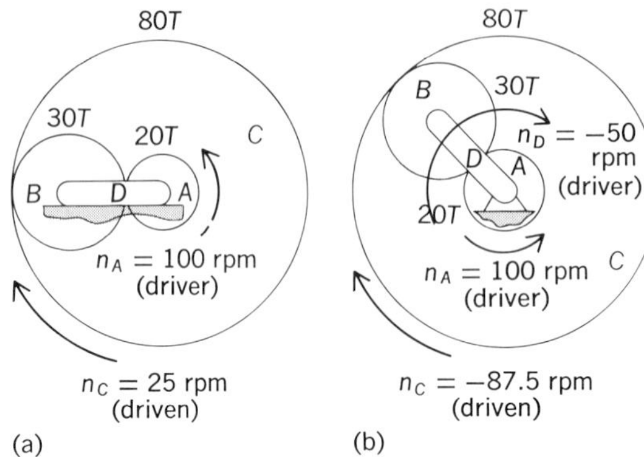

Fig. 1. Gear trains. (*a*) Ordinary. (*b*) Epicyclic.

axis moves relative to the frame. In Fig. 1*b* gear *B*, whose axis is in motion, is called a planet. The gears *A* and *C* are sun gears.

An ordinary gear train is a single degree of freedom mechanism: A single input, such as an input to gear *A* of the train in Fig. 1*a*, suffices to control the motions of the other moving members. But an epicyclic gear train (Fig. 1*b*) has two degrees of freedom: Two inputs are necessary. In the epicyclic train of Fig. 1*b*, the two input members are gear *A* and the planet carrier, link *D*. Only if both these members are controlled by external agencies can the motions of gears *B* and *C* be predicted. Frequently one gear of an epicyclic train is fixed. This then is one of the input members with a velocity of zero revolutions per unit time.

A simple gear train is one in which each gear is fastened to a separate shaft, as in Fig. 1*a*. If at least one shaft has two or more gears fastened to it (Fig. 2), the train is said to be compound. The train of Fig. 2 is also a reverted gear train, because the input and output shafts are in line. (A reverted epicyclic gear train is shown in Fig. 3.) If the input shaft does not line up with the output shaft, the train is said to be nonreverted.

The 1000-hp mill drive and pinion stand in Fig. 4 are an example of a large industrial ordinary gear train. The first speed reduction is with the opposed single helical gears, and the second is through the herringbone gears. This train is compound and nonreverted.

Ratios. For an ordinary gear train, the ratio of the angular velocity of the last driven gear to that of the driving gear is called the train value. The ratio of the driver's velocity to that of the last driven gear is the velocity ratio. Train value and velocity ratio are, according to these definitions, reciprocal quantities. The train value for the ordinary train of Fig. 1*a* is −0.25. The output gear *C* turns at one-fourth the speed of the input gear *A* and in the opposite direction.

For epicyclic trains the situation can be a little more complicated. Since in general the output velocity is dependent on the two input velocities, the term train value is nonspecific. However, if one gear of an epicyclic train is fixed, the output velocity is some multiple of the velocity of the nonfixed input member; the term train value then applies.

Direction of rotation. The direction of rotation of any gear in an ordinary gear train is best determined by inspection, keeping in mind the fact that mating external gears have opposite directions of rotation, while an internal gear has the same direction as its mating gear.

Ordinary train. Train value is by definition Eq. (1) when the angular velocities are measured with

$$E=\frac{\text{angular velocity of the last driven gear}}{\text{angular velocity of the first driving gear}} \tag{1}$$

respect to the frame supporting the gears. Referring to the ordinary train of Fig. 2, the train value is (ignoring sign) Eq. (2), where n is the speed in revo-

$$E=\frac{\text{rpm of shaft 4}}{\text{rpm of shaft 1}}=\frac{n_4}{n_1} \tag{2}$$

lutions per unit time. This can be expanded as an identity for the entire train to Eq. (3). Each of the ratios on the right side of Eq. (3) is the train value

$$E=\frac{n_4}{n_1}=\frac{n_2}{n_1}\times\frac{n_3}{n_2}\times\frac{n_4}{n_3} \tag{3}$$

for a pair of meshing gears. Since for any two gears in mesh the speeds vary inversely as the numbers of teeth, the train value for the whole train must be (again ignoring sign) Eq. (4), where N is the number

$$E=\frac{n_4}{n_1}=\frac{N_A}{N_B}\times\frac{N_B}{N_C}\times\frac{N_D}{N_E} \tag{4}$$

of teeth. In Eq. (4) the number of teeth appearing in the numerator of the expression on the right are those for driving gears, while those appearing in the denominator are the numbers of teeth on the driven gears. Thus, the general expression for the magnitude of the train value for an ordinary gear train is Eq. (5). Because any two gears in mesh

$$E=\frac{\text{product of numbers of teeth on driving gears}}{\text{product of numbers of teeth on driven gears}} \tag{5}$$

must have the same diametral pitch, expressions for the train value can also be written in terms of pitch diameters instead of numbers of teeth.

Note that in the expression for the train value of the train of Fig. 2 the number of teeth on gear *B* cancels out. Gear *B* is an idler; it is both a driver and a driven gear. Its size has no effect on the train value's magnitude. It does affect, however, the sign of the train valve. Idlers also are useful where a relatively large center distance must be spanned.

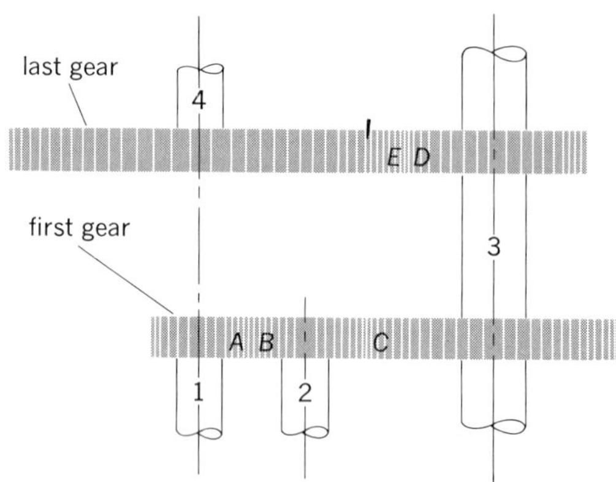

Fig. 2. Reverted gear train.

Transmissions. If a machine must be operated at any one of several output speeds, a multiple-speed gearbox, or transmission, may be used as a component part. Machine tools and motor vehicles are familiar instances of the need for transmissions. The speed of the output shaft of a transmission can be varied by sliding gears in and out of contact or by connecting gears in continual mesh to shafts by means of clutches. A compact 9-speed transmission utilizing 10 gears and 3 shafts is shown in Fig. 5. Gears *A*, *B*, and *C* slide as a unit on the constant-speed shaft, providing three ways in which the constant speed and the intermediate shafts can be connected (*A* and *D*, or *B* and *E*, or *C* and *G*). Similarly gears *H*, *I*, and *J* slide as a unit on the variable-speed shaft, providing three ways in which it can be connected to the intermediate shaft (*D* and *H*, or *E* and *I*, or *F* and *J*). Consequently nine possible combinations are available. For example, the lowest output speed is attained when *A* and *D* mesh and *E* and *I* mesh. The train value is then as in Eq. (6).

$$E = \frac{20 \times 37}{47 \times 57} = 0.276 \qquad (6)$$

The highest speed occurs when *C* and *G* mesh and *D* and *H* mesh. The train value for this combination is shown in Eq. (7). *See* AUTOMOTIVE TRANSMISSION.

$$E = \frac{41 \times 47}{26 \times 47} = 1.577 \qquad (7)$$

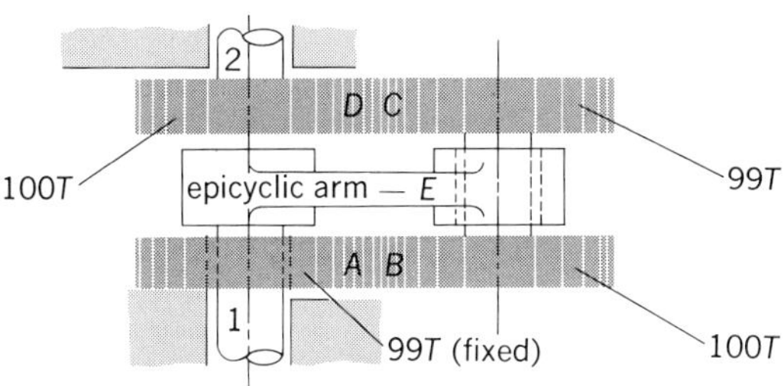

Fig. 3. Reverted epicyclic gear train.

Epicyclic train. An epicyclic train is named for the path described by any point on the pitch circle of a planet gear as it rolls on a sun gear. The term epicyclic refers only to the motion and is not related to the gear tooth form, which may be involute or any other form satisfying the law of gearing. An epicyclic train can be used to obtain a considerably greater velocity ratio than would be possible with an ordinary gear train of the same size. *See* GEAR.

In the compound and reverted epicyclic gear train of Fig. 3, gear *A* is fixed to the frame, but not to shaft 1. The planet carrier, link *E*, is fastened to shaft 1 and carries planet gears *B* and *C*, which are fastened to a common shaft. Planet *B* rolls on sun gear *A*, while planet *C* rolls on sun gear *D*. Gear *A* is one of the input members; it has a velocity of 0 rpm. The other input could be either the carrier or gear *D*. In the example to follow it will be assumed that the arm is the input member.

To determine the train value of an epicyclic train with a fixed gear, the following special but very convenient technique can be used. The net

Fig. 4. Helical gear speed reducer. (*Farrel-Birmingham Co.*)

motion is divided into two parts. In the first, the train is locked with no relative motion between any of its components. For this step gear *A* is released from the frame and locked to gear *B*. The locked train is rotated one revolution in an arbitrary direction. Counterclockwise rotation is assumed positive, clockwise negative. In the second step, the assembly is viewed as an ordinary gear train with the planet carrier regarded as fixed. While the fixed gear is returned to its original position by one revolution in the sense opposite that used in the first step, the consequent rotation of the other gears is noted. For any one gear, the algebraic sum of the revolutions during the two parts is the net motion of that gear for one revolution of the planet

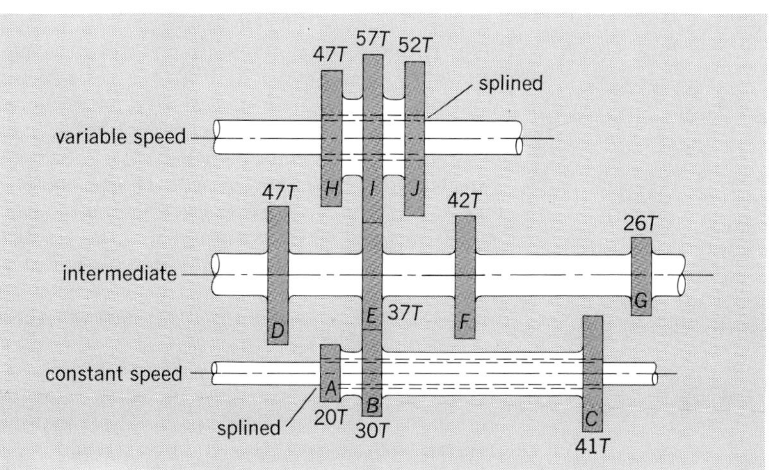

Fig. 5. Compact 9-speed transmission utilizing 10 gears and 3 shafts. (*From J. R. Zimmerman, Elementary Kinematics of Mechanisms, Wiley, 1962*)

Algebraic sum of revolutions for a reverted epicyclic gear train

Category	Gear A	Gear B	Gear C	Gear D	Arm E
Locked train	+1	+1	+1	+1	+1
Locked arm	−1	$+\frac{99}{100}$	$+\frac{99}{100}$	$-\frac{99}{100}\times\frac{99}{100}$	0
Net motion	0	$+\frac{199}{100}$	$+\frac{199}{100}$	$+\frac{199}{10{,}000}$	+1

carrier. To illustrate, consider the table for the train of Fig. 3, with the number of teeth as marked for each gear. The table shows that, for each revolution of the carrier, gear D will turn 199/10,000 of a revolution in the same direction.

For epicyclic trains with no fixed gears, a modification of the two step method is needed. In the first step, the locked train is given the same rotation as the known motion of the planet carrier. In the second step, with the carrier stationary, the gear whose velocity is known is rotated sufficient turns in the proper direction to make its net motion equal to the known value.

Probably the most common epicyclic train with no fixed gear is the automobile differential. When the automobile is moving along a straight path, there is no relative motion between the bevel differential gears fastened individually to the right and left axles. But as the car makes a turn, the differential gears move relative to one another in the manner of an epicyclic bevel gear train with no fixed gears. *See* DIFFERENTIAL; PLANETARY GEAR TRAIN. [JOHN R. ZIMMERMAN]

Bibliography: C. A. Scoles and R. Kirk, *Gear Metrology*, 1969; J. F. Shannon, *Marine Gearing*, 1978; J. R. Zimmerman, *Elementary Kinematics of Mechanisms*, 1962.

Generator

Any machine by which mechanical power is transformed into electric power. Generators fall into two main groups, alternating-current (ac) and direct-current (dc). They may be further classified by their source of mechanical power, called the prime mover. Generators are usually driven by steam turbines, hydraulic turbines, engines, gas turbines, or motors. Small generators are sometimes powered from windmills or through gears, belts, friction, or direct drive from parts of vehicles or other machines. *See* PRIME MOVER.

Theory of operation. The theory of operation of most electric generators is based upon Faraday's law. When the number of webers of magnetic flux linking a coil of wire is caused to change, an electromotive force proportional to the product of the number of turns times the rate of change of flux is generated in the coil. The instantaneous induced voltage is given in Eq. (1), where n is the

$$e = -\,n\,(d\phi/dt)\ \text{volt} \tag{1}$$

number of turns, ϕ is the flux in webers, and t is time in seconds. The minus sign indicates that the induced voltage opposes the effect which produced it. Voltage is induced in the windings of a generator by mechanically driving one member relative to the other, thereby causing the magnetic flux linking one set of coils, called the armature windings, to vary, pulsate, or alternate. The magnetic flux may originate from a permanent magnet, a dc field winding, or an ac source.

Figure 1 shows an elementary generator having a stationary field and a single, rotating, armature coil. It is apparent that the magnetic flux threading the coil reverses direction twice per revolution, thereby generating one cycle of voltage in the armature coil for each revolution. If this variation in flux is expressed as a function of time, the voltage generated in the coil would be given by Eq. (1). For example, let the flux vary as given by Eq. (2),

$$\phi = \Phi_m \cos 2\pi ft \tag{2}$$

where Φ_m is the maximum flux, f is the frequency, and t is time. By differentiating Eq. (2) with respect to time, and substituting in Eq. (1), the instantaneous voltage in the coil is found to be Eq. (3). This

$$e = 2\pi fn\Phi_m \sin 2\pi ft \tag{3}$$

ac voltage can be taken from the armature by brushes on the slip rings shown. If the coil terminals were brought to a two-segment commutator instead of slip rings, a pulsating dc voltage would appear at the brushes. *See* ALTERNATING-CURRENT GENERATOR; DIRECT-CURRENT GENERATOR.

Construction. In practice, permanent-magnet fields are used only in small generators. Large generators, except induction generators, are equipped with dc field windings. The field coils are wound on the stators of most dc generators to permit mounting the armature coils and commutator on the rotor. On ac generators, the field coils are normally located on the rotors. Field coils require only low voltage and power, and only two lead wires. They are more easily insulated and supported against rotational forces and are better suited to sliding contacts than the relatively high-voltage armature windings, which often have six leads brought out.

Any part of the magnetic circuit not subject to changing flux may be of solid steel. This includes the field poles of dc machines and portions or all of the rotating field structure of some ac generators. In machines with small air gaps the poles are frequently of laminated steel, even though their flux may be substantially constant. Laminations help minimize pole-face losses arising from tooth-fre-

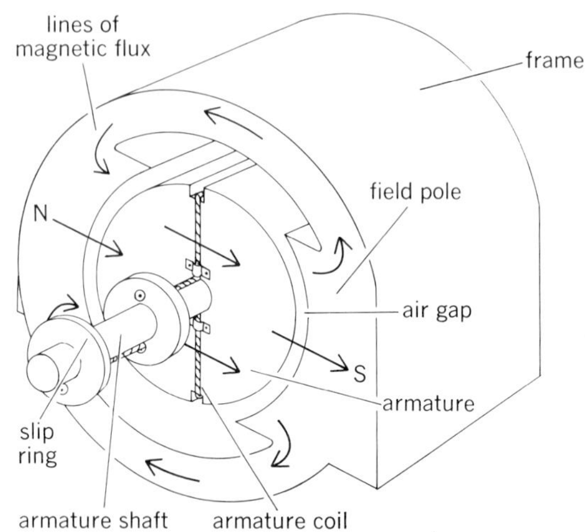

Fig. 1. Elementary generator.

Fig. 2. Typical turbogenerator unit with a direct-connected exciter, rated 22,000 kW. (*Allis-Chalmers*)

quency pulsations. The armature core is almost always composed of thin sheets of high-grade electrical steel to reduce core loss. *See* CORE LOSS.

The windings are insulated from the magnetic structure, and are either embedded in slots distributed around the periphery or mounted to encircle the field poles. The terminals from the stator windings and from the brush holders are usually brought to a convenient terminal block for external wiring connections. *See* WINDINGS IN ELECTRIC MACHINERY.

Turbogenerators. Generators driven by steam or gas turbines are sometimes called turbogenerators. Although in small sizes these may be gear-driven, and some may be dc generators, the term turbogenerator generally means an ac generator driven directly from the shaft of a steam turbine. A typical turbogenerator unit is shown in Fig. 2.

In order to achieve maximum efficiency, the steam turbine must operate at high speed. Consequently, direct-connected turbogenerators are seldom built to operate below 1500 rpm. To minimize windage loss and to keep rotational stresses down to a safe level, turbogenerator rotors are usually long and slender, in some cases five to six times the diameter in length of active iron. Long rotors operate above their first critical speed, and in some cases near or above their second, thereby introducing mechanical problems in balancing and resonance. To shorten the length of turbogenerators, conductor cooling has proved effective. *See* ELECTRIC POWER GENERATION; ELECTRIC ROTATING MACHINERY.

[LEON T. ROSENBERG]

Bibliography: S. L. Dixon, *Fluid Mechanics, Thermodynamics of Turbomachinery*, 2d ed., 1974; D. G. Fink and H. W. Beaty, *Standard Handbook for Electrical Engineers*, 11th ed., 1978; M. G. Say, *Alternating Current Machines*, 4th ed., 1976.

Geothermal power

Thermal or electrical power produced from the thermal energy contained in the Earth (geothermal energy). Use of geothermal energy is based thermodynamically on the temperature difference between a mass of subsurface rock and water and a mass of water or air at the Earth's surface. This temperature difference allows production of thermal energy that can be either used directly or converted to mechanical or electrical energy.

CHARACTERISTICS AND USE

Temperatures in the Earth in general increase with increasing depth, to 200–1000°C at the base of the Earth's crust and to perhaps 3500–4500°C at the center of the Earth. Average conductive geothermal gradients to 10 km (the depth of the deepest wells drilled to date) are shown in Fig. 1 for representative heat-flow provinces of the United States. The heat that produces these gradients comes from two sources: flow of heat from the deep crust and mantle; and thermal energy generated in the upper crust by radioactive decay of isotopes of uranium, thorium, and potassium. The gradients of Fig. 1 represent regions of different conductive heat flow from the mantle or deep

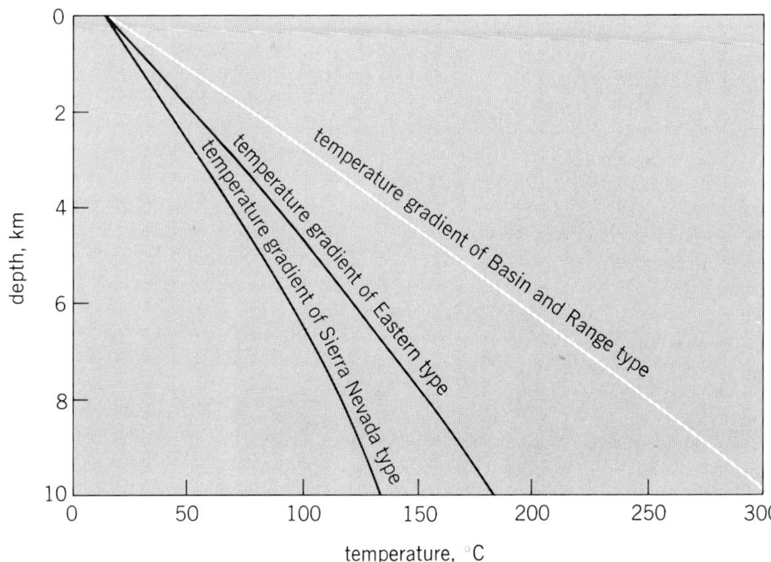

Fig. 1. Calculated average conductive temperature gradients to a depth of 10 km in representative heat-flow provinces of the United States. (*Adapted from D. E. White and D. L. Williams, eds., Assessment of Geothermal Resources of the United States—1975, USGS Circ. 726, 1975*)

crust. Some granitic rocks in the upper crust, however, have abnormally high contents of U and Th and thus produce anomalously great amounts of thermal energy and enhanced flow of heat toward the Earth's surface. Consequently, thermal gradients at shallow levels above these granitic plutons can be somewhat greater than shown on Fig. 1.

The thermal gradients of Fig. 1 are calculated under the assumption that heat moves toward the Earth's surface only by thermal conduction through solid rock. However, thermal energy is also transmitted toward the Earth's surface by movement of molten rock (magma) and by circulation of water through interconnected pores and fractures. These processes are superimposed on the regional conduction-dominated gradients of Fig. 1 and give rise to very high temperatures near the Earth's surface. Areas characterized by such high temperatures are the primary targets for geothermal exploration and development.

Natural geothermal reservoirs. Commercial exploration and development of geothermal energy to date have focused on natural geothermal reservoirs—volumes of rock at high temperature (up to 350°C) and with both high porosity (pore space, usually filled with water) and high permeability (ability to transmit fluid). The thermal energy is tapped by drilling wells into the reservoirs. The thermal energy in the rock is transferred by conduction to the fluid, which subsequently flows to the well and then to the Earth's surface.

Natural geothermal reservoirs, however, make up only a small fraction of the upper 10 km of the Earth's crust. The remainder is rock of relatively low permeability whose thermal energy cannot be produced without fracturing the rock artificially by means of explosives or hydrofracturing. Experiments involving artificial fracturing of hot rock have been performed, and extraction of energy by circulation of water through a network of these artificial fractures may someday prove economically feasible.

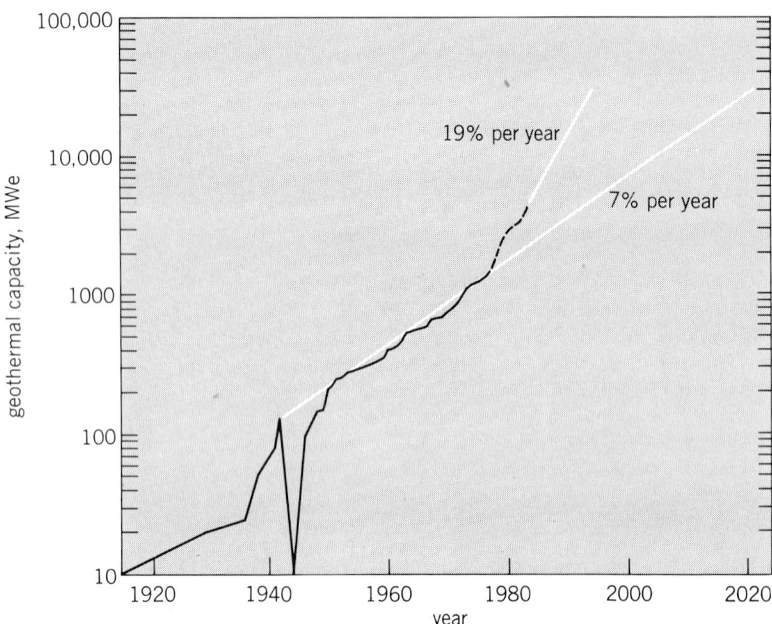

Fig. 2. Graph showing growth of geothermal electrical generating capacity of the world. *(From L. J. P. Muffler, ed., Assessment of Geothermal Resources of the United States—1978, USGS Circ. 790, 1979)*

There are several types of natural geothermal reservoirs. All the reservoirs developed to date for electrical energy are termed hydrothermal convection systems and are characterized by circulation of meteoric (surface) water to depth. The driving force of the convection systems is gravity, effective because of the density difference between cold, downward-moving, recharge water and heated, upward-moving, thermal water. A hydrothermal convection system can be driven either by an underlying young igneous intrusion or by merely deep circulation of water along faults and fractures. Depending on the physical state of the pore fluid, there are two kinds of hydrothermal convection systems: liquid-dominated, in which all the pores and fractures are filled with liquid water that exists at temperatures well above boiling at atmospheric pressure, owing to the pressure of overlying water; and vapor-dominated, in which the larger pores and fractures are filled with steam. Liquid-dominated reservoirs produce either water or a mixture of water and steam, whereas vapor-dominated reservoirs produce only steam, in most cases superheated.

Natural geothermal reservoirs also occur as regional aquifers, such as the Dogger Limestone of the Paris Basin in France and the sandstones of the Pannonian series of central Hungary. In some rapidly subsiding young sedimentary basins such as the northern Gulf of Mexico Basin, porous reservoir sandstones are compartmentalized by growth faults into individual reservoirs that can have fluid pressures exceeding that of a column of water and approaching that of the overlying rock. The pore water is prevented from escaping by the impermeable shale that surrounds the compartmented sandstone. The energy in these geopressured reservoirs consists not only of thermal energy, but also of an equal amount of energy from methane dissolved in the waters plus a small amount of mechanical energy due to the high fluid pressures.

Use of geothermal energy. Although geothermal energy is present everywhere beneath the Earth's surface, its use is possible only when certain conditions are met: (1) The energy must be accessible to drilling, usually at depths of less than 3 km but possibly at depths of 6–7 km in particularly favorable environments (such as in the northern Gulf of Mexico Basin of the United States). (2) Pending demonstration of the technology and economics for fracturing and producing energy from rock of low permeability, the reservoir porosity and permeability must be sufficiently high to allow production of large quantities of thermal water. (3) Since a major cost in geothermal development is drilling and since costs per meter increase with increasing depth, the shallower the concentration of geothermal energy the better. (4) Geothermal fluids can be transported economically by pipeline on the Earth's surface only a few tens of kilometers, and thus any generating or direct-use facility must be located at or near the geothermal anomaly.

Electric power generation. The most conspicuous use of geothermal energy is the generation of electricity. Hot water from a liquid-dominated reservoir is flashed partly to steam at the Earth's surface, and this steam is used to drive a conventional turbine-generator set. In the relatively rare vapor-dominated reservoirs, superheated steam produced by wells can be piped directly to the turbine without need for separation of water. Electricity is most readily produced from reservoirs of 180°C or greater, but reservoirs of 150°C or even lower show promise for electrical generation, either by using steam directly or by transferring its heat to a working fluid of low boiling point such as isobutane or Freon. The increase in worldwide geothermal electrical capacity with time is shown in Fig. 2. The importance of geothermal electricity to a small country is illustrated by El Salvador, where in 1977 the electricity generated from the Ahuachapán geothermal field represented 32% of the total electricity generated in that country. *See* ELECTRIC POWER GENERATION.

Direct use. Equally important worldwide is the direct use of geothermal energy, often at reservoir temperatures less than 100°C. Geothermal energy is used directly in a number of ways: to heat buildings (individual houses, apartment complexes, and even whole communities); to cool buildings (using lithium bromide absorption units); to heat greenhouses and soil; and to provide hot or warm water for domestic use, for product processing (for example, the production of paper), for the culture of shellfish and fish, for swimming pools, and for therapeutic (healing) purposes.

Major localities where geothermal energy is directly used include Iceland (30% of net energy consumption, primarily as domestic heating), the Paris Basin of France (where 60–70°C water is used in district heating systems for the communities of Melun, Creil, and Villeneuve la Garenne), and the Pannonian Basin of Hungary.

Prospects. In any analysis of the possible contribution of geothermal energy to human energy needs, one must keep in mind that the geothermal resource (that is, the potentially usable geothermal energy) is only a fraction of the thermal energy in a subsurface volume of rock and water. For favorable hydrothermal convection systems, this fraction can be 25% or greater, but for systems of restricted permeability the fraction is likely to be far smaller. Only this recoverable energy can be meaningfully compared with the thermal energy equivalent of barrels of recoverable oil, cubic meters of recoverable gas, tons of minable coal, or kilograms of minable uranium.

Use of geothermal energy is likely to increase greatly in many countries as other sources of energy become less abundant and more expensive and as geothermal reservoirs become better defined through systematic exploration and resource assessment. In the United States, the U.S. Geological Survey has estimated that the geothermal resources in identified and undiscovered hydrothermal convection systems are 2400×10^{18} joules, equivalent to the energy in 430×10^9 barrels (68×10^9 m^3) of oil. Geopressured-geothermal resources (both thermal energy and energy from dissolved methane) of the northern Gulf of Mexico Basin are estimated to be between 430 and 4400×10^{18} joules, equivalent to 74 to 780×10^9 barrels (11.8 to 124×10^9 m^3) of oil. Any energy that might be developed in the future from rock of low permeability or form magma would be in addition to this amount. Clearly, geothermal energy can play an important role in the energy economy of the United States as well as in the economies of many other countries throughout the world.

[L. J. PATRICK MUFFLER]

PRODUCTION AND POLLUTION PROBLEMS

The chief problems in producing geothermal power involve mineral deposition, changes in hydrological conditions, and corrosion of equipment. Pollution problems arise in handling geothermal effluents, both water and steam.

Mineral deposition. In some water-dominated fields there may be mineral deposition from boiling geothermal fluid. Silica deposition in wells caused problems in the Salton Sea, California, field; more commonly, calcium carbonate scale formation in wells or in the country rock may limit field developments, for example, in Turkey and the Philippines. Fields with hot waters high in total carbonate are now regarded with suspicion for simple development. In the disposal of hot wastewaters at the surface, silica deposition in flumes and waterways can be troublesome.

Hydrological changes. Extensive production from wells changes the local hydrological conditions. Decreasing aquifer pressures may cause boiling of water in the rocks (leading to changes in well fluid characteristics), encroachment of cool water from the outskirts of the field, or changes in water chemistry through lowered temperatures and gas concentrations. After an extensive withdrawal of hot water from rocks of low strength, localized ground subsidence may occur (up to several meters) and the original natural thermal activity may diminish in intensity. Some changes occur in all fields, and a good understanding of the geology and hydrology of a system is needed so that the well withdrawal rate can be matched to the well's long-term capacity to supply fluid.

Corrosion. Geothermal waters cause an accelerated corrosion of most metal alloys, but this is not a serious utilization problem except, very rarely, in areas where wells tap high-temperature acidic waters (for example, in active volcanic zones). The usual deep geothermal water is of near-neutral pH. The principal metal corrosion effects to be avoided are sulfide and chloride stress corrosion of certain stainless and high-strength steels and the rapid corrosion of copper-based alloys. Hydrogen sulfide, or its oxidation products, also causes a more rapid degradation than normal of building materials, such as concrete, plastics, and paints. *See* CORROSION.

Pollution. A high noise level can arise from unsilenced discharging wells (up to 120 decibels adjusted), and well discharges may spray saline and silica-containing fluids on vegetation and buildings. Good engineering practice can reduce these effects to acceptable levels.

Because of the lower efficiency of geothermal power stations, they emit more water vapor per unit capacity than fossil-fuel stations. Steam from wellhead silencers and power station cooling towers may cause an increasing tendency for local fog and winter ice formation. Geothermal effluent waters liberated into waterways may cause a thermal pollution problem unless diluted by at least 100:1.

Geothermal power stations may have four major effluent streams. Large volumes of hot saline effluent water are produced in liquid-dominated fields. Impure water vapor rises from the station cooling towers, which also produce a condensate stream containing varying concentrations of ammonia, sulfide, carbonate, and boron. Waste gases flow from the gas extraction pump vent.

Pollutants in geothermal steam. Geothermal steam supplies differ widely in gas content (often 0.1–5%). The gas is predominantly carbon dioxide, hydrogen sulfide, methane, and ammonia. Venting of hydrogen sulfide gas may cause local objections if it is not adequately dispersed, and a major geothermal station near communities with a low tolerance to odor may require a sulfur recovery unit (such as the Stretford process unit). Sulfide dispersal effects on trees and plants appear to be small. The low radon concentrations in steam (3–200 nanocuries/kg or 0.1–7.4 kilobecquerels/kg), when dispersed, are unlikely to be of health significance. The mercury in geothermal stream (often 1–10 μg/kg) is finally released into the atmosphere, but the concentrations created are unlikely to be hazardous.

Geothermal waters. The compositions of geothermal waters vary widely. Those in recent volcanic areas are commonly dilute (<0.5%) saline solutions, but waters in sedimentary basins or active volcanic areas range upward to concentrated brines. In comparison with surface waters, most geothermal waters contain exceptional concentrations of boron, fluoride, ammonia, silica, hydrogen sulfide, and arsenic. In the common dilute geothermal waters, the concentrations of heavy metals such as iron, manganese, lead, zinc, cadmium, and thallium seldom exceed the levels permissible in drinking waters. However, the concentrated brines may contain appreciable levels of heavy metals (parts per million or greater).

Because of their composition, effluent geothermal waters or condensates may adversely affect potable or irrigation water supplies and aquatic life. Ammonia can increase weed growth in waterways and promote eutrophication, while the entry of boron to irrigation waters may affect sensitive plants such as citrus. Small quantities of metal sulfide precipitates from waters, containing arsenic, antimony, and mercury, can accumulate in stream sediments and cause fish to derive undesirably high (over 0.5 ppm) mercury concentrations. *See* WATER POLLUTION.

Reinjection. The problem of surface disposal may be avoided by reinjection of wastewaters or condensates back into the countryside through disposal wells. Steam condensate reinjection has few problems and is practiced in Italy and the United States. The much larger volumes of separated waste hot water (about 50 metric tons per megawatt-electric) from water-dominated fields present a more difficult reinjection situation. Silica and carbonate deposition may cause blockages in rock fissures if appropriate temperature, chemical, and hydrological regimes are not met at the disposal depth. In some cases, chemical processing of brines may be necessary before reinjection. Selective reinjection of water into the thermal system may help to retain aquifer pressures and to extract further heat from the rock. A successful water reinjection system has operated for several years at Ahuachapán, El Salvador. [A. J. ELLIS]

Bibliography: H. C. H. Armstead, *Geothermal Energy*, 1978; H. C. H. Armstead (ed.), *Geothermal Energy: Review of Research and Development*, UNESCO, 1973; R. C. Axtmann, *Science*, 187: 795–803, 1975; A. J. Ellis and W. A. J. Mahon, *Chemistry and Geothermal Systems*, 1977; L. J. P. Muffler (ed.), *Assessment of Geothermal Resources of the United States—1978*, USGS Circ. 790, 1979; *Proceedings of the First United Nations Symposium on the Development and Utilization of Geothermal Resources, Pisa, Italy, Sept. 1970*, spec. issue no. 2 of *Geothermics*, 2 vols., 1973; *Proceedings of the Second United Nations Symposium on the Development and Use of Geothermal Resources, San Francisco, May 1975*, U.S. Government Printing Office, 3 vols., 1976; E. F. Wahl, *Geothermal Energy Utilization*, 1977; D. E. White and D. L. Williams (eds.), *Assessment of Geothermal Resources of the United States—1975*, USGS Circ. 726, 1975.

GERT

A procedure for the formulation and evaluation of systems using a network approach. Problem solving with the GERT (graphical evaluation and review technique) procedure utilizes the following steps:

1. Convert a qualitative description of a system or problem to a generalized network similar to the critical path method–PERT type of network.
2. Collect the data necessary to describe the functions ascribed to the branches of a network.
3. Combine the branch functions (the network components) into an equivalent function or functions which describe the network.
4. Convert the equivalent function or functions into performance measures for studying the system or solving the problem for which the network was created. These might include either the average or variance of the time or cost to complete the network.
5. Make inferences based on the performance measures developed in step 4.

Both analytic and simulation approaches have been used to perform step 4 of the procedure. GERTE was developed to analytically evaluate network models of linear systems through an adaptation of signal flow-graph theory. For nonlinear systems, involving complex logic and queuing situations, Q-GERT was developed. In Q-GERT, a simulation of the network is performed in order to obtain statistical estimates of the performance measures of interest.

GERTE. The components of GERTE networks are directed branches and exclusive-or nodes. Two parameters are generally associated with each branch. These are the probability p that the branch is taken, given that the node from which it emanated is realized; and the time t (or cost, profit, and so on) required, if the branch is taken, to accomplish the activity which the branch represents. The time parameter can be a random variable.

For this type of network, the probability and time parameters can be combined into a single parameter, as in the equation below, where $M_t(s)$ is

$$w(s) = pM_t(s)$$

the moment-generating function of the time parameter. The calculation of the equivalent w function is shown in Fig. 1 for three basic types of networks. For general GERTE networks, an equivalent w function can be calculated by using the topology equation or Mason's rule of signal flow–graph theory.

Q-GERT. For nonlinear systems, different node and branch types are required in order to obtain realistic network models. Q-GERT is used for such systems where branches represent activities and nodes are used to model milestones, decision points, and queues. Flowing through the Q-GERT network are items referred to as transactions. Transactions can represent physical objects, information, or a combination of the two. Different types of nodes are included in Q-GERT to allow for the modeling of complex queuing situations and project management systems. For example, activities can be used to represent servers of a queuing system, and Q-GERT networks can be developed to model sequential and parallel service systems.

The symbol to represent an activity is a branch with a syntax, as shown in Fig. 2. The distribution type and parameter set number characterize the time delay involved in the activity. The activity number is a label, and the number of parallel servers specifies a limit on the number of transactions that can proceed through the branch concurrently. The probability or condition specifies when the activity is to be taken and is used in conjunction with a routing type specified for a node. Routing characteristics are the means for directing transactions across activities and hence through the network. The routing types which may be specified for any node are shown in Fig. 3.

There are seven node types included in Q-GERT. These node types are shown in Fig. 4. The BASIC node type is used to create transactions, to terminate the existence of transactions, to accumulate transactions, and to collect statistics. The release requirement associated with the node specifies the number of incoming transactions required before an output is generated from the node. The choice criterion specifies the characteristics to be associated with any transaction leaving the node after it is released. Branching from the BASIC node is done in accordance with one of the routing types shown in Fig. 3. The Q-node provides a means for storing transactions waiting for service. The SELECT node is a means for selecting from among parallel queues and parallel servers. The MATCH node provides a mechanism for matching transactions prior to their continuation through the network.

The last three node types relate to the allocation of resources to transactions. Transactions requiring resources wait in Q-nodes which precede the ALLOCATE node. If resources are available or when they become available, they are allocated to transactions at the ALLOCATE nodes. The number of units and the type of resource required are characteristics of the ALLOCATE node. Since many Q-nodes may precede an ALLOCATE node, a queue selection rule is specified for the ALLOCATE node. After a transaction has completed the activities for which the resources are required, the resources can be made available by routing the transaction through a FREE node. At the FREE node, the number of units of the particular resource to be freed is specified. Also indicated is the order in which ALLOCATE nodes should be polled in the attempt to reassign the freed resources. The last node type is the ALTER node, which allows the capacity of resources to be changed. When the capacity of a resource type is increased, a list of ALLOCATE nodes is appended to the ALTER node to specify where the newly created units are to be considered for allocation.

Applications. Many applications of GERT have been made. GERT networks have been designed, developed, and used to analyze the following situations: claims processing in an insurance company,

network type	graphical representation	paths	loops	equivalent function
series		$w_a\ w_b$	—	$w_a\ w_b$
parallel		w_a, w_b	—	$w_a + w_b$
self-loop		w_a	w_b	$\frac{w_a}{1 - w_b}$

Fig. 1. Calculation of the equivalent w function.

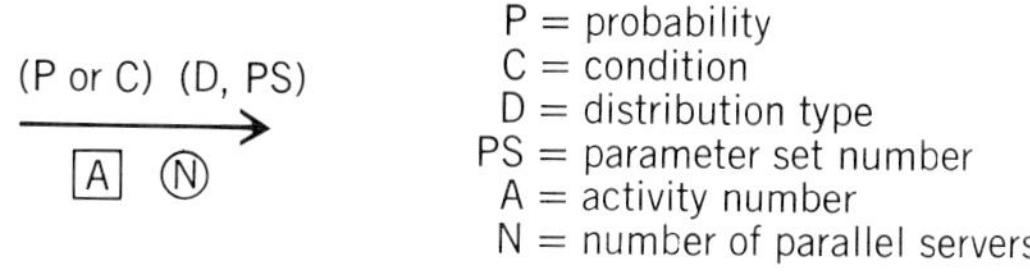

Fig. 2. Symbol and syntax for a Q-GERT activity.

GERT

deterministic

probabilistic

conditional, take-first

conditional, take-all

Fig. 3. Routing types and symbols.

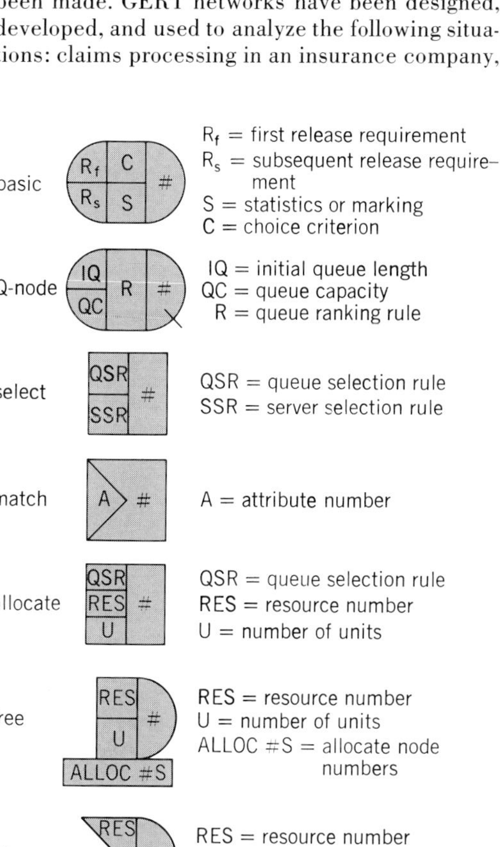

Fig. 4. Symbols and syntax of Q-GERT nodes.

production lines, quality control in manufacturing systems, assessment of job performance aids, burglary resistance of buildings, capacity of air terminal cargo facilities, judicial court system operation, equipment allocation in construction planning, refueling of military airlift forces, planning and control of marketing research, planning for contract negotiations, risk analysis in pipeline construction, effects of funding and administrative strategies on nuclear fusion power plant development, research and development planning, and system reliability. *See* CRITICAL PATH METHOD (CPM); PERT.

[A. ALAN B. PRITSKER]

Bibliography: S. E. Elmaghraby, *Activity Networks*, 1978; L. J. Moore and E. R. Clayton, *Introduction to Systems Analysis with GERT Modeling and Simulation*, 1976; A. A. B. Pritsker, *GERT: Graphical Evaluation and Review Technique*, RM-4973-NASA, 1966; A. A. B. Pritsker, *Modeling and Analysis Using Q-GERT Networks*, 2d ed.; G. E. Whitehouse, *Systems Analysis and Design Using Network Techniques*, 1973.

Governor

A device used to control the speed of a prime mover. A governor protects the prime mover from overspeed and keeps the prime mover speed at or near the desired revolutions per minute. When a prime mover drives an alternator supplying electrical power at a given frequency, such as 60 Hz, a governor must be used to hold the prime mover at a speed that will yield this frequency. An unloaded diesel engine will fly to pieces unless it is under governor control. *See* PRIME MOVER.

Speed control. A governor regulates the speed of a prime mover by properly varying the flow of energy to or from it. In the case of gas and steam turbines and internal combustion engines, the fuel furnishes the energy to the prime mover. For such applications, the governor usually controls the speed of the unit by regulating the rate at which fuel, and hence energy, is furnished to the prime mover. The governor controls the fuel flow so that the speed of the prime mover remains constant regardless of load and other disturbances, or changes in accordance with such operating conditions as changes in speed setting.

For a diesel engine the governor is connected to the rack which controls the amount of fuel injected. A governor on a gas or gasoline engine is attached to the engine throttle. A steam turbine governor strokes the steam valve or valves which regulate the steam flow to the turbine. *See* DIESEL ENGINE; INTERNAL COMBUSTION ENGINE; STEAM TURBINE.

The output mechanism of a gas turbine governor is connected to the fuel value with the stroke normally limited in each direction by the allowable combustion chamber temperature and other factors. If the fuel rate is too low, the fire will go out. Compressor stall must be avoided. To give rapid control, combustion chamber temperature is often computed in the governor by measuring other variables and applying the laws of thermodynamics. *See* GAS TURBINE.

A hydraulic turbine governor regulates the flow of water to the turbine by varying the openings of gates or other components. *See* HYDRAULIC TURBINE.

An aircraft propeller governor varies the pitch of the propeller to keep constant the speed of the engine attached to the propeller. This type of governor varies the load on the engine and thus controls the speed by regulating the energy flow from the prime mover.

Ballhead governor. The speed of a prime mover is usually measured by a ballhead that contains flyweights driven at a speed proportional to the speed of the prime mover. The force from the flyweights is balanced, at least in part, by the force of compression of a speeder spring (Fig. 1). The upper end of this spring is positioned according to the speed setting of the governor.

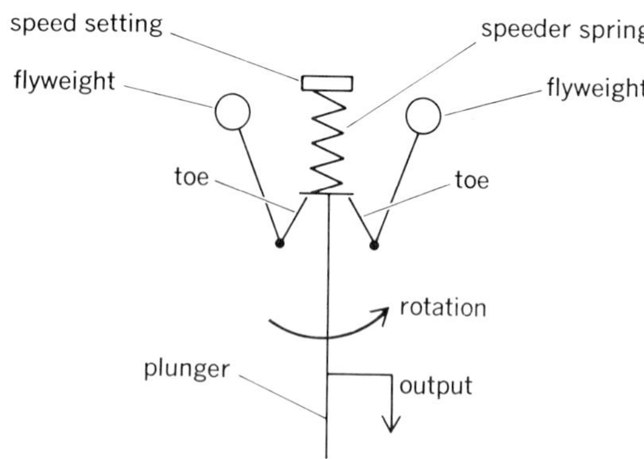

Fig. 1. Ballhead governor.

The ballhead toes press against one end of a plunger. In the simplest version of a governor, the plunger position is a function of the engine speed as a result of the balance between the centrifugal and spring forces. The plunger is directly connected to the throttle or other energy controlling means of the prime mover. Because the governor output power is drawn directly from the speed measuring means, the output power and the precision of such a governor are severely limited.

Governor response. In automatic control theory, the input to the governor is taken to be the difference between the speed setting and prime mover speed. This difference is the speed error. For the simplest governor, the position of the plunger is the output. This governor is proportional since the governor output is proportional to the governor input. In equilibrium, where the engine is running in steady state, the speed of the prime mover depends on both the load and the speed setting. With the mechanical governor and a fixed speed setting, the equilibrium speed decreases as the load increases. A governor with this property is a droop governor, or a governor operating on droop. The governor-prime mover unit is then said to be running on droop.

To increase the power output of a governor, a hydraulic amplifier is often employed. A governor that keeps the speed of a prime mover constant is said to be isochronous. In a simple isochronous governor, the ballhead senses the speed and strokes a pilot valve plunger that regulates the flow of fluid to a servomotor (Fig. 2). Normally, the fluid is oil. The servomotor is a piston in a cylinder; the piston is connected to the engine throttle or equivalent energy controlling mechanism. The position

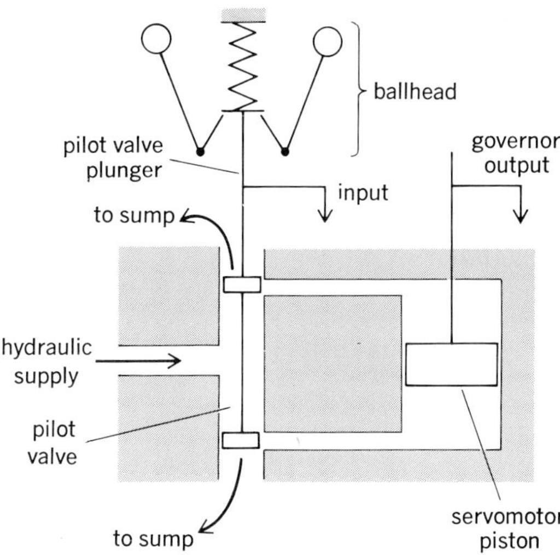

Fig. 2. Isochronous governor.

of the servomotor piston is the output of the governor. This output is made proportional to the integral of the governor input (speed error), and the governor is therefore an integral, or reset, controller. This governor is intended to bring the prime mover speed back to the speed setting after any change in load.

A hydraulic droop governor is obtained from the simple isochronous governor by the introduction of feedback from the governor output to the pilot valve (Fig. 3). This governor behaves like a simple mechanical governor except that a smaller ballhead is generally employed, greatly improving the precision of the governor by reducing hysteresis, dead band, and friction in the ballhead. The power output of the governor is much larger so that the effect of the load on governor performance incurred in moving the throttle or equivalent mechanism, is considerably diminished. A hydraulic governor may be sensitive to speed changes of as little as 1/1000 of 1%. For normal disturbances, prime mover speed error may be kept to 0.1% or better. The power output of such a governor may exceed 1 horsepower (hp). The mechanical version may be insensitive to speed changes of 1%; the output is generally limited to 1/50 hp. Lag in a hydraulic governor is usually below 1/10 sec, but in a mechanical governor 1/2 sec or more is common.

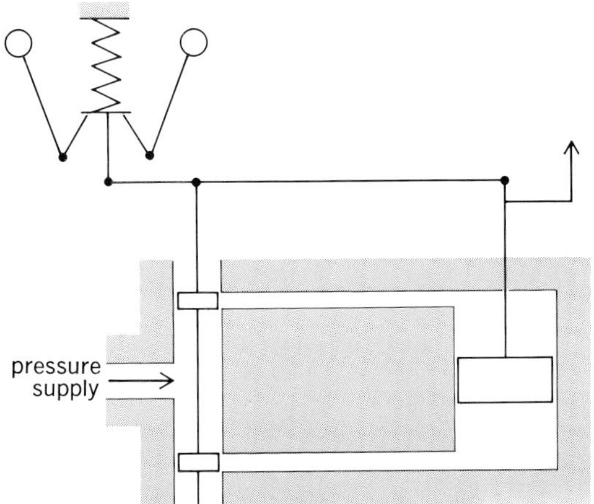

Fig. 3. Hydraulic droop governor.

Use of dashpot. The performance of the simple isochronous governor is often greatly improved by the introduction of a dashpot in the feedback path from the output to the ballhead. If there is little damping in the prime mover, instability often occurs when the simple isochronous governor is used, whereas this instability is removed when the dashpot is incorporated. A system is stable when for each disturbance that dies out, response of the system settles to an equilibrium condition. When instability occurs, the prime mover speed oscillates continuously or increases indefinitely until the unit flies to pieces.

In a dashpot (Fig. 4), deflection of a spring-loaded piston from equilibrium depends on the velocity of the other, or transmitting, piston. In its simplest version this dependence is a proportionality. The spring-loaded piston is termed the receiving piston. The position of the transmitting piston is the dashpot input, whereas the position of the receiving piston is the output. Between input and output there is a time lag. The receiving piston attains its equilibrium when the pressure drop across the dashpot orifice is zero. Oil is used as the dashpot fluid.

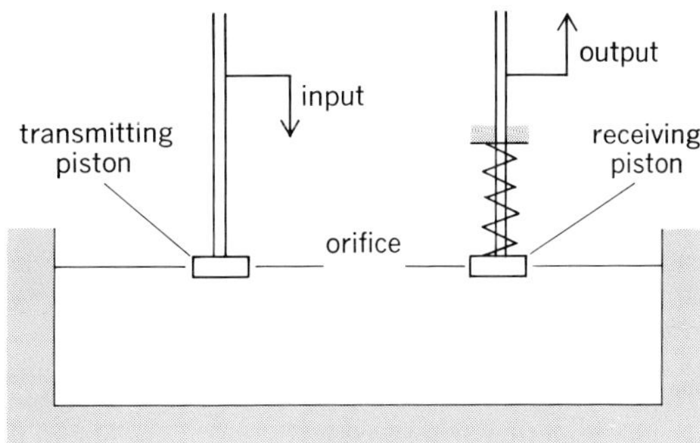

Fig. 4. Dashpot, to improve governor performance.

When a dashpot is incorporated into a governor (Fig. 5), the governor output becomes a function of the speed error and the integral of this error. Such a governor is a proportional plus integral controller. The velocity of the servomotor then depends on prime mover speed and prime mover acceleration. The time lag in the dashpot makes the governor sensitive to prime mover acceleration. As this lag is increased, the response of the governor to prime mover acceleration tends to increase. This increase in lag is accomplished by moving the dashpot needle valve, which controls the orifice area, toward the closed position.

Instead of mechanical feedback from the governor, force feedback is generally preferred (Fig. 6). An isochronous dashpot governor is turned into a droop governor by adding direct mechanical feedback from the servomotor to the ballhead.

Use of flywheel. Acceleration governors are sometimes used in place of governors with dashpots. In such governors a flywheel is employed instead of a dashpot. The prime mover drives the flywheel through a spring. This combination yields a motion proportional, except for a time lag, to the

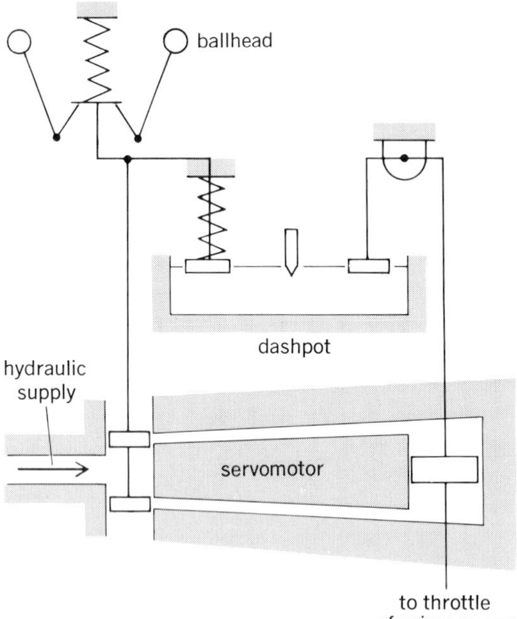

Fig. 5. Isochronous dashpot governor.

acceleration of the prime mover. This motion is added mechanically or otherwise to the output of the governor ballhead, and the result is used to stroke the pilot valve plunger. The pilot valve meters the flow of fluid to the servomotor cylinder. The force from the flywheel is usually amplified hydraulically before the motion is added to the ballhead output. The outputs of acceleration governors tend to jiggle because they are more sensitive to high-frequency variations in the governor ballhead drive.

Governor applications. Governors for large hydraulic turbines require a second stage of hydraulic amplification. The governor servomotor piston is

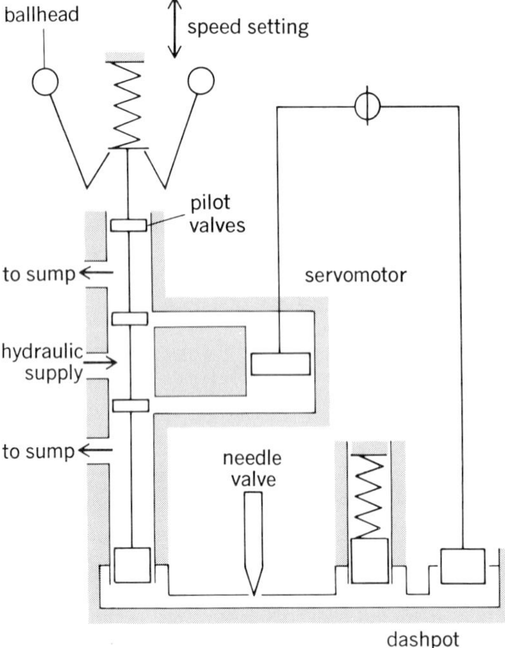

Fig. 6. Governor with force feedback.

connected to the plunger of a relay valve that regulates the flow of oil to the turbine servomotors. The governor servomotor is then termed the controller. Turbine servomotors often require 10 hp or more.

If at all times the driving torque is equal to the load torque, the prime mover will neither accelerate nor decelerate. This principle is employed in load control governors where a measurement of load is employed to rapidly position a load servomotor piston so that the driving and load torques are approximately equal, and speed errors kept small. To the motion of this piston is added the motion of the regular governor piston, which now functions as a vernier control.

The speed setting of a governor is often adjusted by an electric, hydraulic, or pneumatic motor as a function of auxiliary variables. Thus in an electrical power system the deviation of the frequency of the system from the desired value is used to position the governor speed setting so as to bring the frequency to the right value. Adjustment of governor speed settings is required to keep electric clocks on time. Oil and gas pipeline governors control pressure rather than speed. The pressure in the line is measured and the speed setting of the governor is adjusted accordingly. This affects the speed of the prime mover, which drives a compressor, raising or lowering the pressure as required.

Paralleled prime movers. Prime movers may be paralleled to supply power to the same load. In an electrical power system, prime movers drive alternators electrically connected to the system. At most, one of the governors of paralleled prime movers can be isochronous; the rest must be on droop. The prime movers may all be on droop. Rated load on a prime mover is 100% load; rated speed is 100% speed. If, as the load is increased 100%, the prime mover speed falls μ%, the governor is on μ% droop. The amount of droop is determined by the amount of servomotor motion fed back to the governor ballhead. In operation the percentage of droop is usually fixed, while the speed setting is adjusted to make the prime mover take its assigned load. A droop of 5% is common. The droop generally falls in the range of 1–10%.

When two or more identical prime mover-generator units are paralleled, and one is controlled by an isochronous governor while the others are on droop, electrical coupling will force all units to run at the same steady-state speed. The alternators then supply electrical power at the same frequency. The load taken on by a unit with a droop governor is now a function of the speed setting of the governor (Fig. 7). At 100% speed, the speed of the prime movers produces the required system frequency. The load taken on by a unit is determined by the intersection of its load line with the 100% speed line. As the speed setting N_s of the unit is increased, the load line moves up, retaining the same slope but increasing the load taken on by the unit. By adjusting the speed settings of the droop governors, load may be arbitrarily divided among the generating units.

Aircraft engines are synchronized and synchrophased by using one engine as the master, and adjusting the speed settings of the governors on the other engines to make them follow the master.

Governor design. Proper design of a governor involves making the governor characteristics fit those of the prime mover and load so as to give

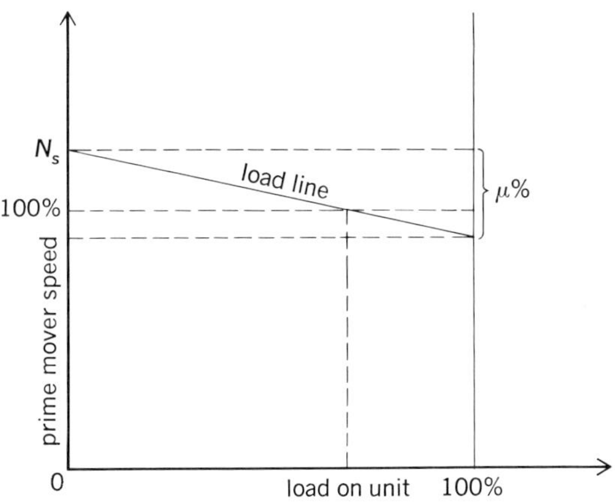

Fig. 7. Droop characteristic.

acceptable overall performance. Desirable and undesirable nonlinearities occur in governors to complicate design calculations. Thus centrifugal force varies as the square of speed. To compensate for this, nonlinear speeder springs are often employed. Bypass is often used in dashpots to limit the range of the dashpot output. When the receiving piston for a dashpot (Fig. 4) is sufficiently far removed from the equilibrium position, passages are uncovered, which permit the fluid to flow from one side of the piston to the other, thus limiting its travel.

The output of a generator driven by the prime mover is sometimes used to obtain a voltage proportional to prime mover speed. This voltage, properly filtered, is fed to an electronic, magnetic, or other circuit, and eventually to a transducer to move the pilot valve. A gear with permanent magnet teeth is often employed with a pickup coil, where the gear is driven at a speed proportional to that of the prime mover. The electrical output of the coil is a high-frequency alternating current which is filtered to obtain a voltage proportional to prime mover speed. This voltage is integrated and sometimes differentiated by classical electrical means with the aid of capacitors and operational amplifiers to obtain electrical equivalents of the computing components of hydraulic governors. Solid-state devices are frequently used to manipulate electrical signals. The output pressure of a hydraulic pump is sometimes employed as a measure of prime mover speed. Other speed-measuring means are utilized.

Limit or topping governors are often employed to meet fail-safe specifications. The limit governor is usually a mechanical governor that takes over control from the main governor to shut the prime mover down if its speed reaches a fixed overspeed above its rated speed. The flyweight force increases with an increase in the distance of the flyweights from the axis of rotation of the ballhead; this flyweight force opposes the speeder spring force. The flyweight force increases as the square of prime mover speed. When it is sufficiently large, the governor gain becomes infinite, so that a small input to the governor causes an unlimited output. This causes a snap action of the governor at the overspeed for which the governor is set.

The natural frequency of a ballhead is generally so high that the ballhead lag can be neglected in first-approximation studies of governor–prime mover system performance. When the ballhead drive involves dominant frequencies near the ballhead resonant frequency, damping is sometimes provided in the ballhead to prevent excessive governor output jiggle.

In the case of a hydraulic governor, the speeder spring force is aided by the hydraulic reaction force tending to close the pilot valve. This reaction force occurs because of a change in momentum of the fluid passing through the valve and is proportional to the opening of the valve for small valve openings. This force behaves like the force of a linear spring. The effect of hydraulic reaction is to diminish the gain of the ballhead, so that a given input causes a smaller output.

In the interests of economy the four-way valve servomotor (Fig. 2) is often replaced by a two-way valve-differential servomotor (Fig. 8). The area on one side of the servomotor is half the area on the other side. The supply pressure on the small area is sometimes replaced by a spring.

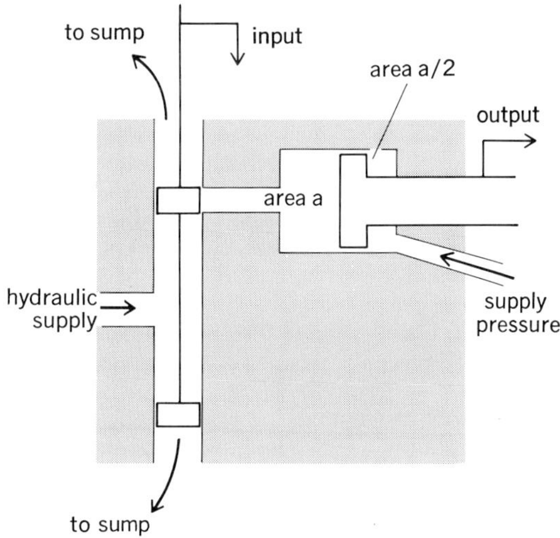

Fig. 8. Two-way valve-differential servomotor.

Governors are normally designed so that the response to a sudden disturbance dies out in 1–20 sec.

A load is termed isolated if the power supplying this load comes from one prime mover only. In the design of a governor for a prime mover supplying power to an isolated load, a prime mover differential equation is obtained relating the input to the prime mover, such as throttle position, to the speed of the prime mover, which is taken to be its output. This equation is obtained by mathematically balancing the torques on the prime mover shaft. The equation also involves a delay between a change in the input to the prime mover and the resulting change in the driving output torque. Part of the delay is a dead time. This dead time is normally 1/100–1/2 sec. The rest of the delay is usually in the order of 1/10 sec.

The nature of a prime mover may introduce other terms in the equation. For example, with hydraulic turbines the equation is complicated because of water hammer; a sudden closing of the gates to decrease the speed of the turbine causes the turbine to speed up initially, resulting in a correction opposite to that desired. As the gates are moved, pressure waves travel up and down the hydraulic conduit from the source of the water to the discharge from the turbine. These waves tend to unstabilize the unit.

The equation of the prime mover is combined with the equation of the governor to yield an equation for overall system response from which the governor response is determined. When full load on a diesel engine is rejected, an overspeed of 3% is often considered acceptable with the response dying out in 1 sec or less. For a hydraulic turbine this overspeed may be 30–40% and the response may endure as much as 10–20 sec. Performance for other prime movers under governor control falls between these extremes.

[RUFUS OLDENBURGER]

Bibliography: T. Baumeister (ed.), *Standard Handbook for Mechanical Engineers*, 8th ed., 1978; H. A. Rothbart, *Mechanical Design and Systems Handbook*, 1964.

Grab bucket

A digging and materials-handling tool with hinged sides or leaves. It is particularly suited for digging deep holes or for raising material to high places. A grab bucket may have two, three, or four movable sides. A two-sided bucket is somewhat rectangular in shape and is called a clamshell. A three- or four-leafed bucket is ball-shaped and is called an orange peel. For hard digging, clamshells are equipped with hardened prongs or teeth; for handling loose material such as sand, cinders, or fine coal, no teeth are used. Standard clamshell capacities range from $\frac{3}{8}$ yd^3 to 6 yd^3 but larger sizes have been made. Orange peel buckets are manufactured with capacities up to 3 yd^3, but this size is unusual. Most orange peels are small; some are only slightly bigger than a large grapefruit. They are used mainly for cleaning sewer catch basins, vertical pipes, caissons, and other small vertical openings. Most grab buckets are still suspended from and activated by cables, although cable-suspended buckets with hydraulic activators (openers and closers) have increased in popularity. On a few specialized machines relatively small buckets (up to $1\frac{1}{2}$ yd^3) are connected to the ends of rigid but movable masts that provide precise bucket placement and a powered downthrust for increased efficiency.

[EDWARD M. YOUNG]

Grinding

The process of removing metal by the cutting action of a solid, rotating, grinding wheel. The abrasive grains of the wheel perform a multitude of minute machining cuts on the workpiece. Although grinding is sometimes used as the complete machining operation on surfaces, it is generally considered a finishing process used to obtain a fine surface finish and extremely accurate dimensional tolerances. From 0.0005 to 0.020 in. of stock is commonly removed, depending on the operation.

Materials. The grinding process is used in machining a wide range of metals in addition to such materials as cemented carbide, marble, stone, and certain ceramics. Because commercial grinding abrasives are many times harder than the metals to be machined, the grinding process may be used on metals too hard to machine otherwise.

Grinding wheels are composed of abrasive grains or grit plus a bonding material. The abrasives most commonly used are silicon carbide, SiC, and aluminum oxide, Al_2O_3. The silicon carbide crystals are very hard and tend to fracture easily. They are used to grind low-tensile-strength materials such as aluminum, brass, and copper. They are also used to cut hard, brittle materials such as hard alloys, cemented carbide, gray iron, marble, and stone. Aluminum oxide crystals are tough and tend to resist fracture. In general they are used on high-tensile-strength materials such as alloyed steels or wrought iron.

The common bonding processes are the vitrified or ceramic, silicate, rubber, resinoid, and shellac.

Wheel types. Wheels are classified according to type of abrasive (silicon or aluminum), grain size, grade or hardness, structure (referring to density or porosity), and type of bonding material.

Coarse-grained wheels are used for rapid removal of stock; wheels with fine grains cut more slowly but give smoother finishes. Usually the harder the material to be ground, the softer the bond in the wheel should be, while the softer the material to be ground, the harder the bond should be. On soft materials the grit cuts in deeply and therefore requires a strong bonding material to hold it. On hard substances the grit dulls rapidly and a soft bond allows it to tear out and expose new cutting edges. The correct wheel for the job should hold the grain in place until it becomes dull.

Operation of a wheel above recommended speeds can be dangerous and may actually hinder cutting.

Wheel care. Grinding wheels are kept in proper condition by truing and dressing. Truing a wheel means cutting it, usually by means of a diamond cutting tool, for the purpose of making it concentric about its axis. Dressing, which opens or reconditions the grinding surface of a wheel to afford maximum cutting qualities, may not necessarily give concentricity. Dressing may be done with steel cutters or abrasive dressing wheels, or by crush dressing. The last is used to shape the face of the wheel by forcing a rotating hard iron wheel of the desired shape into the grinding wheel, and crushing away the grain of the grinding wheel until the two wheels mate.

Grinding fluids or coolants are applied to the grinding point to dissipate the heat generated and also to trap the abrasive dust. Although grinding oils are used, water emulsions containing soda or alkali are more common. Plain water could be used except for the problem of rust. Abrasive dust and chips can be removed by filtering the coolant.

Grinding machines. Grinding machines are designed to hold a workpiece rigidly and to feed it smoothly through the cutting path of a rotating grinding wheel. These machines must be constructed with sufficient accuracy to produce the finely finished surfaces and close dimensional tolerances expected. Where tolerances are less critical, as in removing burrs or in sharpening a cutting tool, an operator may handhold the workpiece against the side or face of a wheel (Fig. 1).

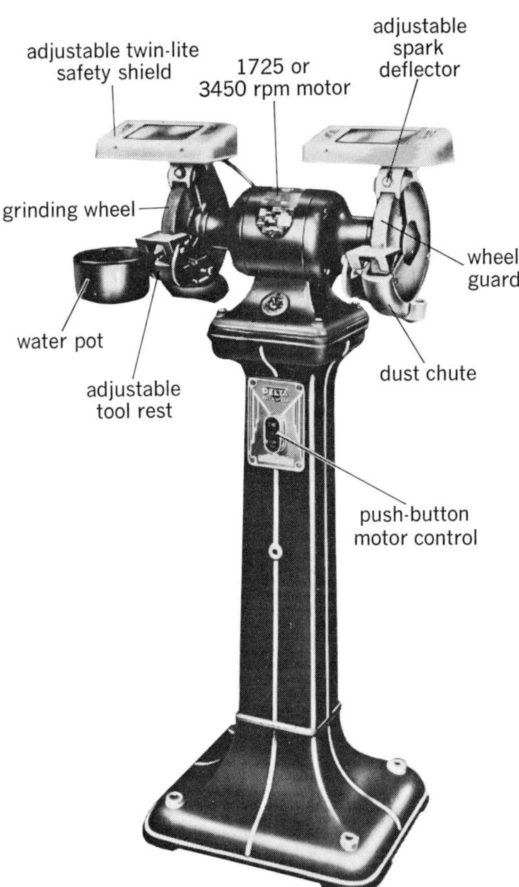

Fig. 1. Power-driven grinding wheels used on handheld workpieces. (*Rockwell Manufacturing Co.*)

Grinding machines may be listed under four main classifications: cylindrical, surface, internal, and special. Grinding machine operations are classified as cylindrical, surface, or internal; most special operations are variations of these.

Cylindrical grinders. Cylindrical grinding is performed on the peripheries or shoulders of workpieces composed of concentric cylindrical or conical shapes. Examples of such workpieces are shafts, cylinders, rolls, and axles.

Cylindrical grinders rotate the workpiece from a power headstock while it is held between centers, gripped in a jawed chuck, or fastened to a faceplate. Usually, the power headstock and tailstock are held on a worktable which in turn is mounted on the main table. The latter may move longitudinally on ways and thus carry the workpiece past the face of the grinding wheel.

When the work and wheel are moved longitudinally past each other to grind a length greater than the width of the wheel, it is called traverse grinding. If the wheel is advanced directly into the work to form shoulders and contours on the piece, or if the length to be ground is less than the width of the wheel, it is referred to as infeed or plunge grinding.

Centerless cylindrical grinders carry the work on a support or blade between two abrasive wheels. One is a regular grinding wheel rotating at normal grinding speed, while the second is usually a rubber-bonded abrasive regulating wheel turning in the same direction as the first wheel but at a much slower speed. The regulating wheel does not grind the work but causes it to rotate against the grinding wheel at a uniform speed. Its distance from the grinding wheel determines the finished size of the workpiece.

Because the workpiece is neither rigidly held nor gripped, centerless grinding machines are ideally suited to mass production (Fig. 2). Centerless grinders are made in sizes to handle both large and small workpieces.

With straight, cylindrical workpieces having no interfering shoulders, the regulating wheel may be set to run at a slight angle to the grinding wheel. This causes the rotating workpiece to move longitudinally past the face of the grinding wheel and is termed through feeding.

Workpieces having shoulders and more than one diameter may be ground by infeeding. An end stop keeps the rotating workpiece from moving longitudinally and the regulating wheel is advanced, forcing the work directly against the grinding wheel until the desired diameter is obtained.

End feeding is used when a taper is desired. The grinding wheel, regulating wheel, and sup-

Fig. 2. Centerless thread grinding operation. (*Landis Machine Co.*)

port are set at the desired angle and the workpiece is fed in from the side until it reaches the point at which the desired taper and size have been ground on it.

Centerless grinders are designed for both automatic and manual feeds.

Surface grinders. Surface grinding is accomplished by holding one or several workpieces on either a rotary or a reciprocating horizontal table and passing them through the cutting path of a rotating grinding wheel. The wheel may be mounted in either a vertical or a horizontal plane.

Surface grinders are classed according to the axis of the grinding wheel as horizontal or vertical. Horizontally mounted wheels grind on their peripheries; vertically mounted wheels grind on their circular faces.

On machines with reciprocating tables, the workpiece is moved back and forth underneath the grinding wheel until the desired dimension and finish are obtained. Where a rotary-table surface grinder is used, enough parts may be mounted on the table to cover its entire working area. As the table rotates, each successive piece passes through the path of the grinding wheel.

Internal grinders: Internal grinding is the process of grinding the surfaces of holes to obtain a close dimensional tolerance or a desired surface finish. Cylindrically shaped holes as well as those with tapered sides may be ground.

Internal grinders vary in the manner in which they hold the workpiece and also in the method of grinding. Some machines may hold the part in a chuck or on a faceplate and rotate it in a fixed position. The rotating grinding wheel reciprocates through the length of the hole and is fed sideways to grind the piece to size. Other types may rotate and also reciprocate the workpiece while the wheel revolves in a fixed position. Large parts, ground on a planetary type grinder, remain stationary with the grinding wheel revolving on its own axis and also rotating around the axis of the hole. Where high production is desired, a centerless-type internal grinder may be employed. *See* HONING; LAPPING; MACHINING OPERATIONS; POLISHING.

[ALAN H. TUTTLE]

Bibliography: F. T. Farago, *Abrasive Methods Engineering*, 1976; R. L. Little, *Metalworking Technology*, 1976.

Grinding mill

A machine that reduces the size of particles of raw material fed into it. The size reduction may be to facilitate removal of valuable constituents from an ore or to prepare the material for industrial use, as in preparing clay for pottery making or coal for furnace firing. Coarse material is first crushed. The moderate-sized crushings may be reduced further by grinding or pulverizing.

Grinding mills are of three principal types, as shown in the illustration. In ring-roller pulverizers, the material is fed past spring-loaded rollers. The rolling surfaces apply a slow large force to the material as the bowl or other container revolves. The fine particles may be swept by an air stream up out of the mill. In tumbling mills the material is fed into a shell or drum that rotates about its horizontal axis. The attrition or abrasion between particles grinds the material. The grinding bodies may be flint pebbles, steel balls, metal rods lying parallel to the axis of the drum, or simply larger pieces of the material itself. In hammer mills, driven swinging hammers reduce the material by sudden impacts. *See* BALL-AND-RACE-TYPE PULVERIZER; PEBBLE MILL; TUMBLING MILL.

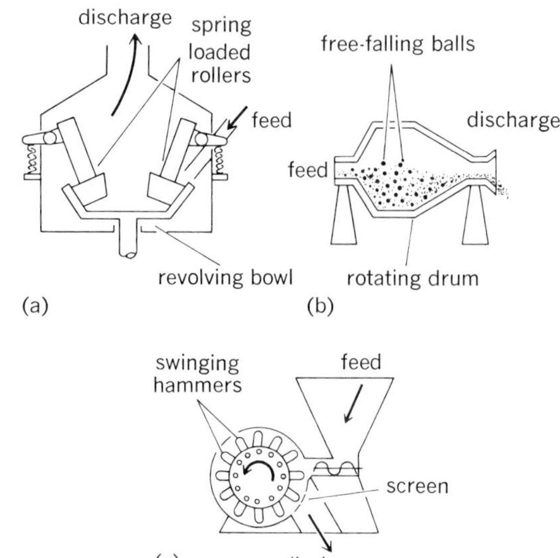

Basic grinding mills. (*a*) Ring-roller mill. (*b*) Tumbling mill. (*c*) Hammer mill.

Depending on the required fineness and uniformity of the finished particles, the discharge may or may not be classified by size. Oversized particles may be returned in a closed circuit to the grinding mill for further reduction. Material may be ground dry or wet, in batches or continuously. *See* CRUSHING AND PULVERIZING.

[RALPH M. HARDGROVE]

Bibliography: F. T. Farago, *Abrasive Methods Engineering*, 1976; J. H. Perry and C. H. Chilton (eds.), *Chemical Engineers' Handbook*, 5th ed., 1973.

Groin

A structure oriented perpendicular to a shoreline and extending into the sea for the purpose of retarding erosion or trapping littoral drift, thus establishing, stabilizing, or widening a beach.

A single structure may suffice, but a series of structures uniformly spaced are usually more efficient. Groins may be permeable or impermeable, high or low, fixed or adjustable, and may be constructed of timber, steel, stone, concrete, and other materials. *See* COASTAL ENGINEERING.

[EDWARD J. QUIRIN]

Grounding

Intentional electrical connections to a reference conducting plane, which may be earth (hence the term ground), but which more generally consists of a specific array of interconnected electrical conductors, referred to as the grounding conductor. The symbol which denotes a connection to the grounding conductor is three parallel horizontal lines, each of the lower two being shorter than the one above it (Fig. 1). The electric system of an airplane or ship observes specific grounding practices with prescribed points of grounding, but no connection to earth is involved. A connection to such a reference grounding conductor which is

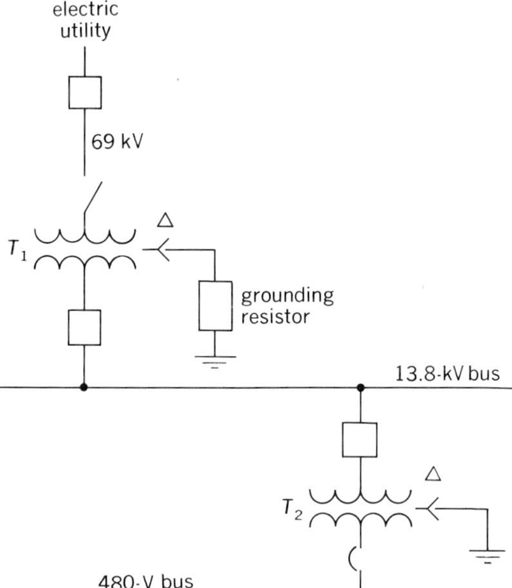

Fig. 1. Each conductively isolated portion of a distribution system requires its ground.

independent of earth is denoted by use of the symbol shown in Fig. 2.

The subject of grounding may be conveniently divided into two categories: system grounding and equipment grounding. System grounding relates to a grounding connection from the electric power system conductors for the purpose of securing superior performance qualities in the electrical system. Equipment grounding relates to a grounding connection from the various electric-machine frames, equipment housings, metal raceways containing energized electrical conductors, and closely adjacent conducting structures judged to be vulnerable to contact by an energized conductor. The purpose of such equipment grounding is to avoid environmental hazards such as electric shock to area occupants, fire ignition hazard to the building or contents, and sparking or arcing between building interior metallic members which may be in loose contact with one another. The design of outdoor open-type installations presents special problems.

System grounding. Appropriately applied, a grounding system can (1) avoid excess voltage stress on electrical insulation within the system, leading to longer apparatus life and less frequent breakdown; (2) improve substantially the operating quality of the overcurrent protection system; and (3) greatly diminish the magnitude of arc fault heat-energy release at an insulation breakdown point, lessening arc burning damage and fire ignition possibility.

Sectionalization. Each voltage transformation point employing an insulating transformer interrupts the continuity of the system grounding circuit. Figure 1 illustrates the point. System grounding connections made on the 69-kV electric service companies' lines extend their influence only to the 69-kV winding of transformer T_1. The grounding connections established at the 13.8-kV winding of transformer T_1 extends its influence only to the 13.8-kV winding of transformer T_2. The grounding connections established at the 480-V secondary winding of transformer T_2 apply to the 480-V conductor system only. Two distinct advantages result. First, the system grounding arrangement of each voltage-level electric system is independent of all others. Second, the type and pattern of system grounding to be used with any individual voltage-level electric system can be tailored to optimize the performance of that particular electric-system section. *See* ELECTRIC POWER SYSTEMS.

It is preferable to locate the grounding connection at the source-point electrical neutral of the particular voltage-level system, and mandatory to do so at the service entrance point if the point of origin is outside the local building.

Common patterns. The great majority of system grounding patterns fall into one of the varieties that are shown in Fig. 3. The most used varieties of system grounding impedance are illustrated in Fig. 4.

The use of solid grounding exclusively for grounding patterns of Fig. 3*a*, *b*, and *d* are influenced by two considerations: (1) Overcurrent

Fig. 2. Symbol to denote connection to a reference ground that is independent of earth.

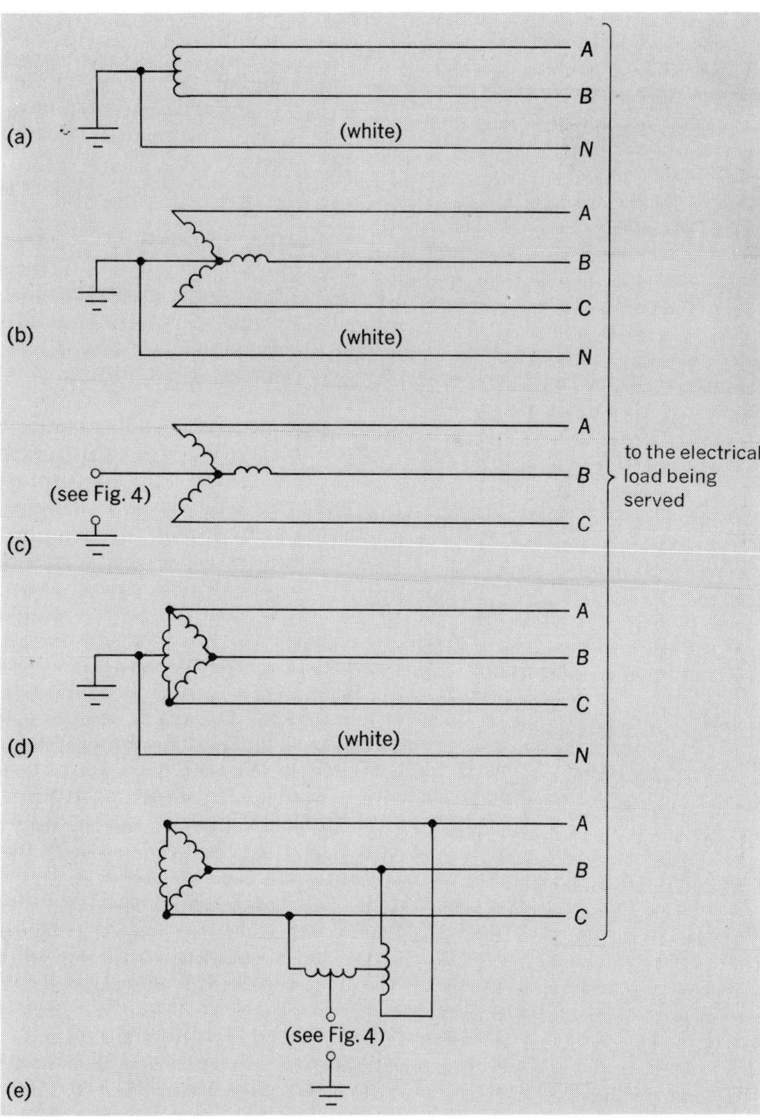

Fig. 3. Commonly used patterns of grounding. (*a*) Single-phase three-wire 240/120-V service. (*b*) Three-phase four-wire 208/120- or 480/277-V service. (*c*) Three-phase three-wire pattern. (*d*) Three-phase four-wire Δ 240-V line-to-line pattern. (*e*) Three-phase Δ with derived neutral.

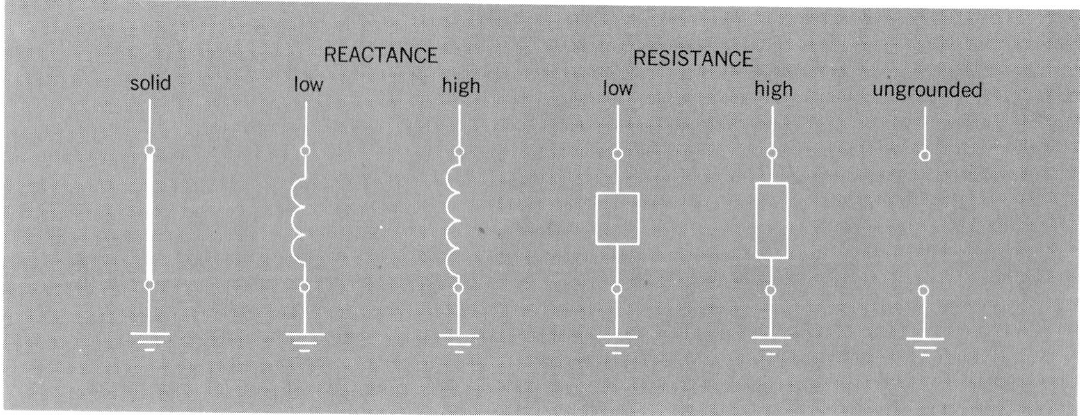

Fig. 4. Varieties of system grounding impedances in common use.

protection is present in only the phase conductor of single-phase, one-side grounded circuits. (2) The National Electrical Code (NEC) requires any electric system that can be solidly grounded in a manner which will avoid a phase-to-ground potential in excess of 150 volts to be so grounded.

Solid grounding is also used almost exclusively in the case of operating voltages of 69 kV and higher, as in Fig. 3c, to achieve the most rigid control of overvoltage stress. Such control allows reduced-rating lightning arresters, which in turn permits the successful use of reduced insulation level on station apparatus.

The high, unrestricted magnitude of short-circuit current created by a line-to-ground (L-G) fault on a solidly grounded system can pose severe design problems with costly solutions. The desire to artificially reduce the magnitude of L-G fault current is the chief reason for the use of other grounding impedances.

A low-reactance (inductive) grounding connection impedance can be used to effect a moderate reduction in the L-G short-circuit current, particularly for the purpose of avoiding excess short-circuit current flow in a phase winding of a rotating machine or of accomplishing a desired distribution of neutral unbalanced load current among source machines. Use of reactance to achieve the reduction in L-G fault current to below about 40% of the three-phase value enters the high-reactance region, which is subject to the generation of damaging transient overvoltages as a result of a ground fault condition, unless appropriate resistive damping circuits are added. Only one special case of high reactance grounding is free of overvoltage trouble: the ground-fault neutralizer case, in which the reactance magnitude is carefully matched with the electric-system L-G capacitance to create a natural frequency of oscillation almost exactly equal to power system operating frequency. It is critical, however, because a small deviation from the resonant value will destroy the overvoltage immunity.

Resistance grounding. By substituting a resistive grounding impedance (a totally dissipative impedance), much greater reductions in the L-G fault current can be intentionally created without danger of transitory overvoltages.

The low-resistance region is characterized by an established level of available ground-fault current well below the three-phase fault value yet ample to properly operate protective devices responsive to ground-fault current flow. Typical current values in use range from a few thousand amperes downward. Present-day protective practices allow the current value to be set at 400 A for general purpose medium-voltage electric systems widely used in industry. The far more critical electric-shock hazards incident to electric power supply to portable excavating machinery have led to the selection of a much lower level of available ground-fault current, typically in the 25–50-A region.

Most electrical breakdowns occur line-to-ground and many remain so throughout the interval of detection and isolation. A summary of the operating advantages achieved by intentional reduction of available ground-fault current is given below and illustrated in Fig. 5. (1) Low heat-energy release at the fault location because of the low current magnitude. (2) No noticeable dip in the system line-to-line voltages, which means no disturbance to the operation of all healthy load circuits. (3) Diminished interrupting duty on the circuit interrupter (low current and high power factor), which contributes to infrequent maintenance requirements. (4) Diminished duty imposed on the equipment grounding conductor network, which allows superior performance achievement at less cost.

High-resistance grounding relates to a mode of operation in which the fault location and subsequent corrective action are undertaken manually by skilled maintenance personnel. It is used prin-

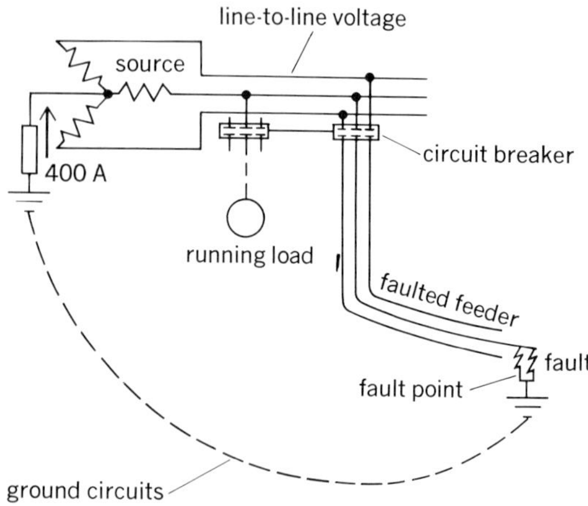

Fig. 5. Resistance grounding. (*General Electric Co.*)

cipally in electric systems that serve critical continuous-process machines.

The available ground-fault current is reduced to the same order as the electric system charging current to ground (generally less than 5 A). This resistive component of fault current is sufficient to arrest the generation of transient overvoltages by L-G fault disturbances and provides a positive signal for identifying the presence of an L-G fault somewhere on the system. Successful results depend upon the presence of a skilled maintenance crew who can respond quickly to the ground-fault alarm, promptly locate the faulty circuit element, and take effective action to remove this faulty circuit from the system before a second insulation failure is induced.

Ungrounded system. Although a system may have no intentional grounding connection, and hence is named an ungrounded system, it is in fact unavoidably capacitively grounded. The layer of insulation surrounding every energized conductor metal constitutes the dielectric film of a minute distributed capacitor between power conductors and ground. In the aggregate this capacitance can amount to a substantial fraction of a microfarad for the complete metallically connected system. Surge voltage suppression filters typically include line-to-ground-connected capacitors which add to the distributed capacitors inherent in the system proper. Unless modified by the stabilizing qualities of grounding connections previously described, L-G fault disturbances can create dangerous overvoltage transients (impressed on system insulation) in about the same ways possible had the grounding connection impedance of Fig. 4 been a capacitor. The equivalent circuit of one phase conductor of an ungrounded three-phase power supply (relative to ground) takes the form illustrated in Fig. 6.

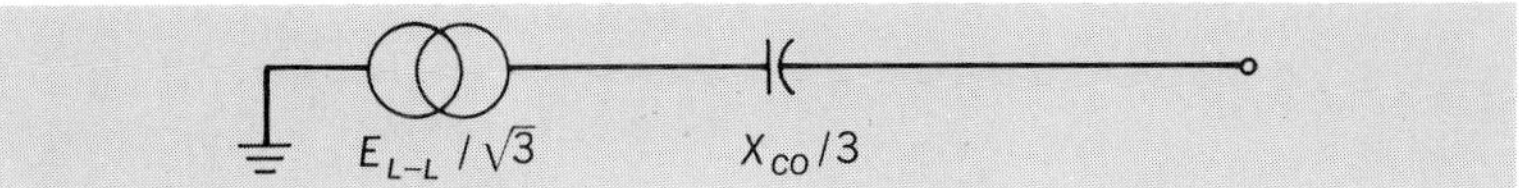

Fig. 6. Equivalent supply circuit to one phase conductor (relative to ground) of an "ungrounded" electric power system. E_{L-L} = line-to-line voltage; $X_{CO}/3$ = capacitive reactance coupling to ground.

Equipment grounding. At each electrical equipment, grounding serves to establish a near-zero potential reference plane (even during L-G fault conditions). This reference extends to the outer reaches of the particular voltage-level electric system to which a solid grounding connection can be made from the metal frames of served electric machines, the metal housings that contain switching equipment or other electric-system apparatus, and the metal enclosures containing energized power conductors; these enclosures may be metal cable sheaths or metal raceways.

The purposes of this interconnected mesh conducting network, drawn as the heavy lines in Fig. 7, are listed below.

1. To avoid electric shock hazard to any occupant of the area who may be making bodily contact with a metallic structure containing energized conductors, one of which has made an electric fault connection to the mentioned structure.

2. To provide an adequately low impedance to the return path of L-G fault current so as not to interfere with the operation of system overcurrent protectors.

3. To provide ample conductivity (cross-sectional area) to carry the possible magnitude of ground-fault current for the duration controlled by the overcurrent protectors in the electric system.

4. To avoid, by installation, with appropriate geometric spacing (relative to phase conductors) dangerous amounts of ground-fault current diverted into paralleling conductive paths.

The fast-growing use of low-signal-level input high-speed electronic computing and data-processing systems places added emphasis on the need to minimize the transmission of stray electrical noise from the electric power circuits to the surrounding space in which these critical equipments are located. The presence of fast-acting solid-state switching devices among the electric-system switching components can aggravate the problem by intensifying the amount of high-frequency disturbance present with the electric-system voltage carried by the power conductors.

The complete metal enclosure of the electric power system, as shown in Fig. 8, contributes immensely to the elimination of electrical noise. The NEC requires that all such metal enclosures be interconnected to form an adequate continuous electrical circuit, and also that they be grounded (with some exceptions). The effect of this construction is to enclose the entire electric power system conductor array within a continuous shell of grounded metal that functions as a Faraday shield to confine the electrostatic and electromagnetic fields associated with the power conductors to the space within the metallic shell. The contribution of electrical noise external to the enclosures is reduced to almost zero.

To avoid a by-pass circuit by which power-con-

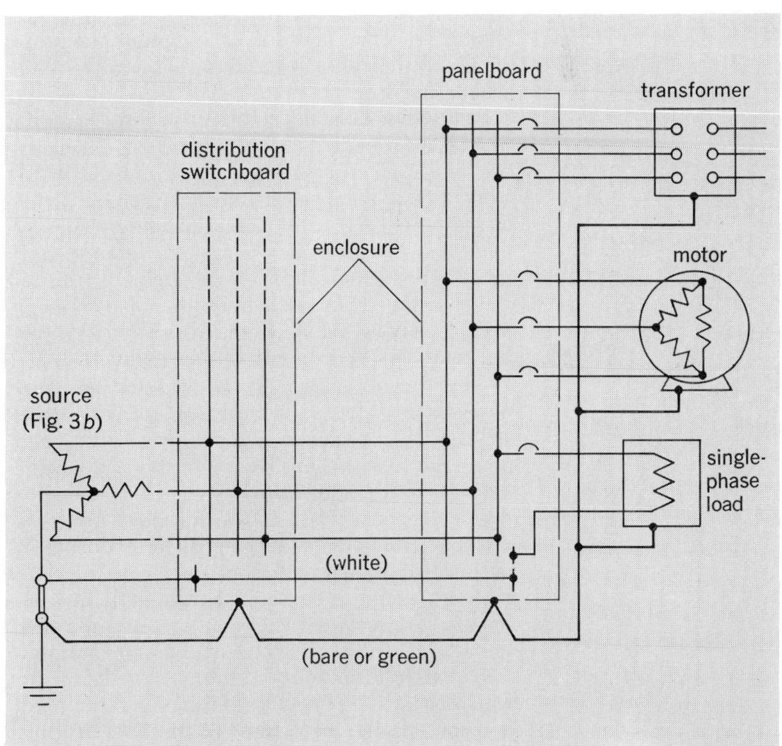

Fig. 7. Heavy line shows typical equipment grounding conductor.

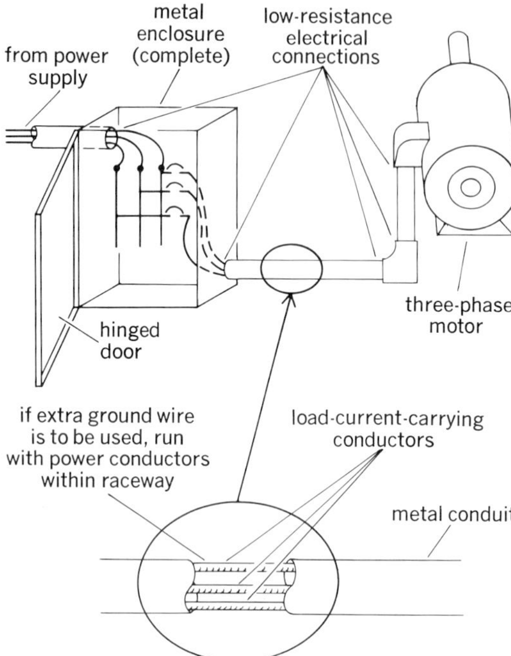

Fig. 8. Complete enclosure of power conductors within a continuous grounded steel shell confines electric and magnetic effects of power currents.

ductor noise voltages might be conductively transmitted to the outside of the metallic enclosure, a careful check of the integrity of the insulation of the grounded power conductor (white wire) throughout the building interior is warranted. The NEC prescribes that the power-system grounded conductor be connected to the grounding conductor at the point of service entrance to the building, and at no other point within the building beyond the service equipment, with some exceptions. Only if a supply feeder extends from one building into another is it permissible to reground the white wire, and then only at the point of entrance to the second building.

Protection considerations. The importance of limited magnitude L-G fault current in easing the problem of electric shock exposure control will be evident from the discussion below. Section 230-95 of the NEC (1978) contains a mandatory requirement for installation of automatic ground-fault-responsive tripping of the power supply at the service equipment for all solidly grounded wye-connected electric services of more than 150 V L-G but not those exceeding 600 V line to line (L-L), for each service disconnecting means rated 1000 A or more. In the interests of assuring superior electric shock protection, section 210-8 of the NEC mandates the installation of ultrasensitive ground-responsive tripping features (type-GFCI personal protectors) on certain 120-V one-sided grounded circuits. Protection is accomplished by deenergizing the supply power to the receptacle if even a minute amount of the circuit current (a few milliamperes) becomes diverted to a return path other than the grounded power conductor of the circuit.

Grounding of outdoor stations. Installations in which earth is used as a reference ground plane present special problems. To design an earth "floor surface" for an outdoor open-type substation which will be free of dangerous electric shock voltage exposure to persons around the station is a difficult task.

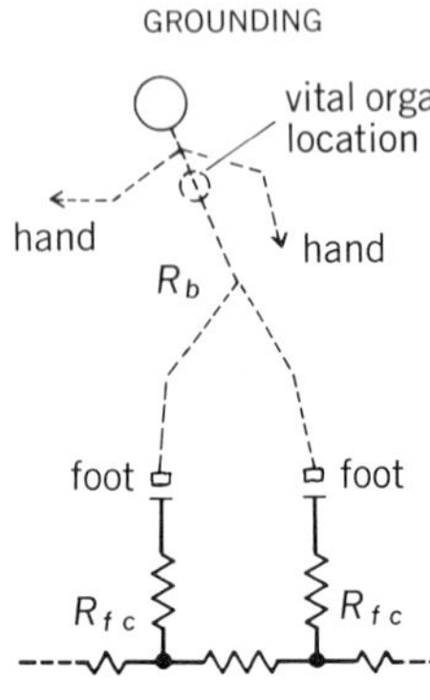

Fig. 9. A simple "person model" to be used in the evaluation of electric shock voltage-exposure intensity.

Personal electric shock danger. A beginning step in dealing with the shock hazard is to establish the magnitude of electric shock exposure which can be accepted by a person without ill effect. The evaluation of dangers from electric shock is aided by the simple "person model" (Fig. 9). The organs which can be vitally disoriented are the heart and lungs. Lung muscles can be propelled to a tight closed spasm, shutting off the respiratory action. However, the lungs resume normal action once the severe shock voltage is removed. On the other hand, a single short-time excess-value electric shock incident may throw the heart muscles into a state of fibrillation from which they are unable to automatically recover normal action, resulting in death within a few minutes unless antifibrillation is accomplished.

Electric shock is communicated to the body via electrical contact to the body extremities (Fig. 9), hand to hand, hand to foot, or foot to foot, each of which creates nearly the same shock intensity at the central chest area. The hand-to-foot case applies to a person standing on the substation floor and touching nearly conducting parts of the station with the extended arms; this applied shock hazard is named the E_{touch} exposure. The foot-to-foot case applies to a person walking across the substation floor during which activity the person's two feet, with considerable separation distance between, contact the substation surface; this shock hazard is named the E_{step} exposure. The effect of a surface layer of higher-resistance material over the active area of the station may be represented by an added resistance element R_{fc} (Fig. 9).

The basic limits of tolerable electric shock exposure are expressed by Eq. (1), where i is the current magnitude through the body and t the elapsed time in seconds, or, in similar form, by Eq. (2).

$$i = \frac{0.116 \text{ ampere}}{\sqrt{t}} \qquad (1)$$

$$i^2t = 0.0134 \text{ ampere}^2 \text{ seconds} \qquad (2)$$

The sources of potential electric shock hazard are commonly expressed in terms of volts (rather than amperes), but the limiting shock exposure tolerance can be converted to terms of voltage and time if the body resistance R_b (Fig. 9) is known. The value of R_b may go up into the megohm region in the absence of excess moisture or perspiration, but drops drastically when these substances are present on the skin. Assuming the lower limit of the body resistance to be 1000 ohms, the maximum allowable impressed body voltage V_s is given by Eq. (3).

$$V_s = i(R_b) = i(1000) \text{ volts} = \frac{116 \text{ volts}}{\sqrt{t}} \qquad (3)$$

Use of ground rods. The voltage exposure values of both E_{touch} and E_{step} depend heavily on the magnitude of voltage gradients along the surface of the outdoor station floor. An outdoor station might receive incoming power at 69 kV and an L-G circuit fault at this terminal would result in the injection of 3000 A into the grounding conductor at the station. During this current flow the station grounding conductor system could rise above mean earth potential by as much as 2500 V or more. The resulting potential distribution pattern across the station floor must be evaluated and tailored to achieve the limiting values of E_{touch} and E_{step}.

As a first step, one may consider driving a standard ground rod (8 ft or 2.4 m long, 3/4 in. or 19 mm in diameter) in the center of the substation floor area. In a typical soil makeup this might be found to establish a 25-ohm connection to earth. It takes only 100 A injected into such a connection to produce a voltage drop of 2500 V, about half of which takes place within a radius of about 2½ ft (0.8 m). As a voltage probe is moved away from the rod on the station surface, the potential drops rapidly in a conical pyramidal fashion.

The earth-connection properties of a single driven ground rod are thus controlled, almost totally, by a small cylinder of earth, about 10 ft (3 m) in diameter, immediately surrounding it. The resistance value is usually substantial (on the order of 25 ohms) and is reduced only slightly by a second ground rod within the influence zone of the first. A low-value-resistance connection to earth therefore requires a multiplicity of distributed ground rods, spaced to be nearly independent of the influence fields of each other.

To meet the electric shock safety requirements in open-type outdoor stations usually requires an elaborate array of metallic grounding conductors buried in the surface soil of the station area. The array is commonly composed of parallel horizontal conductors with perhaps 8 ft (2.4 m) horizontal spacing at a depth of some 1 to 2 ft (0.3 to 0.6 m) below the surface (Fig. 10*a*). The geometry of this conductor system usually matches that of the station structure and makes use of the below-grade concrete reinforcing steel involved in the construction of the station. The ground-surface voltage, within one specific bay, relative to the station grounding conductor potential and above the mean earth potential, resembles that shown in Fig. 10*b*. An artificially controlled level of available ground-fault current at 400 A is common in industrial electric power system design, in contrast to a value of 3000 A, not uncommon elsewhere, representing a 7-to-1 advantage.

Hazards near station boundary. There is the possibility that unacceptable levels of electric shock voltage exposure E_{step} and E_{touch} may be found in some places external to the outdoor station area. To ensure the absence of electric shock danger to persons who may frequent these areas adjoining the substation, it is important to check out suspect spots and institute corrective measures as necessary. [R. H. KAUFMANN]

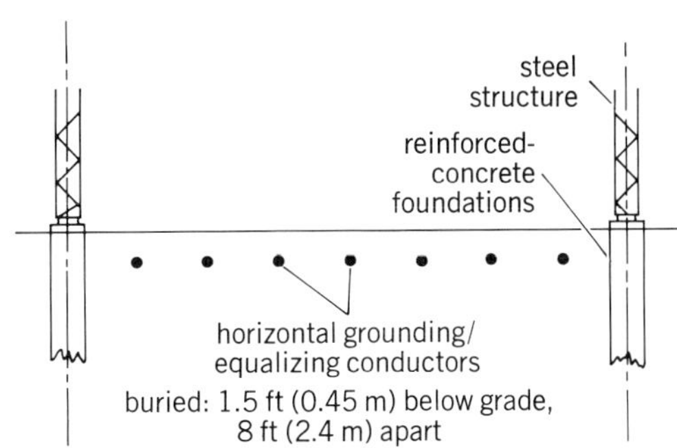

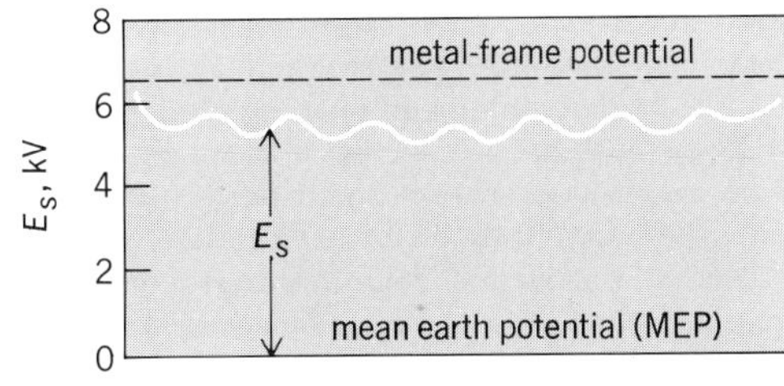

Fig. 10. Surface-potential control in an earth-floor electrical station obtained by a mesh grid of buried grounding/equalizing conductors. (*a*) Physical pattern of conductors in one bay. (*b*) Station surface potential E_s above mean earth potential during a local L-G fault.

Bibliography: L. E. Crawford and M. S. Griffith, A closer look at "the facts of life" in ground mat design, *IEEE Trans., Ind. Appl.*, IA-15:241, 1979; C. F. Dalziel and W. R. Lee, Reevaluation of lethal electric currents, *IEEE Trans. Ind. Gen. Appl.*, IGA-4:467–476, 1968; Institute of Electrical and Electronics Engineers, *IEEE Guide for Safety in AC Substation Grounding*, IEEE Stand. no. 80, 1976; Institute of Electrical and Electronics Engineers, *IEEE Recommended Practice for Electric Power Systems in Commercial Buildings*, IEEE Stand. no. 241, 1974; Institute of Electrical and Electronics Engineers, *IEEE Recommended Practice for Grounding of Industrial and Commercial Power Systems*, IEEE Stand. no. 142, 1972; R. H. Kaufman, Some fundamentals of equipment grounding, *IEEE Trans. Appl. Ind.*, vol. 73, pt. 2, pp. 227–231, 1954; National Fire Protection Association, *National Electrical Code*, NFPA no. 70, 1978; F. J. Shields, The problem of arcing faults in low-voltage power distribution systems, *IEEE Trans. Ind. Gen. Appl.*, IGA-3(1):15, 1967; R. B. West, Grounding for emergency and standby power systems, *IEEE Trans. Ind. Appl.*, IA-15(2):124, 1979.

Gypsum plank

A structural precast unit designed for use as the roof deck of industrial or other steel-frame buildings, and in some cases for the floor system as well. The four edges of the plank are bound with a tongue-and-groove steel form which, when fitted into an adjacent unit, produces a rigid steel I beam at the joint. The gypsum core is reinforced with welded galvanized-steel mesh. The planks are usually 2 in. thick, 15 in. wide, and 8–10 ft in length. Specially designed clips fasten the planks to the supporting purlins and to an asphalt-impregnated felt built-up roofing used to resist the weather. The planks are fire-resistant and lightweight, are easily erected, and may be readily cut or drilled. *See* ROOF CONSTRUCTION.

[CHARLES M. ANTONI]

Hardness scales

Arbitrarily defined measures of the resistance of a material to indentation under static or dynamic load, to scratch, abrasion, or wear, or to cutting or drilling. Standardized tests compare similar materials according to the particular aspect of hardness measured by the test. Widely used tests for metals are Brinell, Rockwell, and Scleroscope tests, with modifications depending upon the size or condition of the material. Indentation tests compare species of wood or flooring materials, and abrasion tests serve as an index of performance of stones and paving materials.

Hardness tests are important in research and are widely used for grading, acceptance, and quality control of manufactured articles. The hardness designation or scale is associated with the test method or instrument used.

Scratch hardness. Resistance to scratching is defined by comparison with 10 selected minerals, which are numbered in the order of increasing hardness. This mineralogical scale, called Mohs scale, is 1, talc; 2, gypsum; 3, calcite; 4, fluorite; 5, apatite; 6, orthoclase; 7, quartz; 8, topaz; 9, corundum; and 10, diamond. Minerals lower in the scale are scratched by those with higher numbers. The scale is extended to provide finer distinction of harder materials by additional minerals: 7, vitreous pure silica; 8, quartz; 9, topaz; 10, garnet; 11, fused zirconia; 12, fused alumina; 13, silicon carbide; 14, boron carbide; and 15, diamond.

File-test hardness. Materials are differentiated qualitatively according to resistance to scratching or cutting by files especially selected for the purpose. Whether or not a visible scratch is produced on the material indicates its hardness in comparison with a sample of desired hardness. The method is used for routine inspection of hardened surfaces in production.

Brinell hardness. Resistance to indentation by a hardened steel or tungsten carbide ball under specified load is the basis for Brinell hardness. Standard procedure uses a 10-mm ball with loads of 3000 kg for hard material, 1500 kg for intermediate, and 500 kg or less for soft materials. Various machines apply and control the specified load. The diameter of the impression is measured with a micrometer microscope. Brinell hardness number (Bhn), expressed in kilograms per square millimeter, is obtained by dividing the load by the spherical surface area of the impression. Different-size balls may be used according to size, thickness, and hardness of the specimen, and give the same hardness number provided the loads are proportional to the square of the ball diameter. Carboloy balls are used for very hard material. Time of load application, minimum thickness, and size of specimen are standardized. A close relation exists between Bhn and ultimate tensile strength.

Vickers hardness. Indentation of a square-based diamond pyramid penetrator with an angle between opposite faces of 136° measures Vickers hardness. Applied load may be varied from 5 to 120 kg in increments of 5 kg according to size of test piece. Vickers hardness number, also called diamond pyramid hardness, is equal to the load divided by the lateral area of the pyramidal impression. The area is computed from measurements of the diagonals of the square impression. Vickers hardness is the most reliable measure for very hard material and is applicable to thin sheets and hardened surfaces.

Rockwell hardness. Depth of indentation of either a steel ball or a 120° conical diamond with rounded point, called a brale, under prescribed load is the basis for Rockwell hardness. The ball is normally 1/16 in. in diameter, but 1/8-, 1/4-, or 1/2-in. balls are used for soft materials. A specially designed machine applies loads of 60, 100, or 150 kg. The depth of impression, referred to the position under an initial minor load, is indicated on a dial whose graduations represent the hardness number. Hardness is designated by a number with a standard system of prefix letters to indicate type of penetrator and load used.

Superficial Rockwell hardness is measured by a special machine differing from the standard Rockwell tester in that it applies lighter loads with a more sensitive depth-measuring dial. It produces a shallow impression and is suitable for thin sheet material and where surface hardness to a limited depth is of interest.

Monotron hardness. The pressure in kilograms per square millimeter required to embed a 0.75-mm hemispherical diamond penetrator to a depth of 0.0018 in., producing an impression 0.36 mm in diameter, is the measure of Monotron hardness. The depth is controlled by a separate dial graduated to 1 kg/unit area. The method is applicable to the entire range of hardness and is suited to thin sheet and case-hardened surfaces.

Shore Scleroscope hardness. Height of rebound of a diamond-tipped weight or hammer falling within a glass tube from a height of 10 in. and striking the specimen surface measures Shore Scleroscope hardness. The standard hammer is 1/4 in. in diameter, 3/4 in. long, and weighs 1/12 oz. The hardness number is the height of rebound referred to an arbitrary scale graduated to 140 divisions within the glass tube. The method is a dynamic load test, and the rebound reflects the size of indentation produced, which determines the energy absorbed by deformation and hence that available for rebound. A recording instrument with a dial indicates the rebound hardness directly. Both instruments are portable and permit rapid determinations.

Herbert pendulum hardness. Resistance to cold working is measured as Herbert pendulum hardness. The apparatus consists of a rocking device, called a pendulum, supported on a 1-mm steel ball in contact with the specimen. A curved level bubble measures amplitude of oscillation. The time hardness number is the time in seconds for 10 complete swings of the pendulum through a small arc. The work-hardening capacity is measured by the maximum time hardness after previously repeated single swings of the pendulum. The scale hardness number is the angular oscillation of a half swing after tilting the pendulum through a definite angle before release. Scale hardness is taken as a measure of resistance to flow as in rolling, drawing, and stamping. The method is applicable to studies of machining and forming of metals.

Microhardness. Resistance to indentation over very small areas (as on small parts, the constituents of metal alloys, or for exploration of hardness variations) is called microhardness. One tester employs the Vickers square-based pyramidal diamond penetrator attached to the end of a vertically guided shaft having a weight of 25 g. An arrangement of microscopes permits centering and measurement of the diagonals of the impression. The hardness number is the pressure intensity in kilograms per square millimeter.

Another procedure employs a Tukon tester

applying loads of 25–3600 g using a Knoop indenter, which is a diamond ground to produce a diamond-shaped impression with ratio of diagonal lengths of 7:1. The location of the indenter and measurement of the diagonals of the impression is accomplished with microscopes. The hardness number is the ratio of the applied load to the projected area of the impression.

Hardness of wood. The load required to embed a 0.444-in.-diameter steel ball to half its diameter expresses the hardness of wood. Used as a means of comparison, the values vary with species and grain characteristics. Hardness values of poplar and Douglas fir are approximately 400 and 900 lb, respectively.

Hardness of paving. Wear or abrasion hardness applies primarily to natural stones, paving, or flooring materials. It is measured by a specified test providing an index of service performance. Hardness of stone is reflected by weight loss of a cylindrical core rubbed on a sand bed. Deval abrasion test determines weight loss of a charge of broken stone tumbled in a cylinder. Similarly, the Los Angeles rattler and the standard rattler test for paving block tumble a charge including steel balls in a drum to determine percentage loss of weight as an index of wear. Special wear tests are applied to floor surfaces.

[WILLIAM J. KREFELD/WALDO G. BOWMAN]

Bibliography: H. E. Davis, G. E. Troxell, and C. T. Wiskocil, *Testing and Inspection of Engineering Materials*, 3d ed., 1964; A. A. Ivan'ko, *Handbook of Hardness Data*, 1971.

Harmonic speed changer

A mechanical-drive system used to transmit rotary, linear, or angular motion at high ratios and with positive motion. In the rotary version, illustrated schematically in Fig. 1, the drive consists of a rigid circular spline, an input wave generator, and a flexible spline. Any one of these can be fixed, used as the input, or used as the output. Any combination (fixed, driver, or driven) may be used. *See* SPLINES.

In Fig. 1 the fixed member is a rigid circular spline with 132 internal teeth. The driven part is a flexible spline, a ring gear with 130 external teeth of same size as on the rigid spline. The driving member is shown as a two-lobed member generating a traveling circular wave on the flexible spline.

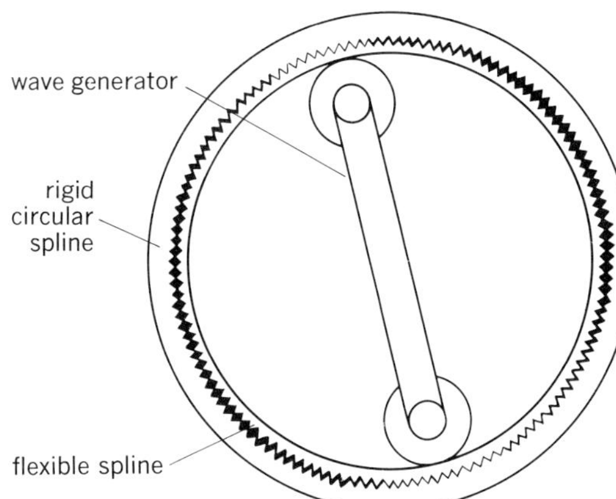

Fig. 1. Rotary-to-rotary harmonic speed changer.

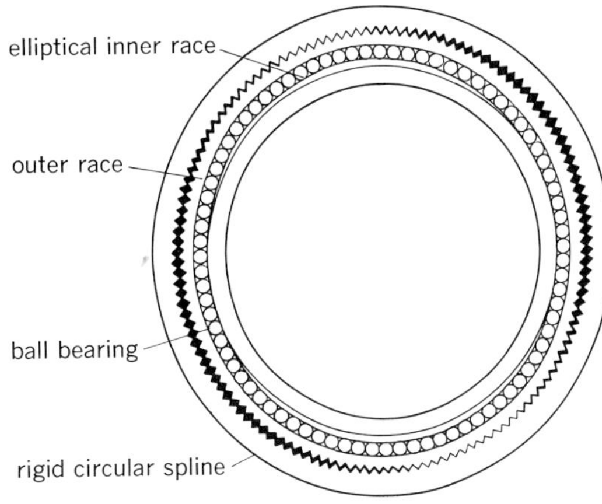

Fig. 2. Diagram of actual construction of a rotary-to-rotary harmonic speed changer.

The flexible spline meshes with the rigid circular spline at two diametrically opposite regions of the spline, which is shown flexed as an ellipse with its major axis nearly vertical. The teeth on the splines clear each other along the nearly horizontal minor axis.

As the wave generator rotates, the axes of the flexible spline correspondingly rotate along with the regions of contact and clearance between the teeth on the splines. Because the number of teeth on the flexible spline is less than the number of teeth on the fixed and rigid circular spline, when the wave generator rotates one turn, the flexible spline will rotate 2/132 parts of a turn in a direction opposite that of the driving wave generator. The speed reduction is therefore 2:132 (or 1:66). By increasing the number of teeth on the two splines, the flexible spline always having two less (with a two-lobed wave generator) than the number on the circular spline, the speed reduction can be increased correspondingly.

In actual construction (Fig. 2), the wave generator is composed of a ball bearing with an elliptical inner race. In rotating, the inner race flexes the outer race, engaging its external teeth along its major axis with the internal teeth of the rigid spline.

The advantages of this drive are (1) high-ratio gearing in small space, (2) speed reduction (or increase) in fixed ratio up to 1000:1 in one unit, (3) negligible wear of teeth, (4) balanced bearing loads, (5) negligible backlash, (6) efficiency of approximately 80% in a gear ratio of 400:1, and (7) adaptability to rotary-to-rotary, rotary-to-linear, and linear-to-linear drives. *See* GEAR DRIVE.

[PAUL H. BLACK]

Bibliography: E. A. Dijksman, *Motion Geometry of Mechanism*, 1976; H. H. Mabie and F. W. Ocvirk, *Mechanisms and Dynamics of Machinery*, 3d ed., 1975.

Heat exchanger

A device used to transfer heat from a fluid flowing on one side of a barrier to another fluid (or fluids) flowing on the other side of the barrier.

When used to accomplish simultaneous heat transfer and mass transfer, heat exchangers become special equipment types, often known by other names. When fired directly by a combustion process, they become furnaces, boilers, heaters, tube-still heaters, and engines. If there is a change in phase in one of the flowing fluids—condensation of steam to water, for example—the equipment may be called a chiller, evaporator, sublimator, distillation-column reboiler, still, condenser, or cooler-condenser.

Heat exchangers may be so designed that chemical reactions or energy-generation processes can be carried out within them. The exchanger then becomes an integral part of the reaction system and may be known, for example, as a nuclear reactor, catalytic reactor, or polymerizer.

Heat exchangers are normally used only for the transfer and useful elimination or recovery of heat without an accompanying phase change. The fluids on either side of the barrier are usually liquids, but they may also be gases such as steam, air, or hydrocarbon vapors; or they may be liquid metals such as sodium or mercury. Fused salts are also used as heat-exchanger fluids in some applications.

With the development and commercial adoption of large, air-cooled heat exchangers, the simplest example of a heat exhanger would now be a tube within which a hot fluid flows and outside of which air is made to flow for the purpose of cooling. By similar reasoning, it might be argued that any container of a fluid immersed in any fluid could serve as a heat exchanger if the flow paths were properly connected, or that any container of a fluid exposed to air becomes a heat exchanger when a temperature differential exists. However, engineers will insist that the true heat exchanger serve some useful purpose, that the heat recovery be meaningful or profitable.

Most often the barrier between the fluids is a metal wall such as that of a tube or pipe. However, it can be fabricated from flat metal plate or from graphite, plastic, or other corrosion-resistant materials of construction. If, as is often the case, the barrier wall is that of a seamless or welded tube, several tubes may be tied together into a tube bundle (see diagram) through which one of the fluids flows distributed within the tubes. The other fluid (or fluids) is directed in its flow in the space outside the tubes through various arrangements of passes. This fluid is contained by the heat-exchanger shell. Discharge from the tube bundle is to the head (heads) and channel of the exchanger. Separation of tube-side and shell-side fluids is accomplished by using a tube sheet (tube sheets).

Applications. Heat exchangers find wide application in the chemical process industries, including petroleum refining and petrochemical processing; in the food industry, for example, for pasteurization of milk and canning of processed foods; in the generation of steam for production of power and electricity; in nuclear reaction systems; in aircraft and space vehicles; and in the field of cryogenics for the low-temperature separation of gases. Heat exchangers are the workhorses of the entire field of heating, ventilating, air-conditioning, and refrigeration.

Classifications. The exchanger type described in general terms above and illustrated by the diagram is the well-known shell-and-tube heat exchanger. Shell-and-tube exchangers are the most numerous, but constitute only one of many types. Exchangers in use range from the simple pipe within a pipe—with a few square feet of heat-transfer surface—up to the complex-surface exchangers that provide thousands of square feet of heat-transfer area.

In between these extremes is a broad field of shell-and-tube exchangers often specifically named by distinguishing design features; for example, U tube, fin tube, fixed tube sheet, floating head, lantern-ring packed floating head, socket-and-gland packed floating head, split-ring internal floating head, pull-through floating head, nonremovable bundle with floating head or U-tube construction, and bayonet type.

Also, varying pass arrangements and baffle-and-shell alignments add to the multiplicity of available designs. Either the shell-side or tube-side fluids, or both, may be designed to pass through the exchanger several times in concurrent, countercurrent, or cross flow to the other fluids.

The concentric pipe within a pipe (double pipe) serves as a simple but efficient heat exchanger. One fluid flows inside the smaller-diameter pipe, and the other flows, either concurrently or countercurrently, in the annular space between the two pipes, with the wall of the larger-diameter pipe serving as the shell of the exchanger.

To solve new processing problems and to find more economical ways of solving old ones, new

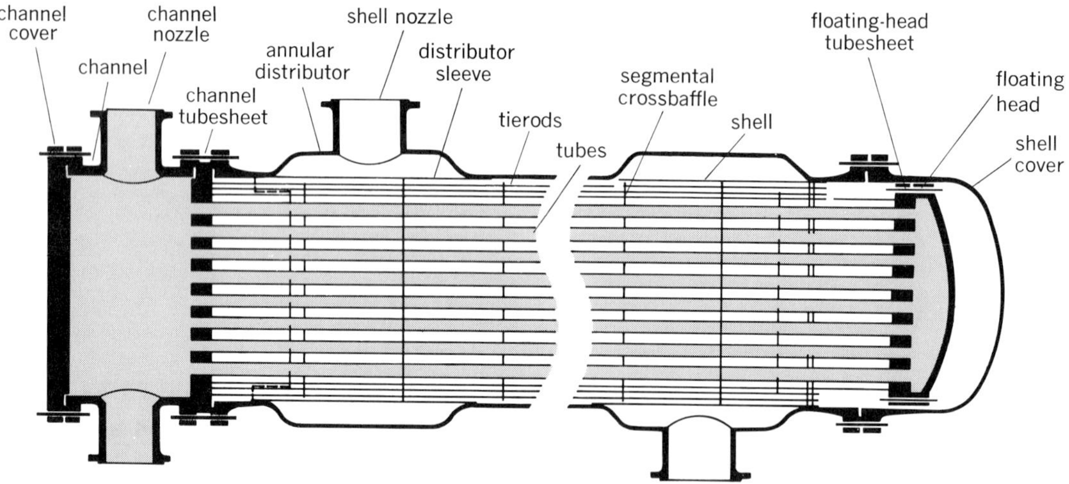

Schematic diagram of heat exchanger.

types of heat exchangers are being developed. There has been much emphasis on cramming more heat-transfer surface into less and less volume. Extended-surface exchangers, such as those built with fin tubes, are finding wide application.

Water shortage has added a new dimension to heat-exchanger design and has led to use of air-cooled exchangers.

Plate-type heat exchangers, long used in the milk industry for pasteurization and skimming, have moved into the chemical and petroleum industries. Coiled tubular exchangers and coiled-plate heat exchangers are winning new assignments. Spiral exchangers offer short cylindrical shells with flat heads, carrying inlets and outlets leading to internal spiral passages. These passages may be made with spiral plates or with spiral banks of tubes. Exchangers with mechanically scraped surfaces are finding favor for use with very viscous and pastelike materials.

A somewhat unusual type of plate heat exchanger is one in which sheets of 16-gage metal, seam- and spot-welded together, are embossed to form transverse internal channels which carry the heat-transfer medium. This type is often used for immersion heating in electroplating and pickling.

Materials of construction. Every metal seems to be a possible candidate as a material of construction in fabrication of heat exchangers. Most often, carbon steels and alloy steels are used because of the strength they offer, especially when the exchanger is to be operated as a pressure vessel. Because of excellent heat conductance, brass and copper find wide use in exchanger manufacture.

Corrosion plays a key role in the selection of exchanger construction materials. Often, a high-priced material will be selected to contain a corrosive tube-side fluid, with a cheaper material being used on the less corrosive shell side.

For special corrosion problems, exchangers are built from graphite, ceramics, glass, bimetallic tubes, tantalum, aluminum, nickel, bronze, silver, and gold. *See* CORROSION.

Problems of use. Each of the fluids and the barrier walls between them offers a resistance to heat transfer. However, another major resistance that must be considered in design is the formation of dirt and scale deposited on either side of the barrier wall. This resistance may become so great that the exchanger will have to be removed from service periodically for cleaning.

Chemical and mechanical methods may be used to remove the dirt and scale. For mechanical cleaning, the exchanger is removed from service and opened up. Perhaps the entire tube bundle is pulled from the exchanger shell if the plant layout has provided space for this to be done. If the deposit is on the inside of straight tubes, cleaning may be accomplished merely by forcing a long worm or wire brush through each tube.

More labor is required to remove deposits on the shell side. After removal of the tube bundle, special cleaning methods such as sandblasting may be necessary.

Much engineering effort has gone into the design of heat exchangers to allow for fouling. However, D. Q. Kern has suggested that methods are available to design heat exchangers that, by accommodating a certain amount of dirt in a thermal design, will allow heat exchangers to run forever without shutdown for cleaning. Commercial units designed in this fashion are operating today.

Another operating problem is allowance for differential thermal expansion of metallic parts. Most operating difficulties arise during the startup or shutdown of equipment. Therefore, M. S. Peters suggests the following general rules:

1. Startup. Always introduce the cooler fluid first. Add the hotter fluid slowly until the unit is up to operating conditions. Be sure the entire unit is filled with fluid and there are no pockets or trapped inert gases. Use a bleed valve to remove trapped gases.

2. Shutdown. Shut off the hot fluid first, but do not allow the unit to cool too rapidly. Drain any materials which might freeze of solidify as the exchanger cools.

3. Steam condensate. Always drain any steam condensate from heat exchangers when starting up or shutting down. This reduces the possibility of water hammer caused by steam forcing the trapped water through the lines at high velocities.

Standardization. Users have requested that heat exchangers be made available at lower prices through the standardization of designs. Organizations active in this work are the Tubular Exchangers Manufacturers' Association, American Petroleum Institute, American Standards Association, and the American Institute of Chemical Engineers. *See* COOLING TOWER; FURNACE CONSTRUCTION; VAPOR CONDENSOR.

[RAYMOND F. FREMED]

Bibliography: W. E. Glausser and J. A. Cortright, How to specify heat exchangers, *Chem. Eng.*, 62(12):203–206, 1955; W. M. Kays and A. L. London, *Compact Heat Exchangers*, 2d ed., 1964; D. Q. Kern, Speculative process design, *Chem. Eng.*, 66(20):127–142, 1959; K. Kornwell, *The Flow of Heat*, 1977; M. S. Peters, *Elementary Chemical Engineering*, 1954; D. J. Portman and E. Ryznar, *An Investigation of Heat Exchange*, 1971; J. C. Smith, Trends in heat exchangers, *Chem. Eng.*, 61(6):232–238, 1953.

Heat pump

The thermodynamic counterpart of the heat engine. A heat pump raises the temperature level of heat by means of work input. In its usual form a compressor takes refrigerant vapor from a low-pressure, low-temperature evaporator and delivers it at high pressure and temperature to a condenser (Fig. 1). The pump cycle is identical with the customary vapor-compression refrigeration system. *See* REFRIGERATION CYCLE.

Application to comfort control. For air-conditioning in the comfort heating and cooling of space, a heat pump uses the same equipment to cool the conditioned space in summer and to heat it in winter, maintaining a generally comfortable temperature at all times (Fig. 2). *See* AIR CONDITIONING; COMFORT HEATING.

This dual purpose is accomplished, in effect, by placing the low-temperature evaporator in the conditioned space during the summer and the high-temperature condenser in the same space during the winter (Fig. 3). Thus, if 70°F is to be maintained in the conditioned space regardless of the season, this would be the theoretical temperature of the evaporating coil in summer and of the

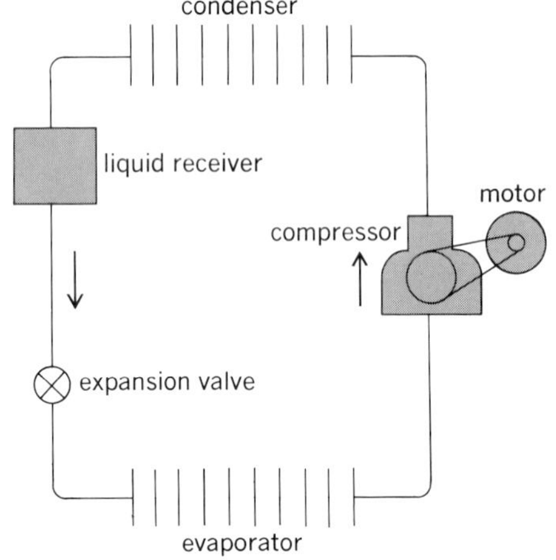

Fig. 1. Basic flow diagram of heat pump with motor-driven compressor. For summer cooling, condenser is outdoors and evaporator indoors; for winter heating, condenser is indoors and evaporator outdoors.

condensing coil in winter. The actual temperatures on the refrigerant side of these coils would need to be below 70°F in summer and above 70°F in winter to permit the necessary transfer of heat through the coil surfaces.

If the average outside temperatures are 100°F in summer and 40°F in winter, the heat pump serves to raise or lower the temperature 30° and to deliver the heat or cold as required. The ultimate ideal cycle for estimating performance is the same Carnot cycle as that for heat engines. The coefficient of performance COP_c as cooling machine is given in Eq. (1), and the coefficient COP_w as a warming machine is given in Eq. (2), where T is temperature in

$$COP_c = \frac{\text{refrigeration}}{\text{work}} = \frac{T_c}{T_h - T_c} \quad (1)$$

$$COP_w = \frac{\text{heat delivered}}{\text{work}} = \frac{T_h}{T_h - T_c} \quad (2)$$

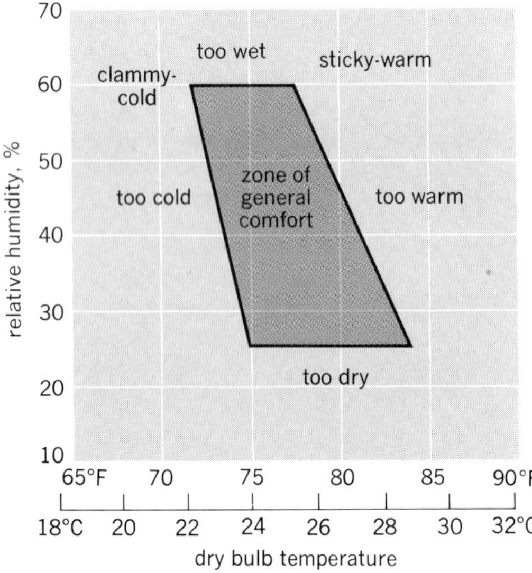

Fig. 2. Indoor climatic conditions acceptable to most people when doing desk work; continuous air motion with 5–8 air changes per hour.

degrees absolute and the subscripts c and h refer to the cold and hot temperatures, respectively.

For the data cited, the theoretical coefficients of performance are as in Eqs. (3). The significance of

$$\begin{aligned} COP_c &= \frac{460+70}{(460+100)-(460+70)} = 17.7 \\ COP_w &= \frac{460+70}{(460+70)-(460+40)} = 17.7 \end{aligned} \quad (3)$$

these coefficients is that ideally for 1 kilowatt-hour (kwhr) of electric energy input to the compressor there will be delivered $3413 \times 17.7 = 60{,}000$ Btu/hr as refrigeration or heating effect as required. This is a great improvement over the alternative use of resistance heating, typically, where 1 kWhr of electric energy would deliver only 3413 Btu. The heat pump uses the second law of thermodynamics to give a much more substantial return for each kilowatt-hour of electric energy input, since the

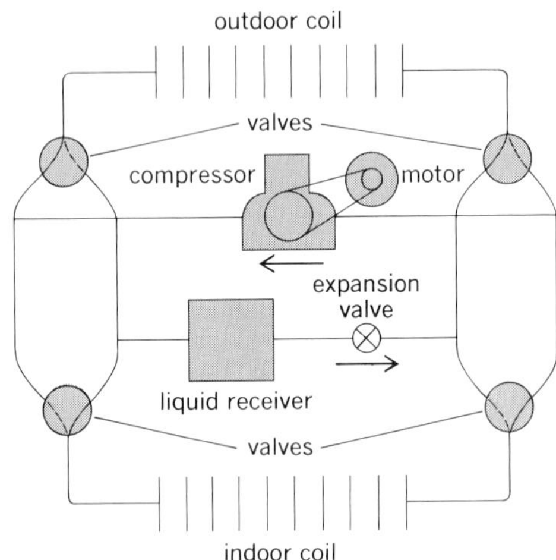

Fig. 3. Air-to-air heat pump installation; fixed air circuit with valves in the summer positions (the broken lines show the winter positions).

electric energy serves to move heat which is already present to a desired location.

Effect of seasonal loads. With the prevalent acceptance of summer comfort cooling of space, it is entirely possible and practical to use the same compressor equipment and coils for winter heating and for summer cooling and to dispense with the need for direct-fired apparatus using oil, gas, coal, or wood fuel.

For an economical installation, equipment must be of correct size for both the summer cooling and winter heating loads. Climatic conditions have a significant influence and can lead to imbalance on sizing. If the heating and cooling loads are equal, the equipment can be selected with minimum investment. However, generally the loads are not balanced; in the temperate zone the heating load is usually greater than the cooling load. This necessitates (1) a large, high horsepower compressor fitted to the heating demand, (2) a supplementary heating system (electrical resistance or fuel), or (3) a heat-storage system.

If well water or the ground serves as the heat

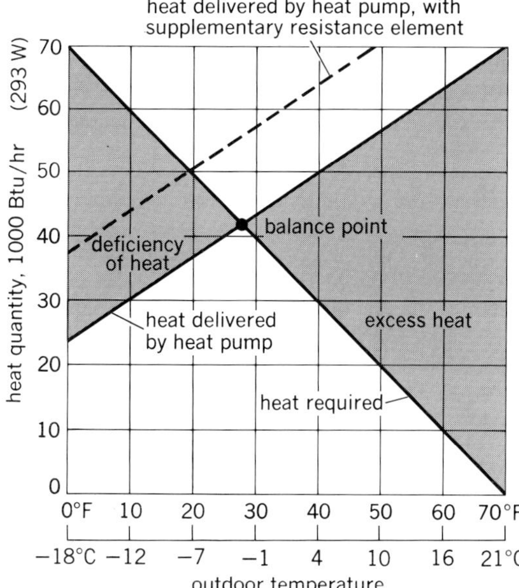

Fig. 4. Performance of air-to-air, self-contained, domestic heat pump on the heating cycle.

source, the imbalance is less severe than when atmospheric air is the source. However, the uncertain heat transfer rates with ground coil, the impurities, quality, quantity, and disposal of water, and the corrosion problems mitigate the use of these sources.

Atmospheric air as a heat source is preferable, particularly with smaller domestic units. A self-contained, packaged unit of this type offers maximum dependability and minimal total investment. The performance of such a unit for the heating cycle is illustrated in Fig. 4. Curves of heat required and heat available show the limitations on capacity. The heat delivered by the pump is less than the heat required at low temperatures, so that there is a deficiency of heat when the outside temperature goes below, in this case, about 28°F (−2°C). The intersection of the two curves is the balance point. There is an area of excess heat to the right and of deficiency of heat to the left of the balance point. Many devices and methods are offered to correct this situation, such as storage systems, supplementary heaters, and compressors operating alternatively in series or in parallel.

In temperate regions heat-pump installations achieve coefficients of performance in the order of 3 on heating loads when all requirements for power, including auxiliary pumps, fans, resistance heaters, defrosters, and controls are taken into account. Automatic defrosting systems, when air is the heat source, are essential for best performance, with the defrost cycle occurring twice a day.

Heat pumps are uneconomical if used for the sole purpose of comfort heating. The direct firing of fuels is generally more attractive from an overall financial viewpoint. The investment in heat-pump equipment is higher than that for the conventional heating system. Unless the price of electric energy is sufficiently low or the price of fuels very high, the heat pump cannot be justified solely as a heating device. However, if there is also need for comfort cooling of the same space in summer, the heat pump, to do both the cooling and heating, becomes attractive. The widespread use of air conditioning will probably lead to an increase in heat pumps.

The heat pump is also used for a wide assortment of industrial and process applications such as low-temperature heating, evaporation, concentration, and distillation. [THEODORE BAUMEISTER]

Solar energy–assisted system. The use of a heat pump for space heating with ambient air as the source is attractive in the temperate regions where ambient temperatures do not go significantly below the freezing point of water for extended periods of time. However, in the colder regions where ambient temperatures remain about (or below) 0°F (−18°C) for extended periods, the heat pump with outside air as a source presents problems. With a decrease in source temperature, both the capacity of the heat pump and the coefficient of performance (COP) fall. Figure 5 illustrates these characteristics for a commercially available heat pump unit. At an outside air temperature of 0°F (−18°C), the capacity of the heat pump unit is only about 43% of the capacity at 40°F (4°C). The COP also falls from 2.92 at 40°F to 1.8 at 0°F. Thus, when the heating load increases at decreasing ambient temperatures, the capacity of the heat pump decreases and the heat pump requires greater electrical energy to run it. Under these conditions there is also the problem of ice buildup on the evaporator coils, necessitating frequent defrosting. For the heat pump to be more attractive in colder regions, the source temperature for the heat pump should be increased. One possible method for achieving this is solar energy.

Solar collectors. For space heating utilizing solar energy, flat-plate collectors are generally used. A typical flat-plate collector contains a metallic plate painted black, with one or two glass covers and the sides and bottom of the collector well insulated. Solar energy is transmitted through the glass, and a significant part of that reaching the metallic plate—about 90%—is absorbed by the plate, increasing its temperature. The energy absorbed by the plate is, in turn, transferred to a working fluid—usually air or water. If air is the working fluid, it passes over the metallic plate; if water, it

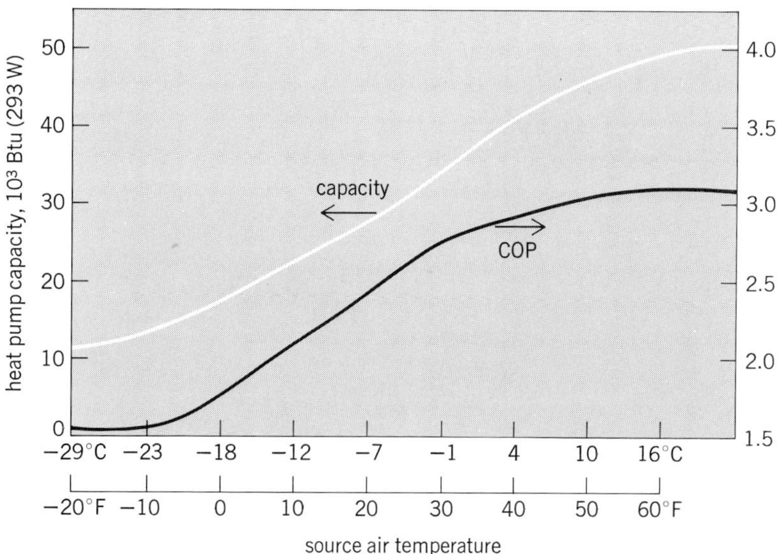

Fig. 5. Heat pump characteristics.

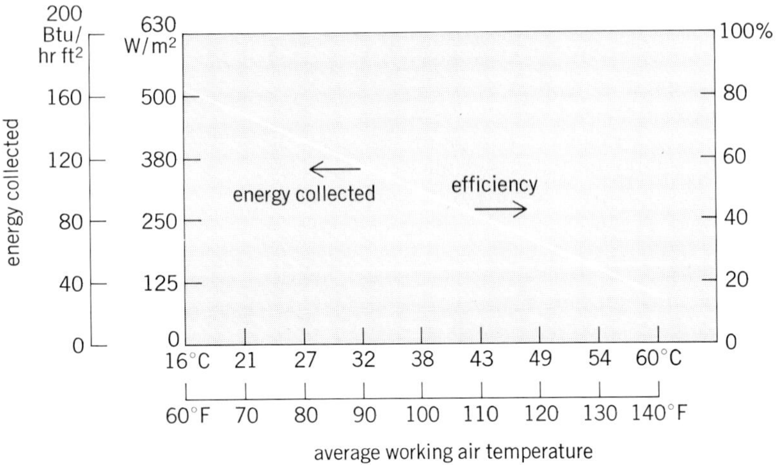

Fig. 6. Solar collector characteristics. An ambient temperature of 20°F (−7°C) and a solar insolation of 200 Btu/hr ft² (630 W/m²) have been assumed.

flows through tubes attached to the metallic plate.

Because of the heat transfer from the collector to the surroundings, not all the radiant energy reaching the collector is transferred to the working fluid. The efficiency of a collector, defined as the ratio of the heat transfer to the working fluid to the radiant energy reaching the collector surface, decreases with an increase in the absorber temperature. The higher the temperature at which the working fluid is operated, the higher is the temperature of the absorber surface and the lower the efficiency. The efficiency of a typical flat-plate collector with one glass cover and air as the working fluid is shown in Fig. 6, where the abscissa represents the average temperature of the working air. An ambient temperature of 20°F (−7°C) and a solar insolation of 200 Btu/hr ft² (630 W/m²) have been assumed. If warm air is to be directly used for space heating, its temperature should be about 110°F (43°C). At this temperature, the useful energy collected is 80 Btu/hr ft² (250 W/m²), corresponding to an efficiency of 40%. But when the temperature of the working air is 60°F (15.5°C), the useful energy collected (other conditions being the same as before) increases to 166 Btu/hr ft² (523 W/m²), corresponding to an efficiency of 83%. To make use of this increased solar energy at 60°F for space heating, a heat pump is needed. *See* SOLAR HEATING AND COOLING.

Advantages. The solar energy–assisted heat pump system (SAHPS) offers several advantages. Most buildings constructed during and since the 1970s are equipped with central air-conditioning systems, and with some modifications the air-conditioning system can serve as the heat pump system. As solar energy can be utilized either directly or through the heat pump at high values of COP during sunny days, electric power consumption during such days is reduced, thus contributing to load leveling at the central generating stations. Since the collectors operate at high efficiencies at low temperatures, the solar energy contribution from a given collector area can be increased. As the absorber surface temperature is lower, only one glazing may be sufficient in the majority of cases.

Working fluid and size. A heat pump system utilizing solar energy for space heating is schematically illustrated in Fig. 7. Air is the working fluid for the solar collectors, and the system uses an air-to-air heat pump. Energy storage is necessary as, during some periods, solar energy may not be available at all, such as during the night, or mav be available in very limited quantities, such as on cloudy days. The system can operate in three different modes as indicated in Fig. 7. In mode 1, the solar energy stored is not at a sufficiently high temperature for direct use but at a temperature higher than ambient temperature—conditions likely to be met on cold, sunny days. The heat pump uses the stored solar energy in this mode. In mode 2, the stored solar energy is at a sufficiently high temperature for direct utilization—probable conditions during sunny days in the fall and spring. The heat pump is then bypassed. In mode 3—during extended periods of cloudy days or very cold weather—the stored solar energy is depleted and outside air is the source for heat pumps.

Sizing the heat pump to meet the entire heating needs at the lowest expected ambient temperature, on the basis of heat pump output–source temperature characteristics, may lead to a heat pump which is significantly oversized for periods when ambient temperatures are higher. To avoid such oversizing, heat pump capacity may be determined on the basis of a source temperature higher than the lowest expected ambient temperatures, and the system may be equipped with an auxiliary heating source which will supply the required amount of energy to raise the temperature of the

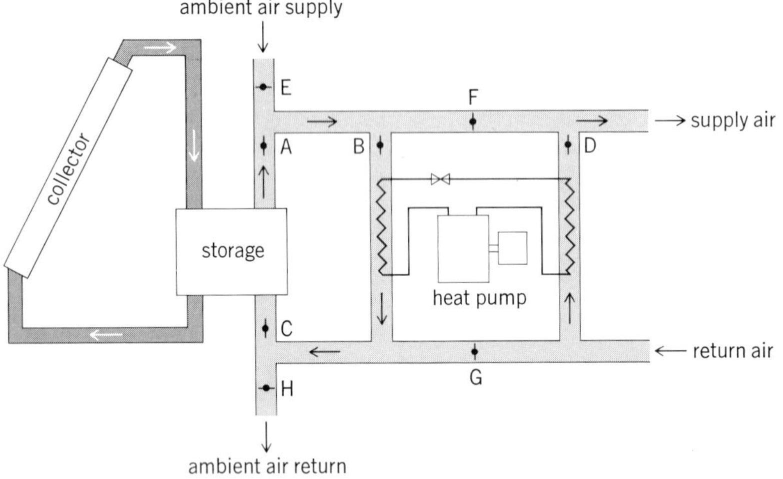

	open	close	mode
1	A,B,C,D	E,F,G,H	temperature of stored energy too low for direct use but higher than ambient air
2	A,C,F,G	E,B,D,H	stored solar energy temperature sufficiently high; direct utilization of solar energy (bypass heat pump)
3	B,D,E,H	A,C,F,G	stored solar energy temperature lower than ambient air temperature; heat pump source is ambient air

Fig. 7. Schematic arrangement of solar-assisted heat pump system.

ambient air to the designated temperature during extremely cold days.

Besides air-to-air heat pumps, water-to-air or water-to-water heat pumps can also be used. Because of the better heat transfer characteristics when water is used (with an antifreeze additive to prevent freezing) as the working fluid, a higher efficiency for the collector and a higher COP for the heat pump can be realized.

Thermal energy storage. As indicated earlier, some form of energy storage is essential for any system utilizing solar energy, since solar energy is available only intermittently. The amount of energy storage to be provided in a given system depends on collector area and weather conditions. Storage plays a significant role in determining the performance of the system. For systems with air as the working fluid in the collector, the commonly used thermal energy storage (TES) material is crushed rocks, which are inexpensive and readily available. With water as the working fluid, water itself is probably the best TES material.

TES can be either on the low-temperature (evaporator) side or high-temperature (condenser) side of the heat pump. High-temperature storage requires that the heat pump operate during periods of sunshine even when there is no need for space heating. This is avoided if thermal storage is on the low-temperature side, thus reducing the operation of the electrically driven heat pump during periods of peak electric demand.

Use. Some interest was shown in solar-assisted heat pump systems in the 1950s but declined as a result of alternate sources of inexpensive energy. Since the mid-1970s, however, there has been a renewal of interest in solar energy utilization for space heating, and several heat pump systems utilizing solar energy are in use in New Mexico, New York, Pennsylvania, Vermont, and other states.

Performance studies. Several performance studies—both analytical and experimental—have been made. Performance tests were conducted over two partial winter heating seasons during 1974–1975 and 1975–1976 on an office building in Albuquerque, NM, that was equipped with a SAHPS and completed in 1956, and was subsequently enlarged to a floor area of 5000 ft^2 (465 m^2). All building heating requirements were met by the SAHPS with negligible use of auxiliary heat. The average COP for the heat pump alone was 4.2, but was somewhat lower if all the power to run the auxiliary equipment was included.

A computational study of a SAHPS for a two-story elementary school building in Boston, MA, with 75,000 ft^2 (6970 m^2) of floor area, a collector area of 7500 ft^2 (697 m^2), and a water storage of 22,500 gal (85 m^3) indicated that about 75% of the heating requirements could be met by the SAHPS. Retrofitting of a block of 19 town houses at the Heron Gate project, Ottawa, Canada, is designed to supply about 40% of the space heating requirements from the SAHPS. Collectors operate in the range of 70–100°F (20–40°C). A water-to-water heat pump is employed, and thermal energy is provided by a 7000-gal (26-m^3) underground storage tank on the low-temperature side, with a collector area of 2300 ft^2 (214 m^2).

Although there are several studies relating to SAHPS, because of the many variables—weather, available solar energy in the specific location, collector area, thermal storage, and so forth—it is not yet possible to easily predict the performance of a particular system; performance predictions can be made only after a suitable simulation employing a computer.

Solar-assisted heat pumps were not competitive at 1977 costs, but with subsequent increases in prices of home heating oil, these systems have become economical under certain conditions in certain areas of the United States. With improvements in heat pump performance—particularly their reliability and ability to operate at relatively high source temperatures—such systems are likely to become attractive in many areas.

[N. V. SURYANARAYANA]

Domestic use. Developments in the energy field have brought about an increase in the use of heat pumps. The greatly increased cost of fossil fuels, as well as the limitations on gas usage in many areas, has resulted in cost justification of the slightly more expensive heat pump over a much larger geographical area of the United States. In addition, new types of heat pump systems and innovations in design have come about as a direct result of the energy picture.

Product applications. A number of factors have changed the application of heating-cooling products in the residential market. For example, the average size of homes has decreased since the mid-1970s. Associated with this, the heat loss per square foot has been decreased even faster. Some of the energy-conserving methods that have been responsible are the increased use of insulation in walls, floors, and ceilings; use of smaller and fewer windows; placement of the larger walls toward the southern exposure; and reduced infiltration due to tighter construction.

The heating-cooling industry has nationally experienced the impact of smaller homes in the capacity mix of unitary product shipments. Until the mid-1970s a nominal 3-ton (10.6-kW) cooling system represented a manufacturer's normal mix of products. This dropped to 2½ tons (8.8 kW) by 1978, and is expected to continue to drop and possibly level out at 2 tons (7 kW) by the mid-1980s.

Probably the most significant breakthrough in heat pump applications has come from the design and marketing of "add-on" heat pumps. These involve a split-system set of components, very similar to a standard air-source split-system heat pump, and are capable of being added to any existing forced-air furnace. The system consists of an outdoor section, indoor coil, and a control box. The indoor coil is mounted on the discharge-air side of the existing furnace, and the control box is the interface between the heat pump and furnace. Various designs have been marketed in the add-on category. The most efficient allows for heat pump operation below the balance point, since a large number of annual heating hours fall below this temperature; higher-efficiency heat pumps give a utilization efficiency higher than the existing furnace in this range. The advantages of this type of system include the following: (1) When mated with a fossil fuel furnace, the seasonal utilization efficiency of the two different fuels is maximized, greatly increasing the net efficiency of the raw resources (minimum of 35%). Thus, nonrenewable energy sources, such as fuel oil and natural gas, can be preserved for longer periods. (2) Application in combination with an existing electric forced-air furnace gives essentially the same utilization efficiency as a conventional heat pump. This permits improvement of the seasonal COP from 1.0 to approximately 2.0, for a 50% energy savings.

New product design considerations. Previous heat pump designs were simply air conditioners which were modified to become reversible refrigeration systems. Today, the heat pump designer must evaluate the cost effectiveness of each component in the system to justify the higher first cost resulting from improved efficiency. Due to the many hours of operating during the cooling cycle and relatively few in the heating cycle in the southern United States, a unit designed for this location would emphasize efficient operation during the cooling season. The reverse would be true for the northern market. A heat pump designed for universal use throughout the United States would compromise to provide equal improvements for both operating cycles, resulting in nearly equal annual operating cost savings regardless of the geographical location. The same high-efficiency unit would provide greater energy savings in the South during the summer and in the North during the winter. Improved heat pump efficiencies are achieved by using larger heat transfer surface; more efficient compressors, fans, and fan motors; and better thermal insulation. Care must be exercised to limit the amount of heat transfer surface so that system reliability will not be significantly affected. Larger surface area also requires more refrigerant charge, and compressor life is inversely proportional to the system charge.

Good product planning will anticipate energy costs in future years to justify higher first cost for equipment in a reasonable payback period. The typical time-span from original conception through development to production is from 2½ to 3 years. Since heat pump sales have now penetrated colder climates, there is a greater need for system reliability under more severe operating conditions. In addition, the high energy cost for standby electric heat during system outages and the increasing labor cost for service have demanded more emphasis on reliable operation.

The system reliability of heat pump designs has been improved in a variety of ways. Compressors have been designed specifically for heat pump duty, where stress conditions and operating hours are more demanding than for cooling-only units. Compressor improvements include heavy-duty connecting rods and wrist pins to withstand high compression ratios encountered in heating operation, nonfoaming lubricants, better motor-winding insulation, and motor overload protectors that are more sensitive to unsafe operating conditions.

Heat pumps have been designed with solid-state controls which can monitor and regulate unit operation more completely than previously was possible with electromechanical controls. Also included are such functions as short-cycling protection; excessive temperature and pressure protection; avoidance of nuisance defrosts due to wind gusts; overriding capability of defrost termination for gusty winds; protection against repeated defrost cycles during snow drifts; and system lockout and visual indication for abnormal operation.

Design testing of heat pumps has become much more extensive so as to better simulate field conditions. Test facilities can operate over a wider range of temperature, humidity, and precipitation conditions. Feedback from malfunctioning field installations and teardown of failed systems have given engineers insights into further improvements.

Until the mid-1970s most heat pumps were built with the time-temperature method of defrost initiation, because of its low cost and simplicity. At outdoor temperatures below about 42°F (6°C), a time clock would initiate a defrost cycle approximately every 60 min of compressor operation, whether the coil surface was frosted or not. Since this wastes energy and penalizes the operating cost, there has been an increase in the use of the true demand-type method of defrost. Here, a defrost cycle is initiated only when there is significant frost buildup on the coil surface. The sensor generally measures an increase in air resistance across the coil due to frost buildup. Field tests have indicated that a time-temperature defrost system results in 10–15 times as many defrost cycles per year as a well-designed demand system. [JERRY E. DEMPSEY]

Bibliography: E. R. Ambrose, *Heat Pumps and Electric Heating*, 1966; T. Baumeister (ed.), *Standard Handbook for Mechanical Engineers*, 8th ed., 1978; E. E. Drucker et al., *Commercial Building Unitary Heat Pump System with Solar Heating*, Rep. NSF/RANN/SE/GI-43895/FR/75/3 to ERDA, October 1975; T. L. Freeman et al., Performance of combined solar heat pump systems, *Solar Energy*, 22:125–135, 1979; S. F. Gilman, M. W. Wildin, and E. R. McLaughlin, Field study of a solar energy assisted heat pump heating system, *Solar Cooling and Heating: A National Forum*, Miami Beach, ERDA Doc. C00-2704-4, Dec. 14, 1976; F. C. McQuiston and J. D. Parker, *Heating, Ventilating and Air Conditioning: Analysis and Design*, 1977; S. Walters (ed.), Ottawa town houses heated by solar energy, *Mech. Eng.*, 101(11):64–65, November 1979.

Heat treatment (metallurgy)

A procedure of heating and cooling a material without melting. The heating and cooling sequence may involve temperatures above, below, and at the ambient. Controlled heating and cooling rates, and a variety of furnace atmospheres and heating media may be used. Plastic deformation may be included in the sequence of heating and cooling steps, thus defining a thermomechanical treatment. Typical objectives of heat treatments are hardening, strengthening, softening, improved formability, improved machinability, stress relief, and improved dimensional stability. Heat treatments are often categorized with special names, such as annealing, normalizing, stress relief anneals, process anneals, hardening, tempering, austempering, martempering, intercritical annealing, carburizing, nitriding, solution anneal, aging, precipitation hardening, and thermomechanical treatment.

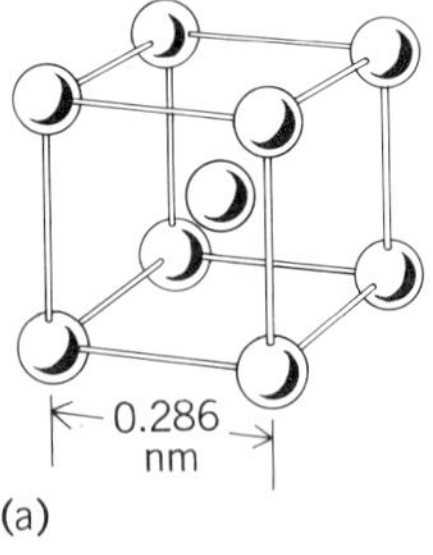

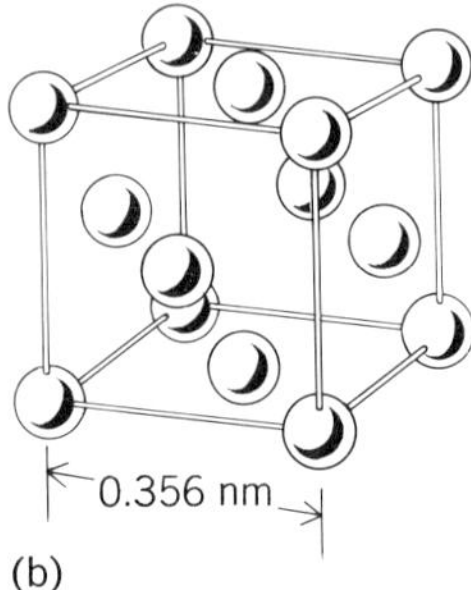

Fig. 1. Crystal structures of iron (*a*) Body-centered cubic α-iron. (*b*) Face-centered cubic γ-iron.

STEELS AND OTHER FERROUS ALLOYS

All metals and alloys in common use are heat-treated at some stage during processing. Iron alloys, however. respond to heat treatments in a unique way because of the multitude of phase changes which can be induced, and it is thus convenient to discuss heat treatments for ferrous and nonferrous metals separately. *See* STEEL.

General principles. At room temperature the equilibrium crystal structure of pure iron is body-centered cubic α-iron (Fig. 1*a*), also known as ferrite. On heating above 910°C (1670°F), α-iron is transformed to face-centered cubic γ-iron (Fig. 1*b*), also called austenite. At 1400°C (2552°F) γ-iron transforms to δ-ferrite, which is also body-centered cubic and structurally similar to α-iron, but δ-ferrite is seldom involved in heat-treating procedures. The addition of carbon to iron influences the transformation from one form of iron to another, and the resultant structures are summarized in the iron-carbon phase diagram shown in Fig. 2. Practical steels and irons contain other elements such as manganese, silicon, sulfur, phosphorus, aluminum, silicon, chromium, molybdenum, and nickel. These alloy elements influence the shape of the iron-carbon diagram, but if the total alloy content is less than 2%, the phase diagram is not affected substantially.

The general principles of the heat treatment of plain-carbon and low-alloy steels may be understood from the basic iron-carbon diagram. The diagram indicates the microconstituents, or phases, which are observed for a given carbon content at each temperature. In addition to ferrite and austenite, the other principal phase is the intermetallic compound, Fe_3C, or cementite (cm). Cementite is, however, not the most stable form of carbon in iron, and at true equilibrium, graphite is formed after prolonged heating. Graphite is a constituent of the common grades of cast iron, but these irons usually contain 2–4% silicon to promote graphitization. From a practical standpoint, cementite is, however, stable in most alloys with less than 2% carbon. Approximate ranges of the carbon contents for irons, steels, and cast iron are shown in Fig. 2.

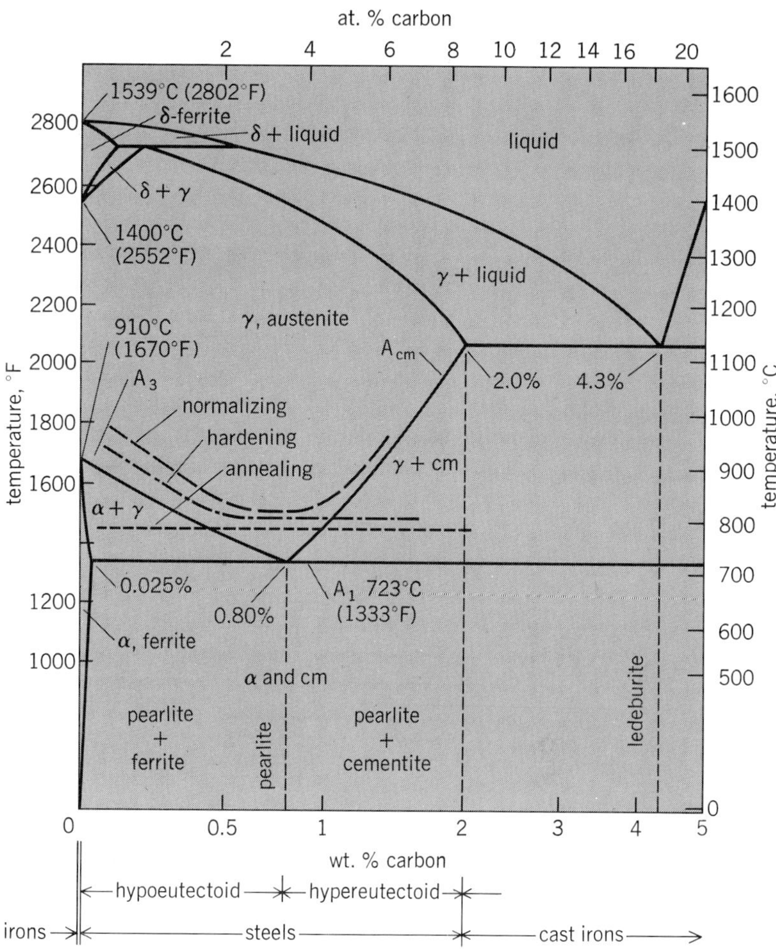

Fig. 2. Iron-carbon phase diagram.

Steels which contain 0.80% carbon are called eutectoid steels, those with less carbon are hypoeutectoid, and those with more are hypereutectoid. The changes which occur on slowly heating a typical hypoeutectoid steel containing 0.40% carbon are the following. At room temperature such a steel contains ferrite and cementite. On heating above the A_1 critical temperature, 723°C (1333°F), the cementite dissolves, austenite is formed, and the structure consists of austenite and ferrite. Above the A_3 temperature, which is 800°C (1472°F) for this alloy, the structure is entirely austenitic. Melting is observed at 1450°C (2640°F), and in practice, heat-treating and hot-forming temperatures are below 1200°C (2200°F). On cooling, the sequence of phase changes is reversed, but the shapes and distribution of the phases depend on the cooling rate. If the cooling is very slow, as in furnace cooling, the carbides will tend to have a spheroidal shape. If the cooling rate is more rapid, as in air cooling, the cementite will form a duplex lamellar structure with the ferrite, called pearlite, and the microstructure will consist of 50% ferrite grains and 50% pearlite colonies as shown in Fig. 3. A eutectoid steel, containing 0.80% carbon, will consist entirely of pearlite on air-cooling from above A_1. Hypereutectoid steels will consist of pearlite colonies and excess cementite. It should be emphasized that pearlite is not a phase, and it is shown on the phase diagram only for convenience.

Annealing heat treatments. Annealing heat treatments are used to soften the steel, to improve the machinability, to relieve internal stresses, to impart dimensional stability, and to refine the grain size. Several typical annealing treatments are discussed below.

Anneal or grain-refining anneal. Hypo- and hypereutectoid steels are heated just above A_1, and cooled moderately slowly to room temperature. Recrystallization of the ferrite grains will occur during the heating, and the material will be softened by the stress relief and the removal of crystallographic imperfections at temperature and during the slow cooling. Anneals of this type are used to improve the machinability of hypereutectoid steels since the blocky pearlite which forms (Fig. 3) aids in the break-up of the chips during machining.

Spheroidizing anneal. In this treatment, steels are heated just above A_1 and cooled very slowly through the critical temperature, using a programmed reduction of the temperature over a period of 10–15 h to 650°C (1200°F), and then cooling more rapidly to room temperature. In other procedures, the temperature is cycled slowly from just above to just below the critical, or the steel may be cooled to just below the critical and held for a long period. The latter are called isothermal anneals. The objective of all of these treatments is to produce spheroidized carbides. Such structures provide superior cold formability in hypoeutectoid steels and good machinability in hypereutectoid steels.

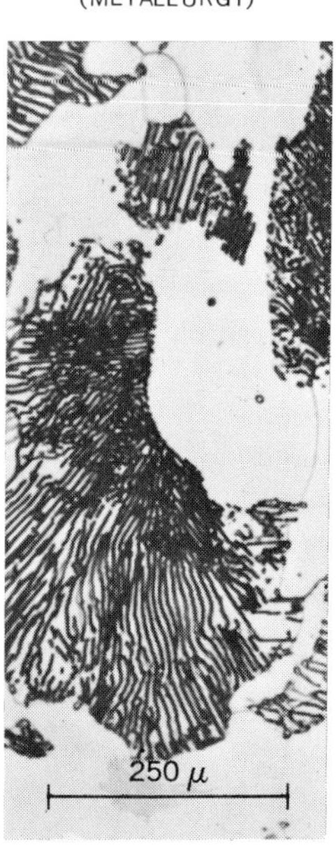

Fig. 3. Microstructure of annealed 0.40% carbon steel, after polishing and etching. White regions are ferrite; lamellar regions, pearlite.

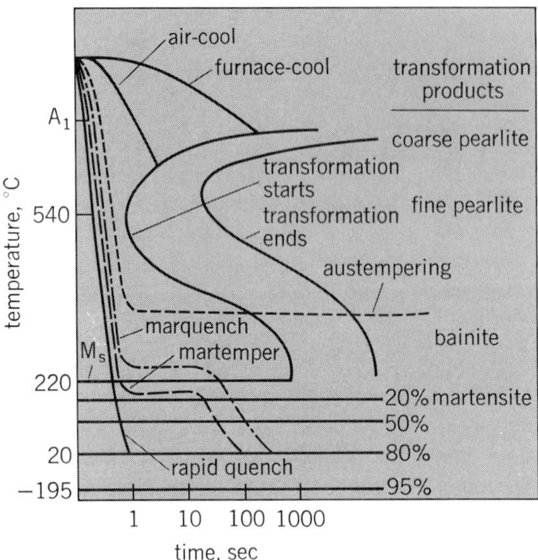

Fig. 4. Schematic isothermal transformation diagram for a eutectoid steel.

Stress-relief anneal. This treatment consists of heating below A_1 and cooling slowly. The objective is to relieve residual stresses, such as those introduced by cold forming and machining. Phase transformations are avoided in order to minimize distortion and this treatment is used frequently to improve the dimensional stability of a complex part.

Normalizing. This treatment involves heating above A_3 or A_{cm} and air-cooling to room temperature. It is used for hypereutectoid steels to remove carbide networks which may have formed in previous processing (such as carburizing), and for hypoeutectoid steels to produce a material of intermediate strength with high ductility and low residual stresses. Typically, such a treatment is specified for large shafts, made of hypoeutectoid steels, after forging. Normalizing is used to improve the machinability of hypoeutectoid steels, because the resultant microstructure contains pearlite, although the pearlite colonies will not be as well developed as those in a steel which has been given a full anneal.

In-process anneal. This is an annealing treatment introduced at the steel mill during the processing of steel bar, wire, sheet, tube, or other product. These may be spheroidizing, softening, normalizing, or stress relief anneals, depending on the ultimate use.

Hardening treatments. Steels are hardened by heating to a temperature at which austenite is formed and then cooling with sufficient rapidity to make the transformation to pearlite and/or ferrite unfavorable. The phase diagram is useful in determining suitable austenitizing, or hardening, temperatures, but the isothermal transformation curve, also called a TTT (time, temperature, transformation) curve is more helpful in assessing the hardenability of a steel. A typical isothermal transformation diagram for a eutectoid steel austenitized at 815°C (1500°F) is shown in Fig. 4. This diagram is obtained experimentally for each combination of steel and austenitizing temperature by austenitizing small pieces, cooling rapidly to a holding bath, typically a lead or salt bath, and quenching into water after various times at the holding temperature. The constituents which form isothermally are determined by metallographic methods and are indicated on the diagram.

The superposition of typical cooling rates used in heat treatments onto the isothermal diagrams will indicate the constituents which will form initially on cooling. A continuous cooling transformation diagram is used to record all of the phases which form at various cooling rates, but such diagrams are difficult to determine and the isothermal curves are frequently used for this purpose. For example, furnace and air-cooling curves intersect the transformation curve above the nose of the C-curve (Fig. 4), and the resultant microstructure will be pearlite. A rapid quench, in water or in some oils, will miss the pearlite nose, and martensite will form.

Martensite has a body-centered tetragonal crystal structure (Fig. 5) with the c and a dimensions dependent on the carbon content. Martensite can be formed only from austenite and contains all of the carbon and alloy elements which are dissolved in the austenite. The hardening of steel occurs by the formation of martensite, and the hardness of the martensite increases with the carbon content, reaching a plateau at 0.80% C.

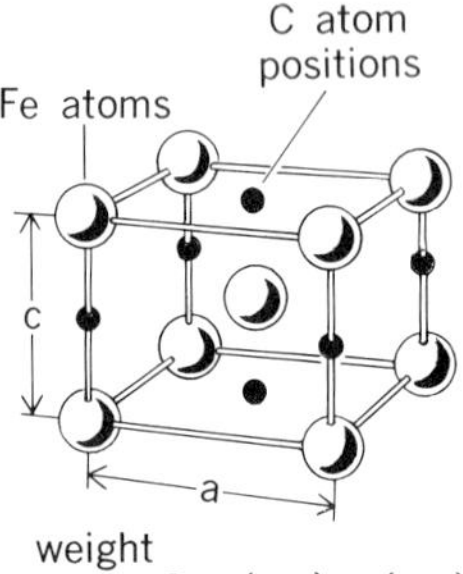

weight percent C	c (nm)	*a* (nm)
0	0.286	0.286
0.20	0.288	0.2858
0.40	0.291	0.2856
0.80	0.295	0.285

Fig. 5. Crystal structure of martensite.

Martensite begins to form at the M_s temperature, which is 270°C (520°F) in the example shown in Fig. 4. The percentage of martensite is shown as a series of horizontal lines on the transformation diagram since very little martensite formation occurs isothermally. At room temperature, such a steel will contain about 80% martensite by volume, with the remainder retained austenite. Some of the retained austenite may be transformed by cooling below room temperature, but the austenite cannot be eliminated solely by refrigeration, even to liquid helium temperatures. The M_s temperature is lowered, and the retained austenite contents increased, by increasing alloy and carbon contents of the austenite.

The position of the nose of the C-curve is dependent on the alloy and carbon contents, and this greatly influences the cooling rates required to harden a steel. In plain-carbon steels the nose is far to the left, and water or rapid oil quenches must be used. For medium-alloy steels (for example AISI 4340, which contains 0.40 C, 1.8 Ni, 0.8 Cr, 0.25 Mo) moderately thick sections may be hardened by oil quenching. For high-alloy steels (for example, type A-2 tool steel, which contains 1.0 C, 5.0 Cr, 1.0 Mo, 0.25 V) air cooling may be sufficient. The hardenability of steel is greater the farther to the right the position of the nose of the C-curve. Hardness should be distinguished from hardenability. Hardness refers to the resistance to indentation. Hardenability refers to the severity of cooling which must be used to produce martensite in a steel. If the steel can be hardened by using a slow cooling rate, such as air cooling, it has high hardenability. If a very rapid cooling rate, such as water quenching, must be used, it has low hardenability.

Freshly quenched martensite containing more than 0.10% C cannot be deformed easily, and a subsequent heating, or tempering, is used to increase the ductility. Microstructural changes occur on tempering, and the sequence of structures in low- and medium-alloy steels follows three stages. In the first, ϵ-carbide, which is hexagonal with the approximate composition Fe_2C, precipitates at temperatures up to 220°C (450°F). In the second stage, 210–310°C (410–590°F), the retained austenite decomposes to bainite, and in the third stage, 210–540°C (590–1000°F), ferrite and cementite are formed. Typically, cutting tools and bearings made of low-alloy steels are tempered through the first and second stages, and high-strength structural steels, such as AISI 4340, are tempered into the third stage in order to obtain sufficient ductility. The tempering sequence is somewhat different for high-alloy tool steels, such as type A-2 or the high-speed tool steels. In these steels, high-alloy carbides precipitate on tempering in the range 490–550°C (900–1000°F), lowering the M_S temperature of the retained austenite. The retained austenite transforms to martensite on cooling, and these steels thus undergo a secondary hardening on tempering in this range.

Austempering. Several variations in the quenching procedure are used to achieve special properties. Austempering involves quenching rapidly to a temperature below the pearlite nose, but above the M_S, as shown in Fig. 4. The objective is the formation of bainite, which is similar in form and has about the hardness of martensite tempered at the same temperature. Bainite, however, has better impact toughness and lower residual stresses than the corresponding martensite.

Marquenching and martempering. Marquenching involves quenching to a temperature just above the M_S and then air-cooling to room temperature. In martempering, the procedure is similar, but the interrupting temperature is just below M_S. The steels are then tempered in the normal fashion. These treatments are used to minimize the distortion which occurs in the direct quench to room temperature.

HSLA and dual-phase steels. High-strength low-alloy (HSLA) steels usually have a low carbon content (typically 0.02–0.10%) and sufficient content of alloy elements such as manganese, silicon, molybdenum, titanium, columbium, chromium, and vanadium to provide enough hardenability for the formation of low-carbon martensite in thin sheets. Such steels may harden on air-cooling from the austenitizing temperatures, although similar results may be obtained by water-quenching thin sheets of plain-carbon steel. Dual-phase (or duplex) steels are HSLA steels for which a particularly high degree of formability has been achieved by a special heat treatment. These steels are heat-treated by austenitizing in the $\alpha + \gamma$ phase region (this is often called an intercritical anneal) and cooling rapidly to room temperature. The temperature is chosen to result in a microstructure with approximately 80% ferrite and 20% martensite-plus-retained-austenite, and this provides a combination of good formability and high strength in the finished part.

Surface treatments. Some heat treatments are used to alter the chemistry at the surface of a steel, usually to achieve preferential hardening of a surface layer. *See* SURFACE HARDENING OF STEEL.

Carburizing. Most carburizing is carried out in an atmosphere of partially combusted natural gas which is further enriched in carbon by additions of natural gas, or some other carbonaceous gas. The starting material is a low-carbon steel, and carbon diffuses into the outer layers, thus producing a duplex structure with a hypereutectoid steel in the case and a hypoeutectoid steel in the core. A hardening treatment may then be used to produce a hard case and a soft core.

Nitriding. Steels, usually containing aluminum, molybdenum, titanium, or chromium, are heated at 540°C (1000°F) in an atmosphere containing undissociated ammonia. Nitrogen diffuses into the surface, forming nitrides and causing a substantial increase in hardness of the surface layer. Additional hardening treatments are not required after the nitriding treatment.

Chromizing. Chromium may be added to the surface by diffusion from a chromium-rich material packed around the steel or dissolved in molten lead. This process is quite slow and must be carried out at an elevated temperature. It is used to improve the corrosion resistance of the surface.

NONFERROUS METALS AND ALLOYS

Many nonferrous metals do not exhibit phase transformations, and it is not possible to harden them by means of simple heating and quenching treatments as in steel. Unlike steels, it is impossible to achieve grain refinement by heat treatment alone, but it is possible to reduce the grain size by a combination of cold-working and annealing treatments. Some nonferrous alloys can be hardened, but the mechanism is one by which a fine precipitate is formed, and the reaction is fundamentally different from the martensitic hardening reaction in steel. There are also certain ferrous alloys that can be precipitation hardened. However this hardening technique is used much more widely in nonferrous than in ferrous alloys. In titanium alloys, the β phase can transform in a martensitic reaction on rapid cooling, and the hardening of these alloys is achieved by methods which are similar to those used for steels.

Annealing treatments. This process involves heating the metal or alloy to an elevated temperature, particularly after cold-working operations such as wire drawing, deep drawing, cold forming, cold extrusion, or heavy machining, and cooling slowly to room temperature. The range of annealing temperatures for copper-base alloys is 350–850°C (660–1560°F), for aluminum alloys 290–430°C (560–800°F), and for nickel-base alloys 820–1200°C (1470–2190°F). The metals are softened by a recrystallization mechanism, which involves the nucleation of a new set of strain-free grains, and the subsequent growth of these strain-free grains at the expense of the cold-worked grains. The temperature at which recrystallization starts is not precisely defined and depends primarily on the amount of cold work, that is, deformation, the material experienced. Recovery from cold work can also be achieved without the formation of new grains by prolonged heating below the recrystallization temperature. Recovery treatments are used in instances where stress relieving is required for dimensional stability but full softening is undesirable because of the accompanying reduction in strength.

Precipitation and age hardening. Certain nonferrous alloys can be hardened by a precipitation-hardening treatment, sometimes called age hardening. In order to be amenable to such a treatment, the solubility of a phase or a compound in the alloy must be such as to be completely dissolved at a high temperature and supersaturated at a lower temperature. For example, one of the well-known precipitation-hardening alloys is an aluminum alloy containing 4% copper, 0.5% manganese, and 0.5% magnesium. It is solution-treated by heating at about 510°C (950°F) and quenching into water. At the solution temperature all of the copper is in solid solution, and the solid solution is retained on quenching to room temperature. The alloy is soft after this treatment (or solution anneal) and may be readily machined or deformed. If the alloy is subsequently aged at room temperature, the strength and hardness gradually increase, reaching a maximum in 5 or 6 days. The aging may be accomplished more rapidly by heating for several hours at 175°C (400°F).

During the age-hardening period, copper-rich regions form in the aluminum lattice and a precipitate which is coherent with the aluminum matrix is gradually developed. The coherency refers to a matching of the lattice spacings in the copper-rich cluster with those in the aluminum, and the resultant local strains produce the hardening of the alloy. If the alloy is overaged, by heating above the precipitation-hardening temperature, the equilibrium precipitate, $CuAl_2$, which is not coherent, will be formed and the hardness will be reduced. Overaging is utilized where dimensional stability is favored over achieving the maximum strength.

Some precipitation-hardening alloys achieve maximum strength only after a cold-working operation is introduced following the solution treatment and prior to the precipitation treatment. These treatments are usually carried out at the mill, and the material is purchased in the hardened condition.

A variety of heat treatments and thermomechanical treatments are used for aluminum alloys, and these are designated by letters and digits which follow the alloy designation. A few of the common designations are given below.

F = as fabricated
O = annealed
H = strain-hardened
W = solution heat-treated
T1 = naturally aged only
T2 = annealed (case products only)
T3 = solution heat-treated, cold-worked, and naturally aged
T4 = solution heat-treated and naturally aged
T5 = artificially aged only
T6 = solution heat-treated and artificially aged
T7 = solution heat-treated and stabilized (overaged)
T8 = solution heat-treated, cold-worked, and artificially aged
T9 = solution heat-treated, artificially aged, and cold-worked
T10 = artificially aged and cold-worked

See ALLOY; METAL, MECHANICAL PROPERTIES OF.

[B. L. AVERBACH]

Bibliography: *Aluminum Standards and Data*, Aluminum Association, 1972; *Atlas of Isothermal Transformation Diagrams*, 2d ed., 1951; *The Making, Shaping and Testing of Steel*, 9th ed., U. S. Steel, 1971; *Metals Handbook*, American Society for Metals, vol. 1, 1961, and vol. 2, 1964.

High-temperature materials

A metal or alloy which serves above about 1000°F (540°C). More specifically, the materials which operate at such temperatures consist principally of some stainless steels, superalloys, refractory metals, and certain ceramic materials. The giant class of alloys called steels usually see service below 1000°F. The most demanding applications for high-temperature materials are found in aircraft jet engines, industrial gas turbines, and nuclear reactors. However, many furnaces, ductings, and electronic and lighting devices operate at such high temperatures.

In order to perform successfully and economically at high temperatures, a material must have at least two essential characteristics: it must be strong, since increasing temperature tends to reduce strength, and it must have resistance to its environment, since oxidation and corrosion attack also increase with temperature.

High-temperature materials, always vital, have acquired an even greater importance because of developing crises in providing society with sufficient energy. The machinery which produces electricity or some other form of power from a heat source operates according to the basic Carnot cycle law, where the efficiency of the device depends on the difference between its highest operating temperature and its lowest temperature. Thus, the greater this difference, the more efficient is the device—a result giving great impetus to create materials that operate at very high temperatures. Following is a discussion of the theory of some of the more significant alloys and a description of several developments. *See* CARNOT CYCLE; EFFICIENCY; MECHANICAL ENGINEERING.

Design theory. Metals used at high temperatures in heat engines are alloys composed of several elements. High-temperature metallurgists, using both theoretical knowledge and application of empirical experimental techniques, depend upon three principal methods for developing and maintaining strength in alloy systems. The alloys, of course, are composed of grains of regular crystalline arrays of atoms and show usable ductility when one plane of atoms slips readily over the next (dislocation movement); excessive dislocation movement leads to weak alloys. Thus, in terms of mechanical properties, alloy design attempts to inhibit but not completely block dislocation movement. The most common technique is that of solid solution strengthening. Foreign atoms of a different size than the parent group cause the crystal lattice to strain. This distortion impedes the tendency of the lattice to slip and thus increases strength. *See* ALLOY.

A second significant mechanism is dispersion strengthening. This is the introduction and dispersion into the alloy lattice of extremely hard and fine foreign particles such as carbide particles or oxide particles. Carbide dispersions can usually be created by a solid-state chemical reaction within the alloy to precipitate the particles, while oxide particles are best added mechanically, as described later. These hard particles impede slippage of the metal lattice simply by intercepting and locking the dislocations in place. Still another strengthening technique is coherent-phase precipi-

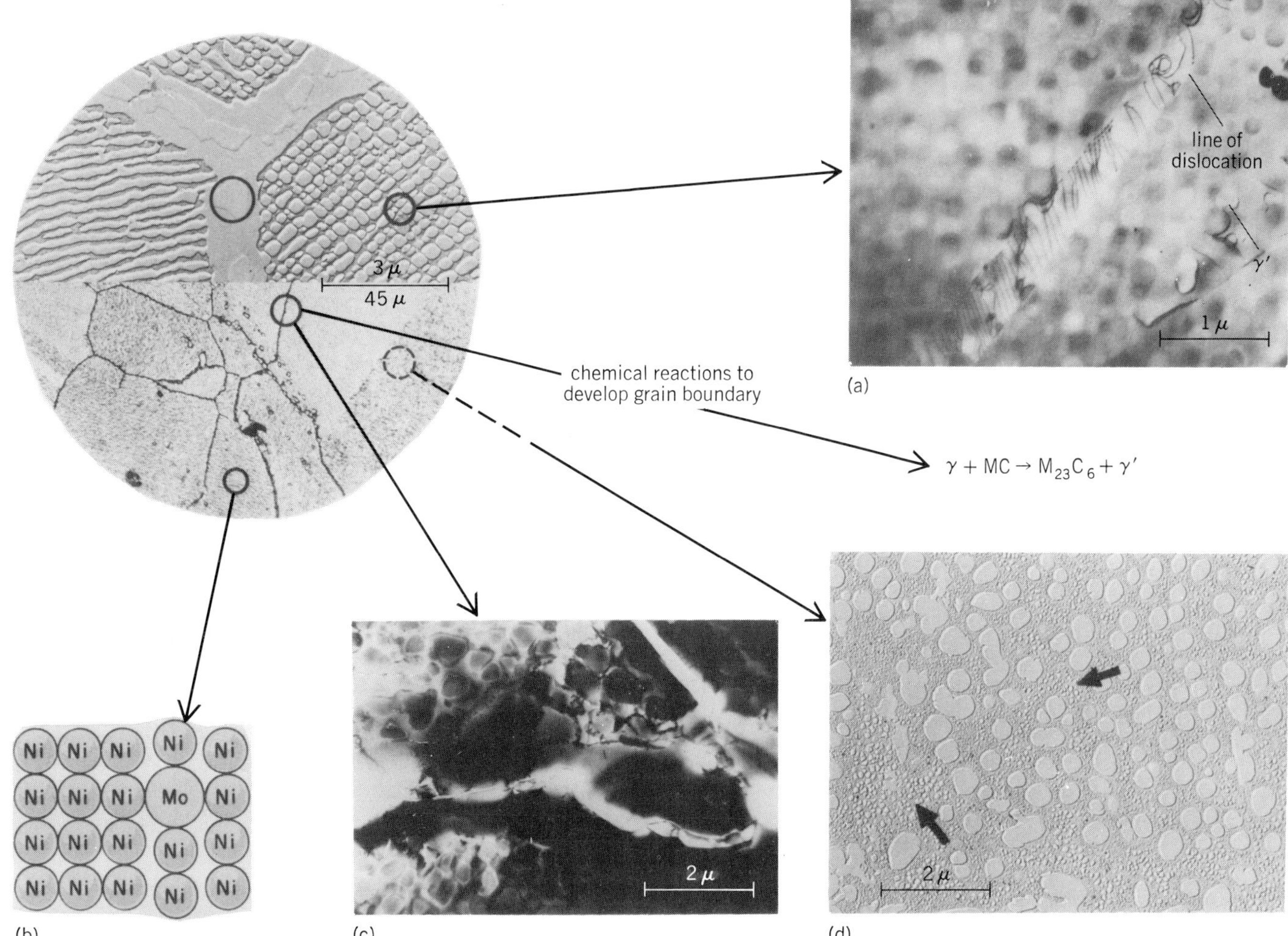

(a) (b) (c) (d)

Fig. 1. Major features for achieving useful strength in nickel-base superalloys. (*a*) Dislocation interaction with γ', a coherent crystalline phase which shows dislocation movement. (*b*) Lattice after solid solution strengthening with molybdenum. (*c*) Carbide strengthening at grain boundaries. (*d*) Fine γ (arrows) for low-temperature strengthening; the big blobs are large γ'. (*From C. T. Sims and W. C. Hagel, eds., The Superalloys, chap. 2, 1973*)

tation; a foreign phase component of similar but modestly differing crystalline structure is introduced into the alloy—always by a chemical solid-state precipitation mechanism. The phase develops significant binding strength with the mother alloy in which it resides (it is "coherent") and, like the other mechanisms, then impedes and controls dislocation flow through the metal lattice. Coherent phase precipitation is a special case within precipitation hardening. *See* HEAT TREATMENT (METALLURGY).

Superalloys. Undoubtedly, the most complex and sophisticated group of high-temperature alloys are the superalloys. A superalloy is defined as an alloy developed for elevated temperature service, usually based on group VIIIa elements, where relatively severe mechanical stressing is encountered and where high surface stability is frequently required. Superalloys, classically, are those utilized in the hottest parts of aircraft jet engines and industrial turbines; in fact, it is the demand of these technically sophisticated applications which created the need for superalloys.

Superalloys are strengthened by all of the methods described above, as well as by some even more subtle ones, such as control of the boundaries between the grains of the crystalline metal by minor additives and other little-understood solid-state chemical reactions. Figure 1 illustrates how these mechanisms combine to make nickel-base superalloys relatively very strong closer to their melting points than any other alloy system. The superalloy René 77 (a nickel-base casting alloy used for turbine blades) is shown in two magnifications as a basis for illustrating some of these strengthening mechanisms. The continuous matrix of the alloy is the face-centered-cubic-phase austenite, referred to as γ. Chemical reactions between the γ and MC (a simple metallic carbide often present in superalloys) create $M_{23}C_6$ (a complex metallic carbide which strengthens the grain boundaries of the alloy) and some additional γ, referred to above. Carbides, precipitated at boundaries between the grains, lock the grain boundaries in place and generate the dislocations. While these alloys were originally processed by melting in air, the large amounts of reactive metals present eventually demanded processing in vacuum to maintain alloy cleanliness; these alloys contain as many as 15 different alloying additions.

Stainless steel. Stainless steels are strengthened principally by carbide precipitations and solid solution strengthening; since they cannot form the γ' coherent precipitate, their use is limited to about 1200°F (650°C), except where strength may not be needed. *See* STAINLESS STEEL.

Refractory alloys. Refractory metal alloys are based on elements which have extremely high melting points—greater than 3000°F (1650°C). These elements—tungsten, tantalum, molybdenum, and columbium—are strengthened principally by a combination of solid solution strengthening, carbide precipitation, and unique metalworking. However, they cannot be used in modern heat engines because no commercially successful method of preventing their extensive reaction with oxidizing environments has ever been found. However, a significant potential use in gas turbines in a completely different environment (helium) is described later.

Oxidation and corrosion. In addition to possessing strength, high-temperature alloys must resist chemical attack from the environment in which they serve. Most commonly this attack is characterized by a simple oxidation of the surface. However, in machines which utilize crude or residual oils or coal or its products for fuel, natural or acquired contaminants can cause severe and complex chemical attack. Involved reactions with sulfur, sodium, potassium, vanadium, and other elements which appear in these fuels can destroy high-temperature metals rapidly—sometimes in a matter of a few hours. This is called hot corrosion, and it is currently the major problem facing otherwise well-suited alloys for service in turbines operating in the combustion products of coal. *See* CORROSION.

Some nuclear reactors operate above 1000°F, and the working fluid is often not an oxidizing gas. For instance, high temperature gas cooled reactors utilize high-temperature, high-pressure helium. However, the helium inevitably contains very low levels of impurities—oxygen, carbon, hydrogen, and others. Since the He atmosphere does not contain enough oxygen to form a protective oxide film on the metals which contain it, impurities enter these alloys readily, reducing their strength by precipitating excessive amounts of oxide and carbide phases and by other deleterious effects.

However, the refractory metal molybdenum is characteristically very resistant to this attack. It does not react significantly with low concentrations of oxygen in helium and appears resistant to carbon as well. Brown-Boveri of Mannheim, Germany (a turbine producer), has reported 100,000-hr (11 1/2-year) tests in which Mo-TZM (a carbide-strengthened molybdenum alloy) showed little or no effect from helium at 1800°F (1000°C) presumably containing some impurities. Under other simulated helium reactor atmospheres, meanwhile, most stainless steels and superalloys have suffered severe attack and significant loss of strength. For gas turbines, this is a new problem, and superalloy metallurgists are working to develop more resistant species for this service. Meanwhile, the long-range future of molybdenum for helium service seems assured.

In other reactor applications such as the liquid-metal fast-breeder reactor, construction metals must resist the high-temperature liquid metal used to transfer heat from hot uranium-plutonium fuel. The liquid metal is usually sodium or potassium. All classes of high-temperature alloys—stainless steels, superalloys, and refractory metals—are under evaluation for this service. As in helium gas, small amounts of impurities are the critical item. They can react with the metal at one temperature and transfer it to a component operating at a lower temperature. Stainless steels remain the prime candidate for these breeder reactors. *See* NUCLEAR REACTOR.

The most significant problem, however, remains that of resistance to oxidation and high-temperature corrosion in present heat engines. Superalloys contain small-to-moderate amounts of highly reactive elements such as chromium and aluminum which react easily with oxygen to form a thin tenacious semiplastic oxide surface film. This prevents further reaction of the aggressive environment with the underlying metal. This is the reason why stainless steels are "stainless"; all contain a minimum of 10% chromium. Superalloys follow approximately the same rule, but often also contain about 5% aluminum, which further enhances oxidation resistance.

This natural protective system works well when oxidation is the only or primary type of attack. However, when the contaminants—vandium and others—are present, they react in a myriad of ways in the 1000–2000°F (540–1100°C) temperature range to destroy the protective oxide and eventually the alloy. Coating is a method used to combat this problem.

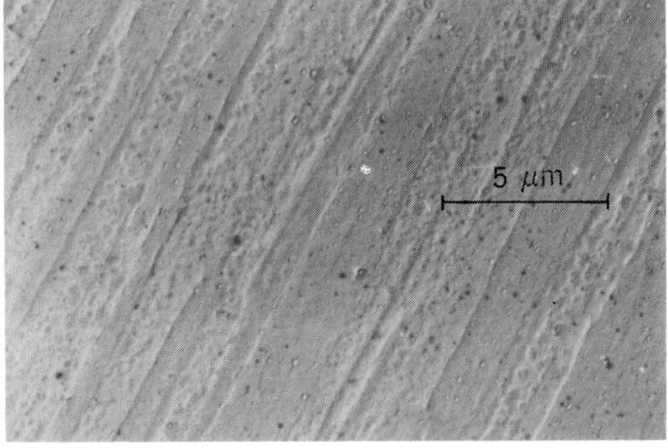

(a)

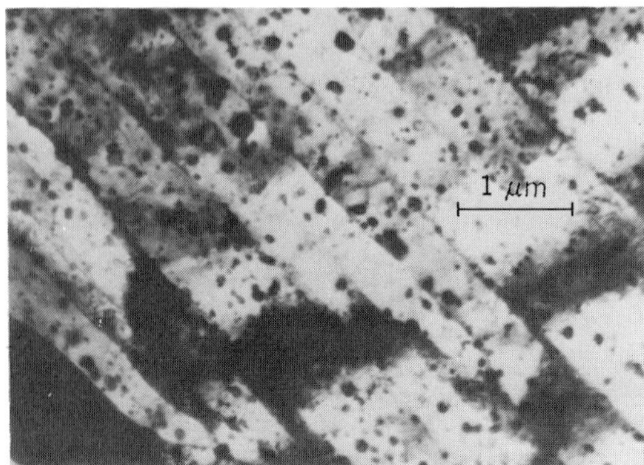

(b)

Fig. 2. Structure of thoria-strengthened nickel showing thoria particles. (*a*) Electron micrograph. (*b*) Transmission electron micrograph. (*From C. T. Sims and W. C. Hagel, eds., The Superalloys, chap. 3, 1973*)

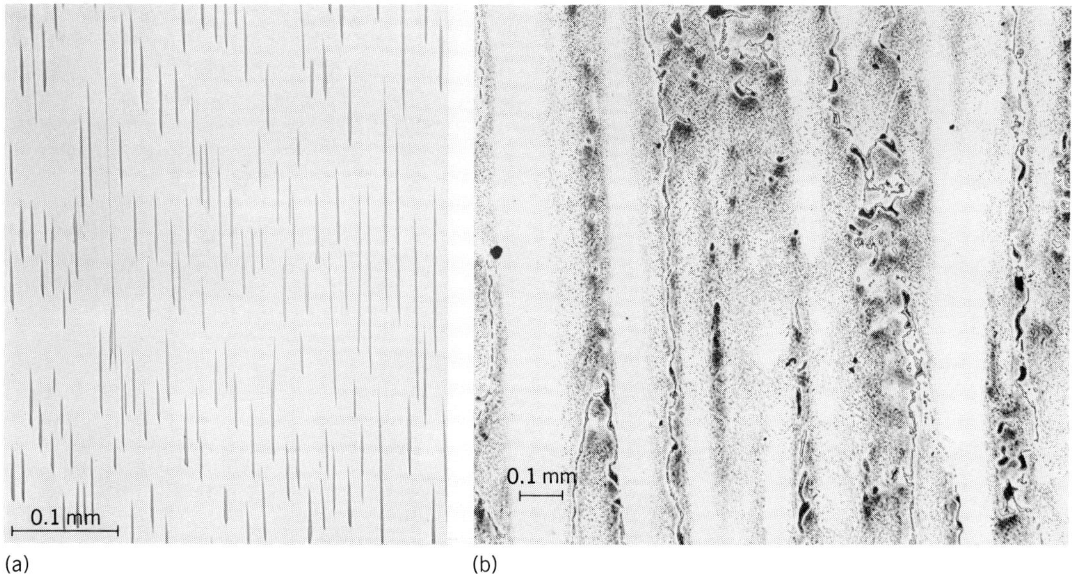

Fig. 3. Structures developed to strengthen superalloys. (*a*) TaC needles in a nickel-chromium matrix. (*b*) Directionally solidified cobalt-base alloy FSX-414 following testing at 1000°C (1800°F).

New superalloy processes. Metallurgists have been active in the superalloy field to increase capability. One technique to generate improved strength is called oxide dispersion strengthening (ODS). The objective is to distribute very fine, uniformly dispersed nonreactive oxide particles throughout the alloy. The processing usually involves starting with very fine particles of the metal itself, to which the oxide is added by a chemical or mechanical process step. The alloy is then consolidated by a mechanical pressing operation and forged into a final useful shape. ODS materials are characterized by unusual creep resistance at very high temperatures; however, intermediate-temperature creep and all tensile properties are mediocre. An example of ODS is shown in Fig. 2.

Success has been obtained in combining some of the classic strengthening factors described previously with ODS. This gives a balance of property enhancement which can be particularly useful to turbine metallurgists—good high-temperature creep strength from ODS and good intermediate-temperature strength from the other mechanisms. In this process, prealloyed superalloy power (carbide- and γ'-strengthened) is hammer-milled in the presence of a very reactive metal, such as yttrium, with oxygen present. The result is a superalloy containing finely dispersed Y_2O_3 particles.

A technique adaptable to investment casting is that of directional solidification, which is particularly suitable to the complex shapes of airfoil parts used in gas turbines. It has been found that by commencing the freezing of molten superalloys (held in a ceramic mold of the shape desired) at the bottom, then allowing the freezing process slowly to proceed up through the shape to be cast, the grains of the structure acquire a long slender shape in the direction of freezing. This significantly increases the ability of the structure to withstand mechanical load in the freezing direction, and particularly increases resistance of the superalloy to a complex phenomenon involving stress and temperature cycling or temperature gradients, called thermal fatigue. Some aircraft engine parts are now made by directional solidification.

A significant further development born from direction solidification is eutectic solidification. Two metallic elements, of course, are related (as one is added to the other) by a series of solid solutions and by new phases of differing crystal structure that can form, separated by two-phase regions. (Actually, three elements can interact to form ternary phases, but four are forbidden.) Metallurgists have discovered that by balancing two or three elements they can create long needlelike strengthening particles formed directly in the castings as they are frozen by the directional solidification process. Structures with extremely high strength can be made. However, the structures produced do not have the best balance with other needs such as oxidation and corrosion resistance, and eutectics are now limited to small parts due to the stringent heat transfer requirements during freezing. A directionally solidified structure and a eutectic structure are shown in Fig. 3. *See* EUTECTICS.

Coatings. Superalloys and stainless steels must possess a balance of properties for high-temperature service. As one might expect from nature, the most oxidation- and corrosion-resistant alloys do not have acceptable strength for most structural applications. Therefore, a viable solution is to utilize strong alloys for airfoil structures, but then coat them to create environmental stability. This is done by adding elements such as chromium and aluminum into the surface of the alloy; these elements react with the aggressive environment to form very protective oxides. Because of the great mechanical stress which occurs in service, the coating must be bonded extremely well to the structural alloy. Accordingly, in one method the coating is applied by chemically reacting the turbine part with an atmosphere containing chromium or aluminum halides at very high temperature so that the active elements diffuse into the surface of the alloy to form the protective layer *in* the alloy. Thus, ultimately, the coating layer is composed

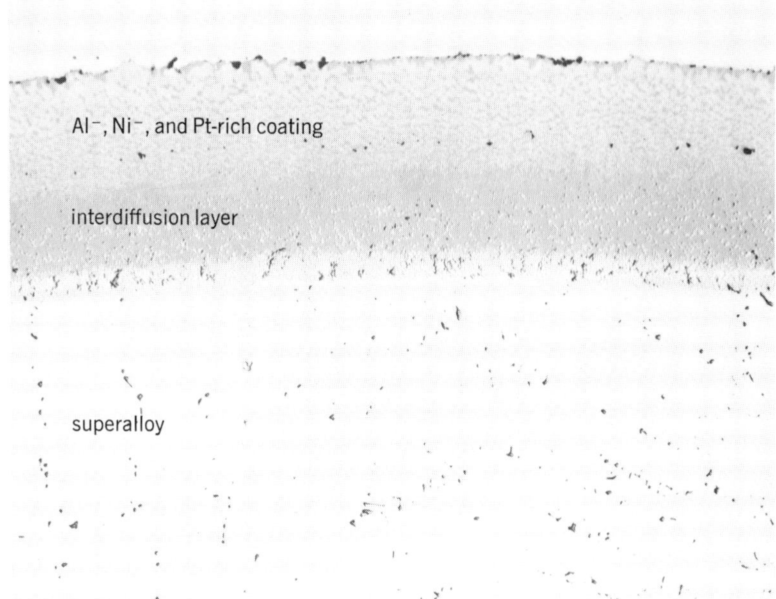

Fig. 4. Aluminide-platinide coating on a nickel-base superalloy. (*Courtesy of E. Buchanan, General Electric Co., Schenectady*)

mainly of nickel from the alloy, and aluminum or chromium or both added in the coating process. Another method is by electron bombardment with needed elements; the metal vapors are directed at the finished parts, forming the coating directly, followed by a heat treatment to cause the necessary interdiffusion. Such aluminum- and chromium-rich coatings can triple the life of industrial gas turbine parts at temperatures such as 1600°F (870°C) in oxidizing atmospheres. *See* METAL COATINGS.

Eventually, however, the coatings fail by oxidizing away or by further interdiffusion with the superalloy underneath. If a very thin layer of platinum is used, the concentration of aluminum in the surface appears to be enhanced, creating an even greater measure of protection. A metallographic picture of a superalloy with a coating made by the electron-beam process, with aluminum increasing in concentration toward the surface, is shown in Fig. 4.

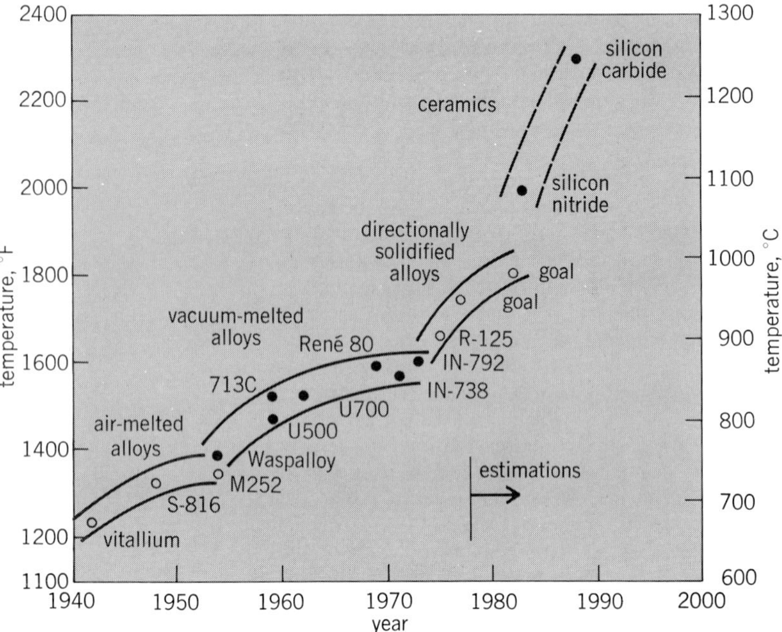

Fig. 5. Temperature capability of superalloys and ceramics shown as a function of the year of first use. Temperature capability shown is the temperature at which the alloy can sustain 25,000 psi (172,500 kPa) for 10,000 hr before rupture.

Ceramics. No discussion of high-temperature materials is complete without mention of the considerable activity now occurring to attempt to adapt ceramics to high-temperature applications, under the extreme and dynamic mechanical loading present in gas turbines. These are ceramics of the covalent-bonded type, such as silicon carbide and silicon nitride. Oxide ceramics, such as Al_2O_3 and ZrO_2, possess ionically bonded structures, and so tend not to possess usable high-temperature creep resistance. Turbine designers and materials engineers are struggling with the problems of utilizing the covalent ceramics, since they possess great strength. However, their complete lack of ductility means that new design techniques are required to prevent early and catastrophic failure. It has also become apparent that these ceramics possess extremely great oxidation and corrosion resistance—features that are expected to be very useful in turbine equipment which must handle high-temperature products of combustion from coal. Data have shown that SiC and Si_3N_4 are attacked to only 2 or 3 mils (0.05–0.08 mm) in depth after 6000–8000 hr exposure in corrosive atmospheres; most high-temperature metals or alloys cannot meet this performance level.

Summary. The capability of high-temperature materials to survive extreme mechanical loading and aggressive environments continues to increase. This is particularly well demonstrated by the continued developments in nickel- and cobalt-base superalloys. New processes, together with advances in classic alloy metallurgy, are generating this progress. The constant improvement over the past decades is shown in Fig. 5, with the potential for ceramics also indicated. *See* METAL, MECHANICAL PROPERTIES OF.

[CHESTER T. SIMS]

Bibliography: Americal Society for Metals, *Metals Handbook*, vol. 1, 1961; W. Betteridge and J. Heslop (eds.), *The Nimonic Alloys*, 2d ed., 1974; J. J. Burke et al., *Ceramics for High Performance Applications*, 1974; C. T. Sims and W. C. Hagel (eds.), *The Superalloys*, 1973.

Highway engineering

Highway planning, location, design, and maintenance. Before the design and construction of a new highway or highway improvement can be undertaken, there must be general planning and consideration of financing. As part of general planning it is decided what the traffic needs of the area will be for a considerable period, generally 20 years, and what construction will meet those needs. To assess traffic needs, the highway engineer collects and analyzes information about the physical features of existing facilities, the volume, distribution, and character of present traffic, and the changes to be expected in these factors. The highway engineer must determine the most suitable location, layout, and capacity of the new routes and structures. Frequently, a preliminary line, or location, and several alternate routes are studied. The detailed design is normally begun only when the preferred location has been chosen.

In selecting the best route careful consideration is given to the traffic requirements, terrain to be traversed, value of land needed for the right-of-way, and estimated cost of construction for the various plans. The photogrammetric method, which makes use of aerial photographs, is used extensively to indicate the character of the terrain on large projects, where it is most economical. On small projects ground-mapping methods are preferred. *See* SURVEYING; TRAFFIC ENGINEERING.

Financing considerations determine whether the project can be carried out at one time or whether construction must be in stages, with each stage initiated as funds become available. In deciding the best method of financing the work, the engineer makes an analysis of whom it will benefit. Improved highways and streets benefit, in varying degrees, three groups: users, owners of adjacent property, and the general public.

Users of improved highways benefit from decreased cost of transportation, greater travel comfort, increased safety, and saving of time (Fig. 1). They also obtain recreational and educational benefits. Owners of abutting or adjacent property may benefit from better access, increased property value, more effective police and fire protection, improved street parking, greater pedestrian traffic safety, and the use of the street right-of-way for the location of public utilities, such as water lines and sewers.

Evaluation of various benefits from highway construction is often difficult but is a most important phase of highway engineering. Some benefits can be measured with accuracy, but the evaluation of others is more speculative. As a result numerous methods are used to finance construction, and much engineering work may be involved in selecting the best procedure.

Environmental evaluation. The environmental impact of constructing highways has received increased attention and importance. Many projects have been delayed and numerous others canceled because of environmental problems. The environmental study or report covers many factors, including noise generation, air pollution, disturbance of areas traversed, destruction of existing housing, and possible alternate routes.

Right-of-way acquisition. Highway engineers must also assist in the acquisition of right-of-way needed for new highway facilities. Acquisition of the land required for construction of expressways leading into the central business areas of cities has proved extremely difficult; the public is demanding that traffic engineers work closely with city planners, architects, sociologists, and all groups interested in beautification and improvement of cities to assure that expressways extending through metropolitan areas be built only after coordinated evaluation of all major questions, including the following:

(1) Is sufficient attention being paid to beautification of the expressway itself? (2) Would a change in location preserve major natural beauties of the city? (3) Could a depressed design be logically substituted for those sections where an elevated expressway is proposed? (4) Can the general design be improved to reduce the noise created by large volumes of traffic? (5) Are some sections of the city being isolated by the proposed location?

Fig. 1. Construction of expressways in metropolitan areas requires careful planning and is costly. Often many bridges are required for an expressway in the heart of a city, as illustrated by this aerial view of St. Paul, MN. At the lower left is the State Capitol.

Because of the large land areas needed for expressways, in several communities air rights above expressways are being sold to permit apartment buildings, parking garages, and similar structures to be built over the expressway, or such areas are being put to public use (Fig. 2).

At other locations, playgrounds are being built beneath elevated highways, as are parking garages. In Honolulu, Hawaii, construction of a new post office to be located under an elevated section of a freeway is being studied.

Detailed design. Detailed design of a highway project includes preparation of drawings or blueprints to be used for construction. These plans show, for example, the location, the dimensions of such elements as roadway width, the final profile for the road, the location and type of drainage facilities, and the quantities of work involved, including earthwork and surfacing.

Soil studies. In planning the grading operations the design engineer considers the type of material to be encountered in excavating or in cutting away the high points along the project and how the material removed can best be utilized for fill or for constructing embankments across low areas elsewhere on the project. For this the engineer must analyze the gradation and physical properties of the soil, determine how the embankments can best be compacted, and calculate the volume of earthwork to be done. Electronic calculating procedures are now sometimes used for the last step. Electronic equipment has also speeded up many other highway engineering calculations.

Powerful and highly mobile earth-moving machines have been developed to permit rapid and economical operations. For example, now in use in the United States is a self-propelled earthmover weighing 125 tons (113 metric tons) and capable of hauling 100 yd^3 (87 m^3) of earth. This unit is powered by a 600-hp (450-kW) diesel electric motor.

Surfacing. Selection of the type and thickness of

HIGHWAY ENGINEERING

Fig. 2. Footbridge over the vehicular roadway crossing the Mississippi River in Minneapolis, MN. (*Minnesota Department of Highways*)

roadway surfacing to be constructed is an important part of design. The type chosen depends upon the maximum loads to be accommodated, the frequency of these loads, and other factors. For some routes, traffic volume may be so low that no surfacing is economically justified and natural soil serves as the roadway. As traffic increases, a surfacing of sandy clay, crushed slag, taconic tailings, crushed stone, caliche, crushed oyster shells, or a combination of these may be applied. If gravel is used, it usually contains sufficient clay and fine material to help stabilize the surfacing. Gravel surfaces may be further stabilized by application of calcium chloride, which also aids in controlling dust. Another surfacing is composed of portland cement and water mixed into the upper few inches of the subgrade and compacted with rollers. This procedure forms a soil-cement base that can be surfaced with bituminous materials. Roadways to carry large volumes of heavy vehicles must be carefully designed and made of considerable thickness. *See* PAVEMENT.

Drainage structures. Much of highway engineering is devoted to the planning and construction of facilities to drain the highway or street and to carry streams across the highway right-of-way.

Removal of surface water from the road or street is known as surface drainage. It is accomplished by constructing the road so that it has a crown and by sloping the shoulders and adjacent areas so as to control the flow of water either toward existing natural drainage, such as open ditches, or into a storm drainage system of catchbasins and underground pipes. If a storm drainage system is used, as it would be with city streets, the design engineer must give consideration to the total area draining onto the street, the maximum rate of runoff expected, the duration of the design storm, the amount of ponding allowable at each catchbasin, and the proposed spacing of the catchbasins along the street. From this information the desired capacity of the individual catchbasin and the size of the underground piping network are calculated.

In designing facilities to carry streams under the highway the engineer must determine the area to be drained, the maximum probable precipitation over the drainage basin, the highest expected runoff rate, and then, using this information, must calculate the required capacity of the drainage structure. Generally designs are made adequate to accommodate not only the largest flow ever recorded for that location but the greatest discharge that might be expected under the most adverse conditions for a given number of years.

Factors considered in calculating the expected flow through a culvert opening include size, length, and shape of the opening, roughness of the walls, shape of the entrance and downstream end of the conduit, maximum allowable height of water at the entrance, and water level at the outlet.

There is a trend to use designs that permit drainage structures to be assembled from standard sections manufactured at a central yard. Such procedures permit better control of the work, quicker construction, and less field work. For example, with precast concrete pipe sections it is often possible to avoid building small box culverts in the field. Circular culverts of large diameter are now also constructed in this way. However, culverts built of corrugated metal are specified for many projects for reasons of economy and to avoid placing of small volumes of concrete at numerous locations.

Numerous small bridges are being designed to permit precast beams or girders to be placed side by side across the bridge opening to form the support for the roadway. These members are frequently of prestressed concrete. When precast members are used, the need for falsework to construct the bridge deck is eliminated, an especially beneficial move if the bridge is being built over a railroad or busy street. *See* PRECAST CONCRETE; PRESTRESSED CONCRETE.

Planning and construction of tunnels is an important part of highway engineering. Many tunnels are being built to avoid removal of important structures aboveground and to preserve existing facilities. Long tunnels are also being built to extend highways through mountains. *See* TUNNEL.

The 8941-ft-long Eisenhower Memorial tunnel 80 mi west of Denver, Colorado, carries Interstate Highway 70 under the Continental Divide. The Eisenhower Memorial is the world's highest highway tunnel.

The 38,280-ft-long Mont Blanc tunnel in the Alps is the world's longest highway tunnel. In Japan a tunnel 28,015 ft long was completed in 1975 to carry a dual-lane highway under the Japan Alps. This tube is the second longest vehicular tunnel in the world.

Rest facilities. Much engineering and construction work has been done to provide rest stops along major expressway routes, especially the national system of interstate highways. These facilities must be carefully located to permit easy and safe exit and return access to the highway. Many units are being built at scenic locations in forested areas to permit picnic grounds and walkways through the forest. These rest areas are especially beneficial to those drivers traveling long distances with few stops (Fig. 3).

Fig. 3. Rest stop along a major highway in Minnesota.

Noise barriers. The control and reduction of noise along busy routes, especially expressways, has become an important part of highway engineering. In many communities high walls have been built along either side of the expressway. Such walls can be costly to construct, but can prove very beneficial. Barriers can reduce overall noise levels by over 50%.

Construction operations. Although much engineering and planning must be done preliminary to it, the actual construction is normally the costliest part of making highway and street improvements.

Staking out. With the award of a construction contract following the preparation of the detailed plans and specifications, engineers go onto the site and lay out the project. As part of this staking out, limits of earthwork are shown, location of drainage structures indicated, and profiles established.

Compaction. Heavy rollers are used to compact the soil or subgrade below the roadway in order to eliminate later settlement. Pneumatic-tired rollers and sheepsfoot rollers (steel cylinders equipped with numerous short steel teeth or feet) are often employed for this operation. Vibratory rollers have been developed and used on some projects in recent years. One type vibrates up to 3400 times/min, compacting the underlying material to an appreciable depth.

Vertical sand drains are sometimes employed to help stabilize fills or embankments constructed across wet and unstable ground (Fig. 4). Holes are

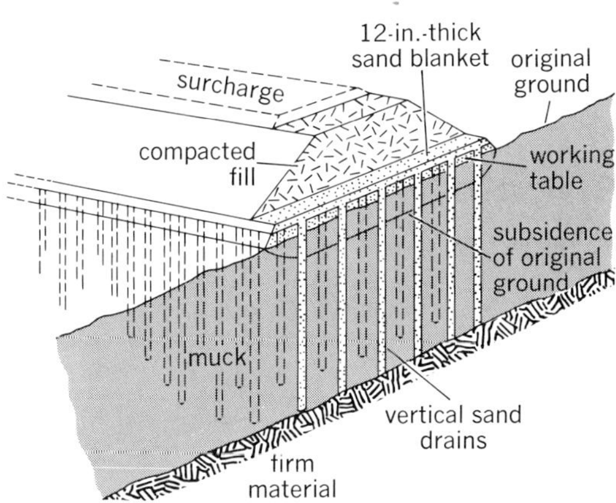

Fig. 4. Cross section of a vertical sand drain installation. Drains are 13–28 in. in diameter and are spaced 9 ft apart. Broken lines represent positions of staggered sand drains. (*From L. I. Hewes and C. H. Oglesby, Highway Engineering, copyright © 1954 by John Wiley and Sons, Inc.; used with permission*)

drilled into the existing material and filled with sand. A horizontal layer of pervious sand or gravel is laid over the network of drains. As the fill is added over the pervious layer and the drains, the undesired water is forced upward through the columns of sand. By allowing the water to move horizontally to the edge of the embankment, the pervious layer prevents collection of water under the roadway.

To obtain quicker compaction of the unstable soil, a surcharge may be placed above the compacted fill. The surcharge material, usually intended for building the compacted fill elsewhere on the project, is removed once the unstable soil has been sufficiently compacted.

Maintenance and operation. Highway maintenance consists of the repair and upkeep of surfacing and shoulders, bridges and drainage facilities, signs, traffic control devices, guard rails, traffic striping on the pavement, retaining walls, and side slopes. Additional operations include ice control and snow removal. Because it is valuable to know why some highway designs give better performance and prove less costly to maintain than others, engineers supervising maintenance can offer valuable guidance to design engineers. Consequently, maintenance and operation are important parts of highway engineering. *See* CIVIL ENGINEERING.

[ARCHIE N. CARTER]

Bibliography: American Association of State Transportation Officials, *Highway Design and Operational Practices Related to Highway Safety*; H. W. Busching and R. Russell, American road building: Fifty years of progress, *J. Constr. Div. ASCE*, 101(CO3):565–581, September 1975; R. W. Cockfield, *A Design Method for the Preparation of a Preliminary Urban Land Use/Transport Plan*, Tech. Rep. Dep. Civil Eng., University of Waterloo, Ontario, May 1970; I. Cook and T. J. Schultz, *Highway Noise and Acoustical Buffer Zones*, 1974; K. A. Godfrey, Jr., Interstate highway system, *Civil Eng.*, p. 51, March 1975; K. A. Godfrey, Jr., Urban freeways: Salvation of cities or their death, *Civil Eng.*, p. 80, May 1975; Institute of Transportation Engineers, *Introduction to Transportation Engineering: Highways and Transit*, 1978; R. S. Mayo, An introduction to fifty years of tunneling, *J. Constr. Div. ASCE*, 101(CO2):259–263, June 1975; J. N. Normann, Improved design of highway culverts, *Civil Eng.*, p. 70, March 1975; D. C. Oliver, Legal liability and highway design and maintenance, *Transp. Eng. J. ASCE*, 101(TE3): 425–435, August 1975; E. L. Priestas and T. E. Mulinazzi, Traffic volume counting recorders, *Trans. Eng. J. ASCE*, 101(TE2):211–223, May 1975; T. N. Tamburri and R. N. Smith, *The Safety Index: A Method for Measuring Safety Benefits*, State of California Department of Transportation, Analytical Studies Branch, November 1973; P. Wang, G. L. Peterson, and J. L. Schofer, Population change: An indicator of freeway impact, *Transp. Eng. J. ASCE*, 101(TE3):491–504, August 1975; P. H. Wright and R. J. Poquette, *Highway Engineering*, 1979.

Hoisting machines

Mechanisms for raising and lowering material with intermittent motion while holding the material freely suspended. Hoisting machines are capable of picking up loads at one location and depositing them at another anywhere within a limited area. In contrast, elevating machines move their loads only in a fixed vertical path, and monorails operate on a fixed horizontal path rather than over a limited area. *See* ELEVATING MACHINES; MONORAIL.

The principal components of hoisting machines are: sheaves and pulleys, for the hoisting mechanism; winches and hoists, for the power units; and derricks and cranes, for the structural elements.

Block and tackle. Sheaves and pulleys or blocks are a means of applying power through a rope, wire, cable, or chain. Sheaves are wheels with a grooved periphery, in appropriate mountings, that change the direction or the point of application of a force transmitted by means of a rope or cable. Pulleys are made up of one or more sheaves mounted in a frame, usually with an attaching swivel hook,

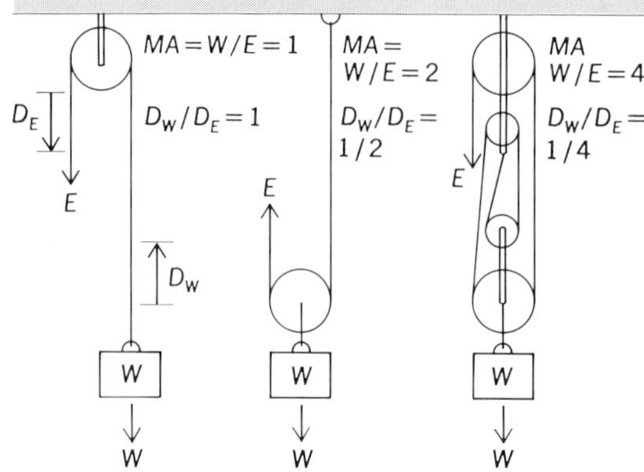

Fig. 1. Mechanical advantage of pulley systems.

eye, or similar device at one or both ends. Pulley systems are a combination of blocks.

The mechanical advantage *MA* of a pulley system is the ratio of weight lifted *W* to input effort exerted *E*. The mechanical advantage, neglecting friction, for any of various arrangements can be determined readily because it equals the number of strands that support the load (Fig. 1). The ratio of distance D_W through which the weight is lifted to distance D_E through which the effort is expended is inversely proportional to the mechanical advantage.

Sometimes used alone, sheaves and pulleys find their most usual application as the hoisting tackle of derricks and cranes.

Winches and hoists. Normally, winches are designed for stationary service, while hoists are mounted so that they can be moved about, for example, on wheel trolleys in connection with overhead crane operations.

A winch is basically a drum or cylinder around which cordage is coiled for hoisting or hauling. The drum may be operated either manually or by power, using a worm gear and worm wheel, or a spur gear arrangement. A ratchet and pawl prevent the load from slipping; large winches are equipped with brakes, usually of the external band type. Industrial applications of winches include use as the power element for derricks and as the elevating mechanism with stackers (Fig. 2).

Floor- and wall-mounted electric hoists are used for many hoisting and hauling jobs from fixed locations in industrial plants and warehouses. Heavy-duty types are standard equipment for powering ship's gear in cargo handling. They are also mounted on over-the-road carriers to facilitate the moving of heavy bulky loads, and serve as the power units of power cranes and shovels. A railroad car puller employs the same principle; however, the drum is mounted vertically and is used for spotting railroad cars in freight yards.

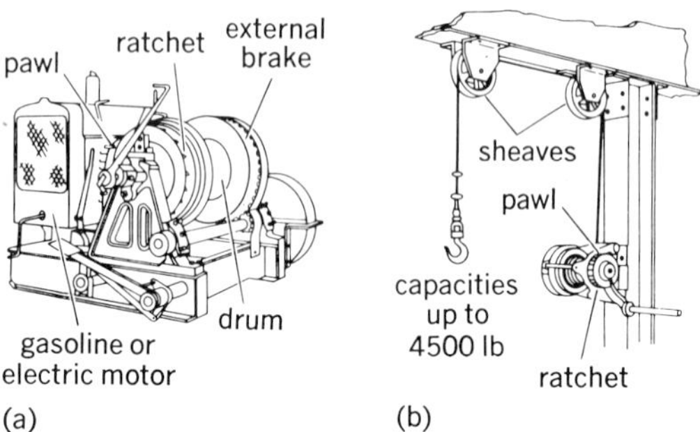

Fig. 2. Powered and hand winches. (*a*) Heavy-duty single-drum winch. (*b*) Wire-rope hand winch.

Hoists are designed to lift from a position directly above their loads and thus require mobile mountings. Hoists are classified by their power source, such as hand, electric, or pneumatic.

Hand hoists are chain operated. There are four types: spur geared, worm geared, differential, and pull-lift or lever (Fig. 3). The last, with its lever for operation and its ratchet for holding, is the simplest and most economical. However, since the operating lever is located on the anchor end of the hoist, it is not as convenient for vertical lifting as it is for horizontal pulling.

The spur-gear type costs the most but is the most economical to operate, with an efficiency as high as 85%. Where hoists are to be used frequently, it is the type most recommended.

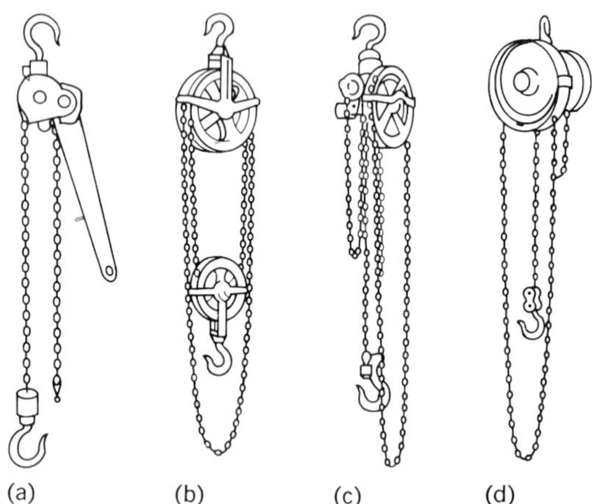

Fig. 3. Chain hand hoists. (*a*) Lever (ratchet). (*b*) Differential. (*c*) Worm gear. (*d*) Spur gear.

Worm-gear hoists transform about 1/3 – 1/2 the input energy to useful work. Offsetting this low efficiency is the locking characteristic of the worm drive: The load cannot turn the mechanism; consequently the load is at all times restrained from running away. In contrast, with a ratchet the load is locked only at positions where the pawl engages a step of the ratchet. Differential hoists use only about one-third the energy input; they too prevent loads from running away during lowering. Spur-gear hoists are more efficient, but require a brake to restrain loads during lowering or holding.

Electric hoists lift their loads by either cable or chain (Fig. 4). The cable type has a drum around which a wire cable is coiled and uncoiled for hoisting and lowering. Chain models have either a roller chain and sprocket or a link chain and pocketed wheel for hoisting and lowering.

There are innumerable below-the-hook attachments, such as slings, hooks, grabs, and highly specialized devices to facilitate practically any handling requirement. Many of these devices are designed to pick up and release their loads auto-

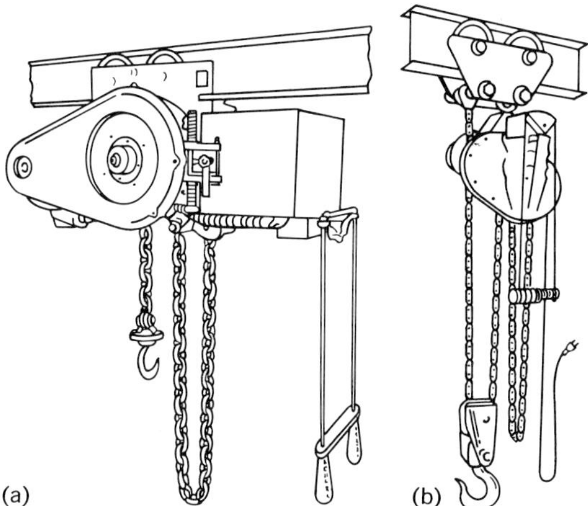

Fig. 4. Hoists. (*a*) Chain. (*b*) Roller chain.

matically. All chain hoists are designed with the lower hooks as the weakest parts; not being interchangeable with the anchor hook, therefore, if the hoist is overloaded, the first indication is the spreading or opening up of the lower hook. If the inside contour of the hook is not a true arc of a circle, this is an indication that the hook has been overloaded.

Pneumatic or air hoists are constructed with cylinders and pistons for reciprocating motion and air motors for rotary motion. Compressed air is the actuating medium in both. Various arrangements admit air to and discharge it from cylinders mounted to operate vertically or horizontally. Pneumatically operated hoists provide smooth action and sensitive response to control; these characteristics account for their wide use in handling fragile materials, such as molds in foundries. In addition, freedom from sparking makes them useful in locations where the presence of explosive mixtures make electrical equipment hazardous.

Derricks and cranes. A derrick is distinguished by a mast in the form of a slanting boom pivoted at its lower end and carrying load-supporting tackle at its outer end. In contrast, jib cranes always have horizontal booms.

Derrick masts are supported by guy lines or stiff legs; some are arranged to rotate 360°. Winches, hand or powered, usually in conjunction with pulleys, do the lifting. Derricks are standard equipment on construction jobs; they are also used on freighters for loading and unloading cargo, and on barges for dredging operations.

Jib cranes, when carried on self-supporting masts, are called pillar cranes; those mounted on walls are called wall-bracket cranes. Cranes with jiblike booms are frequently used in shops, mounted on columns or walls, but have limited coverage. They may have their own running gear or be mounted on trucks (Fig. 5). Mobile types for heavier service are called yard cranes or crane trucks. These may or may not be able to rotate their booms. More powerful machines of this type belong to the power crane and shovel group. *See* BULK-HANDLING MACHINES.

Overhead-traveling and gantry cranes. Hoisting machines with a bridgelike structure spanning the area over which they operate are overhead-traveling or gantry cranes. In the overhead-traveling type, the bridge is carried by, and moves along, overhead trackage which is usually fixed to the building structure itself. The gantry crane is normally supported by fixed structures or arranged for running along tracks on ground level. Gantry cranes are standard equipment in shipside operations. Basic arrangements of overhead-traveling cranes are top running and underhung. In the former, the bridge's end trucks ride on top of the runway rails; in the latter, the end trucks carry the bridge suspended below the rails (Fig. 6). Types for relatively light duty can be made of elements used in the construction of overhead track.

Both overhead-traveling and gantry cranes are called bridge cranes. These cranes span a rather large area and differ among themselves primarily only in the construction of the bridge portion of the crane, and in the method of suspension of the bridge. Where smaller areas are to be spanned, standard beams are used for the bridge structure; however, built-up girders or trusslike bridge structure are used for larger spans. A full gantry crane has both its supporting elements erected on the ground, usually riding on tracks. A half-gantry crane has a supporting structure at one end of the bridge that reaches to the ground, and the other end is carried directly on overhead tracks. Selection of either type depends primarily on building design and the areas in which the crane is to be utilized.

Any one or several of the hoists described earlier may be attached to the bridge, usually suspended from a trolley attached to an I-beam track. The combination of a hoist on a track and the bridge crane itself moving on tracks provides for usable movement of equipment within a rectangular area governed only by the length of the bridge

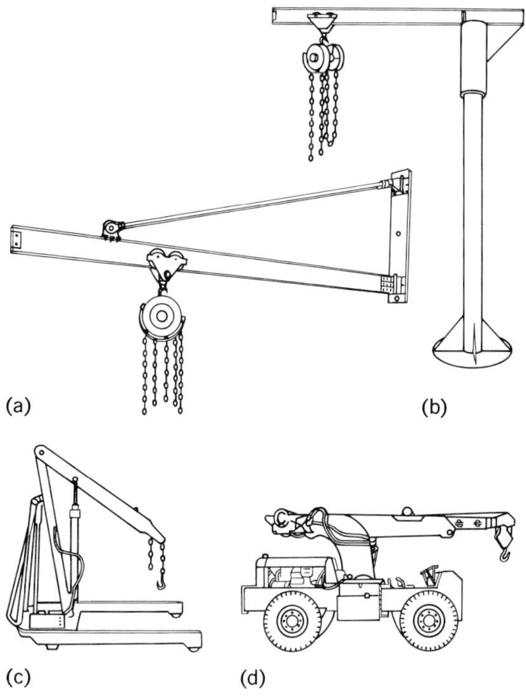

Fig. 5. Four types of jib cranes. (*a*) Wall crane. (*b*) Pillar crane. (*c*) Movable hydraulic crane. (*d*) Crane truck.

and the total horizontal movement of the bridge or gantry crane.

In simpler units the bridges and trolley hoists may be hand-propelled. In heavy-duty units hand operation is not practical and separate electric motors drive each motion. Controls for the motors vary from pendant-type push buttons operated from the ground to remote or automatic control. Pendant controls are satisfactory when a crane has intermittent use during the work day; in larger units, where the crane is in constant use and heavy loads are the rule, an operator may be stationed in a cab mounted to the bridge structure. A more sophisticated extension of this is the operatorless crane, which is worked by means of an electronic control.

A more recent type of lifting mechanism, used on a modified type bridge crane, is a fork-lift attachment suspended from the overhead truck or bridge; it is referred to as a stacker crane. This unit is especially used for such locations as stock rooms, die racks, or finished-goods storage because it has the advantage of being able to handle unit loads in narrow aisles.

[ARTHUR M. PERRIN]

Bibliography: American Society of Mechanical Engineers, *Safety Code For Cranes, Derricks and Hoists*, ANSI B30.2, 1976; J. M. Apple, *Plant Layout and Materials Handling*, 3d ed., 1977; Electric Overhead Crane Institute, *Specifications and General Information for Standard Industrial Service: Electric Overhead Traveling Cranes*; W. E. Rossnagel, *Handbook of Rigging*, 3d ed., 1964; H. Shapiro, *Cranes and Derricks*, 1980.

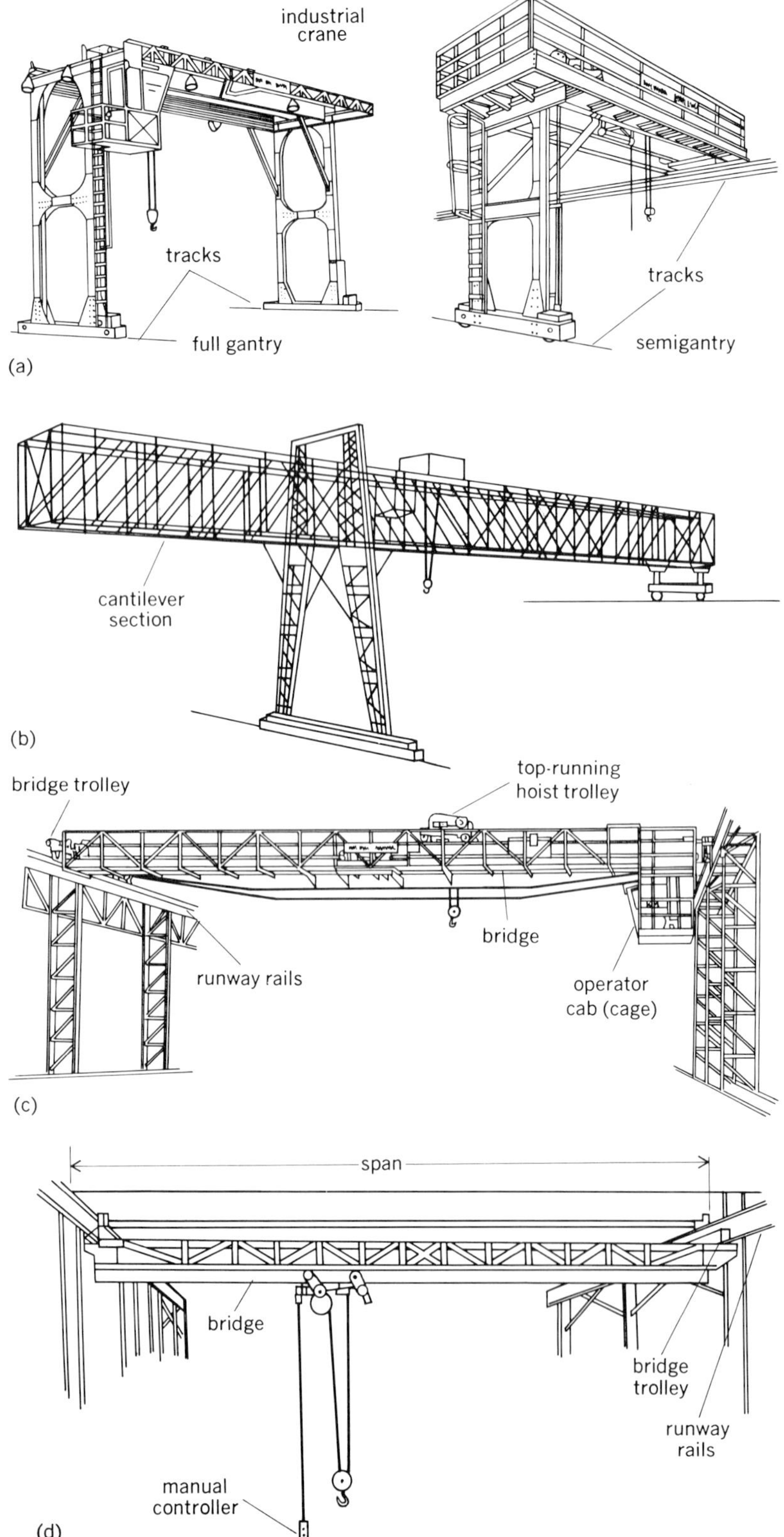

Fig. 6. Basic types of overhead-traveling cranes. (*a*) Traveling gantry cranes. (*b*) Cantilever crane. (*c*) Top-running crane. (*d*) Underhung crane.

Honing

The process of removing a relatively small amount of material from a surface by means of abrasive stones to obtain a desired finish or extremely close dimensional tolerance. Seldom is more than a few thousandths of an inch of stock removed. The abrading action of the fine grit stones occurs on a wide surface area rather than on a line of contact as in grinding. As it applies to machining, honing refers primarily to work done on cylindrical surfaces. In addition to metals and carbides, materials such as plastics, ceramics, and glass may be honed.

A hone consists of a holding device containing several oblong stones arranged in a circular pattern. These may be set at a given diameter or forced against the work by a wedging action. The hone floats in the hole as it rotates and reciprocates. While it will remove high spots and surface inaccuracies, it will not correct the position of a hole or establish its alignment. For external honing, the workpiece usually rotates and reciprocates.

Honing is done with manually operated equipment as well as with vertical and horizontal honing machines. *See* GRINDING; MACHINING OPERATIONS.

[ALAN TUTTLE]

Hooke's law

A generalization applicable to all solid materials, stating that stress is directly proportional to strain and expressed as

$$\frac{\text{Stress}}{\text{Strain}} = \frac{S}{\epsilon} = \text{constant} = E$$

where E is the modulus of elasticity, or Young's modulus, in pounds per square inch. The constant relationship between stress and strain applies only to stress below the proportional limit. For materials having a nonlinear stress-strain diagram, the law is an approximation applicable to low stress values. *See* STRESS AND STRAIN; YOUNG'S MODULUS.

[W. J. KREFELD/W. G. BOWMAN]

Bibliography: R. G. Budynas, *Advanced Strength and Applied Stress Analysis*, 1977; H. E. Davis, G. E. Troxell, and G. T. Wiskocil, *The Testing and Inspection of Engineering Materials*, 3d ed., 1964.

Hot-water heating system

A heating system for a building in which the heat-conveying medium is hot water. Heat transfer in British thermal units (Btu) equals pounds of water circulated times drop in temperature of water. For other liquids, the equation should be modified by the specific heats. The system may also be modified to provide cooling.

A hot-water heating system consists essentially of water-heating or -cooling means and of heat-emitting means such as radiators, convectors, baseboard radiators, or panel coils. A piping system connects the heat source to the various heat-emitting units and includes a method of establishing circulation of the water or other medium and an expansion tank to hold the excess volume of water as it is heated and expands. Radiators and convectors have such different response characteristics that they should not be used in the same system.

Types. In a one-pipe system (Fig. 1), radiation units are bypassed around a one-pipe loop. This type of system should only be used in small installations.

In a two-pipe system (Fig. 2), radiation units are connected to separate flow and return mains, which may run in parallel or preferably on a reverse return loop, with no limit on the size of the system.

In either type of system, circulation may be provided by gravity or pump. In gravity circulation each radiating unit establishes a feeble gravity circulation; hence such a system is slow to start, is unpredictable, and is not suitable for convectors, baseboard radiation, or panel coils because circulating head cannot be established, and circulation cannot be supplied to units below the mains. The pipes must be large in size. For these reasons gravity systems are no longer being used.

In forced circulation a pump is used as the source of motivation. Circulation is positive and units may be above or below the heat source. Smaller pipes are used.

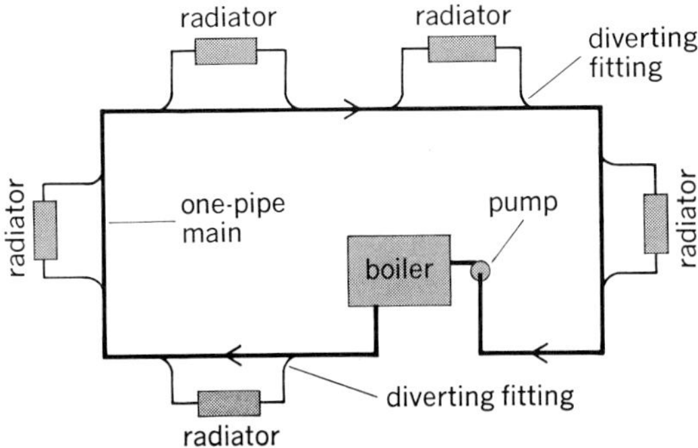

Fig. 1. One-pipe hot-water heating system.

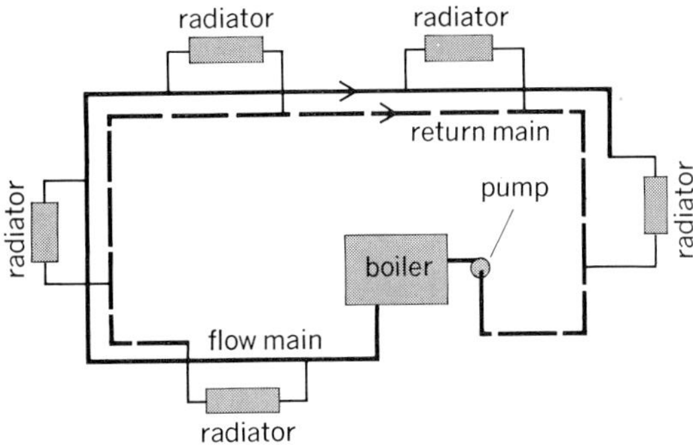

Fig. 2. Two-pipe reverse return system.

Operation. For perfect operation it is imperative that the friction head from the heat source through each unit of radiation and back to the heat source be the same. To achieve this condition usually requires careful balancing after installation and during operation.

Expansion tanks may be open or closed. Open tanks are vented to the atmosphere and are used where the water temperature does not exceed 220°F (at sea level). They provide the safest operation, practically free from explosion hazards. Closed tanks, used for higher water temperatures, must be provided with safety devices to avoid possible explosions.

One outstanding advantage of hot-water systems is the ability to vary the water temperature according to requirements imposed by outdoor weather conditions, with consequent savings in fuel. Radiation units may be above or below water heaters, and piping may run in any direction as long as air is eliminated. The system is practically indestructible. Flue-gas temperatures are low, resulting in fuel savings. The absence of myriads of special steam fittings, which are costly to purchase and to maintain, is also an important advantage. Hot water is admirably adapted to extensive central heating where high temperatures and high pressures are used and also to low-temperature panel-heating and -cooling systems.

Circulating hot-water pumps must be carefully specified and selected. On medium-size installations, it is recommended that two identical pumps be used, each capable of handling the entire load. The pumps operate alternately but never together in parallel. On large installations three or more pumps may be used in parallel, provided they are identical and produced by the same manufacturer. The casings and runners must be cast from the same molds, and the metals and other features that affect their temperature characteristics must be identical. All machined finishes must be identical, and the pumps must be thoroughly shop-tested to operate with identical characteristics.

When the system is in operation, the pump can be disconnected from the boiler by throttling down the valve at the boiler return inlet; it should not be closed completely. This procedure permits the water in all boilers to be at the same temperature so that when a boiler is thrown back into service, the flue gases do not impinge on any cold surfaces, thus producing soot and smoke to further contaminate the outdoor atmosphere. *See* COMFORT HEATING; DISTRICT HEATING; OIL BURNER.

[ERWIN L. WEBER]

Bibliography: American Society of Heating, Refrigerating, and Air Conditioning Engineers, *Guide*, 1962, 1964, 1966, and 1967; American Society of Heating and Ventilating Engineers, *Heating, Ventilating, Air Conditioning Guide*, vol. 37, 1959; F. E. Giesecke, *Hot-water Heating and Radiant Heating and Radiant Cooling*, 1947.

Human-factors engineering

The area of knowledge dealing with the capabilities and limitations of human performance in relation to design of machines, jobs, and other modifications of the physical environment. Human-factors engineering seeks to ensure that humans' tools and environment are best matched to their physical size, strength, and speed and to the capabilities of the senses, memory, cognitive skill, and psychomotor preferences. These objectives are in contrast to forcing humans to conform or adapt to the physical environment.

Human-factors engineering has also been termed human engineering, engineering psychology, applied experimental psychology, ergonomics, and biotechnology. It is related to the field of human-machine systems engineering but is more general, comprehensive, and empirical and not so wedded to formal mathematical models and physical analysis.

Among the problems of human-factors engineering are design of visual displays for ease and speed of interpretation; design of tonal signaling systems and voice communication systems for accuracy of communication; design of knobs and control handles and pedals; design of seats, work places, cockpits, and consoles in terms of humans' physical size, comfort, strength, and visibility. Human-factors engineering addresses itself to problems of physiological stresses arising from such environmental factors as heat and cold, humidity, high and low atmospheric pressure, vibration and acceleration, radiation and toxicity, illumination or lack of it, and acoustic noise. Finally, the field includes psychological stresses of work speed and load and problems of memory, perception, decision-making, and fatigue. *See* HUMAN-MACHINE SYSTEMS.

History. The foundations of human-factors engineering were laid by Fredrick W. Taylor, who demonstrated that, by proper design of work places and procedures, the productivity of workers could be greatly increased. Early systematic studies were made by Frank B. and Lillian M. Gilbreth, whose therblig (an anagram after Gilbreth) system of categorizing hand movements is still a standard motion-and-time analysis technique. With World War II came a great demand for psychologists and physiologists to help engineers design aircraft, ships, tanks, and other weapons to ensure that these devices could be operated under stress by men with relatively little training. All too often it was found that engineers had designed the equipment around themselves and, as a result, large sailors could not fit in the required spaces, small pilots could not reach the required controls, or an appreciable percentage of operators were lacking in sufficient strength or were confused by procedures for operating the complex implements of war. After World War II formal courses in human-factors engineering were introduced in psychology and engineering departments of colleges and professional organizations in the United States and in many European countries.

Early efforts were directed to providing experimental data for the more obvious gaps in human-factors knowledge, such as the variations in the physical dimensions of men and women of different ages, the strength and speed capacities of humans, and the physical conditions under which they could just barely detect or read certain visual images and hear or discriminate certain sounds. As more powerful techniques come to be developed, other sensory capacities of touch, taste, smell, and motion, as well as the more complex

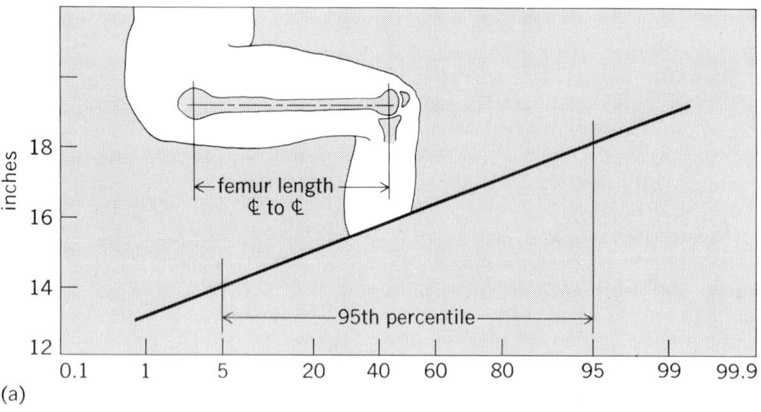

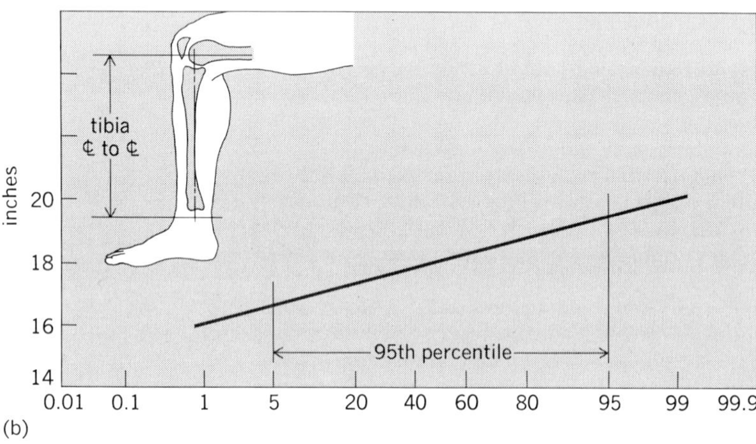

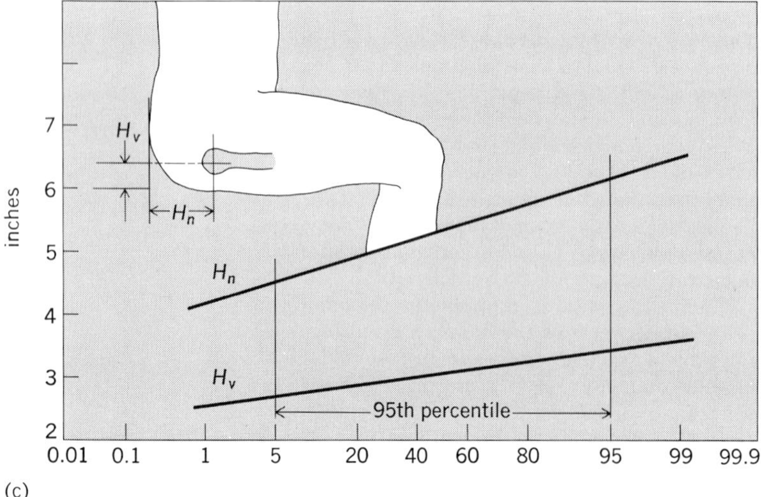

Fig. 1. Variations in the distributions of body dimensions as studied in servicemen. (*a*) Femur length. (*b*) Tibia length. (c) Femur-to-seat distances. (*From L. J. Fogel, Biotechnology, Prentice-Hall, 1963*)

and subtle problems of training, stress, and fatigue, became amenable to laboratory experiment.

Anthropometry. The specialized field dealing with the physical dimensions of the human body is called anthropometry. In dimensioning a seat, console, workplace, or special piece of personal equipment (such as a space suit), it is important that as large a fraction of the intended users as possible be accommodated. But fitting a console to the largest person usually means the smallest person cannot reach the controls unless various features of the equipment are adjustable. Moreover, people are not shaped in simple proportion; the person with the largest head often does not have the longest legs or greatest girth. There is no average human: The person who is average in one dimension is not usually average in another (Fig. 1). To accommodate everyone, many features of a machine need to be adjustable independently. Because this is not always possible, it is important in sizing the equipment to know what percentage of large people and what percentage of small people will not be accommodated, and to weigh the relative costs and advantages of adding adjustments or building the equipment in different sizes.

Displays and controls. Problems of visual displays have been of considerable interest in designing aircraft, spacecraft, ships, submarines, nuclear reactors, electronic equipment, and tools of all kinds. Human-factors engineers have sought to specify the light intensity required for reading different dials and signs in relation to the illumination level of the background. They have recommended, for specific tasks and environments, optimum shape and spacing of numerals and indicator marks on dials and specified what colors give the best contrast and how far away markings of certain size can be read (Fig. 2).

Electronics have made possible a number of new visual display techniques: Self-illuminating electroluminescent numerals and holographic photographs, when illuminated by coherent (laser) beams, appear strikingly three-dimensional. Cathode-ray tubes driven by computer can provide a tremendous variety of images, including letters and numbers, graphs, and television-type pictures. Indeed, in new aircraft and other applications, many of the single-purpose dials and indicators are being replaced by relatively few computer displays, which are completely flexible and can display a great variety of kinds of information. Sometimes these displays can be called up by the operator by keying in certain code letters. Alternatively, a number of different indications, both qualitative and quantitative, can be integrated on one display, such as the integrated aircraft landing display.

With respect to sound signaling systems (auditory displays), human-factors engineers can specify how loud various sounds must be to be heard against background noises of differing loudness. Because consonants, which have relatively higher frequencies and lower loudness, are known to carry much more information than vowels, the fidelity of voice communication systems at these higher frequencies (2000–4000 Hz) is recognized to be more important than the fidelity at lower frequencies. New electronic reading machines and mobility aids for the blind make considerable use of auditory signals to replace visual senses.

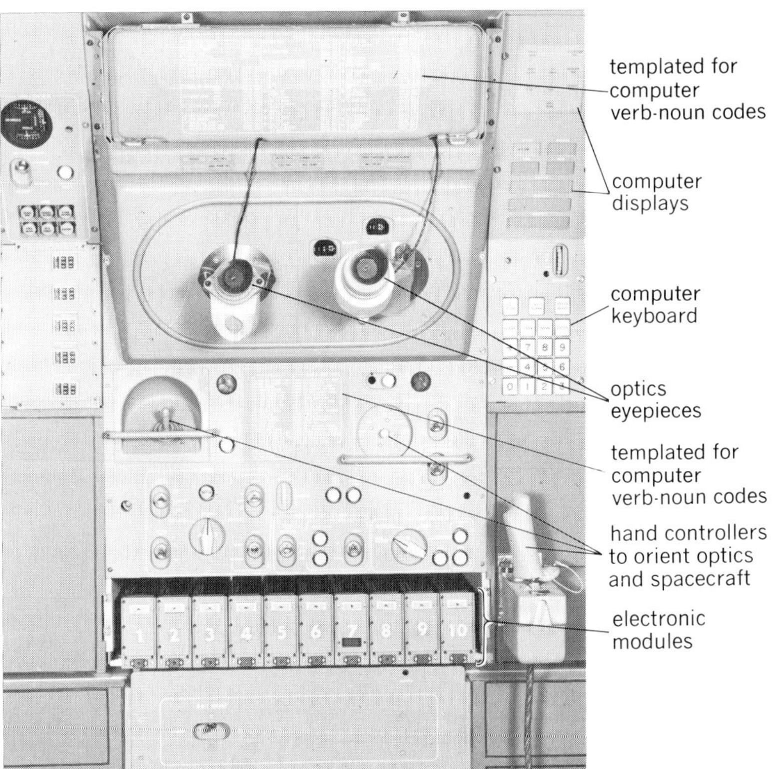

Fig. 2. Design for optimum visual display: astronaut's guidance and navigation control panel in Apollo spacecraft. (*Massachusetts Institute of Technology*)

An important human-factors engineering problem is coordination of displays with the controls by which responses to these displays are made. In practice there are many violations of even simple commonsense design principles such as locating controls adjacent to displays when they are used together, arranging for controls to move in the same direction as the associated display, or having all displays and controls move in the same direction to turn them off. Although the human operator can adapt to a reasonably broad range of forces and sizes of controls, some controls even now are installed that require either more strength or sensitivity than many operators can exercise. Human-factors handbooks are available that summarize various experimental data and give recommendations for display and control design.

Environmental stress. Human-factors engineers are concerned both with the design of the environment to enhance humans' information-processing ability and with the design of the environment to keep the body functioning normally. These concerns require an understanding of human tolerance ranges to heat and cold, to high and low pressures, to varying atmospheric concentrations, and to acceleration and vibration. Engineering solutions can take the form of space suits, diving suits, special clothing for extreme climates, and atmospheric-control units for sealed cabins and tanks. Alternatively, working conditions can be so arranged as to limit exposure to hazard, for example, by determining the least fatiguing distance from the open door of a blast furnace or by recommending the maximum time that can safely be spent at certain altitudes or depths without pressure-breathing equipment.

Bodily comfort is dependent upon a combination of ambient temperature, pressure, humidity, air movement, and amount of clothing—all of which interact to keep the internal body temperature constant at 98.6°F. Proper respiration depends upon the combination of ambient pressure and oxygen concentration that keep the lungs at the equivalent of breathing air (20% oxygen) at sea level. Constant high accelerations, experienced by astronauts in boost and reentry, can drain the blood from eyes and brain and temporarily prevent man's proper functioning; such adverse effects can be prevented by orienting the body properly or using g suits, which keep blood from pooling in the veins of the trunk and legs. Vibration is likely to cause motion sickness if it occurs at certain frequencies at which body parts resonate sympathetically (worst at 2–5 Hz). Levels of thermal, pressure, atmospheric, acceleration, and vibration stresses, which are individually experienced as mild, can have violent effects when they occur in combination.

Safety. Human-factors considerations often determine whether a particular vehicle, tool, or environment is safe. The ability of the human body to withstand various sudden acceleration forces is an important aspect of highway safety. Seat-belt designs for automobiles are based on rocket sled experiments originally conducted to design crash harnesses and parachute harnesses for aircraft pilots.

Human reaction time is also a large factor in accidents of all types. It has been shown that human reaction under the best conditions, such as those that exist when an alert, expectant test subject is required to push a button in response to a light signal, is about 1/4 sec. As the number N of response choices from among which the subject must choose increases, the reaction time increases as the logarithm of N. Reaction time can increase many times as a function of boredom and fatigue, but reliable quantitative predictors for these factors are not available.

Work load. Much recent research has tried to specify the optimum work load for humans. This research is motivated by the gradual disappearance of routine tasks which can be performed at an even pace and somewhat independently of the environment. Humans are used increasingly as a monitor of complex semiautomatic systems where the workload for most of the time is relatively light but where detection of certain low-probability contingencies, or failure of the system, requires rapid and dependable overt action. Examples occur in industrial inspection processes, monitoring of nuclear reactors, piloting of spacecraft, and rapid transit trains.

Early studies on ship's watchkeeping and radar watching indicated that humans could remain vigilant only about 1/2 hr for signals that occur rarely if ever (such as a ship on a collision course with another at night or in fog or the approach of enemy aircraft or missiles). After this interval the chances of rare events being detected diminish markedly. One technique employed to keep watchkeepers and monitors alert is to introduce artificial signals that the observer initially cannot discriminate from the real ones. For example, it was reported that in an experiment one soft-drink-bottle inspector passed fewer dirty or defective bottles when, to keep him alert, cockroaches were added to a small percentage of otherwise clean bottles!

The Manned Spacecraft Center of NASA has conducted simulations of long space flights where not only was the workload negligible but in some cases sound, light, temperature, and gravity sensations were reduced to a minimum. Under such conditions humans become decreasingly alert and even hallucinate to substitute imagined sensation in compensation for the lack of real sensation.

At the other end of the workload scale, human-factors research has determined the upper limits on what humans can do with their sense organs and muscles, both in transmitting information and in performing mechanical work. In terms of the unit of information measure, the bit (the average logarithm of the number of choices or binary decisions made per second), humans can transmit up to 35 bits/sec in such tasks as piano playing and speaking, but for most routine manipulation skills their rate is much lower. As an engine for mechanical work, humans are rather inefficient because, even while operating at peak efficiency (for example, in pedaling a racing bicycle), experiments show that they can produce 1 hp for only about 1 min—and in order to do so their bodies produce several times that amount of energy in wasted heat.

Humans versus machine. A fundamental problem of ever-increasing importance for human-factors engineers is what tasks should be assigned to people and what to machines. It is a fallacy to think that any given whole task can be accomplished best either by a human or a machine without aid of the other, because often some elements of both provide a mixture superior to either alone.

Machines are superior in speed and power; are more reliable for routine tasks, being free of boredom and fatigue; can perform computations at higher rates; and can store and recall specific quantitative facts from memory faster and more dependably. Humans, by contrast, have remarkable sensory capacities which are difficult to duplicate in range, size, and power with artificial instruments (the ratio of the greatest to the least energy which people can either see or hear is about 10,000,000,000,000). Humans' ability to perceive patterns, make relevant associations in the memory, and induce new generalizations from empirical data remains far superior to that of any computer existing or planned. Thus, while people's overt information-processing rate in simple skills is low, their information-processing rate for these pattern recognition and inductive-reasoning capabilities (of which little is understood) appears far greater.

The science of allocating tasks to human and machine, based upon what each can do best, is still in its infancy.

[THOMAS B. SHERIDAN]

Bibliography: A. Chapanis (ed.), *Ethnic Variables in Human Factors Engineering*, 1975; P. M. Fitts and M. I. Posner, *Human Performance*, 1967, reprint 1979; E. J. McCormick, *Human Factors in Engineering and Design*, 1975.

Human-machine systems

The area of knowledge, also known as man-machine systems, dealing quantitatively with the communication and cooperation between people and machines in performing specific tasks. Human-machine systems engineering includes

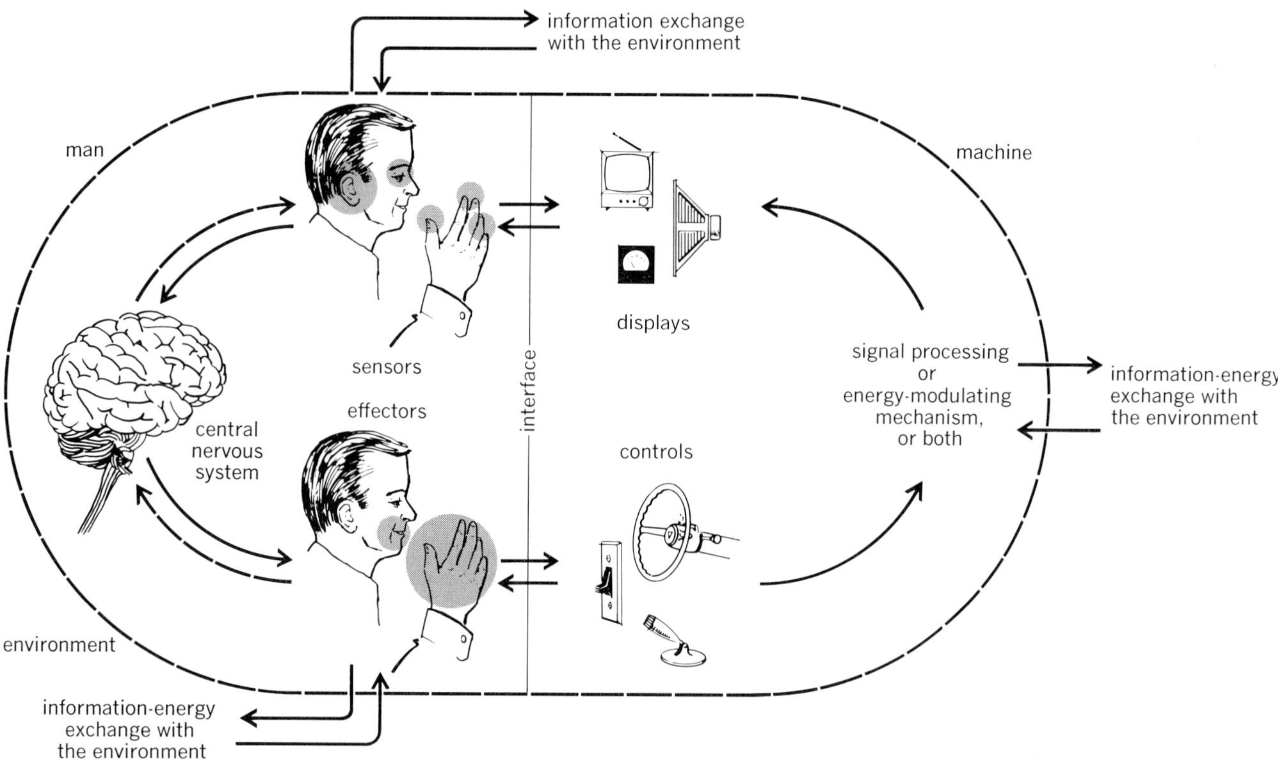

Fig. 1. Schematic diagram of a human-machine system.

the experimentation techniques, the mathematical theory appropriate to handling data derived from experiments, and the engineering methods for applying this knowledge to the design of machines and systems. The pilot and the airplane, the driver and the automobile, the secretary and the typewriter, the talker and the telephone, the programmer and the computer are examples of human-machine systems. *See* HUMAN-FACTORS ENGINEERING.

Though it draws heavily on psychology and physiology, the field uses the language and concepts of engineering to analyze how people receive information through their vision, hearing, touch, and other senses, how they utilize this information to make decisions, and how they implement these decisions with their muscles. Often the response an individual makes in turn affects the new information which is displayed to the senses, thus completing a closed loop of information flow linking the person and machine, as indicated in Fig. 1. The diagram shows only one person in the system, but can easily be expanded to represent two individuals connected, for example, in conversation through a telephone system—or even many individuals simultaneously interacting through separate consoles to control a spaceship or industrial plant.

Human as input-output component. The field of human-machine systems grew out of the engineer's need to specify the behavior of human-operated systems in the same kind of language as is used to specify the operation of electrical and mechanical systems. As with pumps, gears, motors, and electronic circuits, engineers seek to describe the human operator in terms of definite input or stimulus quantities and output or response quantities and to connect these quantities by mathematical equa-

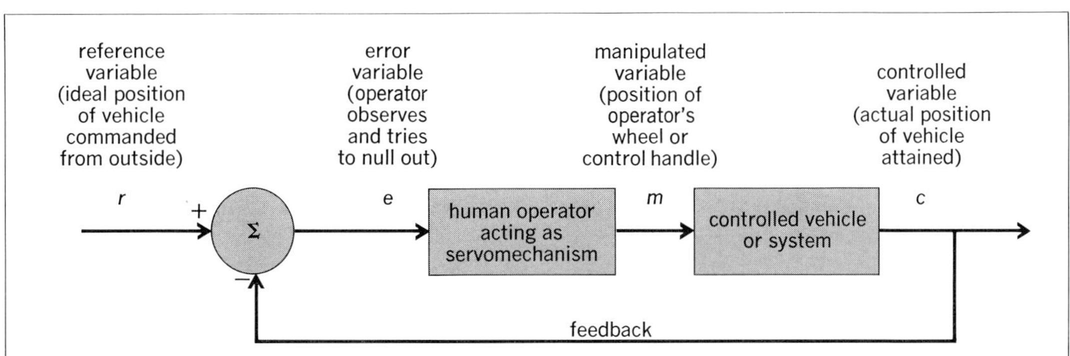

Fig. 2. A closed-loop block diagram of a servomechanism model of a human-controlled system. Although the human (*e* to *m*) relation is largely unknown, the control engineer requires the human-machine combination relation (*e* to *c*) in order to be able to make predictions about the closed-loop *r* to *c* relation.

tions with a view to predicting the behavior o whole human-machine systems. The idea of predicting the behavior of a whole system from a knowledge of how the components are interconnected and what are the input-output equations of these components is at the root of what is commonly called systems theory.

Systems theory does not require understanding the inner workings of the components themselves, that is, why they yield a particular output for a given input; it is only necessary to know what that input-output relation is. This is why systems engineers often refer to their components as "black boxes." In the man-machine systems the engineer likewise need not understand the inner workings of the human nervous system—only what the stimulus-response relationship is in quantitative terms. The mathematical model of man moreover need not correspond to man's anatomy or physiolo-boxes." In the human-machine systems the engineer likewise need not understand the inner workings of the human nervous system—only what the stimulus-response relationship is in quantitative terms. The mathematical model of humans moreover need not correspond to their anatomy or physiology in any way. It is only necessary that the black box representing the individual be connectable to those for the machine. Human-machine systems engineering is closely related to the field called human-factors engineering, which applies psychology and physiology to the design of machines but is not as concerned with compatible input-output equations for both humans and machine. In its origins it goes back to the development of theories of servomechanisms and automatic control in the 1930s, and of the mathematical theory of communication in the 1940s, and to the subsequent application of these theories and techniques to weapons systems during World War II.

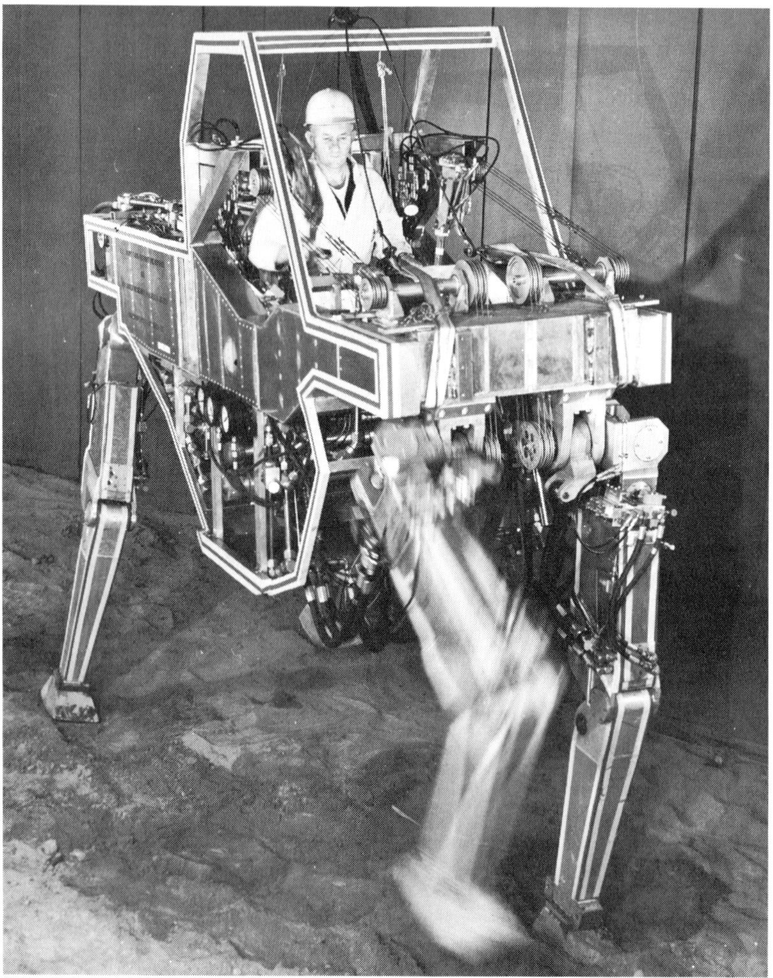

Fig. 3. Research prototype of four-legged walking machine which was developed for use over various types of terrain. Movements of operator's arms control the front legs of the machine, and movements of operator's legs control the rear legs of the machine. (*General Electric Co.*)

The simple manual control system is the type of human-machine system on which most research has been done and which best illustrates how equations for both human operators and machine interact. Figure 2 is a diagram of a vehicle which is continuously steered or otherwise controlled by a human operator to make the vehicle's controlled output (position or direction or other variable) match some given roadway (or ideal path). Researchers have assumed for simplicity that the human operator's steering movements are in some direct relationship to the size of the error e observed between actual and ideal course. Then they have set up simulated aircraft, submarines, spaceships, and automobiles in the laboratory to measure the manual responses of various human subjects. From these measurements, and after the application of suitable statistics, they have derived equations which relate m to e and which best match the given sample of test drivers. These equations are used by engineers to predict pilot or driver response in various kinds of aircraft or automobiles.

In simple tracking situations the individual's manual controller (joystick or steering wheel) position $m(t)$ at time t is given by the differential equation below, where K_1, K_2, K_3, and K_4 are constants

$$K_1\ddot{m}(t) + K_2\dot{m}(t) + m(t) = K_3\,\dot{e}(t-0.15) + K_4\,e(t-0.15) + (\text{noise})$$

which are specified for different task situations; $\dot{m}$ (the dot means first time derivative of m) is the controller velocity and $\ddot{m}$ (the second time derivative) is its acceleration at time t; $\dot{e}(t-0.15)$ is the error velocity measured 0.15 sec (one average human reaction time) before t; and $e(t-0.15)$ is the error position 0.15 sec before t. "Noise" indicates a random signal (not related to the observed input e) which the human adds unavoidably; in simple tasks it is negligible, but for complex tasks it may increase to as much as 50% of the average response motion.

The human operators can modify the nature of their own response equations to adapt to those of the vehicles or processes they are controlling as well as the pattern of the input. An example of such interaction is the walking machine shown in Fig. 3. In the above equation, K_4 is a direct sensitivity coefficient and is normally made as large as possible to prevent excess noise and instability. If the controlled vehicle is very sluggish, the human operator can counteract this effect by increasing K_3. K_1 and K_2 tend to increase the smoothness and evenness of response, and if desirable, these too may be adjusted beyond the smoothness naturally due to the mass and viscosity of the muscle. As a rule of thumb, the human operator adjusts the human response equation coefficients K_1, K_2, K_3, and K_4 to directly compensate for the position of the vehicle, or until the human-plus-machine combination becomes what control engineers call a good servomechanism where $\dot{c}(t) = Ke(t-0.15)$, $c(t)$ being the actual position of the controlled vehicle or system and $\dot{c}(t)$ its velocity. Thus in this instance the person need not be specified separately from the machine, for the individual "becomes" that equation which, together with the machine, yields a common equation for the human-machine combination.

Clearly these equations treat only the more mechanical aspects of human response and do not purport to cope with the broader problems of thinking, motivation, social interaction, and so on, which of course do occur in real human-machine systems. Human-machine systems engineers do not deny the existence of these complex psychological and social variables, but regard such problems as constituting the art rather than the science of the engineer's task.

Figure 4 illustrates a simple laboratory simulator for studying manual control in automobile driving. The driver operates a steering wheel, brake, and accelerator pedal to guide a miniature car around a test course by remote control. A view of the test course as seen by a TV camera mounted on the miniature car is presented to him on a TV screen. With this mockup, emergency situations can be created in the laboratory without danger of actual bodily harm to test drivers: brakes can be made to fail, vision can be obscured, and so on. There are many other types of laboratory driving simulators. In some simulators, models of the roadway and of other vehicles are carried on a moving belt past an optical system through which the driver sees. In others the driver views a moving picture which has been filmed previously from a car driving on a real road. Still others utilize digital computers to generate line drawings of the highway and other vehicles. The armed forces and commercial airlines have used such simulators for many years for pilot training, and soon high schools and other driver-training institutions may use such simulators to teach learner drivers how to cope with skids and other emergencies on the highway.

Fig. 4. Laboratory automobile driving simulator.

Simulators have been used in conjunction with mathematical models not only to understand how the driver or pilot operates the controls but also how the instruments are monitored. Information rates may be assigned to the various instruments in the cockpit on the basis of how rapidly they are likely to change, how accurately they must be read, and how important they are. Then mathematical models are devised to suggest the optimal pattern for visual scanning—where the human operator should move the eyes and for how long to gain the most information from the display. Records of the actual durations and patterns of eye fixations of drivers and pilots obtained by means of special photographic techniques have shown that both in simulators and in actual driving and flying, human operator fixation patterns are fairly close to optimal.

Human-computer interaction. A rapidly growing concern of human-machine systems engineering is the interaction between human programmer and digital computer, especially with so-called on-line interaction, where the individual and computer communicate with each other continually (as distinct from the situation where the human programmer simply feeds in punched cards or tape containing programmed sets of instructions). *See* DIGITAL COMPUTER; ENGINEERING DESIGN.

Through being connected by wire to a large number of electric typewriter consoles scattered throughout a company, a city, or even across a continent, a single large computer may be simultaneously on line to hundreds of operators. The key design feature of most such multiclient computers is time sharing where, because computation speed is so much greater than that of a human, the computer can serve first one master and then another and another in round-robin fashion. But obviously if one human user poses a very difficult task for the machine, less of the computer's capacity is available to the other users who may have to wait several seconds in the course of their typewritten conversations. The strategy by which the user poses problems and the strategy by which the computer allocates its time are important current problems for human-machine systems engineers.

With suitable programming, time-shared computers have proved capable of conversing in what seems to be intelligent English. An example is given in Fig. 5. This program, developed by J. Weizenbaum at Massachusetts Institute of Technology, takes advantage of certain key words and triggers responses already stored in the computer memory. Weizenbaum's experiments are not meant to have any practical use in themselves except to demonstrate the potential of this kind of man-computer interaction. Time-shared computer systems of the type described above are now in wide use in universities, industrial companies, and government agencies and are being developed for use in retail stores and homes.

One very promising use of time-shared computers is computer-aided instruction, an outgrowth of the simple mechanical teaching machine of the 1950s. The human user reads a question from a computer display (cathode-ray tube) and indicates an answer by keying a teletypewriter or pointing to one of several alternative answers presented on the display face by using an electronic pointer that detects the answer the user chooses. Such devices are being developed for teaching spelling to first graders as well as for teaching physics to college students. An especially useful feature of the computer in teaching is its ability to sense patterns of wrong answers and call up a series of questions specifically designed to give more practice or suggest alternative ways of viewing that aspect of the problem giving difficulty. *See* DRAFTING.

Men are all alike.
IN WHAT WAY
They're always bugging us about something or other.
CAN YOU THINK OF A SPECIFIC EXAMPLE
Well, my boyfriend made me come here.
YOUR BOYFRIEND MADE YOU COME HERE
He says I'm depressed much of the time.
I AM SORRY TO HEAR YOU ARE DEPRESSED
It's true. I am unhappy.
DO YOU THINK COMING HERE WILL HELP YOU NOT TO BE UNHAPPY
I need some help, that much seems certain.
WHAT WOULD IT MEAN TO YOU IF YOU GOT SOME HELP
Perhaps I could learn to get along with my mother.
TELL ME MORE ABOUT YOUR FAMILY
My mother takes care of me.
WHO ELSE IN YOUR FAMILY TAKES CARE OF YOU
My father.
YOUR FATHER
WHAT RESEMBLANCE DO YOU SEE
You are not very aggressive but I think you don't want me to notice that.
WHAT MAKES YOU THINK I AM NOT VERY AGGRESSIVE
You don't argue with me.
WHY DO YOU THINK I DON'T ARGUE WITH YOU
You are afraid of me.
DOES IT PLEASE YOU TO BELIEVE I AM AFRAID OF YOU
My father is afraid of everybody.
WHAT ELSE COMES TO MIND WHEN YOU THINK OF YOUR FATHER
Bullies.
DOES THAT HAVE ANYTHING TO DO WITH THE FACT THAT YOUR BOYFRIEND MADE YOU COME HERE

Fig. 5. Typical computer printout of Weizenbaum's ELIZA program for human-computer interaction. Computer's responses are in capital letters. When in conversation with ELIZA, the user types in some statement or set of statements in natural language using normal punctuation and sentence structures. The user's statement is terminated by a double carriage return which serves to turn control over to ELIZA. ELIZA then analyzes the user's statement and generates some response which it types out. Control is then again with the user. (*From Communications of the ACM, vol. 9, no. 1, January 1966*)

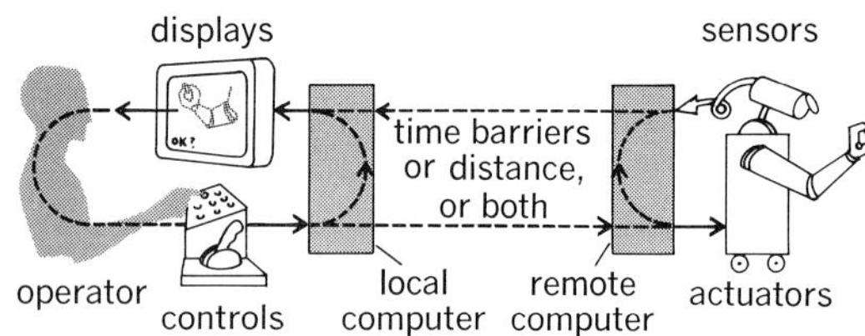

Fig. 6. Supervisory controlled manipulator. (*From E. T. Johnsen and W. R. Corliss, Teleoperators and Human Augmentation, NASA, 1968*)

A second example of human-computer interaction concerns the control of a mechanical hand in space or deep-ocean handling operations. To send radio messages to the Moon takes 1.3 sec and to receive a reply takes another 1.3 sec. The time for sending communications to the planets is many minutes. Such time delays place great restrictions on remote control operations initiated from Earth. Errors cannot be corrected in time, and even if they could, the inch-by-inch steering of the manipulator would require the human to wait after each elemental command before giving the next, a process which would require far too much time for most jobs.

A solution to this problem is to endow the remote mechanical hand with a computer intelligence of its own so that a human operator on Earth need not issue elemental commands one at a time. Instead the operator need only specify subgoals ("Put the nut A on bolt B, move to coordinates x, y and drill a hole z inches deep") and let the computer evolve elemental commands until the subgoal has been achieved or until the computer reports a difficulty with which it cannot cope unaided. This is analogous to the situation of a child performing some elementary manipulative task under parental supervision. Operation of such human-supervised manipulators has been demonstrated in the laboratory. Figure 6 illustrates such a device.

Computer augmentation of expert panel. A final use of computers which shows much promise is accepting input data from experts in the form of subjective judgments about the probability of occurrence of alternative events, and the worth (or utility) of each if it does occur. Consider, for example, a team of geologists prospecting for oil and performing a number of different tests, each of which yields some kind of result. Is there a better method for determining whether there is a good oil reserve at a particular location than reliance on a human judge to simply weigh the evidence and decide? Suppose instead that one were to ask human judges to specify the probabilities (or odds) of each test result having occurred on the assumption that each alternative hypothesis (good oil, poor oil, no oil, and so on) was the true one and then to instruct the computer to pool these likelihood estimates and come up with the best bet. It is true that the suggestion that humans supply component judgment data and that the major decisions be made by computers runs counter to commonly accepted notions of how computers should be used. Yet laboratory experiments conducted by researchers at the University of Michigan indicate that this method is superior, one reason being that human judges do not weigh the evidence properly and are unwilling to commit themselves to one alternative when mathematical optimization theory says they should. The area of research that has opened up here promises to be a very active and fruitful one. [THOMAS B. SHERIDAN]

Bibliography: R. C. Dorf, *Modern Control Systems*, 3d ed., 1980; E. Edwards and F. P. Lees (eds.), *The Human Operator in Process Control*, 1974; H. M. Parsons, *Man-Machine System Experiments*, 1972; A. I. Siegel and J. J. Wolf, *Man-Machine Simulation Models*, 1969, reprint 1979.

Humidistat

A controller that measures and controls relative humidity. A humidistat may be used to control either humidifying or dehumidifying equipment by the regulation of electric or pneumatic switches, valves, or dampers. Most methods for measuring humidity rely upon the swelling and shrinking of materials, such as human hair, silk, horn, goldbeater's skin, and wood, with increases and decreases in relative humidity.

Human hair is most commonly used because of its small diameter, which contributes to rapid absorption and dissipation of moisture. Strands of hair are bunched and several such bunches are combined in a ribbonlike element (Fig. 1).

As the relative humidity of the air decreases, the strands of hair shorten; this movement is transmitted through a suitable lever mechanism to an electric switch or pneumatic valve, which is part of the humidistat.

An electronic humidistat includes a sensing element and a relay amplifier. The sensing element consists of alternate metal conductors on a small, flat plate with a plastic coating (Fig. 2).

An increase or decrease of the relative humidity causes a decrease or increase in the electrical resistance between the two sets of conductors; the change in resistance is measured by the relay amplifier. Small changes in relative humidity can be measured in this way for precise control. See HUMIDITY CONTROL.

[JOHN E. HAINES/RICHARD L. KORAL]

Bibliography: R. W. Maines, *Control Systems for Heating, Ventilating and Air Conditioning*, 2d ed., 1977; A. Wexler, *Humidity and Moisture*, 4 vols., 1964–1965.

Humidity control

Regulation of the degree of saturation (relative humidity) or quantity (absolute humidity) of water vapor in a mixture of air and water vapor. Humidity is commonly mistaken as a quality of air.

When the mixture of air and water vapor is heated at constant pressure, not in the presence of water or ice, the ratio of vapor pressure to saturation pressure decreases; that is, the relative humidity falls, but absolute humidity remains the same. If the warm mixture is brought in contact with water in an insulated system, adiabatic humidification takes place; the warm gases and the bulk of the water are cooled as heat is transferred to that portion of the water which evaporates, until the water vapor reaches its saturation pressure corresponding to the resultant water-air-vapor mixture temperature. Relative humidity is then 100% and absolute humidity has increased. Heating of the mixture and use of the heated mixture to evaporate water is typical of many industrial drying processes, as well as such common domestic applications as hair drying. This same sequence occurs when warm furnace air is passed over wetted, porous surfaces to humidify air for comfort conditioning. *See* AIR CONDITIONING.

To remove moisture from the air-vapor mixture, the mixture is commonly cooled to the required dew point temperature (corresponding to the absolute humidity to be achieved) by passage over refrigerated coils or through an air washer where the mixture is brought in contact with chilled water. The result is a nearly saturated mixture which can be reheated, if required, to achieve the desired relative humidity. *See* DEHUMIDIFIER.

Moisture is also removed without refrigeration by absorption, a process in which the mixture passes through a spray of liquid sorbent that undergoes physical or chemical change as it becomes more dilute. Typical sorbents include lithium and calcium chloride solutions and ethylene glycol.

Another means of dehumidification, by adsorption, uses silica gel or activated bauxite which, through capillary action, reduces the vapor pressure on its surface so that the water vapor in its vicinity, being supersaturated, condenses.

[RICHARD L. KORAL]

Bibliography: N. C. Harris and D. F. Conde, *Modern Air Conditioning Practice*, 2d ed., 1974; R. Havrella, *Heating, Ventilating and Air Conditioning Fundamentals*, 1981; A. Wexler, *Humidity and Moisture*, 4 vols., 1964–1965.

HUMIDISTAT

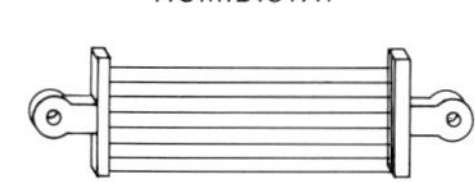

Fig. 1. Human hair-element humidistat. (*Honeywell Inc.*)

HUMIDISTAT

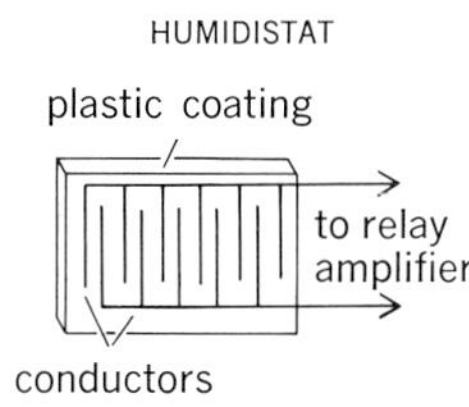

Fig. 2. Electronic humidistat. (*Honeywell Inc.*)

Hydraulic actuator

A cylinder or fluid motor that converts hydraulic power into useful mechanical work. The mechanical motion produced may be linear, rotary, or oscillatory. Operating pressures range from a few pounds per square inch gage (psig) to several thousand, but usually are about 500–5000 psig (3.4–34 MPa). Sizes vary from 0.2-in.^{2} (1.3-cm^2) linear actuators capable of a few pounds of force to extremely large linear or rotary actuators capable of exerting hundreds of tons of force. Operation exhibits high force capability, high power per unit weight and volume, good mechanical stiffness, and high dynamic response. These features lead to wide use in precision control systems and in heavy-

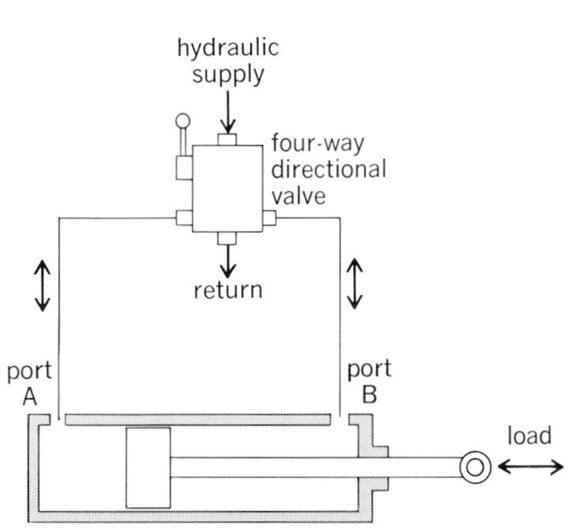

Fig. 1. Function of a hydraulic double-acting cylinder.

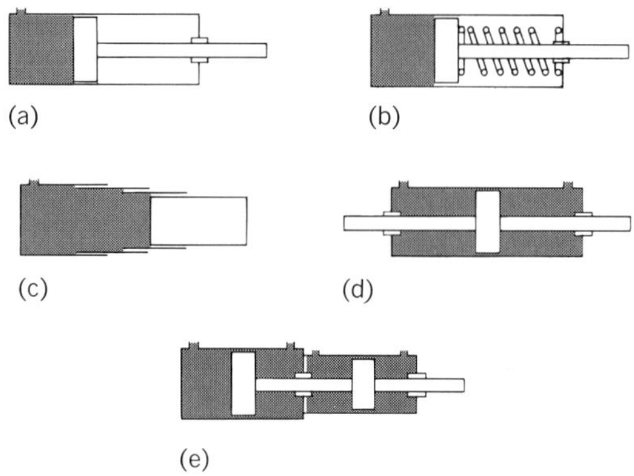

Fig. 2. Types of hydraulic cylinder (linear) actuators. (*a*) Single-acting, external return. (*b*) Single-acting, spring return. (*c*) Telescopic. (*d*) Double-acting double-ended rod. (*e*) Tandem cylinders.

duty machine tool, mobile, marine, and aerospace applications.

Cylinder actuators. To provide a fixed length of straight-line motion, linear cylinder actuators usually consist of a tight-fitting piston moving in a closed cylinder. The piston is attached to a rod that extends from one end of the cylinder to provide the mechanical output. The double-acting cylinder (Fig. 1) has a port at each end of the cylinder to admit or return hydraulic fluid. A four-way directional valve functions to connect one cylinder port to the hydraulic supply and the other to the return, depending on the desired direction of the power stroke.

Some examples of the many possible types of linear hydraulic actuators are shown in Fig. 2. Single-acting types have a power stroke in one direction only, with the return stroke accomplished by some external means or effected by a spring. Telescopic cylinders are used to accomplish extremely long single-acting motions. A double-ended rod is employed when it is necessary to have equal forces and velocities in each direction of travel. Tandem cylinder actuators consist of two or more pistons and rods fastened together so that they operate as a unit. In this actuator a combination of low-force and high-speed piston travel can be obtained, followed by high-force and low-speed piston travel.

Limited-rotation actuators. For lifting, lowering, opening, closing, indexing, and transferring movements, limited-rotation actuators produce limited reciprocating rotary force and motion. Rotary actuators are compact and efficient, and produce high instantaneous torque in either direction.

In the piston-rack type of rotary actuator (Fig. 3), the pinion gear is attached to the load. Hydraulic fluid is applied to either the two end chambers or the central chamber to cause the two pistons to retract or extend simultaneously so that the racks rotate the pinion gear. The single-vane type of actuator (Fig. 4) has a fixed stationary barrier and a rotating vane that forms two variable volume chambers. Hydraulic fluid is ported to one chamber and returned from the other to effect output shaft rotation of about 280° in either direction. A double-vane actuator has two stationary barriers and two rotating vanes, which limit rotation to about 100°. Its advantages are balanced radial loads on the output shaft and a doubling of the torque output over that of a single-vane unit of comparable dimensions.

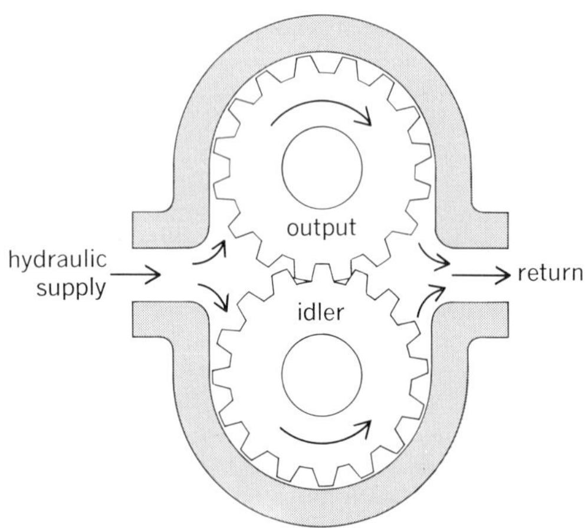

Fig. 5. Gear motor rotary actuator.

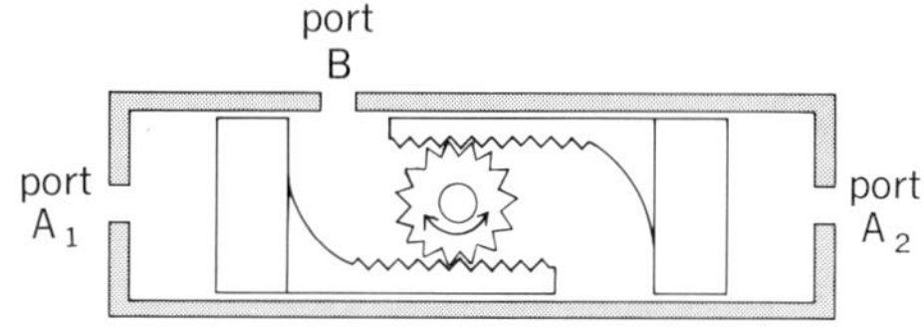

Fig. 3. Piston-rack type rotary actuator.

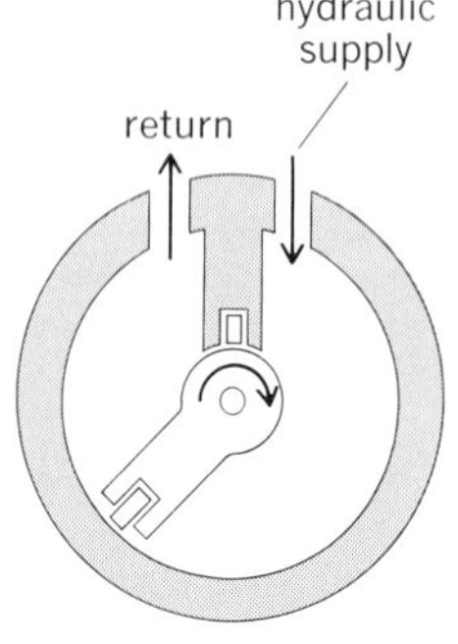

Fig. 4. Single-vane rotary actuator.

Rotary motor actuators. Coupled directly to a rotating load, rotary motor actuators provide excellent control for acceleration, operating speed, deceleration, smooth reversals, and positioning. They allow flexibility in design and eliminate much of the bulk and weight of mechanical and electrical power transmissions.

Motor actuators are generally reversible and are of the gear or vane type. The gear motor actuator (Fig. 5) has one gear connected to the output shaft. Supply fluid enters and flows around the chamber as shown, forcing the gears to rotate. The vane motor actuator (Fig. 6) consists of a rotor with sev-

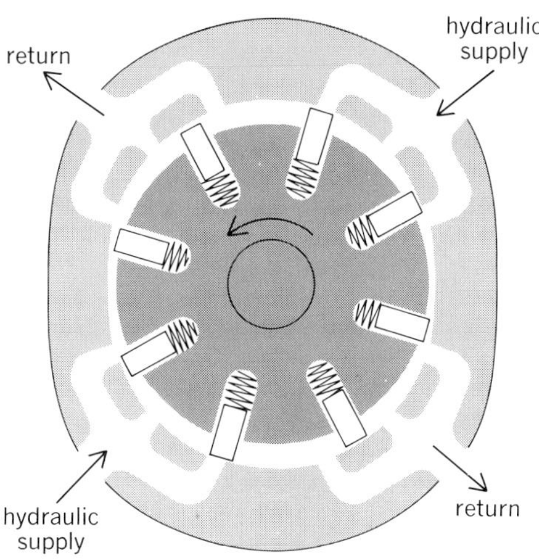

Fig. 6. Vane motor rotary actuator.

eral spring-loaded sliding vanes in an elliptical chamber. Hydraulic fluid enters the chamber and forces the vanes before it as it moves to the outlets. [CHARLES MANGION]

Bibliography: H. E. Merritt, *Hydraulic Control Systems*, 1967; D. A. Pease, *Basic Fluid Power*, 1967; J. Prokes, *Hydraulic Mechanisms in Automation*, 1972; H. L. Stewart and J. M. Storer, *Fluid Power*, 1973; J. Stringer, *Hydraulic Systems Analysis*, 1976; F. D. Yeaple, *Hydraulic and Pneumatic Power and Control*, 1966.

Hydraulic press

A combination of a large and a small cylinder connected by a pipe and filled with a fluid so that the pressure created in the fluid by a small force acting on the piston in the small cylinder will result in a large force on the large piston. The operation

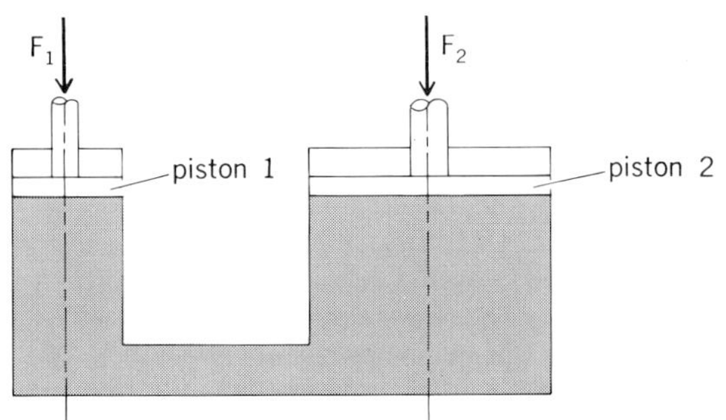

Fig. 1. Principle of hydraulic press.

depends upon Pascal's principle, which states that when a liquid is at rest the addition of a pressure (force per unit area) at one point results in an identical increase in pressure at all points. Therefore, in Fig. 1, the pressure due to the application of force F_1 is shown by Eq. (1), and the equilibrium

$$p=F_1/A_1 \quad (1)$$

force F_2 is shown by Eq. (2), where A_1 and A_2 are

$$F_2=pA_2=F_1\frac{A_2}{A_1} \quad (2)$$

the areas of pistons 1 and 2, respectively. The mechanical advantage is shown by Eq. (3).

$$\mathrm{MA}=\frac{F_2}{F_1}=\frac{A_2}{A_1} \quad (3)$$

The principle of the hydraulic press is used in lift jacks, earth-moving machines, and metal-forming presses (Fig. 2). A comparatively small supply pump creates pressure in the hydraulic fluid. The fluid then acts on a substantially larger piston to produce the action force. In this way forces greater than 15,000 tons are developed over the entire stroke of hydraulic presses. Heavy objects are

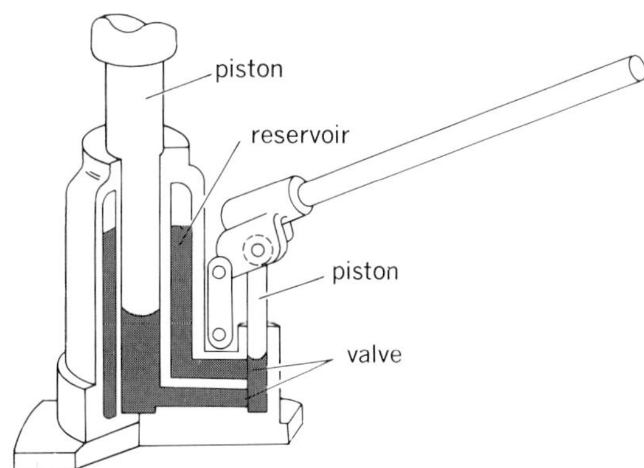

Fig. 2. Hydraulic jack.

accurately weighed on hydraulic scales in which precision ground pistons introduce negligible friction. *See* MECHANICAL ADVANTAGE; SIMPLE MACHINE. [RICHARD M. PHELAN]

Bibliography: J. P. Pippenger and T. G. Hicks, *Industrial Hydraulics*, 2d ed., 1970; W. F. Walker and G. N. Sellers, *A Guide to Industrial Hydraulics*, 1973.

Hydraulic turbine

A machine which converts the energy of an elevated water supply into mechanical energy of a rotating shaft. Most old-style waterwheels utilized the weight effect of the water directly, but all modern hydraulic turbines are a form of fluid dynamic

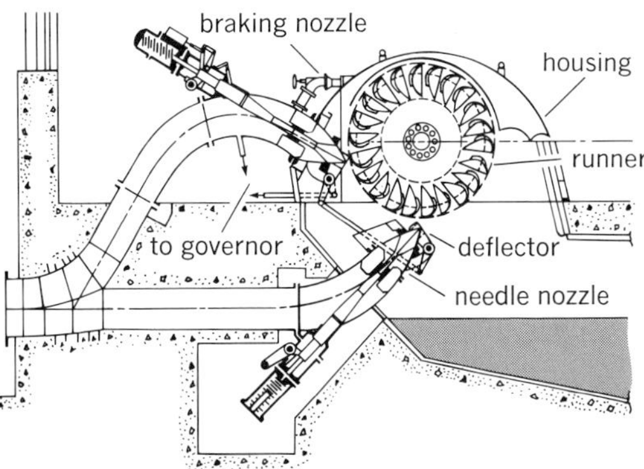

Fig. 1. Cross section of an impulse (Pelton) type of hydraulic-turbine installation.

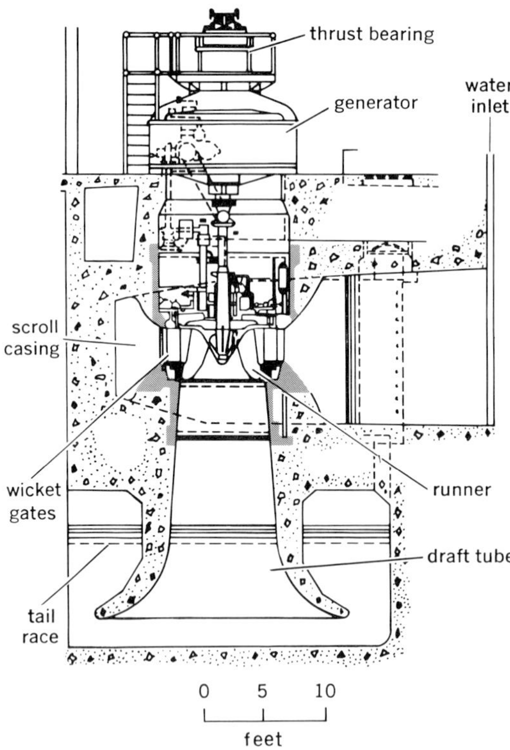

Fig. 2. Cross section of a reaction (Francis) type of hydraulic-turbine installation. 1 ft = 0.3 m.

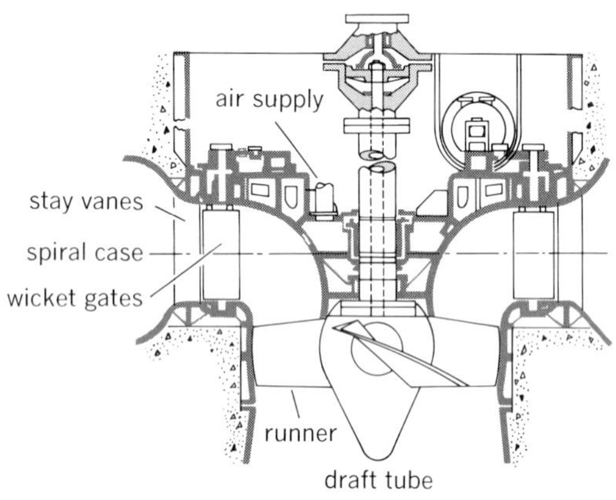

Fig. 3. Cross section of a propeller (Kaplan) type of hydraulic turbine installation.

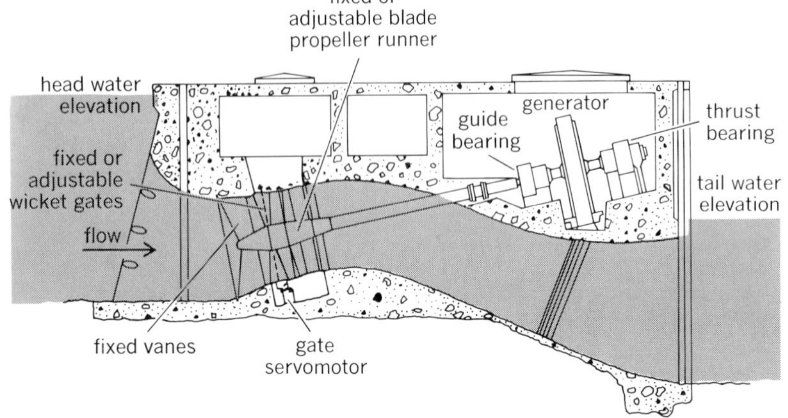

Fig. 4. Axial-flow tube-type hydraulic-turbine installation.

machinery of the jet and vane type operating on the impulse or reaction principle and thus involving the conversion of pressure energy to kinetic energy. The shaft drives an electric generator, and speed must be of an acceptable synchronous value. *See* GENERATOR.

The impulse or Pelton unit has all available energy converted to the kinetic form in a few stationary nozzles and subsequent absorption by reversing buckets mounted on the rim of a wheel (Fig. 1). Reaction units of the Francis or the Kaplan types run full of water, submerged, with a draft tube and a continuous column of water from head race to tail race (Figs. 2 and 3). There is some fluid acceleration in a continuous ring of stationary nozzles with full peripheral admission to the moving nozzles of the runner in which there is further acceleration. The draft tube produces a negative pressure in the runner with the propeller or Kaplan units acting as suction runners; the Francis inward-flow units act as pressure runners. Mixed-flow units give intermediate degrees of rotor pressure drop and fluid acceleration. *See* IMPULSE TURBINE; PELTON WHEEL.

For many years reaction turbines have generally used vertical shafts for better accommodation of the draft tube whereas Pelton units have favored the horizontal shaft since they cannot use a draft tube. Vertical-shaft Pelton units have found increasing acceptance in large sizes because of multiple jets (for example, 4–6) on a single wheel; these provide reduced runner windage and friction losses and, consequently, higher efficiency. Axial-flow (Fig. 4) and diagonal-flow reaction turbines offer improved hydraulic performance and economic powerhouse structures for large-capacity low-head units. Kaplan units employ adjustable propeller blades as well as adjustable stationary nozzles in the gate ring for higher sustained efficiency. Pelton units are preferred for high-head service (1000± ft or 300± m), Francis runners for medium heads (200± ft or 60± m), and propeller or Kaplan units for low heads (50± ft or 15± m).

Hydraulic-turbine performance is rigorously defined by characteristic curves, such as the efficiency characteristic. The proper selection of unit type and size is a technical and economic problem. The data of Fig. 5 are significant because they show synoptically the relationship between unit type and site head as the result of accumulated experience on some satisfactory turbine installations. Specific speed N_s is a criterion or coefficient which is uniquely applicable to a given turbine type and relates head, power, and speed, which are the basic performance data in the selection of any hydraulic turbine. Specific speed is defined in the equation below, where rpm is revolutions per minute, shp is shaft horsepower

$$N_s = \frac{\text{rpm} \times (\text{shp})^{0.5}}{(\text{head})^{1.25}}$$

(1 hp = 0.7457 kW), and head is head on unit in feet (1 ft = 0.3048 m). Specific speed is usually identified for a unit at the point of maximum efficiency. Cavitation must also be carefully scrutinized in any practical selection.

The draft tube (Fig. 2) is a closed conduit which (1) permits the runner to be set safely above tail water level, yet to utilize the full head on the site, and (2) is limited by the atmospheric water column to a height substantially less than 30 ft, and when made flaring in cross section will serve to recover velocity head and to utilize the full site head.

Efficiency of hydraulic turbine installations is always high, more than 85% after all allowances for hydraulic, shock, bearing, friction, generator, and mechanical losses. Material selection is not only a problem of machine design and stress loading from running speeds and hydraulic surges, but is also a matter of fabrication, maintenance, and resistance to erosion, corrosion, and cavitation pitting.

Governing problems are severe, primarily because of the large masses of water involved, their positive and negative acceleration without interruption of the fluid column continuity, and the consequent shock and water-hammer hazards. *See* PRIME MOVER.

Pumped-storage hydro plants have employed various types of equipment to pump water to an elevated storage reservoir during off-peak periods and to generate power during on-peak periods where the water flows from the elevated reservoir through hydraulic turbines. Although separate, single-purpose, motor-pump and turbine-generator sets give the best hydraulic performance, the economic burden of investment has led to the development of reversible pump-turbine units. Components of the conventional turbine are retained, but the modified pump runner gives optimum performance when operating as a turbine. Compromises in hydraulic performance, with some sacrifice in efficiency, are more than offset by the investment savings with the dual-purpose machines. *See* WATERPOWER.

[THEODORE BAUMEISTER]

Bibliography: T. Baumeister (ed.), *Standard Handbook for Mechanical Engineers*, 8th ed., 1978; D. G. Fink and J. M. Carroll, *Standard Handbook for Electrical Engineers*, 11th ed., 1978.

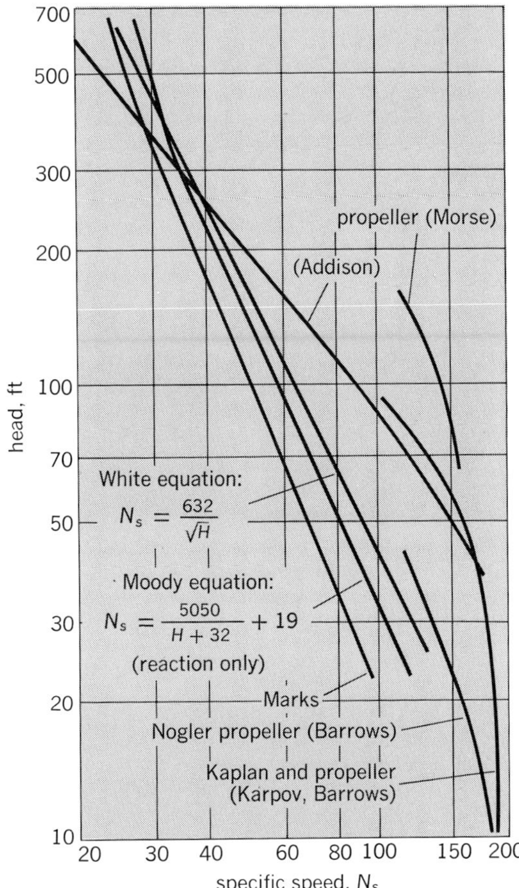

Fig. 5. Hydraulic-turbine experience curves, showing specific speed versus head. 1 ft = 0.3048 m.

Hydraulic valve lifter

A device that eliminates the need for mechanical clearance in the valve train of internal combustion engines.

Clearance is normally required to prevent the valve's being held open and destroyed as the valve train undergoes thermal expansion. However, clearance requires frequent adjustment and is responsible for much operating noise. The hydraulic lifter is a telescoping compression strut in the linkage between cam and valve, consisting of a piston and cylinder (see illustration). When no opening load exists, a weak spring moves the piston, extending the strut and eliminating any clearance. This action sucks oil into the cylinder past a check valve. The trapped oil transmits the valve-opening forces with little deflection. A slight leakage of oil during lift shortens the strut, assuring valve closure. The leakage oil is replaced as the spring again extends the strut at no load. *See* VALVE TRAIN.

[AUGUSTUS R. ROGOWSKI]

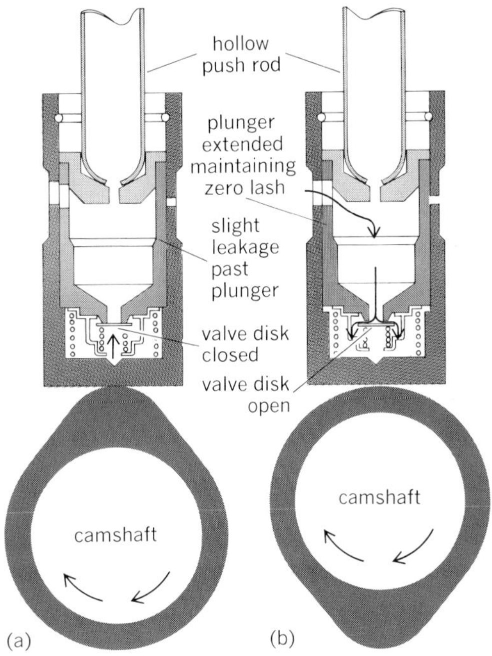

Positions of the hydraulic valve lifter, with engine valve (*a*) open and (*b*) closed. (*Adapted from W. H. Crouse, Automotive Mechanics, 5th ed., McGraw-Hill, 1965*)

Hydraulics

The behavior of water or other liquids, chiefly when in motion. As a part of fluid mechanics, hydraulics deals with the properties, behavior, and effects of all liquids at rest against, or in motion relative to, boundary surfaces or objects. Its laws also apply to gases when changes in density are small, that is, compressibility effects are negligible.

Applications. Common hydraulic applications encountered in civil engineering include the flow of water in pipes, canals, and rivers, in flood control, in land reclamation, and in hydroelectric power projects. Hydraulics also influences appurtenant structures and devices such as dams, dikes, locks, spillways, weirs, piers, ship hulls, gates, and valves. Mechanical and chemical engineering applications include the flow of gases, oils, or other liquids; lubrication; and fluid machinery such as pumps, turbines, propellers, fans, and fluid power transmission and control devices, including servomechanisms.

The physical laws of hydrostatics govern the effects of fluids at rest. Hydrokinematics covers fluid motion wherein consideration of the forces causing the motion is not required. Hydrodynamics treats of fluid motion, the forces involved, and the accompanying energy changes. Characteristics of particular phenomena in hydraulics include fluid properties; the shapes, sizes, relative roughnesses, and relative motions of the surfaces or objects involved; the relative times and distances (short or long); whether flow is closed as in a pipe or open as in a canal; and whether motion is in a relatively extensive fluid as a submarine deeply submerged or whether motion occurs where there is a free surface as a ship at sea.

Physical laws are expressed by mathematical equations; two are required—an equation of condition (continuity) and an equation of motion (momentum or energy). When introduced by methods of dimensional analysis, physical quantities which are known, assumed, or sought experimentally may enter the equations in combinations which are dimensionless numbers or ratios.

Properties. Hydraulics deals with bulk fluids which are regarded as homogeneous; it is not concerned with molecular sizes. As distinct from solids, fluids, such as water, are unable to resist shearing forces while remaining in a state of equilibrium. Among fluids, a liquid mass will have a definite volume, varying only slightly with temperature and pressure, whereas a mass of gas will fill any space available to it, the pressure adjusting itself accordingly.

A common fluid property is density, the mass per unit volume, sometimes expressed dimensionally in slugs per cubic foot, or $[(\text{lb force})(\text{sec})^2/(\text{ft})^4]$. Specific weight, the weight per unit volume in pounds force per cubic foot, is an alternate way of expressing density.

Viscosity, the ability of a fluid in motion to resist shearing forces, is the seat of all fluid resistance. The coefficient of viscosity (absolute viscosity) is the shearing stress divided by the velocity gradient, that is, the rate at which adjacent fluid layers slip past each other; it has the dimension $[(\text{slug})/(\text{ft})(\text{sec})]$ or $[(\text{lb force})(\text{sec})/(\text{ft})^2]$. The slippage rate is greatest adjacent to a solid boundary. There, in the boundary layer, the velocity rises sharply from zero at the wall toward the value in the mainstream. Where viscosity and density enter a problem, the coefficient of viscosity is divided by density to give the kinematic viscosity, which has the dimension square feet per second.

Elasticity denotes the relative compressibility of a fluid under pressure. The volume modulus of elasticity is the change in pressure intensity divided by the corresponding unit volumetric change. This is not important in ordinary applications but plays a significant role when gases flow at high velocity or when, in a pipeline carrying a liquid, a valve is closed or opened rapidly, thereby causing a pressure wave.

Surface tension is a force per unit of length that binds surface molecules of a liquid to one another or to a solid boundary for which they may have an affinity; it is of minor importance except where thin films with a free surface exist or capillary movements occur. *See* DENSITY; ELASTICITY.

[WILLIAM ALLAN]

Hydroelectric generator

An electric rotating machine that transforms mechanical power from a hydraulic turbine or water wheel into electric power (see illustration). Hydroelectric generators may have horizontal or vertical shafts, depending upon the turbine. The most common type in large ratings is the vertical-shaft, synchronous generator. The hydroelectric generator may be arranged to operate as a motor during periods of low power demand. In such a case the turbine serves as a pump, raising water to a high elevation for reuse to generate power at a period of peak load. Such a unit is termed a pump turbine. *See* ELECTRIC POWER GENERATION; GENERATOR; HYDRAULIC TURBINE; TURBINE; WATERPOWER.

[LEON T. ROSENBERG]

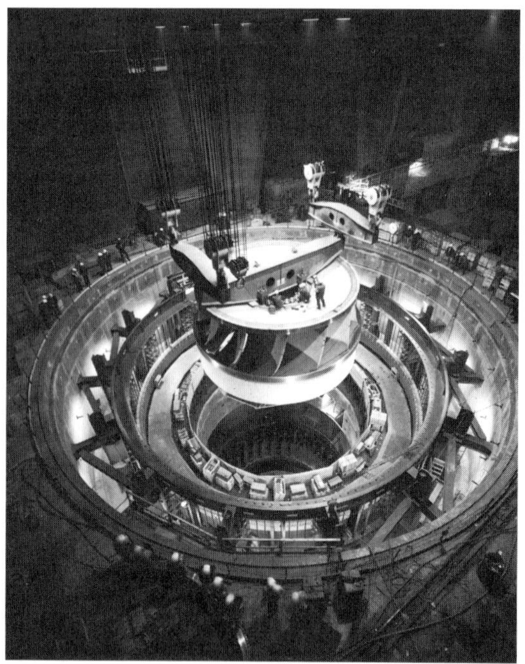

Turbine runner being lowered into hydroelectric generator unit number 19 at the third power plant, Grand Coulee, WA. (*Bureau of Reclamation, U.S. Department of the Interior*)

Hydrometallurgy

The extraction and recovery of metals from their ores by processes in which aqueous solutions play a predominant role. Two distinct processes are involved in hydrometallurgy: putting the metal values in the ore into solution via the operation known as leaching; and recovering the metal values from solution, usually after a suitable solution purification or concentration step, or both. The scope of hydrometallurgy is quite broad and extends beyond the processing of ores to the treatment of metal concentrates, metal scrap and revert materials, and intermediate products in metallurgical processes. Hydrometallurgy enters into the production of practically all nonferrous metals and of metalloids, such as selenium and tellurium.

A generalized metallurgical flow sheet (see illustration) provides an indication of the nature and extent of the role of hydrometallurgy in metal production. It also shows the manner in which hydrometallurgical and pyrometallurgical processes complement each other. *See* PYROMETALLURGY.

The advantages of hydrometallurgy are applicability to low-grade ores (copper, uranium, gold, silver), amenability to the treatment of materials of quite different compositions and concentrations, adaptability to separation of highly similar materials (hafnium from zirconium), flexibility in terms of the scale of operations, simplified materials handling as compared with pyrometallurgy, and good operational and environmental control.

The first commercial hydrometallurgical operation was the application of cyanidation to the treatment of gold ores in 1889, more than 40 years after the discovery that gold can be dissolved in dilute aqueous solutions of sodium cyanide, as shown in reaction (1). During 1890–1910, much of the equip-

$$4Au + 8NaCN + O_2 + 2H_2O \rightarrow 4NaAu(CN)_2 + 4NaOH \quad (1)$$

ment required in modern hydrometallurgy was developed.

Leaching. A major unit process in hydrometallurgy is the leaching of ores and concentrates. A particularly noteworthy and time-honored dissolution process is the caustic pressure leaching of bauxite ore to produce pure alumina. This process, which was developed in the 19th century, is still in use as the principal method of producing alumina. The ever-increasing demand for aluminum and the rising cost of bauxite have stimulated research on the development of processes for leaching nonbauxite alumina materials such as alunite, anorthosite, clays, and coal shales.

In the case of sulfide concentrates, leaching in acid or ammoniacal solution under oxygen pressure provides a number of advantages, including accelerated leaching and the recovery of sulfur in a suitable form, such as ammonium sulfate in the

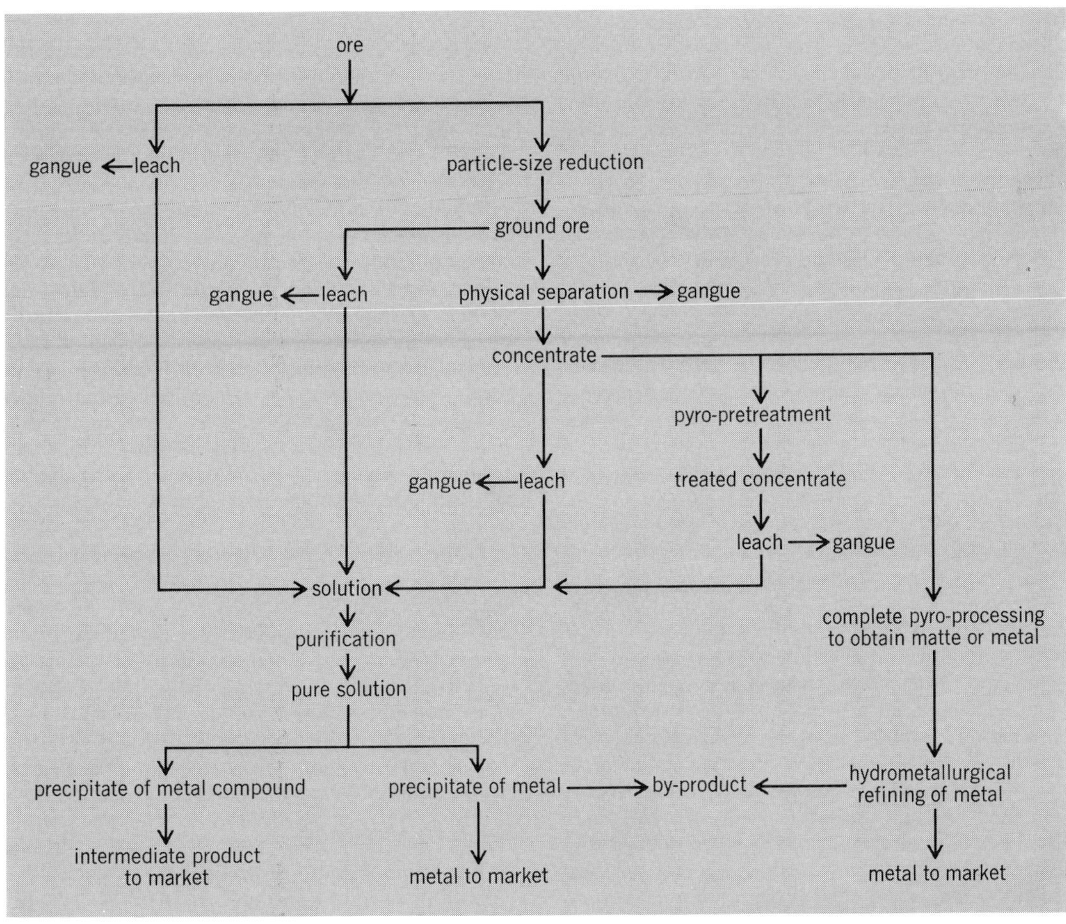

Generalized metallurgical flow sheet.

ammoniacal leaching of nickel concentrates in the Sherritt-Gordon process or as elemental sulfur in the acid pressure leaching of zinc concentrates.

The economic recovery of metals from low-grade ores is becoming increasingly important as mining costs rise and as higher-grade materials are more difficult to find. In many operations, what was at one time waste is now considered as ore to be treated by newer leaching methods such as heap, dump, and in-place leaching. In the United States, copper in excess of 160,000 tons (145,000 metric tons) is extracted annually by such processes. The same technology is applied in the leaching of uranium ores and in the treatment of gold and silver ores. The leaching of low-grade sulfide copper ores and of uranium ores in the presence of pyrite can be facilitated greatly by the action of the bacterium *Thiobacillus ferrooxidans*, and hence the interest in what is known as bacterial or microbiological leaching.

The establishment of much more stringent regulations in many countries concerning sulfur dioxide emissions from smelters has led not only to the adoption of new and improved smelting processes, but also to a great deal of research on hydrometallurgical alternatives to smelting, especially of copper concentrates. In the case of copper concentrates, the principal difficulty has been the realization of a hydrometallurgical process which has sufficiently well-defined economic advantages. From an environmental standpoint, there are definite advantages which are afforded by hydrometallurgy, principally the elimination of sulfur dioxide emissions and the conversion of sulfur to elemental sulfur or a sulfate. The overall economics of copper hydrometallurgy can be improved by the recovery of copper in final form by a process other than electrowinning, such as the hydrogen reduction of cuprous chloride.

Environmental considerations have led to a marked decline in the production of zinc by the carbothermic reduction of zinc calcine at 1200–1300°C in horizontal retorts by reaction (2),

$$ZnO(s) + C(s) \rightarrow Zn(g) + CO(g) \qquad (2)$$

where (s) and (g) denote solid and gas. This process has been replaced by the sulfuric acid leaching of zinc calcine, purification of the resulting leach solution, and the recovery of zinc by electrowinning.

Ion exchange. The recovery of metals from leach solutions is often preceded by the purification or concentration of the metal to be isolated. Two processes which can be used to achieve this objective are ion exchange and solvent extraction. Ion exchange involves the selective adsorption of metal ions onto a synthetic organic ion-exchange resin (commonly a styrene-divinylbenzene copolymer containing suitable functional groups for ion exchange). The process is stoichiometric and reversible, so that the adsorbed ions are subsequently desorbed or eluted from the resin by a solution known as the eluant. The resulting solution or eluate is treated for metal recovery or for production of a metal compound, such as ammonium diurante, $(NH_4)_2V_2O_7$, in uranium extraction. Ion exchange, which is widely used in water purification, is employed extensively in uranium recovery. Here continuous ion-exchange processes are in commercial use, with resultant increases in efficiency and productivity. New resins are being developed, notably chelating resins, which should find increasing use in the separation of metals such as copper, nickel, cobalt, and zinc.

Solvent extraction. As applied in hydrometallurgy, solvent extraction relates to the selective transfer of metal species from an aqueous solution to an immiscible organic solvent with which it is in contact. The solvent itself can be the active reagent in the process, as in the extraction of uranium by tributylphosphate. On the other hand, the solvent can contain an organic extractant which forms a complex or chelation compound with the metal ion from the aqueous phase. The extraction process is reversible, so that the extracted metal can be stripped from the organic solution and returned to the aqueous phase.

With the development of suitable chelating compounds, solvent extraction has become established as an excellent technique for the upgrading of copper leach solutions. The application of solvent extraction is especially suited to the treatment of the dilute solutions resulting from the heap and dump leaching of low-grade copper ores. The selectivity of the organic reagents, notably the hydroxyoximes, is such as to permit an excellent separation from iron. The stripping of the organic phase by spent electrowinning electrolyte leads to a strong solution, from which copper can be electrowon without difficulty.

Solvent extraction is used commercially in the separation of many metals, including copper, nickel, cobalt, uranium, thorium, zirconium, hafnium, molybdenum, tungsten, niobium, tantalum, and beryllium.

Recovery. Metal recovery from aqueous solution can be effected by a number of reduction processes, the most prominent being electrowinning and gaseous reduction using hydrogen. The latter process yields metal powder, whereas the former results in cathodes which are melted and cast to produce various marketable shapes.

Hydrometallurgy occupies an important role in the production of aluminum, copper, nickel, cobalt, zinc, gold, silver, platinum metals, selenium, tellurium, tungsten, molybdenum, uranium, zirconium, and other metals. Considering the versatility of hydrometallurgy and the need to process more complex ores, as well as lower-grade ores and secondary materials, there is no doubt that hydrometallurgical processes will come to play an even greater part in the production of metals in the future. *See* ELECTROMETALLURGY; METALLURGY.

[W. CHARLES COOPER]

Bibliography: J. M. Cigan, T. S. Mackey, and T. J. O'Keefe (eds.), *Lead-Zinc-Tin '80: TMS-AIME World Symposium on Metallurgy and Environmental Control*, 1980; D. J. I. Evans and R. S. Shoemaker (eds.), *AIME International Symposium on Hydrometallurgy*, 1973; F. Habashi, *Principles of Extractive Metallurgy*, vol. 2: *Hydrometallurgy*, 1970; J. C. Yannopoulos and J. C. Agarwal (eds.), *Extractive Metallurgy of Copper*, vol. 2: *Hydrometallurgy and Electrowinning*, 1976.

Hysteresis motor

A synchronous motor without salient poles and without dc excitation which makes use of the hysteresis and eddy-current losses induced in its hardened-steel rotor to produce rotor torque. The stator and stator windings are similar to those of an induction motor and may be polyphase, shaded-pole, or capacitor type. The rotor is usually made up of a number of hardened steel rings on a non-magnetic arbor. The hysteresis motor develops constant torque up to synchronous speed. The motor can, therefore, synchronize any load it can accelerate. These motors are built in small sizes, for instance, for electric clocks. *See* INDUCTION MOTOR; SYNCHRONOUS MOTOR.

[LOYAL V. BEWLEY]

Ignition system

The ignition system of an internal combustion engine initiates the chemical reaction between fuel and air in the cylinder charge. For maximum efficiency it is necessary to ignite the charge in each cylinder shortly before the piston reaches the top of its compression stroke. The best timing is determined experimentally for various engine speeds and loadings, and the ignition system is designed to provide this timing automatically. For smooth running, multicylinder engines are usually built so that the various pistons arrive at their firing top center positions in evenly spaced intervals. The sequence in which the cylinders reach their firing points depends upon the geometrical arrangement of the cylinders and crankshaft, and is called the firing order of the engine. The ignition system distributes the ignition impulse to the cylinders in this order.

A large class of engines, including most automobile engines, operate on a premixed charge of fuel vapor and air. In these engines the charge is ignited by passing a high-voltage electric current between two electrodes in the combustion chamber. The electrodes are incorporated in a removable unit. When a spark of sufficient energy jumps the gap between the electrodes, a self-propagating flame is produced in the fuel-air mixture which spreads rapidly throughout the charge. *See* DISTRIBUTOR; SPARK PLUG.

Battery system. The electric energy for the spark is obtained from a storage battery in practically all spark-ignition engines. The battery, usually 6 or 12 volts, is part of a low-voltage primary circuit, which includes a switch, the primary winding of the ignition coil, a capacitor (or condenser), and breaker points (Fig. 1). When the switch and breaker points are closed, current flows from the battery through the primary winding, through the points, and back to the battery through the engine frame or other ground connection. The coil is wound about a soft iron core. The current in the primary winding induces a magnetic field in and around this core. When the breaker points open as the engine-driven cam rotates, the current, which had been passing through the points, now flows into the capacitor. As the capacitor becomes charged, the primary current falls rapidly, and the magnetic field collapses.

The secondary winding consists of many turns of fine wire wound on the same core with the primary winding. The rapid collapse of the magnetic field in the core induces a very high voltage (10,000 to 20,000 volts) in the secondary winding.

Without the capacitor the primary voltage caused by the rapid collapse of the magnetic field around the primary winding would cause an arc across the breaker points. The arc would burn the points and soon destroy them. The capacitor also

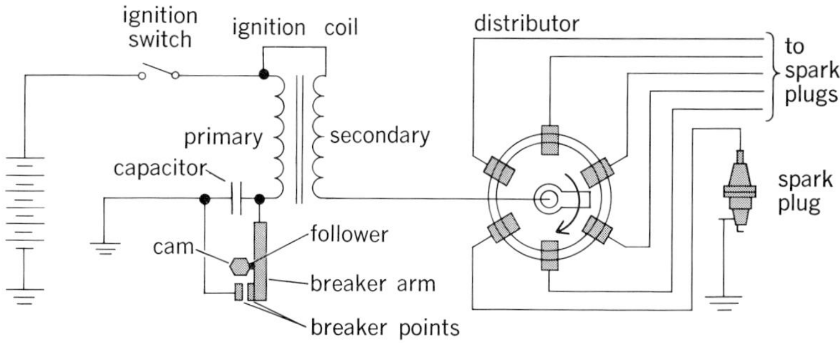

Fig. 1. Diagram of battery ignition system. (*Modified from A. R. Rogowski, Elements of Internal-combustion Engines, McGraw-Hill, 1953*)

assists the rapid drop in the primary current and collapse of the magnetic field, both of which are needed to produce the high secondary voltage.

The secondary voltage is led to the spark plugs in proper sequence by a rotary switch called the distributor, which is driven by the same shaft that drives the breaker point cam. From the head of the distributor, well-insulated wires carry the secondary voltage to the central electrodes of the spark plugs. The discharge which takes place between the central electrode and the grounded electrode inside the combustion chamber ignites the cylinder charge. The timing of the ignition spark is controlled by the opening of the breaker points. The ignition timing may be varied by rotating the plate upon which the breaker points are mounted relative to the cam. Timing may also be varied by changing the angular relationship between the breaker cam and the shaft that drives it.

Failures. Ignition system failures are usually due to breaker points which have become overheated by the primary current or to spark plugs which are fouled by carbon or other electrically conducting cylinder deposits. Ignition systems are being considered in which a solid-state switching device, called a transistor, is used in place of the breaker points. The transistor is capable of interrupting the primary current without wear or burning. Breaker points are still used but only to control the transistor, and the control current is very small. Another type of ignition system that has been found superior in certain applications where plug fouling is common consists of circuitry which charges a capacitor to several hundred volts. When an ignition spark is required, the capacitor is discharged into the primary of the ignition coil. A transistor, controlled by breaker points or other triggering devices, is used to make the connection between the capacitor and the coil. The sudden application of high voltage to the primary coil produces a much more rapid rise in secondary voltage than is obtainable with the conventional ignition system. Because of the rapid voltage rise, the spark jumps the spark plug gap before the electric charge has had time to leak away through the conducting carbon deposits on the plug.

Magneto system. The magneto system is similar to the battery system except that the voltage required to cause a flow of current in the primary winding is generated by the rotation of a set of permanent magnets, instead of being supplied by a battery. The magneto assembly normally includes, within one housing, the rotating magnets, ignition

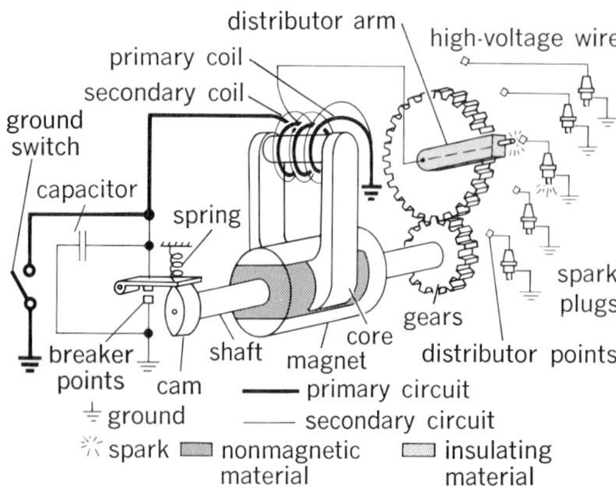

Fig. 2. Diagram of magneto ignition system. (*From C. H. Chatfield, C. F. Taylor, and S. Ober, The Airplane and Its Engine, 5th ed., McGraw-Hill, 1949*)

coil windings, iron core, breaker points, cam, capacitor, and distributor (Fig. 2).

Rotation of the magnet assembly completely reverses the direction of the magnetic flux in the soft iron core about which the primary and secondary coils are wound. This rapid change in flux intensity and direction drives a current through the primary circuit. When the primary current reaches a high value and the contribution to the total flux is principally from the primary current, and not from the permanent magnets, the breaker points open with the same result as in the battery system described. The magneto is often used with small compact installations such as lawnmowers, chain saws, and small outboard engines, where a battery would represent undersirable weight and bulk.

Compression ignition. In the diesel or compression ignition engine, the inducted cylinder air is heated sufficiently by the compression stroke of the piston so that ignition takes place shortly after the fuel is sprayed into the cylinder. Ignition timing is thus controlled by the phasing of the fuel injection pump, while the firing order of the cylinders is established by the geometry of the pump. In compression-ignition engines, the fuel burns as soon after ignition as it is able to find air with which to combine, so that less attention need be paid to automatic injection timing control. When control is used, pump shaft position relative to engine shaft can be changed automatically by centrifugal weights and a mechanical linkage. *See* COMBUSTION CHAMBER.

Sometimes in small engines a premixed charge of fuel and air is ignited near the end of the compression stroke by the temperature of compression. Under these conditions the entire charge, once ignited, burns rapidly, since all parts of the charge are equally heated and prepared for combustion. The result is rough, pounding combustion due to the high reaction rate. No precise way exists to control the exact timing of ignition with this method, so that the combustion may occur too early or too late depending upon such factors as the temperature of the cylinder and the kind of fuel being used; efficiency with such ignition is likely to be poor. [A. R. ROGOWSKI]

Bibliography: American Institute of Physics, *Automobile Ignition Systems*, 1975; C. Brant, *Transistor Ignition Systems*, 1976; T. Baumeister (ed.), *Standard Handbook for Mechanical Engineers*, 8th ed., 1978; W. H. Crouse, *Automotive Mechanics*, 8th ed., 1980; D. G. Fink and H. W. Beaty, *Standard Handbook for Electrical Engineers*, 11th ed., 1978.

Imhoff tank

A sewage treatment tank named after its developer, Karl Imhoff. Imhoff tanks differ from septic tanks in that digestion takes place in a separate compartment from that in which settlement occurs. The tank was introduced in the United States in 1907 and was widely used as a primary treatment process and also in preceding trickling filters. Developments in mechanized equipment have lessened its popularity, but it is still valued as a combination unit for settling sewage and digesting sludge. *See* SEPTIC TANK; SEWAGE.

The Imhoff tank is constructed with the flowing-through chamber on top and the digestion chamber on the bottom (see illustration). The upper chamber is designed according to the principles of a sedimentation unit. Sludge drops to the bottom of the tank and through a slot along its length into the lower chamber. As digestion takes place, scum is formed by rising sludge in which gas is trapped. The scum chamber, or gas vent, is a third section of the tank located above the lower chamber and beside the upper chamber. As gases escape, sludge from the scum chamber returns to the lower chamber. The slot is so constructed that particles cannot rise through it. A triangle or sidewall deflector below the slot prevents vertical rising of gas-laden sludge. Sludge in the lower chamber settles to the bottom, which is in the form of one or more steep-sloped hoppers. At intervals the sludge can be withdrawn. The overall height of the tank is 30–40 ft (9–12 m), and sludge can be expelled under hydraulic pressure of the water in the upper tank. Large tanks are built with means for reversing flow in the upper chamber, thus making it possible to distribute the settled solids more evenly over the digestion chamber.

Design. Detention period in the upper chamber is usually about $2\frac{1}{2}$ hr. The surface settling rate is usually 600 gal/(ft^2)(day) or 25 m/(day). The weir overflow rate is not over 10,000 gal/ft or 125 m^3/m of weir per day. Velocity of flow is held below 1 ft/sec or 0.3 m/sec. Tanks are dimensioned with a length-width ratio of 5:1–3:1 and with depth to slot about equal to width. Multiple units are built rather than one large tank to carry the entire flow. Two flowing-through chambers can be placed above one digester unit. The digestion chamber is normally designed at 3–5 ft^3 (0.085–0.14 m^3) per capita of connected sewage load. When industrial wastes include large quantities of solids, additional allowance must be made. Ordinarily sludge withdrawals are scheduled twice per year. If these are to be less frequent, an increase in capacity is desirable. Some chambers have been provided with up to 6.5 ft^3 (0.184 m^3) per capita. The scum chamber should have a surface area 25–30% of the horizontal surface of the digestion chamber. Vents should be 24 in (0.6 m) wide. Top freeboard should be at least 2 ft (0.6 m) to contain rising scum. Water under pressure must be available to combat foaming and scum.

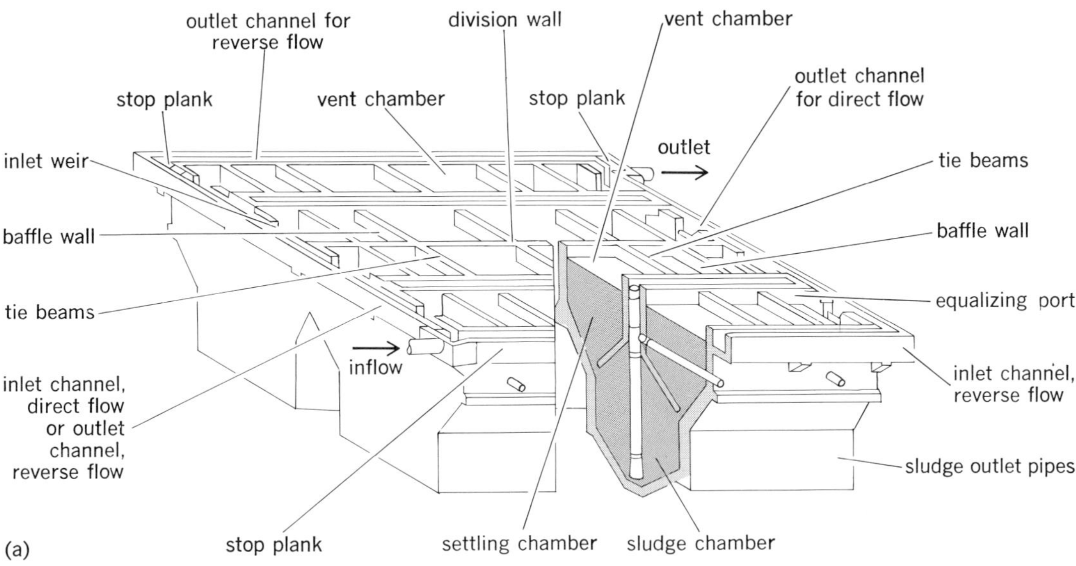

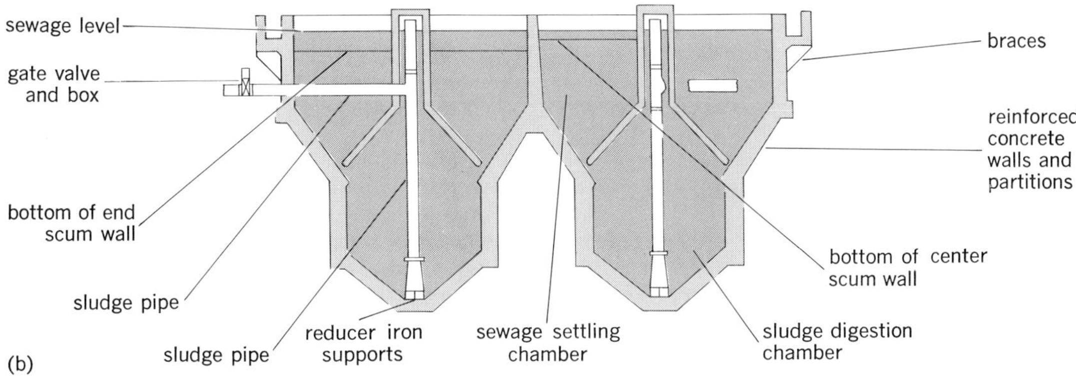

Diagram of typical large Imhoff tank for sewage treatment. (*a*) General arrangement. (*b*) Cross section. (*From H. E. Babbitt and E. R. Baumann, Sewerage and Sewage Treatment, 8th ed., copyright © 1958 by John Wiley and Sons, Inc.; used with permission*)

Efficiency. The efficiency of Imhoff tanks is equivalent to that of plain sedimentation tanks. Effluents are suitable for treatment on trickling filters. The sludge is dense, and when withdrawn it may have a moisture content of 90–95%. Imhoff sludge has a characteristic tarlike odor and a black granular appearance. It dries easily and when dry is comparatively odorless. It is an excellent humus but not a fertilizer. Gas vents may occasionally give off offensive odors. [WILLIAM T. INGRAM]

Impulse generator

An electrical apparatus which produces very short surges of high-voltage, or high-current, power. High impulse voltages are used to test the strength of insulators and of power equipment against lightning and switching surges. High-current impulses are produced by the discharge of capacitors connected in parallel. Such current surges may be used to magnetize permanent magnets or to produce the rising magnetic field in circular particle accelerators.

High-voltage impulse generators commonly employ the principle, originally suggested by E. Marx, of charging capacitors in parallel and discharging them in series. The figure shows a four-stage Marx circuit in which the capacitors are first charged in parallel through charging resistors R, then connected in series and discharged through the test piece by the simultaneous sparkover of spark gaps G. The discharge is precipitated by placing a sufficient voltage on the middle electrode of the three-electrode gap between the first and second capacitor banks.

Although 1,000,000 to 2,000,000 volts (peak) is most common, over 7,500,000 volts to ground has been obtained. The waveform shows a rapid rise followed by a less rapid decline to zero, expressed by $v \propto (e^{-mt} - e^{-nt})$, where v is the instantaneous voltage t seconds after onset of the discharge, and m and n are the exponential decay constants of the circuit. Typical industrial laboratory waveforms are the 0.5–5, the 1–10, and the 1.5–40, in which the first number is the time in microseconds to the peak of the voltage wave, and the second is the time to one-half voltage of the tail of the wave. These discharges simulate the transient voltages induced in electrical conductors by natural lightning.

Impulses of still shorter rise times and duration are now used to produce short intense bursts of ionizing radiation for radiation damage studies. By

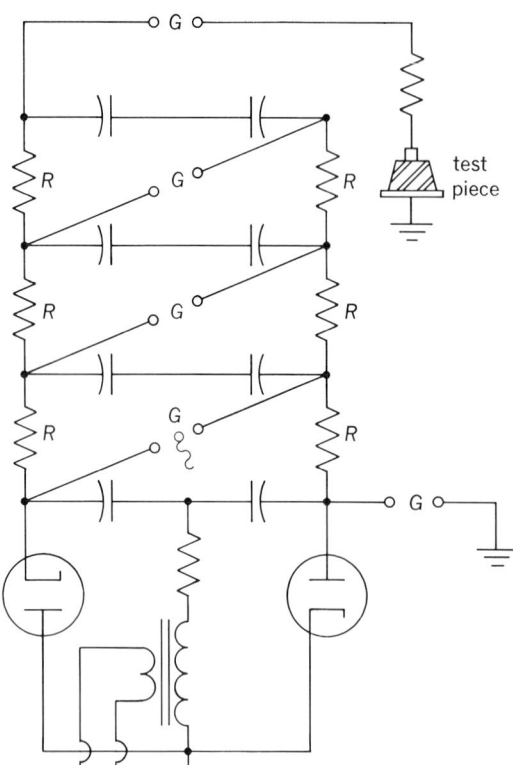

Typical four-stage Marx impulse generator circuit.

the discharge of a single high-voltage condenser, such as the terminal of a Van de Graaff electrostatic generator insulated in compressed gas, voltage pulses which last less than 0.05 μsec have been applied to an electron acceleration tube. Transient electron pulses have been attained with energies of several million volts with currents of more than 25,000 amp.

[JOHN G. TRUMP]

Bibliography: E. Kuffel and M. Abdullah, *High Voltage Engineering*, 1970.

Impulse turbine

A prime mover in which fluid (water, steam, or hot gas) under pressure enters a stationary nozzle where its pressure (potential) energy is converted to velocity (kinetic) energy. The accelerated fluid then impinges on the blades of a rotor, imparting its energy to the blades to produce rotation and overcome the connected rotor resistance. The impulse principle is basic to many turbines.

The impulse principle can be distinguished from the reaction principle by considering the flow of water from a hole near the bottom of a bucket (Fig. 1). The hole is a nozzle that serves to convert potential energy to the kinetic form ΔE; the impulse force F_i in the issuing jet is given by the expression $\Delta E = F_i v/2$, where v is the velocity of the jet.

IMPULSE TURBINE

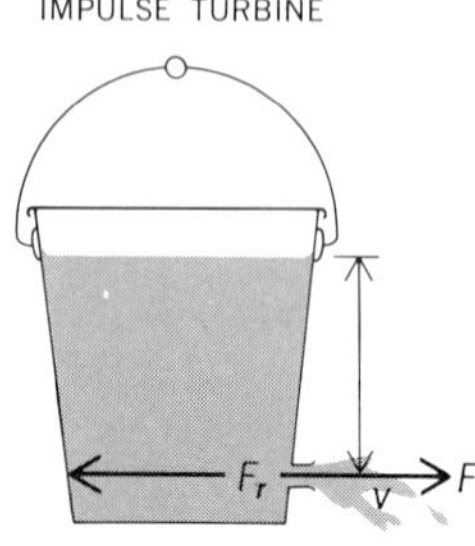

Fig. 1. Water escaping through a nozzle near the base of a bucket, free to swing, illustrating the impulse F_i and reaction F_r forces of the jet issuing with a velocity v.

If the jet is allowed to impinge on a series of vanes mounted on the periphery of a wheel, the impulse force can overcome the resistance connected to the shaft (Fig. 2). The efficiency of the device is ideally dependent upon the vane curvature and the absolute vane velocity. For flat blades (Fig. 2*a*), the efficiency cannot exceed 50%. With curved blades and complete reversal of the jet (Fig. 2*b*), the efficiency will be 100% when the vane velocity u is one-half the jet velocity v.

If the bucket in Fig. 1 is suspended and free to move, there will be, by Newton's third law of motion, a reaction force F_r equal and opposite to the impulse force F_i. For maximum efficiency (100%) the swinging bucket will have to move with an absolute tip velocity u equal to the jet velocity, v. This is the reaction principle and is demonstrated by the Barker's mill (Fig. 3).

The basic difference between the reaction principle and the impulse-turbine principle is determined by the presence or absence of moving nozzles. A nozzle is a throat section device in which there is a drop in pressure with consequent acceleration of the emerging fluid. An impulse turbine has stationary nozzles only. A reaction turbine must have moving nozzles but may have stationary nozzles also so that the fluid can reach the moving nozzle. This is the usual condition for any practical reaction turbine.

The idealized vector diagrams of Fig. 4 demonstrate distinguishing features for the construction which uses a row of blades mounted on the periphery of a wheel and for which flow is axially through the blade passages from one side of the wheel to the other. The theoretical condition further presupposes complete (180°) reversal of the jet and no friction losses. In Fig. 4, v is the absolute velocity of the fluid, u is the absolute velocity of the moving blade, and w is the relative velocity of the fluid with respect to the moving blade. Subscripts 1 and 2 apply to entrance and exit conditions, respectively. The vectors of the illustration demonstrate that maximum efficiency (zero residual absolute fluid velocity v_2) obtains (1) when the

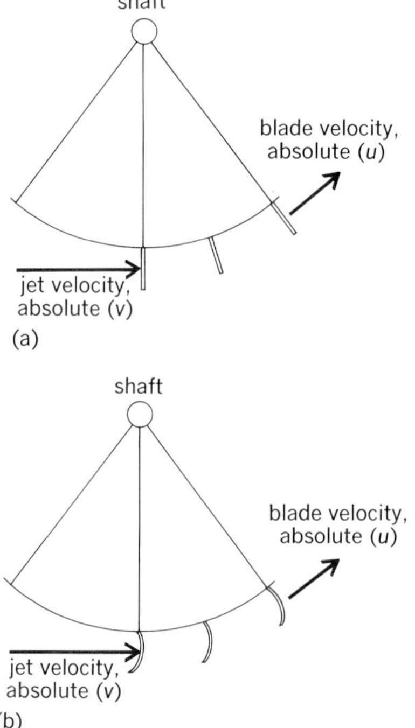

Fig. 2. Diagrams of a jet impinging on a series of (*a*) flat blades and (*b*) curved blades mounted on the periphery of a wheel.

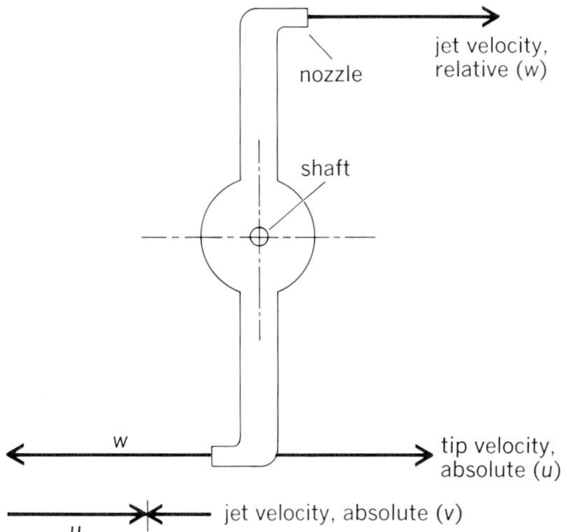

Fig. 3. Barker's mill, a reaction-type jet device, illustrating the application of moving nozzles and the resultant absolute and relative velocities.

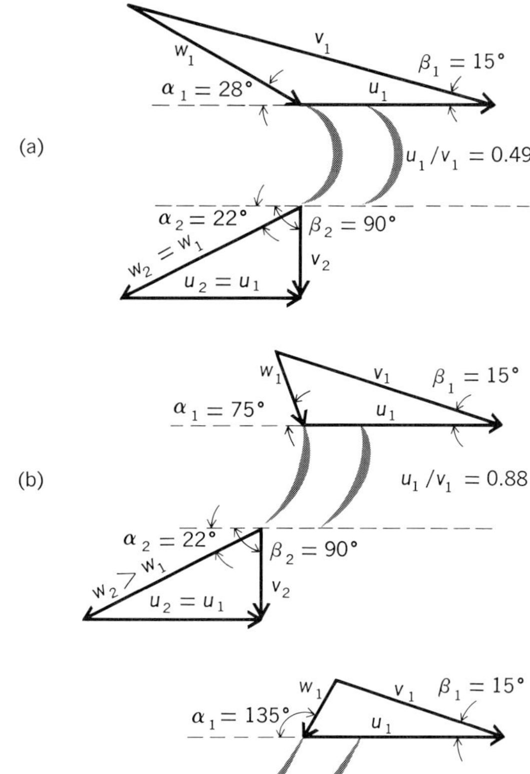

Fig. 5. Vector diagrams for idealized (frictionless) axial fluid flow on (a) impulse-turbine blading and (b, c) reaction-turbine blading; jet reversal less than 180°. Exit vector triangles are identical with consequent equal efficiencies. Resultant speed ratios (u_1/v_1) are 0.49, 0.88, and 1.22, respectively.

blade speed is half the jet speed for impulse turbines and (2) when the blade speed is equal to the jet speed for reaction turbines.

Figure 5 shows the situation more practically as complete reversal of the jet is not realistic. The vector diagrams of Fig. 5 show speed ratios of 0.49, 0.88, and 1.22 for reasonable degrees of jet reversal and no friction losses. By varying the angle of entrance α_1 to the moving blades, a wide range of speed ratios is available to the designer. In each case shown in Fig. 5, the stationary nozzle entrance angle β_1 is fixed at 15°; the exit vector triangle from the moving blades is identical in all cases. The data in Fig. 5*c* demonstrate that it is entirely possible to have speed ratios greater than unity with the reaction principle. This condition is utilized in many designs of hydraulic turbines, for example, propeller and Kaplan units.

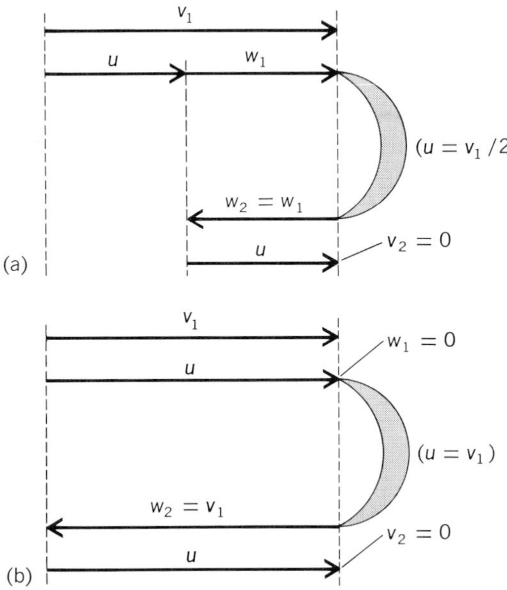

Fig. 4. Velocity vector diagrams for idealized axial fluid flow on (a) impulse-turbine blading and (b) reaction-turbine blading; 180° jet reversal. In *b* the fluid is accelerated to *w* across the moving blades (nozzles). In both cases the leaving absolute fluid velocity *v* is zero, giving maximum efficiency.

The choice of details is at the discretion of the turbine designer. He can determine the extent to which the basic principles of impulse and reaction should be applied for a practical, reliable, economic unit in the fields of hydraulic turbines, gas turbines, and steam turbines. *See* GAS TURBINE; HYDRAULIC TURBINE; PELTON WHEEL; REACTION TURBINE; STEAM TURBINE.

[THEODORE BAUMEISTER]

Bibliography: T. Baumeister (ed.), *Standard Handbook for Mechanical Engineers*, 8th ed., 1978; G. T. Csanady, *Theory of Turbomachines*, 1964; D. G. Shepherd, *Principles of Turbomachinery*, 1956; W. G. Stelz (ed.), *Turbomachinery Developments in Steam and Gas Turbines*, 1977; V. L. Streeter, *Handbook of Fluid Dynamics*, 1961.

Incandescent lamp

An electric lamp that produces light by heating a metallic filament to intense heat by passing an electric current through it. It is designed to produce light in the visible portion (at wavelengths of 380–760 mμ) of the electromagnetic spectrum. The filament is prepared of special materials and is enclosed in either an evacuated enclosure or one filled with an inert gas. In addition to radiation in the visible spectrum, infrared and ultraviolet energy are emitted, thus lowering the luminous efficiency of the lamp. When either of these radiations is accentuated, however, the lamp may be used as a source of that radiant energy. The light-source efficacy, formerly called light-source efficiency, is expressed in lumens per watt (lu/W). Luminous efficiency is the ratio of the luminous flux to the radiant flux.

Lamp construction. The important parts of an incandescent lamp are the lamp enclosure or bulb, the filament, and the base. The parts of a typical incandescent lamp are shown in Fig. 1.

Standard lamps have various bulb shapes, bases, and filament constructions. The bulb may be clear, colored, inside-frosted, or coated with diffusing or reflecting material. Most lamps have soft-glass bulbs; hard glass is used when the lamp will be subjected to sudden and severe temperature changes. In addition, lamps are available with a variety of bulb shapes, base types, and filament structures, as shown in Fig. 2. These vary according to the type of service planned, the need for easy replacement, and other environmental and service conditions.

The efficient design of an incandescent lamp centers about obtaining a high temperature at the filament without the loss of heat or disintegration of the filament. The early selection of carbon, which has the highest melting point of any element (6510°F) was a natural one. Carbon evaporates from its solid phase (sublimates) below this temperature, so carbon filaments must be operated at relatively low temperatures to obtain reasonable life. Two other elements, osmium (mp 4890°F) and tantalum (mp 5250°F), claimed attention because they could be operated at a high temperature with a longer life and less evaporation. With the advent of ductile tungsten a nearly perfect filament material was discovered. Ductile tungsten has a tensile strength four times that of steel, its melting point is high (6120°F), and it has relatively low evaporation. Hot tungsten is an efficient light radiator; it has a continuous spectrum closely following that of a blackbody radiator with a relatively high portion of the radiation in the visible spectrum. Because of its strength, ductility, and workability, it may be formed into coils, these coils again recoiled (for coil-coil filaments), and these again recoiled for cathodes in fluorescent lamps. If tungsten could be held at its melting point, 52 lu/W would be radiated. Because of physical limitations, however, 22 lu/W is the highest practical radiation for general-service lamps; some special lamps reach 35.8 lu/W. The higher the lumens per watt, the higher the filament temperature and the color temperature and the whiter the light.

Because the temperature of the filament controls the life of the lamp and its efficiency, it also controls the economics of lighting. For an economical installation, the factors affecting lamp life are weighed against the cost of the lamp and its installation and the cost of operation. This type of economic study, however, is rarely made of a lighting installation. The usual practice is to select a desired lamp size from the stock of the supplier for the accepted regional voltage.

Vaporization of the filament is reduced as much as possible. A small amount remains, however, and causes blackening of the bulb. The evaporated tungsten particles are carried to the upper part of the bulb by convection currents. With the lamp in a base-up position the blackening is confined to the socket area, and the light output is only slightly affected. In a base-down position the blackening reduces the output a few percent. To reduce blackening, the inner atmosphere of the lamp is kept as clean as possible by use of a getter, which combines chemically with the tungsten particles. In lamps in which a getter is not adequate, tungsten powder is enclosed in the bulb and used to scour the surface by shaking. In some lamps a grid is placed to attract and hold the evaporated tungsten particles.

When cement failures are likely because of the heat, mechanical fastening is used to hold the base to the glass. When large electric currents are present, either mechanical fastening or bipost construction is used. Bipost and prefocus bases allow accurate placement of the filament with respect to the equipment for which it is designed. There are two common failures of the lamp bulb: in projectors, when the filament image is focused upon the glass and the bulb blisters; and when the hot bulb comes in contact with some low-temperature medium and thermal cracks develop. To protect the electric circuit from some lamp failures, a fuse may be placed in the lead-in wire.

Lamp ratings. Lamps are built for various voltage conditions, the most common being 115, 120, and 125 volts. High-voltage lamps are designed for ratings of 220–260 volts and low-voltage lamps for 6–64 volts. Lamps for use in 525–625 volt systems are designed to operate in groups of five in series across the line. Christmas-tree lights are designed to operate in parallel, or in series with eight lamps placed across the line. Street lamps are of the series type operating on 6.6 A, except in series systems where individual lamp transformers are used with a lamp current of 15–20 A. In a series street-lighting system continuity is maintained by a device that short-circuits the lamp when the filament burns out.

Lamp characteristics. Two characteristics of the incandescent lamp of particular interest are operation and color. Figure 3 shows the relative energy in the various color regions for the major types of incandescent lamp. Figure 4 shows the change of the various operating characteristics with a change of voltage.

At times it is necessary to determine lamp characteristics under other than normal conditions. Equations (1)–(5) with exponents given in Table 1 permit solution for actual service conditions as compared to normal operation.

To use Eqs. (1)–(5) it is necessary first to know the normal characteristics of the lamp, such as life, lumens, lumens per watt, volts, amperes, watts, and ohms, if all the characteristics are to be studied under abnormal conditions. By substituting into the equations the specific normal characteristics (upper case letters) and the special conditions (lower case letters) with the proper exponent

INCANDESCENT LAMP

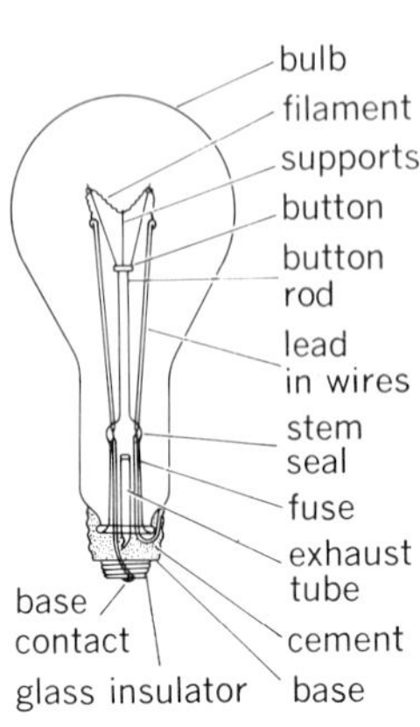

Fig. 1. Parts of an incandescent lamp. (*General Electric Co.*)

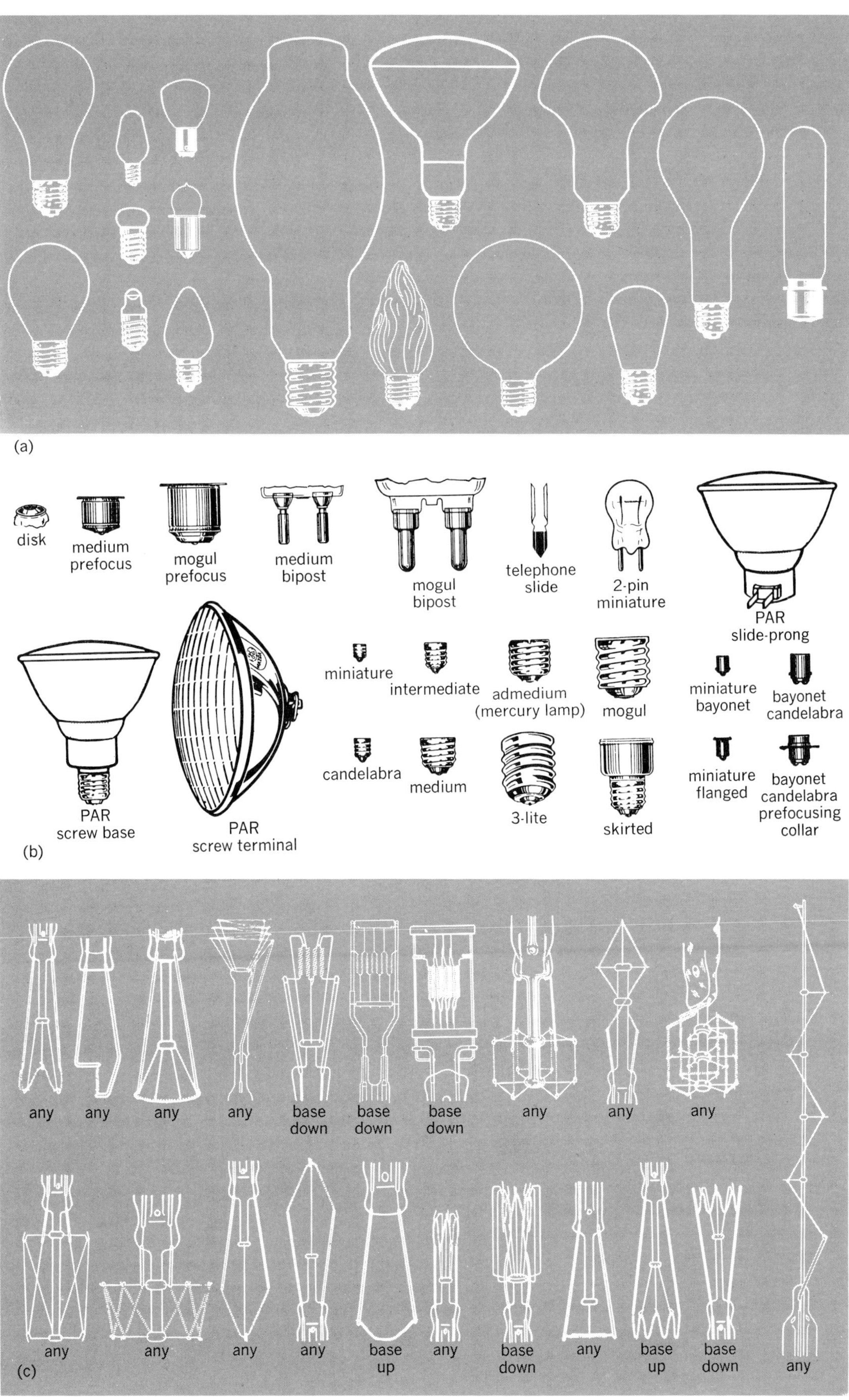

Fig. 2. Variations of incandescent lamps. (*a*) Bulb shapes. (*b*) Types of lamp bases. (*c*) Filament structures (notes designate burning position for each example). (*General Electric Co.*)

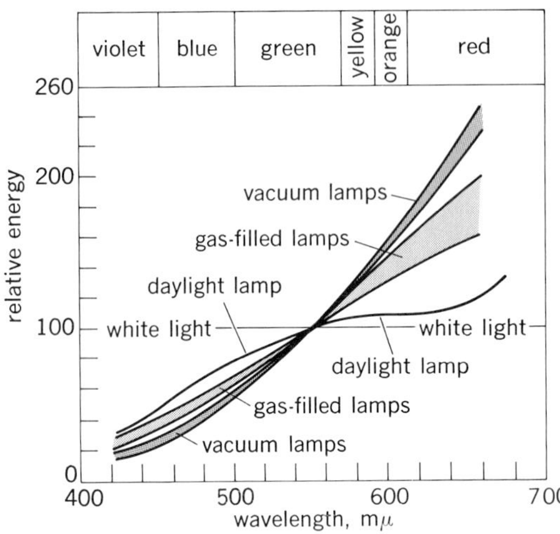

Fig. 3. Spectral energy distribution for important types of incandescent lamps.

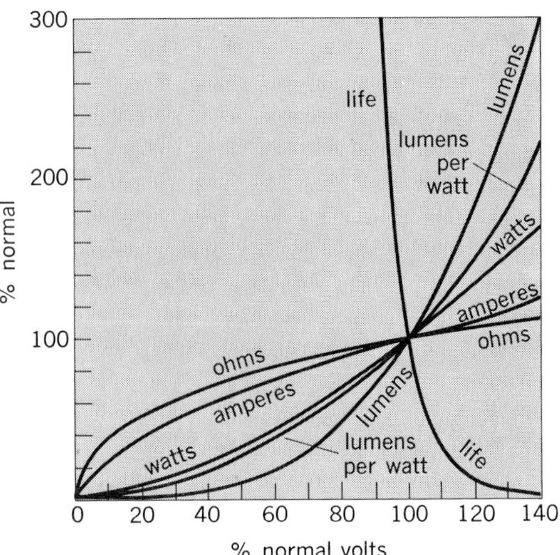

Fig. 4. Characteristic curves for incandescent lamps.

$$\frac{\text{life}}{\text{LIFE}}=\left(\frac{\text{LUMENS}}{\text{lumens}}\right)^a=\left(\frac{\text{LUMENS/WATT}}{\text{lumens/watt}}\right)^b=\left(\frac{\text{VOLTS}}{\text{volts}}\right)^d=\left(\frac{\text{AMPERES}}{\text{amperes}}\right)^u \quad (1)$$

$$\frac{\text{lumens}}{\text{LUMENS}}=\left(\frac{\text{volts}}{\text{VOLTS}}\right)^k=\left(\frac{\text{lumens/watt}}{\text{LUMENS/WATT}}\right)^h=\left(\frac{\text{watts}}{\text{WATTS}}\right)^s=\left(\frac{\text{amperes}}{\text{AMPERES}}\right)^y=\left(\frac{\text{ohms}}{\text{OHMS}}\right)^z \quad (2)$$

$$\frac{\text{LUMENS/WATT}}{\text{lumens/watt}}=\left(\frac{\text{LUMENS}}{\text{lumens}}\right)^f=\left(\frac{\text{VOLTS}}{\text{volts}}\right)^g=\left(\frac{\text{AMPERES}}{\text{amperes}}\right)^j \quad (3)$$

$$\frac{\text{amperes}}{\text{AMPERES}}=\left(\frac{\text{volts}}{\text{VOLTS}}\right)^t \quad (4)$$

$$\frac{\text{watts}}{\text{WATTS}}=\left(\frac{\text{volts}}{\text{VOLTS}}\right)^n \quad (5)$$

Table 1. Exponents for Eqs. (1)–(5) in text

Exponent	Gas-filled	Vacuum
a	3.86	3.85
b	7.1	7.0
d	13.1	13.5
u	24.1	23.3
k	3.38	3.51
h	1.84	1.82
s	2.19	2.22
y	6.25	6.05
z	7.36	8.36
f	9.544	0.550
g	1.84	1.93
j	3.40	3.33
t	0.541	0.580
n	1.54	1.58

from Table 1, the unknown characteristic may be determined. The exponents in Table 1 are changed at times, but for practical purposes and for normal voltage range those given will suffice.

Table 2 gives the effect, in percentage, of operating incandescent lamps below normal voltage. Low voltage increases lamp life, but the economic sacrifice of light advises against such practice.

Applications and special types. Incandescent lamps have been developed for many services. Most common are those used in general service and the miniature lamp. Special types have been developed for rough service applications, bake-oven use, severe vibration applications, show-case lamps, multiple lights (three-way lamp), sign lamps, spotlights, floodlights, and insect-control lamps.

Series street-lighting lamps are designed with heavy filaments. The circuit operates at a constant current. The voltage is adjusted to the number of burning lamps to keep the current constant.

Reflector-type lamps are designed with built-in light control, that is, silvering placed on a portion of the lamp's outside surface. The reflecting surface is effective for the life of the lamp.

Projector-type lamps have molded-glass reflectors, silvered inside the lamp cavity, with either a clear glass cover or a molded control-lens cover. The reflector and the cover are sealed together, forming a lamp with an internal reflector. The parts are of hard glass, and the lamps may be used for outdoor service in floodlight and spotlight installations. This is also the type of lamp used in the sealed-beam headlight for automobiles, locomotives, and airplanes. A high degree of accuracy—just short of optical accuracy—is achieved by molding the contours of the reflector, thereby obtaining accurate beam control. With this sturdy structure the filament can be positioned for the best use of the lens, and the lamp has little depreciation during its life. Being constructed of hard glass, it lends itself to high-wattage use.

Miniature incandescent lamps are used in many fields and in many pieces of equipment, from the ordinary flashlight to the "grain of wheat" lamps used in surgical and dental instruments. These

Table 2. Typical performance of large incandescent lamps burned below rated voltage

Percentage of normal operation			
Voltage	Light output	Watts	Efficiency
100.0	100.0	100.0	100.0
99.2	97.3	98.8	98.2
98.3	94.4	97.4	96.9
97.5	91.8	96.1	95.4
96.7	89.2	95.0	93.8
95.8	86.4	93.6	92.3
95.0	84.0	92.4	90.8
94.2	81.5	91.2	89.4
93.3	78.0	89.8	87.7
92.5	76.6	88.7	86.2
91.7	74.1	87.5	84.8
90.0	69.5	85.0	81.7
88.3	65.0	82.5	79.4
86.7	60.8	80.3	75.8
85.0	56.6	77.9	72.8
83.3	52.0	75.5	69.7

lamps are designed to give the highest efficiency consistent with the nature of the power source employed.

Special picture-projection lamps are designed for accurate filament location in the focal plane of the optical system, with the filament concentrated as much as possible in a single plane and in a small area. These precision lamps use a prefocus base for accurate positioning of the filament with respect to the base. Projector lamps run at high temperatures, and forced ventilation is frequently required.

A special class of lamp is designed for the photographic field, where the chief requirement is actinic quality. Frequently the most important rating is the color temperature, with little regard for economic efficiency or life. Photoflood lamps give high illumination for a short life, obtaining twice the lumens from high filament temperature, and high color temperature with three times the photographic effectiveness. The "daylight blue" photoflood lamp gives a very white light at 4800 K color temperature at 35.8 lu/W.

For other types of incandescent lamps *see* ARC LAMP; INFRARED LAMP.

[JOHN O. KRAEHENBUEHL]

Bibliography: Illuminating Engineering Society, *IES Lighting Handbook*, 6th ed., 1981; also see various manufacturers' publications.

Inclined plane

A plane surface inclined at an angle with the line of action of the force that is to be exerted or overcome. The usual application is illustrated in the diagram. A relatively small force acting parallel to the surface holds an object in place, or moves it at a constant speed.

In the free-body diagram shown here, three forces act on the object when no friction is present. The forces are its weight W, the force F_p parallel to the surface, and a force F_n normal to the surface. The summation of the forces acting in any direction on a body in static equilibrium equals zero; therefore, the summation of forces parallel to and forces normal to the surface are given by Eqs. (1) and (2). A force slightly greater than $W \sin \theta$ moves

$$F_p - W \sin \theta = 0 \tag{1}$$

$$F_n - W \cos \theta = 0 \tag{2}$$

the object up the incline, but the inclined plane supports the greater part of the weight of the object. The principal use of the inclined plane is as ramps for moving goods from one level to another. Wheels may be added to the object to be raised to decrease friction but the principle remains the same. The wedge and screw are closely related to the inclined plane and find wide application. *See* SIMPLE MACHINE.

[RICHARD M. PHELAN]

INCLINED PLANE

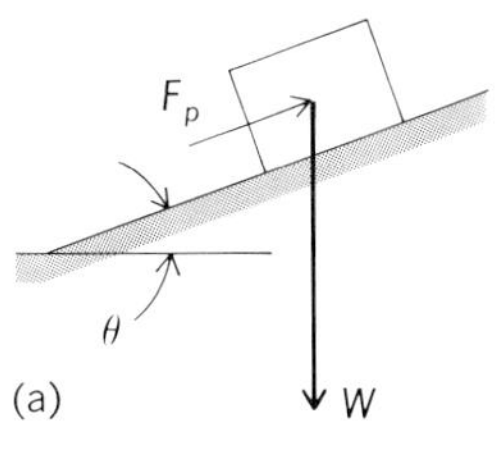

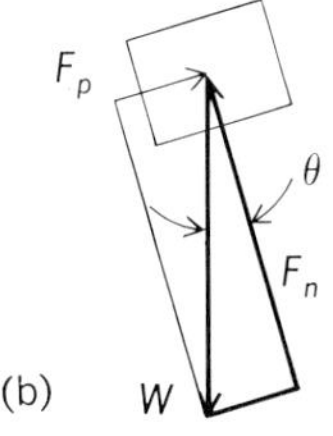

Weight resting on an inclined plane (*a*) with principal forces applied, and (*b*) their resolution into normal force.

Inductance

That property of an electric circuit or of two neighboring circuits whereby an electromotive force is induced (by the process of electromagnetic induction) in one of the circuits by a change of current in either of them. The term inductance coil is sometimes used as a synonym for inductor, a device possessing the property of inductance. *See* ELECTROMOTIVE FORCE (EMF).

Self-inductance. For a given coil, the ratio of the electromotive force of induction to the rate of change of current in the coil is called the self-inductance L of the coil, given in Eq. (1), where e is

$$L = -\frac{e}{dI/dt} \tag{1}$$

the electromotive force at any instant and dI/dt is the rate of change of the current at that instant. The negative sign indicates that the induced electromotive force is opposite in direction to the current when the current is increasing (dI/dt positive) and in the same direction as the current when the current is decreasing (dI/dt negative). The self-inductance is in henrys when the electromotive force is in volts, and the rate of change of current is in amperes per second.

An alternative definition of self-inductance is the number of flux linkages per unit current. Flux linkage is the product of the flux Φ and the number of turns in the coil N. Then Eq. (2) holds. Both

$$L = \frac{N\Phi}{I} \tag{2}$$

sides of Eq. (2) may be multiplied by I to obtain Eq. (3), which may be differentiated with respect to t,

$$LI = N\Phi \tag{3}$$

as in Eqs. (4). Hence the second definition is equivalent to the first.

$$L\frac{dI}{dt} = N\frac{d\Phi}{dt} = -e$$

$$\text{or} \quad L = -\frac{e}{dI/dt} \tag{4}$$

Self-inductance does not affect a circuit in which the current is unchanging; however, it is of great importance when there is a changing current, since there is an induced emf during the time that the change takes place. For example, in an alternating-current circuit, the current is constantly changing and the inductance is an important factor. Also, in transient phenomena at the beginning or end of a steady unidirectional current, the self-inductance plays a part.

Consider a circuit of resistance R and inductance L connected in series to a constant source of potential difference V. The current in the circuit does not reach a final steady value instantly, but rises toward the final value $I=V/R$ in a manner that depends upon R and L. At every instant after the switch is closed the applied potential difference is the sum of the iR drop in potential and the back emf $L\,di/dt$, as in Eq. (5), where i is

$$V=iR+L\frac{di}{dt} \qquad (5)$$

the instantaneous value of the current. Separating the variables i and t, one obtains Eq. (6). The solution of Eq. (6) is given in Eq. (7).

$$\frac{di}{(V/R)-i}=\frac{R}{L}dt \qquad (6)$$

$$i=\frac{V}{R}\left(1-e^{-(R/L)t}\right) \qquad (7)$$

The current rises exponentially to a final steady value V/R. The rate of growth is rapid at first, then less and less rapid as the current approaches the final value.

The power p supplied to the circuit at every instant during the rise of current is given by Eq. (8).

$$p=iV=i^2R+Li\,di/dt \qquad (8)$$

The first term i^2R is the power that goes into heating the circuit. The second term $Li\,di/dt$ is the power that goes into building up the magnetic field in the inductor. The total energy W used in building up the magnetic field is given by Eq. (9). The

$$W=\int_0^t p\,dt=\int_0^t Li\frac{di}{dt}dt=\int_0^I Li\,di=\tfrac{1}{2}LI^2 \qquad (9)$$

energy used in building up the magnetic field remains as energy of the magnetic field. When the switch is opened, the magnetic field collapses and the energy of the field is returned to the circuit, resulting in an induced emf. The arc that is often seen when a switch is opened is a result of this emf, and the energy to maintain the arc is supplied by the decreasing magnetic field.

Mutual inductance. The mutual inductance M of two neighboring circuits A and B is defined as the ratio of the emf induced in one circuit $\mathscr{E}$ to the rate of change of current in the other circuit, as in Eq. (10).

$$M=-\frac{\mathscr{E}_B}{(dI/dt)_A} \qquad (10)$$

The mks unit of mutual inductance is the henry, the same as the unit of self-inductance. The same value is obtained for a pair of coils, regardless of which coil is taken as the starting point.

The mutual inductance of two circuits may also be expressed as the ratio of the flux linkages produced in circuit B by the current in circuit A to the current in circuit A. If Φ_{BA} is the flux threading B as a result of the current in circuit A, Eqs. (11) hold. Integration leads to Eq. (12).

$$\mathscr{E}_B=-N_B\frac{d\Phi_{BA}}{dt}=-M\frac{dI_A}{dt} \qquad (11)$$

or

$$N_B\,d\Phi_{BA}=M\,dI_A$$

$$M=\frac{N_B\Phi_{BA}}{I_A} \qquad (12)$$

[KENNETH V. MANNING]

Induction coil

A device for producing a high-voltage alternating current or high-voltage pulses from a low-voltage direct current. The largest modern use of the induction coil is in the ignition system of internal combustion engines, such as automobile engines. Devices of similar construction, known as vibrators, are used as rectifiers and synchronous inverters. *See* IGNITION SYSTEM.

Figure 1 shows a typical circuit diagram for an induction coil. The primary coil, wound on the iron core, consists of only a few turns. The secondary coil, wound over the primary, consists of a large number of turns.

When the switch S is closed, the iron core becomes magnetized and therefore attracts the armature A. This automatically breaks the circuit to the coil through contact B and the armature. The armature is returned to its initial position by a spring and again makes contact with the contact B, restoring the circuit to the primary coil. The cycle is then repeated rapidly.

While current is flowing in the primary coil, a magnetic field is produced. When the contact between A and B is broken, the magnetic field collapses and induces a high voltage in the secondary coil, similar to transformer action. The self-inductance of the coil must be limited; therefore, the core is a straight bundle of iron wires, which minimize eddy-current losses, rather than a closed iron circuit as is used in a transformer. *See* TRANSFORMER.

The capacitor C is placed across the breaker contacts to reduce the voltage across the contacts at the moment of their opening and thus reduce sparking. Sparking is caused by the induced voltage in the primary winding resulting from the collapsing magnetic field. The capacitor allows some of the energy of the magnetic field to be converted to electrostatic energy in the capacitor, rather than into heat at the contacts.

Induction coils of a different type are used in telephone circuits to step up the voltage from the

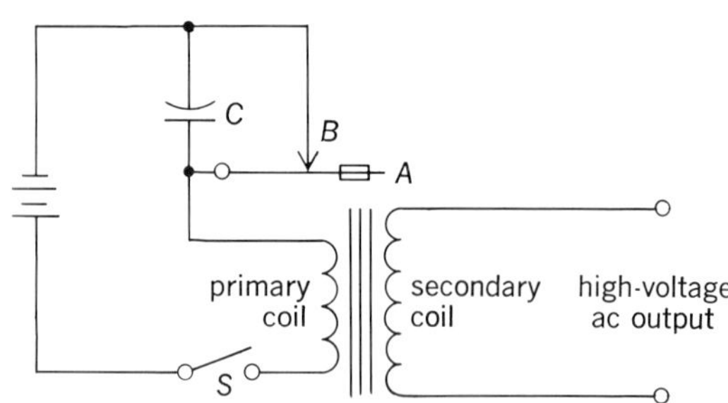

Fig. 1. Typical circuit for an induction coil.

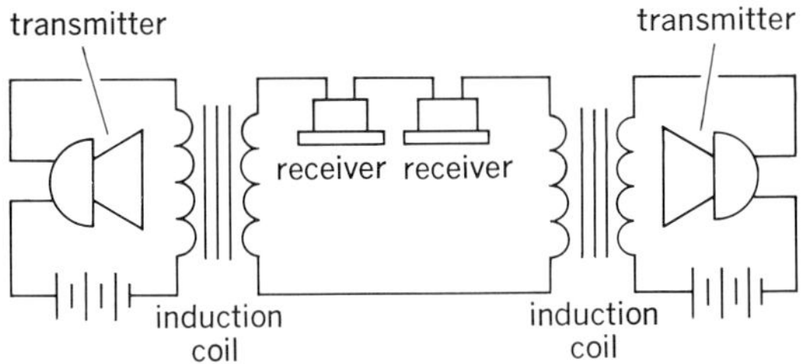

Fig. 2. Induction coils in telephone circuit.

transmitter and match the impedance of the line. The direct current in the circuit varies in magnitude at speech frequencies; therefore, no interrupter contacts are necessary. The battery and primary winding are connected in series with the transmitter as in Fig. 2. The secondary winding and the receiver are connected in series with the line. This circuitry reduces the required battery voltage.

Still another type of induction coil, called a reactor, is really a one-winding transformer designed to produce a definite voltage drop for a given current. *See* REACTOR (ELECTRICITY).

In 1892 Nicola Tesla used a form of induction coil to obtain currents of very high frequencies and high voltages. The oscillatory discharge of a Leyden jar was used as the interrupter. The terminals of the secondary of an induction coil are connected, one to the inner coating and the other to the outer coating of an insulated Leyden jar C_1 (Fig. 3).

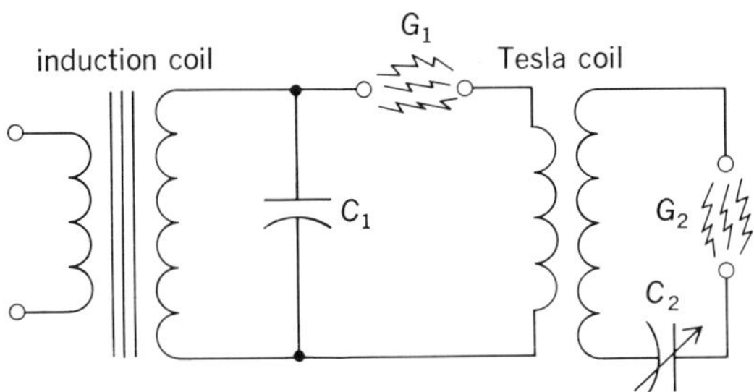

Fig. 3. Circuit diagram of Tesla coil.

The circuit is completed through the primary winding of the Tesla coil, and the primary gap G_1. The coils and the Leyden jar act as a resonant circuit in the production of the high-frequency oscillation.

The primary of the Tesla coil consists of a half-dozen turns of wires wound on a nonmagnetic core. The secondary consists of many turns. The two coils are separated by air or oil as insulation. The alternation from the Leyden jar may have a frequency of several million hertz. Hence, the current induced in the secondary is not only of high voltage but also of high frequency.

[NORMAN R. BELL]

Induction heating

The heating of a nominally electrical conducting material by eddy currents induced by a varying electromagnetic field. The principle of the induction heating process is similar to that of a transformer. In Fig. 1, the inductor coil can be considered the primary winding of a transformer, with the workpiece as a single-turn secondary. When an alternating current flows in the primary coil, secondary currents will be induced in the workpiece. These induced currents are called eddy currents. The current flowing in the workpiece can be considered as the summation of all of the eddy currents.

In the design of conventional electrical apparatus, the losses due to induced eddy currents are minimized because they reduce the overall efficiency. However, in induction heating, their maximum effect is desired. Therefore close spacing is used between the inductor coil and the workpiece, and high-coil currents are used to obtain the maximum induced eddy currents and therefore high heating rates. *See* CORE LOSS.

INDUCTION HEATING

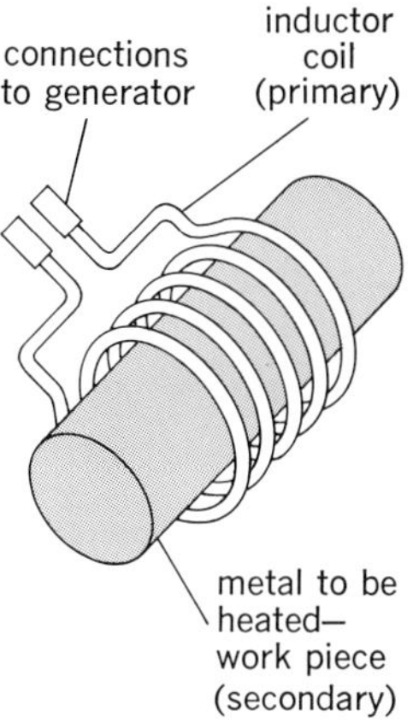

Fig. 1. Basic elements of induction heating.

Applications. Induction heating is widely employed in the metalworking industry for a variety of industrial processes. While carbon steel is by far the most common material heated, induction heating is also used with many other conducting materials such as various grades of stainless steel, aluminum, brass, copper, nickel, and titanium products. Induction heating is widely employed in (1) melting, holding, and superheating of ferrous and nonferrous metals in coreless and channel furnaces; (2) forging, forming, and rolling of slabs, billets, bars, and portions of bars; (3) heat treatment, such as hardening, annealing, and tempering applications; these processes are performed as continuous in-line systems or with automated handling of discrete parts to improve the metal's physical properties; (4) surface conditioning, such as curing of coatings, sintering, and semiconductor processing; (5) metal joining such as welding, brazing, and soldering. *See* BRAZING; HEAT TREATMENT (METALLURGY); SOLDERING; WELDING AND CUTTING OF METALS.

The advantages of induction heating over the conventional processes (like fossil furnace or salt-bath heating) are the following: (1) Heating is induced directly into the material. It is therefore an extremely rapid method of heating. It is not limited by the relatively slow rate of heat diffusion in conventional processes using surface-contact or radiant heating methods. (2) Because of skin effect, the heating is localized and the heated area is easily controlled by the shape and size of the inductor coil. (3) Induction heating is easily controllable, resulting in uniform high quality of the product. (4) It lends itself to automation, in-line processing, and automatic-process cycle control. (5) Startup time is short, and standby losses are low or nonexistent. (6) Working conditions are better because of the absence of noise, fumes, and radiated heat.

Heating process. The induced currents in the workpiece decrease exponentially from the surface toward the center of the material being heated. This current flowing through the material's own resistivity generates heat. The depth of heat penetration is defined as the depth to which 87% of the heat is generated. It is proportional to the square root of the material's resistivity divided by the square root of its permeability and frequency of the ac power applied to the inductor ($\delta_W = \sqrt{p/\mu f}$).

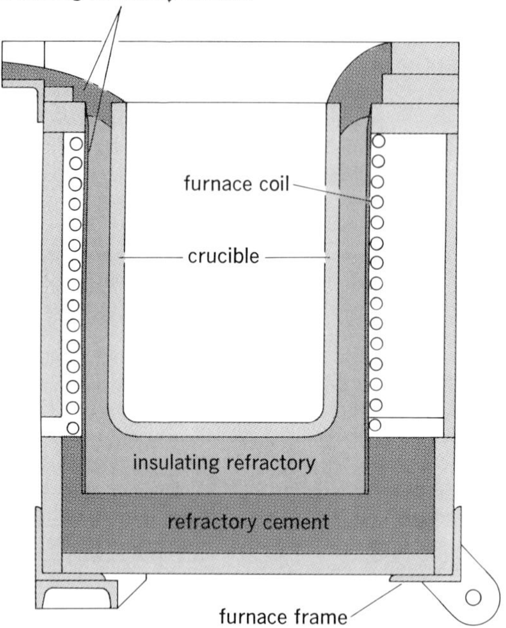

Fig. 2. Cross section of a typical induction melting furnace. (*Inductotherm Corp.*)

The effective depth of heat penetration is therefore greater for lower frequencies and higher-resistivity materials. In magnetic materials such as steel, the depth of current penetration is less below the Curie temperature (approximately 1350°F or 730°C where the steel is magnetic) than it is above the Curie temperature.

The thermal energy required can easily be calculated by the formula below.

$$P_{th} = MC\Delta T \times 10^{-3}\ \text{kW}$$
$$= 17.6M'C'\Delta T' \times 10^{-3}\ \text{kW}$$

P_{th} = rate at which thermal energy is required
M = kilograms of product heated per second
C = specific heat, (J/kg)/°C
ΔT = temperature rise in °C
M' = pounds of product heated per minute
C' = specific heat, (Btu/lb)/°F
$\Delta T'$ = temperature rise in °F

When the proper frequency is employed, the overall heating efficiency averages 75% for magnetic materials, and 30–50% for nonmagnetic materials. For efficient heating, the frequency employed must be high enough so that the depth of current penetration is less than one-third the diameter or cross section of the material being heated. The table shows the range of frequencies used for applications of the induction heating process.

Since the capital equipment cost (in dollars per kilowatt) increases with the frequency of the power supply, the lowest frequency that will properly heat the part is employed. Where high-volume production is involved, as in forging applications, line frequency and dual frequencies with separate induction coils are often employed to maximize production at minimum equipment costs. When the workpieces are small, it is necessary to use higher frequencies to efficiently heat the part. Likewise, higher frequencies must be used when it

Fig. 3. A 1600-kW bar end heating system for feeding an automatic upsetter for forging flanges on automotive axles at a production rate of 720 per hour. (*Westinghouse Electric Corp.*)

Frequencies used in induction heating

Frequency	Power source	Application
60 Hz	Line	Melting
180 Hz – 3 kHz	Solid state	
60 Hz	Line	Forging, forming, rolling
180 Hz – 10kHz	Solid state	
3 – 10 kHz	Solid-state	Heat treatment and surface conditioning
200 – 500 kHz	Vacuum tube oscillators	
10 kHz	Solid-state	Metals joining
200 – 500 kHz	Vacuum tube oscillators	

is necessary to concentrate the heat near the surface, as in surface hardening applications.

Power sources. The equipment used as power sources depends on the frequency required for the application. When line frequencies (generally 60 Hz) are used, suitable transformers, power factor correction capacitors, and control equipment are employed.

For higher frequencies, up to 10,000 Hz, solid-state power supplies have replaced rotating equipment (motor generators). These power supplies utilize high-power thyristor solid-state devices to generate frequencies of 180 Hz to 10 kHz and ratings from 30 to 2000 kW. The conversion efficiencies of these power supplies average 90% or better as compared with the motor generator equipment with conversion efficiencies of only about 75%.

Process use. Induction heating is used for many processes as shown in the table. The construction of a typical melting furnace is shown in Fig. 2. A typical bar-heating system is shown in Fig. 3. Many different types of material-handling equipment such as scanners, indexing fixtures, conveyors, and lift and carry mechanisms are used with automated induction heating systems. *See* FURNACE CONSTRUCTION.

Induction heating differs from other methods in its high speed of heating and in its ability to generate heat from within the metal itself. In applications such as hardening, time and temperature affect the end product. Therefore slightly different heating cycles, temperatures, and materials are selected to provide the desired metallurgical properties in the end product. For other heating methods *see* ELECTRIC HEATING.

[GEORGE F. BOBART]

Bibliography: G. F. Bobart, Electric induction heating for metalworking, *Plant Eng.* 31(25): 137–140, Dec. 8, 1977; D. W. Brown, *Induction Heating Practice*, 1956; P. G. Simpson, *Induction Heating: Coil and System Design*, 1960; C. A. Tudbury, *Basics of Induction Heating*, 1960.

Induction motor

An alternating-current motor in which the currents in the secondary winding (usually the rotor) are created solely by induction. These currents result from voltages induced in the secondary by the magnetic field of the primary winding (usually the stator). An induction motor operates slightly below synchronous speed and is sometimes called an asynchronous (meaning not synchronous) motor.

Induction motors are the most commonly used electric motors because of their simple construction, efficiency, good speed regulation, and low cost. Polyphase induction motors come in all sizes and find wide use where polyphase power is available. Single-phase induction motors are found mainly in fractional-horsepower sizes, and those up to 25 hp are used where only single-phase power is available.

POLYPHASE INDUCTION MOTORS

There are two principal types of polyphase induction motors: squirrel-cage and wound-rotor machines. The differences in these machines is in the construction of the rotor. The stator construction is the same and is also identical to the stator of a synchronous motor. Both squirrel-cage and wound-rotor machines can be designed for two- or three-phase current.

Stator. The stator of a polyphase induction motor produces a rotating magnetic field when supplied with balanced, polyphase voltages (equal in magnitude and 90 electrical degrees apart for two-phase motors, 120 electrical degrees apart for three-phase motors). These voltages are supplied to phase windings, which are identical in all respects. The currents resulting from these voltages produce a magnetomotive force (mmf) of constant magnitude which rotates at synchronous speed. The speed is proportional to the frequency of the supply voltage and inversely proportional to the number of poles constructed on the stator.

Figure 1 is a simplified diagram of a three-phase, two-pole, Y-connected stator supplied with currents I_1, I_2, and I_3. Each stator winding produces a pulsating mmf which varies sinusoidally with time. The resultant mmf of the three windings (Fig. 1*c*) is constant in magnitude and rotates at synchronous speed. Figure 1*b* shows the direction of the mmf in the stator for times t_1, t_2, and t_3 shown in Fig. 1*a* and shows how the resultant mmf rotates. The synchronous speed N_s is shown by Eq. (1), where f is the frequency in hertz and p is the

$$N_s = \frac{120f}{p} \text{ rpm} \tag{1}$$

number of stator poles. For any given frequency of operation, the synchronous speed is determined by the number of poles. For 60-Hz frequency, a two-pole motor has a synchronous speed of 3600 rpm; a four-pole motor, 1800 rpm; and so on. For details of stator windings *see* WINDINGS IN ELECTRIC MACHINERY.

Squirrel-cage rotor. Figure 2 shows the bars, end rings, and cooling fins of a squirrel-cage rotor. The bars are skewed or angled to prevent cogging (operating below synchronous speed) and to reduce noise. The end rings provide paths for currents that result from the voltages induced in the rotor bars by the stator flux. The number of poles on a squirrel-cage rotor is always equal to the number of poles created by the stator winding.

Figure 3 shows how the two motor elements interact. A counterclockwise rotation of the stator flux causes voltages to be induced in the top bars of the rotor in an outward direction and in the bottom bars in an inward direction. Currents will flow in these bars in the same direction. These currents interact with the stator flux and produce a force on the rotor bars in the direction of the rotation of the stator flux.

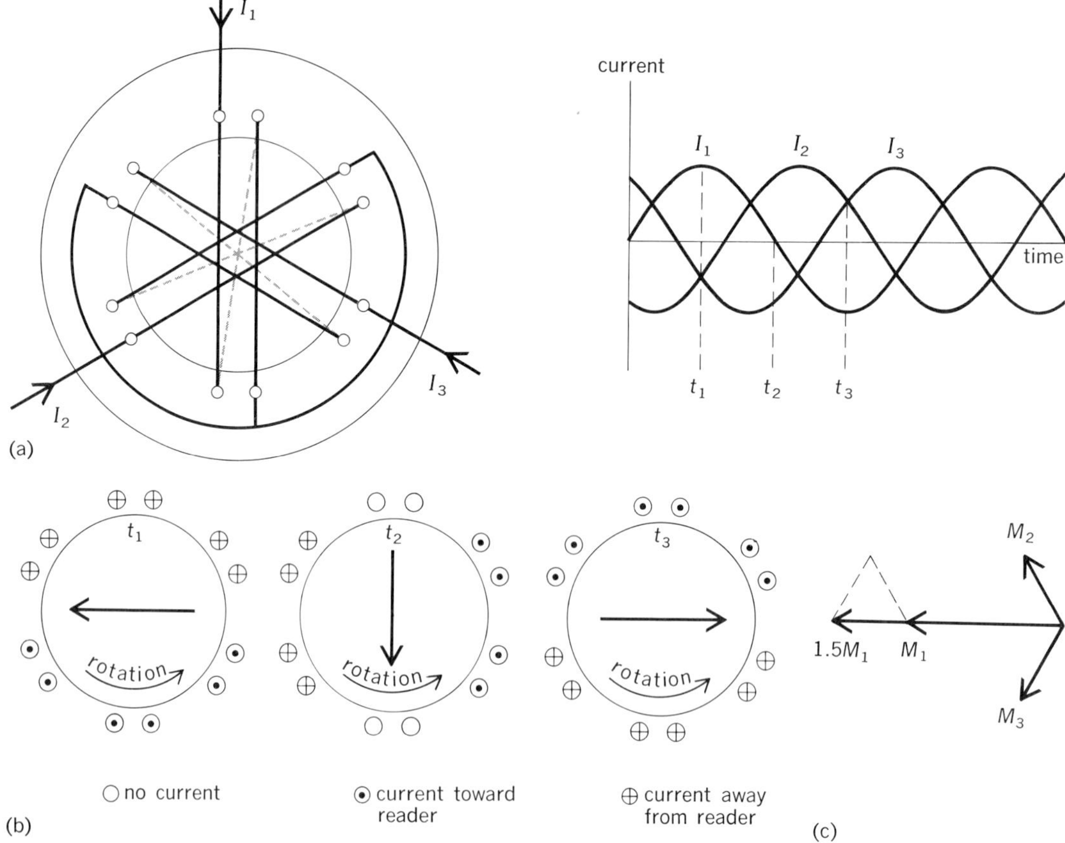

Fig. 1. Three-phase, two-pole, Y-connected stator of induction motor supplied with currents I_1, I_2, and I_3. (a) Stator windings and currents. (b) Rotating field. (c) Magnetomotive forces produced by stator winding.

When not driving a load, the rotor approaches synchronous speed N_s. At this speed there is no motion of the flux with respect to the rotor conductors. As a result, there is no voltage induced in the rotor and no rotor current flows. As load is applied, the rotor speed decreases slightly, causing an increase in rotor voltage and rotor current and a consequent increase in torque developed by the rotor. The reduction in speed is therefore sufficient to develop a torque equal and opposite to that of the load. Light loads require only slight reductions in speed; heavy loads require greater reduction. The difference between the synchronous speed N_s and the operating speed N is the slip speed. Slip s is conveniently expressed as a percentage of synchronous speed, as in Eq. (2).

$$s = \frac{N_s - N}{N_s} \times 100\% \qquad (2)$$

When the rotor is stationary, a large voltage is induced in the rotor. The frequency of this rotor voltage is the same as that of the supply voltage. The frequency f_2 of rotor voltage at any speed is shown by Eq. (3), where f_1 is the frequency of the supply voltage and s is the slip expressed as a decimal. The voltage e_2 induced in the rotor at any speed is shown by Eq. (4), where e_{2s} is the rotor voltage at standstill. The reactance x_2 of the rotor is a function of its standstill reactance x_{2s} and slip, as shown by Eq. (5). The impedance of the rotor at any speed is determined by the reactance x_2 and the rotor resistance r_2. The rotor current i_2 is shown by Eq. (6). In the equation, for small

$$f_2 = f_1 s \qquad (3)$$

$$e_2 = (e_{2s})s \qquad (4)$$

$$x_2 = (x_{2s})s \qquad (5)$$

$$i_2 = \frac{e_2}{\sqrt{r_2^2 + x_2^2}} = \frac{(e_{2s})s}{\sqrt{r_2^2 + (x_{2s})^2 s^2}} = \frac{e_{2s}}{\sqrt{\left(\frac{r_2}{s}\right)^2 + (x_{2s})^2}} \qquad (6)$$

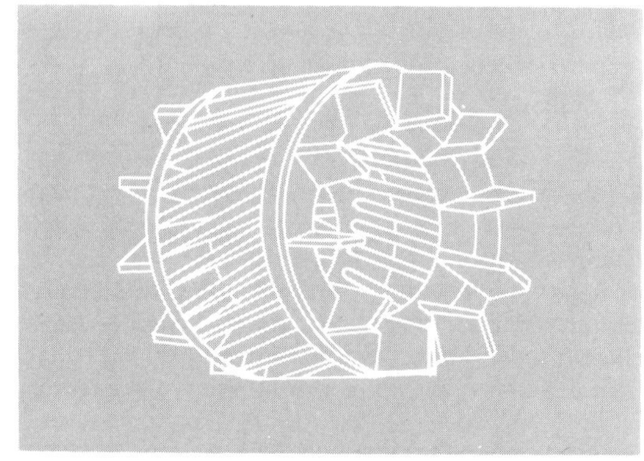

Fig. 2. Bars, end rings, and cooling fins of a squirrel-cage rotor.

values of slip, the rotor current is small and possesses a high power factor. When slip becomes large, the r_2/s term becomes small, current increases, and the current lags the voltage by a large phase angle. Standstill (or starting) current is large and lags the voltage by 50–70°. Only in-phase, or unity-power-factor, rotor currents are in space phase with the air-gap flux and can therefore produce torque. The current i_2 contains both a unity power-factor component i_p and a reactive component i_r. The maximum value of i_p and therefore maximum torque are obtained when slip is of the correct value to make r_2/s equal to x_{2s}. If the value of r_2 is changed, the slip at which maximum torque is developed must also change. If r_2 is doubled and s is doubled, the current i_2 is not changed and the torque is unchanged.

This feature provides a means of changing the speed-torque characteristics of the motor. In Fig. 4, curve 1 shows a typical characteristic curve of an induction motor. If the resistance of the rotor bars were doubled without making any other changes in the motor, it would develop the characteristic of curve 2, which shows twice the slip of curve 1 for any given torque. Further increases in the rotor resistance could result in curve 3. When r_2 is made equal to x_{2s}, maximum torque will be developed at standstill, as in curve 4. These curves show that higher resistance rotors give higher starting torque. However, since the motor's normal operating range is on the upper portion of the curve, the curves also show that a higher-resistance rotor results in more variation in speed from no load to full load (or poorer speed regulation) than the low-resistance rotor. Higher-resistance rotors also reduce motor efficiency. Except for their characteristic low starting torque, low-resistance rotors would be desirable for most applications.

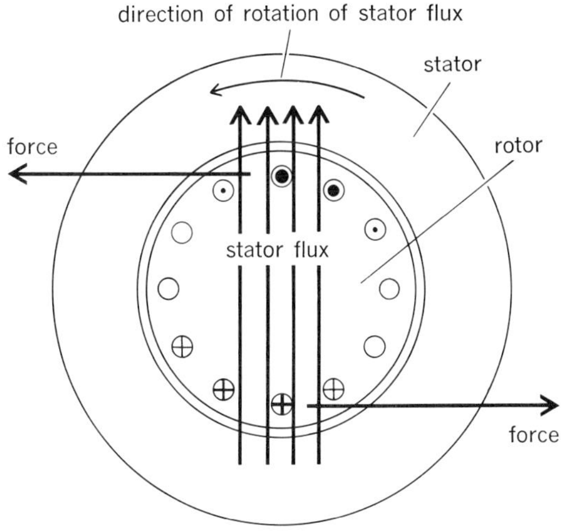

Fig. 3. Forces on the rotor winding.

Wound rotor. A wound-rotor induction motor can provide both high starting torque and good speed regulation. This is accomplished by adding external resistance to the rotor circuit during starting and removing the resistance after speed is attained.

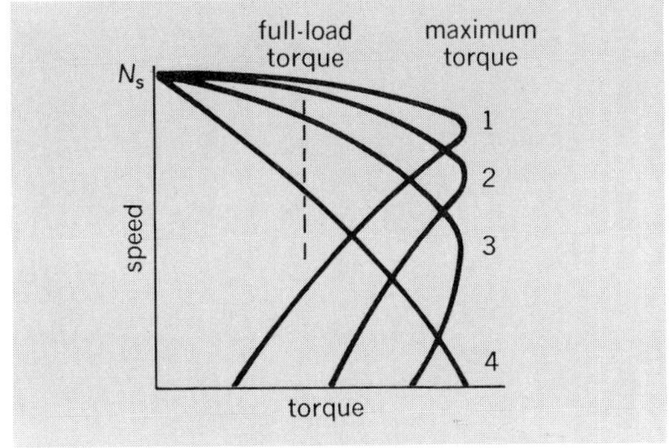

Fig. 4. Speed-torque characteristic of polyphase induction motor.

The wound rotor has a polyphase winding similar to the stator winding and must be wound for the same number of poles. Voltages are induced in these windings just as they are in the squirrel-cage rotor bars. The windings are connected to slip rings so that connections may be made to external impedances, usually resistors, to limit starting currents, improve power factor, or control speed.

Figure 5 shows the connection of a rheostat used to bring a wound-rotor motor up to speed. The rheostat limits the starting current drawn from the supply to a value less than that required by a squirrel-cage motor. The resistance is gradually reduced to bring the motor up to speed. By leaving various portions of the starting resistances in the circuit, some degree of speed control can be obtained, as in Fig. 4. However, this method of speed control is inherently inefficient and converts the motor into a variable-speed motor, rather than an essentially constant-speed motor. For other means of controlling speed of polyphase induction motors and for other types of ac motors *see* ALTERNATING-CURRENT MOTOR.

SINGLE-PHASE INDUCTION MOTORS

Single-phase induction motors display poorer operating characteristics than polyphase machines, but are used where polyphase voltages are not available. They are most common in small sizes (1/2 hp or less) in domestic and industrial applications. Their particular disadvantages are low

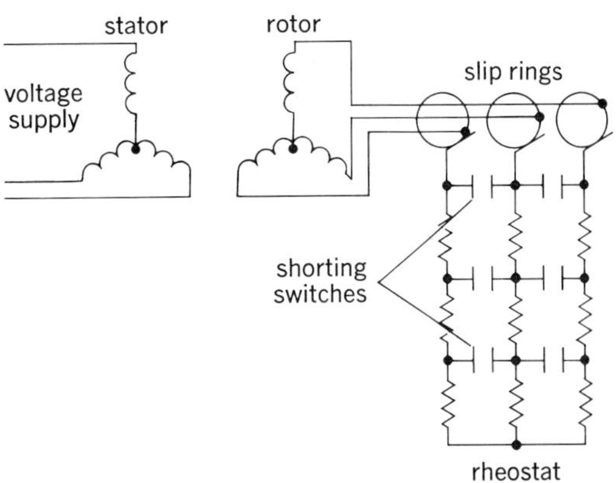

Fig. 5. Connections of wound-rotor induction motor.

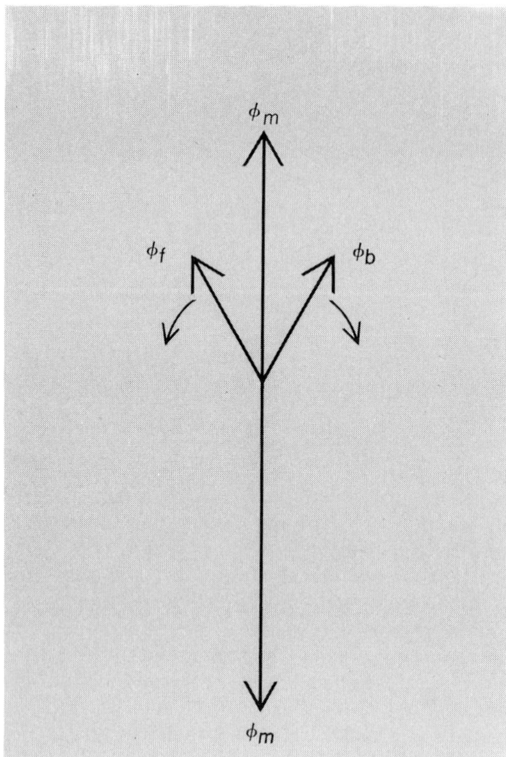

Fig. 6. Fluxes associated with the single-phase induction motor.

power factor, low efficiency, and the need for special starting devices.

The rotor of a single-phase induction motor is of the squirrel-cage type. The stator has a main winding which produces a pulsating field. At standstill, the pulsating field cannot produce rotor currents that will act on the air-gap flux to produce rotor torque. However, once the rotor is turning, it produces a cross flux at right angles in both space and time with the main field and thereby produces a rotating field comparable to that produced by the stator of a two-phase motor.

An explanation of this is based on the concept that a pulsating field is the equivalent of two oppositely rotating fields of one-half the magnitude of the resultant pulsating field. In Fig. 6, ϕ_m is the maximum value of the stator flux ϕ, which is shown only by its two components ϕ_f and ϕ_b, which represent the two oppositely rotating fields of constant equal magnitudes of $\phi_m/2$. Each component ϕ_f and ϕ_b produces a torque T_f and T_b on the rotor. Figure 7 shows that the sum of these torques is zero when speed is zero. However, if started, the sum of the torques is not zero and rotation will be maintained by the resultant torque.

This machine has good performance at high speed. However, to make this motor useful, it must have some way of producing a starting torque. The method by which this starting torque is obtained designates the type of the single-phase induction motor.

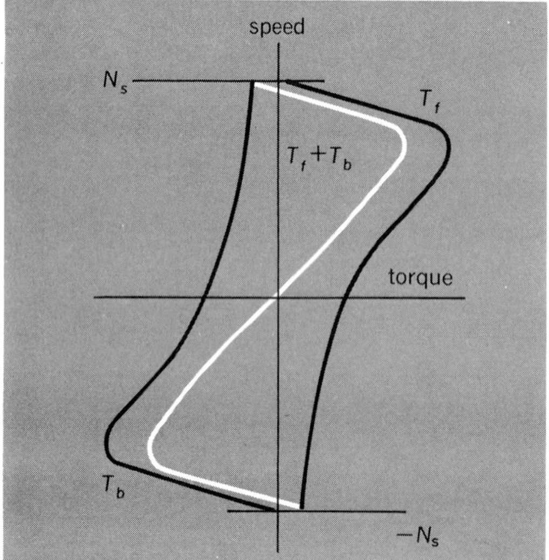

Fig. 7. Torques produced in the single-phase induction motor.

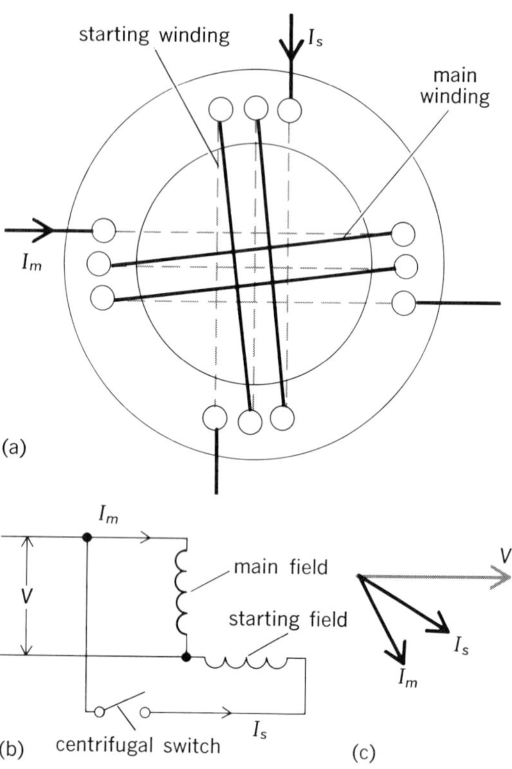

Fig. 8. Split-phase motor. (*a*) Windings. (*b*) Winding connections. (*c*) Vector diagram.

Split-phase motor. This motor has two stator windings, the customary main winding and a starting winding located 90 electrical degrees from the main winding, as in Fig. 8*a*. The starting winding has fewer turns of smaller wire, to give a higher resistance-to-reactance ratio, than the main winding. Therefore their currents I_m (main winding) and I_s (starting winding) are out of time phase, as in Fig. 8*c*, when the windings are supplied by a common voltage V. These currents produce an elliptical field (equivalent to a uniform rotating field superimposed on a pulsating field) which causes a unidirectional torque at standstill. This torque will start the motor. When sufficient speed has been attained, the circuit of the starting winding can be opened by a centrifugal switch and the motor will operate with a characteristic illustrated by the broken-line curve of Fig. 7.

Capacitor motor. The stator windings of this motor are similar to the split-phase motor. However, the starting winding is connected to the supply through a capacitor (Fig. 9*a*). This results in a starting winding current which leads the applied voltage. The motor then has winding currents at standstill which are nearly 90° apart in time, as well as 90° apart in space. High starting torque and high power factor are therefore obtained. The starting winding circuit can be opened by a centrifugal switch when the motor comes up to speed. A typical characteristic is shown in Fig. 9*c*.

In some motors two capacitors are used. When the motor is first connected to the voltage supply, the two capacitors are used in parallel in the starting circuit. At higher speed one capacitor is removed by a centrifugal switch, leaving the other in series with the starting winding. This motor has high starting torque and good power factor.

Shaded-pole motor. This motor is used extensively where large power and large starting torque are not required, as in fans. A squirrel-cage rotor is used with a salient-pole stator excited by the ac supply. Each salient pole is slotted so that a portion of the pole face can be encircled by a short-circuited winding, or shading coil.

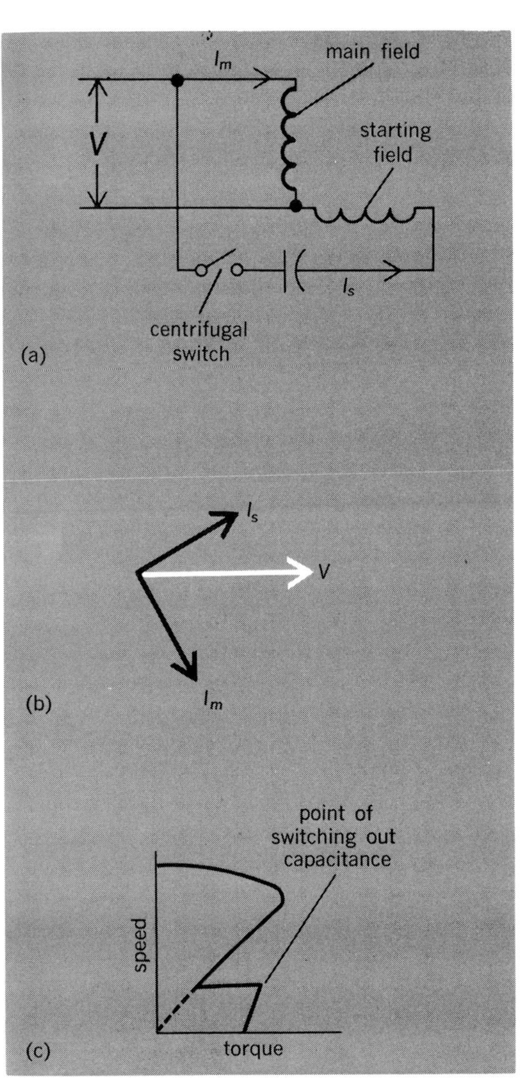

Fig. 9. Capacitor motor. (*a*) Winding connections. (*b*) Vector diagram. (c) Characteristic.

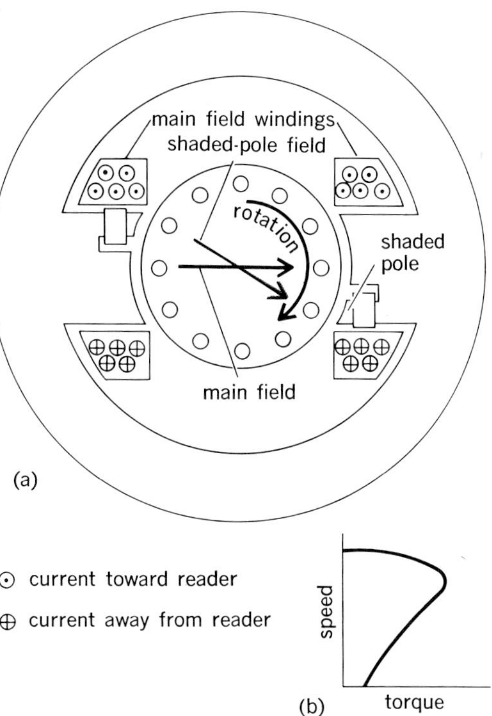

Fig. 10. Shaded-pole motor. (*a*) Cross-sectional view. (*b*) Typical characteristic.

The main winding produces a field between the poles as in Fig. 10. The shading coils act to delay the flux passing through them, so that it lags the flux in the unshaded portions. This gives a sweeping magnetic action across the pole face, and consequently across the rotor bars opposite the pole face, and results in a torque on the rotor. This torque is much smaller than the torque of a split-phase motor, but it is adequate for many operations. A typical characteristic of the motor is shown in Fig. 10*b*.

For other single-phase alternating-current motors *see* REPULSION MOTOR; UNIVERSAL MOTOR. For synchronous motors built for single-phase *see* HYSTERESIS MOTOR; RELUCTANCE MOTOR.

Linear motor. Figure 11 illustrates the arrangements of the elements of the polyphase squirrel-cage induction motor. The squirrel cage (secondary) is embedded in the rotor in a manner to provide a close magnetic coupling with the stator winding (primary). This arrangement provides a small air gap between the stator and the rotor. If the squirrel cage is replaced by a conducting sheet as in Fig. 11*b*, motor action can be obtained. This machine, though inferior to that of Fig. 11*a*, will function as a motor. If the stator windings and iron are unrolled (rectangular laminations instead of circular laminations), the arrangement of the elements will take a form shown in Fig. 11*c*, and the field produced by polyphase excitation of the primary winding will travel in a linear direction instead of a circular direction. This field will produce a force on the conducting sheet that is in the plane of the sheet and at right angles with the stator conductors. A reversal of the phase rotation of the primary voltages will reverse the direction of motion of the air-gap flux and thereby reverse the force on the

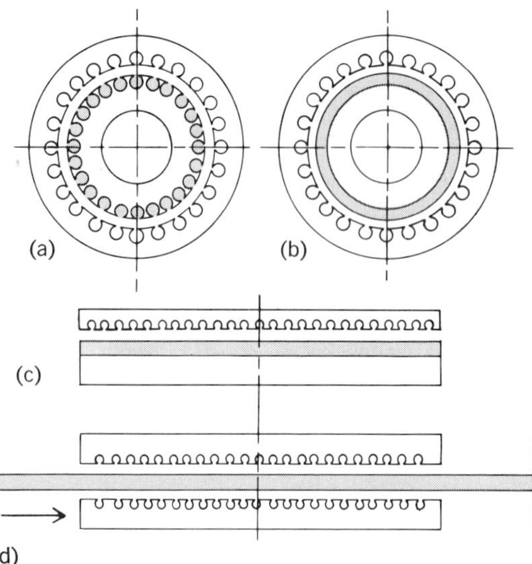

Fig. 11. Evolution of linear induction motor. (*a*) Polyphase squirrel-cage induction motor. (*b*) Conduction sheet motor. (*c*) One-sided long-secondary linear motor with short-flux return yoke. (*d*) Double-sided long-secondary linear motor.

secondary sheet. No load on the motor corresponds to the condition when the secondary sheet is moving at the same speed as the field produced by the primary. For the arrangement of Fig. 11*c*, there is a magnetic attraction between the iron of the stator and the iron of the secondary sheet. For some applications, this can be a serious disadvantage of the one-sided motor of Fig. 11*c*. This disadvantage can be eliminated by use of the double-sided arrangement of Fig. 11*d*. Here the primary iron of the upper and lower sides is held together rigidly, and the forces that are normal to the plane of the sheet do not act on the sheet.

Applications. In conventional transportation systems, traction effort is dependent on the contact of the wheels with the ground. In some cases, locomotives must be provided with heavy weights to keep the wheels from sliding when under heavy loads. This disadvantage can be eliminated by use of the linear motor. Through use of the linear motor with air cushions, friction loss can be reduced and skidding can be eliminated. Figure 12 illustrates the application of the double-sided linear motor for high-speed transportation.

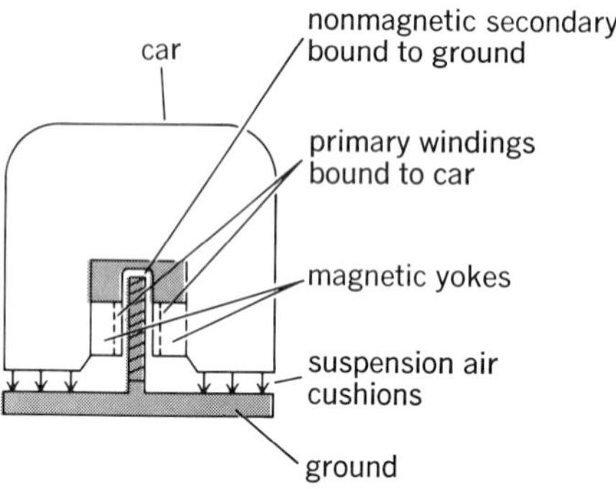

Fig. 12. Drawing of linear induction-motor configuration. Magnetic interaction supplies the forces between the car and the ground. (*Le Moteur Lineaire and Societé de l'Aerotrain*)

Conveying systems that are operated in limited space have been driven with linear motors. Some of these, ranging from 1/2 to 1 mi (805–1609 m) in length, have worked successfully.

Because of the large effective air gap of the linear motor, its magnetizing current is larger than that of the conventional motor. Its efficiency is somewhat lower, and its cost is high. The linear motor is largely in the experimental stage.

[ALBERT G. CONRAD]

Bibliography: P. D. Agarwal and T. C. Wang, Evaluation of fixed and moving primary linear induction motor systems, *Proc. IEEE*, pp. 631–637, May 1973; P. L. Alger, *Induction Machines: Behavior and Uses*, 1970; J. H. Dannon, R. N. Day, and G. P. Kalman, A Linear-induction motor propulsion system for high-speed ground vehicles, *Proc. IEEE*, pp. 621–630, May 1973; E. R. Laithwaite, Linear electric machine: A personal view, *Proc. IEEE*, pp. 220–290, February 1975; M. G. Say, *Alternating Current Machines*, 4th ed., 1976; David Schieber, Principles of operation of linear induction devices, *Proc. IEEE*, pp. 647–656, May 1973.

Industrial cost control

A specific system or procedure for monitoring manufacturing costs to achieve profitability goals. The industrial cost control system should provide the framework for improving product profitability by identifying the variations between actual and target performance, discovering the causes of this variation, and eliminating the detrimental causes of cost variation.

Controlling costs is crucial to a firm's profitability. In a competitive market, price is determined by the demand for the product and the prices of competing goods and services. Profit exists only to the extent that cost is lower than income. To maintain its financial position, a firm must control its costs to avoid deterioration in its level of profits. Effective cost control requires a system that provides current data which can be used to monitor and rapidly correct the manufacturing operation or process. Such a system usually exists outside the cost accounting system because of the need for rapid and specialized reports.

In industrial cost control, performance of the control system is usually measured by the cost control performance of the factors of direct labor, material, and manufacturing burden or overhead. Methods for determining the standard or target cost for each one of these three elements will be discussed separately.

Direct labor control. Direct labor consists of the labor of employees which can be identified with the manufacture of a product or the providing of a service. To control these direct labor costs, a company must control its operations so that the amount of direct labor actually used will be equal to or less than the cost standard.

Historical data. Standard labor costs can be established either through the use of historical data on past performance or through the development of time standards for all operations required to produce the item. Using historical data is the simplest procedure. For example, parts are produced on an injection molding machine at a rate of 240 parts per hour. The worker monitoring the injection molding machine earns $6.00 per hour.

The direct labor cost assignable to each piece is then $6.00 ÷ 240, or $0.025. To extend the example further, assume that a finishing operation is required for each piece after the molding operation. A worker making $5.40 per hour can be expected to finish 60 pieces per hour. This adds $0.090 per piece, $5.40 ÷ 60, to the cost of the item. The cost of the total direct labor assignable to each item in this extended example is $0.025 for the molding operation plus $0.090 for the finishing operation, or a total of $0.115 per piece.

Standard times. Standard direct labor costs can also be established using standard times. A standard time is a unit-time value for the accomplishment of a work task as determined by the proper application of the appropriate work measurement techniques. Time and motion studies could be made of the tasks required to produce the item. Predetermined time standards could also be used to establish the unit-time. This latter method analyzes the basic motions required to perform a job and assigns a fixed time value to these motions. For example, the motions of a finishing operation might be: reach to the part, grasp the part, move the part to a grinding wheel, position the part on the wheel, grind the part, move the part to the conveyor belt, and release the part. Time values for these basic motions have been predetermined, based on factors such as distance traveled. Assigned values are published by sources such as the MTM Association for Standards and Research. The sum of the times for each of the basic motions yields a total time for performance of the operation. This total time is called the normal time. The standard time is the normal time plus the time for desired allowances such as normal work interruption, fatigue, and personal needs.

In summary, standard direct labor costs are established by determining which operations would be performed, how long the operations should take, and what classes of labor should perform the operations.

Variations in costs. To determine the cause of variations in the direct labor costs, it is necessary to know only three things: the standard time, the actual time, and the labor rate of the operator. If the actual time needed to produce the item is equal to or less than the standard time, and if the rate of the operator is equal to or less than that specified, the cost will be under control. If the standard time is exceeded, the target cost will be exceeded and the cost will be out of control. Likewise, if a higher-rated or higher-cost operator is used, even though the actual time equals the standard time, the cost will be out of control. The cost control system provides immediate identification of either cause of cost variation, operator rate, or operations time, so that corrective action may be taken. The most common cause of missing the cost target is exceeding the standard time in one or more of the manufacturing operations.

Cost control. The most obvious way to control direct labor cost is to compare actual time to standard time on a daily basis, establishing a daily rate variance. This can be done for each part, operation, and employee. The foreperson will then be able to identify detrimental variations from standard times and correct the causes of those variations on a daily or even hourly basis. Summary variance reports can also be provided to supervisory personnel. Control in this manner is simple, inexpensive, and effective.

Material control. A material control system must indicate whether or not the standard material cost per unit is being maintained at a level equal to or less than the target value. The standard material cost per unit is calculated by multiplying the standard unit quantity by the standard unit price. The standard unit quantity is the amount of material required to produce a unit of product, including a normal allowance for scrap. The standard unit quantity of material per item can be found by dividing the quantity of materials used in the manufacture of a group of items by the number of items produced. The standard quantity can also be found by examining the product design and material specifications, using the bill of material and a normal scrap allowance to determine the amount of material which should be used to produce an item.

Material usage. Whether or not the operator produces the expected number of parts from a given amount of material is usually a quality control problem. Excessive material usage is identified by examining production count records and scrap reports. These reports, coupled with a material requisition system which would affirm the amount of material used, provide a constant check on the amount of material per unit of production. The amount of material used will vary from item to item due to variations in material composition and quality, operator performance, and so forth. It is a common practice to establish statistical limits of allowable variation for material usage. The process is considered to be out of control when the amount of material used to produce an item exceeds the predetermined allowance. Regular reports, often compiled daily, provide information upon which to take corrective actions if necessary. Such actions would include machine adjustment, replacement of substandard or defective raw material or purchased components, or operator reorientation.

Net cost. The standard price per unit for materials is the total cost of the material up to the point that it is used in manufacturing. This is the net cost after deducting trade and quantity discounts. This cost includes identifiable costs such as freight, receiving, testing, and storage. The importance of including costs such as receiving, testing, and storage in the cost control system is that these costs can usually be controlled by the firm and immediate action can be taken if they exceed a target amount.

Price characteristics. Materials are generally divided into two classes based on price characteristics. The first class is purchased components which exhibit few price changes and are often bought on long-term contracts. Little control is required for these items, only an assurance that the cost paid on the long-term contract is compatible with the target price. The second class usually represents raw materials, such as steel, copper, or lumber, whose prices fluctuate regularly as commodities. Only if the material cost exceeds a prescribed level is there any concern. Then the question to be asked is whether the excessive costs are temporary or permanent. If the cost increases appear to be permanent, the firm should review whether a price increase is possible and justified.

Manufacturing burden. Manufacturing burden includes all costs involved in the manufacturing operation other than direct labor and material. Burden costs include items such as utilities, indirect labor for maintenance, supervision, and depreciation. They are a significant portion of the total cost of the manufacturing item, usually amounting to over 200% of the direct labor cost. These costs deserve, but seldom receive, closer attention than direct labor costs. Manufacturing burden costs are actually management costs, as they result from conscious management actions. As such, they are the costs which can be most effectively reduced and controlled through management action.

Target costs. The first step in controlling manufacturing burden costs is to identify each specific cost and create an account for that cost, showing the variance from the target cost. The majority of burden costs are fixed; that is, they are attributable to the time period involved, not to the number of items produced. Such fixed costs include rent, supervision, and depreciation. For fixed burden costs, the target value would be a constant figure derived from historical records. Other burden costs are variable, or related to some degree to the volume of goods produced. These costs could include utilities, or forms and supplies associated with production records. The relationship between burden costs and production volume or direct labor expense is often found by using simple linear regression. Using this relationship, a target burden cost can be established for a particular level of direct labor cost for control purposes. Target costs are related to comparative bases such as hours of direct labor, time periods, or number of pieces produced.

Burden cost. A correct target burden cost for a particular item must accurately reflect the costs assignable to that item. All products do not have the same overhead. In a multiproduct operation, it is typical for one product to require 75% of the invoicing, 50% of the engineering, and 40% of the time-study effort. Expenses vary with product or process complexity. The total burden cost must be properly allocated to each product. "Average costs" apply only to "average products."

Responsibility. Burden cost accounts must clearly spell out who is responsible for each item of cost. The responsible person must be able to control the cost in question. For example, the foreperson should not be responsible for controlling rent, as the foreperson has no control over the rental contract. After the manufacturing burden cost accounts have been created and target costs established, the individuals responsible for those costs can control the costs against the target standards, as was suggested for direct labor and material costs.

Indirect operations. Indirect labor operations should be studied just as rigorously as direct labor. Incentive plans should be established to improve performance, especially in areas such as maintenance. Reductions in burden costs are achieved by studying the operations contributing to burden costs and by reducing wasted time, duplicative effort, and unnecessary work.

Summary. After creating the control accounts for direct labor, material, and overhead costs and after defining target costs for each account, those costs deviating from the desired standard can be quickly identified. Corrective action can then be taken to bring deviant costs back to the target level. Industrial cost control can be a simple and effective way to achieve goals of continued profitability. *See* INDUSTRIAL ENGINEERING; INSPECTION AND TESTING; INVENTORY CONTROL; MASS PRODUCTION; OPERATIONS RESEARCH; PERFORMANCE RATING; PERT; PROCESS ENGINEERING; PRODUCTION ENGINEERING; PRODUCTION METHODS; QUALITY CONTROL.

[RAYMOND P. LUTZ]

Industrial engineering

As defined by the American Institute of Industrial Engineers, a branch of engineering "concerned with the design, improvement, and installation of integrated systems of people, material, equipment, and energy. It draws upon the specialized knowledge and skills in the mathematical, physical and social sciences together with the principles and methods of engineering analysis and design to specify, predict, and evaluate the results to be obtained from such systems."

Background. The advent of industrial engineering is frequently associated with the Industrial Revolution in the 19th century. It was during this period that household production was replaced by production in factories. Frederick W. Taylor is usually recognized as the father of industrial engineering and scientific management. His pioneering work in the design, measurement, planning, and scheduling of work from 1880 to the time of his death in 1915 was the impetus for the conceptualization and growth of industrial engineering. Taylor introduced the concepts of time study, methods engineering, tool standardization, costing methods, routing, employee job selection, and incentives.

The term classical has been applied to the traditional industrial engineering activities of Taylor and F. B. Gilbreth. These techniques have had great practical value, and for decades have been considered the hallmarks of industrial engineering. Modern industrial engineering techniques address the more quantitative computer-systems approach to the solution of industrial engineering problems.

The classical activities of methods, job standards, plant layout, and costs are vital to effective management; in fact, many of the quantitative systems models need data which are obtained by time study, synthetic time standards, or work sampling. For an industrial engineer to be effective necessarily demands the capability of evaluating and utilizing the appropriate classical and the more sophisticated management science–operations research techniques in solving unstructured, real-life problems.

Activities. Industrial engineering is similar to the other major engineering specializations (civil, mechanical, electrical, and chemical) in that it is concerned with analysis and design, and applying the laws and materials of nature to useful and constructive purposes. It is different from these other fields of engineering in that it is specifically concerned with equipment and systems in which people are an integral part. The industrial engineer must be able to use mathematics, materials, machinery, devices, chemistry, electricity, electronics, and so on, just as all of the other types of engineers. But, unlike them, the industrial engineer must also understand and be able to integrate people into his or her designs, and must know their physical, physiological, psychological, and other characteristics—singly and in groups.

The early, and still major, activities of industrial engineers include work methods analysis and improvement, work measurement and the establishment of standards, job and workplace design, plant layout, materials handling, wage rates and incentives, cost reduction, suggestion evaluation, production planning and scheduling, inventory control, maintenance scheduling, equipment evaluation, assembly line balancing, systems and procedures, overall productivity improvement, and special studies—all done, almost exclusively in the early days, in manufacturing industries. As the technology evolved, additional activities were added to this list. These include machine tool analysis; numerically controlled machine installation and programming; computer analysis, installation, and programming; linear programming; queueing and other operations research techniques; simulations; management information systems; value analysis; human factors engineering; human/machine systems design; ergonomics; biomechanics; and the use of robots and automation. Moreover, it was found that the industrial engineering techniques used successfully in the factory could also be applied in the office, laboratory, classroom, hospital (including the operating room), the government, the military, and other nonindustrial areas.

The industrial engineer of today is using computers more in his or her production and test equipment, controls, systems, and for analyses and special studies. Minicomputers and microprocessors are in wide and growing use.

Computers are already being used to eliminate much of the calculation drudgery of work measurement. The power of the computer also facilitates the synthesis of various methods configurations.

Computer aided manufacturing (CAM) makes use of the computer in an attempt to develop a total systems approach to the manufacturing process. It links together computer-aided design (CAD), automatically programmed tools, work measurement, sequencing, material handling, and inventory control. Computer interactive graphics are also being introduced in the design and analysis functions. *See* COMPUTER-AIDED DESIGN AND MANUFACTURING.

Numerical control (N/C) is the term usually associated with automatically programmed equipment. An N/C machine consists of a reader, a control unit, and the machine tool. Preprogrammed machine instructions on tape are transmitted to the control unit, which interprets the instructions and causes the machine tool to execute. While N/C has been used primarily for metal removal machinery, it has also been applied successfully in material storage, assembly, and packaging.

The capability of N/C can be enhanced by direct numerical control (DNC), which uses an on-line computer to provide the instructions directly to the machine. Further extensions are in the area of adaptive control systems and robots.

The production function is dynamic by its inherent activities, and traditional industrial engineers are frequently called upon to determine the impact of worker power changes, product mix variances, and equipment additions or deletions on facilities arrangement and line balancing in the manufacturing area. This task becomes increasingly more complex today with the additional energy computation and pollution level considerations. This complicated system of interrelated activities is almost impossible to evaluate manually; however, through the use of simulation on a computer, answers can be given to "what if" types of questions which provide significant input to the decision maker.

The age-long inventory control function can now be monitored constantly by a real-time computerized system. The computer greatly facilitates the storage and retrieval of historical and current information on the status of inventory and concomitant purchase orders. Computers can be programmed to evaluate the trade-offs between the savings in inventory investment and the activity required in the planning and control phases. *See* INVENTORY CONTROL.

In the early days of industrial engineering, safety engineering was considered an important element in the educational programs. However, in the 1950s and 1960s safety courses were dropped in many of the curricula as more sophisticated analytical techniques were added. However, the huge costs of industrial accidents, and the regulatory measures established by the Occupational Safety and Health Administration (OSHA) demand that safety be considered in the initial facilities design in a more in-depth manner than previously. It should be recognized that the criteria utilized for plant layout, such as minimizing material handling, minimizing congestion and hazards, and providing for good housekeeping, are compatible with the objectives associated with safety engineering. *See* INDUSTRIAL HEALTH AND SAFETY.

Plant layout is considered a classical industrial engineering activity; however, the traditional approach to facilities design has been supplemented by the use of computers in the design and calculation process of a layout, and also by the application of mathematical modeling and optimization to those problems. It must be emphasized that the basic concepts of plant layout are still important, and the computer and operations research methods must be skillfully blended with the traditional methods in order to obtain better answers to facilities problems which become increasingly more complex—for example, by inclusion of accident risk factors. *See* PLANT FACILITIES (INDUSTRY).

Systems concept. The trend in industrial engineering is from micro- to macrolevel structures, such as a computer-aided manufacturing system, a total information system design of a company, or the complex materials-handling function involving many interrelated departments.

Management of industrial, business, or service activities is growing increasingly complicated as the sciences and humanities become more interdependent. These organizations need not only management personnel, but individuals capable of designing and installing new integrated systems of people, materials, and equipment which will function effectively in a society made more complex by the technological explosion. Today's industrial engineer has a foundation in the basic physical and social sciences, engineering, and computers, and a skill in systems analysis and design which crosses traditional disciplinary lines. The result is a capability to meet the demands imposed by a dynamic social system.

The logical structure of systems study may be outlined as follows:

1. Identification of the components of a complex system. Examples include a production scheduling system, a computer software system, an educational scheduling system, a project planning and control system for underdeveloped countries, a hospital information system, and a space information system.

2. Development of the topological properties of the system, and the generation of mathematical models and analogs, or the adaptation of a heuristic approach when desirable.

3. Establishment of the functional relationships between the variables of the system, together with the necessary feedback required to control the operational system.

4. Selection and evaluation of optimization criteria. Examples of criteria are profit, costs, idle capacity, energy consumption, and information entropy.

5. Analysis and design of the nondeterministic functions to make them amenable to solution of the total system structure.

6. Integration of behavioral patterns in different environments, as required, within the physical framework of the system. Examples are the sociopolitical environment of a country for which the system is designed, the behavioral attitudes of nurses in a hospital, and those of production line workers in a fabrication shop.

7. Design of simulation models which permit a rigorous critique of the parameters of a dynamic system.

8. Economic analysis of the total system and concomitant subsystems with extensions into alternative structures.

Operations research and industrial engineering design. The real-world problems are complex systems configurations which require the use of more sophisticated methods than previously to present meaningful results to the decision maker. In the development of modern systems technology, the pacing factor is management. This is the function which directs, coordinates, and controls the many facets of a system. Technological advancement depends on optimal use of all resources and requires that new discoveries be translated into new products in minimal time. Productivity must be increased. The terms optimal and minimal are characteristics of management science and operations research.

Management science and operations research are often used synonymously since the tenor of their objectives is compatible. Operations research has been referred to as a sharper kind of industrial engineering. Even though no two definitions of operations research are exactly the same, the following four common denominators can be abstracted from most definitions: (1) formulation of objectives—seek to attain goals established by management; (2) systems perspective—this is frequently concerned with the interrelationship among many components; (3) alternatives—choices exist among decision variables; (4) optimization—attempt to make best decision relative to cited objectives, recognizing that in the real world this is not usually achieved in the strict sense.

In the current identification of engineering design and operations research, two apparently diverse disciplines, their components, and processes are found to be very similar. Industrial engineers are involved with both the design and analysis aspects of engineering in solving many of their problems. Engineering design is the process of devising a system, which includes its components and processes, to meet desired needs. It is a decision-making process (often iterative and interactive) in which the basic sciences, mathematics, and engineering sciences are applied to convert resources optimally to meet a stated objective. Among the fundamental elements of the design process are the establishment of objectives and criteria, synthesis, analysis, construction, testing, and evaluation.

The criteria established for engineering design evaluation provide a powerful impetus to utilize operations research foundations in engineering problems. It is important to note that in industrial

Applications of operations research techniques in production*

Application areas	Linear programming	Dynamic programming	Network models	Simulation	Queueing theory	Game theory	Regression analysis	Others
Production scheduling	30 (41.1)	7 (9.6)	6 (8.2)	26 (35.6)	9 (12.3)	0 (0.0)	5 (6.8)	10 (13.7)
Production planning and control	19 (26.0)	3 (4.1)	7 (9.6)	18 (24.7)	4 (5.5)	0 (0.0)	3 (4.1)	3 (4.1)
Project planning and control	10 (13.7)	1 (1.4)	28 (38.4)	9 (12.3)	2 (2.7)	0 (0.0)	0 (0.0)	3 (4.1)
Inventory analysis and control	11 (15.1)	3 (4.1)	3 (4.1)	27 (37.0)	4 (5.5)	1 (1.4)	12 (16.4)	7 (9.6)
Quality control	2 (2.7)	0 (0.0)	1 (4.1)	2 (2.7)	0 (0.0)	0 (0.0)	15 (20.5)	9 (12.3)
Maintenance and repair	0 (0.0)	1 (1.4)	3 (4.1)	8 (11.0)	3 (4.1)	1 (1.4)	4 (5.5)	3 (4.1)
Plant layout	13 (17.8)	0 (0.0)	5 (6.8)	19 (26.0)	5 (6.8)	1 (1.4)	2 (2.7)	3 (4.1)
Equipment acquisition and replacement	4 (5.5)	0 (0.0)	1 (1.4)	11 (15.1)	1 (1.4)	0 (0.0)	0 (0.0)	7 (9.6)
Blending	32 (43.8)	0 (0.0)	1 (1.4)	6 (8.2)	1 (1.4)	0 (0.0)	3 (4.1)	1 (1.4)
Logistics	27 (37.0)	1 (1.4)	8 (11.0)	24 (32.9)	3 (4.1)	2 (2.7)	6 (8.2)	2 (2.7)
Plant location	32 (43.8)	2 (2.7)	8 (11.0)	23 (31.5)	1 (1.4)	0 (0.0)	5 (6.8)	4 (5.5)
Other	7 (9.6)	1 (1.4)	2 (2.7)	7 (9.6)	1 (1.4)	1 (1.4)	3 (4.1)	4 (5.5)

*All data are expressed in terms of numbers (percent). The percentages do not total 100% because many respondents indicated they used more than one technique in a given application area.

SOURCE: From A. N. Ledbetter and J. F. Cox, Are OR techniques being used, *Ind. Eng.*, p. 21, February 1977.

engineering operations research is currently being used in solving the problems. The data in the table provide insights relative to the use of operations research techniques in the production function, which is a major area of activity for industrial engineers. The techniques used most frequently are linear programming and simulation. Network models have been used extensively in project planning and control, and regression analysis in inventory and quality control. *See* OPERATIONS RESEARCH.

Even though the inception of industrial engineering dates back to the latter part of the 19th century, the first official society for industrial engineers, the American Institute of Industrial Engineers (AIIE), was founded in 1948.

[ALBERT G. HOLZMAN]

Bibliography: P. E. Hicks, *Introduction to Industrial Engineering and Management Science*, 1977; H. B. Maynard, *Industrial Engineering Handbook*, 3d ed., 1971.

Industrial health and safety

Industrial health and safety is concerned with freedom from harm, injury and disease in the work environment. Because of this, the safety professional is closely aligned with the hygienist and the industrial medical workers. When a person becomes disabled the problem is of concern to the individual and the employer. In addition to causing pain and suffering, accidents produce economic and social loss, impair individual and group productivity, cause inefficiency, and generally retard progress. (The term "accident" as used here means any unexpected happening that interrupts the work sequence, or a process and that may result in injury, illness, or property damage to the extent that it causes loss.)

Industrial safety and health hazards may mean: conditions that cause legally compensable illnesses; or any conditions in the workplace that impair the health of employees enough to make them lose time from work or to work at less than full efficiency. Both are preventable, and their correction is properly a responsibility of management.

Recognition of hazards involves knowledge and understanding of the environmental stresses of the workplace and the effect of these stresses upon the health of the worker. Control involves the reduction of environmental stresses to values that the worker can tolerate without impairment of health or productivity. Measurement and quantitative evaluation of environmental stress is the essential ingredient that has brought modern industrial hygiene to its high level of value for conserving the health and well-being of the workers.

Safety is a multidisciplinary profession, drawing its workers from many areas, including education, engineering, psychology, medicine, and biophysics. The work of the safety professional follows a pattern. Before taking any steps in the containment of illness or accidents, the safety professional first identifies and appraises all existing safety and health hazards, both immediate and potential. Accurate records are essential in the search for the cause of an illness or an accident and they can help find the means to prevent similar incidents in the future. Upon determining the cause, the safety professional will have a firm basis on which to propose preventive measures and put them into operation. Finally, the safety professional must evaluate the effectiveness of safety and health control measures after they have been put into practice.

Preventive measures are obviously better than corrective measures. The safety professional can thus be of great value in examining (and sometimes in helping to draft) the specifications for materials, job procedures, new machinery and equipment, and new structures from the standpoint of health and safety well before installation or construction. As part of the overall safety and health program, the safety professional should recommend policies, codes, safety standards, and procedures.

The occupational health program requires the services of two main professional disciplines, the physician and the industrial hygienist, each supported by ancillary safety and health professionals, including industrial nurses, industrial toxicologists, and health physicists.

FEDERAL REGULATIONS

The Occupational Safety and Health Act (OSHAct) of 1970 brought a restructuring of programs and activities relating to safeguarding the health of the worker. Uniform occupational health regulations now apply to all businesses engaged in commerce. Nearly every employer is required to implement some element of an industrial hygiene or occupational health program in order to be responsive to OSHAct and its health regulations. The declared congressional purpose of the OSHAct is to "assure so far as possible every working man and woman in the nation safe and

Fig. 1. Grinder in a foundry. When inhaled, silica dust produced by the grinding operation can cause a scarring of the lungs called silicosis.

healthful working conditions and to preserve our human resources." The Federal government is authorized to develop and set mandatory occupational safety and health standards applicable to any business affecting interstate commerce. The responsibility for promulgating and enforcing occupational safety and health standards rests with the Department of Labor.

The OSHAct established the National Institute for Occupational Safety and Health (NIOSH) to conduct research on the health effects of exposures in the work environment; to develop criteria for dealing with toxic materials and harmful agents, including safe levels of exposure; to train an adequate supply of professional personnel to meet the purposes of the OSHAct; and in general, to conduct research and assistance programs for improving protection and maintenance of worker health.

ENVIRONMENTAL FACTORS OR STRESSES

The various environmental factors or stresses that may cause sickness, impaired health, or significant discomfort or inefficiency in workers may be classified as chemical, physical, biological, or ergonomic (Figs. 1–3).

Chemical hazards arise from excessive exposure to airborne concentrations of mists, vapors, gases, or solids that are in the form of dusts or fumes (Fig. 1). In addition to the hazard of inhalation, many of these materials may act as skin irritants or may be toxic when absorbed through the skin.

Physical hazards include excessive levels of radiation or noise and extremes of temperature or pressure.

Biological hazards include contamination of food, drinking water, or sanitary facilities by insects, molds, fungi, and bacteria; and improper removal of industrial waste and sewage.

Ergonomic hazards include improperly designed tools or work areas, improper lifting or reaching, poor visual conditions, or repeated motions in an awkward position. Careful designing of both the tools and the job, and intelligent application of engineering and biomechanical principles are required to eliminate hazards of this kind.

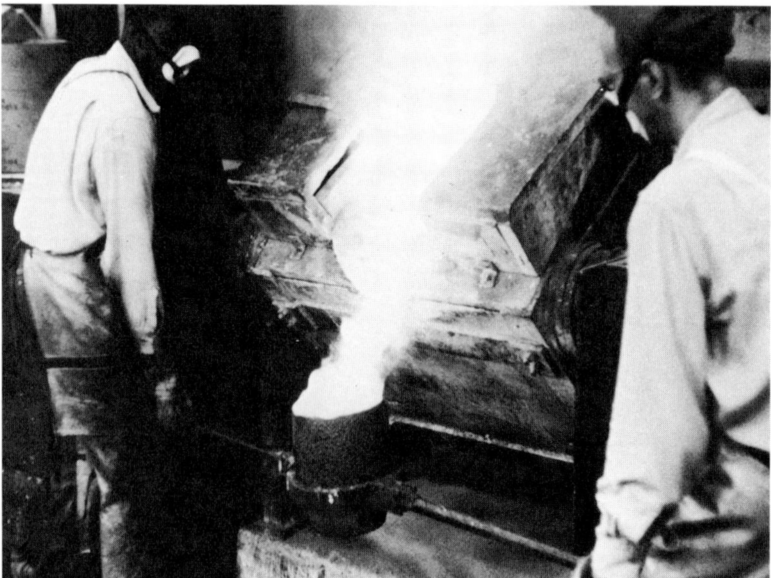

Fig. 2. Workers pouring metal in a foundry. In such molten metal processes, workers may often be exposed to toxic metal fumes.

Exposure to many of these hazards may produce an immediate acute response, due to the intensity of the exposure, or a chronic response, resulting from longer exposure at a lower intensity. In normal circumstances, an employee rarely experiences exposure to a single environmental stress, but rather to an interplay of multiple stresses.

Threshold limit values. Control of the work environment is based on the assumption that, for each substance, there is some safe or tolerable level of exposure, known as the threshold limit value, below which no significantly adverse effect occurs. Threshold limit values specifically refer to limits published by a committee of the American Conference of Governmental Industrial Hygienists (ACGIH). These limits are reviewed and updated each year to assimilate new information. Threshold limit values apply to airborne concentrations of substances and represent conditions under which it is believed that nearly all workers may be repeatedly exposed without adverse effect.

Because individual susceptibility varies widely, an occasional exposure of an individual at (or even below) the threshold limit may not prevent discomfort, aggravation of a preexisting condition, or occupational illness. In addition to the threshold limit values set for chemical compounds, there are limits for physical agents, such as noise, microwaves, and heat stress.

Toxicity versus hazard. The toxicity of a material is its capacity to produce injury or harm. Hazard is the possibility that exposure to a material will cause injury when a specific quantity is used under certain conditions.

The key elements to be considered when evaluating a material health hazard are the following: How much of the material is required to be in contact with a body cell and for how long to produce injury? What is the probability that the material will be absorbed or come in contact with body cells? What is the rate of generation of airborne contaminants? What control measures are in use?

The effects of exposure to a substance are dependent on dose, rate, physical state of the substance, temperature, site of absorption, diet, and general state of the individual's health.

It is important to recognize that toxicity is not necessarily the most important factor in determining the extent of a health hazard associated with the use of that material. The nature of the process in which that material is used or generated, the possibility of reaction with other agents (physical or chemical), the degree of effective ventilation control, the amount of enclosure, and the duration of exposure all relate to the potential hazard. Consideration should also be given to the type and degree of toxic response that the material may elicit in both the average and the hypersusceptible worker.

Chemical stresses. Chemical compounds in the form of liquids, gases, mists, dusts, fumes, and vapors may cause problems by absorption (through direct contact with the skin), by ingestion (eating or drinking), or by inhalation (breathing).

Skin absorption. While absorption through the skin can occur quite rapidly if the skin is cut or abraded, intact skin offers a reasonably good barrier to chemicals. Some substances are absorbed through the hair follicles. Others, such as organic lead compounds, many nitro compounds, and organic phosphate pesticides, dissolve in the fats and oils of the skin.

Ingestion. In the workplace, people may unknowingly eat or drink harmful chemicals. Toxic compounds are capable of being absorbed from the gastrointestinal tract into the blood and can cause serious problems. In this situation, careful and thorough handwashing is required before eating and at the end of every shift.

Inhalation. This type of stress occurs through inhaling those airborne contaminants physically classified as gases, vapors, and particulate matter, including dusts, fumes, smokes, aerosols, and mists. As a route of entry, inhalation is particularly important because of the rapidity with which a toxic material can be absorbed in the lungs, pass into the bloodstream, and reach the brain.

1. Gas and vapor contaminants are classified in six categories as follows.

Inert: These generally do not react with other substances. They create a hazard by displacing air and producing oxygen deficiency. Examples of inert gases are helium, neon, and argon.

Acidic: These are acids or substances that react with water to produce an acid (positively charged hydrogen ions). They taste sour and many are corrosive to tissues. Commonly encountered acids include hydrogen chloride, sulfur dioxide, fluorine, nitrogen dioxide, acetic acid, carbon dioxide, hydrogen sulfide, and hydrogen cyanide.

Alkaline: These are alkalies or substances that react with water to produce an alkali [negatively charged hydroxyl ions (OH^-)]. They taste bitter, and many are corrosive to tissues. Examples include ammonia, amines, phosphine, arsine, and stibine.

Organic: These are carbon compounds, including: saturated hydrocarbons (methane, ethane, butane); unsaturated hydrocarbons (ethylene, acetylene); alcohols (methyl alcohol, propyl alcohol); ethers (methyl ether, ethyl ether); aldehydes (formaldehyde); ketones (dimethyl ketone); organic acids (formic acid, acetic acid); halides (chloroform, carbon tetrachloride); amides (formamide, acetamide); nitriles (acetonitrile); isocyanates (toluene diisocyanate); amines (methylamine); epoxies (epoxyethane, propylene oxide); and aromatics (benzene, toluene, xylene).

Organometallic: These are compounds in which organic groups and metals are bonded, such as ethyl silicate, tetraethyl lead, and organic phosphates.

Hydrides: These are compounds in which hydrogen is bonded to metals and certain other elements, such as diborane and lithium hydride.

2. Particulate contaminants are produced by mechanical means by the disintegration processes of grinding, crushing, drilling, blasting, and spraying; or by the physiochemical reactions such as combustion, vaporization, distillation, sublimation, calcination, and condensation. Particles are classified in six types as follows.

Dust: A solid mechanically produced particle that may vary in size from submicroscopic to visible.

Spray: A liquid, mechanically produced particle generally in the visible range.

Fume: A solid condensation particle generally less than 1 μm in diameter (Fig. 2).

Mist: A liquid condensation particle ranging from submicroscopic to visible.

Fog: A mist of sufficient density to obscure vision.

Smoke: A mass of substances which includes the products of incomplete combustion of organic materials in the form of solid and liquid particles and gaseous products in air, smoke usually obscures vision.

Respiratory hazards. Respiratory hazards may be broken down into two main groups: (1) oxygen deficiency, where the oxygen concentration (or partial pressure of oxygen) is below that considered safe for human exposure; and (2) air that contains harmful or toxic contaminants.

Oxygen-deficient atmospheres. Each living cell in the body requires a constant supply of oxygen. Deficiency of oxygen in the atmosphere of confined spaces can be a problem in industry. For this reason, the oxygen content of any tank or other confined space should be measured with instruments such as an oxygen analyzer before entry is made. Oxygen-deficient atmospheres may occur in tanks, vats, holds of ships, silos, or mines, or in areas where the air may be diluted or displaced by asphyxiating levels of gases or vapors or where the oxygen may have been consumed by chemical or biological reactions.

Airborne contaminants. Dusts are solid particles generated by handling, crushing, grinding, rapid impact, detonation, and decrepitation (breaking apart by heating) of organic or inorganic materials, such as rock, ore, metal, coal, wood, and grain. Dust may enter the air from various sources, for instance, when a dusty material is handled, or when solid materials are reduced to small sizes in

Fig. 3. Welding operation in a steel fabrication plant. This worker could be exposed to toxic metal fumes when welding in confined areas.

processes such as grinding, crushing, blasting, shaking, and drilling.

Proper evaluation of dust exposures requires a knowledge of the chemical composition and particle size of the dust, its concentration in the air, how it is dispersed, and many other factors.

Fumes are formed when the material from a volatilized solid condenses in cool air. The solid particles that are formed make up a fume that is extremely fine—usually less than 1 μm in diameter—and is therefore readily inhaled. In most cases, the hot vapor reacts with the air to form an oxide.

Welding, metalizing, and other operations involving vapors from molten metals may produce fumes that can be harmful under certain conditions. Arc welding, for example, volatilizes metal vapor that condenses, as the metal or its oxide, in the air around the arc. In addition, the rod coating is partly volatilized (Fig. 3).

Gases are formless fluids that expand to occupy the space in which they are confined. Gases are a state of matter in which the molecules are unrestricted by cohesive forces. Examples are welding gases, internal combustion engine exhaust gases, and air.

Vapors are the volatile form of substances that are normally in the solid or liquid state at room temperature and pressure. Evaporation is the process by which a liquid is changed into the vapor state and mixed with the surrounding atmosphere. Gases and vapors are not fumes, although the terms are often mistakenly interchanged.

Physical stresses. Such things as noise, temperature extremes, ionizing and nonionizing radiation, and pressure extremes are physical stresses.

Noise. Noise ("unwanted sound") is a form of vibration that may be conducted through solids, liquids, or gases. The effects of noise on humans include: psychological effects (noise can startle, annoy, and disrupt concentration, sleep, or relaxation); interference with communication by speech, and, as a consequence, interference with job performance and safety; physiological effects (noise-induced loss of hearing, or aural pain when the exposure is severe).

If the ear is subjected to high levels of noise for a significant period of time, some loss of hearing may occur. Because of the complex relationships of noise and exposure time to hearing-threshold shift (reduction in hearing level) and the many possible contributory causes, establishment of criteria for protecting workers against hearing loss presents many difficulties. The criteria for hearing conservation, required by the OSHAct, establishes the permissible levels of harmful noise to which an employee may be subjected. The permissible decibel levels (dBA) and hours (duration per day) are specified. For example, a noise of 90 dBA is permissible for 8 hours; 95 dBA for 4 h, and so on.

There are three nontechnical criteria that can be used to determine if the work area has excessive noise levels: if it is necessary to speak in a very loud voice or shout directly into the ear of a person in order to be understood; if employees say that they hear ringing noises in their ears at the end of the workday; if employees complain that the sounds of speech seem muffled after leaving work but that their hearing is fairly clear in the morning when they return to work.

Extremes of temperature. Probably the most elementary factor of environmental control is the ability to adjust the thermal environment. General experience shows that extremes of temperature affect the amount of work that a person can do and the manner in which it is done. The industrial problem is more often one of high temperatures than low temperatures.

The body is continuously producing heat through its metabolic processes. Since the body processes are designed so that they can operate only within a very narrow range of internal temperature, the body must dissipate this heat as rapidily as it is produced if it is to function efficiently. The body attempts to counteract the effects of high temperature by increasing the rate of heartbeat. The capillaries in the skin also dilate to bring more blood to the surface so that the rate of cooling is increased.

Usually, the answer to a cold work area is to supply heat where possible, except for areas that must be cold, such as food storage areas.

General hypothermia is an acute problem resulting from prolonged cold exposure and heat loss. An individual who becomes fatigued during physical activity will be more prone to heat loss, and as exhaustion approaches, sudden vasodilation (blood vessel dilation) occurs with resultant rapid loss of heat.

Cold stress is proportional to the total thermal gradient between the skin and the environment, because this gradient determines the rate of heat loss from the body by radiation and convection. When blood vessel constriction is no longer adequate to maintain body heat balance, shivering becomes an important mechanism for increasing body temperature by causing metabolic heat production to increase to several times the resting rate.

Ionizing radiation. Radioactive materials that emit x-rays, gamma rays, or neutrons are external hazards. In other words, such materials can be located some distance from the body and emit radiation that will produce ionization (and thus damage) as it passes through the body. Control by limiting exposure time, working at a safe distance, use of barriers, or a combination of all three is required for adequate protection against external radiation hazards.

A radioactive material that emits only alpha particles and remains outside the body is not a hazard. Internally, it is a hazard because of the ionizing ability of alpha particles at very short distances in soft tissue. Alpha-emitting radioactive materials can enter the body by inhalation or ingestion or through an open wound. Those that will concentrate as persisting deposits in specific parts of the body are considered very dangerous.

Beta emitters, although generally viewed as an internal hazard, can also be classed as an external hazard because they may produce burns when in contact with the skin.

Lasers. Lasers emit light beams of coherent radiation with a single wavelength and frequency; that is, the light waves are nearly parallel to each other and all travel in the same direction. Atoms are "pumped" full of energy, and when they are stimulated to fall to a lower energy level, they give off radiation that is directed to produce the coherent laser beam. Since the laser is highly collimated (has a small divergence angle), it can have a large energy density in a narrow beam. Uses for lasers include machining, cutting, welding, measuring, aligning, and inspecting.

The eye is the most vulnerable organ to laser-induced injury. This is because the lens focuses the laser beam on a small spot on the retina. The fact that infrared radiation of certain lasers may not be visible to the naked eye contributes to the potential hazard. Lasers generating in the ultraviolet range of the electromagnetic spectrum often produce corneal burns rather than retinal damage. Direct viewing of a laser source or its reflections should therefore be avoided and the work area should contain no reflective surfaces (such as mirrors or highly polished furniture), since even a reflected laser beam can be hazardous. Suitable shielding to contain the laser beam should be provided. The OSHAct covers protection against laser hazards in its construction regulations.

Extremes of pressure. It has long been recognized that people working under greater than normal (hyperbaric) atmospheric pressure are subject to various job-related ills. Occupational exposure to hyperbaric environments occurs in caisson or tunneling operations, where a compressed-gas environment is used to exclude water or mud and to provide support for structures.

Hyperbaric environments are also encountered by divers. Decompression sickness, commonly known as the bends, results from the release of nitrogen bubbles into the blood and tissues during rapid decompression. If the bubbles lodge at the joints and under muscles, they cause severe cramps. To prevent this trouble, decompression is carried out slowly and by stages so that the nitrogen can be eliminated slowly. Deep-sea divers are supplied with a mixture of helium and oxygen for breathing, and because helium is an inert diluent and less soluble in blood and tissue than is nitrogen, it presents a less formidable decompression problem.

Ergonomic stresses. The term ergonomics literally means the customs, habits, and laws of work. It is the application of human biological science in conjunction with the engineering sciences to achieve the optimum mutual adjustment of individuals and their work. The benefits are measured in terms of human efficiency and well-being. Ergonomic stresses can impair the health and efficiency of a worker just as significantly as the other more commonly recognized environmental stresses.

The human body can endure considerable discomfort and stress and can perform many awkward and unnatural movements for a limited period of time. However, when these conditions or motions are continued for prolonged periods, physiological limitations may be exceeded. To ensure a high level of performance on a continuing basis, work systems must be tailored to human capacities and limitations (Fig. 4). Ergonomics includes the physiological and psychological stresses of the task. The task should not require excessive muscular effort, considering age, sex, and state of health of the worker. The job should not be so easy that boredom and inattention lead to unnecessary errors, scrap material, and accidents.

Fig. 4. An ergonomically designed workbench. Notice that all the part bins are within easy reach of the worker.

The task of the design engineer and safety professional is to find the happy blend between "easy" and "difficult" jobs. With very low levels of physiological and psychological stress, performance is also low; as stress increases, however, performance also increases, to a point. The task is to design jobs that will be centered on optimum performance. In any human-machine system, there are tasks that are better performed by a person than by a machine and, conversely, tasks that are better handled by machines.

Ergonomics deals with the interactions between workers and such traditional environmental elements as atmospheric contaminants, heat, light, sound, tools, and equipment pertaining to the workplace. The modern concept is that the worker is to be considered the monitoring link of a human-machine environment system. Management is interested in the behavior and physical operating characteristics of the worker to the extent that they affect the economic and productive outputs from such systems.

To achieve maximum efficiency, a human-machine system must be designed as a whole, with the worker being complementary to the machine and the machine being complementary to the abilities of the worker. Consideration should be given to the general physical and mental demands of the task, so as not to overload the operator. Because there are many situations which can make excessive demands on an operator, an overload can develop unless information reaches the individual at a speed which can be coped with and in a form which is easily understood. The demands should be properly spaced in time, so that the operator can act upon them without undue mental stress. If a study of the demands of the task suggest that overload is possible, steps should be taken to bring the demands within the expected capacity of the operator and so reduce the load. *See* HUMAN-FACTORS ENGINEERING.

Biological agents. Biological agents may be associated with certain occupations. In the work environment, these agents include viruses, rickettsiae, bacteria, and parasites of various types. Diseases transmitted from animals to humans are common. Infections and parasitic diseases may also result from exposure to contaminated water or to insects.

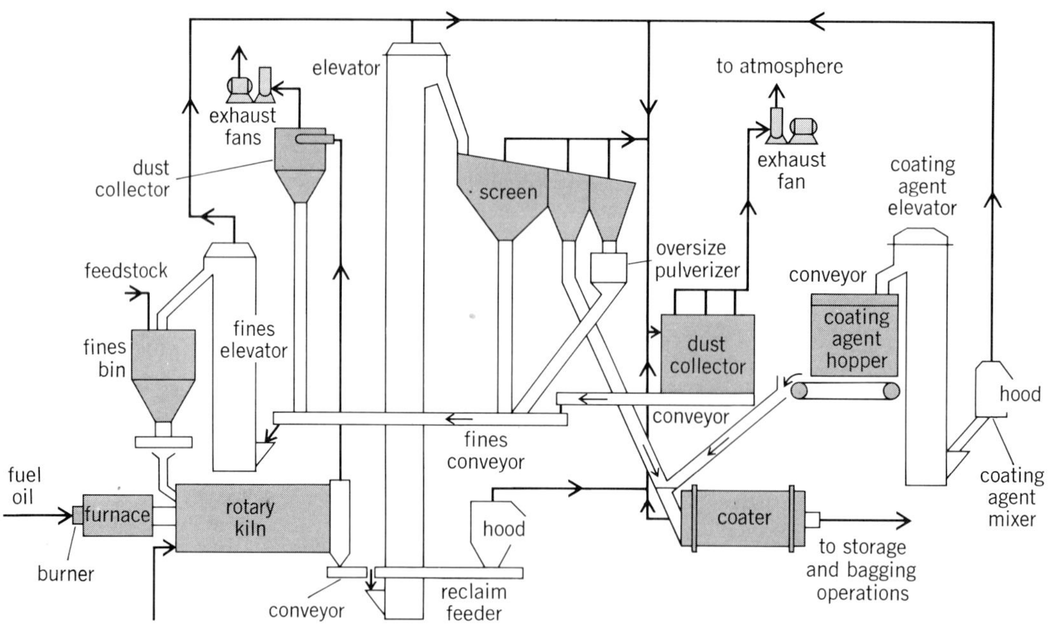

Fig. 5. A simple process flow sheet showing the stepwise introduction of raw material and the product of each step. The extent of chemical or physical hazards that may arise out of any step in the operation should be determined.

U.S. DEPARTMENT OF LABOR
Occupational Safety and Health Administration

Form Approved
OMB No. 44-R1387

MATERIAL SAFETY DATA SHEET

Required under USDL Safety and Health Regulations for Ship Repairing, Shipbuilding, and Shipbreaking (29 CFR 1915, 1916, 1917)

SECTION I

MANUFACTURER'S NAME	EMERGENCY TELEPHONE NO.
ADDRESS *(Number, Street, City, State, and ZIP Code)*	
CHEMICAL NAME AND SYNONYMS	TRADE NAME AND SYNONYMS
CHEMICAL FAMILY	FORMULA

SECTION II - HAZARDOUS INGREDIENTS

PAINTS, PRESERVATIVES, & SOLVENTS	%	TLV (Units)	ALLOYS AND METALLIC COATINGS	%	TLV (Units)
PIGMENTS			BASE METAL		
CATALYST			ALLOYS		
VEHICLE			METALLIC COATINGS		
SOLVENTS			FILLER METAL PLUS COATING OR CORE FLUX		
ADDITIVES			OTHERS		
OTHERS					

HAZARDOUS MIXTURES OF OTHER LIQUIDS, SOLIDS, OR GASES	%	TLV (Units)

SECTION III - PHYSICAL DATA

BOILING POINT (°F.)		SPECIFIC GRAVITY (H_2O=1)	
VAPOR PRESSURE (mm Hg.)		PERCENT, VOLATILE BY VOLUME (%)	
VAPOR DENSITY (AIR=1)		EVAPORATION RATE (_______ =1)	
SOLUBILITY IN WATER			
APPEARANCE AND ODOR			

SECTION IV - FIRE AND EXPLOSION HAZARD DATA

FLASH POINT (Method used)	FLAMMABLE LIMITS	Lel	Uel
EXTINGUISHING MEDIA			
SPECIAL FIRE FIGHTING PROCEDURES			
UNUSUAL FIRE AND EXPLOSION HAZARDS			

SECTION V - HEALTH HAZARD DATA

THRESHOLD LIMIT VALUE

EFFECTS OF OVEREXPOSURE

EMERGENCY AND FIRST AID PROCEDURES

SECTION VI - REACTIVITY DATA

STABILITY	UNSTABLE		CONDITIONS TO AVOID
	STABLE		
INCOMPATABILITY *(Materials to avoid)*			
HAZARDOUS DECOMPOSITION PRODUCTS			
HAZARDOUS POLYMERIZATION	MAY OCCUR		CONDITIONS TO AVOID
	WILL NOT OCCUR		

SECTION VII - SPILL OR LEAK PROCEDURES

STEPS TO BE TAKEN IN CASE MATERIAL IS RELEASED OR SPILLED

WASTE DISPOSAL METHOD

SECTION VIII - SPECIAL PROTECTION INFORMATION

RESPIRATORY PROTECTION *(Specify type)*		
VENTILATION	LOCAL EXHAUST	SPECIAL
	MECHANICAL *(General)*	OTHER
PROTECTIVE GLOVES		EYE PROTECTION
OTHER PROTECTIVE EQUIPMENT		

SECTION IX - SPECIAL PRECAUTIONS

PRECAUTIONS TO BE TAKEN IN HANDLING AND STORING

OTHER PRECAUTIONS

Fig. 6. Sample Material Safety Data Sheet to be used to obtain information from the manufacturer. (*Occupational Safety and Health Administration*)

A number of occupational diseases are common to workers in agriculture and in closely related industrial jobs. For example, grain handlers who inhale grain dust are likely to come in contact with some of the fungi that contaminate grain. Some of these fungi can flourish in human lungs, causing a condition called farmer's lung.

Anthrax is a highly virulent bacterial infection. In spite of considerable effort in quarantining infected animals and in sterilizing imported animal products, this disease remains a problem. Q fever, which is similar to but apparently not identical to "tick fever," has been reported among meat and livestock handlers. There is evidence that Q fever may come from contacting freshly killed carcasses or droppings of infected cattle. In the latter case, the disease is probably transmitted by inhalation of the infectious dust. Prevention undoubtedly depends primarily upon recognition and elimination of the disease in the animal host.

BASIC HAZARD-RECOGNITION PROCEDURES

There is a basic, systematic procedure that can be followed in the recognition and evaluation of environmental health hazards. Almost any work environment has either potential or actual environmental hazards that need to be identified, measured, and monitored. The safety professional must always be alert in order to be aware of these and must use the following steps in evaluating work conditions. First, the raw materials, the degree of hazard to the worker, and the potential of that material to do harm are considered. Next, the safety professional evaluates how these raw materials are modified through intermediate steps. Finally, the finished product is examined to determine any possible harmful effects on the worker. Each step, from raw material to finished product, must be evaluated under normal conditions and also under anticipated emergency conditions (Fig. 5).

The degree of hazard arising from environmental causes depends on four factors: the nature of the environmental factor or stress; the level of exposure; the duration of exposure; and human variability or individual differences.

It is of the utmost importance to obtain a list of all the chemicals used as raw materials and to determine the nature of the products and by-products arising out of industrial operations (Fig. 6).

After the list of chemicals and physical agents to which employees are exposed has been prepared, it is necessary to determine which of the environmental factors or stresses result in hazardous exposures and need further evaluation.

MONITORING

Monitoring is a continuing program of observation, measurement, and judgment, all of which are necessary to recognize the potential health hazards and to evaluate the adequacy of protection. It implies an awareness of potential health hazards and a continuing assessment of the adequacy of control measures being used.

Monitoring is more than a simple sampling of the air to which an employee is exposed, or an examination of the medical status of that employee. It is an entire series of actions that permits a judgment of the adequacy of employee protection (see table).

Four types of monitoring systems generally used in occupational health surveillance are personal, environmental, biological, and medical.

Personal monitoring. This is the measurement of an employee's exposure to airborne contaminants. The measurement device, a dosimeter, is placed as close to the contaminant's portal of entry into the human body as possible. In the case of an air contaminant that is toxic by inhalation, the measurement device is placed close to the breathing zone (Fig. 7). In the case of noise, the device is placed close to the ear.

Environmental monitoring. This is the measurement of contaminant concentrations in the workroom. The measurement device is placed adjacent to the worker's station, and then the air contaminant concentration or physical energy stress is calculated or estimated.

Biological monitoring. This involves measurement of changes in composition of body fluid, tis-

Sampling procedures for industrial hygiene surveys

Chemical hazards (airborne)		*Physical hazards*	
Gases and vapors	May be determined by use of approved calibrated field indicator tubes yielding direct readings May be collected in containers or absorbed on charcoal for laboratory evaluation Organic vapor may be absorbed on charcoal or determined chromatographically	Pressure	May be measured barometrically
		Temperature	May be measured by thermometer, thermocouple, or radiometer; determination of heat stress, however, requires in some form the measurement of evaporation rate; usually heat stress is inferred from humidity and air velocity
Fumes and mists	May be absorbed and measured in the field May be absorbed and evaluated in the laboratory May be collected on filter media and analyzed in the laboratory	Ionizing radiation	May be measured by survey meter, personal dosimetry, or film badge techniques
Dusts	May be collected by a personal air sampler and fractionated into respirable size by a cyclone separator, and the fractions weighed to determine the concentration May be collected on an open-faced filter and weighed May be collected in an appropriate manner and counted	Noise levels	May be measured with sound-level meters or octave-band analyzers; vibration may be determined with additional sound-level equipment
		Nonionizing radiation	May be measured by a number of direct-reading meters

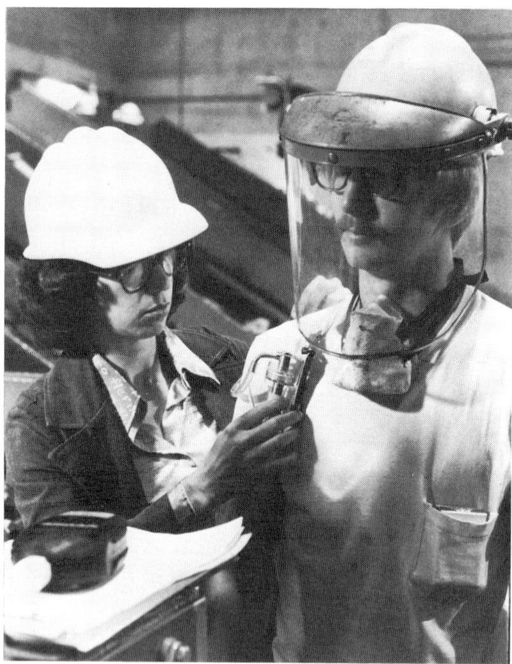

Fig. 7. Safety professional putting air-sampling device on worker. The worker will usually wear the device for 8 h to determine if the threshold value has been exceeded during a normal work shift.

sues, or expired air in order to determine excessive absorption of a contaminant. Examples are the measurement of lead, fluoride, cadmium, mercury, and so forth in blood or urine, or the identification of phenol in urine to determine excessive benzene absorption.

Medical monitoring. This refers to medical surveillance of the worker's response to a contaminant. Biological and medical monitoring provides information only after the fact of absorption of contaminant. Procedures which can be used to assess the adequacy of protective measures as well as the overall health of employees include development of a baseline health inventory, followed by periodic reevaluation. It is important to establish a preplacement baseline for the functioning of organ systems which are known to be affected by materials present in work areas. For a respiratory hazard, the occupational physician would perhaps want to know the status of the employee's chest x-rays. Preplacement medical examinations are also used to detect medical conditions which may predispose an employee to the toxic effects of the agent.

CONTROLS

It is best to introduce control measures to minimize hazards when an industrial facility is being designed so that they can be integrated into the building and operations. The proposed facility's layout and design must include an analysis of construction type, proposed activities in all areas, and possible health and safety hazards. The influence of one work activity on another must be assessed for combined hazards.

The types of control measures depend on the nature of the harmful substance or agent, and its routes of absorption into the body. For example, when air contaminants are created, generated, or released in hazardous concentrations, protection is usually provided by means of ventilation. Exposure to an airborne substance is related to the concentration of contaminants in the breathing zone and the amount of time the employee is exposed to this concentration. Reducing the quantity of contaminant in the employee's breathing zone or the time that the employee spends in the area will reduce the overall exposure. Protection that is inherent in the design of a process is preferable to a control method which depends on continual human involvement. To decrease exposure, it is necessary to determine the contaminant's source and the path it travels, as well as the employee's work pattern, and finally to use protective equipment. The type and extent of control methods will also depend upon the physical, chemical, and toxic properties of the environmental factor involved. The extensive controls needed for lead oxide dust, for example, would not be needed for limestone dust, since much greater quantities of limestone dust are permissible.

General methods. Controlling harmful environmental factors or stresses includes the following general methods: (1) substitution of a less harmful material for one that is hazardous to health; (2) change or alteration of a process to minimize worker contact; (3) isolation or enclosure of a process or work operation to reduce the number of persons exposed; (4) wet methods to reduce generation of dust in operations such as mining and quarrying; (5) local exhaust at the point of generation and dispersion of contaminants; (6) general or dilution ventilation with clean air to provide a healthful atmosphere; (7) personal protective devices, such as special clothing and eye and respiratory protectors; (8) good housekeeping, including cleanliness of the workplace, waste disposal, adequate washing and eating facilities, healthful drinking water, and control of insects and rodents; (9) special control methods for specific hazards, such as reduction of exposure time, monitoring devices, and continuous sampling with preset alarms; (10) medical programs to detect intake of toxic materials; (11) training and education to supplement engineering controls.

Personal protective equipment. When it is not feasible to render the working environment completely safe, it may be necessary to exclude the worker from that environment. Robots have application in unsafe areas. Personal protective equipment must be provided and used: where it is not possible to enclose or isolate the process or equipment or to provide ventilation or other control measures; where there are short exposures to hazardous concentrations of contaminants; or where unavoidable spills may occur.

Administrative controls. When the employee's exposure cannot be reduced to permissible safe levels through engineering controls, then an effort should be made to limit the employee's exposure through administrative controls. Some administrative controls that may be employed are: arrangement of work schedules and the related duration of exposures so that employees are minimally exposed to health hazards; transferring employees who have reached their upper permissible limits of exposure to an environment that is free from exposure; assigning as many individuals to a job as are needed to keep the exposure of each within permissible time limits.

Housekeeping and maintenance. Good housekeeping plays a key role in the control of occupational health hazards. Immediate clean-up of any spills of toxic material is a very important control measure. Dust should be removed before it can become airborne by traffic, vibration, and random air currents. A regular clean-up schedule using vacuum cleaners is an effective method of removing dirt and dust from the work area. An air hose should never be used to remove dust.

Waste disposal. Trained individuals should use established procedures for the safe disposal of toxic residues, other contaminated waste, and containers of chemicals that are no longer needed.

The disposal of hazardous materials in sewage systems can create many problems; the mixing of materials such as acids with arsenic or cyanide compounds, for example, can create serious health hazards. At any given time it would be difficult to know what a neighboring company may be dumping into a sewer with or without permission. Consequently, there can be unplanned mixing of waste chemicals with possibly disastrous results.

JOB SAFETY AND HEALTH ANALYSIS

Job safety and health analysis (JSA) is a procedure used by health and safety professionals to review job methods and uncover hazards that may have (1) been overlooked in the layout of the plant or building or in the design of the machinery, equipment, or processes, (2) developed after production started, or (3) resulted from changes in available work force.

Once the safety and health hazards are known, the proper solutions can be developed. Some solutions may be physical changes that control the hazard, such as enclosures to contain an air contaminant or placing a guard over exposed moving machine parts. Others may be job procedures that eliminate or minimize the hazard, such as safe piling of materials.

SYSTEM SAFETY

As a result of increased discussion and exploration by safety professionals of system approaches to control safety and health hazards, safety professionals have been called on to discover ways of implementing system techniques. Although complete system safety analysis requires specially trained engineers and rather sophisticated mathematics, safety professionals have found that some knowledge of these techniques can help to both codify and direct their programs. A complex process can be greatly clarified with a chart or model that provides a comprehensive, overall view of the process by showing its principal elements and the ways in which they are interrelated.

Many progressive safety professionals have investigated and applied a concept of total accident and illness control based on studies of "near misses" (noninjury accidents) and detailed analyses of hidden as well as direct accident costs. This approach consists of three closely related, logically ordered steps: hazard identification, elimination, and protection.

Hazard identification. To prevent accidents and loss of control, it is first necessary to determine those areas or activities in an operation where losses can occur by identifying all safety and health hazards. This requires studying processes at the research stage, reviewing design during engineering, checking pilot plant operations and startup, and monitoring production regularly.

Hazard elimination. Toxic, flammable, or corrosive chemicals can sometimes be replaced by safer materials. Machines can be redesigned to remove danger points, and plant layouts improved by eradicating blind corners or limited visibility crossings.

Hazard protection. Hazards that cannot be removed must be protected against. There are many familiar examples: mechanical guards to keep fingers from pinch-points, safety shoes to safeguard toes against dropped objects, ventilation systems which control the buildup of air contaminants. The kind of provisions necessary to protect against safety and health hazards will vary from plant to plant, from one toxic material to another, and from process to process.

[JULIAN B. OLISHIFSKI]

Bibliography: *The Industrial Environment: Its Evaluation & Control*, 3d ed., National Institute of Occupational Safety and Health, 1974; F. E. McElroy (ed.), *Accident Prevention Manual for Industrial Operation*, National Safety Council, 1974; J. B. Olishifski (ed.), *Fundamentals of Industrial Hygiene*, 2d ed., National Safety Council, 1979; C. Zenz, *Occupational Medicine: Principles and Practical Applications*, 1975.

Industrial robots

A programmable mechanism designed to move and do work within a certain volume of space. The robot differs from a parts-transfer mechanism in that the action patterns can easily be changed by software changes in a controlling computer or sometimes by adjusting the mechanism. Because of this feature the industrial robot has been called off-the-shelf automation, and fills the gap between hard automation—that is, mechanisms designed for a specific job—and human labor.

The industrial robot was first developed in the United States in the late 1950s. After a slow beginning, its use in industry has accelerated rapidly, and robots of all configurations are presently in extensive use in most mechanized societies. The largest user is the automobile industry, where robots are performing a wide variety of tasks from spot-welding bodies to assembling instrument panels. Other users of robots span the manufacturing industry from woodworking (applying polish to fine furniture) to the aerospace industry.

CLASSIFICATION

Industrial robots may be divided into four classifications based on their capital cost and capabilities: pick-and-place, lightweight electric, heavy duty, and special-purpose. These are described below, and examples of their use given.

Pick-and-place robots. The pick-and-place robot is the simplest and cheapest of the four classifications. Bordering on a transfer mechanism, it is classified as an industrial robot because of its ability to be reconfigured for different tasks with a minimum of effort. A pick-and-place robot is characterized as one that moves between predeter-

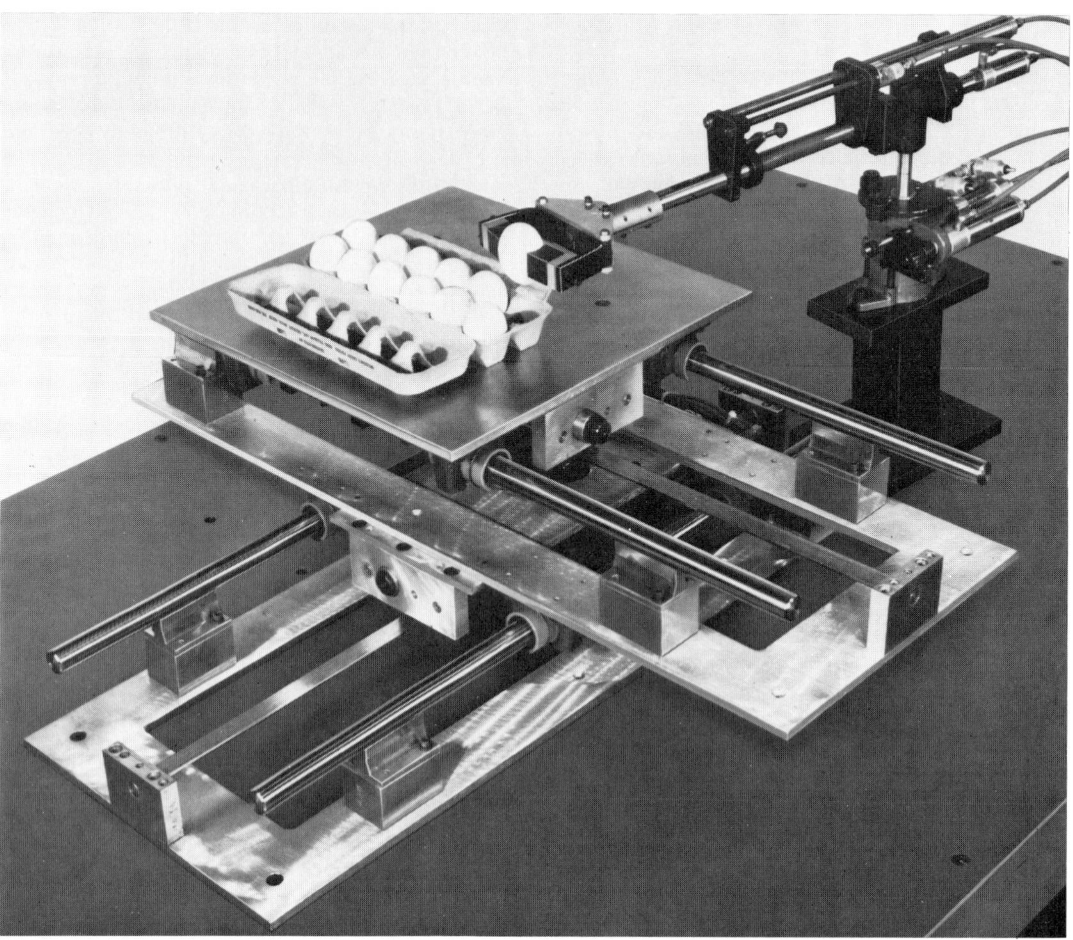

Fig. 1. Pick-and-place robot working in conjunction with an X-Y table. (Auto-Place, Inc.)

Fig. 2. PRAB hydraulic pick-and-place robot unloading machine flanges. *(PRAB Conveyors, Inc.)*

mined limits, which are set by hard stops or by microswitches, and is usually pneumatically powered, although larger models may be hydraulically powered. Because of its simple construction and low basic cost, this robot can be obtained in various configurations, depending upon the nature of the task to be performed. Very simple models may have only one or two independent axes of motion; more complicated models may have up to five or six axes, including rotatable and twisting wrists.

Figure 1 shows a basic pick-and-place robot working in conjunction with an X-Y table. Each independently movable axis or "degree of freedom" is powered by a small air cylinder. Because the limits of motion are determined by preset mechanical stops, there are only two stationary positions for each axis, namely, in or out, up or down. Tailoring the pick-and-place robot to its task involves adjusting the stops so that the end positions are appropriate, then programming the control system to move to these positions in the correct sequence.

To maintain the low-cost concept of the machine, the control system is simple. Its function is to energize valves which admit air or hydraulic fluid to the cylinders of the machine in the proper sequence. In addition, the control system has a limited capability of interacting with the outside world, such as recognizing the closing of microswitches or activating external relays. Any simple programmable sequencer is suited for this purpose; low-cost programmable controllers, drum switches, or pneumatic logic systems have all been used with success.

The pick-and-place robot is used for application where low initial cost is an important consideration. Because of its simple construction and control system, its capabilities are limited, and it is usually used for parts transfer between two predetermined positions.

Although most robots in this classification are designed for lightweight part-transfer applications, the pick-and-place robot can be made much larger and more powerful to enable it to handle heavier loads, while still maintaining the inherent simplicity, low cost, and ease of maintenance of the pick-and-place concept. Figure 2 shows a hydraulically powered pick-and-place robot unloading a machine tool. This is a popular application of this type of robot, because the path to be followed is simple and without variations.

An additional advantage provided by hydraulic power is the ability of stopping at intermediate positions between the hard stops. This is accomplished by placing a microswitch at the desired position to control the hydraulic servo valve. Because hydraulic fluid is incompressible, the robot remains at the intermediate position. This feature, together with the long reach and 150-lb (68-kg) load-handling capacity, make the hydraulically powered pick-and-place robot an attractive option for many jobs.

Lightweight electric robots. The lightweight electric robot resembles a small pick-and-place robot in its size and load-handling capacity. It is, however, a considerably more sophisticated machine, and is also much more expensive. In contrast to the pick-and-place robot, the lightweight electric machine can be driven to any position in its operating space.

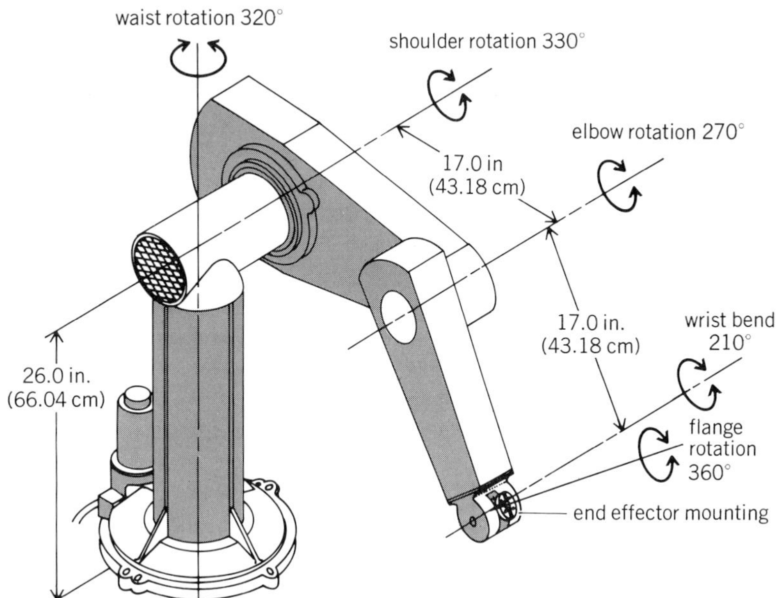

Fig. 3. Most popular configuration of a lightweight electric robot, showing the human-sized proportions. (*Unimation, Inc.*)

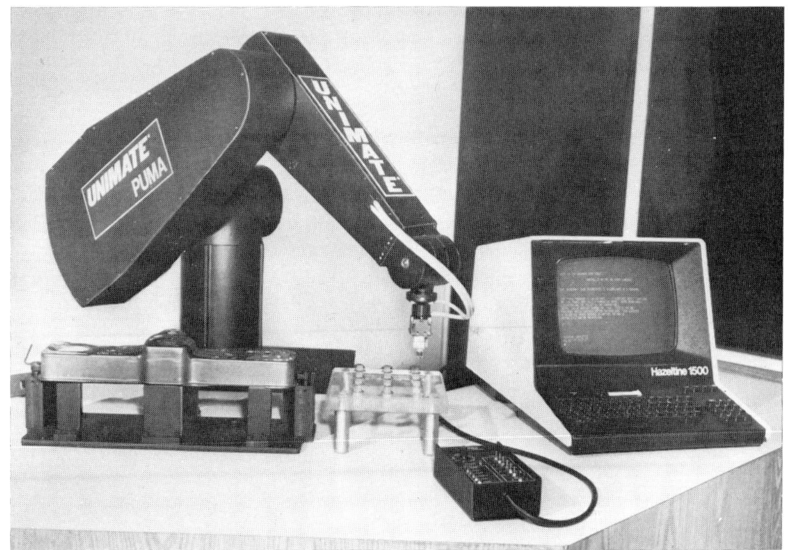

Fig. 4. Unimate PUMA robot installing lights in an automobile panel. (*Unimation, Inc.*)

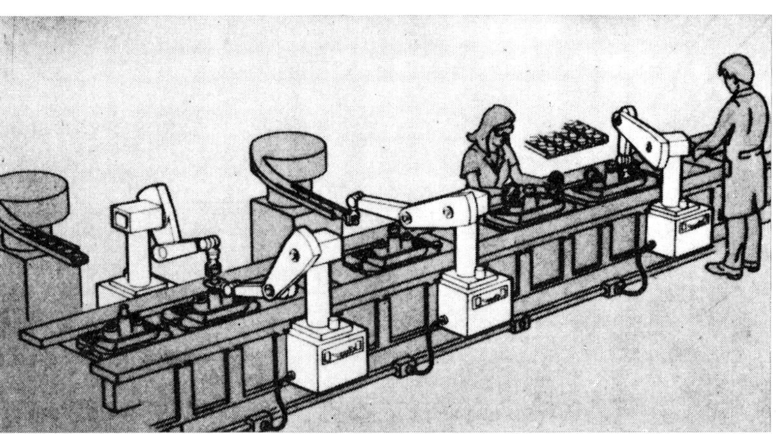

Fig. 5. Assembly line of the future at General Motors. PUMA robots and humans work side by side to assemble components. (*Unimation, Inc.*)

Fig. 6. The PRAB-AMF Versitran robot is an example of a cylindrical-coordinate heavy-duty machine. *(PRAB Conveyors, Inc.)*

The most popular configuration of this robot is shown in Fig. 3. The dimensions of the arm are usually chosen around the dimensions of the human arm. This is a convenient size since it permits many jobs which are traditionally done by human beings to be performed by the robot, and minimizes the space required to operate it. The robot shown in Fig. 3, for example, has dimensions between the major joints, shoulder to elbow and elbow to wrist, of 17 in. (43 cm). This gives it a reasonable working space of 30 in. (76 cm) radius from the base. The machine can also handle loads that a seated operator can conveniently work with—in the neighborhood of 5 lb (2.3 kg).

Theoretically, six independent axes of motion provide all the flexibility required. Three of these axes are needed to position the hand at the appropriate point in space, while the other three axes are used to orient it. However, in many cases only five independent axes are used, three to position the hand and two for orientation. Five axes slightly limit the ability to position an object, but still retain 90% of the capability of the arm while reducing the cost and complexity significantly.

Each joint, or axis, of the robot is controlled by a closed-loop servo system and is equipped with an electric motor, a suitable gear train, some sort of position feedback mechanism, and usually a feedback tachometer and electric brake. The motor, position feedback, and velocity feedback (tachometer) form part of a servo loop which is driven by a control computer. The electric brake locks the joint while it is not in motion.

The robot is "taught" by moving it to the appropriate positions and recording those positions in the computer. The simplest way to do this is to release all the brakes and move it manually. The computer is then instructed to remember the coordinates of the joints for that position. A more sophisticated method is to use a teach box which, when plugged into the controlling computer, enables the operator to position the arm by pressing certain buttons. The teach box also incorporates the ability to inch the arm by small amounts, which aids in precise positioning. Having placed the arm in the desired position, the operator keys a switch which causes the computer to read and store the positions of all the joints. The robot is taught by repeatedly defining a series of stored positions. Upon playback, the computer will move the arm through the taught positions and thus mimic the motion defined by the operator. Advanced methods of teaching computer-controlled robots involve the use of a "high-level language."

The lightweight electric robot costs 5–10 times as much as the basic pick-and-place robot. However, because of its ability to move objects to any place within its working space and the large memory capacity of its control computer, it is a much more versatile machine. Moreover, the computer can be programmed to handle complex decisions, and thus cause the robot to react to the external environment.

The principal market for the lightweight electric robot is in tasks such as assembly of small components typically performed by human operators. Figure 4 shows a lightweight electric robot set up to install lights in an automobile dash panel. Figure 5 is an artist's illustration of a series of lightweight electric robots at work on an assembly line. Each arm occupies approximately the same volume of space as a human operator would, and is designed so that if the robot fails it can be removed from the assembly line and its function performed by a human. Thus the entire assembly line is not dependent upon the robot.

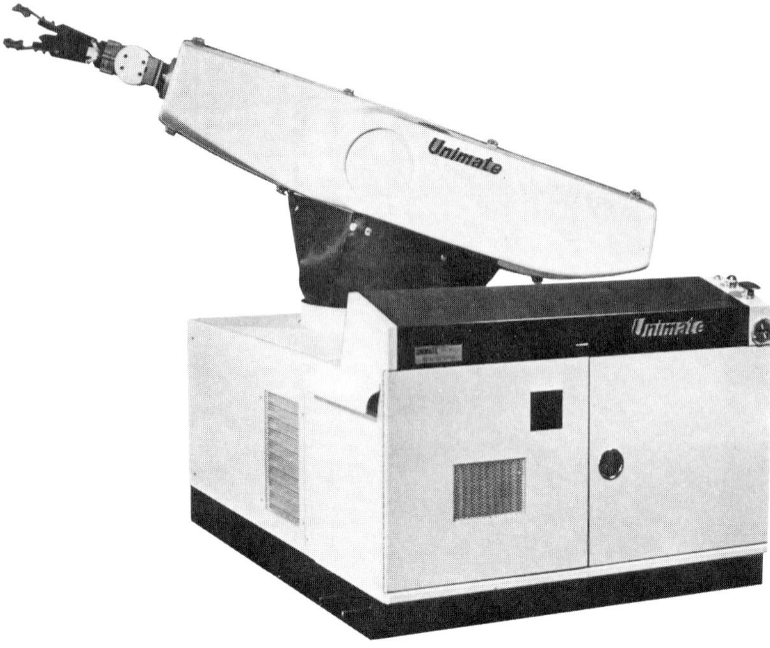

Fig. 7. The spherical-coordinate heavy-duty robot is typified by the Unimate Model 2000. *(Unimation, Inc.)*

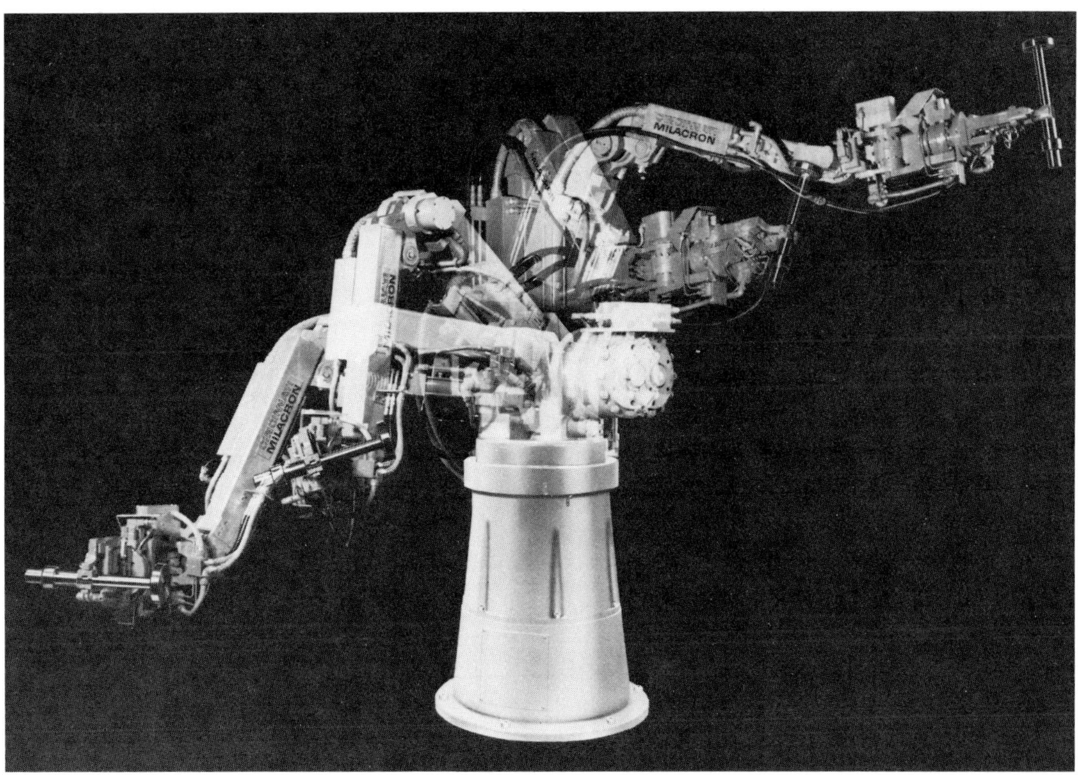

Fig. 8. Cincinnati: Milacron T^3 (The Tomorrow Tool) robot has articulated configuration. (*Cincinnati: Milacron, Inc.*)

Fig. 9. Multi-robot automobile spot-welding line. (*Unimation, Inc.*)

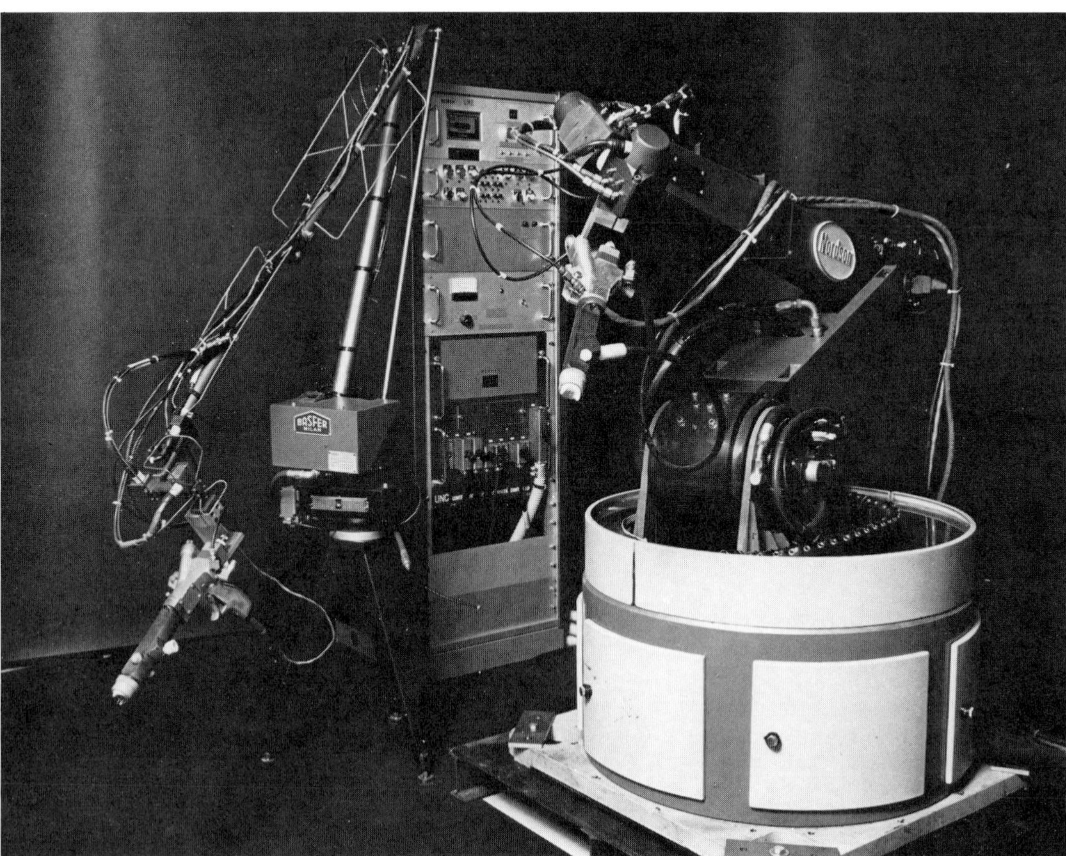

Fig. 10. The Nordson spray-painting robot shown beside its lightweight teaching arm. *(Nordson, Inc.)*

Heavy-duty robots. Presently the heavy-duty robot is the workhorse of the industry. Although at least one model is driven electrically, most heavy-duty robots are hydraulically powered because of their high torque requirements. As the name implies, these robots are capable of handling loads in the 200–500-lb (90–230-kg) range, depending on the model. With a reach of approximately 6 ft (1.8 m), the heavy-duty robot considerably exceeds human capabilites.

Several different configurations of heavy-duty robots are available, as shown in Figs. 6, 7, and 8. The robot in Fig. 6 operates in cylindrical coordinates—the three primary axes to move the wrist are a rotation, a vertical motion, and a radial motion of the boom. The robot of Fig. 7, on the other hand, operates in spherical coordinates. The appropriate three axes are rotation and elevation of the turret and extension of the boom. The third robot, shown in Fig. 8, uses an articulated configuration. There are no significant differences in performance between the three configurations, except that the articulated machine has an advantage in terms of reaching awkward positions.

Hydraulically powered heavy-duty robots are equipped with a hydraulic pump mounted either in the base or at a remote location. Each joint of the robot contains a hydraulic motor and position and velocity feedback elements, as is the case with the lightweight electric robot. A brake, however, is not necessary because the joint can be locked by closing the hydraulic valve.

The robot is controlled by a built-in computer, and all models come equipped with a teach box or similar device for instructing the robot. By means of controls on the teach box, the arm is positioned to appropriate points, and by pressing a button these points are recorded. On playback, the arm moves smoothly through the recorded positions. Alternatively, some models are equipped with sophisticated controls that allow continuous motion for special tasks such as arc welding or working on articles on a moving conveyor belt.

The heavy-duty robot is used for tasks that are hazardous, hot, strenuous, or otherwise difficult for humans to perform. Its usefulness is limited only by imagination; its versatility allows it to perform many different tasks, ranging from loading and unloading machine tools to TIG (tungsten–inert gas) welding of truck bodies. A major air-frame manufacturer is using an articulated heavy-duty robot to drill holes in aircraft wings. In this case, the robot replaces jobs previously done by three or four workers using large templates. A common use is spot-welding auto bodies, where the reprogrammability of the robot allows it to weld several different models in succession. Figure 9 shows the production line of a major car manufacturer with seven heavy-duty robots at work. This line is known as the "turkey farm" because of the appearance and action of the robots.

Special-purpose robots. One of the major advantages of an industrial robot is that it is a general-purpose mechanism which is tailored to a specific task by software changes. The concept of a special-purpose robot would appear to invalidate this advantage. There are, however, some tasks which are sufficiently different from those performed by other robots, and sufficiently broad in scope, that special-purpose robots have been designed and built to fill their requirements. Special-purpose robots presently on the market have been designed for spray painting and for welding.

Figure 10 shows a robot designed expressly for spray painting. It is hydraulically powered, but it differs from the heavy-duty hydraulic robot in its lighter construction and, more importantly, in its control method. In contrast with the control system of other robots, which concentrate on accuracy of the end positions and are unconcerned with the path between those end positions, the control system for a spray-painting robot must accurately follow a predetermined path.

To program this machine, the operator, a skilled spray painter, holds a spray gun attached to the end of a special teaching arm (on some models the spray gun on the robot arm is used for teaching); he then sprays the article in the manner necessary to achieve a good coat of paint, and the computer continuously records the positions of all the robot joints. On playback, it reproduces the exact motions of the skilled painter and so duplicates the spray pattern.

Besides the obvious advantage of reducing labor costs, the spray-painting robot improves throughput and consistency and realizes paint savings of up to 50%. Because the robot repeats spray patterns with high consistency, problems such as heavy edges and sags are eliminated. In addition, it can function in a spray booth with a ventilation level unhealthy for people.

The spray-painting robot is not limited to applications of paint. Any material which is applied by spraying is suitable, and uses range from spraying car bodies to applying ablative coatings to the space shuttle.

Fig. 11. The Unimate Apprentice robot laying a weld on a ship's hull. *(Unimation, Inc.)*

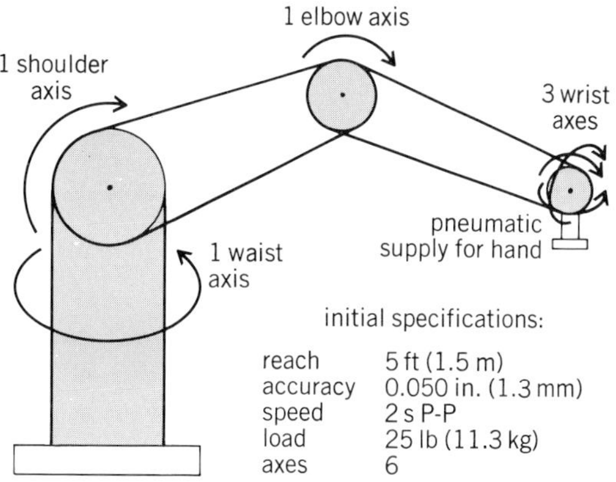

Fig. 12. Initial sketch and specifications for a hypothetical robot.

A second type of special-purpose robot is designed for on-site arc welding, and is a light-weight portable unit which is attached to the job to be welded. Whereas the spray-painting robot is useful for repetitive jobs, the welding robot is designed expressly for single welding jobs in cramped or poorly ventilated environments which are difficult for humans.

Figure 11 shows the robot laying a weld on a ship's hull. The machine has five axes of motion, and is electrically powered. Like the spray-painting robot, its control system follows the path defined by the skilled operator; however, it has additional capabilites which allow the weld to be "weaved" and the welding wire to be automatically fed. Programming is accomplished by using a teaching wheel attachment. The operator attaches the wheel and then rolls it along the seam to be welded. All movements made while the record button is held down are memorized by the robot. The operator then leaves the area, and the robot lays down the weld. Transverse speed, wire speed rate, voltage, and weave conditions are programmed from the control panel.

ANATOMY OF A ROBOT

An industrial robot is an engineering system with many interacting elements. In order to understand the elements and to appreciate their interaction, the design criteria for a hypothetical machine will be reviewed.

Design criteria. The first step in designing a robot is to select the mission that it is to accomplish. From this, certain mechanical parameters are specified, namely, (1) the size and configuration of the robot, (2) the required positioning accuracy, (3) the load that is to be handled, and (4) the speed at which it is to be moved. For the purpose of discussion, let it be assumed that the robot is to have a reach of 5 ft (1.5 m) and have an articulated construction. The required accuracy is to be 0.050 in. (1.27 mm) at the hand, and the machine should be able to handle a 25-lb (11.3-kg) load and be fast enough to move between any two positions in 2 s or less. Figure 12 shows the initial design layout and key parameters.

After determining the configuration, the mass and moments of inertia of the links are calculated.

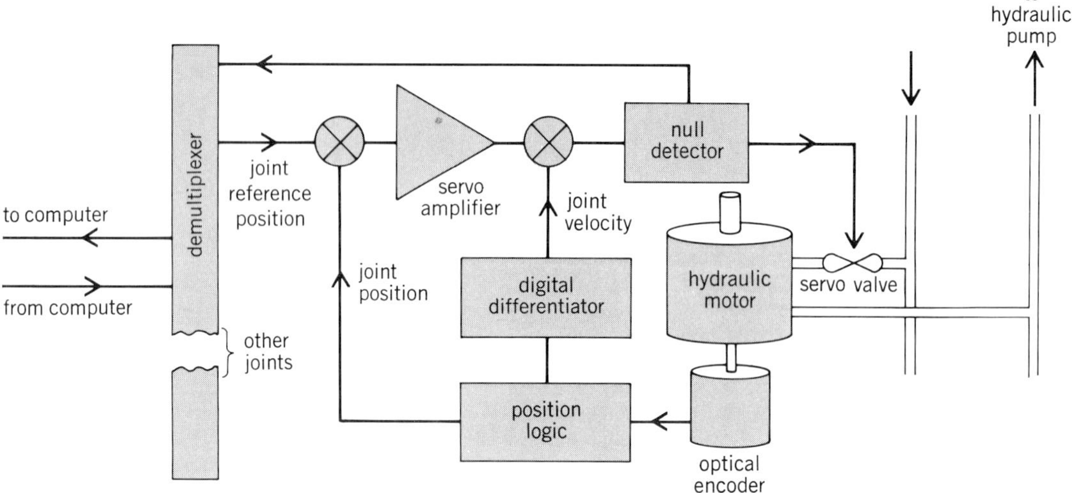

Fig. 13. Interface diagram for one of the six joints of the hypothetical robot.

To do so, the construction method of each link should be determined by considering the size and accuracy requirements. With a reach of 5 ft (1.5 m) and a total accuracy of 0.050 in. (1.27 mm), a rigid construction technique is necessary. On the other hand, a high speed is required, so that the construction cannot be too heavy. Assume that a cast aluminum construction is selected.

From the mass and moments of inertia of the links, it is possible to determine the motor torque required to move them at their rated speeds. The acceleration and maximum velocity of each joint is determined from point-to-point time of 2 s. The torque required to accelerate the joint is then found and added to the torque required to hold the link against gravity. From these numbers the type of motor for each joint is selected. In choosing the motor, attention must also be paid to the total mass of the motor, because three motors on the wrist and one on the elbow are carried on the arm and add to the weight. The motor must also be capable of sufficiently fine control so that the arm can be positioned with the necessary accuracy. Let it be assumed that direct-drive hydraulic motors are the optimum in this case.

Finally the feedback elements for each joint must be chosen. The primary feedback determines the position of the joint; the secondary feedback (tachometer) gives information on the velocity of the joint, and is used for stabilizing the control loop. Three types of position feedback elements are available—precision potentiometers, resolvers, or optical encoders. Because of space and precision requirements and the fact that a digital output is an advantage, incremental optical encoders have been selected.

Velocity feedback is usually provided by a tachometer. However, tachometers, being small generators, must be driven from reasonably high-speed shafts. Because direct-drive hydraulic motors have been selected, a tachometer is not suitable. Velocity feedback will therefore be provided by digitally differentiating the position feedback signal.

The final element to be considered is the hand, or end effector. The hand is usually the only element of the robot that is changed to suit the task. A suitable mounting flange for a custom-made hand and a means of actuating it will therefore be provided. A two-state hand motion is all that is usually required, and so an air supply for a pneumatically operated hand will be available at the end of the wrist.

Interface and control computer. The sophistication of the control system of a robot greatly affects its versatility. With the increasing capabilities and the decreasing cost of microprocessors, systems that would have been prohibitively expensive 10 years ago are now possible.

The control system consists of two parts—the interface and the computer control. The function of the interface is to receive control commands from the computer and convert them into appropriate outputs to drive the motors. It is also responsible for closing the feedback servo loop to make sure that the joints follow the commanded positions.

The data from the computer to the interface are in the form of a stream of position instructions for each joint of the robot arm. Depending on the degree of sophistication of the control system, these position commands can represent either successive end positions for a particular motion or incremental points along a path to be followed.

If the design robot must be capable of complex motions, the second type of control system is selected, which enables path control to be maintained. Position data are provided by the computer, typically 60 times a second. Since there are six joints to be controlled, six position commands are transmitted in sequence every 1/60 s. In the interface, these position commands are separated (demultiplexed) and passed to the appropriate servo controllers.

Figure 13 is a block diagram of the interface. Note that the position feedback is differentiated and used to provide velocity damping. The output from the servo amplifier also has a null detector, which provides information to the computer as to when the final position has been reached.

The function of the control computer is to provide the stream of data defining the incremental

Sample instructions of a robot high-level language

SPEED 50	Run at 50% speed
COARSE	Set coarse tolerances
WAIT 51	Wait for switch 51
MOVE P3	Move to position P3
SET N = N + 1	Increment counter N
IF N = 4 GO TO 30	If counter = 4, branch to instruction 30

position changes of all the joints in the robot. It must also establish, by checking the null detectors on the outputs of the servo amplifiers, that the robot arm has arrived at a particular position before initiating the next sequence of action. The computer also generates all appropriate timing pauses and provides necessary outputs for ancillary equipment, such as incrementing feeding mechanisms or signaling that a task is completed.

The easiest way for a computer to provide this data is to play back a prerecorded sequence of events. The spray-painting robot shown in Fig. 10 is controlled in just such a manner. This method does, however, unnecessarily restrict the utility of a general-purpose robot. The presence of the computer on a robot enables it to be a decision-making, or adaptively controlled, machine. Even simple adaptive control can greatly increase the versatility of the robot. For example, with appropriate input on its position, the robot can track and operate on a moving object such as a conveyor belt. To do this, it must first establish the required position of the hand at every instance and then calculate the appropriate angles for all the joints in real time.

A more sophisticated version of adaptive control enables the robot to generate paths that are not specified by the programmer. One example would be if the machine were equipped with vision (described below). In this case the path to be taken depends upon the input from the vision unit, and cannot be determined in advance. With an adaptive control program, the computer generates its own path and provides the appropriate stream of data to the interface.

FUTURE DEVELOPMENTS

Industrial robot technology is just beginning what will surely be a long and successful career. The necessary mechanical components have been available for a number of years, but it is the introduction of small, very capable computers that has spurred the development of the technology. The advent of the microprocessor and the continuing drop in price of electronics indicate a bright future for all aspects of automation technology.

High-level languages. Side by side with the development of hardware must come the development of suitable software. Since the major purpose of the industrial robot is to provide off-the-shelf automation that can be utilized by a typical engineer in a factory, some form of easy-to-use high-level language must be developed. Several research organizations are presently studying this concept, and there is at least one high-level language on the market. However, there are no universally accepted language and no agreed-upon constraints from which to construct such a language.

Two major schools of thought exist. In one, the language should be small and self-contained, since the robot should not need access to any other computer in order to be instructed or to execute the instructions. In the other school of thought, the computer should be integrated into a large, general-purpose, computer-aided manufacturing (CAM) system. With such integrated control, the robot would be capable of generating its own motions based solely upon the design of the object to be handled, with little or no input from the engineer. The second type of high-level language is considerable interest to the aerospace industry.

The table shows a section of a program written in the first type of high-level language. The arm is instructed to move by commands such as MOVE, DRAW, and GO TO. These commands cause the arm to move in different manners to the appropriate point. The exact coordinates of the points are not specified at the time of writing the program, but may either be defined later or even be made programmable, that is, the computer can calculate the coordinates of the point and insert them into the program.

More sophisticated programming techniques will ultimately enable robots to be programmed by voice in a conversational manner. This technique falls under the classification of artificial intelligence, and is being actively studied in several universities in the United States. Practical implementation is, however, several years into the future.

Assembly aids. A major area of interest for robot manufacturers is automatic assembly of mechanisms such as small electric motors, brake cylinders, and typewriter components. Although the level of sophistication and dexterity required is higher than is available from most present-day robots, several industry- and government-sponsored laboratories are developing techniques to assist the machine.

Fig. 14. The remote center compliance (RCC) unit greatly simplifies tight-tolerance assembly tasks. Diameter is 2 in. plus (50 mm plus). *(Charles Stark Draper Laboratories, Inc.)*

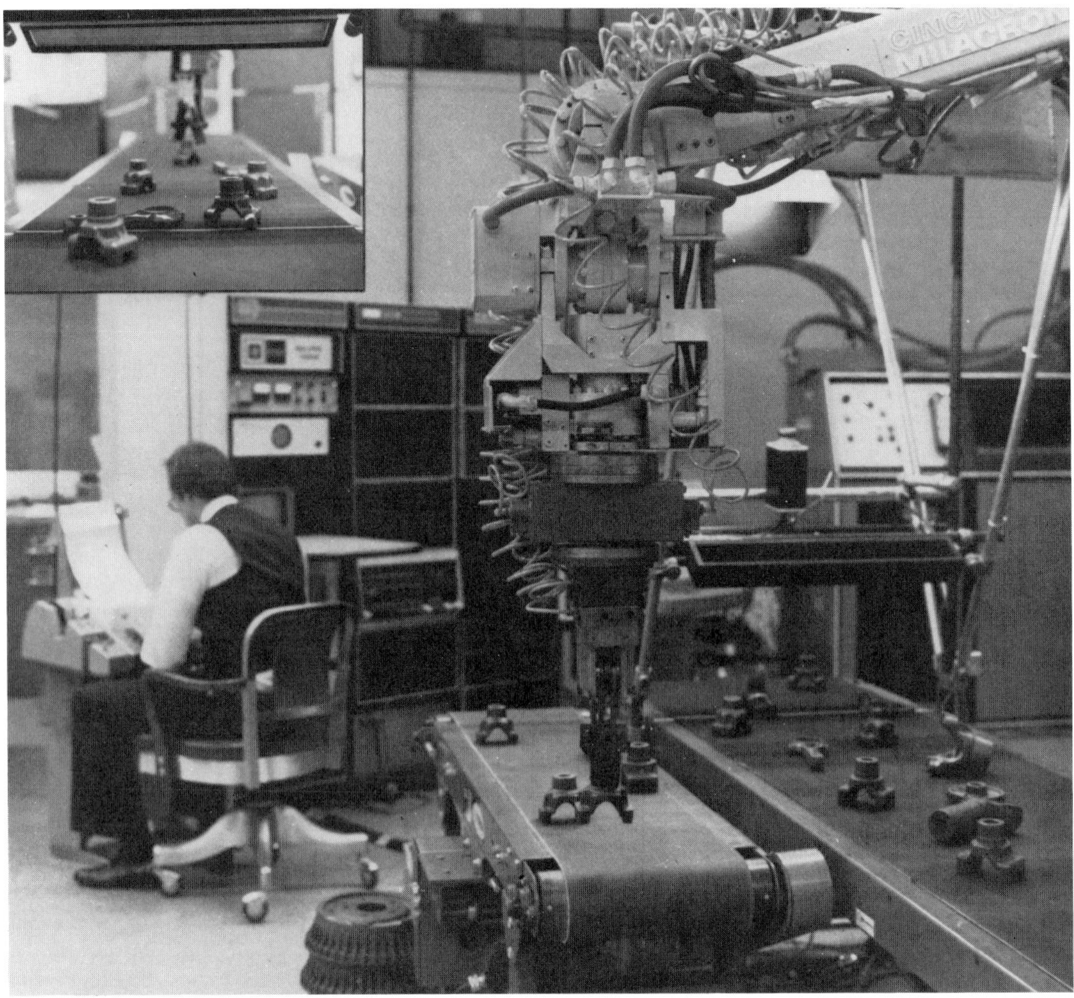

Fig. 15. Using Consight, a robot acquires randomly placed items on a conveyor belt. *(General Motors, Inc.)*

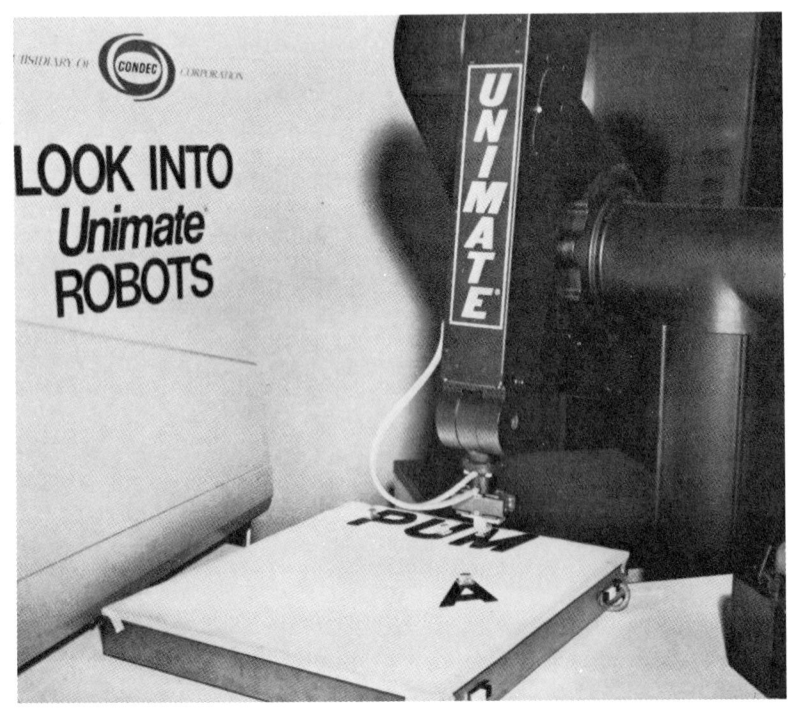

Fig. 16. This Unimate PUMA robot uses vision to acquire the letters that spell out its name, and arranges them in order. *(Unimation, Inc.)*

One fundamental aspect of assembly is the fitting of tight-tolerance objects, for instance, a bearing into its housing. A human fits a bearing by invoking delicate and complicated feedback sensation from the fingertips, which are entirely lacking in the robot. Moreover, the resolution of the robot arm is usually not capable of sufficiently fine adjustments to insert the bearing. Researchers at the Charles – Stark Draper Laboratory have studied this problem and produced an ingenious solution. A device known as a remote center compliance unit, shown in Fig. 14, is attached to the end of the arm, which by means of levers and springs has the effect of changing the apparent point at which the force is applied to the bearing. In this manner, although the bearing is in fact being pushed into the housing, it is mechanically identical to a bearing which is being pulled into the housing from inside. Using this device, robots can insert bearings into housings with 0.0005 in. (12.7 μm) clearances in 0.5 s.

Vision. A limitation of present robots is the fact that they position their hand by deadreckoning. The object to be handled is expected to be in the appropriate position, which often requires elaborate feed mechanisms to dispense and properly place the object. Since parts feeders must be tailored individually for each part, this is not in keeping with the concept of off-the-shelf automation. To overcome this limitation, several research facilities are developing elaborate feedback systems, most of them based upon vision.

Compared with human vision, machine vision is rudimentary and poor in resolution. It is, however, capable of locating an object within a certain known area and of describing the identity, position, and orientation of that object. This information can be passed to the computer controlling the robot, which is then able to acquire the object.

There are several different approaches to machine vision. Figure 15 illustrates one approach, in which a bar of light projected across the conveyor belt illuminates a strip of the object as it passes through it. The outline of the object is matched against known templates, and the object identified by finding the closest match. A second approach is used in the demonstration shown in Fig. 16. Here a television camera (not shown) recognizes objects by establishing certain parameters, such as area and perimeter, associated with each object. In this demonstration the robot is selecting the letters that spell out its name and arranging them in the proper order.

CONCLUSIONS

The advent of low-cost microcomputers has turned the industrial robot from a mechanical novelty into a significant automation aid. The basic mechanism, with relatively simple control programs, has demonstrated its reliability and utility in various industries, and future developments in the fields of tactile and visual feedback systems will generate a considerable impact on the economics of automation. [GORDON I. ROBERTSON]

Industrial trucks

Manually propelled or powered carriers for transporting materials over level, slightly inclined, or slightly declined running surfaces. Some industrial trucks can lift and lower their loads, and others can also tier them. In any event, all such trucks maintain contact with the running surface over which they operate and, except when towed by a chain conveyor, follow variable paths of travel as distinct from conveying machines or monorails. *See* MATERIALS-HANDLING MACHINES.

Running gear. The means employed to support a truck and its load and to provide rolling-friction contact with the running surface is the running gear. Factors in the selection of running gears include load capacity, operating conditions, travel surface, kind of material to be handled, protection of load and machine, economy, and, in the case of hand trucks, ease of manipulation and the reduction of operator fatigue.

Rollers, used in dollies, are of solid or tubular steel with antifriction bearings. Rigid and swivel casters are used with the dollies and with hand and powered trucks. Swivels differ from rigid casters in that they are offset wheels that swivel about their own vertical axis, thereby easing the turning of the hand truck with its load. Steel, solid rubber, semipneumatic, and other wheels fitted with plain or antifriction bearings are designed to meet specific requirements. Industrial wheels for heavier-duty and special automotive-type wheels in a wide selection of tire treads are used with powered trucks and tractors.

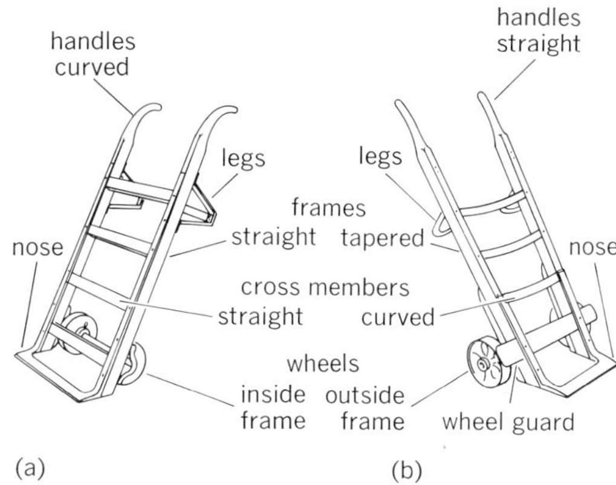

Fig. 1. Two-wheel hand trucks. (*a*) Eastern. (*b*) Western.

Hand trucks. The manually operated, two-wheel hand truck, regarded by some as old fashioned and inefficient, still plays a prominent part in the handling of materials in some of the most modern plants. The hand truck is inexpensive, requires a negligible amount of maintenance, and is convenient for handling light loads for short distances. Its light weight makes it ideally suited for applications where small quantities are handled, with a low frequency of movement, where floor loading restrictions preclude the use of the heavier powered trucks. The hand truck is also easily adapted for use in explosive atmospheres.

The key feature to the hand truck is the running gear. Care must be exercised in selecting the type and size of wheel to suit the carrying load and operating conditions. In many cases, depending on the loads to be carried and the surface of the operating floor, inexpensive cast-iron or steel wheels may be more suitable than higher-priced pneumatic tires or plastic-rimmed wheels.

Basic types of hand trucks and their distinctive features are shown in the table. Two-wheel hand trucks are classified broadly as eastern and western (Fig. 1). Multiwheel hand trucks are produced in many models, but platform types continue to be the most widely used in industry and distribution. Stakes, end-and-side gates, solid panels, and other superstructures add versatility to the basic machines. Low-lift types of hand trucks elevate their loads just sufficiently to clear the running surface (Fig. 2).

Powered trucks. The powered industrial truck can be defined as any self-propelled, power-driven truck or tractor used to carry, push, pull, lift, or

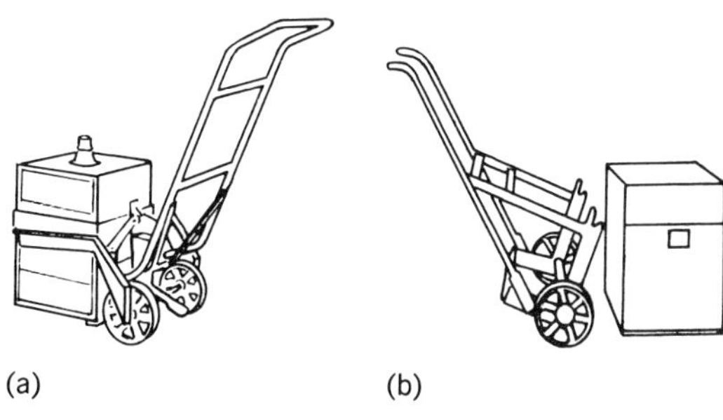

Fig. 2. Low-lift hand trucks elevate loads to clear running surface. (*a*) Carboy. (*b*) Trunnion box.

Handtrucks and their features*

Type	Description	Capacity	Range
Pry	Lever bars with long wooden handles, short steel noses, and two wheels at the fulcrum. Used singly or in pairs for moving and spotting crates and similar heavy articles in freight cars.	Up to 5000 lb each	Very short distances
Dolly	Low platforms or specially shaped carriers mounted on rollers or combinations of fixed and swivel casters. Designated as furniture, milk can, paper roll, and so on to indicate intended uses.	From few pounds up to 80 tons for moving machinery	A few feet
Two-wheel truck	Normally constructed of wood and steel but available in aluminum and magnesium. Stevedore and warehouse models are general purpose, but many special-purpose types are available for egg crates, drums, or paper rolls.	Normally 400–500 lb, but exceptionally up to 1500 lb	Up to 150 ft
Multiwheel platform	Trucks with flat platforms mounted on various combinations of 3 (unstable), 4, or 6 rigid and swivel casters. Addition of stakes, side and end gates, solid panels, and similar accessories increases the versatility of these trucks.	Normally 1500–2000 lb for one-man operation	Weight is an important factor, but normally within a few hundred feet
Specials	A wide variety of trucks with special superstructures and running gear. Examples are box, frames for spools and sheets of metal, frames for plate glass, shelves for small parts and tools, shapes to accommodate bar stock, textile beams, and stock items.	Same as multiwheel platform	Same as multiwheel platform

*1 ft = 0.3 m; 1 lb = 0.45 kg; 1 ton = 0.9 metric ton.

tow loads. Relatively new in the use of industrial trucks are personnel and burden carriers for transporting messengers, watchmen, mail, and blueprints, as well as materials between buildings and in-plant areas covering considerable ground. Shop tractors are used extensively for miscellaneous towing jobs in plants and warehouses (Fig. 3). More recently an electronically controlled tractor was introduced, directed by radio, overhead wires, or wires concealed in the floor.

Whatever the type or function, these trucks depend on compact, high-capacity storage batteries or small internal combustion engines. Sources of power for internal combustion engines are gasoline, diesel fuel, and more recently liquid petroleum gas; some manufacturers offer trucks powered in this fashion as original equipment and conversion kits are available for existing equipment. Characteristics of these sources of power are briefly outlined as follows.

Storage battery. Trucks powered from storage batteries are quiet and clean in operation, have low maintenance cost and smooth acceleration, and when properly charged present no starting problems in cold weather. However, the speed is generally considered too low for such trucks to be used in long-distance travel.

Battery-charging facilities are required at many points for truck travels over a large area, and battery drainage is excessive on steep slopes; additional capital costs are involved in providing batteries for multishift operations.

Diesel engine. Higher speed of operation is provided by diesel-powered trucks. They can withstand severe operating conditions, such as working on inclines or traveling over long distances; refueling is a simple operation; and fuel cost is low. However, operation is noisy and fumes are objectionable though nontoxic. Maintenance cost is higher than for electric-powered trucks.

Gasoline engine. Basically the characteristics are the same as the diesel engine, except that the exhaust fumes are toxic.

Liquid petroleum gas. High octane rating permits higher compression and greater efficiency; clean combustion, even at idling speed, prolongs the life of the engine; and there is less carbon monoxide in exhaust fumes than from a gasoline engine. However, gas is transported and stored in tanks under pressure requiring stringent safety precautions; the fuel is heavier than air and odorless and requires a smell additive to make a leak detectable.

Unit load principle. In industry the principle of handling materials in unit loads has developed in parallel with the increased use of powered industrial trucks, particularly the forklift type. These mobile mechanical handling aids have removed

INDUSTRIAL TRUCKS

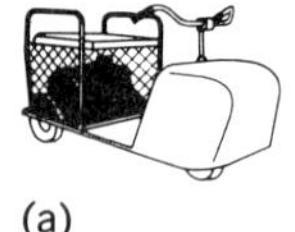

(a)

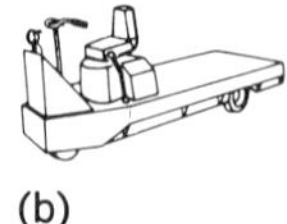

(b)

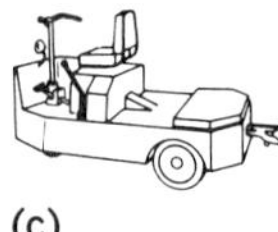

(c)

Fig. 3. Powered nonlift platform trucks. (*a*) Personnel carrier. (*b*) Burden carrier. (*c*) Shop tractor.

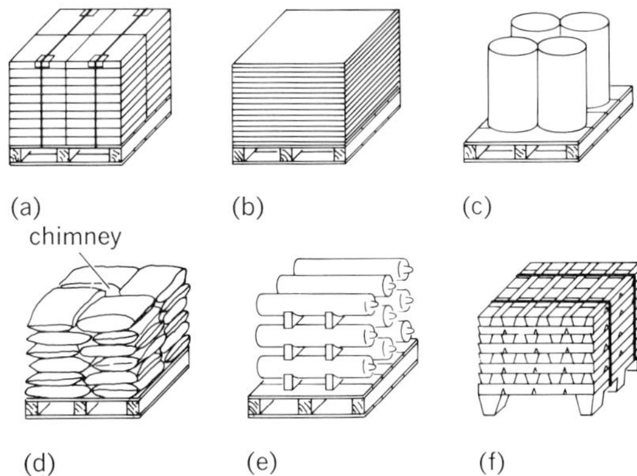

Fig. 4. Loads assembled for handling as units. (*a*) Strapped load. (*b*) Sheet or flat stack. (*c*) Barrels or drums. (*d*) Bags, pinwheel pattern. (*e*) Cylinders with dunnage. (*f*) Ingots with molded legs.

the limitations that existed when the weight and size of a load for movement and stacking depended mainly on the ability of a man to lift it manually. The unit-load principle of materials handling underlies the skid-platform and the pallet-forklift methods of operation. Both methods, especially the latter, have revolutionized handling techniques and equipment and even production equipment.

A unit load can be defined as a quantity of material (bulk or individual items) assembled and, if necessary, restrained to permit handling as a single object. When referred to in connection with industrial trucks, it is implied that a unit load will retain its shape and arrangement when deposited by the truck to allow subsequent movement, if required (Fig. 4).

In order to gain the maximum benefits from unit-load handling, the principle should be applied as far back and as far forward as possible in the manufacturing cycle. In particular, the possibility of introducing unit loads should be thoroughly investigated from raw material receipt through production operations, process storage, and warehousing. Investigation may reveal that modification of production equipment is justified. Some or all of the following benefits can be expected from the correct application of the unit load principle: (1) cheaper direct cost of unit handling because of the movement of more pieces or a larger quantity at a time; (2) maximum use of the cubic space available in storage area; (3) quicker loading and unloading and consequently a faster turn around of transport vehicles; (4) unit load lending itself to an efficient storage system and hence easier and quicker stock control; (5) well-secured unit load deterring pilfering; (6) product damage reduced by the elimination of manual handling of individual pieces; (7) reduction of packaging costs; (8) reduction of materials-handling accidents, particularly injuries, such as damaged fingers or strained backs, which result in lost time; and (9) value of correctly designed load units as a sales aid since most customers welcome receipt of goods in suitable unit loads.

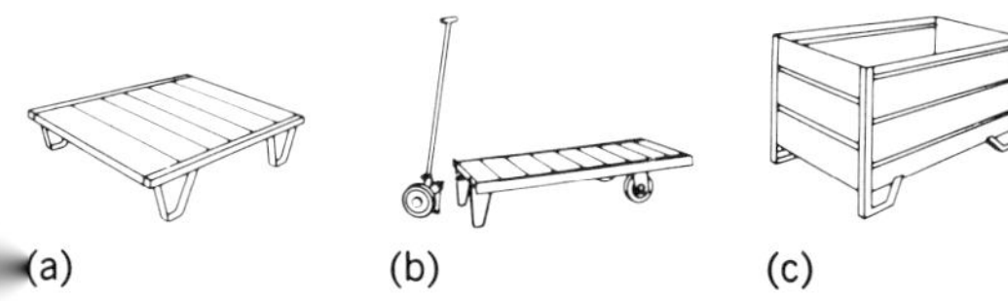

Fig. 5. Skids for load units. (*a*) Dead skid. (*b*) Semilive skid and jack. (c) Corrugated metal box.

Skids are constructed in three basic types: (1) live skids, with fixed and swivel casters, which are too lively for tiering purposes; (2) semilive skids, with two fixed legs in the front and a pair of rigid casters at the rear and made mobile by means of a jack; and (3) dead skids, having either two solid runners or four metal legs (Fig. 5). All wooden dead skids and those made of hardwood platforms with metal edges and metal legs are conventional; others are made of metal such as steel or aluminum and have special superstructures designed with the most efficient configuration for the product being handled.

Pallets differ from skids in having two decks, or faces, separated by two or more lengthwise members called stringers. The National Wooden Pallet Manufacturers Association has prepared and issued specifications for lumber, fasteners, and other pallet components. Different constructions are called by descriptive names (Fig. 6); for exam-

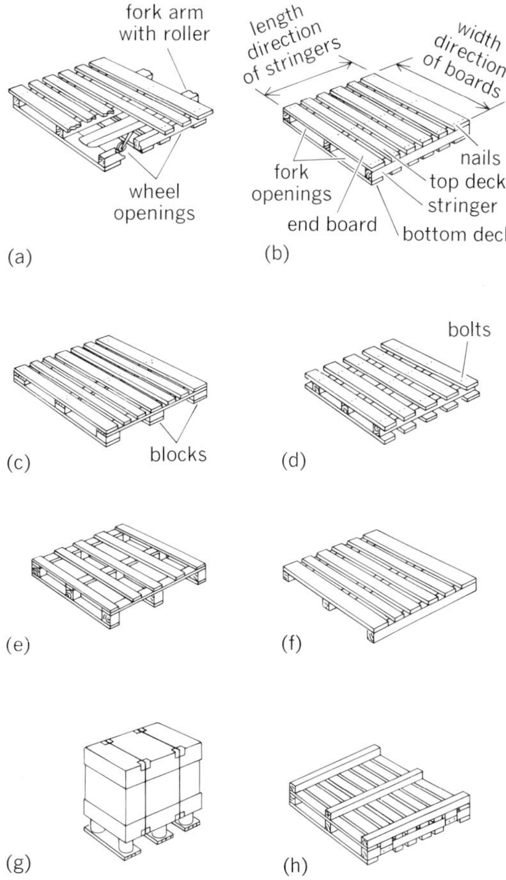

Fig. 6. Standard and special pallets. (*a*) Semiwing, nonreversible, two-way pallet. (*b*) Flush-type, reversible, two-way pallet. (*c*) Four-way pallet. (*d*) Stevedore (double-wing), reversible, two-way pallet. (*e*) Expendable (one-way shipper) pallet. (*f*) Single-faced pallet. (*g*) Fiberboard expendable pallet. (*h*) "Take it or leave it" pallet unit; load rests on stringers.

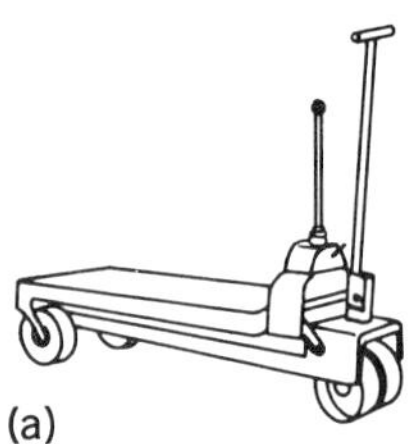
(a)

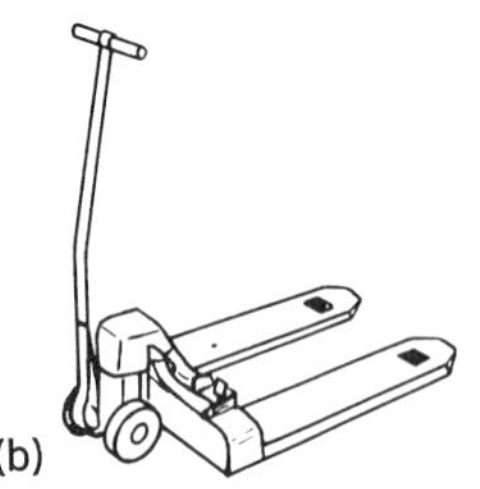
(b)

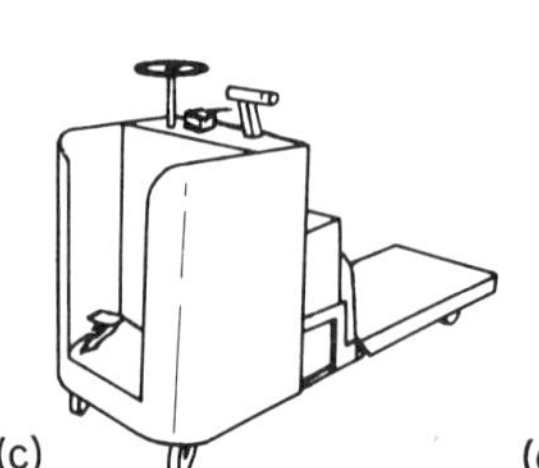
(c)

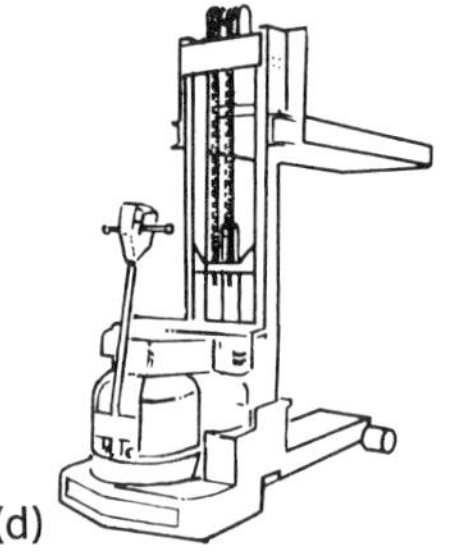
(d)

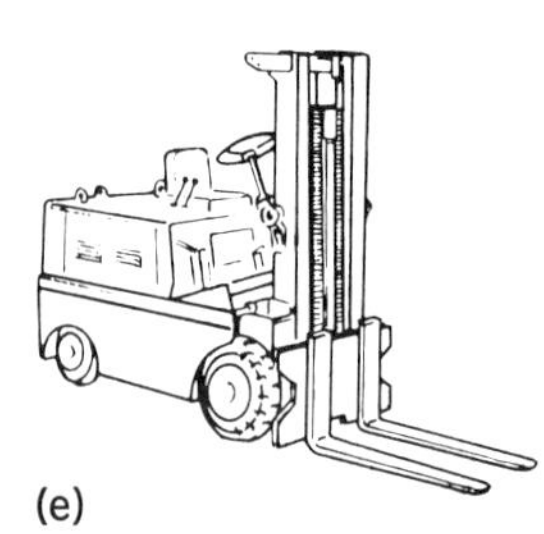
(e)

Fig. 7. Hand-operated, counterbalanced, powered, stacker-type skid and pallet trucks. (*a*) Hand. (*b*) Hand. (*c*) Powered low-lift skid. (*d*) Noncounterbalanced stacker with skid platform. (*e*) High-lift fork.

ple, those with double wings to permit the use of sling bars are the stevedore type. A newer development in pallets is the "take it or leave it" pallet (Fig. 6*h*), which permits the choice of retrieving the pallet and unit load together or of retrieving the unit load and leaving the pallet.

There are nine standard rectangular pallet sizes and three square sizes, which range from 24 × 32 to 88 × 108 in. (60.96 × 81.28 to 223.52 × 274.32 cm). The International Organization for Standardized Sizes recommends the following sizes: 32 × 40, 32 × 48, and 40 × 48 in. (81.28 × 101.6, 81.28 × 121.92, and 101.6 × 121.92 cm), all of which can be carried advantageously in over-the-road trucks and trailers and in box and refrigerated railroad cars. Other sizes may be suitable for the product being handled; however, these sizes should be developed with regard to proper increments to utilize maximum space in motor or rail carriers.

The pallet was developed primarily for use with forklift trucks. If a pallet is to be used with a low-lift truck, the bottom deckboards must be so spaced that openings are left near both ends to permit the rollers in the fork arms of the truck to drop to the floor when the pallet is being elevated. Pallets so constructed are called nonreversible. Special pallets are those made of steel, magnesium, or even plastic; expendable varieties (one-way shippers) are made of fiberboard or inexpensive wooden construction and, as the name implies, are designed for extremely limited service.

Structural differences between skids (Fig. 5), with considerable clearance underneath, and pallets (Fig. 6), with stringers and bottom boards presenting obstructions, account for the use of platform trucks with the former and, of necessity, fork-equipped trucks with the latter. The double-deck feature of pallets makes them more suitable for multiple tiering than the runners or legs of skids, which may damage the supporting surface of the lower load.

Low- and high-lift trucks. Skids and pallets are handled by self-loading machines. The lift trucks pick up, transport, set down, and in the case of high-lift types tier their loads without manual handling. Powered models evolved from prototype hand machines and, because the first of the self-propelled machines were led by the operator, were called "walkies." The name persists, even though most of them are now produced as rider trucks (Fig. 7).

Low-lift trucks, hand and powered, lift their loads sufficiently to make them mobile. The elevating mechanisms of hand types may be operated by

Fig. 8. Industrial high-lift truck raising a load. (*Forest Products Laboratory, Madison, Wis.*)

pulling the handle down one or more times. Others have a lever- or pedal-operated hydraulic mechanism or a powered hydraulic system.

Noncounterbalanced high-lift trucks evolved from hand-propelled stackers (Fig. 8). Fork-equipped types are called outrigger or straddle trucks because of the supports which straddle the pallet. All varieties come with fixed or telescoping masts. These trucks are advantageous when space is at a premium. For example, when piling vertically from aisles, these machines operate in aisles 6 ft wide, compared with the 10-ft aisles required by the counterbalanced trucks of equal capacity. Because they have no counterweight, these machines are lighter than the counterbalanced types and hence are a boon to handling operations in old multistory buildings.

Counterbalanced lift trucks are constructed by extending the wheelbase, adding counterweight, and removing the outriggers. By using a forward- and backward-tilting mast, they are used where outriggers cannot function advantageously, but the entire truck is longer than its noncounterbalanced equivalent.

Forklift trucks. Conventional fork trucks are made with any desired source of power, the selection depending upon the service for which the truck is intended. Elevation of the forks, forward and backward tilt of the mast, and the power for actuating attachments are provided by the hydraulic system. The hydraulic system consists of a pump to draw oil from a reservoir and force it through the control valves, which are manipulated by hand levers selected by the machine operator, into piston actuators.

Free lift is an important factor in some operations. This is the distance through which the forks are elevated before the mast starts to rise. Where headrooms are low, as in boxcars, double-tiering is possible if sufficient free lift is provided. Attachments for use with high-lift trucks are designed to handle specific products singly or in multiple units. Many attachments eliminate the need for a pallet or other supporting carrier, a feature with direct economic advantages.

The rated capacity of a high-lift fork truck is based on the weight of the load in pounds that the machine can lift when the load center is located a specified distance from the vertical faces of the forks. The load center is the center of gravity of a uniform load. The load center is 15 in. (38.1 cm) for trucks up to 2000 lb (907 kg) capacity, 24 in. (60.96 cm) from 2000 to 10,000 lb (907 to 4536 kg), 36 in. (91.44 cm) from 10,000 to 20,000 lb (4536 to 9072 kg), and as designated above 20,000 lb (9072 kg).

Trends in forklift truck design have been toward the replacement of friction clutches by torque converters in gasoline and diesel machines, more complete protection from exhaust fumes through more efficient catalytic mufflers, and changes in mast construction to provide better visibility for the operator.

Trucks and trailers. Other trucks capable of handling unit loads include end-loading (also called gantry and straddle-carries) and side-loading types (Fig. 9). The former type must straddle its load before picking up and transporting. This type unit is most suitable for handling long, unwieldy items such as pipe and lumber. These trucks, however, cannot tier their loads so that, when tiering is essential, they team up with forklift trucks. In contrast, side-loading trucks can transport and tier within the reach of their forks. Furthermore, side-loading trucks can be equipped with crane arms for handling lengths of pipe or lumber and with reel carriers for paying out cable as the truck advances.

Tractor-trailer or trackless trains are motorized versions of industrial (narrow-gage) railroads but have an advantage over the railroads in being able to follow variable paths in travel (Fig. 10). Industrial tractors, powered electrically or by internal-combustion engines, are classified as three- or four-point contact, according to the number of supporting wheels. They are used primarily for service along loading platforms and in yard operations.

Of industrial trailers, those with caster steer have good trailability (that is, they follow closely the path established by the tractor); those with four-wheel steer are rated as having excellent trailability but are difficult to maneuver manually; and those with fifth-wheel (wagon) steering are effective with heavy loads. They are offered in a wide variety of constructions, which makes it possible to select one with proper characteristics to meet given requirements.

Each type of industrial truck is used to the best advantage when carrying the particular load for which it is intended. Several types may be found working singly or in teams. For example, palletized loads may be transported on tractor-trailer trains and then distributed and tiered by forklift trucks. *See* TRUCK.

[ARTHUR M. PERRIN]

Inertia welding

A welding process used to join similar and dissimilar materials at very rapid speed. It is, therefore, a very attractive welding process in mass production

INDUSTRIAL TRUCKS

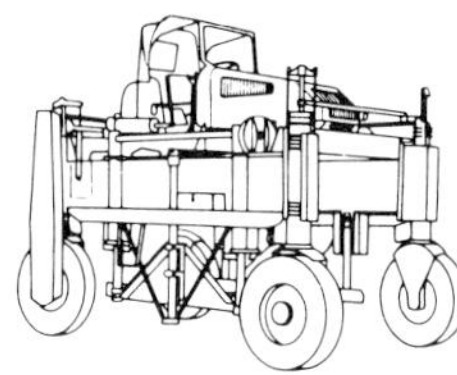

(a)

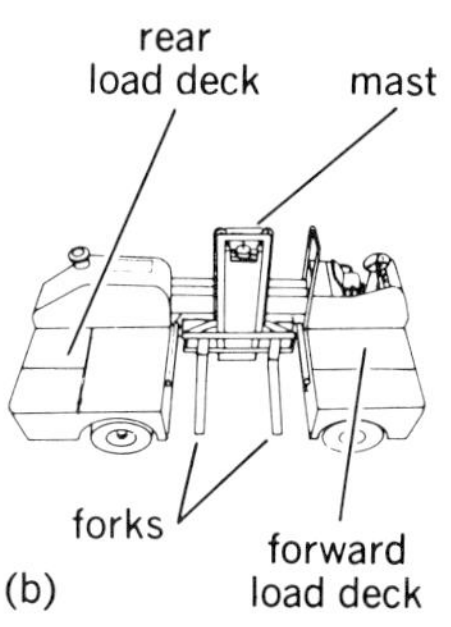

(b)

Fig. 9. Heavy-duty industrial trucks. (*a*) Straddle truck. (*b*) Side-loading truck.

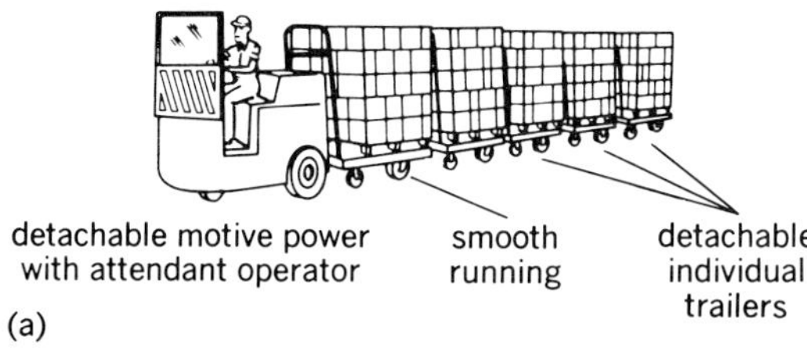

(a)

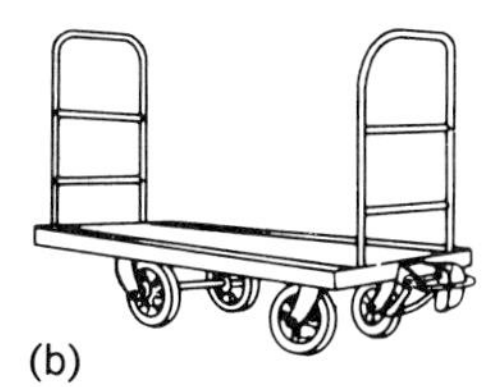

(b)

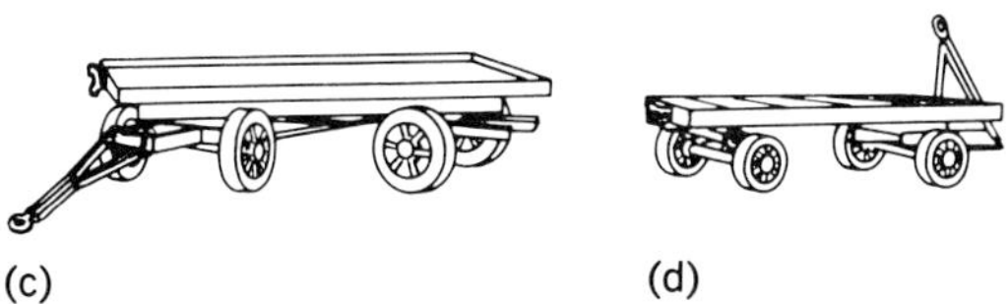

(c) (d)

Fig. 10. Industrial trailers. (*a*) In-plant, tractor-trailer, trackless train. (*b*) Caster steer. (*c*) Four-wheel steer. (*d*) Wagon (fifth-wheel) steer.

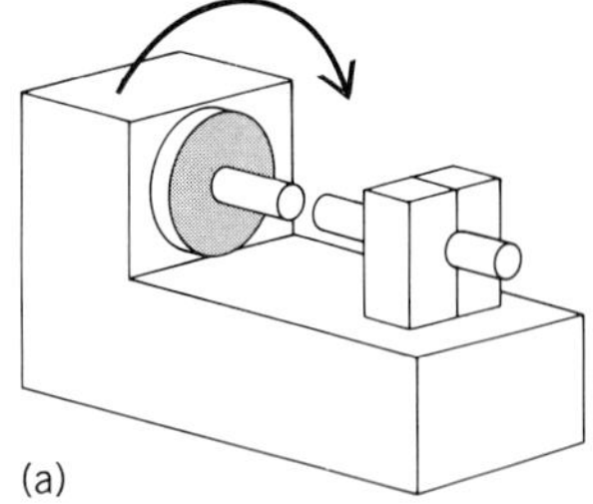

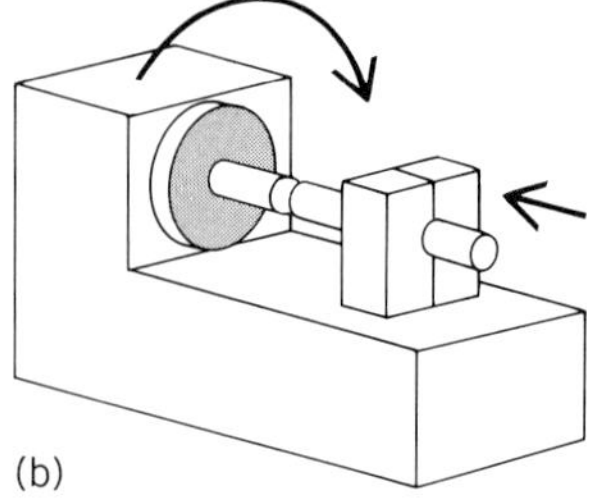

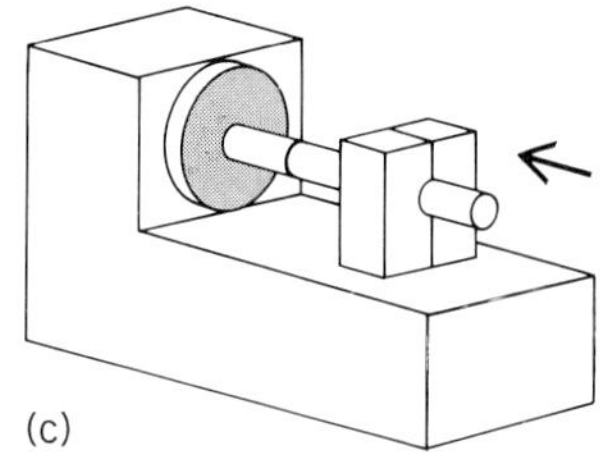

Fig. 1. Work sequence of inertia welding process. (*a*) Acceleration of flywheel and chuck-holding workpiece to preset speed. (*b*) At speed, drive is cut and workpieces are thrust together. (*c*) Workpieces bond together to stop flywheel rotation; weld is complete.

of good-quality welds. The ability to join dissimilar materials provides further flexibility in the design of mechanical components.

Principles. Inertia welding is a type of friction welding which utilizes the frictional heat generated at the rubbing surfaces to raise the temperature to a degree that the two parts can be forged together to form a solid bond. The energy required for inertia welding comes from a rotating flywheel system built into the machine. Figure 1 illustrates the principle of the process. Like an engine lathe, the inertia welding machine has a headstock and a tailstock. One workpiece held in the spindle chuck (usually with an attached flywheel) is accelerated rapidly, while the other is clamped in a stationary holding device of the tailstock. When a predetermined spindle speed is reached, the drive power is cut and the nonrotating part on the tailstock is pushed against the rotating part under high pressure. Friction between the rubbing surfaces quickly brings the spindle to a stop. At the same time the stored kinetic energy in the flywheel is converted into frictional heat which raises the temperature at the interface high enough to forge the two parts together without melting.

During the process the material near the interface is softened and expelled from the original contacting surfaces to form a flash. This brings the nascent sublayers of the material close together to form a metallurgical bond. Therefore, the welding process is autogenous and carried out in solid state without the need of flux or filler material.

Weld characteristics. Inertia welds of similar materials consist of a thin layer of a severely deformed but highly refined grain structure at the weld interface. The heat-affected zone is very narrow as compared with any fusion welding process. For a properly welded joint, the weld strength is significantly higher than that of the base material. This is because during inertia welding the forging process improves the hardness of the material at the weld zone by means of work hardening. An inertia weld of carbon steels can have an average hardness at the weld some 25% higher than that of the base material. A tensile test will result in a fracture occurring at the base material instead of at the weld. If the specimen is annealed after welding, the interface would completely disappear as if the materials were not welded.

One of the principal advantages of inertia welding is its ability to join dissimilar metals which would be unweldable or uneconomical to join by any other means. Figure 2 is a micrograph of a specimen of SAE 1020 steel welded to tungsten carbide. The process is also used to join aluminum to steel and stainless steel, aluminum to copper, copper to steel, copper to titanium, and zirconium alloys to each other. The weldability chart (Fig. 3) shows that some metal combinations can be welded to achieve strength at the weld equal to or better than the strength of the weaker material in the pair. In some cases, however, it may be necessary to perform an appropriate postweld heat treatment to realize the full weld strength. The chart indicates that there are other combinations that can also be welded, but with strengths usually less than the strength of the weaker material.

The mechanisms which cause bonding of dissimilar metals are rather complex. Factors such as physical properties, surface energy, crystalline structure, mutual solubility, and possible formation of intermetallic compounds from the materials being welded may all play a role in the bonding mechanism. Since the key to achieving a metallurgical bond in solid state is to break up and remove the contaminated surface layers, metals with a substantial difference in hardness are generally more difficult to weld. In such cases, the hard material incurs practically no deformation, while the soft one has already been severely deformed or extruded. In general, inertia welding can be applied with steels, aluminum, titanium, cobalt, nickel, and their alloys, copper, brass, and some bronzes, refractory materials, precipitation-hardened metal alloys, sintered materials, and heat-treated materials. The process cannot be successfully applied to cast irons, dry bearing materials, beryllium, and other materials wherever a distinct weak phase is present in the microstructure, such as in graphite, manganese sulfide, or a large amount of free lead or tellurium.

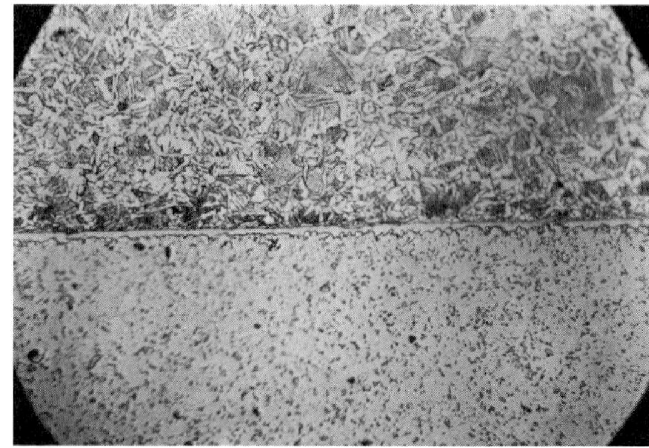

Fig. 2. Micrograph of a joint of SAE 1020 steel welded to tungsten carbide.

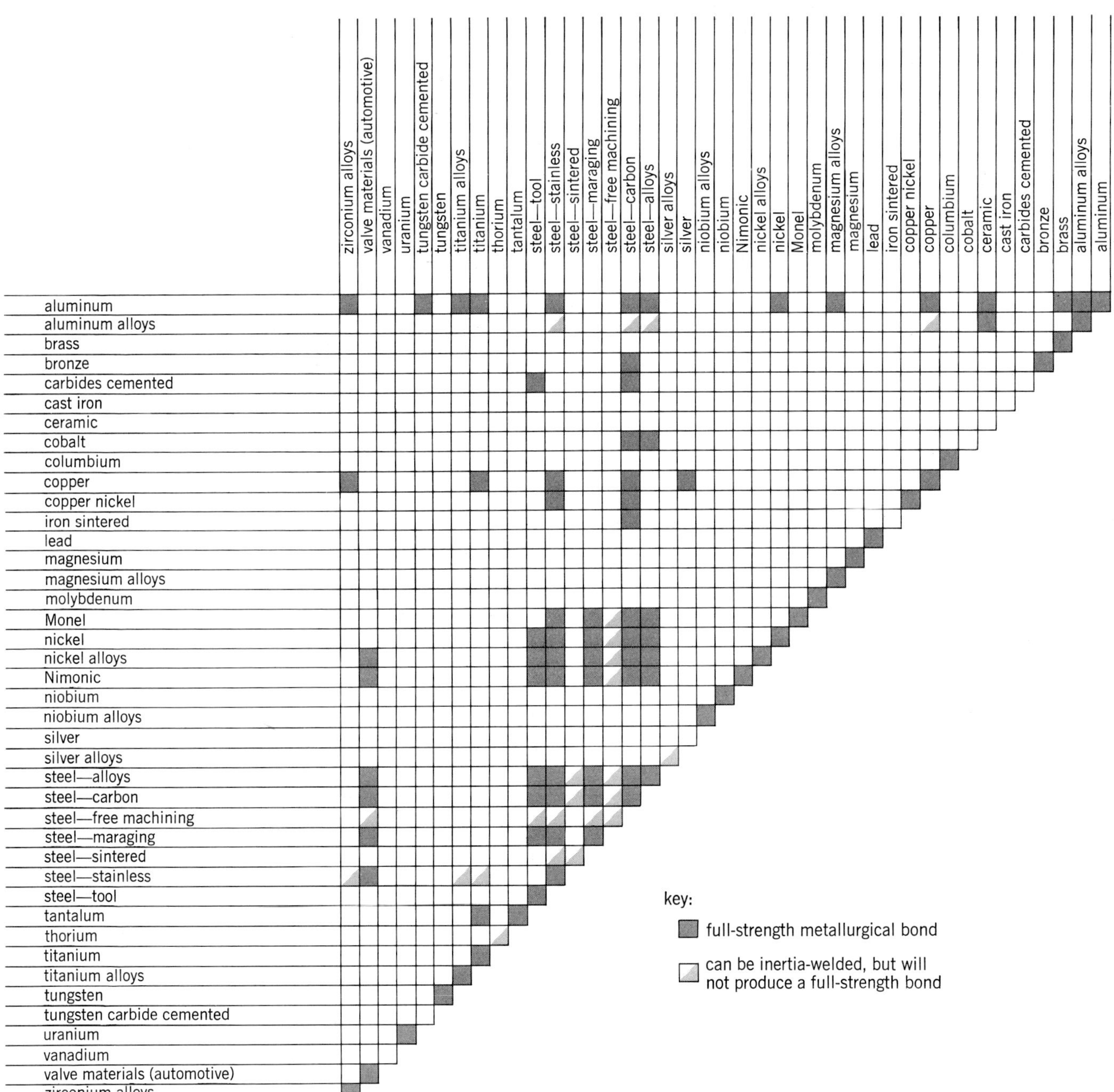

Fig. 3. Weldability between metals by inertia welding.

Applications. In addition to the advantages of high weld strength and short welding-cycle time, inertia welding machines can be readily automated for mass production. The automotive and truck industry is the major user of the process. Typical applications are engine valves (a medium-carbon-steel bar stem welded to a forged heat-resistant-alloy steel head), base-cup shock absorber assemblies, universal joints, axle housings, and hydraulic cylinder components. Auto makers, as well as other manufacturers, are constantly looking for ways to reduce the weight of components without sacrificing strength. An inexpensive way of achieving this goal is to replace standard open-die forgings or castings by weldments of precision closed-die forgings or castings welded to less expensive tubes, bar stocks, or stampings. Inertia welding of a tool pin joint to a drill pipe is one typical application in the oil drill industry. Figure 4 is a closeup view of such a machine showing a tool pin joint in the chuck and the part-feeding mechanisms.

Other areas of application include welding of stainless steel to plain carbon steel for outboard motor shafts or electric motor shafts. Inertia welding is used to replace electron-beam welding for joining low-alloy steel shafts to nickel-base superalloy turbine wheels or turbochargers in the aircraft industry. The process is also economical for making form-cutting tools by welding carbon-steel shanks to tool-steel heads. If one part has a round cross section and the materials are weldable by this process, the number of possible applications is virtually unlimited.

Other developments include design of multiple-

Fig. 4. Inertia welding machine for oil drill pipe sections.

spindle machines and radial inertia welding for joining a sleeve onto a ring in the radial direction. *See* METAL, MECHANICAL PROPERTIES OF; WELDED JOINT; WELDING AND CUTTING OF METALS.

[K. K. WANG]

Bibliography: American Welding Society, *Welding Handbook*, 7th ed., vol. 3, ch. 7, 1980; W. M. Hallett, *A Review of Inertia Welding*, SME Pap. no. AD76-090, 1976; K. K. Wang, *Friction Welding*, Welding Res. Counc. Bull. no. 204, April 1975.

Infrared lamp

A special type of incandescent lamp that is designed to produce energy in the infrared portion of the electromagnetic spectrum. The lamps produce radiant thermal energy which can be used to heat objects that intercept the radiation. All incandescent lamps have radiation in three regions of the electromagnetic spectrum, the infrared, the visible, and the ultraviolet. An infrared lamp with a filament operating at 2500 K will release about 85% of its energy in the form of thermal radiant energy, about 15% as visible light, and a tiny fraction of a percent as ultraviolet energy. The amount of infrared radiation produced by a lamp is a function of the input wattage of the lamp and the temperature of its tungsten filament. For most infrared lamps, the thermal radiation will be about 65–70% of the input wattage. A typical 250-W infrared lamp radiates about 175 W of thermal energy.

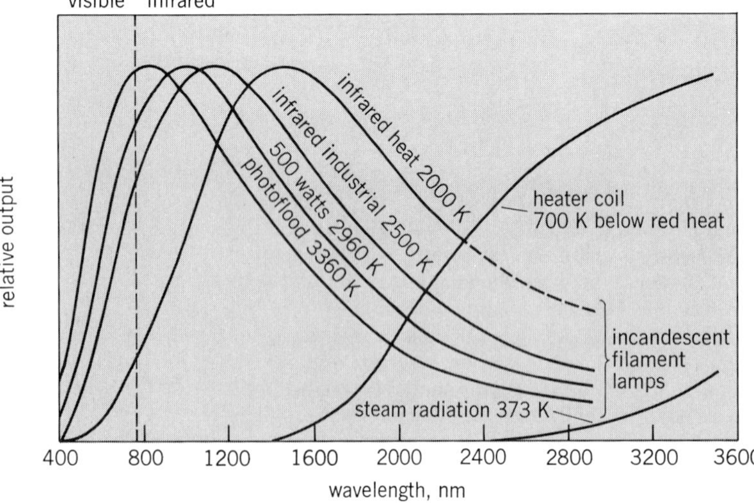

Fig. 1. Relative energy output of four incandescent lamps as a function of wavelength and filament temperature. As temperature increases, more of the lamp's energy shifts to shorter wavelengths. 2000 K = 1727°C = 3140°F; 2500 K = 2227°C = 4041°F; 2960 K = 2687°C = 4869°F; 3360 K = 3087°C = 5589°F. (*From Illuminating Engineering Society, IES Lighting Handbook, 5th ed., 1972*)

Temperature and energy spectrum. The temperature of the incandescent filament determines the portion of the energy output allocated to each of the spectral regions. As the filament temperature increases, the wavelength at which peak energy output occurs shifts toward the visible region of the spectrum (Fig. 1). Lamp filament temperatures are always stated in kelvins (K).

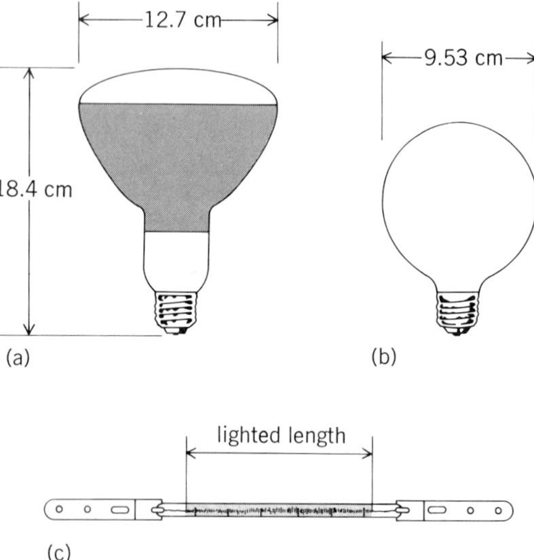

Fig. 2. Shapes of infrared lamps. (*a*) Glass bulb with built-in reflector unit. (*b*) Glass bulb used with separate external reflector. (c) Tubular quartz bulb.

Types. The lamps are supplied in three shapes (Fig. 2). The most common shape, for general use, is that shown in Fig. 2*a*, since the reflector unit is built in and needs only a suitable socket to form an infrared heating system. These lamps are available in 125-, 250-, and 375-W ratings, all in glass bulbs with a nominal reflector diameter of 12.7 cm (5.0 in.) and voltage rating of 115–125 V. In industry, the lamps shown in Fig. 2*b* and *c* are used with separate external reflectors designed to distribute the heat as desired. The lamp in Fig. 2*b* is available in 125-, 250-, 375- and 500-W ratings, all in glass bulbs 9.53 cm (3.75 in.) in diameter, with voltage ratings of 115–125 V. More than 30 sizes of the tubular quartz bulb lamps shown in Fig. 2*c* are available in 375- to 5000-W ratings. Lighted lengths range from 12.7 cm (5 in.) to 127 cm (50 in.), overall lengths (including connecting terminals) from 22.38 cm ($8^{13}/_{16}$ in.) to 138.68 cm, ($53^{13}/_{16}$ in.) and tube diameter is 0.95 cm (0.375 in.). Voltage ratings are a function of length, with nominal values ranging from 120 V for short tubes to 960 V for the longest tubes. Because the lamp bases become quite hot, the lamps in Fig. 2*a* and *b* should be used only in porcelain sockets, rather than in the common brass shell socket which would fail with prolonged use, posing a fire hazard. Some of the lamps in Fig. 2*a* are made with the glass face of the lamp stained red to reduce the amount of light emitted by the lamp. This reduces the undesirable glare in situations where the lamps must be in sight of the workers.

Uses. The major advantage of infrared heating is that it is possible to heat a surface that intercepts the radiation without heating the air or other objects that surround the surface being heated. Infrared lamps have many uses, some of which will be discussed.

Paint drying. Many industrial infrared ovens are used to dry painted surfaces as they pass by on a conveyor belt. Drying times are much shorter than for other methods of drying paint.

Evaporative drying. Infrared lamps can be used to remove moisture or solvents from the surface of objects by speeding the evaporation process. Porous materials with internal moisture require longer exposure, since the entire object must be heated by conducting heat from the surface to the interior.

Farm heating of animals. The lamps can be used to heat brooders and to keep other livestock warm without having to heat a large, poorly insulated shed.

Food. Many R lamps are used in restaurant kitchens to keep food warm while they are waiting to be served to customers.

Comfort heating. Areas under theater marquees can be made more pleasant for patrons waiting in line to buy tickets, and sidewalk cafes and similar areas can be made comfortable in relatively chill weather by installing suitable lamp systems.

Other uses. These include therapeutic heating of portions of the body to relieve muscle strains, tobacco drying, textile and wallpaper drying, and drying of printing inks so that press operation can be speeded up without causing image offset.

Design of heating systems. Most infrared heating ovens or systems are designed by calculating the number and wattage of lamps, plus experimenting with the system after it is constructed. The latter is necessary because the number of variables is large, and precise data on the magnitude of the variables are often difficult to obtain. Figure 3 shows the heat output of both the frosted and red glass-faced 250-W reflector-type lamp shown in Fig. 2*a*. The curves enable the computing of the watts per square centimeter received by a surface perpendicular to the lamp axis at a distance *a* from the face of the lamp.

Voltage and lifetime. Moderate variations in voltage will alter lamp operation somewhat. A 1% increase in voltage will increase both the input and radiated power in watts by 1.5–2%, and it will also increase the filament temperature and the light from the lamp. Most lamps are rated for use over a range of voltage, such as 115–125 V for a lamp intended for use on a 120-V power system. The life of all infrared lamps exceeds 5000 hours. The greatest cause of life reduction is subjecting the lamp to vibration.

Exposure to radiation. Exposure of persons to moderate levels of infrared is not known to be

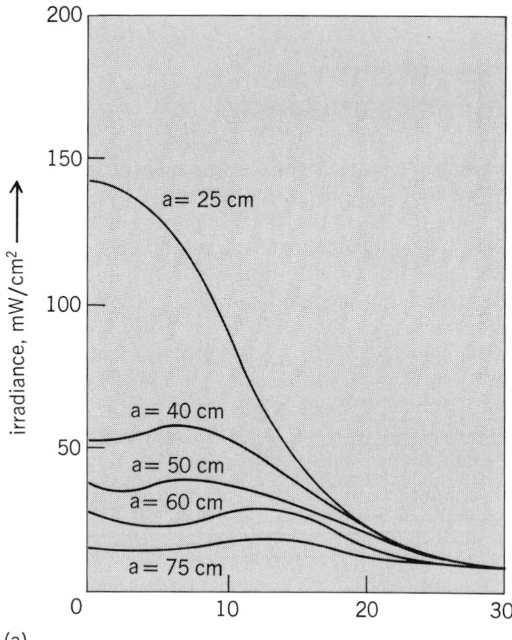

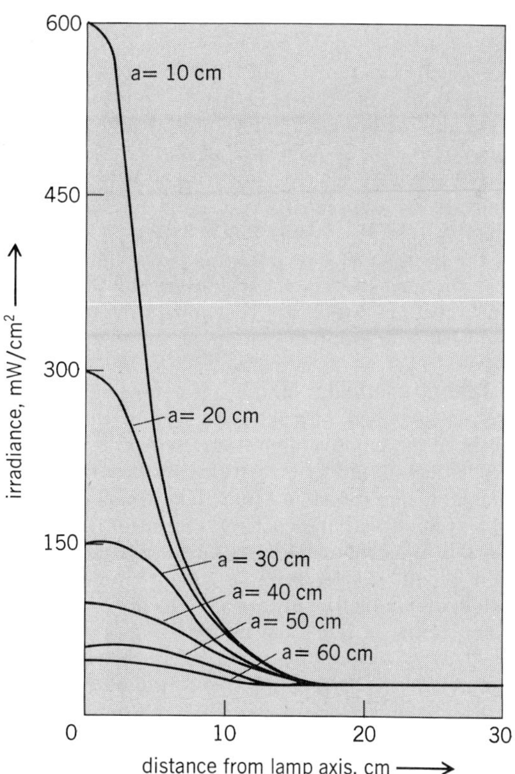

Fig. 3. Irradiance as a function of the distance from the axis of the lamp for various distances *a* from the face of the lamp to the irradiated surface (perpendicular to the axis), for the reflector-type infrared lamp shown in Fig. 2*a*. (*a*) 250-W ruby-stained lamp. (*b*) 250-W frosted lamp. (*From M. La Toison, Infrared and Its Thermal Applications, Philips Technical Library, 1964*)

harmful. High levels could cause burns, but the exposed individual feels the heat and would normally move away before a burn actually occurred. *See* INCANDESCENT LAMP.

[G. R. PEIRCE]

Bibliography: J. D. Hall, *Industrial Applications tions of Infrared*, 1947; Illuminating Engineering Society, *IES Lighting Handbook*, 5th ed., 1972; M. La Toison, *Infrared and Its Thermal Applications*, 1964.

Inspection and testing

Industrial activities which ensure that manufactured products, individual components, and multicomponent systems are adequate for their intended purpose. Complete inspection and testing programs cover evaluation of the product design prior to manufacturing, conformance to the design during manufacturing, and durability during subsequent storage, transport, and use. The terms inspection and testing are used interchangeably and jointly to describe closely related activities, but strictly speaking, inspection per se is the total activity dealing with: interpretation of product specifications; measurement and testing of the characteristics described in the specifications; methods of sampling and statistical analysis of results; and decisions on the disposition of the product evaluated.

Specifications. A specification is the technical description of a product verbally, numerically, and pictorially. It contains target values and permissible limits known as tolerances, together with methods for carrying out the tests. Typical characteristics described may be dimensional, physical, chemical, functional, and safety-related. Resistance to use and abuse, esthetic or sensory qualities, and endurance under normal and extreme conditions may also be spelled out. Inspection instructions may also be given on the number of units to be sampled from a given manufactured quantity, and the quality levels expected in terms of the number of nonconforming units or the degree of nonconformance which can be tolerated. Specifications which are recognized by an industry, nation, or group of nations are known as standards.

Testing methods. Methods for measuring and testing specified characteristics span the whole range of human invention from the simple scale to sophisticated modern instruments such as the scanning electron microscope. The science of metrology deals with dimensional measuring methods capable of extreme degrees of accuracy and precision which are essential in modern high-speed, automated industrial processes. The products evaluated range in size from an electrical generating plant down to subminiature electric circuits and components. Instruments used include simple scales, fixed gages, and jigs for checking size and geometric shape. For greater precision, micrometers and calipers with vernier scales are called upon. Ultimate precision depends on devices which replace the judgment of the human eye and hand by magnifying or amplifying the readings by using electrical, optical, or pneumatic methods. Typical high-precision instruments in common use are microscopes, interferometers, autocollimators, and optical comparators (Fig. 1).

Testing for strength, performance, and safety is carried out by a variety of mechanical and electrical methods, by using familiar laboratory instruments modified where necessary for high-volume throughput. Test samples are sometimes subjected to extremes of environmental, electrical, and mechanical stresses in special ovens, refrigerators, and humidity chambers. Such accelerated or high-stress testing, together with the practical and theoretical study of components and systems performance in service, is known as reliability engineering.

Testing which destroys or impairs the product is known as destructive testing. However, it is possible to do much testing nondestructively by using x-rays, ultrasonics, eddy currents, gamma rays, magnetic particles, penetrating dyes, electrical capacitance, fiber optics, and other methods. *See* NONDESTRUCTIVE TESTING.

There are many subjective properties of products which are virtually impossible to test by instrumentation. Such sensory and esthetic qualities as taste, smell, and visual appearance and many performance characteristics are tested by human panels. Skin sensitivity and allergic reaction are checked on human panels under carefully controlled conditions.

Sampling inspection. Methods of reading inspection and test data from the myriad of instruments in service range from the unaided human eye to electronic digital displays. Some instruments are coupled directly with the production machinery to feed back self-correcting information. In other cases the data must be analyzed to aid human judgment on questions of quality. Due to the natural variation in raw materials and processes, it is rare to find any two manufactured units which are exactly alike; therefore in order to make decisions about the quality of a batch of products, several units from that batch are measured. In special circumstances 100% inspection of the batch is carried out, but it is more usual to test a relatively small random sample of units to provide information on raw materials, goods in process, and finished products. Such procedures are generally known as sampling inspection, and they fall into several categories such as process inspection, incoming goods inspection, and finished goods inspection.

Quality control. The analysis of sampling inspection data and the decisions made from them constitute an important branch of statistics known

Fig. 1. Razor blade samples from sharpening machines are examined under high-power microscopes for defects. (*Tektronix, Inc.*)

Fig. 2. Quality control charts of inspection data are produced by computer graphics. (*Tektronix, Inc.*)

as statistical quality control. This in turn is a topic of total quality control, which embraces all the technical and management activities concerned with inspection and testing, and the actions taken on the results. In most manufacturing operations, inspection and testing departments are part of the quality control department, although similar activities can often be found in the research and development department and engineering department.

The statistical techniques referred to above are designed to make inferences about large quantities of material (populations) from small random samples. Wide use is made of such statistical measures as the average to measure closeness to target value, and standard deviation to measure the amount of scatter of individuals about the target. These values are often plotted on quality control charts, on which statistical limits are drawn to indicate whether process adjustments are necessary (Fig. 2). Histograms also provide at-a-glance summaries of results compared with specification values. *See* QUALITY CONTROL.

Sampling plan. In mass production it is virtually impossible to avoid small quantities of material falling outside the specification tolerances, and this is usually allowed for in contracts between producers and consumers. Such contracts define a percentage of nonconforming material which is acceptable, and a somewhat higher percentage which is not. These are called the acceptable quality level (AQL) and rejectable quality level (RQL), respectively. Sampling plans are then agreed upon which ensure that batches of acceptable quality are passed most of the time, and rejectable quality batches are rejected most of the time. A simple example of a sampling plan follows: Draw a random sample of 80 units from the batch. If two or less nonconforming units are found in the sample, pass the batch. If more than two are found, the batch is failed.

With this plan, batches containing 1.0% nonconforming material (the AQL) will pass 95 times in 100, whereas batches containing 6.5% nonconforming material (the RQL) will pass only 10 times in 100. Tables of plans containing a wide selection of AQLs, RQLs, and sample sizes are available.

Guarantees. The risks of occasionally making the wrong decisions in sampling inspection are an accepted part of industrial practice. The possibility of unsatisfactory product reaching the consuming public under such procedures is covered by guarantees which are now regulated by law. *See* GAGES. [JOHN A. CLEMENTS]

Bibliography: American Society for Quality Control, *J. Qual. Technol.*, *Quality Progress*, *Technometrics*, all periodically; A. V. Feigenbaum, *Total Quality Control*, 1961; E. G. Grant, *Statistical Quality Control*, 4th ed., 1972; J. M. Juran, *Quality Control Handbook*, 3d ed., 1974; C. W. Kennedy, *Inspection and Gaging*, 4th ed., 1967; U.S. Department of Defense, *Sampling Procedures and Tables for Inspection by Attributes*, Military Stand. 105-D.

Instrument transformer

Electric transformers specifically designed for use in measurement and control circuits. The purpose of an instrument transformer is to convert primary voltages or currents to secondary values suitable for use with relays, meters, or other measuring equipment. A second purpose is to isolate the high-voltage primary circuit from the measurement circuit. Thus the construction of measuring devices and the protection of personnel using such devices are greatly simplified by the use of instrument transformers. A sketch showing the basic elements and phasor diagram for an instrument transformer is shown in Fig. 1. For general discussion of transformers *see* TRANSFORMER.

Specialized design techniques and materials are used in instrument transformers to insure a high degree of stability and reliability, because improper operation could result in economic loss in metering applications or power system damage in relaying applications.

There are two general types of instrument transformers: instrument potential transformers, which are used in voltage measurements, and instrument current transformers, which are used in current measurement.

The primary of potential transformers is connected across the line and the current that flows

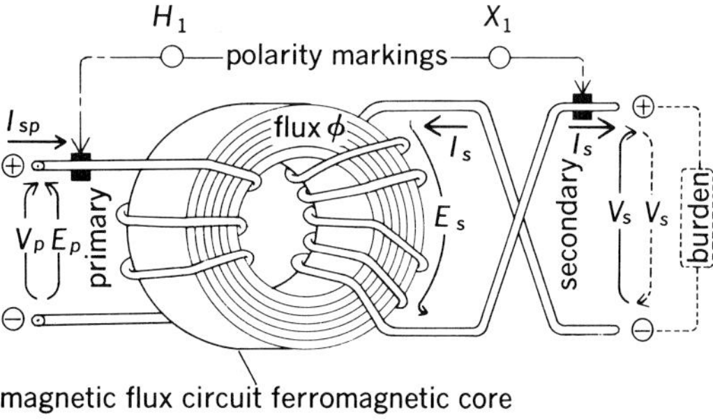

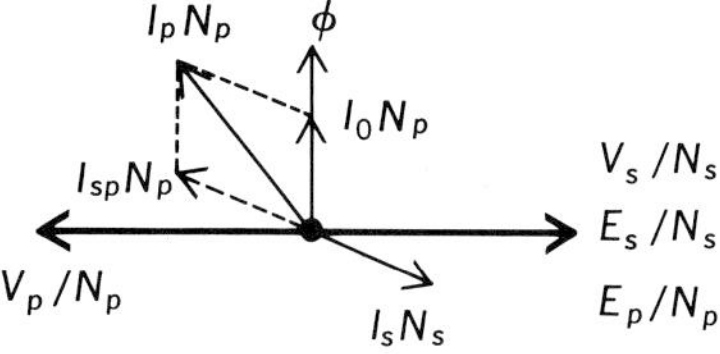

Fig. 1. A simple instrument transformer and its phasor diagram. Core-loss components and impedance drops have been omitted from diagram. (*General Electric Co.*)

through this winding produces a flux in the core. Since the core links both the primary and secondary windings, a voltage is induced in the secondary circuit. The ratio of primary to secondary voltage is roughly in proportion to the number of turns in the primary and secondary windings. Usually this ratio is selected to produce 115 or 120 volts at the secondary terminals when rated voltage is applied to the primary.

The current transformer differs from the potential transformer in that the primary winding is designed for connection in series with the line. The ratio of primary to secondary current is, roughly, inversely proportional to the ratio of primary to secondary turns. Usually 5 amp of secondary current is produced by rated current in the primary.

Instrument transformers may differ considerably, since the physical construction depends upon the intended application. A potential transformer usually has a magnetic core and coil arrangement that is quite similar to that of the conventional power or distribution transformer. Since the core operates continuously at the highest flux density consistent with good magnetic design, the core is the major heat producer. Therefore, shell-type construction usually is employed for effective heat dissipation.

In a current transformer the magnetic core density is a function of line current and, except for occasional transients caused by line faults, is usually very low. The core loss is negligible, and the windings are the major source of heat. Therefore, the core type of construction is generally employed.

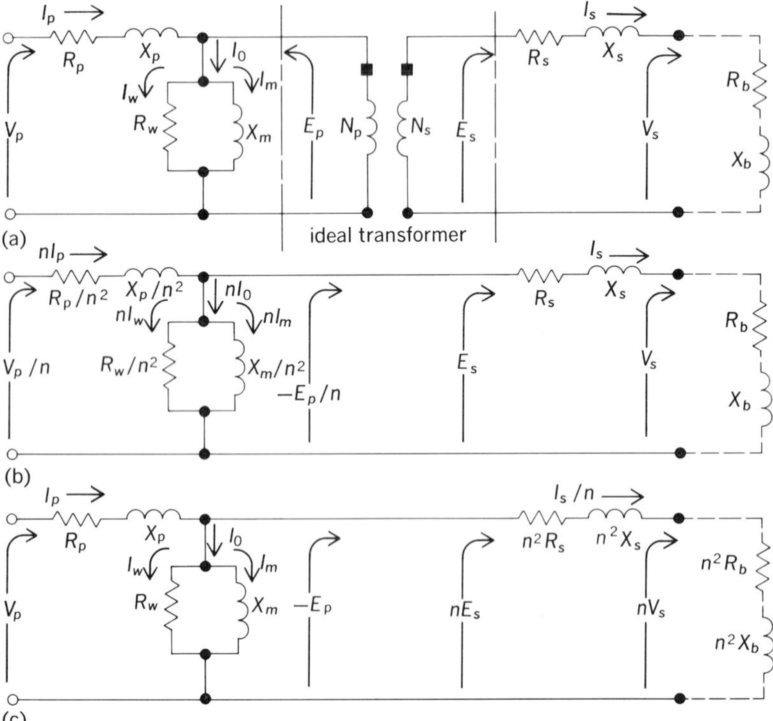

Fig. 2. Instrument transformer equivalent circuits. (*a*) Ideal transformer. (*b*) All quantities are referred to secondary. (*c*) All quantities are referred to primary. (*From I. F. Kinnard, Applied Electrical Measurements, copyright © 1956 by John Wiley and Sons, Inc.; used with permission*)

Special insulations, such as arc-resistant butyl rubber, have been developed for winding insulation to ensure necessary reliability and safety.

Instrument transformers have a marked ratio, which differs from the turns ratio and more nearly represents the true ratio of output to input. However, since the true ratio changes slightly with burden, frequency, temperature changes, and other factors, the marked ratio must be multiplied by a ratio correction factor to give the true ratio under specified operating conditions.

Principles of operation. Instrument transformers approach an ideal transformer, and their principle of operation is explained by a study of the ideal. The practical differences from the ideal can then be incorporated, and equivalent circuits can be established which allow analysis of most instrument transformer problems.

In the ideal instrument transformer the output voltage or current is exactly proportional in magnitude, and exactly opposite in phase, to the corresponding input voltage or current. This means there is no core or winding loss, and no resistance or reactance voltage drops. All the flux links both primary and secondary windings.

An alternating voltage applied to the transformer produces a primary current and a resulting alternating flux ϕ in the core. The same alternating emf e is induced in each primary and secondary turn by the changing flux linkages $d\phi/dt$. For a complete winding the induced voltage equals $-N(d\phi/dt)$, where N is the number of turns.

The instantaneous flux ϕ in an ideal transformer is sinusoidal with a maximum value of ϕ_m when the applied voltage is sinusoidal, as shown by Eq. (1). The emf is shown by Eq. (2). The root-mean-square (rms) value of the voltage is shown by Eq. (3),

$$\phi = \phi_m \sin \omega t \quad (1)$$

$$e = -\phi_m N\omega \cos \omega t \quad (2)$$

$$E = 4.44\phi_m Nf \quad (3)$$

where f is the frequency. This induced voltage is determined only by the applied voltage and the number of turns, since there are no other voltage drops. Since the induced voltage in the primary E_p must cancel the applied voltage V_p, its polarity is determined for all turns including the secondary, as shown by the relations given in Eq. (4). Instru-

$$E_p/E_s = -V_p/V_s = N_p/N_s \quad (4)$$

ment transformers have polarity marks such that, when the current enters the H primary, it instantaneously leaves the X secondary, and if H has a positive potential, X is also positive.

Closing the secondary circuit of an ideal transformer results in a secondary current i_s, with an mmf $\mathcal{F}$ in the core producing a magnetizing force H. Magnetomotive force and magnetizing force can be computed from Eqs. (5) and (6), respectively,

$$\mathcal{F} = N_s i_s \quad (5)$$

$$H = N_s i_s / l \quad (6)$$

where N_s is the number of secondary turns and l is the magnetic circuit length.

This magnetizing force must exactly cancel that of the primary, since no excitation flux is needed for the ideal transformer. This is shown by Eq. (7), from which Eq. (8) can be derived.

$$N_s i_s / l = -N_p i_p / l \tag{7}$$

$$i_p / i_s = N_s / N_p \tag{8}$$

By defining n as equal to $-N_p/N_s$, Eqs. (9) through (11) can be obtained.

$$i_s = i_p n \tag{9}$$

$$E_s = -E_p / n \tag{10}$$

$$V_s = V_p / n \tag{11}$$

When the secondary circuit is closed, the secondary current is V_s/Z_b, where Z_b is the secondary burden (the load and its wiring) composed of resistance R_b and reactance X_b.

An actual instrument transformer can be simulated by the network shown in Fig. 2*a*. Some ampere turns are needed to produce exciting flux in the magnetic circuit. A current I_o through N_p turns produces this flux; it has a magnetizing component I_m and an eddy current and hysteresis watt loss component I_w. These are simulated in the equivalent circuit by a reactance X_m and a resistance R_w respectively. The transformer windings have resistances shown by R_p and R_s, and magnetic leakage produces small reactances shown by X_p and X_s. Equivalent circuits eliminating the ideal transformer can be drawn by referring all values to the secondary, as in Fig. 2*b*, or to the primary, as in Fig. 2*c*.

Rules for referring quantities from the primary to the secondary are (1) multiply I_p by n, (2) divide V_p and E_p by n, and (3) divide Z_p, R_p, and X_p by n^2.

These equivalent circuits are usually satisfactory to predict transformer performance, although in the transformer the magnetization characteristic is nonlinear. Capacitative effects also exist. These should be evaluated at frequencies above 100 Hz, and it is usually necessary to change values of X_m and R_w as a function of frequency.

Accuracies. Potential transformer accuracies depend on the magnitude of primary and secondary voltage drops. With the secondary open circuited the only voltage drop is caused by the exciting current. With a burden it is possible to add drops due to the burden current in both primary and secondary to obtain the overall change in performance. Potential transformers are designed to have low exciting current and low winding impedances.

In a current transformer the difference between the secondary and primary current is the exciting current. Therefore, small errors in both ratio and phase angle exist fundamentally. These errors vary with primary current, and many compensating schemes have been used to increase accuracy. Both the magnetizing and core-loss currents are kept small, special lamination material is used, and special techniques are used to reduce losses in the transformer.

Commercial instrument transformers have extremely good accuracies, and manufacturers furnish typical curves depicting any small error. American Standard for Transformers, Regulators, and Reactors (ANSI C57) specifies the accuracy classes for specific burdens, and errors are kept within these standard accuracies for transformers of a given class. For even greater accuracies test equipment is available for accurately measuring ratios and phase angles by either direct methods or by comparison with standard transformers.

[ISAAC F. KINNARD]

Bibliography: L. Anderson, *Electric Machines and Transformers*, 1980; M. Braccio, *Basic Electrical and Electronic Tests and Measurement*, 1979; D. Richardson, *Rotating Electric Machinery and Transformer Technology*, 1978.

Insulation resistance testing

The testing of the electrical resistance of dielectric materials or insulators. Insulation resistance is measured in megohms. It consists of (1) the volume resistance resulting from the resistance to current flow through the volume of the material, and (2) the surface resistance due to the resistance to current flow over the surface of the material. These two resistances are electrically in parallel, and it is impossible to separate them in a single measurement. If basic insulating properties of a material are being studied, judicious choice of sample size and electrode configuration can emphasize one effect over the other so that reasonable evaluating data of each factor can be obtained.

If the conditions of measurement do not permit such control, as in the checking of insulation resistance of large generator windings, the testing conditions and physical arrangements must be evaluated carefully to determine which, if either, factor is dominant and whether or not the test will produce the desired evaluation of the winding insulation. Often, only the total insulation resistance is required, and individual evaluation of the surface and volume effects need not be made. Except for the simplest of measurements, such as the resistance of a capacitor terminal or the point-to-point resistance of a switch, the use of a guard circuit is essential for precise and accurate results.

Usually, surface-leakage effects may be eliminated by guarding one terminal or a portion of the measuring circuit. A guard is a low-resistance conductor; it is insulated from the guarded terminal and located to maintain the guard and guarded terminal at near-equal potential and to minimize the current flow between them. The guard circuit also conducts leakage current around the measuring circuit to minimize its effect on the measuring circuit.

The determination of insulation resistance is further complicated by two other factors. First, if the capacitance between the test electrodes is high, the time constant of the circuit, *RC*, will be large, and the time required to charge the capacitance will be long. Second, many insulating materials exhibit dielectric absorption which may continue for extended periods of time.

If a measurement is made without guarding and before the two conditions described above are stabilized—an interval of microseconds to minutes for the time constant effect, and seconds to days for dielectric absorption—an apparent insulation resistance will be determined. A guarded measurement made after stabilization will determine volume resistance from which resistivity may be calculated.

It is therefore necessary to completely specify all test conditions, including sample and electrode configuration and condition, test voltage, state of capacitance charge before test, time of measurement after application of test voltage, and temperature and humidity if reproducible results are to be achieved. *See* ELECTRICAL RESISTANCE.

Guarded deflection method. This method for measuring insulation resistance (Fig. 1*a*) uses the voltmeter-ammeter method. The voltage is measured with a voltmeter of suitable range; current is measured by a calibrated galvanometer, with an Ayrton shunt for range extension, or a vacuum-tube dc amplifier calibrated as a picoammeter with appropriate multipliers.

A modification of this method, which permits the comparison of a high resistance standard with the insulation resistance under test, is shown in Fig. 1*b*. In this method, neither the test voltage nor the currents need be known. With a constant test voltage applied, the deflection of the current measuring device is noted, first, with the sample shorted. For this condition, Eq. (1) holds,

$$E = d_1 R_s \qquad (1)$$

where $d_1 \propto I_S$. Next, with the switch open, a new deflection is noted. For this second condition, Eq. (2) holds, where $d_2 \propto I_x$, and, combined with Eq. (1), yields Eq. (3).

In each of these circuits, requisite guarding of test equipment is shown by dotted lines with the

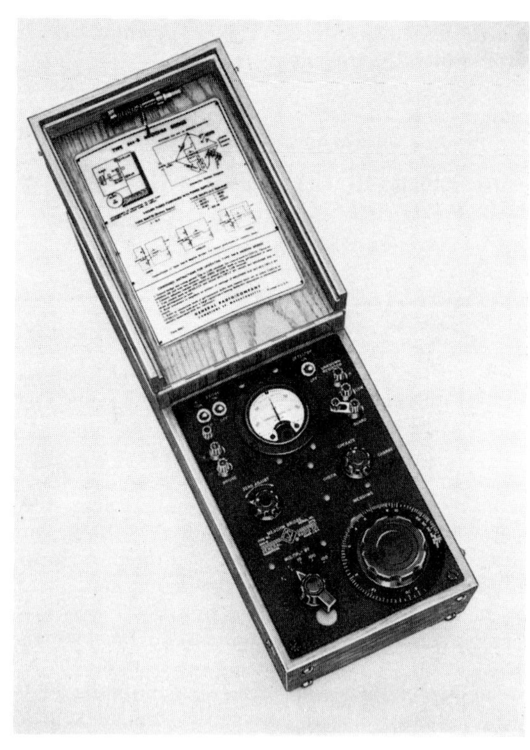

Fig. 2. Megohm bridge for insulation resistance measurement. (*General Radio Co.*)

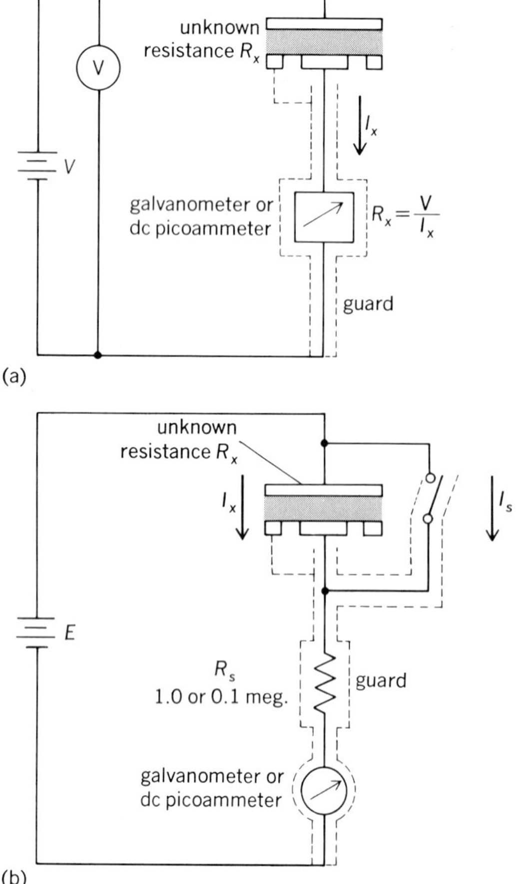

Fig. 1. Circuits for guarded deflection measurement of insulation resistance. (*a*) Voltmeter-ammeter method. (*b*) Comparison method.

$$E = d_2 (R_X + R_S) \qquad (2)$$

$$R_X = R_s (d_1 / d_2 - 1) \qquad (3)$$

guard of the sample connected to the equipment guard. It is essential that all of the equipment shown within the guard lines be mounted on the guard plate or circuit. However, the insulation requirements between guard and measuring circuit are not severe. This insulation shunts the internal resistance of the measuring circuit and therefore need only be 10–100 times this value to minimize the error from this source. If an electronic picoammeter is used, its circuits, including such components as the power supply transformer, must also be guarded.

Guarded Wheatstone bridge. This bridge requires the use of guarded ratio arms, detector, and power supply, and careful attention must be given to the details of construction, particularly with regard to component insulation. Such construction also results in a well-shielded bridge. A bridge of this type may be used for highly accurate and precise measurements even under conditions of high humidity.

Unguarded Wheatstone bridge. An unguarded bridge (Fig. 2) may also be used to measure guarded insulation samples if the guard connections are made as shown in Fig. 3. This connection allows the leakage current to bypass the detector. If the detector input resistance is high and the bridge resistor R_R is no larger than 100,000 ohms, little residual error will result.

Hi-pot testing. In this test, high potentials are applied to electrical materials, components, equipment or systems to determine experimentally the dielectric breakdown characteristics or to proof-test the insulation for manufacture, acceptance or

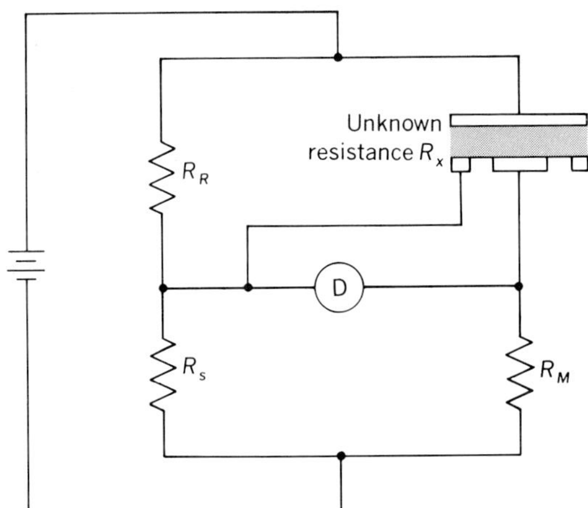

Fig. 3. An unguarded megohm bridge for measuring guarded insulation specimen.

service requirements at voltages higher than operating but lower than expected breakdown potential. This testing may be done with either alternating or direct current. The voltage applied and the time of application are chosen to provide the required information for specific purposes. Interpretation of hi-pot test data is complex; considerable experience is required to evaluate this data properly, particularly when solid dielectrics are involved. [CHARLES E. APPLEGATE]

Bibliography: American Society for Testing and Materials, *AC Loss Characteristics and Dielectric Constants of Solid Electrical Insulating Materials*, ASTM Stand. no. D-150-74, 1974, and *DC Resistance or Conductance of Insulating Materials*, ASTM Stand. no. D-257–66(72), 1966, and *Specific Resistance of Electrical Insulating Liquids*, ASTM Stand. no. D-1169–74, 1974, and *Terms Relating to Electrical Insulation*, ASTM Stand. no. D-1711–74A, 1974, and *Test for Dielectric Breakdown Voltage and Dielectric Strength of Electrical Insulating Materials at Commercial Power Frequencies*, ASTM Stand. no. D-149–64(70), 1964; W. P. Baker, *Electrical Insulation Measurements*, 1966; F. K. Harris, *Electrical Measurements*, 1958; Institute of Electrical and Electronics Engineers, *Recommended Guide for Making Dielectric Measurements in the Field*, IEEE Stand. no. 62, 1958, and *Test Code for Resistance Measurements*, IEEE Stand. no. 118, 1949, and *Recommended Practice for Testing Insulation Resistance of Rotating Machinery*, IEEE Stand. no. 41, 1974; F. A. Laws, *Electrical Measurements*, 2d ed., 1938; H. A. Sauer and W. H. Shirk, Jr., Wheatstone bridge for multi-terohm measurements, *IEEE Trans. Comm. Elec.*, 71:131–136, March, 1964; F. H. Wyeth, J. B. Higley, and W. H. Shirk, Jr., Precision, guarded resistance measuring facility, *Trans. AIEE*, pt. 1, 77(38):471–475, 1958.

Internal combustion engine

A prime mover, the fuel for which is burned within the engine, as contrasted to a steam engine, for example, in which fuel is burned in a separate furnace. *See* ENGINE.

The most numerous of internal combustion engines are the gasoline piston engines used in passenger automobiles, outboard engines for motor boats, small units for lawn mowers, and other such equipment, as well as diesel engines used in trucks, tractors, earth-moving, and similar equipment. This article describes these types of engines. For other types of internal combustion engines *see* GAS TURBINE; ROTARY ENGINE.

The aircraft piston engine is fundamentally the same as that used in automobiles but is engineered for light weight and is usually air cooled.

ENGINE TYPES

Characteristic features common to all commercially successful internal combustion engines include (1) the compression of air, (2) the raising of air temperature by the combustion of fuel in this air at its elevated pressure, (3) the extraction of work from the heated air by expansion to the initial pressure, and (4) exhaust. William Barnett first drew attention to the theoretical advantages of combustion under compression in 1838. In 1862 Beau de Rochas published a treatise that emphasized the value of combustion under pressure and a high ratio of expansion for fuel economy; he proposed the four-stroke engine cycle as a means of accomplishing these conditions in a piston engine (Fig. 1). The engine requires two revolutions of the crankshaft to complete one combustion cycle. The

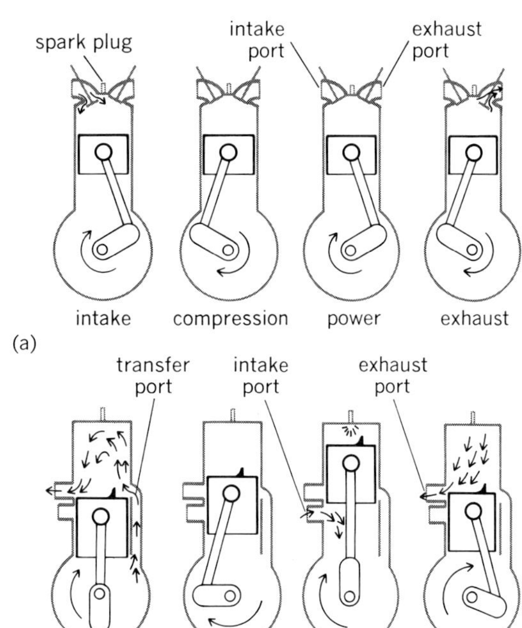

Fig. 1. The four strokes of a modern four-stroke engine cycle. (*a*) Conventional engine. For intake stroke the intake valve (left) has opened and the piston is moving downward, drawing air and gasoline vapor into the cylinder. In compression stroke the intake valve has closed and the piston is moving upward, compressing the mixture. On power stroke the ignition system produces a spark that ignites the mixture. As it burns, high pressure is created, which pushes the piston downward. For exhaust stroke the exhaust valve (right) has opened and the piston is moving upward, forcing the burned gases from the cylinder. (*b*) Three-port two-cycle engine. The same action is accomplished without separate valves and in a single rotation of the crankshaft. (*From M. L. Smith and K. W. Stinson, Fuels and Combustion, McGraw-Hill, 1952*)

first engine to use this cycle successfully was built in 1876 by N. A. Otto. *See* OTTO CYCLE.

Two years later Sir Dougald Clerk developed the two-stroke engine cycle by which a similar combustion cycle required only one revolution of the crankshaft. In this cycle, exhaust ports in the cylinder were uncovered by the piston as it approached the end of its power stroke. A second cylinder then pumped a charge of air to the working cylinder through a check valve when the pump pressure exceeded that in the working cylinder.

In 1891 Joseph Day simplified the two-stroke engine cycle by using the crankcase to pump the required air. The compression stroke of the working piston draws the fresh combustible charge through a check valve into the crankcase, and the next power stroke of the piston compresses this charge. The piston uncovers the exhaust ports near the end of the power stroke and slightly later uncovers intake ports opposite them to admit the compressed charge from the crankcase. A baffle is usually provided on the piston head of small engines to deflect the charge up one side of the cylinder to scavenge the remaining burned gases down the other side and out the exhaust ports with as little mixing as possible.

Engines using this two-stroke cycle today have been further simplified by use of a third cylinder port which dispenses with the crankcase check valve used by Day. Such engines are in wide use for small units where fuel economy is not as important as mechanical simplicity and light weight. They do not need mechanically operated valves and develop one combustion cycle per crankshaft revolution. Nevertheless they do not develop twice the power of four-stroke cycle engines with the same size working cylinders at the same number of revolutions per minute (rpm). The principal reasons for this are (1) the reduction in effective cylinder volume due to the piston movement required to cover the exhaust ports, (2) the appreciable mixing of burned (exhaust) gases with the combustible mixture, and (3) the loss of some combustible mixture through the exhaust ports with the exhaust gases.

Otto's engine, like almost all internal combustion engines developed at that period, burned coal gas mixed in combustible proportions with air prior to being drawn into the cylinder. The engine load was generally controlled by throttling the quantity of charge taken into the cylinder. Ignition was accomplished by a device such as an external flame or an electric spark so that the timing was controllable. These are essential features of what has become known as the Otto or spark-ignition combustion cycle.

Ideal and actual combustion. In the classical presentation of the four-stroke cycle, combustion is idealized as instantaneous and at constant volume. This simplifies thermodynamic analysis, but fortunately combustion takes time, for it is doubtful that an engine could run if the whole charge burned or detonated instantly.

Detonation of a small part of the charge in the cylinder, after most of the charge has burned progressively, causes the knock which limits the compression ratio of an engine with a given fuel. *See* COMBUSTION KNOCK.

The gas pressure of an Otto combustion cycle using the four-stroke engine cycle varies with the piston position as shown by the typical indicator card in Fig. 2*a*. This is a conventional pressure-volume (*PV*) card for an 8.7:1 compression ratio. For simplicity in calculations of engine power, the average net pressure during the working stroke, called the mean effective pressure (mep), is frequently used. It may be obtained from the average net height of the card, which is found by measurement of the area with a planimeter and by division of this area by its length. Similar pressure-volume data may be plotted on logarithmic coordinates as in Fig. 2*b*, which develops expansion and compression relations as approximately straight lines, the slopes of which show the values of exponent n to use in equations for PV relationships.

The rounding of the plots at peak pressure, with

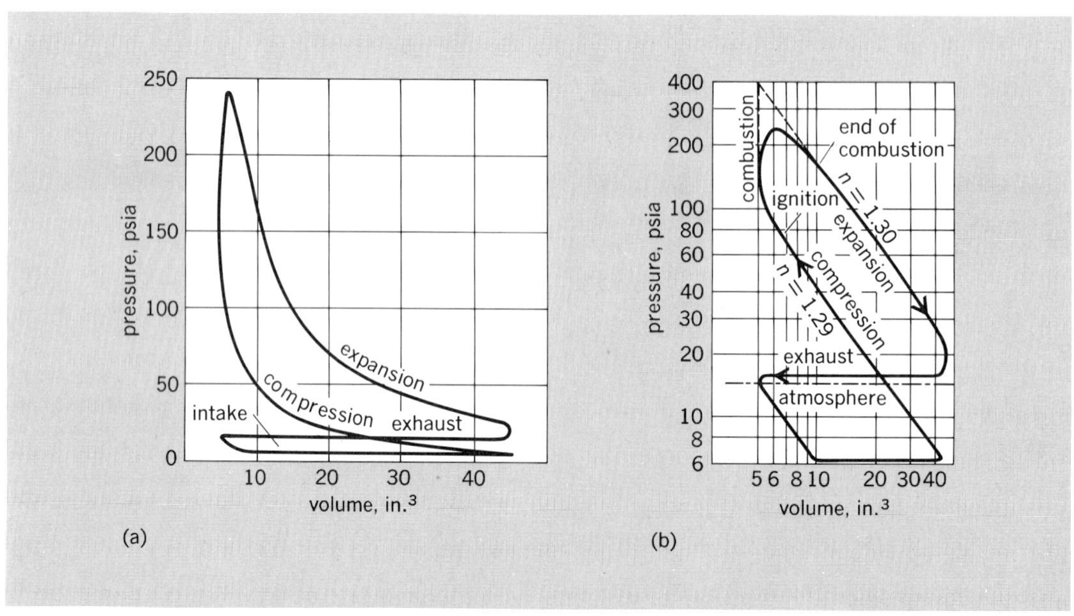

Fig. 2. Typical pressure-volume indicator card plotted on (*a*) rectangular coordinates, (*b*) logarithmic coordinates.

the peak developing after the piston has started its power stroke, even with the spark occurring before the piston reaches the end of the compression stroke, is due to the time required for combustion. The actual time required is more or less under control of the engine designer, because he can alter the design to vary the violence of the turbulence of the charge in the compression space prior to and during combustion. The greater the turbulence the faster the combustion and the lower the antiknock or octane value required of the fuel, or the higher the compression ratio that may be used with a given fuel without knocking. On the other hand, a designer is limited as to the amount he can raise the turbulence by the increased rate of pressure rise, which increases engine roughness. Roughness must not exceed a level acceptable for automobile or other service. *See* COMBUSTION CHAMBER; COMPRESSION RATIO.

Compression ratio. According to classical thermodynamic theory, thermal efficiency η of the Otto combustion cycle is given by Eq. (1), where

$$\eta = 1 - \frac{1}{r^{n-1}} \qquad (1)$$

the compression ratio r_c and expansion ratio r_e are the same ($r_c = r_e = r$). When theory assumes atmospheric air in the cylinder for extreme simplicity, exponent n is 1.4. Efficiencies calculated on this basis are almost twice as high as measured efficiencies. Logarithmic diagrams from experimental data show that n is about 1.3. Even with this value, efficiencies achieved in practice are less than given by Eq. (1). This is not surprising considering the differences found in practice and assumed in theory, such as instantaneous combustion and 100% volumetric efficiency.

Attempts to adjust classical theory to practice by use of variable specific heats and consideration of dissociation of the burning gases at high temperatures have shown that this exponent should vary with the fuel-air mixture ratio, and to some extent with the compression ratio. G. A. Goodenough and J. B. Baker have shown that, for an 8:1 compression ratio, the exponent should vary from about 1.28 for a stoichiometric (chemically correct) mixture to about 1.31 for a lean mixture. Similar calculations by D. R. Pye showed that, at a compression ratio of 5:1, n should be 1.258 for the stoichiometric mixture, increasing with excess air (lean mixture) to about 1.3 for a 20% lean mixture and to 1.4 if extrapolated to 100% air. Actual practice gives thermal efficiencies still lower than these, which might well be expected because of the assumed instantaneous changes in cyclic pressure (during combustion and exhaust) and the disregard of heat losses to the cylinder walls. These theoretical relations between compression ratio and thermal efficiency, as well as some experimental results, are shown in Fig. 3*a*. The data published by C. F. Kettering and D. F. Caris are about 85% and 82%, respectively, of the theoretical relations for the corresponding fuel-air mixtures.

Figure 3*b* gives the theoretical percentage gain in indicated thermal efficiency or power from raising the compression of an engine from a given value. They were plotted from Eq. (1) with $n = 1.3$, but would differ only slightly if obtained from any of the curves shown in Fig. 3*a*. The dotted line crossing these curves shows the diminishing gain

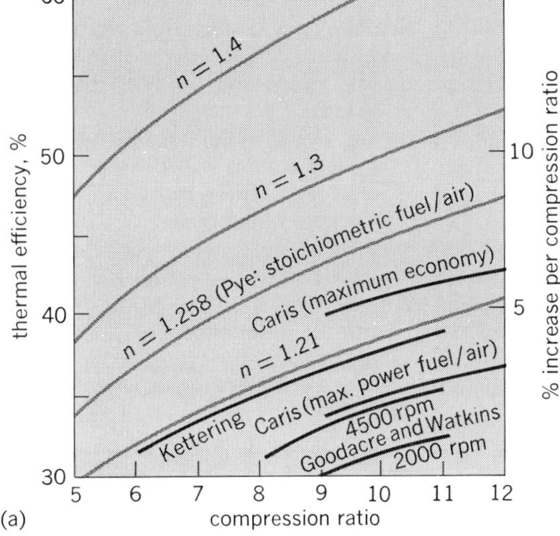

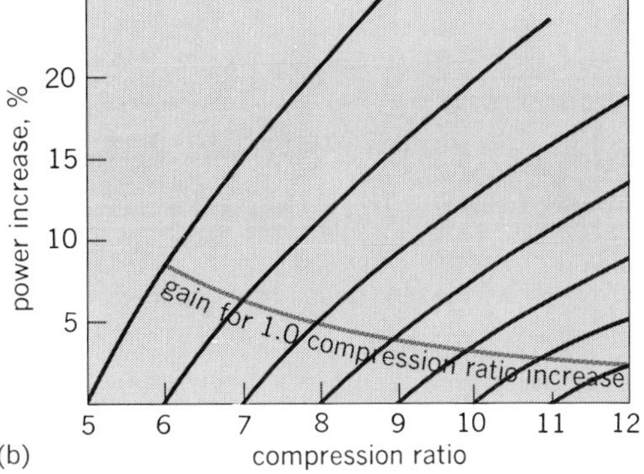

Fig. 3. Effect of compression ratio on thermal efficiency. (*a*) Effect as calculated with different values of *n* and compared with published experimental data. (*b*) Increase in indicated thermal efficiency or power from raising compression ratio as calculated with *n* = 1.3. Percentage values are but little altered by calculating with different values of *n*.

obtainable by raising the compression ratio one unit at the higher compression ratios.

Experimental data indicate that a change in compression ratio does not appreciably change the mechanical efficiency or the volumetric efficiency of the engine. Therefore, any increase in thermal efficiency resulting from an increase in compression ratio will be revealed by a corresponding increase in torque or mep; this is frequently of more practical importance to the engine designer than the actual efficiency increase, which becomes an added bonus.

Compression ratio and octane rating. For years compression ratios of automobile engines have been as high as designers considered possible without danger of too much customer annoyance from detonation or knock with the gasoline on the market at the time. Engine designers continue to raise the compression ratios of their engines as suitable gasolines come on the market.

Little theoretical study has been given to the effect of engine load on indicated thermal efficiency. Experimental evidence reveals that it varies little, if at all, with load, provided that the fuel-air-ratio remains constant and that the ignition time is suitably advanced at reduced loads to compensate for the slower rate of burning which results from dilution of the combustible charge with the larger percentages of burned gases remaining in the combustion space and the reduced turbulence at lower speeds.

Ignition timing. Designers obtain high thermal efficiency from high compression ratios at part loads, where engines normally run at automobile cruising speeds, with optimum spark advance, but avoid knock on available gasolines at wide-open throttle by use of a reduced or compromise spark advance. The tendency of an engine to knock at wide-open throttle is reduced appreciably when the spark timing is reduced 5–10° from optimum, as is shown in Fig. 4. Advancing or retarding the spark timing from optimum results in an increasing loss in mep for any normal engine as shown by the solid curve. The octane requirement falls rapidly as the spark timing is retarded, the actual rate depending on the nature of the gasoline as well as on the design of the combustion chamber. The broken-line curves A and B show the effects on a given gasoline of the use of moderate- and high-turbulence combustion chambers, respectively, with the same compression ratio. Because the mep curve is relatively flat near optimum spark advance, retarding the spark for a 1–2% loss is considered normally acceptable in practice because of the appreciable reduction in octane requirement.

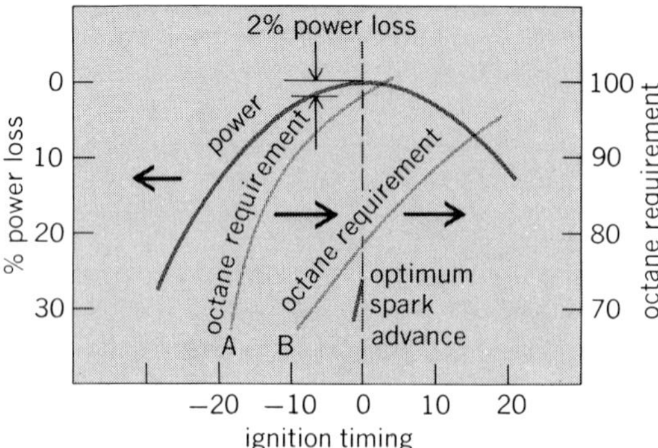

Fig. 4. Effects of advancing or retarding ignition timing from optimum on engine power and resulting octane requirement of fuel in an experimental engine with a combustion chamber having typical turbulence (A) and a highly turbulent design (B) with the same compression ratio. Retarding the spark 7° for a 2% power loss reduced octane requirement from 98 to 93 for design A.

In Fig. 5 similar data are shown by curve A for another engine with changes in mep plotted on a percentage basis against octane requirement as the spark timing was charged. Point a indicates optimum spark timing, where 85 octane was required of the gasoline to avoid knock. By raising the compression ratio, the power and octane requirement were also raised as shown by the broken-line curve B. Although optimum spark required 95 octane (point b), retarding the spark timing and thus reducing the octane requirement to 86 (point c) developed slightly more power than with the original compression ratio at its optimum spark advance. The gain may be negligible at wide-open throttle, but at lower loads where knock does not develop the spark timing may be advanced to optimum (point b), where appreciably more power may be developed by the same amount of fuel.

In addition to the advantages of the higher compression ratio at cruising loads with optimum spark advance, the compromise spark at full load may be advanced toward optimum as higher-octane fuels become available, and a corresponding increase in full-throttle mep enjoyed. Such compromise spark timings have had much to do with the adoption of compression ratios of 10:1 to 13:1.

Fuel-air ratio. A similar line of reasoning shows that a fuel-air mixture richer than that which develops maximum knock-free mep will permit use of higher compression ratios. However, the benefits derived from compromise or superrich mixtures vary so much with mixture temperature and the sensitivity of the octane value of the particular fuel to temperature that it is not generally practical to make much general use of this method. Nevertheless it has been the practice with piston-type aircraft engines to use fuel-air mixture ratios of 0.11 or even higher during takeoff, instead of about 0.08, which normally develops maximum mep in the absence of knock.

Compression-ignition engines. About 20 years after Otto first ran his engine, Rudolf Diesel successfully demonstrated an entirely different method of igniting fuel. Air is compressed to a pressure high enough for the adiabatic temperature to reach or exceed the ignition temperature of the fuel. Because this temperature is in the order of 1000°F, compression ratios of 12:1 to 20:1 are used commercially with compression pressures generally over 600 psi. This engine cycle requires the fuel to be injected after compression at a time and rate suitable to control the rate of combustion. *See* FUEL INJECTION.

Conditions for high efficiency. The classical

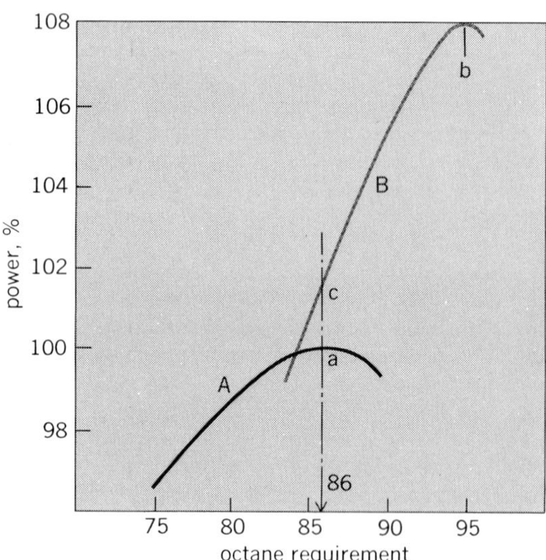

Fig. 5. Effect of raising compression ratio of an experimental engine on the power output and octane requirement at wide-open throttle. While an 86-octane fuel was required for optimum spark advance (maximum power) with the original compression ratio, the same gasoline would be knock-free at the higher compression ratio by suitably retarding the ignition timing.

presentation of the diesel engine cycle assumes combustion at constant pressure. Like the Otto cycle, thermal efficiency increases with compression ratio, but in addition it varies with the amount of heat added (at the constant pressure) up to the cutoff point where the pressure begins to drop from adiabatic expansion. *See* DIESEL CYCLE; DIESEL ENGINE.

Practical attainments. Diesel engines were highly developed in Germany prior to World War I, and made an impressive performance in submarines. Large experimental single-cylinder engines were built in several European countries with cylinder diameters up to 1 m. As an example, the two-stroke Sulzer S100 single-acting engine with a bore of 1 m and a stroke of 1.1 m developed 2050 gross horsepower at 150 rpm. Multiple-cylinder engines developing 15,000 hp are in marine service. Small diesel engines are in wide use also.

Fuel injection. In early diesel engines, air injection of the fuel was used to develop extremely fine atomization and good distribution of the spray. But the need for injection air at pressures in the order of 1500 psi required expensive and bulky multistage air compressors and intercoolers.

A simpler fuel-injection method was introduced by James McKechnie in 1910. He atomized the fuel as it entered the cylinder by use of high fuel pressure and suitable spray nozzles. After considerable development it became possible to atomize the fuel sufficiently to minimize the smoky exhaust which had been characteristic of the early solid-injection engines. By 1930 solid or airless injection had become the generally accepted method of injecting fuel in diesel engines.

Contrast between diesel and Otto engines. There are many characteristics of the diesel engine which are in direct contrast to those of the Otto engine. The higher the compression ratio of a diesel engine, the less the difficulties with ignition time lag. Too great an ignition lag results in a sudden and undesired pressure rise which causes an audible knock. In contrast to an Otto engine, knock in a diesel engine can be reduced by use of a fuel of higher cetane number, which is equivalent to a lower octane number.

The larger the cylinder diameter of a diesel engine, the simpler the development of good combustion. In contrast, the smaller the cylinder diameter of the Otto engine, the less the limitation from detonation of the fuel.

High intake-air temperature and density materially aid combustion in a diesel engine, especially of fuels having low volatility and high viscosity. Some engines have not performed properly on heavy fuel until provided with a super charger. The added compression of the supercharger raised the temperature and, what is more important, the density of the combustion air. For an Otto engine, an increase in either the air temperature or density increases the tendency of the engine to knock and therefore reduces the allowable compression ratio.

Diesel engines develop increasingly higher indicated thermal efficiency at reduced loads because of leaner fuel-air ratios and earlier cutoff. Such mixture ratios may be leaner than will ignite in an Otto engine. Furthermore, the reduction of load in an Otto engine requires throttling, which develops increasing pumping losses in the intake system.

TRENDS IN AUTOMOBILE ENGINES

Cylinder diameters of average American automobile engines prior to 1910 were over $4\frac{1}{4}$ in. (108 mm). By 1917 they had been reduced to only a little over $3\frac{1}{4}$ in. (83 mm) where they stabilized until after 1945. After that, because of the increased demand for more power, with the number of cylinders limited to eight for practical mechanical reasons, the diameters increased until they averaged 4.00 in. (102 mm) in 1970, with a maximum of 4.36 in. (111 mm). After 1975 cylinder diameters decreased, and in 1981 they averaged about 3.7 in. (94 mm).

Stroke-bore ratio. Experimental engines differing only in stroke-bore ratio show that this ratio has no appreciable effect on fuel economy and friction at corresponding piston speeds. Practical advantages which result from the short stroke include (1) the greater rigidity of crankshaft from the shorter crank cheeks, with crankpins sometimes overlapping main bearings, and (2) the narrower as well as lighter cylinder block which is possible. On the other hand, the higher rates of crankshaft rotation for an equivalent piston speed necessitate greater valve forces and require stronger valve springs. Also the smaller depth of the compression space for a given compression ratio increases the surface-to-volume ratio and the proportion of heat lost by radiation during combustion. Nevertheless, stroke-bore ratios have been decreasing over the years.

Cylinder number and arrangement. Engine power may be raised by increasing the number of cylinders as well as the power per cylinder. The minimum number of cylinders has generally been four for four-cycle automobile engines, because this is the smallest number that provides a reasonable balance for the reciprocating pistons. Many early cars had four-cylinder engines. After 1912 six-cylinder in-line engines became popular. They have superior balance of reciprocating forces and more even torque impulses. By 1940 the eight-cylinder 90° V engine had risen in popularity until it about equaled the six-in-line. After 1954, the V-8 dominated the field for American automobile engines. There are several important reasons for this besides the increased power. For example, the V-8 offers appreciably more rigid construction with less bearing deflection at high speeds, provides more uniform distribution of fuel to all cylinders from centrally located downdraft carburetors, and has a short, low engine that fits within the hood demanded by style trends. With the introduction of the smaller "compact" cars in 1959, where the power and cost of eight-cylinder V-type engines were not required, six-cylinder designs increased. The evolution of cylinder arrangements included for a short period the V-12 and even a V-16 cylinder design, but experience showed that in their day there was too much practical difficulty in providing good manifold distribution of fuel, especially when starting cold, and too much difficulty in keeping all spark plugs firing. *See* AUTOMOBILE.

Compression ratio. The considerable increase in power of the average automobile engine over the years is shown in Fig. 6 together with the compression ratios which have had much to do with the increased mep. Such ratios approach practical lim-

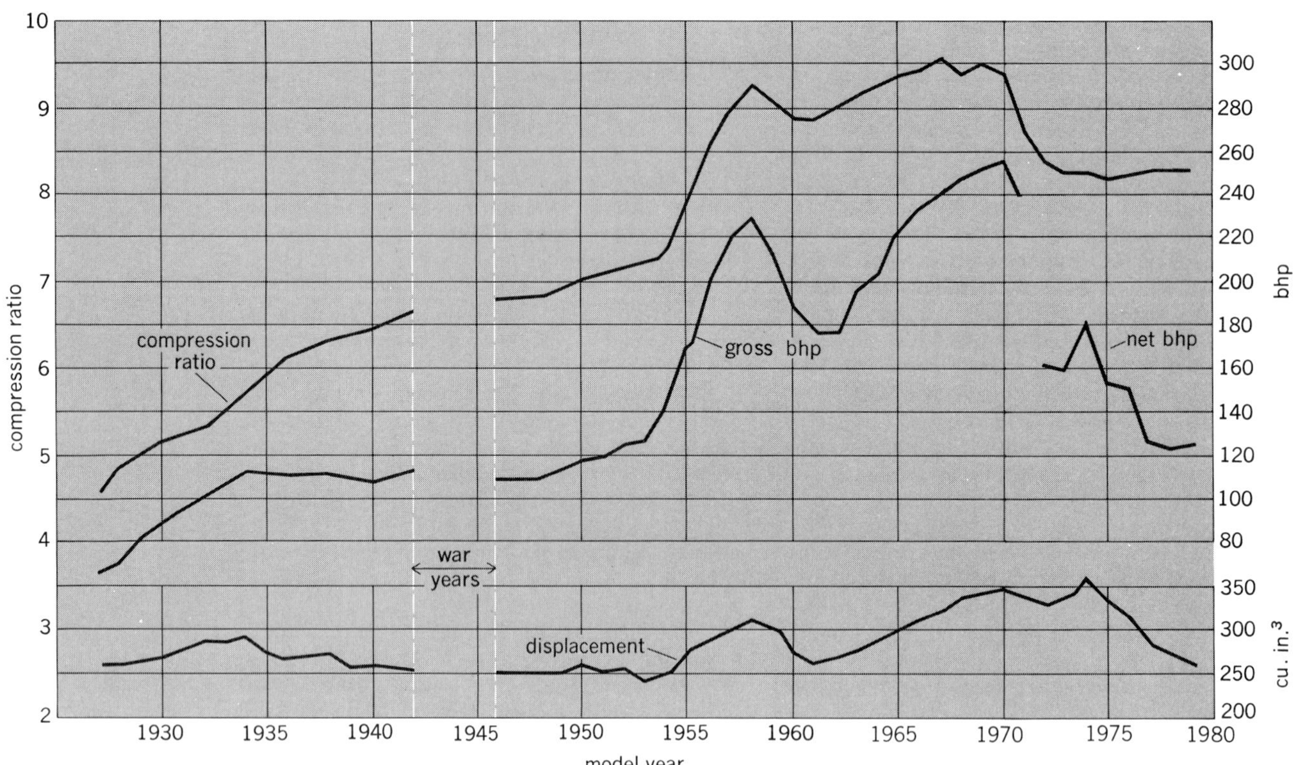

Fig. 6. Trend in compression ratio, power, and displacement of average United States automobile engines. Data for 1927–1970 are weighted for production volume; data for 1971–1979 are weighted on the basis of one engine for each model. Gross brake horsepower (bhp) is given through 1971; net bhp from 1972 on. 1 hp = 0.7457 kW; 1 in.³ = 0.0164 liter. (*Data from Ethyl Corp. and Automotive Industries*)

its imposed by phenomena other than detonation, such as preignition, rumble, and other evidences of undesirable combustion.

The modern trend toward high compression ratios, with their small compression volumes, has dictated the universal use of overhead valves in all American engine designs. High compression ratios also tend to restrict cylinder diameters because the longer flame travel increases the tendency to knock (Fig. 7).

Improved breathing and exhaust. Added power output has been brought about by reducing the pressure drop in the intake system at high speeds and by reducing the back pressure of the exhaust systems (Fig. 8). These results were accomplished by larger valve areas and valve ports, by larger venturi areas, and by more streamlined manifolds. Larger valve areas were achieved by higher lift of the valves and by larger valves. Larger venturi areas in the carburetors were achieved by use of one or more two-stage carburetors; in these, sufficient air velocity was developed to meter the fuel on one venturi at low power; the second venturi was opened for high power. Better streamlining of the manifold passages between carburetors and valves, especially at the cylinder ports, and the use of dual exhausts and mufflers with reduced back pressure have also improved engine breathing and exhaust.

Valve timing. The times of opening and closing of the valves of an engine in relation to the piston position are usually selected to develop maximum power over a desired speed range at wide-open throttle. For convenience the timing of these events is expressed as the number of degrees of crankshaft rotation before or after the piston reaches the end of one of its strokes. Because of the time required for the flow of the burned gas through the exhaust valve at the end of the power or expansion stroke of a piston, it is customary to start opening the valve considerably before the end of the stroke. If the valve should be opened when

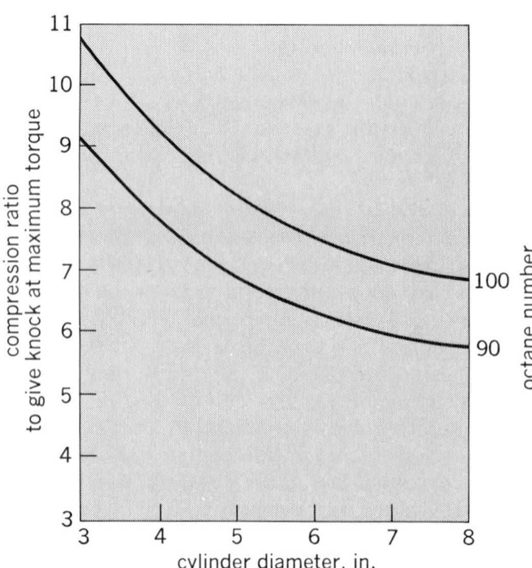

Fig. 7. Relation between cylinder diameter and limiting compression ratio for engines of similar design, one using 90- and the other 100-octane gasoline. 1 in. = 2.54 cm. (*From L. L. Brower, unpublished SAE paper, 243, 1950*)

the piston is nearer the lower end of its stroke, power would be lost at high engine speeds because the piston on its return (exhaust) stroke would have to move against gas pressure remaining in the cylinder. On the other hand, if the valve were opened before necessary, the burned gas would be released while it is still at sufficient pressure to increase the work done on the piston. Thus for any engine there is a time for opening the exhaust valve which will develop the maximum power at some particular speed. Moreover, the power loss at other speeds does not increase rapidly. It is obvious that, when an engine is throttled at part load, there will be less gas to discharge through the exhaust valve and there will be less need for it to be opened as early as at wide-open throttle.

The timing of intake valve events is normally selected to trap the largest possible quantity of air or combustible mixture in the cylinder when the valve closes at some desired engine speed and at wide-open throttle. The intermittent nature of the flow through the valve subjects it to alternate accelerations and retardation which require time. During the suction stroke, the mass of air moving through the pipe leading to the intake valve is given velocity energy which may be converted to a little pressure at the valve when the air mass still in the pipe is stopped by its closure. Advantage of this phenomenon may be obtained at some engine speed to increase the air mass which enters the cylinder. The engine speed at which the maximum volumetric efficiency is developed varies with the relative valve area, closure time, and other factors, including the diameter and particularly the length of this pipe (Fig. 9). These curves reveal the characteristic falling off at high speeds from the inevitable throttling action as air flows at increased velocities through any restriction such as a valve or intake pipe or particularly the venturi of a carburetor.

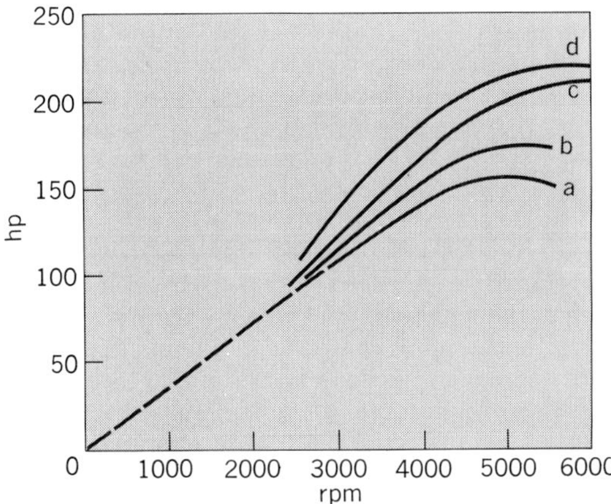

Fig. 8. Increases in peak power of a six-cylinder engine (210 in.³ or 3.44 liters displacement) by successive reductions in resistance to airflow through it. Curve a, power developed by original engine with two $1\frac{3}{4}$ in. (4.45 cm) carburetors; curve b, higher valve-lift valves and dual exhaust systems; curve c, larger valves, smoother valve ports, and two 2-in. (5.08-cm) carburetors; curve d, three double-barrel carburetors. 1 hp = 0.7457 kW. (*Adapted from W. M. Heynes, The Jaguar engine, Inst. Mech. Eng., Automot. Div. Trans., 1952–1953*)

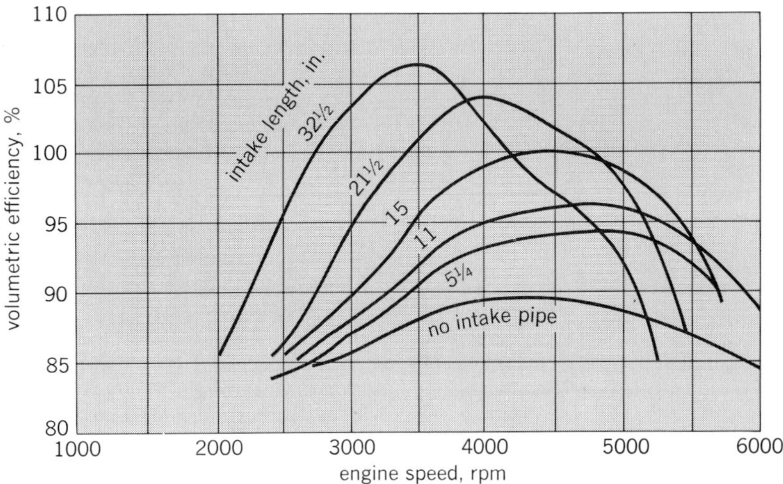

Fig. 9. Effects of intake-pipe length and engine speed on volumetric efficiency of one cylinder of a six-cylinder engine. 1 in. = 2.54 cm. (*From E. W. Downing, Proc. Automot. Div., Inst. Mech. Eng., no. 6, p. 170, 1957–1958*)

The curves shown in Fig. 9 are smoothed averages drawn through data obtained from experiments, but if they had been drawn through data taken at many more speeds, they would have revealed a wavy nature (Fig. 10) due to resonant oscillations in the air column entering the cylinder; these oscillations are somewhat similar to those which develop in an organ pipe.

Volumetric efficiency has a direct effect on the mep developed in a cylinder, on the torque, and on the power that may be realized at a given speed. Since power is a product of speed and torque, the peak power of an engine occurs at a higher speed than for maximum torque, where the rate of torque loss with any further increase in speed will exceed the rate of speed increase. This may be seen in Fig. 11, where the torque and power curves for a typical six-cylinder engine have been plotted. This engine developed its maximum power at a speed about twice that for maximum torque.

It has been shown that the engine speeds at which maximum torque and power are desired in

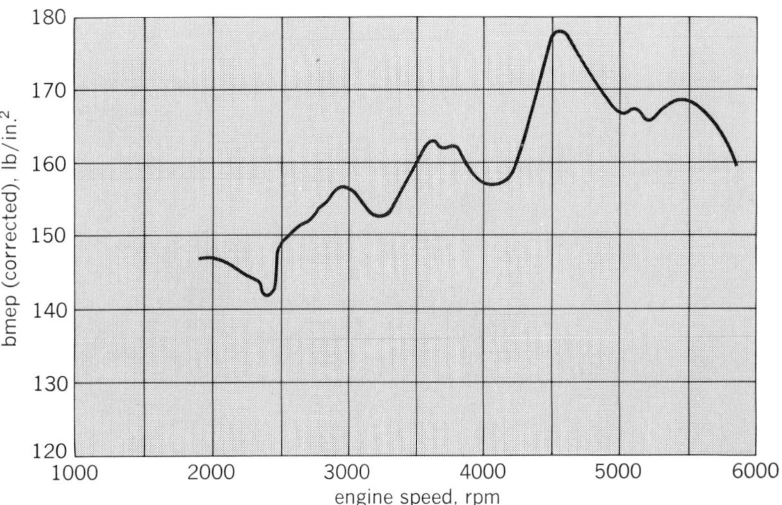

Fig. 10. Evidence of resonant oscillations in the intake pipe to a single cylinder shown by readings taken at small speed increments. 1 lb/in.² = 6.895 kPa. (*From E. W. Downing, Proc. Automot. Div., Inst. Mech. Eng., no. 6, p. 170, 1957–1958*)

practice require closure of the intake valve to be delayed until the piston has traveled almost half the length of the compression stroke. At engine speeds below those where maximum torque is developed by this valve timing, some of the combustible charge which has been drawn into the cylinder on the suction stroke will be driven back through the valve before it closes. This reduces the effective compression ratio at wide-open throttle – thus, the increasing tendency of an engine to develop a fuel knock as the speed and the resulting gas turbulence are reduced.

Another result of this blowback into the intake pipe is the possible reversal of the flow of some of the fuel-air mixture through the carburetor, which will draw fuel each time it passes through the metering system, thereby producing a much richer mixture than if it had passed through the same carburetor only once. The increased fuel supplied to the air reaching the intake valve because of this reversed air flow may be over 100% greater for a single cylinder engine at wide-open throttle than if there were no reversal of flow. Throttling the engine reduces and almost eliminates the blowback through the metering system and the ratio of fuel to air approaches that which would be expected from air flow in one direction only. Manifolding more than one cylinder to the metering system of one carburetor also reduces the blowback through a carburetor because it averages the blowback from individual cylinders. When six cylinders are supplied by a single carburetor, there may be practically no blowback through the carburetor or enrichment of the combustible mixture by this phenomenon. However, when three carburetors are installed on the same six-cylinder engine, with two cylinders supplied by each, the enrichment may be 80 – 90%. When four cylinders are supplied by a single metering system, as with most V-type eight-cylinder engines, the enrichment may be 10 – 50% and varies with

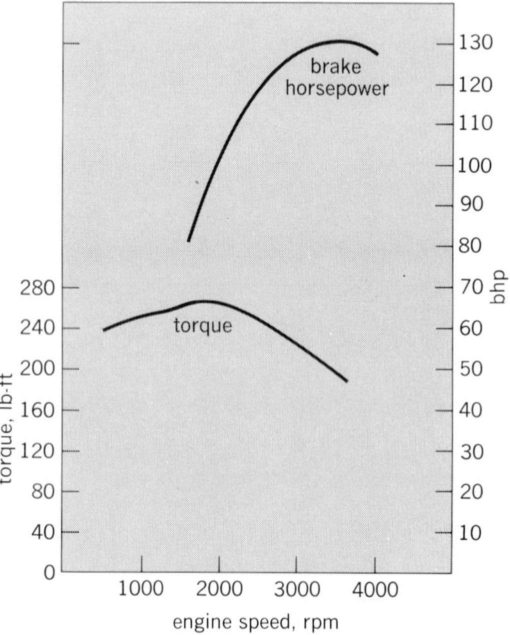

Fig. 11. Typical relation between engine speeds for maximum torque and maximum brake horsepower (bhp). 1 lb-ft = 1.356 N·m; 1 hp = 0.7457 kW.

many factors besides the closing time of the intake valve, such as the exhaust valve events and exhaust back pressure, so that carburetor settings and compensation must usually be made on the engine which it is to supply, and preferably as installed in the car, if optimum power and economy are to be realized. Unfortunately, such settings can not be predicted by simple calculations of fuel flow through an orifice into an airstream flowing at constant velocity.

Intake manifolds. Intake manifolds for multicylinder engines should meet several requirements for the satisfactory performance of spark-ignition engines. They should (1) distribute fuel equally to all cylinders at temperatures where unvaporized fuel is present, as when starting a cold engine or during the warm-up period; (2) supply sufficient heat to vaporize the liquid fuel from the carburetor as soon after starting as possible; (3) distribute the vaporized fuel-air mixture evenly to all cylinders during normal operation and at low speeds; (4) offer minimum restriction to the mixture flow at high power; and (5) provide equal ram or dynamic boost to volumetric efficiency of all cylinders at some desired part of the engine speed range. This requires that each branch from the carburetor to the valve port should be equal in length, as may be inferred from Fig. 10. Accordingly, no cylinder port should be siamesed with another at the end of a leg of the manifold.

For the warming-up period with liquid fuel present, rectangular sections are desirable to impede spiraling of liquid fuel along the walls, and right-angle bends should be sharp, at least at their inner corner, so as to throw the liquid flowing along the inner wall back into the air stream, and there should be an equal number in each branch.

Manifold heat. Intake manifolds of most American automobile engines are heated to the temperature required to vaporize the fuel from the carburetor (120 – 140°F or 50 – 60°C) by exhaust gas passing through a suitable passage in the manifold casting, particularly at the first T beyond the carburetor where the liquid fuel impinges before turning to side branches.

To speed the warm-up process, thermostatically operated valves are generally placed in the engine exhaust system so as to force most of the exhaust gases through the intake manifold heater passages when the engine is cold. After the intake manifold has reached the desired temperature, such valves are intended to open and permit only the necessary small portion of exhaust gases to continue passing through the heater. This is an important feature, for too much heat causes a loss of engine power and aggravates the tendency for the engine toward knock and vapor lock.

On some engines, the intake manifolds are heated by water jackets taking hot water from the engine cooling system. This gives uniform heating over a wide range of operating conditions without danger of the overheating that might result from exhaust gas heat if the thermostatic exhaust valve should fail to open. It has the disadvantage, however, of requiring more time to reach normal manifold temperature, even though the water supply from the cylinder heads is short-circuited through the manifold jacket by a suitable water thermostat during warm-up.

One of the advantages of the V-8 engine is the

excellent intake manifold design permitted by the centrally located carburetor with but small differences in the lengths of the passage between the carburetor and each cylinder, and an equal number of right-angle bends in each, as in a typical intake manifold using a dual carburetor (Fig. 12). With the usual firing order shown in Fig. 19, the firing intervals for each of the lower branches (shown dotted) are evenly spaced 360 crankshaft degrees apart, but for each of the upper branches two cylinders fire 180° and then 540° apart.

Icing. Because gasoline has considerable latent heat of vaporization, it lowers the air temperature as it evaporates. This is true even at the low temperatures, where only a small part of it is vaporized. It is therefore possible for moisture which may be carried by the air to freeze under certain conditions. Ice is most likely to form when the atmosphere is almost saturated with moisture at temperatures slightly above freezing and up to about 40°F (4°C). When ice forms around or near the throttle, it can seriously interfere with the operation of an engine. For this reason small passages have been provided on some engines for jacket water, or exhaust gas from the heating supply for the intake manifold, to warm at least the flange of the carburetor. Here, again, too much heat would produce vapor lock and this would interfere with normal fuel metering. This is one of the reasons for designing some carburetors with separate casting for the throttle bodies which are heated by the manifold through only a thin gasket, while a thick gasket acting as a heat barrier is inserted between it and the float chamber containing the fuel metering systems.

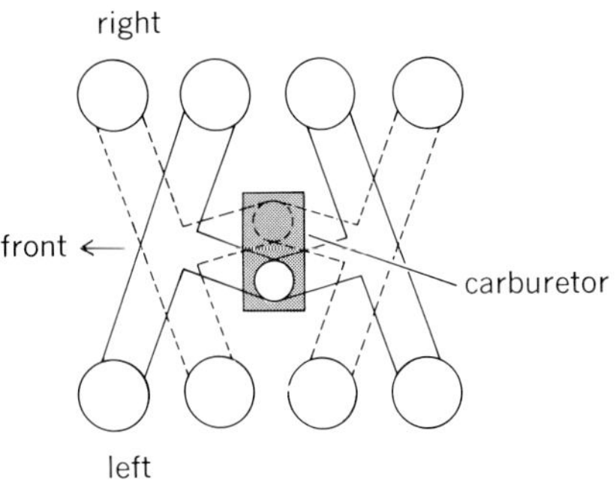

Fig. 12. Schematic of typical intake manifold with dual carburetor on a V-8 engine.

FUEL CONSUMPTION AND SUPERCHARGING

Fuel consumption at loads throughout the operating range of an engine provide insight into such characteristics as friction loss within the engine. Volumetric efficiency of an engine can be increased by use of supercharging.

Part-load fuel economy. When the fuel consumption of a spark-ignition engine is plotted against brake horsepower, straight lines may generally be drawn through the test points at given speeds, as shown in Fig. 13, provided that the tests are run with optimum spark advances and at constant fuel-air ratios. Such lines are similar to the Willans lines long used for the steam consumption of steam engines.

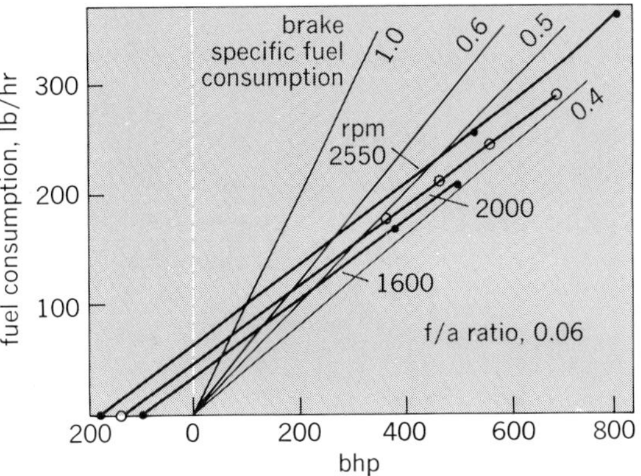

Fig. 13. Fuel consumption of an engine at part loads, plotted as typical Willans lines against brake horsepower (bhp). Data were taken with optimum spark advance and with fuel-air ratio adjusted at each test point. 1 lb/hr = 0.4356 kg/h; 1 hp = 0.7457 kW. Specific fuel consumption is in (lb/hr)/hp, where 1 (lb/hr)/hp = 0.608 (kg/h)/kW.

For practical purposes the lines at various speeds may be considered parallel over a wide range of speeds. The assumption that the negative power indicated by extrapolating these lines to zero fuel consumption reveals the power absorbed by internal friction of the engine would be justified only when the thermal efficiency remains constant over the load range. On these coordinates, lines radiating from the origin represent constant ratios of fuel consumed to power developed and therefore constant specific fuel consumption (sfc). Several such lines are indicated in Fig. 13, from which the sfc at various loads may be read directly where they cross the performance lines at the various speeds.

Similar plots of even greater utility may be drawn on an indicated horsepower basis, as has been done in Fig 14 for the same data. For many engines, a single performance line may be drawn

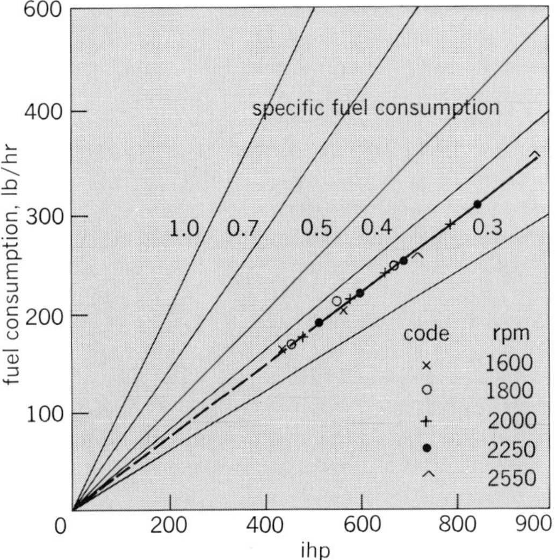

Fig. 14. The fuel consumption data of Fig. 13 plotted against the indicated horsepower (ihp), by which speed differences are neutralized. 1 lb/hr = 0.4536 kg/h; 1 hp = 0.7457 kW. Specific fuel consumption is in (lb/hr)/hp, where 1 (lb/hr)/hp = 0.608 (kg/h)/kW.

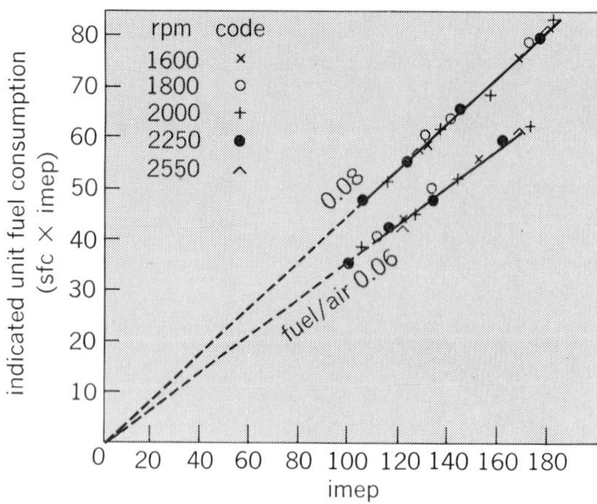

Fig. 15. The fuel consumption data of Fig. 14 plotted on the basis of imep by dividing the fuel scale and the power scale by the same factor (k = ihp/imep). Line for 0.08 fuel-air ratio has been added to show effect of fuel-air ratio on slope of such plots. The imep is in psi, where 1 psi = 6.895 kPa; sfc × imep is in psi · (lb/hr)/hp, where 1 psi · (lb/hr)/hp = 4.194 kPa · (kg/h)/kW.

through all test points at a given fuel-air ratio over a considerable range of speeds. When extrapolated, the performance line passes through the origin as it does for the engine shown in Fig. 14; the indicated sfc and thermal efficiency of the engine remain constant over the load range covered. Frequently a performance line for an engine passes a little to the left of the origin because of conditions causing a decrease in thermal efficiency as the load is reduced, such as insufficient turbulence or too low a manifold velocity. For a more complete picture of the fuel consumption performance of an engine, similar plots may be made on an mep basis. When this is done and both fuel consumption and horsepower are divided by the engine factor which converts horsepower to mep, the slope and nature of the fuel performance line remain unchanged. The fuel consumption scale then becomes equivalent to the product of mep and sfc. Such plots may be on the basis of either indicated mep (imep) or brake mep (bmep). Figure 15 shows the same data as Figs. 13 and 14 plotted on an imep basis for two different fuel-air ratios.

The fuel consumption performance of diesel engines at part loads may be shown on similar bases, but the plots should not be expected to be straight because the effective fuel-air ratio varies with load. This is illustrated in Fig. 16. It is characteristic of most diesel engines that the curvature of the plot generally flattens out at low loads so that it becomes tempting to extrapolate it to zero fuel consumption, and to consider the negative power intercept as friction. Such an intercept would represent friction only if the efficiency did not change with load, as for the engine characteristics shown in Fig. 13. If the thermal efficiency of a diesel engine improves as the load is reduced, as it would in theory for the classical diesel cycle, the zero fuel intercept for a curve such as shown in Fig. 16 would be to the right of the negative power representing engine friction. Although these fuel consumption performance plots are of considerable utility for recording such data for an engine, the fact that they are curved requires three or even more points to fix their location on the plot.

Supercharging spark-ignition engines. Volumetric efficiency and thus the mep of a four-stroke spark-ignition engine may be increased over a part of or the whole speed range by supplying air to the engine intake at higher than atmospheric pressure. This is usually accomplished by a centrifugal or rotary pump. The indicated power of an engine increases directly with the absolute pressure in the intake manifold. Because fuel consumption increases at the same rate, the indicated sfc is generally not altered appreciably by supercharging.

The three principal reasons for supercharging four-cycle spark-ignition engines are (1) to lessen the tapering off of mep at higher engine speed; (2) to prevent loss of power due to diminished atmospheric density, as when an airplane (with piston engines) climbs to high altitudes; and (3) to develop more torque at all speeds.

In a normal engine characteristic, torque rises as speed increases but falls off at higher speeds because of the throttling effects of such parts of the fuel intake system as valves and carburetors. If a supercharger is installed so as to maintain the volumetric efficiency at the higher speeds without increasing it in the middle-speed range, peak horsepower can be increased.

The rapid fall of atmospheric pressure at increased altitudes causes a corresponding decrease in the power of unsupercharged piston-type aircraft engines. For example, at 20,000 ft (6 km) the air density, and thus the absolute manifold pressure and indicated torque of an aircraft engine, would be only about half as great as at sea level. The useful power developed would be still less because of the friction and other mechanical power losses which are not affected appreciably by volumetric efficiency. By the use of superchargers, which are usually of the centrifugal type, sea-level air density may be maintained in the intake manifold up to considerable altitudes. Some aircraft engines drive these superchargers through gearing which may be changed in flight, from about 6.5 to 8.5 times engine speed. The speed change avoids oversupercharging at medium altitudes with corresponding power loss. Supercharged aircraft engines must be throttled at sea level to avoid

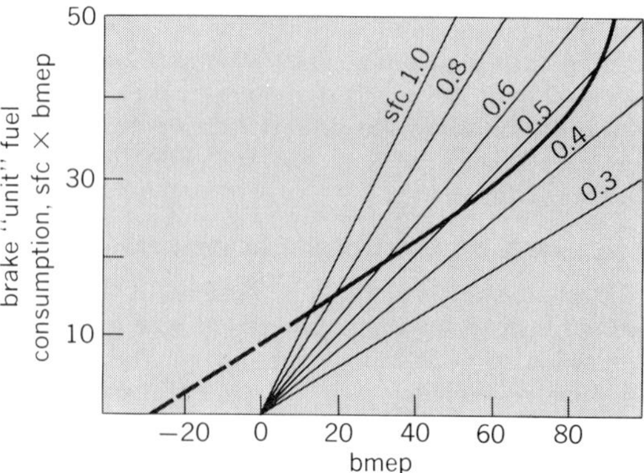

Fig. 16. Fuel consumption of a diesel engine at part loads showing the curvature, typical of such engines on these coordinates, caused by changing effective fuel-air ratios as the loads are increased. The bmep is in psi, where 1 psia = 6.895 kPa; sfc is in (lb/hr)/hp, where 1 (lb/hr)/hp = 0.608 (kg/h)/kW; sfc × bmep is in psi · (lb/hr)/hp, where 1 psi · (lb/hr)/hp = 4.194 kPa · (kg/h)/kW.

damage from detonation or excessive overheating caused by the high mep which would otherwise be developed.

Normally an engine is designed with the highest compression ratio allowable without knock from the fuel expected to be used. This is desirable for the highest attainable mep and fuel economy from an atmospheric air supply. Any increase in the volumetric efficiency of such an engine would cause it to knock unless a fuel of higher octane number were used or the compression ratio were lowered. When the compression ratio is lowered, the knock-limited mep may be raised appreciably by supercharging but at the expense of lowered thermal efficiency. There are engine uses where power is more important than fuel economy, and supercharging becomes a solution. The principle involved is illustrated in Fig. 17 for a given engine.

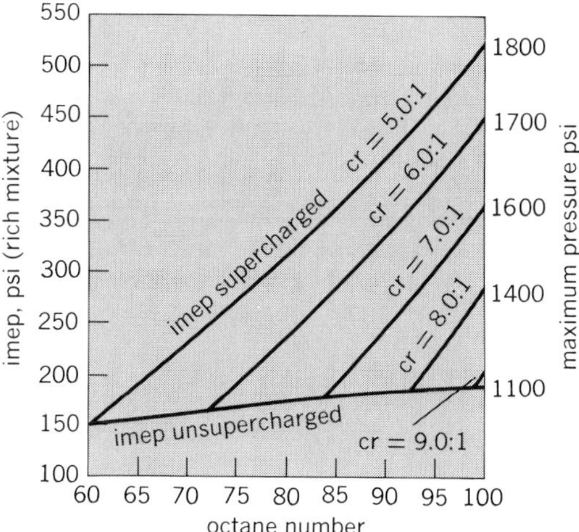

Fig. 17. Graph showing the relationship between compression ratio and knock-limited imep for given octane numbers, obtained by supercharging a laboratory engine. 1 psi = 6.895 kPa. (*From H. R. Ricardo, The High-Speed Internal Combustion Engine, 4th ed., Blackie, 1953*)

With no supercharge this engine, when using 93-octane fuel, developed an imep of 180 psi (1240 kPa) at the border line of knock at 8:1 compression ratio. If the compression ratio were lowered to 7:1, the mep could be raised by supercharging along the 7:1 curve to 275 imep before it would be knock-limited by the same fuel. With a 5:1 compression ratio it could be raised to 435 imep. Thus the imep could be raised until the cylinder became thermally limited by the temperatures of critical parts, particularly of the piston head.

Supercharged diesel engines. Combustion in a four-stroke diesel engine is materially improved by supercharging. In fact, fuels which would smoke badly and misfire at low loads will burn otherwise satisfactorily with supercharging. The imep rises directly with the supercharge pressure, until it is limited by the rate of heat flow from the metal parts surrounding the combustion chamber, and the resulting temperatures. A practical application of this limitation was made on a locomotive built by British Railways where the powers, and thus the heats developed, were held reasonably constant over a considerable speed range by driving the supercharger at constant speed by its own engine. In this way the supercharge pressure varied inversely with the speed of the main engine. The corresponding torque rise at reduced speed dispensed with much gear-shifting which would have been required during acceleration with a conventional engine.

When superchargers of either the centrifugal or positive-displacement type are driven mechanically by the engine, the power required becomes an additional loss to the engine output. Experience shows that there is a degree of supercharge for any engine which develops maximum efficiency; too high a supercharge absorbs more power in the supercharger than is gained by the engine, especially at low loads. Another means of driving the supercharger which is becoming quite general is by an exhaust turbine, which recovers some of the energy that would otherwise be wasted in the engine exhaust. This may be accomplished with so small an increase of back pressure that little power is lost by the engine. This type of drive results in an appreciable increase in efficiency at loads high enough to develop the necessary exhaust pressure.

Supercharging a two-cycle diesel engine requires some means of restricting or throttling the exhaust in order to build up cylinder pressure at the start of the compression stroke, and is used on a few large engines. Most medium and large two-cycle diesel engines are usually equipped with blowers to scavenge the cylinders after the working stroke and to supply the air required for the subsequent cycles. These blowers, in contrast to superchargers, do not build up appreciable pressure in the cylinder at the start of compression. If the capacity of such a blower is greater than the engine displacement, it will scavenge the cylinder of practically all exhaust products, even to the extent of blowing some air out through the exhaust ports. Such blowers, like superchargers, may be driven by the engine or by exhaust turbines.

Engine balance. Rotating masses such as crank pins and the lower half of a connecting rod may be counterbalanced by weights attached to the crankshaft. The vibration which would result from the reciprocating forces of the pistons and their associated masses is usually minimized or eliminated by the arrangement of cylinders in a multicylinder engine so that the reciprocating forces in one cylinder are neutralized by those in another. Where these forces are in different planes, a corresponding pair of cylinders is required to counteract the resulting rocking couple.

If piston motion were truly harmonic, which would require a connecting rod of infinite length, the reciprocating inertia force at each end of the stroke would be as in Eq. (2), where W is the total

$$F = 0.000456WN^2s \qquad (2)$$

weight of the reciprocating parts in one cylinder, N is the rpm, and s is the stroke in inches. Both F and W are in pounds. But the piston motion is not simple harmonic because the connecting rod is not infinite in length, and the piston travels more than half its stroke when the crankpin turns 90° from firing dead center. This distortion of the true harmonic motion is due to the so-called angularity a of the connecting rod, shown by Eq. (3), where r is the

$$a = \frac{r}{l} = \frac{s}{2l} \qquad (3)$$

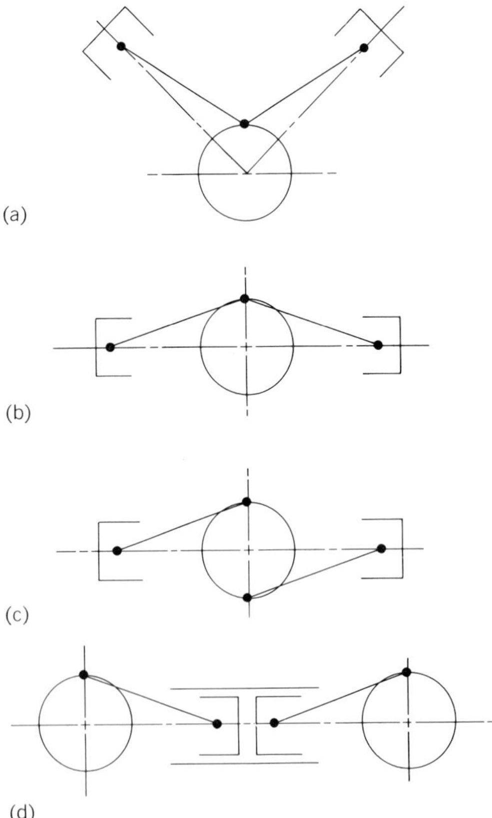

Fig. 18. Arrangements of two cylinders. (*a*) A 90° V formation with connecting rods operating on the same crankpin. (*b*) Opposed cylinders with connecting rods operating on the same crankpin. (*c*) Opposed cylinders with pistons operating on crankpins 180° apart. (*d*) Double-opposed pistons in the same cylinder but with pistons operating on separate crankshafts.

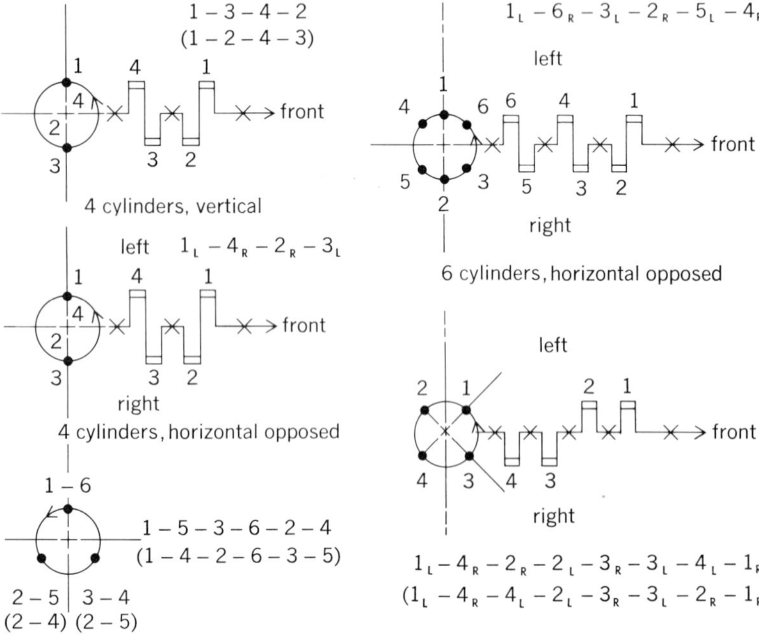

Fig. 19. Typical cylinder arrangements and firing orders.

crank radius, *s* the stroke, and *l* the connecting rod length, all in inches.

Reciprocating inertia forces act in line with the cylinder axis and may be considered as combinations of a primary force—the true harmonic force from Eq. (2)—oscillating at the same frequency as the crankshaft rpm and a secondary force oscillating at twice this frequency having a value of Fa, which is added to the primary at firing dead center and subtracted from it at inner dead center. In reality there is an infinite but rapidly diminishing series of even harmonics at 4, 6, 8, . . . , times crankshaft speed, but above the second harmonic they are so small that they may generally be neglected. Thus, for a connecting rod with the average angularity of the 1969 automobile engines, where $a = 0.291$, the inertia force caused by a piston at firing dead center is about 1.29 times the pure harmonic force, and at inner dead center it is about 0.71 times as large.

Where two pistons act on one crankpin, with the cylinders in 90° V arrangements, as in Fig. 18*a*, the resultant primary force is radial and of constant magnitude and rotates around the crankshaft with the crankpin. Therefore, it may be compensated for by an addition to the weight required to counterbalance the centrifugal force of the revolving crankpin and its associated masses. The resultant of the secondary force of the two pistons is 1.41 times as large as for one cylinder, and reciprocates in a horizontal plane through the crankshaft at twice crankshaft speed.

In engines with opposed cylinders, if the two connecting rods operate on the same crankpin, as in Fig. 18*b*, the primary forces are added and are twice as great as for one piston, but the secondary forces cancel. If the pistons operate on two crankpins 180° apart, as in Fig. 18*c*, all reciprocating forces are balanced. However, as they will be in different planes, a rocking couple will develop unless compensated by an opposing couple from another pair of cylinders. Double-opposed piston pairs, operating in a single cylinder on two crankshafts as in Fig. 18*d* (with a cross shaft to maintain synchronism), are in perfect balance for primary and secondary reciprocating forces as well as for rotating masses and torque reactions.

In the conventional four-cylinder in-line engines with crankpins in the same plane, the primary reciprocating forces of the two inner pistons (2 and 3) cancel those of the two outer pistons (1 and 4), but the secondary forces from all pistons are added. They are thus equivalent to the force resulting from a weight about $4a$ times the weight of one piston and its share of the connecting rod, oscillating parallel to the piston movement, having the same stroke, but moving at twice the frequency. A large a for this type of engine is advantageous. Where the four cylinders are arranged alternately on each side of a similar crankshaft, and in the same plane, both primary and secondary forces are in balance. Six cylinders in line also balance both primary and secondary forces.

Early eight-cylinder 90° engines with four crankpins in the same plane, like those of the early four-cylinder engine, had unbalanced horizontal secondary forces acting through the crankshafts, which were four times as large as those from one pair of cylinders.

In 1927 Cadillac introduced a crank arrangement for its V-8 engines, with the crankpins in two planes 90° apart. Staggering the 1 and 2 crankpins 90° from each other equalizes secondary forces, but the forces are in different planes. The couple thus introduced is cancelled by an opposite couple from the pistons operating on crankpins 3 and 4. This arrangement of crankpins is now universally used on V-8 engines.

Torsion dampers. In addition to vibrational forces from rotating and reciprocating masses, vibration may develop in an engine from torsional resonance of the crankshaft at various critical speeds. The longer the shaft for given bearing diameters, the lower the speeds at which these vibrations may develop. Such vibrations are dampened on most six- and eight-cylinder engines by a vibration damper which is similar to a small flywheel on the crankshaft at the end opposite to the main flywheel but coupled to the shaft only through rubber so arranged as to reduce the torsional resonances. Such vibration dampers are usually combined with the pulley driving the cooling fan and generator. *See* MECHANICAL VIBRATION.

Even though the majority of American automobile engines are now dynamically balanced, it has been general practice for several years to mount them in the chassis frame on rubber blocks. This reduces the transmission of the small-amplitude high-frequency vibrations in torque reaction as well as small unbalance of reciprocating parts in individual cylinders, so that a low noise level is developed in the car from the operation of the engine.

Firing order. Cylinder arrangements are generally selected for even firing intervals and torque impulses, as well as for balance. As a result, the cylinder arrangements and firing orders shown in Fig. 19 may be found in automobile use. It is generally customary to identify cylinder banks as left or right as seen from the driver's seat, and to number the crankpins from front to rear. Manufacturers do not agree on methods of numbering individual cylinders of V-type engines. However, the arrangements and firing orders shown are in general use with the addition in parentheses of alternate arrangements only occasionally used.

Typical American automobile engine. American six- and eight-cylinder engines have many features in common. They generally have overhead valves located in a removable cylinder head and cylinder blocks cast integral with the upper crankcase. The valves are operated from a chain-driven camshaft in the crankcase through push rods and rockers having arm lengths giving a valve lift about 50–75% greater than the cam lift. The valve action is usually silenced by use of hydraulic valve lifters. *See* HYDRAULIC VALVE LIFTER.

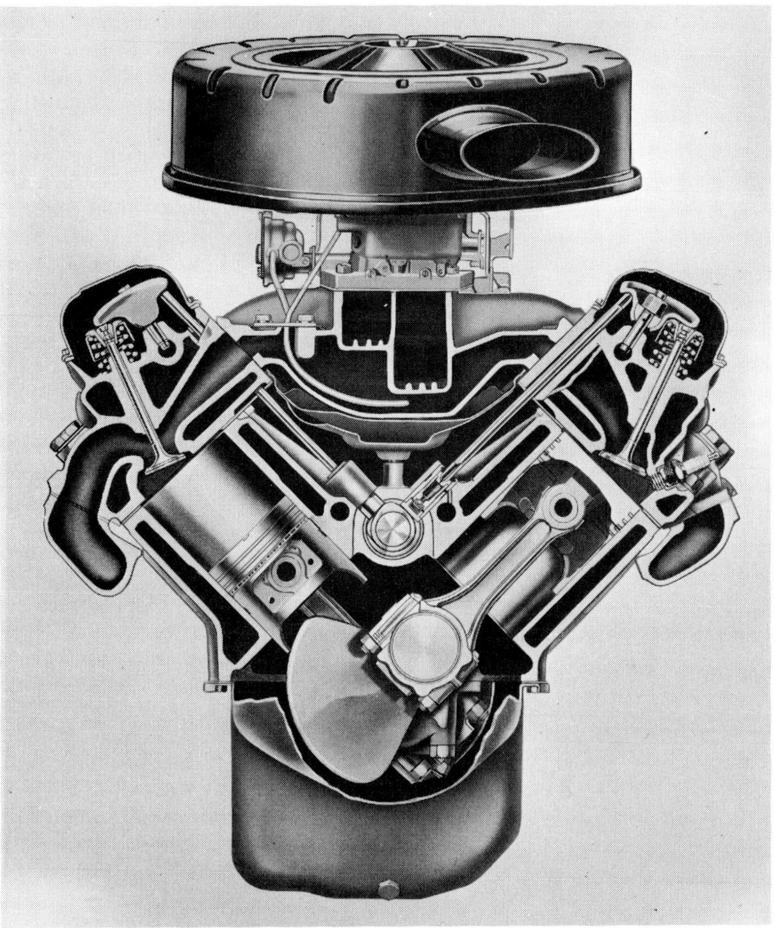

Fig. 20. Cutaway view of a typical V-8 overhead valve automobile engine with air filter. (*Pontiac Div., General Motors Corp.*)

Many designs provide means for rotating at least the exhaust valve to improve valve life. Cylinder barrels are completely surrounded by the

Selected average dimensions of 1981 engines*

Category	Four-cylinder	Six-cylinder	Six-cylinder V	Eight-cylinder V
Cylinder bore, mm	91	89	97	97
Cylinder stroke, mm	81	103	87	82
Stroke-bore ratio	0.89	0.86	0.90	0.84
Displacement, liters	2.1	3.4	3.9	4.9
Compression ratio	8.5	8.4	8.2	9.4
Net brake power, kW/rpm	58/4700	63/3600	87/4000	96/3700
Net torque, N·m/rpm	145/2600	224/1600	259/1900	299/2000

*Adapted from *Automotive Industries*.

jacket cooling water, and most designs extend it the full length of the bore. Main bearings support the crankshaft between each crankpin. Most designs lock the wrist pin to the connecting rod; the others permit it to "float" in both the piston and connecting rod. The use of three piston rings above the wrist pin has become general practice, two being narrow compression rings and the lowest an oil scraper. Compression rings are generally of cast iron about 0.078 in. (1.98 mm) wide and provided with a coating such as chromium or tin to prevent scuffing the surface during the wearing-in period of a new engine and when it is started cold with but little oil on the cylinder wall. Oil-scraper rings are about 3/16 in. (5 mm) wide, are provided with a nonscuffing surface, and have drain holes through the piston for the return of the excess oil scraped from the cylinder wall.

The highest power engines are of the eight-cylinder V-type. A typical engine of this type is shown in cross section in Fig. 20. The right and left banks of cylinders are staggered to enable connecting rods of opposing cylinders to be located side by side on the same crankpin. The V arrangement provides a short and very ridged structure, which is important for high engine speeds because of the minimized deflection of the main bearings. It also makes possible efficient intake-manifold designs and almost symmetrical and equal-length branches to each cylinder port from a centrally located downdraft carburetor. The short length and low height of these engines are also important features for car styling. Some of the principal dimensions and other statistics of these engines have been averaged in the table.

[NEIL MAC COULL]

Bibliography: T. Baumeister (ed.), *Standard Handbook for Mechanical Engineers*, 8th ed., 1978; W. H. Crouse, *Automotive Engines*, 5th ed., 1975; W. H. Crouse, *Automotive Mechanics*, 7th ed., 1975; L. C. Lichty, *Combustion Engine Processes*, 7th ed., 1967; E. F. Obert, *Internal Combustion Engines*, 3d ed., 1968; H. R. Ricardo, *The High-Speed Internal Combustion Engine*, 4th ed., 1953; C. F. Taylor, *The Internal Combustion Engine in Theory and Practice*, vol. 1, 2d ed., vol. 2, 1977.

Inventory control

The process of managing the timing and the quantities of goods to be ordered and stocked, so that demands can be met satisfactorily and economically. Inventories are accumulated commodities waiting to be used to meet anticipated demands. Inventory control policies are decision rules that focus on the trade-off between the costs and benefits of alternative solutions to questions of when and how much to order for each different type of item.

Benefits of carrying inventories. For a firm, some of the possible reasons for carrying inventories are: uncertainty about the size of future demands; uncertainty about the duration of lead time for deliveries; provision for greater assurance of continuing production, using work-in-process inventories as a hedge against the failure of some of the machines feeding other machines; and speculation on future prices of commodities. Some of the other important benefits of carrying inventories are: reduction of ordering costs and production setup costs (these costs are less frequently incurred as the size of the orders are made larger which in turn creates higher inventories); price discounts for ordering large quantities; shipping economies; and maintenance of stable production rates and work-force levels which otherwise could fluctuate excessively due to variations in seasonal demand.

Holding costs. The benefits of carrying inventories have to be compared with the costs of holding them. Holding costs include the following elements: cost of capital for money tied up in inventories; cost of owning or renting the warehouse or other storage spaces; materials handling equipment and labor costs; costs of potential obsolescence, pilferage, and deterioration (these also involve the costs of insurance, security, and protection from natural causes such as humidity or extreme temperatures); property taxes levied on inventories; and cost of installing and operating an inventory control policy. Inventories, when listed with respect to their annual costs, tend to exhibit a similarity to Pareto's law and distribution. A small percentage of the product lines may account for a very large share of the total inventory budget (they are called class A items). Aside from the class A items, and in the opposite direction, there exists a large percentage of product lines which tend to constitute a much smaller portion of the budget (they are called class C items). The remaining 20 to 30% of the items in the middle are called class B items. The ABC analysis may help management direct more of its attention to important issues and may lead to greater cost-effectiveness. For example, if the inventory levels are checked at fixed time intervals, then the status of type A items may be reported weekly, type B items biweekly, and type C items monthly.

Mathematical inventory theory. Mathematical inventory theory dates back to 1915 to the work of F. Harris of the Westinghouse Corporation, who derived the simple lot size formula. For independent items, economic order quantity (EOQ) means the replenishment quantity (Q) which has to be ordered to minimize the sum of the costs of holding (h dollars per item per period) and setups (and/or ordering costs: K dollars per order). If quantity discounts are not available, and if the demand (d items per period) is known and continues at the same rate through time, then the average cost per period, $Kd/Q + hQ/2$, is minimized by setting $Q = \sqrt{2Kd/h}$. This basic model was later extended to allow for various other conditions. For example, if the replenishment of the lot cannot take place instantaneously, but rather happens at a speed of r items per period, then the optimal value of Q becomes $\sqrt{2Kd/[h(1 - d/r)]}$. Other extensions of the EOQ model have been made to cover cases such as: quantity discounts in purchase prices; allowance for back orders at a cost; limited resources or facilities shared by otherwise independent items; inflation; time-varying deterministic demands and production capacities; and multiple echelons.

As for problems where the demands involve uncertainty, substantial research work has been carried out to develop inventory policies which minimize the expected values of various cost functions. Provided that information regarding the probability density function of demand is known, elegant mathematical solutions are now available for independent items on a single echelon. There have also been some findings for certain multiitem, multiechelon problems; however, the amount of further research that needs to be carried out to find optimal solutions for complicated everyday inventory problems is substantial. Inventory control problems of everyday life involve many complications that can include various combinations of sequence-dependent set-up times, multiitems, multiechelons with stochastic lead times, joint orders, dependent probabilistic demands, and situations where adequate information on probability density functions is not available. Even worse, the shape of the unknown probability density function may be time-varying. Regardless, mathematical inventory theory is valuable because by using the insight it provides to simpler problems, good heuristics for more complicated problems of everyday life can be designed.

Simulation. Computer simulation is also used for such purposes. By simulating inventory systems and by analyzing or comparing the performance of different decision policies, further insights can be acquired into the specific problem on hand and a more cost- and service-effective inventory control system can be developed.

Ordering policies. Continuous-review and fixed-interval are two different modes of operation of inventory control systems. The former means the records are updated every time items are withdrawn from stock. When the inventory level drops to a critical level called reorder point (s), a replenishment order is issued. Under fixed-interval policies, the status of the inventory at each point in time does not have to be known. The review is done periodically (every t periods).

Many policies for determining the quantity of replenishment use either fixed-order quantities or maximum-order levels. Under fixed-order quantities for a given product, the size of replenishment lot is always the same (Q). Under maximum-order levels, the lot size is equal to a prespecified order level (S) minus the number of items (of that product) already in the system. Different combinations of the alternatives for timing and lot sizes yield different policies known by abbreviations such as (s,Q), (s,S), (s,t,S), and (t,S). Other variations of the form of inventory control policies include coordination of timing of replenishments to achieve joint orders, and adjustment of lot sizes to the medium of transportation.

Forecasting. Uncertainties of future demand play a major role in the cost of inventories. That is why the ability to better-forecast future demand can substantially reduce the inventory expenditures of a firm. Conversely, using ineffective forecasting methods can lead to excessive shortages of needed items and to high levels of unnecessary ones.

Product design. Careful product design can also reduce inventory costs. Standardization, modularity, introduction of common components for different end products, and extension of the use of interchangeable parts and materials can lead to substantial savings in inventory costs.

MRP system. In the last decade, material requirements planning (MRP) systems (which are production-inventory scheduling softwares that make use of computerized files and data-processing equipment) have received widespread application. MRP systems have not yet made use of mathematical inventory theory. They recognize the implications of dependent demands in multiechelon manufacturing (which includes lumpy production requirements). Integrating the bills of materials, the given production requirements of end products, and the inventory records file, MRP systems generate a complete list of a production-inventory schedule for parts, subassemblies, and end products, taking into account the lead-time requirements. MRP has proved to be a useful tool for manufacturers, especially in assembly operations.

Further developments. An example of developments in computerized inventory control systems is IBM's Communication Oriented Production and Informational Control System (COPICS), which covers a wide scope of inventory-control-related activities, including demand forecasting, materials planning, and even simulation capabilities to test different strategies.

Taking into account the continuing decreases in the costs of computer hardware, the advances in computerized inventory control systems can be expected to receive even wider applications.

[ALI DOGRAMACI]

Bibliography: J. E. Biegel, *Production Control*, 1971; R. G. Brown, *Materials Management Systems*, 1977; L. A. Johnson and D. C. Montgomery, *Operations Research in Production Planning, Scheduling and Inventory Control*, 1974; S. Love, *Inventory Control*, 1979; J. Orlicky, *Material Requirements Planning*, 1975; R. Peterson and E. A. Silver, *Decision Systems for Inventory Management and Production Planning*, 1979; M. K. Starr, *Operations Management*, 1978.

IR drop

That component of the potential drop across a passive element (one which is not a seat of electromotive force) in an electric circuit caused by resistance of the element. This potential drop, by definition, is the product of the resistance R of the element and the current I flowing through it. The IR drop across a resistor is the difference of potential between the two ends of the resistor.

In a simple direct-current circuit containing a battery and a number of resistors, the sum of all the IR drops around the circuit (including that of the internal resistance of the battery itself) is equal to the electromotive force of the battery. This is an important circuit theorem useful in the analytic solution of electrical networks. *See* ELECTRICAL RESISTANCE; KIRCHHOFF'S LAWS OF ELECTRIC CIRCUITS. [JOHN W. STEWART]

Iron alloys

Solid solutions of metals, one metal being iron. A great number of commercial alloys have iron as an intentional constituent. Iron is the major constituent of wrought and cast iron and wrought and cast steel. Alloyed with usually large amounts of silicon, manganese, chromium, vanadium, molybdenum, niobium (columbium), selenium, titanium, phosphorus, or other elements, singly or sometimes in combination, iron forms the large group of materials known as ferroalloys that are important as addition agents in steelmaking. Iron is also a major constituent of many special-purpose alloys developed to have exceptional characteristics with respect to magnetic properties, electrical resistance, heat resistance, corrosion resistance, and thermal expansion. Table 1 lists some of these alloys. *See* ALLOY; FERROALLOY; STEEL.

Because of the enormous number of commercially available materials, this article is limited to some of the better-known types of alloys. Emphasis is on special-purpose alloys; practically all of these contain relatively large amounts of an alloying element or elements referred to in the classification. Alloys containing less than 50% iron are excluded, with a few exceptions.

Iron-aluminum alloys. Although pure iron has ideal magnetic properties in many ways, its low electrical resistivity makes it unsuitable for use in alternating-current (ac) magnetic circuits. Addition of aluminum in fairly large amounts increases the electrical resistivity of iron, making the resulting alloys useful in such circuits.

Three commercial iron-aluminum alloys having moderately high permeability at low field strength and high electrical resistance nominally contain 12% aluminum, 16% aluminum, and 16% aluminum with 3.5% molybdenum, respectively. These three alloys are classified as magnetically soft materials; that is, they become magnetized in a magnetic field but are easily demagnetized when the field is removed.

The addition of more than 8% aluminum to iron results in alloys that are too brittle for many uses because of difficulties in fabrication. However, addition of aluminum to iron markedly increases its resistance to oxidation. One steel containing 6% aluminum possesses good oxidation resistance up to 2300°F.

Iron-carbon alloys. The principal iron-carbon alloys are wrought iron, cast iron, and steel.

Wrought iron of good quality is nearly pure iron; its carbon content seldom exceeds 0.035%. In addition, it contains 0.075–0.15% silicon, 0.10 to less than 0.25% phosphorus, less than 0.02% sulfur, and 0.06–0.10% manganese. Not all of these elements are alloyed with the iron; part of them may be associated with the intermingled slag that is a characteristic of this product. Because of its low carbon content, the properties of wrought iron cannot be altered in any useful way by heat treatment.

Cast iron may contain 2–4% carbon and varying amounts of silicon, manganese, phosphorus, and sulfur to obtain a wide range of physical and mechanical properties. Alloying elements (silicon, nickel, chromium, molybdenum, copper, titanium, and so on) may be added in amounts varying from a few tenths to 30% or more. Many of the alloyed cast irons have proprietary compositions. *See* CAST IRON; CORROSION.

Steel is a generic name for a large group of iron alloys that include the plain carbon and alloy steels. The plain carbon steels represent the most important group of engineering materials known. Although any iron-carbon alloy containing less than about 2% carbon can be considered a steel, the American Iron and Steel Institute (AISI) standard carbon steels embrace a range of carbon contents from 0.06% maximum to about 1%. In the early days of the American steel industry, hundreds of steels with different chemical compositions were produced to meet individual demands of purchasers. Many of these steels differed only slightly from each other in chemical composition. Studies were undertaken to provide a simplified list of fewer steels that would still serve the varied needs of fabricators and users of steel products. The Society of Automotive Engineers (SAE) and the AISI both were prominent in this effort, and both periodically publish lists of steels, called standard steels, classified by chemical composition. These lists are published in the *SAE Handbook* and the AISI's *Steel Products Manuals*. The lists are altered periodically to accommodate new steels and to provide for changes in consumer requirements. There are minor differences between some of the steels listed by the AISI and SAE.

Table 1. Some typical composition percent ranges of iron alloys classified by important uses*

Type	Fe	C	Mn	Si	Cr	Ni	Co	W	Mo	Al	Cu	Ti
Heat-resistant alloy castings	Bal.	0.30–0.50		1–2	8–30	0–7						
	Bal	0.20–0.75		2–2.5	10–30	8–41						
Heat-resistant cast irons	Bal.	1.8–3.0	0.3–1.5	0.5–2.5	15–35	5 max						
	Bal.	1.8–3.0	0.4–1.5	1.0–2.75	1.75–5.5	14–30			1		7	
Corrosion-resistant alloy castings	Bal.	0.15–0.50	1 max	1	11.5–30	0–4		0.5 max				
	Bal.	0.03–0.20	1.5 max	1.5–2.0	18–27	8–31						
Corrosion-resistant cast irons	Bal.	1.2–4.0	0.3–1.5	0.5–3.0	12–35	5 max			4 max		3 max	
	Bal.	1.8–3.0	0.4–1.5	1.0–2.75	1.75–5.5	14–32			1 max		7 max	
Magnetically soft materials	Bal.			0.5–4.5								
	Bal.								3.5	16		
	Bal.									16		
	Bal.									12		
Permanent-magnet materials	Bal.						12		17			
	Bal.						12		20			
	Bal.					20	5			12		
	Bal.					17	12.5			10	6	
	Bal.					25				12		
	Bal.					28	5			12		
	Bal.					14	24			8	3	
	Bal.					15	24			8	3	1.25
Low-expansion alloys	Bal.		0.15	0.33		36						
	Bal.		0.24	0.03		42						
	61–53	0.5–2.0	0.5–2.0	0.5–2.0	4–5	33–35		1–3				

*This table does not include any AISI standard carbon steels, alloy steels, or stainless and heat-resistant steels or plain or alloy cast iron for ordinary engineering uses; it includes only alloys containing at least 50% iron, with a few exceptions. Abbreviation bal. indicates balance percent of composition.

Table 2. AISI standard steel designations

Type	Series designation*	Composition
Carbon steels	10xx	Nonresulfurized carbon steel
	11xx	Resulfurized carbon steel
	12xx	Rephosphorized and resulfurized carbon steel
Constructional alloy steels	13xx	Manganese 1.75%
	23xx	Nickel 3.50%
	25xx	Nickel 5.00%
	31xx.	Nickel 1.25%, chromium 0.65%
	33xx	Nickel 3.50%, chromium 1.55%
	40xx	Molybdenum 0.25%
	41xx	Chromium 0.50 or 0.95%, molybdenum 0.12 or 0.20%
	43xx	Nickel 1.80%, chromium 0.50 or 0.80%, molybdenum 0.25%
	46xx	Nickel 1.55 or 1.80%, molybdenum 0.20 or 0.25%
	47xx	Nickel 1.05%, chromium 0.45%, molybdenum 0.20%
	48xx	Nickel 3.50%, molybdenum 0.25%
	50xx	Chromium 0.28 or 0.40%
	51xx	Chromium 0.80, 0.90, 0.95, 1.00, or 1.05%
	5xxxx	Carbon 1.00%, chromium 0.50, 1.00, or 1.45%
	61xx	Chromium 0.80 or 0.95%, vanadium 0.10 or 0.15% min
	86xx	Nickel 0.55%, chromium 0.50 or 0.65%, molybdenum 0.20%
	87xx	Nickel 0.55%, chromium 0.50%, molybdenum 0.25%
	92xx	Manganese 0.85%, silicon 2.00%
	93xx	Nickel 3.25%, chromium 1.20%, molybdenum 0.12%
	98xx	Nickel 1.00%, chromium 0.80%, molybdenum 0.25%
Stainless and heat-resisting steels	2xx	Chromium-nickel-manganese steels; nonhardenable, austenitic, and nonmagnetic
	3xx	Chromium-nickel steels; nonhardenable, austenitic, and nonmagnetic
	4xx	Chromium steels of two classes: one class hardenable, martensitic, and magnetic; the other nonhardenable, ferritic, and magnetic
	5xx	Chromium steels; low chromium heat resisting

*The "x's" are replaced by actual numerals in defining a steel grade, as explained in the text.

Only the AISI lists will be considered here. The standard steels represent a large percentage of all steel produced and, although considerably fewer in number, have successfully replaced the large number of specialized compositions formerly used.

A numerical system is used to indicate grades of standard steels. Provision also is made to use certain letters of the alphabet to indicate the steelmaking process, certain special additions, and steels that are tentatively standard, but these are not pertinent to this discussion. Table 2 gives the basic numerals for the AISI classification and the corresponding types of steels. In this system the first digit of the series designation indicates the type to which a steel belongs; thus 1 indicates a carbon steel, 2 indicates a nickel steel, and 3 indicates a nickel-chromium steel. In the case of simple alloy steels, the second numeral usually indicates the percentage of the predominating alloying element. Usually, the last two (or three) digits indicate the average carbon content in points, or hundredths of a percent. Thus, 2340 indicates a nickel steel containing about 3% nickel and 0.40% carbon.

All carbon steels contain minor amounts of manganese, silicon, sulfur, phosphorus, and sometimes other elements. At all carbon levels the mechanical properties of carbon steel can be varied to a useful degree by heat treatments that alter its microstructure. Above about 0.25% carbon steel can be hardened by heat treatment. However, most of the carbon steel produced is used without a final heat treatment. *See* HEAT TREATMENT (METALLURGY).

Alloy steels are steels with enhanced properties attributable to the presence of one or more special elements or of larger proportions of manganese or silicon than are present ordinarily in carbon steel. The major classifications of alloy steels are:

High-strength, low-alloy
AISI alloy
Alloy tool
Stainless
Heat-resisting
Electrical
Austenitic manganese

Some of these iron alloys are discussed briefly in this article; for more detailed attention *see* STAINLESS STEEL; STEEL.

Iron-chromium alloys. An important class of iron-chromium alloys is exemplified by the wrought stainless and heat-resisting steels of the type 400 series of the AISI standard steels, all of which contain at least 12% chromium, which is about the minimum chromium content that will confer stainlessness. However, considerably less than 12% chromium will improve the oxidation resistance of steel for service up to 1200°F, as is true of AISI types 501 and 502 steels that nominally contain about 5% chromium and 0.5% molybdenum. A comparable group of heat- and corrosion-resistant alloys, generally similar to the 400 series of the AISI steels, is covered by the Alloy Casting Institute specifications for cast steels.

Corrosion-resistant cast irons alloyed with chromium contain 12–35% of that element and up to 5% nickel. Cast irons classified as heat-resistant contain 15–35% chromium and up to 5% nickel.

Iron-chromium-nickel alloys. The wrought stainless and heat-resisting steels represented by the type 200 and the type 300 series of the AISI standard steels are an important class of iron-chromium-nickel alloys. A comparable series of heat- and corrosion-resistant alloys is covered by specifications of the Alloy Casting Institute. Heat- and corrosion-resistant cast irons contain 15–35% chromium and up to 5% nickel.

Iron-chromium-aluminum alloys. Electrical-resistance heating elements are made of several iron alloys of this type. Nominal compositions are 72% iron, 23% chromium, 5% aluminum; and 55% iron, 37.5% chromium, 7.5% aluminum. The iron-chromium-aluminum alloys (with or without 0.5–2% cobalt) have higher electrical resistivity and lower density than nickel-chromium alloys used for the same purpose. When used as heating elements in furnaces, the iron-chromium-aluminum alloys can be operated at temperatures of 2350°C maximum. These alloys are somewhat brittle after elevated temperature use and have a tendency to grow or increase in length while at temperature, so that heating elements made from them should have additional mechanical support. Addition of niobium (columbium) reduces the tendency to grow.

Because of its high electrical resistance, the 72% iron, 23% chromium, 5% aluminum alloy (with 0.5% cobalt) can be used for semiprecision resistors in, for example, potentiometers and rheostats. *See* ELECTRICAL RESISTANCE; RESISTANCE HEATING.

Iron-cobalt alloys. Magnetically soft iron alloys containing up to 65% cobalt have higher saturation values than pure iron. The cost of cobalt limits the use of these alloys to some extent. The alloys also are characterized by low electrical resistivity and high hysteresis loss. Alloys containing more than 30% cobalt are brittle unless modified by additional alloying and special processing. Two commercial alloys with high permeability at high field strengths (in the annealed condition) contain 49% cobalt with 2% vanadium, and 35% cobalt with 1% chromium. The latter alloy can be cold-rolled to a strip that is sufficiently ductile to permit punching and shearing. In the annealed state, these alloys can be used in either ac or dc applications. The alloy of 49% cobalt with 2% vanadium has been used in pole tips, magnet yokes, telephone diaphragms, special transformers, and ultrasonic equipment. The alloy of 35% cobalt with 1% chromium has been used in high-flux-density motors and transformers as well as in some of the applications listed for the higher-cobalt alloy.

Iron-manganese alloys. The important commercial alloy in this class is an austenitic manganese steel (sometimes called Hadfield manganese steel after its inventor) that nominally contains 1.2% carbon and 12–13% manganese. This steel is highly resistant to abrasion, impact, and shock.

Iron-nickel alloys. The iron-nickel alloys discussed here exhibit a wide range of properties related to their nickel contents.

Nickel content of a group of magnetically soft materials ranges from 40 to 60%; however, the highest saturation value is obtained at about 50%. Alloys with nickel content of 45–50% are characterized by high permeability and low magnetic losses. They are used in such applications as audio transformers, magnetic amplifiers, magnetic shields, coils, relays, contact rectifiers, and choke coils. The properties of the alloys can be altered to meet specific requirements by special processing techniques involving annealing in hydrogen to minimize the effects of impurities, grain-orientation treatments, and so on.

Another group of iron-nickel alloys, those containing about 30% nickel, is used for compensating changes that occur in magnetic circuits due to temperature changes. The permeability of the alloys decreases predictably with increasing temperature.

Low-expansion alloys are so called because they have low thermal coefficients of linear expansion. Consequently, they are valuable for use as standards of length, surveyors' rods and tapes, compensating pendulums, balance wheels in timepieces, glass-to-metal seals, thermostats, jet-engine parts, electronic devices, and similar applications.

The first alloy of this type contained 36% nickel with small amounts of carbon, silicon, and manganese (totaling less than 1%). Subsequently, a 39% nickel alloy with a coefficient of expansion equal to that of low-expansion glasses and a 46% nickel alloy with a coefficient equal to that of platinum were developed. Another important alloy is one containing 42% nickel that can be used to replace platinum as lead-in wire in light bulbs and vacuum tubes by first coating the alloy with copper. An alloy containing 36% nickel and 12% chromium has a constant modulus of elasticity and low expansivity over a broad range of temperatures. Substitution of 5% cobalt for 5% nickel in the 36% nickel alloy decreases its expansivity. Small amounts of other elements affect the coefficient of linear expansion, as do variations in heat treatment, cold-working, and other processing procedures.

A 9% nickel steel is useful in cryogenic and similar applications because of good mechanical properties at low temperatures. Two steels (one containing 10–12% nickel, 3–5% chromium, about 3% molybdenum, and lesser amounts of titanium and aluminum and another with 17–19% nickel, 8–9% cobalt, 3–3.5% molybdenum, and small amounts of titanium and aluminum) have exceptional strength in the heat-treated (aged) condition. These are known as maraging steels.

Cast irons containing 14–30% nickel and 1.75–5.5% chromium possess good resistance to heat and corrosion.

Iron-silicon alloys. There are two types of iron-silicon alloys that are commercially important: the magnetically soft materials designated silicon or electrical steel, and the corrosion-resistant, high-silicon cast irons.

Most silicon steels used in magnetic circuits contain 0.5–5% silicon. Alloys with these amounts of silicon have high permeability, high electrical resistance, and low hysteresis loss compared with relatively pure iron. Most silicon steel is produced in flat-rolled (sheet) form and is used in transformer cores, stators and rotors of motors, and so on that are built up in laminated-sheet form to reduce eddy-current losses. Silicon-steel electrical sheets, as they are called commercially, are made in two general classifications: grain-oriented and non-oriented.

The grain-oriented steels are rolled and heat-treated in special ways to cause the edges of most of the unit cubes of the metal lattice to align themselves in the preferred direction of optimum magnetic properties. Magnetic cores are designed with the main flux path in the preferred direction, thereby taking advantage of the directional properties. The grain-oriented steels contain about 3.25% silicon and are used in the highest efficiency distribution and power transformers and in large turbine generators.

The nonoriented steels may be subdivided into low-, intermediate-, and high-silicon classes. Low-silicon steels contain about 0.5–1.5% silicon and are used principally in rotors and stators of motors and generators; steels containing about 1% silicon are also used for reactors, relays, and small intermittent-duty transformers. Intermediate-silicon steels contain about 2.5–3.5% silicon and are used in motors and generators of average to high efficiency and in small- to medium-size intermittent-duty transformers, reactors, and motors. High-silicon steels contain about 3.75–5% silicon and are used in power transformers and communications equipment and in highest efficiency motors, generators, and transformers.

High-silicon cast irons containing 14–17% silicon and sometimes up to 3.5% molybdenum possess corrosion resistance that makes them useful for acid-handling equipment and for laboratory drain pipes.

Iron-tungsten alloys. Although tungsten is used in several types of relatively complex alloys (including high-speed steels not discussed here), the only commercial alloy made up principally of iron and tungsten was a tungsten steel containing 0.5% chromium in addition to 6% tungsten that was used up to the time of World War I for making permanent magnets.

Hard-facing alloys. Hard-facing consists of welding a layer of metal of special composition on a metal surface to impart some special property not possessed by the original surface. The deposited metal may be more resistant to abrasion, corrosion, heat, or erosion than the metal to which it is applied. A considerable number of hard-facing alloys are available commercially. Many of these would not be considered iron alloys by the 50% iron content criterion adopted for the iron alloys in this article, and they will not be discussed here. Among the iron alloys are low-alloy facing materials containing chromium as the chief alloying element, with smaller amounts of manganese, silicon, molybdenum, vanadium, tungsten, and in some cases nickel to make a total alloy content of up to 12%, with the balance iron. High-alloy ferrous materials containing a total of 12–25% alloying elements form another group of hard-facing alloys; a third group contains 26–50% alloying elements. Chromium, molybdenum, and manganese are the principal alloying elements in the 12–25% group; smaller amounts of molybdenum, vanadium, nickel, and in some cases titanium are present in various proportions. In the 26–50% alloys, chromium (and in some cases tungsten) is the principal alloying element, with manganese, silicon, nickel, molybdenum, vanadium, niobium (columbium), and boron as the elements from which a selection is made to bring the total alloy content within the 26–50% range.

Permanent-magnet alloys. These are magnetically hard ferrous alloys, many of which are too complex to fit the simple compositional classification used above for other iron alloys. As already mentioned in discussing iron-cobalt and iron-tungsten alloys, the high-carbon steels (with or without alloying elements) are now little used for permanent magnets. These have been supplanted by a group of sometimes complex alloys with much higher retentivities. The ones considered here are all proprietary compositions. Two of the alloys contain 12% cobalt and 17% molybdenum and 12% cobalt and 20% molybdenum.

Members of a group of six related alloys contain iron, nickel, aluminum, and with one exception cobalt; in addition, three of the cobalt-containing alloys contain copper and one has copper and titanium. Unlike magnet steels, these alloys resist demagnetization by shock, vibration, or temperature variations. They are used in magnets for radio speakers, watt-hour meters, magnetrons, torque motors, panel and switchboard instruments, and so on, where constancy and degree of magnet strength are important.

[HAROLD E. MC GANNON]

Bibliography: American Chemical Society, *Corrosion Resistance of Metals and Alloys*, 2d ed., 1963; D. Peckner and I. M. Bernstein, *Handbook of Stainless Steels*, 1977; J. P. Saville, *Iron and Steel*, 1977; J. Zotos, *Mathematical Models of the Chemical, Physical, and Mathematical Properties of Engineering Alloys*, 1977.

Isotope (stable) separation

The physical separation of different stable isotopes of an element from one another. Many chemical elements always occur in nature as a mixture of several isotopes. The isotopes of any given element have identical chemical properties, but there are slight differences in their physical properties because of the differences in mass of the individual isotopes. Thus it is possible to separate physically the isotopes of an element to produce material of isotopic composition different from that which occurs in nature. Although these separation processes are all quite difficult and expensive to carry out, they are not inherently different from the usual operations employed in the chemical process industries. *See* ISOTOPE.

The separation of isotopes is particularly important in the nuclear energy field because individual isotopes may have completely different nuclear properties. For example, uranium-235 is used as a fuel for nuclear chain reactors, heavy water (deuterium oxide) is used as a neutron moderator in nuclear chain reactors, and deuterium gas is a possible fuel for thermonuclear reactors. Separated isotopes are also used widely for research on the structure and properties of the nucleus.

The process which is best suited for separating the isotopes of a given element depends upon the mass of the element and the desired quantity of separated material. Research quantities of separated isotopes are best prepared by electromagnetic separation in a mass spectrometer. For example, gram quantities of many separated isotopes have been prepared at Oak Ridge National Laboratory using the large electromagnetic separators which were built during World War II. The electromagnetic process has the advantage that a fairly complete separation of two isotopes can be obtained in one operation.

When moderate quantities of a separated isotope are desired, thermal diffusion may be used. Although thermal diffusion requires a large energy input, this is more than offset by the simplicity of the equipment, absence of moving parts, and high separation obtained in a small volume.

In the large-scale separation of stable isotopes, the best processes are those which have the highest thermodynamic efficiencies. Reversible processes involving distillation and chemical exchange are best for separating the light isotopes such as deuterium. For heavy isotopes such as those of uranium, however, no appreciable separation is obtained by the reversible processes and some type of irreversible process such as gaseous diffusion must be used. Although reversible processes have in general higher efficiencies than irreversible ones, the absolute efficiency of any isotope separation process is very small and the cost is very high in comparison to the usual operations employed in the chemical process industries.

Gaseous diffusion. This process has turned out to be the most economical for the separation of the isotopes of uranium. It is based on the fact that in a mixture of two gases of different molecular weights, molecules of the lighter gas will on the average be traveling at higher velocities than those of the heavier gas. If there is a porous barrier with holes just large enough to permit passage of the individual molecules but without permitting bulk flow of the gas as a whole, the probability of a gas molecule passing through the barrier will be directly proportional to its velocity. From kinetic theory it can be shown that the velocity of a gas molecule is inversely proportional to the square root of its molecular weight, so that the efficiency of gaseous diffusion will depend on the ratio of the square roots of the molecular weights of the two gases present.

The only uranium compound which is a gas at a reasonable temperature and pressure is uranium hexafluoride, UF_6. The two isotopes to be separated are $U^{235}F_6$ and $U^{238}F_6$, and the efficiency of separation depends on the quantity in the equation below. Since this number is close to unity, the

$$\sqrt{U^{238}F_6/U^{235}F_6} = 1.0043$$

separation is very small in any one step of the process.

The separation of the isotopes of uranium in the United States is carried out in the three plants operated for the Atomic Energy Commission which are located at Oak Ridge, Tenn.; Paducah, Ky.; and Portsmouth, Ohio. In each of these installations natural uranium containing 0.71% U^{235} and the balance U^{238} in the form of UF_6 gas is separated into an enriched uranium product containing more than 90% U^{235}, and a waste containing about 0.3% U^{235}. It is also possible to use "depleted" uranium (uranium recovered from plutonium production reactors has lower U^{235} content than the natural) as feed to a gaseous diffusion plant. Britain has a gaseous diffusion plant at Capenhurst, and the Soviet Union has facilities at an undisclosed location.

The success of the gaseous diffusion process is dependent on the performance of the single diffusion stage. In each stage, UF_6 gas is compressed, passed through a cooler to remove the heat of compression, and then admitted to the vessel containing the porous barrier (Fig. 1). About half the gas entering the vessel diffuses through the barrier and passes to the next higher stage. This diffused gas contains slightly more of the U^{235} isotope. The undiffused gas is slightly depleted in the U^{235} isotope, and passes to the next lower stage.

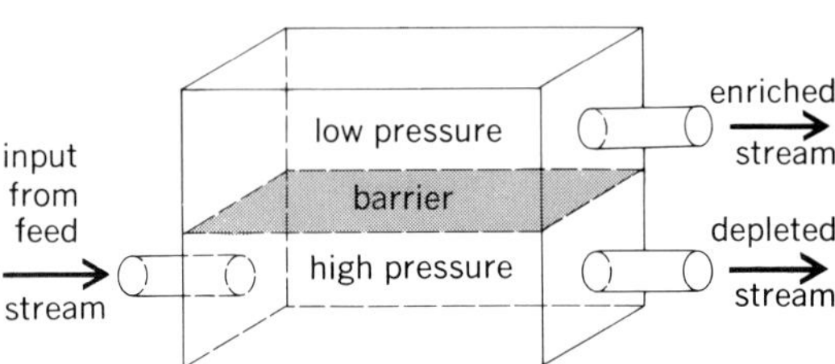

Fig. 1. Gaseous diffusion stage. *(From H. Etherington, ed, Nuclear Engineering Handbook, McGraw-Hill, 1958)*

Several thousand individual stages are required to bring about the necessary overall change in composition. The combination of stages is known as a cascade, and the cascade which brings about the separation with the least work is known as an ideal cascade. The size of the stages varies tremendously; those feeding the natural uranium into the cascade are the largest and the final product stages are the smallest.

Nickel-clad piping and process equipment are used to handle the UF_6 gas. Thousands of pumps, coolers, and control instruments are required. The electric power requirements of the gaseous diffusion process are large; for many years approximately 10% of the total electric power output of the United States was required to operate the three diffusion plants.

The gaseous diffusion process was originally developed as a means of producing highly enriched uranium for atomic bombs. At present, the gaseous diffusion plants are being modified to produce partially enriched uranium to fuel nuclear power reactors. There has also been a substantial decrease in the output from the three plants.

Chemical exchange. The chemical exchange process has proved to be the most efficient for separating isotopes of the lighter elements. This process is based on the fact that if equilibrium is established between, for example, a gas and a liquid phase, the composition of the isotopes will be different in the two phases. Thus, if hydrogen gas is brought into equilibrium with water, it is found that the ratio of heavy hydrogen (deuterium) to light hydrogen is several times greater in the water than in the hydrogen gas. By repeating the process in a suitable cascade, it is possible to effect a substantial separation of the isotopes with a relatively small number of stages.

The chief use of chemical exchange is in the large-scale production of heavy water. A dual-temperature exchange reaction between water, which contains HDO and D_2O molecules as well as H_2O molecules, and hydrogen sulfide (H_2S) gas for the primary separation of heavy water is used at the Savannah River plant. The separation is carried out in a series of hot towers operating at the boiling

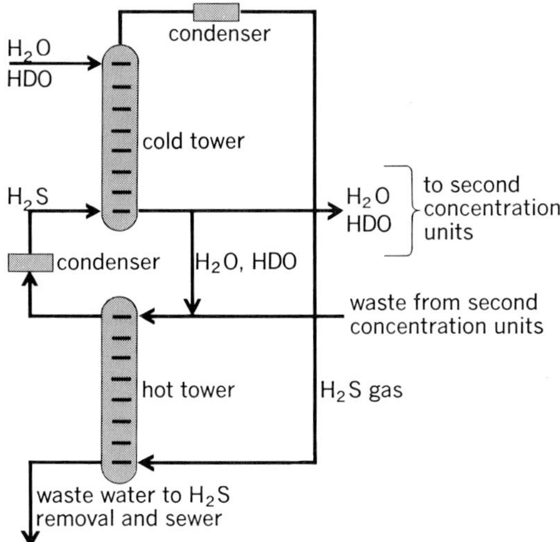

Fig. 2. Dual-temperature H_2S–HDO exchange. The end product is a mixture of H_2O, HDO, and D_2O, with the relative abundance of D_2O increasing in the later stages. (*From R. Stephenson, Introduction to Nuclear Engineering, 2d ed., McGraw-Hill, 1958*)

point and cold towers operating at room temperature (Fig. 2). In each tower liquid water passes downward countercurrent to the rising H_2S gas. The relative distribution of heavy and light water is affected by temperature, and the success of the process is determined by the difference between the concentrations in the hot and the cold towers.

Chemical exchange has also been used for the large-scale separation of other isotopes. For example, the isotopes of boron have been separated by fractional distillation of the boron trifluoride–dimethyl ether complex.

Distillation. The separation of isotopes by distillation is much less efficient than separation by other methods. Distillation was used during World War II to produce heavy water, but the cost was high and the plants are no longer in existence. Fractional distillation has been used at Savannah River to concentrate the product from the dual-temperature process (12–16% D_2O) up to 95–98% D_2O.

Electrolysis. Electrolysis of water is the oldest large-scale method of producing heavy water. Under favorable conditions, the ratio of hydrogen to deuterium in the gas leaving a cell in which water is electrolyzed is eight times the ratio of these isotopes in the liquid. In spite of this high degree of separation, electrolysis can be used only where electricity is very cheap, as in Norway, because of the large power consumption per pound of D_2O produced. Electrolysis is used in the United States only as a finishing step to concentrate to final-product specifications.

Electromagnetic process. Electromagnetic separation was the method which was first used to prove the existence of isotopes. The mass spectrometer and mass spectrograph are still widely used by physicists as a research tool. In the electromagnetic process, vapors of the material to be analyzed are ionized, accelerated in an electric field, and enter a magnetic field which causes the ions to be bent in a circular path. Since the light ions have less momentum than the heavy ions, they will be bent through a circle of smaller radius, and the two isotopes can be separated by placing collectors at the proper location.

During World War II, a large electromagnetic separation plant was built on Oak Ridge to separate the isotopes of uranium. The large mass spectrometers used there were referred to by the code name Calutron, a contraction of California University cyclotron (Fig. 3). The first kilogram quantities of U^{235} were produced in 1944. With the completion of the gaseous-diffusion plant at Oak Ridge, the electromagnetic process was found to be uneconomical and was abandoned in 1946. However, some of the equipment is still being used to produce gram quantities of separated isotopes for research purposes.

Thermal diffusion. The separation of isotopes by thermal diffusion is based on the fact that when a temperature gradient is established in a mixture of uniform composition, one component will concentrate near the hot region and the other near the cold region. Thermal diffusion is carried out in the annular space between two vertical concentric pipes, the inner one heated and the outer one cooled. Because of thermal convection of the fluid, there is a countercurrent flow which greatly increases the separation obtained in simple thermal diffusion and makes possible substantial separations in a reasonable column height.

Thermal diffusion has been used to separate small quantities of isotopes for research purposes. In 1944 a plant was built at Oak Ridge to separate the isotopes of uranium by thermal diffusion. However, the steam consumption was very large, and the plant was dismantled when the gaseous-diffusion facilities were completed.

Centrifugation. The use of a centrifuge to separate isotopes has one major advantage, namely that the separation depends only on the difference in masses of the two isotopes, and not on the ratio of their masses. Thus it is no more difficult to separate the isotopes of uranium than those of the light elements. A disadvantage of the centrifuge method is that a very high speed of rotation is required to obtain any substantial separation in a single unit.

A centrifuge pilot plant was built during World War II, but further work on the process was discontinued because of engineering problems involved in the operation of high-speed rotors, the low capacity of the individual machines, and the large power input required to overcome friction.

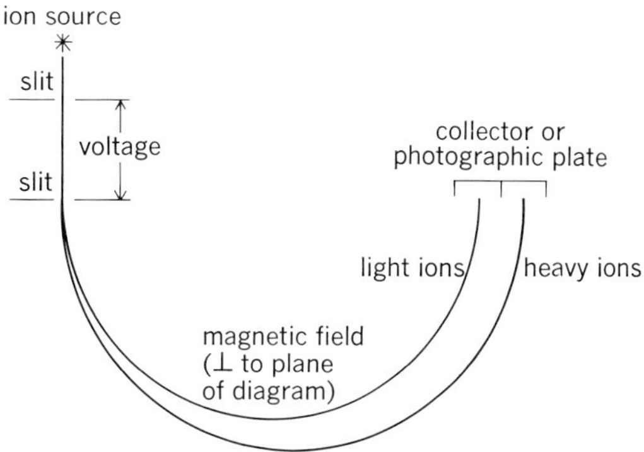

Fig. 3. Diagrammatic representation of Calutron mass spectrometer. (*From R. Stephenson, Introduction to Nuclear Engineering, 2d ed., McGraw-Hill, 1958*)

There has been a renewal of interest in centrifugation, particularly in Europe where several new approaches are being studied. There is some evidence that this will eventually lead to a simple and cheap way to produce enriched U^{235}. Politically, this has been a very sensitive subject since, if successful, it would enable even the smallest nations to manufacture nuclear weapons for possible military use.

Nozzle process. Isotopes can be separated by allowing a gaseous compound to exhaust through a properly shaped nozzle. Preliminary calculations indicate that this relatively new method may be competitive with gaseous diffusion for separating the isotopes of uranium. *See* NUCLEAR FUELS.

Laser methods. Security-classified for many years, laser methods for separating isotopes are under intensive study throughout the world. Most of the effort is directed toward the separation of U^{235}. For example, by the use of a tunable dye laser, U^{235} atoms (or possibly molecules) can be raised to an excited state while the accompanying U^{238} atoms remain unaffected. The U^{235} can then be separated out by conventional electromagnetic means. Laser methods have the advantage of a high rate of separation in one step, but they also have the potential problems of other beam processes such as the obsolete electromagnetic process. [RICHARD M. STEPHENSON]

Bibliography: M. Benedict et al., *Nuclear Chemical Engineering*, 2d ed., 1981; S. Villani, *Isotope Separation*, 2d ed., 1976.

Jetty

A structure extending out from the shore at a river mouth or harbor entrance, located parallel to and at one or both sides of a navigation channel, at sites where littoral drift causes formation of bars across channel entrances. Jetties are used to keep a channel open by preventing littoral drift and by increasing velocity of flow in the channel.

Jetties can be of breakwater-type construction or single lines of sheet piling (timber, concrete, or steel). They can be built in a straight or curved line. A fine example is the 4½-mi (7-km) jetty at the mouth of the Columbia River, Oregon. *See* COASTAL ENGINEERING; RIVER ENGINEERING.

[EDWARD J. QUIRIN]

Jewel bearing

A bearing used in quality timekeeping devices, gyros, and instruments, usually made of synthetic corundum (crystallized Al_2O_3) which is more commonly known as ruby or sapphire. The extensive use of such bearings in the design of precision devices is mainly due to the outstanding qualities of the material. Sapphire's extreme hardness (1520–2200 kg/mm²) imparts to the bearing excellent wear resistance, as well as the ability to withstand heavy loads without deformation of shape or structure. The crystalline nature of sapphire lends itself to very fine polishing and this, combined with the excellent oil- and lubricant-retention ability of the surface, adds to the natural low-friction characteristics of the material. Sapphire is also nonmagnetic and oxidization-resistant, and has a very high melting point (3685°F). Ruby has the same properties as sapphire; the red coloration is due to the introduction of a small amount of chromium oxide. *See* ANTIFRICTION BEARING.

Types. Jewel bearings, classified as either instrument or watch jewels, are also categorized according to their configuration or function. The ring jewel is the most common type. It is basically a journal bearing which supports a cylindrical pivot. The wall of the hole can be either left straight (bar hole) or can be imparted a slight curvature from end to end (olive hole). This last configuration is designed to reduce friction, compensate for slight misalignment, and help lubrication. A large variety of instrument and timing devices are fitted with such bearings, including missile and aircraft guidance systems.

Vee, or V, jewels are used in conjunction with a conical pivot, the bearing surface being a small radius located at the apex of a conical recess. This type of bearing is found primarily in electric measuring instruments.

Cup jewels have a highly polished concave recess mated to a rounded pivot or a steel ball. Typical are compass and electric-meter bearings.

End stone and cap jewels, combined with ring jewels, control the end play of the pivot and support axial thrust. They consist of a disk with highly polished flat or convex ends. Other relatively common jewel bearings are pallet stones and roller pins; both are part of the timekeeping device's escapement.

Dimensions. Minute dimensions are a characteristic of jewel bearings. A typical watch jewel may be .040 in. in diameter with a .004-in. hole, but these dimensions may go down to .015 and .002 in., respectively. Jewels with a diameter of more than 1/16 in. are considered large. It is usual for critical dimensions, such as hole diameter and roundness, to have a tolerance of .0001 in. or less. In some instances these tolerances may be as low as .0000020 in.

Manufacturing. Because of its hardness sapphire can only be worked by diamond, which is consequently the main tool for the production of jewel bearings. Both natural and synthetic diamond are used, mostly under the form of slurry, broaches, and grinding wheels.

The machining of the blanks, small disks, or cylinders of varied diameter and thickness is the first step in the manufacturing process, and is common to most types of jewel bearings. The boules (pear-shaped crystals of synthetic ruby or sapphire) are

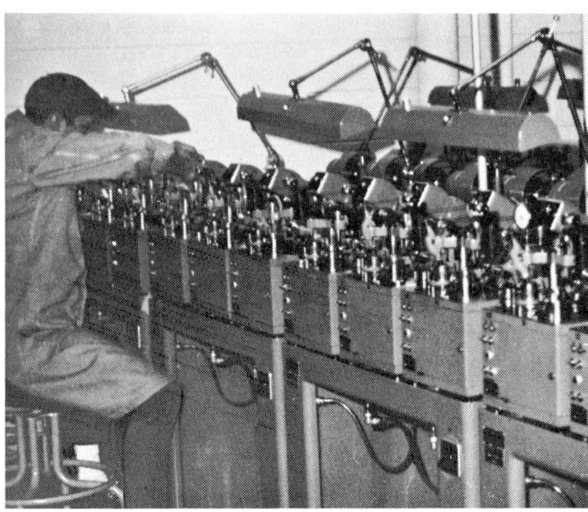

Fig. 1. Automatic cupping machines for the manufacture of jewel bearings. (*Bulova Watch Co.*)

Fig. 2. Individual head of an automatic cupping machine. (*Bulova Watch Co.*)

first oriented according to the optical axis to ensure maximum hardness of the bearing working surface. They are then sliced, diced, and ground flat, and rounded by centerless grinding to the required blank dimensions. From this point on, the process varies considerably, according to the type of bearing.

Ring jewels are drilled with a steel or tungsten wire and coarse diamond slurry, or bored with a small grinding tool. The hole is then enlarged and sized by reciprocating wires of increasingly larger sizes through a string of jewels, until the required hole size is achieved. Fine diamond powder and a very slow rate of material removal permit the respect of strict tolerances and high-finish quality requirements. The jewels, supported by a wire strung through the hole, are then ground in a special centerless-type grinding machine to the desired outside diameter dimension. After the cutting of a concave recess, which functions as an oil reservoir, the top and bottom of the bearing are polished and beveled by lapping and brushing. Finally, the "olive" configuration is obtained by oscillating a diamond charged wire through the hole of the rotating jewel. Between each operation the boiling of the jewels in a bath of sulfuric and nitric acid disposes of remaining slurries and other contaminating products.

The conical recess in vee jewels is first roughly shaped by a pyramidal diamond tool. The wall of the vee and the radius are then polished and blended with an agglomerated diamond broach, and a succession of brushing operations. Lapping of the top of the jewel and brushing a bevel around the upper outside edge conclude the process.

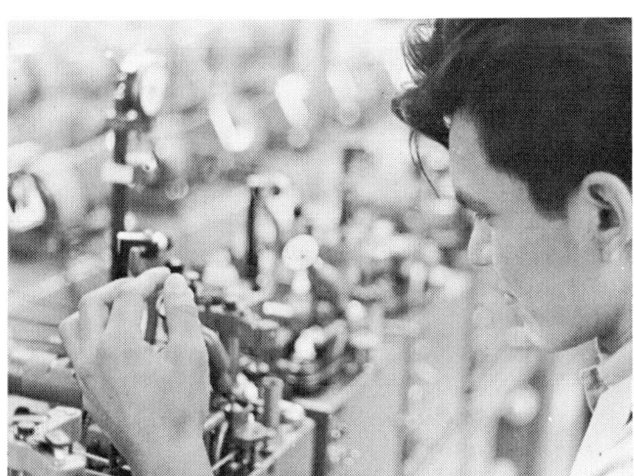

Fig. 3. Technician shown operating an automatic cupping machine. (*Bulova Watch Co.*)

Most other types of jewel bearings, such as end stones and pallet stones, are shaped by a series of grinding, lapping, and brushing operations.

A full line of automatic and semiautomatic high-precision equipment has been developed to handle and machine jewel bearings efficiently, permitting mass production and relatively low cost (Figs. 1–3). Traditionally, a large proportion of the labor involved is devoted to in-process and final quality control (Fig. 4).

History. It was around 1700 that the first handmade jewel bearings were incorporated in watch movements. With the advent of the mass production of watches in the middle of the 19th century, a considerable jewel bearing industry developed in Switzerland. The jewels were then made of natural gemstone, but the development of the Verneuil process for synthesis of corundum in the early 1900s gave further impetus to the industry, and jewel bearings started to be used in electric meters and other instruments. At that time and until the end of World War II, most of the world production was concentrated in Switzerland.

Fig. 4. Quality-control section of William Langer Jewel Bearing Plant. (*Bulova Watch Co.*)

United States production. In the United States some efforts were made during World War I to establish a domestic industry, but these efforts were abandoned after the war. When Swiss imports were cut off in 1940, the United States faced a very serious and dangerous shortage of jewel bearings. The government then turned to the watch industry to develop and build specialized manufacturing equipment. By 1943 the watch companies had succeeded in producing the required quantities of instrument jewels necessary for defense, and they turned their activity to the production of watch jewels, still in short supply. However, at the end of the war jewel bearing production was halted.

New problems developed during the Korean War, and steps were taken to establish a domestic industry to take care of defense requirements. In 1952 a pilot plant was established in Rolla, N. Dak., to train and maintain a nucleus of skilled specialists. This plant, the William Langer Jewel Bearing Plant, operated for the Federal government by the Bulova Watch Co. under a contract with General Services Administration, is the only manufacturing facility in the United States with a high-precision jewel bearing manufacturing capability (Fig. 4). *See* MACHINING OPERATIONS.

[ROBERT M. SCHULTZ]

Jig, fixture, and die design

The planning of tools for use in making production parts. Jigs position parts for machining operations such as drilling, milling, reaming, and boring, and they physically guide the cutting tools. Fixtures position parts for machining, welding, or assembly operations, but permit cutting tools to find their own path. Dies are tools which, when mounted in a press, produce parts by punching and forming.

Jig and fixture construction. Fabricated steel, strain-relieved after welding and with hardened steel inserts at the wear points, has generally superseded castings in jig and fixture construction. Aluminum and magnesium are used for many applications because they are lighter to handle and less costly to fabricate and machine. When large parts are involved, such as those in automobile or aircraft industries, the epoxy resins, reinforced with fiberglass or other suitable materials, can be used in conjunction with hardened steel inserts and bushings. Kirksite and other low-melting-point alloys are used where difficult contours must be duplicated.

Locating methods. The accuracies obtained by jigs or fixtures are dependent upon the accuracy with which parts can be consistently positioned in the first and subsequent jigs or fixtures. The dimensional consistency of the part itself is a factor. One desirable practice is to have the first jig or fixture in a series of operations create locating points on the part in the form of milled surfaces or holes which can be used in subsequent operations. The same locating points should be used for as many operations as possible.

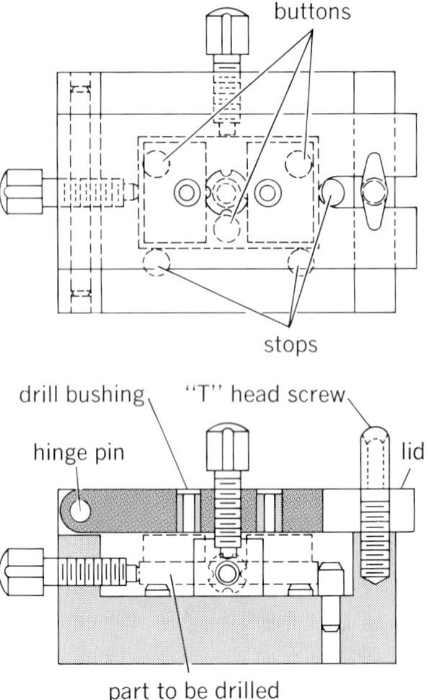

Fig. 1. Buttons and stops locate part in drill jig.

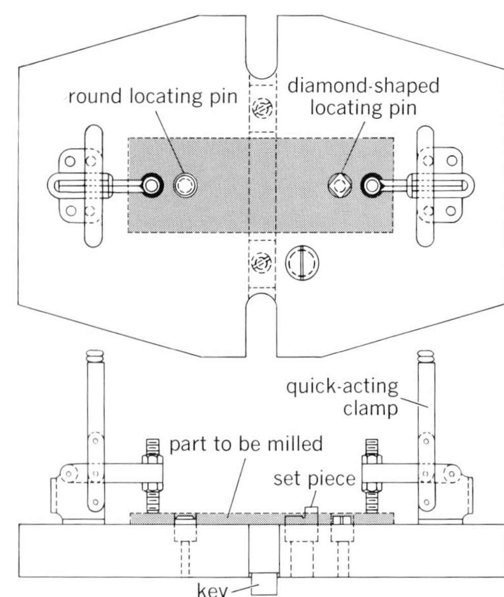

Fig. 2. Round pin paired with diamond pin locates part and allows it to be loaded readily in milling fixture.

Clamping may deform the part if the locating points are not selected advantageously. Locating points should be spaced as far apart as the part design permits and be as small as possible considering the need for adequate support and anticipated wear. Three buttons or small pads determine a plane, and stops or pins determine locations within the plane (Fig. 1).

The use of two or more straight pins is not advisable for locating from holes because a slight dimensional variation between holes will make it difficult to load or unload the part. The most common method is to use one round pin and one diamond or elliptical pin (Fig. 2).

Self-centering devices must be used when it is necessary to locate centrally from two surfaces that can vary. When locating round parts, it is preferable to work from hardened V-shaped locators. The angle of the V should be 90°. Sight holes should be provided to permit the operator to observe visually whether or not the part is down on stops and buttons. Where practical, multiple station jigs and fixtures should be made in such a way that parts can be loaded while others are being machined. The cost of loading or unloading parts is often greater than the machining cost.

Clamping. The clamping arrangement should be simple. Clamping pressure is applied over or between the pads or buttons, and standard vises are adapted as machining fixtures where practical. Machining is done toward the fixed jaw. Standard air cylinders are used. If more than three supports in one plane are necessary, all in excess of three should be adjustable. Where a wrench must be used, all nuts should be standardized. Large hand knobs ensure ease of operation, and cam-type clamps are usually faster and more positive than other methods.

Tool guides. In jigs, tool guides consist of bushings that guide drills, reamers, and boring bars. American Standards Association sizes should be used when possible. Bushings normally extend as close to the part as practical to assure accuracy of hole locations and to eliminate chip tangles within the jig. Use of slip-removable bushings in hardened liners permits secondary operations such as tapping, reaming, and counterboring. Where accuracy demands, reamers may also be guided in slip bushings either by pilots or by their cutting edges.

Tool guides in fixtures are limited to set pieces for establishing a relationship between the cutting tools and the fixture. These are permanently located on the fixture and clearance is allowed between the set pieces and the cutting tools. A feeler gage, the thickness of the clearance, is then used to locate and set the cutting tools.

Chip control. During drilling, chips are either brought out through the bushing by allowing zero clearance between the bushing and the part, or permitted to stay in the jig by allowing sufficient clearance for that purpose. The former method is the more desirable, but if tolerances are not critical and the chips are discontinuous, wear on the bushings can be reduced by the latter method. To compromise between zero and ample clearance creates the disadvantages of both with none of their advantages.

Jigs and fixtures are designed with sufficient openings for chips to be removed by gravity, coolant flow, brush, or air blast. Dust collectors are used to collect dry grinding chips or chips of nonmetallic or toxic materials.

Safety. Practice dictates the use of no. 1/2 to 13 or larger tapped holes for eyebolts in all jigs or fixtures weighing 50 lb or more and the removal of all sharp corners that could injure operators.

Stamping dies. Dies produce parts either by stressing the part material in shear until fracture occurs, or by stressing the part material beyond the yield point where it takes a permanent set but not beyond the rupture point where it breaks. Dies using the former method include blanking, piercing, cutoff, and lancing dies, and those employing the latter method include drawing, bending, embossing, and forming dies. Any practical combination of the above or other operations can be included in one die and can be performed either simultaneously or in steps using a progressive-type die (Fig. 3).

Die construction. Conventional stamping dies are composed of the following main parts: the die set, consisting of top and bottom shoes, guide pins, and bushings; the punch, which is the male cutting or forming member; the die, which is the female cutting or forming member; strippers that remove the part from the punch or die; gages to locate the part or part material; and blank holders that remove wrinkles and cause an even flow of material in drawing (Fig. 4).

Die selection. The die type selected for a job is determined by part design and economic considerations, such as the number of pieces to be produced, cost of alternate methods of production, and maintenance of the tool itself.

Very low activities or liberal part tolerances often justify special types of dies, such as steel rule dies or plow steel dies for cutting operations and plastic or rubber dies for forming operations. Most applications involving moderate to high activities require dies employing hardened tool steels.

Large activities can sometimes justify the high cost of dies made from tungsten carbide. Well-designed and well-built tungsten carbide dies can turn out a large volume of extremely accurate parts at high speeds with little sharpening or repair.

Materials for stampings. Any material softer than the punch and die and which will not shatter with impact may be cut, pierced, or blanked in a die. In drawing, bending, and forming operations, ductility is the prime requisite. Springback and deformations, including changes in material thickness occurring during the forming operation, are taken into consideration. Material is used either in individually sheared pieces or in strip or coil form. Strip or coiled material makes possible the use of various feeding mechanisms with resultant labor economies not usually practical with sheared material unless a fully automated setup is provided.

Stamping tolerances. Obtainable accuracies are broad in range, depending on the part material, thickness of stock being punched, sizes and part

JIG, FIXTURE, AND DIE DESIGN

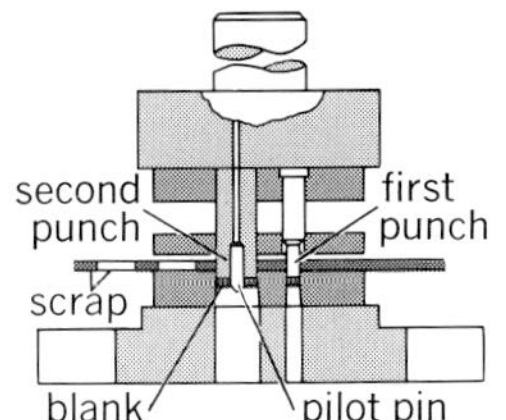

Fig. 3. Progressive die performs several operations sequentially; material advances from first to second punch between strokes.

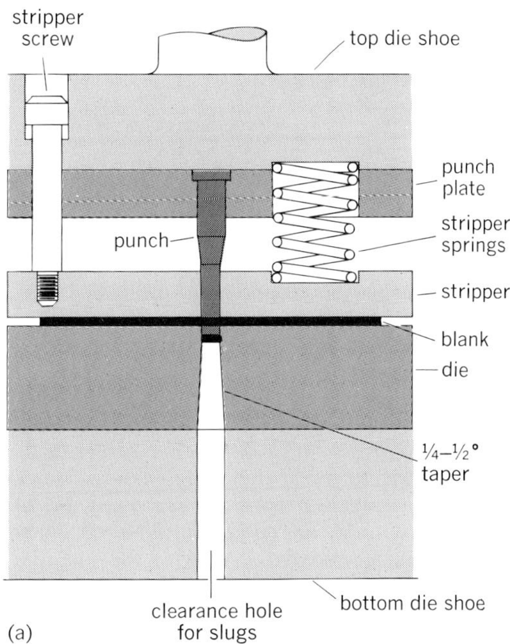

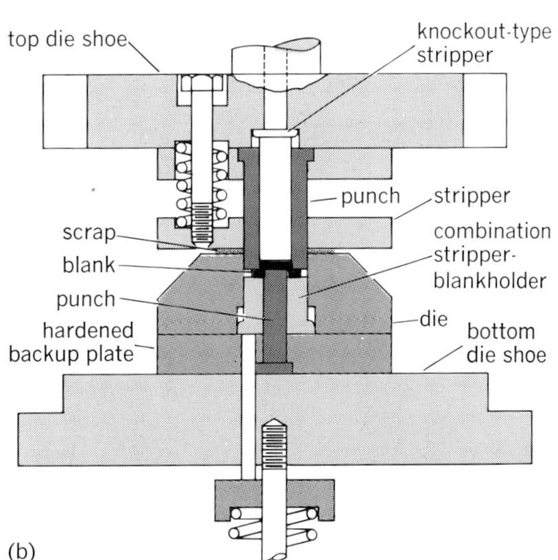

Fig. 4. Main parts of conventional stamping dies. (*a*) For piercing. (*b*) For blanking and drawing.

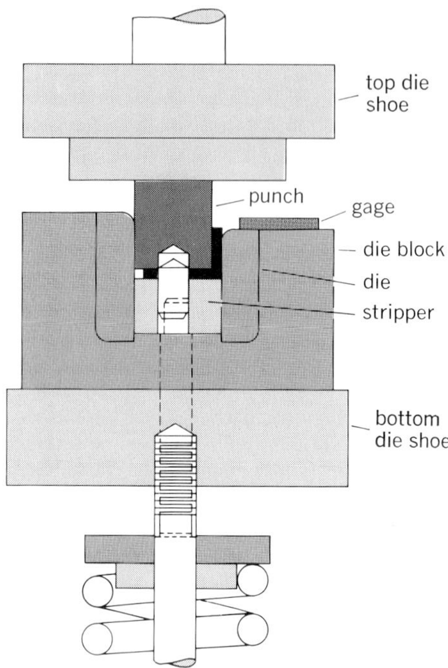

Fig. 5. Bending die, showing adequate clearance between hole and bend.

configuration, and the proximity of bends or holes. Accuracies of ±0.0005 in. are obtainable on thin materials using compound-type dies; however, under normal operating conditions, practical tolerances on light material can be expected to vary from ±0.002 to ±0.010 in. and on heavy material from 0.010 to 0.035 in. To avoid distortion, holes are placed a distance no less than the material thickness from the edge of another hole and no less than three times the material thickness from a bend radius. Hole locations established previous to bending operations are subject to material thickness tolerances and the stretch of the material which occurs during the bending operations (Fig. 5).

General design. A tool lineup is made to determine the sequence of operations and the type and number of dies required to complete the part or blank. A blank layout is made to obtain optimum material usage and to reduce scrap (Fig. 6), and the material is checked to make sure the burr is on an acceptable side of the blank. Material is normally fed into the die from right to left or front to back. When tolerances permit, the finished edge of the stock can be used as a side of the blank.

Tonnage to shear or form is computed and the die is designed for a press that will handle this capacity. It is generally possible to improve die life by using a press rated for twice the required capacity. Work can be performed on presses of less than the computed capacity by staggering the lengths of punches or by adding shear to the punch or die. Safety requires that tapped holes be provided by eyebolts in dies 50 lb or more in weight and that safety bolts be provided to hold the die together when not in use. Guards and undercuts are applied to dies to protect the hands of the operator. Standard punches and die inserts for round, oblong, square, or rectangular holes can be purchased in small increments of size and are readily available for replacement when breakage occurs. *See* MACHINING OPERATIONS; METAL FORMING; SHEET-METAL FORMING. [R. L. CAMMACK]

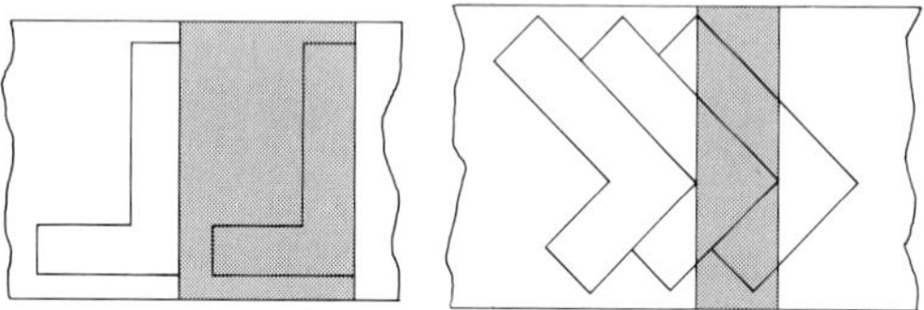

Fig. 6. Blank layout compares the amount of material which is used to produce one blank (shaded portions) by two different die positions.

Bibliography: H. E. Grant, *Jigs and Fixtures*, 1967; F. W. Wilson (ed.), *ASTME Die Design Handbook*, American Society of Tool and Manufacturing Engineers, 1965; F. W. Wilson (ed.), *ASTME Handbook of Fixture Design*, 1962; F. W. Wilson (ed.), *ASTME Tool Engineer's Handbook*, 2d ed., 1959.

Joint (structures)

The surface at which two or more mechanical or structural components are united. Whenever parts of a machine or structure are brought together and fastened into position, a joint is formed. *See* STRUCTURAL CONNECTIONS.

Mechanical joints can be fabricated by a great variety of methods, but all can be classified into two general types, temporary (screw, snap, or clamp, for example), and permanent (brazed, welded, or riveted, for example). The following list includes many of the more common methods of forming joints.

1. Screw threads. Bolt and nut, machine screw and nut, machine screw into tapped hole, threaded parts (rod, pipe), self-tapping screw, lockscrew, studs with nuts, threaded inserts, coiled-wire inserts, drive screws. *See* BOLT; BOLTED JOINT; NUT; SCREW; SCREW FASTENER; WASHER.
2. Rivets. Solid, hollow, explosive and other blind side types. *See* RIVET; RIVETED JOINT.
3. Welding. *See* WELDED JOINT; WELDING AND CUTTING OF METALS.
4. Soldering. *See* SOLDERING.
5. Brazing. *See* BRAZING.
6. Adhesive.
7. Friction-held. Nails, dowels, pins, clamps, clips, keys, shrink and press fits.
8. Interlocking. Twisted tabs, snap ring, twisted wire, crimp.
9. Other. Peening, staking, wiring, stapling, retaining rings, magnetic. Also, pipe joints are made with screw threads, couplings, caulking, and by welding or brazing; masonry joints are made with cement mortar.

[WILLIAM H. CROUSE]

Kinetics (classical mechanics)

That part of classical mechanics which deals with the relation between the motions of material bodies and the forces acting upon them. It is synonymous with dynamics of material bodies.

Basic concepts. Kinetics proceeds by adopting certain intuitively acceptable concepts which are associated with measurable quantities. These essential concepts and the measurable quantities used for their specification are as follows:

1. Space configuration refers to the positions and orientations of bodies in a reference frame adopted by the observer. It is expressed quantitatively by an arbitrarily chosen set of space coordinates, of which cartesian and polar coordinates are examples. All space coordinates rest on the motion of distance measurement.

2. Duration is expressed quantitatively by time measured by a clock or comparable mechanism.

3. Motion refers to change of configuration with time and is expressed by time rates of coordinate change called velocities and time rates of velocity change called accelerations. The classical assumption that coordinates behave as analytic functions of time permits representation of velocities and accelerations as first and second derivatives, respectively, of the space coordinates with respect to time.

4. Inertia is an attribute of bodies implying their capacity to resist changes of motion. A body's inertia with respect to linear motion is denoted by its mass.

5. Momentum is an attribute proportional to both the mass and velocity of a body. Momentum of linear motion is expressed as the product of mass and linear velocity.

6. Force serves to designate the influence exercised upon the motion of a particular body by other bodies, not necessarily specified. A quantitative connection between the motion of a body and the force applied to it is expressed by Newton's second law of motion, which is discussed later.

Distance, time, and mass are commonly regarded as fundamental, all other dynamical quantities being definable in terms of them.

Newton's second law. A primary objective of classical kinetics is the prediction of the behavior of bodies which are subject to known forces when only initial values of the coordinates and momenta are available. This is accomplished by use of a principle first recognized by Isaac Newton. Newton's statement of the principle was restricted to the linear motion of an idealized body called a mass particle, having negligible extension in space.

The basic dynamical law set forth by Newton and known as his second law states that the time rate of change of a particle's linear momentum is proportional to and in the direction of the force applied to the particle. This statement, although special in form, serves as a basis for more comprehensive statements of the principle which have since appeared.

Stated analytically, Newton's second law becomes the differential equation, Eq. (1), in which m

$$\frac{d\,(mv)}{dt} = F \tag{1}$$

represents the particle's mass, v its velocity, F the applied force, and t the time. Equation (1) provides a definition of force and of its units if units of mass, distance, and time have previously been adopted. The classical assumption of constancy of mass permits Eq. (1) to be expressed as Eq. (2), where a

$$ma = F \tag{2}$$

represents the linear acceleration. A particle in physical space requires three cartesian coordinates, x, y, and z, to specify its position. Its linear acceleration is a vector with three cartesian components, the second time derivatives of x, y, and z. Equation (2) therefore equates two vectors, requiring equality of their components expressed in detail by Eqs. (3a)–(3c). These are the Newtonian

$$m\frac{d^2x}{dt^2} = F_x \tag{3a}$$

$$m\frac{d^2y}{dt^2} = F_y \tag{3b}$$

$$m\frac{d^2z}{dt^2} = F_z \tag{3c}$$

equations of motion of an unconstrained particle in space. If the three force components F_x, F_y, and F_z are expressed functions of the coordinates and time, the dependence of each coordinate upon the time is implied and can in favorable cases be found as solutions of the equations of motion in the form of Eqs. (4). The primary objective of kinetics is achieved in the discovery of such functions.

$$x = x(t) \qquad y = y(t) \qquad z = z(t) \tag{4}$$

One-dimensional particle motion. The motion of a particle which remains on the x axis, either because of constraints or initial conditions, is determined by Eq. (3a) alone, whose solution is simplified by the absence of y and z. Such one-dimensional dynamical problems provide an attractively simple introduction to the subject. Examples are the motion of a body falling vertically, subject to gravitational force, and linear harmonic motion.

Two-dimensional particle motion. The motion of a particle remaining in the plane of the x and y axes is determined by Eqs. (3a) and (3b), from which z is absent. Two-dimensional problems are reasonably tractible and include many of physical interest such as the motion of a projectile (exterior ballistics), and of a body attracted toward a central point, as in planetary motion. Solution of a two-dimensional problem is frequently simplified by change of variables which reduces it to a pair of one-dimensional problems.

Three-dimensional particle motion. All three equations of motion, Eqs. (3a)–(3c), apply to an unconstrained particle in space. Complete solutions are possible only when the functions expressing force components are relatively simple in character. Fortunately, many of the solvable cases correspond to important physical examples in which simplicity of the forces allows separation into one- and two-dimensional motions. Three-dimensional projectile motion without friction is an example.

Newton's third law. The behavior of systems composed of two or more interacting particles is treated by Newtonian dynamics augmented by Newton's third law of motion which states that when two bodies interact, the forces they exert on one another are equal and oppositely directed. The important laws of momentum and energy conservation are derivable for such systems (the latter only for forces of special type) and useful in solution of problems. The equations of motion for systems of more than two interacting particles in space are mathematically intractible in the absence of geometrical constraints or special initial conditions, but assumptions approximating the physical situation permit solution of many problems of physical interest. The principles of particle dynamics are transferred to extended bodies by regarding them as systems of particles subject to specified mutual constraints and mutual forces. *See* ACCELERATION; FORCE; MOMENTUM; VELOCITY. [RUSSELL A. FISHER]

Bibliography: H. Goldstein, *Classical Mechanics*, 2d ed., 1980; C. Kittel, W. D. Knight, and M. A. Ruderman, *Mechanics*, Berkeley Physics Course, vol. 1, 2d ed., 1973; K. R. Symon, *Mechanics*, 3d ed., 1971.

Kirchhoff's laws of electric circuits

Fundamental natural laws dealing with the relation of currents at a junction and the voltages around a loop. These laws are commonly used in the analysis and solution of networks. They may be used directly to solve circuit problems, and they form the basis for network theorems used with more complex networks.

In the solution of circuit problems, it is necessary to identify the specific physical principles involved in the problem and, on the basis of them, to write equations expressing the relations among the unknowns. Physically, the analysis of networks is based on Ohm's law giving the branch equations, Kirchhoff's voltage law giving the loop voltage equations, and Kirchhoff's current law giving the node current equations. Mathematically, a network may be solved when it is possible to set up a number of independent equations equal to the number of unknowns. *See* CIRCUIT (ELECTRICITY).

When writing the independent equations, current directions and voltage polarities may be chosen arbitrarily. If the equations are written with due regard for these arbitrary choices, the algebraic signs of current and voltage will take care of themselves.

Kirchhoff's voltage law. One way of stating Kirchhoff's voltage law is: "At each instant of time, the algebraic sum of the voltage rise is equal to the algebraic sum of the voltage drops, both being taken in the same direction around the closed loop."

The application of this law may be illustrated with the circuit in Fig. 1. First, it is desirable to consider the significance of a voltage rise and a voltage drop, in relation to the current arrow. The following definitions are illustrated by Fig. 1.

A voltage rise is encountered if, in going from 1 to 2 in the direction of the current arrow, the polarity is from minus to plus. Thus, E is a voltage rise from 1 to 2.

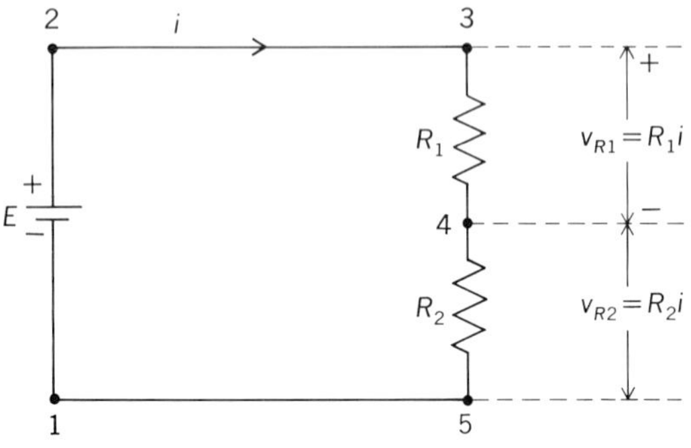

Fig. 1. Simple loop to show Kirchhoff's voltage law.

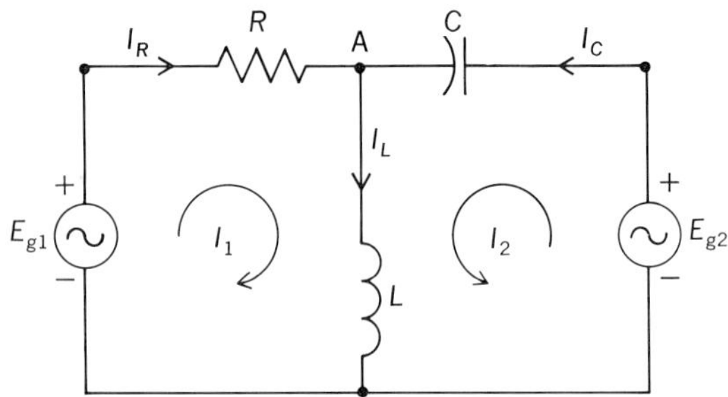

Fig. 2. Two-loop network demonstrating the application of Kirchhoff's voltage law.

A voltage drop is encountered if, in going from 3 to 4 in the direction of the current arrow, the polarity is from plus to minus. Thus, $v_{R1}=R_1i$ is a voltage drop from 3 to 4. The application of Kirchhoff's voltage law gives the loop voltage, Eq. (1).

$$E=v_{R1}+v_{R2}=R_1i+R_2i \tag{1}$$

In the network of Fig. 2 the voltage sources have the same frequency. The positive senses for the branch currents I_R, I_L, and I_C are chosen arbitrarily, as are the loop currents I_1 and I_2. The voltage equations for loops 1 and 2 can be written using instantaneous branch currents, instantaneous loop currents, phasor branch currents, or phasor loop currents.

The loop voltage equations are obtained by applying Kirchhoff's voltage law to each loop as follows.

By using instantaneous branch currents, Eqs. (2)

$$e_{g1}=Ri_R+L\frac{di_L}{dt} \tag{2}$$

and (3) may be obtained. By using instantaneous

$$e_{g2}=\frac{1}{C}\int i_C\,dt+L\frac{di_L}{dt} \tag{3}$$

loop currents, Eqs. (4) and (5) are obtained. Equa-

$$e_{g1}=Ri_1+L\frac{d(i_1+i_2)}{dt} \tag{4}$$

$$e_{g2}=\frac{1}{C}\int i_2\,dt+L\frac{d(i_2+i_1)}{dt} \tag{5}$$

tions (6) and (7) are obtained by using phasor

$$\mathbf{E}_{g1}=R\mathbf{I}_R+j\omega L\mathbf{I}_L \tag{6}$$

$$\mathbf{E}_{g2}=-j\frac{1}{\omega C}\mathbf{I}_C+j\omega L\mathbf{I}_L \tag{7}$$

branch currents. By using phasor loop currents, Eqs. (8) and (9) may be obtained.

$$\mathbf{E}_{g1}=R\mathbf{I}_1+j\omega L(\mathbf{I}_1+\mathbf{I}_2) \tag{8}$$

$$\mathbf{E}_{g2}=-j\frac{1}{\omega C}\mathbf{I}_2+j\omega L(\mathbf{I}_2+\mathbf{I}_1) \tag{9}$$

Kirchhoff's current law. Kirchhoff's current law may be expressed as follows: "At any given instant, the sum of the instantaneous values of all the currents flowing toward a point is equal to the sum of the instantaneous values of all the currents flowing away from the point."

The application of this law may be illustrated

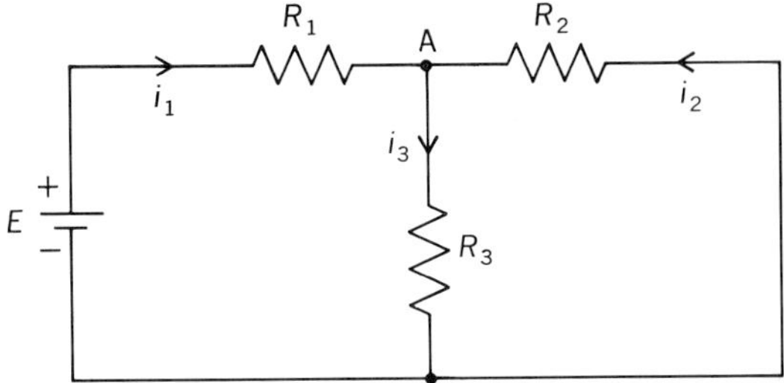

Fig. 3. Circuit demonstrating Kirchhoff's current law.

with the circuit in Fig. 3. At node A in the circuit in Fig. 3, the current is given by Eq. (10).

$$i_1 + i_2 = i_3 \qquad (10)$$

The current equations at node A in Fig. 2 can be written by using instantaneous branch currents or phasor branch currents.

By using instantaneous branch currents, Eq. (11) is obtained.

$$i_R + i_C = i_L \qquad (11)$$

By using phasor branch currents, Eq. (12) is obtained.

$$\mathbf{I}_R + \mathbf{I}_C = \mathbf{I}_L \qquad (12)$$

See DIRECT-CURRENT CIRCUIT THEORY.

[K. Y. TANG/ROBERT T. WEIL]

Lamp

An electric lamp is a device for converting electrical energy into illumination. In the common incandescent lamp a resistance wire, such as tungsten, is heated to incandescence by an electric current. In a vapor lamp the passage of electricity through mercury vapor or sodium vapor serves to ionize the gas and produce a brilliant visible glow discharge. Inert gases may be used in place of vapors to give other colors of light, as in neon lamps and in luminous tubing for advertising signs. A fluorescent lamp is a type of vapor lamp in which the radiation from ionized mercury vapor is converted into a more suitable white light by fluorescent coating on the inside of the glass tubing. In an arc lamp the light is produced by an electric arc passing through the space between two electrodes. *See* ARC LAMP; FLUORESCENT LAMP; INCANDESCENT LAMP; MERCURY-VAPOR LAMP; VAPOR LAMP. [JOHN MARKUS]

Lapping

A precision abrading process used to bring a surface to a desired state of refinement or dimensional tolerance by removal of an extremely small amount of material. Lapping is accomplished by abrading a surface with a fine abrasive grit rubbed about it in a random manner. Usually less than 0.0005 in. of stock is removed.

A loose unbonded grit is used. It is traversed about with a mating piece or lap of a somewhat softer material than the workpiece. The unbonded grit is mixed with a vehicle such as oil, grease, or soap and water compound. When a bonded grit is used, it may be in the form of a bonded abrasive lap or a charged lap such as cast iron or copper with the lapping compound embedded in it. In some cases, abrasive-covered paper laps are used.

Although some lapping is done by hand, most production work is done on a lapping machine. Various types are designed for work on flat, cylindrical, and spherical surfaces. *See* GRINDING; MACHINING OPERATIONS.

[ALLEN H. TUTTLE]

Laser welding

Welding with a laser beam. The primary apparatus being used for both commercial application and industrial research and development is the continuous-wave, convectively cooled CO_2 laser with either oscillator/amplifier (Gaussian output beam) or unstable resonator (hollows output beam) optics. These lasers, available in output powers ranging from approximately 1000 to 15,000 W, have been used to demonstrate specific welding accomplishments in a variety of metals and alloys. In addition, several high-power laser systems have been designed, produced, and delivered for actual commercial applications. Substantial advances in laser technology made possible the production of fully automated multikilowatt industrial laser systems which can be operated on a continuous production basis. These systems can be used for a variety of development programs and on-line production applications.

Welding of HY-130 steel. A significant metallurgical phenomenon was reported following the laser welding of HY-130 alloy steel. After welding with a 7-kW continuous-wave coaxial-flow CO_2 laser with oscillator-amplifier optics (Gaussian mode beam), concurrent increases in the hardness, tensile strength, and impact energy of the welds were noted when compared with base metal properties. Welding was accomplished in the open atmosphere under gas shielding with no preheating, postheating, or filler metal additions. A decrease in the visible inclusion content was noted both on the impact fracture surface (Fig. 1) and in metallographic sections. Chemical analysis revealed significant decreases in the oxygen and nitrogen contents of the weld metal as compared to the base metal. It was concluded that during laser welding of this alloy a purification of the fusion zone with respect to inclusions, and perhaps a change to a more favorable inclusion size distribution, occurs.

Welding of pipeline steel. Single- and dual-pass laser welds have been made in an alloy of the 80,000-psi (5.5×10^8 N/m²) yield strength class. Figure 2 shows cross sections of these welds. The welds were formed with a 12-kW beam from a continuous-wave CO_2 laser operating in the TEM_{00} mode. Welding speeds of 25 and 30 ipm (1.06 and 1.27 cm/sec) and 60 and 65 ipm (2.54 and 2.75 cm/sec) were used for the single- and dual-pass welds respectively. The welds were evaluated by visual inspection, x-ray, metallography, hardness tests, cross-weld tensile tests, transverse root, face, and side bend tests, longitudinal face bend tests, and Charpy impact tests at temperatures ranging from 70 to −100°F (21.1 to −73.3°C). The laser welds exhibited excellent overall mechanical properties and a Charpy shelf energy greater than 264 ft-lb (358 J), which is substantially greater than that of the base material. Dual-pass welds exhibited

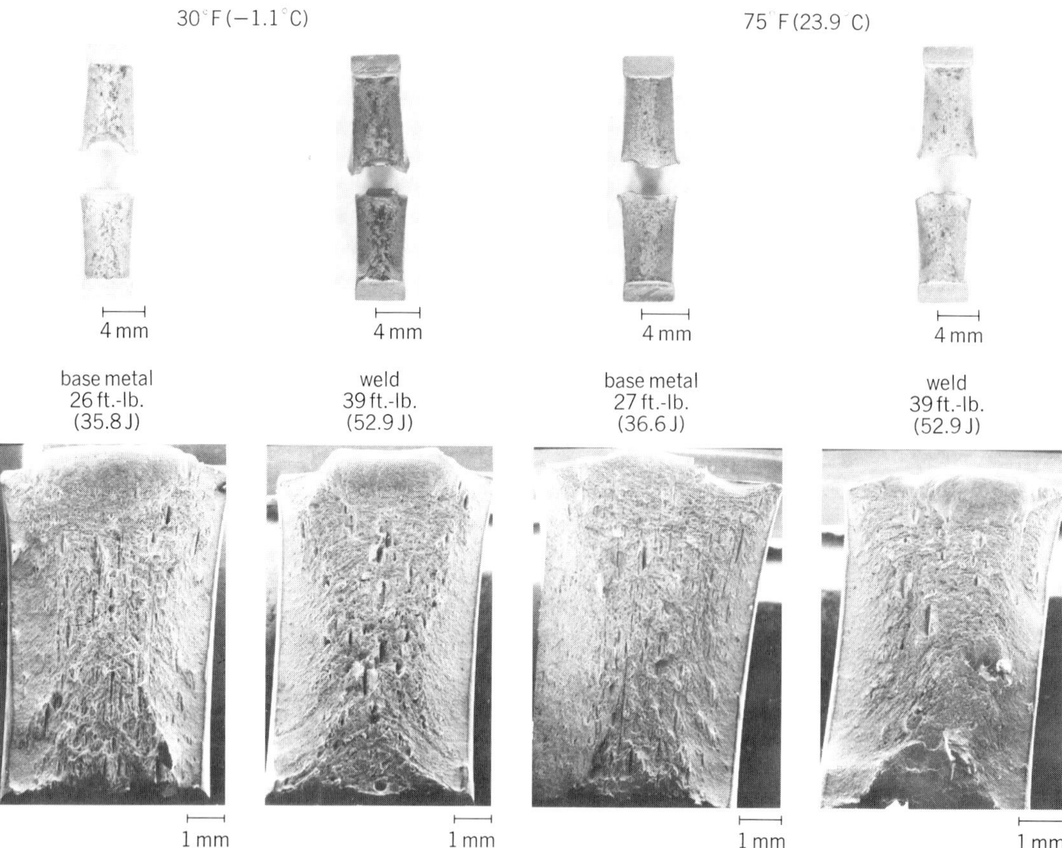

Fig. 1. A change in the shape and distribution of the inclusions on the fracture surfaces of HY-130 weld and base metal impact specimens is shown. The top row comprises low-magnification macrophotographs of both halves of the fractured specimens. The bottom row comprises higher-magnification scanning electron micrographs of one-half of the fracture. Welds were made at 5 kW and a speed of 45 ipm. (*From E. M. Breinan and C. M. Banas, Fusion zone purification during welding with high power CO_2 lasers, Proceedings of the 2d International Symposium of the Japan Welding Society, Osaka, Aug. 24–28, 1975.*

a ductile-to-brittle transition temperature below −60°F (−51.1°C). The increased shelf energy was attributed to a reduction in the visible inclusion content of the fusion zone, while transition temperature was shown to be strongly dependent on grain size. These results indicated that laser welding should be a strong candidate for future pipeline welding in certain alloys.

Titanium alloys. Two in-depth studies involving laser welding of Ti-6Al-4V alloy were conducted. In one study, laser welding was compared with electron-beam and plasma-arc welding. Sound welds were made by all processes, and the tensile properties of these welds exceeded base metal properties. The laser and electron-beam welds, because of their lower specific energies for fabrication, were substantially finer-grained and possessed finer substructures than the plasma-arc welds. Due to the finer microstructure of the beam welds (fine acicular alpha), they exhibited lower K_{IC} (fracture toughness in plane strain) values than the plasma-arc welds, which had a coarser, Widmanstatten-type structure. The finer structure of the laser and electron-beam welds showed potential for good fatigue crack initiation resistance.

A comparison study of the fatigue behavior of laser and plasma-arc welds was also conducted. It was found that the distributions of fatigue lives at the 60,000–80,000 psi ($4.1-5.5 \times 10^8$ N/m²) stress level with $R = 0.1$ were quite similar for the laser welds and the state-of-the-art plasma-arc welds with which they were compared. Both techniques were capable of making weld specimens where fatigue failures initiated in the base metal; however, there were a number of specimens welded by each process in which fractures initiated at pores in the weld which were below the radiographic detectability limit of ~0.006 in. (0.015 cm). While both the laser and plasma-arc processes appear to be capable of making useful welds in Ti-6Al-4V, the process which exhibits the best reproducibility and shows the least tendency to produce fine porosity will ultimately be favored for critical applications.

Aluminum alloys. While successful weld penetrations have been made in some aluminum alloys, and reasonable-looking beads have been produced, no extensive weld-quality studies or mechanical-property studies have been reported. Most weld penetrations in aluminum alloys made to date show some porosity, but it appears that the potential of welding aluminum exists if proper effort is devoted to solving the problems associated with aluminum laser welding. It was expected that considerable effort would be devoted toward laser welding of aluminum alloys during 1975.

Industrial applications. Five multikilowatt laser welding systems are in use for industrial welding

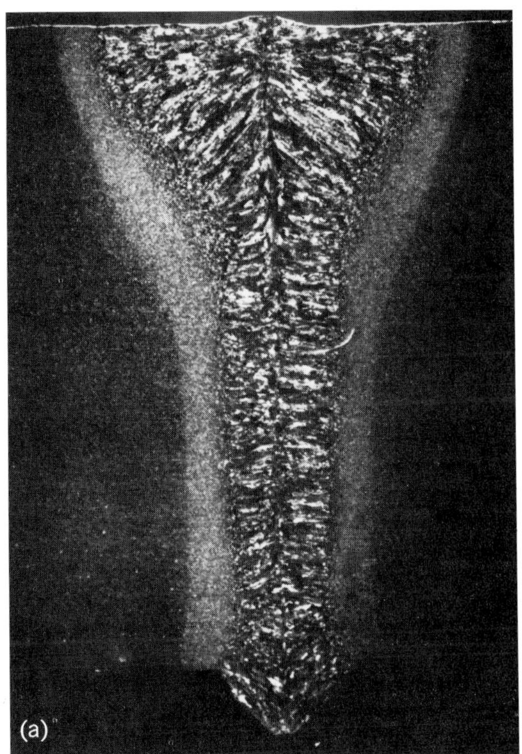

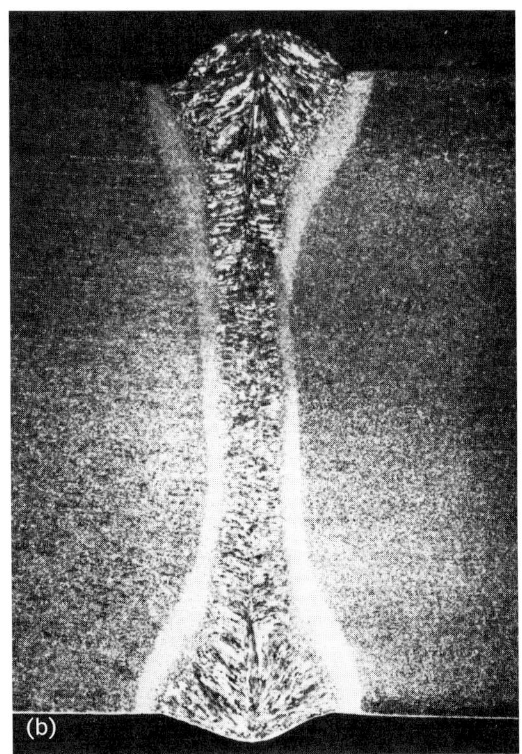

Fig. 2. Cross sections (0.52 in., or 1.32 cm, thick plate) of laser welds in X-80 arctic pipeline steel. (*a*) Single-pass weld. (*b*) Dual-pass weld. (*From E. M. Breinan and C. M. Banas, Weld. Res. Counc. Bull. No. 201, pp. 47–57, December 1974*)

and heat-treating applications. Ford Motor Company purchased an automated underbody welding system powered by a 6-kW laser from Hamilton Standard Division of United Technologies Corporation. Hamilton Standard also sold two 3-kW battery welding systems to the Western Electric Company. Caterpillar Tractor and General Motors Corporation both purchased 10-kW lasers (Model HPL-10) from AVCO Everett Research Laboratories, primarily for use in laboratory development of heat-treating applications.

The system which displays the advantages of laser processing most clearly is the underbody welding system at Ford (Fig. 3). As illustrated, a total of five degrees of freedom are achieved in movement of the focused beam relative to the intricately curved underbody panels. Only one of these, the X direction, is achieved by mechanical actuation of the parts; all the rest are achieved by rotation or translation of water-cooled copper mirrors. A unique 90° off-axis parabolic focusing mirror turns the beam 90° as it focuses the beam, thus allowing easy implementation of the α and θ axes. The focused beam creates a continuous lap weld in the thin steel sheet to a depth of 0.060 in. (0.15 cm) and at a speed of 500 in./min (21 cm/sec). The continuous weld achieves much greater strength than spot welding techniques, and assures leak-tightness of the joint. At that welding speed, four large panels are joined together into an underbody in a total cycle time of 1 min.

Future applications. As the power available in commercial lasers increases, the scope of applications is also increasing. In addition to assembly-line operations, several opportunities are seen in the area of welding of heavy sections for industrial purposes. Autogenous welds have been made at depths up to 1 in., or 2.54 cm (welding from both sides), and further research and development work should increase laser welding capabilities into weld depths of several inches. These capabilities will be of great utility in a number of industries, including shipbuilding, pipeline welding, and general heavy steel fabrication. The use of added filler metal, which has already proved feasible with laser, promises to extend the thickness capability of the laser welding process.

Laser welding has several potential advantages for deep welding tasks. For one thing, high welding speeds are achievable. For example, single-pass butt welds have been made in 0.5-in. (1.27 cm) steel at a speed of 50 ipm (2.1 cm/sec) with a 12-kW beam. In addition, the lack of necessity for filler metal or substantial reduction in the amount of filler required, promises economic advantages for the laser process, particularly when welding arctic pipeline steels and in other applications normally requiring large quantities of expensive filler metal. Reductions in edge preparation cost may also be realized due to the elimination of the edge chamfer required, although some machining of the butt edges is usually necessary to achieve a joint fitup which is suitable for laser welding. Another important factor is the ability of a single laser installation to supply a beam to several different welding stations, by appropriate indexing of turning mirrors. This capability, along with the high welding speed, allows high productivity on a welding installation.

A final advantage to laser welding is in the quality of the joint, apart from the low cost of welding.

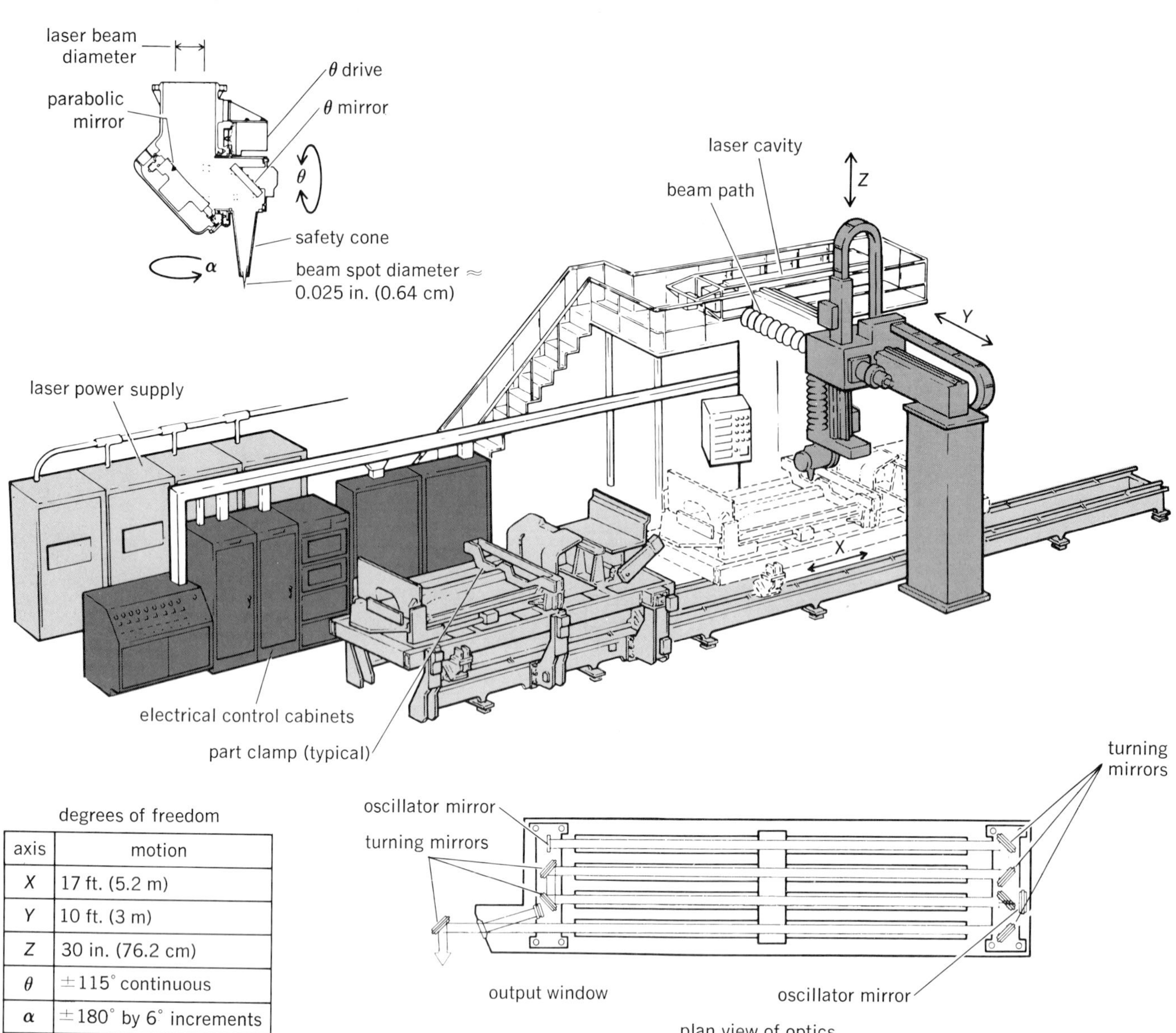

axis	motion
X	17 ft. (5.2 m)
Y	10 ft. (3 m)
Z	30 in. (76.2 cm)
θ	±115° continuous
α	±180° by 6° increments

Fig. 3. Schematic diagram of 6-kW multiaxis laser welding system at Ford Motor Company; arrows indicate direction of motion along various axes. This is the first complete high-power laser welding system designed to perform a high-production-rate industrial fabrication operation. (*Hamilton Standard Division, United Technologies Corporation*)

For quality-critical applications, such as welding of nuclear reactor components or arctic pipeline steel, the joint quality alone may prove to be an important factor in the selection of laser welding. *See* WELDING AND CUTTING OF METALS.

[EDWARD M. BREINAN]

Bibliography: E. M. Breinan and C. M. Banas, *Weld. Res. Counc. Bull. No. 201*, pp. 47–57, December 1974; F. F. Gagliano, *SME Pap. MR74-954*, 1974; E. V. Locke, *SME Pap. MR74-952*, 1974; F. D. Seaman, *SME Pap. MR74-957*, 1974; M. Yessick and D. J. Schmatz, *SME Pap. MR74-962*, 1974.

Lathe

A machine for the removal of metal from a workpiece by gripping it securely in a holding device and rotating it under power against a suitable cutting tool. The tool may be moved radially or longitudinally in respect to the turning axis of the workpiece either manually or by attached power. Forms such as cones, spheres, and related concentric-shaped workpieces as well as true cylinders can be turned on a lathe. Machining operations such as facing, boring, and threading, which are variations of the turning process, can also be performed on a lathe.

Turning equipment may be classed as being either of the horizontal or vertical type referring to the turning axis of the workpiece in the machine. The basic engine lathe is primarily a manually operated machine. Filling the gap between the engine lathe and fully automatic turning equipment is the turret lathe. Fastest and best suited to high-quantity production is the automatic screw machine.

Engine lathe. The versatile engine lathe ranges in size and design from small bench and speed lathes to large floor types. The workpiece may be held between tapered centers and rotated with the power spindle by means of a clamping device, or it may be held in a chuck, or even fixed to a rotating plate. When the work is swung between centers, the tailstock, which holds the stationary tapered center, is clamped firmly in place. Long workpieces may be supported in the middle by either a steady rest or a mechanically driven follow rest, which moves with the cutting tool.

The single-point tool is held in a tool post or block, which is supported on a cross slide and carriage. The tool may be moved radially or longitudinally in relation to the workpiece.

Angular cuts are possible, and longitudinal tapers may be cut by offsetting the tailstock, or by adjustment of a taper attachment on the rear of the lathe which actuates the cross slide.

Turret lathe. The tailstock of the engine lathe can be replaced with a multisided, indexing tool holder or turret designed to hold several tools, the machine becoming a turret lathe. The single-tool post and compound is usually replaced by a four-position indexing tool post. Power cuts may be taken from both of these tool mountings either individually or simultaneously. Turret lathes are constructed in vertical models and also in horizontal models.

Horizontal turret lathes are classed as either bar or chucking machines referring to the manner in which the workpiece is held. The headstock of the bar machine is constructed so that bar stock may be slid through it and the collet on line with the turning axis of the lathe. The collet closes, holding the piece firmly. Chucking-type machines grasp a unit size workpiece in a chuck or jawed device. Chucking devices permit work of relatively large diameters to be machined.

Horizontal machines may be further classed as being either of the ram or saddle type, a designation referring to the manner in which the turret is mounted on the machine. On the ram machine, the turret is mounted on a ram or slide which rides on a saddle. When the turret is indexed for successive operations, the saddle acts as a guide for the ram in its strokes to and from the work. On the saddle-type machine, designed without a ram, the turret mounts directly on the saddle, which slides on the bedways of the lathe (Fig. 1). Ram-type lathes with their short turret travel are generally of lighter construction than the saddle type. Excessive overhang of the ram reduces tool rigidity. Fast in operation, they perform best on small-diameter work and light chucking jobs. The rigid construction and the longer stroke of the saddle-type lathe enable it to handle both longer and heavier bar and chuck work than the ram-type machine.

The vertical turret lathe, similar in principle to the horizontal machine, is capable of handling heavier and bulkier workpieces. The vertical machine is constructed with a rotary, horizontal worktable whose diameter normally designates the capacity of the machine. Machine tables range from 30 to 74 in. in diameter. A crossrail mounted above the table carries a turret, which indexes in a vertical plane with tools that may be fed either across or downward (Fig. 2). The crossrail may also carry a vertical swiveling ram with a nonindexing

Fig. 1. Saddle-type turret lathe. (*Jones and Lamson Machine Co.*)

tool holder which feeds in a manner similar to the turret. Below the crossrail, a side head with an indexing tool holder is sometimes provided. This tool may be fed in horizontally or moved vertically. Tools may be operated simultaneously either manually or by power.

Automatic screw machines. When a high production rate of relatively small turned parts is required, automatic screw machines are used. The screw machine was originally developed from the lathe for the purpose of more economical manufacture of screws and bolts. The name has persisted even though numerous partially or completely

Fig. 2. Vertical turret lathe. (*Bullard Co.*)

finished products are produced on them. Generally these machines are classified as single-spindle or multiple-spindle automatics. The usual machine is of the horizontal type.

The single-spindle machine is constructed to feed bar stock through the hollow machine spindle and collet similar to a turret lathe. When the bar meets a stop, the collet closes on the piece. The manner in which the cutting tools are held may vary. Usually a small five- or six-position turret or drum indexes and feeds tools longitudinally against the end of the rotating workpiece. Turret tools may include drills, reamers, hollow mills, and counterboring tools; at times a single-threading die mounted on line with the spindle is used.

Usually two independent cam-actuated cross slides, front and rear, are provided to hold forming, grooving, or cutoff tools. Turret indexing, actuation of the collet, feeding of stock, and spindle clutch operations are automatic. Machining operations commonly performed include facing, drilling, reaming, forming, and knurling. Special attachments permit such operations as milling, index drilling, or thread chasing to be performed.

Multiple-spindle automatics employ several machine spindles arranged in a circular pattern. Each spindle is equipped to hold stock in a manner similar to a single-spindle machine. In some instances automatically operating chucks capable of holding irregularly shaped workpieces replace the collet-type chucks. Standard machines may have as many as eight rotating spindles. These in turn index in a carrier about a nonrotating turret, which holds a variety of cutting tools. As each spindle indexes and progresses around the carrier, successive machining operations are performed by the turret tools. Simultaneously, tools located in successive positions are performing their respective operations on the various workpieces. Cross slides mounted at right angles to the spindles carry the forming, grooving, or cutoff tools.

The time for the longest cut plus the allowance for such programming actions as indexing and tool traversing sets the time required to produce one finished piece. By dividing long cuts over two or more operations, the cycling time may be reduced. Machining operations performed on these machines are generally the same as those done on single-spindle automatics.

Vertical multiple-spindle machines of the semiautomatic chucking type are constructed with as many as 16 spindles. A horizontal, circular table holding the rotating chucks indexes under the vertical spindles with a different operation being performed at each station. This machine is constructed to handle much larger workpieces than horizontal automatics; on many models the speed, feed, and direction of rotation for each spindle combination may be varied from one position to the next. This permits selection of the correct feed and speed for the operation being performed at each station. *See* MACHINING OPERATIONS.

[ALAN H. TUTTLE]

Bibliography: J. Anderson and E. E. Tatro, *Shop Theory*, 5th ed., 1968; I. Bradley, *The Grinding Machine*, 1973; H. D. Burghardt and A. Axelrod, *Machine Tool Operation*, part 1: *Safety, Measuring Instruments, Bench Work, Drill Press, Lathe, and Forge Work*, 5th ed., 1959.

Layout drawing

A design drawing or graphical statement of the overall form of a component or device, which is usually prepared during the innovative stages of a design. Since it lacks detail and completeness, a layout drawing provides a faithful explanation of the device and its construction only to individuals such as designers and drafters who have been intimately involved in the conceptual stage. In a sense, the layout drawing is a running record of ideas and problems posed as the design evolves. In the layout drawing, for instance, considerations of kinematic design of a mechanical component are explored graphically in incomplete detail, showing only those aspects of the elements and their interrelationships to be considered in the design. In most cases the layout drawing ultimately becomes the primary source of information from which detail drawings and assembly drawings are prepared by other drafters under the guidance of the designer. *See* DRAFTING; ENGINEERING DRAWING.

[ROBERT W. MANN]

Lead alloys

Substances formed by the addition of one or more elements, usually metals, to lead. Lead alloys may exhibit greatly improved mechanical or chemical properties as compared to pure lead. The major alloying additions to lead are antimony and tin. The solubilities of most other elements in lead are small, but even fractional weight percent additions of some of these elements, notably copper and arsenic, can alter properties appreciably.

Cable-sheathing alloys. Lead is used as a sheath over the electrical components to protect power and telephone cable from moisture. Alloys containing 1% antimony are used for telephone cable, and lead-arsenical alloys, containing 0.15% arsenic, 0.1% tin, and 0.1% bismuth, for example, are used for power cable. Aluminum and plastic cable sheathing are replacing lead alloy sheathing in many applications, but improvements in methods of applying a lead sheathing (continuous extrusion) may offset this trend somewhat.

Battery-grid alloys. Lead alloy grids are used in the lead-acid storage battery (the type used in automobiles) to support the active material composing the plates. Lead grid alloys contain 6–12% antimony for strength, small amounts of tin to improve castability, and one or more other minor additions to retard dimensional change in service. No lead alloys capable of replacing the lead-antimony alloys in automobile batteries have been developed, although research in this area has been extensive. An alloy containing 0.03% calcium for use in large stationary batteries has had success.

Chemical-resistant alloys. Lead alloys are used extensively in many applications requiring resistance to water, atmosphere, or chemical corrosion. They are noted for their resistance to attack by sulfuric acid. Alloys most commonly used contain 0.06% copper, or 1–12% antimony, where greater strength is needed. The presence of antimony lowers corrosion-resistance to some degree.

Type metals. Type metals contain $2\frac{1}{2}$–12% tin and $2\frac{1}{2}$–25% antimony. Antimony increases hardness and reduces shrinkage during solidification. Tin improves fluidity and reproduction of detail. Both elements lower the melting temperature of the alloy. Common type metals melt at 460–475°F.

Bearing metals. Lead bearing metals (babbitt metals) contain 10–15% antimony, 5–10% tin, and for some applications, small amounts of arsenic or copper. Tin and antimony combine to form a compound which provides wear-resistance. These alloys find frequent application in cast sleeve bearings, and are used extensively in freight-car journal bearings. In some cast bearing bronzes, the lead content may exceed 25%.

Solders. A large number of lead-base solder compositions have been developed. Most contain large amounts of tin with selected minor additions to provide specific benefits, such as improved wetting characteristics.

Free-machining brasses, bronzes, and steels. Lead is added in amounts from 1 to 25% to brasses and bronzes to improve machining characteristics. Lead remains as discrete particles in these alloys. It is also added to some construction steel products to increase machinability. Only about 0.1% is needed, but the tonnage involved is so large that this forms an important use for lead. *See* ALLOY; SOLDERING; TIN ALLOYS.

[DUDLEY WILLIAMS]

Bibliography: W. Hofmann, *Lead and Lead Alloys: Properties and Technology*, 1970.

Level (surveying)

An instrument for establishing a horizontal line of sight in order to find the difference of elevation between two points upon which a graduated rod is successively held to be sighted. The level usually consists of a telescope and attached spirit level, mounted on a tripod for rotation about a vertical axis. The line of sight is fixed by cross hairs. Stadia hairs may be provided for convenience in keeping foresight and backsight distances balanced. The tubular spirit level vial is long, thus quite sensitive, with its upper interior surface being finely ground to a circular arc. A tangent to the center of the ground arc is parallel to the line of sight; thus the line of sight is horizontal (level) when the spirit bubble is centered. The instrument is leveled by manipulation of foot screws in the mounting.

The engineer's dumpy level has the telescope rigidly attached to the supporting yoke (Fig. 1). Another type, the tilting level, has a circular or bull's-eye bubble for approximate leveling of the instrument; a micrometer screw then inclines the telescope for final centering of the tubular bubble, which is seen through a prism at the instant the rod is read (Fig. 2). A most recent type is the self-leveling or automatic level, which employs a prism suspended on fine wires, or other pendulum-mounted prism, as part of the optical train. After preliminary leveling with foot screws and a circular level, the line of sight remains horizontal for all pointings. For the use of levels in surveying *See* SURVEYING.

[B. AUSTIN BARRY]

Fig. 1. Engineer's dumpy level. (*W. and L. E. Gurley Co.*)

LEVEL (SURVEYING)

Fig. 2. High-precision tilting level. (*Wild Heerbrugg Instruments, Inc.*)

Lever

A pivoted rigid bar used to multiply force or motion, sometimes called the lever and fulcrum (Fig. 1). The lever uses one of the two conditions for static equilibrium, which is that the summation of moments about any point equals zero. The other condition is that the summation of forces acting in any direction through a point equals zero. *See* INCLINED PLANE.

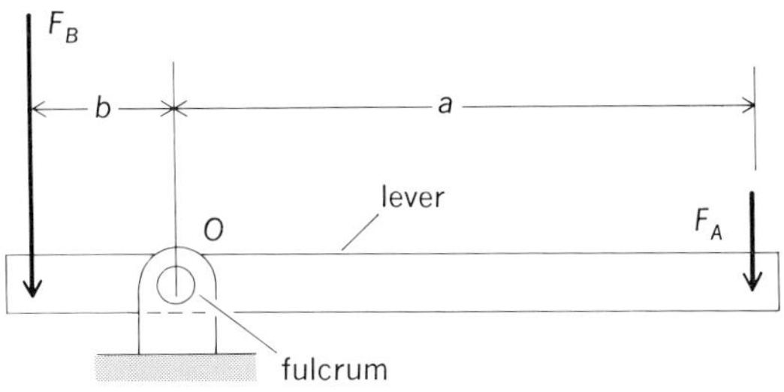

Fig. 1. The lever pivots at the fulcrum.

If moments acting counterclockwise around the fulcrum of a lever are positive, then, for a frictionless lever, $F_Bb - F_Aa = 0$, which may be rearranged to give Eq. (1).

$$F_B = \frac{a}{b} F_A \qquad (1)$$

If F_B represents the output and F_A represents the input, the mechanical advantage, MA, is then given by Eq. (2).

$$\mathrm{MA} = \frac{F_B}{F_A} = \frac{a}{b} \qquad (2)$$

Applications of the lever range from the simple nutcracker and paper punch (Fig. 2) to complex multiple-lever systems found in scales and in testing machines used in the study of properties of materials. *See* SIMPLE MACHINE.

[RICHARD M. PHELAN]

LEVER

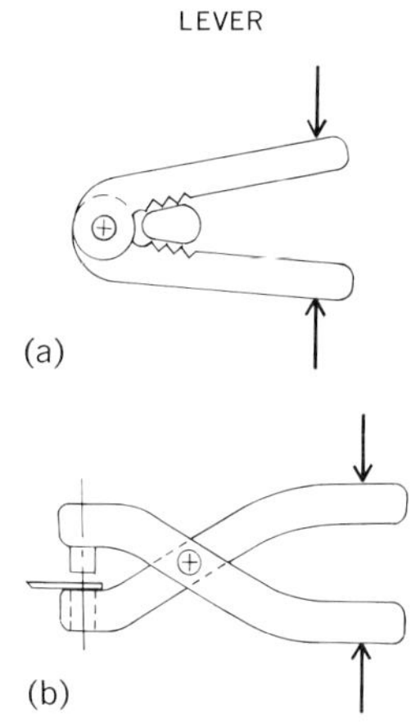

Fig. 2. Two applications of the lever. (*a*) Nutcracker. (*b*) Paper punch.

Lightning and surge protection

Means of protecting electrical systems, buildings, and other property from lightning and other high-voltage surges.

The destructive effects of natural lightning are well known. Studies of lightning and means of either preventing its striking an object or passing the stroke harmlessly to ground have been going on since the days when Franklin first established that lightning is electrical in nature. From these studies, two conclusions emerge: (1) Lightning will not strike an object if it is placed in a grounded metal cage. (2) Lightning tends to strike, in general, the highest objects on the horizon.

One practical approximation of the grounded metal cage is the well-known lightning rod or mast (Fig. 1*a*). The effectiveness of this device is evaluated on the cone-of-protection principle. The protected area is the space enclosed by a cone having the mast top as the apex of the cone and tapering out to the base. Laboratory tests and field experience have shown that if the radius of the base of the cone is equal to the height of the mast, equipment inside this cone will rarely be struck. A radius equal to twice the height of the mast gives a cone of shielding within which an object will be struck occasionally. The cone-of-protection principle is illustrated in Fig. 1*b*.

A building which stands alone, like the Empire State in New York City, is struck many times by lightning during a season. It is protected with a mast and the strokes are passed harmlessly to ground. It is interesting to note, however, that lightning has been observed to strike part way down the side of this building (Fig. 2). This shows that lightning does not always strike the highest object but rather chooses the path having the lowest electrical breakdown.

The probability that an object will be struck by lightning is considerably less if it is located in a valley. Therefore, electric transmission lines which must cross mountain ranges often will be routed through the gaps to avoid the direct exposure of the ridges. *See* ELECTRIC DISTRIBUTION SYSTEMS; ELECTRIC POWER SYSTEMS.

Overhead lines of electric power companies are vulnerable to lightning. Lightning appears on these lines as a transient voltage, which, if of sufficient magnitude, will either flash over or puncture the weakest point in the system insulation.

(a)

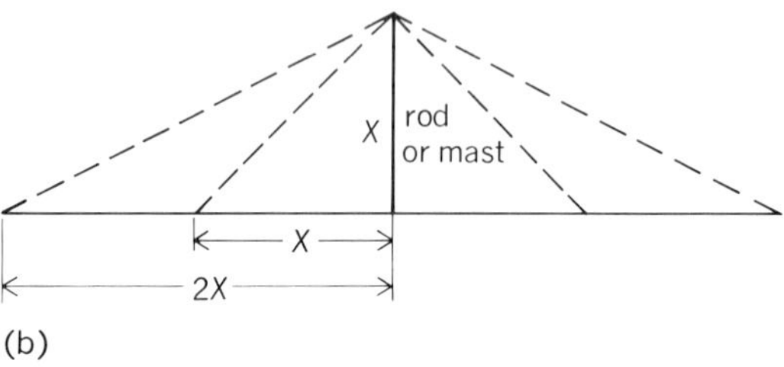

(b)

Fig. 1. Lightning rod cone of protection. (*a*) Configuration of rods on a house. (*b*) Geometry of the principle.

Fig. 2. Multiple lightning stroke A to Empire State and nearby buildings, Aug. 23, 1936. Single, continuing stroke B to Empire State Building, Aug. 24, 1936.

Many of the troubles that cause service interruptions on electrical systems are the results of flashovers of insulation wherein no permanent damage is done at the point of fault, and service can be restored as soon as the cause of the trouble has disappeared. A puncture or failure of the insulation, on the other hand, requires repair work, and damaged apparatus must be removed from service.

There are a number of protective devices to limit or prevent lightning damage to electric power systems and equipment. The word protective is used to connote either one or two functions: the prevention of trouble, or its elimination after it occurs. Various protective means have been devised either to prevent lightning from entering the system or to dissipate it harmlessly if it does.

Overhead ground wires and lightning rods. These devices are used to prevent lightning from striking the electrical system.

The grounded-metal-cage principle is approached by overhead ground wires, preferably two, installed over the transmission phase conductors and grounded at each tower. The ground wires must be properly located with respect to the phase conductors to provide a cone of protection and have adequate clearance from them, both at the towers and throughout the span. If the resistance to ground of the tower is high, the passing of high lightning currents through it may sufficiently elevate the tower in potential from the transmission line conductors so that electrical flashover can occur. Since the magnitude of the lightning current may be defined in terms of probability, the expected frequency of line flashover may be predicted from expected storm and lightning stroke frequency, the current magnitudes versus probability, and the tower footing resistance. Where the expected flashover rate is too high, means to lower the tower footing resistance are employed such as driven ground rods or buried wires connected to the base of the tower (Fig. 3).

The ground wires are often brought in over the

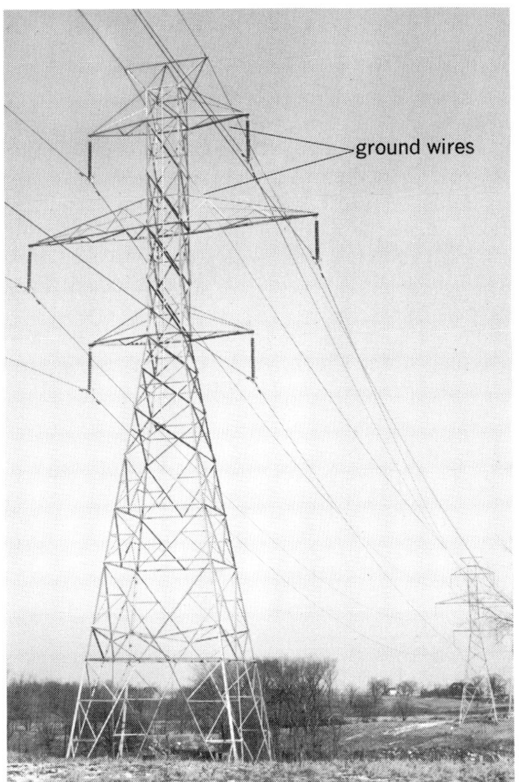

Fig. 3. Double-circuit suspension tower on the Olive–Gooding Grove 345-kV transmission line, with two circuits strung and with two ground wires. (*Indiana and Michigan Electric Co., American Electric Power Co.*)

terminal substations. For additional shielding of the substations, lightning rods or masts are installed. Several rods are usually used to obtain the desired protection. *See* ELECTRIC POWER SUBSTATION.

Lightning arresters. These are protective devices for reducing the transient system overvoltages to levels compatible with the terminal-apparatus insulation. They are connected in parallel with the apparatus to be protected. One end of the arrester is grounded and also connected to the case of the equipment being protected; the other end is connected to the electric conductor (Fig. 4).

Lightning arresters provide a relatively low discharge path to ground for the transient overvoltages, and a relatively high resistance to the power system follow current, so that their operation does not cause a system short circuit.

In selecting an arrester to protect a transformer, for example, the voltage levels that can be maintained by the arrester both on lightning surges and surges resulting from system switching must be coordinated with the withstand strength of the transformers to these surges. Arrester gap sparkover voltages, discharge voltages on various magnitudes of discharge currents, the inductance of connections to the arrester, voltage wave shape, and other factors must all be evaluated. *See* SURGE ARRESTER.

Rod gaps. These also are devices for limiting the magnitude of the transient overvoltages. They usually are formed of two 1/2-in.[2] rods, one of which is grounded and the other connected to the line conductor but may also have the shape of rings or horns. They have no inherent arc-quenching ability, and once conducting, they continue to arc until the system voltage is removed, resulting in a system outage.

These devices are applied on the principle that, if an occasional flashover is to occur in a station, it is best to predetermine the point of flashover so that it will be away from any apparatus that might otherwise be damaged by the short-circuit current and the associated heat.

The flashover characteristics of rod gaps are such that they turn up (increase of breakdown voltages with decreasing time of wavefront) much faster on steep-wavefront surges than the withstand-voltage characteristics of apparatus, with the result that if a gap is set to give a reasonable margin of protection on slow-wavefront surges, there may be little or no protection for steep-wavefront surges. In addition, the gap characteristics may be adversely affected by weather conditions and may result in undesired flashovers.

Immediate reclosure. This is a practice for restoring service after the trouble occurs by immediately reclosing automatically the line power circuit breakers after they had been tripped by a short circuit. The protective devices involved are the power circuit breaker and the fault-detecting and reclosing relays.

This practice is successful because the majority of the short circuits on overhead lines are the result of flashovers of insulators and there is no permanent damage at the point of fault. The fault may be either line-to-ground or between phases. Reclosing relays are available to reenergize the line several times with adjustable time intervals between reclosures.

If the relays go through the full sequence of reclosing and the fault has not cleared, they lock out.

Fig. 4. Installation of three single-pole station lightning arresters rated 276 kV for lightning and surge protection of large 345-kV power transformer. (*General Electric Co.*)

If the fault has cleared after a reclosure, the relays return to normal.

Permanent faults must always be removed from a system and the accepted electric protective devices are power circuit breakers and suitable protective relays. *See* CIRCUIT TESTING (ELECTRICITY); ELECTRIC PROTECTIVE DEVICES.

[GLENN D. BREUER]

Bibliography: R. H. Golde, *Lightning Protection*, 1973; Institute of Electrical and Electronics Engineers, *Surge Arresters for AC Power Circuits*, IEEE Stand. no. 28, 1974.

Link

An element of a mechanical linkage. A link may be a straight bar or a disk, or it may have any other shape, simple or complex. It is assumed, for simple analysis, to be made of unyielding material; that is, its shape does not change. The frame, or fixed member, of a linkage is one of the links, whether the frame is fixed relative to the Earth or relative to a movable body such as the chassis of an automobile. *See* VELOCITY ANALYSIS.

Two links of a kinematic chain meet in a joint, or pair, by which these links are held together. Just as each joint is a pair having two elements, one from each link, so a link in a kinematic chain is the rigid connector of two or more elements belonging to different pairs. *See* SLIDING PAIR.

Each free link has three degrees of freedom in the plane (two translational and one rotational). If the tail of one link is pin-joined to the head of its predecessor two degrees of freedom are lost. If several links are joined so as to form a closed loop, $2g$ degrees of freedom are lost, where g is the number of lower pair joints. If one member of an n-member chain is fixed, all three degrees of freedom of the fixed member are eliminated, the remaining degrees of freedom F being given by Eq. (1). Hence for a triangle $F = 0$ and for a

$$F=3(n-1)-2g \qquad (1)$$

chain of four links called simply a four bar linkage, $F=1$.

Assuming that $F=1$ is stipulated for a mechanism, and investigating n in Eq. (1), it is found that Eq. (2) results. Because 4 is even and $2g$ is even, n

$$3n=4+2g \qquad (2)$$

must always be even. There can be no pin-joint mechanism with five links; the next highest must have six links. *See* FOUR-BAR LINKAGE; LINKAGE (MECHANISM); MECHANISM.

[DOUGLAS P. ADAMS]

Bibliography: J. Bickford, *Mechanism for Intermittent Motion*, 1972; R. S. Hartenberg and J. Denavit, *Kinematic Synthesis of Linkages*, 1964.

Linkage (mechanism)

A set of rigid bodies, called links, joined together at pivots by means of pins or equivalent devices. A body is considered to be rigid if, for practical purposes, the distances between points on the body do not change. Linkages are used to transmit power and information. They may be employed to make a point on the linkage follow a prescribed curve, regardless of the input motions to the linkage. They are used to produce an angular or linear displacement $f(x)$, where $f(x)$ is a given function of a displacement x. *See* LINK.

LINKAGE (MECHANISM)

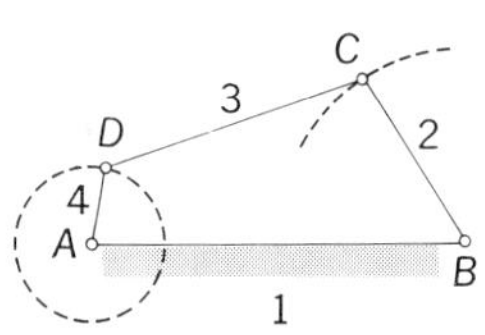

Fig. 1. Four-bar linkage

If the links are bars the linkage is termed a bar linkage. In first approximations a bar may be treated as a straight line segment or a portion of a curve, and a pivot may be treated as a common point on the bars connected at the pivot. A common form of bar linkage is then one for which the bars are restricted to a given plane, such as a four-bar linkage (Fig. 1). A body pivoted to a fixed base and to one or more other links is a crank. As crank 4 in Fig. 1 rotates, link 2 (also a crank) oscillates back and forth. This four-bar linkage thus transforms a rotary motion into an oscillatory one, or vice versa. Link AB, marked 1, is fixed. Link 2 may be replaced by a slider 5 (Fig. 2). As end D of crank

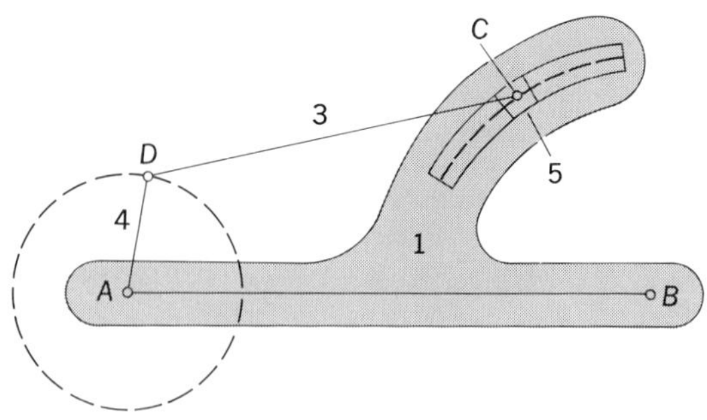

Fig. 2. Equivalent of four-bar linkage.

4 rotates, pivot C describes the same curve as before. The slider moves in a fixed groove. *See* FOUR-BAR LINKAGE.

A commonly occurring variation of the four-bar linkage is the linkage used in reciprocating engines (Fig. 3). Slider C is the piston in a cylinder, link 3 is the connecting rod, and link 4 is the crank.

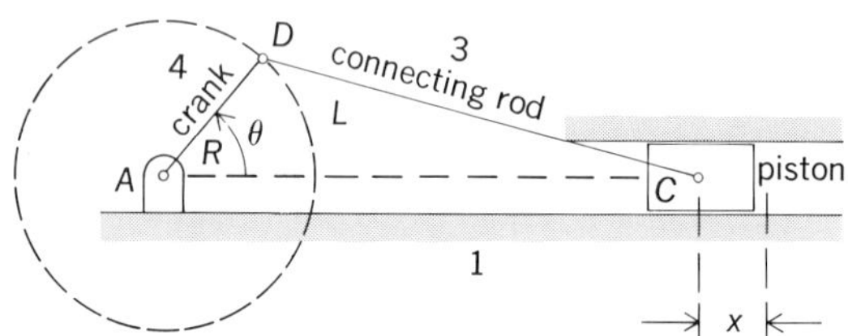

Fig. 3. Slider crank mechanism.

This mechanism transforms a linear into a circular motion, or vice versa. The straight slider in line with the crank center is equivalent to a pivot at the end of an infinitely long link. Let R be the length of the crank, L the length of the connecting rod, while θ denotes the angle of the crank as shown in Fig. 3. Also, let x denote the coordinate of the pivot C measured so that $x=0$ when $\theta=0$. Length L normally dominates radius R, whence the approximate relation in Eq. (1) holds.

$$x=R(1-\cos\theta)+\frac{R^2}{2L}\sin^2\theta \qquad (1)$$

A Scotch yoke is employed to convert a steady rotation into a simple harmonic motion (Fig. 4). The angle of the crank 2 with respect to the horizontal is denoted by θ, and the coordinate of the

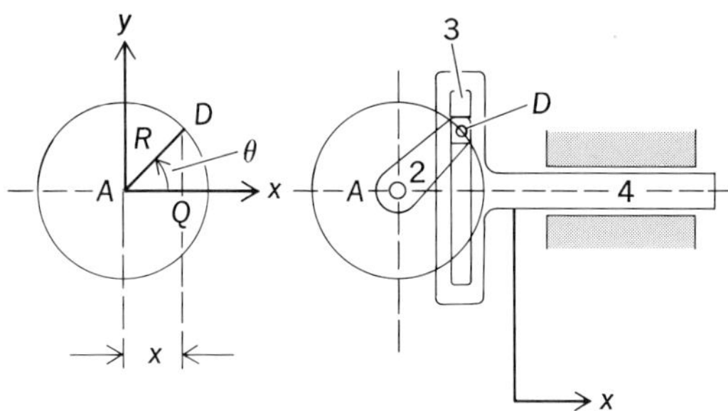

Fig. 4. Scotch yoke.

slider 4 by x. If R again denotes the length of the crank, Eq. (2) holds. In computer mechanisms, slider 3 can be moved along the crank so that the component of a vector can be obtained mechanically. *See* SLIDER-CRANK MECHANISM.

$$x = R\cos\theta \qquad (2)$$

A lever is normally understood to be a bar connected to three other links. A lever is often used for addition (Fig. 5). Point B is at the center of lever AC. It is assumed that x and y are numerically small compared to the length of the lever. Levers with a fixed pivot (fulcrum) are particularly useful for amplifying or attenuating linear displacements and forces. *See* LEVER.

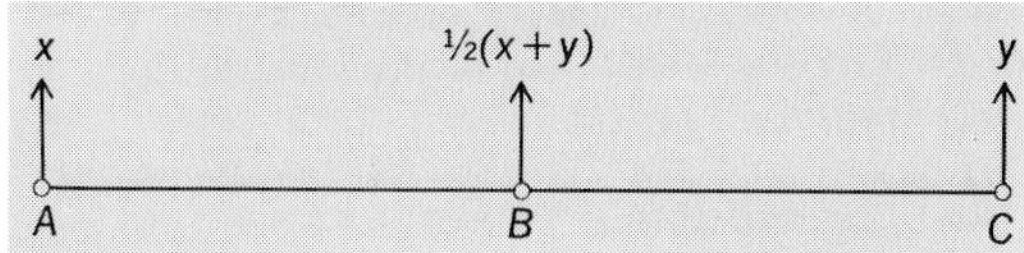

Fig. 5. Addition with a lever.

The pantograph shown in Fig. 6 is a five-link mechanism employed in drafting to magnify or reduce diagrams. As point Q on link 5 describes a curve, point P on link 4 will describe a similar but enlarged curve. Here $ABCD$ is a parallelogram. Pivot A is fixed. *See* PANTOGRAPH.

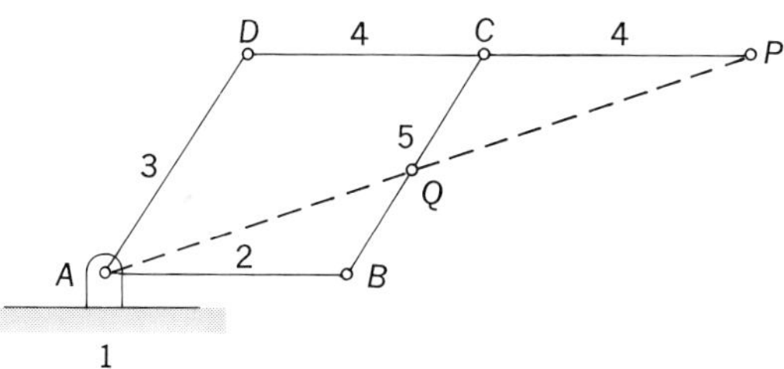

Fig. 6. Pantograph.

A mechanism composed of six links, introduced by Joseph Whitworth (1803–1887), makes the return stroke of a body in oscillatory linear motion faster than the forward stroke. Universal joints for transmitting motion between intersecting axes and escapements permitting rotation about an axis in one direction only are examples of the many kinds of linkages used in machines. However, complicated linkages are becoming obsolete because of the ease with which information and power can be transmitted by electrical, hydraulic, and pneumatic means. *See* MECHANISM.

[RUFUS OLDENBURGER]

Bibliography: N. P. Chironis, *Mechanisms, Linkages, and Mechanical Controls*, 1965; R. S. Hartenburg and J. Denavit, *Kinematic Synthesis of Linkages*, 1964; H. H. Mabie and F. W. Ocvirk, *Mechanisms and Dynamics of Machinery*, 3d ed., 1975; J. E. Shigley and J. J. Viker, *Theory of Machines and Mechanisms*, 1980; C. H. Suh and C. W. Radcliffe, *Kinematics and Mechanisms Design*, 1978.

Loads, dynamic

A force exerted by a moving body on a resisting member is called a dynamic or energy load. Loads suddenly applied with appreciable striking velocity, as when a falling weight strikes another body, are called impact loads. Such loads are produced by moving machine parts, cars on a bridge, airplane landing shock, falling weights, and other nonstationary loading conditions where forces are applied in relatively short time intervals. Dynamic load is expressed in terms of the amount of energy transferred or by an equivalent static load which has the same stress-producing effect on the member.

Impact or load factor is the measure of the increased static load required to produce the same stress as the dynamically applied load. It is thus the ratio of the equivalent static load to the weight of the moving body. Impact affects both the magnitude of the stress and the properties of the material. Increased strain rates associated with impact increase the elastic and ultimate strength.

Elastic behavior. If a material or structure has sufficient elastic energy capacity as a resisting member to absorb completely the energy load through elastic action and recover completely from the deformation, it behaves elastically. Elastic stresses associated with stored strain energy depend on the kind of deformation produced. Part of the kinetic energy is dissipated by deformation of connections, friction, and inertia effects. *See* SHOCK ABSORBER.

Dynamic stresses and deformations can be found by equating externally applied energy to the internal strain energy expressed in terms of stresses and dimensions. Dynamic force produced by weight W falling through height h onto a beam or spring is approximately indicated in the equation shown below, where Δ_{st} is elastic deflection due to gradual application, and c is an inertia-correcting factor. Stresses and deflection are found by expressions of the same form.

$$P_{\text{dyn}} = W + W\sqrt{1 + \frac{2h}{\Delta_{st}}\,(c)}$$

Inelastic behavior. An energy load exceeding the elastic energy capacity of the member produces inelastic deformation. For a known amount of energy, the dynamic load and deformation can be found approximately from the boundaries of the area under a statically determined load deformation curve, which represents absorbed energy. Similarly, the inelastic stress produced by impact or the dimensions necessary to limit the stress can be found from areas in the static stress-strain curve.

Toughness is the energy absorbed per unit volume when the material is stressed to fracture, and is represented by the total area under the stress-strain diagram. Ability to store strain energy is increased by uniformity of stress and ductility. Capacity to dissipate energy overloads by plastic deformations is desirable in members subject to shock or impact. *See* STRESS AND STRAIN.

[W. J. KREFELD/W. G. BOWMAN]

Loads, repeated

Forces reapplied many times and causing varying stresses as the load changes. Repeated loads exist in crankshafts, axles, springs, piston rods, rails, bridge members, and many other machine and structural elements. Repeated loads considerably smaller than similarly applied static loads cause failure by progressive fracture. Designs for repeated loads depend on the maximum value of repeated stress which the material can resist, called the fatigue strength, and the magnitude of localized stress concentrations. *See* STRESS CONCENTRATION.

LOADS, REPEATED

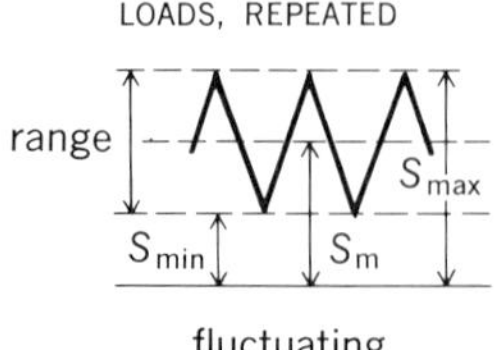

fluctuating

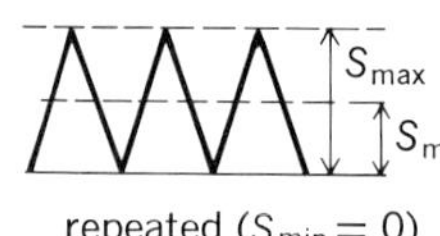

repeated ($S_{min} = 0$)

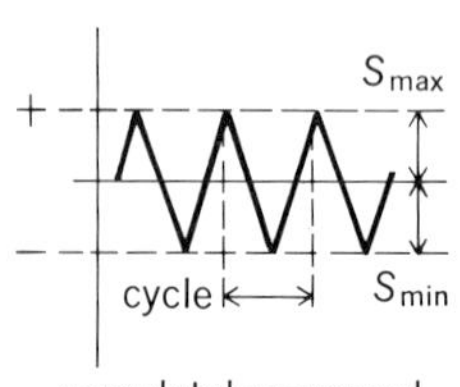

completely reversed ($+S_{max} = -S_{min}$)

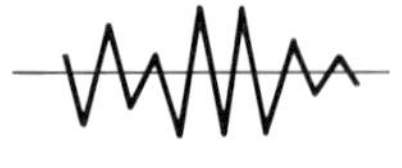
random reversed (range and stress varies)

Types of repeated stress.

Cycle of stress. The variation of stress S during one typical fluctuation when the load is repeatedly applied and removed or when the load fluctuates continuously from tension to compression is a stress cycle (as illustrated). A fluctuating or pulsating stress varies between maximum and minimum values of the same sign; this condition is called repeated if the stress reduces to zero in each cycle. Reversed stress varies between values of opposite signs and is partly reversed when the opposite stresses are unequal, and completely reversed when the opposite stresses are equal.

Stress variation is described either by the maximum stress, its type (tension or compression), and the range of fluctuations, or by the mean or steady stress (S_m in illustration) together with the magnitude of the superimposed alternating stress producing the cyclic variation. Repeated loads superimposed on a constant load, such as dead weight of the structure, may produce fluctuation without reversal. Engine crankshafts and parts, rails, and axles are subjected to the reversed-type cyclic loading, which may involve billions of repetitions in the life of the member.

Stress raisers. Dimensional changes, internal discontinuities, or other conditions causing localized disturbance of the normal stress distribution, with increase of stress over that normally produced, are termed stress raisers. The maximum stress is called a stress concentration. High local stresses exist at bolt threads, abrupt change of shaft diameter, notches, holes, keyways, or fillet wells. Internal discontinuities, such as blowholes, inclusions, seams or cracks, ducts and knots in wood, voids in concrete, and variable stiffness of component constituents, cause stress concentrations. At points of contact of ball or roller bearings, gear teeth, or other local load applications, the stress may be greatly increased. Maximum stress is obtained by multiplying the nominal stress, computed without regard to the modifying effect of the stress raiser, by a stress concentration factor. This factor, found experimentally, depends on the type of discontinuity and the properties of the material under static load. Ductile material relieves most of the stress concentration by plastic yielding. Stress raisers contribute to failure under high stress rates and low temperatures, which tend to inhibit plastic flow, and are of great importance under repeated loads, which produce progressive fracture, called fatigue failure.

[JOHN B. SCALZI]

Loads, transverse

Forces applied perpendicularly to the longitudinal axis of a member. Transverse loading causes the member to bend and deflect from its original position, with internal tensile and compressive strains accompanying change in curvature.

Concentrated loads are applied over areas or lengths which are relatively small compared with the dimensions of the supporting member. Often a single resultant force is used to analyze effects on the member. Examples are a heavy machine occupying limited floor area, wheel loads on a rail, or a tie rod attached locally. Loads may be stationary or they may be moving, as with the carriage of a crane hoist or with truck wheels.

Distributed loads are forces applied continuously over large areas with uniform or nonuniform intensity. Closely stacked contents on warehouse floors, snow, or wind pressures are considered to be uniform loads. An equivalent uniform load may represent multiple closely spaced, concentrated loads. Variably distributed load intensities include foundation soil pressures and hydrostatic pressures.

Bending and shear. Transverse forces produce bending moments and transverse (shearing) forces at every section which must be balanced by an internal couple or resisting moment M and an internal shear force V (Fig. 1*a*). These component forces are evaluated by free-body analysis. The bending moment and resultant shear force can be expressed analytically as functions of the loads and the distance locating the section from a reference origin. The magnitude and distribution of internal stresses associated with the external

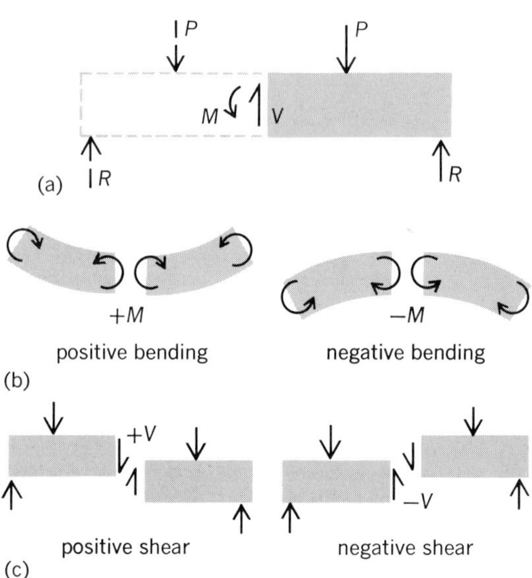

Fig. 1. Beam loading analysis. (*a*) By free-body analysis of beam. (*b*) Conventions of sign for bending moment. (*c*) Conventions of sign for shear.

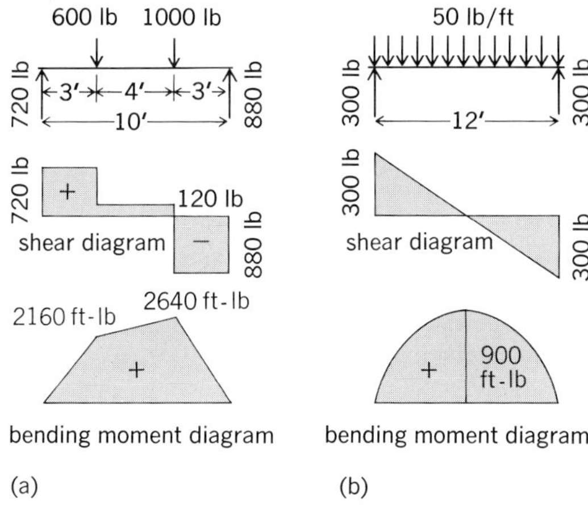

Fig. 2. Typical shear and bending moment diagrams for (*a*) concentrated and (*b*) uniformly distributed loads. 1 ft = 0.3048 m; 1 lb = 4.45 N; 1 ft-lb = 1.36 N-m.

force system are calculated by the theory of flexure.

Bending moment diagram. A graphical representation indicates the bending moment produced at all sections of a member by a specified loading. The bending moment is the sum of the moments of external forces acting to the left of the section, taken about the section. The sign of the curvature is related to the sign of the bending moment (Fig. 1*b*).

Typical bending moment and shear diagrams are shown in Fig. 2 for concentrated and uniformly distributed loads on a simply supported beam. The reactions and moments are calculated for equilibrium. *See* STATICS.

Shear diagram. A graphic representation of the transverse (shearing) force at all sections of a beam produced by specified loading is called a shear diagram. The shear at any section is equal to the sum of the forces on a segment to the left of the section considered (Fig. 1*c*). Positive shears are produced by upward forces (Fig. 2).

Relationships between shear and bending moment that assist in construction of diagrams are: (1) maximum moment occurs where shear is zero; (2) area in the shear diagram equals change of bending moment between sections; (3) ordinates in the shear diagram equal slope of moment diagram; (4) shear is constant between concentrated loads and has constant slope for uniformly distributed loads. Combined loading can be represented by conventional composite moment diagrams or by separate diagrams, called diagrams by parts.

Beams. Members subjected to bending by transverse loads are classed as beams. The span is the unsupported length. Beams may have single or multiple spans and are classified according to type of support, which may permit freedom of rotation or furnish restraint (Fig. 3).

The degree of restraint at supports determines the stresses, curvature, and deflection. A beam is statically determinate when all reaction components can be evaluated by the equations of statics. Two equations are available for transverse loads and only two reaction components can be found. Fixed-end and continuous beams are statically indeterminate, and additional load deformation relationships are required to supplement statics.

Bending stresses are the internal tensile and compressive longitudinal stresses developed in response to curvature induced by external loads. Their magnitudes depend on the bending moment and the properties of the section. The theory of flexure assumes elastic behavior. No stresses act along a longitudinal plane surface within the beam called the neutral surface. A segment subjected to constant bending moment, as when bent by end couples, is in pure bending, and stresses vary linearly across the section which remains plane. Simultaneous shear causes warping and nonlinear stress distribution. The stress is maximum at boundary surfaces of sections where bending moment M is greatest. Maximum stress S at distance c from the neutral axis to the extreme element of a section having a moment of inertia I, is $S = Mc/I = M/Z$, where $Z = I/c$ is called the section modulus, a measure of the section depending upon shape and dimensions. Greatest economy results when the section provides the required section modulus with least area. The common theory of flexure applies only to elastic stresses. Stresses exceeding the elastic limit involve plastic strains producing permanent deflections upon load removal.

Inelastic stresses are first produced at surface elements of sections resisting maximum moment. A distinct yield point in materials, such as structural grade steel, causes stresses near the surface to remain constant, while interior stresses increase with increasing load. Redistribution of stress continues until the entire section behaves plastically. The fully plastic resisting moment of a standard rolled, wide-flange section is about 12% greater than the moment just producing yield point at the surface. Where small increases in deflection can be tolerated, this plastic-hinge moment is used in design.

Shearing stresses are intensities of the shear force distributed over a beam cross section. At any point in the beam, the vertical and longitudinal shear stresses are equal. They are zero at the boundary surfaces and maximum at the neutral axis. The shear stress according to the common theory of flexure is $S_s = VQ/Ib$, where V is total shear force, Q is statical moment about the neutral axis of that area of the beam section between the level of investigation and the top or bottom of the section, I is moment of inertia, and b is width. In

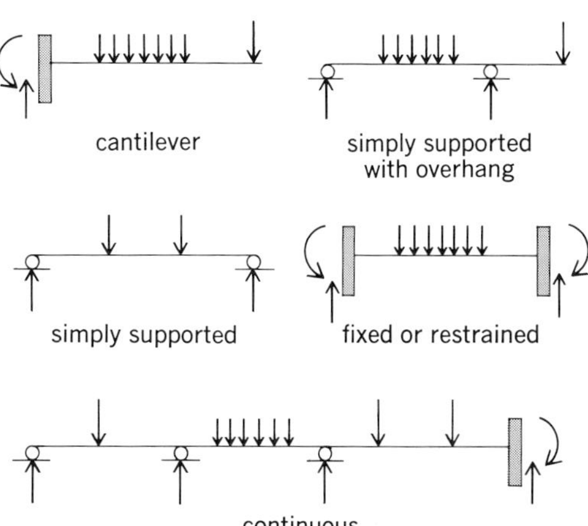

Fig. 3. Types of beams.

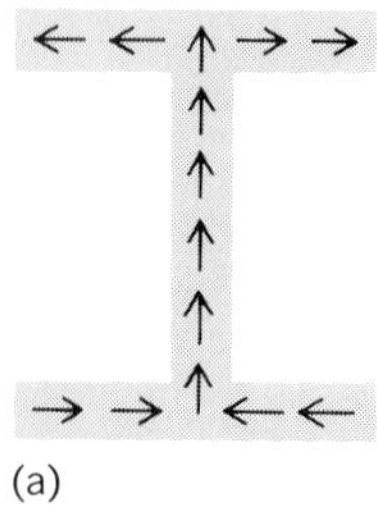

(a)

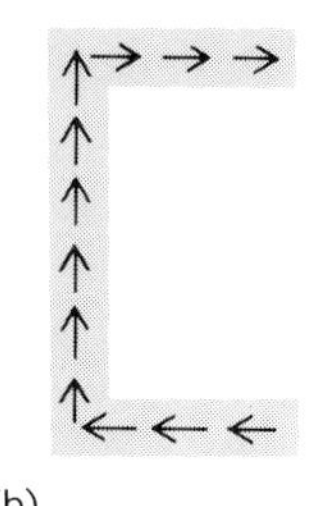

(b)

Fig. 4. Shear flow in (*a*) a wide-flange section and (*b*) a channel section.

rectangular sections the distribution is parabolic, and maximum S_s is one and one-half times the average stress.

Shear flow is the shear force per unit length acting along component areas of the section. The concept of flow is derived from the distribution of unit shear forces represented by vectors pointing in the same direction along the median line of a thin element, and the similarity of the expression for unit shear to the equation for quantity of liquid flowing in a pipe or channel. The direction of unit shear forces acting in typical structural shapes is shown in Fig. 4. *See* TORSION.

Deflection. A beam is said to undergo deflection when the displacement is normal to the original unloaded position and is due to curvature produced by loads. Deflections may be limited to less than structurally safe values in buildings to avoid cracking the ceiling or in machines because of necessary clearances.

Elastic curve is the curved shape of the longitudinal centroidal surface when loads produce elastic stresses only. Deflection at any point is calculated from the equation of this curve. Shear forces contribute additional deflection.

Radius of curvature depends on the moment and stiffness of the beam and is constant, with bending to a circular arc, when the moment is constant. This is simple bending. Curvature varies under variable moment.

Statically indeterminate beam. In the presence of more reaction components than can be determined by the equations of statics alone, a beam is statically indeterminate. The first step in analysis is to find the reactions. Equations of statics are supplemented by additional equations relating external forces to slope and deflection.

The curvature and accompanying stresses depend on the loads and the restraint imposed by the supports. In a simply supported beam, only vertical displacement is prevented at the supports, whereas when the beam is fixed, rotation is also prevented at the ends. Similarly, beams with overhanging ends and beams continuous over multiple supports are partially restrained. Introduction of the boundary conditions describing the conditions imposed by the supports in the solution of the basic differential equation for the elastic curve furnishes relations between the loads, reaction components, and properties of the beam. For elastic behavior these relations, together with the equations of statics, evaluate the indeterminate reactions. Inelastic behavior is caused by plastic deformations at critical sections of high flexural stress when the yield point of the material is exceeded. When a fully plastic condition is reached, the section acts as a plastic hinge, with constant resisting moment during unrestrained rotation. These known moments at supports or intermediate points permit determination of the reactions by the equations of statics. Advantage is taken of this inelastic action in plastic design.

Unsymmetrical bending. Loads that produce bending moments in planes oblique to the principal axes of a section cause unsymmetrical bending. Components of the applied moment produce curvature and deflection in planes perpendicular to these bending axes; the stresses produced by these components acting separately can be superimposed. To avoid torsional twist, the oblique forces must pass through either the geometric center of symmetrical sections or a particular point called the shear center for unsymmetrical sections.

Lateral buckling of beams is caused by instability of the compression flange, which results in lateral deflection and twist of the section. The action corresponds to buckling of long columns. The critical compressive stress producing buckling depends on the sectional dimensions and type of loading, which in turn determine the flange stresses and torsional rigidity of the section. Beams having small moments of inertia about the weak axis with low torsional resistance are susceptible to buckling. For rolled beams, buckling resistance is related approximately to the Ld/A_f ratio, where L is span, d is beam depth, and A_f is the area of the flange.

Buckling of beam webs is caused by excessive diagonal compression induced by the shear stresses in the web or by the combined shear and bending stresses in the web acting as an edge-supported plate. Because the web is relatively thin, it tends to buckle laterally in a series of waves. For deep beams the shear stress must be reduced by increased web thickness or reinforcement in the form of stiffener angles or plates added to increase resistance to lateral buckling.

Variable cross sections make more efficient use of material in beams by providing resisting moment at all sections more nearly equal to the applied moment. If the section modulus varies directly as the bending moment, stress M/Z will be constant at all sections, and the beam is said to have constant strength. Dimensions of the section can be varied by tapering the width and depth or by adding cover plates to a built-up section to vary the section modulus. Common examples of beams with variable section are tapered cantilevers, plate girders with multiple cover plates, forgings, and other machine elements. A leaf spring is a tapered beam in which uniform stress increases energy absorption.

Shear center. A point on a line parallel to the axis of a beam through which any transverse force must be applied to avoid twisting of the section is

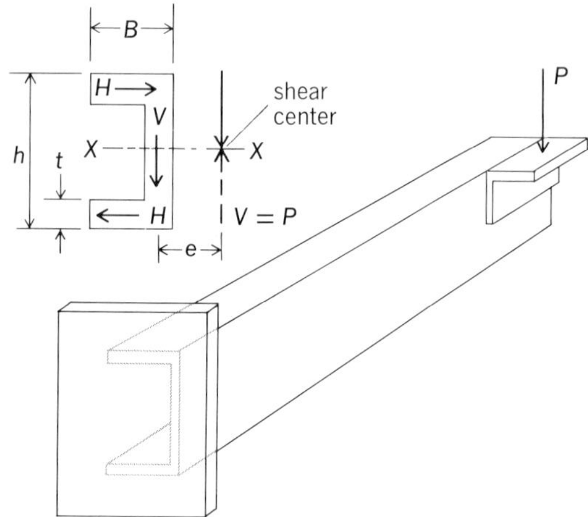

Fig. 5. Shear center under *P* at distance *e* from centerline of web, on symmetrical axis.

called the center of twist or shear center. A beam section twists when the resultant of the internal shearing forces is not in the same plane as the externally applied shear. The shear center can be determined by locating the line of action of the resultant of the internal shear forces. A channel section has one axis of symmetry and is subject to twist unless the external shear passes through a point located e from the center of the web so that $e=B^2h^2t/4I_x$, approximately, when the flange and web thicknesses are small, where B is flange width, h is depth of section, t is thickness, and I_x is moment of inertia about the symmetrical axis (Fig. 5).

Beams on elastic foundation are members subjected to transverse loads while resting either on a continuous supporting foundation or on closely spaced supports which behave elastically. The curvature, deflection, and bending moment depend upon the relative stiffness of the beam and supporting foundation. Beams of this type include railroad rails on ties, a timber resting on level ground, a long pipe supported by closely spaced hanger rods or springs, and a concrete footing on a soil foundation. The unknown reactive force per unit length of beam is assumed to be proportional to the deflection, and the solution of this statically indeterminate problem requires derivation of the equation of the elastic curve whose differential equation is $EI(d^4y/dx^4)=-ky$, where k is the spring constant per unit length of support and ky is the elastic force exerted per unit length of beam. Solution of this equation determines the distribution of reactive forces from which the shear and moments can be calculated. *See* STRENGTH OF MATERIALS.

[JOHN B. SCALZI]

Location fit

Mechanical sizes of mating parts such that, when assembled, they are accurately positioned in relation to each other. Locational fits are intended only to determine the orientation of the parts. For normally stationary parts that require ease of assembly or disassembly, for parts that fit snugly, and for parts that move yet fit closely as in spigots, slight clearance is provided between parts. Where accuracy of location is important, transition fits are used. In these fits the holes and shafts are normally nearly the same diameter. For greater accuracy of location, the shafts are made slightly larger than the holes; such fits are termed location interference fits. *See* ALLOWANCE. [PAUL H. BLACK]

Lubricant

A gas, liquid, or solid used to prevent contact of parts in relative motion, and thereby reduce friction and wear. In many machines, cooling by the lubricant is equally important. The lubricant may also be called upon to prevent rusting and the deposition of solids on close-fitting parts.

Liquid hydrocarbons are the most commonly used lubricants because they are inexpensive, easily applied, and good coolants. In most cases, petroleum fractions are applicable, but for special conditions such as extremes of temperature, select synthetic liquids may be used. At very high temperatures, or in places where renewal of liquid lubricants is impossible, solid lubricants (graphite or molybdenum disulfide) may be used.

Petroleum lubricants. Crude petroleum is an excellent source of lubricants because a very wide range of suitable liquids, varying in molecular weight from 150 to over 1000 and in viscosity from light machine oils to heavy gear oils, can be produced by various refining processes. Studies carried out by the American Petroleum Institute and others have established that crude petroleum fractions consist of saturates (normal, iso-, and cycloparaffins); monoaromatics which may contain saturated rings as well as saturated side chains; substituted polyaromatics; hetero compounds containing sulfur, nitrogen, and oxygen; and asphaltic material made up of polycondensed aromatics and hetero compounds. Of these, the wax-free saturates and monoaromatics are desired in the finished oil in order to obtain the desired viscometric properties and oxidation and thermal stability. So far as possible, the other types of compounds are generally removed. Classically, only those crudes (so-called paraffinic crudes) relatively rich in the desired type of hydrocarbons were used in lubricating oil manufacture. By atmospheric distillation, neutrals were produced as distillates and the residual fractions were treated with activated clays, which removed asphaltic and hetero material to produce bright stock.

In modern refining, vacuum distillation removes the desired hydrocarbons from the asphaltic constituents (which may be present in amounts up to 40% of the total) and gives oils in the required viscosity and boiling range. Extraction of the distilled fractions with solvents such as liquid sulfur dioxide, furfural, and phenol permits the removal of the polyaromatic and hetero compounds to improve viscosity-temperature characteristics (viscosity index, VI) and stability of the oil. Dewaxing removes the high-melting paraffins.

If the viscosity index of the oil is not important for a particular application, the most reactive polyaromatics and hetero compounds may be removed by treatment with concentrated sulfuric acid. In either case, the oil is treated finally with an active clay to remove trace amounts of residual acids and resins (heterocyclics).

Small amounts of heterocyclics, such as substituted benzthiophenes, quinolines, and indoles, may remain in the finished oil. The sulfur-containing compounds serve the useful purpose of acting as natural antioxidants. The nitrogen compounds may be harmful in certain applications because of their propensity to form deposits on hot surfaces and are generally reduced to low concentrations.

The table lists typical specification data for lubricants in several applications. Viscosity is a determining factor in lubricant selection. Machines are generally designed to operate on the lowest practicable viscosity, since the lighter fluids give lower friction and better heat-transfer rates. However, if loads or temperatures are high, more viscous and less volatile lubricants are required. Change of viscosity with temperature is often of considerable practical importance and is customarily expressed in terms of viscosity index, an arbitrarily chosen scale on which an oil from a Pennsylvania crude, high in saturate and monoaromatic content, is assigned a value of 100, and those containing a relatively large amount of cyclohydrocarbons, both paraffinic and aromatic (from

Viscosity of oils for various applications

Application	Viscosity in centistokes at 25°C (77°F)	Primary function
Engine oils		
SAE 10W	60–90	Lubricate piston rings, cylinders, valve gear, bearings; cool piston; prevent deposition on metal surfaces
SAE 20	90–180	
SAE 30	180–280	
SAE 40	280–450	
SAE 50	450–800	
Gear oils		Prevent metal contact and wear of spur gears, hypoid gears, worm gears; cool gear cases
SAE 80	100–400	
SAE 90	400–1000	
SAE 140	1000–2200	
Aviation engine oils	220–700	Same as engine oils
Torque converter fluid	80–140	Lubricate, transmit power
Hydraulic brake fluid	35	Transmit power
Refrigerator oils	30–260	Lubricate compressor pump
Steam-turbine oil	55–300	Lubricate reduction gearing, cool
Steam cylinder oil	1500–3300	Lubricate in presence of steam at high temperatures

naphthenic crudes) are in the 0–50 viscosity-index range. As already described, the refiner can produce oils of high viscosity index from naphthenic stocks, but yields may be low.

Multigrade oils. In order to standardize on nomenclature for oils of differing viscosity, the Society of Automotive Engineers (SAE) has established viscosity ranges for the various SAE designations (see table). By the use of relatively large amounts of additives for improving viscosity index, it is possible to formulate one oil which will fall within the range of more than one SAE viscosity grade, the so-called multigrade oils, illustrated by Fig. 1. Since all viscosity-index improvers also increase viscosity, it is necessary to use a base oil of low viscosity to formulate such oils. Since the viscosity-increasing effect of the additive decreases with increase in shear rate, an artificially thickened oil of this type behaves as a light oil in engine parts under high shear, and thereby friction is low, but acts as a heavier oil in low-shear regions or at higher temperatures. However, volatility and ability to protect high shear parts from rubbing set limits to the use of light oils in these formulations.

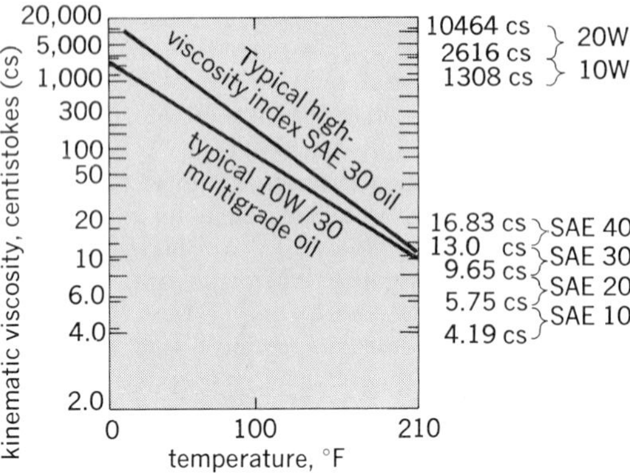

Fig. 1. Viscosity-temperature relationships indicated for additive-thickened multigrade oils.

Additives for lubricating oils. It is often desirable to add various chemicals to lubricating oils to improve their physical properties or to obtain some needed improvement in performance.

Viscosity-index improvers. The fall in viscosity with increase in temperature of oils of a given grade, or viscosity level, can be made less steep by thickening lighter oils with polymeric substances such as polybutenes and copolymers of polymethacrylates. The polymer may increase in solubility as the temperature increases, and correspondingly, the molecules uncoil and thicken the base oil more at high than at low temperatures and thus counteract, to some extent, the natural decrease in viscosity the base oil would undergo (Fig. 1).

Pour-point depressants. The dewaxing process removes the higher melting hydrocarbons, but some components remain which may solidify and gel the oil, and thus reduce its fluidity at low temperatures. Small amounts of chemicals, such as metallic soaps, condensation products of chlorinated wax and alkyl naphthalenes or phenols, polymethacrylates, and a host of others, increase fluidity at low temperatures. The mechanism of their action is still uncertain, but it probably involves adsorption of additive molecules on the surface of the growing wax crystals. The adsorbed layer either reduces intercrystal forces or modifies crystal growth.

Antioxidants. Lubricants are exposed to oxidation by atmospheric oxygen in practically all of their applications. This results in formation of acids and sludges which interfere with the primary function of the lubricant. Substantial increases in service life can be obtained by using small amounts (0.1–1.0%) of antioxidants. For lighter, more highly refined oils such as steam turbine oils, hindered phenols such as dibutyl-*p*-cresol are very effective. In heavy lubricants, as engine oils, amines (phenyl-α-naphthylamine), metal phenates (alkali-earth salts of phenol disulfides), and zinc salts of thiophosphates and carbamates are used.

Antiwear and friction-reducing additives. If sliding surfaces can be completely separated by an oil film, friction and wear will be at a minimum. There are many systems, however, in which the combination of component geometry and operating conditions is such that a continuous oil film cannot be maintained. If no separating film of any kind were interposed, however, complete seizure would result. If the pressures and temperatures between such contacting surfaces are moderate, the provision of a boundary lubricating film will suffice, whereas if, as in some gears, conditions of both temperature and pressure are severe, some form of extreme pressure (EP) lubrication may be necessary.

The form in which wear manifests itself in machine components varies widely with the conditions, from the catastrophic welding together of gear-tooth surfaces to the slow continuous removal of material from an engine cylinder; even within the same mechanical system, it may change profoundly with changes in operating and environmental conditions.

In internal-combustion engines operating at low cylinder temperatures, for instance, condensation of acids from the gaseous combustion products on the cylinder and ring surfaces results in corrosive wear; in such cases, the addition of alkaline-earth phenates to the lubricating oil to neutralize the acids has succeeded in reducing wear rates.

Extreme pressure additives. Certain types of gears, particularly the hypoids used in automotive rear-axle transmissions, operate under such severe conditions of load and sliding speed, with resulting high temperature and pressure, that ordinary lubricants cannot provide complete protection against metal contact; this leads to welding, transfer of surface metal, and ultimate destruction of the gears. Also, in certain machining operations, it is necessary to prevent the chip from welding to the cutting tool. For such applications, lubricants containing sulfur and chlorine compounds are used. At the temperatures developed in the contact, these react chemically with the metal surfaces; the resulting sulfide and chloride films provide penetration-resistant, low-shear-strength films which prevent damage to the surfaces. Care in formulating these lubricants must, however, be exercised to ensure that corrosion of metal at normal temperatures does not occur. Figure 2 shows the damaging results which may occur in the absence of such protective additives.

Dispersants. In internal-combustion engines, some of the products of combustion, which include carbonaceous particles, partially burnt fuel, sulfur acids, and water, enter the oil film on the cylinder walls. These materials may react to form lacquers and sludges which are deposited on the working parts of the engine. The lubricant is called upon to keep engine parts (oil screens, ring grooves) clean, and this is accomplished by the use of so-called detergent oils (Fig. 3).

The most commonly used dispersants are alkaline-earth salts. Of these, the most successful have been the salts of petroleum sulfonic acids, phenates, salicylates, thiophosphates, and oxidized olefinphosphorus pentasulfide reaction products. These materials owe their effectiveness partly to their surface activity, which ensures that foreign particles are kept in suspension in the lubricating oil, and partly to their alkalinity, which enables them to neutralize combustion acids that would otherwise catalyze the formation of lacquers.

Polymeric additives have been introduced for dispersancy. They are copolymers of a long-chain methacrylate, for example, lauryl methacrylate, and a nitrogen-containing olefin, such as vinyl pyrrolidone. These nonash dispersants are superior to the metal-containing additives for low-temperature operation and can be used very effectively in motor oils. Because they are less alkaline than metallic additives, the polymeric additives are less effective in diesel engines whose fuels often lead to combustion products which are much more acid than those from gasoline.

Boundary lubricants. Boundary conditions are encountered in many metal-forming processes in which the pressures required to deform the metal are too high to allow an oil film to form. In such applications, fatty oils, such as palm oil, or lubricants containing fatty materials are employed to reduce the friction and wear; the fatty acids react with the metal surface to form a tenacious soap film which provides lubrication up to temperatures near the melting point of the soap, usually about 120°C. *See* MACHINABILITY OF METALS.

Synthetic oils. During World War II, synthetic lubricants were extensively employed by Germany as substitutes for mineral lubricants which were in short supply. Some of these, such as the ester oils, were in fact superior to mineral lubricants in some applications. Since the war, the use of synthetics for special applications, where their performance justifies the higher costs, has steadily increased. Their main advantage is that they have a greater operating range than a mineral oil (Fig. 4). Esters such as 2-ethylhexyl sebacate, containing oxidation inhibitors and sometimes mild extreme pressure additives, are used as lubricants for aircraft jet engines. Silicones, although ideal with regard to thermal stability and viscosity characteristics, are poor lubricants for steel on steel, and this has restricted their use, although they are invaluable in some applications. Another class of widely used synthetic oils are the polyglycols, such as polypropylene and ethylene oxides. These polymers are available in a wide molecular-weight range, and they vary considerably in solubility in water and hydrocarbons. Thus, water-base lubricants may be formed which are used when a fire hazard exists.

Solid lubricants. The most useful solid lubricants are those with a layer structure in which the molecular platelets will readily slide over each other. Graphite, molybdenum disulfide, talc, and boron nitride possess this property.

The principal difficulty encountered with the use of solid lubricants is that of maintaining an adequate lubricant layer between the sliding metal surfaces. If the solid lubricant is applied as a suspension in a fluid, there is a tendency for the particles to settle out, and they may not reach the region where they are required. If they are applied as a thick paste to overcome the tendency to settle out, it is frequently difficult to force the paste through the narrow clearances between the sliding surfaces. A third method is to pretreat the surfaces with a relatively thick film of solid lubricant suspended in a resin and bake it on. Difficulties may then arise through progressively increasing clearances as the film wears, and there is no method for its continuous renewal. Materials such as graphite and molybdenum disulfide oxidize quite rapidly in air at 400°C. Unless air can be excluded, therefore, alternative solid lubricants will be required at temperatures greatly in excess of this.

A unique type of solid lubricant is provided by the plastic polytetrafluoroethylene (PTFE). At low speeds, PTFE slides on itself or on metals with a coefficient of friction in the lowest range observed for boundary lubrication (about 0.05). The friction rises to values common for other plastics after the sliding speed passes a critical value, and this high friction tends to persist when returning to low speeds.

It is likely that this is a thermal effect, the incidence of which is promoted by the extremely low thermal conductivity of PTFE, and it can be combated to a considerable extent by using the PTFE in a copper matrix. The use of such a matrix also

LUBRICANT

new gear

scoring

plowing

Fig. 2. Gear failures.

LUBRICANT

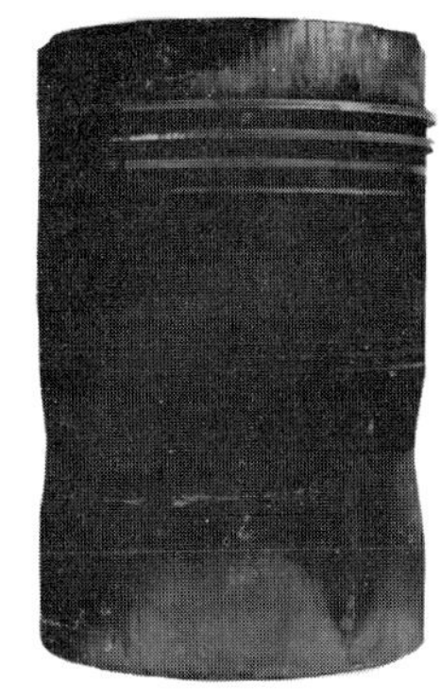

(a)

(b)

Fig. 3. Effect of dispersant on cleanliness. (*a*) Ordinary oil. (*b*) Oil containing dispersant additive.

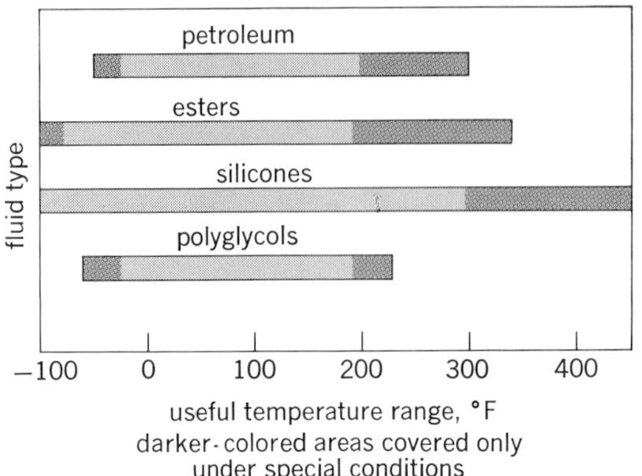

Fig. 4. Useful temperature range of synthetic oils.

reduces the wear which would occur with PTFE alone. This type of structure is used as a self-lubricated bearing material.

Greases. A lubricating grease is a solid or semifluid lubricant comprising a thickening (or gelling) agent in a liquid lubricant. Other ingredients imparting special properties may be included. An important property of a grease is its solid nature; it has a yield value. This enables grease to retain itself in a bearing assembly without the aid of expensive seals, to provide its own seal against the ingress of moisture and dirt, and to remain on vertical surfaces and protect against moisture corrosion, especially during shut-down periods.

The gelling agents used in most greases are fatty acid soaps of lithium, sodium, calcium, and aluminum. Potassium, barium, and lead soaps are used occasionally. The fatty acids employed are oleic, palmitic, stearic, and other carboxylic acids derived from animal, fish, and vegetable oils. The use of soaps in greases is limited by phase changes (or melting points) that occur in the range 50–200°C. Because many applications require operation above this temperature, greases have been introduced which are thickened by high-melting solids such as clays, silica, organic dyes, aromatic amides, and urea derivatives. In these cases, the upper limit of application is set by the volatility or by the oxidation and thermal stability of the base oil. To extend the operating range to extremes of high and low temperatures, synthetic oils may be used as the substrate. Silicones, esters, and fluorocarbons have all been employed for this purpose. As in the case of liquid lubricants, various additives such as oxidation inhibitors, EP agents, and antirust additives are frequently used in greases. Volatility of the base oil is particularly important in greases, because it is constantly exposed as a thin film on the bearing surfaces.

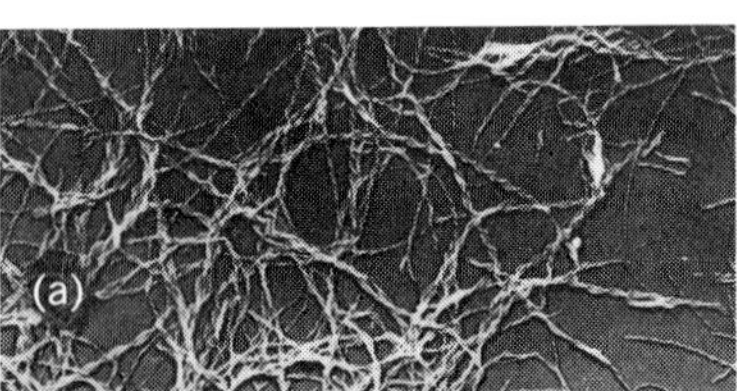

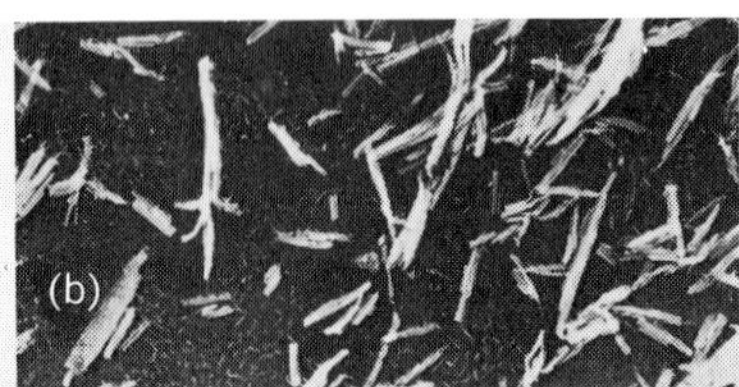

Fig. 5. Electron micrographs showing (*a*) soap-thickened grease and (*b*) dye-thickened grease.

Structure of greases. The gelling agent in most soap-base greases is present as crystalline fibers having lengths of 1–100 μm and diameters 1/10–1/100 of their lengths. The fibers thicken the oil by forming a network or brush-heap structure in which the oil is held by capillary forces. For a given concentration of soap, the larger the length-to-diameter ratio of the fibers, the greater is the probability of network junction formation and the harder is the grease. However, not all soaps form such crystalline fibers; aluminum soap is a notable exception. There is evidence that aluminum soap molecules form a network akin to that formed by polymer molecules.

Nonsoap thickeners are present as small isometric particles much smaller than 1 μm in linear size. Silica particles are round, and some clay particles are plates. Forces of interaction between silica, clay particles, and oil are easily destroyed by water, so it is necessary to waterproof the primary particles by special treatments. Electron micrographs of various types of greases are shown in Fig. 5.

Mechanical properties of greases. When the applied stress exceeds the yield stress, the grease flows and the viscosity falls rapidly with further increase of stress until it reaches a value only a little higher than that of the base oil. This fall in viscosity is largely reversible, since it is caused by the rupture of network junctions which, following the release of stress, can reform. However, continued vigorous shearing usually results in the rupture of fibers and permanent softening of the grease. Here, certain greases such as those derived from lithium salts of hydroxy acids are superior to other types, although the manner of grease preparation exercises considerable influence.

In a major grease application, for example, ball and roller bearings, shearing between the races and rolling elements is extremely severe. The reason that a good bearing grease does not soften and run out lies in the fact that only a very small fraction of the grease put in the bearing is actually subjected to shearing. As soon as a freshly packed bearing is set in motion, most of the grease is redistributed to places where it can remain static in the cover plate recesses and attached to the bore of the cage between the balls or rollers. The small amount of grease remaining on the working surfaces of the bearing is quickly broken down to a soft oily material which lubricates these parts, and because it has only a low viscosity, it develops very little friction.

A number of methods exist for the measurement of yield stress and viscosity at various temperatures. One of the oldest of grease tests, still universally used, is a consistency (hardness) test in which the depth of penetration of a cone of standard weight and dimensions is determined. However, few of the many tests used are helpful in predicting the performance of a grease. Consequently, in development work, recourse is made to actual performance data and to the use of rigs that simulate field conditions. *See* ANTIFRICTION BEARING.

[ROBERT G. LARSEN]

Bibliography: C. J. Boner, *Gear and Transmission Lubricants*, 1964, reprint 1971; E. R. Braithwaite, *Solid Lubricants and Surfaces*, 1963;

H. M. Drew, *Metal-Based Lubricant Compositions*, 1975; M. W. Ranney, *Lubricant Additives: Recent Developments*, 1978; H. H. Zuidema, *Performance of Lubricating Oils*, 2d ed., 1959.

Machinability of metals

The ease and economy with which a metal may be cut under average conditions. Frequently no truly quantitative assessment is made, but rather a rating or an index is established vis-à-vis a reference material. More quantitative comparisons are based on tool life. For example, maximum cutting speeds for a given tool life may be used as a rating of machinability. Alternatively, tool wear rate may be the basis for a machinability rating. Surface finish is sometimes used for assessing machinability.

Machining process. The wide range of metal-cutting processes may be represented, with some oversimplification, by the orthogonal cutting process (Fig. 1). A rectangular metal workpiece is machined by a tool with a face at a rake angle γ_c measured from the normal of the surface to be machined. A clearance angle Θ exists between the flank of the tool and the machined surface of the workpiece. The tool cuts into the workpiece to a depth h_c. However, the emerging chip is thicker than h_c. The chip thickness h_2 reflects intense shearing that takes place at an angle ϕ to the workpiece surface.

Cutting speed. The machining response of ductile metals is very sensitive to cutting speed (Fig. 2). Below about 0.02 m/s, chips form discontinuously, metal chunks are lifted out of the surface, and the surface is scalloped or pockmarked (Fig. 2*a*). When the speed is in the 0.1 m/s range, chips are formed continuously, the shear zone is narrow, and the chip slides on the tool face (Fig. 2*b*). Under these conditions a cutting fluid can lubricate both the rake and flank faces of the tool. If the speed is increased to 0.3 m/s or so, sticking starts to occur at the tool-workpiece interface due to the increased heat generation. The metal begins to shear along a built-up edge of metal which is stuck to the tool face (Fig. 2*c*). The appearance of the built-up edge has several practical consequences. The effective rake angle becomes quite large, and energy consumption drops. The surface finish is poor, however, owing to the ill-defined tool tip and

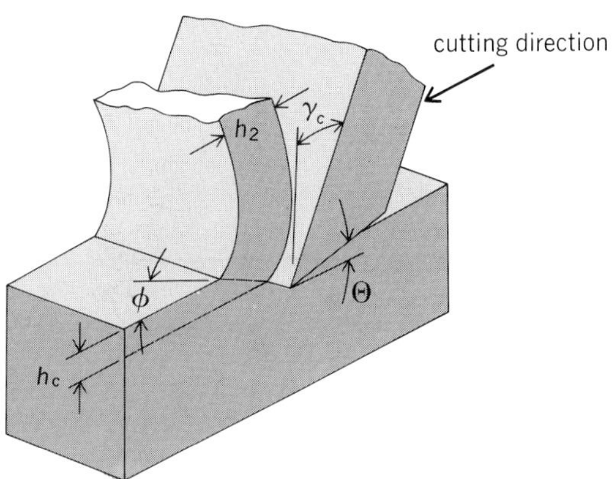

Fig. 1. Orthogonal cutting process: a simplified view of machining. *(From J. A. Schey, Introduction to Manufacturing Processes, McGraw-Hill, 1977)*

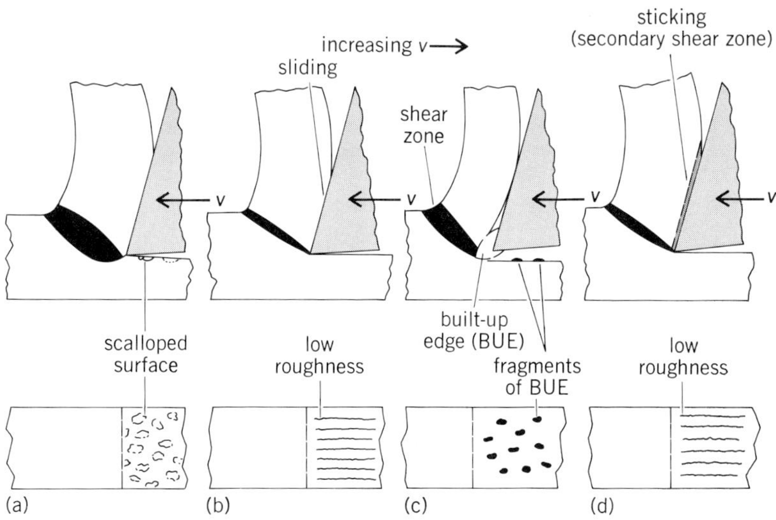

Fig. 2. Changes in chip formation and surface finish with increasing cutting speed: *(a)* 0.02 m/s; *(b)* 0.1 m/s; *(c)* 0.3 m/s; and *(d)* above 1.0 m/s. *(From J. A. Schey, Introduction to Manufacturing Processes, McGraw-Hill, 1977)*

sporadic breakdowns in the built-up edge. At much higher speeds (above 1.0 m/s), heat generation is even more intense, and temperature increases of several hundred degrees Celsius may occur. Under these conditions the built-up edge disappears, and the chip makes full, sticking contact with the tool face (Fig. 2*d*). The chip moves by shearing along the so-called secondary shear zone. With the high temperature, workpiece strength may be so low that power consumption is lower in spite of the extensive shearing. The most serious problem associated with the high temperature may be the possibility of diffusional bonding occurring between the tool and workpiece. If alloying elements of the tool material diffuse into the workpiece, very rapid wear (called crater wear) develops.

Cutting fluid. The interaction of the tool and the workpiece is considerably affected by the presence of cutting fluids. Cutting fluid has two primary functions. First, as long as the cutting speed is slow, the cutting fluid can act as a lubricant between the chip and the tool face. Even at higher speeds some lubricating effect at the flank face may be present. Second, and perhaps more importantly, the cutting fluid serves as a coolant. In most instances the cutting fluid will be an emulsion of a lubricating phase (oil, graphite, and so on) in water, since water is the best heat transfer medium readily available. Beyond lubrication and cooling, the cutting fluid can be used to flush out the cutting zone.

Surface quality and tool wear. Cutting speed thus affects the all-important machinability considerations of surface quality and tool wear. As implied by Fig. 2, surface finish is best with well-lubricated, moderately low-speed operation or high-speed cutting with no built-up edge. The high pressures and temperatures of operation, abetted by shock loading and vibrations, can lead to rapid tool wear. Tool wear is often sufficiently rapid to make tool replacement a major factor in machining economics. Tools must be replaced when they break or when they have worn to the point of producing an unacceptable surface finish or an unacceptable degree of surface heating. Whereas cracking and chipping occur suddenly, other forms of wear occur gradually. Gradual wear is related to rubbing distance and temperature, and thus to speed v. It often follows the Taylor power law, $C = vt^n$, where C is the cutting speed leading specifically to 1 minute of tool life, t is actual tool life in minutes, and n is the Taylor exponent. Both C and n reflect the tool and workpiece combination.

Metal properties. A machining operation can be optimized to effect metal removal with the least energy, or the best surface, or the longest tool life, or a reasonable compromise among these factors. Even so, it remains that the optimum ease and economy of machining some alloys is vastly different from that of others, and some basic attributes of easily machined alloys can be set forth.

Toughness. To ensure that chip separation occurs after minimum sliding, low ductility is required. To minimize cutting force, low strength is desirable. The two features of low strength and low ductility combine to mean low toughness. Toughness is generally defined as energy per unit volume consumed en route to fracture. Ironically, materials of maximized toughness are desirable for most engineering applications and, thus, some of the most attractive alloys, such as austenitic stainless steels, are difficult to machine.

Adhesion. The degree to which the metal adheres to the tool material is important to its machinability. Actually this attribute can work to advantage or disadvantage. If diffusion results, the tool can be weakened, and rapid wear occurs. Otherwise, high adhesion will stabilize the secondary shear zone.

Workpiece second phases. Small particles or inclusions in the metal can have a marked effect on machinability. Hard, sharp oxides, carbides, and certain intermetallic compounds abrade the tool and accelerate tool wear. On the other hand, soft second phases are beneficial because they promote localized shear and chip breakage.

Thermal conductivity. In some cases, workpiece thermal conductivity can be important to machinability. A low thermal conductivity generally results in high shear-zone temperature. This can be advantageous in reducing the strength of the metal or in softening second-phase particles. Of course, if adhesion and diffusional depletion of tool alloy content result, the high temperature is a problem. The workpiece temperature can be managed by cutting-speed and cutting-fluid manipulation.

Alloy systems. Commercial alloys can be grouped into two categories, namely those designed for ease of machining (so-called free machining grades), and the vast majority which are of widely varying but generally less than optimum machinability. Considering these latter, ordinary alloys, it can be shown that their machinability may be considerably improved by metallurgical operations which limit strength or ductility or both. Of course, it is not often possible to simultaneously reduce both strength and ductility. Even so, machinability often can be improved by grossly reducing one or the other property.

Consider the range of plain carbon steels represented in Fig. 3. In each case, the steel can be produced in three relevant forms: fully annealed (pearlitic), spheroidized, or cold-worked. For steel in the 0.20% C range, the annealed and spheroidized conditions involve too much ductility for optimum machinability. Better machinability can be had by reducing the ductility through cold work, even if strength is increased. In the 0.45% C range, the cold-worked material is too strong, and the intermediate strength and ductility of the annealed condition is optimum. In the 0.70% C range, the large quantities of lamellar carbide in the annealed pearlitic form are too abrasive and promote tool wear. Better machinability is possible with the globular carbides and lower strength of the spheroidized condition despite its high ductility. *See* STEEL.

Similar cases may be cited for other alloy systems, and the following principles are fairly general: (1) soft, ductile materials are more machinable when work-hardened; (2) hard materials with hard second phases are best machined in the well-annealed or overaged condition where globular particle shapes are developed; (3) moderate-strength age-hardenable alloys (such as age-hardenable Al alloys) are best machined in the aged and strengthened condition. *See* ALLOY.

Cast iron. Cast iron is a "natural" free-machining material. Gray cast iron with its large graphite flakes produces short chips without extra alloying

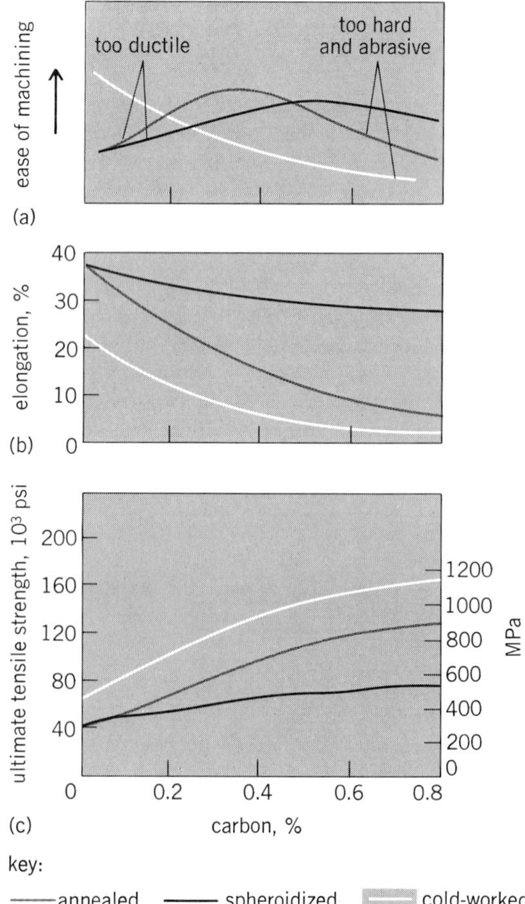

Fig. 3. The ease of machining carbon steels as a function of their metallurgical condition. *(a)* Ease of machining versus carbon content. *(b)* Elongation versus carbon content. *(c)* Ultimate tensile strength versus carbon content. *(From J. A. Schey, Introduction to Manufacturing Processes, McGraw-Hill, 1977)*

or further process control. Moreover, the graphite itself serves as a lubricant. However, the machined surface is somewhat roughened by the breaking out of graphite particles. The machinability of gray iron may be further enhanced with the use of a subcritical anneal (1 h at about 750°C) to graphitize the carbide that exists in the pearlite lamellae. The resultant structure is totally iron and flake graphite.

Lead additive. One of the most universal free-machining additives is lead. Lead is insoluble in iron, copper, aluminum, and alloys thereof. It can be finely dispersed as a soft second phase, which helps to break up the chips and, furthermore, lubricates the tool-workpiece interface. *See* MACHINING OPERATIONS; METAL, MECHANICAL PROPERTIES OF; PLASTIC DEFORMATION OF METAL.

[ROGER N. WRIGHT]

Bibliography: American Society for Metals, *Metals Handbook*, 8th ed., vol. 3: *Machining*, 1967; G. Boothroyd, *Fundamentals of Metal Machining and Machine Tools*, 1975; J. A. Schey, *Introduction to Manufacturing Processes*, chap. 8, 1977; United States Steel Corporation, *The Making, Shaping and Treating of Steel*, chap. 51, 1971.

Machine

A combination of rigid or resistant bodies having definite motions and capable of performing useful work. The term mechanism is closely related but applies only to the physical arrangement that provides for the definite motions of the parts of a machine. For example, a wristwatch is a mechanism, but it does no useful work and thus is not a machine.

Machines vary widely in appearance, function, and complexity from the simple hand-operated paper punch to the ocean liner, which is itself composed of many simple and complex machines. No matter how complicated in appearance, every machine may be broken down into smaller and smaller assemblies, until an analysis of the operation becomes dependent upon an understanding of a few basic concepts, most of which come from elementary physics. *See* SIMPLE MACHINE.

[RICHARD M. PHELAN]

Bibliography: R. M. Phelan, *Fundamentals of Mechanical Design*, 3d ed., 1970; J. E. Shigley, *Mechanical Engineering Design*, 3d ed., 1976.

Machine design

Application of science and invention to the development and construction of machines. An understanding of the basic laws of nature is essential to a proper perspective in the approach to machine design. Knowledge of the past development of machine elements makes possible their effective application. Inventiveness consists of producing new combinations of old elements or, where extreme need arises, of exercising genius either in breaking the bounds of convention, or in evolving new principles not hitherto applied or known.

In machine design, accomplishment takes on two forms: one is the drawings and blueprints, which completely describe the machine, and the other is the assembled product. In addition, most machines go through periods of evolution, and later models may show little outward similarity to the original design.

Machine design consists of the conception of a machine that will meet a specific need. Before constructing a machine to fulfill the need, the designer must thoroughly understand the application, and mentally modify an old machine or devise a new machine as required. He estimates a certain cost for the machine and a probable time for its construction. He envisions the materials required, the equipment necessary for its manufacture and testing, and the final operation in meeting the original need. If the machine is desirable, he converts his thoughts into drawings and materials and follows through to its fabrication. In time the machine may become obsolete due to advances in the technology; it may then be rebuilt or replaced, possibly under the direction of the original designer.

The working tools in machine design are an understanding of the basic elements of machines that have been developed in the past and a thorough knowledge of the mechanical fields of science including mathematics, physics, statics and dynamics, strength of materials, kinematics, mechanisms, and the laboratories associated with them. *See* MECHANISM.

[JAMES J. RYAN]

Bibliography: R. H. Creamer, *Machine Design*, 2d ed., 1976.

Machine elements

Elementary mechanical parts used as building blocks for the construction of most devices, apparatus, and machinery. The gradual development of these building blocks, following the invention of the roller or wheel and the arm or lever in ancient times, brought about the Industrial Revolution, starting with the assembly of James Watt's engine for harnessing the force of steam and proceeding into the advanced mechanization of present automatic control.

The most common example of a machine element is a gear, which, fundamentally, is a combination of the wheel and the lever to form a toothed wheel as illustrated. The rotation of this gear on a hub or shaft drives other gears which may rotate faster or slower, depending upon the number of teeth on the basic wheels. The material from which the gear is made establishes its strength, and the hardness of its surface determines its resistance to wear. Knowledge of the forces on the gear makes possible the determination of its size. Changes in its shape allow modifications in its use. These applications, as in most machine elements, have developed into many standard forms, such as spur, bevel, helical, and worm gears. Each of these forms has required the development of a special technology for its production and use. *See* GEAR.

Other fundamental machine elements have evolved from wheels and levers. A wheel must have a shaft on which it may rotate. The wheel is fastened to the shaft with a key, and the shaft is joined to other shafts with couplings. The shaft must rest in bearings, such as journal bearings, ball bearings, or roller bearings. The shaft may be started by a clutch or stopped with a brake. It may be turned by a pulley with a flat belt, a V belt, or a chain or a rope connecting it to a pulley on a second shaft. The supporting structure may be assembled with bolts or rivets or by welding. Proper application of these machine elements depends upon a knowledge of the forces on the structure and the strength of the materials employed. In the design,

MACHINE ELEMENTS

wheel (succession of levers)
gear tooth (end of lever)
hub (fulcrum)

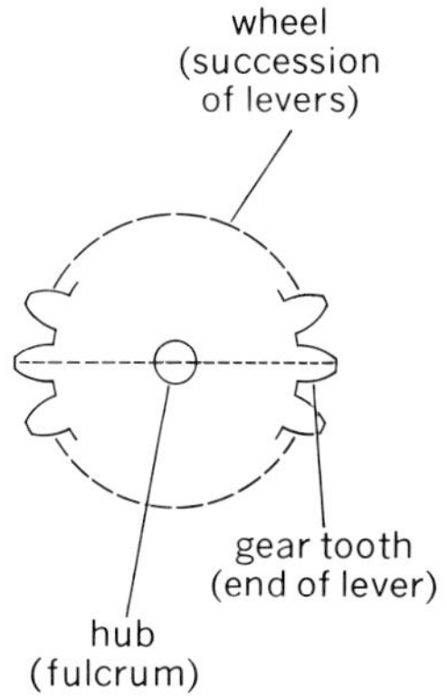

The gear, a machine element which combines features of the wheel and the lever.

calculations must accommodate the forces to the materials in the simplest construction.

Other machine elements have been evolved whose applications are more specific in construction. *See* ANTIFRICTION BEARING; BRAKE; CAM MECHANISM; CLUTCH; COUPLING; FLYWHEEL; FOUR-BAR LINKAGE; GOVERNOR; SCREW FASTENER; SHAFTING; SPRING.

Machine parts which are commonly used have been developed into standardized designs. Manufacturing specialists have concentrated upon the development of standard elements and have mass-produced these parts with a high degree of perfection at reduced cost. Standard elements, as applied in machine design, may be modified as desired, although certain ones, through the hazards to safety under improper use, must be modified only in line with the requirements of codes established by regulating bodies. *See* DESIGN STANDARDS. [JAMES J. RYAN]

Bibliography: E. Oberg et al., *Machinery's Handbook*, 20th rev. ed., 1975.

Machine key

The most common function of a key is to prevent relative rotation of a shaft and the member to which it is connected, such as the hub of a gear, pulley, or crank. Many types of keys are available, and the choice in any installation depends on such factors as power requirements, tightness of fit, stability of connection, and cost. For light power requirements a setscrew may be tightened against the round shaft or against a flat spot on the shaft. For most requirements a positive connection, such as by a key, is necessary. A setscrew is frequently used to seat the key and to prevent axial motion.

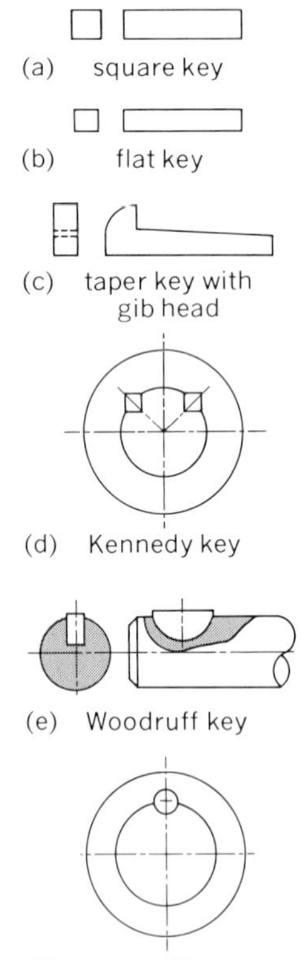

Fig. 1. Types of keys. (*From P. H. Black and O. E. Adams, Jr., Machine Design, 3d ed., 1968*)

Square keys are common in general industrial machinery (Fig. 1*a*). Flat keys are used where added stability of the connection is desired, as in machine tools (Fig. 1*b*). Square or flat keys may be of uniform cross section or they may be tapered. In tapered keys the width is uniform and the height of the key tapers. Tapered keys may have gib heads to facilitate removal (Fig. 1*c*).

The Kennedy key is used for heavy duty and consists of two keys driven 90° apart (Fig. 1*d*). The hub is bored to fit the shaft and is then rebored slightly off center. The keys force the shaft and hub into concentric position.

The Woodruff key requires a key seat formed by a special side-milling cutter (Fig. 1*e*). This key will align itself in the key seat. It has the disadvantage of weakening the shaft more than a straight key.

The round key, or pin, introduces less stress concentration at the key seat in the shaft and is satisfactory except for the necessity of drilling the hole to accommodate the pin after assembly of the hub and shaft (Fig. 1*f*). This may be a disadvantage in production and prevents interchangeability.

A spline fitting is composed of a splined shaft formed by milling and a mating hub with internal splines formed by broaching (Fig. 1*g*). The splines in reality are a number of keys integral with the shaft. Splined fittings are adaptable to mass production and are used where radial space must be conserved.

Straight-side splines are being replaced at an increasing rate by stub involute splines. These splines have the advantages of greater strength, a self-centering feature, and production economy.

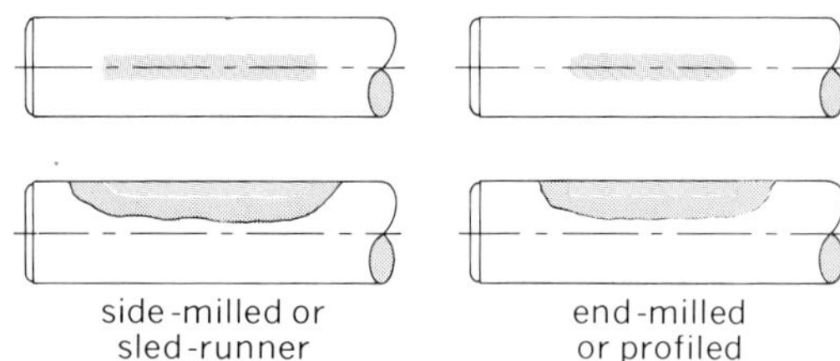

Fig. 2. Types of keyways. (*From P. H. Black and O. E. Adams, Jr., Machine Design, 3d ed., 1968*)

Keyways for straight keys are formed either by a side-milling cutter which forms a sled-runner keyway, or by an end-milling cutter which forms profiled keyways (Fig. 2). The sled-runner keyway requires a longer space between the end of the key and the end of the keyway than does an end-milled keyway. This favors the end-milled keyway in locations near a shoulder. However, the end-milled keyway reduces the endurance limit of a shaft more than does the sled-runner type.

Feather keys are used where it is necessary to slide a keyed gear or pulley along the shaft. The key is generally tight in the shaft and with clearance between key and hub keyway. When a gear or pulley must be moved axially along the shaft while power is being transmitted, it is desirable to provide for a minimum of force necessary to move the hub along the shaft. The use of two feather keys equally spaced reduces the necessary axial force to half that for one key. [PAUL H. BLACK]

Bibliography: P. H. Black and O. E. Adams, Jr., *Machine Design*, 3d ed., 1968; E. Oberg et al., *Machinery's Handbook*, 20th rev. ed., 1975; R. M. Phelan, *Fundamentals of Mechanical Design*, 3d ed., 1970.

Machinery

A group of parts arranged to perform a useful function. Normally some of the parts are capable of motion; others are stationary and provide a frame for the moving parts. The terms machine and machinery are so closely related as to be almost synonymous; however, machinery has a plural implication, suggesting more than one machine. Common examples of machinery include automobiles, clothes washers, and airplanes; machinery differs greatly in number of parts and complexity.

Some machinery simply provides a mechanical advantage for human effort. Other machinery performs functions that no human being can do for long-sustained periods. A jackscrew does nothing until a man pulls on a lever; then he is able to move objects many times his weight. Conversely, an internal combustion engine can run unattended for hours, requiring only the press of a button to start it.

The need for machinery usually stems from a desire to do a job at less cost. Evolution of machinery for a certain function may be gradual or rapid. If only small quantities of a product are needed, it is likely that machinery used in making the product will not change rapidly or possess the highest degree of automation. On the other hand, machinery used in making automobiles has evolved into some of the most complex automatic machines in existence. *See* MACHINE; MACHINE DESIGN; MECHANICAL ENGINEERING; SIMPLE MACHINE.

[ROBERT S. SHERWOOD]

Machining operations

Methods for the removal of material from a workpiece by the use of a cutting medium. Machining operations are usually carried out in a power-driven arrangement which gives the piece a desired shape or finish.

Cutting tools. Single- or multiple-point cutting tools, including broaching tools, milling cutters, honing stones, abrasives, and saw blades, are common forms of cutting devices. Although varied in configuration, the basic cutting action performed by each is much the same. Some tools remove material in large chips, shavings, or pieces; others such as the saw, honing stone, or abrasive grits remove stock by cutting away small particles.

The machines which hold the cutting tools and in some cases the workpieces, and furnish the power for cutting, vary greatly in size and configuration. Machines vary from small hand-held drilling or grinding devices to large, automatic, multioperation machine tools. Machining operations commonly performed are turning and facing, boring, milling, sawing, broaching, shaping, planing, drilling, threading, tapping, the various types of gear cutting, grinding, honing, lapping, superfinishing, buffing, polishing, and nibbling. Some of these overlap each other; operations such as polishing may not be commonly thought of as machining.

Motion between tool and work. In power machining operations there must be a relative movement between cutting tool and workpiece. During some machining operations the workpiece moves; in others only the cutting tool moves; frequently both tool and workpiece are in motion. The various machining operations that employ cutting tools use either a rotating or a traversing motion. For example, if either a turning or boring operation is performed on a lathe, the workpiece rotates while the cutting tool traverses longitudinally along or through the piece. On a milling machine the opposite may be true; the cutting tool, or milling cutter, rotates while the workpiece is clamped to a table and made to traverse the path of the cutter teeth.

In certain cases an operation is accomplished by motion of the tool on one machine and by motion of the workpiece on another machine. An example is a drilling operation performed on a turret lathe; the drill is traversed without rotating, while the workpiece revolves. Had the operation been done on a drill press, the workpiece would have remained stationary while the drill rotated as it advanced into the work.

Basically in any machining operation in which a cutting tool is used, one or more relative motions between the workpiece and the cutting device are necessary.

Many particular machining operations can be performed on a workpiece most easily or most economically by a certain machine tool, although it is frequently possible to perform the operations on more than one type of equipment. Of primary consideration in fitting the job to the machine is the nature of the required operation. Surfaces to be machined may be considered as being flat, flat contoured, curved, externally cylindrical (turned), internally cylindrical (bored), and so on. In addition to these, there are the surfaces or configurations produced by specialized operations such as threading, tapping, and hobbing.

Choice of machine method. Once the machining requirements have been assessed, the job is fitted to the most appropriate machine available.

Basic machine tool functions

No.	Machine type	Tool movement	Work movement	Machined surfaces
1	Drilling machine	Rotate and traverse	Stationary	Internally cylindrical; specialized
2	Horizontal lathe (engine, turret, screw machine)	Traverse (longitudinally and radially)	Rotate	Externally cylindrical; internally cylindrical; shoulders or ends of cylinders; specialized surfaces
3	Vertical lathe	Traverse (longitudinally and radially)	Rotate	Same as horizontal lathe
4	Boring machine (horizontal)	Rotate and traverse	Traverse	Externally cylindrical; internally cylindrical; shoulders or ends of cylinders; flat and flat contoured; specialized surfaces
5	Boring machine (vertical)	Traverse	Rotate	Externally cylindrical; internally cylindrical; shoulders or ends of cylinders
6	Broaching machines (horizontal and vertical)	Traverse and stationary	Traverse and stationary	Flat and flat contoured; curved; cylindrical
7	Grinder (cylindrical)	Rotate and traverse	Rotate (also traverse on centerless type)	Externally cylindrical
8	Milling machine	Rotate	Traverse	Flat and flat contoured; curve contoured
9	Planer (vertical)	Traverse	Traverse	Same as milling machine
10	Shaper	Traverse	Stationary	Same as milling machine
11	Planer (horizontal)	Stationary (planer miller rotates)	Traverse	Same as milling machine
12	Grinder (surface)	Rotate	Traverse	Flat surfaces
13	Nibbling machine	Traverse	Traverse	Flat or curved
14	Saw (circular)	Rotate	Traverse	Flat
15	Saw (band)	Traverse	Traverse	Flat or curved
16	Flame cutter	Traverse	Stationary	Flat or curved

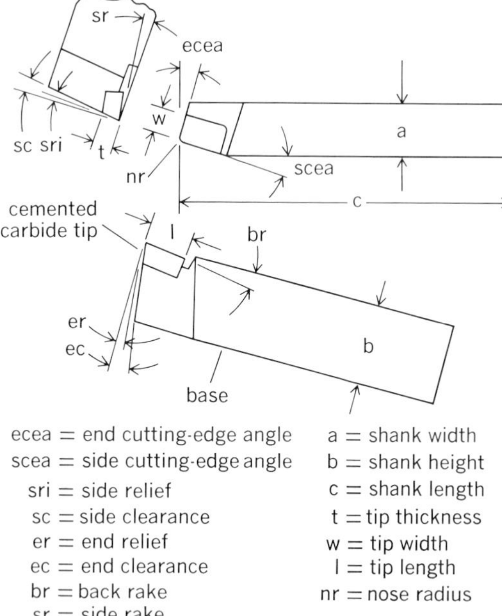

Fig. 2. Single-point tool nomenclature. (*Kennametal Inc.*)

The basic rotary and traverse machining motions provide a primary step in relating the work requirements to the machine. The table relates these motions to many of the common machine tools and also to the various surface configurations that the particular machine is capable of producing.

Machines for cylindrical surfaces. The table shows that machines 1–7 are capable of machining or finishing some type of cylindrical surface and most of them can machine other types of surface as well. Several of these first seven machines are extremely versatile. The horizontal boring mill, in addition to turning and boring, can perform milling operations equally well.

The difference in the work produced by these machines lies mainly in (1) the size of either the bore or the outside diameter of the piece, (2) the physical size of the workpiece handled, and (3) the tool used. The drilling machine, for example, using a twist drill for a tool, is able at best to produce holes only a few inches in diameter. While radial drilling machines are built to handle pieces of considerable size, nevertheless machine drilling usually refers to pieces of small to medium size. By contrast, large vertical boring mills, using single-point cutting tools, are able to bore a hole several feet in diameter in a piece weighing many tons. The drill actually produces a hole while the boring machine finishes an existing one; yet both machine internally cylindrical surfaces or bores.

Machine 7, the cylindrical grinder, is designed to refine or finish externally cylindrical surfaces.

Machines for flat surfaces. Machines 8–12 on the chart are all capable of machining flat or flat contoured surfaces. These machines are usually able to produce curved contoured surfaces by means of contour cutters, two-dimensional movement of either the tool or workpiece, or by special fixtures.

The machines vary in size, capacity, and type of tool. Common milling machines, with rotating cutters, are used primarily for smaller items. Horizontal planers and milling planers, with single-point and rotating face mills, respectively, are built in models designed to handle pieces of all sizes. Versatile machines such as boring mills, broaching machines, and vertical turret lathes can finish flat surfaces. Machine 11, the surface grinder, as in the case of the cylindrical grinder, actually refines or finishes flat surfaces.

Machines for parting or trimming. Machines 13–16 are used for parting stock or removing excess material. The nibbling machine with its punching action is able to cut out irregular shapes from relatively thin sheets of material, while a tool such as the flame cutter can cut through several inches of steel plate. The power saws fall between these two extremes.

The factors that differentiate one machine from another are size, capabilities, workpiece requirements, machine speed, rigidity, and work loads. These and other factors enter into scheduling work on one type of machine in preference to another. *See* PRODUCTION METHODS.

Single-point tool. Basic among the various tool forms used in power machining is the single-point tool (Fig. 1). A form of this tool is used on lathes, boring mills, shapers, and planers, with variations employed on other machines. Single-point tools vary in design, each adapted to the operation for which it is intended. The tool may be straight, bent, offset, or have some other special form that will enable its cutting tip to reach the desired area of machining (Fig. 2). The tool may be made from one solid piece of tool steel with the cutting edge ground on the end, or it may have an insert or tip attached to it to provide a harder cutting surface.

Fig. 1. Inspection of a single-point cutting tool on an optical comparator and measuring machine. (*Jones and Lamson Machine Co.*)

The insert may be a piece of high-quality tool steel or a special material such as sintered carbide and be brazed, welded, or fastened to the tool shank by mechanical means.

Tools may be sharpened or shaped by grinding, and inserts may be replaced or rotated when chipped or worn. Frequently a groove or irregularity is ground in the face of the tool behind the cutting edge. This groove, called a chipbreaker, causes removed stock to break into chips or small curls, thus becoming easier to manage or to dispose of.

Multipoint tool. A second type of cutting device commonly used on such power tools as milling machines, horizontal boring machines, and planer millers is the revolving multitoothed cutter (Fig. 3). This multiple cutting edge tool gives the advantage of much faster cutting than can be obtained using a single-point tool. Like the single-point tool, each cutter tooth has a rake angle and clearance angle. The rake angle aids in chip removal and in most

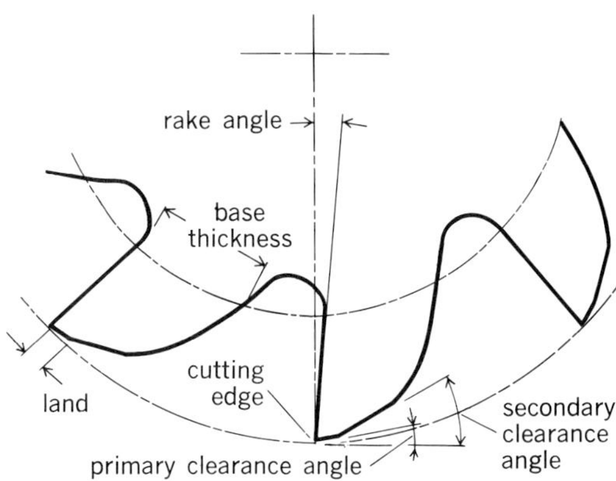

Fig. 3. Typical milling cutter teeth.

cases provides a sharper cutting edge. The clearance angle allows the tooth area behind the cutting edge to clear the work, avoiding drag or friction.

Multitoothed cutters are made in many designs, each to perform efficiently a certain milling operation. Probably the most common cutter is the regular or plain milling cutter. This cylindrical type of cutter is made in varying diameters and face widths. Cylindrical cutters are made in face widths that range from the thin metal slitting saws to slab mills several inches in width. Side-milling cylindrical cutters have teeth on one or both sides. Other cutters may be constructed in the form of a cone to mill angles or to do beveling.

The face-milling cutter, another frequently used tool, has teeth arranged around the edge of a plate or disk. Usually several inches in diameter, these cutters are able to face or mill a large area rapidly by traversing along it.

The end mill is a cylindrical cutter with teeth on its circumferential face and one end. An integral shank on the other end is used for holding and driving. End mills vary in size from a fraction of an inch to several inches in diameter. Regular milling cutters are able to traverse mill at a depth, thus producing a slot or groove. Two-lip end mills are able to cut directly into the work in much the same manner as a drill.

Speed and feed. Two important considerations are present in the relative movement of the cutting tool and workpiece: cutting speed and feed. The cutting speed is the relative surface speed between the work and the tool. For example, in a turning operation, cutting speed is usually measured on the uncut surface of the work ahead of the tool. Cutting speed is commonly given in feet per minute. The feed is the relative amount of motion or travel of the cutting tool into the workpiece per revolution, stroke, or unit of time.

To obtain high machining efficiency, desired surface finish, and appreciable tool life, speed and feed must be closely correlated during any machining operation. Many factors affect this relationship, including the type of tool used, its durability, and the rigidity of the machine and tool, as well as the power of the machine, and the material, hardness, and configuration of the workpiece. No general rule or equation for the correct feed and speed can cover all conditions. Some information comes from publications and handbooks, some from the advice of others, and much from personal experience.

Most manufacturers of machines and cutting tools supply information on recommended feeds and speeds for their particular product. In addition, machine shop handbooks provide adequate information on feeds and speeds for normal conditions encountered in machining.

Cutting fluid. Most metal machining operations require the use of some type of cutting fluid. Machining of metals causes deformation, rubbing, and friction. The resulting temperature rise can warp the work and damage or excessively wear the tool. Also, the metal will tend to expand, causing inaccuracies in the work. Cutting fluids are used mainly for cooling, but they may serve other purposes. In certain instances the surface finish is improved by the use of a cutting fluid. Often the fluid also serves to lubricate the slides on the machine or to protect the machine from corrosion. In many operations the fluid washes away metal chips and particles that could clog or interfere with the tool and the machine.

There are two general types of cutting oils, one based on mixtures with water, the other on oil compounds. In general, if cooling is the primary requirement, an oil-and-water emulsion is used. If lubrication of slides or other parts is required in addition to cooling, a mineral oil compound is usually employed. There are, however, no hard and fast rules, and in many cases only experience or actual test will tell the type of cutting oil to use. *See* BORING; BROACHING; DRILLING MACHINE; GEAR CUTTING; GRINDING; LAPPING; LATHE; MILLING MACHINE; NIBBLING; PLANER; POLISHING; REAMER; SAWING; SHAPER; THREADING.

[ALLAN TUTTLE]

Bibliography: J. L. Feirer, in D. G. Gilmore (ed.), *Machine Tool Metalworking*, 2d ed., 1973; Society of Manufacturing Engineers, *Tool and Manufacturing Engineers Handbook*, 3d ed., 1976.

Magnesium alloys

The most important alloying ingredients used in magnesium alloys are aluminum, zinc, manganese, silicon, zirconium, rare-earth metals, and thorium. The specific gravity of magnesium alloys ranges from 1.74 to 1.83. This low specific gravity has led to a great many structural applications in the aircraft, transportation, materials-handling, and portable-tool and -equipment industries. This article deals exclusively with the structural alloys and uses of magnesium.

Commercial alloys. The great expansion in the use of magnesium came during World War II, and was based almost entirely on magnesium-aluminum-zinc alloys and, to a much more limited extent, on the binary magnesium-manganese alloy, whose use has been abandoned. The Mg-Al-Zn alloys are still the most important. The designations and chemical compositions of the most important of these alloys are given in the table. Magnesium-aluminum-zinc alloys are produced in the form of sand, permanent-mold, and die castings; extrusions; rolled sheet and plate; and forgings. They are also cast by some of the less common methods, including plaster-molding, centrifugal-molding, shell-molding, and investment-molding processes. Their properties may be modified by appropriate heat treatments.

The need for magnesium alloys with improved strength and creep resistance at elevated temperatures has been met by the development of alloys containing rare-earth metals or thorium, or both as the principal alloying constituents. The temperature range over which magnesium alloys exhibit structurally useful mechanical strength has been markedly extended by the use of rare-earth metals and thorium. Magnesium-thorium alloys, for example, are used up to temperatures in the range of 700–900°F (371–482°C) and even higher, depending on the time duration of exposure at the elevated temperature. The alloys containing rare-earth metals and thorium are included in the table.

Chemical composition of common magnesium alloys, in percentage*

Alloy	Al	Mn (min.)	Rare earth	Th	Zn	Zr
AM100A	10.0	0.1			0.3 (max.)	
AM60A	6.0	0.13				
AS41A†	4.2	0.35				
AZ31B	3.0	0.2			1.0	
AZ61A	6.5	0.15			1.0	
AZ63A	6.0	0.15			3.0	
AZ81A	7.5	0.13			0.7	
AZ80A	8.5	0.15			0.5	
AZ91B	9.0	0.13			0.7	
AZ91C	8.7	0.13			0.7	
AZ92A	9.0	0.10			2.0	
EZ33A			3.0‡		2.7	0.6
HK31A				3.0		0.7
HM21A		0.45		2.0		
HM31A		1.2		3.0		
HZ32A				3.0	2.1	0.7
K1A						0.7
QE22A§			2.0¶			0.7
QH21A§			1.0¶	1.1		0.7
ZE41A			1.2‡		4.2	0.7
ZE63A			2.6‡		5.75	0.7
ZH62A				1.8	5.7	0.75
ZK51A					4.6	0.7
ZK60A					5.7	0.55
ZK61A					6.0	0.8

*Balance is magnesium in all cases.
†Alloy also contains 1.0% silicon.
‡Rare-earth metals present as misch metal.
§Alloy also contains 2.5% silver.
¶Rare-earth metals present as didymium.

A little-known advantage of magnesium over other common structural materials is its inherently high damping capacity, that is, its capacity to absorb mechanical vibrations. In general, addition of alloying elements decreases this property. Alloy K1A has been developed offering the best combination of damping capacity, strength, and castability.

Fabricability. In addition to being adaptable to all the primary working operations already mentioned, magnesium alloys can be fabricated by all the common metalworking processses such as stamping, deep and shallow drawing, blanking, coining, spinning, impact extrusion, and forging. For forging operations, both press and hammer equipment are used, but the former is more commonly used because the physical structure of magnesium alloys makes the metals better adapted to the squeezing action of the forging press.

Magnesium alloys exhibit excellent machinability. They can be machined at higher speeds and with larger feeds and depths of cut than is possible with most other commonly used metals. Power requirements needed in machining are the lowest for magnesium among all structural metals.

Magnesium alloy parts can be joined by any of the common methods. Inert-gas, shielded-arc, and electric-resistance welding, adhesive bonding, and riveting are in daily production use. Brazing and gas welding, although not as frequently used as the other methods, are also suitable ways of joining magnesium alloys.

A wide variety of protective and decorative surface-finishing systems can be applied to magnesium alloys. Magnesium alloys can be treated chemically and electrochemically to produce a protective and paint-adherent surface. In addition they can be painted, electroplated, anodized, and clad with plastic sheathing.

Uses. Reviewing the nonstructural uses of magnesium gives an accurate perspective of the entire industry. Nonstructural uses are those in which magnesium is used for its particular chemical, electrochemical, and metallurgical properties. The principal nonstructural uses are as: alloying constituent in other metals (principally aluminum); nodularizing agent for graphite in cast iron; reducing agent in the production of other metals, such as uranium, titanium, zirconium, beryllium, and hafnium; desulfurizing agent in iron and steel production; sacrificial anode in the protection of other metals against corrosion; and anode material in reserve batteries and dry cells. These nonstructural uses account for about 85% of all the magnesium used in the United States. Consumption as an alloying constituent in aluminum alone amounts to 55% of the total. In Europe, on the other hand, magnesium is used more extensively as a structural metal, and consumption for such uses amounts to about 35% of total usage. On a worldwide basis, structural uses account for about 22% of total consumption.

As structural materials, magnesium alloys are best known for their light weight and high strength-to-weight ratio. Accordingly, they are used generally in applications where weight is a critical factor and where high mechanical integrity is needed. Magnesium alloys are most commonly used in the form of die casting. Die castings account for 75% of the total worldwide usage in structural applications. Of all the metalworking and fabricating processes, die casting is the most readily adaptable to the specific characteristics of magnesium. Magnesium has a relatively low melting point, low heat content per unit volume, and low reactivity toward ferrous materials. Coupling these characteristics with the high productivity of the die-casting process results in the most economical technique for the mass production of a large number of complex parts of the same configuration. Until the mid-1970s, all magnesium die castings were produced on cold-chamber die-casting machines. Since then, hot-chamber die casting machines have been developed in Europe, and they have been introduced in the United States. The hot-chamber machine is even more amenable to automation than the cold-chamber machine and is capable of higher production rates and overall lower operating costs.

The development of the hot-chamber machine greatly enhances the potential for expansion of the use of magnesium die castings in the automotive and appliance industries. Magnesium die castings are used extensively on Volkswagen automobiles. The VW Beetle, which was still being produced in Latin America in the early 1980s, contains about 45 lb (20 kg) of magnesium alloys in such parts as the engine crankcase, transmission housing, fan housing, gearbox, and many others. The newer models of the VW produced in Europe no longer have a magnesium crankcase, but they still have 10–15 lb (4.5–7 kg) of magnesium die castings distributed among many smaller parts. The United States automotive industry is showing a renewed interest in magnesium because of the weight-saving potential with the advent of the hot-chamber process and the economic advantages of this manufacturing technique. Some 1980 models contained a few magnesium parts such as the distributor diaphragm, steering column brackets, and a lever cover plate. Several other parts were being considered. Magnesium-alloy die castings are also used on chain saws, portable power tools, cameras and projectors, office and business machines, tape reels, sporting goods, luggage frames, and many other products.

Uses of sand and permanent mold castings are confined largely to aircraft engine and airframe components. Engine parts include gearboxes, compressor housings, diffusers, fan thrust reversers, and miscellaneous brackets. On the airframe, magnesium sand castings are used for leading edge flaps, control pulleys and brackets, entry door gates, and various cockpit components. This is a small market in volume, but one requiring highly technical skills in production.

The use of magnesium wrought products has become rather limited. Magnesium sheet, extrusions, and forgings continue to be used in a few limited airframe applications. The high-temperature alloys are used in at least 20 different missiles (including ICBMs) and on various spacecraft. The Agena is designed almost exclusively in HM21A sheet and forgings and HM31A extrusions. The Titan ICBM has about 2000 lb (900 kg) of magnesium alloys containing thorium in various product forms. In nonmilitary applications, magnesium extrusions (AZ31B) are used for luggage frames and bakery racks. AZ31B plate is used for tooling jigs and fixtures. A special grade of AZ31B (PE alloy) is used for printing plates in the offset printing process. *See* ALLOY; HEAT TREATMENT (METALLURGY); METAL CASTING; METAL FORMING.

[THOMAS E. LEONTIS]

Bibliography: A. V. Beck, *The Technology of Magnesium and Its Alloys*, 1943; E. F. Emley, *The Principles of Magnesium Technology*, 1966; W. H. Gross, *The Story of Magnesium*, 1949; G. V. Raynor, *The Physical Metallurgy of Magnesium and Its Alloys*, 1959; C. S. Roberts, *Magnesium and Its Alloys*, 1960.

Magnet wire

The insulated copper or aluminum wire used in the coils of all types of electromagnetic machines and devices. It is single-strand wire insulated with enamel, varnish, cotton, glass, asbestos, or combinations of these. To meet the immense variety of uses and to gain competitive advantage, a great number of kinds of enamel and of fiber insulations have been developed and are widely available. And to ensure that the many new synthetic materials and combinations are properly applied, many elaborate test procedures have been devised to evaluate heat shock, solubility, blister and abrasion resistance, flexibility, and so on as covered in the National Electrical Manufacturers Association (NEMA) Standards. Most important, the thermal endurance is determined by repeated exposure to high temperature, moisture, and electrical stress, as specified in the American Society for Testing and Materials (ASTM) Standard D2307-64T.

The Institute of Electrical and Electronics Engineers (IEEE) thermal classes of insulation, defined by upper temperature limits at which the untreated insulation will have a life expectancy of at least 20,000 hr, are O (80°C), A (105°C), B (130°C), F (155°C), and H (180°C). In general, materials such as cotton, paper, and silk are class O. Organic materials, such as oleoresinous and Formvar enamels, varnish-treated cotton, paper, and silk are class A, and asbestos, mica, silicone varnishes, and polyimide are class H, while various synthetic enamels fall in the B and F classes. The polyesterimide enamels, however, are capable of withstanding temperatures of 180–200°C.

Almost all magnet wire is insulated soft-drawn electrolytic copper, but aluminum is being used more, especially in times when copper is scarce or high-priced and where space is not limiting. Round aluminum wire, being soft, flattens under pressure, giving a higher space factor in coils than might be expected. At temperatures above 200°C copper oxidizes rapidly; it also becomes brittle when under stress at such temperatures. Its high-temperature strength is increased considerably by adding a small amount of silver (about 30 oz/ton). Anodized aluminum is preferred for higher temperatures, up to 300°C or more, while copper nickel wire with an anodized aluminum coating may be operated at still higher temperatures.

The increase in wire diameter due to a single coat of enamel varies from about 1 mil for a wire diameter of 80 mils to about 5 mils for a wire diameter of 200 mils. The double coating that is frequently called for is not quite twice as thick. The fiber coatings are about three times as thick as the enamel, and they too can be obtained in double thickness. *See* ELECTRICAL INSULATION; MAGNET.

[PHILIP L. ALGER; C. J. HERMAN]

Bibliography: American Society for Testing and Materials, Standard ASTM-D2307-64T, 1964; D. G. Fink and J. M. Carroll, *Standard Handbook for Electrical Engineers*, 10th ed., 1968; National Electrical Manufacturers Association, *NEMA Handbook;* National Electrical Manufacturers Association, *Magnet Wire*, Standard MW 1000-1967.

Magnetic amplifier

A device that employs saturable reactors to modulate the flow of alternating-current electric power to a load in response to a lower energy level direct-current input signal. Magnetic amplifiers are often referred to as power amplifiers, because they are well suited for use in driving electric motors and other output devices.

The concept of a saturable reactor is illustrated by the system shown in Fig. 1. The saturable reactor is much like an ordinary transformer, but it is operated with currents in its windings that can readily saturate the core material. The core is made of one of several special magnetic materials, rather than transformer iron, to achieve a given desired saturation characteristic. One of the most common core materials is Delta Max, which has a sharp saturation characteristic, high permeability, and a nearly rectangular flux density versus magnetizing force characteristic. When the dc control-signal voltage V_C is zero in the system of Fig. 1, the saturable reactor is not quite saturated and it acts as a large inductance in the load circuit, thereby generating a back electromotive force which is nearly equal to and 180° out of phase with the voltage of the ac supply V_S. In other words, the impedance Z_r of the unsaturated reactor is much larger than the impedance Z_L of the load, and acts to block the flow of current from the ac supply through the load circuit. When the dc control-signal voltage is increased, the reactor (which was close to saturation with zero dc control signal) is caused to saturate during the part of each cycle when the magnetizing currents of the control and load windings are acting together. The duration of the period of saturation increases with increasing dc control voltage. Similarly, if the dc control signal had decreased from a zero value to a negative value, the reactor would have started to saturate during the other half-cycle of the ac sinusoid. As soon as the reactor saturates, the impedance of the load winding suddenly drops to a small fraction of its unsaturated value (its back emf falls to a low value). Full ac supply voltage then acts to drive the load, and a pulse of current flows through the load until the reactor is desaturated a short time later as the load current from the ac supply decreases to zero. The shape of the load-current pulse depends on the exact nature of the load impedance.

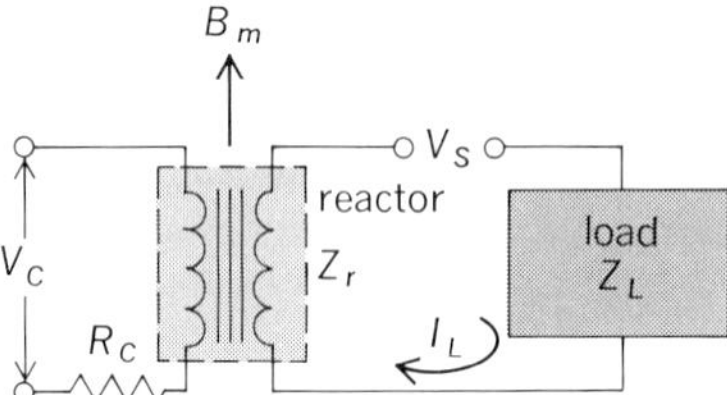

Fig. 1. Single-core saturable reactor circuit.

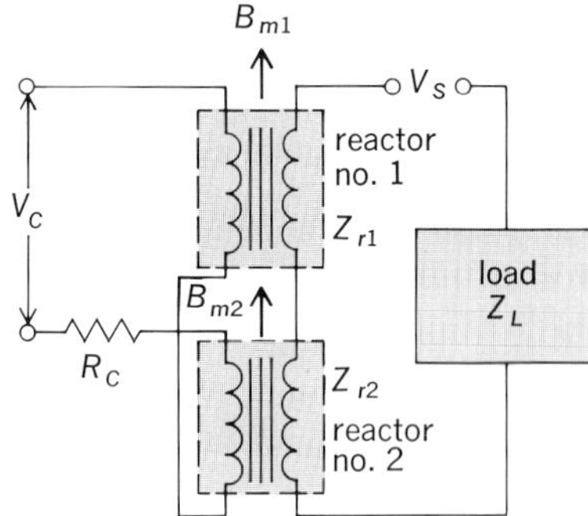

Fig. 2. Series-connected saturable reactors.

The saturable reactor is essentially a synchronous switch which closes for a controlled portion of the ac supply cycle. The duration, and therefore the energy content, of the load-current pulse is a function of the magnitude of the dc control voltage, increasing with increasing dc control voltage.

Magnetic amplifiers often employ two saturable reactors combined to drive a single load from a single dc control signal. The system shown in Fig. 2, a series-connected arrangement employing two saturable reactors, overcomes many of the limitations of the single-reactor system shown in Fig. 1. The series arrangement makes it possible to obtain a controlled pulse during each half-cycle of the ac power sinusoid, and the problem of minimizing the effects of pulses in the control circuit is considerably simplified. Because of the pulses of current induced in the control windings, special care must be exercised in the design of the amplifier (or other device) that provides the dc control signal for a magnetic amplifier. Magnetic amplifiers are very nonlinear in their operation, and detailed analysis of their characteristics requires careful application of the theory of nonlinear magnetic circuits.

The speed of response of magnetic amplifiers is often severely limited by the frequency of the ac power source, because the speed depends on the period of the ac power source. Impedance-matching problems at the input can also make it difficult to provide desired speed of response from the amplifier which drives the magnetic amplifier input. *See* SATURABLE REACTOR.

[J. LOWEN SHEARER]

Magneto

A type of permanent-magnet alternating-current generator frequently used as a source of ignition energy on tractor, marine, industrial, and aviation engines. The higher cost of magneto ignition is not warranted in modern automobiles, where storage batteries are required for other electrically operated equipment. Hand-operated magneto generators were once widely used for signaling from local battery telephone sets. *See* GENERATOR.

Modern induction-type magnetos consist of a permanent-magnet rotor and stationary low- and high-tension windings, also called the primary and

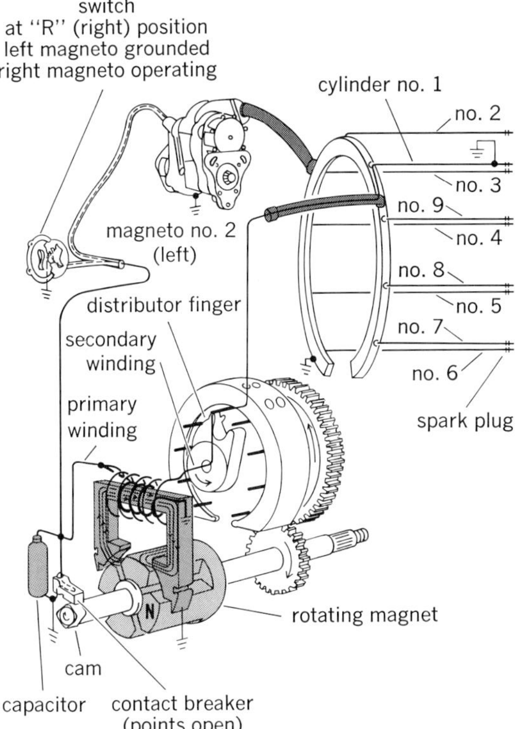

Multipolar aviation magneto in dual-ignition circuit for nine-cylinder radial engine. *(Bendix Aviation Corp.)*

secondary windings. The illustration shows two induction-type magnetos in a dual-ignition circuit for a nine-cylinder aircraft engine. The two magnetos are identical, but only magneto no. 1 is shown schematically. The ignition system may receive energy from either or both magnetos, according to the position of the switch.

The energy output of a magneto is obtained as a result of a rapid rate of change of flux through the stationary windings. The primary winding has comparatively few turns and the secondary winding has many thousand turns of fine wire. One end of the secondary winding is connected to an end of the primary winding and grounded to the frame of the magneto. The primary winding is closed on itself through a breaker mechanism actuated by a cam on the magneto shaft. The breaker is mechanically set to interrupt the primary circuit each time the flux through the winding is changing at its greatest rate. The sudden collapse of the primary current induces a very high voltage in the secondary winding.

Magnetos are always geared to the engine shaft and timed to open the breaker at the proper instant. Magneto speed depends on the number of poles of the magneto and the number of engine cylinders. The distributor finger speed is always one-half engine speed. *See* IGNITION SYSTEM; INTERNAL COMBUSTION ENGINE.

[ROBERT T. WEIL, JR.]

Bibliography: T. Baumeister (ed.), *Standard Handbook for Mechanical Engineers*, 8th ed., 1978; D. Fink and H. W. Beatty (eds.), *Standard Handbook for Electrical Engineers*, 11th ed., 1978; H. Pender and W. A. Del Mar (eds.), *Electrical Engineers' Handbook: Electric Power*, 4th ed., 1949.

Magnetohydrodynamic power generator

A system for the generation of electrical power through the interaction of a flowing, electrically conducting fluid with a magnetic field. As in a conventional electrical generator, the Faraday principle of motional induction is employed, but solid conductors are replaced by an electrically conducting fluid. The interactions between this conducting fluid and the electromagnetic field system through which power is delivered to a circuit are determined by the electrical magnetohydrodynamic (MHD) equations, while the properties of electrically conducting gases or plasmas are established from the appropriate relationships of plasma physics. Major emphasis has been placed on MHD systems utilizing an ionized gas, but an electrically conducting liquid or a two-phase flow can also be employed.

Improvement of the overall thermal efficiency of central station power plants has been the continuing objective of power engineers. Conventional plants based on steam turbine technology are limited to about 40% efficiency, imposed by a combination of working-fluid properties and limits on the operating temperatures of materials. Application of the MHD interaction to electrical power generation removes the restrictions imposed by the blade structure of turbines and enables the working-fluid temperature to be increased substantially. This enables the working fluid to be rendered electrically conducting and yields a conversion process based on a body force of electromagnetic origin. *See* ELECTRIC POWER GENERATION; STEAM TURBINE.

Principle. Electrical conductivity in an MHD generator can be achieved in a number of ways. At the heat-source operating temperatures of MHD systems (1000–3000 K), the working fluids usually considered are gases derived from fossil fuel combustion, noble gases, and alkali metal vapors. In the case of combustion gases, a seed material such as potassium carbonate is added in small amounts, typically about 1% of the total mass flow. The seed material is thermally ionized and yields the electron number density required for adequate electrical conductivity above about 2500 K. With monatomic gases, operation at temperatures down to about 1500 K is possible through the use of cesium as a seed material. In plasmas of this type, the electron temperature can be elevated above that of the gas (nonequilibrium ionization) to provide adequate electrical conductivity at lower temperatures than with thermal ionization. In so-called liquid metal MHD, electrical conductivity is obtained by injecting a liquid metal into a vapor or liquid stream to obtain a continuous liquid phase.

When combined with a steam turbine system to serve as the high-temperature or topping stage of a binary cycle, an MHD generator has the potential for increasing the overall plant thermal efficiency to around 50%, and values higher than 60% have been predicted for advanced systems. This follows from the increased efficiency made available by the higher source temperature. Thus, the MHD generator is a heat engine or electromagnetic turbine which converts thermal energy to a direct electrical output via the intermediate step of the kinetic energy of the flowing working fluid. *See* ENGINE.

The MHD generator itself consists of a channel

or duct in which a plasma flows at about the speed of sound through a magnetic field. The power output per unit volume W_e is given by $W_e = \sigma v^2 B^2 k \cdot (1-k)$, where σ is the electrical conductivity of the gas, v the velocity of the working fluid, B the magnetic flux density, and k the electrical loading factor (terminal voltage/induced emf). Typical values are $\sigma = 10-20$ mhos/m; $B = 5-6$ tesla; $v = 600-1000$ m/s; and $k = 0.7-0.8$. These values yield W_e in the range 25–150 MW/m^3. The high magnetic field strengths required are to be provided by a superconducting magnet, and the development of suitable electrodes to conduct current into and out of the gas is a major development problem. *See* SUPERCONDUCTING DEVICES.

Types. When this generator is embedded in an overall electrical power generation system, a number of alternatives are possible, depending on the heat source and working fluid selected. The temperature range required by MHD can be achieved through the combustion of fossil fuels with oxygen or compressed preheated air. The association of MHD with nuclear heat sources has also been considered, but in this case limitations on the temperature of nuclear fission heat sources with solid fuel elements has thus far precluded any practical scheme being developed where a plasma serves as the working fluid. The possibility of coupling MHD to a fusion reactor has been explored, and it is possible that 21st-century central station power systems will include both a fusion reactor and an MHD energy conversion system. *See* NUCLEAR REACTOR.

Development efforts on MHD are focused on fossil-fired systems, the fossil fuel selected being determined by national energy considerations. In the United States, coal is the obvious candidate, whereas in the Soviet Union, where ample reserves of natural gas are available, this is the preferred fuel. In Japan, major emphasis is on the use of petroleum-based fuels.

MHD power systems are classified into open- or closed-cycle systems, depending respectively on whether the working fluid is utilized on a once-through basis or recirculated via a compressor. For fossil fuels, the open-cycle system offers the inherent advantage of interposing no solid heat exchange surface between the combustor and the MHD generator, thus avoiding any limitation being placed on the cycle by the temperature attainable over a long period of operation by construction materials in the heat exchanger. Closed-cycle systems were originally proposed for nuclear heat sources, and the working fluid can be either a seeded noble gas or a liquid metal–vapor mixture.

Features. The greatest development effort in MHD power generation has been applied to fossil-fired open-cycle systems, but sufficient progress has been made in closed-cycle systems to establish their potential and to identify the engineering problems which must be solved before they can be considered practical. The rest of this article discusses fossil-fired open-cycle systems.

In addition to offering increased power plant efficiencies, MHD power generation also has important potential environmental advantages. These are of special significance when coal is the primary fuel, for it appears that MHD systems can utilize coal directly without the cost and loss of efficiency resulting from the processing of coal into a clean fuel required by competing systems. The use of a seed material to obtain electrical conductivity in the working fluid also places the requirement on the MHD system that a high level of recovery be attained to avoid adverse environmental impact and also to ensure acceptable plant economics. The seed recovery system required by an MHD plant also serves to recover all particulate material in the plant effluent. A further consequence of the use of seed material is its demonstrated ability to remove sulfur from coal combustion products. This occurs because the seed material is completely dissociated in the combustor, and the recombination phenomena downstream of the MHD generator favor formation of potassium sulfate in the presence of sulfur. Accordingly, seed material acts as a built-in vehicle for removal of sulfur. Laboratory experiments have shown that the sulfur dioxide emissions can be reduced to levels below those experienced with natural-gas–fired plants. A further important consideration is the reduction of the emission of oxides of nitrogen through control of combustion and the design of component operating conditions. Laboratory scale work has demonstrated that these emissions can be controlled to the most exacting standards prescribed by the Environmental Protection Administration (EPA).

While not a property of the MHD system in itself, the potential of MHD to operate at higher thermal plant efficiencies has the consequence of substantial deduction in thermal waste discharge, following the relationship that the heat rejected per unit of electricity generated is given by $(1-\eta)/\eta$, where η is the plant efficiency. As the technology of MHD is developed along with that of advanced gas turbines, there also exists the possibility that MHD systems can dispense entirely with the need for large amounts of cooling water for steam condensation through the coupling of MHD generators with closed-cycle gas turbines. *See* WATER POLLUTION.

State of development. The most complete plant constructed is the U-25 installation near Moscow. It is of the open-cycle type and fired with natural gas. It is operated both at its rated power of 20.5 MW and continuously for 250 h at lower power levels, in both cases delivering its output to the Moscow grid. Results from the U-25 are providing the design basis for a 500-MW (electrical) commercial demonstration plant being constructed at Ryazan and scheduled for startup in 1984–1985.

In the United States, important engineering progress has been achieved on coal-fired systems, and 1000 h of operation of an open-cycle channel have been completed. At large experimental generator at the Arnold Engineering Development Center in Tennessee has reached an electrical power above 30 MW.

[WILLIAM D. JACKSON]

Bibliography: International MHD Liaison Group, Nuclear Energy Agency, OECD, Paris, *1976 MHD Status Report*, 1976; W. D. Jackson and P. S. Zygielbaum, Open cycle MHD power generation: Status and engineering development approach, *Proc. Amer. Power Conf.*, 37:1058–1071, 1975; R. J. Rosa, *Magnetohydrodynamic Energy Conversion*, 1968.

Manufacturing engineering

A function of industry that deals with all aspects of the production process.

Function. The function of manufacturing engineering is to develop and optimize the production process. This can best be understood by considering the total process through which a designer's concept becomes a marketable product: (1) The product designer (or product design department) conceptualizes a product. Drawings and one or more prototypes of this product are produced, usually by manual methods, but increasingly by computer-aided design and manufacturing. (2) The finalized prototype and its part drawings are released to the manufacturing engineering department, which starts designing and building an economically justifiable process by which the product will be produced. (3) When the manufacturing process developed by manufacturing engineering has been thoroughly tried and proved workable, it is turned over to the production department, which assumes responsibility for product manufacture.

In brief, the manufacturing engineering department bridges the gap between product design and full production. It discharges responsibilities in a number of subordinate fields of engineering: process engineering, tool engineering, material handling, plant engineering, and standards and methods.

It must be noted that this list of component disciplines, while typical, is not universal. In virtually all manufacturing facilities the composition of subordinate engineering functions is adjusted to meet the needs of the product to be made. As an example, the manufacturing engineering structure in an aerospace plant would be very different from that of an automobile plant. A single company, such as Westinghouse or General Electric, will show marked differences in the composition of manufacturing engineering within its different branches (see illustration). In all cases, however, the objec-

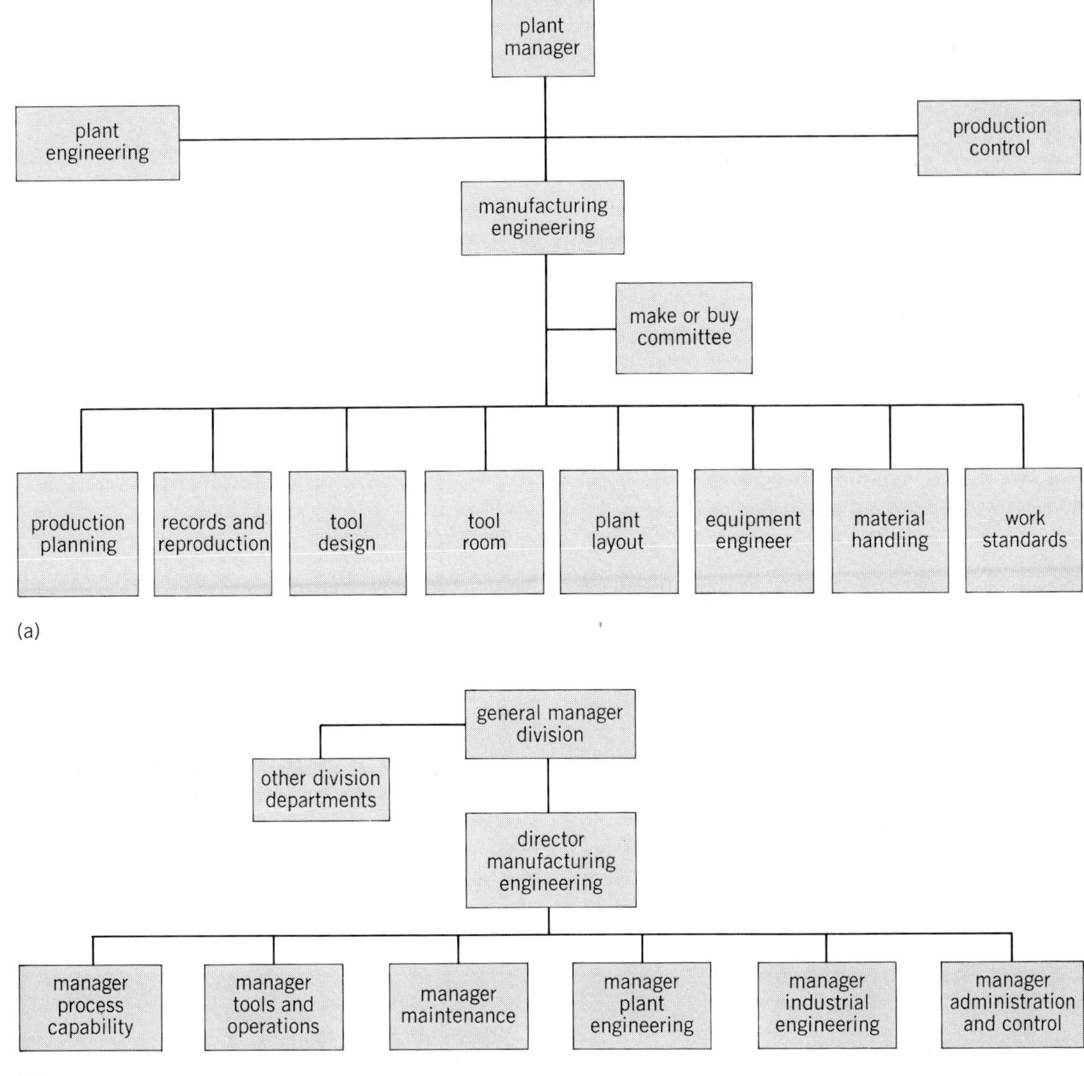

Manufacturing engineering organization varies with the needs of the company, as is seen in two markedly different organizational charts. (*a*) Ford Aeronutronics (*from H. W. Wage, Manufacturing Engineering, McGraw-Hill, 1963*). (*b*) Whirlpool Corporation (*from R. E. Finley and H. R. Ziobro, eds., The Manufacturing Man and His Job, American Management Association, 1966*).

tive of manufacturing engineering is the same: to design and develop the manufacturing process in the most economical manner possible.

Process engineering. The function of this component is to develop a logical sequence of manufacturing operations for each product part. A gear is a typical, extremely simple example. In one of the several methods by which a gear may be made, the gear blank must be sawed from round bar stock; its thickness, outside and inside diameters, and teeth must be machined; it must go through a heat-treatment process to obtain requisite hardness, after which all surfaces, including the teeth, must be ground. It is the function of process engineering to determine the correct sequence of operations, interposing the appropriate number and types of inspection operations between operations. Each operation must be assigned to an appropriate machine; preliminary estimates of the amount of time required for each operation must be made. All of this information is recorded on the process sheet, the most important document used in the entire manufacturing engineering process.

Tool engineering. The process sheets specify the need for tools by which the various operations are performed. In this context, the term tool refers to an array of industrial equipment such as dies, jigs, fixtures, gages, drills, reamers, counterbores, taps, milling cutters, and single-point turning and boring tools. Many of these tools—drills, for example—are standardized. As such, they can easily be procured through commercial outlets. Other tools, such as dies, jigs, and fixtures, are not standard; they must be made to suit the needs of the individual product component. Thus the first assignment of tool engineering is to design the nonstandard tools.

With tool design complete, tool engineering must determine which tools can be produced internally, that is, in the company's own toolroom, and which must be produced externally, in contract tool shops. The decision is usually based on economic considerations (the bid process in contract shops assures lower prices), on toolroom load (most plants do not have sufficient toolroom capacity to handle an entire tooling program), and on union contracts (many contracts place severe constraints on the release of work to outside vendors). Whether the tooling is produced internally or externally, tool engineering bears responsibility for its fabrication according to company standards and for its ability to perform its specified assignments in production.

Material handling. This third component of manufacturing engineering also starts work with the process sheet. The sheet specifies that certain operations will be performed at certain machines or at specified points in the assembly process. This implies the movement of parts in process over extended distances. Part movement from one plant to another is not uncommon, sometimes over widely separated geographical areas. Thus material handling is faced with the problem of moving parts in a time context. A failure in this regard means parts piling up in one area while production comes to a halt in another due to lack of materials. In brief, the flow of materials must be constant and of correct quantity. Material handling must establish the routing and the schedules.

Plant engineering. Closely allied to the material handling function is plant layout or plant engineering. This function involves the correct locating and positioning of machine tools and other pieces of production equipment. The importance of this function can best be appreciated by considering the vast changes made in the automobile plants of the United States during the downsizing operations of the late 1970s and early 1980s. The abrupt change in automobile size, plus significant changes in the materials used, necessitated drastic alteration in the physical layout of each plant. Plant engineering effected these changes.

Smaller changes in plant layout are part of an ongoing process. Many are made in response to changes in product design. But other changes in plant layout are made to optimize the production process. The objective is to reduce manufacturing costs to a minimum. This, above all else, is the corporate objective of manufacturing engineering.

Standards and methods. Every manufacturing concern of any size finds it necessary to establish a standards department. Closely allied to standards is methods engineering. Standards comprise work standards and tool and machine standards. A work standard is the scientifically established amount of time required to perform a given work assignment. It is one of the most socially sensitive aspects of modern engineering, and of critical importance. In mass production the entire system operates by the coordinated efforts of hundreds, perhaps thousands, of persons. Each worker must perform his or her task in an allotted period of time. Accordingly it is imperative that work standards (that is, time standards) be established.

Tool and machine standards pertain to the physical characteristics of plant hardware and material. These standards are especially prevalent in tooling. As an example, stamping dies are built in die sets, which must be adaptable to the company's presses. To assure their adaptability, the standards department develops appropriate dimensions, which the tool engineering department must adhere to in the design of tooling. Similarly experience may have proved that a certain type of steel exhibits superior performance in given tooling applications. The standards department then establishes that steel as a company standard for use in similar applications. This process of standardization reaches into every aspect of production, the objective being to simplify the work and thus reduce production costs.

Through methods engineering, the individual jobs and the entire workplace are made more efficient. This department studies the methods by which work is done. (Does a worker have to reach too far to pick up an incoming workpiece? Could a worker perform more efficiently if the workplace were better lighted? Does the worker have efficient access to tools?) Thus methods engineering and standards engineering work hand in hand to simplify work and reduce its time requirement.

Other functions. Manufacturing engineering is involved in numerous other aspects of the production process, often by committee approach. A key committee is the make-or-buy. Every component must be evaluated as to whether it should be made in-house or purchased from vendors. The economics of each decision is based largely on capabilities and costs. In its make-or-buy role, manufacturing engineering performs a function that cannot be handled via routine purchasing department practice.

Valuable feedback to product design is provided by manufacturing engineering's value analysis committee. It is not uncommon for product design to increase production costs unnecessarily through certain design requirements. After performing a value analysis of each component, manufacturing engineering can often obtain design concessions that lead to production-cost reductions while in no way affecting part performance.

Academic roots. Manufacturing engineering has evolved in response to a need—that of bridging the gap between product design and full production. It is now developing academic roots that extend deeply into the conventional engineering subjects of mathematics, chemistry, physics, and the like. These are followed with intensive study of the newer and more exotic technologies, namely, the laser, robotics, numerical control, and all aspects of electrical machining. Additionally, students are given intensive indoctrination in the use of the computer as the primary tool of manufacturing. These curricula, many of which are still in the formative state, are currently being given impetus through the educational activities of the Society of Manufacturing Engineers (SME) and through the financial support of its Manufacturing Engineering Education Foundation.

Mechanical engineering. This well-established engineering discipline has evolved in response to a more generalized industrialized need. Mechanical engineers are especially proficient in all aspects of machine design, of industrial research, and in the development of various forms of industrial equipment. Many mechanical engineers have gravitated into key posts in manufacturing engineering, where their academic training has enabled them to adapt to the demands of this related profession. Mechanical engineers are given basic courses in mathematics and the sciences, followed by intensive indoctrination in thermodynamics, strength of materials, fluid dynamics, chemistry encompassing qualitative (and sometimes quantitative) analysis, plus advanced courses in electricity.

Industrial engineering. This engineering function was originally established by Frederick Winslow Taylor, a pioneer American engineer and inventor, and the conceptualizer of scientific management. As conceived by Taylor, industrial engineering was concerned with the development of work standards and methods. As with manufacturing engineering, the role of industrial engineering is largely determined by individual companies. In some concerns, manufacturing engineering and industrial engineering perform parallel functions; in others, industrial engineering retains its original function (standards and methods), but has been enlarged to include the plant engineering function. Increasingly, however, industrial engineering is—like tool engineering, process engineering, and the others—a subordinate branch of the more inclusive function and discipline of manufacturing engineering.

[DANIEL B. DALLAS]

Bibliography: D. B. Dallas, *Manufacturing Engineering Defined*, Spec. Rep. Soc. Manuf. Eng., 1978; R. E. Finley and H. R. Ziobro (eds.), *The Manufacturing Man and His Job*, American Management Association, 1966; *The Manufacturing Engineer—Past, Present and Future*, Battelle Memorial Institute, 1979; H. W. Wage, *Manufacturing Engineering*, 1963.

Marine mining

The process of recovering mineral wealth from sea water and from deposits on and under the sea floor. Unknown except to technical specialists before 1960, undersea mining is receiving increasing attention. Frequent references to marine mineral resources and marine minerals legislation by national and international policy makers, increasing activity in marine minerals exploration, and the launching of major new seagoing mining dredges for South Africa and Southeast Asia all indicate the beginnings of a viable and expanding marine mining industry.

There are sound reasons for this sudden emphasis on a previously little-known source of minerals. While the world's demand for mineral commodities is increasing at an alarming rate, most of the developed countries have been thoroughly explored for surface outcroppings of mineral deposits. The mining industry has been required to advance its capabilities for the exploration and exploitation of low-grade and unconventional sources of ore. Corresponding advances in oceanology have highlighted the importance of the ocean as a source of minerals and indicated that the technology required for their exploitation is in some cases already available.

There is a definite realization that the venture into the oceans will require large investments, and the trend toward the consortium approach is very noticeable, not only in exploration and mining activities but in research. Undersea mining has become an important diversification for many major oil and aerospace companies, and a few mining companies appear to be taking an aggressive approach. In the late 1960s over 80 separate exploration activities were reported in coastal areas worldwide. In the 1970s interest turned more to the deep-seabed deposits, which have been the subject of much debate at the United Nations 3d Law of the Sea Conference. Conflicts over the rights of management of these minerals in international waters and the implication of the term "common heritage of mankind" which was applied to them have led to attempted moratoria on deep-seabed mining activities. Anticipating agreement, several countries, including the United States and West Germany, prepared interim legislation which would permit nationals of the respective countries and reciprocating states to mine the deep seabed for nodules containing manganese, copper, nickel, and cobalt. At least five major multinational consortia were active in testing deep-seabed mining systems in the late 1970s, including companies from the United States, Canada, United Kingdom, Australia, West Germany, France, Belgium, Netherlands, and Japan.

While mineral resources to the value of trillions of dollars do exist in and under the oceans, their exploitation is not simple. Many environmental problems must be overcome and many technical advances must be made before the majority of these deposits can be mined in competition with existing land resources.

The marine environment may logically be divided into four significant areas: the waters, the deep sea floor, the continental shelf and slope, and the seacoast. Of these, the waters are the most significant, both for their mineral content and for their unique properties as a mineral overburden. Not only do they cover the ocean floor with a fluid medium quite different from the earth or atmosphere and requiring entirely different concepts of ground survey and exploration, but the constant and often violent movement of the surface waters combined with unusual water depths present formidable deterrents to the use of conventional seagoing techniques in marine mining operations.

The mineral resources of the marine environment are of three basic types: the dissolved minerals of the ocean waters; the unconsolidated mineral deposits of marine beaches, continental shelf, and deep-sea floor; and the consolidated deposits contained within the bedrock underlying the seas. These are described in Table 1, which shows also the subclasses of surficial and in-place deposits, characteristics which have a very great influence on the economics of exploration and mining.

As with land deposits, the initial stages preceding the production of a marketable commodity include discovery, characterization of the deposit to assess its value and exploitability, and mining, including beneficiation of the material to a salable product.

Exploration. Initial requirements of an exploration program on the continental shelves are a thorough study of the known geology of the shelves and adjacent coastal areas and the extrapolation of known metallogenic provinces into the offshore areas. The projection of these provinces, which are characterized by relatively abundant mineralization, generally of one predominant type, has been practiced with some success in the localization of certain mineral commodities, overlain by thick sediments. As a first step, the application of this technique to the continental shelf, overlain by water, is of considerable guidance in localizing more intensive operations. Areas thus delineated are considered to be potentially mineral-bearing and subject to prospecting by geophysical and other methods.

A study of the oceanographic environment may indicate areas favorable to the deposition of authigenic deposits in deep and shallow water. Some deposits may be discovered by chance in the process of other marine activities.

Field exploration prior to or following discovery will involve three major categories of work: ship operation, survey, and sampling.

Ship operation. Conventional seagoing vessels are used for exploration activities with equipment mounted on board to suit the particular type of operation. The use of submersibles will no doubt eventually augment existing techniques but they are not yet advanced sufficiently for normal usage.

One of the most important factors in the location of undersea minerals is accurate navigation. Ore bodies must be relocated after being found and must be accurately delineated and defined. The accuracy of survey required depends upon the phase of operation. Initially, errors of 1000 ft (300 m) or more may be tolerated.

However, once an ore body is believed to exist in a given area, maximum errors of less than 100 ft (30 m) are desirable. These maximum tolerated errors may be further reduced to a few feet in detailed ore body delineation and extraction.

There are a variety of types of electronic navigation systems available for use with accuracies from 3000 ft (900 m) down to approximately 3 ft (1 m). Loran, Lorac, and Decca are permanently installed in various locations throughout the world. Small portable systems are available for local use that provide high accuracy within 30–50-mi (50–80-km) ranges. For deep-ocean survey, navigational satellites have completely revolutionized the capabilities for positioning with high accuracy in any part of the world's oceans.

During sampling and mining operations, the vessel must be held steady over a selected spot on the ocean floor. Two procedures that have been fairly well developed for this purpose are multiple anchoring and dynamic positioning.

A three-point anchoring system is of value for a

Table 1. Marine mineral deposits

Dissolved	Unconsolidated			Consolidated
	Continental shelf, 0–200 m (littoral)	Continental slope, 200–3500 m (bathyal)	Deep sea, 3500–6000 m (abyssal)	
Sea water:	*Nonmetallics:*	*Authigenics:*	*Authigenics:*	*Disseminated, massive, vein, tabular,*
Fresh water	Sand and gravel	Phosphorite	Ferromanganese nodules	*or stratified deposits of:*
Metals and salts of:	Lime sands and shells	Ferromanganese oxides	and associated:	Coal
Magnesium	Silica sand	and associated minerals	Cobalt	Ironstone
Sodium	Semiprecious stones	Metalliferous mud with:	Nickel	Limestone
Calcium	Industrial sands	Zinc	Copper	Sulfur
Bromine	Phosphorite	Copper		Tin
Potassium	Aragonite	Lead	*Sediments:*	Gold
Sulphur	Glauconite	Silver	Red clays	Metallic sulfides
Strontium			Calcareous ooze	Metallic salts
Boron	*Heavy minerals:*		Siliceous ooze	Hydrocarbons
Uranium	Magnetite			
Other elements	Ilmenite			
	Rutile			
Metalliferous brines:	Monazite			
Concentrations of:	Chromite			
Zinc	Zircon			
Copper	Cassiterite			
Lead				
Silver	*Rare and precious minerals:*			
	Diamonds			
	Platinum			
	Gold			
	Native copper			

coring vessel working close to the surf. A series of cores may be obtained along the line of operations by winching in the forward anchors and releasing the stern anchor. Good positive control over the vessel can be obtained with this system, and if conditions warrant, a four-point anchoring system may be used. Increased holding power can be obtained by multiple anchoring at each point.

Dynamic positioning is useful in deeper water, where anchoring may not be practical. The ship is kept in position by use of auxiliary outboard propeller drive units or transverse thrusters. These can be placed both fore and aft to provide excellent maneuverability. Sonar transponders are held submerged at a depth of minimum disturbance, or the system may be tied to shore stations. The auxiliary power units are then controlled manually or by computer to keep the ranges constant.

Survey. The primary aids to exploration for mineral deposits at sea are depth recorders, subbottom profilers, magnetometers, bottom sampling devices, and subbottom sampling systems. Their use is dependent upon the characteristics of the ore being sought.

For the initial topographic survey of the sea floor, and as an aid to navigation, in inshore waters, the depth recorder is indispensable. It is usually carried as standard ship equipment, but precision recorders having a high accuracy are most useful in survey work.

In the search for marine placer deposits of heavy minerals, the subbottom profiler is probably the most useful of all the exploration aids. It is one of several systems utilizing the reflective characteristics of acoustic or shock waves.

Continuous seismic profilers are a development of standard geophysical seismic systems for reflection surveys, used in the oil industry. The normal energy source is explosive, and penetration may be as much as several miles.

Subbottom profilers use a variety of energy sources including electric sparks, compressed air, gas explosions, acoustic transducers, and electromechanical (boomer) transducers. The return signals as recorded show a recognizable section of the subbottom. Shallow layers of sediment, configurations in the bedrock, faults, and other features are clearly displayed and require little interpretation. The maximum theoretical penetration is dependent on the time interval between pulses and the wave velocity in the subbottom. A pulse interval of 1/2 s and an average velocity of 8000 ft/s (2400 m/s) will allow a penetration of 2000 ft (600 m), the reflected wave being recorded before the next transmitted pulse.

Penetration and resolution are widely variable features on most models of wave velocity profiling systems. In general, high frequencies give high resolution with low penetration, while low frequencies give low resolution with high penetration. The general range of frequencies is at the low end of the scale and varies from 150 to 300 Hz, and the general range of pulse energy is 100–25,000 joules for nonexplosive energy sources. The choice of system will depend very much on the requirements of the survey, but for the location of shallow placer deposits on the continental shelf the smaller low-powered models have been used with considerable success.

With the advent of the flux gate, proton precession, and the rubidium vapor magnetometer, all measuring the Earth's total magnetic field to a high degree of accuracy, this technique has become much more useful in the field of mineral exploration.

Anomalies indicative of mineralization such as magnetic bodies, concentrations of magnetic sands, and certain structural features can be detected. Although all three types are adaptable to undersea survey work, the precession magnetometer is more sensitive and more easily handled than the flux gate, and the rubidium vapor type has an extreme degree of sensitivity which enhances its usefulness when used as a gradiometer on the sea surface or submerged.

Once an ore body is indicated by geological, geophysical, or other means, the next step is to sample it in area and in depth.

Sampling. Mineral deposit sampling involves two stages. First, exploratory or qualitative sampling locate mineral values and allow preliminary judgment to be made. For marine deposits this will involve such simple devices as snappers, drop corers, drag dredges, and divers. Accuracy of positioning is not critical at this stage, but of course is dependent on the type of deposit being sampled. Second, the deposit must be characterized in sufficient detail to determine the production technology requirements and to estimate the profitability of its exploitation. This quantitative sampling requires much more sophisticated equipment than does the qualitative type, and for marine work few systems in existence can be considered reliable and accurate. However, in particular cases, systems can be put together using available hardware which will satisfy the need to the accuracy required. Specifically, qualitative sampling of any mineral deposit offshore can be carried out with existing equipment. Quantitative sampling of most alluvial deposits of heavy minerals (specific gravity, less than 8) can be carried out at shallow depths (less than 350 ft or 107 m overall) using existing equipment but cannot be carried out with reliability for the higher-specific-gravity minerals such as gold (specific gravity 19). Quantitative sampling of any consolidated mineral deposit offshore can be carried out within limits.

Any system that will give quantitative samples can be used for qualitative sampling, but in many cases heavy expenses could be avoided by using the simpler equipment.

To obviate the effects of the sea surface environment, the trend is toward the development of fully submerged systems, but it should be noted that the deficiencies in sampling of the heavy placer minerals are not due to the marine environment. Even on land the accuracy of placer deposit evaluation is not high and the controlling factors not well understood. There is a prime need for intensive research in this area.

Evaluation of surficial nodule deposits on the deep seabed at depths of 5000–6000 m may be carried out by using combinations of optical or acoustic imagery and sampling. Except for box corers which are used for geotechnical sampling, devices lowered from a stationary vessel have given way to the use of free-fall (boomerang) samplers equipped with corers, grabs, or cameras, and discharged at preselected points in groups of 8–12 for later retrieval. Also widely utilized are towed

Table 2. Production from dissolved mineral deposits offshore

Mineral	Location	Number of operations	Annual production*
Sodium, NaCl	Worldwide	90+	10,000,000 tons
Magnesium, metal	United States, United	2	221,000 tons
$Mg, MgO, Mg(OH)_2$	Kingdom, Germany, Soviet Union	25+	800,000 tons
Bromine, Br	Worldwide	7	102,000 tons
Fresh water	Middle East, Atlantic region, United States	150+	†
Heavy water	Canada	1	†
Total		275+	

*1 ton = 0.907 metric ton.
†Not reported.

television or multiple-shot photographic cameras which give a fully or nearly continuous coverage along selected tracks. Nodules are commonly assayed aboard ship and the data from all systems analyzed by using onboard computers. A variety of seabed maps showing bathymetry, seabed topography, and nodule distribution and grade may be produced while the vessel is still on site. Ultimately methods for continuous in-place assaying while under way should be perfected, which will allow the requisite data maps to be produced in real time.

Exploitation. Despite the intense interest in undersea mining, new activities have been limited mostly to conceptual studies and exploration. The volume of production has shown little change, and publicity has tended to overemphasize some of the smaller, if more newsworthy, operations. All production comes from nearshore sources, namely, sea water, beach and nearshore placers, and nearshore consolidated deposits.

Minerals dissolved in sea water. Commercial separation techniques for the recovery of minerals dissolved in sea water are limited to chemical precipitation and filtration for magnesium and bromine salts and solar evaporation for common salts and fresh-water production on a limited scale. Other processes developed in the laboratory on pilot plant scale include electrolysis, electrodialysis, adsorption, ion exchange, chelation, oxidation, chlorination, and solvent extraction. The intensive interest in the extraction of fresh water from the sea has permitted much additional research on the recovery of minerals, but successful commercial operations will require continued development of the combination of processes involved for each specific mineral.

As shown in Table 2, three minerals or mineral suites are extracted commercially from sea water: sodium, magnesium, and bromine. Of these, salt evaporites are the most important. Japan's total production of salt products comes from the sea. Magnesium extracted from sea water accounts for 75% of domestic production of this commodity in the United States, and fresh water compares with bromine in total production value.

Unconsolidated deposits. Unconsolidated deposits include all the placer minerals, surficial and in place, as well as the authigenic deposits found at moderate to great depths.

The mining of unconsolidated deposits became widely publicized with the awareness of the potential of manganese nodules as a source of manganese, copper, nickel, and cobalt, and because of the exciting developments in the exploitation of offshore diamonds in South-West Africa in the late 1960s. Despite the fact that there are presently no operations for nodules and the offshore diamond mining operations have been suspended, unconsolidated deposits have for some years presented a major source of exploitable minerals offshore.

So far the methods of recovery which have been

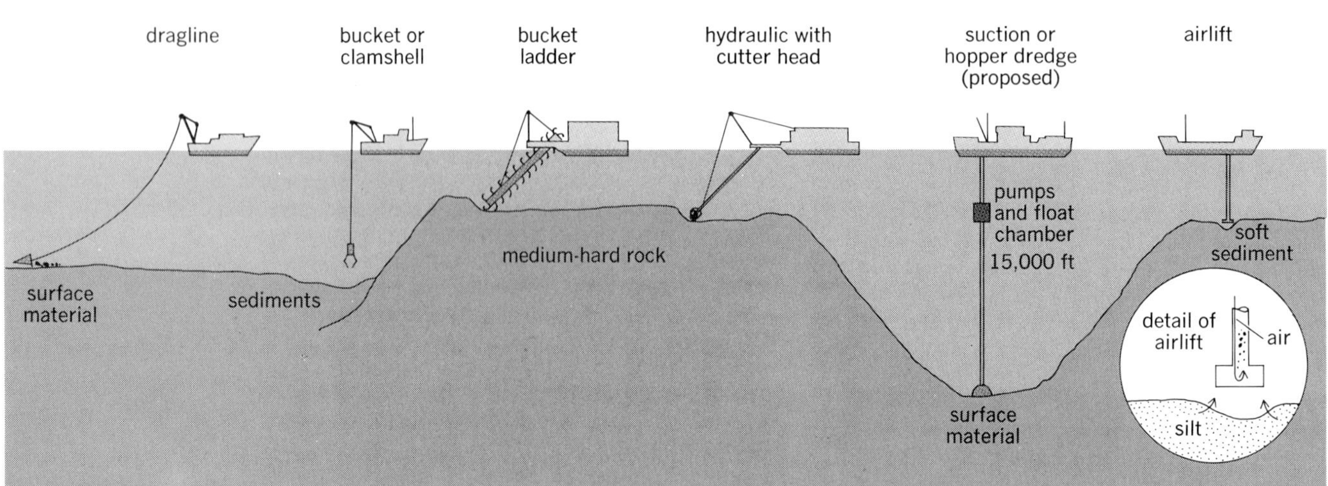

Fig. 1. Methods of dredging used in the exploitation of unconsolidated mineral deposits offshore.

used or proposed have been conventional, namely, by dredging using draglines, clamshells, bucket dredges, hydraulic dredges, or airlifts. All these methods (Fig. 1) have been used in mining to maximum depths of 200 ft (60 m) and hydraulic dredges for digging to 300 ft (90 m) are being built. Extension to depths much greater than this does not appear to present any insurmountable technical difficulties.

More than 70 dredging operations were active in the 1970s, exploiting such diverse products as diamonds, gold, heavy mineral sands, iron sands, tin sands, lime sand, and sand and gravel. The most important of all of these commodities is the least exotic; 60% of world production from marine unconsolidated deposits is involved in dredging and mining operations for sand and gravel. Other major contributors to world production are the operations for heavy mineral sands (ilmenite, rutile, and zircon), mostly in Australia, and the tin operations in Thailand and Indonesia, which account for more than 10% of the world's tin.

Economics of these operations are dictated by many conditions. The spectacular range of costs offshore results from the effects of different environmental conditions. In general, it may be said that offshore operations are more costly than similar operations onshore.

The operations of Marine Diamond Corporation are of considerable historical interest. The first pilot dredging commenced in 1961 with a converted tug, the *Emerson K*, using an 8-in. (20-cm) airlift. The operation expanded until 1963, with the fleet consisting of 3 mining vessels, all using an air- or jet-assisted suction lift, 11 support craft, and 2 aircraft. Production totaled over $1,700,000 of stones during that year from an estimated 322,000 yd^3 (246,000 m^3) of gravel. At that time, the estimation of mining cost was $2.33/yd^3 ($3.05/m^3), showing a profit of nearly $3/yd^3 ($4/m^3). Subsequent unexpected problems, including severe storms, operating difficulties, and loss of one of the mining units, led to a reduction of profits and transfer of company control.

Production operations of Marine Diamond Corporation fluctuated considerably. *Diamantkus*, a vessel designed to produce 7000 yd^3/h (9000 m^3/h), was withdrawn as uneconomic after only 30 months of service. Only two mining units, *Barge III* and *Colpontoon*, were in operation in 1968, both converted pipe-laying barges using combination airlifts and suction dredging equipment. A third and larger unit, the *Pomona*, a multiple-head suction dredge, was commissioned in March 1967, but was damaged by storms on the first trial run. The characteristics of the bedrock, with its many potholes and extremely irregular surface, added to the difficulty in recovering the maximum amount of diamonds. Mining operations ceased in May 1971, but large areas remained to be explored.

Liberal offshore mining laws introduced in 1962 in the state of Alaska resulted in an upsurge of exploration activity for gold, particularly in the Nome area. In 1968 a mining operation was attempted, using a 20-in. (50-cm) hydrojet dredge, in submerged gravels about 60 mi (100 km) east of Nome. No production was reported.

Over 70% of the world's heavy mineral sand production is from beach sand operations in Australia, Sri Lanka, and India. Only two oceangoing dredges are used. The majority being pontoon-mounted hydraulic dredges, or draglines, with separate washing plants.

The Yawata Iron and Steel Co. in Japan used a 10.5-yd^3 (8.0-m^3) barge-mounted grab dredge and a hydraulic cutter dredge to mine iron sand from the floor of Ariake Bay in water depths of 50 ft (15 m). These operations were suspended finally in 1966, the reason being given that the reserves had not been accurately surveyed and the cost of mining had not been competitive with Yawata's alternate sources of supply.

An interesting comparison between a clamshell dredge from Aokam Tin and a bucket-ladder dredge from Tongka Harbor, working offshore on the same deposit in Thailand, was made. The clamshell was set up as an experimental unit using an oil tanker hull. It was designed for a digging depth of 215 ft (65 m), and mobility and seaworthiness were prime factors in its favor. However, in practice, it was never called upon to dig below 140 ft (43 m), its mobility was superfluous, and the ship hull proved very unsatisfactory in terms of usable space. Although it was able to operate in sea states which prevented the operation of the neighboring bucket dredge, its mining recovery factor was low and its operating costs much higher than anticipated. It was withdrawn from service after only 9 years, in favor of the bucket dredge.

Another major operation is run by Indonesian State Mines off the islands of Bangka and Belitung. The operations are as far as 3 mi (5 km) from shore in waters which are normally calm. They do have storms, however, which necessitate delays in the operation and the taking of precautions unnecessary onshore. The operations employ 12 dredges, one of which, the *Belitung I*, constructed in 1979, is the world's largest mining bucket-line dredge. With a maximum digging depth of 150 ft (46 m), it is designed to dig and treat 650,000 yd^3 (500,000 m^3) per month of 600 h and produce over 1500 tons (1350 metric tons) of tin metal per year, depending on the richness of the ground. The dredge is working up to 15 mi (24 km) offshore.

Lime shells are mined as a raw material for portland cement. Two United States operations for oyster shells in San Francisco Bay and Louisiana employed barge-mounted hydraulic cutter dredges of 16- and 18-in. (41- and 46-cm) diameter in 30–50 ft (9–15 m) of water. The Iceland Government Cement Works in Akranes uses a 150-ft (46 m) ship to dredge sea shells from 130-ft (40 m) of water, with a 24-in. (61-cm) hydraulic drag dredge.

In the United Kingdom, hopper dredges are used for mining undersea reserves of sand and gravel. Drag suction dredges up to 38 in. (97 cm) in diameter are most commonly used with the seagoing hopper hulls. Similar deposits have been mined in the United States, and the same type of dredge is employed for the removal of sand for harbor construction or for beach replenishment. Some sand operations use beach-mounted drag lines for removal of material from the surf zone or beyond.

For deep-seabed mining at depths below 5000 m, systems have been tested using airlift or suction lift and a mechanical system using a continuous loop of plastic line from which buckets are suspended every 100 m or so. The gathering devices tested on the bottom include towed and self-propelled systems, the latter incorporating a

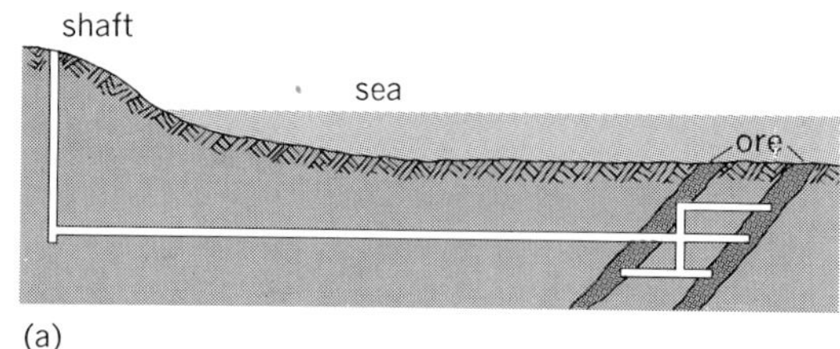

(a)

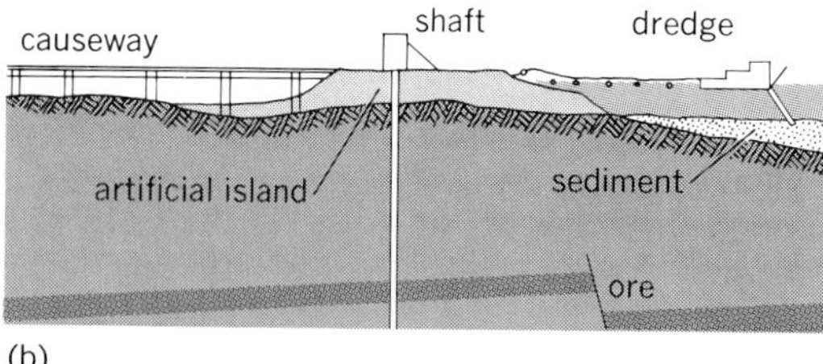

(b)

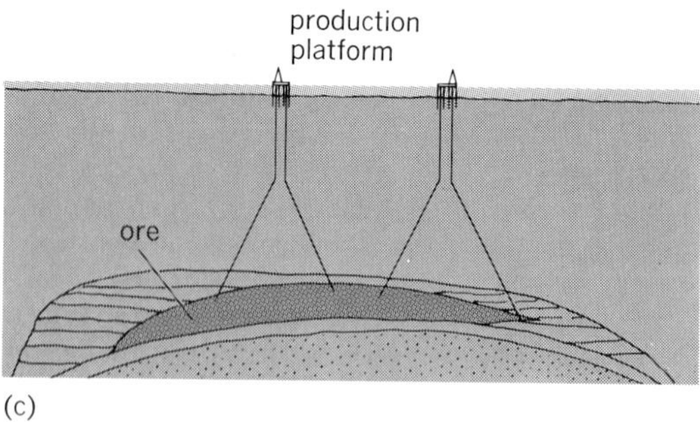

(c)

Fig. 2. Methods of mining for exploitation of consolidated mineral deposits offshore. (*a*) Shaft sunk on land, access by tunnel. (*b*) Shaft sunk at sea on artificial island. (*c*) Offshore drilling and in-place mining.

means to crush and slurry the nodules before pumping. All the tests carried out in the 1970s have been about one-fifth scale for production systems of 1,000,000 to 3,000,000 tons (900,000 to 2,700,000 metric tons) annually. Concurrent tests to identify the environmental effects of mining have not been conclusive, but tend to indicate that adverse effects are minor and can be mitigated.

Consolidated deposits. The third and last area of offshore mineral resources has an equally long history. The production from in-place mineral deposits under the sea is quite substantial, particularly in coal deposits. Undersea coal accounts for almost 30% of the total coal production in Japan and just less than 10% in Great Britain.

Extra costs have been due mainly to exploration, with mining and development being usually conventional. In the development of the Grand Isle sulfur mine off Louisiana, some $8,000,000 of the $30,000,000 expended was estimated to be due to its offshore location. There is no doubt that costs will be greater, generally, but on the other hand, in the initial years of offshore mining as a major industry, the prospects of finding accessible, high-grade deposits will be greater than they are at present on land.

Some of the mining methods are illustrated in Fig. 2. For most of the bedded deposits which extend from shore workings a shaft is sunk on land with access under the sea by tunnel. Massive and vein deposits are also worked in this manner. Normal mining methods are used, but precautions must be taken with regard to overhead cover. Near land and in shallow water a shaft is sunk at sea on an artificial island. The islands are constructed by dredging from the seabed or by transporting fill over causeways. Sinking through the island is accompanied by normal precautions for loose, waterlogged ground, and development and mining are thereafter conventional. The same method is also used in oil drilling. Offshore drilling and in-place mining are used only in the mining of sulfur, but this method has considerable possibilities for the mining of other minerals for which leaching is applicable. Petroleum drilling techniques are used throughout, employing stationary platforms constructed on piles driven into the sea floor or floating drill rigs. *See* OIL AND GAS, OFFSHORE; UNDERGROUND MINING.

In summary, Table 3 shows offshore mining production. Though this is only a fraction of world mineral production, the results of the extensive exploration activity taking place off the shores of all five continents may alter this considerably in the future.

The future. Despite the technical problems which still have to be overcome, the future of the undersea mining industry is without doubt as potent as it is fascinating.

Deposits of hot metalliferous brines and oozes enriched with gold, silver, lead, and copper have been located over a 38.5-mi² (100-km²) area in the middle of the Red Sea at depths of 6000–7000 ft (1800–2100 m). Similar deposits are indicated over vast areas of the East Pacific Rise.

Major problems of dissolved mineral extraction must be solved before their exploitation, and significant advances must be made in the handling of these sometimes corrosive media at such depths and distances from shore.

The mining of unconsolidated deposits will call for the development of bottom-sited equipment to perform the massive earth-moving operations that are carried out by conventional dredges today. The remarkable deposits of Co-Ni-Cu-Mn nodules covering the deep ocean floors will require new

Table 3. Summary of production from mineral deposits offshore

Type	Minerals	Number of operations
Dissolved minerals	Sodium, magnesium, calcium, bromine	275+
Unconsolidated minerals	Diamonds, gold, heavy mineral sands, iron sands, tin sands, lime shells, sand and gravel	71
Consolidated minerals	Iron ore, coal, sulfur	60

concepts in materials handling, and while some initial attempts are being made to mine them from the sea surface, it is almost certain that future operations will include some form of crewed equipment operating on the sea floor.

Consolidated deposits may call for a variety of new mining methods which will be dependent on the type, grade, and chemistry of the deposit, its distance from land, and the depth of water. Some of these methods are illustrated in Fig. 3. The possibility of direct sea floor access at remote sites through shafts drilled in the sea floor has already been given consideration under the U.S. Navy's Rocksite program and will be directly applicable to some undersea mining operations. In relatively shallow water, shafts could be sunk by rotary drilling with caissons. In deeper water the drilling equipment could be placed on the sea floor and the shaft collared on completion. The laying of large-diameter undersea pipelines has been accomplished over distances of 25 mi (40 km) and has been planned for greater distances. Subestuarine road tunnels have been built using prefabricated sections. The sinking of shafts in the sea floor from the extremities of such tunnels should be technically feasible under certain conditions. *See* BORING AND DRILLING (MINERAL).

Submarine ore bodies of massive dimensions and shallow cover could be broken by means of nuclear charges placed in drill holes. The resulting broken rock could then be removed by dredging. Shattering by nuclear blast and solution mining is a method applicable in any depth of water. This method calls for the contained detonation of a nuclear explosive in the ore body, followed by chemical leaching of the valuable mineral. Similar techniques are under study for land deposits.

There are many other government activities which may have a direct bearing on the advancement of undersea mining technology but possibly none as much as the International Decade of Ocean Exploration. The discovery of new deposits brings with it new incentives to overcome the multitude of problems which are encountered in marine mining.

[MICHAEL J. CRUICKSHANK]

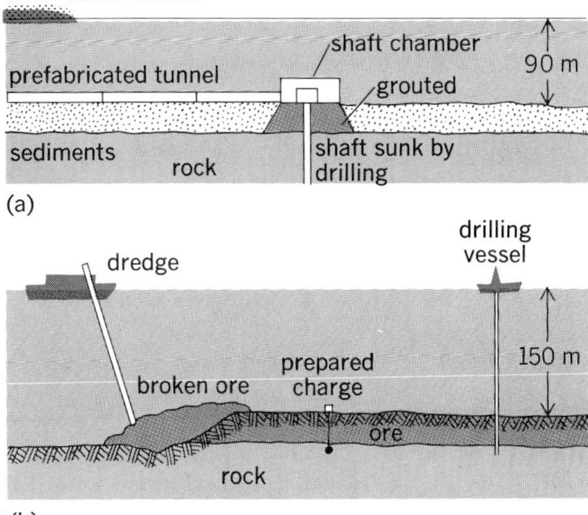

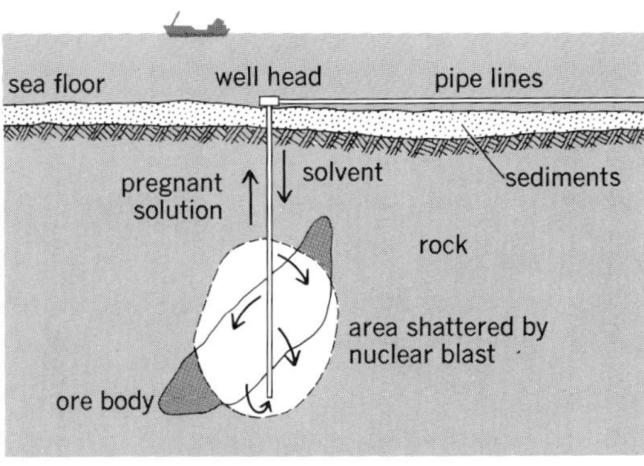

Fig. 3. Possible future methods of mining consolidated mineral deposits under the sea. (*a*) Shaft sinking by rotary drilling from tunnels laid on the sea floor. (*b*) Breaking by nuclear blasting and dredging. (*c*) Shattering by nuclear blast and solution mining.

Bibliography: C. F. Austin. In the rock: A logical approach for undersea mining of resources, *Eng. Mining J.*, August 1967; G. Baker, *Detrital Heavy Minerals in Natural Accumulates*, Australian Institute of Mining and Metallurgy, 1962; H. R. Cooper, *Practical Dredging*, 1958; M. J. Cruickshank, Mineral resources potential of continental margins, in Burke and Drake (eds.), *Geology of Continental Margins*, pp. 965–1000, 1974; M. J. Cruickshank, Mining and mineral recovery, *U.S.T. Handbook Directory*, pp. A/15–A/28, 1973; M. J. Cruickshank, *Technological and Environmental Considerations in the Exploration and Exploitation of Marine Mineral Deposits*, Ph.D. thesis, University of Wisconsin, 1978; M. J. Cruickshank et al., Marine mining, in *Mining Engineering Handbook*, AIME Society of Mining Engineers, sec. 20, pp. 1–200, 1973; M. J. Cruickshank, C. M. Romanowitz, and M. P. Overall, Offshore mining: Present and future, *Eng. Mining J.*, January 1968; R. H. Joynt, R. Greenshields, and R. Hodgen, Advances in sea and beach diamond mining techniques, *S.A. Mining Eng. J.*, p. 25 ff., 1977; J. L. Mero, *The Mineral Resources of the Sea*, 1965; M. A. Meylan et al., *Bibliography and Index to Literature on Manganese Nodules (1874–1975)*, U.S. Department of Commerce, NOAA, 1976; C. G. Welling, and M. J. Cruickshank. Review of available hardware needed for undersea mining, *Transactions of the Second Annual Marine Technology Society Conference*, 1966.

Mass production

The manufacture of products in large quantities. The products may be machine-made individual parts, assembled items, or process items such as chemicals, coal, oil, cement, or cereals.

Eli Whitney, one of the foremost inventors of the early 1800s, is credited with being the father of mass production through his development of the concept of interchangeable manufacture. Up to his time a great deal of handwork had to be done to parts before they could be made to fit together into the final product. Whitney invented machines, tools, gages, jigs, and fixtures so that all parts could be made to close dimensional tolerances.

Today it is difficult to realize that until about 1920 if anyone broke an automobile axle there was usually a wait of at least a week for the new one to be sent out from the factory. Even then it would most likely have to be taken to a local machine shop, where it would be fitted to the particular model car. Long, trouble-free operation of a product is

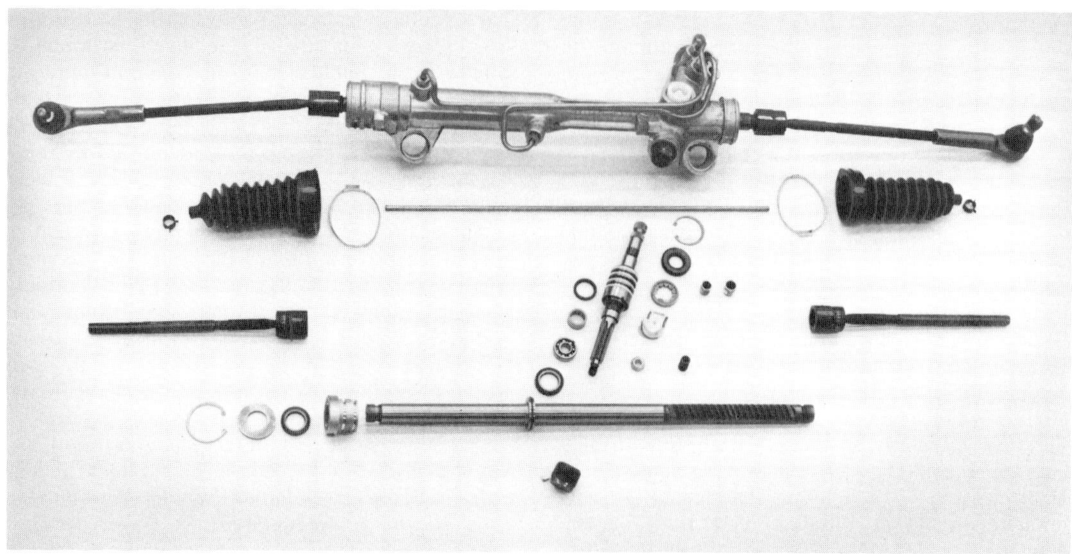

Fig. 1. Parts for the assembly of a power-steering gear.

possible only through more and more exacting tolerances. Parts are now often mass-produced that have tolerances of plus or minus a few thousandths of an inch or even one or two ten-thousandths. Close tolerances are not only necessary for long service life but also for automatic assembly, which is increasingly being done by robots. *See* DESIGN STANDARDS; TOLERANCE.

An example of the type of automatic assembly that makes mass production of complicated parts possible today is the assembly of the 39 parts of a rack-and-pinion power-steering gear (Fig. 1). The

Fig. 2. An assembly machine for a power-steering gear.

65-station assembly machine (Fig. 2), operating at a cycle time of 9 s, produces 400 completed assemblies per hour. Incorporated into the system are inspection stations that assure accurate part placement and continuous verification of critical testing functions, such as pressure-leak tests. *See* QUALITY CONTROL.

In order for an item to warrant mass production, there must be a large, stable market to justify the type of specialized equipment required for economical production. Even this concept is now in the process of evolution to what is being called flexible manufacturing. Flexible manufacturing means changes can be made more easily then in the past. The main reason is that many operations, whether machining or assembly, can be controlled by computers, programmable controllers, or numerical control systems. Robots can be quite easily "trained" to change from the requirements of one design to another. When a robot is trained to spray-paint a car, for example, an expert painter flips the robot into its "learn" mode and then proceeds to paint a sample body exactly as the robot is expected to paint it. The machine will then endlessly duplicate this "lesson" within 0.0625 in. (1.59 mm) for the entire program. *See* INDUSTRIAL ROBOTS.

Figure 3 shows a number of robots being used on an automotive assembly line. As many as 540 welds per car are made by the robots. The welds not only are made faster but have proved to be superior to the welds produced manually.

Robots are increasingly being used to take the drudgery out of many repetitious jobs such as loading and unloading punch presses, drill presses, lathes, forging presses, heat-treat furnaces, and die-casting machines.

Mass-produced products are usually made for a very competitive market. Great care is exercised in the initial design to make certain it does not contain unnecessary cost. As an example, a camera may have a market of 1,000,000 customers. A fraction of a cent shaved off the cost of manufacture can represent a substantial savings.

The effects of mass production are not all technical. In the United States, mass production, particularly of consumer goods, has contributed to a high standard of living. Other countries have large mass-production industries; however, in some nations most of the consumer goods manufactured are exported. In such instances, the local standard of living is not changed appreciably. As was pointed out previously, mass production generally brings with it varying degrees of automation. The automation that lowers the unit cost of a product enables more people to buy the product, creates a need for new maintenance services, calls for more raw materials, and increases shipping and handling jobs both to and from the factory, with consequent overall increases in employment. *See* AUTOMATION; OPERATIONS RESEARCH; PRODUCTION ENGINEERING.

[ROY A. LINDBERG]

Fig. 3. Robots on an automotive assembly line.

Materials handling

The loading, moving, and unloading of materials. The loading, moving, and unloading of ore from a mine to a mill and of garments within a factory are examples of materials handling. There are hundreds of different ways of handling materials. These are generally classified according to the type of equipment used. For example, the International Materials Management Society has classified equipment as (1) conveyor, (2) cranes, elevators, and hoists, (3) positioning, weighing, and control equipment, (4) industrial vehicles, (5) motor vehicles, (6) railroad cars, (7) marine carriers, (8) aircraft, and (9) containers and supports.

Every materials-handling problem starts with the material – its dimensions, its nature, and its characteristics. Engineers who fail to start here usually end up trying to justify equipment rather than achieving safe and economical movement of the material. The quantity to be moved – both in total and in rate of moving desired – is next in importance regarding the type of handling. Then comes the sequence of operations or the routing. Basically, this what, when (how much and how often), and where is the minimum information needed to evaluate or determine any handling system or equipment.

Materials handling is both a planning and an operating activity. These two activities are generally separated in industry; an analytical group designs or selects the system or equipment and the operating group puts it to use. *See* MATERIALS-HANDLING MACHINES; PLANT FACILITIES (INDUSTRY).

[RICHARD MUTHER]

Materials-handling machines

Devices used for handling materials in an industrial distribution activity. The equipment moves products as discrete articles, in suitable containers, or as solid bulk materials which are relatively free-flowing. These machines do not include the means employed to control the flow of fluids.

Many different types of machines result from combinations and permutations of the following factors: (1) The route over which the product is moved may be fixed or variable; (2) the path of travel may be horizontal, inclined, declined, or vertical; (3) motion may be imparted to the product manually, by the force of gravity, by air pressure, by vacuum, by vibration, or by power-actuated components of the machine; (4) the motion may be continuous or intermittent (reciprocating); and (5) the product may be supported or carried suspended during the handling operation. Based upon their most common characteristics, materials-handling machines can be grouped into six broad categories, listed in the cross references. *See* BULK-HANDLING MACHINES; CONVEYING MACHINES; ELEVATING MACHINES; HOISTING MACHINES; INDUSTRIAL TRUCKS; MONORAIL.

Improvements in handling techniques stem from the wide adoption by industry of palleting and of the fork-lift truck. These innovations have produced far-reaching effects. Among these are radical changes in plant layout, elevator design for multistory operations, and the increasing trend to single-story facilities.

Automation, in the sense of feedback control and advanced mechanization in the fabrication and transfer of products from one operation to the next, brings together two major industrial technologies. *See* INDUSTRIAL ENGINEERING; PRODUCTION ENGINEERING; TRANSPORTATION ENGINEERING.

Computer-controlled electronic data-processing devices facilitate the compiling of inventory records and the handling of orders. Use of photoelectric devices for counting and controlling the action of doors, conveyors, and other materials-handling machines is another example of how electronic techniques are being applied. Television and two-way radio improve the communication for materials handling in plants and yards.

[ARTHUR M. PERRIN]

Bibliography: J. M. Apple, *Materials Handling Systems Design*, 1972; J. M. Apple, *Plant Layout and Materials Handling*, 3d ed., 1977; A. P. Boresi et al., *Advanced Mechanics of Materials*, 3d ed., 1978.

Mean effective pressure

MEAN EFFECTIVE PRESSURE

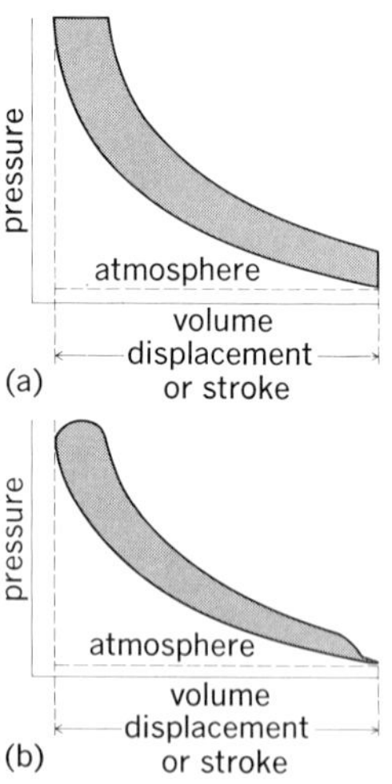

Pressure-volume diagrams (indicator cards) for diesel engine. (*a*) Theoretical (ideal) conditions. (*b*) Actual two-cycle engine card. The colored areas represent the effective work of the cycle.

A term commonly used in the evaluation for positive displacement machinery performance which expresses the average net pressure difference in pounds per square inch (psi) on the two sides of the piston in engines, pumps, and compressors. It is also known as mean pressure and is abbreviated as mep or mp.

In an engine (prime mover) it is the average pressure which urges the piston forward on its stroke. In a pump or compressor it is the average pressure which must be overcome, through the driver, to move the piston against the fluid resistance.

It can be a theoretic value obtained from the pressure-volume diagram of the thermodynamic or fluid dynamic cycle, such as Otto, diesel, or air-compressor cycle, or it may be the actual value, as measured by an instrument (generally called an engine indicator), which traces the real performance cycle diagram as the machine is running. The illustration shows the difference with a two-cycle diesel engine under ideal conditions and under real conditions. The colored area represents the effective work of the cycle. If the area is divided by the length of the diagram (the stroke of the engine), the result is an average height—the mean pressure or mean effective pressure acting throughout the cycle.

The mep can be applied to the cylinder dimensions and to the speed of the engine (or compressor) to give the horsepower. The equation is generally identified as the *plan* equation:

$$\text{Horsepower} = \frac{plan}{33{,}000}$$

p = mep, psi
l = length of stroke, ft
a = net area of piston, in.^{2}
n = number of cycles completed per minute

Inspection of this equation shows that, for an engine cylinder of given size (l and a) and for a given operating speed (n), the higher the mean effective pressure (p) the greater will be the horsepower output, resulting in the most effective utilization of the engine bulk, weight, and investment. The criterion of mep thus is a vitally convenient device for the evaluation of a reciprocating engine, pump, or compressor design as judged by initial cost, space occupied, and deadweight. Some representative values of mep are given in the table. *See* AIR AND GAS COMPRESSOR; DIESEL CYCLE; THERMODYNAMIC CYCLE; VAPOR CYCLE.

[THEODORE BAUMEISTER]

Bibliography: T. Baumeister (ed.), *Standard Handbook for Mechanical Engineers*, 8th ed., 1978.

Some representative values of mean effective pressures at full load

Unit	Value, psi
Aircraft engine	100–200
Automobile engine	100–140
Diesel engine	75–150
Outboard engine	50–100
Steam engine	50–150
Air compressor	50–100

Measured daywork

A tool used primarily in manufacturing facilities as a control device to measure productive output in relation to labor input within a specific time period. The measurement of the work content is accomplished through the use of time standards which are usually the result of a stopwatch time study, predetermined time standards (methods-time measurement, the work-factor system), or some other form of work-measurement technique designed to measure tasks of labor under normal and average conditions. A basic premise of a measured daywork system is simply stated as "a fair day's work for a fair day's pay."

Characteristics. A measured daywork plan shares some characteristics of both incentive pay plans and unmeasured daywork plans, thus becoming a unique plan in itself. It is similar to incentive pay plans inasmuch as in both plans, time standards are used as a device to measure operator performance and also for various forms of management planning. There the similarity ends, for with incentive plans, operator earnings are directly related to, and fluctuate accordingly with, productive output. In a measured daywork plan, worker income is based on a fixed hourly rate established by management, and is usually affected only by job classification, shift premiums, and overtime adjustments. Because of the fixed hourly rate in a measured daywork plan, there is little incentive for a worker to exceed a normal or standard level of performance or productivity. On the other hand, time standards are more readily acceptable, and become less an item of contention to the employee and bargaining unit (union).

The fixed hourly rate of measured daywork is

also characteristic of an unmeasured daywork plan. However, unlike an unmeasured daywork plan, measured daywork does determine worker performance factually. Other advantages of measured daywork compared with unmeasured daywork are that costs can be readily identified for specific jobs or products, accurate estimated costs for new products can be derived if standard data have been compiled, and management planning and control are aided by use of time standards when planning for equipment and worker-power needs and scheduling work through the shop.

Operating principles. As previously stated, the term daywork as used in industry denotes a fixed hourly rate that is not raised or lowered by varying worker performance levels. The hourly rate for a particular job should be a fair one relative to other jobs in the shop, and should also be comparable to rates of pay for similar jobs in the industrial community. In order to ensure an equality of pay rates, a job-evaluation program should exist, and be reviewed and updated as required. Once it is established what a fair rate of pay should be for each job, measurements can be made regarding how long it should take under normal, average conditions to complete the job, task, or operation. Allowances should be made for personal time, unavoidable delays, and, when required, fatigue factors. The result is a time standard as it relates to a specific job, task, or operation.

To evaluate operator performance, a record is made as to how long the operation actually took; this figure is divided into the standard time established for that same operation. As an example, suppose an employee spent 8 h assembling a bicycle and the time standard for that particular operation was 6 h. The calculation to determine employee performance would be as follows: 6 standard hours ÷ 8 actual hours = 75%. It can be said that the employee was 75% productive, efficient, or effective (the semantics of performance vary from industry to industry). On the other hand, if the employee took only 5 h to complete the task, the operator performance level would be 120%. In a standard cost accounting system, anything above or below the standard which represents 100% is usually referred to as negative or positive variance. One way to calculate labor cost (assume $5.00 per hour labor rate) in the aforementioned examples would be as follows: In the first example, the employee was 75% productive; therefore the calculation would be: $5.00 per hour × 6.00 standard hours ÷ 75% =$40.00. In the second example, where the employee exceeded a standard level of performance, the labor cost to assemble the bicycle would be $25.00. These are simplified versions of measurement for both operator performance level and labor cost calculation. However, the same basic principles can be applied to measure not only employees, but also operations, work centers, plants, or even entire divisions. *See* PRODUCTIVITY.

Once measured labor time standards have been established for shop operations, in addition to evaluating operator performance and identifying labor costs, new-product costs can also be determined prior to release to production, worker-power planning and scheduling can be done, equipment capacity requirements can be identified, and planning and make/buy decisions can be facilitated. Most importantly, a system has also been established to aid in identifying those areas, operations, and tasks that offer cost reduction potential.

Program criteria. There are several key criteria that are essential to a sound measured daywork program. First, the labor standards must be realistic and kept current to reflect existing shop conditions. In addition, the labor standards must also be fair and attainable, since "tight standards" will invariably result in apathy and disinterest not only in the direct labor employee ranks, but also with line supervision and other factions of management.

Second, the worker should be made aware of what is the standard or production rate/quota. This in itself will often aid in increasing productivity as most people want to know what is expected of them.

Third, since there is no pay incentive for the worker to exceed standard or normal performance, there is the inherent characteristic of employees in a measured daywork system to fall somewhat short of standard or 100% performance level. Consequently it is of the utmost importance to have strong, qualified, and aggressive first-line supervision, or a foreman who will provide direct supervision to the employee. The supervisor's job should be to motivate the employee to produce a fair day's work, ensure good work habits, evaluate employee performance, take corrective action for subpar performance, and ensure that the shop conditions are kept as optimum as circumstances permit. Prompt performance reports to staff-level management from the supervisor should indicate not only employee performance levels, but also conditions affecting performance outside the control of both the supervisor and the worker. It then becomes the responsibility of management to take prompt corrective action to remedy the situation, so that optimum performance can be attained.

Summary. Measured daywork is a form of wage payment based on the application of time standards to productive operations, with the wages of the employee fixed and independent of employee output. The primary advantage of this method of control is that it establishes a favorable climate for increasing productivity and promotes cost reduction through the introduction of employee measurement, improved management planning, and identifying areas for improved production methods and equipment. In a measured daywork system, the emphasis of management and industrial engineering attention is directed toward evaluating and improving productivity. *See* PERFORMANCE RATING; WAGE INCENTIVES; WORK MEASUREMENT. [DAVE SCOTT]

Bibliography: P. D. O'Donnell, Measured daywork, in H. B. Maynard, *Industrial Engineering Handbook*, 3d ed., 1971.

Mechanical advantage

Ratio of the force exerted by a machine (the output) to the force exerted on the machine, usually by an operator (the input). The term is useful in discussing a simple machine, where it becomes a figure of merit. It is not particularly useful, however, when applied to more complicated machines, where other considerations become more important than a simple ratio of forces. *See* EFFICIENCY; SIMPLE MACHINE. [RICHARD M. PHELAN]

Mechanical engineering

One of several recognized fields of engineering. This distinction begins in college where, after 1 year of common engineering studies, a person can be identified as a student of mechanical engineering. After college, mechanical engineers may enter many industries. Common areas of industrial employment for mechanical engineers are private electric power and machinery manufacturing.

To grasp the meaning of mechanical engineering, it is desirable to take a close look at what engineering really is. The Engineers' Council for Professional Development has defined engineering as the profession in which a knowledge of the mathematical and physical sciences gained by study, experience, and practice is applied with judgment to develop ways to utilize economically the materials and forces of nature for the progressive well-being of mankind. Here is a picture of a profession in which study in mathematics and science is blended with experience and judgment for the production of useful things.

Formal training of a mechanical engineer includes mastery of mathematics through the level of differential equations. Training in physical science embraces chemistry, physics, mechanics of materials, fluid mechanics, thermodynamics, statics, and dynamics. Enhancing these subjects are courses in the humanities: literature, economics, philosophy, and history.

Experience, for the mechanical engineer, is not gained solely by the passage of time. For experience to be added, the time spent in engineering must be meaningful. Projects must become more difficult and the consequences of error greater.

Judgment is the hallmark of any competent engineer in whatever recognized branch of engineering he specializes. The exercise of judgment requires the ability to assess and decide between alternate courses of action. In the application of judgment there is for the engineer constant testing of his decisions against his knowledge of the laws of nature, his sense of right and wrong, and economics.

A further essential of an engineer is that efforts be devoted to matters which improve the well-being of humankind. This purpose is kindred to that of all learned professions and is closely allied to the professions of medicine and law.

In relation to other professions the mechanical engineer differs in one large measure. Whereas most doctors and many lawyers are self-employed, almost all engineers are employed by corporations, colleges, or government. This distinction exists for several reasons. Generally the doctor or lawyer is far more concerned with people and their actions than is the engineer. Furthermore, an engineer dealing with machines requires larger sums of money to finance the equipment and facilities used than does the doctor or lawyer.

From the foregoing, a young person thinking of becoming a mechanical engineer should recognize that there is a long period of preparation necessary. Starting in high school, a full, rigorous academic program should be pursued. Then careful self-examination of results in high school are in order. If high interest and success in mathematics, English, and science are demonstrated, there is room to consider the matter further. Finally, there must be evidence of willingness to work and patience to gain the experience and judgment required in any learned profession. *See* ENGINEERING; MACHINE DESIGN; MACHINERY; MASS PRODUCTION; TECHNOLOGY.

[ROBERT S. SHERWOOD]

Bibliography: American Society of Mechanical Engineers, *Mechanical Engineering*, issued monthly; Engineer's Council for Professional Development (ECPD), *So You Want To Be an Engineer*, Pam. no. EC11; A. P. Johnson (ECPD), *Do I Have Engineering Aptitude?*, Pam. no. EC14; W. E. Wickenden (ECPD), in J. R. Henninger (ed.), *A Professional Guide for Young Engineers*, Pam. no. EC43.

Mechanical vibration

The continuing motion, repetitive and often periodic, of a solid or liquid body within certain spatial limits. Vibration occurs frequently in a variety of natural phenomena such as the tidal motion of the oceans, in rotating and stationary machinery, in structures as varied in nature as buildings and ships, in vehicles, and in combinations of these various elements in larger systems. The sources of vibration and the types of vibratory motion and their propagation are subjects that are complicated and depend a great deal on the particular characteristics of the systems being examined. Further, there is strong coupling between the notions of mechanical vibration and the propagation of vibration and acoustic signals through both the ground and the air so as to create possible sources of discomfort, annoyance, and even physical damage to people and structures adjacent to a source of vibration.

Mass-spring-damper system. Although vibrational phenomena are complex, some basic principles can be recognized in a very simple linear model of a mass-spring-damper system (Fig. 1). Such a system contains a mass M, a spring with spring constant k that serves to restore the mass to a neutral position, and a damping element which opposes the motion of the vibratory response with a force proportional to the velocity of the system, the constant of proportionality being the damping constant c. This damping force is dissipative in nature, and without its presence a response of this mass-spring system would be completely periodic.

Free vibrations. In the absence of an exciting force and of the damping component, the spring-mass system can execute free, periodic vibration at a natural frequency ω_n given by Eq. (1). This is

$$\omega_n = \sqrt{k/M} \tag{1}$$

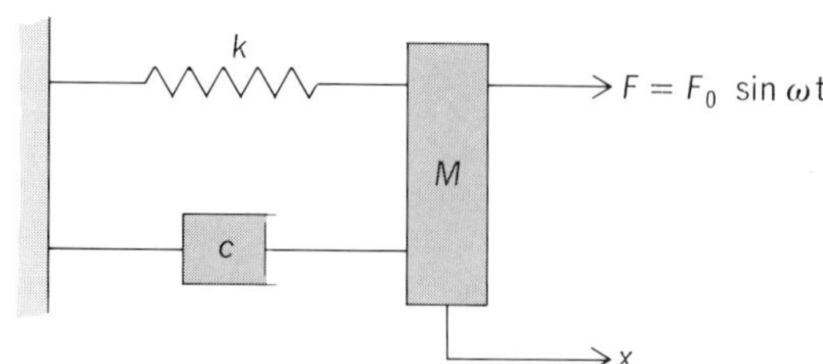

Fig. 1. Vibrating linear system (mass-spring-damper) with one degree of freedom.

the so-called angular frequency and must be divided by 2π to get the actual frequency f_n in hertz (with dimensions time^{-1}). Equation (1) demonstrates that the natural frequency, also called the fundamental or resonant frequency, increases with the stiffness of the system and decreases as the mass of the system is increased. Thus, very stiff systems have high natural frequencies and very massive systems could have very low natural frequencies. In some of the more complex structures referred to above, there are in fact multiple (and even infinite) resonant frequencies, but they all have in common the basic dependence upon the stiffness of the system and the system's mass.

Effect of damping. The inclusion of the damping element with the constant of proportionality c results in a change of the natural frequency of the system and of its response characteristics. If the damping is relatively small in magnitude (the scale against which smallness is to be measured will be given below), then the natural frequency of the system does not change substantially, and the free response of a perturbed mass-spring-damper system is periodic within the confines of a slow exponential decay. However, if the damping constant c becomes large, the motion of the system is not vibratory but is in fact one where the mass creeps back, along an exponential curve, to its initial position without any oscillation. The lowest value of damping for which this loss of vibratory response occurs is called critical damping, and it is defined by a damping constant, Eq. (2). Damping is considered to be small when the ratio c/c_{crit} is small compared to unity.

$$c_{\text{crit}} = 2\sqrt{kM} \tag{2}$$

Impedance and resonance. When the mass-spring-damper system is excited by a periodic force, for example, $F = F_0 \sin \omega t$, where the angular frequency ω $(= 2\pi f)$ can be chosen freely, the response of the system is such that it can be displayed in one of two commonly used forms. One of these (Fig. 2) is a plot of the relative amplitude of vibration (the dimensionless ratio of the actual motion of the perturbed system divided by the corresponding elastic response of the spring alone to the same force) against the normalized (dimensionless) frequency ratio of the forcing frequency divided by the undamped natural frequency. The curves shown in Fig. 2 indicate the response for no damping ($c/c_{\text{crit}} = 0$), small damping, and critical damping. It can be seen that in the undamped case ($c/c_{\text{crit}} = 0$), the amplitude of the response is infinite when the forcing frequency is identically equal to the natural frequency. This is the condition of resonance. It can also be seen that when there is damping, even though it may be small, the system responds so as to peak very near the undamped resonant frequency, but the response is not infinite in amplitude. Finally, if the damping is greater than or equal to c_{crit}, the response is simply the decaying curve that exhibits no peak at the resonant frequency.

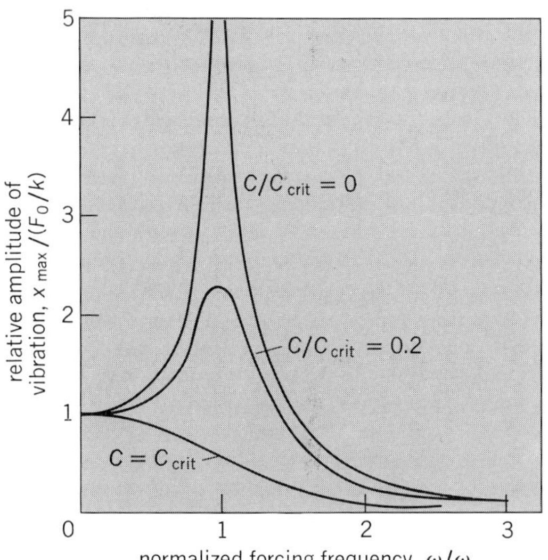

Fig. 2. Variation of amplitude of vibration with forcing frequency and damping. x_{max} = maximum displacement.

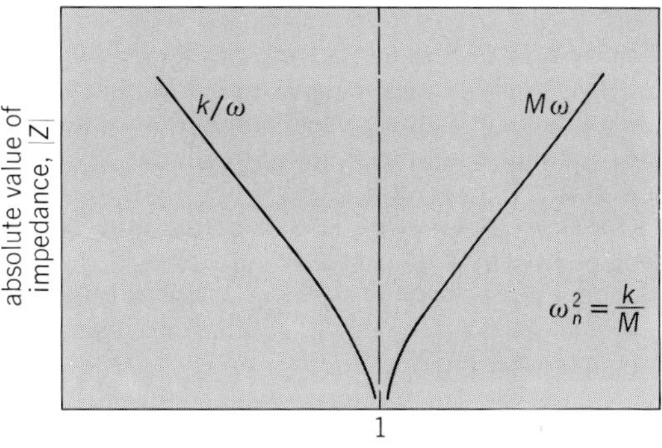

Fig. 3. Impedance of a simple oscillator.

Another way of presenting these results is to look at the quantity called the impedance Z which may be defined here as the ratio of the magnitude of the input force to the magnitude of the velocity of the mass-spring system (Fig. 3). For the case where there is no damping, the impedance can be shown to be in the form of Eq. (3). It follows from Eq. (1), which relates the stiffness k and the mass M to the resonant frequency ω_n, and from the impedance formula that when the forcing frequency is well below the resonant frequency, the impedance can be approximated by Eq. (4). This is considered the stiffness-controlled region wherein the spring-mass system responds largely as a static spring. For frequencies that are significantly above the resonant frequency, the impedance can be written as Eq. (5).

$$Z = k\left(\frac{1}{\omega} - \frac{M\omega}{k}\right) \tag{3}$$

$$Z_k \cong \frac{k}{\omega} \qquad \omega \ll \omega_n \tag{4}$$

$$Z_m \cong -M\omega \qquad \omega \gg \omega_n \tag{5}$$

In this regime the response of the system is said to be mass-controlled, and at high frequencies, well above the resonant frequency, the dynamics of the system are governed by the mass of the system. It follows from Eq. (3) that the impedance vanishes at resonance in the absence of damping. Thus resonance may also be defined as the condition where a very small input force produces an infinite response in the absence of damping. When damping is incorporated in the calculation of impedance, it then turns out that at resonance the impedance is exactly equal to the damping constant; that is, $Z_{res} = c$.

Linear response. One other feature in the simple model is that the response is linear; that is, there is a direct proportionality of the magnitude of the input to the magnitude of the output, and signals of response for different inputs can be superposed to give the total response to a combined signal.

Relation to real systems. This simple model of the response of the single-degree-of-freedom system, as it is commonly called, serves as a model or paradigm for almost all mechanical vibration problems. However, real systems are often extremely complex and involve multiple resonances and multiple sources of excitation. Further, the types of vibrational response that can be seen in different solid bodies are very much dependent not only on the nature of the excitation in terms of its time history, but also on the spatial distribution of the exciting force and the geometry of the system that is responding. There are many different ways in which a solid body can vibrate, and these different ways couple into vibratory and acoustic motions of surrounding media such as air or water in very different ways. Some of the sources of vibration, the types of vibrational response exhibited by solid bodies, and their interactions that result in perceptible, audible, and even dangerous vibration and acoustic signals will be discussed. First, however, some further comments about the general complexity of vibration problems are appropriate in order that the context of the above simple model is more clearly and sharply defined.

Analysis of complex systems. The foregoing model of the linear spring-mass-damper system contains within it a number of simplifications that do not reflect conditions of the real world in any obvious way. These simplifications include the periodicity of both the input and, to some extent, the response; the discrete nature of the input, that is, the assumption that it is temporal in nature with no reference to spatial distribution; and the assumption that only a single resonant frequency and a single set of parameters are required to describe the mass, the stiffness, and the damping. The real world is far more complex. Many sources of vibration are not periodic. These include impulsive forces and shock loading, wherein a force is suddenly applied for a very short time to a system; random excitations, wherein the signal fluctuates in time in such a way that its amplitude at any given instant can be expressed only in terms of a probabilistic expectation; and aperiodic notions, wherein the fluctuation in time may be some prescribed nonperiodic function or some other function that is not readily seen to be periodic.

However, in most cases the notions of periodic response to periodic excitation can be used because it is generally possible to describe the variation in time of a nonperiodic forcing function as a sum of periodic components using the tools of Fourier analysis and Fourier series. In such an approach, the nonperiodic signal is expressed as a sum of periodic signals, each of whose amplitudes and individual periods (or resonant frequencies) is tailored to represent the total nonperiodic signal being described. In such a Fourier analysis of a vibrating system, it is customary to look at the response of the complex system as a sum of responses of linear spring-mass-damper systems to each of the given periodic inputs, and then the simple model outlined earlier can be readily applied.

In any real system there are multiple resonances, each reflecting a different balance of stiffness and mass within a complex elastic system. Again, it is possible in general to describe the response of a complicated system in terms of individual spring-mass oscillators, the stiffness and mass of each of which represents some average of the overall stiffness and mass distributions of the complicated system. Thus, the analysis of complex systems can again be reduced to the superposition of systems of the elementary type. Very often this reduction of the complex to the simple not only involves Fourier analysis in time, but it also represents a Fourier analysis in space, wherein spatial averages are used to reduce spatial variations of both excitation and response to the discrete forms in which they appear in the simple model. Thus, for a continuous system where the variables depend on space as well as time, a Fourier analysis can be used to generate a system that involves multiple degrees of freedom, each degree of freedom being a single discrete spring-mass-damper oscillator.

Sources of vibration. There are many sources of mechanical and structural vibration that the engineer must contend with in both the analysis and the design of engineering systems. The most common form of mechanical vibration problem is motion induced by machinery of varying types, often but not always of the rotating variety; the machinery vibration thus induced can be a problem both for the machinery itself and in terms of vibration and noise propagated to adjacent systems. For example, the vibrations of a compressor in an air-conditioning system in a building can create fatigue and stress problems for the compressor and adjacent piping itself, and can set into motion the floor and walls of the room wherein the compressor is housed. This in turn can set into vibrational motion other walls and components of the building and thus produce vibration and acoustic signals that are both perceptible and audible throughout the building. This simple example could thus run a gamut of problems from structural failure to human annoyance.

Such machinery-induced vibration can also be propagated into adjacent buildings through the coupling of a building foundation with the ground around it, and it is not uncommon for heavy machinery in a factory to produce perceptible vibration at considerable distances from the factory itself. This is more likely to be the case with extremely heavy pieces of equipment, for example a large-scale shredder used in a solid-waste disposal system.

Another source of vibration wherein groundborne propagation is important is vibration due to construction. Construction vibration is a major source of environmental concern, particularly in urban environments, and especially in the construction of large projects within such environments such as urban transportation systems.

Another source of vibration in urban environments is transportation systems themselves and may include vibration from heavy vehicles on conventional pavement as well as vibratory signals from the rail systems common in many metropolitan areas. For such rapid-transit system vibration, the sources include tunnels, at-grade rails, and elevated guideway systems. The propagation characteristics for each of these important vibration sources are different and have been the subject of extensive research.

Other vibrations are induced by natural phenomena, including earthquakes and wind forces. Wave motion is a source of vibration in mechanical and structural systems associated with offshore structures. In the analysis and design of systems that are used in environments where these natural phenomena are important, the basic principles of vibration analysis, such as the elementary model outlined above, remain the same. Thus, although such motion is not in a very strict sense due to ordinary mechanical excitation, it is in terms of practice and design a strongly related phenomenon.

Types of mechanical signals and waves. Due to the varying geometry of both the structural elements involved in a machine or structure and of the loading itself, several different types of vibratory motion may be induced in a solid body. Unlike the acoustic signals that are generated in air and water, wherein all the motion is in the form of waves that are termed body or compressional waves in which the fundamental mechanism is volumetric contraction or expansion, solid bodies support waves of several different varieties and thus produce several different mechanical vibratory signals. These include torsional vibration, longitudinal vibration, and the lateral vibrations of beams, plates, and membranes. The fundamental mechanisms in each case are different in that the elastic restoring forces that are analagous to the spring constant k in the simple model are very different and are produced by different physical effects in these bodies. Each type of vibration will be discussed briefly, although the major emphasis will be placed on vibrations of and in machinery.

Torsional vibration. An engine-driven system (usually diesel or internal combustion powering a generator, a ship's propeller, or other load) has many degrees of freedom and, hence, many natural frequencies of which only the lowest two or three are of practical importance. The installation may, if excited by alternating components in its torque, execute torsional vibrations between its various parts of a magnitude of a fraction of a degree. This relatively small alternating motion is superposed on, and independent of, the continuous rotation of the engine shaft. Whereas the continuous rotation causes no extreme stress in the shaft, the small (quarter degree) angles of vibration wind up the shaft and relax it, causing alternating stresses that on many occasions have caused failure.

The torque on a crankshaft caused by a single cylinder and piston having one explosion for each one or two revolutions has a highly irregular time history containing many harmonics. (Harmonics are integral multiples of the fundamental frequency.) The torques developed by the various cylinders of a multicylinder engine combine in accordance with their times of occurrences. Thus, the exciting torque contains components of the firing frequency (one cycle per one or two revolutions) and multiples of that firing frequency, which are of practical importance up to the 16th multiple or harmonic. Because these 16 exciting or forced frequencies are proportional to the engine speed and the two or three natural frequencies are independent of it, and each exciting frequency can resonate with each natural one, there are as many as 48 resonances or critical speeds to be considered. Many of these will lie outside the habitual running range of the engine.

The designer may shift the severity of the resonances by using flexible couplings or extra flywheels in the engine, or by changing the arrangement of the cylinders, the V-angle of the engine, and the firing order. These suffice to make an installation satisfactory for any one running speed, but it is usually not possible to avoid all dangerous resonances in a wide range of running speeds. The designer must then resort to dampers to keep the amplitude of motion and the stress down. Various types of dampers exist; the most familiar is the pulley in the front of an automobile engine which drives the fan belt. This pulley is usually a small flywheel coupled to the engine shaft through a rubber insert. The assembly serves simultaneously as a torsional spring and as a dashpot damper. The flywheel is so tuned that it holds the principal torsional critical speed of the engine below dangerous torsional stress levels. Fatigue failures of crankshafts and of other shafting due to torsional vibration were common in the past, but such failures are avoidable by proper design and are now rare.

By the principle of action equals reaction, the gas torque acting on the piston-crankshaft assembly is equal and opposite to the gas torque acting on the engine frame. Hence, when the engine frame is rigidly attached to a foundation, that foundation experiences an alternating torque. To protect the foundation and to prevent the vibration from spreading through the structure, the engine is often mounted on metal springs or on rubber bushings. This is now universal practice with automobile and aircraft piston engines.

Longitudinal vibration. A large ship's shafting excited by the propeller blades may come to resonance. The shafting and propeller must be designed to allow for such resonance. This is one cause of vibration in passenger liners. It occurs at a frequency of the shaft revolutions multiplied by the number of propeller blades.

The air or gas column in the suction or discharge lines of internal combustion engines acted upon by the piston motions can come to resonance at certain speeds. This principle has been used to increase the power of the engine by designing those lines so that during the intake more air than usual enters the cylinder, and during exhaust more air than usual goes out, thereby decreasing the backpressure. *See* INTERNAL COMBUSTION ENGINE.

In internal combustion engines the gas pressure exerted on the piston and thence on a crank throw tends to lengthen the crankshaft. Because this effect is periodic, longitudinal vibrations in internal combustion engine shafts have been observed; however, they are of less importance than the torsional ones and become serious only if a longitudinal and a torsional natural frequency fall close together causing coupling oscillations.

Lateral vibrations. The most important of all machinery vibrations are the lateral or bending vibrations of shafts and other rotors caused by the centrifugal force of unbalance. The unbalance consists of a very small deviation of the center of gravity of the rotor from the geometric axis connecting the bearing centers, or of a small angular deviation between that bearing center line and a principal axis of inertia of the rotor. Thus the excitation always has the frequency of the shaft rotation, and when that coincides with one of the natural bending frequencies of the rotor, there is resonance or a critical speed where the bending stresses in the shaft can be a hundred times higher than those caused by the centrifugal force of the unbalance directly.

Slow-speed machines are always designed to run well below the lowest or first critical speed, but for high-speed ones this often is not possible. Steam turbines and electric generators of large power stations have for many years been designed to run between the first and second critical speeds; the newest and largest units are between their second and third criticals. The flexibility of the supporting (nonrotating) bearings is an important factor in the calculation of critical speeds or natural frequencies of rotors, because the flexible bearing decreases the natural frequencies by some 10% on the average below those that would be present if rigid bearings were used. For rotors of a high diameter-length ratio such as in steam turbines with large diameter disks mounted on a comparatively thin shaft, the effect of rotating inertia, sometimes called the gyroscopic effect of the disks, is an important consideration.

Besides the classical unbalance critical speed, which always has a vibration frequency equal t that of the rpm, a number of secondary critical speeds have been observed and explained, in which the frequency of vibration is a multiple of, usually twice, the rpm. The practical importance of these is secondary with respect to the ordinary critical speed.

Beam vibration. Whenever a part of a structure or the entire structure itself has a natural frequency resonating with an alternating excitation, nearby (or sometimes far removed) severe vibrations may result. A typical example is an unbalanced piston machine, such as an air compressor, which frequently causes objectionable vibration in some locations in the building where it is installed. Another example is the vibration of an entire ship in the mode of a free-free (totally unsupported) beam, excited by the propeller blade frequency.

With the advent of jet engines, the effect of high intensity airborne noise on the very light structures of airplanes and missiles has become important. It is characteristic of jet noise, or indeed of all cases of turbulent flow, that the excitation is distributed continuously over a wide band of frequencies. This type of excitation is random and is playing an increasingly important role in the design of aircraft and missiles.

Membrane vibration. A tightly stretched skin, which has negligible bending stiffness like a drumhead, is a membrane. (The diaphragm in a telephone receiver possesses considerable bending stiffness, and is not stretched; hence technically it is a plate, although sometimes it is also called a membrane.) The theory of vibration of a membrane of circular shape has been known for a century and is one of the more beautiful illustrations of the mathematics of Bessel functions. The lowest frequency of vibration corresponds to a shape where the entire membrane bulges in and out, without nodal lines, the periphery remaining fixed. The higher modes of motion possess nodal lines, which are concentric circles or angularly equidistant diameters. The frequency formula for the circular membrane vibration is given in Eq. (6). Here

$$f = \alpha\sqrt{Tg/\gamma r^2} \tag{6}$$

T is the tension in the membrane, g is the acceleration due to gravity, γ is the surface weight density of the membrane, r is the radius, and a is a numerical factor having the values shown in the table.

Numerical factor for membrane frequency

Number of nodal circles	Number of nodal diameters			
	0	1	2	3
0	0.383	0.610	0.819	1.02
1	0.880	1.12	1.34	1.55
2	1.38	1.62	1.85	2.08

Plate vibration. Plate vibrations are very important for a variety of reasons. They occur in large, flat, solid surfaces that are excited in planes normal to that surface. The reason that such vibrations are important in mechanical systems is that they occur frequently in practice and in addition they are very good acoustical radiators; that is, they couple extremely well with the air around them and produce significant quantities of vibratory energy that can be both felt and heard as acoustic signals. The basic mechanism by which a plate operates is similar to a beam in that the action consists of the bending of the plate about its middle surface, except that whereas a beam is a one-dimensional element with bending in only one direction, a plate is a surface that has bending in two orthogonal directions as well as a torsional or twisting type of bending at angles between these two orthogonal directions. Thus, more complicated curvature and surface interaction effects are involved.

Self-excited vibrations. In the cases discussed so far, the existence of an alternating exciting force (or torque) has been assumed which would continue to exist by itself even after the vibratory motion was prevented or stopped. These are called forced vibrations. There is another class of motions, called self-excited, in which the exciting force is generated by the vibrating motion itself and hence disappears with that motion.

A familiar mechanical example of self-excited vibration is the piston of an engine. The back and forth motion (vibration) is maintained by an alternating gas or steam pressure steered by the valve mechanism. The initial source of energy is without alternating properties (the gasoline supply or the boiler steam) but is made alternating by the valves, which are operated by the vibratory piston itself.

A familiar electrical example is the self-oscillating electronic circuit which has a steady source of energy (for example, a battery), and a transistor that plays the role of the valve.

The oldest practical examples of useful self-excited vibrations are musical instruments. The violin operates on the behavior of the friction between the rosined bow and the string, which has aptly been described as stick-slip friction. Vibrations of this type have appeared repeatedly and still are often met within machinery with insufficient lubrication; the most vexing and difficult case is that of the chattering machine tool cutter which leaves a wavy cut instead of a smooth one.

The clarinet operates on the passage of air (from the mouth to the instrument) though a narrow leakage opening whose width varies periodically with the vibration of the reed. Serious vibrations of this character occur in steam, gas, or hydraulic turbines, heat exchangers, and other apparatus in which a fluid or gas passes through narrow passages. Other self-excited vibrations are the pulsating flow sometimes observed in fans and blowers, and the shimmying motion of wheels, which has been a serious problem in automobiles and in the landing gear of aircraft. Self-excited vibrations appear in autopilots of aircraft and missiles, and in general are apt to occur in servomechanisms with high gain.

A class of dangerous vibrations is provided by the various phenomena of flutter, whereby an elastic system becomes self-excited in the stream of air, gas, or fluid of sufficient speed. This aviation problem first arose in airplane wings, but as speeds increased, other parts of airplanes and other machinery, such as turbine blades, displayed flutter. Flutter theory has grown into a subject by itself.

Serious vibrations have occurred in rocket engines and other types of combustion chambers, whereby the combustion becomes unstable and the burning gas mass enters a state of self-excited vibration. These phenomena have been known for a century and still are not completely understood.

Effect of vibrations. The most serious effect of vibration, especially in the case of machinery is that sufficiently high alternating stresses can produce fatigue failure in machine and structural parts. Less serious effects include increased wear of parts, general malfunctioning of apparatus, and the propagation of vibration through foundations and buildings to locations where the vibration or its acoustic realization is intolerable either for human comfort or for the successful operation of sensitive measuring equipment.

Whole-body vibration. Figure 4 shows a comparison of suggested building damage criteria and human threshold-of-perception curves for whole-body vibration. The threshold curves for human perception are based upon tests wherein people are seated on shakers, so the perception is of motion that affects the entire body. There are a variety of ranges of vibration, and these ranges are strongly dependent upon the frequency (expressed here in hertz, equal to $\omega/2\pi$). The ordinate of this graph is expressed as an acceleration level (in decibels) with reference to g (the acceleration of gravity) as the base acceleration level. The decibel notation indicates a logarithmic dependence such that an acceleration level of −20 dB represents an acceleration level that is 0.1 g in magnitude, an acceleration of −40 dB represents a level that is 0.01 g in level, and so on. The magnitudes of vibration that are felt by humans are extremely small; that is, on the order of −60 dB or 0.001 g in magnitude. While identifying the human threshold of perception from whole-body laboratory experiments is very useful, other factors must be considered in practice. Human annoyance may be related to secondary vibratory effects such as vibration-induced noise of the rattling of dishes, visual cues such as the vibration-induced motion of household items such as plants and mirrors, and so on. Thus, human annoyance could depend on the observer's location and activity as well as other suggested factors (Fig. 4). The values for which damage to a building might be realized are significantly higher than those for which vibration is perceptible by humans. In fact, the difference tends to be on the order of two orders of magnitude (a multiplicative factor of 100 in amplitude).

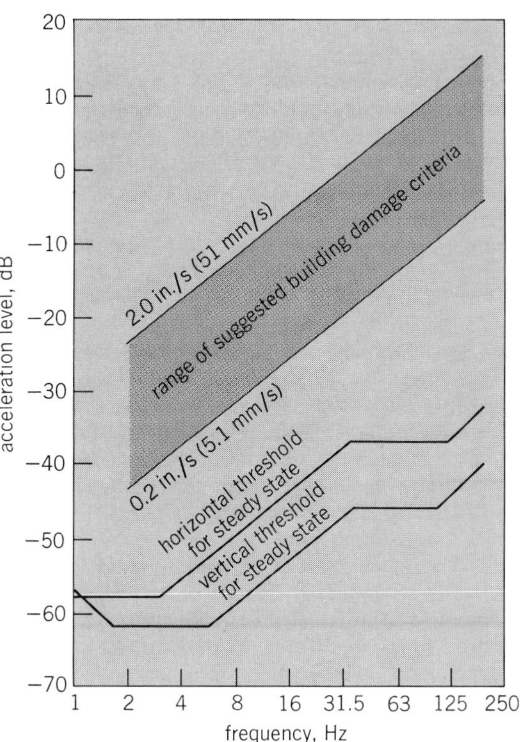

Fig. 4. Comparison of suggested building damage criteria and human threshold of perception curves for whole-body vibration. Acceleration level in dB = 20 log acceleration level/g. Because they are negative logarithms, the decibel values on the vertical axis indicate increasing acceleration levels from bottom to top. *(From T. G. Gutowski, L. E. Wittig, and C. L. Dym, Some aspects of the ground vibration problem, Noise Control Eng., 10(3): 94–100, 1978)*

Generation of acoustic noise. Figure 5 relates vibration magnitudes both to perceptible vibration level and to audible acoustical levels. These data demonstrate that an acoustic signal can be generated by a mechanical vibration and produce auditory noise that can be very uncomfortable. For example, as discussed above, noise can propagate in a building due to an air-conditioning compressor. The curves displayed in Fig. 5 also show so-called NC-equivalent curves used to describe ar-

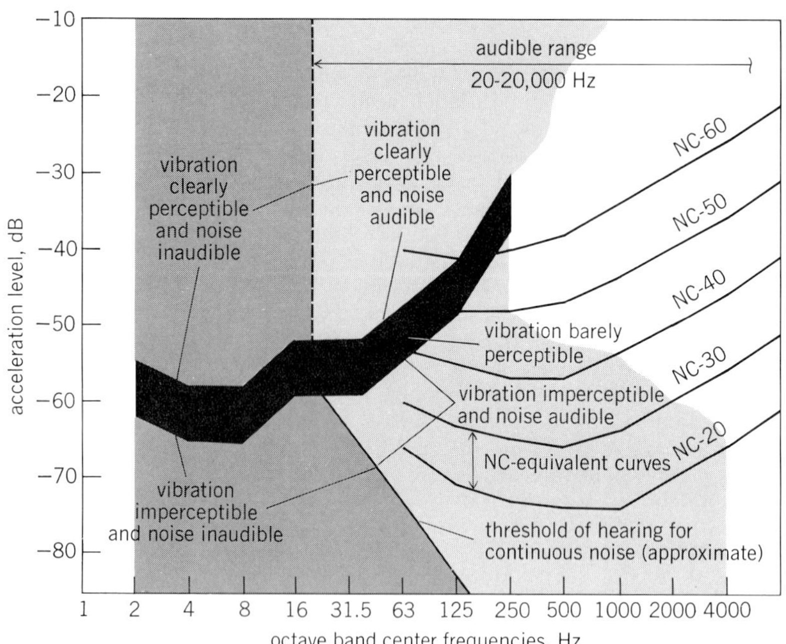

Fig. 5. Levels of perceptibility of vibration and audibility of noise. Acceleration level in dB = 20 log acceleration level/*g*. Because they are negative logarithms, the decibel values on the vertical axis indicate increasing acceleration levels from bottom to top. (*From C. L. Dym and D. Klabin, Architectural implications of structural vibration, Architect. Rec., 157(9):125–127, 1975*)

chitectural levels of quietness in a design environment. Thus an NC-20 equivalent curve is the design standard for a first-class concert hall, while an NC-40 criterion would be adequate for a grade-school classroom. The acoustic signals are normally expressed in logarithmic amplitudes of acoustic pressure, but are expressed here in terms of amplitudes of vibration, that is, acceleration levels. An equivalence between the two is based on the recognition that when a large surface vibrates, it simultaneously sets into motion the acoustic medium around it, whether it be air or water. The bending vibrations of plates, for example, couple extremely well with the compressional waves of a surrounding acoustic medium. Thus when a building wall or floor is set into motion by a mechanical source of vibration, it in turn causes a very effective propagation of an acoustical signal, or noise, in the surrounding air. Another significant example of good coupling is the mechanical excitation of a drumhead by a drumstick. The lateral vibrations of the stretched membrane (the drumhead) also couple well with surrounding air to produce the corresponding tympanic sound of the drum which can be heard at great distances from the drum itself.

There are many factors that enter into the coupling of a vibrating structural element and the surrounding acoustical medium. The principal factors are, however, the surface area of the mechanical source that is exposed to the medium and the nature of the mechanical vibration, that is, whether it is perpendicular to the surface (as with bending vibrations of beams and plates or lateral vibrations of membranes) or along the direction of the surface (as in torsional and longitudinal vibration). It is the perpendicular motion of large surfaces that produces the best acoustical coupling and, hence, the best chance of radiation of noise from vibration.

[CLIVE L. DYM; J. P. DEN HARTOG]

Bibliography: R. E. D. Bishop, *Vibration*, 2d ed., 1979; J. P. Den Hartog, *Mechanical Vibrations*, 4th ed., 1956; C. L. Dym and D. Klabin, Architectural implications of structural vibration, *Architect. Rec.*, 157(9):125–127, September 1975; C. M. Harris and C. E. Crede, *Shock and Vibration Handbook*, 2d ed., 1976.

Mechanism

Classically, a mechanical means for the conversion of motion, the transmission of power, or the control of these. Mechanisms are at the core of the workings of many machines and mechanical devices. In modern usage, mechanisms are not always limited to mechanical means. In addition to mechanical elements, they may include pneumatic, hydraulic, electrical, and electronic elements. In this article, the discussion of mechanism is limited to its classical meaning. *See* MACHINE.

Mechanisms are found in internal combustion engines; compressors; locomotives; agricultural, earth-moving, excavating, mining, packaging, textile, and other machinery; machine tools; printing presses; engraving machines; transmissions; ordnance equipment; washing machines; lawn mowers; sewing machines; projectors; pinspotters; toys; pianos; ski bindings; artificial limbs; door locks; nutcrackers; counters; microswitches; speedometers; and innumerable other devices.

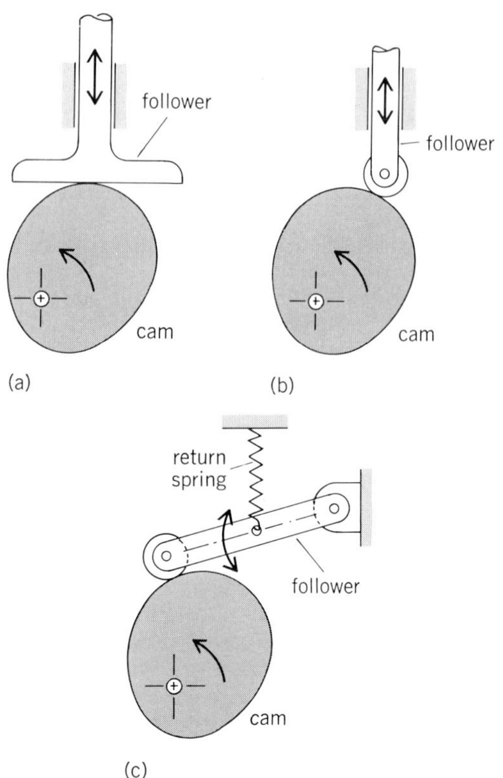

Fig. 1. Cam-follower configurations. (*a*) Disk cam with flat-faced translating follower. (*b*) Disk cam with translating roller follower. (*c*) Disk cam with swinging roller follower.

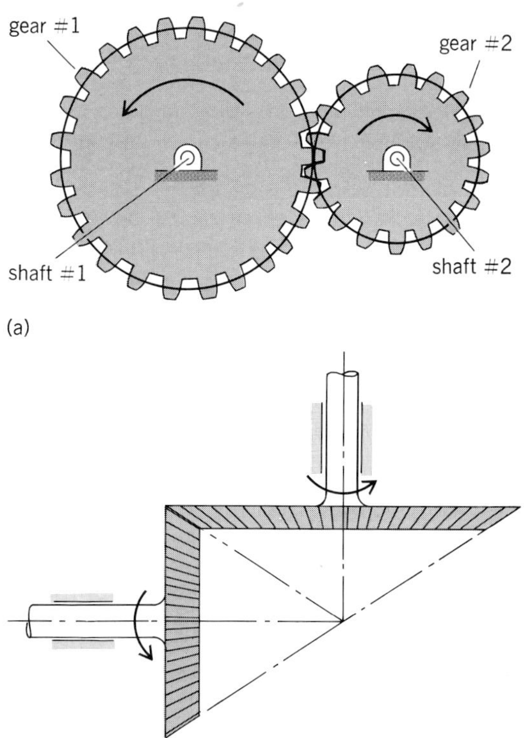

Fig. 2. Gear configurations. *(a)* Spur gears. *(b)* Bevel gears.

Components. Most mechanisms consist of combinations of a relatively small number of basic components. Of these, the most important are cams, gears, links, belts, chains, and logical mechanical elements.

Cams and cam followers. A cam is a specially shaped part designed to impart a prescribed law of motion to a contacting part called the follower. In Fig. 1 the cam is in the form of a disk, the shape of which governs the motion of the follower. Followers may be translating (Fig. 1*a* and *b*) or swinging (Fig. 1*c*) and may be in sliding (Fig. 1*a*) or rolling (Fig. 1*b* and *c*) contact with the cam. The law of motion is the relation between cam rotation and follower displacement. Contact between cam and follower is often (but not necessarily) maintained by a return spring. *See* CAM MECHANISM.

Gears. Gears are toothed wheels that provide a positive connection between rotating shafts. The most familiar are spur gears, which are used to connect parallel shafts. The teeth of straight spur gears (Fig. 2*a*) may be imagined mounted on right circular cylinders. When the shaft axes are intersecting, bevel gears can be used. The teeth of the straight bevel gears (Fig. 2*b*) may be imagined mounted on right circular cones. Gear teeth are so shaped that the angular-velocity ratio of the connected shafts is constant. Other forms of gearing include helical (these may be spur gears or crossed helical gears), worm, and hypoid. The latter two, as well as crossed helical gears, are used to connect shafts with nonparallel, nonintersecting axes. *See* GEAR.

Links. For practical purposes, a link may be defined as a rigid body which is connected to other rigid bodies by joints. The most common joints (Fig. 3) are the pin joint (symbolized by R), the sliding joint (P), the cylindrical joint (C), the ball joint (S), and the screw joint (H). The tooth connection between gears and the contact between cam and follower are also usually regarded as joints. Typical link configurations are shown in Fig. 4. *See* LINKAGE (MECHANISM).

Belts and chains. These flexible connectors include pulleys and flat belts (Fig. 5), timing belts and sprockets, and chain belts and sprockets. The first of these is a friction drive, while the latter two are positive drives. All transmit a practically constant angular-velocity ratio between the connected shafts. *See* BELT DRIVE; CHAIN DRIVE.

Logical mechanical elements. These include ratchets, trips, detents, interlocks, and the like. The most interesting of these is the ratchet. Figure 6 shows a type of ratchet mechanism known as an escapement. The ratchet is the toothed wheel, the rotation of which is controlled by a pawl—a pivoted link which may be in or out of engagement with the ratchet. This arrangement can provide a stop-and-go unidirectional motion, such as that of the hands of a pendulum-driven clock. *See* RATCHET.

Representative mechanisms. Mechanical components can be combined into mechanisms in an infinite variety of ways. For example, in automobiles there is a slider-crank mechanism (Fig. 7), connecting piston to crankshaft; the cam-follower system (Fig. 8) for the actuation of the intake or ex-

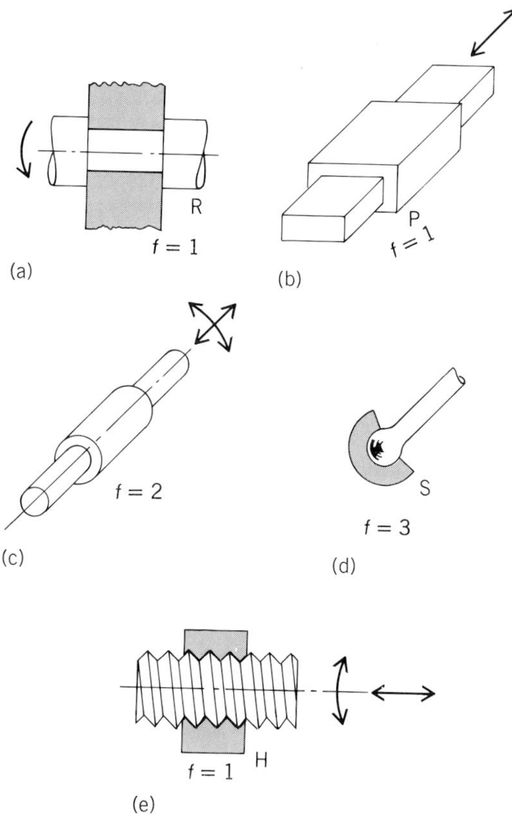

Fig. 3. Common joints, including joint symbols and degree of freedom *(f)* of relative motion at joint. *(a)* Pin joint. *(b)* Sliding joint. *(c)* Cylindrical joint. *(d)* Ball joint. *(e)* Screw joint.

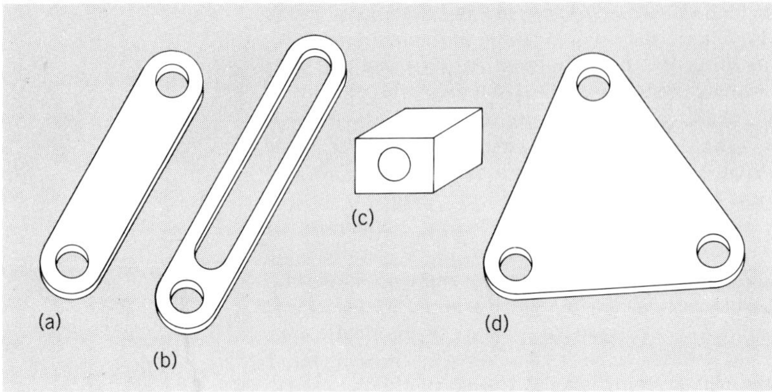

Fig. 4. Typical links. (*a*) Crank. (*b*) Slotted link. (*c*) Slider with provision for pin connection. (*d*) Triangular link.

haust valves in an overhead valve arrangement; and the differential (Fig. 9). The first is a linkage mechanism, the second a cam-follower mechanism, and the third a gear train. The first converts the translation of the piston into crankshaft rotation and transmits power from piston to crankshaft. The second controls the opening and closing of an intake or exhaust valve. Although some power is required to actuate the valve, this is of secondary

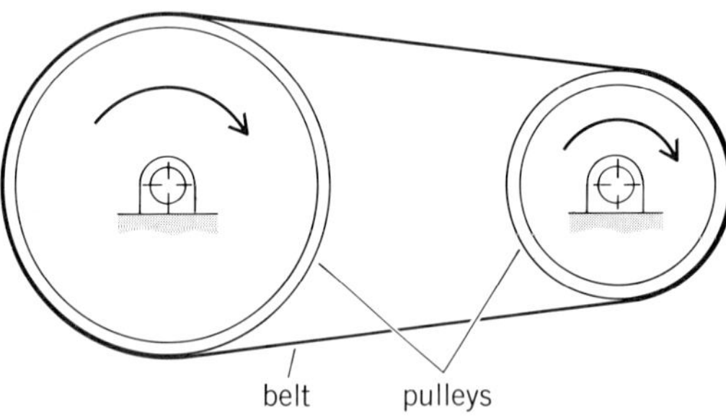

Fig. 5. Belt and pulleys.

significance. The differential is a two-degree-of-freedom gear train, which couples the driveshaft to the tire axles and transmits engine power to the tires. The two degrees of freedom are needed to permit unequal tire speeds when negotiating a turn. In order to understand these and other mechanisms, their degree of freedom, structure, and kinematics must be considered.

Degree of freedom. This is conveniently illustrated for mechanisms with rigid links. The discussion is limited to mechanisms which obey the general degree-of-freedom equation (1), where F =

$$F=\lambda(l-j-1)+\Sigma f_i \qquad (1)$$

degree of freedom of mechanism, l = number of links of mechanism, j = number of joints of mechanism f_i = degree of freedom of relative motion at ith joint (see Fig. 3), Σ = summation symbol—summation over all joints, and λ = mobility number—the most common cases are $\lambda = 3$ for plane mechanisms and $\lambda = 6$ for spatial mechanisms.

There are mechanisms, some of them highly significant, which do not obey the general degree-of-freedom equation. The degree of freedom of

MECHANISM

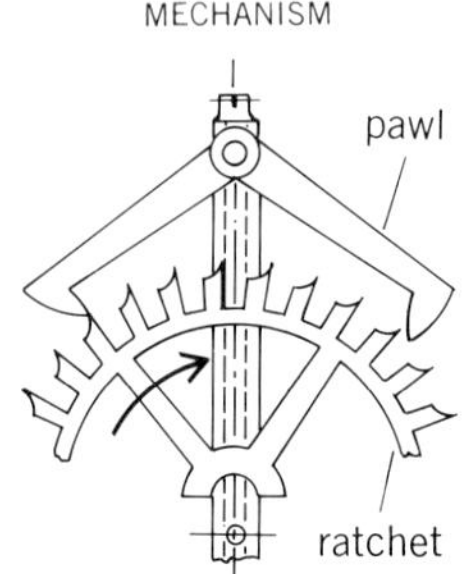

Fig. 6. Escapement.

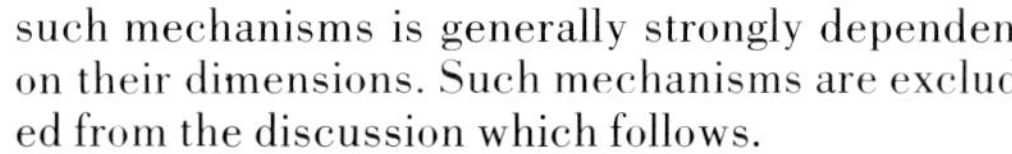
such mechanisms is generally strongly dependent on their dimensions. Such mechanisms are excluded from the discussion which follows.

For plane mechanisms with pin joints, Eq. (2) applies, while for gear trains with j_r pin joints and j_g gear meshes Eqs. (3) apply, where L_{ind} denotes the number of independent loops of the mechanism and is given by Eq. (4) for mechanisms in which the graphs can be drawn without crossing edges.

$$F=3l-2j-3 \qquad (2)$$

$$F=j_r-j_g \qquad j_g=L_{\text{ind}} \qquad l=F+L_{\text{ind}}+1 \qquad (3)$$

$$L_{\text{ind}}=j-l+1 \qquad (4)$$

Structure. The kinematic structure of a mechanism refers to the identification of the joint connection between its links. Just as chemical compounds can be represented by an abstract formula and electric circuits by schematic diagrams, the kinematic structure of mechanisms can be usefully represented by abstract diagrams.

Mathematical representation of kinematic structure. The structure of mechanisms for which each joint connects two links can be represented by a structural diagram, or graph, in which links are denoted by vertices, joints by edges, and in which the edge connection of vertices corresponds to the joint connection of links; edges are labeled according to joint type (and pin axis in the case of gear trains), and the fixed link is identified as well. Thus the graph of the slider-crank mechanism of Fig. 7*a* is as shown in Fig. 7*b*. In this figure the circle around vertex 1 signifies that link 1 is fixed.

Mechanisms can be considered identical or different, depending on whether their graph representations are identical (isomorphic) or different (nonisomorphic).

Classification and creation of mechanisms according to kinematic structure. Mechanisms can be classified and enumerated according to their kinematic structure, with the restrictions on their structure for many mechanisms being given by Eqs. (1)–(4).

It can be shown, for example, that when Eqs.

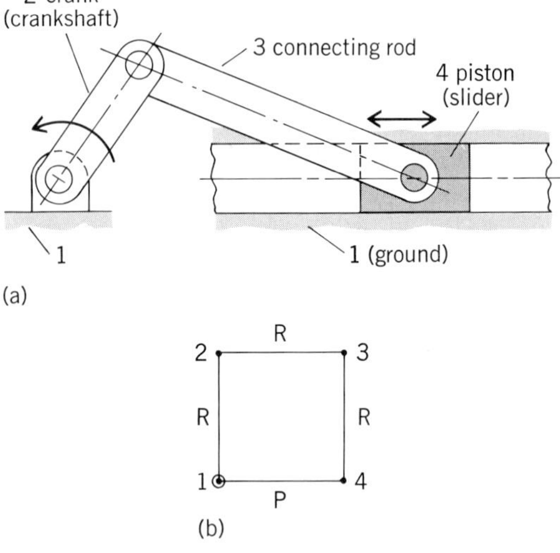

Fig. 7. Slider-crank mechanism. (*a*) Mechanism. (*b*) Graph of mechanism.

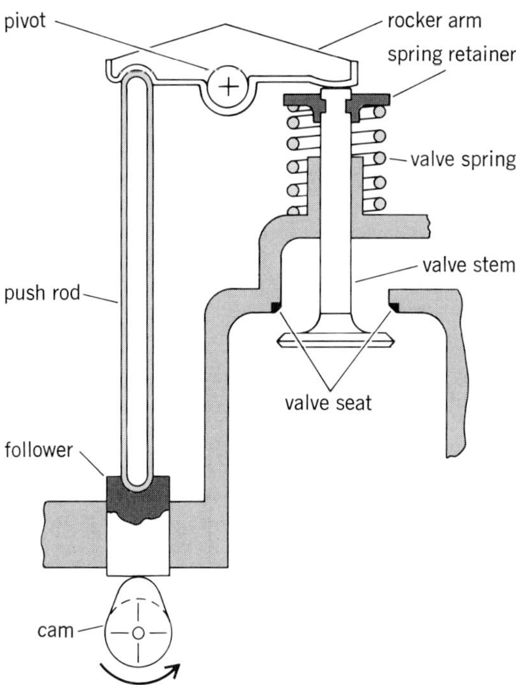

Fig. 8. Valve train of an internal combustion engine. (*Courtesy of Albert Pisano*)

(1)–(4) are applicable, there are seven plane single-degree-of-freedom mechanisms with four links and pin joints and up to two sliding joints. The structures of plane pin-jointed single-degree-of-freedom linkages with up to ten members and the structures of single-degree-of-freedom gear trains with up to six members have been determined as well.

The creation of mechanisms according to kinematic structure is also known as type synthesis or structural synthesis. It is based on the separation of kinematic structure from function. Thus, if, for example, mechanisms are sought which function as shaft couplings, the potential kinematic structures of mechanisms having the required degree of freedom, number of parts, and admissible types of joints are considered. The mechanisms corresponding to these structures are then screened by functional considerations. The result is an unbiased collection of potentially useful shaft-coupling configurations, complete within the specification of the search. This procedure can be a useful tool in the conceptual and inventive phases of mechanical design, supplementing atlases and collections of known mechanisms.

Kinematics. The subject is divided into kinematic analysis (analysis of a mechanism of given dimensions) and synthesis (determination of the proportions of a mechanism for given motion requirements). It includes the investigation of finite as well as infinitesimal displacements, velocities, accelerations and higher accelerations, and curvatures and higher curvatures in plane and three-dimensional motions.

Displacements in mechanisms can be found graphically or analytically from the conditions of closure of the independent loops of the mechanism. Velocities, accelerations, and higher-motion derivatives can be found by differentiation. It can be shown that at any instant in a general, continuous plane motion (the term general excludes translations, for example), there is one point for which the velocity vanishes (the instant center), one point for which the acceleration vanishes (the acceleration center), one point for which the acceleration derivative vanishes, and so on. The theory of instantaneous invariants has proved a powerful tool in such investigations.

For the position guidance of a rigid body through a prescribed sequence of finitely separated coplanar positions, one is interested in the loci of points in the body or the frame which may be used to attach cranks or sliders, so that the body can be guided mechanically. For four arbitrary positions these loci include the circlepoint curve (locus of points in the moving body, the four corresponding positions of which lie on a circle), the centerpoint curve (locus of the centers generated by points of the circlepoint curve), and the Ball point (that point in the moving system, the four corresponding positions of which are collinear). Circlepoints and their corresponding centerpoints can be used for locating moving and fixed crankpins, and the Ball point can be used for the attachment of a slider. In this way one can determine the proportions of linkage mechanisms, which can guide a rigid body through a prescribed sequence of positions.

The circlepoint and centerpoint curves are cubic algebraic curves. These curves can be used in the synthesis of linkage mechanisms for a considerable variety of motion requirements. The theory has been extended to five general coplanar positions and to three-dimensional motions. The optimization of linkage proportions so as to obtain the closest match of a linkage motion to a desired motion has also been the subject of many investigations, especially with the aid of Chebyshev approximations.

Useful concepts, which occur frequently in kine-

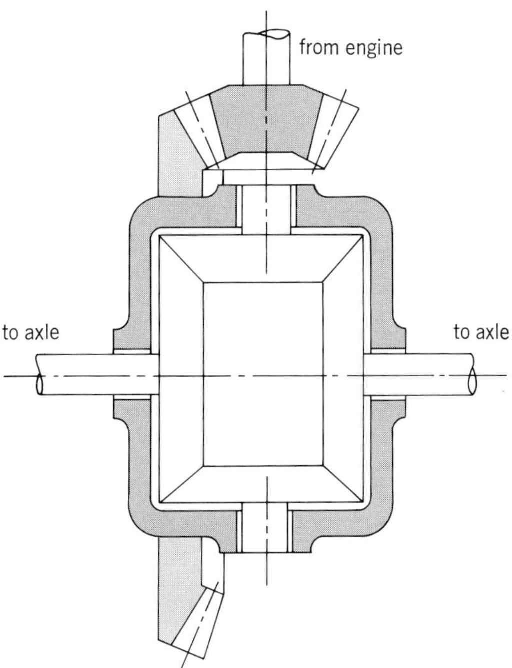

Fig. 9. Automotive differential.

matic analysis, include kinematic inversion (the process of holding different links of a mechanism fixed), relative motion, toggle positions (positions providing a very high mechanical advantage), lock-up positions (positions in which the mechanism cannot be driven from its usual input member), pressure angles and transmission angles (which govern the efficiency of force and power transmission), angular-velocity ratios, and angular-acceleration ratios and shock (the derivative of acceleration with respect to time).

Design. The design of mechanisms involves many factors. These include their structure, kinematics, dynamics, stress analysis, materials, lubrication, wear, tolerances, production considerations, control and actuation, vibrations, critical speeds, reliability, costs, and environmental considerations.

Compared to other areas of engineering design, the design of mechanisms is complicated by their generally high degree of nonlinearity (with the exception of gear trains), the inability in most instances to limit their operation to small displacements, and the great variety of combinations of components and mechanical elements. At the same time this complexity provides a challenge to the creativity of the engineer and inventor.

Current trends in the design of mechanisms emphasize economical design analysis by means of computer-aided design techniques, which offer ever-increasing speeds and more sophisticated performance requirements while maintaining reliability and economy of space, weight, and cost. *See* COMPUTER-AIDED DESIGN AND MANUFACTURING.

[FERDINAND FREUDENSTEIN]

Bibliography: O. Bottema and B. Roth, *Theoretical Kinematics*, 1979; A. Erdman and G. N. Sandor, *Mechanisms Design: Analysis and Synthesis*, 1982; F. D. Jones, H. L. Horton, and J. A. Newell (eds.), *Ingenious Mechanisms for Designers and Inventors*, 4 vols., 1968–1971; S. B. Tuttle, *Mechanisms for Engineering Design*, 1967.

Mercury-vapor lamp

A vapor or gaseous discharge lamp in which the arc discharge takes place in mercury vapor. This lamp is widely used for roadway and all other forms of illumination, and as a source of ultraviolet radiation for industrial applications.

Construction. The arc discharge takes place in a transparent tube of fused silica, or quartz, and this quartz tube is usually mounted inside a larger bulb of glass (see illustration). The outer bulb reduces the ultraviolet radiation of the inner arc tube, encloses and protects the mount structure, and can be coated with phosphors that greatly improve the color of the light emitted.

The arc tube has electrode assemblies mechanically sealed into each end. The main electrodes (one at each end) consist of a molybdenum shank welded to a very thin ribbon of molybdenum and completed by a shank of tungsten wound with tungsten wire and filled with an emission material of metal oxides. At one end of the arc tube, a smaller starting electrode is located close to the main electrode. The starting electrode is connected through a high resistance to the opposite electric polarity of the adjacent main electrode. The arc tube itself is filled with argon gas and a small amount of pure mercury before being sealed.

The outer bulb supports the mount structure built around the arc tube and is sealed to a glass stem with conducting wires. These wires pass through the outer bulb and are connected to the center post and shell of a metal base.

Starting and operation. Mercury lamps, like other vapor lamps, require sizable voltages to start their arcs, after which the lamp current will continue to increase unless limited by a resistor (which severely reduces efficiency) or a magnetic device (a choke or transformer) called a ballast. The ballast, when energized, supplies proper voltage to the lamp base for starting, and with zero current full starting voltage appears between the main and starting electrode at one end of the lamp. Electrons move across a short gap, ionizing some argon gas in the tube. The ionized gas diffuses until an arc strikes between the main electrodes, where the path now contains less resistance than that between the main and starting electrodes. Once the main arc is formed, the heat from the arc vaporizes the mercury droplets, and they become ionized current carriers. As the lamp current increases to its full current rating, the self-limiting design of the ballast simultaneously reduces the supply voltage to maintain stable operation and extinguishes the glow between starter and main electrode at the reduced voltage. Initial light output requires minutes of warm-up and, similarly, if the ballast is deenergized or power is lost, it is necessary for the lamp to cool and the mercury pressure to be reduced before the starting glow can occur again. *See* TRANSFORMER.

Color. Radiation from the mercury arc is confined to four specific wavelengths in the visible portion of the spectrum and several strong lines in the ultraviolet. Careful design of the arc tube can adjust mercury pressure and cause some shifting

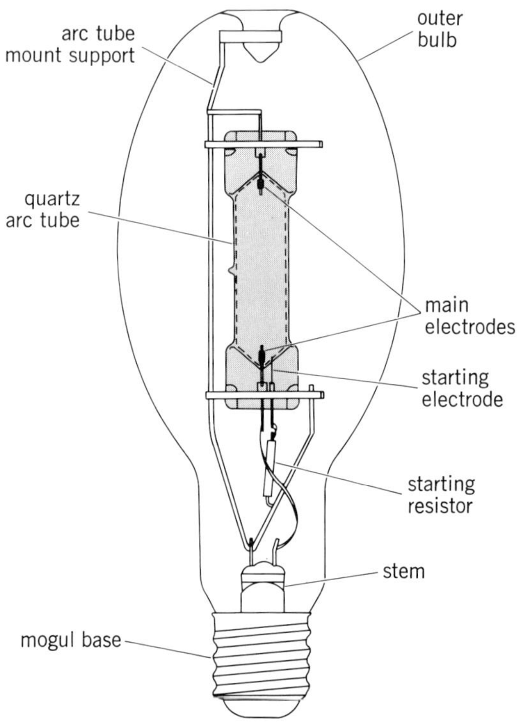

High-pressure mercury-vapor lamp.

of radiaton among these wavelengths, thereby changing slightly the color of the illumination.

The visible radiation from clear mercury lamps is all concentrated in the blue, green, and yellow-green regions of the spectrum. This provides light with a distinct blue-green appearance and poor color rendition. Red objects, for example, look brown or black, and human skin looks unattractive. For this reason, most mercury lamps have a phosphor color-correcting coating, and clear bulbs are used only where appearance of colors is secondary to some gain in efficiency.

The phosphors applied to the inner wall of the outer bulb can provide superior color rendition through conversion of the arc tube's ultraviolet radiation to sizable reradiation in the red part of the spectrum or additional yellow, green, and blue-green radiation. One phosphor used for deluxe white lamps has made their use popular in many commercial lighting applications such as department stores and office areas.

Applications. After the fluorescent lamp, the mercury-vapor lamp is the most popular vapor lamp in use today, with its greatest use in roadway lighting. Many millions of these lamps are produced in the world annually, and despite the fact that other gaseous discharge lamps are more efficient, the long life, reliable performance, and relatively low cost of this lamp and ballast system ensure its use for some years to come.

The sizable radiation of ultraviolet by the bare arc tube suits it well as a sun lamp and as a source for ultraviolet applications in industry, where it is used in reproduction of tracings and to speed chemical reactions such as curing of protective coatings and inks. One wavelength in the near-ultraviolet, not harmful to humans, and transmitted by the bulb glass has the property of causing some dyes and pigments to fluoresce. When the visible light is filtered out and it is used for this fluorescing purpose, it is called a black light.

Consideration should be given to avoiding close proximity of humans to the standard mercury-vapor lamp when the outer bulb is broken since its ultraviolet radiation at close distances can cause erythemal reactions or sunburn, and these standard lamps are now being offered in alternate designs with internal switches or other devices to extinguish the arc if the outer bulb should break. *See* LAMP; ULTRAVIOLET LAMP; VAPOR LAMP.

[T. F. NEUBECKER]

Metal

An electropositive chemical element. Physically, a metal atom in the ground state contains a partially filled band with an empty state close to an occupied state. Chemically, upon going into solution a metal atom releases an electron to become a positive ion. Consequently in biotic systems metal atoms function prominently in ionic transport and electron exchange. In bulk a metal has a high melting point and a correspondingly high boiling temperature; except for mercury, metals are solid at standard conditions. The term metal often refers to a structural material distinct from stone or wood. Direct observation shows a metal to be relatively dense, malleable, ductile, cohesive, highly conductive both electrically and thermally, and lustrous. When crystals of the elements are classified along a scale from plastic to brittle, metals fall toward the plastic end. Furthermore, molten metals mixed with each other over wide ranges of proportions form, upon slowly cooling, homogeneous close-packed crystals. In contrast, a metal mixed with a nonmetal completely combines into a homogeneous crystal only in one or a few discrete stoichiometric proportions.

This article summarizes the characteristics that distinguish a metal from other elements and materials. For detailed discussions of metals, in particular, exceptions to generic behavior, see separate articles on each metal.

Classification of elements. Logically each element might be assigned to either of two mutually exclusive categories: metal or nonmetal. However, properties of elements are not so distinct as this dichotomy implies. Typically, metallic oxides react with water to give basic solutions, but oxides of a few metallic elements, including chromium, yield slightly acidic solutions.

Another subclass, of which antimony is typical, has physical properties of a metal but behaves chemically, depending on conditions, as either a metal losing an electron or as a nonmetal gaining an electron in a reaction with another element.

When metals are arranged by electrical properties, another classification uncertainty arises. Some elements are neither distinctly conductors and hence metals, nor distinctly insulators and hence nonmetals. Between these classes are semimetals, such as germanium. For pure samples of distinctly metallic crystals, electrical conductivity decreases as temperature rises, whereas for naturally occurring semimetals conductivity increases as temperature rises.

Measures of metallic character. No single simple observable property suffices as criterion for listing metals in sequence by the degree to which they are metallic. External properties arise from interaction of several atomic properties, including weight and size as well as outer electrons. However, the energy exchange that accompanies release of the first electron from a metal atom serves as an initial sequencing sieve.

Atomically, the energy necessary for release of an electron, or for ionization of an atom, is a measured in the gaseous state. It is the energy necessary to remove an electron from its ground-state orbital in the atom, leaving behind a positive ion. This ionization energy of free atoms is generally lower for metals than for nonmetals in the same row of the periodic table.

Also important is the number of electrons lost from an atom. If an atom loses one electron, as metals readily do, the atom is monovalent. If the number of electrons it loses varies depending on conditions, the metal is described as polyvalent. In a metallic crystal, depending on lattice, a metal atom can share electrons with several nearest neighbors without violating the Pauli exclusion principle because a metal has fewer valence electrons than valence orbitals.

Electrically, the energy necessary to release an electron from an atom in a crystal and to maintain a current flow depends on the difference in energy levels between the valence band and the conductance band. One measure of this energy is the work function of the metal. For a solid metal crystal in a vacuum environment, this energy is determined experimentally as the threshold frequency, characteristic of each metal, above which the incident electromagnetic wave or photon dislodges an electron.

In a liquid environment, as with marine uses, metal pipes buried in moist ground, or electrodes in a battery, the electrolytic action describes the tendency to release an electron. *See* CORROSION.

Families of metals. Distinctly metallic elements—the ones with low ionization energies relative to their positions in the periodic table—are subdivided into several families.

Metals begin in the lower left-hand corner of the periodic table and extend into the central region. Along the right-hand margin of this region is an indefinite diagonal boundary beyond which are nonmetals in the upper right-hand corner. The metallic character of the elements disappears irregularly along a period with increase in atomic number. In contrast, in a given subgroup, metallic properties tend to appear more strongly with increase in atomic number.

Metal crystals. Observed metallic properties can be related to the structure of atoms as described by quantum mechanics. Spectroscopic analysis of vaporized metals, where each atom is independent of the others, provides information about orbital energies within atoms. Direct visual observation of shapes of single-metal crystals suggests regular arrangements of their constituent atoms. Diffraction of x-rays and of neutrons by metal crystals, where atoms are closely associated with each other in the solid state, provides evidence for the spacing between nuclei and hence, by inference, the sizes of atoms or ions.

Atomic boundaries. In a crystal the wave nature of electrons provides some compression or expansion of the atomic boundary (Fermi level), so atomic size may depend on circumstances, but in the close-packed metal crystal, interatomic contacts are achieved for nearest neighbors in several directions. From these considerations, and for geometric simplicity, a usable first approximation is that atoms are spheres arranged in a regular lattice. This configuration serves as a beginning description of metal crystals that relates structure to characteristics.

Electronic structure. The outer shell of each metal atom contains relatively few filled orbitals. Upon the metal's cooling from the molten state, a valence electron from one atom may, with high probability, overlay an empty orbital in a neighboring atom. Nuclear vibrations distort the overlapping outer orbitals of adjacent atoms with the result that metals predominantly interact with a valence of +1 unless other conditions intervene. Either atom loses an electron so that, when two atoms join, an electron temporarily occupies a vacant orbital in either adjacent atom. With further cooling, the aggregation similarly adds more atoms, so that upon solidification, the first valence electron from any atom is energetically free to migrate to any other close atom. The remaining nucleus of each atom with its inner electrons constitutes a positive core that becomes increasingly fixed in location by mutual electrostatic repulsion between it and other members of this array of cores. The array takes on the regular crystal lattice characteristic of the particular metal. Meanwhile the electrostatic attraction to the negative electrons dispersed about the crystal both holds the cores in place and preserves the local electrical neutrality of the crystal and hence its integrity. Simultaneously the electrons unite the cores into the crystal lattice by their transitory ground-state orbital exchanges that constitute the metallic bond.

These flexible and cohesive actions within the crystal give rise to the directly observable properties that characterize a metal, such as generally high melting and boiling points, low vapor pressures, high tensile and compressive strengths, and high electrical conductivity. Current flows as a result of even a small excitation above the ground state by application of a steady electric field, a changing magnetic field, or a thermal gradient.

Electron mobility. The mobility of the electrons distributed on the surface and throughout the body of the metallic crystal electronically screens incident electromagnetic radiation to render the bulk material opaque and reflective over a frequency range that includes the visible portion of the spectrum (with the noticeable exception of copper and gold). This gives the material its luster and its characteristic reflection of polarized light. When the internal heat energy of the metallic body is sufficient to deliver to a surface electron energy above the work function, the electron leaves the crystal.

Thermal conductance. While the crystal is solid but still hot, each core vibrates about its central location, thus extending the space it occupies. The cores, by their vibrations against each other within the close-packed crystal, and the electrons, by their migrations from core to core, rapidly distribute this heat energy, giving rise to the observed high thermal conductance.

In cooling, the crystal contracts, displaying a characteristic thermal dimensional coefficient. If the metal is one of the naturally occurring ferromagnetic materials, of which iron is the most common, then when it cools below the critical Curie temperature, it spontaneously exhibits its unique property arising from an unpaired internal electron.

Solidification. Because with cooling, the cores remain closer to their neutral sites, electron waves set in motion by an applied electric potential propagate with little dispersion through the crystal. In the sense that electrical conductivity is a metallic characteristic, the metal thus becomes more metallic as it cools.

On the contrary, cooling to low temperatures decreases mechanical metallic properties so that, for example, lead, which in temperate climates is malleable, in arctic temperatures is brittle. Finally, at a sufficiently low temperature to reduce translational vibrations to the zero point energy, and with negligible surface magnetic fields or high applied pressures, electron waves in some metals (one of which is titanium), once set into circulation, continue to propagate about a closed metal loop in the absence of a continuing electromotive force.

Solid solutions. Had the melt initially been a mixture of two metals, because of their common feature of incompletely filled outer shells, the different atoms would have distributed more or less uniformly about the bulk, the greater quantity acting as a solvent to the lesser quantity behaving as a solute. Upon crystallizing, the resulting solid solution would have possessed physical properties of a metal but with different values from those of either parent metal. *See* ALLOY.

Had the proportions of the parent metals and their orbital energies been such that electrons lost by one species filled the empty orbitals of the outer shell of the other species, this sharing of electrons between the two metals would have produced a compound with metallic properties distinctly different from those of the parent metals.

Physical properties of a metal crystal are a function of the crystal structure. Generally in metallic phases, bonds between metal atoms are essentially nondirectional, so that high symmetrical coordination is typical. However, crystal geometry imposes a directional feature on the bonding so that a metallic bond may show some features of a covalent (chemical) bond.

Anistropy arises predominantly from the crystal structure and becomes apparent when a crystal is subject to energy changes as by a strong applied force. Then the crystal slips—a peculiarly metallic behavior—along one direction more readily than along another. In slip, the atoms index past one another, possibly one at a time in a wave of motion propagating along a line of atoms or in a burst of lattice vibrations (phonons). After the slip action, the crystal is again a solid structure, the metallic bonds reestablishing with each atom of a line now opposite a new nearest neighbor. Slip occurs along limited atomic planes and in certain crystal directions, each combination of slip plane and direction constituting a slip system. The number of slip systems is a consequence of the pattern in which nearest-neighbor cores can cluster about a given core.

Malleability and ductility arise from this ability of the cores along each side of a slip plane to glide past each other in the presence of intense compression or tension and to promptly reestablish the metallic bond. *See* METAL FORMING; PLASTIC DEFORMATION OF METAL.

Internally a metal crystal is in force equilibrium. The applied force can be either tension or compression. In either case the spacings within the crystal lattice change so that the electrostatic and bonding forces adjust to equilibrium with the external force. Upon removal of the force, the crystal returns to its initial force equilibrium and dimensions. This elastic behavior, characteristic of all crystals, whether metal or nonmetal, extends over a particularly wide range of applied force for metals. In addition, within a transition region, above a critical high load, metal plastically deforms through a sequence of slips so that, upon removal of the load, the metal permanently retains some of the deformation. The exact value of the load beyond which the metal is observed to no longer behave elastically—the elastic limit—depends on test conditions, such as properties of the testing machine and sensitivity of the measuring instruments. *See* ELASTIC LIMIT.

Role in biotic systems. Other properties than those discussed above take precedence in selection of a metal for use in a biotic environment. Because of their resistance to oxidation, gold and silver have long served in the form of amalgams for a filling material in tooth cavities. Tantalum is inert enough to be used for bone support in surgery. Similarly, platinum among the natural metals is biologically inert and so is used as internal support for fractured bones.

In an entirely different manner, yet with a structural feature, calcium in the form of salts impregnates the fibrils of the bone substrate to impart stiffness to that structural organ. However, because the bonding is to nonmetal atoms rather than to another metal, bone is brittle.

As practical analytic procedures and trace analysis improve, smaller quantities of metals in tissue become measurable. Analytic limits are in the range of $1-5 \times 10^{-6}$ to 1×10^{-9} gram of metal per gram of sample, with sensitivities for detection a further thousandth of these limits. As a consequence of this measurement sensitivity, most metals are now known to serve as foods in small quantities. The behavior of metal atoms used as food begins with the low-energy release of an electron but in a far more varied atomic environment than that described above. Most metals, at least in trace quantities, seem essential to energy exchanges and molecular building in biotic systems. A metal ion assembles reactants in a reversible way into a molecular structure so that metal complexes seem to be agents of stereoselectivity. Iron, copper, sodium, and potassium are examples of metal elements known to have important functions in biotic systems. Beyond their roles in these activities, metals function in molecule building in a wide class of compounds formed between carbon and the metals.

When ingested in controlled quantities, metals function in organic growth, but in excess metals are toxic. The inherent toxicity of a metal is an expression of its electrochemical character: the solubility, stability, and reactivity of its compounds in body fluid and tissue. Selenium, beryllium, and mercury are all highly toxic. The toxicity of mercury is largely the result of the solubility of its compounds; moreover, it does not appear necessary for any biologic function.

Mining and metallurgy bring metals from their natural environment in the lithosphere. There metals occur commonly as compounds, occasionally somewhat concentrated in ore and mineral deposits, with aluminum being both the most abundant and widely distributed.

In their natural state, metals are often confined by their combination with other elements. In releasing them for uses in plows and as fertilizers for the support of the burgeoning world population, some technologies may cause metals to enter the biosphere faster than the biologic adaptation can construct defenses against disruptive effects of excess metals on cellular metabolism or molecular replication. Therefore technologies and possibly new industries are developing for the recycling, storage, and deactivation of metals after they have passed through a primary economic cycle. Waste containment or confined disposal can limit their incidental distribution. After energy has been expended to mine and purify metals, the need for complete duplication of that energy expenditure may be minimized by recycling metals after they enter the economic cycle. *See* METALLURGY.

[FRANK H. ROCKETT]

Bibliography: G. Caglioti, *Atomic Structure and*

Mechanical Properties of Metals, 1976; T. D. Luckey and B. Venugopal, *Metal Toxicity in Mammals*, 1977; W. P. Prarson, *The Crystal Chemistry and Physics of Metals and Alloys*, 1972; Helmut Sigel (ed.), *Metal Ions in Biological Systems*, vol. 7: *Iron in Model and Natural Compounds*, 1978; W. J. McG. Tegart, *Elements of Mechanical Metallurgy*, 1966.

Metal, mechanical properties of

Commonly measured properties of metals (such as tensile strength, hardness, fracture toughness, creep, and fatigue strength) associated with the way metals behave when subjected to various states of stress. The properties are discussed independently of theories of elasticity and plasticity, which refer to the distribution of stress and strain throughout a body subjected to external forces.

Stress states. Stress is defined as the internal resistance, per unit area, of a body subjected to external forces. The forces may be distributed over the surface of a body (surface forces) or may be distributed over the volume (body forces); examples of body forces are gravity, magnetic forces, and centrifugal forces. Forces are generally not uniformly distributed over any cross section of the body upon which they act; a complete description of the state of stress at a point requires the magnitudes and directions of the force intensities on each surface of a vanishingly small body surrounding the point. All forces acting on a point may be resolved into components normal and parallel to faces of the body surrounding the point. When force intensity vectors act perpendicular to the surface of the reference body, they are described as normal stresses. When the force intensity vectors are parallel to the surface, they describe a state of shear stress. Normal stresses are positive, when they act to extend a line (tension). Shear stresses always occur in equal pairs of opposite signs.

A complete description of the state of stress requires knowledge of magnitudes and directions of only three normal stresses, known as principal stresses, acting on reference faces at right angles to each other and constituting the bounding faces of a reference parallelepiped. Three such mutually perpendicular planes may always be found in a body acted upon by both normal and shear forces; along these planes there is no shear stress, but on other planes either shear or shear and tensile forces will exist.

The shear stress is a maximum on a plane bisecting the right angle between the principal planes on which act the largest and smallest (algebraic) principal stresses. The largest normal stress in the body is equal to the greatest principal stress. The magnitude and orientation of the maximum shear stress determine the direction and can control the rate of the inelastic shear processes, such as slip or twinning, which occur in metals. Shear stresses also play a role in crack nucleation and propagation, but the magnitude and direction of the maximum normal stress more often control fracture processes in metals capable only of limited plastic deformation.

It is often useful to characterize stress or strain states under boundary conditions of either plane stress (stresses applied only in the plane of a thin sheet) or plane strain (stresses applied to relatively thick bodies under conditions of zero transverse strain). These two extreme conditions illustrate that strains can occur in the absence of stress in that direction, and vice versa.

Tension and torsion. In simple tension, two of the three principal stresses are reduced to zero, so that there is only one principal stress, and the maximum shear stress is numerically half the maximum normal stress. Because of the symmetry in simple tension, every plane at 45° to the tensile axis is subjected to the maximum shear stress. For other kinds of loading, the relationship between the maximum shear stress and the principal stresses are obtained using the same method, with the results depending upon the loading condition.

For example, in simple torsion, the maximum principal stress is inclined 45° to the axis of the bar being twisted. The least principal stress (algebraically) is perpendicular to this, at 45° to the bar axis, but equal to and opposite in sign to the first principal stress—that is, it is compressive (Fig. 1). Both of these are in a plane perpendicular

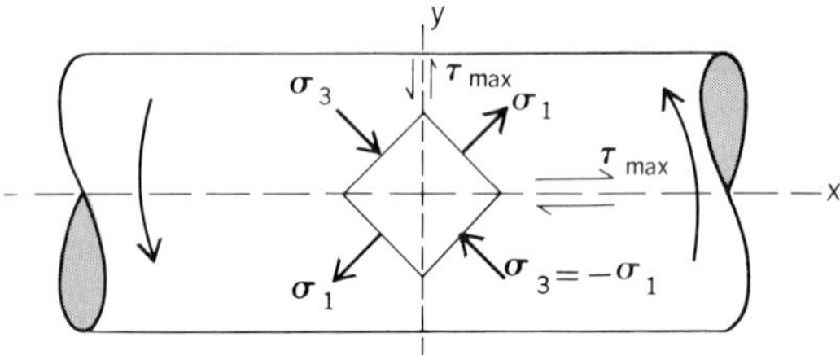

Fig. 1. State of stress in torsion. (*From G. E. Dieter, Mechanical Metallurgy, McGraw-Hill, 1965*)

to the radial direction, the direction of the intermediate principal stress, which in this case has the magnitude zero. Every free external surface of a body is a principal plane on which the principal stress is zero. In torsion, the maximum shear stress occurs on all planes perpendicular to and parallel with the axis of the twisted bar. But because the principal stresses are equal but of opposite sign, the maximum shear stress is numerically equal to the maximum normal stress, instead of to half of it, as in simple tension. This means that in torsion one may expect more ductility (the capacity to deform before fracture) than in tension. Materials that are brittle (exhibiting little capacity for plastic deformation before rupture) in tension may be ductile in torsion. This is because in tension the critical normal stress for fracture may be reached before the critical shear stress for plastic deformation is reached; whereas in torsion, because the maximum shear stress is equal to the maximum normal stress instead of half of it as in tension, the critical shear stress for plastic deformation is reached before the critical maximum normal stress for fracture.

Another important situation arises at the root of a geometrical discontinuity in a loaded body. Such a discontinuity is referred to as a notch. At the root of a notch, immediately beneath but very near the surface, the magnitude of the maximum normal stress may increase to values considerably greater than the applied normal stress; the maximum shear stress, instead of being equal to the maximum normal stress as in torsion, or half of it as in tension, may be a considerably smaller fraction. Also, the critical normal stress for fracture is more easily reached before the critical shear stress for deformation than in torsion and tension, and one may expect notches to have an embrittling effect. This embrittling effect of notches is of very great practical importance and accounts for the majority of service failures in engineering structures. The depth and radius of notches are parameters of overriding importance in determining fracture resistance.

Stress-strain curve. If a metal is strained at a fixed rate, the resistance to deformation increases with straining at a diminishing rate. A plot of the resistance to deformation against the amount of strain is called a stress-strain curve, and the slope of this curve at any point is called the rate of strain hardening. Such a curve is normally obtained by making a tension test and plotting the maximum normal stress against the maximum normal strain. These are identical with the ordinary tensile stress (the tensile force divided by the area of the specimen) and the ordinary tensile strain (the extension per unit of length). The curve produced is known as an engineering stress-strain curve and is widely used to provide design strength information and as an acceptance test for materials. For small strains, the original area and the original gage length may be used without serious error in calculating the stress and strain. Where the strains are large, it is better to use the true average stress, obtained everywhere on the curve by dividing the tensile force by the actual area (which decreases with increasing deformation), and the so-called true strain. The true strain is obtained for every point on the curve by integration of dl/l, between the limits of original length l_0 and actual length l, thus taking into account the changing gage length. The true strain is thus the natural logarithm of the ratio of the final length to the original length as expressed in Eq. (1). The ratio of the final to the original length may be replaced by the ratio of the original area to the final area, because of the constancy of volume in inelastic deformation. True stress-strain curves are more likely to be utilized in circumstances where fundamental information on deformation mechanisms is desired.

$$\int_{l_0}^{l} \frac{dl}{l} = \ln \frac{l}{l_0} \tag{1}$$

For single-phase metals which have not suffered any prior permanent deformation, the stress-strain curve in simple compression is usually nearly identical with the stress-strain curve in simple tension. This is not true, however, for two-phase materials where the second phase is present as dispersed hard particles, nor is it true for materials within which occur residual stresses resulting from, for example, nonuniform thermal contraction from the processing temperature to the test temperature in the coexisting phases. For other states of stress, additional complications arise. In torsion, for instance, the maximum normal stress reaches a value, at corresponding strains, of less than 60% of the value reached in simple tension. A more generally applicable stress-strain relationship is obtained by plotting the von Mises function of the three principal stresses σ_1, σ_2, and σ_3 (the effective stress $\bar{\sigma}$), as in Eq. (2), against an easily derivable function of the three principal strains that may be called the effective strain $\bar{\epsilon}$, given in Eq. (3),

$$\bar{\sigma} = [\tfrac{1}{2}(\sigma_1 - \sigma_2)^2 + (\sigma_2 - \sigma_3)^2 + (\sigma_3 - \sigma_1)^2]^{1/2} \tag{2}$$

$$\bar{\epsilon} = [\tfrac{2}{3}(\epsilon_1{}^2 + \epsilon_2{}^2 + \epsilon_3{}^2)]^{1/2} \tag{3}$$

where $\bar{\epsilon}$ is the effective strain and ϵ_1, ϵ_2, and ϵ_3 are the principal strains. Both the effective stress and the effective strain reduce to the ordinary tensile stress and tensile strain for simple tension testing, so that the tensile stress-strain curve without change in units may be used as the effective stress-strain curve.

Unless some microstructural change occurs in the metal during straining, such as recrystallization or precipitation of a second phase, the effective stress-strain curve usually is describable by the power function $\bar{\sigma} = k\bar{\epsilon}^n$, where k is the effective stress required to produce unit effective strain, and n is an exponent that is always less than unity and commonly has a value between 0.1 and 0.4. The exponent n is called the strain-hardening index because differentiation of $\bar{\sigma} = k\bar{\epsilon}^n$ to find the rate of strain hardening $d\bar{\sigma}/d\bar{\epsilon}$ yields for this expression $n(\bar{\sigma}/\bar{\epsilon})$, so that the slope of the stress-strain curve (the strain-hardening rate) is proportional to n as well as to the ratio of the coordinates of the point on the curve. Higher values of n are associated with annealed pure metals and annealed single-phase alloys; lower values of n are associated with heat-treated alloys such as quenched and tempered steels.

Modern theory makes use of line imperfections (dislocations) to explain the low resistance to initial plastic deformation of annealed metal crystals and polycrystalline aggregates, as well as strain hardening and viscous effects. Imperfection-free crystals deform only elastically and fracture at very high stress levels. This has been demonstrated with metal whiskers of high purity and of extremely small diameter. Larger crystals invariably contain dislocations which move through the crystal lattice at low stress levels, but pile up at barriers such as hard particles or surfaces of contact with other crystals (grain boundaries). The dislocations are surrounded by stress fields which interact with each other, and strain hardening is ascribed to interactions between dislocations moving on different slip systems as well as interactions with other barriers. Viscous effects are associated with the rate at which line imperfections can break loose from their obstacles, usually other line imperfections, with the help of thermal motion. *See* ELASTICITY; PLASTIC DEFORMATION OF METAL; STRESS AND STRAIN.

Tension test. To achieve uniformity of distribution of stress and strain in a tension test requires that the specimen be subjected to no bending moment. This is usually accomplished by providing flexible connections at each end through which the force is applied. The specimen is stretched at a controllable rate, and the force required to deform it is observed with an appropriate dynamometer. The strain is measured by observing the extension between gage marks adequately remote from the ends, or by measuring the diameter and calculating the change in length by using the constancy of volume that characterizes plastic deformation. Diameter measurements are applicable even after necking-down has begun. The elastic properties are seldom determined since these are structure-insensitive. Special refinements are necessary when the elastic stress-strain relationships are in question because the strains are very small and because elastic deformation is not characterized by constancy of volume.

Yield strength. The elastic limit is rarely determined. Metals are seldom if ever ideally elastic, and the value obtained for the elastic limit depends on the sensitivity of strain measurement. The proportional limit, describing the limit of applicability of Hooke's law of linear dependence of stress on strain, is similarly difficult to determine. Modern practice is to determine the stress required to produce a prescribed inelastic strain, which is called the yield strength. The amount of strain used to define the yield strength varies with the application, but is most commonly taken as 0.2% (a unit strain of 0.002 in./in.; 1 in. = 25.4 mm). Because upon unloading the behavior is almost linearly elastic, it is possible to use the offset method of determining the yield strength from a plotted stress-strain curve; a line with a slope equal to the elastic slope is drawn, displaced from the stress-strain curve at low stress levels by the amount of strain used in the definition of the yield strength, and the stress at the intersection of this line with the stress-strain curve is taken as the yield strength (Fig. 2). The elastic slope will not be

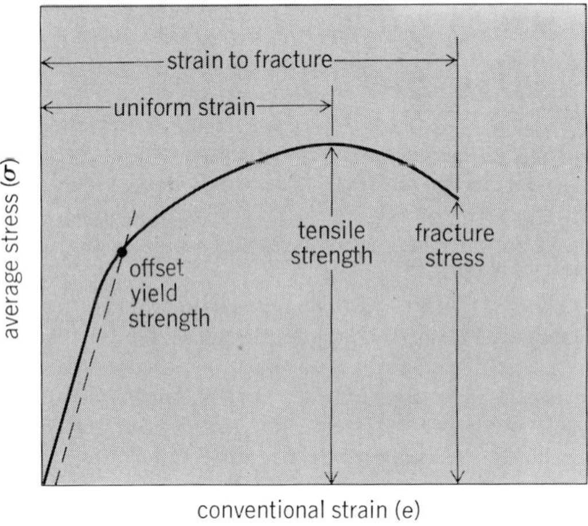

Fig. 2. The engineering stress-strain curve. (*From G. E. Dieter, Mechanical Metallurgy, McGraw-Hill, 1965*)

equal to Young's modulus unless an exceedingly stiff tensile machine is employed or strain gages are attached to the specimen.

The stress-strain curve obtained at room temperature is not appreciably affected by changes in the rate of straining in ordinary tensile tests. At higher temperatures, however, strain rate effects are much more important. A general relation between flow stress and strain rate $\dot{\epsilon}$, at constant temperature and strain, is shown in Eq. (4), where

$$\sigma = C\dot{\epsilon}^m \tag{4}$$

m is the strain-rate sensitivity. For most metals, m increases with both test temperature and strain.

Test temperature at constant strain rate has a very important influence on yield strength. For most metals, strength increases and ductility decreases with decreasing temperature. For face-centered cubic metals such as copper, nickel, and silver, these effects are small, but for iron, steels, and other body-centered cubic metals and alloys, increases in strength of up to 800% can occur between room temperature and 77 K. Ductility also may change rapidly in this temperature range. The temperature dependence of yielding in many metals below room temperature may be influenced by impurities or intentional alloying elements.

Changes in strength with temperature will occur also when microstructural changes such as precipitation, strain aging (clustering of solute atoms around dislocations), or recrystallization occur. Extended exposure at elevated temperatures may cause creep, as discussed in a later section of this article.

The stress-strain properties of metals also may be altered drastically by changes in grain (crystal) size. A linear relation is observed for most metals between yield stress σ_y, and the reciprocal square root of grain size $d^{-1/2}$, as given by Eq. (5), where

$$\sigma_y = \sigma_o + k_y d^{-1/2} \tag{5}$$

σ_o is a measure of the stress to plastically deform a single crystal, while k_y is a parameter which reflects the ease of propagating slip from one grain to another. For metals in which slip is transmitted with difficulty across grain boundaries (for example, hexagonal close-packed metals such as zinc, cadmium, magnesium, and certain face-centered cubic alloys), k_y is relatively large. However, for pure face-centered cubic or body-centered cubic metals, k_y is a small number. Both σ_o and k_y tend to increase with decreasing temperature. Thus grain boundary strengthening is particularly effective in metals such as zinc, cadmium, and magnesium at low test temperatures.

Tensile strength. Tensile strength, usually called the ultimate tensile strength, is calculated by dividing the maximum load by the original cross-sectional area of the specimen. It is, therefore, not the maximum value of the true tensile stress, which increases continuously to fracture and which is always higher than the nominal tensile stress because the area continuously diminishes. For ductile materials the maximum load, upon which the tensile strength is based, is the load at which necking-down begins. Beyond this point, the true tensile stress continues to increase, but the force on the specimen diminishes. This is because the rate of strain hardening has fallen to a value less than the rate at which the stress is increasing because of the diminution of area. If at any time during the extension of a specimen the force required to extend it diminishes, the strain will become localized. The condition for necking down is therefore that the force go through a maximum; that is, $\partial F/\partial l$ (the partial derivative of the force with respect to extension) goes through zero. When force and length are expressed as functions of stress and strain, it is easy to show that this condition is satisfied when the true rate of strain hardening falls to a value numerically equal to the stress; that is, $\partial\sigma/\partial\epsilon = \sigma$. For the stress-strain relationship used above, $\bar{\sigma} = k\bar{\epsilon}^n$, this slope is reached when $\epsilon = n$. Necking-down, therefore, will occur when the tensile strain reaches a value equal to n (usually between 10 and 40% extension), and the

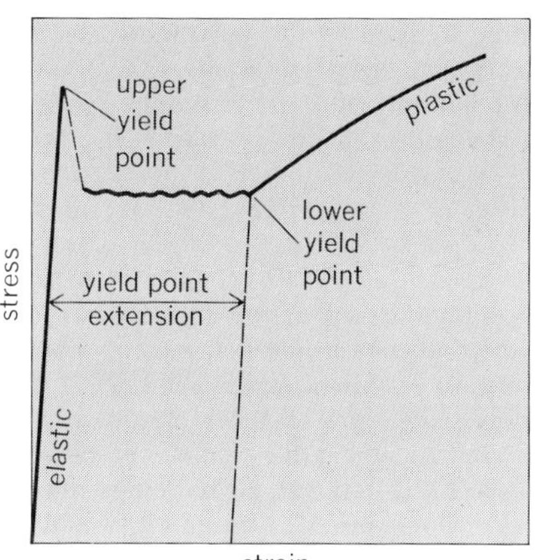

Fig. 3. Yield point, mild steel.

tensile strength for a ductile material has nothing to do with fracture. It describes only the stress required to produce the tensile instability evidenced by necking down. For a relatively brittle material, one which fractures before the strain reaches the necking-down value, the tensile strength reflects the fracture strength, but again is not equal to it.

Yield point. A considerable number of alloys, including those of iron, molybdenum, tungsten, cadmium, zinc, and copper, exhibit a sharp transition between elastic and plastic flow. The stress at which this occurs is known as the upper yield point. A sharp drop in load to the lower yield point accompanies yielding, followed, in ideal circumstances, by a flat region of yield elongation; subsequently, normal strain hardening is observed (Fig. 3). In the case of polycrystalline mild steels as well as concentrated solid solutions of some single-crystal alloys, the yield elongation marks a period in which strain occurs by the movement of a plastic strain wave along the specimen length. The force on the specimen remains constant during the advance of this strain wave because plastic strain occurs only at the advancing edge of the wave. The wave front is visible on the specimen surface because of surface relief; the markings are known as Lüder's bands. When the strain wave has completely traversed the specimen, the extension of the entire specimen is identical to that achieved at the wave front.

Straining of metals beyond the yield point causes the latter to disappear upon subsequent unloading and restraining. However, annealing at an appropriate low temperature causes the yield point to return. This observation has led to the conclusion that yield points in metal single crystals and polycrystals are associated with the locking of dislocations by interstitial or substitutional impurities. Yielding occurs by a process of tearing away dislocations from their impurity "atmospheres," resulting in a sharp drop in load. Annealing of the crystal allows impurities to migrate back to dislocations so that the process may be repeated. It should be noted that other mechanisms for a sharp yield point have been identified as being applicable to metal whiskers and to nonmetallic crystals.

Elongation. The tensile test provides a measure of ductility, by which is meant the capacity to deform by extension. The elongation to the point of necking-down is called the uniform strain or elongation because, until that point on the stress-strain curve, the elongation is uniformly distributed along the gage length (see Fig. 2). The strain to fracture or total elongation includes the extension accompanying local necking. Since the necking extension is a fixed amount, independent of gage length, it is obvious that the total elongation will depend upon the gage length, and will be greater for short gage lengths and less for long gage lengths. The American standard ratio of gage length to diameter for a cylindrical test specimen is 2:0.505 (the standard bar has a 2-in. gage length with a diameter of 0.505 in. corresponding to an area of 0.2 in.^{2}). Specimen configurations have been compiled also for sheet, plate, tube, and wire specimens by the American Society for Testing and Materials. Absolute size does not affect the distribution of deformation. The shape of the cross section has no effect up to ratios of width to thickness of 5:1; rectangular cross sections elongate the same amount as circular cross sections of the same area. The reduction of area at the point of fracture, expressed usually as percentage reduction of area, is independent of gage length, and from it can be calculated the true strain at fracture; this makes the tensile reduction of area a more satisfactory measure of ductility than elongation unless one is interested in the amount that a metal will stretch before it begins to neck down. The reduction of area in the tensile test correlates well with measures of ductility obtained in other ways, such as the strain to fracture on the outside of a bend, provided the strain gradient is not too great, such as in notched-bar testing.

Ductile-to-brittle transition. Many metals and alloys, including iron, zinc, molybdenum, tungsten, chromium, and structural steels, exhibit a transition temperature, below which the metal is brittle and above which it is ductile. The form of the transition for polycrystalline iron and several iron-cobalt alloys is shown in Fig. 4. The transition temperature very clearly is sensitive to alloy content, but it will vary even for the same material, depending upon such external test conditions as stress state and strain rate, and microstructural variables such as purity and grain size. The ductility transition frequently is accompanied by a change in the mechanism of fracture (as in iron and steels or zinc), but this need not be so. In gen-

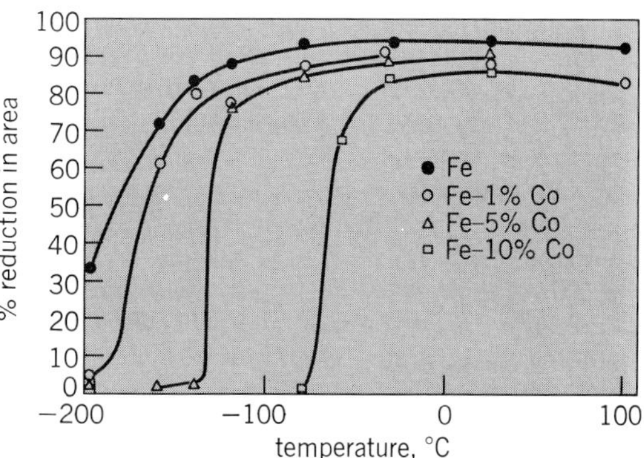

Fig. 4. Ductile-to-brittle transition, iron-cobalt alloys. *(From N. S. Stoloff, R. G. Davies, and R. C. Ku, Low temperature yielding and fracture in Fe-Co and Fe-V alloys, Trans. Metallurg. Soc. AIME, 233:1500–1508, 1965)*

eral, face-centered cubic materials do not exhibit a ductility transition. Among high-purity body-centered cubic (bcc) metals, those in group Va of the periodic table (V, Nb, Ta) normally are ductile to quite low temperatures; similarly, pure hexagonal close-packed (hcp) titanium and zirconium do not exhibit a ductility transition. However, the presence of interstitial impurities or certain substitutional alloying elements may produce a ductile-to-brittle transition in all of these metals. Moveover, it should be noted that ductile-to-brittle transitions can be induced even in face-centered cubic metals by testing in certain aggressive surface environments (for example, alpha brass in mercury) or by adding appropriate small impurities (for example, antimony in copper).

The transition temperature is different for differing states of stress because the ratio of the maximum shear stress to the maximum normal stress is different. The transition temperature is elevated by the presence of a severe notch and is depressed by testing in torsion. In the case of steels, transition temperatures may vary from near 20 K for unnotched low-carbon grades to near 250 K in structures which contain defects such as weld cracks or geometrical stress raisers. All large structures contain such defects, and the relationship between defect sizes and fracture stresses or energies embodies the concept of fracture mechanics (see below). There is also a size effect in steels resulting from inhomogeneities in microstructure or coarser grain sizes resulting from slower cooling rates in large structures. Consequently, thick steel plates are characterized by high transition temperatures.

Among the factors which decrease the transition temperature of metals, the following are of general importance to all susceptible metals: (1) increased purity, particularly for bcc metals in the case of interstitial elements (C, O, N, H) and elements which segregate preferentially at grain boundaries (Sb, P, Sn); (2) reduced grain size; (3) reduced strain rate; (4) introduction of very small (diameter of 1000 A or 100 nm), hard particles.

In addition, alloying elements in solid solution may be utilized to reduce the transition temperature. Nickel and manganese often are added to steels to improve low-temperature toughness (although manganese is effective only in steels containing relatively large amounts of carbon). Similarly, lithium lowers the transition temperature of magnesium, and rhenium (a transition metal of the platinum group) improves low-temperature ductility of chromium and tungsten.

Notch tensile test. Notch sensitivity in metals cannot be detected by the ordinary tension test on smooth bar specimens. Either a notched sample may be used in a tension test or a notched-bar impact test may be conducted. The notch in a tensile bar is characterized by the sharpness $d/2r$ and the depth, defined as $1-d^2/D^2$, where d is the diameter at the root of the notch, D is the original diameter, and r is the notch root radius. Notches produce triaxial stresses under the notch root as tensile forces are applied, thereby decreasing the ratio of shear stress to normal stress and increasing the likelihood of fracture.

Materials are evaluated by a quantity, notch strength, which is the analog of the ultimate tensile strength in an ordinary tensile test. The notch strength is defined as the maximum load divided by the original cross-sectional area at the notch root. For cylindrical bars, a notch ductility also may be defined as in Eq. (6), where $A_{o(n)}$ = original

$$q_n=\frac{A_{o(n)}-A_{f(n)}}{A_{o(n)}} \tag{6}$$

cross-sectional area at notch and $A_{o(f)}$ = cross-sectional area at notch after fracture. This quantity is often very small and is difficult to measure accurately. The ratio of notch tensile strength to ultimate tensile strength is called the notch sensitivity ratio (NSR). If the NSR is less than 1, the material is considered to be notch-sensitive. The notch sensitivity of steels generally increases with increasing tensile strength.

Sharp-notch tension testing is applicable not only to ordinary cylindrical tensile bars but to high-strength sheet materials as well. The method provides a comparative measure of the resistance of sheet materials to unstable fracture resulting from the presence of cracklike defects, and may be utilized in research and development of metals to study effects of composition, processing, and heat treatment. The method also permits comparative evaluation of crack propagation resistance of a number of materials under consideration for a particular application.

Compression test. Very brittle metals, or metals to be utilized in products which are to be formed by compressive loading (rolling, forging) often are tested in compression to obtain yield strength or yield point information. Compression test specimens are generally in the form of solid circular cylinders. The ratio of specimen length to diameter is critical in that high ratios increase the likelihood of buckling during a test, thereby invalidating the test results. Proper specimen alignment is important for the same reason. In addition, care must be taken to lubricate specimen ends to avoid spurious effects from friction between the specimen ends and the testing machine. In the case of a metal which fails in compression by a shattering fracture (for example, cast iron), a quantity known as the compressive strength may be reproducibly obtained by dividing the maximum load carried by the specimen by its cross-sectional area. For materials which do not fail in compression by shattering, the compressive strength is arbitrarily defined as the maximum load at or prior to a specified compressive deformation.

Notched-bar impact testing. Notched-bar impact tests are conducted to estimate the resistance to fracture of structures which may contain defects. The common procedure is to measure the work required to break a standardized specimen, and to express the results in work units, such as foot-pounds or newton-meters. The notched-bar impact test does not provide design information regarding the resistance of a material to crack propagation. Rather, it is a comparative test, useful for preliminary screening of materials or evaluation of processing variables. The notch behavior indicated in a single test applies only to the specimen size, notch geometry, and test conditions involved and is not generally applicable to other specimen sizes and conditions. The test is most useful when conducted over a range of temperatures so that the ductile-to-brittle transition can be

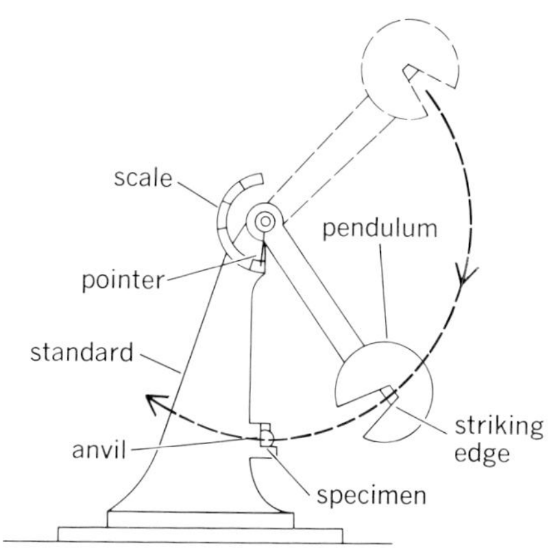

Fig. 5. Diagram of Charpy impact machine.

determined. In some materials, notably pure polycrystalline bcc metals, there is an abrupt change in energy absorbed with temperature, the curve being similar in shape to that for the change in ductility with test temperature shown in Fig. 4, and an unambiguous transition temperature may be defined.

However, most structural alloys, including steels, exhibit a gradual decrease in impact energy absorption with decreasing temperature, so that arbitrary definitions of transition temperature, based upon a fixed energy level, a fixed contraction at the root of the notch, an average energy, or a change in fracture appearance must be devised. In the case of face-centered cubic metals, there is little change in energy absorption and no change in fracture mechanism with temperature, and no transition can be defined.

Notched-bar tests are usually made in either a simple beam (Charpy) or a cantilever beam (Izod) apparatus, in both of which the specimen is broken by a freely swinging pendulum; the work done is obtained by comparing the position of the pendulum before it is released with the position to which it swings after striking and breaking the specimen (Fig. 5). In the Izod test (Fig. 6*a*), the specimen is held in a vise, with the notch at the level of the top of the vise, and broken as a cantilever beam in bending with the notch on the tension side. In the Charpy test (Fig. 6*b*), the specimen is laid loosely on a support in the path of the pendulum and broken as a beam loaded at three points; the tup (striking edge) strikes the middle of the specimen, with the notch opposite the tup, that is, on the tension side. Both tests give substantially the same result with the same specimen unless the material is very ductile, a situation in which there is little interest.

Specimen geometry and test conditions. The choice of specimen depends to some degree upon the nature of the material, therefore several specimen types are recognized. Sharp, deep notches are generally required to compare properties of more ductile materials. Typical Charpy bars (Fig. 6) are 10 mm (0.394 in.) square by 55 mm (2.165 in.) long, with either a 45° V notch, or a parallel-sided notch ending in a 2-mm-diameter hole (keyhole notch). Keyhole notches usually provide a more abrupt transition at lower temperatures than do V notches, as shown in Fig. 7. Notches may also be made with a milling cutter, resulting in a 2-mm-wide U-shaped notch. Izod specimens typically are 10 mm square by 75 mm long, with a 45° V notch 28 mm (1.102 in.) from one end. The transition temperature for a given steel will be different for different tests and specimen geometries.

When testing at temperatures other than room temperature, the Charpy three-point loading arrangement is more convenient than the Izod configuration, in which the specimen is clamped in a vise. It has been established that the temperature of the specimen will not change significantly at the root of the notch for several seconds, so that the specimen may be brought to temperature in a controlled temperature bath and transferred to the testing machine for breaking.

The V-notch transition temperature correlates

METAL, MECHANICAL PROPERTIES OF

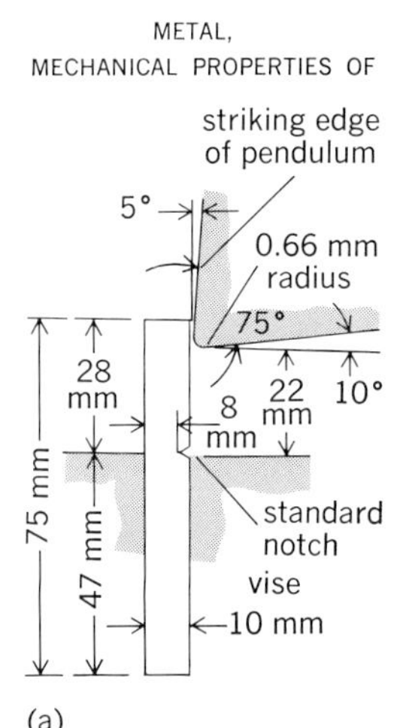

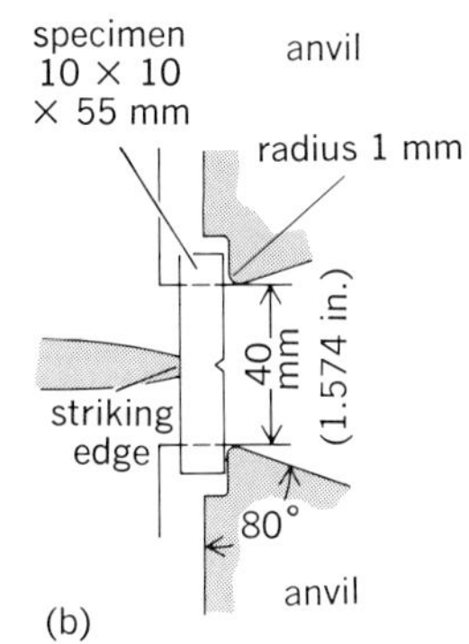

Fig. 6. Mounting of notched-bar test specimens. (*a*) Izod test. (*b*) Charpy test.

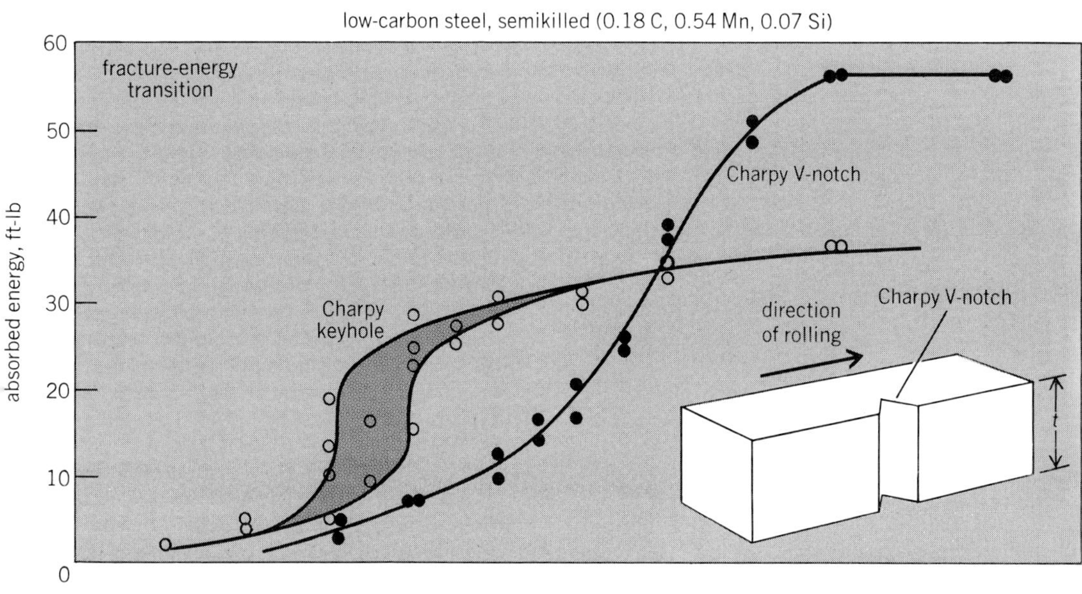

Fig. 7. Fracture energy transition for Charpy keyhole and V notch, showing a more abrupt transition for the keyhole notch. (*From American Society for Metals, Metals Handbook, redrawn from W. S. Pellini, vol. 10, 8th ed., 1975*)

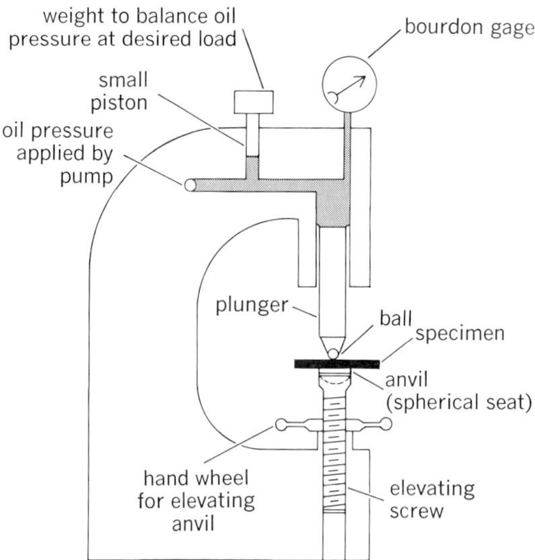

Fig. 8. Features of hydraulic-type Brinell machine.

fairly directly with practical defects in structures, on the basis that brittle fractures in service are seldom observed when a specimen exhibits some minimum level of energy absorbed in the V-notch test. The number of foot-pounds that are needed to ensure freedom from brittle fracture in a structure depends on the material being tested, for the resistance to deformation as well as the ductility enters into the work required to break the specimens, along with the distribution of deformation, which varies from one material to another. For this reason the impact test cannot provide design data.

As in all tests involving brittle behavior, because brittle fracture is concerned with properties in a small region rather than the average throughout a large volume, considerable statistical variation is to be expected in notched-bar testing. Therefore, enough tests must be made to obtain a reliable result by statistical standards, especially when testing in the vicinity of the transition temperature.

Hardness testing. Tension testing provides a complete description of the relationship between stress and strain in plastic deformation, some useful information about ductility, and—in conjunction with the Griffith theory, which relates fracture stress to defect size in an elastically strained body—some information about the resistance to crack propagation under conditions in which the metal is embrittled by the presence of a notch or crack. Often these methods of testing are more time-consuming and more expensive than is necessary, in particular when the only information that is needed is the comparison of the resistance to deformation of a particular sample or lot with a standard material. For such purposes, indentation hardness tests are used. They are relatively inexpensive and fast. They tell nothing about ductility and little about the relationship between stress and strain, for in making the indent the stress and strain are nonuniformly distributed.

In all hardness tests, a standardized load is applied to a standardized indenter, and the dimensions of the indent are measured. This applies to such methods as scratch hardness testing, in which a loaded diamond is dragged across a surface to produce, by plastic deformation, a furrow whose width is measured, and the scleroscope hardness test, in which an indent is produced by dropping a mass with a spherical tup onto a surface. The dimensions of the indent are proportional to the work done in producing it, and the ratio of the height of rebound to the height from which the tup was dropped serves as an indirect measure of the hardness.

Test methods. In the Brinell hardness test, a hardened steel ball is forced into a surface by a force appropriate to the hardness of the material being tested, for a standard time, usually 30 sec (Fig. 8). With steel and other hard materials, a 3000-kg force is applied to a 10-mm-diameter steel or tungsten carbide ball; with softer alloys a 500-kg load is often employed. In every instance, the diameter of the indentation crater is kept between one-fourth and one-half the diameter of the ball. The Brinell hardness number (BHN) is defined, in Eq. (7), as the force applied divided by the area of

$$\mathrm{BHN} = \frac{2P}{\pi D[D - (D^2 - d^2)]^{1/2}} \tag{7}$$

contact between the ball and the test piece after the ball has been removed, where P = applied load, kg; D = diameter of ball, mm; and d = diameter of indentation, mm. The units of Brinell hardness are kg/mm². The surface on which the impression is made should be relatively smooth and free from scale or dirt. The Brinell number will vary with applied load and ball diameter. In order to obtain a constant hardness number, it is necessary to keep d/D = constant; this can be approximated when P/D^2 is maintained constant.

The Meyer hardness number is defined as the force applied divided by the area of the indent projected onto the plane of the original surface of the test piece. This number, just as useful as the Brinell hardness number, is easier to calculate but has not replaced the standard Brinell hardness number, the calculation of which is everywhere facilitated by easily available tables. The Meyer hardness is less sensitive to load than is the Brinell hardness. A Meyer analysis of the Brinell test is carried out by plotting the logarithm of the applied force against the logarithm of the diameter of the indent. This plot is linear and thereby demonstrates the validity of the relationship $F = ad^n$, where F is the force and a and n are the Meyer constants. Meyer's a is the intercept of the line at unit diameter (1 mm, normally), and Meyer's n is the slope of the logarithmic plot; it reflects the strain-hardening characteristics of the metal, provided that a minimum load level is applied.

The indentation hardness number is independent of the force applied to the indenter when a conical or pyramidal indenter is used. A durable diamond indenter is used in the Vickers penetration hardness testing method. It is a square-base diamond pyramid with a dihedral angle between opposite faces of 138°. This is the contact angle in the Brinell test when the indent is three-eighths of the ball diameter, in the middle of the range of indent sizes used in that test. This choice of angle yields hardness numbers in the Vickers penetration test about the same as those obtained in the Brinell test, with the advantage that the hardness number is independent of the load used down to quite small loads and quite small indents. The diamond pyramid hardness number (DPH) or Vickers hardness number (VHN) is defined as the applied load divided by the surface area of the impression; loads typically range from 1 to 120 kg depending upon the hardness of the material under investigation. The length of diagonals of the impression are determined by microscopic measurement, and the DPH is determined from Eq. (8),

$$\mathrm{DPH} = \frac{1.854P}{L^2} \qquad (8)$$

where P is measured in kg and L is the average length of diagonals in mm. In practice, the DPH is read from a chart once the diagonals are known. The Vickers test is widely accepted for research applications because a continuous scale of hardness from 5 DPH for very soft metals to 1000 DPH for very hard metals is achieved. Below loads of about 25 g, departures are observed. The Knoop hardness test is a modification using a diamond pyramid whose base is a rhombus, the diagonals of which are in the ratio of 7:1. The plastic deformation produced by this indenter is nearly plane strain, which can be accomplished in brittle materials without fracture in the adjoining regions under tensile stress, whereas cracks are produced by the nonplane strain produced by symmetrical indenters. Low loads are applied (under 1 kg) and, since surface impressions are about one-half the depth of those for Vickers tests under the same load, the test is particularly suitable for analysis of surface layer properties. For the same reason, it is necessary to prepare surface layers much more carefully in the case of Knoop hardness than for tests involving heavier loads.

Rockwell hardness. The most popular hardness test is the Rockwell test, which is carried out quite rapidly in a convenient machine by sacrificing any attempt to express the hardness as a resistance to deformation in units of force per area. An indenter chosen from two shapes and various sizes is forced into the test-piece surface, first by a minor load, under which the position of the indentor is established as a reference point, and then by a major load which deepens the indent. Upon removing the major load and leaving the minor load applied, the amount by which the indent has been deepened is established. The hardness number is simply the amount of deepening on a linear scale, with the scale reversed so that soft materials having deeper indents are characterized by smaller numbers. The two shapes of indenters are steel spheres of various diameters, for which the starting point on the linear scale for zero deepening is 130; and a conical diamond with a spherical tip for which the starting point on the scale is 100. The hardness number is the number read from the dial gage indicating the depth of the indent on this reversed scale; it must be accompanied by a letter indicating the kind of indenter and load used. For example, for a load of 100 kg applied to a spherical steel indenter of 1/16-in. diameter, the letter B is used, so that a hardness number might be Rockwell B80; for the diamond indenter and a load of 150 kg, the letter C is used, so that a typical hardness number would be Rockwell C55. Tables are available to convert hardness data for steels from one Rockwell scale to another, and to other hardness scales or to tensile strength. *See* HARDNESS SCALES.

Fatigue. Fatigue is a process involving cumulative damage to a material from repeated stress (or strain) applications (cycles), none of which exceed the ultimate tensile strength. The number of cycles required to produce failure decreases as the stress or strain level per cycle is increased. When cyclic stresses are applied, the results are expressed in the form of an *S-N* curve in which stress or logarithm of stress is plotted against the number of cycles to cause failure. The fatigue strength or fatigue limit is defined as the stress amplitude which will cause failure in a specified number of cycles, usually 10^7 cycles. For a few metals, notably steels and titanium alloys, an endurance limit exists, below which it is not possible to produce fatigue failures no matter how often stresses are applied.

It is often desirable to impose fixed cyclic strains on a material, rather than fixed stresses. Under these circumstances, fatigue data are plotted as logarithm of plastic strain per cycle, $\Delta\epsilon_p$, versus logarithm of cycles to failure, N_f. Most engineering metals exhibit a linear dependence of $\Delta\epsilon_p$ on N_f, known as the Coffin-Manson relationship, shown in Eq. (9), where ϵ'_f is a material

$$\frac{\Delta\epsilon_p}{2} = \epsilon'_f N_f^{\alpha} \qquad (9)$$

constant known as the fatigue ductility coefficient, and α is a constant which for most metals tested at room temperature in air is near -0.5; ϵ'_f typically approximates the ductility of the material in monotonic tension, indicating that fatigue life increases with ductility of the material. Equation (9) is obeyed best at lives shorter than 10^4 cycles, the so-called low-cycle fatigue range, as shown for a compilation of several metals tested at a variety of temperatures in Fig. 9.

Fatigue testing in stress control is frequently done with a mean stress of zero, although mean positive stresses are not uncommon. A stress that varies sinusoidally with time is commonly applied, sometimes in direct tension and compression but more usually in bending by utilizing a sample of

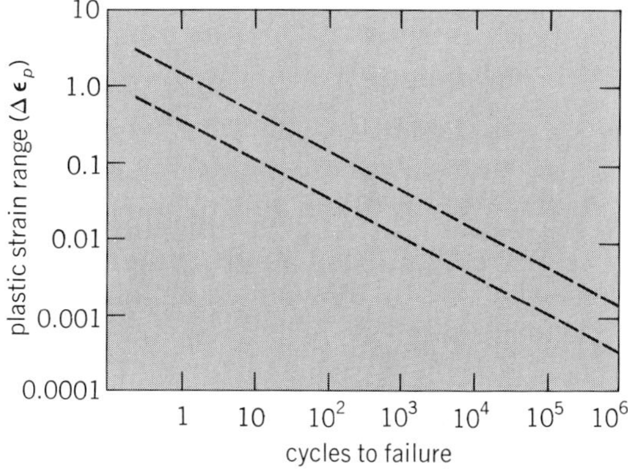

Fig. 9. Low-cycle fatigue data for several alloys. A single band incorporates all data for 11 metals at a variety of test temperatures and test atmospheres. (*From L. F. Coffin, The effect of high vacuum on the low cycle fatigue law, Metallurg. Trans., 3:1777–1788, 1972*)

circular cross section rotated under a constant bending moment. Other waveforms also may be used in tension-compression, most notably triangular or square waves, usually for specialized research applications.

Test frequency is not an important variable at ordinary temperatures, but may assume overriding importance at temperatures sufficiently elevated to permit the introduction of cyclic creep or surface oxidation. Under either of the latter circumstances, low-frequency tests result in shorter lives for the same applied loads. Mean stress superposed upon cyclic stresses can have significant effects, particularly in steel and other alloys sensitive to brittle fracture under tensile loading. In this situation, life decreases as R, the ratio of minimum to maximum cyclic stress, is increased.

Cracks may form very early in a fatigue test, and then grow exceedingly slowly, or, alternatively, crack nucleation may be delayed until very near the total number of cycles to failure. Cracks invariably initiate at surface or near-surface origins such as slip bands, inclusion or precipitate particles, scratches, notches, or other stress raisers, or at material defects such as casting pores. Slip-band cracking is particularly prevalent in pure metals, and under such circumstances the nature of the slip process is a critical factor in determining fatigue life. The greater the multiplicity of active slip systems, the easier it is to initiate a crack; on the other hand, the predominance of one slip system, as in suitably oriented single crystals of zinc and other hcp metals, renders crack nucleation so difficult that it is sometimes impossible to achieve fatigue cracking at all.

Crack propagation occurs at a rate da/dN (crack extension per cycle) which depends upon the change in stress intensity K (stress × square root of crack length) per cycle raised to a power m, as expressed in Eq. (10), where C and m are constants that depend upon material and mean stress as well as environment (see Fig. 10). The value of m has been observed to average near 4 for a wide variety of metals. The fatigue crack continues to grow until the remaining cross-sectional area is insufficient to support the peak stress, fracture then occurring in a single cycle by overload. The appearance of the fracture surface in the fatigue and overload zones differs considerably for ductile materials in that parallel bands known as fatigue striations often occur perpendicular to the direction of fatigue crack propagation. The overload region contains no striations and, for ductile metals, generally comprises microvoids centered about inclusion or precipitate particles. Each striation marks the position of the fatigue crack at the end of a cycle, and the distance between striations serves as a useful measure of crack growth rate. Usually, however, these must be supplemented by microscopic observations of the crack profile with time in order for crack propagation relations such as that of Eq. (10) to be obtained. This is particularly true in

$$\frac{da}{dN} = C(\Delta K)^m \qquad (10)$$

materials where striations are not always well developed, such as structural steels and other high-strength alloys with complex microstructures.

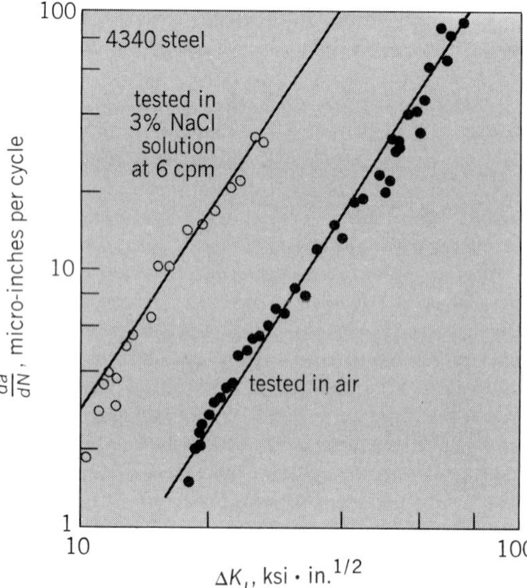

Fig. 10. Crack growth rate in air and in 3% NaCl. *(From American Society for Metals, Metals Handbook, redrawn from Imhof and Barsom, vol. 10, 8th ed., 1975)*

Notches may significantly lower fatigue life, particularly in stronger, less ductile alloys. The notch sensitivity factor q in fatigue is given by Eq. (11),

$$q = \frac{K_f - 1}{K_t - 1} \qquad (11)$$

where K_t is the theoretical stress concentration factor for ideal elastic behavior and K_f is the ratio of fatigue strength of an unnotched specimen divided by the fatigue strength of a notched specimen. Both K_f and K_t are greater than unity always. The value of q varies between 0 and 1 and is highest for strong alloys.

Fatigue strengths of engineering alloys tend to be proportional to tensile strength. Therefore any factor which increases monotonic strength, for example, testing at low temperatures or utilizing samples with finer grain size, tends to increase fatigue life. While for many metals the fatigue strength is on the order of one-half the tensile strength, this ratio is reduced for some precipitation-hardened alloys (nickel-base superalloys, aluminum alloys) to as little as one-fourth the tensile strength. Ratios of fatigue strength to tensile strength approaching 0.8 have been noted, however, in some composite materials; for these materials, cracks may be deflected out of their paths by the presence of weak interfaces between phases (artificial composites such as boron reinforced with aluminum) or due to the inability of cracks to penetrate strong fibers (nickel-base or cobalt-base eutectic composites which contain strong nonmetallic reinforcing fibers).

Fatigue life can also be affected by environmental reactions, as shown in Fig. 10. Corrosion acting simultaneously with cyclic deformation can produce appreciable lowering of fatigue strength (corrosion fatigue). The adsorption of liquid metals on solid metals (such as mercury on copper-aluminum alloys) also can lead to striking decreases in fatigue life. Another important factor is residual stresses; tensile stresses act adversely and compressive stresses favorably. Compressive stresses are deliberately applied to metal surfaces to delay crack initiation and thereby prolong fatigue life.

Fracture toughness. Many modifications of conventional test procedures have arisen from an effort to obtain laboratory information which would be useful to designers of engineering structures. Since such structures often contain macroscopic defects, in particular weld flaws, large inclusion particles, or other manufacturing flaws, most effort has been directed toward studying the properties of notched bars. Unfortunately, neither the notched tensile test nor the several impact tests described above provide sufficiently quantitative relations between defect size, stress state, and the likelihood of brittle or otherwise unstable fracture, particularly in cases where crack propagation can occur below the macroscopic yield stress as measured in a tensile test. In the case of materials which exhibit little or no plastic deformation at fracture, the most successful approach has been that of A. A. Griffith, who in 1929 deduced Eq. (12)

$$\sigma_F{}^2 = \frac{kE\gamma}{\pi c} \qquad (12)$$

from both thermodynamic and stress considerations. Here σ_F is the nominal applied stress to cause fracture, k is a numerical constant which depends upon stress state and crack geometry, E is Young's modulus, c is the depth of an edge crack or the half-width of an internal crack, and γ is the elastic energy necessary to create a new surface. While this equation correctly predicts a lowering of fracture stress as flaw size increases, and in fact provides an accurate measure of elastic work to cause fracture in brittle solids such as glass, several critical problems arise when the equation is applied to metals. When fracture stresses are related to flaw sizes, calculated values of work for fracture, γ_p, are several orders of magnitude higher than those deduced for elastic work by other means. E. Orowan first pointed out that plastic work accompanying fracture must be incorporated into the Griffith equation. This is best done by considering that the total work for fracture, γ_p, is related to γ, the elastic energy, through a proportionality factor which is of magnitude equal to crack tip radius ρ divided by interatomic spacing a. The modified Griffith relation, Eq. (13), is then obtained, where k' is a numerical constant. This

$$\sigma_F{}^2 = \frac{k'E\gamma\rho}{\pi ac} \qquad (13)$$

equation reduces to the Griffith condition, Eq. (12), when ρ approaches a in magnitude, that is, as cracks become very sharp. In practice, ρ is difficult to measure, and a more generally applicable approach has been suggested by G. R. Irwin. A stressed material in which a crack propagates releases strain energy at a rate G. A transition from slow to rapid fracture is denoted by a critical value G_c, which is known as the critical strain energy release rate. Alternatively, G_c may be obtained from a relation between elastic compliance, $1/M$, and crack length c in a series of precracked specimens of differing initial crack lengths as shown in Eq. (14), where $d(1/M)/dc$ is the change in compliance with increasing length of precrack, B is the thickness, and G_c is a measure of work necessary to cause the crack to propagate. G_c is, therefore, equivalent to γ_p, the distinction being that G_c is an experimentally measured parameter which is not dependent upon assumptions as to its origin. G_c is typically of magnitude 100–1000 in.-lb/in.^{2} for metals. Part of this energy may be dissipated in forming new surface, the rest being converted to plastic work or kinetic energy of the crack.

$$G_c = \frac{P^2}{2B}\frac{d(1/M)}{dc} \qquad (14)$$

An alternative and completely complementary approach, based on stress concentrations near crack tips, also has been developed by Irwin, and has been much more widely applied to engineering design situations. Equations for the normal and shear stresses near the crack edge for bodies of various geometries to which forces are applied normal or parallel to the crack edge have been developed. In all cases, the local stresses at the crack edge fall off with distance away from the crack. Fracture of the body occurs when a critical stress distribution, described by the stress intensity factor K is reached. K depends upon applied stress and crack geometry; for most situations it may be defined as Eq. (15), where α is a parameter

$$K = \sigma(\alpha\pi c)^{1/2} \qquad (15)$$

depending upon specimen and crack geometry. For a crack in an infinite plate, subjected to loading perpendicular to the crack edge, $\alpha = 1$. K, which is expressed in units of psi$\sqrt{\text{in.}}$, is determined experimentally by stressing a precracked specimen in bending or in tension (Fig. 11), and noting the critical stress intensity K_c at which unstable (rapid) crack propagation occurs. K and G are related as shown in Eq. (16), where $A = (1-\nu^2)$

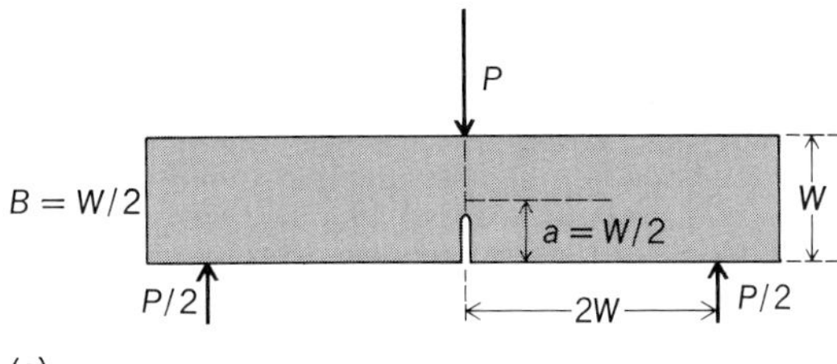

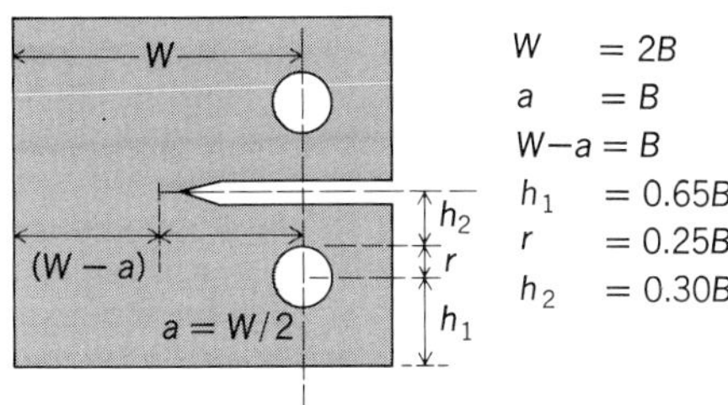

Fig. 11. Standard fracture toughness specimens. (*a*) Edge-notched bend. (*b*) Compact tension. (*From N. H. Polakowski and E. J. Ripley, Strength and Structure of Engineering Materials, Prentice-Hall, 1965*)

$$G = A\frac{K^2}{E} \qquad (16)$$

for plane strain and $A = 1$ for plane stress; ν is Poisson's ratio.

Critical values of stress intensity K_c and strain energy release rate G_c are related similarly. In addition, since K_c corresponds to a fracture condition, Eq. (16) may be rewritten as Eq. (17), which, when combined with Eq. (15), provides the relation given in Eq. (18), which is identical in form to the Griffith condition, Eq. (12).

$$K_c^2 = \alpha\sigma_F^2\pi c \qquad (17)$$

$$\sigma_F^2 = \frac{EG_c}{\Lambda\alpha\pi c} \qquad (18)$$

Unlike the yield stress, which is a reproducible material property for specimens of identical composition and microstructure, K_c and G_c vary with the depth of the precrack relative to the dimensions of the test specimen, particularly width W and thickness B. Specimens with relatively long cracks fracture with appreciable plasticity occurring at the crack tip. The contribution of plastic work is included in G_c or K_c, and is particularly high in thin sections and in low-strength alloys. For high-strength material, and lower-strength alloys in which c/W and c/B are small, K_c and G_c decrease to lower limiting values K_{I_c} and G_{I_c}, respectively (Fig. 12). These values refer to plane-strain conditions, and the K_{I_c} parameter is known as the plane-strain fracture toughness. K_{I_c} appears to be a material constant insofar as specimen-crack geometry is concerned, and therefore is extremely useful for designers who must estimate structural parameters which are necessary to avoid catastrophic failures. However, K_{I_c} may vary with test temperature, strain rate, and microstructure of the metal. In general, fracture toughness decreases as strength increases, so that K_{I_c} tends to be low for quenched and tempered alloy steels, and much higher for medium-strength carbon steels or aluminum alloys.

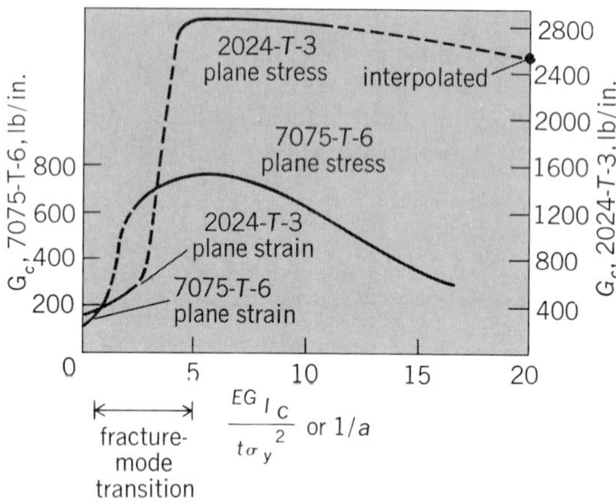

Fig. 12. Critical strain energy release rate as function of thickness. *(From G. R. Irwin, in J. N. Goodier and N. J. Hoff, eds., Structural Mechanics, Pergamon Press, 1960)*

Specimen designs which are currently recommended in the United States and Great Britain are shown in Fig. 11*a* and *b*, although additional designs have been utilized in specific situations. Both are single-edge notched specimens which are precracked in fatigue prior to the fracture toughness determination. The bend specimen in Fig. 11*a* is deformed under three-point loading, while the compact tension specimen in Fig. 11*b* is loaded through pins above and below the crack faces. In order to relate applied loads to stress intensity, test specimens must be calibrated. A compliance coefficient (Y function) is published for each specimen-crack-load geometry, which allows load to be converted directly to a K value. The primary variable in specimen preparation is the ratio of crack length c to specimen width W. Calibrations are presented either in tabular or polynomial form as functions of c/W. Plasticity corrections are necessary for situations in which fracture events are not precisely localized at the crack tip, that is, when plastic flow spreads ahead of an advancing crack. The specification of size of the region in which the stress intensity accurately describes the stress field is a critical aspect of fracture toughness testing. In practice, it is usually estimated to be about 1/50 the crack length. Therefore the plastic zone at fracture must be limited to $0.02c$ or less. This is accomplished by ensuring that c and $B \geq 2.5\ (K_{I_c}/\sigma_y)^2$, where σ_y is the 0.2% offset yield stress of the material. Calculation of K_{I_c} is based upon the lowest load at which significant measurable extension of the crack occurs. The method has been applied with considerable success to high- and medium-strength steels, as well as to high-strength aluminum- and titanium-base alloys. Modifications of the standard fracture toughness test have also been applied to lower-strength materials, but the test procedures have not yet been standardized.

Creep and stress rupture. Time-dependent deformation under constant load or stress is measured in a creep test. Creep tests are those in which the deformation is recorded with time, while stress rupture tests involve the measurement of time for fracture to occur. The two types of test provide complementary information, and indeed both creep rate and time may be recorded in a single test. Closely related are stress relaxation tests, in which the decay of load with time is noted for a body under a fixed state of strain. Although creep deformation may occur under any state of stress, it is most commonly measured in uniaxial tension. Test durations vary from seconds or minutes to tens of thousands of hours. Appreciable deformation occurs in structural materials only at elevated temperatures, while pure metals may creep at temperatures well below room temperature. In general, the temperature at which creep processes become significant exceeds approximately one-half of the melting temperature on the absolute scale. Under constant load conditions, three stages of inelastic deformation are noted, as shown in Fig. 13. These are primary creep (stage I), secondary or steady-state creep (stage II), and tertiary creep (stage III). A considerable portion of primary creep in anelastic and is sometimes referred to as transient creep. The second stage marks a balance between strain hardening of the material and recovery of the microstructure and strength; a constant, minimum creep rate may be noted over most of the life of the specimen, and it is the steady-state creep rate, therefore, which is often determined for structural alloys. Tertiary creep resembles necking in a tensile test in that both are instabilities arising from prefracture events. In this stage, strain hardening cannot compensate for the loss in cross-sectional area during straining as well as the formation of fine cracks. Information usually derived from a series of creep experiments

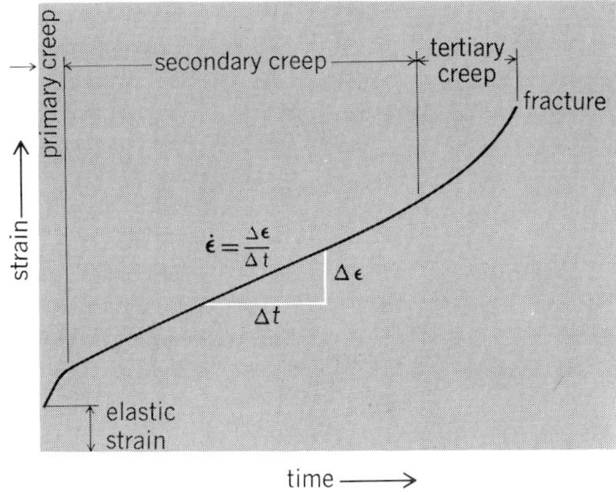

Fig. 13. Creep curve. *(From M. E. Eisenstadt, Introduction to Mechanical Properties of Materials, Macmillan, 1971)*

may include, aside from the steady-state creep rate, data on the stress versus logarithm of time for total deformations of up to 5%, as initial stress versus logarithm of time to rupture, and logarithm of stress versus logarithm of creep rate. It should be noted that since the cross-sectional area of a creep specimen decreases with time, the stress increases under constant load conditions. It is possible to run constant stress tests by providing means for load to be reduced as the specimen elongates.

Since creep deformation and rupture time are temperature- and stress-dependent, it is usually necessary to test a material at several stresses and temperatures in order to establish the creep or stress-rupture properties in adequate detail. Interpolation of creep and rupture data is frequently employed to predict behavior at stresses and temperatures which are not included in the test program. However, no generally accepted methods are available for extrapolating creep and rupture data to temperatures, times, and stresses outside of the tested range. Specimen size has little effect on creep and rupture properties. Consequently, test specimen configurations usually resemble those of flat or round tensile bars with reduced gage sections. Temperature measurement must be precise; the difference between indicated and actual test temperature should not exceed ±2°C up to 1000°C and ±3°C at higher temperatures. Creep strain is measured by an extensometer for tests in which creep rates are to be determined. In the case of stress-rupture tests, dial gages often are employed to provide approximate creep elongations to supplement stress-to-rupture data. *See* METAL; METALLURGY.

[N. S. STOLOFF]

Bibliography: American Society for Metals, *Metals Handbook*, vol. 1: *Properties and Selection*, 8th ed., 1961, and vol. 10: *Failure Analysis and Prevention*, 8th ed., 1975; American Society for Testing and Materials, *Standards A370–72a, E92–72, E10–66, E9–70, E8–69, E399–72, E6–66, E23–72*; G. E. Dieter, *Mechanical Metallurgy*, 1965; M. M. Eisenstadt, *Introduction to Mechanical Behavior of Materials*, 1971; J. F. Knott, *Fundamentals of Fracture Mechanics*, 1973; F. A. McClintock and A. S. Argon, *Mechanical Behavior of Materials*, 1966; N. H. Polakowski and E. J. Ripling, *Strength and Structure of Engineering Materials*, 1966.

Metal casting

The introduction of molten metal into a cavity or mold where, upon solidification, it becomes an object whose shape is determined by mold configuration. Casting offers several advantages over other methods of metal forming: it is adaptable to intricate shapes, to extremely large pieces, and to mass production; it can provide parts with uniform physical and mechanical properties throughout; and, depending on the particular material being cast, the design of the part, and the quantity being produced, its economic advantages can surpass other processes.

CATEGORIES

Two broad categories of metal-casting processes exist: ingot casting (which includes continuous casting) and casting to shape. Ingot castings are produced by pouring molten metal into a permanent or reusable mold. Following solidification these ingots (or bars, slabs, or billets, as the case may be) are then further processed mechanically into many new shapes. Casting to shape involves pouring molten metal into molds in which the cavity provides the final useful shape, followed only by machining or welding for the specific application.

Ingot casting. Ingot castings make up the majority of all metal castings and are separated into three categories: static cast ingots, semicontinuous or direct-chill cast ingots, and continuous cast ingots.

Static cast ingots. Static ingot casting simply involves pouring molten metal into a permanent mold (Fig. 1). After solidification, the ingot is withdrawn from the mold and the mold can be reused. This method is used to produce millions of tons of steel annually. *See* STEEL MANUFACTURE.

Semicontinuous cast ingots. A semicontinuous casting process is employed in the aluminum industry to produce most of the cast alloys from which rod, sheet, strip, and plate configurations are made. In this process molten aluminum is transferred to a water-cooled permanent mold (Fig. 2*a*) which has a movable base mounted on a long piston. After solidification has progressed from the mold surface so that a solid "skin" is formed, the piston is moved down, and more metal continues to fill the reservoir (Fig. 2*b*). Finally the piston will

METAL CASTING

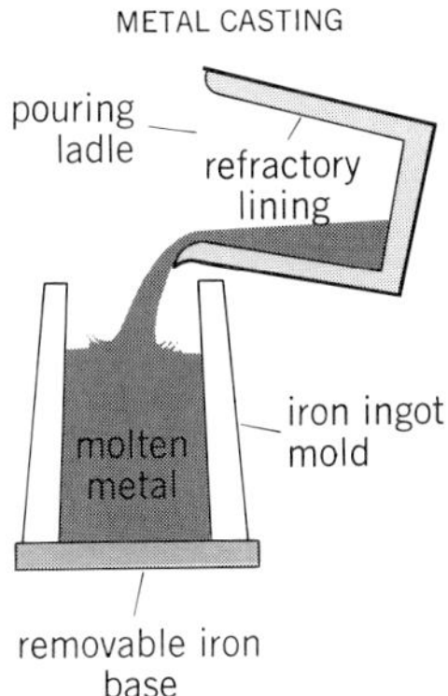

Fig. 1. Static ingot casting.

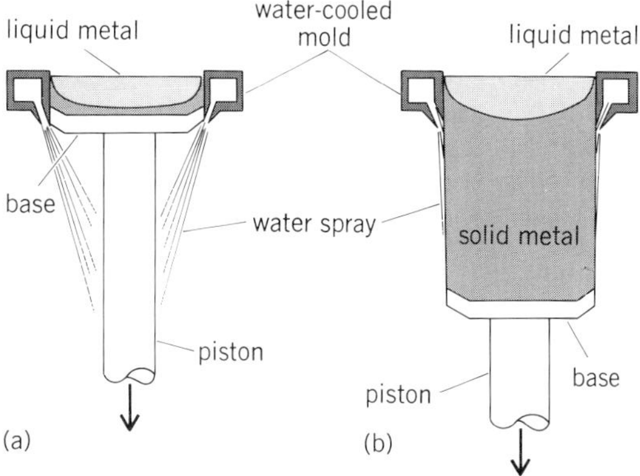

Fig. 2. Semicontinuous (direct-chill) casting. (*a*) Molten aluminum solidifies in a water-cooled mold with a movable base. (*b*) The piston is moved down so more molten metal can be poured into the reservoir.

have moved its entire length, and the process is stopped. Conventional practice in the aluminum industry utilizes suitably lubricated metal molds. However, technological advances have allowed major aluminum alloy producers to replace the metal mold (at least in part) by an electromagnetic field so that molten metal touches the metal mold only briefly, thereby making a product with a much smoother finish than that produced conventionally.

Continuous cast ingots. Continuous casting provides a major source of cast material in the steel and copper industry and is growing rapidly in the aluminum industry. In this process molten metal is delivered to a permanent mold, and the casting begins much in the same way as in semicontinuous casting. However, instead of the process ceasing after a certain length of time, the solidified ingot is continually sheared or cut into lengths and removed during casting. Thus the process is continuous, the solidified bar or strip being removed as rapidly as it is being cast. This method has many economic advantages over the more conventional casting techniques; as a result, all modern steel mills produce continuous cast products.

Casting to shape. Casting to shape is generally classified according to the molding process, molding material, or method of feeding the mold. There are four basic types of these casting processes: sand, permanent-mold, die, and centrifugal.

Sand casting. This is the traditional method which still produces the largest volume of cast-to-shape pieces. It utilizes a mixture of sand grains, water, clay, and other materials to make high-quality molds for use with molten metal. This "green sand" mixture is compacted around a pattern (wood, plaster, or metal), usually by machine, to 20–80% of its bulk density. The basic components of a sand mold and of other molds as well are shown in Fig. 3. The two halves of the mold (the cope and drag) are closed over cores necessary to form internal cavities, and the whole assembly is weighted or clamped to prevent floating of the cope when the metal is poured.

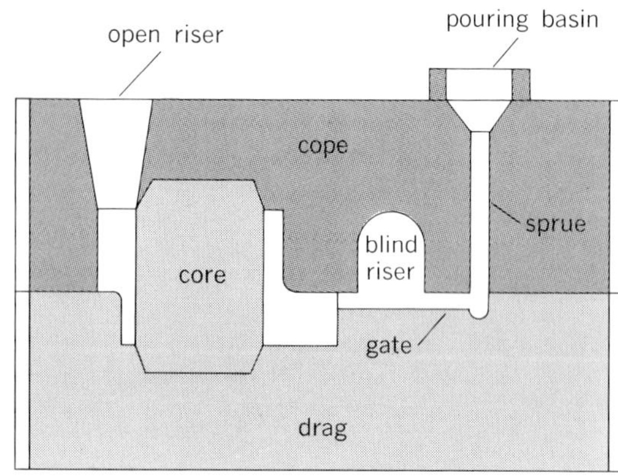

Fig. 3. Section through a sand mold showing the gating system and the risers.

Other casting processes which utilize sands as a basic component are the shell, carbon dioxide, investment casting, ceramic molding, and plaster molding processes (see table). In addition, there are a large number of chemically bonded sands which are becoming increasingly important.

Permanent-mold casting. Many high-quality castings are obtained by pouring molten metal into a mold made of cast iron, steel, or bronze. Semipermanent-mold materials such as aluminum, silicon carbide, and graphite may also be used. The mold cavity and the gating system are machined to the desired dimensions after the mold is cast; the smooth surface from machining thus gives a good surface finish and dimensional accuracy to the casting. To increase mold life and to make ejection of the casting easier, the surface of the mold cavity is usually coated with carbon soot or a refractory

Details of casting processes that utilize sands as a basic component

Process name	Pattern type	Molding aggregate	Type of bond	Current relative cost
Green sand	Wood Plaster Metal	Sand Clay Water	Due to clay plasticity and compaction	Low
Shell	Heated metal	2.5–10% thermosetting resin	Polymerization of resin by pattern heat	Medium
CO_2	Wood Metal	Sand 2–6% sodium silicate	Inorganic bond by chemical reaction of CO_2 with silicate	Medium
Investment	Wax Expendable plastic	Sand Slurry Ceramic binder	Setting ceramic binder	High
V-Process®	Wood Metal	Sand enclosed in plastic with vacuum	Vacuum bond	Medium
No-bake	Wood Metal	Sand Organic compound	Organic bond created by reaction with catalyst	Medium
Full-mold®	Polystyrene coated with thin layer of ceramic	Sand	None	Medium

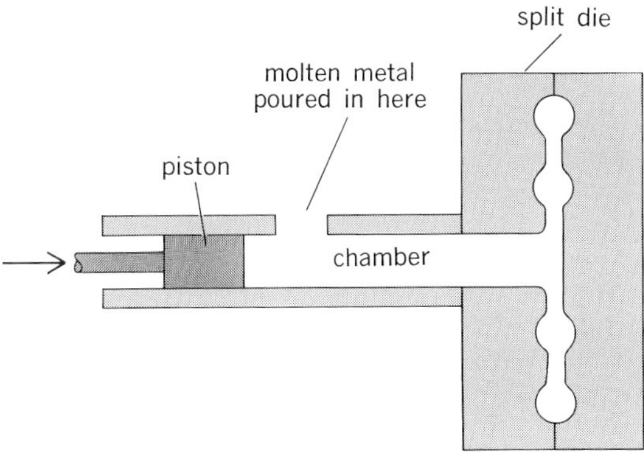

Fig. 4. Cold-chamber die casting machine.

slurry; these also serve as heat barriers and control the rate of cooling of the casting. The process is used for cast iron and nonferrous alloys, with advantages over sand casting such as smoother surface finish, closer tolerances, and higher production rates.

Die casting. A further development of the permanent molding process is die casting. Molten metal is forced into a die cavity under pressures of 100–100,000 psi (0.7–700 MPa). Two basic types of die-casting machines are hot-chamber and cold-chamber. In the hot-chamber machine, a portion of the molten metal is forced into the cavity at pressures up to about 2000 psi (14 MPa). The process is used for casting low-melting-point alloys such as lead, zinc, and tin.

In the cold-chamber process (Fig. 4) the molten metal is ladled into the injection cylinder and forced into the cavity under pressures which are about 10 times those in the hot-chamber process. High-melting-point alloys such as aluminum-, magnesium-, and copper-base alloys are used in this process. Die casting has the advantages of high production rates, high quality and strength, surface finish on the order of 40–100-microinch (1.0–2.5 μm) rms (root mean square), and close tolerances, with thin sections.

Rheocasting is the casting of a mixture of solid and liquid. In this process the alloy to be cast is melted and then allowed to cool until it is about 50% solid and 50% liquid. Vigorous stirring promotes liquidlike properties of this mixture so that it can be injected in a die-casting operation. A major advantage of this type of casting process is expected to be much reduced die erosion due to the lower casting temperatures.

Centrifugal casting. Inertial forces of rotation distribute molten metal into the mold cavities during centrifugal casting, of which there are three categories: true centrifugal casting, semicentrifugal casting, and centrifuging. The first two processes (Fig. 5) produce hollow cylindrical shapes and parts with rotational symmetry, respectively. In the third process, the mold cavities are spun at a certain radius from the axis of rotation; the centrifugal force thus increases the pressure in the mold cavity.

The rotational speed in centrifugal casting is chosen to give between 40 and 60 g acceleration. Dies may be made of forged steel or cast iron. Colloidal graphite is used on the dies to facilitate removal of the casting.

PRINCIPLES AND PRACTICE

Successful operation of any metal-casting process requires careful consideration of mold design and metallurgical factors.

Design considerations. In all types of ingot casting, successful production of ingots depends upon several factors. In the cases of the more technologically complicated semicontinuous or continuous casting processes, the design of cooling systems, choice of lubrication, and rate of ingot movement all play an important role in the final outcome.

In the production of castings which are not subsequently processed mechanically, the construction of a pattern is a most important step toward successful production of castings. The following considerations all play a role in the final design of a usable pattern from which molds can be made.

Gating system. The gating system (Fig. 3), which includes the pouring basin, downsprue, runners, and gates, must be designed to allow the liquid metal to enter the mold cavity with a minimum of turbulence, slag, and temperature loss. The application of Bernoulli's law together with the criterion of conservation of mass allows gas aspiration, turbulence, and slag entrapment to be kept to a minimum.

Riser design. A riser is a volume which is attached to the casting (Fig. 3) and functions to provide metal to feed the shrinkage that occurs when a liquid cools and solidifies. In fact there are three contributions of shrinkage that must be accommodated: shrinkage of the liquid as it cools between the pouring temperature and the melting temperature; shrinkage as the liquid transforms to solid;

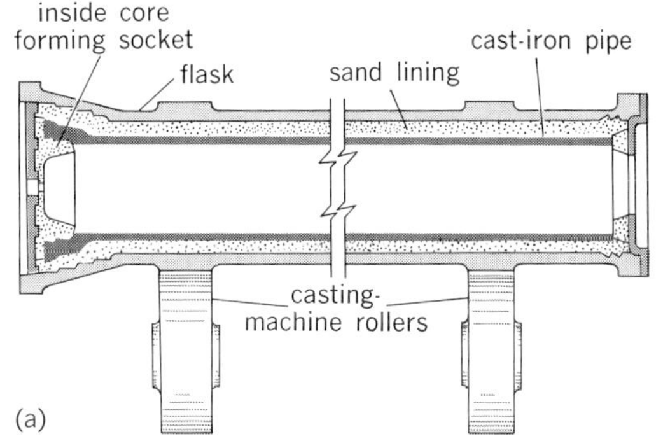

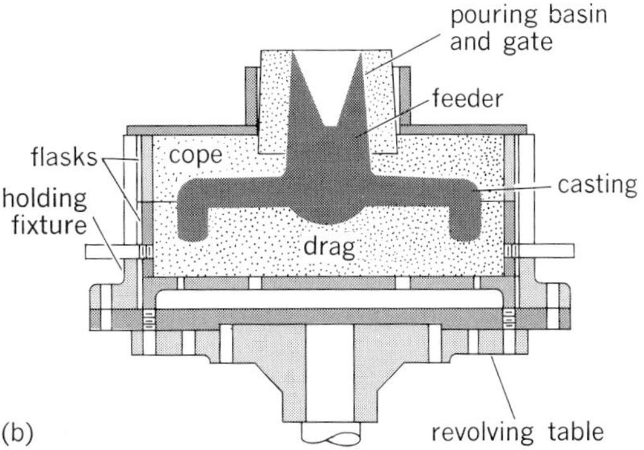

Fig. 5. Centrifugal casting. (*a*) Iron pipe cast in a centrifugal mold. (*b*) Flywheel cast in a semicentrifugal mold.

shrinkage of the solid portion on cooling during the time when some liquid remains. In order to supply liquid metal for the shrinking casting, the riser must necessarily also freeze after the cavity has frozen. It is known that the time of solidification t_s is related directly to the rate of heat transfer in the mold and to the volume-to-surface area ratio (V/A) of the casting. Thus the foundry engineer can ensure the soundness of a casting in situations where the riser and casting have identical mold surroundings by making the V/A ratio of the riser larger than the V/A ratio of the casting. In large bulky castings this means that the riser may actually be larger than the casting, which increases the cost of production. Reducing this cost can sometimes be accomplished by insulating a smaller riser (changing the mold surroundings), thereby reducing the rate of heat transfer and prolonging the time during which the riser is molten. There are several types of risers which are described in detail elsewhere. Two of these are shown in Fig. 3, the open riser and a blind riser.

Ease of pattern extraction. Patterns for sand molding must be designed with a draft of from 1/4 to 2° so that after the mold material has been compacted around the pattern, the pattern can be extracted without damage to the mold. Dies in permanent molding must also be constructed so that after solidification the casting can be ejected without damaging either the casting or the die. The wax patterns of investment castings must be capable of being poured out of the cavity during heating so that no pockets of wax remain.

Machining considerations. In those castings requiring subsequent machining, the patterns must be designed within certain dimensional tolerances so that enough stock is left for machining purposes.

Metallurgical considerations. While design factors are important in producing sound castings with proper dimensions, factors such as the pouring temperature, alloy content, mode of solidification, gas evolution, and segregation of alloying elements control the final structure of the casting and therefore its mechanical and physical properties. Typically, pouring temperatures are selected which are within 100–300°F (60–170°C) of the alloy's melting point. Exceedingly high pouring temperatures can result in excessive mold metal reactions, producing numerous casting defects.

Most metals and alloys pass from the liquid state to the solid state in a predictable manner. As pointed out above, the rate of solidification is a function of the rate at which heat is withdrawn by the mold and of the size of the casting. In the case of continuous casting or semicontinuous casting, new liquid is being added at the same rate as other liquid is changing to solid. Thus, a steady state is established. In the case of static cast ingots or cast-to-shape products, the entire liquid may be poured into the mold before any solidification begins. In either of these two extremes, however, the manner in which solidification occurs is quite similar for most commercial materials.

The relationship between the processing history and the solidification event is illustrated in Fig. 6 for a part which is cast to shape. Solidification begins with the formation of dendrites on the mold walls, which grow with time into the liquid and eventually consume the entire liquid volume. Dendrites are described as treelike or fernlike solids with primary arms similar to tree trunks and secondary arms resembling tree branches. These dendrites form in almost all commercial metals and are responsible to a large extent for the resulting as-cast properties. In pure metals dendrites form if supercooling below the melting point occurs. Supercooling in this sense simply refers to the liquid existing below the temperature where it would normally change to a solid. Supercooling can be accomplished by cooling the metal rapidly and by providing an extremely clean molten bath, thereby hindering the nucleation of solid particles.

A particularly effective way of producing the desired conditions to promote supercooling is through levitation melting, a technique whereby the metal is suspended in a magnetic field so that it touches no foreign object. Thus no mold surface is available upon which the dendrites can grow, and there are no sharp temperature gradients within the metal. This phenomenon of levitation is similar to that mentioned above with regard to the semicontinuous casting of aluminum alloy ingots, where the molten metal is levitated away from the mold surface. It has been suggested that levitation melting would be a natural method for casting in space. In the absence of gravity, large quantities of molten metals could conceivably be processed and handled by levitation techniques. In addition to the potential for casting in the vacuum of space, levitation melting provides exciting possibilities for producing high-purity materials free from gaseous contaminants and their products.

In materials where an element is alloyed with other elements, solidification usually occurs over a range of temperatures. In this instance, dendrites grow in response to constitutional supercooling of the liquid below the liquidus temperature (the highest temperature over which solidification occurs). One would usually desire alloying elements to be distributed uniformly throughout the material after solidification has occurred. Unfortunately the dendritic manner of growth will invariably cause the alloying elements to be segregated and concentrated between the dendrite arms. This high concentration of alloying elements increases with dendrite growth and may ultimately lead to the formation of another solid phase when solidification is complete (Fig. 6). A good example of this type of structure is to be found in a leaded copper alloy containing mostly copper with a small amount of lead. Due to certain atomic forces between copper and lead atoms, nearly pure copper dendrites will form on solidification, leaving all of the lead to form as small spheres between the dendrites. The result is a desirable structure in a bearing material where the lead acts as a built-in lubricant. Many other commercial cast-to-shape materials have properties which depend on the presence of a second phase in the interdendritic volumes; cast iron structures are basically pure graphite particles (plates or spheres) between iron dendrites or dendrite arms; in aluminum silicon alloys, pure silicon particles (plates or rods) are located between aluminum dendrites. These phases are responsible in large part for the desirable properties of the material. However, in many ingot cast alloys, particularly aluminum alloys and steels, the alloying elements that solidify between the dendrite arms are undesirable in this form and must be redistributed to obtain the desired properties. This redistribution, or breaking up of the dendritic structure, is accomplished by various combinations of mechanical working and heat treatment.

In addition to creating segregation, the dendritic form of the solid provides a perfect place to trap gas bubbles which result because of the lower solubility of most gases in the solid than in the liquid state. In addition, the very tightly spaced dendrite arms restrict liquid flow, often resulting in shrinkage porosity between dendrite arms (Fig. 6). Degassing operations, in which certain gases are bubbled through the melt, may be utilized to minimize the gas porosity. As described by Sievert's law, the purging gas bubbles will attract the undesirable gaseous element and remove it from the melt. In the case of shrinkage, the amount of microshrinkage may be reduced by increasing the cooling rate (sometimes called chilling). This action serves to make the dendrites stubbier, thereby providing more free access for metal flow to feed shrinkage. Ideally, with the use of carefully regulated molding materials, solidification can be controlled so that dendritic growth occurs smoothly from the farthest point away from the riser directly to the riser, so that the last volume to solidify is the riser itself. This directional solidification concept is utilized in the production of many unique cast metal parts. Perhaps the ultimate development of this technique is the controlled growth of investment cast turbine blades, where dendrites begin growing from the base of the blade and continue uniformly up to the top. This dendritic configuration provides optimum properties in a critical application.

Contraction of the casting during solidification can cause a defect known as hot tearing. Tensile stresses may be set up in the solidifying casting, which causes tears to open up between the dendrites. These types of defects are quite common in both ingot and cast-to-shape products. If these defects are not filled by liquid metal, then these "cracks" remain in the solidified casting. These defects can usually be avoided by proper mold design (or controlled cooling of ingots) to minimize tensile-stress development. Residual stresses are elastic stresses which may form in solid materials due to uneven heating or cooling. These can cause warping or failure in castings following the solidification event unless steps are taken to ensure uniform cooling.

Other defects which will often plague metal-casting processes are: (1) macrosegregation in ingots, caused by significant interdendritic fluid flow; (2) pinhole porosity, a smooth-walled defect elongated in a direction perpendicular to the mold wall and immediately below the casting skin; (3) metal penetration, resulting from liquid metal penetrating sand, and causing a metal-sand crust attached to the casting;(4) blows, cavities resulting from entrapped gases originating from cores; and (5) cold shuts, seams of oxide or slag resulting from low pouring temperatures.

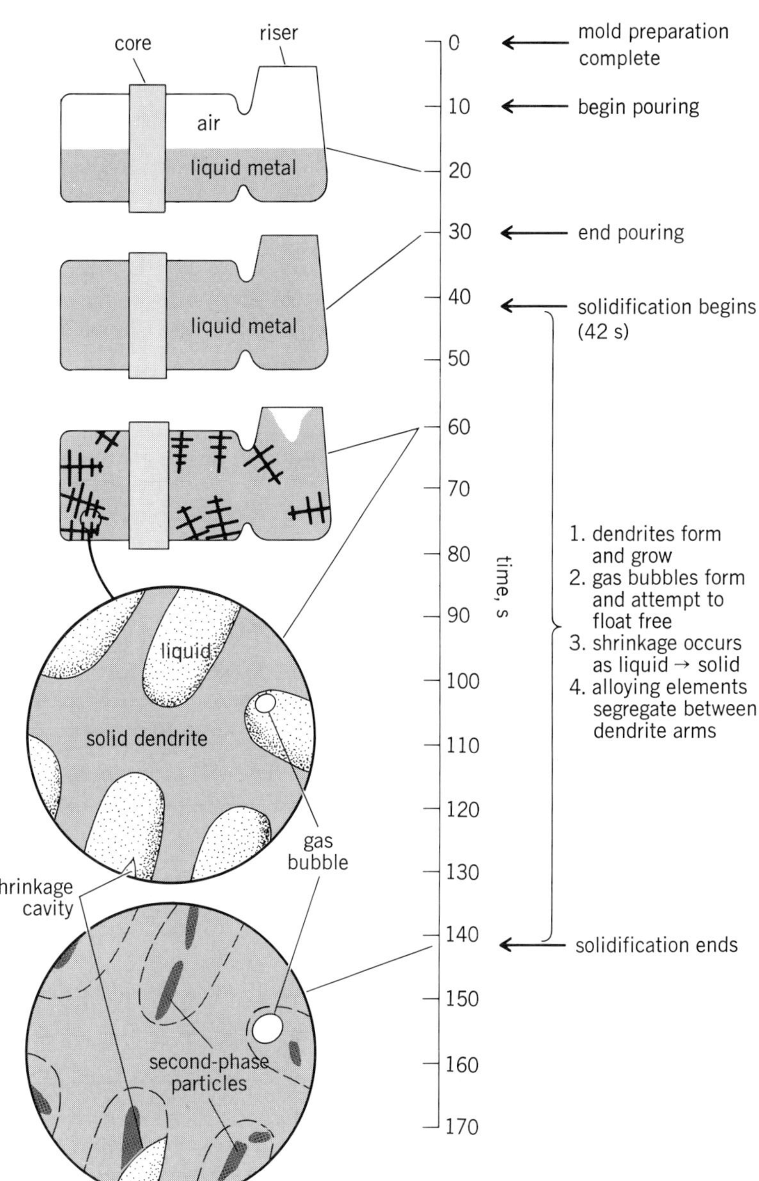

Fig. 6. Solidification and processing history for a cast-to-shape part.

APPLICATION

Almost all metals and alloys which are utilized by engineering specialists have at one time been in the molten state and will have been cast. Metallurgists have in general lumped these materials into two categories, ferrous and nonferrous. Ferrous alloys, steels and cast irons, constitute the largest volume of metals cast. Aluminum-, copper-, zinc-, titanium-, cobalt-, and nickel-base alloys are also cast into many forms, but in much smaller quantity than iron and steel. Selection of a given material for a certain application will depend upon the physical and chemical properties desired, as well as cost, appearance, and other special requirements.

Aluminum alloy castings have advantages such as resistance to corrosion, high electrical conductivity, ease of machining, and architectural and decorative uses. Magnesium alloy castings have the lowest density of all commercial casting alloys. Copper alloy castings, although costly, have advantages such as corrosion resistance, high thermal and electrical conductivity, and wear properties suitable for antifriction-bearing materials. Steel castings have more uniform (isotropic) properties than the same component obtained by mechanical working. On the other hand, because of the high temperatures required, casting of steel is relatively expensive and requires considerable knowledge and experience. Cast irons, which constitute the largest quantity of all metals cast to shape, have properties such as hardness, wear resistance, machinability, and corrosion resistance. Gray iron castings are commonly produced for their low cost, machinability, good damping capacity, and uniformity. Nodular cast irons have significantly higher strengths than gray irons, but are more costly and not as machinable. An intermediate grade of cast iron called CG (compacted graphite) iron has been developed which shares the best properties of both gray and nodular irons. *See* ALLOY; METAL, MECHANICAL PROPERTIES OF; METAL FORMING.

[KARL RUNDMAN]

Bibliography: American Society for Metals, *Casting Metals Handbook*, vol. 5, 8th ed., 1970; J. J. Burke, M. C. Flemings, and A. E. Gorum (eds.), *Solidification Technology*, Army Materials Conference Series, 1974; M. C. Flemings, *Solidification Processing*, 1974; R. Heine, C. R. Loper, and P. C. Rosenthal, *Principles of Metal Casting*, 2d ed., 1967.

Metal coatings

Thin films of material bonded to metals in order to add specific surface properties, such as corrosion or oxidation resistance, color, attractive appearance, wear resistance, optical properties, electrical resistance, or thermal protection. In all cases proper surface preparation is essential to effective bonding between coating and basis metal, so that coated metals can function as duplex materials. This article discusses various methods of applying either metallic coatings (see table), or nonmetallic coatings, such as vitreous enamel and ceramics, and the conversion of surfaces to suitable reaction-product coatings. Other methods for the protection of metal surfaces are described in other articles. *See* CLADDING; ELECTROLESS PLATING; ELECTROPLATING OF METALS.

Hot-dipped coatings. Low-melting metals provide inexpensive protection to the surfaces of a variety of steel articles. To form hot-dipped coatings, thoroughly cleaned work is immersed in a molten bath of the coating metal. The coating consists of a thin alloy layer together with relatively pure coating metal that adheres to the work as it is withdrawn from the bath (Fig. 1). Sheet, strip, and wire are processed on a continuous basis at speeds of several hundred feet per minute. On the other hand, hardware and hollow ware are handled individually or in batches.

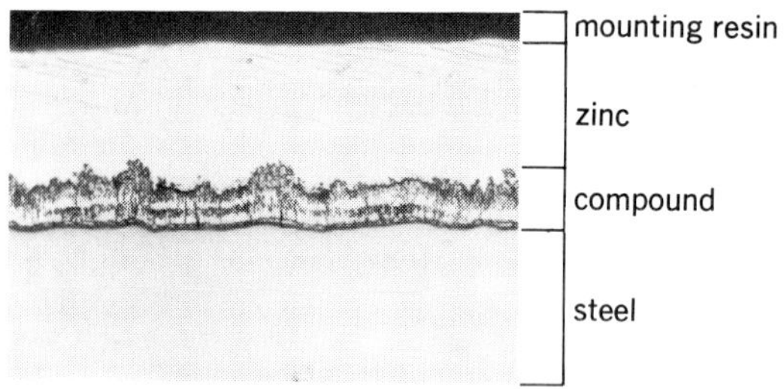

Fig. 1. Cross section of galvanized steel.

Coating weights for zinc coatings are expressed as ounces of coating per square foot (1 oz/ft^2 is equivalent to 0.305 kg/m^2 or 0.0017-in. or 43-μm thickness on hardware or wire, and to one-half this thickness on sheet). Hot-dipped tin and lead-tin (Terne) coatings are expressed as pounds per base box. A base box is 31,360 in.^{2} (20.23 m^2); thus 1 lb (0.4536 kg) per base box is equivalent to 0.000059 in. (1.50 μm) of tin.

Galvanized steel, that is, steel hot-dipped in zinc, is used for roofing, structural shapes, hardware, sheet, strip, and wire products. Coatings normally range up to 3 oz/ft^2 for outdoor service. Hot-dipped tinplate is now largely supplanted by electroplated tinplate for tin cans. Terne plate, with coatings up to 0.0008 in. (20 μm), is used for roofing, chemical cabinets, and gasoline tanks. Hot-dipped aluminum-coated (aluminized) steel with coatings up to 0.004 in. (10 μm) thick are used for oil refinery equipment and furnace and appliance parts where protection at temperatures up to 1000°F (538°C) is required.

Sprayed coatings. A particular advantage of sprayed coatings is that they can be applied with portable equipment. The technique permits the coating of assembled steel structures to obtain corrosion resistance, the building up of worn machine parts for rejuvenation, and the application of

Commonly used methods for applying metallic coatings

Coating metal	Hot dip	Electro-plate	Spray	Cemen-tation	Vapor deposition	Cladding	Immersion
Zinc	X	X	X	X			
Aluminum	X		X	X	X	X	
Tin	X	X					X
Nickel		X	X		X	X	X
Chromium		X		X	X		
Stainless steel			X			X	
Cadmium		X			X		
Copper		X				X	X
Lead	X	X				X	
Gold		X			X	X	X
Silver		X			X	X	X
Platinum metals		X			X	X	X
Refractory metals		X	X			X	

highly refractory coatings with melting points in excess of 3000°F (1650°C).

Nearly any metal or refractory compound can be applied by spraying. Coating material in the form of wire or powder is fed through a specially designed gun, where it is melted and subjected to a high-velocity gas blast that propels the atomized particles against the surface to be coated (Fig. 2). The surface is initially blasted with sharp abrasive to provide a roughened surface to which the particles can adhere. Coatings such as zinc and aluminum are applied with a gun which provides heat by burning acetylene, propane, or hydrogen in oxygen. A compressed air blast atomizes and propels the coating metals onto the surface. Highly refractory coating materials, such as oxides, carbides, and nitrides, can be applied by plasma-arc spraying (Fig. 3). In this process temperatures of 20,000°F (12,000°C) or more may be produced by partially ionizing a gas (nitrogen or argon) in an electric arc and passing the gas through a small orifice to produce a jet of hot gas moving at high velocity. Another variation for applying refractory coatings is detonation-flame plating. In this process a mixture of oxygen and gas-suspended fine particles are fired four times per second by a timed spark.

Sprayed zinc or aluminum coatings up to 0.010 in. (250 μm) are used to protect towers, tanks, and bridges. Such coatings are normally sealed with an organic resin to enhance protection.

Sprayed refractory coatings have been developed for high temperatures experienced in aerospace applications. They are also used for wear resistance, heat resistance, and electrical insulation.

Cementation coatings. These are surface alloys formed by diffusion of the coating metal into the base metal, producing little dimensional change. Parts are heated in contact with powdered coating material that diffuses into the surface to form an alloy coating, whose thickness depends on the time and the temperature of treatment. A zinc alloy coating of 0.001 in. (25 μm) is formed on steel

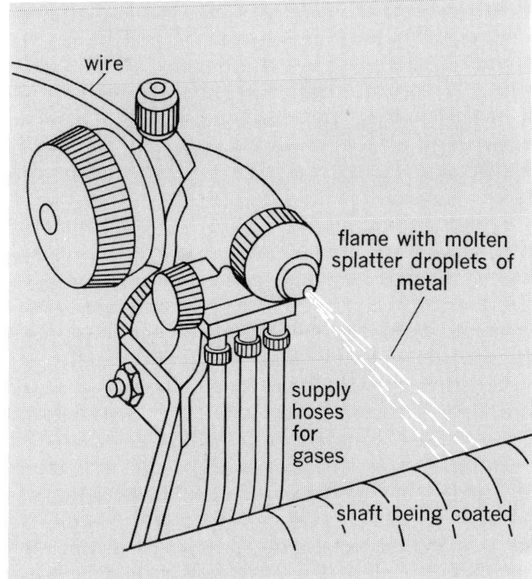

Fig. 2. Schooping gun with the coating material furnished in the form of wire.

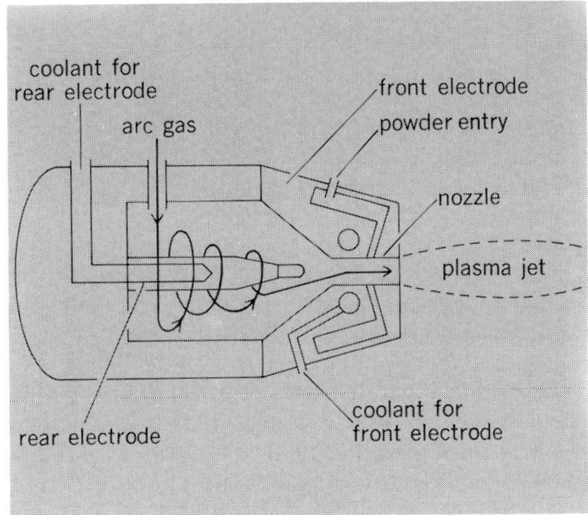

Fig. 3. Cross-sectional diagram of plasma-spray jet.

in 2–3 hr at 375°C (700°F). A chromium alloy (chromized) coating of 0.004 in. (100 μm) is formed in 1 hr at 1000°C (1830°F).

Chromized coatings on steel protect aircraft parts and combustion equipment. Sherardized (zinc-iron alloy) coatings are used in threaded parts and castings. Calorized (aluminum-iron alloy) coatings protect chemical equipment and furnace parts. Diffusion coatings are used to provide oxidation resistance to refractory metals, such as molybdenum and tungsten, in aerospace applications where reentry temperatures may exceed 3000°F (1650°C). In addition to the pack process described above, such coatings may be applied in a fluidized bed. In forming disilicide coatings on molybdenum, the bed consists of silicon particles suspended in a stream of heated argon flowing at 0.5 ft/sec (0.15 m/s), to which a small amount of iodine is added. The hot gases react with the silicon to form SiI_4, which in turn reacts with the molybdenum to form $MoSi_2$.

Vapor deposition. A thin specular coating is formed on metals, plastics, paper, glass, and even fabrics. Coatings form by condensation of metal vapor originating from molten metal, from high-voltage (500–2000 volts) discharge between electrodes (cathode sputtering), or from chemical means such as hydrogen reduction or thermal decomposition (gas plating) of metal halides. Vacuums up to 10^{-6} mm mercury (10^{-4} Pa) often are required.

Aluminum coatings of 0.000005 in. (0.125 μm) are formed on zinc, steel, costume jewelry, plastics, and optical reflectors. Chemical methods are capable of forming relatively thick coatings, up to 0.010 in. (250 μm). The 200-in. (5-m) mirror for the Mount Palomar telescope was prepared by vapor coating with aluminum in a vacuum chamber 19 ft (5.8 m) in diameter and 7 ft (2.1 m) high.

Immersion coatings. Either by direct chemical displacement or for thicker coatings by chemical reduction (electroless coating), metal ions plate out of solution onto the workpiece.

Tin coatings are displaced onto brass and steel notions and on aluminum-alloy pistons as an aid during the breaking-in period. Displacement nickel coatings of 0.00005 in. (1.25 μm) are formed on steel articles. Electroless nickel, involving the reduction of a nickel salt to metallic nickel (actually a nickel-phosphorus alloy), permits the formation of relatively thick uniform coatings up to 0.010 in. (250 μm) on parts with recessed or hidden surfaces difficult to reach by electroplating.

Vitreous enamel coatings. Glassy but noncrystalline coatings for attractive durable service in chemical, atmospheric, or moderately high-temperature environments are provided by enamel or porcelain coating. In wet enameling a slip is prepared of a water suspension of crushed glass, flux, suspending agent, refractory compound, and coloring agents or opacifiers. The slip is applied by dipping or flow coating; it is then fired at a temperature at which it fuses into a continuous vitreous coating. For multiple coats the first or ground coat contains an oxide of cobalt, nickel, or molybdenum to promote adherence.

Dry enameling is used for castings, such as bathtubs. The casting is heated to a high temperature, and then dry enamel powder is sprinkled over the surface, where it fuses.

Firing temperatures for conventional enameling of iron or steel ranges up to 870°C (1600°F). Low-temperature enamels have been developed, permitting the enameling of aluminum and magnesium.

Coatings of 0.003–0.020 in. (75–500 μm) are used for kitchenware, bathroom fixtures, highway signs, and water heaters. Vitreous coatings with crystalline refractory additives can protect stainless steel equipment at temperatures up to 950°C (1740°F).

Ceramic coatings. Essentially crystalline, ceramic coatings are used for high-temperature protection above 1100°C (2000°F). The coatings may be formed by spraying refractory materials such as aluminum oxide or zirconium oxide, or by the cementation processes for coatings of intermetallic compounds such as molybdenum disilicide. Cermets are intimate mixtures of ceramic and metal, such as zirconium boride particles, dispersed throughout an electroplated coating of chromium. *See* CERMET.

Surface-conversion coatings. An insulating barrier of low solubility is formed on steel, zinc, aluminum, or magnesium without electric current. The article to be coated is either immersed in or sprayed with an aqueous solution, which converts the surface into a phosphate, an oxide, or a chromate. Modern solutions react so rapidly that sheet and strip materials can be treated on continuous lines.

Phosphate coatings, equivalent to 100–400 mg/ft^2 (1–4 g/m^2), are applied to bare or galvanized steel and to zinc-base die castings as preparation for painting. The coating enhances paint adhesion and prevents underfilm corrosion. Phosphate coatings, containing up to 4000 mg/ft^2 or 40 g/m^2 (lubricated), serve as an aid in deep-drawing steel and in other friction-producing processes or applications. Iridescent chromate coatings on zinc-coated steel improve appearance and reduce zinc corrosion. Chromate, phosphate, and oxide coatings on aluminum or magnesium are used to prepare the surface before painting.

Anodic coatings. Coatings of protective oxide may be formed on aluminum or magnesium by making them the anode in an electrolytic cell. Anodized coatings on aluminum up to 0.003 in. (75 μm) thick are formed in sulfuric acid to form a porous oxide that may be sealed in boiling water or steam to provide a clear, abrasion-resistant, protective coating. If permanent color is required, the coating is impregnated with a dye prior to sealing. Such coatings are used widely on aluminum furniture, automobile trim, and architectural shapes. Thin, nonporous, electrically resistant coatings are formed on aluminum in a boric acid bath in the production of electrolytic capacitors. Anodized coatings on magnesium are thicker (up to 0.003 in. or 75 μm) and harder than those formed by chemical conversion. Anodic coatings 0.0003 in. (7.5 μm) thick are often used as a paint base. *See* CORROSION; SURFACE HARDENING OF STEEL.

[WILLIAM W. BRADLEY]

Bibliography: A. I. Andrews, *Porcelain Enamel*, 2d ed., 1961; R. M. Burns and W. W. Bradley, *Protective Coatings for Metals*, 3d ed., 1967; J. Huminik (ed.), *High Temperature Inorganic Coatings*, 1963; *Materials in Design Engineering*, Metals Selector Issue, vol. 80, no. 4, September 1974; C. F. Powell, J. H. Oxley, and J. M. Blocher (eds.), *Vapor Deposition*, 1966; H. H. Uhlig, *Corrosion Handbook*, 1948.

Metal forming

Manufacturing processes by which parts or components are fabricated from metal stock. In the specific technical sense, metal forming involves changing the shape of a piece of metal. In general terms, however, it may be classified roughly into five categories: mechanical working, such as forging, extrusion, rolling, drawing, and various sheet-forming processes; casting; powder and fiber metal forming; electroforming; and joining processes. The selection of a process, or combination of processes, requires a knowledge of all possible methods of producing the part if a serviceable part is to be produced at the lowest overall cost.

Influence on service behavior. The service behavior of a part depends on the interrelation of the design, material, processing of the material (such as manufacturing process, heat treatment, and surface treatment), and operating and environmental conditions (such as temperature of operation, loading condition, and corrosive atmosphere). The metal-forming process is a factor in satisfactory service performance in that it affects the microstructure of the metal and its surface finish, it may introduce large residual stresses into the part, and it may affect the final design of the part which, in turn, influences its service behavior.

Process selection. The more important factors to be considered in choosing the optimum process (or combination of processes) include: type of material, metallurgical structure effect inherent in the process, size of the part, shape or complexity of the part, tolerances or finish required, quantity to be produced, cost, and production factors such as availability of equipment, rate of production, and time required to initiate production.

Material. The particular metal or alloy specified for a part is of major importance in the selection of the forming process. Some aluminum or copper alloys may be fabricated by practically any of the manufacturing processes; other alloys may be brittle under cold-working conditions but may be hot-worked. Highly refractory materials, such as tungsten and tungsten carbide, are not suitable for casting and must be fabricated by powder metallurgy methods. This process is also used for making porous metal products, or parts requiring combinations of two materials (not an alloy of the two, however). Alloys that are extremely hard, and therefore unsuitable for machining operations, can be precision cast and then ground if extremely close tolerances are required. Higher-melting-point alloys such as steel can be fabricated by most of the major classes of processes but are not suitable for all of the individual processes within a major class. Where a specific material must be used for the part, the choice of the optimum fabricating process may be definitely limited. The formability of a material may be predicted either from the reduction of area or the percent elongation in a tension test; the higher these values, the better is the formability. Other factors that influence formability are rate of deformation, temperature, type of loading, environment, impurities in the metal, and surface condition of the original stock. *See* ALLOY; METALLURGY; STEEL.

Metallurgical structure. Each manufacturing process has a different effect on the microstructure of the metal and, consequently, on its mechanical properties. Casting processes generally produce a relatively coarse-grained structure and a random orientation of nonmetallic inclusions. The result is isotropic properties, but lower ductility than that of wrought products. Castings may also be porous. Hot-working processes, such as forging, align the inclusions (fiber structure) and thus impart anisotropic properties, with the strength and ductility generally being higher in a direction parallel to that of the inclusions. This orientation may be an advantage or disadvantage, depending upon the direction of the applied loads. Cold-working processes (such as cold rolling) also produce directional properties in the metal because of the tendency of the grains (or crystals) to align in certain directions. In addition the grains become distorted, and the metal becomes harder and stronger but less ductile. Cold-working operations (or any process causing nonuniform deformations) generally leave residual stresses in the part. Residual stresses left in a part algebraically add to the stresses induced by service loading, and in some instances these residual stresses are of major importance.

Size. Metal-forming processes may be limited as to the size of the part they can produce. Among the processes that are limited to relatively small parts are precision casting, die casting, powder metallurgy, and screw machining. Large parts are best produced by sand casting, forging, or building up component sections by welding or other joining processes. Parts with rotational symmetry can be produced by spinning; other large parts, such as domes, made of sheet or plate stock can be formed by explosive-forming techniques developed in the aerospace industries.

Shape. The complexity of a part often dictates the process or combination of processes used in its manufacture. Generally, castings can be more complex than parts made by most other fabrication processes; however, some casting processes (sand and precision investment casting) are capable of producing parts of greater complexity than others. Parts produced by the powder metallurgy process have definite restrictions as to design because of the inability of metal powder to flow like a liquid. In some instances the complexity of a part may require the fabrication of the structure by welding or brazing several sections together. The change from forming a part as a single piece to fabricating it from several sections usually requires design modifications if optimum serviceability is to be attained. The design should be left flexible until all feasible manufacturing processes have been considered.

Tolerances and finish. Parts requiring close tolerances or smooth finishes can be formed directly by precision investment casting, die casting, or such cold-working processes as swaging, drawing, or stamping. If formed by other processes, they can be finished by machining or grinding. Hot-working processes, such as forging, result in relatively rough, oxidized surfaces and relatively low dimensional accuracies. Welding operations generally result in some distortion or dimensional change. The final overall cost often governs whether the desired tolerances should be attained in the original process or obtained in secondary operations. As an example, in hot-working operations, such as rolling, forging, and extrusion, surface roughness ranges from about 100 to as high as 2000 μin., arithmetic average. In cold working, the range is between 10 and 250 μin.; in some machining operations, it is as low as 1 μin., for example, in grinding and honing.

Quantity. The number of parts to be produced is a major factor in determining the method of manufacture, and whether the part should be produced within the plant or subcontracted. Some processes are suitable only for large-quantity production because of high tooling costs; for example, permanent-mold and die casting, certain forging processes, and deep drawing. Processes such as sand casting, spinning, and welding are readily adaptable, but not necessarily restricted, to small-quantity production.

Cost. If the quantity to be produced is large, the overall finished cost of the product is usually a prime consideration in the selection of a process. In many cases cost is the deciding factor in choosing the fabrication process and perhaps the material as well. In determining the overall cost, the replacement cost of the part based on its estimated service life should be included with more immediate factors such as material and tooling costs, labor, and scrap loss.

Production factors. In some instances the time necessary to initiate production may be of significance in selecting the fabrication process. Those methods involving extensive tooling necessarily require a long lead time before production starts. Availability of equipment may be the deciding factor in choosing between two equally feasible processes, especially if it has been decided that the part is to be produced in a certain plant rather than subcontracting it to another organization. Another production factor which may be important is the required rate of production. Processes such as die casting, powder metallurgy, and deep drawing have high production rates. Conversely, sand casting, spinning, hydraulic press forging, and fusion welding are relatively slow processes.

It is thus evident that there are many factors involved in selecting an optimum for material and processes to manufacture a certain part. A very wide range of variables are involved. Strength levels for metals range from as low as 1000 psi to the order of 500,000 psi. Size of parts may range from a thousandth of an inch to a few feet. Rates of deformation can be as high as 40,000 fpm, as in explosive forming, while working temperatures can be as low as cryogenic to the range of 4000°F. Capacities of equipment for forming metal components are as high as 50,000 tons, with a 200,000-ton hydraulic forging press, which is presently in the design stage. *See* DRAWING OF METAL; EXTRUSION; FORGING; MACHINING OPERATIONS; METAL CASTING; METAL ROLLING; POWDER METALLURGY; SHEET-METAL FORMING.

[SEROPE KALPAKJIAN]

Bibliography: J. M. Alexander and R. C. Brewer, *Manufacturing Properties of Materials*, 1963; American Society for Metals, *Forming*, 1969; American Society of Mechanical Engineers, *ASME Handbook: Metals Engineering Processes*, 1958; American Society of Tool and Manufacturing Engineers, *Tool Engineers' Handbook*, 2d ed., 1959; B. W. Niebel and E. N. Baldwin, *Designing for Production*, rev. ed., 1963.

Metal matrix composite

A material consisting of two or more substances, one of which is a metal or metallic alloy, and which exhibits its own distinctive behavior. The latter criterion sets a composite apart from conventional metallic materials such as steels and nonferrous alloys, both of which contain two or more constituents. For example, the stiffness or strength of a composite may be higher than that of the individual components, or radically different from either.

Typically, metal matrix composites consist of wires, filaments, fibers (fine wires or filaments), whiskers, or thin sheets of a strong, relatively brittle material embedded in the weaker but more ductile metallic constituent. This combination and particular distribution of the constituents result in enhanced stiffness and strength coupled with ductility and toughness (the ability to withstand impact or shock loading). The need for a fine-scale distribution of the reinforcement, and the role of the matrix in stress transfer, crack propagation, and protection of the reinforcement are considerations which apply to metal matrix composites. The reinforcement may be either continuous, that is, extending the full length of the composite, or discontinuous, consisting of chopped lengths having a particular ratio of length to diameter (aspect ratio).

A combination of high stiffness in relation to density (relative stiffness) and high strength in relation to density (relative strength) means that the designer has a wide choice of materials. The overall property combinations of metal matrix composites are particularly attractive in aerospace and marine engineering, propulsion, and transportation. Unlike polymer matrix composites, metal matrix composites can be utilized at both ambient temperature and high temperatures. A good example of this is the use of nickel-matrix composites in gas turbine components, such as fan blades, where temperature cycles from ambient to in excess of 1800°F (980°C) are experienced.

This article examines the various classes of metal matrix composites in terms of their structure and associated properties. Attention is also directed to the methods and techniques available for the production and fabrication of these composites.

Composite systems. While several classification modes are possible, it is convenient to subdivide metal matrix composites on the basis of matrix density, the matrix being the continuous phase or constituent in the composite. Low-density matrices utilize Al, Mg, Ti, and their alloys. Materials and their physical form commonly used for reinforcement are listed in Table 1. Representative high-density metal matrices are Co, Ni, Fe, Cb, W, and their alloys. Reinforcement materials and form are listed in Table 2.

Mechanical properties. Composite stiffness and strength are reasonably well predicted from the rule of mixtures, that is, the contributions from each phase are weighted in proportion to the volume fraction of each phase in the composite. For example, with continuous fibers aligned uniaxially, the composite strength in a direction parallel to the fibers is given by the equation below, where S_c is

$$S_c = S_f V_f + \sigma_m (1 - V_f)$$

the strength of the composite, S_f is the average strength of the fibers, V_f is the volume fraction of the fibers, $(1 - V_f) = V_m$ is the volume fraction of the matrix, and σ_m is the average stress in the matrix at fiber fracture.

Clearly, to enhance strength, a reinforcing material having a high intrinsic value of S_f must be selected and V_f must be made as large as possible. Similar considerations apply to stiffness except that now the elastic modulus of the fiber (E_f) is the materials property of importance. To provide a perspective on the strength and stiffness that can be achieved, the strength and modulus of typical metal matrices and reinforcements are listed in Table 3. A comparison of the strength and stiffness of various materials is presented in Fig. 1. Some

Table 1. Reinforcement of metal matrix composites based on Al, Mg, Ti

Wire*	Filament†	Whisker‡	Platelets§
Be	B	Al_2O_3	$CuAl_2$
Mo, Mo alloys	SiC	B_4C	
Stainless steel	Graphite	SiC	
	Al_3Ni	Al_3Ni	

*Diameter $\gtrsim$ 0.005 in. (1 in. = 25.4 mm).
†Diameter $\simeq$ 0.005 in. ‡Diameter $\lesssim$ 0.0001 in.
§Sheets having small through-thickness dimension $\lesssim$ 0.001 in.

Table 2. Reinforcement materials for high-density metal matrices

Wire	Filament	Whisker	Platelets
W	B	Al_2O_3	Ni_3Cb
Mo alloys	SiC	SiC	
Steel	Graphite	TaC	
		$Co,Cr-(Cr,Co)_7C_3$	

Table 3. Strength and modulus of metal matrix and reinforcement

Material	Tensile strength, 10^6 psi* Theoretical	Fiber reinforcement†	Bulk	Young's modulus, 10^6 psi‡
Graphite	5	3 (w)	<0.1	98
SiC	5	3 (w)	<0.1	100
B	2.5	0.5 (f)	0.1	64
		Matrix		
Fe	1.8	~1.8 (f)	0.7	30
Cu	0.9	0.6 (w)	0.2	18
Al	0.5	0.04 (W)	<0.1	10

*1 psi = 6895 Pa.
†w = whisker form; f = filament; W = wire.
‡ = measured values; 1 psi = 6895 Pa.

Table 4. Ambient-temperature strength levels of Al- and Ni-matrix composites

Matrix	Fiber	V_f	S_c, 10^3 psi	S_c/density, 10^3 in.*
Al	B	0.5	110	1690
	Steel	0.25	173	1210
	Be	0.4	80	830
	SiO_2	0.48	118	1340
	Al_2O_3	0.35	161	1425
	Al_3Ni	0.1	48	470
Ni	B	0.75	384	1470
	W	0.09	61	173
	Al_2O_3	0.19	171	600
	Graphite	0.48	50	260

*1 in. = 2.54 cm.

representative strength and specific strength levels of high- and low-density metal matrix composites are assembled in Table 4.

The most recently developed metal matrix composites utilize a nickel- or cobalt-alloy matrix reinforced with metal carbides or Ni_3Cb. These advanced materials have a tensile strength in excess of 150,000 psi (1 psi = 6895 Pa) at ambient temperature. Of primary importance, however, in relation to application in gas turbines is their high-temperature strength; values $\gtrsim$ 60,000 psi have been achieved at 2000°F (1093°C).

In general, metal matrix composites exhibit good creep resistance (deformation under constant stress or load), fatigue properties (response to cyclic loading), and toughness (ability to absorb energy on impact without failure). These properties, in combination with strength and stiffness, are the ones used by the designer in selection of a material for structural applications. Also of importance to the designer is the fact that metal matrix composites are anisotropic; that is, the mechanical properties depend critically on the orientation of the fibers with respect to the applied force and on the nature of the stress (tensile or compressive). Anisotropy can be an advantage or disadvantage; again this depends on the state of stress on the part in service.

Making metal matrix composites. A variety of techniques are available for the production of continuous or discontinuous metal matrix composites. These may be broadly classified as diffusion processes, deposition processes, and liquid processes. Examples of each are illustrated in Fig. 2.

Diffusion processes. These include the pressing and sintering of powder metal matrix and bare or coated fibers, and the cold or hot pressing of the reinforcement between thin foils of the matrix metal or alloy.

Deposition processes. Primary techniques are

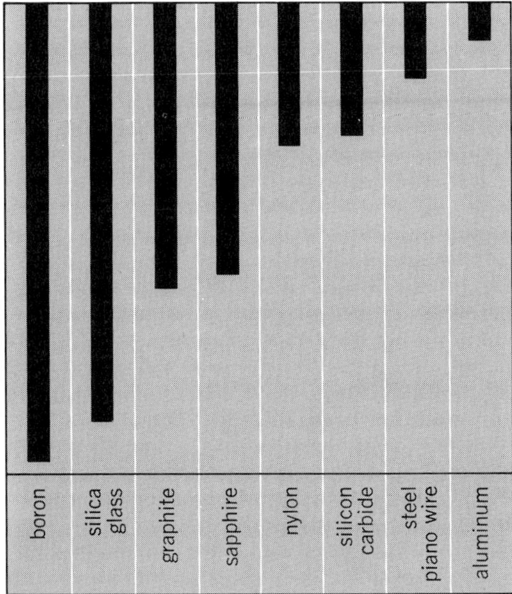

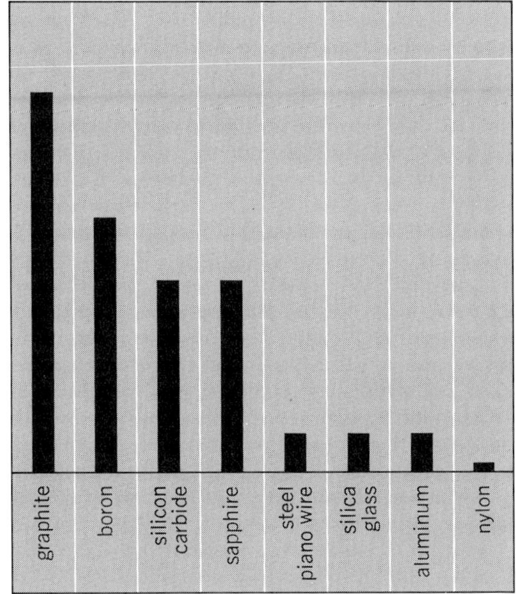

Fig. 1. Strength and stiffness for various materials. (*a*) Strength is represented as greatest free-hanging length; that of boron is 189.4 mils. (*b*) Stiffness is represented on an arbitrary scale indicating relative stiffness per unit weight. (*From Materials: A Scientific American Book, W. H. Freeman and Co., 1967*)

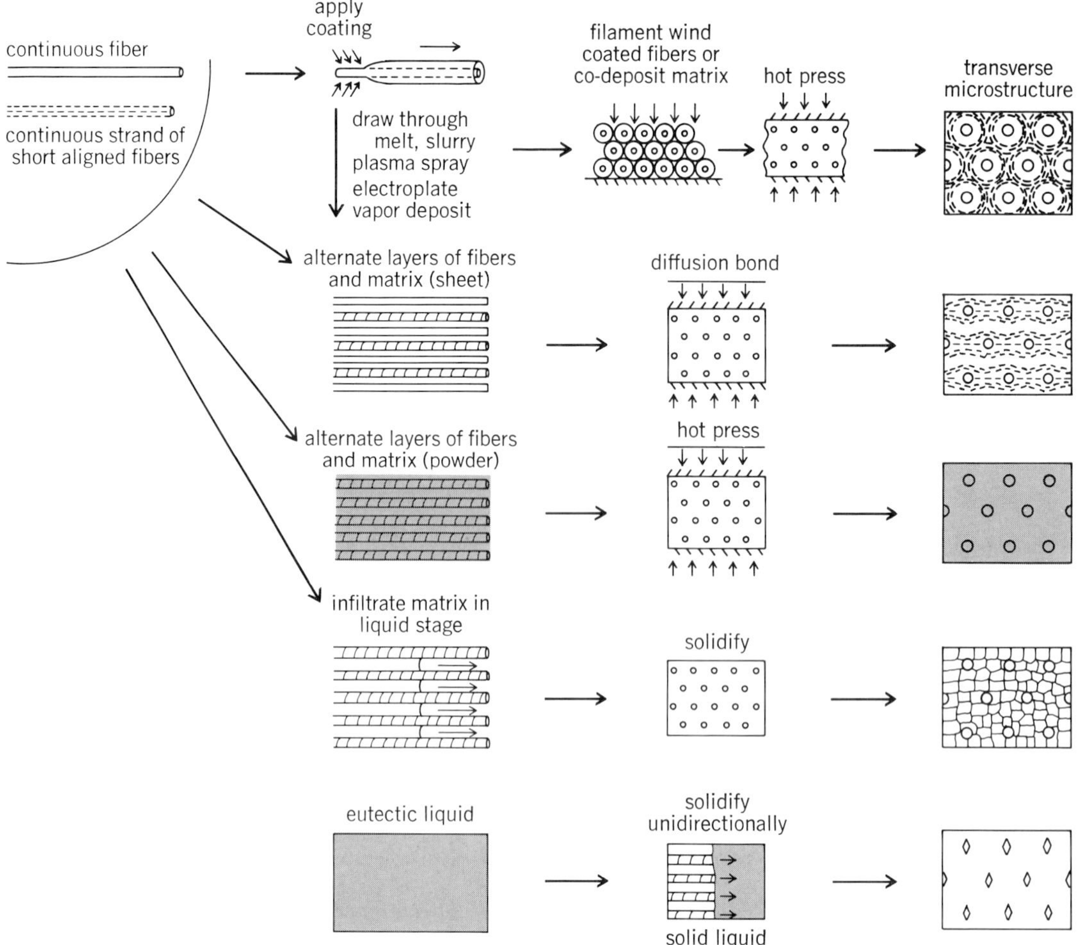

Fig. 2. Examples of the methods which are used to make metal matrix composites. (*Space Sciences Laboratory, General Electric Valley Forge Space Technology Center, King of Prussia, PA, 1967*)

electrodeposition of matrix around fibers, plasma spraying of matrix around fibers, or vacuum deposition of matrix around fibers. In each case, it is usual to complete the process by cold or hot pressing.

Liquid processes. The important techniques in this category are infiltration of the liquid matrix around fibers, the pressing of powdered matrix and fibers at a temperature above the melting point of matrix, and unidirectional solidification of eutectic alloys. *See* EUTECTICS.

Unidirectional solidification of eutectic alloys deserves special mention. Here the constituents of the composite are formed in place as the alloy solidifies. Composites formed in this way are much less susceptible to chemical reaction between matrix and reinforcement than are composites manufactured by the other techniques. This stability of the composite structure is a key advantage when service at elevated temperature is contemplated.

As of 1976, research on metal matrix composites emphasized structural applications. While this trend is expected to continue, the potential of metal matrix composites in nonstructural areas is beginning to be explored. In particular there is now an interest in possible unique electrical and magnetic property combinations; this would make the composites amenable to application in electronic devices. *See* ALLOY; METAL, MECHANICAL PROPERTIES OF.

[ALAN LAWLEY]

Bibliography: L. J. Broutman and R. H. Krock (eds.), *Modern Composite Materials*, 1967; H. Herman (ed.), *Advances in Materials Research*, vol. 5, p. 83, 1971; A. Kelly, The nature of composite materials, in *Materials: A Scientific American Book*, p. 97, 1967; A. Kelly, *Strong Solids*, 1973; F. Lemkey and M. Salkind, *Crystal Growth*, p. 171, 1967; A. G. Metcalfe (ed.), *Composite Materials*, vol. 1: *Interfaces in Metal Matrix Composites*, 1974.

Metal rolling

Reducing or changing the cross-sectional area of a workpiece by the compressive forces exerted by rotating rolls. The original material fed into the rolls is usually an ingot from a foundry. The largest product in hot rolling is called a bloom; by successive hot- and then cold-rolling operations the bloom is reduced to a billet, slab, plate, sheet, strip, and foil, in decreasing order of thickness and size. The initial breakdown of the ingot by rolling changes the coarse-grained, brittle, and porous structure into a wrought structure with greater ductility and finer grain size.

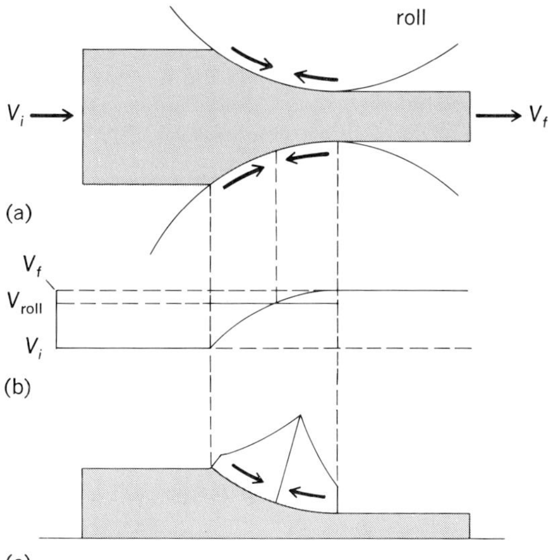

Fig. 1. The rolling process. (*a*) Direction of friction forces in the roll gap. (*b*) Velocity distribution. (*c*) Normal pressure acting on the strip in the roll gap.

Process. A schematic presentation of the rolling process, in which the thickness of the metal is reduced as it passes through the rolls, is shown in Fig. 1*a*. The speed at which the metal moves during rolling changes, as shown in Fig. 1*b*, to keep the volume rate of flow constant throughout the roll gap. Hence, as the thickness decreases, the velocity increases; however, the surface speed of a point on the roll is constant, and there is therefore relative sliding between the roll and the strip. The direction of this relative velocity changes at a point along the contact area, this point being known as the neutral or no-slip point. At the neutral point, roll and strip have the same velocity; to the left of this point (entry side) the strip moves more slowly than the roll; and to the right of this point (exit) it moves faster. Hence, the direction of frictional forces acting on the strip are opposite in these two regions, as shown in Fig. 1*a*. The net frictional force acting on the strip must be in the direction of exit to enable the rolling operation to take place. Although friction is a disadvantage in many metalworking processes, it is a necessity in rolling; without friction the rolls cannot pull the strip through the roll gap. It has been observed that in hot rolling the coefficient of friction may be as much as 0.7, while in cold rolling it generally ranges 0.02–0.3.

The normal pressure distribution on the roll and hence on the strip is of the form shown in Fig. 1*c*. Because of its particular shape this pressure distribution is known as the friction hill. The overall slope of this hill depends on the coefficient of friction and the ratio of roll-strip contact length to the thickness of the strip. The size of the friction hill (area under the curve) represents the magnitude of the roll separating force per unit width that tends to push the rolls apart. This force can be reduced by increasing the workpiece temperature, reducing friction, taking smaller bites, and using smaller roll radii. Another method of reducing this force is by applying tension to the strip; this lowers the apparent compressive flow stress of the material. The tension can be back tension (entry) or front tension (exit) or both. Depending on the magnitude of these tensions, the neutral point shifts from its original position.

The roll-separating force can become so great that it is not possible to reduce further the thickness of the strip during the particular pass. With such a force the rolls deform elastically (roll flattening). In such cases rolling may be accomplished by improving lubrication, rolling a stack of sheets at the same time, or annealing the strip.

The maximum possible draft (the difference between the initial and final thickness of the strip) is a function of the friction coefficient and roll radius; it can be shown analytically that maximum draft is equal to the coefficient of friction squared times the roll radius.

The workpiece may become wider during rolling, this is called spreading. It is found that spreading increases with the thickness-width ratio of the workpiece and with decreasing coefficient of friction and decreasing radius-thickness ratio of the roll.

Practice. A great variety of roll arrangements and equipment are used in rolling. Some basic types are shown in Fig. 2. In another method (Steckel rolling) the strip is pulled through idler

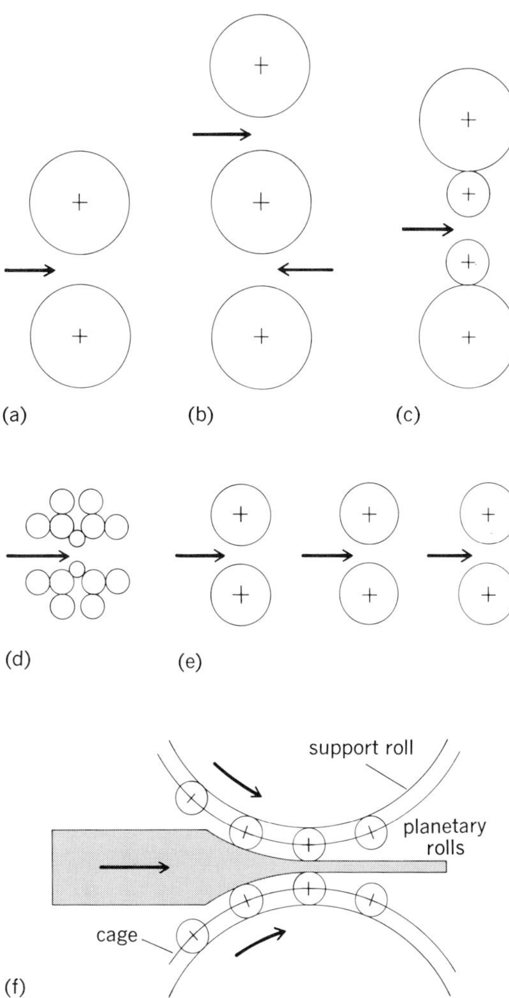

Fig. 2. Basic types of roll arrangement. (*a*) Two-high. (*b*) Three-high. (*c*) Four-high. (*d*) Cluster. (*e*) Tandem rolling with three stands. (*f*) Planetary mill.

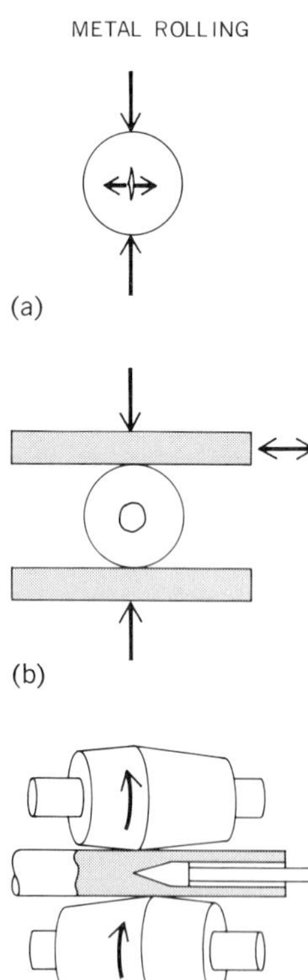

Fig. 3. Development of the Mannesmann process for producing seamless tubing (*a–c* explained in text).

rolls by front tension only. The proper reduction per pass in rolling depends on the type of material and other factors; for soft, nonferrous metals, reductions are usually high, while for high-strength alloys they are small.

Temperatures in the hot rolling of various metals and alloys are similar to those in forging, namely, for aluminum alloys, 750–850°F; copper alloys, 1500°F; alloy steels, 1700–2300°F; titanium alloys, 1400–1800°F; and refractory metal alloys, 1800–3000°F. Rolling speeds range up to about 5000 ft/min. Although hot rolling has the advantage of lowering forces and increasing the ductility of the metal, cold rolling gives smoother finish, closer tolerances, and increased strength.

The thickness of a rolled sheet is identified by gage number: The smaller the number the thicker the sheet. Actual thickness depends on the particular standard used; there are different standards, depending on the type of metal. The terminology for hardness for rolled sheets is shown in the table. A sheet that is reduced in thickness by one gage number from the annealed condition is called quarter hard, by two gage numbers half hard, and so on, as listed in the table. The dead-soft condition is fully annealed.

Successful rolling practice requires a careful balance of factors such as reduction per pass, control of temperature, lubrication, roll size and finish, and intermediate annealing.

Requirements for roll materials are mainly strength and resistance to wear. Common roll materials are cast iron, cast steel, and forged steel. In cold rolling, cemented carbide is also used for small rolls. To avoid variation in strip thickness due to bending of the rolls, rolls are ground to a particular geometry, called camber, whereby the center of the roll has a diameter of a few thousandths of an inch greater than the diameter of its ends.

Hot rolling is usually carried out without a lubricant, although graphite or grease may be used. Cold rolling is carried out with low-viscosity lubricants, paraffin being a suitable lubricant for nonferrous metals. The type of residual stresses in the strip depends on the reduction and also on roll radius. Small rolls or small reductions produce compressive residual stresses on the surface of the strip and tensile stresses in the central portions, whereas large rolls produce the opposite residual stress pattern. A number of defects, such as wavy edges, zipper breaks, edge cracks, and alligatoring, can result during the rolling process. These are usually eliminated or reduced by changes in operating variables or by special techniques.

Hardness terminology for rolled sheets

Terminology	Increase in gage number	Reduction in thickness, %
Annealed (dead-soft)	0	0
1/8 hard	–	6
1/4 hard	1	11
1/2 hard	2	21
3/4 hard	3	29
Hard	4	37
Extra hard	6	50
Spring hard	8	60
Extra spring hard	10	69

Miscellaneous processes. Most seamless tubing is produced by a technique based on the principle that the inside of a round rod is subjected to secondary tensile stresses when compressed radially (Fig. 3*a*). A simple demonstration of this can be made by rolling an eraser, removed from the end of a pencil, between a flat surface and a ruler that is moved back and forth (Fig. 3*b*). In a short time a hole is produced in the center of the eraser. The roll-piercing of round bars to make seamless tubing (Mannesmann process) is based on this principle (Fig. 3*c*). The compressive radial forces are supplied by two rolls with their axes in parallel planes, but at an angle to each other so as to move the workpiece through the rolls. The rolling action causes the center of the billet to rupture. The purpose of the mandrel is to expand the tube and improve the surface finish of the inside. For this process to be successful, the billet material should be of high quality and free of defects.

In the ring-rolling process, a thin, large-diameter ring is produced from a thicker and smaller-diameter ring. This is accomplished by placing the ring between two rotating rolls; the reduction in thickness of the ring is compensated by an increase in diameter; there is little or no change in the width of the ring.

Sections such as railroad tracks and I beams are also rolled by passing the stock through a number of specially designed rolls with their axes placed in different directions. Roll-pass design for such sections requires considerable experience to avoid defective products and to obtain desirable properties. Metal powders are also rolled into sheets or strip by special techniques. *See* METAL COATINGS; POWDER METALLURGY.

[SEROPE KALPAKJIAN]

Bibliography: W. L. Roberts, *Cold Rolling of Steel*, vol. 2, 1978.

Metallic glasses

Alloys having amorphous or glassy structures. A glass is a solid material obtained from a liquid which does not crystallize during cooling. It is therefore an amorphous solid, which means that the atoms are packed in a more or less random fashion similar to that in the liquid state. The word glass is generally associated with the familiar transparent silicate glasses containing mostly silica and other oxides of aluminum, magnesium, sodium, and so on. These glasses are not metallic; they are electrical insulators and do not exhibit ferromagnetism. One obvious answer to obtaining a glass having metallic properties is to start from a melt containing metallic elements instead of oxides. However, liquid metals and alloys crystallize so rapidly on cooling that it was not until 1960 that the first true metallic glass, $Au_{80}Si_{20}$, an alloy containing 80 at. % Au and 20 at. % Si, was obtained by a method capable of achieving rates of cooling of the order of 1,000,000°C per second.

Preparation. The original method used for obtaining the amorphous Au-Si alloy was the gun technique (or splat cooling). It consisted of propelling a small liquid globule by means of a shock tube and spraying it into a thin foil on a copper substrate. The substrate is curved so that centrifugal force promotes good thermal contact between the liquid layer and the substrate. This method ex-

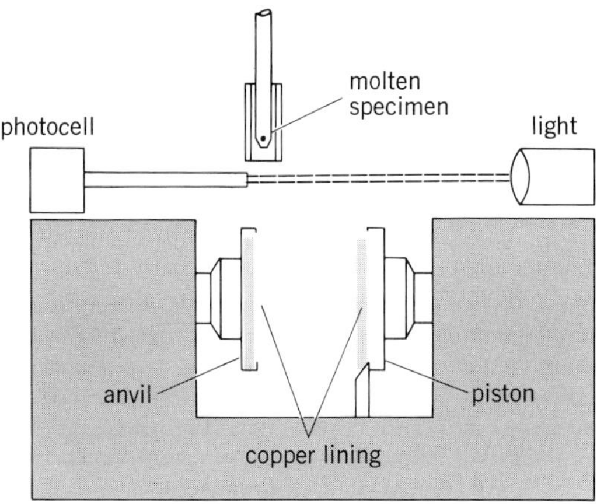

Fig. 1. Piston-and-anvil liquid-quenching device.

tracts the heat from the melt from one side only, a limitation eliminated in the subsequent method in which the liquid globule is squeezed between a fast-moving piston and a fixed anvil (Fig. 1) that are lined with copper to ensure fast heat removal. An advantage of the piston-anvil method is the good parallelism between the two faces of a foil, which is about 25 mm in diameter and 40 μm thick.

These two method are quite satisfactory for preparing laboratory specimens of metallic glasses, but are inadequate for mass production of useful shapes. Because of the foreseen practical applications of metallic glasses, a large effort has been devoted to various methods of rapidly freezing a metallic melt into a continuous ribbon. The principle of this method is to feed a continuous jet of liquid alloy on the outside rim of a rapidly rotating cylinder. The molten alloy solidifies rapidly into a thin ribbon (about 50 μm thick) which is ejected tangentially to the rotating cylinder at rates as high as 2000 m/min. With a single liquid metal jet, the width of the ribbon is limited to a few millimeters; using a multijet design, it is possible to produce strips as wide as 100 mm. Because some metallic glasses have excellent magnetic properties, there is a great incentive for developing advanced techniques for producing large sheets of these materials to be used as transformer cores.

With the available rates of cooling from the liquid state, of the order of 10^6 to 10^{8}°C/s, it is not possible to obtain amorphous pure metals for which a minimum cooling rate through the liquid-solid transition would be at least 10^{14}°C/s. Therefore metallic glasses always contain at least two or more different kinds of atoms, if the term glass is used with its proper meaning of an amorphous solid obtained from the liquid state.

Structure. Experimental methods for studying the structure of crystalline or amorphous solids, as well as that of liquids, are based on the scattering of radiation or particles, giving rise to interference effects.

The most widely used method is x-ray diffraction. The difference between the diffraction patterns of a crystalline and an amorphous solid is striking (Fig. 2). The crystalline pattern consists of a succession of very sharp and intense peaks located at well-defined diffraction angles (Bragg angles). These peaks are not present in the pattern coming from an amorphous solid and are replaced by broad maxima in the intensity curve. Although only one of these maxima is shown in Fig. 2, the complete pattern up to higher diffraction angles may show up to six maxima progressively decreasing intensities.

The diffraction pattern of an amorphous alloy is very similar to that of the same alloy in the liquid state, and the analysis which has been used for a long time to study the atomic arrangement in liquids is now routinely applied to metallic glasses and other amorphous solids. From such an analysis it is possible to determine the number of atoms in the nearest-neighbor shell surrounding any one atom and the average radius of this shell. For most metallic glasses, the number of nearest neighbors, also called the coordination number, is close to 12, just as in a close-packed crystalline structure. To give a quantitative interpretation of the diffraction curve, it is necessary to assume a certain model for the packing of atoms, generally taken as rigid spheres. The dense random-packing model previously used by J. D. Bernal in his studies of liquids provides a reasonably accurate explanation for the observed diffraction experiments.

Predicting favorable alloy compositions. There is no theory capable of predicting the composition of metallic alloys which can be quenched from the liquid state into an amorphous structure. So many metallic glasses have been synthesized since 1960 that it is now possible to formulate some empirical rules, hoping that these will eventually constitute a foundation for establishing a theory of glass formation in alloys.

Shortly after the first metallic glass ($Au_{80}Si_{20}$) was obtained, its composition was considered as being favorable to glass formation for the two following reasons: first, the alloy has a low melting point compared with those of its constituents and is located near a eutectic point in the AuSi phase diagram; and second, the atomic sizes of Au and Si are quite different, Si being much smaller than Au. A very simple justification for the first condition is that a low melting point means that during cooling

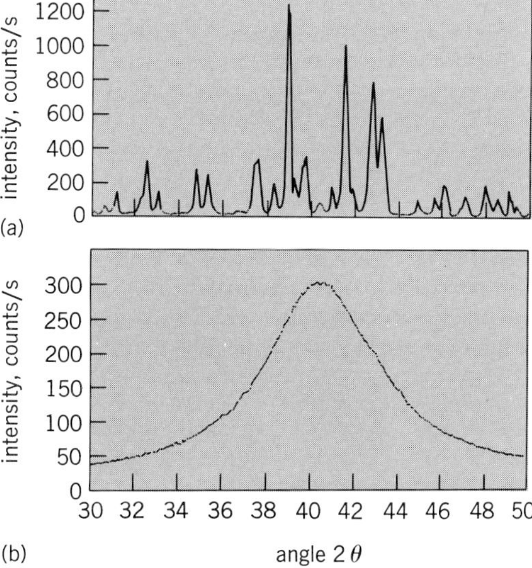

Fig. 2. X-ray diffraction patterns of a $Pd_{80}Si_{20}$ alloy in the (*a*) crystalline and (*b*) amorphous states.

the alloy has a strong tendency to remain liquid and the nucleation of solid phases is relatively slow. It can be visualized that high rates of cooling might prevent nucleation altogether. The second condition, which involves the difference in atomic size, finds its justification in the analysis of the Bernal model of dense random packing of rigid spheres.

A mathematical analysis of this model shows that the volume fraction of voids in the packing is about 19% if the spheres all have the same size. If the model is made of spheres of two different sizes, it may be assumed that the smaller spheres will fit into the voids, and one might expect that the ratio of large atoms to small ones will be in the neighborhood of 20%. It is remarkable that the composition of metallic glasses made of the transition metals such as Ni, Co, and Fe, with the metalloid atoms B, C, Si, and P, is in the range of 75 to 85% transition metals with 25 to 15% metalloids. The 20% rule is particularly applicable to metal-metalloid glasses but fails to explain the concentration in binary alloys containing metals only. A typical example is found in Zr-Cu metallic glasses, which can be obtained over a wide range of compositions extending from about 40 to 60% Cu. The difference in atomic size between Zr and Cu still satisfies the size rule, but the explanation of the small atom, in this case Cu, filling the holes in a random close packing of Zr atoms is not valid. Obviously the model based on a dense random packing of atoms cannot explain the structure of the different types of metallic glasses. In addition to the size of atoms, the type of bonding, which is either metallic only in the case of ZrCu or metallic and covalent in the case of AuSi, must be taken into consideration in formulating a model.

The composition of metallic glasses is the most important factor controlling the temperature above which the glass will transform to the more stable crystalline state. The crystallization from the liquid state is an extremely fast phase change taking place at well-defined temperatures, whereas the glass-to-crystalline transformation is a rate process. Strictly speaking, there is no such a thing as a crystallization temperature, but rather a range of temperatures within which the crystallization rate is relatively high. What is generally called the crystallization temperature refers to measurements made by heating the specimen at a rate of 10°C/min.

One of the best methods for studying the early stage of nucleation and the growth of crystals within the glassy matrix is transmission electron microscopy. Many different types of morphology have been observed. The crystals may be similar to the treelike, so-called dendrites generally observed during solidification of a liquid alloy, or have very sharp, almost straight boundaries, as shown in Fig. 3. The kinetics of crystallization of metallic glasses is a very important problem because it determines the maximum temperature and the maximum length of time that metallic glasses will stand in service without noticeable changes in their physical properties. It is probable that this temperature will never be higher than about 400–500°C.

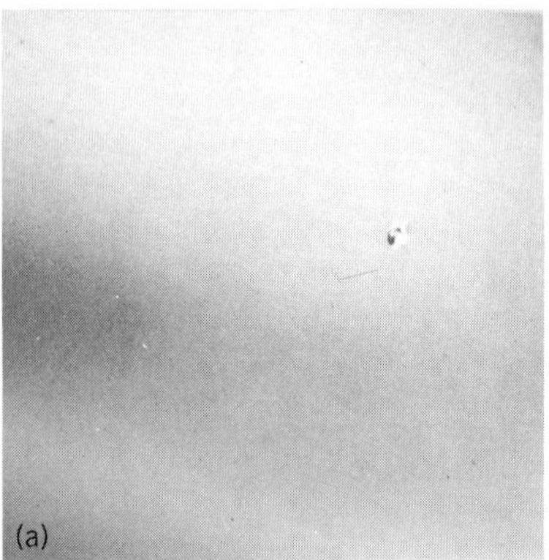

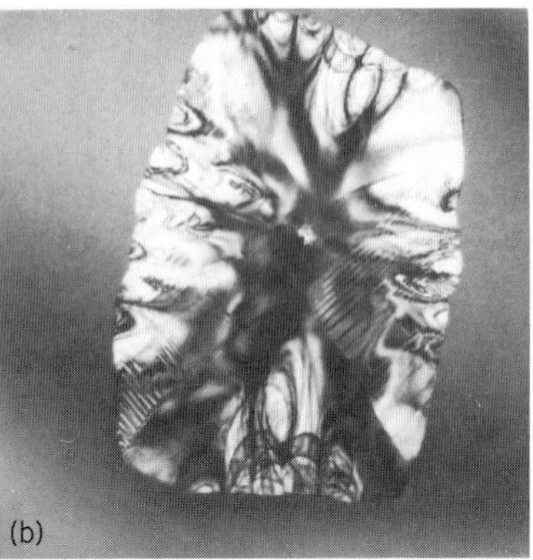

Fig. 3. Transmission electron microscope pictures of crystallization of a $Pt_{77}Si_{23}$ metallic glass. A very small, barely visible crystalline region in *a* grows to an almost rectangular $1 \times 3 \mu m$ region in *b*.

Electrical properties and superconductivity. The electrical properties of crystalline metals and alloys are generally well understood. The absence of a crystal lattice in metallic glasses results in substantial changes in their electrical properties and has theoretical applications in studies of transport properties in solids.

Electrical resistivity. The electrical resistivity of metallic glasses is high, for example 100 $\mu\Omega$-cm and higher, which is in the same range as the familiar nichrome alloys widely used as resistance elements in electric circuits. Another interesting characteristic of the electrical resistivity of metallic glasses is that it does not vary much with temperature. The temperature coefficient is of the order of 10^{-4} K^{-1} and can even be zero or negative, in which case the resistivity decreases with increasing temperature.

Because of their insensitivity to temperature variations, metallic glasses may find applications in electronic circuits for which this property is an essential requirement. In a number of metallic glasses containing an atom having a magnetic moment, there is a resistivity minimum at low temperature. At temperatures below the resistivity minimum there is a very steep increase in resistivity with decreasing temperature. This behavior is similar to the so-called Kondo effect found in some crystalline alloys, but it is much more pronounced, and the temperature at which the minimum occurs is as high as 200°C instead of less than 50 K in crystalline alloys. A metallic glass, $Pd_{80}Si_{20}$, in which Pd is replaced by Cr up to 7 at. %, has been used as a low-temperature thermometer. In this alloy the increase in resistivity with decreasing temperature is so large that resistance-measuring instruments can be used to measure temperature with an accuracy of 1 K or less. This is not as sensitive as the standard germanium thermometer, but while the single crystal of germanium is severely damaged by radiation, the metallic glass can sustain high doses of radiation without a change in its electrical properties. Therefore it will be very useful in fission- or fusion-reactor environments.

Superconductivity. The first superconducting metallic glass was reported in 1975. This was an alloy containing 80 at. % La and 20 at. % Au. The absence of a crystal lattice in amorphous metals does not exclude the existence of superconductivity. In fact, superconductivity was observed below 6 K in an amorphous Bi thin film obtained by vapor deposition in high vacuum and on a liquid helium-cooled substrate. As with all amorphous films deposited on a liquid helium-cooled substrate, the Bi films are very unstable and crystallize below room temperature. Superconducting metallic glasses are much more stable, and some of them do not crystallize at temperatures as high as 500°C.

Some superconducting metallic glasses contain only two metals, such as $Zr_{75}Rh_{25}$, and some are more complex alloys in which there is approximately 20% of metalloid elements, mostly B, Si, or P. The maximum transition temperature reported is 8.71 K for a complex metal-metalloid alloy of composition $(Mo_{80}Re_{20})_{80}P_{10}B_{10}$. This transition temperature is much lower than that of some crystalline superconductors which are as high as 23 K.

The theory of superconductivity does not predict that the absence of a crystalline lattice will automatically lead to a low transition temperature. It is possible, therefore, that metallic glasses with higher transition temperatures will be found. One of the main reasons for continuing research on new superconducting glasses is their projected usefulness in high-field electromagnets which will be required to contain the high-temperature plasma in fusion reactors. The present requirement for these magnets is a field of at least 100,000 G, which is not attainable with conventional copper-wound electromagnets. However, several crystalline superconductors could be used, but it has been shown that these alloys will be seriously damaged by the high radiation field in the vicinity of the reactor. For example, the transition temperature of crystalline Nb_3Sn alloy which could meet the field requirements is reduced by radiation damage from about 18 K to less than 4 K. By contrast, no decrease in the transition temperature of the metallic glass $(Mo_{60}Ru_{40})_{82}B_{18}$ was detected after the same exposure to radiation as that used for Nb_3Sn. This observation is in agreement with previous results, showing that the electrical resistivity of metallic glasses is not affected by radiation.

In spite of these encouraging results, the properties of metallic glasses must be improved before they can meet the requirements for high-field magnets. All metals and alloys lose their superconducting capabilities when they are subjected to a magnetic field more intense than a certain critical value. This is known as the Meissner effect. A similar loss of the superconducting properties of an alloy occurs when the intensity of the current passing through it is greater than a critical value. In metallic glasses both critical field and critical current are much lower than in crystalline alloys. The main reason for this behavior is the amorphous nature of the alloys, which is so homogeneous that the superconducting flux lines are not "pinned" by local defects such as grain boundaries and dislocations in crystals and are expelled from the conductor by the magnetic field. This inherent characteristic of amorphous superconductors might seriously limit their application to superconducting power transmission lines and high-flux magnets. The problem is probably not insoluble. It may be possible, for example, to introduce pinning centers in an amorphous alloy by a suitable thermal treatment, which would result in fine crystalline precipitates having the required critical size and spaced at a critical distance to maximize the flux pinning effect.

Magnetic properties. The ferromagnetic properties of metallic glasses have received a great deal of attention, probably because of the possibility that these materials can be used as transformer cores.

Ferromagnetic amorphous alloys had been prepared before the technique of rapid cooling from the liquid state was developed. Electrolytic deposits of NiP alloys are slightly ferromagnetic for P concentrations less than 17 at. %. Amorphous CoP alloys can be electrodeposited in the amorphous state for P concentrations from 18 to 25 at. % Co and are also ferromagnetic. Ferromagnetism was also measured in alloys of Co with Au in the form of vapor-deposited thin films. These results suggested that it should be possible to obtain a ferromagnetic metallic glass from a liquid alloy containing a high enough percentage of ferromagnetic metals. The choice of alloying elements was guided by trying to satisfy the low-melting-point eutectic composition of the original AuSi glass, and Fe was the most obvious choice for the metal constituent.

The low eutectic condition is satisfied in FeP binary alloys, but the liquid-quenched alloys of eutectic composition were only partially amorphous, in the sense that their structure consisted of a mixture of amorphous regions and microcrystals of the phosphide Fe_3P. When carbon was added to the melt, completely amorphous alloys were obtained; $Fe_{75}P_{15}C_{10}$ became the first strongly ferromagnetic metallic glass and is the prototype for all other ferromagnetic glasses. It is possible to replace Fe in the basic composition of the alloy by other transition metals with *3d* electrons, namely Ni, Co, Cr, and Mn. Similarly, the P and C can be replaced by other metalloids, mostly B and Si.

A general formula for all metallic ferromagnetic glasses of practical interest may be written $T_{70\text{-}85} - M_{30\text{-}15}$ in which T is the sum of the transition metals and M the sum of the metalloid atoms. The compositions of many soft ferromagnetic glasses now produced in the form of ribbons and strips follow this general formula. Among ternary alloys are $Fe_{76}Cr_7B_{17}$, $Ni_{75}P_{15}B_{10}$, and $Co_{73}P_{17}B_{10}$. More complex alloys containing four, five, and as many as six elements are also of interest; a typical example is $Fe_{38.5}Ni_{38.5}P_{14}B_6Al_{1.5}Si_{1.5}$. One very soft ferromagnetic glass has the composition $Co_{72}Fe_3P_{16}B_6Al_3$. After suitable annealing, this alloy has a coercive force which is in the range of the best very soft ferromagnets of the Permalloy type used in high-performance transformers.

With the development of ferromagnetic glasses and improvements in their fabrication techniques, the transformer industry is likely to be the biggest consumer of metallic glasses. In spite of the fact that a transformer is already a very efficient device, the powers involved are so high that even less than 1% improvement in reducing the hysteresis losses in a transformer core represents a large number of kilowatt-hours saved. *See* EUTECTICS.

Mechanical and other properties. The investigations of mechanical properties started to expand very rapidly when foils and continuous ribbons of glassy alloys of uniform thickness became available. Because of their very small permanent deformation before fracture in tension, glassy metals might be classified as brittle materials. This apparent brittleness, however, is not at all comparable to that of silicate glasses. Metallic glasses show evidence of ductility in compression tests. Relatively large specimens of liquid-quenched amorphous $Pd_{77.5}Cu_{16}Si_{16.5}$ can be obtained, and their stress-strain curves in compression are similar to those of ductile crystalline alloys. In addition, most metallic glass foils can be reduced to half or even one-third of their thickness by rolling at room temperature.

The growing interest in the mechanical properties of metallic glasses is motivated by their high rupture strength and toughness. Most of the high-strength metallic glasses are in the same class as those having interesting magnetic properties; that is, they are made of about 80% transition metals and 20% metalloids. It is believed that high concentrations of B are an important factor in achieving high strength levels.

Metallic glasses have superior corrosion resistance, and some metallic glasses compare favorably with stainless steel. The fundamental reasons for the surface of an amorphous alloy to be chemically less active than that of its crystalline equivalent are not yet well understood. Qualitatively it might be argued that an amorphous surface has a low activity because it is very homogeneous and free of crystalline defects such as grain boundaries, vacancies, and dislocations. *See* CORROSION; METAL, MECHANICAL PROPERTIES OF.

Future. Considering the unusual physical and chemical properties of metallic glasses, there is no doubt that they will play an important role as an engineering material in the future. The possible applications of these materials have already been demonstrated on a small scale. The first attractive products are probably thin sheets for power transformer cores. These should eventually be produced in widths of the order of at least 50 cm. The melt-spinning process of producing metallic glasses should also result in substantial savings in labor and energy when compared with the present technology, because it can produce a thin sheet of material by direct casting from the liquid state. By contrast, the starting point in the traditional process is a large ingot weighing as much as several tons, which is forged to small dimensions and then hot-rolled into sheets. This is an energy-consuming, multistep operation requiring many intermediate reheats of the steel before the final sheet thickness is obtained.

Techniques have been studied for producing metallic glasses in forms other than thin sheets. Experiments have shown that a high-power laser can modify the microstructure of a metal surface. Under certain conditions the laser can melt a very thin layer of material (perhaps 50 to 100 μm), and this thin layer is then cooled rapidly by the massive substrate acting as a heat sink. If the alloy is susceptible to solidify into a glass, an amorphous skin can be obtained. If not, suitable alloying elements can be introduced, probably in powder form, to modify the chemical composition of the base material. The obvious applications for this technique are the development of surfaces which have improved resistance to corrosion or wear.

Another development is an application of powder metallurgy methods to the fabrication of amorphous alloys. In the process of "atomizing" liquid alloys into a very fine powder, the rate of cooling may be high enough to produce an amorphous powder. The problem of compacting and sintering this powder into a massive ingot without crystallizing has been solved. By using explosive loading, it is indeed possible to subject the powder to very high pressure and probably some rather high temperature, for a short enough time to avoid crystallization, and achieve near-theoretical densities. Massive ingots of metallic glasses, susceptible of being forged and rolled into various shapes, may become available within the not-too-distant future. *See* ALLOY; POWDER METALLURGY.

[POL E. DUWEZ]

Bibliography: B. Cantor (ed.), *Proceedings of the 3d International Conference on Rapidly Quenched Metals, Metals Society, London*, 1978; P. Duwez, Structure and properties of glassy metals, *Annu. Rev. Mater. Sci.*, 6:83–117, 1976; P. Duwez and S. C. H. Lin, Amorphous ferromagnetic phase in iron-carbon-phosphorus alloys, *J. Appl. Phys.*, 38:4096–4097, 1967; W. L. Johnson, S. J. Poon, and P. Duwez, Amorphous superconducting lanthanum-gold alloys obtained by liquid quenching, *Phys. Rev. B*, 11(1):150–154, 1975.

Metallography

The study of the structure of metals and alloys by various methods, especially optical and electron microscopy. In older usage the term metallography designated the science relating the properties of metals to their structure, but this science is now called physical metallurgy. *See* METALLURGY.

The optical microscopy of metals is conducted with reflected light on surfaces suitably prepared to reveal structural features; a resolution of about 2000 A (200 nm) and a linear magnification of at most 2000× can be obtained. Electron microscopy is generally carried out either with replicas of specimen surfaces or with thin foils prepared from bulk specimens. The replica technique provides a resolution of detail as small as 50 A (5 nm) and magnifications of up to approximately 50,000×. The thin-foil technique provides a resolution of better than 10 A (1 nm) and a useful magnification of 500,000×.

Metallography serves both research and industrial practice. Optical microscopy has long been a standard method for the investigation of the phase constitution of metallic systems and the microstructure of metals. Phase transformations and the mechanisms by which metals deform plastically are becoming better understood through the application of electron microscopy. Metallographic methods are used in industry for the control of production processes, especially heat treatments. These methods are frequently indispensable in the analysis of the causes of service failures of metallic objects. Metallographic methods have been adapted to the examination of nonmetallic materials, especially ceramics, semiconductors, and minerals.

Typical applications. The photomicrographs in Figs. 1, 2, and 3 show typical structures investigated by optical microscopy. They include the size and shape of the grains making up the structure of metals (Figs. 1*a* and 2*a*); the microstructure of al-

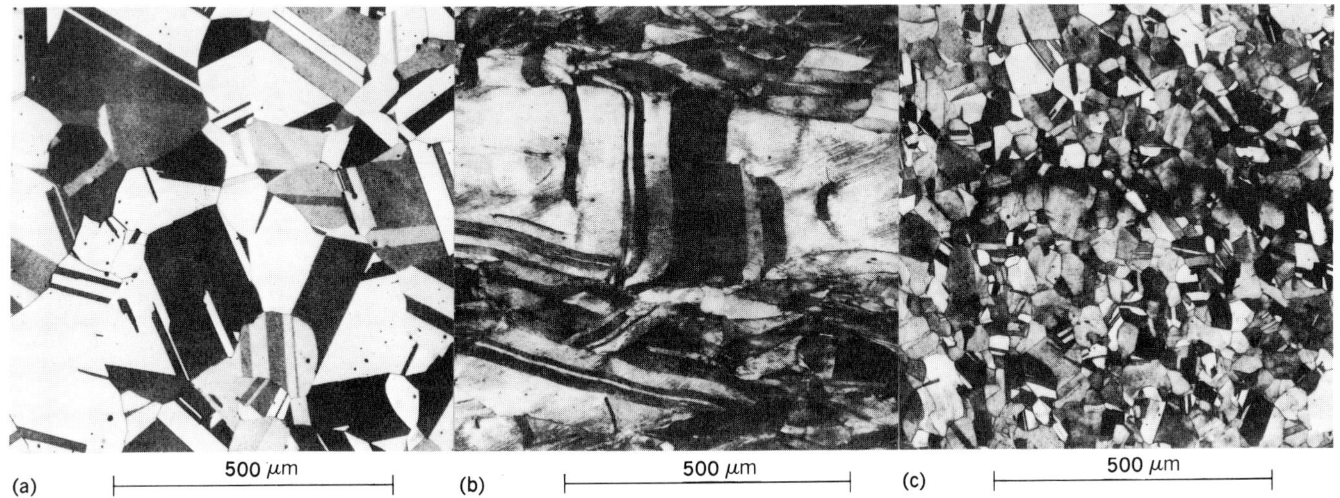

Fig. 1. Photomicrographs of typical structures of brass (70% Cu, 30% Zn). (*a*) Annealed (grains with twin bands). (*b*) Reduced 40% by cold rolling (distorted grains). (*c*) Stress-corroded, with horizontal crack. (*Courtesy of W. R. Johnson*)

loys containing several phases (Fig. 3*a* and *b*) and particularly of steels (Fig. 2*b*, *c*, and *d*); effects of deformation (Fig. 1*b*) and of stress corrosion (Fig. 1*c*); and etch pits revealing the crystal orientation of a grain (Fig. 3*c*). Other structural features investigated by optical microscopy include the size, shape, and arrangement of precipitates; inhomogeneities in the composition of metals (microsegregation); microporosity; strain markings due to deformation; mechanical twins; microcracks; surface layers, such as carburized cases and electroplated metals; nonmetallic inclusions; and heat-treating defects, such as overheated steel or decarburized layers.

Figure 4 shows a two-phase structure as observed by electron microscopy involving the use of a replica. The structure of an alloy observed by transmission electron microscopy is shown in Fig. 5. Typical structural features which can be investigated by electron microscopy are the products of phase transformations and age hardening and the density and arrangement of imperfections, especially dislocations and stacking faults. In an electron microscope, electron diffraction patterns can be obtained from small regions of the specimen; in this manner the crystallographic orientation of features such as precipitates observed by electron microscopy can be determined.

Optical microscopy. The selection of representative specimens for metallographic examination is essential. For example, if a part such as a forged shaft has directional properties, transverse and longitudinal sections should be examined. Sheet, wire, and other small specimens are mounted in fixtures or plastic mounts.

A specimen is first ground on a series of emery papers of decreasing grit size or on laps. It is then polished on one or more cloth-covered wheels with an abrasive such as aluminum oxide, magnesium oxide, or diamond dust. These operations render the specimen surface scratch-free and mirrorlike by means of the progressive removal of surface irregularities; however polishing, even when properly carried out, produces a thin layer of distorted metal.

Electrolytic polishing is an alternative to mechanical polishing. Electrolytic polishing consists of controlled anodic dissolution; it does not distort the surface layer of the metal, and is therefore particularly suitable for soft metals. However, inclusions and compounds present in a specimen may

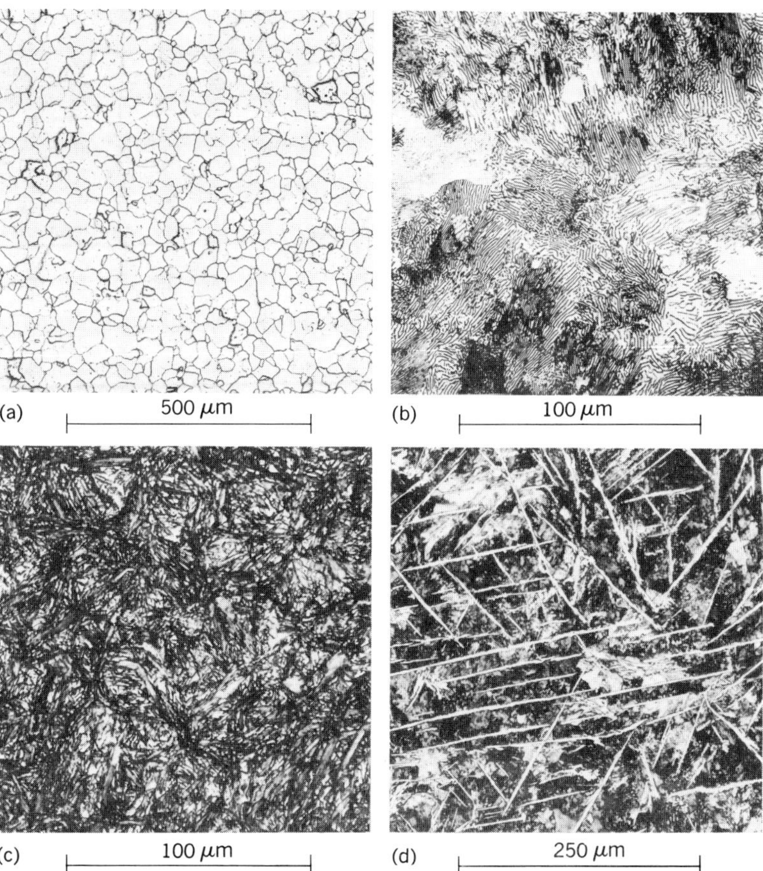

Fig. 2. Photomicrographs of typical structures of iron and steel. (*a*) High-purity iron. (*b*) Steel containing 0.85% C, slowly cooled; pearlite structure is composed of light ferrite and dark cementite. (c) Steel containing 0.80% C, quenched and tempered (dark needles of tempered martensite). (*d*) Steel containing 1.10% C, cooled at moderate rate (a Widmannstätten structure of white cementite plates in a matrix of fine pearlite). (*Courtesy of W. R. Johnson*)

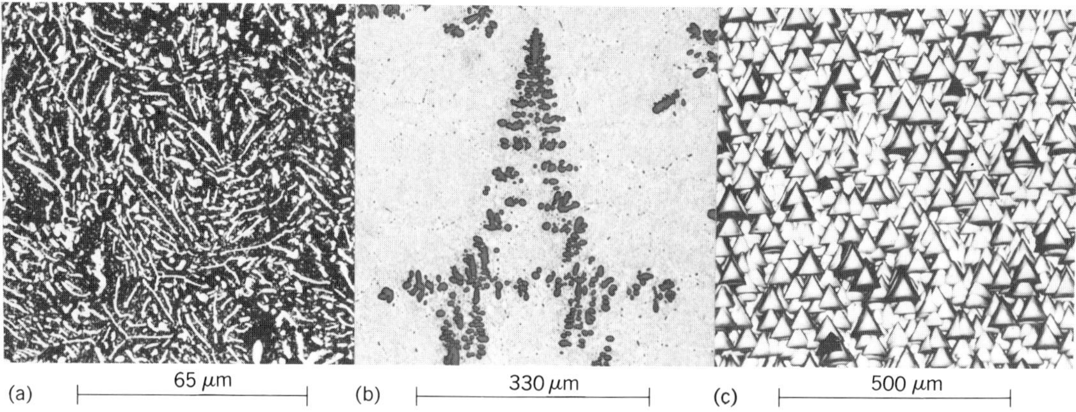

Fig. 3. Photomicrographs of typical structures. (*a*) Eutectic in an alloy of lead and antimony containing 11% Sb and 89% Pb; mixture of two phases solidified together. (*b*) A large dendrite of copper oxide in a matrix of copper—copper oxide eutectic; branches of dark particles have grown from a single trunk. (*c*) Etch pits in germanium; identical orientation of the etch pits indicates that the area of the specimen in the field of view consists of a single grain. (*Courtesy of W. R. Johnson*)

react with the electrolyte. Once the operating conditions have been established, electrolytic polishing is a simple operation, and it prepares a surface which may be superior to a mechanically polished surface.

Polished specimens reveal only a few structural features, such as inclusions, microcracks, and microporosity. Etching with an appropriate reagent is necessary to bring out the microstructures. Etching also removes the distorted metal layer from mechanically polished specimens.

Etching reagents are acids, bases, or complex substances, generally in dilute solution in water, alcohol, or glycerin. Most reagents act by dissolution, a few by selective deposition of reaction products (staining). In a single-phase alloy, etching is effective because it attacks different parts of the structure selectively. Grain-boundary regions dissolve in preference to the body of the grains; the resulting grooves are observed as a dark network, as shown schematically in Fig. 6*a* and illustrated in the photomicrograph in Fig. 2*a*. Adjoining grains may develop facets of different orientation, which reflect differing amounts of light as shown in Fig. 6*b* and illustrated in Fig. 1. In multiphase alloys, the phases are attacked selectively; Fig. 3*a* and *b* shows examples of this.

Etching may also be carried out in an electrolyte by making the specimen the anode. Etching by heating in a vacuum or an oxidizing atmosphere has proved useful. Cathodic etching, in which the specimen is bombarded by ions of an inert gas, is effective with some metals which resist etching by other methods.

The examination of metallographic specimens by optical microscopy depends on the differences in the amount and quality of light reflected by different structural constituents.

In the standard metallographic microscope, a beam of light normal to the surface illuminates the specimen. The objective of the microscope serves as a condensing system for the incident beam and as an image-forming system for the light reflected by the specimen. This is known as bright-field illumination. Oblique illumination by an off-center beam improves contrast by casting shadows. Conical-stop illumination achieves the same effect. In dark-field illumination, a hollow beam of light converges on the specimen; a mirrorlike surface appears dark, whereas a rough surface appears bright.

Polarized light permits identification of some constituents, especially nonmetallic inclusions. It also may bring out contrast between adjoining grains, particularly in anisotropic metals. Ultraviolet light improves resolution, but has found little use in metallography. In phase-contrast microscopy, differences in levels of the specimen surface appear as differences in brightness. Phase contrast is also sensitive to the nature of the reflecting surfaces. Fine structural details (veining, slip lines, and precipitates) have been observed by phase contrast when ordinary methods failed. Interference microscopy permits quantitative investigation of surface contours.

Observation at elevated temperatures allows processes such as recrystallization or precipitation to be followed as they occur. The microscope stage must accommodate a heating device, and most

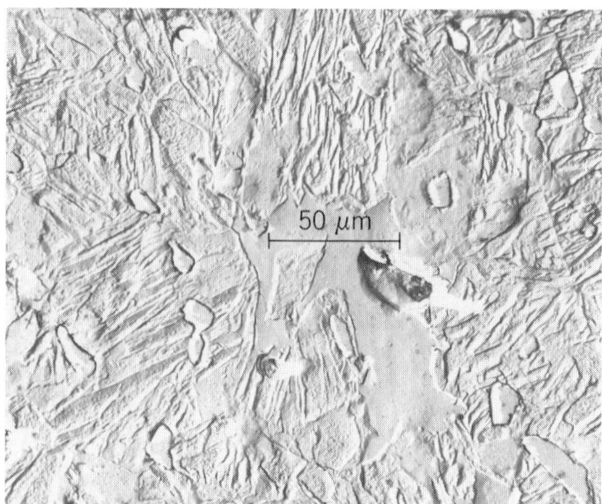

Fig. 4. Electron micrograph of a two-phase structure of a steel containing 0.35% carbon and 0.04% boron, commonly designated 10B35. The sample was quenched from the austenite phase field, producing a structure which consists of martensite (the needlelike component) and ferrite (the smooth, more rounded component). The plastic (parlodion) replica was shadowed with chromium to enhance the contrast.

specimens must be protected from oxidation by a protective atmosphere or vacuum. Also, the objective must be kept cool. Their long working distance makes reflecting objectives particularly suitable for this application. In high-temperature metallography, thermal etching reveals the structural features.

Photomicrographs of metallographic specimens are often taken. For this purpose, special microscopes known as metallographs have been developed. Metallography of highly radioactive specimens calls for equipment operated entirely by remote control.

In microautoradiography, radiations emitted by radioactive tracers are recorded on photographic emulsions, which are then investigated microscopically. Microradiography operates in an analogous manner with the use of x-rays.

Electron microscopy. A major consideration in the application of transmission electron microscopy (TEM) to microstructural observations is that the specimen should not exceed a certain thickness if it is to be transparent to electrons. This thickness depends on the metal and the electron acceleration voltage used; maximum thickness is of the order of 3000 A (300 nm) for 100-keV electrons for an element of medium atomic number. Suitable specimens can be prepared in two ways: replication of the specimen surface and preparation of thin foils from the bulk.

Specimens are prepared for replication by polishing and etching in a manner resembling the preparation of specimens for optical microscopy. Greater care, however, must be exercised in order to avoid artifacts. Only etching reagents which act by dissolution rather than by staining are suitable, and a lighter etch is used. When a representative surface has been prepared, a thin layer of the replica material is applied. The replica material may be a polymer (for example, formvar) dissolved in a volatile solvent or a polymeric film softened in solvent. Alternatively, carbon is deposited by evaporation onto the specimen surface. The contrast to be obtained with a polymer or carbon replica can be enhanced by shadowing, that is, the deposition of a heavy metal, such as gold or chromium on the replica at a glancing angle. The shadowing metal can also be deposited by simultaneous evaporation with the carbon. By a variation of the replica technique an oxide layer can be formed on aluminum alloys. This layer is removed by dissolving the underlying metal.

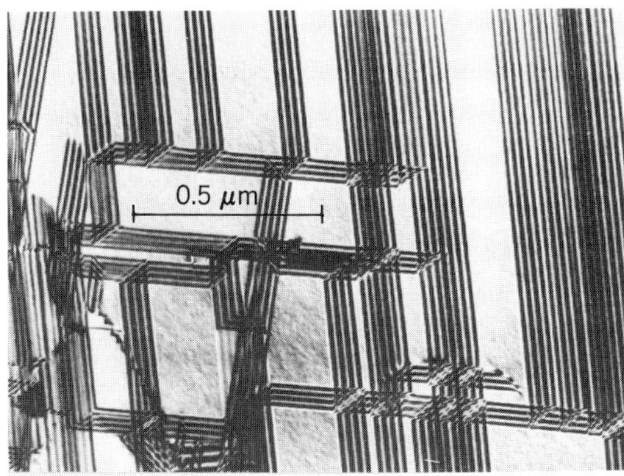

Fig. 5. Electron micrograph of a thin foil of a cobalt-base alloy showing structural imperfections resulting from a phase transformation. The narrow black and white bands are the images produced by planar imperfections in the sample. Many of these planar imperfections intersect one another; this observation has important implications in reference to the mechanical behavior of the alloy.

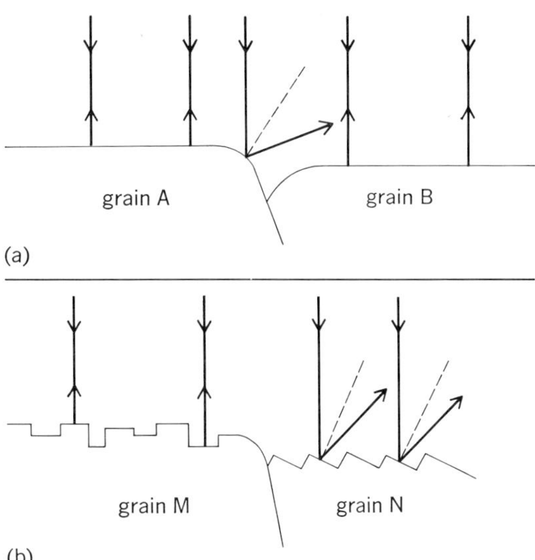

Fig. 6. Schematic diagrams normal to specimen surfaces showing the reflection of light by etched metal surfaces. (*a*) Preferentially etched grain boundary, dark in image. (*b*) Etched facets on two adjoining grains, with differences in light reflected.

The replica produced by these techniques is transferred to a fine screen or grid for support. In the case of most polymer replica techniques, the polymer is dissolved away, usually by acetone, leaving the carbon–heavy-metal film behind. In all replication procedures the specimen to be observed is transparent to electrons in its final state. The grid carrying the specimen is inserted in the specimen holder of the electron microscope.

The advantage obtained by preparing thin foils from the bulk, as compared to replicating the surface structure, is that microstructural features present in the volume can be directly investigated. The thin-foil specimens are prepared by cutting a slice from the material to be investigated. This slice is thinned, usually in the case of metals by chemical or electrolytic dissolution, until a perforation appears, and the ultra-thin fringe of this perforation is examined. Precautions must be taken to ensure that preparation of the thin foil does not alter the structure of the specimen; for example, mishandling can add dislocations to the specimen.

Electron microscopes can be equipped with cold stages and hot stages. Processes can be observed directly in specimens cooled or heated in the microscope. Similarly, specimens may be deformed in the instrument and observed during deformation. Tilting (goniometer) stages permit the diffraction condition to be varied and make possible a quantitative evaluation of the structure. For example, a grain boundary can be tilted to maximize its projected width in order to reveal its structure more fully.

In scanning electron microscopy (SEM), a finely focused beam of electrons is moved over the specimen surface in a sequence of closely spaced paths; the backscattered, or secondary, electrons are collected; and the resulting information is displayed on a cathode-ray tube (scanned synchronously with the electron beam) or otherwise recorded. This basic scheme can be modified in various ways. For instance, the x-rays generated by the electron beam during normal sample observation can be collected and a display produced consisting of x-rays generated by a specific element. In this way, the distribution of this element in the sample can be ascertained. The spatial resolution attainable in the SEM ranges from 1000 A (100 nm) to a few hundred angstroms (tens of nanometers) under best conditions. The two major advantages of the SEM over the optical microscope are greater depth of field and better resolution. The SEM is particularly well suited for the direct examination of rough surfaces such as those of fractured metals (microfractography). It is also employed in the investigation of the microstructure of metals and corrosion products.

Quantitative metallography. Metallographic methods can be used for the quantitative characterization of microstructures. For quantitative metallography, representative microsections are particularly important. The basic problem to be solved is the conversion of measurements on a two-dimensional section into quantities representing the three-dimensional structure of the material. The parameters used for "testing" the microstructure are test points, lines, and areas. The steps needed to carry out a quantitative analysis consist of defining a parameter which characterizes the structural property to be measured, establishing the relation between this parameter and the possible measurements available on the plane of polish, and determining the sampling error and the efficiency of the analysis.

The quantities most frequently measured by quantitative metallography are grain size, relative amounts of several phases in a multiphase alloy, total area of grain boundaries, particle size of a dispersed phase, and spacing of lamellae and dispersed particles. Increasing attention is being devoted to the quantitative characterization of the shapes of grains and dispersed particles.

The determination of grain size is one of the most frequently occurring problems of quantitative metallography. This determination is usually carried out by measuring the average length of intercepts between grain boundaries (average linear intercept) or by counting the number of grains enclosed in a known area at a known magnification. The grain size can be estimated by comparing the grain structure of a microsection with published standards.

The relative amounts of each of several phases present in a multiphase alloy can be measured by several equivalent procedures. The volume fractions are equal to the fractions of the areas occupied by each of the phases in a random section (area analysis); they are also equal to the fractions of superposed random lines or points lying in the various phases (linear analysis and point counting). The total area of internal boundaries (grain boundaries) in a given volume is equal to twice the number of intersections a random line of given length makes in a random section with the traces of the boundaries. The particle sizes of a phase dispersed in a matrix are more difficult to determine; the solution of this problem depends upon the shape of the particles and the distribution of particle sizes.

The classical technique for carrying out quantitative metallography has required direct involvement of the experimenter through the application of test figures, counting of intersections, tabulation of data, and statistical analysis. A minimum of automation was available. The advent of computers, however, has had a major impact on quantitative metallography through the introduction of image-analyzing computers. An image-analyzing computer uses a TV scanner to break an image into picture points. Each picture point is distinguished by its contrast and fed into a module which performs a specific image-analysis function, for instance, volume fraction or some other quantity mentioned above. The image-analyzing computer consists of a series of these mutually compatible modular units. Since the role of the experimenter now becomes the task of operating the instrument, a substantial saving in time, improvement in accuracy, and higher degree of reproducibility result.

[MICHAEL B. BEVER; JOHN B. VANDER SANDE]

Bibliography: R. M. Brick et al., *Structure and Properties of Engineering Materials*, 4th ed., 1979; T. Lyman (ed.), *Metals Handbook, Properties and Selection: Irons and Steels*, American Society of Metals, 1978; T. Lyman (ed.), *Metals Handbook, Properties and Selection: Nonferrous Alloys and Pure Metals*, American Society of Metals, 1978; J. L. McCall and P. M. French (eds.), *Metallography as a Quality Control Tool*, 1980.

Metallurgy

The technology and science of metallic materials. Metallurgy as a branch of engineering is concerned with the production of metals and alloys, their adaptation to use, and their performance in service. As a science, metallurgy is concerned with the chemical reactions involved in the processes by which metals are produced and the chemical, physical, and mechanical behavior of metallic materials.

Metallurgy has played an important role in the history of civilization. Metals were first produced more than 6000 years ago. Because only a few metals, principally gold, silver, copper, and meteoric iron, occur in the uncombined state in nature, and then only in small quantities, primitive metallurgists had to discover ways of extracting metals from their ores. Fairly large-scale production of some metals was carried out with technical competence in early Near Eastern and Mediterranean civilizations and in the Middle Ages in central and northern Europe. Basic metallurgical skills were also developed in other parts of the world.

The winning of metals would have been of little value without the ability to work them. Great craftsmanship in metalworking developed in early times: the objects produced included jewelry, large ornamental and ceremonial objects, tools, and weapons. It may be noted that almost all early materials and techniques that later had important useful applications were discovered and first used in the decorative arts. In the Middle Ages metalworking was in the hands of individual or groups of craftsmen. The scale and capabilities of metalworking developed with the growth of industrial organizations. Today's metallurgical plants supply metals and alloys to the manufacturing and construction industries in many forms, such as beams, plates, sheets, bars, wire, and castings. Rapidly developing technologies such as communications, nuclear power, and space exploration continue to demand new techniques of metal production and processing.

The field of metallurgy may be divided into process metallurgy (production metallurgy, extractive metallurgy) and physical metallurgy. According to another system of classification, metallurgy comprises chemical metallurgy, mechanical metallurgy (metal processing and mechanical behavior in service), and physical metallurgy. The more common division into process metallurgy and physical metallurgy, which is adopted here, classifies metal processing as a part of process metallurgy and the mechanical behavior of metals as a part of physical metallurgy.

Process metallurgy. Process metallurgy, the science and technology used in the production of metals, employs some of the same unit operations and unit processes as chemical engineering. These operations and processes are carried out with ores, concentrates, scrap metals, fuels, fluxes, slags, solvents, and electrolytes. Different metals require different combinations of operations and processes, but typically the production of a metal involves two major steps. The first is the production of an impure metal from ore minerals, commonly oxides or sulfides, and the second is the refining of the reduced impure metal, for example, by selective oxidation of impurities or by electrolysis. Process metallurgy is continually challenged by the demand for metals which have not been produced previously or are difficult to produce; by the depletion of the deposits of the richer and more easily processed ores of the traditional metals; and by the need for metals of greater purity and higher quality. The mining of leaner ores has greatly enhanced the importance of ore dressing methods for enriching raw materials for metal production. Several nonferrous metals are commonly produced from concentrates. Iron ores are also increasingly treated by ore dressing. *See* ELECTROMETALLURGY; HYDROMETALLURGY; PYROMETALLURGY; PYROMETALLURGY, NONFERROUS.

Process metallurgy today mainly involves large-scale operations. A single blast furnace produces crude iron at the rate of 3000–11,000 tons per day. A basic oxygen furnace for steelmaking consumes 800 tons of pure oxygen together with required amounts of crude iron and scrap to produce 12,000 tons of steel per day. Advanced methods of process analysis and control are now being applied to such processing systems. The application of vacuum to extraction and refining processes, the leaching of low-grade ores for the extraction of metals, the use of electrochemical reduction cells, and the refining of reactive metals by processing through the vapor state are other important developments. *See* STEEL MANUFACTURE.

Because the production of metals employs many different chemical reactions, process metallurgy has been closely associated with inorganic chemistry. Techniques for analyzing ores and metallurgical products originated several centuries ago and represented an early stage of analytical chemistry. Application of physical chemistry to equilibria and kinetics of metallurgical reactions has led to great progress in metallurgical chemistry.

Physical metallurgy. Physical metallurgy investigates the effects of composition and treatment on the structure of metals and the relations of the structure to the properties of metals. Physical metallurgy is also concerned with the engineering applications of scientific principles to the fabrication, mechanical treatment, heat treatment, and service behavior of metals. *See* ALLOY; HEAT TREATMENT (METALLURGY).

The structure of metals consists of their crystal structure, which is investigated by x-ray, electron, and neutron diffraction, their microstructure, which is the subject of metallography, and their macrostructure. Crystal imperfections provide mechanisms for processes occurring in solid metals; for example, the movement of dislocations results in plastic deformation. Crystal imperfections are investigated by x-ray diffraction and metallographic methods, especially electron microscopy. The microstructure is determined by the constituent phases and the geometrical arrangement of the microcrystals (grains) formed by those phases. Macrostructure is important in industrial metals. It involves chemical and physical inhomogeneities on a scale larger than microscopic. Examples are flow lines in steel forgings and blowholes in castings. *See* METALLOGRAPHY.

Phase transformations occurring in the solid state underlie many heat-treatment operations. The thermodynamics and kinetics of these transformations are a major concern of physical metallurgy. Physical metallurgy also investigates changes in the structure and properties resulting from mechanical working of metals. *See* PLASTIC DEFORMATION OF METAL.

The composition of metallic objects is often characterized by the presence of impurities, nonuniform distribution (segregation) of solute elements and nonmetallic inclusions, especially in steel. These phenomena, which originate in the production process, can have important effects on the properties of metals and alloys. They illustrate the close relation between process and physical metallurgy.

The applications for which a metal is intended determine the properties that are of practical interest. For use in machinery and construction, mechanical properties including deformation and fracture behavior are of greatest importance. In transportation equipment the strength-to-weight ratio deserves special consideration. In other applications, electrical or thermal conductivity or magnetic properties may be decisive. Resistance to corrosion is a common requirement and accounts for another close link between metallurgy and chemistry, especially electrochemistry and surface chemistry. *See* CORROSION.

General principles. An old distinction between ferrous and nonferrous metallurgy, although still of some practical significance, is no longer considered fundamental. Certain general principles have become established and apply to all metals. In some respects, however, there are great differences between metals, which has led to specialization in research and industrial practice. On the other hand, an underlying structural science is developing to bring all materials—metallic and other inorganic as well as organic materials—within a unified framework.

Metallurgy occupies a position at the juncture of physics, chemistry, mechanical engineering, and chemical engineering. It also borders electrical, civil, aeronautical, and nuclear engineering. Metallurgical knowledge can be relevant to fields that are as removed from engineering as archeology, crime investigation, and orthopedic surgery. In the newly emerging field of materials science and engineering, metallurgy takes its place as one of the oldest and most highly developed disciplines.

The area of concern of metallurgy has widened in recent years. Problems of materials availability are becoming recognized and are drawing attention to the need for new production processes and the recycling of secondary metals. Environmental considerations require new technology for pollution abatement. Energy conservation favors more efficient processes in metals production. Conservation of materials also calls for more effective utilization of metals and the substitution of more plentiful metals for scarce ones. *See* ELECTROPLATING OF METALS; METAL COATINGS; METAL FORMING.

[MICHAEL B. BEVER]

Bibliography: C. S. Barrett and T. B. Massalski, *Structure of Metals*, 3d rev. ed., 1980; R. W. Cahn (ed.), *Physical Metallurgy*, 2d rev. ed., 1971; G. F. Carter, *Principles of Physical and Chemical Metallurgy*, 1979; A. H. Cottrell, *Introduction to Metallurgy*, 2d ed., 1975; G. E. Dieter, *Mechanical Metallurgy*, 2d ed., 1976; R. E. Reed-Hill, *Physical Metallurgy Principles*, 2d ed., 1973.

Methods engineering

A technique used by progressive management to improve productivity and reduce costs in both direct and indirect operations of manufacturing and nonmanufacturing business organizations. Methods engineering is applicable in any enterprise wherever human effort is required. It can be defined as the systematic procedure for subjecting all direct and indirect operations to close scrutiny in order to introduce improvements that will make work easier to perform and will allow work to be done smoother in less time, and with less energy, effort, and fatigue, with less investment per unit. The ultimate objective of methods engineering is profit improvement. *See* OPERATIONS RESEARCH; PRODUCTIVITY.

Activities. Methods engineering includes five activities: planning, methods study, standardization, work measurement, and controls (Fig. 1). Methods engineering, through planning, first identifies the amount of time that should be spent on a project so as to get as much of the potential savings as is practical. Invariably the most profitable jobs to study are those with the most repetition, the highest labor content (human work as distinguished from mechanical or process work), the highest labor cost, or the longest life-span. Next, through methods study, methods are improved by observing what is currently being done and then by developing better ways of doing it. After ideal methods have been devised, standardization of equipment, methods, and working conditions takes place. The standardization phase includes the training of the operator to follow the standard method. After all this is done, the number of standard hours in which operators working with standard performances can do their job is determined by measurement. Finally, the established method is periodically audited, and various management controls are adjusted with the new time data. The system may include a plan for compensating labor that encourages attaining or surpassing a standard performance. *See* WAGE INCENTIVES.

Methods study. Methods engineering entails analysis work at two different times. This analysis work is termed methods study. Initially, the methods study should be used in the development of the best way to perform a job before it has been done, and second, the procedure should be used to improve the way work is being accomplished. Thus methods engineering may be thought of as a continuing activity to assure the economic health of the business or enterprise. The more thorough the methods study made during the preproduction stages, the less the necessity for additional detailed methods studies during the life of the product.

The first step in the planning stage of methods engineering is to get and present the facts. If a new job, not yet in production, is being studied, a practical method of doing the work is visualized. Then better ways of accomplishing the same objective are sought. The procedure is one of starting with an existing method, either actual or visualized, analyzing it so that it is clearly understood in detail, and then searching for better ways of doing parts or all of the work.

In getting the facts for planning the methods study, the use of work sampling is recommended. It is always advisable to record the facts in a form that facilitates the primary approaches to operations analysis as used in methods study. The four process charts that are most used in presenting the

Events recorded on a process chart

Symbol	Event
●	*Operation.* An operation occurs when an object is intentionally changed in any of its physical or chemical characteristics, is assembled or disassembled from another object, or is arranged or prepared for another operation, transportation, inspection, or storage. An operation also occurs when information is given or received, or when planning or calculating takes place.
➡	*Transportation.* A transportation occurs when an object is moved from one place to another, except when such movements are a part of the operation or are caused by the operator at the work station during an operation or an inspection.
■	*Inspection.* An inspection occurs when an object is examined for identification or is verified for quality or quantity in any of its characteristics.
D	*Delay.* A delay occurs to an object when conditions except those which intentionally change the physical or chemical characteristics of the object do not permit or require immediate performance of the next planned action.
▼	*Storage.* A storage occurs when an object is kept and protected against unauthorized removal.

facts are the operation process chart, flow process chart, multiple activity process chart, and operator process chart.

Process chart analysis. A process chart is a graphic representation of events occurring during a series of actions or operations. It is useful for obtaining an initial understanding of a process. Process charts present a clear picture of a given process and often reveal unnecessary work or duplication of effort. The operation process chart and flow process chart are used for the analysis of processes involving a number of events or operations. The multiple activity process chart and operator process chart are used to analyze single operations in detail. The events which are recognized in charting a process and the symbols for them are shown in the table.

An operation process chart represents the points at which materials are introduced into the process and the sequence of inspections and all operations except material handling (Fig. 2). It is particularly useful in connection with plant layout studies.

A flow process chart represents the sequence of all operations, transportations, inspections, delays, and storages occurring during a process or procedure, and includes information considered desirable for analysis, such as time required and distance moved. It is used to follow either material or workers through a process and is an essential tool for material handling studies. Figure 3 shows a flow process chart of the material type.

Operations analysis. There are 10 major points of operation analysis that are applied in methods study. Listed in the order in which they usually are considered, they are: purpose of operation, design of part, tolerances and specifications, material, process of manufacture, setup and tools, material handling, plant layout, working conditions, and motion-and-time study.

Where work has not previously been studied, methods engineers repeatedly find that much of what is being done is entirely unnecessary. Sometimes whole operations are eliminated merely by asking why something is being done, and recognizing that there is no sound answer. The primary technique used to discover such situations is through the analysis of the process charts, giving consideration to the methods study approach of purpose of operation.

The methods analyst studies all major factors which affect a given operation, then conscientiously and open-mindedly applies the questioning attitude and asks why, where, what, when, who, and how. If this is done systematically, worthwhile improvements are almost certain to result.

When the methods analyst considers the design of the part, the objective is to discover manufacturing economics which the design engineer may have overlooked. A consideration of tolerances and specifications will often suggest better ways of obtaining necessary quality and product reliability. Questioning the materials of which the product is made and the way the supply materials are used in performing the operations required to produce the product will frequently lead to economies. There are usually many different ways to perform a given operation. A process that is best today may not be best tomorrow because of improved technology.

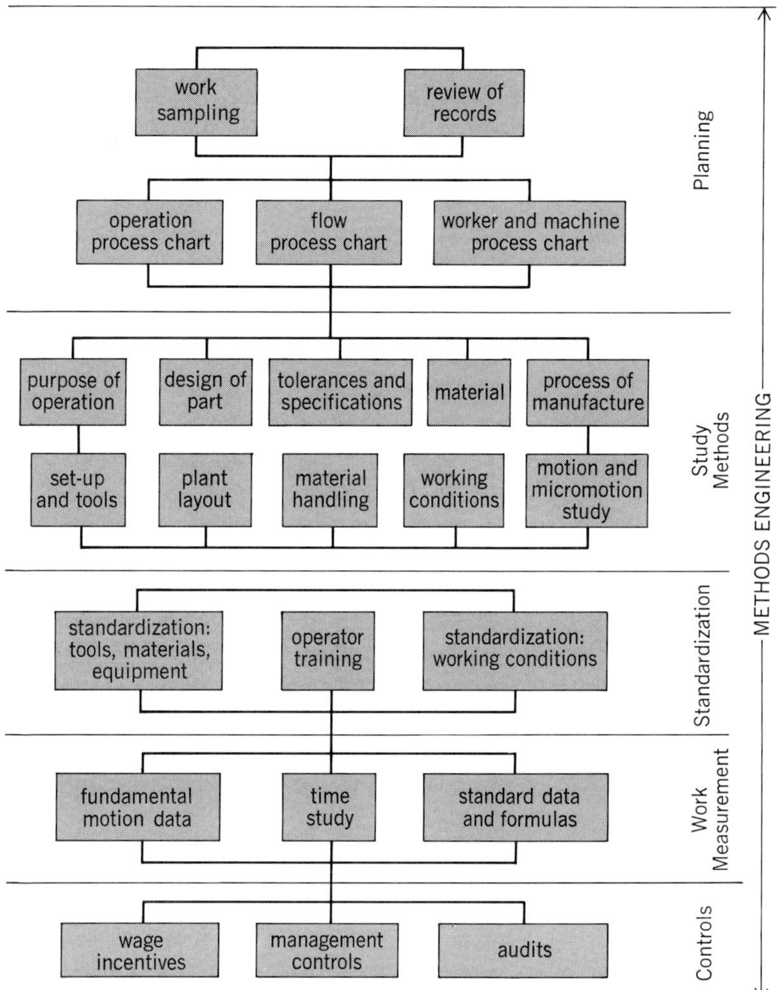

Fig. 1. Graphic analysis of elements of methods engineering.

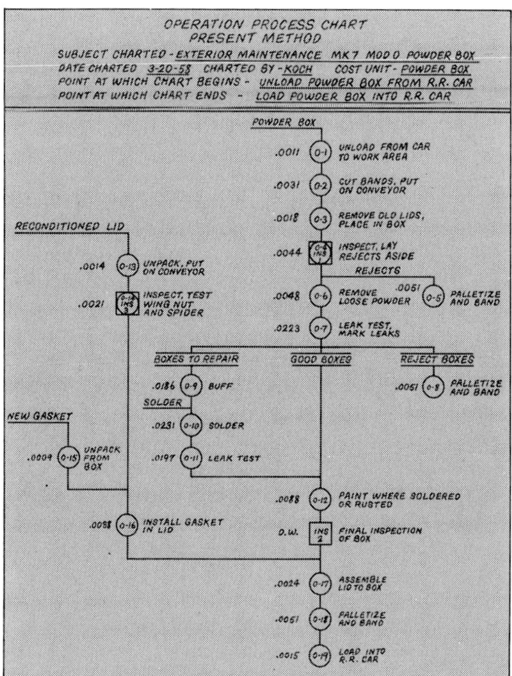

Fig. 2. Typical operation process chart.

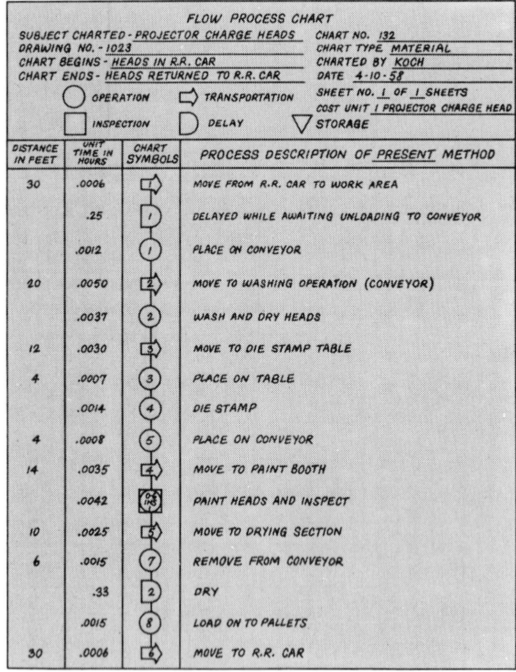

FLOW PROCESS CHART

SUBJECT CHARTED - PROJECTOR CHARGE HEADS | CHART NO. 132
DRAWING NO. - 1023 | CHART TYPE MATERIAL
CHART BEGINS - HEADS IN R.R. CAR | CHARTED BY KOCH
CHART ENDS - HEADS RETURNED TO R.R. CAR | DATE 4-10-58
SHEET NO. 1 OF 1 SHEETS
COST UNIT 1 PROJECTOR CHARGE HEAD

OPERATION; TRANSPORTATION; INSPECTION; DELAY; STORAGE

DISTANCE IN FEET	UNIT TIME IN HOURS	CHART SYMBOLS	PROCESS DESCRIPTION OF PRESENT METHOD
30	.0006	1	MOVE FROM R.R. CAR TO WORK AREA
	.25	1	DELAYED WHILE AWAITING UNLOADING TO CONVEYOR
	.0012	1	PLACE ON CONVEYOR
20	.0050	2	MOVE TO WASHING OPERATION (CONVEYOR)
	.0037	2	WASH AND DRY HEADS
12	.0030	3	MOVE TO DIE STAMP TABLE
4	.0007	3	PLACE ON TABLE
	.0014	4	DIE STAMP
4	.0008	5	PLACE ON CONVEYOR
14	.0035	4	MOVE TO PAINT BOOTH
	.0042		PAINT HEADS AND INSPECT
10	.0025	5	MOVE TO DRYING SECTION
6	.0015	7	REMOVE FROM CONVEYOR
	.33	2	DRY
	.0015	8	LOAD ON TO PALLETS
30	.0006	6	MOVE TO R.R. CAR

Fig. 3. Typical flow process chart, material type.

By questioning the process of manufacture, the methods analyst will consider all alternative processes to ascertain if the best ones are currently being utilized. Just as there are usually several alternative processes to be considered, so there are alternative tooling and setup procedures. The larger the quantity to be produced, the more advanced should be the tooling. As quantity requirements change, existing setup and tooling may become inefficient. Consequently, the methods analyst should always consider this side of operations analysis. *See* ASSEMBLY METHODS; PRODUCTION ENGINEERING.

The seventh consideration, material handling, should always be given serious attention. Within a factory or other enterprise, handling of material adds to the cost of a product without adding to its salable value. Therefore material handling should be eliminated or minimized. The operation and flow process charts are especially useful to give a clear picture of present material handling methods and to point the way to new and improved ones.

The physical layout is an important element of an entire production system. The method analyst will strive to have an effective layout that permits the manufacture of the desired number of products of the desired quality at the least cost. Work stations and facilities, including service facilities, should be arranged to permit the least travel and the most efficient processing of a product with a minimum of handling. *See* PLANT FACILITIES (INDUSTRY).

Analysis of the working conditions of the immediate environment of each work station can be another source of savings. The methods analyst will study the working conditions of each work center shown on the operation process chart in connection with the following common possibilities for job improvement: installing gravity delivery chutes; using drop delivery; comparing methods if more than one operator is working on the job; providing the correct chair for each operator; improving jigs or fixtures by providing ejectors and quick-acting clamps; using foot-operated mechanisms; arranging for two-handed operation; arranging tools and parts within normal working area; changing layout to eliminate backtracking and to permit coupling of machines; and providing correct light, ventilation, temperature, and safety features for the operator. *See* HUMAN-FACTORS ENGINEERING; INDUSTRIAL HEALTH AND SAFETY.

The last of the 10 primary approaches, motion study, involves the study of both manual and eye movements that occur in an operation or work cycle in order to eliminate wasted movements and establish a better sequence and coordination of movements.

In making a motion study, the methods engineer studies each individual motion in detail and tries to shorten it, combine it with others, or eliminate it altogether. Ineffective motions are identified and replaced, if possible, with effective motions. As a result of this intensive study, easier, quicker ways of doing the work are developed.

Motions may be studied individually by direct observation, or they may be observed and timed in groups during the making of a stopwatch time study. Time-study data are useful when studying the relationship between the work done by an employee or groups of employees and the work done by a machine or groups of machines. When these data are properly arranged on a multiple-activity process chart, possibilities for improvement are often revealed.

Effective motion studies can be made by observation with the aid of one of the predetermined elemental time systems. As a tool of motion study, these systems are similar in ease of application to motion study by observation and, in the hands of one trained in their use, approach the results obtained by micromotion study in developing improved methods. *See* WORK MEASUREMENT.

When more intensive motion study is desirable, the methods engineer can turn to micromotion study. A video tape or motion picture of the operation under study is taken, and the resulting film is analyzed frame by frame. When the results of the analysis are recorded on an operator process chart, a clear and permanent record of the method is obtained. Motions are classified into therbligs, also called Gilbreth basic elements or basic divisions of accomplishment. The therbligs (Gilbreth spelled backward) generally recognized are: transport empty or reach; transport loaded or move; grasp; position; disengage; release; do; examine; change direction; pre-position; search; select; plan; balancing delay; hold; avoidable delay; unavoidable delay; and rest to overcome fatigue. The first eight, in general, accomplish necessary work; the next six retard accomplishment; and the last four do not accomplish. The last two groups should be eliminated when possible.

Operator training and standards. Once the best possible method has been developed through methods study, then standardization of the tools, equipment, materials, and working conditions should take place so as to assure the proposed method is followed at all applicable work centers. Highly important in the standardization process is the training of the operator who is to do the work. Depending upon the nature and importance of the job, training can consist of verbal instructions, demonstrations at or away from the workplace, written instruction cards, visual guidance with video tape or motion pictures, or a combination of these and still other training techniques. In some cases, attitude training to develop "want-to" accompanies training in "know-how." *See* OPERATOR TRAINING.

After the method has been made as efficient as is economically justified and after standardization has been accomplished, the job is ready for work measurement. When stopwatch time study is to be used, the methods engineer selects a suitable operator, explains the purpose of the study, and makes observations, usually rating the performance of the operator by judging the skill and effort exhibited or by assessing the speed with which motions are made as compared to a normal working pace. The methods engineer makes the necessary calculations, adds allowances—usually determined from separate work sampling studies—for fatigue and personal, unavoidable, and special delays, and establishes a standard time for doing the work.

If a system of predetermined elemental times has been used during the methods study, a list of the motions required to perform the operation will have been developed. The methods engineer can then measure the work merely by assigning predetermined time values to each motion. When these are totaled and the necessary allowances are added, the standard time is obtained without need for further observation or measurement.

On nonrepetitive work, the cost of establishing standards by individual studies is usually prohibitive. In such cases, the methods engineer develops the time standard from standard data or time formulas. With these compilations of detailed time data as a table, chart, or algebraic formula, the methods engineer only has to identify major variables affecting performance time, such as size, weight, or the like, and select appropriate data from the tables or substitute in the formula to establish a time standard. Although the cost of developing a time formula may be substantial, once developed, the job of establishing accurate and consistent standards becomes routine, often being accomplished in less than 5 min per standard. Time formulas are particularly useful in the measurement and control of indirect labor.

When time standards are available, they may be used for control purposes in connection with production planning and control, standard costs, budgetary control, top management controls, worker-shift controls, machine loading, and operator performance control. They may further be used as the basis for incentive wage payment.

In summary, the procedures shown in Fig. 1 grouped under the category of methods study are used to find the best way to do a job. The procedures grouped under standardization, work measurement, and controls are employed to get the operator to use the best method, once it has been found. *See* INDUSTRIAL ENGINEERING.

[BENJAMIN W. NIEBEL]

Micrometer

A precision instrument used to measure small distances and angles. A common use is on a machinist's caliper, as in the illustration. *See* CALIPER.

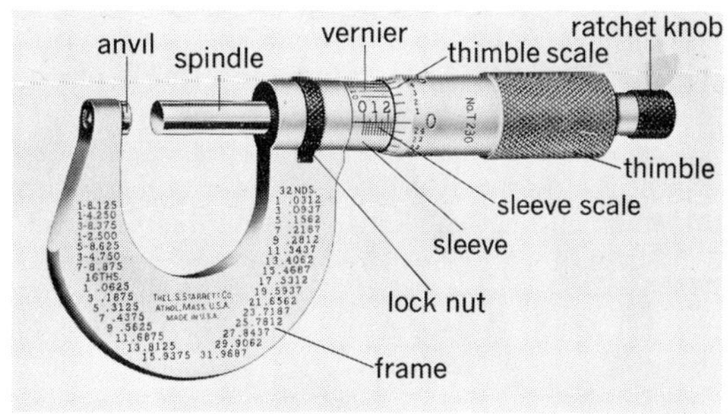

Machinist's outside caliper with micrometer reading 0.250 in. (*L. S. Starrett Co.*)

The spindle of the caliper is an accurately machined screw, which is rotated by the thimble or the ratchet knob until the object to be measured is in contact with both spindle and anvil. The ratchet slips after correct pressure is applied, ensuring consistent, accurate gaging. The number 1 on the sleeve represents 0.1 in.; the smallest divisions are 0.025 in. The thimble makes one complete turn for each 0.025 in. on the sleeve, and the 25 divisions on the thimble allow reading to the nearest 0.001 in. A vernier scale allows accurate reading to 0.0001 in. [FRANK H. ROCKETT]

Milling machine

A machine for the removal of metal by feeding a workpiece through the periphery of a rotating circular cutter. The multitoothed cutter of a milling machine produces a milled surface as each revolving tooth removes a portion of metal from the passing workpiece.

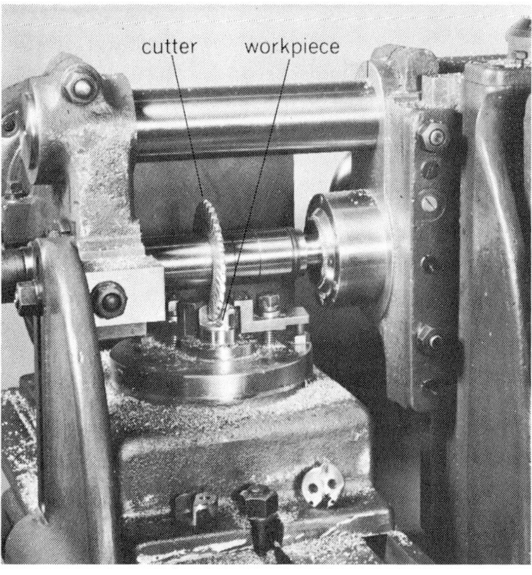

Part of a horizontal milling machine. (*Brown and Sharpe Manufacturing Co.*)

Milling machines may generally be classed as horizontal or vertical, depending on the power spindle's axis of rotation. A movable, horizontal table holds the workpiece and carries it through the path of the cutter teeth (see illustration). On the universal mill, the table may also be rotated horizontally; on machines with a universal head, the cutter may be tilted to the desired angle. Milling machines of considerable size may be made for special applications.

Milling cutters may be made from a single piece of steel or they may be composed of a body with attached cutting elements of a harder material. Various cutters are designed to do specific types of milling operations. *See* MACHINING OPERATIONS.

[ALAN H. TUTTLE]

Mining

The taking of minerals from the earth, including production from surface waters and from wells. Usually the oil and gas industries are regarded as separate from the mining industry. The term mining industry commonly includes such functions as exploration, mineral separation, hydrometallurgy, electrolytic reduction, and smelting and refining, even though these are not actually mining operations. *See* METALLURGY.

The use of mineral materials, and thus mining, dates back to the earliest stages of man's history, as shown by artifacts of stone and pottery and gold ornaments. The products of mining are not only basic to communal living as construction, mechanical, and raw materials, but salt is necessary to life itself, and the fertilizer minerals are required to feed a populous world.

Methods. Mining is broadly divided into three basic methods: opencast, underground, and fluid mining. Opencast, or surface mining, is done either from pits or gouged-out slopes or by strip mining, which involves extraction from a series of successive parallel trenches. Dredging is a type of strip mining, with digging done from barges. Hydraulic mining uses jets of water to excavate material.

Underground mining involves extraction from beneath the surface, from depths as great as 10,000 ft (3 km), by any of several methods.

Fluid mining is extraction from natural brines, lakes, oceans, or underground waters; from solutions made by dissolving underground materials and pumping to the surface; from underground oil or gas pools; by melting underground material with hot water and pumping to the surface; or by driving material from well to well by gas drive, water drive, or combustion. Most fluid mining is done by wells. A recent type of well mining, still experimental, is to wash insoluble material loose by underground jets and pump the slurry to the surface. *See* OPEN-PIT MINING; PLACER MINING; PETROLEUM ENGINEERING; STRIP MINING; UNDERGROUND MINING.

The activities of the mining industry begin with exploration which, since mankind can no longer depend on accidental discoveries or surficially exposed deposits, has become a complicated, expensive, and highly technical task. After suitable deposits have been found and their worth proved, development, or preparation for mining, is necessary. For opencast mining this involves stripping off overburden; and for underground mining the sinking of shafts, driving of adits and various other underground openings, and providing for drainage and ventilation. For mining by wells drilling must be done. For all these cases equipment must be provided for such purposes as blasthole drilling, blasting, loading, transporting, hoisting, power transmission, pumping, ventilation, storage, or casing and connecting wells. Mines may ship their crude products directly to reduction plants, refiners, or consumers, but commonly concentrating mills are provided to separate useful from useless (gangue) minerals. *See* PROSPECTING.

Economics. Mining is done by hand in places where labor is cheap, but in the more industrialized countries it is a highly mechanized operation. Some surface mines use the largest and most expensive machines ever developed—unless a large ship can be called a machine.

There are many small- and medium-sized mines but also a growing number of large ones. The trend to larger mines and particularly to large opencast operations is due to the great demand for mineral products, depletion of high-grade reserves, technological progress, and the need for economies of scale in mining low-grade deposits. The largest open-pit coal mine moves 415,000 tons (375,000 metric tons) of material per day and the largest metal mine, 300,000 tons (270,000 metric tons). The largest underground coal mine produces 17,000 tons (15,000 metric tons). of coal per day, and the largest metal mine 150,000 tons (135,000 metric tons) of ore.

The quality of deposits which can be mined economically depends on the market value of the contained valuable minerals, on the costs of mining and treatment, and on location. Alluvial gold gravels may run as low as 1/350 oz of gold/yd^3 (0.1 g/m^3) and gold from lodes as low as 1/10 oz/ton (3 g per metric ton). Uranium ore containing 1.5 lb of uranium oxide per ton (0.75 kg per metric ton) is mined underground, and copper-molybdenum ore as low grade as 0.35% copper and 0.05% molybdenum is taken from open pits. On the other hand, iron ore is rarely mined below 25% iron, aluminum ore below 30% aluminum, and coal less than 90% pure. Crude petroleum is usually over 95% pure.

Use of natural resources. A unique feature of mining is the circumstance that mineral deposits undergoing extraction are "wasting assets," meaning that they are not renewable as are other natural resources. This depletability of mineral deposits not only requires that mining companies must periodically find new deposits and constantly improve their technology in order to stay in business, but calls for conservational, industrial, and political policies to serve the public interest. Depletion means that the supplies of any particular mineral, except those derived from oceanic brine, must be drawn from ever-lower-grade sources. Consciousness of depletion causes many countries to be possessive about their mineral resources and jealous of their exploitation by foreigners. Depletion also accounts for some controversial attitudes toward conservation. Some observers would reduce the scale of domestic production and increase imports in order to extend the lives of domestic deposits. Their opponents argue that encouragement of mining through tariff protection, subsidies, or import quotas is desirable, on the grounds that only a dynamic industry in being can develop the means of mining low-grade deposits or meet the needs of a national emergency. They point out that protection encourages the extraction of marginal resources that would otherwise be condemned through abandonment.

Despite its essential nature, mining today is being constrained as to where it can operate by wilderness lovers and by rapidly expanding urbanization. Concern over pollution, some of which is caused by mine water, mining wastes, and smelter effluents, has grown rapidly. More and more objection is developing to defacement of landscapes by surface mining, and many states now require restoration of the surface to a cultivable or forested condition after strip mining. People in residential areas usually resist the development of industries near their homes, particularly if such industries employ blasting, produce smoke or fumes, or cause traffic by large vehicles.

In countries with Anglo-Saxon traditions, title to minerals on private lands is vested in the private owners, but in many countries minerals are state property. Where minerals are privately owned, mining is commonly done by purchase of title or under lease and royalty contracts. State-owned minerals are mined under claims acquired through discovery and denouncement, leases, or concessions. For centuries governments have found it desirable to encourage prospecting and mining to lure the adventurous into these challenging, useful arts.

[EVAN JUST]

Bibliography: Engineering Journal Magazine, *Operating Handbook of Mineral Underground Mining*, vol. 3, 1979; *Eng. Min. J.*, issued monthly; I. A. Given et al., *SME Mining Engineering Handbook*, 1973; R. B. Lewis and G. B. Clark, *Elements of Mining*, 3d ed., 1964; Mining Information Services, *Mining Methods and Equipment*, 1980; T. A. Rickard, *Man and Metals*, 2 vols., 1932, reprint 1974; G. Robson, *Economics of Mineral Engineering*, 1977; Society of Mining Engineers, *Surface Mining*, 1968; U.S. Bureau of Mines, *Mineral Facts and Problems*, no. 667, 1976; W. A. Vogely et al., *Economics of the Mineral Industries*, 3d ed., 1976.

Mining excavation

In mining for coal, metallic, and nonmetallic minerals, the process of removing minerals from the Earth. Excavation consists of fragmentation (or in special cases solution) of minerals from their solid state, loading them, and transporting them to the surface. Fragmentation is accomplished by the use of explosives or mechanical means, the former being most commonly applied. Excavation is also involved in establishing mine entries and other development workings in waste rock for access to the minerals. Mechanical fragmentation by means of boring machines (Figs. 1–3) is being introduced at some mines for this purpose.

Hard rock is generally broken with explosives to attain fragmentation. Some moderately soft deposits, such as coal, potash, and borax, are fragmented mechanically by machines without the use of explosives. When fragmentation is by machine, the loading device is commonly an integral part of the machine. In special cases (sulfur, salt, and potash) the mineral may be excavated by solution. When solution mining is used, the mineral-bearing solution is pumped to the surface; thus loading and transport become integral parts of the process.

Explosives. Two general classes of explosives are used for mining, black powder and high explosives or dynamites. Black powder is not used underground, where commercial high explosives commonly known as dynamites are preferred. A special class of explosive, designated "permissible" after being tested and passed by the United States Bureau of Mines, is required in gaseous and dusty coal mines. Permissible explosives are especially designed to produce a flame of small volume, short duration, and low temperature. Dynamites are typed according to properties and further subdivided into grades based on strength. The principal properties are strength, density, sensitiveness, velocity, water resistance, freezing resistance, and fume products of detonation. Strength refers to the energy content of an explosive, and is based on the percent by weight of nitroglycerin or equivalent energy when other strength-imparting ingredients are substituted for nitroglycerin.

High explosives are detonated with blasting caps, small metal tubes closed at one end and charged with a highly heat-sensitive explosive.

Blasting agents not technically classified as explosives are increasingly being used for blasting,

Fig. 1. Raise boring machine (drills pilot holes between levels) in Idaho silver-lead mine. (*Hecla Mining Co.*)

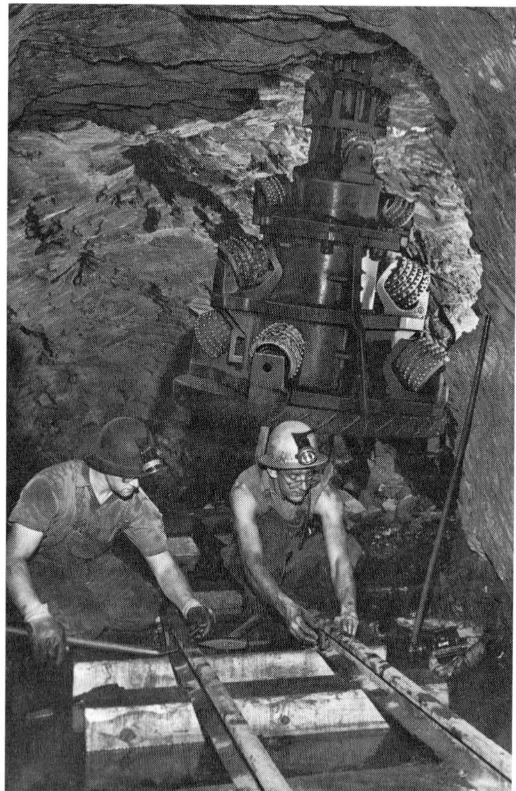

Fig. 2. Raise boring machine set to ream pilot hole to finished dimension of raise. (*Homestake Mining Co.*)

both in surface mines and underground. They are non-cap-sensitive materials or mixtures in which none of the ingredients are classed as explosives but which can be detonated by a high-explosive primer for blasting purposes. A commonly used blasting agent of this type is a mixture of fertilizer-grade ammonium nitrate and fuel oil (ANFO) that has found wide acceptance because of its safety features and low cost.

Fragmentation. Fragmentation is the process of breaking ground with explosives and machines.

Explosives fragmentation. The object of explosives fragmentation is to break the minerals and produce fragments of a size best suited for handling. It is cheaper and generally desirable to break waste rock in coarse sizes. Customarily, coal has been produced as lumps, but now the trend is toward using fine sizes for more economical handling and combustion. Fine sizes are desirable for most metallic and some nonmetallic ores that are processed after mining. Some nonmetallic minerals can be marketed more profitably in medium coarse sizes (1/4 to 2 in. or 6 to 50 mm), so fine material is wasted and must be minimized. The size of mineral fragments can be controlled partly by the amount and strength of the explosive used in blasting and somewhat by the spacing between shot holes and by their depth.

An assemblage of shot holes drilled into the face of a stope, drift, crosscut, shaft, raise, winze, adit, or tunnel and blasted at one time is a "round" (Fig. 4). The pattern of the round contributes to the effectiveness of a blast. A wide variety of patterns is used, depending upon the character of the material to be broken, the size and shape of the desired opening, and the desired size of fragments. These rounds have been given names such as pyramid, triangle, V, and fan, suggestive of the pattern of the drill holes that are blasted to give the initial cut in the center of the face. Others, such as Norwegian and Michigan, have been named according to place of origin. Michigan is also called burned cut because the holes which are drilled close together in a group near the center of the face, when blasted, pulverize the rock and discharge it at a high velocity. *See* BORING AND DRILLING (MINERAL).

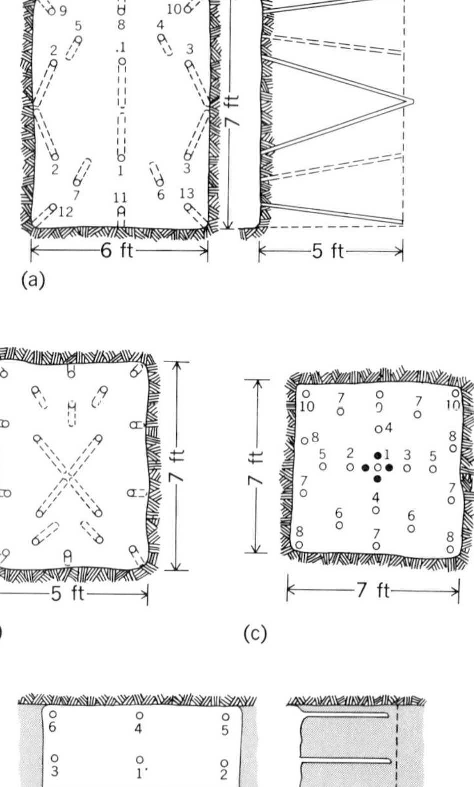

Fig. 4. Typical rounds for explosives fragmentation. (*a*) Horizontal V cut. (*b*) Four-hole pyramid cut. (*c*) Five-hole burn cut. (*d*) Undercut coal face. 1 ft = 0.3 m.

Fig. 3. Tunnel boring machine used in mining. (*Robbins Company*)

To break a round, each hole is charged with explosives. The explosives are in the form of cylindrical cartridges, or in the case of blasting agents, may be in bulk. Cartridges are tamped into place in the drill hole and explosives in bulk form are blown into the hole pneumatically. A detonating cap with attached safety fuse or electric wire is inserted into a cartridge and placed in each hole to form a primer. The round is then blasted by igniting the fuse or sending an electric current through the wiring. The sequence in which holes of a round are blasted is important. Usually a group of holes near the center or along one side is blasted first. Then successive groups of holes are blasted in series. Crude timing between successive blasts is attained by trimming fuses to different lengths. Precision timing is possible with electric blasting caps constructed to vary in detonation by a few thousandths of a second.

Mechanical fragmentation. This method is used primarily in coal mining. The principal machines are the continuous miner, auger, coal plow, and coal cutter. The continuous miner is manufactured in several sizes by both domestic and foreign companies. There are many variations in detail to meet varying conditions in coal seams.

Fig. 5. A continuous miner, finishing cut as it rips overlapping paths in coal seam 18 in. (46 cm) deep by 42 in. (107 cm) wide by 90 in. (229 cm) high. Coal is discharged into waiting shuttle. 20 ft = 6.1 m. (*U.S. Bureau of Mines, and Joy Manufacturing*)

One representative model is shown in Fig. 5. The coal is fragmented by a front-end cutting head comprising a number of continuously revolving chains upon which are mounted hard-metal-tipped cutters or picks. The chains are mounted on a bar which can be rotated horizontally to cut an entry 12–20 ft (3.7–6.1 m) wide and vertically to cut from 6 in. (15 cm) below the bottom of the machine to 7 or 8 ft (2.1–2.4 m) above. The cutter-bar frame is mounted on a carriage which in turn is mounted on caterpillar treads. The broken coal is collected by a conveyor which transports it from the cutting head to the transportation system at the rear of the machine.

Augers up to 5 ft (1.5 m) in diameter (Fig. 6) are used under high banks of abandoned strip mines to recover coal to a boring depth of about 200 ft (60 m). The auger is made in sections, the first one having a hard-metal-faced cutting tool that does the cutting as the auger is rotated. The force providing thrust and rotation to the auger is supplied by a diesel engine at the surface. Some machines are self-propelling and handle and store the auger sections mechanically.

Fig. 6. View from spoil bank showing operation of auger mining and coal being loaded into truck in the Pittsburgh coal bed, West Virginia. (*U.S. Bureau of Mines*)

The coal plow or planer was developed in Germany to mine coal by the longwall system. The plow is drawn back and forth along the face and cuts or slabs off the coal in a series of slices. The planer unit comprises (1) an armored double-chain (Panzer) conveyor which rests on the mine floor and extends the entire length of the longwall face, (2) an electric driving unit at each end of the conveyor, (3) the planer assembly, and (4) pneumatic conveyor shifters. The electric motors move the plow by means of a rotating drum and chain attached to the planer. Broken coal falls to the conveyor, which moves it to entries adjacent to the mining areas. The pneumatic shifters move the conveyor and the plow toward the face as mining advances.

Where coal is fragmented by explosives, it is first undercut by a coal-cutting machine along the bottom of the seam. It either falls by gravity (undercutting and mass falling), or is broken by permissible explosives or other substitutes, shot holes are placed to take advantage of such natural features as the presence of hard and soft bands and the direction of cleavage planes in the coal.

Three classes of machines are used for cutting coal: disk, bar, and chain, classified by the method employed for holding the cutting tools or cutter picks.

Fig. 7. Electrically powered chain coal cutter making shear cut in coal face of West Virginia mine. (*U.S. Bureau of Mines, and Joy Manufacturing*)

Fig. 8. Caterpillar-mounted electric machine loading broken limestone into a diesel-powered haulage truck in a deep limestone mine in Ohio. (*U.S. Bureau of Mines*)

The chain coal cutter (Fig. 7) is in most common use and is manufactured in various sizes. These electrically powered machines are designed to cut in an arc of 180°. The smaller units are maneuvered by power-actuated drums and cable attached to stationary anchors. The larger self-propelled units are mounted on track or use pneumatic tires. The cutting jib can be rotated while cutting without moving the machine. It can also be rotated to cut either vertically or horizontally.

Loading. Since most loading is mechanized, hand shoveling is limited. Underground loaders include the continuous loader, scraper loader, and revolving and overcast shovels. Clamshell loaders are used in sinking shafts. The continuous loader (Fig. 8) is used extensively in underground coal mining and less extensively in other mines. Its essential parts comprise a gathering headframe, or scoop, on the front end that is crowded into the broken material by the machine, and a set of gathering arms to rake the material onto a bar chain conveyor that transfers it to the haulage system. The scraper loader consists of a hoe-type scraper, double-drum electric hoist, and ramp. Fragmented material is scraped up the ramp and discharged into the transportation system. The scraper loader is used in many places in mining to scrape ore and waste short distances up to about 200 ft (60 m). Revolving power shovels are used for excavation in large underground openings. The overcast shovel loader, which runs on rail or caterpillar tracks, is best adapted for use in confined areas. Loading is accomplished by crowding the unit into the broken rock; when the shovel is full, it casts back over the unit into cars or trucks.

Fig. 9. Loading 60 ft³-capacity, Granby-type mine car at uranium ore-pass station on haulage level, San Juan County, Utah. Underswing arc gate is operated by a compressed-air cylinder controlled by motorman helper on right. (*U.S. Bureau of Mines*)

Transportation. Moving or hauling of men, supplies, and broken minerals is one of the most complicated operations in mining. Transportation includes (1) transfer of broken material through mine openings by gravity and scraping, (2) rail haulage on surface and underground, (3) trackless, wheeled haulage on surface and underground, (4) hoisting and cable haulage from open pits and underground mines, (5) movement of broken material by numerous types of conveyors, and (6) pumping broken ore through pipelines.

Gravity and scraping methods are used to gather small quantities of material into larger volume for transportation by some other method. Broken material may fall directly into raises and ore passes where it flows by gravity to a chute or ore pocket.

Underground rail haulage has been adapted from railroad practice, but on a much smaller scale. It is used mostly for collecting broken material from chutes (Fig. 9) and various loaders, and transporting it to the hoist or directly to the surface. Track gage ranges from 18 to 48 in. (46 to 122 cm). Trains are made up of cars ranging from 1 ton (1 ton = 0.9 metric ton) to about 20 tons capacity. Cars are equipped for manual dumping or with some dumping device. Small units are drawn by storage battery locomotives and larger units by electric trolley locomotives.

Trackless, wheeled haulage underground usually is by electric cable reel and diesel-electric shuttle cars with self-contained discharging conveyors (Fig. 10). They are mounted on pneumatic tires and have capacities ranging to 20 tons. Dump trucks are used underground to a limited extent. When so used they are equipped with diesel engines for safety. Most trackless underground haulage equipment negotiates grades up to 10%.

Electric hoisting is employed in deep underground mines. Either vertical or inclined shafts are divided into three to six parallel shaftways—one for a manway, pipe, or power lines, and a minimum of two for hoisting. In large deep shafts, two compartments are commonly used for hoisting broken material, and two others for handling supplies, equipment, and personnel.

The hoisting layout comprises a headframe, usually of timber, concrete, or steel, erected over the collar of the shaft. A sheave wheel is mounted at the top of the headframe for each hoisting compartment. In addition, the headframe contains bins into which hoisted material is discharged.

Most hoists are electrically driven and have two winding drums that can be operated in balance or separately. When four compartments are employed for hoisting, a larger, double-drum unit hoists rock and a smaller one hoists men and supplies. Each drum is wound with a steel wire cable long enough to reach the sump (bottom) of the shaft. Cables range up to about 1½ in. (4 cm) in diameter. Rock is hoisted from various levels of the mine in skips (buckets, baskets, or open cars) ranging up to 10 tons capacity. Miners and supplies and, at some smaller mines, loaded cars are hoisted in cages with one to three decks.

Fig. 10. Loader and a trackless, diesel-electric shuttle car mounted on pneumatic tires in an underground Missouri lead-zinc mine. (*U.S. Bureau of Mines*)

An important safety feature of the skip and the cage is a device that stops them if the hoisting cable fails. When failure occurs, a spring releases a set of safety catches that engage the wooden guides along each side of the skip or cage, thereby stopping it.

Several different types of conveyors are used in transporting coal in underground mines. The most important of these are belt conveyors. In some mines virtually all the coal is transported by a system of belt conveyors from the mining face through the various entries to surface. *See* MINING OPERATING FACILITIES; UNDERGROUND MINING.

[JAMES E. HILL]

Mining machinery

Apparatus used in removing and transporting valuable solid minerals from their place of natural origin to a more accessible location for further processing or transportation. Many of the machines are identical to, or minor adaptations of, those used for excavating in the construction industry. In a wider sense mining machinery could also include all equipment used in finding (exploring and prospecting), removing (mining: developing and exploiting), and improving (processing: ore dressing, milling, concentrating or beneficiating, and refining) valuable minerals; it could even include metallurgical (smelting) and chemical processing equipment used in extracting or purifying the final product for industry. The term is also applied to special equipment for recovery of minerals from beneath the sea. In usual context the term does not include apparatus used principally in the petroleum industry. Perhaps those machines most often considered as uniquely mining machinery are drills, mechanical miners, and specially adapted materials-handling equipment for use in mining underground or on surface (where a large proportion of mines are located). In addition, some unique auxiliary equipment and processing equipment are used in the mining industry.

Design and construction. All components of mining machinery—including primary mechanism, controls, means of powering, and frame—require the following features to a much greater extent than do other machines (with the possible exception of some military, construction, oil well, and marine units).

Ruggedness. Equipment is handled roughly, frequently receiving severe and sudden shock from dropping, striking, and blasting vibration; overloading is common, and long life is demanded by the economics of mining.

Weather resistance. Operations extend over a wide range of climate and altitude.

Abrasion resistance. Minerals include some of the hardest substances known. Dust and fine particles are always present.

Water and corrosion resistance. Moisture and water, often acidic, are common in mining operations.

Infrequent and simple maintenance. Equipment is often widely scattered and in locations with restricted access. Trained mechanics and repair parts are generally limited in availability because of the remoteness of operations.

Easy disassembly and reassembly. Access to the machinery at the site of operation is frequently limited. Also, the working space near it and mechanical aids to moving or lifting it may be limited or nonexistent.

Safety. Mining has had a poor safety record. As a result, most governments have testing bureaus, inspection agencies, and enforcement laws for the approval of mining equipment.

Simplicity of operation, low initial cost, and low operating costs are also desirable features of mining machinery.

Exploration machinery (vehicles, drills, and accessory equipment) is subjected to the same operating conditions as other mining equipment. Mineral-beneficiating equipment must have, above all, abrasion resistance. Smelting equipment has the requirement of heat resistance. Chemical refining process equipment must be highly resistant to corrosion. Reliability of all processing equipment is critical because slurries are commonly handled, and they can cause considerable difficulty in restarting after shutdowns.

Underground requirements. Machinery operated underground must meet special design requirements.

Low-ventilation demand. Quantity and geometry of passageways for air are rather rigidly fixed, so that high air consumption is a problem and noxious gases cannot be readily dispersed. Heat removal is a problem in deep mines.

Compactness. Space is at a premium, especially height, particularly in bituminous coal mines.

Easy visibility. Most operating areas are lighted only by individual cap-mounted or hand-held lamps.

Hand portability. Units or components must frequently be hand-carried into an operating area.

Absence of spark and flame. Equipment is often used in or near explosives, timber supports, and natural or man-made combustible gas or dust. In the presence of hydrocarbons, as well as of certain metal ores such as some sulfides, complete absence of open sparks or flames is a major requirement.

Power source. Mining machinery is very commonly powered by compressed air, but electricity is also widely used and is often the basic source. Compressed air has the advantages of simplicity of transmission and safety under wet conditions. It is especially advantageous underground as an aid to ventilation. Machines powered by compressed air can be easily designed to accommodate overloads or jamming, which is desirable on surface as well as underground. Large central compressors and extensive pipeline distribution systems are common, especially at underground mines.

Electric power, purchased from public sources or locally generated at a large central station, is common in open-pit and strip mines and dredging operations. Underground coal and saline-mineral mines often use electric-powered production machinery, but in other underground mines electricity is normally used only for pumps and transportation systems in relatively dry or permanent locations. Direct-current devices are dominant because of simplicity of speed and power control, but alternating-current apparatus is becoming common. Mobile equipment is often either battery of cable-reel (having a spring-loaded reel of extension power cable mounted on the machine) type. Processing machinery units are almost exclusively powered by individual electric motors.

Diesel engines are popular for generating small quantities of electric power in remote areas and for transportation units. Underground, abundant ventilation is essential, as well as wet scrubbers, chemical oxidizers, and other accessories to aid the removal of noxious and irritant exhaust gases. Hydraulic (oil) control and driving mechanisms are widely used. Transfer of power by wire rope is common, especially for main vertical transportation.

Drills. Drills make openings, of relatively small cross section and long length, which are used to obtain samples of minerals during exploration, to emplace blasting explosives, and to extract natural or artificial solutions or melts of minerals. Exploration holes are generally vertical or inclined steeply downward, less than 6 in. (15 cm) in diameter and up to 10,000 ft (3 km) long. Blastholes range from 3/4 to 12 in. (2 to 30 cm) in diameter and usually are under 50 ft (15 m) long in any direction, with the larger usually downward. Solution wells normally are vertical and 6 to 12 in. (15 to 30 cm) in diameter, sometimes reach depths of several thousand feet, and are equipped with several concentric strings of pipe.

Rock drills. Percussion, rotary, or a combination action of a steel rod or pipe, tipped with a harder metal chisel or rolling gearlike bit, chips out holes up to 12 in. (30 cm) or more in diameter by 125 ft (38 m) or more long from the surface, and 1–3 in. (2.5–8 cm) by 5–200 ft (1.5–60 m) from underground. Crawler or wheeled carriers are used, and the smaller drills are often attached to hydraulically maneuvered booms. Air or liquids flush out the chips.

Diamond drills. Rotation of a pipe tipped with a diamond-studded bit is used in exploration to penetrate the hardest rocks. Large units make holes up to 3 in. (8 cm) in diameter and 5000 ft (1.5 km) or more deep; at the other extreme are units so small that they can be pack-carried. A cylindrical core is usually recovered.

Water-jet drills. For exploration and blasting in loose or weakly bonded materials, a water jet washes out a hole as a wall-supporting pipe is inserted.

Jet flame drills. For economical surface blast holes in hard abrasive quartzitic rock, a high-velocity flame is used to spall out a hole.

Mechanical miners. There are many machines designed to excavate the valuable mineral or the access openings by relatively continuous dislodgement of material without resorting to the more common practice of intermittent blasting in drill holes. These units also frequently transport the mineral a short distance, and when designed for weakly bonded minerals, they often become primarily materials-handling equipment.

Continuous miners. For horizontal openings in coal and saline deposits, toothlike lugs on moving chains, or rotating drums or disks rip material from the face of the opening as the assembly crawls ahead.

Plows or planers. In coal and other mineral deposits of medium hardness, bladelike devices continuously break off a 6-in. (15-cm) layer as they are pulled by various mechanisms along a wall several hundred feet long and 3–5 ft (0.9–1.5 m) high.

Augers. Coal and soft sediments are mechanically mined by augers up to 5 ft (1.5 m) in diameter and 100 ft (30 m) long, usually used horizontally.

Shaft and raise drills and borers. For vertical and inclined openings up to 8 ft (2.4 m) even larger in diameter, various rotary coring and fullface boring equipment is used, both in an upward direction (raising) for several hundred feet or downward (sinking) for several thousand feet. These units usually use many rigid teeth or rolling gearlike bits to chip out the mineral.

Tunneling machines. There are similar rotary boring units for horizontal, or nearly so, openings of any length and up to 35 ft (11 m) in diameter (in soft rock).

Rock saws. To remove large blocks of material, narrow slots or channels are cut by the action of a moving steel band or blade and a slurry of abrasive particles (sometimes diamonds) rather than teeth. Small flame jets are also used.

Hydraulic monitors. Water jets of medium to high pressure (some to 5000 psi or 34 MPa) are used to excavate weakly cemented surface material and brittle hydrocarbons both on the surface and underground.

Special materials-handling equipment. Loose material (muck) is picked up (mucked or loaded) and transported (hauled or hoisted) by a wide variety of equipment.

Excavator loaders. For confined places underground there are various unique grab-bucket shaft muckers and overcasting shovel tunnel muckers, and also gathering-head loading-conveyor units having eccentric arms, lugged chains, and screws or oscillating pans for handling muck in horizontal openings.

Dragline scrapers. Scrapers (slushers) with a flat plowlike blade or partially open bucket pulled by a wire rope are commonly used to move muck up to a couple hundred feet, especially in underground mines.

Dipper shovels and dragline cranes. In surface mining single-bucket loads can handle up to 100 yd^3 (75 m^3) of material. Many of the intermediate size (20–40 yd^3 or 15–30 m^3) units move on unique walking shoes.

Bucketline and bucketwheel excavators. These are for surface use and can dig up to 5000 yd^3/hr (3800 m^3/hr), using a series of buckets on a moving chain or a rotating wheel supported on crawlers, railcars, or floating hulls (dredges).

Suction dredges and pipelines. On surface, up to 3700 yd^3/hr (2800 m^3/hr) of moderately loose mineral up to several inches in diameter can be picked up and moved as a slurry (mix of water and solid material) by pumps mounted on floating hulls.

Trucks. Diesel and electric shuttle cars (short-haul trucks) of unusual design, often having very low profile and conveyor bottoms, are used underground. At surface mines there are diesel-electric- and electric-trolley-type dump trucks of over 100-ton (1 ton = 0.9 metric tons) capacity.

Railroads. Underground locomotives range from 1/2 to 80 tons in weight, with electric (storage battery, cable-reel, or trolley), diesel, and sometimes compressed-air power units. Cars are usually of special design.

Conveyors. Unique movable, self-propelled, sectional and extensible conveyors are used in underground mining.

Wire rope hoists. Hoists or winders of up to 6000 hp (4.5 mW) are used in shafts for vertical or steeply inclined transportation in single lifts of as much as 6000 ft (1.8 km). Of various particular designs, there are two basic types: drum, simply a powered reel of rope; and friction, in which the rope is draped over a powered wheel and a counterweight is attached to one end and a conveyance to the other.

Auxiliary equipment. Drainage pumps handling hundreds of gallons of water per minute at heads of 1000 ft or more are used underground. For underground roof support, primarily in coal, there are mechanically moved jacks of 100-ton capacity. Ventilation fans are capable of moving several hundred thousand cubic feet of air per minute. Crushers can handle pieces of hard rock several feet across in two dimensions. In processing, minerals are sorted by size or density, or both, by a variety of screens, classifiers, and special concentrators, using vibration, fluid flow, centrifugal force, and other principles. Froth flotation and magnetic and electrostatic equipment take advantage of other special properties of minerals. *See* BULK-HANDLING MACHINES; GRINDING MILL; MINING OPERATING FACILITIES.

[LLOYD E. ANTONIDES]

Bibliography: Engineering and Mining Journal, *Mining Guidebook*, annual; Engineering and Mining Journal, *Operating Handbook of Mineral Underground Mining*, vol. 3, 1979; R. S. Lewis and G. B. Clark, *Elements of Mining*, 3d ed., 1964; R. Peele (ed.), *Mining Engineers' Handbook*, 2 vols., 3d ed., 1941; Mining Information Services, *Mining Methods and Equipment*, 1980; Society of Mining Engineers, *SME Mining Engineering Handbook*, 1973.

Mining operating facilities

Physical aids, procedures, and plans widely used to expedite development and production in mining. Several of these are of such outstanding significance that they are the subject of separate articles. *See* BORING AND DRILLING (MINERAL); MINING EXCAVATION.

This article includes discussions of other facilitating aspects of mining: power, ventilation, illumination, drainage, storage and shipping, field sampling and ore estimation, mine evaluation, and the use of computers.

MINING POWER

Power is applied to mining in six distinctive ways: electricity, diesel power, compressed air, hydraulic power, steam power, and hydroelectricity.

Electricity. Both alternating and direct electric current are used in modern mining. Direct current (dc) is used for locomotive haulage and for the major portion of underground coal operations because of the high torque developed in dc series motors under heavy loads and starting. Alternating current (ac) is used less extensively although some mines employ both in a dual system.

In the United States, 6000-, 4160-, and 2300-volt three-phase alternating current is generally received from service companies at the surface substation. In a well-planned mine, all the power is metered at one point and the power factor is adjusted between 90 and 95% by the use of synchronous motors and capacitors. Neoprene-covered cables, type SHD, 15,000 volts, are recommended to take the power underground through boreholes or power shafts. Each of three insulated conductors is covered with copper shielding braiding to eliminate static stresses, and a ground conductor for each power conductor is placed in the interstices of the cable.

Converting ac to dc underground may be accomplished by several methods. A 250- or 500-volt substation can be provided by converters, motor-generator sets, mercury-arc rectifiers including the glass-bulb type, and dry rectifiers made of selenium, germanium, or silicon (Fig. 1). Portable conversion units offer increased convenience and efficiency.

Trolley and feeder lines supply power throughout the main roadways. These distribution lines should not be extended more than 3500–4000 ft from the power source to avoid low voltage at the end of the line. If local laws permit, 500-volt systems are sometimes used to minimize line voltage drop. This supplies twice the load for the same size and length of cable, or conversely, the same load can be supplied at twice the distance. Trailing cables, fastened to the trolley nipping stations or power centers, supply power to machines in sections of the mine where explosive gases may be present.

Fig. 1. A 300-kW Westinghouse ignitron rectifier installed at a mine.

Sectionalization is a method of distributing mine power so that power cables can be isolated for reasons of faults or repairs without shutting off the main supply to several working sections. For main distribution, circuit breakers, disconnect switches, and various overcurrent protective devices are essential for properly installed sectionalization. At face areas, safety circuit centers and associated intrinsically safe circuitry make it possible to connect or disconnect short cables without danger of causing incendiary arcing which could ignite gas.

Alternating-current power for mining is increasing in popularity because its equipment is less costly than dc and its maintenance is simpler. Alternating-current motors, for example, cost one-third to one-half as much as the equivalent dc and are more compact. High voltage is transformed to 440 or 220 volts at underground substations. Portable units are also utilized. Sectionalization is applicable to ac distribution.

Alternating-current power is commonly used in strip mining. High voltage of 33,000 volts is stepped down to 7300, 6600, 4160, or 2300 volts. Permanent substations equipped with lightning arresters, circuit breakers, ground protective equipment, and other protective devices, and semiportable substations are employed for distribution. Power is distributed by pole lines or cable systems, or a combination of the two.

Diesel power. This type is rapidly gaining favor in metal and noncoal mining because of its flexibility. Some states require that underground diesel-powered equipment be approved by the U.S. Bureau of Mines. Details of these standards include explosion-proof housings, mine ventilation necessary to dilute exhaust gases, control of hot exhaust gases and surface temperatures, concentrations of toxic constituents in exhaust gases, and recommendations for the selection, handling, and storage of diesel fuel oil.

Compressed air. As a mine facility, pneumatic power is utilized in a variety of applications, mostly in the metal and noncoal fields. It is used to power drills and hoists; pneumatic tools, such as grinders, drills, riveters, chippers, pneumatic diggers, and spades; air-driven sump pumps; direct-acting and air-lift pumps; pile drivers for shaft sheathing; air pistons for unloading cars; drill-steel sharpeners; air motors; compressed-air locomotives; shank and detachable-bit grinders; mine ventilation; and in supplying air for blowing converters; starting diesel engines; and coal preparation. Compressed air is used in coal mining for blasting. This method works with 9000–10,000 pounds per square inch (psi) and the air is released from a tube which is inserted in a hole in the face of the coal when a metal diaphragm ruptures. The force is released through slots in the tube and breaks up the coal, previously undercut.

Hydraulic power. For mining, hydraulic applications are rather limited in usage and may employ either water or oil as a fluid. Oil types are used in connection with small tools, lifts, and in the intricate operation of continuous mining machines and other equipment. An available waterfall may be directed to power equipment such as air compressors. A unique use of hydraulic power, called jet mining, uses air pressure to force water through 1/4-in. nozzles under 2000 psi pressure. This jet action has been developed to cut a material called gilsonite, a solid hydrocarbon, by use of water at a rate of 300 gal/min.

Steam power. Although formerly used to drive compressors, hoists, generators, and other equipment, steam is now rarely used in mining but has definite, although limited, application in some coal mines in which there is an abundance of waste fuel or unmarketable coal. Parts of certain coal seams contain impure bands that must be rejected, or that are difficult to clean, but they can be burned under boilers with special firing equipment.

Hydroelectricity. As applied to mining, hydroelectricity is used mostly in the electrometallurgical fields, where vast amounts of power are required at a reasonable cost. Some hydroelectricity has been used for normal, electrical, mining power, but mostly it is used for processes beyond the mining operation such as beneficiation, smelting, and refining. In certain cases these are done near the mine mouth in isolated areas in order to reduce bulk before shipment. [ROBERT S. JAMES]

MINE VENTILATION

The purpose of mine ventilation is to provide comfortable, safe, and healthful atmospheric conditions at places where miners work or travel.

Airflow fundamentals. The following points summarize airflow principles for mine ventilation: (1) Airflow is induced by a pressure difference between intake and exhaust; (2) the pressure created must be sufficient to overcome the system resistance and may be either negative or positive; and (3) air flows from the point of higher to lower pressure. Also, mine airflow is considered turbulent and follows the square law relationship between volume and pressure; that is, a doubled volume requires four times the pressure.

The principles of fans may be summarized as follows: (1) Air quantity varies directly as fan speed, and is independent of air density; (2) pressures induced vary directly as the square of the speed, and directly as the air density; (3) the fan power input varies directly as the air density and the cube of the speed; (4) the fan mechanical efficiency is independent of speed and density. *See* FAN.

The amount of air movement induced will be dependent upon the fan characteristic and mine resistance as shown by Fig. 2. The pressure H required to pass a quantity of air Q through a mine or segment is expressed by the formula $H=RQ^2$, where R, the mine resistance factor, may be calculated from known pressure losses, or from the common ventilation formula $R=KlO/5.2A^3$ in

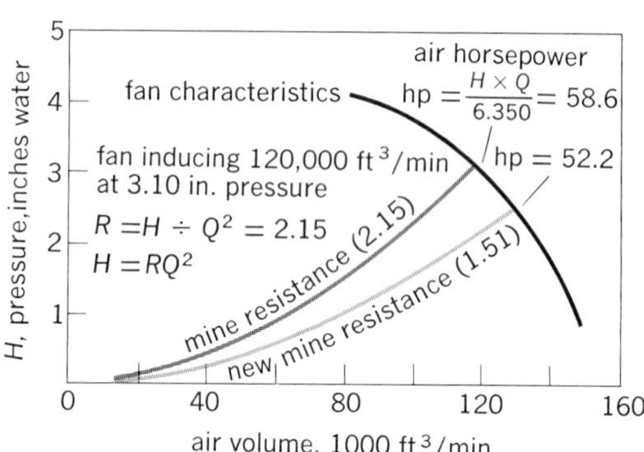

Fig. 2. Fan characteristics versus mine resistance.

which K is the frictional coefficient, l is the length of the airway, O is the perimeter of the airway, and A is the cross-sectional area of the airway. For simplicity of calculation, frictional coefficients should be expressed without decimals, and the quantity Q of air in cubic feet per minute (cfm) should be divided by 100,000 before using. The table gives reasonably frictional coefficients; the table's use may be exemplified by an application to the following case: What pressure is necessary to induce 60,000 cfm through a single airway 6 ft high, 12 ft wide, and 2500 ft long in sedimentary rock? The airway is straight, has average irregularities, and is moderately obstructed. K from table is 70; l is 2500; perimeter is 36; area is 72 ft². Q is 60,000 cfm ÷ 100,000 = 0.60. The pressure H is given by Eq. (1).

$$H = RQ^2 = \frac{KlOQ^2}{5.2A^3} = \frac{70 \times 2500 \times 36 \times (0.6)^2}{5.2 \times (72)^3} = 1.17 \text{ in. water pressure} \quad (1)$$

Parallel air flow can be determined by the square-law relationship. For example, the pressure H required to pass 60,000 cfm through 2500 ft of single entry is 1.17 in. of water. For two identical entries the pressure is given by Eqs. (2).

$$H_1 \left(\frac{1}{2}\right)^2 = H_2$$
$$H_2 = \frac{H_1}{4} = \frac{1.17}{4} = 0.292 \text{ in. water} \quad (2)$$

When resistance factors are known or entries are not identical, the formula is Eq. (3). For example, what is the combined resistance factor of two entries 1000 ft long? One entry is substantially larger, $R_1 = 1.50$, $R_2 = 4.0$. Also, what pressure is required to pass 60,000 cfm through 2500 ft of the combined entries, assuming average conditions throughout? Computation is as follows:

$$\frac{1}{\sqrt{R}} = \frac{1}{\sqrt{R_1}} + \frac{1}{\sqrt{R_2}} + \frac{1}{\sqrt{R_3}} + \cdots + \frac{1}{\sqrt{R_n}} \quad (3)$$

$$\frac{1}{\sqrt{R}} = \frac{1}{\sqrt{1.5}} + \frac{1}{\sqrt{4.0}} = \frac{1}{1.22} + \frac{1}{2.00}$$
$$= 0.82 + 0.50 = 1.32$$

$$\frac{1}{\sqrt{R}} = 1.32 \qquad R = 0.58 \text{ per 1000 ft entry}$$

R for 2500 ft = 2.5 × 0.58 = 1.45

Pressure = 1.45 × (0.60)² = 0.52 in. water

Air quantity requirements. Unless covered by state laws, the common criteria for adequate ventilation are absence of smoke and dust with moderate air temperatures in metal mines and the absence of methane, smoke, and dust in coal mines. Natural conditions of gas, rock temperatures, dust, and operating practices determine requirements. Good quality air is not deficient of oxygen and is free of harmful amounts of physiological or explosive contaminants. *See* MINING SAFETY.

Mine gases. Important contaminants of mine air are carbon dioxide, hydrogen sulfide, methane, carbon monoxide, and sulfur dioxide.

Carbon dioxide, CO_2, specific gravity 1.529, is produced by oxidation and combustion of organic compounds and is occluded in the rock strata of certain mines. It is heavy, colorless, and odorless and is usually found in low, poorly ventilated areas.

Hydrogen sulfide, H_2S, specific gravity 1.191, is the product of decomposition of sulfur compounds. It is colorless, has the odor of rotten eggs, and may be found near areas of stagnant water in poorly ventilated areas.

Methane, CH_4, specific gravity 0.554, is a natural constituent of all coals. It may be occluded in carbonaceous shales and sandstones and may infiltrate into metal mines at contacts with carbonaceous rocks. It is colorless, odorless, and may be found in high, poorly ventilated cavities.

Carbon monoxide, CO, specific gravity 0.967, is not a normal constituent of mine air, but is produced in mines by the incomplete combustion of carbonaceous matter, mine fires, or from gas or dust explosions. It is colorless and odorless.

Sulfur dioxide, SO_2, specific gravity 2.264, is not common, but may be found in sulfur mines and in mines with rich sulfide ores as the result of fires. It is a water-soluble and colorless gas with a suffocating odor.

Blackdamp is a common term applied to oxygen-deficient atmospheres; it is not a specific gas mixture but may contain any of many gases produced by oxidation and processes that use oxygen and liberate carbon dioxide.

Small quantities of air contaminants must be determined by laboratory analysis of air samples. On-the-spot safety determinations for carbon dioxide and oxygen deficiency may be made by flame safety lamp or small portable absorption instruments. Methane can be detected by flame safety lamp and commercial testers; carbon monoxide and hydrogen sulfide by special hand-held colorimetric indicators.

Reasonable frictional coefficients

		Straight			Sinuous or curved: Moderate			Sinuous or curved: High degree		
Type of airway	Irregularities	Clean	Slightly obstructed	Moderately obstructed	Clean	Slightly obstructed	Moderately obstructed	Clean	Slightly obstructed	Moderately obstructed
Smooth-lined	Minimum	10	15	25	25	30	40	35	40	50
	Average	15	20	30	30	35	45	40	45	55
	Maximum	20	25	35	35	40	50	45	50	60
Sedimentary rock or coal	Minimum	30	35	45	45	50	60	55	60	70
	Average	55	60	70	70	75	85	80	85	95
	Maximum	70	75	85	85	95	100	95	100	110
Timbered (5-ft centers)	Minimum	80	85	95	95	100	110	105	110	120
	Average	95	100	110	110	115	125	120	125	135
	Maximum	105	110	120	120	125	135	130	135	145
Igneous rock	Minimum	90	95	105	105	110	120	115	120	130
	Average	145	150	160	160	165	175	170	175	195
	Maximum	195	200	210	210	215	225	220	225	235

Dust and dust hazards. Dust is defined as the solid particulate matter thrown into suspension by mining operations. The size of particles may range upward from less than 1 micron (0.001 mm) diameter (as shown in Fig. 3, the Frank chart); particles larger than 10 μ can usually be seen by the naked eye. The dust hazard may be physiological or explosive or both physiological and explosive as is coal dust. The common physiological diseases are mostly various pneumoconioses. Preventive measures are to suppress dust at the source with water sprays, foam, fog, wetting agents, or dust collectors. Additional protection may be provided through suitable respirators. Accumulations of explosive dust should be removed and inert material, such as rock dust, applied to surface areas. The dust hazard can be determined by systematic sampling of airborne dust at critical points. The impinger, a common instrument used, draws a known volume of air through a liquid or filter to remove the dust. The dust concentrations of the sample can then be determined by count using a microscope or microprojector.

Air temperature and humidity. Temperature rise in mine workings is from (1) heat conducted from surrounding strata because of the thermal gradient (depth per °F rise in temperature); (2) adiabatic compresiion of descending air columns (approximately 5.5°F/1000 ft or 10°C/km); and (3) heat from oxidation of minerals.

The factors that influence humidity are: rise in dry-bulb temperature; volume changes caused by pressure and temperature changes; and moisture picked up from shafts and roadways.

The air temperature approximates the temperature of adjacent walls; seasonal temperature changes are noticeable only short distances underground. Workers' efficiency (Fig. 4) is dependent upon temperature, humidity, and motion velocity of air. A solution to excessive air temperatures is air conditioning by passing the air currents through heat exchangers filled with chilled air.

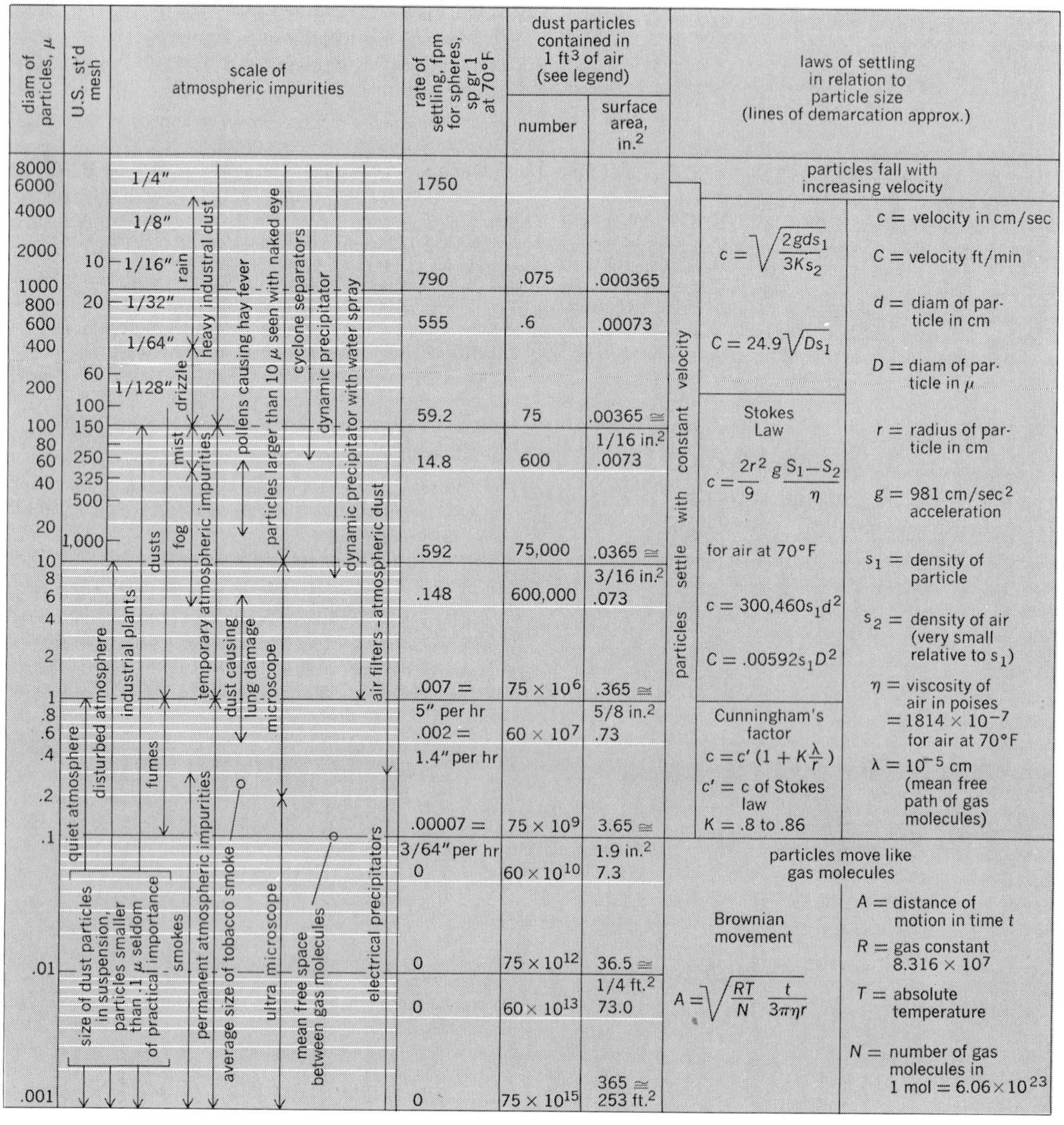

diam of particles, μ	U.S. st'd mesh	rate of settling, fpm for spheres, sp gr 1 at 70°F	dust particles contained in 1 ft³ of air (see legend): number	surface area, in.²
8000–6000		1750		
2000–1000	10	790	.075	.000365
800–600	20	555	.6	.00073
100	150	59.2	75	.00365 ≅ 1/16 in.²
60	250	14.8	600	.0073
10	1,000	.592	75,000	.0365 ≅ 3/16 in.²
6		.148	600,000	.073
1		.007 = 5" per hr	75×10^6	.365 ≅ 5/8 in.²
.6		.002 = 1.4" per hr	60×10^7	.73
.1		.00007 = 3/64" per hr	75×10^9	3.65 ≅ 1.9 in.²
		0	60×10^{10}	7.3
.01		0	75×10^{12}	36.5 ≅ 1/4 ft.²
		0	60×10^{13}	73.0
.001		0	75×10^{15}	365 ≅ 253 ft.²

Diameter scale (μ): 8000, 6000, 4000, 2000, 1000, 800, 600, 400, 200, 100, 80, 60, 40, 20, 10, 8, 6, 4, 2, 1, .8, .6, .4, .2, .1, .01, .001

U.S. st'd mesh: 10, 20, 60, 100, 150, 250, 325, 500, 1,000

Scale of atmospheric impurities: 1/4", 1/8", 1/16", 1/32", 1/64", 1/128"; rain; drizzle; mist; fog; heavy industrial dust; dusts; temporary atmospheric impurities; pollens causing hay fever; particles larger than 10 μ seen with naked eye; cyclone separators; dynamic precipitator; dynamic precipitator with water spray; air filters - atmospheric dust; quiet atmosphere; disturbed atmosphere; industrial plants; fumes; dust causing lung damage; microscope; smokes; permanent atmospheric impurities; average size of tobacco smoke; ultra microscope; mean free space between gas molecules; electrical precipitators; size of dust particles in suspension, particles smaller than .1 μ seldom of practical importance

Laws of settling in relation to particle size (lines of demarcation approx.):

- Particles fall with increasing velocity: $c = \sqrt{\dfrac{2gds_1}{3Ks_2}}$; $C = 24.9\sqrt{Ds_1}$
- Particles settle with constant velocity — Stokes Law: $c = \dfrac{2r^2}{9}\, g\, \dfrac{s_1 - s_2}{\eta}$; for air at 70°F: $c = 300{,}460 s_1 d^2$; $C = .00592 s_1 D^2$
- Cunningham's factor: $c = c'(1 + K\frac{\lambda}{r})$; c' = c of Stokes law; K = .8 to .86
- Particles move like gas molecules — Brownian movement: $A = \sqrt{\dfrac{RT}{N}\,\dfrac{t}{3\pi\eta r}}$

Legend: c = velocity in cm/sec; C = velocity ft/min; d = diam of particle in cm; D = diam of particle in μ; r = radius of particle in cm; g = 981 cm/sec² acceleration; s_1 = density of particle; s_2 = density of air (very small relative to s_1); η = viscosity of air in poises = 1814×10^{-7} for air at 70°F; $\lambda = 10^{-5}$ cm (mean free path of gas molecules); A = distance of motion in time t; R = gas constant 8.316×10^7; T = absolute temperature; N = number of gas molecules in 1 mol = 6.06×10^{23}

Fig. 3. Size and characteristics of airborne solids (compiled by W. G. Frank). It is assumed that particles are of uniform spherical shape having specific gravity 1 and that dust concentration is 0.6 gr/1000 ft³ of air, average for metropolitan districts.

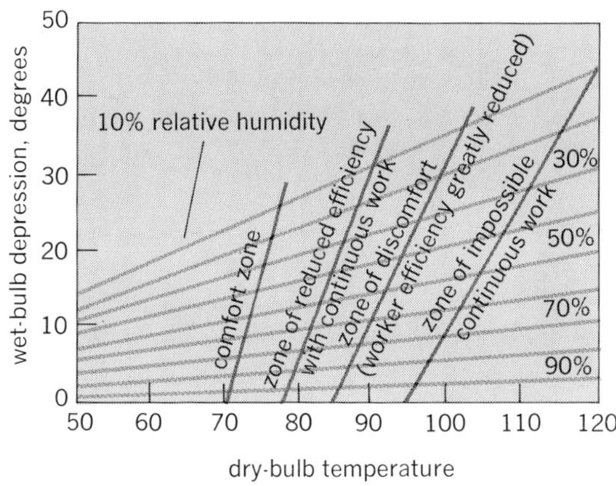

Fig. 4. Graph showing the influence of humidity and temperature on worker efficiency.

Natural ventilation. Natural ventilation is induced by differences of total weights of air columns for the same vertical distance. Natural draft may operate with or against the mechanical draft (Fig. 5) or may be the only source of pressure. Natural draft pressures at standard density can be estimated as 0.03 in. water-gage for each 10°F average temperature difference per 100 ft increment vertical elevation (135 Pa for each 1°C temperature difference per 1 m increment). For accurate determinations the average air densities of influencing air columns must be calculated.

The formula for determining air density is given by Eq. (4), where d is the density in lb/ft^3, T is the

$$d=\frac{1.327}{460+T}(B-0.378\mathrm{VP}) \qquad (4)$$

dry-bulb temperature, B is the barometric pressure, in inches of mercury, and VP is the vapor pressure of water at the dew point, in inches of mercury.

With most calculations, the vapor pressure influence can be ignored. The simplified formula then is Eq. (5).

$$d=\frac{1.327}{460+T}(B) \qquad (5)$$

Auxiliary ventilation. This term applies to booster fans and auxiliary fans. A booster fan is placed underground and operated in series with the main fan to increase ventilating pressure of one or more splits of the ventilating current. The booster fan in effect reduces the mine resistance, thereby increasing the air quantity circulated. An auxiliary fan is a small fan installed in the air current to divert, through air tubing or ducts, a part of the ventilating current to ventilate some particular place or places. In metal mines, they are used to ventilate developing drifts, raises, crosscuts, and stopes. In coal mines, they are used to conduct air to working faces.

[DONALD S. KINGERY]

Ventilation system evaluation. Accurate analysis and evaluation of a ventilation system, for the purpose of initiating improvements or projecting the system, require pressure and quantity measurements from which actual resistances can be determined. Proper utilization of such data permit efficient and economic changes in the present system and the accurate determination of requirements for the projected system. Although friction coefficients from the table enable engineers to design, with good results, ventilation systems for new mines, the continuously changing conditions following mining make it almost impossible to apply these values in evaluating older systems.

The aneroid barometer measures absolute static pressure and is the instrument usually used for mine-pressure surveys. The instrument is rugged and portable but relatively few have sufficient sensitivity and precision for pressure surveys. A variety of graduated scales are provided, the most common being "inches of mercury" and "feet of air." Instruments with the latter scale are termed altimeters. Regardless of scale on the instrument, mine pressure and pressure differentials are generally converted to inches of water for analysis of pressure data.

The surface absolute static pressure (barometer reading) is primarily a function of elevation and air density. Underground, within the ventilation system, absolute static pressure is a function of the same conditions plus the static pressure resulting from pressure generated by the fan. By compensating for elevation differences and atmospheric changes, the ventilating pressure at any point and between points resulting from fan operation can be calculated. By supplementing the pressure deter-

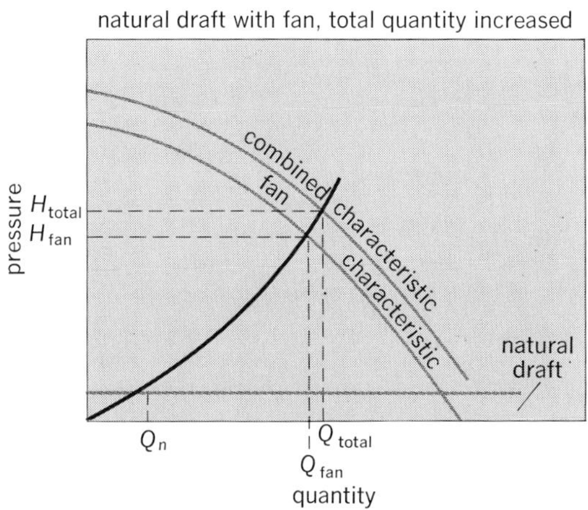

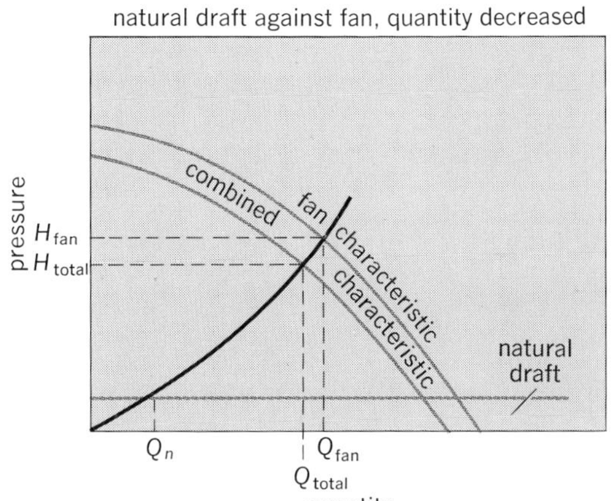

Fig. 5. Graphs of influence of natural draft to mine ventilation.

Fig. 7. The fluid network analyzer used to solve mine ventilation distribution problems. (*U.S. Bureau of Mines*)

minations with water-gage reading across stoppings and regulators, pressure losses for any part of the system can be readily determined.

Analysis of the pressure data by means of a pressure gradient will pinpoint high resistance areas. The pressure gradient (Fig. 6) is constructed by plotting distance of air travel against pressure drop. Portions of the gradient with a steep slope indicate high-resistance airways.

For a complete analysis of any system, pressure data must be combined with extensive quantity determinations. Quantity flow is determined by area and velocity measurements. The more common instruments for velocity measurements include vane anemometer, pitot tube, and smoke tube. Instruments selected will depend on velocity of the airstream.

The quantity survey is useful in locating areas of excessive intake to return leakage, but more important, combined with pressure data, the actual resistance factor for each segment of the system can be calculated. These factors are necessary for accurate analysis of the effect of any major system changes or projections. Results may be determined mathematically or by analog computer.

For the economic solution of problems involving mine ventilation improvement, the U.S. Bureau of Mines employs almost exclusively an electric analog computer designed especially for analyzing mine-ventilation distribution problems (Fig. 7). The heart of the analyzer is a nonlinear resistor known as a Fluistor which closely approximates the square-law resistance characteristics of mine airflow; that is, the voltage drop varies approximately as the square of the current.

In developing problem solution the ventilation circuit is divided into segments and the resistance factor for each segment is calculated. The system is reproduced electrically on the analog, using Fluistors of the proper coefficient to simulate airway resistance. Once the analog network is balanced to mine conditions represented by field data, any changes can be readily made and analyzed. The master meter can be connected directly to any segment in the circuit to measure voltage or current. The meter is a digital voltmeter which reads out pressure and airflow in units of inches of water and cubic feet per minute, respectively.

[E. J. HARRIS]

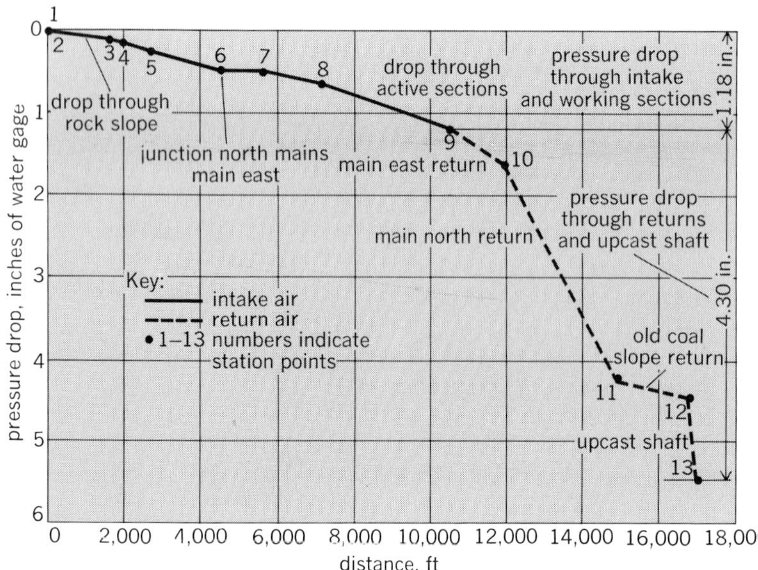

Fig. 6. Ventilation pressure gradient for a small slope mine. (*U.S. Bureau of Mines*)

MINE ILLUMINATION

The lighting of mines is accomplished by use of both movable and stationary lamps.

Mobile illumination. Cap lamps, flashlights, portable hand lamps, trip lamps for haulage cars, and lights in mobile machinery provide mobile lighting. The carbide lamp has been outmoded by the safer and more efficient electric cap lamp operated by a 4- or 5-volt battery. Such a lamp, using a polished reflector, can produce a beam candlepower of 1000 candles in a small spot. Less light with greater spread is obtained with a matte reflector.

Permissible continuous mining machines, shuttle cars, and other mining production equipment are fitted with explosion-proof 150-watt, 120-volt lights, front and rear. All permissible lights for use in gassy or dusty mines in the United States are approved by the U.S. Bureau of Mines.

Stationary or area illumination. This type of lighting is provided by incandescent lamps or dual

Fig. 8. An installation of permissible fluorescent lights in a coal mine. (*U.S. Bureau of Mines*)

15-in. (38-cm) fluorescent lights at intervals of 20–90 ft (6–27 m). Incandescent types, 50–150 watts, dc-operated, may be connected singly or in groups in ventilated haulageways but not in gassy areas. Fluorescent lighting, first approved in 1957 for gassy or dusty mines, employs 14-watt lamps on ac power (Fig. 8).

Fluorescent lights have proved very effective in gassy areas, along beltways and shuttle-car roadways, and have been credited with reduction in accidents, better morale among the workers, and better production records. Three-wire conventional grounding or two-wire isolated ungrounded systems provide shock protection. Special connectors facilitate the disconnection of lights without arcing. The U.S. Bureau of Mines approved a flashlamp for underground photographic use in 1957. [ROBERT S. JAMES]

MINE DRAINAGE

It is necessary to prevent water infiltration into mines from the surface or from sources underground. Control, transportation, and ultimate disposal of such waters constitute mine drainage.

Mine drainage varies in facilities and importance, at different localities, according to conditions of the mine water. There must be a careful study of its occurrence, corrosive and erosive character, volume to be handled, and the handling and disposal facilities that can economically provide efficient drainage. Major facilities consist of flumes, dikes, and storage ponds to prevent or minimize seepage of surface waters, underground ditches and sumps to collect, store, and control mine water, and pumping plants and drainage tunnels for the disposal of the mine water.

Control of stream pollution by mine drainage is important for both esthetic and economic reasons.

Sources and types of mine waters. Physical characteristics of the individual mining areas create distinctive sources and types of mine waters. Water enters directly into open-pit or underground workings, seeps through pervious strata, drains from normal water channels both on the surface and underground, and infiltrates from impounded waters into mine workings through cracked, crushed, and broken formations.

Mine waters vary widely in character in different mines because of the different geological formations. Waters from mines may be classed as alkaline (pH 7–14), neutral (pH 7), and acidic (pH 0–7). They may carry abrasive matter in suspension and be erosive to a degree dependent on the amount of abrasive substances. Although ground waters are normally alkaline, analyses from the majority of mines show considerable variation because the waters contain varying quantities of dissolved sulfate salts. Acidic water formed by such salts in the majority of mines has resulted in tremendous losses by corrosion, especially to pipes and pumping equipment.

Sumps, drains, and tunnels. The infiltration of water into mines, the volume of water to be handled, and pumping facilities vary with the seasons. The storage of mine water until it can be disposed of satisfactorily is of great importance. Ample sumpage where mine water can be stored is necessary in any drainage system. Sumps should be provided near pumping stations, have ample capacity, and be arranged so as to permit easy cleaning. If practicable, they should be designed to provide gravity feed to the pumps.

Drains, flumes, or ditches. These are used on the surface and underground to divert and convey water and thereby to prevent stream pollution and inflow through crop caves, strippings, cracks and fissures, and stream beds. Large flumes are used for those purposes in subsided areas, particularly where flash floods or quick runoff occur. Because ditches are usually made in mine haulage roads, they should be designed to carry the maximum drainage.

Drainage tunnels between different working levels are preferred means of handling and keeping mine water from reaching lower mine workings. Drainage tunnels from a mine or a group of mines to some disposal point situated favorably with respect to surface disposal areas are a means of draining mines economically and saving reserves of minerals that would be lost otherwise. They are advantageous over a long economic life of a mine for handling large quantities of water that sometimes could not be pumped to the surface. Although the initial cost is high, their upkeep is comparatively low.

Pumping. As long as mines are operated, mine drainage will most likely be done in whole or in part by pumps. Because mine water is usually acidic, suitable metals or alloys should be used for mine pumping equipment. The designing engineer should consult with qualified technicians before attempting to solve corrosion problems. Centrifugal pumps are both horizontal and vertical and known as standard, deepwell, and shaft. Displacement pumps, piston and plunger, are being discarded. Mine tailing pumping is sometimes done by means of centrifugal sand pumps. Deep-well pumps in a shaft or bore hole have proved successful for emergency use and for unwatering abandoned mines. Pumping systems and controls must be designed to handle the maximum inflow of water in mine workings. *See* PUMP. [SIMON H. ASH]

STORAGE AND SHIPPING

Although a task of great proportions, at least with respect to coal mining, the operation does not require extensive mine operating facilities. Coal is mined only when market demands for the product have been established and shipment to the consumer is made immediately. Storage facilities at the coal mines seldom provide more than surge capacity and are used to permit continuous operation of mine during temporary shutdown of preparation plant or loading outlets. Where storage of coal is required to provide a continuing fuel supply, necessary storage facilities are provided by the purchaser at the point of consumption.

Ore storage facilities are needed at mines because market cycles and changing seasons prevent steady ore shipments. At mines with mills or washing plants, different operating rates require surge capacity to assure uniform flow of ore to each unit (Fig. 9).

Bulky ores, such as those of iron, limestone, gypsum, sulfur, sand, and gravel, are kept in stockpiles on the ground. Iron mines, for example, store 5,000,000–6,000,000 tons of ore each winter in stockpiles. Common practice is to build the piles about 50 ft (15 m) high by dumping from a trestle or aerial tramway. At open-pit mines, stockpiles

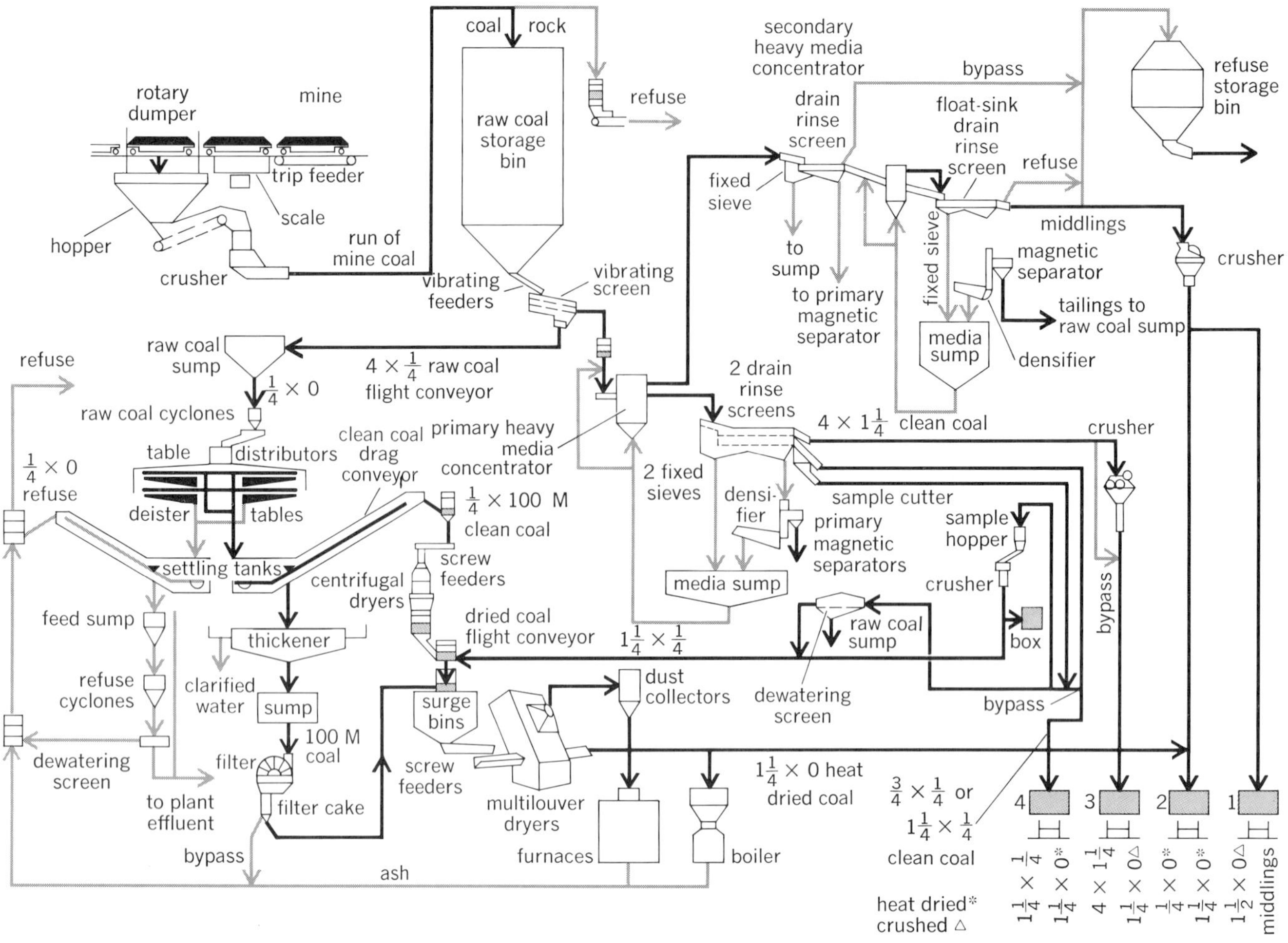

Fig. 9. Flow diagram of a complex coal preparation plant; dimensions in inches or M (mesh). (*Link-Belt Co.*)

are formed with dump trucks or stacker conveyor belts. Ore is withdrawn from stockpiles with mechanical loaders and occasionally through tunnel draw holes.

Underground mines, in themselves, provide a certain amount of ore storage in the stopes and in skip pockets, thus providing uniform flow to mills. At mills and ore-treatment plants, surge bins, with capacities of 50–3000 tons for coarse ore, are used ahead of primary crushers and bins of 5- to 50-ton capacity ahead of fine crushers and screens. Ore concentrate storage facilities range from 10-ton capacity at small mines to a 1,700,000-ton capacity unit, at Silver Bay, Minn., for storing iron ore pellets during the winter months.

Fig. 10. Dense-medium coal washing unit. (*Link-Belt Co.*)

Bulky ores are shipped by truck, in open railroad cars, and in bulk-cargo vessels. Concentrate is shipped short distances uncovered in standard dump trucks, and long distances by rail in closed box cars. Most mineral concentrates are shipped as bulk cargo in ocean-going vessels.

Mine refuse and tailings disposal. More than 20% of the product received by coal preparation plants is mine refuse and tailings. Depending upon the terrain and the availability of disposal areas to the mining operation, refuse is hauled from less than 100 ft (30 m) to more than 2 mi (3.2 km) to the disposal site. Trucks, carryall scrapers, conveyors, aerial trams, and hydraulic systems are the most popular means of transporting mine refuse away from the coal preparation plant.

In the ore industry, whenever possible, mine and mill waste is stored so that valuable mineral constituents can be reclaimed, or it is stocked for possible later use. Coarse tailing is loaded into cars or trucks and stockpiled. Fine tailing is impounded by dams made from the waste material.

Fig. 11. Compact, efficiently designed coal tipple containing automatic machinery for a processing flow similar to that of Fig. 10. (*Link-Belt Co.*)

These dams are formed of the sand portion of tailing slurry transported to the site in launders, or runningwater transporters, arranged to discharge the slime to the pond and the sand to the dam. Flocculants are often added to ensure clear-water overflow. At some underground mines, tailing is deslimed, dewatered, and returned to the mine for ground support.

Sorting, washing, and screening plants. About 60% of the coal produced in the United States is subjected to some form of mechanical cleaning. These plants use a variety of mechanical cleaning methods (Fig. 10), ranging from rudimentary equipment to multimillion dollar installations operating with a high degree of efficiency. Virtually all coal mines provide means of sizing the raw coal loaded from the mine into the size grades best adapted to the uses of the coal. When mechanical cleaning is not provided, the incombustibles are removed by handpicking before the coal is shipped. Even when coal washing facilities are used, the large-size lumps are frequently subjected to visual inspection and the refuse material discarded. Only in isolated cases can hand-picking be justified economically on coal less than 2 in. in size. Early preparation plants used stationary bar screens over which the coal flowed by gravity. The fine material passing through the bars was discarded as waste. With the development of coal combustion units capable of burning small-size coal, shaker and vibrating screens became commonplace in most coal preparation plants.

Coal and mine tipples usually house the washing, screening, and loading facilities of the mine. The early tipples were disproportionately large compared to capacity, but the modern tipple is a masterpiece of efficient design embodying automation to a considerable degree without retarding the ability to supply a satisfactory product covering a range of quality specifications (Fig. 11).

[WILLIAM L. CRENTZ; HORACE T. RENO]

FIELD SAMPLING AND ORE ESTIMATION

These processes and activities are somewhat the field counterparts of laboratory assaying. However, the samples so derived may be sent to the laboratory for more refined analysis.

Sampling. Ore sampling is the process of taking a small portion of ore in such a manner that the portion will be representative of the whole in respect to some quality or characteristic. The quality with which sampling is concerned in mineral deposits is usually the chemical analysis or grade of the material for commercial use. This is also called assay value. When a mineral deposit contains sufficient valuable ingredients to be mineable at a profit over costs, the mineral-bearing material is called ore, and the deposit an ore body. It is the assay value of the ore body which is of interest.

The unit of valuable material varies among ores: for example, for the precious metals (gold, silver, platinum), the assay is given in troy ounces of metal per short ton of ore, for the nonferrous metals (copper, lead, zinc, tin), the percentage of metal in the ore is used, and for other metals (iron, manganese, chromium), a more complete chemical analysis is used which may include the percentage or ratio of other specified ingredients occurring in the ore.

Mineral deposits are seldom exposed to view to any appreciable extent. Outcrops may appear at

MINING OPERATING FACILITIES

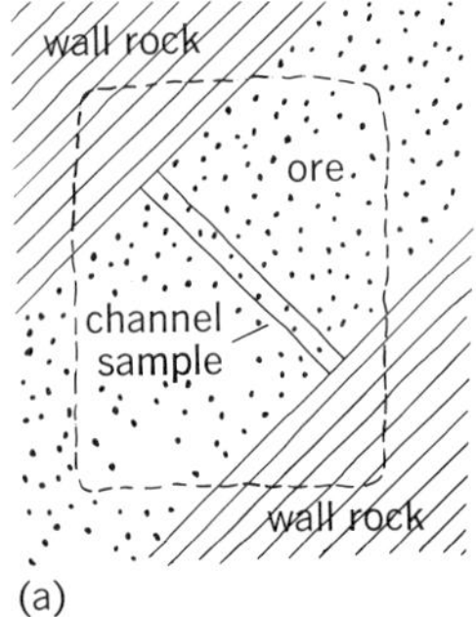

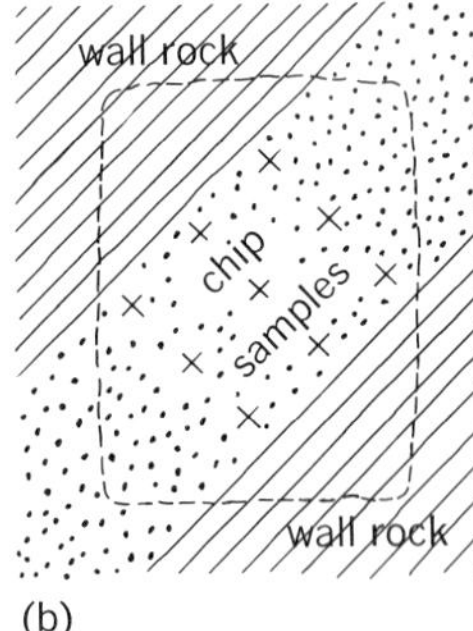

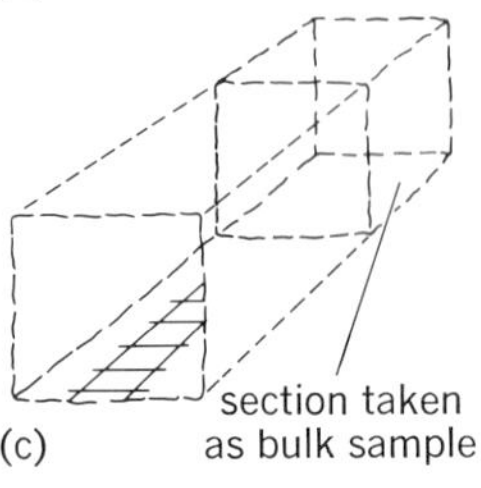

Fig. 12. Diagrams within a mine drift of major field sampling process. (*a*) Channel sampling. (*b*) Chip sampling. (*c*) Bulk sampling.

the surface, and if the deposit is being exploited, there will be limited underground exposures in shafts and other passageways. In addition, drill holes are used to investigate hidden deposits.

Three widely used methods of sampling are referred to as channel, chip, and bulk sampling. They are essentially the same in principle but differ in field procedure.

Channel sampling. In this sampling method, grooves are cut or chiseled across the exposed face at specified intervals (Fig. 12*a*). The grooves are about 4 in. (10 cm) wide by 1 in. (2.5 cm) deep and are preferably cut at right angles to the inclination of the vein or formation. Material cut from the channel is the sample. It is carefully saved for chemical analysis. If there is too much material from the cut, it is mixed thoroughly on the canvas on which it is caught and quartered down to a suitable amount. Long channels (across a thick deposit or lengthwise of a passageway) are often cut in 5- or 10-ft (1.5- or 3-m) sections, each section representing a sample.

Chip sampling. As distinguished from a channel cut, pieces are chipped from the face at designated points in chip sampling (Fig. 12*b*). Grab sampling, used in testing broken ore, such as ore in transit, is essentially the same in principle as chip sampling.

Bulk sampling. A substantial amount (perhaps several tons) of material is taken at each test location (Fig. 12*c*). This bulk is then either crushed and sampled or used for a mill test.

Sample calculations. A mineral deposit is three-dimensional; it has length, breadth, and thickness. Within this three-dimensional mass, each sample constitutes a point (chip sample), or a line (channel), or a unit (bulk cut), the location of which is known.

To determine the average assay value of the whole deposit or some designated part of it, it is necessary to combine the values of all samples located within the selected limits. To accomplish this, the several samples must be related to each other according to the portion of the deposit which each represents and correspondingly weighted for mathematical resolution to an average assay value. In basic terms, each sample is considered to represent the volume of material (ore) surrounding it and extending halfway to each adjacent sample. This volume, or a proportional factor, is the true numerical weight of the sample assay value when combining it with other assays.

When sample assays are used to determine the overall (average) grade of a mining face (either to guide progress of work or preliminary to estimating available ore tonnages), each assay is weighted according to its area of influence (assumed to extend halfway to adjacent samples) because the third dimension (depth or thickness) is considered identical for all points on the face. This procedure does not violate the principle of volume weighting previously noted. Face averages may then be combined on a volume basis to determine the average assay of the related body of ore.

Figure 13 is a diagram of a channel-sampled face to illustrate the principle of weighting. Since only the average grade of the exposed face is desired, no third dimension is involved and since the channels are uniformly spaced, the interval between them (second dimension) may also be omitted providing the face area in question is taken to extend one-half an interval beyond the end channels.

Estimation of ore. In addition to preliminary sampling and calculation of average mineral content, estimation of ore for mining implies recoverable values and hence involves efficiency of mining (mining losses plus ore left in place for support), and milling and smelting losses. To convert volume of ore into tonnage requires testing of the deposit for specific gravity along lines similar to sampling for assay values.

[ROLAND D. PARKS]

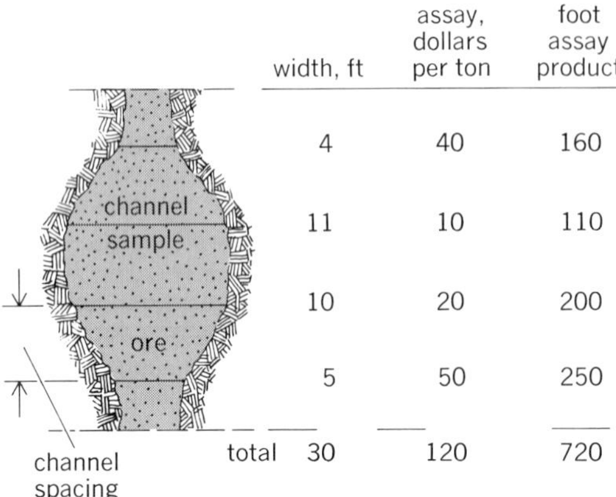

	width, ft	assay, dollars per ton	foot assay product
	4	40	160
	11	10	110
	10	20	200
	5	50	250
total	30	120	720

Fig. 13. Diagram of weighting based on channel-sampled face of ore body. Computations from this case are: Foot assay product ÷ sum of widths = average assay, so 720 ÷ 30 = $24/ton.

MINE EVALUATION

An appraisal of the monetary worth of a workable mineral deposit generally arrived at by experts in the mining industry. Such appraisal weighs the values of recoverable mineral content, operating efficiency when developed to production, and marketability of product. There are two parts to an evaluation: determination of factors, and calculation of present worth.

Determination of factors. Although past records may serve as a guide to the producing ability of the business and to indicate the cost-price relationship that may have obtained, the evaluation, as such, is concerned only with estimated future production and profits.

To estimate future operations the following features must be examined, analyzed, and verified: (1) grade and size of ore body and its remaining recoverable portion; (2) suitable production schedule, methods of mining and treatment, and corresponding plant and equipment requirements; (3) costs of production, treatment, and transportation; (4) grade, market and selling price of product, and operating profits; (5) production life; and (6) interest rates suitable to the business risks involved.

Size, shape, attitude, and quality of the mineral deposit are determined by geologic study (and maps, if available), by inspection of the deposit and sampling of exposures, and by test drilling and sampling so planned as to reveal hidden areas. Quality is expressed as the amount of valuable constituent per commercial unit of material, such as by troy ounces of gold per ton, or by percentage of copper, lead, or zinc, or by full chemical analysis. Structural quality and metallurgically undesirable constituents may also be included factors in the analysis. When the constituent is sufficiently valuable to warrant exploitation, the mineral-bearing material is spoken of as ore and the deposit is called an ore body.

Methods of mining, whether by open-pit or underground excavation, will be determined by the depth, size, and shape of deposit, by the structural strengths of ore and surrounding rock formations plus considerations of output rate, and by the time and investment required for development, plant, and equipment.

Output rate, mechanization, and selection of specific method and practices from applicable alternatives are considered in estimating operating costs. Production life is a function of recoverable reserves and annual output.

In contemplating investment in a mining operation, consideration must be given to business and market risks which will be related to the industry, location of the property, political environment, taxation, problems of climate, transportation, labor force, and probable hazards of the specific operation.

The process of gathering and analyzing data is spoken of as the examination of the property. When complete data are needed, as in the case of an operating property or one with substantial considerations at stake, a formal, or thorough, examination is made. For less-developed properties a preliminary examination may suffice to show whether or not to go ahead with detailed study.

Once the examination of the property is completed and essential factors for evaluation are established, including suitable interest rates on capital to be invested, the remainder of the evaluation consists of mathematical calculations to reduce the estimated future earnings to present worth.

Calculation of present worth. The evaluator, in the second part of the evaluation, faces the unique feature of the mineral industry, namely depletion. Minerals are exhaustible resources. When mined, they cannot be replenished, except through eras of geologic time. Unlike the agricultural industry where production may be regrown next year, a mine can produce the minerals from its ore body only once; the producer must then move on to new ore bodies to continue operations. This feature, depletion, has long been realized as fundamental to evaluation. Courts have recognized the principle of depletion in respect to mineral resources (State of Minnesota, Supreme Court 30723, 1935). Income tax authorities also have assented by allowing percentage depletion as a deduction in calculating taxable profit.

In view of depletion, the purchaser of a mineral deposit or a mining business must recover his capital during the remaining productive life. In evaluating mineral property, capital recovery is provided for in the mathematical calculation of present worth of estimated future profits. This is done by separating annual net profit into two parts, annual return of capital, and annual return on capital. The annual increments of capital return are considered as sinking fund payments which, when accrued at safe interest, will amount to the total invested capital at the end of the productive life.

The remainder of the annual profit is taken by the investor as interest on his capital investment. By this procedure the investor receives interest on his capital during the productive life and at the depletion of the ore body, has intact his full original capital with which to start a new operation. This type of "redemption fund" calculation follows the Hoskold premise for mine valuation.

The Hoskold equation for determining the present value of anticipated uniform annual income may be expressed as Eq. (6), where V_p is the present value, A is the annual income (profit), r is the safe interest rate (for sinking fund), r' is the rate of interest on invested capital, n is the years of productive life, and $R=(1+r)$ or \$1.00 plus one year's interest.

$$V_p = \frac{A}{[r/(R^n-1)]+r'} \qquad (6)$$

When estimated future income results in nonuniform annual profits, a variation of the above equation, embodying the same basic principles, may be used to calculate present value.

Capital outlays for plant, development, and taxes prior to production are deducted from present value of earnings in arriving at purchase or sale valuation. Interest on working capital and taxes are expense items to the operator.

Royalty is a fee paid to the owner for the privilege of working a mineral deposit. Although generally viewed as division of gross profit between fee owner and operator, royalty represents gross profit to the fee owner, or lessor, and a cost of production to the operator, or lessee.

Use of computers. Electronic computers are being applied to statistical analysis of large numbers of samples and to estimation of ore tonnages and grades. They have proved useful in correlating mining operations with ore grades and tonnages to effect maximum efficiency.

These applications do not affect the basic principles fundamental to sampling, ore estimation, and mine evaluation. The next section discusses some applications of computers. [ROLAND D. PARKS]

COMPUTERS IN MINING

Automatic computers are electronic devices which perform mathematical operations with great speed. These mathematical operations, no matter how complex, can always be broken down into the simplest arithmetical operations such as addition, subtraction, and discrimination (determining which of two numbers is the larger). There are two broad classes of computers depending upon how numbers are represented in the device. In the digital computers, the individual digits of a number are represented by separate devices, each device having as many states as the values assumed by the digit, that is, 10 states corresponding with the values 0, 1, 2, 3, 4, 5, 6, 7, 8, and 9 if using the decimal system. In the analog computers, numbers are represented by some physical quantity such as length, angular rotation, or electrical voltage, with the numbers bearing some simple, constant relationship to the physical quantity. In general, digital computers are more accurate but comparatively slower than analog devices. There has evolved a new type called the hybrid computers which is designed to take advantage of the better features of the two principal classes.

General applications. Since the function of computers is to perform calculations rapidly, the use of computers in mining is in connection with computations of interest in mining. The need for complex and lengthy calculations in mining stems from applications of operations research, systems analysis, and statistics to the operational, economic, and management problems of mining. It should be noted that computer uses in mining are relatively unsophisticated and few in number at the present time. However, once this introductory phase is over they may be expected to increase exponentially.

One use of computers for statistical calculations is directed at reducing large amounts of data to two or more numbers. Thus the mean value and the variance may be calculated and used to indicate the central and dispersion tendencies of the data. Calculation of grade or assay of an ore body or production from an ore body is fairly common. Another fairly common application is the use of regression analysis to determine the functional relationship between two variables of interest. For example, the grade of an ore body may be varying in some given direction with length measured from some fixed point of reference. Analysis may indicate a probable linear relationship between grade and position. The "best" value of the constants used in this linear function are then calculated. It should be made clear that it is not the fact that statistical calculations are being performed which necessitates the use of a computer but the volume of the calculations which is the determining factor. As statistics is used more frequently in mining, any and all statistical calculations involving large masses of data may be expected to be performed by computers. One factor stimulating such use is the existence of standard programs for the usual statistical calculations. This greatly simplifies the task of writing a program of instructions to the computer as to the various computational steps it must take to arrive at an answer.

In a very broad sense, all problems arising from the operations, management, and economics of mining call for a decision as to action to be taken. Heretofore the process followed was to collect and assimilate all the information available and bearing on the problem and then, based on experience and judgement, to arrive at some recommendation for action. With the development of operations research, systems analysis, statistical decision theory, and so forth, this process has been made more systematic and amenable to logic. Further, in many cases, the decision can be rated by giving a number which measures the confidence which may be placed in the answer.

It is not possible to enumerate the various applications involving computer use or to discuss in detail even a single application. However, a simple description of areas of application may indicate scope and complexity. B. Mackenzie in his discussion of the theory of the small mining firm considers the risks inherent in exploration with limited capital. Using a computer program, he is able to estimate costs versus expected gains and the probability of success. R. M. Ellis and J. H. Blackwell tackled the problem of an optimal drilling program whose object is to locate ore deposits in space, assuming the number and size of deposits are known. K. T. Marshall has dealt with the problem of sequencing drill holes in the exploration of a deposit.

In mine design, A. Soderberg has attempted to arrive at an open-pit design by minimizing unit production costs. He assumes constant metal prices, and examines future cash flow discounted to present value for different cutoff grades of the ore. By relating to capital investment he arrives at rate of return. M. D. Hassialis and coworkers have developed a computer program for the design of a mining sequence which will deliver ore to the treatment plant within specified limits of grade. The ore deposit is flat-lying and exhibits variation in grade in a vertical but not a horizontal direction. As the ore is mined by room-and-pillar using a heading and benching round, it is necessary to properly mix ores from these rounds to give the desired grade. L. Gibbs, J. Gross, and E. Pfleider have developed a computer program for the comparative evaluation of truck performance over a given road haul. A second program calculates individual cycle times for different truck-and-shovel combinations and the fleet requirements for any desired production rate and estimates comparative costs. This approach typifies the problem of equipment selection. By use of a computer program, C. Margueron has analyzed the international iron ore market and accounted for almost three-quarters of its variability. Inventory-control, transportation, and additional programs developed in other industries have been adapted to mining.

One of the most fruitful areas for application is mine design. As the design proceeds, points are arrived at where the designer has more than one alternative available for choice. These alternatives may differ in kind or in magnitude (for the same kind). This occurs again and again as the design sequence proceeds. The resulting combinations of alternatives multiplies so rapidly that it becomes impossible to compute the results for each combination so that a comparative basis for selection is available. The customary practice is to make a decision at each of these multiple-alternative points and so limit the computation to at most two or three cases. By writing a program which encompasses all or most of these alternatives it is possible to know the consequences in relatively short time and for reasonable cost.

Simulation of operations. One of the most interesting and ultimately most useful applications of computers in mining is for the calculations involved in simulation of operations. Here the object is to simulate the performance of an operation by constructing a mathematical model which depicts as accurately as possible the situation being represented. The model is a set of mathematical equations representing the operation together with a set of constraints which limit the range of applicability of the variables of the mathematical equations. These equations together with the constraints are embodied in a computer program, and the operation of this program gives the performance of the model.

One of the most useful aspects of a simulation program is the ability to incorporate nondeterministic information. Suppose operation of a drill was being simulated. It is known that drill steel breaks; further, past experience has shown that out of 1000 pieces of drill steel 100 have a life of 200 hr, 300 a life of 275 hr, and so forth. From this information, a frequency distribution for steel life can be constructed and made part of the computer program. When the computer needs this information to perform some detail of the calculation, it effectively samples this distribution and selects a particular value. If the calculation was to be repeated, the next time the computer needs steel life it may select another value. Thus, if the calculation was to be repeated many times, the number of times particular values for steel life are used by the computer would be proportional to their frequency. If more than one datum is represented by a frequency distribution, then running the program many times permits the operator to sample the performance of the model under all possible conditions.

This type of experimentation with the model or sampling the model's performance is extremely useful. It permits estimation of the worst, the best, and the average performance. Further, it permits estimation of the probability that performance will lie between certain limits. From this it is possible to estimate certain risks. It must be stressed however that the validity of this information depends completely on the validity of the model, that is, upon how accurately it represents reality. This is the basic weakness of simulation. As more and more detail is added to the model to make it conform closer to reality, the program may become so long and complex that computer costs become excessive.

An early attempt in the simulation area was made by J. Dunlap and H. Jacobs, who created a model for the operation of a dragline in a phosphate mine. M. Sheinkin and D. Julin established a very large simulation program for the Climax Molybdenum Mine in Colorado. Using files of sample data for each working area in the mine, the program simulates the effects on production and grade of different mining plans while maintaining control of the cave surface. The program also gives sampling and development control data. Another program was developed by T. O'Neil and C. Manula to simulate materials handling in open-pit mining. The program cycles trucks between assigned loading and discharge points over a measured haulage route. Performance characteristics of loading and hauling units, service time distributions, and many other details needed for a realistic model are included. P. Czegledy and M. D. Hassialis wrote a program simulating the interaction of several adjacent but different uranium-bearing strata with a core-hole scintillometer to determine the effectiveness of interpreting radiometric logs. The program takes into account the various alternative events which may take place when radiation traveling from some point within the stratum encounters an atom of matter. In mathematical terms the method is known as a "random walk."

To summarize, the use of computers in mining is still very preliminary and relatively unsophisticated. However, as mining engineers trained in operations research, statistical decision theory, and systems analysis become available, it is to be expected that use will increase rapidly with great benefits to the industry. *See* OPERATIONS RESEARCH.

[M. D. HASSIALIS]

Bibliography: American Standards Association, *American Standard Recommended Practice for Drainage of Coal Mines*, U.S. Bur. Mines Bull. no. 570, 1957; S. H. Ash et al, *Corrosive and Erosive Effects of Acid Mine Waters on Metals and Alloys for Mine Pumping Equipment and Drainage Facilities*, U.S. Bur. Mines Bull. no. 555, 1955; S. H. Ash, H. A. Dierks, and P. S. Miller, *Mine Flood Prevention and Control*, U.S. Bur. Mines Bull. no. 562, 1957; R. M. Ellis and J. H. Blackwell, Optimum prospecting plans in mining exploration, *Geophysics*, April, 1959; L. W. Gibbs, J. R. Gross, and E. P. Pfleider, System analysis for truck and shovel selection, *Trans. A.I.M.E.*, December, 1967; E. J. Harris, The fluid network analyzer as an aid in solving mine ventilation distribution problems, *Trans. A.I.M.E.*, vol. 226, 1963; H. L. Hartman, *Mine Ventilation and Air Conditioning*, 1961; D. S. Kingery, *Introduction to Mine Ventilating Principles and Practices*, U.S. Bur. Mines Bull. no. 589, 1960; R. S. Lewis and G. B. Clark, *Elements of Mining*, 3d ed., 1964; E. E. McElroy and D. S. Kingery, *Making Ventilation-Pressure Surveys with Altimeters*, U.S. Bur. Mines Inform. Circ. no. 7809, 1957; R. D. Parks, *Examination and Valuation of Mineral Property*, 4th ed., 1957; R. Peele and J. A. Church (eds.), *Mining Engineers' Handbook*, 3d ed., 2 vols., 1941; T. J. O'Neil and C. B. Manuala, Computer simulation of materials handling in open pit mining, *Trans. S.M.E. of A.I.M.E.*, June, 1967; A. Roberts (ed.), *Mine Ventilation*, 1960; M. Sheinken and D. E. Julin, Computer simulation aids in long-range production planning at climax, *Trans. S.M.E. of A.I.M.E.*, February, 1967; J. Sinclair, *Winding and Transport in Mines*, 1959; J. Sinclair, *Geological Aspects of Mining*, 1958; W. W. Staley, *Introduction to Mine Surveying*, 2d ed., 1964.

Mining safety

The prevention of mine worker injury by precautionary practices in mining operations. This article discusses records of mine safety, methods of fire prevention, government regulations, training programs for mine workers, and mine environment problems causing disease. For other aspects of mine operations *see* COAL MINING; MINING OPERATING FACILITIES; UNDERGROUND MINING.

Safety records. Vigilance is the price of safety in a mine, as it is in any other activity. Fortunately, the presentation of mining hazards in motion pictures and television bears little relation to fact. The Mine Safety Appliances Co. in the late 1930s originated the John T. Ryan trophies, to be awarded annually to the Canadian mines with the best safety records each year. These awards include two national trophies, one for metalliferous mining and one for coal mining, and six regional trophies, two for coal mining and four for metalliferous mining. The awards are based on the number of lost-time accidents during 1,000,000 worker-hours of exposure for metalliferous mines, and for 120,000 worker-hours of exposure in coal mines.

A lost-time accident is defined as one that results in a compensation payment for a total or partial disability that lasts more than 3 days. In Canada, before Aug. 1, 1968, a 3-day waiting period was required in some provinces before an injury became compensable. The records of the workers' compensation boards are used to compile safety records, and all locations of employment are included, such as mine, mill, offices, and shops, but smelters and open-pit or strip-mine operations are not included.

The Canada John T. Ryan trophy for metalliferous mines is usually won by a mine with no lost-time accidents as defined. A mine with four or five accidents is likely to be disqualified from the countrywide competition. A fatal accident eliminates a mine. The coal mining record is not as good as that of the metalliferous mines; for one thing, there are fewer coal mines. The low incidence of accidents is not attained without effort. Supervisors, like traffic police, spend a lot of their time trying to keep people from acting foolishly. In North America governments have regulations regarding the safety of workmen in mines and inspectors to see that they are observed.

Fire hazards. Government regulations for mine operation and the safety of workmen in mines were established in North America during the 19th century, patterned on those of Europe and Great Britain, where there was a longer history of coal mining than in North America. The regulations were first directed to coal mining because it presents the hazard of combustible gas from bituminous coal and flammable dust. Coal is made up of volatile material and fixed carbon. If the ratio

$$\frac{\text{volatile}}{\text{volatile} + \text{fixed carbon}}$$

exceeds 0.12, the dust can burn explosively when mixed with air in dangerous proportions, and the range of dangerous proportions is quite wide. All bituminous coals have a ratio in excess of 0.12. Combustible gas is especially dangerous because it is mobile and if it is ignited the ensuing fire or explosion may affect men remote from the source of trouble. Other accidents, a fall of rock, for instance, do not affect those working outside the location of the accident.

Methane gas, the major component of the combustible gas, is dangerous in two ways. It dilutes the mine air and so reduces the amount of life-sustaining oxygen. Coal mines are classified as gassy, slightly gassy, and nongassy. A nongassy mine has less than 0.05% methane in the air. If methane is in a concentration of 5–13.9%, it may be ignited and will burn explosively. The most violent explosive mixture is 9.4% methane. At some point, as the location of the emission is approached, the explosive ratio will be reached; a single spark can set off an explosion. The resultant turbulence in the air will stir up the coal dust and the fire can spread with explosive violence.

A mine fire has side effects. Men may be caught in the actual fire, but the burning produces carbon monoxide and carbon dioxide which will spread far beyond the fire area. If the roof is sulfurous shale, it may be brought down by the heat.

Prevention of ignition. The propagation of a fire in the dust stirred up by the air turbulence accompanying an explosion may be inhibited by dusting. An incombustible rock dust is spread throughout the workings. When stirred up, it dilutes the mine dust and absorbs heat, so combustion cannot be maintained. The initiation of combustion must be prevented.

Equipment. In the United States only equipment certified by the U.S.B.M. (United States Bureau of Mines) as permissible may be used where gas or combustible dust may be present. The restriction applies to electrical equipment, including cap lamps and machinery. The certification is based on the no-sparking qualities of switches and adequate current-carrying capacities for electrical equipment; and on ensurance that machinery has no operating parts that can heat to the ignition temperature for gas or dust. In Canada a British certification is also accepted.

Explosives. Only permitted explosives may be used. All explosives involve combustion and high temperatures. The U.S.B.M. has certified certain explosives as permissible for use in coal mines. They are compounded to produce a lower temperature and volume of flame than ordinary explosives, and certain salts are included in the formulas to quench the flame rapidly. Permitted explosives must be used as prescribed for prescribed conditions. Special attention is required for stemming to prevent blowouts.

Safety lamp. The biggest single advance in coal mine safety was probably the invention of the Davy safety lamp in 1815. A mantle of wire gauze around the flame dissipated the heat from the flame to below the ignition temperature of methane, which is about 650°C. It provided light and the miner no longer had to work with an open flame. The height and color of the flame gave a measure of the amount of methane in the air. Improved models are still in use, though they are being superseded by no-flame instruments that give a prompt reading or can monitor the methane content continually. The miner is no longer dependent on a flame for light since the electric cap lamp became available.

Ventilation. Coal mine operators have developed ventilation techniques, which until recently, surpassed those used in metal mines. Great care is taken to sweep working places with enough air to prevent dangerous accumulations of gas or of dust. The advent of mechanical coal-cutters and other mechanical equipment has made ventilation more difficult, but the problem has been overcome.

Other mine fires. Methane is not confined to coal mines. It is occasionally released from pockets in metalliferous mines. It is seldom troublesome unless it has accumulated in an unused unventilated working or in a sump at the bottom of a shaft.

Sulfide ores containing certain sulfide minerals may oxidize and generate sulfur dioxide. This may generate enough heat to ignite wood. Pyrrhotite should always be suspected as such an oxidizing sulfide.

Dry, timbered mines in which there is careless smoking are obvious fire hazards. Most metalliferous mines are damp and vigorous fires are not expected, but smoldering fires may be even more hazardous. They may burn for a long time without being detected and the incomplete combustion generates carbon monoxide which is heavier than air, odorless, and lethal. The fire is usually in rubbish accumulated in abandoned places. Good housekeeping is required. Many fires are started in old power cables when the insulation has broken down, and a surprising number start during locomotive battery charging. Sparks from acetylene torch burning are another common source of fire.

Mine rescue teams. The U.S.B.M. officials have accumulated a vast store of fire-fighting knowledge. Most mines have teams of men trained in mine rescue and in the use of U.S.B.M.-approved equipment in accordance with manuals prepared by the U.S.B.M. officials. They work under their own supervisors but U.S.B.M. officials are available for consultation.

Government regulations. Government regulations for safety in mine operations must be enforced. The higher the government authority and the less localized the enforcement agent, the more effective the enforcement will be. The basic enforcement unit in the United States is the state authority, and even obviously good regulations are often hard to enforce. In addition ot the human tendency to resist change, pressure may be brought to bear on the legislatures of the mining states. Some government inspectors may be appointed for political expediency; in the past the company safety inspector was too often an employee given a sinecure in lieu of a pension.

The U.S.B.M. has been assigned responsibility for the health and safety of United States miners. The period of 1907–1913 was formative. States are jealous of their prerogatives, and so it was 1941 before the U.S.B.M. officers had the right to enter a mine. They were chiefly involved by invitation after a disaster. They attained the right to order a coal mine closed for dangerous practices only in 1952. Legislation in 1969 resulted in a safety code to apply to all mines, coal and metalliferous. The initial responsibility continues to be with the state authority, and the U.S.B.M. will interfere only when the state regulations and enforcement do not meet the requirements of the Federal code.

Regulations in Canada are prepared and enforced by the provinces. The Northwest Territories, Yukon, and the Arctic Islands are not provinces and are under the authority of the federal Canadian government. The Ontario metal mining code is the most comprehensive and has been used as a model for the preparation of the codes in other provinces and countries. There is no coal mining in Ontario. The coal mining provinces based their regulations on British codes originally, and these regulations have since been modified in the light of experience in the United States.

Occurrence of accidents. The following discussion is based on data from the *Ontario Department of Mines Inspection Branch Annual Report for 1967*, which gives accident experience for an average of 17,461 men who put in 32,391,000 worker-hours in underground mines and open pits of all sizes and degrees of organization. There were 1887 lost-time accidents. The experience did not vary much from that of the preceding 4- or 5-year period.

The time distribution for lost-time accidents was 1 per 17,100 worker-hours or, using 1850 worker-hours per worker-year, 1 per 9.4 worker-years. Whether or not a worker has had an accident during any one year has no bearing on whether another could take place in the same year or any other year, so this is a Poisson distribution. A worker would have a 0.33 probability of having no lost-time accidents in 9.4 years.

Actually, the mathematical probability is misleading because a worker is more apt to have an accident in the earlier years of employment before becoming mine-wise, or if the worker is employed in a developing mine, before a safety program is well organized.

The 16 fatal accidents had an incidence of 1 per 2,020,000 worker-hours, or 1 per 1090 worker-years. Mathematically a group of 100 miners working 10 years would have about 0.37 probability of having no fatal accident. That is subject to the same reservations as those set out above for the lost-time accidents.

Training. The necessity for training miners to work safely cannot be overstated. Of the 1887 nonfatal accidents in Ontario in 1967 about 34% (641) were personal accidents – fall of persons, or strains while moving or lifting or handling material other than rock or ore. There is reason to suspect laxity in the use of safety equipment, such as gloves and safety belts and boots; ineffective instruction on how to lift; and insufficient emphasis on the importance of good housekeeping in working places. It must be emphasized that the data are from all sizes of mines and from mines that range from those operated under rather primitive conditions to those using modern equipment and management. Some of them would be close runners-up in the Canada John T. Ryan trophy competition, and one of them won the Regional trophy. Others were barely passing the government inspections. One runner-up for the Canada trophy had 2 accidents and the next had 3 accidents in 1967. The winner of the Ontario Regional trophy had 7 accidents per 1,000,000 worker-hours.

The most lethal class of accident is the fall of rock or ore. About 10% (185) of all accidents, including fatal accidents, in Ontario in 1967, were falls of rock or ore, but 31% (5) of the fatal accidents were from that cause. The real cause was probably the failure to recognize an unsafe condition due to neglect or lack of experience. Either cause requires better training of the supervisors and the workers. Nearly half of the accidents occurred during drilling or scaling of loose rock, clearly showing a lack of skill or the neglect of an unsafe condition.

The advent of mechanized mining in large deposits requires greater areas of exposure to provide room to maneuver equipment. Fortunately, it has been accompanied by the extended use of rock bolts as a means for rock reinforcement, and the incidence of falls of rock has been reduced. The illustration shows an experimental stope in a mine where U.S.B.M. engineers tested the effectiveness of rock bolting.

The ultimate in mine safety cannot be achieved without the complete support of top management, but the key person in the achievement is the worker, and in the chain of responsibility it is the supervisor at the lowest level of contact with the worker. Though the worker may have the best of intentions, the supervisor is handicapped because miners work in scattered places, out of sight except for short intervals each day. That condition has improved in mines that work larger deposits with mechanical equipment. There is a concentration of working places and the crews are under less intermittent surveillance.

Most mines have a staff safety engineer responsible only to top management. The safety department usually has no line authority. The most important duty that the safety engineer has is the education of supervisors and workers by lectures and training sessions.

It must be recognized that any accident could be fatal, even a neglected scratch from a nail. Furthermore, an accident need not involve injury to a person. Any undesirable happening at any time or in any place that has not been foreseen is an accident. It may be only by chance that there is no personal injury.

Environmental problems. Instantly recognizable injuries to persons attract more attention, but for ages miners have been subject to pneumocon-

Rock bolting in a Michigan copper mine. (*Photograph by H. R. Rice*)

iosis. This includes all lung diseases, fibrotic or nonfibrotic, caused by breathing in a dusty atmosphere. It takes time to develop, and began to be recognized as something to be eliminated early in the 20th century. Most varieties are not fibrotic and will clear up when the person is out of the dust-laden atmosphere; silicosis and the effects of breathing radioactive dust or gas, however, will not.

Silicosis. South African mine doctors led in the diagnosis of silicosis. About 1920 mine operators in North America became aware of the magnitude of the problem and moved to do something about it. It is caused by inhaling silica dust, and possibly some other mineral dusts. It is not confined to mining. Any industry that produces a silica-dusty atmosphere is dangerous. Furthermore, a clear-looking atmosphere may still be dangerous because the particles that are harmful are less than 0.001 mm in size. At that size they settle slowly and do not show in a beam of light.

The small particles pass the filtering hairs and mucus in the nose and throat. They reach the lungs and by some action, mechanical or chemical, create fibrosis. The useful volume of the lung is reduced. The disease is seldom lethal in itself but the victim is susceptible to tuberculosis and pneumonia.

As soon as the cause was recognized, the operators and the government authorities increased ventilation requirements and insisted on water sprays, particularly after blasting, to knock down the dust. Wet drilling was well established at that time but there have been some improvements in the drills. The dust content of the mine air at working places and elsewhere was measured, especially in the fine sizes because the presence of coarse dust is easily detected. What are thought to be safe, or acceptable, working levels have been established.

All employees were examined by x-ray for fibrosis in the lungs, and for a proneness to develop silicosis. Some lung shapes are more likely to develop it than others are. Employees with developing fibrosis were given work in dust-free locations, or treatments and pensions. Employees are now examined at least yearly to detect the disease in the primary stage so that prompt action may be taken, and the incidence of silicosis is lessening. Aluminum dust sprayed into the air in mine change houses and underground has been found to have a prophylactic, and possibly therapeutic, effect. Ventilation is the only completely effective remedy.

Radioactivity. It was observed that in some mine areas silicosis seemed to be much more virulent than in others. This was thought to be due to some additive to the silica. It was found that coal dust speeded up the development of the disease.

When uranium mining got under way, it was found that the rocks emitted radon gas which broke down in the lungs and produced lung cancer. When the instruments developed for radioactivity research became available, it was learned that many rocks that had a low level of radioactivity emitted small amounts of radon gas, which in some cases were carried into the mine workings in the mine water and released under the reduced pressure. That provided one reason for the virulence of silicosis in some mining areas.

The government authorities moved quickly and established safe levels of radioactivity and safe exposure time in those levels. The regulations are enforced. The acceptable levels are attained by ventilation.

Other considerations. When diesel engines were introduced underground for motive power, they had to be certified for an acceptable level of carbon monoxide and oxides of nitrogen in the fumes. The required levels were attained by engine design and by scrubbers in the exhaust system. An engine must have a U.S.B.M. certificate of approval, and then it is only approved to travel certain routes for which the government inspectors consider the ventilation to be adequate. [A. V. CORLETT]

Bibliography: R. S. Lewis and G. B. Clark, *Elements of Mining*, 3d ed., 1964; R. Peele (ed.), *Mining Engineers' Handbook*, 3d ed., vol. 2, 1941.

Moment of inertia

A relation between the area of a surface or the mass of a body to the position of a line. The analogous positive number quantities, moment of inertia of area and moment of inertia of mass, are involved in the analysis of problems of statics and dynamics respectively.

The moment of inertia of a figure (area or mass) about a line is the sum of the products formed by multiplying the magnitude of each element (of area or of mass) by the square of its distance from the line. The moment of inertia of a figure is the sum of moments of inertia of its parts.

Moment of inertia of area. In practice, only moments of inertia of a plane area about mutually perpendicular axes (lines) in or normal to its plane are useful (Fig. 1).

The moment of inertia of plane area A about the X and Y axes in its plane are respectively $I_X = \int y^2\,dA$ and $I_Y = \int x^2\,dA$. In these, x and y are the coordinate locations of area element dA.

The polar moment of inertia of a plane area is its moment of inertia about an axis normal to the plane of area. The polar moment of area A about the Z axis is $J_Z = \int r_Z^2\,dA$. As referred to a common origin of axes $J_Z = I_X + I_Y$.

Moment of inertia of area is measured in quartic length units, such as ft^4.

Moment of inertia of mass. For a body of mass distributed continuously within volume V, the moment of inertia of the mass about the X axis is given by either $I_X = \int r_X^2\,dm$ or $I_X = \int r_X^2 \rho\,dV$, where dm is the mass included in volume element dV at whose position the mass per unit volume is ρ (Fig. 2). Similarly, $I_Y = \int r_y^2 \rho\,dV$ and $I_Z = \int r_Z^2 \rho\,dV$. Mass moment of inertia is measured in units of mass times length units squared, such as g-cm^2.

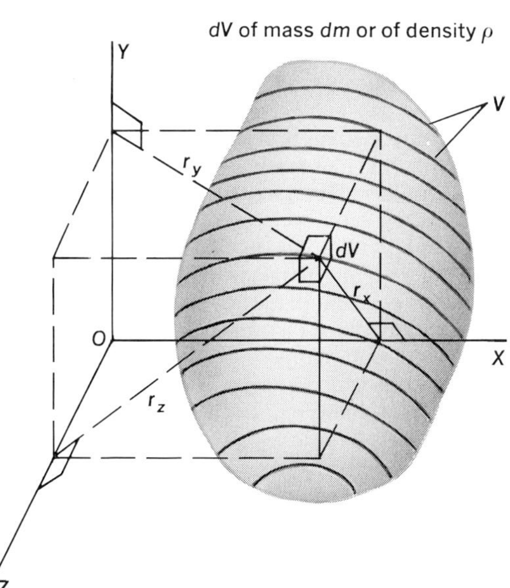

Fig. 2. Moment of inertia of a volume.

Parallel-axis theorem. The moment of inertia of a figure about any axis is the sum of its moment of inertia about a parallel axis containing the centroid of the figure and the product formed by multiplying the magnitude of the figure (its area or mass) by the distance squared between the parallel axes; for area, $I = \bar{I} + AD^2$; for mass, $I = \bar{I} + MD^2$. Accordingly the moment of inertia about a centroidal axis is less than its moment about any parallel axis.

Principal axes of inertia. The moments of inertia of a figure about lines which intersect at a common point are generally unequal. The moment is greatest about one line and least about another line perpendicular to the first one. A set of three orthogonal lines consisting of these two and a line perpendicular to both are the principal axes of inertia of the figure relative to that point. If the point is the figure's centroid, the axes are the central principal axes of inertia. The moments of inertia about principal axes are principal moments of inertia. [NELSON S. FISK]

MOMENT OF INERTIA

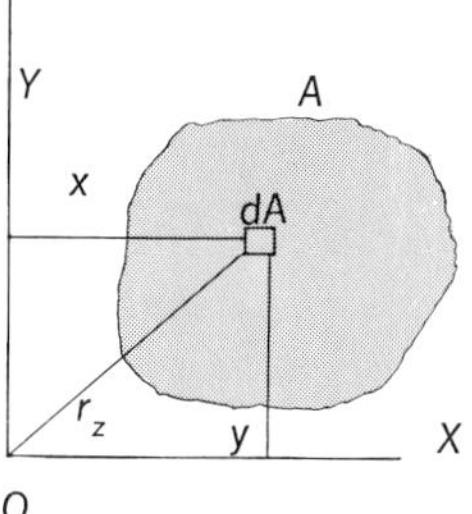

Fig. 1. Moment of inertia of an area.

Momentum

Linear momentum is the product of the mass and the linear velocity of a body. It is defined by Eq. (1),

$$\mathbf{P} = m\mathbf{v} \qquad (1)$$

where m is the mass and $\mathbf{v}$ is the linear velocity. Since linear momentum is the product of a scalar and a vector quantity, it is a vector and hence has both magnitude and direction.

The angular momentum of a body is defined as the product of its moment of inertia and its angular velocity.

No special names are given to the units of linear momentum. The units are gram-centimeters per second, kilogram-meters per second, and slug-feet per second in the centimeter-gram-second, meter-kilogram-second, and British engineering systems of units, respectively.

According to the general statement of Newton's second law, for a force $\mathbf{F}$, a momentum $\mathbf{P}$, and a time t, Eq. (2) holds. Thus Newton's second law

$$\mathbf{F} = d\mathbf{P}/dt \qquad (2)$$

involves the time rate of change of momentum. Usually the mass of a body is constant, and the time rate of change of momentum of a body equals the product of its mass and acceleration. However, under certain conditions the mass can change, as when a rocket moves through space by consuming part of its mass as fuel. Whenever a change in mass occurs, the total time rate of change of momentum must be considered in describing the motion. Changes of momentum are important in collision processes.

When a group of bodies is subject only to forces that members of the group exert on one another, the total momentum of the group remains constant.

[PAUL W. SCHMIDT]

Monorail

A distinctive type of materials-handling machine that provides an overhead, normally horizontal, fixed path of travel in the form of a trackage system and individually propelled hand or powered trolleys which carry their loads suspended freely with an intermittent motion. Because monorails operate over fixed paths rather than over limited areas, they differ from overhead-traveling cranes, and they should not be confused with such overhead conveyors as cableways.

Relatively simple but adequately efficient monorail systems for specialized applications have flat steel bars or galvanized pipes for trackage. However, standard I beams or other similar shapes are used more and more frequently, with the wheels of the trolley bearing directly upon the lower flange of the beam. The latter type are used more for heavy-load service.

Garment manufacturers and cleaning establishments use the simple pipe-rail system for hanging garments on hangers or wheel-equipped trolleys. Switches, crossovers, and other components make setups flexible (see illustration). These systems can be arranged to run onto freight elevators, along loading platforms, and directly into carriers.

Of primary importance in a conventional monorail system are the rails. They are connected with butt or lap joints and must be smooth to allow the trolley to move freely. Clamps or brackets suspend the trackage from ceilings or walls. Because monorail systems do not operate with continuous motion, the tracks need not be arranged in self-closing lines. Spurs can be run into paint booths, cooling rooms, and similar work areas. These paths are selected by switches, such as the tongue or the slide (glider) varieties. Specially constructed sections permit 90° changes in trolley travel; turntables permit articles to be turned through 360°.

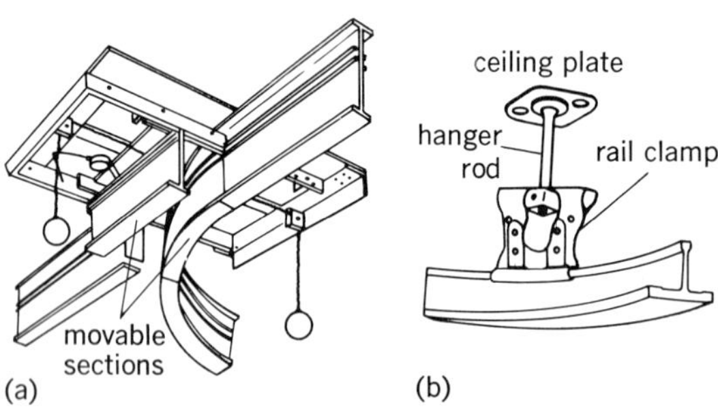

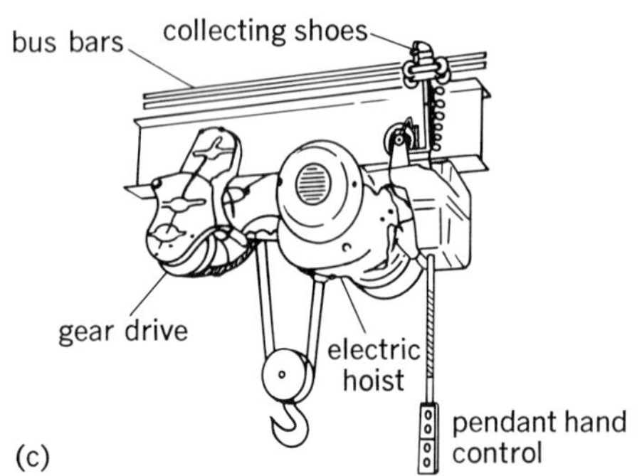

Typical parts of overhead monorails. (*a*) Slide (glider) type. (*b*) I-beam track. (*c*) Powered trolley.

Where monorails pass through fire doorways, lift-out sections permit the door to close in the event of fire. Lift and drop sections shift the flow of traffic from one line to another at a different level, thus eliminating the need for inclined and declined tracks.

Wheels mounted in trolleys ride on the flanges of the track, in most varieties. Two-wheeled trolleys, connected by bars, combine to make up carriers. The number of wheels depends upon the desired load-carrying capacity. In many instances hand or powered hoists are suspended from the carriers. To provide current for powered hoists, special electrification equipment, such as bus bars and collecting shoes, are used. If powered carriers are required, the drive may be secured in one of several ways. Examples are the motor-driven rubber tire which contacts the under flange of the track, or the trolley wheels which are driven through gear trains by an electric motor. Trolleys of the first type operate without slippage and make possible the introduction of sections with slight inclines and declines. Special carriers are used with monorail systems for handling specific products, such as batches in bakeries and the movement of clothes from one process to another in laundries. Scale sections can be included in a trackage line to weigh such products as batches, textile beams, or rolls of paper without causing a delay in traffic flow. Many below-the-hook devices can be used with monorails. *See* HOISTING MACHINES.

Monorail elements are utilized in the construction of overhead-traveling cranes if distribution over an area rather than along a fixed path is required. Such cranes can not accommodate the heavy loads handled by cranes built with structural girders, but they are adequate for many industrial applications. *See* BULK-HANDLING MACHINES; MATERIALS-HANDLING MACHINES.

[ARTHUR M. PERRIN]

Bibliography: J. M. Apple, *Plant Layout and Materials Handling*, 3d ed., 1977.

Motor

An electric rotating machine which converts electric energy into mechanical energy. Because of its many advantages, the electric motor has largely replaced other motive power in industry, transportation, mines, business, farms, and homes. Electric motors are convenient, economical to operate, inexpensive to purchase, safe, free from smoke and odor, and comparatively quiet. They can meet a wide range of service requirements—starting, accelerating, running, braking, holding, and stopping a load. They are available in sizes from a small fraction of a horsepower to many thousands of horsepower, and in a wide range of speeds. The speed may be fixed (or synchronous), constant for given load conditions, adjustable, or variable. Many are self-starting and reversible. For uniformity and interchangeability, motors are standardized in sizes, types, and speeds. *See* ELECTRIC ROTATING MACHINERY.

Electric motors may be alternating-current (ac) or direct-current (dc). There are many types of each. Although ac motors are more common, dc motors are unexcelled for applications requiring simple, inexpensive speed control or sustained high torque under low-voltage conditions.

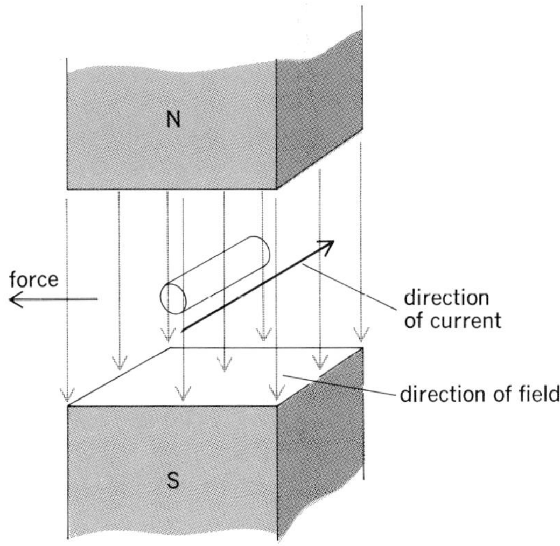

Relative directions of field flux, current, and force.

Motor classification. Motors are classified in many ways. The following classifications show some of the many available variations in types of motors.

1. Size: flea, fractional, or integral horsepower.
2. Application: general purpose, definite purpose, special purpose, or part-winding start. May be further classified as crane, elevator, pump, and so forth.
3. Electrical type: alternating-current induction, synchronous, or series; direct-current series, permanent magnet, shunt, or compound.
4. Mechanical protection and cooling: (*a*) open: dripproof, splashproof, semiguarded, fully guarded, externally ventilated, pipe ventilated, weather protected; (*b*) totally enclosed: nonventilated, fan cooled, explosionproof, dustproof, ignitionproof, waterproof, water cooled, water-air cooled, air-to-air cooled, pipe ventilated, fan cooled guarded.
5. Speed variability: constant speed, varying speed, adjustable speed, adjustable varying speed, multispeed.
6. Mounting: floor, wall, ceiling, face, flange, vertical shaft.

Characteristics. Each electrical type of motor has its own individual characteristics. Each motor is selected to meet the requirements of the job it must perform. For individual motor characteristics *see* DIRECT-CURRENT MOTOR; INDUCTION MOTOR; REPULSION MOTOR; SYNCHRONOUS MOTOR; UNIVERSAL MOTOR. For comparison of all ac motors *see* ALTERNATING-CURRENT MOTOR.

Principles of operation. When a conductor located in a magnetic field carries current, a mechanical force is exerted upon it (see illustration). This force has the value shown in Eq. (1), where i

$$F = Bil \qquad \text{newtons} \tag{1}$$

is the current in amperes, B is the magnetic density in webers per square meter, and l is the conductor length in meters. The illustration shows the relative directions of current, field, and force. The force reverses with either current or field reversal, but not when both are reversed. The torque T is the product of this force and the rotor radius.

If the conductor moves in the direction of F, an emf e is generated which opposes the current (motor action). If the conductor is moved against F, this emf will assist the current (generator action). Its value is shown in Eq. (2), where v= velocity of conductor across the flux in meters per second.

$$e = vBl \qquad \text{volts} \tag{2}$$

The product ei represents the power converted, in watts, shown in Eqs. (3) and (4), which are the

$$\text{Motor output} = ei - \text{rotative loss} \tag{3}$$

$$\text{Generator shaft input} = ei + \text{rotative loss} \tag{4}$$

bases for the emf and output formulas of dynamo machinery. Many machines will operate as either a motor or a generator, but they should be designed for the particular service.

Current may be fed into the field and armature by conduction, as in dc machines and ac series motors, or into the stator by conduction and the rotor by induction, as in ac induction and repulsion motors. [ALBERT F. PUCHSTEIN]

Bibliography: A. E. Fitzgerald and C. Kingsley, *Electric Machinery*, 3d ed., 1971; V. Gourishankar and D. H. Kelly, *Electromechanical Energy Conversion*, 2d ed., 1973; I. L. Kosow, *Electric Machinery and Transformers*, 1972; L. Matsch, *Electromagnetic and Electromechanical Machines*, 2d ed., 1977.

Motor-generator set

A motor and one or more generators, with their shafts mechanically coupled, used to convert an available power source to another desired frequency or voltage. The motor of the set is selected to operate from the available power supply; the generators are designed to provide the desired output. Motor-generator sets are also employed to provide special control features for the output voltage.

The principal advantage of a motor-generator set over other conversion systems is the flexibility offered by the use of separate machines for each function. Assemblies of standard machines may often be employed with a minimum of engineering required. Since a double energy conversion is involved, electrical to mechanical and back to electrical, the efficiency is lower than in most other conversion methods. In a two-unit set the efficiency is the product of the efficiencies of the motor and of the generator.

Motor-generator sets are used for a variety of purposes, such as providing a precisely regulated dc current for a welding application, a high-frequency ac power for an induction-heating application, or a continuously and rapidly adjustable dc voltage to the armature of a dc motor employed in a position control system. *See* GENERATOR; MOTOR. [ARTHUR R. ECKELS]

Nailing

The driving of nails in a manner that will position and hold two or more members, usually of wood, in a desired relationship to each other. The contact pressures between the surfaces of the nails and the surrounding wood fibers hold the nails in position. For some types of nails see illustration.

Strength of a nailed joint. Factors that determine the strength and efficiency of a nailed joint are (1) the type of wood, (2) the nail used, (3) the conditions under which the nailed joint is used, and (4) the number of nails.

Nail
spiral-threaded, insulated siding, face nail
annular-ring, gypsum board, dry-wall nail
asbestos shingle nails: annular-ring, spiral-threaded
annular-ring, plywood roofing nail for applying wood or asphalt shingles over plywood sheathing
annular-ring, plywood siding nail for applying asbestos shingles and shakes over plywood sheathing
spiral-threaded, casing head, wood siding nail
annular-ring roofing nail for asphalt shingles and shakes
spiral-threaded roofing nail for asphalt shingles and shakes
annular-ring roofing nail with neoprene washer
spiral-threaded roofing nail with neoprene washer
insulated siding nail
gypsum lath nail
wood shake nail
wood shingle nail
roofing nail
general-purpose finish nail
sinker head, wood siding nail
casing head, wood siding nail

Special- and general-purpose nails.

In general, hard, dense woods hold nails better than soft woods. The better the resistance of a nail to direct withdrawal from a piece of wood, the tighter the joint will remain. Nails driven into green wood tend to loosen slightly as the wood dries and shrinks. In seasoned material the resistance to withdrawal diminishes only slightly with time, unless moisture affects the wood. Withdrawal resistance is always higher when nails are driven into the side grain than when into the end grain. Because the lighter woods do not usually split as readily as the denser ones, more and larger nails can be used to offset the poorer nail-holding properties of the former. Hardwoods are more difficult to nail; they are sometimes used green or with holes drilled for nailing, to prevent splitting.

The surface condition of a nail affects its holding ability. The withdrawal resistance of a common nail increases directly with the distance it penetrates into the wood and increases almost directly with its surface area. A rusty nail may offer more resistance to withdrawal than a smooth one.

Means to increase withdrawal resistance. To increase resistance to withdrawal or loosening, nails may be coated, etched, spirally grooved, annularly grooved, or barbed, as illustrated. Grooved nails tend to hold well despite a change in moisture content. Coated nails usually provide a greater increase in withdrawal resistance when used in softer woods than when used in the denser woods. The increase in withdrawal resistance tends to decrease, however, with time.

In most cases, nails driven on a slant have more withdrawal resistance than nails driven straight into the wood. If a slant-driven nail is pulled in a direction which is at right angles to the surface, considerable resistance is encountered from the wood fibers on the pressure side. The nail may also progressively bend as it is pulled out. Both of these factors seem to offer continued holding power, even though the wood fibers are not gripping the entire surface of the nail. Nails slant-driven into the end grain of wood seem to gain proportionately more withdrawal resistance than those slant-driven into the side grain.

When members of a nailed joint tend to separate sideways, the nails are subjected to side loads. In this case doubling the diameter of the nail increases its lateral load capacity by nearly three times. This is true, however, only if the nail point has already been driven a suitable distance into the piece of material which is receiving it.

Blunt-pointed nails are often used to prevent the wood from splitting. Using nails of a smaller diameter also tends to prevent splitting but requires a greater number of nails per joint. Beeswax is sometimes applied to nail points to make them drive more easily, but it also reduces the holding power of the nail.

[ALAN H. TUTTLE]

Newton's laws of motion

Three fundamental principles which form the basis of classical, or newtonian, mechanics. They are stated as follows:

First law: A particle not subjected to external forces remains at rest or moves with constant speed in a straight line.

Second law: The acceleration of a particle is directly proportional to the resultant external force acting on the particle and is inversely proportional to the mass of the particle.

Third law: If two particles interact, the force exerted by the first particle on the second particle (called the action force) is equal in magnitude and opposite in direction to the force exerted by the second particle on the first particle (called the reaction force).

The first law, sometimes called Galileo's law of inertia, can now be regarded as contained in the second. At the time of its enunciation, however, it was important as a negation of the Aristotelian doctrines of natural placement and continuing force.

The third law, sometimes called the law of action and reaction, was also to some extent established prior to Newton's statement of it. However, Newton's formulation of the three laws as a mutually consistent set, with the nature of force clearly defined in the second law, provided the basis for classical dynamics.

The newtonian laws have proved valid for all mechanical problems not involving speeds comparable with the speed of light (approximately 300,000 km/sec) and not involving atomic or subatomic particles. The more general classical methods of Lagrange and of Hamilton are elaborations of the newtonian principles. See FORCE; KINETICS (CLASSICAL MECHANICS). [DUDLEY WILLIAMS]

Nibbling

The cutting of material by the action of a reciprocating punch. The nibbler takes repeated small bites as the work is passed beneath it (see illustration). The workpiece must be backed up by a support or die. Ferrous and nonferrous metals as well as some nonmetallic compositions may be cut by nibbling. Cuts may be made in mild steel up to approximately 1/2 in. thick.

Nibbling machine cutting a cam. (*Wilson Mechanical Instrument Division of American Chain and Cable Co.*)

Nibbling machines are constructed with considerable distance, or throat, between the punch and its supporting upright. This distance, plus the use of a round punch which allows the workpiece to be moved about, permits the cutting of irregular shapes. Duplicate pieces may be made by using templates as guides for the punch. Tubing may also be cut. Internal holes must be started from previously made holes. See MACHINING OPERATIONS. [ALAN H. TUTTLE]

Nickel alloys

Combinations of nickel with other metals. Nickel has been used in electroplating since 1843 and as an alloying addition to steels since about 1889. It was first used as a base for alloys with the introduction of Monel nickel-copper alloy in about 1905. The nominal compositions of some of the currently available alloys containing more than 50% nickel are given in Table 1.

Nickel-base alloys may be melted in open-hearth, electric-arc, or induction furnaces in air, under inert gas, or in vacuum. Casting may also be done under these same ambient conditions. Cast shapes are made in sand or investment molds or by shell molding. Ingots for wrought products are cast in metal molds and are hot-worked by forging, rolling, or extruding. In some instances further work may be done cold by rolling or drawing. Nickel-base alloys, made available in this way in bar, rod, wire, plate, strip, sheet, and tubular forms, may be fabricated into finished products by using conventional metalworking and metal-joining techniques.

Alloyed nickels. Nickel 211 and Duranickel alloy 301 are essentially binary alloys with 4.75% manganese and 4.5% aluminum, respectively. Manganese, in the first of these, extends the range of applicability in the presence of sulfur by about 300°F to a limiting temperature in the neighborhood of 1000°F. A characteristic use of this material is as wire for spark-plug electrodes.

Aluminum and titanium confer age-hardening characteristics on Duranickel alloy 301 and a tensile strength in excess of 200,000 psi is attainable in this alloy by appropriate cold work and heat treatment. In this condition it is well suited to the manufacture of springs and diaphragms.

Monel alloy 400. This alloy contains about two-thirds nickel and one-third copper and is the oldest of the commercial nickel-base alloys, dating from about 1905 when it was directly smelted from the copper-nickel matte obtained from Sudbury, Ontario, sulfide ore. The good fabricating characteristics and corrosion resistance of this alloy have made it widely used in marine applications and in the chemical-processing and petroleum industries. It has applicability in the new and expanding field of nuclear propulsion. As with nickel, this alloy can be made age-hardenable by the addition of aluminum and titanium. Monel alloy K-500 has corrosion-resisting characteristics similar to those of the non-age-hardenable composition, and is widely specified for such applications as marine propellers, shafting, valves, pump parts, and springs. A usable tensile strength of about 175,000 psi is obtainable in cold-drawn and age-hardened wire.

Nickel-chromium alloys. Nickel-chromium binary alloys are used primarily in specialty high-temperature service. An 80 nickel-20 chromium is a common high-quality resistance-heating-element material possessing good resistance to oxidation up to about 2100°F, superior to either of its two component elements. The alloy is used both in industrial-furnace and household-appliance heating elements. Chromel P is used with the chromium-free nickel-base Alumel in temperature-sensing devices known as thermocouples. This particular alloy couple has favorable thermoelectric characteristics for applicability in the measurement of temperatures up to 2000°F.

Nickel-chromium and related complex alloys are widely used for structural and general-purpose applications at high temperatures and in certain corrosive environments, particularly where freedom from stress-corrosion cracking is essential. In this latter instance Inconel alloy 600 has applicability in nuclear propulsion units.

This class of alloys encompasses a broad range of high-temperature properties. Nimonic 75 nickel-chromium alloy is widely used as a scale-resistant sheet material. Neither this material nor Inconel alloy 600 responds to age hardening, and hence they are on the low side of the elevated-temperature mechanical property range. By contrast, Inconel alloy X-750, which contains added aluminum, titanium, and columbium, develops greatly

Table 1. Nominal composition of some nickel-base alloys, weight %

Trademark	Ni	Cu	Cr	Co	Mo	Ti	Al	Cb	Fe	Mn	Si	C	Other
Nickel 211	95	–	–	–	–	–	–	–	–	4.75	–	0.08	–
Duranickel alloy 301	93.7	0.05	–	–	–	0.4	4.4	–	0.35	0.3	0.5	0.17	–
Monel alloy 400	66	31.5	–	–	–	–	–	–	1.35	0.9	0.15	0.18	–
Monel alloy K-500	66	29	–	–	–	0.5	2.75	–	0.9	0.75	0.5	0.15	–
Chromel P	90	–	10	–	–	–	–	–	–	–	–	–	–
Nichrome V	80	–	19.5	–	–	–	–	–	–	2.5*	1	0.25	–
Alumel	94	–	–	–	–	–	2	–	–	3	1	–	–
Nimonic 75	Bal	0.5*	19.5	–	–	0.4	–	–	5*	1*	1*	0.12	–
Nimonic 80A	Bal	–	19.5	2*	–	2.2	1.1	–	5*	1*	1*	0.1*	–
Inconel alloy 600	Bal	0.5*	15.5	–	–	–	–	–	8	1*	0.5*	0.15*	–
Inconel alloy X-750	Bal	0.5*	15	–	–	2.5	0.9	0.9	7	0.7	0.4	0.04	–
Inconel alloy 718	53	0.3*	19	1.0*	3	0.9	0.5	5	Bal	0.35*	0.35*	0.08*	–
Alloy 713C	Bal	–	12	–	4	0.5	6	2	5*	1*	1*	0.2*	0.012 B, 0.10 Zr
Udimet 500	Bal	–	17.5	16.5	4	2.9	2.9	–	4*	0.75*	0.75*	0.15*	–
Waspaloy	Bal	–	19	14	3	2.5	1.2	–	2	0.7	0.4	0.05	–
M252	55	–	19	10	10	2.5	0.75	–	2	1	0.7	0.1	–
GMR 235	Bal	–	15.5	–	5	2.5	3	–	10	0.25*	0.6*	0.15	0.06 B
Hastelloy B	61	–	1*	2.5*	27.5	2	–	–	5.5	1*	1	0.05*	0.4 V
Hastelloy C	54	–	15.5	2.5*	15.5	–	–	–	5.5	1*	1*	0.08*	0.35 V*, 4W
Hastelloy D	82	3	1*	1.5*	–	–	–	–	2*	1	9	0.12*	–

*Maximum.

improved high-temperature strength after proper heat treatment. The more highly alloyed Alloy 713C, a nickel-chromium cast alloy, exhibits further strength improvement at the high side of the temperature range of applicability for the complex nickel-chromium alloys. For comparison, 100-hr rupture strengths of these three alloys are listed in Table 2.

The aforementioned materials and a number of similar proprietary alloys such as Udimet 500, M252, Waspaloy, and GMR235, which combine high strength and oxidation resistance, have found application in jet engine and gas turbines for such parts as combustion liners, blades, vanes, and disks.

Inconel alloy 718 is an age-hardenable alloy with a slow aging response, a characteristic which permits welding and annealing without spontaneous hardening. It is suitable for service from cryogenic temperatures (−423°F) to moderately elevated temperatures (1300°F).

In the interest of optimizing properties, these complex nickel-chromium alloys are being produced in increasing quantities by vacuum-melting and vacuum-pouring techniques.

A cast heat-resisting alloy carrying the Alloy Castings Institute designation HW (nominally 60% nickel, 12% chromium, 23% iron) is used principally for furnace parts and heat-treating fixtures. It has good resistance to oxidation and carburization, only modest hot strength, but good thermal shock resistance.

Table 2. Rupture strengths (100 hr) of nickel-chromium alloys, psi

Alloy	1500°F	1700°F
Inconel alloy 600	8,000	3,800
Inconel alloy X-750	25,000 – 30,000	8,000 – 10,000
Alloy 713C	60,000	30,000

Hastelloy alloys. Hastelloy alloys B, C, and D are used primarily in corrosive environments. Hastelloy B is resistant to hydrochloric and sulfuric acids within certain limits of concentration, temperature, and degree of aeration. It is not recommended for service involving strong oxidizing acids or oxidizing salts. Hastelloy C is unusually resistant to oxidizing solutions and to moist chlorine. Hastelloy D has exceptional resistance to hot concentrated sulfuric acid.

Nickel-iron alloys. Alloys containing more than 50% nickel are used in various applications involving controlled thermal expansivity or certain magnetic requirements. In the range 50–52% nickel, the balance iron, the alloys have thermal expansion characteristics useful in making some types of glass-to-metal seals.

In the range 77–80% nickel, with or without about 4% molybdenum, the balance iron, the alloys have very high initial and maximum magnetic permeabilities when properly processed. *See* ALLOY; IRON ALLOYS; STAINLESS STEEL.

[E. N. SKINNER; GAYLORD SMITH]

Nondestructive testing

The use of tests to examine an object or material to detect imperfections, determine properties, or assess quality without changing its usefulness. Nondestructive testing is a term generally applied to certain tests that are utilized in industrial operations, although some of the same technologies are used in the medical field, such as radiography (x-ray) and ultrasonics. Industrial uses of nondestructive tests include monitoring the quality of products from manufacturing processes, and maintenance of the products and of process machinery. Nondestructive testing is commonly employed in shipbuilding, aerospace vehicle and probe construction, automotive manufacturing, metals manufacturing, railroad maintenance, building and bridge construction, and electric power plant construction and maintenance.

Some nondestructive tests involve complex high-technology systems, while others, which employ the same basic test principle, are easily applied by unskilled operators. Nearly every form of energy is used in nondestructive tests, including all wavelengths of the electromagnetic spectrum, as well as a broad range of frequencies of vibrational mechanical energy. Although there are many methods used for nondestructive testing, six are applied widely in industry: visual-optical, liquid-penetrant, magnetic-particle, eddy-current, ultrasonic, and radiographic.

Visual-optical method. This method was the earliest nondestructive test practiced. It can be applied to the examination of surfaces and transparent materials. A visual test might be as simple as an inspector looking at an object to check color or surface quality. At the other extreme, laser beams and complex image-recognition equipment can be used in automated scanning systems to perform an inspection. Endoscopes are rigid or flexible (fiber-optic) devices used to inspect internal surfaces of objects (Fig. 1). They can be very small in diameter (3 mm) and usually include their own illumination source. Effective visual inspection of an object depends on the viewing angle and the illumination of the object. Color, intensity, angle, and character of illumination are important. An imperfection may be detected under diffuse, polarized, collimated, or coherent light.

Fig. 1. Fiber-optic endoscope being used to inspect the interior components of a jet engine without disassembly. (*Olympus Corporation of America*)

Two of the nondestructive test methods, liquid-penetrant and magnetic-particle, are basically aids to visual examination. They permit an inspector to see imperfections that could otherwise go undetected, even with careful inspection aided by optical magnification.

Liquid-penetrant method. Testing with liquid penetrants is useful for locating imperfections that are open to the surface of a wide variety of nonporous materials. Although generally applied to nonmagnetic objects, such as those made from aluminum, penetrant tests are also used on objects made from magnetic materials, especially small objects of complex shape or very large objects. To perform the test, a thin oillike liquid containing a dye is applied to the material by a brush, flow, or dip method; the liquid, chosen to have the proper surface-tension properties, moves into small openings or cracks (Fig. 2) during a penetration time (1–30 min). After the excess liquid on the surface is carefully removed (by wiping or washing) and the object is dried, a developer is applied to the surface. This material is typically a fine powder, such as talc, usually in suspension in a liquid. It acts like a blotter and pulls the liquid penetrant up out of surface imperfections. The liquid tends to spread in the developer, thereby enlarging the indication. The developer and penetrant dye are chosen to have high contrast. For example, visible dye penetrants are often red and developers white. A very sensitive penetrant method makes use of a fluorescent dye that emits light upon excitation with ultraviolet radiation (0.365 nm). In this case the inspection is carried out in a semidarkened booth with ultraviolet lamps. After testing, the penetrant and developer are removed by washing with water, sometimes aided by an emulsifer, or with a solvent. Leak testing can also be accomplished by applying a liquid penetrant on one surface and detecting it on the other one.

Magnetic-particle method. Magnetic-particle testing is used for detecting discontinuities at or near the surface in magnetic materials. Steel and cast iron parts such as those used in automobiles, aircraft, and railroad equipment are subjected to

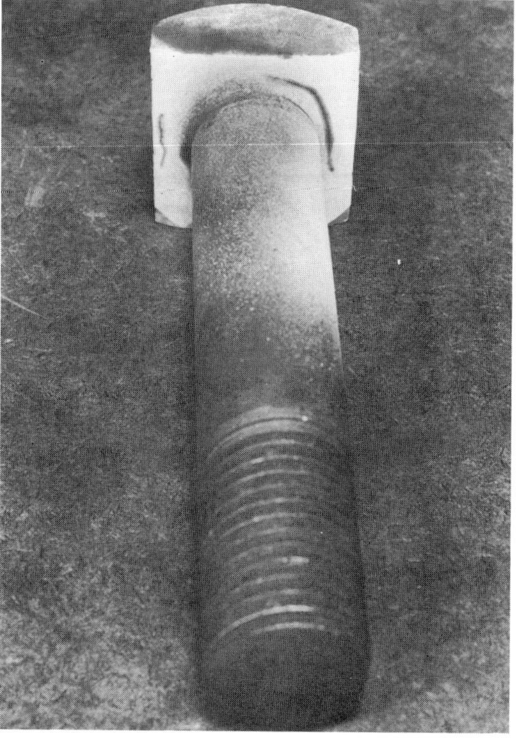

Fig. 2. Fatigue cracks detected in a large bolt head by use of visible-dye liquid penetrant. (*Magnaflux Corp.*)

Fig. 3. Cracks in a gear revealed under ultraviolet light by a fluorescent magnetic-particle test. (*Magnaflux Corp.*)

magnetic-particle testing to detect processing or fatigue cracks. To perform the test, the test object is properly magnetized, and then finely divided magnetic particles are applied to its surface. When the object is properly oriented to the induced magnetic field, a discontinuity creates a leakage field which attracts and holds the particles, forming a visible indication. Magnetic field direction and character are dependent upon how the magnetizing force is applied and upon the type of current used. For best sensitivity, the magnetizing current must flow in a direction parallel to the principal direction of the expected defect. Circular fields, produced by passing current through the obejct, are almost completely contained within the test object. Longitudinal fields, produced by coils or yokes, create external poles and a general leakage field. Alternating, direct, or half-wave direct current may be used for the location of surface defects. Half-wave direct current is most effective for locating subsurface defects.

Magnetic particles may be applied dry or as a wet suspension in a liquid similar to kerosine. Colored dry powders are advantageous when testing for subsurface defects and when testing objects which have rough surfaces, such as castings, forgings, and weldments. Wet particles are preferred for detection of very fine cracks, such as fatigue or grinding cracks. Fluorescent wet particles are used to inspect objects with the aid of ultraviolet light (Fig. 3). Fluorescent inspection is most widely used because of its greater sensitivity. Application of particles while magnetizing current is on (continuous method) produces stronger indications than those obtained if the particles are applied after the current is shut off (residual method). Interpretation of subsurface defect indications requires experience. Demagnetization of the test object after inspection is advisable.

Fig. 4. Portable eddy-current test instrument being used to determine heat effect upon metal in an aircraft jet engine. (*K. J. Law Associates Inc.*)

Eddy-current method. Eddy current tests are based upon correlation between electromagnetic properties and physical or structural properties of a test object. Eddy currents are induced in metals whenever they are brought into an ac magnetic field. These eddy currents create a secondary magnetic field, which opposes the inducing magnetic field. The presence of discontinuities or material variations alters eddy currents, thus changing the apparent impedance of the inducing coil or of a detection coil. Coil impedance indicates the magnitude and phase relationship of the eddy currents to their inducing magnetic field current. This relationship depends on the mass, conductivity, permeability, and structure of the metal, upon the frequency intensity, and upon the distribution of the alternating magnetic field.

To perform the test, a probe coil is placed on an object, or the object is placed within a circular coil, and an indication of object condition is read from a meter. Conditions such as heat treatment, composition, hardness, phase transformation, case depth, cold working, strength, size, thickness, cracks, seams, and inhomogeneities are indicated by eddy-current tests. Thickness measurement of metallic and nonmetallic coatings on metals is performed by using eddy-current principles. Coating thicknesses measured typically range from 0.0001 to 0.100 in. (0.00025 to 0.0025 cm). Correlation data must usually be obtained to determine whether test conditions for desired characteristics of a particular test object can be established. Because of the many factors which cause variation in the electromagnetic properties of metals, care must be taken that the instrument response to the condition of interest is not nullified or duplicated by variations due to other conditions. Multiple-frequency tests often permit greater precision in detection of certain characteristics of an object.

Alternating-current frequencies between 1 and 5,000,000 Hz are used for eddy-current testing. Test frequency determines the depth of current penetration into the test object, because the ac phenomenon of skin effect, varies with conductivity, permeability, and frequency. High-frequency eddy currents are more sensitive to surface defects or conditions, while low-frequency eddy currents are sensitive also to deeper internal defects or conditions.

The coils are of two general types: the circular coil, which surrounds an object; and the probe coil, which is placed on the surface of an object. Coils are further classified as absolute, when testing is conducted without direct comparison with a reference object in another coil, or differential, when comparison is made through use of two coils connected in series opposition. Many variations of these coil types are utilized. Coils may be of the air-core or magnetic-core type.

Instrumentation for the analysis and presentation of electric signals resulting from eddy-current testing includes a variety of means, ranging from meters to oscilloscopes. Instruments vary from the small battery-operated special-purpose type (Fig. 4) to the automatic multiple-instrument system (Fig. 5). Alarm circuits are adjusted to be sensitive only to signals of certain electrical phase and amplitude, so that selected conditions are indicated while others are ignored. Automatic testing is one of the significant advantages of the eddy-current method.

Ultrasonic method. The term ultrasonic is applied to mechanical vibrational waves that cannot be heard by the human ear (greater than 20 kHz). An ultrasonic frequency selected for testing will usually lie between 1 and 15 MHz, depending upon the material, its grain size, its elastic modulus, and the test sensitivity required. Ultrasonic waves can detect and locate structural discontinuities or differences and measure the thickness of a variety of materials. To perform a contact test, a transducer is placed on an object to which a coupling material has been applied, and the results are interpreted from cathode-ray-tube indications. To accomplish this, an electric pulse is generated in a test instrument and transmitted through a cable to a transducer, which converts the electric pulse into mechanical vibrations. These low-energy-level vibrations are transmitted through a coupling material into the test object, where the ultrasonic energy is attenuated, scattered, or reflected to indicate conditions within the object. Reflected or transmitted sound energy is usually reconverted to electrical energy by the same transducer and returned to the test instrument. The received energy is then usually displayed on a cathode-ray tube. The presence, position, and amplitude of echoes indicate conditions of the test-object material.

Materials capable of being tested by ultrasonic energy are those which transmit vibrational energy. Metals are tested in dimensions of up to 30 ft (9 m). Noncellular plastics, ceramics, glass, new concrete, organic materials, and rubber can be tested. Each material has a characteristic sound velocity, which is a function of its density and modulus (elastic or shear).

Material characteristics determinable through ultrasonics include structural discontinuities, such as flaws and unbonds, physical property and metallurgical differences, and thickness (measured from one side). A common application of ultrasonics is the inspection of welds for inclusions, porosity, lack of penetration, and lack of fusion. Other applications include location of unbond in nuclear-fuel elements, location of fatigue cracks in machinery, and medical applications. Automatic testing is frequently performed in manufacturing applications.

Ultrasonic systems are classified as either pulse-echo, in which a single transducer is used, or through transmission, in which separate sending and receiving transducers are used, Pulse-echo systems are much more common. In either system, ultrasonic energy must be transmitted into, and received from, the test object through a coupling medium, since air will not transmit ultrasound of these frequencies. Water, oil, grease, and glycerin are commonly used couplants. Two types of testing are used: contact and immersion. In contact testing, the transducer is placed directly on the test object. In immersion testing, the transducer and test object are separated from one another in a tank filled with water, or by a column of water, or by a liquid-filled wheel. Immersion testing eliminates transducer wear and facilitates scanning of the test object. Scanning systems usually have paper-printing equipment for readout of test information (Fig. 6).

Fig. 5. Multiple-test eddy-current system to monitor material chemistry and heat-treat condition and to detect internal and external longitudinal cracks in automobile bumper shock-absorber tubes at a rate of 1200 parts per hour. (*K. J. Law Associates, Inc.*)

Ultrasonic transducers are piezoelectric units which convert electrical energy into acoustical energy and convert acoustical energy into electrical energy of the same frequency. Quartz, barium titanate, lithium sulfate, lead metaniobate, and lead zirconate titanate are commonly used transducer crystals, and are generally mounted with a damping backing in a housing. Transducers range in size from 0.06 to 5 in. (0.15 to 12.7 cm) and are of circular or rectangular shape. Ultrasonic beams can be focused to improve resolution and definition of imperfections. Transducer characteristics and beam patterns depend on frequency, size, crystal material, and construction.

Radiographic method. Radiography can be used to reveal internal structure in industrial materials and assemblies, just as it does for a doctor or dentist who inspects parts of the body. The most widely used inspection approach involves the transmission of x-rays through an object to produce an image on x-ray film. This inspection method can be applied to many materials and assemblies. Proper placement of components in assemblies, detection of porosity, inclusions, or cracks, and differentiation of different materials, thicknesses, or densities are among the useful results obtained by radiography. X-rays can also be detected by special television systems; the x-ray image (often greatly magnified) can be displayed on a television monitor. Objects can be manipulated and viewed in real time with a television radiographic system.

Fig. 6. Multiple-channel ultrasonic test immersion system used to examine the interior of large components of aircraft, such as forgings or castings. The test instrument and automatic scanning controls are in the center, the immersion tank and transducer are on the right, and a paper printer for graphic test results is on the left. (*Automation Industries, Inc.*)

Advantages and limitations. The table compares these six nondestructive test methods and lists the basic advantages and limitations of each. The methods tend to complement each other. Therefore it is not unusual to find an object inspected by more than one method. Visual-optical and penetrant methods inspect surfaces for surface-connected imperfections. Magnetic-particle and eddy-current techniques can reveal imperfections at and near surfaces, but primarily in ferromagnetic and electrically conducting materials, respectively. Radiography and ultrasonic methods are capable of probing deeply into materials and structures. Radiography offers advantages for determining the component location in complex structures, whereas ultrasonic methods are able to quickly pene-

Comparison of nondestructive tests

Method	Characteristics detected	Advantages	Limitations
Visual-optical	Surface characteristics such as finish, scratches, cracks or color; strain in transparent materials	Often convenient; can be automated; inexpensive	Can be applied only to surfaces, through surface openings, or to transparent material
Liquid penetrants	Surface openings due to cracks, porosity, seams, or folds	Inexpensive; easy to use; readily portable; sensitive to small surface flaws	Flaw must be open to surface; not useful on porous materials
Ultrasonics	Changes in acoustic impedance caused by cracks, non-bonds, inclusions, or interfaces	Can penetrate thick materials; moderate cost; can be automated	Requires coupling to material either by contact to surface or immersion in a fluid such as water
Radiography	Changes in density from voids, inclusions, material variations; placement of internal parts	Can be used to inspect a wide range of materials and thicknesses; versatile; film provides record of inspection	Radiation safety requires precautions; expensive; detection of cracks can be difficult
Magnetic particles	Leakage magnetic flux caused by surface or near-surface cracks, voids, inclusions, material, or geometry changes	Inexpensive; sensitive both to surface and near-surface flaws	Limited to ferromagnetic material; surface preparation and postinspection demagnetization may be required
Eddy currents	Changes in electrical properties caused by material variations, cracks, voids, or inclusions	Readily automated; moderate cost	Limited to electrically conducting materials; limited penetration depth

trate very thick objects without any risk to personnel safety.

Other methods. There are, of course, many methods in addition to these six that are used for nondestructive testing. For example, sonic energy emitted by a material under stress can be detected and analyzed in a method called acoustic emission. Optical holography, which uses coherent light from a laser, can display strain distribution and thereby detect imperfections. The heat-transmission properties of an object can be analyzed by thermal methods and, in some cases, an infrared image can provide very useful test information. Novel methods such as these and adaptations of the six major methods are also used for industrial nondestructive testing.

Indirect data. All of these nondestructive test methods yield data that are often indirectly related to the information desired from the inspection. One may wish to know the yield strength of a particular object, for instance. The nondestructive test may reveal that there is some porosity or a crack in the object, or that it is of some category of hardness or composition. This information, combined with experience from previous destructive tests and analytical procedures, will help relate the nondestructive test result to the necessary performance level. Despite some difficulties in making correlations between test results and serviceability, nondestructive testing offers the significant advantage that every part of an assembly can be tested before being placed in service, and monitored thereafter.

[DONALD D. DODGE]

Bibliography: American Society for Metals, *Metals Handbook*, vol. 11, 1976; C. E. Betz, *Principles of Magnetic Particle Testing*, 1966; C. E. Betz, *Principles of Penetrants*, 1969; Eastman Kodak Co., *Radiography in Modern Industry*, 1969; J. Krautkramer and H. Krautkramer, *Ultrasonic Testing of Materials*, 1969; H. L. Libby, *Introduction to Electromagnetic Nondestructive Test Methods*, 1971; W. J. McGonnagle, *Nondestructive Testing*, 1969; R. C. McMaster (ed.), *Nondestructive Testing Handbook*, 2 vols., 1959; R. S. Sharpe (ed.), *Research Techniques in Nondestructive Testing*, vols. 1 and 2, 1970, 1973.

Nuclear engineering

That branch of engineering that deals with the production and use of nuclear energy. It is concerned with the development, design, construction, and operation of power plants which convert energy produced by fission or fusion to other useful forms such as heat or electrical energy. Development of these unique sources of energy requires novel solutions to difficult mechanical, electrical, and materials problems. Because many of the components and systems operate in the presence of intense high-energy radiation, special problems that are generated by the interaction of radiation with various materials are encountered. Such problems are unique to nuclear engineering. Training of nuclear engineers places special emphasis on this area. *See* NUCLEAR FUELS; NUCLEAR POWER; NUCLEAR REACTOR; RADIATION DAMAGE TO MATERIALS; REACTOR PHYSICS.

Radioactive materials are used in a wide variety of industrial processes and equipment, ranging from nondestructive testing of welds to low-temperature sterilization of pharmaceuticals. Handling and storage of the large quantities of radioactive substances used as reactor fuel, produced as by-product material, or generated as waste introduce problems in the protection of personnel, equipment, and the environment. Such protection from high-energy radiation requires the design and construction of a variety of radiation shields and of shipping and handling equipment. *See* NONDESTRUCTIVE TESTING; NUCLEAR FUEL CYCLE; NUCLEAR FUELS REPROCESSING; RADIATION SHIELDING; RADIOACTIVE WASTE MANAGEMENT.

Nuclear explosives are being investigated for use in large-scale excavation and for stimulation of the production of natural gas. Nuclear reactors are now used for propulsion of a variety of naval vessels. Serious consideration is being given to the use of nuclear power for propulsion of commercial ships.

More than 60 colleges and universities in the United States offer educational programs in nuclear engineering. Undergraduate curricula emphasize design and analysis of fission reactor power plants, industrial applications of radiation and radioactive isotopes, and radiation protection. Graduate programs typically place emphasis on research in fission reactor fuels management, reactor safety, effects of radiation on materials, generation and control of magnetically confined high-temperature plasmas, laser-generated fusion, design of fusion power plants, radiation measuring devices and systems, and medical applications of radiation. Many nuclear engineering departments operate research or training reactors which are used for laboratory instruction and as intense sources of neutron and gamma radiation for research.

[WILLIAM KERR]

Bibliography: J. J. Duderstadt and L. J. Hamilton, *Nuclear Reactor Analysis*, 1976; A. R. Foster and R. L. Wright, *Basic Nuclear Engineering*, 2d ed., 1973; J. R. Lamarsh, *Introduction to Nuclear Engineering*, 1975; R. L. Murray, *Nuclear Energy*, 1975.

Nuclear fuel cycle

The steps by which fissionable (for example ^{233}U, ^{235}U, ^{239}Pu) and fertile (for example, ^{238}U, ^{232}Th) materials are prepared for use in, and recycled or discarded after discharge from, the nuclear reactor. These steps include mining of uranium- or thorium-bearing ore and milling of the ore to form concentrates. The uranium concentrate is converted to the volatile uranium hexafluoride (UF_6) that is used in the separation of isotopes to produce uranium enriched in the fissile ^{235}U. Another part of the fuel cycle is the fabrication of the enriched uranium into fuel assemblies. After the fuel has liberated the desired amount of heat in the reactor, the spent assemblies are reprocessed to separate the remaining fissionable and fertile material (uranium and plutonium) from the nuclear wastes. Other steps in the fuel cycle include the various transportation operations that move materials from one step to another, often connecting plants many hundreds of miles apart. Finally, waste management includes the treatment, storage, and disposal of radioactive wastes from the many other parts of the fuel cycle. The fuel cycle for a thorium-based reactor requires, in addition to the uranium fuel cycle steps outlined above, the mining, milling, and purification of thorium and the reprocessing of thorium-containing fuel into its components, which include unused thorium, uranium in the form of fissionable ^{233}U, and nuclear waste.

Mining and purification. Even though uranium is relatively abundant in the Earth's crust, it is found only at low concentrations in most ores. Concentrations of uranium in typical, high-grade ores in the United States range about 0.2%, and hence the ore, obtained by either open-pit or underground mining, must be treated to concentrate the uranium. For some types of ores, physical processes such as grinding, washing, flotation, and gravity settling will result in separation of the gangue (for example, sandstone, clay, limestone) from uranium-bearing ore. These physical processes are useful only in concentrating the ore to less than 50% uranium content and can be carried out at the mine.

The concentrated ore is treated at a mill to further concentrate and separate the uranium from metallic impurities and the rest of the gangue material. The two chemical methods commonly used are the carbonate leach method and the acid leach method. The choice of method is largely determined by the type of ore to be treated. The product of the purification process is a precipitate, most commonly of ammonium diuranate, that is dried to a powdered oxide containing small amounts of impurities. This material is shipped to conversion facilities, where it may be dissolved and the uranium purified further by solvent extraction methods. The product of this purification is a solution of uranium that is dried and calcined to form U_3O_8.

Conversion to hexafluoride. Conversion of purified uranium to the hexafluoride is done by a multistep process in which uranium oxides are fluorinated with hydrogen fluoride and with fluorine. Purified U_3O_8 is reduced to UO_2 and converted to UF_4 by gaseous HF in a fluidized-bed reaction vessel. Formation of UF_6 requires elemental fluorine and is also carried out in a fluidized bed to simplify removal of heat from the exothermic reaction. Variations exist in the details of the conversion process, including elimination of the initial purification of the uranium concentrate, inclusion of fractional distillation of UF_6 to obtain a pure product, and the use of various types of reaction vessels.

Isotopic enrichment. Natural uranium contains the fissile isotope ^{235}U at a concentration (0.72%) too low to be useful in reactors which use ordinary water as moderator and coolant, that is, light water reactors. The UF_6 is used in the process of separating isotopes by diffusion that increases the ^{235}U content to about 3%, at which concentration it can be conveniently used in reactors. For other purposes such as for fuel used in certain small reactors (for example, research reactors), enrichment of ^{235}U to more than 90% is carried out by the same process.

Fuel fabrication. Fuel for large nuclear reactors producing electric power is fabricated by conversion of the slightly (that is, about 3%) enriched UF_6 to UO_2. Hydrolysis of UF_6 and precipitation of compounds such as $(NH_4)_2U_2O_7$ is followed by calcination to U_3O_8 and reduction with hydrogen to powdered UO_2. The dioxide, selected because of its chemical stability, is compacted into pellets that are sintered at high temperatures. The fuel assembly is made of an array of sealed tubes of a zirconium alloy (Zircaloy) containing the fuel pellets, of end plates, and of other hardware. Many such assemblies are charged into the core of a reactor. *See* NUCLEAR POWER; NUCLEAR REACTOR.

Reprocessing. Following the discharge of fuel assemblies that no longer contribute efficiently to the generation of heat in the reactor, the spent fuel is allowed to dissipate nuclear radiation and the heat generated by it (that is, to "cool") while submerged in water for some time. The spent fuel is processed to recover the residual ^{235}U and the plutonium that was formed while the fuel was in the reactor, and to separate the radioactive fission product wastes for storage. The processing steps for typical power reactor fuel include disassembly; chopping of the Zircaloy tubes to expose the fuel; chemical dissolutions of the uranium, plutonium, and fission products; and separation and purification of the uranium and plutonium by solvent extraction methods. The waste from this step contains more than 99% of the radioactive fission products and must be handled carefully to ensure confinement and isolation from the biosphere.

The uranium product from the processing plant is converted to UF_6 by procedures similar to those used to make UF_6 from natural uranium, except that the slight enrichment of the spent uranium requires attention to nuclear criticality safety. The product UF_6 is reenriched in the isotope separation plants to about 3% concentration of ^{235}U for recycle as fuel. The plutonium product can be combined with natural uranium to make a "mixed oxide" fuel (that is, PuO_2-UO_2) containing about 5% plutonium that can be used in power reactors otherwise using slightly enriched uranium. Alternate use for plutonium includes fuel for fast breeder reactors. *See* NUCLEAR FUELS REPROCESSING.

Waste management. Most of the radioactive wastes from reprocessing plants are in the form of aqueous solutions containing high levels of radioactive materials. These solutions can be temporarily stored in cooled tanks. Following interim storage of the solutions, the wastes are solidified by evaporation of water and acid and are converted to stable oxides. In order to decrease the likelihood of leaching of radioactive wastes by water, the solidified wastes can be incorporated into an inert matrix such as glass and can be held in thick metallic containers.

Other wastes include discarded equipment, trash, filters, and miscellaneous materials, all of which may be contaminated to varying degrees. These wastes are treated to reduce their volume and to limit the dispersibility of the radioactive contamination. Waste from the milling operation, called tailings, is treated and stored near the mill. Final disposal of some of the radioactive waste will take advantage of the stability of selected geologic formations to ensure the necessary and prolonged isolation of the waste. *See* NUCLEAR FUELS; RADIOACTIVE WASTE MANAGEMENT.

[MARTIN J. STEINDLER]

Bibliography: D. M. Elliott and L. E. Weaver (eds.), *Education and Research in the Nuclear Fuel Cycle*, 1972; M. Etherington (ed.), *Nuclear Engineering Handbook*, 1958; J. T. Long, *Engineering for Nuclear Fuel Reprocessing*, 1967.

Nuclear fuels

The fissionable and fertile elements and isotopes used as the sources of energy in nuclear reactors. Although many heavy elements can be made to fission by bombardment with high-energy alpha particles, protons, deuterons, or neutrons, only neutrons can provide a self-sustaining reaction.

The number of neutrons ν released in the fission process varies from one per many fissions for elements just beyond the fission point (silver) to two or more per fission for the heavier elements, such as thorium and uranium. Even in such elements, neutron capture by the nucleus accompanied by the release of excess energy in the form of a gamma ray occurs in many cases, rather than nuclear fission. This reduces the number of neutrons available for further fission. The ratio of neutron capture to neutron fission varies from nucleus to nucleus and changes with the energy of the bombarding neutrons. Only a few isotopes of the heavy elements have a higher probability of fission than capture. These fissionable isotopes, ^{233}U, ^{235}U, and ^{239}Pu, are the only materials that can sustain the fission reaction and are therefore called nuclear fuels. *See* NUCLEAR REACTOR.

Of these isotopes, only ^{235}U occurs in nature as 1 part in 140 of natural uranium, the remainder being ^{238}U. The other two fissionable isotopes must be produced artificially, ^{233}U by neutron capture in ^{232}TH and ^{239}Pu by neutron capture in ^{238}U. The isotopes ^{232}Th and ^{238}U are called fertile materials.

By using a mixture of fissionable and fertile isotopes in a nuclear reactor, it is possible to reduce the rate of depletion of the nuclear fuel, because capture of excess neutrons by the fertile materials replenishes the fissionable material. Thus, ^{235}U can be burned (fissioned) and the surplus neutrons used to produce plutonium from ^{238}U or ^{233}U from thorium. Such nuclear reactors are called converter reactors.

The efficiency of production of new nuclear fuel depends on the extent of neutron losses due to undesirable neutron absorptions in the reactor or to neutron leakage. In some cases these losses can be kept small enough so that more nuclear fuel is produced than burned. Moreover, the ^{233}U and ^{239}Pu can be subsequently used as fuel in place of the original ^{235}U and by this means a large fraction of fertile material can be gradually converted into fissionable material. Reactors that burn ^{233}U and ^{239}Pu and produce as much fuel as is consumed, or more, are called breeders.

The total energy that can be produced from the fissionable ^{235}U in known resources of high-grade uranium ores corresponds to less than 5% of that from economically recoverable fossil fuels. Thus, nuclear fission will not become an important source of power in the long term unless the breeding and conversion fuel cycles are utilized.

Breeding and conversion. The nuclear reactions governing the consumption and production of nuclear fuel in a reactor are listed in Table 1. Also shown are values for the thermal-neutron (0.004 aJ or 0.025-eV neutron) cross section (probability that the reaction will take place) and the half-life for radioactive decay of the relatively unstable isotopes.

In a mixture of ^{235}U and ^{238}U, three competing reactions take place with thermal neutrons: (1) ^{235}U capture, (2) ^{235}U fission, and (3) ^{238}U capture (numbers in parentheses refer to reactions in Table 1). Reaction (3) leads to the production of ^{239}Pu by successive decay of ^{239}U and ^{239}Np, as shown by reactions (4) and (5). The conversion ratio (relative production and consumption of nuclear fuel) of the system is given by the relative probability that reaction (3) will take place as compared to reactions (1) and (2).

Table 1. Nuclear reactions in a thermal-neutron spectrum

Reaction number	Equation	Cross section, barns*	Half-life
(1)	$^{235}U + n \rightarrow {}^{236}U + \gamma$	107	
(2)	$^{235}U + n \rightarrow$ Fission + 2.47 n	582	
(3)	$^{238}U + n \rightarrow {}^{239}U + \gamma$	2.74	
(4)	$^{239}U \rightarrow {}^{239}Np + \beta^-$		23.5 m
(5)	$^{239}Np \rightarrow {}^{239}Pu + \beta^-$		2.33 d
(6)	$^{239}Pu + n \rightarrow {}^{240}Pu + \gamma$	277	
(7)	$^{239}Pu + n \rightarrow$ Fission + 2.88 n	748	
(8)	$^{240}Pu + n \rightarrow {}^{241}Pu + \gamma$	250	
(9)	$^{241}Pu + n \rightarrow {}^{242}Pu + \gamma$	390	
(10)	$^{241}Pu + n \rightarrow$ Fission + 3.06 n	1025	
(11)	$^{242}Pu + n \rightarrow {}^{243}Pu + \gamma$	19	
(12)	$^{243}Pu \rightarrow {}^{243}Am + \beta^-$		4.98 h
(13)	$^{232}Th + n \rightarrow {}^{233}TH + \gamma$	7.3	
(14)	$^{233}TH \rightarrow {}^{233}Pa + \beta^-$		23.3 m
(15)	$^{233}Pa \rightarrow {}^{233}U + \beta^-$		27.4 d
(16)	$^{233}U + n \rightarrow {}^{234}U + \gamma$	52	
(17)	$^{233}U + n \rightarrow$ Fission + 2.51 n	527	
(18)	$^{234}U + n \rightarrow {}^{235}U + \gamma$	90	

*Accepted values for monoenergetic thermal neutrons at 2200 m/s (0.00405 aJ or 0.0253 eV); 1 barn = 10^{-28} m^2.

Similarly, in a mixture of ^{239}Pu and ^{238}U the conversion ratio (breeding ratio) is given by the relative probability of reaction (3) as compared to reactions (6) and (7). In this case, however, the higher isotopes of plutonium that are formed have a long half-life and start to absorb neutrons as their concentration builds up by means of reactions (6), (8), and (9). The ^{243}Pu formed by reaction (11) decays rapidly to americium, as shown, to end the chain effectively. Thus, after long exposure to thermal neutrons in a reactor, a mixture of ^{235}U and ^{238}U will contain appreciable concentrations of ^{236}U, ^{239}Pu, ^{240}Pu, ^{241}Pu, and ^{242}Pu, all of which must be taken into consideration in determining the overall conversion ratio.

Reactions (1), (2), and (13)–(18) represent the reactions taking place in a mixture of ^{235}U and thorium. In this fuel cycle secondary isotopes of importance to the conversion ratio are ^{233}U, ^{234}U, ^{236}U, and ^{233}Pa. For most efficient neutron utilization (capture in thorium), it is important to minimize losses due to neutron absorption in ^{233}Pa by keeping the average neutron flux as low as possible.

Maximizing conversion ratio. When it is desired to maximize the neutron-conversion ratio in a reactor, neutron losses are held to a minimum by suitable selection of the materials making up the reactor system, their arrangement in the reactor, and its operating conditions. For example, neutron leakage is reduced if the reactor is made large; fission-product poisons (neutron absorbers) can be lowered by frequent processing of fuel; and nonfission neutron capture by fuel can be minimized by designing the reactor so that the average energy of the neutrons is optimum for causing fission. How-

Table 2. Capture-to-fission ratio (α) and neutron yield (η) as function of energy

Neutron energy, aJ (eV)	^{233}U α	^{233}U η	^{235}U α	^{235}U η	^{239}Pu α	^{239}Pu η
0.004 (0.025)	0.102	2.28	0.190	2.07	0.380	2.09
0.016 (0.10)	0.08	2.33	0.17	2.11	0.59	1.81
0.048 (0.30)	0.15	2.19	0.25	1.97	0.70	1.70
16 (10^2)			0.52	1.62	0.72	1.67
1.6×10^4 (10^5)			0.18	2.09	0.60	1.80
1.6×10^5 (10^6)	0.03	2.44	0.08	2.28	0.10	2.62

ever, the extent to which these methods of improving neutron utilization can be applied is limited by economic considerations. Thus for any nuclear power application, there is an optimum reactor size and configuration and an optimum fuel-processing cycle. *See* NUCLEAR FUELS REPROCESSING.

The control of neutron losses due to parasitic capture in fuel by varying the relative amounts of neutron-scattering material (moderator) and fuel to give the proper neutron energy is the most important factor in achieving a high conversion ratio. The effect of neutron energy on α (the ratio of neutrons lost by parasitic capture in fuel to those leading to fission) and on η (the number of neutrons emitted per neutron absorbed in fuel), as seen in the equation below, are shown in Table 2.

$$\eta = \frac{\nu}{1+\alpha}$$

Table 2 indicates that the theoretical maximum conversion ratio (given by $\eta - 1$) is above 1.0 for all three fissionable materials as long as the average energy of the neutrons causing fission is either very high (~ 160 fJ or 1 MeV) or very low (~ 0.004 aJ or 0.025 eV). In a practical reactor design, however, both these neutron energy conditions are difficult to achieve, because for any given mixture of fuel and moderator there will exist neutrons moving at all energies, ranging from those for fission neutrons (fast or high-energy neutrons) down to those moving at approximately the same velocities as the moderator atoms. Even in a highly thermalized reactor (high ratio of moderator to fuel), the neutron energy will vary considerably from the mean that is established by the moderator temperature. Because of this, the conversion ratio is affected by the moderator temperature. This is especially true in the case of the ^{235}U, ^{238}U, ^{239}Pu fuel cycle, as shown in Table 3.

In nuclear power reactors that operate with high moderator and coolant temperatures to achieve high thermal efficiencies, it is difficult to get a high conversion ratio because of the effect just described. One solution to this problem is to insulate the moderator thermally from the coolant and to maintain the moderator at a lower temperature. Such a technique cannot be applied in graphite-moderated reactors because of the necessity for keeping the graphite hot to minimize its expansion due to radiation damage and to minimize the build-up of stored energy. *See* RADIATION DAMAGE TO MATERIALS.

Table 3. Effect of moderator temperature on the nuclear properties of ^{235}U and ^{239}Pu

Average moderator temperature, °C	Average neutron energy (kT), aJ (eV)	Fast neutrons produced per thermal neutron absorbed in: ^{235}U	^{239}Pu
75	0.0048 (0.030)	2.083	2.006
200	0.0066 (0.041)	2.094	1.936
350	0.0086 (0.054)	2.102	1.875
600	0.0120 (0.075)	2.103	1.871

Fast reactors. By eliminating the moderator, it is possible to raise the average neutron energy in a reactor to a value close to that of the fission neutrons (320 fJ or 2.0 MeV, average). Coolants, fertile material, and structural material in the core, however, tend to degrade the energy so that the average is normally 96–32 fJ or 0.6–0.2 MeV. Under these conditions, the ratio of parasitic fuel captures to fuel fissions varies from 0.12 to 0.25. Corresponding breeding ratios range from 1.96 to 1.40 for ^{239}Pu-fueled fast (unmoderated) reactors and from 1.34 to 1.08 for ^{235}U-fueled reactors. In all cases the neutron yield from fissions in fuel is increased by fast-neutron fissions in fertile material (^{238}U or ^{232}Th) resulting in a higher breeding ratio than that given simply by $\eta - 1$.

It is evident from the foregoing that considerably higher conversion or breeding ratios are possible in a ^{235}U- or ^{239}Pu-fueled fast reactor than in a thermal reactor. Fast reactors, therefore, provide a means of utilizing a far greater proportion of natural uranium than would be otherwise possible.

In the case of thorium utilization by means of the ^{233}U-thorium cycle, breeding is possible with both fast and thermal neutrons. Here, the difference in breeding ratio between thermal and fast reactors is not as great as for the ^{235}U-plutonium cycle, and the choice depends upon other considerations, such as the amount of fissionable material required for criticality in each case.

In addition to achieving a high conversion ratio in a nuclear power reactor, it is also desirable to have a high thermal efficiency and high material economy (heat output per unit weight of fuel and fertile material). Unfortunately, in most cases these three characteristics cannot be maximized simultaneously. For example, in a boiling water reactor, which generates steam inside the reactor core for power production, an increase in the rate of steam generation increases the neutron losses and decreases the neutron economy. Thus, the optimum design of a boiler reactor and most other reactor types involves a compromise between high power density and high neutron economy.

Fuel requirements and supply. To estimate future fuel requirements for the United States nuclear industry, it is necessary to predict the industry's growth rate, the probable types of nuclear

Table 4. Status of United States nuclear power plants at end of 1978*

Category	Units	Electrical power, MW
Operating license	72	52,396
Construction permits	92	101,148
Limited work authorization	4	4,112
Reactors on order	30	35,082
Total	198	192,738

From E. Gordon, *Uranium 1978*, Atomic Industrial Forum, Inc., 1979.

reactors, and the amount of uranium or thorium needed for each type of reactor. Because of uncertainties in these predictions, it is obvious that the resulting estimate of future nuclear fuel requirements must be very approximate. Nevertheless, it is important to make an approximate estimate and to compare it with estimated resources of low-cost uranium to evaluate whether such resources are sufficient to meet long-term needs. Some facts concerning the present nuclear industry help in such an evaluation. It is now clear that up to some time early in the 21st century the industry will consist primarily of light-water reactors, including both pressurized water and boiling water types. In earlier analyses the optimum fueling characteristics of each type of reactor have been calculated on the basis of anticipated fuel cycle economic conditions, including the costs of fresh uranium and recycled plutonium. However, a national policy decision to delay indefinitely the recycle of plutonium has established a guideline for the industry to focus on optimization of a once-through slightly enriched uranium fuel cycle. Implicit in this policy is the deferral of fast breeder reactors which operate most efficiently with plutonium as fuel.

Estimates of demand. An estimate of future uranium demand can be made on the basis of the number of nuclear power plants already on the line and those forecast to be on the line in the future. In 1978 there were 72 operating nuclear plants in the United States which accounted for 12.5% of the nation's total electrical output. The status of United States nuclear power plants as of the end of 1978 is shown in Table 4. Since the late 1970s the number of domestic nuclear plant order cancellations have exceeded the number of new orders, and it is not clear when the nation's electric utilities will again initiate orders for new plants.

Nevertheless, the U.S. Department of Energy through its Energy Information Administration routinely prepares forecasts of nuclear power usage. These forecasts are based on projections of annual growth in gross national product, total energy requirements, the share of total energy which is electricity, and the share of electricity which is generated by nuclear power. The forecasts are made in each area for low-, mid-, and high-growth cases and result in a projected range of nuclear plant capacities and concomitant uranium ore requirements. As of the end of 1978, the Department of Energy growth forecasts for domestic nuclear power and uranium ore requirements are those shown in Table 5 for the mid-growth-rate scenario. The U_3O_8 demand shown is based on several assumptions, including a lifetime average plant capacity factor of 66.6%, reactor lifetime of 30 years, no plutonium recycle, and current light-water reactor technology for fuel utilization efficiency. Based on these parameters, a nuclear power plant generating 1000 MW of electric power requires about 5600 metric tons of U_3O_8 over its lifetime.

Table 5. Forecast of domestic uranium ore requirements for mid-growth-rate scenario*

Year	Cumulative nuclear power capacity, GW of electric power	Cumulative ore required, 10^3 metric tons of U_3O_8
1979	58	34
1980	66	39
1985	111	139
1990	172	294
1995	250	512
2000	325	795

*From U.S. Department of Energy.

Estimates of resources. Estimates of United States domestic uranium resources are shown in Table 6. The estimated uranium available is categorized on the basis of the evidence for its existence, ranging from proved "reserves" to "speculative" as indicated in the table. The amount available in each category is shown as a function of cost, which is primarily a reflection of the quality of the ore and the difficulty involved in its recovery. While the numbers presented in Table 6 are subject to change with further uranium exploration,

Table 6. Estimated United States uranium resources as of Jan. 1, 1979[a]

Cost category, \$/lb U_3O_8[b]	10^3 metric tons U_3O_8			
	Reserves[c]	Probable[d]	Possible[e]	Speculative[f]
15	—	378	192	69
15–30	—	536	422	204
30	630	915	614	273
30–50	206	408	395	227
50	836	1368	1064	500

[a]From U.S. Department of Energy.
[b]Does not represent market price. 1 lb = 0.4536 kg.
[c]Proved resources.
[d]Estimated to occur in known productive uranium districts.
[e]Estimated to occur in undiscovered or partly defined deposits in geologic settings productive elsewhere within the same geologic province.
[f]Estimated to occur in undiscovered or partly defined deposits in geologic settings not previously productive within a productive geologic province not previously productive.

these are the best estimates available and may be used to assess the potential for long-range nuclear power development.

Comparison of resources and demand. From the forecast of United States nuclear installations through the year 2000 (325 GW of electric power), the lifetime uranium commitment to plants existing at that time would be about 1,830,000 metric tons of U_3O_8. From Table 6 it may be seen that "reserves" recoverable at $50 per pound are less than half of the requirement. The balance of the uranium needed to satisfy the commitment to plants forecast for the year 2000, while it is believed to exist as indicated, will require major investment in exploration effort and resource development.

Projection of the availability of uranium resources to sustain expansion of a light-water nuclear reactor industry into the next century is thus highly uncertain. Therefore, the long-range potential for nuclear power appears to depend on either discovery of much larger uranium resources than those now estimated or on much more efficient utilization of known resources than is permitted by current light-water reactor technology.

In the long range, fast breeder reactors, which produce more fissile fuel than they consume, appear to offer a technical solution to the problem posed by limited uranium resources. Fast breeder technology is being developed in the United States and abroad, but commercialization of this technology is not being pursued in the United States. Establishing a fast breeder reactor industry in the United States would require a national policy decision which must consider economics, safety, environmental acceptability, ability to safeguard nuclear-weapons-grade fissile material, and availability of alternative energy sources. In the absence of commercial fast breeders, the large-scale utilization of nuclear power may be only an interim energy source serving until other renewable energy sources can be developed.

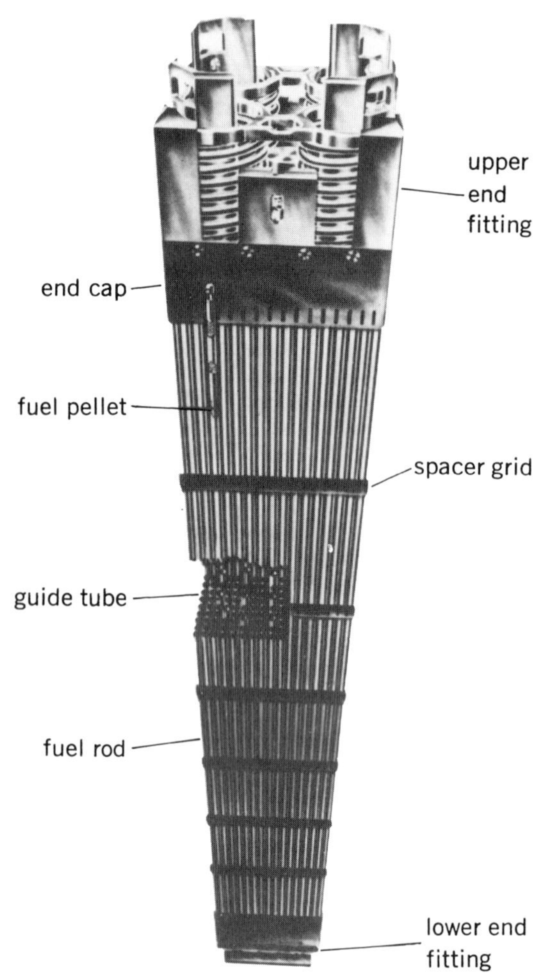

Fig. 2. Pressurized water reactor fuel assembly showing fuel rod and fuel pellet perspective. Total assembly length is about 4 m. *(Babcock & Wilcox Co.)*

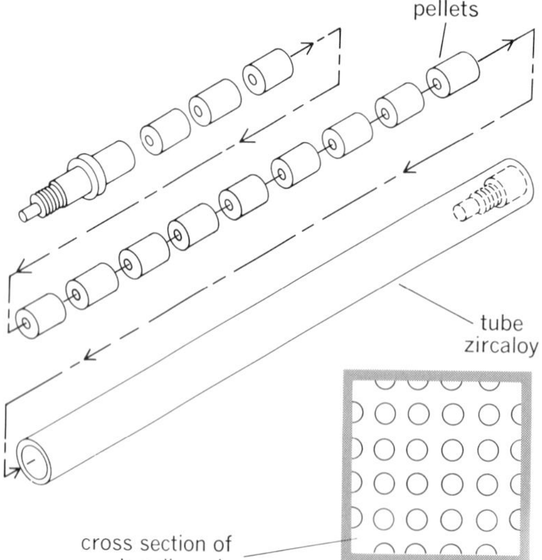

Fig. 1. Cylindrical pellets of UO_2 are pressed to exacting specifications for size and weight. After finishing, pellets are inserted into stainless steel or Zircaloy tubes. Tubes are sealed and welded, then assembled into bundles to form the rod-type element. *(From Nuclear Fuel Elements, General Electric Co.).*

Preparation of uranium fuel. Starting with ore, six major steps are required in the preparation of enriched uranium fuel: (1) recovery of uranium from ore (concentration), (2) purification of crude concentrate, (3) conversion of oxide to UF_6, (4) isotopic enrichment, (5) reduction of enriched UF_6 to UO_2 or metal, and (6) fabrication of the fuel element.

Concentration. Because of the variety of natural sources of uranium, no one concentration method is uniquely suited to all ores. Concentration by gravity methods, for example, is applicable for pitchblende but not for carnotite or autunite, from which uranium is extracted almost exculsively by leaching with acid or alkali carbonate. This is followed by a precipitation process (or by ion exchange or solvent extraction) to recover the uranium from the leach solutions.

Purification. To make uranium suitable for use in a nuclear reactor, it is desirable to reduce the concentration of neutron-absorbing impurities such as boron, cadmium, and the rare earths to levels of 0.1–10 parts per million. This is accomplished either by selective extraction of uranyl nitrate from aqueous solutions by certain oxygenated organic solvents, notably diethyl ether, methyl isobutyl ketone, or tributyl phosphate in kerosine, or by quantitative precipitation or uranium peroxide, $UO_4 \cdot 2H_2O$, from weakly acid solutions of uranyl salts.

Conversion. Conversion of the purified uranyl nitrate or UO_4 to UF_6 is carried out by first calcining the salt to produce UO_3. This is then reduced to UO_2, which is treated with HF to produce green salt, UF_4. The UF_6, which is a gas at temperatures above 56°C, is produced from UF_4 by reaction with fluorine.

Isotope enrichment. Separation of the uranium isotopes, ^{235}U and ^{238}U, depends upon the physical differences arising from the difference in their atomic weights. Gaseous diffusion using UF_6 is the process now employed to enrich the product to its specified ^{235}U content. A centrifuge process is being developed for future additional enrichment capacity because it is much more energy-efficient than gaseous diffusion.

UF_6 reduction. The UF_6 product from the diffusion plant must be reduced to uranium oxide or uranium metal for incorporation into fuel elements. To produce UO_2, which is the fuel used in commercial power reactors, the UF_6 is hydrolyzed to uranyl fluoride, UO_2F_2, reacted with ammonia, NH_3, to produce ammonium diuranate, $(NH_4)_2U_2O_7$, and calcined in hydrogen. Uranium metal, which is used in alloy form as fuel only for test reactors and plutonium production reactors, is obtained by reduction with calcium or magnesium metal.

Fuel element fabrication. For power reactor application, fuel elements consist of UO_2 pellets contained in Zircaloy or stainless steel tubes which are grouped into fuel bundles or assemblies (Figs. 1 and 2). Current light-water reactors use Zircaloy cladding almost exclusively because of its lower neutron absorption relative to stainless steel. Fabrication beginning with the calcined UO_2 described above involves milling and blending, powder granulation, compaction into pellets, sintering to high density, and centerless grinding to obtain the precise pellet diameter needed to maintain close contact with the metal tube for good heat transfer. The UO_2 pellets are inspected, dried, and loaded into the Zircaloy tubes, which are filled with helium and welded shut. The tubes must maintain leak tightness during service to prevent release of radioactive fission products to the reactor coolant circuit. The fuel rods are then inspected and assembled into fuel bundles, which are loaded into the reactor. For test reactor application, fuel elements usually consist of U_3O_8 or uranium alloy metal sandwiched in aluminum plates (Fig. 3). For naval propulsion reactors, the fuel elements are stainless-steel-clad UO_2 dispersed in stainless steel. *See* NUCLEAR FUEL CYCLE; NUCLEAR POWER.

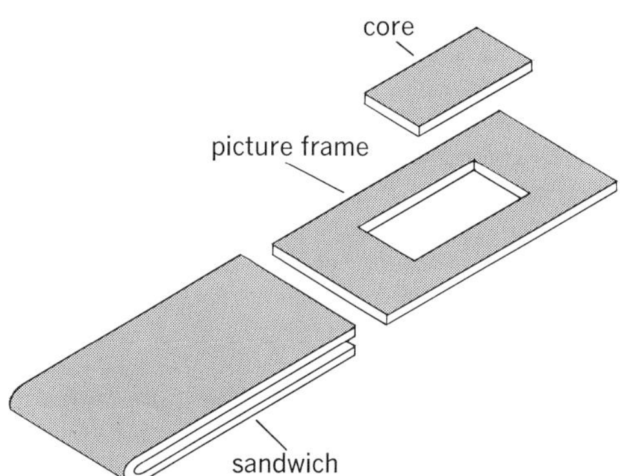

Fig. 3. A fuel plate is assembled with the core, or uranium alloy piece, fitting into a picture frame of aluminum plate. Aluminum plate is placed on either side, and the entire sandwich is hot-rolled to effect bonding. After centering the core by x-ray, the plates are trimmed to size, assembled, and mechanically bonded to the side plates. *(From Nuclear Fuel Elements, General Electric Co.)*

[R. L. BEATTY]

Bibliography: M. Benedict et al. (eds.), *Nuclear Chemical Engineering*, 2d ed., 1981; F. R. Bruce, J. M. Fletcher, and H. H. Hyman (eds.), *Process Chemistry*, in *Progress in Nuclear Energy*, ser. 3, vol. 1, 1958; H. Etherington (ed.), *Nuclear Engineering Handbook*, 1958; S. Glasstone, *Sourcebook on Atomic Energy*, 3d ed., 1967; E. Gordon, *Uranium 1978*, Atomic Industrial Forum, Inc., 1979; C. R. Tipton, *Reactor Handbook*, vol. 1: *Materials*, 1960; W. D. Wilkinson and W. F. Murphy, *Nuclear Reactor Metallurgy*, 1958.

Nuclear fuels reprocessing

The treatment of spent reactor fuel elements to recover fissionable and fertile material. Spent fuel is usually discharged from reactors because of chemical, physical, and nuclear changes that make the fuel no longer efficient for the production of heat, rather than because of the complete depletion of fissionable material. Therefore, discharged fuel usually contains fissionable material in sufficient amounts to make its recovery attractive. In the case of breeder reactors, in order to take advantage of the characteristic of the breeder reactor to produce more fissionable material than is used, reprocessing of fuel must be done to recover the fissile material bred into part of the fuel. If fertile material is also contained in the fuel, it is ordinarily recovered and purified during fuel reprocessing. Purification of the valuable constituents consists of the removal of fission products and extraneous structural material present in the fuel. *See* NUCLEAR FUELS; NUCLEAR REACTOR.

Because of the frequency of fuel discharge and because of the high value of fissionable materials, it is important that the degree of recovery approach 100% as closely as practicable. It is often necessary to reduce the fission product impurity content of discharged fuel by a factor of 10^6 to 10^7 in order to make the recovered material safe to handle during refabrication into new fuel for reuse.

There are several basic steps involved in fuel reprocessing. After fuel has been discharged from a nuclear reactor, it is common practice to store the fuel submerged in 15–20 ft (4.6–6.1 m) of water (for cooling and radiation-shielding purposes) for a period of 50–200 days; this allows the short-lived fission products to decay radioactively. During this period the radioactivity of the fuel decreases rapidly and substantially, so that when reprocessing is commenced, shielding requirements are reduced to practical thicknesses, heat evolution of the spent fuel assemblies is reduced to more easily managed levels, and radiation damage to chemicals or special structural materials in the reprocessing plant can be held to tolerable magnitudes. Following the cooling period, the fuel is mechanically cut or disassembled into convenient sizes. At this point the fuel is ready for chemical treatment to enable recovery and purification of the desired materials.

The specific steps next undertaken depend upon the particular reprocessing method employed to separate the desired products from each other, from fission products, and from extraneous structural materials. Although many separation methods exist, the one based upon solvent extraction principles is most frequently used for fuel reprocessing. Therefore, the discussion of the sequence of steps for recovery and purification will be based on the use of the solvent extraction method. Further details of this process are given in the discussion below.

Dissolution of spent fuel. The fuel assembly has been cut into pieces so that the fuel, normally held in metal tubes called cladding, is exposed. Fuel from reactors other than those using UO_2 contained in Zircaloy tubes is disassembled to allow access of chemical reagents to the fuel or fuel assembly components. The cut-up or disassembled fuel is charged, along with an appropriate aqueous dissolution medium, generally nitric acid, into a vessel. Here the solid fuel is dissolved. Except for a few fission products which are volatilized during dissolution, all the constituents initially in the fuel are retained in the dissolver solution as soluble salts or, in the case of minor constituents, as an insoluble sludge. Generally, the resulting solution must be treated by various means to accommodate its use as a feed solution to the solvent extraction process. Important reasons for such pretreatment are the adjustments of oxidation states, removal of sludge, and the adjustment of concentrations of solution constituents for optimum recovery and purification performance in the solvent extraction process.

Solvent extraction. In the solvent extraction steps, purification of the desired constituents (that is, uranium, plutonium and, if present, thorium) is achieved by the selective transfer of these constituents into an organic solution that is immiscible with the original aqueous solution. Fission products and other impurities remain in the aqueous solution and are discarded as waste. Recovery of the desired constituents is achieved by adjusting conditions to retransfer them into a clean aqueous solution. The most commonly used organic solution contains the extractant tributyl phosphate (TBP) dissolved in purified kerosine or similar materials such as long-chain hydrocarbons (for example, *n*-dodecane). The separation process using TBP with nitric acid as the aqueous medium and salting agent is called the Purex process. This is the process in general use for power reactor fuels. The basic requirements for separation during solvent extraction are immiscibility of the organic solvent with the aqueous solution of irradiated fuel, and appreciable differences with which components, initially present in the aqueous fuel solution, distribute or partition themselves between the organic and the aqueous solutions when the organic solution is first thoroughly mixed with the aqueous solution and is later separated from it.

If a quantity of suitably prepared solution of irradiated fuels is well mixed with a similar quantity of TBP-kerosine solution and is allowed to stand, the following results. The TBP-kerosine solution floats, essentially quantitatively, on the aqueous solution because the organic solution is not miscible with, and is less dense than, the aqueous solution. Analyses of the separated liquids show that a large fraction of the uranium and plutonium (and thorium if it is also present) transfers to the organic solution, but only a minute fraction of the fission products and other impurities transfers.

To enhance the transfer of uranium and plutonium into the organic solution without appreciably influencing the transfer of fission products, it is customary to have large concentrations of certain chemicals called salting agents in the aqueous solution. However, if the organic solvent containing the uranium and plutonium is now brought into contact with a clean aqueous solution wherein salting agents are absent, the uranium and plutonium will retransfer almost quantitatively to the new aqueous solution. The solvent can then be reused. Because only a minute fraction of the fission products was initially transferred to the organic mixture, the new aqueous solution contains recovered uranium and plutonium well separated from fission products. It is also possible to separate the uranium and plutonium from each other. The same organic solvent can be used, taking advantage of the fact that under certain conditions (reduced oxidation state of plutonium) plutonium extraction by the solvent is very small. Thus separation of the two heavy elements is achieved in much the same way that the impurities (for example, fission products) were initially removed from uranium and plutonium.

Separation and purification of uranium and plutonium can be done by repeated contact of the fuel solution with quantities of fresh solvent. This is the batch extraction mode and, while useful in the laboratory, it is seldom employed in large-scale practice. Operation of extraction equipment in the continuous countercurrent mode is normally more convenient, efficient, and economic for the purification of large amounts of fuel. The basic principles of separation are the same in the continuous countercurrent mode as in the batch extraction mode. The continuous operation provides repeated mixing and separation of the organic solvent and the aqueous solution of irradiated fuel from which it is desired to remove all of the valuable products freed of impurities. The continuous nature of operation is also applied in the step wherein the purified products are retransferred to a solution free of salting agents. Continuous separation of plutonium and uranium (or thorium and uranium) can also be performed. The principal advantages of continuous overbatch operations are more uniform product quality, greater ease of process control, and economy.

The countercurrent aspects of the operation are derived from having the organic solvent flow in equipment in a direction opposite to that of the aqueous solution. This allows maximum loading of the organic solution with the components to be extracted because fresh solvent encounters initially low concentrations of these components. Progressively higher concentrations occur as the solvent moves toward the point at which the aqueous solution is introduced. The aqueous solution moves counter to the flow of organic extractant and is thereby depleted to a high degree of the valuable products (for example, uranium and plutonium). Similarly, reextraction (stripping) of the products from the organic phase into a clean aqueous phase, when done in a countercurrent mode, reduces product losses to the organic solvent. Thus, a minimum of solvent is needed for maximum recovery of desired materials with solvent extraction equipment of a given efficiency.

With proper process conditions and suitable equipment, the solvent extraction operation yields nearly complete recovery and purification. Products are usually obtained in the form of dilute aqueous solutions. These solutions are subsequently further processed to give the form of plutonium, uranium, or thorium which is suitable for reuse in nuclear reactors. In the case of uranium which has been depleted in its ^{235}U content, the processing may include ^{235}U isotope reenrichment in gaseous diffusion plants. The fission products and other impurities initially present in the fuel are waste and are also obtained in the form of aqueous solutions. The waste is concentrated by evaporation and then may be introduced into underground tanks for interim storage. *See* RADIOACTIVE WASTE MANAGEMENT.

Processing plants. Plants for fuel reprocessing are large and expensive. They can be a few hundred yards in length, are normally built above and below ground level, and may cost hundreds of millions of dollars. Modern reprocessing plants are usually integrated into facilities that also provide other fuel cycle services such as conversion of uranium to UF_6, solidification of solutions of high-level waste, and conversion of plutonium product solutions to a solid suitable for shipment.

There are many factors that contribute to the high cost of these plants. Some of the more important factors are the large amounts of massive shielding (up to 7 ft or 2.1 m of high-density concrete) used to separate the process equipment from normally occupied work areas; the stringent design criteria for resistance of critical parts of the structure and equipment against natural forces such as earthquakes and tornadoes; the high integrity and rigid manufacturing standards of process equipment; the extensive systems to confine particulate radioactivity; and the barriers interposed between stored wastes and the environment.

Operation of the plants, including sampling for process control, is conducted by remote means. In some plants, even repairs and modifications of equipment in high-radiation zones are made by remote techniques. The additional cost of this type of maintenance is large. For those plants in which maintenance is performed by direct methods, the initial capital cost is reduced. This saving may be offset to some extent by increased operating costs when decontamination is difficult and permissible working time of maintenance personnel is limited. Because of the difficulty and cost of maintenance by either remote or direct methods, more spare equipment and higher standards of design, construction, and installation are necessary in fuel-reprocessing plants than in conventional chemical plants. Special precautions, which also contribute to increased capital and operating costs, must be taken in fuel-reprocessing plants to avoid nuclear criticality accidents from inadvertent accumulation of fissionable materials. This is particularly important when highly enriched fuels are reprocessed.

Thus the gross capital and operating costs are high for a fuel-reprocessing plant. The unit cost of recovered products is also very large because the output of moderately large plants is relatively small. They process only a few tons of uranium per day, for uranium containing 3% or less of the ^{235}U isotope, or as little as 10–20 lb/day (4.5–9 kg/day) when uranium containing greater than 90% of the ^{235}U isotope (highly enriched uranium) is processed. Unit cost can be substantially reduced, however, by increased capacity, because total capital and operating costs do not increase proportionally.

The operating experience to date with fuel-reprocessing plants has shown them to be relatively safe in spite of hazards from radiation and nuclear criticality, as well as other hazards of a more conventional nature.

[MARTIN J. STEINDLER]

Bibliography: M. Benedict et al., *Nuclear Chemical Engineering*, 2d ed., 1981; A. Chayes and W. B. Lewis (eds.), *International Arrangements for Nuclear Fuels Reprocessing*, 1977; H. Etherington (ed.), *Nuclear Engineering Handbook*, 1958; J. T. Long, *Engineering for Nuclear Fuel Reprocessing*, 1978; W. S. Lyon (ed.), *Analytical Chemistry in Nuclear Fuel Reprocessing*, 1978.

Nuclear power

Power derived from fission or fusion nuclear reactions. More conventionally, nuclear power is interpreted as the utilization of the fission reactions in a nuclear power reactor to produce steam for electrical power production, for ship propulsion, or for process heat. Fission reactions involve the breakup of the nucleus of heavy-weight atoms and yield-energy release which is more than a millionfold greater than that obtained from chemical reactions involving the burning of a fuel. Successful control of the nuclear fission reactions provides for the utilization of this intensive source of energy, and with the availability of ample resources of uranium deposits, significantly cheaper fuel costs for electrical power generation are attainable. Safe, clean, economic nuclear power has been the objective both of the Federal government and of industry's programs for research, development, and demonstration. Critics of nuclear power seek a complete ban or at least a moratorium on new commercial plants.

Considerations. Fission reactions provide intensive sources of energy. For example, the fissioning of an atom of uranium yields about 200 MeV, whereas the oxidation of an atom of carbon releases only 4 eV. On a weight basis, the 50,000,000 energy ratio becomes about 2,500,000. Only 0.7% of the uranium found in nature is uranium-235, which is the fissile fuel used. Even with these considerations, including the need to enrich the fuel to several percent uranium-235, the fission reactions are attractive energy sources when coupled with abundant and relatively cheap uranium ores. Although resources of low-cost uranium ores are extensive (see Tables 1 and 2), more explorations in the United States are required to better establish the reserves. Most of the uranium resources in the United States are in New Mexico, Wyoming, Colorado, and Utah. Major foreign sources of uranium are Australia, Canada, South Africa, and southwestern Africa; smaller contributions come from France, Niger, and Gabon; other sources include

Table 1. United States uranium resources, in tons of U_3O_8 as of Jan. 1, 1979*

Production cost, $/lb U_3O_8†	Proved reserves	Potential reserves			Total reserves
		Probable	Possible	Speculative	
15	290,000	415,000	210,000	75,000	990,000
30	690,000	1,005,000	675,000	300,000	2,670,000
50	920,000	1,505,000	1,170,000	550,000	4,145,000

*1 short ton = 0.907 metric ton.
†Each cost category includes all lower-cost resources. $1/lb = $2.20/kg.
SOURCE: *Statistical Data of the Uranium Industry*, U.S. Department of Energy, 1979.

Sweden, Spain, Argentina, Brazil, Denmark, Finland, India, Italy, Japan, Mexico, Portugal, Turkey, Yugoslavia, and Zaire.

Government administration and regulation. The development and promotion of the peaceful uses of nuclear power in the United States was under the direction of the Atomic Energy Commission (AEC), which was created by the Atomic Energy Act of 1946 and functioned through 1974. The Atomic Energy Act of 1954, as amended, provided direction and support for the development of commercial nuclear power. Congressional hearings established the need to assure that the public would have the availability of funds to satisfy liability claims in the unlikely event of a serious nuclear accident, and that the emerging nuclear industry should be protected from the threat of unlimited liability claims. In 1957 the Price-Anderson Act was passed to provide a combination of private insurance and governmental indemnity to a maximum of $560,000,000 for public liability claims. The act was extended in 1965 and again in 1975, each time following congressional hearings which probed the need for such protection and the merits of having nuclear power. The Federal Energy Reorganization Act of 1974 separated the promotional and regulatory functions of the AEC, with the creation of a separate Nuclear Regulatory Commission (NRC) and the formation of the Energy Research and Development Administration (ERDA). In 1977 ERDA was absorbed into the newly formed Department of Energy.

Safety measures. The AEC, overseen by the Joint Committee on Atomic Energy (JCAE), a statutory committee of United States senators and representatives, had sought to encourage the development and use of nuclear power while still maintaining the strong regulatory powers to ensure that the public health and safety were protected. The inherent dangers associated with nuclear power which involves unprecedented quantities of radioactive materials, including possible widescale use of plutonium, were recognized, and extensive programs for safety, ecological, and biomedical studies, research, and testing have been integral with the advancement of the engineering of nuclear power. Safety policies and implementation reflect the premises that any radiation may be harmful and that exposures should be reduced to "as low as reasonably achievable" (ALARA); that neither humans nor their creations are perfect and that suitable allowances should be made for failures of components and systems and human error; and that human knowledge is incomplete and thus designs and operations should be conservatively carried out. The AEC regulations, inspections, and enforcements sought to develop criteria, guides, and improved codes and standards which would enhance safety, starting with design, specification, construction, operation, and maintenance of the nuclear power operations; would separate control and safety functions; would provide redundant and diverse systems for prevention of accidents; and would provide to a reasonable extent for engineered safety features to mitigate the consequences of postulated accidents.

Criteria for siting a nuclear power station involve thorough investigation of the region's geology, seismology, hydrology, meteorology, demography, and nearby industrial, transportation, and military facilities. Also included are emergency plans to cope with fires or explosions, and radiation accidents arising from operational malfunctions, natural disasters, and civil disturbances. The AEC Directorates of Licensing, Regulatory Operations, and Regulatory Standards thus functioned to achieve an extraordinary program of safety to be commensurate with the extraordinary risks involved with nuclear power.

Starting about 1970, regulatory safety measures have been significantly augmented in response to the introduction of nuclear power reactors with larger powers and higher specific power ratings; to improved technology, experiences, and more sophisticated analytical methods; to the National Environmental Policy Act of 1969 (NEPA) and the interpretations of its implementations; and to public participation and criticism. A variety of special assessments, studies, and hearings have been undertaken by such parties as congressional committees other than JCAE, the U.S. General Accounting Office (GAO), the American Physical Society, the National Research Council representing the National Academy of Sciences and the National Academy of Engineering, and by organizations which represent public interests in the environmental impacts of the continued use of nuclear power.

Table 2. Foreign resources of uranium, in tons of U_3O_8 as of Jan. 1, 1979*

Production cost, $/lb U_3O_8†	Reasonably assured	Estimated additional	Total resources
30	1,460,000	870,000	2,330,000
50	2,010,000	1,350,000	3,360,000

*Excluding People's Republic of China, Soviet Union, and associated countries. 1 short ton = 0.907 metric ton.
†Each cost category includes all lower-cost resources. $1/lb = $2.20/kg.

Nuclear Regulatory Commission. The independent NRC is charged solely with the regulation of nuclear activities to protect the public health and safety and the environmental quality, to safeguard nuclear materials and facilities, and to ensure conformity with antitrust laws. The scope of the activities include, in addition to the regulation of the nuclear power plant, most of the steps in the nuclear fuel cycle: milling of source materials; conversion, fabrication, use, reprocessing, and transportation of fuel; and transportation and management of wastes. Not included are uranium mining and operation of the government enrichment facilities. *See* NUCLEAR FUEL CYCLE.

Public issues. Public issues of nuclear power have covered many facets and have undergone some changes in response to changes being effected. Key issues include possible theft of plutonium, with threatening consequences; management of radioactive wastes; whether, under present escalating costs, nuclear power is economic and reliable; and protection of the nuclear industry from unlimited indemnity for catastrophic nuclear accidents.

Special nuclear materials. Guidance for improving industrial security and safeguarding special nuclear materials has been initiated. Scenarios studied include possible action by terrorist groups, and evaluations have been undertaken of effective methods for preventing or deterring thefts and for recovering stolen materials. Loss of plutonium by theft and diversionary tactics could pose serious dangers through threats to disperse toxic plutonium oxide particles in populated areas or to make and use nuclear bombs.

In the commercial nuclear fuel cycle which had been envisioned previous to 1977, the more critical segments in the safeguard program would involve the chemical reprocessing plants where the high-level radioactive wastes would be separated from the uranium and the plutonium, the shipment of the plutonium oxide to the fuel manufacturing plant, and the fuel manufacturing plant where plutonium oxide would be incorporated in the uranium oxide fuel. Plutonium is produced in the normal operation of a nuclear power reactor through the conversion of uranium-238. For each gram of uranium-235 fissioned, about 0.5–0.6 g of plutonium-239 is formed, and about half of this amount is fissioned to contribute to the operation of the power reactor. Reactor operations require refueling at yearly intervals, with about one-fourth to one-third of the irradiated fuel being replaced by new fuel. In the chemical reprocessing, most of the uranium-238 initially present in the fuel would be recovered, and about one-fourth of the uranium-235 initially charged would remain to be recovered along with an almost equal amount of plutonium-239.

The only commercial chemical reprocessing plant, the Nuclear Fuel Services, Inc., facility in West Valley, NY, opened in 1966, recovered uranium and plutonium as nitrates, and stored the high-level wastes in large, underground tanks. The facility was closed in 1972, and in 1976 Nuclear Fuel Services withdrew from nuclear fuels reprocessing because of changing regulatory requirements. The Midwest Fuel Recovery Plant at Morris, IL, was to have begun operations in 1974, but functional pretests revealed that major modifications would have to be undertaken before initiating commercial operations. Maximum use of the facility has been made to accommodate storage of irradiated fuel. A third plant, the Allied General Nuclear Services Barnwell Nuclear Fuel Plant at Barnwell, SC, whose construction was begun in 1971, has not been licensed. Thus, no commercial reprocessing plant is in operation in the United States, and there is only very limited use of test fuel assemblies containing mixed oxides of plutonium and uranium.

In April 1977 the Carter administration decided to defer indefinitely the reprocessing of spent nuclear fuels. This decision reflected a policy of seeking alternative approaches to plutonium recycling and the plutonium breeder reactor for the generation of nuclear power. This policy was motivated by a concern that the use of plutonium in other parts of the world might encourage the use of nuclear weapons.

Prior to the administration's decisions, the NRC had developed a system of reviews, including public participation through hearings and through comments received on draft regulations, to determine whether recycling of plutonium was to be licensed in a generic manner. Consideration has also been given to the possible collocation of chemical reprocessing and fuel manufacturing plants, and whether there are net gains achieved through the concentration of nuclear power reactors and fuel facilities in energy parks.

Some foreign countries have opposed the anti-plutonium policies of the United States, and proceeded with the development of reprocessing plants and plutonium breeder reactors. A small commercial reprocessing plant for oxide fuel began operation in 1976 at La Hague, France, and construction of a much larger facility has been undertaken at this site. Design has begun for an oxide fuel reprocessing plant at Windscale in the United Kingdom to serve overseas markets as well as domestic markets. *See* NUCLEAR FUELS REPROCESSING.

Public participation in licensing. The procedures for licensing a nuclear facility for construction and for operation provide opportunities for meaningful public participation. Unique procedures have evolved from the Atomic Energy Act of 1954, as amended, which are responsive to public and congressional inquiry. The applicant is required to submit to the NRC a set of documents called the Preliminary Safety Analysis Report (PSAR), which must conform to a prescribed and detailed format. In addition, an environmental report is prepared. The docketed materials are available to the public, and with the Freedom of Information Act and the Federal Advisory Committee Act, even more public access to information is available. The NRC carries out an intensive review of the PSAR, extending over a period of about a year, involving meetings with the applicant and its contractors and consultants. Early in the review process, the NRC attempts to identify problems to be resolved, including concerns from citizens in the region involved with the siting of the plant. Formal questions are submitted to the applicant, and the replies are included as amendments to the PSAR.

Environmental Impact Statement. Major Federal actions that significantly affect the quality of human environment require the preparation of an

Environmental Impact Statement (EIS) in accordance with the provisions of NEPA. The EIS presents (1) the environmental impact of the proposed action; (2) any adverse environmental effects which cannot be avoided should the proposal be implemented; (3) alternatives to the proposed action; (4) relationships between short-term uses of the environment and the maintenance and enhancement of long-term productivity; and (5) any irreversible and irretrievable commitments of resources which would be involved in the proposed action should it be implemented. To better achieve these objectives, the NRC staff supplements the applicant's submittal with its own investigations and analyses and issues a draft EIS so as to gain the benefit of comments from Federal, state, and local governmental agencies, and from all interested parties. A final EIS is prepared which reflects consideration of all comments.

Safety Evaluation Report. A second major report issued by the NRC staff is the Safety Evaluation Report (SER), which contains the staff's conclusions on the many detailed safety items, including discussions on site characteristics; design criteria for structures, systems, and components; design of the reactor, fuel, and coolant system; engineered safety features; instrumentation and control; both off-site and on-site power systems; auxiliary systems, including fuel storage and handling, water systems, and fire protection system; radioactive waste management; radiation protection for employees; qualifications of applicant and contractors; training programs; review and audit; industrial security; emergency planning; accident analyses; and quality assurance.

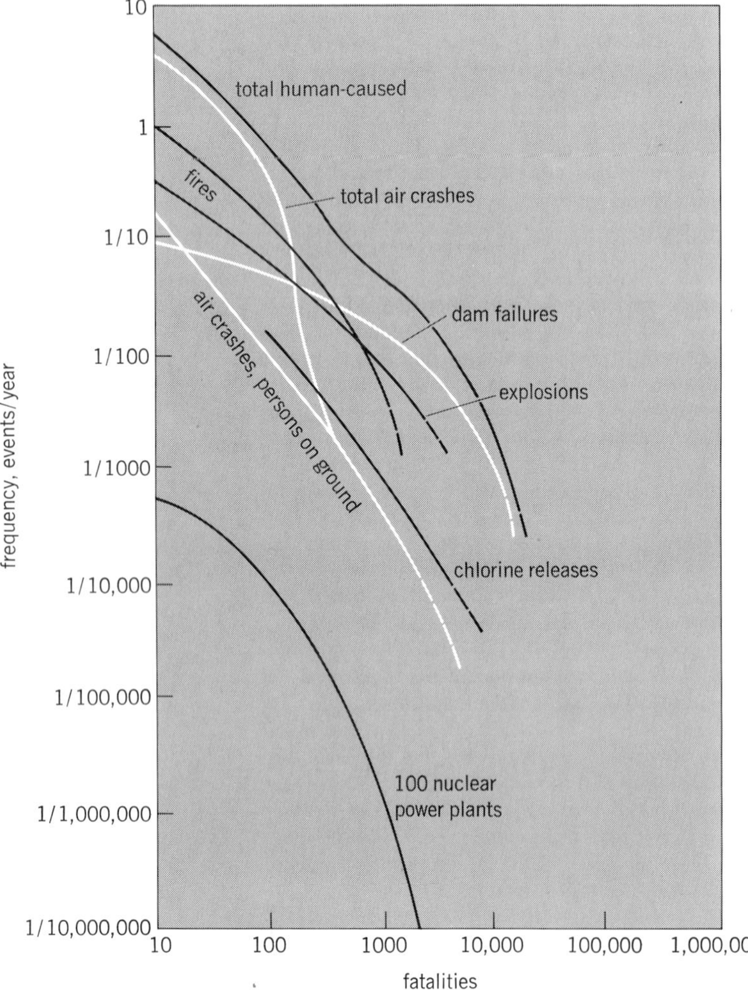

Fig. 1. Estimated frequency of fatalities due to human-caused events.

Independent review. Two additional steps are required before the decision is made regarding the construction license. An independent review on the radiological safety items is made and reported by the Advisory Committee on Reactor Safeguards (ACRS), and a public hearing is held by the Atomic Safety and Licensing Board (ASLB). The ACRS is a statutory committee consisting of a maximum of 15 members, covering a variety of disciplines and expertise. Appointments are made for this part-time activity by the NRC. Members are selected from universities, national laboratories and institutes, and industry, including experienced engineers and scientists who have retired, and, in each case, any possible conflicts-of-interest are carefully evaluated. The ACRS has a full-time staff and has access to more than 90 consultants. The ACRS conducts an independent review on nuclear safety issues and prepares a letter to the chairman of the NRC. Both subcommittee and full committee meetings are held to review the documents available and to discuss the applicant's and NRC staff's views on specific and generic issues.

Public hearing and appeal. Public participation is a major objective in the public hearing conducted by the ASLB. The ASLB is a three-member board, chaired by a lawyer, with usually two technical experts. For each application, a board is chosen from among the members of the Atomic Safety and Licensing Panel. Most members of the panel are part-time, and all members are appointed by the NRC. Prehearing conferences are held by the ASLB to identify parties who may wish to qualify and participate in the public hearing. Attempts are made to improve the understanding of the contentions, to see which contentions can be settled before the hearing, and to agree on the issues to be contested. The hearing may probe the need for additional electrical power, the suitability of the particular site chosen over possible alternative sites, the justification of the choice of nuclear power over alternate energy sources, and special issues regarding environmental impact and safety. The ASLB makes a decision on the construction application and may prescribe conditions to be followed. The decision is reviewed by and may be appealed to the Atomic Safety and Licensing Appeal Board. The appeal board is chosen from a panel completely separate from the ASLB panel. The NRC retains the authority to accept, reject, or modify the decisions. Parties not satisfied by the review process and the decisions rendered can take their case to the courts. In several cases, resort to the U.S. Supreme Court has been utilized.

Authorization. A construction permit license is not issued until the NRC, ACRS, and ASLB reviews have been completed and the application has been approved, including conditions to be met during the construction review phase. Depending upon the justification of need, a Limited Work Authorization may be granted for limited construction activities following satisfactory review of the EIS, but prior to completion of the public hearings. Construction of a nuclear power plant may take 5

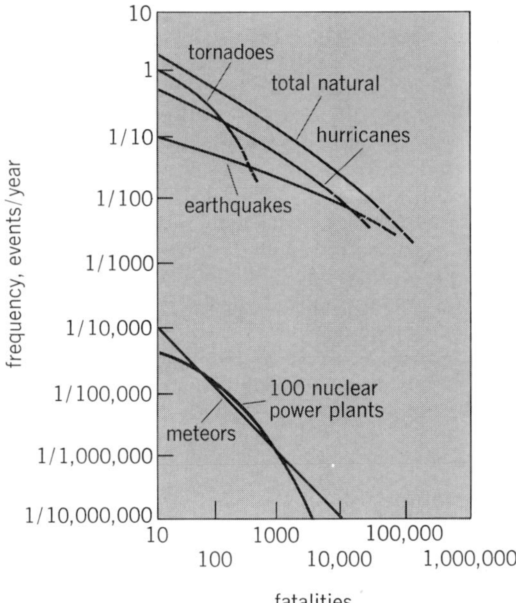

Fig. 2. Estimated frequency of fatalities due to natural events.

to 6 years or more, during which time the NRC Office of Inspection and Enforcement is involved in monitoring and inspection programs. Several years prior to the completion of construction, a Final Safety Analysis Report (FSAR) is submitted by the applicant, and again an intensive review is undertaken by the NRC staff, and later by the ACRS. A detailed Safety Evaluation Report is prepared by the NRC, and a letter is prepared by the ACRS. If all items have been satisfactorily resolved and the construction and preoperational tests have been completed, a license for operation up to the full power is granted by the NRC. Technical specifications accompany the operating license and provide for detailed limits on how the plant may be operated. In some situations where additional information is sought, less than full power is authorized. A public hearing at the operation license stage is not mandatory and would be held only if an intervenor justifies sufficient cause.

Hearings on general matters. Public hearings have also been held on generic matters to establish rules for operation. The two rulemaking hearings conducted by the AEC that have attracted much attention were the Emergency Core Cooling Systems (ECCS) and the "As Low As Practicable" (ALAP) hearings. The hearing on the criteria and conditions for evaluating the effectiveness of the ECCS for a postulated loss of coolant accident lasted from January 1972 to July 1973, and provided more than 22,000 pages of transcript, with probably twice as much additional material in supporting exhibits. EISs on major activities, such as the liquid metal fast breeder reactor (LMFBR) program and the management of commercial high-level and transuranium-contaminated radioactive waste, have provided a process for public interactions and influence.

Reactor Safety Study. A detailed, quantitative assessment of accident risks and consequences in United States commercial nuclear power plants has been carried out (WASH-1400, October 1975). The final report has had the benefit of comments and criticisms on a previous draft from governmental agencies, environmental groups, industry, professional societies, and a broad spectrum of other interested parties. Although the study was initiated by the AEC and continued by the NRC, the ad hoc group directed by N. C. Rasmussen of the Massachusetts Institute of Technology carried out an independent assessment. Aside from the very significant technical advancements made in the risk assessment methodology, the "Reactor Safety Study" represents an approach to deal with one phase of the controversial impact of technology upon society. The study presents estimated risks from accidents with nuclear power reactors and compares them with risks that society faces from both natural events and nonnuclear accidents caused by people. The judgment as to what level of risk may be acceptable for the nuclear risks still remains to be made.

Figures 1 and 2 illustrate that the frequency of human-caused nonnuclear accidents and natural events is about 10,000 times more likely to produce large numbers of fatalities than accidents at nuclear plants. The study examined two representative types of nuclear power reactors from the 50 operating reactors, and has considered that extrapolation to a base for 100 reactors is a reasonable representation. Improvements in design, construction, operation, and maintenance would be expected to reduce the risks for later expansion in nuclear reactor operations. The fatalities shown in Figs. 1 and 2 do not include potential injuries and longer-term health effects from either the nonnuclear or nuclear accidents. For the nuclear accidents, early illness would be about 10 times the fatalities in comparison to about 8,000,000 injuries caused annually by other accidents. The long-term health effects such as cancer and genetic effects are predicted to be smaller than the normal incidence rates, with increases in incidence difficult to detect even for large accidents. Thyroid illnesses, which rarely lead to serious consequences, would begin to approach the normal incidence rates only for large accidents.

The likelihood and dollar value of property dam-

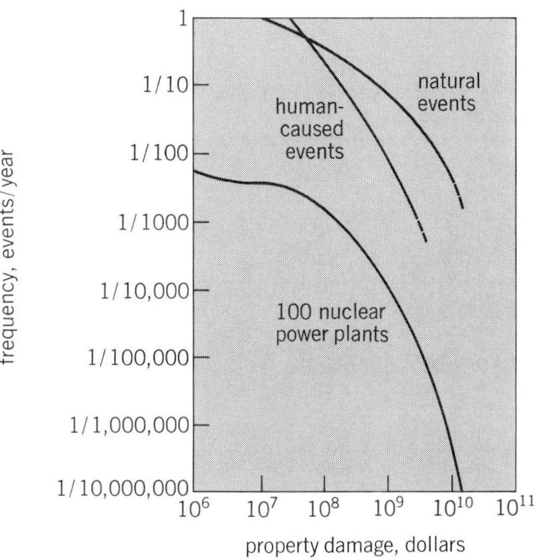

Fig. 3. Estimated frequency of property damage due to natural and human-caused events.

Table 3. Average risk of fatality by various causes

Accident type	Total number	Individual chance per year
Motor vehicle	55,791	1 in 4,000
Falls	17,827	1 in 10,000
Fires and hot substances	7,451	1 in 25,000
Drowning	6,181	1 in 30,000
Firearms	2,309	1 in 100,000
Air travel	1,778	1 in 100,000
Falling objects	1,271	1 in 160,000
Electrocution	1,148	1 in 160,000
Lightning	160	1 in 2,000,000
Tornadoes	91	1 in 2,500,000
Hurricanes	93	1 in 2,500,000
All accidents	111,992	1 in 1,600
Nuclear reactor accidents (100 plants)	–	1 in 5,000,000,000

age arising from nuclear and nonnuclear accidents are illustrated in Fig. 3. Both natural events (tornadoes, hurricanes, earthquakes) and human-caused events (air crashes, fires, dam failures, explosions, hazardous chemicals) might result in property damages in billions of dollars at frequencies up to 1000 times greater than that for accidents arising from the operation of 100 nuclear power plants.

Figures 1, 2, and 3 represent overall risk information. Risk to individuals being fatally injured through various causes is summarized in Table 3. The results of the study indicate that the predicted nuclear accidents are very small compared to other possible causes of fatal injuries.

The probability of an accident leading to the melting of the fuel core was estimated to be one chance per 20,000 reactor-years of operation, or for 100 operating reactors, one chance in 200 per year. The consequences of a core melt depend upon a number of subsequent factors, including additional failures leading to release of radioactivity, type of weather conditions, and population distribution at the particular site. The factors would have to occur in their worst conditions to produce severe consequences. Table 4 illustrates the progression of consequences and the likelihood of occurrence.

There has been considerable controversy concerning the risk estimates given in the Reactor

Table 4. Approximate values of early illness and latent effects for 100 reactors

Chance per year	Consequences: Early illness	Latent cancer fatalities* per year	Thyroid illness* per year	Genetic effects† per year
1 in 200‡	< 1.0	< 1.0	4	< 1.0
1 in 10,000	300	170	1,400	25
1 in 100,000	3,000	460	3,500	60
1 in 1,000,000	14,000	860	6,000	110
1 in 10,000,000	45,000	1500	8,000	170
Normal incidence per year	4×10^5	17,000	8,000	8,000

SOURCE: Nuclear Regulatory Commission, *Reactor Safety Study*, WASH-1400, 1975.

*This rate would occur approximately in the 10–40-yr period after a potential accident.

†This rate would apply to the first generation born after the accident. Subsequent generations would experience effects at decreasing rates.

‡This is the p edicted chance per year of core melt for 100 reactors.

Safety Study as a result of an NRC-sponsored review of them by a panel chaired by H. W. Lewis in 1977–1978. The Lewis panel argued that the Reactor Safety Study had, by its quantitative estimates, suggested a higher estimating precision than was justified by the data and procedures used. The Lewis group suggested that, rather than providing a single numerical estimate for a given risk, a numerical range should be quoted to more correctly reflect the uncertainties involved. However, Lewis also suggested that the Reactor Safety Study estimates were probably overconservative in that the most conservative choice had been made at each branch of the fault tree analyses whereas in an actual incident it would be anticipated that at least some of the branches would involve positive rather than negative choices.

Enrichment facilities. Only 0.7% of the uranium that is found in nature is the isotope uranium-235, which is used for the fuel in the nuclear power reactors. An enrichment of several percent is needed, and the Federal government (ERDA) owns the enrichment facilities. Expansion of the enrichment facilities is needed to meet expected demands. *See* NUCLEAR FUELS.

Breeder reactor program. In a breeder reactor, more fissile fuel is generated than is consumed. For example, in the fissioning of uranium-235, the neutrons released by fission are used both to continue the neutron chain reaction which produces the fission and to react with uranium-238 to produce uranium-239. The uranium-239 in turn decays to neptunium-239 and then to plutonium-239. Uranium-238 is called a fertile fuel, and uranium-235, as well as plutonium-239, is a fissile fuel and can be used in nuclear power reactors. The reactions noted can be used to convert most of the uranium-238 to plutonium-239 and thus provide about a 60-fold extension in the available uranium energy source. Breeder power reactors can be used to generate electrical power and to produce more fissile fuel. Breeding is also possible using the fertile fuel thorium-233, which in turn is converted to the fissile fuel uranium-233. An almost inexhaustible energy source becomes possible with breeder reactors. The breeder reactors would decrease the long-range needs for enrichment and for mining more uranium.

Development. The development of the breeder nuclear power reactors had been initiated by the AEC, and the experimental breeder reactor I (EBR-I) was the first reactor to demonstrate production of electrical energy from nuclear energy (Dec. 20, 1951) and to prove the feasibility of breeding utilizing a fast reactor and a liquid metal coolant. In a special test in 1955, an operator error led to a substantial melting of the fuel, but with no off-site effects. The construction on EBR-II was begun in 1958, and it has been operating since 1963. The EBR-II installation is an integrated nuclear reactor power, breeding, and fuel cycle facility. The first commercial breeder licensed for operation, in 1965, was the Enrico Fermi Fast Breeder Reactor. In 1966, while the reactor was operating at a low power, partial coolant blockage occurred, leading to some melting of fuel and release of some radioactivity within the containment building. Although the operation of the reactor was resumed in 1970, the project was discontinued in 1972, primarily for economic considerations.

Environmental impact studies. The AEC had established that the priority breeder program would utilize the LMFBR concept. In response to a 1973 Court of Appeals ruling requiring an environment impact statement for the total LMFBR research and development program, the AEC issued a seven-volume *Proposed Final Environmental Statement* in December 1974 (WASH-1535), covering environmental, economic, social, and other impacts; alternative technology options; mitigation of adverse environmental impacts; unavoidable adverse environmental impacts; short- and long-term losses; irreversible and irretrievable commitments of resources; cost-benefit analysis; and responses to the many critical comments received during the development of the environmental statements in previous drafts. With due regard to the inherent hazards and the need to carefully manage plutonium, including a vigorous program to strengthen and improve safeguards, the conclusion reached was: "The LMFBR can be developed as a safe, clean, reliable and economic electric power generation system and the advantages of developing the LMFBR as an alternative energy option for the Nation's use far outweigh the attendant disadvantages."

A reexamination of the LMFBR program was undertaken by ERDA with its issuance of the 10-volume *Final Environmental Statement* (ERDA-1535) to cover additional and supporting information for ERDA's findings and responses to the comment letters on WASH-1535. The possible impact of a new technology upon society has never been so thoroughly questioned. The *Final Environmental Statement* addressed major uncertainties in nuclear reactor safety, fuel cycle performance, safeguards, waste management, health effects, and uranium resource availability. The environmental acceptability, technical feasibility, and economic advantages of the LMFBR cannot be ascertained until additional research, development, and demonstration programs are undertaken. The ERDA decision was to proceed with a plan to continue the research and supporting programs so as to provide sufficient data by 1986 to make the decision on commercialization of the LMFBR technology. The plan contemplated the licensing, construction, and operation of the Clinch River Breeder Reactor; the design, procurement, component fabrication and testing, and the licensing for construction of the prototype large breeder reactor; and the planning of a commercial breeder reactor (CBR-I). The heat transport and power generation system for the Clinch River Breeder Reactor is illustrated in Fig. 4.

United States policies. In April 1977 the Carter administration proposed to cancel construction of the Clinch River Breeder Reactor, reflecting the policies on plutonium recycling and breeding discussed above. However, Congress decided to continue funding for this project in fiscal years 1978 and 1979.

Foreign development. Prototype breeder reactors have been constructed in France, England, and the Soviet Union. Construction of a full-scale (1250 megawatts electric power, or MWe) commercial breeder, the Super Phénix, has been undertaken in France.

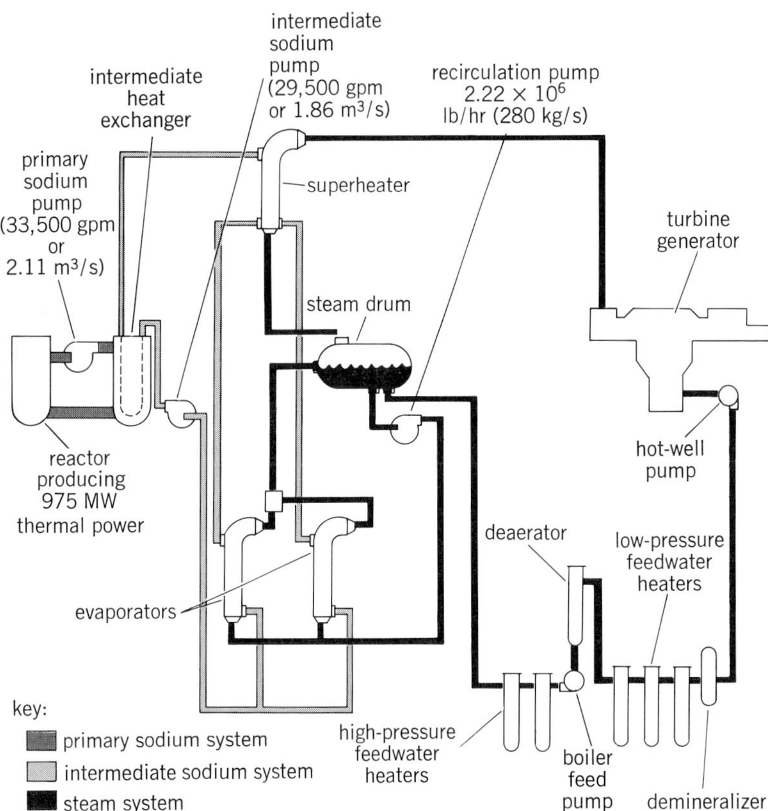

Fig. 4. Heat transport and power generation system for the Clinch River Breeder Reactor.

Radioactive waste management. In 1975 the responsibility for the development of a proposed Federal repository for high-level radioactive wastes was transferred from the AEC to ERDA (absorbed by the Department of Energy in 1977). The AEC had established regulations that the high-level radioactive wastes from the chemical reprocessing of irradiated fuel must be converted to a stable solid within 5 years of its generation at the reprocessing plant, and that the solids be sealed in high-integrity steel canisters and delivered to the AEC for subsequent management within 10 years of its generation at the reprocessing plant. Initially the AEC had sought to develop a salt mine near Lyons, KS, as the repository, but effective public intervention disclosed deficiencies for the site selected and the proposal was withdrawn. Subsequently, a more extensive study was undertaken to review permanent, safe, geologic formations and to locate possible sites. Stable, deep-lying formations could serve to isolate the wastes and dissipate their heat generation without need for maintenance or monitoring. During the period of time that might be used for the demonstration of a repository on a pilot plant scale, a retrievable surface storage facility (RSSF) would be placed in operation. The RSSF would require maintenance and monitoring, and could be engineered using known technology.

Critics of nuclear power consider the radioactive wastes generated by the nuclear industry to be too great a burden for society to bear. Since the high-level wastes will contain highly toxic materials with long half-lives, such as about a few tenths of

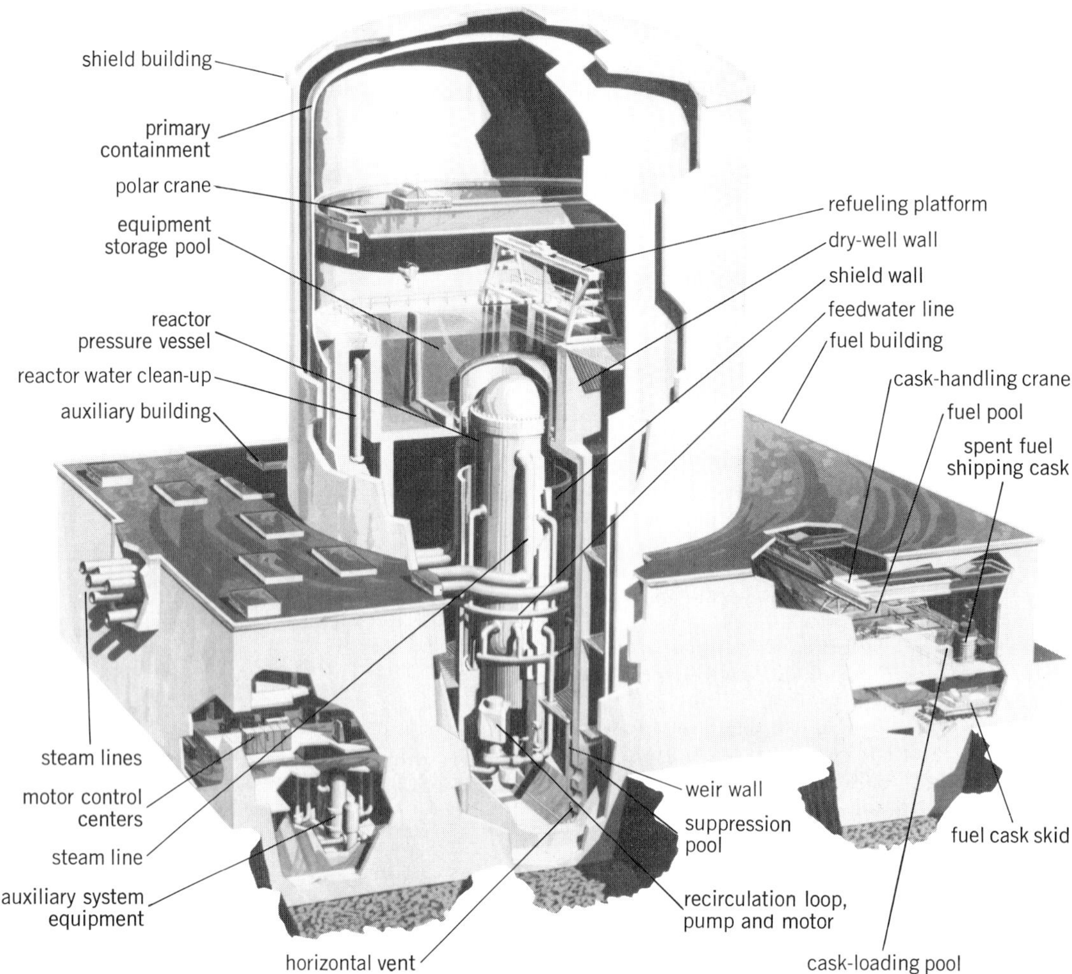

Fig. 5. Mark III containment of a boiling-water reactor, which illustrates safety features designed to minimize the consequences of reactor accidents. (*General Electric Co.*)

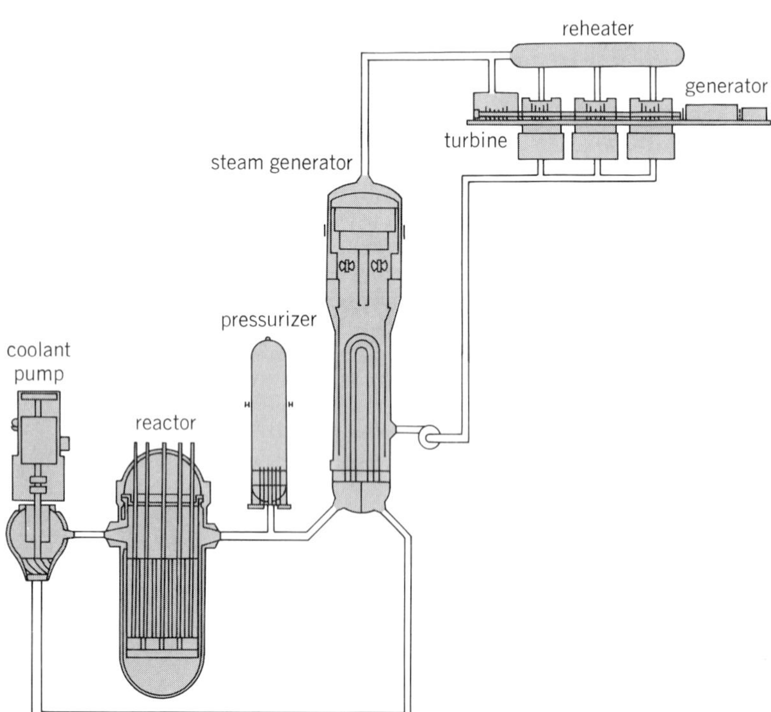

Fig. 6. Closed-cycle PWR. (*Westinghouse Electric Corp.*)

one percent of plutonium that was in the irradiated fuel, the safekeeping of these materials must be assured for time periods longer than social orders have existed in the past. Nuclear proponents answer that the proposed use of bedded salts, for example, found in geologic formations that have prevented access of water and have been undisturbed for millions of years provide assurance that safe storage can be engineered. A relatively small area of several square miles would be needed for disposal of projected wastes.

Research and development has been underway since the 1950s on methods for solidifying the wastes; however, neither ERDA nor the NRC has provided the detailed criteria and guides needed by industry to carry forth the design of the waste solidification portions for the chemical reprocessing plants. Until such time as the chemical reprocessing plants begin operation, irradiated fuel will be retained at the reactor sites and at separate facilities away from reactors in appropriate water storage pools. Spent fuels can be stored in such pools for a period of decades if necessary. The water provides shielding and also carries away residual heat.

The decision by the Carter administration in 1977 to defer indefinitely the reprocessing of spent nuclear reactor fuels forced a reconsideration of the plans and schedules that had been under development for handling spent fuel from reactors in the United States. Utilities have expanded their on-site storage capabilities for spent fuel amid uncertainty about if and when reprocessing would be permitted. With indefinite deferral of reprocessing, some people are calling for geologic disposal of spent fuel assemblies. These assemblies are about the same size as the glass rods of processed waste and could also be placed in permanent geologic storage. However, this concept has drawbacks on a resource basis because it would mean throwing away the remaining fuel value of the uranium and plutonium.

Transportation. Transportation of nuclear wastes has received special attention. With increasing truck and train shipments and increased probabilities for accidents, the protection of the public from radioactive hazards is achieved through regulations and implementation which seek to provide transport packages with multiple barriers to withstand major accidents. For example, the cask used to transport irradiated fuel is designed to withstand severe drop, puncture, fire, and immersion tests.

Low-level wastes. Management of low-level wastes generated by the nuclear energy industry requires use of burial sites for isolation of the wastes and decay to innocuous levels. Operation of the commercial burial sites is subject to regulations by Federal and state agencies.

Routine operations of nuclear power stations result in very small releases of radioactivity in the gaseous and water effluents. The NRC has adopted the principle that all releases should conform to the ALARA standard. Extension of ALARA guidance to other portions of the nuclear fuel cycle has been undertaken. *See* RADIOACTIVE WASTE MANAGEMENT.

Phase-out of governmental indemnity. In 1954 the Atomic Energy Act of 1946 was revised, making possible the possession and use of fissionable materials for industrial uses. As noted above, the Price-Anderson Act of 1957, extended in 1965 and again in 1975, protects the nuclear industry against unlimited liability in the unlikely event of a catastrophic nuclear accident. The total amount was set at $560,000,000, with private liability insurance starting at $60,000,000 in 1957 and increasing to $140,000,000 by 1977, and the government indemnity commensurately decreasing to $420,000,000. About 67% of the premiums paid to the private sector are placed in a reserve fund, and after a 10-year period, approximately 97–98% of this reserve fund has been returned. The smaller premiums paid to the Federal government are not returned. A 1966 amendment to the Price-Anderson Act provided features for no-fault liability and provisions for accelerated payment of claims. The 1975 extension of the act to 1987 provides for phasing out government indemnification, permits the $560,000,000 limit to float upward, and extends coverage to certain nuclear incidents that may occur outside the territorial limits of the United States. Each licensee is assessed a deferred payment, with a maximum level to be set per reactor. For example, in 1978 the licensees of the 66 operating reactors were each liable for a retrospective premium assessment of $5,000,000, making

Fig. 7. Oconee Nuclear Power Station containment structures.

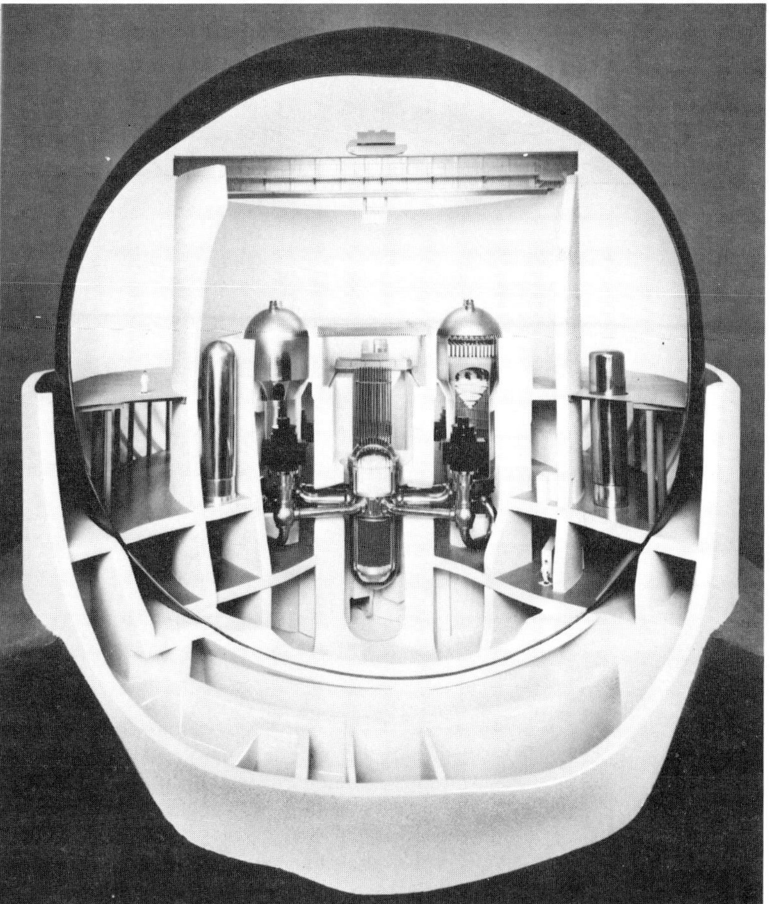

Fig. 8. Spherical containment design for a PWR. *(Combustion Engineering, Inc.)*

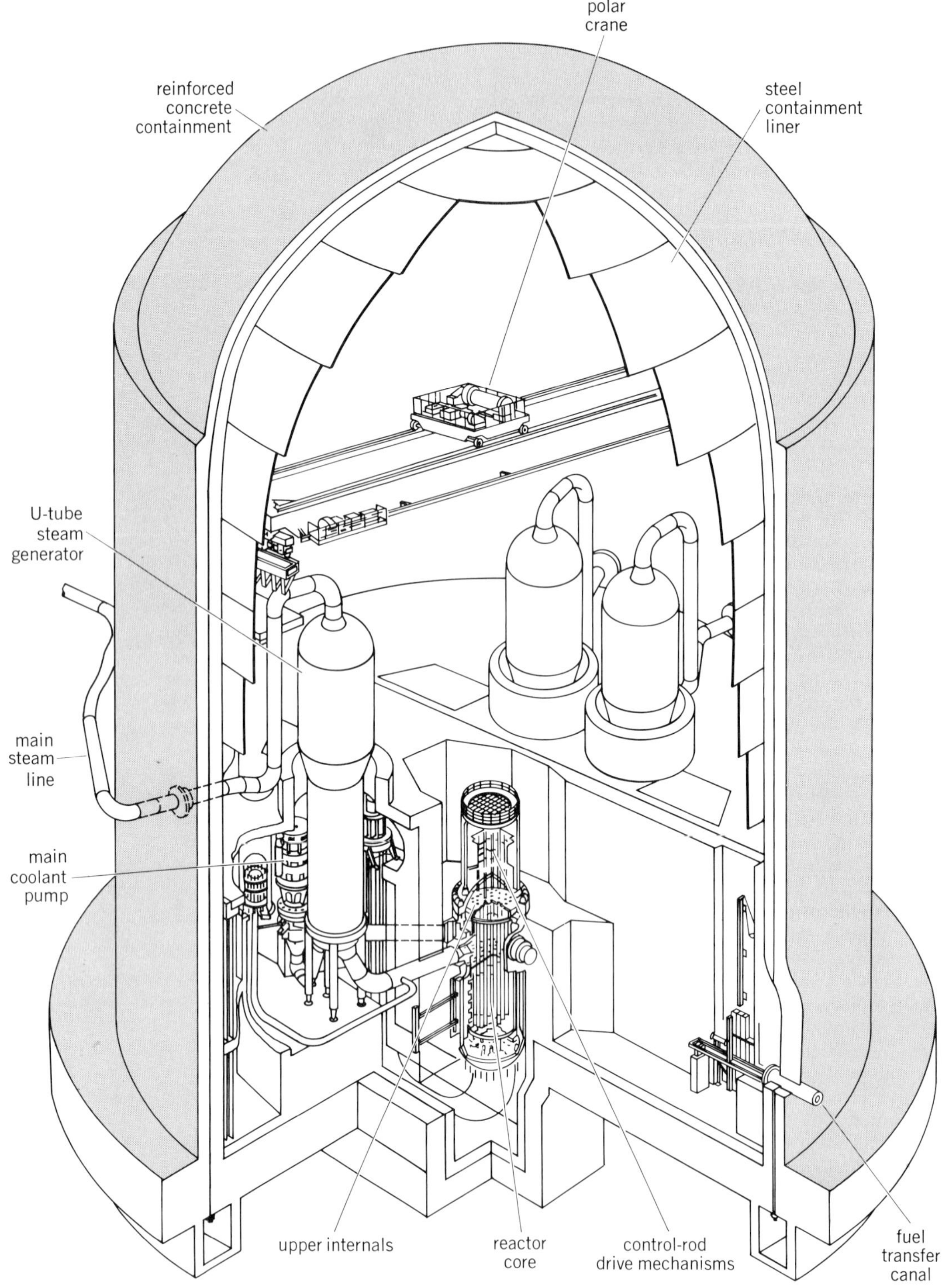

Fig. 9. Cutaway view of containment building of a typical PWR system. (*Westinghouse Electric Corp.*)

$330,000,000 available in addition to the base level of $140,000,000, and leaving $90,000,000 for the government indemnity. As the number of operating reactors increases, the government indemnity would phase out and would permit increases of total indemnity to exceed $560,000,000.

In addition to the indemnity insurance, private pools have provided property damage up to $175,000,000. New nuclear power stations would have capital costs in excess of $1,000,000,000.

Types of power reactors. There are five commercial nuclear power reactor suppliers in the United States: three for pressurized-water reactors, Combustion Engineering, Inc., Babcock and Wilcox, and Westinghouse Electric Corporation; one for boiling-water reactors, General Electric Company; and one for high-temperature gas-cooled reactors, General Atomic Company. Approximately two-thirds of the orders for nuclear power reactors are shared by General Electric and

Westinghouse, and the remaining one-third by Combustion and Babcock and Wilcox. One high-temperature gas-cooled reactor (HTGR), Fort St. Vrain Nuclear, with 330 MW net electrical power, has been licensed for operation, but orders placed for the higher-power units (with up to 1160 MW net electrical power) have been canceled or deferred indefinitely.

Boiling-water reactor (BWR). In a General Electric BWR designed to produce about 3580 MW thermal and 1220 MW net electrical power, the reactor vessel is 238 in. (6.05 m) in inside diameter, 5.7 in. (14.5 cm) thick, and about 71 ft (21.6 m) in height. The active height of the core containing the fuel assemblies is 148 in. (3.76 m). Each fuel assembly contains 63 fuel rods, and 732 fuel assemblies are used. The diameter of the fuel rod is 0.493 in. (12.5 mm). The reactor is controlled by the cruciform-shape control rods moving up from the bottom of the reactor in spaces between the fuel assemblies (177 control rods are provided). The water coolant is circulated up through the fuel assemblies by 20 jet pumps at about 70 atm (7 MPa), and boiling occurs within the core. The steam is fed through four 26-in. diameter (66 cm) steam lines to the turbine. About one-third of the energy released by fission is converted into electrical energy, and the remaining heat is removed in the condenser. The condenser operations are typical of both fossil and nuclear power plants, with heat being removed by the condenser having to be dissipated to the environment. Some limited use of the low-temperature heat source from the condenser is possible. The steam produced in the nuclear system will be at lower temperatures and pressures than that from fossil plants, and thus the efficiency of the nuclear plant in producing electrical power is less, leading to proportionately greater heat rejection to the environment.

Shielding is provided to reduce radiation levels, and pools of water are used for fuel storage and when access to the core is necessary for fuel transfers. Among the engineered safety features to minimize the consequences of reactor accidents is the containment (Fig. 5). The function of the containment is to cope with the energy released by depressurization of the coolant should a failure occur in the primary piping, and to provide a secure enclosure to minimize leakage of radioactive material to the surroundings. The BWR utilizes a pool of water to condense the steam produced by the depressurization of the primary water coolant. Various arrangements have been used for the suppression pool. Other engineered safety features include the emergency core-cooling systems.

Pressurized-water reactor (PWR). Whereas in the BWR a direct cycle is used in which steam from the reactor is fed to the turbine, the PWR employs a closed system, as shown in Fig. 6. The water coolant in the primary system is pumped through the reactor vessel, transports the heat to a steam generator, and is recirculated in a closed primary system. A separate secondary water system is used on the shell side of the steam generator to generate steam, which is fed to the turbine, condensed, and recycled back to the steam generator. A pressurizer is used in the primary system to maintain about 150 atm (15 MPa) pressure to suppress boiling in the primary coolant. One loop is shown in Fig. 6, but up to four have been used.

The reactor pressure vessel is about 44 ft (13.4 m) in height, about 14.5 ft (4.4 m) in inside diameter, and has wall thickness in the core region at least 8.5 in. (22 cm). The active length of the fuel assemblies may range from 12 to 14 ft (3.7 to 4.3 m), and different configurations are used by the manufacturers. For example, one type of fuel assembly contains 264 fuel rods, and 193 fuel assemblies are used for the 3411-MW-thermal, four-loop plant. The outside diameter of the fuel rods is 0.374 in. (9.5 mm). For this arrangement, the control rods are grouped in clusters of 24 rods per cluster, with 61 clusters provided. In the PWR the control rods enter from the top of the core. Control of the reactor operations is carried out by using both the control rods and a system to increase or decrease the boric acid content of the primary coolant.

Figure 7 is an external view of the Oconee Nuclear Power Station with three reactors, each housed in a separate containment building. The prestressed concrete containment buildings are designed for a 4-atm (400 kPa) rise in pressure and have inside dimensions of about 116 ft (35.4 m) in diameter and 208.5 ft (63.6 m) in height. The walls are 45 in. (1.14 m) thick. The containments have cooling and radioactive absorption systems as part of the engineered safety features. Figure 8 is a view of a design for a 3800-MW-thermal nuclear power reactor utilizing a two-loop system placed in a spherical steel containment (about 200, or 60 m, in diameter), surrounded by a reinforced-concrete shield building. A cutaway view of the containment building is shown in Fig. 9.

The instrumentation and control for a nuclear power station involves separation of control and protection systems, redundant and diverse features to enhance the safety of the operations, and ex-core and in-core monitoring systems to ensure safe, reliable, and efficient operations. *See* ELECTRIC POWER GENERATION; ENERGY SOURCES; NUCLEAR REACTOR.

[H. S. ISBIN]

Bibliography: Atomic Energy Commission, *Proposed Final Environmental Statement: Liquid Metal Fast Breeder Reactor Program*, WASH-1535, 1974; Energy Research and Development Administration, *Final Environmental Statement*, ERDA-1535 (including WASH-1535), 1975; International Atomic Energy Agency, *Nuclear Power and Its Fuel Cycle*, vols. 1–7, 1977–1978; H. S. Isbin (coordinator), *Public Issues of Nuclear Power*, 1975; Nuclear Regulatory Commission, *Operating Units Status Reports, such as Licensed Operating Reactors*, NUREG-75/020-12, December 1975; Nuclear Regulatory Commission, *Reactor Safety Study*, WASH-1400, 1975; Subcommittee on Energy and the Environment of the Committee on Interior and Insular Affairs, House of Representatives, 94th Congress, first session, pt. 1: *Overview of the Major Issues*, pt. 2: *Nuclear Breeder Development Program*, 1975; U.S. Congress, Office of Technology Assessment, *Nuclear Proliferation and Safeguards*, 1977.

Nuclear reactor

A system utilizing nuclear fission in a controlled and self-sustaining manner. Neutrons are used to fission the nuclear fuel, and the fission reaction produces not only energy and radiation, but also additional neutrons. Thus a neutron chain reaction ensues. A nuclear reactor provides the assembly of materials to sustain and control the neutron chain reaction, to appropriately transport the heat produced from the fission reactions, and to provide the necessary safety features to cope with the radiation and radioactive materials produced by the operation of the nuclear reactor.

Nuclear reactors are used in a variety of ways as sources for energy, for nuclear radiations, and for special tests and feasibility demonstrations. Since the first demonstration of a nuclear reactor, made beneath the West Stands of Stagg Field at the University of Chicago on Dec. 2, 1942, more than 500 nuclear reactors have been built and operated in the United States. Extreme diversification is possible with the materials available, and reactor dimensions may vary from football size to house size. The rates of energy release for controlled operations may vary from a fraction of a watt to thousands of megawatts. The critical size of a nuclear reactor is governed by the factors affecting the control of the neutron chain reaction, and the thermal output of the reactor is determined by the factors affecting the effectiveness of the coolant in removing the fission energy released.

The generation of electric energy by a nuclear power plant requires the use of heat to produce steam or to heat gases to drive turbogenerators. Direct conversion of the fission energy into useful work is possible, but an efficient process has not yet been realized to accomplish this. Thus, in its operation the nuclear power plant is similar to the conventional coal-fired plant, except that the nuclear reactor is substituted for the conventional boiler.

The rating of a reactor is usually given in kilowatts (kW) or megawatts (MW) thermal, representing the heat generation rate. The net output of electricity of a nuclear plant is about one-third of the thermal output. Significant economical gains have been achieved by building improved nuclear reactors with outputs of about 3000 MW thermal and about 1000 MW electrical. *See* ELECTRIC POWER GENERATION; NUCLEAR POWER.

FUEL AND MODERATOR

The fission neutrons are released at very high energies and are called fast neutrons. The average kinetic energy is 2 MeV, with a corresponding neutron speed of 1/15 the speed of light. Neutrons slow down through collisions with nuclei of the surrounding material. This slowing-down process is made more effective by the introduction of lightweight materials, called moderators, such as heavy water (deuterium oxide), ordinary (light) water, graphite, beryllium, beryllium oxide, hydrides, and organic materials (hydrocarbons). Neutrons that have slowed down to an energy state in equilibrium with the surrounding material are called thermal neutrons. The probability that a neutron will cause the fuel material to fission is greatly enhanced at thermal energies, and thus most reactors utilize a moderator for the conversion of fast neutrons to thermal neutrons.

With suitable concentrations of the fuel material, neutron chain reactions also can be sustained at higher neutron energy levels. The energy range between fast and thermal is designated as intermediate. Fast reactors do not have moderators and are relatively small.

Reactors have been built in all three categories. The first fast reactor was the Los Alamos assembly called Clementine, which operated from 1946 to 1953. The fuel core consisted of nickel-coated rods of pure plutonium metal, contained in a 6-in.-diameter (15 cm) mild (low-carbon) steel pot. Coolants for fast reactors may be steam, gas, or liquid metals. Current fast reactors utilize liquid sodium as the coolant and are being developed for breeding and power. An example of an intermediate reactor was the first propulsion reactor for the submarine USS *Seawolf*. The fuel core consisted of enriched uranium with beryllium as a moderator; the original coolant was sodium, and the reactor operated from 1956 to 1959. Examples of thermal reactors are given later.

Fuel composition. Only three isotopes—uranium-235, uranium-233, and plutonium-239—are feasible as fission fuels, but a wide selection of materials incorporating these isotopes is available.

Uranium-235. Naturally occurring uranium contains only 0.7% of the fissionable isotope uranium-235, the balance being essentially uranium-238. Uranium with higher concentrations of uranium-235 is called enriched uranium.

Uranium metal is susceptible to irradiation damage, which limits its operating life in a reactor. The life expectancy can be improved somewhat by heat treatment, and considerably more by alloying with elements such as zirconium or molybdenum. Uranium oxide exhibits better irradiation damage resistance and, in addition, is corrosion-resistant in oxidizing media. Ceramics such as uranium oxide have a very low thermal conductivity and lower density than metals, which are disadvantageous in certain applications.

Uranium metal can be fabricated by relatively well-established techniques, provided proper care is taken to prevent oxidation. The metal is melted in vacuum furnaces and can be cast by gravity or injection. Ingots can be rolled or extruded, and relatively complicated shapes can be fabricated. Most commonly, fuel elements are in shape of rods or plates and are fabricated by casting, rolling, or extrusion.

Current light-water-cooled nuclear power reactors utilize uranium oxide as a fuel, with an enrichment of several percent uranium-235. Cylindrical rods are the most common fuel-element configuration. They can be fabricated by compacting and sintering cylindrical pellets which are then assembled into metal tubes which are sealed.

Developmental programs for attaining long-lived solid-fuel elements include studies with uranium oxide, uranium carbide, and other refractory uranium compounds.

Plutonium-239. Plutonium-239 is produced by neutron capture in uranium-238. It is a by-product in power reactors and is becoming increasingly available as nuclear power production increases. However, plutonium as a fuel is at a relatively early stage of development and the commercial recycle of plutonium from processed spent fuel was deferred indefinitely in the United States by the Carter administration in April 1977.

Plutonium is extremely hazardous to handle

because of its biological toxicity and must be fabricated in glove boxes to ensure isolation from operating personnel. It can be alloyed with other metals and fabricated into various ceramic compounds. It is normally used in conjunction with uranium-238; alloys of uranium-plutonium, and mixtures of uranium-plutonium oxides and carbides, are of most interest. Except for the additional requirements imposed by plutonium toxicity, much of the uranium technology is applicable to plutonium. For the light-water nuclear power reactors, the oxide fuel pellets are contained in a zirconium alloy tube. Stainless steel tubes are used for containing the oxide fuel for the fast breeder reactors.

Uranium-233. Uranium-233, like plutonium, does not occur naturally, but is produced by neutron absorption in thorium-232, a process similar to that by which plutonium is produced from uranium-238. Interest in uranium-233 arises from its favorable nuclear properties and the abundance of thorium. However, studies of this fuel cycle are at a relatively early stage.

Uranium-233 also imposes special handling problems because of biological toxicity, but it does not introduce new metallurgical problems. Thorium is metallurgically different, but it has very favorable properties both as a metal and as a ceramic. *See* NUCLEAR FUELS.

Fuel distribution. Fuel-moderator assemblies may be homogeneous or heterogeneous. Homogeneous assemblies include the aqueous-solution-type water boilers and molten-salt-solution dispersions, slurries, and suspensions. The few homogeneous reactors built have been used for limited research and for demonstration of the principles and design features. In the heterogeneous assemblies the fuel and moderator form separate solid or liquid phases, such as solid-fuel elements spaced either in a graphite matrix or in a water phase. Most power reactors utilize an arrangement of closely spaced, solid-fuel rods, about 1/2 in. (13 mm) in diameter and 12 ft (3.7 m) long, in water In the arrangement shown in Fig. 1, fuel rods are arranged in a grid pattern to form a fuel assembly, and over 200 fuel assemblies are in turn arranged in a grid pattern in the reactor core.

The first homogeneous reactor was the Los Alamos Water Boiler, which commenced operations in 1944 at 1/20 watt. Various modifications were carried out to upgrade the thermal output. The aqueous solutions of uranium sulfate and later, uranium nitrate, with enrichments of about 17% were contained in a 1-ft-diameter (0.3 m) sphere. Homogeneous reactors which were used to demonstrate the feasibility of producing electrical power include the Homogeneous Reactor Experiment No. 1 (HRE-1) and HRE-2. HRE-1 operated from 1952 to 1954, generated 140 kW net electrical, contained an aqueous homogeneous solution of UO_2SO_4 with an enrichment in excess of 90%, and was self-stabilizing because of its large negative coefficient. HRE-2 operated from 1957 to 1961 and generated 300 kW net electrical.

HEAT REMOVAL

The major portion of the energy released by the fissioning of the fuel is in the form of kinetic energy of the fission fragments, which in turn is converted into heat through the slowing down and stopping of the fragments. For the heterogeneous reactors this heating occurs within the fuel elements. Heating also arises through the release and absorption of the radiations from the fission process and from the radioactive materials formed. The heat generated in a reactor is removed by a

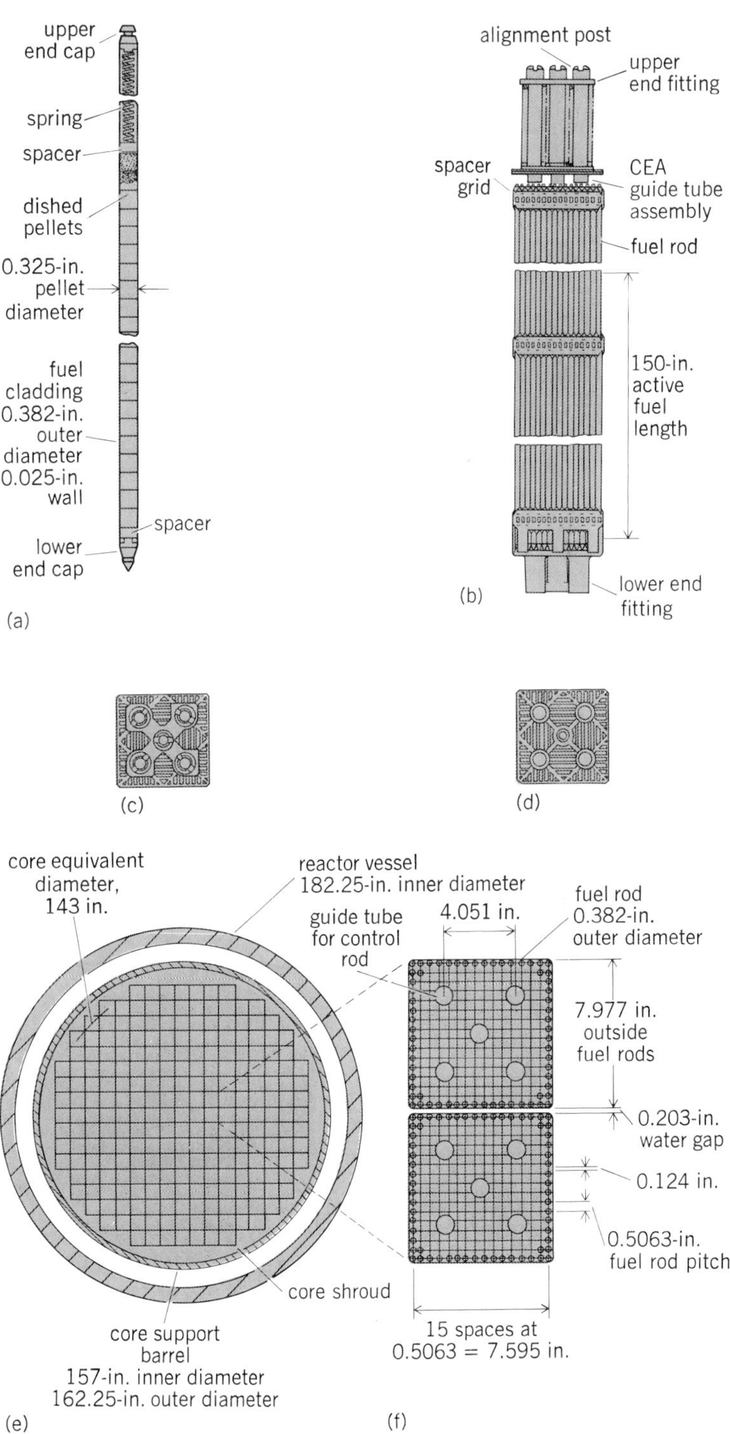

Fig. 1. Arrangement of fuel in the core of a pressurized-water reactor, a typical heterogeneous reactor. (*a*) Fuel rod; (*b*) side view (CEA = control element assembly), (*c*) top view, and (*d*) bottom view of fuel assembly; (*e*) cross section of reactor core showing arrangement of fuel assemblies; (*f*) cross section of two adjacent fuel assemblies, showing arrangement of fuel rods. 1 in. = 25.4 mm. (*Combustion Engineering, Inc.*)

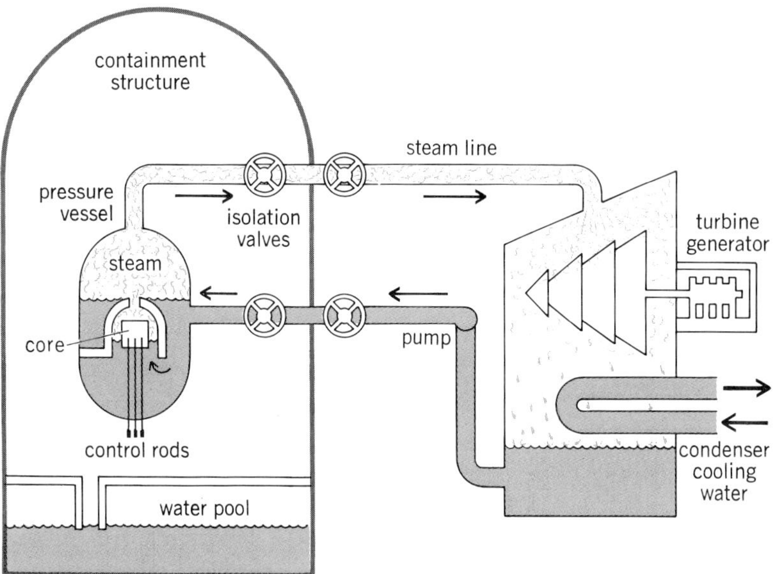

Fig. 2. Boiling-water reactor (BWR). (*Atomic Industrial Forum, Inc.*)

primary coolant flowing through the reactor.

Heat is not generated uniformly in a reactor. The heat flux decreases axially and radially from a peak at the center of the reactor, or near the center if the reactor is not symmetrical in configuration. In addition, local perturbations in heat generation can occur because of inhomogeneities in the reactor structure. These variations impose special considerations in the design of reactor cooling systems, including the need for establishing variations in coolant flow rate through the reactor to achieve uniform temperature rise in the coolant; avoiding local hot-spot conditions; and avoiding local thermal stresses and distortions in the structural members of the reactor.

Nuclear reactors have the unique thermal characteristic that heat generation continues after shutdown because of fission and radioactive decay of fission products. Significant fission heat generation occurs for only a few seconds after shutdown. Radioactive-decay heating varies with the decay characteristics of the fission products.

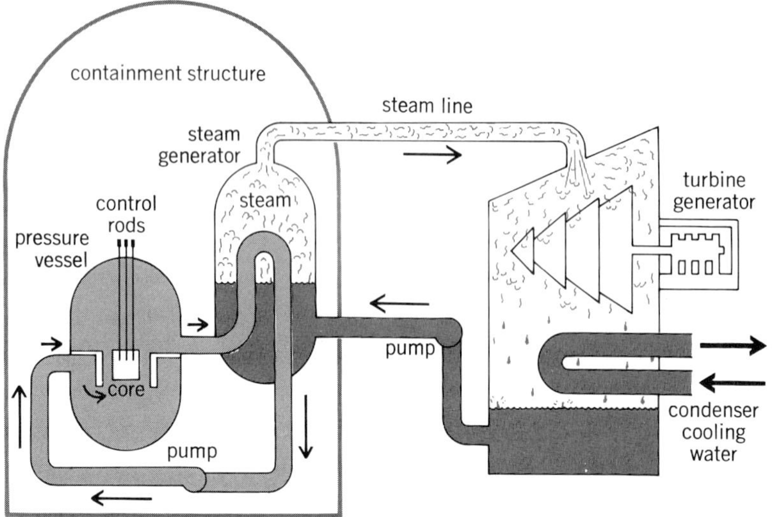

Fig. 3. Pressurized-water reactor (PWR). (*Atomic Industrial Forum, Inc.*)

Accurate analysis of fission heat generation as a function of time immediately after reactor shutdown requires detailed knowledge of the speed and reactivity worth of the control rods. The longer-term fission-product-decay heating depends upon prior reactor operation. Typical values of the total heat generation after shutdown (as percent of operating power) are 10–20% after 1 sec, 5–10% after 10 sec, approximately 2% after 10 min, 1.5% after 1 hr, and 0.7% after 1 day.

Reactor coolants. Coolants are selected for specific applications on the basis of their heat-transfer capability, physical properties, and nuclear properties.

Water. Water has many desirable characteristics. It was employed as the coolant in the first production reactors and most power reactors still utilize water as the coolant. In a boiling-water reactor (BWR; Fig. 2) the water is allowed to boil and form steam that is piped to the turbine. In a pressurized-water reactor (PWR; Fig. 3) the coolant water is kept under increased pressure to prevent boiling, and transfers heat to a separate stream of water in a steam generator, changing that water to steam. Figure 4 shows the relation of the core and heat removal systems to the condenser, electrical power system, and waste management system in the Prairie Island Nuclear Plant, which is typical of plants using pressurized-water reactors. Cool intake water is pumped through hundreds of 1-in.-diameter (25 mm) tubes in the condenser, and the warm water from the condenser is then pumped over cooling towers and returned to the plant. *See* COOLING TOWER; RADIOACTIVE WASTE MANAGEMENT; VAPOR CONDENSER.

For both boiling-water and pressurized-water reactors, the water serves as the moderator as well as the coolant. Both light and heavy water are excellent neutron moderators, although heavy water (deuterium oxide) has a neutron-absorption cross section approximately 1/500 that for light water.

There is no serious neutron-activation problem with pure water; ^{16}N, formed by the (n,p) reaction with ^{16}O (absorption of a neutron followed by emission of a proton), is the major source of activity, but its 7.5-sec half-life minimizes this problem. The most serious limitation of water as a coolant for power reactors is its high vapor pressure. A coolant temperature of 550°F (288°C) requires a system pressure of approximately 1500 psi (10 MPa). This temperature is far below modern power station practice, for which steam temperatures in excess of 1000°F (538°C) have become common. Lower thermal efficiencies result from lower temperatures. Boiling-water reactors operate at about 70 atm (7 MPa), and pressurized-water reactors at 150 atm (15 MPa). The high pressure necessary for water-cooled power reactors imposes severe design problems, which will be discussed later.

Gases. Gases are inherently poor heat-transfer fluids as compared with liquids because of their low density. This situation can be improved by increasing the gas pressure; however, this introduces other problems. Helium is the most attractive gas (it is chemically inert and has good thermodynamic and nuclear properties and has been selected as the coolant for the development of high-temperature gas-cooled reactor (HTGR) systems (Fig. 5), in which the gas transfers heat from

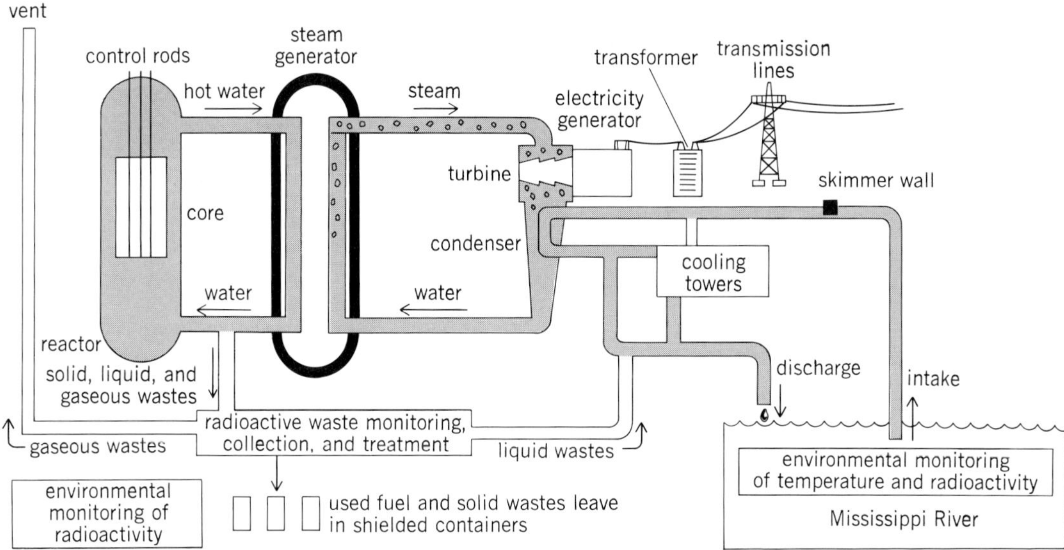

Fig. 4. Prairie Island Nuclear Plant, using pressurized-water reactors. *(Northern States Power Company)*

the reactor core to a steam generator. Gases are capable of operation at extremely high temperature, and they are being considered for special process applications and direct-cycle gas-turbine applications. Hydrogen was used as the coolant for the reactors developed in the Nuclear Engine Rocket Vehicle Application (NERVA) Program, now terminated. Heated gas discharging through the nozzle developed the propulsive thrust.

Organic coolants. Diphenyl and terphenyl possess good neutron-moderating properties and have lower vapor pressures than water. Organic coolants are noncorrosive and relatively inexpensive. Their major disadvantage is dissociation or decomposition under irradiation.

Liquid metals. The alkali metals, in particular, have excellent heat-transfer properties and extremely low vapor pressures at temperatures of interest for power generation. Sodium is the most attractive because of its relatively low melting point (208°F; 98°C) and high heat-transfer coefficient. It is also abundant, commercially available in acceptable purity, and relatively inexpensive. It is not particularly corrosive, provided low oxygen concentration is maintained. Its nuclear properties are excellent for fast reactors. In the liquid metal fast breeder reactor (LMFBR; Fig. 6) sodium in the primary loop collects the heat generated in the core and transfers it to a secondary sodium loop in the heat exchanger, from which it is carried to the steam generator.

Sodium presents an activation problem because ^{24}Na is formed by the absorption of a neutron and is an energetic gamma emitter with a 15-hr half-life. The containing system requires extensive biological shielding, and approximately 2 weeks is required for decay of ^{24}Na activity prior to access to the system for repair or maintenance. Sodium does not decompose, and no makeup is required. Sodium reacts violently with water, imposing severe problems in the design of sodium-to-water steam boilers. The poor lubricating properties of sodium and its reaction with air further complicate the mechanical design of sodium-cooled reactors. The other alkali metals exhibit similar characteristics and appear to be less attractive than sodium. The eutectic alloy of sodium with potassium (NaK), however, has the advantage that it remains liquid at room temperature.

Heavy metals have been considered for use as reactor coolants. Uranium is sufficiently soluble in bismuth at high temperatures to permit a liquid-fuel system. Bismuth also has an extremely small thermal-neutron-absorption cross section. It is a relatively poor heat-transfer fluid and, in addition, the formation of biologically toxic polonium by neutron capture imposes severe leakage restrictions. The high melting point (520°F; 267°C) of bismuth is also a disadvantage. Essentially the same considerations apply to lead-bismuth alloy, except for its more favorable melting point (257°F; 125°C).

Although mercury has seen some application as a heat-transfer fluid, it is not a particularly attractive reactor coolant. As a coolant, mercury has relatively poor heat-transfer and nuclear characteristics and also is toxic and expensive.

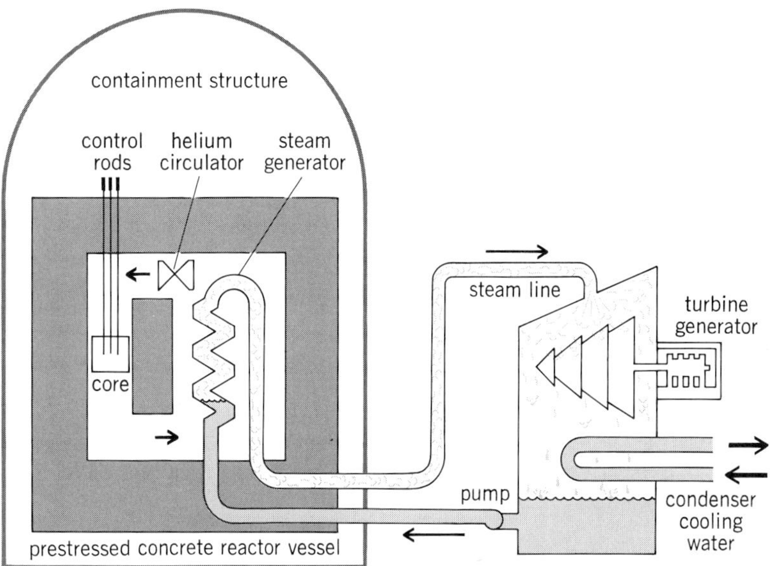

Fig. 5. High-temperature gas-cooled reactor (HTGR). *(Atomic Industrial Forum, Inc.)*

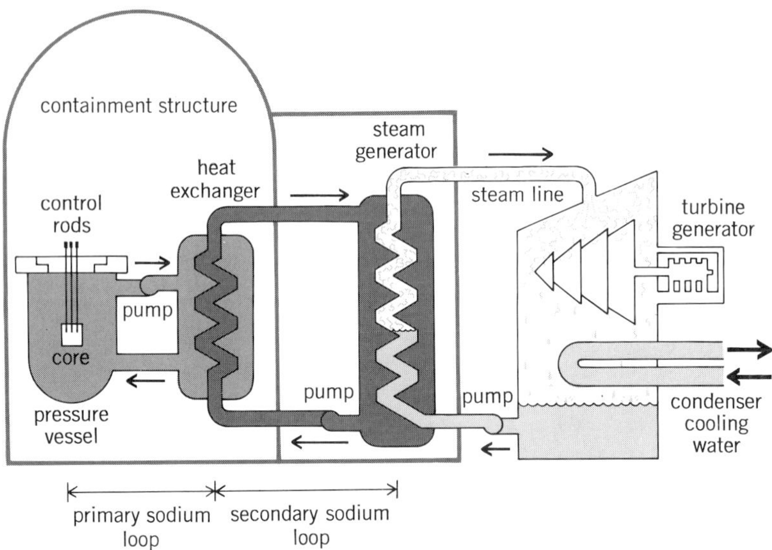

Fig. 6. Liquid metal fast breeder reactor (LMFBR). *(Atomic Industrial Forum, Inc.)*

Molten salts. Molten salts have been used as reactor coolants because they have favorable high-temperature properties and because mixtures of salts containing fuel permit fluid fuel-coolant systems. In one small experimental power reactor of this type, a molten mixture of the fluorides of beryllium, lithium, zirconium, and uranium was pumped through channels in a graphite moderator within a reactor vessel in which the fuel salt forms a critical mass and generates energy by fissioning the uranium. Design studies have been made of a larger reactor of this type. The fuel salt would be heated to 1300°F (704°C) and would deliver its heat to sodium fluoroborate in an intermediate loop that would isolate the steam boilers from the radioactive fuel circuit. Steam would be generated in the boilers at a temperature of 1000°F (538°C) and a pressure of 3500 psi (24 MPa) to achieve a thermal efficiency of about 45%. In addition to producing high thermal efficiency, this method has the potential of achieving high neutron efficiency because fission products can be removed continuously from the fluid fuel, thus reducing the fraction of neutrons lost nonproductively by capture in fission products. However, the pumping and processing of the intensely radioactive liquid fuel impose special requirements in design, fabrication, and operation.

Fluid flow and hydrodynamics. Because heat removal must be accomplished as efficiently as possible, considerable attention must be given to the fluid-flow and hydrodynamic characteristics of the system.

The heat capacity and thermal conductivity of the fluid at the temperature of operation have a fundamental effect upon the design of the reactor system. The heat capacity determines the mass flow of the coolant required. The fluid properties (thermal conductivity, viscosity, density, and specific heat) are important in determining the surface area required for the fuel to permit transfer of the heat generated at reasonable temperature differences. This, in turn, affects the design of the fuel—in particular, the amount and arrangement of the fuel elements. These factors combine to establish the pumping characteristics of the system because pressure drop and coolant-temperature rise are directly related.

Secondary considerations include other physical properties of the coolant, particularly its vapor pressure. If the vapor pressure is high at the operating temperature, local or bulk boiling of the fluid may occur. This in turn must be considered in establishing the heat-transfer coefficient for the fluid.

Because the coolant absorbs and scatters neutrons, variations in coolant density also affect reactor performance. This is particularly significant in reactors in which the coolant exists in two phases, for example, the liquid and vapor phases in boiling systems. Gases, of course, do not undergo phase change, nor do liquids operating at temperatures well below their boiling point; however, the fluid density does change with temperature and may have an important effect upon the reactor.

Power generation and, therefore, the heat-removal rate are not uniform throughout the reactor. If the mass flow rate of the coolant is uniform throughout the reactor, then unequal temperature rise of the coolant results. This becomes particularly significant in power reactors in which it is desired to achieve the highest possible coolant outlet temperature to attain maximum thermal efficiency of the power cycle. The performance limit of the coolant is set by the temperature in the hottest region or channel of the reactor. Unless the coolant flow rate is adjusted in the other regions of the reactor, the coolant will leave these regions at a lower temperature and thus will reduce the average coolant outlet temperature. In high-performance power reactors, this effect is reduced by orificing the flow in each region of the reactor commensurate with its heat generation. This involves very careful design and analysis of the system. In the boiling-type reactor, this effect upon coolant temperature does not occur because the exit temperature of the coolant is at the saturation temperature for the system. However, the variation in power generation in the reactor is reflected by a difference in the amount of steam generated in the various zones, and orificing is still required to achieve most effective use of coolant flow.

In very-high-performance reactors, the flow rate and consequent pressure drop of the coolant are sufficient to create mechanical problems in the system. It is not uncommon for the pressure drop through the fuel assemblies to exceed the weight of the fuel elements in the reactor, with a resulting hydraulic lifting force on the fuel elements. Often this requires a design arrangement to hold the fuel elements down. Although this problem can be overcome by employing downward flow through the system, it is often undesirable to do so because of shutdown-cooling considerations. It is desirable in most systems to accomplish shutdown cooling by natural-convection circulation of the coolant. If downflow is employed for forced circulation, then shutdown cooling by natural-convection circulation requires a flow reversal, which can introduce new problems.

Thermal stress considerations. The temperature of the reactor coolant increases as it circulates through the reactor. This increase in temperature is constant at steady-state conditions. Fluctuations in power level or in coolant flow rate result in variations in the temperature rise. These are reflected as temperature changes in the coolant exit temperature, which in turn result in temperature changes in the coolant system.

A reactor is capable of very rapid changes in power level, particularly, reduction in power level. Reactors are equipped with mechanisms (reactor scram systems) to ensure rapid shutdown of the system in the event of an operational abnormality.

Therefore, reactor-coolant systems must be designed to accommodate the temperature transients that may occur because of rapid power changes. In addition, they must be designed to accommodate temperature transients that might occur as a result of a coolant-system malfunction, such as pump stoppage. The consequent temperature stresses induced in the various parts of the system are superimposed upon the thermal stresses that exist under normal steady-state operations.

In very-high-performance systems, it is not uncommon for the thermal stresses alone to approach the allowable stresses in the materials of construction. In these cases, careful attention must be given to the transient stresses, and thermal shielding is commonly employed in critical sections of the system. Normally, this consists of a thermal barrier, which, by virtue of its heat capacity and resistance to heat transfer, delays the transfer of heat, thereby reducing the rate of change of temperature and protecting critical system components from thermal stresses.

Thermal stresses are also important in the design of reactor fuel elements. Metals that possess dissimilar thermal-expansion coefficients are frequently required. Heating of such systems gives rise to distortions, which in turn can result in flow restrictions in coolant passages. Careful analysis and experimental verification are often required to avoid such circumstances.

Coolant-system components. The development of reactor systems has necessitated concurrent development of special components for reactor coolant systems. These have been required even for systems employing conventional coolants, such as water or air.

Because of the hazard of radioactivity, leak-tight systems and components are a prerequisite to safe, reliable operation and maintenance. Special problems are introduced by many of the fluids employed as reactor coolants.

More extensive component developments have been required for the alkali metals (sodium, NaK, and potassium) which are chemically very active and are extremely poor lubricants. Centrifugal pumps employing unique bearings and seals must be specially designed. Liquid metals are excellent electrical conductors, and, in some special cases, electromagnetic-type pumps have been developed. These pumps are completely sealed, contain no moving parts, and derive their pumping action from electromagnetic forces imposed directly on the fluid.

In addition to the variety of special pumps developed for reactor coolant systems, there is a variety of piping-system components and heat-exchange components. As in all flow systems, flow-regulating devices such as valves are required, as well as flow instrumentation to measure and thereby control the systems. Here again, leak-tightness has necessitated the development of special valves with metallic bellows around the valve stem to ensure system integrity. Measurement of flow and pressure has also required the development of sensing instrumentation that is reliable and leak-tight.

Many of these developments have borrowed from other technologies because toxic or flammable fluids are frequently pumped in other applications. In many cases, however, special equipment has been developed specifically to meet the requirements of the reactor systems. An example of this type of development involves the measurement of flow in liquid-metal piping systems. The simple principle of a moving conductor in a magnetic field is employed by placing a magnet around the pipe and measuring the voltage generated by the moving conductor (coolant) in terms of flow rate. Temperature compensation is required, and calibration is critical.

Although the development of nuclear power reactors has introduced many new technologies, no method has yet displaced the conventional steam cycle for converting thermal energy to mechanical energy. Steam is generated either directly in the reactor (direct-cycle boiling reactor) or auxiliary steam-generation equipment in which steam is generated by transfer of heat to water from the reactor coolant. These steam generators require very special design, particularly when dissimilar fluids are involved. Typical of these problems are the sodium-to-water steam generators in which absolute integrity is essential because of the violent chemical reaction between sodium and water.

CORE DESIGN AND MATERIALS

A typical reactor core for a power reactor consists of the fuel element rods supported by a grid-type structure inside a vessel (Fig. 1).

The primary function of the vessel is to contain the coolant. Its design and materials are determined by such factors as the nature of the coolant (corrosive properties), operating conditions (temperature and pressure), and quantity and configuration of fuel. To complicate vessel design even further, the vessel is pierced by devices which are used for controlling reactor operation, for loading and unloading the fuel, and for coolant entrance and exit.

Design must also take account of thermal stresses caused by temperature differences in the system. Another problem is radioactivity induced in core materials because of neutron absorption during reactor operation. This precludes normal maintenance of the equipment and, in some areas, makes repairs virtually impossible. For this reason, an exceptionally high degree of integrity is demanded of this equipment. Reactors have been designed to permit removal of the internals from the vessel; this is difficult, however, and tends to complicate the design of the system.

Structural materials. Structural materials employed in reactor systems must possess suitable nuclear and physical properties and must be compatible with the reactor coolant under the conditions of operation. Some requirements are especially severe because of secondary effects; for example, corrosion limits may be established by the rate of deposition of coolant-entrained corrosion products on critical surfaces rather than by the rate of corrosion of the base material.

The most common structural materials employed in reactor systems are aluminum, stainless steel, and zirconium alloys. Aluminum and zirconium alloys have favorable nuclear properties, whereas stainless steel has favorable physical properties. Aluminum is widely used in low-temperature reactors; zirconium and stainless steel are used in high-temperature reactors. Zirconium is relatively expensive, and its use if therefore confined to applications where neutron absorption is critical.

The 18–8 series stainless steels have been used for structural members in both water-cooled reactors and sodium-cooled reactors because of their corrosion resistance and favorable physical properties at high temperatures. Type 304 and type 347 stainless steel have been used most extensively because of their weldability, machinability, and physical properties. To reduce cost, heavy-walled pressure vessels are normally fabricated from carbon steels and clad on the internal surfaces with a thin layer of stainless steel to provide the necessary corrosion resistance.

Although pressure vessels have been constructed for other industries to meet even more severe service requirements, the complex requirements for reactors have introduced new design and fabrication problems. Of particular importance is the dimensional precision required and the special nozzles and other appurtenances required.

As the size of power reactors has increased, it has become necessary, in some instances, to field-fabricate reactor vessels. This involves field-welding of heavy wall sections and subsequent stress relieving. Prestressed concrete vessels for gas-cooled reaction are also field-fabricated and have the potential capability of being fabricated in much larger sizes than steel vessels.

Research reactors operating at low temperatures and pressures introduce special experimental considerations. The primary objective is to provide the maximum volume of unperturbed neutron flux for experimentation. It is desirable, therefore, to extend the experimental irradiation facilities beyond the vessel wall. This has introduced the need for vessels constructed of materials having a low cross section for neutron capture. Relatively large aluminum reactor vessels with wall sections as thin as practicable have been manufactured for research reactors. Special problems with respect to dimensional stability have necessitated unique supporting structures. The vessel design is complicated further by the variety of openings that must be provided to accommodate experimental apparatus. It is highly desirable to provide access to the reactor proper for experiments and, in many cases, to have apparatus installed in so-called through holes that penetrate the vessel from side to side.

In some instances, stainless steel vessels have been employed for research and test reactors at the sacrifice of some experimental flexibility. The experimental irradiations are performed within the reactor vessel and limited in use if made of the space external to the reactor vessel.

A special problem is introduced by research reactors employing heavy water as a moderator and light water as a coolant. A calandria-type design has been employed, consisting of an all-aluminum multitube container for the heavy water, with additional aluminum tubes connected to separate coolant headers for circulation of the light-water coolant. This arrangement introduces the special problems associated with the multitudinous welds to contain a system within a system, each being tight with respect to leakage to the atmosphere and to the other system.

Fuel cladding. Heterogeneous reactors maintain a separation of fuel and coolant by cladding the fuel. The cladding material must be compatible with both the fuel and the coolant.

The cladding materials must also have favorable nuclear properties. The neutron-capture cross section is most significant because the parasitic absorption of neutrons by these materials reduces the efficiency of the nuclear fission process. Aluminum is a very desirable material in this respect; however, its physical strength and corrosion resistance in water decrease very rapidly above about 300°F (149°C).

Zirconium has very favorable neutron properties, and in addition can be made reasonably corrosion-resistant in high-temperature water. It has found extensive use for water-cooled power reactors. The technology of zirconium and zirconium-base alloys, Zircaloy, has advanced tremendously under the impetus of the various reactor development programs.

Stainless steel is used for the fuel cladding in fast reactors.

CONTROL AND INSTRUMENTATION

The control of reactors requires the measurement and adjustment of the critical condition. A reactor is critical when the rate of production of neutrons equals the rate of consumption in the system. The neutrons are produced by the fission process and are consumed in a variety of ways, including absorption to cause fission, nonfission capture in fissionable materials, capture in fertile materials, capture in structure or coolant, and leakage from the reactor. A reactor is subcritical (power level decreasing) if the number of neutrons produced is less than the number consumed. The reactor is supercritical (power level increasing) if the number of neutrons produced exceeds the number consumed.

Reactors are controlled by adjusting the balance between neutron production and neutron consumption. Normally, neutron consumption is controlled by varying the absorption or leakage of neutrons; however, the neutron-generation rate can be controlled by varying the amount of fissionable material in the system.

It is essential to orderly control and management of a reactor that the neutron density be sufficiently high to permit reliable measurement. During reactor startup, a source of neutrons is essential, therefore, to the control and instrumentation of reactor systems. Neutrons are obtained from the photo-neutron effect in materials such as beryllium. Neutron sources consist of a photon (γ-ray) source and beryllium, such as antimony-beryllium. Antimony sources are particularly convenient for use in reactors because the antimony is actjvated by the reactor neutrons each time the reactor operates.

Control drives and systems. The reactor control system requires the movement of neutron-absorbing rods (control rods) in the reactor under very exacting conditions. They must be arranged to increase reactivity (increase neutron population) slowly and under absolute control. They must be capable of reducing reactivity, both rapidly and slowly.

Normal operation of the control drives can be accomplished manually by the reactor operator or by automatic control systems. Reactor scram (very rapid reactor shutdown) can be initiated automatically by one or more system scram-safety signals, or it can be started manually by depressing a scram button convenient to the operator in the control room.

Control drives are normally electromechanical devices that impart linear or swinging motion to the control rods. They are usually equipped with a relatively slow-speed reversible drive system for normal operational control. Scram is usually effected by a high-speed overriding drive accompanied by unlatching or disconnecting the main drive system. To enhance reliability of the scram system, its operation is usually initiated by deenergizing appropriate electrical circuits. This also automatically produces reactor scram in the event of a system power failure. Hydraulic or pneumatic drive systems, as well as a variety of electromechanical systems, have also been developed.

In addition to the actuating motions required, control-rod-drive systems must also provide accurate indication of the rod positions at all times. Various types of selsyn drive, as well as arrangements of switches and lighting systems, are employed as position indicators. It is possible to provide control-rod-position indication accurate to a few thousandths of an inch.

Reactor instrumentation. Reactor control requires measurement of the reactor condition. Neutron-sensitive ion chambers are used to measure neutron flux. These neutron detectors may be located outside the reactor core, and the flux measurements from the detectors are combined to measure an average flux that is proportional to the average neutron density in the reactor. The chamber current is calibrated against a thermal power measurement and then applied over a wide range of reactor power level. The neutron-sensitive detector system must respond to the lowest neutron flux in the system produced by the neutron source.

Normally, many channels of instrumentation are required to cover the entire operating range. Several channels are required for low-level operation, beginning at the source level, whereas others are required for the intermediate and high-power-level ranges. Ten channels of detectors are not uncommon in reactor systems, and some systems contain a larger number. The total range to be covered is in the range of 7–10 decades of power level.

The chamber current can be employed as a signal, suitably amplified, to operate automatic control-system devices as well as to actuate reactor scram. In addition to absolute power level, rate of change of power level is also an important measurement which is recorded and employed to actuate various alarm and trip circuits. The normal range for the current ion chambers is approximately 10^{-14} to 10^{-4} A. This current is suitably amplified in logarithmic and period amplifiers, and can be measured directly with a galvanometer.

APPLICATIONS

Reactor applications include production of fissionable fuels (plutonium and uranium-233); mobile, stationary, and packaged power plants; research, testing, teaching-demonstration, and experimental facilities; space and process heat; dual-purpose designs; and special applications. The potential use of reactor radiation for sterilization of food and other products, for chemical processes, and for high-temperature applications has been recognized.

Production reactors. Reactor installations at Hanford, WA, and Savannah River, SC, were designed to produce plutonium-239 from uranium-238. Natural uranium is used as the fuel material. The moderator for the reactors at Hanford is graphite, and heavy water is used as the moderator at Savannah River. Water is used as a coolant in the United States production reactors, whereas in the United Kingdom, gas cooling has been the basis for most designs. The thermal, heterogeneous, natural-uranium, graphite-moderated reactors are representative of the largest reactors. The eight graphite-moderated production reactors at Hanford have been shut down, and the remaining operating production reactor, the N Reactor, is a dual-purpose unit producing special nuclear materials as well as steam for a gross power output of 860 MW electrical.

Breeder reactors. The term "converter" is applied to a reactor that converts a fertile material (for example, uranium-238) to a fissionable material (for example, plutonium). A breeder reactor, strictly speaking, produces the same fissionable material that it consumes (for example, it consumes plutonium fuel and at the same time breeds plutonium). The fuel cycle, of course, could be based on fissionable uranium-233 and fertile thorium-232 rather than uranium-238 and plutonium. In popular usage, however, any reactor that has a conversion ratio of over 100% (that is, produces more fuel than it consumes) is called a breeder, even if the fuels that are consumed and produced are different. The Experimental Breeder Reactor I (EBR-I) operated from 1951 to 1964 and was the first reactor to produce electrical power and to demonstrate the feasibility of breeding. The only operating breeder in the United States in 1976 was EBR-II, which demonstrated the use of integral facility for central station power and a closed remote reprocessing and fabrication system for the fuel cycle. EBR-II is used for fast neutron testing of fuels and materials. Prototype nuclear power breeder reactors are in operation in the United Kingdom, France, and the Soviet Union. Liquid sodium is used for the coolant (Fig. 6). Studies of fast breeder reactors utilizing gas cooling have been undertaken. Breeding with thermal reactors is also possible, and the molten salt breeder reactor (MSBR) concept, for example, has received some consideration. *See* NUCLEAR FUELS REPROCESSING.

Power reactors. Nuclear power reactors are used extensively by the U.S. Navy for propulsion of submarines and surface vessels, and by the nuclear industry for the generation of electrical power. A variety of organizations and public-interest groups have sought to slow or halt the use of the commercial nuclear power reactors.

As of 1975, more than 130 reactors have been operated by the Navy. The total United States military program has involved more than 200 reactors, operable, being built, planned, or shut down. The prototype of the first reactor used for propulsion operated in 1953, and the first reactor-powered submarine, the USS *Nautilus*, was placed in operation in 1955. Water is used as coolant and moderator and is maintained at 2000 psi (14 MPa) to suppress boiling. Two submarines, the USS *Thresher* and the USS *Scorpion*, were lost in the Atlantic in 1963 and 1968, respectively. Pressurized-water reactors are in use and under further development for submarines, cruisers, aircraft carriers, merchant ships, and (in the Soviet Union) icebreakers. The first civilian maritime reactor application (1961) was the nuclear ship *Savannah*, which utilized a pressurized-water reactor rated at 22,000 shaft horsepower (16.4 MW).

The first reactors for central-station power plant prototypes include the pressurized-water reactors—Shippingport Atomic Power Station (Pennsylvania, 231 MW thermal; 60 MW electrical, 1957) and the Atomic Power Station (Obninsk, Soviet Union, 30 MW thermal; 5 MW electrical, 1954); and the gas-cooled reactors—Calder Hall Station (Sellafield, England, originally 180 MW thermal, increased to 210 MW; 35 MW electrical with four reactors, 1956). The Dresden Nuclear Power Station (Morris, IL) is a boiling-water reactor with an output of 700 MW thermal and 208 MW electrical started in 1959. The 175-MW-electrical Yankee plant (Rowe, MA) is a pressurized-water reactor, started in 1960.

As of December 1979, the electrical generating capacity for the commercial nuclear power reactors was 9% of the total United States generating capacity, with nuclear power in some areas furnishing almost 50% of the energy source for the electric power generation. In the United States, as of April 1980, 72 nuclear power plants had operating licenses with a net electric generating capacity of 53,241 MW; 89 with construction permits, 97,762 MW; and 21 plants on order, 24,532 MW. High capital costs and slow downs experienced in use of electrical power, as well as other factors, have contributed to some cancellations and a number of deferrments for 1 to 4 years in nuclear power as well as in fossil power plant constructions. As of December 1979, the status of nuclear power plants outside the United States was 166 plants operable with a net electric generating capacity of 70,200 MW; 156 plants under construction, 125,364 MW; 33 plants on order, 27,472 MW; and 233 plants planned, 224,003 MW. Nuclear power was used in 22 countries with planned operations extending to 28 additional countries.

Research and test reactors. The research-and-development aspects of a nuclear reactor may be considered from two points of view. One is that the reactor provides experimental irradiation facilities, and the other is that the reactor itself may represent a test of a given design.

Research with reactors covers such activities as measurements of the probabilities of nuclear reactions, shielding measurements, studies of the behavior of materials under neutron and γ-irradiation, and other studies in nuclear physics, solid-state physics, and the life sciences. The irradiation facilities are used extensively for production of isotopes. High-neutron-flux reactors, designed specifically for experimental exposures of materials, are called materials-testing reactors. Reactors built to test design features are called experiments or experimental reactor facilities. Several different types of low-cost reactors, which are called teaching-demonstration reactors, have been promoted to accentuate the teaching aspects. *See* NUCLEAR ENGINEERING.

The four major varieties of research reactors are (1) uranium-fueled, graphite-moderated, air-cooled reactors; (2) uranium-fueled, heavy-water-moderated reactors; (3) enriched-fuel, aqueous-solution-type reactors; and (4) water-moderated, enriched-fuel, pool-type, and tank-type reactors. All the reactors are thermal and, with the exception of the third type, heterogeneous. Both natural and enriched uranium are used in the first two types.

The bulk shielding reactor, or BSR (Oak Ridge, TN, 1950), was the first reactor with the core submerged in an open pool of water—hence the term "swimming-pool reactor." The water is the moderator, coolant, and shield. With forced circulation of water, reactor levels of 1000 kW of heat are possible. Some reactor designs involve the use of a tank instead of a pool. Features of other pool- and tank-type reactors include variability of fuel-element design and configuration, fixed and movable cores, and a lightly pressurized (for tank-type), forced-convection water-cooling system.

The materials-testing reactor, or MTR (1952–1970), was a high-flux irradiation facility designed for studying the behavior of materials for use in power reactors. The maximum neutron fluxes available at 40 MW (thermal) were 5.5×10^{14} thermal neutrons/(cm^2)(s) and 3×10^{14} fast neutrons/(cm^2)(s). Nearly 100 experimental and instrument holes or exposure posts were provided. Other test reactors have been built to accommodate the specialized materials development programs necessary for the continued advancement of the nuclear reactor industry. Included are the engineering test reactor (ETR), 175 MW thermal, in operation since 1957, and the advanced test reactor (ATR), 250 MW thermal, completed in 1967. The ATR provides a flux up to 2.5×10^{15} neutrons/(cm^2)(s).

Test facilities for the fast-breeder-reactor physics program included the zero power reactors (ZPR), zero power plutonium reactors (ZPPR), and the Southwest Experimental Fast Oxide Reactor (SEFOR) reactor. The fast-flux test facility (FFTR) is designed to provide fast neutron environments for testing fuel and materials for fast reactors.

Among the many thermal research reactors is the high-flux isotope reactor (HFIR) at the Oak Ridge National Laboratory. A principal use of this reactor is the production of transplutonium elements such as berkelium, californium, einsteinium, and fermium.

Experimental reactors. A variety of reactors have been built to test the feasibility of given reactor designs. Reactors already noted include the experimental breeder reactors and the homogeneous reactor experiments. Several types of reactors have been designed and operated under severe power excursions to study reactor stability. Five boiling-water reactor experiments (Borax-1 to -5) have been carried out to study the behavior of such ractors at atmospheric and at elevated pressures and with different kinds of fuel elements, including nonmetallic fuels. Power-excursion experiments have been performed with the homogeneous aqueous-solution-type reactors. For example, kinetic experiment water boiler (KEWB, Canoga Park, CA) has successfully handled a power excursion of 0–530 MW in less than 1 s.

The use of boiling water as a coolant for power-producing reactors was established by the experimental boiling water reactor (EBWR, Argonne National Laboratory, Lemont, IL, 1956) and the Vallecitos boiling water reactor (VBWR, Vallecitos, CA, 1957).

The use of sodium as a high-temperature coolant for power reactors was demonstrated by the sodium-graphite reactor experiment (SRE, 1957–1964).

The feasibility of organics as coolants or coolant-moderators for reactors was studied in the organic moderated reactor experiment (OMRE, 1957–1963. The organic was a polyphenyl compound.

Test reactors for the nuclear engine for the NERVA Program included the Phoebus, NRX, and Kiwi reactors, ranging up to 4000 MW thermal. The adaptation and further testing of the reactors for space vehicles was completed successfully with ground experimental engines (XE).

Among the many other reactor experiments, two additional ones are noted here. The feasibility of the molten-salt-reactor concept has been successfully demonstrated by the molten-salt-reactor experiment (MSRE) (1965–1969). The ultra-high-temperature reactor experiment (UHTREX) (1968–1970) employed helium as a coolant and was designed to operate at 2400°F (1316°C).

Thermoelectric power. In early 1959 the AEC Los Alamos Laboratory announced the first successful production of electricity directly from a reactor core without the use of a heat-transfer medium or conventional generating equipment. The experimental unit operated by means of a thermoelectric process. The thermoelectric medium was cesium vapor, and the heat source was enriched uranium.

Specialized nuclear power units. Nuclear power units are being developed for small electrical outputs, but with special purpose for land, sea, and space applications. A 500-W reactor, SNAP-10A, was orbited in 1965 and operated successfully for 43 days. SNAP reactors are used to supply power for lunar surface experiments left behind by Apollo astronauts. Other systems for nuclear auxiliary power (SNAP) include SNAP-8, a 600-kW, thermal unit, and a series of odd-numbered units employing radioisotopes, such as plutonium-238, curium-242, polonium-210, and promethium-147, for the energy source. Other isotopes being considered are cobalt-60, strontium-90, and thulium-171. [HERBERT S. ISBIN]

Bibliography: J. M. Harrer and J. B. Beckerley, *Nuclear Power Reactor Instrumentation Systems Handbook*, National Technical Information Service, TID-25952-P1 and -P2, vol. 1, 1973, vol.2, 1974; *Nuclear Reactors Built, Being Built, or Planned in the United States*, National Technical Information Service, TID-8200, printed twice yearly as of June 30 and December 31; A. Sesonske, *Nuclear Power Plant Design Analysis*, National Technical Information Service, TID-26241, 1973; U.S. Atomic Energy Commission, *The Safety of Nuclear Power Reactors (Light Water-Cooled) and Related Facilities*, WASH-1250, 1973; J. Weisman, *Elements of Nuclear Reactor Design*, 1977.

Nucleonics

The technology based on phenomena of the atomic nucleus. These phenomena include radioactivity, fission, and fusion. Thus, nucleonics embraces such devices and fields as nuclear reactors, radioisotope applications, radiation-producing machines (such as cyclotrons and Van de Graaff accelerators), the application of radiation for biological sterilization and for the induction of chemical reactions, and radiation-detection devices. Nucleonics makes use of and serves virtually all other technologies and scientific disciplines. *See* NUCLEAR ENGINEERING.

That part of the industry concerned with nuclear reactors involves a cross section of the entire industrial complex. The chemical industry is concerned with uranium ore refining, fuel and moderator preparation, and fuel reprocessing; the light and heavy metals industry, with fuel fabrication, special component fabrication to withstand environmental conditions including radiation, and containment materials; the machinery industry with control rods, fuel charge and discharge devices, and manipulators; and the instrument industry with control systems. The many applications of nuclear reactors and isotopes also bring the industries making use of them into the field, so that electrical generation, marine propulsion, process heat, special industrial devices, and agriculture, to name a few, are industries participating to some degree in nucleonics. *See* NUCLEAR REACTOR.

A number of service activities such as reactor-design consultation, film-badge reading, special shipping and disposal of radioactive nuclear materials and wastes, and analytical services by such techniques as low-level counting and activation are included in the nucleonics industry. The unique radiation hazards and benefits associated with nuclear technology have also engendered special legal, political, and mercantile aspects. *See* RADIOACTIVE WASTE MANAGEMENT.

[BERNARD I. SPINRAD]

Nut

In mechanical structures, an internally threaded fastener. Plain square and hexagon nuts for bolts and screws are available in three degrees of finish: unfinished, semifinished, and finished. There are two standard weights: regular and heavy. For specific applications, there are other standard forms such as jam nut, castellated nut, slotted nut, cap nut, wing nut, and knurled nut (see illustration).

Hexagon jam nuts are used as locking devices to keep regular nuts from loosening and for holding set screws in position. They are not as thick as plain nuts. Jam nuts are available in semifinished form in both regular and heavy weight.

Castellated and slotted nuts have slots so that a cotter pin or safety wire can hold them in place. They are commonly used in the automotive and allied fields on operating machinery where nuts tend to loosen. The slotted nut is a regular hexagon nut with slots cut across the flats of the hexagon. They are standardized in regular and heavy weights in semifinished hexagon form. Finished thick slotted nuts are available. Castle nuts are hexagonal with a cylindrical portion above, through which slots are cut.

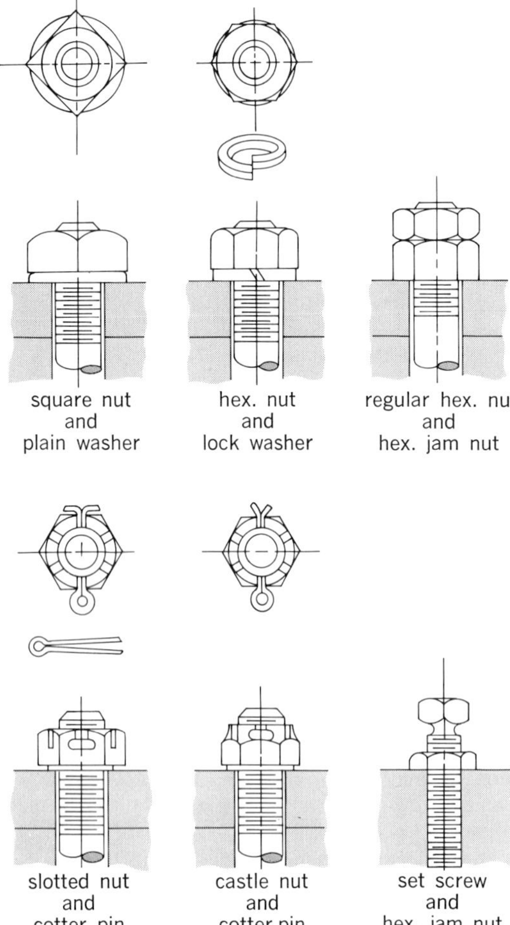

Examples of application of nuts.

Machine screw and stove-bolt nuts may be either square or hexagonal. Hexagon machine-screw nuts may have the top chamfered at 30° with a plain bearing surface, or both top and bearing surfaces may be chamfered. Square nuts have flat surfaces without chamfer. Square nuts are available with a coarse thread; hexagon nuts may be supplied with either coarse or fine thread.

Wing and knurled nuts are designed for applications where a nut is to be tightened or loosened by using finger pressure only. *See* SCREW FASTENER.

[WARREN J. LUZADDER]

Ohm's law

The direct current flowing in an electrical circuit is directly proportional to the voltage applied to the circuit. The constant of proportionality R, called the electrical resistance, is given by the equation below, in which V is the applied voltage and I is the current.

$$V=RI$$

This relationship was first described by Georg Simon Ohm in 1827 and was based on his experiments with metallic conductors. Since that time numerous deviations from this simple, linear relationship have been discovered. *See* ELECTRICAL RESISTANCE.

[CHARLES E. APPLEGATE]

Oil and gas, offshore

Oil and gas prospecting and exploitation on the continental shelves and slopes. Since Mobil Oil Co. drilled what is considered the first offshore well off the coast of Louisiana in 1945, exploration for petroleum and natural gas on the more than 8,000,000 mi² (20.7×10^6 km²) of the world's continental shelves and slopes, lying between the shore and 1000-ft (300-m) water depth, has expanded rapidly to include exploration or drilling or both off the coasts of more than 75 nations. More than 100 national and private petroleum companies have joined in the worldwide offshore search for oil and gas. The construction costs of equipping mobile offshore drilling fleets has risen tremendously due to inflation and the added sophistication of the equipment.

Until early in 1975, most of the offshore rigs were employed in drilling on the United States Outer Continental Shelf. After that time, this changed dramatically, with more than 75% of the offshore drilling being conducted outside the United States waters.

Paralleling the development of drilling vessels able to withstand the rigors of operation in the open sea have been remarkable technological developments in equipment and methods. More than 125 companies in the United States devote a large share of their efforts to the development and manufacture of material and devices in support of offshore oil and gas.

For many years petroleum companies stopped at the water's edge or sought and developed oil and gas accumulations only in the shallow seas bordering onshore producing areas. These activities were usually confined to water depths in which drilling and producing operations could be conducted from platforms, piers, or causeways built upon pilings driven into the sea floor. Major accumulations along the Gulf Coast of the United States, in Lake Maracaibo of Venezuela, in the Persian Gulf, and in the Baku fields of the Caspian Sea were developed from such fixed structures.

Exploration deeper under the sea did not begin in earnest until the world's burgeoning appetite for energy sources, coupled with a lessening return from land drilling, provided the incentives for the huge investments required for drilling in the open sea. There has been a steady annual increase in the number of wells drilled in deep water beyond the continental shelves, although the costs of developing the actual producing systems are enormous.

Geology and the sea. There is a sound geologic basis for the petroleum industry turning to the continental shelves and slopes in search of needed reserves. Favorable sediments and structures exist beneath the present seas of the world, in geologic settings that have proven highly productive onshore. In fact, the subsea geologic similarity—or in some cases superiority—to geologic conditions on land has been a vital factor in the rapid expansion of the free world's investment in offshore exploration and production.

Subsea geologic basins, having sediments considered favorable for petroleum deposits, total approximately 6,000,000 mi² (15.5×10^6 km²) out to a water depth of 1000 ft (300 m), or about 57% of the world's total continental shelf. This 6,000,000-mi^2 area is equivalent to one-third of the 18,000,000 mi^2 (46.6×10^6 km^2) of geologic basins found on land.

Offshore reserves have been estimated at 21% of the world's proved and produced reserves, while estimates of ultimate resources of petroleum offshore range from 30 to 35%, indicating that 40–50% of all future resources will come from offshore. While the exploration potential of the continental shelf is fairly well established, the potential of deep-water areas is relatively unknown, has had little petroleum test drilling, and involves a wide range of estimates (3–25% of future resources). Deep-water areas have not been included in most estimates of future resources. In addition, petroleum data from offshore areas exhibit a change in the dominance of basin types from those onshore, changes in trap types and age and depth of reservoirs, and a reduction in supergiants and average size of giants—particularly from fields partially onshore to fields that are totally offshore. Many fields with low productivity over an extensive area which are commercial onshore might be uneconomic offshore, where a thick productive column and high productivity over a concentrated area are required.

A high percentage of future petroleum resources is expected to come from offshore, where fields smaller than those on land seem to be prevalent. In addition, productive basins on land seem to have smaller fields in the final stages of development. Therefore it is probable that future resources will involve a greater number of reserves from smaller fields than in the past.

Virtually all of the world's continental shelf has received some geological study, with active seismic or drilling exploration of some sort either planned or in effect. Offshore exploration planned or underway includes action in the North Sea; in the English Channel off the coast of France; in the Red Sea bordering Egypt; in the seas to the north, west, and south of Australia; off Sumatra; along the Gulf Coast and east, west, and northwest coast of the continental United States; in the Cook Inlet and off the west coast of Alaska; in the Persian Gulf; off Mexico; off the east and west coasts of Central America and South America; and off Nigeria, North Africa, and West Africa. In addition, offshore exploration has been undertaken in the Caspian and Baltic seas off the coasts of East Germany, Poland, Latvia, and Lithuania; the China seas; Gulf of Thailand: Irish Sea; Arctic Ocean; and Arctic islands. *See* PETROLEUM PROSPECTING.

Successes at sea. The price of success in the offshore drilling is staggering in the amount of capital required, the risks involved, and the time required to achieve a break-even point.

Offshore Louisiana, long the center of much of the world's offshore drilling, has yielded a large number of oil and gas fields. However, offshore discoveries there have been made at a diminishing rate because the shelf in the area of the northern Gulf of Mexico has been thoroughly explored. Areas under exploration off the United States include the Baltimore Canyon, the Southern Georgia Embayment, the Georges Bank (southeast of Cape Cod, MA), and the eastern Gulf of Mexico (off the coast of Florida). There also has been offshore exploration for hydrocarbon deposits off the coasts of California and Alaska.

Major oil and gas strikes have been made in the North Sea, and oil and gas production has been established in the Bass Straits off Australia. These have the potential for changing historical energy sources and the economies of the countries involved. Discoveries made since 1964 in the Cook Inlet of Alaska are making a major impact on the West Coast market of the United States; although they involve between 1.5×10^9 to 2×10^9 bbl (2.4×10^8 to 3.2×10^8 m^3) of oil, the extremely high exploration and development costs will delay the break-even point. The most active offshore areas outside the United States are eastern Canada, the Labrador Sea, the North Sea, Southeast Asian seas, Nigeria, and the Bay of Campeche.

The Okan field of Nigeria reached 45,000 barrels per day (bpd) production in 1966. While offshore drilling in Nigeria has declined, exploration in Africa continues in other areas, extending out into the Atlantic to greater depths. Major activity has centered off West Africa and also in the Gulf of Suez.

The Bay of Campeche off southeastern Mexico had yielded evidence of giant oil and gas reserves, while a large discovery of oil was made in the Norwegian North Sea in 1978.

Mobile drilling platforms. The underwater search has been made possible only by vast improvements in offshore technology. Drillers first took to the sea with land rigs mounted on barges towed to location and anchored or with fixed platforms accompanied by a tender ship (Fig. 1). A wide variety of rig platforms has since evolved, some designed to cope with specific hazards of the sea and others for more general work. All new

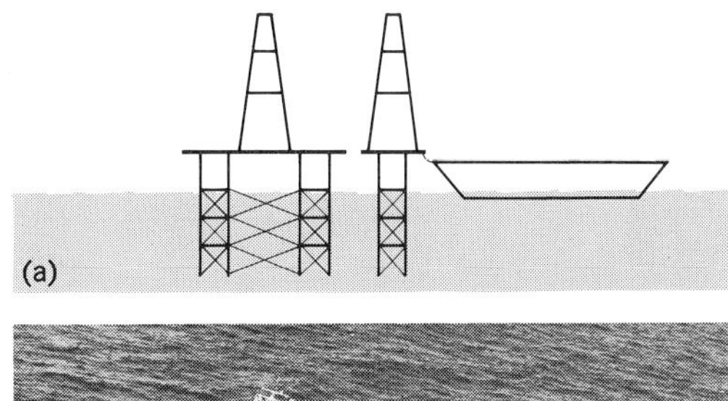

Fig. 1. Offshore fixed drilling platform. (*a*) Underwater design (*World Petroleum*). (*b*) Rig on a drilling site (*Marathon Oil Co., Findlay, Ohio*).

types stress characteristics of mobility and the capability for work in even deeper water.

The world's mobile platform fleet can be divided into four main groupings: self-elevating platforms, submersibles, semisubmersibles, and floating drill ships.

The most widely used mobile platform is the self-elevating, or jack-up, unit (Fig. 2). It is towed to location, where the legs are lowered to the sea floor, and the platform is jacked up above wave height. These self-contained platforms are especially suited to wildcat and delineation drilling. They are best in firmer sea bottoms with a depth limit out to 300 ft (90 m) of water.

The submersible platforms have been developed from earlier submersible barges which were used in shallow inlet drilling along the United States Gulf Coast. The platforms are towed to location and then submerged to the sea bottom. They are very stable and can operate in areas with soft sea floors. Difficulty in towing is a disadvantage, but this is partially offset by the rapidity with which they can be raised or lowered, once on location.

Semisubmersibles (Fig. 3) are a version of submersibles. They can work as bottom-supported units or in deep water as floaters. Their key virtue is the wide range of water depths in which they can operate, plus the fact that, when working as floaters, their primary buoyancy lies below the action of the waves, thus providing great stability. The "semis" are the most recent of the rig-type platforms.

Floating drill ships (Fig. 4) are capable of drilling in 60-ft (18-m) to abyssal depths. They are built as self-propelled ships or with a ship configuration that requires towing. Several twin-hulled versions have been constructed to give a stable catamaran design. Floating drill ships use anchoring or ingenious dynamic positioning systems to stablize their position, the latter being necessary in deeper waters. Floaters cannot be used in waters much shallower than 70 ft because of the special equipment required for drilling from the vessel subject to vertical movement from waves and tidal changes, as well as minor horizontal shifts due to stretch and play in anchor lines. Exploration in deeper waters necessitates building more semisubmersible and floating drill ships. A conventional exploratory hole has been drilled 50 mi (80 km) off the coast of Gabon in 2150 ft (655 m) of water by Shell Deep Water Drilling Company using the Sedco 445 drill ship. The *Glomar Challenger* has drilled stratigraphic holes in the sea floor to a depth of 3334 ft (1016 m) in water depths of 20,000 ft (6 km).

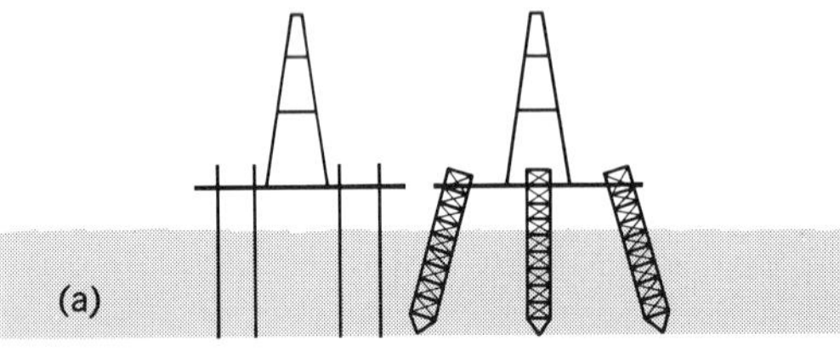

Fig. 2. Offshore self-elevating drilling platform. (*a*) Underwater design (*World Petroleum*). (*b*) Self-elevating drilling platform (*Marathon Oil Co., Findlay, Ohio*).

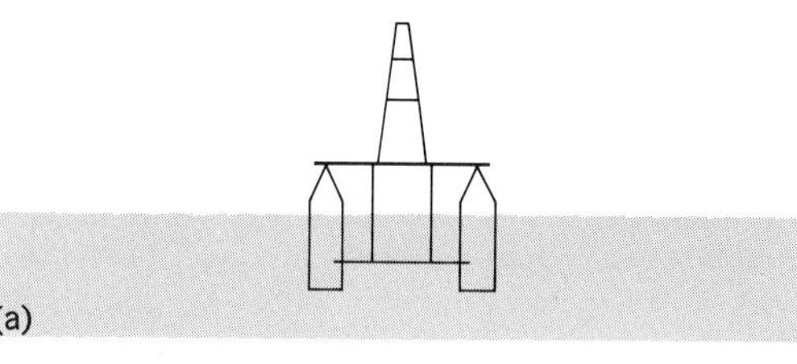

Fig. 3. Offshore semisubmersible drilling platform. (*a*) Underwater design (*World Petroleum*). (*b*) Santa Fe Marine's Blue Water no. 3 drilling rig on a drilling site (*Marathon Oil Co., Findlay, Ohio*).

Production and well completion technology. The move of exploration into the open hostile sea has required not only the development of drilling vessels but a host of auxiliary equipment and techniques. A whole new industrial complex has developed to serve the offshore industry.

Of particular interest is the development of diving techniques and submersible equipment to aid in exploration and the completion of wells. A platform was constructed for installation in 850 ft (260 m) of water in the Santa Barbara Channel. Tentative plans for platforms in 1000 ft (300 m) of water have been announced for the Gulf of Mexico. Economics will soon force sea-bottom completions which require men or robots or both to make the necessary pipe and well connections. Such work is necessary even in the water depths now being developed.

A robot device has been developed which operates from the surface and uses sonar and television for viewing; it can excavate ditches for pipelines and make simple pipe connections and well hookups in water depths to 2500 ft (760 m). Limitations of robot devices are such that the more complex

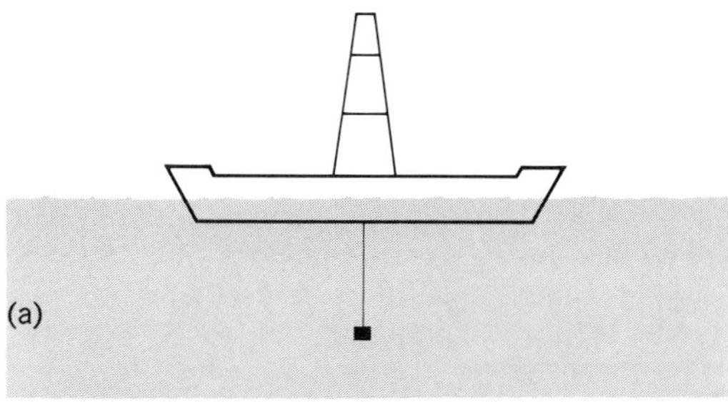

Fig. 4. Floating drill ship. Such ships can drill in depths from 60 to 1000 ft (18 to 300 m) or more. (*a*) Underwater design (*World Petroleum*). (*b*) Floating drill ship on a drilling site (*Marathon Oil Co., Findlay, Ohio*).

needs of well completions and service require the actions of men. To fill this need, diving specialists have been used to sandbag platform bases, recover conductor pipe, survey and remove wreckage, and make pipeline connections and well hookups. Diving depths have been increased to the point where useful work has been performed at depths in excess of 600 ft (180 m). Pressure chambers to take divers to the bottom and to return them to the surface to be decompressed are operational. One has been operating routinely in 425 ft (130 m) of water for Esso Exploration, Norway. This deep diving is made possible by the development of saturation diving, which uses a mixture of oxygen, helium, nitrogen, and argon. This technique has allowed divers to remain below 200 ft (60 m) for 6 days while doing salvage work on a platform. It has also allowed prolonged submergence at 600 ft (180 m) in preparation for actual work on wells at this depth. *See* INDUSTRIAL ROBOTS.

Miniature submarines have taken their place in exploration and completion work allowing the viewing of conditions, the gathering of samples, and simple mechanical tasks. Their depth range is for all practical purposes unlimited.

Technical groups are experimenting with the design of drilling and production units that would be totally enclosed and be set on the sea bottom. Living in and working from these units, personnel would be able to carry out all the necessary oil field operations. In effect, such units would resemble a miniature city on the sea floor, from which a man would need to return to the surface only when his tour of work was completed. *See* OIL AND GAS FIELD EXPLOITATION; OIL AND GAS WELL COMPLETION.

Concomitant with the progress of the petroleum industry in its venture into the open sea has been a vast increase in the knowledge of the sea and its contained wealth. Mining of the sea floor using some of petroleum's technology has started in several areas of the world, and actual farming or ranching of the life in the sea is being planned. *See* MARINE MINING.

Hazards at sea. As the petroleum industry pushed farther into the hostile environment of the sea, it sustained a series of disasters, reflected by the doubling of offshore insurance rates in April 1966. Between 1955 and 1968, for example, 23 offshore units were destroyed by blowout and 6 by hurricane and breakup and collapse at sea. The United States Gulf Coast, where a large percentage of the world's offshore drilling has taken place, was severely hit in 1964 and 1965 by hurricanes, which caused great damage to vessels, fixed platforms, and mobile platforms. In 1980 an offshore oil platform in the North Sea was overturned and severely damaged by near-hurricane winds, with a loss of more than half of the personnel aboard the platform.

Expenses sustained from loss of wells, removal of wrecked equipment, and loss of production are huge. Such liabilities have raised insurance rates on platforms valued at millions of dollars to as much as 10% or more per year, depending on the platform type and location. Much design work is aimed at engineering better safety features for the benefit of both the crews and structures.

Despite the hazards and monumental cost involved in extracting oil and gas from beneath the sea, the world's population explosion and its ever increasing demand for petroleum energy will force the search for new reserves into even deeper waters and more remote corners of the world. In truth, the search is only just beginning.

[G. R. SCHOONMAKER]

Bibliography: F. W. Mansvelt-Beck and K. M. Wiig, *The Economics of Offshore Oil and Gas Supplies*, 1977; L. G. Weeks, Petroleum resources potential of continental margins, in C. A. Burk and C. L. Drake (eds.), *The Geology of Continental Margins*, 1974.

Oil and gas field exploitation

In the petroleum industry, a field is an area underlain without substantial interruption by one or more reservoirs of commercially valuable oil or gas, or both. A single reservoir (or group of reservoirs which cannot be separately produced) is a pool. Several pools separated from one another by barren, impermeable rock may be superimposed one above another within the same field. Pools have variable areal extent. Any sufficiently deep well located within the field should produce from one or more pools. However, each well cannot produce from every pool, because different pools have different areal limits.

DEVELOPMENT

Development of a field includes the location, drilling, completion, and equipment of wells necessary to produce the commercially recoverable oil and gas in the field.

Related oil field conditions. Petroleum is a generic term which, in its broadest meaning, includes all naturally occurring hydrocarbons, whether gaseous, liquid, or solid. By variation of the temperature or pressure, or both, of any hydrocarbon, it becomes gaseous, liquid, or solid. Temperatures in producing horizons vary from approximately 60°F

(16°C) to more than 300°F (149°C), depending chiefly upon the depth of the horizon. A rough approximation is that temperature in the reservoir sand, or pay, equals 60°F (16°C), plus 0.017°F/ft (0.031°C/m) of depth below surface. Pressure on the hydrocarbons varies from atmospheric to more than 11,000 psi (76 MPa). Normal pressure is considered as 0.465 psi/ft (10.5 kPa/m) of depth. Temperatures and pressure vary widely from these average figures. Hydrocarbons, because of wide variations in pressure and temperature and because of mutual solubility in one another, do not necessarily exist underground in the same phases in which they appear at the surface.

Petroleum occurs underground in porous rocks of wide variety. The pore spaces range from microscopic size to rare holes 1 in. or more in diameter. The containing rock is commonly called the sand or the pay, regardless of whether the pay is actually sandstone, limestone, dolomite, unconsolidated sand, or fracture openings in relatively impermeable rock.

Development of field. After discovery of a field containing oil or gas, or both, in commercial quantities, the field must be explored to determine its vertical and horizontal limits and the mechanisms under which the field will produce. Development and exploitation of the field proceed simultaneously. Usually the original development program is repeatedly modified by geologic knowledge acquired during the early stages of development and exploitation of the field.

Ideally, tests should be drilled to the lowest possible producing horizon in order to determine the number of pools existing in the field. Testing and geologic analysis of the first wells sometimes indicates the producing mechanisms, and thus the best development program. Very early in the history of the field, step-out wells will be drilled to determine the areal extent of the pool or pools. Step-out wells give further information regarding the volumes of oil and gas available, the producing mechanisms, and the desirable spacing of wells.

The operator of an oil and gas field endeavors to select a development program which will produce the largest volume of oil and gas at a profit. The program adopted is always a compromise between conflicting objectives. The operator desires (1) to drill the fewest wells which will efficiently produce the recoverable oil and gas; (2) to drill, complete, and equip the wells at the lowest possible cost; (3) to complete production in the shortest practical time to reduce both capital and operating charges; (4) to operate the wells at the lowest possible cost; and (5) to recover the largest possible volume of oil and gas.

Selecting the number of wells. Oil pools are produced by four mechanisms: dissolved gas expansion, gas-cap drive, water drive, and gravity drainage. Commonly, two or more mechanisms operate in a single pool. The type of producing mechanism in each pool influences the decision as to the number of wells to be drilled. Theoretically, a single, perfectly located well in a water-drive pool is capable of producing all of the commercially recoverable oil and gas from that pool. Practically, more than one well is necessary if a pool of more than 80 acres (32 hectares) is to be depleted in a reasonable time. If a pool produces under either gas expansion or gas-cap drive, oil production from the pool will be independent of the number of wells up to a spacing of at least 80 acres per well (1866 ft or 569 m between wells). Gas wells often are spaced a mile or more apart. The operator accordingly selects the widest spacing permitted by field conditions and legal requirements.

Major components of cost. Costs of drilling, completing, and equipping the wells influence development plans. Having determined the number and depths of producing horizons and the producing mechanisms in each horizon, the operator must decide whether he will drill a well at each location to each horizon or whether a single well can produce from two or more horizons at the same location. Clearly, the cost of drilling the field can be sharply reduced if a well can drain two, three, or more horizons. The cost of drilling a well will be higher if several horizons are simultaneously produced, because the dual or triple completion of a well usually requires larger casing. Further, completion and operating costs are higher. However, the increased cost of drilling a well of larger diameter and completing the well in two or more horizons is 20–40% less than the cost of drilling and completing two wells to produce separately from two horizons.

In some cases, the operator may reduce the number of wells by drilling a well to the lowest producible horizon and taking production from that level until the horizon there is commercially exhausted. The well is then plugged back to produce from a higher horizon. Selection of the plan for producing the various horizons obviously affects the cost of drilling and completing individual wells, as well as the number of wells which the operator will drill. If two wells are drilled at approximately the same location, they are referred to as twins, three wells at the same location are triplets, and so on.

Costs and duration of production. The operator wishes to produce as rapidly as possible because the net income from sale of hydrocarbons is obviously reduced as the life of the well is extended. The successful operator must recover from the productive wells the costs of drilling and operating those wells, and in addition must recover all costs involved in geological and geophysical exploration, leasing, scouting, and drilling of dry holes, and occasionally other operations. If profits from production are not sufficient to recover all exploration and production costs and yield a profit in excess of the rate of interest which the operator could secure from a different type of investment, he is discouraged from further exploration.

Most wells cannot operate at full capacity because unlimited production results in physical waste and sharp reduction in ultimate recovery. In many areas, conservation restrictions are enforced to make certain that the operator does not produce in excess of the maximum efficient rate. For example, if an oil well produces at its highest possible rate, a zone promptly develops around the well where production is occurring under gas-expansion drive, the most inefficient producing mechanism. Slower production may permit the petroleum to be produced under gas-cap drive or water drive, in which case ultimate production of oil will be two to four times as great as it would be under gas-expansion drive. Accordingly, the most rapid rate of production generally is not the most efficient rate.

Similarly, the initial exploration of the field may indicate that one or more gas-condensate pools exist, and recycling of gas may be necessary to secure maximim recovery of both condensate and of gas. The decision to recycle will affect the number of wells, the locations of the wells, and the completion methods adopted in the development program.

Further, as soon as the operator determines that secondary oil-recovery methods are desired and expects to inject water, gas, steam, or, rarely, air to provide additional energy to flush or displace oil from the pay, the number and location of wells may be modified to permit the most effective secondary recovery procedures.

Legal and practical restrictions. The preceding discussion has assumed control of an entire field under single ownership by a single operator. In the United States, a single operator rarely controls a large field, and this field is almost never under a single lease. Usually, the field is covered by separate leases owned and operated by different producers. The development program must then be modified in consideration of the lease boundaries and the practices of the other operators.

Oil and gas know no lease boundaries. They move freely underground from areas of high pressure toward lower-pressure situations. The operator of a lease is obligated to locate the wells in such a way as to prevent drainage of the lease by wells on adjoining leases, even though the adjoining leases may be owned by that operator. In the absence of conservation restrictions, an operator must produce petroleum from wells as rapidly as it is produced from wells on adjoining leases. Slow production on one lease results in migration of oil and gas to nearby leases which are more rapidly produced.

The operator's development program must provide for offset wells located as close to the boundary of the lease as are wells on adjoining leases. Further, the operator must equip the wells to produce as rapidly as the offset produces and must produce from the same horizons which are being produced in offset wells. The lessor who sold the lease to the operator is entitled to a share of the recoverable petroleum underlying the land. Negligence by the operator in permitting drainage of a lease makes the operator liable to suit for damages or cancellation of the lease.

A development program acceptable to all operators in the field permits simultaneous development of leases, prevents drainage, and results in maximum ultimate production from the field. Difficulties may arise in agreement upon the best development program for a field. Most states have enacted statutes and have appointed regulatory bodies under which judicial determination can be made of the permissible spacing of the wells, the rates of production, and the application of secondary recovery methods.

Drilling unit. Commonly, small leases or portions of two or more leases are combined to form a drilling unit in whose center a well will be drilled. Unitization may be voluntary, by agreement between the operator or operators and the interested royalty owners, with provision for sharing production from the well between the parties in proportion to their acreage interests. In many states the regulatory body has authority to require unitization of drilling units, which eliminates unnecessary offset wells and protects the interests of a landowner whose acreage holding may be too small to justify the drilling of a single well on that property alone.

Pool unitization. When recycling or some types of secondary recovery are planned, further unitization is adopted. Since oil and gas move freely across lease boundaries, it would be wasteful for an operator to repressure, recycle, or water-drive a lease if the adjoining leases were not similarly operated. Usually an entire pool must be unitized for efficient recycling, or secondary recovery operations. Pool unitization may be accomplished by agreement between operators and royalty owners. In many cases, difference of opinion or ignorance on the part of some parties prevents voluntary pool unitization. Many states authorize the regulatory body to unitize a pool compulsorily on application by a specified percentage of interests of operators and royalty owners. Such compulsory unitization is planned to provide each operator and each royalty owner his fair share of the petroleum products produced from the field regardless of the location of the well or wells through which these products actually reach the surface.

EXPLOITATION—GENERAL CONSIDERATIONS

Oil and gas production necessarily are intimately related, since approximately one-third of the gross gas production in the United States is produced from wells that are classified as oil wells. However, the naturally occurring hydrocarbons of petroleum are not only liquid and gaseous but may even be found in a solid state, such as asphaltite and some asphalts.

Where gas is produced without oil, the production problems are simplified because the product flows naturally throughout the life of the well and does not have to be lifted to the surface. However, there are sometimes problems of water accumulations in gas wells, and it is necessary to pump the water from the wells to maintain maximum, or economical, gas production. The line of demarcation between oil wells and gas wells is not definitely established since oil wells may have gas-oil ratios ranging from a few cubic feet (1 cubic foot $= 2.8 \times 10^{-2}$ m^3) per barrel to many thousand cubic feet of gas per barrel of oil. Most gas wells produce quantities of condensable vapors, such as propane and butane, that may be liquefied and marketed for fuel, and the more stable liquids produced with gas can be utilized as natural gasoline.

Factors of method selection. The method selected for recovering oil from a producing formation depends on many factors, including well depth, well-casing size, oil viscosity, density, water production, gas-oil ratio, porosity and permeability of the producing formation, formation pressure, water content of producing formation, and whether the force-driving the oil into the well from the formation is primarily gas pressure, water pressure, or a combination of the two. Other factors, such as paraffin content and difficulty expected from paraffin deposits, sand production, and corrosivity of the well fluids, also have a decided influence on the most economical method of production.

Special techniques utilized to increase productivity of oil and gas wells include acidizing, hydraulic fracturing of the formation, the setting of screens, and gravel packing or sand packing to increase permeability around the well bore.

Aspects of production rate. Productive rates per well may vary from a fraction of a barrel (1 barrel = 0.1590 m^3) per day to several thousand barrels per day, and it may be necessary to produce a large percentage of water along with the oil.

Field and reservoir conditions. In some cases reservoir conditions are such that some of the wells flow naturally throughout the entire economical life of the oil field. However, in the great majority of cases it is necessary to resort to artificial lifting methods at some time during the life of the field, and often it is necessary to apply artificial lifting means immediately after the well is drilled.

Market and regulatory factors. In some oil-producing states of the United States there are state bodies authorized to regulate oil production from the various oil fields. The allowable production per well is based on various factors, including the market for the particular type of oil available, but very often the allowable production is based on an engineering study of the reservoir to determine the optimum rate of production.

Useful terminology. A few definitions of terms used in petroleum production technology are listed below to assist in an understanding of some of the problems involved.

Porosity. The percentage porosity is defined as the percentage volume of voids per unit total volume. This, of course, represents the total possible volume available for accumulation of fluids in a formation, but only a fraction of this volume may be effective for practical purposes because of possible discontinuities between the individual pores. The smallest pores generally contain water held by capillary forces.

Permeability. Permeability is a measure of the resistance to flow through a porous medium under the influence of a pressure gradient. The unit of permeability commonly employed in petroleum production technology is the darcy. A porous structure has a permeability of 1 darcy if, for a fluid of 1 centipoise (cp) [10^{-3} Pa·s] viscosity, the volume flow is 1 $cm^3/(sec)(cm^2)$ [10^{-2} $m^3/(sec)(m^2)$] under a pressure gradient of 1 atm/cm (1.01325×10^7 Pa/m).

Productivity index. The productivity index is a measure of the capacity of the reservoir to deliver oil to the well bore through the productive formation and any other obstacles that may exist around the well bore. In petroleum production technology, the productivity index is defined as production in barrels per day (1 barrel per day ≅ 0.1590 m^3/day) per pound per square inch (psi = 6.895 kPa) drop in bottom-hole pressure. For example, if a well is closed in at the casinghead, the bottom-hole pressure will equal the formation pressure when equilibrium conditions are established. However, if fluid is removed from the well, either by flowing or pumping, the bottom-hole pressure will drop as a result of the resistance to flow of fluid into the well from the formation to replace the fluid removed from the well. If the closed-in bottom-hole pressure should be 1000 psi, for example, and if this pressure should drop to 900 psi when producing at a rate of 100 bbl/day (a drop of 100 psi), the well would have a productivity index of one.

Barrel. The standard barrel used in the petroleum industry is 42 U.S. gal (approximately 0.1590 m^3).

API gravity. The American Petroleum Institute (API) scale that is in common use for indicating specific gravity, or a rough indication of quality of crude petroleum oils, differs slightly from the Baume scale commonly used for other liquids lighter than water. The table shows the relationship between degrees API and specific gravity referred to water at 60°F (15.6°C) for specific gravities ranging from 0.60 to 1.0.

Viscosity range. Viscosity of crude oils currently produced varies from approximately 1 cp (10^{-3} Pa·s) to values above 1000 cp (1 Pa·s) at temperatures existing at the bottom of the well. In some areas it is necessary to supply heat artificially down the wells or circulate lighter oils to mix with the produced fluid for maintenance of a relatively low viscosity throughout the temperature range to which the product is subjected.

In addition to wells that are classified as gas wells or oil wells, the term gas-condensate well has come into general use to designate a well that produces large volumes of gas with appreciable quantities of light, volatile hydrocarbon fluids. Some of these fluids are liquid at atmospheric pressure and temperature; others, such as propane and butane, are readily condensed under relatively low pressures in gas separators for use as liquid petroleum gas (LPG) fuels or for other uses. The liquid components of the production from gas-condensate wells generally arrive at the surface in the form of small droplets entrained in the high-velocity gas stream and are separated from the gas in a high-pressure gas separator.

PRODUCTION METHODS IN PRODUCING WELLS

The common methods of producing oil wells are (1) natural flow; (2) pumping with sucker rods; (3) gas lift; (4) hydraulic subsurface pumps; (5)

Degrees API corresponding to specific gravities of crude oil at 60°/60°F*

Specific gravity, in tenths	Specific gravity, in hundredths									
	.00	.01	.02	.03	.04	.05	.06	.07	.08	.09
0.60	104.33	100.47	96.73	93.10	89.59	86.19	82.89	79.69	76.59	73.57
0.70	70.64	67.80	65.03	62.34	59.72	57.17	54.68	52.27	49.91	47.61
0.80	45.38	43.19	44.06	38.98	36.95	34.97	33.03	31.14	29.30	27.49
0.90	25.72	23.99	22.30	20.65	19.03	17.45	15.90	14.38	12.89	11.43
1.00	10.00									

*60°F = 15.6°C.

electrically driven centrifugal well pumps; and (6) swabbing.

Numerous other methods, including jet pumps and sonic pumps, have been tried and are used to slight extent. The sonic pump is a development in which the tubing is vibrated longitudinally by a mechanism at the surface and acts as a high-speed pump with an extremely short stroke.

The total number of producing oil wells in the United States at the end of 1977 was reported to be 508,340, while the total number of producing gas wells was 145,453.

A total of 48,384 wells were drilled in the United States during 1978. Of this number, 17,747 were productive oil wells, 12,941 were classified as gas wells, 16,228 were nonproductive (dry holes), and 1468 were service wells. Service wells are utilized for various purposes, such as water injection for water flooding operations, salt-water disposal, and gas recycling.

A discussion of production methods, in approximate order of relative importance, follows.

Natural flow. Natural flow is the most economical method of production and generally is utilized as long as the desired production rate can be maintained by this method. It utilizes the formation energy, which may consist of gas in solution in the oil in the formation; free gas under pressure acting against the liquid and gas-liquid phase to force it toward the well bore; water pressure acting against the oil; or a combination of these three energy sources. In some areas the casinghead pressure may be of the order of 10,000 psi, so it is necessary to provide fittings adequate to withstand such pressures. Adjustable throttle values, or chokes, are utilized to regulate the flow rate to a desired and safe value. With such a high-pressure drop across a throttle valve the life of the valve is likely to be very short. Several such valves are arranged in parallel in the tubing head "Christmas tree" with positive shutoff valves between the chokes and the tubing head so that the wearing parts of the throttle valve, or the entire valve, can be replaced while flow continues through another similar valve.

An additional safeguard that is often used in connection with high-pressure flowing wells is a bottom-hole choke or a bottom-hole flow control valve that limits the rate of flow to a reasonable value, or stops it completely, in case of failure of surface controls. Figure 1 shows a schematic outline of a simple flowing well hookup. The packer is not essential but is often used to reduce the free gas volume in the casing.

Flow rates for United States wells seldom exceed a few hundred barrels per day because of enforced or voluntary restrictions to regulate production rates and to obtain most efficient and economical ultimate recovery. However, in some countries, especially in the Middle East, it is not uncommon for natural flow rates to exceed 10,000 bpd/well [1590 m^3/(day)(well)].

Lifting. Most wells are not self-flowing. The common types of lifting are outlined here.

Pumping with sucker rods. Approximately 90% of the wells made to produce by some artificial lift method in the United States are equipped with sucker-rod–type pumps. In these the pump is installed at the lower end of the tubing string and is actuated by a string of sucker rods extending from the surface to the subsurface pump. The sucker rods are attached to a polished rod at the surface. The polished rod extends through a stuffing box and is attached to the pumping unit, which produces the necessary reciprocating motion to actuate the sucker rods and the subsurface pump. Figure 2 shows a simplified schematic section through a pumping well. The two common variations are mechanical and hydraulic long-stroke pumping.

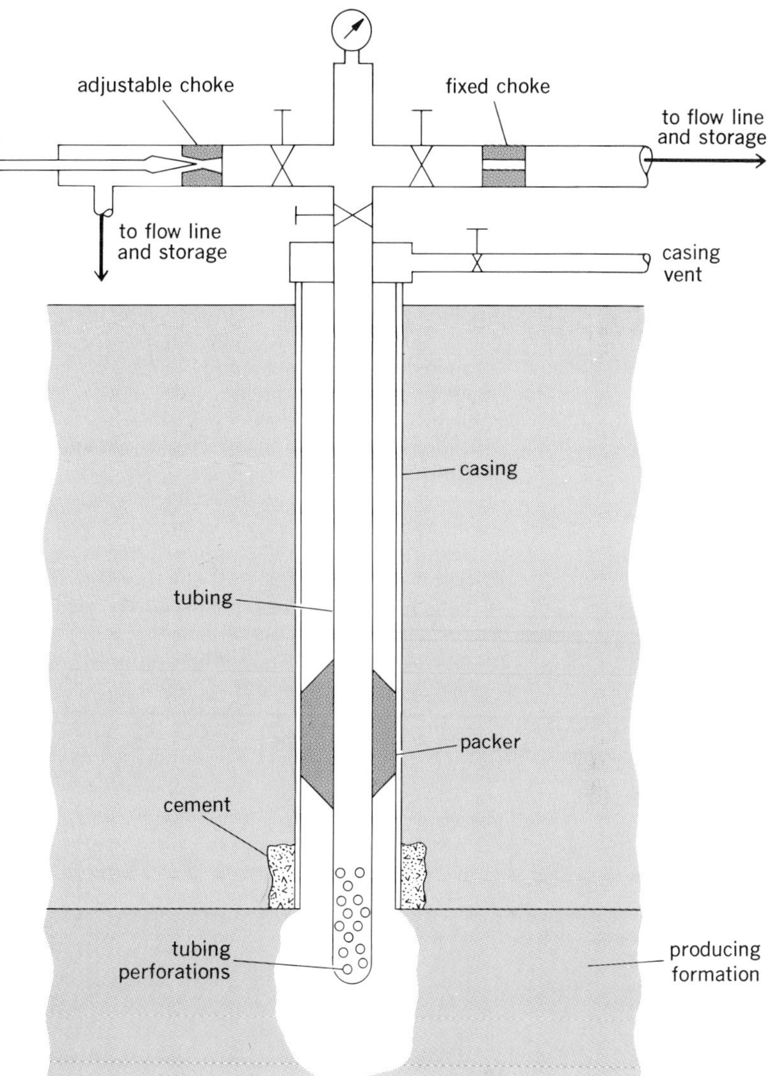

Fig. 1. Schematic view of well equipped for producing by natural flow.

1. Mechanical pumping. The great majority of pumping units are of the mechanical type, consisting of a suitable reduction gear, and crank and pitman arrangement to drive a walking beam to produce the necessary reciprocating motion. A counterbalance is provided to equalize the load on the upstroke and downstroke. Mechanical pumping units of this type vary in load-carrying capacity from about 2000 to about 43,000 lb (900 to 19,500 kg), and the torque rating of the low-speed gear which drives the crank ranges from 6400 in.-lb (720 N·m) in the smallest API standard unit to about 1,500,000 in.-lb (170,000 N·m) for the largest units now in use. Stroke length varies from about 18 to 192 in. (46 to 488 cm). Usual operating speeds are from about 6 to 20 strokes/

min. However, both lower and higher rates of speed are sometimes used. Figure 3 shows a modern pumping unit in operation.

Production rates with sucker-rod–type pumps vary from a fraction of 1 bpd in some areas, with part-time pumping, to approximately 3000 bpd (480 m^3/day) for the largest installations in relatively shallow wells.

2. Hydraulic long-stroke pumping. For this the units consist of a hydraulic lifting cylinder mounted directly over the well head and are designed to produce stroke lengths of as much as 30 ft (9 m). Such long-stroke hydraulic units are usually equipped with a pneumatic counterbalance arrangement which equalizes the power requirement on the upstroke and downstroke.

Hydraulic pumping units also are made without any provision for counterbalance. However, these units are generally limited to relatively small wells, and they are relatively inefficient.

Gas lift. Gas lift in its simplest form consists of initiating or stimulating well flow by injecting gas at some point below the fluid level in the well. With large-volume gas-lift operations the well may be produced through either the casing or the tubing.

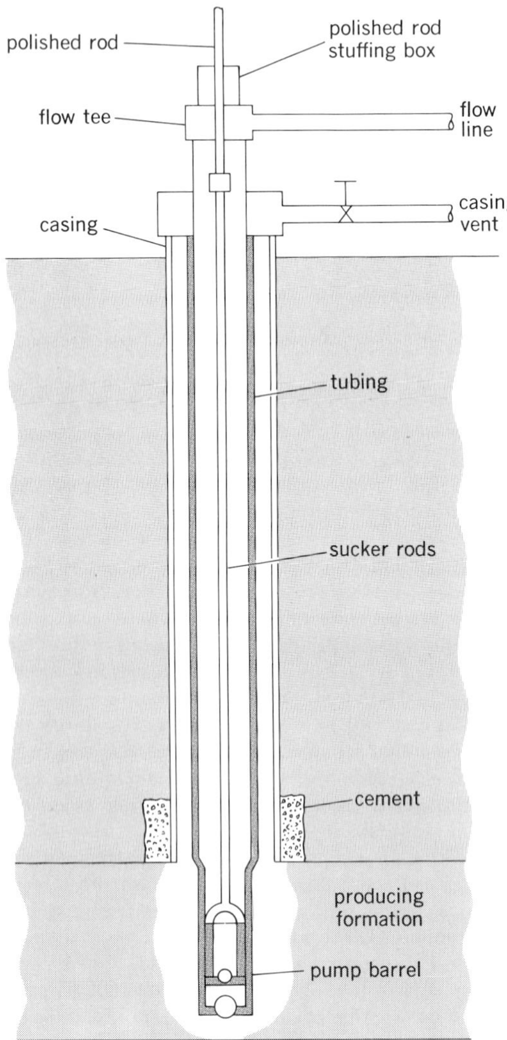

Fig. 2. A schematic view of a well which is equipped for pumping with sucker rods.

Fig. 3. Pumping unit with adjustable rotary counterbalance. (*Oil Well Supply Division, U.S. Steel Corp.*)

In the former case, gas is conducted through the tubing to the point of injection; in the latter, gas may be conducted to the point of injection through the casing or through an auxiliary string of tubing. When gas is injected into the oil column, the weight of the column above the point of injection is reduced as a result of the space occupied by the relatively low-density gas. This lightening of the fluid column is sufficient to permit the formation pressure to initiate flow up the tubing to the surface. Gas injection is often utilized to increase the flow from wells that will flow naturally but will not produce the desired amount by natural flow.

There are many factors determining the advisability of adopting gas lift as a means of production. One of the more important factors is the availability of an adequate supply of gas at suitable pressure and reasonable cost. In a majority of cases gas lift cannot be used economically to produce a reservoir to depletion because the well may be relatively productive with a low back pressure maintained on the formation but will produce very little, if anything, with the back pressure required for gas-lift operation. Therefore, it generally is necessary to resort to some mechanical means of pumping before the well is abandoned, and it may be more economical to adopt the mechanical means initially than to install the gas-lift system while conditions are favorable and later replace it.

This discussion of gas lift has dealt primarily with the simple injection of gas, which may be continuous or intermittent. There are numerous modifications of gas-lift installations, including various designs for flow valves which may be installed in the tubing string to open and admit gas to the tubing from the casing at a predetermined pressure differential between the tubing and casing. When the valve opens, gas is injected into the tubing to initiate and maintain flow until the tubing pressure drops to a predetermined value; and the valve closes before the input gas-oil ratio becomes excessive. This represents an intermittent-flow–type valve. Other types are designed to maintain continuous flow, proper pressure differential, and proper gas injection rate for efficient operation. In some cases several such flow valves are spaced up the tubing string to permit flow to be initiated from various levels as required.

Other modifications of gas lift involve the utilization of displacement chambers. These are installed on the lower end of the well tubing where oil may accumulate, and the oil is displaced up the tubing with gas injection controlled by automatic or mechanical valves.

Hydraulic subsurface pumps. The hydraulic subsurface pump has come into fairly prominent use. The subsurface pump is operated by means of a hydraulic reciprocating motor attached to the pump and installed in the well as a single unit. The hydraulic motor is driven by a supply of hydraulic fluid under pressure that is circulated down a string of tubing and through the motor. Generally the hydraulic fluid consists of crude oil which is discharged into the return line and returns to the surface along with the produced crude oil.

Hydraulically operated subsurface pumps are also arranged for separating the hydraulic power fluid from the produced well fluid. This arrangement is especially desirable where the fluid being produced is corrosive or is contaminated with considerable quantities of sand or other solids that are difficult to separate to condition the fluid for use as satisfactory power oil. This method permits use of water or other nonflammable liquids as hydraulic power fluid to minimize fire hazard in case of a failure of the hydraulic power line at the surface.

Centrifugal well pumps. Electrically driven centrifugal pumps have been used to some extent, especially in large-volume wells of shallow or moderate depths. Both the pump and the motor are restricted in diameter to run down the well casing, leaving sufficient clearance for the flow of fluid around the pump housing. With the restricted diameter of the impellers the discharge head necessary for pumping a relatively deep well can be obtained only by using a large number of stages and operating at a relatively high speed. The usual rotating speed for such units is 3600 rpm, and it is not uncommon for such units to have 50 or more pump stages. The direct-connected electric motor must be provided with a suitable seal to prevent well fluid from entering the motor housing, and electrical leads must be run down the well casing to supply power to the motor.

Swabs. Swabs have been used for lifting oil almost since the beginning of the petroleum industry. They usually consist of a steel tubular body equipped with a check valve which permits oil to flow through the tube as it is lowered down the well with a wire line. The exterior of the steel body is generally fitted with flexible cup-type soft packing that will fall freely but will expand and form a seal with the tubing when pulled upward with a head of fluid above the swab. Swabs are run into the well on a wire line to a point considerably below the fluid level and then lifted back to the surface to deliver the volume of oil above the swab. They are often used for determining the productivity of a well that will not flow naturally and for assisting in cleaning paraffin from well tubing. In some cases swabs are used to stimulate wells to flow by lifting, from the upper portion of the tubing, the relatively dead oil from which most of the gas has separated.

Bailers. Bailers are used to remove fluids from wells and for cleaning out solid material. They are run into the wells on wire lines as in swabbing, but differ from swabs in that they generally are run only in the casing when there is no tubing in the well. The capacity of the bailer itself represents the volume of fluid lifted each time since the bailer does not form a seal with the casing. The bailer is simply a tubular vessel with a check valve in the bottom. This check valve generally is arranged so that it is forced open when the bailer touches bottom in order to assist in picking up solid material for cleaning out a well.

Jet pumps. A jet pump for use in oil wells operates on exactly the same principle as a water-well jet pump. Advantage is taken of the Bernoulli effect to reduce pressure by means of a high-velocity fluid jet. Thus oil is entrained from the well with this high-velocity jet in a venturi tube to accelerate the fluid and assist in lifting it to the surface, along with any assistance from the formation pressure. The application of jet pumps to oil wells has been insignificant.

Sonic pumps. Sonic pumps essentially consist of a string of tubing equipped with a check valve at each joint and mechanical means on the surface to vibrate the tubing string longitudinally. This creates a harmonic condition that will result in several hundred strokes per minute, with the strokes being a small fraction of 1 in. in length. Some of these pumps are being used in relatively shallow wells.

Lease tanks and gas separators. Figure 4 shows a typical lease tank battery consisting of

Fig. 4. Lease tank battery with four tanks and two gas separators. *(Gulf Oil Corp.)*

four 1000-bbl tanks and two gas separators. Such equipment is used for handling production from wells produced by natural flow, gas lift, or pumping. In some pumping wells the gas content may be too low to justify the cost of separators for saving the gas.

Natural gasoline production. An important phase of oil and gas production in many areas is the production of natural gasoline from gas taken from the casinghead of oil wells or separated from the oil and conducted to the natural gasoline plant. The plant consists of facilities for compressing and extracting the liquid components from the gas. The natural gasoline generally is collected by cooling and condensing the vapors after compression or by absorbing in organic liquids having high boiling points from which the volatile liquids are distilled. Many natural gasoline plants utilize a combination of condensing and absorbing techniques. Figure 5 shows an overall view of a natural gasoline plant operating in western Texas.

Fig. 5. Modern natural gasoline plant in western Texas. (*Gulf Oil Corp.*)

PRODUCTION PROBLEMS AND INSTRUMENTS

To maintain production, various problems must be overcome. Numerous instruments have been developed to monitor production and to control production problems.

Corrosion. In many areas the corrosion of production equipment is a major factor in the cost of petroleum production. The following comments on the oil field corrosion problem are taken largely from *Corrosion of Oil- and Gas-Well Equipment* and reproduced by permission of NACE-API.

For practical consideration, corrosion in oil and gas-well production can be classified into four main types.

1. Sweet corrosion occurs as a result of the presence of carbon dioxide and fatty acids. Oxygen and hydrogen sulfide are not present. This type of corrosion occurs in both gas-condensate and oil wells. It is most frequently encountered in the United States in southern Louisiana and Texas, and other scattered areas. At least 20% of all sweet oil production and 45% of condensate production are considered corrosive.

2. Sour corrosion is designated as corrosion in oil and gas wells producing even trace quantities of hydrogen sulfide. These wells may also contain oxygen, carbon dioxide, or organic acids. Sour corrosion occurs in the United States primarily throughout Arbuckle production in Kansas and in the Permian basin of western Texas and New Mexico. About 12% of all sour production is considered corrosive.

3. Oxygen corrosion occurs wherever equipment is exposed to atmospheric oxygen. It occurs most frequently in offshore installations, brine-handling and injection systems, and in shallow producing wells where air is allowed to enter the casing.

4. Electrochemical corrosion is that which occurs when corrosion currents can be readily measured or when corrosion can be mitigated by the application of current, as in soil corrosion.

Corrosion inhibitors are used extensively in both oil and gas wells to reduce corrosion damage to subsurface equipment. Most of the inhibitors used in the oil field are of the so-called polar organic type. All of the major inhibitor suppliers can furnish effective inhibitors for the prevention of sweet corrosion as encountered in most fields. These can be purchased in oil-soluble, water-dispersible, or water-soluble form.

Fig. 6. Two pumping wells with tank battery. (*Oil Well Supply Division, U.S. Steel Corp.*)

Paraffin deposits. In many crude-oil–producing areas paraffin deposits in tubing and flow lines and on sucker rods are a source of considerable trouble and expense. Such deposits build up until the tubing or flow line is partially or completely plugged. It is necessary to remove these deposits to maintain production rates. A variety of methods are used to remove paraffin from the tubing, including the application of heated oil through tubular sucker rods to mix with and transfer heat to the oil being produced and raise the temperature to a point at which the deposited paraffin will be dissolved or melted. Paraffin solvents may also be applied in this manner without the necessity of applying heat.

Mechanical means often are used in which a scraping tool is run on a wire line and paraffin is scraped from the tubing wall as the tool is pulled back to the surface. Mechanical scrapers that attach to sucker rods also are in use. Various types of automatic scrapers have been used in connection with flowing wells. These consist of a form of piston that will drop freely to the bottom when flow is stopped but will rise back to the surface when flow is resumed. Electrical heating methods have been used rather extensively in some areas. The tubing is insulated from the casing and from the flow line, and electric current is transmitted through the tubing for the time necessary to heat the tubing sufficiently to cause the paraffin deposits to melt or go into solution in the oil in the tubing. Plastic coatings have been utilized inside tubing and flow lines to minimize or prevent paraffin deposits. Paraffin does not deposit readily on certain plastic coatings.

A common method for removing paraffin from flow lines is to disconnect the line at the well head and at the tank battery and force live steam through the line to melt the paraffin deposits and flow them out. Various designs of flow-line scrapers have also been used rather extensively and fairly successfully. Paraffin deposits in flow lines are minimized by insulating the lines or by burying the lines to maintain a higher average temperature.

Emulsions. A large percentage of oil wells produce various quantities of salt water along with the oil, and numerous wells are being pumped in which the salt-water production is 90% or more of the total fluid lifted. Turbulence resulting from production methods results in the formation of emulsions of water in oil or oil in water; the commoner type is oil in water. Emulsions are treated with a variety of demulsifying chemicals, with the application of heat, and with a combination of these two treatments. Another method for breaking emulsions is the electrostatic or electrical precipitator type of emulsion treatment. In this method the emulsion to be broken is circulated between electrodes subjected to a high potential difference. The resulting concentrated electric field tends to rupture the oil-water interface and thus breaks the emulsion and permits the water to settle out. Figure 6 shows two pumping wells with a tank battery in the background. This tank battery is equipped with a wash tank, or gun barrel, and a gas-fired heater for emulsion treating and water separation before the oil is admitted to the lease tanks.

Gas conservation. If the quantity of gas produced with crude oil is appreciably greater than that which can be efficiently utilized or marketed, it is necessary to provide facilities for returning the excess gas to the producing formation. Formerly, large quantities of excess gas were disposed of by burning or simply by venting to the atmosphere. This practice is now unlawful. Returning excess gas to the formation not only conserves the gas for future use but also results in greater ultimate recovery of oil from the formation.

Salt-water disposal. The large volumes of salt water produced with the oil in some areas present serious disposal problems. The salt water is generally pumped back to the formation through wells drilled for this purpose. Such salt-water disposal wells are located in areas where the formation already contains water. Thus this practice helps to maintain the formation pressure as well as the productivity of the producing wells.

Offshore production. Offshore wells present additional production problems since the wells must be serviced from barges or boats. Wells of reasonable depth on land locations are seldom equipped with derricks for servicing because it is more economical to set up a portable mast for pulling and installing rods, tubing, and other equipment. However, the use of portable masts is not practical on offshore locations, and a derrick is generally left standing over such wells throughout their productive life to facilitate servicing. There are a considerable number of offshore wells along the Gulf Coast and the Pacific Coast of the United States, but by far the greatest number of offshore wells in a particular region is in Lake Maracaibo in Venezuela. Figure 7 shows a considerable number of derricks in Lake Maracaibo with pumping wells

Fig. 7. Numerous offshore wells located in Lake Maracaibo, Venezuela. (*Creole Petroleum Corp.*)

in the foreground. These wells are pumped by electric power through cables laid on the lake bottom to conduct electricity from power-generating stations onshore. An overwater tank battery is visible at the extreme right. All offshore installations, such as tank batteries, pump stations, and the derricks and pumping equipment, are supported on pilings in water up to 100 ft (30 m) or more in depth. There are approximately 2300 oil derricks in Lake Maracaibo. A growing number of semipermanent platform rigs and even bottom storage facilities are being used in Gulf of Mexico waters at depths of more than 100 ft (30 m).

Instruments. The commoner and more important instruments required in petroleum production operations are included in the following discussion.

1. Gas meters, which are generally of the orifice type, are designed to record the differential pressure across the orifice, and the static pressure.

2. Recording subsurface pressure gages small enough to run down 2-in. ID (inside diameter) tubing are used extensively for measuring pressure gradients down the tubing of flowing wells, recording pressure buildup when the well is closed in, and measuring equilibrium bottom-hole pressures.

3. Subsurface samplers designed to sample well fluids at various levels in the tubing are used to determine physical properties, such as viscosity, gas content, free gas, and dissolved gas at various levels. These instruments may also include a recording thermometer or a maximum reading thermometer, depending upon the information required.

4. Oil meters of various types are utilized to meter crude oil flowing to or from storage.

5. Dynamometers are used to measure polished-rod loads. These instruments are sometimes known as well weighers since they are used to record the polished-rod load throughout a pumping cycle of a sucker-rod—type pump. They are used to determine maximum load on polished rods as well as load variations, to permit accurate counterbalancing of pumping wells, and to assure that pumping units or sucker-rod strings are not seriously overloaded.

6. Liquid-level gages and controllers are used. They are similar to those used in other industries, but with special designs for closed lease tanks.

A wide variety of scientific instruments find application in petroleum production problems. The above outline gives an indication of a few specialized instruments used in this branch of the industry, and there are many more. Special instruments developed by service companies are valued for a wide variety of purposes and include calipers to detect and measure corrosion pits inside tubing and casing and magnetic instruments to detect microscopic cracks in sucker rods.

[ROY L. CHENAULT]

Bibliography: American Petroleum Institute, *History of Petroleum Engineering*, 1961; K. E. Brown, *The Technology of Artificial Lift Methods*, 1977; E. L. DeGolyer (ed.), *Elements of the Petroleum Industry*, 1940; ETA Offshore Seminars, Inc., *The Technology of Offshore Drilling and Production*, 1976; L. L. Farkas, *Management of Technical Field Operations*, 1970; T. C. Frick (ed.), *Petroleum Production Handbook*, vol. 1: *Mathematics and Production Equipment*, 1962; L. M. Harris, *An Introduction Deep Water Floating Drilling Operations*, 1972; M. Muskat, *Physical Principles of Oil Production*, 1949; T. E. W. Nind, *Principles of Oil Well Production*, 1964; L. T. Stanley, *Practical Statistics for Petroleum Engineers*, 1973; L. C. Uren, *Petroleum Production Engineering: Oil Field Development*, 4th ed., 1956; L. C. Uren, *Petroleum Production Engineering: Oil Field Exploitation*, 3d ed., 1953; J. Zaba and W. T. Doherty, *Practical Petroleum Engineering*, 5th ed., 1970.

Oil and gas storage

Crude oil and natural gas, after being produced from their natural reservoirs, are stored in great quantities. Large amounts of refined products are stored as well. Storage is necessary to meet seasonal and other fluctuations in demand and for efficient operation of producing equipment, pipelines, tankers, and refineries. Storage also provides ready reserves for emergency use. According to the U.S. Bureau of Mines, 265×10^6 bbl (1 bbl = 0.1590 m^3) of crude oil were in storage in the United States at the end of 1974. In addition, 808×10^6 bbl in the form of refined products, natural gasoline, plant condensate, and unfinished oils were in storage. The American Gas Association reported 4.788×10^{12} ft^3 (1 ft^3 = 2.832×10^{-2} m^3) of natural gas stored in underground reservoirs in the United States at the end of 1978.

Crude oil and refined products. Oil from producing wells is first collected in welded-steel, bolted-steel, or wooden tanks of 100 bbl or greater capacity located on individual leases. These tanks, upright cylinders with low-pitched conical roofs, provide temporary storage while the oil is awaiting shipment. Several tanks grouped together are a tank battery. Assemblages of large steel tanks, known as tank farms, are used for more permanent storage at pipeline pump stations, points where tankers load and unload, and refineries.

With the trend toward giant tankers, accelerated by the closing of the Suez Canal in 1967, large storage facilities are needed at both the loading and unloading ends of the tanker runs. Some tanks with capacities of 1×10^6 bbl are now in use. Large-capacity excavated reservoirs with concrete linings have been used for many years in California to store both crude and fuel oil. One such reservoir with a fixed roof and elliptical in form, is 780 ft (1 ft = 0.3048 m) long, 467 ft wide, and 23 ft deep. It covers $9\frac{1}{4}$ acres and provides storage for more than 1×10^6 bbl. Another reservoir has a capacity of 4×10^6 bbl and covers 16 acres.

Offshore storage. For offshore producing fields a number of unique storage systems have been designed. In several instances old tankers have been adapted for storage, and barges have been constructed especially for offshore storage use. One underwater installation consists essentially of three giant inverted steel funnels. Each unit is 270 ft in diameter and 205 ft high, weighs 28×10^6 lb, and has a capacity of 0.5×10^6 bbl. The bottom is open, and the unit is anchored to the sea floor by 95-ft pilings. A reinforced concrete installation features a nine-module storage unit with 1×10^6 bbl capacity surrounded by a perforated wall 302 ft in diameter that serves as a breakwater. The outer wall is about 270 ft high and extends about 40 ft above the water surface. A submerged floating storage tank 96 ft in diameter and 305 ft high is held in place by six anchor lines and has a capacity of 300,000 bbl. One relatively small unit consists of a platform with four vertical legs, each holding 4700 bbl, and four horizontal tanks at the bottom holding 1850 bbl each, for a total capacity of 26,200 bbl. A second small unit, utilizing bottom tanks as an anchor, holds 2400 bbl underwater and 600 bbl in a spherical tank above the surface. A third small unit consists of a sea-floor base, connected by a universal joint to a large-diameter vertical cylinder, about 350 ft high, which extends above the surface of the water. One proposed design includes an excavated cavern beneath the sea floor, and nuclear cavities have also been suggested.

Volatility problems. To minimize vaporization losses, lease tanks are sometimes equipped to hold several ounces pressure. At large-capacity storage sites, special tanks are generally used. Tanks with lifter or floating roofs are used to store crude oil, motor gasoline, and less volatile natural gasoline. Motor and natural gasolines are also stored in spheroid containers. Spherical containers are used for more volatile liquids, such as butane. Horizontal cylindrical containers are used for propane and butane storage. Refrigerated insulated tank systems enabling propane to be stored at a lower pressure are also in use. One tank has a capacity of 900,000 bbl.

Underground storage. Large quantities of volatile liquid-petroleum products, including propane and butane, are stored in underground caverns dissolved in salt formations and in mined caverns, gas reservoirs, and water sands. In 1973 the underground storage capacity for liquid-petroleum products in the United States was 255.231×10^6 bbl. Underground storage capacity in Canada was 14.996×10^6 bbl. Liquid-petroleum products are also being stored underground in Belgium, France, Germany, Italy, and the United Kingdom. Caverns are also used for storing crude oil in Sweden, Germany, and France. In Pennsylvania an abandoned quarry with a capacity of 2×10^6 bbl has been equipped with a floating roof for storing fuel oil. Refrigerated propane is also being stored in excavations in frozen earth and in underground concrete tanks. To provide security in the event of another oil embargo, the National Petroleum Council has proposed developing salt cavern storage for 500×10^6 bbl of crude oil.

Natural gas. Natural gas is stored in low-pressure surface holders, buried high-pressure pipe batteries and bottles, depleted or partially depleted oil and gas reservoirs, water sands, and several types of containers at extremely low temperature (−258°F, or −161°C) after liquefaction.

Low-pressure holders, which store relatively small volumes of gas, basically use either a water or a dry seal, and variations of each type exist. With the displacement of manufactured gas by natural gas in the United States, the need for surface holders has greatly diminished and they have disappeared almost entirely.

Underground storage. In the United States gas pipeline and utility companies store large quantities of natural gas in underground reservoirs. In most cases these reservoirs are located near market areas and are used to supplement pipeline supplies during the winter months when the gas demand for residential heating is very high. Since gas can be stored in the summer when the gas demand is low, underground storage permits greatly increased pipeline utilization, resulting in lower transportation costs and reduced gas cost to the consumer. Underground storage is the only economical method of storing large enough quantities of gas to meet the seasonal fluctuations in pipeline loads, and has enabled gas companies to meet market requirements which otherwise could not be satisfied.

Gas was first stored underground in 1915 in a partially depleted gas field in Ontario, Canada. The following year gas was injected into a depleted gas field near Buffalo, NY. At the end of 1978 gas was being stored in 311 reservoirs, which utilized depleted gas and oil fields, water sands, salt caverns, and an abandoned coal mine. These reservoirs are located in 26 states and are operated by 86 different companies. They have a total capacity of 7.330×10^{12} ft³ and, at the end of 1978, held 4.788×10^{12} ft³ of stored gas plus 1.054×10^{12} ft³ of negative gas. The maximum volume of gas in storage during 1978 was 5.301×10^{12} ft³, excluding native gas. During 1978 a total of 2.150×10^{12} ft³ was withdrawn from storage. The total maximum daily output from all of these reservoirs was 28.3×10^9 ft³. Canada has 17 storage reservoirs, 14 in gas and oil reservoirs and 3 in salt caverns, with a total capacity of 325×10^9 ft³. At the end of 1978 these reservoirs held 250×10^9 ft³ of natural gas. Gas is also being stored in underground reservoirs in France, East and West Germany, Austria, Italy, Poland, Rumania, Czechoslovakia and the Soviet Union. In the United Kingdom a salt cavern is being used for gas storage.

Gas storage in water sands was first undertaken in 1952, and this method of storage has steadily increased, especially in areas where no gas or oil fields are available. In the United States at the end of 1974, aquifer-type storages numbered 51 and these reservoirs were operated by 25 companies in 10 states. They had a total capacity of 1.408×10^{12} ft³ and, at the end of 1974, held 849×10^9 ft³. During 1974 the maximum volume in storage was 899×10^9 ft³. The total maximum daily output from these reservoirs was 4.7×10^9 ft³. A cross section of a typical aquifer storage field is shown in the illustra-

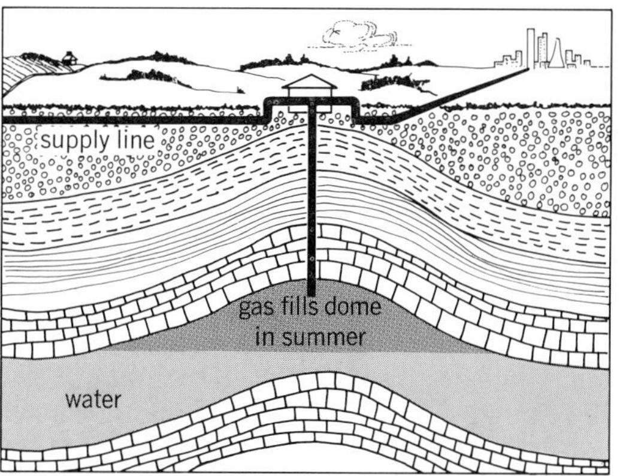

Schematic cross section of typical aquifer gas storage field showing injected gas displacing water. (*Natural Gas Pipeline Company of America*)

tion. A geologic trap having adequate structural closure and a suitable caprock is needed. The storage sand must be porous and thick enough and under sufficient hydrostatic pressure to hold large quantities of gas. The sand must also be sufficiently permeable and continuous over a wide enough area so that water can be pushed back readily to make room for the stored gas. In some cases, water removal wells are also utilized.

Reservoir pressure. In operating storage reservoirs only a portion of the stored gas, called working gas, is normally withdrawn. The remaining gas, called cushion gas, stays in the reservoir to provide the necessary pressure to produce the storage wells at desired rates. In aquifer storages some water returns to help maintain the reservoir pressure. The percentage of cushion gas varies considerably among reservoirs. Based on American Gas Association figures, cushion gas amounted to 59% of the maximum gas in storage, including native gas, in the United States in 1974. In some instances, the original reservoir pressure of the oil and gas field is exceeded in storage operations. This has resulted in storage volumes greater than the original content and has substantially increased well deliverabilities. In aquifer storages the original hydrostatic pressure must be exceeded in order to push the water back.

In the Soviet Union aquifer gas storage is being undertaken in one area where no appreciable structure. In an inconclusive field test, air was injected into a center well with control of the lateral spread of the air bubble attempted by injecting water into surrounding wells. Storage of gas in cavities created by nuclear explosions has been proposed and seriously considered.

Liquefied gas. Storage of liquefied natural gas has rapidly increased throughout the world. Storage is in connection with shipment of liquefied natural gas by tanker, and is located at the loading and unloading ends of the tanker runs as well as at peak sharing facilities operated by gas pipeline and local utility companies. In 1974 the United States and Canada had more than 100 liquefied natural-gas storage installations either operational or under construction. These had a total storage capacity of 22.8×10^6 bbl, or 78.5×10^9 ft^3. England, France, the Netherlands, West Germany, Italy, Spain, Algeria, Libya, and Japan also have installations. Storage is in insulated metal tanks, buried concrete tanks, or frozen earth excavations. In two projects using frozen earth excavations, excessive boil-off of the liquefied gas has led to replacement with insulated metal tanks. See OIL AND GAS FIELD EXPLOITATION; PIPELINE.

[PETER G. BURNETT]

Bibliography: American Gas Association, *The Underground Storage of Gas in the United States and Canada*, Annual Report on Statistics; D. C. Bond, *Underground Storage of Natural Gas*, Illinois State Geological Survey, 1975; G. D. Hobson and W. Pohl, *Modern Petroleum Engineering*, D. L. Katz et al., *Handbook of Natural Gas Engineering*, 1959; D. L. Katz and P. A. Witherspoon, *Underground and Other Storage of Oil Products and Gas*, Proceedings of the 8th World Petroleum Congress, 1971; Stone and Webster Engineering, *Gas Storage at the Point of Use*, Amer. Gas Ass. Proj. PL-56, 1965.

Oil and gas well completion

The operations that prepare for production a well drilled to an oil or gas reservoir. Various problems of well casing during and at the end of drilling are related to modes of completing connection between the proper reservoirs and the surface. Tubing inside the casing and valves and a pumping unit at the surface must deliver reservoir products to the surface at a controlled rate. Variations in reservoir and overlying formations may require special techniques to keep out water or sand or to increase the production rate. See OIL AND GAS FIELD EXPLOITATION; OIL AND GAS WELL DRILLING.

Casing. Oil and gas wells are walled with steel tubing which is cemented in place.

Tubing. Steel tubing is manufactured in various diameters, wall thicknesses, lengths, and steel alloys, selected to satisfy specific needs. These lengths, called joints, are threaded and coupled so that they may be joined together in a continuous string in the well bore. Properly placed and cemented in the hole, casing protects fresh-water reservoirs from contamination, supports unconsolidated rock formations, maintains natural separation of formations, aids in the prevention of blowouts and waste of reservoir energy, and acts as a conduit for receiving pipe of smaller diameters through which the well effluent may be brought to the surface under controlled conditions.

It may be necessary to set many strings of casing in one hole before reaching the objective (Fig. 1). The determining factors are many, such as depth of hole, loss of circulation, high-pressure formations, hole sloughing, and wearing out a

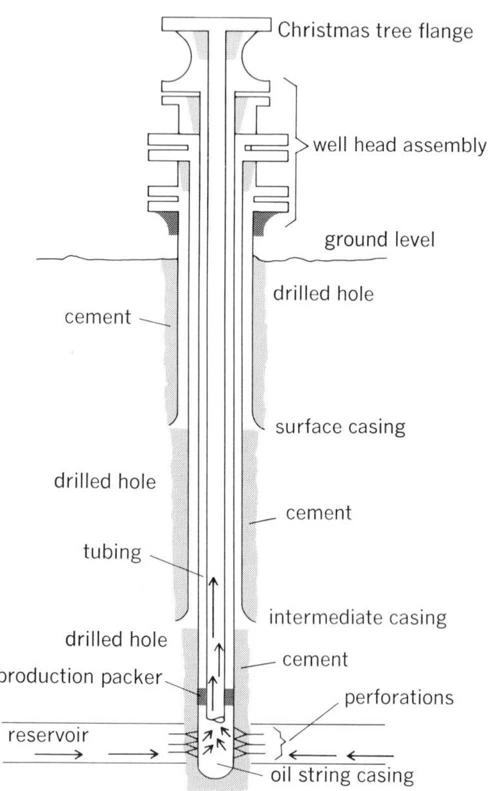

Fig. 1. Casing detail; casing strings in an oil well.

string of casing while rotating drill pipe through it over a long period of time.

Cementing casing. Basically, ordinary portland cement is used in cementing casing. In order to obtain the protection and fulfil the purposes, it is imperative that each string of casing be securely sealed to the walls of the hole for at least some distance up from the bottom of the casing string. After casing is in place, the cement is pumped down the inside and up the outside to a predetermined height to occupy the space between the casing and the walls of the hole, thereby effecting the desired seal. In the pumping and measuring process, plugs are used to separate the cement from other fluids to eliminate contamination; also, the cement inside the casing is displaced with fluid. Oil or gas well cementing is not performed with the drilling equipment; an outside service company equipped with mobile, high-pressure mixing and pumping equipment and accessories operated by trained personnel is employed (Fig. 2).

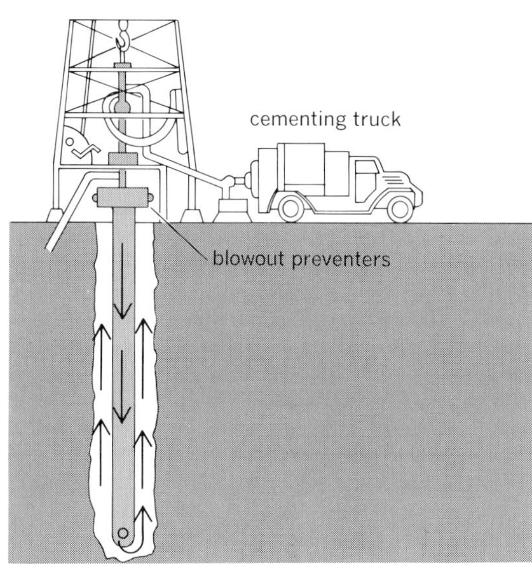

Fig. 2. Diagram of cementing process, showing truck, equipment, and well job.

Well hole–reservoir connection. Reservoir conditions, known to exist or later defined, determine the type of completion technique to be followed: (1) barefoot completion, (2) preperforated liner, or (3) casing set through and perforated.

Barefoot completion. This type of completion is frequently used when the character of the producing rock is such that it does not require supplemental support or screening, for example, in formations such as limestone, dolomite, or hard sandstone. With this method, the production casing is seated above the producing section in the conventional manner. The casing is cleaned out by insertion of fluid-separating plugs and drilled out through the casing and into the producing formation below. The formation contents enter the bore hole from the bare or unlined producing stratum or strata, hence the term barefoot (Fig. 3).

Liner-type completion. This type of completion is similar to a barefoot completion except that the open portion of the hole is cased with a preperforated section of casing called a liner (Fig. 4). This liner is smaller in diameter than the casing previously set in the hole and is usually suspended from the upper casing near the bottom from a liner hanger. The hanger is attached to the top of the liner, and when it is set, it effects a seal between the liner and the casing. The purpose of the liner is to permit gas and liquids to enter the hole and screen out formation particles.

Gun perforating. Gun perforating is a method of forming holes through the casing and into a formation from within a well bore. The two more popular methods are bullet perforating, as with a rifle, and jet perforating, as with a torch.

The gun is fitted around the outside with barrels containing the perforating medium. Each barrel is wired to fire by remote control from the surface. The gun is run into the hole on a wire line from a service company's shooting truck. The wire line serves to lower and raise the gun in and out of the hole and, when the gun is in position to be fired, the operator sends an electric impulse down the line to trigger it. The hole is thus formed through the steel casing, the cement sheath, and some inches into the reservoir rock, creating an entry for the reservoir content into the well bore. Guns of the bullet type are retrieved and reloaded (Fig. 5). Jet-type guns are expendable and disintegrate (Fig. 6).

Production-flow control. A steel tube, the same as casing except that it is smaller in diameter, serves as a production flow line within the well. The tubing is run inside the casing and is either suspended in the hole or set on a production packer at or near the producing interval. The top of the tubing string terminates at the surface in a sealing element in the wellhead assembly to which the so-called Christmas tree is attached. A manifold constructed of steel valves and fittings, placed on top of the casings protruding above the surface, is called a Christmas tree (Fig. 7). Its purpose is to maintain the well under proper control, to receive the formation products under pressure, and to control the rate of daily production from the reservoir

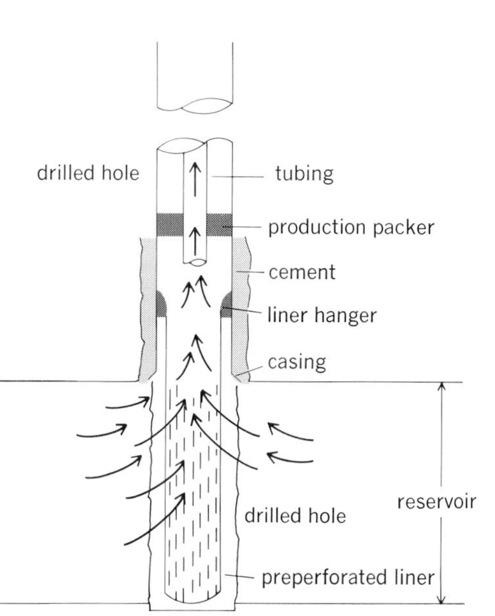

Fig. 4. Liner-type completion; preperforated liner.

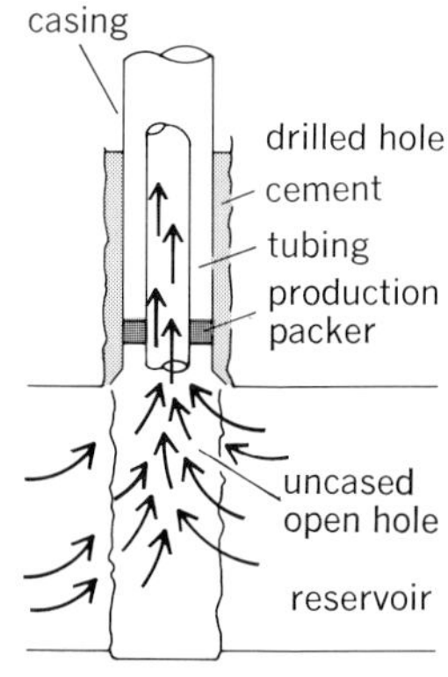

Fig. 3. Diagram of barefoot completion.

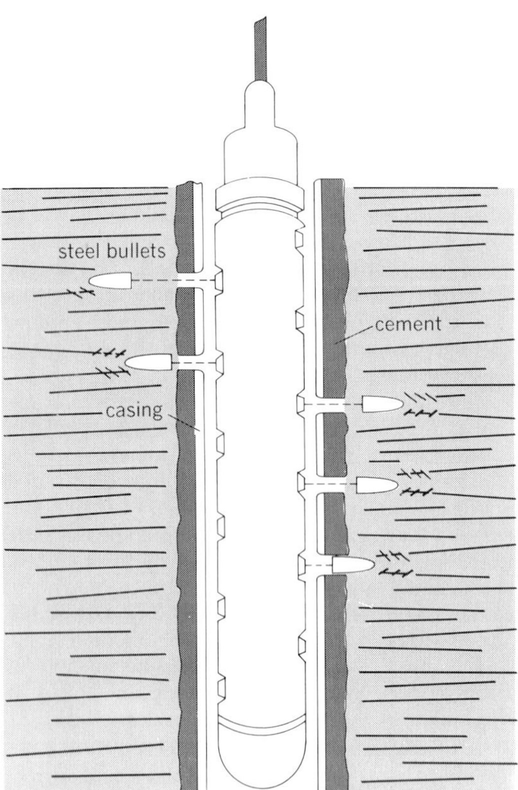

Fig. 5. Gun perforator bullets from bullet-type gun.

and direct it into a pipeline, generally at reduced pressures, to the oil-gathering station.

Pumping unit. Relatively few oil wells flow from natural pressures. Most require secondary means of removing the reservoir product. The most common of several methods is the pumping unit. A walking beam is operated like a seesaw, raising and lowering a plunger-type pump set near the bottom of the hole. The rods between the walking bean and pump are called sucker rods (Fig. 8).

Multiple completion. An oil or gas well from which several separate horizons are individually, separately, and simultaneously produced is called a multiple completion. Such a completion is accomplished by the use of multiple-zone packers and separate tubing strings. The producing zones are separated one from another by proper placement of packers in the well bore. An individual string of tubing is attached to each packer and extended to the surface where each is interconnected with the Christmas tree or flow assembly (Fig. 9).

The several advantages of such a completion from a single well bore are increase in daily production, more efficient and economic utilization of a well bore with multiple reservoirs, increase in ultimate recovery, accurate measurement of product withdrawal from each reservoir, and elimination of mixing of products of different gravities and basic sediment and water content.

Water problems. The production of water in quantity from an oil or gas well renders it uneconomical; means are provided for water exclusion.

Water-exclusion methods. Water exclusion may be effected by the application of cements or various types of plastic. If it is determined that water is entering from the lower portion of a producing sand in a relatively shallow, low-pressure well, a cement plug may be placed in the bottom of the hole so that it will cover the oil-water interface of the reservoir. This technique is called laying in a plug and may be accomplished by placing the cement with a dump-bottom bailer on a wire line or by pumping cement down the drill pipe or tubing. For deeper, higher-pressure, or more troublesome wells, a squeeze method is used. Squeeze cementing is the process of applying hydraulic pressure to force a cementing material into permeable space of an exposed formation or through openings in the casing or liner. In many conditions cement, plastic, or diesel-oil cement may be squeezed into water-, oil-, or gas-bearing portions of a producing zone to eliminate excessive water without sealing off the gas or oil. A few of the applications are: repair of casing leaks; isolation of producing zones prior to perforating for production; remedial or secondary cementing to correct a defective condition, such as channeling or insufficient cement on a primary cement job; sealing off a low-pressure formation that engulfs oil and gas or drilling fluids; and abandonment of depleted producing zones to prevent migration of formation effluent and to reduce possibilities of contaminating other zones or wells.

The squeeze-method tool is a packer-type device designed to isolate the point of entry between or below packing elements. The tool is run into the hole on drill pipe or tubing, and the cementing material is squeezed out between or below these confining elements into the problem area. The well is then recompleted. It may be necessary to drill the cement out of the hole and reperforate, depending upon the outcome of the job performed in the squeeze process.

Water-exclusion plug back. Simple water shutoff jobs in shallow, deep, or high-pressure wells may also be performed in multizone wells in which the lower producing interval is depleted or the remain-

OIL AND GAS WELL COMPLETION

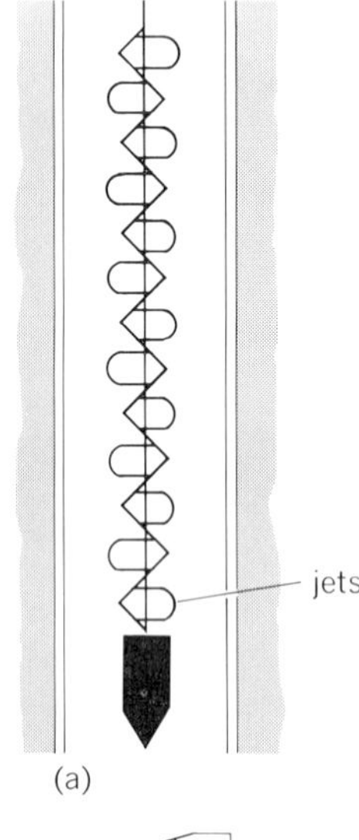

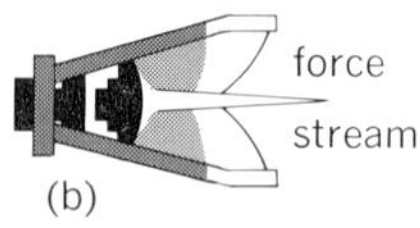

Fig. 6. Jet-type perforator. (*a*) Charges in firing position. (*b*) Side view of a shaped charge.

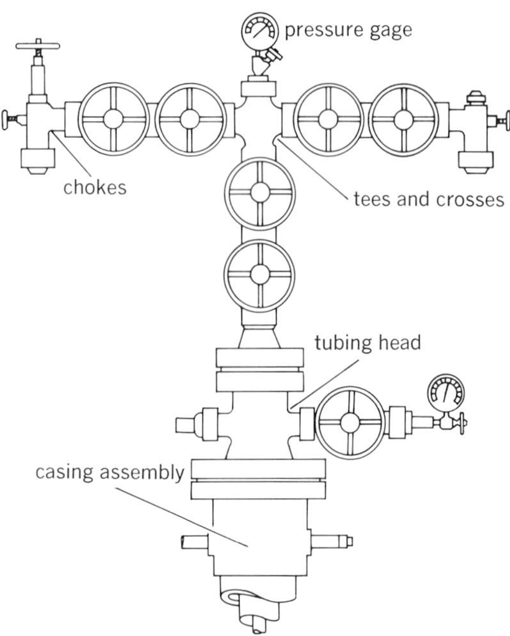

Fig. 7. Typical layout of a Christmas tree.

ing recoverable reserves do not justify rehabilitation.

Here, water may be excluded by placing a packer-type plug (cork) above the interval, then producing formations that are already open or perforating additional intervals that may be present higher up the hole.

Production-stimulation techniques. The initial testing or production history often indicates subnormal production rates, signifying the necessity for remedial action. Any method designed to increase the production rate from a reservoir is defined as production stimulation. Three of the methods used are acidizing, fracturing, and employing explosives.

Acidizing. Varied volumes of hydrochloric acid are used in limestone and dolomite or other acid-soluble formations to dissolve the existing flow-channel walls and enlarge them (Fig. 10). High-pressure equipment, pumps, and wellheads are necessary for satisfactory performance. Fast pumping speeds and acid inhibitors are used to alleviate corrosion of the well equipment.

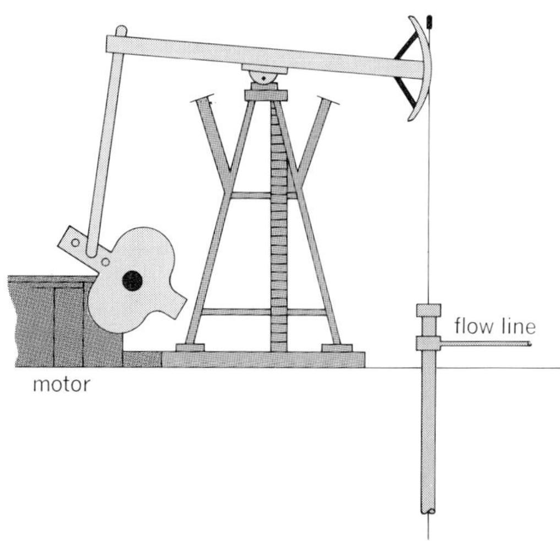

Fig. 8. Pumping unit diagram. These are most common if natural pressure is lacking for well flow.

Fracturing. Formation fracturing is a hydraulic process aimed at the parting of a desired section of formation. Selected grades of sand or particles of other materials are added to the fracturing fluid in varied quantities. These particles pack and fill the fracture, acting as a propping agent to hold it open when the applied pressure is released (Fig. 11). Such fractures increase the flow channels in size and number, improving the fluid-flow characteristics of the reservoir rock. The particle-carrying agent (fluid) is of considerable importance and is varied to fit particular demands. Some of the fluids which are used in this process are crude oil (sand oil fracturing), special refined oils (sand oil fracturing), water (river fracturing), acid (acid fracturing), and oil, water, and chemical emulsion (emulsifracturing).

Explosives. The idea of stimulating production by use of explosives was first used in a well in Pennsylvania on Jan. 21, 1865. The first torpedo consisted of 8 lb of gunpowder, contained in a metal tube, which was lowered into the well and detonated. Its more important function, in this shallow well, was to clear away paraffin, which was accomplished, but a decided increase in production was also accomplished. The method has since been improved with new explosives, firing mechanisms, and procedures, but the basic idea is the same, that is, to remove reservoir-blocking material from the reservoir face and to create fractures in the rock to increase production (Fig. 12).

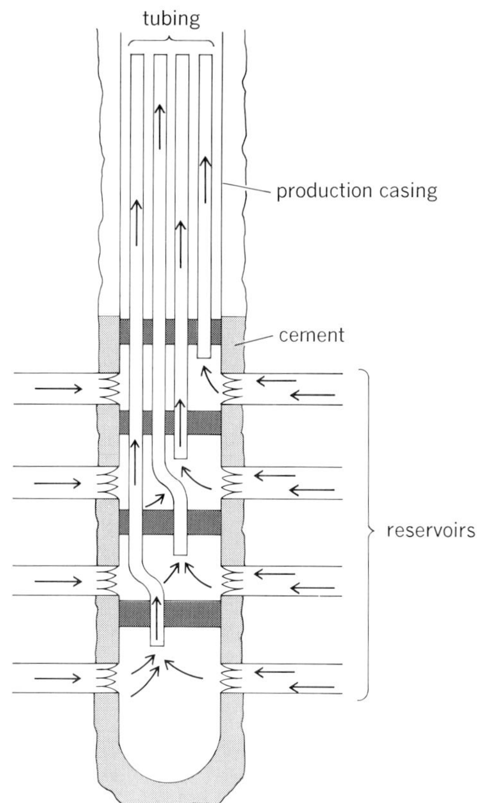

Fig. 9. A representative multiple completion diagram.

Sand exclusion. Some reservoir rock is of an unconsolidated nature similar to beach sands;

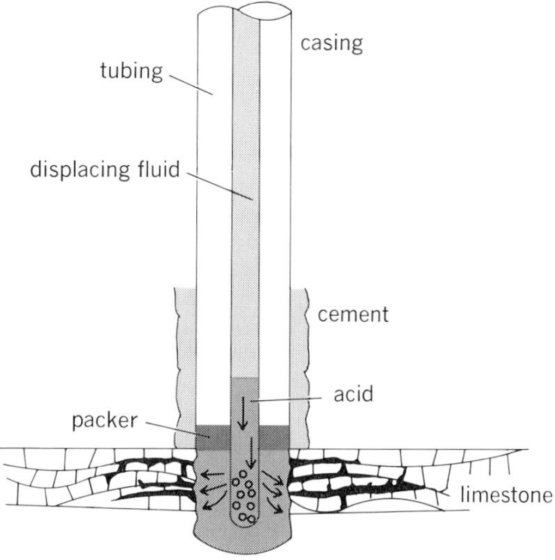

Fig. 10. Outline of the acidizing process.

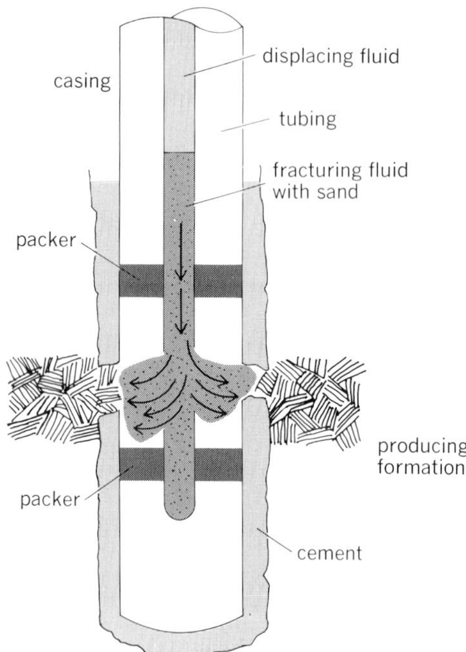

Fig. 11. Outline of the fracturing process.

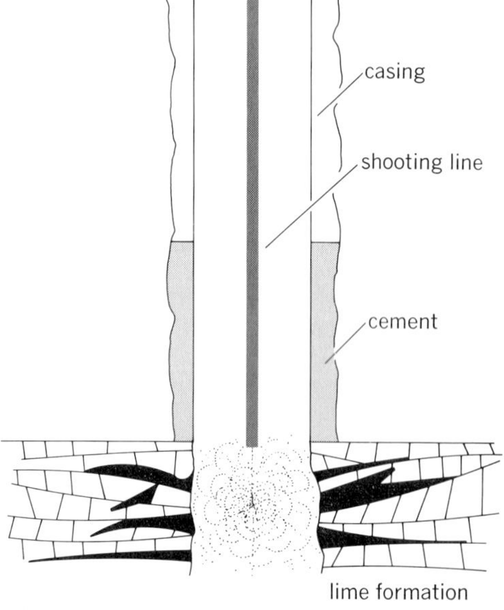

Fig. 12. Stimulating production by use of explosives.

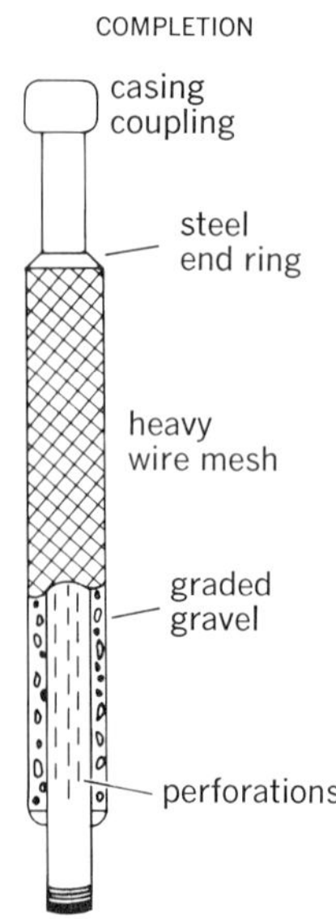

Fig. 13. Prepacked gravel liner.

after having been penetrated with a bore hole, it will slough, if unsupported. This rock also has a tendency to flow with its formation fluids, resulting in plugging of the well bore and restriction or elimination of the entry of formation fluids. Several measures can be used to combat this condition, but no single measure can be used universally.

Screen liner. This type of liner is a segment of preslotted pipe wrapped with wire screen designed to screen out or retain outside the bore hole all except the fine particles that may be brought to the surface with the reservoir fluid. The original design is such that a calculated snug fit is obtained between the liner and bore hole. The coarse screened-out particles form a secondary gravel pack between the liner and the hole, additionally supporting the formation and reducing sand incursion.

Sand consolidation. Sand consolidation is the result of successfully placing a binding material in the producing sand. In effect this glues the sand grains together without completely destroying the porosity and permeability of the sand. The binding material, generally a form of plastic, is forced into the sand through perforations in the casing. The purpose is to consolidate the sand around the well bore to eliminate sloughing and sand incursion.

Prepack gravel liners. This type of liner (Fig. 13) is made by using a perforated section of steel pipe, over which has been fitted a tubular sleeve formed of an inner and outer screen of heavy wire mesh or perforated sheet steel. The pipe section is held in concentric relationship by spacers with gravel of proper size packed between the screens and sealed at both ends to retain the gravel. Set in the hole, it serves as a screen liner. *See* PETROLEUM RESERVOIR ENGINEERING. [HARRY S. BRIGHAM]

Oil and gas well drilling

The drilling of holes for exploration and extraction of crude oil and natural gas. Deep holes and high pressures are characteristics of petroleum drilling not commonly associated with other types of drilling. In general, it becomes more difficult to control the direction of the drilled hole as the depth increases, and additionally, the cost per foot of hole drilled increases rapidly with the depth of the hole. Drilling-fluid pressure must be sufficiently high to prevent blowouts but not high enough to cause fracturing of the bore hole. Formation-fluid pressures are commonly controlled by the use of a high-density clay-water slurry, called drilling mud. The chemicals used in drilling mud can be expensive, but the primary disadvantage in the use of drilling muds is the relatively low drilling rate which normally accompanies high bottom-hole pressure. Drilling rates can often be increased by using water to circulate the cuttings from the hole; when feasible, the use of gas as a drilling fluid can lead to drilling rates as much as 10 times those attained with mud. Drilling research has the objectives of improving the utilization of current drilling technology and the development of improved drilling techniques and tools.

Hole direction. The hole direction must be controlled within permissible limits in order to reach a desired target at depths as great as 25,000 ft. Inclined layers of rocks with different hardnesses tend to cause the direction of drilling to deviate; consequently, deep holes are rarely truly straight and vertical. The drilling rate generally increases as additional drill-collar weight is applied to the bit by adjusting the pipe tension at the surface. However, crooked-hole tendency also increases with higher weight-on-bit. In recent years a so-called packed-hole technique has been used to reduce the tendency to hole deviation. One version of this technique makes use of square drill collars which nearly fill the hole on the diagonals but permit fluid and cuttings to circulate around the sides. This procedure reduces the rate at which the hole direction can change.

Cost factors. Drilling costs depend on the costs of such items as the drilling rig, the bits, and the drilling fluid, as well as on the drilling rate, the time required to replace a worn bit, and bit life. At depths below 15,000 ft (4.6 km), for example, the cost per foot of hole drilled can be relatively modest. Operating costs of a large land rig required for deep holes are about thousands of dollars per day, whereas comparable costs for an offshore drilling platform or floating drilling vessel can vary from several hundred to tens of thousands of dollars per day. Although weak rock at shallow depth can be drilled at rates exceeding 100 ft/hr (30 m/hr), drilling rates often average about 5 ft/hr (1.5 m/hr) in deep holes. Conventional rotary bits have an operating life of 10–20 hr, and 10–20 hr are required for removing the drill pipe, replacing the worn bit, and lowering the pipe back into the hole. Diamond bits may drill for as long as 50–200 hr, but the drilling rate is relatively low and the bits are expensive. For more economical drilling it is desirable to increase both bit life and drilling rate simultaneously.

Drilling fluids. The increased formation or pore-fluid pressures existing at great depths in the Earth's crust adversely affects drilling. Gushers, blowouts, or other uncontrolled pressure conditions are no longer tolerated. High-density drilling fluids maintain control of well pressures. A normal fluid gradient for salt water is about 0.5 psi/ft (11 kPa/m) of depth, and the total stress due to the weight of the overburden increases approximately 1 psi/ft (23 kPa/m). Under most drilling conditions in permeable formations, the well-bore pressure must be kept between these two limiting values. If the mud pressure is too low, the formation fluid can force the mud from the hole, resulting in a blowout; whereas if the mud pressure becomes too high, the rock adjacent to the well may be fractured, resulting in lost circulation. In this latter case the mud and cuttings are lost into the fractured formation.

High drilling-fluid pressure at the bottom of a bore hole impedes the drilling action of the bit. Rock failure strength increases, and the failure becomes more ductile as the pressure acting on the rock is increased. Ideally, cuttings are cleaned from beneath the bit by the drilling-fluid stream; however, relatively low mud pressure tends to hold cuttings in place. In this case mechanical action of the bit is often necessary to dislodge the chips. Regrinding of fractured rock greatly decreases drilling efficiency by lowering the drilling rate and increasing bit wear.

Drilling efficiency can be increased under circumstances where mud can be replaced by water as the drilling fluid. This might be permissible, for example, in a well in which no high-pressure gas zones are present. Hole cleaning is improved with water drilling fluid because the downhole pressure is lower and no clay filter cake is formed on permeable rock surfaces. So-called fast-drilling fluids provide a time delay for filter cake buildup. This delay permits rapid drilling with no filter cake at the bottom of the hole and, at the same time, prevents excessive loss of fluid into permeable zones above.

In portions of wells where no water zones occur, it is frequently possible to drill using air or natural gas to remove the cuttings. Drilling rates with a gas drilling fluid are often 10 times those obtained with mud under similar conditions. Sometimes a detergent foam is used to remove water in order to permit gas drilling in the presence of limited water inflow. In other instances porous formations can be plugged with plastic to permit continued gas drilling. However, in many cases it is necessary to revert to either water or mud drilling when a water zone is encountered. Another possibility is the cementing of steel casing through the zone containing water and then proceeding with gas drilling.

Research. Drilling research includes the study of drilling fluids, the evaluation of rock properties, laboratory simulation of field drilling conditions, and the development of new drilling techniques and tools. Fast-drilling fluids have been developed by selecting drilling-fluid additives which plug the pore spaces very slowly, thereby providing the desired time delay for filter cake buildup. Water-shutoff chemicals have been formulated which can be injected into a porous water-bearing formation in liquid form and then, within a few hours, set to become solid plastics. Well-logging techniques can warn of high-pressure permeable zones so that the change from gas or water drilling fluid to mud can be made before the drill enters the high-pressure zone. Downhole instrumentation can lead to improved drilling operations by providing information for improved bit design and also by feeding information to a computer for optimum control of bit weight and rotary speed. Computer programs can also utilize information from nearby wells to determine the best program for optimum safety and economy in new wells.

Since rock is a very hard, strong, abrasive material, there is a challenge to provide drills which can penetrate rock more efficiently. A better understanding of rock failure can lead to improved use of present equipment and to the development of better tools. Measurements of physical properties of rocks are beginning to be correlated with methods for the theoretical analysis of rock failure by a drill bit. A small 1¼-in. (32-mm) diameter bit, called a microbit, has been used in scale-model drilling experiments in which independent control is provided for bore-hole, formation-fluid, and overburden pressures. These tests permit separation of the effects of the various pressures on drilling rates.

Novel drilling methods which are being explored include studies of rock failure by mechanical, thermal, fusion and vaporization, and chemical means. Some of the new techniques currently have limited practical application while others are still in the experimental stage. For example, jet piercing is widely used for drilling very hard, spallable rocks, such as taconite. Other methods include the use of electric arc, laser, plasma, spark, and ultrasonic drills. Better materials and improved tools can be expected in the future for drilling oil and gas wells to greater depths more economically and with greater safety. *See* BORING AND DRILLING (MINERAL); OIL AND GAS WELL COMPLETION; ROTARY TOOL DRILL; TURBODRILL.

[J. B. CHEATHAM, JR.]

Bibliography: L. W. Ledgerwood, Efforts to develop improved oil-well drilling methods, *J. Petrol. Tech.*, 219:61–74, 1960; W. C. Maurer, *Novel Drilling Techniques*, 1968; A. W. McCray and F. W. Cole, *Oil Well Drilling Technology*, 1976.

Oil burner

A device for converting fuel oil from a liquid state into a combustible mixture. A number of different types of oil burners are in use for domestic heating. These include sleeve burners, natural-draft pot burners, forced-draft pot burners, rotary wall flame burners, and air-atomizing and pressure-atomizing gun burners. The most common and modern type that handles 80% of the burners used to heat United States homes is the pressure-atomizing—type burner shown in Fig. 1.

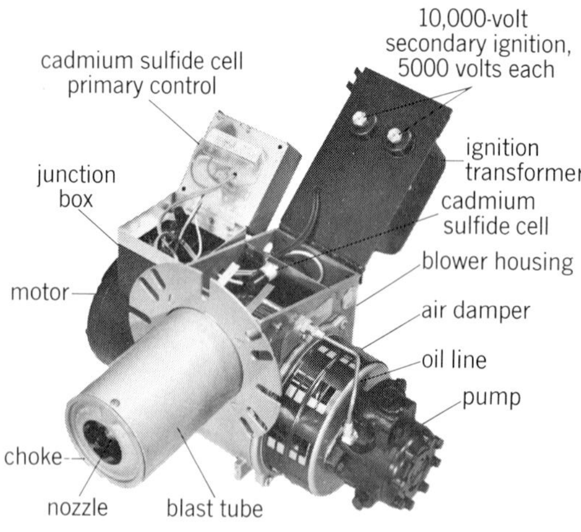

Fig. 1. An oil burner of the pressure-atomizing type. (*Automatic Burner Corp.*)

Characteristics. The sleeve burner, commonly known as a range burner because of its use in kitchen ranges, is the simplest form of vaporizing burner. The natural-draft pot burner relies on the draft developed by the chimney to support combustion. The forced-draft pot burner is a modification of the natural-draft pot burner, since the only significant difference between the two types is the means of supplying combustion air. The forced-draft pot burner supplies its own air for combustion and does not rely totally on the chimney. The rotary wall flame burners have mechanically assisted vaporization. The gun-type burner uses a nozzle to atomize the fuel so that it becomes a vapor, and burns easily when mixed with air.

The most important feature of a high-pressure atomizing gun burner is the method of delivering the air. The most efficient burner is the one which completely burns the oil with the smallest quantity of air. The function of the oil burner is to properly proportion and mix the atomized oil and air required for combustion (Fig. 2).

Efficiency. If a large quantity of excess air is used to attempt to burn the oil, there is a direct loss of usable heat. This air absorbs heat in the heating unit which is then carried away through the stack with the combustion gases. This preheated air causes high stack temperatures which lower the efficiency of the combustion. The higher the CO_2 (carbon dioxide), the less excess air. Over-

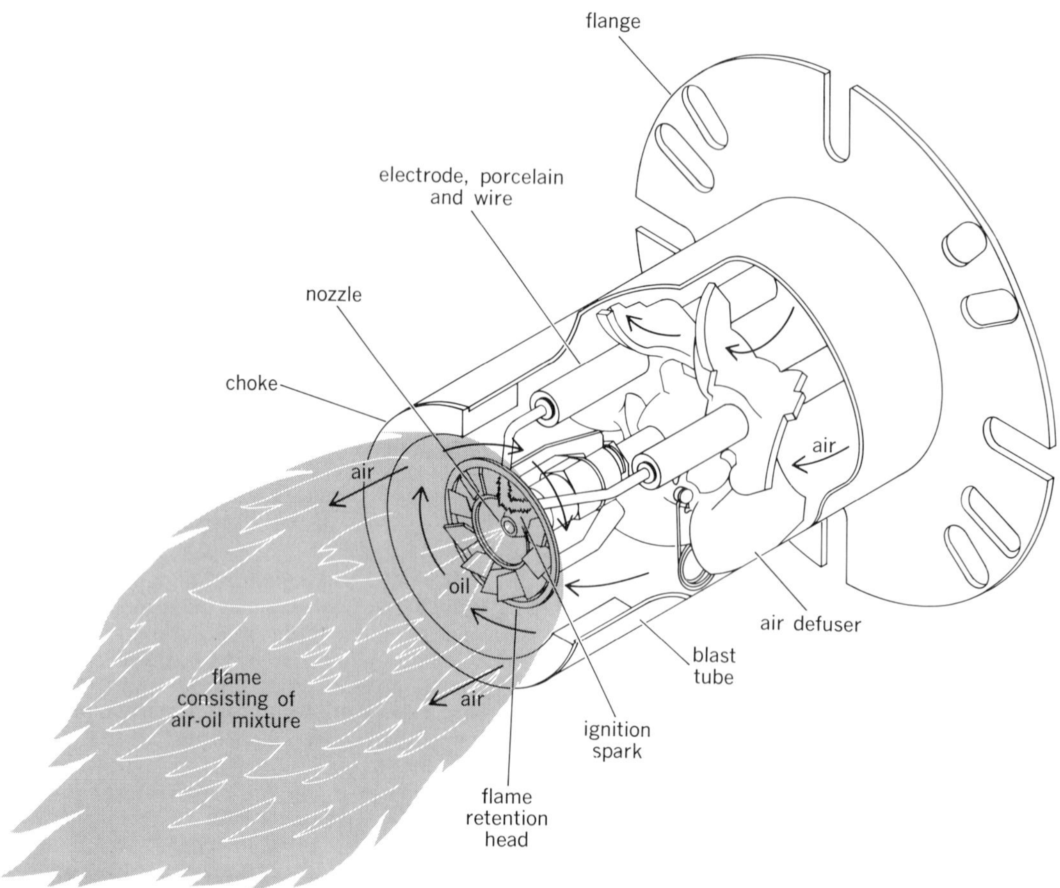

Fig. 2. Flame-retention–type burner, an example of a high-efficiency burner.

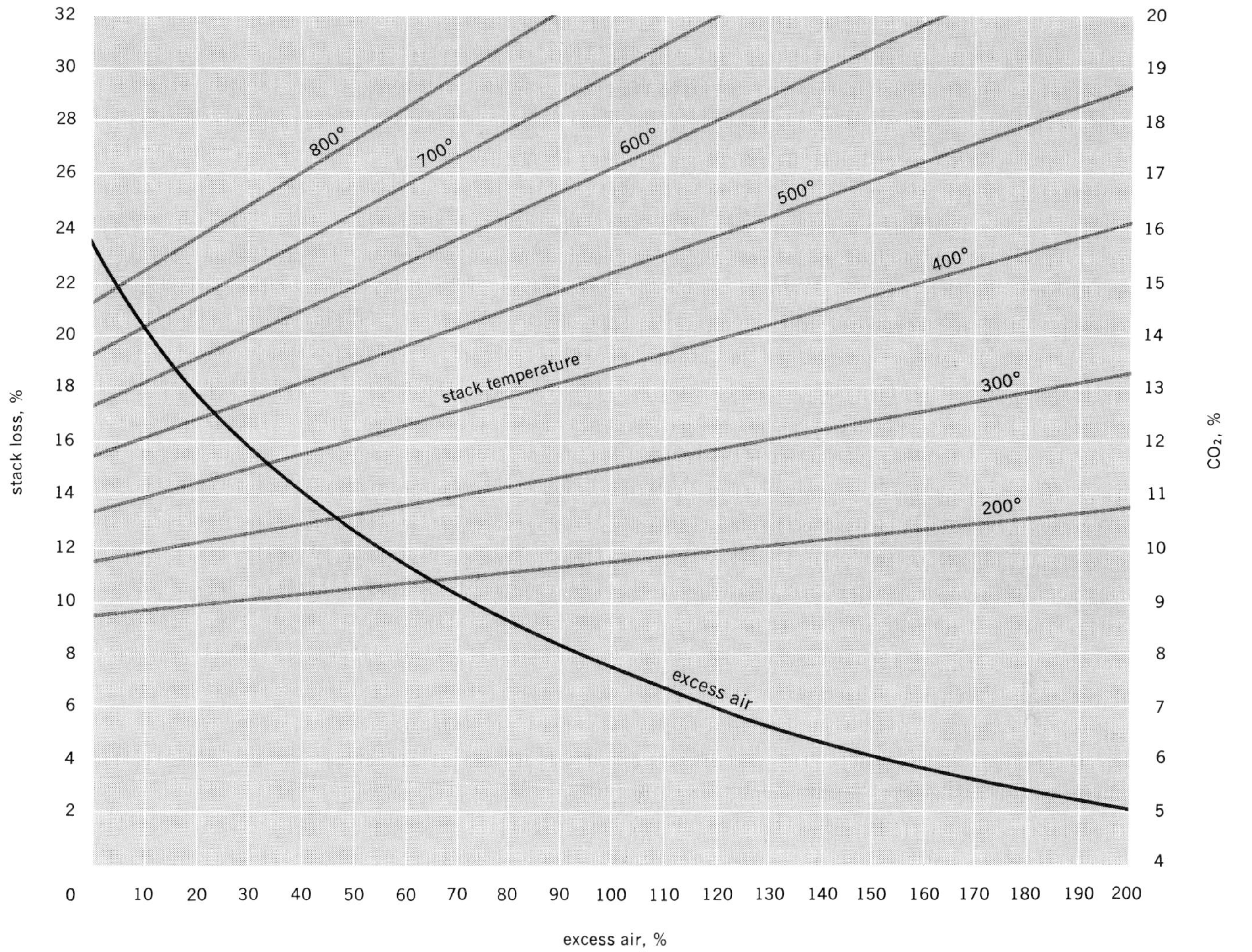

Fig. 3. Typical stack loss chart. To determine stack loss and efficiency, start with correct CO_2 and follow horizontal line to excess air curve, then vertical line to stack temperature, and finally horizontal line to stack loss. Overall efficiency percent = 100 − stack loss.

all efficiencies or stack loss can be estimated by the use of the stack loss chart (Fig. 3).

Increased efficiency can be obtained through the use of devices located near the flame end of the burner. CO_2 is not the ultimate factor in efficiency. Burners producing high CO_2 can also produce high smoke readings. Accumulations of 1/8-in. (3-mm) soot layers on the heating unit surface can increase fuel consumption as much as 8%. An oil burner is always adjusting to start smoothly with the highest CO_2 and not more than a number two smoke on a Bacharach smoke scale. For designs in high-efficiency burners commonly known as flame-retention–type burners see Fig. 2.

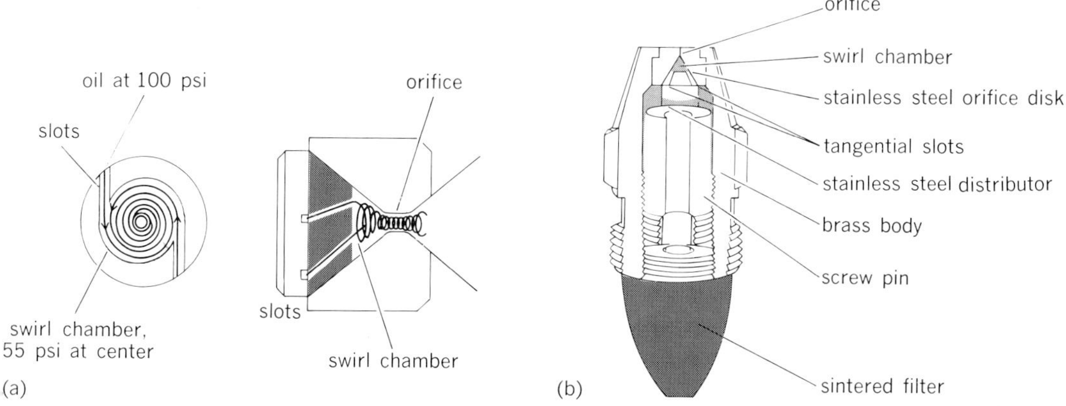

Fig. 4. Nozzle. (*a*) Operation of a nozzle. (*b*) Cutaway of nozzle. (*Delavan Manufacturing Co.*)

Nozzle. The nozzle is made up of two essential parts: the inner body, called the distributor, and the outer body, which contains the orifice that the oil sprays through. Under this high pump pressure of 100–300 psi (700–2100 kPa) the oil is swirled through the distributor and discharged from the orifice as a spray (Fig. 4). The spray is ignited by the spark and combustion is self-sustaining, provided the proper amount of air is supplied by the squirrel cage blower. Air is delivered through the blast tube and moved in such a manner that it mixes well with the oil spray (Fig. 2). This air is controlled by a damper either located on the intake side or the discharge side of the fan (Fig. 1).

The essential parts of the pressure gun burner are the electric motor, squirrel-cage–type blower, housing, fuel pump, electric ignition, atomizing nozzle, and a primary control (Figs. 1 and 5).

The fuel oil is pumped from the tank through the pump gears, and oil pressure is regulated by an internal valve that develops 100–300 psi (700–2100 kPa) at the burner nozzle. The oil is then atomized, ignited, and burned.

The fine droplets of oil that are discharged from the nozzle are electrically ignited by a transformer which raises the voltage from 115 to 10,000 volts. The recently developed electric ignition system called the capacitor discharge system incorporates a capacitor that discharges voltage into a booster transformer developing as much as 14,000 volts. The electric spark is developed at the electrode gap located near the nozzle spray (Figs. 2 and 5).

The high velocity of air produced by the squirrel cage fan helps develop the ignition spark to a point where it will reach out and ignite the oil without the electrode tips actually being in the oil spray. Figure 2 shows a typical gun-type burner inner assembly.

Installations. Fuel oils for domestic burners are distilled from the crude oil after the lighter products have been taken off. Consequently, the oil is nonexplosive at ordinary storage temperatures. Domestic fuel oils are divided into two grades, number one and number two, according to the ASTM specification. There are several different methods of installing an oil tank system with an oil burner (Fig. 6).

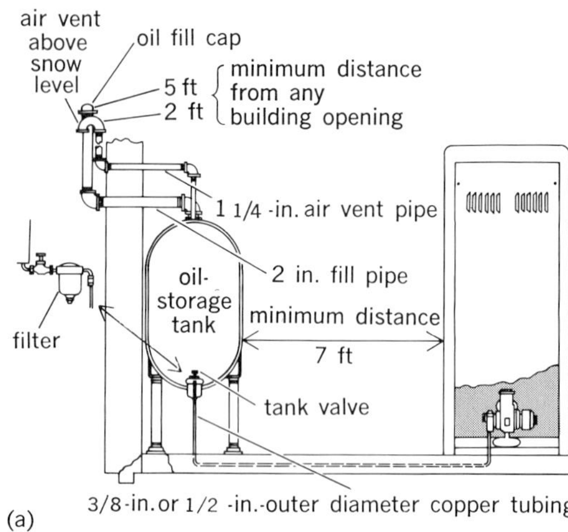

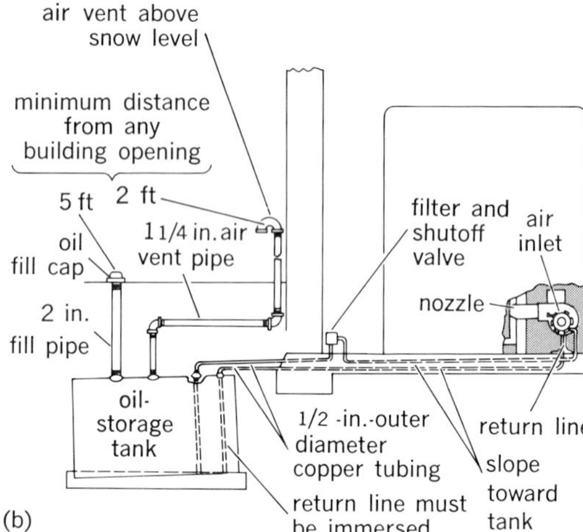

Fig. 6. Two typical installation methods for oil storage tank and oil burner. (*a*) Inside basement. (*b*) Outside underground storage tank. 1 ft = 0.3 m. 1 in. = 2.5 cm.

One of the most important safety items of an automatic oil burner is the primary control. This device will stop the operation when any part of the burner or the heating equipment does not function properly. This control protects against such occurrences as incorrect primary air adjustments, dirt in the atomizing nozzle, inadequate oil supply, and improper combustion. The most modern type of primary control is called the cadmium cell control. The cadmium sulfide flame detection cell is located in a position where it directly views the flame. If any of the above functional problems should occur, the electrical resistance across the face of the cell would increase, causing the primary control to shut the burner off in 70 sec or less.

Draft caused by a chimney or technical means is a very important factor in the operation of domestic oil burners. The majority of burners are designed to fire in a heating unit that has a minus 0.02 in. water column draft over the fire. Because

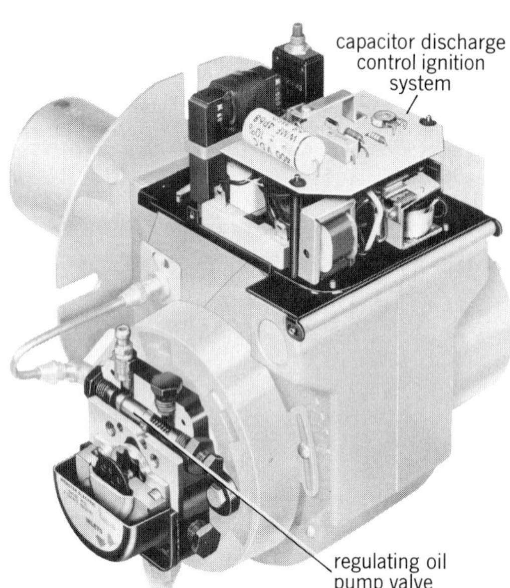

Fig. 5. Oil burner, showing capacitor discharge control ignition system and regulating valve for fuel pump. (*Automatic Burner Corp.*)

the burner develops such low static pressures in the draft tube, over fire drafts are an important factor in satisfactory operation.

If the heating unit is designed to create pressure over the fire, a burner developing high static pressures in the blast tube must be used. These burners develop high static by means of higher rpm on the motor and fan. Burners of this nature are available and are being produced. Flame control is very important. Firing heads similar to Fig. 2 are used on applications of this nature.

The oil burner is also used for a wide assortment of heating, air conditioning, and processing applications. Oil burners heat commercial buildings such as hospitals, schools, and factories. Air conditioners using the absorbtion refrigeration system have been developed and fired with oil burners. Oil burners are used to produce CO_2 in greenhouses to accelerate plant growth. They produce hot water for many commercial and industrial applications. *See* AIR COOLING; COMFORT HEATING; HOT-WATER HEATING SYSTEM; OIL FURNACE.

[ROBERT A. KAPLAN]

Bibliography: American Society of Heating, Refrigerating, and Air-conditioning Engineers (ASHRAE), *ASHRAE Handbook and Product Directory*: *Fundamentals*, 1977, and *Equipment*, 1979; C. H. Burkhardt, *Domestic and Commercial Oil Burners*, 3d ed., 1969; E. M. Field, *Oil Burners*, 3d ed., 1977.

Oil field model

A small-scale and commonly simplified replica of subsurface conditions of interest and value in petroleum prospecting and oil-field development. The term model has been applied by geologists to a simplified diagram and by mathematical physicists to a formal analysis with special boundary conditions and related attributes. These and somewhat more complex physical models have value in transmitting concepts and relationships to the nonspecialist, but they are generally designed by geologists and petroleum engineers to aid investigation of a particular problem. To do so, the model is scaled to size, shape, and like attributes, or according to forces upon it, relative to its archetype.

Size-scaled models. These are commonly models of ore and mineral bodies and of those blocks on which surface topography and surface geology are shown on top and geologic cross sections on the sides, also to scale. Peg models are constructed for oil fields with wells shown as pegs or wires. Cutout block diagrams are also used. Models of this kind are found in many geological museums and in mining companies' offices.

Forces and movement models. This type contains mobile material to simulate by its movement a movement that has taken place or will take place in the prototype. Transparent solid plastics have been incorporated in mobile parts of models for enhanced internal visibility and to obtain photoelastic data. Such models are also used to make graphic representation of changes recorded from observations with geophysical instruments, for example, from microinstruments used to survey physical response of a small mass in an artificial field. Force and movement models have come into use for study of movements of earth materials. By this, past changes resulting in present features can be made as dimensionally credible experiments in periods tremendously short as compared with geologic time. Models with mobility are of three principal types.

Fluid or fluids moving in porous medium. The models for study of fluid movement through a geometrically stable medium have important applications to hydrologic and petroleum engineering. The hydrologic case is mostly of an interface underground between fresh water and salt water caused by extraction or by injection of fresh water through wells. Where the interface owes its position to fresh-water–salt-water patterns found in many coastal zones and along many shorelines, the setting is difficult to describe without a model. Such demonstrations have added to the understanding of water conditions in California's Santa Clara Valley and other critical water areas.

The petroleum engineering application is to movement of fluid in respect to another kind of interface underground. This is between petroleum and natural gas, or both, and salt water, and especially when petroleum is being extracted through wells and salt water simultaneously injected through other wells into the same porous stratum, so that the important flow is radial, that is, two-dimensional.

The model used involves scaling down from the actual distance between wells in the oil field under study, and also the device of an analog for the fluids, whereby the pressure distribution in steady-state porous flow is exactly the same as the potential distribution in an electrical conducting medium.

Uniform high-viscosity materials. The second geologic realm where models are used extensively is that of the movement of large segments of the earth made of materials of essentially uniform high viscosity, comprising nearly all rock species in the Earth's crust. This has little direct bearing on oil-field models.

Adjacent materials of diverse viscosities. Among the movements of the earth which are of interest to geologists are those involving what may be called soft layers between relatively hard ones. This is the realm where diapir structures are inferred to develop and also their subform, the salt dome. Because of the economic importance of salt domes and their associated oil traps, investigations with models have become widespread.

L. L. Nettleton produced a model for salt-dome formation using two viscous fluids of different density to emphasize the role of the lower density of salt to that of adjacent sediments in the process of salt-dome movements. M. B. Dobrin made a parallel mathematical analysis. T. J. Parker and A. N. MacDowell extended salt-dome studies by using materials with higher yield points than Netleson's first model, and their model showed fractures above and around the "salt" analog in their model.

Also in both the prototype and the model, boundary conditions modify vectors nearby. This seems particularly hard to handle in cases of hydrologic model study, such as at the University of California.

Models versus mathematical analysis. There is a possible alternative to the use of physical models—mathematical analysis, which is more attractive now that computing machines are available. Dobrin used this method for the case of Nettleton's model investigation of salt dome. However, mathematical analysis is as yet a subject of controversy among physicists in many simple geologic settings; physical models are presently available and their value is acknowledged.

Physical models are probably more helpful in educating the nonspecialist in geology; they present the earth features in a readily visible fashion easier to grasp than mathematical analysis. For a discussion of mathematical models *see* PETROLEUM RESERVOIR MODELS. [PAUL WEAVER]

Bibliography: J. W. Amyx, D. M. Bass, and R. L. Whiting, *Petroleum Reservoir Engineering*, vol. 1: *Physical Properties*, 1960; L. P. Drake, *Fundamentals of Reservoir Engineering*, 1979; A. Mayer-Gurr, *Petroleum Engineering*, 1976; D. W. Peaceman, *Fundamentals of Numerical Reservoir Simulation*, 1977; A. E. Scheidegger, *Principles of Geodynamics*, 2d ed., 1963; A. E. Scheidegger, *Theoretical Geomorphology*, 2d ed., 1970; D. K. Todd, *Ground Water Hydrology*, 1959.

Oil furnace

A combustion chamber in which oil is the heat-producing fuel. Fuel oils, having from 18,000 to 20,000 Btu/lb (42–47 MJ/kg), which is equivalent to 140,000 to 155,000 Btu/gal (39–43 MJ/liter), are supplied commercially. The lower flash-point grades are used primarily in domestic and other furnaces without preheating. Grades having higher flash points are fired in burners equipped with preheaters.

The ease with which oil is transported, stored, handled, and fired gives it a special advantage in small installations. The fuel burns almost completely so that, especially in a large furnace, combustible losses are negligible. *See* OIL BURNER; STEAM-GENERATING FURNACE.

Domestic oil furnaces with automatic thermostat control usually operate intermittently, being either off or operating at maximum capacity. The heat-absorbing surfaces, especially the convection surface, should therefore be based more on maximum capacity than on average capacity if furnace efficiency is to be high. The combustion chamber should provide at least 1 ft^3 for each 1.5–2 lb (1m^3 for each 24–32 kg) of fuel burned per hour. Gas velocity should be below 40 ft/sec (12m/sec). The shape of the chamber should follow the outline of the flame. [FRANK H. ROCKETT]

Oil mining

The surface or subsurface excavation of petroleum-bearing sediments for subsequent removal of the oil by washing, flotation, or retorting treatments. Oil mining also includes recovery of oil by drainage from reservoir beds to mine shafts or other openings driven into the oil rock, or by drainage from the reservoir rock into mine openings driven outside the oil sand but connected with it by bore holes or mine wells.

Surface mining consists of strip or open-pit mining. It has been used primarily for the removal of oil shale or bituminous sands lying at or near the surface. Strip mining of shale is practiced in Sweden, Manchuria, and South Africa. Strip mining of bituminous sand is conducted in Canada.

Subsurface mining is used for the removal of oil sediments, oil shale, and Gilsonite. It is practiced in several European countries and in the United States. Some authorities consider this the best method to recover oil when oil sediments are involved, because virtually all of the oil is recovered.

European experience. Subsurface oil mining was used in the Pechelbronn oil field in Alsace, France, as early as 1735. This early mining involved the sinking of shafts to the reservoir rock, only 100–200 ft (30–60 m) below the surface, and the excavation of the oil sand in short drifts driven from the shafts. These oil sands were hoisted to the surface and washed with boiling water to release the oil. The drifts were extended as far as natural ventilation permitted. When these limits were reached, the pillars were removed and the openings filled with waste.

This type of mining continued at Pechelbronn until 1866, when it was found that oil could be recovered from deeper, more prolific sands by letting it drain in place through mine openings, without removing the sands to the surface for treatment.

Subsurface mining of oil shale also goes back to the mid-19th century in Scotland and France. It is not so widely practiced now because of its high cost as compared with that of usual oil production, particularly in the prolific fields of the Middle East.

United States oil shale mining. The U.S. Bureau of Mines carried out an experimental mining and processing program at Rifle, Colo., between 1944 and 1955 in an effort to find economically feasible methods of producing oil shale.

One of the more important phases of this experimental program was a large-scale mine dug into what is known as the Mahagony Ledge, a rich oil shale stratum that is flat and strong, making it favorable for mining. This stratum lies under an average of about 1000 ft (300 m) of overburden and is 70–90 ft (21–27 m) thick.

The Bureau of Mines adopted the room-and-pillar system of mining, advancing into the 70-ft ledge face in two benches. The mine roof was supported by 60-ft (18-m) pillars staggered at 60-ft intervals and supplemented by iron roof bolts 6 ft (1.8 m) long.

Multiple rotary drills mounted on trucks made holes in which dynamite was placed to shatter the shale; the shale was then removed from the mine by electric locomotive and cars (Fig. 1).

Fig. 1. Mine locomotive and cars removing shale from U.S. Bureau of Mines Shale Mine at Rifle, Colo. Just visible left of center is the Colorado River, nearly 3000 ft (900 m) below. (*After R. Fleming, U.S. Bureau of Mines*)

Fig. 2. Close-up view of Union Oil Co. of California shale-oil retort near Grand Valley, Colo. Right part of structure is portion of the system for removing oil vapors that would otherwise escape in the gas stream.

The experimental mining program ended in February, 1955, when a roof fall occurred. Despite this occurrence, however, the Bureau is convinced that the room-and-pillar method used in coal, salt, and limestone mines is feasible for shale oil mining in Colorado.

Since 1955, several companies have conducted experimental efforts to produce shale oil. Those continuing to move toward commercialization include companies which plan to use traditional mining techniques combined with some form of surface retorting. One company is testing an on-site process in which both mining and retorting are done underground, and another firm is well along in development work using a method involving a gas-combustion retort similar to that used by the Bureau of Mines in various pilot plants operated during the 1950s and 1960s.

Another possibility is a modified open-cast surface method, which proponents claim has a 95% recovery rate of the minable reserves, compared with a maximum 40% rate by room-and-pillar underground mining and 20% by on-site mining and retorting.

Oil shale does not contain oil, as such. Draining methods, therefore, are not applicable. It does, however, contain an organic substance known as kerogen. This substance decomposes and gives off a heavy, oily vapor when it is heated above 700°F (370°C) in retorts. When condensed, this vapor becomes a viscous black liquid called shale oil, which resembles ordinary crude but has several significant differences.

Colorado's Mahagony Ledge yields an average of about 30 gal of oil per ton (125 liters per metric ton). This means that large amounts of oil shale must be mined, transported, retorted, and discarded for production of commercial quantities of oil. Various types of retort also have been tested in Colorado, but none is in commercial use (Fig. 2).

Gilsonite. Gilsonite is a trade name, registered by the American Gilsonite Co., for a solid hydrocarbon found in the Uinta Basin of eastern Utah and western Colorado. The American Gilsonite Co. uses a subsurface wet-mining technique to extract about 700 tons (635 metric tons) of Gilsonite daily from its mine at Bonanza, Utah.

Conventional mining methods were found unsuitable for mass output of Gilsonite because it is friable and produces fine dust when so mined. This dust can be highly explosive. In the system now being used, tunnels are driven from the main shaft by means of water jetted through a 1/4-in. (6-mm) nozzle under pressure of 2000 psi (14 MPa). The stream of water penetrates tiny fissures and the ore falls to the bottom of the drift. The drifts are cut on a rising grade of about 2.5°. The ore is washed down to the main shaft where it is screened. Particles of sizes smaller than 3/4 in. (19 mm) are pumped to the surface in a water stream; larger pieces are hoisted in buckets. A long rotary drill with carbide-tipped teeth is used to remove ore that cannot be broken with water jets.

Gilsonite is moved through a pipeline in slurry form to a refinery, where it is dried and melted and then heated to about 450°F (232°C). The melted oil is fed to a coker and other processing units to make gasoline and other petroleum products.

[A. L. PONIKVAR]

North American tar sands. The world's only tar sand mining operations, the 45,000-barrel-per-day (bpd, 7150 m^3 per day) synthetic crude complex of Suncor, Inc. (formerly Great Canadian Oil Sands Limited), which is located 34 km north of Fort McMurray, Alberta, and started production in 1967, and the 129,000-bpd (20,500 m^3 per day) project of Syncrude Canada Ltd., which is located approximately 8 km away and started production in 1978, are situated in the minable area of the gigantic Athabasca Tar Sands deposit, the world's largest oil reservoir. The minable area is that portion with an overburden thickness of less than 45 m, embracing about 10% of the estimated 6.24×10^{11} bbl (9.92×10^{10} m^3) of total in-place bitumen reserves, may ultimately support five or more such surface mining plants operating at the same time. In 1979 Suncor commenced an expansion program to increase its plant capacity to 58,000 bpd (9200 m^3 per day) by 1982. Also in 1979, Alsands, a consortium of nine major oil companies, received preliminary regulatory approval to construct, by 1986, a 140,000-bpd (22,300 m^3 per day) project, which will be fashioned after the Syncrude model.

In both the Suncor and the Syncrude projects, about half of the terrain is covered with muskeg, an organic soil resembling peat moss, which ranges from a few centimeters to 7 m in depth. The major part of the overburden, however, consists of Pliestocene glacial drift and Clearwater Formation sands and shales. The total overburden varies from 7 to 40 m in thickness. Underlying is the McMurray Formation (Lower Cretaceous), in which the oil-impregnated sands reside. The sand strata in the region of the Suncor and Syncrude operations, also variable in thickness, average about 45 m, although typically 5–10 m must be discarded because of bitumen content below the economic cutoff grade of 6 wt %.

Composition. The bitumen content of tar sand can range from 0 to 20 wt %, but feed-grade mate-

Fig. 3. Large draglines dig tar sand at the Syncrude project. Note the relative size of the vehicles in the foreground.

rial normally runs between 10 and 12 wt %. The balance of the tar sand is composed, on the average, of 5 wt % water and 84 wt % sand and clay. The bitumen, which has a gravity on the American Petroleum Institute (API) scale of 8 degrees, is heavier than water and very viscous. Tar sand is a competent material, but it can be readily dug in the summer months; during the winter months, however, when temperatures plunge to −45°C, tar sand assumes the consistency of concrete. To maintain an acceptable digging rate in the winter, mining must proceed faster than the rate of frost penetration; if not, supplemental measures, such as blasting, are required.

Overburden removal. For muskeg removal, a series of ditches are dug 2 or 3 years in advance of stripping to permit as much of the water as possible to drain away. Despite this preparation, the spongy nature of the muskeg persists and removal is best accomplished after freeze-up.

Suncor has tried, and rejected, several overburden removal methods including shovels and trucks, scrapers, and front-end loaders and trucks. In 1976 a bucketwheel excavator was purchased to replace seven 14-m^3 front-end loaders and five D9G bulldozers. Through twin chutes, the excavator loads a fleet of 21 140-metric-ton-capacity WABCO trucks. Additional equipment is used for maintaining the haul roads and for spreading and compacting the spoiled material.

Bucketwheel excavators. Mining of tar sand is performed at Suncor mainly by two large bucketwheel excavators of German manufacture, each operating on a separate bench. These units, weigh-

Fig. 4. Bucketwheel excavators get into position to load windrows of tar sand onto the conveyor belts in order to feed the Syncrude plant which is visible in the background.

ing 1600 metric tons, have a 10-m-diameter digging wheel on the end of a long boom. Each wheel has a theoretical capacity of 8600 tons/hr, but the average output from digging is about 4500 tons/hr. Because the availability of these machines is normally 55–60%, the extraction plant has been designed to accept a widely fluctuating feed rate. To facilitate digging of the highly abrasive tar sand and to achieve a reasonable bucket and tooth life, Suncor routinely preblasts the tar sand on a year-round basis. At the rate of 127,000 tons/day, tar sand is transferred from the mine to the plant by a system of 152-cm wide conveyor belts and 183-cm trunk conveyors, operating at 333 m/min. Following extraction by a process using hot water, the bitumen is upgraded by coking and hydrogenation to a high-quality synthetic crude.

Dragline scheme. Syncrude opted for an even more capital-intensive mining scheme. Four large draglines, each equipped with a 70-m^3 bucket at the end of a 111-m boom, are employed to dig both a portion of the overburden, which is free-cast into the mining pit, and the tar sand, which is piled in windrows behind the machines (Fig. 3). Four bucketwheel reclaimers, similar to the Suncor excavators but larger to handle the additional capacity, load the tar sand from the windrows onto conveyor belts which transfer it to the plant (Fig. 4). With a peak tar sand mining rate of 300,000 tons/day, the Syncrude project is the largest mining operation in the world.

The advantages of the dragline scheme lie in its ability to open a new mine faster, to handle certain types of overburden at a lower cost, and to reject with greater selectivity lenses of low-grade tar sand and barren material. The disadvantages include the necessity of rehandling the tar sand and the production of an increased percentage of lumps in the plant feed, which can damage conveyor belting. Certain Clearwater overburden strata, occurring mainly on the west side of Syncrude's mine, are too weak to support the draglines and must be prestripped by other methods.

Several years of comparative operation will be required to firmly establish whether one of the schemes enjoys an economic advantage.

United States tar sands. No United States reservoirs have sufficient resources reachable by surface mining to justify the larger-scale operations used in Canada.

Economic exploitation of surface accumulations of tar sands in the United States appears remote because of the high costs associated with developing such small deposits.

Mine-assisted in-situ production. Underground mining may yet play a significant role in tar sand exploitation. For almost a century, production of heavy oil from the Wietze reservoir in Germany has been increased beyond that available from conventional wells, by digging vertical shafts into the formation and collecting the seepage; more recently, horizontal tunnels have attained a higher seepage rate. At the Yarega reservoir in Siberia, the Soviets have taken the underground recovery method a step further by combining tunnels under the formation with 34,000 closely spaced vertical wells rising from the tunnels. Steam is injected into the Yarega formation from above to raise the temperature of the bitumen and enhance its flow characteristics. Approximately 1.8×10^7 bbl (2.9×10^6 m^3) had been produced in this manner by 1979. A major pilot test is under way to investigate the applicability of the mine-assisted in-situ production (MAISP) method to Alberta's Athabasca deposit. The MAISP method may offer the best hope for recovering reserves too deep for surface mining (more than 80 m deep) but too shallow for in-place techniques (less than 300 m deep).

[G. RONALD GRAY]

Bibliography: W. L. Oliver and G. R. Gray, Technology and economics of oil sands operations, *10th World Petroleum Congress*, Bucharest, Romania, 1979; G. S. Smith and R. M. Butler, Studies on the use of tunnels and horizontal wells for the recovery of heavy crudes, *United Nations Institute of Training and Research (UNITAR) 1st International Conference on the Future of Heavy Crude and Tar Sands*, Edmonton, Alberta, 1979.

Oldham's coupling

A flexible coupling that permits two slightly misaligned shafts to be joined. In the conventional form of Oldham's coupling (see illustration), each shaft end is fitted with a hub having a smooth slot or groove. A floating member with orthogonal ridges or tongues mates with the two hubs. The two degrees of freedom thus provided enable the coupling to accommodate to axis displacement, but not to angular displacement. A similar action is provided by a square block mating with widely slotted hubs. Adequate lubrication is essential to efficient operation. *See* COUPLING; UNIVERSAL JOINT.

[JAMES J. RYAN]

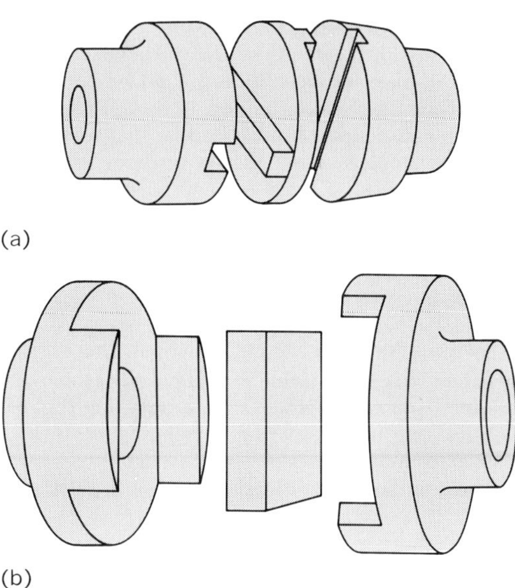

Parallel displaced rotating shafts joined by a floating center member. (*a*) Conventional Oldham's coupling. (*b*) Modern coupling with Oldham principle.

Open circuit

A condition in an electric circuit in which there is no path for current between two points; examples are a broken wire and a switch in the OPEN, or OFF, position. *See* CIRCUIT (ELECTRICITY).

Open-circuit voltage is the potential difference

between two points in a circuit when a branch (current path) between the points is open circuited. Open-circuit voltage is measured by a voltmeter which has a very high resistance (theoretically infinite), such as a vacuum-tube voltmeter.

[CLARENCE F. GOODHEART]

Open-pit mining

The extraction of ores of metals and minerals by surface excavations. This method of mining is applicable for near-surface deposits which do not have a ratio of overburden (waste material that must be removed) to ore that would make the operation uneconomic. Where these criteria are met, large-scale earth-moving equipment can be used to give low unit mining costs. For a discussion of other methods of surface mining *see* COAL MINING; PLACER MINING; STRIP MINING.

Most open-pit mines are developed in the form of an inverted cone, with the base of the cone on the surface. Exceptions are open-pit mines developed in hills or mountains (Fig. 1). The walls of most open pits are terraced with benches to permit shovels, front-end loaders, or bucket-wheel excavators to excavate the rock and to provide access for trucks, trains, or belts to transport the rock out of the pit. The final depth of the pits ranges from less than 51 feet to nearly 3000 ft (15.5 to 914 m) and is dependent on the depth and value of the ore and the cost of mining. The cost of mining usually increases with the depth of the pit because of the increased distance that the ore must be hauled to the surface, the waste hauled to dumps, and the increased amount of overburden that must be removed. Numerous benches or terraces in the pit permit a number of areas to be worked at one time and are necessary for giving slope, typically less than a 2-to-1 slope, to the sides of the pit and for ore blending. Using multiple benches also permits a balanced operation in which much of the overburden is removed from upper benches at the same time that ore is being mined from lower benches. It is undesirable to remove all of the overburden from an ore deposit before mining the ore because the initial cost of developing the ore is then extremely high.

The principal operations in mining are drilling, blasting, loading, and hauling. These operations are usually required for both ore and overburden. Occasionally the ore or the overburden is soft enough that drilling and blasting are not required.

Fig. 1. Roads and rounds that have developed open cuts on Cerro Bolivar in Venezuela. (*U.S. Steel Corp.*)

Rock drilling. Drilling and blasting are interrelated operations. The primary purpose of drilling is to provide an opening in the rock for the placing of explosives. If the ore or overburden is weak enough to be easily excavated without blasting, it is not necessary for it to be drilled. The cuttings from the hole are often used to fill it after the explosive charge has been placed at the bottom. Blast holes drilled in ore serve a secondary function in that the cuttings can be sampled and assayed to determine the mineral content of the ore. Sampling of drill holes is frequently done for ores not readily identified by visual means. *See* BORING AND DRILLING (MINERAL).

The basic methods of drilling rock are rotary, percussion, and jet piercing. The drill holes are usually vertical and range in size from 11/2 to 15 in. (3.8 to 38 cm) in diameter and are drilled to varying depths and spacings as required for the particular type of rock being mined. The drill hole depth ranges from 20 to 50 ft (6 to 15 m), depending on the bench height used in the mine. The drill hole spacing (distance between holes) is governed by the depth of the holes and the hardness of the rock and generally ranges from 12 to 30 ft (3.7 to 9 m). For short holes or hard rock, the hole spacing must be close; and for long holes and soft rock, the spacing may be increased.

Rotary drilling. Drilling is accomplished by rotating a bit under pressure. Compressed air is forced down the hollow drill shank or steel, and is allowed to escape through small holes in the bit for cooling the bit and blowing the rock cuttings out of the hole. Rotary drill holes range from a minimum diameter of 4 in. to a maximum diameter of 15 in. (10 to 38 cm) and are drilled by machines mounted on trucks or a crawler frame for mobility (Fig. 2). The weight of these machines ranges from 30,000 to 300,000 lb (13,600 to 136,000 kg) and, in general, the heavier machines are required for the larger-diameter holes. The rotary method is the most common type of drilling used in open-pit mines because it is the cheapest method of drilling; however, it is restricted to soft and medium-hard rock.

Percussion drilling. Percussion drilling is accomplished with a star-shaped bit which is rotated while being struck with an air hammer operated by high-pressure air. Air is also forced down the drill steel to cool the bit and to blow the cuttings out of the hole. Small amounts of water are frequently added with the air to reduce the dust. The hole diameters range from 11/2 to 9 in. (3.8 to 23 cm), depending upon the type of drill. The drilling machine used for the smaller-diameter holes (1 1/2 to 41/2 in. or 3.8 to 11.4 cm) is small and usually mounted on a lightweight crawler or rubber-tired frame weighing a few thousand pounds (1 lb = 0.45 kg). The air hammer and rotating device are mounted on a boom on the machine, and the hammer blows are transferred by the drill steel to the drill bit. For the larger-diameter and deeper holes, the air hammer is attached directly to the bit, and the drill unit is lowered down the vertical hole because the impact of the hammer blows is dissipated in the larger and longer columns of drill steel.

Fig. 2. Aerial view of rotary drill. *(Kennecott Copper Corp.)*

Percussion drilling is most applicable to brittle rock in the medium-hard to hard range. The depth of the smaller-diameter holes is limited to about 15 to 30 ft (4.6 to 9 m) but the larger holes can be drilled to 40 to 50 ft (12 to 15 m) without significant loss of efficiency.

*Jet piercing.*The jet piercing method was developed to drill the very hard iron ores (taconite). In this method, a hole is drilled by applying a high-temperature flame produced by fuel oil and oxygen to the rock. The holes range in size from 6 to 18 in. (15 to 46 cm), tend to be irregular in diameter, and require careful control of the flame to prevent overenlargement. The drilling machines are integrated units which control the fuel oil and oxygen mixture and lower the jet piercing bit down the hole. The jet piercing method is limited to very hard rock for which other types of drilling are more costly.

Blasting. The type and quantity of explosive are governed by the resistance of the rock to breaking. The primary blasting agents are dynamite and ammonium nitrate, which are detonated by either electric caps or a fuselike detonator called Primacord.

Dynamite is available in varying strengths for use with varying rock conditions. It is commonly used either in cartridge form or as a pulverzied, free-flowing material packed in bags. It can be used for almost any blasting application, including the detonation of ammonium nitrate, which cannot be detonated by the conventional blasting cap.

Commercial, or fertilizer-grade, ammonium nitrate has become a popular blasting medium because of its low cost and because it is safer to handle, store, and transport than most explosives. Granular or prill-size ammonium nitrate is commonly obtained in bulk, stored at the mine site in bins, and transported to the blasting site in hopper trucks that mix the agent and fuel oil as the explosive is discharged into the hole. It may also be obtained packed in paper, textile, or polyethylene bags. The carbon necessary for the proper detonation of ammonium nitrate is usually provided by the addition of the fuel oil. Granular or prilled ammonium nitrate is highly soluble and becomes insensitive to detonation if placed in water, and therefore its use is restricted to water-free holes. A portable blast hole dewatering pump is available for use with wet holes.

During the early 1960s ammonium nitrate slurries were developed by mixing calculated amounts of ammonium nitrate, water, and other ingredients such as TNT and aluminum. These slurries have

Fig. 3. A 27-yd³ (20.5-m³) shovel loading a 170-ton (153-metric ton) haulage truck. *(Kennecott Copper Corp.)*

several advantages when compared with dry ammonium nitrate: They are more powerful because of higher densities and added ingredients, and they can be used in wet holes. Further, the slurries are generally safer and less costly than dynamite, and are used extensively in mines which are wet or have rock difficult to blast with the less powerful ammonium nitrate. Both ammonium nitrate and slurries are detonated by a small charge of dynamite.

Mechanical loading. Ore and waste loading equipment in common use includes power shovels for medium to large pits and tractor-type front-end loaders for the smaller pits, or a combination of these. The use of scrapers is becoming more common for overburden removal and in some cases for ore mining. The loading unit must be selected to fit the transportation system, but because the rock will usually be broken to the largest size that can be handled by the crushing plant, the size and weight of the broken material will also have a significant influence on the type of loading machine. Of equal importance in determining the type of loading equipment is the required production or loading rate, the available working room, and the required operational mobility of the loading equipment.

Power shovels. Shovels in open-pit mines range from small machines equipped with 2-yd³ (1.5-m³) buckets to large machines with 36-yd³ (27.5-m³) buckets. A 6-yd³ (4.6-m³) shovel will, under average conditions, load about 6000 tons (5400 metric tons) per shift, while a 12-yd³ (9.2-m³) shovel will load about 12,000 tons (10,900 metric tons) per shift. However, the nature of the material loaded will have a significant effect on productivity. If the material is soft or finely broken, the shovel productivity will be high; but if the material is hard or poorly broken with a high percentage of large boulders, the shovel production will be adversely affected. Power may be derived from diesel or gasoline engines or diesel electric or electric motors. The use of diesel or gasoline engines for power shovels is usually limited to shovels of up to 4-yd³ (3.1-m³) capacity, and diesel electric drives are not common in shovels of more than 6-yd³ (4.6-m³) capacity. Electric drives are used in shovels ranging up from 4-yd³ (3.1-m³) capacity and are the most widely used power sources in the open-pit mining operations (Figs. 3 and 4).

Draglines. A dragline is similar to a power shovel but uses a much longer boom. A bucket is suspended by a steel cable over a sheave at the end of the boom. The bucket is cast out toward the end of the boom and is pulled back by a hoist to gather a load of material which is deposited in an ore haulage unit or on a waste pile. Draglines range in size from machines with buckets of a few cubic yards' capacity to machines with buckets of 150-yd³ (115-m³) capacity. Draglines are extensively used in the phosphate fields of Florida and North Carolina. The overburden and the ore (called matrix in phosphate operations) are quite soft, and no blasting is required. The phosphate ore is deposited in slurrying pits, where it is mixed with water and transported hydraulically to the concentrating plant.

Front-end loaders. Front-end loaders are tractors, both rubber-tired and track-type, equipped

Fig. 4. An 8-yd^3 (6.2-m^3) shovel loading an ore train. *(Kennecott Copper Corp.)*

with a bucket for excavating and loading material (Fig. 5). The buckets are usually operated hydraulically and range in capacity from 1 to 36 yd^3 (0.76 to 27.5 m^3). Front-end loaders are usually powered by diesel engines and are much more mobile and less costly than power shovels of equal capacity. On the other hand, they are not generally as durable as a power shovel, nor can they efficiently excavate hard or poorly broken material. A front-end loader has about one-half the productivity of a power shovel of equal bucket capacity; that is, a 10-yd^3 (7.6-m^3) front-end loader will load about the same tonnage per shift as a 5-yd^3 (3.8-m^3) shovel. However, the loader is gaining in popularity where mobility is desirable and where the digging is relatively easy. *See* BULK HANDLING MACHINES.

Mechanical haulage. The common modes of transporting ore and waste from open-pit mines are trucks, railroads, and belt conveyors. The application of these methods or combination of them depends upon the size and depth of the pit, the production rate, the length of haul to the crusher or dumping place, the maximum size of the material, and the type of loading equipment used.

Truck haulage. Truck haulage is the most common means of transporting ore and waste from open-pit mines because trucks provide a more versatile haulage system at a lower capital cost than rail or conveyor systems in most pits (Fig. 6). Further, trucks are sometimes used in conjunction with rail and conveyor systems where the haulage distance is greater than 2 or 3 mi. (3 or 5 km). In these cases, trucks haul from the pit to a permanent loading point within the pit or on the surface, where the material is transferred to one or the other system. Trucks range in size from 20- to 350-ton (18- to 318-metric ton) capacity and are powered by diesel engines. Mechanical drives are used almost exclusively in trucks up to 75-ton (68-metric ton) capacity. Between 25- and 100-ton (23- and 91-metric ton) capacity, the drive can be either mechanical or electrical. In the larger trucks, the

Fig. 5. A 3-yd^3 (2.3-m^3) front-end loader loading a haulage truck in a small open-pit mine. *(Eaton Corp.)*

electric drive, in which a diesel engine drives a generator or alternator to provide power for electric motors mounted in the hubs of the wheels, is most common. Locomotive traction motors are used in some trucks of over 200-ton (181-metric ton) capacity. The diesel engines in the conventional drives range from 175 to 1200 hp (130 to 900 kW), while the engines in the diesel electric units range from 700 to 2400 hp (520 to 1800 kW). Most haulage trucks can ascend road grades of 8 to 12% fully loaded and are equipped with various braking devices, including dynamic electrical braking, to permit safe descent on steep roads. Tire cost is a major item in the operating cost of large trucks, and the roads must be well designed and maintained to enable the trucks to operate efficiently at high speed.

The size of trucks used is primarily dependent upon the size of the loading equipment, but the required production rate and the length of haul are also factors of consideration. Usually 20- to 40-ton-capacity (18- to 36-metric ton) trucks are used with 2- to 4-yd³ (1.5- to 3.1-m³) shovels, while the 70- to 100-ton (64- to 91-metric ton) trucks are used with 8- to 10-yd³ (6.1- to 7.6-m³) shovels.

In the early 1960s major advances were made in truck design which put the truck haulage system in a favorable competitive position compared with the other methods of transportation. Trucks are used almost exclusively in small and medium-size pits and the majority of the large pits, but in some cases are used in conjunction with rail or conveyor haulage. Trucks have the advantage of versatility, mobility, and low cost when used on short-haul distances.

Rail haulage. When mine rock must be transported more than 3 or 4 mi (5 or 6 km), rail haulage is generally employed. Since rail haulage requires a larger capital outlay for equipment than other systems, only a large ore reserve justifies the investment. As a rough rule of thumb, the reserve should be large enough to support for 25 years a production rate of 30,000 tons (27,000 metric tons) of ore per day and an equivalent or greater tonnage of stripping. Adverse grades should be limited to a maximum of 3% on the main lines and 4% for short distances on switchbacks. Good track maintenance requires the use of auxiliary equipment such as mechanical tie tampers and track shifters. The latter are required for relocating track on the pit benches as mining progresses and on the waste dumps as the disposed material builds up adjacent to the track. Ground movement in the pit resulting from disturbance of the Earth's crust or settling of the waste dumps makes track maintenance a large part of mining cost.

Locomotives in use range from 50 to 140 tons (45 to 127 metric tons) in weight, with the largest sizes coming into increased use for steeper grades and larger loads. Most mines operate either all electric or diesel electric models. The use of all-electric locomotives creates the problem of electrical distribution in the pit and on the waste dumps, and requires the installation of trolley lines adjacent to all tracks (Fig. 7).

Mine cars range in capacity from 50 to 100 tons (45 to 91 metric tons) of ore, and to 50 yd³ (39 m³) of waste. Ore is transported in various types of cars: solid-bottom, side-dump, or bottom-dump. The solid-bottom car is cheapest to maintain but requires emptying by a rotary dumper. Waste is mostly handled by the side-dump car. Truck haulage has replaced much of the rail haulage in recent years because of advances in truck design and reduction in truck haulage costs.

Conveyors. Rubber belt conveyors may be used to transport crushed material from the pit at slope angles up to 20°. Conveyors are especially useful for transporting large tonnages over rugged terrain and out of pits where ground conditions preclude building of good haulage roads, and where long haul distances are required. Improved belt design is permitting greater loading rates, higher speeds, and the substitution of single-flight for multiple-flight installations. The chief disadvantage of this transport system is that, to protect the belt from damage by large lumps, waste as well as ore must be crushed in the pit before loading on the belt.

Waste disposal problems. To keep costs at a minimum, the dump site must be located as near the pit as possible. However, care must be taken to prevent location of waste dumps above possible future ore reserves. In the case of copper mines, where the waste contains quantities of the metal which can be recovered by leaching, the ground on which such waste is deposited must be impervious to leach water. Where the creation of dumps is necessary, problems of possible stream pollution and the effect on farms and on real estate and land values must all be considered.

Slope stability and bench patterns. In open-cut and open-pit mining, the material ranges from unconsolidated surface debris to competent rock. The slope angle, that is, the angle at which the benches progress from bottom to top, is limited by the strength and characteristics of the material. Faults, joints, bedding planes, and especially groundwater behind the slopes are known to decrease the effective strength of the material and contribute to slides. In practice, slope angles vary from 22 to 60° and under normal conditions are about 45°. Steeper slopes have a greater tendency to fail, but may be economically desirable because

Fig. 6. Copper mine waste is dumped from a 170-ton (153-metric ton) haulage truck. (*Kennecott Copper Corp.*)

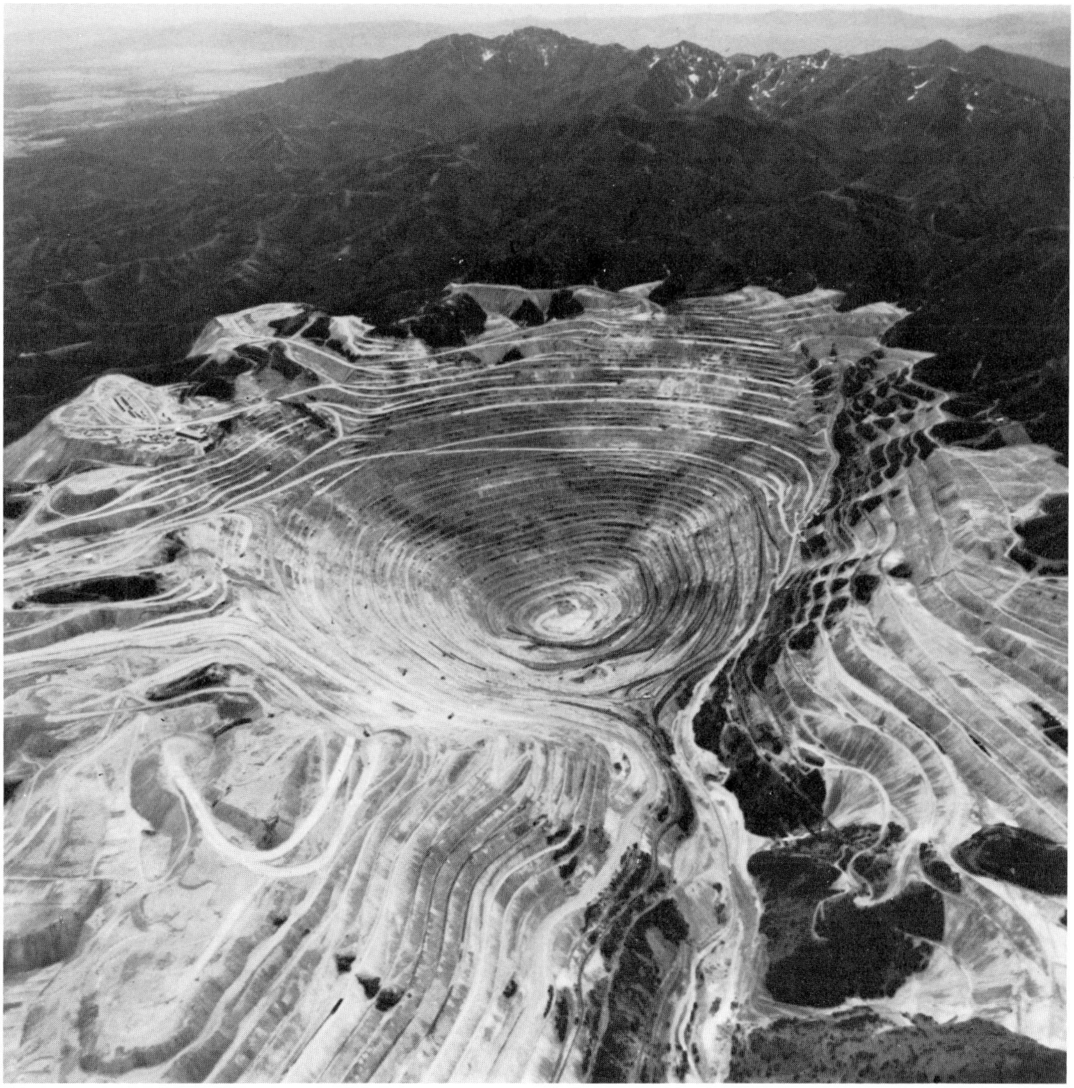

Fig. 7. Aerial view of Kennecott's Bingham Mine. (*Kennecott Copper Corp.*)

they lower the quantity of waste material that must be removed to provide access to the ore.

Technology has been developed that permits an engineering approach to the design of slopes in keeping with measured rock and water conditions, making quantitative estimates of the factor of safety of a given design possible. Precise instrumentation can detect the boundaries of moving rock masses and the rate of their movement. Instrumentation to give warning of impending failure is being used in some pits. *See* ENGINEERING GEOLOGY.

Communication. Efficiency of mining operations, especially loading and hauling, is being improved by the use of communication equipment. Two-way, high-frequency radiophones are proving useful for communicating with haulage and repair crews and shovel operators.

[CARL D. BROADBENT]

Operations research

The application of scientific methods and techniques to decision-making problems. A decision-making problem occurs where there are two or more alternative courses of action, each of which leads to a different and sometimes unknown end result (Fig. 1). Operations research is also used to maximize the utility of limited resources. The objective is to select the best alternative, that is, the one leading to the best result. Often, however, it is not simply a matter of searching a table, as

alternatives	resulting end states θ_1	θ_2	θ_3	θ_4	θ_5
a_1	r_{11}	r_{12}	r_{13}	r_{14}	r_{15}
a_2	r_{21}	r_{22}	r_{23}	r_{24}	r_{25}
a_3	r_{31}	r_{32}	r_{33}	r_{34}	r_{35}
a_4	r_{41}	r_{42}	r_{43}	r_{44}	r_{45}

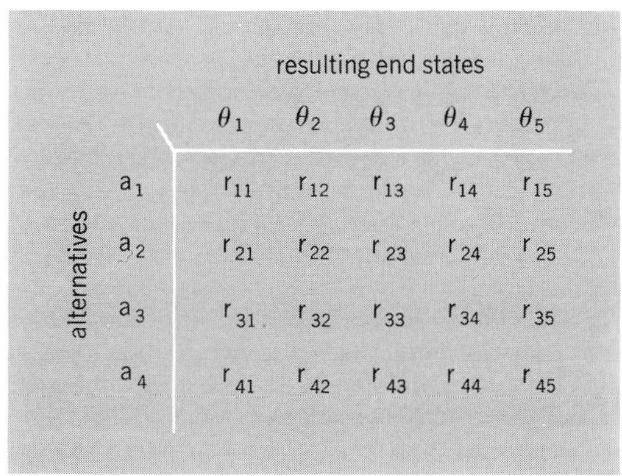

Fig. 1. End states for alternatives in a decision-making process.

there may literally be an infinity of outcomes. More intelligent means are needed to seek out the prime result.

To put these definitions into perspective, the following analogy might be used. In mathematics, when solving a set of simultaneous linear equations, one states that if there are seven unknowns, there must be seven equations. If they are independent and consistent and if it exists, a unique solution to the problem is found. In operations research there are figuratively "seven unknowns and four equations." There may exist a solution space with many feasible solutions which satisfy the equations. Operations research is concerned with establishing the best solution. To do so, some measure of merit, some objective function, must be prescribed.

In the current lexicon there are several terms associated with the subject matter of this program: operations research, management science, systems analysis, operations analysis, and so forth. While there are subtle differences and distinctions, the terms can be considered nearly synonymous.

The field can be divided into two general areas with regard to methods. These are those that can be termed mathematical programming and those associated with stochastic processes. While computers are heavily used to solve problems, the term programming should not be considered in that sense, but rather in the general sense of organizing and planning. Also, the tools of probability and statistics are used to a considerable extent in working with stochastic processes. These areas will be explored in greater detail in a later section. With regard to areas of applications, there are very few fields where the methods of operations research have not been tried and proved successful. Following is a brief history of the field, and then the general approach to solving problems.

HISTORY

While almost every art and every science can reach back into antiquity for its roots, operations research can reach back only a half-century to find its beginnings. During World War I, F. W. Lancaster developed models of combat superiority and victory based on relative and effective firepower. Thomas Edison studied antisubmarine warfare, and in 1915 F. Harris derived the first economic order quantity (EOQ) equation for inventory. Starting in 1905 and continuing into the 1920s, A. Erlang studied the flow of calls into a switchboard and formed the basis of what is now known as queueing theory.

Empiricists. The formal beginning of operations research was in England during World War II, where the term was and still is operational research. Early work concerned air and coastal defense—the coordination of fighter aircraft, antiaircraft guns, barrage balloons, and radar. Typical of the research groups formed was "Blackett's Circus." This interdisciplinary group consisted of three physiologists, two math physicists, an astrophysicist, an army officer, one surveyor, one general physicist, and two mathematicians. The basic mode of operation was to observe the problem area, and then call on the expertise of the various disciplines to apply methods from other sciences to solve the particular problem. In retrospect, this was an era of "applied common sense," and yet it was novel and highly effective.

Pragmatists. Following World War II, operations research continued to exist mainly in the military area. The operations research groups formed during the war stayed together, and a number of civilian-staffed organizations were established—RAND (1946), ORO (1948), and WSEG (1948). The real impetus to this era and to the whole field was the work done by George Dantzig and colleagues at RAND on Project Scope, undertaken for the U.S. Air Force in 1948. In attacking the problem of assigning limited resources to almost limitless demands, they developed the techniques of linear programming. Perhaps no other method is more closely associated with the field. Its use quickly spread from the military to the industrial area, and a new dimension was added to operations research. No longer was it simply "observe, analyze, and try." For the first time the field became "scientific." It could now "optimize" the solutions to problems.

It was during this "dynasty" that formal courses in operations research were first offered. It was also during this time that operations research suffered its first lapse. Linear programming soon was looked to as the cure for too many of industry's ills. Unfortunately, industry's problems were not all linear, and "straightening them out" to fit resulted in many aborted projects and reports that were simply shelved instead of implemented. The early 1950s saw a growth in the number of industrial operations research groups; by the end of the decade, many had disappeared.

Theorists. As if in reaction to the failures of the third dynasty, toward the end of the 1950s a number of highly skilled scientists emerged who made some substantial contributions to the field. Operations research came of age and matured, and its adherents now strove for respectability. In fact, the movement was so far advanced toward developing a sound theoretical base that a new problem arose—the practicality gap.

METHODOLOGY

Operations research today is a maturing science rather than an art—but it has outrun many of the decision makers it purports to assist. The success of operations research, where there has been success, has been the result of the following six simply stated rules:

1. Formulate the problem.
2. Construct a model of the system.
3. Select a solution technique.
4. Obtain a solution to the problem.
5. Establish controls over the system.
6. Implement the solution (Fig. 2).

The first statement of the problem is usually vague, inaccurate, and sometimes not a statement of the problem at all. Rather, it may be a cataloging of observable effects. It is necessary to identify the decision maker, the alternatives, goals, and constraints, and the parameters of the system. At times the goals may be many and conflicting; for example: Our goal is to market a high-quality product for the lowest cost yielding the maximum profit while maintaining or increasing our share of the market through diversifications and acquisitions;

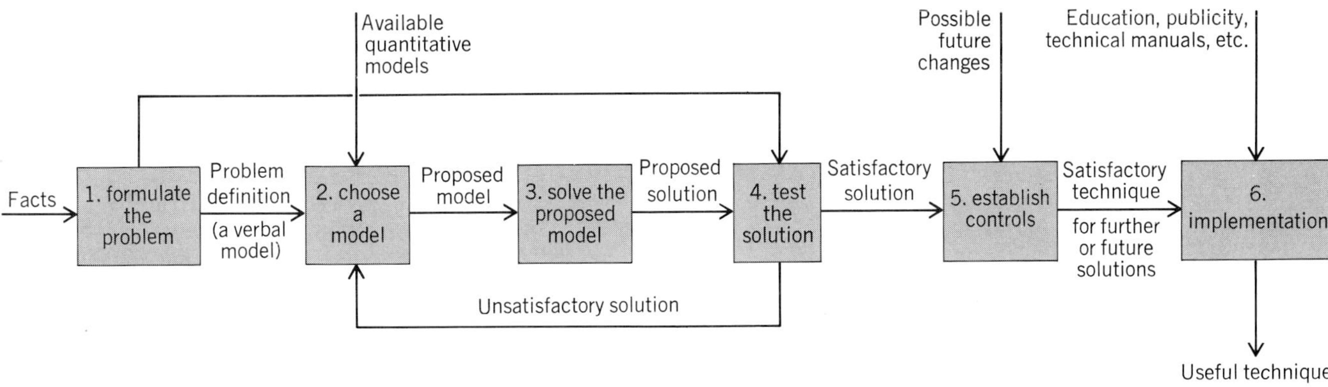

Fig. 2. Operations research approach—the six basic rules of success.

yielding a high dividend to our stockholders while maintaining high worker morale through extensive benefits, without the Justice Department suing us for constraint of trade, competition, price fixing, or just being too big.

More properly, a statement of the problem contains four basic elements that, if correctly identified and articulated, greatly eases the model formulation. These elements can be combined in the following general form: "Given (the system description), the problem is to optimize (the objective function), by choice of the (decision variable), subject to a set of (constraints and restrictions)."

In modeling the system, one usually relies on mathematics, although graphical and analog models are also useful. It is important, however, that the model suggest the solution technique, and not the other way around. Forcing a model to fit a preferred technique led to some of the bad operations research of the third dynasty.

With the first solution obtained, it is often evident that the model and the problem statement must be modified, and the sequence of problem-model-technique-solution-problem may have to be repeated several times. The controls are established by performing sensitivity analysis on the parameters. This also indicates the areas in which the data-collecting effort should be made.

Implementation is perhaps of least interest to the theorists, but in reality it is the most important step. If direct action is not taken to implement the solution, the whole effort may end as a dust-collecting report on a shelf. Given the natural inclination to resist change, it is necessary to win the support of the people who will use the new system. To do this, several ploys may be used. Make a member of the using group also a member of the research team. (This provides liaison and access to needed data.) Educate the users about what the "black box" does—not to the extent of making them experts, but to alleviate any fear of the unknown. Perhaps the major limitation to successful use of operations research lies in this phase.

MATHEMATICAL PROGRAMMING

Probably the one technique most associated with operations research is linear programming. The basic problem that can be modeled by linear programming is the use of limited resources to meet demands for the output of these resources. This type of problem is found mainly in production systems, but definitely is not limited to this area. Since this method is so basic to the operations research approach, its use will be illustrated.

Linear program model. Consider a company that produces two main products—X and Y. For every unit of X it sells, it gets a $10 contribution to profit (selling price minus direct, variable costs), and for Y it gets a $15 contribution. How many should they sell of each? Obviously there must be some limitations—on demand and on productive capacity. Suppose they must, by contract, sell 50 of X and 10 of Y, while at the other end the maximum sales are 120 of X and 90 of Y. Unfortunately they have only 40 total hours of productive capacity per period, and it takes 0.25 h to make an X and 0.4 h to make a Y. What is their optimal strategy? First, the problem must be formally stated: "Given a production system making two products, the problem is to maximize the contribution to profit, by the choice of how many of each product to make, subject to limits on demand (upper and lower) and available production hours."

If X equals the number of first products made and Y the number of the second products made, the system can be modeled as follows:

$$\begin{aligned}
&\text{Maximize} \\
&\quad \text{Profit contribution} = 10X + 15Y \\
&\text{Subject to} \quad X \geq 50 \\
&\qquad X \leq 120 \\
&\qquad Y \geq 10 \\
&\qquad Y \leq 90 \\
&\qquad .25X + .40Y \leq 40
\end{aligned}$$

The first two constraints are the lower and upper bounds of demand on X, the next two are for Y, and the last constraint refers to the productive capacity.

The problem is illustrated in Fig. 3. Any point in the feasible region of solution and the edges will satisfy all the demand and capacity constraints. To pick the best, the objective—"maximize contribution to profit"—is used. Several isoprofit lines, that is, lines where the profit is a constant value, have been added to Fig. 3. Note that the $600 line is below the feasible region, the $1800 line is above it, and the $1200 line runs through it. If other lines were formed between $1200 and $1800, one could graphically find the maximum-profit-level line that still had one point in the feasible region. This

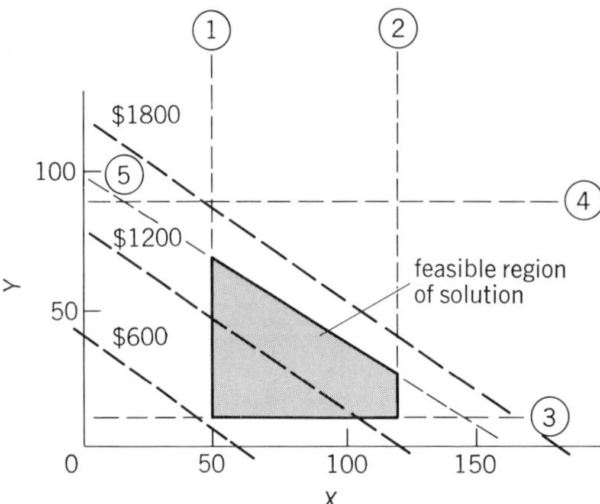

Fig. 3. Graphical presentation of production problem; circled numbers refer to the constraints.

would occur at the intersection of the second and fifth constraints. At this point, the solution is:

$$\begin{aligned} \text{Profit contribution} &= \$1575 \\ X &= 120 \\ Y &= 25 \end{aligned}$$

It is fairly easy to verify that this is the optimal solution, that all the constraints are satisfied, and that no larger profit can be found.

While this simple example did not pose much of a computational problem, actual cases do, in that there may be thousands of variables and thousands of constraints. Problems of scheduling the product mix from a refinery where a blend of different crudes can be used (each with different costs and properties) are in this category. Efficient algorithms and computer codes have been developed to solve these problems.

Other models. There are a number of other linear programming models that relate to specific types of problems. However, advantage is taken of the special structure of the model to develop special and more efficient methods of solution.

1. Transportation problem. Given a set of sources (as, for example, factories) and a set of demands (such as regional warehouses), the problem is to minimize the cost of transporting product to destination, by choice of how much will be supplied from each source to each destination, subject to limits on demand and capacity.

2. Assignment problem. Given a set of tasks to be performed and a set of workers (machines), the problem is to maximize the overall efficiency by choice of assignment of task to worker, subject to the constraint that all tasks must be done and a worker can do at most one task.

Network models. There is another set of linear programming models that fall under the general category of network models:

The shortest-path problem determines the shortest path from one node to another. (The reverse of this is the critical path method used in project management.) *See* CRITICAL PATH METHOD (CPM).

The max flow problem determines the capacity of a network such as a pipeline or highway system.

The min cost problem is a variation of the transportation model.

Integer linear programming. There is a class of problems that have the linear programming form, but in which the variables are limited to be integer values. These represent the most difficult set to solve, especially where the variables take on values of 0 or 1. (This problem occurs, for instance, in project selection—the project is either funded, 1, or not funded, 0.) While a number of algorithms have been developed, there is no assurance that a given problem will be solved in a reasonable amount of computer time. These algorithms fall into two general categories—the cutting plane method, which adds constraints that cut off noninteger solution points, and the branch and bound method, which examines a tree network of solution points.

Nonlinear programming. Another category of problems arise where either the objective function or one or more of the constraints, or both, are nonlinear in form. Again, a series of methods have been developed that have varying degrees of success, depending on the problem structure.

Geometric programming. One of the new techniques developed relates to a certain class of nonlinear models that use the arithmetic-geometric mean inequality relationship between sums and products of positive numbers. Such models result from modeling engineering design problems, and at times can be solved almost by inspection. Because of the ease of solution, a considerable effort has been made to identify the various problems that can be structured as a geometric programming model.

Dynamic programming. This technique is not as structured as linear programming, and more properly should be referred to as a solution philosophy rather than a solution technique. Actually, predecessors of this philosophy have been known for some time under the general classification of calculus of variation methods. Dynamic programming is based on the principle of optimality as expressed by Richard Bellman: "An optimal policy has the property that whatever the initial state and initial decision are, the remaining decision must constitute an optimal policy with regard to the state resulting from the first decision."

The operative result of this principle is to start at the "end" of the problem—the last stage of the decision-making process—and "chain back" to the beginning of the problem, making decisions at each stage that are optimal from that point to the end. To illustrate this process, the fly-away-kit problem (also known as the knapsack problem) will be briefly described: Given a set of components, each with a unit weight and volume, the problem is to maximize the value of units carried to another location, by choice of the number of each to be taken, subject to limitations on volume and weight that can be carried.

In this problem each component will represent a stage at which a decision is to be made (the number to be included in the kit), and the amount of volume and weight left are defined as state variables. The end of the problem, then, is where there is only one more component to consider (Fig. 4*a*).

Usually at this stage the solution is almost trivial. As many of the last components (X_1) are selected as can be within the limits of available volume (V_1) and weight (W_1). In practice, all possible values of these two state variables are solved for, since it is not known how much will be left when the last stage is reached. To this solution is now

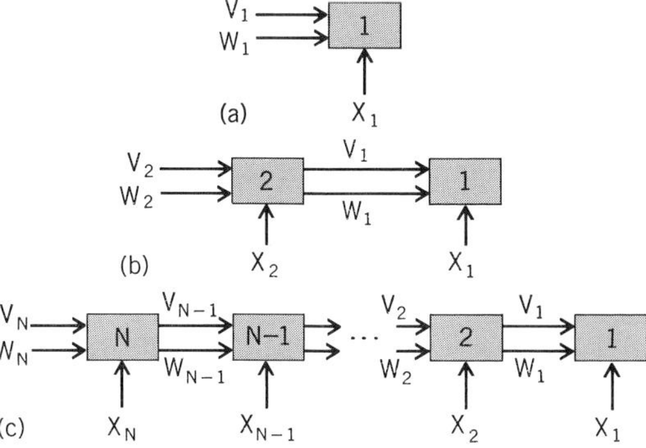

Fig. 4. Dynamic programming problems. (*a*) Final decision stage. (*b*) Two-stage problem. (*c*) "N"-stage problem.

"chained" the problem of how to select the second-last component (Fig. 4*b*).

Here one considers the problem that, for each level of volume and weight, one must choose between taking one or more of component 2 or passing it on to the last stage. Combinations of components 1 and 2 must now be looked at to find the best mix, considering the available resources. In like manner, one "chains back" to the beginning of the problem (Fig. 4*c*).

The measures of value can be in any terms. If the components are simple cargo, they could have a monetary value. If they are spare parts for some system, a reliability measure could be used.

STOCHASTIC PROCESSES

A large class of operations research methods and applications deals with stochastic processes. These processes can be defined as where one or more of the variables take on values according to some, perhaps unknown, probability distribution. These are referred to as random variables, and it takes only one to make the process stochastic.

In contrast to the mathematical programming methods and applications, there are not many optimization techniques. The techniques used tend to be more diagnostic than prognostic; that is, they can be used to describe the "health" of a system, but not necessarily how to "cure" it. This capacity is still very valuable.

Queueing theory. Probably the most studied stochastic process is queueing. A queue or waiting line develops whenever some customer seeks some service that has limited capacity. This occurs in banks, post offices, doctors' offices, supermarkets, airline check-in counters, and so on. But queues can also exist in computer centers, repair garages, planes waiting to land at a busy airport, or at a traffic light.

Queueing theory is the prime example of what can be said about the state of the system but not how to improve it. Fortunately, the improvements can be made by increasing the resources available. For example, another teller opens up a window in a bank, or another check-out stand is opened in the supermarket. Other possible changes are more subtle. For example, the "eight items or less" express lane in a supermarket minimizes the frustration of a customer in line behind another with two full shopping carts. Some post offices and banks have gone to a single queue where the person at the head of the line goes to the next available window.

While it is not possible to generalize the results of queueing analysis, it is possible to provide some general measures. One is the load or traffic factor. This is the ratio of arrival rate to combined service rate, considering the number of service centers in operation. It is possible to plot queueing statistics against this factor as shown in Fig. 5.

When the load factor (ρ) is low, there is excess serving capacity. There may be occasional lines, but not often. As the load factor rises, a fairly linear rise is obtained in any queueing statistics until approximately the point where $\rho = 0.75$. After this there is a sharp and continuous rise in the lengths of line and waits. The system is becoming saturated. If $\rho = 1.00$, the best that can be said is, "the larger the line, the larger the wait."

The mathematics of the probability distributions of arrival rates and service times often define closed-form solutions, that is, being able to directly solve for an answer. In these cases another technique has proved very useful.

Simulation. Simulation is defined as the essence of reality without reality itself. With stochastic processes, values are simulated for the random variables from their known or assumed probability distributions, a simpler model is solved, and the process is repeated a sufficient number of times to obtain statistical confidence in the results.

As a simple example of this, assume that someone makes the statement that average female student at a university is of a certain height. To verify this assumption the actual heights of all female students could be measured, but this may be a long process. As an alternative, a hopefully random sample may be taken, the heights measured, the average determined, and an inference made as to the whole female population.

Many stochastic processes do not lend themselves to such a direct approach. Instead an assumption is made as to what the underlying probability distributions of the random variables are, and these are sampled. This sampling is done with random numbers. True random numbers are difficult to obtain, so pseudorandom numbers are generated by some mathematical relationship. This apparent paradox is justified by the fact that the numbers appear to be random and pass most tests

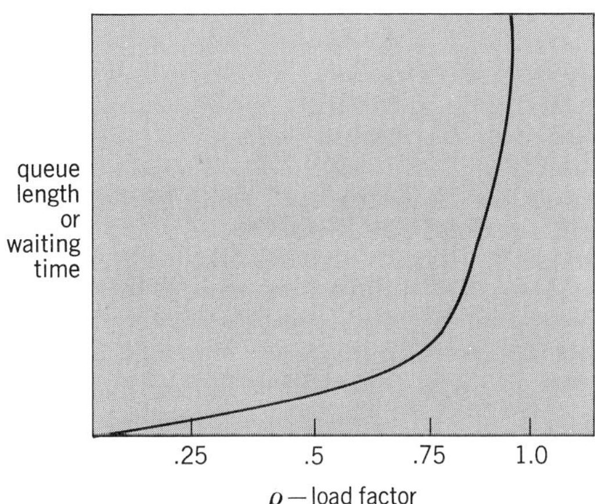

Fig. 5. Plot for general queues relating load factor to serving capacity.

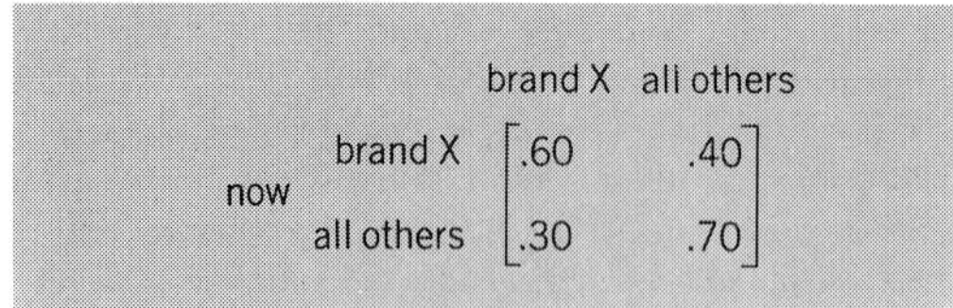

Fig. 6. Probability matrix for brand-switching model.

for randomness. As an example, a queueing system could be modeled by having two wheels of fortune—one with numbers in random order from 1 to 15, and the other from 1 to 20. The first could represent the number of arrivals into the system per 10-min period, and the other the number served during the same period. In this manner a system with a load factor of 0.75 could be simulated. Many simulation computer languages have been developed to ease the work of analyzing queueing systems.

Markov processes. This class of stochastic processes is characterized by a matrix of transition probabilities. These measure the probability of moving from one state to another. Such methods have been used to determine the reliability of a system, movement of a stock in the stock market, aging of doubtful accounts in credit analysis, and brand switching. This last application will be illustrated.

A company is considering marketing a new product. A preliminary market survey indicates that if a customer uses the product, there is a 60% probability that he or she will continue to do so. Likewise, the advertising campaign is such that 70% of those using all other products will be tempted to switch. The question is, what market share will the product be expected to finally obtain? The probability matrix is shown in Fig. 6.

Initially the market share of brand X is 0%, but after one period this will rise to 30%; that is, the advertising campaign will induce 30% to try brand X. In the next period, only 60% of these 30% will continue to purchase brand X, but 30% of the other 70% will switch, for a total market share of 39%.

Period by period, there will be transitions from state to state, but soon the variations will dampen out and a steady state will be achieved. This can be found for this problem by matrix value, and the final market share for brand X is 42.8%.

Decision trees. While not originated by the field of operations research and not strictly in its domain, decision trees are an important tool in the analysis of some stochastic processes. More properly, they are a part of what may be considered statistical decision making. Their use can be illustrated with an example from the oil industry.

An oil company is developing an oil field and has the option to lease the mineral rights on an adjacent block of land for $100,000. An exploratory well can be drilled at a cost of $250,000. If the well is considered moderate or good, it will cost an additional $50,000 to complete before production can start.

For simplicity, one can assume that the well, if good, has either moderate or good production. Also, the present worth of the net returns on all the future production is $1,000,000 and $3,500,000 for these two states. The problem facing the oil company is what they should do.

To analyze this problem, some subjective probabilities must be estimated. Assume the probability that the well is dry is 70%, that the flow is moderate is 20%, and that it is good is 10%. With these percentages, a decision tree can be constructed (Fig. 7).

By a process of "folding back" the values and probabilities, the expected value at each decision point is seen to be represented by a square in Fig. 7. Actually, while there are three sequential decisions—lease, drill, and complete the well—once the decision to lease is made, the sequence is fixed unless the well is discovered to be dry. The expected value of this decision can be calculated as follows:

$$
\begin{aligned}
E\,[\text{lease} - \text{drill}] &= -100{,}000 - 250{,}000 + 0.70\,(0) \\
&\quad + 0.20\,(-50{,}000 + 1{,}000{,}000) \\
&\quad + 0.10\,(-50{,}000 + 3{,}500{,}000) \\
&= \$185{,}000
\end{aligned}
$$

Note that this expected value is slightly positive. In reality one of three things will happen—the well is dry, and the loss is $350,000; the well is moderate, with a net gain of $600,000; or the well is good, with a net gain of $3,100,000. This expected value can be interpreted as follows: if a large number of wells are drilled, 70% will be dry at a loss of $350,000 each, 20% will be moderate, and 10% will be good. The average gain will be $185,000.

SCOPE OF APPLICATION

There are numerous areas where operations research has been applied. The following list is not intended to be all-inclusive, but is mainly to illustrate the scope of applications: optimal depreciation strategies; communication network design; computer network design; simulation of computer time-sharing systems; water resource project selection; demand forecasting; bidding models for offshore oil leases; production planning; assembly line balancing; job shop scheduling; optimal location of offshore drilling platforms; optimal allocation of crude oil using input-output models; classroom size mix to meet student demand; optimizing waste treatment plants; risk analysis in capital budgeting; electric utility fuel management; public utility rate determination; location of ambulances;

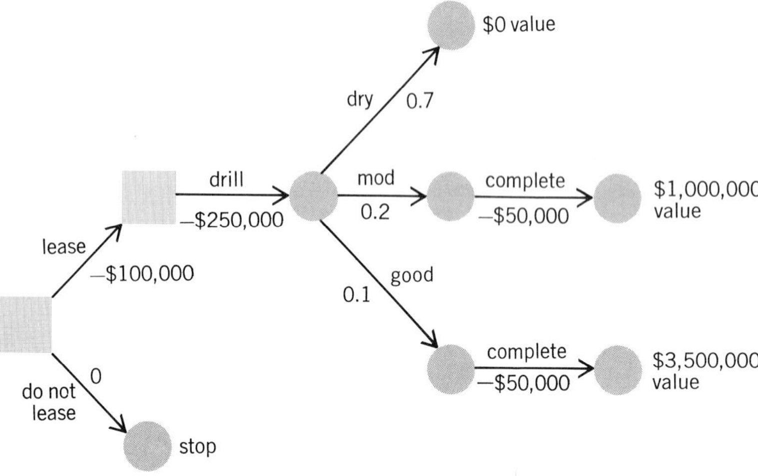

Fig. 7. Decision tree for an oil well problem.

optimal staffing of medical facilities; feedlot optimization; minimizing waste in the steel industry; optimal design of natural-gas pipelines; economic inventory levels; random jury selection; optimal marketing-price strategies; project management with CPM/PERT/GERT; air-traffic-control simulations; optimal strategies in sports; system availability/reliability/maintainability; optimal testing plans for reliability; optimal space trajectories.

It can be seen from this list that there are few facets of society that have not had an application for operations research. *See* GERT; PERT.

[WILLIAM G. LESSO]

Bibliography: R. E. Bellman and S. E. Dreyfus, *Applied Dynamic Programming*, 1962; V. N. Bhat, *Elements of Applied Stochastic Processes*, 1969; F. S. Hillier and G. Lieberman, *Introduction to Operations Research*, 3d ed., 1980.

Operator training

The specialized education of an organization's employees in the general knowledge, specific skills, and overall mental attitude required to do their jobs effectively. It is regarded as so important to the continued soundness of an enterprise that it is not considered as a luxury or fringe benefit but rather as an essential function. As science advances and technology becomes more complex, competent and continuous training increases in importance.

Purposes. The objective of operator training is to enable the operator to perform the job in a manner which is satisfactory to the employer and satisfying to the employee. It should contribute to increased output, productivity, quality, pride of workmanship, and morale and to decreased errors, rejects, rework, waste, accidents, injuries, equipment downtime, unit costs, frustration, absenteeism, and labor turnover.

Scope. Operator training can be of new employees and can include introduction to the company, orientation to the work situation, indoctrination in the rules and regulations of the firm, instruction in specific job skills, knowledge of particular subjects, and attitude improvement. Training can also be directed to existing employees to prepare them for job transfers or promotions, to maintain or upgrade their existing skills levels, to keep them current on new tools, techniques, or technologies, to aid in their personal or professional growth, or to improve their morale and attitudes. Training can also be of employees who have become injured, aged, or infirm in some way so that they can no longer do the job for which they were originally trained. In some cases, a cutback in a company's work force will result in a reshuffling of the jobs of those workers retained, and a certain amount of retraining will be required. In addition, all workers at all levels may be trained in special subjects, such as quality control, safety, first aid, human relations, communications, new products, policies, and other topics of particular importance to the company.

Productive industrial work requires the presence of at least six elements: (1) proper tools, machines, and equipment; (2) proper raw materials and supplies; (3) an operator skilled in the work to be done; (4) an environment suitable for the above three elements to operate; (5) correct instructions clearly communicated to the operator; and (6) motivation on the operator's part to perform.

Generally, the more mechanized or automated a process, the less skill the operator requires, but the more skill the engineers, designers, builders, and maintainers of the machines require. It is important therefore to match the skills required by the process to those possessed by the operator—both in the present and into the forseeable future, since changes are to be expected.

A distinction may be made between training and development. Training may be said to be the imparting to the operator the correct skills, knowledge, and attitude to do the present or immediately prospective job. Development may be said to be the more general education and upgrading of the employee in subjects not as much required in the present job but anticipated to be needed in future work, usually an advancement or a promotion. Human resources development (HRD) is a term used by some firms to encompass these functions.

Training decisions. The following decisions must be made and implemented in order to have a successful operator-training program: job definition, worker identification, needs determination, program design, mode selection, trainer selection, program preparation, trainee preparation, organizational decisions, and administrative decisions.

Job definition. The first thing that must be accomplished as a basis for any subsequent operator training is to analyze the jobs, work, activities, and operations to be done to identify and define the skills and other attributes required. The needed operator abilities must be defined in sufficient detail. Rather than merely stating that the job requires an ability to weld, for instance, the definition should state that the job requires an ability to butt- or lap-weld $1/8$- to $1/2$-in. (3- to 13-mm) thick alloy steel plates using oxyacetylene equipment at a stated speed or output rate and at a stated quality or inspection test level. It is important to note that the job skills being defined may or may not be for an immediate opening. It may be in anticipation of an operator being promoted, released, or retired. It may be in anticipation of a skilled-labor shortage, of an expansion, of a new product line, or of a new technology. It any case, the required skills must be quantified, qualified, and placed in an order of priority and in a time frame of when needed. The required skills to perform a job may be shown graphically (Fig. 1).

The bar chart of Fig. 1 shows that the particular operation requires at least eight essential skills in

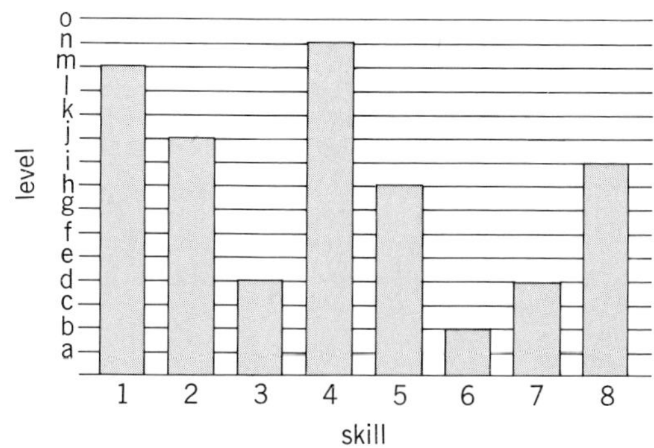

Fig. 1. Skill levels required to perform a given job.

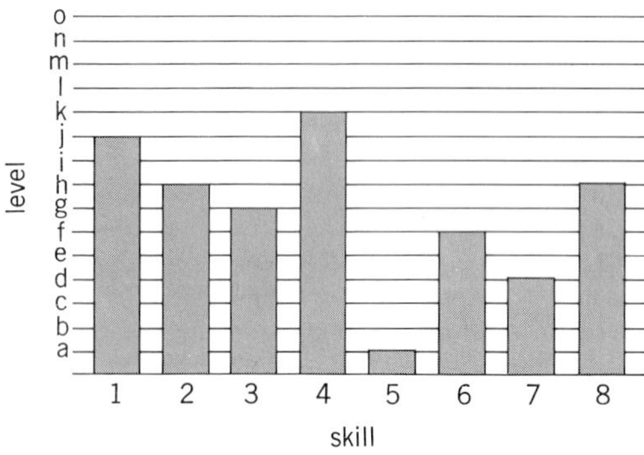

Fig. 2. Skill levels possessed by a candidate operator.

varying degrees of advancement or competence. Priority #1 (bar 1) might be a certain (qualified) level of proficiency in operating a turrent lathe, shown as a column reaching the "m" level; the next skill (#2) might be the ability to use go–no go gages; the next, #3, to read a vernier micrometer; #4 to adjust the machine; #5 to read blueprints; and so on.

Worker identification. Candidates to perform required present or prospective operations include those already doing the operation or a similar operation, those employed by the organization but presently doing unrelated work, and those outsiders not now employed by the firm. Each candidate possesses a greater or lesser degree of each of the skills, knowledge, experience, attitudes, and other attributes deemed necessary to perform the required work satisfactorily. An effort to find someone with precisely the skill and attribute profile desired (Fig. 1) may be a complete success or only a partial success. The perfect new operator will bring all of the requirements to the job, and the operator's personal profile will match the desired job profile. An operator who has only a part of the desired job profile may currently possess skills and other attributes (Fig. 2). The knowledge and experience possessed by a candidate operator may be revealed by means of applications, interviews, reference checks, tests, and work trials.

Needs determination. Naturally an effort is made to locate a person who possesses all of the skills and other requirements of the operation, in which case no training is required–save a brief introduction and orientation to the new job. In practice, a perfect match is rarely achieved, and if the best match (that is, the least mismatch) of required job skills to currently possessed candidate operator skills is the selection of the person with the profile of Fig. 2 to fill the job of Fig. 1, then mismatches exist to the extent shown in Fig. 3.

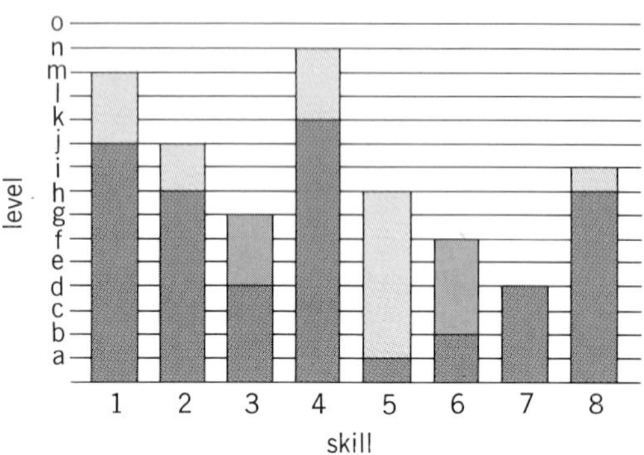

Fig. 3. Differences between the skill levels required to perform a given job and those possessed by a candidate operator, showing both deficiencies and excesses in skill levels of the operator.

Examination of Fig. 3 reveals that there are two types of mismatches: skills and experience that the candidate operator brings to the job which are not presently required; and skills required in the job which the candidate operator presently lacks. In some cases the requirements of the job can be modified to better fit the abilities of the candidate, but for the most part, where needed skills are lacking, the operator must be trained.

Program design. The list of required skills to be taught constitute the contents of the operator-training program designed. Essentially the contents will be subject matter to fill the gaps between what the student knows and what the student should know. From Fig. 3, the program content must be such that it will raise the trainee's skill level in item #1 from 1-j to 1-m, in #2 from 2-h to 2-j, and so on. No training need be done in skills #3, #6, and #7 since the new operator presently possesses enough, or more than enough, experience in them. In addition to skills and knowledge training, new employees must be given an induction and orientation to the company and its history, organization, products, facilities, personnel, policies, and rules and regulations. Also, they must be informed of their job title, supervisor, hours of work, attendance rules, holidays, vacations, insurance coverage, other fringe benefits, wage rates, safety rules, causes for reprimand or termination, job duties, and other responsibilities.

All employees should be educated to improve their morale and attitudes. They should be imbued with feelings of pride, responsibility, importance, security, belongingness, participation, cooperation, and satisfaction.

Mode selection. Operator training may be done informally or formally. Informally the trainee can be turned over to the immediate supervisor, who will instruct, watch, coach, and correct the operator as an ongoing activity; or the supervisor may assign a senior experienced operator to do the foregoing–this arrangement is called the "buddy" system; or the supervisor may assign an assistant to do the training or have the help of a staff-training specialist. All these methods are on-the-job training arrangements. As a more formal method, a vestibule school may be used. This is an off-the-job arrangement in which the work station is simulated in a separate location so that the full attention of a professional trainer may be given to the trainee, and so that the trainee's slow initial pace and high error rates do not upset regular production. Other arrangements include sending the trainee to a local school or training institute for courses in the required subjects, bringing in a training consultant to conduct a special course, having the trainee pursue a correspondence course or some other self-learning approach, setting up a formal apprentice program, or a combination of these modes. A full-scale formal apprentice training program can be for a period of 4 years, or longer, under the tutelage of a master craftsman. It can be composed of a combination of classroom study of the basic subjects of mathematics, science, and technology, on-the-job training, and job rotation, resulting in the apprentice being made a journeyman.

Trainer selection. Depending on the mode of operator training selected, the trainer will be the supervisor, an assistant, a master craftsman, a staff-training specialist, an outside teacher, or a special consultant. Regardless of who is selected to train the operator, it is essential that they possess the technical knowledge of the subjects, a mastery of the techniques of teaching and communicating, patience, and the intangible ability to instill confidence, pride of workmanship, and a positive mental attitude in the trainee.

Program preparation. Each part of the program must be prepared with care. A determination must be made as to specific learning objectives—what the trainee must learn, how well it must be learned, and in what time span. At the end of the instruction, the trainee should be able to perform at a specific quantified performance level. The total program should be broken into individual lessons—each a small dose, a specific subject with a few stressed key points.

Trainee preparation. The trainee must be conditioned to accept, retain, and subsequently correctly use the information and skills taught. The individual must be put at ease, made comfortable, told how the operation to be learned and performed fits into the larger picture, and convinced that the job is important and worth doing well.

Organizational decisions. The length of the program and of each session or lesson must be decided, as must the time of day when the training is to be done. The frequency of training must be determined, that is, the number of sessions per day or week. The size of the group of trainees to be trained at one time must also be decided upon. The training facility must be selected or designed so as to minimize distractions and maximize teaching effectiveness. All audiovisual equipment and material must be selected and tested before use, as must all other teaching aids, samples, books, manuals, and so forth. Provision must be made for review, feedback, testing, and evaluation of the effectiveness of the teaching, the learning, and the entire program.

Administrative decisions. Lastly, some administrative decisions must be made. Will the operator be paid while being trained? How much? What percent of the regular pay rate? Will the training be done during working hours instead of work, or after work in addition to regular work? Will certificates of successful completion or competency be awarded? And so on.

Instruction. With the foregoing accomplished, the actual teaching can begin. A method found effective in training operators in work skills can be called the "tell-show-watch-check" technique, whereby:

1. Tells
   a. Instructor tells what will be taught.
   b. Teaches.
   c. Tells what was just taught.
2. Shows
   a. Instructor does task or operation.
   b. Instructor recites steps while doing them.
   c. Trainee recites steps after instructor does them.
   d. Trainee tells instructor steps before instructor does them.
3. Watches
   a. Instructor recites steps and trainee does them.
   b. Trainee recites steps (and reasons for each) as he or she does them.
4. Checks
   a. Instructor follows up by observing (with increasing time between observations as trainee gains competence and confidence) trainee doing the operation.
   b. Corrects or compliments, as warranted.

It is essential that the trainee comprehend earlier lessons before the instructor proceeds to advanced training which builds upon such knowledge. It is wise, in cases where the total training program is divided into separate lessons, for the instructor to review the key points of prior lessons before teaching the current lesson. Other training guides include teaching in small doses, stressing key points, reviewing frequently, correcting errors promptly, avoiding embarrassing the trainee, encouraging questions, and building confidence in the trainee as he or she gains competence.

In job training the trainer attempts to have the trainee think and act in a particular desired manner, as contrasted to that person's prior inability to perform at all or in a previously different behavior pattern. It is important that the trainee accept the authority and superior knowledge of the trainer and the logic and value of the new information or method. In order to effect a permanent new behavior pattern, the trainee must be convinced that to perform in that way will lead to something desirable or that it is something which must be done in order to avoid something undesirable. *See* PERFORMANCE RATING; PRODUCTIVITY.

[VINCENT M. ALTAMURO]

Optical flat

A disk of high-grade quartz glass approximately 3/4 in. (2 cm) thick, having at least one side ground and polished with a deviation in flatness usually not

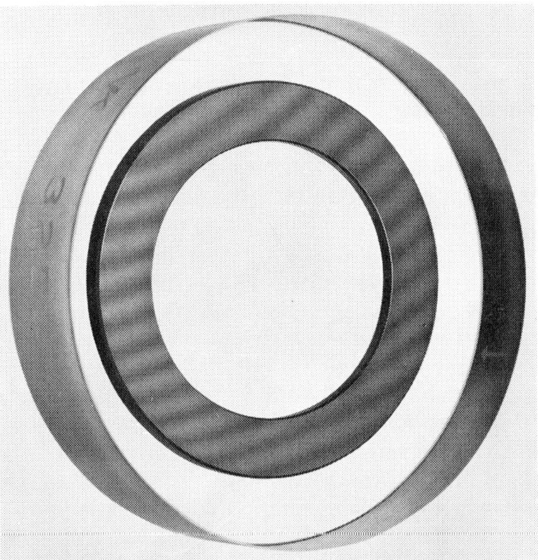

Fig. 1. Optical flat being used to determine flatness of seal ring. Interference bands on seal ring face show lines of constant depth. (*Van Keuren Co.*)

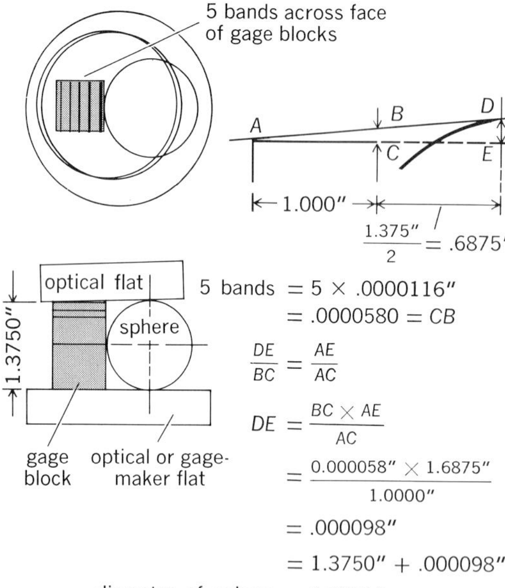

Fig. 2. Measurement of height of sphere determined by means of gage blocks and optical flat. 1″ = 25.4 mm.

exceeding 0.000002 in. (50 nm) all over, and a surface quality of 5 microfinish or less. When two surfaces of this quality are placed lightly together so that the air is not wrung out from between them, they are separated by a film of air and actually touch at only one point. This point is the vertex of a wedge of air separating the two pieces.

If parallel beams of light pass through the flat, part will be reflected against the surface being inspected, while part will be reflected directly back through the flat. Because the distance between the surfaces is constantly increasing along the angle, the beams reflected from the flat and the beams reflected from the workpiece will alternately reinforce and interfere with each other, producing a pattern of alternate light and dark bands (Fig. 1). Each succeeding full band from a point of contact means the distance between surfaces is one wavelength thicker. If the light is relatively monochromatic, the wavelength is known. Red with a wavelength of 0.0000116 in. (295 nm) is commonly used. Thus a definite relationship is established between lineal measurement and light waves. Optical flats are used for two general purposes.

Determination of surface contour. If the surface being inspected is flat, the light bands are parallel. Any deviation from flatness shows as curvature of the lines. The principle in interpreting a pattern is almost identical with the principle in interpreting a topographical map in that the bands connect points of equal distance from the master surface of the optical flat. Deviation from flatness can be reduced to rational figures.

Comparison of lineal measurement. When an optical flat is placed across a gage block or a buildup of blocks and another object, as an end standard, or a cylinder or sphere, both resting on a precisely flat surface, then the angle between the blocks and the flat can be measured, and consequently the difference in length between the gage blocks and the length or height or diameter of the unknown can be determined (Fig. 2). With the high degree of accuracy available in electrical and pneumatic comparators, optical flats are seldom used in this way except for spheres or irregular surfaces where point contact by the comparator is impractical.

The principle of interferometry is the standard method for measurement of gage blocks. However, an interferometer and not the optical flat itself is used. The wavelength of light is the present standard of all lineal measurements. See GAGES.

[RUSH A. BOWMAN]

Otto cycle

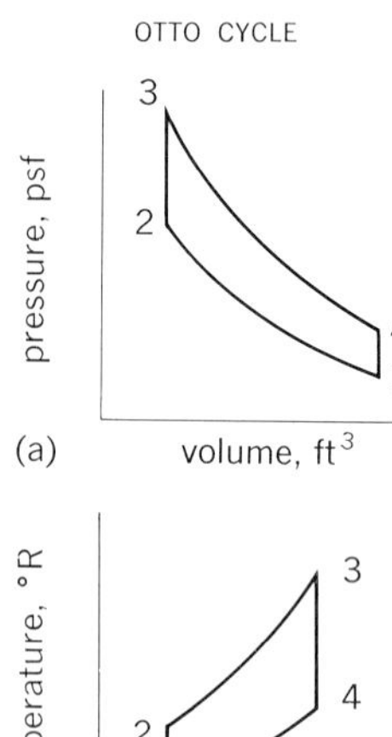

Fig. 1. Diagrams of (*a*) pressure-volume and (*b*) temperature-entropy for Otto cycle; phase 1–2, isentropic compression; phase 2–3, constant-volume heat addition; phase 3–4, isentropic expansion; phase 4–1, constant volume heat rejection.

The basic thermodynamic cycle for the prevalent automotive type of internal combustion engine. The engine uses a volatile liquid fuel (gasoline) or a gaseous fuel to carry out the theoretic cycle illustrated in Fig. 1. The cycle consists of two isentropic (reversible adiabatic) phases interspersed between two constant-volume phases. The theoretic cycle should not be confused with the actual engine built for such service as automobiles, motor boats, aircraft, lawn mowers, and other small (generally $<300 \pm$ hp) self-contained power plants.

The thermodynamic working fluid in the cycle is subjected to isentropic compression, phase 1–2; constant-volume heat addition, phase 2–3; isentropic expansion, phase 3–4; and constant-volume heat rejection (cooling), phase 4–1. The ideal performance of this cycle, predicated on the use of a perfect gas, Eqs. (1), is summarized by Eqs. (2) and (3) for thermal efficiency and power output.

$$V_3/V_2 = V_4/V_1 \qquad T_3/T_2 = T_4/T_1$$
$$\frac{T_2}{T_1} = \frac{T_3}{T_4} = \left(\frac{V_1}{V_2}\right)^{k-1} = \left(\frac{V_4}{V_3}\right)^{k-1} = \left(\frac{P_2}{P_1}\right)^{\frac{k-1}{k}} \qquad (1)$$

$$\text{Thermal eff} = \frac{\text{net work of cycle}}{\text{heat added}} = [1 - (T_1/T_2)] = \left[1 - \left(\frac{1}{r^{k-1}}\right)\right] \qquad (2)$$

$$\begin{aligned}\text{Net work of cycle} &= \text{heat added} - \text{heat rejected}\\ &= \text{heat added} \times \text{thermal eff}\\ &= \text{heat added } [1 - (T_1/T_2)]\\ &= \text{heat added } [1 - (1/r^{k-1})]\end{aligned} \qquad (3)$$

In Eqs. (2) and (3) V is the volume in cubic feet; P is the pressure in pounds per square inch; T is the absolute temperature in degrees Rankine; k is the ratio of specific heats at constant pressure and constant volume, C_p/C_v; and r is the ratio of compression, V_1/V_2.

The most convenient application of Eq. (3) to the positive displacement type of reciprocating engine uses the mean effective pressure and the horsepower equation, Eq. (4), where hp is horsepower;

$$\text{hp} = \frac{\text{mep } Lan}{33{,}000} \qquad (4)$$

mep is mean effective pressure in pounds per square foot; L is stroke in feet; a is piston area in square inches; and n is the number of cycles completed per minute. The mep is derived from Eq. (3) by Eq. (5), where 778 is the mechanical equivalent

$$\text{mep} = \frac{\text{net work of cycle} \times 778}{144\,(V_1 - V_2)} \qquad (5)$$

of heat in foot-pounds per Btu; 144 is the number of square inches in 1 ft²; and $(V_1 - V_2)$ is the vol-

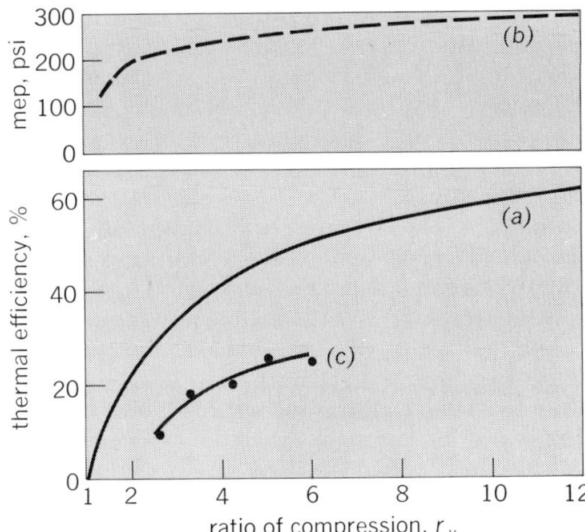

Fig. 2. Effect of compression ratio on thermal efficiency and mean effective pressure of Otto cycle. Curve *a* shows thermal efficiency, air-standard cycle; curve *b* mean effective pressure, air-standard cycle; and curve c thermal efficiency of an actual engine.

ume swept out (displacement) by the piston per stroke in cubic feet. *See* MEAN EFFECTIVE PRESSURE.

Air standard. In the evaluation of theoretical and actual performance of internal combustion engines, it is customary to apply the above equations to the idealized conditions of the air-standard cycle. The working fluid is considered to be a perfect gas with such properties of air as volume at 14.7 psia and 492°R equal to 12.4 ft³/lb, and the ratio of specific heats *k* as equal to 1:4. Figure 2 shows the thermal efficiency for this air-standard cycle as a function of the ratio of compression *r*, and the mep for a heat addition of 1000 Btu/lb of working gases. These curves demonstrate the intrinsic worth of high compression in this thermodynamic cycle.

The table gives a comparison of the important gas-power cycles on the ideal air-standard base for the case of compression ratio = 10 and 1000 Btu added per pound of working gases. The Otto, Brayton, and Carnot cycles show the same thermal efficiency of 60%. The mean effective pressures, however, show that the physical dimensions of the engines will be a minimum with the Otto cycle but hopelessly large with the Carnot cycle. The Brayton cycle ideal mep is only about one-fifth that of the Otto cycle, and it is accordingly at a distinct disadvantage when applied to a positive displacement mechanism. This disadvantage can be offset by use of a free-expansion, gas-turbine mechanism for the Brayton cycle. The Diesel cycle offers a lower themal efficiency than the Otto cycle for the same conditions, for example, 42 versus 60%, and some sacrifice of mep, 200 versus 290 psi. As opposed to the Otto engine, the diesel can utilize a much higher compression ratio without preignition troubles and without excessive peak pressures in the cycle and mechanism. Efficiency and engine weight are thus nicely compromised in the Otto and diesel cycles. *See* GAS TURBINE.

Thermal efficiency, mean effective pressure, and peak pressure of air-standard gas-power cycles*

Cycle	Efficiency	Mep	Peak pressures
Otto	60	290	2100
Diesel	42	200	370
Brayton	60	61	370
Carnot	60	Impossibly small	Impossibly high

*Ratio of compression = 10; heat added = 1000 Btu per pound working gases.

Actual engine process. This reasoning demonstrates some of the valid conclusions that can be drawn from analyses utilizing the ideal air-standard cycles. Those ideals, however, require the modifications of reality for the best design of internal combustion engines. The actual processes of an internal combustion engine depart widely from the air-standard cycle. The actual Otto cycle uses a mixture of air and a complex chemical fuel which is either a volatile liquid or a gas. The rate of the combustion process and the intermediate steps through which it proceeds must be established. The combustion process shifts the analysis of the working gases from one set of chemicals, constituting the incoming explosive mixture, to a new set representing the burned products of combustion. Determination of temperatures and pressures at each point of the periodic sequence of phases (Fig. 1) requires information on such factors as variable specific heats, dissociation, chemical equilibrium, and heat transfer to and from the engine parts.

N. A. Otto (1832–1891) built a highly successful engine that used the sequence of engine operations proposed by Beau de Rochas in 1862. Today the Otto cycle is represented in many millions of engines utilizing either the four-stroke principle or the two-stroke principle. *See* INTERNAL COMBUSTION ENGINE.

The actual Otto engine performance is substantially poorer than the values determined by the theoretic air-standard cycle. An actual engine performance curve *c* is added in Fig. 2, in which the trends are similar and show improved efficiency with higher compression ratios. There is, however, a case of diminishing return if the compression ratio is carried too far. Evidence indicates that actual Otto engines offer peak efficiencies (25 ± %) at compression ratios of 15 ±. Above this ratio, efficiency falls. The most probable explanation is that the extreme pressures associated with high compression cause increasing amounts of dissociation of the combustion products. This dissociation, near the beginning of the expansion stroke, exerts a more deleterious effect on efficiency than the corresponding gain from increasing compression ratio. *See* BRAYTON CYCLE; CARNOT CYCLE; DIESEL CYCLE; THERMODYNAMIC CYCLE.

[THEODORE BAUMEISTER]

Bibliography: T. Baumeister (ed.), *Standard Handbook for Mechanical Engineers*, 8th ed., 1978; J. B. Jones and G. A. Hawkins, *Engineering Thermodynamics*, 1960; J. H. Keenan, *Thermodynamics*, 1970; L. C. Lichty, *Combustion Engine Processes*, 7th ed., 1967; E. F. Obert, *Internal Combustion Engines*, 3d ed., 1973.

Overdrive

An automotive device supplied as special equipment and containing a step-up planetary gear arrangement located between the transmission and propeller shaft. Overdrive permits the propeller shaft to be driven at transmission output shaft speed, or faster than transmission output shaft speed when the overdrive comes into operation.

Purpose. For a given forward speed of the car, overdrive gives lower engine speed, quieter operation, and reduced gasoline consumption. Most overdrives used in the United States reduce engine speed 28% for a given car speed. For example, without overdrive action, a car engine might turn at 2000 rpm to achieve a speed of 40 mph. As overdrive comes into operation, engine speed could drop to about 1400 rpm at the same car speed of 40 mph. This provides quieter operation and reduced gasoline consumption and engine wear.

Operation. The overdrive contains two major components, a planetary gear system and a free-wheeling device (or overrunning clutch). *See* CLUTCH; PLANETARY GEAR TRAIN.

The overrunning clutch has two functions. It locks the transmission mainshaft and the overdrive output shaft together when the planetary gear system is inactive, providing direct drive through the overdrive. The second function is to permit the output shaft to overrun the transmission mainshaft when the planetary gear train is in action, providing overdrive. The overrunning clutch consists of an outer shell attached to the output shaft (see illustration). A circular inner member attached to the transmission shaft carries a series of cams or flats; a hardened steel roller rides on each cam. When the clutch is in action, the rollers wedge between the cams and outer shell so that the cams drive the outer shell. During overrunning, the outer shell rotates faster than the cams and the rollers move from loaded contact with the cams to disengage the clutch.

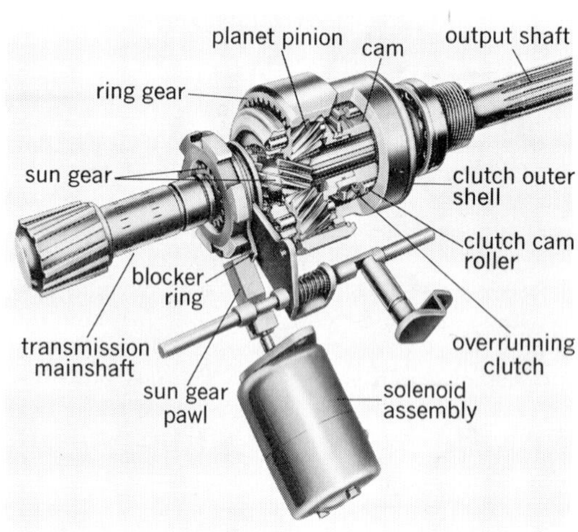

Cutaway view of overdrive mechanism.

The planetary gear system contains a sun gear which floats freely on the transmission mainshaft in direct drive. In overdrive, the sun gear is held stationary by a pawl that enters one of the slots in the sun gear plate which is splined to the sun gear.

With the sun gear stationary, the planet pinions are forced to rotate as they are carried around the sun gear by rotation of the planet pinion cage splined to the transmission mainshaft. Pinion teeth meshed with the sun gear are momentarily at rest. The teeth opposite, which are meshed with the ring gear, are therefore required to move faster than the cage. The ring gear is thus driven faster than the cage and transmission mainshaft. The ring gear is integral with the clutch outer shell and the output shaft. Therefore, with the sun gear stationary, the output shaft is driven faster than the mainshaft.

Controls. The overdrive will not operate at low car speed. However, when operating speed is reached (around 20 mph in many cars), a governor that senses car speed connects the solenoid to the car battery. The solenoid plunger compresses an internal spring which urges the pawl against a ledger on the blocker ring. The blocker ring is frictionally mounted to the sun gear plate, and has only limited freedom of rotary movement.

The driver of the car must make a conscious movement to cause the overdrive to go into operation. This movement is a momentary release of the accelerator pedal. As this takes place, the engine slows down slightly and causes a reversal of rotation of the sun gear. As the sun gear reverses rotation, friction on the blocker ring causes it to turn a few degrees. The ledge on the blocker ring moves out from under the pawl, allowing the pawl to drop into one of the notches in the sun gear plate, locking the plate, and thus the sun gear. With the sun gear stationary, the overdrive goes into action.

When the car speed drops below an established speed, the governor deenergizes the solenoid, permitting withdrawal of the pawl to restore direct drive. When direct drive is desired at higher speeds for additional passing ability, depressing the accelerator pedal past the full-throttle position will interrupt the ignition for a fraction of a second, causing an instantaneous reversal of the sun gear and permitting return to direct drive by the process just described.

In reverse gear, the sun gear is moved rearward, and its teeth mesh with the internal splines in the pinion cage; the overdrive input shaft is thus connected to the overdrive output shaft. Use of the friction torque of the engine in first or second transmission gear for hill descent is accomplished by locking out the overrunning clutch by pulling out a knob under the instrument board.

Automatic transmissions, with torque ratios considerably greater than those of standard layshaft transmissions, have permitted the use of rear-axle ratios equivalent to the overall ratios previously used by overdrives. This fact, together with cost considerations, has resulted in application of the overdrive to standard layshaft transmissions only. The automatic transmissions generally have a means of obtaining increased passing ability by depression of the accelerator pedal past the full-throttle position, similar to that on overdrives. *See* AUTOMOTIVE TRANSMISSION.

[FOREST R. MC FARLAND]

Packing

A seal usually used for high pressure as in steam and hydraulic applications. The motion between parts may be infrequent as in valve stems, or continual as in pump or engine piston rods. There is no sharp dividing line between seals and packing; both are dynamic pressure resistors under motion.

Such diverse materials are used for packing as impregnated fiber, rubber, cork, or asbestos compounds. The form of the packing may be square, in ring or spiral form, trapezoidal, or V-, U-, or O-ring in section. In packings, it is necessary that the surface finish of the contacting metal part be smooth for long life of the material. *See* PRESSURE SEAL.

[PAUL H. BLACK]

Panel heating and cooling

A system in which the heat-emitting and heat-absorbing means is the surface of the ceiling, floor, or wall panels of the space which is to be environmentally conditioned. The heating or cooling medium may be air, water, or other fluid circulated in air spaces, conduits, or pipes within or attached to the panel structure. For heating only, electric current may flow through resistors in or on the panels. *See* ELECTRIC HEATING.

Warm or cold water is circulated in pipes embedded in concrete floors or ceilings or plaster ceilings or attached to metal ceiling panels. The coefficient of linear expansion of concrete is 0.000095; for steel it is 0.000081, or 15% less than for concrete. For copper it is 0.000112, or 20% more than for concrete, and for aluminum it is 0.000154, or 60% more than for concrete. Since the warmest or coolest water is carried on the inside of the pipes and the heat is transmitted to the concrete, only steel pipe should be used for panel heating and cooling systems, except when metal panels are used.

Cracks are very likely to develop in the concrete or plaster, breaking the bonds between the pipes and the concrete or plaster. The pipes move freely, causing scraping noises. An insulating layer of air is formed between the concrete or plaster and this markedly reduces the coefficient of conductivity between the liquid heating medium and the active radiant surfaces.

Heat transfer. Heat energy is transmitted from a warmer to a cooler mass by conduction, convection, and radiation. Radiant heat rays are emitted from all bodies at temperatures above absolute zero. These rays pass through air without appreciably warming it, but are absorbed by liquid or solid masses and increase their sensible temperature and heat content.

The output from heating surfaces comprises both radiation and convection components in varying proportions. In panel heating systems, especially the ceiling type, the radiation component predominates. Heat interchange follows the Stefan-Boltzmann laws of radiation; that is, heat transfer by radiation between surfaces visible to each other varies as the difference between the fourth power of the absolute temperatures of the two surfaces, and is transferred from the surface with the higher temperature to the surface with the lower temperature.

The skin surface temperature of the human body under normal conditions varies from 87 to 95°F and is modified by clothing and rate of metabolism. The presence of radiating surfaces above these temperatures heats the body, whereas those below produce a cooling effect. *See* RADIANT HEATING.

Cooling. When a panel system is used for cooling, the dew-point temperature of the ambient air must remain below the surface temperature of the heat-absorbing panels to avoid condensation of moisture on the panels. In regions where the maximum dew point temperature does not exceed 60°F, or possibly 65°F, as in the Pacific Northwest and the semiarid areas between the Cascade and Rocky mountains, ordinary city water provides radiant comfort cooling. Where higher dew points prevail, it is necessary to dehumidify the ambient air. Panel cooling effectively prevents the disagreeable feeling of cold air blown against the body and minimizes the occurrence of summer colds.

Fuel consumption records show that panel heating systems save 30–50% of the fuel costs of ordinary heating systems. Lower ambient air temperatures produce a comfortable environment, and air temperatures within the room are practically uniform and not considerably higher at the ceiling, as in radiator- and convector-heated interiors. *See* COMFORT HEATING; HOT-WATER HEATING SYSTEMS.

[ERWIN L. WEBER/RICHARD KORAL]

Bibliography: American Society of Heating, Refrigerating, and Air Conditioning Engineers, *ASHRAE Handbook and Product Directory: Systems*, 1976.

Pantograph

A four-bar parallel linkage, with no links fixed, used as a copying device for generating geometrically similar figures, larger or smaller in size, within the limits of the mechanism. In the illustration the curve traced by point T will be similar to that generated by point S. This similarity results because points T and S will always lie on the straight line $\overline{OTS}$; triangles $\overline{OBS}$ and $\overline{TCS}$ are always similar because lengths $\overline{OB}$, $\overline{BS}$, $\overline{CT}$, and $\overline{CS}$ are constant and $\overline{OB}$ is always parallel to $\overline{CT}$. Distance $\overline{OT}$ always maintains a constant proportion to distance $\overline{OS}$ because of the similarity of the above triangles.

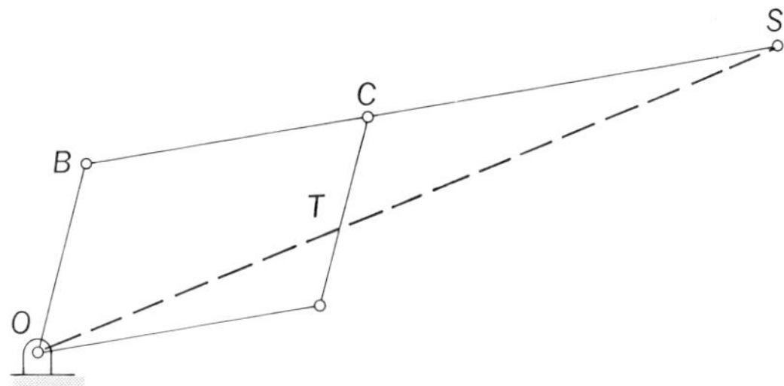

Similar triangles of a pantograph.

Numerous modifications of the pantograph as a copying device have been made. *See* FOUR-BAR LINKAGE.

James Watt applied the pantograph as a reducing motion in his beam engine. The pantograph has also served as a reducing motion for an engine indictator: S is attached to the engine crosshead, while the indicator cord is attached to T. *See* STRAIGHT-LINE MECHANISM.

A second use of the pantograph geometry is seen in the collapsible parallel linkage used on electric locomotives and rail cars to keep a current-collector bar or wheel in contact with an overhead wire. Two such congruent linkages in planes parallel to the train's motion are affixed securely on the top of the locomotive with joining horizontal members perpendicular to each other.

The uppermost member collects the current, and powerful springs thrust the configuration upward with sufficient pressure normally to make low-resistance contact from wire to collector. The inevitable variation of distance from wire to track produces, during high speeds, undesirable dynamic behavior of both wire and collector-pantograph, with oscillations of both that can break the connection between them or cause excessive arcing and wear with attendant loss of transmission. Stitching the wire to a stouter supporting wire can minimize the bounce of both collector and wire and assure more reliable current flow.

[DOUGLAS P. ADAMS]

Bibliography: N. P. Chironis, *Mechanisms, Linkages, and Mechanical Controls*, 1955.

Parallel circuit

An electric circuit in which the elements, branches (elements in series), or components are connected between two points with one of the two ends of each component connected to each point. The illustration shows a simple parallel circuit. In more complicated electric networks one or more branches of the network may be made up of various combinations of series or series-parallel elements.

PARALLEL CIRCUIT

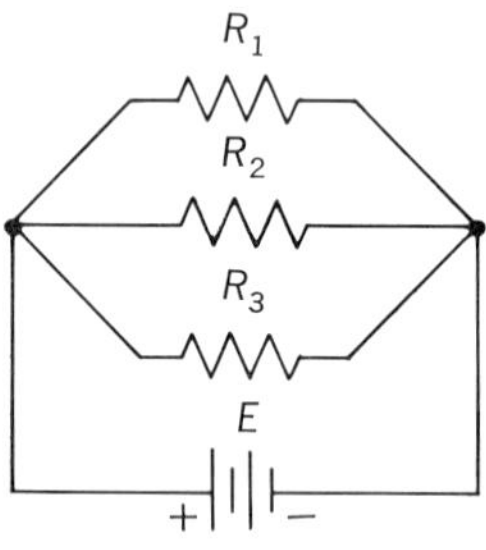

Schematic of a parallel circuit.

In a parallel circuit the potential difference (voltage) across each component is the same. However, the current through each branch of the parallel circuit may be different. For example, the lights and outlets in a house are connected in parallel so that each load will have the same voltage (120 volts), but each load may draw a different current (0.50 amp in a 60-watt lamp and 10 amp in a toaster).

For a discussion of parallel circuits *see* ALTERNATING-CURRENT CIRCUIT THEORY; DIRECT-CURRENT CIRCUIT THEORY.

[CLARENCE F. GOODHEART]

Patent

Common designation for letters patent, which is a certificate of grant by a government of an exclusive right with respect to an invention for a limited period of time. A United States patent confers the right to exclude others from making, using, or selling the patented subject matter in the United States and its territories. Portions of those rights deriving naturally from it may be licensed separately, as the rights to use, to make, to have made, and to lease. Any violation of this right is an infringement.

An essential substantive condition which must be satisfied before a patent will be granted is the presence of patentable invention or discovery. To be patentable, an invention or discovery must relate to a prescribed category of contribution, such as process, machine, manufacture, composition of matter, plant, or design. In the United States there are different classes of patents for different members of these categories.

Utility patents. Utility patents, which include electrical, mechanical, and chemical patents, are the most familiar; they each have a term beginning upon issue and ending 17 years later. This 17-year class of patent is granted (or issued) for an inventive improvement in a process, manufacture, machine, or composition of matter. A process may, for example, be a method of inducing or promoting a chemical reaction or of producing a desired physical result (such as differential specific gravity ore separation by flotation). "Manufacture" means any article of manufacture, and includes such diverse items as waveguides, transistors, fishing reels, hammers, buttons, and corks. "Machines" has its broadest conventional meaning, and "composition of matter" includes anything from drugs to alloys. This class of patent affords protection for the invention and normally a range of equivalents for doing substantially the same thing in substantially the same way. Thus mere changes in form, material, inversion, or rearrangement will not avoid infringement of the utility patent. A straightforward substitution of one active device for another, as a transistor for an electronic tube, is not normally such a change as to avoid infringement of a utility patent.

Design patent. This class of patent is granted for any new, original, and ornamental design for an article of manufacture. To the extent that shape is determined by functional, rather than ornamental, considerations, it is not proper subject matter for a design patent. Unlike the utility patent, the design patent may be avoided by a change in proportions or shape, although the essential function may be retained. The design patent is issued for different periods, depending upon the desires of the applicant for patent and the fee paid by the applicant. These terms are 3½, 7, or 14 years.

Plant patents. This class of patent is granted to one who discovers and asexually reproduces any distinct and new variety of plant other than a tuber-propagated plant. The right of exclusion extends only to the asexual propagation of such a plant. The term of the plant patent is 17 years from the date of issue by the Patent Office.

Procedure. The discussion in the balance of this article is limited to utility patents. In the United States, letters patent are granted upon the making of written application to the Patent and Trademark Office. The essential parts of this application are (1) petition (request that the patent be granted); (2) drawing showing one or more representative embodiments of the invention whenever description of the invention would be aided by its presence; (3a) specification (abstract of the disclosure, detailed written description of the invention and the manner and process of making and using it in such full, clear, concise, and exact terms as to enable any person skilled in the art to which it pertains, or with which it is most nearly connected, to make and use it, and setting forth the best mode contemplated by the inventor for carrying out the invention); (3b) the claims, which define the scope of the requested exclusionary right; (4) an oath in which the applicant swears that he or she is the original and first inventor, that the invention was not known or used by others in the United States or patented or described in a printed publication in this or a foreign country before his or her invention or discovery thereof, or patented or described in a printed publication in this or a foreign country or in public use or on sale in this country more than 1 year prior to the date of his or her application; and (5) the prescribed filing fee.

Owing to the complexities of obtaining proper patent protection, it is customary to retain an attorney who has been registered to practice before the

Patent and Trademark Office for the purpose of preparing the application and pursuing it through that office.

When the application is received, it is examined at the Patent and Trademark Office by an examiner. The examiner searches the prior art, consisting of prior patents and other publications, to determine whether the application and claims comply with the rigorous standards required by the statute for patentability. If it does not, the examiner points this out in a letter of rejection, specifying the particulars of the basis for rejection. If the application, on the other hand, is allowable, the examiner sends the applicant a notice of allowance. Following the date when the Patent Office mails a rejection and within the time period specified by the Patent Office, ordinarily 3 months, but never more than 6 months, the applicant or the attorney must respond. Normally the application would be amended to make it comply with the statutes. If not, the examiner would issue a final rejection, from which the applicant could appeal to the Board of Appeals. If the amended application was found to be in accord with the statute, the examiner would issue a notice of allowance. The patent is issued upon the payment of the issue fee. The entire process requires about 2 years on the average.

Invention. Patents are granted for new inventions or discoveries which are within the statutory classes of patentable subject matter, which were not known in the United States before the invention date and not printed anywhere in the world before that date, which are the subject of applications timely filed, and in which the subject matter is not obvious to one skilled in the art to which the invention relates.

It is generally considered that there are two discernible steps in invention. The first is thinking of the invention; the second is constructing it. The first is termed conception. The second is called reduction to practice.

Conception. The formation in the mind of the inventor of a definite and permanent idea of the complete and operative invention as it is thereafter to be applied in practice is conception. It is not merely a perception of what is done, or considered desirable to do, but, going beyond this, how it is to be done, in terms of a currently realizable instrumentality or group of instrumentalities. Because conception is a mental act, there must be some external, verifying manifestation in the form of impartation to another if the act is to be established in later controversy over priority of invention. Although oral transmission is adequate in theory, the perishability of the human memory favors the unchanging written word, dated, signed by the inventor, and witnessed by the corroborating party.

It is this which makes the keeping of written records by the inventor in the course of his or her work so important, because a failure to have this verified external manifestation of the conception may cost the inventor the patent if the date becomes important and is challenged. Because the record is not a proof, but only a document capable of proof, it is important that the recording be in some permanent form in which undetectable alteration is practically impossible; otherwise the weight of the proof will be diminished. Predating of the document is damaging, although a record, made a few days later, of a previously observed event, is valuable, if appearing on its face as such. The parties involved in a record of conception will be tested for verity by cross examination should the dates ever be challenged.

Frequently the moment of conception is clear only in retrospect, when there is a long sequence of experimental effort directed to attaining the desired result. There is no dramatic thunderclap ushering in the birth of most inventions, and hence the desirability of keeping current dated and witnessed records of all work. Such records tend further to corroborate the conception by providing a clue to the entire thought train, and revealing the completeness of understanding by the skill with which its principles are later applied.

Conception alone does not give the inventor any vested right in the invention.

Reduction to practice. Two forms of reduction to practice are recognized, actual and constructive. The filing of a patent application which does not become abandoned is a constructive reduction to practice.

Actual reduction to practice requires that the invention be carried out in a physical embodiment demonstrating its practicability. In a process or method, this is sufficiently done when the steps are actually carried out to produce the desired result. In a machine or article of manufacture, it is required that there be a construction showing every essential feature of the claimed invention. Practicality is demonstrated by operating the apparatus under the conditions which it is anticipated will be encountered in actual service. For example, the testing of an automobile hot-water heater by using water from the hot-water tap has been held not an actual reduction to practice because some conditions might occur in a motor vehicle which would not be observed in a stationary installation with a heat source of relatively unlimited capacity. Materials are reduced to practice when they are produced, unless utility is not self-evident, for example, in the case of drug compounds, where it is required to establish that the drug is useful for the purpose stated, not merely harmless. To be effective, the reduction to practice must be by the inventor, by one acting as his or her agent, or by one who has acquired rights from the inventor.

A reduction to practice which results from diligent efforts following conception of the invention gives a vested right in the invention, unless followed by an abandonment. If there is a gap in such diligence, the effective date of the right to assert ownership of the invention is only that date which can be connected by a continuous train of diligence with the reduction to practice. Like conception, reduction to practice must be corroborated by a third-party witness, and it is advisable to have a contemporary written record of what was done and what was observed, accompanied by the date of the observations, to refresh the recollection of the witnesses. The witness must have sufficiently acquainted himself or herself with the internal details of any apparatus to be able later to establish the identity between what was demonstrated and what is sought later to be patented. The diligence required is that which is reasonable under all the circumstances, but must be directed to the reduction to practice of the invention, and to collateral factors. The safest course is to make every effort to reduce to practice consistent with the inventor's physical, intellectual, and financial capacity. For example, if reduction to practice would be clearly within the inventor's financial means, alternative attempts to secure financing could not constitute diligence.

Interferences. When two or more persons are claiming substantially the same invention, a contest to determine priority between the two is instituted by the Patent and Trademark Office. This proceeding is termed an interference. That party who can carry his or her reduction to practice back to the earliest date toward and including conception by a continuous train of diligence will prevail in the interference and be awarded the patent, providing those acts have been in the United States. If any or all of the acts were performed outside the United States, the inventor is limited to the date of his or her first efforts in the United States. Exception is made for foreign inventors when they have first filed a corresponding application for patent in a foreign country which is a signatory to the Paris Convention. In such instance the inventor will be credited with a date corresponding to his or her first foreign filing date if the filing in the United States occurs within a year of that date and if the inventor promptly and duly requests this.

An applicant may provoke an interference with an issued patent which claims an invention that the applicant believes should rightly belong to him or her. This is done within 12 months of issue by filing in the Patent and Trademark Office, as part of a pending application for patent, all the claims which it is sought to contest, applying them to his or her disclosure, and requesting the declaration of an interference.

Following declaration of an interference, normally the parties are called upon to file preliminary statements under oath setting forth pertinent data surrounding the genesis of the invention and its disclosure to others. When these preliminary statements have been received and approved by the Patent and Trademark Office, the applicant is notified of the setting of a period of time during which motions may be filed. During this time, access to the adversary's application is permitted, marking one of the few times that the secrecy with which the Patent and Trademark Office surrounds each patent application is penetrated. During the motion period, various requests for modification or termination of the interference, known as motions, may be presented and set for hearing. When an issued patent is involved, there may be no motion for dissolution of the interference upon the ground that the claim in issue is unpatentable. After the motions are disposed of, times for taking the testimony are set, during which the parties may take sworn statements from their witnesses before duly qualified officers, which are then filed with the Patent and Trademark Office, accompanied by any proper exhibits, and used as the basis for the presentation of written arguments, followed by an oral hearing if the applicant so requests. The burden of proof in the interference is on the party who was last to file the application in the Patent and Trademark Office, and he or she accordingly has the right to open and close the written and oral presentations. If the junior party does not take testimony during the time allotted, the interference is terminated in favor of the senior party without any testimony being taken by him or her.

The conduct of an interference proceeding is frequently an arduous and complex matter, being fraught with many technicalities.

Inventor. In the United States only a natural person may be an inventor, as distinguished from a corporation. Inventors may be either sole or joint.

Assignment. Patents and applications for patents have the general attributes of personal property, and interests in them are assignable by instrument in writing. Such assignments will be recorded by the Patent and Trademark Office upon filing of a request accompanied by copy of the assignment and payment of the proper fee. To be good against subsequent purchasers without notice, the assignment must be recorded within 3 months from its date, or before the date when such subsequent transfer of rights was made.

Witnesses. Corroboration witnesses, as required in connection with the establishment of conception and reduction to practice, must be someone other than the inventor. No joint inventor can serve as a corroborating witness for another joint inventor in connection with the invention which they have jointly made. Beyond this, the rules normally governing witnesses apply. A corroborating witness should understand the subject matter involved in his or her testimony.

Enforcement. Enforcement is available against the manufacturer, the user, and the seller. Enforcement of patents is through the Federal judicial system, action being initiated in the Federal judicial district where the defendant resides, or where the defendant has committed the alleged act of infringement and has a regular and established place of business. Damages may be awarded, and an injunction granted, prohibiting further infringement by the defendant. If damages are awarded, there can be recovery for a period not longer than 6 years preceding the filing of the complaint.

When the infringer is the United States government, or a supplier of the government operating with the authorization and consent of the government, the suit must be filed against the government in the United States Court of Claims.

A patent may become unenforceable through improper use, for example, use as a part of an act in violation of the antitrust law.

Under the 6-year statute of limitations, an action may be maintained on a patent up to 6 years after its expiration, the accounting being limited in such instance to damages for infringing activities within the life of the patent.

Licenses. Licenses to operate under a patent may be granted, either nonexclusive or exclusive, and may be in writing or may arise as a necessary implication of other actions of the patentee. Except for an exclusive license, licenses are not ordinarily recorded by the Patent and Trademark Office.

Foreign filing. A United States patent is void if the United States inventor should file an application for the same invention in any country foreign to the United States before 6 months from the date when he or she filed in the United States, unless license to do so be first obtained from the Commissioner of Patents and Trademarks. Because the United States is a party to the Paris Convention, applications filed in foreign countries also members of the Paris Convention within 12 months of the date when the parent case is filed in the United States are accorded an effective filing date which is the same as the date of filing in the United States. The procedural details of foreign filing vary from country to country and from time to time. The Patent Cooperation Treaty, to which the United States, Japan, most European countries, and other countries are party, has greatly facilitated filing of patent applications in foreign countries. In particular it has established a standardized application and claim format acceptable in the patent offices of all member countries.

Nuclear and atomic energy. Patents may not be issued for inventions or discoveries useful solely in the utilization of special nuclear material or atomic energy in an atomic weapon, nor does any patent confer rights upon the patentee with respect to such uses. As a substitute for the patent incentive for disclosure in this field, there is a mandatory provision in the law that requires anyone making an invention or discovery useful in the production of special nuclear material or atomic energy, in the utilization of such special nuclear materials in the production of an atomic weapon, or in the utilization of atomic energy weapon, to report such invention promptly to the Atomic Energy Commission, unless it is earlier described in an application for patent. Awards may then be requested from the Patent Compensation Board of the Atomic Energy Commission.

Aeronautics and astronautics. No patent may be issued for an invention having significant utility in the field of aeronautics or space unless there be filed with the Commissioner of Patents and Trademarks a sworn statement of the facts surrounding the making of the invention and establishing that it was done without any relation to any contract with the National Aeronautics and Space Administration. This is subject to waiver by the administrator, but in the event of waiver the administrator is required by law to retain a license for the United States and foreign governments.

Marking. A patented product may carry a notice of this fact, including the patent number. The affixation of such notice is of advantage in establishing constructive notice and fixing the period for which damages may be collected. In many cases damages may not be collected, in the absence of such marking, without actual notice to the infringer. By statute, false marking is a criminal offense. The marking "Patent pending" or "Patent applied for" gives no substantive rights, but may give rise to sanction under the above-mentioned statute if without foundation in fact.

Foreign patents. The principles guiding most foreign patent systems are essentially the same as those underlying the United States system: the granting of a carefully defined exclusionary right for a limited term of years in return for a laying open of the invention through letters patent. In the socialist countries the grant is sometimes made for innovation as well as invention and establishes a right to compensation. There are some differences in the classes and terms of patent, and in the nature of subject matter which may be patented. The most significant departures are the measuring of the term, in most instances, from the filing date, the imposition of annual taxes increasing each year of the life of the patent, and compulsory licensing.

Generally speaking, countries can be divided into two classes: the examination countries, that is, those which examine the application in relation to the prior art to measure whether patentable invention is present before granting the patent; and the registration countries, that is, those granting the patent as a ministerial act on the basis of fulfillment of certain formal requisites, leaving substantive determinations wholly to the courts. Typical examination countries are the United States, West Germany, and Japan. Typical registration countries are Belgium and Italy.

In some examination countries the mounting magnitude of material to be searched so slowed the rate of disposition of patent applications that the delays and the accumulation of pending patent applications created intolerable uncertainties. Led by the Netherlands and followed by West Germany, the practice of deferred examination has been introduced, according to which the application is given cursory formal examination, and published after 18 months, but examination on the merits is delayed. Examination on the merits is made upon request, but must be made before enforcement of the patent or the end of 7 years, whichever first occurs. The theory was that only a few of the patents are ever brought to litigation, and many are without value after 7 years, so that the load on the examining staff would be reduced to manageable proportions. In fact, however, the system created additional uncertainties for the business community.

To eliminate duplicate searching of the same invention by the different patent offices and to standardize the requirements for patent applications, as well as the form of patent claims, the Patent Cooperation Treaty has, as noted above, been entered into by many countries. Under this treaty the application is examined by a competent searching authority, the criteria for which are set by the treaty. The application, together with the results of that search, are then published. Any country to which the applicant subsequently applies has the benefit of that search, which may be utilized by that country in determining whether the invention is in fact patentable in that country.

In somewhat similar fashion the countries of Europe have by treaty established a European Patent Office in Munich to which application can be made. The European Patent Office conducts the search on behalf of the other countries of Europe who are members of that treaty. As in the case of the Patent Cooperation Treaty, further searching may be done by each country, but need not be.

Developments in United States law. Unfortunately the patentability of inventions relating to, or involving, computer programs is not clear. The cases which have reached the Supreme Court to date have resulted in holdings which do not clearly differentiate between the patentable and the unpatentable in this area. In 1980 the Supreme Court had before it the first case raising the issue of patentability of microorganisms, plants or animals of microscopic size, especially a protozoan or bacterium. This first case involved, as one item, a single-cell microorganism useful in degrading oil spills. The Supreme Court decided that the subject matter was patentable even though "alive."

[DONALD W. BANNER]

Bibliography: Matthew Bender and Co., *Patent Law Annual*, annually; Matthew Bender and Co., *Patent Law Developments: Enforceability of Rights*, 1965; Matthew Bender and Co., *Patent Law Developments: Proprietary Intellectual Rights*, 1964; A. Casalonga, *Traité Technique et Pratique des Brevets d'Invention*, 1949; A. W. Deller, *Deller's Walker on Patents*, 1964; U. Draetta, *Il Regime Internazionale della Properieta Industriale*, 1967; H. Nathan, *Erfinder-und Neuererrecht der Deutschen Demokratischen Republik*, 1968; *N.Y. Law School Law Rev.*, vol. 22, no. 2, 1976, and vol. 22, no. 3, 1977; E. Ridsdale, *Patent Claims*, 1949; E. Ridsdale, *Patent Licenses*, 3d ed., 1958; U.S. Bureau of National Affairs, *Patent Procurement and Exploitation*, 1963; U.S. Department of Commerce, Patent Office, *Manual of Patent Examining Procedures*, 3d ed., 1961; F. L. Vaughan, *The United States Patent System*, 1956.

Pavement

An artificial surface laid over the ground to facilitate travel. In this article only road pavements are discussed. The engineering involved is closely similar to that for airport surfacing, another major type of pavement. A pavement's ability to support loads depends primarily upon the magnitude of the load, how often it is applied, the supporting power of the soil underneath, and the type and thickness of the pavement structure. Before the necessary thickness of a pavement can be calculated, the volume, type, and weight of the traffic (the traffic load) and the physical characteristics of the underlying soil must be determined.

Traffic load. Traffic data can readily be obtained. Traffic is counted to learn the total volume and the proportion of heavy vehicles. Loadometer scales are placed next to the road and trucks are weighed, front and rear wheels separately. Traffic trends are studied. Such data provide a basis for estimating the total volume of traffic to be carried by a pavement during its service life. They also permit an estimate of the magnitude and frequency of the expected load. *See* TRAFFIC ENGINEERING.

Base and surface courses of pavements are designed to withstand many applications of load over a prolonged period, in some cases up to 30 years. In structural design it is also necessary to give consideration to the direction of traffic. For example, a pavement from a mine to a nearby railroad siding may carry a high percentage of heavily loaded trucks on the inbound lanes and a low percentage on the return lanes.

In general, the larger the volume of heavy vehicles on a highway, the greater the structural capacity required in the pavement. But equally as important as the volume of heavy vehicles are the magnitude of the applied loads and the conditions that will influence the effect of those loads on the pavement. Under the action of vehicular traffic, the surface of a pavement is subjected to a series of highly concentrated forces applied through the wheels of the vehicle. These forces exert an influence throughout the depth of the pavement.

The applied loads vary considerably, depending upon the number and spacing of the wheels of each vehicle, the gross weight of the vehicle, and the distribution of that weight among the axles. Two vehicles of the same gross weight may differ widely in the wheel loads applied to the highway. A relatively small truck may cause a load of higher unit stress than a larger vehicle with larger tires and more axles. Consideration of the actual wheel loads has become increasingly important with the large increase in the volume of heavy vehicles on most highways.

All highways, regardless of their design, have some limit to their ability to support the frequent application of a heavy load. A large number of load applications can be supported by a given pavement if the load does not exceed a particular magnitude, but once this magnitude is exceeded, distress and failure of the pavement becomes increasingly evident. Weather also influences the ability of the pavement to support a load. Definite load limitations are imposed by law on many highways. Further limitations are often imposed on specific highways of lighter design, especially during such conditions as spring freezing and thawing.

In the preceding discussion wheel loads were treated as static loads. For adequate design analysis the effects of dynamic loads must be evaluated. The vertical force exerted on the pavement by a moving wheel may be considered to be the sum of the static weight of the wheel and the impact or dynamic force from the wheel's movement over irregularities in the pavement surface. The static load is a constant factor except as the movement of the vehicle along the highway sways or oscillates the load.

An exceedingly variable factor, the dynamic force generated by a moving wheel depends upon (1) the magnitude of the static wheel load; (2) the operating speed of the vehicle; (3) the type, size, and cushioning properties of the tire equipment; and (4) the smoothness of the pavement. An increase in static wheel load or pavement roughness, a decrease in the cushioning qualities of the tires, and, within limits, an increase in vehicle speed all result in increased dynamic forces.

The importance of the foregoing variables and factors has long been recognized by design engineers. The difficulty has been to express the data in terms that could be rationally applied to design formula and then correlated with the performance of foundation soils and materials used to construct the pavement. The problem has not been given a rigorous mathematical solution but rather has depended largely upon field observations under actual operating conditions.

Several methods of load evaluation have been developed and used by various highway agencies and organizations. All of the methods are more or less empirical in approach and thus are subject to revision and adjustment when field observations indicate changing conditions of traffic, climate, or soil performance. No universal method of load evaluation has been developed so far and indeed cannot be until a method is devised that can be readily adjusted when other factors influencing the performance of a pavement structure change.

Methods of evaluation in use include (1) numerical count method, in which the number of vehicles using a particular highway is actually counted and the weight of various vehicles listed as light, medium, heavy, and extra heavy; (2) wheel load method, in which factors based upon the actual weight of the wheel load are used; (3) load frequency method, in which the wheel load weights are combined with the volumetric count of the commercial vehicles; and (4) equivalent wheel load method, in which destructive effects of the actual wheel loads being applied to a pavement are expressed in terms of a standard wheel load.

Evaluation of subgrade. Factors that must be considered in evaluating the ability of the underlying soil or subgrade to support the pavement include (1) type of soil, such as loam or clay; (2) gradation and variation in particle size; (3) strength or bearing value; (4) modulus of deformation; and (5) swell or volume change characteristics and related properties. Measurement of the supporting power of the subgrade presents numerous difficulties. Tests sometimes used include the plate bearing test, the direct shear method, the triaxial compression test, and the bearing ratio procedure.

Some soil types are unsuitable for supporting pavements because they have low bearing values or undergo changes in volume with variation in moisture content. For example, it is desirable to excavate peat and muck along the road and replace it with soil of higher bearing value.

Rigid and flexible pavements. Once the grading operation has been completed and the subgrade compacted, construction of the pavement can begin. Pavements are either flexible or rigid. Flexible pavements have less resistance to bending than do rigid pavements. Both types can be designed to withstand heavy traffic. Selection of the type of pavement depends among other things, upon (1) estimated construction costs; (2) experience of the highway agency doing the work with each of the two types; (3) availability of contractors experienced in building each type; (4) anticipated yearly maintenance costs; and (5) experience of the owner in maintenance of each type.

Flexible pavements. Flexible paving mixtures are composed of aggregate (sand, gravel, or crushed stone) and bituminous material. The latter consists of asphalt products, which are obtained from natural asphalt products or are produced from petroleum; and tar products, which are secured in the manufacture of gas or coke from bituminous coal or in the manufacture of carburetted water-gas from petroleum distillates. Structural strength of a bituminous pavement is almost wholly dependent upon the aggregate, which constitutes a high percentage of the volume of the mixture and forms the structure that carries the wheel load stresses to the base layers. The bituminous material cements the aggregate particles into a compact mass with enough plasticity to absorb shock and jar; it also fills the voids in the aggregate, waterproofing the pavement.

Among the many types of bituminous surface used are surface treatments, penetration macadams, road mixes, and plant mixes, as well as variations of these. With surface treatment a liquid bituminous material is applied over a previously prepared aggregate base.

In building penetration-macadam pavement, a base, usually of crushed stone or gravel, is constructed in layers and firmly compacted by rolling. Often during the rolling, water is applied to make what is termed a water-bound base, or macadam. Then, the keyed and wedged fragments in the upper portion of the base are bonded in place by working alternate applications of bituminous material and choke stone into the surface voids.

With road-mix designs a base is constructed, and a layer of aggregate and bituminous material, mixed on the road with a motor grader, is then uniformly spread and compacted with rollers.

When a plant-mix design is used, construction of a base is also necessary, but the aggregate and bituminous materials are mixed at a central plant, trucked to the job, and then spread or placed with a paving machine and compacted. With the plant-mix procedure more accurate mixing is possible.

For flexible pavements designed to support heavy loads a subbase built of materials similar to those of the base but of poorer quality may also be used. Thickness of the wearing surface, the base, and the subbase depends upon the design load. A typical flexible pavement design for light loads or a 5-ton axle loading is shown in Fig. 1. Where a flexible pavement is designed to carry a large traffic volume, for example, 2000–5000 vehicles per day, including 150–300 heavy commercial vehicles, the gravel subbase would be increased to 10 in., the gravel base to 5 in., and the plant-mixed bituminous surfacing to 4 in. It is assumed that the maximum wheel load would be restricted to 9 tons per axle. Some flexible pavement designs have called for mixing asphalt with the gravel base material to provide greater pavement strength. Bases treated in this manner are termed bituminous stabilized bases.

Rigid pavements. Coarse aggregate, fine aggregate, and portland cement are mixed with clean acid-free water to produce the concrete used for rigid pavements. The coarse aggregate may consist of coarse gravel or crushed stone and the fine aggregate of sand or crushed-stone screenings. The thickness of the pavement slab may vary from 6 in. (15 cm) for light traffic to 18 in. (45 cm) or

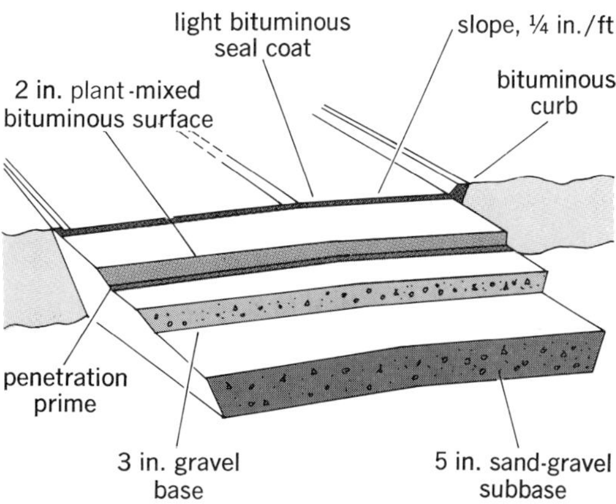

Fig. 1. Flexible pavement design for a city collector street with maximum traffic load of 5 tons (4.5 metric tons) per axle. Right-of-way is 60 ft (18 m) wide and the pavement width is 38 ft (11.6 m). Berms or boulevards at the sides are sloped in order to drain toward the street. 1 in. = 2.5 cm.

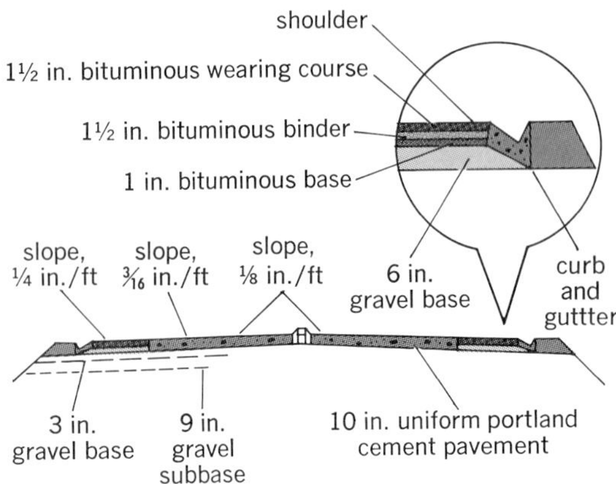

Fig. 2. Rigid main roadway with flexible shoulders. The design is for maximum loads of 9 tons (8 metric tons) per axle. 1 in. = 2.5 cm. 1 in./ft = 8 cm/m.

more for airport pavements accommodating heavy aircraft. A layer of granular material such as sand, sandy gravel, or slag is generally used as a subbase under the concrete slab to prevent frost heave and to increase the supporting power of the underlying soil. A rigid pavement is shown in cross section in Fig. 2.

No steel reinforcement is used with bituminous or flexible pavements. With rigid pavements, especially those designed for heavy loads, reinforcement is often used to strengthen the pavement and to prevent cracking. The reinforcement may consist of welded wire fabric or bar mats assembled by tying transverse and longitudinal steel rods together at their point of intersection (Fig. 3). The reinforcement is usually placed about 2 in. (5 cm) below the upper surface of the concrete slab.

Fig. 3. A rigid pavement reinforced by welded steel wire mesh. Underneath the mesh are 6 in. (15 cm) of concrete. A 3-in. (8-cm) layer of concrete is being laid on top.

In constructing rigid pavement a longitudinal control joint is used between adjacent lanes. Transverse joints, such as expansion and contraction joints to prevent cracking of the pavement when the temperature changes, may also be included. With flexible pavement no joints are used.

Research. Much remains to be learned regarding the performance and length of service of various types of pavement under different traffic, foundation, and climatic conditions. Probably the most comprehensive experimental pavement study ever undertaken was conducted near Ottawa, Ill. Pavements of different types and thicknesses were investigated. This comprehensive study showed that heavy wheel loads and the trip frequency of heavily loaded vehicles were the two major factors in the destruction of both rigid and nonrigid pavements. The tests also provided valuable data for designing residential streets and other routes where traffic volumes are low and truck traffic is a small percentage of the traffic flow. *See* HIGHWAY ENGINEERING. [ARCHIE N. CARTER]

Bibliography: R. G. Ahlvin et al., *Multiple-Wheel Heavy Gear Load Pavement Test*, Tech. Rep. S-71-17, vols. 1–4, U.S. Army Engineer Waterways Experiment Station, Corps of Engineers, Vicksburg, MI, November 1971; *Airfield Flexible Pavements: Air Force*, chap. 2, TM 5-824-2/AFM 88-6, Departments of the Army and the Air Force, February 1969; American Society for Testing and Materials, Standard specification for hot-mixed, hot-laid asphalt paving mixtures, *ASTM Book of Standards*, D1163, pt. II; R. E. Boyer and M. E. Harr, Predicting pavement performance, *Transp. Eng. J. ASCE*, 100(TE2):431–442, May 1974; Y. T. Chou, R. L. Hutchinson, and H. H. Ulery, Jr., Response of flexible pavements to multiple loads, *Transp. Eng. J. ASCE*, 101(TE3):537–551, August 1975; R. H. Ledbetter, H. H. Ulery, Jr., and R. G. Ahlvin, Traffic tests of airfield pavements for the jumbo jets, *Proceedings of the 3d International Conference on the Structural Design of Asphalt Pavements*, London, September 1972; V. F. Nakamura and H. L. Michael, *Serviceability Ratings of Highway Pavements*, Highway Res. Rec. no. 40, Highway Research Board, 1963; Paving train meets strict specification slipforming spacecraft runway, *Eng. News Rec.*, p. 22, Aug. 7, 1975; P. S. Pell and S. F. Brown, The characteristics of materials for the design of flexible pavement structures, *Proceedings of the 3d International Conference on* *havior of Airfield Flexible Pavements under Loads: Theory and Experiments*, Tech. Rep. no. AFWL-TR-72-215, Air Force Weapons Laboratory, Air Force Systems Command, Kirtland Air Force Base, Albuquerque, July 1973; P. Cook, A history of joints in concrete pavements, *J. Constr. Div. ASCE*, 101(CO1):29–36, March 1975; J. A. Epps and C. L. Monosmith, Fatigue of asphalt concrete mixtures: Summary of existing information, *Fatigue of Compacted Bituminous Mixtures*, American Society for Testing and Materials, STP 508, p. 27, 1972; W. H. Highter and M. E. Harr, Cumulative deflection and pavement performance, *Transp. Eng. J. ASCE*, 101(TE2):247–263, May 1975; Y. T. Chou and R. H. Ledbetter, *The Be-*

the Structural Design of Asphalt Pavements, London, pp. 326–342, 1972; W. Witczak and J. F. Shook, Full-depth asphalt airfield pavements, *Transp. Eng. J. ASCE*, 101(TE2):297–309, May 1972.

Pawl

The driving link or holding link of a ratchet mechanism, also called a click or detent. In Fig. 1 the driving pawl at *A*, forced upward by lever *B*, engages the teeth of the ratchet wheel and rotates it counterclockwise. Holding pawl *C* prevents clockwise rotation of the wheel when the pawl at *A* is making its return stroke. Pawl and ratchet are an open, upper pair. *See* SLIDING PAIR.

Driving and holding pawls likewise engage rack teeth on the plunger of a ratchet lifting jack, such as those supplied with automobiles. A ratchet wheel with a holding pawl only, acting as a safety brake, is fastened to the drum of a capstan, winch, or other powered hoisting device.

A double pawl can drive in either direction (Fig. 2*a*) or be easily reversed in holding (Fig. 2*b*). A cam pawl (Fig. 2*c*) prevents the wheel from turning clockwise by a wedging action while permitting free counterclockwise rotation. This technique is used in the automobile hill-holder to prevent the vehicle's rolling backward.

The spacing mechanism of a typewriter, although frequently termed an escapement, is properly a ratchet device in which a holding pawl is withdrawn from a spring-loaded rack to allow movement of the carriage, while an arresting pawl is introduced momentarily to permit the holding pawl to engage the next tooth of the rack.

In designing a pawl, care should be taken that the line of contract has a normal *N* passing between centers *R* and *P* (Fig. 3); otherwise the pawl will ride out of ratchet step *S*. The pallet of an escapement mechanism is closely related to the pawl. *See* RATCHET.

[DOUGLAS P. ADAMS]

Bibliography: H. H. Mabie and F. W. Ocvirk, *Mechanisms and Dynamics of Machinery*, 3d ed., 1978; W. J. Patton, *Kinematics*, 1979.

PAWL

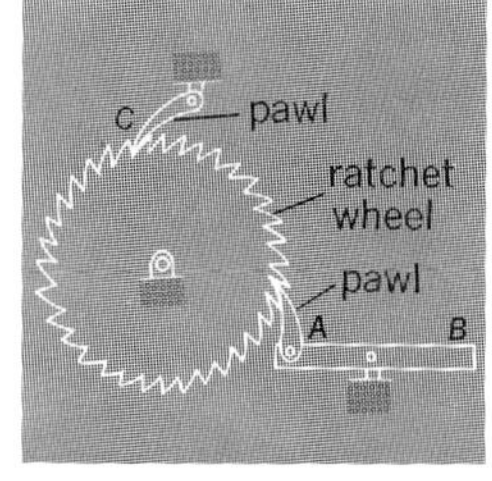

Fig. 1. Holding and driving pawls with a ratchet wheel.

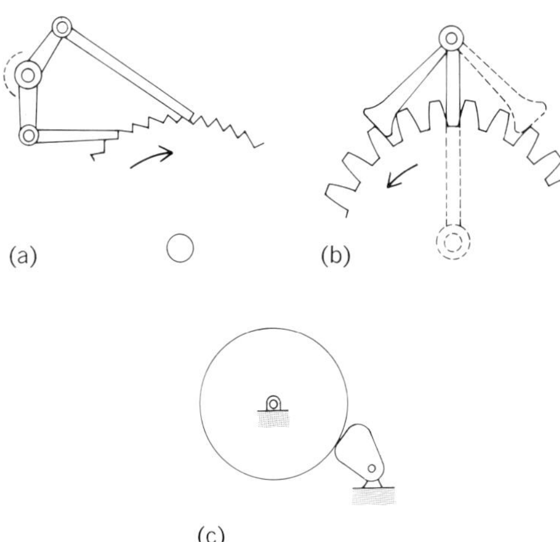

Fig. 2. Some pawl applications. (*a*) Double-acting pawl. (*b*) Reversible pawl. (*c*) Cam pawl.

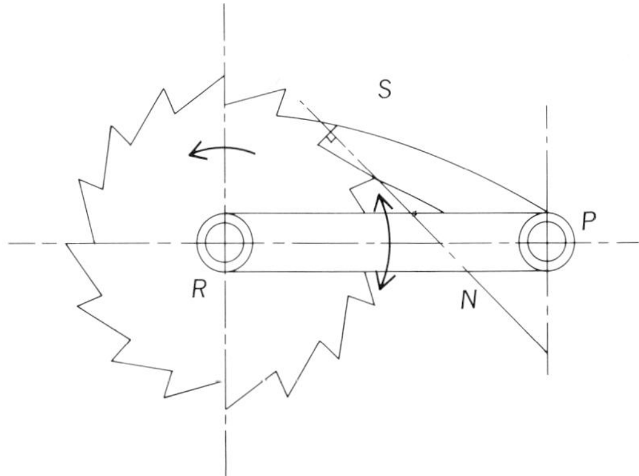

Fig. 3. Essential geometry of pawl design.

Pebble mill

A tumbling mill that grinds or pulverizes materials without contaminating them with iron. Because the pebbles have lower specific gravity than steel balls, the capacity of a given size shell with pebbles is considerably lower than with steel balls.

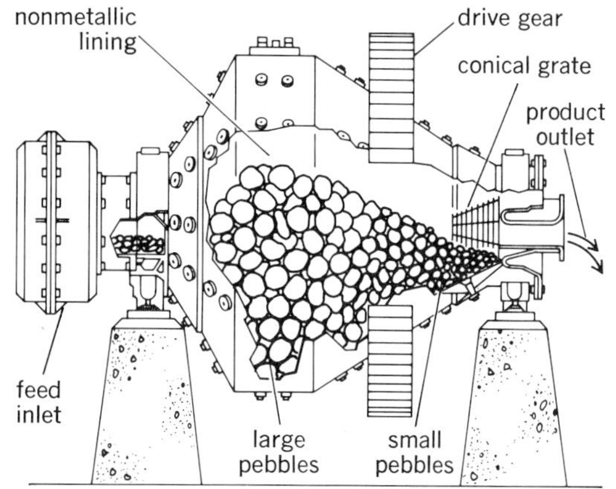

Diagrammatic sketch of a conical pebble mill.

The lower capacity results in lower power consumption. The shell has a nonmetallic lining to further prevent iron contamination, as in pulverizing ceramics or pigments (see illustration). Selected hard pieces of the material being ground can be used as pebbles to further prevent contamination. *See* TUMBLING MILL. [RALPH M. HARDGROVE]

Peening

A metal-finishing operation, also called shot peening, in which small steel shot is thrown against a piece such as a cutting tool. The impact of the shot on the work plastically deforms the surface to a depth of a few thousandths of an inch, producing residual compressive stress. The material is thus made more resistant to fatigue failure. Surface hardness of the material is also increased slightly by the cold working produced by the shot. The shot is hurled at high velocity upon the work by centrifugal force or by an air blast. *See* MACHINING OPERATIONS. [ALAN H. TUTTLE]

Pelton wheel

An impulse type of hydraulic turbine. In the impulse turbine, pressure of the water supply is converted into velocity by a nozzle. The water jet then impinges on the buckets of the turbine wheel or runner. In a Pelton wheel the buckets have a split-

A 12-in. (30 cm) Pelton wheel, a hydraulic turbine. (*Photograph by R. L. Daugherty and P. Kyropoulas*)

ter in the middle to divide the water jet and cause it to flow across the cupped faces of the buckets and emerge at the sides (see illustration). The water jet thus imparts its kinetic energy to the buckets. Pelton wheels are usually operated from high-head sources. *See* HYDRAULIC TURBINE.

[FRANK H. ROCKETT]

Bibliography: O. A. Johnson, *Fluid Power for Industrial Use*: *Hydraulics*, vol. 2, 1980.

Performance rating

A procedure for determining the value for a factor which will adjust the measured time for an observed task performance to a task time that one would expect of a trained operator performing the task, utilizing the approved method and performing at normal pace under specified workplace conditions. Since the introduction of stopwatch time study more than a hundred years ago, there has been a need to adjust an observed operator time to a "normal time." Normal time is the time that a trained worker requires to perform the specified task under defined workplace conditions, employing the assumed philosophy of "a fair day's work for a fair day's pay."

Normal performance must ultimately be subjectively based on the observer's opinion of what represents a fair or equitable pace of activity for a worker for a specified task under specified conditions. It is assumed under the "fair day's work for a fair day's pay" philosophy that skilled operators of average physique for the tasks they are performing, performing widely differing tasks under widely varying conditions, will be equally fatigued at the end of an average day's work. During a normal workday, a specified amount of time is set aside as nonwork time and is called allowances. The performance rating process is concerned with determining normal pace during the work portion of an average day and must, therefore, consider the fatigue recovery aspects of allowance times occurring during the day.

The following two equations relate factors in determining how much time a worker will be allowed per unit of output:

Standard time = normal time + allowances
Normal time = observed time × rating factor

If the observed time for a task is adjusted by the performance rating factor to determine normal time, and allowance time is added for nonwork time, the standard time will represent the allowed time per unit of production. For example, if the observed time is 60 min, and the rating factor is estimated to be 80%, and the allowances are 15% of normal time, then the normal time is $60 \times 0.8 = 48$ min, and the standard time is $48 + 0.15(48) = 55.2$ min. In a typical 8-h shift, the expected production from this worker would be $8(60)/55.2 = 8.7$ units of production.

Two frequently used bench marks of normal pace are walking and dealing playing cards. These bench marks are walking at a pace of 3 mph (1.33 m/s), and dealing a deck of 52 cards into four piles 1 ft (30 cm) apart in 0.5 min. Films of these activities being performed at different speeds are frequently viewed by performance raters so as to have their ratings be more consistent and more standardized.

What has been observed in rating has been variously called speed, effort, tempo, pace, or some other word connoting rate of activity or exhibited effort. In 1927 a technique referred to as leveling, which required the evaluation of four factors—skill, effort, conditions, and consistency—was developed by the Westinghouse Corporation. Evaluation scales for the first two factors, effort and skill (see table) were employed in determining values for these factors. The factors of skill, conditions, and consistency were employed to adjust what was observed to what was intended as standard for the specified task. However, if an average trained operator is observed performing the specified task under specified conditions, there is no underlying theoretical justification for considering these factors. Conditions and consistency, as adjustment factors, are not in common use today in employing this technique.

Leveling factors for performance rating

Effort			Skill		
Category	Code	Value	Category	Code	Value
Excessive	A1	+0.13	Superskill	A1	+0.15
	A2	+0.12		A2	+0.13
Excellent	B1	+0.10	Excellent	B1	+0.11
	B2	+0.08		B2	+0.08
Good	C1	+0.05	Good	C1	+0.05
	C2	+0.02		C2	+0.03
Average	D	0.00	Average	D	0.00
Fair	E1	−0.04	Fair	E1	−0.05
	E2	−0.08		E2	−0.10
Poor	F1	−0.12	Poor	F1	−0.16
	F2	−0.17		F2	−0.22

The word "effort" has been defined in the past in such unfortunate terms as "demonstration of the will to work effectively," which is possibly unobservable and likely to be unmeasurable and irrelevant. Regardless of an employee's will, it is the pace of a worker that the employer has a right to expect (that is, employers do not pay for trying, they pay for accomplishment).

In 1949 the original Westinghouse method was

revised extensively, and the new plan, called the performance rating plan, included three major classifications of factors—dexterity, effectiveness, and physical demand. Each classification contains from two to four factors. As indicated above, there is no theoretical justification for any factor other than pace, if the appropriate employee utilizes the specified method under specified conditions.

If two tasks of widely varying job difficulty are rated without consideration of the inherent relative job difficulty, and the two employees are paid on an incentive basis, the employee with the more difficult task is penalized in terms of potential incentive earnings, because although it may be an equally demanding task to perform at day work pace, the more difficult task is more difficult to increase in pace in comparison to the simpler task. The objective rating technique, first published in 1946, considers job difficulty and, in so doing, adjusts for this problem. Such subfactors as amount of body movement, use of foot pedals, bimanualness, eye-hand coordination, handling requirements, and weight are considered in evaluating the primary factor–job difficulty. Pace is the other primary factor considered.

The most commonly employed rating technique throughout the history of stopwatch time study, including the present, is referred to today as pace rating. A properly trained employee of average skill is time-studied while performing the approved task method under specified work conditions. Rating consists only of determining the relative pace (speed) of the operator in relation to the observer's concept of what normal pace should be for the observed task, including consideration of expected allowances to be applied to the standard. *See* HUMAN-FACTORS ENGINEERING; METHODS ENGINEERING; WORK MEASUREMENT.

[PHILIP E. HICKS]

Bibliography: R. M. Barnes, *Motion and Time Study*, 6th ed., 1968; M. E. Mundel, *Motion and Time Study: Principles and Practices*, 5th ed., 1978; B. W. Niebel, *Motion and Time Study*, 5th ed., 1972.

PERT

An acronym for program evaluation and review technique; a planning, scheduling, and control procedure based upon the use of time-oriented networks which reflect the interrelationships and dependencies among the project tasks (activities). The major objectives of PERT are to give management improved ability to develop a project plan and to properly allocate resources within overall program time and cost limitations; and to control the time and cost performance of the project, and to replan when significant departures from budget occur.

Background. In 1958 the U.S. Navy Special Projects Office, concerned with performance trends on large military development programs, introduced PERT on the Polaris weapons system. Since that time the use of PERT has spread widely throughout the United States and the rest of the industrialized countries. At about the same time that the Navy was developing PERT, the DuPont Company, concerned with the increasing cost and time required to bring new products from research to production, and to overhaul existing plants, introduced a similar technique called the critical path method (CPM). *See* CRITICAL PATH METHOD (CPM).

Requirements. The basic requirements of PERT, in its time or schedule form of application, are:

1. All individual tasks required to complete a given program must be visualized in a clear enough manner to be put down in a network composed of events and activities. An event denotes a specified program accomplishment at a particular instant in time; in effect, it represents a state of the project system. An activity represents the time and resources that are necessary to progress from one event to the next. Emphasis is placed on defining events and activities with sufficient precision so that there is no difficulty in monitoring actual accomplishment as the program proceeds. Figure 1 shows a typical operating-level PERT network from the electronics industry. Events are shown by squares, and activities are designated by arrows leading from predecessor to successor events.

2. Events and activities must be sequenced on the network under a logical set of ground rules. The activity sequencing is not arbitrary, but rather it is based on technological constraints; a foundation must be dug before the concrete can be poured. The network logic is merely the requirement that an event is said to occur when all predecessor activities are completed, and only then can the successor activities begin. The initial event, without predecessors, is self-actuated when the project begins, and the occurrence of the final event (without successor activities) denotes completion of the project. This logic requires that all activities in a network must be completed before the project is complete, and no "looping" of activities in the network is allowed. Another technique, called GERT, relaxes these logic constraints. *See* GERT

3. Time estimates can be made for each activity of the network on a three-way basis (the three numbers shown along the arrows in Fig. 1). Optimistic (minimum), most likely (modal), and pessimistic (maximum) performance time figures are estimated by the person or persons most familiar with the activity involved. The three-time estimates are used as a measure of uncertainty of the eventual activity duration; they represent the approach used in PERT to express the probabilistic nature of many of the tasks in development-oriented and nonrepetitive programs. It is important to note, however, that for the purposes of critical path computation and reporting, the three-time estimates are reduced to a single expected time T_E, and it is used in the same way that CPM employs a single (deterministic) time estimate of activity duration time.

4. Finally, critical path and slack times are computed. The critical path is that sequence of activities and events on the network that will require the greatest expected time to accomplish. Slack time is the difference between the earliest time that an activity may start (or finish) and its latest allowable start (or finish) time, as required to complete the project on schedule. Thus, for any event, it is a measure of the spare time that exists within the total network plan. If total expected activity time along the critical path is greater than the time

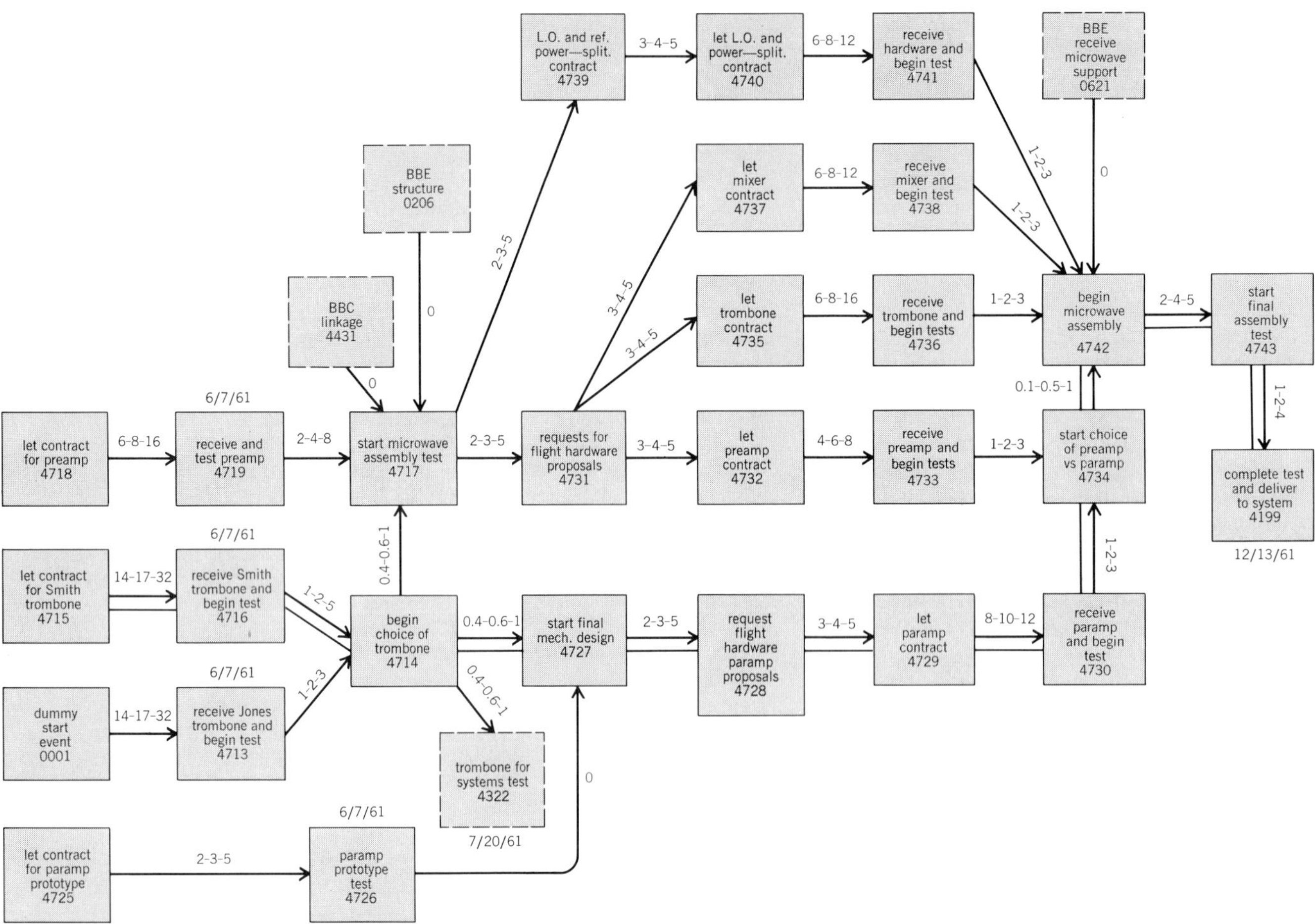

Fig. 1. Typical PERT network of an electronic module development project. Arrows and lines denote the critical path. (*Applied Physics Laboratory, Johns Hopkins University*)

available to complete the project, the program is said to have negative slack time. This figure is a measure of how much acceleration is required to meet the scheduled program completion date.

5. The difference between the pessimistic (b) and optimistic (a) activity performance times is used to compute the standard deviation (σ) of the hypothetical distribution of activity performance times ($\hat{\sigma} = (b - a)/6$). The PERT procedure employs these expected times and standard deviations (σ^2 is called variance) to compute the probability that an event will be on schedule, that is, will occur on or before its scheduled occurrence time. The procedure merely adds the expected time (T_E) and variances (σ^2) of the activities on the critical path to get the mean and variance of the hypothetical distribution of project duration times. [See columns headed EXP TIME and EXP VAR in Fig. 2, and the total values of 28.6 and 1.9, respectively, for the last activity (4004-743 to 4004-199) on the critical path.] The normal distribution is then used to approximate the probability of meeting the project schedule, as the area under the normal distribution curve to the left of (earlier than) the project scheduled completion date.

A computer-prepared analysis of the illustrative network contains data on the critical path (first group of activities in the table) and slack times for the other, shorter network paths. Note that the events (points in time) are labeled in the network, but the computer output is by activities (identified by event numbers) which also have descriptive labels. Under the column heading PROB(ability), note the figure 0.12 for the final activity on the critical path (4743-4199). This analysis indicates that the expected completion time of 12/25/61 results in a low probability of meeting the scheduled time of 12/13/61. This computer output (slack order report) is the most important of a number of outputs provided by most PERT computerized systems. Other reports may give greater or lesser details for different levels of management; they may deal with estimated and actual costs by activities (system called PERT COST); and so forth.

In the actual utilization of PERT, review and action by responsible managers is required, generally on a biweekly basis, concentrating on important critical path activities. Where necessary, effective means of shortening critical path time must be found by applying new resources or additional funds, often obtained from those activities that can afford them because of their slack condition. Alternatively, sequencing of activities along the critical path may be compromised to reduce overall duration. A final alternative may be a change in the scope of the work along the critical path to meet a given program schedule. Utilization of PERT requires constant updating and reanalysis, since the

RUN 1 ENDING EVENT
BY PATHS OF CRITICALITY DATE 06-07-61
CHART AJ LR SN 9 ELECTRONIC MODULE (ILLUSTRATIVE NETWORK) SYSTEM W 034

EVENT PREDECESSOR	SUCCESSOR	NOMENCLATURE	DEP	DATE EXPECTED	ALLOWED	DATE SCHD/ACT	PROB	SLACK	EXP TIME	EXP VAR
4004-715	4004-716	REV DATE (SMITH TROMBONE RECD-BEG TEST)	98		05-26-61	A06-07-61		-1.6	+	
4004-716	4004-714	SMITH TROMBONE TESTED	0146	06-23-61	06-12-61			-1.6	+ 2.3	.4
4004-714	4004-727	TROMBONE CHOSEN-BEGIN MECH DESIGN	0146	06-28-61	06-16-61			-1.6	+ 3.0	.5
4004-727	4004-728	RFP PARAMP FLIGHT HARDWARE	0146	07-20-61	07-08-61			-1.6	+ 6.1	.7
4004-728	4004-729	PARAMP CONTRACT LET	0146	08-17-61	08-05-61			-1.6	+10.1	.8
4004-729	4004-730	PARAMP RECEIVED		10-26-61	10-14-61			-1.6	+20.1	1.3
4004-730	4004-734	PARAMP TESTED	0146	11-09-61	10-28-61			-1.6	+22.1	1.4
4004-734	4004-742	CHOICE BETWEEN PREAMP-PARAMP	0146	11-13-61	11-01-61			-1.6	+22.6	1.4
4004-742	4004-743	COMPL MICROWAVE ASSY	0146	12-09-61	11-28-61			-1.6	+26.5	1.6
4004-743	4004-199	COMPL FINAL TEST MICWAVE ASSY-DELIVERED	0146	12-25-61	12-13-61	12-13-61	.12	-1.6	+28.6	1.9
4000-001	4004-713	REV DATE (JONES TROMBONE RECD-BEG TEST)	99		05-29-61	A06-07-61		-1.3	+	
4004-713	4004-714	JONES TROMBONE TESTED	0146	06-21-61	06-12-61			-1.3	+ 2.0	.1
4004-714	4004-717	TROMBONE CHOSEN-BEGIN MICWAVE ASSY TEST	0146	06-28-61	07-02-61			+ .5	+ 3.0	.5
4004-717	4004-731	RFP FOR FLIGHT HDW-MIXER-TROMB-PREAMP	0146	07-20-61	07-24-61			+ .5	+ 6.1	.7
4004-717	4004-739	COMPL MICWAVE ASSY TEST-RFP LOC OSCIL	0146	07-20-61	07-24-61			+ .5	+ 6.1	.7
4004-731	4004-735	TROMBONE CONTRACT LET	0146	08-17-61	08-21-61			+ .5	+10.1	.8
4004-731	4004-737	MIXER CONTRACT LET	0146	08-17-61	08-21-61			+ .5	+10.1	.8
4004-739	4004-740	CONTRACT LET FOR LOC OSCIL AND PWR SPLT	0146	08-17-61	08-21-61			+ .5	+10.1	.8
4004-735	4004-736	TROMBONE RECEIVED		10-14-61	10-18-61			+ .5	+18.5	1.8
4004-737	4004-738	MIXER RECEIVED		10-14-61	10-18-61			+ .5	+18.5	1.8
4004-740	4004-741	LOC OSC-PWR SPLITTER RECEIVED		10-14-61	10-18-61			+ .5	+18.5	1.8
4004-736	4004-742	TROMBONE TESTED	0146	10-28-61	11-01-61			+ .5	+20.5	1.9
4004-738	4004-742	MIXER TESTED	0146	10-28-61	11-01-61			+ .5	+20.5	1.9
4004-741	4004-742	LOC OSC-PWR SPLITTER TESTED	0146	10-28-61	11-01-61			+ .5	+20.5	1.9

Fig. 2. Typical PERT computer output. First three paths of Fig. 1 are shown here. *(From J. J. Moder and C. R. Phillips, Project Management with CPM and PERT, 2d ed., Van Nostrand–Reinhold, 1970)*

outlook for completion of activities in a complex program is constantly changing. Systematized methods of handling this aspect have been developed.

Advantage. A major advantage of PERT is the kind of planning required to create an initial network. Network development and critical path analysis reveal interdependencies and problem areas before the program begins that are often not obvious or well defined by conventional planning methods. Another advantage, especially where there is a significant amount of uncertainty, is the three-way estimate. If there is a minimum of uncertainty, the single-time approach may be used while retaining the advantages of network analysis.

In summary, it should be stated that while the developments of PERT and CPM were independent, they are both based on the same network logic to represent the project plan. PERT emphasizes the time performance of a project, including a probabilistic treatment of the uncertainty in the activity performance times and scheduled completion dates, while CPM treats time deterministically and addresses the problem of minimizing total (direct plus indirect) project cost as a function of scheduled project duration. The acronym PERT is now used as a generic term for network-based project management schemes that have evolved over the years. Today these schemes are hybrids of both PERT and CPM, but they are most often referred to as PERT. Finally, mention should be made of GERT, which denotes a generalization of the PERT/CPM network logic to complex situations where branching at events and closed loops of activities are required to adequately portray a complex project plan in the form of a network. See INDUSTRIAL COST CONTROL. [JOSEPH J. MODER]

Bibliography: D. G. Malcolm et al., Applications of a technique for R and D program evaluation, *Operations Res.*, 7(5):646–669, 1959; J. J. Moder and S. E. Elmaghraby, *Handbook of Operations Research, Models and Applications*, vol. 2, 1978; J. J. Moder and C. R. Phillips, *Project Management with CPM and PERT*, 2d ed., 1970.

Petroleum engineering

An eclectic discipline comprising the technologies used for the exploitation of crude oil and natural gas reservoirs. It is usually subdivided into the branches of petrophysical, geological, reservoir drilling, production, and construction engineering. After an oil or gas accumulation is discovered, technical supervision of the reservoir is transferred to the petroleum engineering group, although in the exploration phase the drilling and petrophysical engineers have played a role in the completion and evaluation of the discovery.

Petrophysical engineering. The petrophysical engineer is perhaps the first of the petroleum engineering group to become involved in the exploitation of the new discovery. By the use of down-hole logging tools and of laboratory analysis of cores made during the drilling operation, the petrophysicist estimates the porosity, permeability, and oil content of the reservoir rock that has been sampled at the drill site.

Geological engineering. The geological engineer, using the petrophysical data, the seismic surveys conducted during the exploration operations, and an analysis of the regional and environmental geology, develops inferences concerning the lateral continuity and extent of the reservoir. However, this assessment usually cannot be verified until additional wells are drilled and the geological and petrophysical analyses are combined to produce a firm diagnostic concept of the size of the reservoir, the distribution of fluids therein, and the nature of the natural producing mechanism. As the understanding of the reservoir develops with continued drilling and production, the geological engineer, working with the reservoir engineer, selects additional drill sites to further develop and optimize the economic production of oil and gas. *See* SEISMIC EXPLORATION FOR OIL AND GAS.

Reservoir engineering. The reservoir engineer, using the initial studies of the petrophysicist and geological engineers together with the early performance of the wells drilled into the reservoir, attempts to assess the producing rates (barrels of oil or millions of cubic feet of gas per day) that individual wells and the entire reservoir are capable of sustaining. One of the major assignments of the reservoir engineer is to estimate the ultimate production that can be anticipated from both primary and enhanced recovery from the reservoir. The ultimate production is the total amount of oil and gas that can be secured from the reservoir until the economic limit is reached. The economic limit represents that production rate which is just capable of generating sufficient revenue to offset the cost of operating the reservoir. The proved reserves of a reservoir are calculated by subtracting from the ultimate recovery of the reservoir (which can be anticipated using available technology and current economics) the amount of oil or gas that has already been produced. *See* PETROLEUM RESERVES.

Primary recovery operations are those which produce oil and gas without the use of external energy except for that required to drill and complete the wells and lift the fluids to the surface (pumping). Enhanced oil recovery, or supplemental recovery, is the amount of oil that can be recovered over and above that producible by primary operation by the implementation of schemes that require the input of significant quantities of energy. In modern times waterflooding has been almost exclusively the supplementary method used to recover additional quantities of crude oil. However, with the realization that the discovery of new petroleum resources will become an increasingly difficult achievement in the future, the reservoir engineer has been concerned with other enhanced oil recovery processes that promise to increase the recovery efficiency above the average 33% experienced to the United States (which is somewhat above that achieved in the rest of the world). The restrictive factor on such processes is the economic cost of their implementation. *See* PETROLEUM ENHANCED RECOVERY.

In the past the reservoir engineer was confined to making predictions of ultimate recovery by using analytical equations for fluid flow in the framework of an overall definition of the geological and lithological description of the reservoir. Extensive references to reservoirs that are considered to have analogous features, and history matching, such as curve fitting and extrapolation of the declining production, are important tools of the reservoir engineer. The analytical techniques are limited in that reservoir heterogeneity and the competition between various producing mechanisms (solution gas drive, water influx, gravity drainage) cannot be accounted for, nor can one reservoir be matched exactly by another. The reservoir engineer has also been able to use mathematical modeling to simulate the performance of the reservoir. The reservoir is divided mathematically into segments (grid blocks), and the appropriate flow equations and material balances are repeatedly applied to contiguous blocks. Extensive history matching is still required to achieve a reliable predictive model, but this can usually be achieved more quickly and reliably than with the use of analytical solutions and analogy. The cost is high, however, and cannot always be justified. *See* PETROLEUM RESERVOIR ENGINEERING; PETROLEUM RESERVOIR MODELS.

Drilling engineering. The drilling engineer has the responsibility for the efficient penetration of the earth by a well bore, and for cementing of the steel casing from the surface to a depth usually just above the target reservoir. The drilling engineer or another specialist, the mud engineer, is in charge of the fluid that is continuously circulated through the drill pipe and back up to surface in the annulus between the drill pipe and the bore hole. This mud must be formulated so that it can do the following: carry the drill cuttings to the surface, where they are separated on vibrating screens; gel and hold cuttings in suspension if circulation stops; form a filter cake over porous low-pressure intervals of the earth, thus preventing undue fluid loss; and exert sufficient pressure on any gas- or oil-bearing formation so that the fluids do not flow into the well bore prematurely, blowing out at the surface. As drilling has gone deeper and deeper into the earth in the search for additional supplies of oil and gas, higher and higher pressure formations have been encountered. This has required the use of positive-acting blow-out preventers that can firmly and quickly shut off uncontrolled flow due to inadvertent unbalances in the mud system. *See* OIL AND GAS WELL DRILLING; ROTARY TOOL DRILL.

Production engineering. The production engineer, upon consultation with the petrophysical and reservoir engineers, plans the completion procedure for the well. This involves a choice of setting a liner across the formation or perforating a casing that has been extended and cemented across the reservoir, selecting appropriate pumping techniques, and choosing the surface collection, dehydration, and storage facilities. The production engineer also compares the productivity index of the well (barrels per day per pounds per square inch of drawdown around the well bore) with that anticipated from the measured and inferred values of permeability, porosity, and reservoir pressure to determine whether the well has been damaged by the completion procedure. Such comparisons can be supplemented by a knowledge of the rate at which the pressure builds up at the well bore when the well is abruptly shut-in. Using the principles of unsteady state flow, the reservoir engineer can evaluate such a buildup to assess quantitatively the nature and extent of well bore damage. Damaged wells, like wells of low innate productivity, can be stimulated by acidization, hydraulic fracturing, additional performation, or washing with selective solvents and aqueous fluids. *See* OIL AND GAS WELL COMPLETION.

Construction engineering. Major construction projects, such as the design and erection of offshore platforms, require the addition of civil engineers to the staff of petroleum engineering departments, and the design and implementation of natural gasoline and gas processing plants require the addition of chemical engineers.

Summary. The relative importance and numbers of petroleum engineers employed in the industry have increased in recent years because of the increasing value of crude oil and natural gas and the need for more economical recovery of these fluids. The technology has become increasingly sophisticated and demanding with the implementation of new recovery techniques and the expanding frontiers of the industry into the hostile territories of arctic regions and deep oceans. *See* OIL AND GAS, OFFSHORE.

[TODD M. DOSCHER]

Petroleum enhanced recovery

Novel technology designed to enhance the fraction of the original oil in place in a reservoir. Heightened interest in developing enhanced recovery technology has developed as it has become more certain that over two-thirds of the oil discovered in the United States, and a still greater percentage in the rest of the world, will remain unrecovered through the application of conventional primary and secondary (waterflood) operations. Thus there is a strong incentive for the development and implementation of advanced technology to recover some of this oil. *See* PETROLEUM RESERVES.

The problems encountered in developing such technology are very great because of the nature of fluid flow within a subsurface reservoir, and because of the inability of engineers to exercise any intimate degree of control over the flow and distribution of fluids within the reservoir. The only points of contact with the reservoir are at the surface of the producing and injection wells that lead down to the reservoir sands.

Reservoir quality. A subsurface reservoir comprises the interconnected pores of a sandstone whose genesis was the compaction and cementation of sand-rich sediments. Oil is also found in carbonate rocks in which the porosity may have resulted from solution, diagenesis, and fracturing. The flow of fluids through these reservoir rocks is along very tortuous and nonuniform microscopic channels, frequently interrupted by inclusions of shales and clays. Further, a single reservoir is usually composed of several successive beds or layers of rock which have significantly different permeabilities. As a result, there is a great tendency for oil being displaced by encroaching water or an expanding gas to be bypassed by the displacing phase. Since the path of least resistance is always followed, and because of the dendritic nature of the channels that can be delineated through the porous network, the opportunities for bypassing are numerous. The tendency to bypass is exacerbated by the differences in wettability of the reservoir minerals by water and oil. If the encroaching water succeeds in bypassing oil, then pressure gradients, above an attainable level, will be required to effect displacement of the trapped ganglia.

Thus, the occurrence of unrecoverable oil following waterflooding is not unanticipated. The overall recovery of 33% from reservoirs in the United States does not represent the mean of a continuous distribution, but is the volumetric average of discrete values which are quantitatively different from each other. The low-permeability Spraberry reservoir in West Texas is anticipated to yield a recovery of less than 10%, whereas many reservoirs in south Louisiana and east Texas will experience recoveries of well over 60% of the original oil in place. The difference between the two is due to the far better permeability and porosity development and the presence of a strong water drive in the latter reservoirs. The target for enhanced oil recovery would superficially appear to be very large in the Spraberry and less in the other reservoirs; however, the nature of the Spraberry mitigates against a high sweep and contact of the remaining oil by any enhanced recovery scheme, and therefore the realistic prospects for additional recovery are small.

Thermal technology. One exception to this relationship between unrecoverable oil and reservoir quality is the heavy-oil reservoirs of California. Numerous shallow reservoirs in that state are at a low temperature and a relatively low pressure, and have a high porosity that is saturated with medium- to high-viscosity oils. The viscosity is so high and the pressure so low that production rates are virtually uneconomic. Because of the much lower mobility of the viscous oils, injected water will readily finger through the reservoir without displacing a significant quantity of oil. During the late 1950s and early 1960s, following the Suez crisis, the industry developed technology by which the viscosity of the crude within the reservoir could be lowered and a suitable pressure gradient applied to achieve significant recovery of crude from such reservoirs. The methods are the steam soak, or cyclic stimulation, in which oil is produced from the same well into which steam had earlier been injected; the steam drive, which is much like a waterflood but with the substitution of steam for injected water; and wet, or quenched, in-situ combustion in which a steam drive is generated in situ by the combustion of a limited fraction of the oil in place.

Steam soak. Cyclic steam injection is very effective in reservoirs which have some modest degree of natural energy either due to gas in solution or due to a substantial thickness to support gravity drainage. It is an extremely profitable operation because of the rather short deferment of costs prior to the production of oil. However, it is usually limited to the recovery of only 5–20% of the oil in place because of the inability of the reservoir energy to maintain suitable rates of influx of oil into the region around the borehole where effective heating occurs. If wells are drilled closely, which is economically feasible in some very shallow reservoirs, the recovery in thick reservoirs can reach values as high as 50%.

Steam drive. The steam drive is more effective in displacing oil; recoveries in excess of 50% have been achieved, although at a lower thermal efficiency than in the case of the steam soak. The major drawback to the process is its intensive use of energy. About a third of the produced oil must be used as fuel for generating the steam. The extension of the process to the recovery of still more viscous oils (tar sands and bitumens), to reservoirs less than 40 ft (12 m) or so in thickness, and to reservoirs containing a low waterflood residual, less than 1000 bbl per acre-foot (13% by volume), will require the consumption of a still greater fraction of the produced oil. This is due to the fact that, for the conditions cited, the amount of heat that is required to raise the temperature of the reservoir and satisfy the heat losses becomes an increasing fraction of the energy contained in the produced crude.

The use of coal as a substitute for produced oil as boiler fuel and the possible use of solar-powered steam generators would increase the salable oil from a steam drive operation. Severe regulatory obstacles are presently in the way of using coal-fired systems, and there are severe economic hurdles that will have to be overcome for the widespread use of solar power.

In-situ combustion. This method, in which air and water are injected simultaneously in the preferred modification, has not fared as well as steam injection in its level of application. Capital and operating costs for air compressors, corrosion and the production of difficult fluid emulsions, and a lower reservoir sweep efficiency have contributed to its being a second choice even though its theoretical efficiency is higher, in that it avoids the heat losses in surface and well bore facilities and generates heat from residual oil.

Production rates. Over 250,000 bbl (40,000 m^3) of oil a day are being produced by steam drive and cyclic steam operations in California, and this rate will probably increase substantially in the years ahead as a result of developing technology, higher needs for liquid fuels, and higher prices. In Canada, where vast accumulations of heavy oil occur in the Athabasca Bituminous Sands and in the Peace River and Cold Lake districts in Alberta, and in eastern Venezuela along the Orinoco River, the application of steam drives will eventually boost the reserves of these nations significantly.

Other technologies. For most of the residual or unrecoverable oil in the United States, the use of thermal technology to increase recovery is out of the question because of the relatively low energy content of the recoverable oil (even if all the residual is recovered) compared to the energy required to heat the reservoir. Instead, technology must be used which focuses on liberating the oil, trapped by capillary forces, by reducing the interfacial tension between the oil and water in the reservoir. There are two basic schemes being pursued: the injection of aqueous solutions of surfactants that can reduce the interfacial tension to very low values, and the injection of solutes which can dissolve in and swell the residual oil and restore mobility to the trapped oil. The latter process ideally would use a displacing phase that, more than merely dissolving in the oil, would be completely miscible with it.

Micellar/polymer fluids. Aqueous solutions of surfactants have been developed that are capable of recovering virtually 100% of the residual oil in a laboratory experiment. These are known as micellar/polymer fluids since they contain surfactants at concentrations above the critical micellar value and polymers (such as polysaccharides and hydrolyzed polyacrylamides) that develop aqueous-phase viscosities that will assure stable displacement (minimum bypassing). The theoretical foundation for the use of such systems is fundamentally sound: a low interfacial tension means a low displacement pressure.

Field tests of the process, however, have been generally disappointing. This performance can be traced back to the fact that, because of the cost of these systems, the solutions of surfactants and polymers can be injected only as slugs, rather than continuously, if economic recovery of crude oil is to be achieved. The slug size is probably limited to less than 5% of the reservoir pore volume, certainly no more than 10%, and the integrity of the slug is weakened by numerous factors: temperature and shear degradation, precipitation by ions occurring in the connate water or released by ion exchange with the reservoir clays, adsorption on mineral surfaces, cross-flow and diffusion into low-permeability layers, and transfer of the active surfactants into the oil phase. Increasing oil prices do not result in a proportionate increase in profit potential, since the surfactants and polymers that are found to be useful are derived from petroleum or use petroleum products in their manufacture, and their price escalates with the price of crude. The micellar/polymer process, however, is potentially the most widely applicable scheme for enhanced oil recovery, and therefore research and development continue. A successful breakthrough in such research would have inestimable value for the United States and the world in increasing the ultimate recovery of crude oil.

Solvents. Again during the 1950s, liquefied petroleum gas (LPG; propane) was sufficiently inexpensive compared to crude oil that it could be considered as a sacrificial solvent for the tertiary recovery of residual oil, or, in some cases, even as a substitute for water in secondary recovery operations. Laboratory studies indicated that the efficiency of the LPG was impaired by its relatively high mobility compared to that of crude oil and water. The LPG tended to finger through the reservoir, recovering only a small fraction of the residual oil, and this at very high and uneconomic ratios of injected LPG to produced crude. A search for cheaper solvents revealed that carbon dioxide under high pressure was very soluble in many crudes and led to significant volumetric expansion and decreased viscosity of the residual oil. With some crudes there was further evidence that a miscible carbon dioxide–rich phase could be generated by repeated contacts of successively enriched phases. These observations gave rise to the expectation that successful recovery of crude could be achieved despite the very high mobility of the carbon dioxide and carbon dioxide–rich phases.

Field pilots have borne out the expectation that carbon dioxide would recover additional crude oil, but in tertiary (after waterflood) pilots the carbon dioxide/produced oil ratios are significantly higher than anticipated. (In reservoirs which have not been waterflooded, the injection of carbon dioxide has resulted in significant improvements in recovery at what appears to be economically viable ratios.) A particularly attractive province for the use of carbon dioxide is in dipping reservoirs on the flanks of salt domes in south Louisiana and offshore in the Gulf of Mexico. The updip injection of carbon dioxide is anticipated to result in a gravity-stabilized displacement of the residual oil.

Research and development activities on the use of carbon dioxide as an enhanced recovery agent are continuing, with several projects being devoted to reducing the mobility of the reagent by injecting it as a foam or taking advantage of intermediate phases created by mixing crude oil with carbon dioxide and additives. Development studies are under way on securing economic sources of carbon dioxide, both from natural reservoirs and from manufacturing operations.

Future productivity. It is difficult to anticipate the amount of reserves that can be added by successful implementation of enhanced oil recovery operations, and similarly difficult to anticipate the production rates that may be achieved. Several studies have indicated a marked sensitivity of enhanced oil recovery to the real market price of the produced crude oil. Considering the rapid escalation in the value of crude oil, the chief limiting factor on enhanced recovery will be the development and implementation of technology for amenable reservoirs. *See* OIL MINING.

[TODD M. DOSCHER]

Petroleum prospecting

The search for commercially valuable accumulations of petroleum. This search at the one extreme may be carried out in a completely haphazard manner with entire dependence on luck for success or, at the other extreme, it may be a highly organized procedure involving the use of complex precision instruments, skilled and experienced personnel, and advanced scientific reasoning. In either case the final and critical step is always the drilling of an exploratory hole. Moreover, in neither case can the successful outcome of the exploratory hole be assured in advance because no infallible means of detecting the presence of a commercial petroleum accumulation ahead of the drill has yet been devised. Much petroleum has been found both by luck and by the application of scientific methods, but statistics demonstrate that at the present time the success ratio of holes located with the benefit of scientific or technical advice is nearly twice as great as that of those located without such advice.

The classic requisities for petroleum accumulations are (1) source or mother rocks from which petroleum can have originated; (2) carrier and reservoir rocks possessing sufficient permeability to provide avenues of migration as well as sufficient porosity to provide storage space; (3) traps adequate to cause commercial concentration of petroleum at local points in the reservoir beds; and (4) proper time and spatial relations in the development of source, reservoir, and trap. A favorable hydrodynamic condition might also be mentioned as a requisite to initial accumulation as well as to later preservation of a petroleum deposit. *See* PETROLEUM RESERVOIR ENGINEERING.

Scientific petroleum prospecting consists of (1) determination of generally favorable regions with respect to source, reservoir, trap, timing, and hydrodynamic conditions; (2) finding local geological features (anticlines, fault traps, and pinch-outs) within these regions, believed to be suited to the trapping of petroleum; (3) location and programming of exploratory holes to test the presence or absence of commercially significant petroleum accumulations on these local features; and (4) after initial discovery, determination of the extent and character of the accumulation. Prospecting methods are the means employed to gain the information called for in these four steps. Prospecting methods are commonly classified as geological and geophysical, but there is no sharp distinction between the two; all involve geological reasoning and interpretation.

This article outlines two major aspects of petroleum prospecting—geological prospecting and geophysical prospecting.

GEOLOGICAL PETROLEUM PROSPECTING

Nearly all prospecting entails certain preliminary library and cartographic background research. Some mention of base maps is followed in this section by the topics of: surface geology; photogeology; drilling; structure and core drilling; wild-cat wells; subsurface geology; geological laboratory methods; and regional geology.

Base maps. A requisite to petroleum prospecting is accurate base-map control. Horizontal control is necessary for location of property boundaries, physical features, roads, wells, and other cultural features, and for the map location of the points from which geological or geophysical data are obtained. Vertical control is necessary for providing topographic information for operational purposes as well as for adjustment of geological, geophysical, and well data to a common datum. Topography may also be of important geological significance. Aerial photography and electronic and radio positioning systems are largely replacing the theodolite, alidade, and plane table for mapping and geographic control work both on land and over water, and these methods have advantages both in speed and in accuracy.

Surface geology. The examination and study of outcropping rocks as a clue to the structure and stratigraphy of an area is the oldest of petroleum-prospecting methods and one which is still extremely important. The surface geologist maps the topographic expression and distribution of exposed rock units, determines and plots their structural attitude, measures and describes stratigraphic sections, identifies surface structural anomalies such as anticlines or faults which may reflect deeper structures, and prepares cross sections showing the hypothetical distribution of rocks and structure at depth. Study of the rocks exposed at the surface may yield important information on the presence and position of source and reservoir rocks as well as on structural and stratigraphic accumulation traps. Finally, surface geological examination is the only means of acquiring information on the occurrence and location of petroleum seepages. There is no more encouraging indication of a petroliferous province than the presence of actual oil seepages. In difficult terrain, helicopters commonly attached to surface geological parties aid transportation and communication and facilitate geological observation.

Photogeology. The mapping of surface geologic features is frequently best carried out through study of aerial photographs. In addition to greatly expediting the study of the surface geology of any area, the photographic method provides coverage of regions where access on the ground would be prohibitively difficult. It usually provides more complete detail than is possible by surface-mapping methods and has the additional advantage of greater overall perspective. In regions where outcrops are scarce, photogeology is employed as a means of determining geologic structure and distribution of formations indirectly through interpretation of geomorphologic features, vegetation, fracture patterns, and soil characters. Photogeology does not replace surface study, which is always desirable, but it does constitute an extremely valuable supplement.

Drilling. There is no method of petroleum prospecting so effective as the drilling of a hole to the objective horizon, and if the cost of deep drilling were not so great this method would supplant almost all others. Even so, the great bulk of all money spent on petroleum prospecting goes for drilling of exploratory holes. *See* BORING AND DRILLING (MINERAL); OIL AND GAS WELL COMPLETION.

Structure drilling and core drilling. These terms are applied to relatively shallow drilling where the purpose is purely that of securing geological information. Highly portable drilling rigs with depth capacities of a few hundred to a few thousand feet are used, and on the basis of cuttings, cores, or electrical logs, information is obtained on near-surface stratigraphy and structure which may guide deeper drilling for petroleum.

Wildcat wells. Exploratory holes drilled with the aim of discovering new petroleum pools are true wildcat wells. Usually these are programmed and equipped for completion as producers if successful, but the so-called stratigraphic test hole, which penetrates potentially productive horizons, is aimed only at providing geological information and is not equipped for production. Even the drill is not always conclusive in prospecting for petroleum. Many potentially productive wells are abandoned as dry each year because of inefficient testing. Under current methods of drilling, the hole is usually kept filled with heavy mud to prevent caving and to hold back excessive fluid pressures; consequently, potentially productive petroleum horizons may frequently be penetrated by the bit with very little indication of their fluid content. To avoid overlooking such horizons, rock samples cut by the bit and brought up in the circulating mud are carefully and concurrently studied for petroleum indications; instrumental equipment is installed on the drilling rig to analyze the mud automatically for traces of petroleum; and electrical and radioactive devices are run down the hole from time to time to record the properties of the rocks penetrated with respect to the probability of their carrying petroleum. Likewise, cores and side-wall samples are taken from intervals suspected of being productive.

Subsurface geology. Regardless of whether production is obtained, an exploratory hole is usually a valuable contribution to prospecting knowledge. The study of the geological and geophysical data made available through drilling is called subsurface geology. The subsurface geologist stationed at the well constantly watches the cores, cuttings, and drilling fluid for direct traces of petroleum and studies the characteristics of electric logs, radioactive logs, and geothermal logs for indirect indications of petroleum. The lithology, paleontology, and mineralogy of the cores and cuttings also yield clues to the stratigraphic position at which the well is drilling and the remaining depth to objective producing horizons. Determination of the attitude of bedding from cores and from dip-meter surveys gives important evidence as to whether the well is off structure with respect to the fold, fault, or other trap structure on which it is being drilled and also gives a factor for correcting the drilled thickness of a formation to its true thickness.

As more wells are drilled in a region, the steadily increasing background of subsurface geological information becomes progressively more effective as a means of locating new structural or stratigraphic traps for testing. Correlation, the identification and tracing of stratigraphic units from one well to another, allows conclusions to be reached on the relative structural positions of wells, on the probable location of new fold and fault structures, on the presence of unconformities, and on lateral changes in thickness and lithology. These correlation data provide the subsurface geologist with the base for cross sections and various kinds of subsurface maps: structure-contour, isopach, paleogeologic, lithofacies, palinspastic, and others.

Geological laboratory methods. Many of the determinations which can usefully be made on rock samples, either from outcrops or from wells, require such specialized knowledge and equipment that the surface or subsurface geologist sends them to specialists in a geological laboratory. Paleontologic and micropaleontologic studies are valuable in determination of the age or stratigraphic position of samples, in correlation, and in determination of past environments of deposition which may bear on source, reservoir, and stratigraphic trap conclusions. Study of the Foraminifera and Ostracoda has been particularly useful in petroleum prospecting, and spore and pollen studies have recently been increasing in importance.

Laboratory determination of heavy detrital minerals furnishes useful information for correlation and provenance. Among other laboratory methods which may be useful in identification and correlation are analysis for insoluble residues, size and shape analysis, differential thermal analysis, and calcimetry. Refractive-index determinations made on solvent extracts from rock samples provide useful information on the presence and gravity of even minute traces of oil. Computers and data-processing machines are being employed in some geological laboratories to aid in the sorting and analysis of large batches of data from surface geology and wells.

Regional geology. In petroleum prospecting the various contributions of surface geology, subsurface geology, geophysics, and other methods should all be put together and coordinated to give as complete a regional geologic picture as possible. Given adequate information on the character and attitude of the physical rock framework of a region, its geologic history, and the conditions of movement of its fluids, it should be theoretically possible to predict the location of all its petroleum accumulations. This information is of course never fully forthcoming, but the acquisition of as much of it as can be obtained and the imaginative but intelligent extrapolation of the remainder from experience are essential to long-range success in petroleum prospecting.

[HOLLIS D. HEDBERG]

GEOPHYSICAL PROSPECTING FOR PETROLEUM

Geophysical techniques have contributed decisively to the world supply of oil since about 1930. The seismograph technique has accounted for most of this activity. The seismic method has been the most expensive of those available. For this reason the cheaper gravity and magnetic methods have often been used for reconnaissance purposes, and the more limited anomalous areas thus revealed are then subjected to seismic investigation. Means and procedures have been developed which permit seismic operations in coastal waters at unit costs comparable to those of gravity and magnetic surveys under the same conditions. The rate of expenditure is very high for such operations, but the output rate is also high, so that acceptable unit costs are achieved. The seismic, gravity, magnetic, electrical, and various well-logging methods account for all but a tiny fraction of the geophysical work done in the search for oil.

The geophysical measurements made in oil prospecting are to a large extent related to the configuration and properties of the rocks which enclose oil pools. At best, the results of geophysical surveys indicate the presence, position, and nature of a structure which may or may not contain oil. The discovery of the oil itself is made by drilling a hole into the structure. If oil is found, the well serves the purposes of both discovery and exploitation.

Seismic surveys. Contour maps are generally produced to show the elevation of geological horizons with reference to some datum plane, preferably at the general depth level of formations which are known to produce oil elsewhere or which are thought to be potential reservoirs. The data furnish evidence of possible traps for petroleum accumulation, which must be identified and confirmed by drilling.

This method has progressed considerably in the transition to digital methods of processing the field data. Therefore, large masses of data can be analyzed, and use can be made of new communication theory, velocity filtering, and nondynamite sources. This enables exploration of the remaining more difficult geological provinces, for example, the coastal areas and the deep formations of West Texas and New Mexico, and detection of stratigraphic traps. Seismic surveys furnish the most conclusive evidence available from any geophysical technique. *See* SEISMIC EXPLORATION FOR OIL AND GAS.

Electrical methods. Electrical methods, except for drill hole surveys, have been used very little in prospecting for petroleum in the free world. The Soviet Union, however, is finding them useful, particularly for reconnaissance in unexploited areas.

Remote sensing. Improved air photography, using infrared and microwave, is being tested for reconnaissance work and surface indicators of deeper oil accumulations.

Magnetic surveys. In the search for oil, magnetic surveys are now made almost exclusively from aircraft. Oil is found in deep sedimentary basins, and the magnetic anomalies found in such areas arise from the igneous floor beneath the sediments. The depth below surface to the igneous basement rocks usually is thousands of feet. Aeromagnetic surveys have been made over millions of square miles.

The sedimentary structures which are oil-bearing often lie above uplifts or topographic features of the igneous basement surface. Local magnetic anomalies are associated with such features and therefore are a key to the discovery of basement uplifts. Other anomalies are related to differences in the magnetization of the igneous rocks. If anomalies are present in sufficient numbers and well distributed, the configuration of the sedimentary basin can be predicted and the principal structural features in the basement indicated in advance of any drilling.

In areas where the sedimentary structure arises mainly from thrusting (force applied sidewise on rockbeds), magnetic surveys may be of little help.

Gravity surveys. These have been most successful in discovering and detailing salt domes, a great percentage of which have associated oil accumulation. A newly discovered salt dome therefore represents an oil prospect of high potential. A salt dome usually contains one or more cubic miles of salt, which is ordinarily of lower density than most of the surrounding sediments. A gravity minimum is therefore characteristic of a salt-dome structure.

The gravity manifestation of other structural types is generally more complex. If a structure involves the position of dense beds nearer to the surface, a gravity high will be found. Such anomalies are customarily investigated further by seismic techniques before drilling is undertaken.

A great percentage of the potential oil-producing areas of the United States has been covered by gravity surveys.

Well logging. Geophysical techniques of well logging are now applied to practically every well drilled by the oil industry. As in pregeophysical days, geologists prepare a graphical log of the formations through which a drill hole extends, based on visual examination of drill cuttings brought to the surface by the drilling mud and on core samples. Such logs show lithology and fossil distribution with depth. Structure maps result from correlations between well horizons which can be identified as being the same or substantially equivalent in all of them. Geophysical well logging is merely an extension of this procedure to other physical properties which require physical measurements for their determination.

Various geophysical well-logging methods have been developed, electrical, radioactive, acoustic, and gravity.

[G. E. ARCHIE]

Bibliography: M. B. Dobrin, *Introduction to Geophysical Prospecting*, 3d ed., 1976; G. D. Hobson and E. N. Tiratsoo, *Introduction to Petroleum Geology*, 1980; K. K. Landes, *Petroleum Geology*, 1959, reprint 1975; L. W. LeRoy and D. O. LeRoy, *Subsurface Geology in Petroleum Mining Construction*, 4th ed., 1977; A. I. Levorsen, *Geology of Petroleum*, 2d ed., 1967; G. B. Moody (ed.), *Petroleum Exploration Handbook*, 1961; W. E. Wrather and F. H. Lahee (eds.), *Problems of Petroleum Geology*, 2 vols., 1976.

Petroleum reserves

Proved reserves are the estimated quantities of crude oil liquids which with reasonable certainty can be recovered in future years from delineated reservoirs under existing economic and operating conditions. Thus, estimates of crude oil reserves do not include synthetic liquids which at some time in the future may be produced by converting coal or oil shale, nor do reserves include fluids which may be recovered following the future implementation of a supplementary or enhanced recovery scheme.

Indicated reserves are those quantities of petroleum which are believed to be recoverable by already implemented but unproved enhanced oil recovery processes or by the application of enhanced recovery processes to reservoirs similar to those in which such recovery processes have been proved to increase recovery.

Thus, crude oil reserves can be called upon in the future with a high degree of certainty, subject of course to the limitations placed on production rate by fluid flow within the reservoir and the capacity of the individual producing wells and surface facilities to handle the produced fluids. It is important to bear in the mind the distinction between resources and reserves. The former term refers to the total amount of oil that has been discovered in the subsurface, whereas the latter refers to the amount of oil that can be economically recovered in the future. The ratio of the ultimate recovery (the sum of currently proved reserves and past production) to the resource or original oil in place is the anticipated recovery efficiency. *See* PETROLEUM ENHANCED RECOVERY.

Levels. In earlier years crude oil reserves were estimated by first defining the volume of the resources from drilling data, the nature of the natural producing mechanism from the performance of the reservoir, particularly the rate of decline in productivity, and then applying a recovery factor based on analogy with similar reservoirs. Although more sophisticated technology is in use today, earlier rule-of-thumb estimates have proved to be surprisingly valid.

Reserves are increased by the discovery of new reservoirs, by additions to already discovered reservoirs by continued drilling, and by revisions due to a better-than-established anticipated perforformance or implementation of an enhanced recovery project. Reserves are decreased by production, and by negative revisions due to poorer-than-anticipated performance or less-than-projected reservoir volumes. *See* PETROLEUM RESERVOIR MODELS.

In the 15 years between 1954 and 1969, reserves of crude oil in the United States remained relatively stable. In 1969 the reserves jumped to a value of 39×10^9 bbl (6.2×10^9 m^3) as a result of the discovery of the gigantic Prudhoe Bay oil field on the North Slope of the Brooks Range in Alaska. In the 1970s the reserves in the United States fell steadily, reaching at the end of 1978 the lowest level since 1952. Some giant oil fields have been discovered in the Overthrust Belt of the Rocky Mountains, and further exploration in this area is continuing.

Discovery and production. In retrospect, it can be seen that reserves in the United States were sustained during the 1950s and early 1960s as a result of extensions and revisions, the latter due primarily to the installation of waterfloods, rather than as a result of significant new discoveries.The discovery at Prudhoe Bay indicated the necessity of exploration in new frontiers if new reserves were to be added to the United States total. Subsequent exploration failures in other frontier areas, such as the eastern Gulf of Mexico, the Gulf of Alaska, and the Baltimore Canyon, however, again emphasized the limited occurrence of crude oil in the Earth's crust.

The existence of an accumulation of petroleum fluids in the Earth's crust represents the fortuitous sequence of several natural events: the concentration of a large amount of organic material in ancient sediments, sufficient burial and thermal history to cause conversion to mobile fluids, and the the migration and eventual confinement of the fluids in a subsurface geological trap. Although some oil and gas have been produced in 33 of the 50 states, 4 states account for 85% of the proved reserves: Texas and Alaska account for over 60%, with California and Louisiana ranking third and fourth. Further evidence for the spotty concentration of oil in the crust is gleaned from the fact that although there are some 10,000 producing oil fields in the United States, some 60 of them account for over 40% of the productive capacity.

In this context, it was to be anticipated that oil discovery and production would reach a peak followed by a monotonic decline in reserves and productive capacity. Only the date of peaking might be in doubt, but even this was predicted with a high degree of accuracy at least a decade before it occurred in the late 1960s in the United States. Additional oil discoveries will be made in some of America's remaining new frontiers, such as the deep ocean and in the vicinity of the Arctic Circle; however, most estimates of the amount of oil remaining to be found have been steadily decreasing, and it is highly unlikely that the decline will be substantially affected by future events.

Economics. The effect of economics on reserves has received much attention because of the fact that recovery of crude oil in the United States is expected to reach an average value of only 33% of the discovered resource, and because of the steadily rising price of crude oil. Despite a more than fivefold increase in the average price of crude oil since 1975, reserve additions have not increased significantly from their prior rate.

The basic reason for this can be deduced from the illustration, which shows the oil production rate from a reservoir as a function of time. Produc-

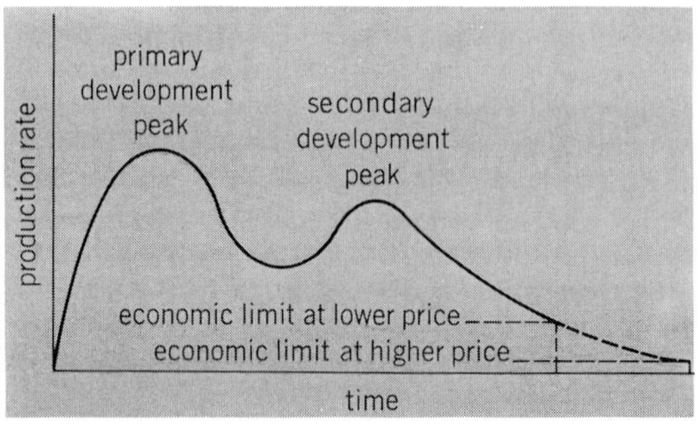

Oil production rate from a reservoir as a function of time.

tion is stopped upon reaching the economic limit, which is the rate of production at which the income from the sale of the produced oil is just sufficient to pay the operating costs, taxes, and overhead encountered in achieving the production. An increase in the sales price (assuming that the associated increase in costs is not completely compensating) will extend the economic limits to a still lower limit. However, the volume of oil produced between any two time points is the area under the curve, and it is obvious that for mature fields the additional production of fluids will be a small percentage of the total producible to the original economic limit. The effect of higher prices for crude oil will provide a stimulant to the exploration for new reserves in hostile, frontier areas and in the development of new technology for enhanced recovery.

Natural gas. Natural gas reserves are estimated in a manner similar to that used for the estimation of crude oil reserves. Some natural gas is found in association with crude oil in the same reservoir, and some is found in reservoirs which do not have a gas-oil contact, that is, nonassociated gas. Gas represents a higher maturation level than that of crude oil, and therefore some geographical areas are more gas-prone than others. The history of reserves of natural gas in the United States parallels that of crude oil, with a peak having been reached in the late 1960s. The ratio of gas reserves to oil reserves in the United States is approximately twice the ratio for the world.

[TODD M. DOSCHER]

Bibliography: Worldwide report, *Oil Gas J.*, published in last issue of each year.

Petroleum reservoir engineering

The technology concerned with the prediction of the optimum economic recovery of oil or gas from hydrocarbon-bearing reservoirs. It is an eclectic technology requiring coordinated application of many disciplines: physics, chemistry, mathematics, geology, and chemical engineering.

Originally, the role of reservoir engineering was exclusively that of counting oil and natural gas reserves. The reserves—the amount of oil or gas that can be economically recovered from the reservoir—are a measure of the wealth available to the owner and operator. It is also necessary to know the reserves in order to make proper decisions concerning the viability of downstream pipeline, refining, and marketing facilities that will rely on the production as feedstocks.

The scope of reservoir engineering has broadened to include the analysis of optimum ways for recovering oil and natural gas, and the study and implementation of enhanced recovery techniques for increasing the recovery above that which can be expected from the use of conventional technology.

Original oil in place. The amount of oil in a reservoir can be estimated volumetrically or by material balance techniques. A reservoir is sampled only at the points at which wells penetrate it. By using logging techniques and core analysis, the porosity and net feet of pay (oil-saturated interval) and the average oil saturation for the interval can be estimated in the immediate vicinity of the well. The oil-saturated interval observed at one location is not identical to that at another because of the inherent heterogeneity of a sedimentary layer. It is therefore necessary to use statistical averaging techniques in order to define the average oil content of the reservoir (usually expressed in barrels per net acre-foot) and the average net pay. The areal extent of the reservoir is inferred from the completion of dry holes beyond the productive limits of the reservoir. The definition of reservoir boundaries can be heightened by study of seismic surveys and analysis of pressure buildups in wells after they have been brought on production.

If the only mobile fluid in the reservoir is crude oil, and the pressure during the production of the crude remains above the bubble point value (at which point the gas begins to come out of solution), the production of crude oil will be due merely to fluid expansion, and the appropriate equation is Eq. (1). Here c is the compressibility of the fluid, V

$$\text{Production} = N_p = (c \cdot V \cdot \Delta P) B_0 \qquad (1)$$

is the volume of crude oil in the reservoir, and ΔP is the decrease in average reservoir pressure associated with the production of N_p barrels of crude oil measured at the surface. The term in parentheses is the expansion measured in reservoir barrels, and B_0 is the formation volume factor (inverse of shrinkage) which corrects subsurface measurements to surface units. Thus, if N_p is plotted as a function of $(c \cdot \Delta P)B_0$, the slope of the resulting straight line will be V, the volume of the original oil is place in the reservoir.

For most reservoirs, the material balance equation is considerably more complicated. Few reservoirs would produce more than 1 or 2% of the oil in place if fluid expansion alone were relied upon. The reasons for the increased complexity of a material balance equation when the pressure falls below the bubble point are as follows: (1) The pore volume occupied by fluids in the reservoir changes with reservoir pressure as does the volume of the immobile (connate) water in the reservoir. (2) As the pressure declines below the bubble point, gas is released from the oil. (3) The volume of gas liberated is greater than the corresponding shrinkage in the oil, and therefore the liberated gas becomes mobile upon establishing a critical gas saturation, and the gas will flow and be produced in parallel with oil production. (4) There may be an initial gas cap above the oil column (above a gas-oil contact), and some of this gas, depending upon the nature of the well completion, may flow immediately upon opening the well to production. (5) Below the oil column, there may be an accumulation of water (below a water-oil contact), and if sufficiently large, the water in this aquifer will expand under the influence of the pressure drop and replace the oil that is produced from the oil column.

Thus the simple material balance equation (1) for fluid expansion must be expanded to take into account all the changes in fluid and spatial volumes and the production of oil and gas. By the suitable manipulation of the various terms in the resulting equation, the material balance can be reduced to the equation of a straight line with the slope and intercept of the line providing information about the magnitude of the original oil in place and the significance of the water encroachment and the magnitude of the gas cap. However, the accuracy of the material balance equation depends strongly on having obtained accurate knowledge about the *PVT* (pressure-volume-temperature) behavior of the reservoir crude, and knowing the average reservoir pressure corresponding to successive levels of reservoir depletion. There are significant limits to acquiring such information.

The material balance equation is not time-dependent; it relates only average reservoir pressure to production. Although it can be extrapolated to provide information on the amount of oil and gas that may be produced in the future when the pressure falls to a given level, it does not provide any information as to when the pressure will fall to such a level. It can be used to estimate the ultimate production from the reservoir if an economically limiting pressure can be specified. This usually can be done since pressure and rate of production are of course explicitly interrelated by the equations for fluid flow, and the economic limit of production is that rate at which the economic income is insufficient to pay for the cost of operation, royalties, taxes, overheads, and other facets of maintaining production.

Fluid flow in crude oil reservoirs. In order to develop some understanding of the future performance of a reservoir on a real-time basis, the reservoir engineer has two options. One is to predict the flow rate history of the reservoir by using the equations for fluid flow, and the second is to extrapolate the already known production history of the reservoir into the future by using empiricism and know-how generated by experience with analogous reservoirs.

The flow of fluids in the reservoir obeys the differential form of the Darcy equation, which for radial flow is Eq. (2), where q is the volumetric flow

$$q=\frac{k\cdot h\cdot 2\pi r(\Delta P/\Delta r)}{\mu} \qquad (2)$$

rate measured in subsurface volumes, r is the radial distance from the well bore to any point in the reservoir, $(\Delta P/\Delta r)$ is the pressure gradient at the corresponding value of the radius, μ is the viscosity of the mobile fluid, and k is the permeability of the reservoir to the mobile fluid.

This equation could be integrated directly if $\Delta P/\Delta r$ and the pressure at the outer limits of the reservoir were constant. However, this condition is true only where there exists a strong natural water drive resulting from a contiguous aquifer. Steady-state conditions then prevail and the flow equation is simply Eq. (3), where q is in barrels per day, k in

$$q=\frac{7.08kh(P_e-P_w)}{\mu\ln\left(\frac{r_e}{r_w}\right)} \qquad (3)$$

darcies, P_e (pressure at the outer boundary) and P_w (pressure at the well bore) are in pounds per square inch, μ in centipoises, and r in feet.

When steady-state conditions do not exist, the Darcy equation must be combined with the general radial diffusivity equation (4), where ϕ, the porosi-

$$\frac{\partial^2 P}{\partial r^2}+\frac{1}{r}\frac{\partial P}{\partial r}=\frac{\phi\mu c}{k}\frac{\partial P}{\partial t} \qquad (4)$$

ty, is the only term not elsewhere defined. Appropriate boundary conditions must be specified in order to obtain an integrated analytical equation for fluid flow in the reservoir. W. Hurst and A. F. Van Everdingen presented solutions for the important two sets of boundary conditions: constant production rate at the well bore, and constant production (well bore) pressure. Solutions are available for both the bounded reservoir in which the pressure declines from its initial value at the physical reservoir boundary, or in the case of multiple wells in a reservoir at the equivalent no-flow boundary between wells, and the infinite reservoir (usually only a transient condition in real reservoirs) in which the pressure has not declined below initial value at the boundary.

However, the use of the solutions to these equations still does not permit prediction of reservoir performance. The solutions are usually presented in terms of dimensionless time and dimensionless pressure or dimensionless cumulative production. The permeability to the mobile fluid and its compressibility enter into these dimensionless parameters, and neither of these functions is constant, nor can they be explicitly calculated. The reasons for this are readily traced.

It has already been noted that as the pressure in the reservoir is reduced, the gas saturation builds up. Accompanying this increase, there is an increase in the permeability to the mobile gas phase, and a decrease in the permeability to oil. The changes are not linear (see illustration). Since the pressure is not uniform throughout the reservoir, neither will the gas saturation be uniform, nor will the corresponding permeabilities to the gas and oil. The compressibility of the reservoir fluids will obviously vary directly as the saturation of the compressible gas. Thus, the integrated forms of the radial diffusivity equation cannot be used with any high degree of accuracy over the entire reservoir. The modern high-speed digital computer has made it possible, however, to represent the entire reservoir by a three-dimensional grid system in which the reservoir is mathematically subdivided, and the fluid flow equations and material balance across each block are iteratively and compatibly solved for adjacent blocks. The use of mathematical simulation techniques requires a detailed specification of the reservoir geometry and lithology,

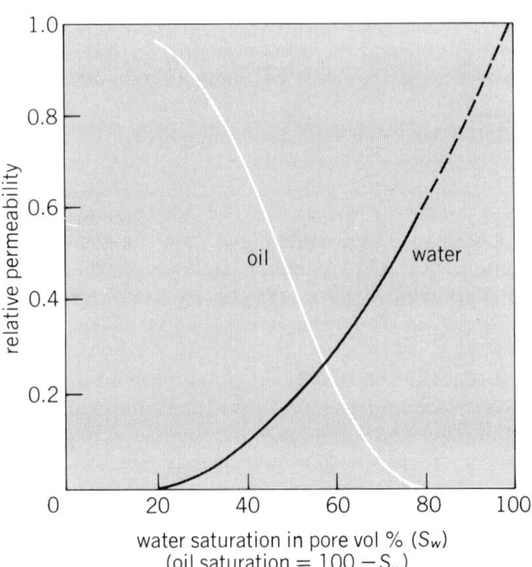

Typical relative permeability relationship for oil and water in a porous medium.

and the relative permeabilities to the mobile fluids under conditions of both diffuse and stratified flow. Again, such knowledge is limited, and as a result it is necessary to calibrate the mathematical simulator by history-matching available production and pressure history for the reservoir. Unique predictions are difficult to achieve and as a result such simulations must be constantly updated. Pressure-transient analyses of individual wells are analyzed by superimposing the flow regimes for the flowing period and the shut-in period to yield data on effective permeability to the flowing fluids and the average reservoir pressure at the time of shutting in the well. Such data are important in calibrating the reservoir simulator for a given reservoir. *See* PETROLEUM RESERVOIR MODELS.

Before the high-speed digital computer was available to the reservoir engineer, and today when the data available for numerical simulation are not available or when the cost of such simulation studies cannot be justified, the reserves may be predicted by decline curve analysis. Depending on the production mechanism, the production rate will decline in keeping with exponential, hyperbolic, or constant-percentage decline rates. The appropriate coefficients and exponents for a particular decline rate are estimated by history matching, and then the decline curves are extrapolated into the future. A variety of nomographs are available for facilitating the history matching and extrapolation.

Reservoir recovery efficiency. The overall recovery of crude oil from a reservoir is a function of the production mechanism, the reservoir and fluid parameters, and the implementation of supplementary recovery techniques. In the United States, recovery shows a geographical pattern because of the differences in subsurface geology in the various producing basins. Many reservoirs in southern Louisiana and eastern Texas, where strong natural water drives, low-viscosity crudes, and highly permeable, uniform reservoirs occur, show a recovery efficiency well over 50% of the original oil in place. Pressure maintenance, the injection of water to maintain the reservoir pressure, is widely used where the natural water influx is not sufficient to maintain the original pressure. In California, on the other hand, strong water drives are virtually absent, the crude oil is generally more viscous than in Louisiana, and the reservoirs, many of which are turbidites, show significant heterogeneity. Both the viscous crude oil and the heterogeneity permit the bypassing of the oil by less-viscous water and gas. Once a volume of oil is bypassed by water in water-wet rocks (the usual wettability), the oil becomes trapped by capillary forces, which then require the imposition of unfeasible pressure gradients or the attainment of very low interfacial tensions for the oil to be released and rendered mobile. As a result, the overall displacement in California reservoirs is relatively poor, and even after the implementation of water flooding, the recovery efficiency in California will probably be only 25–30%.

In general, recovery efficiency is not dependent upon the rate of production except for those reservoirs where gravity segregation is sufficient to permit segregation of the gas, oil, and water. Where gravity drainage is the producing mechanism, which occurs when the oil column in the reservoir is quite thick and the vertical permeability is high and a gas cap is initially present or is developed on producing, the reservoir will also show a significant effect of rate on the production efficiency.

The overall recovery efficiency in the United States is anticipated to be only 32% unless new technology can be developed to effectively increase this recovery. Reservoir engineering expertise is being put to making very detailed studies of the production performance of crude oil reservoirs in an effort to delineate the distribution of residual oil and gas in the reservoir, and to develop the necessary technology to enhance the recovery. Significant strides have been made in at least one area in the United States—California, where the introduction of thermal recovery techniques has increased the recovery from many viscous oil reservoirs from values as low as a few percent to well over 50%. *See* PETROLEUM ENHANCED RECOVERY.

[TODD M. DOSCHER]

Petroleum reservoir models

Physical and computational systems whose behavior is designed to resemble that of actual reservoirs. Early in the petroleum industry's history, there was a great deal of effort invested in the development of very complex physical (laboratory) models, while the computational models were quite simple. Since the explosive evolution of computers in the early 1950s, the computational models have become more and more complex, and there has been a diminishing emphasis on the physical models. Today the emphasis is on combining the best features of the physical and computational models to permit evaluation of the complex behavior of enhanced oil recovery processes.

There are three basic requirements for a useful model: (1) a quantitative description of the relevant properties of a reservoir; (2) a means of describing the mechanics of fluid movements in a reservoir; and (3) a means of combining 1 and 2 by constructing analogous physical systems or by constructing mathematical systems which can be solved by computational techniques.

The purpose of a petroleum reservoir model is to estimate field performance (for instance, oil recovery) under a variety of operating strategies. Whereas the reservoir can be produced only once—and at considerable expense—a model can be produced many times. If this production can be accomplished at low expense and over a short period of time, then observation of the model's performance can give great insight into how to optimally produce a reservoir. Petroleum reservoir models can range in complexity from the intuition and judgment of an engineer to complex mathematical or physical systems describing enhanced oil recovery processes. *See* OIL FIELD MODEL.

Simple models. One of the simplest, yet generally used, computational models is the material balance model. This model, in which the flow of a substance into a region minus the flow of that substance out of the region is equated to the change of mass or volume of that substance in the region [Eq. (1)], is the basic building block for the

$$\mathrm{IN} - \mathrm{OUT} = \mathrm{GAIN} \tag{1}$$

realistic computational models in use today.

Considering the reservoir depicted in Fig. 1, the

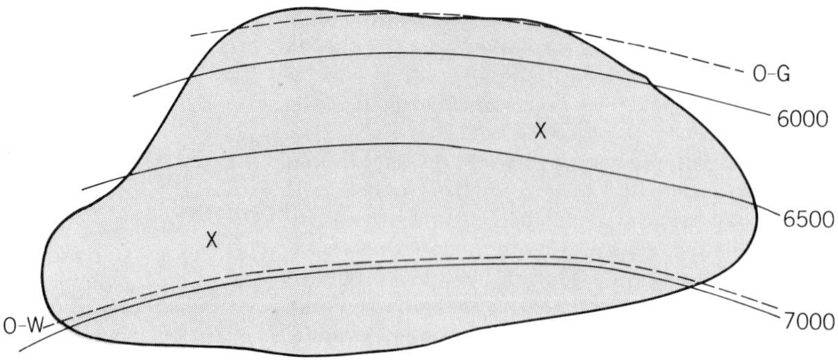

Fig. 1. General material balance. Decrease in oil volume + decrease in gas volume = increase in water volume.

material balance equation will permit relating the fluid withdrawals from this reservoir to the water influx, the gas cap expansion, and the change in pressure. This is an excellent tool for evaluating the primary depletion of a reservoir. However, because the solution of the material balance equation gives only average saturations and pressures for the region, one can tell how much of each fluid is present as a function of time, but cannot tell where the fluids are. Since all enhanced oil recovery (EOR) schemes, including waterflooding, are going after the remaining oil in place, it is imperative to know not only how much oil is present but also where it is. For this reason, present computational models break the region of interest into small pieces by means of a grid (Fig. 2) and solve the material balance equation for each small block in the grid system.

Realistic models. If a small three-dimensional block is removed from the reservoir (Fig. 3), the concept of material balance can be generalized as in Eqs. (2)–(4).

$$\mathrm{IN} = [(F_x)_x \Delta y\, \Delta z + (F_y)_y \Delta x\, \Delta z + (F_z)_z \Delta x\, \Delta y]\, \Delta t \quad (2)$$

$$\begin{aligned}\mathrm{OUT} = q_v\, \Delta x\, \Delta y\, \Delta z\, \Delta t + [(F_x)_{x+\Delta x} \Delta y\, \Delta z \\ + (F_y)_{y+\Delta y} \Delta x\, \Delta z + (F_z)_{z+\Delta z}\, \Delta x \Delta y]\, \Delta t\end{aligned} \quad (3)$$

$$\mathrm{GAIN} = [(C)_{t+\Delta t} - (C)_t]\, \Delta x\, \Delta y\, \Delta z \quad (4)$$

where F is a flux, that is, lb of fluid per ft^2 per day
C is a general concentration, lb per ft^3
q_v is a source term, lb per ft^3 per day

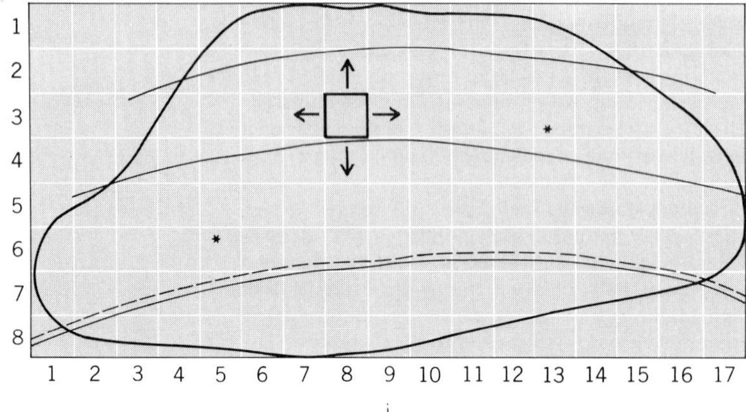

Fig. 2. Grid diagram for a computational model. The material balance equation is solved for each small block in the system.

Even though the mechanics of fluid flow are complex, it has been found in the laboratory that the flow of a single fluid through the pores of a rock can be described by Darcy's law, Eq. (5).

$$u_x = \frac{k_x}{\mu}\left[\left(\frac{p_{x+\Delta x} - p_x}{\Delta x}\right) - \gamma\left(\frac{Z_{x+\Delta x} - Z_x}{\Delta x}\right)\right] \quad (5)$$

where u_x = volumetric velocity, ft^3/(ft^2 of area normal to flow)(day)
k_x = rock permeability, millidarcies (md) × 0.00633
μ = fluid viscosity, centipoises (cp)
x = distance, ft
p = fluid pressure, psi
γ = specific weight, psi/ft (g/g_c)
Z = elevation (vertical position) measured positively downward, ft

For multiphase fluid flow, the velocity u_f of each fluid is found from Eq. (5), modified by relative permeability k_r to result in Eq. (6), where the subscript f refers to phase f, and $k_{r,f}$ is the relative permeability of phase f and is a function of the relative amounts of the phases present (the saturations S_f). Equations (5) and (6) are empirical relationships which have been developed by laboratory experiments.

$$u_{x,f} = -\frac{k_x k_{r,f}}{\mu_f} \cdot \left[\left(\frac{p_{x+\Delta x,f} - p_{x,f}}{\Delta x}\right) - \gamma_f\left(\frac{Z_{x+\Delta x} - Z_x}{\Delta x}\right)\right] \quad (6)$$

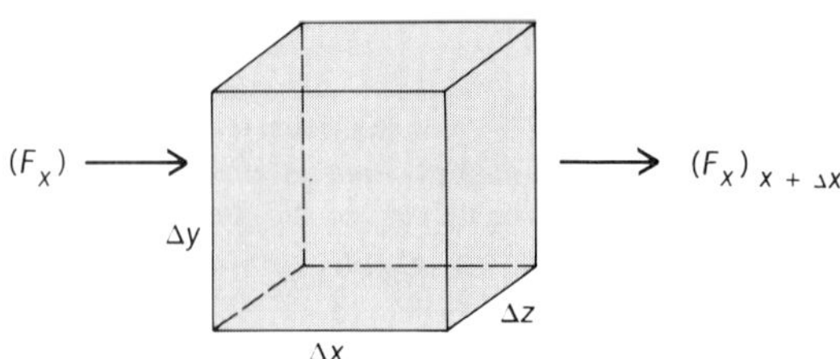

Fig. 3. Material balance in a three-dimensional block from the reservoir.

For a three-phase black oil system where the phases are oil, water, and gas, the three fluxes are given by Eqs. (7), and the concentrations of the phases are given by Eqs. (8).

$$\left.\begin{aligned} F_{x,w} &= b_w u_{x,w} \\ F_{x,o} &= b_o u_{x,o} \end{aligned}\right\} \quad \frac{\text{stock tank barrel}}{\text{ft}^2\text{-day}} \quad (7)$$

$$F_{x,g} = b_g u_{x,g} + b_o R_s u_{x,o} \qquad 10^3\ \text{ft}^3/\text{ft}^2\text{-day}$$

$$\begin{aligned} C_w &= \phi b_w S_w \\ C_o &= \phi b_o S_o \\ C_g &= \phi(b_o R_s S_o + b_g S_g) \end{aligned} \quad (8)$$

where b_f is the formation volume factor in stock tank barrels (STB) per reservoir barrel for phase f
ϕ is the porosity of the rock
R_x is a solution gas/oil ratio

The fluxes are defined in this manner in order

to put them in a consistent set of units, so that all quantities measured are expressed in the standard conditions of 60°F (15.6°C) and atmospheric pressure (101.325 kPa). In all of these equations (7), the material balance model is related to some constant set of conditions. Actually, the conditions in the reservoirs are themselves constantly changing, and the terms are defined in such a way as to represent the amount of oil filling the barrels under standard conditions.

Combining Eqs. (1), (4), (7), and (8) results in the oil material balance equation in two dimensions for the grid shown in Fig. 4. The terms of the oil balance equation are given separately in Eqs. (9), (10), and (11).

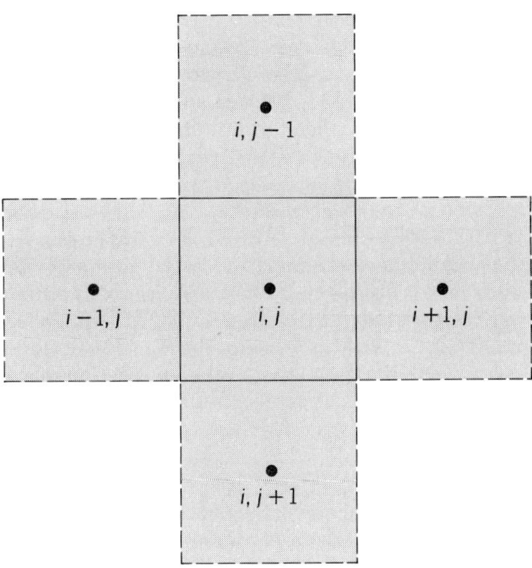

Fig. 4. Oil material balance in two dimensions.

$$\text{IN} = [(b_o u_{x,o})_{i-1/2,\,j,\,n+1/2}\, h_{i-1/2,\,j} \Delta y_j + (b_o u_{y,\,o})_{i,\,j-1/2,\,n+1/2}\, h_{i,\,j-1/2}\, \Delta x_i]\, \Delta t \quad (9)$$

The above term includes the flow from the blocks labeled $i-1$, j, and i, $j-1$ to the block labeled i, j. The variable h refers to vertical thickness.

$$\text{OUT} = [(b_o u_{x,\,o})_{i+1/2,\,j,\,n+1/2}\, h_{i+1/2,\,j}\, \Delta y_j + (b_o u_{y,\,o})_{i,\,j+1/2,\,n+1/2}\, h_{i,\,j+1/2}\, \Delta x_i + q_{o,\,i,\,j,\,k,\,n+1/2}]\, \Delta t \quad (10)$$

The above term includes the flow from the block labeled i, j to blocks $i+1$, j and i, $j+1$. Here q_o is an oil injection or production rate.

$$\text{GAIN} = (\phi b_o S_o)_{i,\,j,\,n+1} - (\phi b_o S_o)_{i,\,j,\,n}\, h_{i,\,j}\, \Delta x_i\, \Delta y_j \quad (11)$$

The above term is the net change of fluid in block i, j. If Eq. (6) is used in Eq. (9), expression (12) results, where $t = n$, $n+1/2$, $n+1$, . . ., and $\ell_{i-1/2,\,j} = (\Delta x_i + \Delta x_{i-1})/2$.

$$\left[\frac{b_o k_{ro}}{\mu_o} k_x h\right]_{i-1/2,\,j,\,n+1/2} \Delta y_j \cdot \left[\left(\frac{p_{o,\,i,\,j,\,n+1/2} - p_{o,\,i-1,\,j,\,n+1/2}}{i-1/2,j}\right) - \gamma_{o,\,i-1/2,\,j,\,n+1/2}\left(\frac{Z_{i,j} - Z_{i-1,j}}{\ell_{i-1/2,\,j}}\right)\right] \quad (12)$$

If material balance equations for the other two phases are written and definition (13) is included,

$$f_{n+1/2} = \frac{f_n + f_{n+1}}{2} \quad (13)$$

where f may represent any variable, a system of three equations at each grid block (i, j) and each time t for the unknowns p_o, p_w, p_g, S_o, S_w, S_g is developed. Since there are more unknowns than equations, two capillary pressure relationships and one constraint at each grid block and at each point in time are required. These are given by Eqs. (14) and (15), where p_{cow} represents water-oil capillary pressure and p_{cog} represents gas-oil capillary pressure.

$$p_{cow}(S_w) = p_o - p_w \qquad p_{cog}(S_g) = p_o - p_g \quad (14)$$

$$S_o + S_w + S_g = 1 \quad (15)$$

At this stage of the development, there is a deterministic system of equations which can be solved for the average pressures and saturations in each grid block with the aid of large-scale digital computers, provided the required reservoir rock and fluid data are available.

For black oil systems, a great deal of experimental work has been done, and functional relationships (16) are generally accepted. All of these

$$\begin{gathered} k_{ro}(S_w, S_g),\ k_{rw}(S_w),\ k_{rg}(S_g) \\ b_o(p_o),\ b_g(p_g),\ b_w(p_w) \\ \mu_o(p_o),\ \mu_g(p_g),\ \mu_w(p_w) \\ \gamma_o(p_o),\ \gamma_g(p_g),\ \gamma_w(p_w),\ R_s(p_o) \end{gathered} \quad (16)$$

data can be measured by using a fluid sample and some cores from the reservoir. In addition, through the use of geology, petrophysics, well testing, and tracer and pulse tests, as well as history-matching production data using the model itself, the important rock characteristics $h_{i,j}$, $Z_{i,j}$, $k_{x,i,j}$, $k_{y,i,j}$, and $\phi_{i,j}(p_o)$ can be determined.

Once a comprehensive laboratory program is combined with some field tests or production data, a computational model can be used to adequately predict the future performance of the reservoir under various operating strategies. This technology is complete and is currently being used throughout the oil and gas industry for dry gas and black oil reservoirs.

Enhanced oil recovery. The reservoir models for enhanced oil recovery (EOR) processes are not nearly as advanced as the black oil model. With the possible exception of steamflooding, for which the first model was developed in 1973, even the fluid mechanics of the other EOR processes are uncertain. Today models exist for simulating CO_2 flooding, combustion, and chemical flooding; however, none of these models has been used extensively enough to have reached the level of sophistication of black oil models.

Because the EOR processes are so complex, one of the major uses of the mathematical models is to explain the behavior of the physical models. By using the computational model to simulate laboratory experiments, it can be determined if the model is truly representing the physics of the process. In this manner, insight is gained into the important aspects of the physical processes, and when the model is capable of modeling the laboratory work, it can be used to scale these results to the field.

Reservoir models require both complex physical models as well as computational models. Each reservoir study requires a comprehensive laboratory program under which actual floods are carried out so that the physics of the EOR process can be properly input to the computational model. In adition to running core floods, a great deal of physical property data regarding the fluids and the reservoir rock must be generated, as is done in the black oil case.

In addition to requiring a comprehensive laboratory program, the EOR models need to be more accurate than the black oil models. This is because small changes in a component's concentration can dramatically affect its ability to mobilize oil, and so the computational model must be able to generate accurate results. Also, because many EOR processes inject slugs of material into the reservoir, the model needs to be capable of computing results on a scale much smaller than is required in black oil models. Finally, all of these complexities lead to requiring a more accurate reservoir characterization than is needed for black oil systems.

In the late 1960s and early 1970s, the need for more accurate computational models was recognized and more accurate approximations were being considered. The most promising of these high-order-accuracy models were the finite-element models; however, at this time (1980) no truly successful finite element model is being used for EOR. Faster computers (such as the Star and the Cray and conformed grid transformations) have enabled the industry to continue using second-order-correct finite-difference approximations. However, the mathematical models of the complex EOR processes are limited to modeling the behavior of symmetry elements selected from patterns and are not capable of simulating a full field's performance. This limitation, of course, puts a greater burden on the engineers who use these models and requires that the engineers of the future be better trained than their counterparts of the 1960s.

The final chapter on modeling is not being written here. As more and more complex physical processes are modeled, there will be better understanding of which aspects are important and which are not. In that way it will be possible to make reasonable and justifiable assumptions which will simplify the models. At the same time, computers will continue to get faster and less expensive, and numerical methods will continue to improve. While an ever-expanding collection of modeling techniques will become available, it is almost certain that the needs will continue to outnumber the means. Much work remains to make the reservoir models respond more and more like the real reservoirs. However, since more and more complex processes will be used to recover oil, the development of reservoir models will not end until oil is replaced with another form of energy. *See* OIL AND GAS FIELD EXPLOITATION; PETROLEUM ENHANCED RECOVERY; PETROLEUM RESERVOIR ENGINEERING.

[HARVEY S. PRICE]

Bibliography: J. C. Cavendish, H. S. Price, and R. S. Varga, Galerkin methods for the numerical solution of boundary value problems, *Soc. Petrol. Eng. J.*, 246:204–220, June 1969; K. H. Coats, In-situ combustion model, *SPE-AIME 54th Annual Meeting*, Las Vegas, SPE Pap. 8394, Sept. 23–26, 1979; K. H. Coats et al., Three-dimensional simulation of steamflooding, *SPE-AIME 48th Annual Meeting*, Las Vegas, SPE Pap. 4500, Sept. 30–Oct. 3, 1973; H. J. Morel-Seytoux, Analytical-numerical method in waterflooding predictions, *Soc. Petrol. Eng. J.*, 9:247–258, 1965; H. J. Morel-Seytoux, Unit mobility ratio displacement calculations for pattern floods in homogeneous medium, *Soc. Petrol. Eng. J.*, 9:217–227, 1966; P. M. O'-Dell, Use of numerical simulation to improve thermal recovery performance in the Mount Paso Field, California, *SPE-AIME Symposium on Improved Oil Recovery*, Tulsa, SPE Pap. 7078, Apr. 16–19, 1978; A. Settari, H. S. Price, and T. Dupont, Development and application of variational methods for simulation of miscible displacements in porous media, *Soc. Petrol. Eng. J.*, pp. 228–246, June 1976; A. Spivak, H. S. Price, and A. Settari, Solution of the equations for multidimensional, two phase, immiscible flow by variational methods, *SPE-AIME 4th Symposium on Numerical Simulation of Reservoir Performance*, Los Angeles, SPE Pap. 5723, Feb. 19–20, 1976; M. R. Todd and C. A. Chase, A numerical simulator for predicting chemical flood performance, *SPE-AIME 5th Symposium on Reservoir Simulation*, Denver, SPE Pap. 7689, Feb. 1–2, 1979; M. R. Todd and W. J. Longstaff, The development, testing, and application of a numerical simulator for predicting miscible flood performance, *J. Petrol. Technol.*, p. 874, July 1973.

Pewter

A tarnish-resistant alloy of lead and tin always containing appreciably more than 63% tin. Other metals are sometimes used with or in place of the lead; among them are copper, antimony, and zinc. Pewter is commonly worked by spinning and it polishes to a characteristic luster. Because pewter work-hardens only slightly, pewter products can be finished without intermediate annealing. Early pewter, with high lead content, darkened with age. With less than 35% lead, pewter was used for decanters, mugs, tankards, bowls, dishes, candlesticks, and canisters. The lead remained in solid solution with the tin so that the alloy was resistant to the weak acids in foods.

Addition of copper increases ductility; addition of antimony increases hardness. Pewter high in tin (91% tin and 9% antimony or antimony and copper, for example) has been used for ceremonial objects, such as religious communion plates and chalices, and for cruets, civic symbolic cups, and flagons. *See* SOLDERING, TIN ALLOYS.

[FRANK H. ROCKETT]

Pictorial drawing

A pictorial drawing shows a view of an object (actual or imagined) as it would be seen by an observer who looks at the object either in a chosen direction or from a selected point of view. One such view often suffices to give the reader a clear picture of the shape and details of the object. Pictorial sketches often are more readily made and more clearly understood than are front, top, and side views of an object. Pictorial drawings, either sketched freehand or made with drawing instruments, are frequently used by engineers and architects to convey ideas to their assistants and

Fig. 1. Perspective drawing of a residence. *(Home Planners Inc.)*

clients. *See* ENGINEERING DRAWING.

In making a pictorial drawing, it is important to select the viewing direction that shows the object and its details to the best advantage. The resultant drawing is *orthographic* if the viewing rays are considered as parallel, or *perspective* if the rays are considered as meeting at the eye of the observer. Making perspective drawings with instruments is time-consuming and requires considerable knowledge and skill. There are, however, commercially available devices which make this chore easier and quicker than is the case when conventional instruments are used. Perspective drawings provide the most realistic, and usually the most pleasing, likeness when compared with other types of pictorial views (Fig. 1).

Several types of nonperspective pictorial views can be sketched, or drawn with instruments. Although each type has some distortion, all provide a good picture of what the object looks like. They are easier and quicker to make than perspective drawings. The isometric pictorial is especially popular because of the direction of its axes and the fact that all measurements along these axes are made with one scale (Fig. 2). In addition to isometric representation, two other forms—dimetric and trimetric—are sometimes used

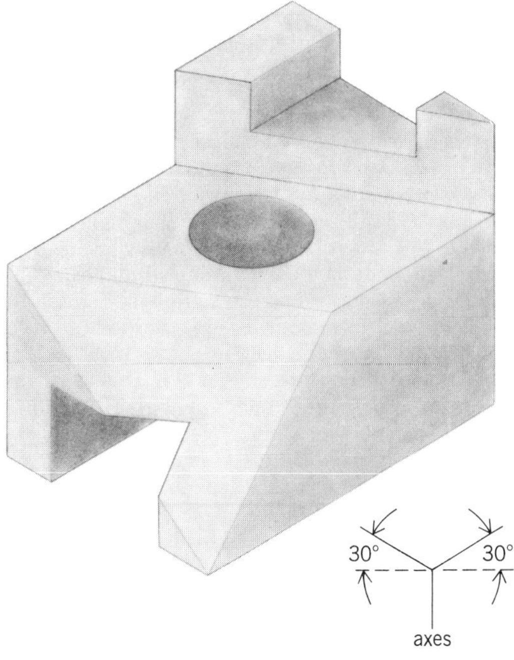

Fig. 2. Isometric drawing; measurements along each axis are made with the same scale.

Oblique pictorial drawings, while not true orthographic views, offer a convenient method for drawing circles and other curves in their true shape (Fig. 3). In order to reduce the distortion in an oblique drawing, measurements along the receding axis may be foreshortened. When they are halved, the method is called cabinet drawing.

An effective freehand sketch of an object can be made if proper attention is given to viewing direction, proportions, orientation of ellipses, and location of tangent points. Shading is sometimes added to enhance the pictorial drawing.

Shaded exploded-view production illustrations greatly facilitate the learning process in assembly of machines and devices (Fig. 4). When this type of illustration is used, the initial assembly of parts

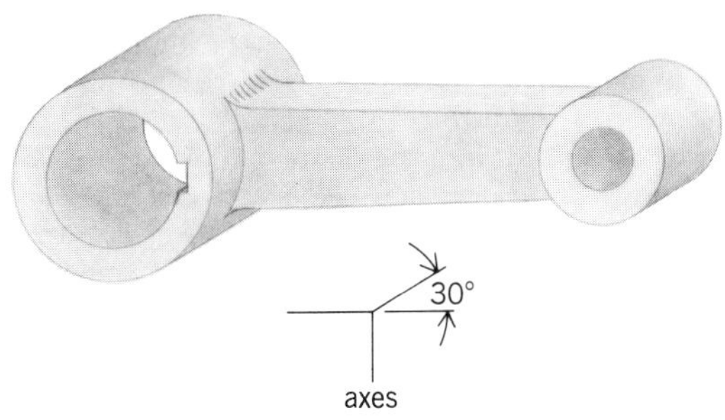

Fig. 3. Oblique pictorial drawing.

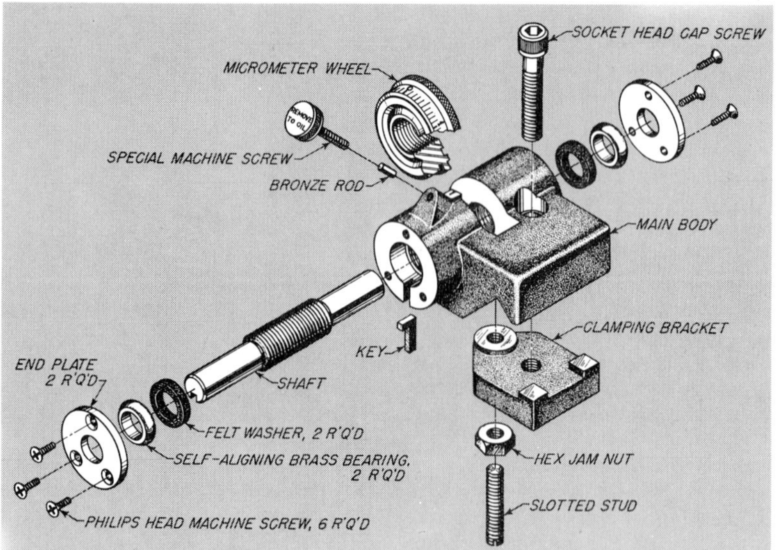

Fig. 4. Exploded-view production illustration; this drawing is isometric.

into a machine has been found to be three or four times faster than if a conventional assembly drawing is used. Photodrawings can be used to achieve the same visual results.

[CHARLES J. BAER]

Pier

A fingerlike structure projecting from shore and providing berths for ships to load and discharge passengers and cargo. Construction usually takes the form of a pile-supported platform using steel, timber, and concrete materials, thus allowing free water flow underneath. Sometimes bulkhead construction is used around the periphery with earth fill inside and pavement topping it. Piers may be housed over with sheds and may have special cargo-handling equipment, such as that used for loading iron ore. Uses of piers sometime extend to recreational purposes, such as fishing piers, or to community purposes, such as car parking. Spaces between adjacent piers are called slips. *See* COASTAL ENGINEERING; WHARF.

[EDWARD J. QUIRIN]

Pile foundation

A part of a foundation system which transfers loads from a superstructure to the soil or rock below the surface by means of structural elements called piles. A pile foundation is used when the upper layers of soil are too soft and compressible to support the loads of the superstructure in a safe and economical manner by means of footings or mat foundations. A pile foundation is often used to support structures over water or adjacent to water, in addition to supporting structures which are constructed over soft soils.

Piles, the structural elements which are installed into the soil or rock, may be wood, concrete, or steel; or concrete and wood or concrete and steel may be combined to form composite piles. Occasionally, wood piles may be reinforced with steel bands.

Wood piles. Wood or timber piles are the first types of piles used to support structural loads. They are the trunks of trees with the limbs removed, and are installed into the soil by driving or other methods. The tip of a wood pile is the lowest portion of the pile after installation in the soil; the butt of the pile is then at the top. Probably the most common type of timber piles used in the United States is southern pine. The other types of trees which are used are Douglas fir, oak, cypress, hickory, spruce, and so forth.

Timber piles last indefinitely as long as the wood is continuously saturated with water. However, if a portion of the wood is above the water table, this portion is subject to deterioration through dry rot or attack by termites (if they exist in the area). The problems of deterioration of wood piles above the water table can be mitigated by pressure treatment of the piles with creosote prior to installation.

Wood piles in a marine environment are subject to attack by organisms such as teredo or limnoria. These organisms bore into the wood of the piles and destroy the structural integrity. Chemical treatment of the wood prior to installation is not always a successful preventive measure against marine borer attacks. Encasement with concrete of the portion of the pile exposed to the marine environment may be a helpful preventive measure.

Concrete piles. Concrete piles are widely used throughout the world. A variety of concrete piles are available for use in supporting structural loads. The types of concrete piles can be divided into two categories: precast and cast-in-place. Precast concrete piles are those which are made in a casting yard either at the site at which they are to be installed or more often at another location. The concrete may be reinforced with steel or prestressed. Precast concrete piles are made in a variety of shapes and sizes. The lower section of the pile may be tapered to allow easier penetration into the soil, or the entire length may be tapered. The cross section of the pile may be square, hexagonal, octagonal, or circular. One innovative precast concrete pile is shaped like a wood screw or an auger and can be installed into the soil by rotation. The area of the cross section of a precast pile varies depending on the load the pile is to support. Precast piles are usually monolithic, but they also may be cast in sections which are connected together in the field during installation. Problems often develop during the installation of monolithic precast piles. If the lengths of the piles have been underestimated, it is difficult to make splices; if the lengths have been overestimated, it is expensive to cut them off. *See* PRECAST CONCRETE.

Cast-in-place concrete piles are those in which the concrete is placed in the field during pile installation. There are a large variety of cast-in-place concrete piles: (1) An auger may be used to drill a hole into the soil, and upon removing the auger the hole is filled with concrete. (2) An auger with a hollow pipe in its center may be used; after the desired penetration is reached, concrete is poured through the pipe as the auger is being removed. (3) A thin steel shell may be driven into the soil by using a mandrel; after reaching the required penetration, the mandrel is removed and the steel is filled with concrete. (4) A heavy steel shell may be driven into the soil, and concrete is poured into the shell as the shell is being removed. (5) A steel pipe may be driven into the soil and filled with concrete. (6) A bulb or bell may be installed at the base of a

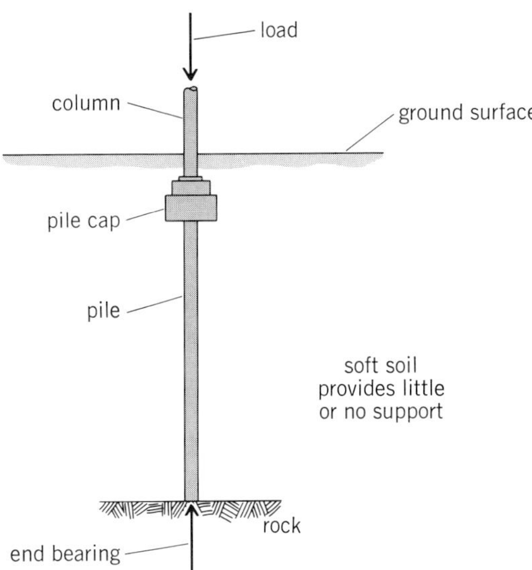

Fig. 1. End-bearing pile.

cast-in-place pile to increase the cross-sectional area of the pile. Reinforcing steel can be placed in some types of cast-in-place concrete piles before the concrete is poured if such steel is required.

Steel piles. Steel piles are either steel pipe or steel H beams installed into the earth to support structural loads. As mentioned, steel-pipe piles may be filled with concrete after installation.

Supporting capacity. Piles obtain supporting capacity from the earth through end bearing or side friction or more generally through a combination of both (Figs. 1 and 2). Piles installed through very soft soils to rock or another firm stratum derive their capacity to support structural loads from the strength of the firm stratum on which the end of the pile is bearing. In such a situation the shaft of the pile acts as a column, and must have sufficient strength to transfer the imposed load from the top to the bottom of the pile. However, even the softest soils provide some lateral restraint, and it is very unusual for an end-bearing pile to fail because of buckling.

Soil is not always underlain by rock or another firm stratum within a depth which can be reached economically by piles. In this condition, the soil in which the pile is embedded and the soil below the pile tip provide the support for the pile. As the pile is subjected to load, it tends to move downward. The tendency to move downward is resisted by the friction between the sides of the pile and the soil in which the pile is embedded, or the shearing resistance of the soil immediately adjacent to the sides of the pile. In addition, as the tip of the pile tends to move downward, the soil below the tip resists this tendency due to its bearing strength.

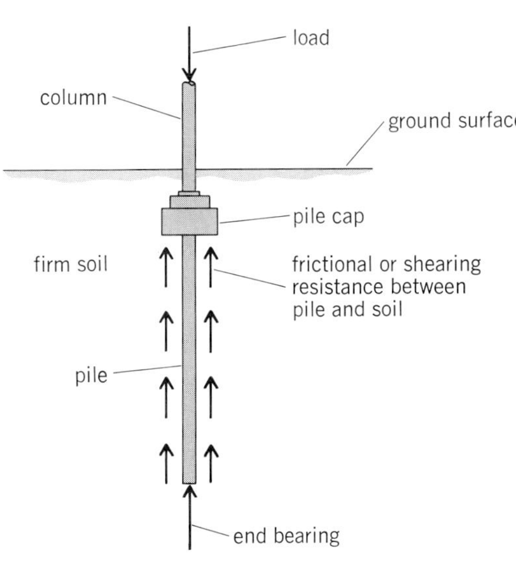

Fig. 2. Friction pile with some support from end bearing.

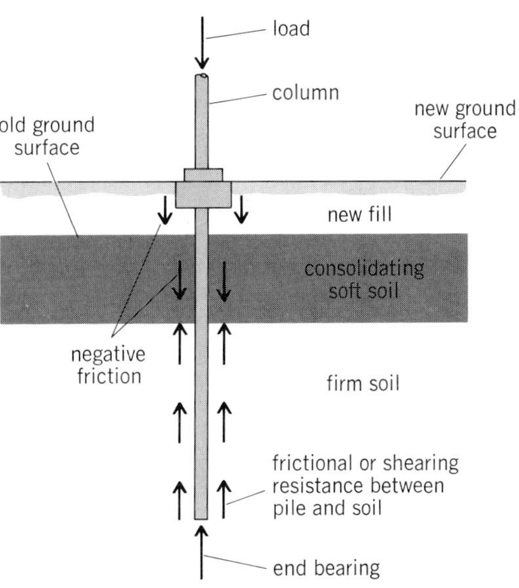

Fig. 3. Pile subjected to negative friction or downdrag. The pile must penetrate a sufficient distance in firm soil to support the external load plus the negative frictional forces.

In certain conditions, the soil in which a pile is embedded may not provide upward resistance to pile movement. In fact, if the soil in which a portion of the pile is embedded is undergoing consolidation, the pile is subjected to a downward force which is in addition to the force imposed from the structural load. A downward force imposed by the soil on the sides of a pile is called negative friction or downdrag. Negative friction occurs when piles are installed through a new fill overlying a layer of soft soils. The soft soils consolidate due to the weight of the fill above them and tend to pull the piles downward. The piles must be installed to a sufficient depth below the layer of soft soils to resist both the structural loads and the negative friction (Fig. 3).

A light column load from a structure may be supported by a single pile; however, piles are more often installed in groups of two or more. The stresses imposed on the surrounding and underlying soil by a single pile are relatively simple to analyze. However, when piles are installed in groups, the stresses from the piles being imposed on the soils may overlap depending on the pile spacing. The supporting capacity of a pile group may be different than the summation of the capacity of single piles within the group.

Pile installation. Most piles are installed by driving the piles into the soil. The equipment used consists of a steel-pile-driving hammer which operates between vertical leads. The leads are attached to a mobile crane. The simplest type of hammer is a steel weight raised above the top of the pile and allowed to fall by gravity. The energy generated by dropping the hammer on the top of the pile causes the tip of the pile to penetrate into the soil. The hammer may be raised by a cable or rope connected to a drum on a crane and allowed to drop. Variations of this process involve the use of steam or compressed air to raise the hammer or ram above the top of the pile. If the hammer, after being raised by steam or air is allowed to fall by gravity, it is called a single-acting hammer. If the steam or compressed air is also used to push the hammer downward with a force in addition to that generated by gravity, it is called a double-acting or differential hammer.

Another type of hammer used is a diesel hammer. In this type, the ram is lifted by an explosion of diesel fuel in a combustion chamber. When the ram reaches a predetermined height above the top of the pile, it drops by gravity; however, the next fuel explosion for lifting occurs at the moment of impact and the top of the pile receives additional energy from the explosion. A second type of diesel hammer is one which has an air chamber above the ram. As the ram is raised by the explosion in the combustion chamber, the air in the air chamber is compressed. As the ram drops, the compressed air adds downward force to the ram.

Pile-driving hammers of the type discussed are rated on the amount of energy delivered per blow. For a hammer which falls by gravity, the energy which it generates is the weight of the hammer multiplied by the height of the fall less frictional losses in the leads. If energy, other than that developed by gravity, is imposed on the hammer, this energy is added to the energy obtained from gravity to determine the rated energy.

Certain cast-in-place piles are installed by driving a thin steel shell into the soil by using one of the pile-driving hammers described and a mandrel within the shell. The driving impact is on the mandrel, which prevents the shell from collapsing. After the desired penetration is reached, the mandrel is removed and the shell is filled with concrete.

A further development in pile installation is the use of vibratory pile drivers. Such a pile driver imparts a vibration into the pile to be installed, while also imposing a static downward force on the pile. The vibrating mechanism is attached to the head of the pile, while the static downward load is separated from the vibrating force by springs. As the pile is vibrated, the soil adjacent to the sides and immediately below the tip of the pile also vibrates and loses its resistance to penetration. The pile then moves downward under its own weight and the weight of the static load from the machine. On some vibratory pile drivers, the frequency of the vibrations transmitted to the pile can be varied so that the soil, pile, and pile driver are in resonance. In this condition, the pile easily penetrates the soil. Vibratory drivers are also used to extract piles from the ground by applying an upward force with a crane while the pile is being vibrated.

Methods of installing piles other than using pile-driving equipment involve drilling a hole into the soil with an auger or a rotating bucket and filling the hole with concrete.

In the installation of piles, certain problems can develop. Wood piles driven into the soil may strike obstructions and break, or the tips of the piles may be splintered because of hard driving. Precast concrete piles may be broken due to hard driving. Steel H piles may strike obstructions and bend so that they are not plumb. Water may infiltrate an open hole into which concrete is poured to form a cast-in-place pile. If the concrete falls through water during installation, the concrete will segregate and will not become a hard solid material.

Determining pile load-supporting capacity. The purpose in installing piles is to support structural loads which cannot be supported by other types of foundations. The designer must select the type of pile most suitable and most economical for the subsurface conditions at the construction site. In addition, the type of pile must be appropriate for the loads to be supported. The subsurface conditions can be evaluated through borings at the site, from which samples of the materials penetrated are extracted. The samples can be subjected to laboratory tests to determine their physical characteristics. By using data from the borings and laboratory tests, analyses can be made to determine the frictional or shear properties of the soils to be penetrated by the piles. In addition, the bearing values of the soils can be determined at various levels to which the pile tips may penetrate. With these data, the supporting capacities of various types of piles installed can be estimated. However, subsurface conditions are variable, and if piles are installed over a large area, there may be changes in the conditions which are not disclosed by the borings. Thus penetrations and supporting capacities computed on the basis of the borings alone may not be applicable for all of the piles to be installed. Monitoring of the pile installation process is always advisable, so that design changes can be made if conditions vary.

Often test piles are installed prior to the installation of production piles. If the piles are to be installed with a pile-driving hammer, the test piles give an indication of the driving conditions which will be encountered. In addition, one or more of the test piles may be subjected to load tests as a check on the design capacity which has been selected. Load tests are performed by applying incremental loads to the tops of test piles by using hydraulic jacks. It is necessary to construct reaction frames to offset the downward loads being applied to the piles being tested. Movements of the piles under the incremental loads are measured and recorded. Generally the piles subjected to load tests are loaded to twice the design load. If failure occurs before twice the design load is reached, it is necessary to investigate the cause of failure and perhaps redesign the pile foundation.

For many years engineers have used pile-driving

formulas to determine the supporting capacities of driven piles. Most of these formulas are based on the energy delivered by the hammer, the work done by the pile as it penetrates a certain distance, and the resistance provided by the soil to the penetration. These factors can be placed into a formula wherein the resistance to penetration is the supporting capacity of the pile with an appropriate factor of safety. Many modifications are made to incorporate other factors, such as hammer efficiency and other losses of energy due to the pile-driving process. All of these formulas are based on dynamic pile-driving conditions. However, the pile, when it is used for its intended purpose, normally is subjected to a static load. The conditions in the soil supporting the pile during a static situation may be much different than they are in the dynamic situation which occurs during pile driving. The static conditions are time-dependent. During pile driving, the soil is displaced by the penetration of the pile. In addition, if the soil is below the groundwater level, the water in the soil may be subjected to transitory pressures which subsequently dissipate under static conditions. When the soil is displaced and the water is subjected to increased pressure, time must pass before the soil and water achieve a new permanent state. The new permanent state of the soil and water provides the support for the pile in the static state.

Another development in relating the driving of piles to the supporting capacity of the driven pile is based on the wave equation theory. In applying this theory to pile driving, the pile is considered to behave as an elastic bar and not as a concentrated mass. When the bar (pile) is struck at one end by an impact force, the stress wave travels axially along the pile at a velocity which is based on the modulus of elasticity and the mass density of the material in the pile. Although the wave velocity is constant for any homogeneous elastic material, the velocity of a point or particle of the elastic material is dependent on its position or location in relation to the impact force and to time. If a point on the pile moves downward, there is a velocity at that point in the pile and also a stress. The stress must be sufficient to overcome the supporting capacity of the soil or the pile will not move downward. The characteristics of the pile material, the energy and efficiency of the pile-driving hammer, the stress-transmission characteristics of any cushion blocks to be used, the general characteristics of the soil conditions, and an estimate of the soil resistance to be obtained from the sides and the tip of the pile are combined in a wave equation which, on solution, gives the ultimate capacity of the pile related to blows of the hammer per unit of penetration. In addition, the stresses in the piling material for blows per unit of penetration are also obtained.

By analyzing the pile-driving operation through the use of the wave equation, the designer can determine if the energy to be imposed on the top of the pile will reach the tip of the pile and cause it to move downward without overstressing the pile material.

A pile foundation is a viable way of supporting structural loads when other types of foundations cannot be used. The designer has many choices in selecting the type of pile to be used and determining the supporting capacities. A pile foundation is usually more expensive than a foundation which is installed without piles. However, if the piles are installed properly, the settlements of the pile foundation are usually minimal. *See* CAISSON FOUNDATION; COFFERDAM; FOUNDATIONS; RETAINING WALL. [GARDNER M. REYNOLDS]

Bibliography: *See* FOUNDATIONS.

Pilot production

The production of a product, process, or piece of equipment on a simulated factory basis. In mass-production industries where complicated products, processes, or equipment are being developed, a pilot plant often leads to the presentation of a better product to the customer, lower development and manufacturing costs, more efficient factory operations, and earlier introduction of the product. Following the engineering development of a product, process, or complicated piece of equipment and its one-of-a-kind fabrication in the model shop, it becomes desirable and necessary to "prove out" the development on a simulated factory basis.

Facilities. To accomplish this aim, the pilot plant, an intermediate step between development laboratory and production factory, is established. It is provided with personnel and facilities that duplicate as nearly as possible actual manufacturing conditions but are free of the day-to-day necessity of meeting delivery schedules.

This requires personnel who are highly skilled in their field but are not necessarily experts. They must understand the thinking and objectives of development engineers and must also appreciate the day-to-day problems that arise in factory operation with unskilled or semiskilled operators.

Adequate equipment and services should be provided to accommodate the full range of factory requirements. Sometimes, however, it may be impractical to include certain facilities in a pilot plant because of floor space limitations, or because of specialized services required. In such cases, arrangements should be made to perform these pilot-plant functions on regular production equipment in the factory but under pilot plant supervision and financial control.

Objectives. Some of the objectives of a pilot plant operation are as follows.

Quality control. Pilot plant operation demonstrates the ability of processes or equipment to function consistently at the desired quality level under factory operating conditions.

Material usage. To demonstrate economical usage of materials during the pilot operation, the materials used should be truly representative and contain the full range of variability permitted by material purchase specifications.

Process reliability. Pilot operation demonstrates whether the processes involved are practical and realistic and whether operating procedures can be followed in their entirety by the type of personnel used in the factory.

Equipment reliability. To demonstrate the capability of the equipment to produce at speeds and machine efficiencies, which will become standard in the factory, data are recorded during pilot production which show whether the equipment will perform at a satisfactory rate of defectives. Pilot plant experience establishes a list of spare parts which will be initially provided in the factory.

Personnel requirements. Pilot experience establishes the amount and labor grade of production and maintenance personnel required to operate the process or equipment. This is done with the cooperation of industrial engineering and industrial relations departments.

Safety. During pilot runs the safety department has an opportunity to review the process or equipment for compliance with safety requirements.

Factory acceptance. Perhaps the most important objective of a pilot plant is to afford representatives of the factory to which the process or equipment will be transferred an opportunity to witness the operation and to agree that it is ready for the factory. Thus, criticism of the development is minimized and its introduction into regular factory operation accomplished more quickly.

Manufacturing costs. Pilot plant operation provides the opportunity to verify engineering estimates of manufacturing cost and may indicate the necessity for pricing changes or for taking further steps to reduce the cost of product in order to reach a satisfactory level of profit.

Development costs. The cost of designing and developing processes or equipment and of instructing personnel in their use is often only a fraction of the total cost involved in producing a final efficient process or machine. The debugging or working out of unforeseen problems can consume a great deal of time and add a large amount of expense before the process or machine is ready for factory operation. The pilot plant relieves the factory of the expense of a trial run both in direct cost and in lost production of established products. The close proximity of the pilot operation to engineering and other skilled personnel permits first-hand observation of operation and quicker decisions on required changes. It may also save the expense and time of such personnel traveling to factory locations and the return of equipment from the factory to its point of origin.

The financing of a pilot plant should be on a budgetary basis. Funds for its facilities, personnel, and operation should be provided as part of a manufacturing development budget. The pilot plant should not be expected to finance itself through manufactured products; otherwise management tends to keep the project in the pilot plant longer than necessary in order to create funds for its operation. The pilot plant then loses its value as a development function, and the desire and interest of its personnel to introduce improvements and to tackle new projects become subordinate to that of meeting production schedules. On the other hand, arrangements should be made to credit the operation with a fair value of the product manufactured so that the net amount remaining will represent development cost more accurately. *See* INDUSTRIAL COST CONTROL; MASS PRODUCTION; PRODUCT DESIGN; PRODUCTION ENGINEERING; QUALITY CONTROL. [JAMES E. WOODALL]

Pipeline

A line of piping and the associated pumps, valves, and equipment necessary for the transportation of a fluid. Major uses of pipelines are for the transportation of petroleum, water (including sewage), chemicals, foodstuffs, pulverized coal, and gases such as natural gas, steam, and compressed air. Pipelines must be leakproof and must permit the application of whatever pressure is required to force conveyed substances through the lines. Pipe is made of a variety of materials and in diameters from a fraction of an inch up to 30 ft. Principal materials are steel, wrought and cast iron, concrete, clay products, aluminum, copper, brass, cement and asbestos (called cement-asbestos), plastics, and wood.

Pipe is described as pressure and nonpressure pipe. In many pressure lines, such as long oil and gas lines, pumps force substances through the pipelines at required velocities. Pressure may be developed also by gravity head, as for example in city water mains fed from elevated tanks or reservoirs.

Nonpressure pipe is used for gravity flow where the gradient is nominal and without major irregularities, as in sewer lines, culverts, and certain types of irrigation distribution systems.

Design of pipelines considers such factors as required capacity, internal and external pressures, water- or airtightness, expansion characteristics of the pipe material, chemical activity of the liquid or gas being conveyed, and corrosion.

Most pipe is jointed, although some concrete pipe is monolithically cast in place. The length of the individual sections of pipe and the method of joining them depend upon the pipe material, diameter, weight, and requirements of use. Steel pipe sections are usually joined by welding, couplings, or riveting. Cast-iron pipe may be joined by couplings or, in the case of bell-and-spigot pipe, by filling the space between the bell and the spigot with calked or melted metal such as lead. Flexible-type joints with rubber gaskets are also used for joining cast iron pipe. The rubber gasket is contained in grooves and is ordinarily the sole element making the joint watertight.

Cement-mortar-filled or lead-filled rigid-type bell-and-spigot joints are usually used for joining concrete or vitrified clay sewer pipe. Tongue-and-groove rigid-type mortar-filled joints are often used for concrete pipe in low-pressure installations. The flexible-type joints are most frequently used for asbestos-cement pipe and concrete pipe under higher pressures.

[LESLIE N. MC CLELLAN]

Placer mining

Placer mining is the working of deposits of sand, gravel, and other alluvium and eluvium containing concentrations of metals or minerals of economic importance. For many years gold has been the most important product obtained, although considerable platinum, cassiterite (tin mineral), phosphate, monazite, columbite, ilmenite, zircon, diamond, sapphire, and other gems have been produced. Other valuables recovered include native bismuth, native copper, native silver, cinnabar, and other heavy weather-resistant metals or minerals.

In addition to onshore placer mining, offshore mining for gold, diamonds, iron ore, lime sand, and oyster shells is being done. Future possibilities exist for mining of manganese nodules containing manganese, iron, copper, nickel, and cobalt and phosphorite nodules which overlie large areas of deep oceans floors. Other apparent concentrations of metals and minerals exist in vast tonnages of clays, muds, and oozes on the ocean floors.

Mining claims. In the United States placer mining claims are initially obtained by complying with mining and leasing laws of the Federal or state government or both. Concessions for mining ground in foreign countries can be obtained by following appropriate procedures of the country concerned. Specific information pertaining to acquiring mining claims in the United States can be obtained through offices of the U.S. Bureau of Land Management and the state land office having jurisdiction over the area in which the mining ground is located.

Prospecting, sampling, and valuation. A uniform grid pattern is best for sampling placers. This generally consists of equally spaced sample points in lines approximately perpendicular to the longest direction of the deposit. The distance between lines may be five times the distance between holes. If the apparent deposit seems equidistant in all directions, a grid of equally spaced samples is often used. Prior to actual laying out a prospecting grid, geophysical prospecting techniques may be used to determine approximate depth to bedrock and to outline probable old stream drainage systems. This information will allow more judicious planning of a sample point pattern. Samples are usually obtained by one or several of the following methods: shaft or caisson sinking, churn, and other drilling, and dozer or backhoe opencut trenches. Drilling in areas covered by water is accomplished by placing equipment on a barge or, in the case of the Northland, on ice. The information obtained allows the value of the deposit to be calculated and gives descriptions of the physical characteristics of the material in the deposit. The latter information is valuable for designing the mining method. Plotting unit values, obtained from sampling, on a map will indicate whether a concentration of mineral exists and the location of a pay streak. Calculations will indicate the total value and yardage in the pay streak. A decision can then be made as to whether or not the deposit can be mined on an economic basis. *See* BORING AND DRILLING (MINERAL); MINING OPERATING FACILITIES; PROSPECTING.

Recovery methods. A form of sluice box is the type of unit most often used to separate and recover the valuable metals or minerals. Such a unit may vary in width from 12 to 60 in. (0.3 to 1.5 m), 30 to 60 in. (0.75 to 1.5 m) being common, and in lengths from 40 to several hundred feet. The sluice box is placed on a grade of about $1\frac{1}{2}$ in. (4 cm) vertical to 12 in. (30 cm) horizontal, but adjusted for the particular conditions. Recovery units used in the sluice include the following.

Riffles. Rocks 3–6-in. (7.5–15 cm), wood blocks, pole riffles, and Hungarian riffles (Figs. 1 and 2) form pockets in which valuable heavy particles will settle and be retained.

Amalgamation. Mercury is sprinkled at the head end of the sluice box to combine with and hold gold in various types of riffles and amalgam traps. Copper plates, mercury-coated and protected from coarse gravel by suitable screen, aid in collecting gold under some conditions.

Undercurrents. These are auxiliary sluices, parallel or at right angles to the main sluice. A variety of grizzly plates, screens, and mattings are used to separate various sizes of material and to catch small particles of concentrates that may be lost when using only a conventional sluice box.

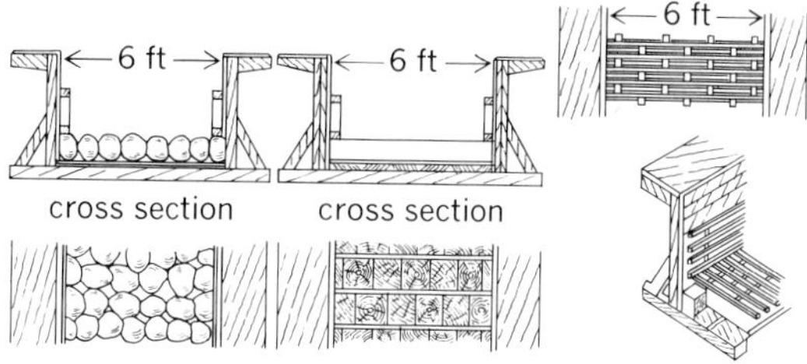

Fig. 1. Cross-sectional diagrams illustrating construction of sluices. 6 ft = 1.8 m. (*From G. J. Young, Elements of Mining, 4th ed., McGraw-Hill, 1946*)

Jigs. This is essentially a box with a screen top upon which rests a 3–6-in. (7.5–15 cm) bed, usually of steel shot, through which pulsating water currents act. Feed to the jig comes onto the bed in a water stream, and the valuable heavy particles pass through the pulsating bed and are recovered as a concentrate. Lighter waterborne material passes over the bed and into other recovery units or emerges as a finished tailing.

Other. In some plants, units such as cyclones, spirals, and grease plates are used to effect recovery. More sophisticated equipment may be used to further concentrate and separate products. Details of design and application of various recovery units appear in technical literature.

Water supply. A large volume of water is essential to nearly all types of placer mining. Water may be brought into the mining area by a ditch, in which case the water is often under sufficient head to give ample gravity pressure for the mining operation. In other cases, water is pumped from its source. Where the water supply is limited, the water is collected after initial use in a settling pond and then pumped back for reuse. Water rights are obtained in accordance with Federal and state regulations in the United States.

Mining methods. A number of mining methods exist, but all have elements of preparing the ground, excavating the pay material, separating and recovering the valuable products, and stacking the tailings. A general classification of present placer mining methods is small-scale hand methods, hydraulicking, mechanical methods, dredging, and drift mining.

Small-scale hand methods. Perhaps most widely known because of gold rush notoriety are methods using pans, rockers, long toms, and other equipment that are responsible for a very small percent of all placer production. Other small-scale methods include shoveling into boxes, ground sluicing, booming, and the use of dip boxes, puddling boxes, dry washers, surf washers, and small-scale mechanical placer machines.

Pan and batea. Panning currently is mostly used for prospecting and recovering valuable material from concentrates. The pan is a circular metal dish that varies in diameter from 6 to 18 in., 16 in. being quite common. Many such pans are 2 or 3 in. deep and have 30–40° sloping sides. The pan with the mineral-bearing gravel, immersed in water, is shaken to cause the heavy material to settle toward the bottom of the pan, while the light surface

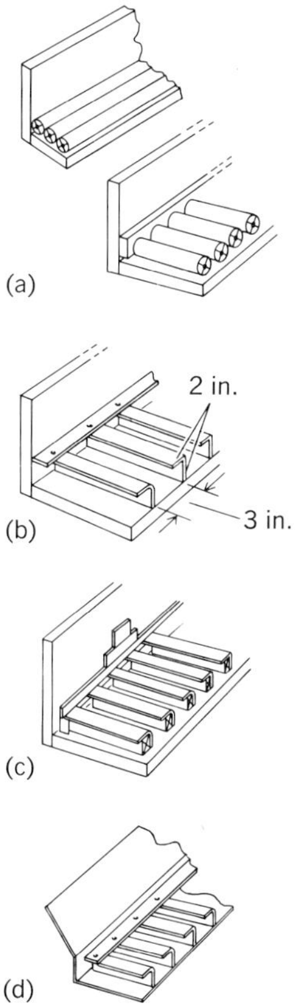

Fig. 2. Types of riffles in partial plan views. (*a*) Pole riffles. (*b*) Hungarian riffle. (1 in. = 2.5 cm.) (*c*) Oroville Hungarian riffle. (*d*) All-steel sluice. (*From G. J. Young, Elements of Mining, 4th ed., McGraw-Hill, 1946*)

material is washed away by swirling and overflowing water. These actions are repeated until only the heavy concentrates remain.

In some countries a conical-shaped wood unit called a batea (12–30-in. or 30–75-cm diameter with about 150° apex angle) is used to recover valuable metals from river channels and bars.

Rockers. Rockers are used to sample placer deposits or to mine high-grade areas (Nome, Alaska, beaches during the gold rush) when installation of larger equipment is not justified (Fig. 3). At present, various types of engine-operated mechanical panning machines and mechanical sluices are used in concentrating samples.

Long tom. A long tom is essentially a small sluice box with various combinations of riffles, matting, and expanded metal screens and, in the case of gold, amalgamating plates.

Hydraulicking. Hydraulic mining utilizes water under pressure, forced through nozzles, to break and transport the placer gravel to the recovery plant (often sluice box), where it is washed. The valuable material is separated and retained in the sluice, and the tailings pass through and are stacked by water under pressure. Hydraulicking is a low-cost method of mining if a cheap, plentiful supply of water is available and streams are not objectionably polluted.

Mechanical equipment. In this category, equipment such as bulldozers, draglines, front-end loaders, pumps, pipelines, hydraulic elevators, and recovery units are used in a number of different combinations depending on the physical characteristics of placer deposits. A typical example is to use bulldozers to prepare the ground and push the placer material to the head end of a sluice box, from which the gravel is washed through the sluice box. The tailings are stacked with a dragline, and a pump is used to return the water for reuse.

Dredging. Large flat-lying areas are best suited for dredging because huge volumes of gravel can be handled fairly cheaply. Prior to dredging a deposit, the ground must be prepared. This preparation may consist of removing trees and other vegetation and overburden such as the frozen loess (muck) found in many parts of Alaska by use of equipment and water under pressure (stripping). Barren gravel may also be removed. If the placer gravel is frozen, the deposit may be thawed by the cold-water method, as has been done in northern Canada (Yukon Territory) and Alaska.

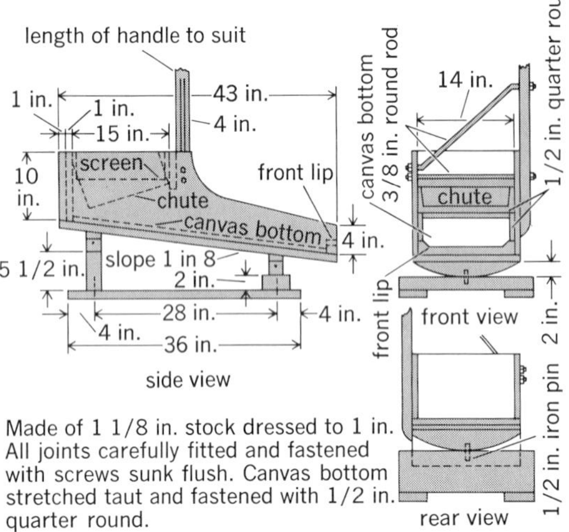

Fig. 3. Diagram of plan details of a rocker. 1 in = 2.54 cm. (*Modified from a drawing in Eng. Mining J., in R. S. Lewis Elements of Mining, 2d ed., copyright © 1941 by John Wiley and Sons, Inc.; used with permission*)

Fig. 4. Yuba bucket-ladder dredge in operation near Marysville, Calif. (*Yuba Manufacturing Co.*)

Bucket-ladder dredges (California type). For onshore placer mining, this mass-handling method now largely supersedes others, such as the chain-bucket dredge and one-bucket dredge for mining placer gold and other heavy placer material. Some large modern dredges can dig to depths of 150 ft and handle 10,000 yd^3 or more/24 hr.

Modern dredges are for the most part steel hulls of the compartment or pontoon types. On the hull is mounted the necessary machinery to cause an endless bucketline to revolve and dig the placer gravels as the dredge swings from side to side in a pond (Fig. 4). At the end of each swing, the buckets on the ladder are dropped and the cycle repeated. When the bucketline reaches the bottom of the pay streak, the ladder with buckets is raised, the dredge is moved ahead or into a parallel cut, and the cycle is repeated. The buckets discharge their load into a hopper, where it is washed with water into a long inclined revolving cylindrical screen (trommel) having holes that commonly vary in diameter from $\frac{1}{4}$ in. (6 mm) to as much as $\frac{3}{4}$ in. (19 mm). In this revolving screen, a stationary manifold supplies water (common pressure 60 psi or 400 kPa) to nozzles spaced equally in a line the length of the screen. Water from these nozzles washes the sand and gravel and causes the fines and valuable minerals and metals to work through the holes in the revolving trommel and fall on the tables (series of parallel sluices), which may consist of various recovery units, such as mercury traps, jigs, rifles, matting, expanded metal screens, and undercurrents. The placer concentrates are held in the recovery units until cleanup time (the end of a day to 4 weeks, but commonly 2 weeks). The accompanying sand passes over the tables and into the dredge pond. The coarse material that will not pass through the holes in the screen is discharged at the lower end of the trommel screen onto an endless conveyor belt (stacker) which causes the coarse tailings to be deposited in back of the dredge. Overflow muddy pond water passes from the mining area through a drain (Fig. 5).

In the case of gold mining, during cleanup time, the gold amalgam is further concentrated by paddling in the sluices, then picked up, cleaned by panning, retorted to separate the gold from the

Fig. 5. Aerial oblique view up-valley over the site of a placer dredging, in which successive cuts, tailing patterns, and dredge pond are shown. The dredge appears in the left foreground. (*Pacific Aerial Survey, Inc.*)

mercury, and the remaining gold is cast into bricks for appropriate sale. The mercury is reused.

Dragline dredges. A washing and recovery plant, mounted on pontoons or crawler treads, is fed directly or by a conveyor from a shovel or dragline unit (Fig. 6). Flow of material and recovery of products are essentially the same as for a bucketline dredge. Such dragline dredge installation is mostly used on deposits too small to justify the installation of a bucket dredge.

Offshore suction dredges. Exploration for offshore mineral deposits in the unconsolidated material of the ocean's floors is receiving a considerable amount of attention with the ultimate hope of increasing the efficiency of mining for gold, tin, diamonds, heavy mineral sands, iron sands, lime shells, and sand and gravel (also excavation for channel excavation). In addition to bucket-ladder dredges, grab dredges (clamshell buckets), hydraulic (suction) dredges, and airlift dredges are being used. In hydraulic dredges water under pressure is released near the intake and, helped by a suction created with pumps, returns to the surface, carrying with it sand and gravel from the ocean floor. In airlift dredges, air replaces water as described for the hydraulic dredge. Flow of material and recovery of products are essentially the same as for a bucketline dredge.

Small-scale venturi suction dredges have been used by skin divers to recover gold from river bottoms.

In the future, development of more sophisticated equipment for offshore mining of minerals can be anticipated.

Drift mining. Because of relatively high costs, drift mining of placer deposits is not used to a large

Fig. 6. Dragline dredge and pond. (*Bucyrus-Erie Co.*)

extent. It consists of sinking a shaft to bedrock, driving drifts up- and downstream in the valley deposits for 250–300 ft (75–90 m), and then extending crosscuts the width of the pay streak. In thawed ground, timbering is necessary, whereas in frozen ground a minimum of timber is required and steam thawing plants are used to thaw the gravels before mining is done. Usually, the thawed gravel is hoisted to the surface and then washed to recover the valuable material. Research utilizing modern technique and equipment should make this method more economically competitive. *See* UNDERGROUND MINING.

Water pollution. Proposed regulations on both state and Federal levels tend to place more restrictions on discharge of water carrying sediments into stream drainages. Eventually, treatment of some type may be necessary to remove such sediments, and proposed new operations will no doubt consider this factor.

Tailing disposal. Increased restriction on tailings in placer mining is being proposed in various areas. Such regulations may require smoothing of tailing piles and in some cases resoiling. In Alaska, in many cases, dredge tailings have added to the value of the ground by removing permafrost and offering a solid foundation for road and building construction in place of the original swampland surface.

Power. Electric power for operation of equipment may be generated by several methods, such as utilizing coal, oil, gas, or water (hydro). Units may be permanently located or may be diesel power units aboard dredges or movable shore plants for dredges.

Developments in technology. In past years a number of improvements have been made to increase the efficiency of placer mining. These may be summarized as follows.

1. Development of the diesel engine allowed diesel-powered pumps, tractors, draglines, dredges, and the like to be manufactured and gave much more compactness, economy, and flexibility to placer mining methods. This in turn resulted in greater yardage at lower unit costs.

2. The adaptation and use of jigs in placer recovery plants have increased the recovery of valuable minerals and metals with less work and more efficiency of compact recovery units.

3. The use of the sluice plate in mechanical mining methods has been an important improvement in recent years. This method does away with the conventional nozzle setup in front of the sluice box and the accompanying labor expense. Gravel is pushed directly into the head of the sluice into a stream of free-flowing water. This results in a fast system of coordinated operation, especially for mining placer gravels that contain a minimum of clay and cemented gravels.

4. Development of more efficient transportation carriers has allowed relatively inaccessible areas to become accessible. These include large airplanes to haul heavy equipment and fuel and land machines for crossing swampy and difficult terrain.

5. Use of rippers on tractors to aid in the mining process is becoming successful.

6. Electronic devices and television cameras for control units further increase efficiency of operation.

7. Modern lightweight pipe with the "snap-on" type of couplings and continuously improving design of equipment such as the automatic moving nozzle (Intelligiant) tend to increase the efficiency of mining.

[EARL H. BEISTLINE]

Bibliography: American Institute of Mining, Metallurgical and Petroleum Engineers, *Surface Mining*, 1968; Institution of Mining and Metallurgy, London, *Opencast Mining, Quarrying and Alluvial Mining*, 1965; C. F. Jackson and J. B. Knaebel, *Small-Scale Placer-Mining Methods*, USBM Inform. Circ. no. 6611, 1932; R. S. Lewis and G. B. Clark, *Elements of Mining*, 3d ed., 1964.

Planer

A machine for the shaping of long, flat, or flat contoured surfaces by reciprocating the workpiece under a stationary single-point tool or tools. Usually the workpiece is too large to be handled on a shaper.

Planers are built in two general types, open-side or double-housing. The former is constructed with one upright or housing to support the crossrail and tools. The double-housing type has an upright on either side of the reciprocating table connected by an arch at the top.

Saddles on the crossrail carry the tools which feed across the work. A hinged clapper box, free to tilt, provides tool relief on the return stroke of the table. A variation is the milling planer; it uses a rotary cutter rather than single-point tools. *See* MACHINING COPERATIONS; SHAPER.

[ALAN H. TUTTLE]

Planetary gear train

An assembly of meshed gears consisting of a central gear, a coaxial internal or ring gear, and one or more intermediate pinions supported on a revolving carrier. Sometimes the term planetary gear train is used broadly as a synonym for epicyclic gear train, or narrowly to indicate that the ring gear is the fixed member. In a simple planetary gear train the pinions mesh simultaneously with the two coaxial gears (Fig. 1). With the central gear fixed, a pinion rotates about it as a planet rotates about its sun, and the gears are named accordingly: The central gear is the sun, the pinions the planets. Figure 2 shows a construction typical in industrial planetary trains.

In operation, input power drives one member of a planetary gear train, the second member is driven to provide the output, and the third member is fixed. If the third member is not fixed, no power is

Table 1. Speed ratios for simple planetary train

Fixed member	Input member	Output member	Overall ratio*	Range of ratios normally used
Ring	Sun	Cage	$\frac{N_R}{N_S}+1$	3:1–12:1
Cage	Sun	Ring	$\frac{N_R}{N_S}$	2:1–11:1
Sun	Ring	Cage	$\frac{N_S}{N_R}+1$	1.2:1–1.7:1

*N_S = number of sun teeth, N_R = number of ring (annulus) teeth.

delivered. This characteristic provides a convenient clutch action. A brake band about the intermediate member and fixed to the gear housing serves to lock or free the third member; the band itself does not enter into the power path.

Any one of these three elements can be fixed: the central sun gear, the carrier, or the ring gear. Either of the two remaining elements can be driven and the other one used to deliver the output. Thus there are six possible combinations, although three of these provide velocity ratios that are reciprocals of the other three. The principal ratios for a simple planetary gear train are given in Table 1. The ratios are entirely independent of the number of teeth on each pinion. However, to assemble the unit, Eq. (1), which does involve the number of

$$N_R = N_S + 2N_P \tag{1}$$

teeth on each pinion, must be satisfied. Typically more than one pinion is used to distribute the load through more than one mesh and to achieve dynamic balance. If the several pinions are to be equally spaced, assembly can be made only if Eq. (2) is followed.

$$\frac{N_R + N_S}{\text{Number of planets}} = \text{an integer} \tag{2}$$

In a compound planetary train, two planet gears are attached together on a common shaft. One planet meshes only with the central sun gear, the other only with the ring gear. As in simple planetary trains, there can be several of these planet pairs distributed around the train to distribute the load and achieve balance. Table 2 shows the ratios obtained for the three principal arrangements.

To assemble the train, the pitch diameters of the gears must agree with Eq. (3), where D_R and D_P

$$D_R = D_S + D_{P_1} + D_{P_2} \tag{3}$$

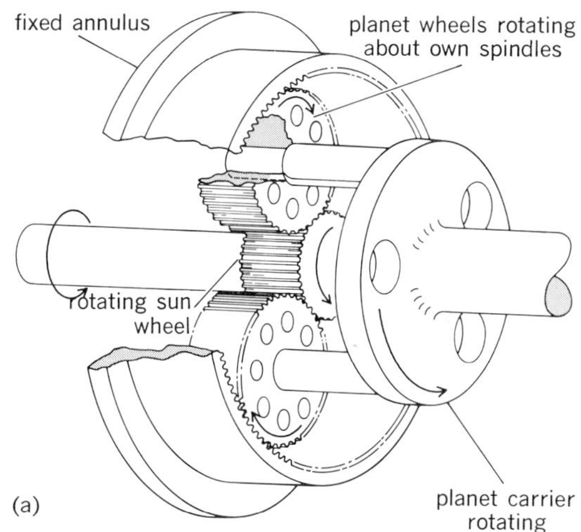

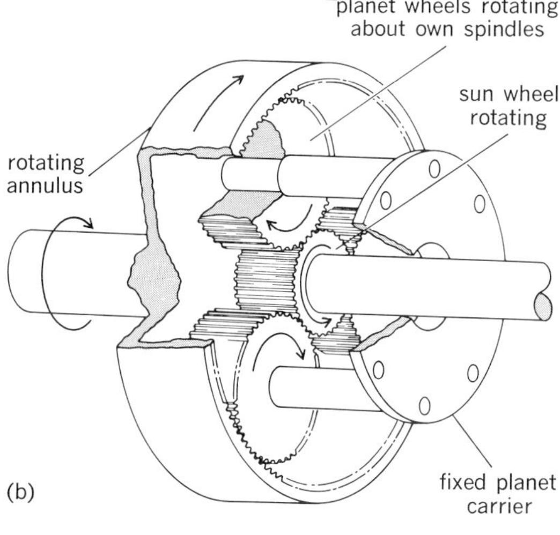

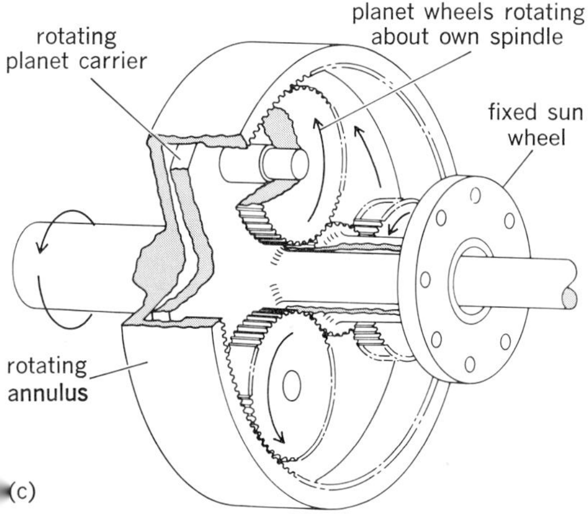

Fig. 1. Cutaway isometrics of modes of operation for simple planetary gear train. (*a*) Annulus locked. (*b*) Planet carrier locked. (*c*) Sun gear locked. (*D. W. Dudley, ed., Gear Handbook, McGraw-Hill, 1962*)

Fig. 2. Stoeckticht planetary gear train. (*DeLaval Steam Turbine Co.*)

Table 2. Speed ratios for compound planetary train

Fixed member	Input member	Output member	Overall ratio*	Range of ratios normally used
Ring	Sun	Cage	$\frac{N_R N_{P1}}{N_S N_{P2}} + 1$	6:1 – 25:1
Cage	Sun	Ring	$\frac{N_{P1} N_R}{N_S N_{P2}}$	5:1 – 24:1
Sun	Ring	Cage	$\frac{N_S N_{P2}}{N_R N_{P1}} + 1$	1.05:1 – 2.2:1

*N_S = number of sun teeth; N_{P1} = number of first-reduction planet teeth; N_{P2} = number of second-reduction planet teeth; N_R = number of ring teeth.

Fig. 3. Six compound planetary pinions which transmit high power from the aircraft engine shaft to the propeller hub with about 2:1 speed reduction. (*Foote Brothers Gear and Machine Corp.*)

designate ring gear and planet diameters. If the planets are to be equally spaced about the train, the number of teeth must agree with Eq. (4).

$$\frac{N_{P_1}N_R + N_{P_2}N_S}{\text{Number of planets}} = \text{an integer} \qquad (4)$$

The aircraft engine drives the outer ring gear of Fig. 3; the central sun gear is fixed; the planet carrier arm, omitted in Fig. 3 to show the compound pinions, has stub shafts that carry the planet pinions in the manner of the cage of a ball bearing. The main shaft of the planet carrier passes through the opening in the sun gear to drive the propeller. The use of an internal gear accounts for much of the power capacity and smooth operation of the gear train. *See* GEAR TRAIN.

[JOHN R. ZIMMERMAN]

Bibliography: N. P. Chironis, *Gear Design and Application*, 1967; D. W. Dudley (ed.), *Gear Handbook*, 1962; J. E. Shigley and J. J. Vicker, *Theory of Machines and Mechanisms*, 1980.

Plant facilities (industry)

The physical properties owned and used by industry. These include land and site improvements, such as roads, rail extensions, parking lots, and fencing; buildings, such as factory, warehouse, and office areas; other structures, such as docks, liquid-storage tanks, incinerators, and gas generating systems; and machinery and equipment used to produce or condition the products of the plant or to support the producing (or conditioning) processes.

Location. An industrial plant is often located merely by preference of one or a few owners of the business. Well-managed companies make a detailed study before deciding on locations for new plants. The decision is based on many factors; most of them are economic and relate to the costs of transportation (for incoming materials and outgoing product), of materials, of labor, and of taxes. Many industries depend directly upon their source of raw materials and the market area for their products. Other industries consider first the availability of satisfactory features such as space, freedom from adjacent dirt and fumes, trained labor supply, water, and utilities. Other factors are community climate or the attitude of the people and their leaders, and impact on the environment by the company itself.

A thorough plant-location study includes evaluating the availability, suitability, and long-range cost of landsite, raw materials, power, fuel, water, utilities, market area, transportation facilities, labor supply, community conditions, taxes, public relations value, laws or regulations, and external hazards.

Plant layout. The layout of an industrial plant embraces the arrangements and orientation of the physical facilities, including storage and supporting services, used by the company in the production of its products. Planning a new or rearranged layout involves four basic phases.

1. Determining the *location* of the area to be laid out is not necessarily a plant-location problem. More often it is one of analyzing whether the new department or expansion should go on the north side, on the third floor, in a separate building, and so on.

2. Establishing the general *overall layout* involves relating the major activities or departments to each other and allocating the necessary area to each. At this phase, major features of the producing machinery, materials-handling equipment, utilities, and the building itself should be incorporated.

3. Planning the *detailed layout* includes the locating and orienting of each machine, working operator, material-in-process container, and supporting service.

4. *Installing* the layout plan involves coaching or training personnel in the procedures and methods on which the proper functioning of the layout has been planned, and moving, placing, and hooking up the machinery and equipment (see illustration).

These four phases fall into chronological sequence but should overlap each other. Also, these four phases make logical check points or organizational divisions for the supervisor of plant layout work.

Effective layouts are based on flow of material to allow sequential movement of the material being produced or conditioned. As the number and diversity of parts or products increase, the complexity of flow analysis grows, and the methods of analysis change. Relationships other than flow also call for activities to be close to each other.

Diagramming the relationships and assigning to the diagram the space required by the activities are steps that follow the determination of the relationships. These steps lead to a physical arrangement of space. Integration of supporting services with the flow pattern and modification of the arrangement in accordance with practical limitations of the handling and storage facilities, machine utilization, building features, flexibility, opportunity for expansion, and personnel needs and conveniences lead to the desired layout. Alternative plans are usually developed. By means of comparative evaluation the most effective plan is selected.

Layouts are most readily visualized by making a three-dimensional model of the proposed plan. Bu

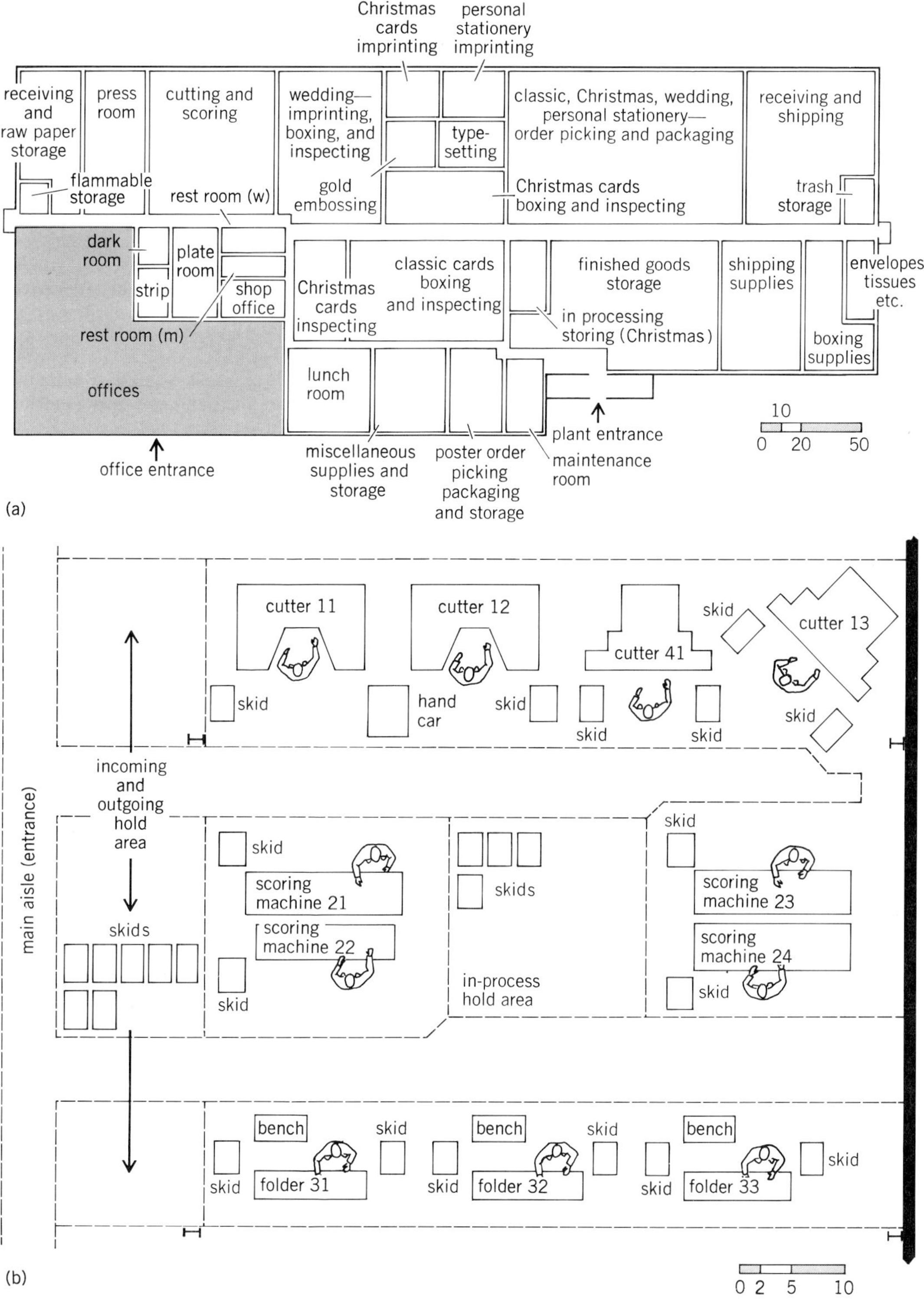

Sample plant layout. (*a*) Overall (block) layout. (*b*) Detailed (equipment) layout of one departmental block.

this is not always necessary. Two-dimensional scale templates representing the space—and the individual pieces of machinery and equipment in detail layouts—are almost always sufficient. With modern equipment and materials, many different layout plans can be made and reproduced with a minimum of draftsmanship and drawing time.

Classical plant layouts center about either the product or the process when forming or treating of materials is involved. Layout by *product* arranges all facilities required for one product together; layout by *process* locates all similar processing equipment together. The former is better for high-volume work; the latter for high variety of work.

In assembly, the choice is between moving the major component progressively to points where other parts are assembled to it—line production—or fixing the location of the major component and bringing other parts to it. Here again volume and variety essentially determine the choice.

In-plant transportation. Materials handling and plant layout go hand in hand. It is seldom possible to plan or change one without affecting the other.

Materials handling is a universal production problem. To form, condition, or assemble the materials, they must be moved to and from the point of operation.

Planning effective materials handling involves the selection of a basic handling method. Individual handling equipment and containers are then fitted into this system. The particular equipment may be hand trucks, overhead cranes, or conveyors. An integrated materials-handling operation depends on the containers and attachments for the handling equipment (clamps, hooks, brackets, and the like) being planned for interchangeable use.

Although improvements in equipment are constantly being adopted, the basic handling system of a company should remain relatively stable. Industries with high initial investment in facilities and with relatively fixed products, processes, and equipment establish a system of handling around which the plant is constructed. Factories with frequent model changes remain flexible and are constantly examining their handling methods for improvement. *See* MATERIALS HANDLING; MATERIALS-HANDLING MACHINES.

Production lines. A production line is an arrangement of work places in the sequence of operations. In its optimum form, a production line moves the material through a series of balanced operations smoothly and continuously at a uniform rate of flow with each operation being located immediately adjacent to the ones which precede and follow it. *See* MASS PRODUCTION.

The savings that come from using line production include reduced material handling; ease of production, control, and supervision; improved work-area methods; ease of training workers; reduced inventory in process; and shorter production time. On the other hand, production lines require a substantial volume of a reasonably standardized—or at least similar—product. As a result, the marketing and product design of a company greatly influence the possibility and nature of its production line. Other limitations include delays due to breakdowns or interruptions, idleness of workers due to unbalanced operation times, and greater investment in machinery because its utilization is seldom as great as in a layout by process.

In some cases, the physical ability and economic practicality to move the product from one operation to the next limits the application of a production line. For these reasons, production lines for assembly are easier to set up than those for forming or fabrication.

There are many variations of the line concept, with the ultimate seldom proving practical. Flexibility of the line to accommodate changing products, materials, processes, schedules, and personnel is too important to many companies to permit them to go all the way to the completely synchronized and automatic production line.

A production line is scheduled and operated as a unit, not as individual operations. Its workers generally need be trained to do but a single operation. These two facts can cause considerable change in the type of employee and the procedures of operating an organization when it converts its operations to production lines.

Types of plant space. The space occupied by plant facilities is typically divided by function. First it is divided into the open/yard space and underroof space. Then each is subdivided by its occupancy. This helps facilities planners keep their space needs in balance and affords a way to approximate the cost of facilities before finite plans are made. Typical underroof classes of space are: primary operating areas (to form, machine, mix, and so forth); secondary operating areas (to assemble, fill, pack, and so forth); storage areas, including receiving and shipping; service areas for personnel, plant, and production, including test and control areas; and offices and office-type areas, often including laboratories. Note that the latter three classes of space serve or support the operating areas; these service areas become increasingly important as the complexities of business operation increase.

Supporting services. Industrial facilities generally include four classes of plant services:

Plant and utility services include the generation and distribution of electricity, heating, air conditioning, ventilation, compressed air, process steam, process water, sanitary sewage, storm sewer, fuel storage, fire protection, and so forth.

Personnel or employee services include food services, medical and health care, locker rooms, rest rooms, personnel/employment offices, recreational facilities, parking lots, storm shelters, payroll/cashier functions, credit union, paging equipment, and so forth.

Production or converting services involve rail, truck, and barge access, weigh scales, freight elevators, maintenance services, toolroom, supplies warehousing, scrap collection and disposition, trash handling, quality control, test facilities, and the like.

Miscellaneous services include conference rooms, special office areas, telephone systems, visitor reception areas, training center, print shop, and so on.

In facilities that require high capital investment, the plant services tend to become a dominant factor. Some special projects require the whole facility to be designed around a process service. An example is an automobile assembly-plant paint shop laid out to conform with water, paint, gas, ventilating, air-cleaning, and fume-removal systems.

Highly labor-intensive companies would suggest facilities with ready access to locker rooms, lunchrooms, first-aid stations, rest areas, and the personnel office.

Utility support systems typically divide into primary distribution (which includes high-voltage, main sewer, and water lines) and secondary distribution (which includes detailed wiring, piping, and duct work within departmental areas). This division is similar to the second and third phases of layout planning.

Maintenance. Vital to any low-cost industrial operation is good maintenance of its facilities. This is particularly so as industry continues to substitute power operations for hand labor. The breakdown of but one key operation can sometimes cause the shutdown of an entire plant because of the synchronization of this operation with subsequent operations.

Preventive maintenance such as cleaning, adjusting, exchanging, and lubricating on a programmed basis eliminates or substantially reduces shutdowns due to machine failure. This is perhaps the most important function of a plant's maintenance group. Other functions may include inspection, repair, overhaul, reconstruction, salvage, waste disposal, plant protection, and storekeeping.

So diverse are the problems of maintenance that one person must be assigned its responsibility. Maintenance should be planned, scheduled, and efficiently executed, but its costs must be subject in part to the availability of funds. Its accomplishment should be measured by performance rather than the amount spent, for frequently not enough money is spent on maintenance of machinery, equipment, and plant for efficient overall operations.

Safety and fire prevention. The cost to industry each year of interruptions due to accidents and fires is substantial, purely aside from noneconomic considerations.

Insurance is one way of reducing losses; fire extinguishers and first-aid stations are another way. However, elimination of the causes and opportunities for an accident to occur is a more direct and rewarding approach.

Safety rules—established for the specific plant involved and consistently enforced—are well worth the effort. Detailed rules for specific equipment, located where they will be read before the machine is used, are another precaution.

Industry is held responsible for the safety of its employees. In the United States the Occupational Safety and Health Act of 1971 established national standards for compliance and an inspection-citation procedure. This covers noise and exposure regulations as well as mechanical and other hazards.

For effective fire control, every plant needs a fire-detection and alarm system, sprinklers, mains, hydrants, and extinguishers. Specific equipment depends on the nature of the industry's products, processes, and plant. *See* INDUSTRIAL ENGINEERING; INDUSTRIAL HEALTH AND SAFETY; PROCESS ENGINEERING; PRODUCTION ENGINEERING.

[RICHARD MUTHER]

Bibliography: J. M. Apple, *Plant Layout and Materials Handling*, 1977; B. T. Lewis and J. P. Marron (eds.), *Facilities and Plant Engineering Handbook*, 1973; J. M. Moore, *Plant Layout and Design*, 1962; R. Muther, *Systematic Planning of Industrial Facilities*, 1979.

Plastic deformation of metal

The permanent change in shape of a metallic body as the result of forces acting on its surface. The plasticity of a metal permits it to be shaped into various useful forms that are retained after the forming pressures have been removed. Plastic deformation is terminated either by removal or reduction of the imposed forces or by fracture of the metal. For a discussion of conditions under which changes in shape are not permanent *see* ELASTICITY.

Complete comprehension of plastic deformation of metals requires an understanding of three areas: (1) the mechanisms by which plastic deformation occurs in metals; (2) the way in which different metals respond to a variety of imposed external or environmental conditions; and (3) the relation between the internal structure of a metal and its ability to plastically deform under a given set of conditions.

Crystal structure. Pure metals are crystalline solids, or mixtures of crystalline solids in the case of some alloys. Most metals and alloys that can undergo significant amounts of plastic deformation have their atoms orderly packed in one of three types of crystal structure: hexagonal close-packed, face-centered cubic, or body-centered cubic, or slight variations thereof. *See* ALLOY.

For any type of atomic packing, as the crystal is viewed from different directions, the atoms can be visualized as lying on differently oriented planes in space. Within each plane the atoms are in a regular array, and certain directions are equivalent with respect to the distance between atoms and the location of their neighbors. It has been known experimentally for many years that the primary step in the plastic deformation of a metal crystal is the translation, or slip, of one part of the crystal with respect to the other across one of a set of crystallographically equivalent planes and in one of several possible crystallographically equivalent directions. These are known as the slip plane and slip direction, respectively. The particular direction and plane orientation differ from one metal to another, depending principally on the type of atom packing and the temperature of plastic deformation. Metals with equivalent crystal structures tend to exhibit a similar plastic response to stresses even though the actual strength and temperature range of such a like response will differ from metal to metal.

When a metal consists of a single crystal, it deforms anisotropically when stressed, depending on the orientation of the operative slip system. These translations leave linear traces on the surface called slip lines which are observable under a light microscope. The earliest knowledge of slip systems was obtained through the analysis of slip-band orientations on polished crystal surfaces.

As normally produced, however, metals are polycrystalline; that is, they are composed of a multitude of tiny crystals or grains, all with identical packing but with each crystal having its principal slip planes or directions oriented differently from its neighbors. On a gross scale this permits a metal when stressed to act as an isotropic body even though each grain, if isolated, would behave in an anisotropic manner that would depend on both its orientation with respect to the stress imposed on it and the particular crystal structure of the metal of which it is a part. One structural factor that the metallurgist can control to alter the properties of a metal is grain size and shape.

Most substances are weak relative to the strength that is theoretically calculated for them on the basis of the strength of the bonds between atoms in the crystal and the interatomic spacing. This strength is estimated to be in the neighborhood of one-tenth of the elastic modulus of the particular metal. The observed maximum strengths of metals, moreover, are more like one-tenth of this calculated strength, and the stress under which plastic deformation begins is often several times lower than the observed maximum strength. The reason for this discrepancy between the predicted and observed strengths of metal has been explained to be caused by submicroscopic defects called dislocations. These defects permit metals to be plastically deformed even though their presence also reduces the maximum attainable strength of the metals to the observed value. Understanding the nature and behavior of individual dislocations and their interactions forms the modern basis for understanding the various phenomena associated with plastic deformation in metals. Only a much over-simplified picture of dislocations and their behavior can be given here.

Dislocations are line defects; that is, they have length but only a negligible width. The simplest type to consider is an edge dislocation, seen in section in Fig. 1. The plane of atoms shown is actually repeated hundreds of thousands of times in a direction perpendicular to the plane of the paper. Hence what appears to be an extra vertical row of atoms crowded into an otherwise perfect crystal lattice is actually an extra half-plane of atoms. The line of atoms that makes up the edge of this half-plane is called an edge dislocation. The horizontal plane in which it lies establishes the macroscopically determined slip plane of the crystal previously discussed.

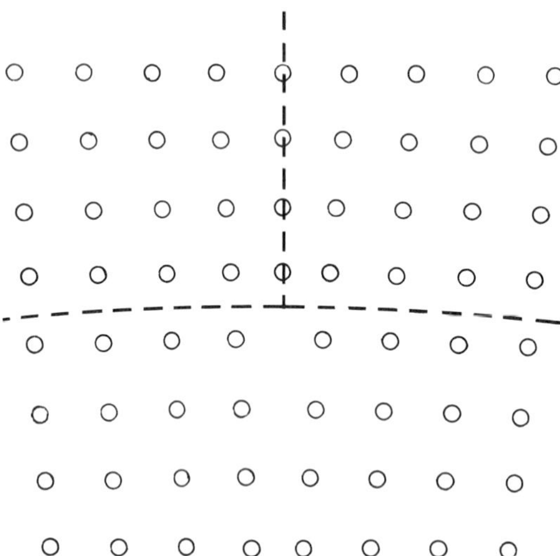

Fig. 1. Edge dislocation in a simple cubic lattice.

The behavior of a perfect metal crystal containing no dislocations may be compared with that of a crystal with one dislocation in the lattice. When shearing forces are applied to the perfect crystal so that the part of the crystal above the slip plane tends to move to the right with respect to that part of the crystal below the slip plane, all the atoms above the slip plane are displaced slightly to the right. The applied force is opposed by the bonds between atoms which are being stretched beyond their normal lowest-energy configuration. This action is like that of stretching springs: The more they are stretched beyond their equilibrium length, the greater the force necessary to stretch them further.

However, when each atom in the upper plane has been translated to a point midway between its counterpart on the plane below and the right-hand neighbor of that atom, it then becomes energetically favorable to break its bond with the former atom to which it was closely attached and link up with the one on the right, thereby moving to a new equilibrium position above it. This movement is aided by the interatomic bonding forces after the midposition has been reached. Since each of the atoms above the plane has now been displaced one interatomic distance to the right, the crystal is said to have slipped such an amount. Because this movement requires the simultaneous cooperative motion of all the atoms above the slip plane, the forces required to slip a perfect crystal are very high.

On the other hand, if a dislocation is present the behavior is considerably different. When no external forces are applied, the extra half-plane of atoms that is crowded into the crystal is symmetrically located with respect to the atoms below the slip plane. Although a considerable number of atoms are out of registry, equal numbers are displaced to the left and to the right producing a balance of nonequilibrium forces. When a shearing force is applied to the upper half of the crystal tending to displace it to the right, this force balance is disturbed. Because the atoms just to the left of the dislocation center are nearly in their "new" equilibrium positions, their interatomic bond forces aid the applied force in shifting those atoms just to the right of the dislocation center farther from their equilibrium position and stretching their interatomic bonds still further. Because these internal forces are nearly balanced, only a very small applied force is needed to cause the dislocation to move. A consideration of the atom movements will show that if all the atoms in the neighborhood of the original dislocation center are shifted slightly to the right, the center of the disturbance in the normal crystal structure (the dislocation) will be reproduced one atom spacing to the right of its unstressed location. Continued application of a low shearing force causes this lattice disturbance to continue to move to the right. When the dislocation reaches the right-hand surface of the crystal, all the atom planes can freely assume their normal equilibrium spacing. Thus the dislocation vanishes and its extra half-plane appears as a step on the surface, the upper half of the crystal having slipped one interatomic distance with respect to the lower. The result is the same as that produced by slipping a perfect crystal; however, only a small fraction of the applied force is needed when the dislocation is present.

A simple analogy may be useful in understanding the tremendous reduction in the level of stress needed to cause plastic deformation through dislocation motion. In the laying of a rubber-backed rug in a room, positioning is difficult because of the nonskid backing of the rug. However, if the back end of the rug is moved forward to the desired final position, a wrinkle is created in the rug at that end. If the back edge is held in position, then the wrinkle may be moved easily along the length of the rug. When it gets to the far end of the room, the extra material composing the wrinkle is flattened out, extending the entire rug to the desired location. The rug has thus been shifted by an amount equal to the excess material in the wrinkle but with a minimum expenditure of force. Further shifting of the rug requires the generation and propagation of additional wrinkles.

Dislocations were originally postulated to explain why metals deformed plastically under stresses so much lower than theoretically predicted. The existence of dislocations has since been confirmed and much of their behavior observed with the aid of the electron microscope and the development of thin-foil electron transmission microscopy. A greatly enlarged image of the foil structure, as seen by the electrons, is displayed on a phosphor-coated screen that can be photographed. Dislocations, because they are line defects in the periodic structure of the metal lattice, appear as images that sharply contrast with the image of the perfect metal matrix. Normally, the dislocations or dislocation networks appear black against a light background such as in Fig. 2. The extremely high magnification of the electron microscope and therefore the small volume of metal under observation permit individual dislocations to be studied in spite of the high density of dislocations in metals. This high density normally precludes a direct study of the effect of an individual dislocation on mechanical properties.

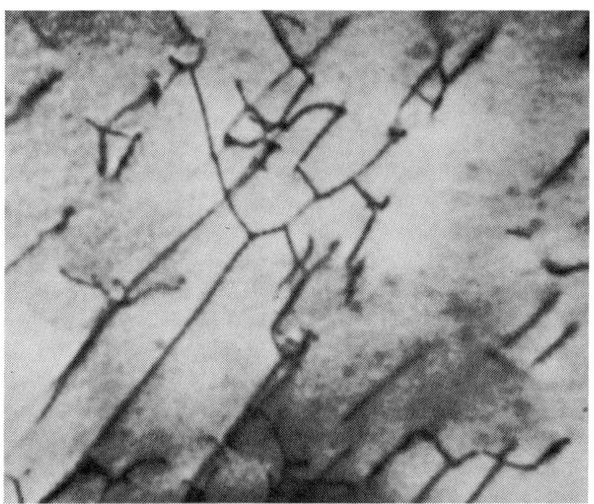

Fig. 2. Dislocation network in iron revealed by electron transmission microscopy; ×28,000. *(From D. McLean, Mechanical Properties of Metals, Wiley, New York, 1962)*

The presence of dislocations has also been revealed and some of their properties studied by the use of chemical etchants on the surface of solids. Certain etchants that are peculiar to an individual metal or other crystalline solid will reveal as etch pits any dislocations that intersect the surface. An example of the appearance of slip bands in lithium fluoride, revealed as a series of dislocation etch pits, is shown in Fig. 3.

Because the movement of one dislocation across a metal crystal results in a slip offset of only about an atom spacing ($\approx 2 \times 10^{-8}$ cm), it is obvious that significant macroscopic shape changes require the movement of billions of such defects. Measurements have shown that the dislocation density in unworked metals is about 10^6 per cubic centimeter, while in highly strained metals it is about 10^{12} per cubic centimeter. In other words, on the average, each initial dislocation has multiplied to form a million more! This inherent ability is required to provide the number of mobile dislocations needed for significant deformation. There are several mechanisms which have been proposed to explain the generation and interaction of dislocations.

Fig. 3. Dislocations in slip bands revealed by etch pits on the surface of lithium fluoride; ×25. *(From J. J. Gilman, Micromechanics of Flow in Solids, McGraw-Hill, 1969)*

Although an understanding of the mechanisms of plastic deformation in metals has been attained only relatively recently and is still receiving vigorous attention by metal scientists, the phenomenology of metal behavior has been explored and documented by metalworkers and metallurgical engineers for centuries. This information has been vital to the design and manufacture or construction of metal objects from tin cans to complex gas turbines. The properties of metals that are associated with plastic deformation are ductility (the ability of a metal to be deformed considerably before breaking), behavior in creep (the time-dependent deformation of metal under stress), and the response to fatigue (conditions where the stresses are applied in a cyclic fashion rather than steadily). *See* METAL, MECHANICAL PROPERTIES OF; METAL FORMING; METALLOGRAPHY.

Ductility. A substance is brittle if it breaks without deforming; it is ductile if it can deform considerably before breaking. Ductility is important in metals for two reasons. It permits a metal to be formed into commercially desirable shapes without breaking, and it permits a metal to absorb shocks and blows in service that could break a stronger but more brittle material. One simple measure of ductility is the reduction in cross-sectional area that a metal sample will show when pulled apart in simple uniaxial tension. Some metals are as brittle as glass, showing virtually no reduction in cross-sectional area in a tensile test, whereas others under the same conditions will be as ductile as taffy, stretching out in a tensile test until the cross section of the sample has contracted to a mere filament. Ductility is not a fixed property of a given metal but will depend upon such factors as the temperature, the speed with which the metal is deformed, the size and shape of the metal, impurities that may be present in the metal, the environment in which the metal is located, and the manner in which the metal deforms prior to breaking.

A load versus elongation curve for a typical metal deformed in tension is shown in Fig. 4. At low loads and small amounts of deformation the metal behaves elastically or nearly so. The exact point at which the deformation becomes plastic, the yield strength, is uncertain, and its measured value depends on the sensitivity of the measuring device.

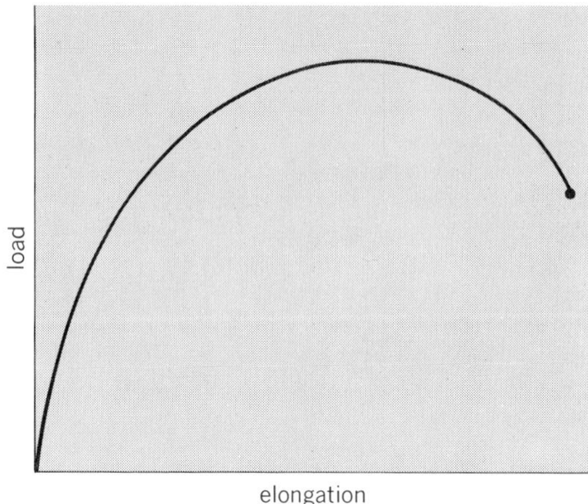

Fig. 4. Typical load-elongation curve for a polycrystalline metal. Exact load at which the deformation becomes plastic is uncertain.

Two things shown by the curve are important: (1) Increasing plastic deformation requires an increasing load; this is known as strain hardening; (2) there exists a maximum load in the curve; this defines the ultimate strength or load-carrying capacity of a metal bar with a given cross section. Beyond this strain the load begins to drop, continuing until the metal fractures.

Strain hardening is associated with the increase in the number and degree of entanglement of dislocations in the metal. Dislocation multiplication is a dynamic process that occurs during plastic deformation, that is, when the majority of dislocations are in motion. Since they are generally moving on different slip planes, many of which intersect, the dislocations will interact with each other forming a tangled network. This impedes the motion of other dislocations, thus requiring higher loads to support further plastic deformation. This is the mechanism of strain hardening. However, because any incremental increase in the number of dislocations is relatively less significant as the total dislocation density increases, the observed rate of strain hardening decreases with increasing strain.

This change in internal structure of a metal bar after plastic deformation is accompanied by a change in external shape. In tension, for example, the bar is lengthened and its cross-sectional area reduced in such proportion that the volume of the bar remains nearly constant. The increased strength of the metal that results from strain hardening permits a bar with an ever decreasing cross sectional area to support a larger and larger load. The load reaches a maximum, however, when the rate of strain hardening decreases to the point where it just compensates for the rate of decrease in cross-sectional area.

Up to the point of maximum load the cross-sectional area is approximately the same at any point along the length of the bar; any tendency for one particular area to reduce further than the others is resisted by the locally greater strain hardening that results. Hence the plastic deformation is said to be uniform. A quantitative measure of this strain is the ratio of the cross-sectional areas of the bar before (a_o) and after (a_f) deformation. Since the range of this ratio is large, the natural logarithm of the ratio is usually plotted. This is termed the true strain. Beyond the point of maximum load, strain hardening can no longer compensate for loss of cross-sectional area. Any local reduction in area is followed by continued reduction in the same location until fracture occurs. This is referred to as the necking of a tensile specimen resulting from plastic instability.

Plastic instability rather than fracture itself limits many important metal-forming operations. An example is the stretch forming of sheet metal to form complex shapes such as automobile fenders. As soon as the strain localizes, little further overall deformation of the sheet can be obtained without fracture in the region of strain localization regardless of its ductility with respect to fracture per se. This leads to localized tearing in stampings, particularly when the part into which the metal is being shaped is complex so that high strains are imposed in a few areas.

Many other metalworking processes are limited instead by the fracture of the metal being deformed. Failure of metals after extensive plastic deformation occurs by the generation and growth of holes within the metal. In a tensile bar the major amount of hole growth takes place in the region near the smallest cross section of the neck. Because of the complex nature of the stresses there, the holes are most likely to grow first and largest on the axis of the bar. An example of a copper bar, sectioned through the neck prior to complete fracture, is shown in Fig. 5. The large central holes are obvious; the many smaller holes are less evident. Continued plastic deformation causes the coalescence of the holes into a central crack which grows by further hole generation and linkage. The ductile fracture surfaces have a well-known silky appearance which results because the surface consists of a myriad of tiny holes. Present thinking is that the holes initiate at the boundary between

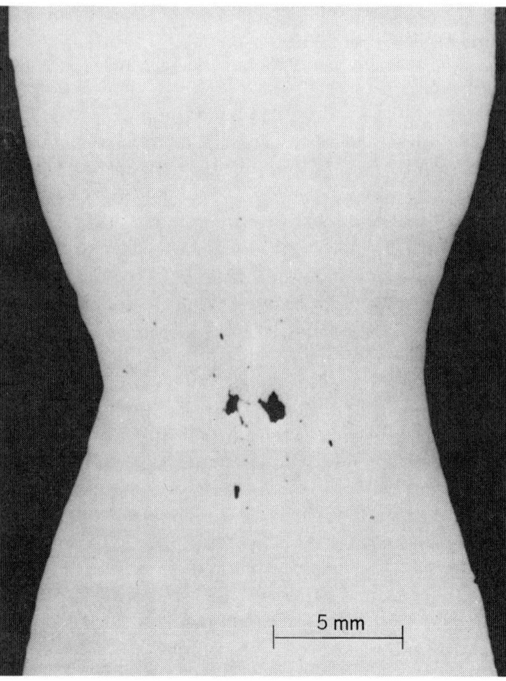

Fig. 5. Section through center of necked region of copper tensile specimen showing large voids that form prior to complete failure. *(From H. C. Rogers, The tensile fracture of ductile metals, Trans. AIME, 218:498, 1960)*

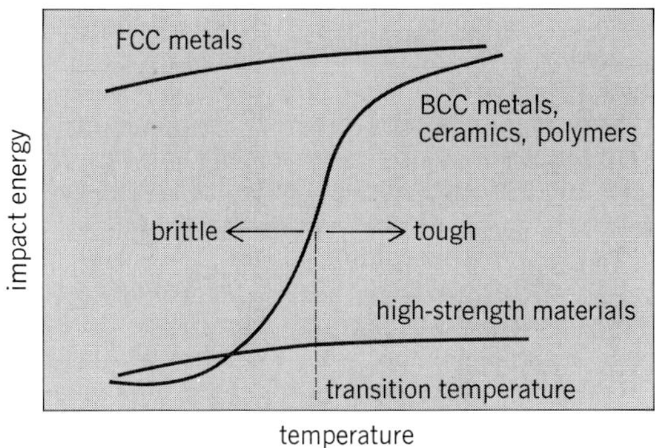

Fig. 6. Illustration of variation of impact energy with temperature for different materials. (*From C. R. Barrett, W. D. Nix, and A. S. Tetelman, The Principles of Engineering Materials, Prentice-Hall, 1973*)

particles in the metal and the surrounding metal matrix or in the particles themselves because they are normally brittle and cannot withstand the plastic deformation of the metal about them without cracking. These particles may be either part of the alloy structure itself or may be tramp impurity particles that remain as a result of the entire processing history of the metal.

In a tensile test, metals crystallizing in the face-centered cubic system (for example, copper, aluminum, nickel, and lead) are generally ductile at either low or moderate temperatures and low rates of deformation, although even these metals (especially those with large grain size) show some brittleness in certain combinations of high temperature and low deformation rates.

Certain of the close-packed hexagonal metals such as zinc and most of the common body-centered cubic metals (notably iron, chromium, tungsten, and molybdenum, but not vanadium or tantalum) and their alloys become brittle at low temperatures and at high deformation rates (Fig. 6). This phenomenon is the basic cause for the dramatic failure of steel structures in cold weather under shock, for example, the breaking in half of the ocean freighter *Flying Enterprise* in the North Atlantic. Metals that behave in a brittle manner at low temperatures and a ductile manner at higher temperature are said to have a ductile to brittle transition temperature. Above this temperature such metals are ductile, absorb large amounts of energy during fracture, and fracture in the manner described above. Below this temperature such metals are nearly as brittle as glass, fracturing along specific crystallographic planes by the severing of atomic bonds across these planes. This brittle fracture process is known as cleavage. With high-purity metals the temperature range over which this transition in behavior occurs is usually very narrow; with most commercial metals, particularly steels, the temperature range may be quite broad. In this case, the center of the range is usually called the transition temperature. Ductility is only one way to define the transition behavior. Others are change in energy absorption or the appearance of the fracture surface, because a cleavage fracture exhibits bright crystalline facets. The transition temperature is not a fixed property of a metal but is sensitive to the alloy content, the cleanliness of the metal in terms of included particles, the grain size, and the amount of plastic deformation the metal has undergone previously. Of major importance also is the rate of stressing and the geometry of a part in service or of a test piece being studied. As the rate at which the part is stressed is increased, the transition temperature is raised, all other variables being constant.

Notches on the surface, abrupt changes in cross section, or large holes or flaws within metal specimens under tension will affect their ductility in two ways. They will lower the general level of ductility of all metals, and they will shift the transition behavior upward in temperature or lower in strain rate or both. Both effects become greater as the area of cross section occupied by defects increases. These considerations are important because a given steel, for example, may be quite ductile, when tested as a smooth bar in a simple tension test, yet when fabricated into a matching part containing notches or holes may be dangerously brittle. The embrittling effects of blowholes from improper casting or voids introduced by improper annealing (for example, the hydrogen annealing of oxygen-bearing copper) stem from this same cause. Undissolved impurity particles whose hardnesses are markedly different from that of the host metal (either harder or softer) are embrittling because they effectively behave as holes (discontinuities) in the matrix.

The transition behavior of steel is of great technological importance because the range of temperatures over which this drastic change in ductility may occur includes "normal" ambient temperatures for many environmental and material conditions. A designer of steel structures or devices, aware of this possibility in the behavior of steel, usually consults a steel producer or other materials expert to ascertain whether the particular steel under consideration and its conditions will permit this product to operate under the service conditions to which it will be exposed. If the tank or tower or the like is exposed to unexpectedly low temperatures or high rates of stress such as in an explosion, or has suffered some unexpected deterioration from the environment or from some undetected damage during fabrication, then it may fail unexpectedly in a brittle and catastrophic manner. From the viewpoint of ductility, then, a material safety factor in addition to a mechanical one must be built into the design of steel structures.

One aspect of ductility that is currently receiving great emphasis is fracture toughness, that is, the ability of a metal to resist the growth of an existing crack, because it is recognized that large structures, especially containing weldments, are difficult to produce without flaws. Most structures are designed to behave elastically under load; however, if they contain a sharp crack, the design loads may cause a plastically deformed zone to form at the crack tip. Catastrophic failure occurs when the crack grows unstably. The larger the crack size when unstable growth begins and the larger the plastic zone size at the crack tip, the tougher or more fracture-resistant is the metal under a given stress system. The fracture toughness of iron and its alloys, as well as other metals of similar structure, decreases as the temperature drops.

Temperature plays a vital role in the plastic deformation of all metals as blacksmiths have known for centuries. In the production of metal parts and structures, plastic deformation at ambient temperatures causes considerable strain hardening, as discussed above, and hence is usually desirable because of the increased strength and stiffness of the resulting part. If, however, the fabrication of the desired shape requires considerable plastic deformation, undesirable effects, such as plastic instability or fracture, may result during processing. These can be prevented by interrupting the metalworking at a point best determined by experience, then exposing the metal to a high temperature. This is called annealing. The choice of temperature depends primarily on the metal or alloy and the amount of prior cold work (that is, deformation at temperatures where substantial strain hardening occurs). The high temperature allows the metal structure to recover because atomic processes will occur very rapidly. Dislocations, when introduced into a metal by plastic deformation, increase its elastic stored energy. It is therefore energetically favorable for the dislocation density to be reduced to that of the undeformed metal, but this process is normally hindered by the kinetics. At high temperatures, however, the dislocations can move about easily and quickly without any externally applied stress. Many interact with each other and, because of the nature of their structure, can annihilate each other in pairs. Furthermore, if the temperature is appropriately high, new grains containing a density of dislocations equivalent to an undeformed metal can form and grow at the expense of the deformed grains until the entire metal piece has a dislocation density equivalent to that of an unworked metal. After annealing, the partly shaped part will have lost the strength and tendency to failure that it gained from strain hardening. Further shaping again causes strain hardening and strengthening. It may be necessary to repeat this intermediate annealing treatment several times before the desired shape has been obtained.

The deleterious effects of strain hardening during plastic deformation may be minimized by the use of temperature in another way. Instead of carrying out the fabrication at ambient temperatures, the metal is actually shaped at high temperatures where, in essence, annealing is occurring simultaneously with plastic deformation. This is referred to as hot working. Since both the yield strength and rate of strain hardening are much lower, the metal is considerably easier to deform. This means that less massive equipment is needed to shape a given-size metal piece. This is particularly important when huge masses of metal weighing many tons such as shafts for giant steam turbines or ship plates are being shaped. At lower strength levels there is also greater ductility; therefore many of the primary working processes used to reduce and homogenize rather fracture-sensitive as-cast metals are carried out hot. Because completely hot-worked metals compare to annealed metals in their reduced strength, in production metal objects are frequently hot-worked during the majority of the fabrication operations, but the final shaping is done cold to impart the strength and stiffness obtainable by cold work.

Creep. High temperature affects metal parts in service in a manner related to that described above, but the result is usually undesirable. Equipment or machinery is built to perform a function; on carrying out this function, the many parts are normally subjected to a variety of loading and environmental conditions. When these include high temperatures and steady loads, the shape of the parts tends to change very slowly, elongating in the direction of tensile stresses and shortening under compressive loads. This causes close-tolerance parts to rub and seize. An example is a turbine wheel in an aircraft gas turbine. The centrifugal force of the high-speed rotor combined with a high operating temperature tends to elongate the buckets, closing the narrow gap between the ends of the buckets and the turbine shell. The change in shape is said to be the result of creep deformation.

Creep is defined as the slow, continuing, or time-dependent, plastic deformation that a metal or any substance evidences when put under stress. It is unlike normal plastic deformation, which is fixed (unchanging with time) unless the stress is raised, and unlike elastic deformation, which is fixed but transient (being recovered after the stress is removed). Creep or time-dependent deformation is the predominant reaction of metals to stress when temperatures are high, that is, in the upper half of the temperature range from absolute zero to the melting point of the metal, and when stresses are low. As the temperature nears the melting point of the metal, creep approximates a truly viscous phenomenon since the deformation rate tends to become constant under a constant stress. The higher the applied stress s, the greater the deformation rate E will be according to the formula $E=As^n$, where A and n are constants, n varying from about 3 to 6. At lower temperatures creep becomes more complex. After an immediate elastic or plastic response to an applied load, the metal will deform with time at a rate that slowly falls off. The deformation may settle down to what appears to be a steady rate, but ultimately, even in tests where the tensile stress is held constant, will accelerate at an ever faster pace until the metal breaks. The terms primary, secondary, and tertiary creep are

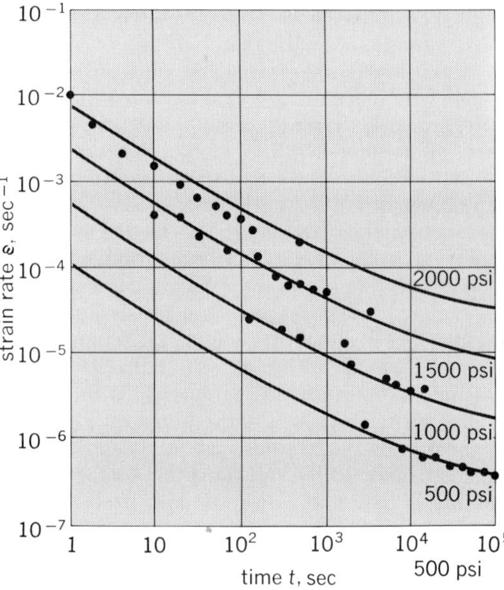

Fig. 7. Creep rate of aluminum for four different stresses plotted against time; in these examples temperature is held constant at 277°C.

used to describe the stages when the creep rate is falling off, remaining constant, or accelerating, respectively.

Mechanisms of plastic deformation at high temperatures differ significantly from those at ambient temperatures, as discussed previously. At low stresses and long times of creep, mass transport by atomic motion (diffusion) plays a significant role. The disordered grain boundaries are weaker than the crystalline grains; neighboring grains slide and rotate relative to each other. This produces voids at grain-boundary defects, weakening the structure. The accelerating rate in tertiary creep is ascribed to deterioration of the structure as voids grow and coalesce to form cracks.

Although the creep rate varies with time in the more complex forms of creep (at temperatures not close to the melting point of the metal), it can be predicted with fair success for a given temperature or stress if it is known over a given period of time at another temperature and stress. Figure 7, for example, shows the logarithm of the creep rate of aluminum plotted against the logarithm of time for four different applied stresses, the temperature being held constant. It is obvious that increasing the stress merely shifts the curve upward without distorting significantly. Similarly Fig. 8 shows the logarithm of the creep rate of aluminum plotted against the logarithm of time for three different temperatures, the stress being held constant. Increasing the temperature in two 50° steps shifts the curve bodily upward and to the left in equal amounts. These relations are expressed by Eq. (1). Here $\dot{\varepsilon}$ is the strain rate in in./in./sec, s is the ap-

$$\dot{\varepsilon} = s^n f\,(te^{-Q/RT}) \qquad (1)$$

plied stress, n is a constant (equal to 3 in the case of aluminum taken for illustration), f is the functional relation between creep rate $\dot{\varepsilon}$ and time t, R is the gas constant, T is the absolute temperature, and Q is a constant equal to the activation energy for creep of the metal in question. Through this and similar relations the effect of changes in stress and temperature can be predicted. The values of n and Q are not necessarily constant even for a given metal. Naturally, the observed strain rate will vary with each metal and it depends also upon such factors as temper and grain size.

Fatigue. Fracture under cyclic loading conditions was named fatigue many years ago because of the apparent suddenness of the failure of many components of operating equipment without any prior changes in shape that would lead to a suspicion that failure might be anticipated shortly. The metal somehow became "tired" after some period of service. Typical components that might be likely to fail in fatigue are gears, rotating shafts in engines and turbines, and parts that vibrate or flutter continuously such as the wing of an airplane or a flat leaf spring in an automobile.

In what is now termed high-cycle fatigue there is little gross shape change associated with a fatigue failure and hence the fracture is sometimes termed brittle; nevertheless, the mechanism of fracture is a ductile one in contrast to the cleavage fracture of metals below the transition temperatures or the brittle fracture of solids such as glass. The apparent brittleness in high-cycle fatigue is the result of an extreme localization of the region that is undergoing intense plastic deformation

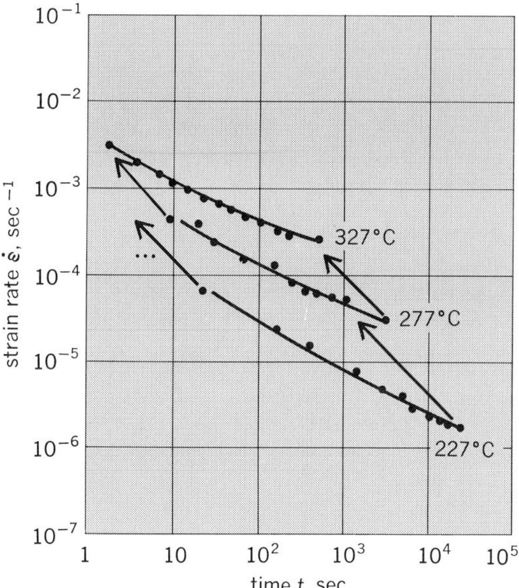

Fig. 8. Creep rate of aluminum at three different temperatures plotted against time; in these examples stress is held constant at 1500 psi.

leading to fracture. Historically, it was rapidly recognized that although the final failure in fatigue occurs suddenly, this is only the result of such a great diminution of cross-sectional area of a load-carrying part that the remaining metal cannot sustain the load without fracturing in tension. The actual fatigue failure itself begins considerably earlier, the fatigue crack growing slowly at the expense of the remaining metal cross section until the catastrophic failure occurs.

Failures in fatigue are characterized quantitatively by the number of stress cycles that a metal part or test specimen can withstand before catastrophic failure at a given level of stress. This is called its fatigue life. Metal behavior in fatigue is usually analyzed by running a series of tests at different stress levels S and determining the resulting fatigue life in terms of the number of cycles to failure N at each stress level. This is usually plotted on a semilogarithmic chart and is referred to as the S-N curve. An example is shown in Fig. 9. It is obvious, both intuitively and from the curve, that as the maximum cyclic tensile stress increases, the life expectancy will decrease sharply. Conversely, as the maximum stress is decreased, the rate of increase of fatigue life for a fixed incremental decrease in stress gets larger. In fact, for steels and a few other metals, a stress is reached below which fatigue failure apparently does not occur for any number of cycles. This is called the endurance limit and is indicated by the horizontal line in Fig. 9. For most other metals, although not possessing a true endurance limit, the rate of increase in life with decreasing stress level becomes so great that it is a simple matter to determine the stress level below which the number of cycles to failure of a metal will far exceed any number that could be expected during its normal operating lifetime.

From the standpoint of service, particularly where safety may be an important factor, it is imperative that the expected number of cycles to failure be known within quite well-defined limits. This

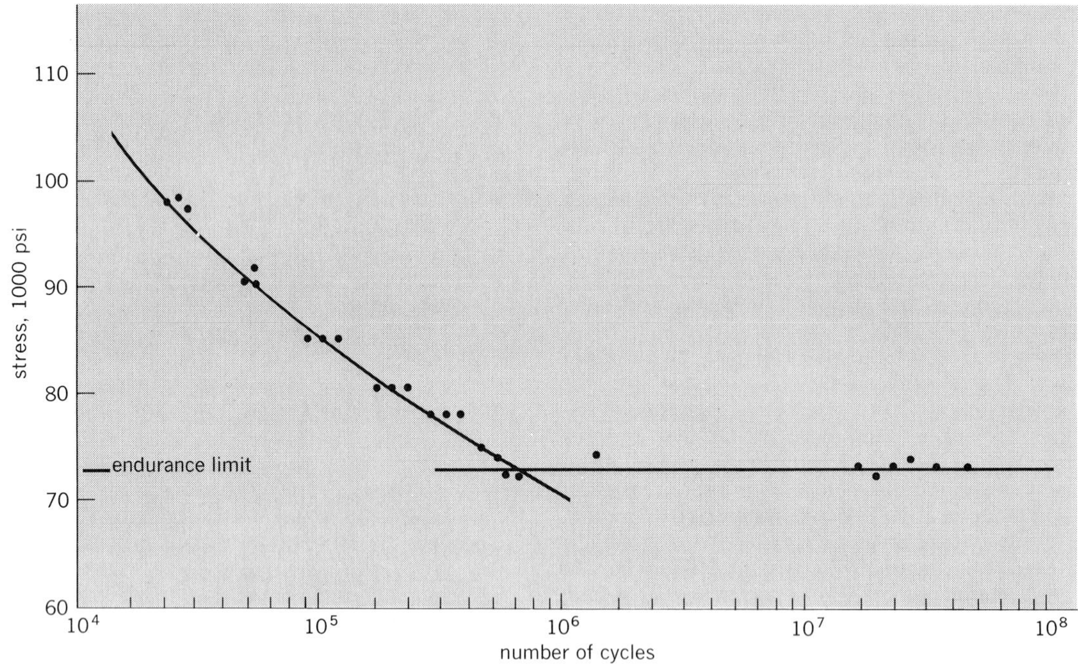

Fig. 9. Semilogarithmic chart of number of cycles to failure as a function of stress for AISI 4340 steel (*S-N* curve). (*From L. H. Van Vlack, Material Science for Engineers, Addison-Wesley, 1970*)

requires a considerably more sophisticated knowledge of the influence of the mean cyclic stress, the frequency of stressing, the temperature, the regularity or randomness of the stress pulses, and the distribution of the magnitudes of the peak stresses if they are not regular, the shape of the stress pulses, and similar external or environmental conditions of stressing than is necessary for an appreciation of the fundamental nature of the fatigue phenomenon. This must be coupled with a complete knowledge of the composition and condition of the metal under stress for an accurate prediction of fatigue behavior.

Plastic deformation is intimately linked with the initiation and propagation of a fatigue crack. Factors that affect plastic deformation in general will similarly affect localized plastic deformation associated with fatigue. Unless initiated at an obvious crack or flaw, fatigue cracks generally start at metal surfaces, usually at a minute surface discontinuity such as a fine scratch. This acts to raise the stress locally causing excessive local plastic deformation. The effect of cyclic stresses on this surface discontinuity is to increase its severity through geometrical changes as well as by changes in the surrounding metal itself resulting from cold work. The condition is continually aggravated until an actual surface crack has formed. Under cyclic stressing this crack grows in increments by a ductile crack growth mechanism during that portion of the cycle when the cracks tend to open up furthest under the applied cyclic stress. The magnitude of the plastic deformation in the local region at the crack tip depends on the variables discussed previously. The incremental local plastic deformation at the tip of the propagating crack associated with the periodic opening and closing of the crack leaves a series of marks or lines on the fracture surface of the metal called fatigue striations. Under normal circumstances one striation is created per cycle. The striation spacing is directly related to the stress level, increasing as the stress level increases. These and related markings on a part fractured in service are useful in locating the origin of the failure as well as the direction and rate which it propagated.

In addition to high-cycle fatigue, another area of fatigue increasingly recognized as important is that which occurs at the high stress end of the *S-N* curve. Because of the high cyclic stresses, reversed plastic deformation occurs throughout a large region of the metal. This produces observable changes in shape and failure will occur after a relatively few cycles; hence it is termed low-cycle fatigue. Everyone is familiar with some simple form of low-cycle fatigue. Anyone who has passed the time by bending a paper clip back and forth causing it to crack has produced a low-cycle fatigue failure. It is also easy when sawing through a metal bar, for example, to stop part-way, then bend the bar back and forth at the saw cut until it breaks. This low-cycle fatigue failure saves considerable time and effort if the resulting rough fracture surface is tolerable.

The principles of low-cycle fatigue are similar to those of high-cycle fatigue except that the plastic strain per cycle is much larger. This minimizes the effect of minor scratches and surface defects, and the surface changes produced by the plastic deformation itself may start to concentrate the strain, eventually leading to fracture. Under controlled conditions whereby the strain increment per cycle is held constant, the fatigue life of a large number of metals at room temperature can be predicted from the Coffin-Manson law, Eq. (2), where N is

$$(N^b)(\Delta\epsilon_p) = C \tag{2}$$

the number of cycles to failure, $\Delta\epsilon_p$ is the strain range imposed per cycle, and C and b are constants that depend on the particular metal being fatigued. The value of b is around 0.5.

A particularly important form of low-cycle fatigue failure is known as thermal fatigue. It occurs when the length of some structural component is held fixed while the metal is subjected to temperature fluctuations. These may be extreme, for example, in the starting up and cooling off of a gas turbine. Because most materials expand and contract with temperature, the maintenance of a fixed length causes the component to be subjected to large stresses, often well into the plastic range. Continued repetition of the temperature cycling may then lead to shape changes and a fatigue failure. [HARRY C. ROGERS]

Bibliography: M. F. Ashby and D. R. Jones, *Engineering Materials: An Introduction to Their Properties and Applications*, 1980; D. R. Axelrad, *Micromechanics of Solids*, 1978; C. R. Barrett, W. D. Nix, and A. S. Tetelman, *The Principles of Engineering Materials*, 1973; D. Hull, *Introduction to Dislocations*, 2d ed., 1976; L. H. Van Vlack, *Materials Science for Engineers*, 1970.

Plasticity

The property of a solid body whereby it undergoes a permanent change in shape or size when subjected to a stress exceeding a particular value, called the yield value. Many solid materials obey Hooke's law at low stresses. As the stress is increased, however, departures from Hooke's law occur, and some plastic flow takes place; that is, the material does not completely recover its original shape or size when the stress is released.

Plastic behavior is often accompanied by time-dependent effects such as creep (the increase in strain with time at constant stress), stress relaxation (the decay of stress with time at constant strain), and elastic aftereffect or recovery (the gradual decrease to a limiting permanent strain when the stress is removed). The study of these phenomena in all their manifestations is the science of rheology. *See* ELASTICITY; HOOKE'S LAW; PLASTIC DEFORMATION OF METAL; STRESS AND STRAIN.

[R. F. S. HEARMON]

Plate girder

A beam built up of steel plates and shapes which may be welded or bolted together to form a deep beam larger than can be produced by a rolling mill (see illustration *a* and *b*). As such, it is capable of supporting greater loads on longer spans. The typical welded plate girder consists of flange plates welded to a deep web plate. A bolted configuration consists of flanges built of angles and cover plates bolted to the web plate. Both types may have vertical stiffeners connected to the web plate, and both may have additional cover plates on the flanges to increase the load capacity of the member. Box girders consist of common flanges connected to two web plates, forming a closed section.

In general, the depth of plate girders is one-tenth to one-twelfth of the span length, varying slightly for heavier or lighter loads. On occasion, the depth may be controlled by architectural considerations.

Webs and flanges. The design of the web is essentially the determination of the required thickness for a selected depth of girder. A minimum thickness is usually prescribed because of corrosion, welding, or bolting limitations. The selection of the thickness is based upon two factors: providing sufficient cross-sectional area to resist the maximum shear; and providing sufficient depth-to-thickness ratio to resist buckling due to the combined shear and bending stresses.

A limiting ratio of web depth to web thickness is expressed below, where h is the clear depth of

$$\frac{h}{t} \leq \frac{14{,}000}{\sqrt{F_y(F_y + 16.5)}}$$

the web in inches, t is the web thickness in inches, F_y is the yield stress of the material in kips per square inch, and the constant 16.5 accounts for the residual stresses due to fabrication. Web buckling may be prevented by the addition of vertical or horizontal stiffeners.

The design of a plate girder proceeds by trial sections determined from the shears and bending

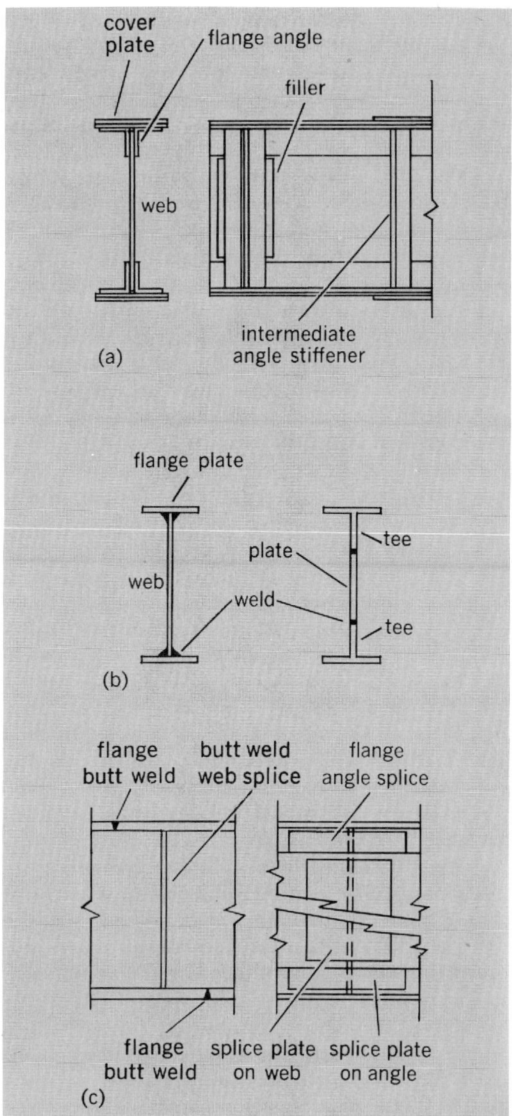

Plate girder configurations: (*a*) bolted type, (*b*) welded type, and (*c*) web splices.

stresses. The capacity of the section is computed and compared with that required by the applied loads, and the section is modified until the desired capacity is achieved. The required section is determined from the elastic theory of bending, such that f equals M/S and is within the allowable stress limit; here f is the bending stress at the extreme fiber (psi), M is the applied moment (inch-kips), and S is the section modulus (cubic inches), Lateral buckling must be considered in proportioning the flange widths and thicknesses.

For economy, cover plates may extend only as far as required by the bending stresses. However, the cost of fabrication should be a consideration in determining the lengths to be used for each cover plate and the amount of welding required to attach them. Welds and bolts transfer the shear from the flange to the web. For a bolted girder, when several cover plates are used, one plate must extend the entire length of the girder, while others may be cut off when no longer required by the bending stresses. However, for a smooth transition of bending stresses, the cut-off cover plates must extend beyond the theoretical termination point to permit a connection to develop the full strength of the plate. For a welded girder, multiple cover plates are not economical because of the additional longitudinal welding, but the flange may consist of butt-welded plates of variable thickness or widths. *See* LOADS, TRANSVERSE.

Stiffeners. Stiffeners, plates or angles, may be attached to the girder web by welding or bolting to increase the buckling resistance of the web. Stiffeners are also required to transfer the concentrated forces of applied loads and reactions to the web without producing local buckling. Intermediate vertical stiffeners are used to increase the critical buckling stress of the web plate. Stiffeners are added to the web when the shearing stresses exceed the allowable shearbuckling stress of the unstiffened web plate. Stiffeners supporting concentrated loads are placed on each side of the web in contact with the flanges. Intermediate stiffeners are not required when the ratio of web depth to thickness (h/t) is less than 260 and the maximum shear stress is less than the allowable stress for the web without stiffeners. A limiting ratio of the stiffener spacing to depth (a/h) is 3.0 for all cases. Stiffeners must be sufficiently rigid to prevent their buckling the web.

The bearing stiffeners reinforce the web at reactions and at concentrated loads. Their function is to prevent local buckling of the web due to the concentrated loads and to transmit the loads to the web. These stiffeners are designed as compression members.

Longitudinal stiffeners are plates welded to the girder web on the compression side of the neutral axis. They serve a double function, increasing the resistance of the web to buckling caused by bending stresses, and contributing to the bending resistance of the flanges.

Splices. Splices are required for webs and flanges when full lengths of plates are not available from the mills or when shorter lengths are more readily fabricated. Splices provide the necessary continuity required in the web and flanges. Welded girders have the web and flange plates butt-welded together. Bolted girders require splice materials in the form of plates or angles. Plates are used for the web splices, one on each side of the web. Plates or angles may be used to splice a flange made up of angles. It has become advisable to stagger the splices in the flanges and the web in order to avoid a concentration of splices at one location of the girder (see illustration *c*).

[JOHN B. SCALZI]

Bibliography: B. Bresler, T. Y. Lin, and J. B. Scalzi, *Design of Steel Structures*, 3d ed., 1968; E. H. Gaylord and C. N. Gaylord, *Design of Steel Structures*, 2d ed., 1972; W. McGuire, *Steel Structures*, 1968.

Polishing

The smoothing of a surface by the cutting action of abrasive grit either glued to or impregnated in a flexible wheel or belt. Polishing, not a precision process, removes stock until the desired surface condition is obtained. Wheels are built up from layers of soft materials, such as wood or leather.

When a considerable amount of stock must be removed, operations may start with a coarse grit. Then a finer-grit wheel may be used for finishing. Common polishing machines may be either bench or floor-mounted. Usually they are lathe-type machines with wheels at either end of a power spindle, or one end may drive an abrasive belt. Various types of semiautomatic polishing machines are used for quantity production. These are designed to move the workpieces through the cutting paths of one or more polishing wheels. *See* GRINDING.

For a mirrorlike surface, a precision abrading process called superfinishing is used. Superfinishing removes minute flaws or inequalities. While it is not primarily intended to remove stock, generally a dimensional change of 0.0001 – 0.0002 in. in diameter occurs. Superfinishing is performed with an extremely-fine-grit abrasive stone, shaped to match and cover a large portion of the work surface. As the workpiece turns, the stone reciprocates across the work under a flood of lubrication. The process is usually performed on symmetrical pieces. *See* LAPPING; MACHINING OPERATIONS.

[ALLAN H. TUTTLE]

Powder metallurgy

A process for making a wide range of components and shapes from a variety of metals and alloys in the form of powder. The process is automated and uses pressure and heat to form precision metal parts into net or near-net shapes, requiring a minimum amount of secondary finishing. Modern powder metallurgy began in the early 1900s when incandescent lamp filaments were fabricated from tungsten powder. Other important products followed, such as cemented tungsten carbide cutting tools, friction materials, and self-lubricating bearings.

Metal powders. The most common metals available in powder form are iron, tin, nickel, copper, aluminum, and titanium, and refractory metals such as tungsten, molybdenum, tantalum, and niobium (columbium). Prealloyed powders such as low-alloy steels, bronze, brass, and stainless steel are produced in which each particle is itself an alloy. Also available in powder form are nickel-cobalt-base superalloys and tool steels.

Significant research and development effort has been devoted to designing processes for producing metal powders. Powder particles are not merely

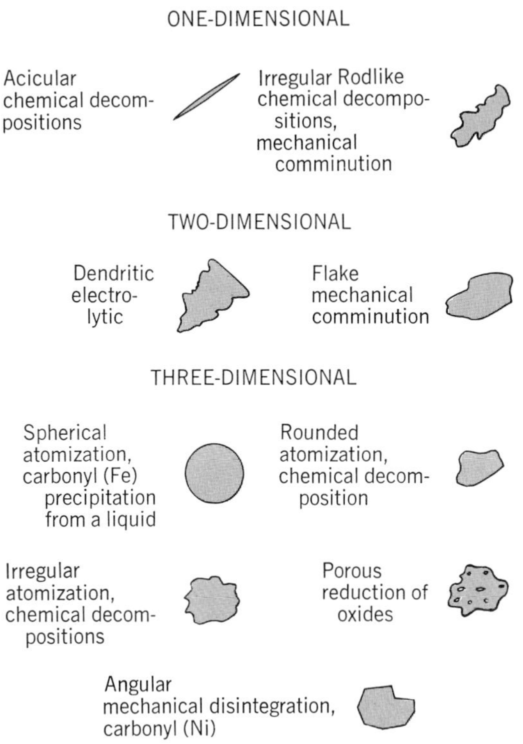

ONE-DIMENSIONAL

Acicular chemical decompositions

Irregular Rodlike chemical decompositions, mechanical comminution

TWO-DIMENSIONAL

Dendritic electrolytic

Flake mechanical comminution

THREE-DIMENSIONAL

Spherical atomization, carbonyl (Fe) precipitation from a liquid

Rounded atomization, chemical decomposition

Irregular atomization, chemical decompositions

Porous reduction of oxides

Angular mechanical disintegration, carbonyl (Ni)

Fig. 1. A simple system of particle shape characterization.

ground-up chips or scraps of metal. They are highly engineered materials of specific shapes and sizes. Metal powders range from 0.1 to 1000 μm in size with shapes that are spherical, acicular, irregular, dendritic, flake, angular, or fragmented (Fig. 1). Particle shape has a direct influence on the density, surface area, permeability, and flow characteristics of a powder.

Porosity of the powder particle varies with the method of production and influences the density of the particle and the final product. While size, shape, density, and structure are the major physical characteristics, chemical characteristics are also important. Commercial powders are available in many grades of purity. Usually, the base metal has a minimum purity of 97.5% and for most applications ranges from 98.5 to 99.5% or higher.

Powder production. Metal powders are produced by three major methods: physical, chemical, and mechanical.

Physical. The most important physical method of metal powder production is atomization, in which a stream of molten metal is broken up into droplets that freeze into powder particles. In most atomizing processes, a stream of liquid, usually water, or a gas impinges upon the liquid metal stream to break it into droplets. Iron and steel, aluminum, copper, stainless steel, brass, tin, bronze, zinc, and high-alloy powders are made this way.

Chemical. Chemical methods of powder production are those in which a metal powder is produced by chemical decomposition of a compound of the metal. Oxide in the form of finely divided solid powder particles may be reduced. Typical are the reduction of tungsten oxide to tungsten powder and of copper oxide to copper powder with hydrogen, and the reduction of iron oxide to iron powder with carbon monoxide. The deposition of a metal by an electric current from an electrolyte containing a metal salt (for example, the electrodeposition of copper powder from an aqueous copper sulfate solution) can be treated as a special kind of reduction process.

Mechanical. Certain metal powders can also be produced by mechanical comminution. Materials include iron, iron-aluminum alloys, and ferrosilicon and ferrophosphorus powders.

Basic process. The powder metallurgy process involves applying intense pressures ranging from 138 to >827 MPa (10 tons/in.^{2} to >60 tons/in.^{2}) to metal powders at room temperature (Fig. 2). After the part has been ejected, it is heated in a controlled-atmosphere furnace to bond the particles into a strong shape. In powder metallurgy the parts are shaped directly from powders. This is in contrast to other metalformed parts: castings are formed from molten metal, and wrought parts are shaped by deformation of hot or cold metal, or by machining.

Special processes. Although most powder metallurgy parts and products are pressed in mechanical or hydraulic compacting presses, other techniques such as cold and hot isostatic pressing, powder metallurgy forging, and direct powder rolling are used to make a variety of important end products.

Isostatic compacting. This is a process by which pressure (gas or fluid) is applied uniformly to a flexible mold or container holding the metal powder to be compacted. Simple or complex shapes requiring uniform distribution of properties are made by the process. Larger parts and parts having a large length-to-diameter ratio are also made this way. Because pressure from all directions is applied to the powder mass, it is possible to obtain a very uniform unsintered density and high uniformity of properties.

Hot isostatic pressing. In this process, metal powder is placed in a sealed mold and then subjected to isostatic pressure at elevated temperatures. These processes are used to make filters, carbides, tool steels, and superalloy parts.

Forging. Powder metallurgy forging or hot forming produces parts which are stronger than conventional powder metallurgy parts, providing toler-

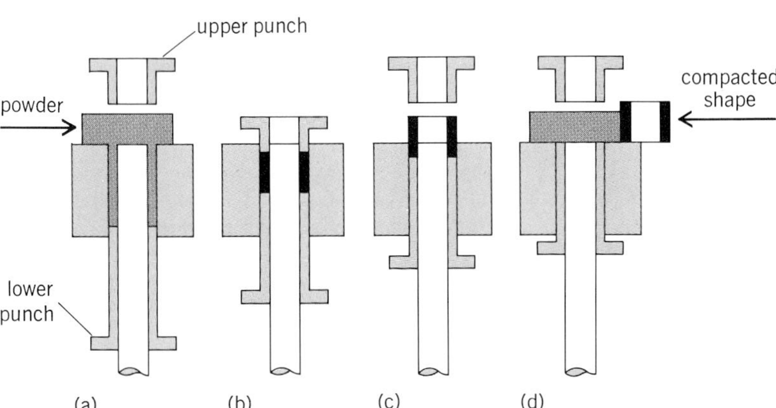

Fig. 2. Diagram of the basic powder metallurgy process. *(a)* Empty die cavity filled with powder. *(b)* Both upper and lower punches simultaneously pressing metal powder in die. *(c)* Upper punch withdrawn and green (unsintered) compact being ejected from die by lower punch. *(d)* Green compact pushed out of pressing area to make ready for another operating cycle.

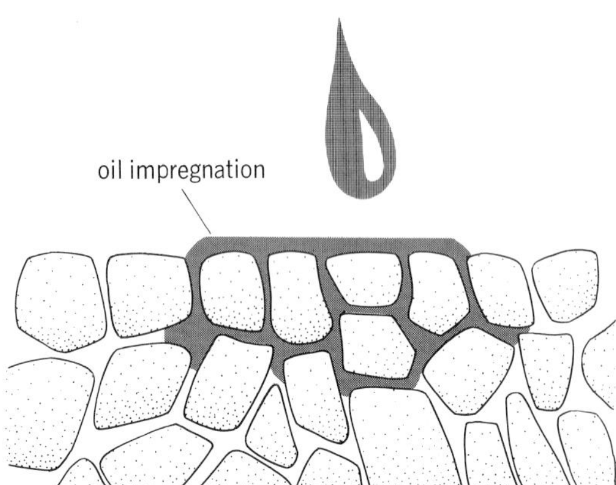

Fig. 3. Interconnecting pores on the surface of a powder metallurgy part.

ances and finishes that, in most cases, require little or no subsequent machining. Preforms are compacted on a conventional compacting press and then sintered. After sintering, the preform is fed automatically through a heating unit which heats the preform to a controlled temperature. The preform is then transferred into the die area of a forging press and formed in a closed die to its final shape with one stroke of the press. The formed part is transferred from the die to a conveyor or container and then to final heat treatment and machining, if required. *See* SINTERING.

Powder rolling. In this process, sometimes called roll compacting, metal powder is fed from a hopper into the gap of a rolling mill and emerges as a continuous compacted strip or sheet. Rolling speeds of up to 100 ft/min (0.5 m/s) are feasible. Important products made by this process include nickel-cobalt alloy strip for electronic tube components, nickel cadmium batteries, and glass-to-metal seals in electronic tubes, transistors, and auto headlights. Coinage strip is also made by this process.

Advantages. Powder metallurgy has environmental advantages in terms of both materials and energy conservation. Instead of generating scrap, the process utilizes scrap from other metal processes and recycles it into useful powders. Energy savings as high as 50% can be realized using lower-density powder metallurgy parts in place of steel parts machined from bar stock that require extensive machining. The process provides a net or near-net shape. Only the amount of raw material needed for the finished part is used.

The single most important feature of powder metallurgy materials is controlled density. In liquid metallurgy, density (or specific gravity) is a fixed property. In powder metallurgy, it can range from the porosity of filters and bearings to the high density of various structural components.

Porosity is the percentage of void volume in a part. A part which exhibits 85% of theoretical density will have 15% porosity. Porosity in powder metallurgy parts can be present as a network of interconnected pores that extend to the surface like a sponge or as a number of closed holes within the part (Fig. 3). Interconnected porosity is important to the performance of a self-lubricating bearing and is part of the specification for this unique powder metallurgy product. Such bearings can hold from 10 to 30% oil by volume when the interconnected voids are filled. Generally, additive-free oils (nonautomotive engine oils) are used. Impregnation is done by soaking the parts in heated oil for a period of time or by vacuum techniques. During use, when the part heats up from friction, the oil expands and flows to the bearing surface. Upon cooling, the oil returns into the pores by capillary action. Oil-impregnated bronze bearings have been used in automobiles since the late 1920s and serve as important components in most appliances and small machines.

Nonstructural applications. Iron powder is used in welding electrodes, in iron-enriched cereals and breads, and in magnetic applications such as in xerography. Aluminum powder is used in outdoor paints such as those on bridges, and in mining explosives, solid rocket fuel, and deodorants. Brass powder is used in the form of flakes in printing inks and finishes. Seed growers also use iron powder to separate damaged or weed seeds from good ones. Since the weed seeds are "hairy" and the damaged ones are sticky, iron powder will cling to them but not to the smooth, dry, good seeds, allowing magnetic separation. Lightweight concrete can be made by incorporating aluminum powder and an alkali in the cement mixture. Reaction of the powder in the alkali causes formation of gas, accompanied by an increase in temperature and production of a foamy structure. Copper, silver, gold, platinum, and palladium flake powders are used for printed circuits, contact strips, and integrated circuits. Atomized aluminum powder is used in solid rocket fuel. The booster rockets on the *Columbia* space shuttle contained about 400,000 lb (180,000 kg) of aluminum powder in addition to an oxidizer and binders. Aluminum has a very high energy output at ignition which provides the needed thrust at launching.

Parts applications. A wide range of products utilize powder metallurgy parts. Examples include gears, bearings, and plates in hand-held power drills; turbine hubs and gears in automobile automatic transmissions; gears, levers, and cams in lawn tractors and garden appliances; gears, bearings, and sprockets in office copying machines and word-processing equipment; and rod guides, pistons, and valve plates in automobile shock absorbers. An automatic washer transmission contains a powder metallurgy steel drive-gear rack used in a rack-and-pinion system for reciprocating agitation action (Fig. 4). This drive-gear rack replaced a number of components that were machined from cast iron blanks. Powder metallurgy permits a design configuration that is technically and economically impossible with other metalforming techniques. A valve seat and valve cap made from stainless steel powder have been developed for a vacuum pump used in industrial equipment (Fig. 5). These parts were redesigned from a machined investment casting. Critical dimensions of the parts include flatness and squareness of through holes to faces and the relationship of the valve cap holes to each other. A steel powder metallurgy helical gear with a bevel pinion is used in a kitchen food processor (Fig. 6). This part replaced two parts that were hobbed and machined from leaded steel. Powder metallurgy was able to offer a manufacturing cost savings of more than 50% over the former method of manufacture.

POWDER METALLURGY

Fig. 4. Powder metallurgy drive-gear rack.

Filters made from powder metallurgy porous metals are used to separate most combinations of solids, liquids, and gases, for example, automotive fuel filters and oil burner fuel filters. Surge dampeners and flame arresters are examples of powder metallurgy porous products for use with gases and liquids. Such porous metal parts have high strength, corrosion resistance, and durability. They have controlled porosity to achieve specified filtration properties and controlled permeability to meet flow specifications. They offer a wide range of design possibilities that permit filtration from 0.5 to 200 μm and can provide flow rates that range from microleaks to high volumes. Porous metal parts can be made by compacting/sintering, gravity sintering, or isostatic compacting.

Developments. It is anticipated that powder metallurgy will have a strong impact on a diverse mixture of markets and materials such as steel components in automobiles, titanium in prosthetics and aerospace hardware, high-temperature superalloys in aircraft and auto turbine engines, tool steels and magnetic applications, high-alloy strip, and isostatically pressed parts. Powder metallurgy provides exceptional cost savings in the manufacture of titanium components. Materials utilization in the production of conventional titanium-wrought products is very poor, yielding about 20% of finished product per ton of raw material. However, powder metallurgy offers a materials-yield range of 60–80%, because components can be formed automatically to precise shapes with little or no secondary finishing. Typical applications include fasteners and components for the chemical industry; external orthopedic devices; porous sheet for filters and electrodes; and fasteners used in aerospace hardware.

The use of powder metallurgy superalloy parts in military and civilian aircraft turbine engines will also increase. Substantial cost savings are achieved compared with conventionally forged superalloy products. Near-net shape-making capabilities coupled with improved properties such as higher tensile strength and higher creep rupture strengths provide more efficient engine operation at higher temperatures, permitting higher rotational speeds of engine components, which in turn yield improved performance at lower fuel consumption rates.

A dispersion-strengthened high-purity copper powder metallurgy material is finding an increasing number of applications as lead wires in light bulbs, as spot welding electrodes for welding automobile frames, and in eyeglass frames and electrical connectors. The material contains submicroscopic particles of insoluble aluminum oxide finely distributed throughout a copper matrix. Through an internal oxidation process, the dispersed phase particles interfere with dislocation movement, thus increasing strength without appreciably decreasing electrical conductivity, especially at elevated temperatures.

Fig. 5. Powder metallurgy valve seat and valve cap (foreground).

Fig. 6. Powder metallurgy helical gear with bevel pinion.

The fabrication of tool steels from metal powder either by cold or hot isostatic pressing or by conventional powder metallurgy compacting is entering a growth phase. Applications include spade drills, knife blades, slotting cutters, insert blades for gear cutters, watch cases, and components for diesel engines.

The future of the powder metallurgy industry is very favorable, as engineers seek to increase manufacturing productivity, conserve critical materials and energy, and reduce costs. *See* METALLURGY.

[PETER K. JOHNSON]

Bibliography: F. V. Lenel, *Powder Metallurgy: Principles and Applications*, 1980; Metal Powder Industries Federation, *Porous Metal Design Guidebook*, 1980; Metal Powder Industries Federation, *Powder Metallurgy Design Guidebook*, 1974.

Power

The time rate of doing work. Like work, power is a scalar quantity, that is, a quantity which has magnitude but no direction. Some units often used for the measurement of power are the watt (1 joule of work per second) and the horsepower (550 ft-lb of work per second). *See* WORK.

Usefulness of the concept. Power is a concept which can be used to describe the operation of any system or device in which a flow of energy occurs. In many problems of apparatus design, the power, rather than the total work to be done, determines the size of the component used. Any device can do a large amount of work by performing for a long time at a low rate of power, that is, by doing work slowly. However, if a large amount of work must be done rapidly, a high-power device is needed. High-power machines are usually larger, more complicated, and more expensive than equipment which need operate only at low power. A motor which must lift a certain weight will have to be larger and more powerful if it lifts the weight rapidly than if it raises it slowly. An electrical resistor must be large in size if it is to convert electrical energy into heat at a high rate without being damaged.

Electrical power. The power P developed in a direct-current electric circuit is $P=VI$, where V is the applied potential difference and I is the current. The power is given in watts if V is in volts and I in amperes. In an alternating-current circuit, $P=VI\cos\phi$, where V and I are the effective values of the voltage and current and ϕ is the phase angle between the current and the voltage. *See* ALTERNATING CURRENT.

Power in mechanics. Consider a force F which does work W on a particle. Let the motion be restricted to one dimension, with the displacement in this dimension given by x. Then by definition the power at time t will be given by Eq. (1). In this

$$P=dW/dt \quad (1)$$

equation W can be considered as a function of either t or x. Treating W as a function of x gives Eq. (2). Now dx/dt represents the velocity v of the

$$P=\frac{dW}{dt}=\frac{dW}{dx}\frac{dx}{dt} \quad (2)$$

particle, and dW/dx is equal to the force F, according to the definition of work. Thus Eq. (3) holds.

$$P=Fv \quad (3)$$

This often convenient expression for power can be generalized to three-dimensional motion. In this case, if ϕ is the angle between the force $\mathbf{F}$ and the velocity $\mathbf{v}$, which have magnitudes F and v, respectively, Eq. (4) expresses quantitatively the observa-

$$P=\mathbf{F}\cdot\mathbf{v}=Fv\cos\phi \quad (4)$$

tion that if a machine is to be powerful, it must run fast, exert a large force, or do both.

[PAUL W. SCHMIDT]

Power factor

The ratio of watts average (or active) power to the apparent power of an alternating-current circuit. By definition, and of general application, the equation below holds, which is the ratio of instrument

$$\text{Power factor (pf)}=\frac{\text{watts average power}}{\text{rms volts}\times\text{rms amperes}}$$

readings. A watt-meter indicates average power and electrodynamometer or iron-vane instruments show rms voltage and current. For the steady-state ac circuit under sinusoidal voltage and current, pf $=\cos\theta$, where θ is the phase angle between the voltage and current. This definition is restricted to sine waves of the same frequency. *See* ALTERNATING-CURRENT CIRCUIT THEORY.

[BURTIS L. ROBERTSON]

Power plant

A means for converting stored energy into work. Stationary power plants such as electric generating stations are located near sources of stored energy, such as coal fields or river dams, or are located near the places where the work is to be performed, as in cities or industrial sites. Mobile power plants for transportation service are located in vehicles, as the gasoline engines in automobiles and diesel locomotives for railroads. Power plants range in capacity from a fraction of a horsepower (hp) to over 10^6 kilowatts (kW) in a single unit (Table 1). Large power plants are assembled, erected, and constructed on location from equipment and systems made by different manufacturers. Smaller units are produced in manufacturing facilities.

Most power plants convert part of the stored raw energy of fossil fuels into kinetic energy of a spinning shaft. Some power plants harness nuclear energy. Elevated water supply or run-of-the-river energy is used in hydroelectric power plants. For transportation, the plant may produce a propulsive jet, as in some aircraft, instead of the rotary motion of a shaft. Other sources of energy, such as winds, tides, waves, geothermal sources, ocean thermal, nuclear fusion, and solar radiation, have been of negligible commercial significance in the generation of power despite their magnitudes.

Table 2 shows the scope of United States power-plant capacity. About a third of the world's electric energy (in kilowatt-hours, kWh) is generated by the United States public utility systems (Fig. 1), and the installed generating capacity (in kW) of the United States (including Alaska and Hawaii) is about the same as the total of the next four countries (Fig 2). Figure 3 shows the declining importance of hydroelectric power in the United States,

Table 1. Representative design and performance data on power plants

Type	Unit size range, kW	Fuel*	Plant weight, lb/kW†	Plant volume, ft³/kW‡	Heat rate, Btu/kWh§
Central station					
Hydro	10,000–700,000				
Steam (fossil-fuel–fired)	10,000–1,300,000	CO		20–50	8,500–15,000
Steam (nuclear)	500,000–1,200,000	N			10,000–12,000
Diesel	1,000–5,000	DG			10,000–15,000
Combustion turbine	5,000–10,000	D′G			11,000–15,000
Industrial (by-product) steam	1,000–25,000	COGW		50–75	4,500–6,000
Diesel locomotive	1,000–5,000	D	100–200	2–3	10,000–15,000
Automobile	25–300	G′	5–10	0.1	15,000–20,000
Outboard motor	1–50	G′	2–5	0.1–0.5	15,000–20,000
Truck	50–500	D	10–20		12,000–18,000
Merchant ship, diesel	5,000–20,000	D	300–500		10,000–12,000
Naval vessel, steam	25,000–100,000	DON	25–50		12,000–18,000
Airplane, reciprocating engine	1,000–3,000	G′	1–3	0.05–0.10	12,000–15,000
Airplane, turbojet	3,000–10,000	D′	0.2–1		13,000–18,000

*C, coal; D, diesel fuel; D′, distillate; G, gas; G′, gasoline; N, nuclear; O, fuel oil (residuum); W, waste.
†1 lb/kW = 0.45 kg/kW. ‡1 ft³/kW = 2.83×10^{-2} m³/kW. §1 Btu/kWh = 1.055 kJ/kWh = 2.93×10^{-4} J(heat)/J(output).

Table 2. Approximate 1981 installed capacity of United States power plants.

Plant type	Capability, 10^6 kW
Electric central stations	557
Industrial	40
Agricultural	60
Railroad	70
Marine, civilian	40
Aircraft, civilian	70
Military establishment	2,000±
Automotive	10,000±
Total	13,000±

where the kWh output from hydro has dropped from 30 to 9% of the total electric generation in 28 years. These data, coupled with the data of Table 2, reflect the dominant position of thermal power both for stationary service and for the propulsion of land-, water-, and air-borne vehicles. *See* ENERGY SOURCES.

Rudimentary flow- or heat-balance diagrams for important types of practical power plants are shown in Fig. 4. Figure 5 is a diagram for a by-product-type industrial steam plant (also known as cogeneration) which has the double purpose of generating electric power and simultaneously delivering heating steam by extraction or exhaust from the prime mover.

Plant load. There is no practical way of storing the mechanical or electrical output of a power plant in the magnitudes encountered in power plant applications, although several small-scale concepts are in the research and development stage. As of now, however, the output must be generated at the instant of its use. This results in wide variations in the loads imposed upon a plant. The capacity, measured in kW or hp, must be available when the load is imposed. Much of the capacity may be idle during extended periods when there is no demand for output. Hence much of the potential output, measured as kWh or hp-h, cannot be generated because there is no demand for output. This greatly complicates the design and confuses the economics of power plants. Kilowatts cannot be traded for kilowatt-hours, and vice versa. *See* ENERGY STORAGE.

The ratios of average load to rated capacity or to peak load are expressed as the capacity factor and the load factor, Eqs. (1) and (2), respectively. The

$$\text{Capacity factor} = \frac{\text{average load for the period}}{\text{rated or installed capacity}} \quad (1)$$

$$\text{Load factor} = \frac{\text{average load for the period}}{\text{peak load in the period}} \quad (2)$$

Table 3. Range of capacity factors for selected power plants

Power plants	Factor, %
Public-utility systems, in general	50–70
Chemical or metallurgical plant, three-shift operation	80–90
Seagoing ships, long voyages	70–80
Seagoing ships, short voyages	30–40
Airplanes, commercial	20–30
Private passenger cars	1–3
Main-line locomotives	30–40
Interurban buses and trucks	5–10

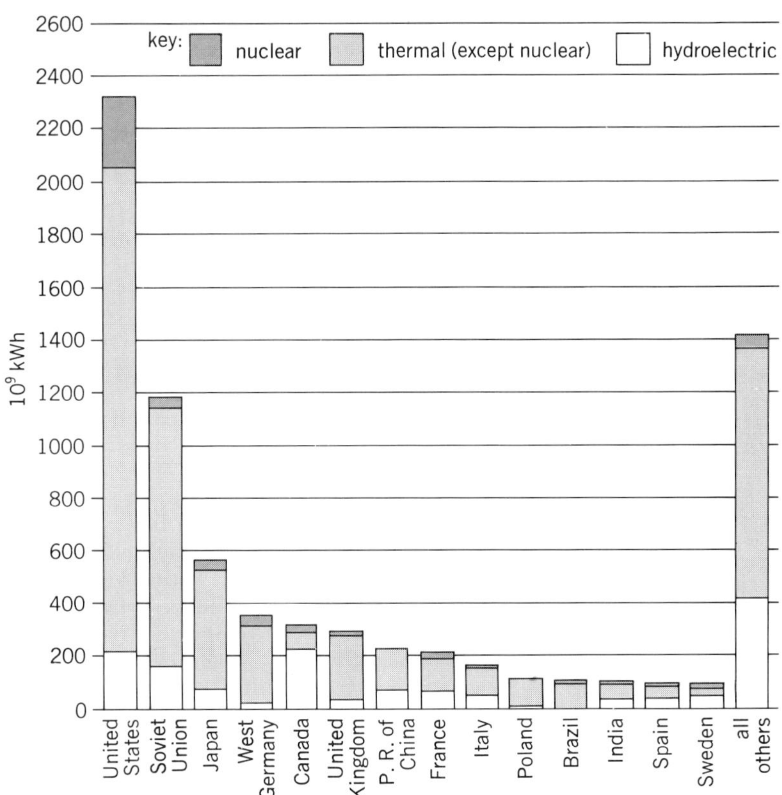

Fig. 1. Electric energy production by type in the 14 highest-producing countries, 1978. Nearly 30% of world electrical energy is generated in United States plants. *(United Nations)*

range of capacity factors experienced for various types of power plants is given in Table 3.

Variations in loads can be conveniently shown on graphical bases as in Figs. 6 and 7 for public

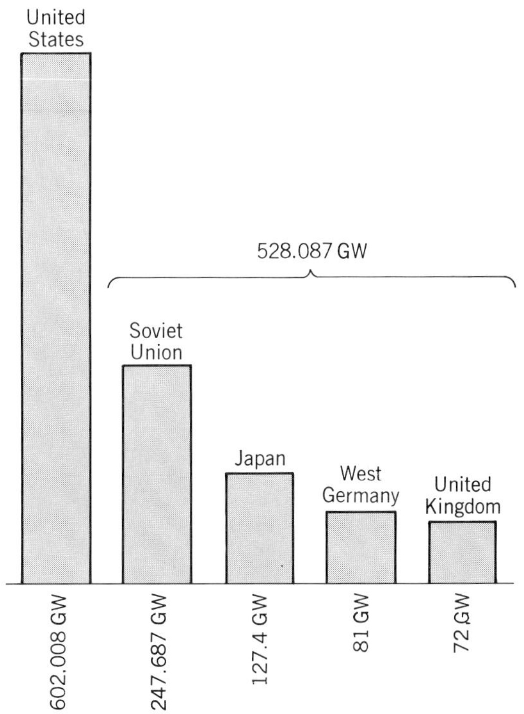

Fig. 2. The five countries with the highest capacities to generate electric power, 1978. *(United Nations)*

utilities and in Fig. 8 for air and marine propulsion. Rigorous definition of load factor is not possible with vehicles like tractors or automobiles because of variations in the character and condition of the running surface. In propulsion applications, power output may be of secondary import: performance may be based on tractive effort, drawbar pull, thrust, climb, and acceleration.

Plant efficiency. The efficiency of energy conversion is vital in most power plant installations. The theoretical power of a hydro plant in kW is $QH/11.8$, where Q is the flow in cubic feet per second and H is the head (height of water intake above discharge level) at the site in feet. In metric units, the theoretical power of the plant in kW is $9.8\ Q'H'$, where Q' is the flow in cubic meters per second and H' is the head in meters. Losses in headworks, penstocks, turbines, draft tubes, tailrace, bearings, generators, and auxiliaries will reduce the salable output 15–20% below the theoretical in modern installations. The selection of a particular type of waterwheel depends on experience with wheels at the planned speed and on the lowest water pressure in the water path. Runners of the reaction type (high specific speed) are suited to low heads (below 500 ft or 150 m) and the impulse type (low specific speed) to high head service (about 1000 ft or 300 m). The lowest heads (below 100 ft or 300 m) are best accommodated by reaction runners of the propeller or the adjustable blade types. Mixed-pressure runners are favored for the intermediate heads (50–500 ft or 15–150 m). Draft tubes, which permit the unit to be placed safely above flood water and without sacrifice of site head, are essential parts of reaction unit installations. *See* HYDRAULIC TURBINE; WATERPOWER.

With thermal power plants the basic limitations of thermodynamics fix the efficiency of converting heat into work. The cyclic standards of Carnot, Rankine, Otto, Diesel, and Brayton are the usual

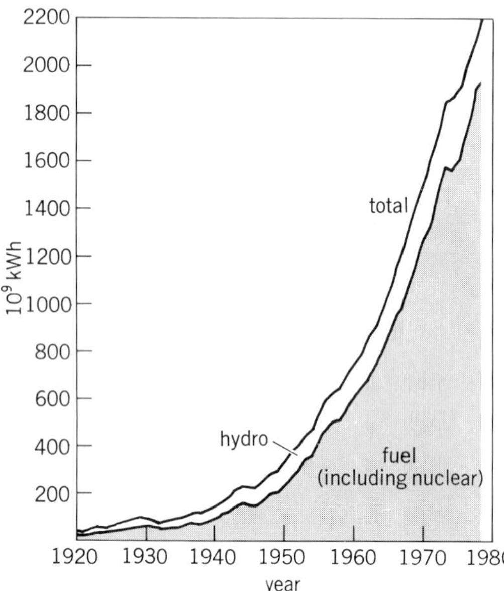

Fig. 3. Total electrical energy generation in the United States utility industry (including Alaska and Hawaii since 1963), by type of prime mover. (*Edison Electric Institute*)

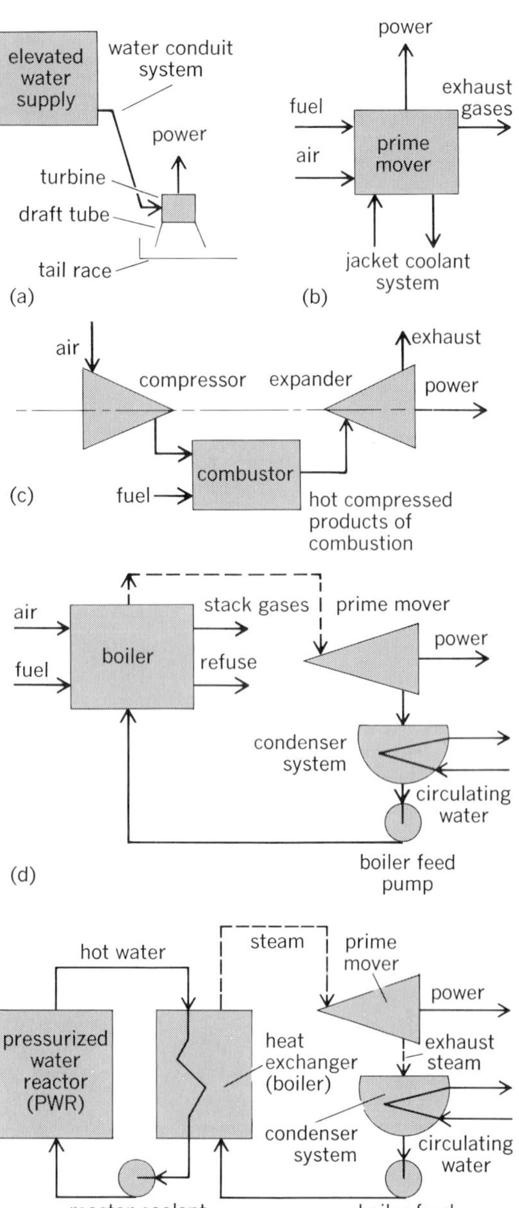

Fig. 4. Rudimentary flow- or heat-balance diagrams for power plants. (*a*) Hydro. (*b*) Internal combustion. (*c*) Gas-turbine. (*d*) Fossil-fuel-fired. (*e*) Nuclear steam (pressurized water reactor, PWR).

criteria on which heat-power operations are variously judged. Performance of an assembled power plant, from fuel to net salable or usable output, may be expressed as thermal efficiency (%); fuel consumption (lb, pt, or gal per hp-h or per kWh); or heat rate (Btu supplied in fuel per hp-h or per kWh). American practice uses high or gross calorific value of the fuel for measuring heat rate or thermal efficiency and differs in this respect from European practice, which prefers the low or net calorific value.

Tables 1 and 3 give performances for selected operations. Figures 9 and 10 reflect the improvement in fuel utilization of the United States electric power industry since 1900, although there have been some minor decreases since 1970. Fig-

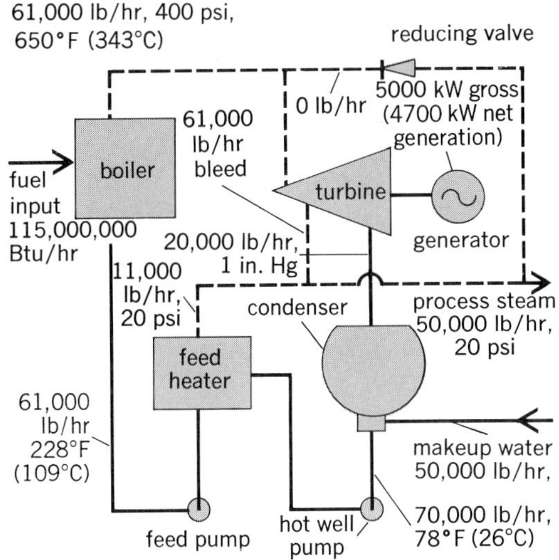

Fig. 5. Heat balance for a by-product industrial power plant delivering both electrical energy and process steam. 1 lb/hr = 1.26×10^{-4} kg/sec. 1 Btu/hr = 0.293 W. 1 psi = 6.895 kPa. 1 in. Hg = 3.386 kPa.

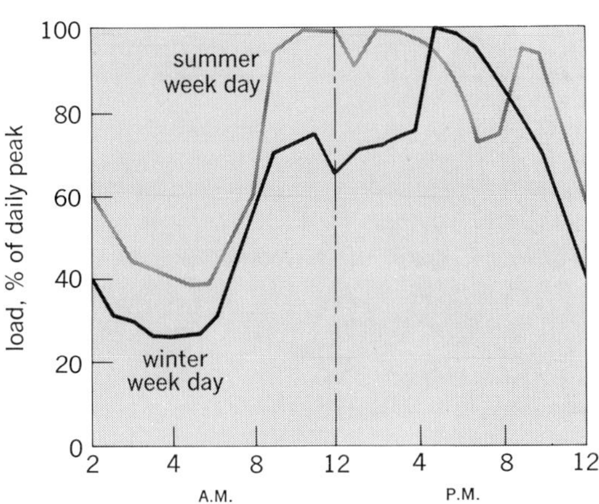

Fig. 6. Daily-load curves for urban utility plant.

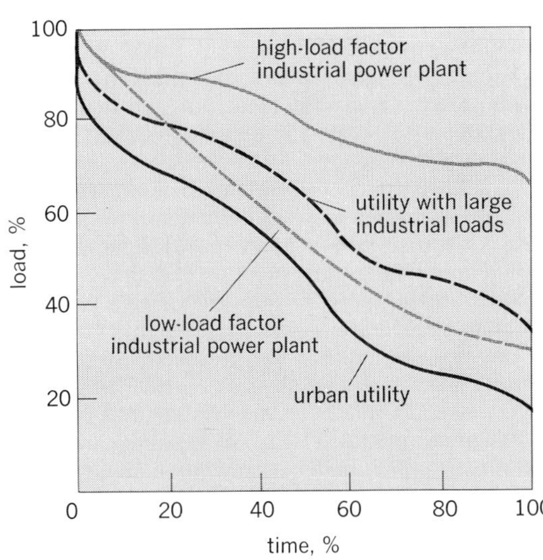

Fig. 7. Annual load-duration curves for selected stationary public utility power plants.

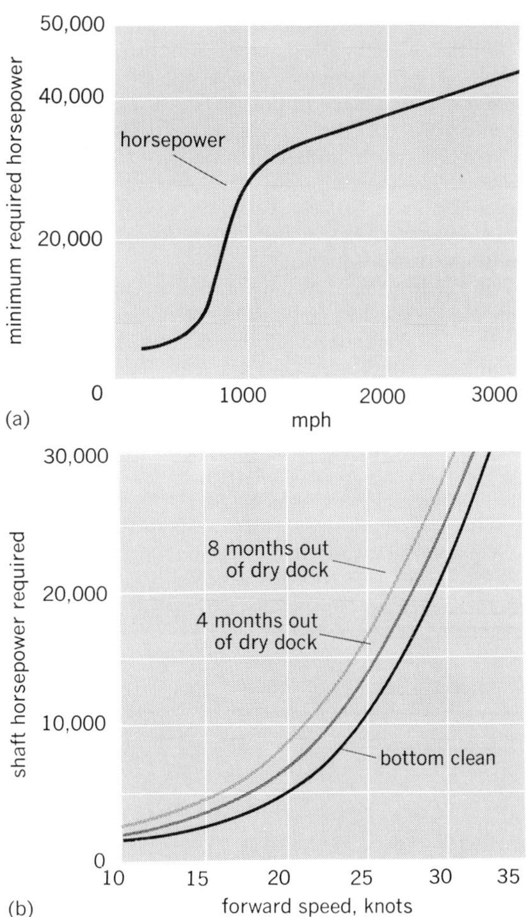

Fig. 8. Air and marine power. *(a)* Minimum power required to drive a 50-ton (45-metric-ton) well-designed airplane in straight level flight at 30,000 ft (9.1 km) altitude. *(b)* Power required to drive a ship, showing effect of fouling. 1 hp = 746 W; 1 mph = 0.447 m/sec; 1 knot = 0.514 m/sec.

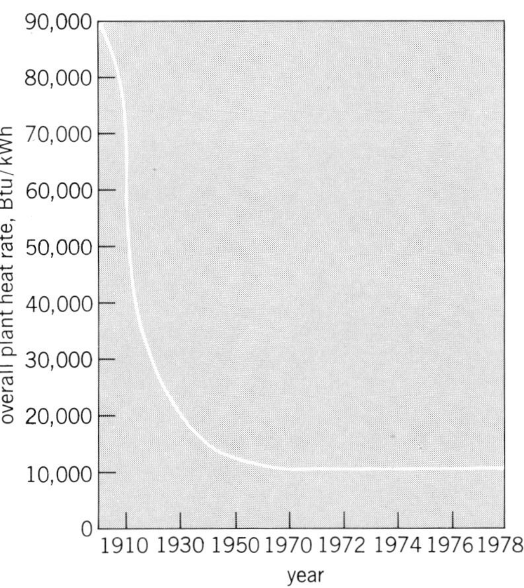

Fig. 9. Thermal performance of fuel-burning electric utility power plants in the United States. 1 Btu/kWh = 1.055 kJ/kWh = 2.93×10^{-4} J(heat)/J(output).

ure 10 is especially significant, as it shows graphically the impact of technological improvements on the cost of producing electrical energy despite the harassing increases in the costs of fuel during the same period. Figure 11 illustrates the variation in thermal performance as a function of load for an assortment of stationary and marine propulsion power plants. *See* BRAYTON CYCLE; CARNOT CYCLE; DIESEL CYCLE; OTTO CYCLE; RANKINE CYCLE; THERMODYNAMIC CYCLE.

In scrutinizing data on thermal performance, it should be recalled that the mechanical equivalent of heat (100% thermal efficiency) is 2545 Btu/hp-h and 3413 Btu/kWh (3.6 MJ/kWh). Modern steam plants in large sizes (75,000–1,300,000 kW units) and internal combustion plants in modest sizes (1000–5000 kW) have little difficulty in delivering a kWh for less than 10,000 Btu (10.55 MJ) in fuel (34% thermal efficiency). Lowest fuel consumptions per unit output (8500–9000 Btu/kWh or 9.0–9.5 MJ/kWh) are obtained in condensing steam plants with the best vacua, regenerative-reheat cycles using eight stages of extraction feed heating, two stages of resuperheat, primary pressures of 3500 psi or 24 MPa (supercritical) and temperatures of 1150°F (620°C). An industrial plant generating electric power as a by-product of the process steam load is capable of having a thermal efficiency of 5000 Btu/kWh (5.3 MJ/kWh).

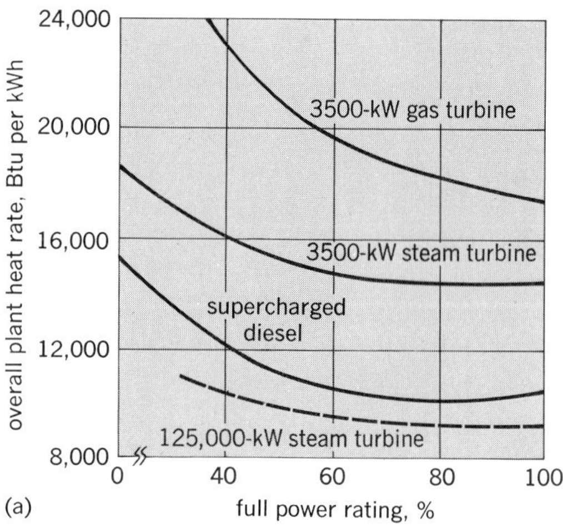

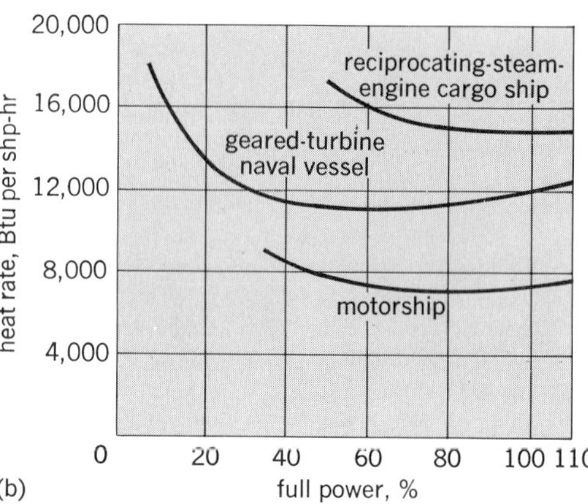

Fig. 11. Comparison of heat rates. (*a*) Stationary power plants. (*b*) Marine propulsion plants. 1 Btu/kWh = 1.055 kJ/kWh = 2.93×10^{-4} J(heat)/J(output). 1 Btu per hp-h = 1.415 kJ/kWh = 3.93×10^{-4} J(heat)/J(output).

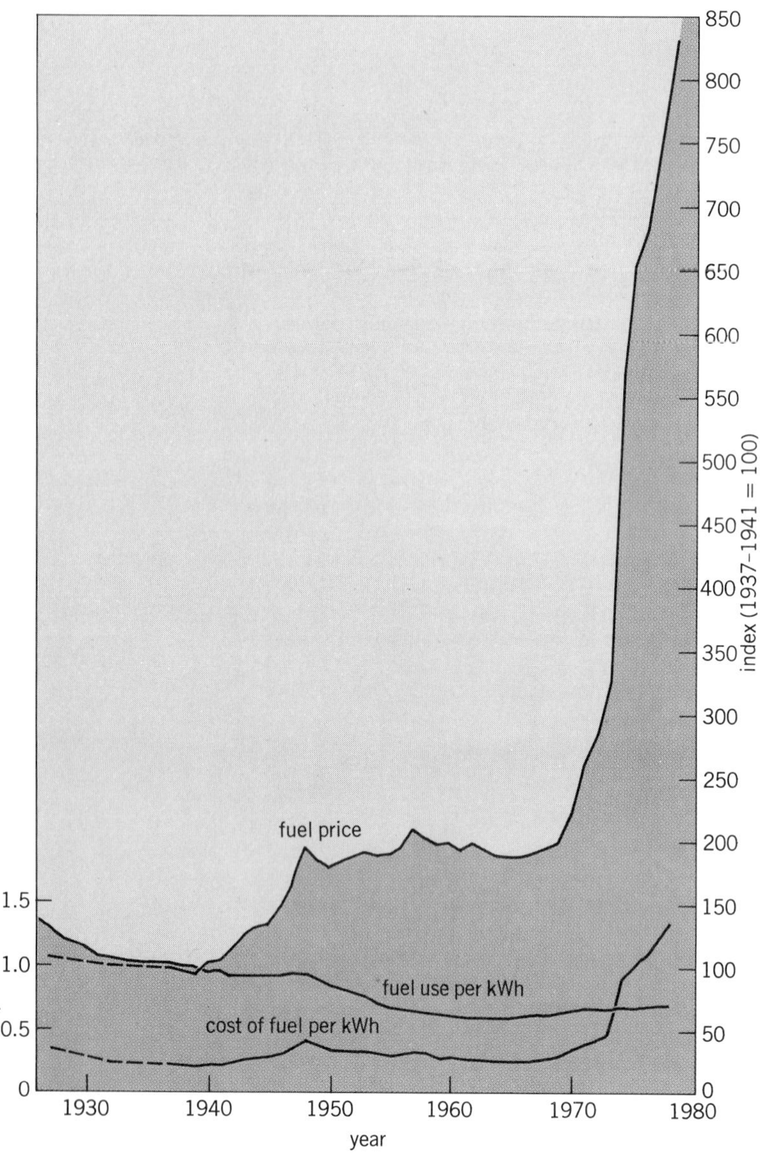

Fig. 10. Effect of fuel price and efficiency of use upon cost of fuel per kilowatt-hour generated in the United States utility industry (including Alaska and Hawaii since 1963). (*Edison Electric Institute*)

The nuclear power plant substitutes the heat of fission for the heat of combustion, and the consequent plant differs only in the method of preparing the thermodynamic fluid. It is otherwise similar to the usual thermal power plant. Low reactor temperatures lead to the overwhelming preference for steam-turbine rather than gas-turbine cycles. When fluid temperatures can be had above 1200°F (650°C), the gas-power cycle will receive more favorable consideration. Otherwise the nuclear power plant is essentially a low-pressure, low-temperature steam operation (less than 1000 psi or 6.9 MPa and 600°F or 320°C). *See* NUCLEAR REACTOR.

Power economy. Costs are a significant, and often controlling, factor in any commercial power plant application. Average costs have little significance because of the many variables, especially load factor. Some plants are short-lived and others long-lived. For example, in most automobiles, which have short-lived power plants, 100,000 mi (160,000 km) and 3000–4000 hr constitute the approximate operating life; diesel locomotives, which run 20,000 mi (32,000 km) a month with complete overhauls every few years, and large seagoing ships, which register 1,000,000 mi (1,600,000 km) of travel and still give excellent service after 20 years of operation, have long-lived plants; electric central stations of the hydro type can remain in service 50 years or longer; and steam plants run round the clock and upward of 8000 hr a year with complete reliability even when 25 years old. Such figures greatly influence costs. Furthermore, costs are open to wide differences of interpretation.

In the effort to minimize cost of electric power to the consumer it is essential to recognize the difference between investment and operating costs, and the difference between average and incremental costs. Plants with high investment (fixed) costs per kW should run at high load factors to spread the burden. Plants with high operating costs (such as fuel) should be run only for the shortest periods to meet peak loads or emergencies. To meet these short operating periods various types of peaking plants have been built. Combustion (gas) turbines and pumped-storage plants serve this requirement. In the latter a hydro installation is operated off-peak to pump water from a lower reservoir to an elevated reservoir. On-peak the operation is reversed with water flowing downhill through the prime movers and returning electrical energy to the transmission system. High head sites (for example, 1000 ft or 300 m), proximity to transmission lines, and low incremental cost producers (such as nuclear or efficient fossil-fuel-fired plants) are necessary. If 2 kWh can thus be returned on-peak to the system, for an imput of 3 kWh off-peak, a pumped-storage installation is generally justifiable. *See* GAS TURBINE.

In any consideration of such power plant installations and operations it is imperative to recognize (1) the requirements of reliability of service and (2) the difference between average and incremental costs. Reliability entails the selection and operation of the proper number and capacity of redundant systems and components and of their location on the system network. Emergencies, breakdowns, and tripouts are bound to occur on the best systems. The demand for maximum continuity of electrical service in modern civilization dictates the clear recognition of the need to provide reserve capacity in all components making up the power system.

Within that framework the minimum cost to the consumer will be met by the incremental loading of equipment. Incremental loading dictates, typically, that any increase in load should be met by supplying that load with the unit then in service, which will give the minimum increase in out-of-pocket operating cost. Conversely, for any decrease in load, the unit with the highest incremental production cost should drop that decrease in load. This is a complex technical, economic, and management problem calling for the highest degree of professional competence for its proper solution. *See* ELECTRIC POWER GENERATION.

[THEODORE BAUMEISTER; LEONARD M. OLMSTED; KENNETH A. ROE]

Bibliography: Babcock and Wilcox Co., *Steam: Its Generation and Use*, rev. 39th ed., 1978; T. Baumeister (ed.), *Standard Handbook for Mechanical Engineers*, 8th ed., 1978; Combustion Engineering, Inc., *Combustion Engineering*, rev. ed., 1966; W. T. Creager and J. D. Justin, *Hydroelectric Handbook*, 1950; Diesel Engine Manufacturing Association, *Standard Practices*, 1972; D. G. Fink and H. W. Beaty, *Standard Handbook for Electrical Engineers*, 11th ed., 1978; G. D. Friedlander, 21st Steam Station Cost Survey, *Elec. World*, 192(10):55–70, Nov. 15, 1979; R. L. Harrington, *Marine Engineering*, 1971; L. C. Lichty, *Combustion Engine Processes*, 7th ed., 1967.

Power shovel

A digging machine, usually self-propelled on crawler, rubber tire, or sometimes rail mountings. It is equipped with a shovel boom, dipper attached to the front end of a dipper stick, dipper trip mechanism, padlock (dipper sheave block), crowding mechanism, and cables all carried on a fully revolving superstructure. Commercial sizes have capacity ratings from $\frac{1}{4}$ to $2\frac{1}{2}$ yd^3; special sizes up to 40 yd^3 have been produced (see illustration). *See* BULK-HANDLING MACHINES; CONSTRUCTION EQUIPMENT.

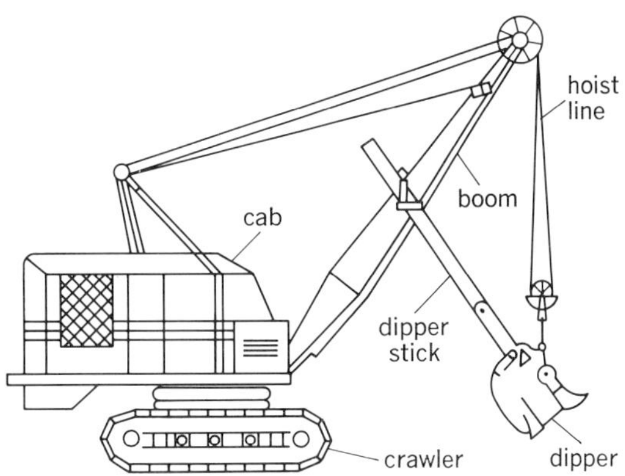

Crawler shovel. (*Cummins Engine Co., Ltd.*)

[ARTHUR M. PERRIN]

Bibliography: R. L. Peurifoy, *Construction Planning, Equipment and Methods*, 1956; Power Crane and Shovel Association (PCSA), *Mobile Hydraulic Crane Standards*, PCSA Stand. no. 2, 1968; Power Crane and Shovel Association (PCSA), *Mobile Power Crane and Excavator Standards*, PCSA Stand. no. 1, 1968; Power Crane and Shovel Association (PCSA), *Proper Sizing of Excavators and Hauling Equipment*, Tech. Bull. no. 3, 1966; U.S. Department of Commerce, *Power Cranes and Shovels*, Commer. Stand. no. CS90, amended 1961.

Power steering

A steering control system for a propelled vehicle in which an auxiliary power source assists the driver by providing the major force required to direct the road wheels. The term power steering commonly applies to any system with a variety of power booster arrangements. The principal components of the power booster are the control valve, the power actuator, and the power source. Dependent upon the type of system, these components normally operate in conjunction with a hand steering wheel, linkage system, and steered wheels. (Most systems provide for manual steering gear backup.) As a vehicle follows a driver-steered course, the valve senses any deviation between the course-set

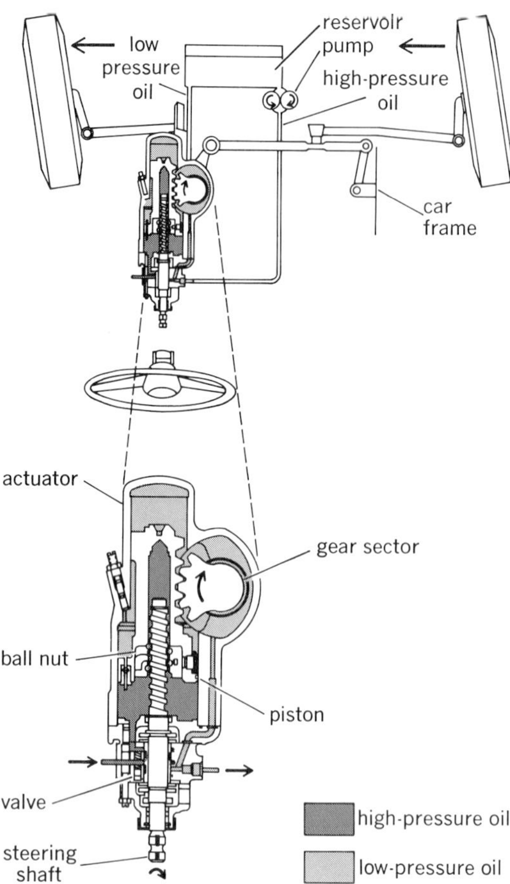

Fig. 1. Integral power steering (*TRW Inc.*)

position of the steering wheel and the corresponding position of the steered wheels and releases power to the actuator until the error is corrected. *See* AUTOMOTIVE STEERING.

History and application. Power steering was developed in 1926 by Francis W. Davis, an engineer with Pierce Arrow. Because the Depression soon followed, the device was never mass-produced; the Cadillac Motor Division of General Motors had planned to offer it as an option in 1933. Not until World War II was power steering adopted, then on military vehicles. Following the war, the servo-power steering arrangement was applied to the commercial vehicle market. The automotive industry lagged until 1952, when modern power steering was introduced by the Chrysler Corporation on their new models as an equipment option.

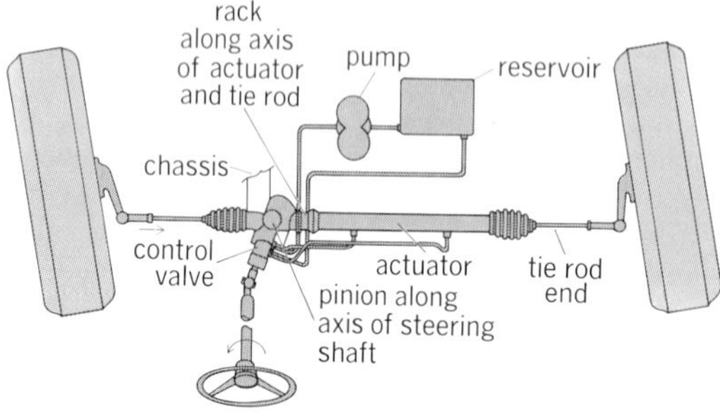

Fig. 2. Power rack-and-pinion steering gear serves to replace the conventional tie rod. (*TRW Inc.*)

Power-assisted steering in commercial vehicles such as buses, trucks, and tractors has increased productivity by a claimed 20%. Heavy-duty dump trucks, wheel loaders, motor graders, and so on demand power steering to meet the steering efforts required by the vehicular load and terrain conditions. Passenger-car application has led to greater safety and convenience by relieving driver fatigue.

Hydraulic servo system. The power steering control system is a hydraulic servomechanism. As the driver's desired course is set at the steering wheel, the system (servoloop) compares the desired course with a measurement of the actual and "signals" a measure of the error. The error signal proportionately actuates the power to assist the steering maneuver. This servoloop essentially controls the power assist function of the steering system. Normally the actual servoloop is added to a manual steering arrangement; however, manual backup steering is not mandatory for off-highway vehicles. The main components of the system that maintain the servo function are the control valve and actuator. This combination can be introduced in various arrangements to provide five principal hydraulic power systems: (1) the integral, (2) the linkage, (3) the in-line linkage, (4) the hybrid, and (5) the hydrostatic.

In each system an engine-driven pump with a reservoir acts as the power source. The pump delivers hydraulic fluid oil under a working pressure to the controller valve. The controller valve meters the working pressure or pump delivery to the actuator, a hydraulic cylinder. The controlling valve responds to a torque from the steering wheel and to reactive shock from the road wheels and directs oil as required to the actuator. In the stroking action the hydraulic cylinder repositions the steered wheels to maintain the vehicle course set by the driver at the steering wheel. In order for the driver to "feel" the road at the steering wheel, the control valve is desensitized to low steering torque efforts, and then manual steer results.

Integral. The servoloop components are combined in a single power steering gear assembly (Fig. 1). The steering shaft carries a valve spool that moves axially to perform its controlling function of metering high-pressure oil. Springs in the valve body retain the valve neutral during normal conditions. To steer in a right turn, the clockwise rotation of the steering wheel is resisted by high forces required to turn the road wheels. Then this resistance causes the shaft to climb down at the ball nut. This movement shifts the valve down, and the valve admits high-pressure oil to one side of the actuator piston and allows oil from the adjacent side to return to the reservoir. The piston-gear sector combination then moves the rest of the system in the corresponding direction to relieve the thrust on the steering shaft. The valve regains its neutral position, thereby curtailing oil from the pump.

A deviation from the conventional integral-type steering gear is the power rack-and-pinion gear illustrated in Fig. 2. Rack-and-pinion steering has been historically utilized on the lightweight sports cars and is used currently on subcompact cars. Not only are the power-steering function valve and actuator combined, but also the steering gear structurally replaces the traditional tie rod required by the conventional rotary-type output integral gear. The lower steered vehicle weight and lower mechanical gear ratios permit the translational rack output. Operation is basically the same as operation of integral-type gear.

Linkage. Components of the linkage system connect to the linkage portion of the steering system. The actuator or booster cylinder anchors to the chassis. Figure 3 shows one such arrangement adapted to a cross steering linkage. The valve consists of a movable spool inside a sealed case. The valve is connected through conduits to the high-pressure pump, the low-pressure reservoir, and to each side of the booster piston. With the spool in its neutral position, annular grooves on the spool connect oil at low pressure to both sides of the piston. The steering system is at rest. Steering effort or road reaction deflects the valve spool relative to the valve case. The spool grooves then connect the high-pressure oil line to one or the other side of the booster cylinder so that oil forces the steering linkages in the direction that will return the valve to its neutral position.

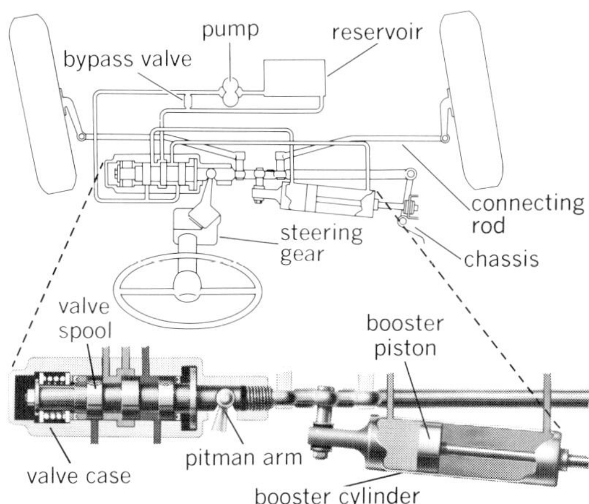

Fig. 3. The linkage power steering reacts between the chassis and steering linkages. (*Ford*)

In-line linkage. Control valve and actuator are combined in a single assembly for in-line linkage power steering. The system of Fig. 4 connects between drag link and idler arm of a fore-and-aft steering linkage. Motion of the drag link operates a valve similar to that used in other systems. The valve energizes the actuator, which drives the linkage to follow the steering motion or to resist road reaction. When the actuator has reduced the linkage forces, springs in the valve return it to its neutral position until unbalanced forces again set the hydraulic assist into action.

Hybrid. A valve in the manual reserve steering gear and a separate actuator in any other part of the steering system, as in Fig. 5, constitute hybrid power steering. Valve and actuator are similar to the corresponding components of other systems, and the operation is basically the same.

Hydrostatic. Hydrostatic steering requires no mechanical linkage to provide position control of the steered vehicle wheels and the driver's steering wheel. This servoloop function, as previously described, relies entirely on hydraulic oil pressure and volume carried by rigid or flexible hoses connecting the control valve to the actuator. Since

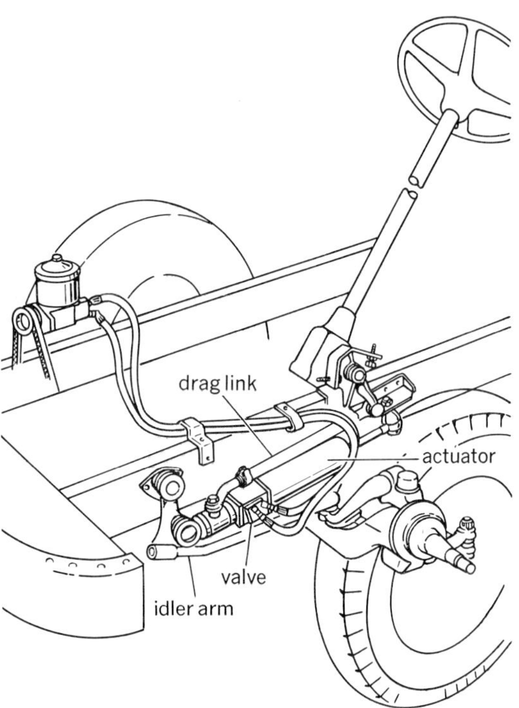

Fig. 4. In-line linkage power steering. (*Bendix*)

no mechanical interconnection exists between the steering wheel and actuator, a small pumping section is required to meter high-pressure oil passed by the control valve. The metering function provided by the pumping element simply "measures" a precise volume of hydraulic fluid, which is proportional to the amount of steering

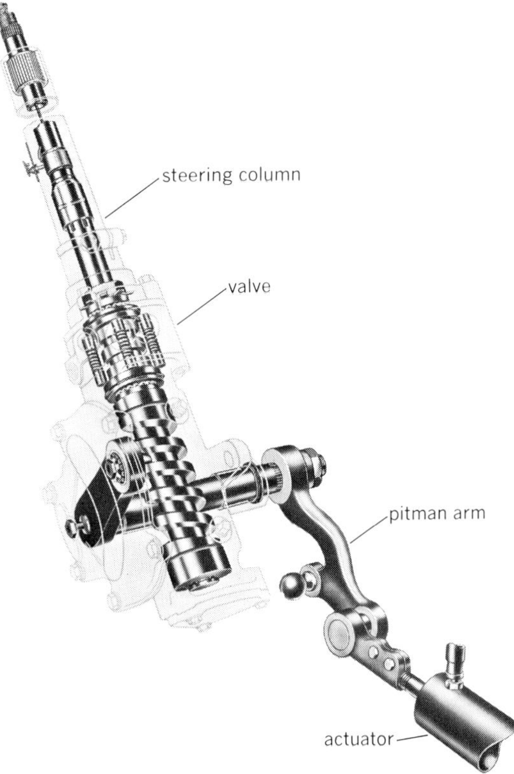

Fig. 5. Hybrid power steering. (*TRW Inc.*)

wheel rotation, to the actuator to effect the appropriate extent of steered wheel movement. Hence, as the steering wheel is turned right, the valve spool shifts forward from neutral position, and high-pressure hydraulic oil is diverted to the working side of the actuator. Return oil is admitted by the valve draining to the reservoir. The valve spool shifting is accomplished by the resistive interaction of the pumping section which is connected to the loaded power actuator. When the steering motion has been completed, the valve returns to neutral.

[JAMES L. RAU]

Bibliography: H. E. Ellinger and R. H. Hathaway, *Automotive Suspension, Steering and Brakes*, 1980.

Precast concrete

Concrete that has been cast into a form which is later incorporated into a structure. A concrete structure may be constructed by casting the concrete in place on the site, by building it of components cast elsewhere, or by a combination of the two. Concrete cast in other than its final position is called precast.

In contrast with cast-in-place concrete construction, in which columns, beams, girders, and slabs are cast integrally or bonded together by successive pours, precast concrete requires field connections to tie the structure together. These connections can be a major design problem.

Form costs are much less with precast concrete because the forms do not have to be supported on falsework in the structure. They may be set on the ground in a convenient position. Furthermore, a thin wall is difficult to concrete if it must be cast vertically because the concrete has to be placed in the narrow opening at the top of the form. Such a wall is easily precast flat on the ground. Moreover, the large side forms are eliminated, as are the braces needed to keep a vertical form in place.

With some types of precast concrete construction, no time is lost in waiting for concrete to gain strength at one level of a structure before the next level can be placed; such delays are common with cast-in-place construction. Frequently, the permanent precast units can be used as a working platform, eliminating the need for a temporary deck.

Precast units can be standardized. Savings can then result from repeated reuse of forms and assembly-line production. Furthermore, high quality can be maintained because of the controls that can be kept on production under plant conditions. However, there is always the possibility that transportation, handling, and erection costs for the precast units will offset the savings. *See* CONCRETE SLAG; PRESTRESSED CONCRETE.

Floor and roof systems. Precast concrete floor and roof systems may be similar to what is generally used for cast-in-place construction. The components may be bolted together, seated on each other or on brackets, held by friction devices, prestressed, or the reinforcing of adjoining members welded and the gap between them filled with cast-in-place concrete. Also, certain systems peculiar to precast construction may be used:

1. I-beam type with cast-in-place or precast slab
2. Hollow-core-type joists
3. Assembled concrete-block type
   a. With contact faces between units ground to provide a slight camber to the assembly
   b. With contact faces parallel but with a tension in the lower moment bars sufficient to align and hold the assembly together, and to provide a slight camber
4. Precast inverted T-beam joists with precast fillers between
5. Integrally precast slab and T-beam joists

Tilt-up construction. Originally, tilt-up construction was the name given to a method of precasting walls in which the units were cast on the ground at the place where they were to be erected, then tilted up to the vertical and anchored when they had gained sufficient strength. Later it became customary to refer to all types of precast wall construction as tilt-up construction.

Generally, the wall is concreted on a casting platform. Only side forms are needed. Sometimes it is advantageous, not only for walls but for floor and roof panels as well, to cast successive units one on top of the other. A bond-breaking agent is applied to the surface of the casting platform and between successive units.

Inserts usually are cast in the panels to facilitate lifting. The precast units may be lifted with a crane or A frame, often equipped with a strongback, or frame, to distribute the uplift forces evenly.

Lift-slab or Youtz-Slick method. In one type of precast construction for buildings, popularly known as lift-slab but sometimes called Youtz-Slick after the developers of the method, floor and roof slabs for a multistory building are cast on the ground around the columns. The slabs are cast one on top of the other, with a bond-breaking agent between them. Jacks atop the columns lift them to their final position, where they are anchored.

[FREDERICK S. MERRITT]

Press fit

A force fit that has negative allowance; that is, the bore in the fitted member is smaller than the shaft which is pressed into the bore. Tight fits have slight negative allowance so that light pressure is required to assemble the parts; they are used for gears, pulleys, cranks, and rocker arms. Medium force fits have somewhat greater negative allowance and require considerable pressure for assembly; they are used for fastening locomotive wheels, car wheels, and motor armatures. *See* ALLOWANCE; FORCE FIT; SHRINK FIT.

[PAUL H. BLACK]

Pressure

The ratio of force to area. The force per unit area at the interior of the Sun is estimated to be 3×10^{17} dynes/cm^2. In interstellar space, pressure approaches zero. Atmospheric pressure at the surface of Earth is in the vicinity of 14 lb/in^2. Pressures in enclosed containers less than this value are spoken of as vacuum pressures; for example, the vacuum pressure inside a cathode-ray tube is 10^{-8} mm of Hg, meaning that the pressure is equal to the pressure that would be produced by a column of mercury, with no force acting above it, that is 10^{-8} mm high. This is absolute pressure measured above zero pressure as a reference level. Inside a steam boiler, the pressure may be 800 lb/in.^{2} or higher. Such pressure, measured above atmospheric pressure as a reference level, is gage pressure, designated psig.

[FRANK H. ROCKETT]

Pressure seal

A seal is used to make pressureproof the interface (contacting surfaces) between two parts that have frequent or continual relative rotational or translational motion; such seals are known as dynamic seals, as compared with static seals. While the pressure in seals is lower than that in gaskets, the motion hinders their effectiveness so that there are more types of seals than gaskets, each type attempting to serve its environment. The materials are leather, rubber, cotton, and flax, and for piston rings, cast iron. The forms of nonmetallic seals are rectangular, V-ring, and O-ring. Cartridge seals are available for rolling-contact bearings. Special seals include carbon ring and labyrinth seals for turbines and mechanical seals for pumps. *See* GASKET; STEAM TURBINE.

[PAUL H. BLACK]

Pressure vessel

A cylindrical or spherical metal container capable of withstanding pressures exerted by the material enclosed. Pressure vessels are important because many liquids and gases must be stored under high pressure. Special emphasis is placed upon the strength of the vessel to prevent explosions as a result of rupture, which would be dangerous to life and property. Codes for the safety of such vessels have been developed that specify the design of the container for specified conditions.

Construction. Most pressure vessels are required to carry only low pressures and thus are constructed of tubes and sheets rolled to form cylinders. Some pressure vessels must carry high pressures, however, and the thickness of the vessel walls must increase in order to provide adequate strength. Hydraulic and pneumatic cylinders are machine elements that are forms of pressure vessels.

Fabrication methods depend upon the vessel diameter, its wall thickness, and the type of cylinder ends employed. For extreme strength, heavy forgings may be welded together; for most normal vessels, rolled sheet and formed ends are fastened together with rivets.

Shell stress in pressure vessels is further dependent upon the type of end construction of the cylinder or vessel, whether welded, riveted or cast; the kind of material, whether ductile or brittle; and the conditions of operation, including pressure and temperature variations and limitations. In choosing the safe allowable stress, these variables are considered.

Although most pressure vessels contain an internal pressure, occasionally an external pressure is applied, which if excessive could buckle the sides and ends of the vessel. Such conditions depend on the elasticity of the material and result in buckling under critical pressures in a manner similar to the critical loads on columns. *See* COLUMN.

Design. Thin-walled pressure vessels are assumed to have uniform stresses through the wall thickness t if the diameter d is 10 or more times as great (see illustration). For a pressure p, the shell stresses s_t are maximum in the circumferential direction, and this stress has the value $s_t = pd/2t$. In the axial direction, the stresses are half as great.

Thick-walled pressure vessels have a hyperbolic stress distribution through the wall thickness if the diameter is less than 10 times the thickness, with the maximum stress s_t at the inside surface. The stress is given in the equation below, where d_o is

$$s_{t(\max)} = \frac{p(d_o^2 + d_i^2)}{(d_o^2 - d_i^2)}$$

the outside and d_i is the inside diameter.

[JAMES J. RYAN]

PRESSURE VESSEL

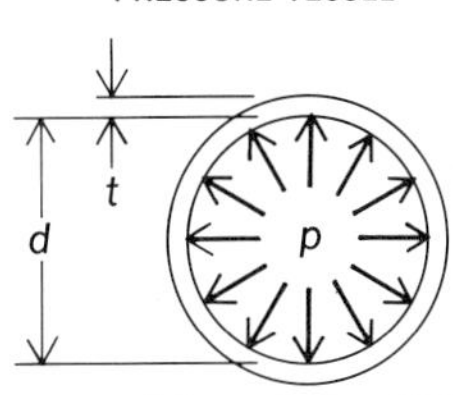

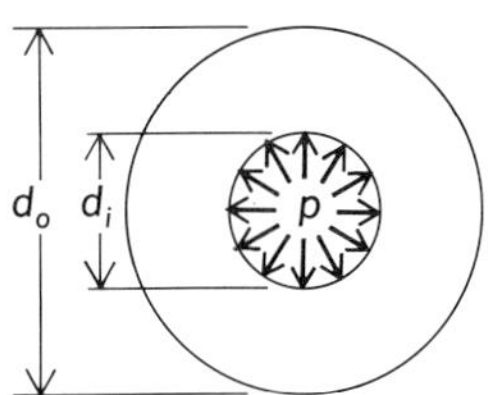

Pressure vessels for moderate and for high pressures.

Pressurized blast furnace

A blast furnace operated under higher than normal pressure. The pressure is obtained by throttling the off-gas line, which permits a greater volume of air to be passed through the furnace at lower velocity and results in an increasing smelting rate. *See* PYROMETALLURGY, NONFERROUS.

The process has advanced markedly since its introduction in the United States in 1943, and is now utilized on all modern blast furnaces in the

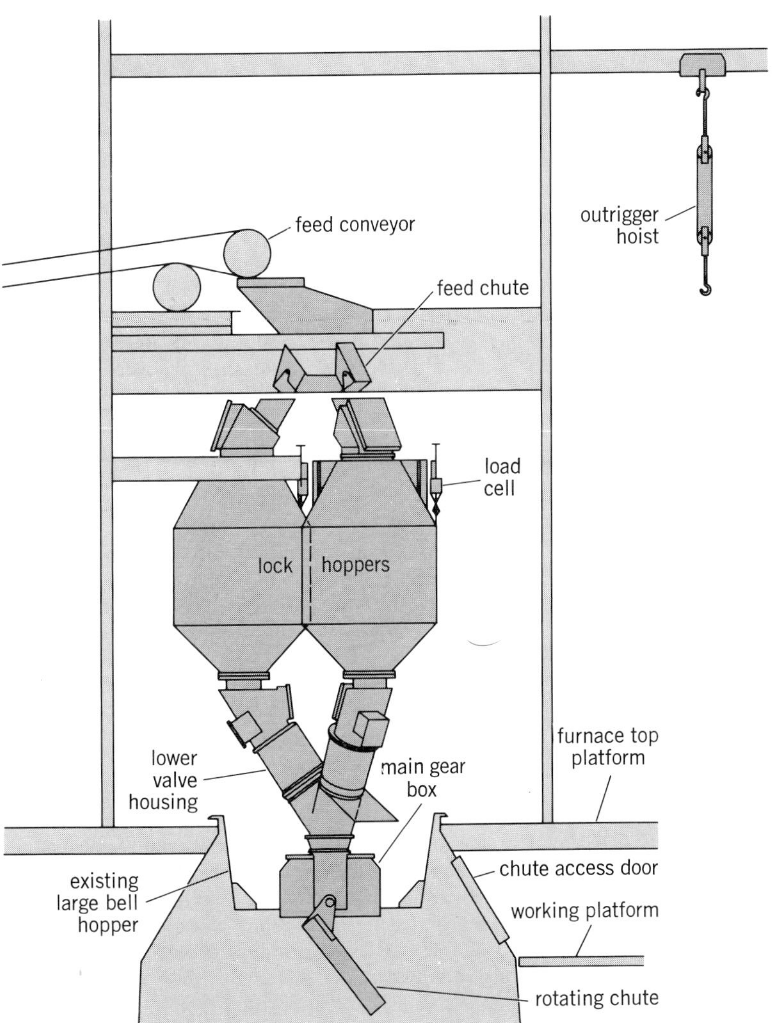

Diagram of a pressurized blast furnace with a Wurth top. (*From R. A. Powell, The blast furnace top of the future, Iron Steel Eng., 50(1):86–90, January 1973*)

world. High-top-pressure operation has made possible the design of mammoth blast furnaces having over 4000 m³ inner volume and hearth diameters of over 14 m; they produce over 10,000 metric tons/day compared with 1400 tons/day in 1943. The process permits large increases in the weight of high-temperature air blown into the bottom of the furnace at lower gas velocities, thus increasing the rate of smelting and decreasing the rate of coke consumption, and also permitting smoother operation with less flue dust production through decreased pressure drop between bottom and top pressures.

Many improvements have also been made in the mechanical features of the flow diagram shown in the illustration. Novel charging mechanisms, such as the Wurth top (lockhoppers with rotating shute), have essentially eliminated wear maintenance problems and improved the distribution of ore, coke, and limestone charged into the furnace. Also, at top pressures of 3 kg/cm², the throttling valve has been replaced by an expansion turbine to recover power from the top gasses. Instrumentation and controls have also been vastly improved. *See* FURNACE CONSTRUCTION. [BRUCE S. OLD]

Bibliography: M. Higuchi, M. Ilzuka, and T. Shibuya, High top pressure operation of blast furnaces at Nippon Kokan K. K., *J. Iron Steel Inst.*, September 1973; B. S. Old, E. L. Pepper, and E. R. Poor, Progress in high pressure operations of blast furnaces, *Iron Steel Eng.*, 25(5):37–43, 1948; O. D. Rice, New blast furnace design boosts iron yields, *Iron Age*, 185(4):96–98, 1960; J. H. Slater, Operation of the iron blast furnace at high pressure, *Yearb. Amer. Iron Steel Inst.*, pp. 125–200, 1947; R. J. Wilson, No. 7 Blast Furnace—Inland Steel Company, *Ironmaking Proc., AIME*, vol. 37, 19XX.

Prestressed concrete

Concrete with stresses induced in it before use so as to counteract stresses that will be produced by loads. Prestress is most effective with concrete, which is weak in tension, when the stresses induced are compressive. One way to produce compressive prestress is to place a concrete member between two abutments, with jacks between its ends and the abutments, and to apply pressure with the jacks. The most common way is to stretch steel bars or wires, called tendons, and to anchor them to the concrete; when they try to regain their initial length, the concrete resists and is prestressed. The tendons may be stretched with jacks or by electrical heating.

Prestressed concrete is particularly advantageous for beams. It permits steel to be used at stresses several times larger than those permitted for reinforcing bars. It permits high-strength concrete to be used economically, for in designing a member with reinforced concrete, all concrete below the neutral axis is considered to be in tension and cracked, and therefore ineffective, whereas the full cross section of a prestressed concrete beam is effective in bending. *See* CONCRETE BEAM.

An especially desirable characteristic of prestressed concrete is that as long as the material is maintained in compression it cannot crack. If cracks should appear under small overload, they generally will close when the load is removed. Sometimes concrete is prestressed principally to prevent cracking.

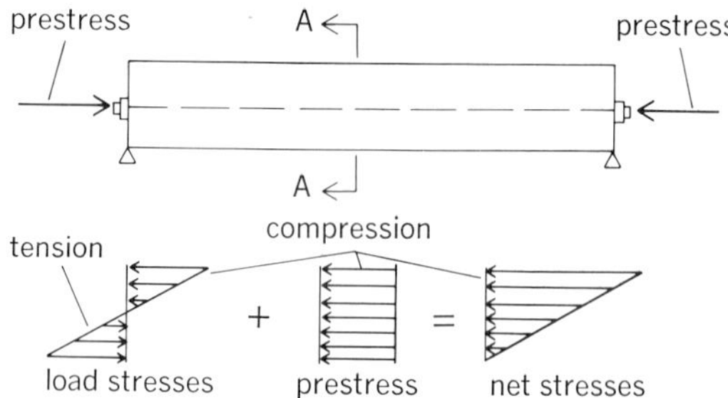

Fig. 1. Diagram showing stresses at section A-A of a concrete beam with uniform prestress.

Basic principles. The effect of compressive prestress may be likened to picking up a group of books by applying pressure to the end pair. As long as the pressure is large enough, none of the books will slip out.

If this concept were applied to a concrete beam in actual practice, steel tendons would be tensioned and placed along the centroidal axis of the beam. The resulting prestress would result in a uniform compression at every section (Fig. 1). Loads would produce both tensile and compressive stresses at the middle of the span. The prestress would combine with these to increase the compression and cancel out the tension. The whole concrete section would be effective in resisting bending, and there would be no cracks.

In practice, however, tendons are rarely placed along the centroidal axis. A smaller prestressing force is required, and therefore less steel for the tendons, if the steel is placed below the centroidal axis of the beam. With the eccentric prestress, stresses at each section of the unloaded beam may vary from tension at the top of compression at the bottom (Fig. 2).

When loads are applied to the beam, they produce both tensile and compressive stresses at the middle of the span. At the top of the beam they cause compressive stresses, which are reduced by the tensile prestress there. Elsewhere, the tensile

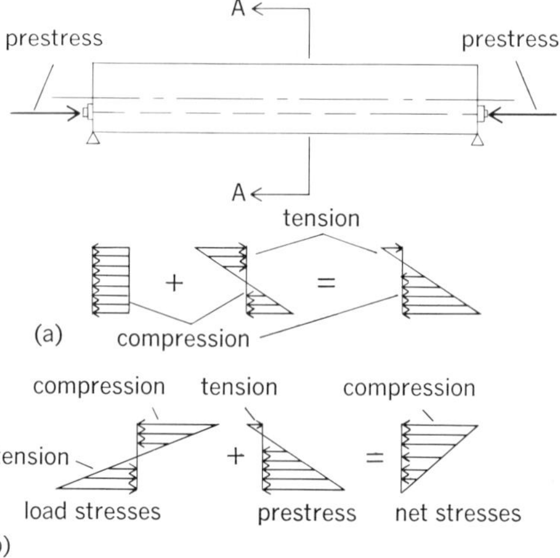

Fig. 2. Stresses at section A-A of a beam with eccentric prestress. (*a*) In unloaded beam simple-bending stress component is largely counteracted by uniform compression component. (*b*) In loaded beam tension components are counteracted and only compression remains.

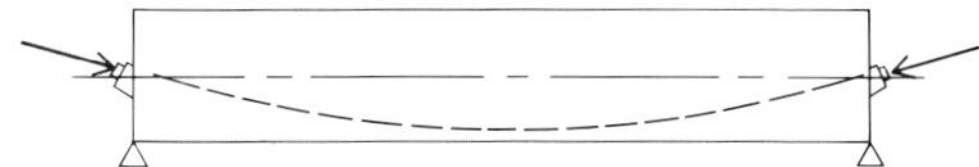

Fig. 3. Beam with tendons draped in a vertical curve.

stresses produced by the loads are counteracted by the compressive prestress.

With this arrangment of the tendons, there is a possibility that near the ends of the beam the tensile prestress may exceed the compressive stresses produced by the loads. The net tension may be undesirable, even though very small. To avoid this condition, the tendons may be draped in a vertical curve (Fig. 3). The distribution of prestress at any section of a beam so prestressed is similar to that for straight tendons applying an eccentric prestress except that the stresses decrease from midspan toward the ends, as do the bending stresses due to the loads. The draped arrangement of the tendons also is advantageous in counteracting diagonal tension near the ends of the beam.

Continuous beams may be prestressed in a similar manner. The tendons may be placed near the bottom of the beams near midspan and near the top over supports.

Tendons. Tendons generally are made of high-strength steel so that they can serve at high working stresses. The reason for this is that losses in stress due to shrinkage and plastic flow of the concrete are relatively high. If the tendons were given a small tension, they might lose nearly all the prestress in a few months. But at a high tension, the loss might be only about 15%, because the increase in stress loss is smaller than the increase in prestress.

Wires are used for prestressing much more frequently than bars because of their greater strength. Wires may be used singly, in pairs, in cables composed of several parallel wires, or in strands. They may be stretched by electric heating, but by far the most common method of tensioning is with jacks. Various devices are used for gripping or anchoring tendons, including swaged fittings on strands, bolts threaded on bars, wedges, and buttonheads on wires.

Prestress losses. One reason for loss of prestress is elastic shortening of concrete and steel due to compression of the concrete. Another reason is creep, or plastic flow, of concrete. It is an inelastic deformation dependent on time that occurs under constant stress.

Concrete shrinks when it dries and chemical changes take place. Shrinkage is dependent on time but not on stresses due to external loading.

Other possible losses in prestress that should be considered include those due to creep of the steel and to friction when the tendons rub against the concrete.

Procedures for estimating the loss of prestress due to shrinkage, creep, friction, and other causes are given in design codes, such as the American Concrete Institute *Building Code Requirements for Reinforced Concrete*, ACI 318, and the American Association of State Highway and Transportation Officials *Standard Specifications for Highway Bridges*.

Pretensioning and posttensioning. Two methods are used in fabricating prestressed beams. In one method, the concrete is bonded to the stretched steel before the prestress is applied. This is called pretensioning. In the other method, posttensioning, the prestress is applied initially through end anchorages and the concrete may or may not be bonded later to the steel.

In pretensioning, the steel is laid through the beam forms and stretched between external abutments. Next, concrete is placed in the forms and allowed to set. When it has gained sufficient strength, the external pull on the tendons is relieved, transferring the prestress to the concrete through bond. This method can be used on casting beds several hundred feet long to mass-produce many beams simultaneously.

In posttensioning, the tendons are prevented initially from bonding to the concrete, usually by encasement in sheaths. The concrete is placed in the beams forms around the sheathed tendons and allowed to set. When it has gained sufficient strength, jacks are used to tension the tendons, and in so doing, the jacks react against the ends of the beam. The tendons then are anchored to the concrete to apply the prestress, and the jacks are released and removed.

Frequently, grout is forced into the sheaths to bond the tendons to the beam concrete. This gives the prestressed beam greater reserve strength and better crack control under overload. Posttensioning appears to be most advantageous for long-span beams and for assembling precast beam components in the field.

Circular prestress. Circular tanks, pipe, or the ring girder of domes may be prestressed, in contrast to the linear prestressing used for beams, by wrapping with steel bars or wires under high tension. Special machines have been developed for rapid circular prestressing with wire. *See* STORAGE TANK.

[FREDERICK S. MERRITT]

Bibliography: M. Fintel, *Handbook of Concrete Engineering*, 1974; B. C. Gerwick, *Construction of Prestressed Concrete Structures*, 1971; A. H. Nilson, *Design of Prestressed Concrete*, 1978; Prestressed Concrete Institute, *PCI Design Handbook: Precast and Prestressed Concrete*, 1979.

Prime mover

The component of a power plant that transforms energy from the thermal or the pressure form to the mechanical form. Mechanical energy may be

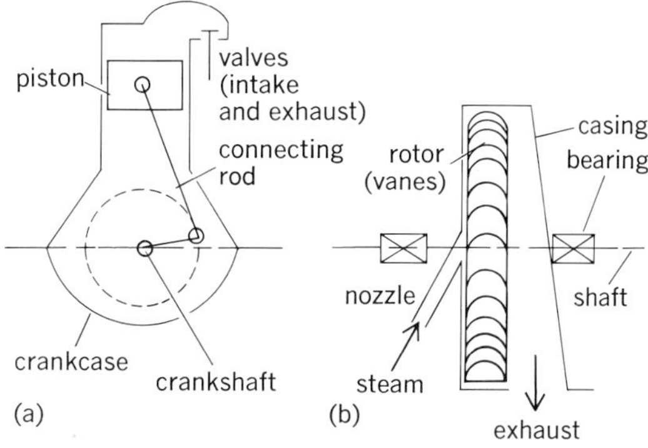

Fig. 1. Representative prime movers. (*a*) Single-acting four-cycle, automotive-type internal combustion engine. (*b*) Single-stage, impulse-type steam turbine.

Table 1. Dimensional and performance criteria of some selected fluid displacement-type prime movers

Type	Size, hp[a]	rpm	Stroke, in.[b]	Bore: stroke ratio	Piston speed, ft/min[c]	Brake mep, psi[d]	Diagram factor, or engine efficiency
Steam engine	25–500	100–300	6–24	0.8–1.2	400–600	50–100	0.6–0.8[e]
Automobile engine	10–300	2000–4000	3–5	0.9–1.1	1000–2000	50–100	0.4–0.6[f]
Aircraft engine	100–3000	2500–3500	4–7	0.8–1.1	1500–3000	100–230	0.4–0.6[f]
Diesel, low-speed	100–5000	100–300	10–24	0.8–1.0	500–1000	40–80	0.4–0.7[f]
Diesel, high-speed	25–1000	1500–2000	3–6	0.8–1.0	800–1500	50–100	0.4–0.6[f]

[a]1 hp = 0.75 kW. [b]1 in. = 25 mm. [c]1 ft/min = 5×10^{-3} m/s. [d]1 psi = 6.9 kPa. [e]Logarithmic standard. [f]Air-card standard.

in the form of a rotating or a reciprocating shaft, or a jet for thrust or propulsion. The prime mover is frequently called an engine or turbine and is represented by such machines as waterwheels, hydraulic turbines, steam engines, steam turbines, windmills, gas turbines, internal combustion engines, and jet engines. These prime movers operate by either of two principles (Fig. 1): (1) balanced expansion, positive displacement, intermittent flow of a working fluid into and out of a piston and cylinder mechanism so that by pressure difference on the opposite sides of the piston, or its equivalent, there is relative motion of the machine parts; or (2) free continuous flow through a nozzle where fluid acceleration in a jet (and vane) mechanism gives relative motion to the machine parts by impulse, reaction, or both. *See* IMPULSE TURBINE; INTERNAL COMBUSTION ENGINE; REACTION TURBINE; STEAM ENGINE.

Displacement prime mover. Power output of a fluid-displacement prime mover is conveniently determined by pressure-volume measurement recorded on an indicator card (Fig. 2). The area of the indicator card divided by its length is the mean effective pressure (mep) in pounds per square inch, and horsepower of the prime mover is given by Eq. (1), where L is stroke in feet, a is piston area

$$\text{Horsepower} = \frac{\text{mep} \times Lan}{33{,}000} \quad (1)$$

in square inches, and n is number of cycles completed per minute. Actual mep is smaller than the theoretical mep and may be related to the theoretical value by diagram factor or engine efficiency (Table 1). *See* MEAN EFFECTIVE PRESSURE.

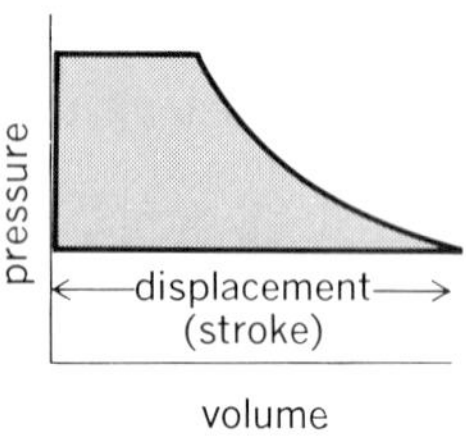

Fig. 2. Pressure-volume diagram (indicator card) for ideal, no-clearance, fluid-displacement types of prime mover.

Acceleration prime mover. Performance of fluid acceleration (hydraulic) prime movers is given by Eq. (2), where Q is water flow rate in cubic

$$\text{Horsepower} = \frac{QH}{8.8} \times \text{efficiency} \quad (2)$$

feet per second, and H is head in feet. For heat-power prime movers of the fluid acceleration type, actual properties of the thermodynamic fluid, as given in tables and graphs, especially the Mollier chart, permit the rapid evaluation of the work or power output from the general energy equation which resolves to the form of Eq. (3), where h is the

$$\Delta W, \text{Btu/lb of fluid} = h_{\text{inlet}} - h_{\text{exhaust}} \quad (3)$$

enthalpy in Btu/lb, and the inlet and exhaust conditions can be connected by an isentropic expansion for ideal conditions, or modified for irreversibility to a lesser difference by engine efficiency

Table 2. Dimensional and performance criteria of some selected fluid acceleration-type prime movers

Type	Rating, kW	Number of stages	Head, ft;[a] or pressure, psi[b]	Temperature, °F (°C)	Exhaust pressure in. Hg abs[c]	rpm	Tip speed, ft/s[d]	Efficiency
Pelton water wheel	1000–200,000	1	500–5000 ft	Ambient	atm	100–1200	100–250	0.75–0.85
Francis hydraulic turbine	1000–200,000	1	50–1000 ft	Ambient	atm[e]	72–360	50–200	0.8–0.9
Propeller (and Kaplan) hydraulic turbine	5000–200,000	1	20–100 ft	Ambient	atm[e]	72–180	70–150	0.8–0.9
Small condensing steam turbine	100–5000	1–12	100–400 psi	400–700 (200–370)	1–5	1800–10,000	200–800	0.5–0.8
Large condensing steam turbine	100,000–1,000,000	20–50	1400–4000 psi	900–1100 (480–590)	1–3	1800–3600	500–1500	0.8–0.9
Gas turbine	500–20,000	10–20	70–100 psi	1200–1500 (650–820)	atm	3600–10,000	500–1500	0.8–0.9

[a]1 ft = 0.3 m. [b]1 psi = 6.9 kPa. [c]1 in. Hg = 3.4 kPa. [d]1 ft/s = 0.3 m/s. [e]Draft tube gives negative pressure on discharge side of runner.

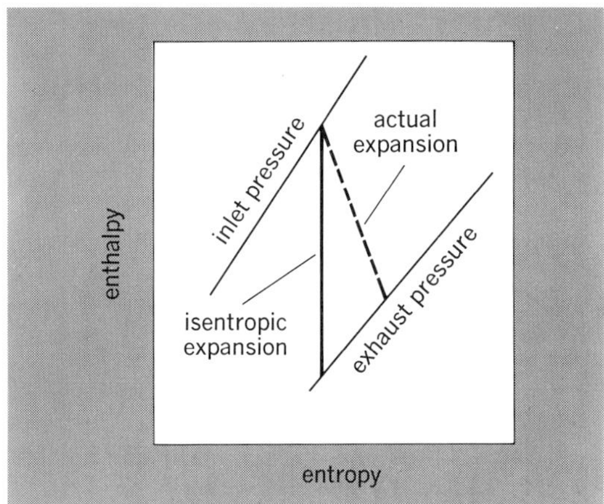

Fig. 3. Enthalpy-entropy (Mollier) chart of performance of steam- or gas-turbine type of prime mover.

(Fig. 3 and Table 2). Fluid consumption follows from Eq. (4) or Eq. (5).

$$\text{Fluid consumption, lb per hphr} = 2545/\Delta W \quad (4)$$

$$\text{Pounds per kWhr} = 3413/\Delta W \quad (5)$$

In the fluid acceleration type of prime mover, jet velocities experienced in the nozzles can be found in feet per second, for nonexpansive fluids, by Eq. (6). For expansive fluids, they may be found by Eq. (7), where H and ΔW are as given above and C is

$$\text{Jet velocity} = C\sqrt{2gH} = 8.02\, C\sqrt{H} \quad (6)$$

$$\text{Jet velocity} = C\sqrt{2g\Delta W} = 223.7\, C\sqrt{\Delta W} \quad (7)$$

the velocity coefficient, seldom less than 0.95 and usually from 0.98 to 0.99.

Selected representative performance values of some prime movers are presented in Tables 1 and 2. *See* GAS TURBINE; HYDRAULIC TURBINE; STEAM; STEAM TURBINE; TURBINE.

[THEODORE BAUMEISTER]

Bibliography: T. Baumeister (ed.), *Standard Handbook for Mechanical Engineers*, 8th ed., 1978.

Process engineering

A branch of engineering in which a process effects chemical and mechanical transformations of matter, conducted continuously or repeatedly on a substantial scale. Process engineering constitutes the specification, optimization, realization, and adjustment of the process applied to manufacture of bulk products or discrete products. Bulk products are those which are homogeneous throughout and uniform in properties, are in gaseous, liquid, or solid form, and are made in separate batches or continuously. Examples of bulk product processes include petroleum refining, municipal water purification, the manufacture of penicillin by fermentation or synthesis, the forming of paper from wood pulp, the separation and crystallization of various salts from brine, the production of liquid oxygen and nitrogen from air, the electrolytic beneficiation of aluminum, and the manufacture of paint, whiskey, plastic resin, and so on. Discrete products are those which are separate and individual, although they may be identical or very nearly so. Examples of discrete product processes include the casting, molding, forging, shaping, forming, joining, and surface finishing of the component piece parts of end products or of the end products themselves. Processes are chemical when one or more essential steps involve chemical reaction. Almost no chemical process occurs without many accompanying mechanical steps such as pumping and conveying, size reduction of particles, classification of particles and their separation from fluid streams, evaporation and distillation with attendant boiling and condensation, absorption, extraction, membrane separations, and mixing.

BULK PRODUCTS

Process engineering begins with a need for a product and some sense of the required scale of production. The end product is always defined; frequently the desired raw materials can also be stipulated at this stage, although sometimes it is very important to consider wholly different sources (for example, motor fuel from coal, petroleum, or renewable resources) and the degree of vertical integration desired—for example, should the process begin with natural products or purchased intermediates. Then one or more process flow sheets are drawn, depicting the series of steps through which the raw materials must pass in the course of transformation into product. On the flow sheet all lines represent movement of material from one process step to another, and all lines must show the intended direction of flow. Each processing step is represented by an enclosed area (circle, rectangle, and so on) on whose edges the lines terminate (unless only heat exchange is intended, in which case at least one line goes through the area unbroken to indicate thermal contact without mixing). Each process flow sheet envisions a definite sequence of operations, including the selection of separation processes and choice of chemical reactions. The sequence may contain subsequences that are serial or parallel or enclosed by a recycle stream.

Batch versus continuous operation. In designing almost any process, an early choice between batch and continuous operation must be made. Small amounts of material can be taken through a series of process steps over time, often using the same apparatus (for example, a thermally jacketed stirred tank) for several or even all steps. Commercial production in batches is feasible and often preferred when small amounts are to be manufactured, when a sequence of many similar reaction steps is envisioned, or when it is feasible for multiple products to be manufactured in the same equipment. However, batch operations are labor-intensive and more difficult to control, analyze, and design (if done with the same care given to continuous processes). Continuous operation is easier to analyze, design, and control, but the plant is expensive, unique to the product, and difficult to operate effectively at other than design rate.

Material and energy balances. Once the process flow sheet is drawn, it is possible—for either batch or continuous processes—to generate material and energy balances. These are drawn for the overall process and for each step on the basis of conservation and rate laws as well as certain assumptions, some of which are only temporary. For example, in a mechanical step, such as dewatering of a fiber slurry during paper manufacture, a dilute stream containing 1% dry fiber and 99% water may be presumed to enter a separator at the rate of 10,000 kg/min, to be separated into two product streams, one containing 99% of the fibers fed at a concentration of 5% fibers and the other, a recycle water stream. The proportioning of the fibers between the streams and the composition of the one outlet stream are assumptions that may well be adjusted at later stages. In this example, the conservation laws require that water and fiber be separately conserved since no reaction is envisioned, and it is immediately possible to calculate that the concentrated fiber stream emerges at the rate of 1980 kg/min, while the balance of the 10,000 kg/min leaves as a stream containing about 0.0125% fiber. In a complex process, many hundreds of such steps must be assessed and matched to each other. (In the case of the process step just considered, accepting the 10,000 kg/min stream constitutes matching this step to a predecessor; and two subsequent steps must be matched to the two exit streams just considered.) When the assessment and matching of individual steps are complete, the overall result must be considered. Is the production rate correct? Is the yield of product satisfactory, or has too much material been turned into by-products or lost in waste streams? Has too much water or heat been used? Is the product pure enough? Frequently one or more of these targets are not met. Then assumptions must be revised and the computations repeated. Sometimes the flow sheet itself must be changed: additional steps added, a new recycle stream provided, one kind of operation replaced by another. Because of the complexity of this process, it is often executed with the aid of computers. Once a flow sheet has been fixed and consistent heat and material balances reckoned, the duty of each process unit is established, and their individual design can be commenced.

Process units and reactors. For many years chemical engineers have recognized that the design of process units was much more dependent upon the kind of operation under consideration (for example, heat transfer, or fluid flow, or distillation) and much less dependent on what was being processed or where in the sequence of process steps a particular unit occurred. Thus evolved the concept of unit operations and a set of principles and rules for design of process units according to their type of unit operation rather than according to the process itself. Notable among unit operations that have been extensively developed are the membrane processes (reverse osmosis, dialysis, electrodialysis, and electroosmosis), techniques for separating solids from fluids, and methods of controlling transport across phase boundaries such as freeze drying.

Reactor design. In completing the engineering of a process, many features peculiar to it must be considered, such as the selection of pipes and tanks strong enough to withstand the temperatures and pressures of operation and the choice of materials that will not be corroded by the materials being processed. Because the chemical reaction is the most critical step in most chemical processes, reactor engineering is a very important component of process engineering. Chemical reactions may be carried out either with a single phase (homogeneously) or near phase boundaries and interfaces (heterogeneously) with or without a catalyst. Process reactors may be stirred tanks containing one fluid, one fluid dispersed in another, or a suspension. Some reactors are fluidized beds: fine particles which may be either reactant or catalyst kept suspended and moving by a flowing gas or liquid. Tubular reactors are used frequently, filled only with fluid, or packed, or equipped with complex so-called internals. Many reactors are jacketed or otherwise equipped for removing heat. Some reactions are conducted in flames or in illuminated chambers. However, in spite of this great geometric variety, general principles of reactor engineering have been developed, based on expressions for the rates of chemical reactions and the transport of molecules and heat as well as the laws of conservation for mass and energy.

Integration and optimization. The widespread availability of computers has led both to higher integration of processes and to much more nearly optimal design. Integration refers to the increased use of recycle streams to recover heat or material that would otherwise be discarded. It is motivated both by possible economies and by its ability to decrease the burdens imposed on the environment by a process. Integration makes design more difficult, usually results in a higher capital cost for the plant, and may make it more difficult to control. Modern petroleum refineries tend to be particularly highly integrated. Optimization is a relative term. In practice it denotes computation of the best split of streams and duties among units and selection of the best operating conditions for each unit to optimize some overall quantity (often plant profitability) subject to constraints (such as environmental restrictions and plant size).

Extreme conditions. Extremes of high temperature and high and low pressure have been extended, so that reactions can be carried out in plasmas at temperatures of 10,000 K, while pressures may go as high as several hundred thousands of psi (10^5 psi $\cong$ 1 GPa) and as low as 10^{-10} torr (10^{-8} Pa). Extension of extremes of feasible operation has been greatly facilitated by development of new materials of construction.

Economic considerations. Process engineering is an activity supported almost entirely by industry, and is thus greatly influenced by economic considerations, much augmented however by regulatory considerations designed to minimize environmental impacts, protect workers, guarantee the safety and efficacy of products, and carry out governmental policies for conservation of resources, especially energy. Still, economics figures heavily in process design: in choosing to make a product, in specifying the degree of vertical integration, in fixing the scale of operations, and in balancing capital costs against operating costs. Other economically significant decisions include estimating the desirable operating lifetime of the process and plant and assessing the degree of risk to be taken in using attractive but not fully proved new technology in the process realization. Thus process engineering utilizes not only principles from the chemical and physical sciences but also detailed assessments of the state of commerce and the economy [EDWARD F. LEONARD]

DISCRETE PRODUCTS

The process engineer must consider all available processes capable of changing the shape of a raw material to the geometrical configuration specified on an engineering drawing. Thus, the process engineer must consider not only those processes within his or her own plant but also those in vendor or subcontractor plants. Usually, several ways exist to produce any part, one being best for a given set of conditions. The process engineer has the responsibility of determining which is best. The principal constraints to be considered in the selection of a basic process in order to bring raw material more closely to the specifications of a functional design are: (1) type of raw material; (2) size of the raw material that the equipment involved can handle; (3) geometric configurations that equipment characteristic of the process is capable of imparting to the raw material; (4) tolerance and surface finish capabilities of the equipment; (5) quantity of finished parts needed and their delivery requirements; and (6) economics of the process. Similar constraints apply in the selection of the most favorable secondary operations. *See* PRODUCTION ENGINEERING.

Choice of process. To identify the most favorable basic processes and secondary processes to be used to produce a given functional design, the process engineer may develop a selector guide to assist in the decision-making process.

Selector guide sheets are usually developed for a given size classification and are shown in part in Tables 1–3. In using these guide sheets, an ana-

Table 1. Part of guide sheet for geometry and other primary factors

Process	Applicable geometry*	Intricacy of geometry	Applicable materials†	Minimum lot for which process is economical	Secondary operation cost	Decision equation for primary process	Expected rate of production per hour	Expected time in weeks to get into production
Die casting	1,2,3,4,5 6,7,8,9	X	t	3000	1	$D_t = 1.623N + 1515$	200–500 injections	12
Investment casting	1,2,3,4,5 6,7,8,9	X	s, t	100	5	$D_t = 2.370N + 320$ $D_s = 1.000N + 320$	10–20 molds	5
Permanent mold	1,2,3,4,5 6,7,8	Y	s, t	500	9	$D_t = 0.770N + 465$ $D_s = 0.335N + 465$	20–30 molds	10
Plaster mold	1,2,3,4,5 6,7,8,9	X	t	100	5	$D_t = 2.160N + 320$	15–28 molds	5
Shell mold	1,2,3,4,5 6,7,8,9	X	s, t	300	6	$D_t = 1.965N + 320$ $D_s = 0.595N + 320$	35–45 molds	5

*Numbers refer to the classifications in text. †S = ferrous; t = nonferrous.

Table 2. Part of guide sheet for raw materials and manufacturing costs

Process	Raw material form	Raw material cost	Tooling cost	Equipment cost	Setup cost	Production cost	Scrap cost	Expected tolerance, mm	Expected surface finish, micrometers
Die casting	Pig or liquid	2	10	4	4	1	3	±.08 except .18 across parting line	.191
Investment casting	Pig	2	4	6	6	8	4	±.13	.318
Permanent mold	Pig	2	6	6	3	4	4	±.38	.381
Plaster mold	Pig	2	3	6	6	7	3	±.13	.254
Shell mold	Pig	2	4	6	6	4	3	±.13 except ±.22 across parting	.318

Table 3. Part of guide sheet for quality considerations

Process	Expected reliability	Directional flow lines	Tool marks	Remarks
Die casting	A for Zn B for Al C for brass	No	Yes	Ejection pins may cause tool marks; surface finish depends on surface finish of die; casting of ferrous alloys generally not economically feasible
Investment casting	B	No	No	Best for complicated design; no parting line; frozen mercury results in better finish than wax or plastic
Permanent mold	C	No	No	Lightest wall thickness 1/8 in. (3 mm); limited complexity because of rigid molds; not suitable for class 9 geometry
Plaster mold	A	No	No	Little finishing required
Shell mold	B	No	No	

lyst refers to the correct sheets for the size of the part under study. There may be several size classifications to handle all products being made by a given company. Next the analyst determines the geometry that best describes the component. Table 1 indicates nine geometric configurations in connection with the five casting processes listed. These nine classifications may be defined and illustrated to facilitate the classification of the geometry of the product being studied. For example:

Class 1 geometry—Solid or partly hollow rounds involving one or more diameters along one axis. The depth of hollow not to be more than two-thirds the diameter of the end containing the hollow.

Class 2 geometry—Hollow or partly hollow rounds involving one or more outside diameters and one or more inside diameters along one axis. The depth of hollow is more than two-thirds the diameter of the end containing the hollow.

Class 3 geometry—Solid or partly hollow shapes other than rounds such as square, triangular, octagonal or irregular, including one or more cross-sectional areas along one axis. The depth of hollow is not more than two-thirds the distance of the major diameter.

Class 4 geometry—Hollow or partly hollow shapes other than rounds involving one or more cross-sectional areas along one axis. The depth of hollow is more than two-thirds the major diameter of the hollow.

Class 5 geometry—Bowl-shaped concentric.

Class 6 geometry—Dish-shaped nonconcentric.

Class 7 geometry—Flats with or without configuration.

Class 8 geometry—Flanged.

Class 9 geometry—Complex characterized by unsymmetrical shapes.

The next parameter that is considered is the material used. Table 1 illustrates only two material considerations, ferrous (s) and nonferrous (t). The required number of parts needed is estimated. The constraint associated with quantity is the minimum number of parts required before it would become economical to provide tooling for the process under study. Decision equations may be developed for typical manufacturing environments of a company. The final step for the analyst is to solve the decision equations for processes still under consideration; most will already have been eliminated by this time because of the constraints of geometry, material, and quantity. The process that provides the smallest value based on the solution of the decision equation will be considered favorably for use. If Table 1 were given in full, it would be seen that two decision equations are provided for most of the processes. The D_s equation is used in conjunction with ferrous materials, and the D_t equation in conjunction with nonferrous materials.

To do a thorough job of process planning, use of the digital computer is recommended. Manual process selection may not give adequate consideration to all possible alternatives. The computer can handle the mathematical manipulation of solving the various decision equations and the final selection problem. The equation used to develop the computer program based on the parameters shown in Table 1 is shown below, where A is the

$$D_t = A_{ijk} \times N + B_{ij}$$

coefficient of N, the number of units to be processed; i is the process to be considered, j is the size classification of the part being process planned, and k is the material specified. B_{ij} is the fixed element.

Sequence of secondary operations. In determining the sequence of operations to be performed, the process engineer considers the logical process order and the geometrical and dimensional control that can be maintained with each processing operation. *See* MACHINING OPERATIONS.

In the logical process order, basic operations are performed first and final inspection near the end of the process, just prior to a product's packaging for shipment. Holes that require reaming would necessarily be drilled before being reamed, and similarly, threaded inside diameters would necessarily be drilled prior to tapping.

In general, the final finishing of internal work is done in advance of external finishing. The principal reason for this sequence is that internal surfaces are less likely to be damaged in material handling and subsequent processes so that their surfaces can be completed earlier in the processing. For internal work, the logical sequence of operations in drilling, boring, recessing, reaming, and tapping. The logical sequence of external work is turning, grooving, forming, and threading.

Rough work involving heavy cuts and liberal tolerances should be performed early in a sequence. Heavy cuts will reveal defects in castings or forgings more readily than light cuts, and it is advantageous to identify defective raw material as soon as possible. Because both tolerance and finish of external surfaces can be adversely affected by subsequent material handling and clamping, close-tolerance operations should be performed late in the processing sequence.

In considering the geometrical and dimensional control that can be maintained, the process engineer should keep in mind those operations that establish locating or mastering points in order to schedule them early in the process. For example, it is usually easier to maintain control from a large, plane surface than from either a curved, irregular, or small surface; therefore, the process engineer establishes a surface which is best qualified for the location of critical features, and schedules it for finishing at a point early in the sequence.

Specifications concerning tools. It is the responsibility of the process engineer not only to specify how a piece should be made but also to order all special tools to be used in conjunction with the specified manufacturing processes. Special tools refer to those jigs, fixtures, dies, and gages that are required to hold work, guide the cutting tool, and inspect results. The requisition for the special tools should specify: (1) those areas or points best suited for locating the workpiece while it is being processed; (2) that portion of the workpiece suited for supporting or holding it while it is being processed; and (3) that portion or area best suited for clamping so that the workpiece is securely held during the processing. *See* JIG, FIXTURE, AND DIE DESIGN.

In specifying the location and number of locators to contact the work, the analyst must keep in mind that three locators are needed to locate a plane, two are needed to determine a line, and one will determine a point. A workpiece can move in either of two opposed directions along three per-

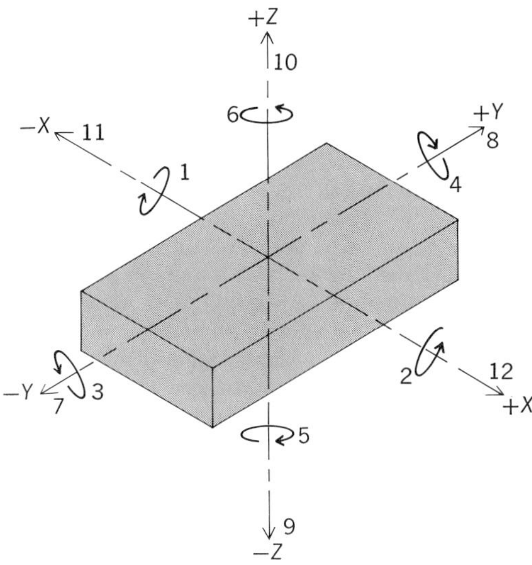

Fig. 1. Twelve degrees of freedom of a workpiece.

pendicular axes (X, Y, and Z). Also, the work may rotate either clockwise or counterclockwise around each of these three axes. Each of these possible movements is a degree of freedom; hence 12 degrees of freedom exist. These 12 degrees of freedom are illustrated in Fig. 1.

Work can be located positively by six points of contact in the tooling. These six points include three points on one plane. For example, in Fig. 2 the three locations A, B, and C on the bottom of the block prevent the work from moving downward and from rotating about the X and Y axes. By adding two locating points D and E on a plane parallel to the plane containing the X and Z axes, the work is prevented from rotating about the Z axis and also from moving negatively along the Y axis. When the sixth and final locating point F is added on a plane parallel to the Y and Z axes, movement upward is prevented. Thus the first three locators prevented movements 1, 2, 3, 4, and 9, as shown in Fig. 1. The next locators prevent movements 5, 6, and 7, and the final locator prevents movement 11. This 3-2-1 locating procedure has prevented movement in 9 of the 12 possibilities. The three remaining degrees of freedom (8, 10, and 12) must not be restricted because they are needed to provide clearance to load and unload the tooling.

When requisitioning special tooling, the process engineer should indicate how the work is to be supported and specify the position of the locators. To assure correct position throughout the process, there must be rigid support so that the work does not deflect because of its static weight or the holding and tool forces applied to it, or both. Adding a support to avoid deflection during processing does not mean adding another locator. The support should not provide a point of location, and the workpiece should never contact the support until the tool or tooling forces are applied; the only purpose of the support is to avoid or limit deflection and distortion.

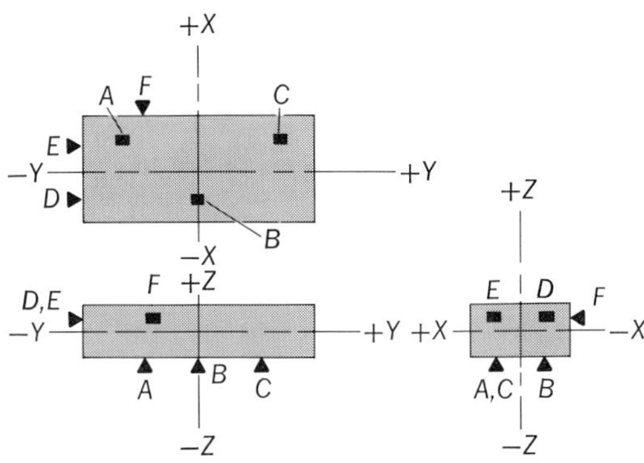

Fig. 2. Three views of workpiece.

The holding force must be of sufficient magnitude to allow all locators to contact the workpiece during the processing cycle, but it is also important not to have holding forces so large that the work becomes marred. It is usually advantageous to place holding forces directly opposite locators. A nonrigid workpiece may require holding forces at several locations to hold the work against all locators.

[BENJAMIN W. NIEBEL]

Bibliography: E. P. DeGarmo, *Materials and Processes in Manufacturing*, 5th ed., 1979; D. F. Eary and G. E. Johnson, *Process Engineering for Manufacturing*, 1962; H. V. Johnson, *Manufacturing Processes Metals and Plastics*, 1973; R. A. Lindberg, *Materials and Manufacturing Technology*, 1968; B. W. Niebel, *Selector Guide for Primary Forming Processes*, U.S. Army Materiel Command, AMCP no. 706; B. W. Niebel and E. N. Baldwin, *Designing for Production*, 1963; B. W. Niebel and A. B. Draper, *Product Design and Process Engineering*, 1974; H. W. Yankee, *Manufacturing Processes*, 1979.

Product design

The determination and specification of the parts of a product and their interrelationship so that they become a unified whole. The design must satisfy a broad array of requirements in a condition of balanced effectiveness. A product is designed to perform a particular function or set of functions effectively and reliably, to be economically manufacturable, to be profitably salable, to suit the purposes and the attitudes of the consumer, and to be durable, safe, and economical to operate. For instance, the design must take into consideration the particular manufacturing facilities, available materials, know-how, and economic resources of the manufacturer. The product may need to be packaged; usually it will also need to be shipped so that it should be light in weight and sturdy of construction. The product should appear significant, effective, compatible with the culture, and appear to be worth more than the price. The emphasis may differ with the instance. Durability in a paper napkin is different from durability in a power shovel.

To determine whether a design is well adjusted to the gross array, criteria are needed. Some are objective and measurable, such as clearances and efficiency, whereas others are quite subtle and even subjective. In a way, product design is an industrial art.

Ultimately the purpose of product design is to ascertain that the product will satisfy human wants and wishes either directly as consumer goods, or indirectly as capital equipment or components.

Except in the case of basic inventions, product design is a redesign to suit changed conditions, criteria, or enlightenment. Change may appear capricious, as in some fashions and toys, may be the result of technological progress, or may result from a change in attitude toward the product or its function. In some areas trends can be discovered, but future preferences of buyers must be predicted, gambled on, or forced to occur.

There are various steps in product design which are not necessarily in particular order. They are analytical studies, creative synthesis, drawings and models for appearance, function and specifications, plus calculations, experiments, and tests. *See* PROCESS ENGINEERING; PRODUCTION ENGINEERING; PRODUCTION PLANNING.

[RICHARD I. FELVER]

Production engineering

The planning and control of the mechanical means of changing the shape, condition, and relationship of materials within industry toward greater effectiveness and value. Production engineering is a relatively new term applied to some aspects of planning and control of manufacturing; it is a service function to the production department.

As industry and technology evolve to greater levels of sophistication, complexity, and specialization, the broad area of figuring out what to do becomes more involved and at the same time better understood. By this process some of what had been originally performed by either the production department or the industrial engineer becomes a separate activity with its own background of knowledge, principles, and techniques.

Planning and purpose. Production engineering as a planning activity takes place between product design and the planning of the overall manufacturing process. Overall manufacturing planning is usually considered within the profession of industrial engineering. But in attitudes of greater specialization, production engineering may be considered a separate profession closely allied to industrial engineering. *See* INDUSTRIAL ENGINEERING.

The purpose of production engineering is to refine and adjust the design of the product (preferably with the product designer) to the problems involved in its proposed manufacture. Conversely it should solve certain problems, mainly mechanical, such as those involved in processing, tools, dies, and new or special equipment necessary to manufacture the product efficiently and according to the established specifications.

Position in the organization. Product design, production engineering, and industrial engineering overlap variously according to the situation, policy, and organization. The techniques of production engineering are mainly in the field of mechanical engineering, but some are closely related in concept and performance to, if not directly derived from industrial engineering. *See* PRODUCT DESIGN.

Intelligent activity in production engineering requires a comprehensive understanding of both the intention and meaning of the product design and the means and principles of industrial engineering. The production engineer often acts as liaison between product design and industrial engineering.

The product design department specifies what is wanted, usually making only a general statement of how such specifications are to be met. The particular means are the problem of the production engineer. Product design takes into consideration performance, life, safety, and other functional requirements, usually fully testing models of various types for these features. In many products, appearance and other sensory qualities have been adjusted. These too must be maintained.

Initial phase. After studying the overall product, the product engineer examines every detail of each operation for forming each piece. Various lines of inquiry are followed: Is the specific shape the most economical in material, labor, and equipment? Is it compatible with either present or obtainable equipment and know-how? Are components readily obtainable if they are to be purchased? *See* INDUSTRIAL COST CONTROL.

Usually the product designer has valid reasons for each detail, but he may not know of the benefits of alternate means; therefore, the production engineer may often make diplomatic inquiry into any details that appear difficult, expensive, or superfluous. Prudent organizations avoid the condition in which the production engineer accepts the design specifications as absolute and final. The opposite situation, where the production engineering staff can make any changes at all to suit easier processes or methods, should also be avoided. There is usually a common-sense ground in between that can be found.

The first operations in production engineering are to examine every detail in relation to its feasibility and economy with respect to the peculiar situation. The first question most likely to arise is whether to manufacture the part or purchase it. This is not always an easy question to answer. It depends upon many factors, from overall utilization of facilities and labor to company policy. For instance, many industries have expanded, by vertical diversification, into being their own suppliers. Like much of production engineering, the answer to this lies within the policy of top management and their vision of where they are going in the light of competition, economics, technology, and cultural change. Peculiarly, here production engineering must cooperate with sales, finance, and even basic research.

Development of overall production. The first work is analytical inquiry. After most of these details are answered, if only tentatively, the major job of designing the total process begins. The whole manufacturing process usually lies within the realm of industrial engineering. Production engineering deals with the mechanical aspects of manufacturing: processing equipment; tools and dies; auxiliary equipment such as fixtures, gauges, and the like; it is also concerned with specially designed equipment such as conveyors, transfer equipment, automatic assembly machines, and inspection equipment. Tool design and machine design are often more specialized functions in their own right. Likewise instrumentation and control design are other special fields. *See* PROCESS ENGINEERING.

After the analytical inquiry into the detailed means, the order of performance or operation sequence is determined. This sequential study ends in a flow diagram, which is similar or supplementary to the flow-process chart and preliminary to plant layout if changes in layout are necessary.

There is a great divergence in industry. Production engineering may require nothing more than set-up and scheduling as in automatic screw machines or wire forming machines. In other instances the creation of whole new factories with special equipment may be required.

Coordination of humans and machine. At this phase the edges of production engineering and industrial engineering diffuse with each other. Industrial engineering is concerned with methods, labor costs, and standards; but where methods become highly mechanized, dominate manual work, or are automatic, the problems become mainly ones of machine design and therefore are in the area of production engineering. From industrial engineering time data there may be indications to a mechanically oriented engineer that it would be feasible, at least economically, to search for a more mechanical means to perform certain operations. These are shifts from purely industrial engineering to an attitude toward production engineering. Many devices perform such operations as transfer, orient, differentiate, and assemble. These operations definitely pertain to industrial engineering, and the industrial engineer may indicate their use. Their installation and adjustment to the total production procedure, however, usually is a mechanical engineering problem; hence industrial engineering utilizes the mechanical bias of the production engineer.

For a long time the design, installation, use, and control of conveyors and automatic equipment have aided manufacturing, and here the mechanical production engineer and the industrial engineer have worked together. The greater the automation the more difficult it is to separate the machine designer and the production engineer from the industrial engineer in their functional position to manufacturing. It is probably safe to assign to machine design certain individual details and components and to industrial engineering the overall plan. Production engineering then falls between and touches both.

If it is found desirable to design special equipment for production, and if the equipment is partially operated manually, principles of human engineering indicate how the operator can most effectively operate it. The equipment is, in a sense, an extension of the people. The force, distance, speed, accuracy, and the understanding of the operator are the limiting dimensions and factors of the equipment. The efficiencies of people in these operations determine the efficiency of the equipment. The equipment should be designed so that the operator is most efficient. *See* HUMAN-FACTORS ENGINEERING; HUMAN-MACHINE SYSTEMS.

Introduction of automation. In newly designed equipment, especially automatic or complex machines, models are often built to find and eliminate "bugs" and to obtain data concerning time and accuracy. Often new products require new processes, which must be developed in laboratories or on prototype models. Also, unless automatic equipment is completely reliable, space should be allowed in the line for substitution of manual labor in case of a breakdown, which would be an expensive bottleneck and could close a line for days.

In industries where automation is well developed, whole lines are built as prototypes, operated to secure synchronization, to develop special skills, to study the process for refinement of elements and their effective interaction, and for optimum application. In some industries the whole production setup is one machine and must be studied and created as a vast interacting unit. The total machine has mechanical, electrical, electronic, hydraulic, and pneumatic components operating together automatically. The production characteristics of such a machine cannot be determined from the characteristics of these separate parts. These characteristics are found by operating models, often in full scale and complete in detail.

Improvement and coordination. The frontier of production engineering is in devising more and simpler automatic machines. The more mechanized a process becomes, the more it is freed from the limitations of human operators, and thus the more possible it becomes to mechanize it further. Because innovations reflect the individualistic approach of the engineers, several different processes or machines may be developed that do the same job equally well, that is, equally fast, efficiently, and economically.

Close control of the use of machines and detailed records of their performance and service can indicate where improvements will be most effective in increasing productivity. As technology increases in complexity, the activities that enter into it become more specialized, and the need for coordination and cooperation grows. To perform his special function efficiently, the production engineer needs to participate in conferences and planning sessions with product-development engineers and factory production managers.

Forces toward obsolescence, from new materials to new viewpoints, render equipment inefficient before it wears out in the physical sense. Production engineering strives to use advance technology wherever it provides an economic advantage. *See* JIG, FIXTURE, AND DIE DESIGN; PERT; PILOT PRODUCTION; PRODUCTION METHODS; QUALITY CONTROL.

[RICHARD I. FELVER]

Bibliography: E. S. Buffa, *Modern Production Management*, 2 vols., 1975; E. S. Buffa and J. C. Miller, *Production Inventory Systems: Planning and Control*, 3d ed., 1979; R. W. Schmenner, *Productions Operations Management*, 1980.

Production methods

In product engineering, the processes that are used to obtain a given product. Basically all production processes and methods can be classified as one of two types. Analytic industrial processes break down a given material into several products, as in an ore-reducing plant. Synthetic processes create one product from several different materials, as in a blast furnace or an automobile assembly plant. Processes may be a combination analytic-synthetic (wood-furniture factory) or synthetic-analytic (feed mill).

Industry frequently classifies processes into those that (1) change the shape, called forming, including cutting, molding, bending, dissolving, and machining; (2) change the chemical or internal characteristics, called treating, including mixing, blending, heat treating, and refining; (3) change the external surface, called finishing, including rinsing, coating, drying, and painting; and (4) add other pieces, called assembling, including attaching, joining, packaging, fitting, and fastening.

Most industries use a combination of at least two basic processes. In fact, some processes can be placed logically in more than one class, for example, metal plating.

Processes have also been classified into continuous or process-type operations, as in an oil refinery, and intermittent (or repetitive) or manufacturing-type operations.

Almost all production processes or methods change the form or condition of some material, or add or deduct other materials, aided by men or machinery, or both, with the end objective being a product which has greater utility by nature of its new form or characteristics than the initial material.

The end product and the start material are of fundamental importance. Together, these are termed the product-material factor; this factor is the chief influencing feature in the choice of production methods. A change in the end product or in the characteristics of the start material may cause or allow significant changes in the production processes or methods. *See* PROCESS ENGINEERING.

Design and specifications. Design is inherent in the production of any product, even if it is not formally recognized and recorded in prints, photos, or other specifications. But generally, product designs are established prior to production and are frequently the result of several years of research, experiment, and development.

Design of products generally calls for specialists of various kinds: scientific, engineering, stylist or artistic, market research or public relations, and production engineers or manufacturing planners. The more these specialists (together with the sales, purchasing, production, and financial departments) can integrate their views and ideas, the more effective will be the product's design for the overall company position.

Product designs may take several forms: formulae in chemical plants, blends in food products, and performance standards or drawings and specifications in manufacturing plants. *See* PRODUCT DESIGN.

The term production design is frequently used for that design which has been engineered for ease and economy of production. It is much more than a design which merely requires functioning of the product; it involves refinements and modification of functional design based on the processes, production equipment, and personnel planned to produce the item.

The design engineer specifies what is to be made and how well it is to be made. The specifications take the form of (1) parts or materials lists describing the elements, (2) required characteristics or dimensions, (3) drawings, photos, blueprints or models, and (4) performance of finished product or test specifications. These are aimed at so describing the product and its elements that the equipment to produce it can be readily planned, the purchased materials can be correctly obtained, and the components can be made and assembled according to how the product has been engineered.

Other specifications pertain to the manufacturing process and the methods. Process specifications are set by process engineers (as distinguished from product engineers) and cover just how processes are to be controlled. Methods instruction is generally set by methods engineers and covers how work area and machinery are to be arranged.

Product specifications usually include tolerances, because nothing can be manufactured exactly. A tolerance is a permissible variation. *See* PRODUCTION ENGINEERING; PRODUCTION PLANNING.

Standard materials and parts. While most products require different components and even the same products require many variations, ranges and preferred choices of product dimensions or other characteristics can be established. Whenever materials and parts can be graded, classified, or otherwise standardized, great savings in time and cost result for designers, purchasers, producers and users.

For example, standard electrical current, horsepower, and dimension for electric motors allow the engineer to detail his overall machine readily, the buyer to specify and buy by numbers and code, the producer to tool up for substantial quantities of standard sizes, and the user of the machine to obtain a replacement quickly should the original motor burn out.

Standards have been established for practically all materials, from lubricating oil to paper, from lumber to metallurgical specifications of sand castings. Standard gages of sheet metal, sizes of screws and nuts, and diameters of bearings are examples of standard parts taken for granted every day. These standards are set by industry associations, the government Bureau of Standards, professional societies, or leading manufacturers. *See* DESIGN STANDARDS.

Standard materials and parts and standardized components lead to interchangeable manufacturing. Each component of a product is made both to fit with its mating parts and to meet a given specification, and the specification is set so that any part so made can be interchanged with the original. *See* INSPECTION AND TESTING; QUALITY CONTROL.

Interchangeable manufacture is the underlying principle that permits mass production. Eli Whitney showed that by making muskets from interchangeable parts a greater quantity of an article can be produced, its quality improved, and its price reduced. *See* MASS PRODUCTION.

Interchangeable manufacture does not necessarily result in a standardized product. Standardized parts and components can be assembled in a variety of different combinations. Current models of a popular automobile offer so many options, for example, that the manufacturer could hardly produce in the life of the model all the combinations that are numerically possible.

Production equipment. Machinery that actually changes the shape or characteristics of the starting material is the production equipment. Although the product design and materials generally dictate what process is to be used, the availability, suitability, and cost of the equipment to execute the process definitely affects the decision. In the final analysis, the choice of production equipment to do the job is dependent on the process selected. *See* MACHINING OPERATIONS; METAL FORMING.

Production equipment includes machinery, which covers the actual mechanical devices working on the product, and equipment, which covers items such as paint booths, ovens, tanks, conveyors, pressure vessels, and others used directly in conjunction with operations performed on the materials. In addition, there is great variety in accessory, utility, or service equipment in any industrial facility.

Production equipment may be classed as general-purpose or special-purpose. The former is universally applied to many materials or parts; the latter is designed to do one specific job, usually on one particular part.

Capacity of production equipment frequently limits a plant's operations. This capacity, along with the ability to keep equipment utilized or in operation an optimum amount of time, determines its productivity.

Tools and accessories. Smaller, easily detachable pieces of machinery or equipment used directly in production or in conjunction with the production equipment are tools and accessories. Hand tools, manual or powered, jigs, fixtures, attachments, controls, hoppers, pumps, and the like, fall into this group of items. Inspection or working gages are also frequently included in the general term tools.

Just as standard parts can be assembled into a number of specific end products, so general-purpose machinery and equipment can be made to do a special job or operation by fitting it with the proper tools or accessory equipment. *See* JIG, FIXTURE, AND DIE DESIGN.

Selection of equipment. To a great extent the selection of equipment is dictated by the production method. A degree of latitude is nearly always available within a given production method.

Much equipment, especially that not directly used in forming or assembling operations, may be common to many production methods. In selecting this equipment, consideration must be given to the following factors:

1. Demand for the product: short- or long-term.
2. Permanency of the product. Is it likely to remain the same or will technological changes force major changes in its design?
3. Risk of equipment obsolescence.
4. Competitive advantages or disadvantages established by choice of equipment.
5. Integration with other available or on-hand equipment.
6. Suitability of equipment in relation to the product.
7. Effect on quality of the product.
8. Skill and availability of labor to operate equipment.
9. Cost of operating and maintaining the equipment.
10. Cost source, and availability of capital.

To resolve the first five factors requires opinions or estimates. Decisions for the second five can be obtained by evaluating facts.

A piece of equipment may be selected from standard available implements or may be built specially for the job. To decide if specially built apparatus is appropriate, additional factors must be evaluated:

1. The rate at which the product is to be made.
2. The volume or quantity of the product to be manufactured.
3. The worker-hours required.
4. The floor space available for the equipment.
5. The adaptability of standard machines.
6. The cost and depreciation charges of the special equipment compared with standard equipment.

To arrive at a sound decision in the selection of equipment, all phases of production under present conditions and anticipated future conditions must be carefully studied and analyzed by people experienced in the fields covered by the various factors.

Production equipment classed as general purpose does not require the extensive financial review that must be given to special equipment. The most difficult factor that must be determined for the financial appraisal of special equipment is that of product life as related to the time necessary to recover the capital invested in the equipment.

In the final analysis, the objective which must be realized in selecting production equipment is to bring about a fair return on the money invested in that equipment. This criterion alone is most frequently used by industrial managers to decide which equipment to select and whether to buy the equipment or put the money into other investments. *See* INDUSTRIAL COST CONTROL.

Supporting services. In addition to the machinery, equipment, and tools used for production, every factory, plant, or mill must have certain supporting services. These take various forms.

Services dealing with materials or product include (1) production control, such as planning, scheduling, machine-loading, dispatching, and recording; (2) material control, including requisitioning, receiving, storing, transporting, and inventorying; (3) quality control, which includes quality levels, inspection, complaints, and specifications release; (4) waste control, dealing with rejects, salvage, scrap, or rework; and (5) warehousing and shipping.

Services relating to machinery and equipment include (1) maintenance, both preventive and repair and overhaul; (2) tool storage and tool conditioning; (3) auxiliary or utility lines, such as water, electricity, heating and ventilating, compressed air or vacuum, lubricating or cutting oil, gas, exhaust, fuel, drains, sewage, and the like.

Services relating to personnel include (1) offices; (2) restrooms, lockers, and showers; (3) eating facilities; (4) parking lots and access ways; and (5) time clocks, drinking fountains, first-aid, bulletin boards, telephones, and so on.

Most supporting services are necessary for a modern production facility. They must be planned into the facility and integrated with the materials, machinery, workers, and building structure. Effective arrangements and organization of these supporting facilities often account for the efficiency of the production methods established. As a result, they should not be overlooked when new or revised production methods are being planned. *See* PLANT FACILITIES (INDUSTRY).

[RICHARD MUTHER]

Production planning

The function of a manufacturing enterprise responsible for the efficient planning, scheduling, and coordination of all production activities. The planning phase involves forecasting demand and translating the demand forecast into a production plan that optimizes the company's objective, which is usually to maximize profit while in some way optimizing customer satisfaction. The twin objectives of maximum profit and maximum customer satisfaction are not always synonomous since the customer may request a product which must be made under high unit costs. During the scheduling phase the production plan is translated into a detailed, usually day-by-day, schedule of products to be made. During the coordination phase actual product output is compared with scheduled product output, and this information is used to adjust production plans and production schedules. *See* INDUSTRIAL COST CONTROL.

If the production or manufacturing process is viewed as an input-output process, then the production planning function can be viewed as a control process with feedback (see illustration). The control is in the form of schedules and plans, while the feedback results from the comparison of the production reports with the production schedules.

Constraints. Production planning generally has several constraints: Demands must be met within reasonable limits; inventories of raw materials, in-process materials, and finished goods must not be excessively large or small; variations in manpower are usually limited; and the available production facilities are limited. A customer will not wait indefinitely for a product to be delivered. The length of time that is acceptable is dependent on the industry and the product. Inventories represent an investment in material that is not producing immediate revenue and therefore must be limited. Some inventories are necessary for efficient manufacturing in lot sizes and for balancing manpower and equipment requirements with customer demand. Production planning must always be cognizant of the inventory balance relationship, which states that the inventory on hand today is equal to the inventory on hand at some prior time plus the product produced in the intervening time minus the demand during the same period. It is because of this relationship that the production process can be called an input-output process. An analogy that explains the concept of inventory and the inventory-production-demand relationship is a bank account. The balance of the account is the "inventory" of funds available, the deposits are the "production," and the withdrawals are the "demands." *See* INVENTORY CONTROL; PRODUCTION ENGINEERING; PRODUCTION METHODS.

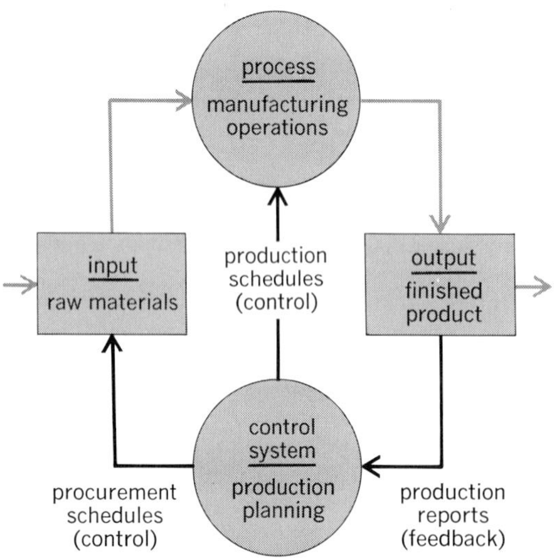

The production process as an input-output process.

Applications. The functions of production planning are dependent upon the type of industry and type of product. If the manufacturing process is of an assembly line nature, the planning is primarily of outputs and inputs and the avoidance of bottlenecks. If the manufacturing process is of the job-shop type, that is, if the product is made to the customer's order, then the planning must involve a detailed schedule of individual parts through the manufacturing process. In both cases, the planning process starts with a forecast of outputs or demands and proceeds to determine manpower and material inputs consistent with demands and company policy. Some of the readily identifiable tasks within the production planning function are planning, scheduling, dispatching, machine loading, expediting, inventory record keeping, and material requisitioning. All of these tasks involve the collection, analysis, interpretation, and dissemination of information and data.

The sophistication of the data system is highly variable from company to company and also within departments or branches of a company.

The computer has greatly changed the data-processing systems of many companies. Its full impact has probably not really been felt as yet. The range of sophistication of electronic data-processing (EDP) methods in production planning varies from plants with no EDP system to those in which all plans, schedules, and the like are made by the EDP system. The computer provides information faster, more accurately, and in greater quantity if the right information is fed into the system, if the program is correct, and if the input information is accurate. The generation of accurate input and output information requires a long development and trial period. The ultimate goal of an EDP system in production planning should be to replace all manual collection and analysis of data in the control system. Current technology could permit a completely automated production planning system. *See* DIGITAL COMPUTER.

The production planning function does not exist independently of other functions within the company. Its activities must be coordinated with purchasing, sales, finance, production, tool engineering, design engineering, maintenance, and so on. Its primary responsibility is to coordinate the manufacture of the product to best satisfy the customer and the constraints imposed by the company management. *See* OPERATIONS RESEARCH.

[JOHN E. BIEGEL]

Bibliography: E. S. Buffa and J. C. Miller, *Production Inventory Systems: Planning and Control*, 3d ed., 1979; F. G. Moore and R. Jablonski, *Production Control*, 3d ed., 1979; J. L. Riggs, *Production Systems: Planning, Analysis, and Control*, 2d ed., 1976.

Productivity

Output in relation to input. Financial measures such as profit-and-loss statements characteristically state the relation of money-valued outputs of an enterprise in relation to costs, which are money-valued inputs. However, productivity refers to measures of output and input that are cast primarily in physical units, for example, the number of units of product in relation to an input factor such as production worker labor-hours.

Measures. Measures of productivity have long been of interest to managers of industrial firms and their engineers, as well as to economists. Each group has a rather different interest in productivity measures. Engineers are typically interested in comparison of output in relation to worker-hour input, or machine-hour input. By successive measures of this kind, engineers attempt to gage the effect on output in relation to input associated with changes in production methods and production organization. The managers of enterprises characteristically have an interest in comparing output in relation to input for similar plants in a multiplant enterprise, or for evaluating the performance of a firm in relation to its industry as a whole.

The interest of economists in measures of productivity is closely associated with measuring the ability of a production system to deliver goods and services for consumption and for production. Accordingly, changes in average output per worker-hour, as in manufacturing as a whole, are an important measure of the ability of a manufacturing system in a given economy to supply the material goods desired by that population.

Productivity measures do not have universal validity. The units to be used for measuring output are characteristically a function of the problem which is being addressed. The output of a single product is of interest to the industrial engineer, whose task is to gage the change in efficiency that is traceable to the installation of a new piece of equipment or a new set of operations for a complete production system. By contrast, the economist is typically interested in output measures that require the calculation of composite representation of output change in the form of an index number.

Neither is there a "universally valid" measure of input. Physical output in relation to labor input has been the productivity measure of greatest interest for more than a century. That has been owing to the continuing importance, in cost terms, of the labor factor. It is also the case that alternative methods of production have normally been available by which the labor input could be traded off in favor of machinery inputs. However, the preferred input factor is a function of the condition of the firm and the production system. Thus the fuel input dominates in the cost of steam central stations. Accordingly, the measure of productivity that is most widely used in the electricity industry is the "heat rate"—the number of Btus required in the fuel input to generate a kilowatt-hour of electricity output. In the case of aluminum reduction the input of primary interest is electric power, since the cost of electric power dominates cost in that industry.

Measures of productivity must be tailor-made in accordance with the problem to be addressed and the nature of the production system that is involved.

Weighting factors. The problems in relation to gaging productivity are usually concentrated on the measure of output, owing to the requirement of measuring a "mix" of products that not only vary in physical characteristics at a given moment, but also vary in proportion through time. In that case it is necessary to "weight" the measure of output, thereby assigning differential importance to its components. This requirement gives rise to the index number problem. The selection of weighting factors and their variation through time are important aspects of the weighting issue. Owing to the inescapable arbitrariness of assigning weights to components of output, the index number problem has no generally valid solution. Nevertheless, the selection of weights and the mode of index number computation can be justified in terms of preferred practice in relation to the problem being addressed for productivity measure.

Measurement base and period. Other technical problems that are involved in the formulation of productivity measures, especially through time, concern the selection of a base year against which to measure relative change in output in relation to input. There is also the issue of the measurement period. Thus a measure of output in relation to input must be congruent with the length of the production cycle for the products in question. An annual measure of productivity is indicated for the output of large motors and generators and for oceangoing vessels.

Attribution problem. The interpretation of productivity measures requires careful avoidance of the attribution problem. The measure of output in relation to input offers no basis from which to infer that the change of output is uniquely caused by a concurrent change in the input factor. In industrial and other operations it is ordinarily feasible to know the cost of a particular input factor. It is never feasible to discover the "contribution" of a particular input factor to the worth of the final product. Modern industrial production is the joint product of many input factors. While their cost is known, their comparative importance for the final product is never known. Thus, while the prices or the wages or the salaries paid for industrial inputs appear to imply an evaluation of relative importance, the fact is that all that is known is the socially determined money value that is assigned to these input factors.

Growth. In the history of the United States economy the growth of productivity has been a major asset. From the middle of the 19th century to about 1965, the long-term growth of output per worker-hour in the manufacturing industries proceeded to grow at about 3% per year. This growth yielded a progressive multiplication of output in relation to input, and gave American industrial managers a powerful capability for offsetting cost increases of whatever sort, since the increase in the output per unit of labor input has typically denoted a parallel increase in the output per unit of capital input.

The growth of United States industrial productivity proceeded until 1965 under the influence of a pervasive pattern of cost minimizing among the managers of industrial and other firms. Production costs in particular were minimized by American managers, the better to maximize profits. The minimization of production cost was made possible by progressive mechanization of work, which in turn was strongly encouraged by the average tendency of machinery prices to rise less rapidly than the wages of labor. This critically important pattern was made possible by the fact that the managers and engineers of the machinery-producing industries themselves strove to minimize the cost of their operations. As they proceeded to offset the rise in the costs of their inputs, the result was that the prices of their machinery products did not parallel the rise in the wages of labor in the industrial system as a whole.

The pattern of cost minimizing that prevailed in the United States economy shifted soon after 1965 in favor of a cost-pass-along type of microeconomy. Machinery prices proceeded to rise as rapidly as or more rapidly than wages of labor from the mid-1950s on, and the classic opportunity for cost minimizing was no longer present for the host of machinery users. As old equipment was retained, the rate of growth in output per worker-hour declined; and so, from 1965 to 1970, the average annual rate of growth in output per worker-hour in United States manufacturing went to 2.1% and then to an average rate of 1.8% from 1970 to 1975. These were immediately recognized as the lowest rates of productivity growth of any Western industrialized country, as well as the lowest long-term rates of productivity growth in American industrial experience.

At the side of the shift to cost pass-along, about 20,000 firms in the United States introduced patterns of "cost maximizing and subsidy maximizing" in the course of functioning as prime contractors to the Department of Defense and NASA. The shift away from traditional cost minimizing had the further effect of diminishing incentives and opportunity for the kinds of cost reduction through mechanization that had classically yielded productivity growth in the United States.

On the average, the pattern of United States productivity growth has been a derived effect of the operation of cost-minimizing management in relation to production. For a given firm that minimizes its production expense through appropriate mechanization, the actual level of productivity achieved has been a function of the host of other factors concerning the organization of production which yielded stability or instability of operations. The optimum productivity achievable from given means of production has been attained under conditions of stable operation of the production system—output variation within predictable and acceptable limits. By contrast, lower levels of productivity with given means of production are obtained as production systems are operated in an unstable fashion. *See* INDUSTRIAL COST CONTROL; PRODUCTION PLANNING.

[SEYMOUR MELMAN]

Bibliography: A. Dogramaci and N. Adam (eds.), *Aggregate and Industry Level Productivity Analysis*, 1981; A. Dogramaci and N. Adam (eds.), *Productivity Analysis: A Range of Perspectives*, 1980; A. Dogramaci and N. Adam (eds.), *Productivity Analysis at the Organization Level*; L. J. Dumas, Productivity and the roots of stagflation, *Proceedings of the American Institute of Industrial Engineers*, May 1979; H. B. Maynard (ed.), *Industrial Engineering Handbook*, 1974; S. Melman, *Decision-Making and Productivity*, 1958; S. Melman, *Dynamic Factors in Industrial Productivity*, 1956; S. Melman, Inflation and unemployment as products of war economy, *Bull. Peace Proposals*, 9(4):359–374, 1979; U.S. Bureau of Labor Statistics, *Bull. no. 1514*, 1966, *Bull. no. 1776*, 1973, *Bull. no. 1933*, 1977, *Rep. no. 559*, 1979.

Prony brake

An absorption dynamometer that applies a friction load to the output shaft by means of wood blocks, flexible band, or other friction surface. The prony brake (see illustration) provides means for measuring the torque T, developed as in the equation below, where L is the distance shown on the drawing,

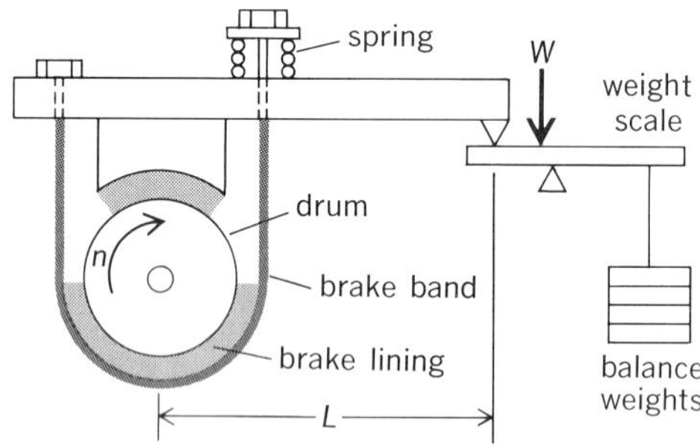

Diagram of a prony brake. *(From T. Baumeister, ed., Standard Handbook for Mechanical Engineers, 7th ed., McGraw-Hill, 1967)*

$$T = L(W - W_0)$$

W is the scale weight with the brake operating, and W_0 is the scale weight with the brake free. When the shaft speed as measured by a tachometer is n rpm, brake horsepower is $2\pi nL(W - W_0)/33{,}000$. Small prony brakes are air-cooled; the drum of larger ones may be filled with water to absorb the heat during operation. *See* DYNAMOMETER.

[FRANK H. ROCKETT]

Prospecting

The search for mineral deposits that can be worked at a profit. A prospect is an occurrence of minerals of potential value before that value has been determined by exploration and development. Mineral deposits include those containing metallic elements, such as gold, copper, lead, zinc, or iron; nonmetallic materials, such as asbestos, clay, phosphates, potash, or sulfur; and mineral fuels, such as coal or petroleum. Deposits worked for their aggregate of materials, such as sand, gravel, or dimension stone, are usually considered deposits of the rock itself. For a discussion of methods of search for petroleum deposits *see* PETROLEUM PROSPECTING.

General mineral prospecting. Much of the world has been intensively prospected for the common metals and nonmetals by simple methods so that in most countries the more obvious deposits have been found. Hope for future discovery, therefore, lies mainly in the buried deposits, which at best may give only a subtle indication of their presence at the surface.

Not only are the older field and library methods

being systematically improved, but the need to prospect for buried deposits has stimulated the use of indirect methods. In these methods geophysical, geochemical, and botanical evidence of subsurface conditions supplement surface and aerial evidence. Because of increased complexity and capital requirements, an increasing proportion of prospecting is done by groups of specialists working for mining companies or government agencies. Such a group may consist of a geologist, a geophysicist, a mining engineer, and assistants as required for traversing and sampling. Specialists in photogeology and in geophysical and geochemical prospecting may be called in as needed. Initial prospecting is followed where warranted by trenching, drilling, underground work, and further sampling. For principles of mine evaluation and field sampling *see* MINING OPERATING FACILITIES.

The individual prospector, independent or employed by mining companies, remains important, but probably less so than in the past. More and more he needs a working knowledge of geology, of the use of geologic maps, and of the mineralogical and petrological and structural relationship of mineral deposits. In addition he must be able to apply the principles of photogeology, to make mineralogical and geochemical tests in the field, to understand the use of the simpler geophysical instruments, and to make use of vegetation, rock staining, and float as guides to mineralization. Deposits of nonmetallic minerals, particularly of the bedded type, are becoming increasingly important.

Outline of methods. Mineral prospecting normally proceeds from the general to the specific from consideration of large regions to smaller favorable areas within the region, and finally to individual prospects. Following a preliminary investigation, including a library search and a study of available maps and reports, prospecting may often be concentrated on smaller areas immediately. Study of the geology will indicate what minerals might be found in a given area. Prospecting methods may be subdivided into direct and indirect methods. Direct methods include geologic and photogeologic mapping, study of guides to ore, and field examination of the surface, supplemented by panning, trenching, pitting, drilling, and sampling. Indirect methods are of two kinds: (1) Geophysical methods include magnetic, electromagnetic, spectral, radioactivity, and infrared surveys from the air; and magnetic, electromagnetic, radioactivity, induced polarization, electrical resistivity, self-potential, gravimetric, and seismic surveys on the surface. Electrical resistivity, self-potential, radioactivity, sonic, and temperature surveys are made in boreholes. (2) Geochemical and botanical surveys are the other category.

Where there is surface evidence of minerals, examination and sampling may be all that is necessary to determine if further exploration is warranted. In little-known regions, or where the deposits are deeply oxidized, leached, or buried, prospecting must be based on geologic inference, and indirect methods and exploration becomes more complex.

Exploration of offshore continental shelves for minerals as well as for petroleum is becoming increasingly widespread and intense. Methods of exploration are basically the same as those used on land. Boring techniques for taking samples from the ocean floors are being continually improved. Comprehensive offshore exploration requires specially equipped ships with sophisticated instrumentation, the high cost of which is beyond the average prospector. *See* MARINE MINING; OIL AND GAS, OFFSHORE.

New tools for mineral exploration are evolving with advances in space age technology. Earth orbital photography, particularly in color, from satellites is the best medium for data collection. This is being complemented by a wide range of remote sensors under continuing development.

DIRECT METHODS

A map of some sort is required to prospect large areas systematically. Government, state, or other published geologic and topographical maps, if available, are suitable bases for plotting mineral occurrence or guides to minerals. Aerial photographs are excellent for such purposes, though ground control is necessary to produce maps from them. Photogeologic studies combined with ground checking may indicate such guides as soil staining, alteration, or structures in deeply weathered areas that might otherwise be missed. Recent work has shown that in some areas aerial observation and color photographs may be useful. Glacial debris and deep weathering mask deposits.

Ore guides. These are mostly associations of geologic and other regional factors. Recognition of ore guides is highly important, though not all guides are valid everywhere. The ore guides that are significant depend on the characteristic rock associations and distribution of the ore in any given region or locality. For example, mineral stream gravel or placer deposits generally indicate veins or lodes in the country rock of the backland.

Among regional ore guides of general rather than specific application are large igneous intrusions with which ore is known to be associated. Examples are the Sierra Nevada batholith of California and the Coast Range batholith of British Columbia. Copper, tungsten, tin, and molybdenum are characteristically associated with granitic rocks; lead and zinc with limestones and dolomites as in the Mississippi Valley; nickel with mafic rocks, especially norite; chromium, nickel, and platinum with ultrabasic igneous rocks; beryllium with pegmatites; and uranium, vanadium, and selenium with terrestrial and shallow-water beds of sandstone and shale. Bedded deposits such as salt, potash, phosphates, and some iron and copper ores are related to local sedimentation.

Faulting and rock weathering offer clues to mineralization. Major zones of faulting may be valid regional guides, although ore deposits are more commonly found in subsidiary faults related to major fault zones. The Mother Lode of California, a mineralized fault and shear zone extending 120 mi along the western side of the Sierra Nevada batholith, is the best-known example in the United States. Bauxite, manganese, barite, and lateritic iron and nickel ores are found in deeply weathered rocks. The porphyry copper deposits of the southwestern United States and western South America occur where arid climate and deep water level have favored oxidation of the original material and preservation of zones of enrichment. Deep weathering and subsequent active erosion favor concentration of gold, platinum, ilmenite, cassiterite, columbite, tantalite, diamonds, zircon, monazite, and some rare-earth minerals in placers.

Ore deposits also are characteristically associated with so-called metallogenetic geological epochs, when conditions favored introduction, concentration, and deposition of minerals. Ore minerals deposited during Precambrian metallogenetic epochs are important in eastern North America, especially in the Canadian Shield; in the shields of Brazil, Scandinavia, Finland, and southern Siberia; and in ancient rocks in Africa; and in other parts of the world. Minerals deposited during Cretaceous and Tertiary epochs are important in western North America, including most of Mexico and Alaska and in western South America.

Local ore guides include rock alteration and channel ways such as fractures, faults, contacts between dissimilar rocks, and breccia pipes. Ore shoots are likely to occur where mineralizing solutions encounter easily replaced rock such as limestone, where veins cross contacts, at intersections of veins and faults, at rolls in faults, and along veins wherever there is a change in physical or chemical environment.

Any departure from normal structure, topography, rock color, or vegetation should be investigated. Old mine workings, dumps, prospect pits, and burrows of animals may yield information on mineralization and on the size and trend of deposits.

The gangue minerals in veins almost invariably extend far beyond the limits of ore and thus constitute important guides. Alteration halos, in materials adjoining veins, are also important, though alteration may be so widespread that it serves merely as a regional guide. Pyrite, white mica, and clay are common and easily recognized alteration minerals, but microscopic study may be necessary before others can be identified. Rock staining may result from weathering of sulfides of iron, manganese, copper, cobalt, and nickel and oxides of uranium and vanadium. Each leaves a characteristic color: red-brown to pale yellow from iron; black from manganese; green and blue from copper; pink to red from cobalt; apple green from nickel; bright yellow, orange, or green from uranium; and yellow from vanadium.

Gossan (mainly hydrated iron oxides) may be the surface residue of an ore deposit from which sulfides of copper, lead, or zinc have been leached. Base-metal sulfides commonly leave cellular pseudomorphs when leached from the primary ore, so that the structure of the gossan may tell much about the original mineral content. Native gold, silver, copper, and bismuth and, locally, oxides of copper, lead, zinc, and manganese tend to remain in gossan. The gossan content of these metals may, in fact, be higher than the underlying primary deposits.

The topography of a region may furnish useful clues, though no general statement can be made concerning the relation between mineral deposits and topography. Outcrops of ore, gossan, or even sparsely mineralized gangue are of course directly useful. Geological maps (called photogeological) can be made from aerial photographs of the surface aided by some ground control as the geology is often closely related to the topography.

Placers are formed in stream channels by weathering, erosion, and concentration of resistant heavy metals and minerals, such as gold, platinum, cassiterite, rutile, and diamonds and other gem minerals. An understanding of the processes involved will therefore aid in recognizing ancient channels, which may bear no relation to the present drainage. In areas where rock outcrops are scarce, assemblages and shapes of minerals in stream channels are indicative of the nature of the parent rocks and of the possible occurrence of economic minerals.

Float (detached fragments of mineralized rock or vein material) indicates a bedrock source at some higher point which may be found by systematic search. Finer material in residual overburden or in alluvium may be traced to its source by panning. Most sulfide minerals are destroyed rapidly by weathering, so their source is likely to be close if they are found in float or by panning. The more resistant materials such as vein quartz, gold, cassiterite, chromite, columbite, tantalite, rutile, zircon, and diamonds and other precious stones may be carried far from their sources.

Ore minerals are genetically associated with minerals that are of no economic importance but may be relatively abundant or resistant and thus indicate the presence of the valuable minerals. As examples, chromite and titaniferous magnetite are associated with metals of the platinum group; pyrite, pyrrhotite, chalcopyrite, and vein quartz with gold; and ilmenite, magnetite, chromite, and pyrope (deep red garnet, often a gemstone) with diamonds. Similarly, fine gold in residual soil may indicate a bedrock source of copper farther uphill, as copper sulfides decompose readily, leaving free gold in the residue. Float tracing and panning work best in unglaciated country.

Testing and sampling. At the point where the mineral indications go beneath the surface, trenches or test pits must be dug or bulldozed. Ground sluicing or hydraulicking may be used to advantage where water is abundant, slopes are moderately steep, and stream pollution is unobjectionable. Drilling is sometimes necessary in prospecting areas covered by swamp, water, or rock or where deep overburden makes trenching or pitting unduly expensive. Drilling is useful in searching for bedded deposits, buried placers, and extensions of known ore bodies and for locating faults or faulted segments of ore bodies. Drilling is not well suited to prospecting for narrow veins, friable ores, or small, irregular ore bodies. *See* BORING AND DRILLING (MINERAL).

Prospects are sampled to determine their composition. Methods for representative sampling depend on the size, character, and accessibility of the project. Exposed prospects are mostly sampled by chip, grab, or channel samples. Buried deposits are commonly sampled by trenching or pitting or by drilling methods. Placers are sampled by pan or rocker.

INDIRECT METHODS

In a large proportion of indirect prospecting geophysical methods are used, involving the measurement of physical quantities associated with buried mineral deposits and geological structures. The measurements are interpreted in terms of geology. Plausible assumptions are made about the subsurface, and the physical effects of assumed structures or deposits are computed or estimated

Fig. 1. Airborne magnetometer survey over rough terrain. (*USGS*)

and compared with the geophysical measurements. The assumptions are modified until there is reasonable agreement between the computations and observed data.

Selection of a suitable geophysical method is facilitated by information on the geologic habits and environment of the deposits sought and on their physical properties compared with the surrounding rocks. Magnetic methods and the various electrical methods are commonly used in prospecting for metallic minerals, because these minerals usually have magnetic and electrical properties that contrast with those of the surrounding rocks. Seismic and gravitational methods have been less used because measurable contrasts in elasticity and density are less common, but they have been indirectly useful in locating buried structures. Little information on depth is theoretically obtainable with magnetic, gravitational, self-potential, and thermal methods, where physical effects are produced by the bodies or structures themselves. Information on depth can usually be obtained with resistivity or seismic methods, where effects are produced by transmitting electrical or seismic energy through the ground.

In the past geophysics has been used less in prospecting for metallic and nonmetallic minerals than for petroleum because of the small size of most ore bodies compared with oil structures and the great variety in their shape, physical properties, and geologic environment. It is now being used increasingly in mineral prospecting, because of the increased emphasis on the search for buried deposits. The development of airborne magnetic, electromagnetic, spectral, and radioactivity methods makes it possible to cover large and otherwise inaccessible areas at comparatively low cost. All three types of surveys may be made simultaneously from the same aircraft, at little more than the cost of a survey by a single method.

Magnetic methods. Airborne and ground magnetic surveys have been used extensively to prospect for magnetic deposits of iron (magnetite), nickel (nickeliferous pyrrhotite), and titanium (titaniferous magnetite) and for nonmagnetic deposits which either contain magnetic gangue minerals or are associated with structures or bodies that have magnetic expression (Fig. 1). Contact metamorphic and replacement deposits commonly contain magnetic gangue minerals and may thus be indicated by magnetic surveys. Faults, dikes, contacts of igneous intrusions, lava beds, and magnetite-bearing sedimentary and metamorphosed rocks, which may control the occurrence of ore, in many cases also produce measurable magnetic anomalies.

In aeromagnetic surveys, flying elevation and spacing of flight lines are governed by terrain and by the geologic habits and associations of the deposits sought. Flights may vary from 1/8 mi apart when prospecting for discontinuous magnetic deposits to 1 mi or more in tracing a continuous magnetic formation that may serve as a guide to ore. In general a small, shallow target requires flying as low as a few hundred feet above the surface. Mineral surveys are made at low levels on closely spaced lines (quarter-mile or less) in order to obtain the maximum resolution for interpretation of the magnetic field. Petroleum surveys, where igneous rocks at great depth are the source of the magnetic field variations, are flown at barometric level and on widely spaced lines.

Ground magnetic surveys are used to locate and detail small areas often chosen on the basis of aeromagnetic surveys. Surveying procedures are determined by the amplitude and extent of the anomalies sought. Dip needle and magnetic declination surveys are adequate for delineating high-amplitude anomalies, such as those associated with near-surface magnetic iron deposits. Surveys with more sensitive and costly equipment, such as the vertical magnetic field balance or the more recently developed nuclear spin magnetometer, are generally required to outline anomalies associated with weakly magnetic deposits, deeply buried deposits of high magnetism, igneous rocks, and geologic structures.

The applicability of magnetic methods may usually be determined by a knowledge of the habits or arrangement of the deposits sought and by determinations of the magnetization of the pertinent rocks and minerals. Hand specimens of strongly magnetic material will deflect a surveyor's compass or dip needle. The magnetization of more weakly magnetic material can be estimated by its effect on a sensitive magnetometer or by crushing specimens and separating the magnetic minerals through the use of a hand magnet.

Electrical methods. Electromagnetic and electrical resistivity surveys are used to locate deposits of metallic sulfides which, except for sphalerite, are good electrical conductors. They may also be used in prospecting for nonmetallic deposits, which are generally poor conductors, provided there is structural or stratigraphic control and a measurable contrast in conductivity. The applications are diverse, as each type of deposit presents its own problem.

A variety of instruments has been used in electromagnetic prospecting, but all measure anomalies in the electromagnetic field set up by induced ground currents rather than in the potential distribution as in resistivity methods. The frequency of the energizing current should be 250–5000 Hz. Too low frequencies reduce the strength of the induced field, but too high frequencies lack depth penetration.

Another technique uses natural, random, alternating magnetic fields of audio and subaudio frequency to locate subsurface electrical conductors, such as sulfide ore bodies. Distortions of the fields caused by variations in conductivity are measured at several frequencies by use of search coil detectors. The method may be used for airborne or ground surveys. Although a high degree of amplification is required, theoretically greater lateral and vertical range is possible than with most electromagnetic methods.

Airborne and ground electromagnetic surveys have been useful mainly in areas where unaltered sulfide deposits are fairly close to the surface, as in the Precambrian Canadian Shield. Other conductors, such as fault gouge, graphite, and some water-bearing formations, may likewise be indicated by electromagnetic surveys. Airborne electromagnetic surveys are normally flown at low levels.

Electrical resistivity surveys require a large field crew and are relatively expensive. Depth to flat-lying deposits or beds may often be determined fairly accurately, using specialized mathematical methods of interpretation.

Electrical and electromagnetic surveys have been used successfully in boreholes to gain information on nearby conducting ore bodies. Electrical surveys using electrodes in boreholes and on the surface, and between two or more boreholes, have also been successful in locating conducting ore bodies. The principles are identical with those governing surface surveys.

Induced polarization (IP) is used in areas of disseminated metallic lustered minerals. A discontinuous pulse is injected into the ground, and the received pulse is distorted by slow current drainage due to disseminated minerals.

A spontaneous polarization or so-called self-potential as large as a volt or so is set up when sulfide ore bodies are oxidized by downward-percolating groundwater. The pattern of currents thus induced may be inexpensively measured by traverses with a potentiometer connected to two nonpolarizable electrodes. The accuracy of measurements is reduced by variations in soil type and soil solutions, roots of trees, and topography, which may set up potentials larger than those associated with some ore bodies. Local spurious potentials can be eliminated largely by repeating the observations with the electrodes in different positions.

Self-potential surveys are useless where the deposits are deeper than 100 ft or where they are below the zone of oxidation. In the latter case good results have been obtained when the water table is lowered during long, dry periods and the sulfide minerals are subject to renewed oxidation.

Gravity, seismic, and thermal methods. Most mineral deposits are too small, or have insufficient density contrast, to permit use of gravity methods. Gravity methods have been highly successful in locating buried salt domes, with their associated sulfur deposits on the Gulf Coast, and massive concentrations of iron, manganese, and chromite sulfides. They have also been used to a minor extent in prospecting for relatively dense deposits, such as magnetite, chromite, and barite.

Seismic surveys are used extensively to locate salt domes and have been used to a limited extent to prospect for buried channels favorable for gold placers or for deposits of uranium. Refraction seismic surveys are being used extensively to determine the thickness of overburden in areas of deep weathering rather than the slower and more expensive drilling methods. Such surveys are also carried out on building sites, dam sites, open pit mines, and highway rights of way for determination of depth to bedrock and bearing characteristics of rock. Costs of shallow seismic surveys are comparable with those of electrical resistivity surveys. *See* SEISMIC EXPLORATION FOR OIL AND GAS.

Oxidation of sulfide ore bodies, or radioactivity associated with extremely rich and large uranium deposits, produces heat which may raise the temperature of the surrounding rocks. Air temperature measurements from boreholes would therefore seem to offer a useful prospecting method. In practice they have been applied in only a few instances, because thermal effects resulting from groundwater circulation or conductivity differences in the rocks are usually larger than those associated with ore bodies.

Infrared surveys are used in airborne surveys to locate hot spots due to mineralization and cool or cold spots due to underground waters. Infrared wavelengths to which air is transparent are used. The surface is scanned with the infrared detector, and an image of the surface is later reconstituted on a photograph by buildup of individual image points from the infrared scanning (similar to television). Abnormal light or dark image points due to variations of surface temperature can be distinguished from those normal for the area.

Spectral photography is another new aid in airborne surveys. Narrow-band-pass filters and spectral film are used to accentuate minor color effects caused by mineralization and alteration which would be undetectable by broad-band normal photography.

Radioactivity methods. Airborne and ground radioactivity methods, which measure the γ-radiation of radioactive elements, are used predominantly to prospect for uranium and thorium (Fig. 2). They may also be used to prospect for marine phosphate deposits, which contain small amounts of uranium, and for thorium- and uranium-bearing placers and beach sands, some of which are mined mainly for titanium minerals. Radioactivity logging of boreholes has aided in prospecting for uranium deposits in sandstone by detecting weakly radioactive halos that sometimes occur around the ore and for potash deposits in salt beds and in correlating stratigraphic horizons.

Airborne and ground prospecting is effective only in areas where the radioactive material is exposed at the surface, as the γ-radiation is effectively masked by a foot or so of barren overburden or a few feet of snow. Airborne surveys must be flown within a few hundred feet of the surface because the air similarly absorbs γ-rays.

Airborne radioactivity surveys could be used to aid in geologic mapping of areas covered by residual soil, which usually produces radiation patterns characteristic of the underlying parent rocks. In this connection they would be of indirect value in prospecting for nonradioactive minerals. There is also evidence that uranium and other metals tend to deposit during the last stages of the crystallization of large granitic intrusions. This suggests that radioactivity highs over the roofs of large intrusions would be an indirect guide to nonradioactive mineral deposits.

GEOCHEMICAL AND BOTANICAL METHODS

Geochemical prospecting involves the analysis of elements in the soils, rocks, surface and underground waters, organisms, and vegetation for the purpose of defining areas, where the anomalous distribution of the elements indicates the presence of ore. Botanical prospecting involves analysis of elements found in deep-rooted plants or plants commonly associated with certain elements or mineral deposits and comparative study of morphological differences in plants growing in mineralized and nonmineralized areas.

Fig. 2. Car-mounted radioactivity logging equipment. (*C. M. Bunker, USGS*)

A variety of methods are used singly or in combinations. Analysis of residual soils for traces of metals is the most widely used geochemical method, as the dispersion pattern corresponds closely with the primary dispersion of metals in the underlying rocks. Analyses of metals in vegetation and of the distribution of indicator plants are used in areas where there is little residual soil, but the relationship between the metal contents of plants and soil is seldom simple. Analysis of water in streams is used for preliminary examinations of large areas. Springs usually give information on areas of limited size.

The choice of elements for analysis depends on mineral associations, relative mobility of the elements, and the availability of suitable analytical methods. Zinc and copper are useful indicators not only for zinc and copper deposits but for other types of metallic deposits. The rare elements in pegmatites and the tin-tungsten-columbium association in high-temperature veins are also useful. Associated noneconomic minerals may serve as a guide where the ore metals do not form a recognizable dispersion pattern. Immobile metals usually form clear-cut dispersion patterns in residual soils, whereas mobile metals are more widely dispersed by groundwater and vegetation. Thus it may be effective to test for the more mobile elements in reconnaissance surveys and for the less mobile elements in detailed surveys.

Kits for use in the field are available commercially for the detection of most elements, including portable radioisotope x-ray fluorescence analyzers for field analyses.

Certain bacteria and other organisms concentrate elements such as sulfur, selenium, boron, cobalt, copper, uranium, and manganese. Prospecting for these elements may eventually be aided by additional knowledge of the processes involved and environments required for their concentration.

In some areas the type of vegetation is related to mineralization or bedded deposits.

[A. F. BANFIELD]

Bibliography: M. B. Dobrin, *Introduction to Geophysical Prospecting*, 3d ed., 1976; M. Kusvart and M. Bohmer, *Prospecting and Exploration for Mineral Deposits*, 1978; C. A. Lamey, *Metallic and Industrial Mineral Deposits*, 1966; D. S. Parasnis, *Mining Geophysics*, 2d ed., 1973; R. M. Pearl, *Handbook for Prospectors*, 5th ed., 1973; Society of Mining Engineers, *Industrial Minerals and Rocks*, 4th ed., 1975.

Pulley

A wheel with a flat, crowned, or grooved rim (Fig. 1) used with a flat belt, V-belt, or a rope to transmit motion and energy. Pulleys for use with V-belt and rope drives have grooved surfaces and are usually called sheaves. A combination of ropes, pulleys, and pulley blocks arranged to gain a mechanical advantage, as for hoisting a load, is referred to as block and tackle. *See* BELT DRIVE; BLOCK AND TACKLE.

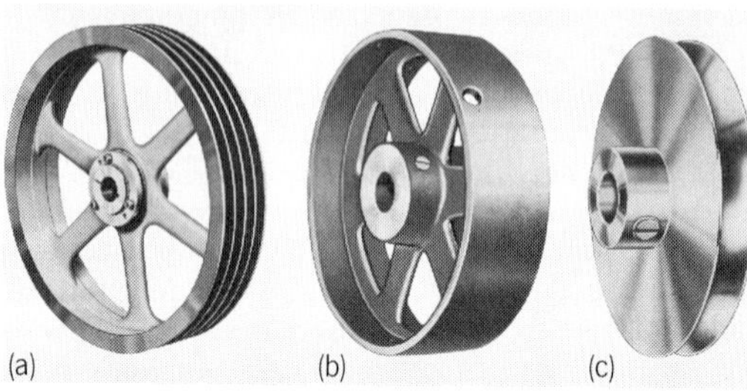

Fig. 1. Some typical examples of pulleys. (*a*) Multi-groove V-belt sheave. (*b*) Solid-hub crown face pulley. (*c*) Brass pulley for round belt (*Boston Gear, INCOM International, Inc.*)

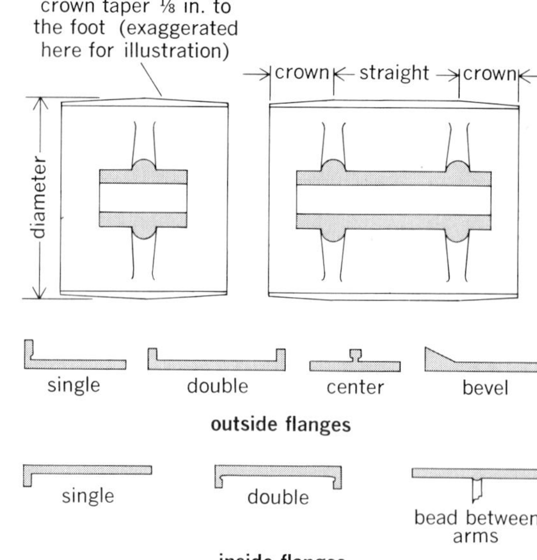

Fig. 2. Diagrams of crowning and flanging on flat-belt pulleys. ⅛ in. to the foot ≅ 1 cm to the m. (*From D. C. Greenwood, Mechanical Power Transmission, McGraw-Hill, 1962*)

Flat-belt pulleys. Pulleys for flat belts are made of cast iron, fabricated steel, wood, and paper. A particular pulley design must be based on such considerations as the ability to resist shock, to conduct heat, and to resist corrosive environments. The face must be smooth enough to minimize belt wear; yet there must be adequate friction between belt and pulley face to carry the load.

Cast-iron pulleys are made in sizes from 3 to 72 in. (7.6 to 183 cm) and are particularly well adapted for service where dampness or acid fumes are present, in contrast to wood or steel pulleys. Because of their greater weight they are useful wherever a flywheel effect is wanted. Available in a wide variety of hub and spoke designs, they can also be furnished in either flat, crowned, or flanged rims. Pulleys are crowned (Fig. 2) to keep the belt on the pulley. The amount of crowning seldom exceeds a taper of 1/8 in./ft (1 cm/m); excessive crown is harmful to the belt and may cause the belt to be thrown off the pulley as quickly as if there were no crown at all. Outside flanging (Fig. 2) helps hold the belt on the pulley; inside flanging stiffens the rim.

The arms on cast-iron pulleys are typically elliptical in cross section to achieve both a pleasing appearance and low windage loss, even though an H-section arm can provide equal strength for less material. Commercial cast-iron pulleys are balanced for rim speeds of 3500–4000 ft/min (18–20 m/s). For higher speeds the pulleys have to be specially balanced. Stock pulleys are made to withstand belt tensions of up to 65 lb/in. (115 N/cm) of width, though they can be furnished for tensions up to 210 lb/in. (370 N/cm).

The majority of pulleys used in flat-belt drives are steel split pulleys. They are low in cost, light in weight, and easily available, and are made with interchangeable bushings which permit their use on any shaft diameter up to the bore size of the pulley itself. Standard sizes range from 3-in. (7.6-cm) diameter by 2-in. (5.1-cm) width up to 144-in. (366-cm) diameter by 36-in. (91-cm) width.

Where lightness is important and moisture is not present, wood pulleys can be used. The resilience of wood makes these pulleys able to withstand shock loads very well. Layers of soft wood are glued together and then compressed under high pressure to increase the density. A ribbed cast-iron hub is forced into an undersize hole in the wood block to complete the construction. Belt speeds of up to 6000 ft/min (30 m/s) are common.

Paper pulleys are made of impregnated paper board and are used particularly for motor pulleys. The recommended rim speed is 5000 ft/min (25 m/s).

If a belt drive must be able to produce several output speeds from a single input speed, stepped pulleys (Fig. 3) can be used. It is good practice to select the several output speeds so that they form a geometric progression. For a crossed-belt drive, the sum of the pulley diameters must be the same for each step. In practice a pair of stepped pulleys is often designed so that they have the same dimensions (Fig. 3). As shown in the equation below,

$$N = (n_3 n_7)^{1/2}$$

the driving shaft speed N is then the geometric mean of the output speeds of the two steps adjacent to the middle one.

PULLEY

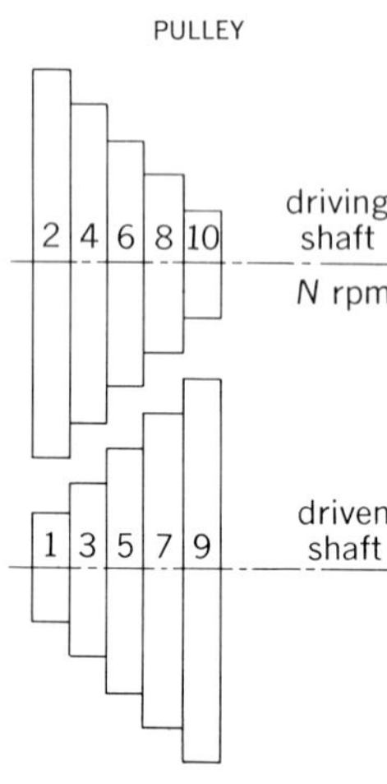

Fig. 3. Equal-stepped pulleys.

V-belt pulleys. The two common types of V-belt pulley are pressed-steel and cast-iron. The pressed-steel pulleys are suitable for single-belt drives. For multiple-belt drives, or in single-belt drives where pulley mass should be high to get a flywheel effect, cast-iron pulleys are used (Fig. 1*a*). Generally the pressed-steel pulleys have noninterchangeable hubs, but the cast-iron pulleys have interchangeable hubs. Interchangeability of hubs reduces the number of pulleys that need be kept in stock: A single pulley can be made to fit a range of shaft diameters. Small pulleys are designed with a groove angle of 30° to increase the wedging action of the belt in the groove. For pulley diameters of 5.4–8 in. (13.7–20 cm) a 34° angle is recommended. Over 8 in. (20 cm) a 38° angle is used.

Variable-pitch pulleys (Fig. 4) are made for drives where it is necessary to vary the speed ratio of the drive continuously over a range. Made for capacities up to 5 hp (3.7 kW), they can accommodate one to six belts.

Rope sheaves. Rope drives are much less efficient than V-belt drives, and so V-belt drives have virtually replaced rope drives. But in cranes and power shovels wire-rope drives are still in use. If a relatively soft material such as gray cast iron is used for the sheave, the wire rope will leave its imprint in the groove. A cast-steel sheave will stand up satisfactorily under the rope pressures experienced in typical installations, and heat-treated alloy steel, hardened to about 350 Brinell, will stand very high loads.

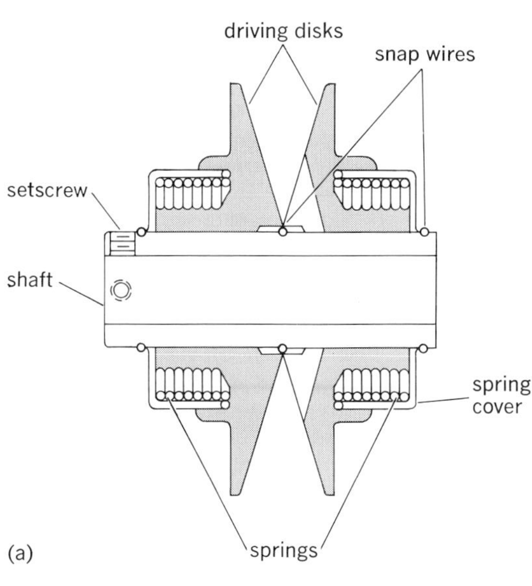

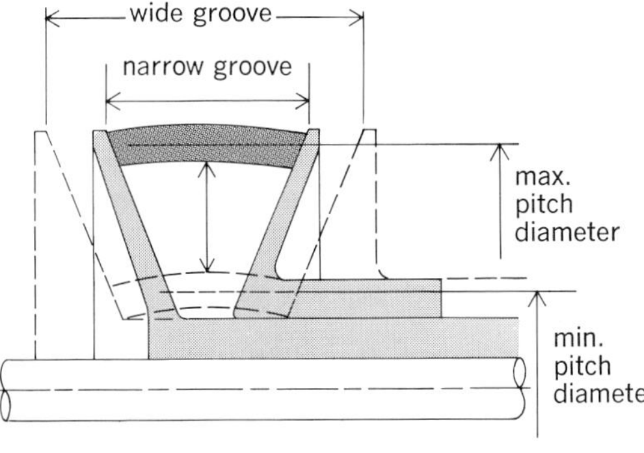

Fig. 4. Typical variable-pitch V-belt pulley (*a*) for standard-section V-belts and (*b*) for wide V-belts. Pulley can accommodate up to six belts. (*Lovejoy, Inc.*)

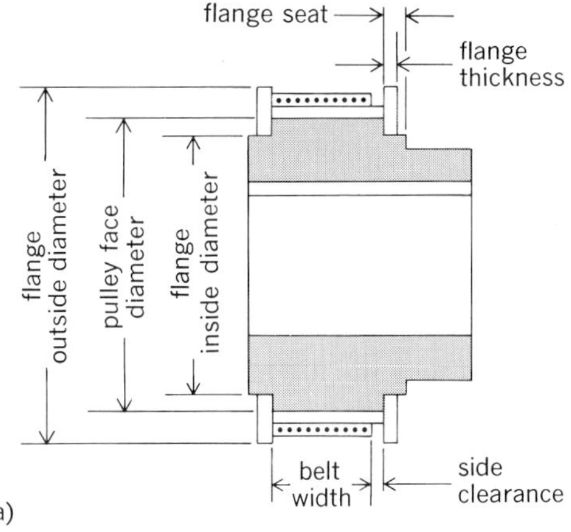

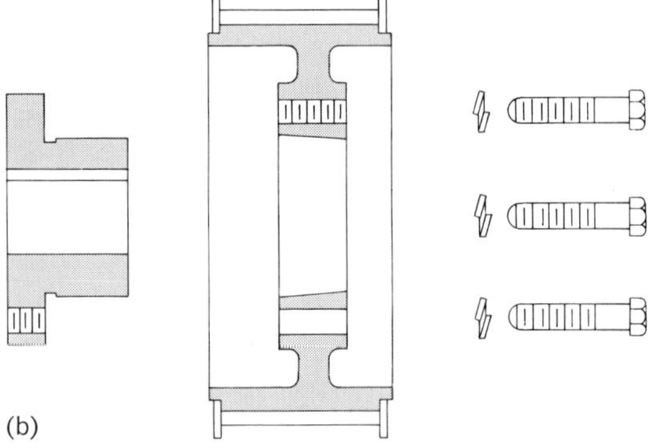

Fig. 5. Timing-belt pulley. (*a*) External plate-type flanges are used to retain belt. (*b*) Standard mounting arrangement permits various pulleys to be used on same shaft diameter. (*T. B. Wood's Sons Co.*)

Timing-belt pulleys. A timing-belt pulley (Fig. 5) is rather like an uncrowned flat-belt pulley, except that the grooves for the belt's teeth are cut in the pulley's face parallel to the axis, using a milling cutter, a gear shaper, or a hob. Most of the materials used in the manufacture of gears can be used as well in making timing-belt pulleys. Steel and cast iron are the most common materials, particularly for heavy loads. But for light loads aluminum alloys, brass, plastics, and zinc or aluminum die castings are used. External flanges are used to keep the belt on the pulley.

[JOHN R. ZIMMERMAN]

Bibliography: W. Kent, *Mechanical Engineer's Handbook*, pt. 1: *Design and Production*, 12th ed., 1950; E. Oberg et al., *Machinery's Handbook*, 20th ed., 1975; W. Patton, *Mechanical Power Transmission*, 1979; Trade and Technical Press Editors, *Handbook of Mechanical Power Drives*, 1977.

Pump

A machine that draws a fluid into itself through an entrance port and forces the fluid out through an exhaust port (see illustration). A pump may serve to move liquid, as in a cross-country pipeline; to lift liquid, as from a well or to the top of a tall building; or to put fluid under pressure, as in a hydraulic brake. These applications depend pre-

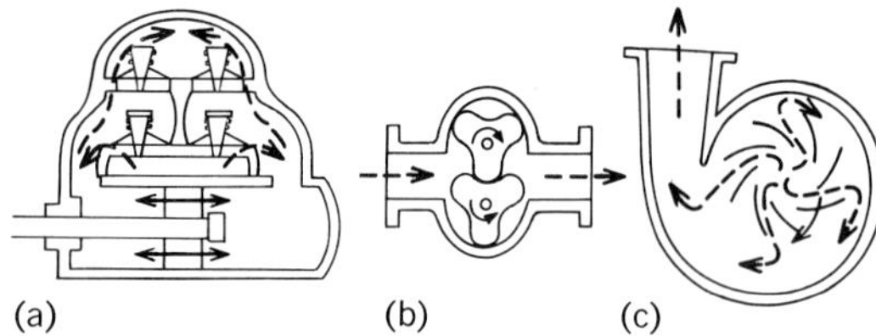

Pumps. (*a*) Reciprocating. (*b*) Rotary. (*c*) Centrifugal.

dominantly upon the discharge characteristic of the pump. A pump may also serve to empty a container, as in a vacuum pump or a sump pump, in which case the application depends primarily on its intake characteristic. *See* CENTRIFUGAL PUMP; COMPRESSOR; DISPLACEMENT PUMP; FAN; FUEL PUMP; PUMPING MACHINERY; VACUUM PUMP. [ELLIOTT F. WRIGHT]

Bibliography: Institute for Power System, *Pump Users Handbook*, 2d ed., 1979; Institute for Power System, *Pumping Manual*, 5th ed., 1979; T. G. Hicks, *Pump Operation and Maintenance*, 1958; A. J. Stepanoff, *Centrifugal and Axial Flow Pumps: Theory, Design and Application*, 2d ed., 1957; H. L. Stewart, *Pumps*, 3d ed., 1978; R. H. Warring, *Pumps Selection, Systems and Applications*, 1979.

Pumping machinery

Devices which convey fluids, chiefly liquids, from a lower to a higher elevation or from a region of lower pressure to one of higher pressure. Pumping machinery may be broadly classified as mechanical or as electromagnetic.

Mechanical pumps. In mechanical pumps the fluid is conveyed by direct contact with a moving part of the pumping machinery. The two basic types are (1) velocity machines, centrifugal or turbine pumps, which impart energy to the fluid primarily by increasing its velocity, then converting part of this energy into pressure or head, and (2) displacement machines with plungers, pistons, cams, or other confining forms which act directly on the fluid, forcing it to flow against a higher pressure. A pump deep in a well may raise water or oil to the surface. At a ground level location a pump may deliver fluid to a nearby elevated reservoir or, through long pipe lines, to a location at similar or different elevation. In a power plant, pumps circulate cooling water or oil at low pressures and transfer water from heaters at moderate pressure to steam generators at pressures of several thousand pounds per square inch. In chemical plants and refineries pumps transfer a great variety of fluids or charge them into reactors at higher pressure. In hydraulic systems, pumps supply energy to a moving stream of oil or water, which is readily controlled, to move a piston or press platen or to rotate a shaft as required by the specific process. *See* CENTRIFUGAL PUMP; DISPLACEMENT PUMP.

Electromagnetic pumps. Where direct contact between the fluid and the pumping machinery is undesirable, as in atomic energy power plants for circulating liquid metals used as reactor coolants or as solvents for reactor fuels, electromagnetic pumps are used. There are no moving parts in these pumps; no shaft seals are required. The liquid metal passing through the pump becomes, in effect, the rotor circuit of an electric motor. The two basic types are conduction and induction.

The conduction type which can be used with either direct or alternating current, confines the liquid metal in a narrow passage between the field magnets. Electrodes on each side of this channel apply current through the liquid metal at right angles to the magnetic field and the direction of flow. This current path is like the flow of current in the armature winding of a motor.

The induction or traveling field type operates only on polyphase alternating current. The liquid metal is confined in a thin rectangular or annular passage thermally insulated from the slotted stator. This stator with its windings is similar to the stator of a squirrel-cage motor cut through on one side and rolled out flat. The traveling field induces currents in the liquid metal similar to the currents in a motor armature. *See* ELECTROMAGNETIC PUMPS. [ELLIOTT F. WRIGHT]

Bibliography: *See* PUMP.

Pyrometallurgy

Processes employing chemical reactions at elevated temperatures for the extraction of metals from ores and concentrates. The use of heat to cause reduction of copper ores by charcoal dates from before 3000 B.C. The techniques of pyrometallurgy have been gradually perfected as knowledge of chemistry has grown and as sources of controlled heating and materials of construction for use at high temperature have become available. Pyrometallurgy is the principal means of metal production.

The advantages of high temperature for metallurgical processing are several: Chemical reaction rates are rapid, reaction equilibria change so that processes impossible at low temperature become spontaneous at higher temperature, and production of the metal as a liquid or a gas facilitates physical separation of metal from residue.

The processes of pyrometallurgy may be divided into preparation processes which convert the raw material to a form suitable for further processing (for example, roasting to convert sulfides to oxides), reduction processes which reduce metallic compounds to metal (the blast furnace which reduces iron oxide to pig iron), and refining processes which remove impurities from crude metal (fractional distillation to remove iron, lead, and cadmium from crude zinc).

The complete production scheme, from ore to refined metal, may employ pyrometallurgical processes (steel, lead, tin, zinc), or only the primary extraction processes may be pyrometallurgical, with other methods used for refining (copper, nickel). In some cases (uranium, tungsten, molybdenum), isolated pyrometallurgical processes are used in a treatment scheme which is predominately nonpyrometallurgical. *See* METALLURGY; PYROMETALLURGY, NONFERROUS.

[HERBERT H. KELLOGG]

Pyrometallurgy, nonferrous

The extraction of nonferrous metals from ores and concentrates based on chemical reactions carried out at high temperatures. Pyrometallurgy for many years was the only method available for the recovery of metals from their ores, and dates back to as early as 3000 B.C. It was not until the late 19th century that hydrometallurgy and electrometallurgy were introduced. Although hydrometallurgy has come to play an increasingly important role in the production of nonferrous metals, numerous improvements in nonferrous pyrometallurgy since the mid-1960s have enabled it to retain its prominence in spite of the much more stringent regulations regarding sulfur dioxide emissions and rapidly rising energy costs. *See* ELECTROMETALLURGY; HYDROMETALLURGY.

Although the basic chemistry of pyrometallurgical processes has remained the same, the practice of pyrometallurgy has undergone major improvements due to a better understanding of the high-temperature chemistry, the introduction of improved refractories and other materials of construction, the availability and use of low-cost oxygen, the development of flash smelting and continuous smelting processes, the application of control devices, including computer process control, and the reduction in gaseous emissions to the atmosphere. The processes of nonferrous pyrometallurgy which have come into prominence in more recent years include fluid-bed roasting, flash smelting, and continuous smelting of copper concentrates using oxygen; shaft furnace melting of copper cathodes, and continuous casting of copper rod, cakes, and billets; and the zinc-lead blast furnace process.

The advantages of metal extraction at high temperatures are related not only to the favorable reaction kinetics but also to the nature of the equilibria which are realized for these reactions at the operating temperature. Thus the conversion of white metal, Cu_2S, to blister copper according to reaction (1) is favored at 1200°C, whereas at 700°C

$$Cu_2S + 2Cu_2O \underset{700°C}{\overset{1200°C}{\rightleftharpoons}} 6Cu + SO_2 \qquad (1)$$

the equilibrium is shifted to the left and conversion is no longer possible (at SO_2 partial pressures close to 1 atm or 100 kPa). The importance of higher temperatures is also evidenced in the reduction of zinc oxide by carbon at atmospheric pressure. The equilibrium in this reaction is such that the reduction of zinc oxide is impossible at temperatures below 900°C, but occurs spontaneously at higher temperatures.

The separation of the waste material as molten slag from the metal values, which are in the form of matte or liquid metal, is favored by the appropriate elevated temperature. The gangue minerals appear in unreduced form in the slag, usually as oxides or silicates (SiO_2, Al_2O_3, CaO, Fe_2O_3, Fe_3O_4).

The high temperature favors the reduction of the metal species to the elemental form and its recovery as the liquid or, in some cases, as the condensed vapor. The reducing agents employed in pyrometallurgy, such as carbon, carbon monoxide, and sulfur (from the metal sulfide), are relatively inexpensive, with the sulfur from the corresponding metal sulfide entailing no cost due to its presence in the concentrate.

An important advantage of pyrometallurgy is the high throughput of metal per unit volume of equipment. For example, in copper smelting, the average concentration (matte plus slag) is approximately 1000 g of copper per liter, whereas in copper hydrometallurgy the concentration of copper in aqueous solution seldom exceeds 50 g of copper per liter, and is often much less than this value.

PYROMETALLURGICAL PROCESSES

The processes in pyrometallurgy can be classified under three headings: preparatory, reduction, and refining processes. These different pyrometallurgical processes are illustrated in Fig. 1 for the production of nickel, copper, and other metals.

Preparatory processes. In preparatory processes the ore or concentrate is converted by chemical reaction to a form suitable for further processing. The most common of these processes are roasting, sintering, and chlorination.

In some cases the preparatory process is carried out to provide a material which is amenable to treatment by hydrometallurgical processing, for example, the roasting of zinc concentrates to produce zinc calcine, which is leached with sulfuric acid solution, and the thermal pretreatment of copper concentrate to convert chalcopyrite, $CuFeS_2$, to bornite, Cu_5FeS_4, prior to acid leaching.

Roasting of sulfides. The roasting of metal sulfides in air or oxygen converts the sulfides to the corresponding metal oxides, sulfates, or a mixture of oxide and sulfate with the production of sulfur dioxide. The objectives in roasting can be to produce an oxide which is subsequently acid-leached to give a solution from which the metal values can be recovered; to provide a partial desulfurization of a concentrate to facilitate the smelting operation, that is, to ensure that the sulfur content is such as to produce a matte of the desired composition; and to produce sulfur dioxide and subsequently sulfuric acid to limit the amount of sulfur which is released to the atmosphere as SO_2. The production of sulfuric acid may not be economically viable. Nevertheless, its production may be necessary in order to meet environmental regulations. The sulfur dioxide gas could be liquefied or even converted to elemental sulfur in a Claus reactor, although the latter process may be too costly to justify its incorporation in the metallurgical flow sheet.

The principal reaction in roasting is shown in reaction (2), where *(s)* and *(g)* indicate the solid and

$$MS(s) + \tfrac{3}{2}O_2(g) \rightarrow MO(s) + SO_2(g) \qquad (2)$$

gas phases, respectively. For all of the common metal sulfides (MS), the reaction is highly exothermic and proceeds spontaneously in the direction of metal oxide (MO) formation. The roasting temperature is usually in the range 650 – 1000°C.

The roaster gas usually contains some sulfur trioxide due to reaction (3). Here the equilibrium

$$SO_2 + \tfrac{1}{2}O_2 \rightarrow SO_3 \qquad (3)$$

shifts to the left as the temperature is raised, so that at 1000°C the roaster gas contains but a little SO_3. The formation of sulfates during roasting is favored by the use of a lower temperature, resulting in a corresponding high concentration of SO_3 in the off-gas.

Three main types of roasting furnaces are employed in metallurgical operations: multiple-hearth, fluid-bed, and flash roasters. The roaster feed is finely divided material of a size −10 mesh or

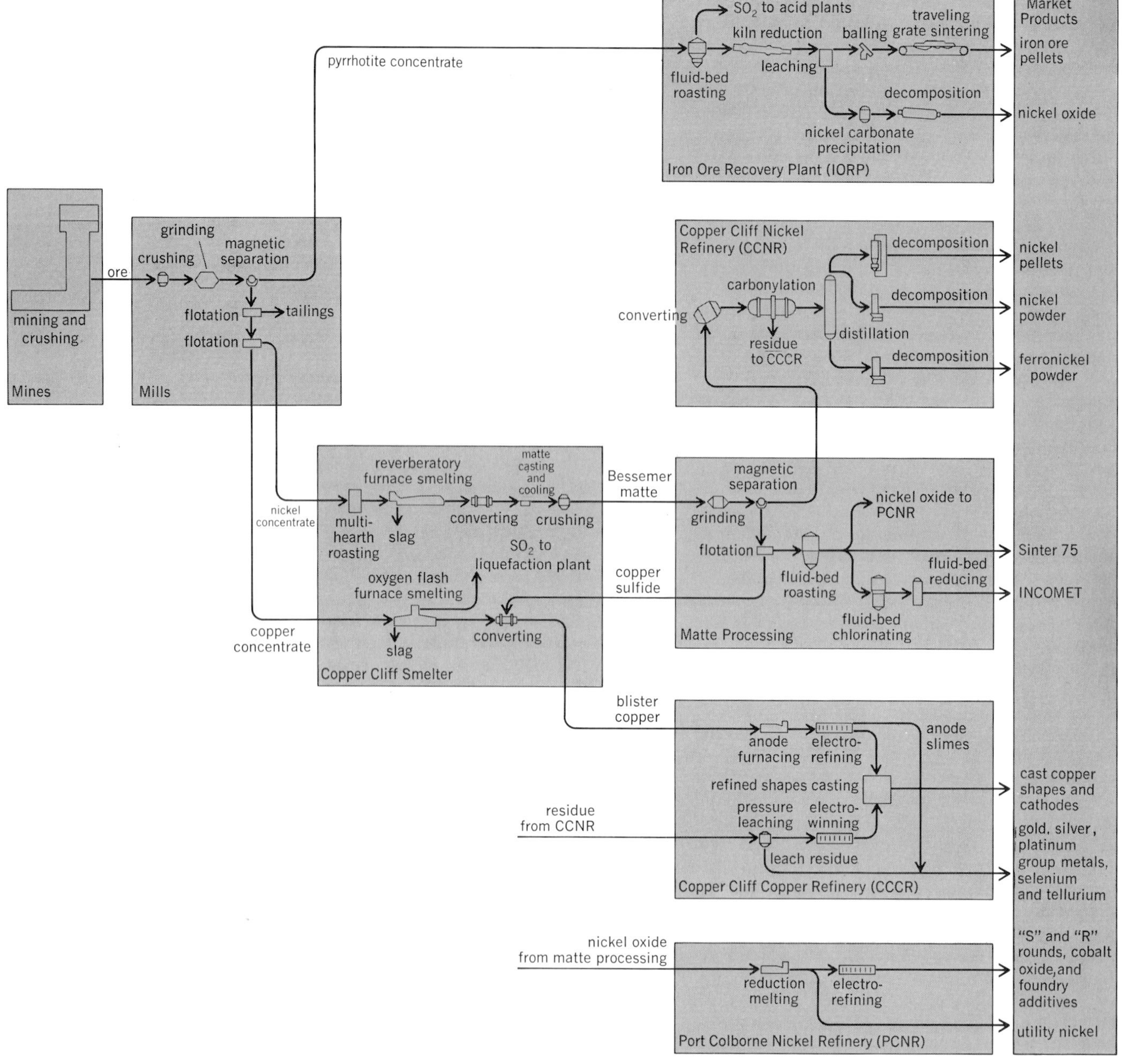

Fig. 1. Flow sheet showing different pyrometallurgical processes associated with the production of nickel, copper, and various by-products from nickel-copper sulfide ores. (*INCO Metals Co.*)

finer, as is the calcine product. Consequently the roaster gases carry considerable dust, which is collected, usually by cyclones, and returned to the primary calcine.

The efficiency of roasting has been improved greatly by the fluid-bed roaster. Although the fluid-bed reactor dates back to a U.S. patent granted to C. E. Robinson in 1879, its first commercial application was in Canada in 1946 for the roasting of arseniferous gold ore concentrates. Since that time the fluid-bed reactor has found numerous applications for roasting and for other solid-gas reactions.

In the operation of a fluid-bed roaster, air is pumped upward through the perforated hearth at a rate sufficient to expand the bed of solid particles and cause a mixing action known as fluidization. Only very fine particles are carried out by the gas stream, since the gas velocity is relatively low above the bed. Sulfide concentrates can be fed as a slurry directly into a hot roaster. Calcine leaves by overflow from a pipe above the fluid bed. The circulation within the bed results in a temperature and gas composition which are highly homogeneous, with the requisite heat being supplied by the strongly exothermic reaction. The fluid-bed reactor has a number of distinct advantages over other types of roasters, including the close control which can be exercised over temperature, gas composition, and product quality. Unlike the multiple-hearth roaster, the fluid-bed roaster is full of the final product and not a mixture of raw feed plus intermediate and final products. The fluid-bed reactor is basically a single-stage reactor in which the reactants in the bed are usually at or near equilibrium. This means that in the roasting of sulfide concentrates in such a reactor, the oxygen efficiency is basically 100%. It should be noted that the fluid-bed roaster is unsuitable for very fine feed material (less than −100 mesh) or for feed in which the particles tend to agglomerate on heating.

In the multiple-hearth roaster the roasting is carried out in a number of stages, with the material passing downward through the roaster from one revolving refractory hearth to another until the roasting is completed. In the flash roaster, the pulverized concentrate is mixed with a stream of air and injected into a hot combustion chamber.

Sintering. In certain pyrometallurgical operations it is preferable to convert finely divided solids into relatively dense aggregates. This material is a more suitable feed for the blast furnace, as well as for other reduction processes. The sintering operation is commonly carried out on a sintering machine of the Dwight-Lloyd type (Fig. 2). The feed, which contains a certain amount of solid fuel, such as sulfide minerals or fine coal, is spread continuously as a 4- to 8-inch (10–20-cm) deep bed on the surface of a modified pan conveyor. Air passes through the bed when the pans pass over a suction box located below the conveyor. The bed passes under a fuel-fired ignition furnace, whereupon the bed material is ignited. The ignition zone passes gradually down through the bed, and in the process the temperature is such as to bring about a partial fusion or sintering of the products. *See* SINTERING.

Chlorination. Although chlorination reactions have been studied for some time as a means of extracting metals from ores and concentrates, and for the removal of impurities, such reactions are used to a limited extent in the processing of the more common nonferrous metals, for example, fluid-bed chlorination of roasted nickel concentrate to reduce the copper content of the nickel oxide (Fig. 1). In the production of refractory metals such as titanium and zirconium, the direct reduction of the oxides of these metals is difficult. On the other hand, the corresponding chlorides are reduced readily by either metallic sodium or magnesium.

The chlorination of titanium is carried out at 800–900°C in the presence of a reducing agent, such as carbon, as in reaction (4).

$$TiO_2(s) + 2C(s) + 2Cl_2(g) \rightarrow TiCl_4(g) + 2CO(g) \quad (4)$$

When the feed material is present as fine particles, the chlorination can be effected by using a fluid bed. An alternative process involves the sintering or briquetting of the metal oxide and carbon, followed by chlorination in a shaft furnace. When the metal chloride is a liquid rather than a gas, the shaft furnace is employed, with the liquid product being tapped from the bottom of the furnace.

Drying and calcination. The removal of water (dehydration) and the decomposition of carbonates to the corresponding oxides are common pyrometallurgical operations which include the decomposition of aluminum hydroxide to alumina for the electrolytic production of aluminum, the calcination of limestone ($CaCO_3$) to CaO. Also of importance is the calcination of dolomite [$CaMg(CO_3)_2$] to CaO·MgO, which material constitutes the feed to the Pidgeon process for the production of magnesium. The rotary kiln is widely used in drying and calcination. Other furnaces include the fluid-bed, the shaft kiln, and the multiple-hearth. Since the reactions are endothermic, heat must be supplied to maintain the required temperature. Alumina calcination using a fluid flash calciner has been found to provide improved fuel utilization, lower operating and maintenance costs, improved purity and uniformity of product, and a lower capital cost per annual metric ton of capacity. *See* FURNACE CONSTRUCTION.

REDUCTION PROCESSES

These processes effect the reduction of a metal compound to the metal and its separation from the residue, and are represented by reaction (5), where

$$MX + Y \rightarrow M + YX \quad (5)$$

MX is the metal compound, Y the reducing agent,

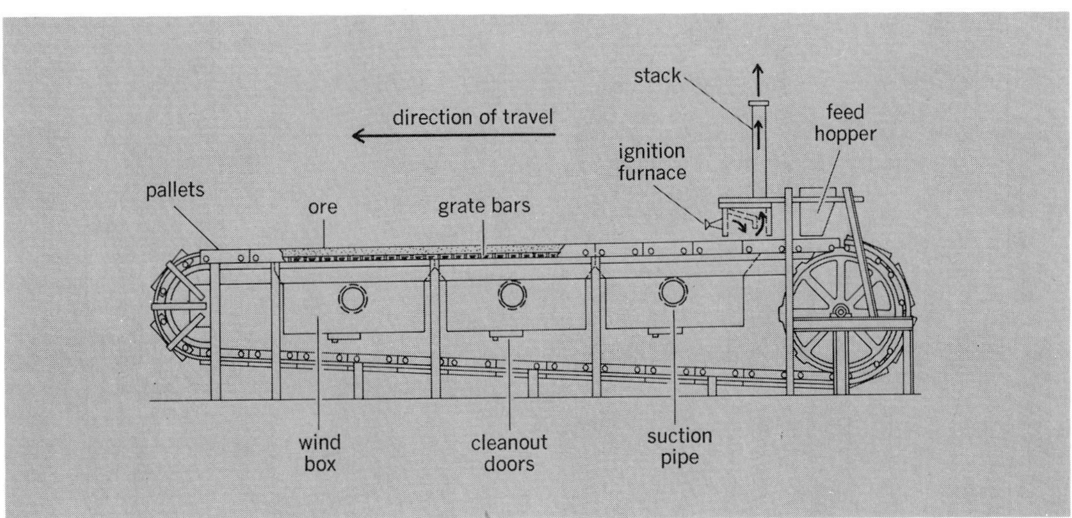

Fig. 2. A schematic diagram of a Dwight-Lloyd (downdraft) sintering machine, a common sintering device. (*From R. E. Kirk and D. E. Othmer eds., Encyclopedia of Chemical Technology, Interscience, 1947–1956*)

Table 1. Reducing agents in nonferrous pyrometallurgy

Reducing agent	Reducing strength for*			Application to production of
	Oxides	Sulfides	Halides	
Carbon (coke)	St	N	N	Zn, Ni, Co, Sn
CO (from coke)	M	N	N	Pb, Cu, Sn
S (from MS)	W	N	N	Cu, Pb, Hg
H_2 (from natural gas)	M	W	M	Mo, W
Fe (scrap)	M	M	M	Cu, Pb, Sb
Si (ferrosilicon)	St	M	M	Mg
Al	VSt	M	St	Ca
Na	St	St	VSt	Ti, Zr
Mg	VSt	St	VSt	Ti, Zr, Hf, Be, U

*VSt, very strong; St, strong; M, moderately strong; W, weak; N, little or none.

and M the metal. The reducing agent, reaction conditions (for example, temperature and pressure), and concentrations of reactants and products are selected so as to achieve a spontaneous reduction reaction.

The reducing agent should be inexpensive relative to the value of the metal produced, and the product YX should be readily separable from the metal. Reducing agents commonly used in nonferrous pyrometallurgy are shown in Table 1, together with the types of reduction for which they are suitable.

Reduction processes can be denoted by the particular physical form of the resulting metal—gas, liquid, or solid.

Reduction to gaseous metal. This particular process is restricted to the readily volatile metals. Of such metals, Zn, Hg, Mg, and Ca can be obtained in industrial practice by this method. The main advantage of the method is found in the simplicity of the reaction and the completeness of the separation of the product from the residue. The gaseous metal can be condensed as a liquid or as a solid in a condenser well removed from the reactants and residue.

The carbothermic reduction of zinc oxide can be carried out at 1200–1300°C (boiling point of Zn, 906°C) in retorts in the absence of air, as in reaction (6). The long-practiced small-batch process

$$ZnO(s) + C(s) \rightarrow Zn(g) + CO(g) \quad (6)$$

involving the use of horizontal retorts (the so-called Belgian retort) has fallen into disfavor on both environmental and economic grounds. Considerable process inefficiency as well as environmental problems are associated with the manual servicing of the numerous small retorts (50–100 pounds or 23–46 kg of zinc per day per retort) heated by a single fuel-fired furnace. Also the difficulty of excluding air gives rise to inefficient condensation of the zinc product.

Continuous vertical retorts have been developed which are definitely superior to the Belgian retort in terms of both their metallurgical efficiency and labor requirements. Their capacity is of the order of 25 tons or 23 metric tons of zinc per day per retort.

The silicothermic process developed just prior to World War II for the production of magnesium involves the reaction of calcined dolomite with ferrosilicon in a vacuum retort. The gaseous magnesium, which is produced according to reaction (7),

$$2(CaO\cdot MgO)(s) + Fe_xSi(s) = 2Mg(g) + (CaO)_2\cdot SiO_2(s) + xFe(s) \quad (7)$$

is captured as solid magnesium in the condenser in the retort. The condenser is a removable steel sleeve that is inserted in cool end of the retort which protrudes from the furnace. This process, although capable of producing a high-quality product, suffers the disadvantages of being a labor- and energy-intensive batch operation. However, the charging and discharging of numerous small retorts (10-inches or 25 cm inside diameter and 12-feet or 3.7-m length) have been improved greatly. Thus the spent residue at a temperature above 1200°C can be removed completely and transported efficiently by using a vacuum pneumatic system. Also, the manual ramming of the 350-pound (159-kg) briquette charge into the retort has been eliminated.

In the French Magnétherm process, dolomite is reduced by ferrosilicon in an electric furnace in the presence of alumina. The alumina serves to

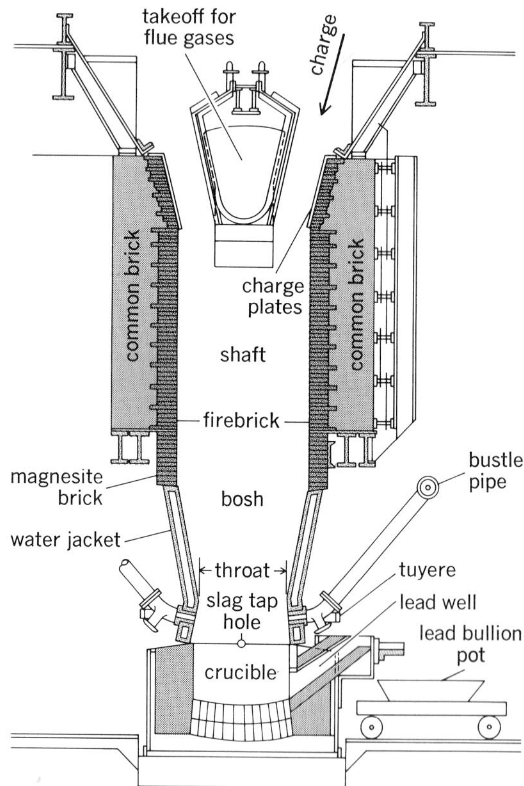

Fig. 3. Cross section of lead blast furnace. (*From R. E. Kirk and D. E. Othmer, eds., Encyclopedia of Chemical Technology, Interscience, 1947–1956*)

produce a calcium aluminum silicate slag which is liquid at the operating temperature of 1500°C. In the absence of alumina, the reaction product would be $(CaO)_2 \cdot SiO_2$ or Ca_2SiO_4, having a melting point of 2100°C. The reduction may be written as shown in reaction (8).

$$2(CaO \cdot MgO)(s) + Fe_xSi(s) + nAl_2O_3(s) \rightarrow 2Mg(g)(CaO)_2 \cdot SiO_2 \cdot nAl_2O_3(l) + {}_xFe(s) \quad (8)$$

Reduction to liquid metal. This process, which is one of the most common procedures for metal reduction, is applied in the case of nonvolatile metals of moderate melting point, such as copper, nickel, cobalt, lead, bismuth, tin, and silver. In the reduction process the practice is to add substances (fluxes) to the furnace charge so that a second liquid phase or slag phase is formed which contains the residue. The pure metal, which now constitutes the other liquid phase, can be separated readily from the impurities. Furnaces commonly used for production of liquid metal include the blast furnace, electric-arc furnace, and the converter.

In the operation of the lead blast furnace (Fig. 3) the charge, consisting principally of roasted and sintered lead concentrates, fluxes, and coke, is added intermittently at the top of the shaft. Air is blown into the furnace through water-cooled nozzles (tuyeres) situated along each side of the furnace near the bottom. Upon combustion the coke provides the requisite heat to the furnace as well as carbon monoxide, which acts as a reducing agent. The reaction is shown in (9), where ΔH is

$$C(s) + \tfrac{1}{2}O_2(g) \rightarrow CO(g) \qquad \Delta H = -26{,}400 \text{ cal/mole} = -110{,}500 \text{ J/mole} \quad (9)$$

the heat of reaction. Upon rising through the charge, the hot carbon monoxide gas reduces the lead oxide to molten lead, reaction (10), which

$$PbO(s) + CO(g) \rightarrow Pb(l) + CO_2(g) \quad (10)$$

passes downward to the crucible at the bottom of the furnace. The slag, containing principally FeO, CaO, and SiO_2, is liquefied when it reaches the tuyere zone of the furnace, where the temperature is highest. The liquid slag, being less dense than the molten lead (3.5–4.0 g/cm^3 in contrast to 11 g/cm^3), appears on the surface of the lead and is tapped periodically through a hole in the furnace wall. The molten lead is removed via the lead well as shown in Fig. 3. Blast furnaces are used in the separation of zinc and lead, and in the production of tin. However, the copper blast furnace has been replaced by the more modern, more efficient flash smelting furnaces.

The reverberatory smelting furnace, long used in the processing of copper concentrates to matte or impure Cu_2S, is now being replaced by flash furnaces such as the Outokumpu furnace (Fig. 4) or by continuous smelting furnaces such as the Noranda reactor (Fig. 5) and the Mitsubishi smelting system (Fig. 6). The efficiency and productivity of these newer furnaces are enhanced by the use of oxygen or air enriched with oxygen. The Outokumpu flash furnace produces a copper matte which is processed to copper metal in a cylindrical horizontal furnace known as a converter. The Noranda reactor can be employed to produce a high-grade matte containing around 70% copper, which

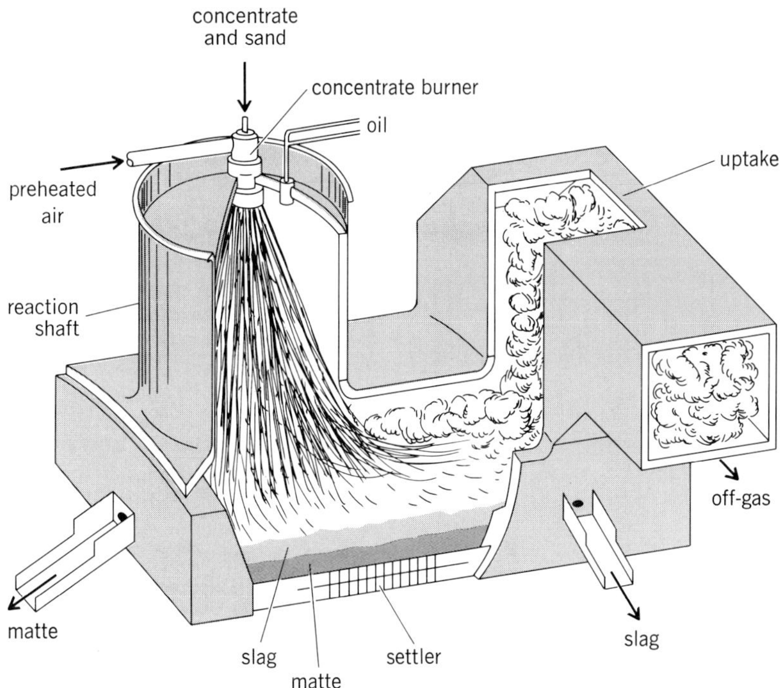

Fig. 4. Cutaway view of Outokumpu flash smelting furnace. (*From A. K. Biswas and W. C. Davenport, The Extractive Metallurgy of Copper, 1976*)

is then transferred to a converter for the production of copper metal. Alternatively the Noranda reactor can be used to yield copper directly. The selected mode of operation depends on the composition of the original concentrate, and in both cases the copper product is melted, fire-refined, and cast into anodes for electrolytic refining. The slag which is produced in the Outokumpu and Noranda furnaces contains significant amounts of copper which can be recovered by the slow cooling of the slag, crushing, grinding, and flotation. The Mitsubishi system incorporates a furnace in which the copper is recovered by liquid-liquid extraction in a slag cleaning step.

Reduction to solid metal. This process is used when the melting point of the metal is very high (W, m.p. 3400°C; Mo, m.p. 2620°C) or when there are no refractories which can be used to contain the liquid metal (for example, Ti, Zr). Compared with other reduction processes, it is expensive, cumbersome, and inefficient, as demonstrated by the Kroll process for producing titanium metal.

All known refractories react with liquid titanium to yield an impure product. Consequently the metal must be obtained as a solid by reduction at lower temperatures. In the Kroll process, magnesium ingots are charged to a steel reaction vessel, which is sealed with a gas-tight cover. The vessel is flushed with argon gas to remove gases such as N_2, O_2, and CO_2, all of which react with titanium, and then heated to about 750°C in a furnace. Following the melting of the magnesium, pure liquid $TiCl_4$ is introduced through a port in the cover. The reduction proceeds according to reaction (11).

$$TiCl_4(g) + 2Mg(l) \rightarrow Ti(s) + 2MgCl_2(l) \quad (11)$$

Upon completion of the reaction, the bulk of the liquid $MgCl_2$ is removed via a tap hole in the bottom of the reactor. The vessel is then cooled to

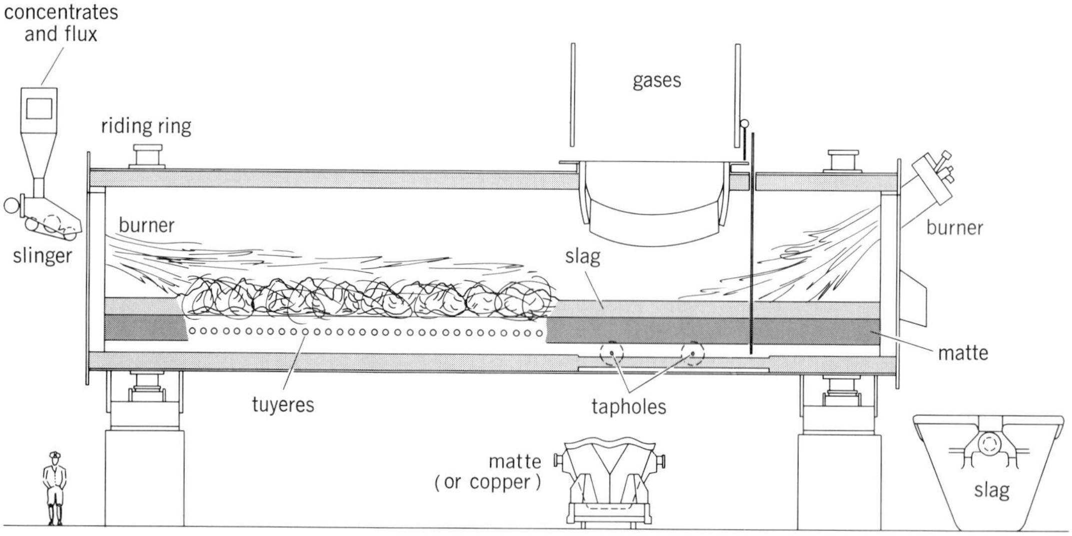

Fig. 5. Noranda process reactor. Copper is present in the reactor only during copper production. *(From J. B. W. Bailey and A. C. Storey, Noranda Mines Limited, Quebec, paper presented at CIM Conference of Metallurgists, Sudbury, Ontario, August 1979)*

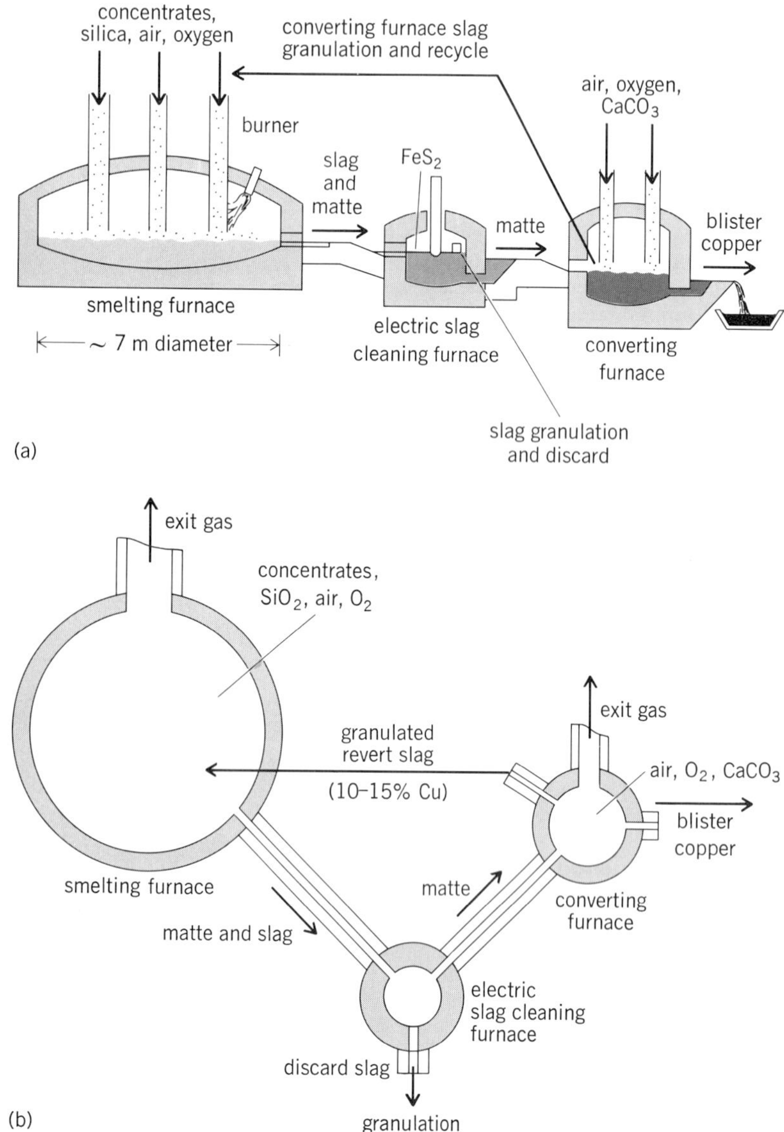

Fig. 6. Mitsubishi continuous smelting system: (*a*) elevation and (*b*) schematic views. *(From A. K. Biswas and W. G. Davenport, The Extractive Metallurgy of Copper, 1976)*

room temperature. The solidified mass, comprising Ti, $MgCl_2$, and any residual Mg, is removed by a boring machine. The $MgCl_2$ and Mg are separated from the titanium by vacuum distillation or by an acid-leaching procedure. The batch size for this process ranges between 700 and 1400 kg.

REFINING PROCESSES

In these processes, impurities are removed to yield a metal product which meets product specifications. The processes can be classified according to the type of separation employed: volatilization (separation of metal or metal compound as a gas from a liquid or solid); drossing and precipitation (separation of metal or impurities as a solid from a liquid melt); and slag refining (separation of metal or impurities by their extraction from one liquid into a second immiscible liquid phase).

Volatilization processes. The metal being refined or the impurities may be separated by volatilization or distillation. Thus zinc is separated from iron and lead impurities by fractional distillation. Crude lead containing zinc as its principal impurity can be refined by a vacuum distillation process in which the zinc is volatilized.

Slag-fuming processes are used to recover metals in metallic form or as volatile compounds. In the case of tin slag fuming, the volatile SnO_2 is returned to the smelting furnace following gas cleaning.

The production of pure nickel via the formation and decomposition of nickel carbonyl, $Ni(CO)_4$, reaction (12), dates back to 1889, when the process

$$Ni(s) + 4CO(g) \underset{232^\circ C}{\overset{49^\circ C}{\rightleftharpoons}} Ni(CO)_4(g) \qquad (12)$$

was discovered by Carl Langer and Ludwig Mond. Nickel carbonyl (b.p. 42°C) is readily separated from iron and cobalt, which carbonylate slowly to yield carbonyls of much lower volatility than $Ni(CO)_4$. Nickel carbonyl is also decomposed by an elevated pressure process which operates at 121°C and 300 psig (2.07 MPa gage pressure).

Highly purified nonvolatile metals such as titanium, zirconium, hafnium, and silicon can be prepared by the formation and decomposition of the corresponding iodides (van Arkel–de Boer process). For titanium the reaction proceeds as shown in reaction (13).

$$Ti(s) + 2I_2(g) \underset{1300°C}{\overset{175°C}{\rightleftharpoons}} TiI_4(g) \qquad (13)$$

Pure titanium is obtained by contacting the TiI_4 vapor with an electrically heated wire at 1300–1400°C. In the process, impurities are retained in the solid residue remaining after the formation of the volatile iodide.

Drossing. In liquid-metal melts the temperature dependence of the solubility of impurities is used to precipitate impurities. The precipitated impurities appear in solid form, and can be removed by skimming (drossing) the liquid surface if the solid is less dense than the liquid. Examples of drossing are the removal of iron from tin and of copper from lead.

Similar to drossing is refining by the addition of a precipitating agent to the liquid metal. A particularly noteworthy application of this type of refining is the Parkes process, in which zinc is added to molten lead to recover small amounts of silver and gold by precipitation of intermetallic compounds of zinc with the precious metals.

Slag refining. This process can be considered as a liquid-liquid extraction in which there is a mass transfer between two immiscible phases. The two phases might be liquid metal or matte in contact with liquid slag. Thus impurities present in a metal can be oxidized by blowing the liquid melt, whereupon the oxidized impurities are taken into the slag phase. The process is particularly applicable to the removal of impurities which are more readily oxidized than the metal itself. In the fire refining of copper, blowing the charge of liquid copper with air permits the slagging of impurities such as iron, lead, zinc, tin, and antimony. More noble elements, such as gold and silver, remain with the copper.

In copper smelting, a significant part of the copper content of the slags produced during the smelting and converting of copper concentrates is present as dissolved copper sulfide, with the remainder as globules of matte. The slag can be cleaned of entrained matte and dissolved copper sulfide by a liquid-liquid extraction process. In this process the bulk of the slag is contacted with a large quantity of molten iron pyrite or molten matte having a copper content lower than the equilibrium value corresponding to the metal content of the slag being cleaned. Upon vigorous agitation of the two phases, a substantial fraction of the copper in the slag is transported to the pyrite or matte phase. The extracted metal is then recovered from the matte by conventional means. *See* METALLURGY.

Energy requirements in smelting. The increasing cost of fuels used in nonferrous smelting has stimulated the adoption of new and improved smelting technology, especially the use of oxygen. The enrichment of the combustion air with oxygen can reduce the fuel consumption, increase the furnace temperature and throughput, reduce the N_2 in the gases, and increase the SO_2 content of the off-gas in the processing of sulfide concentrates, thereby facilitating the capture of SO_2 and the production of sulfuric acid.

Table 2. Energy used by various copper smelting processes

Furnace	Type of operation	Hydrocarbon fuel energy, kJ/metric ton of charge	Electrical energy, kJ/metric ton of charge
Flash			
Inco	Oxygen flash	4×10^5	3×10^5
Outokumpu	Autogenous, O_2 enriched	4×10^5	4×10^5
Outokumpu	Hot air blast	23×10^5	1×10^5
Electric	Dry concentrate	4×10^5	14×10^5
	Hot calcine		9×10^5
Noranda	31% O_2, blister copper product	13×10^5	2×10^5
Mitsubishi	30% O_2, blister copper	16×10^5	2×10^5
KIVCET	Pure O_2, electric zinc reduction	4×10^5	12×10^5

In comparing the use of electricity and fuel for heating, it should be noted that well-designed fuel-fired processes produce fuel efficiencies in the range of 45 to 85%, whereas the overall efficiency of delivered thermally generated electrical energy is around 30%. Thus 1 kWh used at a metallurgical plant may represent approximately 12×10^6 joules of fuel consumed.

The energy used in different copper smelting processes is indicated in Table 2.

Sulfur dioxide containment. A major factor in modern sulfide smelting is the extent to which sulfur dioxide emissions are being contained or reduced. In roasters, flash furnaces, and continuous smelting reactors, the effluent SO_2 is produced at a steady rate and at a concentration (5–80%) which permits its collection and conversion into sulfuric acid. The capture of SO_2 is especially efficient and effective in Japanese copper smelters. *See* PYROMETALLURGY.

[W. CHARLES COOPER]

Bibliography: J. N. Anderson and P. E. Queneau (eds.), *Pyrometallurgical Processes in Nonferrous Metallurgy*, 1967; A. K. Biswas and W. G. Davenport, *The Extractive Metallurgy of Copper*, 2d ed., 1980; J. R. Boldt, Jr., and P. E. Queneau, *The Winning of Nickel*, 1967; J. M. Cigan, T. S. Mackey, and T. J. O'Keefe (eds.), *Lead-Zinc-Tin '80: TMS-AIME World Symposium on Metallurgy and Environmental Control*, 1980.

Q (electricity)

Often called the quality factor of a circuit, Q is defined in various ways, depending upon the particular application. In the simple RL and RC series circuits, Q is the ratio of reactance to resistance, as in Eqs. (1), where X_L is the inductive reactance,

$$Q = X_L/R \quad Q = X_C/R \quad \text{(a numerical value)} \qquad (1)$$

X_C is the capacitive reactance, and R is the resistance. An important application lies in the dissipation factor or loss angle when the constants of a coil or capacitor are measured by means of the alternating-current bridge.

Q has greater practical significance with respect to the resonant circuit, and a basic definition is given by Eq. (2), where Q_0 means evaluation at resonance. For certain circuits, such as cavity resonators, this is the only meaning Q can have.

$$Q_0 = 2\pi \frac{\text{max stored energy per cycle}}{\text{energy lost per cycle}} \qquad (2)$$

For the *RLC* series resonant circuit with resonant frequency f_0, Eq. (3) holds, where R is the total circuit resistance, L is the inductance, and C is the capacitance. Q_0 is the Q of the coil if it contains practically the total resistance R. The greater the value of Q_0, the sharper will be the resonance peak.

$$Q_0 = 2\pi f_0 L/R = 1/2\pi f_0 CR \qquad (3)$$

The practical case of a coil of high Q_0 in parallel with a capacitor also leads to $Q_0 = 2\pi f_0 L/R$. R is the total series resistance of the loop, although the capacitor branch usually has negligible resistance.

In terms of the resonance curve, Eq. (4) holds,

$$Q_0 = f_0/(f_2 - f_1) \qquad (4)$$

where f_0 is the frequency at resonance, and f_1 and f_2 are the frequencies at the half-power points. *See* RESONANCE (ALTERNATING-CURRENT CIRCUITS).

[BURTIS L. ROBERTSON]

Quadrature

The condition in which the phase angle between two alternating quantities is 90°, corresponding to one-quarter of an electrical cycle. The electric and magnetic fields of electromagnetic radiation are in space quadrature, which means that they are at right angles in space.

The current and voltage of a perfect coil are in quadrature because the coil current lags behind the coil voltage by exactly 90°. The current and voltage of a perfect capacitor are also in quadrature, but here the current leads the voltage by 90°. In these last two cases the current and voltage are in time quadrature. *See* ALTERNATING-CURRENT CIRCUIT THEORY.

[JOHN MARKUS]

Quality control

A system of inspection, analysis, and action applied to a manufacturing operation so that, by inspecting a small portion of the product currently produced, an estimate of the overall quality of the product can be made to determine what, if any, changes must be made in the operation in order to achieve or maintain the required level of quality.

The term is usually applied in the context of manufacturing operations. The manufacturer makes one or more items, each having a set of specifications for various characteristics, which must be met to satisfy the needs of the customer. The specifications may be set either by the customer in view of his intended use of the product or by the manufacturing organization in view of its understanding of the customer's intended use, or by both. Quality control is the set of activities in a manufacturing operation which undertakes to ensure that the specifications of the finished product are satisfied. More general terms such as quality assurance and total quality control embrace all the technical and management aspects of product quality and safety during the design, specification, development, manufacturing, and use stages. Periodic reviews of quality standards, and quality operations are also undertaken to ensure that products remain satisfactory to the consumer and competitive in the market place. This can involve comparative testing of competitive products and the undertaking of quality improvement programs. Since most manufacturing organizations have a profit objective, quality control performs its tasks with the objective of minimizing costs, that is, with obtaining the specified product quality at the lowest possible cost.

Inspection and testing. The cost objective requires that the inspection and testing of the product (to ascertain that it meets specifications) must be done at various points in the production process, since if all measurements of specification conformance were done at the last operation, the costs of all preceding operations might be wasted. There are several key points in most manufacturing operations where inspection and testing are essential. One easily identifiable point in the process at which measurements are made is on material purchased by the manufacturer from a vendor. This action attempts to establish that the manufacturer's operations will be performed on material that will enable him to produce acceptable material for his customers. This activity is usually known as incoming-material control and is accomplished by means of acceptance sampling plans, which are discussed later. *See* GAGES; INSPECTION AND TESTING.

Another easily identifiable point in the manufacturing process at which control is exercised is after all operations have been performed. This action attempts to establish that the material going from the manufacturer to his customers meets their requirements. This activity is usually known as finished goods control and is also most often accomplished by some acceptance sampling plan.

The selection of the number and the location of the quality-control points between the incoming-material-control point and the quality-assurance point is a highly involved procedure. The objective is to see that the operations satisfactorily affect the specified characteristics. This activity is usually known as in-process quality control and is often accomplished by means of quality-control charts. *See* CONTROL CHART.

Quality engineering. A quality-control system that adequately covers the areas of incoming materials, in-process, and finished goods control will handle the more routine aspects of assuring that the product meets customers' requirements. Occasionally, however, problems arise that are not routine, such as installation of new equipment, rebuilding of old equipment, selection of a new vendor or a new raw material, and increase in a customer's complaints about some characteristic of the product. These and many other problems have an impact on the process of generating a product which meets the customers' requirements. In each problem, data must be collected and analyzed to evaluate the impact on quality and to make recommendations concerning appropriate action. This activity which undertakes special quality studies and provides statistical and engineering support to other quality control and manufacturing areas is known as quality engineering. Some of the more widely used techniques are briefly discussed in later paragraphs.

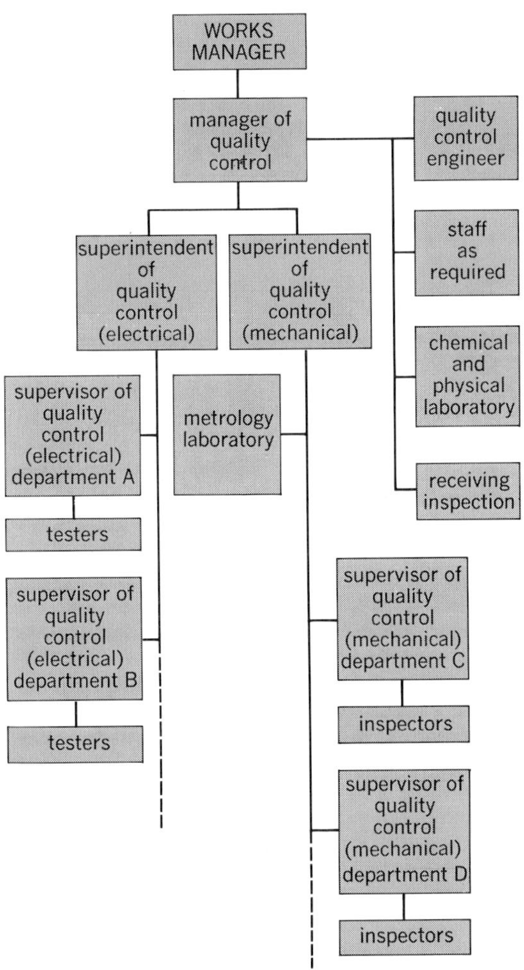

Fig. 1. Typical organization chart of a quality-control department in a plant employing 2000–3000 workers.

Quality management. The work of quality control is incorporated into the overall organization in several ways. The quality-control department should be organized according to accepted management principles. Experience shows that quality control operates most effectively when its manager reports in a staff capacity to the top executive responsible for manufacturing, generally the works manager. The works manager is responsible for both quantity and quality of production. A typical organization chart for a plant employing 2000–3000 productive workers is shown in Fig. 1.

The functions of a quality-control department directly associated with getting out a product can be pictured as an endless chain (Fig. 2). The responsibility of a quality-control department does not end with the shipment of the product. It follows product performance in the field and, by making specific recommendations, motivates the engineering department to redesign the product to improve its quality and performance.

Sampling plans. A sampling plan is a rule used to determine, on the basis of sample results, which of a series of lots of materials should be accepted and which should be rejected, or screened to remove all defective units. The rule is made up of three segments.

1. The sample size is the portion of each lot of material that will be measured. If this portion is 100% of the lot, the inspection is referred to as detailed inspection. If this portion is less than 100% (and greater than 0%), the inspection is referred to as sampling inspection.

2. The decision variable is the number (or numbers) calculated from the sample measurements and used to determine the acceptability or rejectability of the lot.

3. The acceptance values are that set of values of the decision variable, in the sample, for which the lot will be accepted.

Sampling plans can be classified as single, double, multiple, or sequential. In single sampling the decision is made on the basis of the results of a single portion of the lot. In double sampling the lot gets a second chance, so to speak. The first sample results in a decision to accept, reject, or take a second sample; that is, the evidence from the first sample may be such that there is no question as to the disposition of the lot, or it may be inconclusive and lead to the taking of a second sample. The decision based on the results of the second sample, however, is either to accept or reject. The term multiple sampling pertains to sampling plans in which the lot gets more than two chances, but no more than a preselected finite number of chances, to be accepted. The term sequential sampling pertains to sampling plans in which the decision on each sample is to accept, reject, or take a further sample. Thus, in sequential sampling, the maximum number of samples is not specified.

Sampling plans are also classified by the nature of the decision variable. If the decision variable is the result of a counting process, the data are said to be discrete, and the sampling plan is referred to

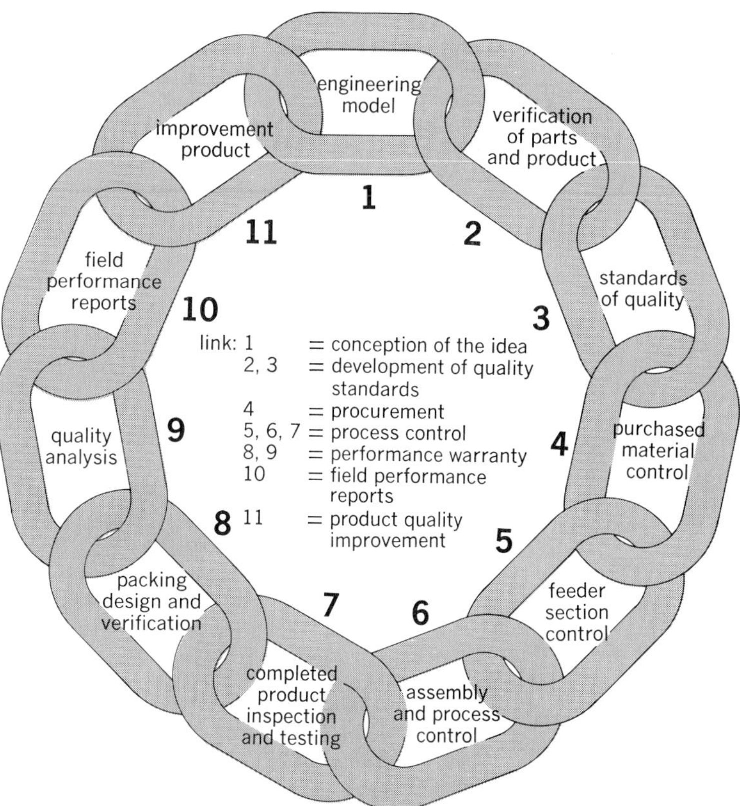

Fig. 2. Quality-control chain showing functions of quality-control department.

as an attribute sampling plan. If, on the other hand, the decision variable is the result of a comparison with some standard unit of measurement, the data are said to be continuous, and the sampling plan is referred to as a variables sampling plan.

All sampling inspection inherently involves taking risks. (All detailed inspection does too, since several studies have shown that only by repeated 100% inspection can all possible defective units be eliminated.) The risks of a sampling plan can be described in terms of an operating characteristic (OC) curve. The OC curve is a plot of the probability that a lot will be accepted (using the selected sampling plan) versus the set of possible quality levels of the lot. An example of the OC curve for a single sample acceptance plan is shown in Fig. 3, where π is the percent defective of the lot; P_A is the probability that the lot will be accepted, using the given sampling plan; n is the sample size; and Ac is the acceptance number (when H, the number of defectives in the sample, exceeds this number the lot is rejected).

There are two points on the OC curve which are widely used to summarize the performance of a sampling plan. One is a high point on the curve at the intersection of the acceptable quality level (AQL), and the producers risk (PR); and the other is a low point at the intersection of the rejectable quality level (RQL), and the consumers risk (CR). The RQL is sometimes referred to as the lot tolerance percent defective (LTPD).

The AQL is the level of quality agreed to as being acceptable to the consumer and achievable by the producer. The sampling plan chosen will give a high value of P_A so that PR which is equal to $1-P_A$ will be low. This will mean that the producer runs only a small risk of having an acceptable batch rejected, and even higher levels of quality will result in fewer rejections. In Fig. 3, AQL = 0.6%, $P_A =$ 0.95, PR = 0.05.

The RQL is the level of quality which is unacceptable to the consumer. The plan chosen will give a low value of P_A at this point so that CR which is equal to P_A will also be low. In Fig. 3, RQL = 3.5%, $P_A = 0.1$, CR = 0.1.

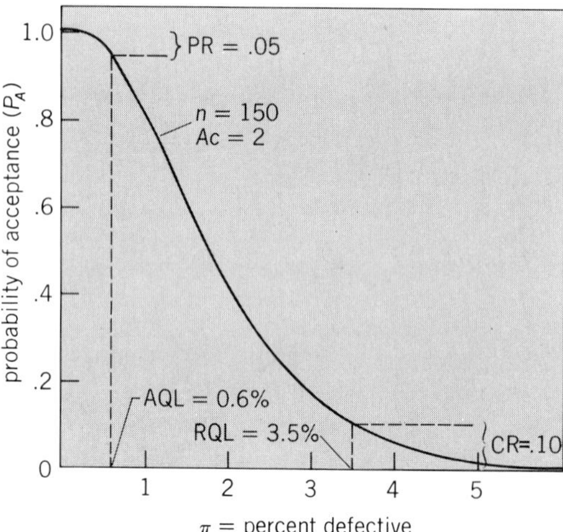

Fig. 3. Operating characteristic (OC) curve for a single sample attribute sampling plan.

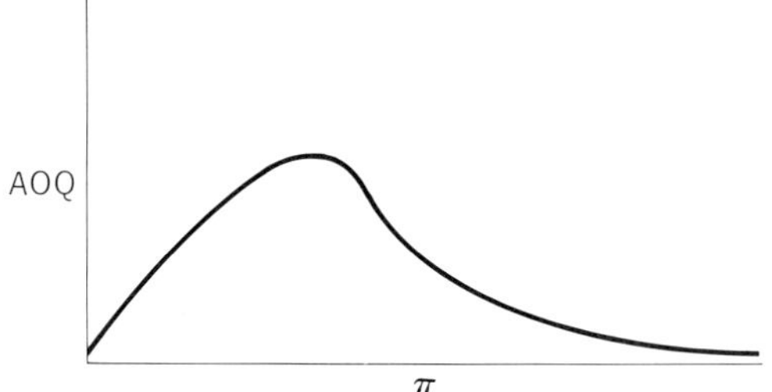

Fig. 4. Average outgoing quality, AOQ, curve shown for a sampling plan.

The values of sample size and acceptance number for a given sampling plan may be calculated by statistical methods if the two points defined above are decided upon. The complete OC curve for the plan may then be calculated.

Another measure of the sampling plan is the average outgoing quality (AOQ). It is assumed that all rejected lots are detailed and that all defective items in the rejected lots are replaced with effective items. With these assumptions, the AOQ curve can be drawn for a given sampling plan and will have the general form illustrated in Fig. 4. The high point on the curve is referred to as the average outgoing quality limit (AOQL) since, if the assumptions above are met, the worst possible average outgoing quality is no greater than that value.

A typical set of sampling plans by attributes is shown in Table 1. A sample of the tabulated normal number of units n is inspected; the size of the sample depends on the size of the lot from which it is taken and on the AQL. The lot is accepted if the number of defectives in the sample does not exceed the corresponding tabulated acceptance number *(Ac)*.

Variables plans are available to control changes in the percent defective or changes in the average value. In order to design a plan to control changes in average value, it is necessary to specify the amount of change which is regarded as important to detect. For example, consider a process with a target value *(T)* equal to 60 mm, and a standard deviation (σ) of 12 mm, for which a shift *(d)* of 5 mm in either direction is considered important. If the producers risk is set at 0.05, and the consumers risk at 0.10, the required sample size *(n)* is given by

$$n = (3.25\sigma/d)^2 = (3.25 \times 12 \div 5)^2 = 61$$

where 3.25 is a factor determined from the stated producer and consumer risks.

The acceptance values A_c are given by

$$\begin{aligned} A_c &= T \pm 1.96\sigma/\sqrt{n} = 60 \pm 1.96 \times 12 \div 7.8 \\ &= 60 \pm 3 \\ &= 57 \text{ and } 63 \end{aligned}$$

where 1.96 is a factor determined from the stated producers risk.

To apply this plan, a random sample of 61 pieces would be drawn from the lot, each unit would be measured, and the average value calculated. If the average value fell between 57 and 63, the lot would be accepted, otherwise it would be rejected. An OC curve for the plan shown in Fig. 5 gives the

Table 1. Single sampling plans by attributes*

Sample size code letter	Sample size	Acceptable quality levels (normal inspection)																									
		0.010	0.015	0.025	0.040	0.065	0.10	0.15	0.25	0.40	0.65	1.0	1.5	2.5	4.0	6.5	10	15	25	40	65	100	150	250	400	650	1000
		Ac Re	Ac Re	Ac Re	Ac Re	Ac Re	Ac Re	Ac Re	Ac Re	Ac Re	Ac Re	Ac Re	Ac Re	Ac Re	Ac Re	Ac Re	Ac Re	Ac Re	Ac Re	Ac Re	Ac Re	Ac Re	Ac Re	Ac Re	Ac Re	Ac Re	Ac Re
A	2	↓	↓	↓	↓	↓	↓	↓	↓	↓	↓	↓	↓	↓	↓	0 1	↓	↓	1 2	2 3	3 4	5 6	7 8	10 11	14 15	21 22	30 31
B	3	↓	↓	↓	↓	↓	↓	↓	↓	↓	↓	↓	↓	↓	0 1	↑	↓	1 2	2 3	3 4	5 6	7 8	10 11	14 15	21 22	30 31	44 45
C	5	↓	↓	↓	↓	↓	↓	↓	↓	↓	↓	↓	↓	0 1	↑	↓	1 2	2 3	3 4	5 6	7 8	10 11	14 15	21 22	30 31	44 45	↑
D	8	↓	↓	↓	↓	↓	↓	↓	↓	↓	↓	↓	0 1	↑	↓	1 2	2 3	3 4	5 6	7 8	10 11	14 15	21 22	30 31	44 45	↑	↑
E	13	↓	↓	↓	↓	↓	↓	↓	↓	↓	↓	0 1	↑	↓	1 2	2 3	3 4	5 6	7 8	10 11	14 15	21 22	30 31	44 45	↑	↑	↑
F	20	↓	↓	↓	↓	↓	↓	↓	↓	↓	0 1	↑	↓	1 2	2 3	3 4	5 6	7 8	10 11	14 15	21 22	↑	↑	↑	↑	↑	↑
G	32	↓	↓	↓	↓	↓	↓	↓	↓	0 1	↑	↓	1 2	2 3	3 4	5 6	7 8	10 11	14 15	21 22	↑	↑	↑	↑	↑	↑	↑
H	50	↓	↓	↓	↓	↓	↓	↓	0 1	↑	↓	1 2	2 3	3 4	5 6	7 8	10 11	14 15	21 22	↑	↑	↑	↑	↑	↑	↑	↑
J	80	↓	↓	↓	↓	↓	↓	0 1	↑	↓	1 2	2 3	3 4	5 6	7 8	10 11	14 15	21 22	↑	↑	↑	↑	↑	↑	↑	↑	↑
K	125	↓	↓	↓	↓	↓	0 1	↑	↓	1 2	2 3	3 4	5 6	7 8	10 11	14 15	21 22	↑	↑	↑	↑	↑	↑	↑	↑	↑	↑
L	200	↓	↓	↓	↓	0 1	↑	↓	1 2	2 3	3 4	5 6	7 8	10 11	14 15	21 22	↑	↑	↑	↑	↑	↑	↑	↑	↑	↑	↑
M	315	↓	↓	↓	0 1	↑	↓	1 2	2 3	3 4	5 6	7 8	10 11	14 15	21 22	↑	↑	↑	↑	↑	↑	↑	↑	↑	↑	↑	↑
N	500	↓	↓	0 1	↑	↓	1 2	2 3	3 4	5 6	7 8	10 11	14 15	21 22	↑	↑	↑	↑	↑	↑	↑	↑	↑	↑	↑	↑	↑
P	800	↓	0 1	↑	↓	1 2	2 3	3 4	5 6	7 8	10 11	14 15	21 22	↑	↑	↑	↑	↑	↑	↑	↑	↑	↑	↑	↑	↑	↑
Q	1250	0 1	↑	↓	1 2	2 3	3 4	5 6	7 8	10 11	14 15	21 22	↑	↑	↑	↑	↑	↑	↑	↑	↑	↑	↑	↑	↑	↑	↑
R	2000	↑	↑	1 2	2 3	3 4	5 6	7 8	10 11	14 15	21 22	↑	↑	↑	↑	↑	↑	↑	↑	↑	↑	↑	↑	↑	↑	↑	↑

*Based on Mil-Srd-105D single sampling plans for normal inspection (master table).

↓ Use first sampling plan below arrow. If sample size equals, or exceeds, lot or batch size, do 100% inspection.

↑ Use first sampling plan above arrow.

Ac = Acceptance number.
Re = Rejection number.

probability of accepting batches with various average values.

Random samples. The foregoing discussion of sampling plans depends heavily on the requirement that samples be drawn at random from the lots being inspected. Put simply, a random sample is one in which each unit in the lot has an equal chance of being included in the sample. Care must be exercised to ensure that samples are not drawn from just one section of the lot, for example, the top of a case or the end of a roll. Such biased samples can give rise to erroneous decisions.

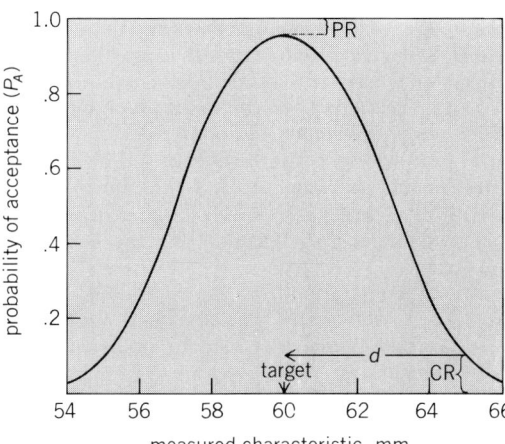

Fig. 5. OC curve for a variables sampling plan sample size = 61. Reject if sample mean is over 63 mm or under 57 mm.

Special quality studies. As one example of a special study, it is often necessary to compare two sample means $\bar{X}_1$ and $\bar{X}_2$, obtained from a pair of random samples of n_1 and n_2 observations with a population in which the standard deviation is σ. From these samples an estimate S of σ can be computed by Eq. (1), where Σ_1 means to sum the

$$S=\sqrt{\frac{\Sigma_1(X-\bar{X}_1)^2+\Sigma_2(X-\bar{X}_2)^2}{n_1+n_2-2}} \qquad (1)$$

deviations of all observations in the first sample and Σ_2 means to sum the deviations of all observations that have been made in the second sample.

Equation (2) is used to determine whether the

$$t=\frac{\bar{X}_1-\bar{X}_2}{S\sqrt{\dfrac{1}{n_1}+\dfrac{1}{n_2}}} \qquad (2)$$

observed difference $(\bar{X}_1-\bar{X}_2)$ could have occurred by chance as a result of the inherent variation in the population.

If t is no larger than the t found in the list for n_1+n_2-2 degrees of freedom, the difference between $\bar{X}_1$ and $\bar{X}_2$ is not significant. If, on the other hand, t is larger than the listed value, the difference is significant. The confidence level of this test is 95% because a significant t (one which is greater than the tabled value) occurs by chance alone only 1 time in 20. A list of t values for indicat-

Table 2. Values of F for indicated degrees of freedom and probability of 0.05

ν_2 \ ν_1	1	2	3	4	5	6	7	8	9	10	12	15	20	24	30	40	60	120	∞
1	161.	200.	216.	225.	230.	234.	237.	239.	241.	242.	244.	246.	248.	249.	250.	251.	252.	253.	254.
2	18.5	19.0	19.2	19.2	19.3	19.3	19.4	19.4	19.4	19.4	19.4	19.4	19.5	19.4	19.5	19.5	19.5	19.5	19.5
3	10.1	9.55	9.28	9.12	9.01	8.94	8.89	8.84	8.81	8.79	8.74	8.70	8.66	8.64	8.62	8.59	8.57	8.55	8.53
4	7.71	6.94	6.59	6.39	6.26	6.16	6.09	6.04	6.00	5.96	5.91	5.86	5.80	5.77	5.74	5.72	5.69	5.66	5.63
5	6.61	5.79	5.41	5.19	5.05	4.95	4.88	4.82	4.77	4.74	4.68	4.62	4.56	4.53	4.50	4.46	4.43	4.40	4.36
6	5.99	5.14	4.76	4.53	4.39	4.28	4.21	4.15	4.10	4.06	4.00	3.94	3.87	3.84	3.81	3.77	3.74	3.76	3.66
7	5.59	4.74	4.35	4.12	3.97	3.87	3.79	3.72	3.68	3.63	3.57	3.51	3.44	3.41	3.38	3.34	3.30	3.27	3.23
8	5.32	4.46	4.07	3.84	3.69	3.58	3.50	3.44	3.39	3.35	3.28	3.22	3.15	3.12	3.08	3.04	3.00	3.97	2.93
9	5.12	4.26	3.86	3.63	3.48	3.37	3.29	3.23	3.18	3.14	3.07	3.01	2.94	2.90	2.86	2.82	2.79	2.75	2.71
10	4.96	4.10	3.71	3.48	3.32	3.22	3.14	3.07	3.02	2.98	2.91	2.84	2.77	2.74	2.70	2.66	2.62	2.58	2.54
12	4.75	3.88	3.49	3.26	3.11	3.00	2.91	2.85	2.80	2.75	2.69	2.62	2.54	2.51	2.47	2.43	2.38	2.34	2.30
15	4.54	3.68	3.29	3.06	2.90	2.79	2.71	2.64	2.59	2.54	2.48	2.40	2.33	2.29	2.25	2.20	2.16	2.11	2.07
20	4.35	3.49	3.10	2.86	2.71	2.60	2.51	2.45	2.39	2.35	2.28	2.20	2.12	2.08	2.04	1.99	1.95	1.90	1.84
24	4.26	3.40	3.01	2.78	2.62	2.51	2.42	2.36	2.30	2.25	2.18	2.11	2.03	1.98	1.94	1.89	1.84	1.79	1.73
30	4.17	3.32	2.92	2.69	2.53	2.42	2.33	2.27	2.21	2.16	2.09	2.01	1.93	1.89	1.84	1.79	1.74	1.68	1.62
40	4.08	3.23	2.84	2.61	2.45	2.34	2.25	2.18	2.12	2.08	2.00	1.92	1.84	1.79	1.74	1.69	1.64	1.58	1.51
60	4.00	3.15	2.76	2.52	2.37	2.25	2.17	2.10	2.04	1.99	1.92	1.84	1.75	1.70	1.65	1.59	1.53	1.47	1.39
120	3.92	3.07	2.68	2.45	2.29	2.16	2.09	2.02	1.96	1.91	1.83	1.75	1.66	1.61	1.55	1.49	1.43	1.35	1.25
∞	3.84	3.00	2.60	2.37	2.21	2.10	2.01	1.94	1.88	1.93	1.75	1.67	1.57	1.52	1.46	1.39	1.32	1.22	1.00

ed degrees of freedom for a confidence level of 95% follows.

Degrees of freedom	*t values*	*Degrees of freedom*	*t values*
1	12.71	15	2.13
2	4.30	16	2.12
3	3.18	17	2.11
4	2.78	18	2.10
5	2.57	20	2.09
6	2.45	22	2.07
7	2.36	24	2.06
8	2.31	26	2.06
9	2.26	28	2.05
10	2.23	30	2.04
11	2.20	40	2.02
12	2.18	60	2.00
13	2.16	120	1.98
14	2.14	∞	1.96

Another example of a special study is the analysis required to determine whether two processes have similar standard deviations (the test is actually run on the sample variances which are squares of the standard deviations), S_1^2 and S_2^2, obtained from a pair of random samples of n_1 and n_2 observations from a population. A sample variance for each sample is computed by the formulations in Eq. (3).

$$S^2 = \frac{\Sigma(X-\overline{X})^2}{n-1} = \frac{\Sigma X^2 - (\Sigma X)^2/n}{n-1} \qquad (3)$$

To determine whether one sample variance is significantly larger than the other, an F ratio of the larger variance to the smaller is formed, Eq. (4). If

$$F = \frac{S_1^2}{S_2^2} \qquad (4)$$

F is not larger than the F value found in Table 2 for $n_1 - 1 = \nu_1$ and $n_2 - 1 = \nu_2$ degrees of freedom, the two variances are not significantly different.

If, on the other hand, F is larger than the tabled value, the variances are significantly different because such a difference can occur as a result of the inherent sampling variation only 1 time in 20 by chance alone. *See* INDUSTRIAL COST CONTROL; INDUSTRIAL ENGINEERING; PRODUCTION ENGINEERING. [JOHN A. CLEMENTS]

Bibliography: D. H. Besterfield, *Quality Control: A Practical Approach*, 1979; A. Bowker and G. Lieberman, *Engineering Statistics*, 2d ed., 1972; I. Burr, *Statistical Quality Control Methods*, 1976; F. Caplan, The *Quality System: A Sourcebook for Managers and Engineers*, 1980; H. Goldberg, *Extending the Limits of Reliability Theory*, 1981; E. L. Grant and R. Leavenworth, *Statistical Quality Control*, 5th ed., 1979; W. C. Guenther, *Scientific Sampling for Statistical Quality Control*, 1977; J. M. Juran (ed.), *Quality Control Handbook*, rev. 3d ed., 1974; I. Miller and J. Freund, *Probability and Statistics for Engineers*, 2d ed., 1977.

Quarrying

The process of extracting stone for commercial use from natural rock deposits. The industry has two major branches: a dimension-stone branch, involving preparation of blocks of various sizes and shapes for use as building stone, monumental stone, paving stone, curbing, and flagging; and a crushed-stone branch, involving preparation of crushed and broken stone for use as a basic construction, chemical, and metallurgical raw material.

Dimension stone. Methods for quarrying dimension stone vary according to the type of rock, depth of deposit, and ultimate use. The major problem is to secure large, sound, and relatively flawless blocks of stone of attractive color and texture. The use of explosives, which tends to shatter the stone, is avoided wherever possible. The usual procedure is to cut the stone from the quarry face into large blocks by some means, undercut the stone at floor level, and break it free by wedging. The mass of stone is then cut into blocks of the desired size, usually 10 × 4 × 4 ft (3 × 1.2 × 1.2 m), by drilling and wedging and is hoisted from the quarry by derricks.

Primary separation from the rock ledge in granite (Fig. 1) and structural sandstone quarries is made by broaching, wire sawing, or jet piercing. Broaching consists of drilling a row of holes close together, usually with detachable tungsten carbide bits, and removing the webs between the holes with broaching tools. A modern method is to employ wire saws (Fig. 2) which consist of one- and three-strand wire cables, up to 16,000 ft long (4900 m), running over pulleys as a belt. When fed with a slurry of sand and water and held against the

Fig. 1. New England granite quarry. (*H. E. Fletcher Granite Co.*)

rock by tension, the wire saw cuts by abrasion. The cables descend at a rate of about 2 in./hr (5 cm/h) in hard granite. Cuts are about 1/4 in. (1¼ cm) wide and may be 50–70 ft (15–21 m) deep. In some quarries a condition of initial strain may exist in the rock which interferes with the broaching operation. In such cases the granite may expand while the holes are being drilled and may pinch the drill bit and render it immovable. Initial separation in such cases is obtained by the use of a jet-piercing drill, which cuts a channel of about 8 in. (20 cm) by utilizing combustion of oxygen and fuel oil fed through a nozzle to burn its way through the rock by disintegrating it into fragments.

Primary separation from the rock ledge in marble (Fig. 3), limestone, and soft sandstone quarries is usually accomplished with electrically powered channeling machines that operate by a chopping action upon three to five steel chisels while traveling back and forth on a track. The channels are about 2 in. (5 cm) wide, 10–12 ft (3.0–3.6 m) deep, and usually 4 ft (1.2 m) apart.

If no horizontal open seams are present in a rock formation, the masses of stone, outlined by the primary separation, are set free at the quarry floor by driving wedges into horizontal drill holes. Subdivision into smaller blocks is usually made by the plug-and-feather method. The feathers are elongated soft-iron plates, used in pairs down each side of a drill hole; the plug is a steel wedge placed between a pair of feathers. When they are inserted in a row of shallow drill holes and sledged lightly in succession, a fracture is made.

The rough blocks are fashioned into finished products by sawing, planing, rubbing, and polishing in mills equipped wih a great variety of stoneworking machines.

Crushed stone. Limestone is the most important rock, accounting for 70% of the total crushed-stone production. Basalt, granite, and quartzite or sandstone are also quarried. Crushed-stone quarries employ a variety of methods and equipment. The operation involves the steps of stripping, drilling, blasting, loading, and conveying to crusher and mill.

Rows of holes are drilled in the quarry bench with percussive or rotary drills, capable of drilling to depths in excess of 100 ft (30 m). The holes, usually 4–12 in. (10–30 cm) in diameter, are loaded with explosive charges and detonated simultaneously. The charges, some reaching several tons,

Fig. 2. Four parallel wire saws in a granite quarry. The four wires appear emerging from their cuts at the lower right. (*H. E. Fletcher Granite Co.*)

Fig. 3. Marble quarry, showing a block ready for hoisting, and electrically powered channeling machines in background. (*Vermont Marble Co.*)

shatter the rock, possibly thousands of cubic yards, and throw it to the quarry floor. This is known as the primary blast. If some masses of rock are too large for loading, they are reduced in size by secondary breakage with small charges of dynamite inserted in holes drilled with a jackhammer or by the use of a drop ball. A steel ball, weighing several tons, is hoisted with a boom and allowed to fall on the stone, shattering it by impact.

The broken stone is usually loaded into dump cars or trucks by crawler-tread electric shovels, whose dipper capacities range from 1 to 15 yd^3 (0.8 to 11.5 m^3), depending upon the size of the operation.

The preparation of crushed-stone products involves a series of crushing, screening, and classification operations. The primary crushers are usually of the jaw or gyratory type. The crushed stone is graded by inclined vibrating screens, revolving screens, or shaking screens. Crushed and ground materials are usually transported through the plant by belt conveyors and bucket elevators, with some drag, screw, and pneumatic conveyors also in occasional use.

The crushed-stone industry exists in every state of the United States and is expected to grow at the same rate as the national population and economy, paralleling the gross national product and the value of construction. [STEFAN H. BOSHKOV]

Bibliography: Society of Mining Engineers, *Industrial Minerals and Rocks*, 4th ed., 1975; U.S. Bureau of Mines, *Bureau of Mines Mineral Yearbook*, issued annually.

Radiant heating

Any system of space heating in which the heat-producing means is a surface that emits heat to the surroundings by radiation rather than by conduction or convection. The surfaces may be radiators such as baseboard radiators or convectors, or they may be the panel surfaces of the space to be heated. *See* PANEL HEATING AND COOLING.

The heat derived from the Sun is radiant energy. Radiant rays pass through gases without warming them appreciably, but they increase the sensible temperature of liquid or solid objects upon which they impinge. The same principle applies to all forms of radiant-heating systems, except that convection currents are established in enclosed spaces and a portion of the space heating is produced by convection. The radiation component of convectors can be increased by providing a reflective surface on the wall side of the convector and painting the inside of the enclosure a dead black to absorb heat and transmit it through the enclosure, thus increasing the temperature of that side of the convector exposed to the space to be heated.

Any radiant-heating system using a fluid heat conveyor may be employed as a cooling system by substituting cold water or other cold fluid. This cannot be done with electric resistance-type radiant-heating systems. Thermoelectric couples will emit or absorb heat, depending upon the polarity of the direct current applied to them. However, the technique is practical only for very special and small-scale heating and cooling applications, certainly not on the scale required for comfort control of an occupied space.

[ERWIN L. WEBER/RICHARD KORAL]

Bibliography: American Society of Heating, Refrigerating, and Air Conditioning Engineers, *ASHRAE Handbook, Fundamentals Volume*, 1977.

Radiation damage to materials

Harmful changes in the properties of liquids, gases, and solids, caused by interaction with nuclear radiations.

The interaction of radiation with materials often leads to changes in the properties of the irradiated material. These changes are usually considered harmful. For example, a ductile metal may become brittle. However, sometimes the interaction may result in beneficial effects. For example, cross-linking may be induced in polymers by electron irradiation leading to a higher temperature stability than could be obtained otherwise.

Radiation damage is usually associated with materials of construction that must function in an environment of intense high-energy radiation from a nuclear reactor. Materials that are an integral part of the fuel element or cladding and nearby structural components are subject to such intense nuclear radiation that a decrease in the useful lifetime of these components can result. *See* NUCLEAR REACTOR.

Radiation damage will also be a factor in thermonuclear reactors. The deuterium-tritium (D-T) fusion in thermonuclear reactors will lead to the production of intense fluxes of 14-MeV neutrons that will cause damage per neutron of magnitude two to four times greater than damage done by 1–2 MeV neutrons in operating reactors. Charged particles from the plasma will be prevented from reaching the containment vessel by magnetic fields, but uncharged particles and neutrons will bombard the containment wall, leading to damage as well as sputtering of the container material surface which not only will cause degradation of the wall but can contaminate the plasma with consequent quenching.

Superconductors are also sensitive to neutron irradiation, hence the magnetic confinement of the plasma may be affected adversely. Damage to electrical insulators will be serious. Electronic components are extremely sensitive to even moderate radiation fields. Transistors malfunction because of defect trapping of charge carriers. Ferroelectrics such as $BaTiO_3$ fail because of induced isotrophy; quartz oscillators change frequency and ultimately become amorphous. High-permeability magnetic materials deteriorate because of hardening; thermocouples lose calibration because of transmutation effects. In this latter case, innovations in Johnson noise thermometry promise freedom from radiation damage in the area of temperature measurement. Plastics used for electrical insulation rapidly deteriorate. Radiation damage is thus a challenge to reactor designers, materials engineers, and scientists to find the means to alleviate radiation damage or to develop more radiation-resistant materials.

Damage mechanisms. There are several mechanisms that function on an atomic and nuclear scale to produce radiation damage in a material if the radiation is sufficiently energetic, whether it be electrons, protons, neutrons, x-rays, fission fragments, or other charged particles.

Electronic excitation and ionization. This type of damage is most severe in liquids and organic compounds and appears in a variety of forms such as gassing, decomposition, viscosity changes, and polymerization in liquids. Rapid deterioration of the mechanical properties of plastics takes place either by softening or by embrittlement, while rubber suffers severe elasticity changes at low fluxes. Cross-linking, scission, free-radical formation, and polymerization are the most important reactions.

The alkali halides are also subject to this type of damage since ionization plays a role in causing displated atoms and darkening of transparent crystals due to the formation of color centers.

Transmutation. In an environment of neutrons, transmutation effects may be important. An extreme case is illustrated by reaction (1). The ^{6}Li

$$^{6}\mathrm{Li} + n \rightarrow {}^{4}\mathrm{He} + {}^{3}\mathrm{H} + 4.8\ \mathrm{MeV} \qquad (1)$$

isotope is approximately 7.5% abundant in natural lithium and has a thermal neutron cross section of 950 barns (1 barn $= 1 \times 10^{-24}$ cm^2). Hence, copious quantities of tritium and helium will be formed. (In addition, the kinetic energy of the reaction products creates many defects.). Lithium alloys or compounds are consequently subject to severe radiation damage. On the other hand, reaction (1) is crucial to success of thermonuclear reactors utilizing the D-T reaction since it regenerates the tritium consumed. The lithium or lithium-containing compounds might best be used in the liquid state.

Even materials that have a low cross section such as aluminum can show an appreciable accumulation of impurity atoms from transmutations. The capture cross section of ^{27}Al (100% abundant) is only 0.25×10^{-24} cm^2. Still the reaction $^{27}\mathrm{Al} + n \rightarrow {}^{28}\mathrm{Al} \xrightarrow[\ ]{\beta^-\ 2.3\ \mathrm{min}} {}^{28}\mathrm{Si}$ will yield several percent of silicon after neutron exposures at fluences of 10^{23} n/cm^2.

The elements boron and europium have very large cross sections and are used in control rods. Damage to the rods is severe in boron-containing materials because of the ^{10}B(n,α) reaction. Europium decay products do not yield any gaseous elements. At high thermal fluences the reaction $^{58}\mathrm{Ni} + n \rightarrow {}^{59}\mathrm{Ni} + n \rightarrow {}^{56}\mathrm{Fe} + \alpha$ is most important in nickel-containing materials. The reaction $(n,n') \rightarrow \alpha$ at 14 MeV takes place in most materials under consideration for structural use. Thus, in many instances transmutation effects can be a problem of great importance.

Displaced atoms. This mechanism is the most important source of radiation damage in nuclear reactors outside the fuel element. It is a consequence of the ability of the energetic neutrons born in the fission process to knock atoms from their equilibrium position in their crystal lattice, displacing them many atomic distances away into interstitial positions and leaving behind vacant lattice sites. The interaction is between the neutron and the nucleus of the atom only, since the neutron carries no charge. The maximum kinetic energy ΔE that can be acquired by a displaced atom is given by Eq. (2), where M is mass of the primary

$$\Delta E = \frac{4Mm}{(M+m)^2} \cdot E_N \qquad (2)$$

knocked-on atom (PKA), m is the mass of the neutron, and E_N is the energy of the neutron.

The energy acquired by each PKA is often high enough to displace additional atoms from their equilibrium position; thus a cascade of vacancies and interstitial atoms is created in the wake of the PKA transit through the matrix material. Collision of the PKA and a neighbor atom takes place within a few atomic spacings or less because the charge on the PKA results in screened coulombic-type repulsive interactions. The original neutron, on the other hand, may travel centimeters between collisions. Thus regions of high disorder are dispersed along the path of the neutron. These disordered regions are created in the order of 10^{-12} s. The energy deposition is so intense in these regions that it may be visualized as a temporary thermal spike.

Not all of the energy transferred is available for displacing atoms. Inelastic energy losses (electronic excitation in metals and alloys and excitation plus ionization in nonmetals) drain an appreciable fraction of the energy of the knocked-on atom even at low energies, particularly at the beginning of its flight through the matrix material. The greater the initial energy of the PKA, the greater is the inelastic energy loss; however, near the end of its range most of the interactions result in displacements. Figure 1 is a schematic representation of the various mechanisms of radiation damage that take place in a solid.

A minimum energy is required to displace an atom from its equilibrium position. This energy ranges from 25 to 40 eV for a typical metal such as iron; the mass of the atom and its orientation in the crystal influence this value. When appropriate calculations are made to compensate for the excitation energy loss of the PKA and factor in the minimum energy for displacement, it is found that approximately 500 stable vacancy-interstitial pairs are formed, on the average, for a PKA in iron resulting from a 1-MeV neutron collision. By multiplying this value by the flux of neutrons [$10^{14\text{-}15}$ n/(cm^2)(s)] times the exposure time [3×10^7 s/yr] one can easily calculate that in a few years each atom in the iron will have been displaced several times.

In the regions of high damage created by the PKA, most of the vacancies and interstitials will recombine. However, many of the interstitials,

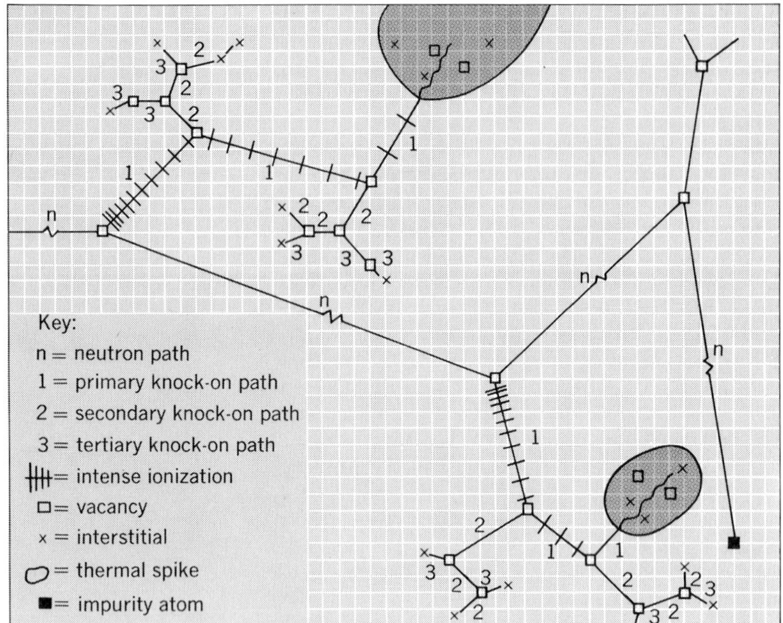

Fig. 1. The five principal mechanisms of radiation damage are ionization, vacancies, interstitials, impurity atoms, and thermal spikes. Diagram shows how a neutron might give rise to each in copper. Grid-line intersections are equilibrium positions for atoms. (*After D. S. Billington, Nucleonics, 14:54–57, 1956*)

being more mobile than the vacancies, will escape and then may eventually be trapped at grain boundaries, impurity atom sites, or dislocations. Sometimes they will agglomerate to form platelets or interstitial dislocation loops. The vacancies left behind may also be trapped in a similar fashion, or they may agglomerate into clusters called voids.

Effect of fission fragments. The fission reaction in uranium or plutonium yielding the energetic neutrons that subsequently act as a source of radiation damage also creates two fission fragments that carry most of the energy released in the fission process. This energy, approximately 160 MeV, is shared by the two highly charged fragments. In the space of a few micrometers all of this energy is deposited, mostly in the form of heat, but a significant fraction goes into radiation damage of the surrounding fuel. The damage takes the form of swelling and distortion of the fuel. These effects may be so severe that the fuel element must be removed for reprocessing in advance of burn-up expectation, thus affecting the economy of reactor operation. However, fuel elements are meant to be ultimately replaced, so that in many respects the damage is not as serious a problem as damage to structural components of the permanent structure whose replacement would force an extended shut down or even reconstruction of the reactor. *See* NUCLEAR FUELS.

Damage in cladding. Swelling of the fuel cladding is a potentially severe problem in breeder reactor design. The spacing between fuel elements is minimized to obtain maximum heat transfer and optimum neutron efficiency, so that diminishing the space for heat transfer by swelling would lead to overheating of the fuel element, while increasing the spacing to allow for the swelling would result in lower efficiencies. A possible solution appears to be in the development of low-swelling alloys.

Damage in engineering materials. Most of the engineering properties of materials of interest for reactor design and construction are sensitive to defects in their crystal lattice. The properties of structural materials that are of most significance are yield strength and tensile strength, ductility, creep, hardness, dimensional stability, impact resistance, and thermal conductivity. Metals and alloys are chosen for their fabricability, ductility, reasonable strength at high temperatures, and ability to tolerate static and dynamic stress loads. Refractory oxides are chosen for high-temperature stability and for use as insulators. Figure 2 shows the relative sensitivity of various types of materials to radiation damage. Several factors that enter into susceptibility to radiation damage will be discussed. *See* METAL, MECHANICAL PROPERTIES OF.

Temperature of irradiation. Nuclear irradiations performed at low temperatures (4 K) result in the maximum retention of radiation-produced defects. As the temperature of irradiation is raised, many of the defects are mobile and some annihilation may take place at 0.3 to 0.55 of the absolute melting point T_m. The increased mobility, particularly of vacancies and vacancy agglomerates, may lead to acceleration of solid-state reactions, such as precipitation, short- and long-range ordering, and phase changes. These reactions may lead to unde-

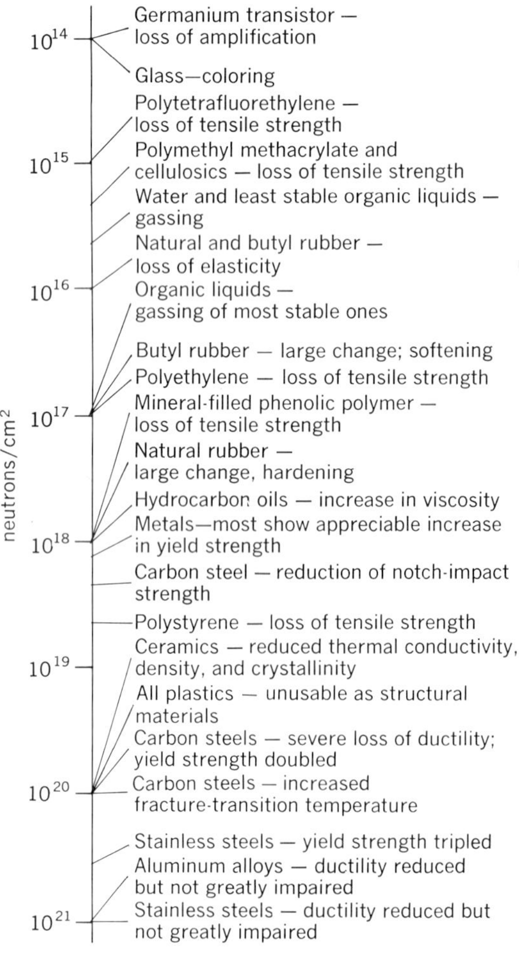

Fig. 2. Sensitivity of engineering materials to radiation. Levels are approximate and subject to variation. Changes are in most cases at least 10%. (*After O. Sisman and J. C. Wilson, Nucleonics, 14:58–65, 1956*)

sirable property changes. In the absence of irradiation many alloys are metastable, but the diffusion rates are so low at this temperature that no significant reaction is observed. The excess vacancies above the equilibrium value of vacancies at a given temperature allow the reaction to proceed as though the temperature were higher. In a narrow temperature region vacancy-controlled diffusion reactions become temperature-independent. When the temperature of irradiation is above 0.55 T_m, most of the defects anneal quickly and the temperature-dependent vacancy concentration becomes overwhelmingly larger than the radiation-induced vacancy concentration. However, in this higher temperature region serious problems may arise from transmutation-produced helium. This gas tends to migrate to grain boundaries and leads to enhanced intergranular fracture, thereby limiting the use of many conventional alloys.

Nuclear properties. Materials of construction with high nuclear-capture cross sections are to be avoided because each neutron that is captured in the structural components is lost for purposes of causing additional fissioning and breeding. The exception is in control rods as discussed earlier. Moderator materials, in particular, need to have low capture cross sections but high scattering cross sections. Low atomic weight is an important feature since moderation of fast neutrons to thermal energies is best done by those elements that maximize the slowing-down process. [See Eq. (2).] Beryllium and graphite are excellent moderators and have been used extensively in elemental form. Both elements suffer radiation damage, and their use under high-stress conditions is to be avoided.

Fluence. The total integrated exposure to radiation (flux × time) is called fluence. It is most important in determining radiation damage. Rate effects (flux) do not appear to be significant. The threshold fluence for a specific property change induced by radiation is a function of the composition and microstructure. One of the most important examples is the appearance of voids in metals and alloys. This defect does not show up in the microstructure of irradiated metals or alloys until a fluence of 10^{19} n/cm^2 or greater has been achieved. Consequently, there was no way to anticipate its appearance and the pronounced effect in causing swelling in structural components of a reactor. This and other examples point to the importance of lifetime studies in order to establish the appearance or absence of any unexpected phenomenon during this time.

Lifetime studies in reactors are time-consuming and are virtually impossible if anticipated fluences far exceed the anticipated lifetime of operating test reactors. A technique to overcome this impasse is to use charged-particle accelerators to simulate reactor irradiation conditions. For example, nickel ions can be used to bombard nickel samples. The bombarding ions at 5 to 10 MeV then simulate primary knocked-on atoms directly and create high-density damage in the thickness of a few micrometers. Accelerators are capable of produced beam currents of several $\mu A/cm^2$; hence in time periods of a few hours to a few days ion bombardment is equivalent to years of neutron bombardment. Correlation experiments have established that the type of damage is similar to neutron damage. Moreover, helium can be injected to approximate n,α damage when these reactions do not occur in accelerator bombardments. However, careful experimentation is required to obtain correlation between results obtained on thin samples and thicker, more massive samples used in neutron studies.

Pretreatment and microstructure. Dislocations play a key role in determining the plastic flow properties of metals and alloys such as ductility, elongation, and creep. The yield, ultimate and impact strength properties, and hardness are also expressions of dislocation behavior. If a radiation-produced defect impedes the motion of a dislocation, strengthening and reduced ductility may result. On the other hand, during irradiation point defects may enhance mobility by promoting dislocation climb over barriers by creating jogs in the dislocation so that it is free to move in a barrier-free area. Moreover, dislocations may act as trapping sites for interstitials and gas atoms, as well as nucleation sites for precipitate formation. Thus the number and disposition of dislocations in the metal alloy may strongly influence its behavior upon irradiation.

Heat treatment prior to irradiation determines the retention of both major alloying components and impurities in solid solution in metastable alloys. It also affects the number and disposition of dislocations. Thus heat treatment is an important variable in determining subsequent radiation behavior.

Impurities and minor alloying elements. The presence of small amounts of impurities may profoundly affect the behavior of engineering alloys in a radiation field. It has been observed that helium concentrations as low as 10^{-9} seriously reduce the high-temperature ductility of a stainless steel. Concentrations of helium greater than 10^{-3} may conceivably be introduced by the n,α reaction in the nickel component of the stainless steel or by boron contamination introduced inadvertently during alloy preparation. The boron also reacts with neutrons via the n,α reaction to produce helium. The addition of a small amount of Ti (0.2%) raises the temperature at which intergranular fracture takes place so that ductility is maintained at operating temperatures.

Small amounts of copper, phosphorus, and nitrogen have a strong influence on the increase in the ductile-brittle transition temperature of pressure vessel steels under irradiation. Normally these carbon steels exhibit brittle failure below room temperature. Under irradiation, with copper content above 0.08% the temperature at which the material fails in a brittle fashion increases. Therefore it is necessary to control the copper content as well as the phosphorus and nitrogen during the manufacture and heat treatment of these steels to keep the transition temperature at a suitably low level. A development of a similar nature has been observed in the swelling of type 316 stainless steel. It has been learned that carefully controlling the concentration of silicon and titanium in these alloys drastically reduces the void swelling. This is an important technical and economic contribution to the fast breeder reactor program.

Beneficial effects. Radiation, under carefully controlled conditions, can be used to alter the course of solid-state reactions that take place in a wide variety of solids. For example, it may be used to promote enhanced diffusion and nucleation, it can speed up both short- and long-range order-disorder reactions, initiate phase changes, stabilize high-temperature phases, induce magnetic property changes, retard diffusionless phase changes, cause re-solution of precipitate particles in some systems while speeding precipitation in other systems, cause lattice parameter changes, and speed up thermal decomposition of chemical compounds. The effect of radiation on these reactions and the other property changes caused by radiation are of great interest and value to research in solid-state physics and metallurgy.

Radiation damage is usually viewed as an unfortunate variable that adds a new dimension to the problem of reactor designers since it places severe restraints on the choice of materials that can be employed in design and construction. In addition, it places restraints on the ease of observation and manipulation because of the radioactivity involved. However, radiation damage is also a valuable research technique that permits materials scientists and engineers to introduce impurities and defects into a solid in a well-controlled fashion.

[DOUGLAS S. BILLINGTON]

Bibliography: American Society for Testing and Materials (ASTM), *Effects of Radiation on Structural Materials*, STP 683, 1979; ASTM, *Irradiation Effects on the Microstructure on Properties of Metals*, STP 611, 1976; ASTM, *Properties of Reactor Structural Alloys After Neutron or Particle Irradiation*, STP 570, 1976; D. S. Billington and J. H. Crawford, Jr., *Radiation Damage in Solids*, 1961; E. E. Bloom et al., Austenitic stainless steels with improved resistance to radiation-induced swelling, *Scripta Met.*, 10:303, 1976; C. J. Borokowski and T. V. Blalock, A new method of Johnson noise thermometry, *Rev. Sci. Inst.*, 45:151–162, 1974; G. J. Dienes (ed.), *Studies in Radiation Effects in Solids*, 4 vols., 1964–1975; International Atomic Energy Agency, Vienna, *Interaction of Radiation with Condensed Matter*, 2 vols., 1977, 1978; International Atomic Energy Agency, Vienna, *Radiation Damage in Reactor Materials*, vol. 1, 1969; J. F. Kircher and R. E. Bowman (eds.), *Effects of Radiation on Materials and Components*, 1964; C. Lehmann, *Interaction of Radiation with Solids and Elementary Defect Production*, 1977; N. L. Peterson and S. Harkness (eds.), *Radiation Damage in Metals*, 1976; L. E. Steel, *Neutron Embrittlement of Reactor Pressure Vessels*, Tech. Publ. 163, International Atomic Energy Agency, Vienna, 1975; Surface effects in controlled fusion, *J. Nucl. Mater.*, 53:1–357, 1974; V. S. Vavilov and N. A. Uklin (eds.), *Radiation Effects in Semiconductors and Semiconducting Devices*, 1977.

Radiation shielding

Physical barriers designed to provide protection from the effects of ionizing radiation; also, the technology of providing such protection. Major sources of radiation are nuclear reactors and associated facilities, medical and industrial x-ray and radioisotope facilities, charged-particle accelerators, and cosmic rays. Types of radiation are directly ionizing (charged particles) and indirectly ionizing (neutrons, gamma rays, and x-rays). In most instances, protection of human life is the goal of radiation shielding. In other instances, protection may be required for structural materials which would otherwise be exposed to high-intensity radiation, or for radiation-sensitive materials such as photographic film and certain electronic components. For the effects on inanimate materials *see* RADIATION DAMAGE TO MATERIALS.

Radiation sources. Nuclear-electric generating stations present a variety of shielding requirements. In the nuclear fission process, ionizing radiation is released not only at the instant of fission but also through radioactive decay of fission products and products of neutron absorption. Of the energy released in the fission process, about 2.5% is carried by prompt fission neutrons and 3% by delayed gamma rays. About 2.5 neutrons are released per fission with average energy 2 MeV. About 90% of the prompt neutrons have energies less than 4 MeV, but because of their greater penetrating ability the remaining 10% are of greater concern in radiation shielding. Neutrons are distributed in energy approximately as shown in Fig. 1, which gives the distribution for fission of uranium-235. The distribution for other fissionable isotopes is similar. *See* NUCLEAR REACTOR; REACTOR PHYSICS.

Fission of one atom releases about 7 MeV of energy in the form of prompt gamma rays. This energy is distributed over some 10 gamma-ray photons with the energy spectrum as shown in Fig. 2. Evaluation of the intensity of delayed gamma-ray sources within a nuclear reactor requires knowledge of the operating history of the reactor. Absorption of neutrons in structural or shielding materials results in the emission of capture gamma rays. Likewise, gamma rays result from inelastic

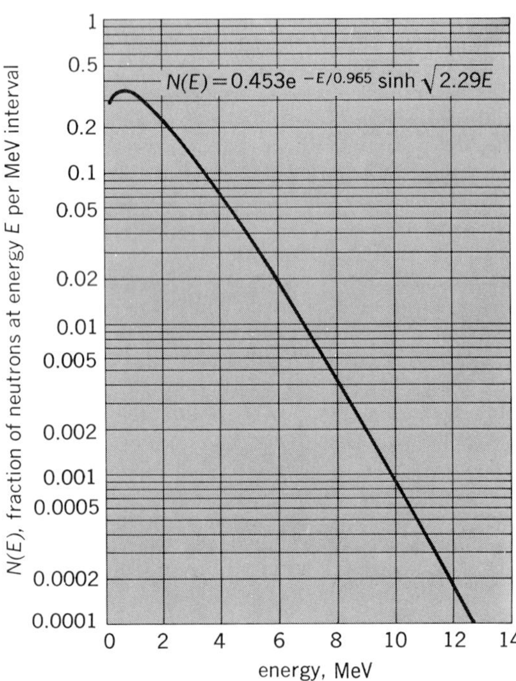

Fig. 1. Neutron spectrum as produced from the fission of U^{235} by thermal neutrons. (*From L. Canberg et al., Phys. Rev. 103:662–670, 1956*)

scattering of neutrons. Because these gamma rays (especially capture gamma rays) are generally of higher energy than prompt fission gamma rays and because they may be released deep within a radiation shield, they require careful consideration in shielding design for nuclear reactors.

Controlled thermonuclear reactors, deriving energy from the nuclear fusion of deuterium and tritium, present radiation shielding requirements similar in kind to those of fission reactors. Highly penetrating (14-MeV) neutrons are released in the fusion process along with charged particles and photons. Capture gamma rays again require careful consideration.

X-ray generators vary widely in characteristics. Typical units release x-rays with maximum energies to 250 KeV, but high-voltage units are in use with x-ray energies as high as tens of MeV. The dominant nature of x-ray energy spectra is that of bremsstrahlung. X-ray generators are but one of many types of charged-particle accelerators and, in most cases, the governing shielding requirement is protection from x-rays. For very-high-energy accelerators, protection from neutrons and mesons produced in beam targets may govern the shield design.

Although many different radionuclides find use in medical diagnosis and therapy as well as in research laboratories and industry, radiation shielding requirements are of special importance for gamma-ray and neutron sources. Alpha and beta particles from radionuclide sources are not highly penetrating, and shielding requirements are minimal. Space vehicles are subjected to bombardment by radiation, chiefly very-high-energy charged particles. In design of the space vehicle and in planning of missions, due consideration must be given to radiation shielding for protection of crew and equipment.

Attenuation processes. Charged particles lose energy and are thus attenuated and stopped primarily as a result of coulombic interactions with electrons of the stopping medium. For heavy charged particles (protons, alpha particles, and such), paths are nearly straight and ranges well defined. Electrons may suffer appreciable angular deflections on collision and may lose substantial energy radiatively. Very-high-energy charged particles may lose energy through nuclear interactions, resulting in fragmentation of the target nuclei and production of a wide variety of secondary radiations.

Gamma-ray and x-ray photons lose energy principally by three types of interactions: photoemission, Compton scattering, and pair production. In photoemission, or the photoelectric effect, the photon transfers all its energy to an atom, and an electron is emitted with kinetic energy equal to the original energy of the photon less the binding energy of the electron in the atom. In Compton scattering, the photon is deflected from its original course by, and transfers a portion of its energy to, an electron. In pair production, the gamma ray is converted to a positron-electron pair. At least 1.02 MeV of gamma-ray energy is required for the rest mass of the pair, and any excess appears as kinetic energy. Ultimately, the positron and an electron recombine and, in annihilation, release two 0.505-MeV gamma rays. Photoemission is especially important for low-energy photons and for stopping media of high atomic number. Compton scattering usually dominates at intermediate photon energies, and pair production at high photon energies.

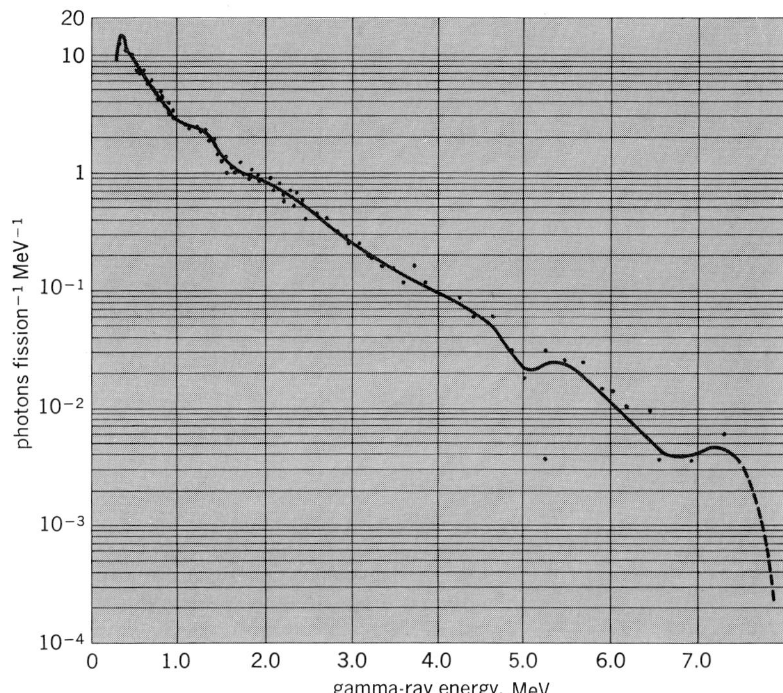

Fig. 2. Energy spectrum of gamma-rays observed within 10^{-7} sec after fission.

Neutrons lose energy in shields by elastic or inelastic scattering. Elastic scattering is more effective with shield materials of low atomic mass, notably hydrogenous materials, but both processes are important, and an efficient neutron shield is made of materials of both high and low atomic mass. The fate of the neutron, after slowing down as a result of scattering interactions, is absorption frequently accompanied by emission of capture gamma rays. Suppression of capture gamma rays may be effected by incorporating elements such as boron or lithium in the shield material. The isotopes ^{10}B and ^{6}Li have large cross sections for neutron capture without gamma-ray emission.

Shielding concepts. The cross section of an atom or electron for interaction with radiation is the effective "target" area presented for the interaction. It is usually given the symbol σ and the units cm^2 or barns (1 barn $= 10^{-24}$ cm^2). The cross section depends on the type of interaction and is a function of the energy of the radiation. When the cross section is multiplied by the number of atoms or electrons per unit volume, the product, identified as the linear attenuation coefficient μ or macroscopic cross section Σ, has the units of reciprocal length and may be interpreted as the probability per unit distance of travel that the radiation experiences in an interaction of a given type. The total attenuation coefficient for radiation of a given energy is the sum of attenuation coefficients for all types of interactions in the shielding medium. The quotient of the linear attenuation coefficient and the density, μ/ρ, is called the mass attenuation coefficient. If μ/ρ for gamma rays is weight-averaged by the fraction of the gama-ray energy

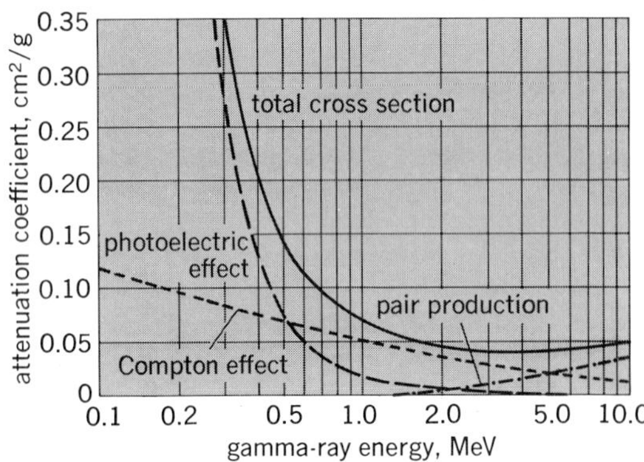

Fig. 3. Gamma-ray attenuation coefficients for lead. (*After G. W. Grodstein*)

locally dissipated subsequent to an interaction, the average μ_a/ρ, is called the mass energy absorption coefficient. The total and component parts of the mass attenuation coefficients for lead are illustrated in Fig. 3. In Fig. 4 are shown total mass attenuation coefficients for all elements at several photon energies.

The flux density, or fluence rate, characterizes the intensity of radiation. It may be thought of as the path length traveled by radiation per unit volume per unit time. It is usually given the symbol ϕ and has units $cm^{-2}\ s^{-1}$.

To illustrate these concepts, consider the flux density at a distance r from a point source isotropically emitting S monoenergetic gamma rays of energy E per second in a uniform medium with total attenuation coefficient μ. If μ were zero, the flux density would be determined just by the inverse square of the distance from the source. If μ were not zero, attenuation of the gamma rays would also be exponential with distance. Thus the equation shown below would hold. This is the flux

$$\phi = \frac{S}{4\pi r^2} e^{-\mu r}$$

density of gamma rays which have traveled distance r without having experienced any interactions in the stopping medium. The energy locally dissipated per unit mass per unit time due to these gamma rays as they experience their first interactions is the product $E(\mu_a/\rho)\phi$. However, some gamma rays reach distance r already having experienced scattering interactions. To account for these secondary gamma rays, a buildup factor B is employed. B is a function of the energy of the source gamma rays and the product μr and depends on the attenuating medium. The total energy locally dissipated per unit mass per unit time is thus the product $BE(\mu_a/\rho)\phi$. Similar concepts apply to neutron shielding; however, the treatment of scattered neutrons is considerably more complicated, and the buildup-factor concept is not well established. In the final phases of shielding design, digital computers are usually employed to carry out the required calculations. It is also common practice to base shielding design, at least in part, on measurements made on a prototype.

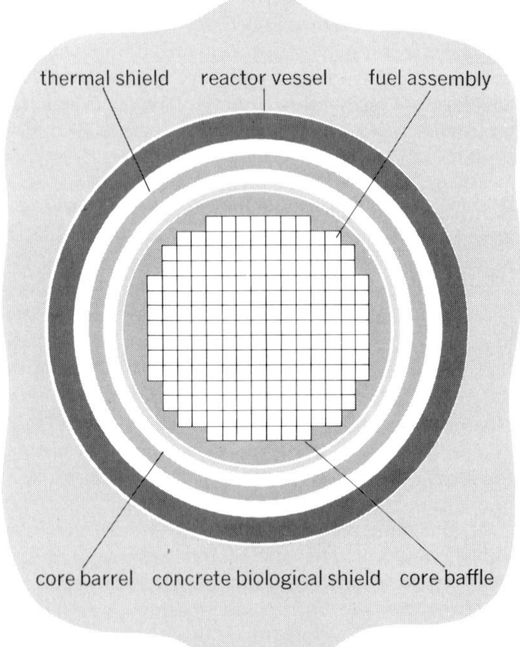

Fig. 5. Typical power reactor shield configuration.

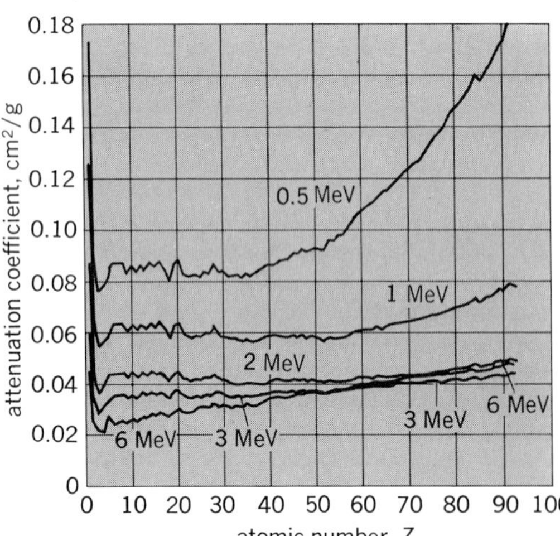

Fig. 4. Gamma-ray attenuation coefficients versus atomic number of absorbing material for various energies.

Shielding materials. The most common criteria for selecting shielding materials are radiation attenuation, ease of heat removal, resistance to radiation damage, economy, and structural strength.

For neutron attenuation, the lightest shields are usually hydrogenous, and the thinnest shields contain a high proportion of iron or other dense material. For gamma-ray attenuation, the high-atomic-number elements are generally the best. For heat removal, particularly from the inner layers of a shield, there may be a requirement for external cooling with the attendant requirement for shielding the coolant to provide protection from induced radioactivity.

Metals are resistant to radiation damage, although there is some change in their mechanical properties. Concretes, frequently used because of their relatively low cost, hold up well; however, if heated they lose water of crystallization, becoming somewhat weaker and less effective in neutron attenuation.

If shielding cost is important, cost of materials must be balanced against the effect of shield size on other parts of the facility, for example, building size and support structure. If conditions warrant, concrete can be loaded with locally available material such as natural minerals (magnetite or barytes), scrap steel, water, or even earth.

Typical shields. Radiation shields vary with application. The overall thickness of material is chosen to reduce radiation intensities outside the shield to levels well within prescribed limits for occupational exposure or for exposure of the general public. The reactor shield is usually considered to consist of two regions, the biological shield and the thermal shield. The thermal shield, located next to the reactor core, is designed to absorb most of the energy of the escaping radiation and thus to protect the steel reactor vessel from radiation damage. It is often made of steel and is cooled by the primary coolant. The biological shield is added outside to reduce the external dose rate to a tolerable level (Fig. 5). [RICHARD E. FAW]

Bibliography: E. P. Blizard, Nuclear radiation shielding, in H. Etherington (ed.), *Nuclear Engineering Handbook*, 1958; E. P. Blizard and L. S. Abbot (eds.), *Reactor Handbook*, vol. 3, pt. B: *Shielding*, 1962; U. Fano, L. V. Spencer and M. J. Berger, *Penetration and Diffusion of X Rays*, vol. 38/2 *of Handbuch der Physik*, 1959; H. Goldstein, *Fundamental Aspects of Reactor Shielding*, 1959; R. G. Jaeger et al. (eds.), *Engineering Compendium on Radiation Shielding*, vol. 1, 1968, vol. 2, 1975, and vol. 3, 1970; A. E. Profio, *Radiation Shielding and Dosimetry*, 1979; R. W. Roussin et al. (eds.), *Nuclear Reactor Shielding*: *Proceedings of 5th International Conference on Reactor Shielding*, 1977.

Radiator

Any of numerous devices, units, or surfaces that emit heat, mainly by radiation, to objects in the space in which they are installed. Because their heating is usually radiant, radiators are of necessity exposed to view. They often also heat by conduction to the adjacent thermally circulated air.

Radiators are usually classified as cast-iron (or steel) or nonferrous. They may be directly fired by wood, coal, charcoal, oil, or gas (such as stoves, ranges, and unit space heaters). The heating medium may be steam, derived from a steam boiler, or hot water, derived from a water heater, circulated through the heat-emitting units.

Cast-iron radiators are made in sections of varying widths and heights and are assembled in the required number by top and bottom nipples. Some are made in a flat panel. They are set on legs or similar supports or affixed to walls by adjustable hangers. The preferred location is under windows.

When windows are low, cast-iron baseboard radiators, less than a foot high, are available for use under the windows. Because of the limited heat from such a radiator, it is frequently extended along the nonwindow baseboard. This radiator is assembled with valve and end sections, extension blocks, and inverted and projecting corner covers. Cast-iron convectors may be used also.

Finned-tube radiators consist of a pipe or tube with affixed fins of steel, copper, or aluminum. They are available with or without enclosures, and usually are wall-mounted. They are more compact than cast-iron radiators, having greater heating area per volume.

Mass production was responsible for the introduction of convectors, which, like finned-tube radiators, consist of steel, copper, or aluminum tubes with extended fin surfaces affixed to them. They have almost completely displaced cast-iron radiators, for the fact that they are lightweight causes freight costs to be lower. They also require less labor to install. They cannot, however, be used as replacements in an existing hot-water heating system with cast-iron radiators for heating surfaces, because the heat-dissipating characteristics of convectors are completely different from those of cast-iron radiators with changes in hot-water temperatures.

Convectors are customarily made of nonferrous finned tubes in a wide variety of enclosures for freestanding, recessed, or wall-hung installation. Air is circulated over the elements by natural convection or by fans, in which case they are known as unit ventilators. Convectors and unit ventilators are not true radiators because they emit most of the heat by conduction to the circulated air.

A limited amount of cooling may be produced by passing chilled water through convectors or radiators, but care must be exercised to dispose of moisture condensation in humid weather.

Electric heating elements may be substituted for fluid heating elements in all types of radiators, convectors, and unit ventilators. *See* HOT-WATER HEATING SYSTEM; RADIANT HEATING; STEAM HEATING. [ERWIN L. WEBER/RICHARD KORAL]

Bibliography: American Society of Heating, Refrigerating and Air Conditioning Engineers, *ASHRAE Handbook and Product Directory*: *Equipment*, 1979.

Radioactive waste management

The treatment and containment of radioactive wastes. Radioactive waste management is required to some degree in all operations associated with the use of nuclear energy for national defense or peaceful purposes. Liquid, solid, and gaseous radioactive wastes are produced in the mining of ore, production of reactor fuel materials, reactor operation, processing of irradiated reactor fuels, and numerous related operations. Wastes also result from the use of radioactive materials, for example, in research laboratories, industrial operations, and medical treatment. The magnitude of waste management operations will increase as the nuclear energy program further extends and diversifies and as a large, widespread nuclear power industry develops. The particular problems associated with the decommissioning of a nuclear facility are discussed in the second part of this article.

ASPECTS OF MANAGEMENT

The chief aim in the safe handling and containment of radioactive wastes is the prevention of radiation damage to humans and the environment by controlling the dispersion of radioactive materials. Harm to humans may result from irradiation by external sources or from the intake (by ingestion, by inhalation, or through the skin) of radioactive materials, their passage through the respiratory and gastrointestinal tract, and their partial incorporation into the body. Radioactive waste contaminants in air, water, food, and other elements of the human environment must be kept below specified concentrations, which differ according to the particular radionuclide or mixture of radionuclides which is present. Liquid or solid waste products containing significant quantities of the more toxic radioactive materials require isolation and permanent containment in media from which any potential escape into the human environment would be at tolerable levels. The radioactive materials of major concern are those that may be readily incorporated into the body and those that have relatively long half-lives, ranging from a few years to thousands of years.

Table 1. Typical materials in high-level liquid waste

Material[b]	Grams per metric ton from various reactor types[a]: Light water reactor[c]	High-temperature gas-cooled reactor[d]	Liquid metal fast breeder reactor[e]
Reprocessing chemicals			
Hydrogen	400	3800	1300
Iron	1100	1500	26,200
Nickel	100	400	3300
Chromium	200	300	6900
Silicon	–	200	–
Lithium	–	200	–
Boron	–	1000	–
Molybdenum	–	40	–
Aluminum	–	6400	–
Copper	–	40	–
Borate	–	–	98,000
Nitrate	65,800	435,000	244,000
Phosphate	900	–	–
Sulfate	–	1100	–
Fluoride	–	1900	–
SUBTOTAL	68,500	452,000	380,000
Fuel product losses[f,g]			
Uranium	4800	250	4300
Thorium	–	4200	–
Plutonium	40	1000	500
SUBTOTAL	4840	5450	4800
Transuranic elements[g]			
Neptunium	480	1400	260
Americium	140	30	1250
Curium	40	10	50
SUBTOTAL	660	1440	1560
Other actinides[g]	<0.001	20	<0.001
Total fission products[h]	28,800	79,400	33,000
TOTAL	103,000	538,000	419,000

SOURCE: From K. J. Schneider and A. M. Platt (eds.), *Advanced Waste Management Studies: High-Level Radioactive Waste Disposal Alternatives*, USAEC Rep. BNWL-1900, May 1974.

[a]Water content is not shown; all quantities are rounded.

[b]Most constituents are present in soluble, ionic form.

[c]U-235 enriched pressurized water reactor (PWR), using 378 liters of aqueous waste per metric ton, 33,000 MWd/MT exposure. (Integrated reactor power is expressed in megawatt-days [MWd] per unit of fuel in metric tons [MT].)

[d]Combined waste from separate reprocessing of "fresh" fuel and fertile particles, using 3785 liters of aqueous waste per metric ton, 94,200 MWd/MT exposure.

[e]Mixed core and blanket, with boron as soluble poison, 10% of cladding dissolved, 1249 liters per metric ton, 37,100 MWd/MT average exposure.

[f]0.5% product loss to waste.

[g]At time of reprocessing.

[h]Volatile fission products (tritium, noble gases, iodine, and bromine) excluded.

Waste management is focused on those radioisotopes which originate in nuclear reactors. Here the fission products (chemical elements formed by nuclear fragmentation of actinide elements such as uranium or plutonium) accumulate in the nuclear fuel, along with plutonium and other transuranic nuclides. (Transuranic elements, also called actinide elements, are those higher than uranium on the periodic table of chemical elements.) The concentrations of plutonium are substantially higher than those found in nature; they range from 10 to 20 kg per metric ton (1000 kg) of uranium, compared to a high of 17 g per metric ton of uranium in minerals from fumarole areas.

Reprocessing. If fuel discharged from the nuclear reactor is reprocessed, uranium and plutonium are recovered after chemical dissolution. During recovery, the favored treatment processes produce high-level waste in the form of an acidic aqueous stream. Other processes are being considered that would produce high-level waste in different forms. The high-level waste contains most of the reactor-produced fission products and actinides, as well as slight residues of uranium and plutonium (see Table 1). These waste products emit large amounts of potentially hazardous ionizing radiation and generate sufficient heat to require substantial cooling. Because the reprocessing step normally does not dissolve much of the nuclear fuel cladding, high-level waste normally contains only a small amount of the radionuclides formed as activation products within the cladding. The cladding hull waste is managed as a separate solid waste stream, as are several other auxiliary waste streams from reprocessing plants.

The nuclear industry can reuse the recovered uranium and plutonium by reconstituting it into nuclear fuels in which plutonium, instead of uranium-235, is the fissile material. Since these fuels contain both uranium and plutonium mixed oxides (MOX), their fabrication generates additional plutonium-containing wastes.

Policy and treatment. The policy of the United States is to assume custody of all commercial high-level radioactive wastes and to provide containment and isolation of them in perpetuity. Regulations require that the high-level wastes from nuclear fuel reprocessing plants be solidified within 5 years after reprocessing and shipped to a Federal repository within 10 years after reprocessing.

Because of the expected increase in the quantities of waste-containing materials or those contaminated with transuranic elements, and because of the long half-life and specific radiotoxicity of these elements, it has also been proposed that all transuranic wastes be solidified and transferred to the United States government as soon as practicable, but at most within 5 years after generation.

Both of these policies require that high-level, cladding, and other transuranic wastes be converted to solid form. A variety of technologies exist for this conversion, including calcination, vitrification, oxidation, and metallurgical smelting, depending on the primary waste.

A typical solidification process, chiefly for high-level waste, is spray calcination-vitrification (Fig.

Table 2. Light-water reactor nuclear wastes from 1 GWe-yr of operation, after 10-year cooling

Unreprocessed spent fuel	
Volume	25 m^3
U	36.4 Mg
Pu	0.31 Mg
Radioactivity	13.0 MCi*
Conditioning waste	
Volume	9 m^3
Radioactivity	0.2 kCi
Reprocessing wastes	
Vitrified high-level waste	
Volume	2.8 m^3
Contained Pu	3.4 kg
Radioactivity	10.4 MCi
Intermediate-level waste	
Volume (in concrete)	52 m^3
Contained Pu	1.7 kg
Radioactivity	0.07 MCi
Hulls and spacers	
Volume (compacted)	7.4 m^3
Contained Pu	0.24 kg
Radioactivity	0.03 MCi
Mixed oxide fuel waste	
Volume (in concrete)	18 m^3
Contained Pu	0.65 kg

*1 Ci = 3.7×10^{10} disintegrations per second = 3.7×10^{10} becquerels.

1). In this process, atomized droplets of waste fall through a heated chamber where flash evaporation results in solid oxide particles. Glassmaking solid frit or phosphoric acid can be added to provide for melting and glass formation, either in a continuous melter or in the vessel that will serve as the waste canister. The molten glass or ceramic is cooled and solidified. At the present time, however, there are no commercial reprocessing plants operating in the United States, and spent fuel is simply being accumulated in water basins. *See* FRIT.

Quantities of waste. Civilian nuclear electrical generating capacity in the United States is projected to increase to about 175,000 to 200,000 megawatts (MW) by the year 2000. Assuming the latter figure, a cumulative total of 50,000 metric tons (or megagrams) of heavy metal (MTHM) will have been discharged from reactors as a result of the production of some 2500 gigawatt-years (or 21.9×10^{12} kilowatt-hours) of electricity.

The Department of Energy has calculated material balances and the amount of waste expected from operation of light-water reactor fuel cycles, either with spent fuel as the primary waste or with reprocessing and plutonium recycle. The characteristics of wastes requiring disposition in a repository are shown in Table 2. The key point is that reprocessing doubles the volume of waste but reduces the actinide content by some fiftyfold.

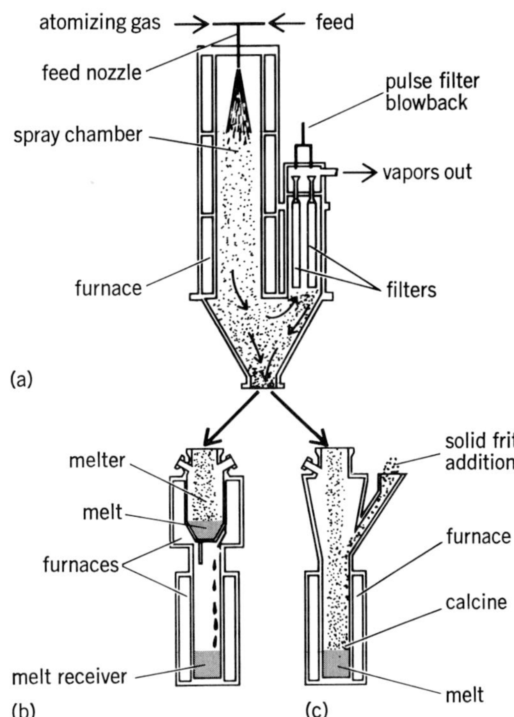

Fig. 1. Spray calcination-vitrification process. (*a*) Spray calciner, producing calcine that drops into either (*b*) continuous silicate glass melter, or (*c*) directly into waste canister vessel for for in-pot melting.

If spent fuel were the primary form of waste, the anticipated packaged waste through the year 2000 would be 62,000 m^3. If this were stacked as a solid cube, it would measure nearly 40 m on a side. About 38,000 megacuries of radioactivity (1.4×10^{21} disintegrations per second) and 175 MW of heat would be associated with this quantity of waste.

In addition to the wastes from peaceful uses of nuclear energy, defense activities in the United States have generated substantial nuclear waste. The 1978 quantities of high-level defense wastes are shown in Table 3, as is the volume which would be occupied by the waste existing in 1990 if it was to be solidified as a glass. The location and the quantities of existing defense-activities transuran-

Table 3. United States' high-level defense wastes

Source site	Present form	Radioactivity, 10^6 Ci*	Volume, 10^3 m^3†	Solidified volume, 10^3 m^3‡
Hanford Plants	Alkaline salt/sludge/liquor	190	173	13
	Separated ^{90}Sr/Y, ^{137}Cs/Ba	210		–
Idaho Chemical Processing Plant	Acid calcine	70 (calcine and liquid combined)	1.5	8 (calcine and liquid combined)
	Acid liquid		10	
Savannah River Plant	Alkaline salt/sludge	560	82	6

*1 Ci = 3.7×10^{10} disintegrations per second = 3.7×10^{10} becquerels.
†As of 1978.
‡Volume if waste existing in 1990 were solidified as a glass.

Table 4. Existing United States' defense transuranic waste, in $10^3 m^3$*

Storage site	Buried	Retrievably stored	Total
Hanford, WA	153	8	161
Idaho National Engineering Laboratory, ID	65	36	101
Los Alamos Scientific Laboratory, NM	116	2	118
Oak Ridge National Laboratory, TN	5	1	7
Savannah River Laboratory, SC	28	2	30
Nevada Test Site	<1	<1	<1
Total	369	49	418
Transuranic content, kg	700	374	1100

*As of 1978.

ic wastes as of 1978 are estimated in Table 4.

Before 1970, the Atomic Energy Commission allowed direct burial of transuranic wastes at selected sites. Since that date, policy has directed that the wastes be packaged and stored in such a way that they can be retrieved easily.

Spent fuel packaging. Probably the most comprehensive study on packaging spent fuel has been conducted by the Swedish project Kaernbraenslesaekerhet (KBS) set up in early 1977. The KBS plan calls for 40 years of water pool storage of spent fuel in a granite cavern some 30 m underground.

After the 40-year storage to let the heat dissipate, groups of 500 fuel rods (1.5 MTHM) would be loaded in copper canisters (Fig. 2). The canisters would then be filled with lead and copper covers welded onto the tops. Each canister would weigh about 20 MT.

The canisters would then be transferred to a final storage/disposal location in granite some 500 m underground. Here they would be placed in holes some 7.7 m deep and 1.5 m in diameter, lined (sides, top, and bottom) with 40 cm of isostatically compressed bentonite.

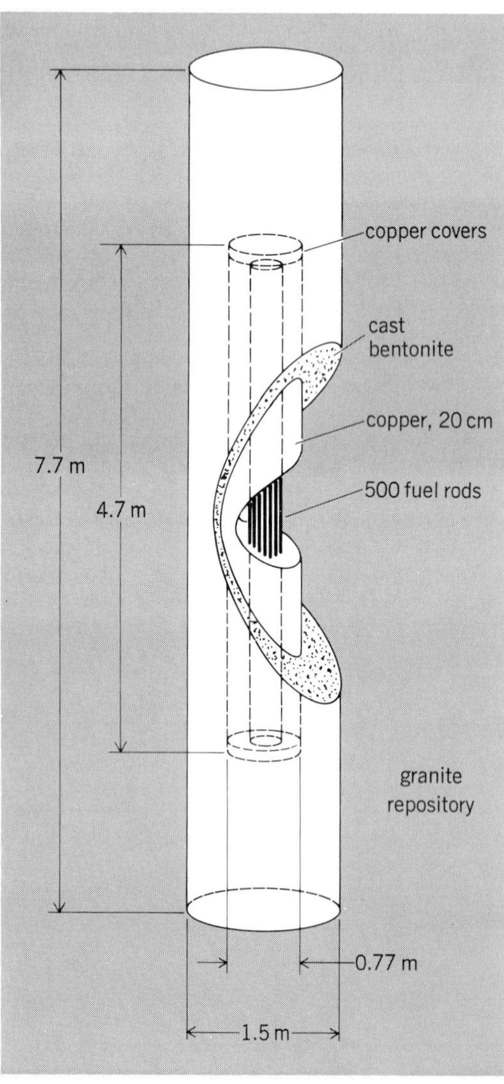

Fig. 2. KBS spent fuel concept.

Early investigation by the KBS of three geologic study areas in granite showed the possibility of water flows of 0.1 to 0.2 liter/m^2/yr. This, in part, was the motivation for the sophisticated packaging system.

In the United States, several options are being investigated for encapsulating the spent fuel. The baseline option is placement of spent fuel in canisters with only an inert-gas fill. Other, more advanced methods are under study as technical alternatives. These include a metal matrix fill, sand fill, other glassy or ceramic materials, and multiple-barrier encapsulation of the spent fuel and canister at the time it is declared a waste for disposal. Experimental packaging and dry storage of spent fuel using facilities previously associated with the nuclear rocket program in Nevada have been demonstrated at the Engine Maintenance and Disassembly (EMAD) facility.

High-level waste packaging. If the United States reprocessed spent fuel, the high-level waste would be solidified as a glass, ceramic, or metal matrix material, encapsulated in a metallic container, and possibly surrounded in the repository by additional barriers. Table 5 gives some characteristics of a typical canister. After 1000 years the canister would generate only 20 W, and the gamma dose rate would be only 1.6 roentgens/hr (1.147×10^{-7} coulomb/kg·s) at the centerline of the cannister 30 cm from the surface.

Repositories and disposal. The major worldwide thrust has narrowed from the many waste management concepts previously described to the investigation of the sub-seabed and mined cavities in geologic formations as the principal alternatives. Both continental and sub-seabed geologic formations exist that have been physically and chemically stable for millions of years.

The basic requirement for acceptable final storage or disposal of radioactive waste is the capability to contain and isolate the waste safely until decay has reduced the radioactivity to nonhazardous levels—or at least to levels found in nature.

Exactly how long near-total containment and isolation of nuclear waste must be maintained is much debated. However, two characteristics of waste change radically in a few hundred years. First, the heat generation rate decreases an order of magnitude in the period of 10 to 100 years and another order of magnitude in the period of 100 to 1000 years.

Second, as shown in Fig. 3, the toxicity (expressed as the quantity of water required to di-

Table 5. Typical high-level waste canister

Decay time, years	Heat generation rate, kW	Cumulative dose, alpha particles per gram	Dose rate 30 cm from surface, roentgens/hr*
1	22	1.0×10^{17}	1.1×10^{6}
10	3.1	2.5×10^{17}	6.2×10^{4}
100	0.36	7.1×10^{17}	5.8×10^{3}
1000	0.02	1.5×10^{18}	1.6
10,000	0.006	3.0×10^{18}	1.3
100,000	0.003	6.1×10^{18}	0.6

Length: 3 m
Diameter: 30 cm
Can material: 304L stainless steel
Volume: 0.21 m^3
Contents: 2.5 metric tons uranium equivalent

*1 roentgen/hr = 2.58×10^{-4} C/kg · hr = 0.717×10^{-7} C/kg · s.

lute to drinking-water tolerances) of the high-level waste needed to produce 1 GW-yr of electric power decreases about three orders of magnitude in the first 300 to 400 years due to decay of short-lived fission products. It is then at levels comparable with those of an equal volume of average ores of common toxic elements. After this time, toxicity diminishes slowly, a million years being required for another two orders of magnitude. Thus it is clear that the first 300 to 400 years of the disposal period are the most critical.

Mined cavity disposal. In this disposal option, radioactive wastes would be emplaced in the walls or floor of deep (600-m) tunnels created by conventional mining techniques. Although there are numerous geologic media that could be considered, only salt, granite, basalt, and shale have been studied extensively. Since the repository should provide both containment and isolation, site selection

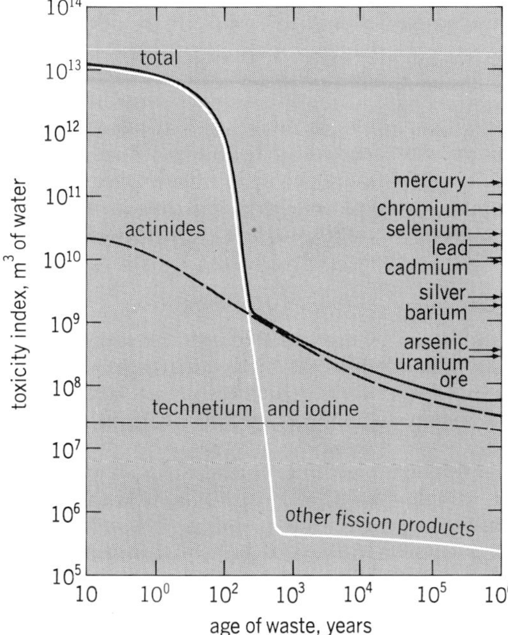

Fig. 3. Toxicity index of high-level waste resulting from production of a gigawatt-year of electric power, and of volumes of average ores of common toxic elements equal to repository volume required by the waste.

will involve consideration of the properties, dimensions, and characteristics of the host rock, the hydrologic properties of the site, its tectonic stability, its resource potential, and the capability of the site geohydrology to provide natural barriers to the movement of waste.

These natural barriers will be further augmented, as previously described, by a solid waste form, by canisters, and possibly by engineered barriers such as absorption backfill and overpack materials.

Another prime design consideration is the initial and long-term heating of the repository site by the radioactive decay process. Preliminary designs are for initial heating from 100 to 350 kW per hectare. Thus a repository area of 1000 to 3500 hectares would be required to accommodate the nuclear wastes which are expected to be generated through the year 2000.

Sub-seabed geologic disposal. Analysis of the ocean regimes has shown that the most appropriate areas for sub-seabed disposal are abyssal hill regions in the centers of sub-ocean tectonic plates underlying large ocean-surface currents known as gyres. These abyssal hill regions are vast, are remote from human activities, have few known natural resources, are biologically unproductive, have weak and variable bottom currents, and are covered with red clays to a depth of 50 to 100 m.

These clay sediments are currently the prime sub-seabed geologic media under consideration for radioactive waste containment. They are soft and pliable near the sediment-water interface and become increasingly rigid and impermeable with depth. Tests have shown that these sediments have high sorption coefficients (radionuclide retention) and low natural pore-water movement. They are found by surface acoustic profiling to be uniformly distributed over large areas (tens of thousands of square kilometers) of the ocean floor. Core analysis has shown that deposition in these areas has been continuous and undisturbed for millions of years, so that they can confidently be predicted to remain stable long enough for radionuclides to decay to innocuous levels.

The multiple barriers in this option consist of the waste form and canister for short-term containment and the sediments for long-term radionuclide retention. Predisposal waste treatments (for example, dilution and longer surface storage) will decrease the thermal-related problems during the early disposal period. The main technical problem yet to be addressed is the response of the geologic medium (red clay) to the heat given off by reprocessed high-level waste or spent fuel in the first 500 years.

One emplacement method being studied is the encasement of the wastes or spent fuel in a needle-shaped projectile which, propelled from a ship, will penetrate into a predetermined location in the sediments. It can thereafter be monitored for location, attitude, condition, and temperature through self-contained instrumentation. Laboratory tests have led to the conclusion that the red clay, thus dynamically penetrated, will reseal itself above a buried canister in a comparatively short time. Only about 0.006% of the available abyssal hill area in the central North Pacific area, for instance, would be needed to dispose of all high-level wastes generated in the United States through the year 2040.

[ALLISON M. PLATT]

DECOMMISSIONING OF A NUCLEAR FACILITY

Because of the buildup of radioactive contamination in a nuclear facility such as a nuclear reactor or reprocessing plant, detailed plans must be made, beginning with the design and construction stages of the facility, to provide adequate protection from radiation exposure during the final decommissioning stage. Some of the radionuclides constituting this contamination have half-lives of many thousands of years, and the levels of contamination in the tanks, equipment, plumbing, instruments, hot cells, and the building itself may reach levels of millions of curies (1 curie $= 3.7 \times 10^{10}$ disintegrations per second $= 3.7 \times 10^{10}$ becquerels). Thus in many cases it is necessary to use special techniques, remote-control equipment, and large shields to remove or isolate this contamination in a safe manner from the human environment for hundreds of thousands of years. First, all high-level radioactive materials, such as spent reactor fuel, equipment for reprocessing fuel, and underground tanks for radioactive liquid waste, are removed to an approved site for disposal of such waste. Then surfaces are scoured with remotely operated equipment using water, steam, jets of sand, and various chemicals. Sometimes the equipment and the facility are completely demolished by demolition balls and explosives, but in that case extreme care must be exercised to prevent spread of the radioactive contamination to the environment.

There are four principal methods of decommissioning in use: layaway, mothballing, dismantlement, and entombment. Layaway and mothballing are temporary measures to postpone final action until some of the radioactivity has decayed. They involve removing the easily removable high-level radioactive objects. With layaway the facility might be put back into use at some future date, but this is unlikely. Both layaway and mothballing require security measures, such as locked or welded doors and guards on constant duty, to prevent the entry of unauthorized persons who might be exposed to high-level radiation. The barriers to reentry of contaminated areas are more secure and rigid with mothballing than with layaway, so that in the former case fewer guards are required and some uncontaminated areas of the plant may be released to public use. Both of these methods require dismantlement or entombment at a later date unless the levels and half-lives of the residual contamination are very low. Entombment, as the name implies, is making the radioactive contamination inaccessible by the use of demolition techniques and covering the residue with reinforced concrete. The choice of decommissioning method depends on the levels and half-lives of the residual contamination, and on cost, environmental factors (rain, wind, hydrology), population density, and many other factors. In any case, the objective should be to take those measures that will result in minimum radiation exposure and minimum consequent radiation damage to persons now living and those to be born in the future. *See* NUCLEAR FUEL CYCLE; NUCLEAR FUELS REPROCESSING; NUCLEAR REACTOR.

[KARL Z. MORGAN]

Bibliography: Department of Energy, *Draft Environmental Impact Statement: Management of Commercially Generated Radioactive Waste*, DOE/EIS0046-D, April 1979; H. W. Dickson, *Standards and Guidelines Pertinent to the Development of Decommissioning Criteria for Sites Contaminated with Radioactive Material*, Oak Ridge National Laboratory, ORNL/DEPA-4, August 1978; Nuclear Regulatory Commission, *Technology, Safety and Costs of Decommissioning a Reference Nuclear Fuel Reprocessing Plant*, NUREG-0278, October 1977; A. M. Platt and J. L. McElroy, *Management of High-Level Nuclear Wastes*, Pacific Northwest Laboratory, PNL-SA-7072, September 1979; K. J. Schneider and A. M. Platt (eds.), *Advanced Waste Management Studies: High-Level Radioactive Waste Disposal Alernatives*, USAEC Rep. BNWL-1900, May 1974; R. I. Smith, G. J. Konzek, and W. E. Kennedy, Jr., *Technology, Safety and Costs of Decommissioning a Reference Pressurized Water Reactor Power Station*, vols. 1 and 2, Nuclear Regulatory Commission, NURE G/eR-0130, 1978.

Railroad engineering

That part of transportation engineering involved in the planning, design, development, construction, maintenance, and use of facilities for the transportation of goods and people in wheeled units of rolling stock running on, and guided by, rails normally supported on cross ties and held to fixed alignment.

UNITED STATES EXPERIENCE

Railroads in the United States have undergone many changes since their beginnings in 1830. The late 1800s saw rapid building of the rail network throughout the country and the standardization of track gage, couplers, and braking systems, thus allowing the many individual companies to operate as a single system. At the turn of the century, railroad mileage was still expanding, with miles of line peaking in 1916 at 254,037 (408,834 km) and miles of track reaching a high in 1930 (Table 1). Consolidations and mergers, as well as abandonment of excessive branch line trackage, have occurred in the years since. Lines constructed since World War II have most often been extensions to industrial developments and to sites of large mineral

Table 1. Trackage trends

Year ended	Amount of operating railroads, mi*	Number of operating railroads
June 30, 1900	192,556	1224
June 30, 1905	216,974	1380
June 30, 1910	240,831	1306
June 30, 1915	257,569	1260
Dec. 31, 1920	259,941	1085
Dec. 31, 1925	258,631	947
Dec. 31, 1930	260,440	775
Dec. 31, 1940	245,740	574
Dec. 31, 1950	236,857	471
Dec. 31, 1960	230,169	407
Dec. 31, 1965	226,015	372
Dec. 31, 1970	220,107	351
Dec. 31, 1975	215,105	340
Dec. 31, 1977	209,332	320
Dec. 31, 1979	184,700	320
Dec. 31, 1980†	179,000	NA

*1 mi = 1.6 km.
†Estimated.

resources and relocations to improve gradients and alignments.

Merger movement. In 1907 there were 1564 line-haul railroad companies, in 1981 about 300. Since the last half of the 19th century, railroads have been undergoing recurrent cycles of consolidation. The depression of 1884–1885 was a major factor in this early consolidation movement. Between 1880 and 1888, 25% of the nation's railroad companies were brought under the control of other railroads. A more recent cycle of consolidations began in 1955 and, by the end of 1980, had resulted in 51 unifications and several other major consolidations pending before the Interstate Commerce Commission. The largest corporate reorganization in history resulted from the Regional Rail Reorganization Act of 1973, which provided for the establishment of the Consolidated Rail Corporation (Conrail), a Federally subsidized, for-profit entity which, on April 1, 1976, commenced operation of some 17,000 miles (27,000 km) of line formerly owned by six bankrupt lines in the northeast and midwest sections of the country.

Fig. 1. Railroad cars loaded with highway trailers or containers. (*Southern Pacific Transportation Co.*)

Fig. 2. At a major piggyback yard, a gantry crane lifts trailers from the railroad flat cars and places them for pickup by trucks. The wheelless containers in the foreground move in international service and may be loaded on a ship deck, railroad flatcar, and flatbed truck at various times during their journeys. (*Santa Fe Railway*)

Traffic trends. From an engineering viewpoint, it is of special interest to note that since the mid-1950s there has been an upward trend in the number of railroad cars loaded with highway trailers or containers (Fig. 1) and in the number of assembled automobiles and trucks moving by rail. Piggyback traffic, combining long-haul fuel economy of rail shipment with the terminal pickup and distribution flexibility of truck transportation is a major innovation for handling freight. This has generated the design and construction of many new loading and unloading facilities at terminal

Fig. 3. A coal-exporting facility. (*Chessie System*)

points (Fig. 2). With the worldwide demand for coal and the resultant increases in coal traffic, it is expected that the 1980s will see continued upgrading of the railroads' coal-hauling lines and major expansion of coal export facilities (Fig. 3). New lines will be constructed to some major coal deposits.

[ANNE O. BENNOF]

FREIGHT HANDLING SYSTEMS

In the United States and in most other countries the part of railroad engineering concerned with fixed properties, including tracks, roadbeds, bridges, buildings, yards, and terminals, is considered to be a part of the broad discipline of civil engineering. In most railroad organizations these functions are performed in an engineering department headed by a chief engineer, although in some larger organizations the department is designated engineering and maintenance of way. In a like manner the part of railroad engineering concerned with rolling stock, including locomotives, cars, car ferries, and other such equipment, is considered to be a part of the broad discipline of mechanical engineering. In most railroad organizations these functions are performed in a mechanical department. A signals and communications department handles the engineering functions involved in the construction and maintenance of such facilities. In many organizations this is a subdepartment of the engineering department. In addition to civil engineers and mechanical engineers, the staffs of railroad organizations include many architects and chemical, electrical, and industrial engineers.

Motive power. Probably the most important transformation that has taken place on railroads in the United States in recent times was the change from steam to diesel power, which was started prior to World War II and was almost complete by 1960. Electric locomotives are used on some 700 mi (1100 km) of Conrail lines, with only a scattering of such power at other locations (see Table 2). *See* DIESEL ENGINE.

Diesel-electric locomotives are designed and built in certain sizes and types by the manufacturers to suit the generalized needs of the railroad industry. Diesel locomotives have much greater availability for service than steam-powered locomotives. Standardized units are comparatively economical to build, and diesels require less time in the shop for repairs. They are properly termed diesel-electric locomotives, and their power comes from oil-burning internal combustion engines driving electric generators which supply power to electric traction motors.

Fig. 4. Low-profile piggyback car.

Diesel locomotive units have been developed with an output of 6000 hp (4.5 MW). Units up to 3600 hp (2.7 MW) are in common use for line-haul movement. Small units are used in switching service. Multiple units of two or more are used with one locomotive crew. During 1967 a 500-car train hauled 48,000 tons (43,500 metric tons) of coal, using six diesel units aggregating 21,600 hp (1.61 MW). Three of the units were on the head end of the train, and three slave units were farther back—all controlled from the head end.

It is the function of the professional engineer in the mechanical department of the railroad to suggest to manufacturers the desired characteristics of the locomotive units and to recommend to railroad management the particular sizes and

Table 2. Locomotives on class I railroads*

Year (average)	Diesel-electric locomotive units	Steam locomotives	Electric locomotive units	Other locomotives
1926–1930	28	58,979	537	9
1931–1935	89	49,998	745	22
1936–1940	396	42,316	833	29
1941–1945	2,386	39,475	857	19
1946–1950	8,647	32,035	817	19
1951–1955	21,761	12,849	704	24
1956–1960	27,457	1,705	556	36
1961–1965	27,888	52	419	43
1966–1970	27,063	20	303	27
1971–1975	27,444	12	234	20
1976–1978	27,463	12	208	
1980	28,483	12	168	

*By Interstate Commerce Commission designation, class I railroads are those companies having operating revenues of $50,000,000 or more per year.

types that should be purchased. These railroad engineers in the mechanical department are also responsible for the servicing, maintenance, and repair of locomotives and other rolling stock.

Rolling stock. Many different kinds of freight cars have been developed. Specially equipped long flat cars with tie-down devices are used for handling trailers in piggyback service; similar cars are equipped to carry containers, which are like trailer bodies but are loaded on rail cars without any roadway wheels. Figure 4 shows an example of a unit designed to carry truck trailers on low-profile, lightweight cars consisting of only a center sill and platforms for the trailer wheels and gear rather than a complete flat car deck. This innovation has resulted in significant fuel savings. Other innovations include designs to carry double stacks of containers and low-profile piggyback cars which carry trailers in a well to allow clearance through eastern tunnels. Other long flat cars are equipped with fully-enclosed bilevel and trilevel racks for handling assembled automobiles and trucks (Fig. 5). High cube cars, longer and higher than ordinary boxcars, are used in the shipment of automobile parts. Large open-top cars are used in hauling wood chips. Covered hopper cars, some of large capacity, are used for shipment of cement, grain, flour, and other dry bulk commodities (Fig. 6). Large-capacity tank cars are in demand for liquid bulk shipments. Most refrigerator cars (Fig. 7) are equipped with gasoline engine – driven mechanical refrigeration units instead of the older ice-filled types.

Fig. 6. Covered hopper cars are used for transporting grain, flour, cement, and other dry bulk commodities. (*Norfolk and Western Railway*)

Fig. 5. Rack cars, bilevel or trilevel, for transporting new automobiles and trucks from factories to central distribution centers. (*Santa Fe Railway*)

Signaling and control. Main lines are equipped with signal systems to aid in the expedition and safety of train movements. Increasingly the signaling facilities are augmented by two-way communication with train crews.

Signal systems. Automatic wayside signals located on poles adjacent to the tracks or on bridge supports over the tracks are generally used for automatic block signal systems to control and protect train movements. Cab signals located inside the locomotives are used on some lines. Centralized traffic control (CTC) is used on a constantly increasing number of lines to permit control from a single location of signals and passing track switches over long distances. A single main track with CTC approaches the traffic-handling capability of a double-track line.

Design, installation, and maintenance of signal systems are functions of the railroad signal engineer. Signal engineers also design, install, and maintain signals and automatic gates used for warning motorists at highway-railroad grade crossings of approaching trains.

Communications. The railroad communications engineer is responsible for telephone, telegraph, teletype, telex, microwave, radio, and closed-circuit television facilities used in railroad operation. The more sophisticated of these methods of communication are rapidly increasing in use. In some railroad organizations, signals and communications are the responsibility of a single department. In other organizations these facilities are the re-

Fig. 7. Three 2500-hp (1.86-MW) diesel freight locomotives coupled to a train of loaded refrigerator cars for a fast run to get perishable products from produce farms to consumer markets in the fastest possible time. (*Missouri Pacific Railroad*)

sponsibility of the engineering department, and the people who are directly concerned with them constitute a subdepartment.

Automation. Electronic computers are widely used by railroads for data processing and for storage and retrieval of information. There are two major programs which function systemwide under centralized control from Washington, DC: Universal Machine Language Equipment Register (UMLER) enables railroads to determine instantly, by computer inquiry, the physical characteristics of any freight car in North America. TRAIN II, the modern, expanded version of TeleRail Automated Information Network (TRAIN), processes reports received through computers from all major railroads and makes available to the Association of American Railroads and to individual railroads, upon inquiry at any time, the location, consist, and destination of every freight car in service.

Another widely used automation device is the hotbox detector, which locates and records overheated car axle bearings in moving trains, thus permitting the defective car to be set out before complete journal failure and possible derailment. The use of these detectors, coupled with the continually greater use of roller bearings on freight cars, reduced hotboxes by more than 90% over a period of 15 years.

Fixed properties. In handling the design, construction, and maintenance of the fixed properties of a railroad, the engineering department occupies an important place in the headquarters staff and also has people located in division and field offices at other important locations on line. The variety of fixed facilities requires some subdivision of the duties of the engineering staff into specialty units. The extent of this subdivision depends upon the size of the railroad.

A study of the committee structure of the American Railway Engineering Association (AREA) shows the general nature of these specialties. The objective of AREA is "the advancement of knowledge pertaining to the scientific and economic location, construction, operation and maintenance of railways." The AREA *Manual of Recommended Practice* is an important guide to railroad engineers responsible for fixed properties. The manual, under continuing review by 21 standing committees, is an excellent reference document.

Railway. Roadway and track constitute a major unit of railroad fixed properties and require a large portion of maintenance expenditures. AREA has six committees with primary interest in these facilities: Roadway and Ballast, Ties and Wood Preservation, Rail, Track, and Concrete Ties.

Fig. 8. Extended lengths of rail with welded joints being placed in position. Welded joints reduce maintenance costs and provide a smoother riding track. (*Chesapeake and Ohio Railway*)

Most of the new rail placed in tracks has welded joints. The elimination of bolted joints makes smoother, quieter riding track, reduces maintenance, and increases the service life of the rail. Standard 39-ft (11.9-m) lengths of rail are welded at a centrally located plant into continuous lengths of 1200 ft (365 m) or more. These lengths are sometimes field-welded together after placement in the track (Fig. 8).

Bridges and buildings. Another important unit of fixed property facilities consists of bridges, buildings, and other structures. AREA committees concerned with these facilities are Buildings, Timber Structures, Concrete Structures and Foundations, and Steel Structures. *See* BRIDGE.

Yards and terminals. The Yards and Terminal Committee of AREA is concerned with an increasingly important group of facilities, particularly the modern automated gravity-operated hump classification yards, which expedite the movement of freight cars through terminal areas with a minimum use of locomotive power. These yards use automation to achieve high efficiency. As entire trains are moved continuously over the hump, the cars are uncoupled at the crest and allowed to move by gravity into their assigned classification tracks, about 40 or 50 in number, for movement to other destinations (Fig. 9). The destination of each car is selected by number, and automatic devices throw the proper switches for the route; weigh the car (if record weight is desired); sense its rollability as determined by its weight, its inherent rolling characteristics, and the prevailing wind and temperature; and adjust its speed by automatic car retarders; the retarders apply the necessary amount of external braking to the wheels so that it moves at a safe speed, which is determined by the distance it must roll in the receiving track to couple with the car ahead of it and which will not damage its lading.

Other interests. The further scope of interest of the railroad engineer is illustrated by the other committees of AREA. The Committee on Engineering Education is concerned with the education of engineers for railroad staff service. Emphasis on economics is highlighted by two committees: Economics of Plant, Equipment and Operations, and Economics of Railway Construction and Maintenance. The Committee on High-Railway Programs is concerned primarily with the construction and maintenance of highway-railway grade separations and traffic control devices at grade crossings.

The titles of other AREA committees indicate rather clearly their scope of interest: Clearances, Electrical Energy Utilization, Engineering Records and Property Accounting, Environmental Engineering, Maintenance of Way Work Equipment, Scales, and Systems Engineering.

Physical features. Some important physical features of United States railroads are discussed in the following paragraphs.

Track gage. Almost all roads use the standard gage of 4 ft $8\frac{1}{2}$ in. (1.4351 m). A few lines in mountainous country, known as narrow-gage lines, use

Fig. 9. Automated hump classification yard at Bellevue, Ohio, receives freight trains at top center. The hump (just below the highway bridge in this view) is the highest point in the yard. From the hump, cars move by gravity at controlled speeds into predetermined tracks. Five groups of eight tracks each (at the viewer's left) are now in use; the space at right is for future expansion. Car repair tracks are along center of yard, and the main tracks extend around the outside. (*Norfolk and Western Railway*)

a 3-ft (0.9144-m) gage. Some urban area transit lines that do not interchange traffic with other railroads use other gages, generally wider than standard.

Alignment. Alignment of tracks and speed of operation must be correlated. On high-speed main lines, maximum curvature of 0°30′ (rad 11,459 ft or 3493 m) allows operation at speeds up to 160 mph (257 km/h). For more normal main tracks, maximum speeds of 100 mph (161 km/h) can be used on curves that are no sharper than 1°15′ (rad 4584 ft or 1397 m). For 80 mph (129 km/h), curves may be 2°00′ (rad 2865 ft or 873 m). Freight lines with speeds of 60 mph (97 km/h) maximum may have curves of 2°30′ (rad 2292 ft or 699 m). Sharper curves require lower speeds. On slow-speed yard tracks, curvature limitations are determined by the kind of equipment operated over them. Where piggyback cars, automobile rack cars, and high cube cars—all approximately 90 ft (27 m) in length—are handled, curves should be no sharper than 15° (rad 382 ft or 116.5 m). In modern industrial track layouts it is desirable to have curves no sharper than 12° (rad 477 ft or 145 m). Diesel locomotives can negotiate curves of 20–40° (rad 286–143 ft or 87–43.5 m), depending on the particular type of wheel and truck arrangement.

Grades. The gradient should be designed for the traffic to be handled and must take into account the topography of the area and the economics of building the line with a light grade in comparison with the economics of using more power when the grades are steeper. The maximum grade on a particular locomotive district between major terminals determines the amount of power required. Usually a grade not in excess of 0.5% is considered satisfactory for main-line operations. However, in generally flat country maximum grades of 0.3% are sometimes used. In mountainous territory main-line grades of 1.5% or even over 2.0% are sometimes justified.

Clearances. Minimum vertical clearances of 22 ft (6.7 m) above the top of the rail are desirable and are required for most structures. Desired minimum lateral clearances are $8\frac{1}{2}$ ft (2.6 m), measured from the center line of track. Lateral clearance along curves depends on radius of curvature and maximum length of rolling stock.

Rail. Much of the original construction of railroads in the United States was with 56-lb and 63-lb rail (or 28 and 31 kg per meter). Rail sections (specified in pounds per yard) have been increased and modified many times in the past and are now up to a maximum of 155 lb (or 77 kg per meter).

Crossties. Most crossties used by United States railroads are of pressure-creosoted wood. Cross-section dimensions vary from 6 by 6 in. (15 by 15 cm) to 7 by 9 in. (18 by 23 cm). Tie lengths are 8 ft 6 in. (2.6 m) or 9 ft (2.7 m). Prestressed concrete crossties have been used to a limited extent since 1960. Concrete crossties cost more per tie, but they are spaced somewhat farther apart in the track. Whether concrete ties are economically competitive depends upon their service life experience, which is yet to be determined. *See* TRANSPORTATION ENGINEERING.

[WILLIAM J. HEDLEY]

Bibliography: American Railway Engineering Association, *Manual of Recommended Practice*, 1957, supplemented annually; *American Railway Signaling Principles and Practices*, Association of American Railroads, 1962; William W. Hay, *Railroad Engineering*, 1953.

PASSENGER SYSTEMS

Experience has shown that, as a country develops, its need for a reliable and efficient railway system grows. There are many distinct characteristics and functions of the guided rail transport system that cannot be achieved as efficiently, as economically, or as acceptably within environmental constraints by other modes of transport. This is particularly true of the passenger services.

Transportation systems in industrialized nations are subject to increasing pressures from population expansion, congestion and overcrowding, energy shortages, and rising costs. Increased utilization of the railroads to bring more balance into the transportation sector could be important in the solution of these problems.

Japan. The post–World War II Japanese experience has been a model of development combined with a dedicated drive for technological excellence. By 1957 the mainline corridor between Tokyo and Osaka had become overcrowded, and forecasts of transport demand for both motor vehicle and rail movement predicted a condition beyond the capability of either system. Construction of Japan's first high-speed toll road started between those regions in 1957. In that same year plans were completed for a separate standard-gage rail line in this same corridor. Construction of the highway and the rail line progressed virtually side by side. The high-speed rail passenger system, Shinkansen, opened in 1964. During the ensuing 17 years this system changed the history of railroading throughout the world. It demonstrates that a railroad is a whole system, and to be commercially and operationally successful, all components must be developed and operated together. The Shinkansen system is considered one of the finest technological achievements of the 20th century. It is very profitable and reliable, operating comfortable trains that run frequently. It has an excellent safety record and has transported more than 1,600,000,000 passengers during its years of service without a single casualty.

France. TGV (Très Grande Vitesse), the French high-speed passenger system, began operations in September 1981. Serving the Paris-Lyon route, it covers 426 km (265 mi), a distance comparable to the Washington, DC–New York City route in the United States. It has a new roadbed that takes advantage of flat or rolling terrain by following the natural contours, with no tunnels. This roadbed has the general appearance of a slowly curving motorway. However, while it is less than half the width of a four-lane highway, it has three times the passenger-carrying capacity per hour.

Avoiding tunnels provided significant economies in the initial construction costs and also removed the possibility of aerodynamic problems that can be encountered when very-high-speed trains pass each other underground. However, such a design requires that the French system accept 3.5% gradients, while the accepted international standard is 1.5%. The TGV electric power system, rated at 6300 kW (nearly 8500 rated horsepower), easily overcomes aerodynamic drag at high speed and is available for these steep grades. A TGV trainset climbing at 260 km/h (162 mph) on full power will run for 3.5 km (2.17 mi) and rise 122 m (400 ft) vertically as its top speed falls to an acceptable 220 km/h (137 mph). To keep the rise and fall of the roadbed within the 3.5% gradient limits, there are land-fill embankments and deep cuts, the deepest being 40 m (131 ft).

While the original design permitted a 30% reduction in initial construction costs, the operating costs of the TGV system are high because of the power required to maintain high speeds on the steep gradients. It is anticipated that high fuel costs eventually will require a modification of the design of the roadbed, with construction of tunnels and smoothing of the excessive gradients.

United States. After World War II the United States railroads began to cast their lot in favor of their freight services to the great detriment of passenger operations. At first there were subtle, almost undetectable, but significant changes. With the availability of more powerful locomotives and the invention of the Janney coupler, along with the air brake, it became possible to make up trains of great length. By 1945 trains of 120 to 130 cars were common. These long trains did appear to reduce operating costs, but it was later discovered that track damage exceeded the saving achieved by purchasing larger vehicles. The very slow speed of operation created another handicap. Whereas the average number of cars per train had been 47.6 in 1929, by 1967 this had reached 70.5, an increase of 48%. The length of the average car also increased, from 38 ft (11.6 m) to 55 ft (16.8 m), so that the average train length increased by 110% (from 1802 ft to 3808 ft) between 1929 and 1970. These developments have been most important.

As trains get longer and heavier, they must get slower. The average freight train speed in the United States (calculated by dividing total train miles by total train hours) is 20.1 mph (32.3 km/h). Such a train must have a flat track, which is unsuitable for passenger operations, since it is slow, uncomfortable, costly, and potentially dangerous.

The ideal railroad track is designed to link two points by as straight a line as possible. The international standard for high-speed rail passenger services calls for a gradient not to exceed 15 units in a thousand and for curvatures with a radius of no less than 4000 m. Then the speed regime for the route is established in consideration of competitive market requirements, the terrain, and the locomotive power available. It is this speed parameter developed from a desired elapsed time between any two points along a rail right of way, that is, origin and destination, which then determines the geometric design and configuration of the track structure between those two points. Once the track has been constructed, all trains must be operated over that segment of track within the design speed regime, that is, neither too fast nor too slow. Thus the design speed produces a condition of equilibrium, with equal weight on all wheels, of each car through all curves. Once the speed parameter is established and built into the system, there must also be a weight limit stipulation.

Heavy freight cars in very long, slow trains do tremendous damage to track structure. In order to guard against this excessive wear, an upper weight limit is set, somewhere in the vicinity of 25 tons (23 metric tons), on the axleload of all freight cars. Consideration of the passenger service problems in the United States—which are very serious on track that has been subjected to this deterioration—requires understanding of the effect of very slow speeds and heavy freight car operation on track. Because the freight-hauling operators have reduced the design of their track to accommodate on average the speed of 20.1 mph (32.3 km/h), the passenger services of the United States average just over 40 mph (64 km/h) as a result of a steady decline in speed limits since 1971, and even at 40 mph the ride is rough.

As the trains were forced to slow for flattened curves, they ran late consistently, and their schedules were lengthened as a consequence. Slower track carrying heavier freight became uncomfortable and unreliable. Clientele who preferred train travel were forced to turn to alternatives just when the nation had embarked on an interstate highway program and the airlines were converting to the above-the-weather jet-propelled aircraft.

The passenger services were all but immobilized by the heavy freight, slow speed, and flat track. Orders for new passenger cars were discontinued. Car manufacturers closed their production lines and turned to other business. These primary car manufacturers depended upon secondary subcomponent manufacturers for nearly 50% of each car and for much of the research and development necessary to keep rolling stock up to date. With no orders the entire industry went out of business. By 1970 the track design in the United States had reached a point where rail passenger services could no longer compete in the market place. The rolling stock had been reduced from a post–World War II peak of 46,500 cars to about 3000 well-worn relics. By 1970 there was no industry that could supply high-technology passenger car rolling stock in the United States. In 1970 Congress passed the Rail Passenger Service Act, and in compliance with the provisions of this Act the National Railroad Passenger Corporation (AMTRAK) was created on May 1, 1971. The intent of this Federal law was clear. However, the law made no provision for improved, passenger-type track or for a rolling stock manufacturing industry. Also, it did not recognize the absolute incompatibility between the design of the network of existing track and the requirements of track for passenger services.

During the next 10 years Amtrak made some superficial progress. First, the Secretary of Transportation selected and designated the Amtrak nationwide network. This would be all the trains saved (approximately 180 per day out of a former 20,000 per day). The incorporators established the routes over certain participating railroads. One of the few assets Amtrak possessed as a result of the law was the stipulation that each participating railroad was required to maintain, until 1996, its track at the level of utility that existed in 1971. This placed a mandate upon the railroads to maintain their track at least as it was in 1971, even though that was not desirable in terms of passenger train design standards.

Amtrak sorted through about 3000 passenger cars, acquiring about 2000 of the best for its use. In 1971 these 2000 cars averaged 19 years in age and in heavy service. Most of them had to be totally renovated before they could be put back in service. By 1981 all of Amtrak's passenger car fleet were essentially new cars or totally rebuilt cars.

By 1975 Amtrak had ordered several sets of a French-designed Turbotrain, nearly 500 coaches, and 284 bilevel Superliners. Since old cars were being retired as new ones arrived, the new cars did not enlarge the fleet, and the total of revenue-producing cars remained at approximately 1200 into 1981. With this small number of cars, only around 250 trains were available in 1981 for United States passenger service. In contrast, the railroads of the European Economic Community had 81,000 passenger cars to serve a population of 274,000,000 in an area about one-third the size of the United States. England, France, and West Germany each had nearly 20,000 rail passenger cars.

In the early 1980s Amtrak's network served all but 4 of the top 50 metropolitan areas. However, in many of them the amount of service was no more than one train a day, with many scheduled times at inconvenient hours. In many countries there are trains every hour. The Japanese Shinkansen "Bullet" trains run a 663-mi (1067-km) route very frequently. During rush hours a train leaves Tokyo bound for Osaka every 7 min. Even international trains such as the Copenhagen-to-Frankfurt train, a 600-mi (966-km), 10.5-h trip, offers eight trains per day in each direction. With over 60% of its system running no more than once a day, Amtrak cannot even offer a reasonable round trip connection to a very large segment of its prospective passengers.

[L. FLETCHER PROUTY]

Rankine cycle

A thermodynamic cycle used as an ideal standard for the comparative performance of heat-engine and heat-pump installations operating with a condensable vapor as the working fluid. Applied typically to a steam power plant, as shown in the illustration, the cycle has four phases: (1) heat addition *bcde* in a boiler at constant pressure p_1 changing water at *b* to superheated steam at *e*, (2) isentropic expansion *ef* in a prime mover from initial pressure p_1 to back pressure p_2, (3) heat rejection *fa* in a condenser at constant pressure p_2 with wet steam at *f* converted to saturated liquid at *a*, and (4) isentropic compression *ab* of water in a feed pump from pressure p_2 to pressure p_1.

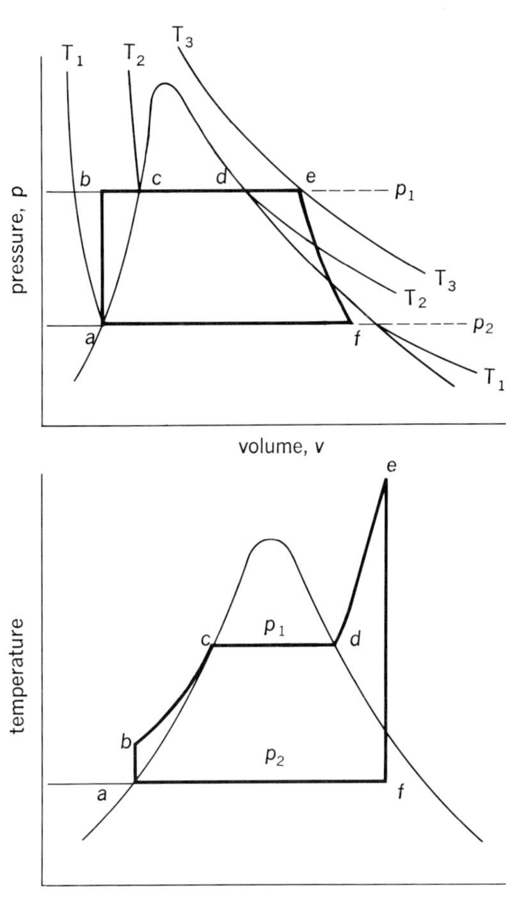

Rankine-cycle diagrams (pressure-volume and temperature-entropy) for a steam power plant using superheated steam. Typically, cycle has four phases.

This cycle more closely approximates the operations in a real steam power plant than does the Carnot cycle. Between given temperature limits it offers a lower ideal thermal efficiency for the conversion of heat into work than does the Carnot standard. Losses from irreversibility, in turn, make the conversion efficiency in an actual plant less than the Rankine cycle standard. *See* CARNOT CYCLE; REFRIGERATION CYCLE; THERMODYNAMIC CYCLE; VAPOR CYCLE. [THEODORE BAUMEISTER]

Bibliography: T. Baumeister (ed.), *Standard Handbook for Mechanical Engineers*, 8th ed., 1978; J. B. Jones and G. A. Hawkins, *Thermodynamics*, 1960; J. H. Keenan, *Thermodynamics*, 1970.

Ratchet

RATCHET

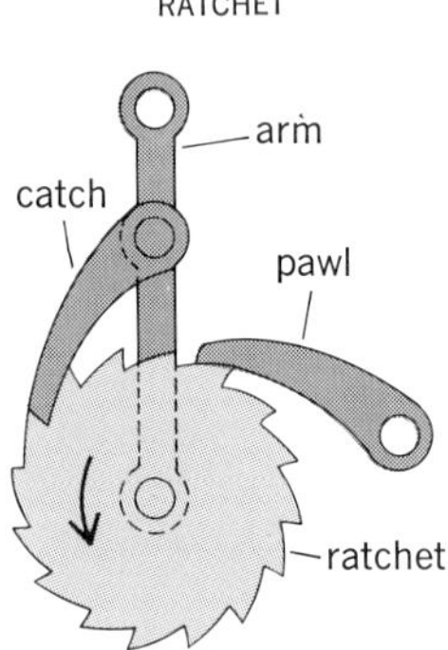

toothed ratchet

roller rachet

roller

driver

follower

Toothed ratchet is driven by catch when arm moves to left; pawl holds ratchet during return stroke of catch. In roller ratchet, rollers become wedged between driver and follower when driver turns faster than follower in direction of arrow.

A wheel, usually toothed, operating with a catch or a pawl so as to rotate in a single direction (see illustration). A ratchet and pawl mechanism locks a machine such as a hoisting winch so that it does not slip. The locking action may serve to produce rotation in a desired direction and to disengage in the undesired direction as in a drill brace. A further adaptation is to drive the catch in a to-and-fro motion against the ratchet to produce intermittent circular motion. The catch or pawl may be of various shapes such as an eccentrically mounted disk or ball bearing. Gravity, a spring, or centrifugal force (with the catch mounted internal to the ratchet) are commonly used to hold the pawl against the ratchet. A ratchet and pawl provides an arresting action, whereas an escapement provides an arresting action followed by a self-initiated momentary release. In high-speed machines, the abrupt action of a toothed ratchet produces severe shock. In such situations, a continuously variable yet directionally sensitive action is achieved by wedging rollers or specially shaped sprags between the input and output members. *See* BRAKE; PAWL.

[FRANK H. ROCKETT]

Bibliography: P. H. Black and O. E. Adams, *Machine Design*, 3d ed., 1968; R. H. Creamer, *Machine Design*, 2d ed., 1976.

Raw water

Water obtained from natural sources such as streams, reservoirs, and wells. Natural water always contains impurities in the form of suspended or dissolved mineral or organic matter and as dissolved gases acquired from contact with earth and atmosphere. Industrial or municipal wastes may also contaminate raw water. *See* WATER POLLUTION.

If admitted to a steam-generating unit, such contaminants may corrode metals or form insulating deposits of sediments or scale on heat-transfer surfaces, with resultant overheating and possible failure of pressure parts.

Principal scale-forming impurities are compounds of calcium and magnesium, or silica. Principal corrosive agents are dissolved oxygen and carbon dioxide. In some localities, raw water has a mineral acidity. Oil and grease impair wetting and heat removal from the steam-generating surfaces and may also form corrosive scale or sludge. Certain organic materials or a high concentration of dissolved solids in the boiler water may cause foaming which contaminates the steam.

Raw water can be treated to remove objectionable impurities or to convert them to forms that can be tolerated. For steam generation, suspended solids are removed by settling or filtration. Scale-forming hardness is diminished by chemical treatment to produce insoluble precipitates that are removable by filtration, or soluble compounds that do not form scale. Essentially complete purification is achieved by demineralizing treatment or evaporation. Demineralization consists of passing the water through beds of synthetic ion-exchange resin particles. Certain of these exchange hydrogen for metallic cations; others exchange hydroxyl for sulfate, chloride, or other anions in solution. The hydrogen and hydroxyl ions combine to form water. The resins may be used in separate or mixed-bed arrangements. They require periodic regeneration by acid and alkaline solutions, respectively; the mixed resins can be separated for such treatment by virtue of the fact that they differ in specific gravity. *See* WATER TREATMENT.

Evaporation requires expenditure of heat for complete vaporization of the water. The vapor is subsequently condensed and collected as purified distillate. Low-pressure steam is used as a heat source; multiple-effect heat exchange provides thermal economy. *See* FEEDWATER. [F. G. ELY]

Reactance

The opposition that inductance and capacitance offer to alternating current through the effect of frequency. Reactance alters the magnitude of current and also changes the circuit phase angle.

Inductive reactance X_L equals $2\pi fL$, where f is the frequency in hertz and L is the self-inductance in henrys. The voltage E across an inductance reaches its peak 90° before the current I reaches its peak, and $I = E/X_L$ amperes. Capacitive reactance X_C equals $1/(2\pi fC)$, where C is the capacitance in farads. The voltage E across a capacitance reaches its peak 90° after the current reaches its peak, and $I = E/X_C$ amperes.

Reactances are components of impedance which, in general, includes resistance R and reactance. Impedance is given by Eq. (1) for the series

$$Z = \sqrt{R^2 + (X_L - X_C)^2} \text{ ohms} \quad (1)$$

RLC circuit. In terms of complex quantities, Eq. (2) holds. Both reactances have magnitude and

$$Z = R + jX_L - jX_C = R + j(X_L - X_C) \quad \text{ohms} \quad (2)$$

angle: $+j$ means $+90°$ for X_L, and $-j$ means $-90°$ for X_C, the angles by which the voltages across them lead, or lag, the current. The resulting phase angle between voltage and current is given by Eq. (3), and current lags, or leads, the voltage de-

$$\theta = \arctan\,[(X_L - X_C)/R] \quad (3)$$

pending upon whether $X_L - X_C$ is positive or negative. *See* ALTERNATING-CURRENT CIRCUIT THEORY.

[BURTIS L. ROBERTSON]

Reaction turbine

A power-generation prime mover utilizing the steady-flow principle of fluid acceleration, where nozzles are mounted on the moving element. The rotor is turned by the reaction of the issuing fluid jet and is utilized in varying degrees in steam, gas, and hydraulic turbines. All turbines contain nozzles; the distinction between the impulse and reaction principles rests in the fact that impulse turbines use only stationary nozzles, while reaction turbines must incorporate moving nozzles. A nozzle is defined as a fluid dynamic device containing a throat where the pressure of the fluid drops and potential energy is converted to the kinetic form with consequent acceleration of the fluid. For details of the two basic principles of impulse and reaction as applied to turbine design *see* IMPULSE TURBINE. [THEODORE BAUMEISTER]

Reactor (electricity)

A device for introducing an inductive reactance into a circuit. Inductive reactance x is a function of the product of frequency f and inductance L; thus, $x = 2\pi fL$. For this reason, a reactor is also called an inductor. Since a voltage drop across a reactor increases with frequency of applied currents, a reactor is sometimes called a choke. All three terms describe a coil of insulated wire.

According to their construction, reactors can be divided into those that employ iron cores and those where no magnetic material is used within the windings. The first type consists of a coil encircling a circuit of iron which usually contains an air gap or a series of air gaps. The air gaps are used to attenuate the effects of saturation of the iron core. The second type, called an air-core reactor, is a simple circular coil, wound around a cylinder constructed of nonmagnetic material for greater mechanical strength. This strength is necessary for the coil to withstand the electromagnetic forces acting on each conductor. These forces become very large with heavy current flow, and their direction tends to compress the coil into less space: radial forces tend to elongate internal conductors in the coil and to compress the external ones while the axial forces press the end sections toward the center of the coil.

Both iron-core and air-core reactors may be of the air-cooled dry type or immersed in oil or a similar cooling fluid. Both types of reactors are normally wound with stranded wire in order to reduce losses due to eddy currents and skin effect. In addition, it is important to avoid formation of short-circuited metal loops when building supporting structures for air-core reactors since these reactors usually produce large magnetic fields external to the coil. If these fields penetrate through closed-loop metal structures, induced currents will flow, causing both losses and heating of the structures. Which of these two reactor types should be used depends on the particular application, which also provides a reactor designation.

Ballast reactor. A ballast reactor consists of a coil wound on an iron core and connected in series with a fluorescent or vapor lamp. The coil compensates for a negative-resistance characteristic of the lamp by providing an increased voltage drop as the current through the lamp is increased. Normally, two coils are magnetically coupled (wound on the same core) to ensure an equal current distribution among two fluorescent lamps operating in parallel. *See* FLUORESCENT LAMP; VAPOR LAMP.

Commutating reactor. The commutating reactor is found primarily in silicon controlled rectifier (SCR) converters, where it is connected in series with a commutation capacitor to form a highly efficient resonant circuit. This circuit is used to cause a current oscillation that turns off (commutates) the conducting SCR. The inductor is normally made of litz wire to reduce losses caused by skin effect and is usually of air-core type.

Current-equalizing reactor. This type is used to achieve desired division of current between several circuits operating in parallel.

Figure 1 shows two interphase reactors connected between two parallel SCR converters. The reactor provides balanced system operation when both converters are conducting by acting as an inductive voltage divider. Since the net direct current through the reactor is zero, the reactor supports only the harmonic voltage generated by SCR switchings.

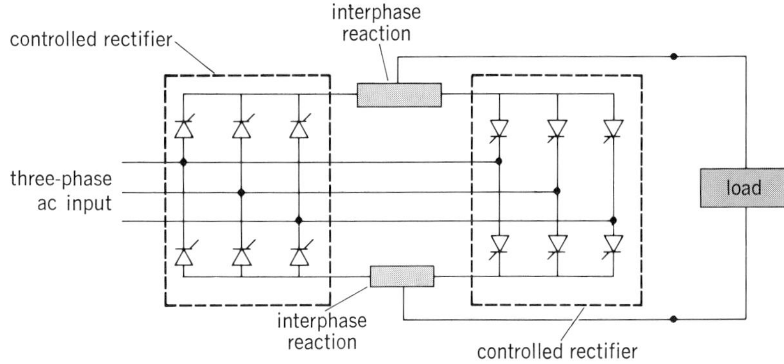

Fig. 1. Two interphase reactors that enable simultaneous (circulating current) operation of two controlled rectifiers.

Figure 2 shows two magnetically coupled reactors used to force equal currents through two parallel conducting paths. The voltage induced across each inductor is proportional to the difference between currents through the two diodes and with a polarity which tends to equalize the two currents. Since the reactors should be as linear as possible and have minimum hysteresis, magnetic materials are not used in the cores.

Current-limiting reactor. This type is used for protection against excessively large currents under short-circuit or transient conditions. By increasing the circuit's time constant (defined as L/R, where R is the total resistance), an inductor decreases the rate at which the current can change. In this way, a reactor provides more time for various protection devices (such as circuit breakers) to act before the current reaches a dangerous level. *See* ELECTRIC PROTECTIVE DEVICES.

Current-limiting reactors are always connected in series with a circuit which they are protecting, are of the air-core type, and can be divided into the following groups.

Bus reactor. This is an air-core inductor connected between two buses or two sections of the same bus in order to limit the effects of voltage transients on either bus. During normal operation both buses are at approximately the same voltage, and a free exchange of current takes place. As a fault occurs, the voltage on that bus abruptly changes and a heavy current starts to flow from one to the other bus. At that instant, a significant voltage drop is developed across the reactor, the magnitude of which depends on the rate of change of the current and on the reactor size (inductance). This voltage drop limits the current flow, minimizing the effect of the fault on the rest of the system and allowing a circuit breaker to disconnect the faulted section.

The di/dt reactor. This is a small inductor connected in series with a silicon controlled rectifier in order to limit the rate of rise of the SCR current. In small static converters, the *di/dt* reactor consists of an air-core inductor. In larger units, a saturable reactor is used, with a saturation point being just below the SCR rated current.

Feeder reactor. This is essentially the same type (but usually of higher power rating) used as a bus

REACTOR (ELECTRICITY)

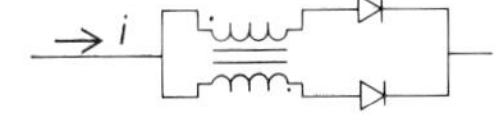

Fig. 2. Two magnetically coupled reactors used to force equal currents through two parallel diodes.

reactor. It is connected in series with a feeder circuit in order to limit and localize disturbances due to faults on the feeder. The principle of operation is the same as for bus reactors.

Generator reactor. This is the same as the feeder reactor but is connected between power plant generators and the rest of the network.

Starting reactor. This type is normally used to limit starting current of electric motors. A starting reactor usually consists of iron-core inductors connected in series with the machine stator winding. The starting reactor is short-circuited by a switch once the motor reaches a predetermined speed.

Filter reactor. This type is used to attenuate (filter) high-frequency currents. Since the voltage drop across an inductor is equal to a product between inductance, current, and frequency, the inductor opposes the flow of high-frequency currents, while a direct current is limited only by the resistance of the wire used to make the reactor. Filter reactors are connected in series and are usually wound on an iron core with an air gap to prevent saturation. Filter reactors are used extensively in both alternating-current and direct-current circuits, especially in conjunction with static power converters.

Grounding reactor. This type is used in grounded power systems and connected between the system's neutral point and a ground. The main purpose of a grounding reactor is to limit disturbances due to ground faults and atmospheric discharges on high-voltage transmission lines. A special grounding reactor, tuned to the rest of the network, which helps to extinguish a line-to-ground fault current, is called a Petersen coil. All grounding reactors are made of heavy cable wound on an iron core. *See* GROUNDING.

Saturable reactor. This is an iron-core reactor, adjusted to saturate at a predetermined current (Fig. 3), and used primarily in alternating-current circuits. Since the slope of the voltage-current curve (Fig. 3) is proportional to the reactor inductance, saturation substantially decreases this inductance, meaning that any further increase in the reactor current will result in very little change in the voltage across the reactor. Saturable reactors are used in a wide variety of circuits, whenever a nonlinear inductance is needed. By placing an additional, direct-current-supplied, control winding

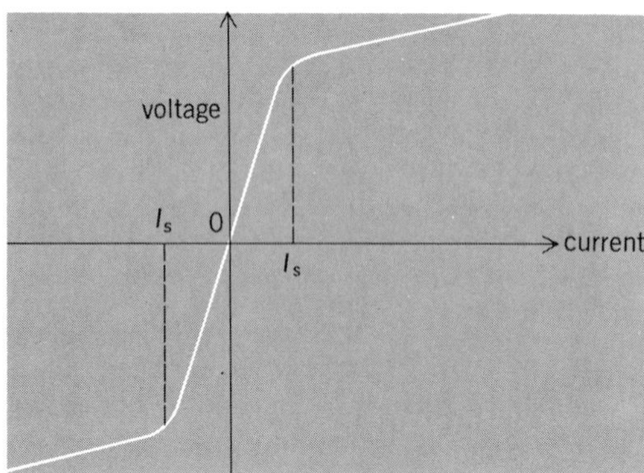

Fig. 3. Voltage-current characteristic of a saturable reactor; I_s is the current level at which the inductor saturates.

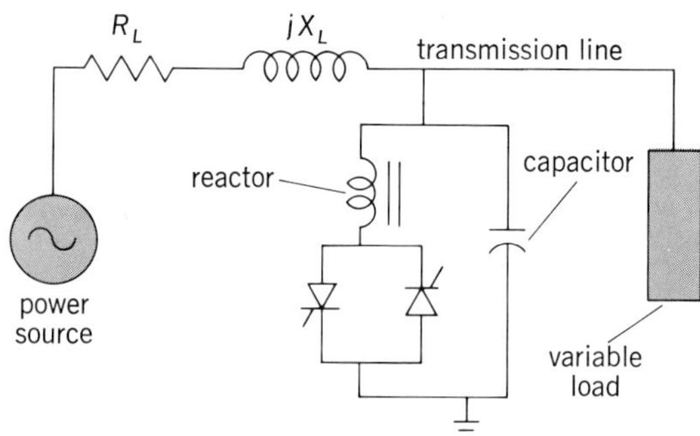

Fig. 4. Simplified single line diagram of a static reactive power system. Silicon controlled rectifiers (SCRs) are used to regulate the current through the reactor. R_L and X_L are the equivalent resistance and reactance of the transmission line.

on the reactor core, one obtains a magnetic amplifier (amplistat or magamp). The direct current is used to control the point at which the reactor saturates and thus the voltage drop across the reactor. Magnetic amplifiers are often used in conjunction with diode rectifiers to regulate dc power flow, but are now being replaced by semiconductor (thyristor or silicon controlled rectifier) converters. *See* MAGNETIC AMPLIFIER; SATURABLE REACTOR.

Shunt reactor. A shunt reactor consists of an iron-core inductor and is used to compensate for capacitance in a transmission line. For this reason, it is also called a compensating reactor. Shunt reactors are normally connected at selected points between a transmission line and a ground. Examples of its application are:

1. Compensating reactors, connected at predetermined points along an underground transmission line to compensate the leading (capacitive) characteristic of an underground cable.

2. The Pupin coil, connected at regular intervals on long telephone lines to compensate line capacitance and to provide a constant line characteristic (impedance) at lower frequencies.

3. A parallel reactor in a static reactive power system (Fig. 4), used in conjunction with silicon controlled rectifiers to regulate voltages on an extra-high-voltage (EHV) overhead transmission line. The regulation is achieved by controlling the amount of current flow through the reactor and is used to prevent voltage variations caused by switching loads.

Tesla reactor. Also known as a tesla coil, this consists of two magnetically coupled coils wound on an air core. The primary winding is made of a few turns of heavy wire, while the secondary consists of fine wire with a very large number of turns. By circulating a high-frequency current in the primary, very high voltage is induced in the secondary. The tesla coil is used in radio and television receivers as well as for producing high discharge voltages. *See* INDUCTION COIL; TRANSFORMER.

[VICTOR R. STEFANOVIC]

Bibliography: S. B. Dewan and A. Straughen, *Power Semiconductor Circuits*, 1975; *Electrical Transmission and Distribution Reference Book*, Westinghouse Electric Corp., 4th ed., 1964; D. R. Grafham, *SCR Manual*, General Electric Co., 6th ed., 1979.

Reactor physics

The science of the interaction of the elementary particles and radiations characteristic of nuclear reactors with matter in bulk. These particles and radiations include neutrons, beta (β) rays, and gamma (γ) rays of energies between zero and about 10^7 electron volts (ev).

The study of the interaction of β- and γ-radiations with matter is, within the field of reactor physics, undertaken primarily to understand the absorption and penetration of energy through reactor structures and shields. For a discussion of problems of reactor shielding *see* RADIATION SHIELDING.

With this exception, reactor physics is the study of those processes pertinent to the chain reaction involving neutron-induced nuclear fission with consequent neutron generation. Reactor physics is differentiated from nuclear physics, which is concerned primarily with nuclear structure. Reactor physics makes direct use of the phenomenology of nuclear reactions. Neutron physics is concerned primarily with interactions between neutrons and individual nuclei or with the use of neutron beams as analytical devices, whereas reactor physics considers neutrons primarily as fission-producing agents. In the hierarchy of professional classification, neutron physics and reactor physics are both ranked as subfields of the more generalized area of nuclear physics.

Reactor physics borrows most of its basic concepts from other fields. From nuclear physics comes the concept of the nuclear cross section for neutron interaction, defined as the effective target area of a nucleus for interaction with a neutron beam. The total interaction is the sum of interactions by a number of potential processes, and the probability of each of them multiplied by the total cross section is designated as a partial cross section. Thus, a given nucleus is characterized by cross sections for capture, fission, elastic and inelastic scattering, and such reactions as (n,p), (n,α) and $(n,2n)$. An outgrowth of this is the definition of macroscopic cross section, which is the product of cross section (termed microscopic, for specificity) with atomic density of the nuclear species involved. The symbols $N\sigma_i^{\,j}$ or $\Sigma_i^{\,j}$ are used for macroscopic cross section, the subscript referring to the nuclear reaction involved, and the superscript to the isotope. The dimensions of Σ are cm^{-1}, and Σ_t (total cross section) is usually of the order of magnitude of unity.

Cross sections vary with energy according to the laws of nuclear structure. In reactor physics this variation is accepted as input data to be assimilated into a description of neutron behavior. Common aspects of cross section dependence, such as variation of absorption cross section inversely as the square root of neutron energy, or the approximate regularity of resonance structure, form the basis of most simplified descriptions of reactor processes in terms of mathematical or logical models.

The concept of neutron flux is related to that of macroscopic cross section. This may be defined as the product of neutron density and neutron speed, or as the rate at which neutrons will traverse the outer surface of a sphere embedded in the medium, per unit of spherical cross-sectional area. The units of flux are neutrons/(cm^2)(sec). The product of flux and macroscopic cross section yields the reaction rate per unit volume and time. The use of these variables is conventional in reactor physics.

The chain reaction, a concept derived from chemical kinetics, is the basis of a physical description of the reactor process. The source of energy in a nuclear reactor is the fission of certain isotopes of heavy elements (thorium, uranium, and plutonium, in particular) when they absorb neutrons. Fission splits the elements into two highly radioactive fragments, which carry away most of the energy liberated by the fission process (about 160 Mev out of about 200 Mev per fission), the bulk of which is rapidly transformed into heat. Neutrons of high energy are also liberated, so that a chain of events alternating between neutron production in fission and neutron absorption causing fission may be initiated. Because more than one neutron is liberated per fission, this chain reaction may rapidly branch out to produce an increasing reaction rate, or divergent reaction; or if the arrangement of materials is such that only a small fraction of the neutrons will ultimately produce fission, the chain will be broken in a convergent reaction.

Criticality. The critical condition is what occurs when the arrangement of materials in a reactor allows, on the average, exactly one neutron of those liberated in one nuclear fission to cause one additional nuclear fission. If a reactor is critical, it will have fissions occurring in it at a steady rate. This desirable condition is achieved by balancing the probability of occurrence of three competing events: fission, neutron capture which does not cause fission, and leakage of neutrons from the system. If ν is the average number of neutrons liberated per fission, then criticality is the condition under which the probability of a neutron causing fission is $1/\nu$. Generally, the degree of approach to criticality is evaluated by computing k_{eff}, the ratio of fissions in successive links of the chain, as a product of probabilities of successive processes.

Reactor constants are the parameters used in determining the probabilities of the various processes which together define k_{eff}. They comprise two sets, those used to characterize nuclear events, and those used to characterize leakage. In the former group are fast effect ε, resonance escape probability p, thermal utilization factor f, and neutrons emitted per fuel absorption η. In the latter group are neutron age or slowing-down area τ or $L_s^{\,2}$, migration area M^2, thermal diffusion area L^2, diffusion coefficient D, and buckling B^2.

Fast effect. The fast effect occurs in thermal reactors containing significant quantities of U^{238} or Th^{232}. These reactors comprise the bulk of plutonium production and civilian power reactors. The isotopes mentioned can undergo fission only when struck by very energetic neutrons; only a little more than one-half of the neutrons born in fission can cause fission in them. Moreover, these fast neutrons (energy greater than about 1.4 Mev) are subject to energy degradation by the competing reaction of inelastic scattering and by collision with moderator. In consequence, only a small number of fast fissions will occur.

The fast effect is characterized by a ratio R of fast to nonfast fissions. From this ratio, a quantity ε, the fast effect, is derived, which determines the neutrons available to the chain reaction after the convergent chain of fast fissions has been completed, per neutron born of nonfast fission.

The common magnitude of ε varies between 1.02 and 1.05. R varies between 0 and 0.15 commonly. When $R < 0.01$, the fast effect is usually omitted from consideration.

Resonance escape probability. The resonance escape probability p is a significant parameter for the same group of reactors that have a significant fast effect and is defined as the probability that a neutron, in the course of being moderated, will escape capture by U^{238} (or Th^{232}) at any of the many energies at which the capture cross section is unusually high (resonance energies). The existence of this resonance absorption prevents most reactors homogeneously fueled with natural uranium from going critical, because resonance absorption, coupled with other losses, causes neutron depletion below the requirement for criticality. Therefore, low-enrichment reactors are heterogeneous, the fuel being disposed in lumps. The lumping of fuel, originally proposed by E. Fermi and E. P. Wigner, increases the probability of resonance escape. Spatial isolation of resonance absorber from moderator increases the number of neutrons which are slowed down without making any collisions with the absorber. Also, because absorption is very probable at resonance energies, the neutrons can travel only very short distances into the fuel lump before being absorbed, so that the interior of the lump is hardly exposed to resonant neutrons. Another effect of lumping is the removal of excess neutron scattering near the absorber so that there is a lesser probability of absorption following multiple collision.

With slight enrichment, homogeneous assemblies can be made critical. The homogeneous problem also provides the formalism by which p can be calculated. Use is made of the resonance integral RI, defined as the absorption probability per absorbing nucleon per neutron slowed down from infinite source energy in a moderator of unit slowing-down power. In a highly dilute system the resonance interval is given by Eq. (1), where σ_a

$$RI = \int_{E_0}^{\infty} \sigma_a \, dE/E \qquad (1)$$

is the neutron absorption cross section of the absorber as a function of energy, and E_0 is an energy taken as the lower limit of the resonance region. The slowing-down power of a moderator nucleus is $\xi\sigma_s$, where ξ is the mean increase in lethargy of the neutron per scattering, and σ_s is the moderator scattering cross section. Thus the resonance absorption probability of a system would appear to be given by Eq. (2), where N_r and N_m are, respectively,

$$1 - p = N_r\,(RI)/N_m\xi\sigma_s \qquad (2)$$

ly, atomic concentrations of absorber and moderator in the reactor volume. However, because the probabilities of escaping capture by successive nuclei must be multiplied, Eq. (3) is better.

$$p = \exp\{-N_r(RI)/N_m\xi\sigma_s\} \qquad (3)$$

The value of E_0 used in the definition of RI follows one of two conventions: it is either taken as some defined thermal cut-off energy between about 0.3 and 1 ev, or as an energy just below the lowest resonance, about 6 ev for U and 20 ev for Th. In the latter case, RI is spoken of as $1/v$ corrected.

Thermal utilization factor. Thermal utilization factor f is the fraction of neutrons which, once thermalized, are absorbed in fuel. In a homogeneous array, f may be calculated from the atomic densities (that is, the number of atoms per cubic centimeter) and thermal-absorption cross sections of the various constituents of the reactor. The problem becomes more complex in a highly absorbing system and in the presence of nuclei (such as Pu or Cd) whose absorption cross sections do not vary with energy in the usual $1/v$ fashion. In this case, it becomes necessary to evaluate the neutron spectrum and average the absorption cross sections over this spectrum in order to obtain reaction ratios.

For low absorptions, the neutron spectrum is given by Eq. (4), the Maxwellian expression, where

$$N(E)\,dE = [2\pi/(\pi kT)^{3/2}]E^{1/2}\exp - \{E/kT\}\,dE \qquad (4)$$

k is the Boltzmann constant and T the absolute temperature of the medium. Deviations from this shape caused by absorption were first formulated by Wigner and J. E. Wilkins. This problem has become very significant with the advent of highly absorbing systems operating with considerable quantities of Pu.

In heterogeneous systems the problem is further complicated by the spatial nonuniformity of the neutron flux. It is therefore necessary to calculate reaction rates in various regions of the lattice by multiplying local absorption cross sections and atomic densities by local neutron fluxes; or, in effect the same thing, by introducing flux weights into the cross sections.

The calculation of neutron flux may be performed by diffusion theory or by more exact and elaborate methods for solving the neutron-transport problem. The problem may be further complicated by spatial effects on spectrum.

The term disadvantage factor, applied in simple systems originally to describe the ratio of mean moderator to mean fuel flux, has fallen into disfavor because of vague and local definitions. The symbol F, called the fuel disadvantage factor, is still in use to describe the ratio of surface to mean volume flux in a fuel lump with isoperimetric flux.

Fission neutrons per fuel absorption. This constant, η, is a characteristic of the fuel and of the neutron spectrum, but not of the spatial configuration of the system. Pure fissionable materials do not always undergo fission when they absorb neutrons; sometimes they lose their energy of excitation by emission of a gamma ray. The number of neutrons per fission ν varies only slightly with incident neutron energy (a few percent per million electron volts), but η can fluctuate considerably even within a fraction of an electron volt. Consequently, the specification of η is dependent upon a good evaluation of neutron spectrum.

For unirradiated, low-enrichment assemblies, the custom of defining fuel as all uranium, U^{235} and U^{238}, persists. Thus, f considers total uranium captures, and η is lowered from the value for pure U^{235} by the fractional absorption rate of U^{235} in uranium. This custom leads to excessive complexity as plutonium builds into the fuel and is therefore declining in use.

Infinite multiplication constant. The infinite multiplication constant, k_∞, is the ratio of neutrons in successive generations of the chain in the absence of leakage. In the formalism just described, the chain is taken from thermal neutron through fission and back to thermal neutron and from the definition of terms, $k_\infty = \eta\varepsilon pf$. This formula is known as the four-factor equation.

For reactors other than weakly absorbing thermal systems, the simplified description breaks down. Thus in a fast reactor significant fission and capture occur at all neutron energies, and no moderator is present; in a very strongly absorbing thermal system, an appreciable fraction of neutrons react at energies between the thermal region and the lowest U^{238} resonance. For such systems, the definition of the neutron chain is usually made in terms of a total time-dependent fission-rate expression, and the parametric representation of k_∞ becomes appreciably more complex. Generally, the spectrum is broken up into energy groups, and k_∞ is defined as the sum of the fission neutron production rate over all groups divided by the sum of absorption rates over all groups.

Neutron age. The neutron age τ is a reactor parameter defined in various ways, all related to the probability of leakage of a fast neutron from a reactor system. The basic definition is that 6τ measures the mean square distance of travel between injection of a neutron at one energy or energy spectrum and its absorption at some other energy in an infinite system. The term age is used because in weakly moderating systems the equation describing neutron slowing down in space and energy has the same form as the time-dependent heat-conduction equation, with τ substituted for time. The British and Canadian usage is L_s^2 for slowing-down length (squared), which more accurately describes the physical parameter.

The common injection spectrum is a fission spectrum, and the common points of measurement or application are absorption at the indium resonance energy, 1.4 ev, or at some arbitrarily defined thermalization energy. The ages of these energies are denoted as τ_{In} or τ_{Th}. When a source other than the fission spectrum is considered, other subscripts are used.

A. M. Weinberg has pointed out that the shape of the spatial distribution in an infinite system for which τ is the second moment can be closely correlated with the fast leakage of a bare finite system. If the finite system has a source and sink distribution which is a solution of Laplace's equation, $\nabla^2\phi + B^2\phi = 0$, then the Fourier transform of the τ distribution for given B will yield a quantity $P_\infty(B^2)$, which is almost exactly the nonleakage probability during slowing down. Two particularly significant cases are those for which the τ distribution is a Gaussian or an exponential curve. In the former case, $P_\infty(B^2) = e^{-\tau B^2}$; in the latter, $P_\infty(B^2) = 1/(1 + \tau B^2)$. The Gaussian distribution is experimentally and theoretically verified for moderators as heavy as Be or heavier. The exponential distribution is crudely applicable to H_2O-moderated systems, and D_2O has a definitely mixed distribution.

In some cases, τ is used in a synthetic way to describe the leakage under some simple approximation to the slowing-down distribution. Thus, τ_{2G} is a number which is used in a two-energy group neutron model to give correct fast nonleakage probability as $P(B^2) = 1/(1 + \tau_{2G}B^2)$. When this model is not a good approximation, $6\tau_{2G}$ is not the second moment of the τ distribution.

For a heavy moderator, the age between two energies is given by the simple approximation of Eq. (5), where D is the diffusion coefficient, N is

$$\tau_{E_1 \to E_2} = \int_{E_2}^{E_1} D(E)/3(N\xi\sigma_s)\, dE/E \qquad (5)$$

atomic density, σ_s is scattering cross section, and ξ is mean lethargy (logarithmic energy) gain/collision.

Thermal diffusion area. The thermal diffusion area L^2 is one-sixth the mean square distance of travel between thermalization and absorption. It is thus the analog of τ for thermal neutrons. Because thermal migration is usually well represented by a diffusion equation, which has an exponential absorption distribution, thermal nonleakage probability is given by $1/(1 + L^2B^2)$. L^2 is defined by $L^2 = D_{\text{Th}}/N\sigma_{a,\text{Th}}$, where D and σ_a are spectrum-averaged diffusion coefficient and absorption cross section, respectively.

Migration area. The migration area M^2 is one-sixth the mean square distance of travel from birth to death of a neutron. For large, small-leakage systems, $1/(1 + M^2B^2)$ is an excellent approximation to the total nonleakage probability. When only thermal absorption exists, $M^2 = \tau + L^2$.

Diffusion coefficient. The diffusion coefficient D is essentially a scaling factor applied to validate Fick's law, a relation between neutron flux and current (the latter being defined as a vector describing net rate of flow of neutron density). Fick's law is expressed by Eq. (6), where $\mathbf{j}$ is current, $\nabla\phi$

$$\mathbf{j} = -D\nabla\phi \qquad (6)$$

flux gradient, and D diffusion coefficient. For generalized systems, D is a tensor, but in regions where $|\nabla\phi|/\phi$ is small, D is approximated by a scalar of magnitude $1/(3\Sigma)$.

Buckling, B^2, is mathematically defined as $\nabla^2\phi/\phi$ in any region of a reactor where this quantity is constant over an appreciable volume. The name is derived from the relationship between force and deflection in a mechanical system with constrained boundaries. In a bare reactor, the buckling is constant except within one neutron mean free path ($\lambda \equiv 1/\Sigma$) of the boundary. For a slab of width t, $B^2 = (\pi/t)^2$; for an infinitely high cylinder of radius a, $B^2 = (2.404/a)^2$; for a sphere of radius a, $B^2 = (\pi/a)^2$; and for other geometries, it is again a geometrical constant. Because B^2 is a definite eigenvalue of Laplace's equation, it is the appropriate number to be used in the formulations of $P(B^2)$ previously described.

Effective multiplication k_{eff} is the ratio of neutron production in successive generations of the chain reaction. It is given by the product of k_∞ and $P(B^2)$, where $P(B^2)$ is itself the product of fast and thermal (or equivalent) nonleakage probabilities.

Reactivity. Reactivity is a measure of the deviation of a reactor from the critical state at any frozen instant of time. The term reactivity is qualitative, because three sets of units are in current use to describe it.

Percent k and millikay are absolute units describing the imbalance of the system from criticality per fission generation. Because $k_{eff}=1$ describes a critical system, one says that it is 1% super- or subcritical, respectively, if each generation produces 1.01 or 0.99 times as many neutrons as the preceding one. Millikay are units of 0.1% k, and are given plus sign for supercriticality and minus sign for subcriticality. In both cases, k_{eff} is the base, as in Eqs. (7). These units are used primarily in design and analysis of control rods.

$$\begin{aligned} \%k &= 100|k_{eff}-1|/k_{eff} \\ \text{Millikay} &= 1000\,(k_{eff}-1)/k_{eff} \end{aligned} \qquad (7)$$

Dollars describe reactivity relative to the mean fraction of delayed neutrons per fission. Because the delayed neutrons are the primary agents for permitting control of the reaction, supercriticalities of less than 1 dollar are considered manageable in most cases. Thus, there is 1 dollar to "spend" in maneuvering power level. The dollar is subdivided into 100 cents. Because the delayed neutron fraction is a function of both neutron energy and fissionable material, the conversion rate between dollars and $\%k$ varies among reactors.

Inhours are reactivity units based on rate of change of power level in low-power reactors. If a low-power reactor is given enough reactivity so that its level would steadily increase by a factor of e per hour (which is also known as a 1-hr period), that much reactivity is 1 inhour (from inverse hour). It is only for very small reactivities, however, that reactivity in inhours may be obtained from reciprocal periods.

Reactivity is measured in inhours primarily by operators of steady-state reactors, in which only small reactivities are normally encountered.

Reflectors. Reflectors are bodies of material placed beyond the chain-reacting zone of a reactor, whose function is to return to the active zone (or core) neutrons which might otherwise leak. Reflector worth can be crudely measured in terms of the albedo, or probability that a neutron passing from core to reflector will return again to the core.

Good reflectors are materials with high scattering cross sections and low absorption cross sections. The first requirement ensures that neutrons will not easily diffuse through the reflector, and the second, that they will not easily be captured in diffusing back to the core.

Beryllium is the outstanding reflector material in terms of neutronic performance. Water, graphite, D_2O, iron, lead, and U^{238} are also good reflectors. The use of Be, H_2O, C, and D_2O as reflectors permits conversion of neutrons leaking at high energy into thermal neutrons diffusing back to the core. Because the reverse flow of neutrons is always accompanied by a neutron-flux gradient, these reflectors show characteristic thermal flux peaks outside the core. They are, therefore, desirable materials for research reactors, in which these flux peaks are useful for experimental purposes.

The usual measurement of reflector worth is in terms of reflector savings, defined as the difference in the reflected dimension between the actual core and one which would be critical without reflector. Reflector savings are close to reflector dimensions for thin reflectors and approach an asymptotic value dependent upon core size and reflector constitution as thickness increases.

Reactor dynamics. Reactor dynamics is concerned with the temporal sequence of events when neutron flux, power, or reactivity varies. The inclusive term takes into account sequential events, not necessarily concerned with nuclear processes, which may affect these parameters. There are basically three ways in which a reactor may be affected so as to change reactivity. A control element, absorbing rod, or piece of fuel may be externally actuated to start up, shut down, or change reactivity or power level; depletion of fuel and poison, buildup of neutron-absorbing fission fragments, and production of new fissionable material from the fertile isotopes Th^{232}, U^{234}, U^{238}, and Pu^{240} make reactivity depend upon the irradiation history of the system; and changes in power level may produce temperature changes in the system, leading to thermal expansion, changes in neutron cross sections, and mechanical changes with consequent change of reactivity.

Reactor control physics. Reactor control physics is the study of the effect of control devices on reactivity and power level. As such, it includes a number of problems in reactor statics, because the primary question is to determine the absorption of the control elements in competition with the other neutronic processes. It is, however, a problem in dynamics, given the above information, to determine what motions of the control devices will lead to stable changes in reactor output.

Particular problems occurring in the statics of reactor control stem from the particular nature of control devices. Many control rods are so heavily absorbing for thermal neutrons that elegant refinements of neutron-transport theory are needed to estimate their absorption. Other types of control rods include isotopes with heavy resonance absorption, and the interaction of such absorbers with U^{238} resonances must be examined. The motion of control rods changes the material balance of reactor regions, and with water-moderated reactors, peaks in the fission rate occur near empty rod channels. By virtue of high absorption of their constituent isotopes, some absorbing materials (for example, cadmium and boron) burn out in the reactor, and a rod made of these materials loses absorbing strength with time. As a final example, the motion of a control rod may change the shape of the power pattern in the reactor so as to bring secondary pseudostatic effects into play. *See* NUCLEAR REACTOR.

Reactivity changes. Long-term reactivity changes may represent a limiting factor in the burning of nuclear fuel without costly reprocessing and refabrication. As the chain reaction proceeds, the original fissionable material is depleted, and the system would become subcritical if some form of slow addition of reactivity were not available. This is the function of shim rods in a typical reactor. The reactor is originally loaded with enough fuel to be critical with the rods completely inserted. As the fuel burns out, the rods are withdrawn to compensate.

In order to decrease requirements on the shim system, many devices to overcome reactivity loss may be used. A burnable poison may be incorporated in the system. This is an absorbing isotope which will burn out at a rate comparable to or greater than the fuel. Burnable poisons are therefore limited to isotopes with very high effective neutron-absorption cross sections. Combinations of poisons, and the use of self-shielding of poisons can, in principle, make the close compensation of considerable reactivity possible without major control rod motions, but the technological problems in their use are formidable.

A more popular method for compensating reactivity losses is the incorporation of fertile isotopes into the fuel. This is desirable because the neutrons captured in the fertile material are not wasted, but used to manufacture new fissionable material; and also because (as with U^{238} in U^{235} reactors or Pu^{240} in Pu^{239} systems) the fertile material is normally found mixed with the fissionable, and isotopic separation may be circumvented or minimized. Depending on the conversion ratio (new fissionable atoms formed per old fissionable atom burned) and the fission parameters of the materials, a reactor so fueled loses reactivity relatively slowly, and in some cases, may show a temporary reactivity increase. The various isotopes produced by successive neutron capture in uranium and plutonium must be considered in this problem, the higher isotopes becoming prominent at very long exposures.

A final consideration of long-term reactivity is the extra parasitic absorption of the fission products as formed. At long exposure, this absorption becomes significant because of the relatively high absorption of many of the fission products. At shorter exposures, isotopes of very high absorption, mainly Sm^{149} and Xe^{135}, are more prominent. These materials have such high cross sections that they reach a steady-state concentration relatively quickly, burning out by neutron capture as rapidly as they are formed in fission.

Xe^{135} is particularly interesting because its cross section is abnormally large, its fission yield is high, it is preceded by an isotope of low cross section and approximately 7-hr half-life (I^{135}), and it undergoes β-decay to the low absorption Cs^{135} with a half-life of about 10 hr. This combination of properties gives several interesting effects. Chief of these is that, at high flux, a reservoir of I^{135} is formed which continues to decay to Xe^{135} even after the reactor is shut down. Because Xe^{135} is maintained at steady state during operation by a balance of buildup against burnout, the shutdown also removes the chief mode of Xe^{135} destruction. Hence, the Xe^{135} concentration increases immediately after shutdown. The reservoir of I^{135} is so large that reactivity is rapidly lost as Xe^{135} builds up; and in some cases, it may be impossible to restart the reactor after a short shutdown. The operator must then wait almost 2 days for the Xe^{135} to disappear by radioactive decay. At very high fluxes, this effect becomes so severe that even a small temporary reduction in power may lead to ultimate subcriticality of the reactor.

Another effect caused by Xe^{135} in high-flux reactors is that, if the reactor is large enough, the Xe^{135} may force the power pattern into oscillations of 1- or 2-day periods. Although this is not a serious dynamic problem, it does emphasize the necessity of monitoring not only the total power, but also the power pattern, so that appropriate countermeasures may be taken. *See* NUCLEAR FUELS; NUCLEAR FUELS REPROCESSING.

Reactor kinetics. This is the study of the short-term aspects of reactor dynamics with respect to stability, safety against power excursion, and design of the control system. Control is possible because increases in reactor power often reduce reactivity to zero (the critical value) and also because there is a time lapse between successive fissions in a chain resulting from the finite velocity of the neutrons and the number of scattering and moderating events intervening, and because a fraction of the neutrons is delayed.

Prompt-neutron lifetime. This is the mean time between successive fissions in a chain, and it is the basic quality which determines the time scale within which controlling effects must be operable if a reactivity excursion is touched off. In thermal reactors the controlling feature is the time between thermalization and capture, because fast neutrons spend less time between collisions, and the total time for moderation of a neutron is at most a few microseconds (μsec). Prompt-neutron lifetimes vary from 10 to a few hundred microseconds for light-water reactors, the shorter times being found in poorly reflected, highly absorbing systems, and the longer in well-reflected systems, in which time spent in the reflector is the dominating factor. Other thermal reactors have longer lifetimes, with some heavy-water reactors having lifetimes as long as a few milliseconds. Fast reactors have lifetimes of the order of 0.01–0.1 μsec, the controlling factor being the amount of scattering material used as diluent.

Delayed neutrons. Delayed neutrons are important because a complete fission generation is not achieved until these neutrons have been emitted by their precursors. A slightly supercritical reactor must wait until the delayed neutrons appear, and this delay allows time for the system to be brought under control. The fraction of delayed neutrons β ranges from about 1/3% for thermal fission of Pu^{239} and U^{233}, to about 3/4% for thermal fission of U^{235}, to several percent for some fast fission events. In these latter cases, however, the extra delayed neutrons have such short lifetimes that they are of only slight extra utility.

In any case, the influence of delayed neutrons is felt only to the extent that they are needed to maintain criticality. When the system is supercritical enough that the delayed neutrons are not needed to complete the critical chain, it is known as prompt critical. Prompt criticality represents in a qualitative sense the threshold between externally controllable and uncontrollable excursions, and it is for this reason that the dollar unit is popular in excursion analysis (a prompt critical system has a reactivity of 1 dollar).

Although it has now been established that a larger number of fission products emit delayed neutrons, the distribution in time after fission of the delayed neutron emission rate is accurately represented for all purposes by a sum of six negative exponentials. For many purposes, however, a three, two, or one group approximation is adequate.

Reactors moderated by D_2O and Be have additional delayed neutrons contributed by photoneutron reactions between the moderator and fission product γ-rays. Although not a large fraction, this effect makes such reactors unresponsive to small reactivity fluctuations and gives them unusual operational smoothness.

Reactor period. This is the asymptotic time required for a reactor at constant reactivity to increase its power by a factor *e*. When a critical reactor is given extra reactivity, its power will rise. At first, the power production rate has a complex shape on a time plot, but ultimately the power will rise exponentially. The period is the measure of this exponential rate.

The relation between the reactor period and the reactivity is known as the inhour equation. If l is the reactor lifetime in seconds, β_i the fraction of delayed neutrons in group i, λ_i the delay constant of group i delayed neutrons in sec^{-1}, S reactor period in seconds and ρ reactivity in thousands of millikay, then Eq. (8) is the inhour equation. The

$$\rho = \frac{l/S + \sum_i [\beta_i/(1+\lambda_i S)]}{1 - \sum_i [\beta_i/(1+\lambda_i S)]} \tag{8}$$

equation has $I+1$ solutions of S for a given ρ, I being the number of groups; and there is always a real value of S with a higher value than the real part of any other solution. This highest S is the period. For very small values of ρ, that is, for very large periods, this value is approximately that given by notation (9). For very large ρ, and therefore small

$$\left(l + \sum_i \frac{\beta_i}{\lambda_i}\right) \Big/ \rho \tag{9}$$

S, the period is approximately $l/(\rho - \beta)$. This same result would be found if the delayed neutrons were thrown away completely.

Reactivity coefficients. There are several functions relating changes in reactivity to changes in the physical state of the reactor. The power coefficient is the change in reactivity per unit change in reactor power; the temperature coefficient relates reactivity to temperature change, and is often broken down into fuel, moderator, and coolant coefficients; for low-power graphite reactors there exists a barometric coefficient; one may define also coolant circulation rate coefficients and void coefficients.

Because the reactivity is commonly a complicated function of all the pertinent variables, the reactivity coefficient generally is the coefficient of the first term in a series expansion of the reactivity about the operating point. This in turn describes a linear theory of reactor dynamics. The theory may be extended to reactivity effects of arbitrary type by considering reactivity coefficients as functionals.

The basic problem of reactor dynamics is the specification of the power coefficient of reactivity. The chain, power affects reactivity which affects power, is thereby analyzable, using the power coefficient functional together with the reactor kinetic equations. The power coefficient is, however, predictable only in terms of changes in temperature and flow resulting from power changes in the system. Thus the analysis of power coefficient implies exhaustive knowledge of system behavior.

Reactivity coefficients may be prompt or delayed, and most delayed effects can be characterized as either of decay or transport type. An example of the decay type of coefficient is the contribution of coolant temperature change to power coefficient. Here, a power pulse gives a thermal effect on the coolant which is instantaneously observable, and which decreases exponentially with a time constant imposed by the heat-transfer equations. An example of a transport type of delay is the delay attributable to coolant circuit times. Here, a finite time lapse exists between the cause and the observable response. Effects due to fuel heating are examples of prompt effects.

Some types of power coefficient yield dangerous or unstable situations. Thus a power coefficient may contain a prompt positive (autocatalytic) term and a larger delayed negative term. Even though such a system may be stable against slow power-level increases, it will undergo a violent excursion whenever power is raised rapidly enough to outstrip delayed effects. Again, a system with prominent delayed effects of the transport type is always unstable beyond some critical power, even if the effect opposes the power shift; here, there is a possibility of phase instability.

The dynamic behavior of a reactor is usually analyzed by techniques common to all feedback systems.

[BERNARD I. SPINRAD]

Bibliography: G. I. Bell and S. Glasstone, *Nuclear Reactor Theory*, 1970; A. R. Foster and R. L. Wright, Jr., *Basic Nuclear Engineering*, 2d ed., 1973; A. D. Galanin, *Thermal Reactor Theory*, 1960; S. Glasstone and A. Sesonske, *Nuclear Reactor Engineering*, 1963; D. L. Hetrick, *Dynamics of Nuclear Reactors*, 1971; J. R. Lamarsh, *Introduction to Nuclear Engineering*, 1975; J. R. Lamarsh, *Nuclear Reactor Theory*, 1966; R. V. Meghreblian and D. K. Holmes, *Reactor Analysis*, 1960; *Nuclear Energy Glossary*, International Organization for Standardization, ISO-921, 1972; *Reactor Physics Constants*, Argonne National Laboratory, ANL-5800, 2d ed., 1963; A. M. Weinberg and E. P. Wigner, *The Physical Theory of Neutron Chain Reactions*, 1958; P. F. Zweifel, *Reactor Physics*, 1975.

Reamer

A multiple-cutting-edge tool designed to enlarge or accurately size and finish an existing hole in solid material by removal of a small amount of stock. The cutting edges may be ground on the apexes between longitudinal flutes or grooves, or cutting may take place on chamfered edges at the end of the reamer.

Reaming is performed either manually or by machine. Hand reamers may be gripped with a wrench on their square-ended tangs, while machine reamer shanks are made either tapered or round to adapt to machine tools. Hand reamers are fluted with a slight end taper to aid in starting. Fluted machine reamers for finishing remove metal by both end and side cutting, while the heavier rose reamers cut only on their chamfered ends. *See* MACHINING OPERATIONS. [ALAN H. TUTTLE]

Refrigeration

The cooling of a space or substance below the environmental temperature. The art was known to the ancient Egyptians and people of India, who

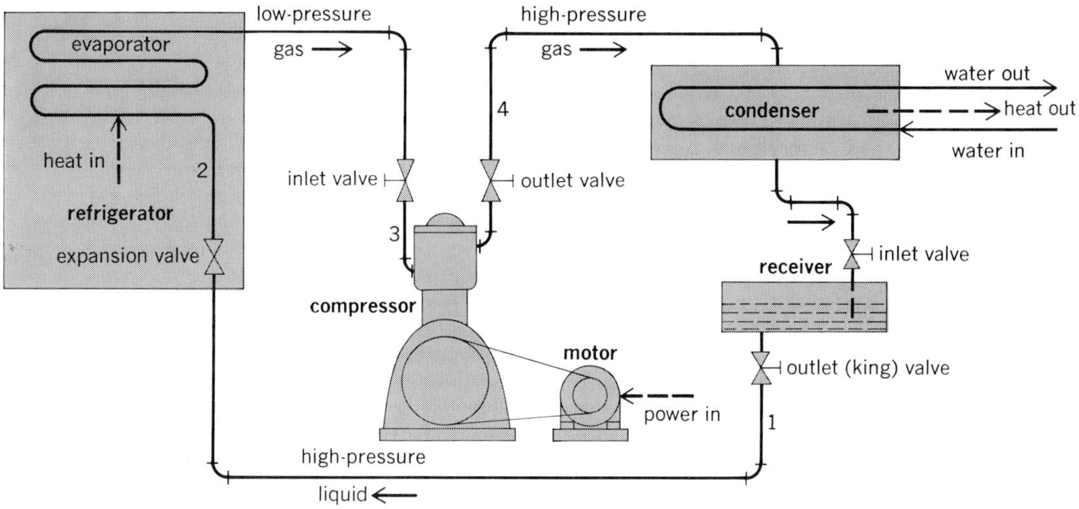

Fig. 1. Diagram of the vapor-compression cycle.

used evaporation to cool liquids in porous earthen jars exposed to dry night air; and to the early Chinese, Greeks, and Romans, who used natural ice or snow stored in underground pits for cooling wine and other delicacies. In the late 18th and early 19th centuries natural ice cut from lakes and ponds in winter was stored underground for use in summer. The technique of mechanical refrigeration began with the invention of machines for making artificial ice. Great strides have been made in the 20th century in the application of mechanical refrigeration to fields other than ice making, including the direct cooling and freezing of perishable foods and air conditioning for industry and human comfort.

Mechanical refrigeration is primarily an application of thermodynamics wherein the cooling medium, or refrigerant, goes through a cycle so that it can be recovered for reuse. The commonly used basic cycles, in order of importance, are vapor-compression, absorption, steam-jet or steam-ejector, and air. Each cycle operates between two pressure levels, and all except the air cycle use a two-phase working medium which alternates cyclically between the liquid and vapor phases.

Vapor-compression cycle. The vapor-compression cycle (Fig. 1) consists of an evaporator in which the liquid refrigerant boils at low temperature to produce cooling, a compressor to raise the pressure and temperature of the gaseous refrigerant, a condenser in which the refrigerant discharges its heat to the environment, usually a receiver for storing the liquid condensed in the condenser, and an expansion valve through which the liquid expands from the high-pressure level in the condenser to the low-pressure level in the evaporator. This cycle may also be used for heating if the useful energy is taken off at the condenser level instead of at the evaporator level. *See* HEAT PUMP.

The theoretical vapor-compression cycle can best be analyzed on the pressure-enthalpy or temperature-entropy coordinates for a two-phase fluid (Fig. 2). Enthalpy is a parameter that replaces heat content. It equals internal energy plus the product of the pressure and the volume divided by 778 and is expressed in units of Btu per pound. Entropy is a parameter that is obtained by dividing the heat flow by the average absolute temperature during the change. It is expressed in units of Btu per pound per degree Rankine (°R = °F + 460). In Fig. 2, process 1–2 represents adiabatic (constant enthalpy) expansion; 2–3′, constant temperature (and pressure) evaporation; 3′–3, suction superheating at constant pressure; 3–4, ideal frictionless adiabatic (constant entropy) compression; 4–4′, removal of discharge superheat at constant pressure; 4′–1′, condensation at constant pressure (and temperature); and 1′–1, liquid subcooling at constant pressure.

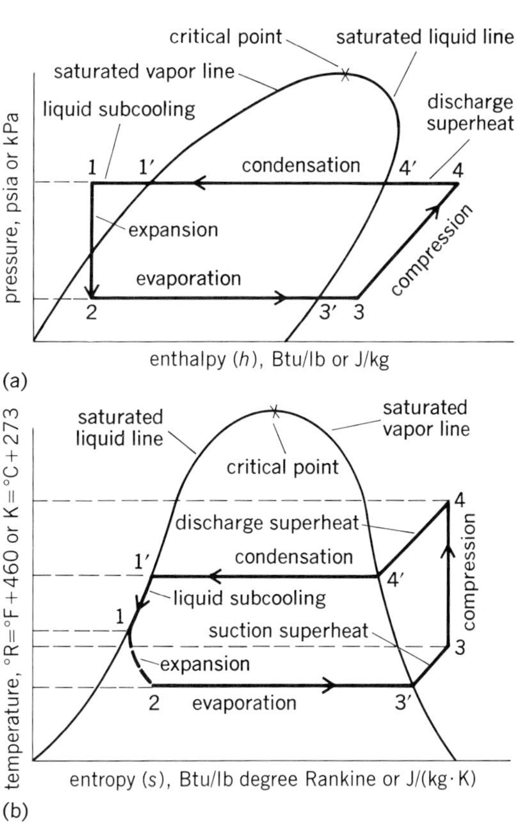

Fig. 2. Vapor-compression cycle. (*a*) Pressure-enthalpy diagram. (*b*) Temperature-entropy diagram.

The efficiency of a heat-power cycle is defined as the ratio of useful output to energy input. For a heat engine, efficiency is less than unity. Efficiency is not very meaningful for the refrigeration and heat-pump cycles, for which the term coefficient of performance (CP) is used instead. Referring to the theoretical cycle (Fig. 2), the refrigeration CP is the ratio of cooling effect in evaporator 2–3 to compressor energy input 3–4 and the heat-pump CP is the ratio of heating effect in condenser 4–1 to compressor energy input 3–4. The coefficient of performance may be considerably greater than unity, and the theoretical heat-pump CP is 1 plus refrigeration CP. For theoretical cycles operating between the same temperature levels, the heat-pump CP is the reciprocal of the heat-engine efficiency.

Absorption cycle. The absorption cycle accomplishes compression by using a secondary fluid to absorb the refrigerant gas, which leaves the evaporator at low temperature and pressure. Heat is applied, by means such as steam or gas flame, to distill the refrigerant at high temperature and pressure. The most-used refrigerant in the basic cycle (Fig. 3) is ammonia; the secondary fluid is then water. The condenser, receiver, expansion valve, and evaporator are essentially the same as in any vapor-compression cycle. The compressor is replaced by an absorber, generator, pump, heat exchanger, and controlling-pressure reducing valve.

The operation of the cycle is based on the principle that the vapor pressure of a refrigerant is lowered by the addition of an absorbent having a lower vapor pressure; and the greater the quantity of absorbent used, the more the depression of the vapor pressure of the refrigerant.

By maintaining the solution in the absorber at the proper temperature and concentration, the vapor pressure of the solution can be kept lower than that of the refrigerant in the evaporator. Spraying the weak solution in the absorber then causes the refrigerant vapor to flow from the evaporator to the absorber. The strong solution thus formed in the absorber is then pumped through a heat exchanger to the generator, where heat is applied to release the refrigerant vapor. Then follows condensation, expansion, and evaporation, as in the standard vapor-compression cycle. Except for small units, an indirect system is used wherein brine is cooled and circulated to the actual refrigeration load.

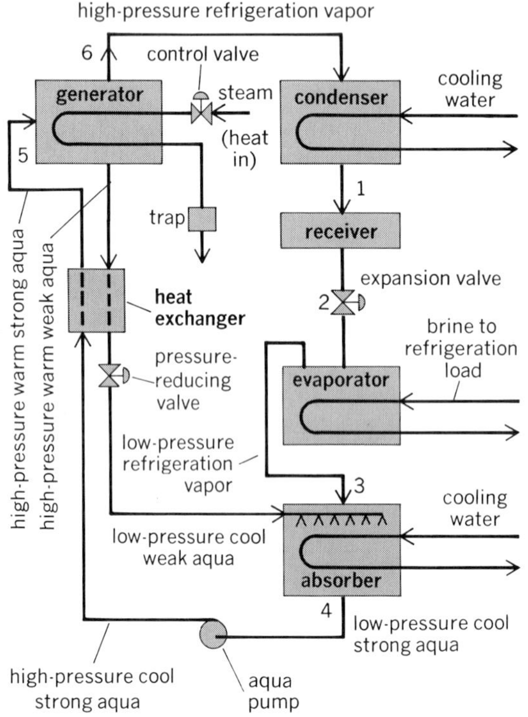

Fig. 3. Basic absorption cycle.

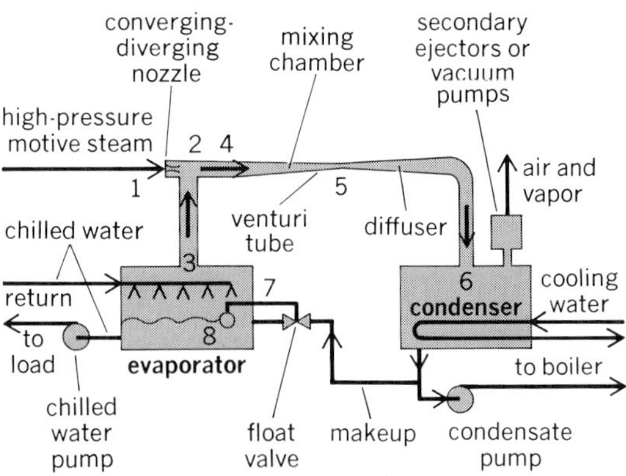

Fig. 4. Steam-jet water-vapor cycle.

For air conditioning, water is the refrigerant and lithium bromide is the absorbent. In terms of the basic absorption cycle (Fig. 3), from 1 to 2 the high-pressure liquid refrigerant is expanded into the evaporator where brine is usually cooled, from 2 to 3 the low-pressure refrigerant vapor is drawn into the absorber, from 3 to 4 the low-pressure refrigerant vapor is absorbed in the weak solution, from 4 to 5 the low-pressure strong solution is pumped through the heat exchanger to the high-pressure generator, and from 5 to 6 heat is applied to drive off the refrigerant vapor and force it into the condenser. The hot weak solution drains back to the absorber through the heat exchanger and pressure-reducing valve.

Steam-jet cycle. The steam-jet cycle uses water as the refrigerant. High-velocity steam jets provide a high vacuum in the evaporator, causing the water to boil at low temperature and at the same time compressing the flashed vapor up to the condenser pressure level. Its use is limited to air conditioning and other applications for temperatures above 32°F (0°C).

The basic steam-jet or ejector cycle (Fig. 4) is usually analyzed on temperature-entropy coordinates (Fig. 5). High-pressure motive steam (Fig. 4) at 1 is expanded to a low absolute pressure at 2 through a converging-diverging nozzle. Path 1–2 (Fig. 5) is the ideal expansion and 1–2′ the actual expansion allowing for nozzle friction. Water vapor in the evaporator at 3 is entrained by the motive steam at 4, the steam having lost some of its energy from 2′ to 4 because the entrainment efficiency is rather low. The motive steam at 4 plus the entrained moisture at 3 are forced through the venturi tube, in which the velocity of the incoming mixture is reduced and converted into pressure head in the condenser at 6. Path 5–6 is the ideal compression and 5–6′ the actual compression, taking into account compression efficiency. Typical efficiencies are nozzle 88%, entrainment 65%, and compression 80%.

In the evaporator or flash chamber, part of the water is evaporated to cool the rest of the water, which is circulated to the cooling load. Makeup water must be added from 7 to 8. Typical operating conditions are 100°F or 38°C (2 in. Hg or 6.8 kPa absolute pressure) in the condenser and 40°F or 4°C (0.25 in. Hg or 850 Pa absolute pressure) chilled water in the evaporator. The condensate from the condenser is pumped back to the boiler; secondary ejectors, or vacuum pumps, are required to remove the air and maintain the high vacuum. Considerably more water is required for condensing than for a vapor-compression system of the same capacity.

Air cycle. The air cycle, used primarily in airplane air conditioning, differs from the other cycles in that the working fluid, air, remains as a gas throughout the cycle. Air coolers replace the condenser, and the useful cooling effect is obtained by a refrigerator instead of by an evaporator. A compressor is used, but the expansion valve is replaced by an expansion engine or turbine which recovers the work of expansion. Systems may be open or closed. In the closed system, the refrigerant air is completely contained within the piping and components, and is continuously reused. In the open system, the refrigerator is replaced by the space to be cooled, the refrigerant air being expanded directly into the space rather than through a cooling coil.

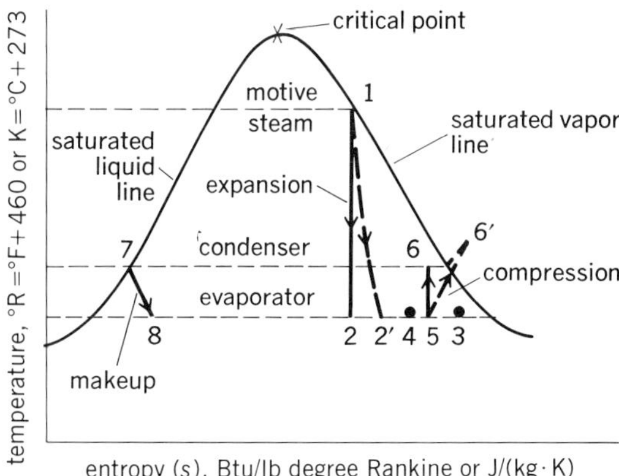

Fig. 5. Temperature-entropy coordinates for analyzing basic steam-jet or ejector cycle.

One of the typical open air-cycle systems used on airplanes is called the "bootstrap" system (Fig. 6). It may be analyzed theoretically on the temperature-entropy coordinates (Fig. 7). From 1 to 2 ambient air is compressed ideally in the engine or supercharger of the airplane. Part of this high-pressure air is bled through a primary heat exchanger, where it is cooled from 2 to 3 by ram air, that is, ambient air compressed by the forward motion of the airplane. From 3 to 4 this air is further increased in pressure by the compressor of the refrigeration machine; from 4 to 5 the air is cooled by ram air in the cooler or secondary heat exchanger; and from 5 to 6 the air is further cooled in the expansion turbine, ideally without moisture. However, there is entrained moisture, about 70% of which is removed by the water separator, resulting in an approximate path 5–6′ as the air is warmed by the heat given up by the condensation and removal of this moisture. The balance of the moisture evaporates in the cabin and contributes to the cooling effect 6′–7, where 6′ is the dry-air rated temperature of the air as it enters the cabin. Dry-air rated temperature is the temperature which would be attained by the expansion of the refrigerant air in the absence of any moisture condensation. Refrigerant air at 7, having absorbed the heat load in the cabin, is released overboard through a pressure-relief device. The equipment is proportioned so that the work of expansion is recovered and is sufficient to drive the compressor and the cooler fan to approximate constant entropy expansion and compression. However, compressor and turbine efficiencies and heat-exchanger pressure drops, neglected in this analysis, reduce the ideal performance.

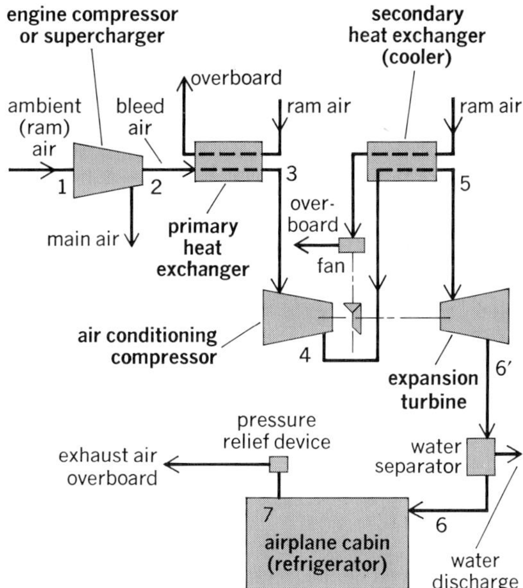

Fig. 6. Open air-cycle bootstrap system for airplanes.

As is well known, the effect of heat application

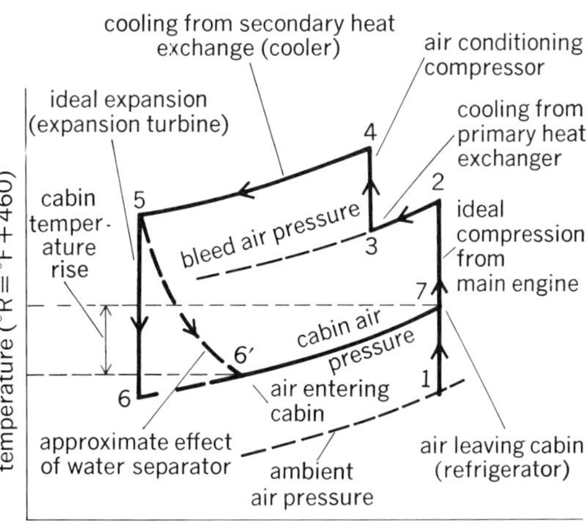

Fig. 7. Temperature-entropy coordinates for analyzing theoretically the air-cycle bootstrap system.

to one junction of a couple composed of two dissimilar conductors is to produce a thermoelectric voltage on the circuit (Seebeck effect). Inversely, this is matched by a cooling effect at one junction of an equivalent couple when a direct-current electrical flow is imposed on the circuit. This is known as the Peltier effect, discovered in 1834. The amount of the refrigerating effect is multiplied by the number of couples in parallel; the heat absorption at the cold junction is rejected at the second junction of each couple (that is, heat-pump effect). With ordinary materials generally available prior to World War II, the Peltier effect, though recognized, was not considered particularly practical. Since the war and with the advent of the transistor and other semiconductor devices leading to more practical and efficient equipment, there has been considerable worldwide interest in the possibilities of this system. Its application is to special-purpose processes, since its low efficiency has to be compensated by its special features, such as elimination of moving parts and noiseless operation.

Refrigerants and equipment. The working fluid in a two-phase refrigeration cycle is called a refrigerant. Commonly used refrigerants are listed in the table. Ammonia and Freon-22 are most important for industrial refrigeration, Freon-11 and Freon-12 being used most often for commercial and air conditioning work in which nontoxic refrigerants are necessary. A secondary cooling liquid that does not change from the liquid phase is called a brine. Solutions of sodium chloride or calcium chloride in water are frequently used as circulating brines in refrigeration systems.

Compressors. Refrigeration compressors may be positive-displacement of the reciprocating, rotary, or gear type for high- and medium-pressure differentials, or of the centrifugal type for low-pressure differentials. Early ammonia compressors were horizontal, double-acting, slow-speed units built like steam engines. Modern reciprocating compressors are vertical, single-acting, multicylinder, high-speed units built like automobile engines. Ammonia and other large compressors require water jacketing, whereas most Freon compressor cylinders are air-cooled.

Condensers. Refrigeration condensers may be air-cooled for small and medium capacities; or water-cooled, as in the case of the shell-and-tube, shell-and-coil, or double-pipe types. Because of the large quantities of condensing water required, cooling towers or spray ponds are commonly used to recool the water for reuse. An evaporative condenser is a device combining a condensing coil and a forced-draft cooling tower in a single unit.

Evaporators. Refrigerant evaporators are the cooling units placed in the room or fluid to be cooled. Plain pipe coils or finned coils, with or without forced circulation of the fluid being cooled, are commonly used. Shell-and-tube coolers, or tanks with wetted or submerged cooling coils, are frequently used where water or brine is circulated as a secondary cooling medium.

Expansion valve. The main flow control in a vapor-compression system is the expansion valve. It permits the liquid refrigerant to expand from the high pressure in the condenser to the lower pressure in the evaporator. The expansion causes part of the liquid to evaporate and thereby to cool the remainder to the evaporator temperature. A float valve with a throttling orifice is often used instead of an expansion valve to provide flooded control and maintain a fixed liquid level in the evaporator. In domestic refrigerators, a capillary tube restricts flow from condenser to evaporator. For completely automatic operation, additional controls are required to maintain the desired temperature in the evaporator, to regulate the compressor operation and the flow of the condensing medium, and to provide safety protection. *See* AIR CONDITIONING; AIR COOLING; COOLING TOWER; REFRIGERATOR.

[CARL F. KAYAN]

Bibliography: Air Conditioning and Refrigeration Institute, *Refrigeration and Air Conditioning*, 1979; W. Stoecker, *Refrigeration and Air Conditioning*, 1958.

Refrigeration cycle

A sequence of thermodynamic processes whereby heat is withdrawn from a cold body and expelled to a hot body. Theoretical thermodynamic cycles consist of nondissipative and frictionless processes. For this reason, a thermodynamic cycle can be operated in the forward direction to produce mechanical power from heat energy, or it can be operated in the reverse direction to produce heat energy from mechanical power. The reversed cycle is used primarily for the cooling effect that it produces during a portion of the cycle and so is called a refrigeration cycle. It may also be used for the heating effect, as in the comfort warming of space during the cold season of the year. *See* HEAT PUMP.

In the refrigeration cycle a substance, called the refrigerant, is compressed, cooled, and then expanded. In expanding, the refrigerant absorbs heat from its surroundings to provide refrigeration. After the refrigerant absorbs heat from such a source, the cycle is repeated. Compression raises the temperature of the refrigerant above that of its natural surroundings so that it can give up its heat in a heat exchanger to a heat sink such as air or water. Expansion lowers the refrigerant temperature below the temperature that is to be produced inside the cold compartment or refrigerator. The sequence of processes performed by the refrigerant constitutes the refrigeration cycle. When the refrigerant is compressed mechanically, the refrig-

Table of common refrigerants

Refrigerant	ASHRAE† standard refrigerant number	Chemical formula	Boiling point at atmospheric pressure, °F (°C)
Air	729		−318 (−194)
Ammonia	717	NH_3	−28 (−33)
Carbon dioxide*	744	CO_2	−109 (−78; sublimes)
Freon-11	11	CCl_3F	74.8 (23.8)
Freon-12	12	CCl_2F_2	−21.6 (−29.8)
Freon-21	21	$CHCl_2F$	48.1 (8.9)
Freon-22	22	$CHClF_2$	−41.4 (−40.8)
Freon-114	114	$C_2Cl_2F_4$	38.4 (3.6)
Methyl chloride*	40	CH_3CL	−10.8 (−23.8)
Methylene chloride*	30	CH_2Cl_2	105.2 (40.7)
Sulfur dioxide*	764	SO_2	14 (−10)
Water	718	H_2O	212 (100)

*Seldom used for new installations in the United States.
†American Society of Heating, Refrigerating and Air-Conditioning Engineers.

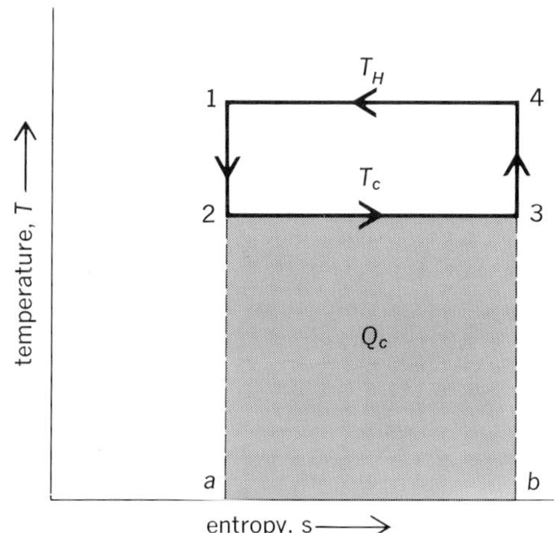

Fig. 1. Reverse Carnot cycle.

erative action is called mechanical refrigeration.

There are many methods by which cooling can be produced. The methods include the noncyclic melting of ice, or the evaporation of volatile liquids, as in local anesthetics; the Joule-Thomson effect, which is used to liquefy gases; the reverse Peltier effect, which produces heat flow from the cold to the hot junction of a bimetallic thermocouple when an external emf is imposed; and the paramagnetic effect, which is used to reach extremely low temperatures. However, large-scale refrigeration or cooling, in general, calls for mechanical refrigeration acting in a closed system. *See* REFRIGERATION.

Reverse Carnot cycle. The purpose of a refrigerator is to extract as much heat from the cold body as possible with the expenditure of as little work as possible. The yardstick in measuring the performance of a refrigeration cycle is the coefficient of performance, defined as the ratio of the heat removed to the work expended. The coefficient of performance of the reverse Carnot cycle is the maximum obtainable for stated temperatures of source and sink. Figure 1 depicts the reverse Carnot cycle on the T-s plane. *See* CARNOT CYCLE.

The appearance of the cycle in Fig. 1 is the same as that of the power cycle, but the order of the cyclic processes is reversed. Starting from state 1 of the figure, with the fluid at the temperature T_H of the hot body, the order of cyclic events is as follows:

1. Isentropic expansion, 1–2, of the refrigerant fluid to the temperature T_c of the cold body.
2. Isothermal expansion, 2–3, at the temperature T_c of the cold body during which the cold body gives up heat to the refrigerant fluid in the amount Q_c, represented by the area 2–3–b–a.
3. Isentropic compression, 3–4, of the fluid to the temperature T_H of the hot body.
4. Isothermal compression, 4–1, at the temperature T_H of the hot body. During this process, the hot body receives heat from the refrigerant fluid in the amount Q_H represented by the area 1–4–b–a. The difference $Q_H - Q_c$ represented by area 1–2–3–4 is the net work which must be supplied to the cycle by the external system.

Figure 1 indicates that Q_c and the net work rectangles each have areas in proportion to their vertical heights. Thus the coefficient of performance, defined as the ratio of Q_c to net work, is $T_c/(T_H - T_c)$.

The reverse Carnot cycle does not lend itself to practical adaptation because it requires both an expanding engine and a compressor. Nevertheless, its performance is a limiting ideal to which actual refrigeration equipment can be compared.

Modifications to reverse Carnot cycle. One change from the Carnot cycle which is always made in real vapor-compression plants is the substitution of an expansion valve for the expansion engine. Even if isentropic expansion were possible, the work delivered by the expansion engine would be very small and the irreversibilities present in any real operations would further reduce the work delivered by the expanding engine. The substitution of an expansion valve, or throttling orifice, with constant enthalpy expansion, changes the theoretical performance but little, and greatly simplifies the apparatus. A typical vapor-compression refrigeration cycle is shown in Fig. 2; it is essentially a reverse Rankine cycle. The irreversible adiabatic expansion 1–2 differs only slightly from the vertical isentropic expansion.

Another practical change from the ideal Carnot cycle substitutes dry compression 3–4 for wet compression e-d in Fig. 2, placing state 4 in the superheat region above ambient temperature; the process is called dry compression in contrast to the wet compression of the Carnot cycle. Dry compression introduces a second irreversibility by

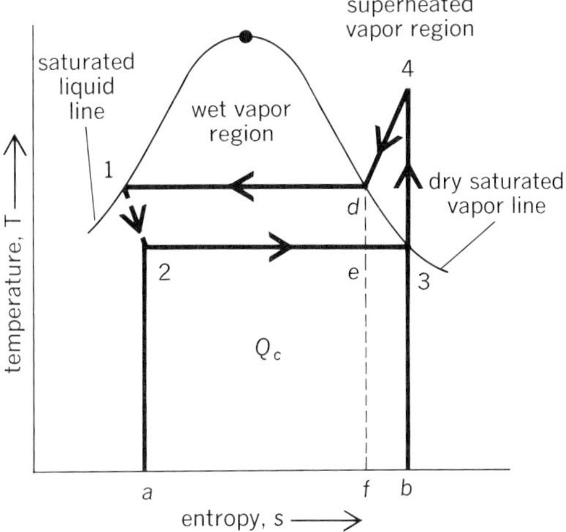

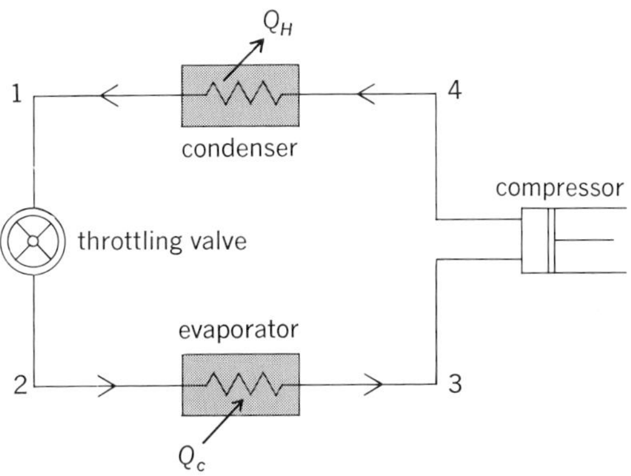

Fig. 2. Vapor-compression refrigeration cycle substitutes valve for expansion engine.

Ideal refrigeration cycle performance[a]

Refrigerant	Saturation pressure, psia[b] Evaporator	Saturation pressure, psia[b] Condenser	Refrigerating effect, Btu/lb[c]	Refrigerant required per ton refrigeration lb/min[d]	Refrigerant required per ton refrigeration cfm[e]	Compressor horsepower per ton refrigeration[f]	Coefficient of performance
Carnot—any fluid							6.6
Freon, F12	47	151	49	4.1	3.6	0.89	5.31
Ammonia	66	247	455	0.44	1.9	0.85	5.55
Sulfur dioxide	24	99	135	1.5	4.8	0.83	5.67
Free air	15	75	32	6.2	77	2.7	1.75
Dense air	50	250	32	6.2	23	2.7	1.75
Steam	0.1	1.275	1000	0.2	590	0.91	5.18

[a]Performance is based on evaporator temperature = 35°F (2°C), condenser temperature = 110°F (44°C). In vapor-compression cycles, the pressures and temperatures are for saturation conditions; vapor enters the compressor dry and saturated; no subcooling in the condenser. [b]1 psia = 6.895 kPa. [c]1 Btu/lb = 2326 J/kg. [d]1 (lb/min)/ton refrigeration = 2.15×10^{-6} (kg/s)/W. [e]1 cfm/ton refrigeration = 1.34×10^{-7} (m^3/s)/W. [f]1 compressor horsepower per ton of refrigeration = 0.212 watt of compressor power per watt of refrigeration.

exceeding the ambient temperature, thus reducing the coefficient of performance. Dry compression is usually preferred, however, because it simplifies the operation and control of a real machine. Vapor gives no readily observable signal as it approaches and passes point *e* in the course of its evaporation, but it would undergo a temperature rise if it accepted heat beyond point 3. This cycle, using dry compression, is the one which has won overwhelming acceptance for refrigeration work.

Reverse Brayton cycle. The reverse Brayton cycle constitutes another possible refrigeration cycle; it was one of the first cycles used for mechanical refrigeration. Before Freon and other condensable fluids were developed for the vapor-compression cycle, refrigerators operated on the Brayton cycle, using air as their working substance. Figure 3 presents the schematic arrangement of this cycle. Air undergoes isentropic compression, followed by reversible constant-pressure cooling. The high-pressure air next expands reversibly in the engine and exhausts at low temperature. The cooled air passes through the cold storage chamber, picks up heat at constant pressure, and finally returns to the suction side of the compressor. See BRAYTON CYCLE.

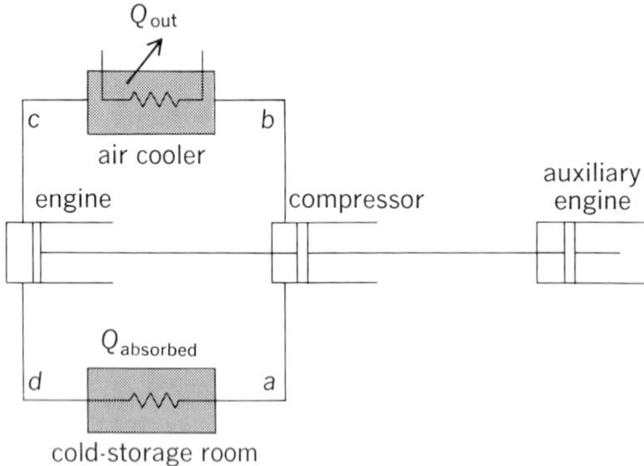

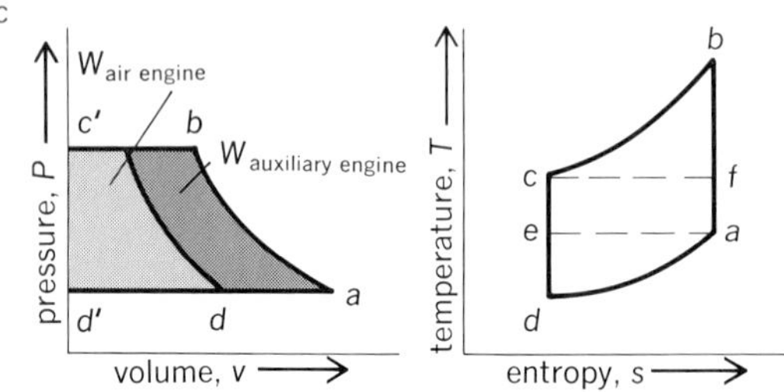

Fig. 3. Schematic arrangement of reverse Brayton, or dense-air, refrigeration cycle.

The temperature-entropy diagram, Fig. 3, points up the disadvantage of the dense-air cycle. If the temperature at *c* represents the ambient, then the only way that air can reject a significant quantity of heat along the line *b-c* is for *b* to be considerably higher than *c*. Correspondingly, if the cold body service temperature is *a*, the air must be at a much lower temperature in order to accept heat along path *d-a*. If a reverse Carnot cycle were used with a working substance undergoing changes in state, the fluid would traverse path *a-f-c-e* instead of path *a-b-c-d*. The reverse Carnot cycle would accept more heat along path *e-a* than the reverse Brayton cycle removes from the cold body along path *d-a*. Also, since the work area required by the reverse Carnot cycle is much smaller than the corresponding area for the reverse Brayton cycle, the vapor-compression cycle is preferred in refrigeration practice. See THERMODYNAMIC CYCLE.

Comparative performance of refrigerants. The table gives the significant theoretic performance data for a selected group of refrigerants when used in the ideal cycles as outlined above. These data include not only the requisite pressures in the evaporator and condenser for saturation temperatures of 35°F (2°C) and 110°F (44°C), respectively, but also the refrigerating effect, the weight and volume of refrigerant to be circulated, the horsepower required, and the coefficient of performance. The vapor compression cycles presuppose that the refrigerant leaves the evaporator dry and saturated (point 3 in Fig. 2) and leaves the condenser without subcooling (point 1 in Fig. 2). The heat absorbed, Q_c, in the evaporator is called the refrigerating effect and is measured in Btu/lb or J/kg of refrigerant. The removal of heat in the evaporator at the rate of 200 Btu/min (3.517 kW) is defined as a ton refrigeration capacity. The weight and volume of refrigerant, measured at the compressor suction, and the theoretic horsepower required to drive the compressor per ton refrigeration capacity are also given in the table. These data show a wide diversity of numerical values. Each refrigerant has its advantages and disadvantages. The selection of the most acceptable refrigerant for a specific application is consequently a practical compromise among such divergent data.

[THEODORE BAUMEISTER]

Bibliography: American Society of Heating, Refrigerating and Air-Conditioning Engineers, *Heating, Ventilating, and Air Conditioning Guide*, revised periodically; T. Baumeister (ed.), *Standard Handbook for Mechanical Engineers*, 8th ed., 1978.

Refrigerator

An insulated, cooled compartment. If it is large enough for the entry of a person, it is termed a walk-in box; otherwise it is called a reach-in refrigerator. Cooling may be by mechanical or gas refrigeration, by water or dry ice, or by brine circulation. Temperatures maintained depend upon the requirements of the product stored, generally varying from 55°F (13°C) down to 0°F (−18°C), and sometimes lower.

A household or domestic refrigerator is a factory-built, self-contained cabinet. The range of storage capacities is wide and varies among manufacturers. Modern designs have a main compartment for holding food above freezing, a second compartment for storage below freezing, and trays for the freezing of ice cubes. The cabinets are usually all metal with 2–3½ in (5–9 cm) of insulation. The refrigeration unit is usually driven by electric motor, but gas refrigerators motivated by the thermal energy of fuel gas have been used extensively where cheap natural gas is available. Low-temperature household refrigerators, or home freezers, for the storage of frozen foods are manufactured in both the chest and the upright, or vertical, types.

A commercial refrigerator is any factory-built refrigerated fixture, cabinet, or room that can be assembled and disassembled readily, in contrast to a built-in refrigerator. Commercial or built-in refrigerators are used in restaurants, markets, hospitals, hotels, and schools for the storage of food and other perishables. A meat cooler is a refrigerator held at about 33°F (1°C) for the storage of fresh meats. Refrigerators for lower temperatures down to 0°F (−18°C) and below are called freezer boxes. Insulation thicknesses vary from 3–8 in. (8–20 cm), depending upon the service. In markets and stores, commercial display refrigerators may be of the self-service type from which customers help themselves. Both vertical types with glass doors, which the customer opens, and chest types with open tops are used. Electric refrigeration units may be built into each fixture, or they may be remotely located. *See* REFRIGERATION.

[CARL F. KAYAN]

Regenerative braking

A system of dynamic braking in which the electric drive motors are used as generators and return the kinetic energy of the motor armature and load to the electric supply system. This method is employed when the load is losing a large amount of potential energy, as in the case of an electric train descending a long grade or a pumped storage reservoir being emptied. The potential energy of the load is first converted to kinetic energy in the motor armature and load and then to electrical energy. The method is applicable to dc motors, to induction motors (with negative slip), and to synchronous motors. *See* DYNAMIC BRAKING; INDUCTION MOTOR; SYNCHRONOUS MOTOR; WATERPOWER.

[ARTHUR R. ECKELS]

Reheating

The addition of heat to steam of reduced pressure after the steam has given up some of its energy by expansion through the high-pressure stages of a turbine. The reheater tube banks are arranged within the setting of the steam-generating unit in such relation to the gas flow that the steam is restored to a high temperature. Under suitable conditions of initially high steam pressure and superheat, one or two stages of reheat can be advantageously employed to improve thermodynamic efficiency of the cycle. *See* STEAM-GENERATING UNIT; STEAM TURBINE; SUPERHEATER; VAPOR CYCLE.

[R. A. MILLER]

Relay

An electromechanical or solid-state device operated by variations in the input which, in turn, operate or control other devices connected to the output. They are used in a wide variety of applications throughout industry, such as in telephone exchanges, digital computers, motor and sequencing controls, and automation systems. Highly sophisticated relays are utilized to protect electric power systems against trouble and power blackouts as well as to regulate and control the generation and distribution of power. In the home, relays are used in refrigerators, automatic washers and dishwashers, and heat and air-conditioning controls. Although relays are generally associated with electrical circuitry, there are many other types, such as pneumatic and hydraulic. Input may be electrical and output directly mechanical, or vice-versa.

Relays using discrete solid-state components, operational amplifiers, or microprocessors can provide more sophisticated designs. Their use is increasing, particularly in applications where the relay and associated equipment are packaged together. The basic operation may be complex, but frequently is similar or equivalent to the units described.

Basic classifications. A basic, simplified block diagram of a relay is shown in Fig. 1. The actual physical embodiment of the blocks varies widely, and one or more blocks may not exist or may be combined in practice. Classifying relays by function and somewhat in the order of increasing complexity there are (1) auxiliary, (2) monitoring, (3) regulating, (4) programming, and (5) protective relays.

An auxiliary relay (telephone type) is shown in Fig. 2. The sensing unit is the electric coil; when the applied current or voltage exceeds a threshold value the coil activates the armature, which operates either to close the open contacts or to open the closed contacts. In Fig. 1 the contacts belong to the amplifier block. Therefore relays of this class generally are used as contact multipliers. Other applications are to isolate circuits or to use a low-power input to control higher-power input. The timing-integrating block for these relays is in the armature and springs. Some time delay is inherent, and a number of techniques are used to add

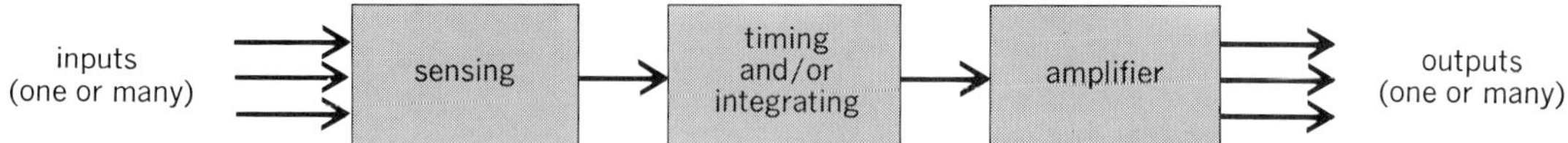

Fig. 1. Basic block diagram of a relay.

more delay or to reduce time of operation to a minimum.

The balanced polar unit of Fig. 3*a* is used as a monitoring relay. The unbalanced type is shown in Fig. 3*b*. A permanent magnet normally polarizes the pole faces on either side of the armature, and direct current is applied to the coil that is around the armature. At nominal values of current, the armature can be adjusted to float in the air gap between the pole pieces by the magnetic shunts, but when this current decreases, or increases significantly, the armature is magnetized and moves to one of the pole pieces to close the contacts. Thus this relay monitors the current level in an electric circuit. The sensing, integrating, and amplifying functions indicated in the block diagram of Fig. 1 are combined in the single mechanical element and its contacts for this polar unit.

The relay unit of Fig. 4, known as the induction-disk type, is widely used in regulating and protective relays. Alternating current or voltage applied to the coil produces mechanical torque to rotate the disk, which is of the order of 3 in. (8 cm) in diameter. Contacts are attached to the disk shaft with a spiral spring to reset the contacts to a backstop or normally closed contacts. A permanent magnet provides damping. By design of the coils, damping magnet, and spring, a variety of inverse time overcurrent and under- or overvoltage characteristics are produced. With reference to Fig. 1, the sensing unit is approximated by the coil, by the timing unit, and the contacts by the amplifier.

For regulating, the output is arranged to adjust or regulate the input within prescribed limits; for monitoring, the output is used to alarm or to shut down the input equipment.

Programming relays provide an automatic se-

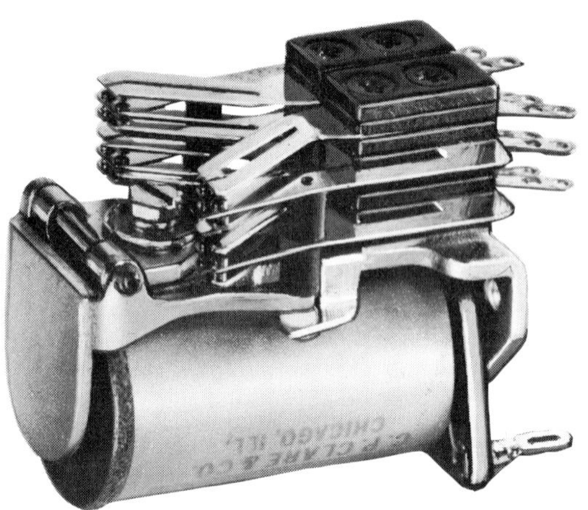

Fig. 2. A multicontrol telephone-type auxiliary relay. (*C.P. Clare and Co.*)

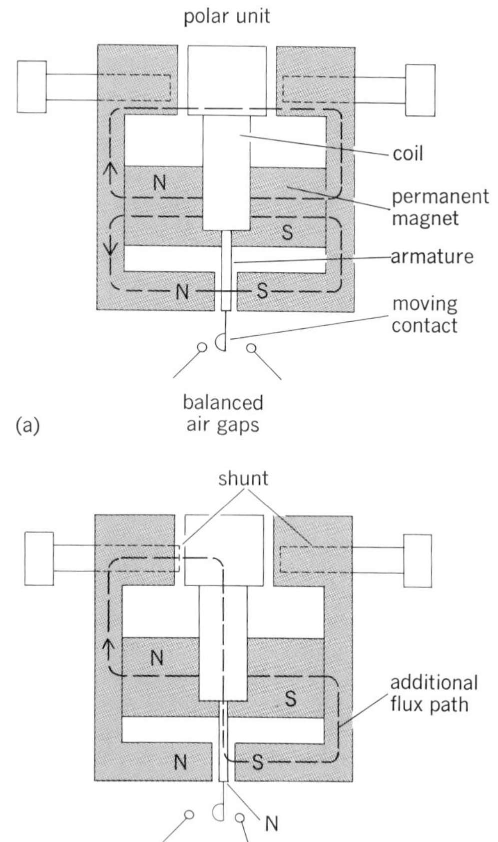

Fig. 3. Polar units. (*a*) Balanced. (*b*) Unbalanced.

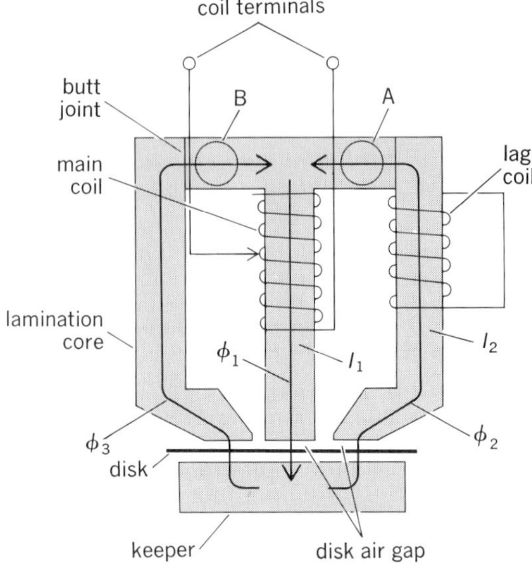

Fig. 4. Schematic of the induction-disk type relay. I_1 = current in main coil; I_2 = current in lag coil; ϕ_1 = flux in main coil; ϕ_2 = flux in lag coil; ϕ_3 = differential induction flux; A = flux field about ϕ_3; B = flux field about ϕ_2.

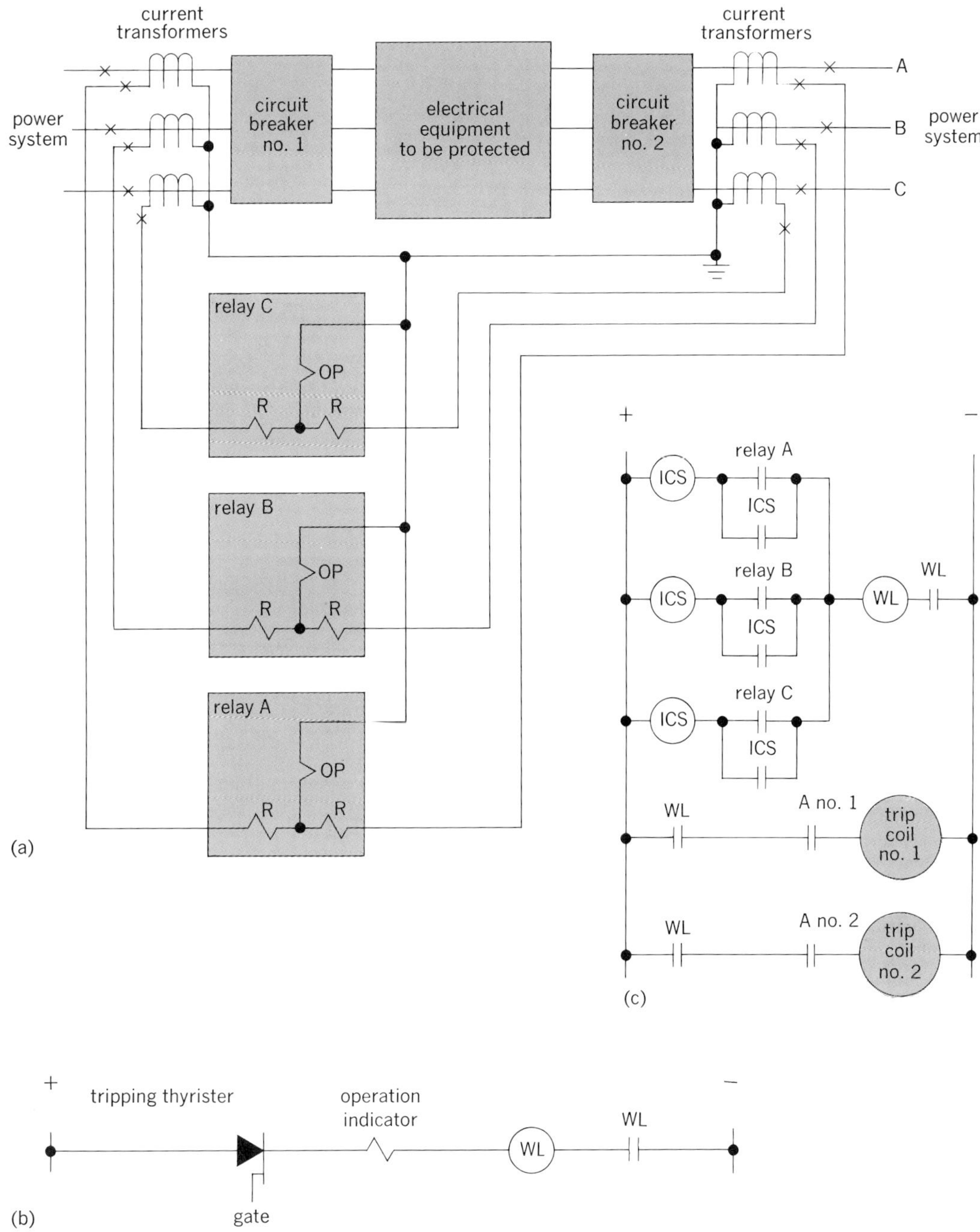

Fig. 5. Differential protection. (*a*) Typical arrangement for electrical equipment. Relays A, B, and C are differential relays. (*b*) Simplified alternate trip circuit (one phase) for solid-state relay. Gate is from differential relay integrating circuitry. (*c*) Trip circuit. A = Auxiliary out-off breaker; ICS = operation indicator and circuit seal-in auxiliary relay; OP = operating coil; R = restraining coil; WL = auxiliary hand-reset lock-out relay.

quence of operations. Good examples of these are the timer control relays on automatic washers and dishwashers, which provide the necessary program in the various washing operations.

Protective relays. Protective relays are compact analog networks, connected to various points of an electrical system, to detect abnormal conditions occurring within their assigned areas. They initiate disconnection of the trouble area by circuit breakers. These relays range from the simple overload unit on house circuit breakers to complex systems used to protect extra-high-voltage power transmission lines. Heavy-duty protective relay systems detect all intolerable system conditions, such as faults caused by lightning and equipment insulation failure, and initiate tripping of power circuit breakers within 6–10 milliseconds. *See* CIRCUIT BREAKER; LIGHTNING AND SURGE PROTECTION.

Fault detection. Fault detection is accomplished by a number of techniques. Some of the common methods are the detection of changes in electric current or voltage levels, power direction, ratio of voltage to current, temperature, and comparison of the electrical quantities flowing into a protected area with the quantities flowing out. The last-mentioned is known as differential protection and is illustrated by Fig. 5. Differential relays are applied to protect a piece of electrical apparatus. The inputs to the relays are currents from a current transformer. Current through the relay (nominally 5 amperes) is proportional to the high-power current of the main circuit. For load through the equipment or for faults either to the right or left of the current transformers, the secondary current flows through the relay restraining coils, and little or no current flows through the operating coils so that operation is prevented. For a fault between the current transformers, secondary current flows through the restraining windings (with reversed phase in one of the coils) and in the operating coils to operate the relay, trip the two circuit breakers, and isolate the fault or damaged equipment from the rest of the power system.

For transmission lines, in which there are considerable miles between the current transformers, the same principle is used, but a set of relays is used at each end. The intelligence to compare the direction of power flow or phase angle of the currents between the terminals is transmitted over a radio-frequency channel superimposed on the power line, by a telephone pair, or by microwave between the terminals. *See* ELECTROMAGNET; SOLENOID (ELECTRICITY); TRANSMISSION LINES.

[J. L. BLACKBURN]

Bibliography: D. Beeman, *Industrial Power System Handbook*, 1955; C. R. Mason, *The Art and Science of Protective Relaying*, 1956; National Association of Relay Manufacturers, *Engineers Relay Handbook*, 1966; R. L. Peek, Jr., and H. N. Wagar, *Switching Relay Design*, 1955; A. R. van C. Warrington, *Protective Relays*, 1962; Westinghouse Electric Corp., *Applied Protective Relaying*, 1957.

Reluctance motor

A synchronous motor which starts as an induction motor and, upon nearing full speed, locks into step with the rotating field and runs at synchronous speed. The stator and rotor windings are similar to those of an induction motor. The rotor is of squirrel-cage construction, to allow induction-motor starting, and has salient-pole projections which provide synchronous operation at full speed. The reluctance motor is built only in small sizes and for situations in which low cost and simplicity are mandatory and efficiency is of little concern. It can be polyphase, but is usually a single-phase motor with a split-phase or capacitor winding for starting. *See* INDUCTION MOTOR; SYNCHRONOUS MOTOR.

[LOYAL V. BEWLEY]

Repulsion motor

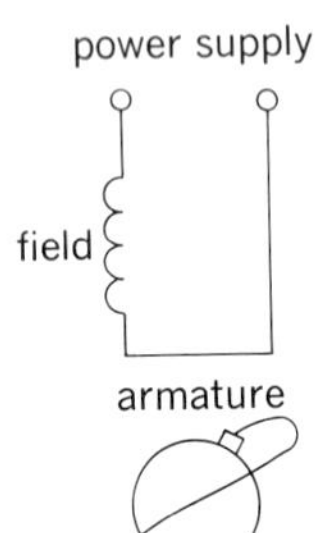

Fig. 1. Schematic of a repulsion motor

An alternating-current (ac) commutator motor designed for single-phase operation. The chief distinction between the repulsion motor and the single-phase series motors is the way in which the armature receives its power. In the series motor the armature power is supplied by conduction from the line power supply. In the repulsion motor, however, armature power is supplied by induction (transformer action) from the field of the stator winding. For discussion of the ac series motor *see* UNIVERSAL MOTOR. *See also* ALTERNATING-CURRENT MOTOR.

The repulsion motor primary or stationary field winding is connected to the power supply. The secondary or armature winding is mounted on the motor shaft and rotates with it. The terminals of the armature winding are short-circuited through a commutator and brushes. There is no electrical contact between the stationary field and rotating armature (Fig. 1). *See* WINDINGS IN ELECTRIC MACHINERY.

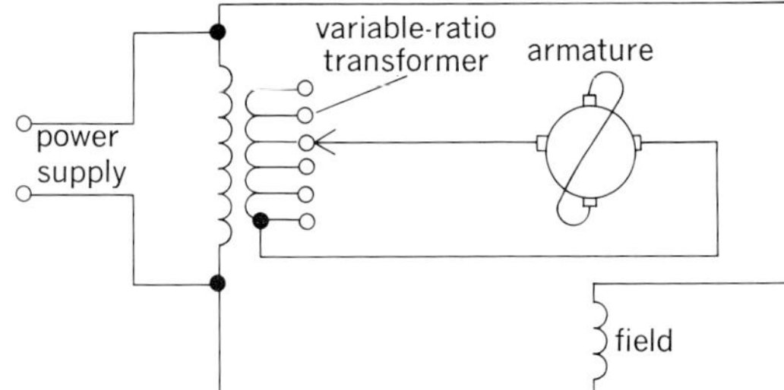

Fig. 2. Doubly excited repulsion motor.

If the motor is at rest and the field coils are energized from an outside ac source, a current is induced in the armature, just as in a static transformer. If the brushes are in line with the neutral axis of the magnetic field, there is no torque, or tendency to rotate. However, if they are set at a proper angle (generally 15–25° from the neutral plane), the motor will rotate.

Repulsion motors may be started with external resistance in series with the motor field, as is done with dc series motors. A more common method is to start the motor with reduced field voltage and increase the voltage as the motor increases speed. This can be done conveniently with a transformer having an adjustable tapped secondary or a variable autotransformer.

It is also possible to doubly excite the motor; that is, the armature may receive its power not only by induction from the stator winding but also by conduction from a transformer with adjustable taps, as shown in Fig. 2.

Repulsion-start, induction-run motor. This motor (Fig. 3*a*) possesses the characteristics of the repulsion motor at low speeds and those of the induction motor at high speed. It starts as a repulsion motor. At a predetermined speed (generally at about two-thirds of synchronous speed) a centrifugal device lifts the brushes from the commutator and short-circuits the armature coils, producing a squirrel-cage rotor. The motor then runs as an induction motor. In Fig. 3*b* the curve *AB* represents the characteristics of an induction motor and curve

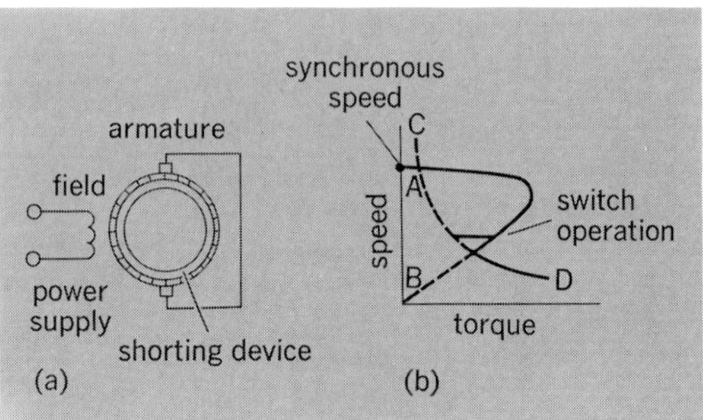

Fig. 3. Repulsion-start, induction-run motor. (*a*) Schematic diagram. (*b*) Speed-torque characteristic.

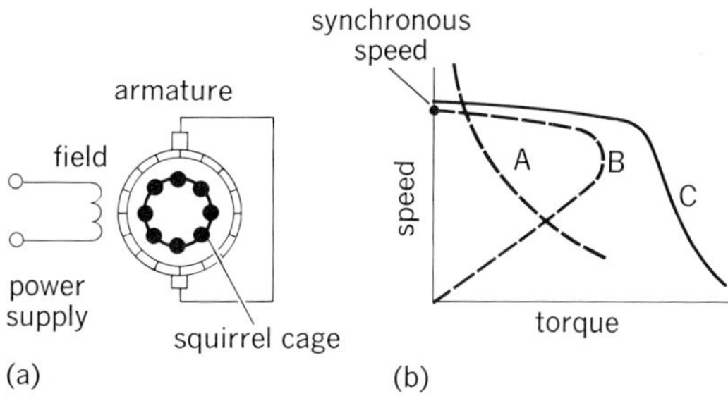

Fig. 4. Repulsion-induction motor. (*a*) Schematic diagram. (*b*) Speed-torque characteristic.

CD a repulsion motor. The solid curve *AD* is the combined characteristic of a repulsion-start, induction-run motor. *See* INDUCTION MOTOR.

Repulsion-induction motor. This motor (Fig. 4*a*) is very similar to the standard repulsion motor in construction, except for the addition of a second separate high-resistance squirrel cage on the rotor. Both rotor windings are torque-producing, and the total torque produced is the sum of the individual torques developed in these two windings. In Fig. 4*b*, curve *A* is the characteristic of the repulsion-motor torque developed in this motor. Curve *B* represents the induction-motor torque. Curve *C* is the combined total torque of the motor. The advantages of this machine are its high starting torque and good speed regulation. Its disadvantages are its poor commutation and high initial cost.

[IRVING L. KOSOW]

Bibliography: I. L. Kosow, *Electric Machinery and Transformers*, 1972; G. McPherson, *An Introduction to Electrical Machines and Transformers*, 1981; M. G. Say, *Alternating-Current Machines*, 4th ed., 1976.

Reservoir

A pond or lake built for the storage of water, usually by the construction of a dam across a river. The size of reservoir needed is a function of the water demands, the natural flows of the river impounded, and the extent of droughts to be encountered. In areas of moderate rainfall, storage may be required during only a few days or weeks in a year. In arid and semiarid areas, storage capacity may be needed to supplement low stream flows over periods of many months or even years. In some instances water can be diverted by gravity flow or by pumping from one river to a reservoir on another stream. Water supplies may be taken directly from the reservoirs, or water may be released from the reservoirs to augment river flows past a water supply intake downstream. *See* DAM; WATER SUPPLY ENGINEERING.

[RICHARD HAZEN]

Resistance heating

The generation of heat by electric conductors carrying current. The degree of heating for a given current is proportional to the electrical resistance of the conductor. If the resistance is high, a large amount of heat is generated, and the material is used as a resistor rather than as a conductor. *See* ELECTRICAL RESISTANCE.

Resistor materials. In addition to having high resistivity, heating elements must be able to withstand high temperatures without deteriorating or sagging. Other desirable characteristics are low temperature coefficient of resistance, low cost, formability, and availability of materials. Most commercial resistance alloys contain chromium or aluminum or both, since a protective coating of chrome oxide or aluminum oxide forms on the surface upon heating and inhibits or retards further oxidation. Some commercial resistor materials are listed in Table 1.

Heating element forms. Since heat is transmitted by radiation, convection, or conduction or combinations of these, the form of element is designed for the major mode of transmission. The simplest form is the helix, using a round wire resistor, with the pitch of the helix approximately three wire diameters. This form is adapted to radiation and convection and is generally used for room or air heating. It is also used in industrial furnaces, utilizing forced convection up to about 1200°F. Such helixes are stretched over grooved high-alumina refractory insulators and are otherwise open and unrestricted. These helixes are suitable for mounting in air ducts or enclosed chambers, where there is no danger of human contact. *See* ELECTRIC FURNACE; INDUCTION HEATING.

For such applications as water heating, electric range units, and die heating, where complete electrical isolation is necessary, the helix is embedded in magnesium oxide inside a metal tube, after which the tube is swaged to a smaller diameter to compact the oxide and increase its thermal conductivity. Such units can then be formed and flattened to desired shapes. The metal tubing is usually copper for water heaters and stainless steel for radiant elements, such as range units. In some cases the tubes may be cast into finned aluminum housings, or fins may be brazed directly to the tubing to increase surface area for convection heating.

Modification of the helix for high-temperature furnaces involves supporting each turn in a grooved refractory insulator, the insulators being strung on stainless alloy rods. Wire sizes for such elements are 3/16 in. (5 mm) in diameter or larger, or they may be edge-wound strap. Such elements may be used up to 1800°F (980°C) furnace temperature.

Table 1. Electric furnace resistor materials and temperature ranges

Material, major elements	Maximum resistor temperature, °F (°C) In air	Maximum resistor temperature, °F (°C) In reducing or neutral atmosphere
*35% nickel, 20% chromium	1900 (1040)	2100 (1150)
*60% nickel, 16% chromium	1800 (980)	2000 (1090)
*68% nickel, 20% chromium, 1% cobalt	2250 (1230)	2250 (1230)
78% nickel, 20% chromium	2250 (1230)	2250 (1230)
*15% chromium, 4.6% aluminum	2050 (1120)	Not used
*22.5% chromium, 4.6% aluminum	2150 (1180)	Not used
*22.5% chromium, 5.5% aluminum	2450 (1340)	Not used
Silicon carbide	2800 (1540)	2500 (1370)
Platinum	2900 (1590)	2900 (1590)
†Molybdenum	Not used	3400 (1870)
†Tungsten	Not used	3700 (2040)
†Graphite	Not used	5000 (2760)

*Balance is largely iron, with 0.5–1.5% silicon.
†Usable only in pure hydrogen, nitrogen, helium, or oxygen or in vacuum because of inability to form protective oxide.

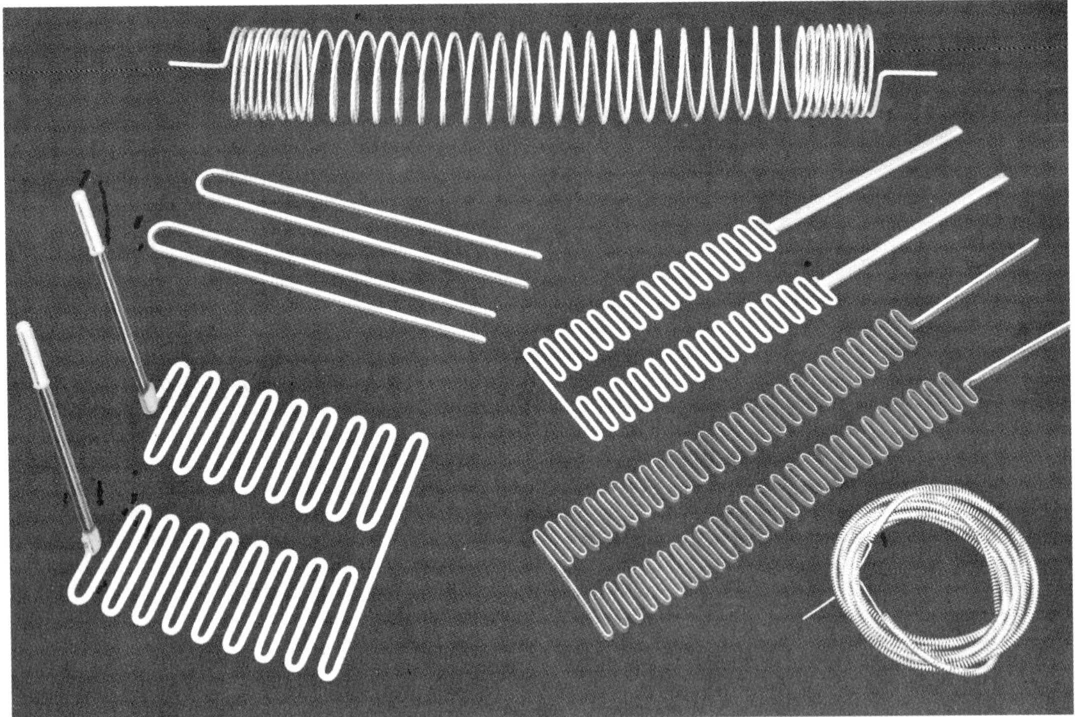

Fig. 1. Metallic resistor forms.

Another form of furnace heating element is the sinuous grid element, made of heavy wire or strap or casting and suspended from refractory or stainless supports built into the furnace walls, floor, and roof. Some of these forms are shown in Fig. 1.

Silicon carbide elements are in rod form, with low-resistance integral terminals extending through the furnace walls, as shown in Fig. 2.

Direct heating. When heating metal strip or wire continuously, the supporting rolls can be used as electrodes, and the strip or wire can be used as the resistor.

In Fig. 3 the electric current passes from roll A through the strip or wire to roll B. Heating by this method can be very rapid. Disadvantages are that the electric currents are large, and uniform contact between the strip and the rolls is difficult to maintain, since both surfaces must be clean and free of oxides. For these reasons, direct heating is not used extensively.

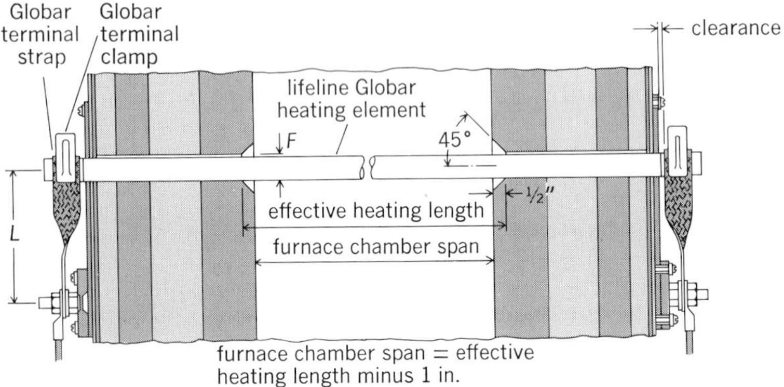

Fig. 2. Silicon carbide heating element. 1 in. = 2.5 cm.

Molten salts. The electrical resistance of molten salts between immersed electrodes can be used to generate heat. Limiting temperatures are dependent on decomposition or evaporization temperatures of the salt. Parts to be heated are immersed in the salt. Heating is rapid and, since there is no exposure to air, oxidation is largely prevented. Disadvantages are the personnel hazards and discomfort of working close to molten salts.

Major applications. A major application of resistance heating is in electric home appliances, including electric ranges, clothes dryers, water heaters, coffee percolators, portable radiant heaters, and hair dryers. Resistance heating is also finding increasing application in home or space heating; new homes are being designed with suitable thermal insulation to make electric heating practicable. A general rule for such insulation is that heat loss be restricted to 10 W/ft² (108 W/m²) of floor living area at 75°F (42°C) temperature difference, inside to outside. This applies to areas having 5000–7500 degree days annually. The degree day is the difference between the 24-hr average outdoor temperature and 65°F (18.3°C). For instance, if the 24-hr average were 40°F (4.4°C), there would be 25 degree days for that day. This includes also a three-quarter air change per hour, except in the basement, where one-quarter air change per hour is calculated. The values in Table 2 are applied to determine overall heat loss.

Heating system capacity. For rooms normally kept warm, the installed capacity should be 20% higher than the calculated losses. For rooms intermittently heated or for entries with frequent door openings, the installed capacity should be 50% more than the calculated losses.

A variety of electric heating units may be used for home heating. The simplest is a grid of helical

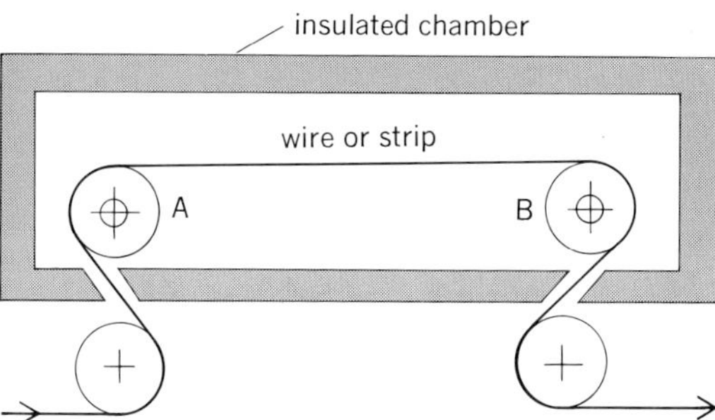

Fig. 3. Direct heating of metal strip or wire.

resistance coils mounted in the central heating unit, with a blower to circulate the warm air through the rooms, and adjustable louvers in each room to control the temperature. Another type is the baseboard heater, consisting of finned sheathed helical resistance coils, with individual room control (thermostat in each room). Still another room unit is the glass or ceramic panel in which resistance wires are embedded. Such panels are designed to radiate infrared wavelengths and are particularly suitable for glassed-in areas or bathrooms. Mounted usually in the ceiling, such panels neutralize the chilling effect of large glass exposures, thus preventing drafts. *See* THERMOSTAT.

For isolated rooms or work areas, integrated units which combine resistors, circulating fans, and magnetic contactors operated by thermostats open and close the power circuits to the heating elements to maintain desired temperature. Where central heating units involving several kilowatts of energy are used, the contactors are arranged in multiple circuits and close in sequence to minimize the transient voltage effects of sudden large energy demands.

Ovens and furnaces. If the resistor is located in a thermally insulated chamber, most of the heat generated is conserved and can be applied to a wide variety of heating processes. Such insulated chambers are called ovens or furnaces, depending on the temperature range and use.

The term oven is generally applied to units which operate up to approximately 800°F (430°C). Ovens use rock wool or glass wool between the inner and outer steel casings for thermal insulation. Typical

Table 2. Maximum recommended heat loss

Structure	Heat loss, W/ft²/°F [W/(m² · C)]
Ceiling	0.015 (0.29)
Wall	0.021 (0.41)
Floor	0.021 (0.41)
Floor slab perimeter	12.0/lineal foot (71/lineal meter)
Basement walls	0.045 (0.77)
Basement walls, 25% of wall aboveground	0.026 (0.50)
Requisites:	
Windows – Double-glazed or storm	
Doors – Storm doors required	
Attic – Ventilating area of 1 ft²/150 ft² or 1 m²/150 m² ceiling area	

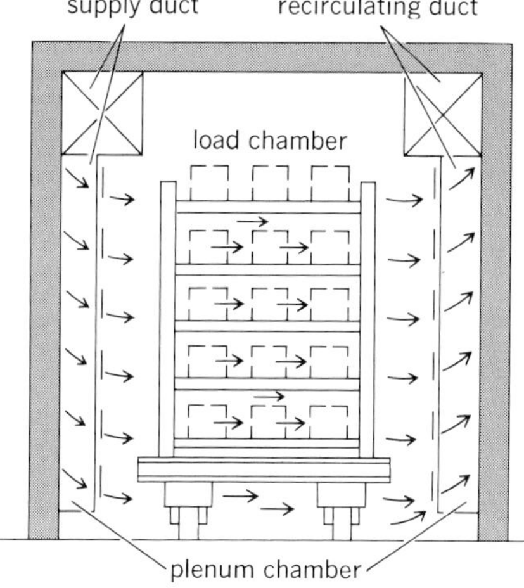

Fig. 4. Typical oven for maximum temperature of 800°F (430°C); heating is by forced convection. (*Carl Mayer Corp.*)

uses are for baking or roasting foods, drying paints and organic enamels, baking foundry cores, and low-temperature treatments of metals.

The term furnace generally applies to units operating above 1200°F (650°C). In these the thermal insulation is made up of an inner wall of fireclay,

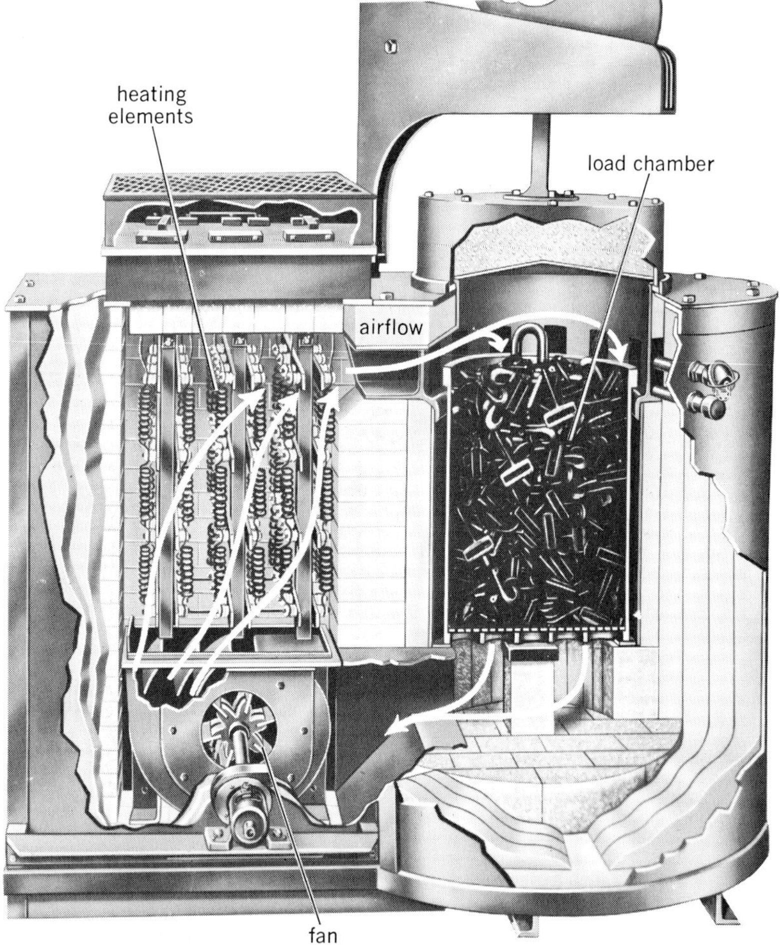

Fig. 5. Low-temperature pit-type furnace, using forced convection and operating at temperatures of 250–1400°F (120–760°C). (*Lindberg Industrial Corp.*)

kaolin, or high-alumina or zirconia brick, depending on the temperature, with secondary insulating blocks made from such base materials as rock wool, asbestos fiber, and diatomaceous earth. Typical uses of furnaces are for heat treatment or melting of metals, for vitrification and glazing of ceramic wares, for annealing of glass, and for roasting and calcining of ores. *See* ELECTRIC HEATING.

Actually, ovens and furnaces overlap in temperature range, with ovens being used at temperatures as high as 1000°F (540°C), and furnaces as low as 250°F (120°C). Electrically heated ovens and furnaces have advantages over fuel-fired units. These advantages often compensate for the generally higher cost of electric energy. The main advantages are (1) ease of distributing resistors, or heating elements, to obtain a uniform temperature in the product being heated; (2) ease of operation, since adjustments by operators are usually unnecessary; (3) cleanliness; (4) comfort, since heat losses are low and there are no waste fuel products; (5) adaptability to the use of controlled furnace atmospheres or vacuum, and (6) high temperatures beyond the range attainable with commercial fuels.

Heat transfer in ovens and furnaces. Heating elements mounted in ovens and furnaces may be located to radiate directly to the parts being heated or may be located behind baffles or walls so that direct radiation cannot take place. The heat then is transferred by circulating the furnace air or gas. Determination of which method or combination of methods to use is based on temperature uniformity, speed of heating, and high-temperature strength limitations of fans or blowers. At temperatures below 1200°F (650°C) radiation is slow, and virtually all ovens and furnaces in this temperature range use forced convection or circulation. From 1200 to 1500°F (650 to 820°C), radiation is increasingly effective, while reduced gas density and lower fan or blower speeds (because of reduced strength at high temperatures) make forced convection less effective. Therefore in this temperature range, combinations of radiation and forced convection are used. Above 1500°F forced convection is employed only when direct radiation cannot reach all parts of the loads being heated. An example of this is a container filled with bolts or a rack filled with gears.

Fig. 6. Interior of high-temperature pit-type carburizing furnace, using radiant heating. (*Lindberg Industrial Corp.*)

Obviously, in vacuum furnaces, heat transfer can be only by radiation and conduction.

Ovens. Figure 4 shows a typical oven, for a maximum temperature of 800°F (430°C); heating is by forced convection. The heated air enters through the supply duct, passing downward into the plenum chamber and distributing louvers into and across the load chamber and returning through the right-side plenum to the recirculating duct. The external heater and blower, which complete the circuit, are not shown.

Low-temperature furnace. Figure 5 shows a pit-type furnace used for tempering steel. It operates at temperatures of 250–1400°F (120–760°C) and uses forced convection.

The steel parts to be tempered are placed in the load chamber at the right. The electrical heating elements are in the heating chamber at the upper left. The fan, at the lower left, circulates air upward over the heating elements and into the load chamber. The hot air is forced down through the load and back to the fan. The heating chamber is thermally insulated from the load chamber so there is no direct radiation. Such furnaces are designed commercially to hold the load temperature within 10° of the control temperature, and can be designed for closer control if desired.

High-temperature furnace. Figure 6 shows the interior of a pit-type carburizing furnace, which uses radiant-heating elements in the form of corrugated metal bands mounted on the inside of the brick walls. The work basket (not shown) rests on the load support at the bottom. These elements are designed to operate at low voltage (approximately 30 volts) so that soot deposits from the carburizing gases will not cause short circuits. *See* FURNACE CONSTRUCTION.

Electrical input. Electric ovens and furnaces are rated in kilowatts (kw). The electrical input is determined from the energy absorption Q of the load and the thermal losses.

The average rate of energy flow into load during the heating period is Q/t, where Q is in kilowatt-hours and t is the heating time in hours.

However, the rate of energy flow into the load is high when the load is cold and decreases as the load temperature approaches furnace temperature. Therefore the energy input must be high enough to take care of the high initial heating rate. This leads to the following approximate formulas for the input, in which L is the thermal losses in kilowatts at the operating temperature:

For batch furnaces, input in kW $= L + 1.5\ Q/t$
In continuous furnaces, input in kW $= L + 1.25\ Q/t$

Operating voltage for heating elements. Heating elements are usually designed to operate at standard service voltages of 115, 230, or 460 volts, if two conditions can be satisfied. These are, first, that the heating element is sufficiently heavy in cross section to avoid sagging or deformation in service, and second, that there is no appreciable electrical leakage through the furnace refractories tending to short-circuit the heating elements. The latter consideration generally limits voltages to 260 volts at 2100°F (1150°C) and to approximately 50 volts at 3100°F (1700°C), because the refractory walls of the furnace become increasingly better conductors at higher temperatures.

In vacuum furnaces, voltages are limited to 230 volts by the tendency to break down into glow discharge at low pressure.

Temperature control. Almost all commercial electric ovens and furnaces have automatic temperature control. The simplest control uses magnetic contactors which open and close the circuit to the heating elements in response to temperature signals from control thermostats or thermoelectric pyrometers. A refinement of this ON-and-OFF control is to modulate it with a timer, with the ON period becoming a progressively smaller percentage as the control temperature is approached. This prevents the overshooting which results from thermal lag in the control thermocouple and the furnace.

True proportioning control is achieved through the use of saturable reactors in series with the heating elements. Figure 7 shows schematically the arrangement used.

The pyrometer, through the amplifier, controls the direct current to the control winding of the reactor. When the direct current is maximum, the reactor offers virtually no impedance to the alternating current flowing through the heating element, and the normal amount of heat is generated. As control temperature is approached, the direct current is decreased, increasing the impedance of the reactor and reducing the current to the heating elements until equilibrium is reached.

A device which is beginning to be used as a heating contactor is the solid-state silicon-controlled rectifier. While more expensive than the magnetic contactor, it has no moving parts, lower maintenance costs, and silent, vibration-free operation. Moreover, it is directly adaptable to proportioning control, obviating the need for saturable reactors. In this latter case it is cost-competitive.

Snow and ice melting. Low-temperature resistors with moisture-proof insulation are embedded in concrete sidewalks or strung in roof gutters to melt snow and ice as needed. Such cables also are used to prevent freezing of exposed water pipes. Analogously, very thin resistance wires may be cast in automobile windows to melt snow and ice.

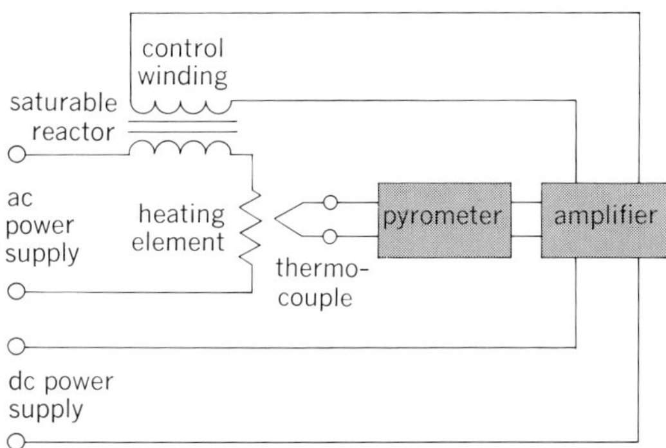

Fig. 7. Temperature control using saturable reactor.

Metal-sheathed units are used at railroad switches and dam locks to keep the moving parts free in freezing temperatures.

Die and platen heating. Metal-sheathed elements are embedded in grooves or holes in dies or platens to maintain these parts at desired temperatures for hot processing or molding of plastics.

Cryogenics. Metal-sheathed immersion-type units are used to supply heat of evaporation for liquefied gases such as argon, helium, nitrogen, hydrogen, and oxygen. Metal sheathing is usually of stainless steel or aluminum to withstand low temperatures. Gaskets are of Teflon for the same reason. [WILLARD ROTH]

Bibliography: American Society of Heating, Refrigeration, and Air Conditioning Engineers, *ASHRAE Handbook and Product Directory: Systems*, 1980; Chemical Rubber Publishing Co., *Handbook of Chemistry and Physics*, published annually; D. G. Fink and H. W. Beatty, *Standard Handbook for Electrical Engineers*, 11th ed., 1978; W. H. McAdams, *Heat Transmission*, 3d ed., 1954; A. G. Robiette, *Electric Melting Practice*, 1972; W. Trinks, *Industrial Furnaces*, vol. 1, 5th ed., 1961, and vol. 2, 4th ed., 1967.

Resistance welding

A process in which the heat for producing the weld is generated by the resistance to the flow of current through the parts to be joined. The application of external force is required; however, no fluxes, filler metals, or external heat sources are necessary. Most metals and their alloys can be successfully joined by resistance welding processes.

Methods. Several methods are classified as resistance welding processes: spot, roll-spot, seam, projection, upset, flash, and percussion (Fig. 1).

Resistance spot welding. In this process, coalescence at the faying surfaces is produced in one spot by the heat obtained from the resistance to electric current through the work parts held together under pressure by electrodes. The size and shape of the individually formed welds are limited primarily by the size and contour of the electrodes (Fig. 2*a*). *See* SPOT WELDING.

Roll resistance spot welding. Separated resistance spot welds are made with one or more rotating circular electrodes. The rotation of the electrodes may or may not be stopped during the making of a weld.

Resistance seam welding. Coalescence at the faying surfaces is produced by the heat obtained from resistance to electric current through the work parts held together under pressure by electrodes. The resulting weld is a series of overlapping resistance spot welds made progressively along a joint by rotating the electrodes (Fig. 2*b*).

Projection welding. Coalescence is produced by the heat obtained from resistance to electric current through the work parts held together under pressure by electrodes. The resulting welds are localized at predetermined points by projections, embossments, or intersections (Fig. 2*c*).

Upset welding. Coalescence is produced, simultaneously over the entire area of abutting surfaces

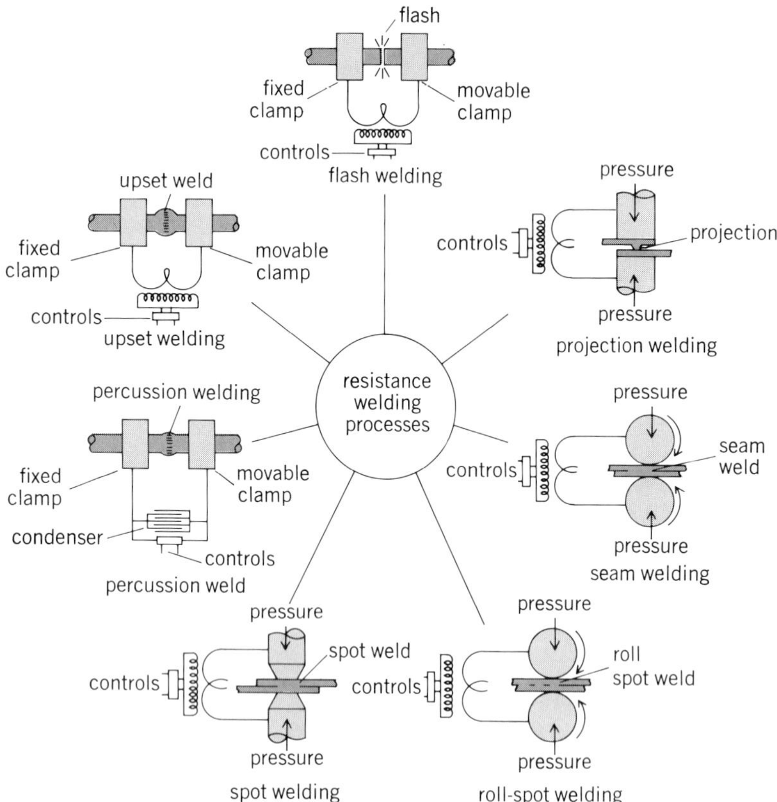

Fig. 1. Comparison of basic resistance welding processes. (*From B. E. Rossi, Welding Engineering, McGraw-Hill, 1954*)

or progressively along a joint, by the heat obtained from resistance to electric current through the area of contact of those surfaces. Pressure is applied before heating is started and is maintained throughout the heating period.

Flash welding. Coalescence is produced simultaneously over the entire area of abutting surfaces by the heat obtained from resistance to electric current between the two surfaces and by the application of pressure after heating is substantially completed. Flash and upsetting are accompanied by expulsion of the metal from the joint. *See* FLASH WELDING.

Percussion welding. Coalescence is produced simultaneously over the entire abutting surfaces by the heat obtained from an arc produced by a rapid discharge of electrical energy with pressure percussively applied during or immediately following the electrical discharge.

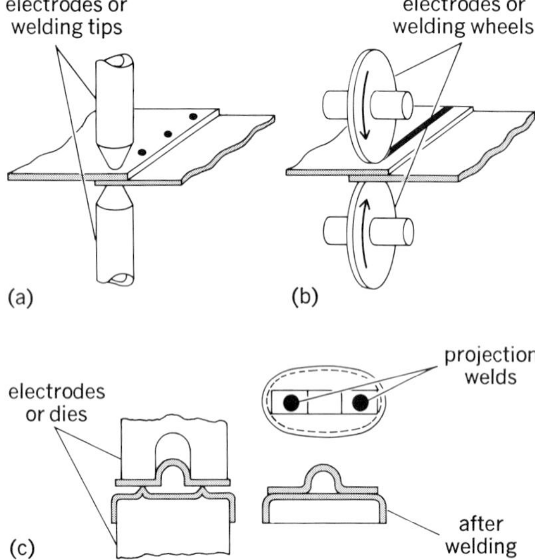

Fig. 2. Three typical resistance welding methods. (*a*) Spot welding. (*b*) Seam welding. (*c*) Projection welding. (*From W. H. Kearns, ed., Welding Handbook, 7th ed., vol. 3, American Welding Society, 1976*)

Fundamentals of process. In all resistance welding processes, heat is generated at a concentrated position and is controlled by adjusting the current magnitude and the duration of current flow. The spot welding process will be used to illustrate the fundamentals involved.

The welding current from the secondary of the transformer flows through a flexible conductor, an electrode, and the workpieces, and back through an electrode and a flexible conductor to the transformer. The application of force to the electrodes is used to position the electrodes in intimate contact with the workpieces, thus completing the electrical circuit. The amount of heat generated during the current flow through the resistance of the parts may be expressed as $H = I^2Rt$, where H is the heat generated in the workpieces in joules (watt-seconds), I is the current in amperes, R is the resistance of the workpieces in ohms measured between the electrodes, and t is the time of current flow in seconds. In a typical spot welding application for joining two sheets of 1.0-mm-thick (0.04-in.) steel, the current is high (typically 10,000 A), the resistance is low (typically 0.0001 ohm), and the time is short (typically 0.15 s). In this case the heat generated in the parts, as calculated by the formula, is 1500 J.

The values of current and time are selected such that sufficient base metal is melted at the desired position to produce a weld of the size and strength appropriate for the application.

Electrode force. The electrode force, which is used to provide intimate contact between the workpieces, reduces porosity and cracking in the fusion zone of the weld. However, when the electrode force is excessive, the weld region becomes severely indented by the electrodes.

Welding current. In the majority of cases, the current during resistance welding is in the range from 5000 to 50,000 A. Because of the high values of current required, resistance welding machines are designed as low-voltage sources, usually in the range from 1 to 25 V. These low voltages are obtained from a step-down transformer which usually has a single- or two-turn, cast-copper, water-cooled secondary. Power supplied to the primary of the transformer is usually obtained from public utility single-phase, alternating-current sources at 440 V and 60 Hz. The step-down ratio of the transformer is approximately 100; thus the current requirements of the primary are lowered to reasonable values, which generally range from 50 to 500 A.

Power sources. Alternating current of 60 Hz is used in about 90% of the installations, although three-phase frequency converters are used to supply 3- to 25-Hz voltage to single-phase transformers. Welding current may also be supplied by a direct-current or stored-energy source. Direct current may be obtained from various low-voltage sources, such as rectifiers, homopolar generators, or storage batteries. The energy may be stored during a relatively long period and released suddenly from capacitors, magnetic fields, storage batteries, or heavy flywheels on homopolar generators. These types of power supplies eliminate large transient loads on power lines.

Welding time. The duration of the welding time is short, generally in the range from 1/2 cycle (1/120 of a second) to a few minutes. In the great majority of applications, the time is in the range of 5–120 cycles of a 60-Hz source (1/12 to 2 s).

Welding electrodes. Resistance welding electrodes serve several purposes in the process: the application of the force required to bring the workpieces into alignment and into intimate contact; the production of proper contact resistance between electrodes and the workpieces; the conduction of welding current; the prevention of weld porosity and cracking; the prevention of "spitting," which is liquid-metal expulsion during welding; dissipation of heat developed at the electrode-workpiece interface; and dissipation of heat developed in the workpieces after the weld is made.

Electrodes are made of copper alloys which provide a satisfactory combination of mechanical properties and electrical and thermal conductivities. Electrodes generally are water-cooled to dissipate the heat developed, to minimize deformation of the electrode tips, and to prevent them from sticking to the workpieces.

Controls. The functions of these controls are to initiate and terminate the welding current, to control the magnitude of the current, and to sequence the operations. Controls are classified into two groups: synchronous and nonsynchronous.

In synchronous controls the timer initiates and terminates each half-cycle of the input current to the primary of the welding transformer at specific times with respect to the input voltage wave. Thus each half-cycle of welding current is identical, and the power delivered for welding is the same for consecutive welds.

With nonsynchronous controls, the timer initiates the current to the welding transformer at random points with respect to the input voltage wave. In the worst case the nonsynchronous closure of the circuit causes a current transient with a duration of about 2 cycles and a magnitude of current twice that of the steady-stage value. The differences in current and power for nonsynchronous and synchronous controls are shown in Fig. 3. The variations in timing and current, which become less important as the time of welding increases, can be neglected for times greater than about 20 cycles. However, when the welding time is less than 10 cycles, synchronous controls must be used for producing consistent welds.

The control of the magnitude of the welding current is accomplished, in coarse steps, by changing the taps on the welding transformer and, in a continuous manner, by adjustment of an electronic heat control. Synchronous controls are provided with electronic heat controls; however, these heat controls may be added to nonsynchronous controls. When a spot or projection weld is made by more than one impulse of current, the process is termed multiple-impulse welding. Multiple-impulse welding is also used to produce overlapping

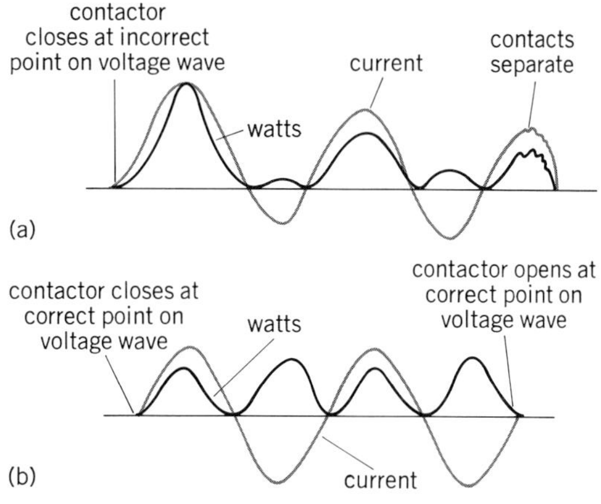

Fig. 3. Curves showing difference in current and power waves when circuit is closed nonsynchronously and synchronously. (*a*) Closure of circuit at incorrect point, with resultant decaying transient. The opening is also nonsynchronous. (*b*) Synchronous opening and closing of circuit at point of zero current corresponding to minimum transient. (*From A. L. Phillips, ed., Welding Handbook, 6th ed., American Welding Society, 1969*)

welds in the seam welding process.

Various levels of impulses can be used to preheat the metal, make the weld, refine the grain structure of the weld, and temper the hardened weld zone.

Machines. Welding machines are selected for use on the basis of the type of weld to be produced, the required weld quality, and production schedules. Simple, standard equipment is often utilized for production operations. However, because of economic considerations, complex resistance welding machines are used in the automotive and appliance industries, where high hourly production rates with minimum labor costs are essential.

Modern machines are designed with low-inertia systems which enable the electrodes to maintain the proper force on the workpieces at all times. If the electrodes fail to maintain adequate force during the welding operation, increased contact resistance will occur at the electrode-workpiece interface and at the workpiece interface. This increased resistance may cause excessive heating and result in deformed electrodes and expulsion of molten metal.

The clamping force of the electrodes is obtained by manual, mechanical, hydraulic, or pneumatic means. Prior to welding, electrodes should approach rapidly but in a controlled manner, so that they are not deformed by impacting the workpieces.

Surface preparation. Preweld surface preparation is usually required to control the magnitude and the variation in magnitude of the contact resistance at the electrode-workpiece interfaces and at the workpiece interface. Thus surface preparation, which removes dirt, oil, and oxide films by mechanical or chemical methods, provides for the production of consistent welds.

Metallurgical considerations. Most metals and alloys can be resistance-welded to themselves and to each other. The weld properties are determined by the metal and by the resultant alloys which form during the welding process. Stronger metals and alloys require higher electrode forces, and poor electrical conductors require less current. Copper, silver, and gold, which are excellent electrical conductors, are very difficult to weld because they require high current densities to compensate for their low resistance. Medium- and high-carbon steels, which are hardened and embrittled during the normal welding process, must be tempered by multiple impulses. *See* WELDING AND CUTTING OF METALS.

[ERNEST F. NIPPES]

Bibliography: W. H. Kearns (ed.), *Welding Handbook*, 7th ed., American Welding Society, 1976; Resistance Welders Manufacturers Association, *Resistance Welding Manual*, 3d ed., 1969.

Resonance (alternating-current circuits)

A condition in a circuit characterized by relatively unimpeded oscillation of energy from a potential to a kinetic form. In an electrical network there is oscillation between the potential energy of charge on capacitance and the kinetic energy of current in inductance. This is analogous to the mechanical resonance seen in a pendulum.

Three kinds of resonant frequency in circuits are officially defined. Phase resonance is the frequency at which the phase angle between sinusoidal current entering a circuit and sinusoidal voltage applied to the terminals of the circuit is zero. Amplitude resonance is the frequency at which a given sinusoidal excitation (voltage or current) produces the maximum oscillation of electric charge in the resonant circuit. Natural resonance is the natural frequency of oscillation of the resonant circuit in the absence of any forcing excitation. These three frequencies are so nearly equal in low-loss circuits that they do not often have to be distinguished.

Phase resonance is perhaps the most useful in many practical situations, as well as being slightly simpler mathematically. The following discussion considers phase resonance in passive, linear, two-terminal networks.

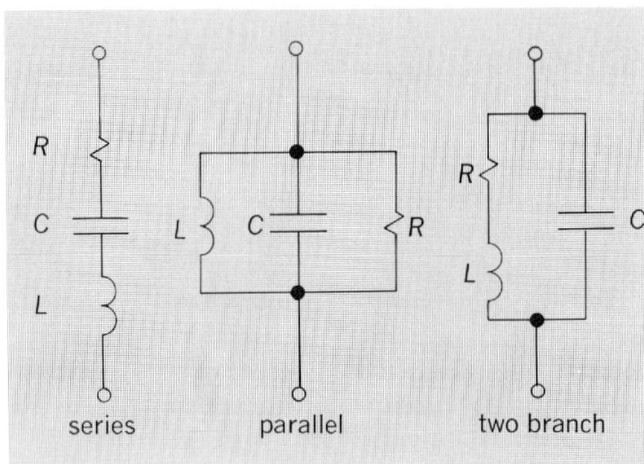

Fig. 1. Resonant circuits.

Resonance can appear in two-terminal networks of any degree of complication, but the three circuits shown in Fig. 1 are simple and typical. The first illustrates series resonance and the second, parallel resonance; the third is a series-parallel resonant circuit of two branches (sometimes referred to as antiresonance). Series resonance is highly practical for providing low impedance at the resonant frequency. Parallel resonance is the dual of series resonance, but it is not practical because it assumes an inductive element with no resistance. The third example, however, shows an eminently practical means of providing the typical characteristic of parallel resonance, which is high impedance at the resonant frequency.

Use. Resonance is of great importance in communications, permitting certain frequencies to be passed and others to be rejected. Thus a pair of telephone wires can carry many messages at the same time, each modulating a different carrier frequency, and each being separated from the others at the receiving end of the line by an appropriate arrangement of resonant filters. A radio or television receiver uses much the same principle to accept a desired signal and to reject all the undesired signals that arrive concurrently at its antenna; tuning a receiver means adjusting a circuit to be resonant at a desired frequency.

Many frequency-sensitive circuits are not truly resonant, and oscillations of a certain frequency can be produced or enhanced by networks that do not involve inductance. It is difficult and expensive to provide inductance with integrated circuits, but frequency selection can be provided by the use of capacitance and resistance, a large amount of amplification being obtained from the semiconductor material employed.

Series resonance. Figure 2 shows a phasor diagram of the voltage, resulting from a given current (steady alternating current) in a series-resonant circuit, such as shown in Fig. 1. The component voltages across the three circuit elements add to give the total applied voltage V, as shown for a frequency slightly above resonance, for the resonant frequency, and for a frequency below resonance. It is of course possible in low-loss (high Q) circuits for the voltage across the capacitance and the voltage across the inductance each to be many times greater than the applied voltage.

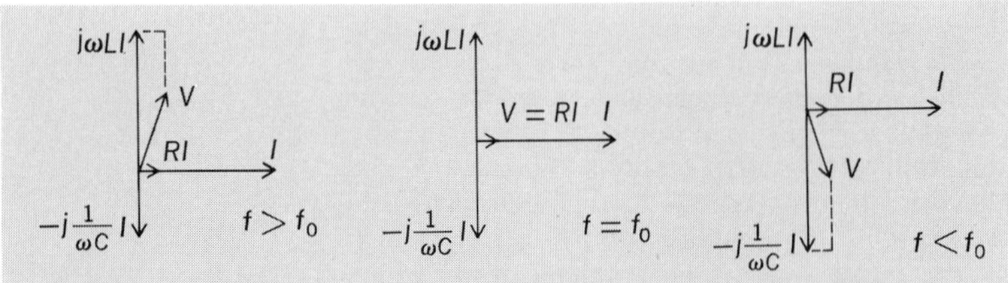

Fig. 2. Phasor diagrams at frequencies near resonance (Q= 5).

Analytically, the impedance of the series-resonant circuit is given by Eq. (1). The resonant frequency f_0 is the frequency at which Z is purely real (phase resonance), so $\omega_0 L = 1/\omega_0 C$, or $2\pi f_0 L = 1/2\pi f_0 C$, from which Eq. (2) obtains.

$$Z = R + j\omega L + \frac{1}{j\omega C} = R + j\left(\omega L - \frac{1}{\omega C}\right) \qquad (1)$$

A more convenient notation is expressed by Eq. (3).

$$f_0 = \frac{1}{2\pi\sqrt{LC}} \qquad (2)$$

$$Z = R_0\left(\frac{R}{R_0} + jQ_0\delta\frac{2+\delta}{1+\delta}\right) \qquad (3)$$

In Eqs. (1)–(3):

Z = impedance at the terminals of the series-resonant circuit

R, L, C = the three circuit parameters
R_0 = resistance (effective) at resonant frequency
Q_0 = $\omega_0 L/R_0$
δ = $(\omega - \omega_0)/\omega_0$
ω = $2\pi f$, where f is frequency (hertz or cps)
ω_0 = $2\pi f_0$, where f_0 is resonant frequency

Equation (3) is true for all series-resonant circuits, but interest is mainly in circuits for which Q_0, the quality factor at the resonant frequency, is high (20 or more) and for which δ, the fractional detuning, is low (perhaps less than 0.1). Assuming high Q_0 and low δ, which means a low-loss circuit and a frequency near resonance, Eq. (4) is very nearly the relative admittance of the series-resonant circuit.

$$\frac{Y}{Y_0} = \frac{Z_0}{Z} = \frac{1}{1 + j2Q_0\delta} \qquad (4)$$

Universal resonance curve. The magnitude and the real and imaginary components of Eq. (4) are usefully plotted in the universal resonance curve of Fig. 3. Since Y/Y_0 is plotted as a function of $Q_0\delta$, this curve can be applied to all series-resonant circuits. (If $Q_0 = 20$, the error in Y barely exceeds 1% of Y_0 for any δ, and is less for small δ.)

Moreover, because of the duality of the network, the curve can also be applied to any parallel-resonant circuit (Fig. 1) provided Q_0 is now interpreted as $Q_0 = R_0/\omega_0 L$. When used for a parallel-resonant circuit, the curve of Fig. 3 gives not Y/Y_0 but the relative input impedance Z/Z_0.

Finally, the universal resonance curve of Fig. 3 can also be applied (with the same slight approximations) to the two-branch resonant circuit of Fig. 1. For this purpose the curve shows Z/Z_0 (as for the three-branch parallel-resonant circuit), but the value of Q to be used is $Q_0 = \omega_0 L/R_0$, exactly as with the series-resonant circuit. Note that Z_0 for this circuit is given by Eq. (5) (instead of being

$$Z_0 = (\omega_0 L)Q_0 = R_0 Q_0^2 \qquad (5)$$

equal to R_0 as it is in the other two circuits of Fig. 1).

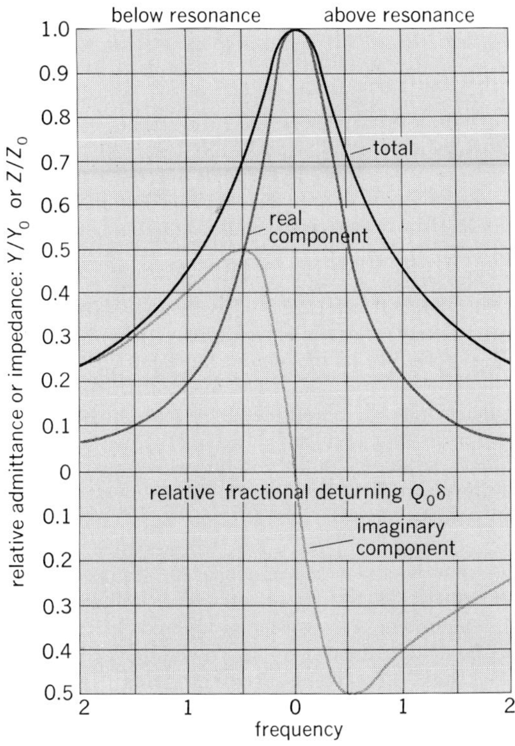

Fig. 3. Universal resonance curve. *(From H. H. Skilling, Electrical Engineering Circuits, 2d ed., copyright © 1965 by John Wiley and Sons, Inc.; used with permission)*

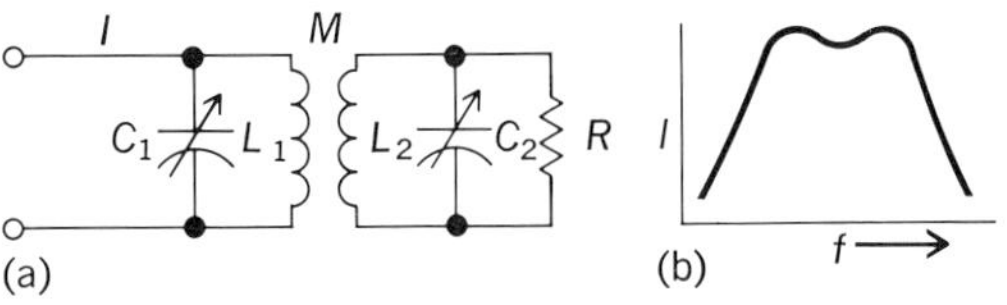

Fig. 4. (*a*) Resonance in a double-tuned network. (*b*) Current in *R* as function of frequency.

Multiple resonance. If two or more coupled circuits are resonant at slightly different frequencies, many valuable characteristics can be obtained. Figure 4 shows a double-tuned network and a typical curve of current in R, the load, as a function of frequency. *See* ALTERNATING-CURRENT CIRCUIT THEORY.

[HUGH HILDRETH SKILLING]

Bibliography: D. Bell, *Fundamentals of Electric Circuits*, 2d ed., 1981; J. R. Duff and M. Kauffman, *Alternating Current Fundamentals*, 1980; Institute of Electrical and Electronics Engineers, *IEEE Standard Dictionary of Electrical and Electronics Terms*, 2d ed., 1977.

Retaining wall

A wall designed to maintain differences in ground elevations by holding back a bank of material. Sometimes a retaining wall also serves as a foundation wall.

Material that exerts pressure on the back of the wall is called backfill. It includes material taken from the excavation and placed behind the wall. The load applied on backfill above the top of the wall level is called surcharge.

Earth pressure against the back of a wall is known as active pressure and acts at an oblique angle; the precise angle depends upon the character of the backfill, slope of the ground surface, presence of surcharge, and groundwater level. Earth pressure against the front of a wall is known as passive pressure and may be of greater unit intensity than active pressure, although the soil must be somewhat compressed before this force develops.

External stability is achieved when a wall is proportioned so that it will neither rotate nor slide under all dead-load and applied forces. In addition to earth pressure, the following forces must be considered.

The weight of the wall acts in a vertical direction through its center of gravity. Surcharge loads may consist of inclined embankments and any live load they carry, for instance, trucks or cranes. Lateral forces may be caused by ice thrust or frost action from repeated freezing and thawing of poorly drained soil. If ice layers can form behind a wall, pressures may be larger than a wall can be designed to resist. Thrust action from clay soils may result from repeated changes in water content, for when the soil drys, it cracks and the cracks fill with loose soil, leaving too little space for expansion in place when the next wetting occurs. Earthquakes may cause lateral forces. Vibrations from machinery or traffic may increase the effects of earth pressure. Uplift under the footing will occur if the water level is above that point or if water is not drained from behind the wall. Uplift may also occur in bedrock if it is seamed or slightly porous. Retaining walls that also serve as foundation walls or abutments may in addition experience lateral forces from cranes, wind, building framing, or tractive effort. Bridge spans cause vertical, longitudinal, and transverse lateral forces on retaining walls used as abutments. *See* BRIDGE.

Vertical-resisting forces are supplied by reactions of the soil or rock under the footing and are known as bearing pressures. Horizontal-resisting forces are supplied by friction under the footing or by shear keys extending below the footing. Passive pressure in front of the wall is sometimes omitted from stability analyses because of the uncertainty as to its magnitude and the amount of movement necessary to bring it into action.

Drainage. A retaining wall is drained primarily to prevent accumulation of water in the backfill, thus avoiding hydrostatic pressure, formation of ice that may cause thrust, swelling of cohesive backfills, and decrease in the stability of the soil. Highly permeable material, such as gravel or crushed stone, should be located in the backfill to collect groundwater. Surface water also should be drained, but its entrance into the backfill should be minimized by use of relatively impervious topsoil or paving.

Water should be led away through weep holes built into the wall or through corrosion-resistant pipe having open joints or perforations. Pocket drains—pockets of stone or gravel at weep holes—may be used if the backfill is moderately permeable. Blanket drains consist of a layer of stone or gravel against the entire back surface of the wall. Water discharge should be carried away by gravity or pumped from collection sumps.

Types of wall. The illustration shows the major types of retaining wall. Gravity walls are of massive, solid construction, proportioned so that tensile stresses are avoided or kept to low values at the toe and along the back. They are more durable than walls with thin, reinforced-concrete sections, and partial disintegration is not as serious since stability depends on weight. Gravity walls are usually low. Unit stresses in the concrete are very low.

Semigravity walls are constructed with narrower stems. A few tension bars are built into the back and toe. Considerable concrete is saved by using small amounts of steel.

A cantilever wall is formed of three cantilever beams: stem, toe projection, and heel projection. Reinforcing steel is required in all members. The simplest form is used for low walls. For higher walls a fillet may be an economical reinforcement. Sliding resistance may be improved by a wall design with a key projecting downward into the soil. This type is the most common. Members can be so designed that concrete and steel unit stresses equal critical values of these materials.

A counterfort wall is a thin, reinforced concrete-face slab backed up by deep vertical cantilever stems, or counterforts. The heel slab of the wall footing carries the backfill load horizontally to the counterforts. The wall may be constructed with a key to increase sliding resistance. The counterfort wall is often the most economical type of high retaining wall. *See* FOUNDATIONS.

[ROBERT D. CHELLIS]

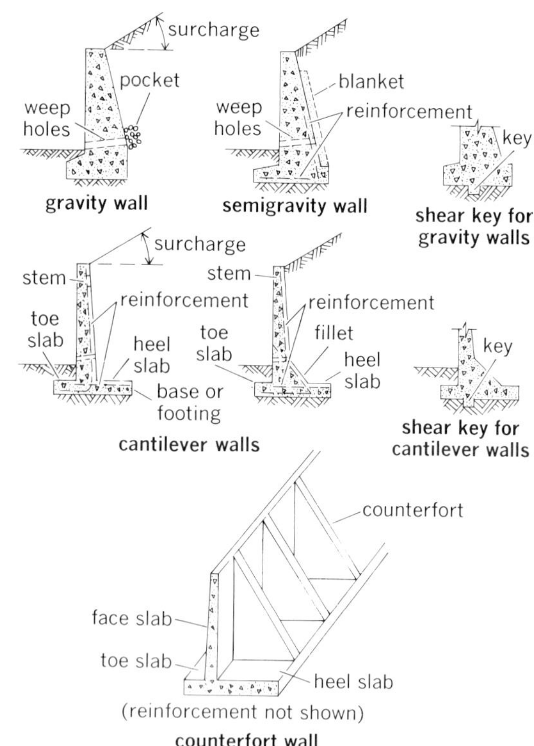

The major types of retaining wall.

Revetment

A means of protecting river banks against erosion. In some cases the revetment must act as a solid barrier; however, as a basic principle, it should be designed to induce the high-velocity thread of the current to move away from the threatened area. This is most readily accomplished when (1) the revetment is initiated at a point upstream from the point of attack and at an angle of not more than about 15° to the current, and (2) the surface of the revetment is rough.

The rough surface generates a zone of turbulence which, in effect, acts as a cushion, causing the high-velocity current to move away from the bank. The extent of this movement has been observed to be approximately 50 ft (15 m), and in numerous cases where the revetment toe is less than about 15 ft (5 m) below the normal low-water elevation, scour depths of 30–40 ft (9–12 m) have occurred without undercutting the revetment. In deeper water or where the angle of attack is greater than 30°, the benefit of the rough surface is diminished. The revetment should be extended to the thalweg of the stream.

Alignment of revetments should be smooth, without holes or projections; and sharp breaks in alignment should be avoided, since they may cause the current to cross abruptly to attack the opposite bank, or may cause eddy action along the bank to scour or undermine the revetment.

Revetments may consist of the following types: (1) rock paving with a heavy rock toe base (Fig. 1);

Fig. 1. Stone revetment of the rock-paving type.

(2) rock, asphaltic, or concrete paving with brush, lumber, or concrete mattress for the underwater portions; and (3) pile dikes or other permeable fences (Fig. 2). Rock paving has the advantage of flexibility in settling to heal breaches and permits easy repair; however, an effective filter should be provided by a gravel blanket or by proper dumping of quarry-run rock to prevent leaching of the underlying bank material. Solid paving has the disadvantage of bridging over small breaks until a major breach develops; and, frequently, it has the added disadvantage of peeling off in large sheets once it is broken.

On streams such as the Missouri and Arkansas rivers, where the depth of water at low-water levels seldom exceeds 15–25 ft (5–8 m), the use of pile dikes and rock paving is effective and economical. Quarry-run rock is used, and its size must be such that at least 30% of the rock is large enough to resist movement by the current. Quarry-run rock is well graded, and placement by dumping provides an inverse filter and the desired rough surface.

On deeper channels such as the Mississippi, where depths may exceed 100 ft (30 m), articulated concrete or reinforced asphaltic mattresses are used.

Pile dikes or other permeable fence-type revetments placed parallel to the bank generate local turbulence which induces a shifting of the current but still permits deposition of sediment behind the revetment, thereby building up the bank. Pile dikes are generally effective if the angle of attack of the current does not exceed 30°. For permanence, particularly in streams with heavy ice or drift runs, pile dikes should be protected with rock to a height which causes enough accretion to support willow growth.

Fig. 2. Permeable fence-type, pile-dike revetment.

Groins are short dikes, approximately perpendicular to the bank, which are sometimes used in lieu of revetment, particularly in areas where the bed is relatively stable. They are usually of solid construction, either sheet piling or dumped materials. For other information on control of rivers *see* RIVER ENGINEERING. *See also* GROIN; RETAINING WALL. [WENDELL E. JOHNSON]

Rheostat

A variable resistor constructed so that its resistance value may be changed without interrupting the circuit to which it is connected. It is used to vary the current in a circuit. The resistive element of a rheostat may be a metal wire or ribbon, carbon disks, or a conducting liquid.

The metallic rheostat is the most common. The wire or ribbon is constructed in a coil or a grid, and taps are brought out from different sections of the element to a multicontact switch which can short-circuit any section of the resistor or switch it out of the circuit. For more continuous control, as is needed for laboratory rheostats, a sliding-contact finger bears directly on closely wound coils of resistive wire.

The carbon-disk type is used only for small currents. The resistive element is varied by changing the pressure on the carbon disks. The advantage of this type is its capacity for fine adjustment.

The electrolytic type is ideally suited to large currents. This type consists of a tank of conducting liquid in which electrodes are placed. The variation of resistance is obtained by changing the distance between the electrodes, the depth of immersion of the electrodes, or the resistivity of the solution. The electrolytic rheostat, also called a water rheostat, has perfectly continuous adjustment.

Rheostats are used whenever it is desired to vary resistance or adjust current. Typical applications are for starting or controlling the speed of motors, for adjusting generator characteristics, for controlling storage-battery charging, for dimming lights, and for imposing artificial loads on electrical equipment during tests.

[FRANK H. ROCKETT]

Rheostatic braking

A system of dynamic braking in which direct-current drive motors are used as generators and convert the kinetic energy of the motor rotor and connected load to electrical energy, which in turn is dissipated as heat in a braking rheostat connected to the armature. This is accomplished by disconnecting the motor armature from the supply lines and connecting them to a braking rheostat, while the field circuit remains connected to the supply line. The polarity of the armature remains unchanged, but the direction of the armature current reverses, resulting in a negative torque. The torque can be controlled by means of the braking rheostat. *See* DIRECT-CURRENT MOTOR; DYNAMIC BRAKING.

[ARTHUR R. ECKELS]

Rigid body

An idealized extended solid whose size and shape are definitely fixed and remain unaltered when forces are applied. Treatment of the motion of a rigid body in terms of Newton's laws of motion leads to an understanding of certain important aspects of the translational and rotational motion of real bodies without the necessity of considering the complications involved when changes in size and shape occur. Many of the principles used to treat the motion of rigid bodies apply in good approximation to the motion of real elastic solids.

[DUDLEY WILLIAMS]

River engineering

A branch of civil engineering concerned with the improvement and stabilization of the channels of rivers (particularly channels in erodible alluvium) to better serve the needs of people.

Application. Throughout history, river valleys and river channels have been important to most civilizations. However, the characteristics of some rivers in their natural states have deterred or presented hazards to development of their valleys or beneficial uses of their channels. In many major river systems there is a need to improve channels in order to provide greater flood flow capacity, provide navigable waterways, improve water supplies, and stabilize the channels to permit development of adjacent valley areas (Fig. 1). This has led to development of extensive and costly river control or management schemes.

Channel improvement and stabilization works have been integral parts of river control plans where meandering, shifting channels or inappropriate channel alignments and cross sections are problems. Large-scale channel improvement and stabilization plans have been carried out on many of the world's major rivers. In the United States the most extensive examples of river engineering are found in the Mississippi Basin, along such streams as the Missouri River, the Arkansas River, the Red River, the Atchafalaya River, and the Mississippi River downstream from St. Louis, MO. Important objectives of these works are the development of navigable waterways, the improvement of flood flow capacities, and the protection of adjacent flood control levees and valley lands from the disastrous effects of major channel changes. River channel improvement and stabilization works often supplement other types of river management development. *See* CANAL; DAM.

Potomology. In essence, river engineering works seek to direct or work with the natural processes involved in a stream with erodible bed and banks (Fig. 2) to arrive at a stable channel of desired cross section and alignment. It has been recognized that the characteristics of such a stream are governed by complex interrelationships among factors such as the geologic history of the area, the volume and variability of streamflow, stream gradients, erosional characteristics of the contributing area and the channel itself, and the type and amounts of sediment load carried by the stream (both as suspended sediments and as bed load). Even the temperature of the water in the stream has been found to have an effect on the depositional forms in a stream bed and the flow capacity of the channel. The study of the factors affecting river channels has led to development of a branch of engineering science called potomology, which aims to systematize field and laboratory investigation of river channels and provide rational bases for predictions of the effects on channel characteristics of proposed engineering works. These systematic studies have provided the civil engineer with a much better understanding of the processes involved in river channels. For example, some early channel improvement projects attempted to straighten meandering channels, but the streams would develop new patterns of erosion that would reestablish meander patterns. Now stream improvement projects are planned to continue the general gradients and patterns of sinuosity that are characteristic of the natural streams.

Types of works. A number of types of construction have been developed for use in improving and stabilizing river channels. Each type seeks to achieve one or more of the following objectives: direct improvement of channel geometry by excavation and filling; armoring of bed or banks of a

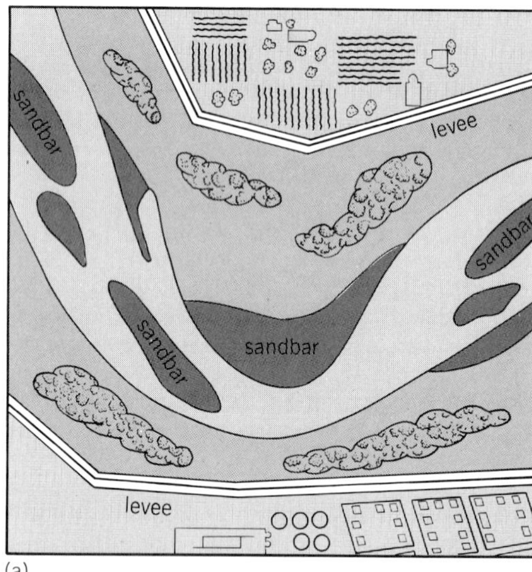

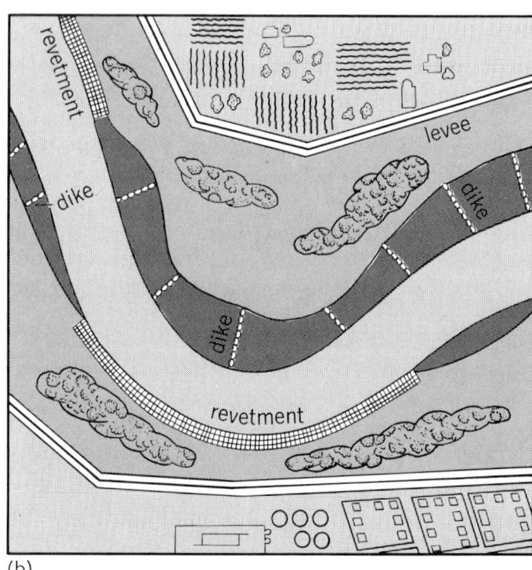

Fig. 1. Effects of river engineering. (*a*) Before: an uncontrolled river (with sharp bends, divided channel, clogging shifting sandbars, and caving banks) threatening flood control levees, cities, and industry and difficult to navigate. (*b*) After: an improved and stabilized river, with a single channel of gentle bends and controlled width for easy navigation, and the threat from caving banks removed.

Fig. 2. Caving Mississippi River bank subject to active erosion.

stream to resist erosion; deflection of river currents from an area subject to undesired erosion; and stilling of river currents in an area of the stream to induce deposition of sediments.

Excavation and filling. Excavation of pilot cutoff channels across meander loops can utilize the erosive forces of river flow to make major changes in stream alignment at relatively low cost. Because of the shorter length of the cutoff channel, as compared with the original channel around the meander bend, and the consequent steeper hydraulic gradient through the cutoff, the new channel through the cutoff can develop relatively rapidly by erosion. In the 1930s an extensive program of cutoff construction on the Lower Mississippi River shortened the river's length by 152 mi (245 km). Excavation by dredging of shoals or sandbars within a river channel is a commonly used method for improving and maintaining project depths in inland waterways.

Armoring. A great many types of materials and methods of construction have been used to armor river beds and banks to resist erosion, including: planting of willows or other vegetation on banks that are submerged infrequently; mats woven of trunks of willow or similar trees or of sawed timbers sunk in the area to be protected and weighted with stones; dumped or hand-placed rock riprap; gabions (or baskets with covers) made of such materials as wire mesh and filled with stones; asphaltic concrete paving (on upper bank areas); and articulated concrete mats or mattresses. Economy and availability of materials and labor, depths and velocities of water in the areas to be protected, size of those areas, and long-term effectiveness of the protection provided have been factors in the selection among types of erosion protection. High labor costs and scarcity of materials have discouraged use of willow or timber mats in recent decades. Blowouts of impervious asphaltic concrete pavement caused by pressure of water trapped behind the pavement during falling river stages have discouraged the use of that type of bank protection.

There is need to safeguard an erosion protection system against loss of underlying foundation soils caused by pumping action of waves or changes in river stages. To prevent such loss of foundation material, a sand and gravel filter layer or especially prepared filter cloth is often placed between the foundation and the erosion-resistant materials forming the armor layer.

Deflection and stilling of currents. Systems of dikes are used to deflect currents and to provide stilling action to induce deposition in an area that otherwise would be subject to undesirable erosion (Fig. 3). Dikes constructed of one or more lines of timber piles driven into the river bottom with timbers connecting the piles to form a fencelike, permeable structure have been used extensively. Rock or rubble mound dikes formed of quarried stone have been used exclusively in recent years on the Lower Mississippi River. One type of deflection structure consists of structural steel sections welded together to produce a frame re-

Fig. 3. Constructing stone fill dike on the Lower Mississippi River.

Fig. 4. Placing articulated concrete mattress revetment for protection against scour.

sembling a very large version of the jacks with which children play. Lines of these frames connected by steel cables and anchored in place have been used to protect stream areas under heavy erosion attack.

Examples of projects. In nature many reaches of the Missouri River were meandering and unstable with sharp bends and shallow, often divided or braided, low-water channels. Since 1950 the Missouri River, from its junction with the Mississippi River near St. Louis, MO, to Sioux City, IA, 750 mi (1200 km) upstream, has been transformed into an inland waterway for 9-ft (2.7-m) draft barge navigation. This has been accomplished primarily through the use of pile dikes to contract, deepen, and realign the low-water channel to provide adequate depths and channel geometry suitable for barge navigation.

On the Arkansas River and the Red River in Arkansas and Louisiana, systems of navigation locks and dams provide waterways for barge navigation, but the projects on both rivers also include extensive bank and channel stabilization works.

Fig. 5. Stone upper bank paving used in conjunction with articulated concrete mattress on the subaqueous bank.

In nature, the Mississippi River below Cape Girardeau, MO, was a wild, ever-changing, unpredictable stream. Throughout the length of the alluvial valley of the Mississippi, from Cape Girardeau to the Gulf of Mexico, numerous ancient oxbow lakes, abandoned river channels, and various types of alluvial deposits mark the past wide meandering of this mighty river as it carried runoff from 41% of what is now the contiguous United States. In 1928 the Mississippi River and Tributaries (MR &T) Project was authorized as a Federal undertaking to provide flood protection and dependable water transportation. The MR&T Project includes such features as extensive levee systems, floodways, and tributary improvements, but one central feature of the basin plan is the improvement and stabilization of the Mississippi River channel. After a half-century of work, dramatic improvement in the river channel has been made, but work to bring this channel under control is still in progress.

As noted above, the channel improvement plan for the MR&T Project included a number of cutoffs to straighten and shorten reaches of the river. Since the cutoffs were made, revetment of banks to provide erosion protection, dikes, and dredging have been the principal means for stabilizing the alignment of the 1000-mi-long (1600-km) channel. The methods of stabilization now in use have evolved over the years, and research and development efforts to improve the methods are continuing. Perhaps the most notable development in the MR&T channel stabilization work has been the articulated concrete mattress revetment currently used for protection against scour (Figs. 4 and 5). This type of mattress is composed of concrete blocks 3 in. (76 mm) thick, each block being approximately 14 in. (36 cm) wide by 4 ft (1.2 m) long. Twenty such blocks are cast as a section approximately 4 ft (1.2 m) wide by 25 ft (7.5 m) long held together by a noncorrosive wire fabric located at the central plane of each block. By a complex and highly mechanized process, these mat sections are manufactured at several casting fields located at strategic points along the Mississippi, hauled by barge to a sinking plant at the mat-laying locations,

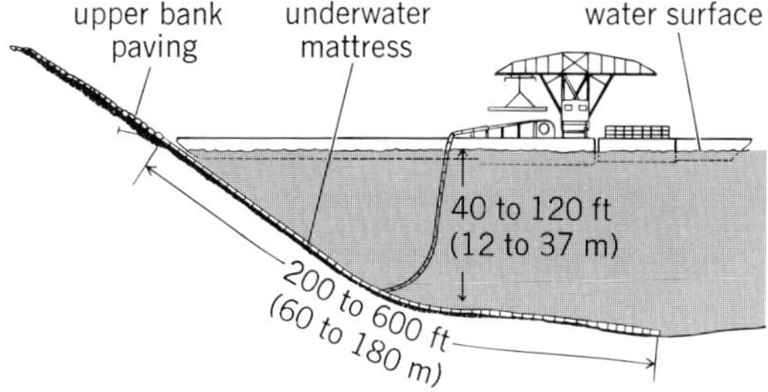

Fig. 6. Cross section of revetment operation.

assembled into mattresses up to 156 ft (47.5 m) in width (held together by corrosion-resisting cables and wire ties), and launched over the side of the sinking plant to rest on the bottom of the river in the area where revetment protection is needed. Placement of each mattress section starts near the water's edge and, as the mattress is assembled and tied together on the sinking plant, the sinking plant is moved out away from the river bank and the mattress is played out into successively deeper water as much as 600 ft (180 m) from the shore (Fig. 6). Each mattress section is tied by wire cables to anchors buried in the river bank at its shoreward end. Mat placement proceeds upstream along the river bank, and each successive mattress overlaps the previously placed downstream mattress.

Prior to placement of the underwater articulated concrete mattress, the river bank is graded to a uniform stable slope from the top of the bank to below the water surface. After mattress placement the graded upper bank is paved with graded riprap.

Dikes are employed extensively in the MR&T channel stabilization work to deflect currents, to trap sand, and to close off back channels at islands and sandbars. Wood piling dikes were used in the past, but to secure more permanence and more immediate effectiveness, dumped quarried-rock dikes have been constructed in recent years.

Excavation of underwater deposits by floating pipeline dredges is employed in development and maintenance of navigable depths on the Mississippi River. The period of falling river stages following a flood, when reduced current velocities accelerate deposition of sediments, is critical in the maintenance of navigable depths in many sections of the river. As planned dike and revetment construction is completed, it is anticipated that maintenance dredging requirements will lessen.

The channel improvement and stabilization work on the MR&T Project has been carried out during a period when there have been large increases in the unit costs of all construction. All experienced costs and all estimates of future costs have been greatly affected by this change in value of the construction dollar. However, the $1,093,000,000 cost of the MR&T channel stabilization program through September 1979 gives some idea of the immensity of this program.

[HOMER B. WILLIS; STAFF OF THE MISSISSIPPI RIVER COMMISSION, U.S. ARMY CORPS OF ENGINEERS]

Bibliography: N. R. Moore, *Improvement of the Lower Mississippi River and Tributaries*, 1972; Office of the President, Mississippi River Commission, *Channel Improvement Feature, Flood Control, Mississippi River and Tributaries Project*, 1977; Public Affairs Office, Mississippi River Commission and U.S. Army Engineer Division, Lower Mississippi Valley, U.S. Army Corps of Engineers, *Flood Control in the Lower Mississippi River Valley*, 1976; U.S. Army Engineer District, Omaha, Corps of Engineers, *Potamology Investigation, Historical Records Research*, 1976; U.S. Army Engineer District, Omaha, Corps of Engineers, *Velocity Trends*, 1971.

Rivet

A short rod with a head formed on one end. A rivet is inserted through aligned holes in two or more parts to be joined; then by pressing the protruding end, a second head is formed to hold the parts together permanently. The first head is called the manufactured head and the second one the point. In forming the point, a hold-on or dolly bar is used to back up the manufactured head and the rivet is driven, preferably by a machine riveter. For high-grade work such as boiler-joint riveting, the rivet holes are drilled and reamed to size, and the rivet is driven to fill the hole completely. Structural riveting uses punched holes.

Small rivets (7/16 in. or 11 mm and under) are used for general-purpose work with head forms as follows: flat, countersunk, button, pan, and truss (Fig. 1). These rivets are commonly made of rivet steel, although aluminum and copper are used for some applications. The fillet under the head may be up to 1/32 in. or 0.8 mm in radius.

Large rivets (1/2 in. or 13 mm and over) are used for structural work and in boiler and ship construction with heads as follows: roundtop countersunk, button (most common), high button or acorn, pan, cone (truncated), and flattop countersunk.

Boiler rivets have heads similar to large rivets with steeple (conical) added but have different proportions from large rivet heads in some cases.

Special-purpose rivets are tinners' rivets, which have flatheads for use in sheet-metal work; coopers' rivets, that are used for riveting hoops for barrels, casks, and kegs; and belt rivets, used for joining belt ends.

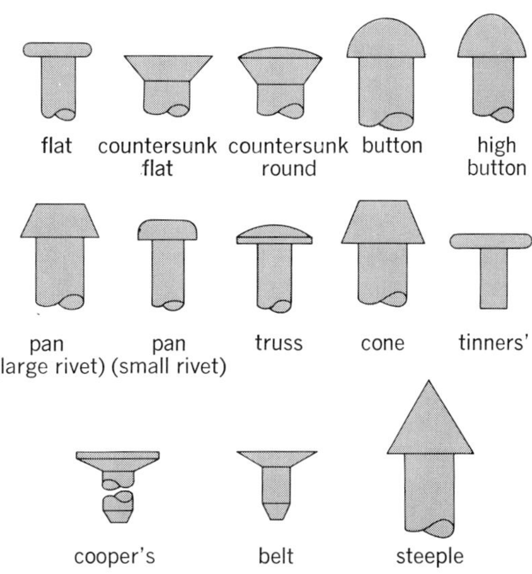

Fig. 1. Standard rivet heads.

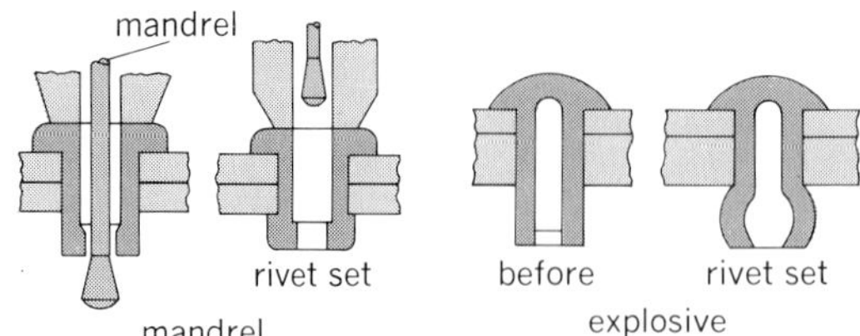

Fig. 2. Two types of blind rivet.

Blind rivets are special rivets that can be set without access to the point. They are available in many designs but are of three general types: screw, mandrel, and explosive (Fig. 2). In the mandrel type the rivet is set as the mandrel is pulled through. In the explosive type an explosive charge in the point is set off by a special hot iron; the explosion expands the point and sets the rivet.

Standard material for rivets is open-hearth steel (containing Mn, P, S) with tensile strength 45,000–55,000 psi or 310–380 MPa. Standards include acceptance tests for cold and hot ductility and hardness. Materials for some special-purpose rivets are aluminum and copper. *See* RIVETED JOINT.

[PAUL H. BLACK]

Bibliography: American National Standards Institute (ANSI), *Semitubular Rivets, Full Tubular Rivets, Split Rivets and Rivet Caps*, B18.7–1972; ANSI, *Small Solid Rivets 7/16 Inch Nominal Diameter and Smaller*, B18.1.1–1972; T. Baumeister (ed.), *Standard Handbook for Mechanical Engineers*, 8th ed., 1978; Institute for Power System, *Handbook of Industrial Fasteners*, 1979.

Riveted joint

The permanent joining of two or more machine or structural members, usually plates, by means of rivets. The plates may be lapped or butted. In the butt joint one or more cover plates must be used to accomplish the joining. One of these cover plates is often made wider than the other (Fig. 1). The rivets in the joint may be disposed in several ways to form single or multiple rows in a regular or staggered arrangement. *See* RIVET.

Terminology. Riveted joints are described in terms of their dimensions (Fig. 1). Pitch p is the distance between adjacent rivets along the gage line. Where different pitches are used on adjacent gage lines, the largest is the pitch for the joint. Gage line is the line through the centers of the rivets parallel to the edge of the plate. A unit strip or length is equal in width to the pitch. The distance p_t between two adjacent gage lines in the same plate is the back pitch or transverse pitch. Diagonal pitch p_d is the distance between adjacent rivets on adjacent gage lines. Margin m is the distance between a gage line and the edge of the plate. The efficiency of a joint is the ratio of the weakest section of a unit strip to the strength of the same width of unperforated plate.

Stresses. Lap joints are subject to eccentric loading, which brings about a bending of the joint (Fig. 2). This in turn complicates the stress pattern on the various components in the joint, and the stresses calculated by straightforward assumptions of simple shear, bearing, or tensile loads must be increased by substantial factors to give adequate design stresses.

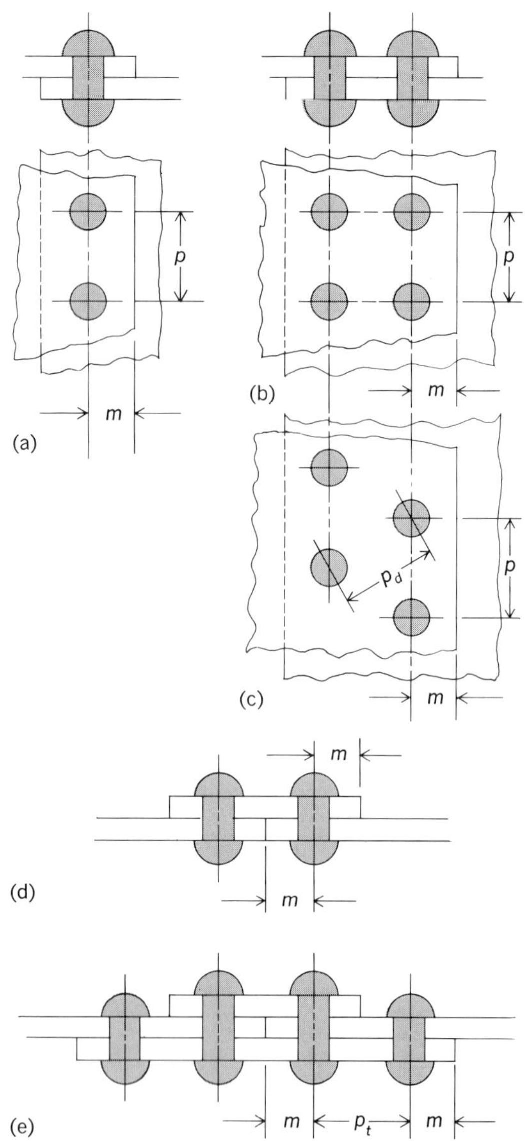

Fig. 1. Diagrams of (*a–c*) lapped joints and (*d, e*) butt joints.

A riveted joint in tension may fail in one of several ways (Fig. 3). In butt joints with two or more plates and two or more rows of rivets, the joint strength calculation is complicated by the uncertainty of the division of the load between the various rivets. Although the strength of a joint is usually considered to be a function only of the strength of the rivets and plate, friction between the plates accounts for a large but indeterminate amount of load capacity.

Strength. The force F that a lapped riveted joint (Fig. 4) can sustain depends on the stress S that can be withstood by the materials of which it is built and their dimensions. Thus, the strength of the solid plate is $F=phS$. Tensile strength F_t at the outer gage line is $F_t=(p-d)hS$. Similarly, the shear strength F_s of all rivets is given by Eq. (1), in

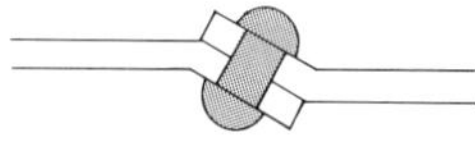
Fig. 2. Eccentric loading bends a lap joint.

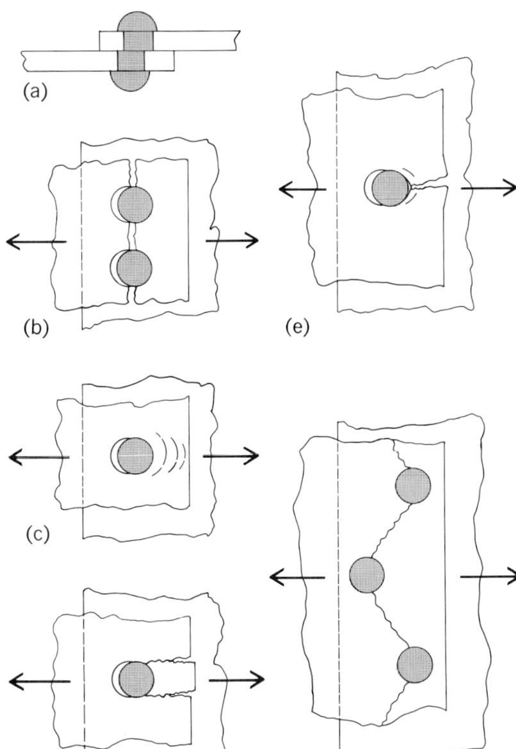

Fig. 3. A riveted joint may fail from various causes: (a) shear, (b) rupture, (c) crushing of rivet or plate, (d) double shear through the margin, (e) rupture of the margin, or (f) zig-zag tension.

$$F_s = (2n_2 - n_1)\pi d^2 S_s/4 \qquad (1)$$

which n_1 is the number of rivets in single shear, n_2 is the number of rivets in double shear, and d is the rivet diameter, assuming all rivets to be the same size. The crushing strength of rivets is given by Eq. (2), where h_2 is thickness of wider strap.

$$F_c = (n_2 h + n_1 h_2) d S_c \qquad (2)$$

See STRENGTH OF MATERIALS.

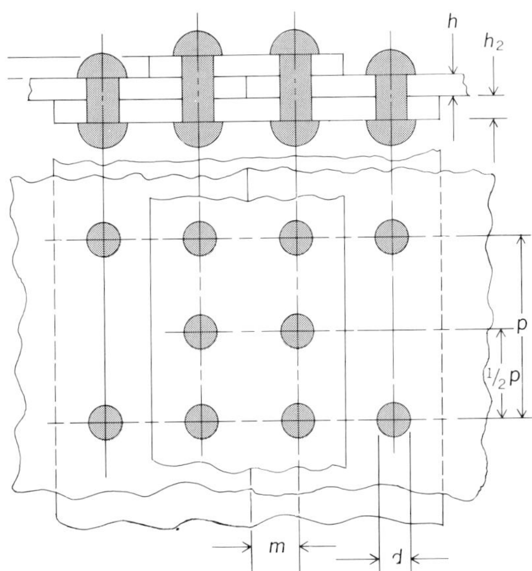

Fig. 4. Dimensions of a rivet joint.

For maximum efficiency e of a joint, $F_t = F_c = F_s$. Under this optimum condition, the maximum efficiency e_{max} is given by Eq. (3).

$$e_{max} = \frac{[n_2 + (n_1 h_2/h)]S_c}{[n_2 + (n_1 h_2/h)]S_c + S} \qquad (3)$$

Resistance in shear does not appear in the equation for maximum efficiency. *See* BOLTED JOINT; JOINT (STRUCTURES); STRUCTURAL CONNECTIONS; WELDED JOINT. [L. SIGFRED LINDEROTH, JR.]

Rock burst

A sudden and violent rock failure around a mining excavation on a sufficiently large scale to be considered a hazard endangering the existence of mine openings, equipment, and personnel. It has been estimated that the energy released in some big bursts was equivalent to that released in exploding 200 tons of TNT. Such bursts resemble small earthquakes and may be detected several hundred miles away.

Rock bursts are related to the fracture of rock in place and require two conditions for their occurrence: a stress in the rock mass sufficiently high to exceed its strength, and physical characteristics of the rock which enable it to store energy up to the threshold value for sudden rupture. Rocks which yield gradually in plastic strain when under load usually do not generate rock bursts.

The state of stress in a mine derives from three main sources:

1. The weight of the superincumbent ground. Since stress is directly related to depth, it follows that the incidence of rock bursts in deep mines is higher than in shallow mines.

2. The existence of tectonic residual stresses "locked" in the rock as a result of its long geologic history. The relief of such stresses, the magnitude of which is independent of depth below the surface, is the cause of rock bursts in surface quarries and open pit mines.

3. The modifications in the stress field due to the conduct of mining operations. The excavation of ground during mining results in the removal of support for the superincumbent rocks, a shift of the weight toward the abutments, and a concentration of stress near the ends of the opening. As the system of mine openings grows complex, the interplay of shifting loads during the mining operations results in a very complicated stress field.

Whereas the first two causes mentioned above are dictated by nature, the third is partially under the control of the mine operator. By careful planning of the sequence of mining operations, the occurrence and severity of the bursts may be minimized. Research and operational control programs usually involve the installation of underground instrumentation to aid in the location of high-stress areas, a testing program to gain insight into the physical properties of the rocks, and the installation of a monitoring system to enable the engineers to predict the burst before it occurs.

Such warning systems may consist of a number of seismographs whose probes are located in boreholes drilled into zones of suspected high stress. It is believed that a burst is preceded by increased internal cracking activity which can be sensed by the seismic system. A critical level of such preburst activity is then used as the warning signal.

The lead time of such warning may be from several hours to several days or weeks. Experience has shown, however, that such systems are not completely reliable, since in numerous recorded cases a burst was preceded by a decrease in the seismic activity.

A remedial mining procedure, designed to reduce the damage potential of rock bursts, has been developed in South Africa. It consists of drilling ahead of mining faces and around existing openings and setting off dynamite charges in the bottoms of the drill holes. This fractures the rock and transforms it from a brittle medium to a semifractured material, thus relieving the concentration of the stress in the vicinity and shifting it away from the zones to be protected. The ensuing rock burst is then contained by the semifractured mass of rock and registers as only a contained bump.

The rock burst problem exercises an overriding constraint on the extension of mines in depth. It is further complicated by time-dependency of the phenomenon and the multiplicity of local geologic factors. Therefore, the existence and severity of the problem as well as the remedies attempted and used vary from one mine to another. *See* UNDERGROUND MINING.

[STEFAN H. BOSHKOV]

Bibliography: R. E. Goodman, *Introduction to Rock Mechanics*, 1980; H. C. Heard et al., (eds.), *Flow and Fracture of Rocks*, 1972.

Roll mill

A series of rolls operating at different speeds. Roll mills are used to grind paint or to mill flour. In paint grinding, a paste is fed between two low-speed rolls running toward each other at different speeds. Because the next roll in the mill is turning faster, it develops shear in the paste and draws the paste through the mill. The film is scraped from the last high-speed roll. For grinding flour, rolls are operated in pairs, rolls in each pair running toward each other at different speeds. Grooved rolls crush the grain; smooth rolls mill the flour to the desired fineness. The term roller mill is applied to a ring-roll mill. *See* GRINDING MILL.

[RALPH M. HARDGROVE]

Rolling contact

Contact between bodies such that the relative velocity of the two contacting surfaces at the point of contact is zero. Common applications of rolling contact are the friction gearing of phonograph turntables, speed changers, and wheels on roadways. Rolling contact mechanisms are, generally speaking, a special variety of cam mechanisms. An understanding of rolling contact is essential in the study of antifriction bearings. The concepts are also useful in the study of the behavior of toothed gearing. *See* ANTIFRICTION BEARING; CAM MECHANISM; GEAR.

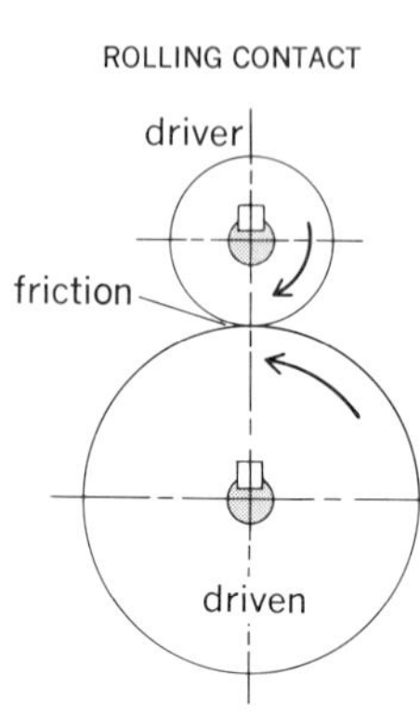

Fig. 1. Rolling friction disks.

Pure rolling contact can exist between two cylinders rotating about their centers, with either external or internal contact. Two friction disks (Fig. 1) have external rolling contact if no slipping occurs between them. The rotational speeds of the disks are then inversely proportional to their radii.

The general case of pure rolling between contacting bodies is illustrated in Fig. 2. Particle Q on body C and particle P on body B coincide at the moment. The velocities of these particles relative to ground frame A must be the same if there is to be no slip. It can be shown, using the theory of kinematics of mechanisms, that for two bodies to have rolling contact at an instant the point of contact must lie on the line joining the two centers of rotation. *See* MECHANISM; VELOCITY ANALYSIS.

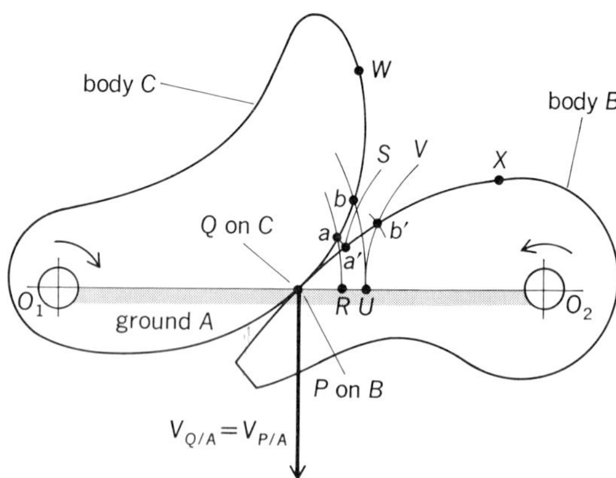

Fig. 2. Pure rolling between bodies in contact. (*From H. A. Rothbart, Cams, copyright © 1956 by John Wiley and Sons, Inc.; used with permission*)

Figure 2 shows that if the contacting particles Q and P were anywhere except on the line of centers O_1O_2 their velocities would necessarily be different, for they would have different directions. Because the centers of rotation are fixed, the sum of the distances from the centers of rotation to the point of contact must equal the center distance. Suppose that, in Fig. 2, W and X are two points that ultimately will come into contact. The bodies must therefore be designed to satisfy Eq. (1).

$$O_1W + O_2X = O_1O_2 \tag{1}$$

Finally, because there is to be no slip between the contacting surfaces, the length of the arc of action on each body must be the same. In Fig. 2, for example, arc length QW must equal arc length PX.

At an instant the angular velocities of the two links are inversely proportional to the distances from the centers of rotation to the point of contact, stated mathematically as Eq. (2).

$$\frac{\omega_C}{\omega_B} = \frac{O_2P}{O_1Q} \tag{2}$$

A graphical technique for designing the shape of body B so that it will have pure rolling contact with given body C is illustrated in Fig. 2. Select a sequence of points along the profile of body C such as a, b, and so on. Then the corresponding points a', b', and so on of body B must be found. Revolve point a about O_1 to the line of centers, giving point R. With O_2 as center and O_2R as radius, draw an arc RS of indefinite length. Because arc length Pa' must equal arc length Qa, the location of a' can be found approximately by using chord Qa as a compass distance and swinging an arc from R to intersect arc RS at a'. The same construction is used for finding b'. By repeating this procedure for a number of closely spaced points, the profile of body B can be determined.

Pure rolling occurs between the bodies with profiles that are the basic plane curves, that is, logarithmic spirals, ellipses, parabolas, and hyperbolas. To achieve pure rolling between these curves, the centers of rotation must be at the foci of the curves; the curves must be properly proportioned and positioned with respect to one another (Fig. 3).

Logarithmic spirals of equal obliquity are shown pivoting about their foci in Fig. 3*a* and *b*. In the first instance the links turn in opposite directions; in the second they turn in the same direction. The latter arrangement is the more compact. A logarithmic spiral rolling against a straight-sided follower is shown in Fig. 3*c*. Here the follower has a reciprocating motion. For the pair of ellipses shown in Fig. 3*d*, pure rolling contact can occur for complete rotations of both members. Rolling parabolas, as in Fig. 3*e*, provide a reciprocating output motion. Where space is limited, the equal hyperbolas of Fig. 3*f* are an especially effective choice.

A difficulty with most of these arrangments is that only partial rotations are possible. The lobe wheels of Fig. 3*g*, on the other hand, make possible continuous action. The lobes are made up of sectors of logarithmic spirals or ellipses.

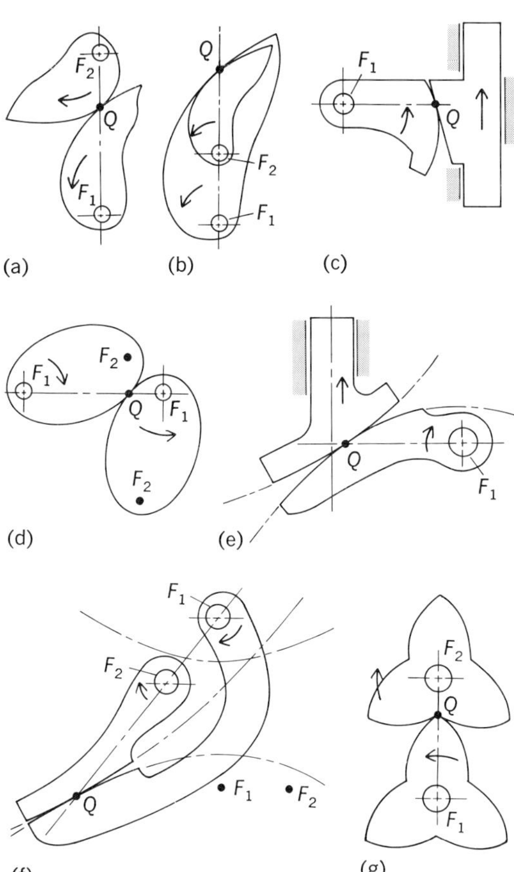

Fig. 3. Pure rolling between bodies in contact. (*a*) Externally rolling logarithmic spirals. (*b*) Internally rolling logarithmic spirals. (*c*) Logarithmic spiral and straight-sided follower. (*d*) Ellipses. (*e*) Equal parabolas. (*f*) Equal hyperbolas. (*g*) Lobe wheels contoured either from sectors of ellipses or logarithmic spirals. (*From H. A. Rothbart, Cams, copyright © 1956 by John Wiley and Sons, Inc.; used with permission*)

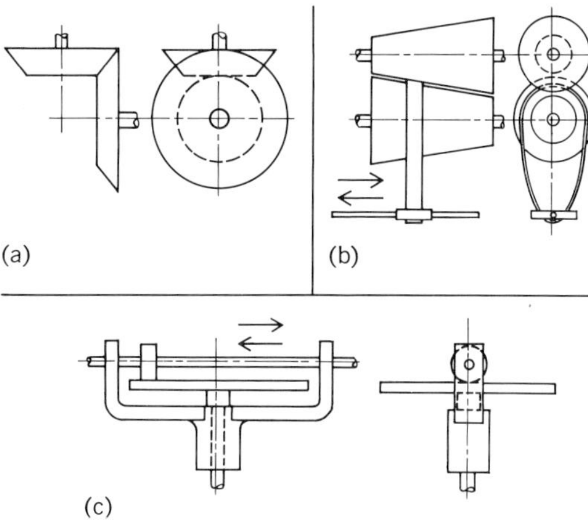

Fig. 4. Friction gears. (*a*) Cones. (*b*) Variable speed. (*c*) Mechanical integrator. (*International Textbook Co.*)

Friction gearing consists mainly of rolling cylinders, cones, and disks. The limited areas of surface contact permit only small quantities of power to be transmitted. Slippage can be often reduced by the use of spring-loaded bearings. Friction devices include friction cones (Fig. 4*a*); Evan's friction cones, which permit variations in output speed by shifting the belt position (Fig. 4*b*); and the Bush wheel and plate arrangement (Fig. 4*c*) used as an integrator for the original Vannevar Bush analog computer. The Bush arrangement permits not only variations in output speed but also reversal of the direction of the output shaft.

[JOHN R. ZIMMERMAN]

Bibliography: F. T. Barwell, *Bearing Systems: Principles and Practice*, 1979; V. M. Faires and R. M. Keown, *Mechanism*, 5th ed., 1960, reprint 1980; H. B. Kepler, *Basic Graphical Kinematics*, 2d ed., 1973; C. H. Suh and C. W. Radcliffe, *Kinematics and Mechanisms Design*, 1978.

Roof construction

The structural form used to provide the cover for homes, buildings, arenas, stadiums, and other activity areas. Whatever form the roof construction takes, it must be capable of resisting the loads to which it will be subjected, carrying them satisfac-

Fig. 1. Intersecting barrel shell roof of the Administration Building, St. Louis Airport. The exposed surface of the concrete is covered with an insulated plywood deck topped with copper sheeting; the interior surface is covered with acoustic tile.

Fig. 2. Thin-shell concrete used to form a folded-plate roof with cantilevered ends.

Fig. 3. Bullock's of Northern California Department Store, San Jose. (*Bill Apton, Photographer*)

torily to the walls or other supports, and providing protection from the elements. Principal roof loads are created by high winds and, in the northern climates, by snow and ice.

Conventional. The most common type of roof is encountered in the construction of wood-framed structures used as dwelling units. This type of construction supports the roof surface, which consists of composition shingles or tile placed on sheathing or plywood on rafters running from the peak of the roof to its eaves. When the width of the area to be covered exceeds approximately 30 ft (9 m), wooden roof trusses replace rafters, and additional support may be furnished through the use of longitudinal members known as purlins that span between trusses. Industrial buildings use essentially the same construction, except that the trusses, purlins, and roofing itself are generally fabricated from steel. When the design calls for a flat roof, precast planks of concrete or gypsum are used on I-beam stringers or floor beams. The roof is made waterproof, although at considerable increase in fire risk, by using from three to five layers of asphalt-impregnated roofing felt, bound together with hot asphalt.

Architects are finding imaginative uses of concrete, reinforced plastics, and steel in roofing large and small areas. Thin reinforced concrete sections have been used for shell roofs of cylindrical or spherical shape, or in the form of hyperbolic paraboloids (Fig. 1). Another form of equilibrium roof is a series of flat plates meeting at various angles to provide support one for the other (Fig. 2). The same configurations have been duplicated in steel, using lightweight sections formed with corrugated material. Structural steel is also used to form domed structures, with a series of radial ribs or a space framework to support the roofing material. The Houston (TX) Astrodome, Buckminster Fuller's geodesic dome covered with plastic panels, and a number of university field houses represent the modern use of one or more of these techniques.

Air-supported. The emergence of tennis as a widely popular sport has sparked the development of a radically new type of structure whose support is provided by differential air pressure. These structures consist of a durable fabric or film supported by cables or, alternatively, make use of a double membrane which provides its own support. In either case, the interior of the structure is maintained at a slightly higher air pressure than the ambient atmospheric pressure, and in this way the shape of the structure is maintained. When the membrane is reinforced, steel cables are used in a network configuration. The membrane or film, whether cable-supported or used doubled, may consist of a vinyl- or neoprene-coated fabric or, more recently, a Teflon-coated fiber-glass fabric has been introduced which is expected to increase the life of these structures from the current 7 years to approximately 20 years.

For tennis courts, storage buildings, and similar types of applications, the air-supported structure not only provides a roof covering, but also constitutes the entire structure. It is a relatively economical way to provide protection from the elements for a large surface area with no internal supports.

The concept is being used also to roof over small-span canopies for department stores (Fig. 3) as well as large stadium areas spanning more than 700 ft (213 m). In the use of air-supported structures for a stadium, the walls are erected and the seating and playing area are covered with one of the membrane materials, which in turn is supported by the slight differential air pressure at which the interior is maintained. This type of structure is exemplified by the stadium at Pontiac, MI, and at Syracuse University, in New York State.

All such modern roof structures result in a pleasing architectural effect at the same time a lightweight roof free from intermediate supports is created, providing large unencumbered areas protected from the weather. *See* GYPSUM PLANK; TRUSS.

[CHARLES M. ANTONI]

Bibliography: American Society for Testing and Materials, *Roofing Systems*, 1976; Task Committee on Air-Supported Structures of the Committee on Metals of the Structural Division, American Society of Civil Engineers, *State-of-the-Art Report of Air Supported Structures*, 1979; J. A. Watson, *Roofing Systems: Materials and Applications*, 1979; G. Winter et al., *Design of Concrete Structures*, 7th ed., 1964; L. Zetlin et al., *Roofs of the Future*, American Institute of Steel Construction, 1964.

Rotary engine

Internal combustion engine that duplicates in some fashion the intermittent cycle of the piston engine, consisting of the intake-compression-power-exhaust cycle, wherein the form of the power output is directly rotational.

Four general categories of rotary engines can be considered: (1) cat-and-mouse (or scissor) engines, which are analogs of the reciprocating piston engine, except that the pistons travel in a circular path; (2) eccentric-rotor engines, wherein motion is imparted to a shaft by a principal rotating part, or rotor, that is eccentric to the shaft; (3) multiple-rotor engines, which are based on simple rotary motion of two or more rotors; and (4) revolving-block engines, which combine reciprocating piston and rotary motion. Some of the more interesting engines of each type are discussed in this article.

Cat-and-mouse engines. Typical of this class is the engine developed by T. Tschudi, the initial design of which goes back to 1927. The pistons, which are sections of a torus, travel around a toroidal cylinder. The operation of the engine can be visualized with the aid of Fig. 1, where piston A operates with piston C, and B with D. In chamber 1 a fresh fuel-air mixture is initially injected while pistons C and D are closest together. During the intake stroke the rotor attached to pistons B and D

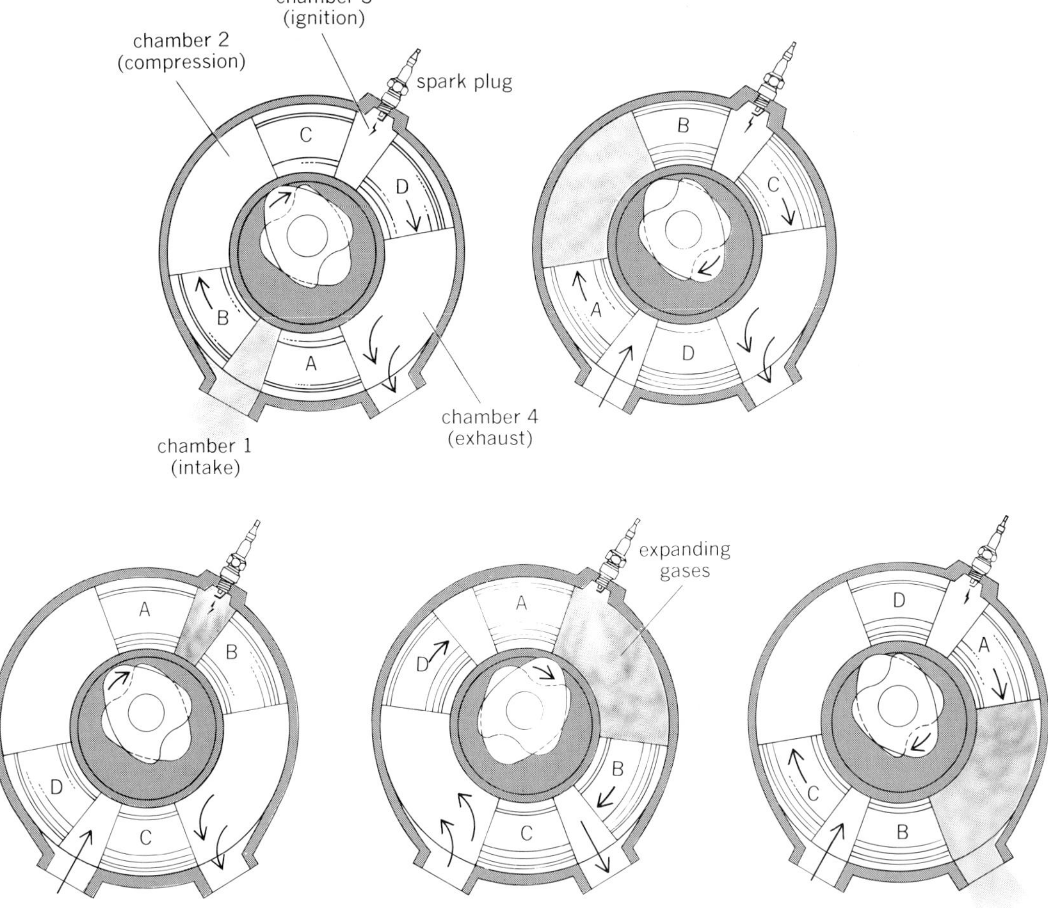

Fig. 1. The Tschudi engine.

rotates, thereby increasing the volume of chamber 1. During this time the A-C rotor is stationary. When piston D reaches its topmost position, the B-D rotor becomes stationary, and A and C rotate so that the volume of chamber 1 decreases and the fuel-air mixture is compressed.

When the volume of chamber 1 is again minimal, both rotors move to locate chamber 1 under the spark plug, which is fired. The power stroke finds piston D moving away from piston C, with the A-C rotor again locked during most of the power stroke. Finally, when piston D has reached bottom, the B-D rotor locks, the exhaust port has been exposed, and the movement of piston C forces out the combustion product gas. Note that four chambers exist at any time, so that at each instant all the processes making up the four-stroke cycle (intake-compression-power-exhaust) are occurring.

The motion of the rotors, and hence the pistons, is controlled by two cams which bear against rollers attached to the rotors. The cams and rollers associated with one of the rotors disengage when it is desired to stop the motion of that rotor. The shock loads associated with starting and stopping the rotors at high speeds may be a problem with this engine, and lubrication and sealing problems are characteristic of virtually all the engines discussed herein. However, the problem of fabricating toroidal pistons does not appear to be as formidable as was once believed.

An engine similar to the Tschudi in operation is that developed by E. Kauertz. In this case, however, the pistons are vanes which are sections of a right circular cylinder. Another difference is that while one set of pistons is attached to one rotor so that these two pistons rotate with a constant angular velocity, the motion of the second set of pistons is controlled by a complex gear-and-crank arrangement so that the angular velocity of this second set varies. In this manner, the chambers between the pistons can be made to vary in volume in a prescribed manner. Hence, the standard piston-engine cycle can be duplicated. Kauertz tested a prototype which was found to run smoothly and to deliver 213 hp (158·90 kW) at 4000 rpm. Here again, however, the varying angular velocity of the second set of pistons must produce inertia effects that will be absorbed by the gear-and-crank system. At high speeds over extended periods, problems with this system are likely to be encountered.

An advanced version of the cat-and-mouse concept called the SODRIC engine has been developed by K. Chahrouri. Unlike the Tschudi engine, in which the four processes of intake-compression-power-exhaust are distributed over 360° of arc, the SODRIC engine performs these same four processes in 60° of arc. Hence, six power strokes per revolution are achieved, resulting in very substantial improvements in engine performance parameters. For example, Chahrouri has estimated that 225 hp (167·85 kW) can be achieved at only 1000 rpm using an engine having a toroid radius of

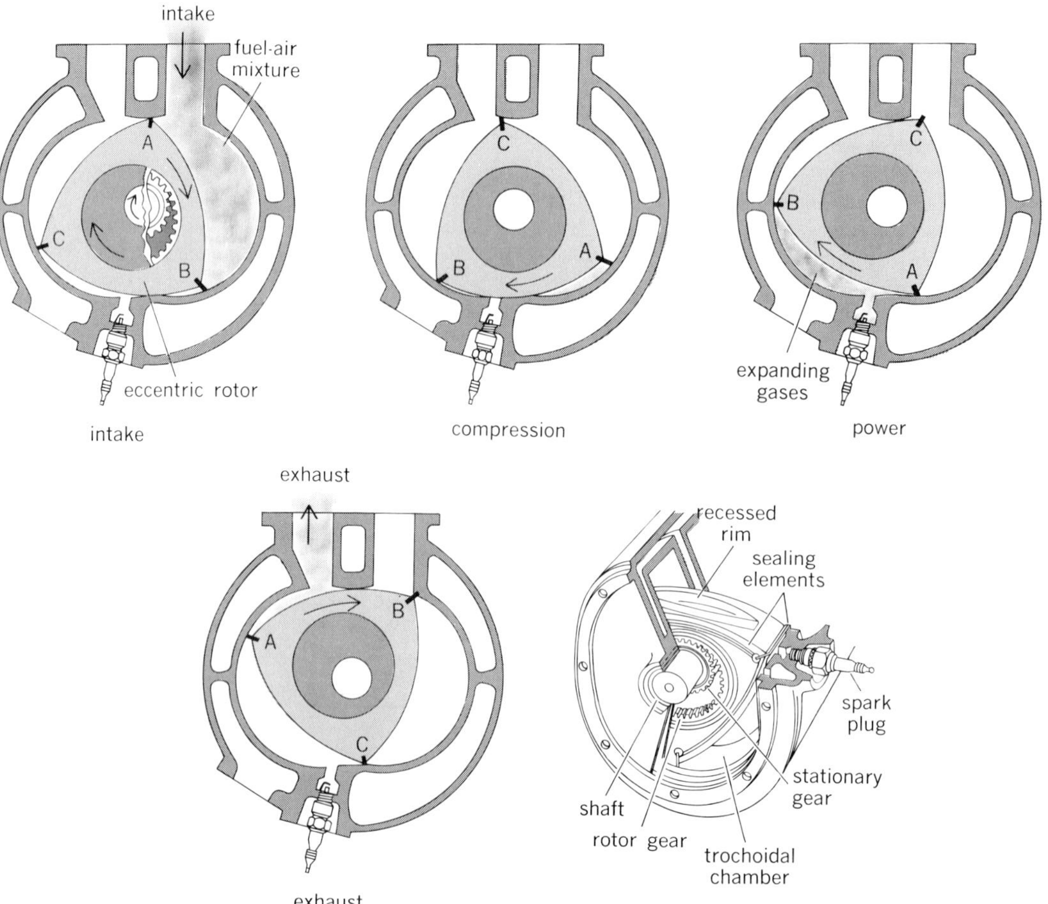

Fig. 2. The Wankel engine.

8 in. (0.2032 m) and 1-in. (0.0254-m-) radius pistons. Chahrouri has also improved upon the method by which alternate acceleration-deceleration of the pistons is achieved, and power is transmitted to the output shaft by using noncircular gears.

The "cat-and-mouse" and "scissors" characterizations of these engines should be clear once the picture of pistons alternately running away from, and catching up to, each other is firmly in mind. Other engines of this type, including the Maier, Rayment, and Virmel designs, differ principally in the system used to achieve the cat-and-mouse effect.

It should be noted that since the length of the power stroke is readily controlled in these engines, good combustion efficiencies (close to complete combustion) should be attainable.

Eccentric-rotor engines. The rotary engine which has received by far the greatest development to date is the Wankel engine, an eccentric-rotor type. The basic engine components are pictured in Fig. 2. Only two primary moving parts are present: the rotor and the eccentric shaft. The rotor moves in one direction around the trochoidal chamber, which contains peripheral intake and exhaust ports.

The operation of this engine can be visualized with the aid of Fig. 2. The rotor divides the inner volume into three chambers, with each chamber the analog of the cylinder in the standard piston engine. Initially, chamber AB is terminating the intake phase and commencing its compression phase, while chamber BC is terminating its compression phase and chamber CA is commencing its exhaust phase. As the rotor moves clockwise, the volume of chamber AB approaches a minimum. When the volume of chamber AB is minimal, the spark plug fires, initiating the combustion phase in that chamber. As combustion continues, the point is reached where the exhaust port is exposed, and the products of combustion are expelled from chamber AB.

To increase the chamber volumes, each segment of the rotor rim is recessed (Fig. 2). During the combustion-expansion phase unburned gas tends to flow at high velocity away from the combustion zone toward the opposite corner. As a result, this engine has a tendency to leave a portion of the charge unburned, similar to the problem encountered in ordinary piston engines. In addition to reducing the engine performance, this unburned gas is a source of air pollution. Efforts have been directed toward increasing the turbulence in each chamber, thereby improving the mixing between the burned and unburned gases, leading to better combustion efficiency.

On the other hand, Wankel engines have demonstrated a number of impressive advantages when contrasted with standard engines. Some of these advantages are listed below.

1. It has superior power-to-weight ratio; that is, the Wankel generally produces more or at least comparable horsepower per pound of engine weight when compared with conventional piston engines.

2. To increase power output, additional rotor-trochoidal chamber assemblies can readily be added, which occupy relatively little space and add little weight. In piston engines, cylinder vol-

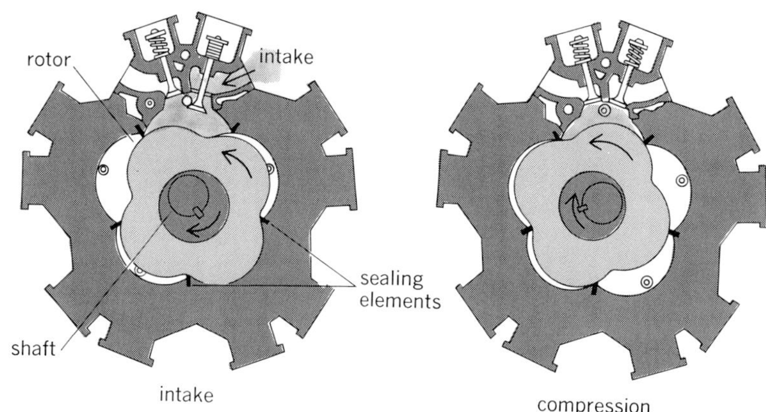

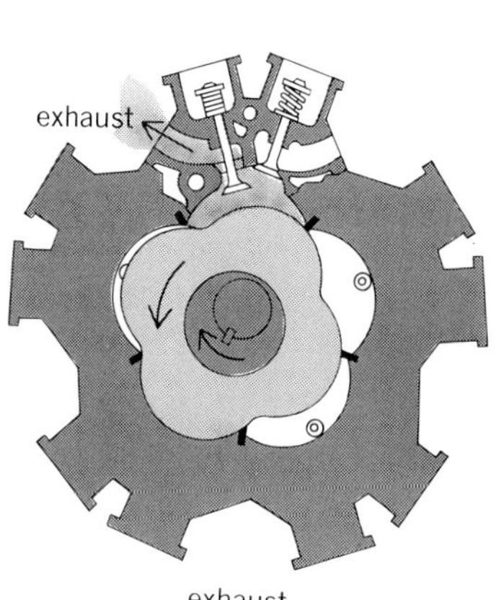

Fig. 3. The Renault-Rambler engine.

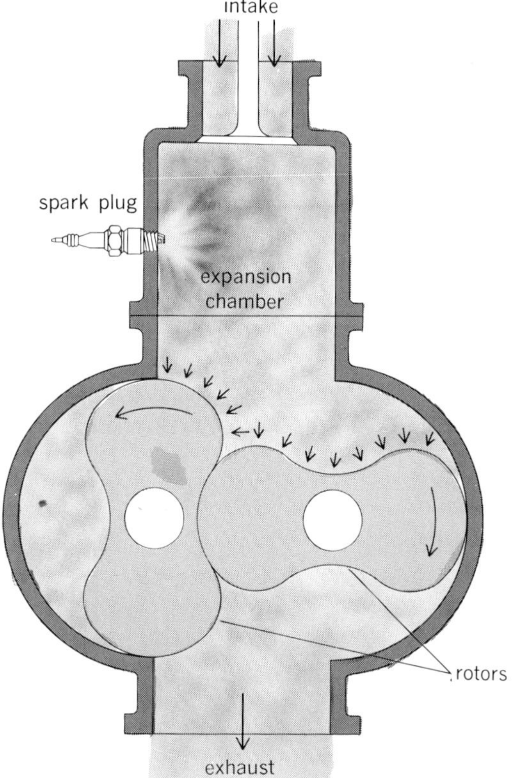

Fig. 4. A simple multirotor engine.

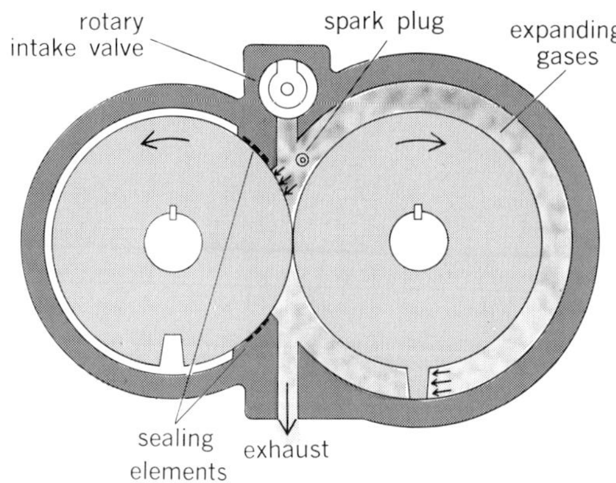

Fig. 5. The Unsin engine.

umes must be increased, leading to substantial increases in weight and installation space.

3. The rotor and eccentric shaft assembly can be completely balanced; since they usually rotate at constant velocity in one direction, vibration is almost completely eliminated and noise levels are markedly reduced.

4. As with the cat-and-mouse engines, the intake and exhaust ports always remain open, that is, gas flow into and from the engine is never stopped, so that surging phenomena and problems associated with valves which open and close are eliminated.

5. Tests indicate that Wankel engines can run on a wide variety of fuels, including ordinary gasoline and cheaper fuels as well.

6. After considerable development, reasonably effective sealing between the chambers has been achieved, and springs maintain a light pressure against the trochoidal surface.

7. The Wankel has so few parts, relative to a piston engine, that in the long run it will probably be cheaper to manufacture.

The initial application of the Wankel engine as an automotive power plant occurred in the NSU Spider. In the early 1970s, however, the Japanese automobile manufacturer Mazda began to use Wankel engines exclusively. However, relatively high pollutant emissions, coupled with rather low gasoline mileage for automobiles of this size and weight, resulted in poor sales in the United States. Mazda ceased marketing Wankel-powered automobiles in the United States in the mid-1970s. In addition, several American automobile manufacturers have experimented with Wankel-powered prototypes, but no production vehicles have emerged.

The Wankel engine is being used as a marine engine and in engine–electric generator installations, where its overall weight and fuel consumption have proved to be superior to those of a diesel engine or gas turbine generating equivalent power. Other projected applications include lawnmower and chainsaw engines. This wide range of applications is made possible by the fact that almost any size of Wankel engine is feasible.

An engine conceptually equivalent to the Wankel was developed jointly by Renault, Inc., and the American Motors Corp. It is sometimes called the Renault-Rambler engine. In this case, however, the rotor consists of a four-lobe arrangement, operating in a five-lobe chamber (Fig. 3). When a lobe moves into a cavity, which is analogous to the upward motion of the piston in the cylinder, the gas volume decreases, resulting in a compression process. The operation of this engine is detailed in Fig. 3. The fact that each cavity has two valves (intake and exhaust) represents a significant drawback. However, sealing between chambers may be simpler than in the Wankel; since each cavity acts as a combustion chamber, heat is evenly distributed around the housing, resulting in little thermal distortion.

It can be concluded that engines of the eccentric-rotor type are an integral part of the internal combustion engine scene. Their inherent simplicity, coupled with their advanced state of development, make them attractive alternatives to the piston engine in a number of applications.

Multirotor engines. These engines operate on some form of simple rotary motion. A typical design, shown in Fig. 4, operates as follows. A fuel-air mixture enters the combustion chamber through some type of valve. No compression takes place; rather, a spark plug ignites the mixture which burns in the combustion chamber, with a consequent increase in temperature and pressure. The hot gas expands by pushing against the two trochoidal rotors. The eccentric force on the left-hand rotor forces the rotors to rotate in the direction shown. Eventually, the combustion gases find their way out the exhaust.

The problems associated with all engines of this type are principally twofold: The absence of a compression phase leads to low engine efficiency, and sealing between the rotors is an enormously difficult problem. One theoretical estimate of the amount of work produced per unit of heat energy put into the engine (by the combustion process), called the thermal efficiency, is only 4%.

The Unsin engine (Fig. 5) replaces the trochoidal rotors with two circular rotors, one of which has a single gear tooth upon which the gas pressure acts. The second rotor has a slot which accepts the gear tooth. The two rotors are in constant frictional contact, and in a small prototype engine sealing apparently was adequate. The recommendation of its inventor was that some compression of the intake charge be provided externally for larger engines.

The Walley and Scheffel engines employ the principle of the engine in Fig. 4, except that in the former, four approximately elliptical rotors are used, while in the latter, nine are used. In both cases the rotors turn in the same clockwise sense,

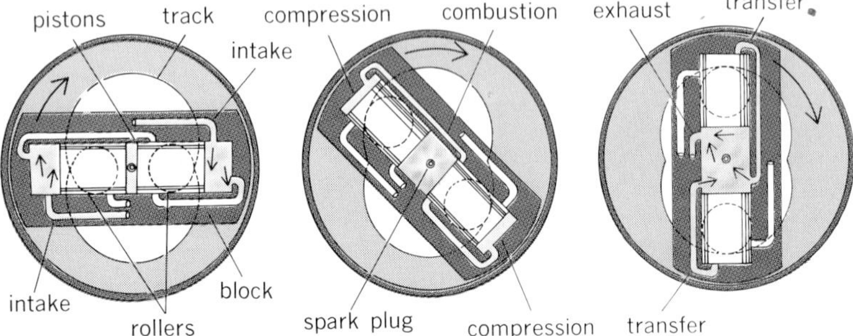

Fig. 6. The Mercer engine.

which leads to excessively high rubbing velocities. (The rotors are in contact to prevent leakage.) The Walter engine uses two different-sized elliptical rotors.

Revolving-block engines. These engines combine reciprocating piston motion with rotational motion of the entire engine block. One engine of this type is the Mercer (Fig. 6). In this case two opposing pistons operate in a single cylinder. Attached to each piston are two rollers which run on a track that consists of two circular arcs. When the pistons are closest together, the intake ports to the chambers behind the pistons are uncovered, admitting a fresh charge. At this moment a charge contained between the pistons has achieved maximum compression, and the spark plug fires. The pistons separate as combustion takes place between them, which results in a compression of the gases behind the pistons. However, the pistons moving apart force the rollers to move outward as well. This latter motion can only occur if the rollers run on their circular track, which consequently forces the entire engine block to rotate. When the pistons are farthest apart, the exhaust ports are uncovered and the combustion gases purged. At this same time the compressed fresh charge behind the pistons is transferred to the region between the pistons to prepare for its recompression and combustion, which must occur because of the continuing rotation of the block.

No doubt some of the fresh charge is lost to the exhaust during the transfer process. In addition, stresses on the roller assembly and cylinder walls are likely to be quite high, which poses some design problems. Cooling is a further problem, since cooling of the pistons is difficult to achieve in this arrangement.

On the other hand, the reciprocating piston motion is converted directly to rotary motion, in contrast with the connecting rod-crank arrangement in the conventional piston engine. Also, no flywheel should be necessary since the entire rotating block acts to sustain the rotary inertia. Vibration will also be minimal.

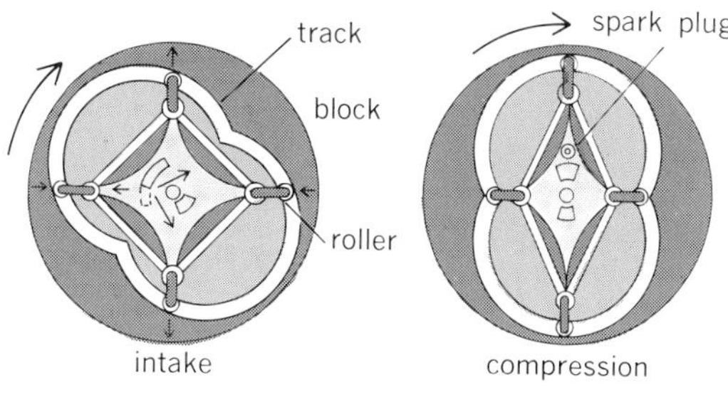

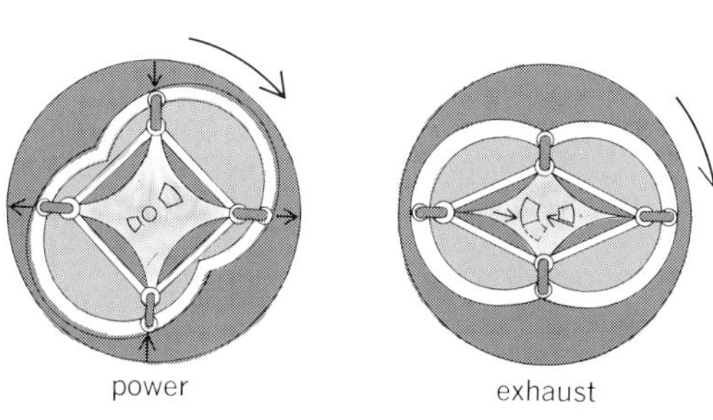

Fig. 7. The Rajakaruna engine.

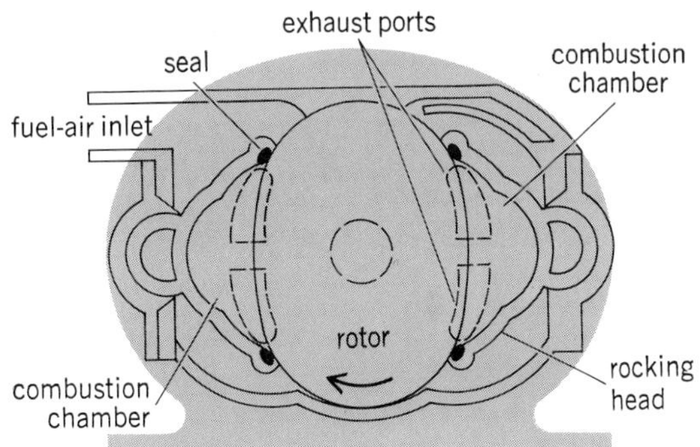

Fig. 8. The Walker engine.

The Selwood engine is similar in operation, except that two curved pistons 180° opposed run in toroidal tracks. This design recalls the Tschudi cat-and-mouse engine, except that the pistons only travel through 30° of toroidal track. This motion forces the entire block to rotate. The Leath engine has a square rotor with four pistons, each 90° apart, with a roller connected to each. As in the Mercer engine, the reciprocating motion of the pistons forces the rollers to run around a trochoidal track which causes the entire block to rotate. The Porshe engine uses a four-cylinder cruciform block. Again, rollers are attached to each of the four pistons. In this arrangement power is achieved on the inward strokes of the piston. Finally, the Rajakaruna engine (Fig. 7) uses a combustion chamber whose sides are pin-jointed together at their ends. Volume changes result from distortion of the four-sided chamber as the surrounding housing, which contains a trochoidal track, rotates. The huge pins are forced against the track. As usual, cooling and lubrication problems will be encountered with this engine, as will excessive wear of the hinge pins and track.

The Ma-Ho engine, invented by G. Hofmann, uses four cylinders welded concentrically around a central shaft. As the pistons oscillate in the cylinders as a result of the intake-compression-power-exhaust processes undergone by the fuel-air mixture, the pistons rotate a barrel cam to which they are connected by cam followers. Hence, the entire block, including the central shaft, is forced to rotate, recalling the Mercer engine.

Engines of other types. Although the vast majority of rotary engines fall into one of the categories discussed above, several ingenious designs which do not are worthy of mention.

The Walker engine (Fig. 8) involves an elliptical rotor which rotates inside a casing containing two C-shaped rocking heads. The fuel-air mixture is drawn into combustion chambers on each side of the rotor; mixture cut-off occurs when the rotor is in the vertical position. As the rotor turns, the mixture undergoes compression, and combustion is initiated by spark plugs. Rotor momentum coupled with the expansion of the combustion gases forces the rotor to continue turning. Compensation for seal wear is made by adjusting the rocking heads closer to the rotor.

The Heydrich engine (Fig. 9) is a vane-type rotary engine which utilizes a small hole to "store"

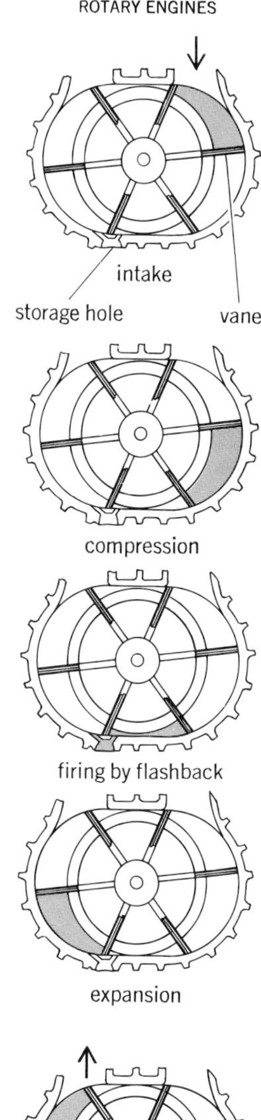

Fig. 9. The Heydrich engine.

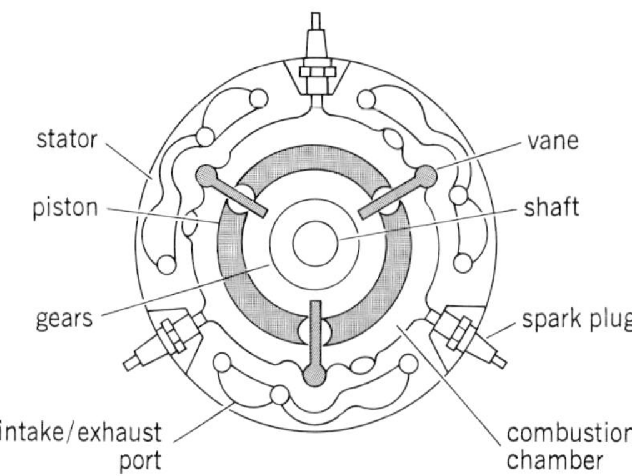

Fig. 10. The Leclerc-Edmon-Benstead engine.

a quantity of high-temperature combustion gases from a previous firing. This gas is then used to ignite the subsequent fuel-air charge. The first charge is ignited by a glow plug. Floating seals make contact with the chamber wall.

The Leclerc-Edmon-Benstead engine (Fig. 10) has three combustion chambers defined by stationary vanes, a cylindrical stator, and end flanges. An output shaft passing through the chamber is geared to a circular, eccentrically mounted piston. As the chambers are fired sequentially, the movement of the piston forces the shaft to rotate. Slots in the flange control porting of the intake and exhaust gases. *See* AUTOMOBILE; COMBUSTION CHAMBER; DIESEL CYCLE; DIESEL ENGINE; GAS TURBINE; INTERNAL COMBUSTION ENGINE; OTTO CYCLE. [WALLACE CHINITZ]

Bibliography: R. F. Ansdale, Rotary combustion engines, *Auto. Eng.*, vol. 53, no. 13, 1963, and vol. 54, nos. 1 and 2, 1965; W. Chinitz, Rotary engines, *Sci. Amer.*, vol. 220, no. 2, 1969; W. Froede, *The Rotary Engine and the NSU Spider*, Soc. Automot. Eng. Pap. no. 650722, October, 1965; R. Wakefield, Revolutionary engines, *Road Track*, vol. 18, no. 3, 1966; F. Wankel, in R. F. Ansdale (ed.), *Rotary Piston Machines: Classification of Design Principles for Engines, Pumps and Compressors*, 1965.

Rotary tool drill

A bit and shaft used for drilling wells. A turntable on the derrick floor rotates a string of hollow steel drill pipe at the bottom of which is a steel bit. The bit grinds the rock. A drilling fluid is pumped down through the drill pipe; the fluid flushes out the rock cuttings and returns up the space between drill string and hole side (see illustration).

The drilling fluid may be air, water, or, most commonly, mud (a mixture of various clays and chemicals, each having a special function). The mud cools and lubricates the bit, removes cuttings from the hole, and cakes the wall of the hole to prevent caving before steel casing is set. The hydrostatic pressure exerted by the column of mud in the hole prevents blowouts which may result when the bit penetrates a high-pressure oil or gas zone. When the mud reaches the surface, it passes over a vibrating screen to filter out large cuttings. The mud then passes on to a settling tank where smaller particles settle out. The cuttings are examined to determine the type of formation being

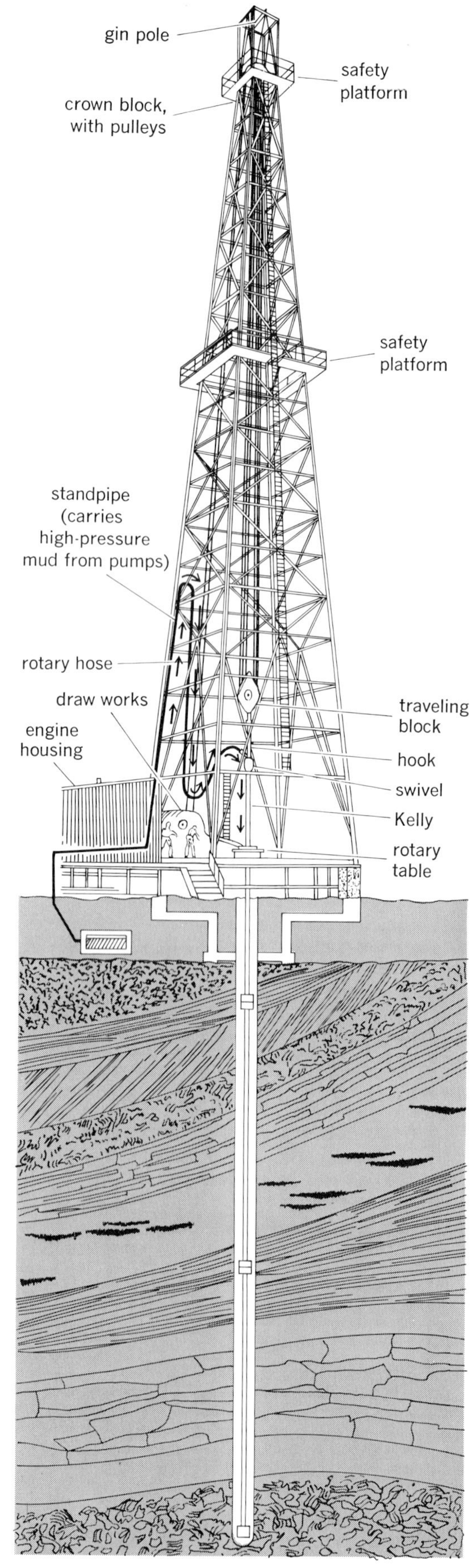

Bit is rotated at bottom of well by connected sections of pipe driven from a rotary table atop well.

drilled and for possibilities of oil or gas production. The mud mixture is sucked up from the pit and recirculated by a high-pressure pump. The viscosity, weight, and filtration properties of the mud are altered as drilling proceeds by changing the proportion of its constituents.

Power is transmitted from an engine to a draw works—a winch which drives the rotary table on the derrick floor and also applies power for hoisting or lowering the drill string as shown in the illustration. The string of drill pipe is topped at the surface by a square-sided length of heavy pipe called the kelly. The square shape permits the rotary table to grip and rotate the kelly, and hence the entire drill string, and yet have sufficient freedom so that it can slip vertically through the table as drilling goes deeper. Rotation speeds range from 40 rpm to 500 rpm or more, depending primarily upon the character of the formation being drilled. The drill string usually consists of 30-ft (9-m) lengths of drill pipe coupled together. On the lower end are heavier-walled lengths of pipe, called drill collars, which help regulate weight on the bit.

The drill string is attached to a swivel suspended from a hook which is connected to a traveling block, or pulley, encased in a frame. The drilling cable runs from the draw works over a crown block at the top of the derrick and down to the traveling block. The mud is pumped through a hose attached to the swivel. An opening in the center of the swivel permits the mud to pass down through the attached drill string.

When the bit has penetrated the distance of a pipe section, drilling is stopped, the string is pulled up to expose the top joint, the kelly is disconnected, a new section added, the kelly attached, the string lowered, and drilling resumed. This process continues until the bit becomes worn out, at which time the entire drill string must be pulled. Pipe is usually disconnected in thribbles, or 90-ft (27-m) sections of pipe, and stacked in the derrick. The height of the derrick determines whether doubles, thribbles, or fourbles can be stacked. The process continues until the bit reaches the surface. A new bit is attached, and the drilling string reassembled and lowered into the hole. Such round trips may take up to two-thirds of total rig-operating time, depending upon the depth of the hole. In hoisting or lowering the drill string, the swivel is disengaged from the hook. Elevators, or clamps, which grip the pipe securely, are attached. The elevators are also used when the hole is lined with steel casing. In lowering drill pipe or casing, each new section of pipe is lifted from the derrick floor and suspended on the elevator until it is screwed to the preceding joint, just above the hole opening; the entire column is then lowered into the hole. While new sections of pipe or casing are being attached to the elevators, the pipe in the hole is supported in the rotary table by slips, or gripping devices.

Derricks can be skid-, truck-, or trailer-mounted, but larger units used in very deep drilling are assembled on the site. Derricks usually range in height from 66 ft (20 m) to nearly 200 ft (60 m). The derrick floor is set 7–20 ft (2.1–6 m) or more above the ground to provide a basement for control devices, such as blowout preventers, below the rotary table. *See* OIL AND GAS WELL DRILLING; TURBODRILL. [A. L. PONIKVAR]

Rotational motion

The motion of a rigid body which takes place in such a way that all of its particles move in circles about an axis with a common angular velocity; also, the rotation of a particle about a fixed point in space. Rotational motion is illustrated by (1) the fixed speed of rotation of the Earth about its axis; (2) the varying speed of rotation of the flywheel of a sewing machine; (3) the rotation of a satellite about a planet, in which both the speed of rotation and the distance from the center of rotation may vary; (4) the motion of an ion in a cyclotron, where the angular speed of rotation remains constant, but the radius of the circular motion increases; and (5) the motion of a pendulum, in which case the particles describe harmonic motion along a circular arc.

This discussion of rotational motion is limited to circular motion such as is exhibited by the first and second examples.

Circular motion is a rotational motion in which each particle of the rotating body moves in a circular path about an axis. The motion may be uniform, that is, with constant angular velocity, or nonuniform, with changing angular velocity.

Uniform circular motion. The speed of rotation, or angular velocity, remains constant in uniform circular motion. In this case, the angular displacement θ experienced by the particle or rotating body in a time t is $\theta = \omega t$, where ω is the constant angular velocity.

Nonuniform circular motion. A special case of circular motion occurs when the rotating body moves with constant angular acceleration. If a body is moving in a circle with an angular acceleration of α radians/sec^2, and if at a certain instant it has an angular velocity ω_0, then at a time t sec later, the angular velocity may be expressed as $\omega = \omega_0 + \alpha t$, and the angular displacement as $\theta = \omega_0 t + \frac{1}{2}\alpha t^2$. *See* ACCELERATION; VELOCITY.

Banking of curves. When a car travels around a horizontal curve on a highway, the path is a circular arc of radius R, where R is the radius of curvature of the roadway. In order to have the car move in this circular arc, a horizontal external force must be applied to give the car an acceleration perpendicular to its path, that is, toward the center of rotation. This force must equal Mv^2/R, where M is the mass of the car and v its speed. This centripetal force is supplied by the friction between the tires and the road. If the force of friction is not great enough to produce this acceleration, the inertia of the car will tend to make it continue with its speed in a straight line, tangent to the road rather than around the curve, and this will cause the car to slide off the road.

To reduce the probability of skidding, roadways are customarily banked as illustrated in Fig. 1, which shows a car of mass M going away from the reader with a speed v and making a right-hand turn along an arc of radius R. The roadway must exert a vertical force F_W upward, equal and opposite to the weight $W = Mg$ of the car (g is the acceleration of gravity), and a horizontal centripetal force $F_C = Mv^2/R$ to make the car move in a circular arc. The net force N of the road on the car is the vector sum of these two forces.

From the diagram, it can be seen that the angle θ' which N makes with the vertical is given by Eq.

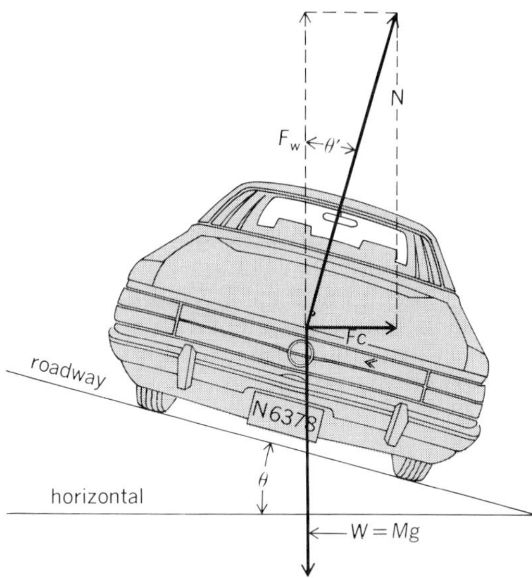

Fig. 1. Banking of a curve.

(1). If this angle θ' is equal to the bank angle θ of

$$\tan\theta' = \frac{F_C}{F_W} = \frac{Mv^2/R}{Mg} = \frac{v^2}{gR} \qquad (1)$$

the road, the force N of the road on the car is perpendicular to the roadway, and there will be no tendency to skid. Equation (1) shows that the correct bank angle is proportional to the square of the speed and inversely proportional to the radius of the curve. For a given curve, there is no correct bank angle for all speeds; thus roadways are banked for the average speed of traffic. Bank angle enters into the design of railroads and into the banking of an airplane when it executes a turn.

Work and power relations. A rotating body possesses kinetic energy of rotation which may be expressed as $T_{\mathrm{rot.}} = \frac{1}{2}I\omega^2$, where ω is the magnitude of the angular velocity of the rotating body and I is the moment of inertia, which is a measure of the opposition of the body to angular acceleration. The moment of inertia of a body depends on the mass of a body and the distribution of the mass relative to the axis of rotation. For example, the moment of inertia of a solid cylinder of mass M and radius R about its axis of symmetry is $\frac{1}{2}MR^2$.

To impart kinetic energy to a rotating body, work must be done. In Fig. 2 there is represented a solid cylinder of mass M and radius R, capable of rotation without friction about an axis perpendicular to the plane of the page through O. By means of a cord wrapped around the cylinder, a constant force F is applied, thus imparting angular acceleration to the cylinder. If the cylinder is originally at rest and the force F acts through a distance $s = R\theta$ equal to the arc PP', thus rotating the cylinder through the angle θ, the work W done is $W = Fs = FR\theta = L\theta$, where $L = FR$ is called the torque or moment of force. The action of this torque L is to produce an angular acceleration α according to Eq. (2), where $I\omega$, the product of moment of inertia

$$L = I\alpha = I\frac{d\omega}{dt} = \frac{d}{dt}(I\omega) \qquad (2)$$

and angular velocity, is called the angular momentum of the rotating body. This equation points out that the angular momentum $I\omega$ of a rotating body,

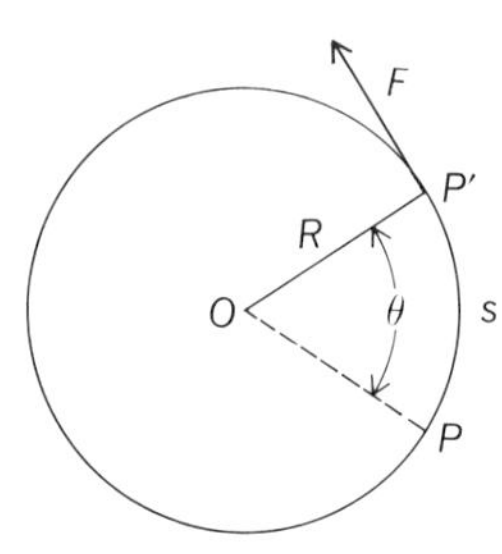

Fig. 2. Work in angular motion.

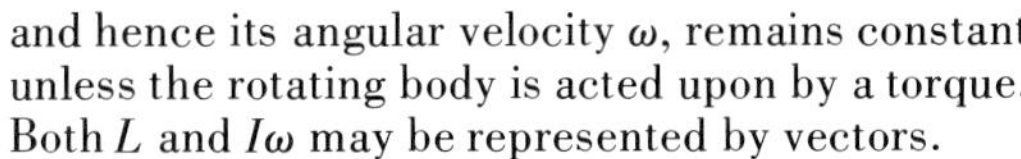

and hence its angular velocity ω, remains constant unless the rotating body is acted upon by a torque. Both L and $I\omega$ may be represented by vectors.

It is readily shown by Eq. (3) that the work done

$$W = L\theta = I\alpha\theta = I\alpha^1\!/_2(\alpha t^2) = {}^1\!/_2 I(\alpha t)^2 = {}^1\!/_2 I\omega^2 \qquad (3)$$

by the torque L acting through an angle θ on a rotating body originally at rest is exactly equal to the kinetic energy of rotation, because, for the case at hand, $\theta = \frac{1}{2}(\alpha t^2)$ and $\omega = \alpha t$.

Power is defined as rate of doing work, and the power P in rotational motion is given by Eq. (4).

$$P = \frac{dW}{dt} = \frac{d}{dt}(L\theta) = L\frac{d\theta}{dt} = L\omega \qquad (4)$$

See MOMENT OF INERTIA; POWER; TORQUE; WORK. [CARL E. HOWE/R. J. STEPHENSON]

Running fit

The intentional difference in dimensions of mating mechanical parts that permits them to move relative to each other. A free running fit has liberal allowance; it is used on high-speed rotating journals or shafts. A medium fit has less allowance; it is used on low-speed rotating shafts and for sliding parts. Running fits are affected markedly by their surface finish and the effectiveness of lubrication. *See* ALLOWANCE.

Running and sliding fits are standardized into nine classes. Close sliding fits accurately locate parts with some sacrifice in free motion; they permit no perceptible play. Sliding fits permit the parts to move but are not intended for freely running parts or moving parts subject to appreciable temperature change. Precision running fits permit parts to run freely at low speeds and at light journal pressures, provided temperature differences are limited. Close, medium, and free running fits are intended for progressively higher surface speeds, journal pressures, and temperature ranges. Loose running fits are for use with cold-rolled shafting and tubing made to commercial tolerances. [PAUL H. BLACK]

Rural electrification

The generation, distribution, and utilization of electricity in nonurban areas, beyond the confines of incorporated cities, villages, and towns. Electric service is now provided to more than 99% of the farms in the United States. This near-universal availability of electricity has contributed greatly to the magnitude and efficiency of farming operations and has been a contributing factor in the location of nonfarm residential, commercial, and industrial developments in rural areas.

Development. The first central station service was provided by the Pearl Street Station in New York City in 1882. Street lighting and service to some commercial establishments constituted the primary loads of the fledgling industry. However, as early as 1898, electricity was first used on a farm—a 5-hp (3.7-kW) electric motor and irrigation pump was installed by a fruit farmer in northern California.

Rural electrification grew slowly. The problems of low population density and undeveloped uses for electricity on the farm presented many economic and technical problems. Some of these problems were eventually solved, and about 10% of the farms in the United States were electrified by 1930 (Fig. 1). The Depression slowed this effort, and it was not until low-interest loans became available through the Rural Electrification Administration (REA) – an agency created by the Federal government in 1935 – that the process of rural electrification again picked up speed.

The technical and economic problems of providing service to rural areas have been solved, and farmers use electricity for many of their operations. In addition to these traditional rural loads, an increasing number of residences and commercial and industrial establishments are found in rural areas.

Uses. The homes found in rural areas, both farm and nonfarm, are as modern and up to date as those found in towns and cities. In some areas, a large percentage of rural homes are heated by electricity due to the unavailability of lower-cost natural gas. This combination of urban comfort in a rural setting has led to more and more nonfarm residential developments. Such developments constitute a major portion of rural loads, with about 21% of the total nonfarm residences in the United States located in rural areas according to the 1970 census.

Modern farming operations, in terms of electrical load, may be equivalent to a small factory. Large livestock feeding and confinement systems, broiler and egg-laying operations, dairying, and other similar livestock operations require large quantities of reliable electrical energy. Modern harvesting equipment and farming methods depend on drying of crops by artificial means. Loads in some rural areas actually peak during the harvesting season due to the many electric motors required to drive the fans associated with grain-drying equipment. In some areas, irrigation constitutes a major portion of the rural load. Irrigation pumps ranging in size from 30 to 250 hp (22 to 186 kW) dot the countryside, requiring high-capacity distribution lines and much electrical energy (Fig. 2). The relative uses of electrical energy on the farm by category are shown in Fig. 3. In 1978 direct use of electrical energy in agriculture totaled 31.7×10^9 kWh, or about 1.6% of total electric utility industry sales.

The processing of agricultural products and other related agribusiness industries have developed

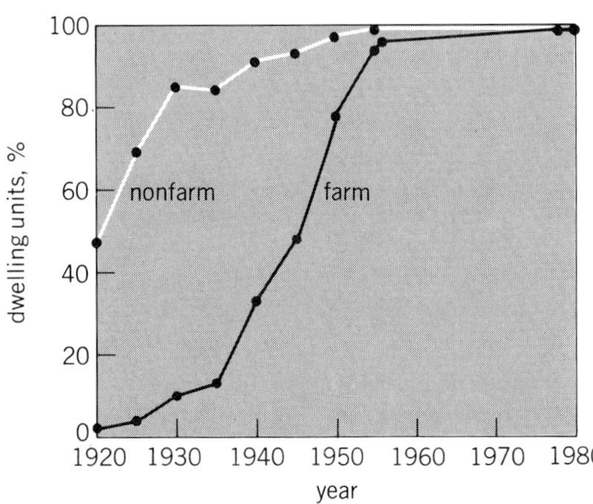

Fig. 1. Percent of United States dwelling units with electric service.

Fig. 2. A center pivot irrigation system irrigating corn. A single unit can water as much as 500 acres (200 hectares) or more. *(Eastern Iowa Light and Power Cooperative)*

in rural areas. These include: grain handling and storage facilities, livestock feed processing plants, canning and freezing plants, fertilizer and farm chemical complexes, farm machinery manufacturing and maintenance, fuel storage and handling facilities, and transportation depots – all related to the agriculture of the area.

Not all loads in rural areas are related to agriculture. Many other types of industries find it desirable to locate in rural areas because of the availability of land, transportation, labor, raw materials, or other incentives. Recreational developments have also become more numerous.

Many rural areas have a broad, prosperous, economic base – much of which is highly dependent on the availability of electrical energy.

Rural utility systems. Rural areas are served by electric utility systems which may be rural electric cooperatives or extensions of investor-owned or public utility systems serving both urban and rural areas. Generally speaking, rural electric cooperatives serve rural areas only, and statistics relating to them are readily available. Other electric utilities, serving a combination of urban and rural areas, usually do not report separate statistics for urban and rural loads on a regular basis. Rural electric cooperative statistics are believed to be representative of several aspects of rural loads and are used here for illustration.

The rural electric cooperatives and the public power utilities may or may not generate their own electrical energy. They may buy from other utilities, be a joint owner in a generation and transmission utility, or buy from some Federal power agency (Fig. 4).

Load density. Regardless of ownership, the primary difference between a rural electric system and an urban electric system is load density; typical area load densities are given in the table. This is also illustrated by a comparison of consumers served per mile of line. Rural electric cooperatives average about 4 consumers per mile compared to an estimated average of 30–40 consumers per mile for the total electric utility industry, or 8–10

RURAL ELECTRIFICATION

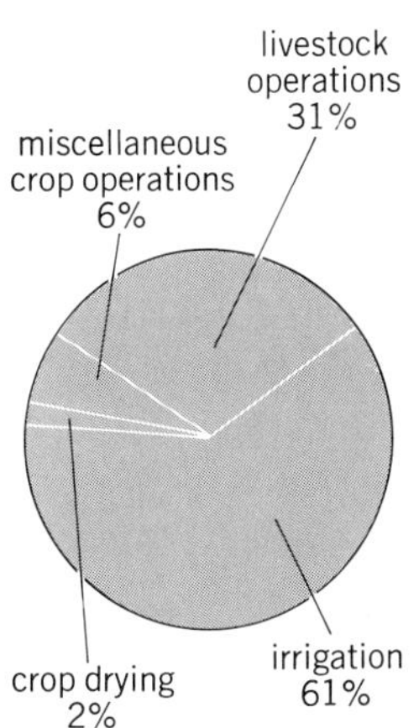

Fig. 3. Major categories of direct electric energy use in United States agriculture, 1978.

Fig. 4. A 55,000-kW-capacity coal-fired steam-electric generating plant which is owned by a rural electric utility in the Midwest. (*Eastern Iowa Light and Power Cooperative*)

times as many consumers per mile as for the rural electric cooperatives. This lower load density is the major factor contributing to the design differences between rural and urban systems.

Distribution system. Rural systems are integrated with the other electric utility systems in the region through transmission system interconnections and reliability coordinating councils. Supply to rural areas is usually provided by a subtransmission system whose voltage is in the range of 33 to 230 kV. This subtransmission system delivers power to many distribution substations, some as small as 1000 kVA, others as large as 15,000 to 25,000 kVA. A typical distribution substation consists of a high-voltage fuse and switch connecting the stepdown transformer to the subtransmission system. The low-voltage side of the transformer is usually connected to voltage regulators—one three-phase or three single-phase—and serves two to six distribution circuits protected by circuit breakers or oil circuit reclosers.

Rural distribution lines may extend up to 50 mi (80 km) or more from a substation, and voltages of 15 or 25 kV are commonly required to provide adequate service. Conductor sizes will range from as small as No. 6 copper equivalent to as large as 350,000 circular mils (1.7735 cm^2) copper equivalent. The voltage drop along a line is roughly proportional to the product of distance from substation times load current. This voltage drop is usually the primary factor in determining the capacity of a rural distribution circuit, whereas conductor cur-

Range of load densities

Type of area	Load density, kVA/mi^2 (kVA/km^2)	Remarks
Residential, low-density (rural area)	10–300 (4–120)	Number of farms or residences, and demand per farm, can cover a broad range of values, for example, 1 farm at 10 kVA to 150 farms at 2 kVA average each, does not include impact of irrigation load
Residential, medium-density (rural and suburban areas)	300–1,200 (120–460)	Based on home saturation on 70 by 100 ft (21 by 30 m) plots, 20% of total area with average diversified kVA per house from 0.5 to 2
Residential, high-density (urban area)	1,200–4,800 (460–1,900)	Based on home saturation on 70 by 100 ft (21 by 30 m) plots, 80% of total area with average diversified kVA per house from 0.5 to 2
Residential, extra-high-density (all-electric area)	15,000–20,000 (6,000–8,000)	An estimated upper limit for total electric homes with heating and air conditioning, also high saturation of homes in total area
Commercial	10,000–300,000 (4,000–120,000)	Type of area and load includes such variations so that more definitive breakdown is not practical; this range of values covers range of small shopping centers and commercial areas to downtown commercial areas of large cities

Fig. 5. Installation of underground rural distribution line. A crawler-type tractor pulls a cable plow and cable reel. The plow opens a furrow and lays the cable in one operation. *(Eastern Iowa Light and Power Cooperative)*

rent-carrying capacity usually determines the capacity of distribution circuits in higher-load-density urban areas.

Most rural distribution lines are overhead. Circuits are three-phase leaving the substations, but may be either three-phase, two-phase, or single-phase at other locations. Single-phase lines provide satisfactory service in many rural areas due to the relatively low load density and small loads.

In some areas, overhead lines are very susceptible to storm damage, particularly from ice storms. A major ice storm can cause havoc in any area, but rural areas are especially susceptible due to the many miles of line required to serve a relatively few customers. Demands for higher service reliability and the high cost of replacing storm-damaged overhead lines have led some utilities to place distribution lines underground. This is especially practical in relatively open areas where underground circuits can be installed by cable plows (Fig. 5).

Rural distribution circuits are usually radial. Because of the relatively low load density, it is seldom practical to provide two-way feed in many rural areas. Most urban distribution systems tend to be "looped," providing two-way feed to most loads. Rural distribution circuits may employ voltage regulators, sectionalizing equipment such as oil circuit reclosers, cutouts, and fuses, and power-factor correction capacitors in order to provide more efficient use of the distribution circuits.

The construction of rural overhead lines differs somewhat from that of urban lines in several ways. Rural lines utilize longer spans and fewer accessories to minimize costs (Fig. 6). The use of common poles by different utilities, such as telephone, cable television, and electric, is not as common in rural areas. In rural areas a small distribution transformer (5–15 kVA) typically serves a single consumer, whereas in urban areas a large transformer (25–50 kVA) may serve several small businesses or residences.

Rural lines are built according to safety standards similar to those for urban lines, with the National Electrical Safety Code usually providing the

Fig. 6. A typical rural single-phase distribution line. The absence of crossarms and longer spans makes lower-cost rural lines possible.

governing design criteria. In some instances, the clearance of rural lines must be increased because of irrigation and other farm equipment. The REA provides design standards and guides for use by rural electric cooperatives. *See* ELECTRIC DISTRIBUTION SYSTEMS; TRANSMISSION LINES.

Decentralization of services. Due to the large geographic area typically covered by a rural utility, district offices and service crews are often established to provide faster and more efficient service to the rural consumers. An urban utility may have a single centralized headquarters facility in each metropolitan area, compared to the rural utility which may have several district or division offices.

Impact on United States. The availability of reliable electrical energy to virtually all areas of the United States has had a major impact on the economy of rural America and the life-style there. Farm output per worker-hour increased by over two-thirds between 1960 and 1980. Automation, mechanization, chemical fertilizers, better management practices, and electrification are credited with this change. Many former urbanites have discovered that living in rural areas is as comfortable as in urban areas. Both agriculture-related and nonagricultural industry have found rural areas desirable places to locate.

The unqualified success of rural electrification programs in the United States has led to the exportation of the concept to developing countries around the world. For example, rural electric cooperatives have been formed in several countries in Latin America and the Far East.

Future. Inflation, energy shortages, and environmental issues affect all electric utilities. Rural areas are particularly vulnerable because a rural utility tends to be more capital-intensive than an urban utility. However, electricity still is a very desirable form of energy for use by farms, rural industries, and agribusinesses, and rural areas will continue to attract city dwellers to the quieter life of rural-residential areas.

Nonfarm developments in rural areas have caused the rural loads in the United States to grow at a much faster pace than for the average utility industry. In many areas, small farms are consolidated into larger operations, resulting in the loss of farmsteads and farm homes. This is offset to some extent by the new, larger farming operation requiring even more energy than the total of the small ones it replaced. The many uses of energy on the farm have also contributed to the increased electrical loads in rural areas.

The increasing cost of energy and efforts to conserve energy are also evident in rural areas. Efforts are being made to use electricity more efficiently in farming operations. However, many farm homes have typically been heated by fuel oil, and diesel engines have been used to drive irrigation pumps. Electricity may be replacing petroleum fuels in many instances, possibly offsetting the impact of conservation efforts.

The electric utility industry has essentially completed the task of making electric service universally available, regardless of the location. The rural customer—who once was desperate to receive electric service of any kind—has become more and more dependent on electric service for the home, farm, or other business. The task now is to continue to provide electrical energy in a safe, economical, reliable, environmentally acceptable manner to the rural consumer. *See* ELECTRIC POWER SYSTEMS.

[C. MAXWELL STANLEY; RONALD D. BROWN]

Bibliography: Edison Electric Institute, *Statistics of the Electric Utility Industry*; National Rural Electric Cooperative Association, *Facts about America's Rural Electric Systems*, 1979; U.S. Department of Agriculture, *Agricultural Statistics 1978*; USDA and Federal Energy Administration, *Energy and Agriculture, 1974 Data Base*; USDA, Economics, statistics and cooperatives service, *Structure Issues of American Agriculture*, Agric. Econ. Rep. 438, 1979; USDA, Rural Electrification Administration, *Annual Statistical Report: Rural Electric Borrowers.*

Safety factor

An empirical number by which the strength of a material is divided to obtain a conservative design stress. A safety factor is used because of uncertainties in operating conditions that may be encountered, nonuniformities in materials, simplifying design assumptions, effects of aging such as corrosion, and strains introduced inadvertently during fabrication and transportation and because of the seriousness of failure. Safety factors vary widely depending on the material, consequences of failure, and operating conditions. For ductile materials, safety factors applied to yield strength are often 1.5–4. For brittle materials that fracture with no prior evidence of incipient failure, factors of 5–8 may be appropriate. *See* STRESS AND STRAIN.

[WILLIAM J. KREFELD/WALDO G. BOWMAN]

Safety valve

A relief valve set to open at a pressure safely below the bursting pressure of a container, such as a boiler or compressed air receiver. Typically, a disk is held against a seat by a spring; excessive pressure forces the disk open (see illustration). Construction is such that when the valve opens slightly, the opening force builds up to open it fully and to hold the valve open until the pressure drops a predetermined amount, such as 2–4% of the opening pressure. This differential or blow-down pressure and the initial relieving pressure are adjustable.

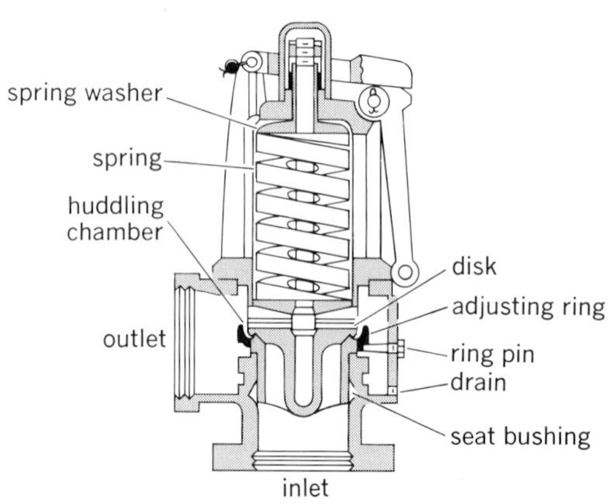

Diagram of a typical safety valve.

Adjustments must be set by licensed operators, and settings must be tamperproof. The ASME Boiler Construction Code gives typical requirements for safety valves. *See* VALVE.

[THEODORE BAUMEISTER]

Sanitary engineering

A specialty field generally developed in civil engineering but not limited to that branch. The National Research Council defines the sanitary engineer as "a graduate of a full 4-year, or longer, course leading to a Bachelor's, or higher, degree at an educational institution of recognized standing with major study in engineering, who has fitted himself by suitable specialized training, study, and experience (1) to conceive, design, appraise, direct and manage engineering works and projects developed, as a whole or in part, for the protection and promotion of the public health, particularly as it relates to improvement of the environment, and (2) to investigate and correct engineering works and other projects that are capable of injury to the public health by being or becoming faulty in conception, design, direction, or management."

Sanitary engineering practice includes surveys, reports, designs, reviews, management, operation and investigation of works or programs for (1) water supply, treatment and distribution; (2) sewage collection, treatment and disposal; (3) control of pollution in surface and underground waters; (4) collection, treatment, and disposal of refuse; (5) sanitary handling of milk and food; (6) housing and institutional sanitation; (7) rodent and insect control; (8) recreational place sanitation; (9) control of atmospheric pollution and air quality in both the general air of communities and in industrial work spaces; (10) control of radiation hazards exposure; and (11) other environmental factors affecting health, comfort, safety, and well-being of people.

Sanitary engineers engage in research in engineering sciences and such related sciences as chemistry, physics, and microbiology and apply these in development of works for protection of humans and control of their environment. *See* SEWAGE; SEWAGE COLLECTION SYSTEMS; SEWAGE DISPOSAL; SEWAGE TREATMENT; WATER POLLUTION; WATER SUPPLY ENGINEERING.

[WILLIAM T. INGRAM]

Bibliography: H. E. Babbitt and E. R. Baumann, *Sewerage and Sewage Treatment*, 8th ed., 1958; H. Blatz (ed.), *Radiation Hygiene Handbook*, 1959; V. M. Ehlers and E. W. Steel, *Municipal and Rural Sanitation*, 6th ed., 1965; G. M. Fair, J. C. Geyer, and D. A. Okun, *Water and Wastewater Engineering*, vol. 2: *Water Purification and Wastewater Treatment and Disposal*, 1968; W. C. L. Hemeon, *Plant and Process Ventilation*, 1955; R. K. Linsley, Jr., and J. B. Franzini, *Elements of Hydraulic Engineering*, 1955; F. S. Merritt (ed.), *Standard Handbook for Civil Engineers*, sect. 22, Sanitary Engineering, 1968; *Municipal Refuse Disposal*, Commission on Refuse Disposal, APWA Research Foundation Project 104, 2d ed., 1966; F. A. Patty et al. (eds.), *Industrial Hygiene and Toxicology*, vol. 2, 2d ed., 1963; *Refuse Collection Practice*, APWA Commission on Solid Wastes, 3d ed., 1966; P. A. Sartwell, *Maxcy-Rosenau Preventive Medicine and Public Health*, 10th ed., 1973; E. W. Steel, *Water Supply and Sewerage*, 5th ed., 1979; H. H. Uhlig, *Corrosion and Corrosion Control*, 1971; M. Willrich and R. K. Lester, *Radioactive Waste: Management and Regulation*, 1977.

Saturable reactor

An iron-core inductor in which the effective inductance is changed by varying the permeability of the core. Saturable-core reactors are used to control large alternating currents where rheostats are impractical. Theater light dimmers often employ saturable reactors.

In the illustration of two types of saturable-core reactors, illustration *a* shows two separate cores, while in illustration *b* a three-legged core is formed by placing two two-legged cores together. The load winding, connected in series with the load, carries the alternating current and acts as an inductive element. The control winding carries a direct current of adjustable magnitude, which can saturate the magnetic core. When the core is saturated by the direct current, the effective inductance (and therefore the reactance) of the coils in the ac circuit is small. Little voltage can therefore be induced in the coil to oppose the applied voltage and therefore reduce the voltage that is supplied to the load.

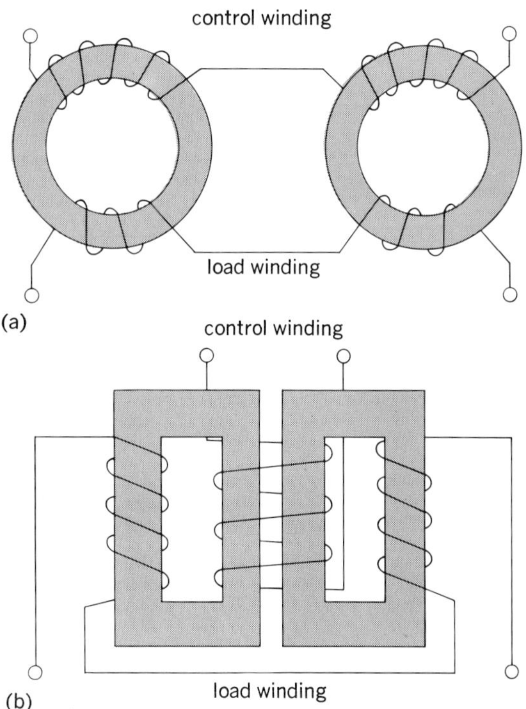

Typical construction of saturable-core reactors. (*a*) Two separate cores. (*b*) Three-legged cores.

Reducing the magnitude of the control current reduces the intensity of saturation. This increases the reactance of the load winding. As the reactance increases, the voltage drop in the load winding increases and causes a reduction in the magnitude of the voltage applied to the load. *See* INDUCTANCE.

[WILSON S. PRITCHETT]

Bibliography: T. Croft, J. Watt, and W. I. Summers, *American Electricians' Handbook*, 10th ed., 1981; D. G. Fink and H. W. Beaty, *Standard Handbook for Electrical Engineers*, 11th ed., 1978.

Sawing

The parting of material by using metal disks, blades, bands, or abrasive disks as the cutting tools. Sawing a piece from stock for further machining is called cutoff sawing, while shaping or forming a piece is referred to as contour sawing.

Machine sawing of metal is performed by five types of saws or processes: hacksawing, band sawing, cold sawing, friction sawing, and abrasive sawing.

Hacksaws are used principally as cutoff tools. The toothed blade, held in tension, is reciprocated across the workpiece. A vise holds the stock in position. The blade is fed into the work by gravity or springs. Sometimes a mechanical or hydraulic feed is used. Automatic machines, handling bar-length stock, are used for continuous production. (Fig. 1).

Fig. 1. Horizontal cutoff saw. (*DoAll Co.*)

Fig. 2. Vertical band saw. (*DoAll Co.*)

Band saws cut rapidly and are suited for either cutoff or contour sawing. The plane in which the blade operates classifies the machine as being either vertical or horizontal. Band saws are basically a flexible endless band of steel running over pulleys or wheels. The band has teeth on one side and is operated under tension. Guides keep it running true. The frame of the horizontal type is pivoted to allow positioning of the workpiece in the vise. Horizontal machines are used for either straight or angular cuts. A table that supports the workpiece and the wide throat between the upright portions of the blade makes the vertical band saw ideal for contour work (Fig. 2). Band saws operating at high speed are frequently used as friction saws.

Cold sawing is principally a cutoff operation. The blade is a circular disk with cutting teeth on its periphery. Blades range in size from a few inches to several feet in diameter. The cutting teeth may be cut into the periphery of the disk or they may be inserts of a harder material. The blade moves into the stock with a positive feed. Stock is positioned manually in some cold-sawing machines, while other models are equipped for automatic cycle sawing.

Friction sawing is a rapid process used to cut steel as well as certain plastics. This process is not satisfactory for cast iron and nonferrous metals. Cutting is done as the high-speed blade wipes the metal from the kerf after softening it with frictional heat. Circular alloy-steel blades perform cutoff work, while frictional band saws do both cutoff and contour sawing. Circular blades are frequently cooled by water or air. Circular blades are advanced into the work, while thick workpieces require power-table feed when friction-cut on a band saw.

Abrasive sawing is a cutoff process using thin rubber or bakelite bonded abrasive disks. In addition to steel, other materials such as nonferrous metals, ceramics, glass, certain plastics, and hard rubber are cut by this method. Cutting is done by the abrasive action of the grit in the disk.

Abrasive disks are operated either wet or dry. For heavy cutting a cooling agent is generally used. The workpiece is firmly held while the wheel traverses through it. Machines are made in manually operated and automatic models. *See* MACHINING OPERATIONS.

[ALAN H. TUTTLE]

Schematic drawing

Concise, graphical symbolism whereby the engineer communicates to others the functional relationship of the parts in a component and, in turn, of the components in a system. The symbols do not attempt to describe in complete detail the characteristics or physical form of the elements, but they do suggest the functional form which the ensemble

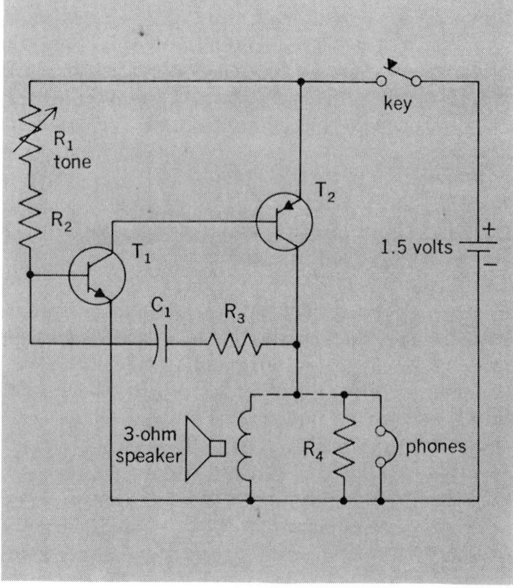

Fig. 1. Simple transistorized code practice oscillator, using standard symbols. (*Adapted from J. Markus, Sourcebook of Electronic Circuits, McGraw-Hill, 1968*).

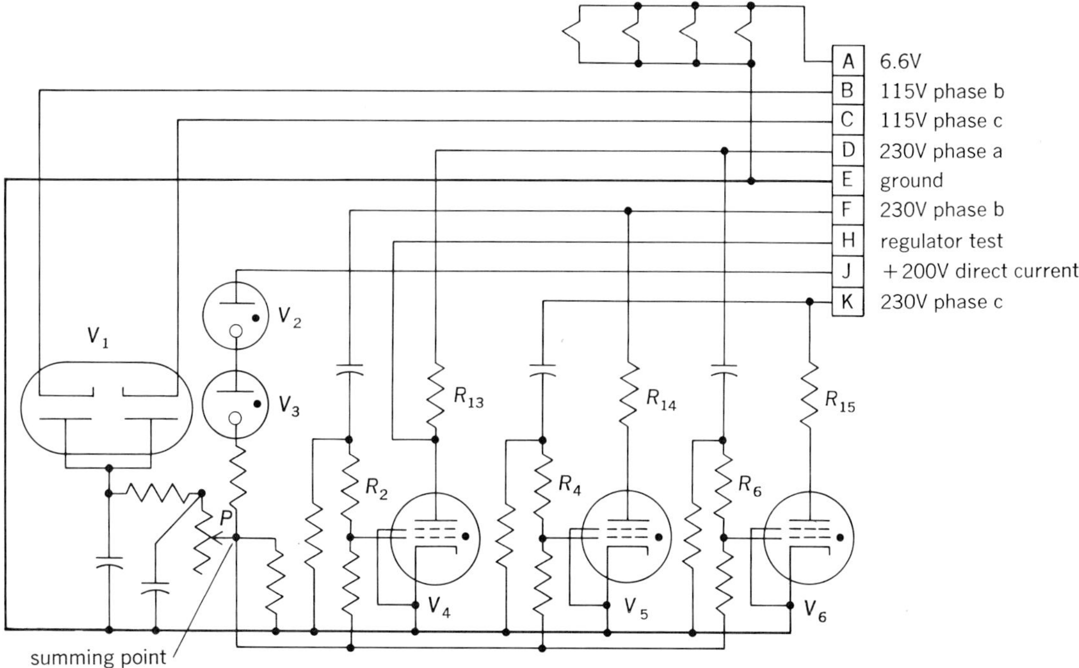

Fig. 2. Electrical schematic of voltage regulator.

of elements will take in satisfying the functional requirements of the component. They are different from a block diagram in that schematics describe more specifically the physical process by which the functional specifications of a block diagram are satisfied. Rather than expressing a mathematical relationship between, for example, an input and an output variable as in a block diagram, a schematic illustrates the physical principles and techniques by which the mathematical requirements of the element are realized. For instance, the schematic indicates whether electrical, hydraulic, mechanical or pneumatic techniques are employed, and suitable symbols indicate the appropriate elements, such as batteries, resistors, valves, gearing, vacuum tubes, and motors.

Electrical schematic. An electrical schematic is a functional schematic which defines the interrelationship of the electrical elements in a circuit, equipment, or system. The symbols describing the electrical elements are stylized, simplified, and standardized to the point of universal acceptance (Figs. 1 and 2).

The simple character of the element symbol makes it possible to represent in a small area the interrelationship of the electrical elements in complex systems. This has the double advantage of economy of space and an increased facility of understanding, because one experienced in the symbolism can easily follow the various functional paths in the electrical schematic. The tracing of a signal path through an electrical schematic is considerably enhanced by the existence of more or less accepted rules with regard to the arrangement of the symbols and of the interconnections between the symbols, all contrived to make more lucid the functional interrelationship of the elements.

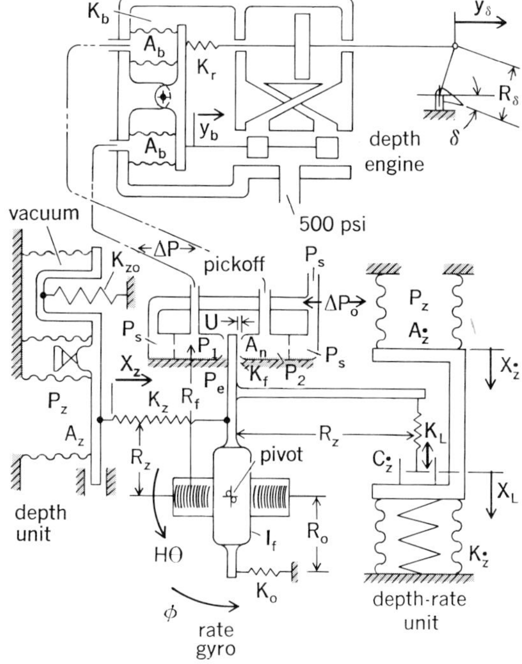

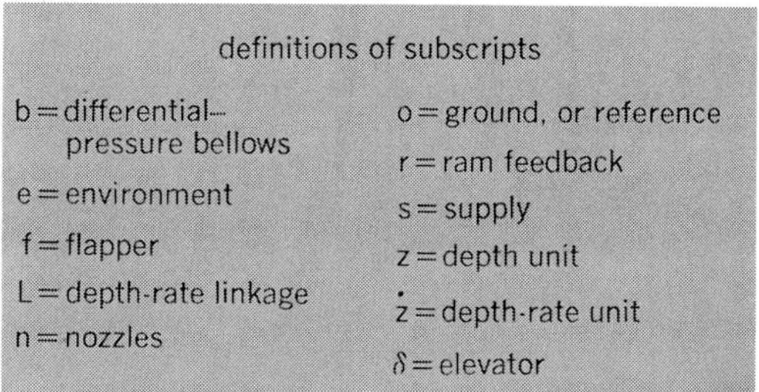

definitions of subscripts
b = differential-pressure bellows
e = environment
f = flapper
L = depth-rate linkage
n = nozzles
o = ground, or reference
r = ram feedback
s = supply
z = depth unit
ż = depth-rate unit
δ = elevator

Fig. 3. Mechanical schematic of the depth-control mechanism of a torpedo.

Mechanical schematic. A mechanical schematic is also a functional schematic. The graphical descriptions of elements of a mechanical system are more complex and more intimately interrelated than the symbolism of an electrical system and so the graphical characterizations are not nearly as well standardized or simplified (Fig. 3). However, a mechanical schematic illustrates such features as components, acceleration, velocity, position force sensing, and viscous damping devices. The symbols are arranged in such a manner and with such simplification so as to economize on space and to facilitate an understanding of the functional interrelationship of the various components which comprise the system. *See* DRAFTING; ENGINEERING DRAWING.

[ROBERT W. MANN]

Screw

A cylindrical body with a helical groove cut into its surface. For practical purposes a screw may be considered to be a wedge wound in the form of a helix so that the input motion is a rotation while the output remains translation. The screw is to the wedge much the same as the wheel and axle is to the lever in that it permits the exertion of force through a greatly increased distance.

Figure 1 shows a frictionless screw with square threads mounted on a ball thrust bearing being used to raise a load Q. The load and other forces may be considered to be concentrated at an effective radius called the pitch radius. The force diagram is similar to the diagram for a wedge, except that wedge angle θ is replaced by screw lead angle λ, which is the angle between the thread of the helix and a plane perpendicular to the axis of rotation. Subscript t is added to indicate that force F is applied tangent to a circle, giving Eq. (1). Because F_t acts at the pitch radius R, the torque on the screw is given by Eq. (2).

$$F_t = Q \tan \lambda \tag{1}$$

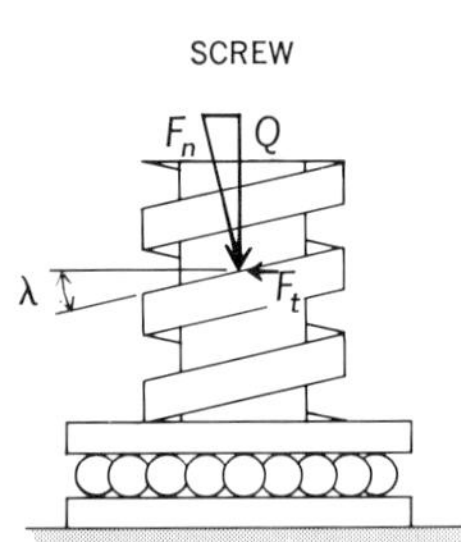

Fig. 1. Diagram of screw. See text for symbols.

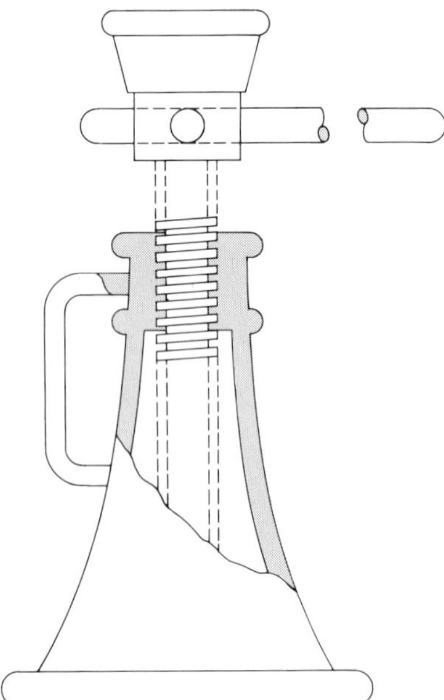

Fig. 2. Screw jack, used to raise heavy objects.

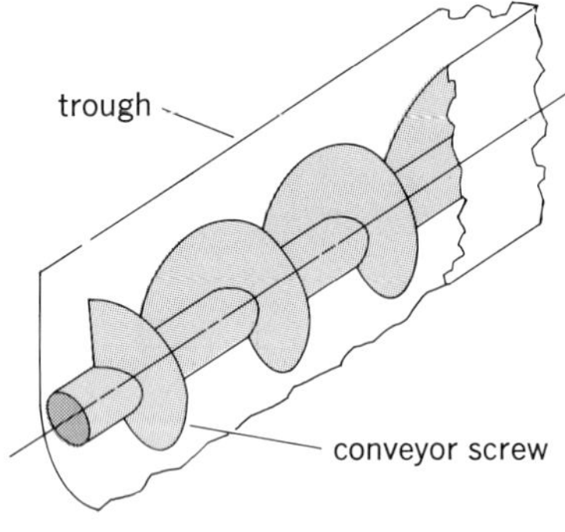

Fig. 3. Screw conveyor, used to move bulk objects.

$$T = F_t R = QR \tan \lambda \tag{2}$$

When a practical screw is considered, friction becomes important, and the torque on a screw with square threads is given by Eq. (3), and the efficiency of a screw with square threads by Eq. (4), where μ is the coefficient of friction.

$$T = QR \frac{\tan \lambda + \mu}{1 - \mu \tan \lambda} \tag{3}$$

$$\eta = \tan \lambda \frac{1 - \mu \tan \lambda}{\tan \lambda + \mu} \tag{4}$$

If a screw jack (Fig. 2) is used to raise a heavy object, such as a house or machine, it is normally desirable for the screw to be self-locking, that is, for the screw not to rotate and lower the load when the torque is removed with the load remaining on the jack. *See* SCREW JACK.

For a screw to be self-locking, the coefficient of friction must be greater than the tangent of the lead angle: $\mu > \tan \lambda$. *See* SCREW FASTENER.

The screw is by far the most useful form of inclined plane or wedge and finds application in the bolts and nuts used to fasten parts together; in lead and feed screws used to advance cutting tools or parts in machine tools; in screw jacks used to lift such objects as automobiles, houses, and heavy machinery; in screw-type conveyors (Fig. 3) used to move bulk materials; and in propellers for airplanes and ships. *See* SCREW THREADS; SIMPLE MACHINE.

[RICHARD M. PHELAN]

Screw fastener

A threaded machine part used to join parts of a machine or structure. Screw fasteners are used when a connection that can be disassembled and reconnected and that must resist tension and shear is required. A nut and bolt is a common screw fastener. Bolt material is chosen to have an extended stress-strain characteristic free from a pronounced yield point. Nut material is chosen for slight plastic flow.

The nut is tightened on the bolt to produce a preload tension in the bolt, as illustrated. This preload has several advantageous effects. It places the bolt under sufficient tension so that during vibration the relative stress change is slight with consequent improved fatigue resistance and locking of the nut. Preloading also increases the friction between bearing surfaces of the joined members so that shear loads are carried by the friction forces rather than by the bolt.

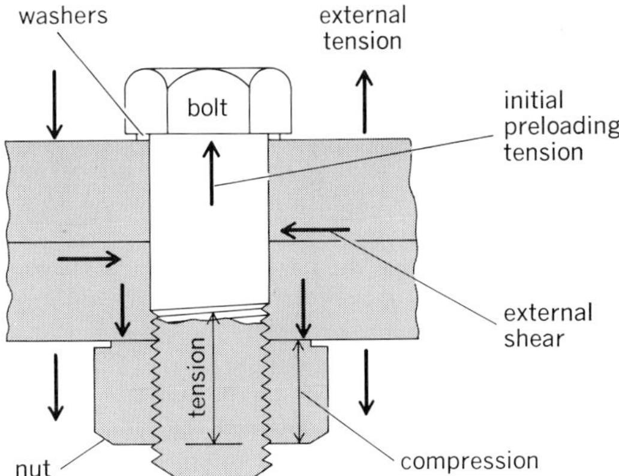

Forces on a bolt fastener under preloading.

The tightened nut is under compression; the bolt is under tension. The deformation that accompanies these forces tends to place the entire preload on the thread nearest the bearing surface. Thus, concentration of loading is counteracted if the nut is slightly plastic so as to set under load. This yielding may be achieved by choice of material or by special shape. If a soft gasket is used in the assembly, preloading is less effective; the bolt may carry the full tension and shear loads. *See* BOLT; JOINT (STRUCTURES); NUT; SCREW THREADS.

[FRANK H. ROCKETT]

Bibliography: American Society for Testing and Materials, *Standards on Fasteners*, 1978; Institute for Power System, *Handbook of Industrial Fasteners*, 1979; F. M. Keely (ed.), *Miscellaneous Fasteners*, 1977.

Screw jack

A mechanism for lifting and supporting loads, usually of large size. A screw jack mechanism consists of a thrust collar and a nut which rides on a bolt; the threads between the nut and bolt normally have a square shape. A standard form of screw jack has a heavy metal base with a central threaded hole into which fits a bolt capable of rotation under a collar thrusting against the load. Screw jacks are also used for positioning mechanical parts on machine tools in order to carry out manufacturing processes. These jacks are small in size and have standard V threads. Load and stress calculations may be performed on screw jacks in the same manner as in power screws and screw fastenings. *See* SCREW; SIMPLE MACHINE.

[JAMES J. RYAN]

Bibliography: N. Chironis, *Mechanisms, Linkages and Mechanical Controls*, 1965; V. M. Faires and R. M. Keown, *Mechanism*, 5th ed., 1960, reprint 1980.

Screw threads

Continuous helical ribs on a cylindrical shank. Screw threads are used principally for fastening, adjusting, and transmitting power. To perform these specific functions, various thread forms have been developed. A thread on the outside of a cylinder or cone is an external (male) thread; a thread on the inside of a member is an internal (female) thread (Fig. 1). *See* SCREW.

Types of thread. A thread may be either right-hand or left-hand. A right-hand thread on an external member advances into an internal thread when turned clockwise; a left-hand thread advances when turned counterclockwise. If a single helical groove is cut or formed on a cylinder, it is called a single-thread screw. Should the helix angle be increased sufficiently for a second thread to be cut between the grooves of the first thread, a double thread will be formed on the screw. Double, triple, and even quadruple threads are used whenever a rapid advance is desired, as on fountain pens and valves. The helices on a double thread start 180° apart, while those on a triple begin 120° apart. A multiple thread produces a rapid advance without resort to a coarse thread.

Pitch and major diameter designate a thread. Lead is the distance advanced parallel to the axis when the screw is turned one revolution. For a single thread, lead is equal to the pitch; for a double thread, lead is twice the pitch. For a straight thread, the pitch diameter is the diameter of an imaginary coaxial cylinder that would cut the thread forms at a height where the width of the thread and groove would be equal.

Thread forms have been developed to satisfy particular requirements (Fig. 2). Those employed on fasteners and couplings and those used for making adjustments are generally of the modified 60°V type (unified and ISO metric). Where strength is required for the transmission of power and motion, a thread having faces that are more nearly perpendicular to the axis is preferred, such as the modified square and the acme. These threads, with their strong thread sections, transmit power nearly parallel to the axis of the screw. The sharp V, formerly found on set screws, is now rarely used because of the difficulty of cutting sharp roots and crests in quantity production.

Thread forms. The unified thread form accompanied by the unified thread standards has in the past fulfilled the necessary requirements for interchangeability of threaded products between the United States, Great Britain, and Canada. This modified thread form represented two compromises between British Whitworth and United States National thread forms. The British accepted the 60° thread angle of the American thread, and the Americans accepted a rounded root and crest similar to the Whitworth form. The United States National unified form for external threads has a rounded root and may have either a rounded or flat

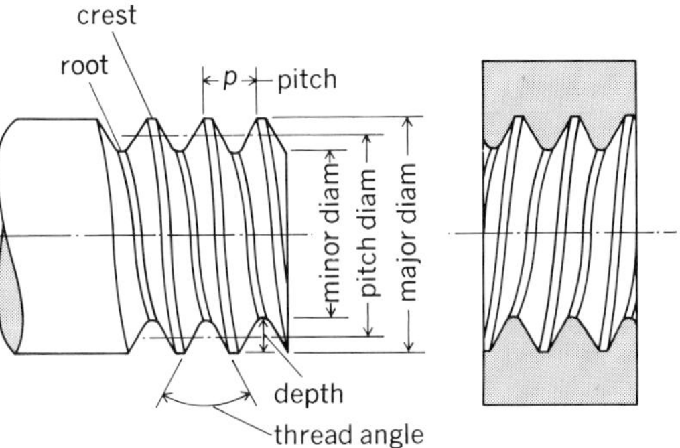

Fig. 1. Screw thread nomenclature.

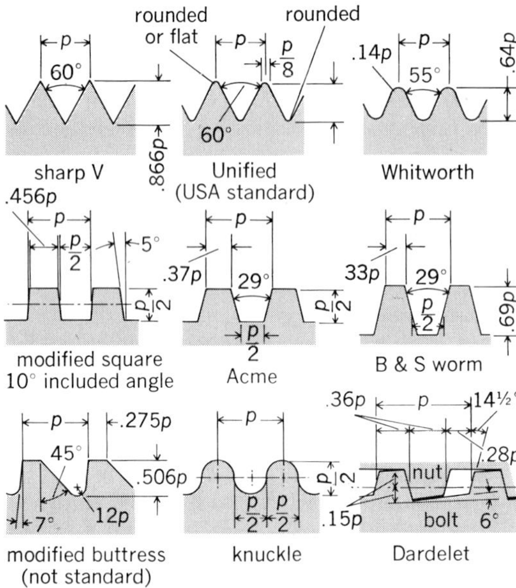

Fig. 2. Screw threads, for particular needs. (*From W. J. Luzadder, Fundamentals of Engineering, 8th ed., Prentice-Hall, 1977*)

crest. The unified thread contours provide advantages over both the earlier forms. The design provided for greater fatigue strength and permitted longer wear of cutting tools. The unified thread is a general-purpose thread that is particularly suited for fasteners.

Threads on bolts and screws may be formed by rolling, cutting, or extruding and rolling. Cut threads frequently produce a tighter fit than do rolled threads because of irregularities and burrs. To facilitate assembly, cut threads should have a finished point. Rolled threads ordinarily can be manufactured at less cost, but the range of sizes of bolts that can be produced is limited.

The Whitworth 55° thread, similar in many respects to the United States standard form, is the standard V form used in Great Britain. The unified form, identical with the United States standard form, is also standard in Great Britain.

A modified square thread, with an angle of 10° between the sides, is sometimes used for jacks and vises where strength is needed for the transmission of power in the direction of the thread axis.

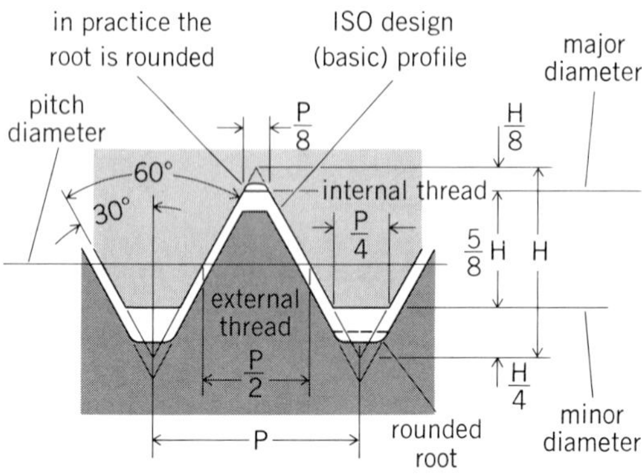

Fig. 3. ISO metric internal and external thread design profiles. (*From W. J. Luzadder, Fundamentals of Engineering Drawing, 8th ed., Prentice-Hall, 1981*)

When originally formulated, the acme thread was intended as an alternate to the square thread and other thread forms that were being used to produce transverse motion on machines. Used for a variety of purposes, the acme thread is easier to produce than the square thread, and its design permits the use of a split nut to provide on-off engagement. A modification of the acme is the 29° stub acme. Because the basic depth of the thread is reduced to 0.30 of the pitch, the thread section is strong and well suited to power applications where space limitations make a shallow thread desirable. A still further modification of the acme is the 60° stub having a 60° angle between the sides of the thread. This thread has a basic depth of 0.433 of the pitch. The Brown and Sharp worm thread, a modified form of the acme thread, is used for transmitting power to a worm wheel.

The buttress or breech-block thread is designed for transmitting an exceptionally high pressure in one direction only. In its original form the pressure flank was perpendicular to the thread axis, and the trailing flank sloped at 45°. To simplify the cutting of the face, modern practice is to give the pressure flank a 7° slope. This form of thread is applicable for assembled tubular members because of the small radial thrust. Buttress threads are used on breech mechanisms of large guns and for airplane propeller hubs.

The knuckle rolled thread is found on sheet-metal shells of lamp bases, fuse plugs, and in electric sockets. The thread form, consisting of circular segments forming crests and roots tangent to each other, has been standardized. The thread is usually rolled, but it may be molded or cast. The Dardelet thread is self-locking in assembly.

SI (ISO) metric thread system. The ISO (International Organization for Standardization) thread form (Fig. 3) has essentially the same basic profile as the unified thread and, when the changeover to the metric system has been completed in the United States, it will supplant the unified form. However, before the metric thread system is fully accepted in the United States, minor modifications of thread characteristics may be made to satisfy industry.

The principal differences between the SI metric thread system and the unified inch system are in the basic sizes, the magnitude and application of allowances and tolerances, and the method for designating and specifying threads.

The basic designation (Fig. 4) consists of the letter M followed by the nominal size (basic major diameter in millimeters) and the pitch. The nominal size and pitch are separated by the sign × as in M8 × 1 (ISO designation) and M20 × 1.5.

For coarse series threads, the indication of the pitch may be omitted. The complete designation of an ISO metric thread includes the basic designation followed by the specification for tolerance class. The tolerance class designation includes the symbol for the pitch-diameter tolerance followed by the symbol for the crest-diameter tolerance. Each of these symbols consists of a number indicating the tolerance grade followed by a letter indicating the tolerance position. A dash separates the tolerance-class designation from the basic designation. When the pitch-diameter and crest-diameter tolerance symbols are identical, the symbol is given only once.

The ISO standard provides a number of tolerance grades that reflect various magnitudes. Basically, three metric tolerance grades are recommended: 4, 6, and 8. Grade 6 is commonly used for general-purpose threads. This is the closest ISO grade to the unified 2A and 2B fits. The numbers of the tolerance grades reflect the size of the tolerance. Tolerances below grade 6 are smaller and are specified for fine-quality requirements. Tolerances above grade 6 are larger and are recommended for coarse-quality and long lengths of engagement.

Tolerance positions establish the maximum material limits of the pitch and crest diameters for both internal and external threads. The series of tolerance-position symbols to reflect varying amounts of allowance are:

For external threads	*For internal threads*
e = large allowance	G = small allowance
g = small allowance	H = no allowance
h = no allowance	

A desired fit between mating threads may be specified by giving the internal thread tolerance-class designation followed by the external thread tolerance class. The two designations are separated by a slash, as in M6 × 1 – 6H/6g.

ISO metric thread series are selections of diameter-pitch combinations that are distinguished from each other by the different pitch applied to a specific diameter (M10 × 1.5 coarse series or M10 × 1.25 fine series). Depending upon design requirements, the thread for a particular application might be selected from the coarse-thread series, fine-thread series, or one of the constant-pitch series.

Unified and standard thread series. Unified and American National screw thread standards for screws, bolts, nuts, and other threaded parts consist of six series of threads and a selection of special threads that cover nonstandard combinations of diameter and pitch. Each series of standard threads has a specific number of threads per inch for a particular diameter. In general, the unified screw threads are limited to three series: coarse (UNC), fine (UNF), and extrafine (UNEF). A $\frac{1}{4}$-in.-diameter thread in the UNC series has 20 threads/in., while in the UNF series it has 28.

In the unified and American National standards, the coarse thread series (UNC and NC) is recommended for general industrial use on threaded machine parts and for screws and bolts. Because a coarse thread has fewer threads per inch than a fine thread of the same diameter, the helix angle is greater, and the thread travels farther in one turn. A coarse thread is preferred when rapidity and ease of assembly are desired, and where strength and clamping power are not of prime importance.

The fine thread series (UNF and NF) is recommended for general use in the automotive and aircraft fields for threads subject to strong vibration. In general, a fine thread should be used only where close adjustment, extra strength, or increased resistance to loosening is a factor.

The extrafine thread series (UNEF and NEF) is used principally in aeronautical structures where an extremely shallow thread is needed for thin-walled material and where a maximum practicable number of threads is required for a given length.

The 8-thread series (8N) with 8 threads/in. for all diameters is used on bolts for high-pressure flanges, cylinder-head studs, and other fasteners against pressure, where it is required that an initial tension be set up so that the joint will not open when steam or other pressure is applied. This series has come into general use for many types of engineering work. It is sometimes used as a substitute for the coarse thread series for diameters greater than 1 in.

The 12-thread series (12UN or 12N) is a uniform pitch series that is widely used in industry for thin nuts on shafts and sleeves. This series is considered to be a continuation of the fine-thread series for diameters greater than $1\frac{1}{2}$ in.

The 16-thread series (16UN or 16N) is a uniform pitch series for applications that require a very fine thread, as in threaded adjusting collars and bearing retaining nuts. This series is used as a continuation of the extrafine thread series for diameters greater than 2 in.

The manufacturing tolerance and allowance permitted distinguishes one class of thread from another under the unified thread system. Because allowance is an intentional difference between correlated size dimensions of mating threads, and tolerance is the difference between limits of size, a thread class may be considered to control the looseness or tightness between mating threaded parts. The American National standard for unified and National screw threads provides classes 1A, 2A, and 3A for external threads, classes 1B, 2B,

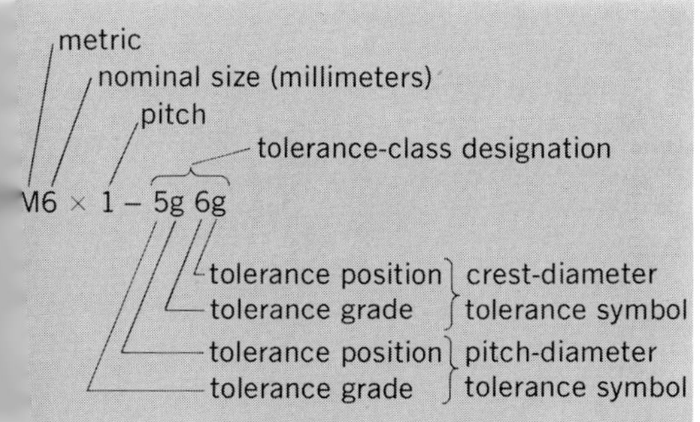

Fig. 4. Metric thread system designations.

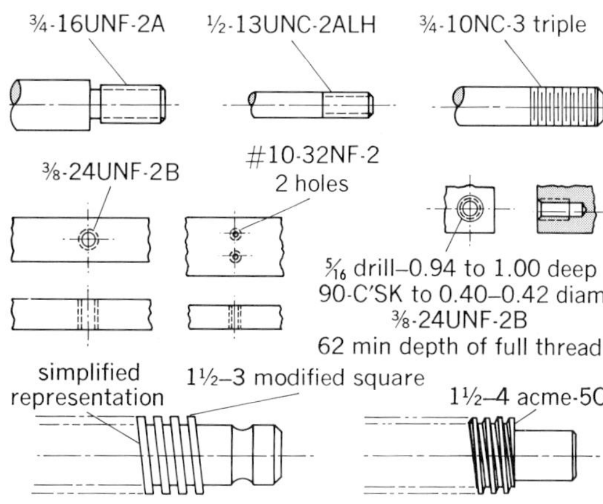

Fig. 5. Thread identification by standardized symbols. (*From W. J. Luzadder, Fundamentals of Engineering Drawing, 8th ed., Prentice-Hall, 1981*)

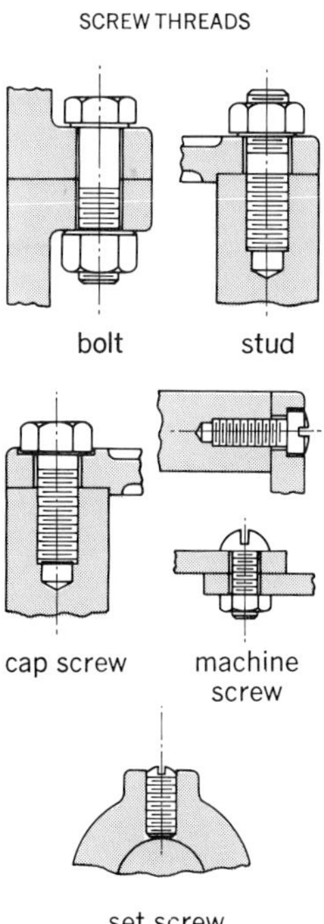

Fig. 6. Common types of fasteners. (*From T. E. French and C. J. Vierck, Engineering Drawing and Graphic Technology, 12th ed., McGraw-Hill, 1978*)

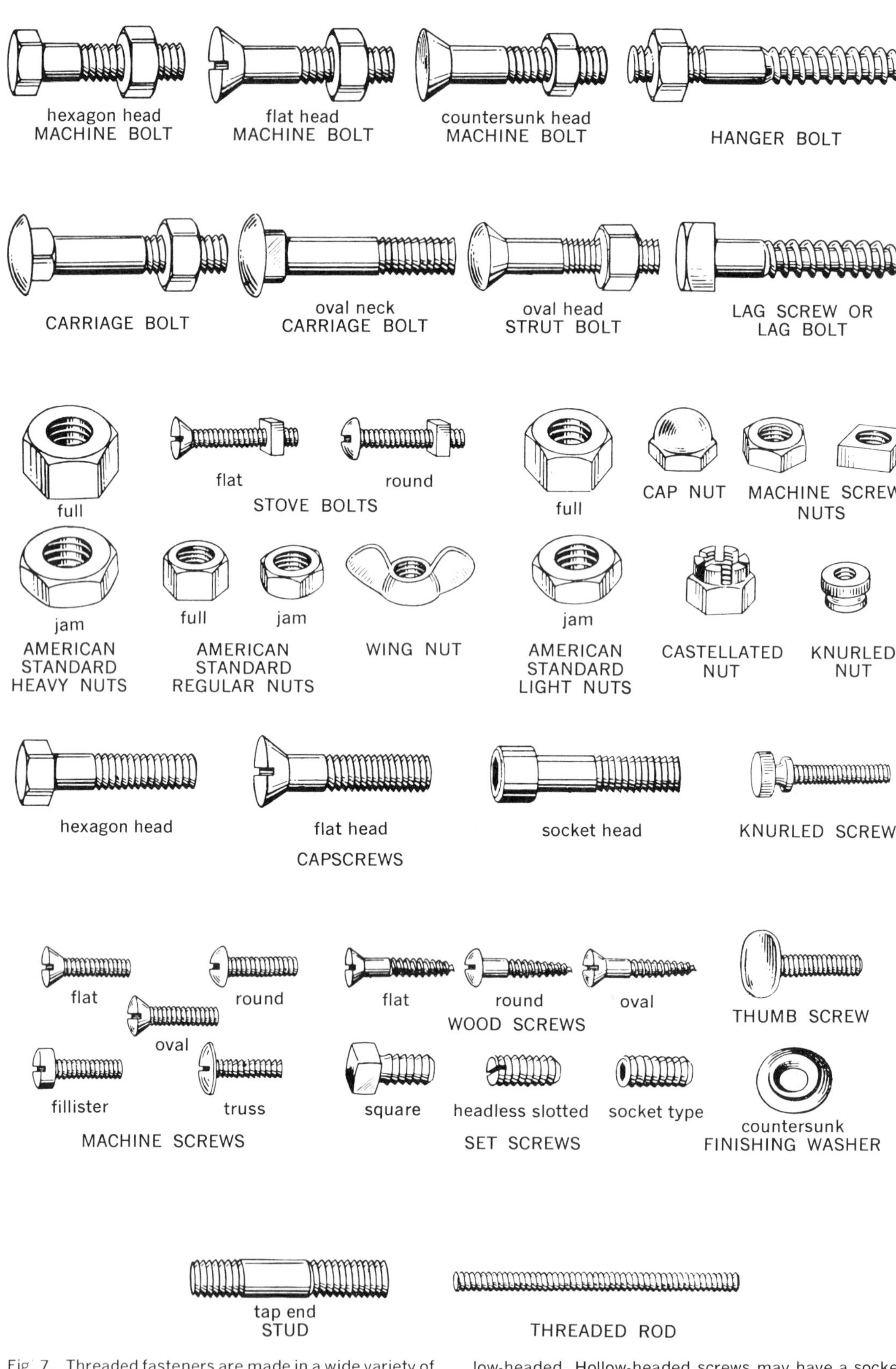

Fig. 7. Threaded fasteners are made in a wide variety of types and used for various purposes. Some of the most common are shown here. Screws may be slotted or hollow-headed. Hollow-headed screws may have a socket or Phillips head or any of a number of special contours for patented turning systems. *(Reynolds Metals Co.)*

and 3B for internal threads, and classes 2 and 3 for both internal and external threads. Classes 1A and 1B are for ordnance and other special uses. Classes 2A and 2B are the recognized standards for the bulk of screw-thread work and for the normal production of threads on screws, bolts, and nuts. Classes 3A and 3B are for applications where smaller tolerances than those afforded by class 2A and 2B are justified and where closeness of fit between mating threads is important. Class 3A has no allowance.

The screw-thread fit needed for a specific appli

cation can be obtained by combining suitable classes of thread. For example, a class 2A external thread might be used with a class 3B internal thread to meet particular requirements.

Designation of unified, square, and acme threads. Threads are designated by standardized notes (Fig. 5). Under the unified system, threads are specified by giving in order the nominal diameter, number of threads per inch, the initial letters such as UNC or UNF that identify the series, and class of thread (1A, 2A, and 3A or 1B, 2B, and 3B). Threads are considered to be right-hand and single unless otherwise noted; therefore the specification for a left-hand thread must have the letters LH included. To indicate multiplicity, the word DOUBLE, TRIPLE, and so on must follow the class symbol.

Thread fastener uses. Most threaded fasteners are a threaded cylindrical rod with some form of head on one end. Various types of threaded fasteners are available; some, such as bolts, cap screws, and machine screws, have been standardized; others are special designs. The use of removable threaded fasteners is necessary on machines and structures for holding together those parts that must be frequently disassembled and reassembled. Because standard threaded fasteners are mass produced at relatively low cost and are uniform and interchangeable, they are used whenever possible.

Fasteners are identified by names that are descriptive of either their form or application: set screw, shoulder screw, self-tapping screw, thumb screw, and eyebolt. Of the many forms available, five types meet most requirements for threaded fasteners and are used for the bulk of production work: bolt, stud, cap screw, machine screw, and set screw (Fig. 6). Bolts and screws can be obtained with varied heads and points (Fig. 7).

A bolt is generally used for drawing two parts together. Having a threaded end and an integral head formed in manufacture, it is passed through aligned clearance holes in the two parts and a nut is applied.

A stud is a rod threaded on both ends. Studs are used for parts that must be removed frequently and for applications where bolts would be impractical. They are first screwed more or less permanently into one part before the removable member with corresponding clearance holes is placed into position. Nuts are used on the projecting ends to draw the parts together.

Cap screws (plated or unplated) are widely used in machine tools and for assembling parts in automotive and aeronautical equipment. They are available in four standard heads: hexagon, flat, round, and fillister. Cap screws may be steel, brass, bronze, or aluminum alloy. Flathead and roundhead screws have slotted heads (Fig. 8). Fillister-head screws have either slotted or socket heads. When mating parts are assembled, the cap screws pass through clear holes in one member and screw into threaded holes in the other. The head, an integral part of the fastener along with the thread, holds the parts together. In the automotive industry, the hexagon-head cap screw in combination with a nut is often used as an automotive hexagon-head bolt.

Machine screws, which are similar to cap screws and fulfill the same purpose, are employed principally in the numbered diameter sizes on small work having thin sections or with a machine nut to function as a small bolt.

Set screws made of hardened steel are used to hold parts in a position relative to one another. Normally, their purpose is to prevent rotary motion between two parts, such as would occur in the case of a rotating pulley and shaft combination. Set screws can be purchased with any one of six types of points in combination with any style of head (Figs. 7 and 9). In the application of set screws, a flat surface may be formed on a shaft to provide a seat for a flat point. A cone point fits into a conical spot.

Wood screws, lag screws, and hanger bolts are used in wood (Fig. 7). Wood screws may have either slotted or recessed heads.

Self-tapping screws (Fig. 10) have a specially hardened thread that makes it possible for the screws to form their own internal thread in sheet metal and soft materials when driven into a hole that has been drilled, punched, or punched and reamed. The use of self-tapping screws eliminates costly tapping operations and saves time in assembling parts.

Metric fasteners (ISO). Existing standards for the production of threaded metric fasteners in the United States are in general agreement with ISO standards, with minor differences. When these differences have been resolved, final standards bearing the approval of both ISO and ANSI (American National Standards Institute) will become available for worldwide use. These approved metric fastener standards will assure international interchangeability. Being more restrictive in the choices of sizes and styles than the standards of the past, the new metric standards will make possible reduced inventories and permit bulk pricing advantages.

Metric threaded fasteners are available in only one series of diameter-pitch combinations that range from M1.6 × 0.35 to M100 × 6. The thread pitches are between those of the coarse thread and fine thread series of the present unified (inch) threads. Except for metric socket screws, all threaded fastener products have one tolerance grade, 6g for bolts, machine screws, and so forth, and 6H for nuts. The 6g/6H threads closely match the unified 2A/2B threads for fasteners, in that there is an allowance on the external thread and no allowance on the internal thread.

The standard lengths for short fasteners are 8, 12, 14, 16, and 20 mm. Fastener lengths from 20 to 100 mm increase in length by 5-mm incre-

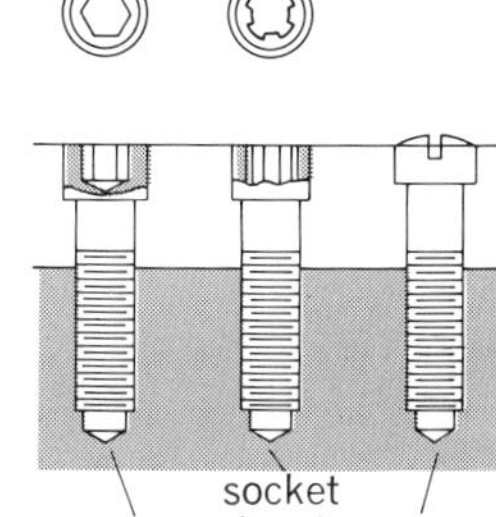

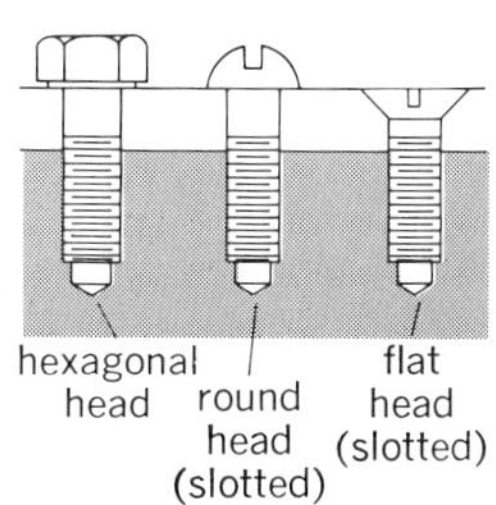

Fig. 8. Cap screws. *(From W. J. Luzadder, Fundamentals of Engineering Drawing, 8th ed., Prentice-Hall, 1981)*

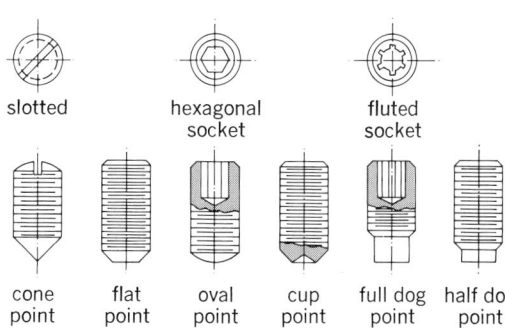

slotted
hexagonal socket
fluted socket
cone point
flat point
oval point
cup point
full dog point
half dog point

Fig. 9. Set screws. *(From W. J. Luzadder, Fundamentals of Engineering Drawing, 8th ed., Prentice-Hall, 1981)*

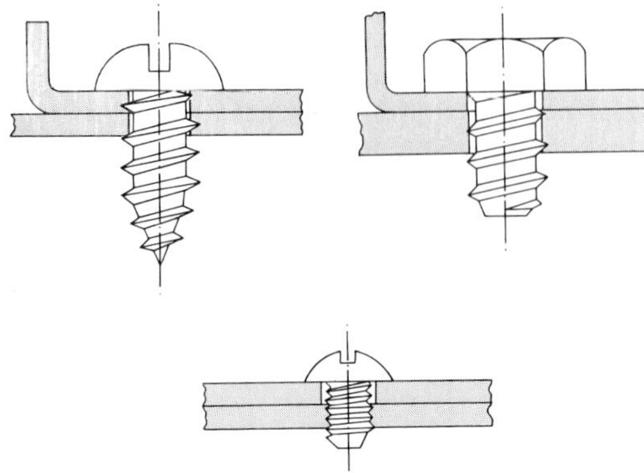

Fig. 10. Three applications of self-tapping screws.

ments. Metric fasteners are specified by giving in sequence the nominal size in millimeters, thread pitch, nominal length, product name (including type of head), material, and protective finish, if needed. Examples are M24 × 3 × 50 STAINLESS STEEL HEX BOLT; M14 × 2 × 80 HEX CAP SCREW CADMIUM PLATED; M8 × 1.25 × 14 HEX SOCKET FLAT POINT SET SCREW.

Pipe thread. Standard (Briggs) taper pipe threads (Fig. 11) are cut (on a cone) to a taper of $\frac{1}{16}$ in./in. to ensure a tight joint. Although a normal connection employs a taper external and a taper internal thread, an American National standard straight pipe thread having the same pitch angle and depth of thread as the corresponding taper pipe thread is used for pressure-tight joints for couplings, pressure-tight joints for fuel- and oil-line fittings, and for loose-fitting and free-fitting mechanical joints. Assemblies made with taper external threads and straight internal threads are frequently preferred to assemblies employing all taper threads; the assumption is made that relatively soft or ductile metals will adjust to the taper external pipe thread. A modified pipe thread, the United States standard Dryseal pipe thread (taper and internal straight), is used for pressure-tight connections that are to be assembled without lubricant or sealer. Except for a difference in the truncation at the roots and crests, the general form and dimensions of this thread are the same as those of the standard taper pipe thread. The principal uses for the Dryseal thread are in refrigerant pipes, automotive and aircraft fuel-line fittings, and for gas and chemical shells (ordnance). For railing joints, where a rigid mechanical thread joint is needed, external and internal taper threads are used. Basically, the external thread is the same as the standard taper pipe thread, except that the length of the thread is shortened at the end to permit use of the larger-diameter portion of the thread. [WARREN J. LUZADDER]

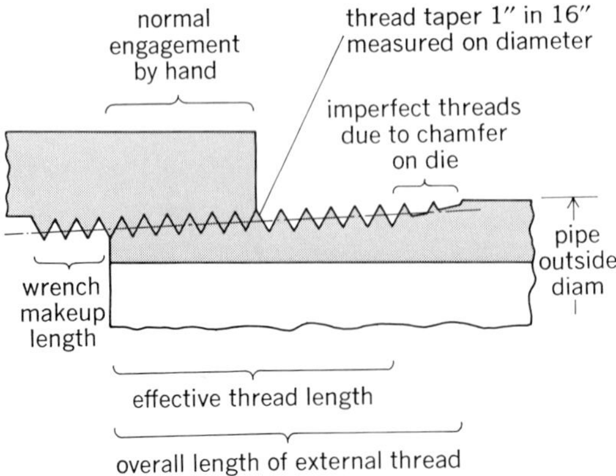

Fig. 11. Standard (Briggs) taper pipe thread.

Bibliography: American National Standards Institute, *Acme Screw Threads*, ANSI B1.5 and B1.8, 1952; American National Standards Institute, *Unified and American Screw Threads Standards*, B1.1, 1960; T. E. French and C. J. Vierck, *Engineering Drawing and Graphic Technology*, 12th ed., 1978; Industrial Fasteners Institute, *Metric Fastener Standards*, 1976; International Organization for Standardization, Publications ISO68, ISO261, and ISO965; W. J. Luzadder, *Fundamentals of Engineering Drawing*, 8th ed., 1981.

Sea wall

A structure at the water's edge to resist encroachment of the sea and to retain the natural soil or deposited fill behind the wall. A usual type of sea wall has a sloped or inclined face toward the sea to deflect and dissipate the wave forces. Sometimes it resembles a breakwater. Timber, steel, and concrete are used in a variety of forms. Failures sometimes occur because of scouring and eventual undermining at the outshore face or because of wave overtopping during highest tides. *See* COASTAL ENGINEERING. [EDWARD J. QUIRIN]

Seismic exploration for oil and gas

Exploration seismology is a geophysical method of determining subsurface geologic structure with man-made elastic waves. Energy, usually in the form of an impulse from an explosion, is introduced into the ground at or near the surface. Spreading out from the source, the energy encounters discontinuities in the physical properties of the rocks and is partially reflected back to the surface, detected, and recorded. Discontinuities of exploration interest are interfaces between different types of rock. The time required for the reflected energy to return indicates the depth of a reflector. Plotting this time for each detected signal while moving along the surface, a geophysicist assembles a picture of the rock layers below from shallow depths to the deepest interface from which returning energy is measurable. Getting such a picture is the goal of the exploration geophysicist. There is no way of detecting oil or gas directly for, with negligible exceptions, the ways that oil-, gas-, and water-saturated rocks respond tc elastic waves are indistinguishable. Even with thi limitation reflection seismology is the primary too of petroleum geophysics. It is the only methoc which combines penetration many thousand fee into the earth with sufficient resolution to delineat changes in the rock layering. *See* PETROLEU PROSPECTING.

Reflection seismology. Figure 1 illustrates th basis of the reflection seismic method. Anothe technique, refraction seismology, preceded th reflection method but is rarely used today. common seismic source for reflection seismolog is dynamite at the bottom of a water-filled ho drilled through the weathered layer. Th weathered layer, soil and rock that are poorly co solidated, absorbs the energy of explosions and is avoided whenever possible. An explosion generates in solids both compressional waves in which the motion is in the direction of travel, and shear waves in which the motion is perpendicular to the wave travel direction. In fluids, only compressional waves (that is, sound) travel. Exploration seismology uses compressional waves alone. These are reflected by interfaces separating rocks with different impedances. (Impedance is the product of rock density and compressional wave velocity.) The amplitude of a reflected wave is proportional to the difference of the impedances across an interface.

Reflection coefficients of 0.01 to 0.1 are common for rocks but higher and lower values occur, the unity coefficient of the air-water interface being particularly important in seismic exploration of water-covered areas.

Seismic signals. Detectors for land exploration, called geophones, measure the vertical component of particle velocity, that is, the time derivative of the vertical displacement of the surface. For marine exploration, hydrophones replace geophones. Hydrophones respond to the excess pressure generated by sound waves in the water. In land or marine work, signals from the detectors are in the range of a few microvolts to several millivolts and must be amplified by circuitry that has gains up to several million. Frequencies of interest for petroleum exploration are 10–100 Hz, and the usual signals are pulses that include a fair portion of the frequency range. Twenty-four detector stations are standard but more are used.

In Fig. 1*b* the recorded signals appear as pulses with positive and negative excursions corresponding to positive and negative excursions of the surface from its undisturbed position. The farther a geophone is from the source, the longer the time required for a reflection from an interface to reach it. For reflectors that are plane interfaces, a plot of arrival time against distance along the surface from the source is hyperbolic. If the reflector is a horizontal plane, the time-distance equation takes the simple form shown below, where t is the arrival

$$t^2 = \left(\frac{2D}{v}\right)^2 + \left(\frac{X}{v}\right)^2$$

time of the reflection, X is the source to detector separation, v is the average velocity with which the wave travels to and from the reflector, and D is the corresponding depth. This is the fundamental equation of exploration seismology. It is only an approximation, for there is no consideration of the bending of rays at each interface in accordance with Snell's law of refraction, or of the fact that a reflector is sometimes tilted. Nonetheless, examination of the equation can expose much of important mathematics of the method.

Geophone and hydrophone arrays. Seismic surveys are conducted along lines called profiles. In terms of Fig. 1*a*, another hole would be drilled beyond the most distant geophone, and the geophones would be moved so that the picture would look identical but moved along the profile line. The process would be repeated again and again until the entire line had been surveyed. Source and geophone configurations may be either symmetric (split-spread) or, as shown here, nonsymmetric (single-ended spread). Whatever the arrangement,

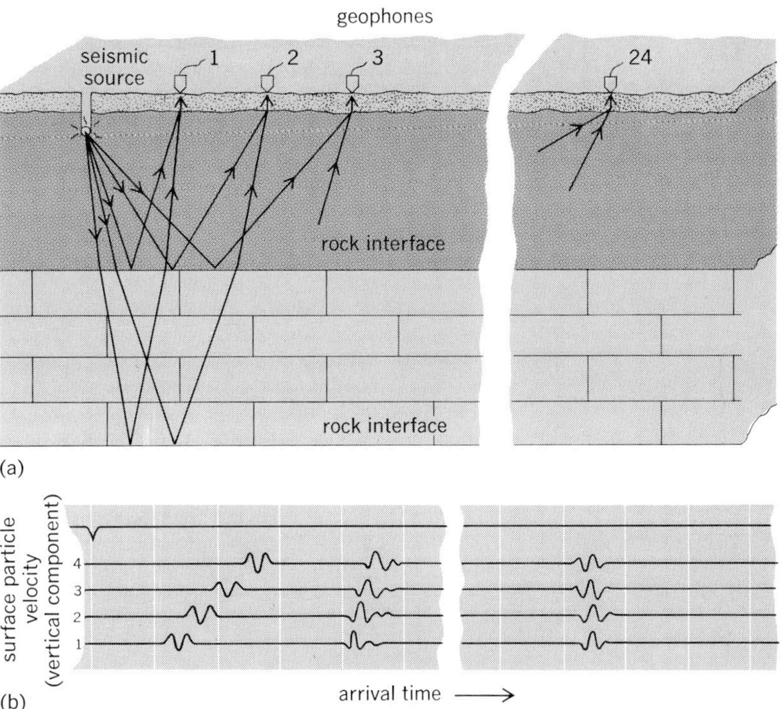

Fig. 1. The basic steps in the reflection seismic method. (*a*) Seismic waves from the source reflect off the rock interfaces and return to the detectors. (*b*) The signals are recorded for each detector.

reflection points along an interface are discrete but without major gaps. This is referred to as continuous subsurface coverage. Nearest and most distant reflection points are separated by half the distance between corresponding geophones, just as a mirror required for full-length viewing needs to be only half the viewer's height. To get an areal picture of each reflector, a geophysicist uses data from a grid of profile lines.

Figure 1 represents World War II technology. Advances have included surface sources of impactor or vibrator types for land exploration and of gases exploded in elastic sleeves or air bubbles released into water for marine work, the replacement of single geophones by linear or areal arrays of many geophones, and particularly the recording of seismic data on magnetic tape. Surface sources have become popular on land. They permit rapid, relatively inexpensive seismic surveying, although generally with reduced resolution. Replacement of high explosives by other sources in offshore surveys has resulted in operational economies and reduces hazards to marine life and the men involved. A single ship, moving at 4–6 knots (2–3 m/s), carries sources activated at 10–30-sec intervals and pulls a cable containing 24 hydrophone stations spaced out to 10,000 ft (3000 m) from the ship. Shore-based electronic equipment locates the ship and defines the profile lines to be followed, but the location methods fail far from shore. Combinations of several methods, including satellite fixes, are expected soon to locate geophysical ships anywhere in the world with an accuracy of, perhaps, 600 ft (180 m). A typical marine seismic crew averages 50 mi (80 km) of a line a day.

Each of the 24 stations in a marine cable has as its output the summed signals of 10–40 hydrophones arranged as a linear array. Arrays discriminate against energy traveling along the detector cable and favor reflections that reach the cable from below. On land, arrays are formed by placing geophones along a line or spreading them out in an areal pattern. The purpose is the same: to emphasize the reflections which arrive very nearly vertically from below. Arrays of up to 150 geophones have been used, but arrays of 12–36 geophones are more common. They are designed to reduce noise generated by the source and traveling mostly in the weathered layer. The strongest noise of this type is called ground roll by the geophysicist. In addition to limiting ground roll, geophone arrays also reduce random noise by a process of statistical addition of the noise.

Recording and processing data. Recording seismic data in analog form on magnetic tape revolutionized exploration seismology in the 1950s; the advent of recording in digital form has produced a second revolution. Array outputs are sampled every 2 or 4 milliseconds and the samples are recorded as digital numbers for a period of 6–10 sec. One complete record is called a seismogram; a seismogram consists of 24 traces, one from each array output. Processing of the magnetic tapes occurs in centers that include one or more large digital computers and the associated programmers, inputters, and data interpreters.

The first processing step is to compensate for the drop in signal amplitude with record time. Spreading of the wave from the source and attenuation within the earth result in a rapid decrease of signal strength that must be reversed during processing. Next is compensation for the varying thickness of weathered material underlying different geophone stations and for variations in source depths. Then a normal moveout (NMO) correction is applied. This correction flattens the hyperbolic time-distance plots by subtracting from the arrival time the quadratic X term of the equation. Since the velocity v changes from reflector to reflector, usually increasing as depth increases, the NMO correction changes with time along the seismogram. After application of a NMO correction, each seismic trace is the one that would have been produced by a coincident source and geophone. The effect of detectors at different distances from the source has been eliminated. Accompanying the NMO process are procedures for determining v from the curvature of the uncorrected time-distance hyperbolas.

Concurrently with NMO correction or preceding it, the data are filtered to emphasize reflections and discriminate against noise. The usual filters pass that band of frequencies over which reflections are strongest and eliminate frequencies for which noise predominates. Other types of filters find use also. These include filters that remove from marine seismograms signals generated by energy bouncing between the surface and bottom of the water. Another important filtering process is common depth point stacking (Fig. 2). Three to twenty-four signals from the same reflection points but with different source to detector separations are, after NMO correction, summed into a single signal. Common depth point stacking both reduces noise and tends to cancel multiple reflections. These reflections, energy that has been reflected more than once from an interface, interfere with the reflections used for subsurface mapping.

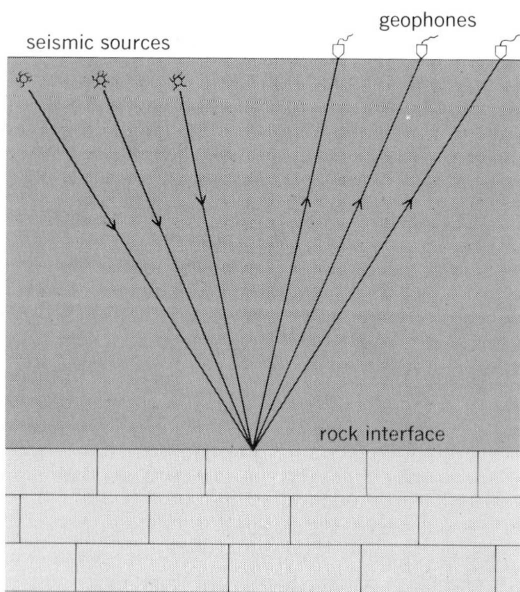

Fig. 2. Seismic signal filtering by the method of common depth point stacking.

The product of processing is a record section. Each trace is displayed on photographic paper as a constant-width strip with the signal amplitude shown in variable intensity, variable area, wiggly trace form, or some combination of these. Laying the traces side by side without a gap between the strips generates a two-dimensional representation of a three-dimensional surface. The abscissa is distance along the profile line from an arbitrary starting point; the ordinate, record time. Amplitude is the dependent variable. To replace record time by depth below the surface requires that depth be computed from a knowledge of v and reflection times. Fully processed record sections present to the geophysicist the configuration of rock layers down to all depths that drills can reach, albeit with resolution no better than 150 or 200 ft (45 or 60 m). Lines many miles long may be viewed.

Although a record section resembles the subsurface, it is really a picture drawn by elastic waves and may differ from the subsurface in detail. For example, rock layers that end abruptly often seem to continue, as diffracted energy extends the reflections. Reflections from interfaces that are not horizontal appear at positions on a record section that do not correspond to their locations in the earth and must be returned to their true position by a process called migration. When all these difficulties and others not listed here have been admitted, the position of reflection seismology as the most valuable tool of petroleum geophysicists is indisputable. No other method approaches the combination of penetration and resolution that is required for a choice of drilling sites.

[FRANKLIN K. LEVIN]

Bibliography: A. Ben-Menahem and S. Singh, *Seismic Waves and Sources*, 1980; M. B. Dobrin, *Introduction to Geophysical Prospecting*, 3d ed., 1976; F. S. Grant and G. F. West, *Interpretation*

Theory in Applied Geophysics, 1965; R. McQuillin et al., *An Introduction to Seismic Interpretation*, 1979; D. S. Parasnis, *Principles of Applied Geophysics*, 3d ed., 1979.

Septic tank

A single-story settling tank in which settled sludge is in immediate contact with sewage flowing through the tank while solids are being decomposed by anaerobic bacterial action. Such tanks have limited use in municipal treatment, but are the primary resource for the treatment of sewage from individual residences. There are probably well over 4,000,000 septic tanks in use in home disposal systems in the United States. Septic tanks are also used by isolated schools and institutions and at small industrial plants.

Home disposal units. Septic tanks have a capacity of approximately 1 day's flow. Since sludge is collected in the same unit, additional capacity is provided for sludge. One formula for sludge storage that has been used is $Q = 17 + 7.5y$, where Q is the volume of sludge and scum in gallons per capita per year, and y is the number of years of service without cleaning. About one-half of a 500-gal (1900-liter) tank is occupied by sludge in 5 years in an ordinary household installation. The majority of states require a minimum capacity of 500 gal (1900 liters) in a single tank. Some states require a second compartment of 300-gal (1100-liters) capacity. Single- and double-compartment tanks are shown in Figs. 1 and 2. Such units are buried in the ground and are not serviced until the system gives trouble because of clogging or overflow. Commercial scavenger companies are available in most areas. A tank truck equipped with pumps is brought to the premises, and the tank content is pumped out and taken to a sewer manhole or a treatment plant for disposal. In rural areas the sludge may be buried in an isolated place.

Municipal and institutional units. These are designed to hold 12–24 hours' flow, with additional sludge capacity provided. Provision is made for sludge withdrawal about once a year. Desirable features of design are (1) watertight and corrosion-resistant material (concrete and well-protected metal have been used); (2) a vented tank; (3) manhole openings in the roof of the tank to permit inspection; (4) baffles at the inlet and the outlet to a depth below the probable scum line, usually 18–24 in. below the water surface; (5) sludge draw-off lines—although seldom used, they should be designed so that they can be rodded or unplugged by some positive mechanism; (6) hoppers or sloped bottoms so that digested sludge can be withdrawn as required; (7) provision for safe handling of septic tank effluent by disposal underground or by chlorination before discharge to a stream, or both.

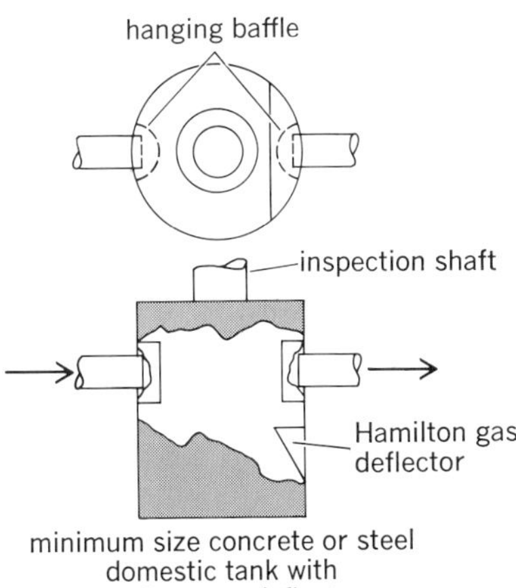

Fig. 1. Circular household septic tank. (*From H. E. Babbitt and E. R. Baumann, Sewerage and Sewage Treatment, 8th ed., copyright © 1958 by John Wiley and Sons, Inc.; used with permission*)

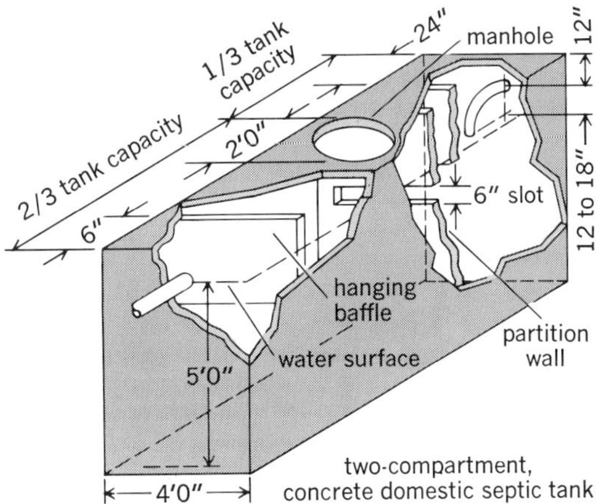

Fig. 2. Two-compartment rectangular household septic tank. 1″ = 2.54 cm; 1′ = 30.48 cm. (*From H. E. Babbitt and E. R. Baumann, Sewerage and Sewage Treatment, 8th ed., copyright © 1958 by John Wiley and Sons, Inc.; used with permission*)

Tank efficiency. Septic tank effluent is dangerous and odorous. It will contain pathogenic bacteria and sewage solids. Particles of sludge and scum are trapped in the flow and will cause nuisance at the point of discharge unless properly handled. Efficiency in removal of solids is less than that for plain sedimentation. While 60% suspended solids removal is used theoretically, it is seldom obtained in practice. Improvement is noted when tanks are built with two compartments. Shallow tanks give somewhat better results than very deep tanks. *See* SEWAGE TREATMENT.

[WILLIAM T. INGRAM]

Bibliography: U.S. Public Health Service, *Manual of Septic Tank Practice*, 1960; P. Warshall, *Septic Tank Practices*, 1979.

Series circuit

An electric circuit in which the principal circuit elements have their terminals joined in sequence so that a common current flows through all the elements.

The circuit may consist of any number of passive and active elements, such as resistors, inductors, capacitors, electron tubes, and transistors.

The algebraic sum of the voltage drops across each of the circuit elements of the series circuit must equal the algebraic sum of the applied voltages. This rule is known as Kirchhoff's second law and is of fundamental importance in electric circuit theory. *See* KIRCHHOFF'S LAWS OF ELECTRIC CIRCUITS.

When time-varying voltages and currents are involved, it is necessary to employ differential or integral equations to express the summation of voltages about a series circuit. If the voltages and currents vary sinusoidally with time, functions of a complex variable are used in place of the calculus. *See* ALTERNATING-CURRENT CIRCUIT THEORY; CIRCUIT (ELECTRICITY); DIRECT-CURRENT CIRCUIT THEORY.

[ROBERT LEE RAMEY]

Sewage

A combination of (1) liquid wastes conducted away from residences, institutions, and business buildings and (2) the liquid wastes from industrial establishments with (3) such surface, ground, and storm water as may find its way or be admitted into the sewers. Category 1 is known as sanitary or domestic sewage; 2 is usually referred to as industrial waste; 3 is known as storm sewage.

Relation to water consumption. Sewage is the waste water reaching the sewer after use; hence it is related in quantity and in flow fluctuation to water use. The quantity of sewage is generally less than the water consumption since some portion of water used for fire fighting, lawn irrigation, street washing, industrial processing, and leakage does not reach the sewer. These losses are compensated for partly by the addition of water from private wells, groundwater infiltration, and illegal connections from roof drains. Water consumption increases with size of community served and many other community characteristics. Characteristics of each city must be studied and analyzed for specific information. As a general average estimate, communities with population under 1000 use about 60 gal per capita per day (gcd) (230 liters per capita per day), while communities of 100,000 use about 140 gcd (530 liters per capita per day). In a study of large cities of the United States, the median consumption was 154 gcd (583 liters per capita per day) and the median population was 658,000

Rates of sewage flow from various sources*

Character of district	Gal per capita per day	Gal per acre per day	Source of sewage	Gal per capita per day
Domestic			Trailer courts	50†
Average	100		Motels	53†
High-cost dwellings	150	7,500	State prisons	
Medium-cost dwellings	100	8,000	Maximum	280‡
Low-cost dwellings	80	16,000	Average	176
Commercial			Minimum	104
Hotels, stores, and office buildings		60,000	Mental hospitals	
			Maximum	216†
Markets, warehouses, wholesale districts		15,000	Average	123
			Minimum	38
Industrial			Grade school	4.4§
Light industry		14,000	High school	3.9§

SOURCE: From H. E. Babbitt and E. R. Baumann, *Sewerage and Sewage Treatment*, 8th ed., copyright 1958 by John Wiley & Sons, Inc.

*A gal = 3.785 liters, 1 gal per acre = 9.35 liters per hectare.

‡From J. C. Frederick, *Public Works*, p. 112, April, 1957.

§Average of 4.4 gal per day per pupil between 7:30 A.M. and 5:30 P.M. The average for the high school is spread over more hours per day. From C. H. Coberly, *Public Works*, p. 143, May, 1957.

(Fig. 1). An accepted unit flow for domestic sewage as shown in the table is 100 gcd (380 liters per capita per day).

Infiltration of groundwater should be held to a minimum. It may be expected to be equal to or less than 30,000 gal/(day)(mile) [70 m^3/(day)(km)] of sewer including house connections. Much depends on the quality of sewer construction. Water may enter through poorly made joints and, in quantity, through poorly constructed, leaky manholes and illegal and abandoned sewers. Sewers in wet ground with a high water table will have more infiltration. Sewers under pressure may have infiltration or leakage to the surrounding ground. The danger of groundwater pollution from leaky sewers should be avoided.

Fluctuations in sewage flow are related to water use characteristics but tend to dampen out since there is a time lag from the time of use to appearance in the sewer mains and trunks (Fig. 2). Hourly, daily, and seasonal fluctuations affect design of sewers, pumping stations, and treatment plants.

Daily and seasonal variations depend largely on community characteristics. Weekend flows may be lower than weekday. Industrial operations of seasonal nature influence the seasonal average. The seasonal average and annual average are about equal in May and June. The seasonal average is about 124% in late summer and may drop to about 87% at the end of winter. Peak flows may reach 200% of average at the treatment plant and may be more than 300% of average in the laterals. Laterals are designed for 400 gcd (1500 liters per capita per day) and mains and trunks for 250 gcd (950 liters per capita per day).

Design periods. These are dependent on the proposed sewer construction. Lateral sewers may be designed for ultimate flow of the area to be sewered. Mains may be designed for periods of 10–40 years. Trunk sewers may be planned for long periods with provision made in design for parallel or separate routings of trunks of smaller size to be constructed as the need arises. Economics, available funds, and engineering judgment affect selection of the design periods. Appurtenances may have a different life, since replacement of mechanical equipment will be necessary. A span of 20–25 years is often selected and a timetable of additions

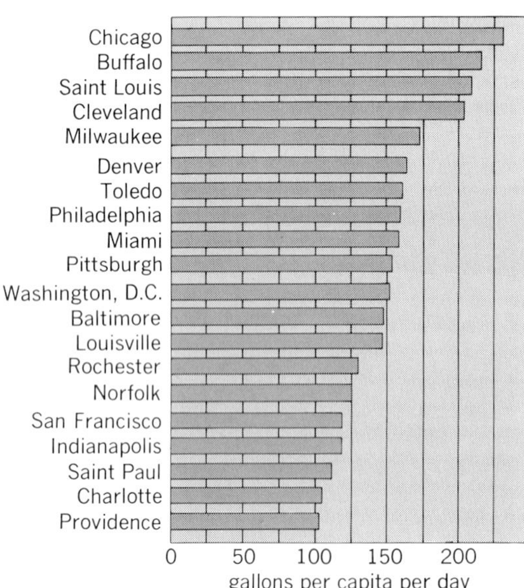

Fig. 1. Estimated water use in 20 major cities of the United States. (*Research Division, New York University College of Engineering*)

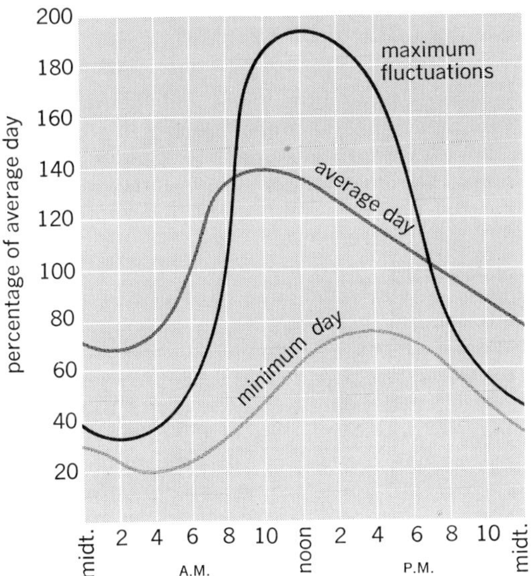

Fig. 2. Typical hourly variations in sewage flow. (*From H. E. Babbitt and E. R. Baumann, Sewerage and Sewage Treatment, 8th ed., copyright © 1958 by John Wiley and Sons, Inc.; used with permission*)

during that period is then scheduled in the overall improvement plan.

Storm sewage. Storm sewage is liquid flowing into sewers during or following a period of rainfall and resulting therefrom. An estimate of the quantity of storm sewage is necessary in sewage design.

Estimating quantity of storm sewage requires a knowledge of intensity and duration of storms, distance water travels before reaching sewer, permeability and slope characteristics of the surface over which water flows to sewer inlet, and shape and amount of area to be drained to inlet. These general considerations are included in the equation $Q=CAIR$ expressing the runoff from a watershed having no retention or storage of water. Q is expressed in cubic feet per second (cfs), A is area, I is the relative imperviousness of the surface expressed as a decimal, and R is the rate of rainfall in inches per hour. C is a coefficient permitting the expression of the factors in convenient units; in the above units it may be taken as 1 so that the equation becomes $Q=AIR$.

Time of concentration is a combination of the theoretical time required for a drop to run from the most distant point to the inlet and time from sewer inlet to point of concentration. The inlet time may range from 3 min for a steep slope on an impervious area to 20 min on a city block. The time of flow is assumed to be the velocity in the full flowing sewer divided by the length of sewer from inlet to point of concentration. Flood crest and storage time while the sewer is filling are usually neglected, the effect being that of assuming a larger rate of flow and so providing a safety factor in design.

Values of I, the runoff coefficient, range from 0.01 in wooded areas to 0.95 on roofed surfaces. A common value used in residential areas with considerable land in lawn, garden, and shrubbery is 0.30–0.40. In built-up areas, values of 0.70–0.90 may be used.

Rainfall intensity values are selected on the basis of frequency and duration of storms. In some sewer design it is necessary to select a value for the expected occurrence of maximum runoff. This is done by using one of the several formulas which will allow a prediction of R for 5, 10, or 15 years. The element of calculated risk is combined with engineering judgment in deciding which R to choose. For lesser structures in residential areas a 5-year frequency may be used with reasonable safety. Where failure would endanger property, the 10-, 15-, or 25-year frequency of occurrence provides a more conservative design basis; 50-year frequency may be selected where flooding could cause lasting damage and disrupt facilities. In such instances cost-benefit studies may be made to guide the selection of a suitable frequency.

Pumping sewage. Not all sewage will flow by gravity without unnecessary expense in circuitous routing or deep excavation; therefore, pumping stations may be advantageous. Pumping stations may be required in the basements of large buildings. Pumping stations are provided with two or more pumps of sufficient capacity so that with one unit out of service the remaining unit or units will pump the maximum flow. Motive power is required from at least two sources, usually electric motors and auxiliary fuel-fed motor drive. Care must be taken to have motive power above flood level and protected from the elements. Screening is usually required ahead of pumping stations, unless the pumps themselves are self-cleaning. Many states require that pumping units be installed in a dry well and that sewage be confined to a separate wet well. Buildings above ground should fit the surroundings. Small pumping stations are often one unit and made fully automatic so that minimum attendance is required. Safety measures must be considered. Centrifugal pumps are used almost exclusively in larger stations. Air ejector units may be installed in smaller stations.

Examination of sewage. Sewage is actually water with a small amount of impurity in it. Examination of sewage is required to know the effects of these impurities. Various tests are used to aid in determining the characteristics, composition, and condition of sewage. These include physical examination, solids determination, tests for determining the oxygen requirement of organic matter, chemical and bacteriological tests, and examination under the microscope.

Physical tests for turbidity, odor, color, and temperature are made. Normal fresh sewage is gray and somewhat opaque, has little odor, and has a temperature slightly higher than the water supply. Decomposition of organic matter darkens the sewage, and odors are characteristic of stale or septic sewage.

Tests for residue or solid matter provide an indication of the types of solids, the strength of the sewage, and the physical state of the solids. Total solids determinations measure both suspended and dissolved solids. A sample of the sewage is filtered. The suspended solids can be determined by drying the material recovered on the filter. The dissolved solids can be determined by evaporation of the filtered portion. Heating the solids residue until organic matter gasifies separates volatile solids from fixed solids or inorganic ash. Loss on ignition represents the volatile or organic fraction and is a good measure of sewage strength.

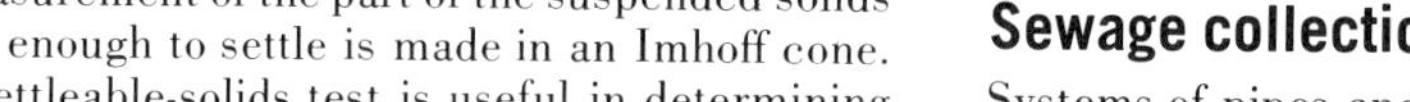

Measurement of the part of the suspended solids heavy enough to settle is made in an Imhoff cone. The settleable-solids test is useful in determining the sludge-producing characteristics of sewage.

Tests for organic matter are made principally to determine the oxygen requirement of sewage. These tests include the biochemical oxygen demand test (BOD), the chemical oxygen demand test, the oxygen consumed test, and the relative stability test. Organisms in sewage require oxygen for growth, and the BOD measures the amount of dissolved oxygen required for decomposition of organic solids for a measured time at a constant temperature. The standard measurement is made for 5 days at 20°C and is a good measure of sewage strength. Since the BOD measurement includes both biological and chemical oxygen requirement, another test, the chemical oxygen demand, is sometimes used to measure the chemical oxygen requirement. Sewage is heated in the presence of an oxidizing agent such as potassium dichromate. The oxygen requirement is that of chemical digestion since all organisms have been killed. This test is increasing in use. The oxygen-consumed test uses potassium dichromate as the oxidizing agent. The result offers some index of the readily oxidizable carbonaceous material. The relative stability test indicates when the oxygen present in plant effluent or polluted water is exhausted. The data express as a percentage the approximate amount of oxygen available in water in relation to the amount required for complete stability. The test is a color test using methylene blue. Reducing agents, precipitation of color, concentration of dye, amount of dissolved oxygen in the sample, and other factors affect the reliability of this test, and it is considered generally as a rough or screening test of the condition of plant effluent. Tests for nitrogen include those for free ammonia, albuminoid ammonia, organic nitrogen, nitrites, and nitrates. The latter are indications of oxidation change and stabilization and are used in checking condition of plant effluent.

Bacteriological tests are made primarily to determine the presence of organisms of the coliform group. The organisms exist in the intestines of warm-blooded animals and are used as an index of the presence of fecal material. The coliform test on chlorinated effluents determines efficiency of chlorination. Occasionally other bacteriological determinations are made to determine the presence of organisms of the *Salmonella* group or dysentery group in polluted water and sewage.

Microscopic tests are not normally made on raw sewage. They are used as part of plant operator control in treatment processes. Examinations for the presence of algae, protozoa, bacteria, fungi, rotifers, and worms are made when necessary.

[WILLIAM T. INGRAM]

Bibliography: APHA-AWWA-WPCA Joint Committee, *Standard Methods for the Examination of Water, Sewage, and Industrial Wastes*, 15th ed., 1981; R. E. Bartlett, *Public Health Engineering: Sewerage*, 2d ed., 1979; G. M. Fair, J. C. Geyer, and D. A. Okun, *Elements of Water Supply and Waste-water Disposal*, 2d ed., 1971; M. Parker, *Wastewater Systems Engineering*, 1975; E. W. Steel and T. McGhee, *Water Supply and Sewerage*, 5th ed., 1979; D. W. Sundstrom and H. E. Klei, *Wastewater Treatment*, 1979.

Sewage collection systems

Systems of pipes and conduits, together with control devices, pumping stations, and appurtenances used for the collection and transfer of waste waters. Waste waters may include sanitary sewage (that is, the liquid wastes of residential premises), industrial wastes, storm waters, and groundwater infiltration. *See* SEWAGE.

Sewer pipe. Sewer pipe is manufactured from a number of materials, such as vitrified clay, concrete, asbestos cement, corrugated iron, cast iron, and steel. Plastics, bituminous wood fiber, and wood stave pipe also are used. All sewer pipes may be surcharged or filled at some time and must be capable of withstanding some hydraulic pressure. Pressure lines connected to pump discharge, or lines carried under roads or streams, that is, inverted siphons, are designed for the particular condition, and materials which withstand pressure are selected.

SEWAGE COLLECTION SYSTEMS

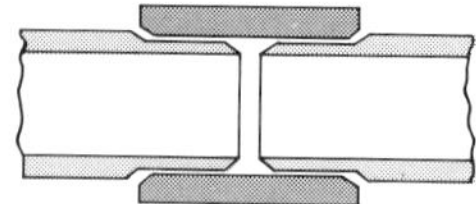

Fig. 1. Joint for asbestos-cement pipe. (*From H. E. Babbitt and E. R. Baumann, Sewerage and Sewage Treatment, 8th ed., copyright © 1958 by John Wiley and Sons, Inc.; used with permission*)

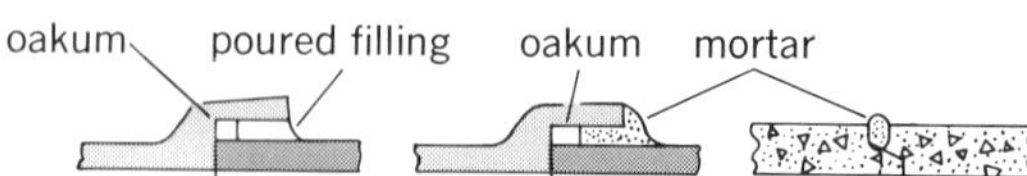

Fig. 2. Types of joints for bell-and-spigot pipes. (*From H. E. Babbitt and E. R. Baumann, Sewerage and Sewage Treatment, 8th ed., copyright © 1958 by John Wiley and Sons, Inc.; used with permission*)

Sewers are laid underground and must be able to withstand external pressures such as those caused by soil, water, and the extra weight of traffic. Large-diameter reinforced concrete sewers or conduits may be constructed in place.

Pipes are made in various lengths, and joints are of many types. An asbestos-cement pipe joint is shown in Fig. 1. The bell-and-spigot and ring types are common (Fig. 2). Jointing materials include cement, mortar, asphalt, plastics, sulfur, rubber, rings, and plastic gasket.

Clay sewer pipe is manufactured in strengths and with dimensions as provided in American Society for Testing and Materials (ASTM) Specifi-

Table 1. Crushing strength requirements for standard-strength clay sewer pipe

	Average strength, min, lb per linear ft†	
Size, in.*	Three-edge-bearing method	Sand-bearing method
4	1000	1500
6	1100	1650
8	1300	1950
10	1400	2100
12	1500	2250
15	1750	2625
18	2000	3000
21	2200	3300
24	2400	3600
27	2750	4125
30	3200	4800
33	3500	5250
36	3900	5850

*1 in. = 2.54 cm. †1 lb/ft = 1.49 kg/m.

Table 2. Crushing strength requirements for extra-strength clay sewer pipe

Nominal size, in.*	Average strength, min, lb per linear ft†	
	Three-edge-bearing method	Sand-bearing method
6	2000	3000
8	2000	3000
10	2000	3000
12	2250	3375
15	2750	4125
18	3300	4950
21	3850	5775
24	4400	6600
27	4700	7050
30	5000	7500
33	5500	8250
36	6000	9000

*1 in. = 2.54 cm. †1 lb/ft = 1.49 kg/m.

cations C13, C261, C200, and C268. Safe supporting strengths are specified in Tables 1 and 2.

Concrete sewer pipe is manufactured as provided in ASTM Specifications C14, C76, and C362. Asbestos-cement pipe is required to meet Federal Specifications SS-P-331A for gravity flow and ASTM Specification C296 where pressure pipe is required. *See* STRENGTH OF MATERIALS.

Precast concrete pipe is made by spinning or tamping semidry concrete against a mold. The centrifugal process was introduced in the United States in the 1920s. Reinforced concrete pipe is also made by the centrifugal method. Reinforcing is placed as a single, double, or elliptical cage (Fig. 3). It may also be a steel plate cylinder.

Special conditions of soil, construction, geology, pressure, and capacity may require that sewers be built in place. Concrete construction of circular, elliptical, egg shape, and horseshoe shape requires special forming of wood or metal (Fig. 4). Machine tamping and vibration equipment are used to produce a dense, impervious concrete shell. Concrete in pipe is designed to withstand 3000–4000 psi (20–27 MPa). Required thickness of shell and amount of reinforcing are designated in specifications.

Both internal and external corrosion of sewer pipe can occur. Metals are attacked unless corrosion-resistant alloys are added. Organic materials in sewage are attacked by bacteria and other microorganisms and form acids which attack concrete and metals. Protective coatings of asphalt compounds and plastics and epoxy resins have been applied to interior pipe surface. Exterior corrosion may occur due to electrolysis, bimetallic corrosion, and electrochemical and bacterial attack. *See* CORROSION.

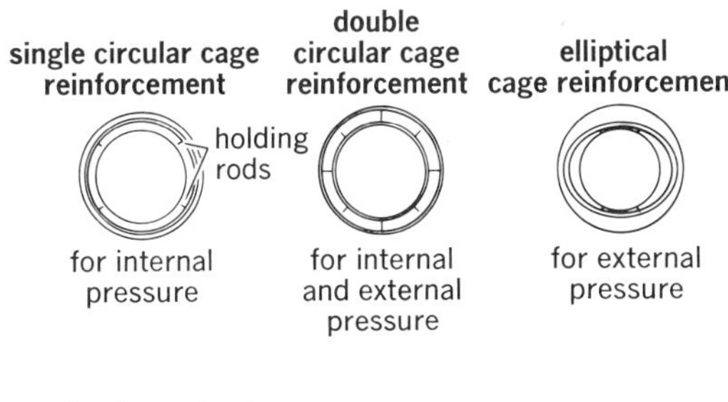

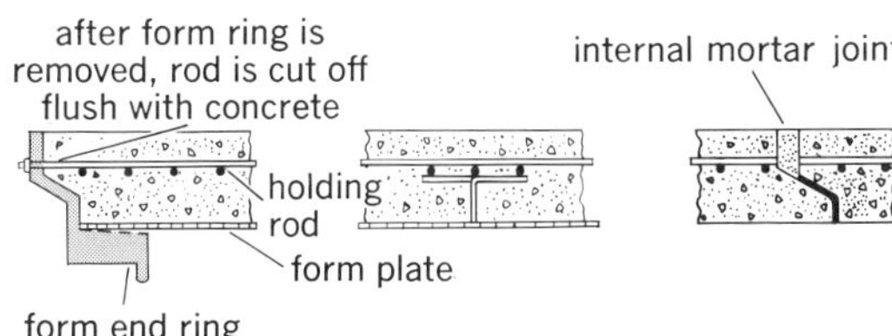

Fig. 3. Methods of reinforcing concrete pipe. (*From H. E. Babbitt and E. R. Baumann, Sewerage and Sewage Treatment, 8th ed., copyright © 1958 by John Wiley and Sons, Inc.; used with permission*)

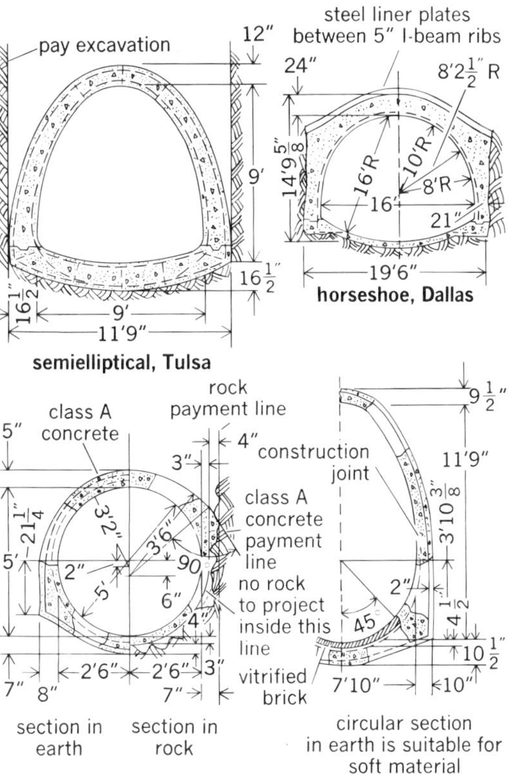

Fig. 4. Sewer shapes. 1" = 2.54 cm; 1' = 30.48 cm. (*From E. W. Steel, Water Supply and Sewerage, 3d ed., McGraw-Hill, 1953*)

Sewage flow. This must be known or estimated before sewer design can be completed. Storm water from roof drainage or ground and street surfaces is excluded.

Velocity. Velocity of flow must be maintained at a rate sufficient to carry contained sewage solids. For sanitary sewage the minimum velocity is 2 ft/sec (0.6 m/sec); for storm water the minimum velocity is 2.5 ft/sec (0.75 m/sec). In a circular pipe the velocity is the same whether flowing at half or full capacity; at one-fourth depth it is about two-thirds that when flowing full. The maximum flow

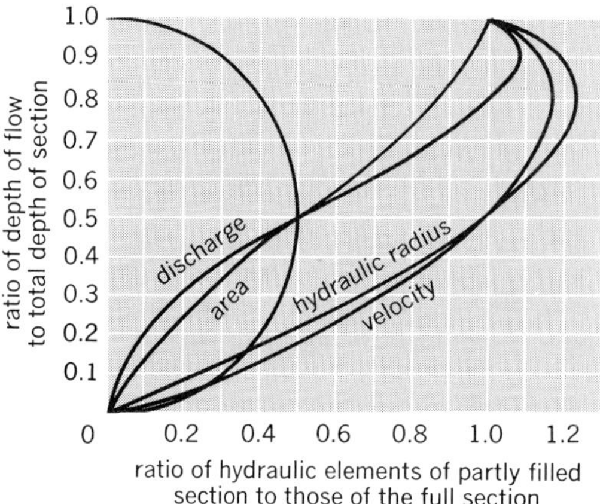

Fig. 5. Graphic representation of the sewage flow in a circular pipe. (*From E. W. Steel, Water Supply and Sewerage, 3d ed., McGraw-Hill, 1953*)

occurs at about 0.9 depth, and the maximum velocity occurs at about 0.8 depth (Fig. 5). A pipe carrying design flow when flowing full provides about 8% safety factor in handling peak flows. Tables, charts, and monographs have been constructed to aid in the solution of pipe flow problems (Fig. 6).

Transitions. Changes in direction, grade, elevation, and pipe size, and the union of two or more sewers into a common trunk are carried out at manholes and junction points. Inlet and exit losses introduced in such structures must be included in computation of the hydraulic grade line. At the end of a sewer pipe or at a marked change in slope, the flow of liquid is no longer uniform. Velocity may be decreasing, as in the case of a reduced slope, or increasing with increase in slope or free fall at end of line. These points of change are called transition points. Design of sewers must provide for transition changes in hydraulic gradient.

Critical depth. Energy is minimal for a given cross section and discharge. Under critical flow condition (Fig. 7) there is only one depth, velocity, and hydraulic gradient that will satisfy the energy equation of flow in open channels, expressed in terms of depth. This relationship may be solved graphically or by nomograph when channels are irregular in shape or when the hydraulic radius and area are not conveniently expressed in terms of depth.

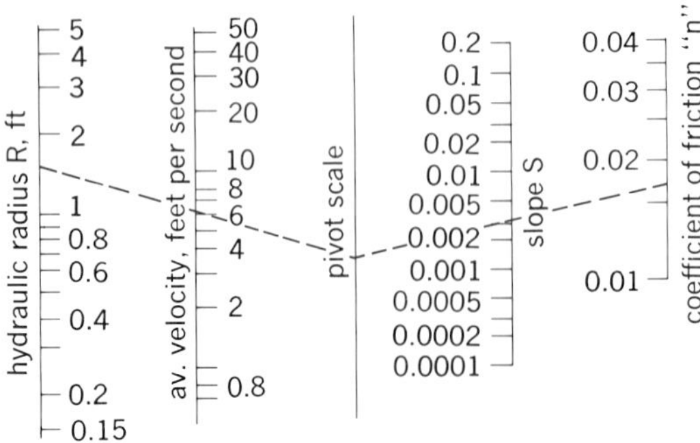

Fig. 6. A nomogram which is based upon the Manning formula. 1 ft = 0.3048 m; 1 ft/sec = 0.3048 m/sec. (*From E. W. Steel, Water Supply and Sewerage, 3d ed., McGraw-Hill, 1953*)

At conditions of flow other than critical, alternate stages at which the same flow may take place are possible. These stages are the normal depth or upper-stage, and lower depth or lower-stage flow. Variations in velocity occur at dropdown curves, backwater curves, and hydraulic jumps. The dropdown curve occurs near the free outlet end of a sewer where the velocity of flow is increasing. The backwater curve is caused by an obstruction in the sewer such as a dam or by discharge into a body of water whose surface is above the normal level of flow in the sewer. A backwater curve will also result from flattened grade. The velocity of flow decreases in the backwater curve. Dropdown and backwater curves are illustrated in Fig. 8.

The hydraulic jump occurs when the depth of flow changes abruptly from the lower stage to the upper stage (Fig. 9). For most sewer shapes the length of transition for the several conditions mentioned is calculated by trial and error, since the formulas are complex.

Sewer appurtenances. Sewer appurtenances are manholes, inlets, regulators, inverted siphons, and outfalls. A manhole is an opening at the ground or street surface permitting a man to enter

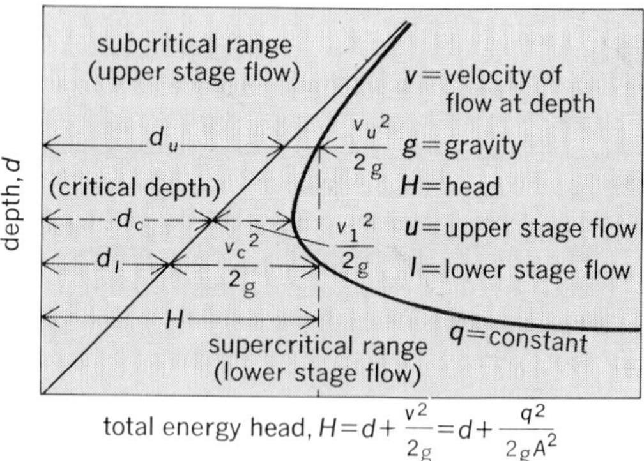

Fig. 7. Depth of flow as related to the total energy head in a sewage channel. (*American Society of Civil Engineers, Manual of Design and Construction of Sanitary and Storm Sewers, no. 37, 1959*)

to make examinations and repairs (Fig. 10). Manholes are spaced along the sewer at 300–500-ft (90–150 m) intervals or placed at any other point where access is necessary.

Inlets. An inlet provides an opening for storm water and is usually placed at the curb line of the street. The structure below the inlet is called a catch basin (Fig. 11). A short length of sewer connects the inlet or catch basin to a manhole on the sewer.

The capacity of inlets is determined by complicated analytical methods. The length of opening is a function of the amount of storm water, gutter shape, and depth. *See* HIGHWAY ENGINEERING; HYDRAULICS.

Regulators. A regulator is a device designed to divert sewage flow from one sewer to another channel. It is most frequently used in combined sewer systems to regulate the amount of storm water permitted to flow to a sewage treatment plant. Types of regulators include side flow weirs, leaping weirs, and float-operated gates and valves. Practice varies, but regulators are usually designed to divert twice or three times the dry weather flow.

Inverted siphons. An inverted siphon is a length of sewer set below hydraulic grade line so that it flows under pressure between an inlet chamber and an outlet chamber (Fig. 12). It is usually constructed with two or more pipes of smaller diameter to regulate velocity at minimum and maximum flow and to avoid clogging and reduce cleaning.

Outfall. The outfall is a structure designed to admit treated or untreated sewage to a receiving body of water (Fig. 13). It may be submerged or partially submerged. Storm waters may flow entirely above water level. The simplest form is a headwall supporting a pipe end equipped in some instances with a tide gate to prevent backflow under high water conditions. Treatment plant outfalls frequently are carried into deep water and the piping ends in a series of outlets, rosette-shaped or in fingers, to provide dispersion of the effluent flow. Openings may be turned upward to prevent clogging by shifting bottom deposits.

Sewer system design. A comprehensive study of the community or area to be sewered is made to estimate the flow that must be handled by the system at some future period of time, such as 10 or 20 years, or the period of ultimate development. Decisions must be made concerning the type of system to be constructed, separate or combined.

Investigations. Major factors affecting the quantity and flow patterns of sanitary sewage are (1) population and population increase; (2) population density and density change; (3) water use, demand, and consumption; (4) industrial requirements, present and future; (5) commercial requirements, present and future; (6) expansion of service geographically; (7) groundwater geology of area; and (8) topography of area.

A preliminary layout of sewers and tentative selection of sizes, grades, and location follows. Physical characteristics of the areas to be sewered are determined with attention given to elevation and plan location of roads, streets, water courses, buildings, basements, underground utilities, and geology. The preliminary report includes a plan of the proposed system together with an estimate of its cost. After the preliminary design is accepted final design begins. Field work is required to establish location and elevation of all existing structures that may affect the design. Borings are made, if necessary, to determine soil and foundation characteristics along the route of sewers and system structures. Final plans and profiles, specifications, and cost estimates are prepared. The project is then ready for the letting of bids and construction. Plans for construction of sewers usually require review and approval by a supervising state agency such as the health department. The engineer must become familiar with specific regulations and legal requirements applicable to the approval of plans for sewers in the state in which work is to be done.

Sanitary sewer system design. The completed design includes a general map of the whole area, showing location of all sewers and structures and the drainage areas; detailed plans and profiles of sewers, showing ground levels, size of pipe, and slope and location of appurtenances; detailed plans of all appurtenances and structures; a complete narrative report with necessary charts, graphs, and tables to make clear the exact nature of the project; complete specifications; and a confidential estimate of costs made available to the authority or owner.

Extensive plans require tabulation of data begin-

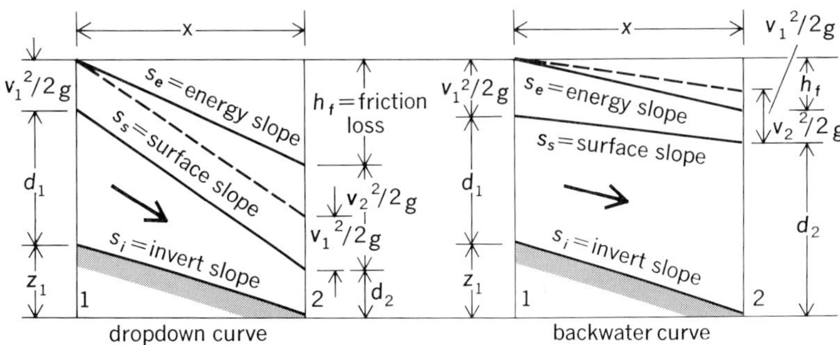

Fig. 8. Dropdown and backwater curves. (*From H. E. Babbitt and E. R. Baumann, Sewerage and Sewage Treatment, 8th ed., copyright © 1958 by John Wiley and Sons, Inc.; used with permission*)

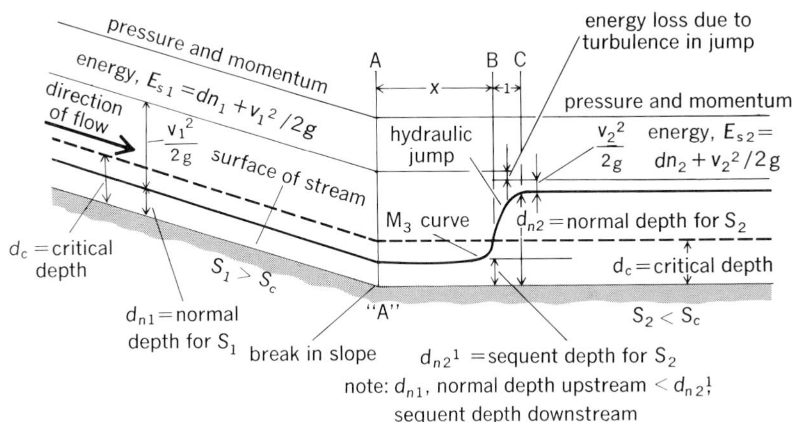

Fig. 9. Free hydraulic jump. (*From H. E. Babbitt and E. R. Baumann, Sewerage and Sewage Treatment, 8th ed., copyright © 1958 by John Wiley and Sons, Inc.; used with permission*)

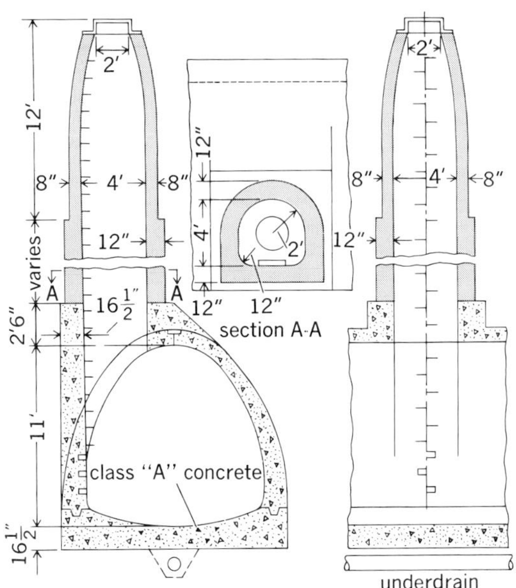

Fig. 10. Cross sections of a deep manhole with access to large sewer. 1" = 2.54 cm; 1' = 30.48 cm. (*From E. W. Steel, Water Supply and Sewerage, 3d ed., McGraw-Hill, 1953*)

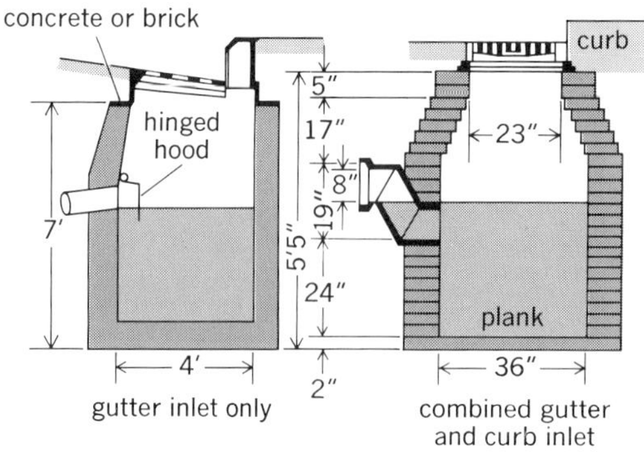

Fig. 11. Types of catch basins. 1″ = 2.54 cm; 1′ = 30.48 cm. (*From E. W. Steel, Water Supply and Sewerage, 3d ed., McGraw-Hill, 1953*)

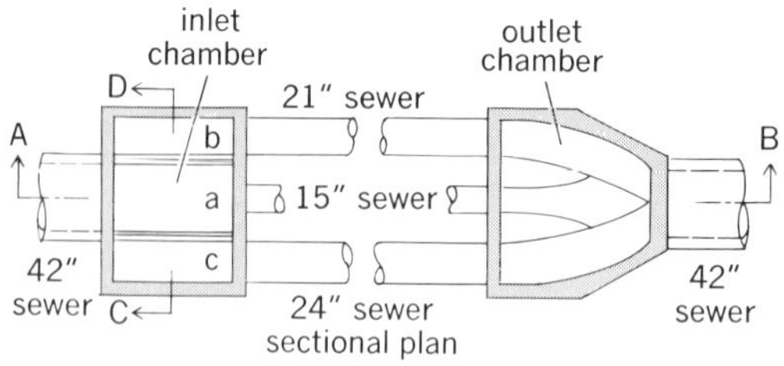

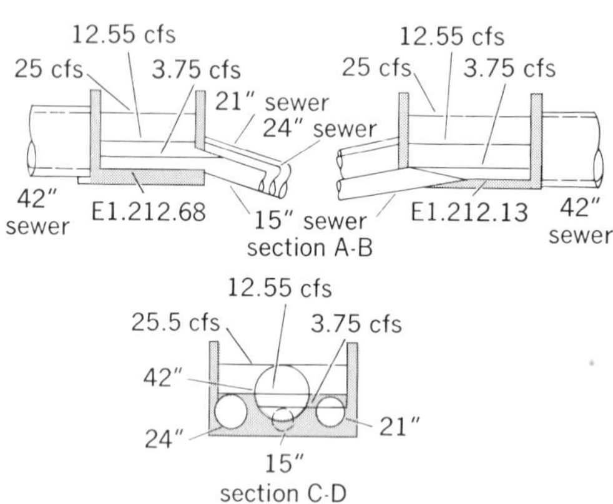

Fig. 12. Inverted siphon. 1″ = 2.54 cm; 1 cfs = 3.785 liters/sec. (*From E. W. Stell, Water Supply and Sewerage, 3d ed., McGraw-Hill, 1953*)

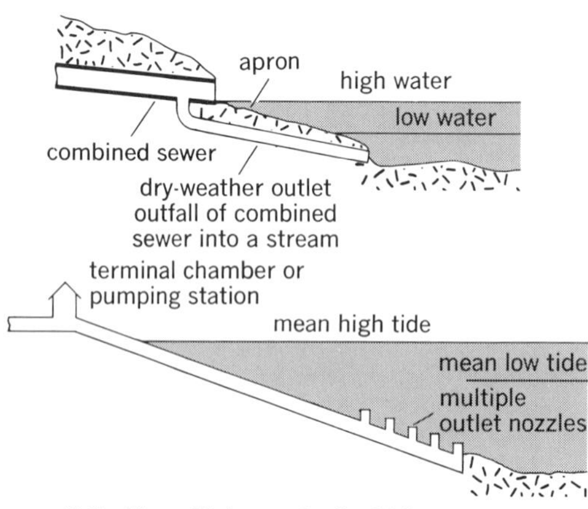

Fig. 13. Two types of sewage outfalls. (*From G. M. Fair and J. C. Geyer, Elements of Water Supply and Waste-Water Disposal, copyright © 1958 by John Wiley and Sons, Inc.; used with permission*)

ning at the upper end of the system and proceeding downstream from manhole to manhole, bringing into the computation each addition to flow from connecting sewers. Details of such tabulations appear in standard textbooks.

Storm sewer system design. As in sanitary sewer design an estimate must be made of the amount of water to be carried by the system. Factors which must be considered include topography, surface permeability, rainfall intensity and duration, time of runoff, and time of concentration in sewers. The method of analysis is discussed in textbooks on hydrology, hydraulics, and sewerage. The procedure for design and report is essentially that described for sanitary sewers.

Combined systems. Combined system design considers factors for both sanitary and storm sewer flow. Provision must be made for handling dry weather or sanitary flow at proper velocity in sewers that may carry large quantities of water following rainfall. Design is complicated by the need for diversion of waters not flowing to the sewage treatment plant. Structures for diversion are located at or near appropriate water courses, and the effects of discharging polluted water, a combination of sanitary and storm waters, must be fully investigated.

Sewerage system construction. Many features of sewer construction are no different from other types of construction. Excavation of trenches and laying of pipe in trenches and tunnels require some variation in construction methods. To construct a sewer it is necessary to remove pavement and ground overburden; use sheeting and bracing to support vertical side walls or the sides and crown of tunnels; dewater the trench; protect all adjacent pipes and structures both in and on the ground; backfill, tamp, and settle soil over finished pipe; and replace pavement. *See* CONSTRUCTION ENGINEERING.

Maintenance. Roots and accumulations of debris must be removed from sewers periodically (Fig. 14). Special equipment is used to clean out sections of pipe between manholes. An instrument designed to cut roots and scrape sand is installed

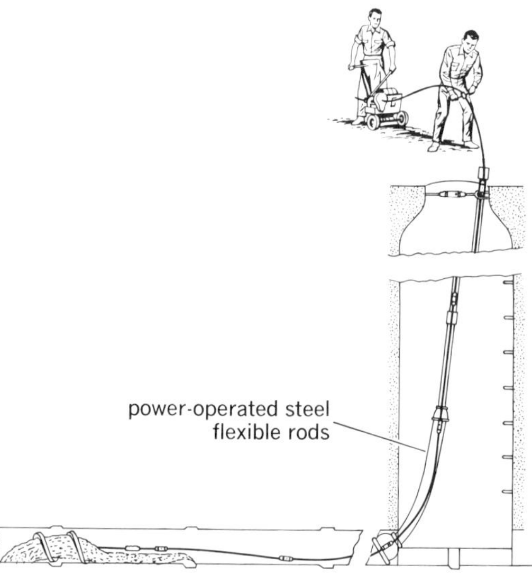

Fig. 14. Sewer-cleaning operation. (*From E. W. Steel, Water Supply and Sewerage, 3d ed., McGraw-Hill, 1953*)

at a manhole and is attached to rods or cable placed in the sewer. This is pushed or pulled forward. Accumulations are raised to the surface by bucket or similar equipment. Safety precautions must be taken by inspectors and workmen to avoid the hazards of sewer gases and insufficient oxygen. *See* SEWAGE DISPOSAL.

[WILLIAM T. INGRAM]

Bibliography: ASCE-FSIWA Joint Committee, *Design and Construction of Sanitary and Storm Sewers*: H. E. Babbitt and E. R. Baumann, *Sewerage and Sewage Treatment*, 8th ed., 1958; R. L. Bolton and L. Klein, *Sewage Treatment: Basic Principles and Trends*, 1961; C. V. Davis (ed.), *Handbook of Applied Hydraulics*, 2d ed., 1952; G. M. Fair and J. C. Geyer, *Elements of Water Supply and Waste-Water Disposal*, 1958; M. Romanoff, *Underground Corrosion*, Nat. Bur. Stand. Circ. no. 579, 1957; H. Rouse (ed.), *Engineering Hydraulics*, 1950; Storm Drainage Research Committee, *The Design of Storm-Water Inlets*, 1956; Texas Water and Sewage Works Association, *Manual for Sewage Plant Operators*, 3d ed., 1964.

Sewage disposal

The discharge of waste waters into surface-water or groundwater courses, which constitute the natural drainage of an area. Most waste waters contain offensive and potentially dangerous substances, which can cause pollution and contamination of the receiving water bodies. Contamination is defined as the impairment of water quality to a degree that creates a hazard to public health. Pollution refers to the adverse effects on water quality that interfere with its proper and beneficial use.

In the past, the dilution afforded by the receiving water body was usually great enough to render waste substances innocuous. Since the turn of the century, however, the dilution of many rivers has been inadequate to absorb the greater waste discharges caused by the increase in population and expansion of industry.

The principal sources of pollution are domestic sewage and industrial wastes. The former includes the used water from dwellings, commercial establishments, and street washings. Industrial wastes constitute acids, chemicals, oils, and animal and vegetable matter carried by cleaning or used process waters from factories and plants. For a discussion of sources of wastes *see* SEWAGE.

Regulation of water pollution. This is primarily a responsibility of the state, in cooperation with the Federal and local governments. The health departments of many states are given statutory power and responsibility for the control of water pollution, and they have established specific water quality standards. There are two basic types of standards—stream standards, dealing with the quality of the receiving water, and effluent standards, referring to strength of wastes discharged. Both types are based on the capacity of the receiving waters to absorb waste substances and on the beneficial uses made of the water.

The self-purification capacity is determined by the available dilution, the biophysical environment of the stream, and the strength and characteristics of the wastes. Beneficial uses include drinking, bathing, recreation, fish culture, irrigation, industrial uses, and disposal of wastes without creation of pollution.

Adjustment of these conflicting interests and equitable distribution of water resources is complex from the technical, economic, and political viewpoints. These considerations have led to the establishment of interstate commissions, which provide a means of coordinated control of the larger rivers.

Water-quality criteria deal with the physical, chemical, and biological parameters of pollution. The most common standards are concerned with physical appearance, odor production, dissolved-oxygen concentration, pathogenic contamination, and potentially toxic or harmful chemicals. The allowable quantity and concentration of these characteristics and substances vary with the water usage.

Absence of odor and unsightliness, and the presence of some dissolved oxygen are common minimum standards. Preliminary or primary treatment of waste waters is usually required for the maintenance of these standards. Highest-quality waters require clarity, oxygen saturation, low bacteriological counts, and absence of harmful substances. In these cases, intermediate or complete treatment may be required. *See* SEWAGE TREATMENT; WATER POLLUTION.

Stream pollution. Biological, or bacteriological, pollution is indicated by the presence of the coliform group of organisms. While nonpathogenic itself, this group is a measure of the potential presence of contaminating organisms. Because of temperature, food supply, and predators, the environment provided by natural bodies of water is not favorable to the growth of pathogenic and coliform organisms. Physical factors, such as flocculation and sedimentation, also help remove bacteria. Any combination of these factors provides the basis for the biological self-purification capacity of natural water bodies.

When subjected to a disinfectant such as chlorine, bacterial die-away is usually defined by Chick's law, which states that the number of organisms destroyed per unit of time is proportional to the number of organisms remaining. This law cannot be directly applied in natural streams because of the variety of factors affecting the removal and death rates in this environment. The die-away is rapid in shallow, turbulent streams of low dilution, and slow in deep, sluggish streams with a high dilution factor. In both cases, higher temperatures increase the rate of removal.

The concentration of many physical characteristics and chemical substances may be calculated directly if the relative volumes of the waste stream and river flow are known. Chlorides and mineral solids fall into this category. Some substances in waste discharges are chemically or biologically unstable, and their rates of decrease can be predicted or measured directly. Sulfites, nitrites, some phenolic compounds, and organic matter are examples of this type of waste.

These simple relationships, however, do not apply to the concentration of dissolved oxygen. This factor depends not only on the relative dilutions, but also upon the rate of oxidation of the organic material and the rate of reaeration of the stream.

Nonpolluted natural waters are usually saturated with dissolved oxygen. They may even be supersaturated because of the oxygen released by green water plants under the influence of sunlight. When an organic waste is discharged into a stream, the dissolved oxygen is utilized by the bacteria in their metabolic processes to oxidize the organic matter. The oxygen is replaced by reaeration through the water surface exposed to the atmosphere. This replenishment permits the bacteria to continue the oxidative process in an aerobic environment. In this state, reasonably clean appearance, freedom from odors, and normal animal and plant life are maintained.

An increase in the concentration of organic matter stimulates the growth of bacteria and increases the rates of oxidation and oxygen utilization. If the concentration of the organic pollutant is so great that the bacteria use oxygen more rapidly than it can be replaced, only anaerobic bacteria can survive and the stabilization of organic matter is accomplished in the absence of oxygen. Under these conditions, the water becomes unsightly and malodorous, and the normal flora and fauna are destroyed. Furthermore, anaerobic decomposition proceeds at a slower rate than aerobic. For maintenance of satisfactory conditions, minimal dissolved oxygen concentrations in receiving streams are of primary importance.

Figure 1 shows the effect of municipal sewage and industrial wastes on the oxygen content of a stream. Cooling water, used in some industrial processes, is characterized by high temperatures, which reduce the capacity of water to hold oxygen in solution. Thermal pollution, however, is significant only when large quantities are concentrated in relatively small flows. Municipal sewage requires oxygen for its stabilization by bacteria. Oxygen is utilized more rapidly than it is replaced by reaeration, resulting in the death of the normal aquatic life. Further downstream, as the oxygen demands are satisfied, reaeration replenishes the oxygen supply.

Any organic industrial waste produces a similar pattern in the concentration of dissolved oxygen. Certain chemical wastes have high oxygen demands which may be exerted quickly, producing a sudden drop in the dissolved oxygen content. Other chemical wastes may be toxic and destroy the biological activity in the stream. Strong acids and alkalies make the water corrosive, and dyes, oils, and floating solids render the stream unsightly. Suspended solids, such as mineral tailings, may settle to the bed of the stream, smother purifying microorganisms, and destroy breeding places. Although these latter factors may not deplete the oxygen, the polutional effects may still be serious.

Deoxygenation. Polluted waters are deprived of oxygen by the exertion of the biochemical oxygen demand (BOD), which is defined as the quantity of oxygen required by the bacteria to oxidize the organic matter. The rate of this reaction is assumed to be proportional to the concentration of the remaining organic matter, measured in terms of oxygen. This reaction may be expressed as Eq. (1),

$$\frac{dL}{dt} = -K_1 L \qquad (1)$$

which integrates to give Eq. (2) or Eq. (3), in which

$$L_t = L_0 e^{-K_1 t} \qquad (2)$$

$$y = L_0 (1 - e^{-K_1 t}) \qquad (3)$$

L_t is BOD remaining at any time t, L_0 is ultimate BOD, y is BOD exerted at end of t, and K_1 is coefficient defining the reaction velocity. The coefficient is a function of temperature given by Eq. (4), in which T is temperature in degrees Cel-

$$K_T = K_{20} \cdot 1.047^{T-20} \qquad (4)$$

sius, K_T is value of the coefficient at T, and K_{20} is value of the coefficient at 20°C.

The BOD of a waste is determined by a standard laboratory procedure and is reported in terms of the 5-day value at 20°C. From a set of BOD values determined for any time sequence, the reaction velocity constant K_1 may be calculated. Knowledge of this coefficient permits determination of the ultimate BOD from the 5-day value in accordance with the above equations. For municipal sewages and many industrial wastes the value of K_1 at 20°C is between 0.15 and 0.75 per day. A common value for sewage is 0.4 per day.

The coefficient determined from laboratory BOD data may be significantly different from that calculated for stream BOD data. The determination of the stream rate may be made from a reexpression of Eqs. (1)–(4) in the form of Eq. (5),

$$K_r = \frac{1}{t} \log \frac{L_A}{L_B} \qquad (5)$$

where L_A is the BOD measured at an upstream station, L_B is the BOD at a station downstream from A, and t is the time of flow between the two stations. Values of K_r range from 0.10 to 3.0 per day. The difference between the laboratory rate K_1 and the stream rate K_r is due to the turbulence of the stream flow, biological growths on the stream bed, insufficient nutrients, and inadequate bacteria in the river water. These factors influence the rate of

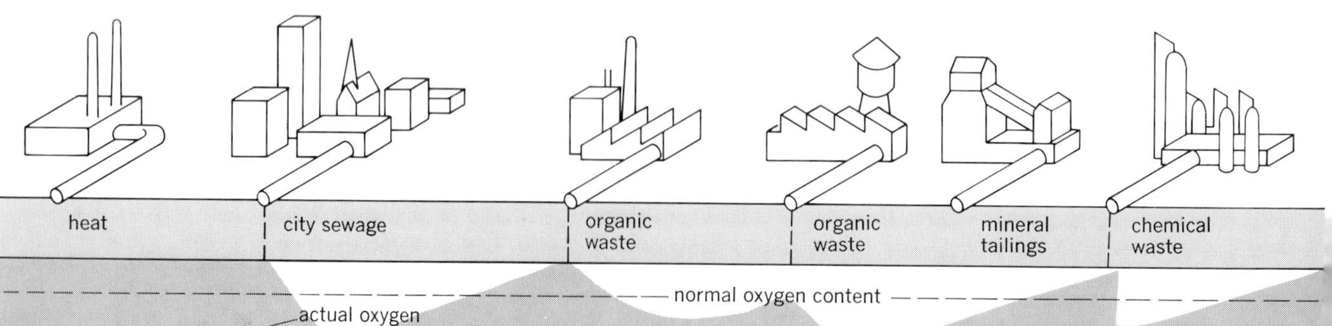

Fig. 1. Variation of oxygen content of polluted stream.

oxidation in the stream as well as the removal of organic matter. Such processes as flocculation, sedimentation, and scour of the organic material in the river affect the removal rate but do not necessarily influence the rate of oxidation and the associated dissolved oxygen concentration. Field surveys are usually required to determine the pollution assimilation capacity of a stream.

When a significant portion of the waste is in the suspended state, settling of the solids in a slow-moving stream is probable. The organic fraction of the sludge deposits decomposes anaerobically, except for the thin surface layer which is subjected to aerobic decomposition due to the dissolved oxygen in the overlying waters. In warm weather, when the anaerobic decomposition proceeds at a more rapid rate, gaseous end products, usually carbon dioxide and methane, rise through the supernatant waters. The evolution of the gas bubbles may raise sludge particles to the water surface. Although this phenomenon may occur while the water contains some dissolved oxygen, the more intense action during the summer usually results in depletion of dissolved oxygen.

Reoxygenation. Water may absorb oxygen from the atmosphere when the oxygen in solution falls below saturation. Dissolved oxygen for receiving waters is also derived from two other sources: that in the receiving water and the waste flow at the point of discharge, and that given off by green plants. The latter source is restricted to daylight hours and the warmer seasons of the year and, therefore, is not usually used in any engineering analysis of stream capacity.

Unpolluted water maintains in solution the maximum quantity of dissolved oxygen. The saturation value is a function of temperature and the concentration of dissolved substances, such as chlorides. When oxygen is removed from solution, the deficiency is made up by the atmospheric oxygen, which is absorbed at the water surface and passes into solution. The rate at which oxygen is absorbed, or the rate of reaeration, is proportional to the degree of undersaturation and may be expressed as in Eq. (6), in which D is dissolved oxygen deficit, t is time, and K_2 is reaeration coefficient.

$$\frac{dD}{dt} = -K_2 D \tag{6}$$

The reaeration coefficient depends upon the ratio of the volume to the surface area and the intensity of fluid turbulence. An approximate value of the coefficient may be obtained from Eq. (7), in which D_L is coefficient of molecular diffusion of oxygen in water, U is average velocity of the river flow, and H is average depth of the river section.

$$K_2 = \frac{D_L U^{1/2}}{H^{3/2}} \tag{7}$$

The effect of temperature on this coefficient is identical with its effect on the deoxygenation coefficient. A common range of K_2 is from 0.20 to 5.0 per day. Many waste constituents, such as surface-active substances, interfere with the molecular diffusion of oxygen and reduce the value of the reaeration rate from that of pure water. Winds, waves, rapids, and tidal mixing are factors which create circulation and surface renewal and enhance reaeration.

Oxygen balance. The oxygen balance in a stream is determined by the concentration of organic matter and its rate of oxidation, and by the dissolved oxygen concentration and the rate of reaeration. The simultaneous action of deoxygenation and reaeration produces a pattern in the dissolved oxygen concentration known as the dissolved oxygen sag. The differential equation describing the combined action of deoxygenation and reaeration is given in Eq. (8), which states that the rate of change in the dissolved oxygen deficit D is the result of two independent rates. The first is that of oxygen utilization in the oxidation of organic matter. This reaction increases the dissolved oxygen deficit at a rate that is proportional to the concentration or organic matter L. The second rate is that of reaeration, which replenishes the oxygen utilized by the first reaction and decreases the deficit. Integration of this equation yields Eq. (9), where L_0 and D_0 are the initial biochemical oxygen demand and the initial dissolved oxygen deficit, respectively, and D_t is the deficit at time t. The proportionality constants K_1 and K_2 represent the coefficients of deaeration and reaeration, respectively, and K_r the coefficient of BOD removal in the stream.

$$\frac{dD}{dt} = K_1 L - K_2 D \tag{8}$$

$$D_t = \frac{K_1 L_0}{K_2 - K_r}\left(e^{-K_r t} - e^{-K_2 t}\right) + D_0 e^{-K_2 t} \tag{9}$$

Figure 2 shows a typical dissolved oxygen sag curve resulting from a pollution of amount L_0 at $t = 0$. The sag curve is shown to result from the deoxygenation curve and the reaeration curve. A point of particular significance on the sag curve is that of minimum dissolved oxygen concentration, or maximum deficit. At this location, the rate of change of the deficit is zero, which results in the numerical equality of the opposing rates of deoxygenation and reoxygenation. The balance at this critical point may be written as Eq. (10), where the BOD at the critical point has been replaced by its equivalent at zero time (the location of the waste discharge). The value of the time t_c may be calculated from Eq. (11).

$$K_2 D_c = K_1 L = K_1 L_0 e^{-K_r t_c} \tag{10}$$

$$t_c = \frac{1}{K_2 - K_r} \log \frac{K_2}{K_r}\left[1 - \frac{D_0(K_2 - K_r)}{K_1 L_0}\right] \tag{11}$$

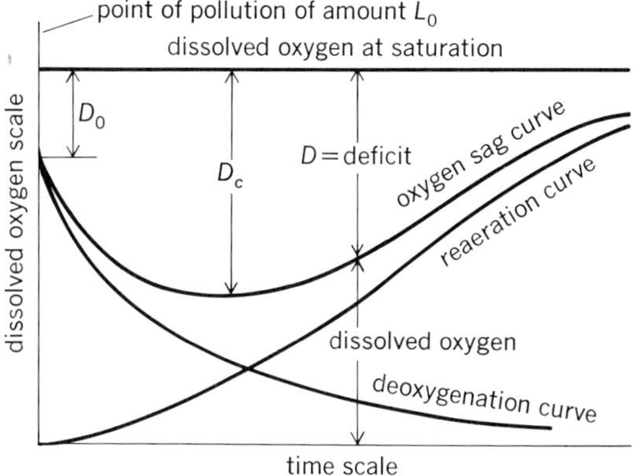

Fig. 2. Dissolved-oxygen sag curve and its components.

Allowable pollutional load. The pollutional load L_0 that a stream may absorb is a function of the dissolved oxygen deficit D_c, the coefficients K_1, K_r, and K_2, and the initial deficit D_0. The dissolved oxygen deficit is usually established by water pollution standards of the health agency, and the initial deficit is determined by upstream pollution. The engineering problem is usually associated with the assignment of representative values of the coefficients K_1, K_r, and K_2 for a given flow and temperature condition.

Seasonal temperatures influence the saturation of oxygen and the rates of deaeration and reaeration. Variation in stream flow with the seasons affects the dilution factor. The most critical conditions occur during the summer when the stream runoff is low and the temperatures are high.

Pollution in lakes and estuaries. In lakes self-purification is slower than in streams because of the low rates of dispersion of the waste waters. There is no turbulence characteristic as in flowing rivers, and mixing depends primarily on winds, waves, and currents. Waste-water outfalls are designed to take advantage of the dispersion induced by these factors and to prevent the development of concentrated sewage fields.

In estuaries, the dispersion of waste waters is complicated by the tides, which carry various portions of the pollutant back and forth over many cycles, and by the difference of density in fresh water, waste water, and salt water. The equation defining the oxygen balance must be modified to allow for the greater time that an average particle of pollution is detained within the estuary; the flushing mechanism of such bodies is therefore of primary concern.

Each estuary presents problems of density currents, configuration, and exchange that distinguish it from others. Field measurements of salinities, currents, and cross sections, in addition to the measurement of physical, chemical, and biological characteristics, are necessary to evaluate the pollution capacity of these watercourses. Dilution and dispersion in ocean waters is complicated by many of the same factors as in estuaries. The death rates of the coliform bacteria are greater in sea water. The outfalls must be designed and located to promote effective dispersion and to prevent the accumulation of sewage fields.

Oxidation ponds and land disposal. The forces of natural purification are utilized in shallow ponds, called oxidation ponds. Successful operation of these basins usually requires relatively high temperatures and sunshine. Carbon dioxide is released by means of the bacterial decomposition of the organic matter. Algae growth develops, consuming the carbon dioxide, ammonia, and other waste products and releasing oxygen under proper climatic conditions.

Oxidation ponds are efficient and relatively economical. Instead of relying on the algae as a primary source of oxygen, mechanical aeration of the pond contents may be employed. Lagoons aerated in this manner are not as susceptible to climatic conditions as the oxidation ponds.

Land disposal of sewage is occasionally practiced by surface or flood irrigation. The former is the discharge of sewage upon the ground, from which it evaporates and through which it percolates. However, a significant portion remains which must be collected in surface drainage channels. Although this method is not particularly efficient for domestic sewage, a modification of it, spray irrigation, has been successfully employed in the treatment of a few industrial wastes. In flood irrigation, all the sewage is permitted to seep through the ground and is usually collected in underdrains. This method takes advantage of the mechanical filtration and biological purification afforded by the soil. Unless the sewage is treated before irrigation, odors and clogging usually occur and possible contamination of ground or surface water can result. *See* SANITARY ENGINEERING; SEWAGE COLLECTION SYSTEMS.

[DONALD J. O'CONNOR]

Bibliography: G. M. Fair, J. C. Geyer, and D. A. Okun, *Elements of Water Supply and Waste-Water Disposal*, 2d ed., 1971; J. Kucharski; and R. G. Rosehart, *Industrial Waste Water Treatment and Disposal*, 1978; Metcalf Eddy Inc. and G. Tchobanglous, *Wastewater Engineering: Treatment, Disposal and Reuse*, 1978; D. A. Okun and G. Ponghis, *Community Wastewater Collection and Disposal*, 1975.

Sewage solids

A semiliquid mass, called sludge, removed from the liquid flow of sewage.

Solids treatment. This depends on the source of the solid and its characteristics. Solids are removed as screenings, grit, primary sludge, secondary sludge, and scum.

Screenings. Screenings are putrescible and offensive. They are either ground and returned to the sewage, ground and transferred to the digestor, incinerated, or buried. The quantity of screenings is variable and is dependent on sewage characteristics. Coarse screenings will vary from 0.3 to 5 ft^3/1,000,000 gal (2.2 to 37 liters/1,000,000 liters). Fine screenings will range from 5 to 35 ft^3/1,000,000 gal (37 to 262 liters/1,000,000 liters). Grit collection has a wide variation. Normally the volume will be between 1 and 10 ft^3/1,000,000 gal (7.5 and 75 liters/1,000,000 liters). *See* SEWAGE.

Sludge. Sludge will vary in amount and characteristics with the characteristics of sewage and plant operations. The following refers to a reasonable normal for each source of sludge. Primary sludge is composed of gray, viscous, identifiable solids, putrescible, odorous, 2500 gal 95% moisture content per 1,000,000 gal (or 2500 liters per 1,000,000 liters). Trickling filter sludge is black, dark brown, granular or flocculent, partially decomposed, not highly odorous when fresh, 500 gal 92.5% moisture content sludge per 1,000,000 gal (or 500 liters per 1,000,000 liters). Activated sludge is dark to golden brown, partially decomposed, granular, flocculent, earthy odor when fresh, 13,500 gal 98% moisture content per 1,000,000 gal (or 13,500 liters per 1,000,000 liters).

Digestion. Digestion is the anaerobic decomposition of organic matter resulting in partial gasification, liquefaction, and mineralization. Sludges (except chemical sludges) from treatment processes can be digested provided there are no substances such as cyanides and chromium, toxic to organisms present in the sludge. Sludges are transferred to separate digestion tanks except where Imhoff-type tanks or septic tanks are in use.

The digestion process is a progressive decompo-

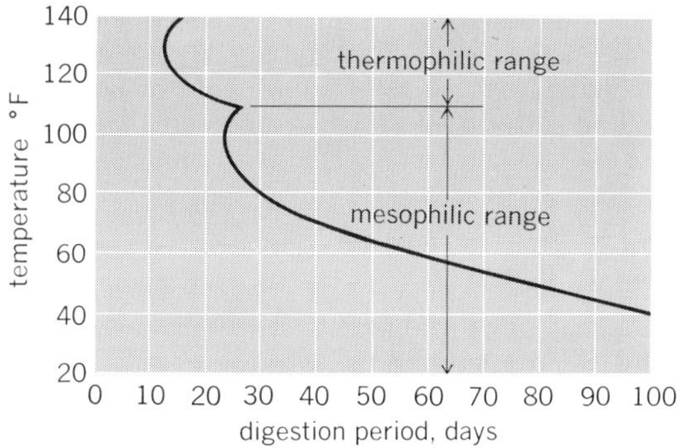

Fig. 1. Effect of temperature on time of digestion of seeded sludge. °C = 5/9 (°F −32). *(From E. W. Steel, Water Supply and Sewerage, 4th ed., McGraw-Hill, 1960)*

sition of organic matter, which makes up about 70% of the total sludge weight. Carbohydrates are attacked first and organic acids are formed. This stage is known as acid fermentation. Organisms living in acid environment continue digestion during a second stage, known as acid digestion, when organic acids and nitrogenous materials are attacked. During the third stage—a period of digestion, stabilization, and gasification—proteins and amino acids are subject to bacterial action. Volatile acids are reduced and pH rises. The final stage is referred to as alkaline digestion. The principal gas produced during this stage is methane.

In a single tank all three stages proceed simultaneously. Fresh solids mixed with well-digested materials providing balance and holding the pH above 7.0 offer a fairly ideal condition. Liquefied materials, excess liquor (or supernatant liquor), and digested solids are removed, making room for fresh material. After the balance has been obtained, it is possible to continue the operation if fresh solids are held to less than 4% of the tank solids measured on a dry-weight basis.

Sludge-digestion tanks are circular or rectangular, heated or unheated units. Most states have established schedules of capacity requirements for different types of sludge on heated and unheated bases. Imhoff tanks are unheated and require sludge capacity of 3–4 ft³ (80–110 liters) per capita. Primary sludges require 2–3 ft³ (60–80 liters) heated and 4–6 ft³ (110–170 liters) per capita unheated. Filter and primary sludge mixed runs from 3–5 ft³ (80–140 liters) heated and 6–10 ft³ (170–280 liters) per capita unheated. Activated sludge requires 4–6 ft³ (110–170 liters) heated and 8–12 ft³ (230–340 liters) per capita unheated. Heated tanks provide controlled temperature for thermophilic (110–140°F or 43–60°C) or mesophilic digestion. Temperatures around 100°F (38°C) are optimum (Fig. 1).

Provision is made for manipulation of the sludge, and the system may include preheater and heater equipment, recirculation pumps with sludge suction at several levels, supernatant liquor drawoff at several levels, gas dome or collector, stirring mechanism, sludge rakes, and drawoff. Covers may be fixed or floating (Fig. 2).

Multistage digestion occurs when two digestors or more are placed in series, the sludge drawoff of the first being connected to the second and continuing. In this system flexibility in manipulating and mixing sludges and in controlling supernatant liquor is possible. *See* IMHOFF TANK.

Supernatant liquor. This is the liquid fraction in a digestor. It is offensive in odor and high in solids and BOD (biochemical oxygen demand). It is discharged to the incoming sewage and treated in the primary sedimentation unit. It is withdrawn in small quantities from a level having fewer solids. Activated sludge with high moisture content is sometimes settled in a preliminary operation before discharge to digestor. The volume may be reduced as much as 50% and the decant liquor is less objectionable than the supernatant.

Sludge gas. Sludge-gas production under good operating conditions is about 12 ft³/lb (0.75 m³/kg) of volatiles destroyed. The gas is 60–70% methane and 20–30% carbon dioxide with minor amounts of impurities such as hydrogen sulfide. Gas has a fuel value of 600–700 Btu/ft³ (22–26 MJ/m³) and is used at the plant to operate auxiliary engines and

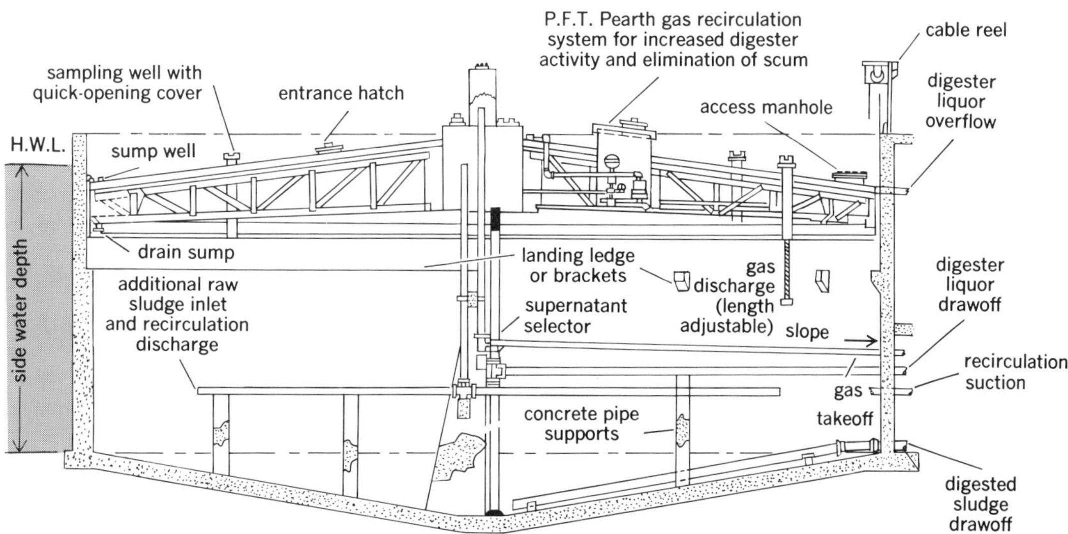

Fig. 2. Floating cover digestor with gas recirculation. *(Pacific Flush Tank Co., Chicago)*

provide heat for sludge-heating systems. Excess gas is burned.

Sludge-drying beds. These are provided at smaller plants to handle sludge removed from digestors without further treatment. Drying beds should have an area of 2–3 ft^2 (0.19–0.28 m^2) per capita. Covered beds require about three-quarters of that area. Beds consist of up to 12 in. (30 cm) of coarse sand over 12 in. (30 cm) of gravel packed around underdrains. Sludge is drawn to a depth of 9–12 in. (23–30 cm) and allowed to drain and dry. A well-digested granular sludge drains easily and reduces to a depth of 3–4 in. (8–10 cm) when dry (60–70% moisture content). Sludge is removed from the bed and eventually may be used as humus material. It has little or no odor.

Sludge processing. Sludge processing may be required if the sludge is to be disposed of by other methods. Elutriation, or washing of sludge with plant effluent, removes undesirable amino-ammonia nitrogen and reduces or eliminates the need for conditioning chemicals. Lime or ferric chloride or polyelectrolytes may be used to prepare sludge for vacuum filtration. Filter cake containing 70–80% moisture is more easily handled. In some plants raw sludge is conditioned and processed on various filters without digestion. Such sludge is offensive and is handled like screenings.

Filter rates expressed as lb/(ft^2)(hr) [1 lb/(ft^2)(hr) = 4.88 kg/(m^2)(hr)] may be taken generally at 3.5 and will range from 2.5 for fresh activated sludge to 8 for primary digested sludge. Filters are revolving drums covered with wire, plastic, or cotton cloth, or flexible metal springlike coils. Drums revolve at 1.5–9 min per revolution, passing through a basin of sludge. Vacuum within the drum picks up sludge against the media and separates water from the solids. The filtrate is returned to sewage flow or to elutriators.

Solids disposal. Sewage solids must be disposed of without nuisance or hazard to health. Burial, incineration, and drying for use as fertilizer are means of final disposal. Lagooning may be used; in seacoast cities sludge may be taken out to sea.

Burial. Screenings are handled by hand in small plants. At one time special screening pits were prepared and the material was placed and covered until it had composted. The recovery does not justify the work, and generally screenings are burned or placed in a sanitary landfill, either on the plant premises or as a part of municipal refuse disposal.

Incineration. Incineration of sludge has developed as a means of disposal in larger plants. Incineration introduces problems of air pollution. Incineration of sludge requires auxiliary heat because the moisture content is high. Gas or oil or digestor gas may be used as fuel. Incinerators used to burn sludge are generally multiple hearth. Sludge is fed to the top hearth and as it dries it is dropped down to the next hearth by agitator arms. Water is driven off and volatile gases are released by the heat. The gases are ignited by the furnace temperature. To avoid excessive odors, the temperature should be maintained at 1200–1400°F (650–760°C). Ash residue is inert and may be used for fill or cover on sanitary landfill. Since the volume of treatment-plant sludge is small, plant incinerators are not operated daily. Auxiliary fuel is required to preheat the combustion chamber. If there is no digestor gas fuel available, the costs of sludge incineration can be excessive. Sludge cake can be mixed with refuse and burned in municipal incinerators, if the two facilities are adjoining.

Drying. Drying of sludge is substituted for incineration in some plants. The dried sludge can be used for fertilizer by enriching it with chemicals, particularly potash, which is lacking in most digested sludges. In this process water is driven off without burning the material. Sludges from drying beds may be stockpiled. After a year or so these sludges are earthlike and may be used as a soil conditioner or as a soil builder when preparing new land areas over sanitary landfill and on sand. Sludges from beds and filter cake may be put into sanitary landfill. The fill is compact but burnable.

Flash driers operate by mixing a portion of dried sludge with the incoming wet sludge cake. A high-velocity, high-temperature gas stream evaporates the water. The dried material is then passed through a cyclone separator and is carried to storage, which may be at a fertilizer plant. Municipalities having refuse incinerators at the sewage treatment site provide a ready source of heat that can be used for sludge drying.

Spray driers have some usefulness in handling liquid sludges. The sludge suspension is ejected under high pressure into a heated chamber. The sudden pressure release atomizes the suspension and water is quickly driven off as the material falls to the bottom. The material is removed by a separator. Heat requirements are high and the method has the limitations of fuel cost if no source of waste heat is available.

Land disposal. Disposal on land has limited application. There are a few locations where sludge may be taken to fields and plowed under. Occasionally liquid sludge may be applied to gardens and lawns around a treatment plant. Land disposal has certain public-health dangers and must be closely supervised. [WILLIAM T. INGRAM]

Bibliography: H. E. Babbitt and E. R. Baumann, *Sewerage and Sewage Treatment*, 8th ed., 1958; J. Borchardt and W. Redman, *Sludge and Its Ultimate Disposal*, 1980; G. M. Fair and J. C. Geyer, *Elements of Water Supply and Waste-Water Disposal*, 2d ed., 1971; R. W. James, *Sewage Sludge Treatment and Disposal*, 1979; E. W. Steel and T. McGhee, *Water Supply and Sewerage*, 5th ed., 1979; G. Tchobanglous, R. Eliassen, and H. Thiesen, *Solid Wastes: Engineering and Management Issues*, 1977; P. A. Vesilind, *Treatment and Disposal of Wastewater Sludges*, 1979.

Sewage treatment

Any process to which sewage is subjected in order to remove or alter its objectionable constituents and thus render it less offensive or dangerous. These processes may be classified as preliminary, primary, secondary, or complete, depending on the degree of treatment accomplished. Preliminary treatment may be the conditioning of industrial waste prior to discharge to remove or to neutralize substances injurious to sewers and treatment processes, or it may be unit operations which prepare the water for major treatment. Primary treatment is the first and sometimes the only treatment of sewage. It is the removal of floating solids and coarse and fine suspended solids. Secondary treatment utilizes biological methods, that is, oxidation

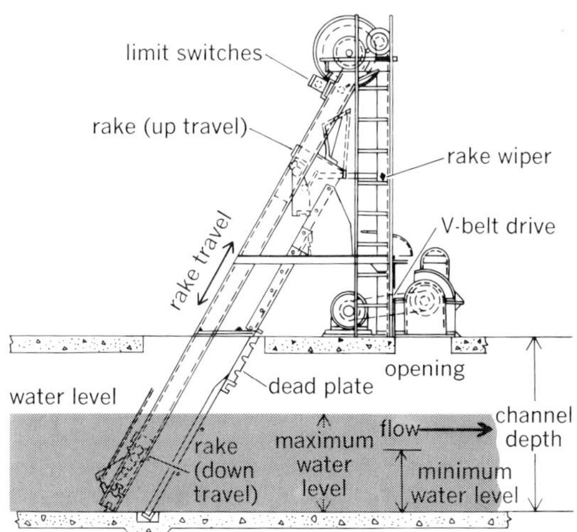

Fig. 1. Mechanically cleaned bar screen. (*From H. E. Babbitt and E. R. Baumann, Sewerage and Sewage Treatment, 8th ed., copyright © 1958 by John Wiley and Sons, Inc.; used with permission*)

processes following primary treatment by sedimentation. Complete treatment removes much of the suspended, colloidal, and organic matter.

Septic tanks and Imhoff tanks are considered secondary treatment methods because sedimentation is combined with biological digestion of the sludge. *See* IMHOFF TANK; SEPTIC TANK.

Coarse solids removal. This is accomplished by means of racks, screens, grit chambers, and skimming tanks. Racks are fixed screens composed of parallel bars placed in the waterway to catch debris. The bars are usually spaced 1 in. (25 mm) or more apart. Screens are devices with openings usually of uniform size 1 in. (25 mm) or less placed in the line of flow. Screens may be fixed or movable and vary in construction as bar screens, band screens, or cage screens. Such screens are hand cleaned or mechanically cleaned (Fig. 1). Grit chambers remove inorganic solids but may also trap heavier particles of organic nature such as seeds (Fig. 2). Grit chambers are designed so that the flow in the chamber is at 1 ft/s (0.3 m/s) or more. At less than that velocity, organic material also settles. Removal of grit is done either by hand or mechanically. Devices are added to mechanically cleaned units which wash most of the organic material out of the grit. Skimming chambers are devices for removing floating solids and grease. Air has been used to coagulate greases which then float and are skimmed off mechanically or by hand.

Fine solids removal. This is accomplished by screens with openings 1/16 or 1/32 in. (1.5875 or 0.79375 mm) wide, by sedimentation, or by both.

Fine screens are set in the line of flow and are operated mechanically. Band screens, drum screens, plate screens, and vibratory screens are in use and the finer particles of floating solids are removed as well as coarse solids passing a rack. In some plants screenings pass through a grinder and return to the flow so that they settle out in the sedimentation tank. Another device, the comminutor, barminutor, or griductor, has high-speed rotating edges working in the sewage flow (Fig. 3). The blades cut, chop, and shred the solids, which then go to the sedimentation unit.

Sedimentation. Sedimentation has one objective, the removal of settleable solids. Some floating materials are also removed by skimming devices, called clarifiers, built into sedimentation units. The basins are either circular or rectangular. In the circular unit sewage flows in at the center and

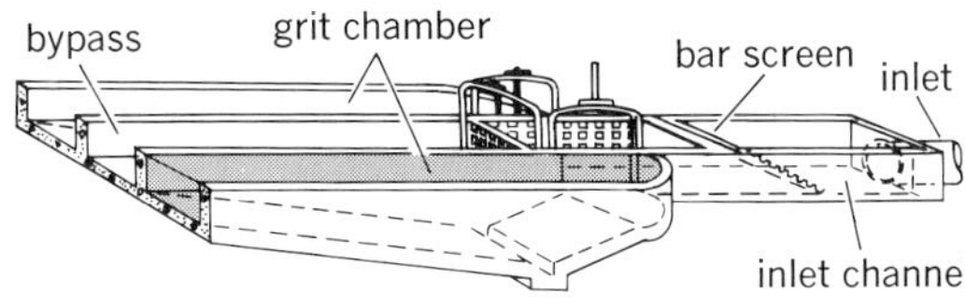

Fig. 2. Grit chamber. (*From Engineering Extension Department of Iowa State College, Bull. no. 58, 1953*)

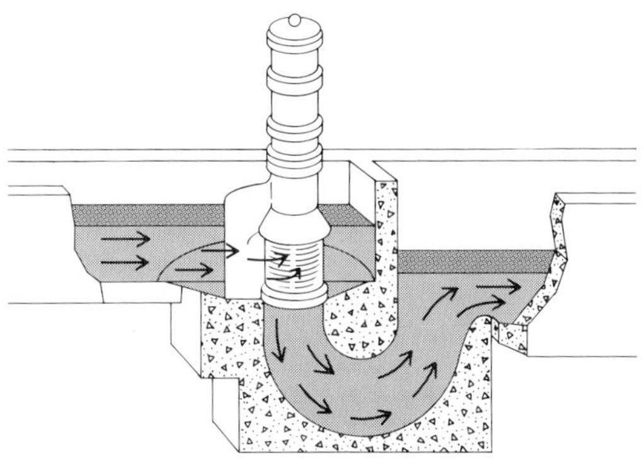

Fig. 3. A comminutor in place. (*Chicago Pump Co.*)

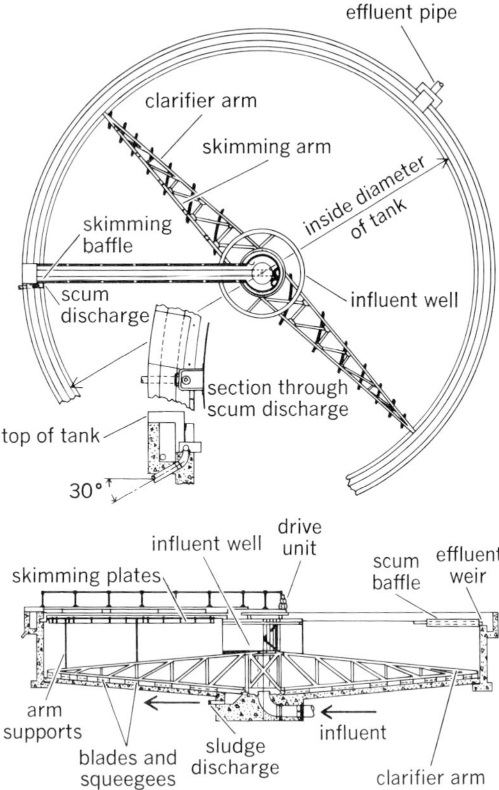

Fig. 4. Typical circular clarifier, a skimming device. (*From H. E. Babbitt and E. R. Baumann, Sewerage and Sewage Treatment, 8th ed., copyright 1958 by John Wiley and Sons, Inc.; used with permission*)

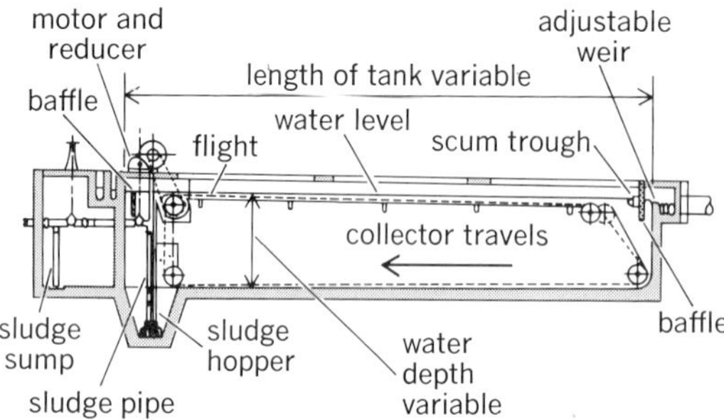

Fig. 5. Longitudinal section of typical rectangular clarifier. (*From E. W. Steel, Water Supply and Sewerage 4th ed., McGraw-Hill, 1960*)

out over weirs along the circumference (Fig. 4). In the rectangular tanks sewage flows into one end and out the other (Fig. 5).

The efficiency of a settling basin is dependent on a number of factors other than particle size, specific gravity, and settling velocity. Concentration of suspended matter, temperature, retention period, depth and shape of basin, baffling, total length of flow, wind, and biological effects all have an effect on solids removal. Density currents and short-circuiting may negate theoretical detention computations. Improper baffling may have the effect of reducing the effective surface area and creating dead or nonflow areas in the tank. A settling tank of good design with surface settling rates of 600 gal/(ft²)(day) [24.5 m³/(m²)(day)] and a 2-hr detention period will remove 50–60% of the suspended solids and at the same time remove 30–35% of the biochemical oxygen demand (Fig. 6).

The settling velocity of a particle is a function of specific gravity of the particle, specific gravity of water, viscosity of liquid, and particle diameter. Settling rates of particles larger than 0.1 mm are determined empirically. Sizes less than 0.1 mm settle in accordance with Stokes' law. Theoretically, if the forward motion of the water is less than the vertical settling rate of the particle, a particle

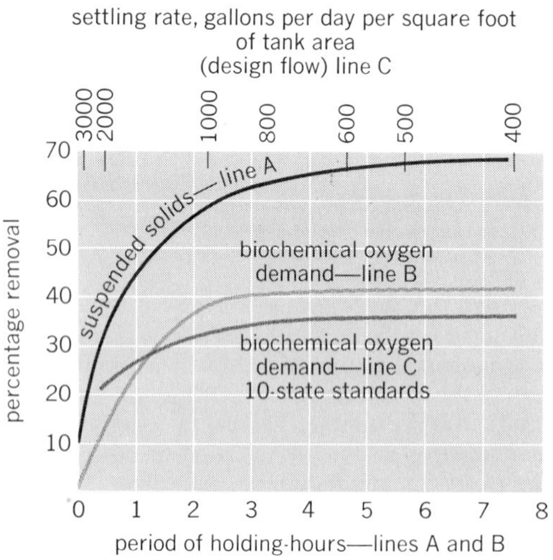

Fig. 6. Probable performance of sedimentation basins. 1 gal/(ft²) (day) = 4.07×10^{-2} m³/(m²) day. (*From H. E. Babbitt and E. R. Baumann, Sewerage and Sewage Treatment, 8th ed., copyright © 1958 by John Wiley and Sons, Inc.; used with permission*)

at the surface will settle some distance below the surface in a given time interval. After that time interval the surface layer of water could be removed and it would contain no solids. The term surface settling rate is introduced as a practical measure of the rate of flow through the basin, if the rate of flow is equal to the surface area times the settling velocity of the smallest particle to be removed. Hence the selection of an overflow or surface settling rate expressed as gallons per day per square foot of surface area (or cubic meters per day per square meter of surface area) establishes a relationship between flow and area.

Flocculent suspensions have little or no settling velocity. These may occur in raw sewage but occur more frequently in secondary settling of effluents from activated sludge units. Such suspensions may be removed by passing the inflowing water upward through a blanket of the material (Fig. 7). Theoretically there is a mechanical sweeping action in which smaller particles are attached to larger particles which then have sufficient weight to

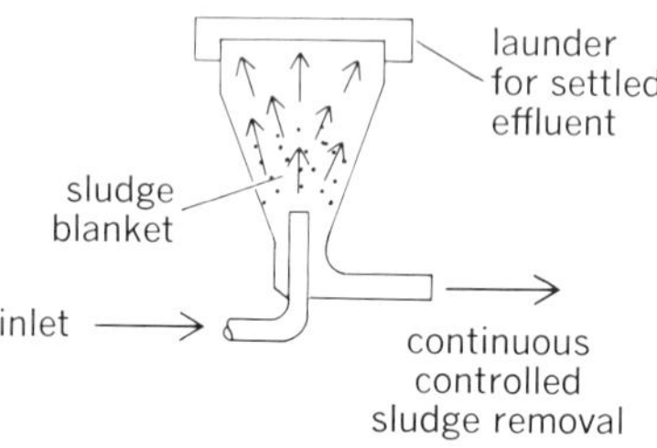

Fig. 7. Diagram of a vertical-flow sedimentation tank. (*From H. E. Babbitt and E. R. Baumann, Sewerage and Sewage Treatment, 8th ed., copyright © 1958 by John Wiley and Sons, Inc.; used with permission*)

settle. Another type of treatment for such material is provided by an inner chamber equipped with baffles which rotate and stir the liquor and aid the formation of larger and heavier floc (Fig. 8). The same purpose is also achieved by agitation with air. Some of the settled sludge is raised by airlift and mixed in with the material, thus forming a mixture with improved settling characteristics.

Sedimentation basin design. Practical considerations and engineering judgment must be applied in designing sedimentation basins. Depth is usually held at 10-ft (3-m) sidewall depth or less. The surface area requirement is usually 600 gal/(ft²) (day) [24.5 m³/(m²)(day)] for primary treatment alone and 800–1000 gal/(ft²)(day) [32.6–40.7 m³/(m²)(day)] for all other tanks. The detention period is normally 2 hr. These three parameters of design must be adjusted since each is dependent on the other for a given design flow (average daily flow at a plant). When mechanical sludge-removal equipment is used, the tank dimensions are usually sized to a conventional equipment specification. Rectangular tanks are built-in units with common walls between units and unit width up to 25 ft (7.6 m). The length-width ratio, frequently determined by economical design dimension, should not be greater than 5:1. The minimum length should be 10 ft (3.0 m). Final sizing may be fitted to convenient equipment dimensions.

Sludge removal on a regular schedule is manda-

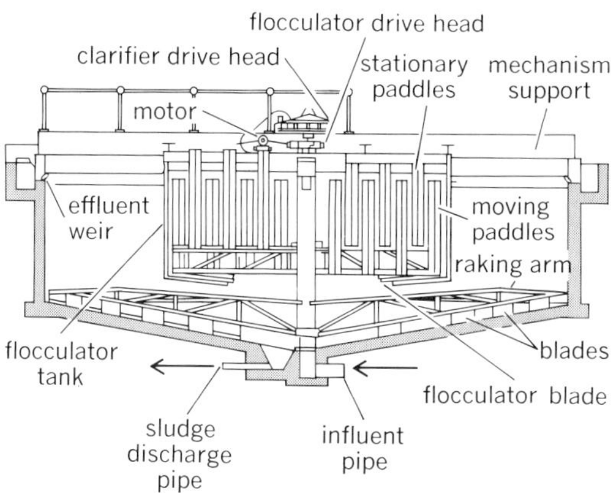

Fig. 8. The Dorr clariflocculator. *(From E. W. Steel, Water Supply and Sewerage, 4th ed., McGraw-Hill, 1960)*

tory in separate sedimentation tanks. If sludge is not removed, gasification occurs and large blocks of sludge begin to appear on the surface. These must then be removed by scum-removal mechanism or broken up so that they will settle. In circular tanks radial blades move the sludge to a center sludge hopper. In rectangular tanks the hopper is located at the inlet end and blades on a traveling chain move sludge in reverse of sewage flow. The heavier solids settle at the inlet and have a short travel path. These same blades may rise to the surface and move scum with the sewage flow to the outlet end where it is held by a baffle and removed by some form of scum-removal device. Sludge-removal mechanisms are often operated intermittently by time-clock relay mechanisms.

Appurtenances in the form of skimmers, scrapers, and other mechanical devices are many. Manufacturers have variants to offer, and competition is keen. Manufacturers' literature should be studied carefully and specifications should be carefully written to procure equipment meeting the requirements of engineering design.

Detention periods are theoretical. The actual flowthrough time is influenced by the inlet and outlet construction. On circular tanks inlets are submerged. Water rises inside a baffle extending downward to still the currents. Rectangular tank inlets may be submerged or, more commonly, sewage is brought to a trough which has a weir extending the width of the tank. The flow then moves forward with less short-circuiting. The outlet device on circular tanks is nearly always a circumferential weir adjusted to level after installation. The weir may be sharp-edged and level or provided with a sawtoothlike series of V-notches. On rectangular tanks, in order to provide enough weir length, a device known as a launder is used. A launder is a series of fingerlike shallow conduits set to water level and receiving flow from both sides of the conduit. Each of the fingers is connected to a common exit trough. The normal weir loading should not exceed 10,000 gal/linear ft (124 m^3/linear m) of weir per day in small plants, or 15,000 gal/linear ft (186 m^3/linear m) in units handling more than 1,000,000 gal per day (1.0 mgd) or 3785 m^3/day.

Chemical precipitation. Many attempts have been made to utilize chemical coagulants in the flocculation of sewage. The process, if used, is similar to that used in water treatment. The cost of chemicals and the somewhat intermediate treatment obtained with chemicals have kept this process out of general use. Its principal use today is in the preparation of sludge for filtration. Various steps in chemical precipitation are shown in Fig. 9. Alum, ferric sulfate, ferric chloride, and lime are used to form an insoluble precipitate which adsorbs colloidal and suspended solids. The entire floc settles and is removed as sludge. The Guggenheim process employs ferric chloride and aeration. The Scott-Darcy process employs ferric chloride made by treating scrap iron with chlorine solution.

Oxidation processes. These are secondary treatment processes, although a few activated-sludge plants have been built without primary sedimentation. Oxidation process methods are (1) filtration by intermittent sand filters, contact filters, and trickling filters; (2) aeration by the activated-sludge process or by contact aerators; and (3) oxidation ponds. There are three basic oxidation methods, all depending on biological growth. Each provides a method of bringing organic matter in suspension or solution in sewage into immediate

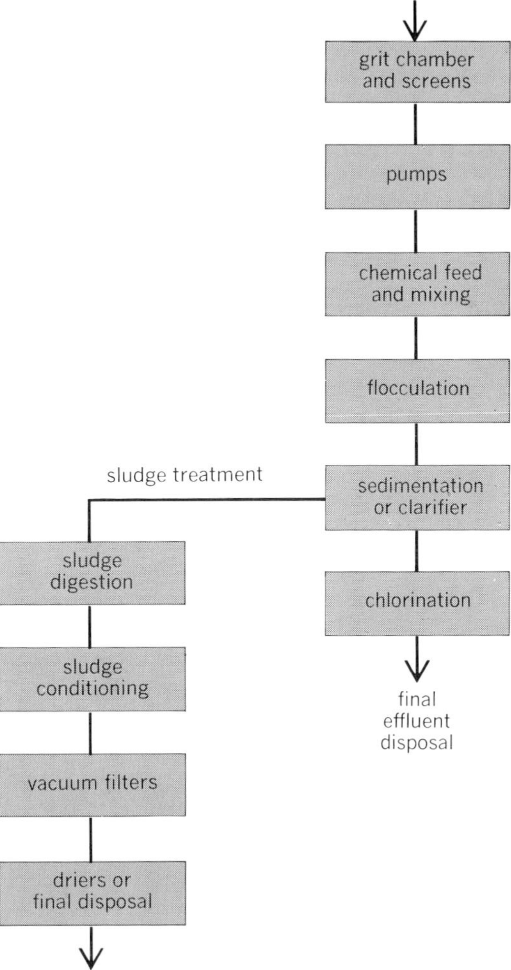

Fig. 9. Flow-through diagram of a chemical treatment plant. *(From H. E. Babbitt and E. R. Baumann, Sewerage and Sewage Treatment, 8th ed., copyright © 1958 by John Wiley and Sons, Inc.; used with permission)*

contact with a population of microorganisms living under aerobic conditions. The processes are called filtration, activated sludge, and contact aeration.

Filtration. Intermittent sand filters are sand beds provided with underdrains. Sewage is dosed intermittently by siphon or by pump, at rates from 20,000 gal per acre per day (gad) [187 m^3 per hectare per day] to a maximum of 125,000 gad (1.17×10^3 m^3 per hectare per day) when operated as a secondary treatment process. Rates may go to 500,000 gad (or 0.5 mgad) [4.7×10^3 m^3 per hectare per day] when operated as a tertiary process. Beds are usually 2 1/2–3 ft (0.76–0.91 m) deep and are constructed with 6–12 in. (15–30 cm) of gravel at the bottom. The sand is sized to a uniformity coefficient of 5.0 or less (3.5 preferred), with effective size of 0.2–0.5 mm. The uniformity coefficient is the ratio between the sieve size that will pass 60% and the effective size. The effective size is the sieve size in millimeters that permits 10% of the sand by weight to pass. A mat of solids is formed in the surface layer of sand and must be removed periodically. The dry surface mat can be scraped clean, but periodically the top 6 in. (15 cm) or so of mat must be removed and replaced. Plants with sand filters operate at better than 95% removal of biochemical oxygen demand (BOD).

Trickling filters are beds of media, usually rock, over which settled sewage is sprayed. Microorganisms form a slime layer on the media surface and the water passes down over the surface in a thin film. Nutrients from the sewage are adsorbed in the slime layer and absorbed as food by organisms. Filters are ventilated through the underdrainage system or by other means, and thus oxygen, sewage, and organisms are brought together. Plants with trickling filtration have been operated at 90–95% efficiency of BOD removal.

Filter media include various materials such as stone, crushed rock, ceramic shapes, slag, and plastics. Preferred media are stone and crushed rock which do not fragment, flour, or soften on exposure to sewage. Rock sizes range from 1 to 6 in. (2.5 to 15 cm); however, current practice employs sizes between 2- and 4-in. (5- and 10-cm) nominal diameter. Plastic corrugated sheets have been employed on very deep filters. Pretreatment of sewage is normally required. When the waste contains a concentration of dissolved solids, as with milk waste, without any great concentration of settleable solids, the waste may be applied directly to the filter. Some advantage is gained by preaeration so that the waste applied to the filter has some dissolved oxygen.

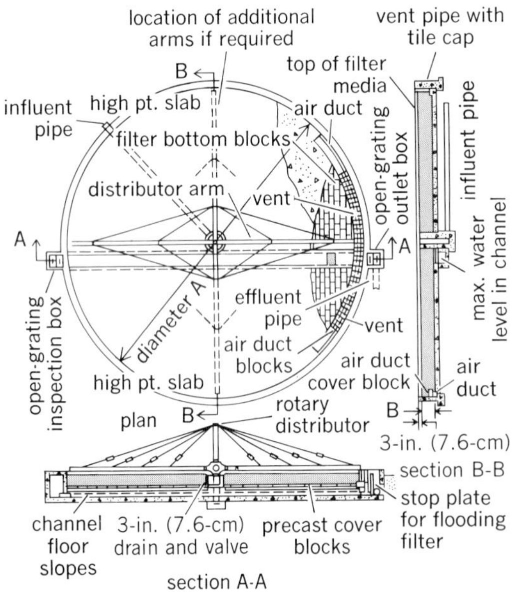

Fig. 10. Circular trickling filter. (*Link-Belt Co.*)

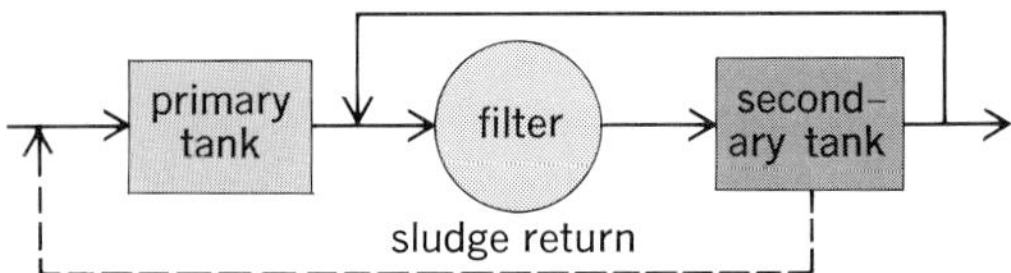

Fig. 11. Flow diagram of a single-stage high-rate trickling filter plant. (*From ASCE-FSIWA Joint Committee, Sewage Treatment Plant Design, 1959*)

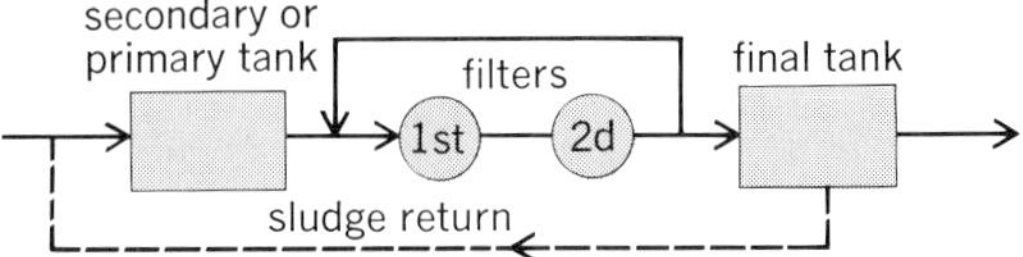

Fig. 12. Flow diagram of a two-stage high-rate trickling filter plant. (*From ASCE-FSIWA Joint Committee, Sewage Treatment Plant Design, 1959*)

Filters are classified as standard or low-rate filters, high-rate filters, and controlled filters. The filter introduced in the United States early in the 20th century was a bed of stone 6–8 ft (1.8–2.4 m) deep with a distribution system of fixed nozzles. This type of filter is called a standard or low-rate filter. The allowable organic loading is about one-third that of a high-rate filter having 3- to 6-ft (0.9- to 1.8-m) depth introduced during 1930–1940 and developed with many variations of recirculation and application of sewage since that time. In 1956 controlled filtration on sectionalized units composing a deep filter was introduced. The loading rate with no recirculation on such filters is 10–12 times that of low-rate filters.

Low-rate filters are dosed at a rate of 1–4 mgad ($9–37 \times 10^3$ m^3 per hectare per day) by siphon through nozzles so spaced that water reaches every part of the filter surface during a dosing cycle. The application of water by this method is intermittent. The rotary distributor (Fig. 10) may also be operated by siphon. This type of distributor has two or four radial arms supported on a center pedestal. Hydraulic force of water passing through the nozzles fixed to the arm causes the arm to rotate. The distributor may be operated in continuous rotation by feeding from a weir box. In either case the filter is sprayed as the arm passes over a given section and the dosing is intermittent with a short time interval between doses. With the fixed-nozzle method the interval may be 5 min, but with the rotary distributor the dosing interval may be no more than 15 sec.

High-rate filters depend on recirculation. The hydraulic loading rate is about 20 mgad (187×10^3 m^3 per hectare per day) with a range of 9–44 mgad ($84–412 \times 10^3$ m^3 per hectare per day). Rotary distributors are used. Pumps pick up settled effluent and return it. Filters are often set up as primary and secondary filters with recirculation of water to each. Several alternative flow arrangements are demonstrated in Figs. 11 and 12. Recirculation ratios range from 1:1 to about 5:1. Final sedimentation is required for both low- and high-rate filters as filter slime and organic debris are washed free.

Aeration. Aeration is accomplished in tanks in which compressed air is diffused in liquid by various devices: filter plates, filter tubes, ejectors, and jets; or in which air is mixed with liquid by mechanical agitation. The high degree of treatment possible with conventional activated sludge, 95–98% BOD removal, has made it a popular method of treatment. Sewage organisms seeded in sludge which has passed through treatment are returned to incoming sewage and mixed thoroughly with the liquor. In this way the biota, oxygen supply, and sewage are brought together. Contact aeration utilizes air diffusion to keep a biota suspension thoroughly mixed; however, the biota are also maintained in active growth on plates of impervious material such as cement-asbestos suspended in the mixed liquor of the aeration tank. Slime growth forms on the plates, and liquid passing by them furnishes the plate biota with nutrients.

Activated-sludge process, the conventional process, requires an aeration period of 4–8 hr. Much of the oxidation takes place in the first 3 hr of detention. Aeration tanks are usually long, narrow, rectangular tanks with porous plates or diffusers along the length to keep the liquor well agitated throughout (Fig. 13). Widths are 15–30 ft (4.6–9.1 m) and depths about 15 ft (4.6 m). Length-width ratio is about 5:1.

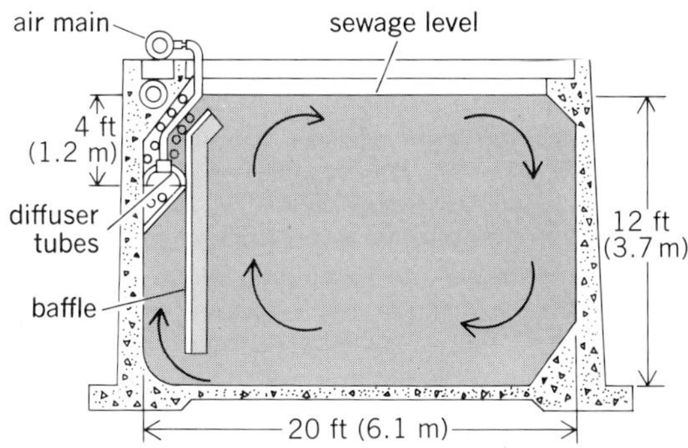

Fig. 13. Cross section of a spiral-flow activated-sludge tank with cylindrical diffusers. (*From E. W. Steel, Water Supply and Sewerage, 4th ed., McGraw-Hill, 1960*)

Air requirements are 0.2–1.5 ft^3 air/gal (1.5–11.2 m^3 air/m^3) of sewage treated. It is necessary to maintain dissolved oxygen (DO) levels at 2 ppm or higher.

Mechanical aeration is done in square or rectangular aeration tanks, depending on the mechanism. In the Simplex method liquor is drawn by impeller up a draft tube and expelled over the tank surface (Fig. 14). In the Link-Belt unit, brushes introduce a spiral motion with considerable agitation. The period of aeration may be up to 8 hr with this method (Fig. 15). Modifications of the aeration process include modified aeration, step aeration, tapered aeration, stage aeration, biosorption, bioactivation, dual aeration, and others.

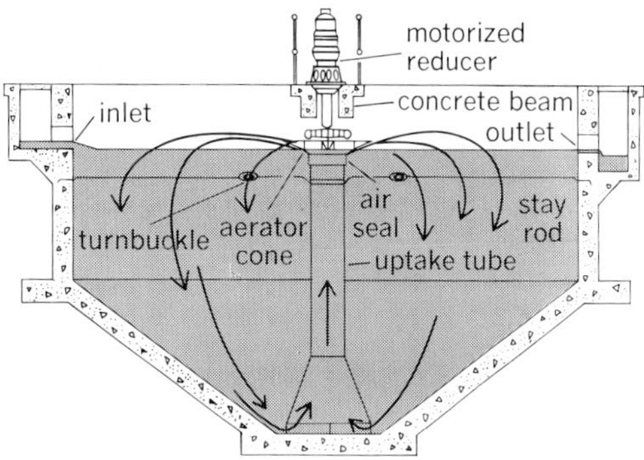

Fig. 14. Simplex aerator. (*From E. W. Steel, Water Supply and Sewerage, 4th ed., McGraw-Hill, 1960*)

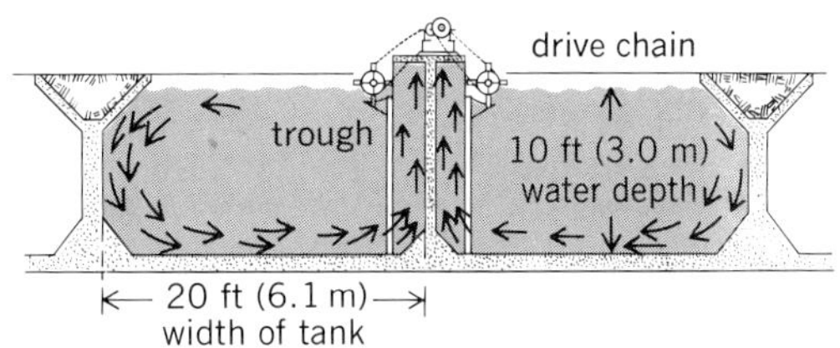

Fig. 15. Cross section of a Link-Belt mechanical aerator. (*From E. W. Steel, Water Supply and Sewerage, 4th ed., McGraw-Hill, 1960*)

Recirculation of sludge is one of the essentials of the process. About 25–35% of the sludge settled in the final sedimentation tank is returned to the aeration tank (Fig. 16). Concentration of solids in mixed liquor may be about 3000 mg/liter in diffused air units and a little less in mechanical aeration units. The ratio of sludge volume settled to suspended solids is the Mohlmann index:

$$\text{Mohlmann index} = \frac{\text{volume of sludge settled in 30 min, \%}}{\text{suspended solids, \%}}$$

A good settling sludge has an index below 100.

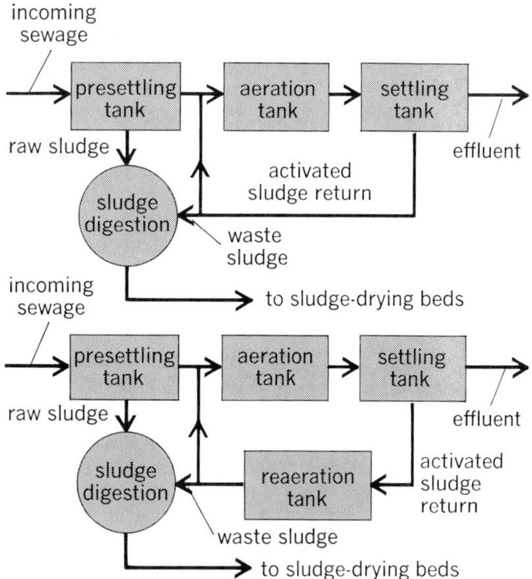

Fig. 16. Flow diagrams of typical activated-sludge plants. (*From E. W. Steel, Water Supply and Sewerage, 4th ed., McGraw-Hill, 1960*)

Sludge age, another important factor, is the average time that a particle of suspended solids remains under aeration and is the ratio of the dry weight of sludge in the tank in pounds to the suspended solids load in pounds per day.

Contact aerators provide an aeration period of 5 hr or more. Aeration is usually preceded by preaeration of the raw sewage before primary settling. The preaeration lasts 1 hr. Loadings are based on two factors: pounds per day per 1000 ft² of contact surface (6.0 or less, or 29.3 kg or less per day per 1000 m² of contact surface), and pounds per day per 1000 ft² per hour of aeration (1.2 or less, or 5.9 kg per day per 1000 m² per hour of aeration). Air supply of 1.5 ft³/gal (11.2 m³/m³) of flow is required. The process has an overall plant efficiency of about 90% BOD removal.

Chlorination. Chlorination of treated sewage has one major purpose: to reduce the coliform group of organisms. Sufficient chlorine to satisfy demand and provide a residual of 2.0 mg/liter should be added. The following magnitude of dosage is possible: primary effluent, 20 mg/liter; trickling filter plant effluent, 15 mg/liter; activated sludge plant effluent, 8 mg/liter; sand filter effluent, 6 mg/liter. The contact period should be at least 15 min at peak hourly flow.

Oxidation ponds. These are ponds 2–4 ft (0.6–1.2 m) in depth designed to allow the growth of algae under suitable conditions in sewage media. Oxygen is absorbed from the air, but the conversion of CO_2 to O_2 by *Chlorella pyrenoidosa* and other algae provides an additional source of oxygen of great value. Oxidation ponds should be preceded by primary treatment. A loading figure of 50 lb BOD/acre (56 kg BOD/hectare) is recommended. BOD removal efficiency may range from 40 to 70%. [WILLIAM T. INGRAM]

Bibliography: R. W. James, *Sewage Sludge Treatment and Disposal*, 1977; D. Sundstrom and H. E. Klei, *Wastewater Treatment*, 1979; G. Tchobanoglous, *Wastewater Engineering: Collection, Treatment, and Disposal*, 1979.

Shaft balancing

The process (often referred to as rotor balancing) of redistributing the mass attached to a rotating body in order to reduce vibrations arising from centrifugal force.

Static versus dynamic balancing. A rotating shaft supported by coaxial bearings (for example, ball bearings together with any attached mass, such as a turbine disk or motor armature) is called a rotor. If the center of mass (CM) of a rotor is not located exactly on the bearing axis, a centrifugal force will be transmitted via the bearings to the foundation. The horizontal and vertical components of this force are periodic shaking forces that can travel through the foundation to create serious vibration problems in neighboring components. The magnitude of the shaking force is $F = me\omega^2$, where m = rotor mass, ω = angular speed, and e = distance (called eccentricity) of the CM from the rotation axis. The product me, called the unbalance, depends only on the mass distribution, and is usually nonzero because of unavoidable manufacturing tolerances, thermal distortion, and so on. When the CM lies exactly on the axis of rotation ($e = 0$), no net shaking force occurs and the rotor is said to be in static balance. Static balance is achieved in practice by adding or subtracting balancing weights at any convenient radius until the rotor shows no tendency to turn about its axis, starting from any given initial orientation. Static balancing is adequate for relatively thin disks or short rotors (such as automobile wheels). However, a long rotor (for example, a turbogenerator) may be in static balance and still exert considerable forces upon individual bearing supports. Suppose that a long shaft is supported on two bearings, and that the unbalance m_1e_1 of the left-hand half of the shaft is equal in magnitude to the unbalance m_2e_2 of the right-hand side, but e_1 and e_2 have opposite signs. In this case, the rotor is in static balance; however, equal but oppositely directed centrifugal forces act on each of the two bearings, transmitting a so-called shaking couple to the foundation. This condition is called dynamic unbalance.

It may be shown that any rigid shaft may be dynamically balanced (that is, the net shaking force and the shaking couple can be simultaneously eliminated) by adding or subtracting a definite amount of mass at any convenient radius in each of two arbitrary transverse cross sections of the rotor. The so-called balancing planes selected for this purpose are usually located near the ends of the rotor, where suitable shoulders or balancing rings have been machined to permit the convenient addition of mass (lead weights, calibrated bolts, and so on) or the removal of mass (by drilling or grinding). Long rotors, running at high speeds, may undergo appreciable elastic deformations. For such flexible rotors it is necessary to utilize more than two balancing planes.

Balancing machines and procedures. The most common types of rotor balancing machines consist of bearings held in pedestals that support the rotor, which is spun at constant speed (as by a belt drive or a compressed air stream). Electromechanical transducers sense the unbalance forces (or associated vibrations) transmitted to the pedestals, and electric circuits automatically perform the calculations necessary to predict the location and amount of balancing weight to be added or subtracted in preselected balancing planes.

For very large rotors, or for rotors which drive several auxiliary devices, commercial balancing machines may not be convenient. The field balancing procedures for such installations may involve the use of accelerometers on the bearing housings, along with vibration meters and phase discriminators (possibly utilizing stroboscopy) to determine the proper location and amount of balance weight.

Committees of the International Standards Organization (ISO) and the American National Standards Institute (ANSI) have formulated recommendations for the allowable quality grade $G = e\omega$ for various classes of machines. A typical value of G (for fans, machine tools, and so on) is 6.3 mm/s, but values as high as 1600 mm/s (for crankshaft assemblies of large two-cycle engines) and as low as 0.4 mm/s (for gyroscopes and precision grinders) have been recommended. *See* MACHINING OPERATIONS; MECHANICAL VIBRATION.

[BURTON PAUL]

Bibliography: J. P. Den Hartog, *Mechanical Vibrations*, 1956; F. Fujisaw et. al., Experimental investigation of multi-span rotor balancing using least squares method, *J. Mech. Des., Trans. ASME*, 102:589–596, 1980; D. Muster and D. G. Stadelbauer, Balancing of rotating machinery, in C. M. Harris and C. E. Crede (eds.), *Shock and Vibrations Handbook*, 1976; B. Paul, *Kinematics and Dynamics of Planar Machinery*, 1979.

Shafting

The machine element that supports a roller and wheel so that they can perform their basic functions of rotation. Shafting, made from round metal bars of various lengths and machined to dimension the surface, is used in a great variety of shapes and applications. Because shafts carry loads and transmit power, they are subject to the stresses and strains of operating machine parts. Standardized procedures have been evolved for determining the material characteristics and size requirements for safe and economical construction and operation.

Types. Most shafting is rigid and carries bending loads without appreciable deflection. Some shafting is highly flexible; it is used to transmit motion around corners.

Solid shafting. The normal form of shafting is a solid bar. Solid shafting is obtainable commercially in round bar stock up to 6 in. (15 cm) in diameter; it is produced by hot-rolling and cold-drawing or by machine-finishing with diameters in increments of 1/4 in. (6 mm) or less. For larger sizes, special rolling procedures are required, and for extremely large shafts, billets are forged to the proper shape. Particularly in solid shafting, the shaft is stepped to allow greater strength in the middle portion with minimum diameter on the ends at the bearings. The steps allow shoulders for positioning the various parts pressed onto the shaft during the rotor assembly.

Hollow shafting. To minimize weights, solid shafting is bored out or drilled, or hollow pipes and tubing are used. Hollow shafts also allow internal support or permit other shafting to operate through the interior. The main shaft between the air compressor and the gas turbine in a jet aircraft engine is hollow to permit an internal speed reduction shaft with the minimum requirement of space and weight. A hollow shaft, to have the same strength in bending and torsion, has a larger diameter than a solid shaft, but its weight is less. The center of large shafts made from ingots are often bored out to remove imperfections and also to allow visual inspection for forging cracks.

Functions. Shafts used in special ways are given specific names, although fundamentally all applications involve transmission of torque.

Axle. The primary shafting connection between a wheel and a housing is an axle. It may simply be the extension of a round member from each side of the rear of a wagon, and on the end of each the hub of a wagon wheel rotates. Similarly, railroad car axles are large, round bars of steel spanning between the car wheels, supporting the car frame with bearings on the axle outside the wheels. Axles normally carry only transverse loads, as in the examples above, but occasionally, as in rear automobile housings, they also transmit torsion loads.

Spindle. A short shaft is a spindle. It may be slender or tapered. A spindle is capable of rotation or of having a body rotate upon it. It is similar to an arbor or a mandrel, and usage defines the small drive shaft of a lathe as a live spindle. The term originated from the round tapering stick on a spinning wheel on which the thread is twisted.

Head. A short stub shaft mounted as part of a motor or engine or extending directly therefrom is a head shaft. An example is the power takeoff shaft on a tractor.

Countershaft. A secondary shaft that is driven by a main shaft and from which power is supplied to a machine part is called a countershaft. Often the countershaft is driven by gears, and thus rotates counter to the direction of the main shaft. Countershafts are used in gear transmissions to obtain speed and torque changes in transmitting power from one shaft to another.

Jackshaft. A countershaft, especially when used as an auxiliary shaft between two other shafts, is termed a jackshaft.

Line shafting. One or more pieces of shafting joined by couplings is used to transmit power from, for example, an engine to a remotely located machine. A single engine can drive many lines of shafting which, in turn, connect in multiple fashion to process equipment machines. Belts operate on pulleys to transmit the torque from one line to another and from the shafting to the machines. Clutches and couplings control the transfer of power from the shafting.

The delivery of power to the machines in a shop has generally been converted from line shafting to individual electric motor drives for each machine. Thus, in a modern processing plant, line shafting is obsolete. *See* BELT DRIVE; PULLEY.

[JAMES J. RYAN]

Shaper

A machine tool for cutting flat or flat, contoured surfaces by reciprocating a single-point tool across the workpiece. A shaper is usually used for small pieces requiring a short cutter stroke rather than

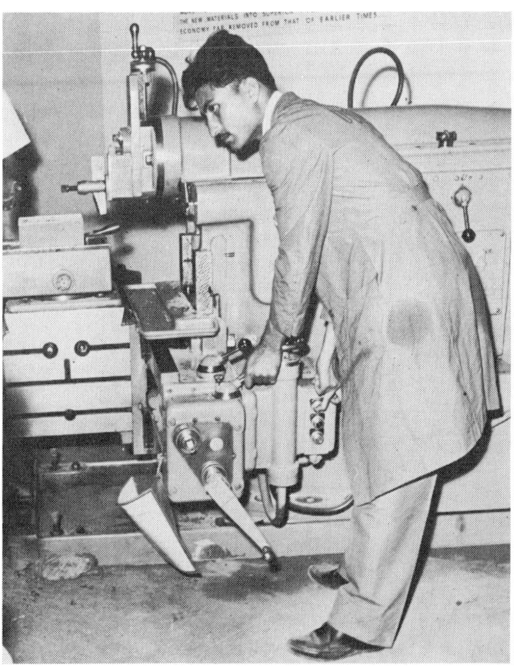

Horizontal ram-type shaper. (*Rockford Machine Tool Co.*)

for large pieces requiring longer strokes provided by a planer.

Shapers are classified as either horizontal or vertical depending on the plane of motion of the reciprocating ram. The vertical type is often referred to as a slotter. The maximum stroke provided by the ram designates the size of the machine.

On horizontal shapers, a movable tool head mounted on the end of the ram permits angular cuts as illustrated. A hinged clapper box provides tool relief on the return stroke, while a movable table holds the workpiece. Vertical machines hold the tool firmly on the ram while the workpiece is fed into the tool by moving the table. *See* MACHINING OPERATIONS. [ALAN H. TUTTLE]

Shear

A straining action wherein applied forces produce a sliding or skewing type of deformation. A shearing force acts parallel to a plane as distinguished from tensile or compressive forces, which act normal to a plane. Examples of force systems producing shearing action are forces transmitted from one plate to another by a rivet that tend to shear the rivet, forces in a beam that tend to displace adjacent segments by transverse shear, and forces acting on the cross section of a bar that tend to twist it by torsional shear (Fig. 1). Shear forces are usually accompanied by normal forces produced by tension, thrust, or bending.

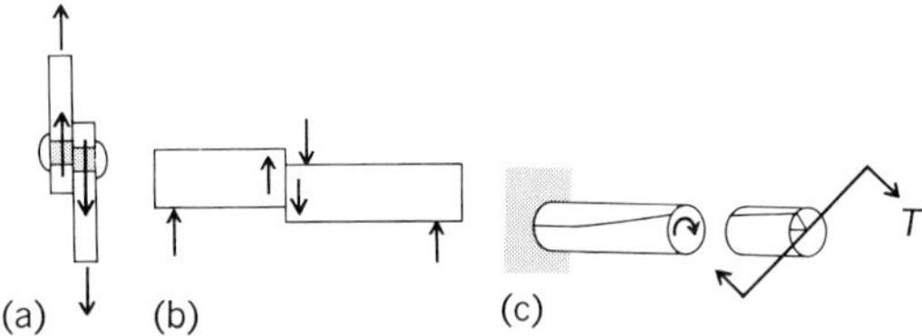

Fig. 1. Shearing actions. (*a*) Single shear on rivet. (*b*) Transverse shear in beam. (c) Torsion.

Shearing stress is the intensity of distributed force expressed as force per unit area. Shear stresses on mutually perpendicular planes at a point in a stressed body are equal. When no normal stresses exist, the state of stress is pure shear, which induces both normal and shear stresses on oblique planes. Elements oriented 45° to planes of pure shear are subjected to biaxial tension and compression equal to the shear stress (Fig. 2*a*). Similarly, pure shear is induced by equal and opposite biaxial stresses (Fig. 2*b*). Pure shear is the maximum shear stress at the point. Under combined stress, the shear stress is found by principal stress analysis.

Shearing strain is the displacement ϵ_s of two parallel planes, unit distance apart, which accompanies shear stresses acting on these planes. Shearing distortion is visualized as sliding without separation or contraction of all planes parallel to the shear forces, like cards in a pack (Fig. 3*a*). For planes unit distances apart, the relationship is shown in Eq. (1), where ϕ is the small angle of distortion.

$$\epsilon_s = \tan\phi \approx \phi \text{ radians} \quad (1)$$

Modulus of rigidity, designated E_s or G, is the shearing modulus of elasticity, which according to Hooke's law is the constant of proportionality between shearing stress S_s and shearing strain S_s/ϕ during elastic behavior.

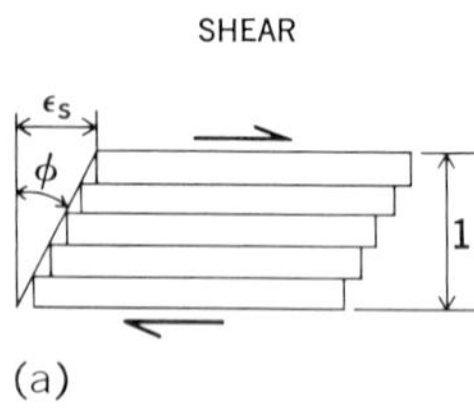

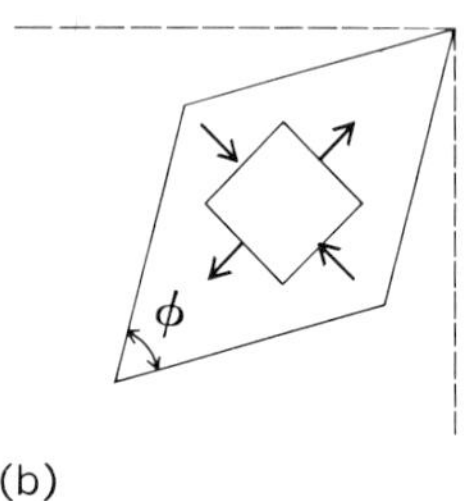

Fig. 3. Shear (*a*) strain and (*b*) distortion.

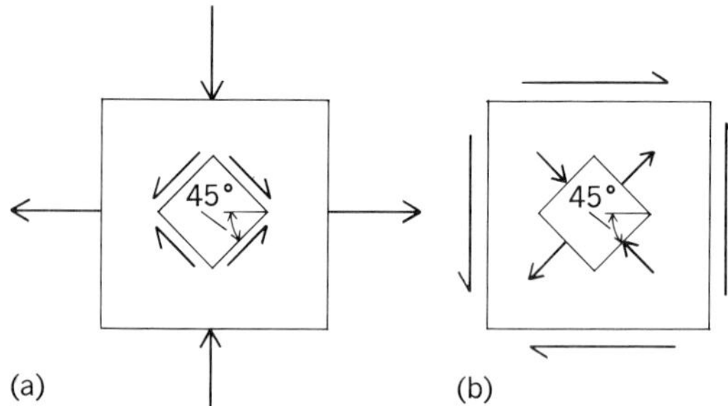

Fig. 2. State of pure shear. (*a*) Shear produces biaxial tension and compression. (*b*) Biaxial stresses induce shear.

After shear distortion, the angles of a rectangular element are altered by the shear strain ϕ (Fig. 3*b*). Changes in length of diagonals must be consistent with the biaxial strains, which leads to the relationship shown in Eq. (2), where E is Young's modulus and μ is Poisson's ratio.

$$E_s = \frac{E}{2(1+\mu)} \quad (2)$$

The value of μ can be found when E_s and E are experimentally determined. *See* STRESS AND STRAIN.

[JOHN B. SCALZI]

Shear center

The point in the plane through a section of a structural member at which a shear force can be applied without producing a twist of that section. The

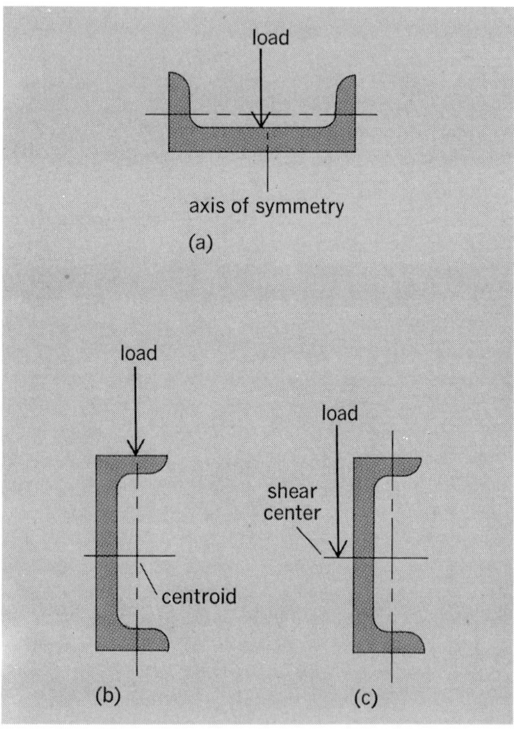

Three conditions of channel section under load. Channel (*a*) resists bending, (*b*) twists, and (*c*) resists twisting.

shear center of a section through a member, such as a rolled wide-flange beam, that has two planes of symmetry coincides with the geometric center or centroid of the section. When such a member is loaded transversely, it bends without twisting. However, the structural member may be unsymmetrical or may have but one plane of symmetry and may be loaded elsewhere than through that plane, as in an open section made of thin material for resisting bending in aircraft construction; then the load force and the reaction force of the structural member constitute a couple that, in general, causes the member to twist. The load can be applied through a unique point in the plane of the section so that the moments in the plane of the section are balanced and the beam will not twist. This unique point is called the shear center. In the illustration, in *a* the channel bends when the load is along the axis of symmetry; in *b* the channel twists, even with the load through the centroid if the load is not along the axis of symmetry; in *c* the channel resists twisting when the load is through the shear center.

Location of the shear center for a section depends only on the section dimensions. The member is subject to twisting only when subjected to shear. Shear center is of no significance for a beam in pure bending. *See* BEAM; LOADS, TRANSVERSE; SHEAR.

[JOHN B. SCALZI]

Bibliography: B. Bresler, T. Y. Lin, and J. B. Scalzi, *Design of Steel Structures*, 2d ed., 1972.

Sheet-metal forming

The shaping of thin sheets of metal (usually less than 1/4 in. or 6 mm) by applying pressure through male or female dies or both. Parts formed of sheet metal have such diverse geometries that it is difficult to classify them. In all sheet-forming processes, excluding shearing, the metal is subjected to primarily tensile or compressive stresses or both. Sheet forming is accomplished basically by processes such as stretching, bending, deep drawing, embossing, bulging, flanging, roll forming, and spinning. In most of these operations there are no intentional major changes in the thickness of the sheet metal. *See* METAL FORMING.

There are certain basic considerations which are common to all sheet forming. Grain size of the metal is important in that too large a grain produces a rough appearance when formed, a condition known as orange peel. For general forming an ASTM no. 7 grain size (average grain diameter 0.00125 in. or 32 μm) is recommended. Another type of surface irregularity observed in materials such as low carbon steel is the phenomenon of yield-point elongation that results in stretcher strains or Lueder's bands, which are elongated depressions on the surface of the sheet. This is usually avoided by cold-rolling the original sheet with a reduction of only 1–2% (temper rolling.) Since yield-point elongation reappears after some time, because of aging, the material should be formed within this time limit. Another defect is season cracking (stress cracking, stress corrosion cracking) which occurs when the formed part is in a corrosive environment for some time. The susceptibility of metals to season cracking depends on factors such as type of metal, degree of deformation, magnitude of residual stresses in the formed part, and environment. *See* METAL, MECHANICAL PROPERTIES OF.

Anisotropy or directionality of the sheet metal is also important because the behavior of the material depends on the direction of deformation. Anisotropy is of two kinds: one in the direction of the sheet plane, and the other in the thickness direction. These aspects are important, particularly in deep drawing.

Formability of sheet metals is of great interest, even though it is difficult to define this term because of the large number of variables involved. Failure in sheet forming usually occurs by localized necking or buckling or both, such as wrinkling or folding. For a simple tension-test specimen the true (natural) necking strain is numerically equal to the strain-hardening exponent of the material; thus, for instance, commercially pure annealed aluminum or common 304 stainless steel stretches more than cold-worked steel before it begins to neck. However, because of the complex stress systems in most forming operations, the maximum strain before necking is difficult to determine, although some theoretical solutions are available for rather simple geometrics. *See* STAINLESS STEEL.

Considerable effort has been expended to simulate sheet-forming operations by simple tests. In addition to bend or tear tests, cupping tests have also been commonly used, such as the Swift, Olsen, and Erichsen tests. Although these tests are practical to perform and give some indication of the formability of the sheet metal, they generally cannot reproduce the exact conditions to be encountered in actual forming operations.

Stretch forming. In this process the sheet metal is clamped between jaws and stretched over a form block. The process is used primarily in the aerospace industry to form large panels with varying curvatures. Stretch forming has the advantages of low die cost, small residual stresses, and virtual elimination of wrinkles in the formed part. *See* JIG, FIXTURE, AND DIE DESIGN.

Bending. This is one of the most common processes in sheet forming. The part may be bent not only along a straight line, but also along a curved path (stretching, flanging). The minimum bend radius, measured to the inside surface of the bend, is important and determines the limit at which the material cracks either on the outer surface of the bend or at the edges of the part. This radius, which is usually expressed in terms of multiples of the sheet thickness, depends on the ductility of the material, width of the part, and its edge conditions. Experimental studies have indicated

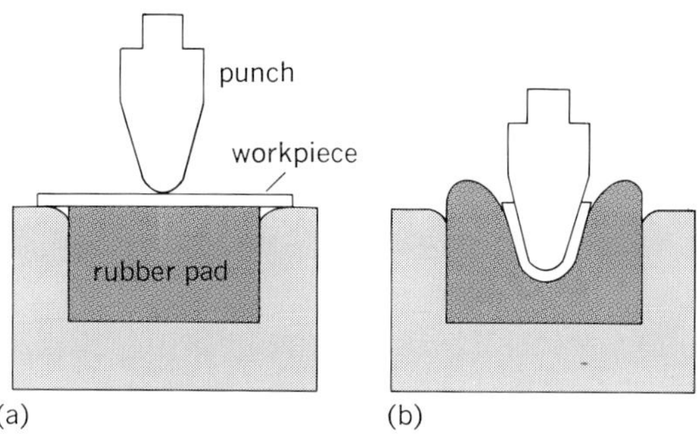

Fig. 1. Bending process with a rubber pad. (*a*) Before forming. (*b*) After forming.

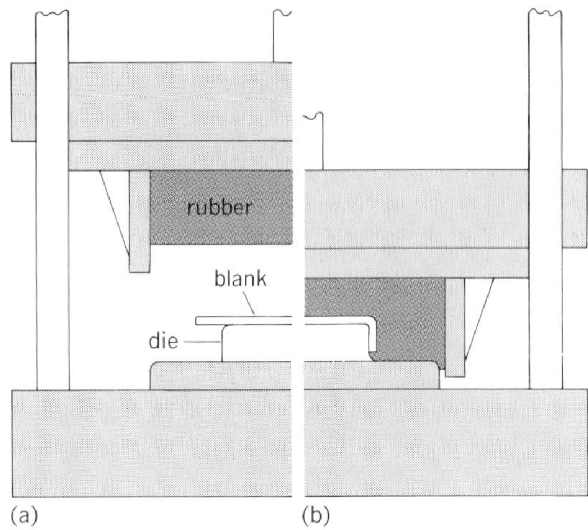

Fig. 2. The Guerin process, the simplest rubber forming process. (*a*) Before forming. (*b*) After forming.

that if bending is carried out in a pressurized environment, with pressures as high as 500,000 psi (3.4 GPa), the minimum bend radius decreases substantially. This observation is important in forming brittle materials. Heating the workpiece is another method of improving bendability.

Springback in bending and other sheet-forming operations is due to the elastic recovery of the metal after it is deformed. Determination of springback is usually done in actual tests. Compensation for springback in practice is generally accomplished by overbending the part; adjustable tools are sometimes used for this purpose.

In addition to male and female dies used in most bending operations, the female die can be replaced by a rubber pad (Fig. 1). In this way die cost is reduced and the bottom surface of the part is protected from scratches by a metal tool. The roll-forming process replaces the vertical motion of the dies by the rotary motion of rolls with various profiles. Each successive roll bends the strip a little further than the preceding roll. The process is economical for forming long sections in large quantities.

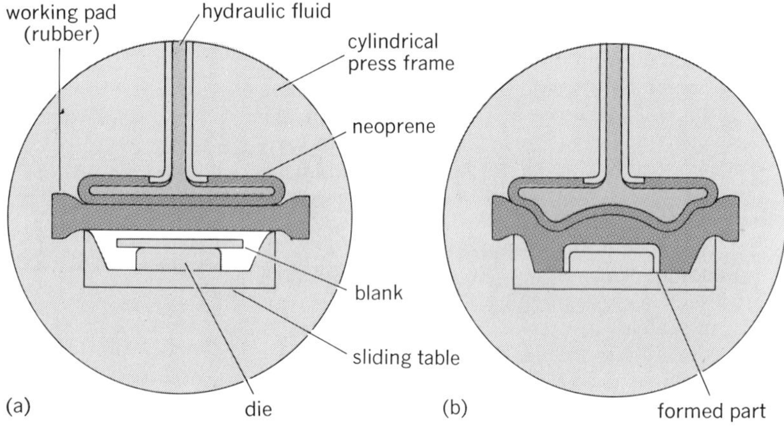

Fig. 3. The Verson-Wheelon process. (*a*) Before forming. (*b*) After forming.

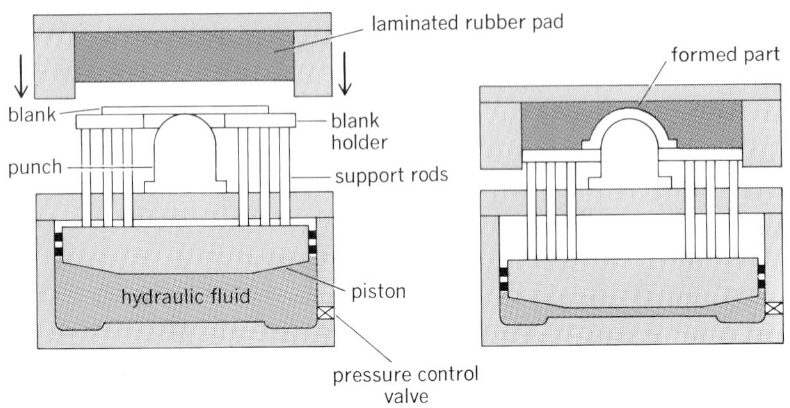

Fig. 4. The Marform process.

Rubber forming. While many sheet-forming processes are carried out in a press with male and female dies usually made of metal, there are four basic processes which utilize rubber to replace one of the dies. Rubber is a very effective material because of its flexibility and low compressibility. In addition, it is low in cost, is easy to fabricate into desired shapes, has a generally low wear rate, and also protects the workpiece surface from damage.

The simplest of these processes is the Guerin process (Fig. 2). Auxiliary devices are also used in forming more complicated shapes. In the Verson-Wheelon process (Fig. 3) hydraulic pressure is confined in a rubber bag, the pressure being about five times greater than that in the Guerin process. For deeper draws the Marform process is used (Fig. 4). This equipment is a packaged unit that can be installed easily into a hydraulic press. In deep drawing of critical parts the Hydroform process (Fig. 5) is quite suitable, where pressure in the dome is as high as 15,000 psi (100 MPa). A particular advantage of this process is that the formed portions of the part travel with the punch, thus lowering tensile stresses which can eventually cause failure.

Bulging of tubular components, such as coffee pots, is also carried out with the use of a rubber pad placed inside the workpiece; the part is then expanded into a split female die for easy removal.

Deep drawing. The basic components of a deep-drawing operation are shown in Fig. 6. A great variety of parts are formed by this process, the successful operation of which requires a careful control of factors such as blank-holder pressure, lubrication, clearance, material properties, and die geometry. Depending on many factors, the maximum ratio of blank diameter to punch diameter ranges from about 1.6 to 2.3. *See* DRAWING OF METAL.

This process has been extensively studied, and the results show that two important material properties for deep drawability are the strain-hardening exponent and the strain ratio (anisotropy ratio) of the metal. The former property becomes dominant when the material undergoes stretching, while the latter is more pertinent for pure radial drawing. The strain ratio is defined as the ratio of the true strain in the width direction to the true strain in the thickness direction of a strip of the sheet metal. The greater this ratio, the greater is the ability of the metal to undergo change in its width direction while resisting thinning.

Anisotropy in the sheet plane results in earing, the appearance of wavy edges on drawn cups. Clearance between the punch and the die is another factor in this process; this is normally set at a value of not more than 1.4 times the thickness of the sheet. Too large a clearance produces a cup whose thickness increases toward the top, whereas correct clearance produces a cup of uniform thickness by ironing. Also, if the blank-holder pressure is too low, the flange wrinkles; if it is too high, the bottom of the cup will be punched out because of the increased frictional resistance of the flange. For relatively thick sheets it is possible to draw parts without a blank holder by special die designs.

Miscellaneous processes. Many parts require one or more additional processes; some of these are described briefly here. Embossing consists of forming a pattern on the sheet by shallow drawing.

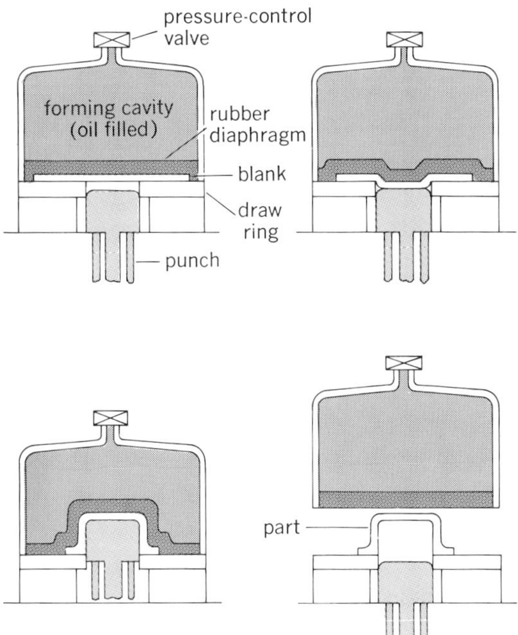

Fig. 5. The Hydroform process.

Coining consists of putting impressions on the surface by a process that is essentially forging, the best example being the two faces of a coin. Coining pressures are quite high, and control of lubrication is essential in order to bring out all the fine detail in a design. Shearing is separation of the material by the cutting action of a pair of sharp tools, similar to a pair of scissors. The clearance in shearing is important in order to obtain a clean cut. A variety of operations based on shearing are punching, blanking, perforating, slitting, notching, and trimming.

Sheet-forming equipment. The most common equipment is a mechanical or hydraulic press. Presses may be single, double, or triple acting, depending on the number of slides to do a certain portion of the operation. Common mechanical presses are the eccentric, crank, knuckle joint, and toggle. The design of dies requires considerable experience because of the great number of

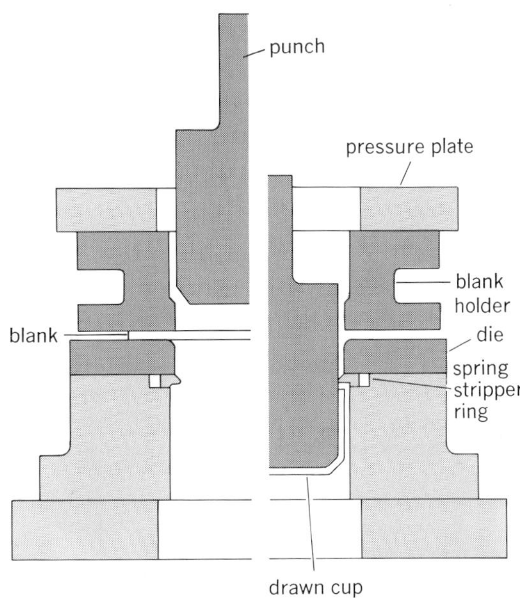

Fig. 6. The deep drawing process.

problems that may arise in production. Tool and die materials for sheet forming are generally made of cast irons, cast alloys, die steels, and cemented carbides for high-production work. Nonmetallic materials such as rubber, plastics, and hardwood are also used as die materials. The selection of the proper lubricant depends on many factors, such as die and workpiece materials, and severity of the operation. A great variety of lubricants are commercially available, such as drawing compounds, fatty acids, mineral oils, and soap solutions.

Pressures in sheet-metal forming generally range between 1000 and 8000 psi (7 and 55 MPa) (normal to the plane of the sheet); most parts require about 1500 psi (10 MPa).

Spinning. This process forms parts with rotational symmetry over a mandrel with the use of a tool or roller. There are two basic types of spinning: conventional or manual spinning, and shear spinning (Fig. 7). The conventional spinning process forms the material over a rotating mandrel with little or no change in the thickness of the original blank. Parts can be as large as 20 ft (6 m) in diameter. The operation may be carried out at room temperature or higher for materials with low ductility or great thickness. Success in manual

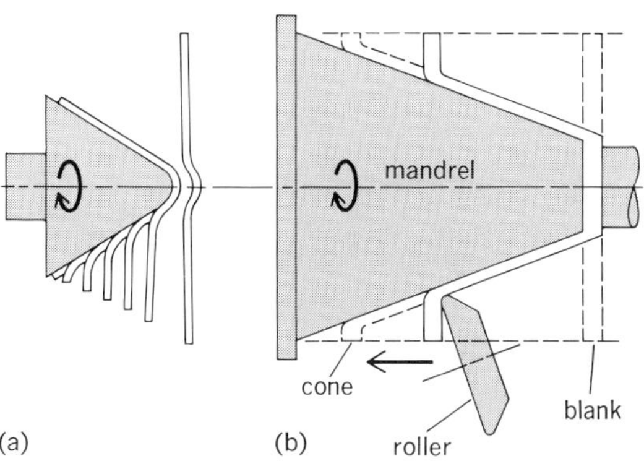

Fig. 7. Spinning processes of (*a*) conventional and (*b*) shear type.

spinning depends largely on the skill of the operator. The process can be economically competitive with drawing; if a part can be made by both processes, spinning may be more economical than drawing for small quantities. *See* SPINNING (METALS).

In shear spinning (hydrospinning, floturning) the deformation is carried out with a roller in such a manner that the diameter of the original blank does not change but the thickness of the part decreases by an amount dependent on the mandrel angle. The spinnability of a metal is related to its tensile reduction of area. For metals with a reduction of area of 50% or greater, it is possible to spin a flat blank to a cone of an included angle of 30° in one operation. Shear spinning produces parts with various shapes (conical, curvilinear, and also tubular by tube spinning on a cylindrical mandrel) with good surface finish, close tolerances, and improved mechanical properties. *See* MACHINING OPERATIONS; METAL COATINGS.

[SEROPE KALPAKJIAN]

Bibliography: D. F. Eary and E. A. Reed, *Techniques of Pressworking Sheet Metal: An Engineering Approach to Die Design*, 2d ed., 1974; D. P. Koistinen and N.-M. Wang (eds.), *Mechanics of Sheet Metal Forming*, 1979; G. Sachs and H. E. Voegeli, *Principles and Methods of Sheet-Metal Fabricating*, 2d ed., 1966.

Shock absorber

Effectively a spring, a dashpot, or a combination of the two, arranged to minimize the acceleration of the mass of a mechanism or portion thereof with respect to its frame or support.

The spring type of shock absorber (Fig. 1) is generally used to protect delicate mechanisms, such as instruments, from direct impact or instantaneously applied loads. Such springs are often made of rubber or similar elastic material. The design of the spring in relation to the natural frequency of the supported system and the forcing frequency of the applied load is most important. *See* SHOCK ISOLATION.

The dashpot type of shock absorber is best illustrated by the direct-acting shock absorber in an automotive spring suspension system (Fig. 2).Here the device is used to dampen and control a spring movement. The energy of the mass in motion is converted to heat by forcing a fluid through a restriction, and the heat is dissipated by radiation and conduction from the shock absorber.

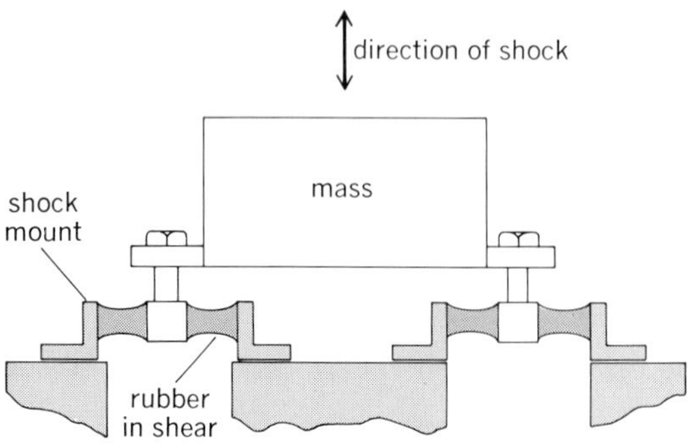

Fig. 1. A spring-type shock absorber.

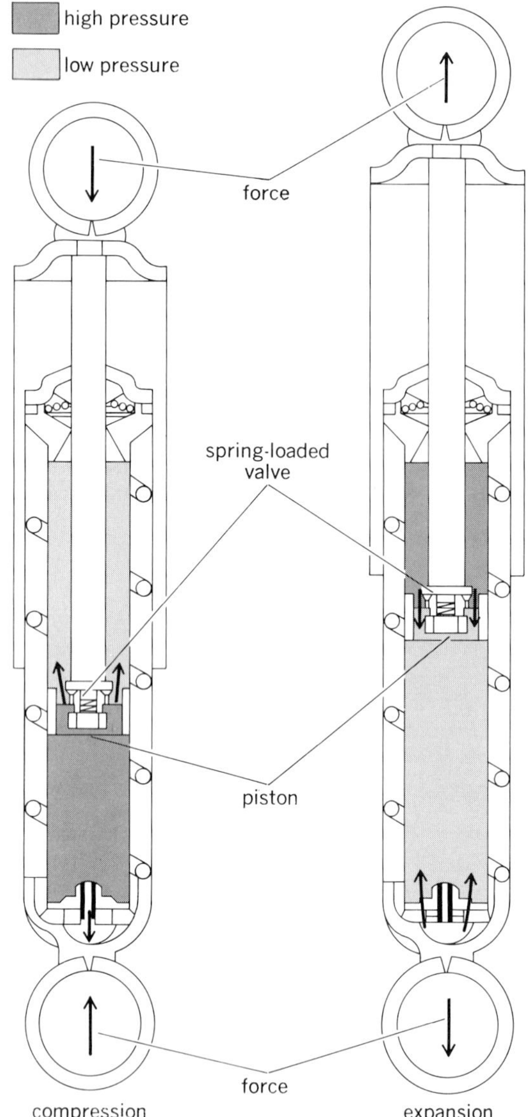

Fig. 2. A dashpot-type shock absorber. (*Plymouth Division, Chrysler Corp.*)

There are also devices available which combine springs and viscous damping (dashpots) in the same unit. They use elastic solids (such as rubber or metal), compressed gas (usually nitrogen), or both for the spring. A flat-viscosity hydraulic fluid is used for the viscous damping.

[L. SIGFRED LINDEROTH, JR.]

Shock isolation

The application of isolators to alleviate the effects of shock on a mechanical device or system. Although the term shock has no universally accepted definition in engineering, it generally denotes suddenness, either in the application of a force or in the inception of a motion.

Shock isolation is accomplished by storing energy in a resilient medium (isolator, cushion, and so on) and releasing it at a slower rate. The effectiveness of an isolator depends upon the duration of the shock impact. An isolator may be effective in one case where there is a high G loading with a short duration, 0.001 millisecond (msec) or less, but may magnify the shock where there is a lower G loading but a longer duration (0.001 – 0.015 msec). The quantity G is equal to the so-called limit acceleration a divided by the acceleration of gravity g and is discussed later. Most shock isolators, also known as shock mounts or shock absorbers, that are available commercially are effective for the 0.001-msec or less interval.

Rubber is the most common material used in commercial shock isolators. Rubber isolators are generally used where the shock forces are created through small displacements. For larger displacement shock forces, such as those experienced by shipping containers in rough handling conditions, thick cushions of felt, rubberized hair, sponge rubber, cork, or foam plastics are used. Shock isolation systems which use the various cushion materials are generally custom designed to the particular application and cannot be considered from the standpoint of standardized isolators, but rather from the standpoint of the basic principles involved.

Absorption of shock. The shock load must be divided between the case, the shock cushion, and the equipment. The case, since it must withstand effects of rough handling such as sliding and dropping, is by necessity rigid. The more rigid the case the closer to a 1:1 ratio will be the transfer of the shock from outside to inside. The absorption of the shock is primarily between the cushion and the equipment.

The dissipation of the energy of a 1-ft (0.3-m) drop with a cushion having a linear spring rate would require a thickness of 2 ft (0.6 m) of cushion. Since cushions of such thickness are not feasible, the equipment itsèlf must withstand part of the shock. The cushion that is needed to dissipate the energy from various heights of drop with equipment sharing part of the load may be determined by consideration of the principles which are involved.

Limit acceleration. When a body moving with velocity v has to be stopped to complete rest, a deceleration (negative acceleration) must be applied. In order to make the stopping process smooth, a maximum value for the acceleration a is prescribed as a limit. Usually the full amount of the limit acceleration cannot be attained for the entire duration of the stopping process. However, an ideal process can be imagined with constant limit acceleration a. Deviation from this ideal case will be considered later.

Such uniformly decelerated motion is exactly the reverse process of uniformly accelerated motion. The same formulas apply.

The total distance traveled S is given by Eq. (1),

$$S=\frac{v^2}{2a} \tag{1}$$

where v is the initial and final velocity. When the velocity to be stopped is produced by a free fall, the necessary height of drop H must be as given by Eq. (2), where g is the gravitational accelera-

$$H=\frac{v^2}{2g} \tag{2}$$

tion. From these formulas Eq. (3) is obtained.

$$v^2=2aS=2gH \tag{3}$$

Since the velocity at the end of the free fall is the same as it was at the beginning of the stopping process, Eq. (4) is obtained. It is customary to give

$$2aS=2gH$$
$$S=H\frac{g}{a} \tag{4}$$

the limit acceleration in the form of a G value, defined by $G=a/g$. With this notation, the stopping distance can be written in the form given in Eq. (5).

$$S=\frac{H}{G} \tag{5}$$

Force-distance diagrams. It is useful to analyze the ideal case again from another aspect. The energy E stored in the falling body after fall from height H is $E=WH$, where W is the weight of the moving body. The same amount of energy must be consumed during the stopping period. Therefore, Eq. (6) is obtained. The active force is GW in this

$$E=GWS \tag{6}$$

stopping phase. Equating the last two equations for the energy yields Eq. (1). As represented in Fig. 1, the energy can be visualized as the area under the curve in a force-distance diagram. The curve is a horizontal line in the case of constant deceleration.

Whatever device is used for checking the velocity, it is hardly possible to obtain a constant deceleration of exactly the limit value. Therefore, allowances have to be made for practical considerations.

If the opposing force is provided by an ordinary spring, no constant force is produced. The force F is built up gradually with the compression of the spring, as shown in Fig. 2. The area under the force-displacement curve is in this case a triangle, representing the energy given in Eq. (7). The kinet-

$$E=\tfrac{1}{2}GWS_1 \tag{7}$$

ic energy from the fall is the same as before, $E=WH$. Equation (8), the stopping distance for the

$$S_1=\frac{2H}{G} \tag{8}$$

spring of Fig. 2, is found by equating Eqs. (6) and (7). This means a distance twice as long as that for the ideal case of Fig. 1 is needed. This less favorable condition is caused by the fact that the permitted maximum force is used only at the end of the process. The diagram of Fig. 2 is not "filled up" completely; it is only half-filled.

Preloaded springs have a better-filled diagram,

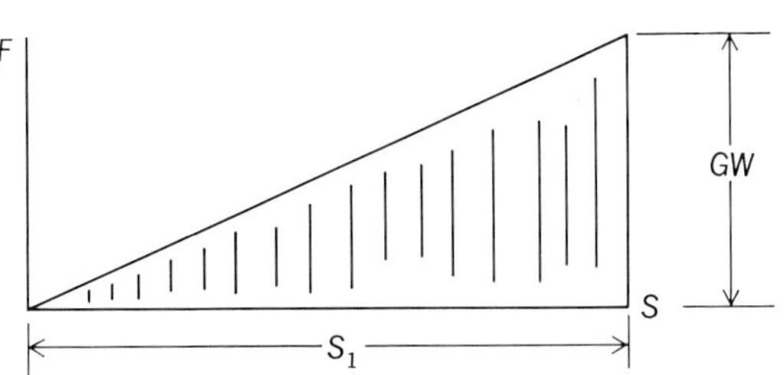

Fig. 2. Force-distance diagram for ordinary spring.

SHOCK ISOLATION

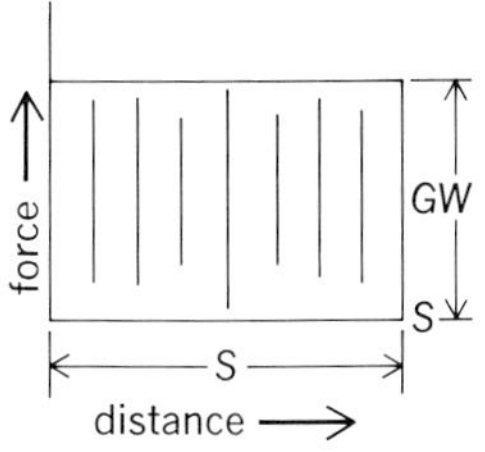

Fig. 1. Force-distance diagram for ideal case.

as shown in Fig. 3. An inherent disadvantage of the better-filled-up spring diagrams is the steep slope at the beginning, which means that the spring system must be designed for a predetermined shock. This design does not help in isolating small shocks. It has not been proved which shocks do the most damage—the large shock that happens occasionally or the small, repeated shock that occurs almost continuously. From the isolation standpoint, the spring characteristic of an isolator should start up with a moderate slope and continue with gradually sharper increase of force. This is why a system representing a completely filled diagram cannot be applied, and therefore a certain deviation from the ideal condition represented by Eq. (1) is unavoidable.

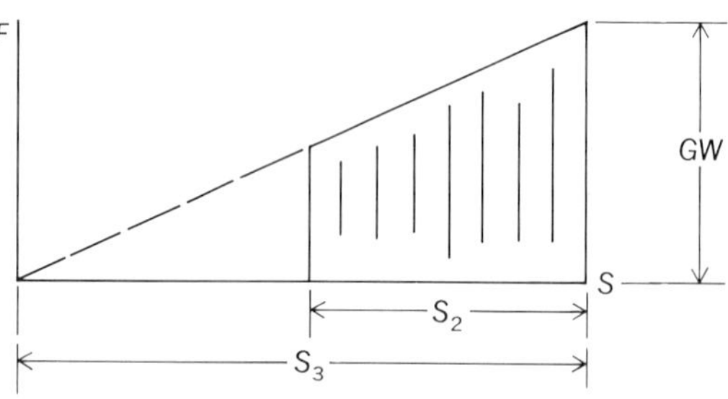

Fig. 3. Force-distance diagram for preloaded spring.

Damping forces are helpful for attaining longer deceleration force at the beginning of the stopping process. Pure viscous damping alone would result in an elliptical diagram like the one shown in Fig. 4.

SHOCK ISOLATION

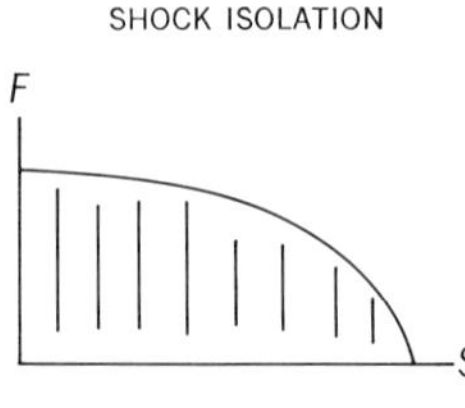

Fig. 4. Force-distance diagram for pure viscous damping.

Since the high velocity at the beginning produces large opposing forces, spring force and damping effects can be combined to make a well-filled diagram, as shown in Fig. 5.

Practical shock absorbers usually fall between the two cases of completely filled diagrams and half-filled diagrams. The stopping distance provided therefore lies between Eqs. (5) and (8). The value given by Eq. (8) is considered conservative and therefore is recommended, since a certain tolerance is necessary.

Using Eq. (8) and assuming that the equipment can withstand $30G$, the operating deflection of the cushion under a 24-in. (60-cm) drop must be as given by Eq. (9). Since S is the operating distance,

$$S = \frac{2 \times 24}{30} = 1.6 \text{ in. (4 cm)} \qquad (9)$$

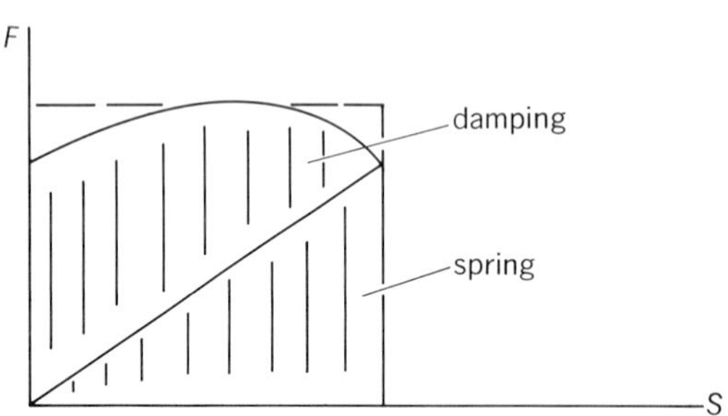

Fig. 5. Force-distance diagram for combined effects of spring force and damping.

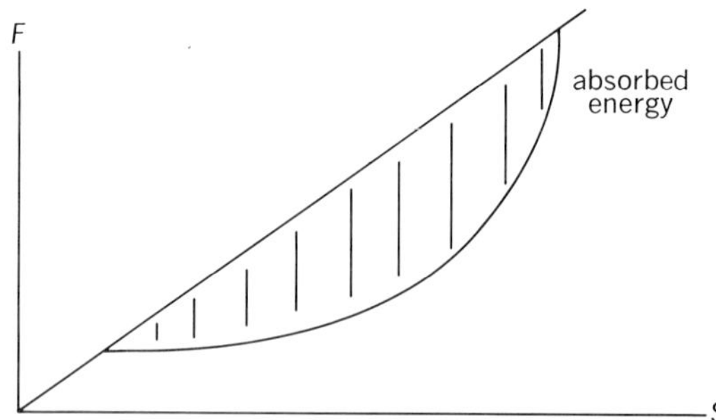

Fig. 6. Diagram of absorbed energy for ideal spring system with damping.

the total thickness of cushion will depend on the compression ratio of the material.

The preceding equation provides the operating thickness needed to bring a given mass to rest with a linear spring cushion, leaving the equipment to share $30G$ of the impact load. It must be remembered that cushions, springs, and so forth store energy and, depending upon their inherent friction, return this energy. Thus, in an ideal spring system, no energy is absorbed. However, by adding damping or by using a material with inherent damping, there is provided a resilient medium that possesses characteristics as shown in Fig. 6. The shaded area describes the absorbed energy.

To accomplish such a system requires engineering information on the cushioning materials under dynamic conditions. Once their behavior is known, the materials that have the desired force-displacement characteristics can be selected, and the damping forces may be added to provide a shock-absorbing system approaching the effectiveness of the one shown in Fig. 6. *See* SHOCK ABSORBER; SPRING; VIBRATION DAMPING.

[K. W. JOHNSON]

Bibliography: C. E. Crede and C. M. Harris (eds.), *Shock and Vibration Control Handbook*, rev. ed., 1976.

Short circuit

A term commonly used to describe an electrical connection of negligible impedance connected across a pair of terminals.

Common use implies an undesirable condition arising from electrical insulation failure due to improper operation, mechanical damage, or damage by natural causes such as lightning and rain. Protection against such short circuits is a major field in electrical power engineering. The location of short circuits, or faults as they are often called, is made by electric bridge measurements. *See* CIRCUIT TESTING (ELECTRICITY); ELECTRIC PROTECTIVE DEVICES.

Although short circuits are undesirable on power transmission lines, they are often used to advantage on high-frequency lines. For instance, a stub transmission line, one-quarter wavelength long and short-circuited at one end, acts as an insulator at the opposite end and therefore is used as a support for a high-frequency transmission line. In a similar manner shorting bars are used to tune transmission lines. *See* CIRCUIT (ELECTRICITY); TRANSMISSION LINES.

[ROBERT LEE RAMEY]

Shrink fit

A fit that has considerable negative allowance so that the diameter of a hole is less than the diameter of a shaft that is to pass through the hole, also called a heavy force fit. Shrink fits are used for permanent assembly of steel external members, as on locomotive wheels. The difference between a shrink fit and a force fit is in method of assembly. Locomotive tires, for instance, would be difficult to assemble by force, whereas a shaft and hub assembly would be convenient for force fit by a hydraulic press. In shrink fits, the outer member is heated, or the inner part is cooled, or both, as required. The parts are then assembled and returned to the same temperature. *See* ALLOWANCE; FORCE FIT.

[PAUL H. BLACK]

Simple machine

Any of several elementary machines, one or more of which is found in practically every machine. The group of simple machines usually includes only the lever, wheel and axle, pulley (or block and tackle), inclined plane, wedge, and screw. However, the gear drive and hydraulic press may also be considered as simple machines. The principles of operation and typical applications of simple machines depend on several closely related concepts. *See* EFFICIENCY; FRICTION; MECHANICAL ADVANTAGE; POWER; WORK.

Two conditions for static equilibrium are used in analyzing the action of a simple machine. The first condition is that the sum of forces in any direction through their common point of action is zero. The second condition is that the summation of torques about a common axis of rotation is zero. Corresponding to these two conditions are two ways of measuring work. In machines with translation, work is the product of force and distance. In machines with rotation, work is the product of torque and angle of rotation. *See* BLOCK AND TACKLE; GEAR DRIVE; HYDRAULIC PRESS; INCLINED PLANE; LEVER; SCREW; WEDGE; WHEEL AND AXLE.

Work is the product of a force and the distance through which it moves. For example, the work done in raising a 10-lb object 15 ft is 150 ft-lb. In this example the work done on the weight goes into increasing the potential energy of the object. Work and energy, both potential and kinetic, have the same units, and in general the purpose of a machine is to convert energy into work.

For rotating machines, it is more convenient to consider torque and angular displacements than force and distance. Work is then expressed as the product of the torque and the angle (in radians) through which the object rotates while acted on by the torque. Torque, in turn, is the force exerted at a given radius from an axis of rotation. Thus, a 10-lb force at the end of a 15-ft crank exerts a torque of 150 lb-ft.

Power is the rate of doing work. For example, one horsepower is arbitrarily defined as 550 ft-lb per second, or 33,000 ft-lb per minute.

[RICHARD M. PHELAN]

Sintering

The welding together and growth of contact area between two or more initially distinct particles at temperatures below the melting point, but above one-half of the melting point in degrees Kelvin. Since the rate of sintering is greater with smaller than with larger particles, the process is most important with powders, as in powder metallurgy and in firing of ceramic oxides.

Powder. Although sintering does occur in loose powders, it is greatly enhanced by compacting the powder, and most commercial sintering is done on compacts. Compacting is generally done at room temperature, and the resulting compact is subsequently sintered at elevated temperature without application of pressure. For special applications, the powders may be compacted at elevated temperatures and therefore simultaneously pressed and sintered. This is called hot pressing or sintering under pressure.

Sintering is observed as an increase in mechanical properties (strength, ductility, and so on) and in many physical properties (for example, electrical and thermal conductivity). In many, but by no means all, sintering processes, the density of the compact increases, that is, the dimensions of the compacts become smaller (the compacts "shrink"), but the shape of the compact is generally preserved. The final density of the sintered compact depends upon the pressure with which the powder is compacted—the higher the pressure, the greater the density—and upon the shrinkage of the compact during sintering. Compacts from a single component powder, for example, a powder of a pure metal or a pure oxide, must be sintered below the melting point of the component, that is, without a liquid phase. Certain compacts from a mixture of different component powders may be sintered under conditions where a limited amount of liquid, generally less than 25 vol %, is formed at the sintering temperature. This is called liquid-phase sintering, important in certain powder-metallurgy and ceramic applications.

Mechanism. The driving force in sintering is surface energy, which decreases because the total surface area decreases as sintering proceeds. When two spheres originally in tangential contact sinter, the area of contact between the spheres increases and a neck is formed. In amorphous materials such as glasses, the process causing neck formation is viscous flow (as when two drops of water coalesce to form one). This type of viscous flow is not possible in crystalline solids. Here the most important material transport process is self-diffusion. Because the material in the neck surface has a highly convex curvature, the number of defects in its crystal structure (vacant lattice sites) is considerably higher than on a flat or concave surface. These defects move by self-diffusion from the convex neck surface to the adjacent flat surface, which means material moves in the opposite direction, that is, the neck grows (illustration *a*). Instead of coming from the adjacent flat surface, the material forming the neck may also move by self-diffusion from the grain boundary between the two spheres to the convex neck surface (illustration *b*). This latter type of movement explains why compacts shrink, because in this case the centers of the spheres approach each other during sintering. Densification (shrinkage of compacts) is inhibited by rapid grain growth during sintering. During grain growth the grain boundaries, which are the sources of the material which is transported into the convex neck surface, that is, into the pores of the compact, are swept out of the compact.

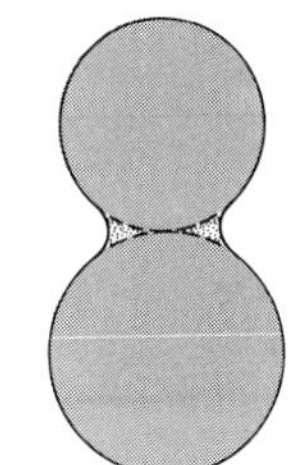

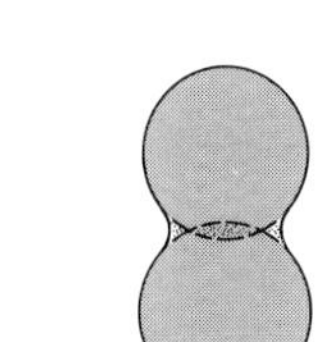

Neck formation. (*a*) Neck growth through movement of defects from the neck surface to the adjacent flat surface. (*b*) Movement of neck-forming material from the grain boundary to the neck surface. (*From G. H. Gessinger et al., Continuous observation of the sintering of silver particles in the electron microscope, ASM Trans. Quart., 61(3):598–604, 1968*)

Liquid. The mechanism of liquid-phase sintering is more complicated. It involves viscous flow of the liquid phase, but also solution of the solid phase in the liquid phase and its reprecipitation in such a way so as to make the compact more dense. As in solid-phase sintering, the driving force in liquid-phase sintering is a decrease in surface energy. *See* CERMET; POWDER METALLURGY.

[F. V. LENEL]

Bibliography: W. D. Kingery, *Introduction to Ceramics*, 2d ed., 1976; G. C. Kuczynski (ed.), *Sintering*, 1980; M. M. Ristic (ed.), *Sintering—New Developments*, 1979; L. H. Yaverbaum (ed.), *Technology of Metal Powders: Recent Developments*, 1980.

Slider-crank mechanism

A four-bar linkage with output crank and ground member of infinite length. A slider crank (Fig. 1) is most widely used to convert reciprocating to rotary motion (as in an engine) or to convert rotary to re-

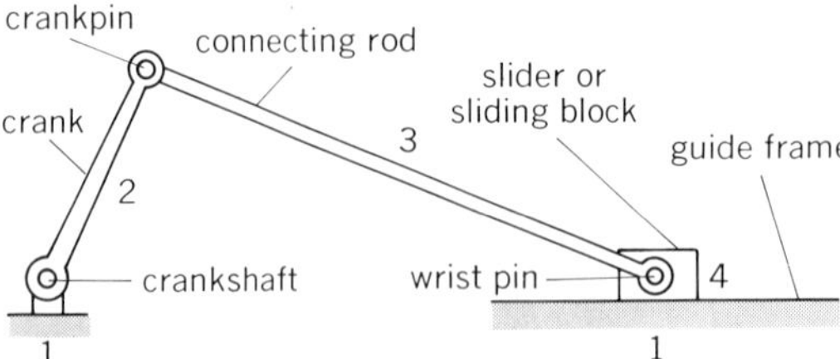

Fig. 1. Principal parts of slider-crank mechanism.

ciprocating motion (as in pumps), but it has numerous other applications. Positions at which slider motion reverses are called dead centers. When crank and connecting rod are extended in a straight line and the slider is at its maximum distance from the axis of the crankshaft, the position is top dead center (TDC); when the slider is at its minimum distance from the axis of the crankshaft, the position is bottom dead center (BDC).

For a given crank throw, or radius, the action of the slider approaches simple harmonic motion as the length of the connecting rod is increased. Maximum accelerations occur at reversal of the slider. For constant angular velocity of the crank, slider acceleration at TDC is somewhat greater, and at BDC somewhat less, than accelerations that would occur if the motion were simple harmonic.

Although the idea of combining a crank with a connecting rod is quite old (15th century), the crank was not successfully applied to a steam engine until 1780; the completion of the linkage to include a slider had to wait for satisfactory lubrication.

Many attempts were made between 1780 and 1830 to produce rotary motion directly and thus eliminate the need for the slider-crank mechanism in prime movers. Hundreds of different rotary engine designs (many of which employed slider-crank linkage in a form not recognized by the inventor) were proposed but, for a prime mover that depends for its operation upon such a fluid as steam or air within a chamber, no arrangement has been found superior to the conventional one described here.

SLIDER-CRANK MECHANISM

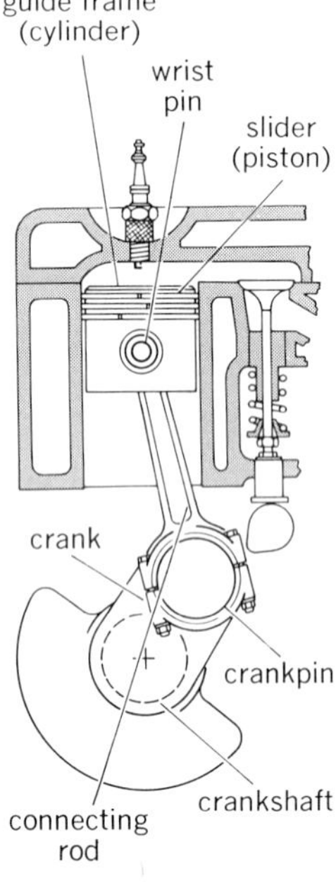

Fig. 2. Slider crank applied to internal combustion engine.

Internal combustion engine. The conventional internal combustion engine employs a piston arrangement in which the piston becomes the slider of the slider-crank mechanism (Fig. 2). Although satisfactory for engines of moderate life-span, the reversal of side thrust on the piston twice during each revolution complicates the problem of keeping the piston tight enough in the cylinder to contain the working medium in the combustion space. Because of angularity of the connecting rod, most wear occurs at the lower end of the cylinder. In some large, low-speed engines, a crosshead, differing in arrangement but similar in principle to the crosshead of a steam engine, is used to reduce cylinder wear.

Radial engines for aircraft employ a single master connecting rod to reduce the length of the crankshaft (Fig. 3). The master rod, which is connected to the wrist pin in a piston, is part of a conventional slider-crank mechanism. The other pistons are joined by their connecting rods to pins on the master connecting rod.

Reciprocating compressors and other applications. To convert rotary motion into reciprocating motion, the slider crank is part of a wide range of machines, typically pumps and compressors. The crankshaft may be driven through belting or by an electric motor. Portable, reciprocating compressors use pistons; stationary compressors generally employ crossheads and guides. *See* COMPRESSOR.

Another use of the slider crank is in toggle mechanisms, also called knuckle joints. The driving force is applied at the crankpin (Fig. 1) so that, at TDC, a much larger force is developed at the slider. This is limited by practicality to the forces at which the links and pins yield. *See* FOUR-BAR LINKAGE.

Other forms that the slider crank can take can be studied by inversion (Fig. 4). Compared to a conventional four-bar linkage (Fig. 4*a*) where link 1 is the fixed member, the slider crank replaces finite moving link 4 with the slider (Fig. 4*b*); in this form link 1 remains the fixed member. Although moving member 4 loses its separate identity, comparison of angles ABO_B in Fig. 4*a* and *b*, and comparison of angles BO_BO_A in the two diagrams,

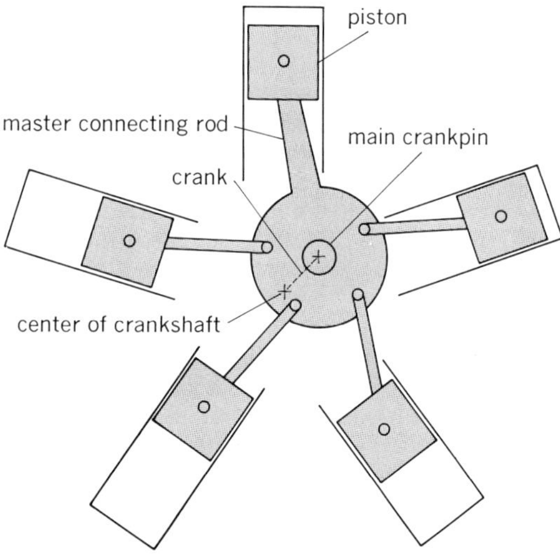

Fig. 3. Multiple-crank mechanism in radial engine.

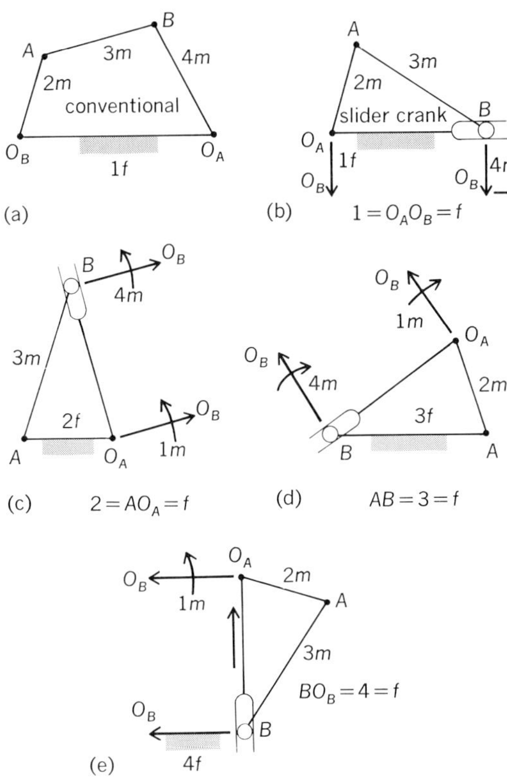

Fig. 4. Inversion of slider-crank mechanism. (*a*) Conventional four-bar linkage of which slider crank is a special case. (*b*) Link 4 made infinitely long, and link 1, which carries the slider guide, held fixed, produce the usual slider-crank mechanism. (*c*) Intermediate link 2 held fixed. (*d*) Link 3, carrying the slider head, held fixed. (*e*) Virtual link 4, to which slider *B* is pinned, held fixed. Links marked *f* are fixed; those marked *m* move.

shows the actions of the two mechanisms to involve the same number of links and turning pairs within their closed loops. By successively holding each link fixed, the alternative forms of Fig. 4*c*–*e* are obtained. *See* SLIDING PAIR. [DOUGLAS P. ADAMS]

Bibliography: C. W. Ham, E. J. Crane, and W. L. Rogers, *Mechanics of Machinery*, 4th ed., 1958; H. H. Mabie and F. W. Ocvirk, *Mechanics and Dynamics of Machinery*, 3d ed., 1975.

Sliding pair

Two adjacent links, one of which is constrained to move in a particular path with respect to the other. A lower, or closed, pair is completely constrained by the design of the links of the pair. A turning pair is always a lower pair, and the connection between links is equivalent to a pin joint in which the pin is encircled by a properly fitted bushing, the two links thus turning about each other. A closed sliding pair has the sliding block of one link constrained by a rod or guide on the other link (Fig. 1). A higher, or open, pair requires an auxiliary force in the mechanism to maintain contact between links. The force may be that of gravity or a spring; in the case of gearing, conjugate rolling surfaces are constrained by fixed centers of rotation (Fig. 2).

The relative constraint of one member of a pair of linkage members upon another was classified and described as above by Franz Reuleaux (1829–1905). R. S. Hartenberg and J. Denavit list six lower pair joints, or constraints of pairings with their degrees of freedom: line slide, 1; surface slide, 3; turn, 1; turn slide, 2; screw, 1; and ball, 3. A surface slide differs from Fig. 2*a* by an upper as well as a lower horizontal surface constraint (in place of gravity). As a consequence, each member can slide in two directions and twist with respect to one another.

Although area contact usually characterizes a lower pair and line or point contact an upper pair, contact is not the basic difference. What distinguishes lower from higher pairs is the mutuality, or its absence, in the roles of the permitted motion. Members of a lower closed pair restrain each other in reciprocal fashion, impose similar constraint, and yield similar motions, as turn-slide or screw. Members of a higher pair, as cam or roller, clearly exert no reciprocity of motion upon each other. *See* MECHANISM. [DOUGLAS P. ADAMS]

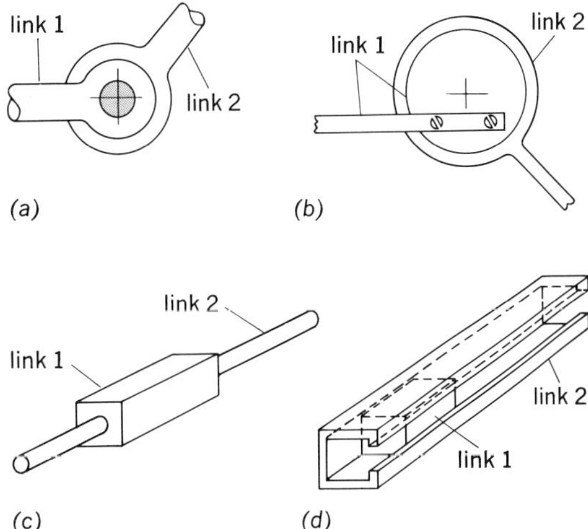

Fig. 1. Lower pairs of mechanical links. (*a*) Turning pair. (*b*) Off-center turning pair. (*c*) Outside sliding pair. (*d*) Inside sliding pair.

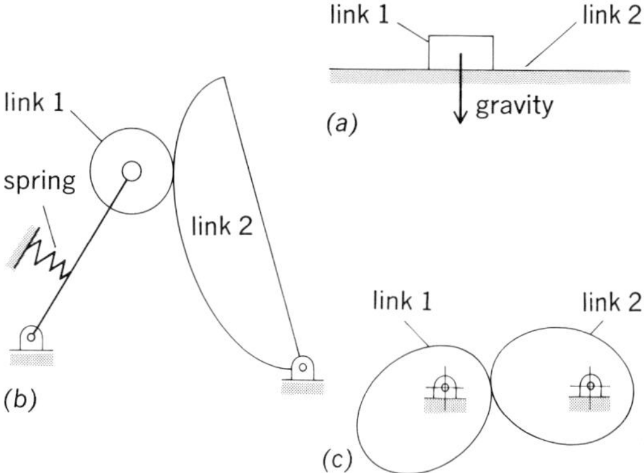

Fig. 2. Higher pairs of mechanical links. (*a*) A sliding pair constrained by gravity. (*b*) A rolling pair constrained by a spring. (*c*) A rolling pair constrained by fixed spacing between the centers.

Bibliography: R. S. Hartenberg and J. Denavit, *Kinematic Synthesis of Linkages*, 1964; A. Ramous, *Applied Kinematics*, 1972.

Slip

The difference between the operating speed of an induction motor and its synchronous speed (the speed of the rotating field). Slip s is usually expressed as a decimal fraction of synchronous speed n_s as in the equation below, where n is the

$$s = \frac{n_s - n}{n_s}$$

rotor, or operating, speed. *See* INDUCTION MOTOR.

[WILLIAM W. SEIFERT]

Slip rings

Electromechanical components which, in combination with brushes, provide a continuous electrical connection between rotating and stationary conductors. Typical applications of slip rings are in electric rotating machinery, synchros, and gyroscopes. Slip rings are also employed in large assemblies where a number of circuits must be established between a rotating device, such as a radar antenna, and stationary equipment.

Electric rotating machines. Slip rings are used in wound-rotor induction motors, synchronous motors and alternators, and rotary converters to connect the rotor to stationary external circuits. These slip rings are usually constructed of steel with the cylindrical outer surface concentric with the axis of rotation. Insulated mountings insulate the rings from the shaft and from each other. Conducting brushes are arranged about the circumference of the slip rings and held in contact with the surface of the rings by spring tension. A typical assembly is shown in Fig. 1. Other arrangements, such as concentric slip rings mounted on the face of an insulating disk, may be employed in special cases. Alternating-current windings normally require one slip ring per phase except that a single-phase winding requires two slip rings. Two slip rings are required for rotating dc field windings. *See* ELECTRIC ROTATING MACHINERY; GENERATOR; MOTOR. [ARTHUR R. ECKELS]

Slip-ring assemblies. These integral mechanical structures contain a plurality of slip rings which, in combination with self-contained brushes, provide continuous electrical connection between electric and electronic equipment mounted on stationary and rotating platforms.

Slip-ring assemblies are designed for a wide range of electric circuits. The same assembly may have circuits for power up to several hundred kilowatts; high voltage to 50 kV; power pulses for radar transmitters, and data signals, including radio frequencies to 100 MHz; strain-gage signals and thermocouple signals. Ring surface speeds range from a few feet per minute to over 15,000 ft/min (75 m/s).

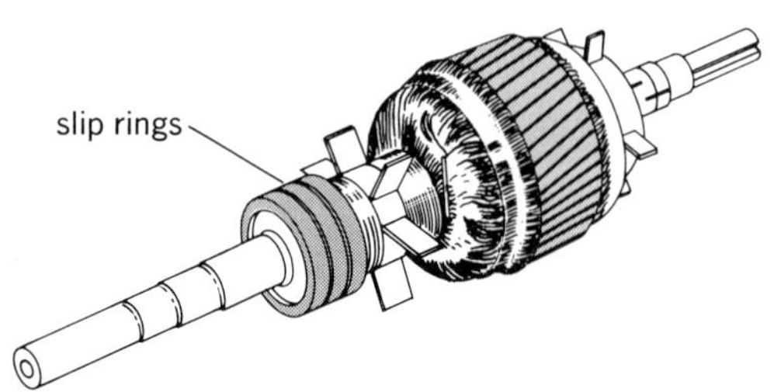

Fig. 1. Rotor of electric rotating machine.

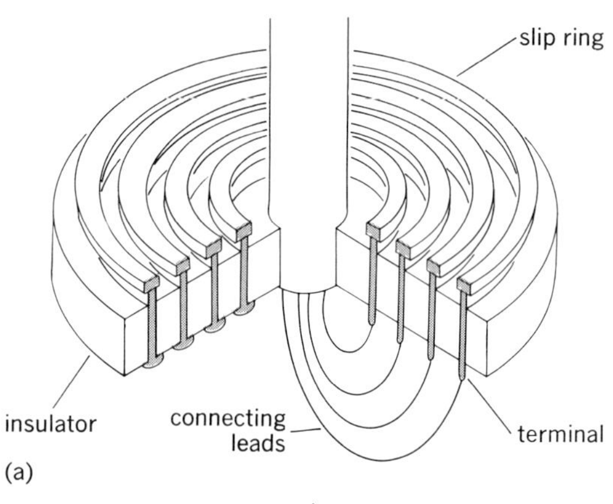

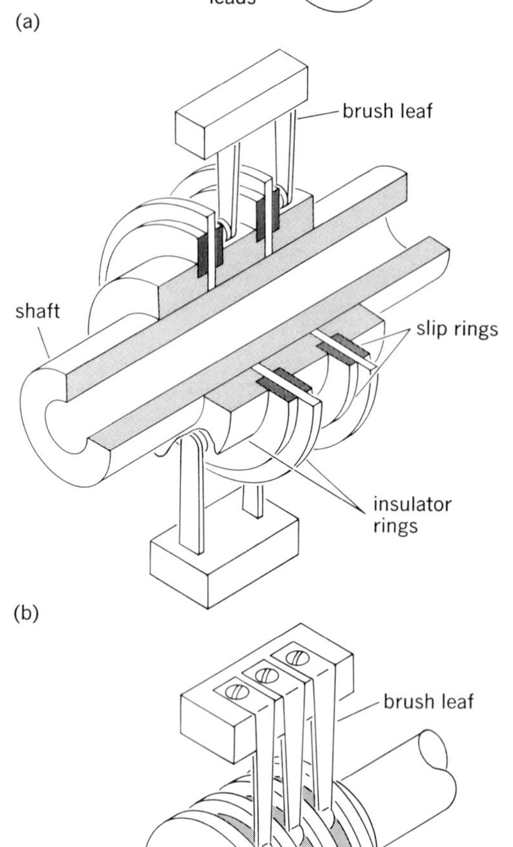

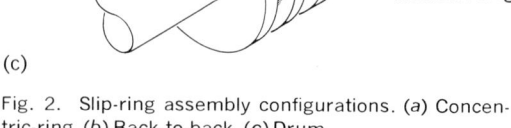

Fig. 2. Slip-ring assembly configurations. (*a*) Concentric-ring. (*b*) Back-to-back. (*c*) Drum.

Slip rings for slip-ring assemblies are made usually of coin silver, stamped from hard rolled sheet or cut from drawn tubing, or made of strip silver overlay on copper, formed into a ring, and silver brazed. Fine silver rings may be electroformed onto the insulating material. Surface finish is machined or mill rolled from 4 to 16 μin. (100 to 400 nm). Brushes for slip-ring assemblies are graphite combined with copper or silver in proportions suitable for the application, and may be welded or brazed to a spring-temper leaf. Leaf brush pressure is from $1\frac{1}{2}$ to 3 oz (43 to 85 g).

Large assemblies, such as those used with radar

antennas, are fabricated from individual insulators and conducting materials arranged in either the concentric-ring configuration (Fig. 2*a*) or the back-to-back ring configuration (Fig. 2*b*). Small assemblies, such as those used to transmit signals through gimbals or high-speed turbines, may be fabricated in the back-to-back ring configuration or the drum configuration (Fig. 2*c*).

Other manufacturing methods employ casting the individual rings into filled epoxy resins, electroforming fine silver into grooves machined on filled epoxy resin tubes, or molding the individual rings with electrical grades of thermosetting resins. The casting, electroforming, and molding methods have the advantage of low tolerance buildup, which is important for the synchro sizes.

Background noise for strain-gage signals should not be greater than a few microvolts; intercircuit interference (crosstalk) should not be greater than 70 dB down at 30 MHz, insertion loss at 30 MHz no greater than 0.5 dB, and brush contact resistance approximately 0.005 ohm.

Synchro slip rings are made from fine silver or gold alloy. Brushes are also made from precious-metal alloys, usually in the form of hard-drawn, spring-temper wires. Surface speeds are usually low.

[WILLARD F. MASON]

Snap ring

A form of spring used principally as a fastener. Piston rings are a form of snap ring used as seals. The ring is elastically deformed, put in place, and allowed to snap back toward its unstressed position into a groove or recess. The snap ring may be used externally to provide a shoulder that retains a wheel or a bearing race on a shaft, or it may be used internally to provide a construction that confines a bearing race in the bore of a machine frame, as illustrated. The size of the ring and its

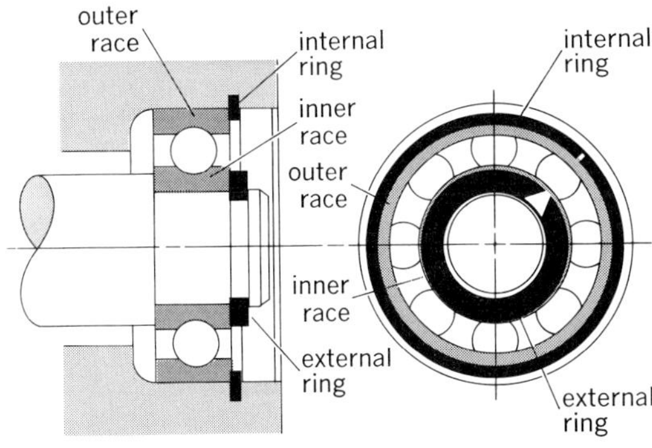

Snap rings hold ball-bearing race in place. Internal ring supports axial thrust of axle. External ring aligns inner race against shoulder of shaft.

recess determines its strength under load. Sufficient clearance and play is needed in the machine so that the ring can be inserted and seated. *See* ANTIFRICTION BEARING; INTERNAL COMBUSTION ENGINE.

[L. SIGFRED LINDEROTH, JR.]

Bibliography: H. Carlson, *Spring Designers Handbook*, 1978.

Sodium-vapor lamp

A low-pressure electric discharge lamp of monochromatic yellow light. The construction of a sodium-vapor lamp is shown in the illustration. The arc tube, in which the arc occurs, is made of glass.

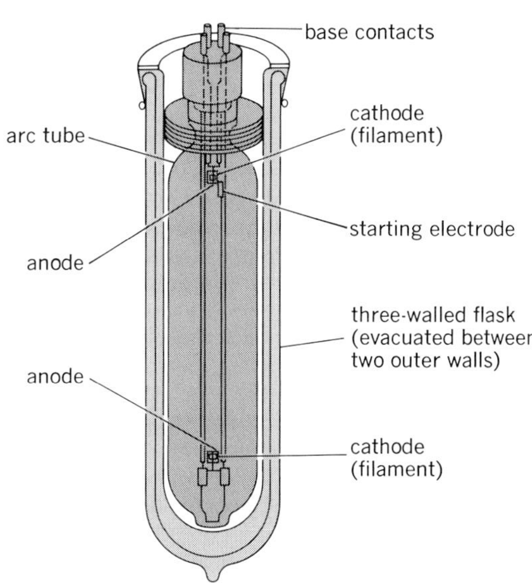

A 10,000-lumen sodium-vapor lamp. (*From Illuminating Engineering Society, IES Lighting Handbook, 3d ed., 1968*)

The outer bulb has three walls for thermal insulation. The electrodes, coils of tungsten, can be preheated before starting by passing an electric current through each coil. The arc tube contains a small amount of sodium and some neon gas to facilitate starting.

During the starting period, the lamp color is essentially red, the characteristic color of the neon glow. As the lamp warms up and the sodium vaporizes, lamp color gradually shifts to the intensely yellow color characteristic of sodium radiation. Sodium radiation is confined to a single band in the yellow region of the spectrum. This results in high luminous efficiency, but extremely poor color rendition limits low-pressure sodium lamps to tasks where color is unimportant and efficiency is a primary consideration. The narrow band of radiation from these lamps suits them ideally to optical work requiring monochromatic light. A special sodium laboratory arc lamp is available for such application.

The high-pressure sodium lamp that is used in the United States utilizes an amalgam of sodium and mercury inside an arc tube surrounded by an evacuated glass jacket. This arc tube is made of translucent aluminum oxide that withstands extremely high temperatures and the chemical attack of alkali metals. Operation of the arc at a higher temperature and pressure produces a continuous radiation pattern rather than the line pattern of yellow light associated with the earlier low-pressure sodium lamp. The light produced by this newer lamp has been described as golden white. Its high luminous efficiency (over 100 lumens per watt), combined with good color rendition, has brought increasing use of the high-pressure sodium lamp in many general lighting applications. *See* LAMP; VAPOR LAMP.

[V. G. MC CLUSKY]

Bibliography: J. E. Kaufman and J. F. Christensen (eds.), *Lighting Handbook*, 5th ed., 1972.

Solar heating and cooling

The use of solar energy to produce heating or cooling for technological purposes. When the Sun's short-wave radiation impinges upon a blackened surface, much of the incoming radiant energy can be absorbed and converted into heat. The temperature that results is determined by: the intensity of the solar irradiance; the ability of the surface to absorb the incident radiation; and the rate at which the resulting heat is removed. By covering the absorbing surface with a material such as glass, which is highly transparent to the Sun's short-wave radiation but is opaque to the long-wave radiation emitted by the Sun-warmed surface, the effectiveness of the collection process can be greatly enhanced. The energy which is collected can be put to beneficial use at many different temperature levels to accomplish: distillation of sea water to produce salt or potable water; heating of swimming pools; space heating; heating of water for domestic, commercial, and industrial purposes; cooling by absorption or compression refrigeration; cooking; and power generation by thermal or photovoltaic means.

DISTILLATION

The oldest of these applications dates back to a 50,000-ft^2 (4600-m^2) installation built in Chile in 1872 to distill saline water and make it potable. Figure 1 shows in cross section the type of glass-roofed solar still used more than a century ago in Chile and used again in modern times on the Greek islands in the eastern Mediterranean and in central Australia. When the Sun shines through the glass cover into the salt or brackish water contained within the concrete channel, the water is warmed and some evaporates to be condensed on the underside of the glazing. The condensate runs down into the scuppers and then into a suitable container. More than 6000 gal (23,000 liters) of pure drinking water were produced on each sunny day by the Chilean still, and the larger version now operating at Coober Pedy, Australia, produces nearly twice as much.

Production of salt from the sea has been accomplished for hundreds of years by trapping ocean water in shallow ponds at high tide and simply allowing the water to evaporate under the influence of the Sun. The residue contains all of the compounds that were present in the sea water, and it is sufficiently pure for use in many industrial applications.

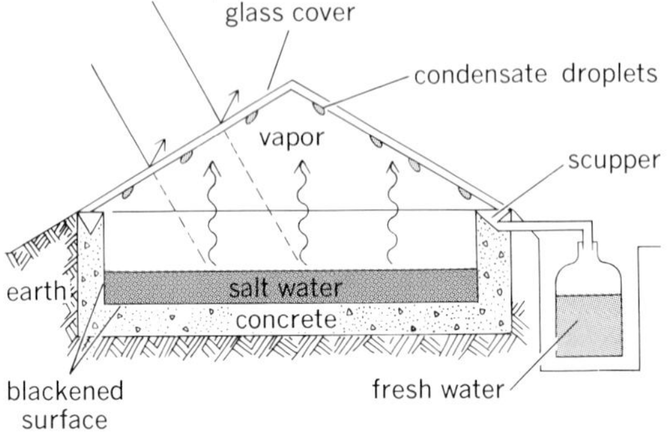

Fig. 1. Roof-type solar still.

SWIMMING POOL HEATING

Swimming pool heating is a moderate-temperature application which, under suitable weather conditions, can be accomplished with a simple unglazed and uninsulated collector similar to the black polyethylene extrusion shown in Fig. 2. When the water to be heated is at almost the same temperature as the surrounding air, little or no heat will be lost from the absorber and so there is no need for either glazing or insulation. Such collectors are usually designed to empty themselves when the circulating pump is shut off so they can avoid the danger of freezing at night. For applications where a significant temperature difference exists between the fluid within the collector passages and the ambient air, both glazing and insulation are essential.

SPACE HEATING

Space heating can be carried out by active systems which use separate collection, distribution, and storage subsystems (Figs. 3 and 4), or by passive designs which use components of a building to admit, store, and distribute the heat resulting from absorbing the incoming solar radiation within the building itself.

Passive systems. Passive systems can be classified as direct-gain when they admit solar radiant energy directly into the structure through large south-facing windows (Fig. 5), or as indirect-gain when a wall (Fig. 6) or a roof (Fig. 7) absorbs the solar radiation, stores the resulting heat, and then transfers it into the building. If the absorber and the storage components are not a part of the building fabric but are separate subsystems which operate by the natural circulation of warmed or cooled air, the term isolated-gain is generally used (Fig. 8).

Passive systems are generally effective where the number of hours of sunshine during the winter months is relatively high, where moderate indoor temperature fluctuations can be tolerated, and where the need for summer cooling and dehumidification is moderate or nonexistent.

Direct-gain systems. Most passive systems make use of the fact that, in winter, whenever the Sun is above the horizon it is in the southern part of the sky, and its altitude above the horizontal plane at noon is relatively low compared to the much higher position which it will attain in summer. This means that in winter when heat is needed, vertical south-facing windows admit solar radiation freely as long as the Sun is shining. The use of double glazing reduces the transmission of the incoming solar radiation by about 16%, but it can halve the thermal loss due to the indoor-outdoor temperature difference. The use of movable insulation which can be placed over the windows at night and removed during the sunlit hours of the day can also greatly improve the performance of direct-gain systems. Window area on surfaces other than those facing south should be kept to the minimum permissible under local building codes. The building components (walls and floors) on which the solar radiation falls should possess as much mass as possible so that the excess solar heat gained during the day can be stored for use at night.

Indirect-gain systems. The indirect-gain concepts shown by Figs. 6 and 7 interpose a thermal mass between the incoming solar radiation and the space to be heated. The system shown in Fig. 6 uses a south-facing glazed wall of concrete or masonry with an air space between the wall's outer surface and the single or double glazing. This is known as a Trombe wall. Vents with dampers are provided near the floor and at the ceiling level so that cool air can be drawn by chimney action from the room into the space between the glass and the wall. The air is heated by contact with the wall, and rises to reenter the room at the ceiling level. At night, the chimney action stops and the dampers are closed to prevent the downflow of cooled air back into the room.

As the Sun's rays warm the outer surface of the concrete, a wave of heat begins to move slowly (at about 5 cm or 2 in. per hour) through the wall. The thickness of the wall is chosen to delay the arrival of the wave of warmth at the indoor surface until after sunset, when the long-wave radiation emitted by the wall will be welcome. Heat also flows outward from the wall to the glazing and thence to the outdoor environment, but this can be minimized by the use of movable insulation within the air space. Windows may be introduced into the Trombe wall, and the building need not be restricted to one story.

Another indirect-gain system, shown in Fig. 7, uses enclosed bags of water, called thermoponds, which are supported by a heat-conducting roof-ceiling and covered by horizontally movable insulating panels. This system collects solar energy during winter days by rolling the insulation away to a storage area, thus admitting Sun's rays into the thermoponds. At night, the insulating panels are rolled back to provide the insulation needed to retain the collected heat.

In summer, the operation is reversed, and the insulating panels are rolled back at night to expose the thermoponds to the sky and thus to enable them to dissipate heat that has been absorbed from the building during the day. Convection to cool night air, when it is available, and radiation to the sky on clear nights are two of the natural processes by which heat can be rejected. The third process, and the most potent, is evaporation, which can occur when the ponds are provided with exposed water surfaces by flooding or spraying them.

Another indirect-gain system makes use of a greenhouse attached to the south side of a building to gather heat during the day. The wall between the glazed space and the building is warmed by the Sun's rays and by contact with the warm air in the greenhouse. Additional heat storage can be transferred into the house by opening windows or vents. The warmth which is gathered even during cold winter days can thus be used to aid in heating the residence as well as to provide an environment in which plants can grow.

Isolated-gain systems. The isolated-gain system shown in Fig. 8 uses a thermosyphon heater to warm air, which rises as it becomes hot, until it encounters the entrance to the rock-bed thermal storage component. The rocks abstract heat from the air which moves downward as it becomes cooler. The air eventually descends to the bottom of the rock bed and thence to the inlet of the heater. The cycle continues as long as the Sun shines on the collector, and heat is stored throughout most of the day. At night, when the occupants of the house

(a)

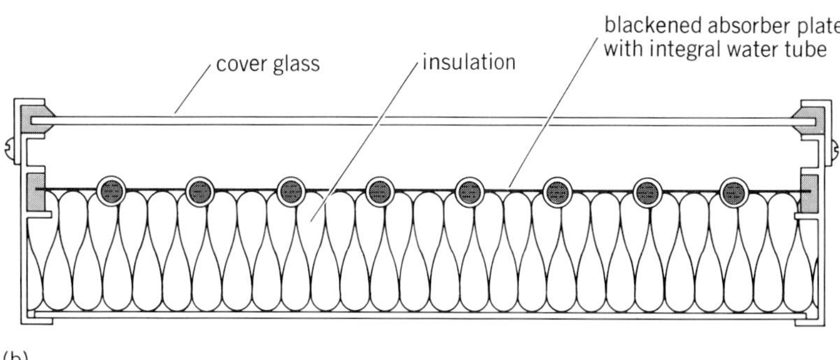

(b)

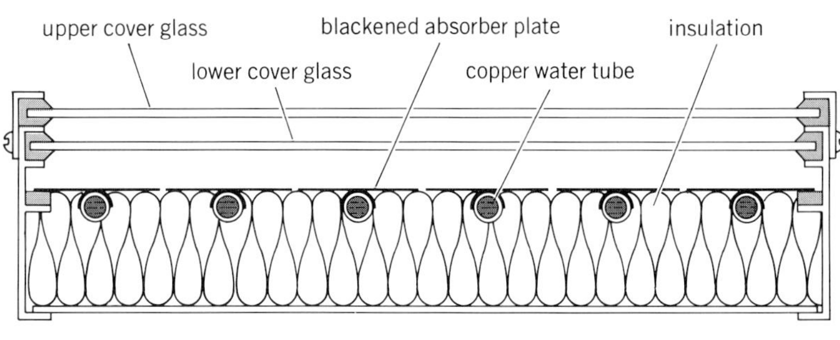

(c)

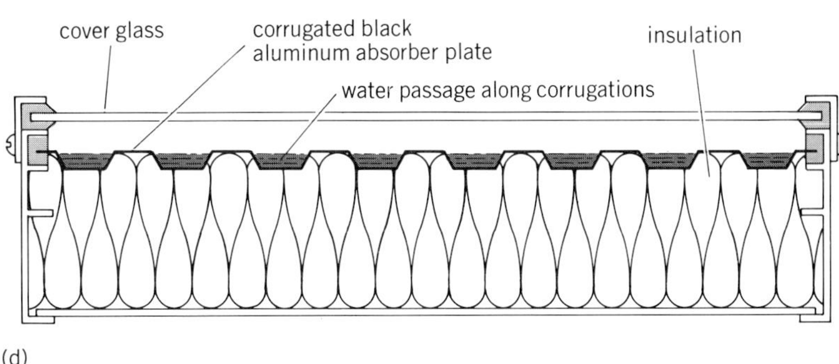

(d)

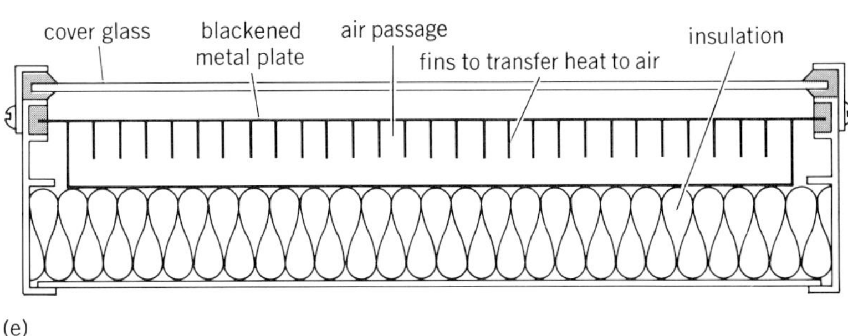

(e)

Fig. 2. Flat-plate collector types. *(a)* Extruded black plastic swimming pool heater. *(b)* Single-glazed collector with tube-in-sheet. *(c)* Double-glazed collector with extruded aluminum absorber element and copper water tube. *(d)* Open-flow collector used in the Thomason "Solaris" system. *(e)* Air heater using finned aluminum heat absorber.

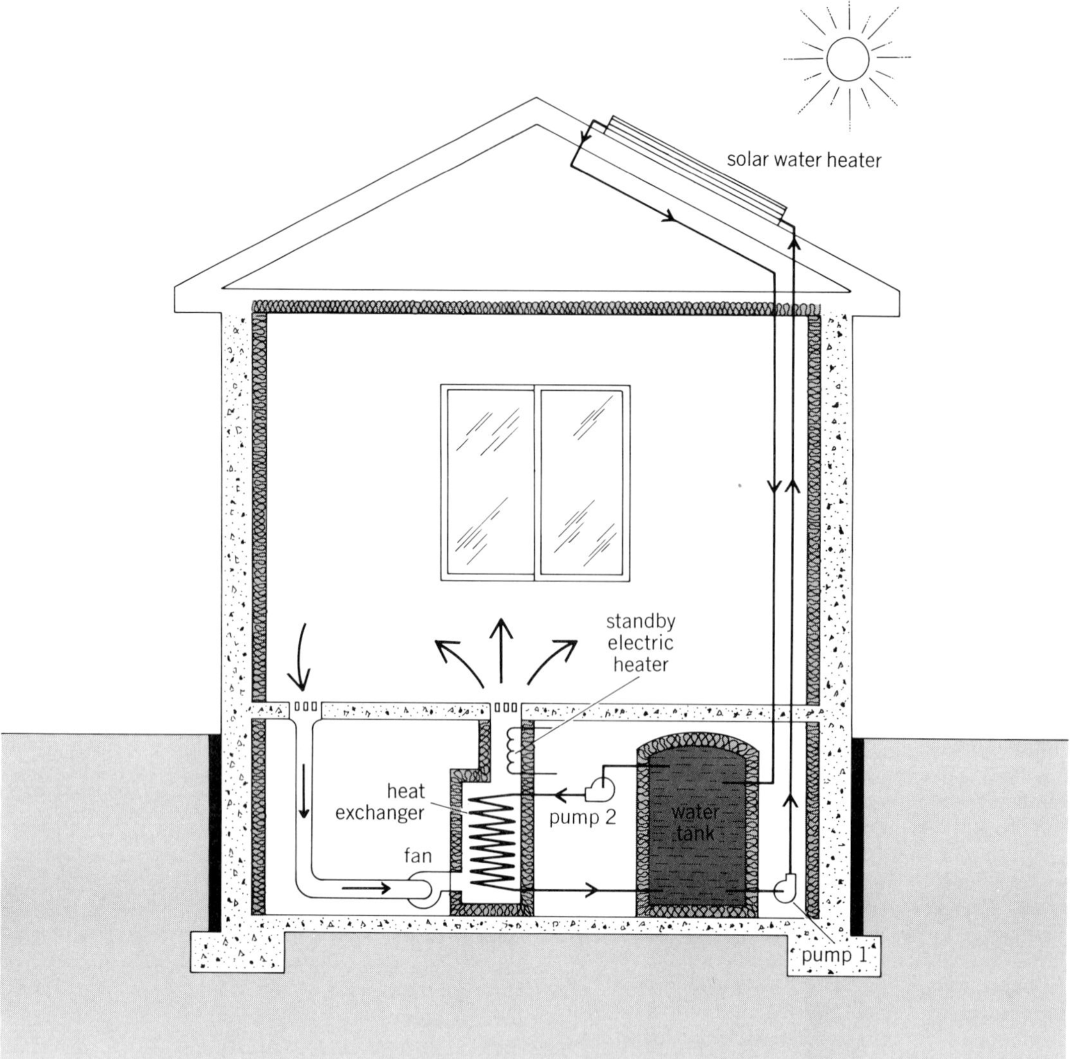

Fig. 3. Active system using roof-mounted solar water heater with storage tank in basement.

sense the need for heat, the damper settings are changed and warm air from the top of the rock bed is admitted to the living space. At the same time, cool air from the north wall of the house is allowed to flow downward through ducts to enter the base of the rock bed. There the air is warmed, and the heat which has been stored during the day is used to warm the house at night.

Active systems. Active systems may use either water (Fig. 3) or air (Fig. 4) to transport heat from roof-mounted south-facing collectors to storage in rock beds or water tanks. The stored heat may be withdrawn and used directly when air is the transfer fluid. When the heat is collected and stored as hot water, fan-coil units are generally used to transfer the heat to air which is then circulated through the warmed space. Each system has both advantages and disadvantages.

Air cannot freeze or cause corrosion, and leakage is not a serious problem. Water requires relatively small pipes compared with the ducts needed to transport the same amount of heat in the form of warm air. Water tanks can store more than three times as much heat as rocks in a given volume per degree of temperature change, but rock beds are considerably lower in cost than water tanks, and rocks can tolerate virtually any temperature while water will boil at 212°F (100°C) unless its pressure is raised above atmospheric. The choice between water and air systems must be made carefully, taking into account the many features of each.

Heat storage. Heat storage can be accomplished with specific-heat materials such as water or rocks, which can store and discharge heat by simply undergoing a change in temperature. A different process is involved in the heat-of-fusion materials, which, like water at 32°F (0°C), can freeze when heat is removed and melt again when the same amount of heat is returned to them. Ice has the unique property of increasing in volume by 7% compared with the water from which it is formed, but virtually all other substances become denser when they solidify and so their solid components sink instead of floating. A number of materials are available which, like the well-known Glauber's salt, change their state from liquid to solid and

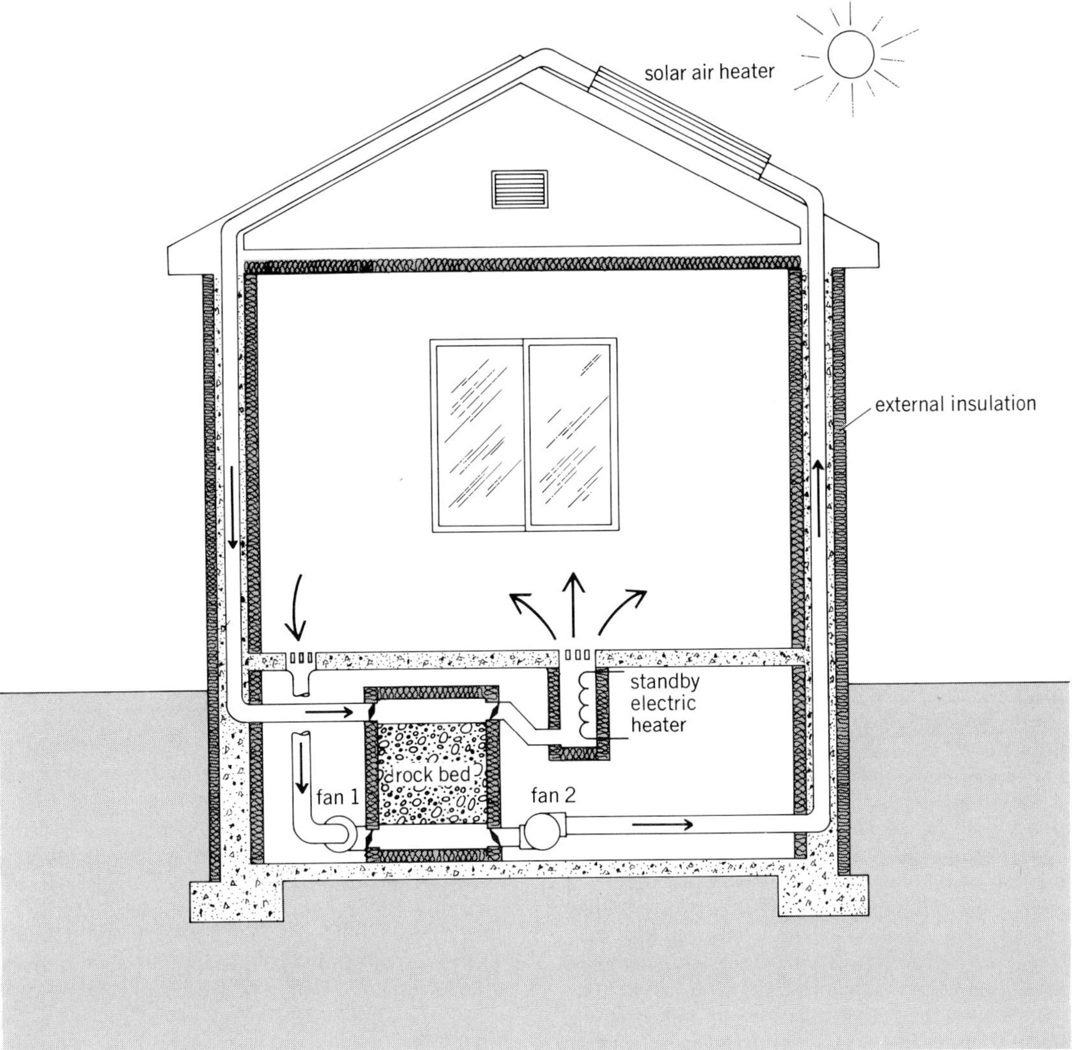

Fig. 4. Active system using air heater with rock-bed heat storage in basement.

back again at temperatures which are useful for either heating or cooling. None has yet been found, however, which possesses all of the attributes required for successful application to solar heating and cooling systems.

Standby electrical sources. Standby electrical sources are shown in Figs. 3 and 4, since some method of providing warmth must be included for use when the Sun's radiant energy is inadequate for long periods of time. The standby heater may be something as simple as a wood-burning stove or fireplace, or as complex as an electrically powered heat pump. Simple electrical resistance heaters are frequently chosen for this service because of their low first cost, but their operating cost can become excessive in applications where they must be used for long periods of time. The rapidly escalating cost of electricity must be borne in mind when the standby energy source is selected. *See* COMFORT HEATING; HEAT PUMP.

SERVICE WATER HEATING

Solar water heating for domestic, commercial, or industrial purposes is one of the oldest and most successful applications of solar-thermal technology. The most widely used water heater, and one that is suitable for use in relatively warm climates where freezing is a minor problem, is the thermosyphon type (Fig. 9). A flat-plate collector of one of the types shown in Fig. 2 is generally used with a storage tank which is mounted above the collector. A source of water is connected near the bottom of the tank, and the hot water outlet is connected to its top. A downcomer pipe leads from the bottom of the tank to the inlet of the collector, and an insulated return line runs from the top of the collector to the upper part of the storage tank which is also insulated.

The system is filled with water, and when the Sun shines on the collector, the water in the tubes is heated. It then becomes less dense than the water in the downcomer, and the heated water rises by thermosyphon action into the storage tank. It is replaced by cool water from the bottom of the tank, and this action continues as long as the Sun shines on the collector with adequate intensity. When the Sun moves away, the thermosyphon action stops, but the glazing on the collector min-

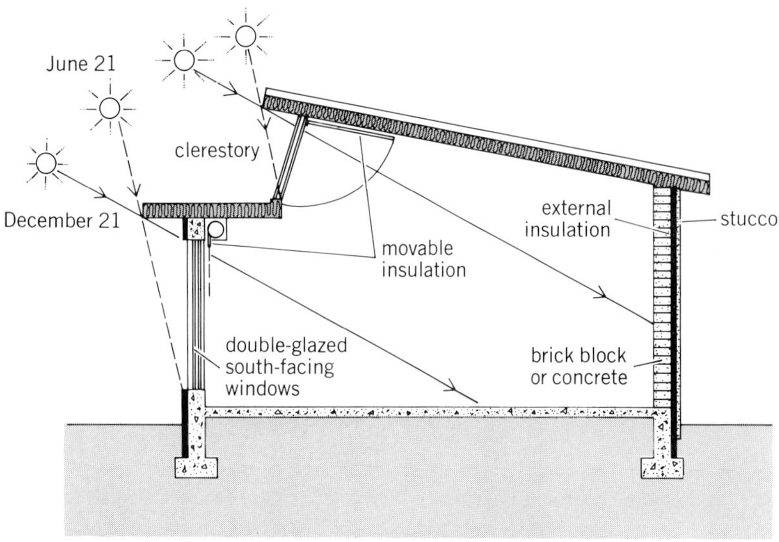

Fig. 5. Direct-gain passive system using large south-facing windows.

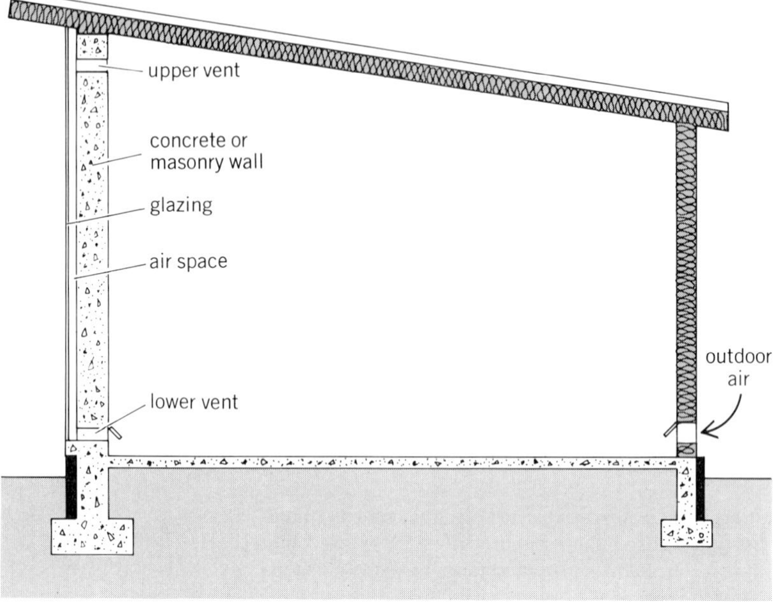

Fig. 6. Indirect-gain passive system using Trombe wall.

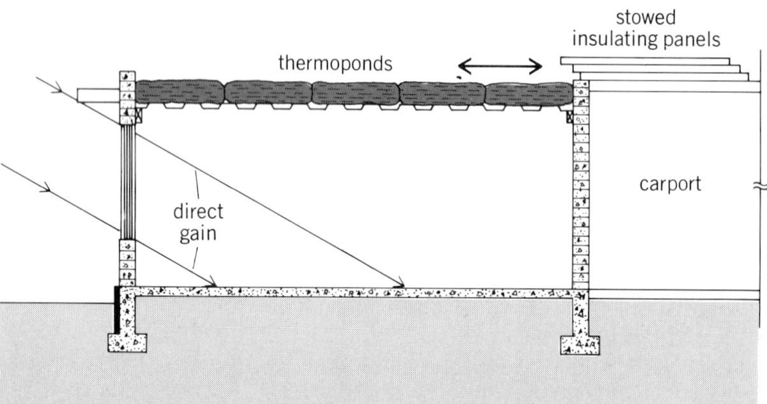

Fig. 7. Indirect-gain "Skytherm" system with movable insulation.

imizes heat loss from this component and the insulation around the storage tank enables it to retain the collected heat until needed. The system must be prevented from freezing, and the components must be strong enough to withstand operating water pressure.

For applications where the elevated storage tank is undesirable or where very large quantities of hot water are needed, the tank is placed at ground level. A small pump circulates the water in response to a signal from a controller which senses the temperatures of the collector and the water near the bottom of the tank. Heat exchangers may also be used with water at operating pressure within the tubes of the exchanger and the collector water outside to eliminate the necessity of using high-pressure collectors. Antifreeze substances may be added to the collector water to eliminate the freezing problem, but care must be taken to prevent any possibility of cross-flow between the collector fluid and the potable water. Many plumbing codes require special heat exchangers which interpose two metallic walls between the two fluid circuits. *See* HOT-WATER HEATING SYSTEM.

COOLING

Cooling can be provided by both active and passive systems.

Active cooling systems. The two feasible types of active cooling systems are Rankine cycle and absorption. The Rankine cycle system uses solar collectors to produce a vapor (steam or one of the fluorocarbons generally known as Freon) to drive an engine or turbine. A condenser must be used to condense the spent vapor so it can be pumped back through the vaporizer. The engine or turbine drives a conventional refrigeration compressor which produces cooling in the usual manner. *See* RANKINE CYCLE; REFRIGERATION.

The absorption system uses heat at relatively high temperature (180 to 200°F or 82 to 93°C) and employs a hygroscopic solution of lithium bromide to absorb water vapor at a very low pressure. The evaporating water is cooled to a temperature low enough (about 45°F or 7°C) to provide the chilled water needed to produce air conditioning. The heat introduced into this cycle by the high-temperature activating water and the low-temperature chilling water is removed by a third stream of water which has been cooled to below the outdoor air temperature by passing through an evaporative cooling tower. The cycle, simple in principle but much more complex in practice, is in wide use in large buildings where the necessary hot activating water is available either from solar collectors or as waste heat from other processes. Use of either Rankine cycle or absorption solar cooling systems for residences is restricted to a few experimental installations, but both hold promise for future development.

Passive cooling systems. Passive cooling systems make use of three natural processes: convection cooling with night air; radiative cooling by heat rejection to the sky on clear nights; and evaporative cooling from water surfaces exposed to the atmosphere. The effectiveness of each of these processes depends upon local climatic conditions. In the hot and dry deserts of the southwestern United States, radiation and evaporation are the two

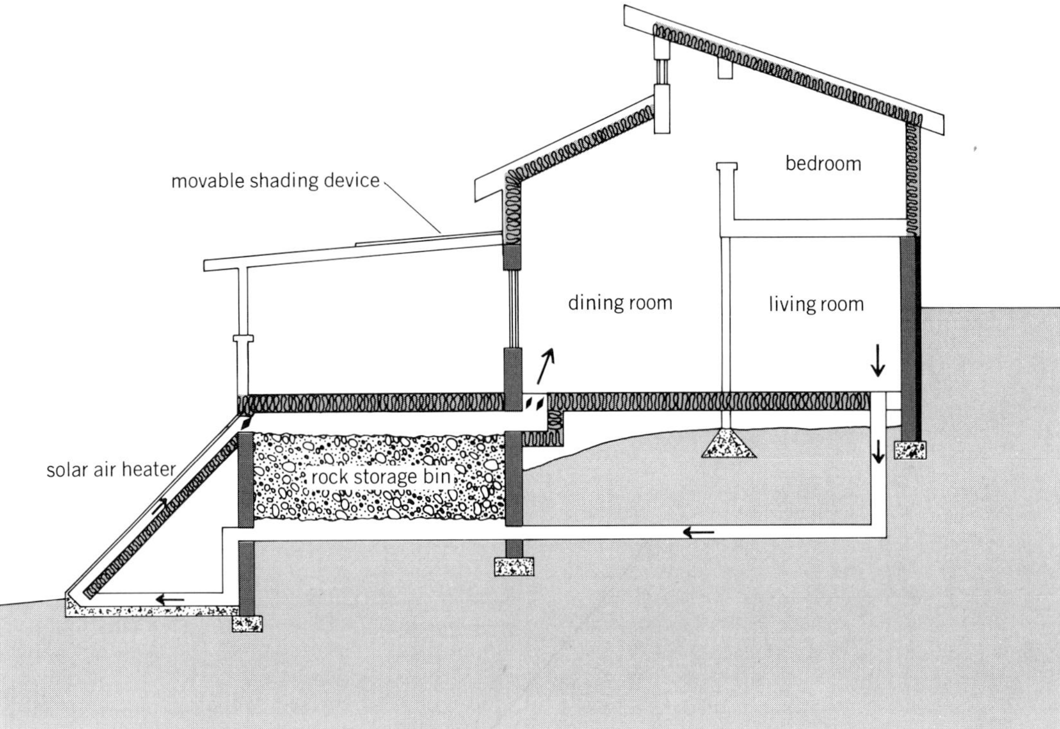

Fig. 8. Isolated-gain system with thermosyphon air heater and rock bed.

most useful processes. In hot and humid regions, other processes must be employed which involve more complex equipment that can first dehumidify and then cool the air.

POWER GENERATION

Electric power can be generated from solar energy by two processes, the first of which uses the Rankine cycle described previously. In order to attain vapor temperatures high enough to make the cycle operate efficiently, concentrating collectors must be used. For installations of moderate size, parabolic troughs (Fig. 10*a*) are adequate, but for very large utility-type plants the "power tower" concept (Fig. 10*b*) is preferred. Concentrators can use only the radiation which comes directly from the solar disk, so some means of tracking must be provided and energy must be stored during the daylight hours to ensure continuity of operation during intermittent cloudy periods. Storage systems for night-long operation are not yet available. Concentrating systems cannot make effective use of the diffuse radiation that comes from the sky, and so their use at the present stage of development of solar technology is limited to areas where bright sunshine is generally available.

Electric power can also be produced from solar radiation, without need for a thermal cycle, by photovoltaic cells which convert radiant energy directly into electricity. Originally used primarily in the United States and Soviet space programs, silicon solar batteries can now be produced at a cost low enough to justify their use in special circumstances for terrestrial applications. One development uses concentrators similar to those in Fig. 10*a* to enable a single solar cell to do the work of ten; the heat which must be removed to make the cells function efficiently can be used for many purposes requiring hot water or hot air. Solar cells produce direct current which can be stored in ordinary automobile-type batteries; the direct current can be converted to alternating current at the standard frequency, 60 Hz in the United States, and at any necessary voltage. One of the goals of the Department of Energy is the reduction of the cost of photovoltaic cells to the point where residential rooftop units become economically feasible.

CONCLUSION

Heating of space and service hot water is currently practical and economical. Passive heating and cooling are coming into wide use as the cost of alternative fuels continues to rise. Active cooling

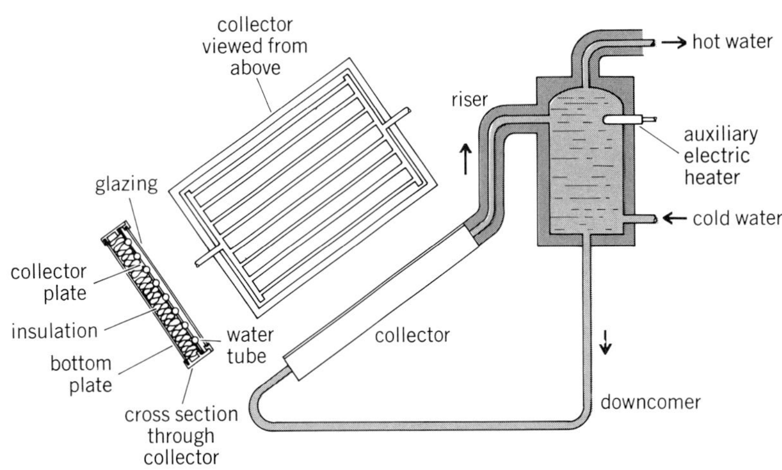

Fig. 9. Passive domestic water heater using thermosyphon system.

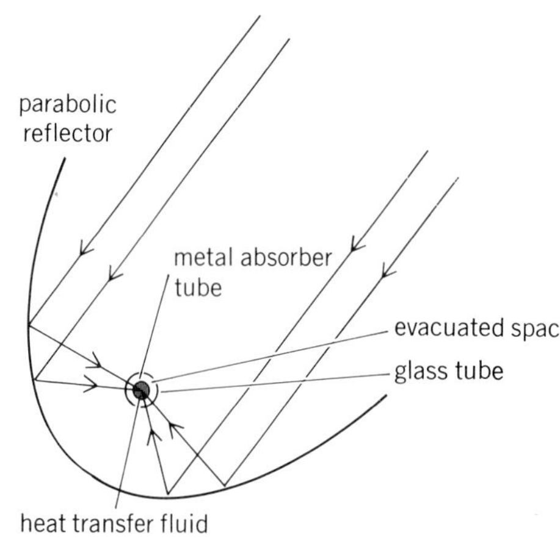

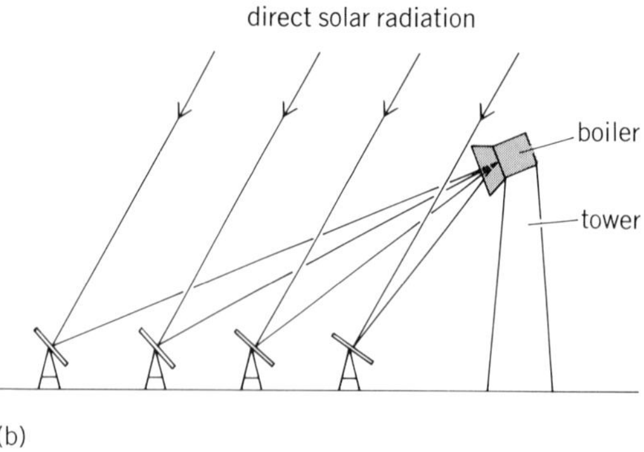

Fig. 10. Concentrating collectors. (*a*) Parabolic trough, for use in installations of moderate size. (*b*) Power tower, for large utility-type plants.

and power generation are too expensive for general use, but it is expected that the cost of these systems will diminish in the future. The major unsolved problem is how to store large quantities of energy for long periods of time. *See* ENERGY STORAGE.

[JOHN I. YELLOTT]

Bibliography: AIA Research Corporation, *A Survey of Passive Systems*, 1979; American Society of Heating, Refrigerating and Air Conditioning Engineers, *ASHRAE Handbook of Fundamentals*, chap. 26, 1977; Copper Development Association, Inc., *Solar Energy Systems*, 1979; F. Daniels, *Direct Use of the Sun's Energy*, 1964; B. Liu and R. C. Jordan (eds.), *Solar Energy Utilization for Heating and Cooling of Buildings*, ASHRAE Pub. GRP 170, 1977; E. Mazria, *The Passive Solar Energy Book*, 1979; A. A. M. Sayigh (ed.), *Solar Energy Engineering*, 1977; D. Watson (ed.), *Energy Conservation through Building Design*, 1979; J. I. Yellott, in Portola Institute, *Energy Primer*, chap. 1, 1975; J. I. Yellott, Solar energy utilization for heating and cooling, in *ASHRAE Handbook of Applications*, chap. 59, 1978.

Soldering

The joining of metals by causing a lower-melting-point metal to wet or alloy with the joint surfaces and then freeze in place. A solder is defined as a joining material that melts below 427°C (800°F); brazing alloys melt above this temperature. Solders are used to establish reliable electrical connections, to make a liquid- or gas-tight joint, and to hold parts together physically. Most soldered joints will sustain loads of only 150–250 psi for long periods. The usual practice is to rivet, crimp, or otherwise support the load and to seal the space with solder.

Solder alloys. The most commonly used solders are alloys of tin and lead that melt below the melting point of tin. Antimony, bismuth, cadmium, silver, and arsenic are sometimes added to improve strength, wetting qualities, or grain size, or to produce alloys having desired melting ranges.

The melting range and other features of all tin-lead compositions are shown in Fig. 1. An alloy of composition E, the eutectic, freezes uniquely at a single temperature, 183°C. All other compositions freeze over a temperature range in which liquid and solid coexist. Thus, an alloy near 38% tin gives a mushy mixture which can be manipulated with cloth pads to wipe joints, as in plumbing and in lead-cable splicing.

The microstructure of a 38% tin solder frozen in contact with copper is shown in Fig. 2. The tin-copper compound is exaggerated for illustration by prolonging the freezing time. The two-stage nature of freezing is apparent.

In order that surfaces will accept solder readily, they and the solder must be free from oxide or other obstructing films. When necessary, parts are cleaned chemically or by abrasion. Also, readily solderable coatings such as gold, silver, or tin may be applied. Even so, fluxes are usually used.

Fluxes. Fluxes range from very mild substances to those of extreme chemical activity. For centuries rosin, a pine product, has been known as an effective and practically harmless flux. It is used widely for electrical connections in which utmost reliability, freedom from corrosion, and absence of electrical leakage are essential. When less stringent requirements exist and when less carefully prepared surfaces are to be soldered, rosin is mixed with chemically active agents that aid materially in soldering. The activated rosin fluxes are offered by most leading solder manufacturers. The rosin-type fluxes may be incorporated as the core of wire solders or dissolved in various solvents for direct application to joints prior to soldering.

Inorganic salts are widely used where stronger fluxes are needed. Zinc chloride and ammonium chloride, separately or in combination, are most common. They may also be obtained as so-called acid-core solder wire or in petroleum jelly as paste flux. Many special-purpose salt mixtures are on the

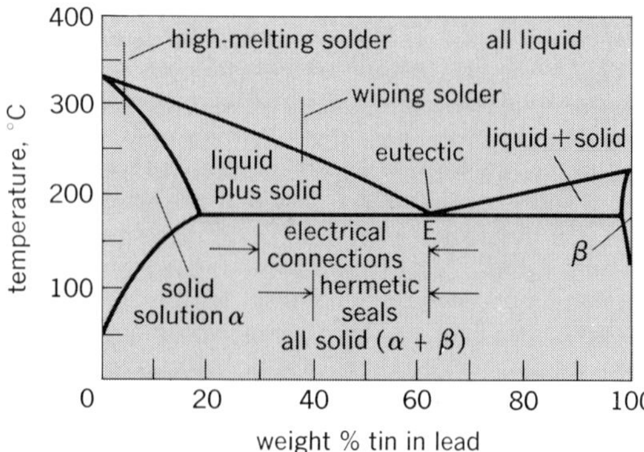

Fig. 1. Constitution diagram of tin-lead alloys.

solid solution first to freeze

laminated eutectic $\alpha + \beta$

copper-tin compound

Fig. 2. Photomicrograph of 38% tin–62% lead solder frozen in contact with copper.

market. All of the salt-type fluxes leave residues after soldering that may be a corrosion hazard. Washing with ample water accompanied by brushing is generally wise.

Application. Under the usual favorable conditions, solders wet the joint surfaces well enough to be drawn into fine crevices or capillaries by surface tension. Joints such as lap seams or wires wound on electric terminals are designed for this.

In applying solders, joints are heated by soldering irons, torches, induction heaters, or furnaces, or by immersion in molten solder. In the first two instances, solder is fed to the joints by hand. In the third and fourth, preformed shapes are placed close to the joints before fluxing and heating.

Gold, copper, silver, tin, lead, and brass are examples of readily solderable materials; iron, nickel, and rhodium are moderately difficult. Stainless steel, nichrome, and germanium are typical materials requiring the strongest fluxes. Materials such as tungsten usually must be electroplated first with more solderable metals to permit solder attachments. *See* BRAZING; WELDING AND CUTTING OF METALS.

[GEORGE M. BOUTON]

Solenoid (electricity)

An electrically energized coil of insulated wire which produces a magnetic field within the coil. If the magnetic field produced by the coil is used to magnetize and thus attract a plunger or armature to a position within the coil, the device may be considered to be a special form of electromagnet and in this sense the words solenoid and electromagnet are synonymous. In a wider scientific sense the solenoid may be used to produce a uniform magnetic field for various investigations. So long as the length of the coil is much greater than its diameter (20 or more times), the magnetic field at the center of the coil is sensibly uniform, and the field intensity is almost exactly that given by the equation for a solenoid of infinite length.

When used as an electromagnet of the plunger type, the solenoid usually has an iron or steel casing. The casing increases the mechanical force on the plunger and also serves to constrain the magnetic field. The addition of a butt or stop at one end of the solenoid greatly increases the force on the plunger when the distance between the plunger and the stop is small. The illustration shows a steel-clad solenoid with plunger and plunger stop. The relation of force versus distance with and without the stop is also shown.

The force for the solenoid rapidly increases as the plunger enters the coil because of the rapid rate of change of the reluctance for the magnetic path. For the solenoid shown in the illustration, the force on the plunger is given approximately by the equation below, where I is the coil current

$$F = ANI\left(\frac{\mu_0 NI}{2(c+x)^2} + \frac{1}{CL}\right) \text{ (newtons)}$$

(in amperes), N is the number of turns, $A = \pi r^2$ is the cross-sectional area of the plunger (in square meters), μ_0 is the permeability of free space ($4\pi \times 10^{-7}$ henry/m), r, a, b, x, and L are as shown in the illustration (in meters), and $c = ra/2b$. The value of the constant C depends upon the dimensions of the solenoid and the degree of plunger magnetization. For normal designs, C has a value in the range 0.6 to 2.9. For a conservative design, one may use the largest value for C or neglect the second term entirely. This term is due to leakage fluxes and becomes negligible as the plunger closes the air gap x.

In principle a solenoid works with either ac or dc excitation. In the dc solenoid the flux is always at its maximum value for a fixed plunger position. In an ac solenoid the force varies at twice the frequency of the supply voltage. The variation, which

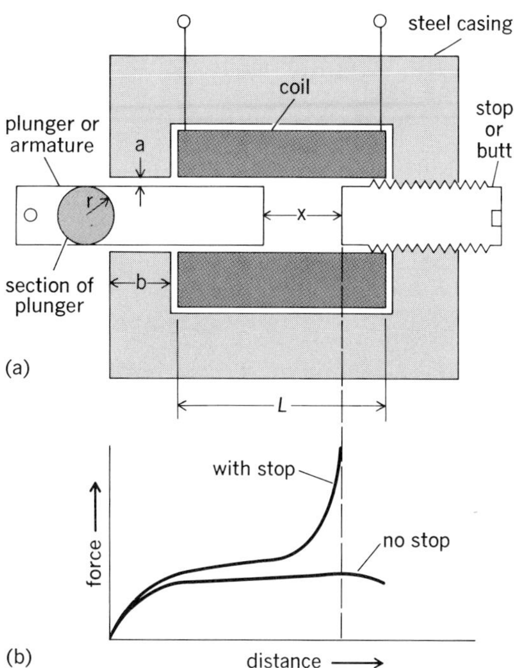

Steel-clad solenoid. (*a*) Cross-sectional view. (*b*) Relation of the force acting on the armature to the displacement of the armature.

is caused by the variation of the magnetic flux, gives rise to excessive chattering or vibration unless a shading coil is embedded in the face of the plunger stop. The shading coil acts to smooth the variation of flux and thereby smoothes the attractive force.

When direct current is used, only the resistance of the coil wire limits the final value of current, while with ac excitation the inductance of the coil must also be considered. It is difficult to calculate the proper value of inductance, because it is a function of the position of the plunger. As the plunger moves into the solenoid, the inductance, or flux per ampere, becomes much greater because the reluctance of the flux path is much less. The current drawn by the coil for constant applied ac voltage becomes smaller as the plunger moves into the solenoid. It is common practice to laminate the plunger, stop, and casing in ac service to reduce eddy current losses. *See* ELECTROMAGNET.

[JEROME MEISEL]

Bibliography: D. G. Fink and H. W. Beaty (eds.), *Standard Handbook for Electrical Engineers*, 11th ed., 1978; B. W. Jones, *Relays and Electromagnets*, 1935; H. C. Roters, *Electromagnetic Devices*, 1941; M. G. Say, *The Electrical Engineers Reference Book*, 13th ed., 1973.

Solids pump

A device used to move solids upward through a chamber or conduit. It is able to overcome the large dynamic forces at the base of a solids bed and cause the entire bed to move upward.

Solids pumps are used to cause motion of solids in process-type equipment in which treatment of solids under special conditions of temperature, oxidation, and reduction can be combined with upward motion and discharge of the spent solids overhead from the reacting vessel.

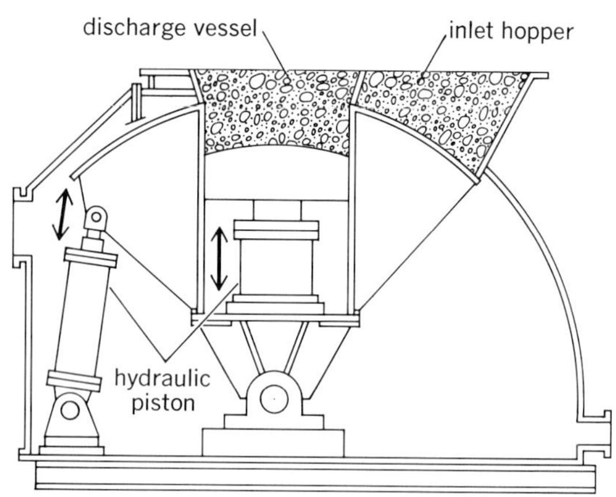

Fig. 2. Hydraulically operated solids pump.

Solids pumps are inherently of the positive displacement type. One practical method uses a reciprocating piston mounted on a trunnion permitting it to swing into an inclined position for filling and then to swing back into vertical position for discharge. Figure 1 shows a mechanically driven solids pump in four positions through its cycle of operation. Figure 2 shows a large hydraulically operated pump used in units of capacity exceeding 1000 tons/day (900 metric tons/day). Hydraulic activation permits very precise control of the feeder mechanism and good efficiency in operation.

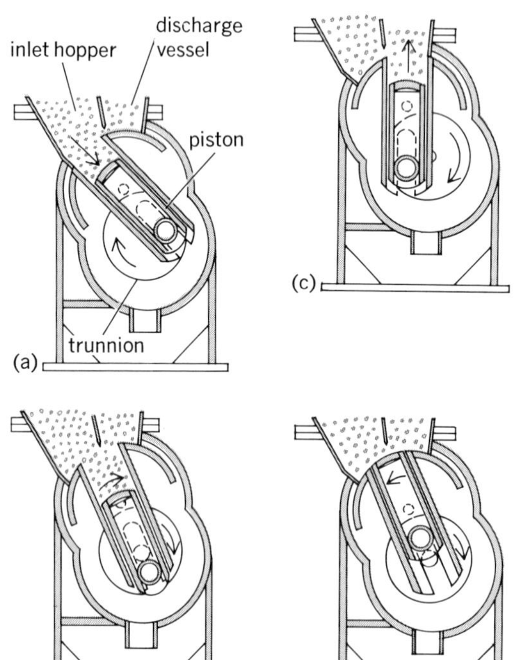

Fig. 1. Operation cycle of mechanically driven solids pump. (*a*) Filling with solids from inlet hopper. (*b*) Piston rotating on a trunnion toward its discharge position. (*c*) The discharge position, with piston pushing charge of solids upward. (*d*) Piston rotating back toward original filling position.

The solids pump has found its principal application in the operation of oil-shale retorts. Here it is used to feed crushed shale into the bottom of a conical vessel and as the shale moves upward through this vessel, air is drawn downward countercurrent. At the top of the retort, the air burns the residual carbon on the shale ash. The hot flue gas so produced contacts shale in the midpoint of the reactor, educting the shale oil. These vapors together with the flue gas are cooled, and the oil condensed on the shale at the bottom of the retort. The oil flows out the bottom countercurrent to the upgoing bed of shale. *See* BULK-HANDLING MACHINES.

[CLYDE BERG]

Spark plug

A device that screws into the cylinder of an internal combustion engine to provide a pair of electrodes between which an electrical discharge is passed to ignite the combustible mixture. It consists of an outer steel casing that is electrically grounded to the engine and a ceramic insulator, sealed into the casing, through which a central electrode passes. The high-tension current jumps the gap between this electrode and a similar one fixed to the outer casing. The electrodes are made of alloys that resist electrical and chemical erosion. The parts exposed to the combustion gases are designed to operate at temperatures hot enough to prevent electrically conducting deposits, but cool enough to avoid ignition of the mixture before the spark occurs. *See* IGNITION SYSTEM.

[AUGUSTUS R. ROGOWSKI]

Speed

The time rate of change of position of a body without regard to direction. It is the numerical magnitude only of a velocity and hence is a scaler quantity. Linear speed is commonly measured in such units as meters per second, miles per hour, or feet per second. It is the most frequently mentioned attribute of motion.

Average linear speed is the ratio of the length of the path Δs traversed by a body to the elapsed time Δt during which the body moved through that path, as in Eq. (1), where s_0 and t_0 are the initial

$$\text{Speed (average)} = \frac{s_f - s_0}{t_f - t_0} = \frac{\Delta s}{\Delta t} \qquad (1)$$

position and time, respectively, s_f and t_f are the final position and time and Δ stands for "the change in."

Instantaneous speed, defined by Eq. (2), is the limiting value of the foregoing ratio as the elapsed time approaches zero.

$$\text{Speed (instantaneous)} = \lim_{\Delta t \to 0} \frac{\Delta s}{\Delta t} = \frac{ds}{dt} \qquad (2)$$

See VELOCITY. [ROGERS D. RUSK]

Speed regulation

The change in steady-state speed of a machine, expressed in percent of rated speed, when the machine load is reduced from rated load to zero. The definition of regulation is usually taken to mean the net change in a steady-state characteristic, and does not include any transient deviation or oscillation that may occur prior to reaching the new operation point. This same definition is used for stating the speed regulation of electric motors as well as for certain drive systems, such as steam turbines.

In specifying the speed regulation of dc motors, for example, one would compute the steady-state regulation R by Eq. (1), where N_R is the speed (in

$$R = \frac{N_0 - N_R}{N_R}(100) \qquad \text{percent} \qquad (1)$$

any convenient units) at rated load and N_0 is the speed (in the same units used to specify N_R) after the load is removed and a new steady-state value is reached. Regulation is usually stated as a positive quantity, even though the regulation characteristic of speed versus load, when plotted as a straight line, would have a negative slope.

For small load changes about an operating point, an incremental steady-state regulation is also defined by Eq. (2), where both the speed N and power

$$R = \frac{\Delta N}{\Delta P}(100) \qquad (2)$$

P are normalized by using rated values as base quantities. Equation (2) recognizes the fact that the speed-versus-power regulation characteristic is not necessarily a straight line. In computing the regulation it should also be specified that all adjustments to the speed-changing mechanism should remain unchanged throughout the test. *See* MOTOR.

The speed regulation characteristic of steam turbines driving synchronous generators is an important parameter in the control of frequency for large power systems. Here the speed regulation, often called the droop characteristic, determines not only the overall systems frequency response characteristic to load changes but also how any load change is shared by the connected turbine-generator units. The droop characteristic is important in estimating steady-state response, but this parameter enters prominently into the determination of transient behavior as well. *See* GENERATOR.

The definition of steady-state regulation (1) is also used in the specification of automatic control system performance, where it is used as the performance measure of a controlled variable, such as speed.

[PAUL M. ANDERSON]

Bibliography: American National Standards Institute, *Terminology for Automatic Control*, ANSI Std. C85.1–1963; N. Cohn, *Control of Generation and Power Flow on Interconnected Power Systems*, 1966; O. I. Elgerd, *Electric Energy Systems Theory: An Introduction*, 1971; A. E. Fitzgerald, C. Kingsley, Jr., and A. Kusko, *Electric Machinery*, 3d ed., 1971; IEEE Power Engineering Society, System Control Subcommittee, *IEEE Standard Definitions of Terms for Automatic Generation Control on Electric Power Systems*, IEEE Std. 94, 1970; *IEEE Standard Dictionary of Electrical and Electronics Terms*, 2d ed., IEEE Std. 100–1977, 1977; Joint IEEE-ASME Committee, *Recommended Specification for Speed-Governing of Steam Turbines Intended to Drive Electric Generators Rated 500 kW and Larger*, IEEE Std. 600, 1959; L. K. Kirchmayer, *Economic Control of Interconnected Systems*, 1959.

Spinning (metals)

A production technique for shaping and finishing metal. In the spinning of metal, a sheet is rotated and worked by a round-ended tool, controlled manually or mechanically. The sheet is formed over a mandrel.

Spinning operations are usually carried out on a special rigid lathe fitted only with a driving headstock, tail spindle, and tool rest. Surface speeds of 500–5000 ft/min (2.5–25 m/s), are used, depending on material and diameter. The work is rubbed with soap, lard, or a similar lubricant during working. The operation can be set up quickly, and thus is desirable for short runs or for experimental units subject to change. Spinning may serve to smooth wrinkles in drawn parts, provide a fine finish, or complete a forming operation as in curling an edge of a deep-drawn part. Other operations include smoothing, necking, bulging, burnishing, beading, and trimming. Spun products range from precision reflectors and nose cones to kitchen utensils. Such materials as steel, aluminum, copper, and their softer alloys are spun in thicknesses up to 1/8 in. (3 mm). It may be necessary to anneal the metal during the spinning. *See* SHEET-METAL FORMING. [RALPH L. FREEMAN]

Splines

A series of projections and slots used instead of a key to prevent relative rotation of cylindrically fitted machine parts. Splines are several projections machined on the shaft; the shaft fits into a

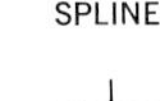

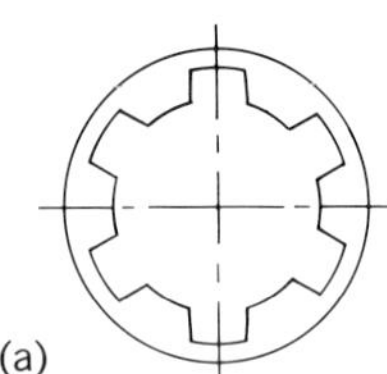

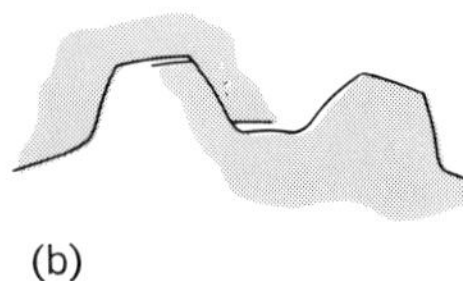

Diagrams of (*a*) square spline and (*b*) involute spline profile.

mating bore called a spline fitting. Splines are made in two forms, square and involute, as illustrated. Since there are several projections (integral keys) to share the force in transmitting power, the splines can be shallow, thereby not weakening the shaft as much as a standard key.

Square splines have 4, 6, 10, or 16 splines. The external part (shaft) may have the splines formed by milling and the internal part (bore) by broaching. Three classes of fits are used: sliding (as for gear shifting) under load, sliding when not loaded, and permanent fit. Square splines have been used extensively for machine parts. In the automotive industry, square splines have been replaced generally by involute splines, which cost less to make accurately for good fit and interchangeability.

Involute splines are used to prevent relative rotation of cylindrically fitted machine parts and have the same functional characteristics as square splines. The involute spline, however, is like an involute gear, and the spline fitting (internal part) is like a mating internal gear. Profiles are the same as for gear teeth of the stub (fractional pitch) form with 30° pressure angle.

Involute splines on the shaft are generated by a hob or a gear shaper, and internal splines are formed by a broach or a gear shaper. Three classes of fit are standard: sliding, close, and press.

Involute serrations are similar to involute splines except that the pressure angle is 45° and, while there are three standard fits (loose, close, and press), serrations are usually press-fitted and used for permanent assembly. They are used for both parallel and tapered shafts. *See* MACHINE KEY.

[PAUL H. BLACK]

Bibliography: American National Standards Institute (ANSI), *Involute Splines and Inspections, Inch Version*, B92.1-1970; ANSI, *Involute Splines and Inspections, Metric Version*, B92.1M-1970; ANSI, *Involute Splines, Metric Modules*, B92.2M-1981; C. W. Ham et al., *Mechanics of Machinery*, 4th ed., 1958; J. E. Shigley, *Mechanical Engineering Design*, 3d ed., 1976.

Spot welding

A resistance-welding process in which coalescence is produced by the flow of electric current through the resistance of metals held together under pressure. A low-voltage, high-current energy source is required (Fig. 1). Usually the upper electrode moves and applies the clamping force. Pressure must be maintained at all times during the

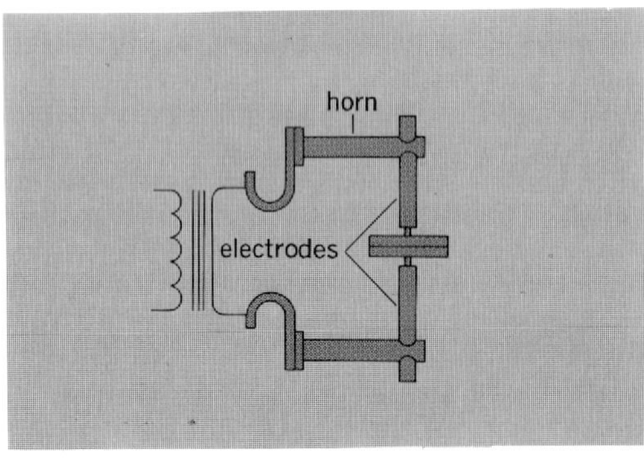

Fig. 1. Spot-welding circuit. When electrodes are closed on the workpiece, the circuit is completed for welding.

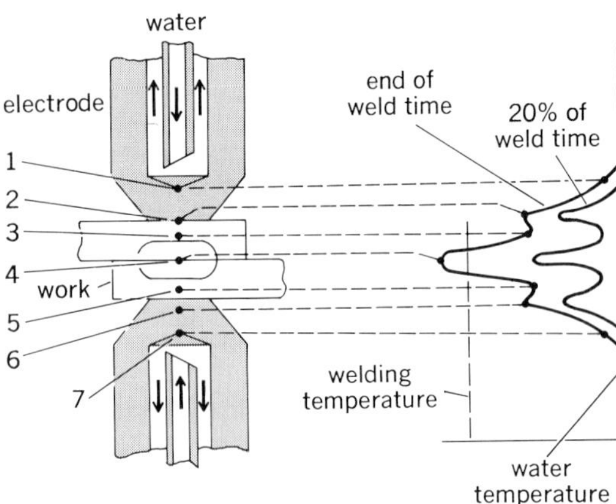

Fig. 2. Distribution of temperature in local (numbered) elements of a spot-welding operation.

heating cycle to prevent flashing at the electrode faces. Electrodes are water-cooled and are made of copper alloys because pure copper is soft and deforms under pressure. *See* RESISTANCE WELDING.

The electric current flows through at least seven resistances connected in series for any one weld. They are (1) upper electrode, (2) contact between upper electrode and upper sheet, (3) body of upper sheet, (4) contact between interfaces of sheets, (5) body of lower sheet, (6) contact between lower sheet and electrode, and (7) lower electrode (Fig. 2). Heat generated in each of the seven sections will be in proportion to the resistance of each. The greatest resistance is at the interfaces in 4, and heat is most rapidly developed there. The liquid-cooled electrodes 1 and 7 rapidly dissipate the heat generated at the contact between electrodes and sheets 2 and 6 and thus contain the metal that is heated to fusion temperature at the interfaces.

After the metals have been fused together, the electrodes usually remain in place sufficiently long to cool the weld. An exception is in welding quench-sensitive metals, where it is desirable to remove the electrodes as soon as possible to allow the heat to be conducted in the surrounding metal, preventing steep quench gradients. *See* WELDING AND CUTTING OF METALS.

[EUGENE J. LIMPEL]

Spring

A machine element for storing energy as a function of displacement. The flywheel, in contrast, is a means for storing energy as a function of angular velocity. Force applied to a spring member causes it to deflect through a certain displacement thus absorbing energy.

A spring may have any shape and may be made from any elastic material. Even fluids can behave as compression springs and do so in fluid pressure systems. Most mechanical springs take on specific and familiar shapes such as helix, flat, or leaf springs. All mechanical elements behave to some extent as springs because of the elastic properties of engineering materials.

Uses. Energy may be stored in a spring for many uses: to be released later, to be absorbed at the instant the energy first appears, and so on.

Motive power. One of the early and still the most frequent uses of springs is to supply motive power in a mechanism. Common examples are clock and watch springs, toy motors, and valve springs in auto engines. In these, energy is supplied to, and stored in, the spring by applying a force through a suitable mechanism to deflect or deform the spring The energy is released from the spring by allowing it to push (as in the valve) or twist (as in a clock) a mechanism through a required displacement.

Return motion. A special case of the spring as a source of motive power is its use for returning displaced mechanisms to their original positions, as in the door-closing device, the spring on the cam follower for an open cam, and the spring as a counterbalance. To a certain extent the springs in vehicles of transportation are in this category. They are designed to keep the car at a certain level with respect to the road or rails, returning the vehicle to this position if displaced by applied forces.

Shock absorbers. Frequently a spring in the form of a block of very elastic material such as rubber absorbs shock in a mechanism. For example, the four legs of a punch press rest on four blocks of rubber. The rubber pads prevent the die-closing inertia forces of the press from transferring down through the legs to the floor with impact or hammer blow proportions. With the rubber pads under the press legs, the force on the floor builds up relatively slowly and no shock is evident. As the acceleration of the die block goes to zero, the inertia force goes to zero and the rubber pad springs, which were deflected by the press blow, are relaxed and ready for the next stroke of the press die. The whole press moves up and down relative to the floor, but by proper selection of the rubber pad the elastic constant is such that this motion is small. *See* SHOCK ABSORBER.

Vibration control. Springs serve an important function in vibration control by supplying the necessary flexibility in the support of the vibrating mechanism and the required opposing forces as a result of their deflection. In controlling vibration, the body or mass of the mechanism must be freely supported so as to generate forces opposing the vibrating forces. These opposing forces tend to

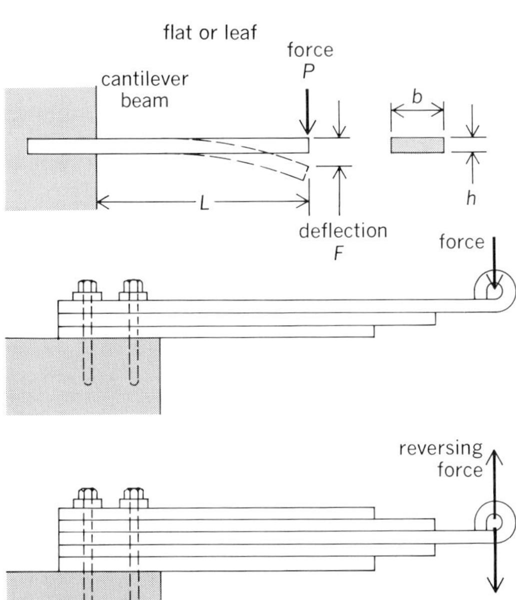

Fig. 1. Flat or leaf spring responds to perpendicular forces in either direction.

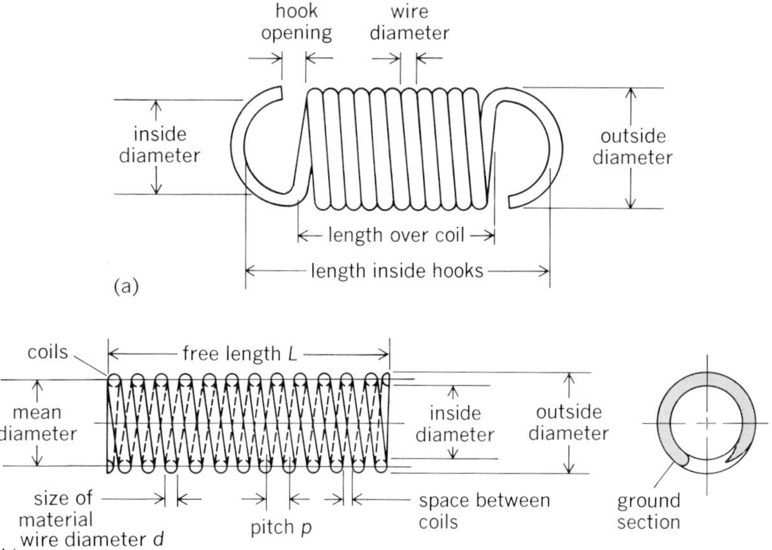

Fig. 2. Helical spring. (*a*) Wound tight to extend under axial tension. (*b*) Wound loose to contract under axial compression.

bring the sum of the forces on the vibrating body to zero. Vibration absorbing mounts are available in a wide range of sizes and spring constants to meet most requirements. These do not prevent vibration, which is a function of speed and the balance of the mechanism, but rather they minimize its effect on the machine frame or the mounting. *See* SHOCK ISOLATION.

Force measurement. Springs have long been used in simple weighing devices. Accurate weighing is usually associated with dead-weight devices or balances, but modern spring scales have received wide acceptance and certification for commercial use. Extremely accurate springs for heavy loads are used to calibrate testing machines, and they are employed in scales over crane hooks for weighing material as it is hoisted. Carefully calibrated springs are used in instruments such as electric meters and pressure gages.

Retaining rings. A relatively modern machine part in which the spring function is used as a holding means is the retaining or snap ring. This device is a split ring of square, rectangular, or special cross section. It fits in a groove on a cylindrical surface or in a bore and stays in place by spring force. A retaining ring prevents or restrains relative axial motion between a shaft or bore and the components on the shaft or in the bore.

Types. Springs may be classified into six major types according to their shape. These are flat or leaf, helical, spiral, torsion bar, disk, and constant force springs.

Flat or leaf spring. A leaf spring is a beam of cantilever design with a deliberately large deflection under a load (Fig. 1). One end of a leaf spring is usually firmly anchored to the frame of the machine and the other end is linked to the moving machine elements by a two-force (pin-ended) link. Force may be tension or compression with no modification of the design. This push-pull feature is the great advantage of a leaf spring, plus the fact that a relatively large amount of energy can be stored in a small space.

Helical spring. The helical spring consists essentially of a bar or wire of uniform cross section wound into a helix. The last turn or two at each end of the spring is modified to a plane surface perpendicular to the helix axis, and force can then be applied to put the helix in compression. The ends of the spring helix may be modified into hooks or eyes so that a force may be applied in tension. In general it is necessary to design helical springs so that force may be transmitted either in tension or in compression but not both ways for the same spring (Fig. 2). Where reversing forces occur in a spring mechanism, it is better to use a bar or leaf spring and in some cases a disk spring.

Spiral spring. In a spiral spring, the spring bar or wire is wound in an Archimedes spiral in a plane. Each end of the spiral is fastened to the force-applying link in the mechanism.

A spiral spring is unique in that it may be deflected in one of two ways or a combination of both of them (Fig. 3). If the ends of the spiral are deflected by forces perpendicular to the spiral plane, the spiral is distorted into a conical helix. For better stability and ease of applying the forces, the spring is often made as a conical helix to start with and is deflected into a plane spiral.

A spiral spring may also have the forces act tangent to the spiral as in a clock spring. The spiral is wound quite open and the tangential force in tension on the spiral tends to close the gaps between successive turns. The spring behaves like a beam, bending to a shorter radius of curvature and thus storing energy.

Torsion bar. A torsion bar spring consists essentially of a shaft or bar of uniform section. It stores energy when one end is rotated relative to the other. It is used in the spring system of the chassis of modern motor cars. *See* TORSION BAR.

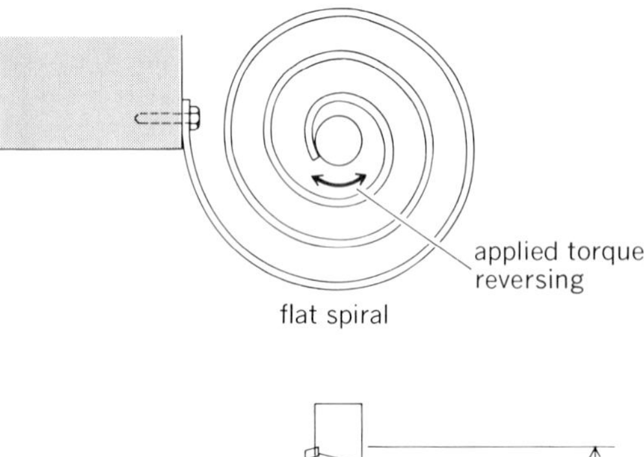

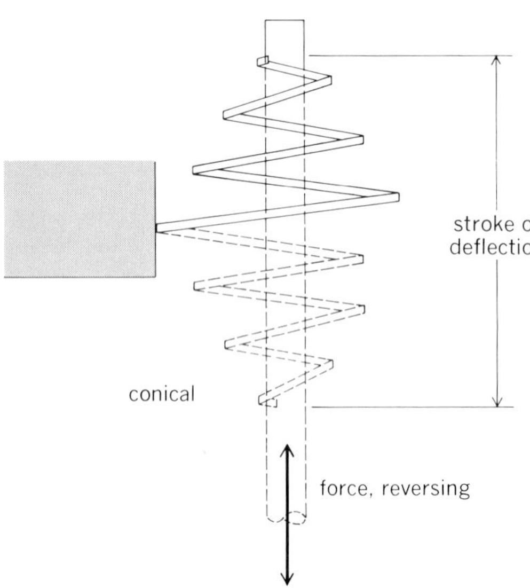

Fig. 3. Spiral spring is unique in responding to torsional or translational forces.

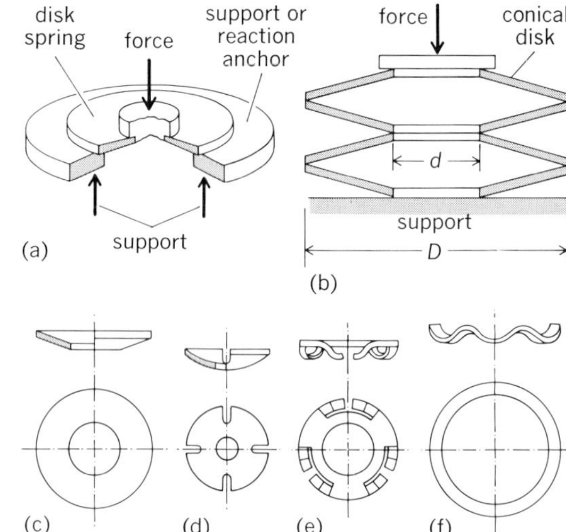

Fig. 4. Disk spring. (*a*) Single tapered disk; (*b*) multiple disk; (*c*) Belleville spring; (*d*) slotted curved washer; (*e*) finger washer; (*f*) wave spring washer.

Disk spring. Where large forces are present and space is at a premium, the disk spring may be used, although it is usually expensive to design and build. It consists essentially of a disk or washer supported at the outer periphery by one force (distributed by a suitable chuck or holder) and an opposing force on the center or hub of the disk (Fig. 4). For greater deflections several disks may be stacked with the forces transmitted from one to the next by inner and outer collars.

Constant-force spring. Many mechanisms require that a constant force be applied regardless of displacement. The counterbalancing of vertically moving masses against the force of gravity is a typical example. The Neg'ator spring of Hunter Spring Co. provides such a constant force; it uses a tight coil of flat steel spring stock. When the outer free end is extended and the coil allowed to rotate on its shaft or pintle, the spring presents a constant restoring force.

Design. Each type spring has its special features and design refinements. Common to all forms of springs are the basic properties of elastic materials. Within the elastic range of a material, the ratio of applied force to resulting deflection is constant. Spring systems can be designed to have a variable ratio. The ratio is the spring rate or scale and has the dimension of force per unit length. *See* HOOKE'S LAW.

In a helical spring the elastic action stresses the wire in torsion. The following variables are subject to the designer's action and decision:

free length	L	modules of elasticity in torsion	G
mean diameter	D	number of active turns	N
wire diameter	d	helix angle or coil pitch	P
allowable stress	S	deflection at load P	F
applied load	P		
spring rate	k		

These variables are related by Eq. (1), which neglects the effect of coil curvature on the stress, and by Eq. (2). For springs subject to frequent cycling, as a valve spring, the correction factor for wire curvature must be included. The generally used correction is the factor proposed by A. M. Wahl. Introducing this factor changes Eq. (1) to the form of Eq. (3), in which $C=D/d$. The portion

$$S=8PD/\pi d^3 \quad (1)$$

$$F=8PD^3N/Gd^4 \quad (2)$$

$$S_{max}=(8PD/\pi d^3)\left[\frac{4C-1}{4C-4}+\frac{0.615}{C}\right] \quad (3)$$

in brackets is Wahl's factor K whence maximum stress is K times the torsional stress (Fig. 5).

Various combinations of spring dimensions for the variables listed above will produce an acceptable spring. Usually the design is chiefly limited by allowable stress of materials and space.

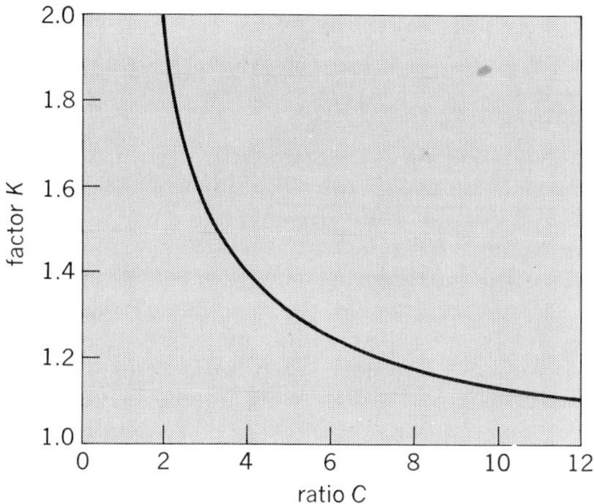

Fig. 5. Correction factor for curvature and shear. (*A. M. Wahl, Mechanical Springs, Penton, 1944*)

Where compression and tension springs are cycled at very high frequencies, surging may cause high local stress and result in early fatigue failure. Surging is the inability of all parts of the spring to deflect at the same rate due to the inherent inertia in the coils. This phenomenon is closely associated with the natural frequency of the spring; springs should be designed so that their natural frequency and cyclic rate are as far apart as practical.

[L. SIGFRED LINDEROTH, JR.]

Bibliography: Associated Spring Corp., *Handbook of Mechanical Spring Design*, rev. ed., 1981; H. Carlson, *Spring Designers Handbook*, 1978; H. Carlson, *Springs: Trouble-shooting and Failure Analysis*, 1980; E. T. Gross et al. (eds.), *Coil Spring Making*, 2d ed., 1974; Spring Research and Manufacturers Association, *Strip Spring Making and Forming*, 1977.

Stadia

A surveying distance-measuring method in which the interval intercepted on a vertically held rod by two cross hairs in the alidade or transit telescope is convertible to the horizontal distance from instrument to rod. The rod is called a stadia rod; the cross hairs are called stadia hairs and are fixed equidistant above and below the horizontal cross hair. Their distance apart, i, is a fixed ratio (conveniently 1:100, or 0.3:100 in levels) of the telescope's focal length. Hence the distance D from the focal point to an object is basically proportional to the interval R observed on the rod.

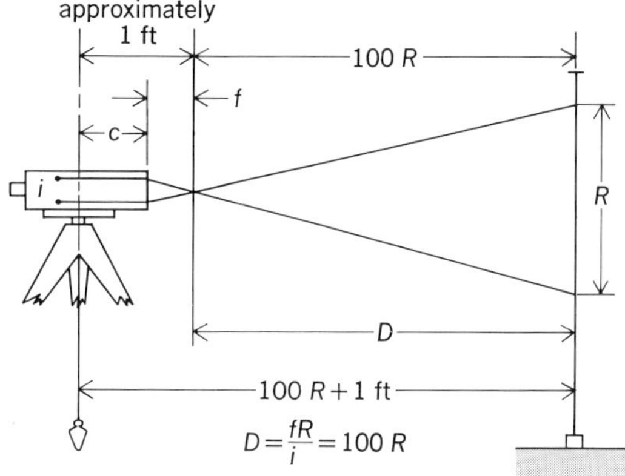

Stadia measurement of distance. 1 ft = 30 cm.

For horizontal stadia sighting, the small distance from the focal point to the instrument center is added to the distance obtained by reading the rod intercept as shown in the illustration. This distance is about 1 ft (30 cm) for externally focusing instruments; it is negligible for internally focusing telescopes. Further conversions are made for inclined sightings to find the correct horizontal component of the slope distance and to obtain the vertical component or difference of elevation between the instrument point and the rod point.

A refined stadia instrument, the reduction tacheometer, incorporates an etched glass reticle that, by rotating for inclined sights, causes the stadia hair separation to vary correctly with the increase or decrease of vertical angle, so as to maintain a proper ratio for direct conversion to horizontal distance. In all of these stadia instruments, the figures and graduations on the stadia rod are read through the instrument telescope. *See* SURVEYING.

[B. AUSTIN BARRY]

Stainless steel

The generic name commonly used for that entire group of iron-base alloys which exhibit phenomenal resistance to rusting and corrosion because of chromium content. The metallic element chromium (Cr) has been used in small amounts to strengthen steel since the famed Eads Bridge spanned the Mississippi at St. Louis, Mo., in 1872; but only in the present century was it discovered that contents of Cr exceeding 10%, with carbon (C) held suitably low, make iron effectively rustproof. Onset of the property is striking (Fig. 1), and the Cr can then be increased to about 27% with even further chemical advantages before the metal becomes structurally useless.

Other alloy elements, notably nickel (Ni) and molybdenum (Mo), can also be added to the basic stainless composition to produce both variety and improvement of properties. Over 100 different stainless steels are produced commercially, about half as standardized grades. Some are more properly classed as stainless irons since they do not harden as steel; others are true steels to which corrosion resistance becomes an added feature. Still others that are neither properly steels nor irons introduce totally new classes of materials,

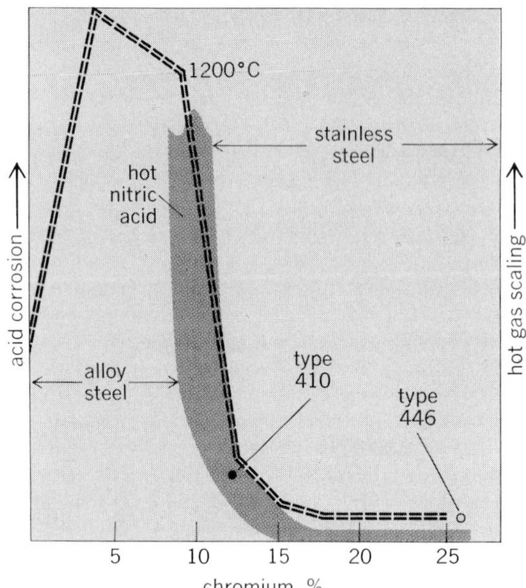

Fig. 1. Passivity phenomenon in stainless steel.

from both mechanical and chemical standpoints. *See* ALLOY; STEEL.

In fact, the great modern developments of heat-resisting superalloys and refractory metals stand as extreme extensions of the two heat-resisting families of stainless steels. That same resistance to rusting at ordinary temperatures shows as resistance to oxidation at elevated temperatures (Fig. 1), producing a family of oxidation-resistant high-temperature alloys; and with this basic feature contributed by the Cr content, further additions of Ni and other elements contribute remarkable properties of workability and strength, which extend not only to high temperatures but to cryogenic temperatures as well. The Ni-base heat-resisting alloys, the most prominent in that field, project from stainless steel in the sense that Ni increasingly replaces the iron (Fe). Similarly, the important titanium (Ti) and cobalt (Co) alloy systems, such refractory metals as niobium (Nb) and tungsten (W), and also Cr and Mo, represent the far extremes for use of these alloy elements in stainless steel.

The shaded band in Fig. 1 covers typical values for corrosion penetration of iron alloys with increasing Cr content, the actual data referring to hot nitric acid. The width of the band is due to differences in temperature and concentration. Attack almost ceases near 12% Cr, the alloy type 410; and it becomes scarcely measurable as Cr increases to type 446. The double-dash boundary in Fig. 1 is a plot of actual data for hot-gas scaling at 1200°C (2190°F). The sudden drop occurs in nearly the same Cr range as for aqueous corrosion, type 446 again being the most resistant alloy. The reversal at the far left is also similar for aqueous corrosion.

In service, the stainless steels cover three major fields: corrosion resistance at ordinary temperatures, scaling resistance at high temperatures, and mechanical strength over the entire temperature range of metallurgical engineering.

Types. Metallurgically, the alloys fall within four groups: ferritic, martensitic, austenitic, and precipitation-hardening.

Ferritic. When Cr is added to ordinary iron to develop corrosion resistance without steel-type hardening, ferritic alloys are obtained. *See* IRON ALLOYS.

The typical alloys in the table are standards listed by the American Iron and Steel Institute (AISI). The first is the least expensive, and a little aluminum (Al) is usually added to help the Cr suppress steel-type hardening and ensure the ferritic condition, since this avoids welding problems from stray martensite. The second is the most popular commercial ferritic grade, widely used for automotive trim; and some fractional martensitic hardening is deliberately courted for additional strength. Type 446 is primarily used for oxidation resistance, for above this Cr content the engineering value of the Cr-Fe alloys rapidly vanishes.

Martensitic. When Cr is added to steels with carefully chosen C contents, the martensitic alloys result (see table). Again, these cover a range from the least expensive and most widely used type 410, through the more select type 420, ideal for cutlery, and on to the highly specialized type 440-C, whose great hardness makes it suitable for ball bearings and surgical instruments.

Austenitic. Nickel becomes the next most important alloying element. Its commonest use is in the renowned "18–8" containing 18% Cr and 8% Ni. Unlike the atomic structure of ferrite, which has the form of a neatly body-centered cube, or the martensite with its atomic disarrangement responsible for strength and hardness, the austenite of this third class of stainless steel has a face-centered cubic form made stable by the Ni. Completely new mechanical properties appear, outstanding for workability, notch toughness, high-temperature strength, and a frequently superior corrosion resistance. These are generally the most prized and popular of all stainless steels and also the most expensive. Only a few of the score of standards appear in the table. The parenthetical C for type 302 refers to a strange sensitivity to C content, such that 18–8 today is more commonly type 304 with C held below 0.08, and even type 304 with

Typical compositions of standard stainless steels

AISI no.	Cr, %	C, %	Other elements,%
Ferritic			
405	13	0.08	
430	17	0.12	
446	26	0.20	
Martensitic			
410	12	0.12	
420	13	0.25	
440-C	17	1.00	
	Cr, %	Ni,%	Other elements,%
Austenitic			
302	18	8	(C = 0.08)
310	25	20	C = 0.20
316	16	13	2.75 Mo
347	18	11	0.50 Nb
Precipitation-hardening			
630 (17-4 PH)	17	4	Cu = 3.2, Nb = 0.25
633 (AM-350)	16.5	4.3	Mo = 2.75, N = 0.10
– (17-4 Cu Mo)	17	14	Cu = 3.0, Mo = 2.5 Nb = 0.50, Ti = 0.25

C held below 0.03. Type 310 lies well toward the Ni-base alloys and is accordingly notable for its heat resistance. The Mo in type 316 confers a special resistance to pitting corrosion, while the Nb in type 347 is a highly specialized addition correcting the C problem in welding 18–8.

While these alloys are called steels, no martensite develops on cooling; neither can the alloys be hardened by the usual quenching and tempering. Contrarily, they are put into their softest state by producing austenite at high temperature, with subsequent quenching to retain it. Strengthening is developed instead by cold working, and by adjusting the 18–8 composition slightly downward toward 17–7. The austenite then tends to decompose inside the cold steel; the martensite crowds the slip planes loosened by cold working, locking them against further movement and thereby increasing the strength (Fig. 2). Hence, ferritic and martensitic steels are little strengthened by cold working, while austenitic steels are greatly strengthened by it, notably type 301.

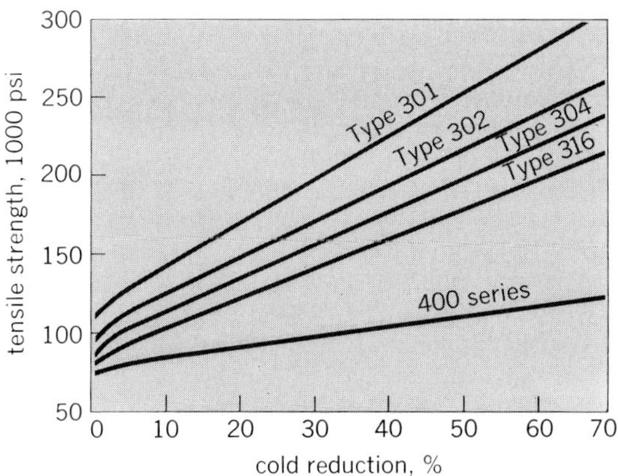

Fig. 2. Graph of the effect of cold-work on tensile strength of stainless steels.

Precipitation-hardening. Exploiting this approach to strengthening, and following the example of duralumin in a neighboring metallurgical field, metallurgists have developed the fourth group of stainless steels known as precipitation-hardening. A score of alloys became available within two decades of their commercial acceptance, and trade names rather than generic designations are still commonly used. The first listed in the table is subclassed martensitic because the metal behaves primarily as martensite which further experiences internal precipitation; the second is typical of what are called semiaustenitic alloys for the reason that they rather closely follow the instability pattern of 17–7; and the third belongs to the austenitic subgroup, with hardening essentially due to precipitation within austenite. The special properties are developed through carefully controlled heat treatments that always begin with a "solution anneal" at some high temperature and are followed by internal precipitation at some lower temperature. The latter may have multiple stages, ranging down to subzero treatments, and effects of cold-work are often also exploited, particularly for the semiaustenitic alloys. These precipitation-hardening stainless steels are the hallmark of the third quarter of the 20th century, currently bringing mechanical properties to levels unmatched by any other metal or alloy. Tensile strengths are actually exceeding the phenomenal figure of 300,000 psi (2.1 GPa).

Fabrication. As with steels in general, the so-called wrought stainless steels come from the melting furnaces in the form of either ingot or continuously cast slabs. Ingots require a roughing or primary hot working, which the other form commonly bypasses. All then go through fabricating and finishing operations such as welding, hot and cold forming, rolling, machining, spinning, and polishing. No stainless steel is excluded from any of the common industrial processes because of its special properties; yet all stainless steels require attention to certain modifications of technique.

Hot working. Hot working is influenced by the fact that many of the stainless steels are heat-resisting alloys. They are stronger at elevated temperatures than ordinary steel. Therefore they require greater roll and forge pressure, and perhaps lesser reductions per pass or per blow. The austenitic steels are particularly heat-resistant.

Welding. Welding is influenced by another aspect of high-temperature resistance of these metals—the resistance to scaling. Oxidation during service at high temperatures does not become catastrophic with stainless steel because the steel immediately forms a hard and protective scale. But this in turn means that welding must be conducted under conditions which protect the metal from such reactions with the environment. This can be done with specially prepared coatings on electrodes, under cover of fluxes, or in vacuum; the first two techniques are particularly prominent. Inert-gas shielding also characterizes widely used processes, among which at least a score are now numbered. As for weld cracking, care must be taken to prevent hydrogen absorption in the martensitic grades and martensite in the ferritic grades, whereas a small proportion of ferrite is almost a necessity in the austenitic grades. Metallurgical "phase balance" is an important aspect of welding the stainless steels because of these complications from a two-phase structure. Thus a minor austenite fraction in ferritic stainless can cause martensitic cracking, while a minor ferrite fraction in austenite can prevent hot cracking. However, the most dangerous aspect of welding austenitic stainless steel is the potential "sensitization" affecting subsequent corrosion, to be discussed later.

Machining and forming. These processes adapt to all grades, with these major precautions: First, the stainless steels are generally stronger and tougher than carbon steel, such that more power and rigidity are needed in tooling. Second, the powerful work-hardening effect (Fig. 2) gives the austenitic grades the property of being instantaneously strengthened upon the first touch of the tool or pass of the roll. Machine tools must therefore bite surely and securely, with care taken not to "ride" the piece. Difficult forming operations warrant careful attention to variations in grade that are available, also in heat treatment, for accomplishing end purposes without unnecessary work problems. Spinning, for example, has a type 305

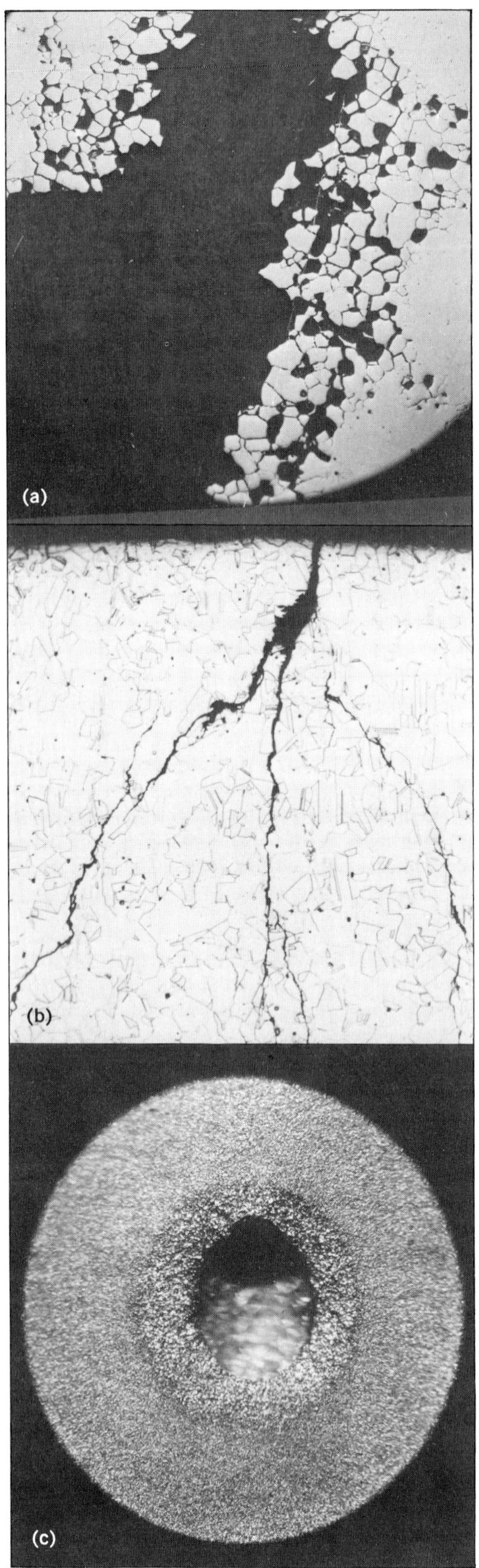

Fig. 3. Three great corrosion effects on stainless steel. (*a*) Sensitization. (*b*) Stress-corrosion cracking. (*c*) Pit corrosion.

modification of 18-8 which greatly favors the operation; while at the opposite extreme is type 301 or 17-7, for those who wish to take advantage of the strengthening due to cold working.

Finishing. These operations produce their best effects with stainless steels. No metal takes a more beautiful polish, and none holds it so long or so well. Stainlessness is not just skin-deep, but body-through. And, of course, coatings are rendered entirely unnecessary. At this point the higher initial cost of these alloys rapidly falls behind the long-range cost of other metals, often making them the cheapest buy on the market.

Service. Service of the stainless steels covers almost every facet of modern metallurgical engineering. They began in the chemical industry for reasons of their corrosion resistance, then graduated into high-temperature service for reasons of resistance to both surfacial sealing and internal creep. They have finally proliferated in every field of mechanical operation, ranging from high strength and hardness in combination with good ductility, at elevated as well as cryogenic temperatures, through cutlery and spring and nonmagnetic materials, to the purely ornamental applications of modern architecture.

However, when stainless steel does break down, the action is usually dramatic, and nothing exceeds the phenomenology of corrosion in the chemical industry. The situation can be understood by recognizing that the passivity of the stainless steels—that which prevents or restrains their reaction with environmental chemicals—is not the nonactivity or nobleness which characterizes the noble metals of the platinum group. It comes as a surprise to the nonscientist that the chromium content makes stainless steel even more disposed to rust than plain iron. But it is this very tendency which immediately fixes the oxygen or rust so tightly upon the surface that a protective coating results. The same quality applies to titanium and aluminum and their alloys. If something disturbs the quite special conditions needed for this passivity, corrosion takes place. The damage then usually takes one of the three major forms illustrated in Fig. 3.

Sensitization. Some phenomenon, not yet fully understood, causes chromium carbide to precipitate in the grain boundaries of the austenitic grades when heated even for short periods of time in the range 800–1500°F (427–816°C); these grain boundaries thereupon become so susceptible to chemical attack that a piece of formerly solid steel can fall apart in the hand like granulated sugar. The most workable explanation is that the carbon has depleted the chromium content in forming the carbide, thus leaving the metal locally nonstainless. Because welding necessarily produces a heat zone alongside the bead which somewhere must cover this critical range, welded steels often are susceptible to sensitization when later subjected to chemical service. The principal cures thus far are: (1) The carbon is lowered to values where carbide cannot form; this has been accomplished with the introduction of type 304L. (2) Other and even more powerful carbide formers—notably titanium (Ti) and niobium (Nb)—have been added to the alloys to tie up carbon at high temperatures before chromium carbide forms. Nevertheless, the problem persists as the principal concern in the chemical service of austenitic stainless steel.

Stress-corrosion cracking. Known for a hundred or more years in the brass industry as season cracking, the same phenomenon seems to have been discovered in stainless steel. In marked contrast to the intergranular nature of sensitization, stress-corrosion cracking cuts right through the body of the grain; and although the austenitic steels are by far the most susceptible, with the martensitic and ferritic following in that order, it appears that any metal or alloy can probably be made to fail in this manner if subjected to the right chemical under appropriate conditions of stress. Present corrections or preventions lie principally in directions of reducing stress levels, both internal and external, and avoiding exposure to chemicals known to be particularly dangerous. A metallurgical breakthrough indicates a close approach to immunity through use of highly purifying melting processes in steelmaking.

Pit and crevice corrosion. These are forms specifically following from oxygen "starvation" in local areas. When the steel can no longer rust, the passive condition converts to an active one; and to worsen the situation, all of the remaining surface then takes on the role of cathode in a galvanic cell which concentrates all the anodic or dissolving activity right at the tiny point of breakdown. If this is just a local spot, the attack takes the form of pitting, but if two flat surfaces come close together as in bolt-and-washer assemblies, the thin film of entrapped liquid can run out of oxygen and cause a general breakdown called crevice corrosion. *See* CORROSION.

Production. Stainless steel is undergoing revolutionary changes in its melting technology. There are two major problems in the melting of stainless steel which distinguish it markedly from melting ordinary steel: chromium must be brought to high levels, and carbon levels must be held low. With minor exceptions in the high-carbon martensitic grades, stainless cannot be produced in ordinary fuel-fired furnaces such as the open hearth. The technology arose solely because of the closer controls possible with electrically heated furnaces, specifically the arc furnace where tonnage production is concerned. Thus stainless melting arose historically out of ordinary steelmaking by graduating into specialty furnaces. Since World War II the technological demands upon stainless steel have pressed the technology into still greater specialization, such that steelmaking in this field now displays the most brilliant technical achievements in the history of metallurgy.

Ordinary steelmaking is concerned only with removal of carbon from the liquid metal, and this is easily accomplished by the addition of iron ore—the carbon seizes the oxygen and escapes as a gas. But chromium also reacts strongly with oxygen, thereby removing itself from the metal to join the slag. Therefore, this branch of steelmaking had to develop a two-stage process: melt down ordinary steel and drive off the carbon, and add the chromium after the oxidizing of carbon was completed. This in turn brought two severe handicaps: scrap stainless steel—the cheapest chromium carrier—could not be used in the first-stage charge because the chromium would be lost; and an expensive chromium-rich alloy with low carbon content would have to be added in the second stage—so-called low-carbon ferrochrome. *See* STEEL MANUFACTURE.

In the 1930s Alexander Feild developed the rustless process in Baltimore, based upon the discovery that the competition between chromium and carbon for oxygen shifts strongly in favor of carbon at very high temperatures. By using chromium oxide itself for the hearth lining and then carefully pressing the arc heating some hundreds of degrees above normal operating temperatures, Feild found that the first-stage charge could be loaded not only with scrap but also with high-carbon ferrochrome, which is much cheaper than the low-carbon grade. Minimal chromium losses to the slag were then "kicked back" with a shower of ferrosilicon, this powerful deoxidizer easily taking the oxygen away from the chromium and leaving silicates in the slag, while returning metallic chromium to the bath.

In the 1940s an abundance of liquid oxygen became available from developments in war industries, and E. J. Chelius developed the idea of the "oxygen lance." A long, insulated steel pipe was plunged beneath the surface of the molten steel, churning the metal with pure oxygen. The difference between iron oxide and gaseous oxygen, with respect to oxidizing a steel bath, is that the latter is a gas, and so is carbon oxide. The bubbling of the one aids the escape of the other, while to chromium oxide it makes no difference. Furthermore, the exceedingly high temperatures developed locally around the lance brought the advantage of the Feild effect in removing carbon while conserving chromium. The 1950s then witnessed a virtual conversion of the entire industry to the oxygen lance, spreading into the open hearth as well as the furnaces used for making stainless steel. At last all stainless could be made with high scrap additions in the charge, right up to 100%. As for carbon contamination and the use of high-carbon ferrochrome, the higher the first-stage carbon the better. It served as a fuel, and a convenient source of heat during lancing.

Scarcely had these achievements come into widespread use when successful development of a porous refractory plug permitted blowing a cleansing gas through the bottom of a ladle. At first this was applied only to the use of an inert gas, such as argon, for removing damaging quantities of hydrogen often dissolved in steels of all kinds. But it shortly diverted to the use of oxygen as a substitute for lancing. However, the violence of the reaction, from entry so deep in the liquid, made it necessary to dilute the oxygen with an inert gas. Again argon was chosen, and the argon-oxygen process was born. New achievements were electrifying: (1) Almost the entire electric furnace processing was dispensed with, the furnace being needed only to melt down the charge, with all further work done in the ladle. The expensive furnace time was cut by factors of three and four, or more. (2) The vastly greater facilities for quality control in a ladle, as compared to a furnace, shortly permitted the further bridging of this technology with the long-developed and highly sophisticated vacuum techniques of physics and laboratory furnaces. The argon-oxygen process then developed variations using vacuum chambers and reduced pressures instead of positive pressures, and in some cases substituting nitrogen or other gases for argon.

Concurrently two major developments entered the field, this time from directions of those highly specialized heat-resistant metals and alloys known as the superalloys and the refractory metals. The superalloys generally comprise systems based upon nickel, cobalt, or titanium, as well as highly complex stainless steels, while the term "refractory metals" typically refers to wolfram (tungsten), tantalum, molybdenum, and niobium (columbium). Frequently the superalloys have quality demands so high as to require remelting a cast form in order to purify it a second time. One process arose naturally out of the vacuum technology just mentioned, and the more critical grades of stainless steel are now often subjected to vacuum-arc remelting. The original melt is cast in the form of a huge electrode which, as with a consumable welding electrode, melts off at the tip and falls into a water-cooled mold in a vacuum chamber. The metal is thus purified, and its solidification structure is very favorable to good mechanical properties. The second process also arose out of a welding concept, namely the use of a slag cover to protect the metal from environmental reactions while molten. First conceived by R. E. Hopkins in America several decades ago, the process was developed in the Soviet Union, then captured worldwide interest in the 1960s. Again the metal in electrode form is remelted electrically, but under a carefully prepared slag blanket instead of a vacuum. Solidification is again well controlled, and the metal is highly purified. *See* HIGH-TEMPERATURE MATERIALS.

Technological advances are continuing at a rapid rate. The 1970s witnessed such sophisticated developments as melting with electron beams, plasma arcs, and lasers—even levitation melting, where crucibles or containers are no longer necessary. The metal just suspends itself in mid-air! While levitation melting remains in curiosity stages and lasers still serve principally for punching holes and welding, the plasma-arc technique has entered commercial usage in welding, with beginnings in steelmaking. The electron-beam furnace has also come of age and has led to remarkable accomplishments in producing commercial quantities of certain ferritic stainless steels whose properties were previously believed impossible.

Most stainless is cast as ingots, although continuous casting is rapidly becoming popular. Virtually all wrought grades have their casting counterparts in the foundry industry, separately standardized by the Alloy Casting Institute (ACI). The metallurgy of the cast alloys is essentially the same as that for the wrought, with the principal exception that certain liberties can be taken in alloy composition because the castings are not hot-worked. In designing casting alloys, one can forget about mechanical problems of forming the part and concentrate on the chemical challenges of corrosion resistance and heat resistance.

[CARL A. ZAPFFE]

Bibliography: American Society for Metals, *Metals Handbook, Properties and Selection: Irons and Steels*, 1978; American Society for Metals, *Source Book on Stainless Steels*, 1976; D. Peckner and I. M. Berstein, *Handbook of Stainless Steels*, 1977; F. B. Pickering (ed.), *The Metallurgical Evolution of Stainless Steels*, 1979; J. A. Sedriks, *Corrosion of Stainless Steels*, 1979; L. L. Shreir (ed.), *Corrosion*, vols. 1 and 2, 2d ed., 1976.

Statics

The branch of mechanics that describes bodies which are acted upon by balanced forces and torques so that they remain at rest or in uniform motion. This includes point particles, rigid bodies, fluids, and deformable solids in general. Static point particles, however, are not very interesting, and special branches of mechanics are devoted to fluids and deformable solids. For example, hydrostatics is the study of static fluids, and elasticity and plasticity are two branches devoted to deformable bodies. Therefore this article will be limited to the discussion of the statics of rigid bodies in two- and three-space dimensions. *See* ELASTICITY.

Mechanics is the study of motions in terms of mass, length, time, and forces. In statics the bodies being studied are in equilibrium. Positions of points in space, velocities, forces, and torques are all vector quantities, since each has direction, magnitude, and units. Beams, bridges, machine parts, and so on are not really rigid, but whenever the largest change in length of a portion of a body Δl is much smaller than that length l, that is, $l \gg \Delta l$, then it is a satisfactory approximation to treat that body as a rigid body. Many of the objects used in architecture, engineering, and physics can be satisfactorily idealized as rigid bodies, and much of building, machine, bridge, and dam design is based upon the study of statics.

The equilibrium conditions are very similar in the planar, or two-dimensional, and the three-dimensional rigid body statics. These are the vector sum of all forces acting upon the body must be zero; and the resultant of all torques about any point must be zero. Thus it is necessary to understand the vector sums of forces and torques.

Force as a vector. The physical effect of a force is a push or pull on an object at its point of application. The effect of a force is to change the velocity of the body. The SI units of force are newtons (1 N = 1 kg-m-s^{-2}). The vector nature of forces is expressed by their direction in space. This direction is the direction of the push or pull exerted by the force.

The notation for a vector quantity is an overarrow, for example, $\vec{F}$ or $\vec{P}$. In Fig. 1 a two-dimensional vector $\vec{P}$ is shown as a directed line segment $\overrightarrow{AB}$ in the plane of the paper. A three-dimensional vector $\vec{P}$ is shown in Fig. 2 as the directed line segment $\overrightarrow{OD}$. A complete description of a vector is

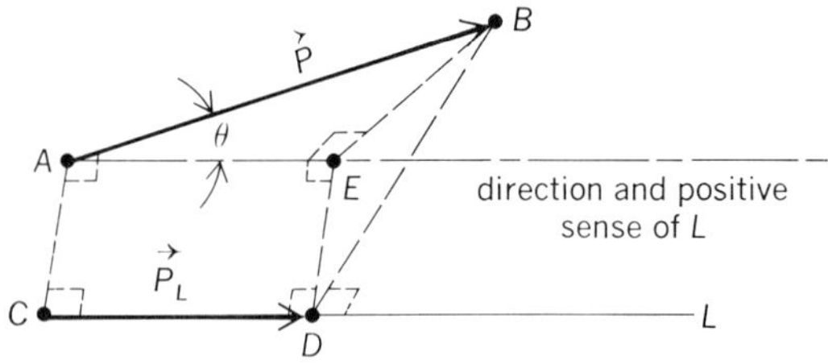

Fig. 1. Directed line AB or $\vec{P}$ is a vector quantity in the plane.

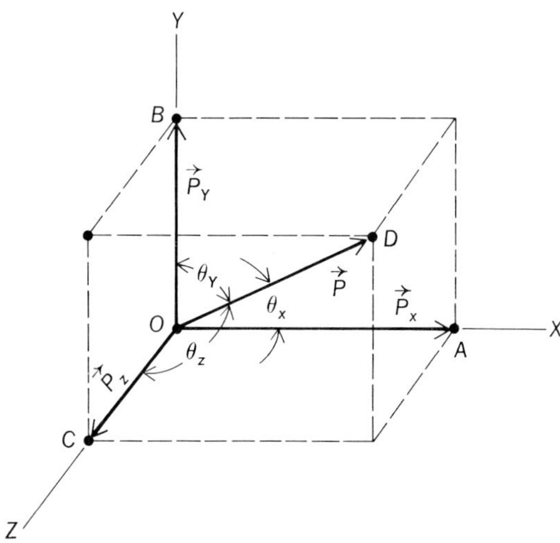

Fig. 2. Rectangular components of a force vector $\vec{P}$ in three-dimensional space.

obtained by specifying a coordinate system and giving its components in that coordinate system. The sum of any two vectors $\vec{A} + \vec{B}$ is the sum of their components, so that $\vec{A}$ and $\vec{B}$ must have the same number of components.

Superposition and transmissibility. In studying statics problems (and general mechanics problems), two principles, superposition and transmissibility, are used repeatedly on force vectors. They are applicable to all vectors, but specifically to forces and torques. (first moments of forces).

1. The principle of superposition of d-dimensional vectors is that the sum of any two d-vectors is another d-vector. Of course, some or all components can be zero. This principle is illustrated in Fig. 3. The two vectors labeled Q, which can be shown applied to a rigid structural member D, add up to zero, and so do the two labeled R. Superposition applies to all sums of vectors not just sums which vanish.

2. The principle of transmissibility of a force applied to a rigid body is that the same mechanical effect is produced by any shift of the application of the force. The principle of transmissibility is illustrated in Fig. 3. The solid rectangle C represents a solid structural member and the line AB is the line of action for the equivalent forces P. Since these two forces have the same magnitudes and direction and are applied along the same line, by transmissibility, all of their mechanical effects are equivalent. To use the superposition principle to add two vectors, the principle of transmissibility is used to move some vectors along their line of action in order to add to their components.

Components of a force. Figure 1, the two-dimensional example, shows the construction by

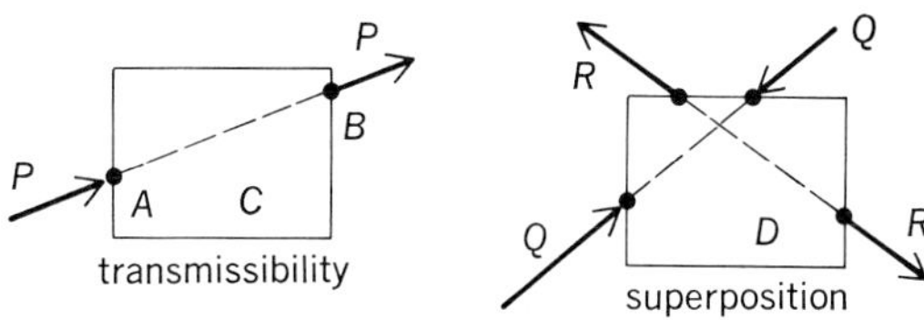

Fig. 3. Illustration of two theorems of statics.

which orthogonal force components are defined. In this figure $\vec{P}$ or $\overrightarrow{AB}$ is a force vector, and L a directed line whose positive sense is toward its labeled end. Construction lines AC and BD are in planes (not shown) normal to L, and θ is the direction angle of $\vec{P}$ relative to L; it is a plane angle between the directions in the positive senses of $\vec{P}$ and L; further, $0 \leq \theta \leq 180°$.

The orthogonal vector component of force $\vec{P}$ on directed line L is a force of direction and magnitude given by P_L or CD where P_L is in the direction of L. Its magnitude is given by Eq. (1).

$$P_L = \overline{CD} = \overline{AE} = \overline{AB}|\cos\theta| = P|\cos\theta| \qquad (1)$$

The component of $\vec{P}$ on L is $P_L = P\cos\theta$, where P_L is positive if $0° \geq \theta < 90°$ and negative if $90° < \theta \leq 180°$. The absolute magnitude of P_L is designated $|P_L|$.

The rectangular components of a force are its components on mutually perpendicular lines.

In Fig. 2 $\vec{P}_X$ or $\overrightarrow{OA}$, $\vec{P}_Y$ or $\overrightarrow{OB}$, and $\vec{P}_Z$ or $\overrightarrow{OC}$ are the rectangular vector components of $\vec{P}$ or $\overrightarrow{OD}$ in the directions of lines (axes) X, Y, and Z, respectively.

The corresponding components have the magnitudes $P_X = P\cos\theta_X$, $P_Y = P\cos\theta_Y$, and $P_Z = P\cos\theta_Z$, and are in the X, Y, Z directions, respectively. The magnitude of the three-dimensional vector $\vec{P}$ is given by Eq. (2).

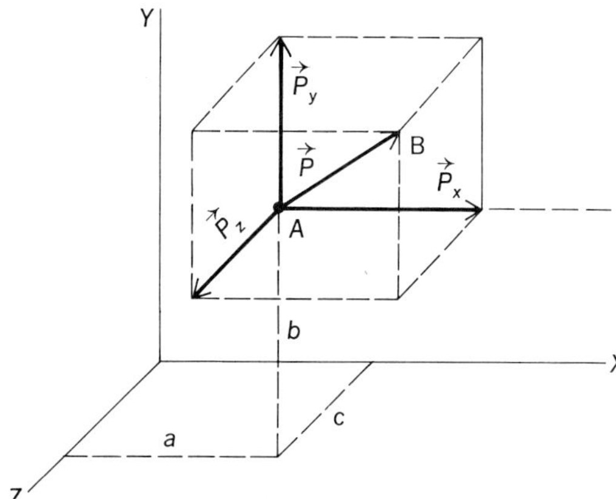

Fig. 4. Moments about axis in rectangular coordinates.

$$P = \sqrt{P_X^2 + P_Y^2 + P_Z^2} \qquad (2)$$

Moment of a force. The moment of a force about a directed line is a signed number whose value can be obtained by applying these two rules:

1. The moment of a force about a line parallel to the force is zero.

2. The moment of a force about a line normal to a plane containing the force is the product of the magnitude of the force and the least distance from the line to the line of the force. Conventionally, the moment is positive if the force points counterclockwise about the line as viewed from the positive end of the line.

In Fig. 4 the moments of $\vec{P}$ about the X, Y, and Z coordinate axes are $M_X = bP_Z - cP_Y$ about the X axis; $M_Y = cP_X = aP_Z$ about the Y axis; and $M_Z =$

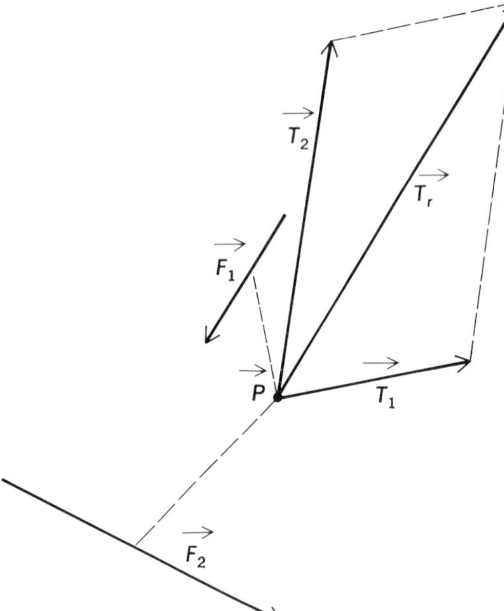

Fig. 5. Resultant of two torques.

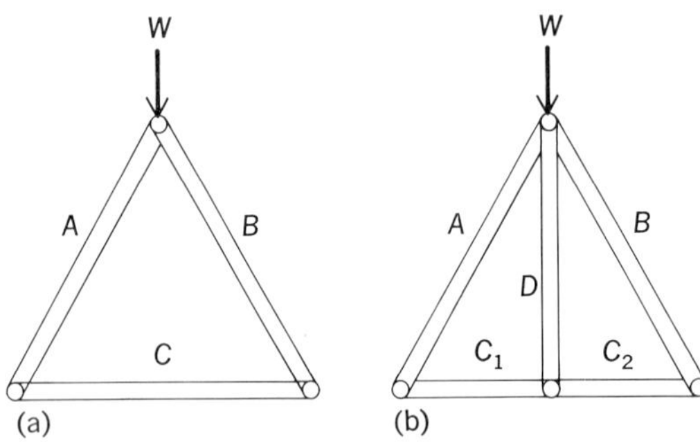

Fig. 6. Statically determinate and indeterminate structures. (*a*) The forces in members *A* and *B* are independent of their strengths (for small deformations). (*b*) The relative strength of the added member *D* will determine how much of the load *W* it supports.

$aP_Y - bP_X$ about the *Z* axis. The resultant of the two torques or moments of force, $\vec{T}_1$ and $\vec{T}_2$ about a point *P* is shown as $\vec{T}_r$ in Fig. 5. The components of these torques are due to forces $\vec{F}_1$ and $\vec{F}_2$ oriented with respect to $\vec{P}$ as shown.

Figure 6 illustrates cases where the forces in a structure may or may not depend on the relative strength of its parts. Case 6*a* is called statically determinate, and case 6*b* statically indeterminate. The terms are actually misnomers, for the forces can be found in both cases. Better terminology would be rigidly determinable and indeterminable. In case 6*b*, if each member were rigid (infinitely strong), the forces in each could not be determined.

In summary, the statics of rigid bodies in statically determinate structures is carried out by summing vector forces and vector torques and setting the resultants equal to zero. *See* COUPLE; FORCE; TORQUE.

[BRIAN DE FACIO]

Bibliography: F. J. Beer and E. Russell Johnston, *Mechanics for Engineers: Statics*, S. Timoshenko and D. H. Young, *Engineering Mechanics: Statics*, 1937.

Steam

Water vapor, or water in its gaseous state. Steam is the most widely used working fluid in external combustion engine cycles, where it will utilize practically any source of heat, that is, coal, oil, gas, nuclear fuel (uranium and thorium), waste fuel, and waste heat. It is also extensively used as a thermal transport fluid in the process industries and in the comfort heating and cooling of space. The universality of its availability and its highly acceptable, well-defined physical and chemical properties also contribute to the usefulness of steam. The engineering profession has developed so much evaporative and condensing equipment for changing the state of water that its properties can be utilized from pressures ranging from 0.18 in. Hg or 0.61 kPa (ice point) to supercritical (>3208.2 psia or 22.120 MPa absolute pressure and 705.47°F or 374.15°C). These temperature ranges are readily adaptable for use with ferrous and nonferrous metal structures at costs which dominate the design, selection, and use of many applications.

The temperature at which steam forms depends on the pressure in the boiler. The steam formed in the boiler (and conversely steam condensed in a condenser) is in temperature equilibrium with the water. The relation is rigorously specified by the vapor tension curve (Fig. 1). (The table gives a few selected values in the first two columns.) Under these conditions, with steam and water in contact and at the same temperature, the steam is termed saturated. Steam can be entirely vapor when it is 100% dry, or it can carry entrained moisture and be wet. It may also be contaminated, as in all practical and engineering applications, with other gases, especially air, in which case it is a mixture and Dalton's law prevails. After the steam is removed from contact with the liquid phase, the steam can be further heated without changing its

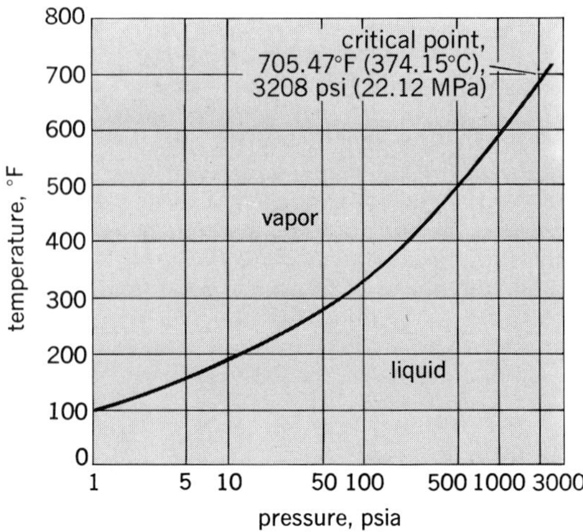

Fig. 1. Vapor-tension curve for water. 1 psia = 6.895 kPa, absolute pressure. Temperature °C = 5/9 (temperature °F − 32).

Thermal properties of steam (approximate)*

Pressure, psia (kPa)	Temperature, °F (°C)	Enthalpy, Btu/lb‡ Saturated liquid†	Latent heat	Saturated vapor†
1 (6.895)	102 (39)	69.7	1036	1106
10 (68.95)	193 (89)	161	982	1143
14.7 (101.3)	212 (100)	180	970	1150
100 (689.5)	328 (164)	298	889	1187
1000 (6895)	545 (285)	542	649	1192
3208 (22,120)	705 (374)	906	0	906

*For more exact values see J. H. Keenan and F. G. Keyes, *Thermodynamic Properties of Steam*, Wiley, 1936, or American Society of Mechanical Engineers, *Thermodynamic and Transport Properties of Steam* (Tables and Charts), 1967.

†Enthalpy is measured from saturated liquid at 32°F (0°C).

‡1 Btu/lb = 2326 J/kg.

pressure. If initially wet, the additional heat will first dry it and then raise it above its saturation temperature. This is a sensible heat addition, and the steam is said to be superheated.

The heat energy in steam can be divided into three parts: (1) enthalpy of the liquid required to raise the water from its initial temperature (usually considered to be 32°F or 0°C) to the boiling temperature, (2) enthalpy of vaporization required to convert the water to steam at the boiling temperature, and (3) enthalpy of superheat that raises the steam to its final temperature. As the steam performs its thermodynamic function of giving up its heat energy, it loses its superheat, becomes wet, and finally condenses to hot water. While wet, the steam has a quality that decreases as the percent of dry, saturated steam present in the wet steam decreases. Dry steam has a quality of 100% (see the table and Fig. 2).

Properties as a gas. Superheated steam at temperatures well above the boiling temperature for the existing steam pressure follows closely the laws of a perfect gas. Thus, $pv \approx 85.8T$ and $k \approx 1.3$, where p is pressure in lb/ft^2, v is volume in ft^3/lb, T is absolute temperature in °R, and k is the ratio of specific heat at constant pressure to specific heat at constant volume. (In SI units, $p'v' \approx 461.5\ T$; where p' is pressure in Pa, v' is volume in m^3/kg, and T' is absolute temperature in K.) However, the behavior of dry, saturated steam departs from that of a perfect gas, and wet steam is a mixture. Therefore, the properties of steam near its vaporization temperature are determined experimentally.

Data can be presented in tabular form, giving pressure, volume, entropy, enthalpy, and temperature for saturated liquid (water at the boiling point), saturated steam, and steam vapor at various temperatures of superheat. Alternatively, thermodynamic properties of steam can be presented diagrammatically. For analysis of thermodynamic cycles, the temperature-entropy chart is widely used (Fig. 3). Area on this chart is proportional to heat energy. At the critical point, steam condenses directly into water without releasing energy. In the uncharted area the pressure is higher than is usually encountered in commercial practice.

For engineering application the Mollier diagram presents steam data in a convenient form (Fig. 4).

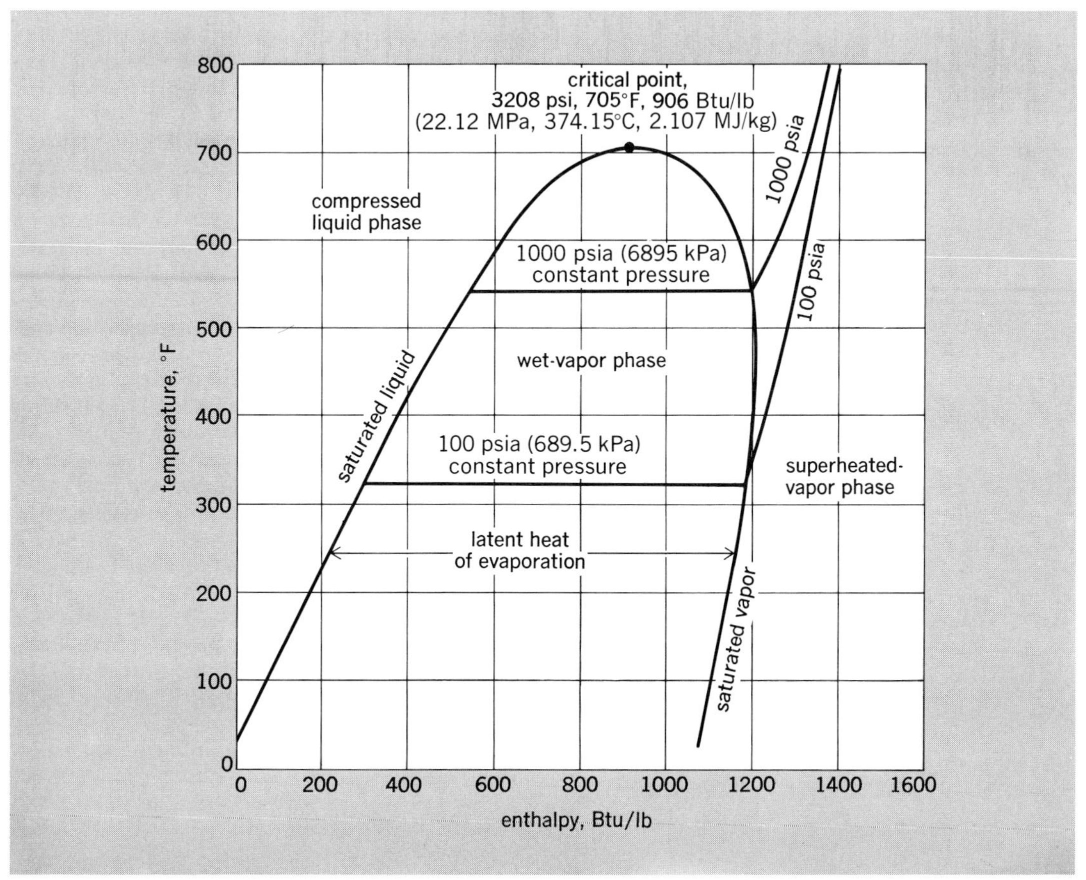

Fig. 2. Temperature-enthalpy chart for steam. 1 Btu/lb = 2326 J/kg. Temperature °C = 5/9 (temperature °F − 32).

In the wet region, lines of constant temperature and of constant pressure are necessarily straight and coincide with each other. Total enthalpies above 32°F (0°C) are plotted as ordinates and total entropies as abscissas.

The specific volume of steam varies widely with temperature and pressure. At 1 psia (6.895 kPa absolute pressure) the specific volume of saturated steam is 334 ft³/lb (20.85 m³/kg), at atmospheric pressure it is 26.8 (1.29), at 1000 psia (6.895 MPa) it is 0.45 (28×10^{-3}), at 2000 psia (13.79 MPa) it is 0.19 (12×10^{-3}), and at the critical (3208 psia or 22.12 MPa) it is 0.05 (3×10^{-3}). Superheating substantially increases values of specific volume of a steam at saturation conditions.

Viscosity of steam increases with temperature and pressure. Saturated steam at atmospheric pressure has a viscosity of about 2.6×10^7 (lb)(sec)/ft² (1.2×10^9 Pa·s). At 1000°F (538°C) and 2000 lb/in.² (13.79 MPa) pressure it is 13.4×10^7 (6.4×10^9 Pa·s). Similarly, the heat conductivity of steam increases with temperature and pressure. The thermal conductivities for the conditions for which viscosities are given are about 14, 37, and 109 Btu/(hr)(ft)(°F) or 24, 64 and 189 W/(m)(°C).

Application. Chiefly because of its availability, but also because of its nontoxicity, steam is widely used as the working medium in thermodynamic processes. It has a uniquely high latent heat of vaporization: 1049 Btu/lb (2.440 MJ/kg) at 1 in. Hg abs (3.39 kPa) and 79°F (26°C), 970 (2.256 MJ/kg) at 14.7 psia (101 kPa) and 212°F (100°C), 889 (2.068 MJ/kg) at 1000 psia (6.895 MPa) and 328°F (164°C), 649 (1.510 MJ/kg) at 1000 psia (6.895 MPa) and 545°F (285°C), and 463 (1.077 MJ/kg) at 2000 psia (13.79 MPa) and 636°F (336°C). Steam has a specific heat in the vicinity of half that of water. For comparison, its specific heat is about twice that of air and comparable to that of ammonia. Except for a few gases such as hydrogen, with a specific heat seven times that of steam, the specific heat of steam is relatively high so that it can carry more thermal energy at practical temperatures than can other usable gases. See BOILER; STEAM ENGINE; STEAM-GENERATING UNIT; STEAM HEATING; STEAM TURBINE; THERMODYNAMIC CYCLE.

[THEODORE BAUMEISTER]

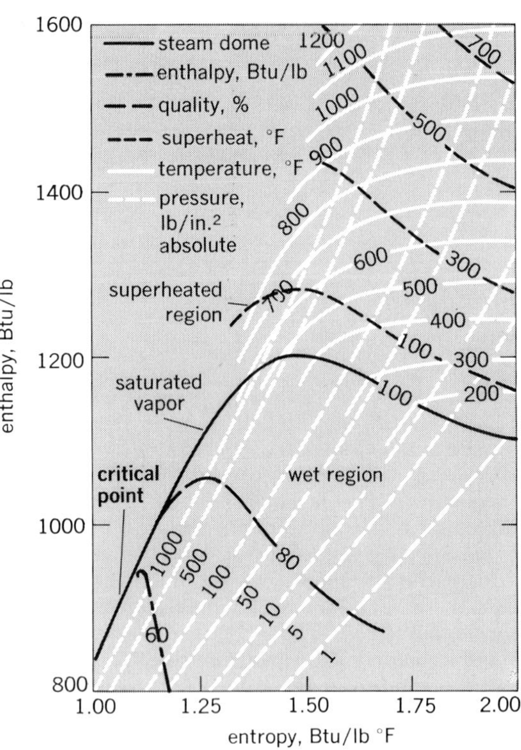

Fig. 4. Mollier diagram for steam. 1 Btu/lb = 2326 J/kg; 1 Btu/lb °F = 4186.8 J/kg °C; temperature °C = 5/9 (temperature °F −32); 1 lb/in. = 6.895 kPa. (*From R. Mollier, Mollier's Steam Tables and Diagrams, Pitman, 1927*)

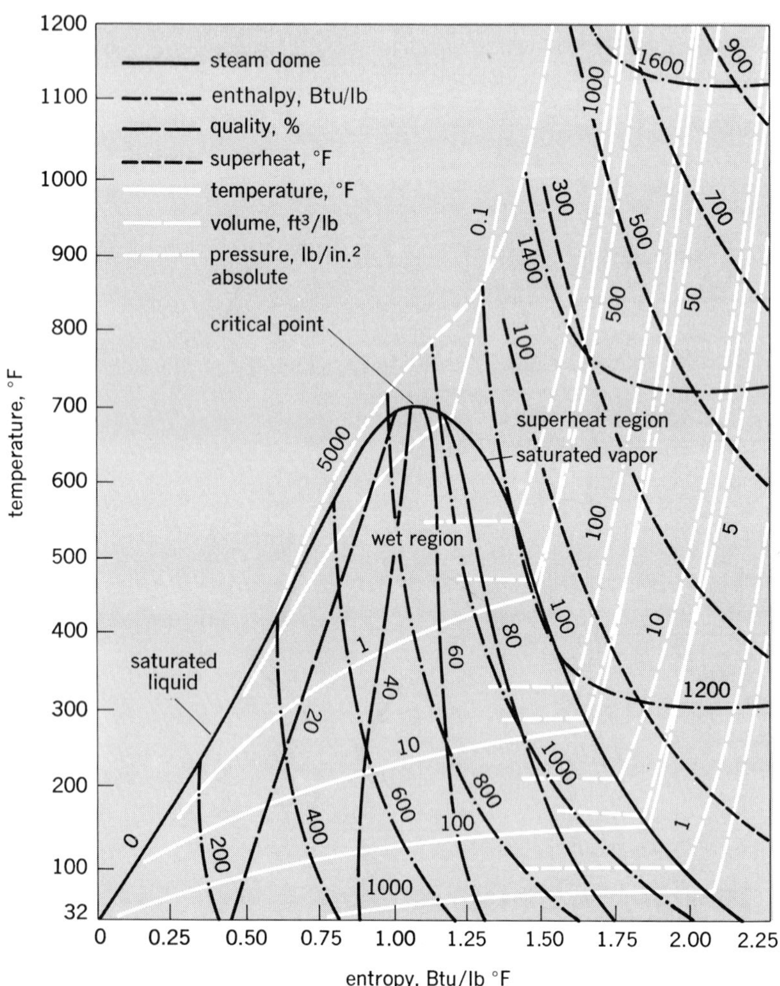

Fig. 3. Temperature-entropy chart for steam. Temperature °C = 5/9 (temperature °F−32); 1 Btu/lb °F = 4186.8 J/kg °C; 1 Btu/lb = 2326 J/kg; 1 ft³/lb = 62.4×10^{-3} m³/kg; 1 lb/in.² = 6.895 kPa. (*From J. H. Keenan and F. S. Keyes, Thermodynamic Properties of Steam, copyright © 1936 by John Wiley and Sons, Inc.; used with permission*)

Bibliography: American Society of Heating, Refrigeration, and Air-Conditioning Engineers, *Guide and Data Book*, 1964, 1965; T. Baumeister (ed.), *Standard Handbook for Mechanical Engineers*, 8th ed., 1978; J. H. Keenan et al., *Steam Tables: Thermodynamic Properties of Water Including Vapor, Liquid, and Solid Phases*, 1978; J. H. Keenan and F. G. Keyes, *Thermodynamics*, 1970; E. Schmidt, *Properties of Water and Steam*, 1979.

Steam condenser

A heat-transfer device used for condensing steam to water by removal of the latent heat of steam and its subsequent absorption in a heat-receiving fluid, usually water, but on occasion air or a process fluid. Steam condensers may be classified as contact or surface condensers.

In the contact condenser, the condensing takes place in a chamber in which the steam and cooling water mix. The direct contact surface is provided by sprays, baffles, or void-effecting fill such as Rashig rings. In the surface condenser, the condensing takes place separated from the cooling water or other heat-receiving fluid (or heat sink). A metal wall, or walls, provides the means for separation and forms the condensing surface.

Both contact and surface condensers are used for process systems and for power generation serving engines and turbines. Modern practice has confined the use of contact condensers almost entirely to such process systems as those involving vacuum pans, evaporators, or dryers, and to condensing and dehumidification processes inherent in vacuum-producing equipment such as steam jet ejectors and vacuum pumps. The steam surface condenser is used chiefly in power generation but is also used in process systems, especially in those in which condensate recovery is important. Air-cooled surface condensers are used increasingly in process systems and in power generation when the availability of cooling water is limited.

Water-cooled surface condensers used for power generation are of the high-vacuum type. Their main purpose is to effect a back pressure at the turbine exhaust. To achieve an economical station heat rate and fixed cost, these condensers must be designed for high heat-transfer rates, minimum steam-side-pressure loss, and effective air removal. The usual power plant steam condenser incorporates a single steam-condensing chamber into which the steam turbine exhausts. Multipressure condensers, having two or more steam-condensing chambers operating at different pressures, have been found to be more economical than single-pressure condensers for large installations, which generally use cooling towers as the final heat sink. Most multipressure condensers are compartmented internally, on the steam-condensing side, with the cooling water arranged to flow through the tubes in one direction from inlet to outlet. The mean cooling water temperature progressively increases as the water flows from compartment to compartment, with back pressure on the turbine increasing proportionately. Economical performance is effected with multiple-exhaust pressure turbines when each exhaust pressure section of the turbine is designed for the back pressure produced in its respective pressure section in the condenser. A rise in water temperature in the order of 20–30°F (11–17°C) is necessary to effect required pressure differences in the turbine.

Air-cooled surface condensers used for power generation are designed to operate at higher back pressures than water-cooled condensers. This is consistent with their lower overall heat-transfer rate, which is of the order of 10–12 Btu/(°F)(ft^2)(hr) [57–68 W/(°C)(m^2)] in contrast with the normal 600–700 Btu/(°F)(ft^2)(hr) [3.4–4.0 kW/(°C)(m^2)] characteristic of water-cooled steam condensers. To compensate for these large differences in heat transfer, greater temperature differences are required between the condensing steam and the cooling air than those needed with conventional water cooling. Steam is condensed on the inside of the heat-transfer tubes, and airflow is usually produced by fans across their outside surface. The tubes are of the extended-surface type with fins on the outside. The normal ratio of outside to inside surface is in the order of 10 or 12 to 1. The use of air as the cooling medium for steam condensers used in smaller industrial steam power plants is increasing rapidly. Air-cooled steam condensers for the production of electric power have been confined to smaller plants; however, with the decreasing availability of water for cooling, the application of air-cooled steam condensers to the larger steam electric plants may be anticipated.

Condenser sizes have increased with the increase in size of turbine generators. The largest, in a single shell, is approximately 485,000 ft^2 (45,000 m^2); the steam-condensing space occupies a volume of 80,000 ft^3 (2,265 m^3). Large units have been built with the surface divided equally among two or three shells. Power plant condensers require 35–100 lb of cooling water to condense 1 lb (35–100 kg for 2 kg) of steam, or, if air-cooled, 2500–5000 ft^3 of air for 1 lb (150–300 m^3 for 1 kg) of steam. Normally, about 0.5 ft^2 (0.05 m^2) of surface is required for water-cooled condensers, and about 10 ft^2 (0.9 m^2) of inside tube surface for air-cooled condensers is required for each kilowatt of generating capacity. *See* STEAM; STEAM TURBINE; VAPOR CONDENSER. [JOSEPH F. SEBALD]

Bibliography: *See* VAPOR CONDENSER.

Steam engine

A machine for converting the heat energy in steam to mechanical energy of a moving mechanism, for example, a shaft. The steam engine dominated the industrial revolution and made available a practical source of power for application to stationary or transportation services. The steam power plant could be placed almost anywhere, whereas other means of power generation were more restricted, experiencing such site limitations as an elevated water supply, wind, animal labor, and so on. The steam engine can utilize any source of heat in the form of steam from a boiler. It was developed in sizes which ranged from that of children's toys to 25,000 hp (18.6 MW), and it was adaptable to pressures up to 200 psi (1.4 MPa). It reached its zenith in the 19th century in stationary services such as drives for pumping plants; drives for air compressor and refrigeration units; power supply for factory operations with shafting for machine shops, rolling mills, and sawmills; and drives for electric generators as electrical supply systems were perfected. Its adaptability to portable and transportation services rested largely on its development of full-rated torque at any speed from rest to full throttle; its speed variability at the will of the operator; and its reversibility, flexibility, and dependability under the realities of stringent service requirements. These same features favored its use for many stationary services such as rolling mills and mine hoists, but the steam engine's great contribution was in the propulsion of small and large ships, both naval and merchant. Also, in the form of the steam locomotive, the engine made the railroad the pratical way of land transport. Most machine elements known today had their origin in the steam engine: cylinders, pistons, piston rings, valves and valve gear crossheads, wrist pins, connecting rods, crankshafts, governors, and reversing gears.

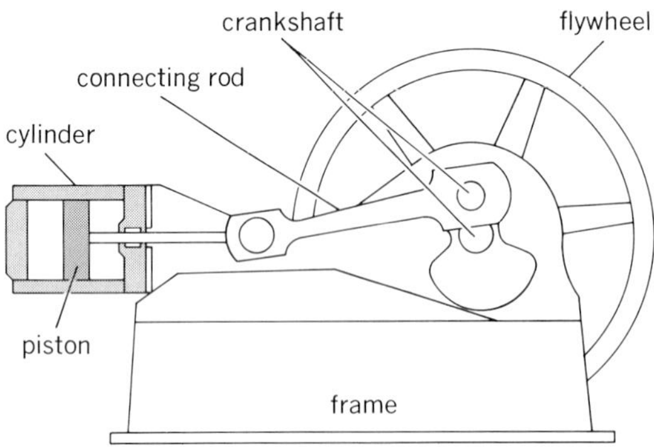

Fig. 1. Principal parts of horizontal steam engine.

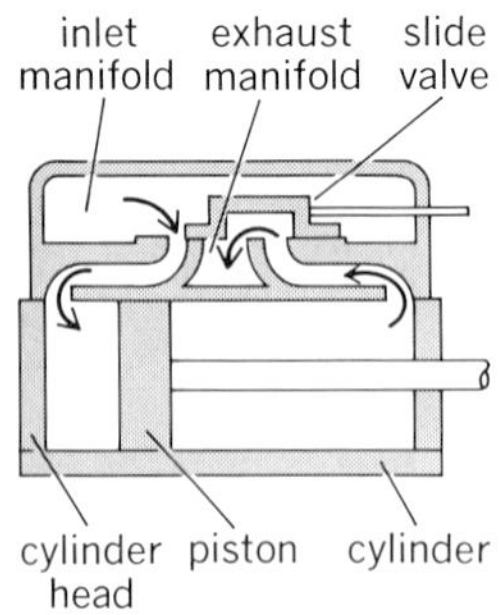

Fig. 2. Single-ported slide valve on counterflow double-acting cylinder.

The 20th century saw the practical end of the steam engine. The steam turbine with (1) its high speed (for example, 3600± rpm); (2) its utilization of maximum steam pressures (2000–5000 psi or 14–34 MPa), maximum steam temperatures (1100±°F or 600±°C), and highest vacuum (29± in. Hg or 98± kPa); and (3) its large size (1,000,000± kW) led to such favorable weight, bulk, efficiency, and cost features that it replaced the steam engine as the major prime mover for electric generating stations. The internal combustion engine, especially the high-speed automotive types which burn volatile (gasoline) or nonvolatile (diesel) liquid fuel, offers a self-contained, flexible, low-weight, low-bulk power plant with high thermal efficiency that has completely displaced the steam locomotive with the diesel locomotive and marine steam engines with the motorship and motorboat. It is the heart of the automotive industry, which produces in a year 10,000,000± vehicles that are powered by engines smaller than 1000 hp (750 kW). Because of the steam engine's weight and speed limitations, it was excluded from the aviation field, which has become the exclusive preserve of the internal combustion piston engine or the gas turbine. *See* DIESEL ENGINE; GAS TURBINE; INTERNAL COMBUSTION ENGINE; STEAM TURBINE; TURBINE.

Cylinder action. A typical steam reciprocating engine consists of a cylinder fitted with a piston (Fig. 1). A connecting rod and crankshaft convert the piston's to-and-fro motion into rotary motion. A flywheel tends to maintain a constant-output angular velocity in the presence of the cyclically changing steam pressure on the piston face. A D slide valve admits high-pressure steam to the cylinder and allows the spent steam to escape (Fig. 2). The power developed by the engine depends upon the pressure and quantity of steam admitted per unit time to the cylinder.

Indicator card. The combined action of valves and piston is most conveniently studied by means of a pressure-volume diagram or indicator card (Fig. 3). The pressure-volume diagram is a thermodynamic analytical method which traces the sequence of phases in the cycle. It may be an idealized operation (Fig. 3*a*), or it may be an actual picture of the phenomena within the cylinder (Fig. 3*b*) as obtained with an instrument commonly known as a steam engine indicator. This instrument, in effect, gives a graphic picture of the pressure and volume for all phases of steam admission, cutoff, expansion, release, exhaust, and compression. It is obtained as the engine is running and shows the conditions which prevail at any instant within the cylinder. The indicator is a useful instrument not only for studying thermodynamic performance, but for the equally important operating knowledge of inlet and exhaust valve leakage and losses, piston ring tightness, and timing correctness.

The net area of the indicator card shows, thermodynamically, the work done in the engine cylinder. By introducing the proper dimensional quantities, power output can be measured. Thus if the net area within the card is divided by the length, the consequent equivalent mean height is the average pressure difference on the piston during the cycle and is generally called the mean effective pressure p, usually expressed in lb/in.^{2} With a cylinder dimension of piston area a (in in.^{2}), length of piston stroke l (in ft), and n equal to the number of cycles completed per minute, the equation given below holds. It has often been referred to as

$$\text{Indicated horsepower} = \frac{plan}{33{,}000}$$

the most important equation in mechanical engineering.

Engine types. Engines are classified as single- or double-acting, and as horizontal (Fig. 1) or vertical depending on the direction of piston motion. If the steam does not fully expand in one cylinder, it can

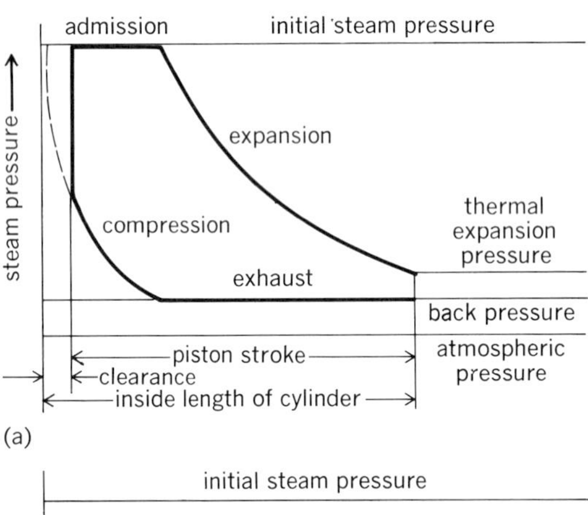

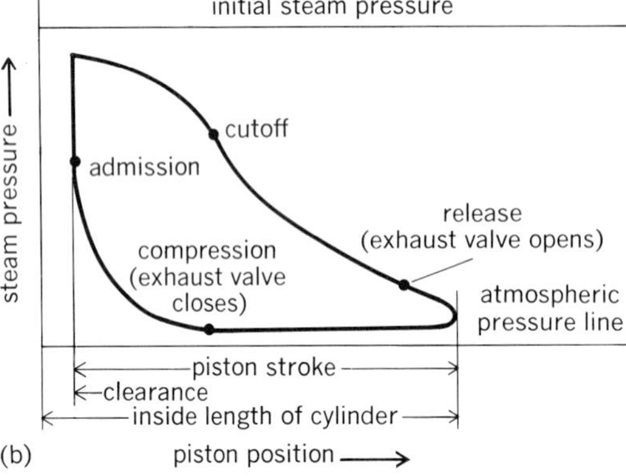

Fig. 3. Events during one cycle of a piston operation. (*a*) In ideal engine. (*b*) As depicted on the indicator card of a noncondensing steam engine.

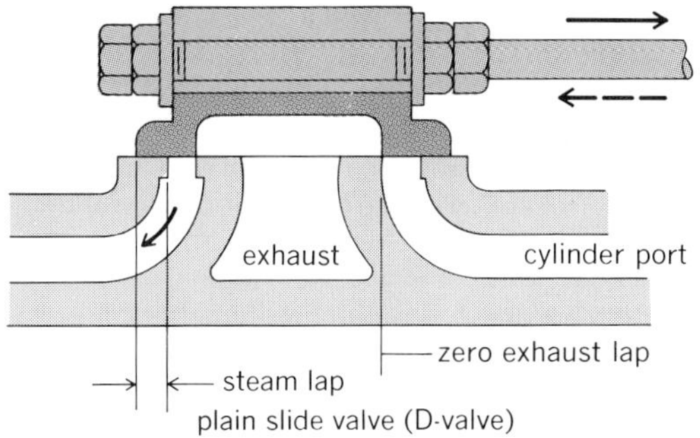

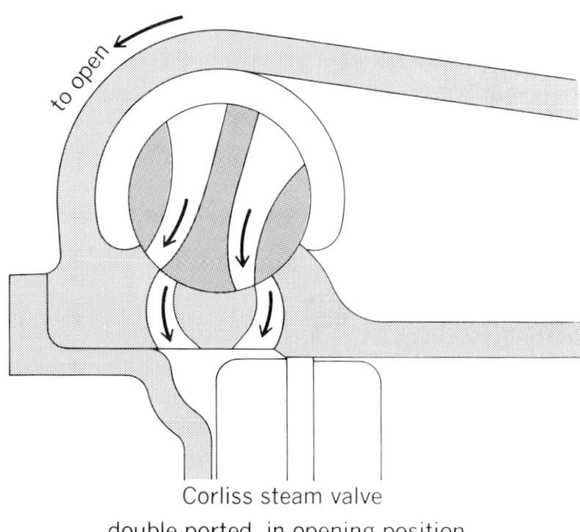

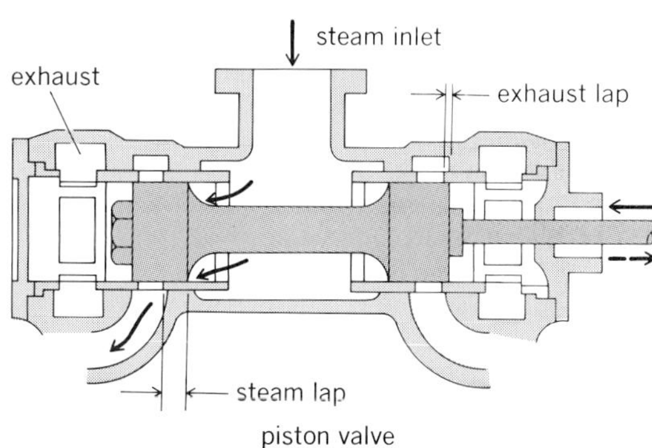

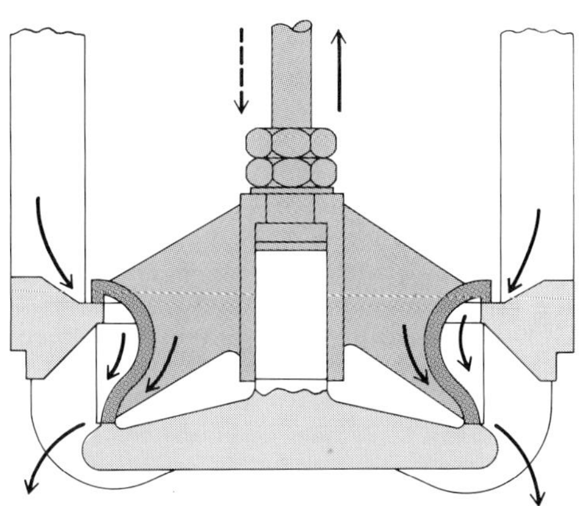

Fig. 4. Steam engine valves in closed positions. The arrows indicate the path steam will travel when the valves are open.

be exhausted into a second, larger cylinder to expand further and give up a greater part of its initial energy. Thus, an engine can be compounded for double or triple expansion. In counterflow engines, steam enters and leaves at the same end of the cylinder; in uniflow engines, steam enters at the end of the cylinder and exhausts at the middle.

Steam engines can also be classed by functions, and are built to optimize the characteristics most desired in each application. Stationary engines drive electric generators, in which constant speed is important, or pumps and compressors, in which constant torque is important. Governors acting through the valves hold the desired characteristic constant. Marine engines require a high order of safety and dependability.

Valves. The extent to which an actual steam piston engine approaches the performance of an ideal engine depends largely on the effectiveness of its valves. The valves alternately admit steam to the cylinder, seal the cylinder while the steam expands against the piston, and exhaust steam from the cylinder. The many forms of valves can be grouped as sliding valves and lifting valves (Fig. 4). *See* CARNOT CYCLE; THERMODYNAMIC CYCLE.

D valves (Fig. 2) are typical sliding valves where admission and exhaust are combined. A common sliding valve is the rocking Corliss valve; it is driven from an eccentric on the main shaft like other valves but has separate rods for each valve on the engine. After a Corliss valve is opened, a latch automatically disengages the rod and a separate dashpot abruptly closes the valve. Exhaust valves are closed by rods, as are other sliding valves.

Lifting valves are more suitable for use with high-temperature steam. They, too, are of numerous forms. The poppet valve is representative.

Valves are driven through a crank or eccentric on the main crankshaft. The crank angle is set to open the steam port near dead center, when the piston is at its extreme position in the cylinder. The angle between valve crank and connecting-rod crank is slightly greater than 90°, the excess being the angle of advance. So that the valves will open and close quickly, they are driven at high velocity with consequently greater travel than is necessary to open and close the ports. The additional travel of a sliding valve is the steam lap and the exhaust lap. The greater the lap, the greater the angle of advance to obtain the proper timing of the valve action.

Engine power is usually controlled by varying the period during which steam is admitted. A shifting eccentric accomplishes this function or, in releasing Corliss and in poppet valves, the eccentric is fixed and cutoff is controlled through a governor to the kickoff cams or to the latch that allows the valves to be closed by their dashpots.

For high engine efficiency, the ratio of cylinder volume after expansion to volume before expansion should be high. The volume before expansion into which the steam is admitted is the volumetric clearance. It may be determined by valve design and other structural features. For this reason, valves and ports are located so as not to necessitate excessive volumetric clearance. *See* BOILER; STEAM. [THEODORE BAUMEISTER]

Bibliography: T. Baumeister (ed.), *Standard Handbook for Mechanical Engineers*, 8th ed., 1978.

Steam-generating furnace

An enclosed space provided for the combustion of fuel to generate steam. The closure confines the products of combustion and is capable of withstanding the high temperatures developed and the pressures used. Its dimensions and geometry are adapted to the rate of heat release, to the type of fuel, and to the method of firing so as to promote complete burning of the combustible and suitable disposal of ash. In water-cooled furnaces the heat absorbed materially affects the temperature of gases at the furnace outlet and contributes directly to the generation of steam.

Furnace walls. Prior to 1925, most furnaces were constructed of firebrick. As the capacity and the physical size of boiler units increased and the suspension burning of pulverized coal was developed, limitations were imposed by the height of the refractory walls that could be made self-supporting at high temperatures and by the inability of refractories to resist the fluxing action of molten fuel ash. Thus refractory is used primarily for special-purpose furnaces, for the burner walls of small boilers, or for the arches of stoker-fired boilers to promote the ignition of the fuel.

The limitations of refractory constructions can be extended somewhat by cooling the brickwork with air flowing through channels in the structure, or by sectionalizing the wall into panels and transferring the structural load to external air-cooled steel or cast-iron supporting members. The heat absorbed can be recovered by using the cooling air for combustion, and the low rate of heat transfer through the refractory helps to maintain high furnace temperatures, thus accelerating ignition and the burning of the fuel.

The low tensile strength of refractories restricts the geometrical shapes that can be built, making it difficult to provide overhanging contours or roof closures. Thus sprung arches or shaped tiles suspended from steel must be used. Many refractory mixtures having air-setting or hydraulic-setting properties are available, and they can be used to form monolithic structures by ramming, guniting, or pouring in forms.

Water-cooling of all or most of the furnace walls is almost universally used in boiler units of all capacities and for all types of fuel and methods of firing. The water-cooling of the furnace walls reduces the transfer of heat to the structural members and, consequently, their temperatures are within the limits that provide satisfactory strength and resistance to oxidation. Water-cooled tube constructions facilitate large furnace dimensions and optimum arrangements of the roof, hopper, and arch, as well as the mountings for the burners and the provision for screens, platens, or division walls to increase the amount of heat-absorbing surface exposed in the combustion zone. External heat losses are small and are further reduced by the use of insulation.

Furnace wall tubes are spaced on close centers to obtain maximum heat absorption and to facilitate ease of ash removal (see illustration). The so-called tangent tube construction, wherein adjacent tubes are almost touching with only a small clearance provided for erection purposes, has been frequently used. However, most new boiler units use membrane-tube walls, in which a steel bar or membrane is welded between adjacent tubes. This construction facilitates the fabrication of water-cooled walls in large shop-assembled tube panels. Less effective cooling is obtained when the tubes are wider spaced and extended metal surfaces, in the form of flat studs, are welded to the tubes; if even less cooling is desired, the tube spacing can be increased and refractory installed between or behind the tubes to form the wall enclosure.

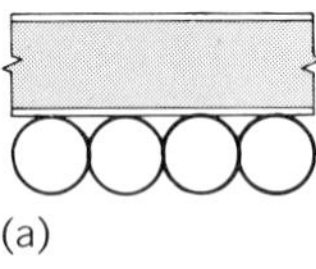

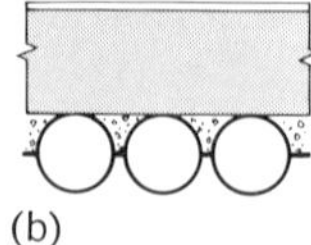

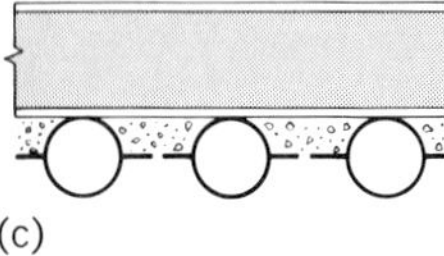

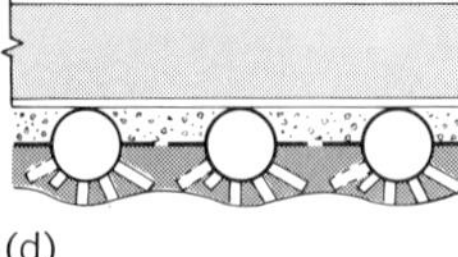

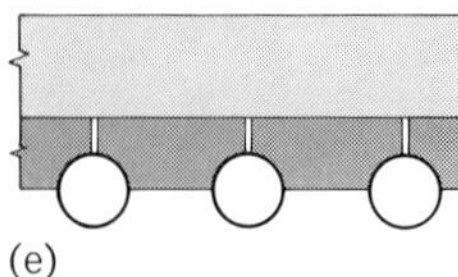

Water-cooled furnace-wall constructions. (*a*) Tangent tube. (*b*) Welded membrane and tube. (*c*) Flat stud and tube. (*d*) Full stud and refractory-covered tube. (*e*) Tube and tile. (*f*) Tubes backed by refractory. (*From T. Baumeister, ed., Marks' Standard Handbook for Mechanical Engineers, 7th ed., McGraw-Hill, 1967*)

Furnace walls must be adequately supported, with provision for thermal expansion and with reinforcing buckstays to withstand the lateral forces resulting from the difference between the furnace pressure and the surrounding atmosphere. The furnace enclosure walls must prevent air infiltration when the furnace is operated under suction, and must prevent gas leakage when the furnace is run at pressures above atmospheric.

Additional furnace cooling, in the form of tubular platens, division walls, or wide-spaced screens, often is used; in high heat input zones the tubes may be protected by refractory coverings anchored to the tubes by studs.

Heat transfer. Heat-absorbing surfaces in the furnace receive heat from the products of combustion and, consequently, lower the furnace gas temperature. The principal mechanisms of heat transfer take place simultaneously and include intersolid radiation from the fuel bed or fuel particles, nonluminous radiation from the products of combustion, convection from the furnace gases, and conduction through any deposits and the tube metal. The absorption effectiveness of the surface is greatly influenced by the deposits of ash or slag.

Analytical solutions to the transfer of heat in the furnaces of steam-generating units are extremely complex, and it is most difficult, if not impossible, to calculate furnace-outlet gas temperatures by theoretical methods. Nevertheless, the furnace-outlet gas temperature must be predicted accurately because this temperature determines the design of the remainder of the unit, particularly that of the superheater. Calculations therefore are based upon test results supplemented by data accumulated from operating experience and judgment predicated upon knowledge of the principles of heat transfer and the characteristics of fuels and slags. *See* STEAM-GENERATING UNIT.

[GEORGE W. KESSLER]

Bibliography: T. Baumeister (ed.), *Marks' Standard Handbook for Mechanical Engineers*, 8th ed., 1978; M. W. Thring, *The Science of Flames and Furnaces*, 1962.

Steam-generating unit

The wide diversity of parts, appurtenances, and functions needed to release and utilize a source of heat for the practical production of steam at pressures to 5000 psi (34 MPa) and temperatures to 1100°F (600°C), often referred to as a "steam boiler for brevity.

The essential steps of the steam-generating process include (1) a furnace for the combustion of fuel, or a nuclear reactor for the release of heat by fission, or a waste heat system; (2) a pressure vessel in which feedwater is raised to the boiling temperature evaporated into steam, and generally superheated beyond the saturation temperature; and (3) in many modern central station units, a reheat section or sections for resuperheating

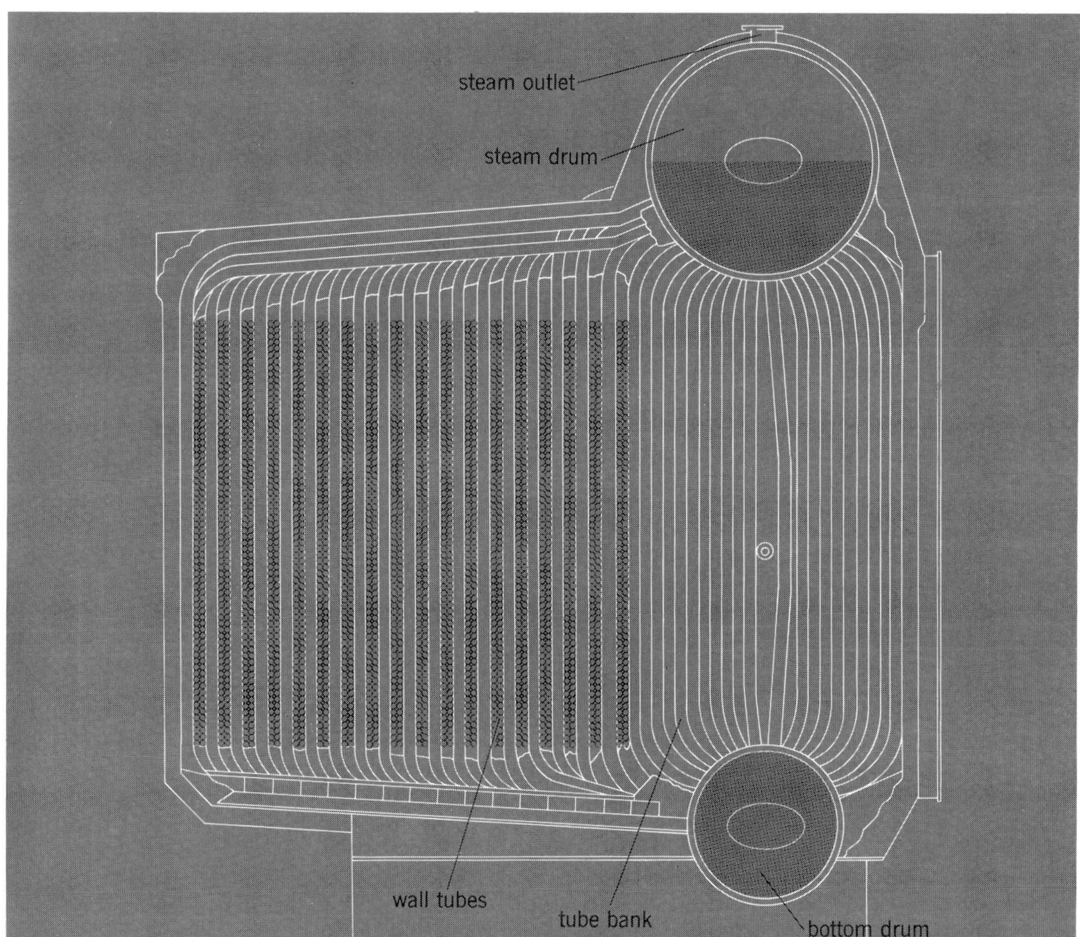

Fig. 1. Package-type water-tube boiler, single gas pass, oil- or gas-fired, 50,000 lb/hr (6.3 kg/sec) steam output for industrial service. (*From T. Baumeister, ed., Standard Handbook for Mechanical Engineers, 7th ed., 1967*)

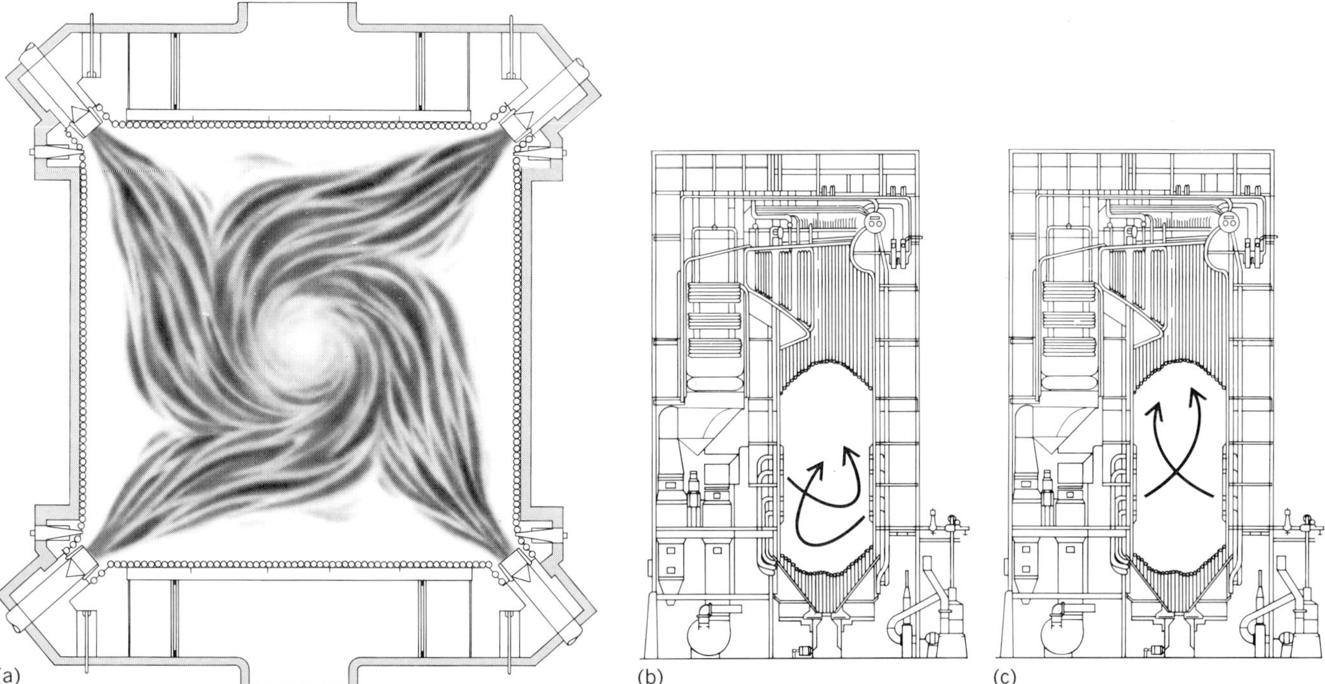

Fig. 2. Large, central-station steam-generating unit, with pump-assisted circulation, corner-fired with tangential, tilting burners for pulverized coal, oil, or gas. (*a*) Plan; (*b*) burners tilted down; (*c*) burners tilted up. Output 2,000,000 lb stream/hr (250 kg/sec) at 2400 psi (16.5 MPa), 1000°F (540°C), and 1000°F reheat. (*Combustion Engineering, Inc.*)

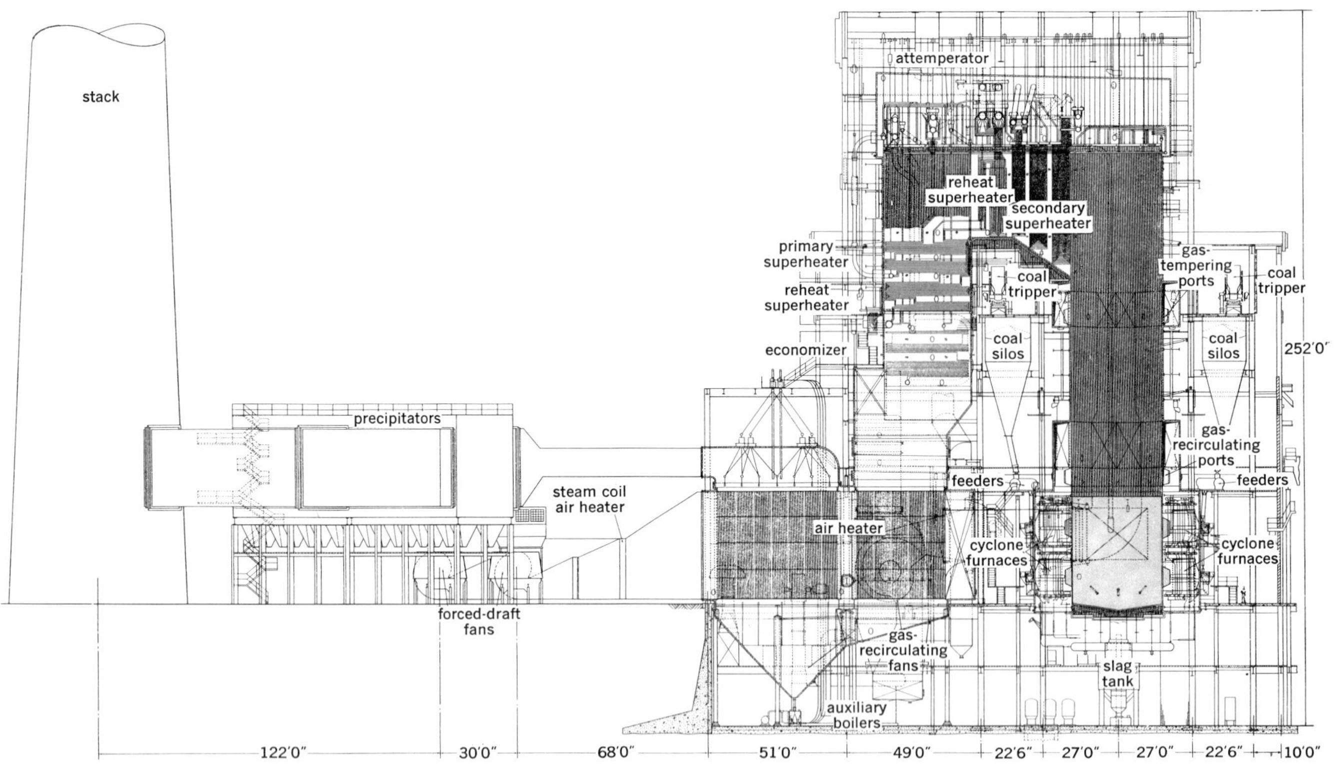

Fig. 3. Supercritical central-station, steam-generating unit with pressurized cyclone-furnace coal firing. Output 5,000,000 lb steam/hr (630 kg/sec) at 3600 psi (24.8 MPa), 1000°F (540°C), and 1000°F reheat. 1 ft = 0.3048 m. (*Babcock and Wilcox Co.*)

steam after it has been partially expanded in a turbine. This aggregation of functions requires a wide assortment of components, which may be variously employed in the interests, primarily, of capacity and efficiency in the steam-production process. Their selection, design, operation, and maintenance constitute a complex process which, in the limit, calls for the highest technical skill.

Service requirements. The wide diversity of commercial steam-generating equipment stems from attempts to accommodate design to the dictates of the imposed service conditions.

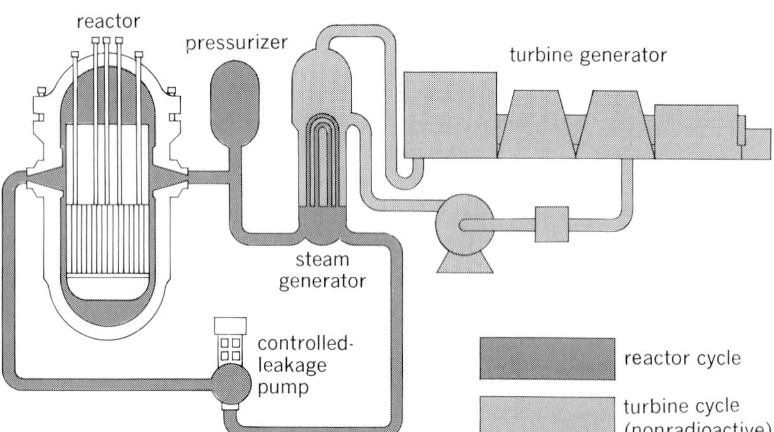

Fig. 4. Pressurized water-cooled reactor (PWR) and steam generator for a large central-station electric-generating plant illustrating the essential components of reactor, pressurizer, steam generator, and circulating pump. Typical performance 6,000,000 lb steam/hr (750 kg/sec) at 700 psi (4.8 MPa), saturated steam. (*Westinghouse Electric Corp.*)

Safety. The pressure parts are a potential bursting hazard and must be designed, maintained, and operated under the most stringent codes, such as the ASME Boiler and Pressure Vessel Code. The codes and regulations are backed by the police power of the state.

Shape. Physical shape must fit the limits imposed. The differences between marine and stationary applications are an example.

Bulk. Space occupied has various degrees of value, and the equipment must best comply with these dictates.

Weight. Frequently, weight is the lowest common denominator in determining portability and economic suitability.

Setting. Confinement of heat source operations and the ensuing transport function will vary the details of the setting, for example, the use of metal or of refractory enclosure.

Character of labor. Highly skilled labor, as in atomic power plants, can utilize design details that would be prohibitive in areas where the labor is essentially primitive.

Cleanability. Impurities on the heat-transfer and heat-release surfaces must not impair safe or efficient operation; surfaces must be cleanable to give acceptable reliability and performance.

Life. Short or long life and short or long operating periods between overhauls are vital to the selection of a preferred design.

Efficiency. Inherent high efficiency can be built into a design, but economics will dictate the efficiency chosen; the full realization of built-in high efficiency requires the services of skilled operators and firemen.

Cost. The overall cost of steam in cents per 1000

lb or 1000 kg is the ultimate criterion, reflecting investment, operating, and maintenance expenses in commercial operations.

Adaptation to use. These service requirements differ in each installation so that a wide variety of designs is available in the commercial markets. There is no one best design for universal application, but propriety rests in the proper selection. A sampling of the diversity of acceptable steam-generating units is represented in Figs. 1–5. These designs illustrate the various degrees of applicability of component parts such as steam drums and tubes; fuel-burning equipment and furnaces; draft systems for air supply and flue gas removal; blowers, fans, and stacks; heat traps such as economizers and air preheaters; structural steel for the support of parts with ample provision for expansion; a casing or setting with possible utilization of water walls, refractory, and insulation; pumps for boiler feed and for water circulation; fuel- and refuse-handling systems; feedwater purification systems; blowdown systems and soot-blowing equipment; and a wide assortment of accessories and instruments such as pressure gages, safety valves, water-level indicators, and sophisticated automatic control with its elaborate interlocks for foolproof, efficient operation and complete computerization.

The maintenance of a safe working temperature for metal parts requires ample circulation of water over the steam-generating parts and ample circulation of steam over the superheating and reheating parts. Water circulation in the generating sections may be by natural convection processes where ample physical height gives the circulatory pressure difference between a column of water in the downcomer and a column of steam mixed with water in the riser parts. With high operating steam pressures, such as 1400 psi (9.7 MPa), pump-assisted circulation is often selected. With supercritical operation (above 3200 psi or 22.1 MPa) there is no difference in density between liquid and vapor so that forced-flow, once-through principles supersede water recirculation practice (Fig. 3). Such supercritical steam generators have no steam or separating drums and, like nuclear steam generators (Figs. 4 and 5), require the highest purity of

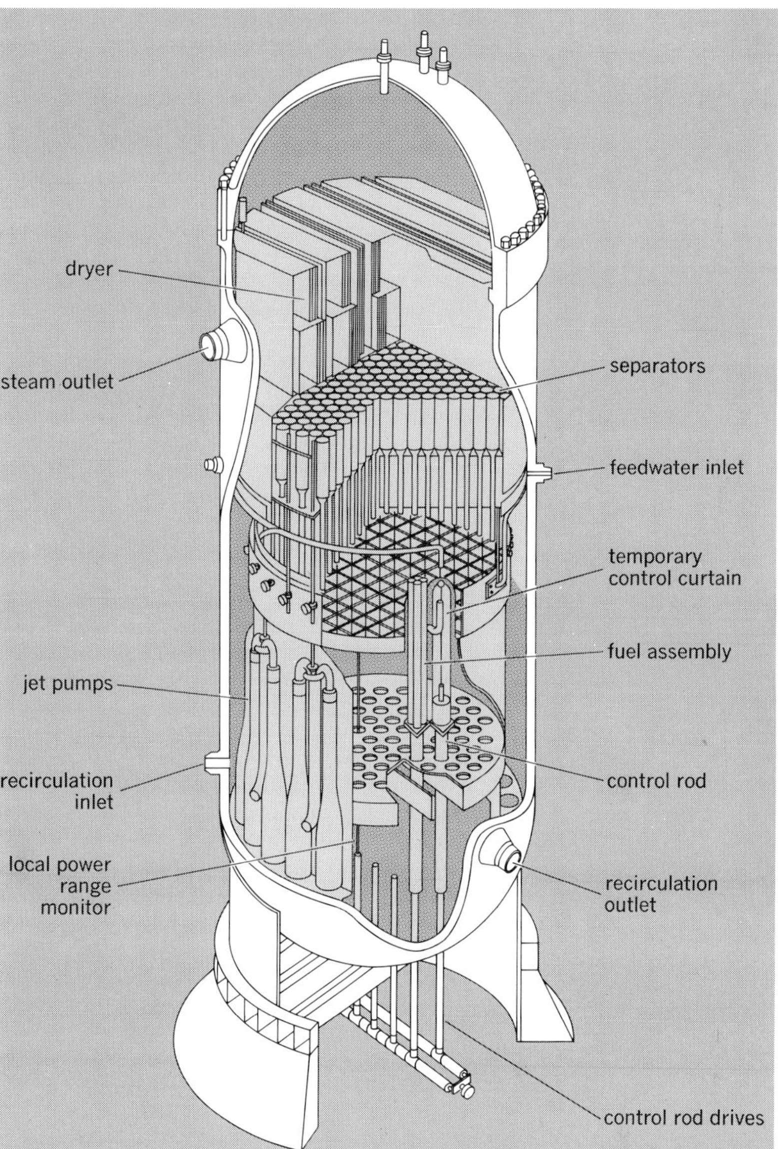

Fig. 5. Boiling-water reactor (BWR) for a large central-station electric-generating plant showing essential components. Typical performance 6,000,000 lb/hr (750 kg/sec) at 1000 psi (6.9 MPa), saturated steam. (*General Electric Co.*)

Performance characteristics of some steam-generating units

Boiler type	Fuel (potential)	Steam output, lb/hr (kg/sec)	Steam pressure, psi (MPa)	Steam temperature, °F (°C)	Boiler efficiency, %
Horizontal return tubular (HRT), fire-tube	Coal, oil, gas	10,000 (1.3)	150 (1.0)	Dry and saturated	70
Large industrial, natural-circulation, water-tube	Coal, oil, gas	300,000 (38)	1400 (9.7)	900 (480)	88
Large central station, natural-circulation	Coal, oil, gas	2,000,000 (250)	2700 (18.6)	1050/1050 (570/570)	90
Large central station, supercritical, once-through	Coal, oil, gas	5,000,000 (630)	3600 (24.8)	1000/1000/1000 (540/540/540)	90
Boiling-water reactor (BWR)	Uranium	6,000,000 (750)	1000 (6.9)	Dry and saturated (550±°F or 290±°C)	98±
Pressurized-water reactor (PWR), hot water, 4 loops	Uranium	6,000,000 (750)	700 (4.8)	Dry and saturated (500±°F or 260±°C)	98±
Electric boiler, small	200± volts	2,000 (0.25)	150 (1.0)	Dry and saturated	98±

feed and boiler water (typically 10^{-9} part of impurities) to avoid deposits in heated circuits or the transport of solids to the turbine.

The illustrations give an indication of the diversity of problems met by a selected group of commercial steam generators. The table gives some performance data for an assorted group of representative units. *See* BOILER; BOILER ECONOMIZER; BOILER FEEDWATER REGULATION; FEEDWATER; FIRE-TUBE BOILER; RAW WATER; REHEATING; STEAM SEPARATOR; STEAM TEMPERATURE CONTROL; SUPERHEATER; WATER-TUBE BOILER.

[THEODORE BAUMEISTER]

Bibliography: American Society of Mechanical Engineers, *Boiler and Pressure Vessel Code*, 1980; Babcock and Wilcox Co., *Steam: Its Generation and Use*, 1955; T. Baumeister (ed.), *Standard Handbook for Mechanical Engineers*, 8th ed., 1978; Combustion Engineering, Inc., *Combustion Engineering*, 1966; J. J. Jackson, *Steam Boiler Operation*, 1980.

Steam heating

A heating system that uses steam generated from a boiler. The steam heating system conveys steam through pipes to heat exchangers, such as radiators, convectors, baseboard units, radiant panels, or fan-driven heaters, and returns the resulting condensed water to the boiler. Such systems normally operate at pressure not exceeding 15 pounds per square inch gage (psig) or 103 kPa gage, and in many designs the condensed steam returns to the boiler by gravity because of the static head of water in the return piping. With utilization of available operating and safety control devices, these systems can be designed to operate automatically with minimal maintenance and attention.

One-pipe system. In a one-pipe steam heating system, a single main serves the dual purpose of supplying steam to the heat exchanger and conveying condensate from it. Ordinarily, there is but one connection to the radiator or heat exchanger, and this connection serves as both the supply and return; separate supply and return connections are sometimes used. Because steam cannot flow through the piping or into the heat exchanger until all the air is expelled, it is important to provide automatic air-venting valves on all exchangers and at the ends of all mains. These valves may be of a type which closes whenever steam or water comes in contact with the operating element but which also permits air to flow back into the system as the pressure drops. A vacuum valve closes against subatmospheric pressure to prevent return of air.

Two-pipe system. A two-pipe system is provided with two connections from each heat exchanger, and in this system steam and condensate flow in separate mains and branches (Fig. 1). A vapor two-pipe system operates at a few ounces above atmospheric pressure, and in this system a thermostatic trap is located at the discharge connection from the heat exchanger which prevents steam passage, but permits air and condensation to flow into the return piping.

When the steam condensate cannot be returned by gravity to the boiler in a two-pipe system, an alternating return lifting trap, condensate return pump, or vacuum return pump must be used to force the condensate back into the boiler. In a condensate return-pump arrangement, the return

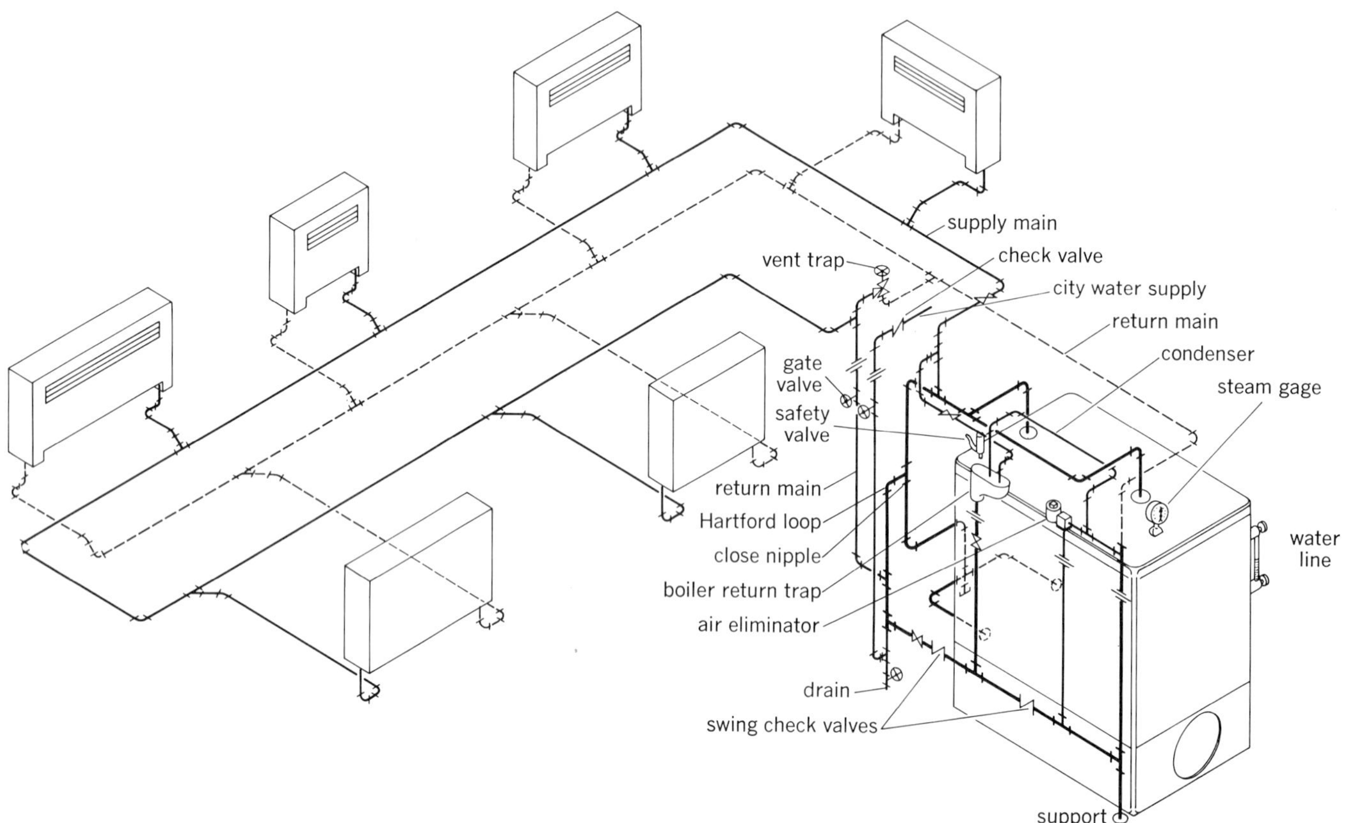

Fig. 1. Two-pipe up-feed system with automatic return trap.

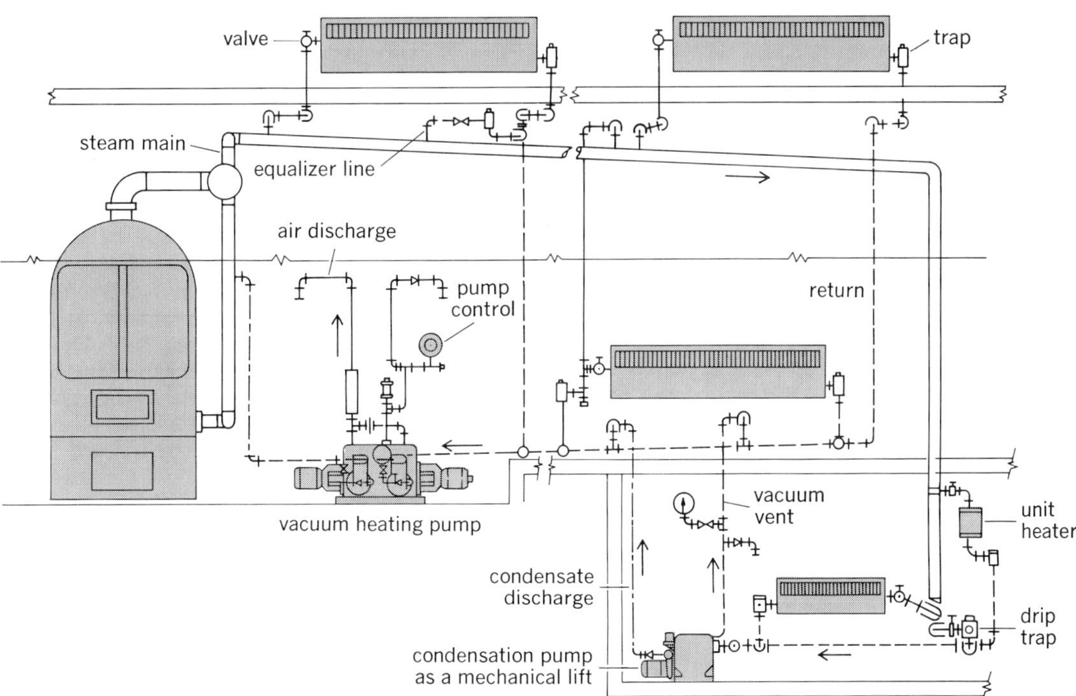

Fig. 2. Layout of a vacuum heating system with condensation and vacuum pumps.

piping is arranged for the water to flow by gravity into a collecting receiver or tank, which may be located below the steam-boiler waterline. A motor-driven pump controlled from the boiler water level then forces the condensate back to the boiler.

In large buildings extending over a considerable area, it is difficult to locate all heat exchangers above the boiler water level or return piping. For these systems a vacuum pump is used that maintains a suction below atmosphere up to 25± in. (max) of mercury in the return piping, thus creating a positive return flow of air and condensate back to the pumping unit. Subatmospheric systems are similar to vacuum systems, but in contrast provide a means of partial vacuum control on both the supply and return piping so that the steam temperature can be regulated to vary the heat emission from the heat exchanger in direct proportion to the heat loss from the structure.

Figure 2 depicts a two-pipe vacuum heating system which uses a condensation pump as a mechanical lift for systems where a part of the heating system is below the boiler room. Note that the low section of the system is maintained under the same vacuum conditions as the remainder of the system. This is accomplished by connecting the vent from the receiver and pump discharge to the return pipe located above the vacuum heating pump.

With the wide acceptance of all-year air conditioning, low-pressure steam boilers have been used to produce cooling from absorption refrigeration equipment. With this system the boiler may be used for primary or supplementary steam heating as diagrammed in Fig. 3.

Exhaust from gas- or oil-driven turbines or engines may be used in waste heat boilers or separators, along with a standby boiler to produce steam for a heating system as indicated in Fig. 4.

Another source for steam for heating is from a high-temperature water source (350–450°F or

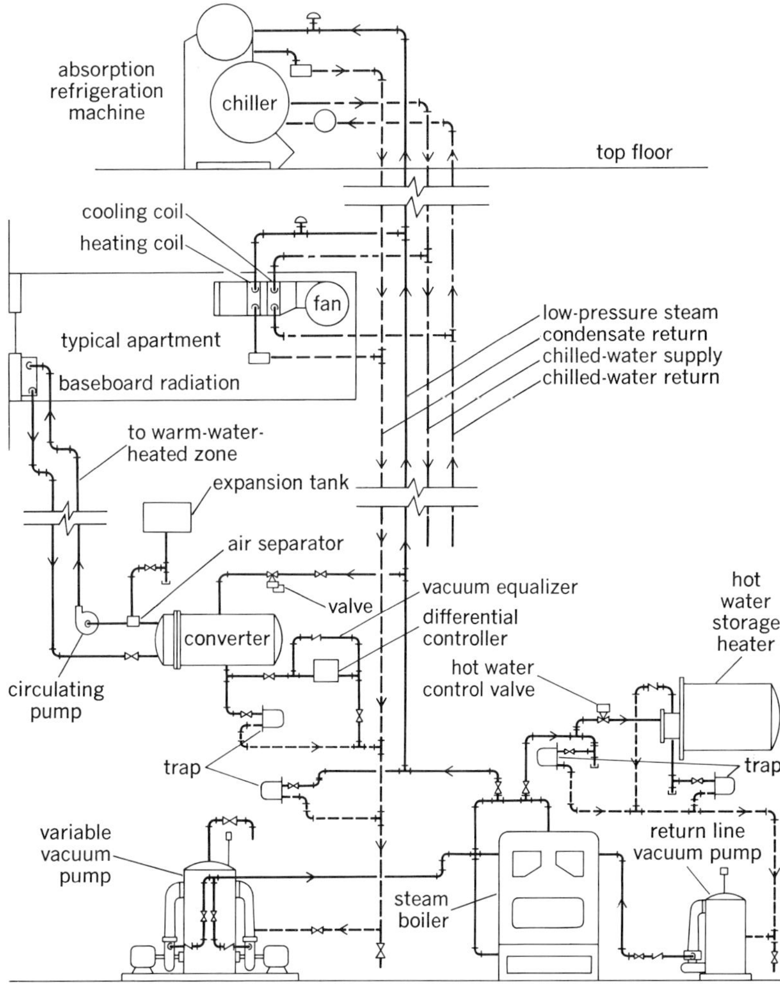

Fig. 3. Diagram of a heating and cooling system for an apartment building employing a low-pressure steam boiler.

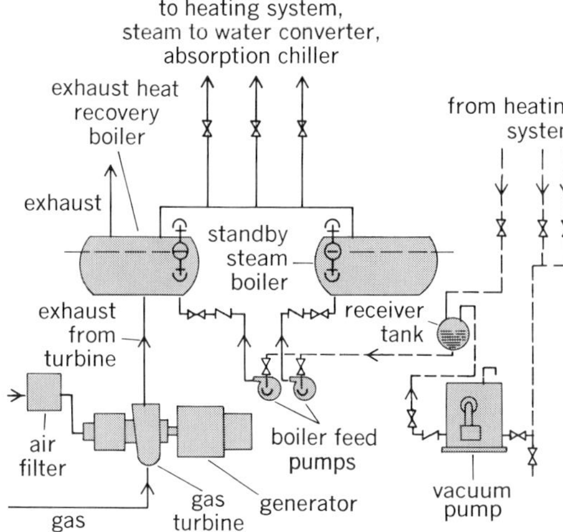

Fig. 4. Steam system using waste heat.

180–230°C) using a high-pressure water to low-pressure steam heat exchanger. *See* COMFORT HEATING; OIL BURNER; STEAM-GENERATING FURNACE. [JOHN W. JAMES]

Bibliography: R. H. Emerick, *Heating Handbook*, 1964; E. B. Woodruff and H. B. Lammers, *Steam-Plant Operation*, 4th ed., 1976.

Steam jet ejector

A steam-actuated device for pumping compressible fluids, usually from subatmospheric suction pressure to atmospheric discharge pressure. A steam jet ejector is most frequently used for maintaining vacuum in process equipment in which evaporation or condensation takes place. Because of its simplicity, compactness, reliability, and generally low first cost, it is often preferred to a mechanical vacuum pump for removing air from condensers serving steam turbines, especially for marine service.

Principle. Compression of the pumped fluid is accomplished in one or more stages, depending upon the total compression required. Each stage consists of a converging-diverging steam nozzle, a suction chamber, and a venturi-shaped diffuser. Steam is supplied to the nozzle at pressures in the range 100–250 psig (0.7–1.7 MPa). A portion of the enthalpy of the steam is converted to kinetic energy by expanding it through the nozzle at substantially constant entropy to ejector suction pressure, where it reaches velocities of 3000–4500 ft/sec (900–1400 m/sec). The air or gas, with its vapor of saturation, which is to be pumped and compressed is entrained, primarily by friction in the high-velocity steam jet. The impulse of the steam produces a change in the momentum of the air or gas vapor mixture as it mixes with the motive steam and travels into the converging section of the diffuser. In the throat or most restricted area of the diffuser, the energy transfer is completed and the final mixture of gases and vapor enters the diverging section of the diffuser, where its velocity is progressively reduced. Here a portion of the kinetic energy of the mixture is reconverted to pressure with a corresponding increase in enthalpy. Thus the air or gas is compressed to a higher pressure than its entrance pressure to the ejector. The compression ratios selected for each stage of a steam jet ejector usually vary from about 4 to 7.

Application. Two or more stages may be arranged in series, depending upon the total compression ratio required (Fig. 1). Two or more sets of series stages may be arranged in parallel to accommodate variations in capacity.

Vapor condensers are usually interposed between the compression stages of multistage steam jet ejectors to condense and remove a significant portion of the motive steam and other condensable vapors (Fig. 2). This action reduces the amount of fluid to be compressed by the next higher stage and results in a reduction in the motive steam required. Both surface and contact types of vapor condensers are used for this purpose. *See* VAPOR CONDENSER.

Ejectors used as air pumps for steam condensers that serve turbines are usually two-stage and are equipped with inter- and aftercondensers of the surface type. The steam condensed is drained through traps to the main condenser and returned to the boiler feed system. Ejectors used as vacuum pumps in process systems may be equipped with either surface condensers or contact condensers of

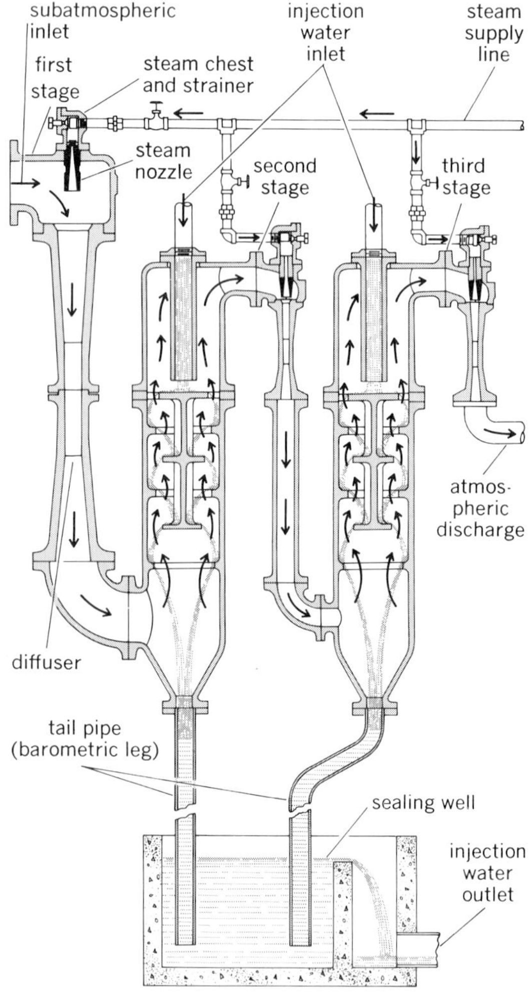

Fig. 1. Typical multistage steam jet ejector with contact barometric condensers; first and second stages condensing and third stage noncondensing.

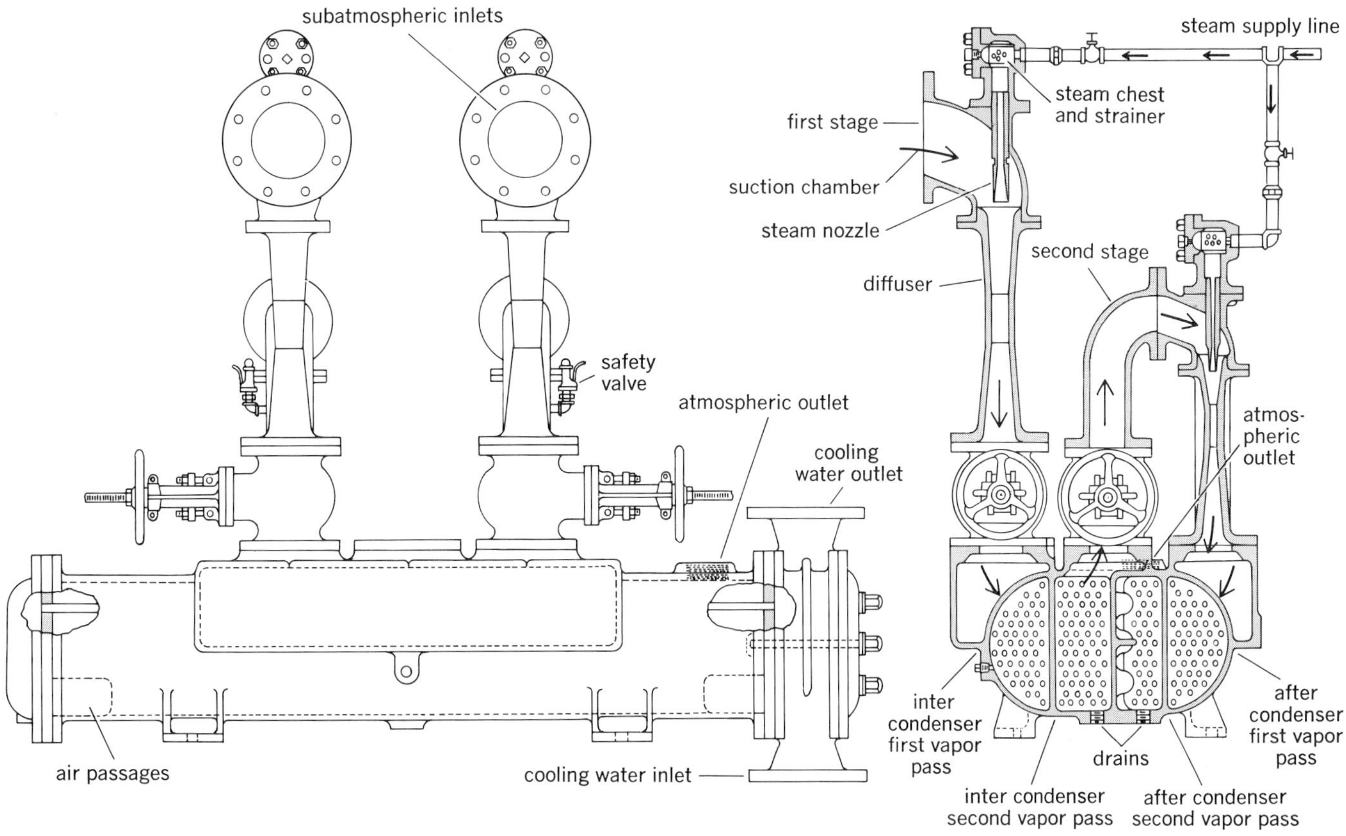

Fig. 2. Multistage steam jet ejector with surface inter- and aftercondenser; first and second stages condensing.

the barometric type between or after stages or both. They may be single or multistage machines. High-vacuum process ejectors with as many as seven stages in series have been built. Industrial or process ejectors are frequently used instead of mechanical vacuum pumps to pump corrosive vapors because they can be manufactured economically from almost any corrosion-resistant material.

The number of stages of compression usually used for various suction pressures with atmospheric discharge pressure is as follows:

No. of ejector stages	*Range of suction pressure*
1	3–30 in. Hg (10–102 kPa) abs
2	0.4–4 in. Hg (1.4–14 kPa) abs
3	1–25 mm Hg (0.13–3.3 kPa) abs
4	0.15–3 mm Hg (20–400 Pa) abs
5	20–300 μm (2.7–40 Pa)
6	5–20 μm (0.7–2.7 Pa)
7	1–5 μm (0.13–0.7 Pa)

Ejectors are also made which use air or other gases instead of steam as the energy source.

[JOSEPH F. SEBALD]

Steam separator

A device for separating a mixture of the liquid and vapor phases of water. Steam separators are used in most boilers and may also be used in saturated steam lines to separate and remove the moisture formed because of heat loss.

In boilers, steam separators must perform efficiently because both steam-free water and water-free steam are required. The force promoting water circulation in a natural-circulation boiler is derived from the density difference between the water in the downcomer tubes and the steam-water mixture in the riser tubes. Solid (steam-free) water in the downcomers maximizes this force. Steam-free water is equally important in a forced-circulation boiler to prevent cavitation in the pump.

Water-free steam is quite necessary because dissolved and suspended solids entering the boiler with the feedwater concentrate in the boiler water, and such solids contaminate the steam if all the water is not separated. Impurity concentrations as low as a few parts per billion can cause solid deposits to form in turbines or can cause product contamination if the steam is used in direct contact with a product. Solid deposits in turbines decrease turbine capacity and efficiency and may even cause unbalance of the rotor in extreme cases. Consequently, the sampling of steam and the measurement of impurities in steam have received much attention.

Many low-pressure boilers operate satisfactorily without steam separators because their capacity ratings are quite conservative. Steam separation in such boilers may be compared with a pan of water boiling on a stove at a moderate rate. However, it is known that if the fire under the pan is turned too high or if the water contains certain solids (like milk, for example), separation becomes inadequate and the pan boils over. Similar phenomena take place in natural separation boilers if the load is too high or the concentration of solids in the water is too great.

Operating pressure also influences natural separation. The force effecting separation results

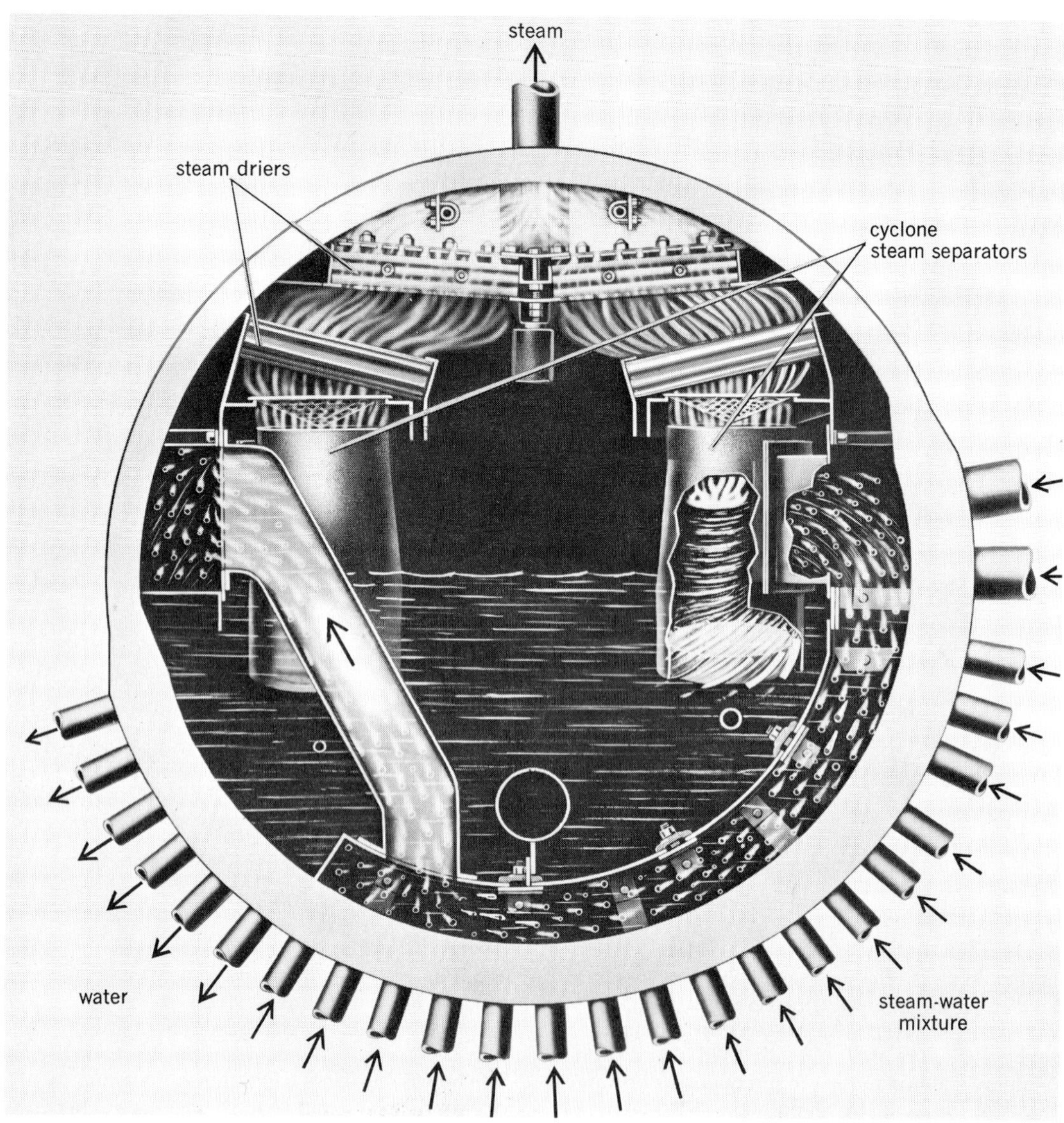

Steam dryers and cyclone steam separators separate steam and water in high-capacity boiler unit.

from the difference in density between water and steam. This difference is quite great at low pressures, but as pressure is increased, the density of water decreases while that of steam increases. resulting in a steadily declining density difference.

Because of these effects, steam separators are used on the majority of boilers. Steam separators have many forms and may be as fundamental as a simple baffle that utilizes inertia caused by a change of direction. Most modern, high-capacity boilers use a combination of cyclone separators and steam dryers, as shown in the figure.

Cyclone separators have two major advantages. First, they slice the steam-water mixture into thin streams so that the steam bubbles need travel only short distances through the mixture to become disengaged. Second, they whirl the mixture in a circular path, creating a centrifugal force many times greater than the force of gravity.

Cyclone separators remove practically all of the steam from the water, producing the steam-free water needed for the boiler downcomer tubes. They also remove the major portion of the water from the steam. After leaving the cyclone separators, the steam is passed through the dryers, where the last traces of moisture are removed. Steam dryers remove small droplets of water from the steam by providing a series of changes in direction and a large surface area to intercept the droplets.

Because of the efficient operation of steam separators, steam is one of the purest mass-produced commodities made. *See* BOILER WATER; STEAM; STEAM TURBINE.

[EARL E. COULTER]

Bibliography: American Society for Testing and Materials, *Methods of Sampling Steam*, ASTM Designation D1066; ASTM, *Methods of Test for Deposit-Forming Impurities in Steam*, ASTM Designation D2186; ASTM, *ASTM Standards*, pt. 23: *Water: Atmospheric Analysis*, annually; P. M. Brister, F. G. Raynor, and E. A. Pirsh, *The Influence of Boiler Design and Operating Conditions on Steam Contamination*, ASME Pap. no. 51–A–95, M. J. Moore and C. H. Sieverding, *Two-Phase Steam Flow in Turbines and Separators: Theory, Instrumentation, Engineering*, 1976.

Steam temperature control

Means for regulating the operation of a steam-generating unit to produce steam at the required temperature. The temperature of steam is affected by the change in the relative heat absorption as load varies, by changes in ash or slag deposits on the heat-absorbing surface, by changes in fuel, by changes in the proportioning of fuel and combustion air, or by changes in feedwater temperature. Low steam temperature lowers the efficiency of the thermal cycle. However, high steam temperature, which increases thermal efficiency, is restricted by the strength and durability of materials used in superheaters. Control of steam temperature is, therefore, a matter of primary concern in the design of modern steam-generating units.

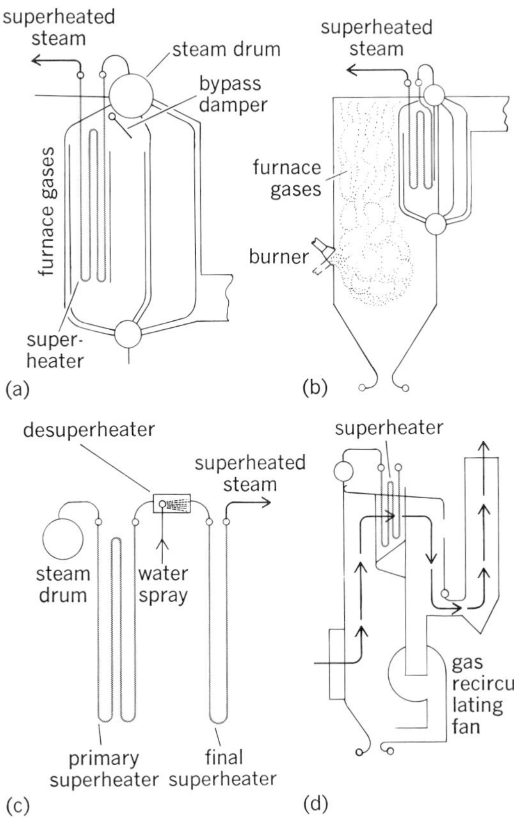

Methods of controlling steam temperature. (*a*) Bypass dampers. (*b*) Tilting burners. (*c*) Spray attemperation. (*d*) Gas recirculation.

Steam temperature can be controlled by one or more of several methods (see illustration). These include (1) the damper control of gases to the superheater, to the reheater, or to both, thus changing the heat input; (2) the recirculation of low-temperature flue gas to the furnace, thus changing the relative amounts of heat absorbed in the furnace and in the superheater, reheater, or both; (3) the selective use of burners at different elevations in the furnace or the use of tilting burners, thus changing the location of the combustion zone with respect to the furnace heat-absorbing surface; (4) the attemperation, or controlled cooling, of the steam by the injection of spray water or by the passage of a portion of the steam through a heat exchanger submerged in the boiler water; (5) the control of the firing rate in divided furnaces; and (6) the control of the firing rate relative to the pumping rate of the feedwater to forced-flow once-through boilers. Generally, these various controls are adjusted automatically. *See* STEAM-GENERATING UNIT. [GEORGE W. KESSLER]

Steam turbine

A machine for generating mechanical power in rotary motion from the energy of steam at temperature and pressure above that of an available sink. By far the most widely used and most powerful turbines are those driven by steam. In the United States well over 85% of the electrical energy consumed is produced by steam-turbine-driven generators. By the mid-1970s, over 25,000 MW (1 MW = 1341 hp) of steam turbine capacity for electrical power generation was shipped in the United States in a single typical year. Individual turbine ratings historically have tended to follow the increasing capacity trend but are now reaching limits imposed by material and machine design considerations. The largest unit shipped during the 1950s was rated 500 MW. Units rated about 1100 MW were in service by the close of the 1960s, and ratings up to 1300 MW saw frequent application in the 1970s and early 1980s. Units of all sizes, from a few horsepower to the largest, have their applications. Manufacturers of steam turbines are located in every industrial country.

Until the 1960s essentially all steam used in turbine cycles was raised in boilers burning fossil fuels (coal, oil, and gas) or, in minor quantities, certain waste products. The 1960s marked the beginning of the introduction of commercial nuclear power. About 50% of the steam turbine capacity ordered from 1965 to 1975 was designed for steam from nuclear reactor steam supplies.

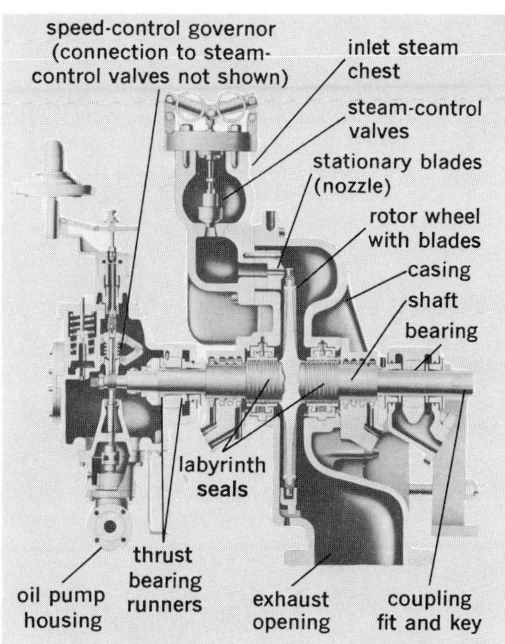

Fig. 1. Cutaway of small, single-stage steam turbine. (*General Electric Co.*)

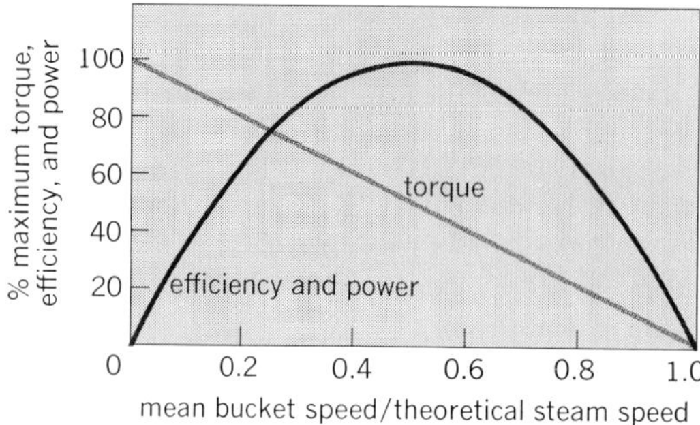

Fig. 2. Illustrative stage performance versus speed.

Approximately 10% of the power generated in 1975 was from nuclear steam plants, about 75% from fossil fuel-fired steam plants, and the balance from other sources.

Turbine parts. Figure 1 shows a small, simple mechanical-drive turbine of a few horsepower. It illustrates the essential parts for all steam turbines regardless of rating or complexity: (1) a casing, or shell, usually divided at the horizontal center line, with the halves bolted together for ease of assembly and disassembly; it contains the stationary blade system; (2) a rotor carrying the moving buckets (blades or vanes) either on wheels or drums, with bearing journals on the ends of the rotor; (3) a set of bearings attached to the casing to support the shaft; (4) a governor and valve system for regulating the speed and power of the turbine by controlling the steam flow, and an oil system for lubrication of the bearings and, on all but the smallest machines, for operating the control valves by a relay system connected with the governor; (5) a coupling to connect with the driven machine; and (6) pipe connections to the steam supply at the inlet and to an exhaust system at the outlet of the casing or shell.

Applications. Steam turbines are ideal prime movers for driving machines requiring rotational mechanical input power. They can deliver constant or variable speed and are capable of close speed control. Drive applications include centrifugal pumps, compressors, ship propellers, and, most important, electric generators.

The turbine shown in Fig. 1 is a small mechanical-drive unit. Units of this general type provide 10–1000 hp (7.5–750 kW) with steam at 100–600 pounds per square inch (0.7–4.1 MPa) gage (psig) inlet pressure and temperatures to 800°F (427°C). These and larger multistage machines drive small electric generators, pumps, blowers, air and gas compressors, and paper machines. A useful feature is that the turbine can be equipped with an adjustable-speed governor and thus be made capable of producing power over a wide range of rotational speeds. In such applications efficiency varies with speed (Fig. 2), being 0 when the rotor stalls at maximum torque and also 0 at the runaway speed at which the output torque is 0. Maximum efficiency and power occur where the product of speed and torque is the maximum.

Many industries need steam at one or more pressures (and consequently temperatures) for heating and process work. Frequently it is more economical to raise steam at high pressure, expand it partially through a turbine, and then extract it for process, than it would be to use a separate boiler at the process steam pressure. Figure 3 is a cross section through an industrial automatic extraction turbine. The left set of valves admits steam from the boiler at the flow rate to provide the desired electrical load. The steam flows through five stages to the controlled extraction point. The second set of valves acts to maintain the desired extraction pressure by varying the flow through the remaining 12 stages. Opening these internal valves increases the flow to the

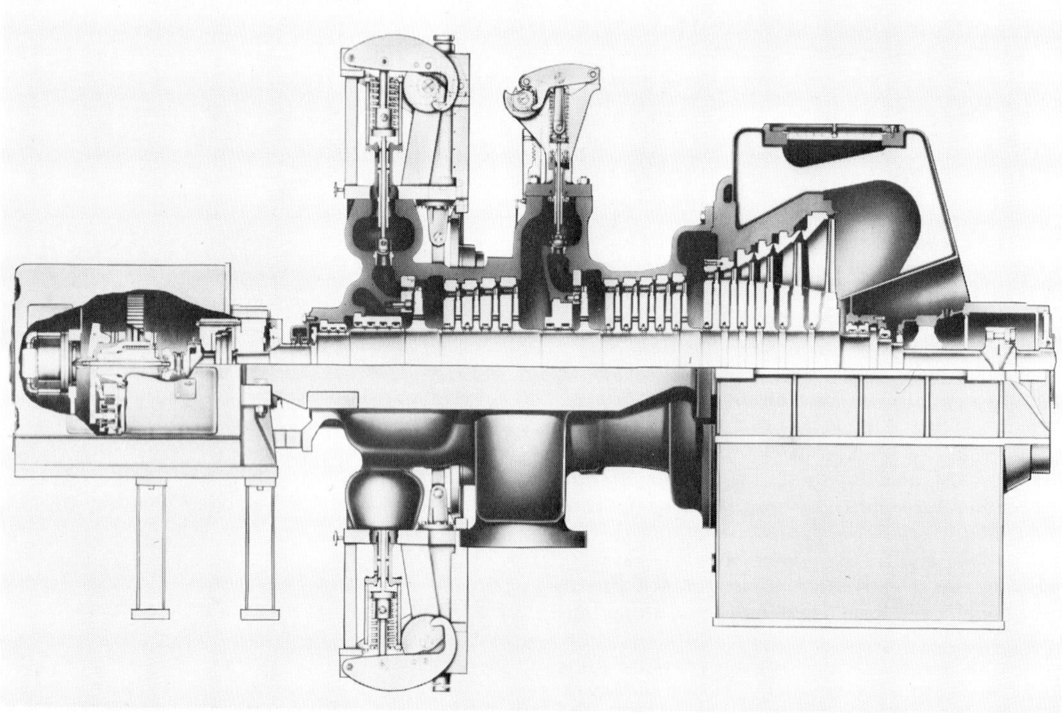

Fig. 3. Cross-section view of single-automatic-extraction condensing steam turbine. (*General Electric Co.*)

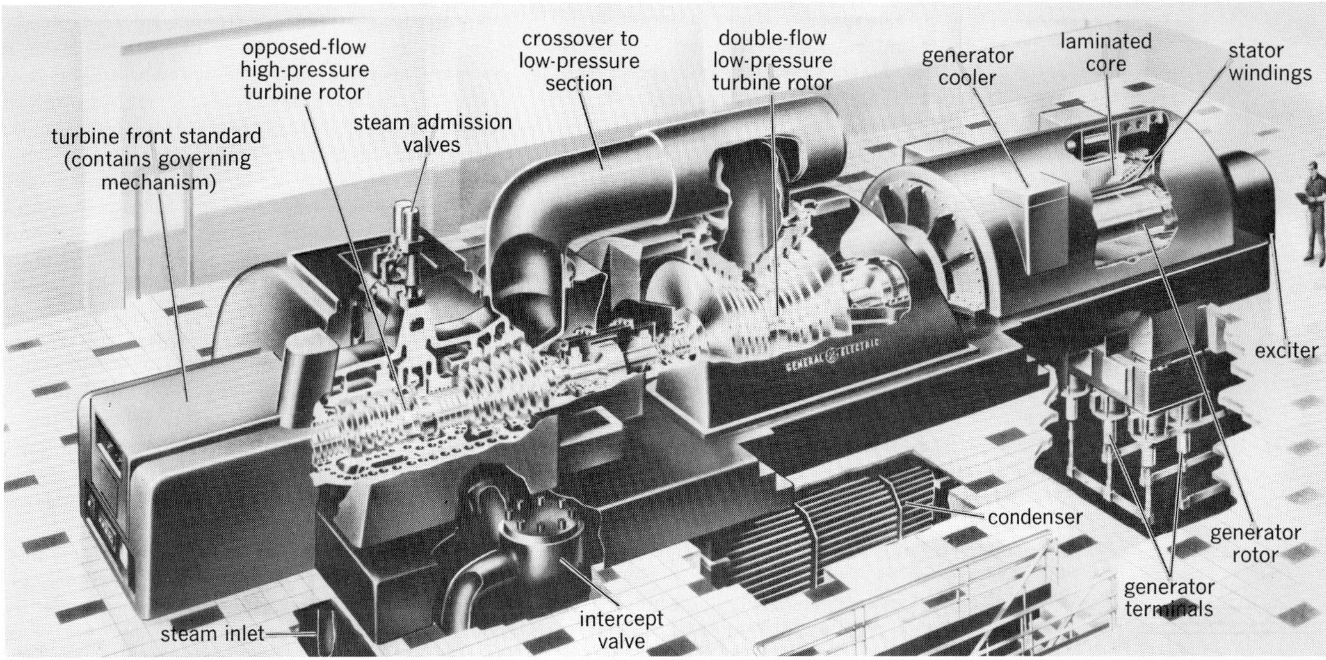

Fig. 4. Partial cutaway view of 3600-rpm fossil-fuel turbine generator. (*General Electric Co.*)

condenser and lowers the controlled extraction pressure.

Industrial turbines are custom-built in a wide variety of ratings for steam pressures to 2000-psig (14 MPa), for temperatures to 1000°F (538°C), and in various combinations of nonextracting, single and double automatic extraction, noncondensing and condensing. Turbines exhausting at or above atmospheric pressure are classed as noncondensing regardless of what is done with the steam after it leaves the turbine. If the pressure at the exhaust flange is less than atmospheric, the turbine is classed as condensing.

Turbines in sizes to about 75,000 hp (56 MW) are used for ship propulsion. The drive is always through reduction gearing (either mechanical or electrical) because the turbine speed is in the range of 4000–10,000 rpm, while 50–200 rpm is desirable for ship propellers. Modern propulsion plants are designed for steam conditions to 1450 psig (10.0 MPa) and 950°F (510°C) with resuperheating to 950°F (510°C). Fuel consumption rates as low as 0.4 lb of oil per shaft-horsepower-hour (0.07 kg/MJ) are achieved.

Central station generation of electric power provides the largest and most important single application of steam turbines. Ratings smaller than 50 MW are seldom employed today; newer units are rated as large as 1300 MW. Large turbines for electric power production are designed for the efficient use of steam in a heat cycle that involves extraction of steam for feedwater heating, resuperheating of the main steam flow (in fossil-fuel cycles), and exhausting at the lowest possible pressure economically consistent with the temperature of the available condenser cooling water.

In fossil-fuel-fired cycles, steam pressures are usually in the range of 1800–3500 psig (12.4–24.1 MPa) and tend to increase with rating. Temperatures of 950–1050°F (510–570°C) are used, with 1000°F (540°C) the most common. Single resuperheat of the steam to 950–1050°F (510–570°C) is almost universal. A second resuperheating is occasionally employed. Figure 4 shows a typical unit designed for fossil-fuel steam conditions. Tandem-compound double-flow machines of this general arrangement are applied over the rating range of 100–400 MW. Initial steam flows through the steam admission valves and passes to the left through the high-pressure portion of the opposed flow rotor. After resuperheating in the boiler it is readmitted through the intercept valves and flows to the right through the reheat stages, then crosses over to the double-flow low-pressure rotor, and exhausts downward to the condenser.

The water-cooled nuclear reactor systems common in the United States provide steam at pressures of about 1000 psig (6.9 MPa), with little or no initial superheat. Temperatures higher than about 600°F (315°C) are not available. Further, reactor containment and heat-exchanger considerations preclude the practical use of resuperheating at the reactor. The boiling-water reactor, for example, provides steam to the turbine cycle at 950-psig (6.55-MPa) pressure and the saturation temperature of 540°F (282°C). Such low steam conditions mean that each unit of steam flow through the turbine produces less power than in a fossil-fuel cycle. Fewer stages in series are needed but more total flow must be accommodated for a given output. Nuclear steam conditions often produce a turbine expansion with water of condensation present throughout the entire stream path. Provisions must be made to control the adverse effects of water: erosion, corrosion, and efficiency loss. In consequence of the differences in stream conditions, the design for a nuclear turbine differs considerably from that of a turbine for fossil fuel application. The former tend to be larger, heavier, and more costly. For example, at 800 MW, a typical fossil-fuel turbine can be built to run at 3600 rpm, is about 90 ft (27 m) long, and weighs about 1000 tons (900 metric tons). A comparable unit for a water-cooled nuclear reactor requires 1800 rpm, is about 125 ft (38 m) long, and weighs about 2500 tons (2300 metric tons). *See* NUCLEAR REACTOR.

Figure 5 represents a large nuclear turbine generator suitable for ratings of 1000 to 1300 MW. Steam from the reactor is admitted to a double-flow high-pressure section, at the left, through four parallel pairs of stop and control valves, not shown. The stop valves are normally fully open and are tripped shut to prevent dangerous overspeed of the unit in the event of loss of electrical load combines with control malfunction. The control valves regulate output by varying the steam flow rate. The steam exhausting from the high-pressure section is at 150 to 200 psia (1.0 to 1.4 MPa absolute pressure) and contains about 13% liquid water by weight. The horizontal cylinder alongside the foundation is one of a pair of symmetrically disposed vessels performing three functions. A moisture separator removes most of the water in the entering steam. Two steam-to-steam reheaters follow. Each is a U-tube bundle which condenses heating steam within the tubes and superheats the main steam flow on the shell side. The first stage uses heating steam extracted from the high-pressure turbine. The final stage employs reactor steam which permits reheating to near initial temperature. Alternate cycles employ reheat with initial steam only of moisture separation alone with no reheat. Reheat enhances cycle efficiency at the expense of increased investment and complexity. The final choice is economic, with two-stage steam reheat, as shown, selected most frequently.

Reheated steam is admitted to the three double-flow low-pressure turbine sections through six combined stop-and-intercept intermediate valves. The intermediate valves are normally wide open but provide two lines of overspeed defense in the event of load loss. Exhaust steam from the low-pressure sections passes downward to the condenser, not shown in Fig. 5.

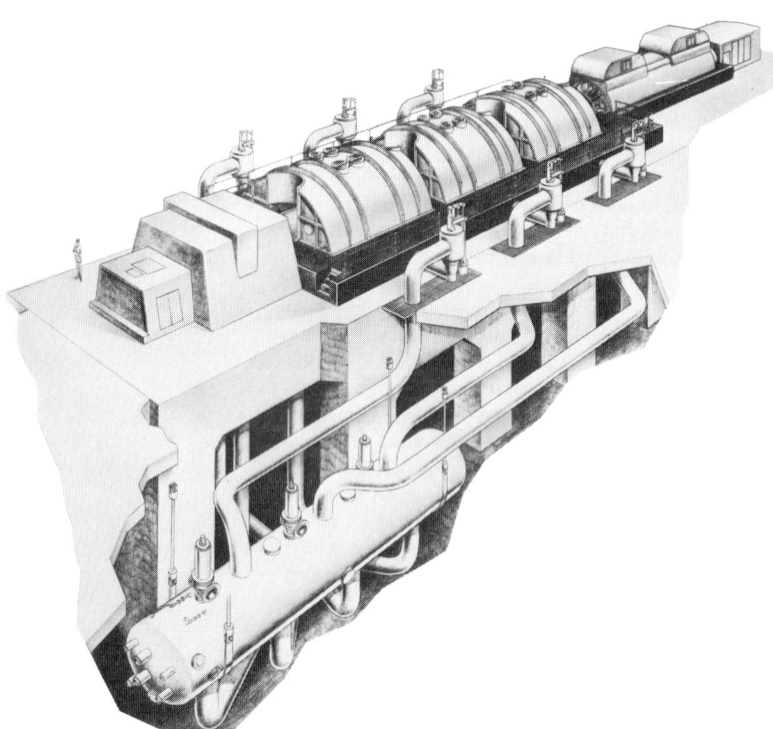

Fig. 5. An 1800-rpm nuclear turbine generator with combined moisture separator and two-stage steam reheater. (*General Electric Co.*)

Turbine cycles and performance. Figures 6 and 7 are representative fossil-fuel and nuclear turbine thermodynamic cycle diagrams, frequently called heat balances. Heat balance calculations establish turbine performance guarantees, provide data for sizing the steam supply and other cycle components, and are the basis for designing the turbine generator.

The fossil-fuel cycle (Fig. 6) assumes a unit rated 500 MW, employing the standard steam conditions of 2400 psig (2415 psia or 16.65 MPa absolute) and 1000°F (538°C), with resuperheat to 1000°F (538°C). As can be seen in the upper left corner, the inlet conditions correspond to a total heat content, or enthalpy, of 1461 Btu/lb (3.398 MJ/kg) of steam flow. A flow rate of 3,390,000 lb/hr (427.1 kg/s) is needed for the desired output of 500 MW (500,000 kW). For efficiency considerations the regenerative feedwater heating cycle is used. Eight heaters in series are employed so that water is returned to the boiler at 475°F (246°C) and 459 Btu/lb (1.068 MJ/kg) enthalpy, rather than at the condenser temperature of 121°F (49°C). Because of the higher feedwater temperature, the boiler adds heat to the cycle at a higher average temperature, more closely approaching the ideal Carnot cycle, in which all heat is added at the highest cycle temperature. The high-pressure turbine section exhausts to the resuperheater at 530 psia (3.65 MPa absolute) pressure and 1306 Btu/lb (3.038 MJ/kg) enthalpy. The reheat flow of 3,031,000 lb/hr (381.9 kg/s) returns to the reheat or intermediate turbine section at 490 psia (3.38 MPa absolute) pressure and 1520 Btu/lb (3.536 MJ/kg) enthalpy. These data are sufficient to calculate the tubine heat rate, or unit heat charged against the turbine cycle. The units are Btu of heat added in the boiler per hour per kilowatt of generator output. Considering both the initial and reheat steam, the heat rate is given by Eq. (1).

Turbine heat rate

$$= \frac{3{,}390{,}000\ (1461 - 459) + 3{,}031{,}000\ (1520 - 1306)}{500{,}000}$$

$$= 8090 \text{ Btu/kWh} = 2.37/\text{kW of heat/kW of output.} \quad (1)$$

The typical power plant net heat rate is poorer than the turbine heat rate because of auxiliary power required throughout the plant and because of boiler losses. Assuming 3% auxiliary power (beyond the boiler-feed pump power given in the turbine cycle in Fig. 6) and 90% boiler efficiency, the net plant heat rate is given by Eq. (2).

Net plant heat rate

$$= \frac{3{,}390{,}000\ (1461 - 459) + 3{,}031{,}000\ (1520 - 1306)}{500{,}000\,((100-3)/100)(90/100)}$$

$$= 9270 \text{ Btu/kWh} = 2.717 \text{ kW of heat/kW of output} \quad (2)$$

The heat rates of modern fossil-fuel plants fall in the range of 8600–10,000 Btu/kWh (2.52–2.93 kW of heat/kW of input). Considering that the heat-energy equivalent of 1 kWh is 3412 Btu, the

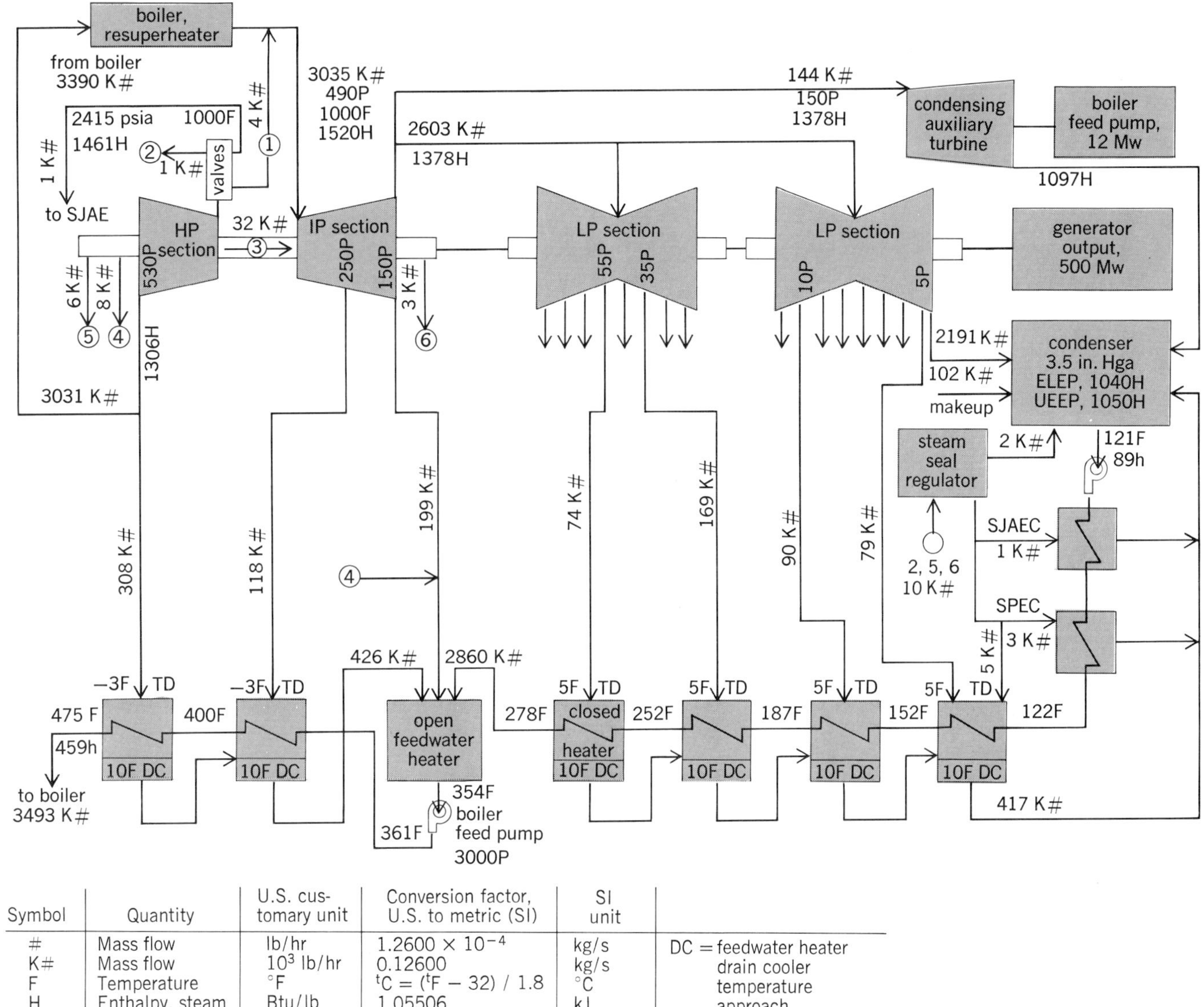

Symbol	Quantity	U.S. customary unit	Conversion factor, U.S. to metric (SI)	SI unit	
#	Mass flow	lb/hr	1.2600×10^{-4}	kg/s	DC = feedwater heater
K#	Mass flow	10^3 lb/hr	0.12600	kg/s	drain cooler
F	Temperature	°F	$^{t}C = (^{t}F - 32) / 1.8$	°C	temperature
H	Enthalpy, steam	Btu/lb	1.05506	kJ	approach
h	Enthalpy, water	Btu/lb	1.05506	kJ	TD = feedwater heater
P	Pressure	psia	6.8948	kPa	terminal temperature
	Pressure	in. Hga	3.3864	kPa	difference
	Heat rate	Btu/kwhr	2.9307×10^{-4}	kJ/(kw·s)	

Fig. 6. Typical fossil-fuel steam turbine cycle.

thermal efficiency of the example is given by Eq. (3).

$$\eta_t = (3412/9270)\,100 = 37\% \qquad (3)$$

Figure 6 shows 2,191,000 lb/hr (276.1 kg/s) of steam exhausting from the main unit to the condenser at an exhaust pressure of 3.5 in. of mercury (11.9 kPa) absolute. The theoretical exhaust enthalpy (ELEP), without considerations of velocity energy loss and friction loss between the last turbine stage and the condenser, is 1040 Btu/lb (2.419 MJ/kg). The actual used energy end point (UEEP) is 1050 Btu/lb (2.442 MJ/kg). The exhaust heat at the condenser pressure is thermodynamically unavailable and is rejected as waste heat to the plant's surroundings. The exhaust steam is condensed at a constant 121°F (49°C) and leaves the condenser as water at 89 Btu/lb (0.207 MJ/kg) enthalpy.

On a heat-rate basis, this cycle rejects heat to the condenser at the approximate rate given by Eq. (4).

Net station condenser heat rejection rate

$$= \frac{2{,}191{,}000\,(1050-89) + 144{,}000\,(1097-89) + 417{,}000\,(100-89)}{500{,}000\,(0.97)}$$

$$= 4650 \text{ Btu/kWh} = 1.363 \text{ kW of heat/kW of output} \qquad (4)$$

If evaporating cooling towers are used, each pound of water provides about 1040 Btu (each kg provides about 2.419 MJ) cooling capacity, which is equivalent to a required minimum cooling-water flow rate of 4.5 lb/kWh (0.57 kg/MJ). The cooling-water needs of a large thermal plant are a most important consideration in plant site selection.

The nuclear cycle (Fig. 7) assumes a unit rated 1210 MW and the steam conditions of the boiling-

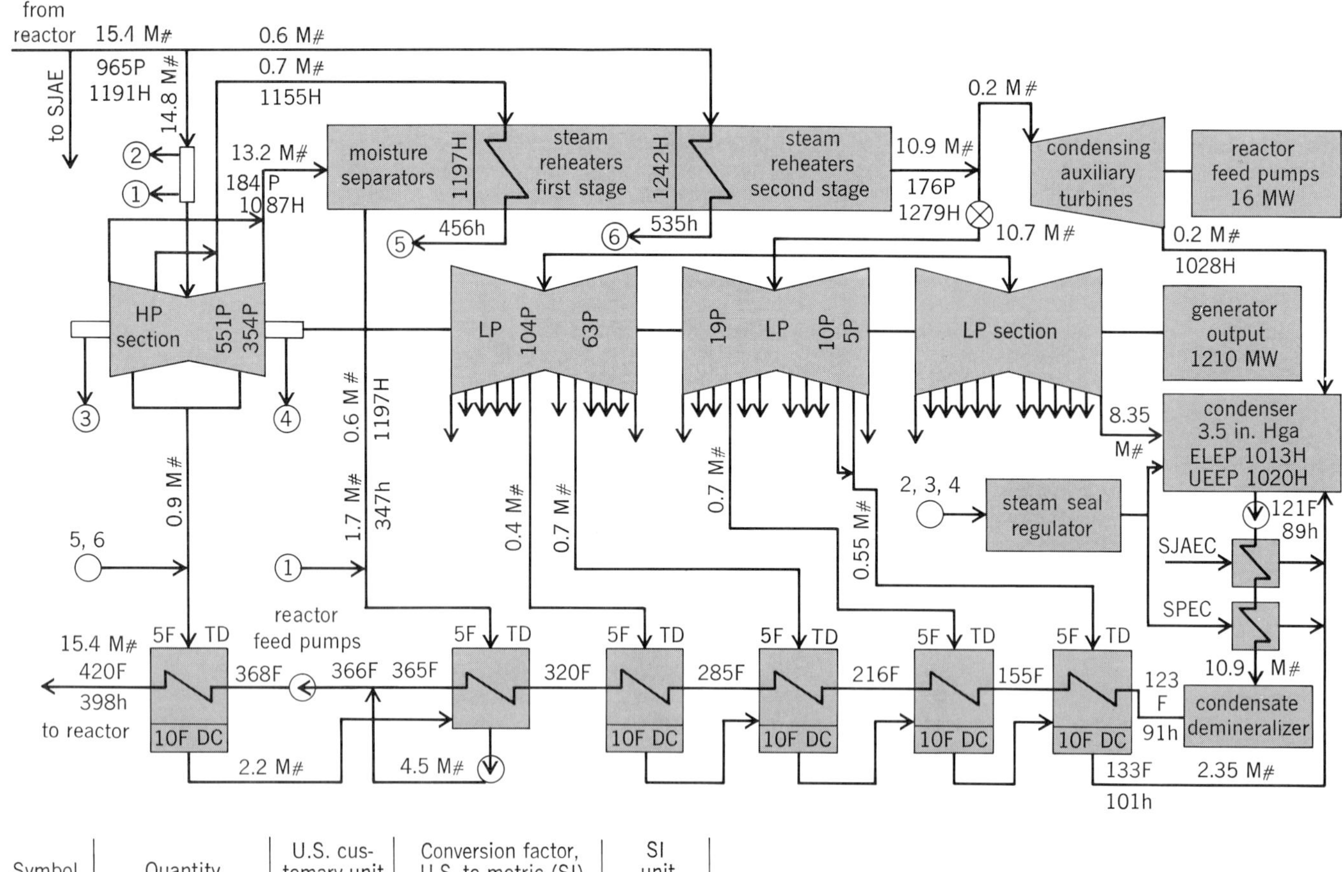

Symbol	Quantity	U.S. customary unit	Conversion factor, U.S. to metric (SI)	SI unit	
#	Mass flow	lb/hr	1.2600×10^{-4}	kg/s	DC = feedwater heater
M#	Mass flow	10^6 lb/hr	0.12600	kg/s	drain cooler
F	Temperature	°F	$^tC = (^tF - 32) / 1.8$	°C	temperature
H	Enthalpy, steam	Btu/lb	1.05506	kJ	approach
h	Enthalpy, water	Btu/lb	1.05506	kJ	TD = feedwater heater
P	Pressure	psia	6.8948	kPa	terminal temperature
	Pressure	in. Hga	3.3864	kPa	difference
	Heat rate	Btu/kwhr	2.9307×10^{-4}	kJ/(kw·s)	

Fig. 7. Typical nuclear steam turbine cycle.

water reactor. Many similarities can be seen to Fig. 6. The major differences include moisture separation and steam reheating and the lack of need for an intermediate pressure element. The low steam conditions are apparent. The consequent turbine heat rate is given in Eq. (5).

Turbine heat rate

$$= \frac{15{,}400{,}000\ (1191 - 398)}{1{,}210{,}000}$$

$$= 10{,}090 \text{ Btu/kWh} = 2.957 \text{ kW of heat/kW of output} \qquad (5)$$

A typical nuclear plant also requires about 3% auxiliary power beyond the reactor feed-pump power already included in Fig. 7. The equivalent boiler efficiency approaches 100% however, and leads to the equivalent net plant heat rate given by Eq. (6).

Net plant heat rate

$$= \frac{15{,}400{,}000\ (1191 - 398)}{1{,}210{,}000\ [(100-3)/100]}$$

$$= 10{,}400 \text{ Btu/kWh} = 3.048 \text{ kW of heat/kW of output} \qquad (6)$$

The corresponding thermal efficiency is given by Eq. (7).

$$\eta_t = (3412/10{,}400)\ 100 = 33\% \qquad (7)$$

Heat is rejected at the condenser at a rate given approximately by Eq. (8).

Net station condenser heat rejection rate

$$= \frac{8{,}350{,}000\ (1020-89) + 200{,}000\ (1028-89) + 2{,}350{,}000\ (101-89)}{1{,}210{,}000\ (0.97)}$$

$$= 6810 \text{ Btu/kWh} = 1.996 \text{ kW of heat/kW of output} \qquad (8)$$

Comparison of the heat rates shows that the nuclear cycle requires about 12% more input heat than does the fossil-fuel cycle and rejects about 46% more heat to the condenser, thus requiring a correspondingly larger supply of cooling water. It consumes heat priced at about half that from coal or one quarter that from oil, and is essentially free of rejection to the atmosphere of heat and combustion products from the steam supply.

Turbine classification. Steam turbines are classified (1) by mechanical arrangement, as single-casing, cross-compound (more than one shaft side by side), or tandem-compound (more than one casing with a single shaft); (2) by steam flow direction (axial for most, but radial for a few); (3) by steam cycle, whether condensing, noncondensing, automatic extraction, reheat, fossil fuel, or nuclear; and (4) by number of exhaust flows of a condensing unit, as single, double, triple flow, and so on. Units with as many as eight exhaust flows are in use.

Often a machine will be described by a combination of several of these terms.

The least demanding applications are satisfied by the simple single-stage turbine of Fig. 1. For large power output and for the high inlet pressures and temperatures and low exhaust pressures which are required for high thermal efficiency, a single stage is not adequate. Steam under such conditions has high available energy, and for its efficient utilization the turbine must have many stages in series, where each takes its share of the total energy and contributes its share of the total output. Also, under these conditions the exhaust volume flow becomes large, and it is necessary to have more than one exhaust stage to avoid a high velocity upon leaving and consequent high kinetic energy loss. Figure 5 is an example of a large nuclear turbine generator which has six exhaust stages in parallel.

Machine considerations. Steam turbines are high-speed machines whose rotating parts must be designed for high centrifugal stress. Difficult stress problems are found in long last-stage blading, hot inlet blading, wheels, and rotor bodies.

Casing or shell stresses. The casings or shells at the high-pressure inlet end must be high-strength pressure vessels to contain the internal steam pressure. The design is made more difficult by the need for a casing split at the horizontal center line for assembly. The horizontal flange and bolt design must be leakproof. Shell design problems lead to the use of small-diameter, high-speed turbines at high pressure, and the use of double shell construction (Fig. 4).

Rotor buckets or blades. Turbine buckets must be strong enough to withstand high centrifugal, steam bending, and vibration forces. Buckets must be designed so that their resonant natural frequencies avoid the vibration stimulus frequencies of the steam forces, or are strong enough to withstand the vibrations.

Sealing against leakage. It is necessary to minimize to the greatest possible extent the wasteful leakage of steam along the shaft both at the ends and between stages. The high peripheral velocities between the shaft and stationary members preclude the use of direct-contact seals. Seals in the form of labyrinths with thin, sharp teeth on at least one of the members are utilized. In normal operation these seals do not touch one another, but run at close clearance. In the case of accidental contact, the sharp teeth can rub away without distorting the shaft.

Vibration and alignment. Shaft and bearings should be free of critical speeds in the turbine operating range. The shaft must be stable and remain in balance.

Governing. Turbines usually have two governors, one to control speed and a second, emergency governor to limit possible destructive overspeed. The speed signal is usually mechanical or electrical. A power relay control, usually hydraulic, converts speed signals to steam valve position. Great reliability is required. *See* GOVERNOR.

Lubrication. The turbine shaft runs at high surface speed; consequently its bearings must be continuously supplied with oil. At least two oil pumps, a main pump and a standby driven by a separate power source, are usually provided on all but the smallest machines. A common supply of oil is often shared between the governing hydraulic system and the lubrication system.

Aerodynamic design. The vane design for highest efficiency, especially for the larger sizes of turbines, draws upon modern aerodynamic theory. Classic forms of impulse and reaction buckets merge in the three-dimensional design required by modern concepts of loss-free fluid flow. To meet the theoretical steam flow requirements, vane sections change in shape along the bucket. To minimize centrifugal forces on the vanes and their attachments, long turbine buckets are tapered toward their tips. *See* CARNOT CYCLE; STEAM-GENERATING UNIT; TURBINE.

[FREDERICK G. BAILY]

Bibliography: F. G. Baily, Steam turbines, in T. Baumeister (ed.), *Standard Handbook for Mechanical Engineers*, sec. 9, pp. 38–54, 8th ed., 1978; F. G. Baily, K. C. Cotton, and R. C. Spencer, Predicting the performance of large steam turbine-generators operating with saturated and low superheat steam conditions, *Combustion*, 3(3):8–13, 1967; R. L. Bartlett, *Steam Turbine Performance and Economics*, 1958; J. K. Salisbury, *Steam Turbines and Their Cycles*, 1950; B. G. A. Strotzki, Steam turbines, *Power*, 106(6):S1–S40, 1962.

Steel

Any of a great number of alloys that contain the element iron as the major component and small amounts of carbon as the major alloying element. These alloys are more properly referred to as car-

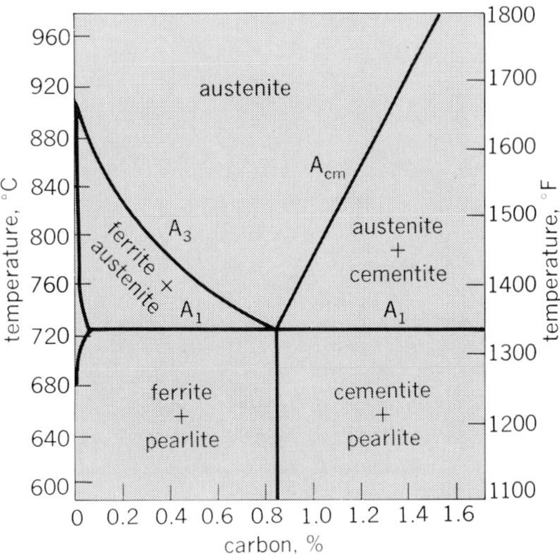

Fig. 1. Critical temperatures in plain carbon steels and constituents which are present in iron-carbon alloys upon slow cooling.

bon steels, and make up well over 90% of the tonnage of steels produced throughout the world. Small amounts, generally on the order of a few percent, of other elements such as manganese, silicon, chromium, molybdenum, and nickel may also be present in carbon steels. However, when large amounts of alloying elements are added to iron to achieve special properties, other designations are used to describe the alloys. For example, large amounts of chromium, over 12%, are added to produce the important groups of alloys known as stainless steels.

Phases. There are three unique thermodynamically stable arrangements of iron and carbon atoms in carbon steels. The arrangements are crystalline and referred to as phases. Ferrite is the phase with a body-centered cubic unit cell, austenite is the phase with a face-centered cubic unit cell, and cementite is the phase with an orthorhombic unit cell. The phase or combination of phases that actually exist in a given steel depends on temperature and composition. Figure 1 is a portion of the phase diagram for the Fe-C system. The diagram is a type of map that shows where the different phases exist under equilibrium conditions. Equilibrium assumes that a steel has been held long enough at a temperature to ensure that the most stable arrangement of atoms and phases has developed.

Figure 1 shows that austenite is stable only at high temperatures and that it is replaced at low temperatures by other phases. This characteristic of iron-carbon alloys is the fundamental basis for the processing and great versatility of steels. When a steel is heated in the austenite phase field, it becomes quite ductile, and heavy sections can be hot-rolled or forged to smaller sizes and more complex shapes. On cooling after hot work, the austenite transforms to structures of higher strength. Another important use of the austenite phase field is in the heat treatment of steels. By heating a finished steel part into the austenite phase, that is, by heating above the critical temperatures identified by A_3 and A_{cm} in Fig. 1, the microstructure existing at room temperature is replaced by austenite. The resulting austenite may then be cooled at various rates to form new arrangements of the phases with a variety of mechanical properties. Cooling at very high rates or quenching produces the hardest structure in

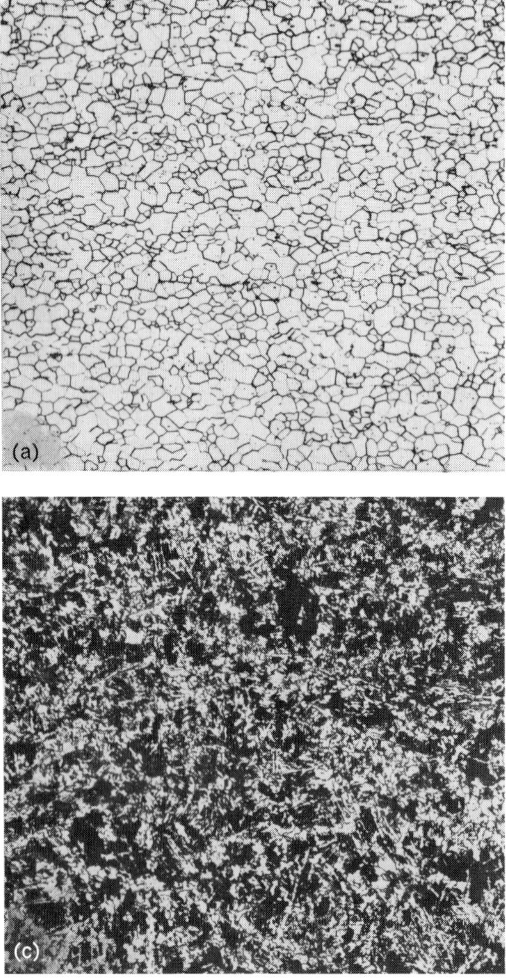

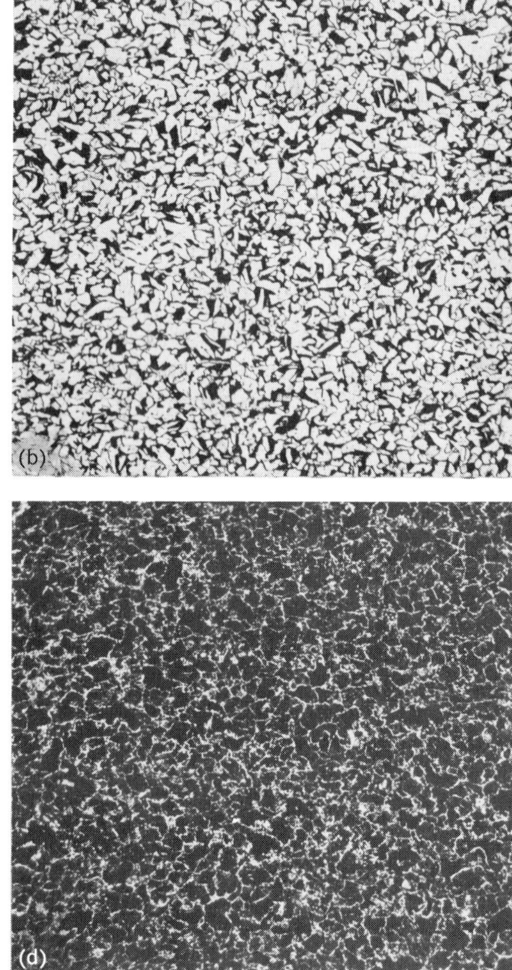

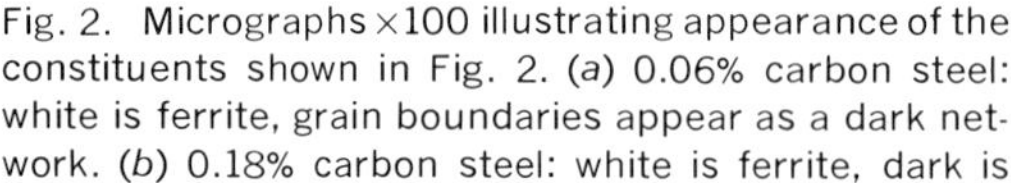

Fig. 2. Micrographs ×100 illustrating appearance of the constituents shown in Fig. 2. (*a*) 0.06% carbon steel: white is ferrite, grain boundaries appear as a dark network. (*b*) 0.18% carbon steel: white is ferrite, dark is pearlite. (*c*) 0.47% carbon steel: same as *b*, but larger amount of pearlite in this higher-carbon steel. (*d*) 1.2% carbon steel: dark is pearlite, white grain boundaries are cementite.

steels, and is discussed in the sections on medium- and high-carbon steels.

Microstructure and properties. When steels are cooled slowly from the austenite phase field, the austenite transforms to microstructures of ferrite and cementite. In steels with carbon content below 0.8%, the austenite first transforms to ferrite, as indicated by the ferrite-plus-austenite field of Fig. 1. Steels with carbon content above 0.8% first form cementite. At or below the A_1 temperature, any austenite not transformed to ferrite or cementite transforms to a mixture of ferrite and cementite referred to as pearlite. The ferrite and cementite form as parallel plates or lamellae, in pearlite. Thus, when a steel reaches room temperature, its microstructure consists of either ferrite and pearlite or cementite and pearlite, depending on its carbon content, as shown in Fig. 1.

Figure 2 shows the microstructures typical of steels cooled very slowly in a furnace, a process referred to as full annealing, or cooled in air, a process referred to as normalizing. The 0.06% carbon steel consists almost entirely of ferrite in the form of fine crystals or grains. The higher-carbon steels contain increasingly greater amounts of pearlite which, at the low magnifications of the micrographs in Fig. 2, appears dark because the lamellae are too close together to be resolvable. Figure 2 *d* shows how cementite has formed on the grain boundaries of the austenite to produce a network of cementite around the pearlite.

The mechanical properties of annealed, normalized, or hot-rolled steels are directly dependent on the amounts of pearlite present in the microstructure. Cementite is a hard carbide phase and, when present in the pearlite, significantly strengthens a steel. Figure 3 shows how the yield and tensile strengths increase with increasing carbon content or increasing pearlite content of steels containing up to 0.8% carbon. However, as strength increases, ductility falls and the steels become less formable, as shown by the decrease in reduction of area and elongation in Fig. 3. The change in mechanical properties with carbon content separates steels into low-, medium-, and high-carbon classifications that are related to the need for formability in manufacturing and strength and toughness for good performance in service.

Low-carbon steels. These steels, sometimes referred to as mild steels, usually contain less than 0.25% carbon, and therefore have microstructures that consist mostly of ferrite and small amounts of pearlite. These steels are easily hot-worked in the austenite condition and are produced in large tonnages for beams and other structural applications. The relatively low strength and high ductility (Fig. 3) of the low-carbon steels make it possible also to cold-work these steels in the ferritic condition. Cold-rolled low-carbon steels are extensively used for sheet applications in the appliance and automotive industries. Cold-rolled steels have excellent surface finishes, and both hot- and cold-worked mild steels are readily welded.

The low-carbon ferritic steels have undergone interesting developments in order to meet the challenges of the energy crisis. Although mild steels have excellent formability, their low strength requires relatively thick sections for a given load-bearing application. New compositions of ferritic

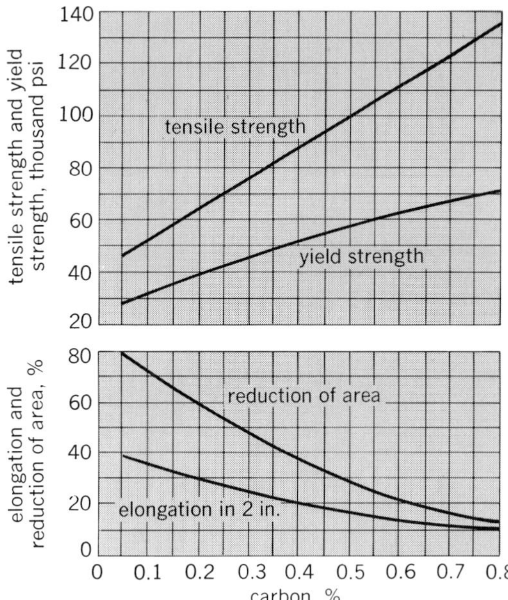

Fig. 3. Effect of carbon content on the tensile properties of hot-worked carbon steels. 10^3 psi = 6.895 MPa; 2 in. = 5 cm.

steels combined with new processing have significantly raised the strength of low-carbon steels, and therefore made possible significant reductions in vehicle weight by permitting the use of thinner sections.

One approach to improved low-carbon ferritic steels has led to the development of a class of steels referred to as high-strength low-alloy (HSLA). HSLA steels are alloyed with very small amounts of strong carbide-forming elements, such as vanadium and niobium. Very fine carbides are formed when the microalloyed steels are finish-hot-worked low in the austenite phase field. The alloy carbides limit the growth of the austenite grains, and therefore the size of the ferrite grains that form from the austenite. The resulting very fine ferrite grain size, together with the fine-alloy carbides, significantly increases strength. Special attention is also paid to inclusion content and shape in HSLA steels, and as a result of inclusion control and the very fine ferritic grain sizes, HSLA steels have excellent toughness for critical applications.

Another approach to producing high strength and good formability in low-carbon steel is to heat a steel with a ferrite-pearlite microstructure into the ferrite-plus-austenite phase field (Fig. 1). Such intercritically annealed steels are referred to as dual-phase steels. The intercritical annealing treatment converts the pearlite areas into pools of austenite and, by controlling cooling rate or alloying the austenite, or both, transforms to structures other than pearlite on cooling. Good formability and high strength are achieved by the elimination of the pearlite and by the introduction of dislocations into the ferrite matrix surrounding the transformed austenite. The dislocations are crystal defects that make possible the shaping of metals. Figure 4 shows the microstructure of a niobium-microalloyed dual-phase steel. The white area containing an internal dark structure (Fig. 4*a*) was aus-

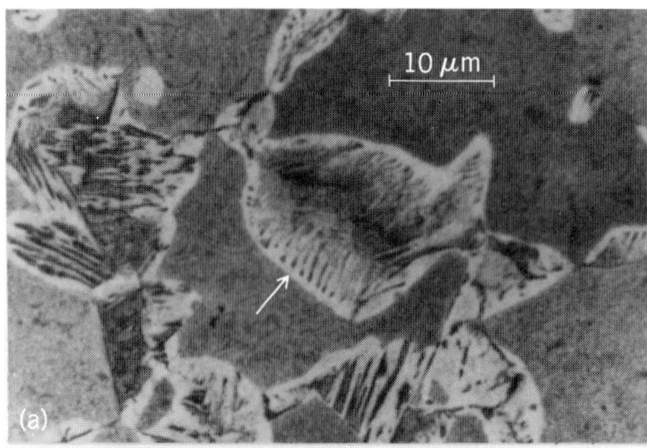

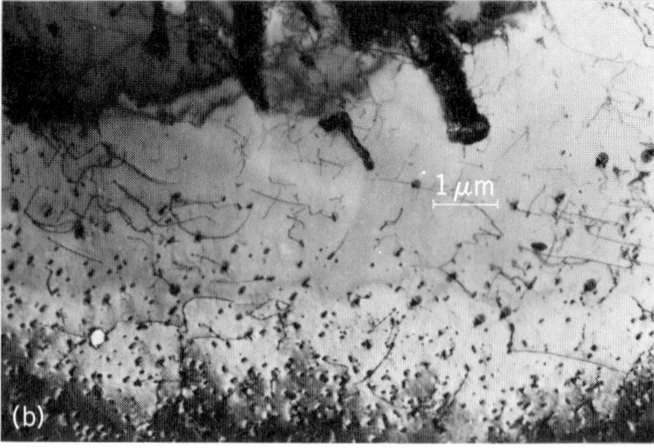

Fig. 4. Niobium-microalloyed dual-phase steel. (*a*) Light micrograph; structure within white areas was produced by transformation of austenite formed during intercritical annealing at 810°C. (*b*) Transmission electron micrograph of area similar to that indicated by arrow in *a*, showing dislocations (fine dark linear features) and alloy carbide particles (fine black spots). (*Courtesy of R. D. Lawson and M. D. Geib, Colorado School of Mines*)

tenite at the intercritical annealing temperature, and has subsequently transformed to ferrite and other phases.

Figure 5 compares the stress-strain curves for plain-carbon mild steels; the HSLA steels SAE 950X and 980X, with yield strengths of 50 kips per square inch (345 MPa) and 80 ksi (552 MPA), respectively; and the dual-phase steel GM 980X. The substantial increase in strength over mild steel is apparent in the stress-strain curves for the HSLA and dual-phase steels. Moreover, the dual-phase steel not only reaches the same maximum strength as the HSLA SAE 980X steel, but also shows a significant increase in ductility and the absence of discontinuous yielding at low strains.

Medium-carbon steels. The medium-carbon steels contain between 0.25 and 0.70% carbon, and are most frequently used in the heat-treated condition for machine components that require high strength and good fatigue resistance (resistance to cyclic stressing). The heat treating consists of austenitizing, cooling at a high enough rate to produce a new phase, martensite, and then tempering at a temperature below the A_1. The martensite is essentially a supersaturated ferrite with carbon atoms trapped between the iron atoms. This structure is very hard and strong. Figure 6 shows that the hardness and strength of martensite increase rapidly with the carbon content of the steel and that the greatest benefits of forming martensite occur above 0.3% carbon content.

However, the same structural factors that make martensite very hard also make it very brittle. Therefore, martensitic steels are tempered to increase their toughness. The tempering essentially makes it possible for the carbon atoms to diffuse from the martensitic structure to form independent carbide particles. Thus the microstructure changes to a mixture of ferrite and carbides on tempering. The extent of this process is controlled primarily by temperature, and a large range of hardness and toughness combinations can be produced.

Martensite can be formed only when austenite transformation to ferrite or pearlite is suppressed. The ferrite and pearlite phases require the diffusion of carbon, a time-dependent process, for their formation. In plain-carbon steels the diffusion can be suppressed only by quenching or cooling very quickly in brine or water. Even when martensite is formed on the surface of a bar of plain-carbon steel, however, the center of the bar may cool too slowly to form martensite.

Small amounts of alloying elements such as chromium, nickel, and molybdenum reduce the rate at which ferrite or pearlite form, and therefore make it possible to form martensite in heavier sections or at slower cooling rates. Slower cooling reduces the tendency to distortion or cracking that sometimes accompanies the high residual stresses introduced by severe quenching. Hardenability is the term used to define the ease of martensite formation relative to cooling rates, section sizes, and steel composition.

High-carbon steels. Steels containing more than 0.7% carbon are in a special category because of their high hardness and low toughness. This combination of properties makes the high-carbon steels ideal for bearing applications where wear resistance is important and the compressive loading minimizes brittle fracture that might develop on tensile loading.

Two types of microstructure are produced in high-carbon steels. One consists entirely of a pearlite with a very fine lamellar spacing produced in

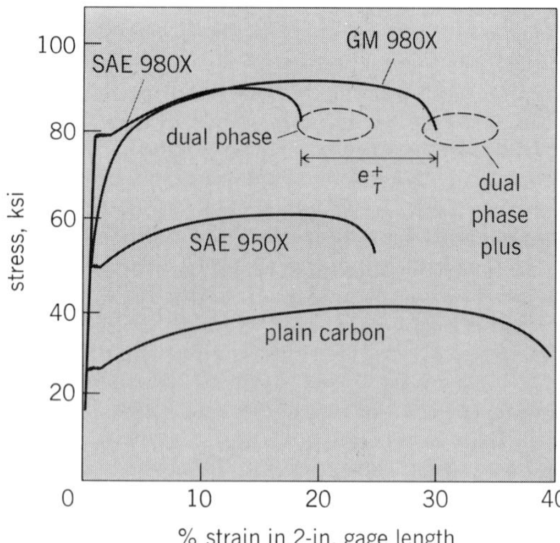

Fig. 5. Schematic stress-strain curves for various steels. Carbon content in all steels is about 0.10%; SAE 950X contains 0.12% titanium, and SAE 980X contains about 0.11% vanadium. The extra elongation due to dual-phase steel processing of GM 908X is indicated as e_T^+; 1 ksi = 6.9 MPa; 2 in. = 5 cm. (*From M. S. Rashid, GM 980X-Potential Applications and Review, Res. Pub. GMR-232, General Motors Corp., Warren MI, 1977*)

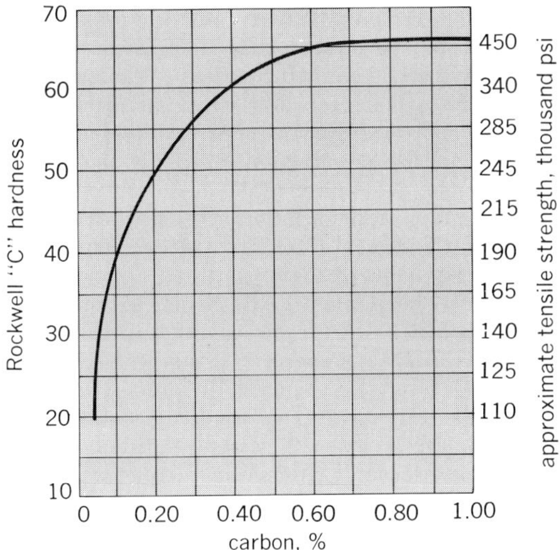

Fig. 6. Relation of hardness and tensile strength to the carbon content in steels which are rapidly quenched from above critical temperature. 10^3 psi = 6.9 MPa.

steels with about 0.80% carbon. Figure 1 shows that no ferrite or cementite form above A_1 at this carbon content. An important application of pearlitic 0.8% carbon steels is in railroad rail. Another application in which pearlitic microstructure is important is high-strength wire for cables and wire rope. In a process referred to as patenting, rods of 0.8% carbon steel are transformed to very fine pearlite. The rods are then drawn to wire to produce an aligned pearlite with strengths up to 350 ksi (2415 MPa).

The other type of microstructure produced in high-carbon steels is tempered martensite. The most common bearing steel is AISI/SAE 52100, which contains 1% carbon and about 1.5% chromium. This steel is oil-quenched to martensite and tempered to retain as high a hardness as possible. Austenitizing is performed in the austenite-cementite phase field. As a result, the microstructure contains very fine cementite particles which not only contribute to wear resistance but also help to maintain a fine austenite grain size. *See* CAST IRON; FERROALLOY; HEAT TREATMENT (METALLURGY); IRON ALLOYS; METAL, MECHANICAL PROPERTIES OF; STEEL MANUFACTURE.

[GEORGE KRAUSS]

Bibliography: American Society for Metals, *Metals Handbook*, vol. 1, 9th ed., 1978; U.S. Steel Corporation, *The Making, Shaping and Treating of Steel*, 9th ed., 1971.

Steel manufacture

The sequence through which blast-furnace iron, scrap iron, and alloying additions are processed to combine and to separate them into desired compositions of steel and residual slag. The manufacture of steel is a tremendous industry with extensive ramifications resulting from the complex interrelation of technical and economic considerations. The United States has been the largest single producer since 1890.

Raw materials and products. Strict chemical usage designates the pure element iron as Fe, but in the ferrous industry "iron" generally refers to the product of the iron blast furnace. Iron, in this sense, is a complex alloy containing about 6% of other common elements whose distribution depends on the raw materials and operation of the particular blast furnace. A typical iron analysis is carbon (C) 4.5%, silicon (Si) 1.5%, manganese (Mn) 0.8%, phosphorus (P) 1.0%, sulfur (S) 0.03%, and many minor impurities. The variation of the Si and P, in particular, makes this product more or less suitable for a specific steelmaking process.

Because this molten product comes from the blast furnace at a temperature of about 1400°C (2600°F), it is called hot metal and is sent directly to the steelmaking operations to be described. If the product is solidified in small molds for convenient handling, it is called pig iron; if solidified in useful shapes such as car wheels, it is called cast iron. Cast iron is relatively brittle and has only limited commercial applications that account for less than 10% of all the iron made. Thus, the steelmaking processes consume most of the iron made, and nearly all of this is in the form of hot metal. Companies using their blast-furnace product in this way are referred to as integrated companies.

Blast-furnace iron provides only about half of the metallic raw material for the steel industry. The other half is obtained from scrap produced in subsequent manufacturing operations and from obsolete steel products. Because the iron units in this scrap are cheaper than those from the blast furnace, the ability of a steelmaking process to use more or less scrap advantageously is an important characteristic of the process.

The steels designated as plain carbon steels generally contain less than 0.8% C, minimum amounts of P and S, and suitably adjusted amounts of Si and Mn. Low-alloy steels, with small additions of other elements such as nickel (Ni), chromium (Cr), and molybdenum (Mo) ordinarily totaling less than 5%, are usually made by the same processes as those for plain carbon steels. The high-alloy steels, such as stainless steel containing 18% Cr and 8% Ni, require special processes. *See* IRON ALLOYS.

A special group of materials known as ferroalloys constitutes the major source of alloy additions for all grades of steel. However, in many cases, these alloying metals are also used in relatively pure forms, such as electrolytic nickel. *See* FERROALLOY.

The melting point of pure iron is 1535°C. This is modified somewhat by the alloying elements present in commercial steels, but to provide proper pouring conditions for both ingots and castings, in all steelmaking processes the melt is generally raised to temperatures of 1550–1650°C. These high temperatures place serious limitations on the refractories that can be used. They also require the most efficient use of heat, whether it is derived from combustion of fuel, electricity, or thermochemical reaction. Iron and steel products have been used by man since antiquity, but steels could not be made in tonnage quantities before 1850 because of the lack of understanding of construction materials, heat generation, and chemistry at these temperatures.

Thus, various modern steelmaking processes have been developed because of their combined attributes for the use of available raw materials, the production of specific grades of steel, and the efficient use of energy and refractory combinations. These features will become evident in the description of the major processes.

Classification of processes. Chemically, all steelmaking processes may be classified as acid or basic, depending upon the refractory and slag combination; each process has particular attributes with regard to the refining it can accomplish.

Acid processes use silica (SiO_2) refractories throughout and are able to accommodate slags that become saturated with this component under operating conditions. These acid systems can be used to eliminate C, Mn, and Si from the charge, but require select raw materials within final steel specifications for P and S.

Basic processes use magnesite (MgO) or equivalent refractories in the portions of the furnace that contact molten slag and metal, and accommodate lower silica slags with compensating amounts of lime (CaO). These systems can eliminate C, Mn, and Si as effectively as can the acid systems; they will also eliminate P and appreciable amounts of S. Basic systems have a decided advantage in flexibility with regard to raw materials consumed and grades of steel produced.

Technologically, processes can be grouped into three main types: (1) pneumatic, (2) open-hearth, and (3) electric. In pneumatic processes all heat is derived from the initial heat content of the charge materials, principally molten, and the thermochemical balance of the refining reactions; the selective oxidation of the refining is accomplished by blowing air or commercial oxygen. In open-hearth processes the major source of heat is the combustion of fuel (usually gas or oil); this in turn depends for its success on the regenerative principle of preheating air to attain steelmaking temperatures efficiently. In electric processes the major source of heat is electric current (arc or resistance or both). Because this heat can be produced in the presence or absence of oxygen, electric furnaces can operate in a neutral or nonoxidizing atmosphere or vacuum, and thus are either preferred or required for alloys with significant amounts of easily oxidized elements and for other special grades where improved control of gases is required.

To use available raw materials and heat sources effectively for particular grades of steel, steelmaking processes of the pneumatic, open-hearth, and electric types exist, and any of these may be either acid or basic in their chemistry. In many cases, there is an overlapping or combination of the above principles in a single process, but the indicated classification is followed in describing the process and its product.

The relative importance of the main commercial processes according to tonnage is given in the table, which shows the rapid changes in the industry. In 1957, steel made by oxygen processes was so small as to be included with crucible steel. By 1967 Bessemer steel had practically vanished as a commercial process, and the previously dominant open-hearth processes were being replaced rapidly by the oxygen and electric processes. The principal characteristics of the individual processes will be described.

PNEUMATIC PROCESSES

In the United States, a limited amount of low-P ore is available. This type ore produces a blast-furnace product suitable for the process developed in England by Henry Bessemer in the years following his original patent in 1857. The same process was proved in the United States by William Kelly after his original idea in 1847. This is an acid process, and was historically the first high-tonnage process for steel production. It will be described first because, in addition to this historical position, it is the simplest major process and provides a reference for appraisal of other processes and modern developments of all types.

From the beginning, the Bessemer process was able to produce steel from hot metal in 10- to 25-ton heats at an average rate of nearly 1 ton/min. Although no scrap is required, the process normally uses 12–15% scrap. This low scrap requirement is an advantage when the supply is limited, and was a major reason for the initial dominance of the process. Suitable hot metal for the process should have a content of 4.00–4.50% C, 1.10–1.50% Si, 0.40–0.70% Mn, 0.09% maximum P, and 0.03% maximum S.

Converters. The typical vessel or converter is a refractory-lined steel shell with tuyeres in the bottom and open at the top (Fig. 1). The vessel is mounted on trunnions and is provided with mechanical means for tilting. Air for the process is blown through one hollow trunnion and is distributed to the tuyeres through the wind box at a pressure of about 30 psi.

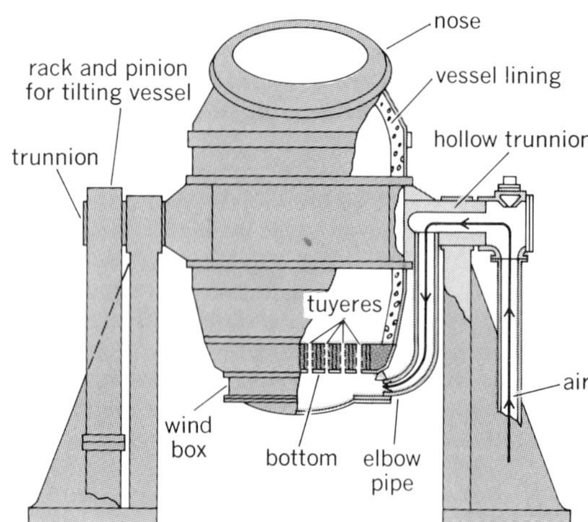

Fig. 1. Schematic cutaway view of Bessemer converter. (*From United States Steel Corp., The Making, Shaping, and Treating of Steel, 7th ed., 1957*)

In a typical blow, the hot metal is charged while the vessel is tilted to the horizontal position to keep the tuyeres clear. The blast is turned on as the vessel is righted, and the air bubbling through the melt, about 18–24 in. deep, provides the oxygen required for refining reactions (1)–(4). The

$$2\,Fe + O_2 \rightarrow 2FeO \quad (slag) \qquad (1)$$

$$2\,Mn + O_2 \rightarrow 2MnO \quad (slag) \qquad (2)$$

$$Si + O_2 \rightarrow SiO_2 \quad (slag) \qquad (3)$$

$$2\,C + O_2 \rightarrow 2CO \qquad (4)$$

reactions occur approximately in the order listed, although there is considerable overlapping. They are exothermic and provide enough heat to raise the temperature from that of the charge (about 1350°C) to that of the product (about 1600°C). This temperature rise can be controlled by regulation of the hot-metal analysis (Si content being the most important) and the amount of scrap added during the blow.

The flame emitted from the mouth of the vessel changes its color and luminosity in a manner which allows the operator (blower) to judge the progress of the refining reactions, to stop the operation at a suitable end point, and to pour the heat into a transfer ladle by tilting the vessel.

Restrictions. The oxides formed in the first three reactions combine to form a silica-saturated slag. This will contain approximately 50% SiO_2 and 50% (MnO + FeO). The silica requirement in excess of that provided by the Si from the hot metal is necessarily obtained from the refractory lining of the vessel. The entire blowing time is only about 15 min, so for success the process depends upon the blower's experience and judgment. The time is too short to allow for control by sampling and analysis.

In addition to the restrictions on raw materials, the process is typically limited to plain carbon steels with less than 0.30% C and with nitrogen contents that make the product unsuitable for many applications.

Additional operations of deoxidation, alloy additions, and teeming (pouring from a transfer ladle into ingot molds) are required to complete the process. These operations are essentially the same as those to be described in connection with the more important basic open-hearth process.

The basic Bessemer or Thomas process is important in Europe, and was developed to use iron from high-P ores, which cannot be refined by acid slags. The refractory lining is made of magnesite (MgO), and an afterblow eliminates the P by reactions (5) and (6). The CaO required for this purpose

$$4P + 5O_2 \rightarrow 2P_2O_5 \qquad (5)$$

$$P_2O_5 + 4CaO\,(slag) \rightarrow 4CaO\cdot P_2O_5\,(slag) \qquad (6)$$

is added with the charge and is dissolved in the slag throughout the blow so that the final slag has the necessary CaO/SiO_2 ratio of about 3. Some S control is accomplished and can be represented by such a reaction as (7). The combined influences of

$$S_{(Fe)} + CaO_{(slag)} \rightleftharpoons CaS_{(slag)} + O_{(Fe)} \qquad (7)$$

temperature, slag gas compositions, and kinetic factors, all of which govern this partition, are complex. The important differences between the acid and basic processes are with regard to the raw-material requirements rather than the type of product. Most recent oxygen steelmaking developments, to be described later, may be related to the original Bessemer and Thomas processes.

BASIC OPEN-HEARTH PROCESS

The basic open-hearth process has been the dominant tonnage steel producer in the United States since the beginning of the 20th century, but most new facilities involve some variation of oxygen steelmaking or electric furnaces so that the classical open hearth may not survive. The conventional reverberatory-type furnace was unable to generate sufficiently high temperatures for proper refining of steels, and only the limited (500 lb or less per day) production of the special grade known as wrought iron was possible.

In 1858, Karl Siemens built an experimental furnace by making use of his regenerative principle, and together with his brother, Frederick, developed it within the next few years into the essential design of modern open-hearth furnaces. Thus, both the pneumatic and the open-hearth processes, which form the basis of the modern era of low-cost high-tonnage steel, were established within a few years of each other.

Initial rapid development of the Bessemer (pneumatic) process resulted from its inherent simplicity. The later dominance of the open-hearth process is the result of its ability to use scrap advantageously, its greater flexibility in use of other raw materials, and its ability to make a wider range of steels. It can be operated as either an acid or basic process by selection of refractories; in either case, a wide variation in practices is possible.

Furnaces. The features of furnace construction incorporate the regenerative principle (Fig. 2). The hearth proper is lined either with magnesite refractory for a basic furnace or with silica for an acid furnace. The roof is made of silica brick with sprung-arch construction, or basic brick with suspended-arch construction. Selection of roof refractories is independent of hearth construction, because these refractories do not contact molten slag. Refractories used in the remainder of the system vary greatly, but they will not be considered here because they do not relate directly to the steelmaking process.

Burners are provided at both ends of the furnace and are operated alternately on a controlled cycle, usually 10–15 min. The usual fuels are natural gas or oil, but this selection depends on availability and economy and is not otherwise limited. To attain useful temperatures in the furnace (above 1650°C) with minimum fuel cost, the air for combustion is preheated by the regenerative system. For example, when fuel is burned at one end, the dampers are adjusted so the air for combustion first passes through the hot checkers at this end. The checkers are an open latticework of refractory brick. The flame and products of combustion pass over the hearth area, through the checker system on the opposite end, and out the stack. The dampers are then reversed and fuel is burned at the opposite end. Thus, by selection of a suitable cycle, high temperatures are achieved with minimum fuel consumption. Typical furnaces have a capacity of 200–300 tons per heat.

The making of a typical batch or heat of plain carbon steel by the basic open-hearth process will be outlined. The furnaces are operated continuously to conserve heat, minimize spalling of refractories, and produce maximum tonnage. After the previous heat has been drained from the hearth, grain magnesite is blown or shoveled into the areas where erosion has occurred, and then the materials for the next heat can be charged. The solids are previously weighed into charging boxes and brought into the furnace area. These boxes can be lifted, thrust through the furnace door, and dumped by the charging machine. The mechanization of these operations is an essential feature of high tonnage rates.

Materials and reactions. The cold charge materials are limestone ($CaCO_3$), ore (Fe_2O_3), and scrap steel, and they are piled into the furnace in this order. The scrap constitutes about one-half the total iron units required for the heat. It will be a mix of light, voluminous and heavy, dense material and is distributed to achieve rapid heat absorption and economy of space and to hold the lime and ore on the bottom as long as possible. The amounts of lime and ore are calculated to give the correct analysis of metal and slag at the end of the heat, and will thus vary according to the type of scrap available and the grade of steel being made. As this charge is heated and melted, additional iron oxide is formed, because the furnace atmosphere must contain an excess of oxygen to maintain efficient combustion. Thus, as the scrap melts, it will be refined by oxidizing reactions (1)–(4) already indicated for the pneumatic processes, and the slag formed will be highly oxidizing. When this scrap is partially melted, the remaining units of iron are added as hot metal by pouring them from a transfer ladle through a trough placed in one of the charging doors.

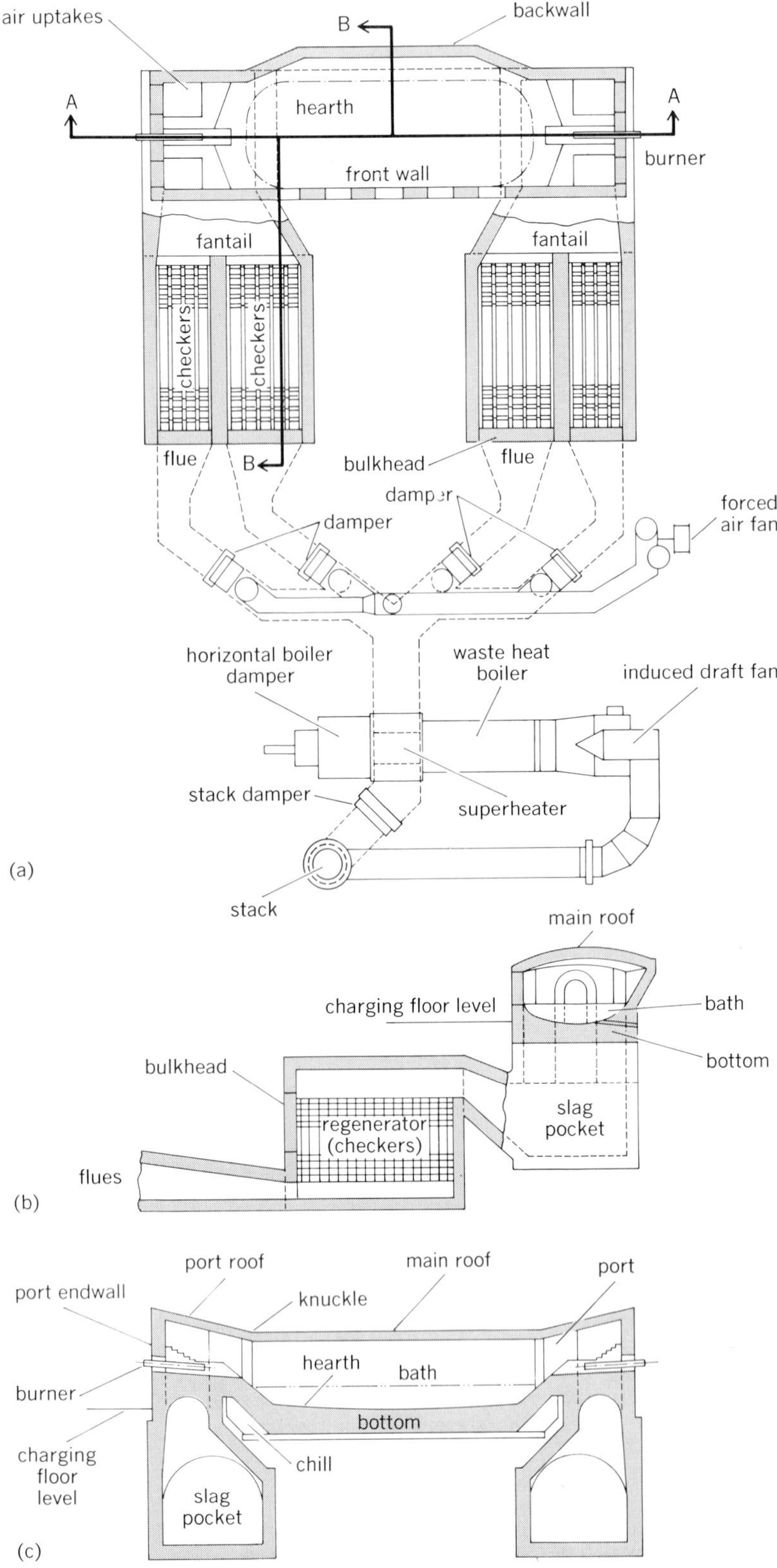

Fig. 2. Schematic sections of liquid-fuel-fired open-hearth furnace. (*a*) Sectioned schematic plan. (*b*) Vertical section across B-B. (*c*) Vertical section across A-A. (*From United States Steel Corp., The Making, Shaping, and Treating of Steel, 7th ed., 1957*)

Details of the remainder of the process depend on the chemical composition required in the end product. To make the description here applicable to all grades of steel and all processes, some important relations will be generalized. These can be made quantitative for any particular situation, but it should be recognized that steelmaking is an intricate and complicated undertaking that involves heterogeneous reactions (gas-metal, gas-slag, and slag-metal) between phases which are not necessarily in equilibrium with each other with regard to either temperature or chemistry.

The various stages and types of steelmaking involve the selective control of oxidizing reactions. The usual sources of oxygen are from the air or products of combustion (CO_2 is oxidizing toward Fe); slag which contains large amounts of iron oxide, usually represented as FeO; or ore represented as Fe_2O_3. Commercial oxygen may be substituted for air to achieve greater fuel efficiency or for ore to give greater speed and control of refining. This combination results in auxiliary improvements in design and flexibility which are still being developed. Whatever the source, there will be an effective oxygen pressure and the molten bath of metal will dissolve an amount of oxygen, represented as $O_{(Fe)}$, corresponding to its temperature and composition. This dissolved $O_{(Fe)}$ will in turn react with the other elements in the metal such as $C_{(Fe)}$ as represented by reaction (8).

$$C_{(Fe)} + O_{(Fe)} \rightarrow CO \quad \text{(gas)} \tag{8}$$

Similar reactions occur with the Si, Mn, and other elements in the molten bath, as already shown under the Bessemer process. The equilibria for all of these reactions are interrelated; however, for present purposes, it is useful to consider the carbon-oxygen reaction in the bath as the dominant one during the refining period of the steelmaking process. The relations between the carbon and oxygen dissolved in the bath are shown schematically in Fig. 3, which indicates that for a given carbon content, the amount of oxygen in the bath is directly related to temperature (that is, will increase with temperature), whereas for a given temperature, the amounts of carbon and oxygen in the bath are inversely related.

Slag and steel. The above information may be applied to a heat under way in the basic open-hearth furnace. Pure iron melts at 1535°C, whereas the hot metal melts at about 1130°C and is probably no hotter than 1300–1350°C. Thus, the low-C steel scrap has a higher melting point than the hot

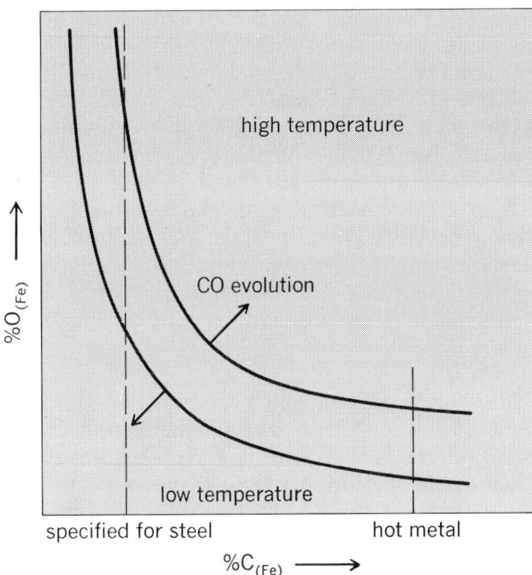

Fig. 3. Generalized relations between carbon and oxygen dissolved in molten iron.

metal (high-C) when most of the scrap has been melted, and therefore this metal bath and the slag which has formed will be high in oxygen in relation to the hot metal being added. This results in the rapid refining of the Si, Mn, P, and C in the system. In particular, the bubbling of CO causes the slag to become foamy and voluminous, and a large amount of it is flushed out of the furnace through either a slag notch at the back or an open door in the front, or both. Thus, this flush slag removes from the system important amounts of the Si, Mn, and P oxides which result from the refining reactions, thereby reducing the weight of slag remaining in the furnace and facilitating the heat transfer and slag control required in the rest of the process.

After melting of the scrap begins, a balance must be maintained between the temperature increase of the bath and its C content so that the bath remains molten at all times. The stirring of the bath by CO evolution is important in achieving the necessary heat transfer, which is often the limiting rate of the process. The reactions during this flush period may become dangerously violent, and their control by proper timing of the hot-metal addition requires great skill by the melter.

During the melt-down period, the limestone on the hearth is calcined, as shown by reaction (9),

$$CaCO_3 \rightarrow CaO + CO_2 \tag{9}$$

and as the last heavy scrap melts, this stone floats up into the slag, where it must be dissolved for the final slag to have some predetermined CaO content. Depending again on the quality of the starting materials and the specifications for the steel being made, this exact composition will vary, but it will usually result in a final ratio of CaO/SiO_2 in the range of 2.5–3. This ratio is a simple measure of slag basicity. This high-lime content is required to achieve the dephosphorization and S control, both characteristic of all basic steelmaking. The major sources of S are the hot metal and fuel.

The equilibria are such that when the required slag basicity and carbon content of the metal bath are reached, the heat is ready to be taken out of the furnace through the taphole at the back where a refractory-lined runner directs it into a transfer ladle. As the metal runs into the ladle, additions are made to adjust the final chemistry with regard to carbon and other elements, as well as oxygen control, according to the type of steel being made.

Compared with the Bessemer process, the open-hearth is a slow process with a total charge-to-tap time of 5–10 hr. The refining rate at the end of the heat can be controlled for sampling and analysis prior to tapping, thus permitting close adjustment to carbon or other specification, providing final solid additions are small enough to avoid chilling the heat in the ladle. In general, additions of up to 10% alloy content can be made.

ACID OPEN-HEARTH PROCESS

In relation to tonnage, the acid open-hearth process is relatively unimportant. It is subject to the restrictions on P and S in the charge, already discussed for acid processes. It has been used largely for the production of high-quality steels in the higher C ranges for steel rolls, castings, and forgings. In these cases, high production rates are unimportant, and a high price for high-quality scrap and other charge materials is tolerable. The process is normally operated on an all-solid charge with the necessary mix of ore, pig iron, and scrap to achieve the specified analysis.

Although there are many important details related to acid open-hearth steelmaking, the essential features can be deduced from the descriptions already given for acid processes in general, the acid Bessemer process, and the principles of deoxidation outlined in connection with the basic open-hearth process. The normal product is fully deoxidized, can be made with low inclusion content, and is believed to be particularly free from the defects associated with hydrogen.

ELECTRIC FURNACE STEEL PRODUCTION

Electric melting began to exert its influence in the industry about 1900 (50 years later than the pneumatic and open-hearth processes, both of which depend on an oxidizing atmosphere for their success). The arc furnace develops the necessary temperatures without the requirement of oxygen in the atmosphere, and has therefore occupied a unique position as a process suitable for making many grades of steel with large amounts of such oxidizable alloys as Cr, vanadium (V), and tungsten (W). Beginning with the original design of Paul L. T. Héroult, the three-electrode furnace has been developed to the point where it is competing with the open-hearth process for all grades of steel in many situations, and it appears destined for a continued increase in importance. The features of the furnace are shown in Fig. 4, which also indicates the differences in refractories for basic or acid practice. The roof and sidewall refractories are subject to wide variation. The distinctions between these practices are the same as those described for other processes, and, to take better advantage of fluctuating sources of raw materials, there is a strong trend in favor of the basic practice. Commercial furnace sizes vary from a few hundred pounds to 200 tons. As a rule, the smaller sizes are used by the castings industry, and the larger sizes are used for the production of ingots. An all-solid charge is normal, but the use of hot

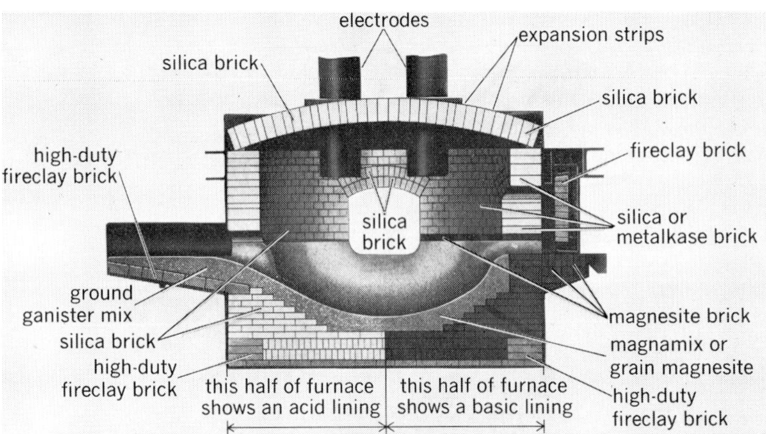

Fig. 4. Section of electric-arc furnace showing refractories for either acid or basic operation. (*Modern Refractory Practice, Harbison-Walker Refractories Co.*)

metal is also feasible. The charging of larger furnaces is a critical rate-determining factor; common practice is to swing the top aside and to place the entire charge in the hearth at once from a previously loaded drop bucket.

The making of a hypothetical heat which illustrates the unique features and versatility of the basic electric-arc process will be described. In most cases, raw materials for the charge could be selected to eliminate or minimize some of the steps included in this example.

Raw materials and reactions. As in the open hearth, the solid charge of ore, limestone, scrap, and pig iron is placed in the hearth after the previous heat has been drained and the bottom repaired. The amounts of these materials are proportioned to achieve the desired specifications when the charge has been melted and brought to temperature; that is, enough oxygen is provided to refine the Si, Mn, P, and C present in the charge, and enough lime is included to give the slag basicity required to retain the P in the slag. These items can be adjusted, if necessary, throughout the melting and refining period, and most of the oxygen is supplied directly by lance.

Up to this point, the process and reactions are comparable to the basic open-hearth process, with the exception that a relatively small part of the oxygen requirement has come from the furnace atmosphere. Air within the furnace is in contact with the graphite electrodes, and all oxygen is rapidly burned to the $CO-CO_2$ equilibrium for the furnace temperature. The basic oxidizing slag will contain the P refined from the charge, so this slag can now be removed from the system, the furnace being tilted to assist in this operation. If the charge materials were sufficiently low in P, as is frequently the case, this slag can be left in the furnace to allow the recovery of such valuable alloying metals as Cr which entered the slag during the oxidizing period.

If the oxidizing slag was removed, and the steel specification requires that large amounts of alloy be added, the next step is to make a new, reducing slag by adding calcined lime (CaO), C (as crushed electrodes or coke), and flux (CaF_2) as well as small amounts of Al_2O_3 or SiO_2. The exact composition of this slag depends a great deal on the type of alloy being made, but it will be highly basic and often carbidic, that is, contain detectable amounts of CaC_2. To be effective, this slag must contain negligible amounts of FeO. So that this condition is reached rapidly, the steel bath is simultaneously deoxidized in the furnace by addition of ferrosilicon or aluminum (Al) or both; the resulting deoxidation products become a part of the reducing slag. In fact, they are the major source of these oxides in this slag. The molten metal bath is thus protected by a nonoxidizing slag, and the atmosphere is kept nonoxidizing by reaction with the electrodes.

Power can be supplied to achieve any desired temperature and to melt any amount of alloy needed to meet the steel specification. This basic reducing slag performs another important useful function. It has much more favorable chemistry for the removal of S than any other treatment in the entire iron and steelmaking cycle, and specifications in the order of 0.002% S can be met.

If a P-free oxidizing slag is left on the bath, it is reduced by the same general methods (addition of ferrosilicon, Al, or C), and the corresponding amount of Cr or other alloy reverts to the bath.

In either case, the bath can be sampled and alloys can be adjusted until they fall within the specified range, and the temperature can be adjusted until the heat is ready to pour. Although not as drastic as in the other steelmaking processes, some final ladle deoxidation is still required.

Product steels. The major tonnage steels for which the electric furnace is required are the stainless steels, such as 18–8 (18% Cr, 8% Ni), but all the special high-alloy grades are made by this process, as are many of the lower-alloy grades with 10% or less of total alloy content, which could be made by the other processes. The latter are made in the electric furnace when they must have the highest cleanliness for critical service requirements or for smaller special orders.

Electric furnace steels are appreciably higher in nitrogen content than the same grades made in the open-hearth process, and this may be desirable or not according to the application. Special procedures are also required to make low-C steels (less than about 0.10% C). In general, the electric furnace is the most versatile of all steelmaking processes, because it can be operated as either an acid or basic, oxidizing or reducing process and thus accommodate any combination of raw materials. A major advantage is its capacity for rapid melting of solid scrap.

SPECIAL PROCESSES

The production data cited in the table show how the major traditional processes dominate the steel industry. The fundamental features of these processes have changed little since their inception, although there have been continual increases in size of the units, and many engineering improvements have led to remarkable modern production rates. Tremendous investments in major facilities preclude sudden changes in structure of such an industry. Since the early 1950s these changes have been rapid because of the combination of inadequate capacity in many areas and improved technology, especially the availability of low-cost commercial oxygen on a tonnage basis.

Use of oxygen. In the conventional steelmaking processes, the use of oxygen is now well established. Oxygen can be used for combustion in the open-hearth process to produce higher flame temperatures and develop heat faster, or to maintain an existing temperature pattern with lower fuel consumption. When coupled with the higher operating temperatures possible with basic roof construction, this offers many possible advantages, which are yet to be fully exploited, such as faster melting and reduced checker chamber capacity.

The use of oxygen for refining by injection into the bath has been important in both the open-hearth and electric furnace. In these applications oxygen may be considered as replacing ore with the advantage of more precise control and a more favorable heat balance, resulting in higher production rates. For this purpose it is common to insert an iron pipe (about 1 in. in diameter) through the charging door into the bath and to blow oxygen fast enough to cool the pipe and avoid excessive melting of the lance. Roof lances are also used. In the electric furnace, the extra heat obtained with the oxygen lance makes possible the recovery of large amounts of Cr from stainless steel scrap.

In the pneumatic processes oxygen can enrich the air blown through the tuyeres to refine the bath more rapidly and to develop the heat required to melt more scrap. When combined with lime injection, this technique also gives better slag control.

Basic-oxygen steelmaking. The most spectacular application of oxygen has been in what originated as the Linnz and Donnewitz (L-D) process, now generally referred to as basic-oxygen steelmaking. The tuyeres of the Bessemer vessel have been eliminated and pure oxygen is blown through a water-cooled lance, usually with multiple openings, onto the top of the bath in a pear-shaped, basic-lined furnace, shown schematically in Fig. 5.

All of the previously described refining reactions of C, Si, Mn, P, and S are operative, and the absence of bulk nitrogen from the air gives a low-nitrogen product as well as the capacity to melt more scrap than in the Thomas process. This lower gas volume also facilitates fume control. Heat sizes comparable to those of the open hearth are made at refining rates of the Bessemer, and there appear to be no restrictions as to the types of plain carbon steel which can be made. The procedure should also be suitable for some high-alloy types where the thermochemistry is favorable.

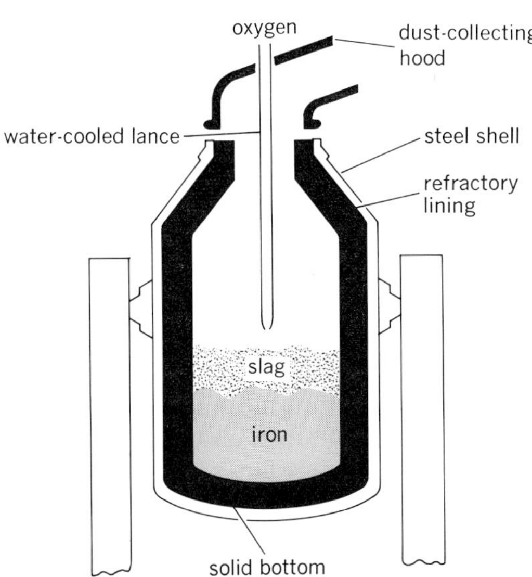

Fig. 5. Schematic section of L-D process vessel. (*From Steel, 43:27, October, 1958*)

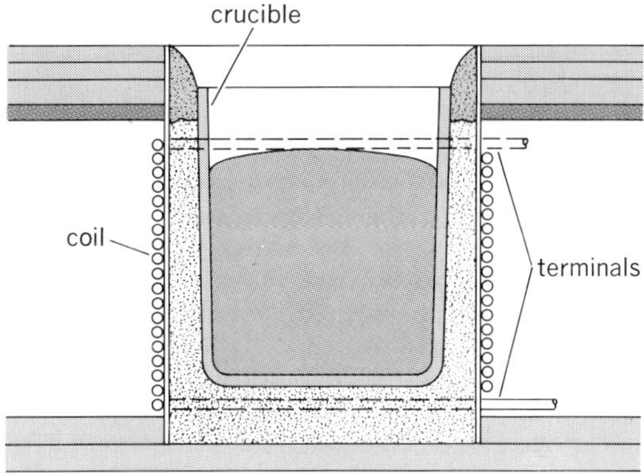

Fig. 6. Section of crucible which is used for high-frequency-induction melting of steel. (*From Ajax Electrothermic Corp., Bull. no. 11-B*)

The requirement for about 75% hot metal is the most important raw material restriction. Nonintegrated plants (those lacking blast furnace facilities) overcome this difficulty by melting pig iron or scrap in cupolas. The single source of oxygen makes the process ideal for automation.

Other new processes use the oxygen lance with a rotating hearth, horizontal or tilting, to obtain mixing of slag and metal, and heat is supplied by an auxiliary burner. None of these variations is as widely used as the basic-oxygen process.

Induction melting. Induction melting has been used to replace the conventional arc furnace in cases where small heats of special alloys are desired, usually for castings. High-frequency power, usually from a motor generator, is supplied to the outer water-cooled coil of the typical furnace shown in Fig. 6. This coil surrounds the crucible containing the charge which acts as the susceptor. The resulting eddy currents in the charge produce the heat required for melting.

Induction melting is usually a straightforward process in which the proper amounts of high-purity raw materials are placed in the crucible, a minimum of slag is formed, and no refining is attempted. Allowance must be made for reactions with air during the meltdown, or a neutral atmosphere is provided.

Vacuum processes. All of the processes described thus far require contact of the molten steel with the air or a furnace atmosphere. It has been indicated that these processes necessarily result in a finished steel containing certain amounts of oxygen, nitrogen, and hydrogen. In turn these elements confer special properties, often undesirable, on the product.

Special practices such as raw material, slag, and temperature control have been developed to minimize these gas problems, but a more complete solution requires the use of inert atmospheres or a vacuum. If the large heat size of an open hearth is required, the most effective current method is to place the mold inside a chamber connected with high-speed pumping equipment, and to use a transfer ladle so that the steel is poured through the evacuated space before reaching the mold. This arrangement is particularly effective for the removal of hydrogen.

When smaller heats are required, more effective control is obtained by conducting the entire melting and casting procedure within the evacuated space, as is now accomplished for heats of several tons. The evacuated chamber is divided into a series of interlocking sections so that a new solid charge can be introduced without admitting air to the crucible or pouring sections. Similarly, ingots or castings can be removed through another vacuum lock. As in normal induction melting, only high-purity raw materials are charged and no ordinary refining is attempted.

In a recently developed vacuum process, the steel passes from the melting furnace to the mold along a "continuous" hearth, heated by a battery of electron guns. This process allows a more thorough exposure of the steel to the vacuum.

The vacuum induction-melted product is low in all of the gaseous elements and is the usual starting material for another type of vacuum melting called the consumable-electrode method. This is also operated in an evacuated chamber which contains a water-cooled copper mold. The mold holds a small starting block of the alloy to be melted; the block acts as one electrode. The other electrode is a long section of the same alloy. The arc generates enough heat to melt the consumable section which drips onto the mold section. Proper control of the arc and cooling rate preserves a small pool of molten metal at the top of the mold. This freezes progressively at the same rate as new melt falls in. Thus, a continuous casting is developed which is free of the shrinkage defects described for conventional ingots and is also low in all gaseous impurities.

These vacuum processes are needed for making special alloys containing large amounts of easily oxidized metals. They can also be used for the melting of more conventional alloy steel compositions where exceptionally low gas and inclusion content is required. *See* VACUUM METALLURGY.

STEEL MANUFACTURE

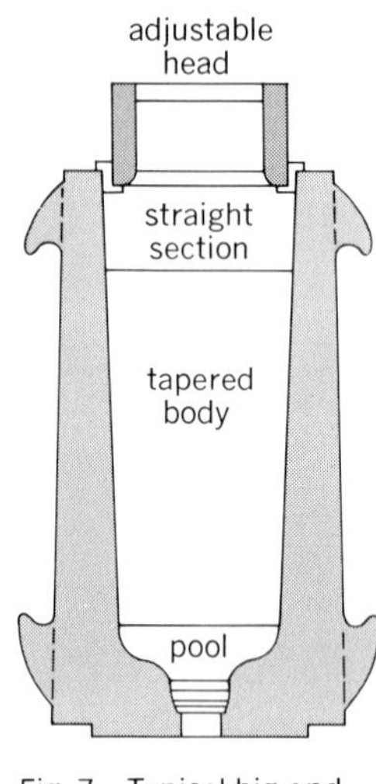

Fig. 7. Typical big-end-up mold with hot top, used in killed steel. *(From J. L. Bray, Ferrous Process Metallurgy, Wiley, 1954)*

OXYGEN CONTROL

Final control of the oxygen content is critical, and this section can be applied, with only minor modification, to the finishing of any heat of steel made by any process. Figure 3 shows that, as the steel with a given carbon composition cools, the $O_{(Fe)}$ required to achieve this carbon is greater than that in equilibrium with the same carbon at a lower temperature. Unless remedial measures are taken, this lowering of $O_{(Fe)}$ content will occur by two processes: (1) formation of an oxide and (2) evolution of CO.

Some oxygen may separate into droplets of an oxide high in Fe whose exact composition will depend on the other elements in the steel. Depending upon the conditions of solidification, these droplets may separate as a slag or be trapped during freezing, in which case they are called inclusions or deoxidation products. These inclusions are more or less harmful, according to their composition and size and the use to be made of the steel.

The evolution of CO bubbles may accompany the adjustment of the $C_{(Fe)}-O_{(Fe)}$ reaction to its temperature equilibrium. Particularly in low-C steels ($C < 0.20\%$), this evolution would generally be so violent as to cause the steel to boil out of the molds during solidification, to cause damage to facilities, and to produce an unsound, worthless product.

According to the final type of oxygen control, steels are classified as fully deoxidized or killed; as rimmed or open, which involves the minimum practical amount of deoxidation; or as semikilled, which covers a range of conditions between killed and rimmed steels.

Killed steel. For sound internal structure, killed steel is made by adding in the ladle some other element such as Al whose equilibrium with oxygen is such that the $O_{(Fe)}$ is lowered to a value that prevents the formation of CO. Other common deoxidizers, used separately or in combination, usually as ferroalloys, are Mn, Si, and titanium (Ti). The killed steel is then teemed into a big-end-up mold, as indicated in Fig. 7. The molds are made of cast iron and provided with a hot top (an insulated upper section) designed to ensure that the steel in it will be the last region to freeze.

The volume of the molten steel at the freezing point is about 8% greater than that of the solid. The loss of heat during solidification is through the mold wall and by radiation from the open top. Thus, if the ingot mold had straight sides and no hot top, a rigid outer skin would freeze early in the process, and the difference in volume between liquid and solid would appear as a shrinkage cavity or pipe along the center line of the ingot. The tapered mold with the big end up and hot top is designed to ensure progressive freezing from bottom to top so that liquid steel can continue to feed into the center shrinkage zone as long as possible. A small pipe in the hot top area is inevitable, but this can be cropped and most of the ingot used as sound metal. Thus, killed steels eliminate the problem of CO evolution during freezing and minimize the objectionable features of the resultant pipe by suitable mold design. The treatment required to accomplish these objectives creates deoxidation products that determine the final steel quality.

All of the elements used as deoxidizers influence the solubility of oxygen in the molten steel in a manner similar to that indicated for carbon in Fig. 3. To be effective, they must lower the $O_{(Fe)}$ below that of the specified carbon. If a solid deoxidation product such as Al_2O_3 is formed when the deoxidizer is added to the ladle, it may have a chance to float out of the metal before it freezes. However, the equilibrium is continually changing with temperature so that more Al_2O_3 will be forming as the steel cools until freezing is completed. These oxides form as a dispersion throughout the metal phase, and it is inevitable that those oxides formed in the later stages of the process must be trapped in the metal. It is thus impossible to make absolutely clean steel by any of the conventional processes requiring final deoxidation to prevent the CO evolution. The foregoing presentation is highly simplified; in any real case, all of the elements present must interact in such a way that the deoxidation products are usually complex associations of the oxides of Fe, Mn, Si, and Al. Similarly, these elements all interact with the S in the steel, and this element too is included in the complex. The final size, shape, and distribution of these inclusions in the steel are important factors in its performance in many applications, and a great deal is done in steelmaking to influence these factors.

Rimmed steel. For good surface characteristics when internal soundness is not critical, rimmed steel is produced. The ladle treatment is not designed to stop the formation of CO but to control it to obtain the following freezing characteristics. Molds for this purpose are described as big-end-down and are placed on a heavy bottom section called a stool, so that the mold can be stripped from the solid ingot by lifting the verticle section from the stool. No hot top is used. The properly deoxidized steel is teemed into the mold, and as freezing begins, because of the chilling action of the mold wall, the $C_{(Fe)} - O_{(Fe)}$ equilibrium is shifted enough to cause CO evolution at the solid-liquid interface. This causes a slight rise of the liquid level in the mold, as well as a stirring of the remaining liquid, which is described as the rimming action. This action sweeps any deoxidation products to the top and also prevents the early freezing-over of the top surface by maintaining a uniform temperature throughout the liquid, hence the alternate term of open steel. Thus, a skin of sound, inclusion-free metal several inches thick is formed. Finally, the top must freeze over, and if the deoxidation has been proper, the remaining gas evolution is just right to compensate for the difference in volume between the remaining liquid and solid, and the ingot freezes with a nearly flat top. The cavity corresponding to the pipe in killed steel is distributed throughout a large region of the ingot as small trapped CO bubbles. No cropping of the pipe is required, so the yield of finished product is high. Most of the small cavities are collapsed by subsequent rolling operations.

Rimmed steels are cheaper because no hot-topping facilities are required, and yields are greater. They have the best surface for many forming operations because they are inclusion-free, and they can be used in many products, such as auto bodies, where an absolutely sound internal structure is not important.

Semikilled steels. As the name implies, semikilled steels are a compromise between the features of killed and of rimmed products; the degree of compromise is subject to considerable control and extensive variation. The rimming action can be stopped mechanically by placing a heavy plate over the top of the mold, or chemically by adding Al as additional deoxidizer. Either method is subject to extensive control and variation. Semikilled steels are used for many applications, such as plate and structural shapes, where the economies gained are not inconsistent with satisfactory performance.

CONTINUOUS STEELMAKING

All of the processes described thus far may be characterized as batch operations with regard to both the chemical refining and the solidification of the steel product as ingots or castings. The chemical industry has demonstrated the advantages in utilization of equipment, energy, and manpower to be derived from continuous operations, for example, in petroleum refining. Steelmakers no longer regard this as an unreasonable objective, and the areas in which rapid progress is being made can be understood readily in terms of the foregoing.

One approach is to use finely divided raw materials; they provide a large surface for rapid reaction with a suitable gaseous atmosphere. Powdered high-grade ore can be reduced directly to iron by hydrogen or CO at low enough temperatures (about 1000°C) to avoid carburization or melting, thus eliminating both the blast furnace and steelmaking. Similarly, hot metal (the molten, high-C blast-furnace product) can be dispersed into fine droplets in a jet of oxygen and powdered lime to accomplish all of the refining of basic oxygen steelmaking. Finally, instead of casting into large ingots, the molten steel can be fed through a water-cooled, reciprocating mold and a solid rod, bar, or slab extracted by pinch rolls. This technique not only eliminates the scrap from cropping of ingots but also avoids the use of expensive processing equipment such as soaking pits and blooming mills. Variations of these processes can be coupled in many ways, and progress has been sufficient to indicate that some will become commercial in the near future. *See* PYROMETALLURGY; STAINLESS STEEL; STEEL.

[GERHARD DERGE]

As another example, one experimental process would replace the large blast furnace by a smaller, continuously operating reactor. The ore-lime mixture would enter through a rotating kiln, with controlled conditions, especially at the high-temperature end, to keep the charge from sticking to the wall. The partially reduced, continuously moving charge would then enter a high-temperature rotary reactor lined with a long-life refractory. Lances, opening into the reactor above the level of the melt, would supply oxygen (or oxygen-fuel burners would supply an oxygen-rich atmosphere); water-cooled lances opening below the surface of the melt would inject a liquid or gaseous hydrocarbon. Experiments with this one-step process have shown that the hydrocarbon cracks to provide the highly reducing conditions necessary for the final stage. The gases escaping from the bath burn partially to heat the reactor and pass out through the kiln to provide the reducing gases and preheat temperature. In its present state of development, the process requires an ore containing more than one-half iron and yields a steel almost saturated with hydrogen. However, the experiment does demonstrate, first, that thermodynamic requirements can be met in a straight-through process, and second, that much of the design can be accomplished by computer modeling with useful agreement between predictions and results.

[FRANK H. ROCKETT]

Bibliography: T. S. Harrison, *Handbook of Analytical Control of Iron and Steel Production*, 1979; Physical Chemistry of Steelmaking Committee, AIME, *Basic Open Hearth Steelmaking*, 3d ed., 1964; Physical Chemistry of Steelmaking Committee, AIME, *Electric Furnace Steelmaking*, vols. 1 and 2, 1963; C. S. Russell and W. J. Vaughan, *Steel Production: Processes, Products, and Residuals*, 1976; United States Steel Corp., *The Making, Shaping, and Treating of Steel*, 9th ed., 1969.

Stirling engine

An engine in which work is performed by the expansion of a gas at high temperature to which heat is supplied through a wall. Like the internal combustion engine, a Stirling engine provides work by means of a cycle in which a piston compresses gas at a low temperature and allows it to expand at a high temperature. In the former case the heat is provided by the internal combustion of fuel in the cylinder, but in the Stirling engine the heat (obtained from externally burning fuel) is supplied to the gas through the wall of the cylinder (Fig. 1). *See* INTERNAL COMBUSTION ENGINE.

The rapid changes desired in the gas temperature are achieved by means of a second piston in the cylinder, called a displacer, which in moving up and down transfers the gas back and forth between two spaces, one at a fixed high temperature and the other at a fixed low temperature. When the displacer in Fig. 2 is raised, the gas will flow from the hot space via the heater and cooler tubes into the cold space. When it is moved downward, the gas will return to the hot space along the same path. During the first transfer stroke the gas has to yield up a large amount of heat to the cooler; an equal quantity of heat has to be taken up from the heater during the second stroke.

The regenerator shown in Figs. 1 and 2 has been inserted between the heater tube and cooler tube in order to prevent unnecessary waste of this heat. It is a space filled with porous material to which

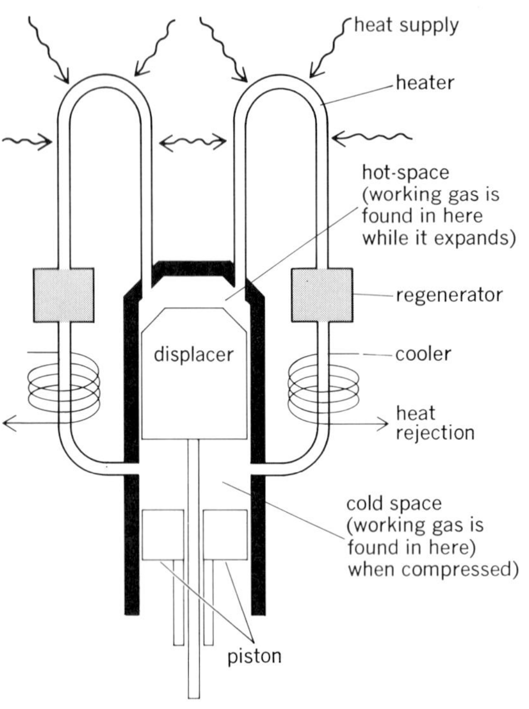

Fig. 1. Principle of Stirling engine, displacer type.

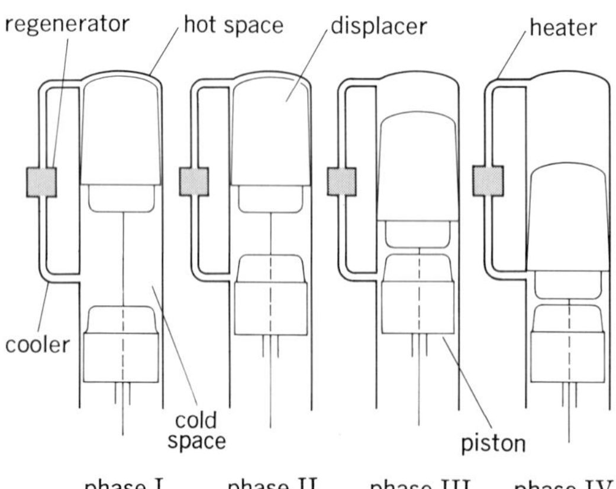

Fig. 2. Stylized Stirling process.

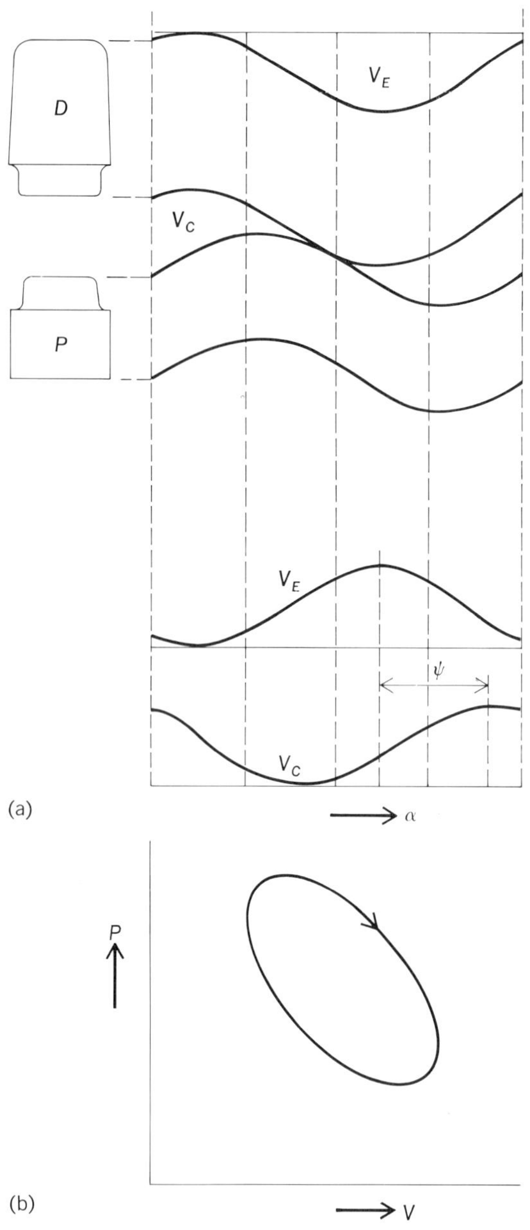

Fig. 3. Piston and displacer movement for a Stirling engine. (*a*) The continuous movement of the piston and displacer shown as a function of crank angle α. There is no clear-cut division between the various phases of the cycle. The variations in volume of the hot space V_E and the cold space V_C are plotted separately in the lower part of the diagram. (*b*) The $p-V$ diagram of the cycle.

the hot gas gives up its heat before entering the cooler; the cooled gas recovers the stored heat on its way back to the heater.

The displacer system serves to heat and cool the gas periodically; associated with it is a piston which compresses the gas while it is in the cold space of the cylinder and allows it to expand while in the hot space. Since expansion takes place at a higher temperature than compression, it produces a surplus of work over that required for the compression.

Stirling cycle. Any practical version of the engine will embody some kind of crank and connecting rod mechanism, in consequence of which there will be no sharp transitions between the successive phases indicated in Fig. 2; but this will not alter the principle of the cycle (nor detract from its efficiency). If, for the sake of simplicity, the piston and displacer are assumed to move discontinuously, the cycle can be divided into the following four phases. (1) The piston is in its lowest position, the displacer in its highest. All the gas is in the cold space. (2) The displacer is still in its highest position; the piston has compressed the gas at low temperature. (3) The piston is still in its highest position; the displacer has moved down and transferred the compressed gas from the cold to the hot space. (4) The hot gas has expanded, pushing the pistons, followed by the displacer, to their lowest positions. The displacer is about to rise and return the gas to the cold space, the piston remaining where it is, to give phase 1 again.

The actual piston and displacer movements might be as indicated in Fig. 3, which shows that the only essential condition for obtaining a surplus of work is that the maximum volume of the hot space occur before that of the cold space. This condition shows that more configurations with pistons and cylinders are possible than just the type with displacer in order to get a Stirling cycle. One of the most compact systems is shown in Fig. 4, which is the system known as the double-acting engine. In the double-acting engine there is a hot space (expansion space) at the top and a cold one (compression space) at the bottom of each of the four cylinders shown. The hot space of a cylinder is connected to the cold one through a heater, a regenerator, and a cooler. The pistons P_n of the cylinders move with a suitable phase shift between them. In the case of four cylinders, as shown in Fig. 4, this shift is 90°.

The theory of an actual engine is very complicated. In order to provide an understanding of the quantities that play a role, the formulas of the power and efficiency of an engine of highly idealized form will be given. On the assumption that the volumes of the hot (expansion) space V_E and that of the cold (compression) space V_C (Fig. 3) vary with the crank angle in a purely sinusoidal way, that both expansion and compression are isothermal at respectively T_E and T_C, and that all kinds of losses caused by flow resistances in the tubes and regenerator, heat losses in the regenerator, and so on are ignored, the power P can be expressed as in Eq. (1). Here $\tau = T_C/T_E$, n is the speed in rpm, p_m is

$$P = (1-\tau)\frac{\pi n}{60} V_{E_{max}} \cdot p_m \frac{\delta}{1+\sqrt{1-\delta^2}} \sin\Theta \qquad (1)$$

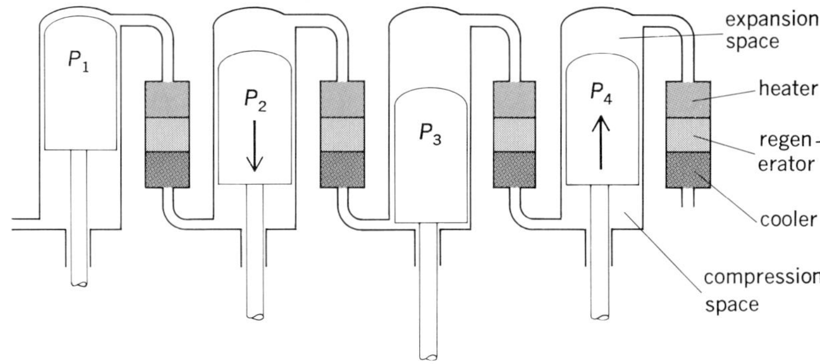

Fig. 4. Principle of the double-acting engine.

the mean pressure of the pressure variation, and δ and Θ are functions of the temperature ratio, swept volume ratio, dead spaces, and phase angle between the two swept volumes. Under these conditions the thermal efficiency η is, of course, that of the Carnot cycle, and is expressed in Eq. (2). *See* CARNOT CYCLE; THERMODYNAMIC CYCLE.

$$\eta = \frac{T_E - T_C}{T_E} = 1 - \tau \qquad (2)$$

Engine design. Actual engines have been built in the Philips Laboratories at Eindhoven, Netherlands, as prototypes in the range of 10–500 hp (7.5–375 kW) per cylinder. After years of research on the Stirling engine, actual thermal effi-

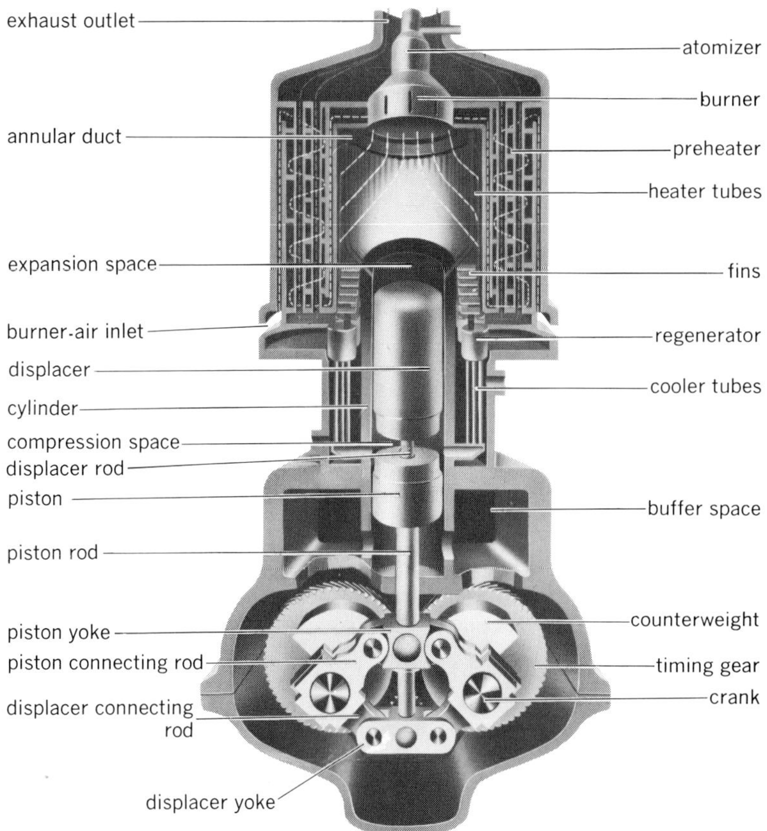

Fig. 5. Philips Stirling engine in cross section. *(Philips Gloeilampenfabrieken)*

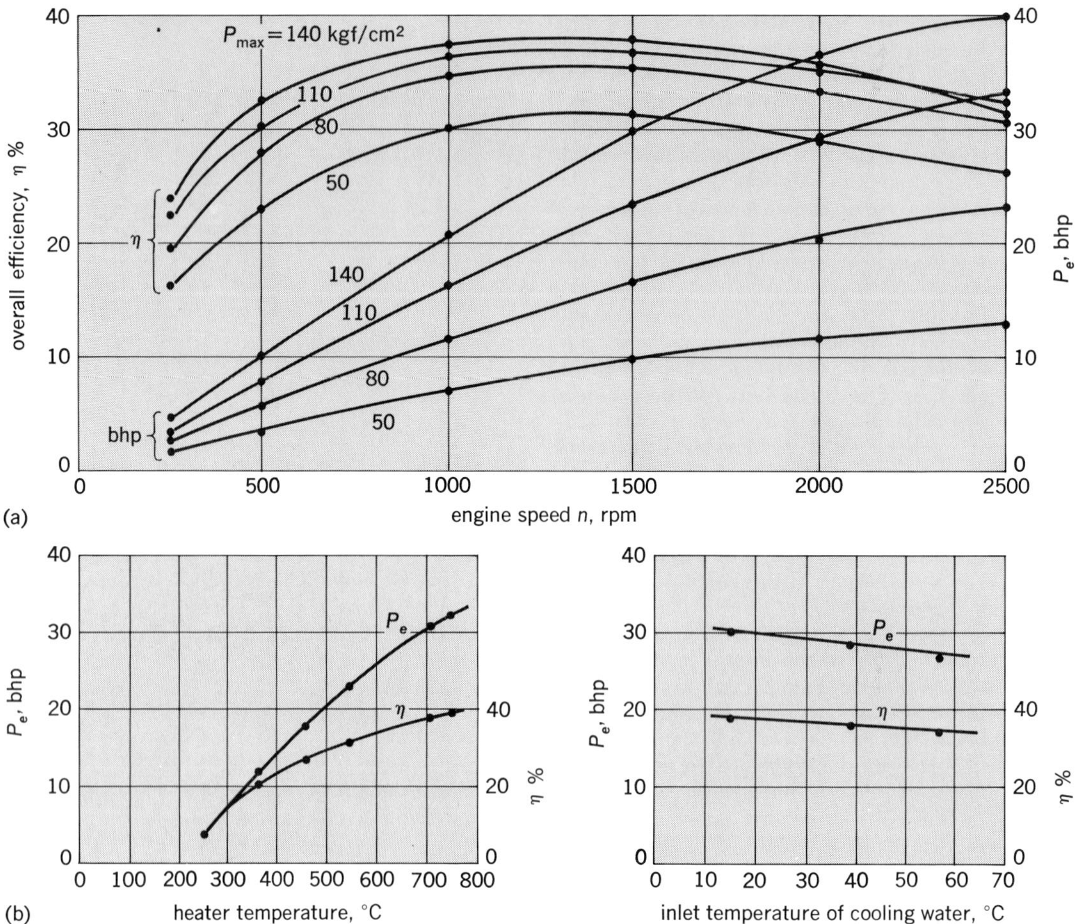

Fig. 6. Test measurements for the Stirling engine. (*a*) Measured shaft power and efficiency of a 40-hp (30-kW) single-cylinder test engine with rhombic drive, plotted as a function of the engine speed *n*, at different values of the maximum pressure of the working fluid in the engine (p_{max}). (*b*) Power and efficiency of the 40-hp (30-kW) test engine, given as a function of the heater temperature and as a function of the inlet temperature of the cooling water. The curves apply to n = 1500 rpm and p_{max} = 140 kgf/cm² = 13.7 MPa. 1 hp = 0.75 kW.

ciencies of 30–45% (depending on the specific output and temperature ratio) and a specific power of 115 hp per liter swept volume of the piston were obtained with the displacer type of engine equipped with rhombic drive, as shown in Fig. 5. In the figure the piston and displacer drive concentric rods, which are coupled to the rhombic drive turning twin timing gears. The cooler, regenerator, and heater are arranged as annular systems around the cylindrical working space. The heater tubes surround the combustion chamber. In the preheater the gas at 800°C from the heater is cooled to 150–200°C while heating the combustion air to about 650°C.

The rhombic drive mechanism allows complete balancing even of a single-cylinder engine and of a separate buffer space, thus avoiding heavy forces acting on the drive. The results of measurements on the first engine of this type, with hydrogen as the working fluid, are shown in Fig. 6. An approximate heat balance is output, 40%; exhaust and radiation, 10%; and heat rejection by cooling water, 50%. Control of engine output is by regulation of the pressure of the working fluid in the engine, while the temperature of the heater is being kept constant by a thermostat; hence the efficiency shows little dependence on the load.

The closed system of the Stirling engine endows this engine with many advantages and also some shortcomings. The continuous external heating of the closed system makes it possible to burn various kinds of liquid fuels and gases, without any modification whatsoever. This multifuel facility can be demonstrated with a 10-hp or 7.5-kW (at 3000 rpm) generator set (Fig. 7). The engine can operate on alcohol, various lead-containing gasolines, diesel fuel, lubricating oil, olive oil, salad oil, crude oil, propane, butane, and natural gas. Furthermore, it allows combustion to take place in such a way that air pollution is some orders of magnitude less than that due to internal combustion engines. Through the intermediary of a suitable heat transport system (for example, heat pipes) any heat source at a sufficiently high temperature can be used for this engine—radioisotopes, a nuclear reactor, heat storage, solar heat, or even the burning of coal or wood.

The almost sinusoidal cylinder pressure variation and continuous heating make the Stirling engine very quiet in operation. An engine having four or more cylinders gives a virtually constant torque per revolution, as well as a constant dynamometer torque over a wide speed range, which is particularly valuable for traction purposes. The present

Fig. 7. Philips Stirling engine with generator to demonstrate its multifuel capacity. (*Philips Gloeilampenfabrieken*)

Fig. 8. Three-kilowatt Stirling engine generator set. (*General Motors Corp.*)

configuration makes complete balancing possible, thus eliminating vibrations. There is no oil consumption and virtually no contamination because a new type of seal for the reciprocating rods shuts off the cycle hermetically from the drive mechanism. Figure 8 shows an engine of this configuration.

Where direct or indirect air cooling is required, the closed cycle has the drawback that more heat has to be removed from the cooler than in comparable engines with open systems, where a greater quantity of heat inevitably escapes through the exhaust.

If it is envisaged as someday taking the place of existing engines, the Stirling might be ideal as a propulsion engine in yachts and passenger ships, and in road vehicles, such as city buses, where a large radiator is acceptable. The system of continuous external heating is also able to open fields of application inaccessible to internal combustion engines. [ROELOF J. MEIJER]

Bibliography: M. J. Collie, *Stirling Engine Design and Feasibility for Automotive Use*, 1979; G. Walker, The Stirling engine, *Sci. Amer.*, 229(2): 80–87, Aug. 1973; G. Walker, *Stirling Engines*, 1980.

Stoker

A mechanical means for feeding coal into, and for burning coal in, a furnace. There are three basic types of stokers (see illustration). Chain or traveling-grate stokers have a moving grate on which the coal burns; they carry the coal from a hopper into the furnace and move the ash out. Spreader stokers mechanically or pneumatically distribute the coal from a hopper at the furnace front wall and move it onto the grate which usually moves continuously to dispose of the ash after the coal is burned. Underfeed stokers are arranged to force fresh coal from the hopper to the bottom of the burning coal bed, usually by means of a screw conveyor. The ash is forced off the edges of the retort peripherally to the ashpit or is removed by hand. *See* STEAM-GENERATING FURNACE.

[GEORGE W. KESSLER]

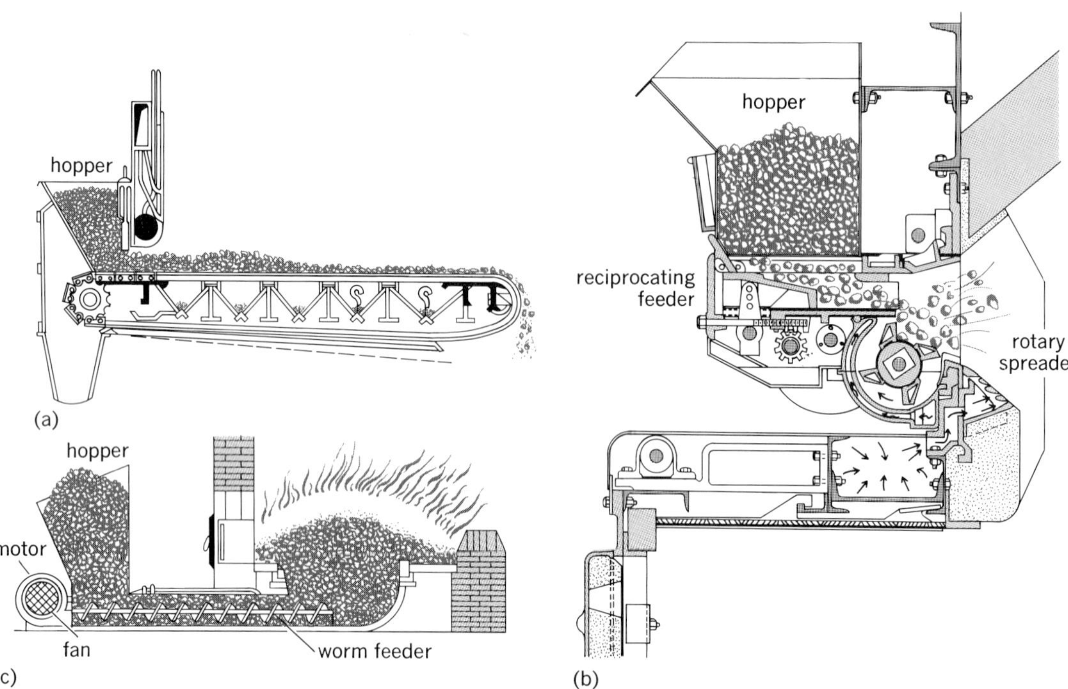

Principal types of stokers. (*a*) Chain grate. (*b*) Spreader. (*c*) Underfeed.

Storage tank

A container for storing liquids or gases. A tank may be constructed of ferrous or nonferrous metals or alloys, reinforced concrete, wood, or filament-wound plastics, depending upon its use. Tanks resting on the ground have flat bottoms; those supported on towers have either flat or curved bottoms. Standpipes, which are usually cylindrical shells of steel or reinforced concrete resting on the ground, are frequently of great height and comparatively small diameter. They are built to contain water for a distribution system, and height is required to maintain pressure in the system. Tanks for other liquids and for gases, where storage is more important than pressure, are generally lower and of greater diameter.

Steel is usually the preferred material for standpipes because of the difficulty of securing watertightness in concrete shells with relatively high head. Unless painted regularly, however, steel standpipes are subject to corrosion. Well-constructed concrete tanks need no surface treatment and are easily maintained. *See* STEEL.

Distribution-system pressure requirements limit the allowable fluctuation in water level in a standpipe or elevated tank to 25–30 ft (7.5–9 m). Therefore, unless a standpipe is located on high ground, only that volume of storage above the elevation required to give the necessary pressure is available for use. Elevated tanks become economical when the cost of the tower is less than the cost of the supporting portion of the standpipe below its useful head.

Elevated tanks for water storage with capacities up to 2,000,000 gal (7500 m^3) are built of either steel or aluminum. The higher initial cost of an all-aluminum tank may be offset by the need for less maintenance.

Elevated tanks with diameters less than about 50 ft (15 m) usually have hemispherical bottoms. Bottoms of ellipsoidal or radial-cone shape are used on tanks of larger diameter. The roof may be conical or domed.

Prestressed-concrete tanks maintain the concrete in compression even when the tank is filled with water. This construction process is used extensively in the United States. Wire is wrapped around a cylindrical concrete tank wall by a wire-winding machine suspended from a track placed on top of the tank wall. The wires are stressed to approximately 140,000 psi (965 MPa) and are coated with a thick layer of pneumatically applied mortar for the final finish. Prestressed-concrete tanks are watertight and as a result require very little maintenance.

Molded plastic tanks in cylindrical and rectangular shapes and plastic liners for metal tanks are widely used in the chemical industry. The aircraft and space industries have developed high-pressure plastic tanks that are wound with a thin filament of fiber glass, boron, or other high-strength material by specially designed machines. For open-ended tanks, the filaments are wound around the circumference; for closed pressure tanks, or for those supporting longitudinal forces, the filament winding is carried out at an angle to the longitudinal axis dependent on the ratio of longitudinal to circumferential force. Tanks of this type have a high strength-to-weight ratio, but are quite expensive. *See* PRESTRESSED CONCRETE; STRUCTURAL MATERIALS.

[CHARLES M. ANTONI]

Bibliography: Prestressing builds up liquified gas storage, *Eng. News Rec.*, 180(19):40, 1968.

Straight-line mechanism

A mechanism that produces a straight-line (or nearly so) output motion from an input element that rotates, oscillates, or moves in a straight line. Common machine elements, such as linkages, gears, and cams, are often used in ingenious ways to produce the required controlled motion. The more elegant designs use the properties of special points on one of the links of a four-bar linkage.

Linkages. Four-bar linkages that generate approximate straight lines are not new. In 1784 James Watt applied the concept to the vertical-cylinder beam engine. Prior to the use of this linkage, the piston of the steam engine was guided with a chain attached to a centrally pivoted circular sector of a "walking beam." Other similar applications used the properties of slider cranks and inverted slider cranks to generate the required approximate straight-line output motion. Examples of some of the classical mechanisms are shown in Fig. 1. *See* SLIDER-CRANK MECHANISM.

By selecting the appropriate link lengths, the

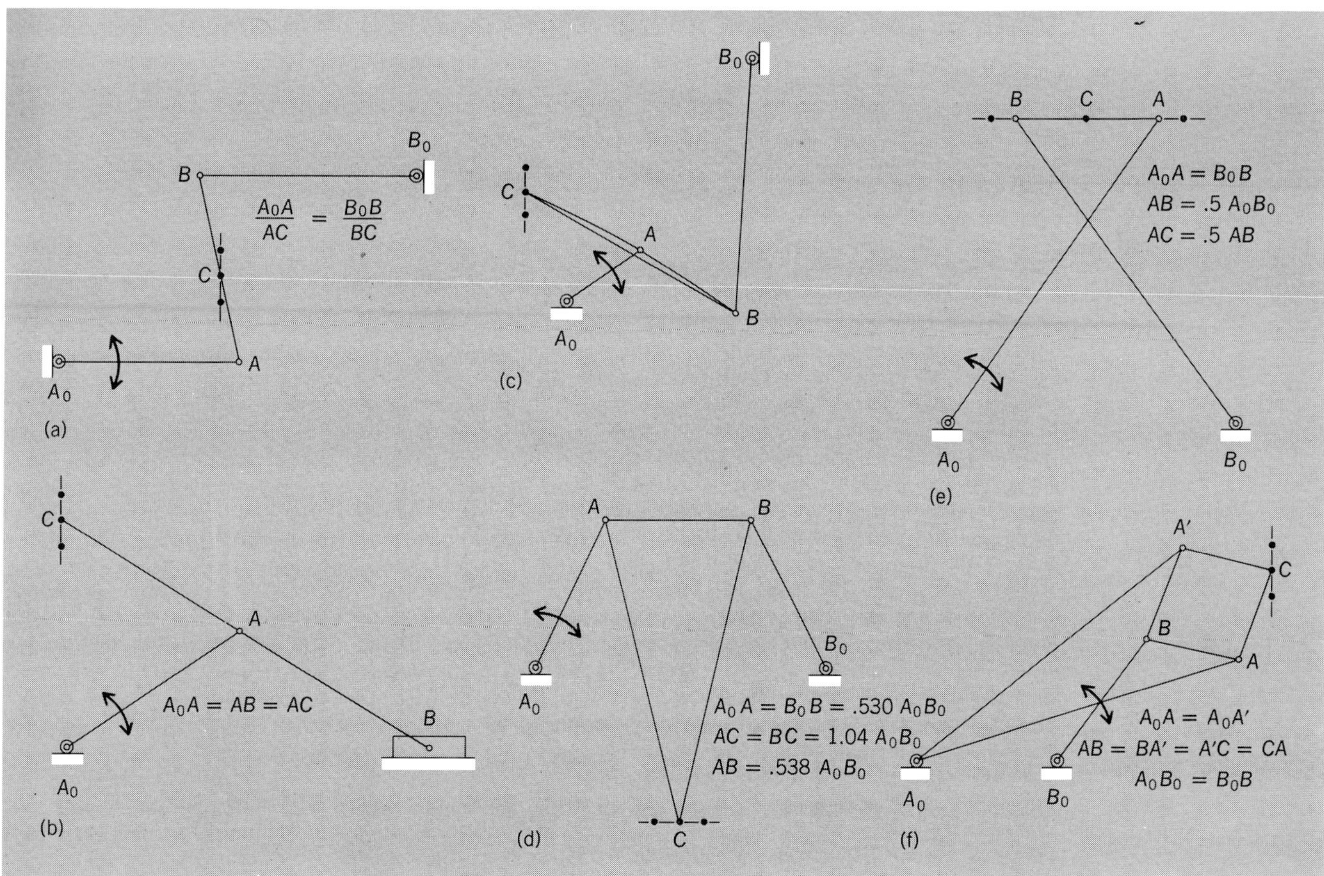

Fig. 1. Classical four-bar linkage straight-line mechanisms: (*a*) Watt, (*b*) Russell, (*c*) Evans, (*d*) Roberts, (*e*) Chebyshev, (*f*) Peaucellier.

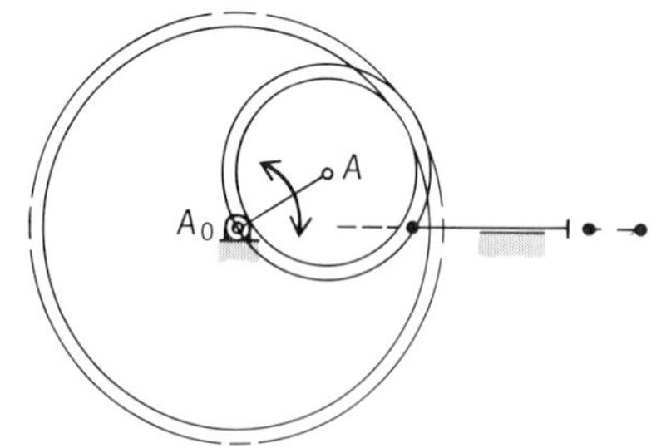

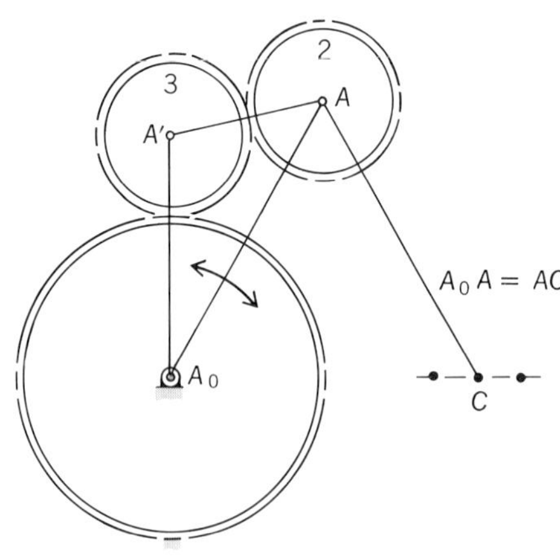

Fig. 2. Geared straight-line mechanisms. (*a*) Cardan circle pair; small-gear diameter = 0.5 large-gear diameter. (*b*) Standard gear; gear diameter 2 = 0.5 gear diameter 1; gear diameters 3 and 4 are arbitrary.

designer can easily develop a mechanism with a high-quality approximate straight line. Contemporary kinematicians have contributed to more comprehensive studies of the properties of the mechanisms that generate approximate straight lines. The work not only describes the various classical mechanisms, but also provides design information on the quality (the amount of deviation from a straight line) and the length of the straight-line output. Subsequently design data were provided on those mechanisms that have a controlled velocity ratio (output velocity/input velocity) along the generated straight-line path. *See* FOUR-BAR LINKAGE.

Gears. Gears can also be used to generate straight-line motions. The most common combination would be a rack-and-pinion gear; other combinations are the Cardan circle pairs (Fig. 2*a*). The unique property of this mechanism is that a point on the pitch circle of the smaller gear moves in a straight line with simple harmonic motion as the smaller gears rotates within the larger one. A number of Cardan pairs can also be combined to produce paths with special output velocity and acceleration specifications. Also, standard gears can be combined to achieve the same result (Fig. 2*b*). Although the mechanism is more complex, its main advantages are that the internal gear is eliminated and the stroke is considerably increased. *See* GEAR.

Cams. Cam mechanisms are generally not classified as straight-line motion generators, but translating followers easily fall into the classical definition. Cam mechanisms have kinematic instantaneously equivalent linkage representations (Fig. 3).

STRAIGHT-LINE MECHANISM

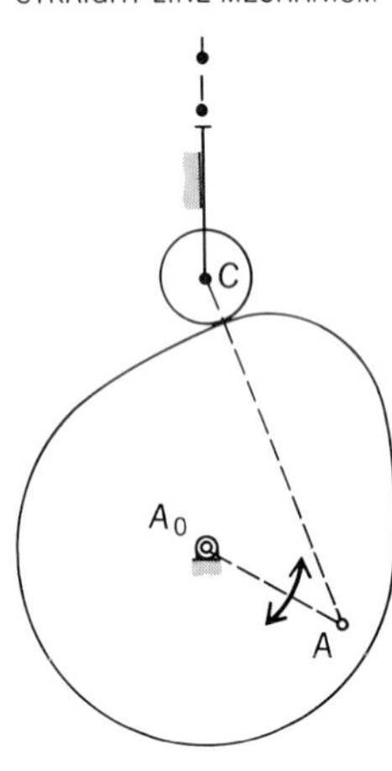

Fig. 3. Translating follower cam mechanism with an instantaneously equivalent slider-crank representation.

However, there is no single equivalent mechanism that can represent the cam motion for the full operational mode. Cam mechanisms have an intrinsic property not possessed by the linkage or gear combinations. As higher-order properties (velocity and acceleration) are assigned to the straight-line portion of the generated path, fewer four-bar linkages are available to a designer. Cam mechanisms are not as restricted by the limitations, and the designer can select the velocity or acceleration properties for the particular application. *See* CAM MECHANISM.

Other elements. Other combinations of gears, belts, chains, and so forth have been combined into useful approximate straight-line motion generators. One example is shown in Fig. 4, where the two sprockets and chain/belt combination move the output link. The end of the link traces the approximate straight-line path; such a mechanism is successfully used in a pen recorder instrument. *See* BELT DRIVE; CHAIN DRIVE.

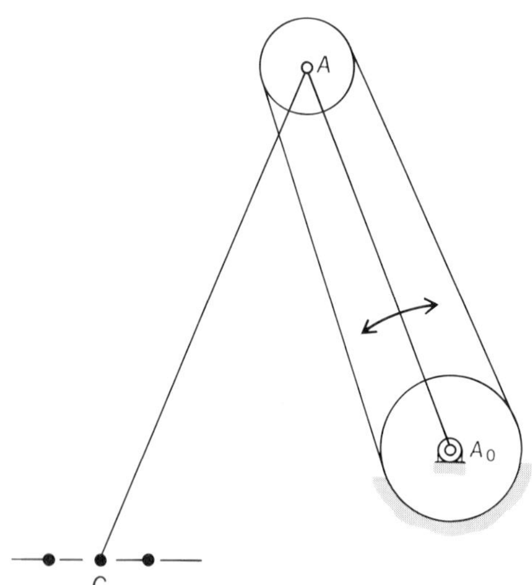

Fig. 4. Chain/belt–sprocket combination.

Numerous mechanisms that can generate approximate straight-line output paths have found useful applications in industry. Kinematic properties can be easily satisfied, but the dynamic response might dictate the usefulness of the mechanism. Clearances in pivot joints, high dynamic loads on the cam surface, and space allocations for the configuration may restrict the designer to only one or two possible choices. *See* MECHANISM.

[JOHN A. SMITH]

Bibliography: F. Freudenstein, Kinematic synthesis of a rotary-to-linear recording mechanism, *J. Franklin Inst.*, vol. 279, no. 5, 1965; A. Myklebust and D. Tesar, *Application of Design Equations for Synthesis of Coupler Mechanisms for Combinations of Higher Order Time and Geometric Derivative Specifications*, ASME Pap. 74-DET-77, October 1974; J. A. Smith, A harmonic series mechanism to approximate a velocity or acceleration specification, *Oklahoma State University Applied Mechanisms Conference*, 1969; J. P. Vidosic, D. Tesar, and H. L. Johnson, *Theoretical Analysis of Four-Bar Mechanisms*, G. Inst. Technol. Final Rep., N.S.F. Grant GP-2748, 1966.

Strain

Deformation or change in shape of a material as a consequence of applied forces. Strain is directly measurable, and from such measurement, within the elastic limit, the internal stress that accompanied the strain can be determined. The strain-producing action sets up stresses in the material, and these stresses cause a deformation or strain; that is, an initial strain is always accompanied by a stress. For this reason, the two phenomena are usually dealt with together. *See* ELASTIC LIMIT; HOOKE'S LAW; STRAIN GAGE; STRESS AND STRAIN; YOUNG'S MODULUS.

[FRANK H. ROCKETT]

Strain gage

A device which measures mechanical deformation. Normally it is attached to a structural element, and uses the change of electrical resistance of a wire or semiconductor under tension. Capacity, inductance, and reluctance are also used.

The strain gage converts a small mechanical motion to an electrical signal by virtue of the fact that when a metal (wire or foil) or semiconductor is stretched, its resistance is increased. The change in resistance is a measure of the mechanical motion. In addition to their use in strain measurement, these gages are used in sensors for measuring the load on a mechanical member, forces due to acceleration on a mass, or stress on a diaphragm or bellows.

Common gages are made of wire, foil, or semiconductor material. Wire and foil gages are constructed by cementing high-resistance metal to a backing of paper, epoxy, or Bakelite, which is then cemented to the structural element. To obtain higher resistance, the conductor is often folded in a zigzag pattern, as shown in the illustration. The terms bonded and unbonded refer to whether the wire is left free-floating or is entirely cemented to the backing.

Unfortunately, most wires and foils which have desirable characteristics as strain-gage material are also sensitive to temperature; that is, temperature changes alter the wire resistance. For tests of short duration this is not necessarily significant, but for long-duration tests this effect must be minimized. It is common therefore to include in the electrical circuit a resistor which is subjected to any changes in temperature but not to strain changes. Often this compensator is placed close to the active area.

Some semiconductor strain gages are made of a piezoresistive material, usually silicon. The sensitive element is a single crystal of silicon bonded to an epoxy base. Leads are attached to deposited gold at each end of the crystal. Semiconductor strain gages are also made by diffusing a conducting pattern on the surface of the semiconductor crystal and attaching electrical leads to the pattern. To maximize the output (especially when used as part of a sensor), the gage often consists of multiple resistors, half of which are in tension and half under compression when attached to an element exhibiting strain. The temperature-compensating element is located next to the diffused pattern whenever possible.

The principle advantage of a semiconductor gage over wire and foil gages is its greater sensitivity to strain. However, the characteristics of wire and foil gages are more predictable because of the high quality of the metals used.

A strain gage is often used in a bridge circuit, such as a Wheatstone bridge. When pressure is applied, the resistance of the strain gage is altered and the output voltage of the bridge changes. The output voltage is usually measured by a voltmeter, and may drive a recording instrument for a continuous record.

[JOHN H. ZIFCAK]

STRAIN GAGE

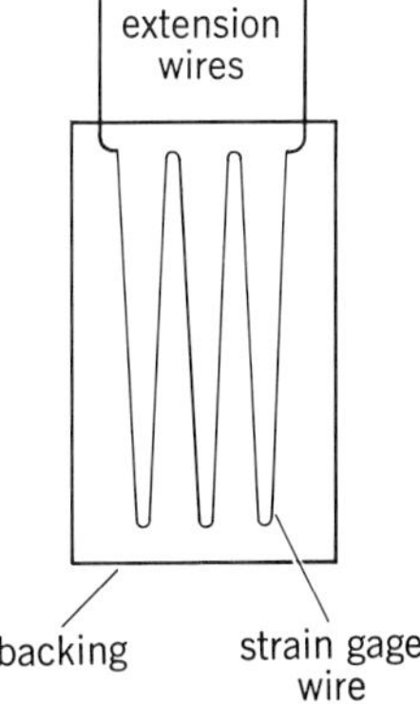

Bonded strain gage.

Strain rosette

A pattern of intersecting gage lines on a surface along which linear strains are measured to determine stresses at a point. A rosette gage is an assembly of strain-measuring components arranged to measure strains in the directions of the respective gage lines. To facilitate computations, three lines are usually oriented to form a rectangular or 45° rosette, or to form an equiangular rosette, also called the delta or 60° rosette (see illustration).

Electrical resistance-type gages are usually employed. With suitable instrumentation, strains are recorded in microinches per inch. Strains in the selected directions can also be measured by mechanical gages. The three measured strains permit evaluation of strains in any direction. The maximum and minimum strains, called principal strains, can be found analytically or by Mohr's circle of strain, and the corresponding stresses by the generalized Hooke's law. This procedure for finding the maximum stresses and their directions from linear strain measurements is called the rosette method. *See* STRAIN GAGE; STRESS AND STRAIN.

[JOHN B. SCALZI]

Bibliography: M. Dean and R. D. Douglas, *Semiconductor and Conventional Strain Gages*, 1962; M. I. Hetenyi (ed.), *Handbook of Experimental Stress Analysis*, 1950; C. C. Perry and H. R. Lissner, *Strain Gage Primer*, 2d ed., 1962.

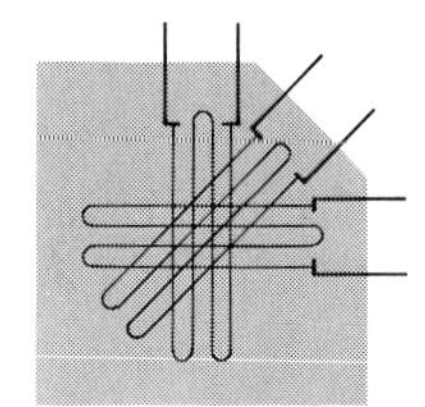

(a)

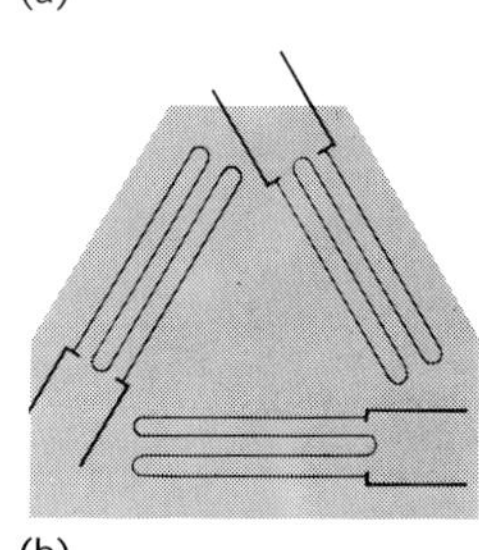

(b)

Strain rosettes. (*a*) 45° type; (*b*) 60° type.

Strength of materials

A branch of applied mechanics concerned with the behavior of materials under load, relationships between externally applied loads and internal resisting forces, and associated deformations. Knowledge of the properties of materials and analysis of the forces involved are fundamental to the investigation and design of structures and machine elements. Mathematical application of principles of mechanics is supplemented by experimentally determined properties of materials and other empirical constants. *See* MACHINE DESIGN; STRUCTURAL MATERIALS.

Investigation of the resistance of a member, dealing with internal forces, is called free-body analysis. In it, principles of statics are applied to selected isolated segments of the loaded member. Determination of the distribution and intensity of the internal forces and the associated deformations is called stress analysis. *See* STATICS; STRESS AND STRAIN.

Internal reactive forces developed in response to straining actions depend on the magnitude and nature of the loads. The possible straining actions are (1) tension or compression, which lengthens or shortens the member; (2) shearing, which produces sliding or angular distortion along the plane of applied forces; (3) bending, in which couples or bending moments produce change in curvature; and (4) torsion, in which couples acting normal to the axis twist the member. *See* BENDING MOMENT; SHEAR; TORSION.

A material offers resistance to external load only insofar as the component elements can furnish cohesive strength, resistance to compaction, and resistance to sliding. The relations developed in strength of materials analysis evaluate the tensile, compressive, and shear stresses that a material is called upon to resist. The most important factors in determining the suitability of a structural or machine element for a particular application are strength and stiffness.

Applications of fundamentals can be broadly classified as (1) investigation of members with known dimensions and materials to determine their ability to resist prescribed loads without excessive deformation, instability, or fracture, and (2) selection of suitable materials and determination of shape and dimensions of a member to perform a prescribed function involving known or estimated external loads. Design is the prediction of suitability for a prescribed function.

[JOHN B. SCALZI]

Stress and strain

Related terms used to define the intensity of internal reactive forces in a deformed body and associated unit changes of dimension, shape, or volume caused by externally applied forces.

Stress is a measure of the internal reaction between elementary particles of a material in resisting separation, compaction, or sliding that tend to be induced by external forces. Total internal resisting forces are resultants of continuously distributed normal and parallel forces that are of varying magnitude and direction and are acting on elementary areas throughout the material. These forces may be distributed uniformly or nonuniformly.

Stresses are identified as tensile, compressive, or shearing, according to the straining action.

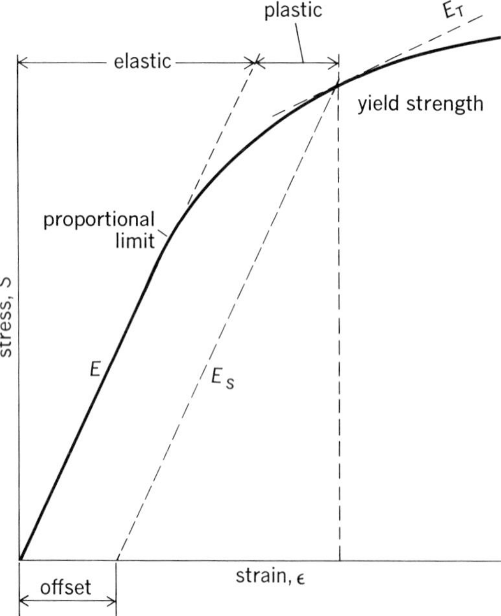

Fig. 1. Stress-strain diagram for an aluminum alloy.

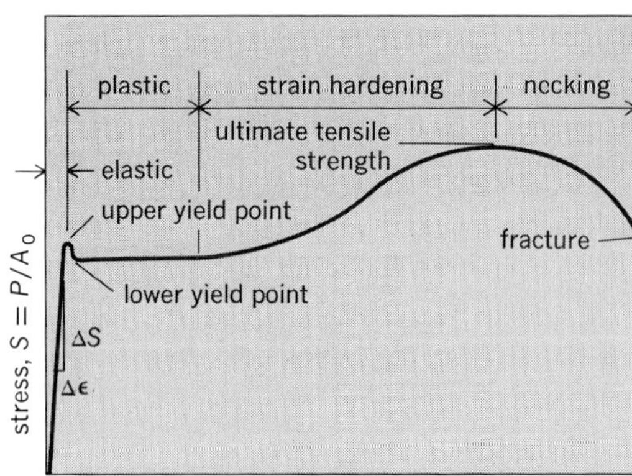

Fig. 2. Stress-strain diagram for a low-carbon steel.

Strain is a measure of deformation such as (1) linear strain, the change of length per unit of linear dimensions; (2) shear strain, the angular rotation in radians of an element undergoing change of shape by shearing forces; or (3) volumetric strain, the change of volume per unit of volume. The strains associated with stress are characteristic of the material.

Strains completely recoverable on removal of stress are called elastic strains. Above a critical stress, both elastic and plastic strains exist, and that part remaining after unloading represents plastic deformation called inelastic strain. Inelastic strain reflects internal changes in the crystalline structure of the metal. Increase of resistance to continued plastic deformation due to more favorable rearrangement of the atomic structure is called strain hardening.

Stress-strain diagram. A graphical representation of simultaneous values of a stress and strain observed in tests indicates material properties associated with both elastic and inelastic behavior. The diagram indicates significant values of stress-accompanying changes produced in the internal structure.

Properties of metals are usually determined by tension or torsion tests. Materials such as wood, concrete, and ceramics that are used to resist compressive loads are tested under compression.

Tension tests are made on suitably machined bars in a testing machine which elongates the specimen and records the resisting force. The stress is the force per unit of original unstrained area of the cross section. Strain is the average unit elongation in the direction of the applied force, determined from total elongation of a selected gage length.

A tension stress-strain diagram for an aluminum alloy is shown in Fig. 1. The diagram for low-carbon steel, which exhibits the unique phenomenon of yielding at a critical stress, is shown in Fig. 2. Diagrams with stresses based on the original sectional area are called engineering or nominal stress-strain curves. A true stress-strain curve is based on the reduced sectional area accompanying an applied force. The interpretation of these diagrams establishes characteristic properties of the material.

Deformation is any alteration of shape or dimensions of a body caused by stresses, thermal expansion or contraction, chemical or metallurgical transformations, or shrinkage and expansions due to moisture change. Deformation is measured by changes in linear dimensions, angular rotation, or volume. In axial tension, it is expressed by percent elongation or percent reduction of area to fracture and is taken as a measure of ductility.

Stress-strain characteristics. Several terms are used to describe the strain behavior of materials in the presence of stress. Figures 1 and 2 describe graphically some of these characteristics.

Yield strength is the stress accompanying a specified permanent plastic strain, which is considered as not having impaired useful elastic behavior and which represents the practical elastic strength for materials having a gradual knee in the stress-strain curve. The offset or total extension methods utilize the stress-strain curve to evaluate the stress at which a specified plastic strain (usually 0.2%) or a specified total strain under load (as 0.5%) has developed. *See* SAFETY FACTOR.

Proportional limit is the greatest stress a material can sustain without departure from linear proportionality of stress and strain. It is indicated on the stress-strain diagram where the curve ceases to follow the slope of the initial straight segment.

Yield point is the stress at which abrupt increase of strain occurs without increase of stress. Only materials such as low-carbon steel exhibit the unique phenomenon of sudden yielding. The stress at which yield initiates is called the upper yield point. The lower yield point is the constant stress while yielding progresses, and is taken as a characteristic property of the material.

Ultimate strength defines the maximum resistance to tensile, compressive, or shearing forces, expressed either as a total load-producing fracture as in the case of a rope or cable, the maximum stress developed prior to fracture, or the stress accompanying some limiting deformation. Ultimate strength in engineering application refers to stress at maximum load resisted. For brittle materials it is the breaking stress. *See* BRITTLENESS.

Ultimate tensile strength is the maximum nominal tensile stress developed during increasing load application, calculated from maximum applied load and original unstrained sectional area. Those materials developing large elongation reach maximum load resistance prior to fracture, which occurs at a locally reduced section. *See* STRESS CONCENTRATION.

Ultimate compressive strength has meaning only when maximum load produces fracture as in brittle materials. Metals that develop large deformation increase in sectional area, thus increasing resistance to load; hence they do not fracture.

Modulus of rupture is a measure of strength in bending or torsion, expressed as a maximum tensile, compressive, or shear stress computed from the maximum load to fracture and the dimensions of the member. Its evaluation incorrectly assumes elastic behavior to fracture and therefore is not the true strength of the material. It serves as an empirical measure of rupture strength, useful in comparing quality of materials such as wood, concrete, and cast iron subjected to standard tests.

Combined stresses. Simultaneous action of stresses produced by independent straining actions such as tension and torsion or bending and thrust produce combined stresses. The state of stress is defined by the magnitude and direction of normal and shearing stresses acting on an element of a structural member, the element having known orientation with the axes of the member. Stresses are three-dimensional when normal and shearing stresses act simultaneously on each face of a cubical element, and are two-dimensional or coplanar when stresses act in mutually perpendicular directions in the same plane. Stresses at a boundary free from surface stress are coplanar.

STRESS AND STRAIN

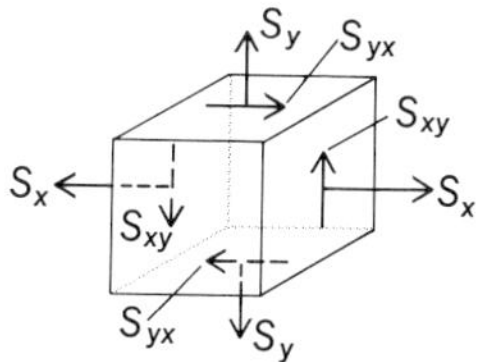

Fig. 3. Combined stresses.

The coplanar stress system shown in Fig. 3 has normal stresses S_x and S_y, and complementary shear stresses of equal intensity forming equilibrating couples. Stresses on planes of reference are found by conventional formulas for independent straining actions. Stresses on other planes are found by statics applied to a free body isolated from the element and subjected to combined stresses. Planes on which maximum and minimum normal stresses occur are called principal planes, and these wholly normal stresses are principal stresses. No shearing stresses exist on principal planes. Maximum shearing stress occurs on planes at 45° to the principal planes. An element subjected only to principal stresses is in a state of biaxial stress; when one of these normal stresses is zero, the condition is uniaxial stress.

Expressed in terms of known normal stresses S_x and S_y and a shear stress S_{xy}, the principal stresses are as shown in Eq. (1). Planes on which principal

$$S_{\substack{\max\\ \min}} = \tfrac{1}{2}(S_x + S_y) \pm \sqrt{[\tfrac{1}{2}(S_x + S_y)]^2 + S_{xy}^{\,2}} \quad (1)$$

stresses act are such that the relationship is as shown in Eq. (2), where θ is the angle with the reference axis of the element.

$$\tan 2\theta = \frac{S_{xy}}{\tfrac{1}{2}(S_x - S_y)} \quad (2)$$

Mohr's circle. A graphical construction called Mohr's circle determines the simultaneous combinations of coplanar normal and shearing stresses on any plane perpendicular to the plane of stress through a given point in a stressed body. The construction solves combined stress problems. For the state of stress shown in Fig. 4*a*, stresses on other planes are found by the construction in Fig. 4*b*. Referring normal and shear stresses to $+S_n$, S_s coordinate axes, OA $= S_x$ and OB $= S_y$. Verticals AD and BE represent the equal shear stresses on faces having normal stresses S_x and S_y. Line ED is the diameter of Mohr's stress circle with center at C. Coordinates of a point on the stress circle represent corresponding normal and shearing stresses on a particular plane. Distances OG and OF are principal stresses. The angle which the maximum principal stress makes with the X reference axis is one-half angle DCF. Ordinate CH is the maximum shear stress acting on a plane defined by angle DCH $= 2\theta_S$. Stress components on any plane whose normal makes an angle θ with the X axis can be found from the stress circle.

Bulk modulus is a constant designated K or E_v; it is a ratio of stress to accompanying volumetric strain, when the material behaves elastically. The volumetric strain ϵ_v is the change of volume per unit volume, $\epsilon_v = \Delta V/V$. According to Hooke's law, $K = E_v = S/\epsilon_v$ under stress S. For mutually perpendicular triaxial stresses, the volumetric strain is equal to the sum of the three linear strains and for

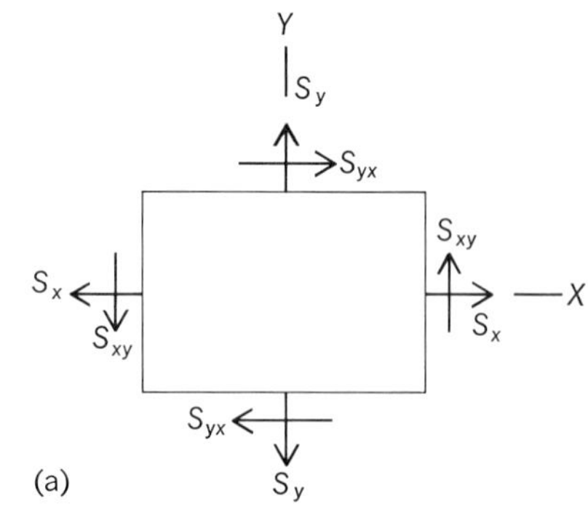

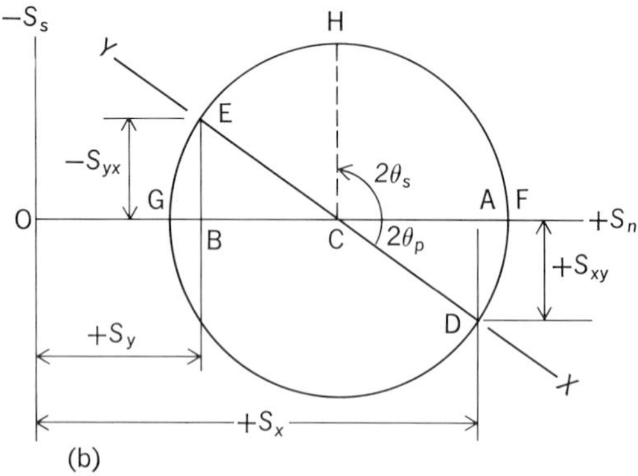

Fig. 4. Mohr's construction. (*a*) Combined stresses. (*b*) Graphical construction.

STRESS AND STRAIN

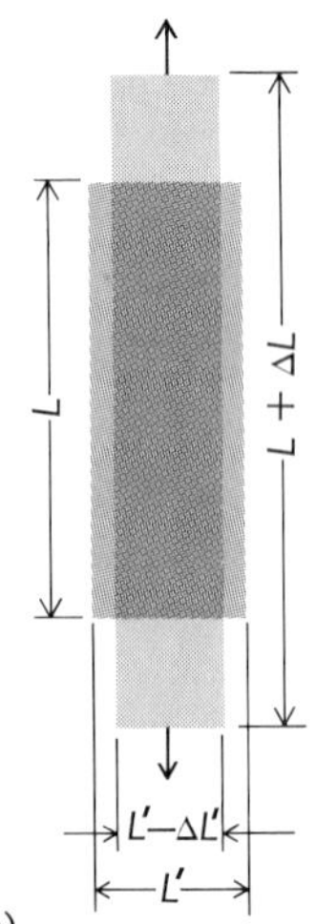

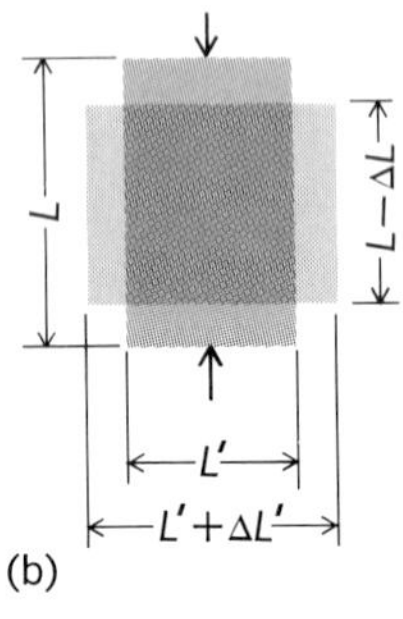

ϵ longitudinal $= \Delta L/L$
ϵ lateral $= \Delta L'/L'$

Fig. 5. Diagram of (*a*) longitudinal and (*b*) lateral strains.

hydrostatic stress $\epsilon_v = 3\epsilon$. For a body subjected to hydrostatic pressure, the relationships are shown by Eq. (3), where $E_v = E/3(1-2\mu)$, in which E is Young's modulus and μ is Poisson's ratio.

$$\epsilon_v = \frac{S}{E_v} + 3\left[\frac{S}{E}(1-2\mu)\right] \qquad (3)$$

Poisson's ratio μ is a material constant expressing the ratio of lateral strain to longitudinal strain (Fig. 5). For uniaxial stress the ratio is expressed by Eq. (4). The ratio varies for different materials,

$$\mu = \frac{\text{lateral strain}}{\text{longitudinal strain}} \qquad (4)$$

usually ranging from 0.25 to 0.35. A maximum value of 0.5 represents plastic behavior.

The generalized Hooke's law for a body subjected to mutually perpendicular normal stresses (including the Poisson's ratio effect) is, for strain in the x direction, as shown by Eq. (5).

$$\epsilon_x = \frac{s_x}{E} - \mu\frac{s_y}{E} - \mu\frac{S_z}{E} \qquad (5)$$

Impact strength. Resistance of a material to dynamically applied loads, expressed as the capacity to absorb energy, in units of inch-pounds per cubic inch, is impact strength. An important dynamic load is one which is applied suddenly, as by the impact of a moving mass. The kinetic energy of the moving mass is transferred to the resisting body in the form of strain energy. The stresses produced by impact will depend upon the amount of energy transferred, type of straining action produced, and the characteristics of the material.

The strain energy per unit volume stored in the material when fractured is represented by the total area under the stress-strain curve and is called the toughness. The strain energy stored while the strains are wholly elastic is called resilience. Maximum elastic strain energy per unit volume is called proof resilience and is a property of the material depending upon the elastic limit and the modulus of elasticity. *See* ELASTIC LIMIT; ELASTICITY.

The capacity to absorb energy under impact load is also measured by special tests in which the energy of a swinging pendulum or falling weight producing fracture is measured. *See* STRENGTH OF MATERIALS. [JOHN B. SCALZI]

Bibliography: J. O. Almen and P. H. Black, *Residual Stresses and Fatigue in Metals*, 1963; A. W. Hendry, *Elements of Experimental Stress Analysis*, 1964; A. J. Kennedy, *Processes of Creep and Fatigue in Metals*, 1963; C. Lipson and R. C. Juvinall, *Handbook of Stress and Strength: Design and Material Applications*, 1963.

Stress concentration

A condition in which a stress distribution has high localized stresses. A stress concentration is usually induced by an abrupt change in shape of a member. In the vicinity of notches, holes, changes in diameter of a shaft, or application points of concentrated loads, maximum stress is several times greater than where there is no geometrical discontinuity. Local stress disturbance is rapidly dissipated and effectively disappears at distances from the discontinuity equal to the major dimension of the section.

The tensile stress distribution in a plate reduced by circular notches is shown qualitatively in the illustration. The stress at the root of the notch is about three times the stress at the center of the plate. Load concentrated on the end of a bar produces nonuniformly distributed normal stresses on adjacent sections with the variation decreasing at more remote sections, as shown.

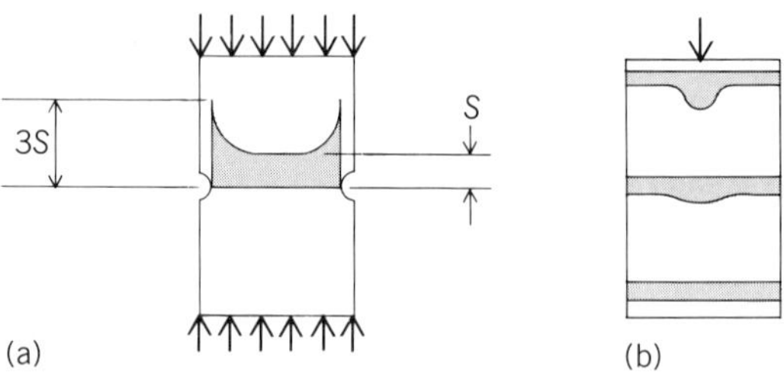

Stress concentrations. (*a*) Plate under uniform tension. (*b*) Bar under concentrated end load.

Stress concentrations are usually determined experimentally by photoelastic methods. The peak stress is expressed as a multiple of the average or nominal stress computed without regard to the concentration, and thus $S_{max} = KS_{nom}$, where K is the stress concentration factor, which depends only on the geometry, without regard to load magnitude. Factors have been evaluated for common types of discontinuity.

Where loads are steady and the material ductile, stress concentrations are relieved by plastic yielding and are not important. Under cyclic loads, fatigue cracks are initiated at points of high local stress, where most failures in machines occur. *See* STRESS AND STRAIN. [JOHN B. SCALZI]

Strip mining

A surface method of mining by removing the material overlying the bed and loading the uncovered mineral, usually coal. It is safer than underground mining because neither the workers nor the equipment is subjected to such hazards as roof falls and explosions caused by gas or dust ignitions. Coal near the outcrop or at shallow depth can be stripped not only more cheaply but more completely than by deep mining, and the need for leaving pillars of coal to support the mine roof is eliminated. The roof over coal at shallow depth is weak and difficult to support in underground workings by conventional methods, yet this same weakness, of cover and of coal seam, making stripping less difficult.

Stripping techniques. Power shovels, draglines, bulldozers, and other types of earth-moving equipment slice a cut through the overburden down to the coal. The cut ranges from 40 to 150 ft (12 to 46 m) wide, depending on the type and size of equipment used. The stripped overburden (spoil) is stacked in a long ridge (spoil bank) parallel with the cut and as far as possible from undisturbed overburden (high wall). The slope of a spoil bank is approximately 1.4:1 and that of a high wall under average conditions is 0.3:1. The uncovered coal (berm) is then fragmented, loaded, and transported from the pit. Spoil from each succeeding cut is stacked overlapping and parallel with the previous ridge and also fills the space left by the coal removed (Fig. 1).

Techniques of stripping methods are similar, but the size of equipment used depends on whether the mine is in prairie or hill country. In prairie areas the thickness of overburden is nearly uniform, the coal bed is extensive, and equipment can be used for years at one mine without dismantling and moving to another location. Large-capacity shovels requiring many months to erect on the site are used at prairie mines. A unit of this type is the 60-yd^3 (46-m^3) rig shown in Fig. 2.

Most coal underlying hills is mined by underground methods, but where the working approaches the outcrop and the overburden is thin, the roof becomes difficult and expensive to support. The coal between the actual or potential underground workings and the outcrop is then more suitable for stripping. Usually, only two or three cuts 40–50 ft (12–15 m) wide can be made on the contour of the coal bed, after which the shovel has to be moved to another site. Thus, in contour stripping, mobility of shovels up to 5-yd^3 (3.8-m^3) capacity is more important than those of larger capacity.

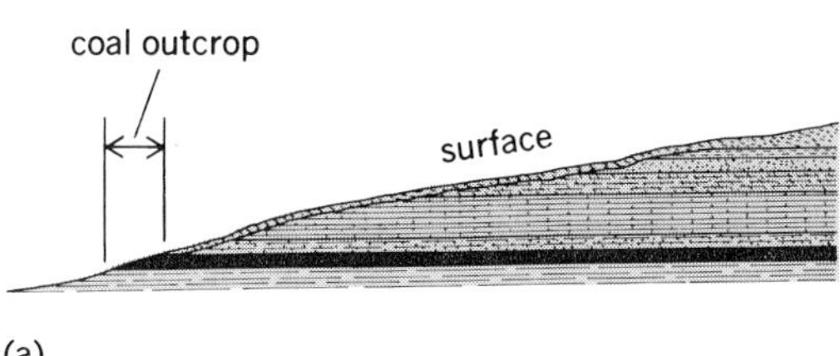

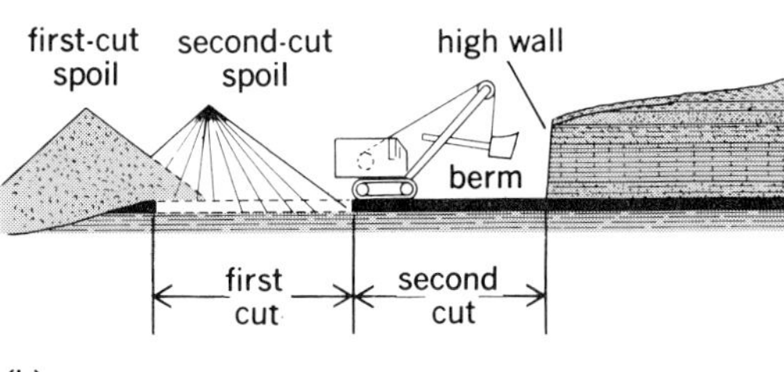

Fig. 1. Representative cross-section profile diagrams of contour strip mining of coal. (*a*) Section before stripping. (*b*) Section after second cut.

Fig. 2. Large electric shovel of the type used in prairie regions, high wall (right), and spoil bank (left) at Hanna Coal Co.'s Georgetown mine, eastern Ohio. (*U.S. Bureau of Mines and Marion Power Shovel Co.*)

Large draglines are used instead of shovels to strip pitching beds of anthracite to depths surpassing 400 ft; however, this use of large draglines could more properly be classed as open pit. *See* OPEN-PIT MINING.

Removal of unconsolidated overburden by hydraulic monitoring is a technique used especially in Alaska. Water under a high-pressure head is directed through a nozzle against the overburden to wash it into deep valleys where swift streams carry it away.

Although the character of the overburden determines the thickness of overburden that can be stripped, the maximum for shovels up to 5-yd^3 (3.8-m^3) capacity is about 50 ft (15 m) and for the largest equipment about 110 ft (33.5 m). To reach these goals it frequently is necessary to use a dragline, carryall, or bulldozer on the high wall or a dragline in tandem with the shovel on the berm to strip the upper few feet of the overburden.

Digging equipment known as the wheel can be used ahead of a large power shovel to remove the upper 20–40 ft (6–12 m) of unconsolidated soil, clay, or weathered strata. This spoil is discharged onto a belt conveyor, then onto a stacker, and finally deposited several hundred feet from the high wall. Overburden thus removed improves the shovel productivity rate materially. Also, coal reserves can be mined that would have been too deep for the power shovel alone to handle.

Rocks overlying coal beds present some diversity of conditions for removal. Materials generally comprise shale, sandstone, and limestone with shale predominating. Proper fragmentation before stripping may be necessary to produce sizes that are smaller than the shovel dipper. Probably more research has been done on overburden drilling and blasting than on any other phase of stripping. The diameter, depth, and spacing of drill holes, the type of drill (whether vertical or horizontal), and the amount and type of explosive for each blast hole are the variables that must be determined for optimum production. Truck-mounted rotary drills have replaced churn drills for drilling vertical holes, and for horizontal drilling, auger drills are used. Package explosives, Airdox, Cardox, and commercial ammonium nitrate mixed with diesel fuel are used for blasting. The ammonium nitrate–diesel fuel mixture is one of the cheaper explosives and has gained favor rapidly. Equipment for mixing this explosive and automatically injecting it through a plastic tube into horizontal drill holes is being tested. If perfected, it will mechanize the only manual operation remaining in the stripping cycle.

Before coal is loaded, spoil remaining on the berm is removed by bulldozer, grader and rotary brooms, or at small mines by hand brooms. Small-diameter vertical holes are augered into the bed and blasted with small charges of explosive to crack the coal. A ripper pulled by a bulldozer is effective in replacing coal drilling and blasting. Broken coal is loaded by $\frac{1}{2}$–5-yd^3 (0.38–3.8-m^3) capacity shovels into trucks of 5–80-ton (4.5–73-metric ton) capacity and transported from the stripping.

Land reclamation. Surface mining for coal can create many problems. The drastic disturbance of the overburden severely changes the chemical and physical properties of the resulting spoils. These altered properties often create a hostile environment for seed germination and subsequent plant growth. Unless vegetative cover is established almost immediately, the denuded areas are subject to both wind and water erosion that pollute surrounding streams with sediment.

Topsoil. Stripmining removes the developed soils and vegetative cover and creates a heterogeneous mass of topsoil, subsoil, and substrata rock fragments. The 1977 United States stripmine law requires that topsoil be removed and reapplied on the spoil surface during regrading and reclamation. Even when topsoil is reapplied over the spoil surface, an organic mulch is generally required for good seed germination and development. These mulching materials alter the surface microclimate and help to conserve soil moisture during the critical seedling establishment period.

The removal of topsoil before mining and its replacement on the spoil surface after final grading have aided materially in the reclamation process. Many of the chemical and physical limitations previously associated with mine spoils have been alleviated or eliminated. Surface grading techniques and seedbed preparation are very important in obtaining good vegetative cover for erosion and sedimentation control. Research has indicated that the surface should be left rough—preferably with small contour furrrows perpendicular to the slope. These furrows catch seed and fertilizer carried in runoff water, thus increasing germination and seedling development and decreasing soil and fertilizer loss from erosion.

Plant species adaptation. Many plant species of economic importance can be used to produce hay, pasture, various horticultural crops, and major row crops. Commercial varieties of grasses that have shown promise on United States eastern stripmine spoils with moderate pH (5.0–6.0) include orchard grass, tall fescue, bromegrass, ryegrass, and timothy. Other grasses that have shown promise on low-pH (4.5 or less) stripmine spoils are weeping lovegrass, bermuda grass, switch grass, bent grass, deer's-tongue, and redtop. Legumes that have shown promise on eastern acid stripmine spoil areas include alfalfa, white clover, crimson clover, bird's-foot trefoil, lespedeza, red clover, crown vetch, hairy vetch, flat pea, kura clover, zigzag clover, and white and yellow sweet clovers.

In the western United States a number of species have been tested. In general, several of the wheatgrass varieties, green needle grasses, side oats grama grass, smooth brome, wheat, and wild rye species have been used with varying degrees of success. Legumes species tested include bird's-foot trefoil, sweet clover, alfalfa, flat pea, and crown vetch. Most spoils in the western United States are returned to perennial grasses for eventual use by grazing livestock and wildlife.

Several woody species can be used on stripmine areas where rainfall is adequate. Tree species should be planted only with or after herbaceous species, such as grasses and legumes, that have stabilized the soil. Almost any species of trees can be grown on stripmine areas as long as their nutrient and environmental needs are met.

Many of the high-organic waste materials such as sewage sludge and composted sewage and garbage materials, are highly beneficial for establishing plant material on stripmine areas, especially with a low pH. These materials can contribute significantly to the plant nutrition on stripmine spoils for the production of vegetative cover.

[ORUS L. BENNETT]

Bibliography: W. S. Doyle, *Strip Mining of Coal: Environmental Solutions*, 1976; R. F. Munn, *Strip Mining: An Annotated Bibliography*, 1973; R. Peele, *Mining Engineers' Handbook*, 3d ed., 1948; U.S. Department of the Interior, *Surface Mining and Our Environment*, 1967; R. A. Wright (ed.), *The Reclamation of Disturbed Arid Lands*, 1978.

Stroboscope

An instrument for observing moving bodies by making them visible intermittently and thereby giving them the optical illusion of being stationary. A stroboscope may operate by illuminating the object with brilliant flashes of light or by imposing an intermittent shutter between the viewer and the object. The rate and duration of the visible periods are adjustable.

Stroboscopes are used to measure the speed of rotation or frequency of vibration of a mechanical part or system. They have the advantage over other instruments of not loading or disturbing the equipment under test. Mechanical equipment may be observed under actual operating conditions with the aid of stroboscopes. Parasitic oscillations, flaws, and unwanted distortion at high speeds are readily detected. It is more economical to obtain this information with a stroboscope than with high-speed motion pictures, and the results are immediately available. Stroboscopes have been employed with balancing machines to locate the lack of balance in lightweight rotating equipment.

By adjusting the rate of viewing to coincide with a multiple of the rate the moving object returns to the same position, parts with a limited range of motion, as in rotation or vibration, may be made to appear stationary (see illustration). The object under view is seen only when it is in one position in space and is freed of the blur normally associated with motion. If the viewing rate is slightly greater than the repetition rate of the motion being observed, the object will appear to move slowly in reverse direction. Similarly, a slightly slower viewing rate makes the object appear to move slowly in the direction of actual motion. Stroboscopes have also been put to a number of nontechnical uses, such as creating the impression of jerking motion when people are dancing.

Stroboscope being used in classroom study of wave motion. (*Editorial Photocolor Archives, Inc.*)

The flashing-light stroboscopes employ gas discharge tubes to provide a brilliant light source of very short duration. A brilliant source is required because of the short period of illumination. Background illumination should be kept as low as possible. Tubes may vary from neon glow lamps, when very little light output is required, to special stroboscope tubes capable of producing flashes of several hundred thousand candlepower with a duration of only a few millionths of a second. *See* VAPOR LAMP.

Tubes may be fired by voltage peaking transformers or by electronic control circuits. When peaking transformers are employed, the flashes are synchronized with the supply line. Electronic controls provide adjustable or synchronous flashing frequencies. Calibrating circuits give direct indication of the frequency of the flashes.

[ARTHUR R. ECKELS]

Structural analysis

The determination of stresses and strains in a given structure. All structures must be designed to carry loads without danger of overall collapse or failure of their components. One way to assure structural safety is to ascertain that stresses and strains produced by loads are less than those allowed by established design codes. This determination of stresses and strains in proposed new structures is a primary objective of structural analysis. *See* STRESS AND STRAIN.

Generally, analysis begins with a check of the overall stability of a structure. This involves first the determination of the forces exerted by the structure against its supports. If the supports are adequate to withstand these forces, they in turn react with equal but opposite forces against the structure.

If computation shows that the reactions balance the loads (weight of structure, occupants, stored materials, vehicles, wind, and earthquake forces), the structure is in static equilibrium. *See* STATICS.

The next step is determination of internal forces and unit stresses in the components of the structure. Finally, if necessary, the deformation of the structure as a whole and of its components may be calculated. These steps are facilitated by use of principles and concepts such as the law of equilibrium. Many of these tools are based on the assumption that the structure is elastic under loads; that is, stress is proportional to strain. *See* STRUCTURAL DEFLECTIONS.

Basic principles. The law of equilibrium is basic in structural analysis. It is useful in computing external reactions of beams, trusses, frames, arches and other structures, as well as internal stresses. For example, the structures in Fig. 1 are acted upon by coplanar, nonconcurrent force systems, or loads and reactions. These must balance if the structures are stable.

When in equilibrium, the structures must satisfy three conditions: (1) The sum of the horizontal components of all the forces must equal zero; (2) the sum of the vertical components must equal zero; and (3) the sum of the moments about any axis normal to the plane must equal zero.

The three independent equations permit three unknowns to be found. Thus, the equilibrium equations can be used directly to compute the reactions, bending moments, shears, and therefore the stresses. When these equations are satisfied, a structure is said to be statically determinate. Figure 1*a* is an example of such a structure; there are three unknowns—vertical and horizontal components of the reaction at the left end and a vertical reaction at the right end.

The structures in Fig. 1*b*, with no horizontal reaction to balance the horizontal load, and Fig. 1*c*, with no balancing support moment, do not satisfy the law of equilibrium; they are unstable.

STRUCTURAL ANALYSIS

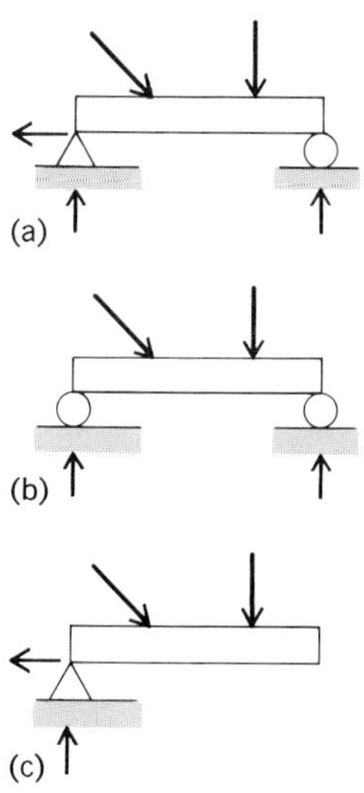

Fig. 1. Structures acted on by force systems. (*a*) Stable structure. (*b*, *c*) Unstable structures. The triangle indicates that the structure is attached, and the circle that the structure, while supported, can move.

Another basic tool of structural analysis is the principle of superposition. It states that the total moments, shears, stresses, and deflections caused by a group of loads are equal to the sum of the effects of the separate loads if the combined effects do not stress the material beyond the elastic limit.

Statically indeterminate structures. When the number of reaction components exceeds the num-

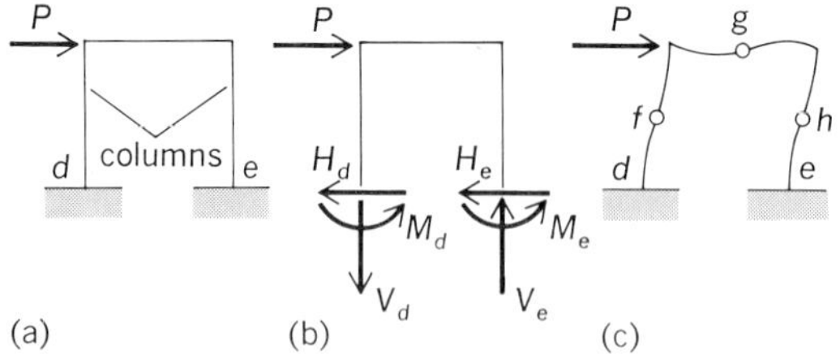

Fig. 2. A statically indeterminate structure. (*a*) Configuration of members. (*b*) Reaction components H_d and H_e are the horizontal reactions to P at points d and e, V_d and V_e are the vertical reactions, and M_d and M_e are the moments. (*c*) Deformation. The points of contraflexure or changes of curvature are f, g, and h.

ber of independent equations that must be satisfied by the loads and reactions when equilibrium exists, the structure is statically indeterminate. For example, if the columns of the rigid frame in Fig. 2*a* are fixed at their bases d and e, there will be six reaction components, as indicated in Fig. 2*b*. This structure will be indeterminate to the third degree; there are six unknowns, whereas the law of equilibrium yields only three equations.

Approximate methods are available for the analysis of statically indeterminate structures. These methods are based on the results of exact analysis or examination of models constructed of flexible materials. Such studies show that in a rigid frame restrained against rotation at the base, the members change curvature somewhere near their midpoints, as indicated in Fig. 2*c*. Bending moment is zero at such points. Thus, these points of contraflexure are equivalent to pin joints.

If their locations are assumed in the columns and the girder of the frame in Fig. 2, and if it is also assumed that column shears are each equal to one-half the lateral load, the reactions may be found from the equations of equilibrium. Similar approximate methods (described in the following subsection) have been used to estimate wind stresses in building frames (Fig. 3).

Approximate methods. The portal method of computing wind stresses in multistory frames assumes that there are points of contraflexure at the midpoints of columns and girders and that each interior column will receive twice as much shear as an exterior column. The resulting structure is statically determinate.

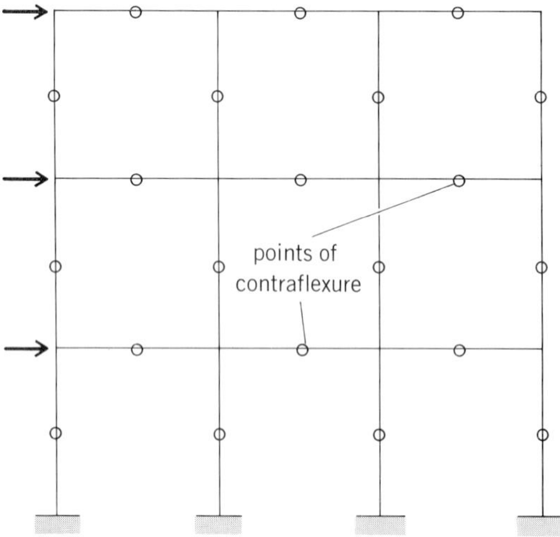

Fig. 3. Diagrammatic representation of the frame of a building using approximate method of estimating wind stresses by placing points of contraflexure at midpoints of columns and beams.

The cantilever method also assumes that there are points of contraflexure at the midpoints of columns and girders. However, it further assumes that the direct stresses in the columns vary as the distance of the columns from the neutral axis, the center of gravity of the columns. As with the portal method, a statically determinate structure results.

Other approximate methods for determining the effects of horizontal loads on multistory frames take into account the stiffness of the members. These include the Witmer K-percentage method, the factor method, and the C method.

Many high buildings were designed with the aid of the portal or cantilever method at a time when the more theoretical (exact) methods available were the laborious, time-consuming Castigliano's theorem and the method of least work. Now with additional analytical methods and electronic computers available, the approximate methods are used for preliminary estimates and spot-checking computer solutions.

Exact methods. The advent of the electronic computer has introduced new methods of analysis for structural systems referred to as the flexibility or force method and the stiffness or displacement method. These two methods consider the behavior of a structure in the linear elastic range of the material, and are applied to those structures which are statically indeterminate.

The essence of the flexibility method is the superposition of displacements which are expressed in terms of the statically determinate force systems. The magnitudes of the redundant forces are determined from the known compatible displacement conditions of the structure. In order to satisfy the compatibility requirements, it is necessary to solve n simultaneous linear equations, where n is the number of redundants in the structural system. An equation is written for each known condition of displacement in the loaded structure. Displacements must be calculated for $(n + 1)$ loading conditions, one equation for the applied loads and n equations for the effects of each of the redundants of the structure.

The stiffness method involves the consideration of linear and angular displacements as the unknown quantities in the analysis; otherwise the method is similar to the flexibility method. Depending upon the number of unknowns, both methods may involve hand calculations for structures with only a few redundants, or for those structures which are more complex a matrix algebra format may be required for the solution.

A comparison of the two methods indicates that each includes the inversion of a matrix. In the flexibility method the matrix is n by n the number of redundants. This method requires more matrix multiplications than the stiffness method. In the stiffness method n is the number of possible movements of the joints which include displacements and rotations. The choice of the better method in any application depends upon the size of the matrix to be inverted.

Although most structures are analyzed for linear elastic behavior, certain extreme loading conditions, such as earthquake effects, require the analysis to be performed by taking into account the nonlinear mechanical properties of the material and the nonlinear geometrical changes caused by the varying load on the structure.

Many of the fundamental concepts of structural analysis serve as background for the analysis methods used with computers. Several of these methods are discussed below.

Castigliano's theorem. This theorem may be applied to continuous beams and to frames of the type illustrated in Fig. 3. The theorem states that the first partial derivative of the strain energy with respect to any particular force (or moment) is equal to the displacement (or rotation) of the point of application of that force (or moment) in the line of its application. Castigliano's theorem is the basis for the method of least work, which states that in a loaded statically indeterminate structure the total internal work is the minimum consistent with equilibrium. For example, in Fig. 2 assume that the structure is fixed at the base of the right column. The unknown reactions are H_d, V_d, and M_d. The total work in the structure is expressed in terms of these reactions and the applied load P. Partial derivatives with respect to the three redundant reactions are set equal to zero. Solving these three equations will yield the values of the unknown reactions.

A frame of the type illustrated in Fig. 3 has nine redundants per story, and is therefore statically indeterminate to the twenty-seventh degree. The solution of the 27 equations required by the method of least work is cumbersome without the aid of computers. In expressing the equation for internal work of a member under direct stress and moment, respectively, use is made of Eqs. (1) and (2), where W is work, S is the direct stress in the

$$W = \frac{S^2 L}{2AE} \qquad (1)$$

$$W = \frac{M^2\,dx}{2EI} \qquad (2)$$

member, L is the length of the member, A is the cross section, E is Young's modulus of elasticity, M is the moment due to the applied loads, and unknowns, expressed in terms of x, and I is the moment of inertia of the member. In many frames the work due to the direct stress is small and may be neglected.

Slope-deflection method. This method is based on Eq. (3), where M_{ab} is the moment at the a end of

$$M_{ab} = \frac{2EI}{L}\,(2\theta_a + \theta_b - 3R) \pm M_{F_{ab}} \qquad (3)$$

member ab; E is the modulus of elasticity of the material; I is the moment of inertia of a cross section of ab; L is the length of ab; θ_a is the rotation in radians of the a end of ab; θ_b is the rotation in radians of the b end of ab; R is an angle in radians equal to the movement of b relative to a divided by L; and $M_{F_{ab}}$ is the fixed-end moment that would occur at a if ab were a fixed beam carrying the actual transverse loads on that member.

In Fig. 4 the moments could be written for the ends of the members if θ_d, θ_e, and R (the horizontal movement of de relative to points c and f, divided by the length of ef) were known. Equations (4) are

$$\begin{aligned} M_{dc} - M_{de} &= 0 \\ M_{ed} - M_{ef} &= 0 \\ M_{cd} + M_{dc} + M_{ef} + M_{fe} + P_1 L_{cd} &= 0 \end{aligned} \qquad (4)$$

the three slope-deflection equations for Fig. 4*b* which may be written to determine these values. Equations (4) are solved to yield the unknowns.

The solution of the forces and stresses in a frame such as that in Fig. 3 would require the solution of 15 simultaneous equations. The unknowns are the rotations of the 12 joints and an R term obtained from the shear in the columns of each story.

Three-moment equation. The three-moment equation permits analysis of beams that are continuous over more than one span. Such beams are statically indeterminate because there are insufficient equilibrium equations for determination of reactions. They differ from simply supported beams because they are subjected to bending moments at supports.

The three-moment equation expresses a relation between the moments over three adjacent supports (one or two may be fixed ends). It is derived by equating the slopes of the two spans that meet over an intermediate support. With the slope-deflection equations, the slopes may be expressed in terms of the moments at the supports and the loads on the spans. The three-moment equation is used as many times as there are unknowns.

Moment-distribution method. The moment-distribution method for continuous beams and frames is based on the fact that the moments in the structure in Fig. 5*a* may be determined by adding the moments from Fig. 5*b* and *c*. In Fig. 5*b*, *de* is regarded as a fixed beam with a fixed-end moment $M_{F_{de}}$. In Fig. 5*c* the moments at joint d will be resisted at d by the four members in the ratio of their stiffness (that is, the I/L of each member). Also there will be induced at the far end of each member a moment equal to one-half of the moment at the d end, if each member is prismatic.

In the analysis of a continuous beam or frame by moment distribution, the members framing into each joint are considered a fixed-end frame of the type shown in Fig. 5*a*, and the moments are distributed as just described. Each such distribution yields an approximate solution. The analyst proceeds from joint to joint a number of times. For example, for the continuous beam in Fig. 6 the distribution might be started at the center support and then proceed to the supports on both sides, after which the procedure would be repeated. The

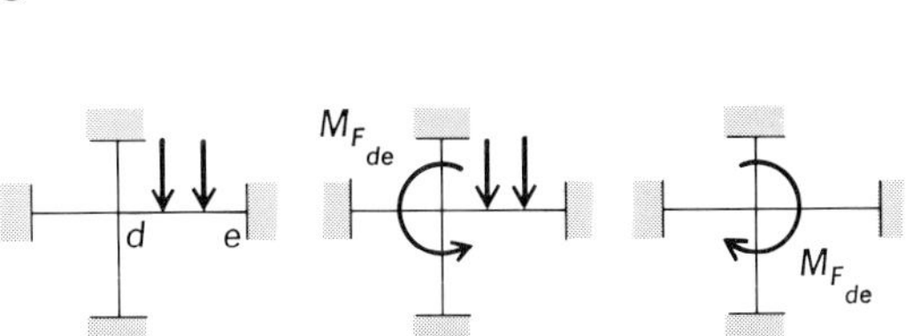

Fig. 5. Illustration of the moment-distribution method for continuous beams and frames. Moments in (*a*) may be determined by adding moments in (*b*) and (*c*).

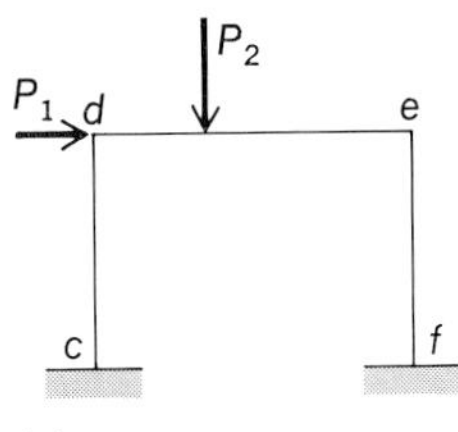

(a)

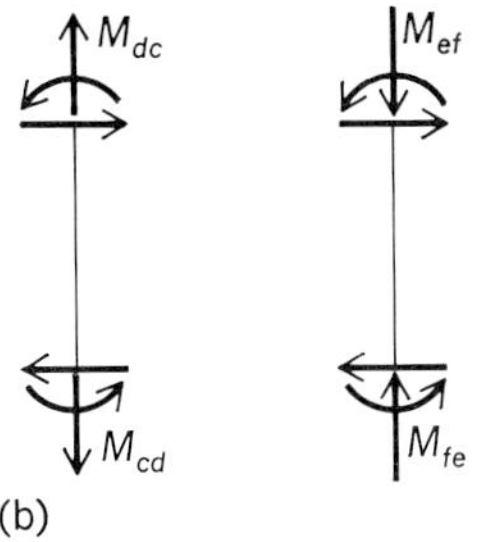

(b)

Fig. 4. Slope-deflection method for statically indeterminate structure. (*a*) Configuration of members. (*b*) Moments at ends of members.

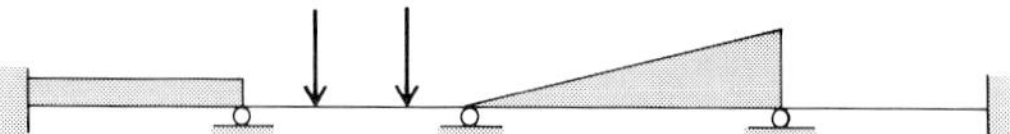

Fig. 6. Moment distribution in a continuous beam.

corrections become smaller and smaller in each cycle.

When, as in Figs. 2 or 3, one end of a member moves relative to the other, sidesway occurs. This complicates the solution, but it can be treated in the same manner. Equations can be written in terms of horizontal shears, or converging approximations can be applied, to determine the moments resulting from sidesway.

Indeterminate trusses. Trusses may also be indeterminate. They may have redundant members, that is, members in excess of the number needed in a statically determinate structure. For example, the double diagonals in the panels of the truss in Fig. 7 may be redundant if they are capable of carrying tension and compression. Like continuous beams, trusses may have redundant reactions. For example, any reaction in Fig. 8*a* (or bar *f*, since its removal will result in a determinate structure).

Types of indeterminate trusses may be solved by the method of least work. Also, the truss in Fig. 8 may be analyzed by any method of computing deflections. The deflection at *e* may be calculated with the center support removed as in Fig. 8*b*, and

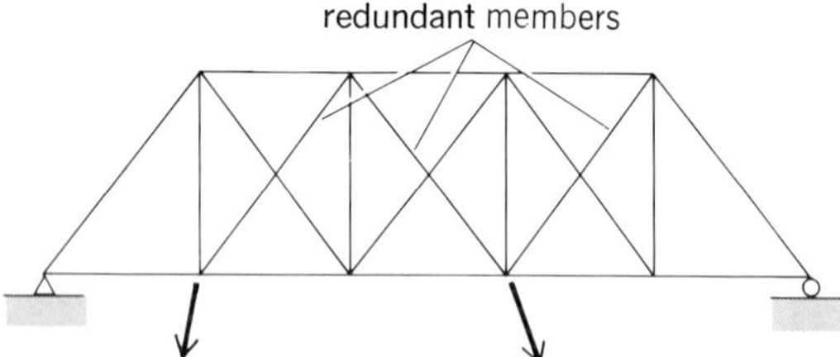

Fig. 7. Redundant members in a statistically indeterminate truss.

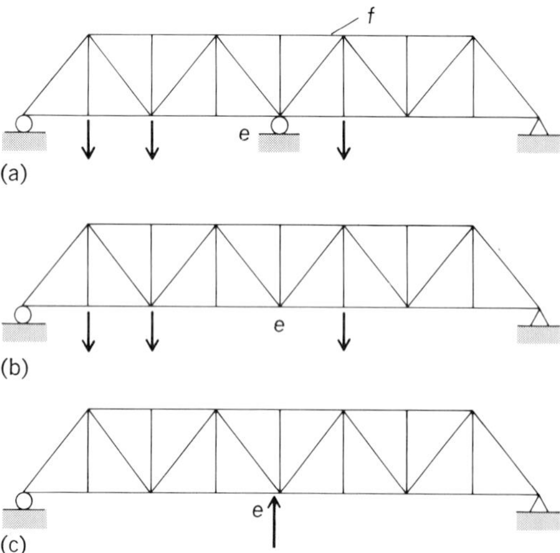

Fig. 8. Computing deflections in an indeterminate truss by methods *a–c*.

the load may be determined that will cause an equal but opposite deflection at *e* as indicated in Fig. 8*c*. The addition of the resulting stresses will be equivalent to those in the truss of Fig. 8*a* in which the deflection at *e* equals zero.

Influence lines. For structures subject to changing or moving load systems, influence lines facilitate analysis. These curves may be plotted for such data as reactions, bending moments, shears, deflections, and stresses.

An influence line for bending moment at a point in a beam is a curve that shows the variation of moment at the point as a unit load passes over the structure. For example, M_c in Fig. 9 is the influence line for moment at *c*.

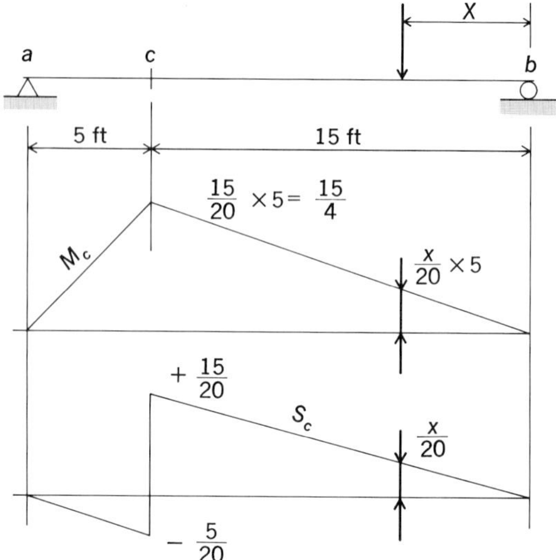

Fig. 9. Influence lines for bending moment M_c at point c and for shear S_c at point c.

A unit load in the position *x* as shown in Fig. 9 will cause at *c* a moment of $5x/20$. An ordinate of this amount is plotted under the load. As *x* varies from zero to 15, the moment at *c* varies from zero to 15/4. Also, as the load passes from *c* to *a* the moment reduces to zero. This influence curve shows that maximum moment at *c* due to a concentrated load *P* will occur when the load is at *c* and will equal $15P/4$. The midpoint of the beam has the highest influence-line ordinate for moment $L/4$ (where *L* is the span).

The influence line for shear at *c* (S_c in Fig. 9) shows that for maximum shear a single load must be placed adjacent to *c* and on the longer of the segments into which *c* divides the beam. A uniform load must be placed only on the longer segment for maximum shear. The highest shear influence-line ordinate occurs for the point adjacent to the support. This ordinate equals one; that is, the shear equals the load.

If an indeterminate structure such as that in Fig. 8 must be analyzed for moving loads, an influence line for the reaction at *e* may be constructed. Plotting this curve can be simplified by applying the principle that a deflection diagram to an appropriate scale is an influence line. For the truss in Fig. 8 with the support at *e* removed, the deflection curve

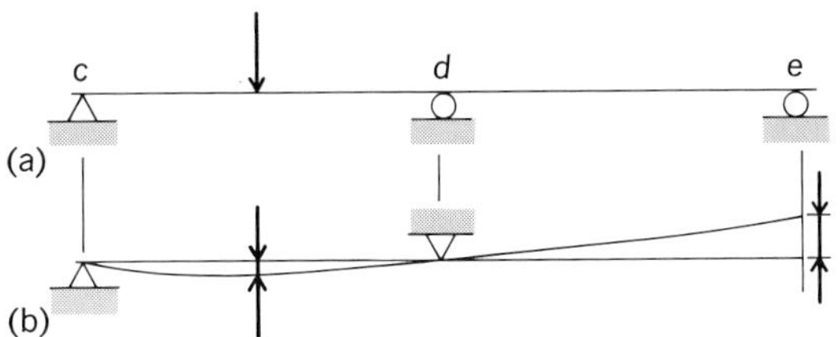

Fig. 10. Example of model analysis. (*a*) Continuous beam. (*b*) Influence line.

due to a load at *e* is, to a different scale, the influence line for reaction at *e*.

Model analysis. The relation between deflection curves and influence lines has led to stress analysis through the use of models. Reactions in a continuous beam, such as *cde* in Fig. 10*a*, may be determined by constructing influence lines with the aid of a spline. For example, if deflection is prevented at *c* and *d* when the spline is forced out of position a unit distance at *e*, as indicated in Fig. 10*b*, the curve formed is the influence line for reaction at *e*. Similarly, if the frame in Fig. 2 is restrained against any motion at *e*, and point *d* is given a unit horizontal motion (with vertical translation and rotation prevented at *d*), the distorted frame will be the influence line for H_d, shear at *d*, when the ordinates are measured normal to the original axis of any member. Similarly, influence lines for axial stress V_d and moment M_d may be constructed.

The Beggs method of model analysis utilizes gages that force a model (usually of celluloid) to make small translations or rotation at a cut section. At target points that have been set on the model, movement is measured with a microscope. The ratio of target movement to movement at the cut section is the same as the ratio of reaction at the cut section to load at the target point. The model need not be composed of prismatic members; they may be tapered, haunched, or curved.

Similitude deals with the effects that are introduced because lengths, transverse dimensions, and moduli of elasticity are different in the model and in the actual structure. These differences will not alter the influence lines for shear and direct stress. However, influence lines for moment must be multiplied by the ratio of lengths in the actual structure to corresponding lengths in the model.

[JOHN B. SCALZI]

Bibliography: B. Bresler, T. Y. Lin, and J. B. Scalzi, *Design of Steel Structures*, 2d ed., 1968; J. W. Fisher and J. H. A. Struik, *Guide to Design Criteria for Bolted and Rived Joints*, 1974; W. McGuire, *Steel Structures*, 1968; C. H. Norris, J. B. Wilbur, and S. Utku, *Elementary Structural Analysis*, 3d ed., 1976; R. J. Roark, *Formulas for Stress and Strain*, 1965; R. N. White, P. Gergely, and R. G. Sexsmith, *Structural Engineering*, vol. 2: *Indeterminate Structures*, 1972.

Structural connections

Methods of joining the individual members of a structure to form a complete assembly. The connections furnish supporting reactions and transfer loads from one member to another. Loads are transferred by fasteners (rivets, bolts) or welding supplemented by suitable arrangements of plates, angles, or other structural shapes. When the end of a member must be free to rotate, a pinned connection is used.

The suitability of a connection depends on its deformational characteristics as well as its strength. Rotational flexibility or complete rigidity must be provided according to the degree of end restraint assumed in the design. A rigid connection maintains the original angles between connected members virtually unchanged after loading. Flexible or nonrestraining connections permit rotation approximately equal to that at the ends of a simply supported beam. Intermediate degrees of restraint are called semirigid.

Framed web connections. A commonly used form of connection for rolled-beam sections, called a web connection, consists of two angles attached to opposite sides of a member and which are in turn connected to the web of a supporting beam, girder, column, or framing at right angles. A shelf angle may be added to facilitate erection.

Riveted or bolted web. The angles and the rivets that fasten them are designed to transmit shear force only as a simple beam connection. The rotation results from the flexibility of the outstanding angle legs (Fig. 1). For rotation approximating that of a simple beam, the angle legs may become permanently bent. A small end moment is developed by the forces necessary to bend the angles. For building construction, web connections have been standardized for varying beam sizes and are listed in handbooks.

Welded web. Flexible welded connections are made with web framing angles attached to the supported beam by fillet welds along the length and ends of the angle legs. These legs are narrow for economy, and the size of weld is such as to resist combined bending and shear without exceeding limiting shear stresses. The legs connected to the supporting member are attached by fillet welds along their outer edges (Fig. 2). To prevent tearing, the edge fillet welds are returned a short distance around the top corners where the stress is greatest. The end rotation of the connected beam results from the flexing of the outstanding legs. Erection bolts are placed near the bottom of these legs in order not to interfere with the bending.

Seat connections. A bracket or seat on which the end of the beam rests is a seat connection; it is intended to furnish the end reaction of the supported beam. The bracket may be attached to the support by bolts or welding. Two general types

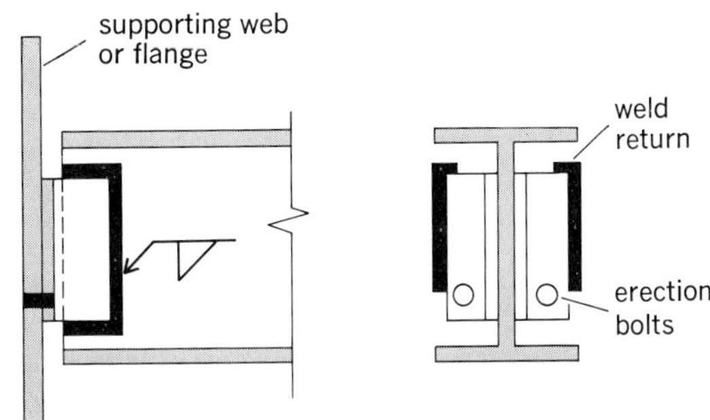

Fig. 2. Welded web connection.

STRUCTURAL CONNECTIONS

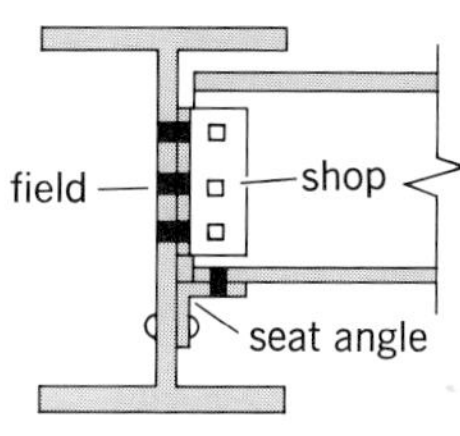

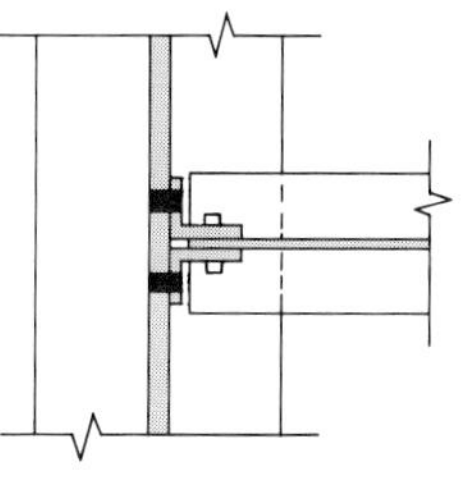

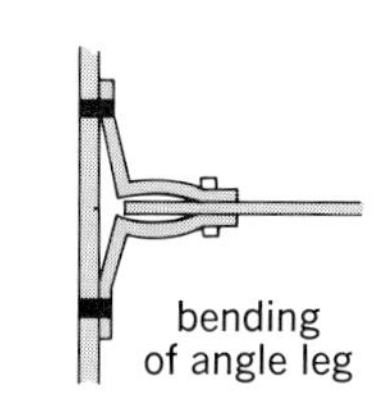

Fig. 1. Riveted or bolted web connections.

are used: The unstiffened seat provides bearing for the beam by a projecting plate or angle leg which offers resistance only by its own flexural strength; the stiffened seat is supported by a vertical plate or angle which transfers the reaction force to the supporting member without flexural distortion of the outstanding seat.

Unstiffened seat. A simple form of unstiffened seat connection consists of an angle sufficiently long to engage the flange width, with its vertical leg attached to the support and the outstanding leg serving as the end bearing for the beam (Fig. 3). The unstiffened angle acting as a cantilever is suitable for small loads.

In bolted angle seats, the size of the angle must be such as to permit sufficient bolts in the vertical leg. The outstanding leg must project enough to distribute the beam reaction and thus avoid crippling of the beam web. Thick angles tend to concentrate the reaction near the outer edges as the beam rotates. The distribution is greater for thinner angles which bend with beam rotation (Fig. 3). The uncertainty of the reaction eccentricity and the effect of connecting the beam flange to the seat require approximations in the analysis. To provide lateral stability, an angle is usually connected to the upper flange of the beam, permitting rotation without shear resistance.

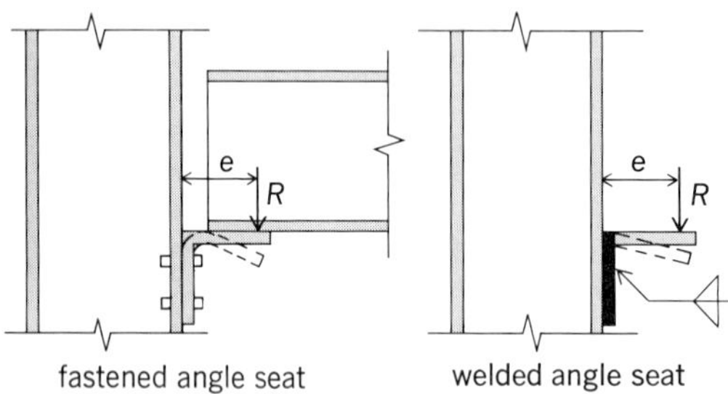

Fig. 3. Unstiffened seat connections.

In welded angle seats, the angle forming the seat is attached to the support by vertical fillet welds along the edges with a short return to the top. These welds are subjected to shear and bending forces due to eccentricity of load. The beam flange is attached to the seat by short edge welds (Fig. 4). The top angle is attached by fillet welds along the outer edges of both legs. Bending of the seat and flexing of the top angle provide a dependable amount of rotation of the beam.

Stiffened seat. To provide sufficient bolts to resist large reactions and also to reinforce the seat against bending, one or two vertical angles, attached to the supporting member, are fitted tightly against the underside of the seat angle leg. The legs of the stiffener angles extend to the outer edge of the seat. The thickness of these angles provides contact area with the seat to limit the compression stress and avoid local buckling. A filler plate is placed back of the extended stiffener angles (Fig. 5*a*). The rivets connecting the stiffeners resist vertical and transverse shear. Tension is produced in the upper rivets, and bearing pressures develop at the bottom of the stiffener because of the moment of the eccentric load.

Welded stiffened seats are made with a length of

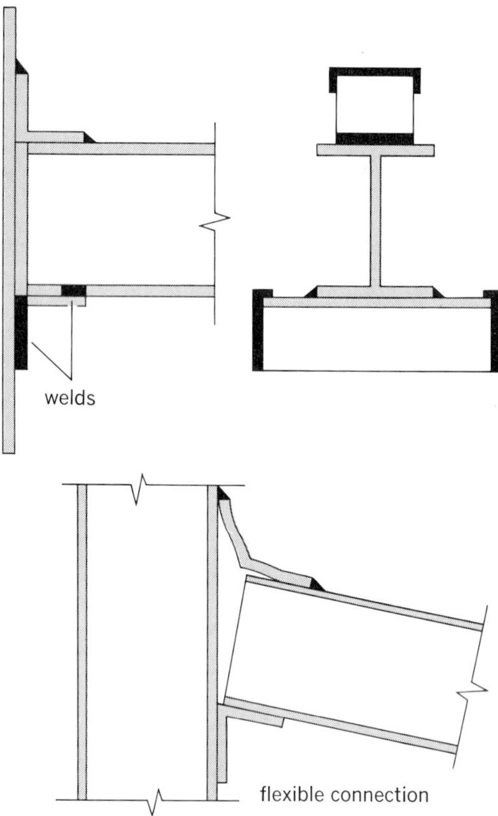

Fig. 4. Welded top angle and seat connection.

structural T, attached to the support by vertical fillet welds on both sides of the stem and additional welds along the underside of the flange to provide torsional stiffness (Fig. 5*b*). A short weld attaches the beam flange to the seat. A top angle is welded along the outer edges of both legs, forming a flexible connection. The stiffness of the seat tends to concentrate the reaction force near the outer edge of the seat. The vertical welds resist shear and moment.

Eccentrically loaded connections. When the action line of a transferred force does not pass through the centroid of the connecting fastener group or welds, the connection is subjected to rotational moment which produces additional shearing stresses in the connectors. The load transmitted by diagonal bracing to a supporting column flange through a gusset plate is eccentric with reference

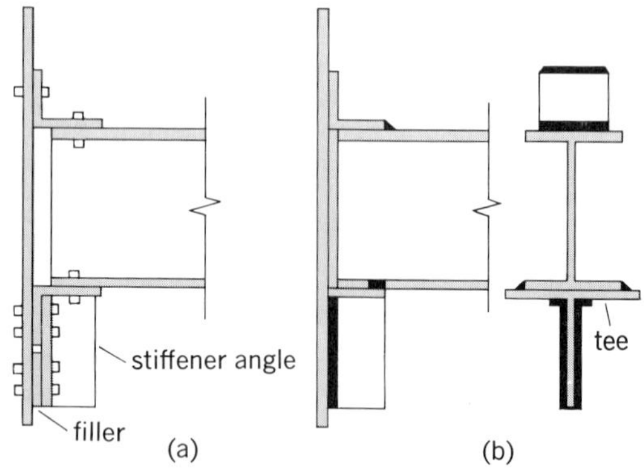

Fig. 5. Stiffened seat connections. (*a*) Bolted stiffened seat. (*b*) Welded stiffened seat.

to the connecting fastener group; as a consequence, moment Pe tends to rotate the plate (Fig. 6*a*). Each fastener must resist its share of P plus the force induced by the moment, which is proportional to the distance from the centroid. These forces are combined vectorially to find the resultant force (Fig. 6*b*). Similar conditions exist in a column bracket or connections required to transmit moments caused by lateral forces on a building frame (Fig. 6*c*). In welded connections the maximum stress in the weld is also the effect of the combined shear and moment.

Fasteners in tension. In beam-to-column connections and stiffened seat connections or when members transfer loads to columns by a gusset plate or a bracket, the fasteners are subjected to tension forces caused by the eccentric connection. Analysis of the tension in the fastener indicates that although there are initial tensions in the fasteners, the final tension is not appreciably greater than the initial tension and the connected parts do not separate under the usual working loads. This results from a balancing of the internal forces with the externally applied load.

When fasteners are subjected to both shear and tension forces, the limiting combined stresses are determined by an empirical procedure which is based on an interaction relationship. The general equation, expressed in working stress terms, is given below. Here F_t and F_v are the allowable work-

$$\left(\frac{f_t}{F_t}\right)^2 + \left(\frac{f_v}{F_v}\right)^2 = 1.0$$

ing stress tensile and shearing unit stresses, respectively, when each is considered by itself, and f_t and f_v are the nominal stress components, calculated from the combined loading. Variations of this equation are included in specifications for building and bridge designs. *See* STRESS AND STRAIN.

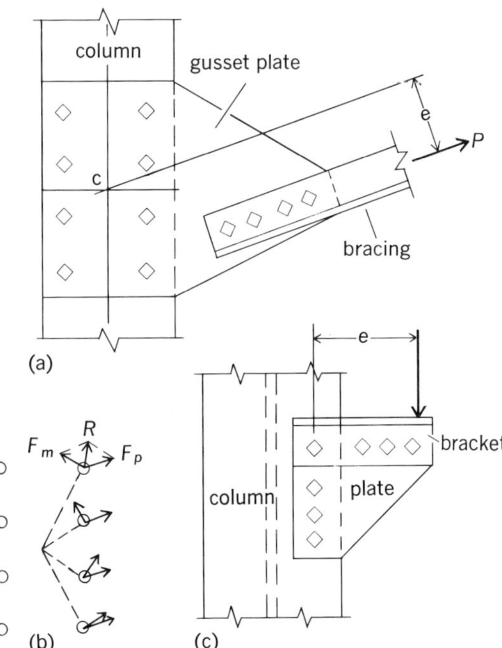

Fig. 6. Eccentrically loaded connections. (*a*) Diagonal bracing. (*b*) Resultant bolt forces. (*c*) Lateral forces to column from frame producing eccentric loading.

Moment-resisting connections. Rigidity and moment resistance are necessary at the ends of beams forming part of a continuous framework which must resist lateral and vertical loads. Wind pressures tend to distort a building frame, producing bending in the beams and columns which must be suitably connected to transfer moment and shear. The resisting moment can be furnished by various forms of angle T for fasteners or welded or bracket connections.

A fastener connection consisting of web framing angles supplemented by angles connected to both top and bottom flanges is suitable for small end moments. The web angles are assumed to transfer the shear, while the flange angles connected to the column apply forces to the beam flanges which form the resisting couple. The end moment is limited by the number of fasteners which can be provided in the legs of the flange angles and by the flexibility of the angles. A similar connection uses web angles and structural T sections for connection of both beam flanges, or a stiffened beam seat connecting the bottom flange with a T attached to the top flange (Fig. 7*a*). The tensile and compressive force required to furnish the resisting couple may be reduced by increasing the lever arm; the increase is accomplished by attaching short lengths of beams called beam stubs to one or both flanges, thus in effect deepening the connected beam and increasing the arm between the connecting T sections (Fig. 7*b*).

Welded moment-resisting connections are commonly made with a stiffened seat and a plate fillet welded to the top flange and butt-welded to the column flange. The column flanges are reinforced against bending by stiffener plates welded between the column flanges (Fig. 7*c*).

Pinned connections. Where appreciable angular change between members is expected, and in special cases where a hinge support without moment resistance is desired, connections are pinned. Many bridge trusses and large girder spans have pin supports. Plates connected to the adjacent members or the webs of the members themselves engage the pin in much the same manner as a bolted connection. As with riveted or bolted joints, the factors in design include bearing, shear, and bending of the pins, the net section of the plates or connected members, and edge tearout. Dishing or buckling of the plates must be avoided. The size of the pin is usually determined by its bending resistance. Reinforcing pin plates may be required to provide sufficient bearing.

High-strength bolts. These are made by heat-treating steel to produce a high-yield and high-tensile-strength material suitable for bolts. There are two grades designated by ASTM as A 325 and A 490. Each type is used in connections of structural members of comparable strength. The mechanical properties for the two grades are:

Bolt	Yield strength	Tensile strength
A 325	81,000 psi (560 MPa)	105,000 psi (725 MPa)
A 490	130,000 psi (895 MPa)	150,000 – 170,000 psi (1035 – 1170 MPa)

The bolts are installed by tightening with a calibrated torque wrench, by the turn of the nut meth-

STRUCTURAL CONNECTIONS

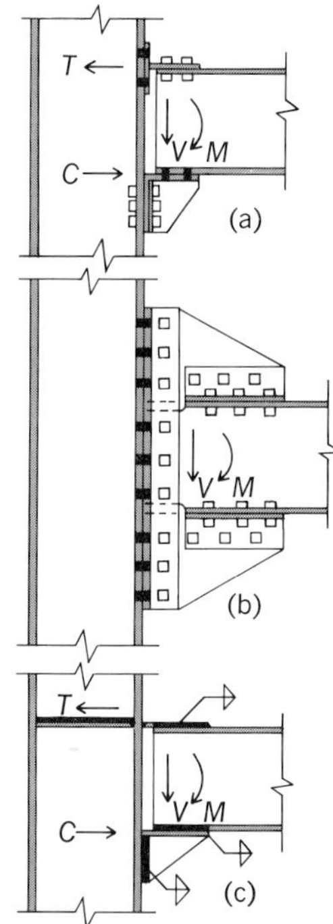

Fig. 7. Moment-resisting connections. (*a*) Seat and *T* or two *T*s. (*b*) Bracket or stiffened beam seat. (*c*) Welded type.

od, or by use of direct tension indicators. Hardened washers are used under the head or unit which is being turned in the tightening process.

The bolted joint transfers the load by friction between the plates (similar to a rivet) which is developed by the high internal tension in the bolt. Connections using these bolts have different capacities depending upon the type of connection, such as friction type or bearing type. The allowable strength of the bolts varies in accordance with use.

Tests have indicated that connections using high-strength bolts have a higher fatigue strength than rivets for the same loading condition. As a result, they are used in bridge construction and other applications where vibratory loads are present. They have also become the conventional method for building construction. *See* BOLT; JOINT (STRUCTURES); RIVET; WELDING AND CUTTING OF METALS. [JOHN B. SCALZI]

Structural deflections

The deformations or movements of a structure and its flexural members, such as beams and trusses, from their original positions. It is as important for the designer to determine deflections and strains as it is to know the stresses caused by loads. *See* STRESS AND STRAIN.

Deflections may be computed by any of several methods. Generally the computation is based on the assumption that stress is proportional to strain. As a result, deflection equations involve the modulus of elasticity E which is a measure of the stiffness of a material.

The relation between deflections at different parts of a structure is indicated by Maxwell's law of reciprocal deflections. This states that if a load P is applied at any point A in any direction a and causes a shift of another point B in direction b, the same load applied at B in direction b will cause an equal shift of A in direction a (see illustration). The

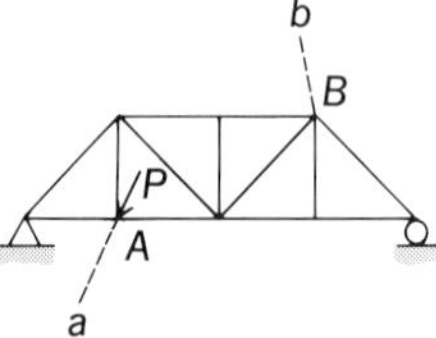

Example of Maxwell's law of reciprocal deflections.

law is used in a number of ways such as in simplifying deflection calculations, checking the accuracy of computations, and producing influence lines. *See* STRUCTURAL ANALYSIS.

Beam and truss deflections usually are computed by similar methods, except that integration is used for beam equations and summation for trusses. Beam deflection equations involve bending moments and moments of inertia. Truss deflection equations are based on the stresses and cross-sectional areas of chords and web members. For example, the virtual work equation takes the form of Eq. (1) for beams, where δ is the deflection,

$$\delta = \int \frac{Mm\,dx}{EI} \quad (1)$$

M is the bending moment due to applied loads, m is the moment due to the virtual unit load applied at the point where the deflection is to be obtained, E is the modulus of elasticity, and I is the moment of inertia of a cross section of the beam. For trusses it takes the form of Eq. (2), where δ is the

$$\delta = \sum \frac{SuL}{AE} \quad (2)$$

deflection, S is the stress in pounds in any member due to the actual load, u is the stress in the member due to the virtual unit load applied where the deflection is to be obtained, L is the length of the member, and A is its area.

Deflections also may be determined graphically. The Williot-Mohr diagram, for example, is used often for trusses. It requires preliminary computation of stresses S and length changes SL/AE for truss members and plotting of these deformations to scale.

[JOHN B. SCALZI]

Structural materials

Construction materials which, because of their ability to withstand external forces, are considered in the design of a structural framework. Materials used primarily for decoration, insulation, or other than structural purposes are not included in this group.

Clay products. The principal products in this class are the solid masonry units such as brick and the hollow masonry units such as clay tile or terracotta.

Brick is the oldest of all artificial building materials. It is classified as face brick, common brick, and glazed brick. Face brick is used on the exterior of a wall and varies in color, texture, and mechanical perfection. Common brick consists of the kiln run of brick and is used principally as backup masonry behind whatever facing material is employed. It provides the necessary wall thickness and additional structural strength. Glazed brick is employed largely for interiors where beauty, ease of cleaning, and sanitation are primary considerations.

Structural clay tiles are burned-clay masonry units having interior hollow spaces termed cells. Such tile is widely used because of its strength, light weight, and insulating and fire-protection qualities. Its size varies with the intended use.

Load-bearing tile is used in walls that support, in addition to their own weight, loads that frame into them, for instance, floors and the roof. Tiles manufactured for use as partition walls, for furring, and for fireproofing steel beams and columns are classed as non-load-bearing tile. Special units are manufactured for floor construction; some are used with reinforced-concrete joists; and others with the steel beams in flat-arch and segmental-arch construction.

Architectural terra-cotta is a burned-clay material used for decorative purposes. The shapes are molded either by hand in plaster-of-paris molds or by machine, using the stiff-mud process.

Building stones. Building stones generally used are limestone, sandstone, granite, and marble. Until the advent of steel and concrete, stone was the most important building material. Its principal use now is as a decorative material because of its beauty, dignity, and durability.

Concrete. Concrete is a mixture of cement, mineral aggregate, and water, which, if combined in proper proportions, form a plastic mixture capable of being placed in forms and of hardening through the hydration of the cement. *See* PRESTRESSED CONCRETE.

Wood. The cellular structure of wood is largely responsible for its basic characteristics, unique among the common structural materials. The strength of wood depends on the thickness of the cell walls. Its tensile strength is generally greater than its compressive strength. The ratio of its strength to its stiffness is much higher than that of steel or concrete; therefore, it is important that deflection be carefully considered in the design of a wooden floor system.

When cut into lumber, a tree provides a wide range of material which is classified according to use as yard lumber, factory or shop lumber, and structural lumber. Timber is lumber that is 5 in. or larger in its least dimension.

Laminated structural lumber is formed by gluing together two or more layers of wood with the grain of all layers parallel to the length of the member. Both laminated lumber and plywood make use of modern gluing techniques to produce a greatly improved product. The principal advantages derived from lamination are the ease with which large members are fabricated and the greater strength of built-up members. Laminated lumber is used for beams, columns, arch ribs, chord members, and other structural members.

Plywood, while laminated also, is formed from three or more thin layers of wood cemented or bonded together, with the grain of the several layers alternately perpendicular and parallel to each other.

Plywood is generally used as a replacement for sheathing, or as form lumber for reinforced concrete structures. Both laminated structural lumber and plywood have the advantage of minimizing the effects of knots, shakes, and other lumber defects by preventing them from occurring in more than one lamination at a given cross section.

Structural metals. Of importance in this group are the structural steels, steel castings, aluminum alloys, magnesium alloys, and cast and wrought iron. *See* CAST IRON; MAGNESIUM ALLOYS; STEEL; STRUCTURAL STEEL.

Steel castings are used for rocker bearings under the ends of large bridges. Shoes and bearing plates are usually cast in carbon steel, but rollers are often cast in stainless steel.

Aluminum alloys are strong, lightweight, and resistant to corrosion. The alloys most frequently used are comparable with the structural steels in strength. However, because aluminum alloys have a modulus of elasticity one-third that of steel, the danger of local buckling is likely to determine the design of aluminum compression members. Also, the accepted ratios of depth-to-span for bridges must be increased to reduce deflections and to give maximum economy of material. Because the alloy is approximately 35% the weight of steel, considerable savings in weight may be achieved in long-span structures. Additional savings occur in machinery, counterweights, and towers of bascule and lift bridges.

Magnesium alloys are produced as extruded shapes, rolled plate, and forgings. The principal structural applications are in aircraft, truck bodies, and portable scaffolding. The alloy weighs approximately 110 lb/ft^3 (1760 kg/m^3) as compared to 490 lb/ft^3 (7850 kg/m^3) for steel.

Gray cast iron is used as a structural material for columns and column bases, bearing plates, stair treads, and railings. Malleable cast iron has few structural applications.

Wrought iron is used extensively because of its ability to resist corrosion. It is used for blast plates to protect bridges, for solid decks to support ballasted roadways, and for trash racks for dams.

Composite materials. Composite material is a combination of two materials which has its own distinctive properties. In terms of strength or other desirable quality, it is better than, or radically different from, either component alone. The concept is leading to the manufacture of a new range of materials that may result in significant changes in their engineering use. Already they have made important contributions to the design of naval, air, and space vehicles and structures.

The most familar composite is formed using filaments or woven cloth of fiber glass embedded in a polyester or epoxy resin base. The resultant material, fiber glass reinforced plastic (FRP), has a set of strength and stiffness properties totally different from either constituent. In their dependence on the orientation of load application to the glass fibers, FRPs closely resemble wood. Other composites are being formed using fibers of graphite, boron, or tungsten in both resin and metallic matrixes.

The FRPs lack the stiffness required for bridge structures. The graphite and boron filaments provide materials with high stiffness-to-weight ratios, but in all cases the composites are not in general use for the common load-carrying structures. In all composites, the strength of the brittle filament is used through the transmittal of load by the resin or other matrix even though the filament is fractured or used in short lengths. New experimental composites are being developed by growing single-crystal whiskers in an appropriate matrix. In one such experimental material, niobium carbide whiskers grown in a niobium matrix demonstrated high strength at a temperature over 1500°C. *See* CERMET.

[CHARLES M. ANTONI]

Bibliography: Building Research Establishment, *Building Materials*, 2d ed., 1972; J. Dubois and F. W. John, *Plastics*, 5th ed., 1974; W. C. Huntington and R. E. Mickadeit, *Building Construction Materials and Types of Construction*, 5th ed., 1980; R. C. Smith, *Materials of Construction*, 3d ed., 1979.

Structural plate

A flat plate or slab which is supported along the edges, such as the bottom or cover of a tank, cylinder head, bulkhead, or floor panel. A structural plate bends when subjected to forces applied normal to its surface. The bending produces curvature in all planes normal to the plate; this dishing action differs from bending of a beam where curvature occurs in a single plane under symmetrical loads. As for a beam, boundary conditions include various degrees of restraint at the edges, from simple supports to complete fixity against rotation. The analysis of stresses and deflections in a plate, while based on assumptions essentially the same as those used for a beam, is more complicated

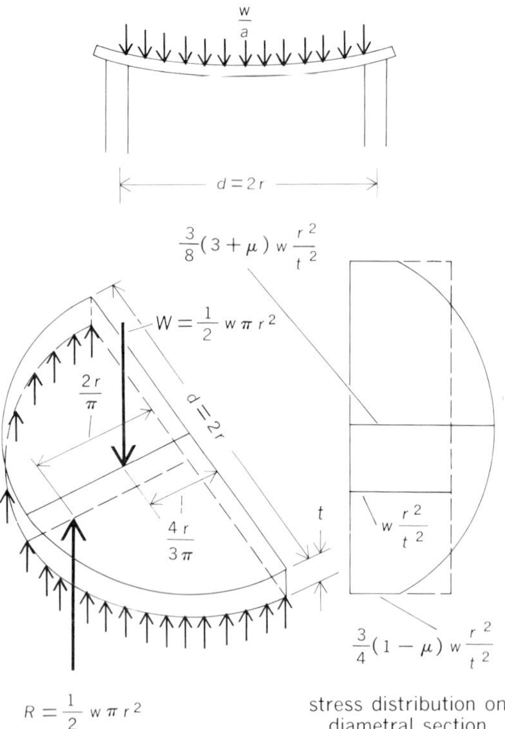

Fig. 1. External forces on the half plate considered in analysis of uniformly loaded circular plate.

because of simultaneous bending in all planes. Simplified approximate analyses predict stresses and deflections useful in design. *See* BEAM.

Bases for design. The relative importance of the straining actions involved depends on the thickness and whether the behavior is elastic or inelastic. For thick plates, as for short deep beams, shearing stresses must be considered, whereas in plates of medium thickness shear stresses can be neglected and only bending action is important. In thin plates, deflection is relatively larger and direct tension due to suspension action is appreciable. For the extreme case of a thin membrane, resistance depends almost entirely on direct tension. Elastic deflection is caused primarily by bending, but during plastic behavior deflections are increased and direct tension resists a progressively larger part of the loads.

The flexure theory for plates, based on assumptions similar to those used for beams, considers the equilibrium of a differential element of a symmetrically loaded plate, included between vertical circumferential and meridional planes, referred to axes of symmetry. The internal stresses and moments are evaluated in terms of angular and linear coordinates of any point of the plate. Application of the general expressions to particular loading and support conditions determines the maximum bending moment. Other methods involve differentiation of an expression for deflection. The stresses S are calculated by the flexure formula $S = Mc/I$, where M is bending moment, c is the distance from the neutral axis, and I is the moment of inertia.

The theory assumes elastic action, neglecting shear and direct tension. A simplified approximate approach evaluates the total bending moment on a critical section by statics and the average bending stress by the flexure formula. Maximum stresses are then found by applying correction factors which are determined by comparison with more exact analysis or experimental results. These analyses assume small deflections and bending as the primary action.

Circular plate. A simply edge-supported circular plate can be analyzed approximately by considering the bending of half of the plate about a diametral axis of symmetry.

Uniform load. With uniform load, the analysis determines the average bending stress on a diametral section (Fig. 1). The resultant total bending moment on the diametral section due to the load and reaction is $wr^3/3$. The section modulus is $rt^3/3$ and the average bending stress at the surfaces is $S = wr^2/t^2$, where the symbols are defined as shown in Fig. 1.

According to the general theory of flexure of plates, the stress reaches a maximum value at the center equal to $(3/8)(3+\mu)wr^2/t^2$, which, for steel with a Poisson's ratio $\mu = 0.30$, is 24% greater than the average. Redistribution after local yielding at the center tends to uniformity of stress, and tests have shown that the average stress can be taken as the significant stress.

Elastic deflection at the center is small, except for very thin plates. For a circular plate, simply edge-supported, deflection y is shown by Eq. (1). For $\mu = 0.30$, $y = (11/16)wr^4/Et^3$ approximately.

$$y = \frac{3}{16}(1-\mu)(5+\mu)\frac{wr^4}{Et^3} \qquad (1)$$

Edges clamped. Circular plates fixed in direction at the edges and uniformly loaded are analogous to fixed-end beams where negative moments at the ends reduce stress and deflection at the center. The moment is greatest at the edges. The maximum radial bending stress at the edges is $3wr^2/4t^2$. For thin plates, the elastic deflection at the center is shown by Eq. (2). For $\mu = 0.30$, this is only 24.5% of the deflection of a simply supported plate.

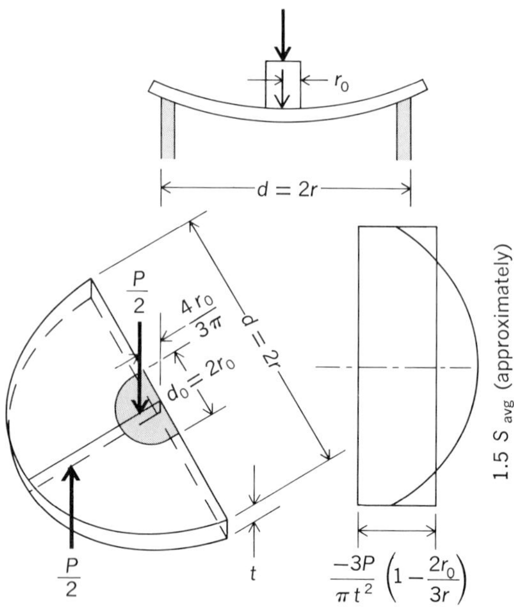

Fig. 2. Central load on simply supported circular plate.

$$y = \frac{3}{16}(1-\mu^2)\frac{wr^4}{Et^3} \qquad (2)$$

For thicker plates with $t/r > 0.1$, the above value is multiplied by a factor $C = 1 + 5.72(t/r)^2$. Complete edge fixity is an ideal condition. Because edge rotation is small for simply supported plates, only a small relaxation of clamping or local yielding will eliminate the greatest part of the fixedness, and behavior approaches that of a simply supported plate.

Central load. If the central load is distributed over a circular area of radius r_0, the external forces on a semicircular segment of a simply supported circular plate are the reactions at the rim and the load on the area of application (Fig. 2). The bending moment M on the diametral section is shown by Eq. (3), and the average stress is $S_{avg} = 3P/\pi t^2$.

$$M = \frac{Pr}{\pi}\left(1 - \frac{2r_0}{3r}\right) \qquad (3)$$

As the load application area decreases, r_0 approaches zero and the average stress approaches $3P/\pi t^2$.

The coefficient to be applied to the average stress to obtain the theoretical maximum stress ($\mu = 1/3$) approaches 1.25 for large values of t_0/r and is nearly 2.20 for $r_0/r = 1/10$.

The theoretical center deflection is shown by Eq. (4).

$$y = \frac{3(1-\mu)(3+\mu)}{4\pi}\frac{Pr^2}{Et^3} \qquad (4)$$

With the edges clamped, the theoretical maximum stress at the center for $r > 1.7r_0$ is shown by Eq. (5). Yielding at the fixed edges by ductile materials relieves the local stress, and the stresses after redistribution approach those of the simply supported plate.

$$S = \frac{3(1+\mu)P}{2\pi t^2}\left(\ln\frac{r}{r_0} + \frac{r_0^2}{4r^2}\right) \qquad (5)$$

When r_0/r is small, the center deflection is shown by Eq. (6). This deflection increases with yielding at supports.

$$y = \frac{3(1-\mu^2)Pr^2}{4\pi Et^3} \qquad (6)$$

Center support. A circular plate, supported at its center, with uniform load has a maximum theoretical stress at the center, as shown by Eq. (7). For r_0/r small, the term $r_0^2/2$ is negligible. Under this condition, the center deflection is shown by Eq. (8).

$$S_{max} = \frac{3wr^2}{2t}\left[(1+\mu)\ln\frac{r}{r_0} + \frac{1}{4}(1-\mu)\left(1 - \frac{r_0^2}{r^2}\right)\right] \qquad (7)$$

$$y_{max} = \frac{3}{16}(1-\mu)(7+3\mu)\frac{wr^4}{Et^3} \qquad (8)$$

When $\mu = 1/3$, this reduces to $y_{max} = wr^4/Et^3$ which is three-fifths of the deflection for the same central load on an edge supported plate.

Elliptical plate. In an elliptical plate simply supported at its edges and with uniform load, the maximum bending stress and curvature occur in the direction of the minor axis (Fig. 3). If the ellipse is elongated with a very much greater than b, the plate action approaches that of a simply supported beam with a span of $2b$. A central strip of unit width along the minor axis resists a maximum moment of $wb^2/2$ and the maximum stress in the plate is $3w\,b^2/t^2$. If a equals b, the plate is circular and the maximum average stress is wb^2/t^2. For intermediate dimensions the coefficient of wb^2/t^2 varies between 1 and 3. An approximate expression for maximum stress, in the direction of the minor axis, is shown in Eq. (9).

$$S_{max} = \frac{3a-2b}{a}\frac{wb^2}{t^2} \qquad (9)$$

STRUCTURAL PLATE

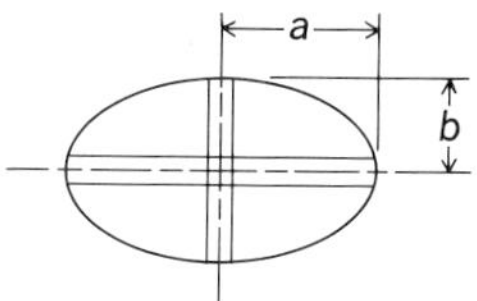

Fig. 3. Elliptical plate.

Square plate. When simply supported at all edges, with uniformly distributed load, a square plate can be analyzed as follows. The critical section is taken along the diagonal of the square, about which maximum stress occurs. The bending moment on a diagonal section is found by considering the forces on the triangular half plate (Fig. 4). The resultants of the applied load and reactions and their locations determine the bending moment on the diagonal section which is $M = wb^3/12\sqrt{2}$. The average stress on the diagonal section is $wb^2/4t^2$ which is the same as for a circular plate with diameter equal to the side of the square.

With fixed edge supports, the average stress on the diagonal section is taken as $wb^2/5t^2$.

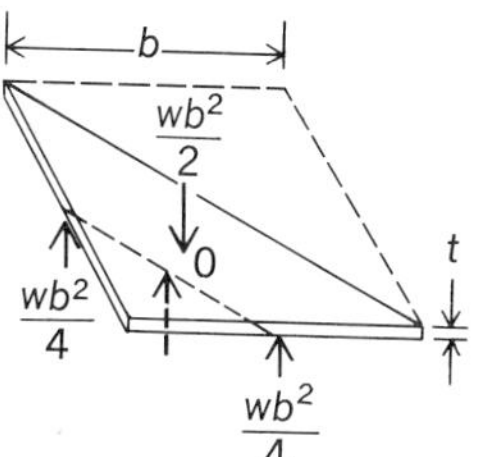

Fig. 4. Forces on triangular half plate serving to analyze uniformly loaded square plate.

Rectangular plate. A simply supported rectangular plate with uniformly distributed load again has the critical section along the diagonal. Because of symmetry, the resultant edge reactions are at the centers of the sides (Fig. 5). On the half rectangle and applying under static equilibrium, the average stress across the diagonal section is shown in Eq. (10). For $a = b$, the expression reduces to that for a square plate. For a rectangle, the stress is always greater in the direction of the shorter span. For a/b very large, the average stress across a midsection parallel to the sides approaches that of a simply supported plate with span $= b$ and $S_{avg} = 3wb^2/4t^2$.

$$S_{avg} = \frac{1}{2}\frac{a^2}{a^2+b^2}\frac{wb^2}{t^2} \qquad (10)$$

Concentrated load at center. A simply supported

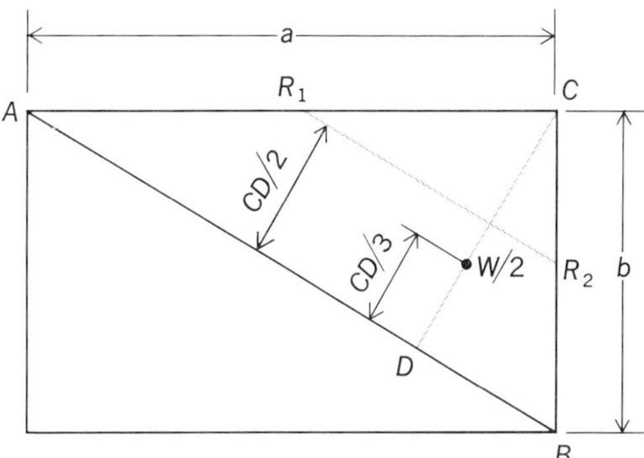

Fig. 5. Rectangular plate simply supported and uniformly loaded, with critical section along diagonal.

square plate with a concentrated load P applied to a central area of diameter d_0 has an average bending stress on the diagonal section $S = 3P/4t^2$. However, the maximum stress at the center calculated from the theory of flexure of plates by H. M. Westergaard is considerably larger, $S = 2.64\ P/t^2$ for $\mu = 1/3$. The higher stresses are localized and plastic yielding, which causes redistribution, does not greatly affect the plate as a whole. For brittle materials or cylic loading involving fatigue, the high localized stress is important.

Continuous plates. In some cases the plate extends beyond its supports. Important examples are flat floor slabs supported by equally spaced columns and "stayed" flat steel plates used as boiler heads.

STRUCTURAL PLATE

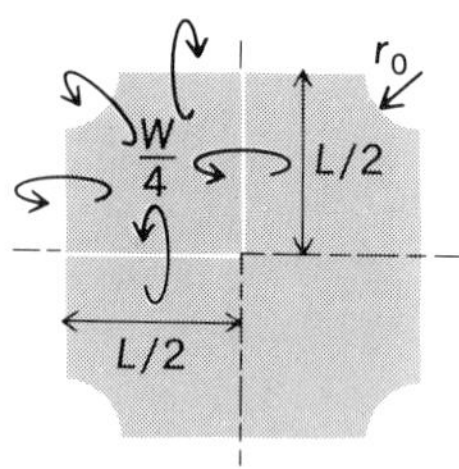

Fig. 6. Continuous plate.

For plates supported by rods or stays attached so as to divide the plate into equal square panels, the maximum bending stress, according to W. C. Unwin, is shown by Eq. (11), where t is thickness,

$$S = \frac{2}{9} w \frac{a^2}{t^2} \tag{11}$$

a is distance between supports (side of the panel) and w is uniformly distributed load per unit area. All units are in inches.

A floor slab may be supported by circular columns L distance apart and dividing the continuous plate into square panels (Fig. 6). The analysis involves moments at five edges of a quarter panel. The distribution of moments along the exterior edges is unknown and the problem is statically indeterminate. From the theory of flexure, Westergaard found expressions for the moment per unit width at various locations in the panel in terms of the average moment per unit width along one entire side of the panel. The expressions take the form shown in Eq. (12), where d_0 is diameter of the

$$M/\text{unit width} = K(w/8)(L - 2d_0/3)^2 \tag{12}$$

column and K has different values according to the location.

Buckling of flat plates. Thin plates subjected to compression loads in the plane of the plate become unstable at a critical elastic stress and suddenly deflect laterally or develop local wrinkling. This susceptibility to buckling depends upon the material constants (E and μ), the ratio of thickness to width or length of the plate, and the edge restraint. *See* COLUMN.

STRUCTURAL PLATE

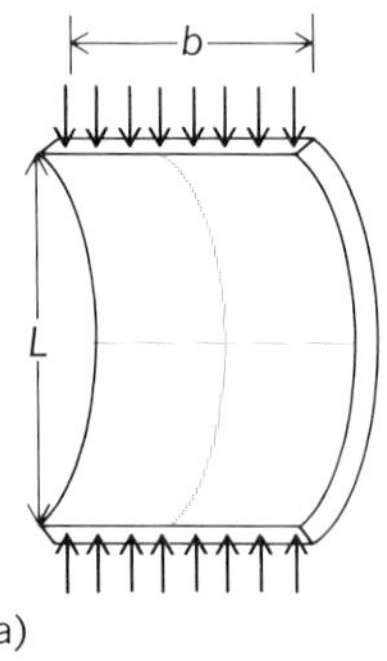

(a)

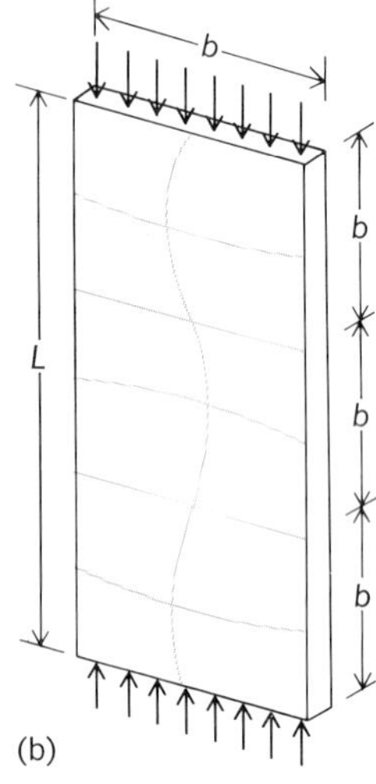

(b)

Fig. 7. Buckling of plates due to distributed loads on ends. (*a*) The edges free and (*b*) the edges simply supported.

A wide thin plate with load uniformly distributed along opposite ends, free to rotate, buckles in primary bending about the longitudinal axis as a long column (Fig. 7). The critical stress initiating buckling is shown in Eq. (13). In the inelastic range, E is replaced by the tangent modulus E'.

$$S_{cr} = \frac{\pi^2 E t^2}{12(1-\mu^2)L^2} \tag{13}$$

A square plate loaded along opposite ends, with edges simply supported so as to be free to rotate but maintained straight, buckles with both lateral and longitudinal curvature. The elastic buckling stress for a square panel with $b = L$ is shown in Eq. (14).

$$S_{cr} = \frac{\pi^2 E}{3(1-\mu^2)} \left(\frac{t}{b}\right)^2 L \tag{14}$$

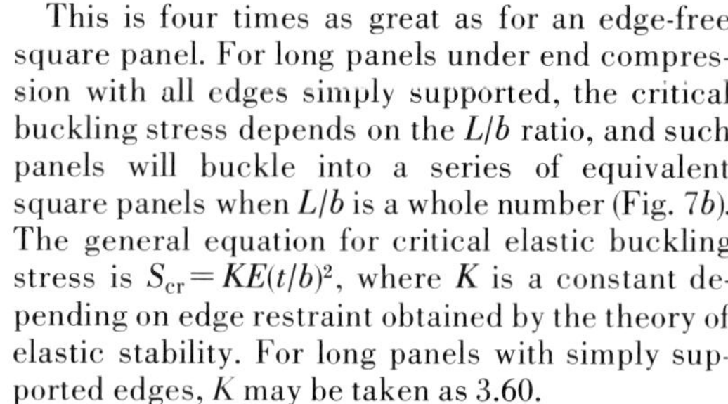

This is four times as great as for an edge-free square panel. For long panels under end compression with all edges simply supported, the critical buckling stress depends on the L/b ratio, and such panels will buckle into a series of equivalent square panels when L/b is a whole number (Fig. 7*b*). The general equation for critical elastic buckling stress is $S_{cr} = KE(t/b)^2$, where K is a constant depending on edge restraint obtained by the theory of elastic stability. For long panels with simply supported edges, K may be taken as 3.60.

For inelastic buckling, an effective modulus replaces E in the elastic formulas such as $\sqrt{EE_t}$, where E_t is the tangent modulus.

[JOHN B. SCALZI]

Bibliography: B. G. Johnston (ed.), *Guide to Stability Design Criteria for Metal Structures*, 3d ed., 1976.

Structural steel

Steel used in engineering structures, usually manufactured by either the open-hearth or the electric-furnace process. The exception is carbon-steel plates and shapes whose thickness is 7/16 in. (11 mm) or less and which are used in structures subject to static loads only. These products may be made from acid-Bessemer steel. The physical properties and chemical composition are governed by standard specifications of the American Society for Testing and Materials (ASTM). Structural steel can be fabricated into numerous shapes for various construction purposes (see illustration).

ASTM-A36 is the specification covering the basic grade of carbon steel used for structures. It has a yield point of 36,000 psi (248 MPa) and a tensile strength of 58,000–80,000 psi (400–552 MPa). An important characteristic is its weldability.

ASTM-A242 is the basic alloy steel in the corrosion-resistant category. It has a yield point of 50,000 psi (345 MPa), tensile strength of 63,000–70,000 psi (434–483 MPa), and is weldable. This steel is given different designations by different

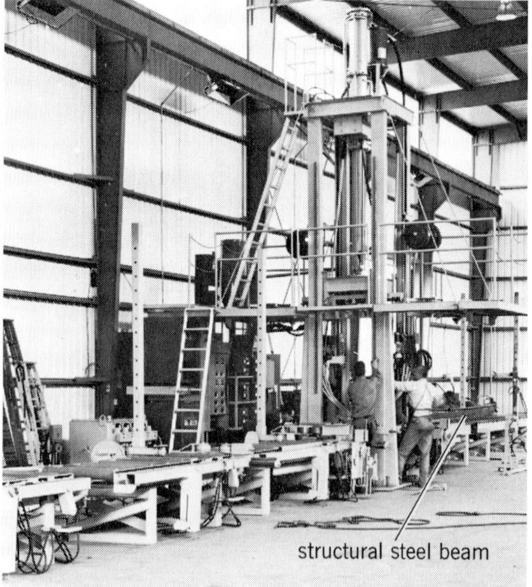

Structural steel beams being fabricated by automatic beam-welding machine. (*Conrac Corp.*)

mills, depending upon the degree of alloying. It contains manganese and copper.

ASTM-441 is alloyed with vanadium, in addition to manganese, copper, and silicon, has a 50,000-psi (345-MPa) yield point [for thicknesses under ¾ in. (19 mm), decreasing to 40,000 psi (276 MPa) for up to 8 in. (20 cm) in thickness], is weldable, and also is widely used for both riveted and bolted work. Another characteristic is that, left unpainted, it weathers to a rich brown color. Another of these weathering steels is ASTM-588. If these steels are used unpainted, precautions must be taken to avoid staining supporting concrete piers with the iron oxide that is washed off in the weathering process.

ASTM-A440 is a high-manganese, corrosion-resistant structural steel intended primarily for riveted and bolted, rather than welded, construction up to a thickness of 4 in. (10 cm). It has a yield point of 50,000 psi (345 MPa) and tensile strength of 63,000–70,000 psi (434–483 MPa).

ASTM-A572 covers a range of low-alloy columbium-vanadium steels with yield points ranging from 42,000 to 65,000 psi (290 to 448 MPa). It can be riveted and bolted, but is not recommended for welding if the yield point exceeds 55,000 psi (379 MPa).

For the highest yield strengths, about 100,000 psi (690 MPa), it is necessary to go to heat-treated steels, which have an ASTM designation of A514. These steels have a low carbon content and may be welded if special procedures are used.

Soft carbon-steel rivets (ASTM-A502-Grade 1) are very satisfactory for carbon-steel structures. Specification ASTM-A502-Grade 2 covers high-strength carbon-manganese rivets which, because of their hardness, are sometimes difficult to drive. ASTM-A406 high-strength rivet steel is free of the driving difficulties and was developed to be used with A242 steels, where corrosion resistance is required.

Many structural connections are made with high-strength bolts, instead of by rivets or welds, and for this purpose ASTM-A325 steel is suitable for bolts up to 1½-in. (3.8-cm) diameter in structures built of A36 steel. For bolts to be used on structures of high-strength steels such as ASTM-A588, and in diameters up to 4 in. (10 cm), ASTM-A490 steel is recommended. *See* STEEL; STRUCTURAL MATERIALS. [WALDO G. BOWMAN]

Structures

Definite arrangements of related elements or members joined together to enable given sets of loads to be supported in a prescribed position. The principal structures designed by civil engineers are bridges, buildings, dams, docks, retaining walls, tanks and bins for storage, transmission towers, radio and television towers, highway pavements, and aircraft landing strips. As man expands his exploration of space and of the ocean depths, engineering structures differing in type and function from those now in use will be necessary. *See* BRIDGE; BUILDINGS; COASTAL ENGINEERING; DAM; FOUNDATIONS; HIGHWAY ENGINEERING; PAVEMENT; STORAGE TANK; TOWER.

A structure should be useful, safe, economical, and esthetically attractive. Primary emphasis must be placed upon safety, and attention once given to economy as the secondary design consideration is giving way to national concern with structural appearance and esthetics. To meet these diverse requirements, the design is usually conceived in four stages, which are carried forward more or less simultaneously.

Functional requirements. The first phase of a design is the development of a general layout that not only will satisfy the functional requirements, but also will permit a structure that will fit attractively into the site chosen. Usually several solutions are prepared for study before one is selected that provides the most satisfactory blend of requirements. *See* CONSTRUCTION ENGINEERING.

Structural scheme. The second major step in design procedure is the development of a structural scheme, or arrangement of members to support the applied loads. Since the functional plan may be greatly influenced by the need to eliminate potential structural difficulties, as well as by the choice of materials and span lengths, the structural scheme is often profitably developed during the functional planning stage. Tentative cost estimates of several possible structural layouts will suggest the most economical scheme. Materials are selected not only on the basis of their structural properties and availability, but also on the availability of the skilled labor needed to work with them. Materials costs and wage scales are also considered. The type of structure and its environment often determine the choice of materials. Steel, aluminum, wood, concrete, masonry, and reinforced and unreinforced plastics have their own distinctive characteristics, and each is particularly suited to a form of construction. *See* STRUCTURAL MATERIALS.

Stress analysis. Analysis of the structural scheme for bending moments, shears, and axial forces caused by applied loads is carried through in the third stage of design. The analysis is based on a sound theoretical background in the laws of statics, the theory of deflections, the principles of statically indeterminate structures, and the art of making the necessary simplifying assumptions. *See* STRUCTURAL ANALYSIS.

Internal forces. In the final phase of the design the individual members of the structure and their connections are proportioned to resist safely the internal forces computed in the structural analysis. The designer must also be proficient in predicting inelastic behavior under combined loadings, in designing for load repetitions, and in predicting buckling loads for inelastic eccentric columns. The foundation for the solution of these problems is a sound knowledge of strength of materials. *See* STRENGTH OF MATERIALS.

These four stages in the design of a structure represent an ideal goal toward which the engineer continually strives but rarely attains. Modern methods of computation, using both analog and digital computers, are gradually providing the assistance that will enable more attention to be paid to alternate designs and layouts than has been possible in the past. Computation programs are already available that permit the analysis of complicated structures, previously avoided. Detailed designs, cost estimates, and even construction drawings are carried through quickly by electronic computation. These advantages remove much of the routine, raise the professional level of the design function, and permit more time to be spent on creative planning for esthetics, function, and structural layout. *See* DIGITAL COMPUTER.

[CHARLES M. ANTONI]

Superconducting devices

Devices that perform functions in the superconducting state that would be difficult or impossible at room temperature, or that contain components which perform such functions. The superconducting state involves a loss of electrical resistance and occurs in many metals and alloys at temperatures near absolute zero. Superconducting devices may be conveniently divided into two categories: small-scale, electronic devices used in measuring instruments and computers, most of which involve Josephson tunneling; and large-scale devices which employ zero-resistance superconducting windings made of type II superconducting materials, and whose applications include high-energy physics research, power generation and transport, motors for marine propulsion, and levitated trains.

Large-scale superconducting devices comprise magnets, motors, generators, and cables using zero resistance superconducting windings. In his original article on the discovery of superconductivity in mercury (1911), H. Kammerlingh Onnes mentioned the possibility of obtaining very high magnetic fields with superconducting coils, since the absence of ohmic losses permitted the use of very high current densities. His original attempts to achieve this failed because of the low critical field of mercury. The superconducting coil of the bubble chamber at the European Commission for Nuclear Research (CERN; Fig. 1) gives an idea of present large-scale applications of superconductivity.

Type II superconductivity. The application of superconductivity in the construction of high-field coils for high-energy physics, motors, generators, levitated trains, and other applications was made possible by the discovery of type II superconductivity: superconducting alloys composed of a superconducting element with impurities in solution (such as bismuth in lead) have a much higher critical field than that of the pure superconductor. In the mid-1960s practical superconducting wires became commercially available with critical fields of 5 to 10 T, and only then did projects for large-scale applications of superconductivity receive serious attention.

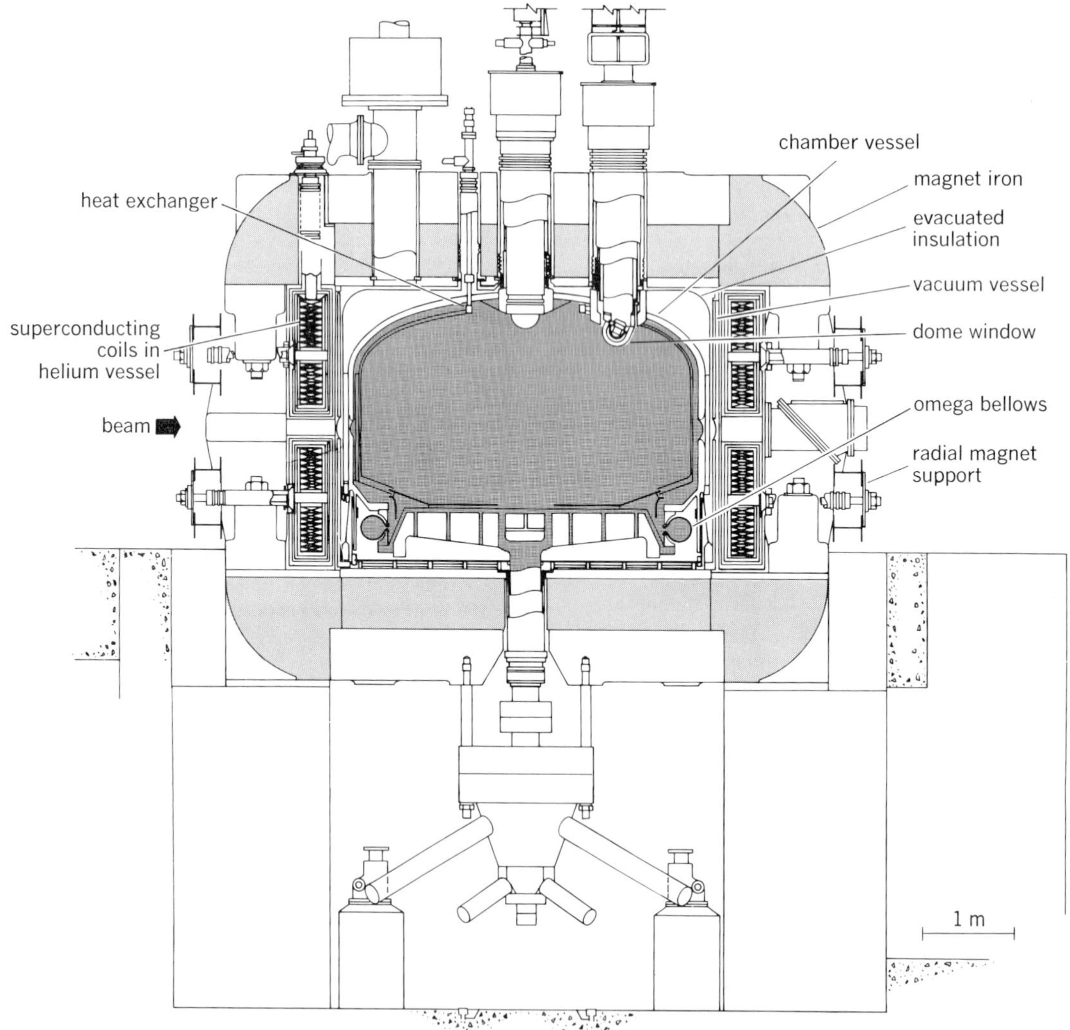

Fig. 1. Hydrogen bubble chamber at CERN. (*Proceedings of the 1967 CERN Conference on Bubble Chambers*).

Higher critical temperatures. Progress in finding alloys with higher critical temperatures has been slow, and superconductivity remains more or less a low-temperature phenomenon: the highest critical temperature known, that of the niobium-germanium alloy Nb_3Ge, is equal to 23 K, and present commercial wires have to be operated below 10 K. The conservative opinion is that superconductors will remain essentially in their present form, although the discovery of an alloy that would superconduct at, say, 30 K can certainly not be excluded. There have been reports of signs of superconductivity in a metastable form of cadmium sulphide (CdS) at 77 K, but these have not been confirmed.

Materials for large-scale applications. Typical values of current densities used in superconducting coils are 10^9 A/m^2 or more—about two orders of magnitude higher than values used in normal conductors, resulting in a large reduction in size and weight as well as in power consumption.

Niobium-titanium alloy. The most commonly used superconducting alloy is the niobium-titanium alloy NbTi. This consists of titanium impurities in solution in niobium, the effect of the titanium impurities being to transform the niobium into a strong type II superconductor. Commercial niobium-titanium wires are composed of thin niobium-titanium filaments, typically 25 μm in diameter, embedded in a copper matrix. The purpose of the copper matrix is to stabilize the superconducting wire: it provides a parallel low-resistance conducting path for the case where a normal spot would appear accidentally along one of the niobium-titanium filaments, thus avoiding an avalanche effect that could quench the whole superconducting coil. The cross section of a typical niobium-titanium-alloy–copper wire is shown in Fig. 2, and its superconducting characteristics are given in Fig. 3.

Niobium-tin alloy. The niobium-tin alloy Nb_3Sn belongs to a different family of alloys, called the A15 compounds, after their crystallographic classification (Nb_3Ge belongs to that same family). They have the highest known critical temperatures and critical fields. Manufacture of A15 wire is difficult because A15 compounds are very brittle. However, filamentary niobium-tin alloy in a copper-tin matrix is used for very-high-field applications. Nb_3Sn can also be used at much higher temperatures than NbTi due to its high critical temperature (18 K). The current-carrying capability of Nb_3Sn at 10 K is about the same as that of NbTi at 4 K. This advantage may be very important for applications where the cost of refrigeration is crucial, such as superconducting cables, and for applications where a large margin of safety is required.

Large-scale magnets. Magnetic fields are used extensively in high-energy physics to accelerate, bend, focus, and store particle beams and to detect and identify elementary particles. Superconducting magnets allow the utilization of higher fields in larger volumes at a lower cost in capital investment and energy expenses. As an example, the bubble chamber magnet at CERN would require a power of 70 MW with conventional coils, while the superconducting version consumes less than 1 MW. It gives a field of 35,000 G (3.5 T) inside a coil about 4.7 m in diameter and 4.4 m in height which is cooled in a liquid helium bath. The energy stored in the coil approaches 1 GJ. Programs employing superconducting magnets that have been undertaken include the construction of a 1000-GeV

Fig. 2. Cross section of NbTi wire consisting of NbTi filaments (the smaller components) embedded in copper matrix. Current capacity is 20 kA at 5 T and 4.2 K. *(From G. Bogner, Large-scale applications of superconductivity, in B. B. Schwartz and S. Foner, eds., Superconductor Applications: SQUIDS and Machines, chap. 20, pp. 547–719, 1977)*

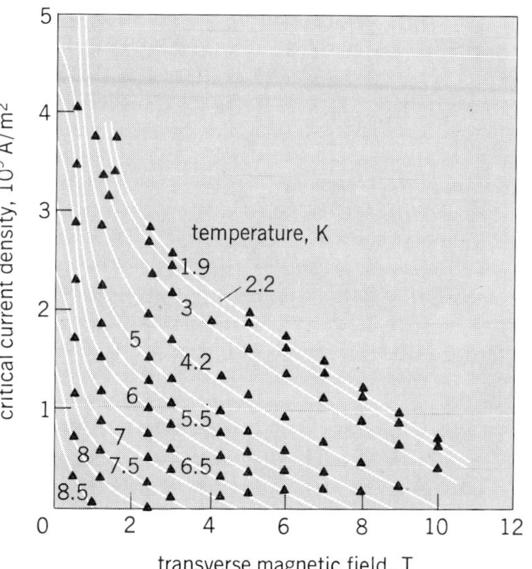

Fig. 3. Critical current density of NbTi as function of transverse magnetic field in the temperature range of 1.9–8.5 K. *(From G. Bogner, Large-scale applications of superconductivity, in B. B. Schwartz and S. Foner, eds., Superconductor Applications: SQUIDS and Machines, chap. 20, pp. 547–719, 1977)*

proton accelerator at the Fermi National Laboratory in Batavia, IL (for which 40,000 kg of superconducting wire is required), and the colliding-beam facility at Brookhaven National Laboratory with two superimposed rings each 2700 m long.

Power generation and transport. Electric power generation has emerged as one of the most promising areas of application of superconductivity.

Superconducting ac generators. Conventional ac generators are composed of a rotating multipole electromagnet (the rotor) and a fixed armature (the stator). Power is produced in the stator windings as the magnetic flux lines produced by the rotor cut them periodically. As the machine rotates at a fixed speed, the power output is determined by the amplitude of the magnetic field produced by the rotor. Because of the losses occurring in a superconductor submitted to a high-intensity ac field, superconducting ac generators already built or currently being developed have a superconducting rotor and a normal armature.

The advantages of the superconducting machine over conventional generators derive from the absence of electrical losses in the rotor and from the higher magnetic field intensity provided by the superconducting coils. The reduced losses and correspondingly higher overall efficiency of the generator (by about 0.5%) result in savings that could pay for the generator over its lifetime (30 to 40 years). The higher field leads to a number of favorable characteristics: reduced weight and size (by about a factor of 2), and therefore reduced installation costs; elimination of the iron from the rotor and armature teeth, hence lower machine reactances and improved transient handling and overall network stability; high-voltage armature feasibility. The table summarizes the projected comparison between the superconducting and the conventional generators at the 1200-MVA rating level.

A 5-MVA machine has been built and tested in the United States, and testing of a 20-MVA ma-

Comparison of superconducting and conventional 1200-MVA generators*

Characteristic	Superconducting	Conventional
Phase to phase voltage, kV	26–500	26
Line current, kA	26.6–1.4	26.6
Active length, m	2.5–3.5	6–7
Total length, m	10–12	17–20
Stator outer diameter, m	2.6	2.7
Rotor diameter, m	1	1
Rotor length, m	4	8–10
Synchronous reactance, pu	0.2–0.5	1.7–1.9
Transient reactance, pu	0.15–0.3	0.3–0.4
Subtransient reactance, pu	0.1–0.2	0.3
Field exciter power, kW, continuous	6	5000
Generator weight, tons†	160–300	600–700
Total losses, MW	5–7	10–15

*From M. Rabinowitz, Cryogenic power generation, *Cryogenics*, 17(6):319–330, 1977. †1 short ton = 0.9 metric ton.

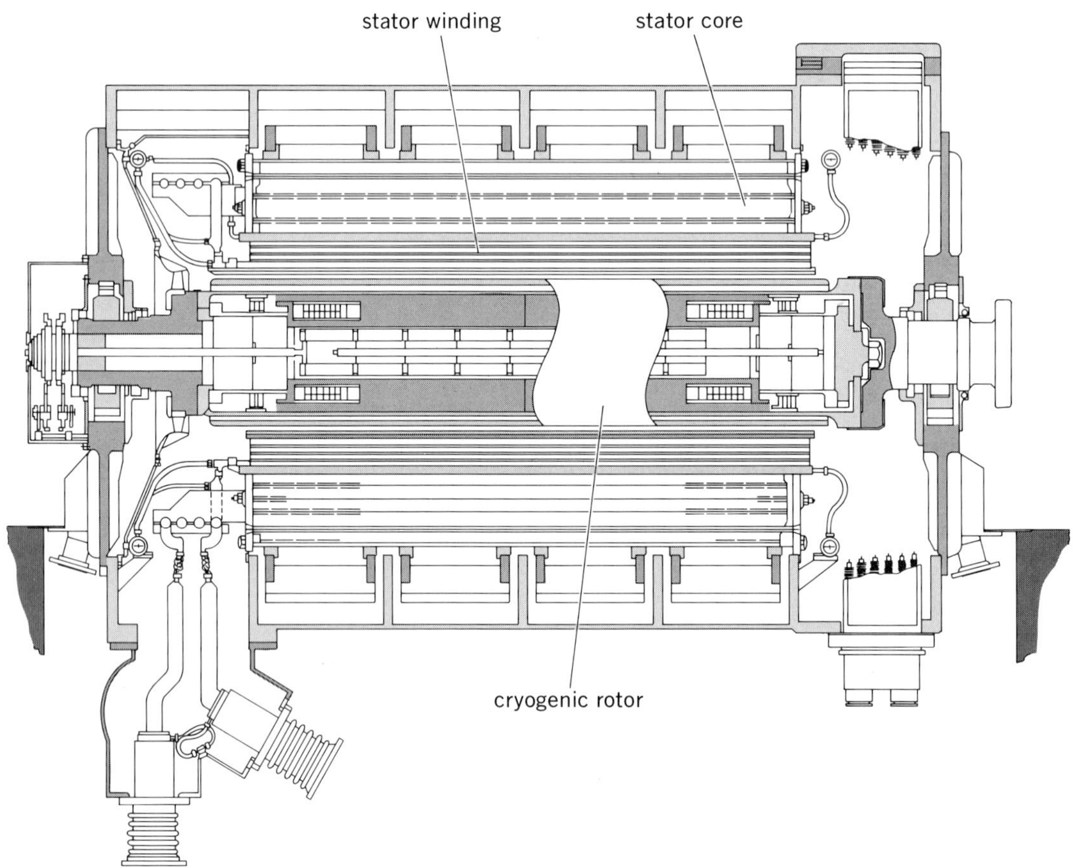

Fig. 4. Configuration of 300-MVA superconducting generator. (*From C. Flick et al., General design aspects of a 300 MVA superconducting generator for utility application, IEEE Trans. Magnet., MAG-17(1):873–879, 1981*)

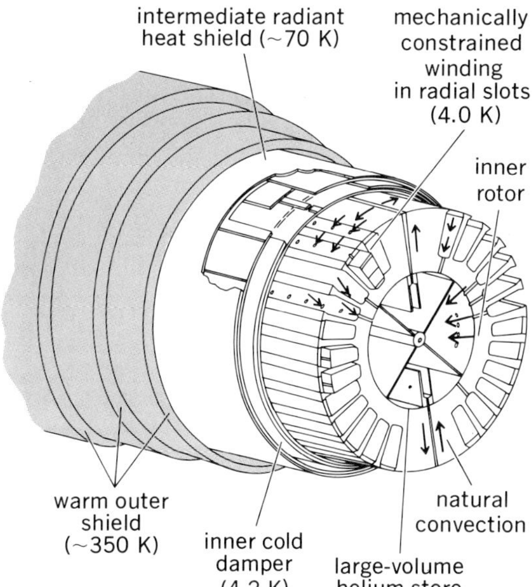

Fig. 5. Cutaway view of rotor of 300-MVA superconducting generator. (*From C. Flick et al., General design aspects of a 300 MVA superconducting generator for utility application, IEEE Trans. Magnet., MAG-17(1):873–879, 1981*)

chine and construction a 300-MVA machine with a 5-T rotor field (Figs. 4 and 5) have been undertaken. The last should be in use in a power station by 1984–1985. If the project is successful, it will signal the first major industrial application of superconductivity. *See* ALTERNATING-CURRENT GENERATOR; GENERATOR.

Superconducting power cables. Short sections of flexible niobium-based superconducting ac cables have been successfully tested. However, due to large refrigeration costs, it is only at ratings of the order of 5 GVA that they could compete with conventional forced-cooling cables. The economics of dc cables using the high-critical-temperature niobium-tin or niobium-germanium alloys could be quite different, and perhaps they could be used in conjunction with magnetohydrodynamic (MHD) power plants, discussed below. *See* DIRECT-CURRENT TRANSMISSION; TRANSMISSION LINES.

Very-large-scale superconducting magnets. Very large superconducting magnets with stored energies of the order of 10 GJ or more have several areas of application.

MHD power plants. In an MHD power plant an electrically conducting expanding fluid moves across magnetic flux lines, and electricity is produced in much the same way as in a conventional generator. Corrosion in the hot expansion chamber and the high intensity of the magnetic field (about 5 T) required in its large volume (approaching 100 m^3) have been the major problems of MHD development. There has been renewed interest in MHD plants because they can burn coal with very high efficiency (50% as compared to 40% in a conventional coal-fired plant). Many of the corrosion problems have been solved, and it is thought that the large-scale superconducting magnets necessary for economical operation can be built (Fig. 6). The first 25-MW MHD plant was built in Moscow, and testing of a superconducting magnet built by Argonne National Laboratory has been undertaken in the Soviet plant. The next step is the construction of a 500-MW commercial-size plant, which may take place at the end of the 1980s. *See* MAGNETOHYDRODYNAMIC POWER GENERATOR.

Controlled fusion. In a controlled thermonuclear reaction by magnetic confinement, large-scale coils are needed to produce the field. The United States fusion-confinement program calls for the construction of a 100-MW power reactor in 1985, a 500-MW reactor in 1990, and a 5000-MW demonstration commercial power plant by the end of the 20th century. The Japanese program also provides for a 2000-MW plant by the year 2000. A similar tokamak program, the JET (Joint European Toroidal Magnetic System) program, has been agreed

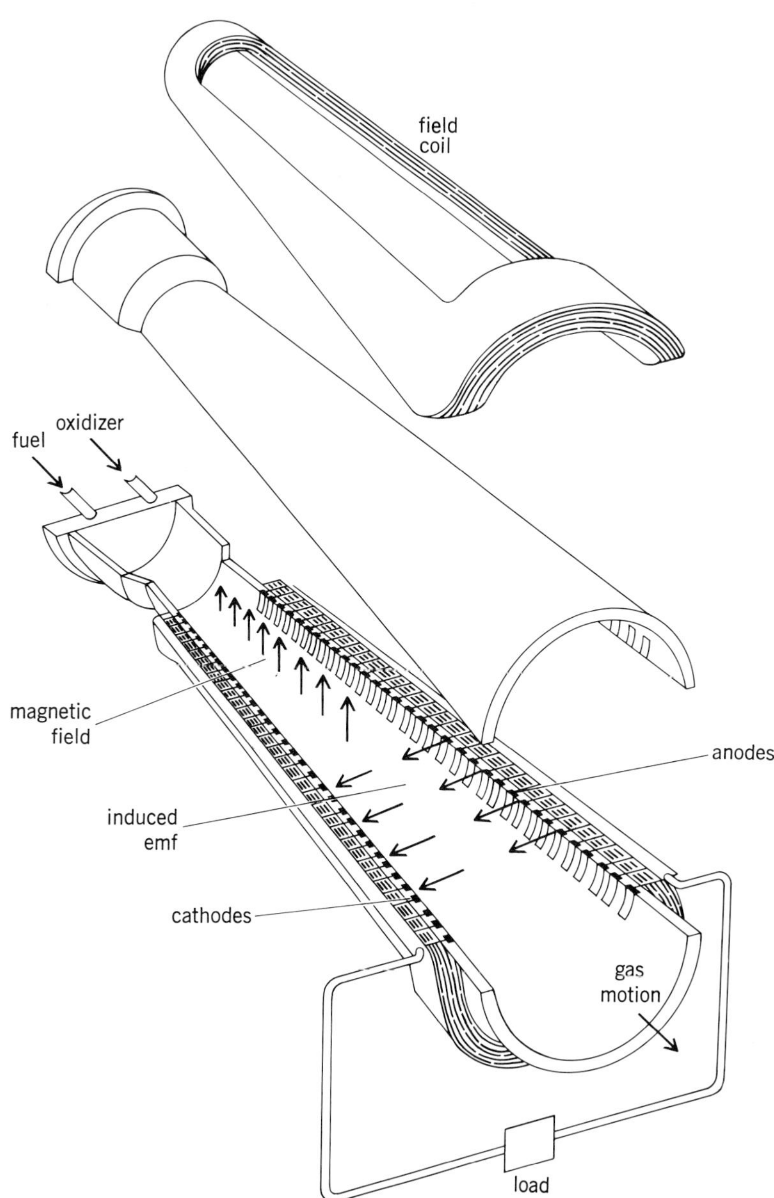

Fig. 6. Exploded view of an MHD expansion chamber and magnet. (*From G. Bogner, Large-scale applications of superconductivity, in B. B. Schwartz and S. Foner, eds., Superconductor Applications: SQUIDS and Machines, chap. 20, pp. 547–719, 1977*)

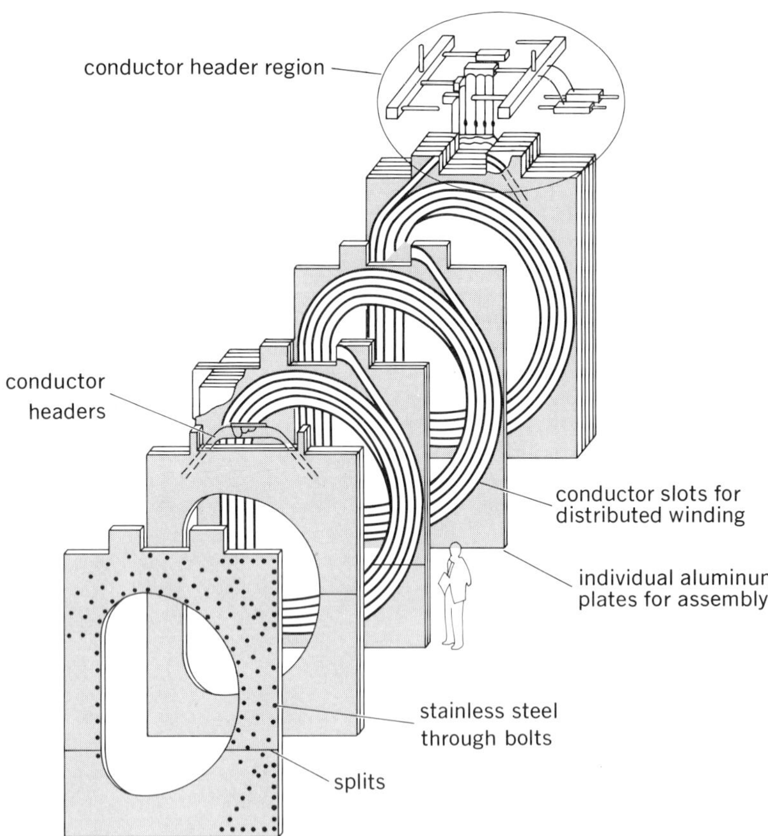

Fig. 7. Exploded view of superconducting coil for fusion reactor. (*From P. N. Haubenreich, Superconducting magnets for toroidal fusion reactors, IEEE Trans. Magnet., MAG-17(1):31–37, 1981*)

upon in Europe; the reactor will be built at Culham, in Great Britain. All these programs will use superconducting coils ranging in diameter from about 1 to 20 m as larger plants are built (Fig. 7).

Energy storage. Giant superconducting coils have been proposed for utilities energy storage (peak leveling), but smaller-size coils with energy storage on the order of 100 MJ are of immediate interest for power oscillation damping and suppression of subsynchronous resonance. *See* ENERGY STORAGE.

Transportation. Large-scale magnets have potential applications for sea and land transportation.

Motors for marine propulsion. Among the most promising applications for superconducting magnets for the short term are motors for marine propulsion. The much smaller weight of the superconducting motor makes electrical propulsion practical. The complete propulsion unit is then composed of a prime mover (and a superconducting generator located where convenient) and a small superconducting motor next to the propeller.

Levitated trains. One of the most spectacular applications of superconducting magnets is for levitated trains for high-speed transportation (Fig. 8). Their principle of operation is quite simple: When a magnet moves over an electrically conducting medium (such as a metallic sheet), the excited eddy currents are equivalent to an image magnet of opposite polarity located beneath the conducting sheet. Repulsion between the magnet and its image results in levitation of the magnet, provided the field intensity and the speed of motion of the magnet are high enough.

Fig. 8. Superconducting test vehicle and track at Erlangen, West Germany. (*Siemens Research Laboratory*)

The Japanese National Railway in collaboration with the Fuji, Hitachi, Mitsubishi, and Toshiba electrical companies has achieved leadership in this field. An experimental vehicle has reached a speed of 520 km/h. The superconducting magnets, working in the persistent mode, are refrigerated by an on-board cooling system. The track is the active element of the propulsion system, which does not require any physical (mechanical) contact with the vehicle. A traveling electromagnetic wave is excited in the track, and the vehicle essentially rides along with the wave in a sort of electromagnetic "surf" mode. Only a portion of the track is excited at a time. Clearance between the vehicle and the track is of the order of 100 mm or more at speeds exceeding 120 km/h. *See* RAILROAD ENGINEERING.

[GUY DEUTSCHER]

Bibliography: S. Foner and B. B. Schwartz (eds.), *Superconducting Machines and Devices: Large-Scale Systems Applications*, 1974; P. G. de Gennes, *Superconductivity of Metals and Alloys*, 1966; J. Matisoo, The superconducting computer, *Sci. Amer.*, 242(5):50–65, May 1980; Proceedings of the 1980 Applied Superconductivity Conference, *IEEE Trans. Magn.*, vol. 17, no. 1, January 1981; B. B. Schwartz and S. Foner (eds.), *Superconductor Applications: SQUIDs and Machines*, 1977; Special issue of Josephson computer technology, *IBM J. Res. Develop.*, 24(2):107–252, 1980; Special issue on superconducting devices, *IEEE Trans. Elec. Devices*, ED-27(10):1855–2042, 1980.

Superheater

A component of a steam-generating unit in which steam, after it has left the boiler drum, is heated above its saturation temperature. The amount of superheat added to the steam is influenced by the location, arrangement, and amount of superheater surface installed, as well as the rating of the boiler. The superheater may consist of one or more stages of tube banks arranged to effectively transfer heat from the products of combustion.

The primary, or initial, stage of a superheater usually consists of a tube bank which is swept by gases that already have given up some of their heat to the final or outlet stage of the superheater. The transfer of heat from the gases to the steam is predominantly by convection; thus this section of the superheater usually is referred to as a convection superheater.

To obtain the highest outlet steam temperature, the final stage of the superheater generally is located in a zone close to the furnace. In some designs the tubes actually form a part of the furnace enclosure or project into the furnace in the form of loops or platens on wide lateral spacings. In these designs the transfer of heat from the products of combustion to the steam is primarily by radiation; such arrangements are known as radiant superheaters.

The temperature characteristics of convection and radiant superheaters differ greatly as the operating rate of the boiler unit is varied. Thus it frequently is advantageous to combine the two types in series to obtain a uniform or relatively constant steam temperature over a wide load range (see illustration). Similar characteristics may be obtained when firing conditions allow superheater arrangements within the boiler tube bank which effectively utilize both radiation and convection heat transfer. *See* STEAM-GENERATING UNIT; STEAM TEMPERATURE CONTROL.

[GEORGE W. KESSLER]

SUPERHEATER

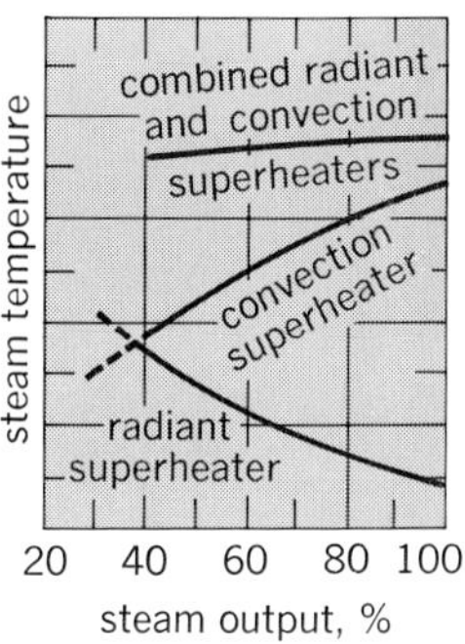

Characteristics of radiant and convection superheaters. (*From T. Baumeister, ed., Standard Handbook for Mechanical Engineers, 7th ed., McGraw-Hill, 1967*)

Superplasticity

The unusual ability of some metals and alloys to elongate uniformly thousands of percent at elevated temperatures, much like hot polymers or glasses. Under normal creep conditions, conventional alloys do not stretch uniformly, but form a necked-down region and then fracture after elongations of only 100% or less.

The earliest superplastic alloys were the zinc–22% aluminum eutectoid and several eutectic alloys. Since 1962, superplasticity has been demonstrated in many commercial alloys as well as ceramics. The most important requirements for obtaining superplastic behavior include a very small metal grain size, a well-rounded (equiaxed) grain shape, a deformation temperature greater than one-half the melting point, and a slow deformation rate. *See* ALLOY; EUTECTICS.

Metal forming. Superplasticity is important to technology primarily because large amounts of deformation can be produced under low loads. Thus, conventional metal-shaping processes (for example, rolling, forging, and extrusion) can be conducted with smaller, and cheaper, equipment. Nonconventional forming methods can also be used; for instance, vacuum-forming techniques, borrowed from the plastics industry, have been applied to sheet metal to form car panels, refrigerator door linings, and TV chassis parts. *See* METAL FORMING.

Stress and strain. The stress applied during superplastic deformation is markedly dependent on the strain rate. This dependence is usually evaluated in terms of the constant m in the equation $\sigma = K\dot{\epsilon}^m$, where σ is the applied stress, K is a constant, $\dot{\epsilon}$ is the strain rate, and m is called the strain-rate sensitivity index. Figure 1*a* depicts a typical curve of stress-versus-strain rate, plotted on log-log coordinates. The slope of the curve at any point gives the value of m at that particular strain rate, and the maximum m occurs at the point of inflection of the curve (Fig. 1*b*).

The values of m in normal and superplastic alloys are significantly different. Normal alloys have low values of m (0.2 or less), while some superplastic alloys attain a maximum m of about 0.8. For values of m that approach closer to 1, the material acts increasingly like a Newtonian-viscous solid, in which elongation in tension is uniform and free of the necking tendencies that limit the maximum elongation of normal metals and alloys.

Although m is a constant for a specified set of conditions, it is a function of the temperature T, the mean grain size $\bar{L}$, and the strain rate $\dot{\epsilon}$. Experimentally it is found that m increases with increasing T or decreasing $\bar{L}$, but goes through a maximum when plotted versus $\dot{\epsilon}$. Since the maximum value of m corresponds closest to the superplastic condition, researchers are also interested in the relationship of m_{max} to these variables. In general, m_{max} increases with decreasing $\bar{L}$, but decreases with increasing T or ϵ. For most metal-forming operations, of course, it is desirable to use a lower

SUPERPLASTICITY

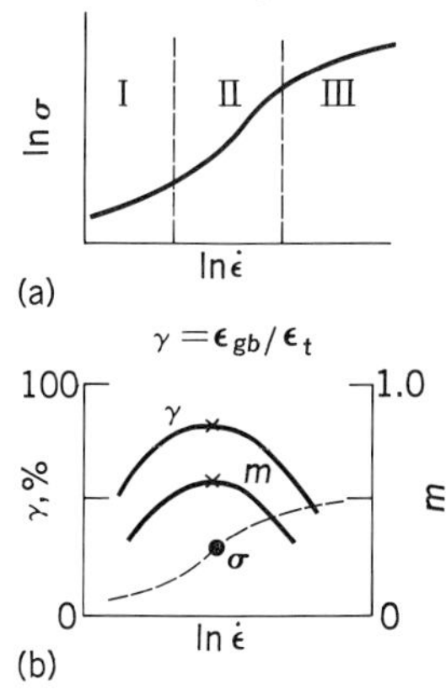

Fig. 1. Stress and strain rate relations for superplastic alloys. (*a*) Typical curve. (*b*) Variations of γ and m with $\ln \dot{\epsilon}$.

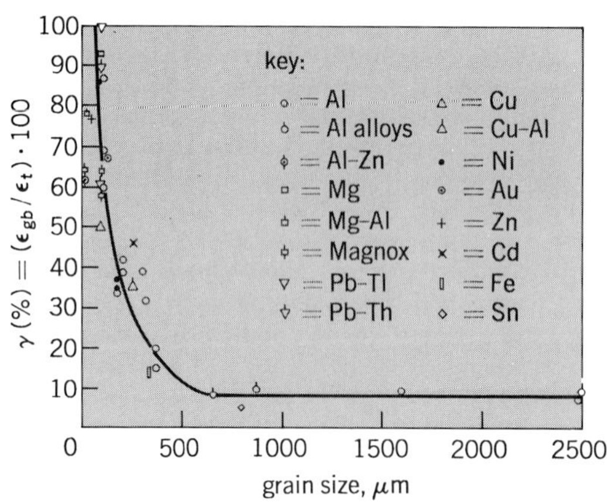

Fig. 2. Grain boundary sliding as a function of grain size (T, high; σ, low; grain size, small).

T, higher $\dot{\epsilon}$, and larger $\bar{L}$, along with the highest possible m value.

Three regions are identified on the plot of Fig. 1*a*. At the lowest stresses and strain rates (region I), diffusion creep predominates, while at the higher stresses and strain rates (region III), dislocation creep and slip processes are controlling. In region II superplastic behavior is observed and grain boundary sliding (GBS) is at a maximum. However, there is some overlapping of regions, so "pure" superplasticity is seldom observed. Although GBS is the predominant mode of deformation, there may be some contribution from either diffusion creep or dislocation creep.

When m is a maximum, there is also a maximum in GBS (Fig. 1*b*). The fraction of the total strain (ϵ_t) due to GBS (ϵ_{gb}) is denoted by γ, which equals ϵ_{gb}/ϵ_t. The extent of GBS in high-temperature deformation is shown in Fig. 2, which plots γ versus grain size for low-load, high-temperature conditions. It is apparent that the total strain depends increasingly on GBS as the grain size becomes smaller, regardless of the type of metal or alloy. All current theories of superplasticity are based on GBS as the primary deformation mechanism. However, an accommodation process, such as the diffusion of atoms along grain boundaries, is necessary in order to smooth the boundaries and grain corners so that sliding can occur readily. *See* PLASTICITY.

[ERVIN E. UNDERWOOD]

Bibliography: T. H. Alden, Review topics in superplasticity, in R. J. Arsenault (ed.), *Treatise on Materials Science and Technology*, vol. 6, 1975; T. J. Headley, D. Kalish, and E. E. Underwood, The current status of applied superplasticity, in J. J. Burke and V. Weiss (eds.), *Ultrafine-Grain Metals*, 1970; C. Massonnet et al., *Plasticity in Structural Engineering: Fundamentals and Applications*, 1980; R. B. Nicholson, The role of metallographic techniques in the understanding and use of superplasticity, in G. Thomas et al. (eds.), *Electron Microscopy and Structure of Materials*, 1972.

Surface condenser

A heat-transfer device used to condense a vapor, usually steam, by absorbing its latent heat in a cooling fluid, ordinarily water. Most surface condensers consist of a chamber containing a large number of 0.5- to 1-in. (1.25- to 2.5-cm) diameter corrosion-resisting alloy tubes through which cooling water flows. The vapor contacts the outside surface of the tubes and is condensed on them. The tubes are arranged so that the cooling water passes through the vapor space one or more times. About 90% of the surface is used for condensing vapor and the remaining 10% for cooling noncondensable gases. Air coolers are normally an integral part of the condenser but may be separate and external to it. The condensate is removed by a condensate pump and the noncondensables by a vacuum pump. *See* STEAM CONDENSER; VAPOR CONDENSER.

[JOSEPH F. SEBALD]

Surface hardening of steel

The selective hardening of the surface layer of a steel product by one of several processes which involve changes in microstructure with or without changes in composition. Surface hardening imparts a combination of properties to the finished product not produced by bulk heat treatment alone. Among these properties are high wear resistance and good toughness or impact properties, increased resistance to failure by fatigue resulting from cyclic loading, and resistance to surface indentation by localized loads. The use of surface hardening frequently is also favored by lower costs and greater flexibility in manufacturing.

Hardening processes. The principal surface hardening processes are: (1) carburizing, (2) the modified carburizing processes of carbonitriding, cyaniding, and liquid carburizing, (3) nitriding, (4) flame hardening and induction hardening, and (5) surface working.

Carburizing introduces carbon into the surface layer of low-carbon steel parts and converts that layer into high-carbon steel, which can be quench-hardened by appropriate heat treatment. Carbonitriding, cyaniding, and liquid carburizing, in addition to supplying carbon, introduce nitrogen into the surface layer; this element permits lower case-hardening temperatures and has a beneficial effect on the subsequent heat treatment. In nitriding, only nitrogen is supplied, and reacts with special alloy elements present in the steel.

Whereas the foregoing processes change the composition of the surface layer, flame hardening and induction hardening depend on a heat treatment applied selectively to the surface layer of a medium-carbon steel. Surface working by shot peening, surface rolling, or prestressing improves fatigue resistance by producing a stronger case, compressive stresses, and a smoother surface.

Table 1 lists the surface-hardening processes for

Table 1. Surface-hardening processes for steel

Process	Elements added	Hardening mechanism
Carburizing	Carbon	Formation of martensite
Carbonitriding	Carbon and nitrogen	Formation of martensite
Cyaniding	Carbon and nitrogen	Formation of martensite
Liquid carburizing	Carbon and nitrogen	Formation of martensite
Nitriding	Nitrogen	Precipitation of alloy nitrides
Flame hardening	None	Formation of martensite
Induction hardening	None	Formation of martensite
Surface working	None	Work hardening

steel and their major characteristics. Processes related to the surface hardening of steel are hardfacing, metal spraying, electroplating, and various diffusion processes involving elements such as aluminum, silicon, and chromium. *See* CLADDING; ELECTROPLATING OF METALS; METAL COATINGS.

Carburizing. The oldest method of surface hardening steel, carburizing, introduces carbon into the surface layer of a low-carbon steel by heating above the transformation range in contact with a carbonaceous material. The carbon diffuses into the steel from the surface and thus converts the outer layer into high-carbon steel. The composite is then heat-treated by the procedures generally applicable to steels. In particular, it must be cooled from above the transformation temperature at a rate sufficiently fast to transform the high-carbon surface layer into a hard martensitic case while the low-carbon core remains tough and shock-resistant. The quench is usually followed by a low-temperature stress-relief anneal. *See* HEAT TREATMENT (METALLURGY).

Pack carburizing. In pack carburizing, carbon is supplied by charcoal or coke to which carbonates or organic materials are added; the mixture is known as the carburizing compound. Parts to be hardened are packed in a steel box with the carburizing compound and heated to the carburizing temperature, usually 1700–1750°F (925–955°C). Carbon is transferred to the steel by the formation of carbon monoxide at the compound surface and by its decomposition to carbon and carbon dioxide at the steel surface according to the reaction given below. The carbonates or organic materials in the

$$C + CO_2 \rightleftharpoons 2CO$$

carburizing compound decompose and increase the concentration of carbon oxide gases in the box required for the transfer of carbon to the steel. The depth of penetration of the carbon depends upon the time and temperature at which the treatment is carried out.

The principal advantage of pack carburizing is its simplicity; no expensive equipment is required. The results are almost certain to be satisfactory with proper temperature control and hardening practice. The hardening treatment usually requires that the parts, after removal from the box, be heated again above the transformation temperature, and that this process be followed by quenching.

Gas carburizing. In gas carburizing the parts are heated in contact with carbon-bearing gases, commonly carbon monoxide and hydrocarbons. The hydrocarbons may be methane, propane, butane, or vaporized hydrocarbon fluids. They are usually diluted with an inert carrier gas to control the amount of carbon supplied to the steel surface and to prevent the formation of soot.

The carbon monoxide and hydrocarbons are decomposed at the steel surfaces, the carbon thus liberated being absorbed by the steel. Close control of gas composition is required because the rate at which carbon is supplied to the steel surface controls the concentration of carbon in the carburized case. This control is an important advantage of gas carburizing. The process is cleaner and entails lower labor costs than pack carburizing. Direct quenching from the furnace is possible in gas carburizing, hence the process is particularly well suited to volume output for the mass production industries. In many installations large continuous furnaces with attached quenching and tempering equipment are used; the parts are charged into one end and leave at the other in the carburized and hardened state. Batch furnaces are also used for gas carburizing.

Steels for carburizing. The selection of steels for carburizing primarily concerns grain growth characteristics, carbon and alloy content, machinability, and cost. Steels that retain a fine-grained structure at the case-hardening temperature are desirable because they permit simple heat-treatment procedures, in particular, hardening by a direct quench. A fine-grained structure in the finished product is essential for maximum shock resistance.

Plain carbon steels are satisfactory for many applications in which low distortion is not a critical requirement and for which optimum core properties are not required. The most common carbon contents are 0.20% in plain carbon steels and 0.08–0.20% in alloy steels. Steels with relatively high sulfur content are frequently used for improved machinability.

The most common alloy elements in carburizing steels are nickel (0.5–3.5%), chromium (0.5–1.6%), and molybdenum (0.1–0.25%). The nickel-molybdenum steels are particularly popular for strength and toughness of the core and toughness of the case. Alloy steels have less tendency to develop coarse-grained structures at the carburizing temperature. They also permit slower quenching rates in hardening, thus reducing distortion and the tendency to crack during quenching. Their higher cost is frequently more than offset by lower finishing costs due to reduced distortion. Improved fatigue resistance is another consideration in the selection of alloy steels.

Typical parts surface-hardened by carburization, as described above, or by modified carburizing processes, described below, include gears, ball and roller bearings, piston pins, sprockets and shafts, clutch plates, and cams.

Modified carburizing processes. Alloying the steel with nitrogen lowers the transformation temperature and reduces the transformation rate. Thus modification of the carburizing process by the diffusing of both carbon and nitrogen into the surface layer of the steel enables the process to be carried out at lower temperatures than with carbon alone. The sources of carbon and nitrogen distinguish the several processes from each other.

Carbonitriding. The carbonitriding process is the same as gas carburizing except that ammonia is added to the furnace atmosphere to provide the source of nitrogen. The amount of nitrogen absorbed by the steel can be controlled by the concentration of ammonia in the furnace atmosphere and the temperature. Thus the nitrogen content of carbonitrided cases may vary from small concentrations to the relatively high level characteristic of cyaniding.

Both because carbonitriding is conducted at lower temperatures than carburizing and because the slower transformation rate permits slower quenching rates, less distortion results from the hardening process. Consequently, plain carbon

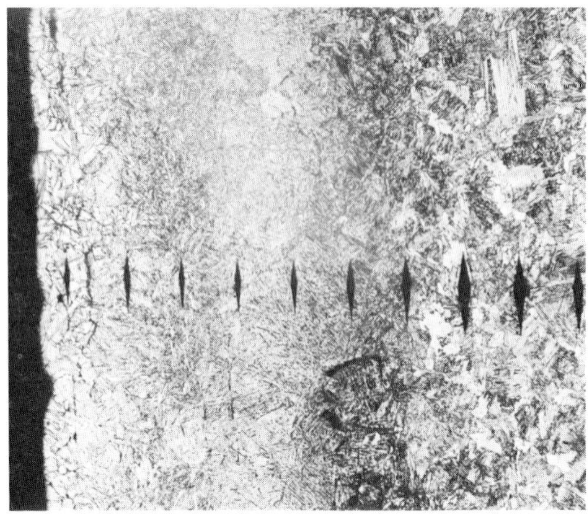

Fig. 1. Microstructure of typical carbonitrided steel. Dark border is specimen mount. Martensite needles (gray) and retained austenite (white) can be seen near surface. Microhardness indentations (black) are smaller in the case than in the core, indicating the greater hardness of the case. *(A. J. Gregor)*

steel can frequently be substituted for alloy steels in carbonitrided parts.

Figure 1 shows a typical carbonitrided case. The microstructure is similar to a carburized case except that it contains a larger fraction of retained austenite near the surface.

Cyaniding and liquid carburizing. If the parts are immersed into molten baths consisting of solutions of cyanides and other salts, the cyanides supply both carbon and nitrogen to the steel. Thus the results from cyaniding and liquid carburizing are similar to those from carbonitriding. By controlling the composition and temperature of the bath, the amounts of carbon and nitrogen absorbed by the steel can be controlled within limits. The term cyaniding is usually applied to processing at temperatures of 1400–1550°F (760–845°C). Liquid carburizing is carried out at 1600–1750°F (870–955°C); at these temperatures the nitrogen absorption is lower so that the process, as far as the product is concerned, approaches pack and gas carburizing.

Nitrogen in cyanided cases results in the same advantages as those obtained from nitrogen in carbonitrided cases, but the control of nitrogen concentration is less precise. Cyaniding and liquid carburizing have the advantage of rapid heating of the charge because of the good heat transfer from the bath to the steel. These processes have considerable flexibility in that selective hardening may be accomplished by partial immersion and that different parts may be treated in the same bath for different times.

One disadvantage of these processes is that the case for a given treatment time is less deep when low operating temperatures are used. The size of salt baths is limited by the necessity to obtain uniform temperatures throughout. The baths are, therefore, less well adapted to quantity production than furnaces employing gaseous atmospheres. The nitrogen dissolved in the steel may cause an appreciable fraction of the steel to remain untransformed as retained austenite in the final product, particularly in the cyaniding process, but also in carbonitriding, especially at low temperatures.

Nitriding. This process is carried out by heating steels of suitable composition in contact with a source of active nitrogen at temperatures of 925–1100°F (495–595°C) for periods of 1–100 hr, depending upon the steel being treated and the depth of case desired. Under these conditions, nitrides form if the steel contains alloying elements such as aluminum, chromium, molybdenum, vanadium, and tungsten. The formation of alloy nitrides at the nitriding temperature accounts for the hardened case. The microstructure of a typical nitrided steel is shown in Fig. 2.

The usual source of active nitrogen for nitriding is ammonia. However, mixtures of molten cyanide salts are also used, with similar advantages and disadvantages as in cyaniding and liquid carburizing. A typical nitriding installation using ammonia consists of a reservoir of ammonia, a furnace, a retort containing the parts to be case hardened, and equipment to control the temperature and gas flow. At the nitriding temperature, part of the ammonia decomposes at the surface of the steel, liberating active nitrogen, some of which diffuses into the steel. The remainder passes into the molecular form, which is inert. For successful nitriding it is therefore necessary to control the gas flow in such a way as to continuously supply fresh ammonia to all steel surfaces. However, an oversupply of ammonia results in the formation of an excessively thick iron nitride layer on the surface. This so-called white layer can be controlled by regulation of the degree of ammonia dissociation, or it must be removed by grinding if it exceeds a depth of 0.0005 in. (12.7 μm).

Steels that produce the hardest nitrided cases consistent with optimum case depth contain about 1% aluminum, 1.5% chromium, and 0.3% molybdenum. Alloy steels containing only chromium and molybdenum as nitride-forming elements are also popular. Stainless steels and high-speed cutting steels are nitrided for improved wear resistance. Among the parts hardened by nitriding are camshafts, fuel injection pump parts, gears, cylinder barrels, boring bars, spindles, splicers, sprockets, valve stems, and milling cutters.

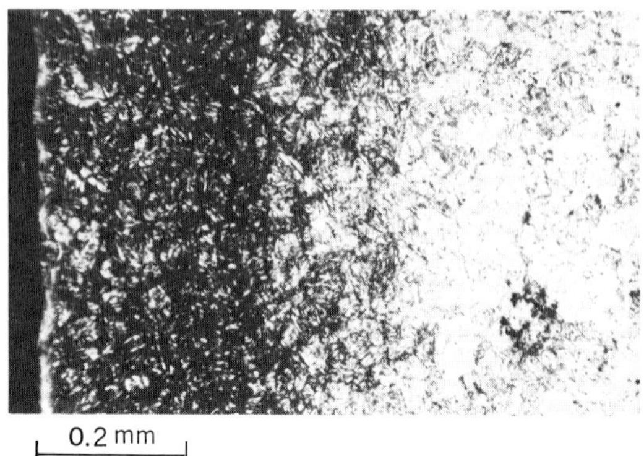

Fig. 2. Microstructure of a nitrided steel, which was heated to 1700°F (925°C) for 1 hr, oil-quenched, tempered at 1250°F (675°C) for 2 hr, air-cooled, and nitrided for 48 hr at 975°F (525°C). Dark border is specimen mount.

Induction and flame hardening. The principle of both flame and induction hardening is to heat quickly the surface of the steel to above the transformation range followed by quenching rapidly. In induction hardening, heat is generated within the part by electromagnetic induction. The part (such as a crankshaft, camshaft, axle, gear, or piston rod) is usually placed inside a copper coil or coils through which a rapidly alternating current is flowing. High-frequency currents are used because they confine the induced currents to the surface of the part to be heated; the higher the frequency, the shallower the case. Short heating cycles minimize conduction of heat to the interior and thus further restrict the heating to the surface layer. *See* INDUCTION HEATING.

In flame hardening, the steel to be surface-hardened is heated by direct impingement of a high-temperature gas flame. The surface layer is quickly heated to a temperature above the transformation range, followed by a quench. In both induction and flame hardening the quenching action is a combination of heat extraction by the cold metal beneath the case and by an external quenching medium.

Induction and flame hardening require close control of time and temperature of heating. In general, induction heating is used when large numbers of symmetrically shaped parts are to be processed. Flame hardening can be readily applied to large parts, such as large gears or lathe ways, and to parts of intricate design, such as camshafts. It is more economical than induction hardening when only a few parts are to be treated.

Steels for induction or flame hardening usually contain 0.4–0.75% carbon. Because no change in composition is involved, the steel is selected for both case and core properties. During hardening the core is not affected and, consequently, the core properties must be developed by proper heat treatment before surface hardening. Cast irons are also induction- or flame-hardened for certain applications. *See* CAST IRON.

Surface-working processes. The selective cold working of the surface layer of parts of steel and other metals increases surface hardness. The working requires a force exceeding the compressive yield strength of the material. Such a force can be applied by various methods, for example, hammer peening, mechanical peening, shot peening, and surface rolling. The surfaces must be accessible to the peening or rolling operation, but the processes can be applied to selected critical areas, such as the fillets of shafts. Prestressing involves stressing a part beyond its yield strength.

Surface-working processes result in substantial improvement of the properties, especially the fatigue resistance. The processes are applied to coil and leaf springs, shafts, gears, and steering knuckles. They have the advantage of being comparatively inexpensive. They must be carefully controlled to avoid overworking or underworking of the surface, both of which fail to give the desired improvement in the fatigue properties.

Properties. The characteristic properties of surface-hardened steels depend on the properties of both the case and the core. Case properties are determined mainly by composition, microstructure, and case depth. Core properties of carburized steels depend primarily on the transformation characteristics of the core during the hardening of the case. In nitriding and in induction and flame hardening, the core properties are developed before the case hardening treatment.

Microstructure. The microstructures of the case and core are controlled by their composition and heat treatment. The structure of carburized, carbonitrided, and cyanided cases is typical of heat-treated high-carbon steel, and that of induction- and flame-hardened cases is typical of medium-carbon steel. The desired microconstituent in each instance is martensite, which usually is tempered. The retained austenite, which is promoted by nitrogen in carbonitrided and cyanided cases, and by some other alloy elements, is generally undesirable. Austenite lowers the hardness and its subsequent transformation in service may cause brittleness and dimensional changes of the case-hardened part. *See* TEMPERING.

The microstructure of nitrided cases consists of finely dispersed alloy nitrides precipitated in a preexisting hardened and tempered structure. In induction- and flame-hardened cases, the fast rate of cooling from the austenitizing temperature results in the formation of martensite. The characteristic feature of surface-worked cases is their cold-worked structure and resulting increase in hardness. *See* PLASTIC DEFORMATION OF METAL.

Case depth. Typical case depths of surface-hardened steels are shown in Table 2. There has been a trend toward thinner cases, which are acceptable because of stronger cores and less distortion during processing. However, for some types of service, especially those involving contact of two loaded parts, the critical stress occurs below the surface; the case depth should preferably be sufficient to allow for this condition. Thick cases are essential in some wear applications.

Table 2. Typical case properties

Process	Typical case depth, in.		Typical hardness
Carburizing	< 0.020	Shallow	55–65 RC
	0.020–0.040	Medium	
	0.040–0.060	Heavy	
	> 0.060	Extra deep	
Carbonitriding	0.003–0.020		55–62 RC
Cyaniding	0.001–0.010		
Nitriding	0.005–0.025		85–95 R-15N
	0.001–0.003	High-speed steel	
Induction hardening	0.010–0.25		50–60 RC
Flame hardening	0.030–0.25		50–60 RC
Surface working	0.020–0.040		

Hardness and other properties. Typical hardness ranges for the several types of cases are listed in Table 2. Other important properties of cases are their frictional and antiseizing characteristics. The latter are determined to a considerable extent by the degree to which the surface softens under frictional heat. Nitrided cases are superior in this respect because their resistance to tempering is greater than that of cases containing martensite.

Residual stresses. Compressive stresses at the surface are desirable because they provide protection against fatigue failure. The expansion accompanying the formation of martensite in carburizing and, similarly, that resulting from the precipitation reactions in nitriding cause these residual compressive stresses. Shot peening or other surface-working processes also cause residual compressive stresses at the surface.

Dimensional changes. In heat treating carburized steels, distortion occurs primarily as a result of uneven quenching, which causes different parts of the surface to transform at different times. In nitriding, dimensional changes result primarily from the increase in volume of the case as the alloy nitrides are precipitated.

[MICHAEL B. BEVER; CARL F. FLOE]

Bibliography: American Society for Metals, *Metals Handbook, Properties and Selection: Irons and Steels*, 1978; C. R. Brooks, *Heat Treatment of Ferrous Alloys*, 1979; D. R. Gabe, *Principles of Metal Surface Treatment and Protection*, 2d ed., 1978.

Surge arrester

A protective device designed primarily for connection between a conductor of an electrical system and gound to limit the magnitude of transient overvoltages on equipment.

The valve arrester consists of a single gap or multiple gaps in series with current-limiting elements (see illustration). The gaps between spaced electrodes prevent the flow of current through the arrester except when the voltage across them exceeds the critical gap flashover. They reseal by developing voltage in the gaps sufficiently to interrupt current flow during normal system voltage, and also for temporary overvoltages. The current-limiting element is a nonlinear resistor whose resistance decreases substantially as the voltage across it increases.

Other types of arresters, called expulsion arresters, have spaced electrodes in an interrupting chamber which contains gas-evolving material.

System overvoltages may be of either external or internal origin, that is, lightning or switching. The arrester is unable to determine the origin of the overvoltage and must attempt to limit the magnitude of all abnormal voltages above the gap sparkover voltage. Hence a lightning arrester is really a voltage-surge arrester.

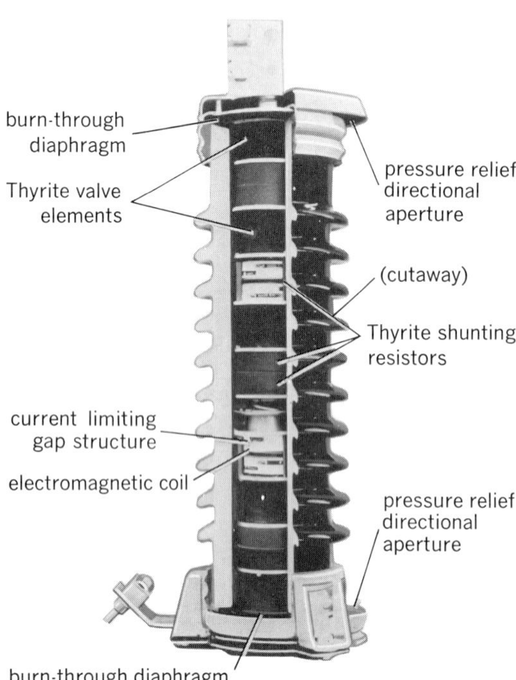

Thirty-kV General Electric Alugard station surge arrester, of valve type. (*General Electric Co.*)

Important application characteristics of arresters are (1) gap sparkover (normal frequency and impulse), (2) gap reseal voltage (maximum permissible normal-frequency or transient voltage across the arrester), (3) discharge voltage across the arrester (valve type) during passage of current, and (4) maximum current discharge capability.

The gap sparkover voltage and the discharge voltage characteristic determine the maximum transient voltage level permitted by the arrester. They are the measures by which the protective efficiency of arresters is determined. The discharge capability is the measure of the arrester endurance to severe lightning and switching surges.

Arresters are classified as station, intermediate, or distribution. Station-class arresters are valve arresters used for the protection of apparatus in important substations where the highest level of protection is desired. Intermediate-class arresters, also valve-type, are used for the protection of apparatus in small- and medium-size substations where economic considerations do not justify the use of station-type arresters. Distribution-class arresters may be either valve or expulsion arresters and are used for the protection of pole-type distribution apparatus, principally distribution transformers. *See* LIGHTNING AND SURGE PROTECTION.

[GLENN D. BREUER]

Bibliography: *Surge Arresters for Alternating Current Power Circuits*, IEEE Stand. 28, ANSI C67.1, 1975.

Surveying

The measurement of dimensional relationships among points, lines, and physical features on or near the Earth's surface. Basically, surveying determines horizontal distances, elevation differences, directions, and angles. These basic determinations are applied further to the computation of areas and volumes and to the establishment of locations with respect to some coordinate system.

Surveying is typically used to locate and measure property lines; to lay out buildings, bridges, channels, highways, sewers, and pipelines for construction; to locate stations for launching and tracking satellites; and to obtain topographic information for mapping and charting.

Horizontal distances are usually assumed to be parallel to a common plane. Each measurement has both length and direction. Length is expressed in feet or in meters. Direction is expressed as a bearing of the azimuthal angle relationship to a reference meridian, which is the north-south direction. It can be the true meridian, a grid meridian, or some other assumed meridian. The degree-minute-second system of angular expression is standard in the United States.

Reference, or control, is a concept that applies to the positions of lines as well as to their directions. In its simplest form, the position control is an identifiable or understood point of origin for the lines of a survey. Conveniently, most coordinate systems have the origin placed west and south of the area to be surveyed so that all coordinates are positive and in the north-east quadrant.

Coordinate systems may be assumed, or established coordinate systems may be used. The most widely used systems in the United States are the various state plane coordinate systems established by the U.S. Coast and Geodetic Survey (now the National Geodetic Survey), which approximate in each case a small portion of the Earth's spherical form.

Vertical measurement adds the third dimension to an object's position. This dimension is expressed as the distance above some reference surface, usually mean sea level, called a datum. Mean sea level is determined by averaging high and low tides during a lunar month, but for certainty this must be carried out for as long as 19+ years.

Precision. Surveying devices and procedures possess individual limitations for accuracy of measurement. The term precision expresses the notion of degree of accuracy.

The choice of methods depends on the precision required for the result. Surveys of major installations or primary control surveys demand higher orders of precision than, for example, a preliminary highway profile or some lesser layout.

Instrumental surveys are designated first-order, second-order, and third-order. First-order is maximum precision, the others gradually less. Established criteria exist to differentiate the levels of precision.

Horizontal control. The main framework, or control, of a survey is laid out by traverse, triangulation, or trilateration. Some success has been achieved in locating control points from Doppler measurements of passing satellites, from aerial phototriangulation, from satellites photographed against a star background, and from inertial guidance systems.

In traverse, adopted for most ordinary surveying, a line or series of lines is established by directly measuring lengths and angles. In triangulation, used mainly for large areas, angles are again directly measured, but distances are computed trigonometrically. This necessitates triangular patterns of lines connecting intervisible points and starting from a baseline of known length. New baselines are measured at intervals. Trigonometric methods are also used in trilateration, but lengths, rather than angles, are measured. The development of electronic distance measurement (EDM) instruments has brought trilateration into significant use.

Traversing. Traverse lines are usually laid out as a closed polygon. This makes it possible to check the accuracy of the field measurements by calculating how nearly the figure closes mathematically. Commonly there are unavoidable random or accidental errors in angular and in distance measurement that tend to coalesce and militate against exact mathematical closure. Angular error is checked by geometric theory: the sum of all interior angles in any polygon equals the number of angles, less two, times 180°. The distance misclosure is determined by resolving the field-measured distance for each line into its north-south component (its latitude) and its east-west component (its departure). For the figure to close, components bearing north must equal in total length those bearing south, and components bearing east must equal in total length those bearing west. The net misclosure in the sum of latitudes and the net misclosure in the sum of departures are the two components of the linear error of closure of the traverse. If the total error of closure is within the prescribed error limit for the order of precision chosen for the survey, the total error is apportioned among the several angles and lengths of the traverse.

Triangulation. This begins with the selection of points whose connecting sight lines form one triangle or a series of triangles. In a series, each triangle has at least one side common to each adjacent triangle. An initial side length, the baseline, is measured, as are all the angles. By successive application of the law of sines, each successive side length is computed. The directions (azimuths) of the sides are also carried forward, so that the relative positions of all points can be determined.

Trilateration. In trilateration the figures are selected as for triangulation. An initial direction is determined and all side lengths are measured. Interior angles are computed by oblique-triangle formulas to obtain geometric checks on distances and to establish triangle-side directions. Ensuing latitude and departure computations yield the relative positions of the angle points of the figures.

Distance measurement. Traverse distances are usually measured with a tape or by EDM, but also may sometimes be measured by stadia, subtense, or trig-traverse.

Taping. The surveyor's tape, a steel ribbon graduated throughout its length, is usually 100 ft (30.48 m) long. However, 200-ft (60.96-m) and 300-ft (91.44-m) tapes are common.

Whether on sloping or level ground, it is horizontal distances that must be measured. Horizontal components of hillside distances are measured by raising the downhill end of the tape to the level of the uphill end. On steep ground this technique is used with shorter sections of the tape. The raised end is positioned over the ground point with the aid of a plumb bob.

Where slope distances are taped along the ground, the slope angle can be measured with the clinometer. The desired horizontal distance can then be computed. Precision in tape measuring is increased by refinements such as standardization, sag correction, tension control, and thermal-expansion correction.

Electronic distance measurement. In EDM the time a signal requires to travel from an emitter to a receiver or reflector and back to the sender is converted to a distance readout. It utilizes the precisely known speed of radio waves or of light, as well as a crystal-controlled chopped or sinusoidal variation in the emitted signal to measure distance in terms of number of wavelengths and a fractional part thereof. Microwave EDM requires a master instrument and a slave instrument, but has a capability of measuring up to 20 mi (32 km) or more. Shorter distances, from a few feet to about 2 mi (3.2 km), are best measured with ordinary light, laser light, or infrared EDM equipment. These require only a single instrument for emitting the signal and a simple retro-reflector or a reflector prism (corner cube) for returning the signal.

The great advantage of electronic distance measuring is its unprecedented precision, speed, and convenience. Further, if mounted directly onto a

Fig. 1. Instrument combines EDM and transit. (*Keuffel & Esser Co.*)

theodolite, and especially if incorporated into it and electronically coupled to it, the EDM with an internal computer can in seconds measure distance (even slope distance) and direction, then compute the coordinates of the sighted point with all the accuracy required for high-order surveying. The development of EDM in recent decades is a great breakthrough, changing the nature and methods of a great part of modern surveying. Accuracies of 1 part in 1,000,000 are now assuredly within reach (Fig. 1).

Stadia. The stadia technique also requires no tape, although it is rather an imprecise distance-measuring method. A graduated stadia rod is held upright on a point and sighted through a transit telescope set up over another point. The distance between the two points is determined from the length of rod intercepted between two horizontal wires in the telescope. *See* STADIA; TRANSIT.

Subtense. In the subtense technique the transit angle subtended by a horizontal bar of fixed length (usually 2.00000 m) enables computation of the transit-to-bar distance (Fig. 2). With the 2-m bar and a theodolite to measure the angle (α) to 1 second, precise distance (D) can be calculated up to 200 m by the formula $D = 1.00000 \cot \frac{1}{2}\alpha$ m. The subtense method can replace taping across a busy highway, canal, or ravine, although it is slower than EDM. For short to medium distances it is more precise than stadia.

Trig-traverse. In trig-traverse the subtense bar is replaced by a measured baseline extending at a right angle from the survey line whose distance is desired. The baseline is at least long enough relative to the desired distance to assure the order of precision desired. The distance calculated in either subtense or trig-traverse is automatically the horizontal distance and so needs no correction because of difference of elevation between theodolite and baseline or subtense bar.

Fig. 2. Subtense bar. (*Lockwood, Kessler, and Bartlett Inc.*)

Angular measurement. The most common instrument for measuring angles is the transit or theodolite. It is essentially a telescope that can be rotated a measurable amount about a vertical axis and a horizontal axis. Carefully graduated metal or glass circles concentric with each axis are used to measure the angles. Glass circles permit much greater magnification and more precise readings because concentrated light is passed through the portions of the circle being read. Further, superimposing the two opposite sides of the circle eliminates eccentricity error. The transit is centered over a point with the aid of either a plumb bob suspended by a string from the vertical axis or (on some theodolites) an optical plummet, which enables the operator to sight along the instrument's vertical axis to the ground through a right-angle prism.

Horizontal angles. To measure a horizontal angle between two intersecting lines, the transit is set up over the intersection. The telescope is sighted along one of the lines, the graduated circle is clamped against rotation, and the telescope is rotated to sight along the other line. The angle is indicated on the horizontal circle by another concentric circular plate, inscribed with an index and a vernier, that rotates with the telescope. Glass-circle transits and theodolites use optical micrometers instead of verniers, giving a much more precise reading. In either case, the angle can be read directly if the initial reading has been set at zero. Otherwise, the angle is found as the difference between the initial and final readings.

To lay off a predetermined angle from some reference line, the initial sight is taken along the line, the telescope is rotated through the angle desired, and a stake or other marker is set on the new line.

The special case of laying off a 180° angle is simply the extension of a straight line. It is done by backsighting along the reference line and rotating the telescope about its horizontal or elevation axis (transiting) for sighting ahead.

Vertical angles. These are measured by rotating the telescope about its elevation axis and reading the angle on the vertical circle. Vertical angles are measured from a horizontal reference usually, though sometimes from the zenith or from the nadir.

Angular precision. Transit precision is denoted by the smallest angular distinction or resolution of which the instrument is capable. A 1-minute transit is so called because its circle is graduated to half-degrees, with the vernier measuring a thirtieth of the graduation, to enable an angular reading to 1 minute. Other transits read to 30 or 20 seconds. Theodolites with glass circles usually give readings to 0.1 minute (6 seconds); higher-order theodolites read to 1 second, even to a few hundredths of a second.

Elevation differences. Elevations may be measured trigonometrically in conjunction with reduction of slope measurements to horizontal distances, but the resulting elevation differences are of low precision.

Differential leveling. Most third-order and all second- and first-order measurements are made by differential leveling, wherein a horizontal line of sight of known elevation is sighted on a graduated rod held vertically on the point being checked. The transit telescope, leveled, may establish the sight line, but more often a specialized leveling instrument is used. For approximate results a hand level may be used. *See* LEVEL (SURVEYING).

In differential leveling, the rod is held on a bench mark, a point of known or assumed elevation. The level is set up and sighted for a reading on the rod. This reading, called the backsight or plus sight, is added to the bench-mark elevation to establish the height of the instrument. With the level remaining where it is, the rod is then moved to a forward point (turning point). The reading to that point, called the foresight or minus sight, is subtracted from the height of the instrument to yield the elevation of the turning point. Distances to foresight and backsight, in pairs, are kept about equal so as to balance small instrumental errors and effects of Earth curvature and refraction of the line of sight. These are shown in exaggerated form in Fig. 3.

The foregoing process is duplicated successively as necessary to obtain the desired new elevation or elevations. Verification of the work is obtained by closing back on the original point by a rerun or by closing on another point of known elevation. Total error, if within allowable limits, is distributed along the level survey.

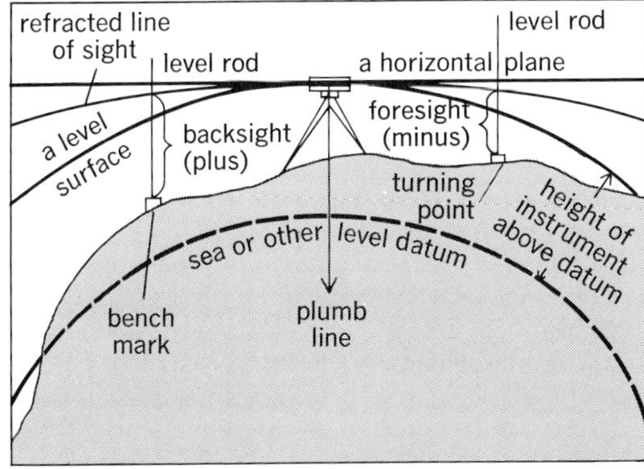

Fig. 3. Theory of differential leveling.

Reciprocal leveling, a variant of straight differential leveling, is used where the level cannot be set up for a long distance, as in crossing a ravine or a river. Here a short backsight and a long foresight (across the inaccessible distance) to a turning point on the opposite side are taken. Then the level is moved to the opposite side so that a long backsight is made to the original point of known elevation, and a short foresight is taken to the turning point. Since any errors of curvature and refraction in the long forward direction are canceled by those of the long backward sight, the average of the two elevation differences so obtained will be the actual difference.

Trigonometric leveling. Elevation difference can be determined also by trigonometric means, as in measuring the height of a mast, or even the elevation difference between two survey points. If the horizontal distance and the vertical angle are measured, the height of the sighted object above the instrument can be calculated by right-triangle methods. In making long sights in this circumstance, a correction may be required for curvature and atmospheric refraction. An altitude observation (elevation angle) on a star or on the Sun will require refraction and parallax corrections.

Barometric leveling. Approximate elevation differences may be measured with the aid of a barometer. This method is particularly useful for reconnaissance of a substantial area.

Airborne profiling. A radar altimeter can give quite accurate elevation of an aircraft above the surface (of the ocean, for instance) by bouncing a signal from the surface back to the aircraft. Using it to give the altitude by bouncing a signal from any surface whose elevation is known can give the exact altitude as well. Thus by continuously recording a terrain profile on an overflight, the correct ground elevations can be obtained if a sufficient number of known ground elevations are scanned to provide control information. The technique is known as airborne profiling and renders fairly accurate results, adequate for a number of less demanding purposes.

Astronomical observations. To determine meridian direction and geographic latitude, observations are made by a theodolite or transit on Polaris, the Sun, or other stars. Direction of the meridian (geographic north-south line) is needed for direction control purposes; latitude is needed where maps and other sources are insufficient.

Meridian determination. The simplest meridian determination is made by sighting Polaris at its elongation, as the star is rounding the easterly or westerly extremity of its apparent orbit. At these times Polaris appears to be moving up or down the transit's cross hair, and there is ample time for assuring an accurate sighting. An angular correction is applied to the direction of sighting, which is referenced to a line on the ground. The correction value is found in an ephemeris.

The ephemeris also gives the noon declination of the Sun at Greenwich throughout the year. Declination is the angle of inclination of the Sun with respect to the equatorial plane. For direct solar observation the direction from the observer to the Sun is found from the equation below, where

$$\cos z = \frac{\sin D}{\cos h \quad \cos L} - \tan h \quad \tan L$$

z is the aximuth angle east or west from the meridian to the Sun; D is the declination at the instant of observation; h is the vertical angle to the Sun, corrected for refraction and parallax; and L is the observer's latitude.

A nonastronomic device for meridian determination is the portable gyroscopic azimuth instrument, a newer development based on missile and aircraft inertial guidance systems.

Latitude determination. Latitude of a position may be determined directly at night by vertical-angle sighting on Polaris at upper or lower culmination (the northerly or southerly extremity of the star's apparent orbit) and application of the suitable vertical-angle correction from the ephemeris. In another method, the vertical angle to the Sun is observed at noon, and the Sun's declination is subtracted algebraically to yield a net angle, which is subtracted from 90° to give the latitude.

Longitude and time observations are made occasionally by noting the passage of the Sun across an established meridian. If Greenwich time is known, longitude can be computed.

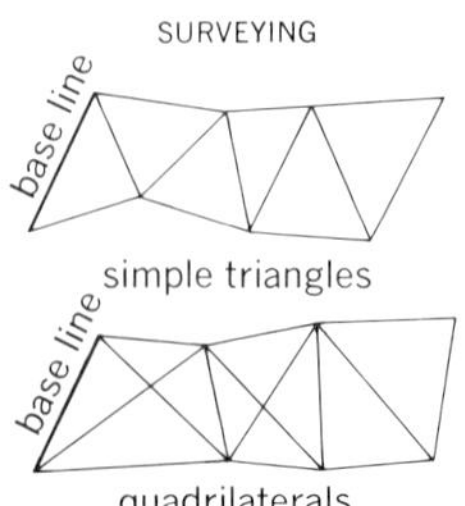

Fig. 4. Triangulation figures, used in geodetic surveys.

Geoid. Astronomic observations from any station yield geographic latitude and longitude for that station; a great number of these stations connected by a network of triangles form a control network. An ellipsoid chosen most nearly to fit the geoidal shape of the Earth affords a base to calculate the interrelationship of the atronomic stations. Lack of conformity between measured distances and calculated distances between stations indicates disparity between astronomical and geoidal (geodetic) latitudes and longitudes, which in turn is traceable to deflection of the astronomic vertical from the ellipsoid vertical at the station or stations. These deflections of the vertical are due to undulations of the geoidal surface and its lack of conformity to the ellipsoid. Adjustments over a wide area (a continent) are usually made. The initial choice of a best ellipsoid renders the conformity closest, so that geodetic (ellipsoidal) computations will best fit the true geoid.

Several ellipsoids have been in use for different continents, but massive accumulations of measurements of gravity and deflections of the vertical are generating the basis for a new ellipsoid that will more closely fit the geoid worldwide. Due in 1983, it will be known as the World Geodetic System Ellipsoid, the first of its kind. Once initiated, it will permit the major geodetic networks of the world to be unified and the coordinates of points anywhere on Earth to be compatible.

Types of surveys. These include geodetic control surveys, route surveys, construction surveys, and cadastral (property) surveys.

Geodetic control surveys. Control surveys provide the reference framework for lesser surveys. A traverse, with elevations of its points, may be the control for mapping a limited area. The broadest control surveys are the geodetic networks established by the National Geodetic Survey (NGS) in the United States and by corresponding agencies in other countries, wherein the horizontal and vertical positions of points are established with first- or second-order accuracy.

In geodetic surveys, Earth curvature is taken into account. Coordinate positions are established in terms of latitude and longitude. Geodetic coordinates are convertible into state plane coordinates.

The traditional method of extending horizontal geodetic control is triangulation. Trilateration and traverse are supplementary techniques.

In geodetic surveys, refinements are added to the simple chain-of-triangles scheme. The most notable refinement is use of the quadrilateral as a geometric unit (Fig. 4). Diagonal directions as well as the sides are observed, so that the quadrilateral includes two pairs of overlapping triangles, one being redundant although valuable as a check. In this way additional angular checks are available, and four separate calculations can be made for the entering side length of the next quadrilateral in the chain. The shapes of the triangles are important, small angles being avoided to preserve high strength of figure. A strength of figure analysis for the four different computational routes reveals the one that will best carry the length forward.

For precision and economy, geodetic triangulation stations are located as far apart as possible, consistent with the requirement of intervisibility. On the other hand, a precise control network for a built-up area may require numerous close-spaced auxiliary stations within quadrilateral figures.

Angles are measured by a precise theodolite, all angles around the horizon being observed several separate times for determination of the statistically best value of each angle. Long sightings are usually made on lighted station signals at night.

Because of the Earth's spheroidal shape, the interior angles of large triangles add up to more than 180°. A 75-mi^2 (200-km^2) triangle, for example, has 1 second more than the 180° total of interior angles. The extra second is called spherical excess.

Large triangulation chains of a primary network will have widely spaced ground stations of calculated position, which are of value as control to only a limited ground area. Thus, secondary triangulation networks are run to connect points on the primary. Then, to densify the control and make it available virtually everywhere, tertiary networks of triangulation or traverse are employed. Precise control for serving a city requires a very heavy saturation of auxiliary stations within the smallest triangulation figures.

Over-ocean geodetic control extension, formerly infeasible, now is accomplished by systems of simultaneous electronic distance measurements from ground stations to airborne signals, applying trilateration principles. Doppler observations of passing satellites have proved an astounding advance, enabling the location of ground stations to within less than a meter virtually anywhere on the Earth's surface.

In another method, simultaneously photographing a satellite against a star background from three camera stations (one, say, on an island) affords measurements to calculate the island station's position, again conquering the overwater problem. Offshore drilling platforms are located by one or several of these now feasible methods.

Each precisely determined distance between geodetic control stations is reduced to its sea-level projection, that is, the distance between the stations if projected vertically to sea level.

Geodetic data convert to state plane coordinate data with the aid of tables published by the NGS. Each plane coordinate system is expressed as a projection of the curved surface onto a plane intersecting either two standard parallels (latitudes) or two meridians. In both cases the meridian and parallel projections are at right angles to each other. At the standard meridians or parallels, scale is exact; elsewhere a known variable scale factor must be applied.

Advantages of state plane coordinate surveys include the ability to initiate a long route survey several places at once, and to interrelate survey points anywhere on the system. Proposed land data systems or a national cadastre would depend on the coordinate systems.

The vertical control system of the NGS consists of a first-order level network with supplementary second-order lines. In most parts of the United States, bench-mark elevations are available within a few miles of any point.

The geodetic leveling procedure is a refinement of that previously described, using more precise instruments and methods. The level instrument is built to rigid specifications. Rod graduations are on an Invar strip (Invar varies only negligibly under temperature change). Special care is taken to equalize foresight and backsight distances, to assure stable turning points, and to protect the instrument from minor stresses.

Route surveys. Surveys for the design and construction of linear works, such as roads, canals, pipelines, or railways, are called route surveys. They begin with reconnaissance and continue through preliminary, location, and construction surveys.

Reconnaissance for a new highway, for example, may be accomplished by study of existing maps together with a visual appraisal of field conditions, or even quick low-order horizontal and vertical field measurements. Controlling points, such as favorable ridge and river crossings, are found, and a preliminary line is selected. This line traditionally is laid out by transit and tape or EDM, being a linear traverse. It is profiled, that is, levels are run along the traverse line to find elevations. Transverse profiles (cross sections) are made at needed intervals. Structures and natural objects that would affect the final location are fixed by side shots. A side shot may be a direction and distance from a transit point, the intersection of two directions or two distances, or a perpendicular offset from a traverse line. Transverse profiles are made by hand level, tape, and level rod.

The result of the preliminary survey is a strip topographic map of sufficient precision to permit preliminary design of the final location, including approximate determination of earthwork quantities. More and more preliminary maps and designs are executed from topographic maps constructed by airborne photographic (photogrammetric) methods, with need for only limited ground surveying prior to construction layout.

The location survey line is conducted with at least third-order precision. Traverse procedures are followed, but curves are laid out and stationed. The result is a staked centerline for the route to be constructed. Profile leveling and cross-sectioning for earthwork quantities also are a part of the location survey.

Contract plans normally are based on the location survey. The construction stakeout is usually conducted later, but sometimes is carried out at the same time as the location survey. A prerequisite is final grade selection so that cuts and fills can be marked on centerline stakes and slope stakes can be set. The slope stake indicates the lateral limits of cut or fill at a given cross section. It is marked with the distance, plus or minus, from existing ground level to the proposed centerline grade. Being at the edge of earthwork, the slope stake supposedly is available for reference throughout construction.

A horizontal curve provides for change in direction for traffic. The directions, called tangents, connect at an angle point, called the point of intersection (P.I.). The intersection angle I is the deflection angle, right or left, from the forward projection of the tangent into the angle point (Fig. 5).

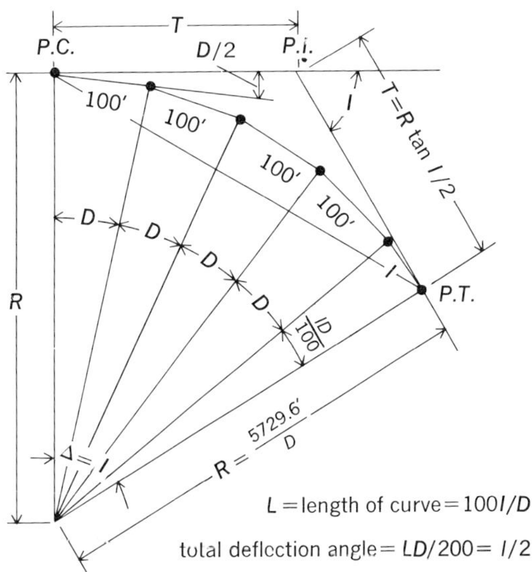

Fig. 5. Circular curve theory.

Any circular curve or arc, connecting two intersecting tangents, has a central angle, equal to the angle I, formed by perpendiculars (radii) to the two tangents, one at the point of curvature (P.C.) and the other at the point of tangency (P.T.). The greater the radius, the more gradual the curve. The sharpness of the curve also is expressed as degree of curvature D, the central-angle increment that subtends 100 ft (30.48 m) of arc (highway definition) or 100 ft of chord (railway definition). A 1° curve has a radius of approximately 5729.6 ft (1746.4 m) under either definition.

Circular curves are laid out by the following procedure, assuming D, and consequently radius R, have been chosen:

1. The tangent distance is measured back from the P.I. to establish the P.C.

2. The transit is set up at P.C., and a series of deflection angles, each equal to $D/2$, is turned in the direction of curvature to establish a point on

the curve for each 100 ft (30.48 m) of chord or arc. For chords or arcs shorter than 100 ft the turned deflection angle is proportionally smaller.

Vertical curves, connecting different grades, usually are parabolic. The parabola has an inherent transition, and it is easy to lay out with tape, level, and level rod after computation of ordinates for vertical-curve grade points along the centerline.

The traditional route survey procedures described above are economical for narrow routes in moderate terrain. Major dual highways with broad rights of way are located more efficiently on large-scale aerial topographic maps. On such maps, "paper" location surveys can be made, complete with earthwork quantity takeoff precise enough for contract bidding. Furthermore, digitized terrain data, directly available from some photogrammetric procedures, permit rapid electronic computation of quantities. With this approach it may be feasible to defer all field staking until construction time, the staking points being selected in advance by scaling or computation of map-coordinate relationships with points established during the small amount of ground survey required to control the photogrammetric survey.

Construction surveys. Surveys for construction layout establish systems of reference points that are not likely to be disturbed by the work. Slope stakes are earthwork references. Buildings, bridge abutments, sewers, and many other structures traditionally are controlled by batter boards, horizontal boards fastened to two uprights. The top of a batter board is set at the elevation of the line to be established, and the horizontal position of the line is indicated by a mark or nail. Across the site another batter board is set up for the given line, and a string or wire stretched between is the line. The line may be a building face at first-floor level, or it may be a reference line. In trench work the line between batter boards may run at a fixed distance above the invert centerline.

In lieu of strings or wires, line of sight may be used. A transit is set up on lines outside the work area, and points of the line are sighted as required. Further, a low-power laser, available as an attachment to the transit or as a separate special-purpose instrument, can project a visible reference line or plane from an unattended position. A means of locating critical construction points, such as those for anchor bolts, is to compute their positions in a coordinate grid, then compute directions and distances from a reference point for their location by transit sighting and tape measurement.

Elevation controls are provided by bench marks near the construction area. The foregoing construction techniques are adaptable into industrial plants for building large mock-ups and jigs as well as for the alignment of parts. Transits and levels on stable mounts may be used; however, related specialized equipment called optical tooling instruments are more readily applied. These include a transit or level telescope with an optical micrometer on the objective, to move the line of sight to a locus parallel with the initial sighting, and flat self-reflecting mirrors for use as targets. Angular sight lines are established by a mirror mounted vertically on a transit base.

Underground surveys. Mine and tunnel surveys impose a few modifications on normal surveying techniques: Repeated independent measurements are made because normal checks (such as closed traverses) are not available; cross hairs are illuminated because work is performed in relative darkness; vertical tape measurements and trigonometric levels, instead of differential levels, frequently must be relied upon; and in adits and tunnels, survey points are placed overhead, rather than underneath, to save them from disturbance by traffic. On the mining transit, an auxiliary telescope outside the trunnion bracket facilitates steep sightings.

Traditionally the most exacting underground survey process has been the transferring of a direction from the surface. In shallow shafts, steep (but not vertical) sights may transfer the direction. Another technique is to hang two weighted wires down the shaft, observe the direction between them on the surface, and use this direction as a control below. The relative shortness of distance between wires makes the transfer geometrically weak, but procedural care enables satisfactory results. An alternative procedure applies inertial navigation principles to underground surveying. A north-seeking gyro mounted on a transit gives azimuth accuracy to about 6 seconds from a plumbed position at shaft bottom and carries tunnel lines forward without the angle-error accumulation in ordinary transit traverse lines (Fig. 6).

Hydrographic surveys. Data for navigation charts and underwater construction are provided by hydrographic surveys. The horizontal locations of depth measurements must be referenced to recognizable controls. Where the shoreline is visible, it is mapped and a system of triangulation stations is established on shore. Transits at two triangulation stations can be used to observe directions to the sounding vessel whenever it signals that a depth measurement is made. A check angle may be obtained by sextant observation of shore points from the deck of the vessel, or a third transit on

Fig. 6. North-seeking gyro eases underground survey tasks. (*Wild Heerbrugg Instruments, Inc.*)

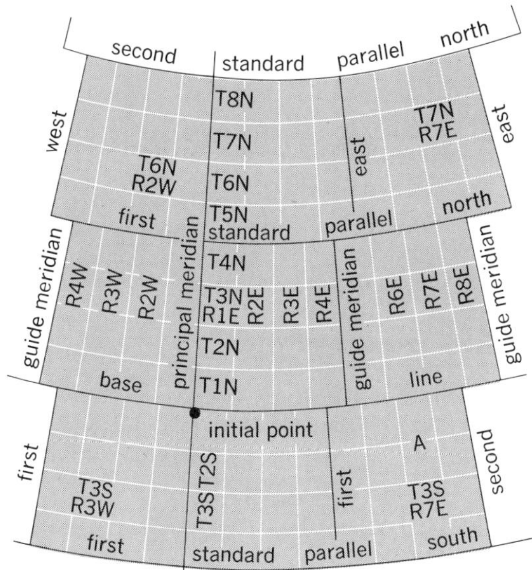

Fig. 7. Rectangular land layout. (*From P. Kissam, Surveying for Civil Engineers, McGraw-Hill, 1956*)

shore may be used to provide a check intersection. Angular observations from a shore point, coupled with EDM lengths to the sounding vessel reflector or transponder, will also fix the sounding location. In another procedure, a sending microwave unit on the vessel can find distances from slave stations set on shore points, to be recorded (even plotted) instantly along with the depth sounded.

Cadastral surveys. To establish property boundary lines, cadastral surveys are made. Descriptions based on horizontal surveys are essential parts of any document denoting ownership or conveyance of land. The basic rule of property lines and corners is that they shall remain in their original positions as established on the ground. This basic rule is important because most land surveys are resurveys. They may follow the original description, but this description is merely an aid to the discovery of the originally established lines and corners. Substantial discrepancies are frequently found in original descriptions because low-order surveying devices such as the compass or the link chain were once used.

Surveys in the original 13 states, plus Tennessee, Kentucky, and parts of others, are conducted on the metes and bounds principle. In the so-called public land states and in Texas, the basic subdivisions are rectangular.

If the boundaries to be described border an irregular line, like a winding stream, for instance, a good mathematical description may be impossible. Such a line can, however, be located by a series of closely spaced perpendicular offsets from an auxiliary straight line.

In the rectangular system, the land parcels of a region are described by their relationship to an initial point. In public-land states the initial point is the intersection of a meridian (principal meridian) and a latitude (baseline), as in Fig. 7. Townships, normally 36 mi^2 (93.24 km^2), are designated by their position with respect to the initial point—the number of tiers north or south of a given baseline and the number of ranges east or west of the corresponding principal meridian.

Within the township each square mile, or section, has a number, from 1 to 36. Sections are subdivided into quarter sections (160 acres or 0.6475 km^2), and they may be subdivided further. North-south section lines are meridians originating at 1-mi (1.609-km) intervals along the baseline and along standard parallels of latitude (correction lines) generally spaced 24 mi (38.62 km) apart. The respacing of meridional lines every 24 mi reduces the effect of meridian convergence on the size of sections.

Resurveys become difficult where the corners are obliterated or lost. An obliterated corner is one for which visible evidence of the previous surveyor's work has disappeared, but whose original position can be established from other physical evidence and testimony. A corner is deemed lost when no sufficient evidence of its position can be found. Restoration requires a faithful rerun of the recorded original survey lines from adjacent points, distance discrepancies being adjusted proportionately. *See* CIVIL ENGINEERING.

[B. AUSTIN BARRY]

Bibliography: B. A. Barry, *Construction Measurements*, 1973; R. C. Brinker and P. R. Wolf, *Elementary Surveying*, 1977; C. M. Brown, *Boundary Control and Legal Principles*, 1969; F. H. Moffitt and H. Bouchard, *Surveying*, 1975.

Synchronous capacitor

A rotating ac machine running with no mechanical load. It is often called a synchronous condenser. These machines generally have salient poles, eight poles being typical. However, two-pole generators are sometimes disconnected from the turbine and operated as synchronous condensers. The purpose of these machines is to provide reactive power to the system or to receive some reactive power. The amount and sign of the reactive power depends on the field current supplied to the rotor. This is an effective method of controlling the voltage at that point in the power system. *See* ELECTRIC POWER MEASUREMENT.

Field excitation above the no-load value tends to raise the voltage and causes reactive current to be supplied to the system. This current is directly demagnetizing, like a zero power factor load on a generator, and requires more excitation for a given apparent power than a normal generator. The machine operating at low excitation draws reactive power from the system, generally limited to 50% of maximum overexcited apparent power. For control of rapidly changing voltage such as that caused by arc furnaces, the reactance of the machine is kept low and a fast electronic exciter is often used. *See* ARC HEATING.

For feeder systems, a small condenser can be used to correct voltage dips. An 80%-power factor synchronous motor running at no load can also be used for this purpose.

Much larger synchronous condensers up to 345 megavolt-amperes overexcited and 290 MVA underexcited have been used on large power systems to improve stability. For example, a synchronous condenser near the center of a long line can add to the maximum power transmitted an amount equal to 0.35 times the peak reactive power supplied. If a synchronous capacitor is regulated so as to increase apparent power supplied in phase with the velocity with which the two ends of a system are swinging apart, an appreciable damping of system oscillations can be obtained.

Since the cost of these rotating machines is high (about 20% more than a generator of the same rating), a solid-state reactive power generator has been developed which uses shunt capacitors in parallel with reactors whose reactive power is controlled by solid-state switches. These can react even faster than the rotating machine, but have a fixed upper current limit, so for some applications requiring high short-time peaks, the rotating machine may be most economical.

Rotating synchronous condensers are sometimes used at the terminals of dc transmission lines where it has been found necessary to have a connected rotating capacity about 30% of the converter terminal rating. *See* DIRECT-CURRENT TRANSMISSION; SYNCHRONOUS MOTOR.

[LEE A. KILGORE]

Bibliography: J. A. Oliver, B. J. Ware, and R. C. Carruth, *345 MVA Fully Water-Cooled Synchronous Condenser for Dumont Station*, IEEE Pap. 71 TP 210 Pwr, 1971.

Synchronous motor

An alternating-current (ac) motor which operates at a fixed synchronous speed proportional to the frequency of the applied ac power. A synchronous machine may operate as a generator, motor, or capacitor depending only on its applied shaft torque (whether positive, negative, or zero) and its excitation. There is no fundamental difference in the theory, design, or construction of a machine intended for any of these roles, although certain design features are stressed for each of them. In use, the machine may change its role from instant to instant. For these reasons it is preferable not to set up separate theories for synchronous generators, motors, and capacitors. It is better to establish a general theory which is applicable to all three and in which the distinction between them is merely a difference in the direction of the currents and the sign of the torque angles. *See* ALTERNATING-CURRENT GENERATOR; ALTERNATING-CURRENT MOTOR; SYNCHRONOUS CAPACITOR.

SYNCHRONOUS MOTOR

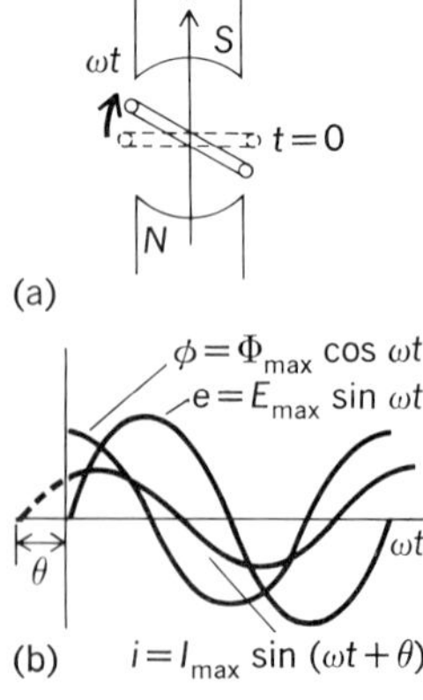

Fig. 1. Single-phase, two-pole synchronous machine.

Basic theory. A single-phase, two-pole synchronous machine is shown in Fig. 1. The coil is on the pole axis at time $t=0$, and the sinusoidally distributed flux ϕ linked with the coil at any instant is given by Eq. (1), where ωt is the angular displacement of the coil and $\Phi_{\max}$ is the maximum value of the flux. This flux will induce in a coil of N turns an instantaneous voltage e, given by Eq. (2). The

$$\phi=\Phi_{\max}\cos\omega t \qquad (1)$$

$$e=-N\frac{d\phi}{dt}=\omega N\Phi_{\max}\sin\omega t=E_{\max}\sin\omega t \qquad (2)$$

effective (rms) value E of this voltage is given by Eq. (3).

$$E=\frac{E_{\max}}{\sqrt{2}}=\sqrt{2}\,\pi f N\Phi_{\max}=4.44\,fN\Phi_{\max} \qquad (3)$$

If the impedance of the coil and its external circuit of resistance R_t and reactance X is given by Eq. (4), there will flow a current, with a value given

$$Z=R_t\pm jX=Z\angle{\pm\theta} \qquad (4)$$

by Eq. (5), in which the phase angle θ is taken pos-

$$\mathbf{I}=\frac{\mathbf{E}}{\mathbf{Z}}=\frac{E}{Z}\angle{\mp\theta} \qquad (5)$$

itive for a leading current. This current will develop a sinusoidal space distribution of armature reaction as in Eq. (6). If this single-phase mmf is

$$A=0.8NI_{\max}\sin(\omega t+\theta) \qquad (6)$$

expressed as a space vector and resolved into direct (in line with the pole axis) A_d and quadrature A_q components, it is given by Eq. (7).

$$\begin{aligned}\mathbf{A}&=A_d+jA_q\\&=0.4\,NI_{\max}\{[\sin\theta+\sin(2\omega t+\theta)]\\&\qquad+j[\cos\theta-\cos(2\omega t+\theta)]\}\end{aligned} \qquad (7)$$

In a three-phase machine with balanced currents, the phase currents are given by Eqs. (8).

$$\begin{aligned}i_a&=I_{\max}\sin(\omega t+\theta)\\i_b&=I_{\max}\sin(\omega t+\theta-120°)\\i_c&=I_{\max}\sin(\omega t+\theta-240°)\end{aligned} \qquad (8)$$

Upon writing Eq. (7) for ωt, $\omega t-120°$, and $\omega t-240°$, respectively, and adding, Eq. (9) results for the polyphase armature reaction.

$$\mathbf{A}=A_d+jA_q=1.2NI_{\max}(\sin\theta+j\cos\theta) \qquad (9)$$

Equation (10) gives the three-phase power of the

$$P=3EI\cos\theta \qquad (10)$$

machine, and Eq. (11) gives the developed torque.

$$T=\frac{P}{\omega}=\frac{3}{\omega}EI\cos\theta \qquad (11)$$

The above equations constitute the essential description of the synchronous generator. The same equations apply for a motor if the currents are reversed, that is, by changing the sign of the current I. They may also be interpreted in the form of phasor diagrams, and show the two cases of a smooth-rotor and a salient-pole machine.

Smooth-rotor synchronous machine. In the smooth-rotor machine, the reluctance of the magnetic path is essentially the same in either the direct or quadrature axes. In Fig. 2*a* let the flux Φ be selected as reference phasor and drawn vertically. Then comparing Eqs. (1) and (2) it is seen that the induced voltage E_f lags the flux by 90°. By Eq. (5) the current I lags the voltage by an angle θ for an inductive circuit, and by Eq. (9) causes a constant mmf of armature reaction A in phase with the current. This armature reaction causes a flux ϕ_a, stationary in space with respect to the field poles, which in turn induces a voltage E_a lagging it by 90°. The two induced voltages E_f (due to the field flux Φ) and E_a (due to the armature reaction flux ϕ_a) combine vectorially to give the resultant voltage E'. But the terminal voltage V is less than E' by the resistance and reactance drops, $R\mathbf{I}$ and $jx_l\mathbf{I}$ in the winding, and Eq. (12) applies.

$$\mathbf{V}=\mathbf{E}'-(R+jx_l)\mathbf{I} \qquad (12)$$

The leakage reactance drop $jx_l\mathbf{I}$ lags the current

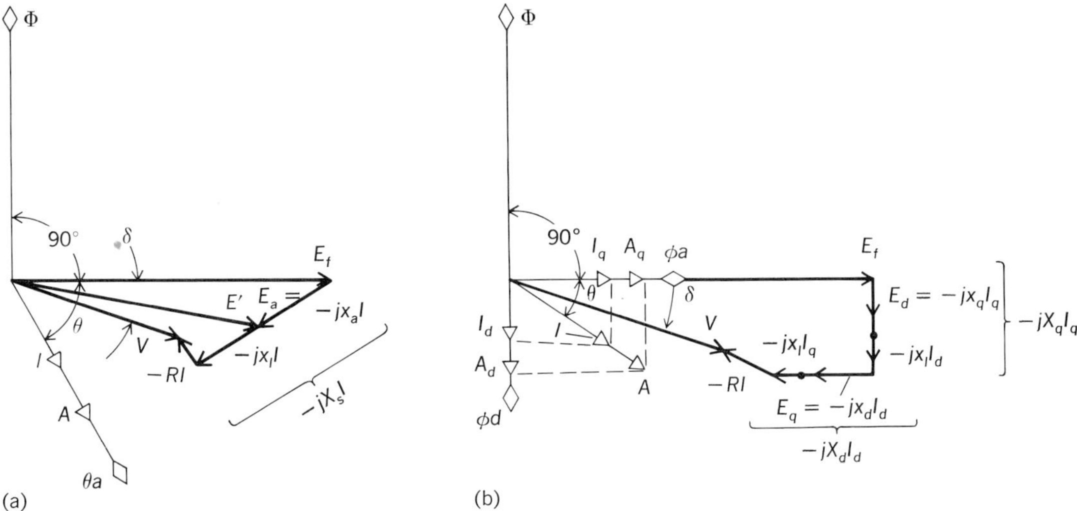

Fig. 2. Vector diagrams of synchronous generators. (*a*) Smooth-rotor machine. (*b*) Salient-pole machine.

by 90° as does the armature reaction voltage E_a. If a fictitious reactance of armature reaction x_a is introduced to account for E_a, it is obvious that Eq. (12) may be rewritten to give Eqs. (13), in which

$$\begin{aligned} \mathbf{V} &= \mathbf{E}_f - jx_a\mathbf{I} - (R + jx_l)\mathbf{I} \\ &= \mathbf{E}_f - R\mathbf{I} - j(x_a + x_l)\mathbf{I} \\ &= \mathbf{E}_f - (R + jX_s)\mathbf{I} \end{aligned} \quad (13)$$

$X_s = x_a + x_l$ is called the synchronous reactance of the machine.

Salient-pole synchronous machine. In a similar fashion the phasor diagram for a salient-pole machine, Fig. 2*b*, may be set up. Here the effects of saliency result in proportionately different armature reaction fluxes in the direct and quadrature axes, thereby necessitating corresponding direct, X_d, and quadrature, X_q, components of the synchronous reactance. The angle δ in Fig. 2 is called the torque angle. It is the angle between the field-induced voltage E_f and the terminal voltage V and is positive when E_f is ahead of V.

The foregoing equations and phasor diagrams were established for a generator. A motor may be regarded as a generator in which the power component of the current is reversed 180°, that is, becomes an input instead of an output current. The motor vector diagram is shown in Fig. 3. Here the torque angle δ is reversed, since V is ahead of E_f in a motor (it was behind in the generator). Therefore a motor differs from a generator in two essential respects: (1) The currents are reversed, and (2) the torque angle has changed sign. As a result the power input, Eq. (10), for a motor is negative, or has become a power output, and the torque is reversed in sign.

When the current **I** is 90° out of phase with the terminal voltage V the torque angle δ is nearly zero, being just sufficient to account for the power lost in the resistance.

Therefore, a synchronous machine is a generator, motor, or capacitor, depending on whether its torque angle δ is positive, negative, or zero. For these conditions the output current is, respectively, at an angle in the first or fourth quadrant, second or third quadrant, or essentially ±90° with respect to the terminal voltage at zero degrees. For any given power input or output, the machine can be made to operate at either leading or lagging power factor by changing the magnitude of the field current producing the flux Φ.

Synchronous capacitor. A synchronous capacitor can be made to draw a leading current and to behave like a capacitance by overexciting its field. Or, it will draw a lagging current on underexcitation. This characteristic thus presents the possibility of power-factor correction of a power system by adjusting the field excitation. A machine so employed at the end of a transmission line permits a wide range of voltage regulation for the line. One used in a factory permits the power factor of the load to be corrected. Of course a synchronous motor can also be used for power-factor correction, but since it must also carry the load current, its power-factor correction capabilities are more limited than for the synchronous condenser.

Power equations. The power output P_o and reactance power output Q_o of a round-rotor synchronous machine are given by Eq. (14), in which $\tan\alpha = R/X_s$ and $\mathbf{Z}_s = R + jX_s = Z_s < 90° - \alpha$.

$$P_o + jQ_o = \left(\frac{VE_f}{Z_s}\sin(\delta+\alpha) - \frac{RV^2}{Z_s^2}\right) + j\left(\frac{VE_f}{Z_s}\cos(\delta+\alpha) + \frac{X_sV^2}{Z_s^2}\right) \quad (14)$$

For a round-rotor motor, both the torque angle δ and the electrical power output are negative.

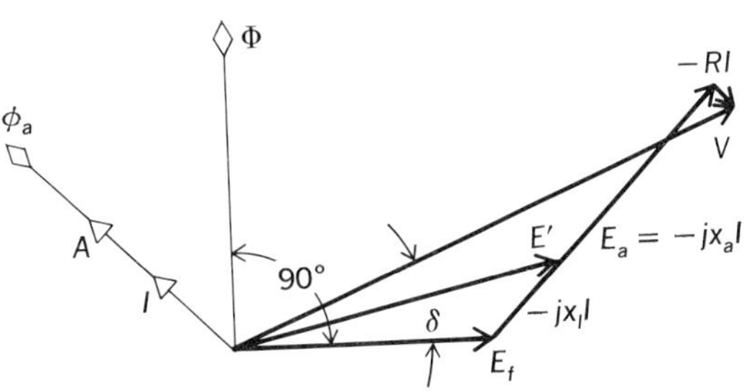

Fig. 3. Phasor diagram of synchronous motor.

For a salient-pole machine, neglecting resistance, Z_s is equal to the direct-axis reactance X_d, α is zero, and P_o is given by Eq. (15). Thus the

$$P_o = \frac{VE_f}{X_d} \sin \delta + V^2 \frac{X_d - X_q}{2X_d X_q} \sin 2\delta \qquad (15)$$

power or torque depends essentially on the product of the terminal and induced voltages and sine of the torque angle δ; but in the case of the salient-pole machine there is also a second harmonic term which is independent of the excitation voltage E_f. This term, the so-called reluctance power, vanishes for nonsaliency when $X_d = X_q$. The small synchronous motors used in some electric clocks and other low-torque applications depend solely on this reluctance torque.

Excitation characteristics. The so-called V curves of a synchronous motor are curves of armature current plotted against field current with power output as parameter. Usually a second set of curves with input power factor (pf) as parameter is superimposed on the same plot. Such curves (Fig. 4), where armature current is plotted against generated voltage, can be determined from design calculations or from test; they yield a considerable amount of data on the performance of the motor. Thus, given any two of the four variables E_f, I, pf, mechanical power P_m, the remaining two may be easily determined, as well as the conditions of maximum power, constant pf, minimum excitation, stability limit, and so forth.

Circle diagrams. Voltage equation (13) and current equation (5) can be combined in such a fashion as to yield Eq. (16), which is the equation of

$$I^2 = \frac{V^2}{Z_s^2} + \frac{E_f^2}{Z_s^2} - 2\frac{V}{Z_s}\frac{E_f}{Z_s} \cos \delta \qquad (16)$$

a set of circles with offset center and with different radii (E_f/Z_s). The locus of these circles is the current.

A companion set of circles can be developed giving the locus of I as a function of its pf angle for different values of constant developed mechanical power.

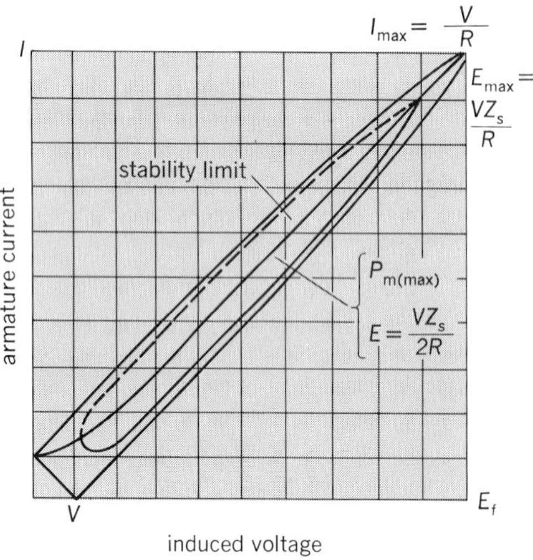

Fig. 4. V curves (armature current versus induced voltage) of synchronous motor.

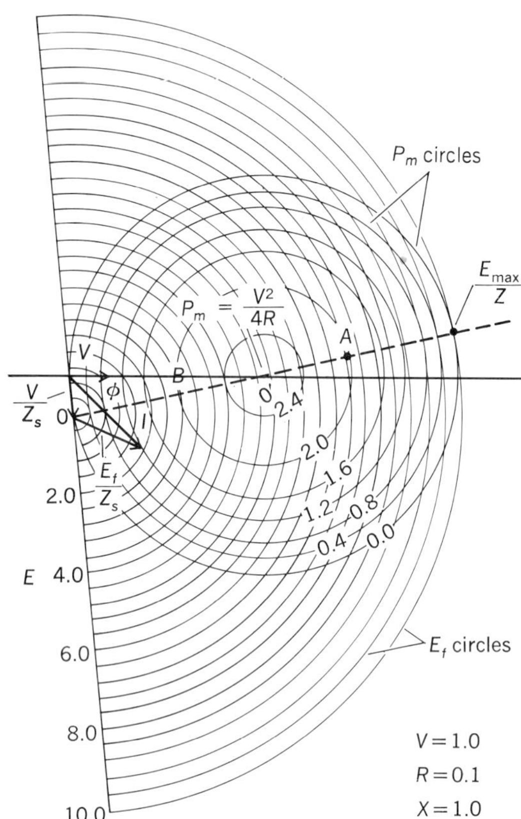

Fig. 5. Circle diagram of synchronous motor.

These two sets of circles are shown in Fig. 5. Such circle diagrams relate the mechanical power, pf angle, armature current, torque angle, and excitation.

Losses and efficiency. The losses in a synchronous motor comprise the copper losses in the field, armature, and amortisseur windings; the exciter and rheostat losses of the excitation system; the core loss due to hysteresis and eddy currents in the armature core and teeth and in the pole face; the stray loss due to skin effect in conductors; and the mechanical losses due to windage and friction. The efficiency of the motor is then given by Eq. (17).

$$\text{Eff} = \frac{\text{output}}{\text{input}} = \frac{\text{output}}{\text{output} + \text{losses}} \qquad (17)$$

Mechanical oscillations. A synchronous motor subjected to sudden changes of load, or when driving a load having a variable torque (for example, a reciprocating compressor), may oscillate about its mean synchronous speed. Under these conditions the torque angle δ does not remain fixed, but varies. As a result the four separate torques expressed in Eq. (18) act on the machine rotor. The

$$\begin{pmatrix}\text{Synchronous}\\ \text{motor torque}\\ \text{Eq. (15)}\end{pmatrix} + \begin{pmatrix}\text{induction motor}\\ \text{torque of}\\ \text{amortisseur}\end{pmatrix} = \begin{pmatrix}\text{torque to}\\ \text{overcome}\\ \text{inertia}\end{pmatrix} + \begin{pmatrix}\text{torque}\\ \text{required}\\ \text{by the load}\end{pmatrix} \qquad (18)$$

possibility exists that cumulative oscillations will build up and cause the motor to fall out of step.

Starting of synchronous motors. Synchronous motors are provided with an amortisseur (squirrel-cage) winding embedded in the face of the field poles. This winding serves the double purpose of starting the motor and limiting the oscillations or hunting. During starting, the field winding is either closed through a resistance, short-circuited, or opened at several points to avoid dangerous induced voltages. The amortisseur winding acts exactly as the squirrel-cage winding in an induction motor and accelerates the motor to nearly synchronous speed. When near synchronous speed, the field is excited, and the synchronous torque pulls the motor into synchronism. During starting, Eq. (18) applies, since all four types of torque may be present. Of course, up to the instant when the field is excited the portion of the synchronous motor torque depending on E_f does not exist, although the reluctance torque will be active.

Other methods of starting have been used. If the exciter is direct-connected and a dc source of power is available, it may be used to start the synchronous motor. In the so-called supersynchronous motor the stator is able to rotate in bearings of its own, and is provided with a brake band. For stator allowed to come up to nearly synchronous speed by virtue of the amortisseur windings; the field is then excited and the stator brought to synchronous speed, the rotor remaining stationary. Then as the brake band is tightened, the torque on the rotor causes it to accelerate while the speed of the stator correspondingly slackens; finally the stator comes to rest and is locked by the brake band. In this way maximum synchronous motor torque is made available for acceleration of the load. For other types of synchronous motors *see* HYSTERESIS MOTOR; RELUCTANCE MOTOR.

[RICHARD T. SMITH]

Bibliography: D. G. Fink and H. W. Beaty (eds.), *Standard Handbook for Electrical Engineers*, 11th ed., 1978; A. E. Fitzgerald, C. Kingsley, Jr., and A. Kusko, *Electric Machinery*, 3d ed., 1971; L. W. Matsch, *Electromagnetic and Electromechanical Machines*, 2d ed., 1977.

Synchroscope

An instrument used for indicating whether two alternating-current (ac) generators or other ac voltage sources are synchronized in time phase with each other. In one type, for example, the position of a continuously rotatable pointer indicates the instantaneous phase difference between the two sources at each instant; the speed of rotation of the pointer corresponds to the frequency difference between the sources, while the direction of rotation indicates which source is higher in frequency. In more modern synchroscopes, a cathode-ray tube serves as the indicating means.

The term synchroscope is also applied to a special type of cathode-ray oscilloscope designed for observing extremely short pulses, using fast sweeps synchronized with the signal to be observed. *See* ELECTRIC POWER GENERATION.

[JOHN MARKUS]

Taper pin

A tapered self-holding pin used to connect parts together. Standard taper pins have a diametral taper 1/4 in. in 12 in. (0.6 cm in 30 cm) and are driven in holes drilled and reamed to fit. The pins are made of soft steel or are cyanide-hardened. They are sometimes used to connect a hub or collar to a shaft (see illustration). Taper pins are frequently used to maintain the location of one surface with respect to another. A disadvantage of the taper pin is that the holes must be drilled and reamed after assembly of the connected parts; hence they are not interchangeable. *See* COTTER PIN.

[PAUL H. BLACK]

Bibliography: D. Lent, *Analysis and Design of Mechanisms*, 2d ed., 1970; *Society of Automotive Engineers Handbook*, revised annually.

TAPER PIN

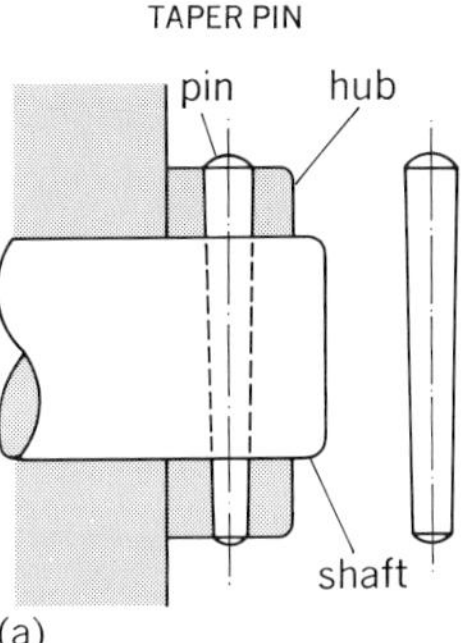

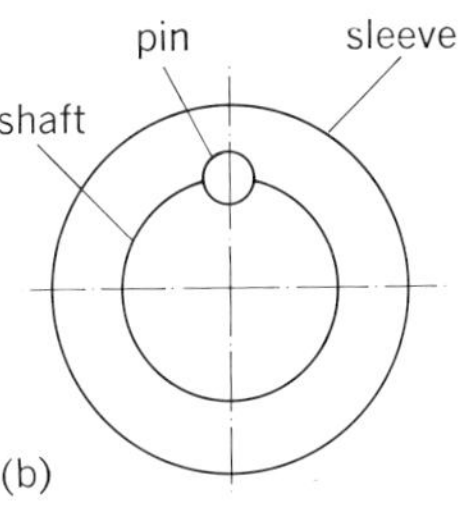

Taper pins. (*a*) Connecting hub or collar to shaft. (*b*) Connecting sleeve to shaft.

Technology

Systematic knowledge and action, usually of industrial processes but applicable to any recurrent activity. Technology is closely related to science and to engineering. Science deals with humans' understanding of the real world about them—the inherent properties of space, matter, energy, and their interactions. Engineering is the application of objective knowledge to the creation of plans, designs, and means for achieving desired objectives. Technology deals with the tools and techniques for carrying out the plans.

For example, certain manufactured parts may need to be thoroughly clean. The technological approach is to use more detergent and softener in the wash water, to use more wash cycles, to rinse and rerinse, and to blow the parts dry with a stronger, warmer air blast. Often such refinements provide an adequate action. However, if they do not suffice, the basic technique may need to be changed. Thus, in this example, science might contribute the knowledge that ultrasonically produced cavitation counteracts surface tension between immiscible liquids and adhesion between clinging dirt and the surface to be cleaned, and thereby produces emulsions. Engineering could then plan an ultrasonic generator and a conveyor to carry the parts through a bath tank in which the ultrasonic energy could clean them. The scientist may use ultrasonic techniques to determine properties of materials. The engineer may design other types of devices that employ ultrasonics to perform other functions. These specialists enlarge their knowledge of ultrasonics and their skill in using this technique not for its own sake but rather for its value in their work. The technologist is the specialist who carries out the technique for the purpose of accomplishing a specified function, and extends knowledge and skill of ultrasonic cleaning by refinement and perfection of the technique for use on various materials soiled in different ways. Technological advances improve and extend the application to cleaning other parts under other conditions. *See* CIVIL ENGINEERING; ELECTRICAL ENGINEERING; ENGINEERING; ENGINEERING, SOCIAL IMPLICATIONS OF; ENGINEERING DESIGN; HUMAN-FACTORS ENGINEERING; INDUSTRIAL ENGINEERING; MECHANICAL ENGINEERING; NUCLEAR ENGINEERING; PRODUCTION ENGINEERING.

[ROBERT S. SHERWOOD/HAROLD B. MAYNARD]

Tempering

The reheating of previously quenched alloy to a predetermined temperature below the critical range, holding the alloy for a specified time at that temperature, and then cooling it at a controlled rate, usually by immediate rapid quenching, to room temperature. Tempering has a long history, beginning as an empirical process to impart toughness. Consequently, the term is broadly applied to any process that toughens a material.

In alloys, if the composition is such that cooling produces a supersaturated solid solution, the resulting material is brittle. Heating the alloy to a temperature only high enough to allow the excess solute to precipitate out and then rapidly cooling the saturated solution fast enough to prevent further precipitation or grain growth result in a microstructure combining hardness and toughness.

With steel, precipitation of carbide out of the supersaturated unstable martensite or interstitial solution of carbon in iron, results in improved ductility. To be most successful, the tempering must be carried out by slow heating to avoid steep temperature gradients, stress relief being one of the objectives. Properties produced by tempering depend on the temperature to which the steel is raised and on its alloy composition. For example, if hardness is to be retained, molybdenum or tungsten is used in the alloy.

For tool steel, to obtain great hardness, the finished or nearly finished tool may be heated to a critical temperature dependent on the particular alloy. While at this temperature below the transformation range, the steel is refined; that is, it forms a solid solution with small grain size. After being held at the refining temperature long enough to reach thermal equilibrium throughout, the steel is rapidly cooled or quenched so as to pass quickly through temperature regions in which grain growth can occur. For steel with 0.83% carbon, the refining temperature is slightly above 700°C; the faster this alloy is cooled the harder it becomes. *See* HEAT TREATMENT (METALLURGY).

[FRANK H. ROCKETT]

Bibliography: American Society for Testing and Materials, *Temper Embrittlement of Alloy Steels*; L. Coudurier and I. Wilkomirsky, *Fundamentals of Metallurgical Processes*, 1978.

Thermodynamic cycle

A procedure or arrangement in which one form of energy, such as heat at an elevated temperature from combustion of a fuel, is in part converted to another form, such as mechanical energy on a shaft, and the remainder is rejected to a lower temperature sink as low-grade heat.

Common features of cycles. A thermodynamic cycle requires, in addition to the supply of incoming energy, (1) a working substance, usually a gas or vapor; (2) a mechanism in which the processes or phases can be carried through sequentially; and (3) a thermodynamic sink to which the residual heat can be rejected. The cycle itself is a repetitive series of operations.

There is a basic pattern of processes common to power-producing cycles. There is a compression process wherein the working substance undergoes an increase in pressure and therefore density. There is an addition of thermal energy from a source such as a fossil fuel, a fissile fuel, or solar radiation. There is an expansion process during which work is done by the system on the surroundings. There is a rejection process where thermal energy is transferred to the surroundings. The algebraic sum of the energy additions and abstractions is such that some thermal energy is converted into mechanical work.

A steam cycle that embraces a boiler, a prime mover, a condenser, and a feed pump is typical of the cyclic arrangement in which the thermodynamic fluid, steam, is used over and over again. An alternative procedure, after the net work flows from the system, is to employ a change of mass within the system boundaries, the spent working substance being replaced by a fresh charge ready to repeat the cyclic events. The automotive engine and the gas turbine illustrate this arrangement of the cyclic processes, called an open cycle because new mass enters the system boundaries and the spent exhaust leaves it.

The basic processes of the cycle, either open or closed, are heat addition, heat rejection, expansion, and compression. These processes are always present in a cycle even though there may be differences in working substance, the individual processes, pressure ranges, temperature ranges, mechanisms, and heat transfer arrangements.

Air-standard cycle. It is convenient to study the various power cycles by using an ideal system such as the air-standard cycle. This is an ideal, frictionless mechanism enveloping the system, with a permanent unit charge of air behaving in accordance with the perfect gas relationships.

The unit air charge is assumed to have an initial state at the start of the cycle to be analyzed. Each process is assumed to be perfectly reversible, and all effects between the system and the surroundings are described as either a heat transfer or a mechanical work term. At the end of a series of processes, the state of the system is the same as it was initially. Because no chemical changes take place within the system, the same unit air charge is conceivably capable of going through the cyclic processes repeatedly.

Whereas this air-standard cycle is an idealization of an actual cycle, it provides an amenable method for the introductory evaluation of any power cycle. Its analysis defines the upper limits of performance toward which the actual cycle performance may approach. It defines trends, if not absolute values, for both ideal and actual cycles. The air-standard cycle can be used to examine such cycles as the Carnot and those applicable to

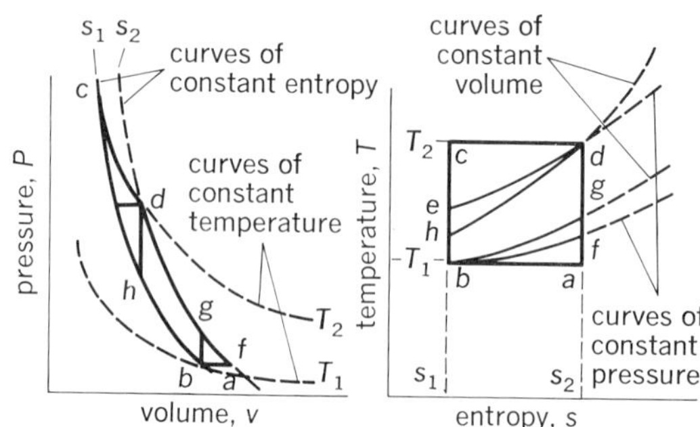

cycles are, in the order of decreasing efficiency:
Carnot cycle (*a-b-c-d-a*)
Brayton cycle (*b-e-d-f-b*)
Diesel cycle (*b-e-d-g-b*)
Otto cycle (*b-h-d-g-b*)

Comparison of principal thermodynamic cycles

the automobile engine, the diesel engine, the gas turbine, and the jet engine.

Cyclic standards. Many cyclic arrangements, using various combinations of phases but all seeking to convert heat into work, have been proposed by many investigators whose names are attached to their proposals, for example, the Diesel, Otto, Rankine, Brayton, Stirling, Ericsson, and Atkinson cycles (see illustration). All proposals are not equally efficient in the conversion of heat into work. However, they may offer other advantages which have led to their practical development for various applications. Nevertheless, there is one overriding limitation on efficiency. It is set by the dictates of the Carnot cycle, which states that no thermodynamic cycle can be projected whose thermal efficiency exceeds that of the Carnot cycle between specified temperature levels for the heat source and the heat sink. Many cycles may approach and even equal this limit, but none can exceed it. This is the uniqueness of the Carnot principle and is basic to the second law of thermodynamics on the conversion of heat into work. *See* BRAYTON CYCLE; CARNOT CYCLE; DIESEL CYCLE; OTTO CYCLE; STIRLING ENGINE.

[THEODORE BAUMEISTER]

Bibliography: T. Baumeister (ed.), *Standard Handbook for Mechanical Engineers*, 8th ed., 1978; J. H. Keenan, *Thermodynamics*, 1970; W. C. Reynolds and H. C. Perkins, *Engineering Thermodynamics*, 2d ed., 1977; M. W. Zemansky, *Heat and Thermodynamics*, 6th ed., 1981.

Thermostat

An instrument which directly or indirectly controls one or more sources of heating and cooling to maintain a desired temperature. To perform this function a thermostat must have a sensing element and a transducer. The sensing element measures changes in the temperature and produces a desired effect on the transducer. The transducer converts the effect produced by the sensing element into a suitable control of the device or devices which affect the temperature.

The most commonly used principles for sensing changes in temperature are (1) unequal rate of expansion of two dissimilar metals bonded together (bimetals), (2) unequal expansion of two dissimilar metals (rod and tube), (3) liquid expansion (sealed diaphragm and remote bulb or sealed bellows with or without a remote bulb), (4) saturation pressure of a liquid-vapor system (bellows), and (5) temperature-sensitive resistance element.

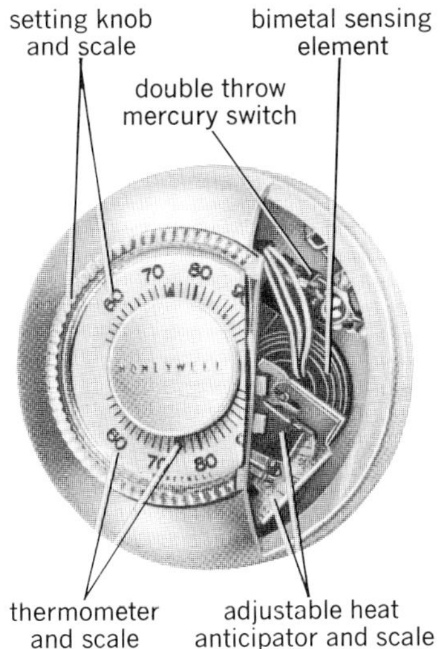

Fig. 1. Typical heat-cool thermostat. (*Honeywell Inc.*)

The most commonly used transducers are (1) switches that make or break an electric circuit, (2) potentiometer with a wiper that is moved by the sensing element, (3) electronic amplifier, and (4) pneumatic actuator.

The most common thermostat application is for room temperature control. Figure 1 shows a typical on-off heating-cooling room thermostat. In a typical application the thermostat controls a gas valve, oil burner control, electric heat control, cooling compressor control, or damper actuator.

To reduce room temperature swings, high-performance on-off thermostats commonly include a means for heat anticipation. The temperature swing becomes excessive if thermostats without heat anticipation are used because of the switch differential (the temperature change required to go from the break to the make of the switch), the time lag of the sensing element (due to the mass of the thermostat) in sensing a change in room temperature, and the inability of the heating system to respond immediately to a signal from the thermostat.

To reduce this swing, a heater element (heat anticipator) is energized during the on period. This

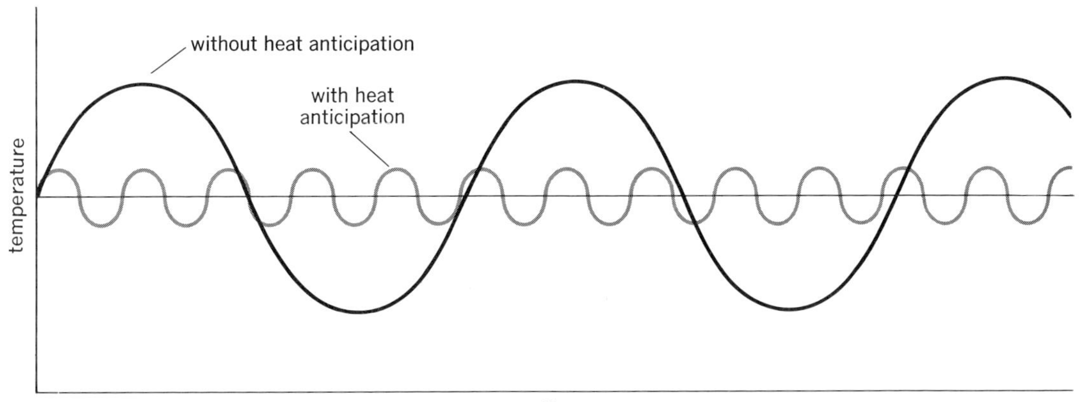

Fig. 2. Comparison of temperature variations using a timed on-off thermostat with and without heat anticipation.

causes the thermostat to break prematurely. Figure 2 shows a comparison of the room temperature variations when a thermostat with and without heat anticipation is used.

The same anticipation action can be obtained on cooling thermostats by energizing a heater (cool anticipator) during the off period of the thermostat. Room thermostats may be used to provide a variety of control functions, such as heat only; heat-cool; day-night, in which the night temperature is controlled at a lower level; and multistage, in which there may be one or more stages of heating, or one or more stages of cooling, or a combination of heating and cooling stages.

Thermostats are also used extensively in safety and limit application. Thermostats are generally of the following types: insertion types that are mounted on ducts with the sensing element extending into a duct, immersion types that control a liquid in a pipe or tank with the sensing element extending into the liquid, and surface types in which the sensing element is mounted on a pipe or similar surface. *See* COMFORT HEATING; OIL BURNER.

[NATHANIEL ROBBINS, JR.]

Bibliography: American Society of Heating, Refrigeration, and Air-Conditioning Engineers, *ASHRAE Handbook and Product Directory: Systems*, 1976; J. E. Haines, *Automatic Control of Heating and Air Conditioning*, 2d ed., 1961; V. C. Miles, *Thermostatic Control*, 2d ed., 1974.

Threading

The forming of a ridge and valley of uniform cross section which spiral about the inner or outer diameter of a cylinder or cone in an even and continuing manner. The work must be produced with sufficient uniformity and accuracy so that the resulting threaded part will accomplish its intended purpose of fastening, transmitting motion or power, or measuring.

Screw threads are classed as either external or internal. Thread chasing by cutting dies and single-point turning plus milling, hobbing, grinding, and rolling are the usual means of producing external threads. Most internal threading is done by tapping plus internal single-point turning, hobbing, grinding, and milling.

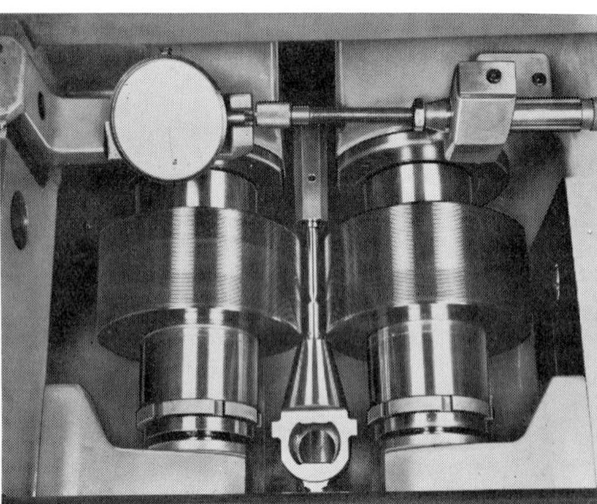

Fig. 1. Thread-rolling machine set up to thread a steel aircraft jet engine mount, with minimum hardness of 40 Rockwell C. Workpiece rotates at 1320 rpm; thread-rolling dies contact the work for only 0.24 sec, producing nearly perfect concentricity and a pitch variation of less than 0.0005 in. (12.5 μm). (*Landis Machine Co.*)

Fig. 2. Threading machine with heat-treated die head set up to cut 3 pitch extra deep thread at 7.6 surface feet (2.3 m) per minute in adjusting screw for construction machine. (*Landis Machine Co.*)

Although machine threading is done on various machine tools, special threading machines may be used for quantity production (Figs. 1 and 2). *See* MACHINING OPERATIONS; SCREW FASTENER; SCREW THREADS.

[ALAN H. TUTTLE]

Tie rod

A tie rod or tie bar, usually circular in cross section, is used in structural parts of machines to tie together or brace connected members, or in moving parts of machines or mechanisms it may connect arms or parts to transmit motion. In the first use the rod ends are usually a threaded fastening, while in the latter they are usually forged into an eye for a pin connection.

In steering systems of automotive vehicles, the rod connects the arms of steering knuckles of each wheel. The connection between the rod and arms is a ball and socket joint. *See* AUTOMOTIVE STEERING.

In pressure piping, large forces are produced between connected parts. The pipes or parts are constrained by tie rods that may be rectangular in cross section, with pinned ends. *See* PIPELINE.

[PAUL H. BLACK]

Tin alloys

Solid solutions of tin and some other metal or metals. Alloys cover a wide composition range and many applications because tin alloys readily with nearly all metals.

Soft solders constitute one of the most widely used and indispensable series of tin-containing alloys. Common solder is an alloy of tin and lead, usually containing 20–70% tin. It is made easily by melting the two metals together. With 63% tin, a eutectic alloy melting sharply at 361°F (169°C) is formed. This is much used in the electrical industry. A more general-purpose solder, containing equal parts tin and lead, has a melting range of 56°F (13°C). With less tin, the melting range is increased further, and wiping joints such as plumbers make can be produced. Lead-free solders for special uses include tin containing up to 5% of either silver or antimony for use at temperatures somewhat higher than those for tin-lead solders, and tin-zinc base solders often used in soldering aluminum. *See* SOLDERING.

Bronzes are among the most ancient of alloys and still form an important group of structural metals. Of the true copper-tin bronzes, up to 10% tin is used in wrought phosphor bronzes, and from 5 to 10% tin in the most common cast bronzes. Many brasses, which are basically copper-zinc alloys, contain 0.75–1.0% tin for additional corrosion resistance in such wrought alloys as Admiralty Metal and Naval brass, and up to 4% tin in cast leaded brasses. Among special cast bronzes are bell metal, historically 20–24% tin for best tonal quality, and speculum, a white bronze containing 33% tin that gained fame for high reflectivity before glass mirrors were invented. *See* COPPER ALLOYS.

Babbitt or bearing metal for forming or lining a sleeve bearing is one of the most useful tin alloys. It is tin containing 4–8% each of copper and antimony to give compressive strength and a structure desired for good bearing properties. An advantage of this alloy is the ease with which castings can be made or bearing shells relined with simple equipment and under emergency conditions. Aluminum-tin alloys are used in bearing applications that require higher loads than can be handled with conventional babbitt alloys. *See* ANTIFRICTION BEARING.

Pewter is an easily formed tin-base alloy that originally contained considerable lead. Thus, because Colonial pewter darkened and because of potential toxicity effects, its use was discouraged. Modern pewter is lead-free. The most favorable composition, Britannia Metal, contains about 7% antimony and 2% copper. This has desired hardness and luster retention, yet it can be readily cast, spun, and hammered.

Type metals are lead-base alloys containing 3–15% tin and a somewhat larger proportion of antimony. As with most tin-bearing alloys, these are used and remelted repeatedly with little loss of constituents. Tin adds fluidity, reduces brittleness, and gives a structure that reproduces fine detail.

Flake and nodular gray iron castings are improved by adding 0.1% tin to give a fully pearlitic matrix with attendant higher hardness, heat stability, and improved strength and machinability.

Tin is commonly an ingredient in die castings hardened with antimony and copper for applications requiring close tolerances, thin walls, and bearing or nontoxic properties; and in low-melting alloys for safety appliances. *See* ALLOY.

[BRUCE W. GONSER]

Bibliography: *Conference on Tin Consumption*, International Tin Council, London, 1972; E. S. Hedges, *Tin and Its Alloys*, 1960.

Toggle

Any of a wide variety of mechanisms, many used to open or close electrical contacts abruptly and all characterized by the control of a large force by a small one. The basic action of a toggle mechanism is shown in illustration *a*. When $\alpha = 90°$ forces P and Q are independent of each other. Again, when $\alpha = 0°$ the forces are isolated, force Q being sustained entirely by the frame, and force P serving only to hold the link in position. At $\alpha = 45°$ from the symmetry $|P| = |Q|$, the mechanism serves to transfer the direction of forces to achieve equilibrium. If frame, connecting rod, and slider blocks with their pivots are sufficiently strong to support the maximum force encountered and if the coefficients of friction at the pin and slider joints are negligible, at intermediate positions the forces are related by the equation written below. The

$$P = Q \tan \alpha$$

essential characteristic of the mechanism arises from this tangent relation between forces. As $\alpha \rightarrow 0$ a small force at P can overcome a large force at Q. *See* COUPLING.

The mechanism finds application in such machines as printing presses, embossing machines, friction brakes and clutches, and rock crushers. In such devices, the frame sustains the output force, the drive supplying only a small force through a sufficient distance to bring the linkage into position. Arranged to drive slightly past dead center,

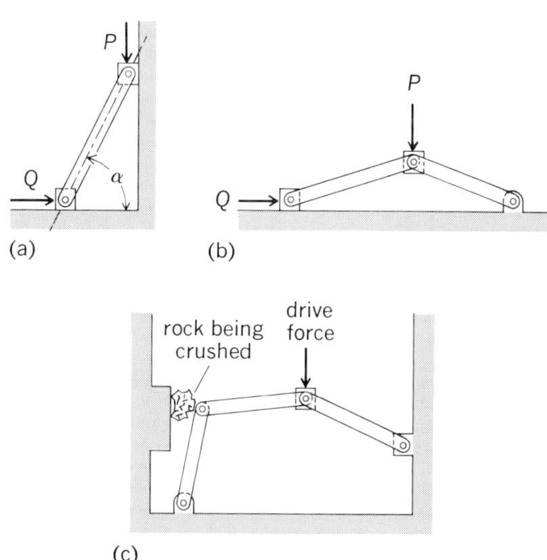

Toggle mechanism. (*a*) Simple structure. (*b*) Traditional configuration. (*c*) Typical application.

the mechanism serves as a clamp that can be released by a small reverse force at P, yet will hold against a large force at Q. Such action finds application in workpiece clamps, tool holders, and electric switches and circuit breakers. A spring is generally added to assist the release once the mechanism has been returned through dead center.

Because the simple configuration of illustration *a* requires low-friction sliders, it is impractical. A more useful structure replaces the vertical slider with a second link pinned to the frame (illustration *b*), in which case input P sets up forces in both links. A further modification (illustration *c*) replaces the other slider with a link. *See* FOUR-BAR LINKAGE; LINKAGE (MECHANISM).

[FRANK H. ROCKETT]

Tolerance

Amount of variation permitted or "tolerated" in the size of a machine part. Manufacturing variables make it impossible to produce a part of exact dimensions; hence the designer must be satisfied with manufactured parts that are between a maximum size and a minimum size. Tolerance is the difference between maximum and minimum limits of a basic dimension. For instance, in a shaft and hole fit, when the hole is a minimum size and the

TOLERANCE

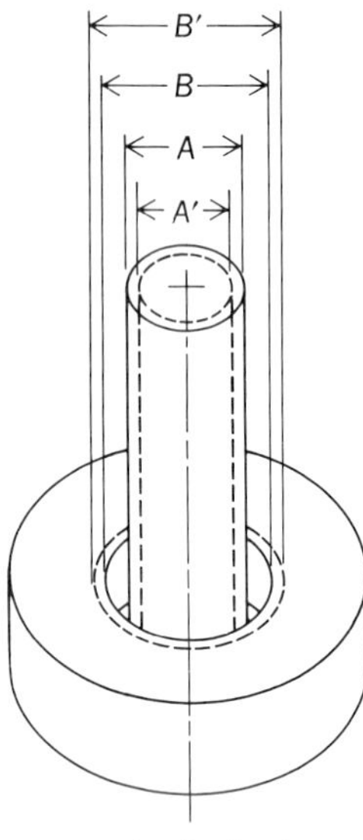

——— size for basic dimension
– – – – size with tolerance
A = basic diameter of shaft
B = basic diameter of hole
B − A = allowance
B′ − A′ = maximum clearance
A − A′ = tolerance on shaft
B − B′ = tolerance on hole

Shaft and hole dimensions.

shaft is a maximum, the clearance will be the smallest, and when the hole is the maximum size and the shaft the minimum, the clearance will be the largest (see illustration).

If the initial dimension placed on the drawing represents the size of the part that would be used if it could be made exactly to size, then a consideration of the operating conditions of the pair of mating surfaces shows that a variation in one direction from the ideal would be more dangerous than a variation in the opposite direction. The dimensional tolerance should be in the less dangerous direction. This method of stating tolerance is called unilateral tolerance and has largely displaced bilateral tolerance, in which variations are given from a basic line in plus and minus values.

As an example, for a 1.5-in. shaft and hole for a free fit the standard allowance is 0.002 in., and the tolerance for hole and shaft is 0.001 in.

$$\begin{aligned}\text{Max. shaft diameter} &= \text{nominal size} - \text{allowance}\\ &= 1.500 - 0.002\\ &= 1.498 \text{ in.}\end{aligned}$$

In the unilateral method for stating tolerance, the shaft diameter is $1.498^{+0.000}_{-0.001}$, or between 1.498 and 1.497 in. The diameter of the hole is $1.500^{+0.001}_{-0.000}$, or between 1.500 and 1.501 in. The maximum clearance is 1.501 minus 1.497, or 0.004 in., and the minimum clearance is 0.002 in.

[PAUL H. BLACK]

Bibliography: D. A. Madsen, *Geometric Dimensioning and Tolerancing*, 1977; E. Oberg and F. D. Jones, *Machinery's Handbook*, 21st rev. ed., 1979.

Ton of refrigeration

A rate of cooling that is equivalent to the removal of heat at 200 Btu/min, 12,000 Btu/hr, or 288,000 Btu/day. This unit of measure stems from the original use of ice for refrigeration. One pound of ice, in melting at 32°F, absorbs as latent heat approximately 144 Btu/lb, and 1 ton (2000 lb) of ice, in melting in 24 hr, absorbs 288,000 Btu/day. In Europe, where the metric system is used, the equivalent cooling unit is the frigorie, which is a kilogram calorie, or 3.96 Btu. Thus 3000 frigories/hr is approximately 1 ton of refrigeration. A standard ton of refrigeration is one developed at standard rating conditions of 5°F evaporator and 86°F condenser temperatures, with 9°F liquid subcooling and 9°F suction superheat. *See* REFRIGERATION.

[CARL F. KAYAN]

Tooling

Auxiliary devices used in manufacturing operations to supplement basic standard and special machine tools. Tooling is used, for example, on drill presses, milling machines, planers, and shapers to facilitate the actions of workers and to adapt machine tools to production of specific parts.

Tooling nomenclature. The overall category of tooling encompasses a broad spectrum of manufacturing accessories; tooling is further subdivided into classifications such as jigs, fixtures, dies, gages, devices, gadgets, and other less universal designations.

There are no universally accepted definitions for differentiating between these subdivisions of tooling. Nomenclature for similar tools varies from one industry to another; it also varies among similar

Types of tooling

Machine type	Prevailing tool nomenclature
Drill press	Drill jig
Boring machine (horizontal or vertical) cutting tool rotates	Fixture
Turning machine (horizontal or vertical) workpiece rotates	Fixture
Milling	Fixture
Welding	Jig
Assembly operations	Jig
Inspection operations	Gage
Presses (stamping, forming, and drawing)	Dies

industries using similar devices. In shop practice, tooling subdivision nomenclature loosely follows the generalities of the table.

Common hand tools and devices such as hammers, wrenches, pliers, screwdrivers, measuring scales, calipers, and standard micrometers are normally excluded from the tooling category.

Standard perishable tools (consumed as a result of production operation) such as drills, reamers, milling cutters, and carbide tool bits are similarly excluded from the tooling category.

Special perishable tools, for example, special-size drills, step drills, form cutters, and special tool holders, are normally regarded as tooling for a specific operation and are usually referred to as special cutting tools.

Auxiliary tooling. The option of physical separation of a machine tool and its tooling is not necessarily directly related to whether the machine tool is standard, semispecial, or special. There are valid reasons why all machine tools to a varying degree utilize removable tooling.

It is usually less costly and more practical to make the machine tool and the tooling (jigs, fixtures, dies) as entities rather than to have these tooling features built directly into the body of the machine tool. For example, it may be most advantageous to make the body of the machine tool from cast iron; it may be more economical or mechanically necessary to make the tooling from a weldment; or the tooling device may best be a bolted assembly. The machining of complex details into a bulky machine body may be extremely awkward as compared to the same detail machining (tool making) of a relatively small tooling component.

By removing one set of tooling and replacing it with another (usually referred to as setup) the usefulness of a machine tool (standard, semistandard, or special) can be extended so as to perform useful work on a variety of parts. Thus removable tooling is often a prerequisite for justifying procurement of, and for improving use of, a machine tool.

The use of removable tooling is a form of insurance against machine obsolescence due to product change. Existing tooling can be modified or new tooling created to cover unforeseen new product requirements.

Tooling can be modified or the effects of damage or wear repaired while the basic machine tool is doing other useful work without loss of productive capacity other than the changeover interval.

OPERATIONS

A tooling operation may consist of a single function or a series of functions.

Functional uses. In operation, most tooling simultaneously performs some combination of a variety of purposes; the following represent only a few of the many purposes for which tooling is used: locating, clamping, positioning, cutter guiding, and others.

There are usually three basic locating problems: concentric locating, plane locating, and radial locating.

Locating. The locating function is accomplished by designing and constructing the tooling device so as to bring together the proper contact points or contact surfaces between the workpiece and the tooling, and by defining a direction of clamping force so that the work will infallibly assume the desired relationship to the tooling, and in turn thus have the desired relationship to the body of the machine tool and to the cutter.

Specific designs may require solution of only one, or simultaneous solution of two or sometimes of all three, of these locating problems. Locating problems may be most practically understood by considering a simple analogy, that of locating a record on a phonograph turntable.

The hole in the record is placed down over the center pin of the turntable. This is known as concentric locating. The center line of the record then coincides with the center line of the turntable. In this instance the record is the female member and the turntable pin is the male member of the concentric engagement. An actual tooling device may be the counterpart of either the record or the turntable; for example, the workpiece may be contained within the body of the fixture or, conversely, a relatively small drill jig may be secured to a side of, or secured inside of, a larger workpiece.

The record is now properly concentrically located, but it may still wobble in a horizontal plane unless it comes home against the face of the turntable. This is known as plane locating. In this instance the clamp is gravity, which keeps the record in both plane and concentric relationship to the table.

A record does not require radial locating; however, suppose that the record must assume an exact radial relationship to the turntable, like the relationship of the winning number on a wheel of chance to the pointer finger. The record must then be so rotated about the concentric locating pin that the desired radial position is achieved; a mechanical arrangement must be provided to locate the record in this relative position. This is known as radial locating.

Empirical procedures in the designing of a tooling device are used because of the conflict among the many factors involved. At the specific stage of manufacture that requires a tooling device, the work may already have different degrees of dimensional variations; for example, it may be rough, semifinished, or finished.

Work may be rough (such as castings, forgings, or weldments) and thus have considerable inherent dimensional variation.

Work may be semifinished; that is, it may be rough on some surfaces but accurately machined

Fig. 1. Trunion-type drill jig in a radial drill press. The work is manually indexed so as to properly position the work to the drill. (*American Tool Works*)

Fig. 2. Drill jig in which a hardened bushing guides the drill bit. After first of two in-line holes is machined, jig is indexed 180° and slip-ring bushing moved to opposite liner. (*American Tool Works*)

Fig. 3. Quick-change rigid tool holder for plunge drilling without guide bushing. For maximum rigidity, drill bit is stub-held to project only minimum required distance. (*Reliance Electric and Engineering Co.*)

Fig. 4. A plunge-drilling operation. In this instance the work-holding device is mounted on a master index table which permits drilling of various combinations of radial holes into the tubular workpiece. The horizontal drill spindle is adjustable for different height settings. (*Reliance Electric and Engineering Co.*)

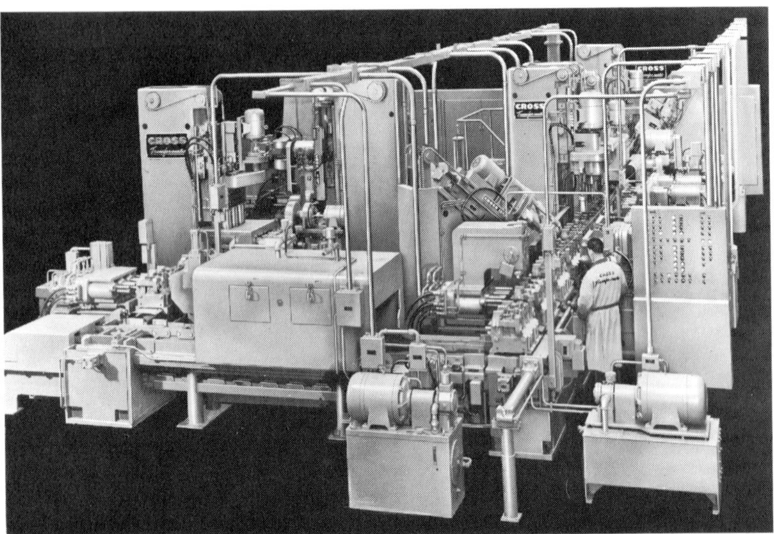

Fig. 5. In this automatic transfer machine, pallets hold three parts each and shuttle through successive machine work stations. Operator manually loads machine; unloading station at left is automatic. Washer automatically cleans pallets before they return to loading station. (*Cross Co.*)

on others. There may be considerable variation between the rough and finished surfaces due to chucking variances in the previous machining operations.

Work may be finished all over, with negligible dimensional variation, or it may have appreciable variations between the various finished surfaces, particularly if they were machined in different independent operations.

Locating mechanisms that are best suited to rough locating are usually quite different to locating mechanisms needed to properly locate semifinished or finished workpieces and vice versa. It is not a problem of degree, but sometimes a basically different locating means for these different conditions is required.

Thus the combination of requirement for concentric locating, plane locating, and radial locating, as well as the variances of rough, semifinished, and finished surfaces, and cost, time, and volume considerations make locating a procedure requiring intimate firsthand knowledge of manufacturing operations. Sometimes it is essential to have a knowledge of the end use of the products, for this often inflicts restrictions or requirements that are not or perhaps cannot be practically expressed in the form of dimensional tolerances on the product drawing.

It may assist the understanding of simultaneously locating in three axes to relate this to the fact that three axes of position are used to describe the relative position of a spacecraft; in that instance, they are commonly termed roll, pitch, and yaw.

Clamping. After the work is located it is clamped; this is a subsequent and cumulative problem. Clamps must act in the required direction with a proper degree of holding force.

In some respects a tooling device acts as a vise or as an anvil. The tooling device must hold the work securely so as to resist the machining forces which may be considerable, as reflected in the amount of torque and pressure that is required to perform the necessary work such as cutting or forming.

In holding the work securely, however, the clamp must not in itself deform the work to an objectionable degree. The locating and clamping schemes should thus be contrived so that the operational forces do not act directly against the clamp (which might then require a powerful clamp) but rather so that the cutting forces are absorbed between the engagements of the workpiece and the tooling, then requiring only a nominal amount of clamping pressure to properly hold together the part and the tooling.

Positioning. The positioning functions of a tooling device may have a variety of interpretations and a variety of means to accomplish these ends.

The tooling may be required to index, tilt, slide, or otherwise manipulate the work in relationship to the working tools (Fig. 1).

The tooling may be required to position different components relative to each other for the purpose of assembly or welding. In welding, the tooling may also be required to manipulate the work so as to achieve downhand welding on different sides.

The positioning function of the tooling device may be essentially a work-handling one where the work is heavy or otherwise impractical to handle by hand. The work itself may be in the hand-handling weight range, but the combination of the work and the necessary tooling may require mechanical assistance in the form of leverage, balance, or power actuation.

Guiding. Another of the many functions of tooling is to guide the cutting tool so as to achieve the desired end result with regard to dimensional accuracy of the finished part.

Drill-press spindles ordinarily do not have side-thrust rigidity. Drill spindles are usually built to absorb the drill-point pressure (feed) but are essentially loose in response to sideways pressures. This is particularly true when long drills or drill extension holders are used. Consequently it is conventional practice to provide drill-guide bushings in a drill jig so as to achieve the necessary degree of dimensional control.

A common example is the use of a drill bushing to guide a drill or a reamer. A drill (or a reamer) cuts on its end rather than on its sides. Thus the engagement of the hard drill bushing with the side of the drill bit does not dull the drill and does not unduly chaff the hardened drill bushing (Fig. 2).

Cutters which cut on their cylindrical peripheries (such as end mills and toothed milling cutters) cannot be so guided because the resultant friction would be detrimental to the cutting-tool edges and to the intended contact guide. Milling machines achieve the desired dimensional relation between cutter and work through the rigidity built into the machine spindle and in the strength of the machine ways along their movements.

Many drilling operations are accomplished without the use of guiding bushing by using the so-called plunge-drilling method. This is done by using a rigid milling type of spindle and holding the drill in a rigid quick-change drill holder (Fig. 3). For maximum rigidity, the drill bit is stub-held and projects only as far as necessary to produce a hole of the required depth. Thus the relationship between the work and the drill is a function of the machine and work movements. Plunge drilling makes possible a variety of work in an efficient manner with simple tooling for the job (Fig. 4). Spindle construction similar to that in milling machines provides the necessary rigidity. Many tape-controlled drilling machines are built around the basic plunge-drilling method.

One of the principal economic justifications for the procurement of a numerically controlled machine tool is the minimizing of the cost of tooling (fixturing) required to produce parts.

Modern numerically controlled machines do offer this potential saving. However, the saving is not due entirely to the flexibility and versatility of numerical control, but also to the fact that plunge drilling (no drill guiding bushings are used) was developed simultaneously with numerical control. The use of drill bushings would require a fixture which encloses the workpiece and supports the drill bushings in space around the workpiece. Stub-type, rigid tool holders (often collet type) were simultaneously developed and are now offered as catalog shelf items. Plunge drilling was also dependent on the development of new techniques in drill pointing (and drill resharpening), which causes the drill point to tend to drill on its own center line rather than to crawl. For example, when using a properly sharpened (so-called self-centering) drill point, it is possible to plunge drill a

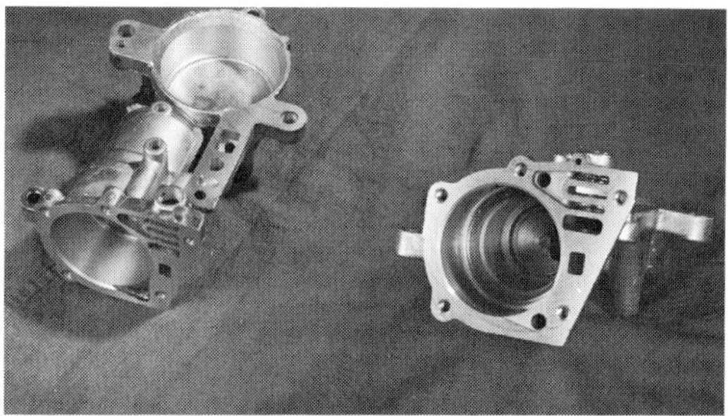

Fig. 6. Workpiece produced in machine of Fig. 5. Results of various milling, boring, drilling, chamfering, and tapping operations are shown. (*Cross Co.*)

hole in a steel ball in a deliberately slightly off-center position. This is virtually impossible to do with the old, conventional, chisel-pointed drill bit unless a drill guided bushing is used.

Plunge drilling, however, has inherent limitations that should be recognized; these are usually governed by the drill diameter, the depth of hole to be drilled, the tolerance of the hole location, the type of surface to be drilled, and the kind of material to be drilled. For products that are bolted together, requiring the drilling and tapping of holes in one member and bolt clearance holes in the other member, plunge drilling is usually sufficiently accurate for the dimensional needs.

The potential savings due to minimizing of fixturing is also enhanced by the fact that many numerically controlled machines are offered as standard items with built-in numerically controlled index tables. This machine feature is thus available to eliminate certain work-positioning features that would otherwise be designed and built into the auxiliary tooling.

Numerical control (NC) machine tools which do not have built-in index tables commonly have built-in connection terminals to facilitate the integration of auxiliary NC indexing heads (horizontal or vertical) into the automatic machining cycle. This is especially true regarding the types of NC machines intended for use on drilling, milling, and boring operations.

Multiple tool functions. Automation, which may be defined as continuous automatic production, although not basically new, achieved a new peak of adaptation during the 1950s. It has brought about many new machine-tool and tooling problems and innovations.

The trend in machine tools for automation is to design and produce basic machine-tool building blocks, each of which is a separate machine tool, that can be linked with others (with or without special units) to form an automated production line that performs a series of milling, drilling, and similar machining operations (Fig. 5). In this way basic machine-tool building blocks perform a variety of jobs (Fig. 6). The standard machine-tool building block also offers considerable protection (through retooling or salvage) against obsolescence of an expensive automatic transfer machine.

The tooling for such operations often follows a pattern in which the work is located and secured to a pallet which is then automatically transferred from one operation to another (Fig. 7). The pallet has master locating surfaces so that at each work

Fig. 7. Work-holding pallet used in conjunction with the continuous automatic production (automation) machine (transfer machine) of Fig. 5. (*Cross Co.*)

station the necessary basic alignment is reestablished and the pallet automatically secured into place. This procedure requires one pallet for each work station and such pallets as are required at the loading and inspection stations. *See* AUTOMATION; JIG, FIXTURE, AND DIE DESIGN; MACHINING OPERATIONS. [JOSEPH I. KARASH]

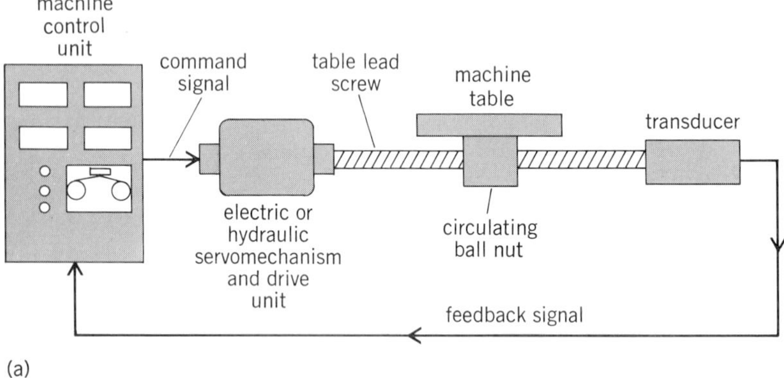

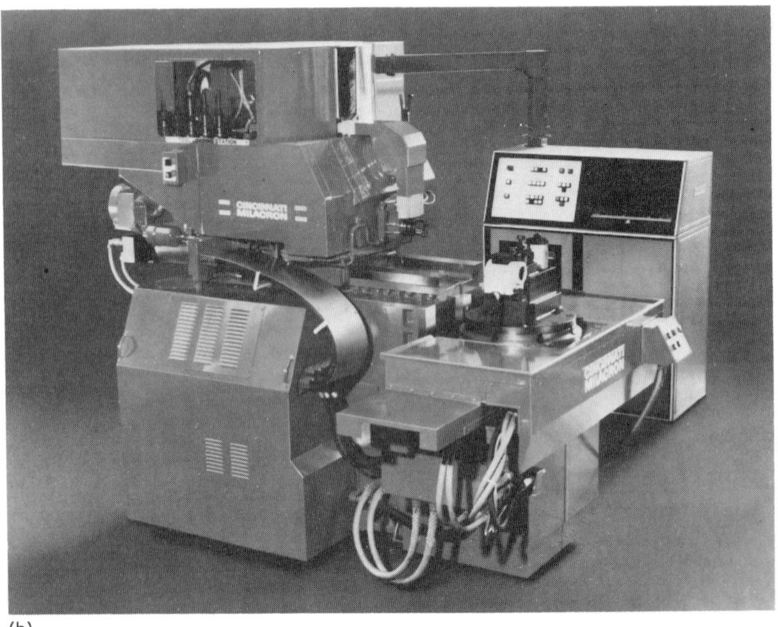

Fig. 8. Numerically controlled machine tool. (*a*) Schematic of an NC one-axis, closed-loop numerical control system. (*b*) Typical NC milling machine and control unit.

NUMERICAL CONTROL

Numerical control (NC), in a broad sense, is an extremely versatile method of automatically operating machines by means of discrete numerical values introduced to the machine through some form of stored-input medium, such as punched cards, punched tape, magnetic tape, or direct control by a digital computer.

History. The concept of controlling equipment from punched cards or paper rolls is not new. In 1725 English knitting machines were produced which to this day operate on the principle of punched program cards. The player piano, introduced in 1863, is another example.

In 1948 the U.S. Air Force contracted for research work on what is now properly called numerical control. The first successful demonstration of this effort occurred in 1952 at the Massachusetts Institute of Technology. That first numerical control, a complex three-axes contouring type which took information from computer-generated magnetic tape input, was retrofitted to a large vertical milling machine that had previously been tracer-controlled.

Following this breakthrough, the Air Force, recognizing the need for its contractors to produce ever more complex parts, awarded contracts for the production of large, numerically controlled, multiaxes contour milling machines costing approximately $500,000 each. As a result, NC undeservedly gained the reputation of being an expensive, complex method for contour-machining large aircraft parts; it was considered beyond the financial reach and technical competence of most metalworking companies, and at that time (1955–1960) this was probably true.

It was not until machine tool and control builders decided to concentrate development efforts on less complex and lower-cost point-to-point equipment for both milling and drilling that NC began to find ever wider applications in all branches of the metalworking industry.

NC commands. By strict definition, numerical control (and certain other processing machines) is the directing of a series of coded instructions for machine operation which is composed largely of numbers and other symbols, including letters of the alphabet and the comma, dash, dollar sign, asterisk, and ampersand.

NC commands for machine tools may range from positioning the spindle, tool, or workpiece to auxiliary functions such as selecting a tool station on a turret, controlling the direction, speed, and feed of a spindle, or turning the coolant off and on.

The commands put together and logically orga-

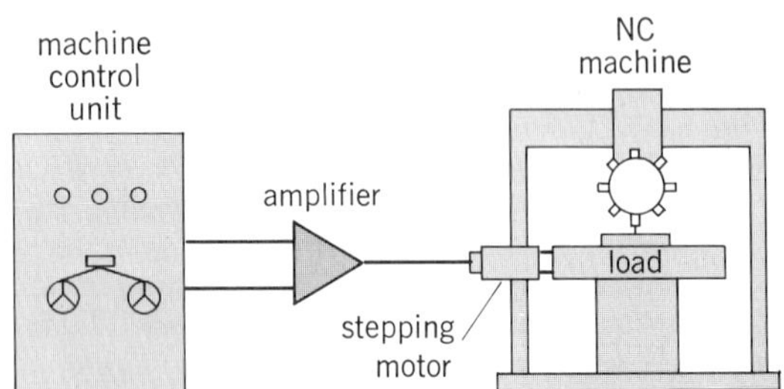

Fig. 9. Schematic of an open-loop NC system.

nized can often be used to program all the operations required to completely machine a workpiece in one setting. This tremendous versatility has caused NC machines to be widely accepted in ever-increasing numbers. NC machines are not generally used for mass production, but rather for limited or medium production where setup times are relatively short. Shifting from one part to another can be done with a minimum of delay once the machines have been programmed for the required workpieces. Tapes or other programmed media can be used over and over again.

NC operational outline. The sequence of events in NC machining starts with the programmer (engineer, shop foreperson, methods person, programming specialist, or anyone who has the ability and inclination to assume a programming function) who studies the part drawing. The programmer must visualize all the machine motions required to make the particular part. Once conceived, they must be documented in logical order on a programming manuscript. The manuscript data are then converted to a medium (often perforated tape) that can be acted upon by the machine control unit.

A schematic closed-loop feedback system is often used in NC machine control of a typical machine tool (Fig. 8). The machine control unit (MCU) accepts instructions from the tape reader, converts them into command signals, and transmits the signals to the machine as required. The command signals, which may be in the form of pulses, can produce machine movements of as little as 0.0002 in. (5 μm) to a pulse. Alternatively, command signals may be of analog (continuous) nature, and a degree of accuracy as required.

The MCU is not generally regarded as a computer, although most NC systems incorporate some computer circuits. The more sophisticated contouring systems have machine control units or directors which are essentially process computers.

Closed-loop control. The signals from the tape or computer go directly to the axis drive motor and servomechanism of the machine. The servomechanism amplifies the incoming signal and provides power to move the machine table or other required mechanical movement. Usually the servomotors drive the slides through low-friction lead screws matched with circulating ball nuts that have zero backlash.

The machine motion provided by the servodrive is recorded or monitored by a feedback system. In this type, typically an encoder (a photoelectric-actuated pulse-generating device) is mounted on the end of the lead screw or is driven by a gear engaging a precision rack mounted on the machine slide. The encoder feeds back to the MCU the pulses which were counted and used to establish the distance traveled, which is then compared with the command or programmed signals. Corrections are automatically and continuously made by the MCU through a null-point arrangement. Systems with feedback signals are classified as closed-loop.

Open-loop systems. Open-loop NC is based on the principle of the stepping motor. Each pulse of electric current from the control system exerts a given torque on the stepping motor and makes it rotate a given amount (Fig. 9). For example, if each pulse produces 1.8° rotation, the motor might be connected to a lead screw that advances 0.001 in. (25 μm) for that 1.8°. To get 1 in. (25 mm) of travel,

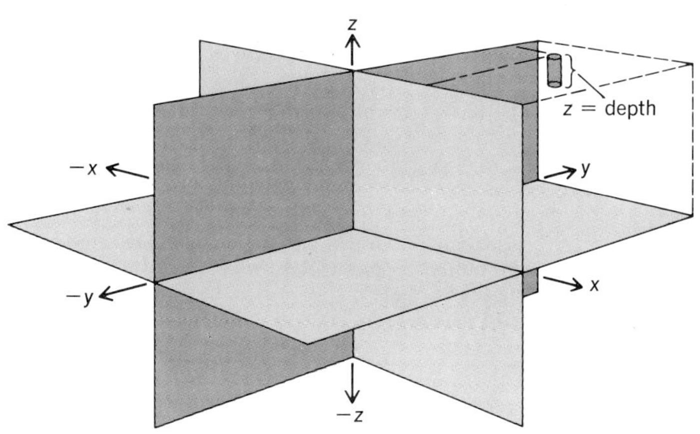

Fig. 10. Three-dimensional cartesian system of coordinates.

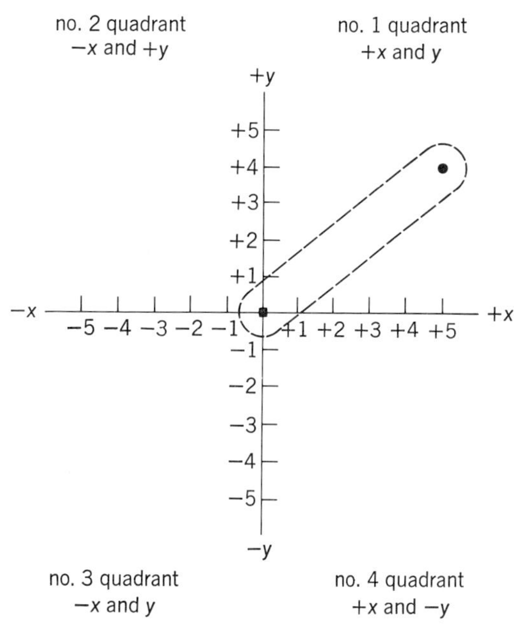

Fig. 11. Rectangular two-dimensional coordinates that represent the movement of an NC machine.

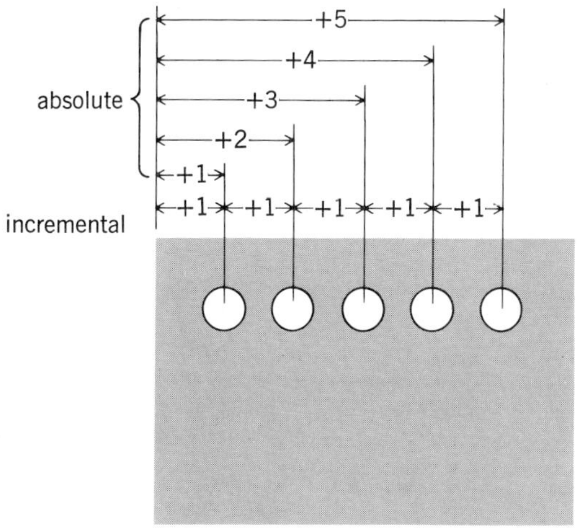

Fig. 12. Difference between absolute and incremental dimensioning.

the control system then puts out 1000 pulses, and the positioning is accurate to 0.001 in (25 μm). However, there is no verification that the pulse motor has actually moved the table the required distance.

The open-loop system is the simplest type of NC. The overall accuracy depends on the accuracy of the lead screw, rigid machine structure, and adequate power in the motors to ensure that they faithfully execute the commands of the controller. The stepping motor is usually electrohydraulic, which is designed to extend the load range of the

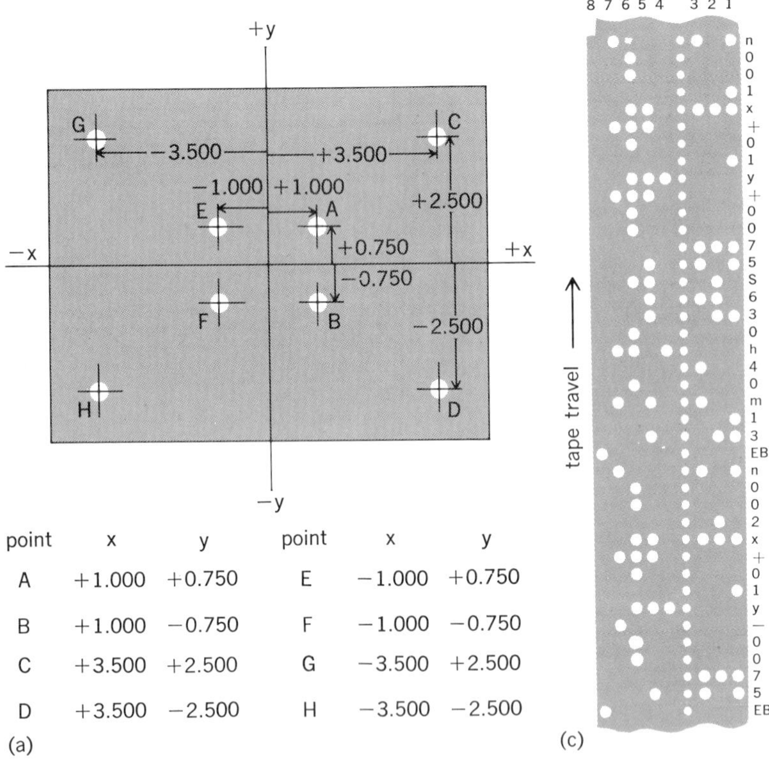

MANUSCRIPT
PTP
XXXX MACHINE

Part no. 1243
Part name: Plate
Date: X/X/XX
Prep. by: ABC
Checked by: DEF

sequence no. (n)	x		y		Spindle speed (s)	Feed (depth) (h)	Aux. code (m)	Comments
001	+	1.000	+	.750	630	40	M13	Use ½" HSS
002	+	1.000	−	.750				Drill, set
003	+	3.500	+	2.500				depth @ 0.35"
004	+	3.500	−	2.500				
005	−	1.000	+	.750				
006	−	1.000	−	.750				
007	−	3.500	+	2.500				
008	−	3.500	−	2.500			M30	Tape will automatically rewind after last operation

(b)

Fig. 13. Programming NC. (*a*) Pattern desired. (*b*) Program manuscript for pattern in *a*. (*c*) Coded tape instructions describing a section of the program in *b*. (*J. J. Childs, Principles of Numerical Control, Industrial Press, 1965*)

open-loop system. This motor has an electric stepper positioned by electric pulses. The pulses position hydraulic valves which drive a hydraulic motor that has more power available than an all-electric stepping motor.

Programming. With NC the role of the part programmer is crucial. He or she must be fully competent in all phases of machine tool operation, such as work-holding methods and correct speeds, feeds, and depth of cut for a wide range of materials, and must know the operation and programming characteristics of each NC machine. To understand how the programmer converts information from part drawings into numerical codes for punching onto the tape, it is necessary to understand the cartesian, or rectangular, coordinate system on which NC is based. In Fig. 10 the origin (or starting point) is placed at the intersection of the planes *X-Y*, *X-Z*, and *Y-Z*. Movements from this point in any direction can be stated in distance along each axis, *x*, *y*, and *z*. For example, the dimensions for two points p_1 and p_2 would be as follows: p_1: $x = 2.350$ in. (59.69 mm), $y = 1.520$ in. (38.61 mm), $z = 3.125$ in. (79.38 mm); p_2: $x = 1.500$ in. (38.10 mm), $y = 3.250$ in. (82.55 mm), $z = -2.500$ in. (−63.50 mm).

Most two-axes drilling and milling machines operate in the *X-Y* plane. The remaining discussion will be simplified by confining the examples to the *X-Y* plane.

Rectangular coordinates. Shown in Fig. 11 are the rectangular two-dimensional coordinates that represent the movement of an NC machine. At the intersection of the *x-y* coordinates is the origin or zero dimension. Each quadrant is given plus or minus values. On the right side, *x* values are all plus; negative values are to the left, and likewise below the horizontal line. The *u* values above the origin are positive, and those below are negative. If a hole is to be drilled at $+5x+4y$ location, the center of the drill will be directed to that programmed point. If a slot is to be milled, as represented by the broken line, the center point of the cutting tool is programmed to go from the point of origin to $+5x+4y$. In this case the MCU would have to simultaneously give 5 units of x motion to 4 units of y motion.

At one time it was thought that all the programming could be done in the all-plus number one quadrant, so that minus dimensions could be avoided. Then came the incremental method of programming where the minus and plus signs did not indicate values, but travel direction along an axis.

Absolute and incremental programming. When a program is prepared in absolute dimensioning, all positions are stated as being measured from a fixed zero or origin point. The point may be a fixed point on the machine or a specific point on the workpiece. Some MCUs are equipped with a floating zero or zero shift which allows the programmer to establish the fixed zero in the most convenient location.

All incremental program locations are given in terms of distance and direction from the immediately preceding point. Thus the plus and minus signs do not refer to the quadrant but to the tool movement. A "plus x" would mean the tool would move to the right of its present location. Likewise a "minus x" command would direct the tool to the left. Absolute and incremental dimensions are compared in Fig. 12.

Figure 13*a* shows a simple part to be machined. The part programmer has listed all the points, A through H, and their corresponding coordinate locations. The information is then written in manuscript form (Fig. 13*b*). The information on the manuscript is converted via a keyboard into punched tape (Fig. 13*c*). It could also be converted to punched cards or magnetic tape, or it may be transferred through the keyboard directly to a computer. An NC system can operate from any one of these input media; however, the advantages of using any particular medium depend upon the type of operation as well as the program-generating equipment available.

Input media. In recent years magnetic tapes have made a comeback via the tape cassette that utilizes the ¼-in.-wide (6.35-mm) magnetic tape.

Diskettes (often called floppy disks) and storage or memory within the control unit itself are rapidly growing types of media to contain workpiece program information. Floppy disks are very similar in appearance to 45-rpm records. Instructions may also be stored in a computer memory remote from the machine tool and control unit.

There are many variations of the keyboard unit. Some are designed solely to punch tape or tabulating cards. Others have multiple-purpose capability, and punch tape and communicate with a computer. Sometimes the data entry is a two-stage process. For example, the unit may be used to generate a tape, which is checked and verified by a printout also known as the hard copy. The tape may then be run through a reader to transfer the information to the memory of a computer or a MCU.

Voice control terminals are also being used to speak directly to a computer. Units now on the market recognize 60 different voice codes. This covers numbers, letters, and some miscellaneous codes.

The punched tape or other medium is placed in the tape reader of the MCU by the machine operator, who also checks to see that the appropriate tools are available and the part is properly positioned and clamped on the machine.

The MCU, with its tape reader, acts upon the coded instructions originated by the part programmer and develops output signals which control the electric, hydraulic, or other type of servomechanism used to physically drive and direct the machine tool.

Tape formats. Two common code-on-tape formats have been developed (Fig. 14). They are called EIA (Electronic Industry Association) and ASCII (American Standard Code for Information Interchange; ASCII is also becoming known as ANSCII, American National Standards Code for Information Interchange). Once the initial shock of having two formats passed, it was found that the difference between the two was almost insignificant.

The standard 1-in. (25.4-mm), eight-channel tape has only six channels that are used (Fig. 14*a*. Channel five is a parity check. Not counting sprocket holes, there must always be an uneven number of holes for any individual code character. Channel eight indicates the end of the block. The

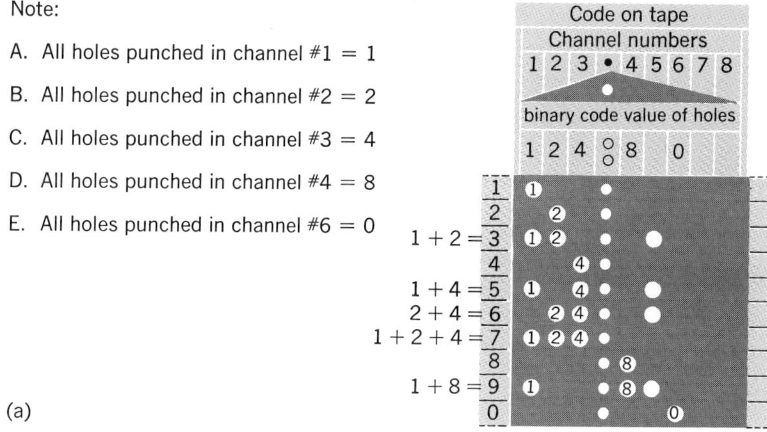

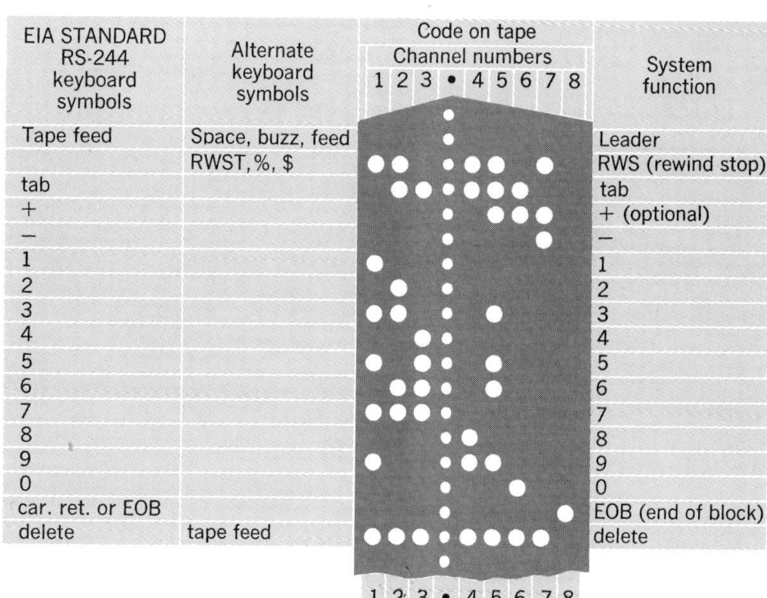

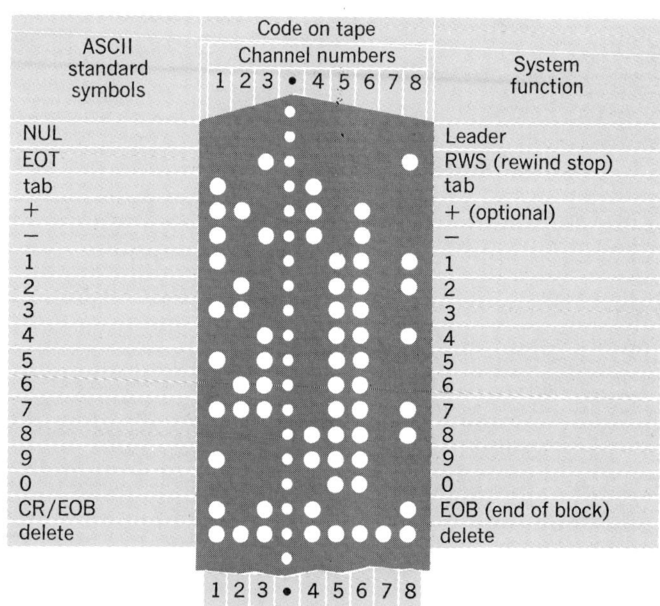

Fig. 14. Code-on-tape formats. (*a*) Binary number code. (*b*) EIA format. (*c*) ASCII format. (*Conway, Bringman, and Doane, Handbook for Slo-Syn Numerical Control, Superior Electric Co.*)

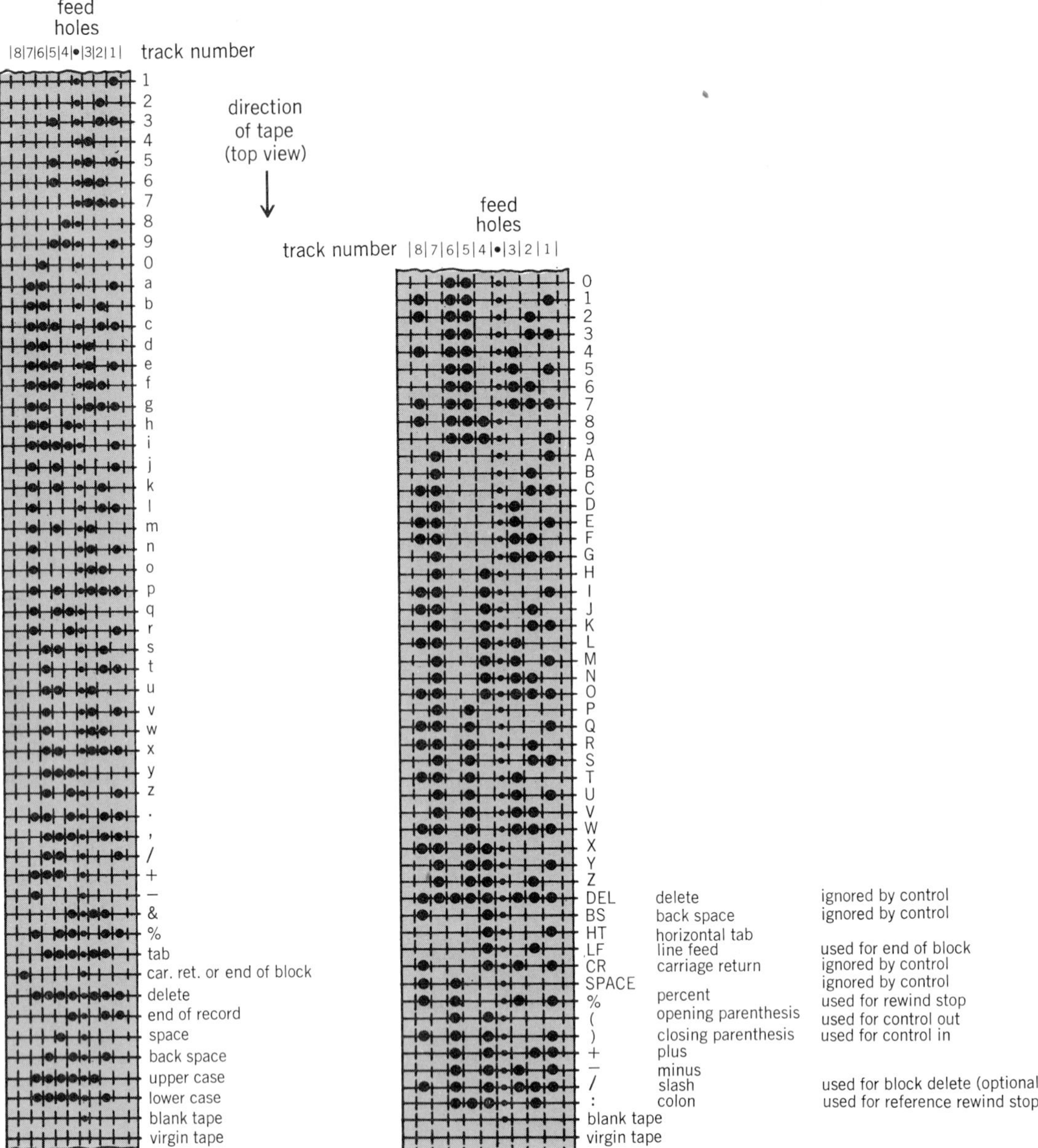

Fig. 15. Original standard EIA tape at left, and revised tape at right. Every level on the revised tape has an odd number of holes for a parity check. Most of the newer MCUs can handle either EIA or ASCII coding. (*Modern Machine Shop, 1979*)

remaining six channels are available at any level to offer hole positions for a single symbol code which may be a letter, number sign, or symbol. There is only one combination that can occupy all six holes, and that is an error-delete sign.

Punched tape is used for much more than numerical control. Computer, telephone, telegraph, government, and other organizations which process data with eight-channel tape want a system that gives maximum capacity. Upper- and lowercase letters account for a 52-code requirement (Fig. 15). For NC work only, lowercase letters are used. Many of the new computer numerical control (CNC) machines accept either EIA or ASCII code and discern which it is through a parity check.

Binary coded decimal. A tape feature that is virtually 100% accepted is the binary coded decimal (BCD) method of individual coding. Five of the channels in the standard eight-channel tape are assigned values 0, 1, 2, 4, and 8 so that any numerical quantity from 0 to 9 can be designated on one horizontal level of the tape (Fig. 14*a*). Any numerical digit or letter or symbol code has its own combination of holes in a single row on the tape. The decimal value of any digit is determined by its relative position within a given numerical quantity.

The binary numbers on which the tape system is built are based on only two digits, normally zero and one. Binary notations are used because electric circuits are stable in either of two conditions: charged or discharged, on or off. Thus in a computer logic system a binary zero might be a positive charge, and a binary one a negative charge. Numerical values expressed as binary notations can be added, subtracted, multiplied, and divided in the same way as the familiar 10-digit system.

The computer programmer does not have to make the conversion from arabic to binary numbers; this is all done by the computer. Also, the final data output is not in binary form, but converted back to the arabic system.

Positioning. NC programming falls into two conceptual realms, point-to-point and continuous-path. This demarcation, once sharp, is now dulled, since many units today are built with a mixture of positioning and contouring capabilities.

Point-to-point programming. Point-to-point positioning is typified by drilling, tapping, reaming, boring, counterboring, and so on, in which any operation is completed at a fixed workpiece location in terms of a two-axis coordinated position. While the tool is moving from one location to another, it is not in contact with the workpiece. Thus, whether the tool moves first on the *Y* axis or the *X* axis is of no significant difference. To save time, most control units generate motions simultaneously along both axes.

Continuous-path programming. The continuous-path machine is capable of having a tool follow a continuous path as programmed. The most common example of this is where lathe tools or milling cutters are used to machine a desired profile. NC contouring may also apply to flame cutting, sawing, welding, grinding, and even adhesive application.

Interpolation. In both point-to-point and continuous-path programming, there must always be rectangular coordinates that are the reference points relating to the machine axis. In interpolation, the method of getting from one program point to the next is done by approximation. For example, a circle is defined as an infinite number of lines joined end to end to form a closed plane. Each line is equidistant from the center point. Since digital control is finite, it can only approximate a curve of any kind. A sufficient number of points have to be described within the framework of rectangular coordinates. Thus a continuous-path machine can produce intricate sculpture if enough coordiate points are described for the desired surface. The computer is often used to approximate a sufficiently large number of points so that the cut appears relatively smooth.

Progress has been made in reducing the complexity of many contouring NC systems to the point where even simple open-loop (nonfeedback) contouring controls can be retrofitted to small vertical milling machines. These newer contouring controls do not achieve a contouring capability by reading long point-to-point tapes faster by means of higher-speed readers; rather they incorporate computer functions within the NC console that permit short-form tapes containing generalized commands to produce a large number of machine movements. These advanced features typically include: (1) buffer storage, which is an information-storage section in the NC console that receives a new block of data from the tape reader while the machine tool is still responding to the previous command of the console; (2) linear interpolation, which is a feature that automatically generates straight-line cuts between programmed points regardless of the angle of cut, and does it with only one command block (Fig. 16*a*); (3) circular interpolation, which is a feature that automatically generates radii between two points in a single command block (Fig. 16*b*).

Editing. No matter how good the programmer or how flawless the operator, there will always be errors. Several editing systems have been developed to check out a program before it is committed to the machine tool loaded with the workpiece.

One type is a graphic plotting unit. Here the program tape (or perhaps data direct from the computer) directs the plotting unit, which is usually a flat piece of paper mounted on a table. Over the paper is a two-axis driving mechanism carrying a pen. The program driving the pen duplicates the cutting tool path. Since the paper is a two-axis surface, there is often provision for selecting the two axes to be portrayed: *XY*, *XZ*, or *YZ*. Some plotters provide an isometric view. This type of unit quickly discloses several types of errors, such as when a supposedly 1-in. (25-mm) move becomes a 10-in. (250-mm) move because the programmer displaced the decimal point.

A similar unit shows the cutter path with a cathode-ray tube very similar to a television tube. Some units have the advantage of being able to portray cutting tool paths in three standard planes plus an isometric view.

Editing units have been developed that are primarily used at the machine tool to make any revisions that become desirable after the program has been put into actual use. A typical shop floor editor contains a keyboard, printout, tape reader, and tape punch which connects directly to the MCU. Data may be taken from the actual program tape and stored in the editing unit. Changes can be made, and a revised tape is punched right at the machine tool.

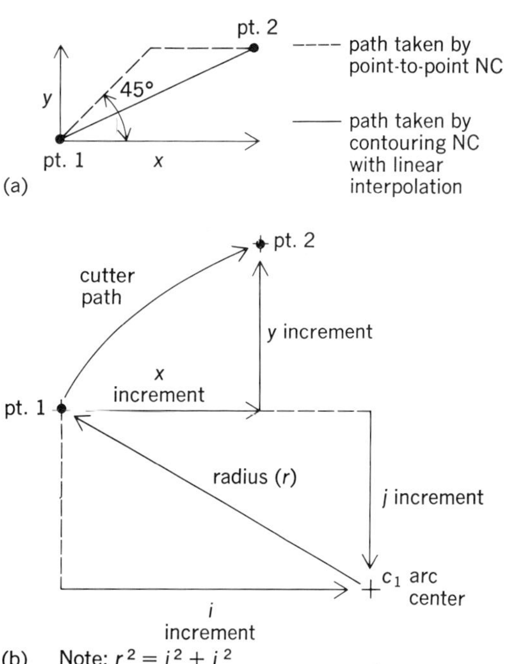

Fig. 16. Contouring operations with point-to-point systems. (*a*) Linear interpolation. (*b*) Circular interpolation.

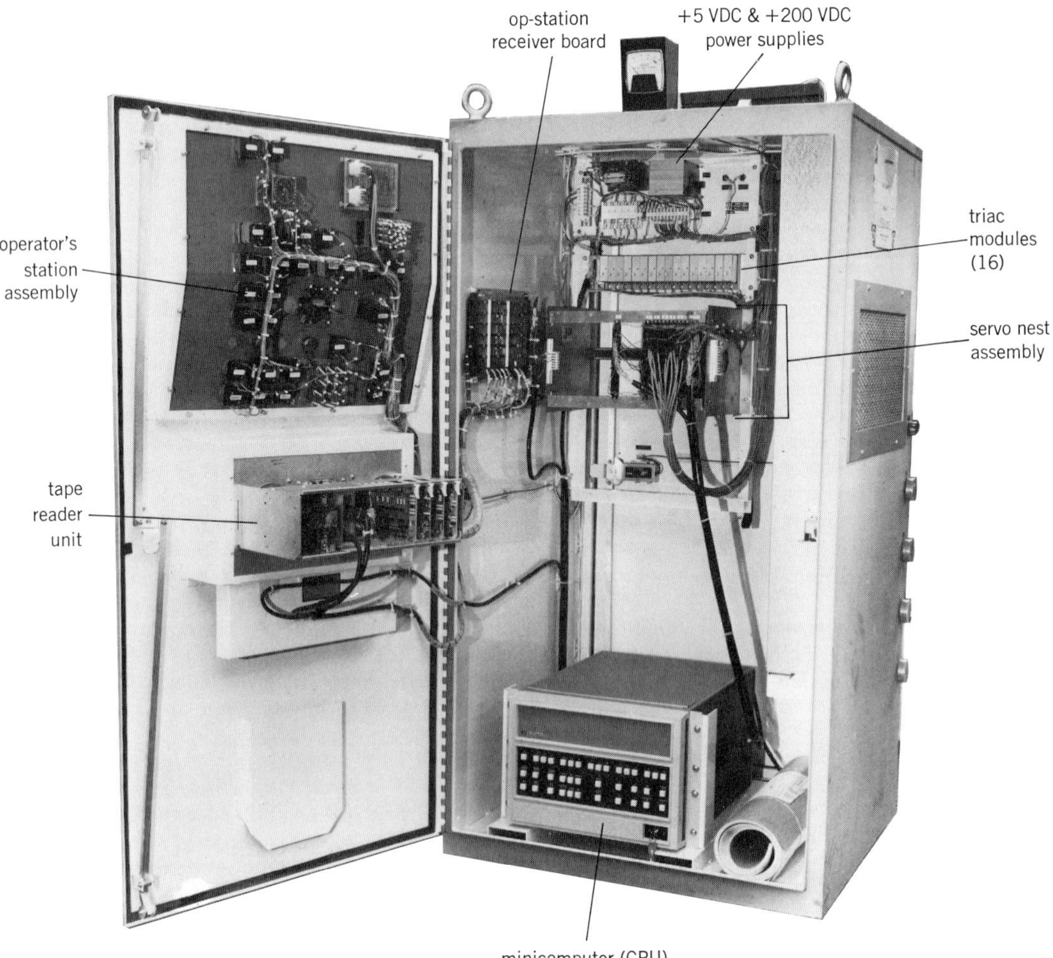

Fig. 17. Inside of MCU as used in CNC. (*Giddings and Lewis*)

The line separating tape-preparation units, editing units, and computational resources is becoming less sharp as microprocessor technology makes the combination of such functions less costly and more reliable.

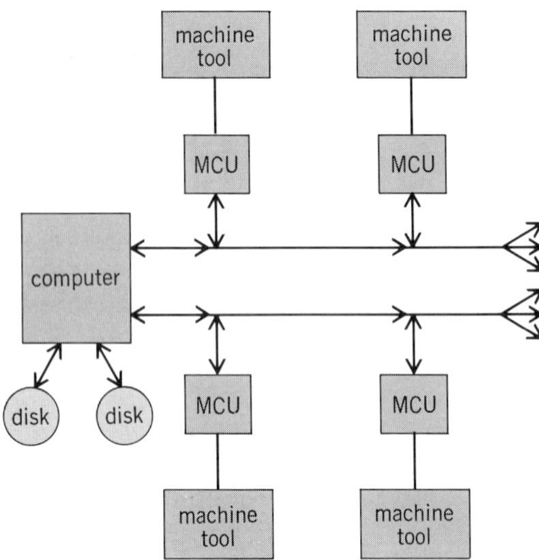

Fig. 18. Schematic diagram to show how one computer is used to control a hierarchy of NC machines as in DNC. (*Modern Machine Shop, 1980*)

Advantages of NC machining. Many of the advantages of NC machining are obvious, such as simpler setups, reduced cutting-tool changing time, less operator skill requirements, less scrap, much greater design flexibility, and better accuracy. However, one of the most important advantages is the increase in production that NC affords over conventional machining. Traditionally, it has been observed that parts in the shop spend 95% of the time moving and waiting, with only 5% of the time on the machine. Of the 5% of the time spent on the machine, less than 30% is spent as actual metalworking time. Through the increased use of NC and computer control, the disproportionate nonproductive time can be greatly reduced.

CNC, DNC, and CAM. The first NC machine generation was classified as hard-wired. That is, the responses to data input, data-handling sequence, and control functions were by a fixed committed circuit with interconnections of discrete decision elements and storage devices. Changes in the response, sequence, or functions could be made only by changing the connections. NC machines have now progressed to the soft-wired type, where the control can be adapted to several machine tool configurations.

Computer numerical control (CNC). CNC machines have a minicomputer installed in the control unit (Fig. 17). When the control unit is connected to the machine tool, the interface is completed

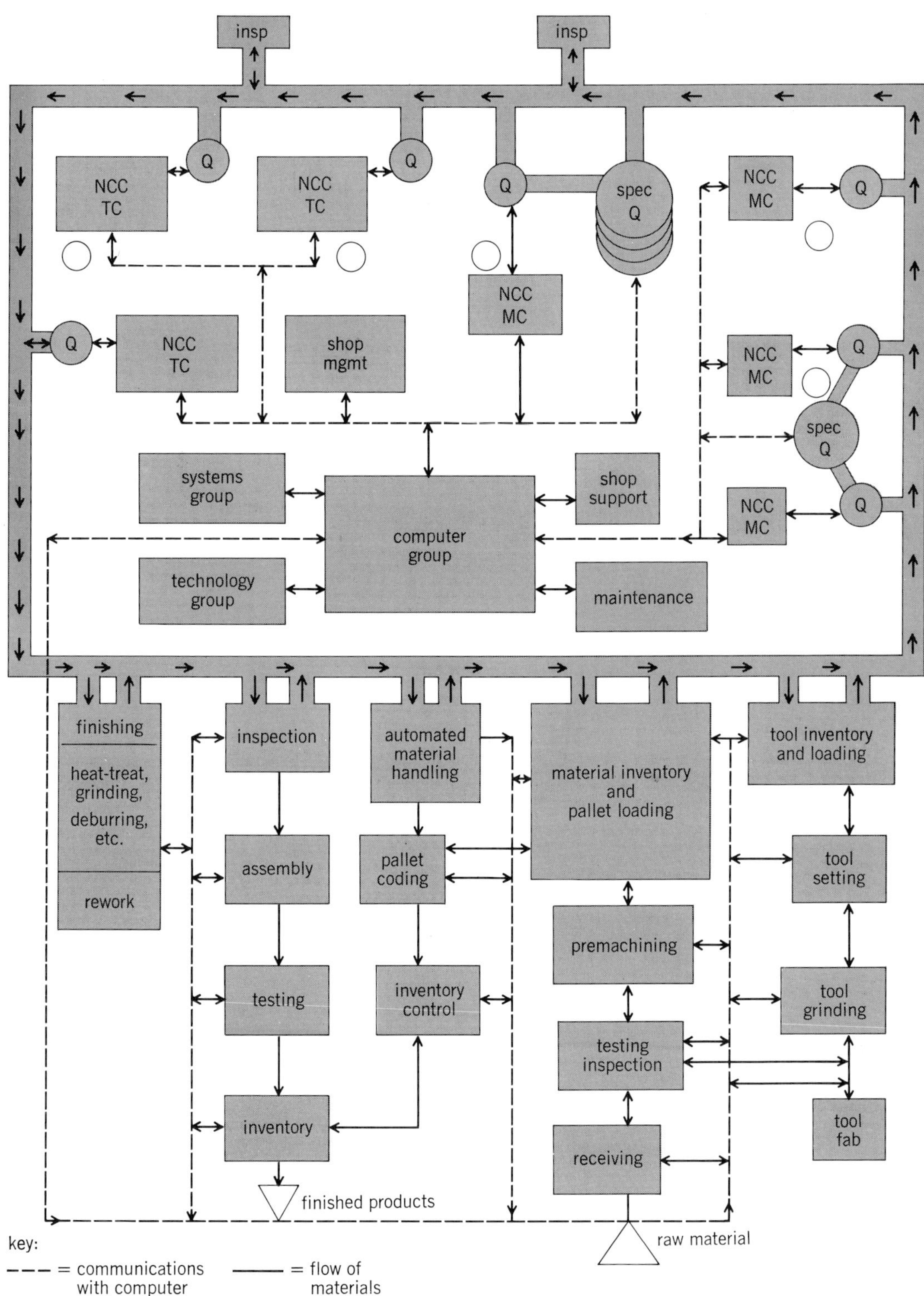

Fig. 19. Schematic diagram showing how an entire manufacturing operation can be controlled by computers when coupled with automatic material-handling equipment. (*Modern Machine Shop, 1980*)

by programming the computer—not by hard-wiring. Interface refers to linking two pieces of electrical equipment having separate functions, for example, a tape reader to a tape processor, or a control to a machine. Thus CNC is referred to as soft-wired control. With the large memory of the minicomputer, it became possible to add many features to the controls, such as the following:

1. When the machine tool is being checked out, some error in the lead screw is noted. This causes an error in the table position. Since this occurs every time at the same spot, corrective action can be programmed into the minicomputer.

2. If the spindle is run at high speed over an extended period of time, it heats up and expands. The minicomputer can be programmed to compensate for the spindle extension.

3. Previously, if 40 workpieces had to be run, the

NC tape would have to be run 40 times. With CNC the tape is run once and stored in the minicomputer memory, which is then used to produce the workpieces. With the tape information in the memory, it is possible to edit tapes at the machine. While the tape information is being proved on the first part, immediate corrections and improvements can be made via the control console. It is possible to get a corrected tape immediately by attaching a tape punching unit to the control.

4. CNC consoles display the command for the current operation, and many units allow operators to look both ahead and behind the command. Thus the operator knows what is happening in the present and can look forward to the next operation as well.

CNC applications continue to grow as new ideas and advantages are added.

Direct numerical control (DNC). Direct numerical control was the next development after CNC. With DNC a number of NC or CNC machines are connected together to a remote computer (Fig. 18). The number of machines connected could be as small as 2 or as great as 256.

With DNC the MCUs are usually NC or CNC units, and instead of carrying a program tape to each machine, the workpiece program tape is fed to the control via a communication line. The tape reader is bypassed or eliminated, with its attendant maintenance.

In both NC and CNC operations, the MCU must read the workpiece program commands and translate them into specific machine motions. With DNC it is possible to have the program commands interpolated by the main computer and sent directly to the machine control.

A big advantage offered by DNC is that the communication between the MCU and the computer is two-way. That is, the machine operator can ask for specific part information or communicate with other departments easily and quickly. For example, if a cutter is broken, the information can be immediately given to the tooling group. Thus, quick corrective action can greatly expedite machining operations and reduce machine downtime.

As more and more programs are added in conventional NC, the number of tapes required with attendant storage and cataloging can become a problem. With DNC disk files (thin, circular metal plates coated on both sides with magnetizable material), the computer can be used to store the equivalent of tens of thousands of punched tapes in a very small area.

The growth of the DNC hierarchy of machine tools and controls led naturally to complete computer-aided manufacturing.

Computer-aided manufacturing (CAM). In DNC the computer can supply information to several machines for individual operation. CAM goes a step further in that all pertinent information is contained in the computer, so that it serves as a communication network for the number of pieces machined at any given station, the amount of uptime and downtime, the production rate, and any other pertinent management data. A schematic CAM plant layout is shown in Fig. 19. The incoming raw material undergoes succeeding steps of inspection, premachining, and so on until it reaches the material inventory and pallet-loading station. As each job order is received from the computer, the pallets are set up with the blanks, cutting tools, and other information. The pallets are coded so that as they go down the line, their paths are controlled by the computer. Depending upon the priority (set by the manager via the computer), the pallets move in queues. The queues may be stacked vertically to take advantage of plant space. The main computer is notified when the individual operations, parts, job lots, or inspections are completed. With input from shop personnel, the main computer can produce an up-to-the-minute status report of everything in the production area.

The schematic diagram in Fig. 18 shows how the concept of CAM can be approached for short-run production items. The configuration would vary considerably for each plant. Although the concept may seem somewhat formidable, most plants are doing all the operations, but with massive human endeavor. The main task in adapting CAM is that of computerizing all operations, including material handling. The CAM concept is one step closer to the ultimate goal of manufacturing, the workerless factory. *See* COMPUTER-AIDED DESIGN AND MANUFACTURING.

[ROY A. LINDBERG]

Bibliography: American Society of Manufacturing Engineers, *Tool and Manufacturing Engineer's Handbook*, 3d ed., 1976; American Society of Manufacturing Engineers, *Tooling for Aircraft and Missile Manufacture*, 1964; J. J. Childs, *Numerical Control Practice and Application*, 1972; J. J. Childs, *Principles of Numerical Control*, 1965; R. Dyke, *Numerical Control*, 1967; Getting the handle on NC, *Mod. Mach. Shop*, April 1979–February 1980; W. C. Leone, *Production Automation and Numerical Control*, 1967; H. E. Linsley, *Broaching: Tooling and Practice*, 1961; D. McCrae, *Optical Tooling in Industry*, 1964; A. Merdinger, *Numerical Control*, 1975; Modern Machine Shop, *NC Guidebook*, 1979–1980; Numeridex, Inc., *NC Glossary*, 1978, 1979; D. E. Ostergaard, *Basic Diemaking*, 1963; W. J. Patton, *Numerical Control Practice and Appreciation*, 1972; D. S. Puckle and J. R. Arrowsmith, *Introduction to the Numerical Control of Machine Tools*, 1964; A. D. Roberts and R. C. Prentice, *Programming for Numerical Control Machines*, 1968.

Torch

A gas-mixing and burning tool that produces a hot flame for the welding or cutting of metal. The torch usually delivers acetylene and commercially pure oxygen producing a flame temperature of 5000–6000°F (2750–3300°C), sufficient to melt the metal locally. The torch thoroughly mixes the two gases and permits adjustment and regulation of the flame. Acetylene requires 2.5 times its volume of oxygen for complete combustion and, being an endothermic compound of carbon and hydrogen, can produce a higher flame temperature than other fuel gases. *See* WELDING AND CUTTING OF METALS.

Torches are of two types: low-pressure and high-pressure. In a low-pressure, or injector, torch, acetylene enters a mixing chamber, where it meets a jet of high-pressure oxygen (Fig. 1). The amount of acetylene drawn into the flame is controlled by the velocity of this oxygen jet. In a high-

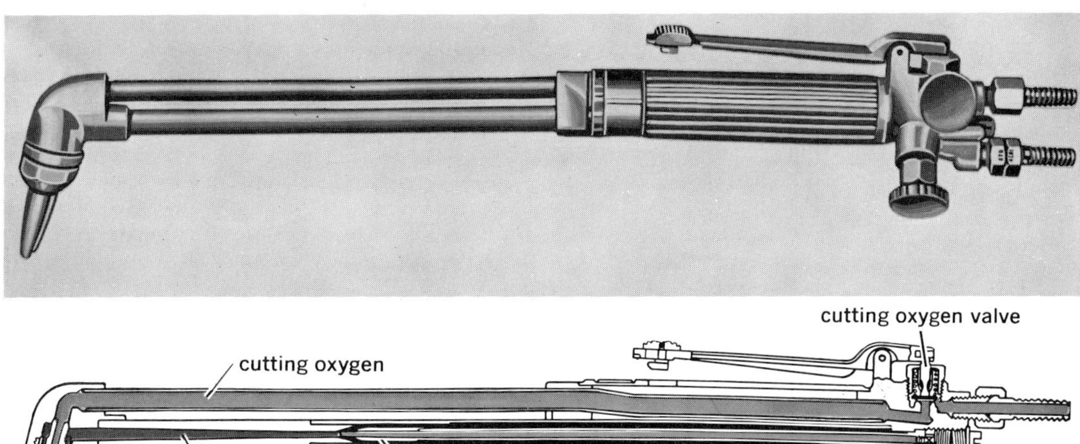

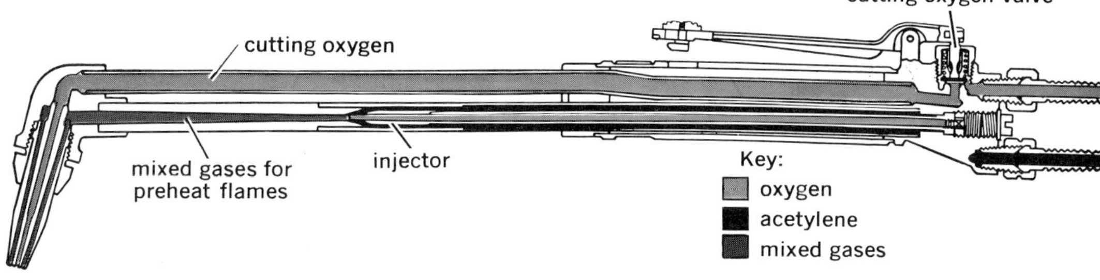

Fig. 1. Low-pressure injector cutting torch. (*Linde Co.*)

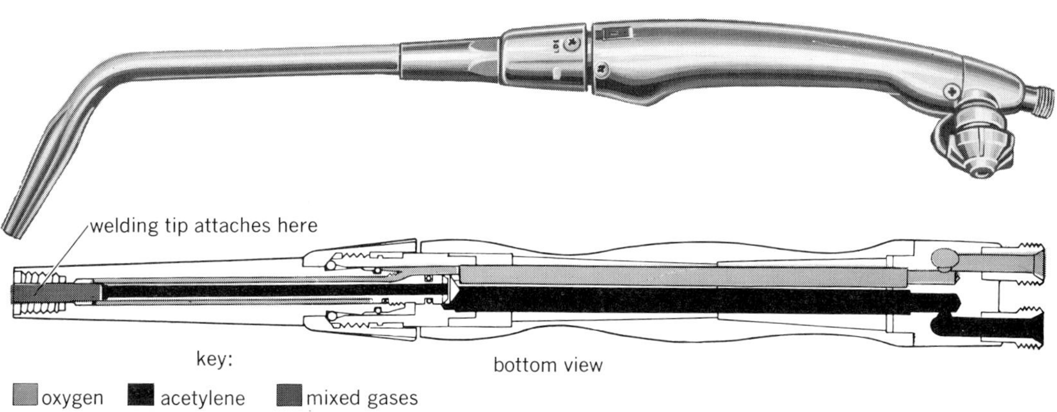

Fig. 2. Welding torch with cartridge mixer operates over wide pressure range. (*Linde Co.*)

pressure torch both gases are delivered under pressure (Fig. 2). Heat developed at the work is controlled principally by the size of the nozzle or tip fitted to the torch. The larger the tip the greater the required gas pressure. Small flames are used with thin-gage metals; large flames are necessary for thick metal parts.

A welding torch mixes the fuel and gas internally and well ahead of the flame (Fig. 1). For cutting, the torch delivers an additional jet of pure oxygen to the center of the flame. The oxyacetylene flame produced by the internally mixed gases raises the metal to its ignition temperature. The central oxygen jet oxidizes the metal, the oxide being blown away by the velocity of the gas jet to leave a narrow slit or kerf. In the case of iron, the oxides fuse at a lower temperature than the iron or steel so that the oxides form, melt, and blow away before the adjacent metal fuses. The temperature for the cutting action, once initiated, is maintained by the oxidization of the iron. Intricate shapes are accurately cut in low-carbon steel by torches automatically guided, such precision cutting being called flame machining.

[FRANK H. ROCKETT]

Torque

The product of a force and its perpendicular distance to a point of turning, also called the moment of the force. Torque produces torsion and tends to produce rotation. Torque arises from a force or forces acting tangentially to a cylinder or from any force or force system acting about a point. A couple, consisting of two equal, parallel, and oppositely directed forces, produces a torque or moment about the central point. A prime mover such as a turbine exerts a twisting effort on its output shaft, measured as torque. In structures, torque appears as the sum of moments of torsional shear forces acting on a transverse section of a shaft or beam. *See* COUPLE; TORSION.

[NELSON S. FISK]

Torque converter

A device for changing the torque-speed ratio or mechanical advantage between an input shaft and an output shaft. A pair of gears is a mechanical torque converter. A hydraulic torque converter, with which this article deals, is an automatically and continuously variable torque converter, in

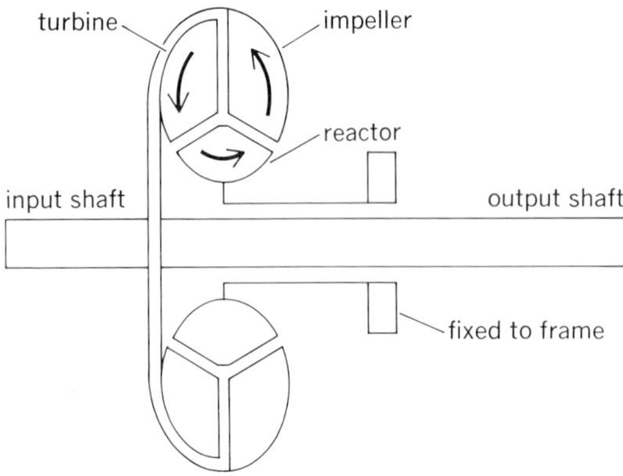

Fig. 1. Elementary hydraulic torque converter.

contrast to a gear shift, whose torque ratio is changed in steps by an external control. *See* AUTOMOTIVE TRANSMISSION; GEAR DRIVE.

Converter characteristics. A mechanical torque converter transmits power with only incidental losses; thus, the power, which is the product of torque T and rotational speed N, at input I is substantially equal to the power at output O of a mechanical torque converter, or $T_I N_I = k T_O N_O$, where k is the efficiency of the gear train. This equal-power characteristic is in contrast to that of a fluid coupling in which input and output torques are equal during steady-state operations. *See* FLUID COUPLING.

In a hydraulic torque converter, efficiency depends intimately on the angles at which the fluid enters and leaves the blades of the several parts.

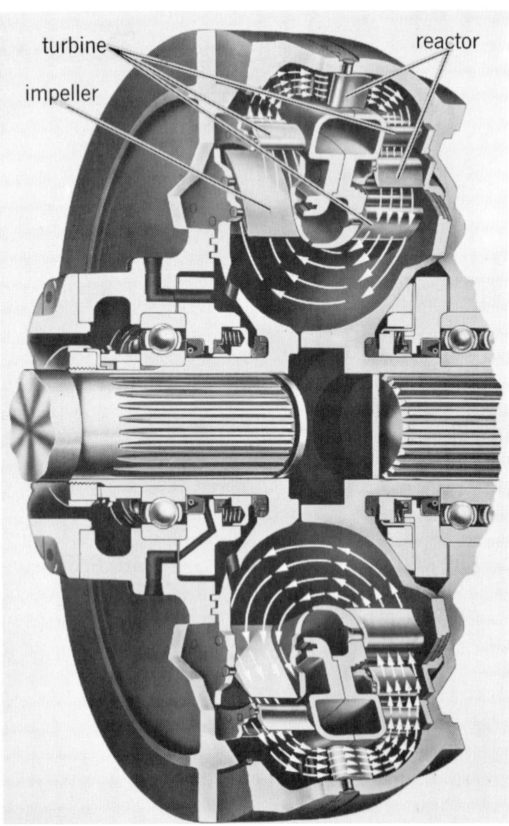

Fig. 2. Three-stage converter showing simplified fluid flow around torus. (*Twin Disc, Inc.*)

Because these angles change appreciably over the operating range, k varies, being by definition zero when the output is stalled, although output torque at stall may be three times engine torque for a single-stage converter and five times engine torque for a three-stage converter. Depending on its input absorption characteristics, the hydraulic torque converter tends to pull down the engine speed toward the speed at which the engine develops maximum torque when the load pulls down the converter output speed toward stall.

Converter power efficiency is highest (80–90%) at a design speed, usually 40–80% of maximum engine speed, and falls toward zero as shaft speed approaches engine speed. Because of this characteristic, the mode of operation may be modified to change from torque conversion to simple fluid coupling or to direct mechanical drive at high speed.

Hydraulic action. These characteristics are achieved by the exchange of momentum between the solid parts of the converter and the fluid (Fig. 1). A vaned impeller on the input shaft pumps the fluid from near the axis of rotation to the outer rim. Fluid momentum increases because of the greater radius and the influence of the vanes. The high-energy fluid leaves the impeller and impinges on the blades of a turbine, giving up its energy to drive the turbine, which is connected to the output shaft. The fluid discharges from the turbine into a bladed reactor. The reactor blades are fixed to the frame; they deflect the fluid flow and redirect it into the impeller. This change in flow direction produced by the stationary reactor is equivalent to an increasing change in momentum which adds to the momentum imparted by the impeller to give a torque increase at the output of the converter (Fig. 2).

In a typical converter, as the output shaft comes up to the speed of the input shaft, efficiency decreases. Therefore, the reaction member may be mounted on a freewheel unit so that it rotates with the fluid at high speed ratio when torque multiplication is no longer possible. In addition, splitting the reaction member to give a four-element polyphase converter gives even more uniform efficiency.

[HENRY J. WIRRY]

Bibliography: G. Raczkowski, *Principles of Machine Dynamics*, 1979; H. F. Tucker, *Automatic Transmissions*, 1980.

Torsion

A straining action produced by couples that act normal to the axis of a member. Torsion is identified by a twisting deformation.

In practice, torsion is often accompanied by bending or axial thrust as in the case of line shafting driving gears or pulleys, or propeller shafts for ship propulsion. Other important examples include springs and machine mechanisms usually having circular sections, either solid or tubular. Members with noncircular sections are of interest in special applications, such as structural members subjected to unsymmetrical bending loads that twist and buckle beams.

When subjected only to torque, the member is in pure torsion, which produces pure shear stresses. The shear properties of materials are determined by a torsion test. *See* SHEAR.

Cylindrical bars. The twist of a bar due to torque can be visualized as the accumulated rotational displacements of imaginary disks cut by transverse sections on which tangential forces operate. Shearing forces vary across the section and together furnish the internal resisting torque.

Torsional angle, designated θ, is the total relative rotation of the ends of a straight cylindrical bar of length L, when subjected to torque (Fig. 1).

Helical angle, designated ϕ, is the angular displacement of a longitudinal element, originally straight on the surface of the untwisted bar, which becomes helical after twisting (Fig. 1). Angle ϕ is the shear strain. For small twist, torsional and helical angles are related by geometry $\phi = R\theta/L$, where R is the radius of the bar.

Elastic shear stress. Within the elastic limit, shear stress S_s is found by Hooke's law, $S_s/\phi = E_s$, and is expressed in terms of the torsional angle as $S_s = (R/L)E_s\theta$, where E_s is the modulus of rigidity. *See* HOOKE'S LAW.

The shear stress varies linearly across the section, being maximum at the surface and zero at the center. For a circular section the maximum shear stress acting perpendicular to the radius at the extreme distance $R = D/2$ from the neutral axis is $S_{max} = 16T/\pi D^3$, where T is the externally applied twisting moment.

Tangential shear stresses on the section are accompanied by longitudinal shear stresses along the bar. These complementary stresses induce tensile and compressive stresses, equal to the shear intensity, at 45° to the shear stresses. The longitudinal stresses are important in laminated materials, wood, or metals with seams. Brittle materials, low in tensile strength, fracture on a 45° helicoidal surface; ductile materials fracture on transverse sections after large twist.

Resisting torque equal to the applied torque is the moment of the internal shear forces about the neutral axis expressed in terms of the sectional dimensions and the stresses. A general expression for resisting torque is $T = S_{max}J/R$, where J is the polar moment of inertia of the section. This relation is applicable to both solid and tubular circular sections which are differentiated by J. In terms of torque T, torsional angle θ is TL/E_sJ. Torsional angle per unit of length is a measure of torsional stiffness, which may limit the required dimensions of a shaft. In power transmission the torque associated with horsepower is found from hp $= TN/63{,}000$, where T is expressed in inch-pounds of moment and N is the rotation of the shaft in revolutions per minute.

Inelastic behavior in torsion. Strains exceeding the elastic limit are not completely recoverable after unloading and the behavior is inelastic. Torsional strains very linearly from the center of the bar during elastic and plastic deformation, and the corresponding shear stresses reflect the stress-stain curve for the material (Fig. 2). After the extreme element reaches the yield point, continued twisting produces inelastic strains at increasing distances from the surface while the stress remains constant. When the action is fully plastic, the stresses are constant, equal to the yield point over the entire section. The fully plastic resisting torque is shown by Eq. (1), which is 1.33

$$T_p = \frac{4}{3}\frac{S_{yp}J}{R} \qquad (1)$$

times that required to just produce surface yielding. Torsional resistance increases because of strain hardening but is of interest only where large deformation can be tolerated. Elastic analysis is applicable to designs where permanent deformation must be avoided and where endurance (fatigue) properties limit the stresses.

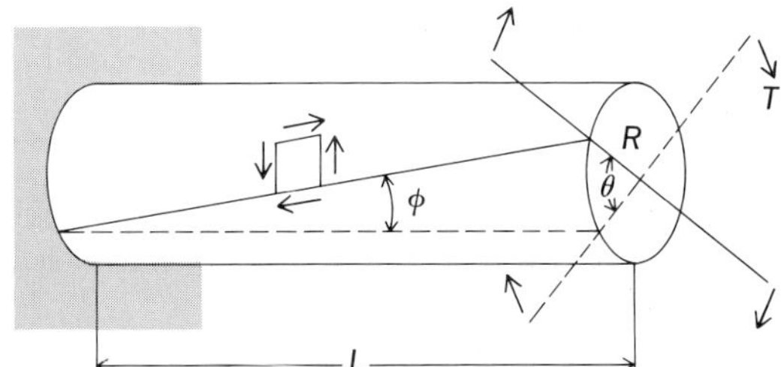

Fig. 1. Cylindrical bar in torsion.

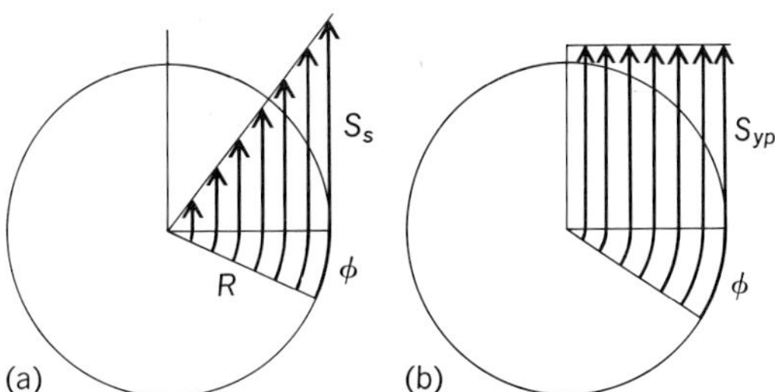

Fig. 2. Stress distribution. (*a*) Elastic torsional stress. (*b*) Fully plastic torsional stress.

Thin-walled tubes. Thin tubular members find application particularly in aircraft. Shear stresses are assumed uniform over the wall thickness, when a thin-walled tube of any shape is subjected to torque at the ends. Shear force q per unit length of perimeter is constant.

Shear flow is the constant shear force q acting along the median line of the wall and is equal to the product of shear stress S times thickness t at any point; thus $q = St$ is constant. The concept of flow is drawn from the similarity of the expression for constant shear force with the constant quantity Q of a liquid passing variable sections of a channel having area A and velocity V, $Q = AV$. Resisting torque T is the summation of moments of shear forces on unit lengths ds of the wall perimeter about the center of rotation $T = 2Aq$, where A is the area enclosed by the center line of the tube wall (Fig. 3). The stress at any point where thickness is t is $S = q/t = T/2At$. The torsional angle produced by applied torque T is found from Eq. (2),

$$\theta = \frac{TL}{4A^2E}\int_0^S \frac{ds}{t} \qquad (2)$$

where S is the length of the perimeter and t is the variable thickness. For constant thickness, $\theta = TLS/4A^2Et$, where S is peripheral length of the center line.

Solid noncircular sections. When a solid member with noncircular section is twisted, the sections become warped and the stresses do not vary linearly as in the case of circular sections. Evaluation of stresses and torsional twist requires the rigorous procedures of the theory of elasticity. If a grid is scribed on the surface of a square or rectan-

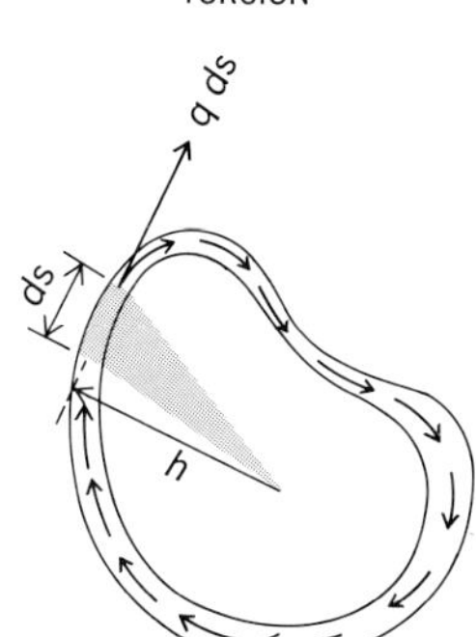

Fig. 3. Shear flow.

gular bar and the bar is twisted, distortions of the grid indicate that maximum shear stress is at a boundary nearest the center. Contrary to theory applicable to circular sections, the stress is zero at the corners, which are the most remote elements. The location of maximum stress is indicated by points of initial plastic yielding as shown by the macrographs of a square and a round bar (Fig. 4). Sections were etched after yielding, thus differentiating the darker plastic zones. Formulas for maximum shear stress and torsional angle for common noncircular sections are presented in Fig. 5.

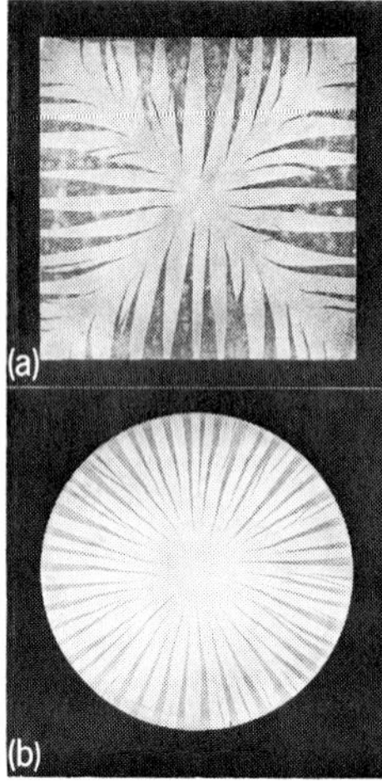

Fig. 4. Plastic strain in torsion. (*a*) Square bar. (*b*) Round bar.

Helical springs subjected to axial loads involve all four possible straining actions: direct stress, transverse shear, bending, and torsional shear. For small obliquity of the coils, as in close-coiled springs, torsional shear is the most important action. When stresses and deflection are determined by formulas applicable to straight bars, a correction is necessary to account for the effect of curvature of the coils. *See* SPRING.

Membrane analogy. Shearing stresses in sections which cannot be conveniently analyzed mathematically are determined experimentally by membrane analogy. The analogy presented by Ludwig Prandtl (1903) is based on the similarity of the equilibrium equation for a membrane with pressure on one side and the differential equation for torsional stresses. In application, a thin membrane, such as a soap film, is placed over an opening in a plate having the same geometrical shape as the section under investigation. Slight air pressure on one side deflects the film, and micrometer measurements determine the contours of equal deflection. The slope at any point and the volume enclosed by the deflected membrane can be found from these measurements. If a bar having this section is twisted, the torsional shearing stress at any point is proportional to the slope of the membrane, the stress direction is tangent to the contour, and the torque is proportional to the volume enclosed by the deflected membrane.

The method is a valuable qualitative aid in locating points of maximum stress by visualizing or observing points of maximum slope of the deflected film. The high stress at a reentrant corner, such as at a fillet of a structural angle or channel section, is indicated by a steep slope of the film. [JOHN B. SCALZI]

Bibliography: R. J. Roark, *Formulas for Stress and Strain*, 4th ed., 1965.

Torsion bar

A spring flexed by twisting about its axis. Design of a torsion bar spring is primarily based on the relationships between the torque applied in twisting the spring, the angle through which the torsion bar twists, and the physical dimensions and material (modulus of elasticity in shear) from which the torsion bar is made. The illustration shows the elements of a simple torsion bar and the important dimensions involved in its design. Equation (1) relates these dimensions.

$$\theta = \frac{32Fal}{\pi D^4 G} \qquad (1)$$

Here θ is angle of twist in radians, F is force in pounds, a is radius arm of force in inches, l is length of torsion bar in inches, D is diameter of torsion bar in inches, and G is modulus of elasticity in shear in pounds per square inch.

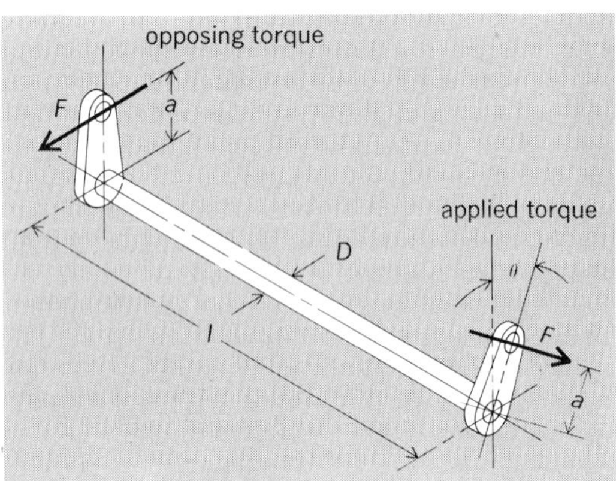

Diagram of torsion bar.

If the deflection or twist of the spring θ is large, force F must change direction if a is to remain constant. For this reason Eq. (1) is frequently written as Eq. (2), in which τ is torque in inch-pounds.

$$\theta = \frac{32\tau l}{\pi D^4 G} \qquad (2)$$

Torsion bar springs are found in the spring sus-

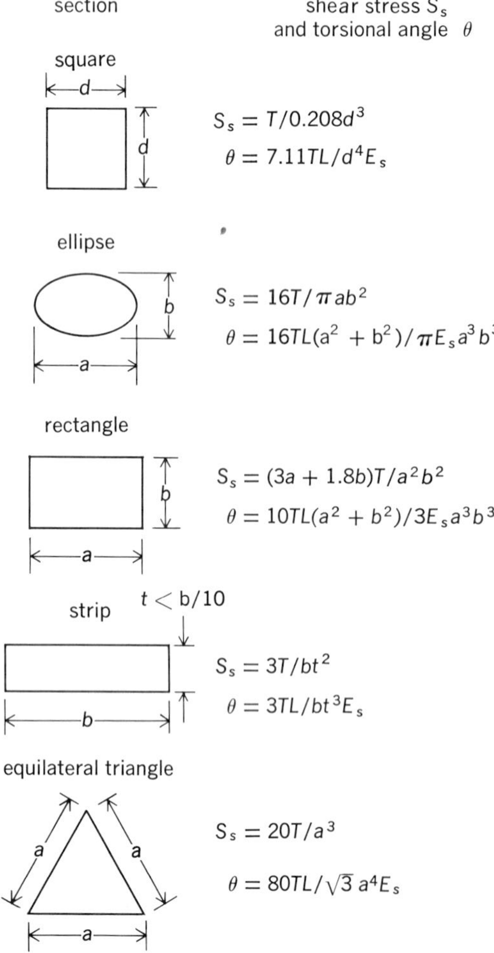

section
shear stress S_s and torsional angle θ
square
d
d
$S_s = T/0.208d^3$
$\theta = 7.11TL/d^4E_s$
ellipse
b
a
$S_s = 16T/\pi ab^2$
$\theta = 16TL(a^2 + b^2)/\pi E_s a^3 b^3$
rectangle
b
a
$S_s = (3a + 1.8b)T/a^2b^2$
$\theta = 10TL(a^2 + b^2)/3E_s a^3 b^3$
strip
$t < b/10$
b
$S_s = 3T/bt^2$
$\theta = 3TL/bt^3E_s$
equilateral triangle
a
a
a
$S_s = 20T/a^3$
$\theta = 80TL/\sqrt{3}\, a^4E_s$

Fig. 5. Formulas for maximum shear stress and torsional angle for common noncircular sections.

pension of truck and passenger car wheels, in production machines where space limitations are critical, and in high-speed mechanisms where inertia forces must be kept to a minimum. *See* SPRING.

[L. SIGFRED LINDEROTH, JR.]

Tower

A concrete, metal, or timber structure that is relatively high for its length and width. Towers are constructed for many purposes, including the support of electric power transmission lines, radio and television antennas, and rockets and missiles prior to launching.

Transmission towers. These towers are rectangular in plan and are not steadied by guy wires. A transmission tower is subjected to a number of forces: its own weight, the pull of the cables at the top of the tower, the effect of wind and ice on the cable, and the effect of wind on the tower itself. Torsional forces on a tower caused by breakage of the cables in one span on one side of the tower must also be considered.

Radio and television towers. Such towers are either guyed or freestanding. Freestanding towers are usually rectangular in plan. In addition to their own weight, freestanding towers support the weight of the antenna and accessories and the weight of ice, unless a deicing circuit is installed. Wind forces must also be carefully considered.

Guyed towers are usually triangular in plan, with the main structural members, or legs, at the vertexes of the triangle. The legs are usually solid round steel bars. All members are galvanized and primed before erection. Television reception requires towers to be as high as possible to increase the area of coverage, and results in the use of the new high-strength steels to produce lighter-weight and taller towers. Field connections are made with galvanized bolts and lock-type nuts. The original design usually provides for increasing the height of the tower, when this is permitted by the Federal Communications Commission.

Two television towers in North Dakota, each 2069 ft high, have the distinction of being the tallest structures in the United States. *See* TRANSMISSION LINES.

[CHARLES M. ANTONI]

Bibliography: D. A. Firmage, *Fundamental Theory of Structures*, 1963, reprint 1971.

Traffic engineering

The determination of the required capacity and layout of highway and street facilities that can safely and economically serve vehicular movements between given points.

Major traffic arteries in urban areas are arterial, secondary, and local streets. Arterial routes may be further identified as either major streets or expressways; the latter are also known as freeways. Secondary streets are sometimes described as collectors. Roads in rural areas are usually designated as primary, secondary, or tertiary, depending upon their importance.

In planning improvement in traffic arteries, consideration must be given to many factors, including traffic volumes, desired speeds of travel, vehicular dimensions, driving habits, vehicular performance, type of terrain traversed, and especially expected future changes in all these factors.

Study of traffic movements. The traffic engineer must know both the pattern and volume of vehicular movements.

Traffic recorders. Mechanical counters or recorders are employed to determine traffic movements on an existing route. Four general types of counters are the air-impulse counter, magnetic detector, photoelectric counter, and radar detector. These devices are used to record hourly variations and total daily volumes of traffic at a given point.

Weigh stations are employed primarily to check against overloaded commercial vehicles. Operation of the stations also provides data on the average weight of the vehicles checked.

Origin-and-destination studies. Information obtained by traffic counters and weigh stations is utilized in planning improvement, and as the widening or strengthening of an existing facility. When planning improvement of a whole network of roads or the construction of an entirely new system, additional engineering tools and procedures may be employed to evaluate traffic requirements. This more comprehensive investigation generally includes an origin-and-destination study. In this investigation house-to-house interviews with drivers may be conducted. To reduce this work and its cost only the residents of a previously selected percentage of the homes may be interviewed.

An origin-and-destination study also may be made by the cordon method. With this procedure, on a selected day and for a specific period of time, all drivers are stopped and interviewed at the fringes of the area under investigation. Detailed data relating to time of trip, purpose of trip, route traveled, means of parking at destination, and desired improvements are secured.

Design volume. Traffic elements evaluated in planning important highway projects include: (1) current average daily traffic using the present facility, or that volume which would use a facility if it existed; (2) average daily traffic expected for a specified future year, which is generally 20 years or more in the future; (3) anticipated directional distribution of the traffic at predicted peak hours; (4) expected ratio of trucks and other commercial vehicles to passenger cars; and (5) a design hour volume. Because the movement of traffic is not uniform throughout the day and night, street and highway improvements must be designed for peak flows, and the interval of time generally used for the design period is 1 hr. Experience has proved that if a system has adequate capacity to handle maximum flows for 1 hr the system will generally have sufficient capacity for longer periods.

Capacity calculations. With uninterrupted flow of traffic, a single lane 12 ft (3.6 m) wide is assumed capable of handling a maximum of 1200 vehicles/hr, although for some conditions a higher figure may be used. Next, those factors that may reduce the 1200 figure are evaluated, and this rate is scaled downward. By using the adjusted figure and knowning the total design hour volume, the desired number of lanes is determined.

Capacity of a street or traffic lane at safe speeds is dependent upon many factors. These include available sight distances, curvature, maximum grades, width and character of roadway, shoulder construction, frequency of intersections, spacing of bus stops, clearances at bridges, location and frequency of loading and unloading zones, type of traffic control devices at intersections, whether or not left-turning movements are permitted at intersections, ratio of trucks and other commercial vehicles to passenger car volume, adequacy of directional signs, parking movements, type of lighting facilities employed, weather conditions, and whether or not traffic traveling in opposite directions is separated by a median or island.

Capacity is particularly influenced by the frequency of intersections of other arteries at grade. If only small volumes of traffic are carried on all streets, stop signs may not be needed at intersections. As traffic increases, stop signs must be provided on minor streets, and with still greater volumes power-operated control devices or signals must be installed at intersections. These signals are of two types: (1) fixed-time, which produce a consistent and regularly repeated sequence of signal indications; and (2) traffic-actuated control devices, in which the intervals of travel are varied in accordance with demands of traffic on the different streets as registered by the actuation of detectors.

Safe speeds. The frequency of side streets and the interference due to parking operations generally limit safe vehicular speeds on city routes to about 30 mph (48 km/hr). In contrast, speeds up to 75 mph (120 km/hr) are permissible on express highways. These higher speeds are possible because grades are gradual, wide traffic lanes and good alignment are provided, sight distances are adequate, access to the highway is controlled, grade separations are used where other routes are crossed, and a traffic interchange is employed where the expressway intersects another important artery.

Design standards. Where a high-speed major highway across flat open country is being planned, traffic engineering indicates that major benefits will follow if uphill grades of the main roadways are limited to 3%, each traffic lane is made at least 12 ft (3.6 m) wide, maximum horizontal curvature is restricted to 5° to provide long sight distances, and vertical curves are at least 600 ft (182 m) in length to assist in securing adequate sight distances. In rolling and mountainous country economy may dictate that these standards be reduced.

Access control. Control of access limits the number of points of entrance and departure from the expressway, thereby increasing its capacity. Access is by entrance ramps incorporating an acceleration lane that permits the entering vehicle to reach the speed of those already upon the expressway before moving onto the main roadway. Exit from the expressway is by an off-ramp incorporating a deceleration lane. Experience of highway engineers has indicated that a distance of a mile between ramps is desirable to provide adequate distances for the streams of traffic to intermingle. This frequent spacing is especially desirable where there are three or more lanes of traffic in each direction and where the weaving problem is more critical than where there are only two lanes of travel in each direction.

Grade separations are bridges or structures used to separate vertically two intersecting roadways, thus permitting traffic on the one road to cross traffic on the other road without interference. They are also called overpasses and underpasses.

Traffic interchanges. A traffic interchange is a system of interconnecting roadways in conjunction with a grade separation or separations providing for the interchange of traffic between two or more roadways or highways on different levels (Fig. 1). Major types of traffic interchanges include directional, cloverleaf, diamond, rotary, trumpet, and variations of these basic types (Fig. 2). An advantage of the simple directional Y interchange is that high speeds of travel may be permitted on all roadways, while with more complicated directional layouts and with cloverleaf interchanges reduced speeds on the ramps are usually necessary. Left-turning movements and need for control signals are disadvantages of the diamond interchange, but this type has the advantage of occupying a minimum area. The rotary type lends itself for use when several streets intersect, but it requires considerable weaving of vehicles, which slows traffic movements and reduces capacity of the interchange.

Factors influencing the type of interchange selected include the expected volumes of traffic to be accommodated, the number of roads or streets involved, the distribution of the total traffic on the various roads, the land available for the interchange, the topography, and the estimated construction cost. Frequently major limitations on the financing dictate a compromise layout.

Safety of operation. When an express highway is built with adequate control of access, good grade separations, and properly planned traffic interchanges, traffic can move over the route much

Fig. 1. One of the most difficult highway engineering problems is the planning, financing, and construction of adequate expressways in large metropolitan areas. This Los Angeles, Calif., freeway cost several million dollars a mile.

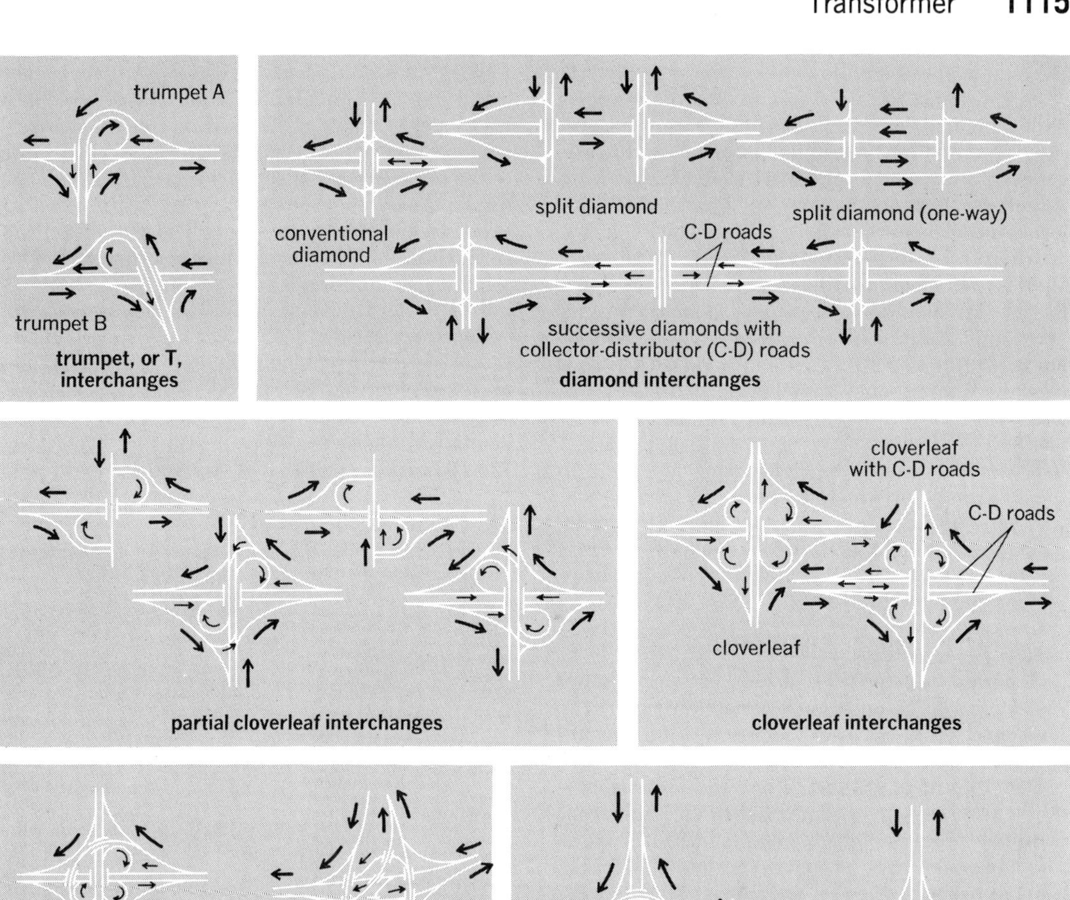

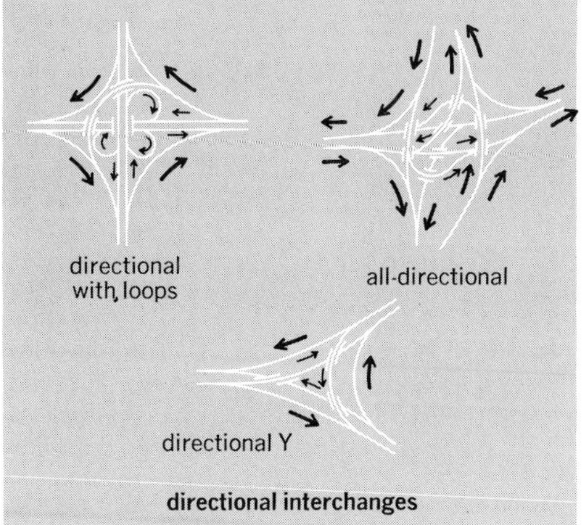

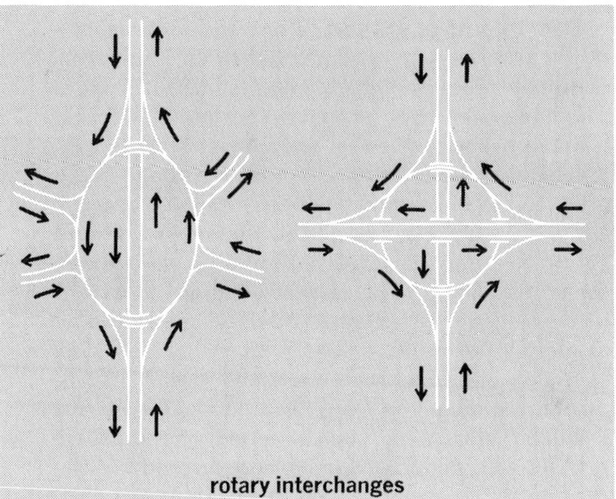

Fig. 2. Diagrams of types of interchanges.

more safely than on arteries without freeway features. For example, during 1 year of the operation of the New York Thruway a fatality rate of only 0.88 deaths for each 100,000,000 mi (160,000,000 km) of travel was established. This rate was far lower than that on nearby roads not designed as freeways.

In addition to providing greater safety and more economical vehicle operation, control of access preserves the capacity of the road from future encroachments, thus safeguarding the investment in the expressway.

Controlled access also assists in the orderly development of the community traversed by providing major transportation arteries that are permanent in character.

Parking. Traffic engineering also deals with parking problems. Investigations relating to parking include determinations as to the desired size, type, and location of off-street facilities, plus studies relating to movements into and out of parking areas. Off-street parking garages adjacent to major streets may involve access connections requiring careful design. In the central section of a city, flow to and from the off-street parking areas may be so highly concentrated in short morning and evening periods that feasible access connections between principal streets and parking structures may become major problems. *See* HIGHWAY ENGINEERING; PAVEMENT; TRANSPORTATION ENGINEERING. [ARCHIE N. CARTER]

Transformer

An electrical component used to transfer electric energy from one alternating-current (ac) circuit to another by magnetic coupling. Essentially, it consists of two or more multiturn coils of wire placed in close proximity to cause the magnetic field of one to link the other. In general, the transformer accomplishes one or more of the following between two circuits: (1) a difference in voltage magnitude, (2) a difference in current magnitude, (3) a difference in phase angle, (4) a difference in impedance level, and (5) a difference in voltage insulation level, either between the two circuits or to ground.

Transformers are used to meet a wide range of requirements. Pole-type distribution transformers supply relatively small amounts of power to residences. Power transformers are used at generating stations to step up the generated voltage to high levels for transmission. The transmission voltages are then stepped down by transformers at the substations for local distribution. Instrument transformers are used to measure voltage and currents accurately. Audio- and video-frequency transformers must function over a broad band of frequencies. Radio-frequency transformers transfer energy in narrow frequency bands from one circuit to another. *See* INSTRUMENT TRANSFORMER.

Transformers are often classified according to the frequency for which they are designed. Power transformers are for power-frequency circuits, audio transformers for audio-frequency circuits, and so forth. Of course, many of the basic principles of operation apply to all.

POWER TRANSFORMERS

A power transformer consists of two or more multiturn coils wound on a laminated iron core. At least one of these coils serves as the primary winding.

Principle of operation. When the primary of a power transformer is connected to an alternating voltage, it produces an alternating flux in the core. The flux generates a primary electromotive force, which is essentially equal and opposite to the voltage supplied to it. It also generates a voltage in the other coil or coils, one of which is called a secondary. This voltage generated in the secondary will supply alternating current to a circuit connected to the terminals of the secondary winding. A current in the secondary winding requires an additional current in the primary. The primary current is essentially self-regulated to meet the power (or volt-ampere) demand of the load connected to the secondary terminals. Thus in normal operation, energy (or volt-amperes) can be transferred from the primary to the secondary electromagnetically.

Figure 1 shows a transformer with a primary of N_1 turns and a secondary of N_2 turns. A primary voltage V_1 causes a current I_1 to flow through the coil. Since all quantities shown are alternating, the arrows indicate only instantaneous polarities.

The magnetic flux ϕ set up by the primary consists of two components. One part passes completely around the magnetic circuit defined by the iron core, thus linking the secondary coil. This is the mutual flux ϕ_m. The second part is a smaller component of flux that links only the primary coil.

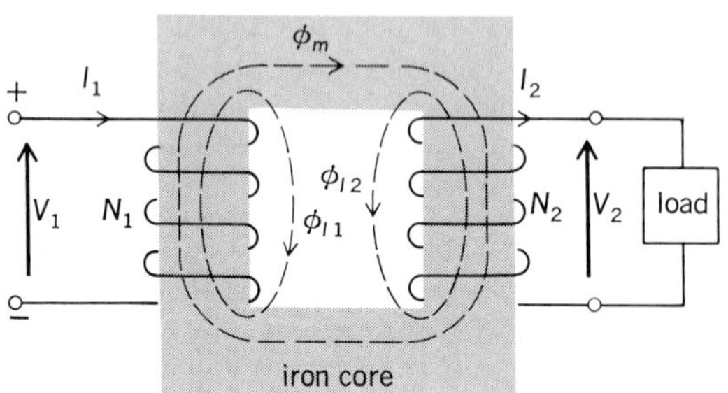

Fig. 1. Basic transformer.

This is the primary leakage flux ϕ_{l1}. If the secondary circuit is completed through a load, a secondary current I_2 flows and in turn creates a secondary leakage flux ϕ_{l2}. These leakage fluxes contribute to the impedance of the transformer. If the leakage flux is small, the coupling between primary and secondary is said to be close. The use of an iron core decreases the leakage flux by providing a low-reluctance path for the flux.

In a power transformer the voltage drops due to winding resistance and leakage are small; therefore V_1 and V_2 are essentially in phase (or 180° out of phase, depending on the choice of polarity). Since the no-load current is small, I_1 and I_2 are essentially in phase (or 180° out of phase). Therefore, Eq. (1) applies, and the voltage ratio is expressed

$$V_i I_1 \approx V_2 I_2 \tag{1}$$

by Eq. (2), in which a is the transformation ratio.

$$\frac{V_1}{V_2} \simeq \mathrm{a} \tag{2}$$

Substituting Eq. (2) into Eq. (1) demonstrates that the current ratio is inversely proportional to the transformation ratio, as in Eq. (3). A transform-

$$\frac{I_1}{I_2} \simeq \frac{1}{a} \tag{3}$$

er therefore may be used to step up or down a voltage from a level V_1 to a level V_2 according to the transformation ratio a. Simultaneously the current will be transformed inversely proportional to a.

Equation (1) may be rewritten in the form of Eq. (4).

$$I_1^2 \frac{V_1}{I_1} \simeq I_2^2 \frac{V_2}{I_2} \tag{4}$$

Since V_2/I_2 is the impedance Z_2 of the load on the secondary and V_1/I_1 is the impedance Z_1 of the load as measured on the primary, Eq. (5) applies.

$$I_1^2 Z_1 \simeq I_2^2 Z_2 \tag{5}$$

Equation (5) may be rewritten in the form of Eq. (6).

$$\frac{Z_1}{Z_2} \simeq \left(\frac{I_2}{I_1}\right)^2 \simeq a^2 \tag{6}$$

The transformer is thus capable of transforming circuit impedance levels according to the square of the transformation ratio; this property is used in telephone, radio, television, and audio systems.

The transmission of power from primary coil to secondary coil is via the magnetic flux. The flux is proportional to the ampere turns in either coil. Since the power in each coil is nearly the same, Eqs. (7) and (8) are obtained. The transformation

$$N_1 I_1 \simeq N_2 I_2 \tag{7}$$

$$\frac{N_1}{N_2} \simeq \frac{I_2}{I_1} \simeq a \tag{8}$$

ratio is therefore approximately equal to the turns ratio.

Construction. Transformer cores are made of special alloy steels rolled to approximately 0.014 in. thick. These thin sheets, or laminations, are stacked to form the transformer core, each sheet being insulated from the others to reduce unwanted eddy-current loss. The steel is heat-treated to obtain low hysteresis loss, low exciting current, and low sound level. *See* CORE LOSS.

Copper conductors are used almost universally. Conductor wires are round in smaller transformers, and rectangular in larger ones.

The conductors are insulated with special paper or cotton covering, with enamel, or with a combination of both. Large outdoor transformers are immersed in oils to obtain good electrical insulation within small spacings and to provide a cooling medium. When lightweight or nonflammable materials are important, transformers may be made with compressed gases as the insulating and cooling medium. Increasing the pressure raises the dielectric strength. The gas is pumped through the transformer and through a gas-to-air heat exchanger for cooling.

The low-voltage (LV) winding is usually in the form of a cylinder next to the core. The high-voltage (HV) winding, also cylindrical, surrounds the LV windings as in Fig. 2*a*. These windings are often described as concentric windings. The number of turns N may be obtained from Eq. (9), where

$$E = \frac{fBAN}{22{,}500} \tag{9}$$

E is the rms voltage, f is the frequency in cps, B is the maximum flux density in kilolines/in.^{2}, and A is the cross-sectional area of the iron core in square inches.

Some manufacturers use a winding arrangement having coils adjacent to each other along the core leg as in Fig. 2*b*. The coils are wound in the form of a disk, with a group of disks for the LV winding stacked alternately with a group of disks for the HV windings. This construction is referred to as interleaved windings.

The core sheets are stacked sheet by sheet to form the desired cross-sectional area. The closed magnetic circuit typically has joints between adjacent sheets, but cores of moderate cross section may be made with a long continuous sheet which has been coiled up to give the required cross section. Passages may be provided between groups of sheets for circulation of the cooling oil.

For single-phase transformers (Fig. 3), the HV and LV coils may be on one leg of a core, with the return path in one, two, or more other legs. The total area of the return legs is equal to that of the main leg. An alternative construction has two legs, each with half of the primary windings and half of the secondary windings.

Figure 4 shows a typical three-phase transformer core with coils. A typical three-phase core has three legs, with the HV and LV windings for one phase on each leg. The yokes of the core connect between the two outer legs and the middle leg on top and bottom. This core-type construction is shown in Fig. 5*a*. The iron in another construction that is sometimes used (shell type) is as shown in Fig. 5*b*. Either concentric windings or interleaved windings may be used with either core.

The core and coils are placed in a steel tank with openings for the electrical connections to the windings, and for the cooling equipment.

Cooling. Small transformers are self-cooled. Radiation, conduction, and convection from the tank or from radiating surfaces remove the heat generated by the power losses of the transformer. On larger units, fans are sometimes added to the radiating surfaces. A transformer may have one rating with a basic method of cooling and a higher rating with supplemental cooling. Pumps may be added to give further cooling. An oil-to-air heat exchanger with finned tubes is used on the very large units. This equipment has a pump for circulating oil and fans for forcing the air against the heat exchanger. Water cooling may be used with cooling coils or with an oil-to-water heat exchanger having an oil pump.

Characteristics. The service conditions for a particular transformer are considered by the designer in choosing materials and the arrangement of parts.

The final design then may be measured by test with respect to a number of characteristics.

No-load loss. The sum of the hysteresis and eddy loss in the iron core is the no-load loss.

Exciting current. The exciting current is that supplied to the transformer at no load when oper-

TRANSFORMER

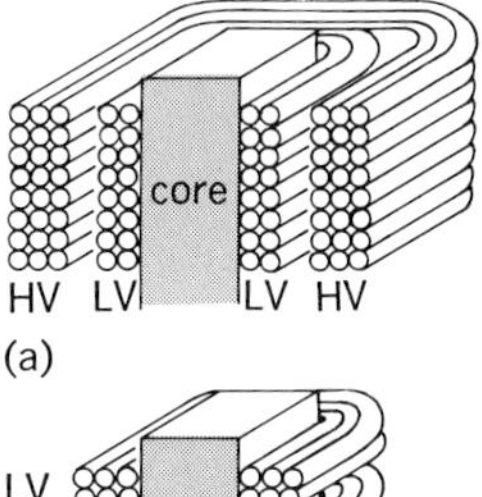

(a)

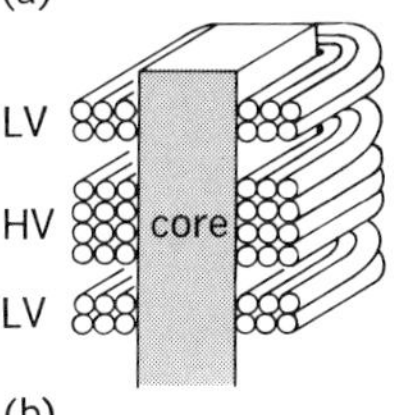

(b)

Fig. 2. Winding arrangements. (*a*) Concentric. (*b*) Interleaved.

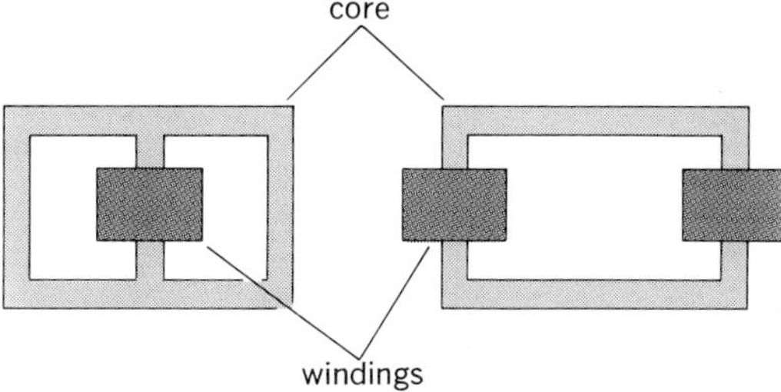

Fig. 3. Location of windings in single-phase cores.

Fig. 4. Three-phase core and coils, rated at 50,000 kVA, 115,000 volts.

TRANSFORMER

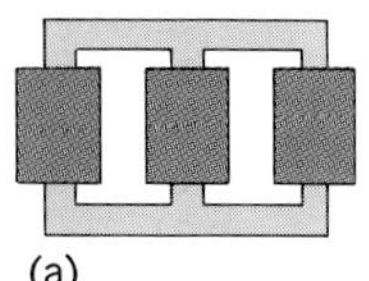

(a)

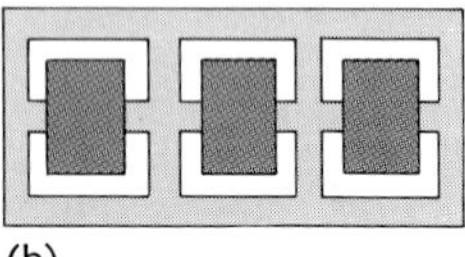

(b)

Fig. 5. Typical three-phase cores showing location of windings. (*a*) Core type. (*b*) Shell type.

ating at rated voltage. This current energizes the core and supplies the no-load loss. Owing to the characteristic shape of the *B-H* curve of iron, the current is not a true sine wave, but has higher frequency harmonics. In a typical power transformer the exciting current is so small (usually less than 1%) that I_2 is approximately $(N_1/N_2)I_1$. In this sense the ampere-turns in the two windings are said to balance.

Load loss. This is the sum of the copper loss, due to the resistance of the windings (I^2R loss), plus the eddy-current loss in the winding, plus the stray loss (loss due to flux in metallic parts of the transformer adjacent to the windings, the flux resulting from current in the windings).

Total loss and efficiency. The total loss in a transformer is the sum of the no-load and full-load losses. Representative values for a 20,000-kVa, three-phase, 115-kV power transformer are no-load loss, 42 kW; load loss, 85 kW; and total loss, 127 kW. Equation (10) expresses the efficiency of a

$$\text{Efficiency} = \frac{\text{output in kW}}{\text{input in kW}} = \frac{\text{output}}{\text{output} + \text{losses}} \qquad (10)$$

transformer. For this transformer the efficiency is 20,000/20,127, or 99.37%.

Voltage ratio. This is the ratio of voltage on one winding to the voltage on another winding at no load. It is the same as the turns ratio.

Impedance. Consider a transformer having equal turns in the primary and secondary windings. If one side is connected to a generator and the other side to a typical power system load, the voltage measured on the load side will be less than that on the generator side, by the amount of the impedance drop through the transformer. *See* ELECTRICAL IMPEDANCE.

Impedance is measured by connecting the secondary terminals together (short-circuited) and applying sufficient voltage to the primary terminals to cause rated current to flow in the primary winding. The transformer impedance in ohms equals the primary voltage divided by the primary current. Impedance is usually referred to the transformer kVA and kV base and given as percent impedance, as in Eq. (11). Percent reactance is usually close in value to percent impedance, since the percent resistance, given by Eq. (12), is small.

$$\% \text{ impedance} = \frac{1}{10} \frac{\text{kVA}}{(\text{kV})^2} \times \text{ohms} \qquad (11)$$

$$\% \text{ resistance} = \frac{\text{load loss in kVA}}{\text{kVA rating}} \times 100 \qquad (12)$$

Typical values for a 20,000-kVa, three-phase, 115-kV self-cooled power transformer are resistance, 0.4% and impedance, 7.5%.

Regulation. Regulation is the change in output (secondary) voltage that occurs when the load is reduced from rated value to zero, with the primary impressed terminal voltage maintained constant. This is usually expressed as a percent of rated output voltage at full load (E_{FL}), as in Eq. (13),

$$\% \text{ regulation} = \frac{E_{NL} - E_{FL}}{E_{FL}} \times 100 \qquad (13)$$

where E_{NL} is the output voltage at no load. When a transformer supplies a capacitive load, the power factor may cause a higher full-load voltage than no-load voltage.

Cooling. Temperature tests (heat run tests) are made by operating the transformer with total losses until the temperatures are constant. In the United States the standard winding rise is 55°C over a 30°C air ambient.

Insulation. Sufficient insulation strength must be built into a transformer so that it can withstand normal operation at its rated voltage and system voltage transients due to lightning and switching surges.

Audio sound. The iron core lengthens and shortens because of magnetostriction during each voltage cycle, giving rise to a hum having a frequency twice that of the voltage. This and other frequencies may cause mechanical vibrations in different parts of the transformer due to resonance.

Taps. The application of a transformer to a power system involves a correct choice of turns ratio for average operating conditions, and the selection of proper taps to obtain improved voltage levels when average conditions do not prevail.

Tap changers are frequently used in the HV winding to give plus or minus two 2½% taps (5% above and 5% below rated voltage). These taps may be changed only when the transformer is deenergized, that is, when the service is interrupted.

A special motor-driven tap changer is used to permit tap changing when the transformer is energized and carrying full load. One of its simpler forms is shown in Fig. 6. The transformer taps are

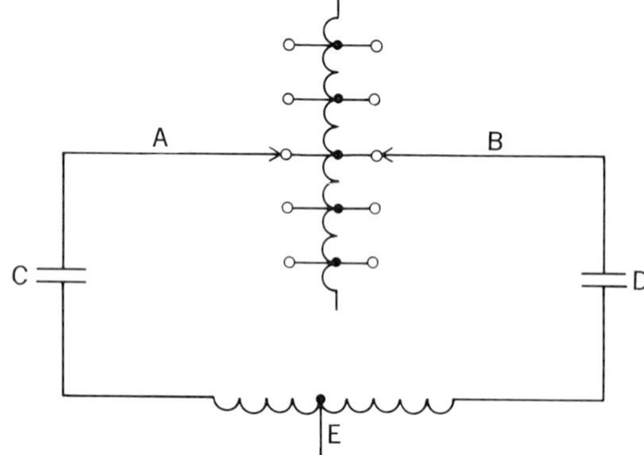

Fig. 6. Typical circuit for tap changing under load.

brought to a tap changer having two sets of fingers A and B. Initially these are on the same tap. When a change is required, a contactor C opens, and A moves to the next lower tap. C now closes. Next D opens and B moves down to the same tap as A. The current, which initially divided half-and-half through A and B, has changed, first to be all in B, then partly in A and partly in B, then all in A, and finally half-and-half in A and B. E is a center-tapped reactor, which limits the current when A and B are not on a common tap.

This equipment is essential where a constant voltage is required under changing loads. It is frequently applied with a tap range of plus or minus 10% of rated voltage. It may be made to operate automatically, maintaining a specified voltage at a predetermined point remote from the transformer.

Tap changing under load equipment is used on power transformers supplying residential loads, where variations in voltage would adversely affect the use of lights and appliances. It is also used for chemical and industrial processes, such as on pot lines for the manufacture of aluminum.

Parallel operation. Two transformers may be operated in parallel (primaries connected to the same source and secondaries connected to the same load) if their turns ratios and per unit impedances are essentially equal. A slight difference in turns ratio would cause a relatively large out-of-phase circulating current between the two units and result in power losses and possible overheating.

Phase transformation. Polyphase power may be changed from 3-phase to 6-phase, 3-phase to 12-phase, and so forth, by means of transformers. This is of value in the power supply to rectifiers, where the greater number of phases results in a smoother dc voltage wave. *See* ALTERNATING-CURRENT CIRCUIT THEORY.

Overloads. Transformers have a capacity for loading above their rating. Such factors as low ambient temperature and type of load carried may be used to increase the continuous load possible on a given transformer. In emergencies it is possible to increase the load further for short times with a calculable loss of transformer life. Such a load would permit, for instance, a 50% overload for 2 hr following full load.

[J. R. SUTHERLAND]

AUDIO AND RADIO-FREQUENCY TRANSFORMERS

Audio or video (broad-band) transformers are used to transfer complex signals containing energy at a large number of frequencies from one circuit to another. Radio-frequency (rf) and intermediate-frequency (i-f) transformers are used to transfer energy in narrow frequency bands from one circuit to another. Audio and video transformers are required to respond uniformly to signal voltages over a frequency range three to five or more decades wide (for example, from 10 to 100,000 Hz), and consequently must be designed so that very nearly all of the magnetic flux threading through one coil also passes through the other. These units are designed to have a coupling coefficient k, given in Eq. (14), nearly equal to one. Here L_p and L_s are the

$$k = M/\sqrt{L_p L_s} \qquad (14)$$

primary and secondary inductances, respectively, and M is the mutual inductance (Fig. 7). The high coupling coefficient is obtained by the use of interleaved windings and a high-permeability iron core, which concentrates the flux. Typical values of k for highest quality video transformers may be as high as 0.9998; that for power transformers need not be greater than 0.98.

The rf and i-f transformers are built from individual inductors whose magnetic fields are loosely

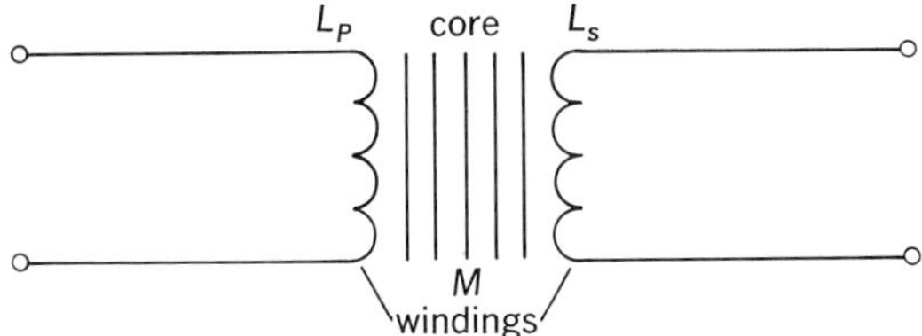

Fig. 7. Schematic of a transformer with symbols.

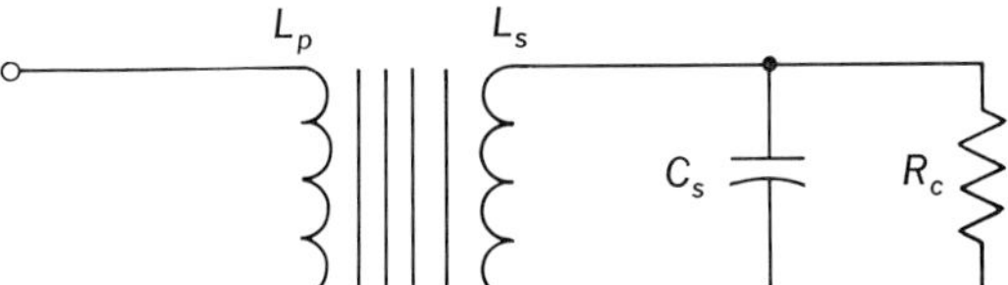

Fig. 8. Circuit of loaded transformer.

coupled together, $k < 0.30$; each inductor is resonated with a capacitor to make efficient energy transfer possible near the resonant frequency. *See* RESONANCE (ALTERNATING-CURRENT CIRCUITS).

The audio and video transformers. Audio and video transformers have two resonances (caused by existing stray capacitances) just as many tuned transformers do. One resonance point is near the low-signal-frequency limit; the other is near the high limit. As the coefficient of coupling in a transformer is reduced appreciably below unity by removal of core material and separation of the windings, tuning capacitors are added to provide efficient transfer of energy. The two resonant frequencies combine to one when the coupling is reduced to the value known as critical coupling, then stay relatively fixed as the coupling is further reduced.

All transformers are devices for transferring energy from one circuit to another. The energy transferred is absorbed either in the circuits themselves or in an external load circuit. For this reason, proper termination is essential for achieving optimum behavior in circuits containing transformers.

Audio and video transformers have a minimum operating frequency at which the open-circuit reactance of the primary is approximately twice its effective loaded impedance. As with wide-band *RC* amplifiers, gain may be traded for bandwidth with transformer-coupled amplifiers. The reduction of the terminating resistance across the secondary of the transformer reduces the minimum operating frequency f_1 and, in the presence of output capacitance, raises the maximum frequency f_2. The approximate values of the minimum and maximum frequencies and the resonant frequency f_r are given by Eqs. (15), where $L_{ss} = L_s - M^2/L_p = L_s(1-k^2)$.

$$\begin{aligned} f_1 &= R_c(N_p/N_s)^2/\pi L_p \\ f_2 &= 1/2\pi R_c C_s \\ f_r &= 1/2\pi\sqrt{L_{ss}C_s} \end{aligned} \qquad (15)$$

This L_{ss} is the secondary inductance with the primary short-circuited; C_s is the output capacitance, both external and internal, on the transformer; and R_c is the load resistance (Fig. 8). The resonant frequency f_r should be larger than f_2 for best performance.

A transformer used to activate terminating circuitry is called an output transformer; one to activate an input circuit is an input transformer; and others are called interstage transformers.

Distortion. The distortion introduced into the amplified signal by a transformer is caused primarily by its hysteresis loss. This loss may be minimized by proper loading on the secondary. The load component of current then is large compared to the magnetizing current. In addition, a resistive load keeps the amplification uniform as a function

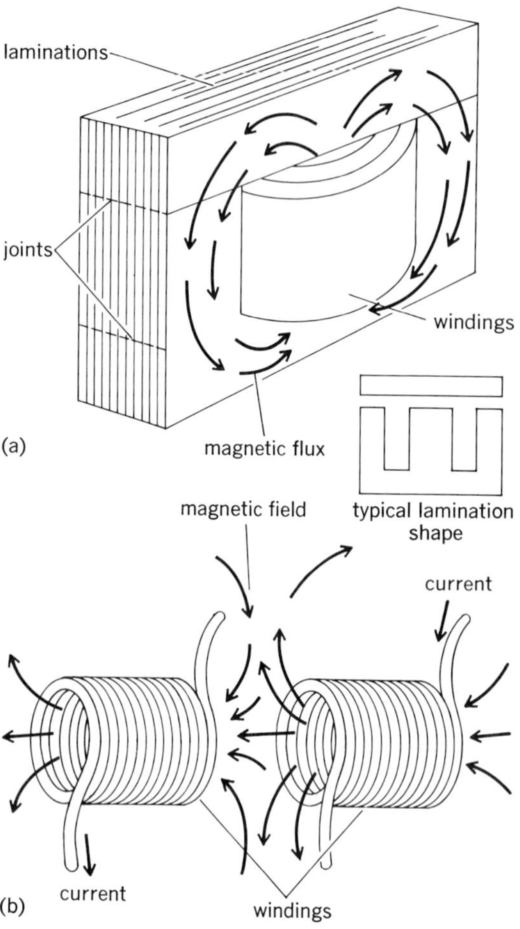

Fig. 9. Audio and rf transformers. (*a*) Iron-core audio transformer. (*b*) Air-core rf transformer.

of frequency, and keeps the phase distortion to a minimum.

The magnetic core in an audio or video transformer is subject to two kinds of saturation, that due to applied direct current in the windings, and that due to excessively large signal currents. The direct current in the windings may make the hysteresis loop of the iron core nonsymmetrical, necessitating the use of a larger core having a built-in air gap. Both large signal amplitudes and low frequencies can cause signal saturation to occur in the core. The structure of audio and rf transformers is shown in Fig. 9.

The rf and i-f transformers. These use two or more inductors, loosely coupled together, to limit the band of operating frequencies. Efficient transfer of energy is obtained by resonating one or more of the inductors. By using higher than critical coupling, a wider bandwidth than that from the individual tuned circuits is obtained, while the attenuation of side frequencies is as rapid as with the individual circuits isolated from one another.

The tuning of the primary, the secondary, or both may be accomplished either by the variation of the tuning capacitor or by an adjustable magnetic or conducting slug that varies the inductance of the inductor (Fig. 10).

The operating impedance of a tuned circuit of an rf transformer is a function of its Q and its tuning capacitance. In general, high-power circuits require a high capacitance for energy storage, and therefore have low values of impedance. In any application, the impedance level must be kept sufficiently small to prevent instability and oscillation.

[KEATS A. PULLEN]

Bibliography: American Institute of Physics, *The Transformer*, 1976; L. Anderson, *Electric Machines and Transformers*, 1980; R. Feinberg, *Modern Power Transformer Practice*, 1979; C. McLyman, *Transformer and Inductor Design Handbook*, 1978; G. McPherson, *An Introduction to Electrical Machines and Transformers*, 1981; A. J. Pansini, *Basic Electrical Power Transformers*, 1976; D. Richardson, *Rotating Electric Machinery and Transformer Technology*, 1978.

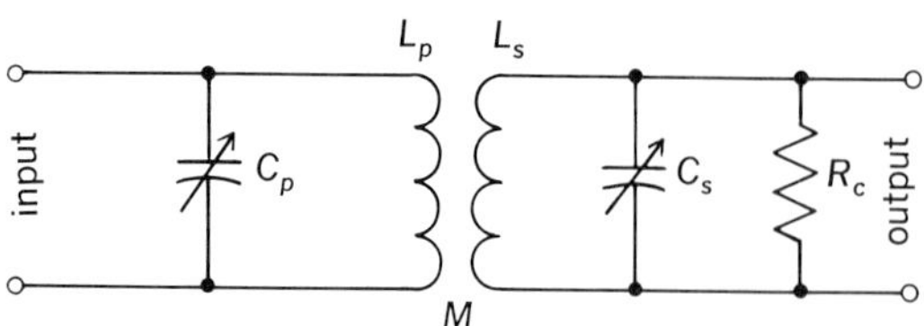

Fig. 10. Tuned rf transformer.

Transit

A surveying instrument for measuring horizontal and vertical angles. It is also used for prolonging a straight line, establishing a level line of sight, and measuring distances by the stadia method. *See* STADIA.

In the disassembled view (see illustration), three main elements are seen to support the telescope

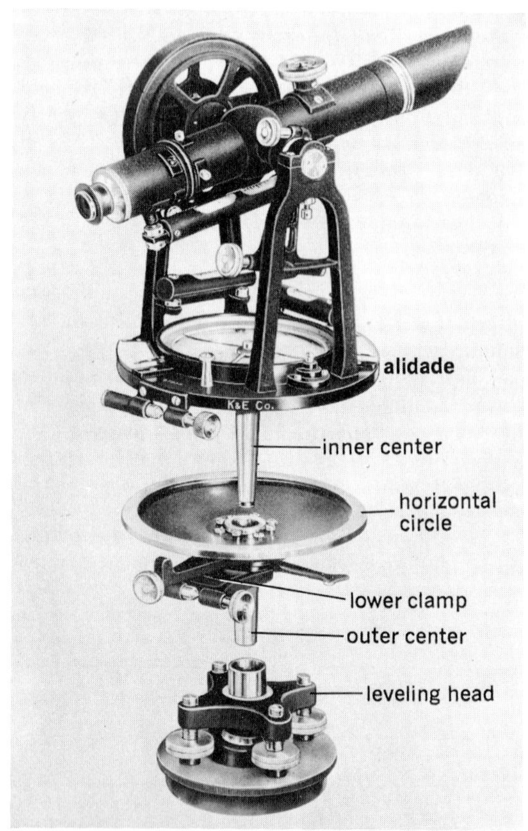

Disassembled transit. (*Keuffel and Esser Co.*)

assembly: the alidade or standard with its solid spindle (inner center), which rotates within the hollow spindle (outer center) and graduated circle, which in turn rotates within the leveling head atop its tripod. These are the elements of the azimuth axis. Clamps are used to prevent one rotation while the other is used. This permits an angle to be turned about the azimuth axis and to be measured on the horizontal circle, even several times incrementally for greater precision. Vertical angular measurement is achieved by rotating the telescope about the elevation axis. The leveling screws on the leveling head enable the azimuth axis to be made vertical. Lateral motion on the leveling head permits the instrument to be centered exactly over a ground station by means of a plumb bob on a string or by use of an optical plummet built into the instrument. *See* SURVEYING.

[B. AUSTIN BARRY]

Transmission lines

A system of conductors suitable for conducting electric power or signals between two or more termini. For example, commercial-frequency electric power transmission lines connect electric generating plants, substations, and their loads. Telephone transmission lines interconnect telephone subscribers and telephone exchanges. Radio-frequency transmission lines transmit high-frequency electric signals between antennas and transmitters or receivers. In this article the theory of transmission lines is considered first, followed by its application to power transmission lines.

Although only a short cord is needed to connect an electric lamp to a wall outlet, the cord is, properly speaking, a transmission line. However, in the electrical industry the term transmission line is applied only when both voltage and current at one line terminus may differ appreciably from those at another terminus. Transmission lines are described either as electrically short if the difference between terminal conditions is attributable simply to the effects of conductor series resistance and inductance, or to the effects of a shunt leakage resistance and capacitance, or to both; or as electrically long when the properties of the line result from traveling-wave phenomena.

TRANSMISSION-LINE THEORY

Depending on the configuration and number of conductors and the electric and magnetic fields about the conductors, transmission lines are described as open-wire transmission lines, coaxial transmission lines, cables, or waveguide transmission lines.

Open-wire transmission lines. Open-wire lines may comprise a single wire with an earth (ground) return or two or more conductors. The conductors are supported at more or less evenly spaced points along the line by insulators, with the spacing between conductors maintained as nearly uniform as feasible, except in special-purpose tapered transmission lines, discussed later in this section.

Open-wire construction is used for communication or power transmission whenever practical and permitted, as in open country and where not prohibited by ordinances.

Open-wire lines are economical to construct and maintain and have relatively low losses at low and medium frequencies. Difficulties arise from electromagnetic radiation losses at very high frequencies and from inductive interference, or crosstalk, resulting from the electric and magnetic field coupling between adjacent lines accompanying the characteristic field configuration (Fig. 1).

Coaxial transmission lines. A coaxial transmission line comprises a conducting cylindrical shell, solid tape, or braided conductor surrounding an isolated, concentric, inner conductor which is solid, stranded, or (in certain video cables and delay cables) helically wound on a plastic or ferrite core. The inner conductor is supported by ceramic or plastic beads or washers in air- or gas-dielectric lines, or by a solid polyethylene or polystyrene dielectric.

The purpose of this construction is to have the shell prevent radiation losses and interference from external sources. The electric and magnetic fields shown in Fig. 1*b* are nominally confined to the space inside the outer conductor. Some external fields exist, but may be reduced by a second outer sheath.

Coaxial lines are widely used in radio, radar, television, and similar applications.

Sheathed cables. Also termed shielded cables, these comprise two or more conductors surrounded by a conducting cylindrical sheath, commonly supported by a continuous solid dielectric. The sheath provides both shielding and mechanical protection.

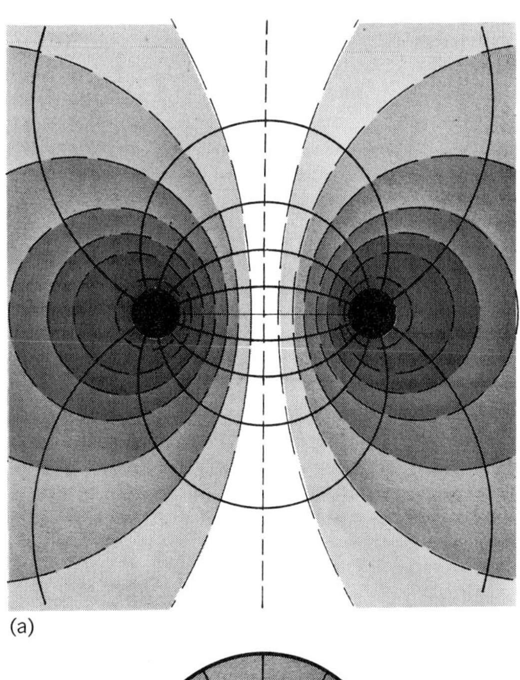

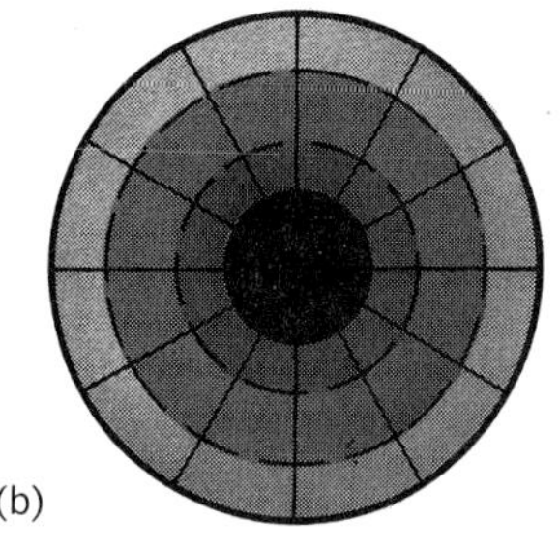

Fig. 1. Electric (solid lines) and magnetic (dashed lines) fields about two-conductor (*a*) open-wire and (*b*) coaxial transmission lines in a plane normal to the conductors, for continuous and low-frequency currents.

Coaxial lines, sheathed cables, or shielded cables are often termed simply cables. For cable assemblies of coaxial lines and other circuits *see* COAXIAL CABLE.

Traveling waves. When electric power is applied at a terminus of a transmission line, electromagnetic waves are launched and guided along the line. The steady-state and transient electrical properties of transmission lines result from the superposition of such waves, termed direct waves, and the reflected waves which may appear at line discontinuities or at load terminals.

Principal mode. When the electric and magnetic field vectors are perpendicular to one another and transverse to the direction of the transmission line, this condition is called the principal mode or the transverse electromagnetic (TEM) mode. The principal-mode electric- and magnetic-field configurations about the conductors are essentially those of Fig. 1. Modes other than the principal mode may exist at any frequency for which conductor spacing exceeds one-half of the wavelength of an electromagnetic wave in the medium separating the conductors. Such high-frequency modes are called waveguide transmission modes.

In a uniform (nontapered) transmission line, the voltage or current applied at a sending terminal determines the shape of the initial voltage or current wave. In a line with negligible losses the transmitted shape remains unchanged. When losses are present, the shape, unless sinusoidal, is altered, because the phase velocity and attenuation vary with frequency.

If a wave shape is sinusoidal, the voltage and current decay exponentially as a wave progresses. The voltage or current, at a distance x from the sending end, is decreased in magnitude by a factor of $\epsilon^{-\alpha x}$, where ϵ is the Napierian base (2.718), and α is called the attenuation constant. The voltage or current at that point lags behind the voltage or current at the sending end by the phase angle βx, where β is called the phase constant.

The attenuation constant α and the phase constant β depend on the distributed parameters of the transmission line, which are (1) resistance per unit length r, the series resistance of a unit length of both going and returning conductors; (2) conductance per unit length g, the leakage conductance of the insulators, conductance due to dielectric losses, or both; (3) inductance per unit length l, determined as flux linkages per unit length of a line of infinite extent carrying a constant direct current; and (4) capacitance per unit length c, determined from charge per unit length of a line of infinite extent with constant voltage applied.

The values of α and β may be found from complex equation (1), where j is the notation for the

$$\alpha+j\beta=\sqrt{(r+j2\pi fl)(g+j2\pi fc)} \qquad (1)$$

imaginary number $\sqrt{-1}$, and f is the frequency of the alternating voltage and current. The complex quantity $\alpha+j\beta$ is often called the propagation constant γ. Since $r+2\pi fl$ is the impedance z per unit length of line, and $g+2\pi fc$ is the admittance y per unit length of line, the equation for the propagation constant is often written in the form of Eq. (2). The velocity at which a point of constant phase

$$\gamma=\sqrt{zy} \qquad (2)$$

is propagated is called the phase velocity v, and is equal to $2\pi f/\beta$. For negligible losses in the line (when r and g are approximately zero) the phase velocity is $1/\sqrt{lc}$, which is also the velocity of electromagnetic waves in the medium surrounding the transmission-line conductors.

The distributed inductance and resistance of the lines may be modified from their dc values because of skin effect in the conductors. This effect, which increases with frequency and conductor size, is usually, but not always, negligible at power frequencies.

Characteristic impedance. The ratio of the voltage to the current in either the forward or the reflected wave is the complex quantity Z_0, called the characteristic impedance.

When line losses are relatively low, that is, when relationships (3) apply, the characteristic imped-

$$\begin{aligned} r &\ll 2\pi fl \\ g &\ll 2\pi fc \end{aligned} \qquad (3)$$

ance is given by Eq. (4), and is a quantity nearly

$$Z_0=\sqrt{l/c} \qquad (4)$$

independent of frequency (but not exactly so since both l and c may be somewhat frequency-dependent). The magnitude of Z_0 is used widely, at high frequencies, to identify a type of transmission line such as 50-ohm line, 200-ohm line, and the popular 300-ohm antenna lead-in line used with television antennas. *See* ELECTRICAL IMPEDANCE.

Distortionless line. Transmission lines used for communications purposes should be as free as possible of signal waveshape distortion. Two types of distortion occur. One is a form of amplitude distortion due to line attenuation, which varies with the signal frequency. The other, delay distortion, occurs when the component frequencies of a signal arrive at the receiving end at different instants of time. This occurs because the velocity of propagation along the line is a function of the frequency.

Theoretically, a distortionless line can be devised if the line parameters are adjusted so that $r/g=l/c$. In practice this is approached by employing loading circuits. Under these conditions the propagation constant is given by Eq. (5).

$$\gamma=\alpha+j\beta=\sqrt{r/g}\,(g+j2\pi fc) \qquad (5)$$

The attenuation constant α is $\sqrt{rg}$, which is independent of frequency f. Therefore, there will be no frequency distortion.

The phase constant β is $2\pi f\sqrt{lc}$ which depends upon frequency. The velocity of propagation along any transmission line is $2\pi f/\beta$, and for the distortionless line this becomes $1/\sqrt{lc}$. Thus the velocity of propagation is independent of frequency, and there will be no delay distortion.

Transmission-line equations. The principal-mode properties of the transmission-line equations are described by Eqs. (6) and (7), in which e and i

$$\frac{\partial e}{\partial x}=-\left(ri+l\frac{\partial i}{\partial t}\right) \qquad (6)$$

$$\frac{\partial i}{\partial x}=-\left(ge+c\frac{\partial e}{\partial t}\right) \qquad (7)$$

are instantaneous values of voltage and current, respectively, x is distance from the sending terminals, and t is time.

For steady-state sinusoidal conditions, the solutions of these equations are given by Eqs. (8) and (9) for voltage E and current I at a distance x from

$$E = E_s \cosh \gamma x - I_s Z_0 \sinh \gamma x \quad (8)$$

$$I = I_s \cosh \gamma x - \frac{E_s}{Z_0} \sinh \gamma x \quad (9)$$

the sending end in terms of voltage E_s and current I_s at the sending end. In Eqs. (8) and (9) $Z_0 = \sqrt{(r+j2\pi fc)/(g+j2\pi fc)}$. All values of current and voltage in these and the following equations are complex.

In terms of receiving-end voltage E_r and current I_r, these solutions are given by Eqs. (10) and (11), where x is now the distance from the receiving end.

$$E = E_r \cosh \gamma x + I_r Z_0 \sinh \gamma x \quad (10)$$

$$I = I_r \cosh \gamma x + \frac{E_r}{Z_0} \sinh \gamma x \quad (11)$$

Reflection coefficient. If the load at the receiving end has an impedance Z_r, the ratio of reflected voltage to direct voltage, known as the reflection coefficient ρ, is given by Eq. (12).

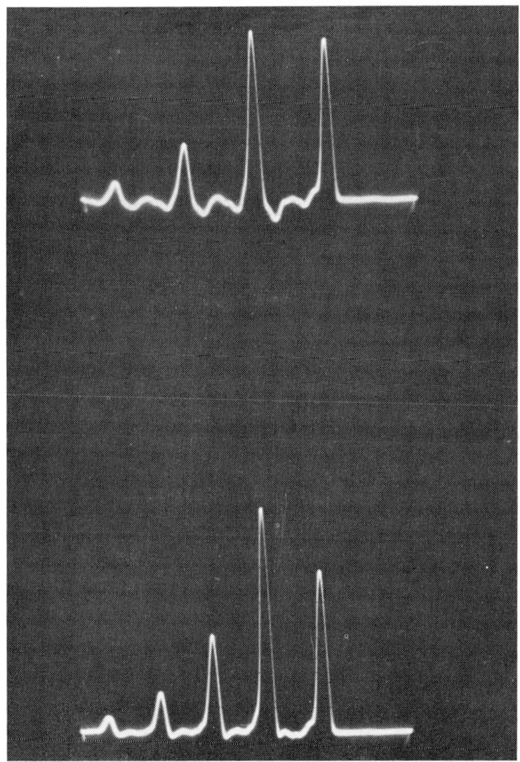

Fig. 2. Typical transient phenomena in a transmission line. These are oscillographic recordings of voltage as a function of time at the sending end of a 300-m transmission line with the receiving end open-circuited. Time increases from right to left; the first (right-hand) pulse is delivered by a generator, equivalent to an open circuit, so that a new forward wave results from each reflected wave arriving at the sending end. At the end of each 2-μsec interval, an echo arrives from the receiving end. In the upper trace, minor discontinuities in the line at intermediate points result in intermediate echos. Intermediate discontinuities are minimized in lower trace.

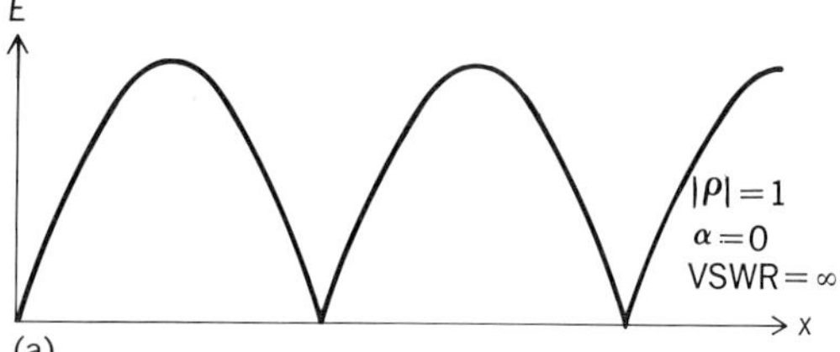

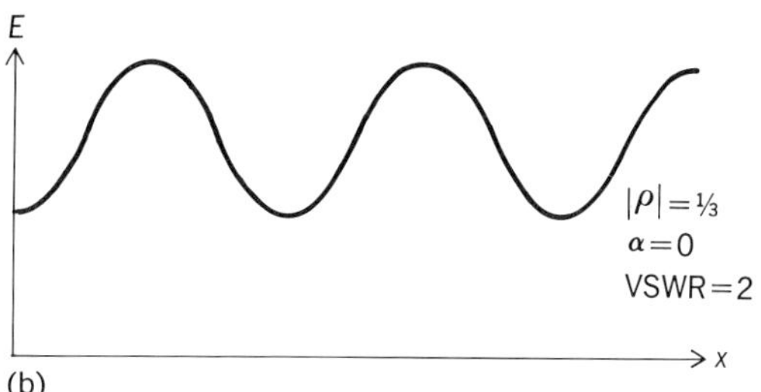

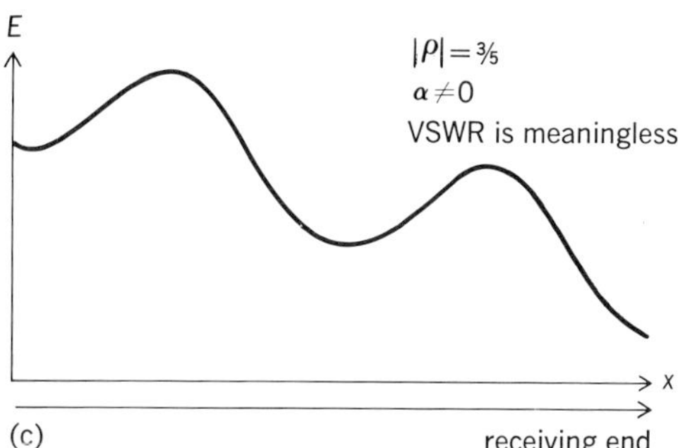

Fig. 3. Voltage distribution under sinusoidal steady-state conditions on a section of transmission line, illustrating three standing-wave conditions: (*a*) line with negligible losses, reflection coefficient of unity, (*b*) line with negligible losses, reflection coefficient of one-third, and (*c*) line with finite losses, reflection coefficient of three-fifths. Position of the voltage wave in each case is dependent on angle of phasor value of reflection coefficient. In each case a current maximum (not shown) appears at a voltage minimum in the wavelength.

$$\rho = \frac{Z_r - Z_0}{Z_r + Z_0} \quad (12)$$

When the load impedance is equal to Z_0, the reflection coefficient is zero. Under this condition the line is said to be matched.

Pulse transients. The transient solutions of Eqs. (6) and (7) are dependent on the particular problem involved. Typical physical phenomena with pulse transients are shown in Fig. 2. The characteristic time delay in transmission is often advantageously employed in radar and other pulse-signal systems.

Standing waves. The superposition of direct and reflected waves under sinusoidal conditions in an unmatched line results in standing waves (Fig. 3).

Voltage standing-wave ratio. When losses are negligible, successive maxima are approximately equal; under this condition a quantity, the voltage standing-wave ratio, abbreviated VSWR, is defined by Eq. (13).

$$\text{VSWR} = \frac{V_{\max}}{V_{\min}} \quad (13)$$

Power standing-wave ratio. This quantity, abbreviated PSWR, is equal to $(\text{VSWR})^2$. Measurements of voltage magnitude and distribution on a line of known characteristic impedance Z_0 can be used to determine the magnitude and phase angle of an unknown impedance connected at its receiving end. Lines adapted for such impedance measurements, known as standing-wave lines, are widely used.

Transmission-line circuit elements. The impedance Z_s at the sending end of a loss-free section of transmission line that has a length d, in terms of its receiving-end impedance Z_r, is given by Eq. (14).

$$Z_s = \frac{Z_r \cos \beta d + jZ_0 \sin \beta d}{\cos \beta d + j(Z_r/Z_0) \sin \beta d} \qquad (14)$$

This equation describes the property of a length of line which transforms an impedance Z_r to a new impedance Z_s. In the simple cases, in which Z_r is a short circuit or open circuit, Z_s is a reactance. Various lengths of line may be used to replace more conventional capacitors or inductors. These properties are widely applied at high frequencies, where suitable values of βx require only physically short lengths of line.

Tapered transmission lines. Transmission lines with progressively increasing or decreasing spacing are used as impedance transformers at very high frequencies and as pulse transformers for pulses of millimicrosecond duration. Although tapers designed to produce exponential-varying parameters, as in the exponential line, are most common, a number of other tapers are useful.

[EVERARD M. WILLIAMS]

POWER TRANSMISSION LINES

In an electric power system the facility used to transfer large amounts of power from one location to a distant location is termed a power transmission line. Techniques of power transmission are presented in this section.

Power transmission lines are distinguished from subtransmission and distribution lines by their higher voltages, greater power capabilities, and greater lengths. With the exception of a few high-voltage dc lines for satisfying special requirements, power transmission lines employ three-phase alternating currents. Such lines require three conductors. The standard frequency in the United States is 60 hertz (Hz). In Europe it is 50 Hz, while in the rest of the world both of these frequencies are used. For transmitting large amounts of power over long distances, high voltages are necessary. Standard transmission voltages in the United States are 69, 115, 138, 161, 230, 345, 500, and 765 kilovolts (kV). These figures refer to the nominal effective voltages between any two of the three conductors. The line conductors are usually placed overhead, supported by poles or towers; however, they may form part of an underground or underwater cable. *See* ELECTRIC DISTRIBUTION SYSTEMS; ELECTRIC POWER SYSTEMS.

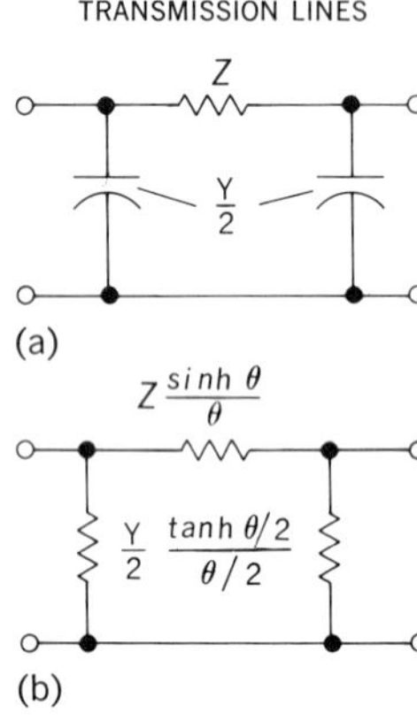

Fig. 4. Lumped-constant representations of a single-phase transmission line or of one phase of a three-phase line. (*a*) Nominal π. (*b*) Equivalent π.

Requirements of transmission. Power transmission systems must be reliable, have good voltage regulation and adequate power capability, and be capable of economical operation.

Reliability. This requirement is met by sturdy construction, by protection against overvoltages, by rapid automatic disconnection of accidentally short-circuited lines, by suitable transmission layouts, and by automatic rapid reconnection of lines experiencing only transitory faults.

Good voltage regulation. When the load voltage does not vary appreciably as the load increases from no load to full load, the regulation is said to be good. The inherent voltage regulation depends mainly on the inductive reactance of the line and the power factor of the load. If the inherent regulation is unsatisfactory, the voltage can be controlled by switched shunt capacitors or synchronous capacitors connected at the load.

Power capability. The maximum power that can be transmitted, with due regard to limitations imposed by losses, temperature of the conductors, voltage regulation, and system stability, is the power capability of the line. It varies approximately as the square of the voltage.

Economy. Fulfillment of this requirement depends on a balance between low first cost and low operating cost, including cost of power loss. The principal loss is the I^2R loss in the conductors.

Constants. From a knowledge of the size and type of conductors and the spacing between them, one can obtain the values of series resistance r and inductive reactance x per phase per unit length of line and of shunt capacitive susceptance b and leakage conductance g per phase per unit length of line. All of these values are multiplied by the length of the line, giving constants R, X, B, and G, respectively. These are then combined to give the complex impedance, admittance, hyperbolic angle, and characteristic impedance, Eqs. (15)–(18), respectively.

$$Z = R + jX \qquad (15)$$

$$Y = G + jB \qquad (16)$$

$$\theta = \sqrt{ZY} \qquad (17)$$

$$Z_0 = \sqrt{\frac{Z}{Y}} \qquad (18)$$

The approximate value of characteristic impedance for a line with low losses, given by $Z_0 = \sqrt{l/c}$, is often called the surge impedance and is real. The power carried by a transmission line is often expressed in terms of its natural power or surge-impedance loading (SIL), which is defined by Eq. (19), where E_n is the nominal voltage, and Z_0 is the surge impedance. Units of kilovolts for E_n, ohms for Z_0, and megawatts for P_n are convenient for power lines. If E_n is the voltage between conductors, P_n is the three-phase power.

$$P_n = \left| \frac{E_n^2}{Z_0} \right| \qquad (19)$$

If a line is operating at its surge-impedance loading, there are no standing waves (Fig. 3), but the graph of voltage magnitude versus distance along the line is flat. In addition, the reactive power E^2B produced by the shunt capacitance is balanced by the reactive power I^2X consumed by the series inductance, and at every point of the line the current is in phase with the voltage.

Equivalent circuit. This circuit indicates lumped values which represent values distributed along the line. A short line can be represented adequately by its nominal π circuit shown in Fig.

4*a*. Here the shunt admittance *Y*, actually distributed uniformly along the line, is assumed to be lumped and divided into two equal parts, one at each end of the line. For a long line the theoretically exact equivalent π circuit should be used. Each of its branch impedances is calculated by multiplying the corresponding branch impedance of the nominal π by a correction factor given in Fig. 4*b*.

By use of equivalent circuits, the steady-state electrical performance of a line can be calculated by the ordinary theory of ac circuits with lumped constants. These circuits can be combined with circuits representing series capacitors, shunt reactors, transformers, and loads. If a complicated network is to be studied, it can be represented by a low-power model in a network analyzer wherein each line is represented by its π circuit or can be solved on a digital computer by use of a suitable power-flow program.

Alternating-current overhead lines. An overhead transmission line consists of a set of conductors, usually bare, which are supported at a specified distance apart and with specified clearances from the ground and from the supporting structures.

Routes. Lower-voltage transmission lines are usually built along highways, whereas higher-voltage lines are put on a special right of way, cleared of trees and brush. Such routes are often chosen from results of aerial surveys.

Supporting structures. Lower-voltage overhead lines are usually supported by wooden poles and higher-voltage lines by wooden H frames or steel towers. Rigid steel towers give the greatest strength and reliability. The higher the voltage, the greater must be the spacing between conductors and the clearance from conductor to ground. The farther apart the towers are placed, the greater is the sag of the conductors and the taller and stronger the towers must be. Figure 5 shows some typical structures. *See* ELECTRICAL INSULATION.

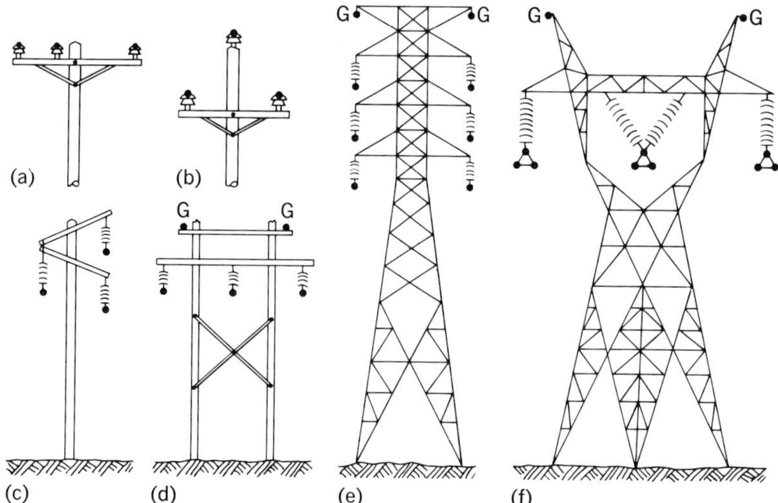

Fig. 5. Supporting structures for electric power transmission lines: (*a-d*) wood poles; (*e, f*) steel towers. Structures in *a–d* and *f* are for single-circuit lines; *e* is for double-circuit lines. In *a* and *b*, pin-type insulators are used; in *c–f*, suspension insulators are used. Structures in *d–f* have ground wires (G) above the line conductors.

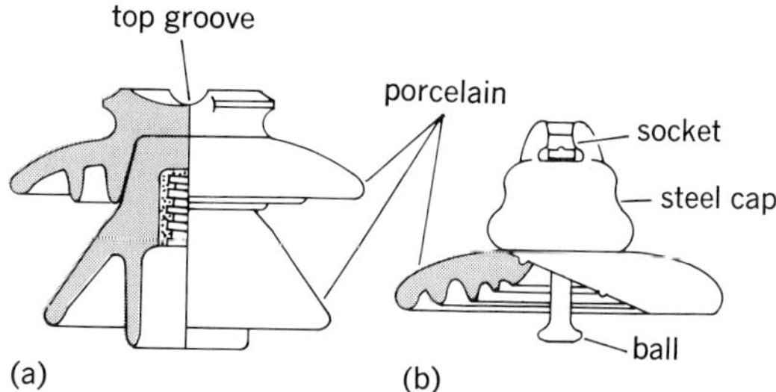

Fig. 6. Insulators used on transmission lines. (*a*) Large two-piece pin-type insulator. (*b*) One unit of a suspension insulator. (*From H. Pender and W. A. Del Mar, Electrical Engineers' Handbook, vol. 1, 4th ed., copyright © 1949 by John Wiley and Sons, Inc.; used with permission*)

The towers shown with vertical strings of suspension insulators are tangent towers or suspension towers. Dead-end towers, used at the ends of a line, and angle towers, used at large angles in the line, have almost horizontal strings of insulators. The center conductor of Fig. 5*f* is supported by V strings, which prevent the conductor from swinging sideways in a cross wind, keeping it from the grounded tower.

Insulators. Conductor supports, or insulators (Fig. 6), are generally made of glazed porcelain or of glass. On lower-voltage lines, they are usually of the pin or post type. On higher-voltage lines, they are of the suspension type, consisting of several units connected by swivel joints. The number of units per string depends on the desired impulse flashover voltage, but is not proportional to it, because the voltage does not divide equally between the several units.

Insulators exposed to industrial dust deposits or to salt spray will, when moist from fog, carry leakage currents which may lead to flashovers at normal operating voltage. Remedies are the use of special fog-type insulators with deeper corrugations on their lower sides for increasing the length of leakage paths, occasional washing of insulators by a stream of water drops from a nozzle, and coating of insulators with silicone grease.

Conductors. For overhead lines, conductors are usually bare, multilayered, concentrically stranded aluminum cables. Adjacent layers of strands are spiraled in opposite directions. Additional tensile strength is provided usually by a core of steel strands (Fig. 7*a*) or sometimes by inclusion of strands of a strong aluminum alloy. These two types of conductor are known by the abbreviations ACSR (aluminum cable, steel reinforced) and ACAR (aluminum cable, alloy reinforced), respectively. If the conductor diameter required for low corona loss is greater than that of an ordinary stranded conductor having the cross-sectional area required for the desired resistance, special "expanded" conductors are used. Some of these have a paper filler between the steel core and the outer layers of aluminum strands; others have two filler layers each of four large aluminum strands (Fig. 7*b*).

Many very high voltage lines (400 kV and above) use multiple, or bundle, conductors, each consisting of two, three, or four stranded subconductors connected one to another through metal spacers and hung from the same insulators (Fig. 5*f*). Bundle conductors have the advantages of lower inductive reactance; higher natural power; lower corona loss, radio interference, and audible noise; and

TRANSMISSION LINES

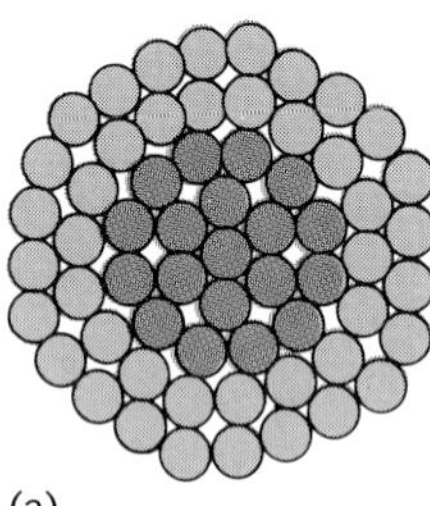

(a)

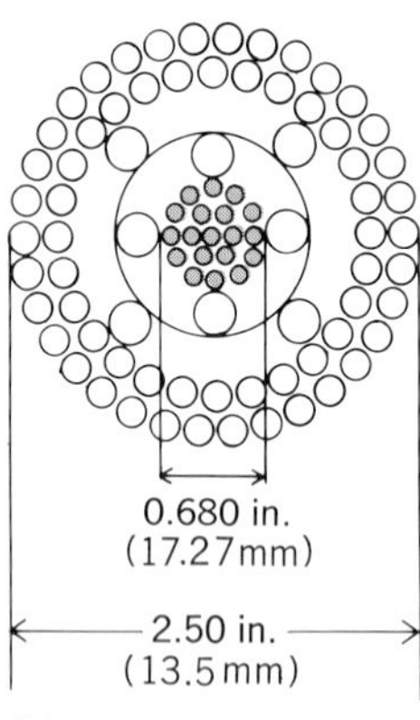

(b)

Fig. 7. Cross sections of typical overhead conductors. (*a*) Steel-reinforced aluminum cable (ACSR) with 19 steel and 42 aluminum strands. (*b*) Expanded ACSR. (*Alcoa Conductor Products Company*)

better cooling than single conductors of the same total cross-sectional area.

Splices in large conductors are usually made with metal sleeves squeezed over the butted ends of the conductor by hydraulic jacks.

Sag and tension. Conductors between adjacent supports hang in a curve called a catenary (Fig. 8). For a given length of span, the greater the tension in the conductor, the smaller is the sag. High mechanical tension is desirable to reduce sag and thus to permit use of longer spans or shorter towers, while maintaining adequate ground clearance. However, the tension must not exceed the tensile yield strength of the conductor under the worst condition, which occurs under a combination of low temperature (causing shortening) simultaneously with the thickest coating of ice on the conductor and the strongest wind.

Vibration. At times the wind causes the conductors to vibrate with low amplitude and audible frequency. This vibration bends the conductor where it is clamped to the insulators and eventually may produce fatigue breakage. A device with duplex weights is used on some lines to reduce conductor vibration. One or more of these dampers are fastened to the conductors several feet from the insulator clamp (Fig. 9).

Sleet. An ice-covered conductor acts as an air foil and is lifted by the wind, so that "dancing" occurs, of such amplitude that one conductor may strike another, producing a short circuit. Conductors should be located so that contact will not be made. Formation of ice is prevented if the current heats the conductor sufficiently. Some power companies make a practice of periodically taking endangered lines out of service and sending high currents through them.

Corona. When the voltage gradient, or electric field strength, at the surface of the conductor exceeds the breakdown gradient of air, the air near the conductor surface becomes ionized. This condition, called corona, is evidenced by a visible glow at night and by a buzzing noise.

Corona results in a loss of power, interference with radio reception, and audible noise, all of which increase rapidly with voltage. Transmission lines are normally operated at a voltage near that at which corona becomes appreciable. The larger the conductor diameter and the greater the number of subconductors, the higher the operating voltage may be.

Inductive coordination. If a telephone line runs near and parallel to a power line for some distance, the high currents and voltages in the power line

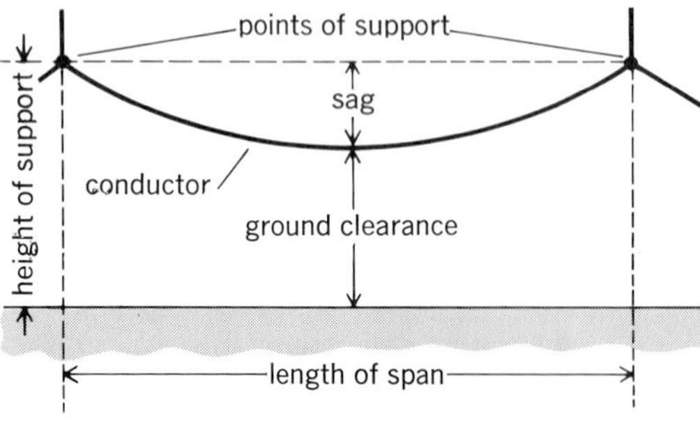

Fig. 8. Catenary curve assumed by one span of a flexible conductor with supports at equal elevations.

Fig. 9. Stockbridge dampers on transmission-line conductor. (*From H. Pender and W. A. Del Mar, Electrical Engineers' Handbook, vol. 1, 4th ed., copyright © 1949 by John Wiley and Sons, Inc.; used with permission*)

may induce currents and voltages in the telephone line. These signals may be comparable in strength to the telephone signals and thus produce objectionable noise in telephone receivers. The worst noise is produced by magnetic coupling from harmonic currents having a ground-return path. The coupling between the two lines can be reduced by greater physical separation, by transposition of the telephone wires, and by shielding, such as that provided by grounded cable sheaths. *See* GROUNDING.

Inspection and fault location. Transmission lines should have both periodic general inspections and special immediate inspections of points where short circuits have occurred, to detect damage, such as broken insulators, which might impair the reliability of the line. Faults can be located approximately by electrical measurements made from the ends of the line and then exactly by visual patrol of the vicinity. *See* CIRCUIT TESTING (ELECTRICITY).

Lightning protection. Lightning is the most detrimental factor affecting the reliability of electric power service, but its damaging effects have been greatly reduced by proper design. Lightning striking a transmission line momentarily impresses a very high voltage on the line, causing spark-over to ground, usually at an insulator. Power current then follows the spark path, producing an arc, which constitutes a short circuit and which can be extinguished only by disconnecting the faulted line from the rest of the power network. Lines built where severe thunderstorms are prevalent are equipped with overhead ground wires (Fig. 5*d*–*f*) for intercepting the lightning stroke and leading it to ground at the nearest tower. *See* LIGHTNING AND SURGE PROTECTION.

Switching surges. Another source of overvoltage, which has become important enough on extra-high-voltage lines (500 kV and above) to determine their insulation levels, is switching. The transient overvoltages caused by reenergization of a line which still has trapped charges left from a recent deenergization may be as high as 3.5 times normal line-to-ground crest voltage. By use of circuit breakers which energize the line through one step of series resistance before making direct connection from the power source to the line, the overvoltage can be reduced to about twice normal. By use of two or three steps of decreasing resistance, the overvoltage can be limited to a still lower value, say 1.5. The values cited are representative, but actual overvoltages vary with the length of line section, the time during which the resistors are inserted, the time span from the closure of the first pole of the breaker to the last pole, and other factors.

Another source of transient overvoltage similar to a switching surge is a short circuit from one line conductor to ground. The overvoltage, having a crest value up to 2.0 times normal crest voltage, appears on the conductor whose voltage phase leads that of the faulted conductor.

Overhead power line constants. Overhead 60-Hz power transmission lines with single conductors per phase have series inductive reactance of about 0.8 ohm/mi (0.5 Ω/km) shunt capacitive susceptance of about 5 micromhos/mi (3 μs/km), and surge impedance of about 400 ohms. For lines with two-conductor bundles per phase these become, respectively, 0.6 ohm/mi (0.37 Ω/km), 6.7 micromhos/mi (4.2 μs/km), and 300 ohms. For any 60-Hz open-wire line, the phase constant β is about 0.0020 radian/mi.

Power capability. The economic loading of an overhead power line is usually in the range of 1.0 to 1.5 times its natural power (SIL loading), depending on conductor size. At much higher loadings, the losses become excessive. The table gives typical values of natural power and economic loading of single-circuit, three-phase, 60-Hz lines.

Natural and economic loading of single-circuit, three-phase, 60-Hz lines

Voltage, kV	Surge impedance, ohms	Natural loading, MW	Economic loading, MW
69	400	12	15
138	400	47	60
230	400	130	170
345	350	340	430
500	300	830	1000
700	300	1600	2000
1000	250	4000	5000

Cost of transmission. The cost of building a transmission line is very nearly proportional to the voltage and to the length of the line. Its power capability is almost proportional to the square of the voltage. Consequently, the cost per unit of power varies directly as the distance and inversely as the square root of the power. If the amount of power to be transmitted is quadrupled, it can be transmitted twice as far for the same unit cost. This explains why it is not economical to transmit for a long distance unless a large quantity of power is involved.

Reactive compensation. There are two kinds of reactive compensation: shunt and series. In shunt compensation the distributed shunt capacitive susceptance B_C of the transmission line is partially or entirely compensated by the addition of lumped inductive susceptance B_L in the form of shunt reactors. The net shunt susceptance is $B=B_C-B_L$. The degree of shunt compensation is the ratio B_L/B_C. The net charging current is reduced in proportion of 1 minus this ratio. Shunt compensation is needed principally on extra-high-voltage overhead lines (500 kV and above) when they are lightly loaded and on long cables.

In series compensation the distributed series inductive reactance X_L of the transmission line itself is partially compensated by the addition of lumped series capacitive reactance X_C in the form of series capacitors. The net series reactance is $X=X_L-X_C$. The degree of series compensation is the ratio X_C/X_L.

Series compensation is used for several different purposes. The first is to reduce the voltage fluctuation (flicker) due to rapidly changing loads, as in electric furnaces and the starting of large motors. The second purpose is to obtain proper division of current between transmission lines connected in parallel. If two such lines have conductors of different cross section, the total losses can be reduced by connecting a capacitor in series with the line having the larger conductors. The capacitor should have such capacitive reactance that the ratio of net reactance to resistance is the same for both lines.

The third purpose of series compensation is to increase the distance that a given amount of power can be transmitted, to increase the power that can be transmitted over a given distance, or, more generally, to increase the possible product of power and distance. Analysis of the equivalent circuits of Fig. 4 for a line with negligible losses shows that the power P transmitted for terminal voltages of fixed magnitudes E_s and E_r but of variable phase difference δ varies according to Eq. (20), where X

$$P=\frac{E_sE_r}{X}\sin\delta \qquad (20)$$

is the series inductive reactance of the line (Z in Fig. 4*a*) or, more exactly, for a long line, of the horizontal branch of the equivalent π of Fig. 4*b*. This equation shows that maximum power is obtained at $\delta=90°$, for which $\sin\delta=1$. Considerations of stability require that the line be operated with δ less than 90°, and experience indicates that it can be operated prudently at $\delta=30°$, $\sin\delta=0.5$. With any fixed values of the terminal voltages and fixed maximum value of $\sin\delta$, there is a maximum value of PX, the product of power and series reactance. Since the frequency is fixed in practice, the series reactance of an overhead line of given frequency is directly proportional to the length of line d, and hence there is a limit to the product Pd (in megawatt-miles) of an uncompensated line. Assuming the economic power to be 1.25 P_n, $f=60$ Hz, and $\sin\delta=0.5$, the limiting distance for that power is about 200 mi (320 km).

In the past, this limitation was not seriously felt because transmission for distances much greater than 200 mi (320 km) between points of maintained voltage was generally not economical in comparison with supplying power from generating stations nearer to the load centers. However, the use of electric power has grown to a point where transmission of large amounts of power over greater distances (perhaps up to 1000 mi or 1600 km) is economical in some circumstances. Transmission of power over such distances becomes technically feasible through series compensation.

Underground ac lines. Insulated cables are used in congested areas where the cost of right of way for overhead lines would be excessive, in city streets where overhead lines would be too unsightly or hazardous, in and around power stations, and for crossing wide bodies of water. About 1% of the total transmission mileage of the United States is underground, located mostly in congested urban areas. Because the cost of underground transmission is so much higher than that of overhead transmission, it is used only where necessary. The cost of constructing an underground line ranges from 8 times (at 69 kV) to 20 times (at 500 kV) that of an overhead line of the same length and power capability.

Cables are now made for alternating voltages up to 500 kV, and cables for 700 kV are being developed.

The conductors of most cables are of stranded copper, insulated by wrapping with layers of paper tape saturated with mineral oil. The thickness of the insulation depends on the voltage, and varies from 0.285 in. (0.7 cm) in 69-kV cables to 1.34 in. (3.4 cm) for 500-kV cables. The dielectric constant of this insulation is from 3.5 to 3.7.

Solid and oil-filled cables. Low-voltage cables are of the solid type, in which the only oil is that put in the paper during manufacture. The disadvantage of this type of cable for high voltages is that small voids may form between layers of paper and that corona in such voids causes the insulation to deteriorate, leading eventually to puncture. Therefore, high-voltage cables are provided with oil under pressure.

In the low-pressure oil-filled cable, there are oil channels in the center of the conductor of single-conductor cables or between the insulated conductors of three-conductor cables. Oil reservoirs, connected to the cable at intervals, keep the cable full of oil at a pressure of from 3 to 20 pounds per square inch (psi) or 20 to 140 kPa in spite of contraction and expansion caused by changes in cable temperature.

The solid and low-pressure oil-filled types of cables have a lead sheath surrounding the insulation to keep moisture out of the cable and the oil in. Such cables are pulled into concrete or fiber ducts which give mechanical protection. At intervals of 600 ft (180 m) or less, these ducts terminate in underground chambers called manholes, where sections of cables are spliced together.

Pipe-type cables. In these, three single-conductor, paper-insulated cables are pulled into a buried steel pipe, which is later filled with oil at high pressure (200 psi or 1.4 MPa). Manholes may be spaced as far apart as 1/2 mi (0.8 km). Oil reservoirs and pumps are required. Power capability ranges from about 225 MVA at 138 kV to 650 MVA at 550 kV.

Compressed-gas-insulated cables. Each of the three isolated phases consists of two coaxial aluminum tubes separated mechanically by epoxy insulating spacers and electrically by these spacers and the compressed gas. The inner tube is the main conductor, while the outer tube, which is grounded, normally carries little current but serves as a sheath, contains the gas, and gives some mechanical protection. Expansion joints incorporated in the central tube allow differential thermal expansion. The gas cable is shipped in factory-assembled sections about 40 ft (12 m) long. These are then welded together in the field. The outer surface of the sheath is provided with a protective coating to prevent corrosion. Additional protection against corrosion may be given by a cathodic protection such as that used on long pipes for oil or gas.

The compressed gas currently used is sulfur hexafluoride (SF_6) at pressures of 45 to 60 psi (300–400 kPa). It combines excellent chemical stability with high electric breakdown strength, good heat-transfer characteristics, and low dielectric constant compared to paper and oil. Cables employing SF_6 can be designed to withstand rated voltage even with the gas pressure reduced to normal atmospheric pressure. Research is under way to find other gases (or mixtures) as good as SF_6 but less expensive. Power ratings range from 600 MVA at 230 kV to 11,000 MVA at 765 kV.

Low-temperature (cryogenic) cables. Cryogenic cables are being investigated. The resistance of aluminum or copper decreases as the temperature decreases. If a cable is cooled by circulating liquid nitrogen at temperature 77 K through a hollow conductor, for example, the power loss (I^2R) in the conductor is greatly reduced. However, the power required for refrigeration of the coolant partly offsets the saving in I^2R loss. Good thermal insulation of such a cable is obviously necessary.

Certain special alloys become superconductive at very low temperatures (below 18.2 K for niobium-tin), and their resistance to direct current becomes zero, leaving only the power for refrigeration as the loss of the cable. With alternating current, there is also some loss in the conductors but much less than that at or above room temperature. If the current or the temperature should exceed its critical value, the conduction would become resistive. For this reason, the superconductive alloy is either plated onto the surface of an ordinary metal or is made in fine filaments embedded in the ordinary metal. Thus, if the alloy should lose its superconductivity, the ordinary metal could carry the current for a long enough time to permit reduction of the current before melting of the conductors could occur.

Terminations. Connections of cables to overhead lines or to substations are made through potheads which provide oil seals and longitudinal as well as radial insulation. High-voltage potheads are encased in porcelain with a corrugated outer surface and are similar in appearance to the insulating bushings used on transformers.

Splices in high-voltage cables must be made with great care and require many hours of skilled labor. The distance between splices depends largely upon the length of cable that can be wound on one shipping reel.

Losses and power capability. The power capability of a cable is limited by the rise of temperature in the conductor and adjacent insulation, because too high a temperature will char the insulation and cause its breakdown. The temperature rise depends on the power losses in the cable and on the rate of conduction of heat from the cable into the surrounding soil. Whereas the loss in the conductor is the only important loss in an overhead line, the cable has also a considerable dielectric loss, which increases with voltage because of the increasing thickness of the insulation. Also, whereas the overhead conductor is bare and directly exposed to the cooling air, the heat produced by losses in a cable must pass through the insulation, the duct walls (if ducts are used) and a considerable thickness of earth before reaching cool earth or air. Thus the permissible current in a cable is less than that in an overhead conductor of equal resistance, and the permissible power for a given voltage is correspondingly less. This is the apparent power, $S = \sqrt{P^2 + Q^2}$, where P is the active power and Q is the reactive power.

Improved cooling. Since the power capability of cables is limited by their temperature rise, capability can be increased by improved heat removal.

Among the means that have been used—but so far only to limited extent—are (1) backfilling the trench in which cable or cable ducts are buried with material of better heat conductivity than that of the original soil, (2) burying cooling pipes in the earth near the cables and circulating water or other fluid through these pipes, and (3) circulating the oil of pipe-type cables through heat exchangers.

Charging current and critical length. Because of the closer spacing between conductors and the higher dielectric constant of the oil-paper insulation, cables have a much higher shunt capacitance per unit length than do overhead lines. Hence for a given voltage the charging current and the charging reactive power of cables are correspondingly higher. Compressed-gas-insulated cables have an advantage in this respect.

The critical length of a cable is that length for which the charging current equals the rated current. A cable longer than the critical length would be overloaded at the sending end even if nothing were connected to the receiving end.

Shunt compensation. The limitation of length or active-power capability due to charging current can be raised by connecting inductive reactors in shunt with the cable at its terminals and at intermediate points. The total shunt current taken by the reactors should be approximately equal in magnitude to the charging current of the whole length of the cable but in phase opposition to it. The economic spacing of shunt reactors on a 60-Hz cable would be between 5 and 10 mi (8 and 16 km).

Shunt compensation of power cables is seldom, if ever, used. Most underground cables are too short to require it.

Submarine ac cables. These are used in crossing rivers, bays, or straits too wide for overhead spans and to transmit power to offshore islands. They are mostly of the solid type. Water pressure prevents formation of voids and the natural cooling is good. Length of uncompensated cables is limited by charging current to about 25 mi (40 km), and shunt compensation is deemed impractical because of the additional complications in laying the cables and in retrieving them when repairs are needed. The lead sheath is protected by an armor of steel wires, sometimes covered with jute. Submarine cables are liable to damage by trawling and by dragging of ship's anchors. Shore sections, used in shallow water and across beaches, usually differ in diameter and amount of armor from the deep-water sections.

Direct-current lines. Although most electric power transmission is by alternating current, there is an increasing number of direct-current transmission lines. These require converter stations at both ends to connect the line to an ac system.

Overhead lines. Bipolar overhead lines are similar in construction to overhead three-phase lines except that they have only two conductors instead of three. For the same conductor size and insulation level, a dc line can carry the same power on two conductors that a three-phase line can carry on three conductors; and the cost of the dc line is about two-thirds the cost of the corresponding ac line. In some places, two monopolar lines spaced well apart (about 1½ mi or 2.4 km) are used instead of one bipolar line for the sake of improved reliability.

Since internal overvoltages are somewhat less on dc lines than on ac lines, lower insulation levels are used for the same crest voltage to ground. Under these conditions leakage currents at normal operating voltage become more important, especially when insulators are dirty from industrial wastes and moist from fog. For this reason, insulators for dc lines are usually of special design, having a higher ratio of length of leakage path to flashover distance.

Cable lines. Most of the dc lines built before 1969 where wholly or partly submarine cables. Since that date the trend has been toward overhead lines, although several cases have been considered for underground cables to bring power to metropolitan areas. Direct-current cables have no charging current and therefore are not subject to the limitation on their length that applies to ac cables. In addition, a dc cable has no dielectric loss and can safely withstand a higher direct voltage than root-mean-square alternating voltage. As a result, a dc cable can carry about six times as much power as the rated apparent power when the same cable is used for alternating current. Single-conductor solid cables are used.

Ground return. The resistance of the ground to direct current is very much lower than to alternating current, being essentially only that in the vicinity of the ground electrodes. The use of ground or sea return for a monopolar line saves most of the cost of one conductor and of its power loss. A bipolar dc line can operate with one pole and ground return while there is a fault on the other pole of the line or of the terminal equipment.

[EDWARD W. KIMBARK]

Bibliography: American Radio Relay League, *ARRL Antenna Book*, revised periodically; L. N. Dworsky, *Modern Transmission Line Theory and Applications*, 1979; D. G. Fink and H. W. Beaty (eds.), *Standard Handbook for Electrical Engineers*, 11th ed., 1978; P. Graneau, *Underground Power Transmission*, 1979; E. W. Kimbark, *Direct Current Transmission*, vol. 1, 1971; W. D. Stevenson, Jr., *Elements of Power System Analysis*, 3d ed., 1975; E. Uhlmann, *Power Transmission by Direct Current*, 1975; B. M. Weedy, *Electric Power Systems*, 3d ed., 1979.

Transportation engineering

That branch of engineering relating to the movements of goods and people. Major types of transportation are highway, water, rail, subways, air, and pipeline.

Highway transportation. Highway transportation engineering deals with the planning, construction, and operation of roads, streets, bridges, and parking facilities. *See* HIGHWAY ENGINEERING.

Important aspects of highway engineering include:

1. Traffic engineering, which relates to the volumes of traffic to be handled, the methods to accommodate these flows, and the general layout of highways. *See* TRAFFIC ENGINEERING.

2. Engineering pertaining to pavements and roadway surfaces. *See* PAVEMENT.

3. Design and construction of highway bridges, structures, tunnels, and similar facilities. *See* BRIDGE; TUNNEL.

Highway transportation engineering has been distinguished by the development of planning and construction techniques that have made possible express highways to accommodate large flows of traffic at high speeds. Express routes built in the United States, Canada, Mexico, Europe, South America, Africa, Japan, and other areas have resulted from the enormous growth of motor vehicle transportation.

Water transportation. Planning and construction of canals, channels, harbor facilities, navigation aids such as lighthouses, and navigation locks and dams are important concerns of water transportation engineers. Another important field of water transportation relates to the design and production of launches, barges, tugs, ferry boats, and other ships. *See* CANAL; COASTAL ENGINEERING; RIVER ENGINEERING.

Water transportation engineering in recent years has been distinguished by the emphasis on larger equipment requiring increased waterway dimension and by changes in the source of power for propelling the bigger ships. A good example of projects to provide deeper inland shipping facilities is the St. Lawrence Seaway, completed jointly in 1959 by Canada and the United States to permit oceangoing ships of 27-ft (8.2-m) draft to travel from the Atlantic Ocean as far west as Duluth, Minn.

Oil-burning equipment provides power for most larger vessels, but several nations have been involved in a program to produce many nuclear-powered vessels for commercial shipping. The NS (Nuclear Ship) *Savannah*, a large merchant ship, was christened by the United States government in 1959.

Despite the great growth in the transportation of commercial products by water, passenger travel across the world's oceans is primarily by air. This fact is strikingly illustrated by the fact that Great Britain sold the two huge ocean liners *Queen Elizabeth* and *Queen Mary* to interests in the United States for uses other than transportation of passengers across the Atlantic Ocean.

Railway transportation. Engineering relating to rail transportation includes the planning and construction of terminals, switchyards, loading and unloading facilities, trackage, bridges, traffic-control and maintenance facilities, and the hauling equipment itself—locomotives and other rolling stock. The railroad industry has continuing programs to develop safer and quicker methods of loading, unloading, and shifting of cars, and of operating trains. These facilities include large marshaling yards where the movement of hundreds of railroad cars is controlled by electronic equipment. *See* RAILROAD ENGINEERING.

One major factor in bringing about these changes was the merging of numerous railway systems, which resulted in quicker and much more economical service.

Subway transportation. Engineering work relating to planning, general layout, detailed design, construction, and operation of subways is an important part of transportation engineering. Subways are a major element in mass transit, and as metropolitan areas increase in size and population and vehicular traffic becomes more congested, subways are being given increased consideration as possible part of the transportation system of a city.

Because of downtown congestion and the difficulty of providing adequate parking, many United States cities have been constructing rapid transit systems. One of the newer subway systems is the 98-mi (156-km) network serving the Washington, DC, metropolitan area. Extensive auto parking facilities near the several terminals is an integral part of the project. In Atlanta, Ga, a transit system has been started. Some 54 mi (86 km) of the system will be rail transportation, but many miles of bus routes are also planned. Many foreign cities also are building new subway systems or extending old facilities, including London, Paris, Brussels, West Berlin, Prague, Rio de Janeiro, Vienna and Santiago.

Engineering planning related to subways includes comprehensive studies to determine whether a subway system is economically feasible. This work involves extensive analysis to evaluate expected construction costs, anticipated passenger volumes, feasible passenger fares and possible income, expected operating costs, depreciation and maintenance of equipment, and passenger safety.

Engineering pertaining to railway transportation has been briefly described in this article. Most of the various types of engineering work relating to railways are required in building a subway, for construction of a subway involves putting a railway of special design underground. However, subway work also includes: (1) extensive costly tunnel construction, with the tunnels often at sufficient depth to pass under rivers and other bodies of water; (2) the building of passenger terminals underground, (3) elaborate ventilation facilities, (4) lengthy underground electric power distribution systems, (5) lighting facilities, (6) escalators for transporting passengers to and from street level to the subways, (7) noise control, and (8) safe and reliable signaling facilities.

Air transportation. The planning, design, and construction of runways, terminals, aircraft, and navigation aids are the major branches of civilian air transportation engineering. Width, length, strength, and layout of the runways are dependent upon the type of aircraft to be accommodated and the frequency of landings and departures. In like manner, design of the airport terminal building is determined by the volume of passenger and air express traffic expected. In the larger air terminals, great emphasis is placed on systems for moving of passengers and baggage, including numerous long conveyor belts at the Charles de Gaulle airport north of Paris and the small railway system at the Dallas–Fort Worth Airport. In planning both airfield and terminal facilities, consideration is also given to helicopter traffic.

The use of jet-powered planes and continued emphasis on larger and faster aircraft are important factors in air transport engineering.

Pipeline transportation. Years ago pipelines were employed primarily for municipal services, such as water supply, sewage disposal, and gas distribution. Today pipelines are used to transport natural gas and petroleum products great distances. In addition, at least one line more than 100 mi (160 km) long is used to move powdered coal. Construction of the 48-in.-diameter (120-cm) Trans-Alaska pipeline to transport oil from near Prudhoe Bay on the Arctic Ocean to the city of Valdez on the Gulf of Alaska has been completed.

Totaling 798 mi (1276 km) in length and costing over $6,000,000,000, the work is considered the world's largest pipeline project to date. Construction was made difficult by Arctic weather, the isolated and mountainous terrain traversed, difficult foundation conditions due to permafrost, and the numerous river and stream crossings. In Brazil a 20-in.-diameter (50-cm) pipeline 250 mi (400 km) long has been built to carry iron ore slurry from the mine to the processing facilities.

Pipeline engineering is concerned with planning and construction of the pipeline, pumping stations to move the material through the line, and any needed storage facilities. Frequently, large bridge structures are required to carry the pipeline across major streams or other barriers. In designing the pipeline, consideration must be given to possible alternate routes, topography, right-of-way needs, foundation conditions, the desired pipe size, thickness of pipe, methods of laying, type of material to be transported, measures to protect the pipe once in place, forces on the conduit due to temperature changes and pumping operations, and adequate maintenance procedures. *See* PIPELINE.

Planning of pumping facilities requires study of power requirements for different types of material moved, standby facilities, spacing of the pumping stations, and related factors.

Bikeway facilities. The increasing cost of gasoline and other petroleum products has resulted in major increases in bicycle travel in many suburban and rural areas. As a result, many highway agencies have established special bikeway divisions, and many miles of special bikeways have been built in numerous states.

[ARCHIE N. CARTER]

Bibliography: M. M. Akins, Washington Metro access facilities, *Civil Eng.*, p. 63, July 1975; P. Braaksma and J. Shortreed, Method for designing airport terminal concepts, *Transp. Eng. J. ASCE*, 101(TE2):321–355, May 1975; A. Chatterjee and C. Sinha, Distribution of benefits of public transit projects, *Transp. Eng. J. ASCE*, 101(TE3):505–519, August 1975; A. Chatterjee and K. C. Sinha, Mode choice estimation for small areas, *Transp. Eng. J. ASCE*, 101(TE2):265–278, May 1975; D. A. Day and B. P. Boisen, Fifty-year highlights of tunneling equipment, *J. Constr. Div. ASCE*, 101(CO2): 265–280, June 1975; E. E. Gilcrease, Jr., W. Kudlick, and M. Padron, Rail transit operating cost guidelines, *Transp. Eng. J. ASCE*, 101(TE2):365–381, May 1975; R. Herman and T. Lam, Carpools at large suburban technical center, *Transp. Eng. J. ASCE*, 101(TE2):311–319, May 1975; A. B. MacPherson and P. E. Egilsrud, Mall tunnel under nation's Capitol, *Civil Eng.*, p. 64, February 1975; *Maintenance and Operating Costs of Urban Rapid Transit Systems*, Institute for Rapid Transit, Chicago, February 1968; T. R. Mongan, N. J. Nielsen, and J. R. Formby, Measuring transportation system performance, *Transp. Eng. J. ASCE*, 101(TE3): 437–454, August 1975; R. E. Paaswell and J. Pafka-Gerbig, Community role in modal choice for transit system planning, *J. Urban Plann. Develop. Div. ASCE*, 101(UP1):35–47, May 1975; F. Palmer and K. C. Roberts, Developments in trench-type tunnel construction, *J. Constr. Div. ASCE*, 101-(CO1):37–49, March 1975; R. B. Peck, A. J. Hendron, and B. Mohraz, State-of-the-art of soft-ground tunnelings, *Proceedings of the 1st North American Rapid Excavation and Tunneling Conference, ASCE-AIME*, Chicago, pp. 259–286, 1972; G. Y. Sebastyan, The new Montreal International Airport, *J. Constr. Div. ASCE*, 101(CO2): 317–334, June 1975; *Standard Specifications for Construction of Airports*, Federal Aviation Administration; F. J. Stastny, Pipeline corridor selection model concept, *Transp. Eng. J. ASCE*, 101(TE2): 337–344, May 1975; Trans-Alaska oil pipeline finally under way, *Eng. News Rec.*, p. 20, May 8, 1975; E. W. Walbridge, Multilane passenger conveyors, *Trans. Eng. J. ASCE*, 101(TE3):463–477, August 1975; T. Zakaria, Analysis of urban transportation criteria, *Trans. Eng. J. ASCE*, 101(TE3): 521–536, August 1975.

Trestle

A succession of towers of steel, timber, or reinforced concrete supporting the horizontal beams of a roadway, bridge, or other structure. Little distinction can be made between a trestle and a viaduct, and the terms are used interchangeably by many engineers. A viaduct is defined as a long bridge consisting of a series of short concrete or masonry spans supported on piers or towers, and is used to carry a road or railroad over a valley, a gorge, another roadway, or across an arm of the sea. A viaduct may also be constructed of steel girders and towers. It is even more difficult to draw a distinction between a viaduct and a bridge than it is between a viaduct and a trestle. *See* BRIDGE.

A trestle or a viaduct usually consists of alternate tower spans and spans between towers. For low trestles the spans may be supported on bents, each composed of two columns adequately braced in a transverse direction. A pair of bents braced longitudinally forms a tower. The columns of one bent of the tower are supported on planed base plates or movable shoes to allow horizontal movement in the longitudinal direction of the trestle. Struts connect the column bases and force the movable shoes to slide. The width of the base of a bent is usually not less than one-third the height of the bent. This width is sufficient to prevent excessive uplift at windward columns when the trestle is unloaded. *See* STRUCTURES; TOWER.

[CHARLES M. ANTONI]

Truck

A wheeled, trackless, self-propelled vehicle for land transportation of commodities.

Every truck is designed to do a specific job: to haul parcels, miscellaneous commodities, or bulk materials in solid or liquid form, in closely figured amounts, over known terrain. Accordingly, there are many models, each representing a combination of components designed to create a unit best suited for the work it is intended to do. A truck is similar to a passenger car in many basic aspects, but its construction is heavier throughout and lower transmission and rear-axle ratios are used to cope with hilly terrain (Fig. 1).

A truck is rated by its gross vehicle weight (gvw), the combined weight of the vehicle and load. This weight ranges from about 4900 to 80,000 lb (2200 to 36,000 kg). Trucks are classified according to gvw as follows: less than 9000 lb (4100 kg), light; 9000–16,000 lb (4100–7300 kg), medium; 16,000–24,000 lb (7300–10,900 kg), light-heavy; above 24,000 lb (10,900 kg), heavy.

Fig. 1. Conventional truck chassis before installation of special body. (*White Motor Co.*)

By definition there are cab-forward-of-engine (CFE), cab-over-engine (COE), and cab-beside-engine types. The cab may be in fixed position or it may tilt forward in order to give access to the engine.

Other designations are four-wheel and six-wheel. The four-wheel type drives through the rear wheels only or through all four wheels. The six-wheel type has a tandem rear axle with the drive through one or both axles.

A truck-tractor is a vehicle of short wheelbase for hauling semitrailers. It carries a swiveling mount, known as the fifth wheel, above the rear axle to support the front end of the semitrailer. If two axles are used, the drive is through the rear axle, but types with tandem rear axles take the drive through one axle, with one trailing, or through both axles.

A semitrailer has one or two axles at the rear; the load is carried on these axles and on the fifth wheel of the tractor. The tractor-semitrailer combination permits the use of longer bodies with greater carrying capacity and better maneuverability than is possible with a conventional truck. Full trailers to be drawn behind semitrailers have a front axle and one or two rear axles. *See* TRUCK AXLE.

The forward positioning of the cab, the short wheelbase of the tractor, and the multiplicity of axles reflect engineering effort to get maximum payloads and operating economy in the face of restriction on overall length imposed by some states, and regulations limiting the weight carried on a single axle.

European manufacturers produce a special class of 1-ton payload trucks which have the engine mounted at the rear with rear-wheel drive, mounted ahead of the front axle with front-wheel drive, or placed under the driver's seat with front-wheel drive (Fig. 2).

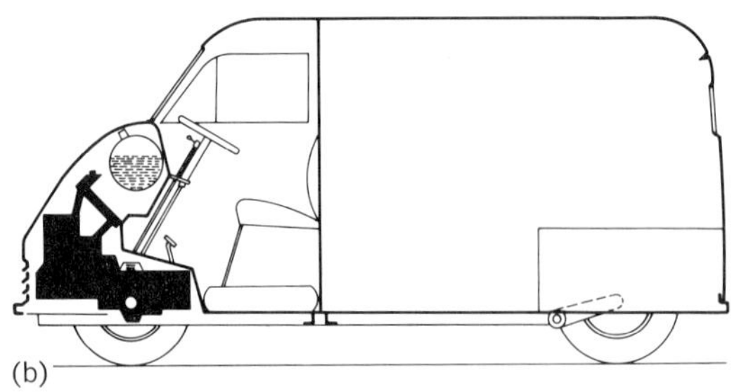

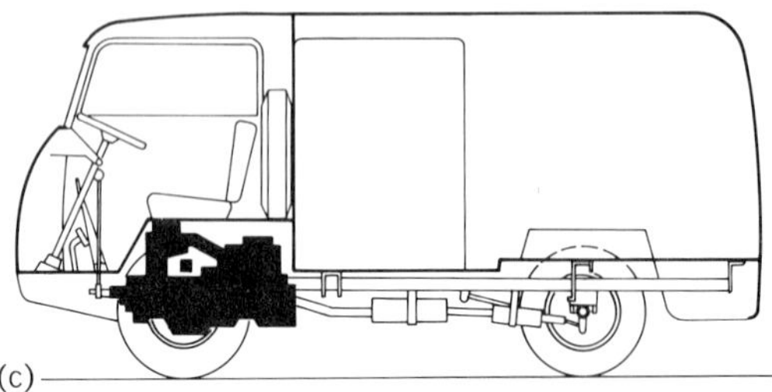

Fig. 2. Usual engine locations arranged for low loading platforms. (*a*) Engine at rear with rear drive. (*b*) Engine in front with front-wheel drive. (*c*) Engine under driver's seat with front-wheel drive.

Engines are in-line, V-type, or pancake and have 4, 6, 8, or 12 cylinders. They may operate on gasoline, LP (liquid petroleum) gas, or diesel fuel. The diesel engines operate on a two-stroke or four-stroke cycle and are water- or air-cooled. Brake horsepower ranges from about 47 at 2800 rpm for a light truck to 356 at 2200 rpm for the heaviest vehicle. Supercharging may be used to develop more horsepower from an engine of given size. *See* DIESEL ENGINE; INTERNAL COMBUSTION ENGINE.

Transmissions have 3, 4, 5, or 7 forward speeds and 1 or 2 reverse speeds. Overdrives are sometimes used. By employing a 5-speed transmission in combination with a 2-speed auxiliary transmission, 10 forward speeds are provided. If a truck has a single axle with two speeds, providing two gear ratios within the axle, as many as 12 gear ratios are afforded. *See* TRUCK TRANSMISSION.

When there is more than one driving axle, an additional gear ratio is provided when needed by an auxiliary transmission known as a transfer case. This is mounted behind the transmission with an output shaft for each of the driven axles.

Two types of semiautomatic transmission are used, a straight mechanical and a conventional transmission with hydraulic torque converter. The semiautomatic leaves the driver free to select any gear ratio he chooses and the power shifting spares him the labor of gear changing. Declutching devices, which permit all forward shifts to be made without depressing the clutch pedal, also help the driver. With these the clutch pedal is used only for standing starts.

Alloy steels and aluminum are used in frames. Cabs and bodies are framed with steel or aluminum and enclosed with sheet steel, aluminum, or fiber-glass reinforced plastics.

Leaf springs are widely used in suspensions. Front wheels may or may not be individually sprung. Air springs are employed to some degree when constant frame height and axle articulation, regardless of load, are important considerations. *See* AUTOMOTIVE VEHICLE; BULK-HANDLING MACHINES.

[PHILIP H. SMITH]

Truck axle

A supporting member carrying the weight of a truck and its payload, and mounting at either end the wheels on which the truck rolls. Drive axles transmit power from an input shaft to the wheels, forcing them to rotate; nondriving axles, often referred to as dead axles, do not power the wheels, merely allowing them to rotate freely. A steering function may be provided on either type by including means to pivot the wheel end portion. Both types may include service, parking, and emergency brakes.

Power is transmitted from an input shaft at the center of the axle to a primary right-angle gear reduction, then to a differential mechanism integral with the gear reduction and through the axle shaft to the wheels. Axles for off-highway trucks usually include additional reduction obtained with a planetary gear set at each wheel.

Axles for industrial-type lift trucks employ spur gear reduction sets at the wheel end in order to offset wheel and axle centerlines. For very large mining trucks, power transmission often includes electric motor and planetary gear reduction in the wheel design.

Highway truck front axles (Fig. 1) are typically nondriving steering type, using a forged steel I-beam cross section between pivot centers. Front axles for off-highway trucks are often of drive steer type.

Construction. Housings for highway trucks (Fig. 2) are of lightest possible design, generally of forged steel with integral spindle. Maximum load ratings of 18,000 and 23,000 lb (8200 and 10,400 kg) are common, consistent with legal maximum single-axle loads in many states. Axles for 1/2- and 3/4-ton (1 ton = 0.9 metric ton) trucks are often of semifloating design, the wheel being supported by the housing rather than the axle shaft as in the former design. Off-highway axles have housings of cast steel or ductile iron and forged steel spindles attached with bolts (Fig. 3). Housings incorporate or provide for spring seats or suspension brackets, torque arm, brake flange, power steer cylinder attachment, and so forth, as required in specific use. *See* AXLE.

Gear reduction ratios. Axle gear reduction ratios are carefully selected to provide a balance between speed and tractive effort requirements in a given truck application. The primary reduction in general use is either a hypoid or spiral bevel gear set. Many axles include a second gear reduction integral with the primary gear reduction unit, thus increasing reduction ratio. Ratios used for highway axles range from 3:1 to 10:1. Axles with planetary gear reduction at wheel end commonly have total ratios between 8:1 and 22:1 with very large trucks or, in unusual applications, even higher.

Double-reduction-type axles have spur, helical, or planetary gear sets in the secondary reduction. Two-speed axles include either an alternate second reduction set, or power flow path if of planetary design, thus permitting operation in two speed ranges. Two-axle tandem or three-axle tridem arrangements are powered by extending the input shaft through the first driving axle and coupling to the second, or in like fashion from second to third driving axles. In one design a three-speed tandem axle is achieved by operating one axle in high range and the other in low, the two units being coupled with an interaxle differential such that an intermediate ratio is obtained.

Differentials. Differential action, which allows one wheel to rotate faster than the other in a turn,

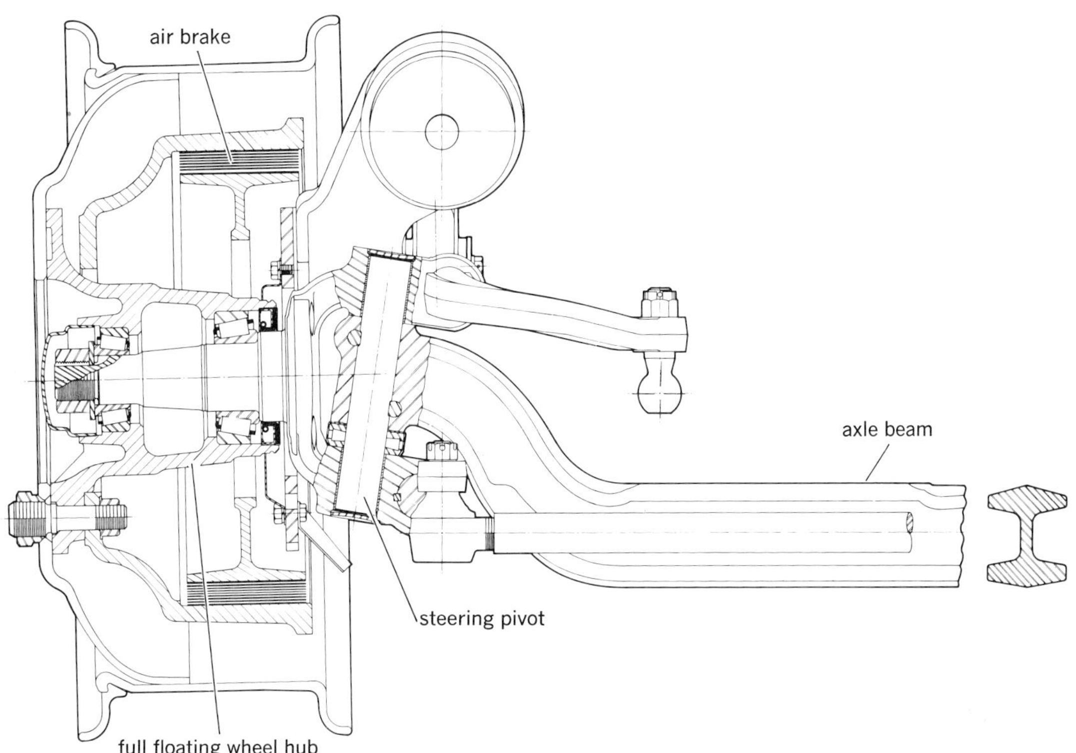

Fig. 1. Typical truck front nondrive steer axle.

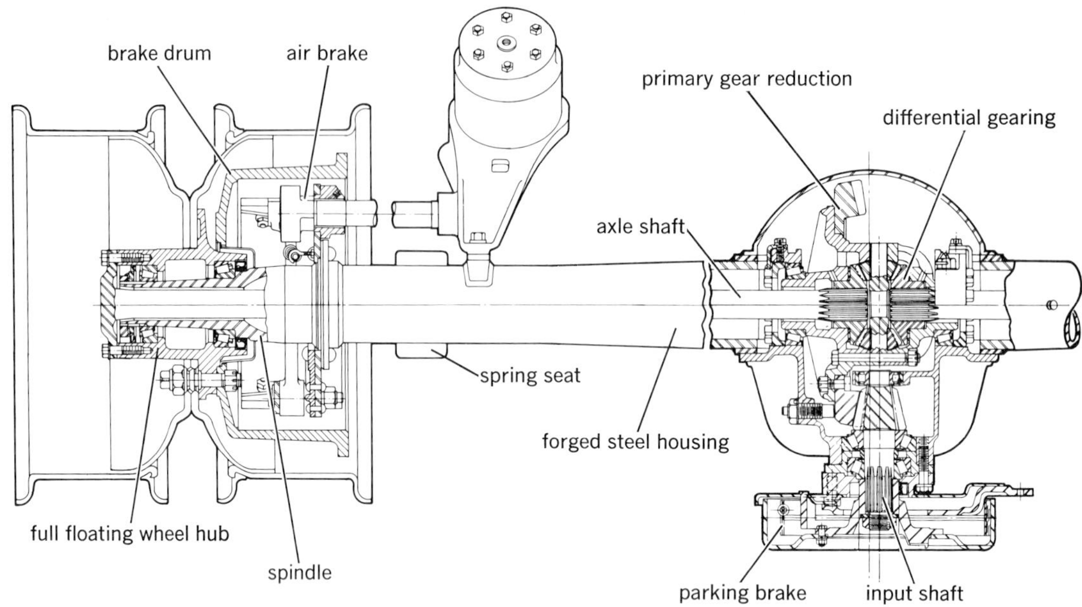

Fig. 2. Typical highway truck rear axle.

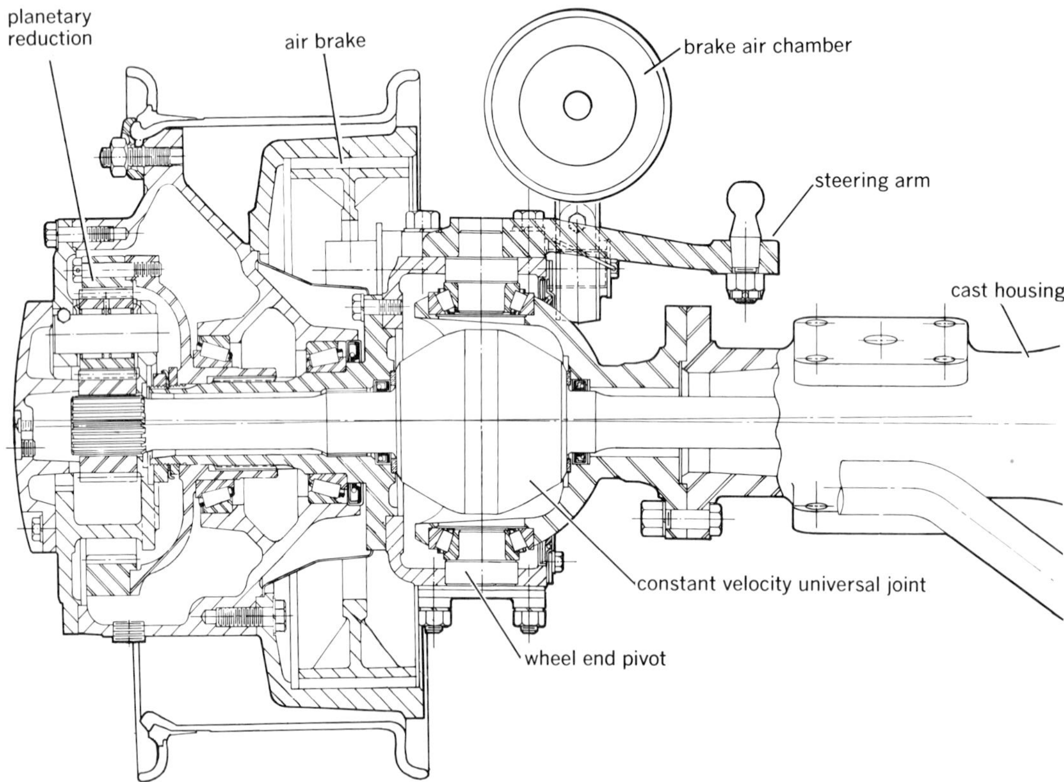

Fig. 3. Typical off-highway planetary drive-steer axle.

is essential for steering control and tire life. Standard types divide torque equally to each wheel and may in poor traction conditions render a vehicle with single-axle drive immobile. There are special features in many differentials which help to overcome this problem.

Limited slip types, also known as torque bias or torque proportioning differentials, enable a greater torque to be absorbed by one wheel than by the other. This is accomplished in some types by using slip-clutch devices to restrain relative motion of the two axle shafts and in others by introducing a deliberate inefficiency in the differential gearing. Another common device substitutes the usual differential gearing with an overrunning jaw-clutch design in which either wheel may overrun in either

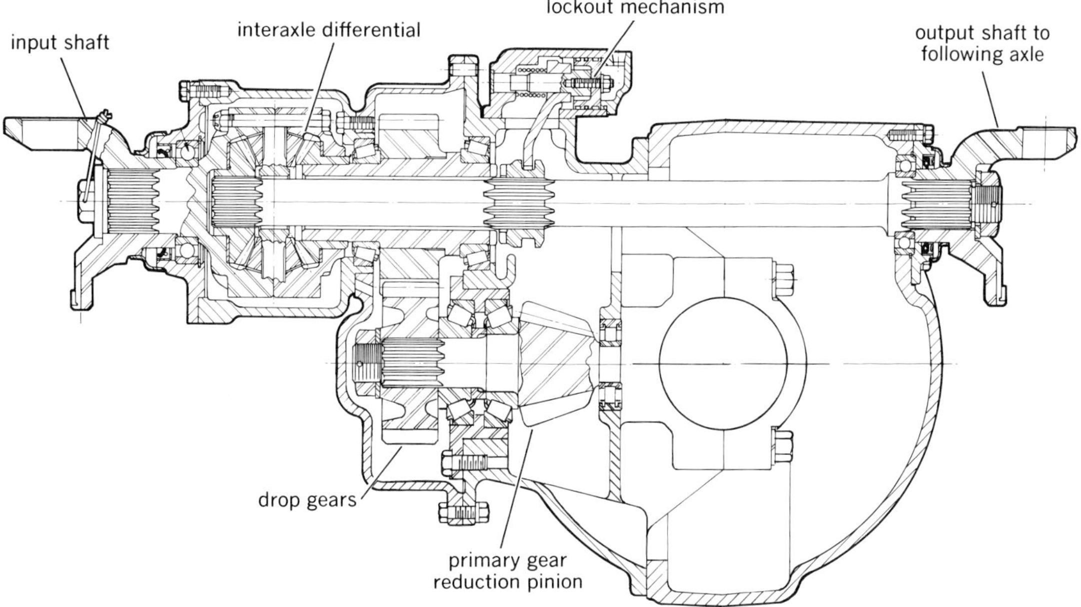

Fig. 4. Differential drive-through arrangement, showing interaxle differential.

direction while total available torque is delivered to the other wheel. A lock-up feature on some differentials enables the operator to eliminate differential action when necessary to obtain required traction and to operate normally in other conditions.

An interaxle differential (Fig. 4) is used in tandem or tridem arrangements to avoid internal torque buildup that could result from operation at slightly differing axle speeds, such as is obtained with unmatched tires. Trucks used in off-highway service may operate satisfactorily without an interaxle differential, especially if in poor traction conditions.

Brakes, suspensions, and lubricants. Service brakes, and commonly emergency or parking brakes, are integral with or mounted on the axle. Air systems prevail because of their adaptability for tractor-trailer systems and their ability to operate at the high temperatures generated in the brake area during severe service.

Suspensions mount the axle to the truck frame. They include leaf-type springs, air or rubber cushions to reduce shock and vibration transmission to the truck, and on tandem or tridem arrangements the necessary beams and torque arms to distribute load to each axle.

Axle lubricants are universally SAE-90 mineral oil modified with extreme-pressure additives because of the high specific loading and sliding tooth contact inherent in hypoid gear sets. *See* TRUCK; TRUCK TRANSMISSION.

[R. H. BOLSTER/W. E. GREEN]

Bibliography: Society for Automotive Engineers, *SAE Handbook*, revised annually.

Truck transmission

A gear unit or a combination of gear units for providing a plurality of gear ratio changes between the engine and rear axle of a commercial automotive vehicle. The primary purpose of a truck transmission system is to permit the vehicle to be operated through a wide range of speed and power requirements while the engine is operated in the most advantageous range for the given conditions.

Functional types. The requirements of a truck transmission system are many and varied and, so there are numerous different transmission arrangements. Transmissions are identified according to their primary function in the vehicle. Types include the main or unit power transmission, which is normally the first transmission behind the engine; the auxiliary transmission, which is a separate unit behind the main transmission to provide additional gear ratios; a transfer case, used behind the transmission to redirect engine power to the front axle for all-wheel drive; and power takeoff transmissions, which are small auxiliary units usually attached to one of the above types to provide power for auxiliary equipment such as pumps and winches.

Main transmissions. Most trucks use mechanical manual-shift transmissions having 3–20 forward speeds and 1–4 reverse speeds.

In a limited number of applications automatic transmissions are used; when these are employed they are torque-converter-type automatics with 4–6 forward speeds behind the converter. The automatics in use are of a planetary-gear design similar to automobile torque-converter transmissions. *See* AUTOMOTIVE TRANSMISSION.

The majority of main transmissions are of the countershaft design (Fig. 1). In their simplest form they have 3–5 speeds forward and usually one speed reverse. The low-gear ratio is normally from 6:1 to 8:1, and the balance of the gear ratios are set to match either the engine operating conditions or other supplemental gearing such as auxiliary transmissions, transfer cases, or multiple speed axles. Synchronizing clutches, universally used in passenger car transmissions, are widely used in light- and medium-duty unit power truck transmis-

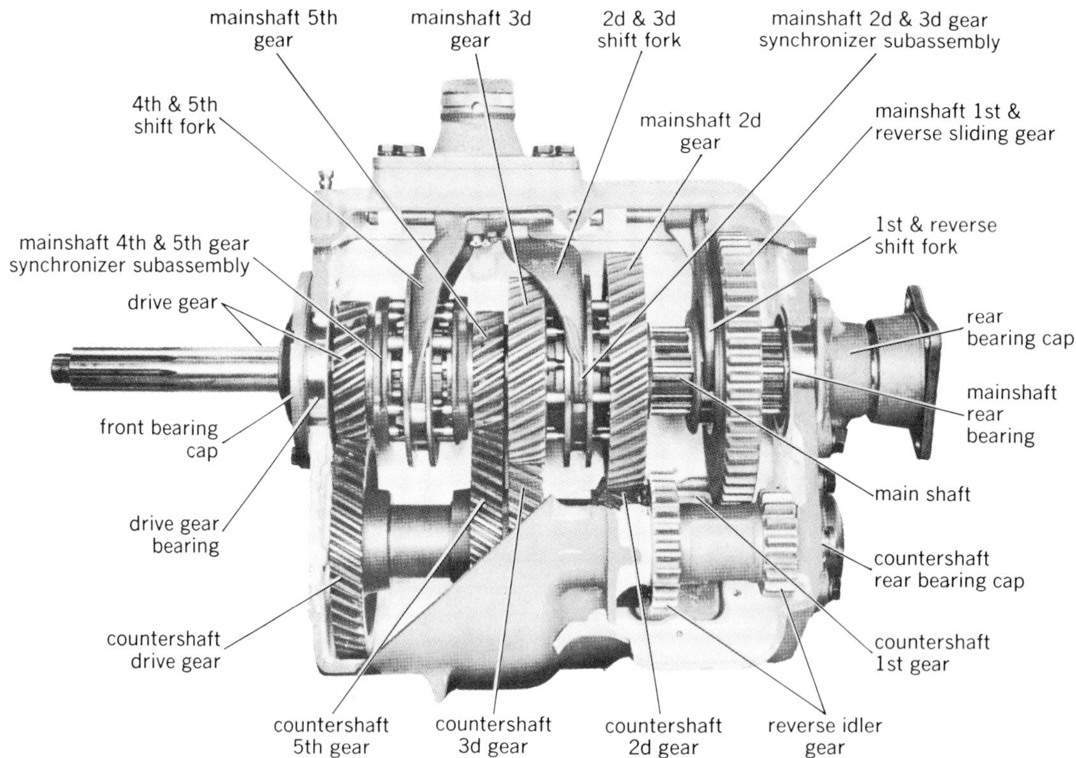

Fig. 1. Five-speed unit power transmission, in neutral position. (*Dana Corp.*)

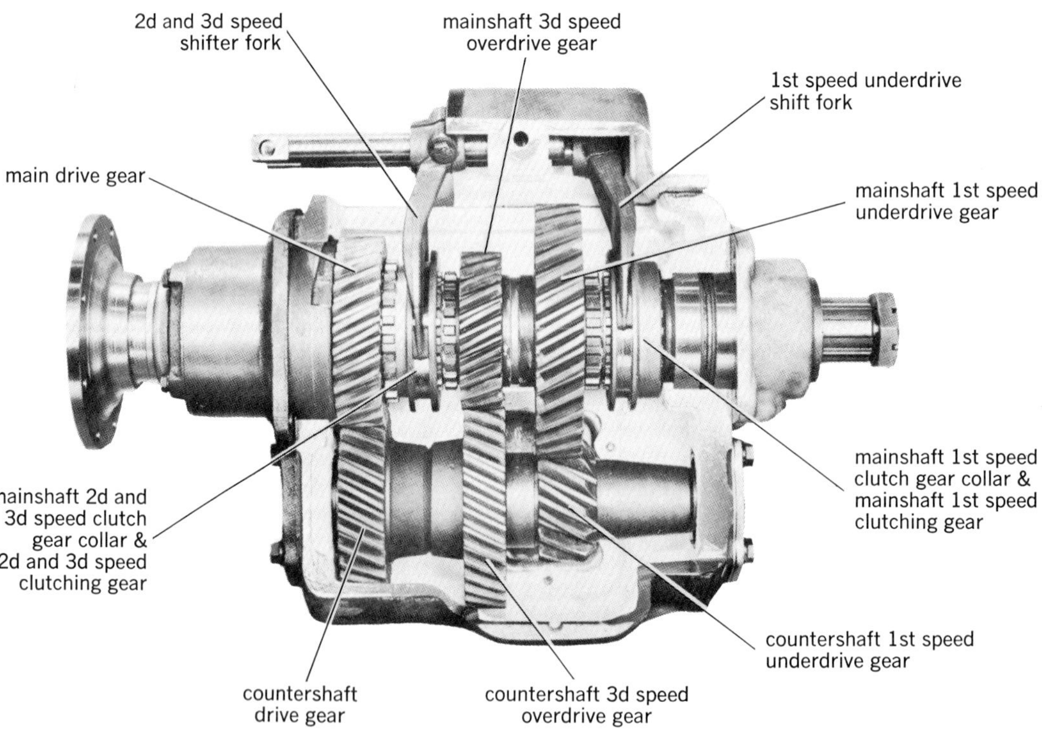

Fig. 2. Three-speed auxiliary transmission. (*Dana Corp.*)

sions but are rarely used in heavy-duty main transmissions or auxiliaries.

Auxiliary transmissions. Auxiliary transmissions are usually of the countershaft design (Fig. 2). They are common in both three- and four-speed units.

Main transmission design. In truck designs it is important to keep the power unit and drive train as short and light as possible so that maximum space and weight can be reserved for the payload. This requirement has led to the development of several unique transmission designs.

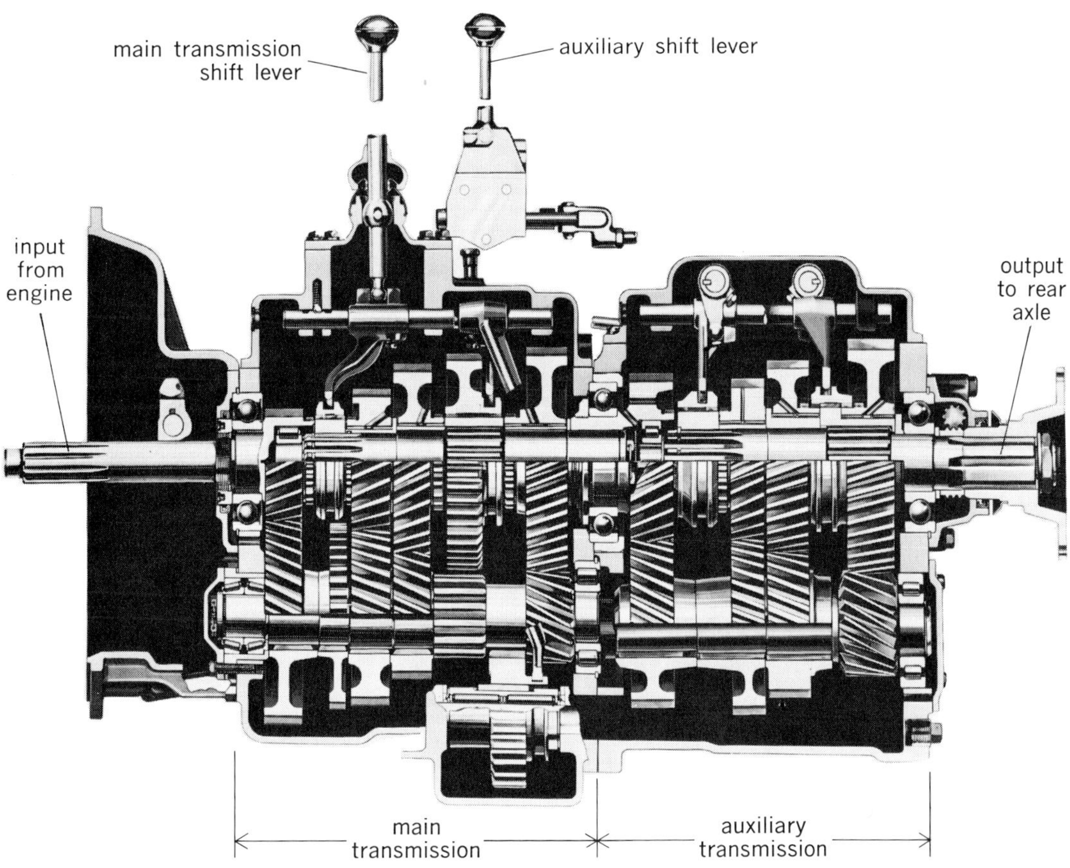

Fig. 3. Compound transmission provides 4 ratios in main unit and auxiliary portion for 16 speed ratios. (*Dana Corp.*)

One method of shortening the transmission combination is the compound transmission (Fig. 3). This unit consists of a four-speed main transmission and a four-speed auxiliary transmission integrally mounted together but functioning in exactly the same manner as separate units. Transmissions of the compound design are available with a four-speed main and two-speed auxiliary (4 × 2, or 8-speed) as well as 4 × 3, or 12-speed; 5 × 2, or 10-speed; 5 × 3, or 15-speed; and 5 × 4, or 20-speed. They are frequently shifted by means of two shift levers, one controlling the main transmission and the other controlling the auxiliary section; in some cases, the auxiliary section can be shifted by a driver-controlled air-shifting means.

When the auxiliary section of a compound transmission has two speeds it is usually operated as a range unit with ratios of 1:1 and approximately 3.5:1. When used with a typical five-speed main transmission having forward-gear ratios of 1.00, 1.28, 1.64, 2.1, and 2.69, the 5 × 2 compound combination provides 10 forward ratios, as shown in Table 1.

When the auxiliary section of a compound transmission has three or four speeds as in a 4 × 3 or 5 × 4, it is usually used as a splitter unit. Typical ratios of the main and auxiliary of a 4 × 3 compound are as shown in Table 2. Twelve forward-speed ratios are available with this combination.

A four-speed auxiliary section usually has three splitter ratios and a low-gear ratio of about 2.5:1, which is used only with first gear in the main transmission to provide a creeper gear.

Numerous other gear combinations are made available by compounding transmissions, but the above are typical of present arrangements.

As a further refinement to reduce the length and weight of truck transmissions, the multiple countershaft transmission design is being employed for

Table 1. Speed ratios available from a typical 2-speed auxiliary in combination with a typical 5-speed main transmission

Speed	Main ratio	Auxiliary ratio	Overall ratio
10	1.00	1.00	1.00
9	1.28	1.00	1.28
8	1.64	1.00	1.64
7	2.10	1.00	2.10
6	2.69	1.00	2.69
5	1.00	3.5	3.50
4	1.28	3.5	4.48
3	1.64	3.5	5.74
2	2.10	3.5	7.35
1	2.69	3.5	9.41

Table 2. Typical ratios of a 4-speed main and 3-speed auxiliary transmission giving overall ratios from 0.81 to 7.72 in 12 steps

Main transmission		Auxiliary transmission	
4th	1.00	Overdrive	0.81
3d	1.85	Direct	1.00
2d	3.42	Underdrive	1.22
1st	6.33		

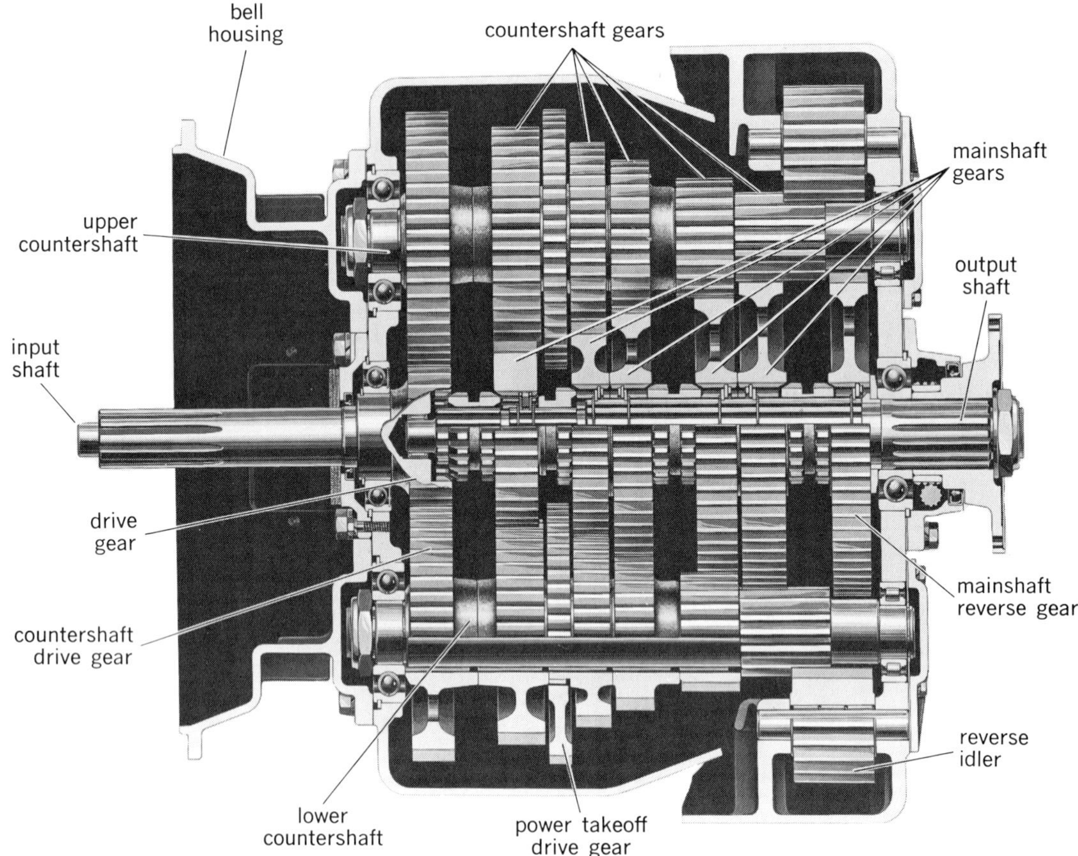

Fig. 4. Two countershafts, by dividing load, permit reduction in total size of six-speed transmission. (*Dana Corp.*)

larger transmissions (Fig. 4). This design incorporates two countershafts spaced 180° apart on opposite sides of the mainshaft. Operation and function of the transmission is identical to that of a single countershaft unit but, because the input torque is divided equally between the two countershafts, the gears, shafts, and bearings can all be made smaller. This design provides high-torque carrying ability with a minimum of length and weight.

In transmissions of this multiple countershaft design, the mainshaft gears are not supported by the mainshaft but float between the two countershaft gears. The amount of float is controlled by careful indexing of the countershaft gears on the countershaft. Some transmissions of this basic design incorporate three countershafts equally spaced around the mainshaft, thus saving even more length but at the cost of added weight.

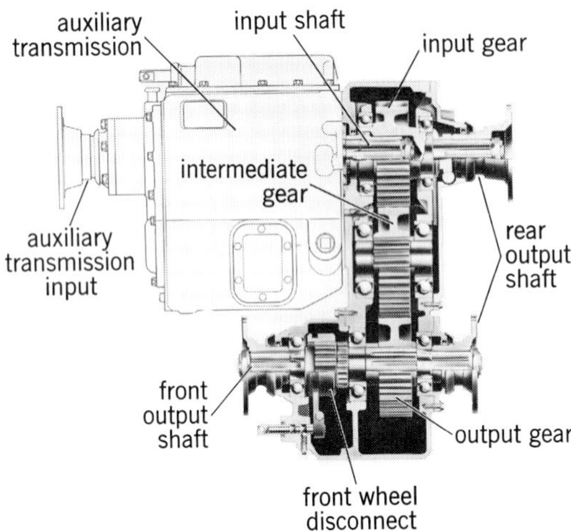

Fig. 5. Typical single-speed transfer case with 1:1 ratio from upper shaft to lower shaft. (*Dana Corp.*)

Transfer case. Figure 5 shows a typical transfer case for which the input is supplied through a propeller shaft or directly from the mainshaft of a transmission. Power is then transmitted through a gear train to some other position in the vehicle. Either the upper or lower rear-output shaft can be used for the universal joint drive to the rear axle. The propeller shaft drive for the front axle always comes from the lower shaft. The front-wheel-drive output shaft is normally equipped with a disconnecting member so that the vehicle can be operated in conventional rear-drive only when front-wheel drive is not required.

Some transfer cases have two different gear ratios and these cases function as a combination two-speed auxiliary and transfer case. Transfer cases are usually used in military vehicles and other trucks expected to engage in considerable off-highway operation.

Power takeoff. Many trucks need auxiliary power for other than the main drive train of the vehicle. This auxiliary power, which is utilized to drive truck-mounted equipment, is normally supplied from a power takeoff. Most transmissions have openings provided in the side of the case for mounting a power takeoff. This opening is posi-

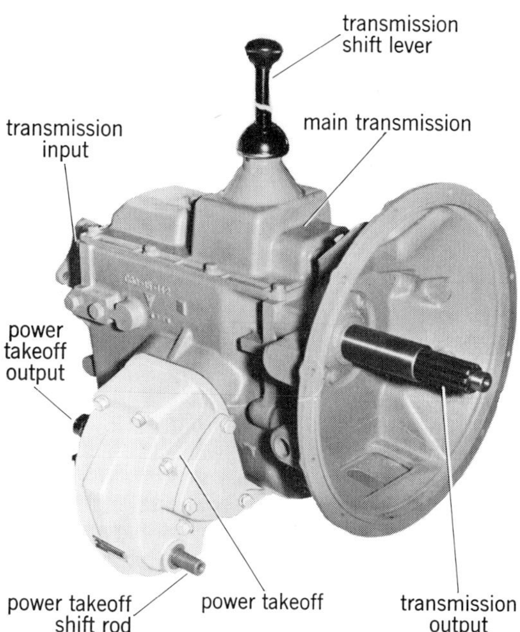

Fig. 6. External view of main transmission with its side-mounted power takeoff. (*Dana Corp.*)

tioned on the case so that when a power takeoff is installed, the input gear of the power takeoff will engage one of the gears in the transmission. Power takeoffs are normally engaged with a countershaft gear so that the transmission can be operated in neutral with the vehicle stationary and still provide power to the power takeoff unit.

Power takeoffs, available in a variety of speeds and normally used in the range of 50 hp or less, are mounted to the sides of transmissions (Fig. 6). Full engine power takeoffs, available for mounting on auxiliary transmissions, normally incorporate one or two forward speeds and one reverse speed and are used for power requirements of up to 250 hp.

[R. E. FLETCHER]

Bibliography: S. L. Abbott, *Automotive Transmissions*, 1980; W. H. Crouse and D. L. Anglin, *Automotive Transmissions and Power Trains*, 5th ed., 1976; Society of Automotive Engineers, *SAE Handbook*, revised annually.

Truss

A system of structural members lying in a single plane and joined at their ends to form a stable framework. A truss is used like a beam, particularly for bridge and roof construction. But because a truss can be made deeper than a beam with solid web and yet not weigh more, it is more economical for long spans and heavy loads. *See* BRIDGE; ROOF CONSTRUCTION.

The simplest truss is a triangle composed of three bars with ends pinned together. If small changes in the lengths of the bars are neglected, the relative positions of the joints do not change when loads are applied in the plane of the triangle at the apexes.

Types. Such simple trusses as a triangle, perhaps with the addition of a vertical bar in the middle, are sometimes used to support peaked roofs of houses and other narrow structures. For longer spans, flat roofs, or bridges many triangles are combined to form a truss, as can be seen in the illustrations.

In metal trusses, connections may be riveted, bolted, welded, or pinned; in wood trusses, they may be bolted, nailed, or glued. Because of long spans, provision must be made to permit movement at one support due to loads and temperature changes; rollers, rockers, or sliding plates generally are used for this purpose.

The top members of a truss are called the upper chord; the bottom members, the lower chord; and the verticals and diagonals, the web members.

Framing to be carried by a truss usually is arranged so that it brings loads to bear on the truss at the intersections of a chord and web members. As a result, truss members are subjected only to direct stress—tension or compression—and can be made of less material than if they also had to resist bending stresses. (In the illustrations members in tension are drawn in light lines, members in compression in heavy lines.)

Roof trusses carry the weight of roof deck and framing and wind loads on the upper chord. They may also support a ceiling or other loads on the lower chord. An example is the Fink truss (Fig. 1). On the other hand, bridge trusses may carry loads on either chord. Deck trusses support loads on the upper chord (Fig. 2); through trusses, on the lower chord (Fig. 3).

To maintain stability of truss construction, bracing must be used normal to the planes of the trusses. Usually framing is inserted between the trusses. For roofs, trussed bracing should be placed in the plane of either the top chord or the bottom chord. For bridges, bracing must be inserted in the planes of both top and bottom chords, because of the greater need for stability under heavy moving loads.

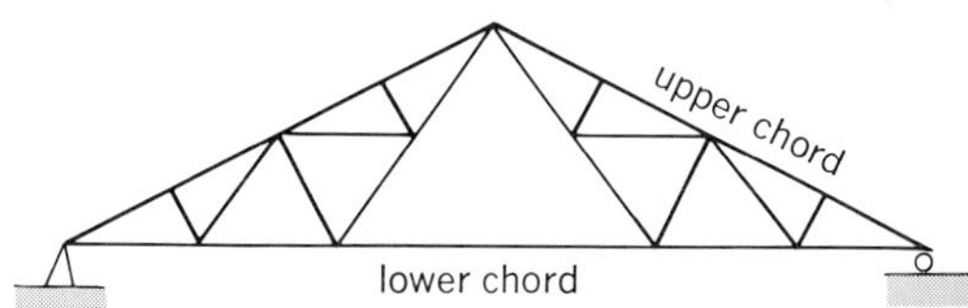

Fig. 1. A Fink truss, used in roof construction.

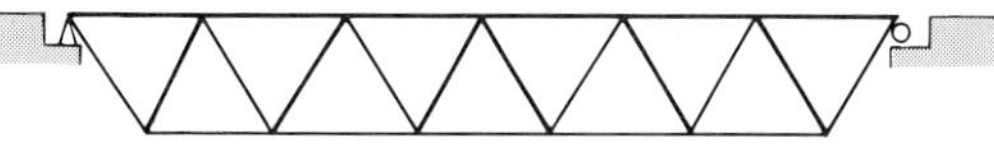

Fig. 2. A deck Warren truss, used in bridge construction. The load is on the upper chord.

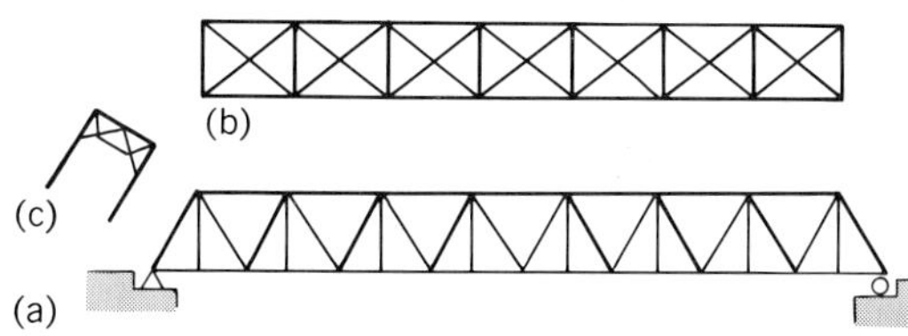

Fig. 3. Loads in through trusses are borne on the lower chord. (*a*) A through Warren truss with verticals. (*b*) Top-chord bracing as seen from above. (*c*) End-on view showing portal bracing.

Computing stresses. Primary stresses in truss members are computed on the assumption that connections at joints are made with frictionless pins. With loads applied at joints, each truss member or bar is subjected to pure tension or compression.

Since the bars change length under load, the angles of each triangle constituting the truss tend to change. But this change is resisted, since pins are not frictionless, and since rivets, bolts, or welds offer restraint. Consequently, members bend slightly, the bending moments creating secondary stresses.

At a truss joint, the primary stresses and loads form a coplanar, concurrent force system in equilibrium. This system satisfies two conditions: The sums of the horizontal and vertical components both equal zero. These equations are used in computing stresses by the "method of joints." In this method, joints with two unknowns are selected in succession and the two equilibrium equations are applied to them to determine the stresses.

A section may be passed through the truss to cut three bars with unknown stresses. These, together with bars with known stresses that are cut and the loads on the part of the structure on either side of the section, constitute a coplanar, noncurrent force system in equilibrium. This system satisfies the two previous conditions; but in addition, the sum of the moments of the forces about any axis normal to the plane equals zero. With these three equations the three unknowns can be determined. However, the unknowns also can be found by the "method of moments," in which two of the unknowns are eliminated by taking the moment axis at their point of intersection, and the third is found by equating the sum of the moments to zero. The "method of shears" is used to determine one force when the other two unknown forces are both normal to the shearing force, for example, for finding the stresses in the diagonals of parallel-chord Warren, Pratt, and Howe trusses.

If n is the number of joints in a truss, stresses can be found by the methods of joints, moments, or shears when the number of bars equals $2n-3$. If a truss is composed of fewer bars, it is unstable; if of more bars, statically indeterminate.

Influence lines are useful in determining the stresses in bridge trusses, because the live load is a moving load. Influence lines can be drawn to show the variation in any function—stress, shear, moment, deflection—as a unit load moves along the truss. *See* STRUCTURAL ANALYSIS.

[HARRY L. BOWMAN/WALDO G. BOWMAN]

Tumbling mill

A grinding and pulverizing machine consisting of a shell or drum rotating on a horizontal axis. The material to be reduced in size is fed into one end of the mill. The mill is also charged with grinding material such as iron balls. As the mill rotates, the material and grinding balls tumble against each other, the material being broken chiefly by attrition.

Tumbling mills are variously classified as pebble, ball, or rod depending on the grinding material, and as cylindrical, conical, or tube depending on the shell shape. *See* CRUSHING AND PULVERIZING; GRINDING MILL; PEBBLE MILL.

[RALPH M. HARDGROVE]

Tunnel

A general term for subterranean passages. Tunnels are used for aqueducts and sewers; for carrying railroad and vehicular traffic under rivers, through mountains, and below cities; and for specific kinds of underground installations, such as hydroelectric plants.

The construction of a tunnel begins with excavation of earth or rock. Excavating methods vary with the character of the material to be removed. Rock tunnels require blasting. If the rock is structurally poor, supports must often be placed under the tunnel ceiling, or arch, as the tunnel is being driven. Tunnels driven through earth do not require blasting, but arch supports are almost always necessary. Tunnels under rivers in soft silts are usually driven with the aid of a short cylinder called a shield, which is pushed through the silt ahead of the excavation to provide advance support of the arch. After excavation has been completed, the exposed tunnel wall is usually lined with concrete. Vehicular tunnel construction often requires the added installation of tile walls and of lighting and ventilation equipment.

Excavation. This is the most costly and hazardous operation in tunneling. Because the nature of the material through which the tunnel is to be driven determines driving methods, the need for supports, and therefore the costs, it is standard procedure to make a thorough geological survey of the subterranean strata to be encountered. This consists of a study of structural characteristics, mineral composition, and hazards. Structural characteristics, such as the frequency and direction of rock joints and planes, determine the need for and type of arch supports. The mineral composition is significant in forecasting behavior of the tunnel faces after excavation. For example, anhydrite when exposed to water during tunneling changes into gypsum with a resulting swelling of the rock and deformation of the tunnel. Hazards such as faults indicate the presence of crushed rock fragments or water channels. The former situation requires extensive tunnel supports; the latter, pumping and possible advance grouting. For a discussion of grouting *see* CONSTRUCTION METHODS.

The required geological information is obtained from a review of geological literature and maps of the area and from field explorations. Seismic and electrical soundings are taken to establish the depth of rock from the surface. Drilled core holes provide samples of the material to be encountered during driving of the tunnel. Topographic surveys made on the ground or from the air are sometimes used to forecast the underground rock contours and character.

Such studies, while invaluable in selecting a tunnel route bypassing certain localized hazards, can forecast exactly neither all such hazards nor the pressures to be developed on the tunnel face during excavation. Exact information can be obtained only by driving small pilot tunnels at the proposed site.

Excavation of tunnels through rock proceeds by repeating the following cycle of operations for the length of the tunnel: drilling blast holes, loading the holes with explosives, setting off the explosives, exhausting the blast fumes, and mucking, or removing blasted rock.

Fig. 1. Drill jumbo is used for drilling blast holes. (*Gardner-Denver Co.*)

Drilling. Blast holes are drilled by compressed-air-operated rock drills mounted on a carriage called a jumbo (Fig. 1). This carriage may consist of several platforms supported on a tractor, truck, or track-riding gantry.

Each drill makes several holes which, together with those made by other drills, form a blast hole pattern. The blast hole pattern consists of an inner group, called cut holes, an intermediate group, called relief holes, and, finally, an outer row called trim holes. Delay caps are placed so that the explosives placed in the cut holes will detonate first, followed by the relief holes, and finally the trim holes. This detonating sequence permits the most economical blasting charge. *See* BLASTING.

Ventilation equipment is required to provide fresh air for the crews at the tunnel face and to eliminate blast fumes. Blowers, set up at the entrance of the tunnel, force fresh air to the tunnel face through large-diameter ducts. These are temporarily reversed to exhaust the blast fumes.

Mucking. The rock or earth is removed by special conveyor loaders or power shovels equipped with short booms. Rock is hauled from the face by means of special trucks or mine cars.

Roof supports. Means must often be provided to support loose rock at the roof. Long wedging bolts are used to anchor large blocks of rock to sound rock. Arch supports for scaly or heavily fragmented rock consist of steel or timber ribs with steel or timber members, called lagging, which span the area between the ribs.

The design of arch supports requires an estimate of the vertical and lateral pressures expected on the support. These vary with the material encountered. Because a natural arch forms above the excavated tunnel, the vertical load never is as great as the depth of rock above it. In moderately jointed rock, the load is equal to a height of rock which is about one-half the width of the tunnel. In squeezing or swelling rock the load may be many times greater. In such extreme cases, the tunnel cross section is made circular and circular ribs are used for wall support.

To save time, it is preferable to erect arch supports at the same time as drilling blastholes for a new round or advance. However, when the exposed roof cannot support itself until drilling has been started for a new round, supports must be set as soon as blast fumes are exhausted. If the rock cannot support itself even that long, a small portion of the face must be excavated at a time to permit more rapid setting of supports.

Tunnel-driving methods. The amount of face that can be taken out at one time determines which of several tunnel-driving methods to select. These include the full-face, the top-heading, the heading and bench, and the drift methods (Fig. 2).

With the full-face method, the most common

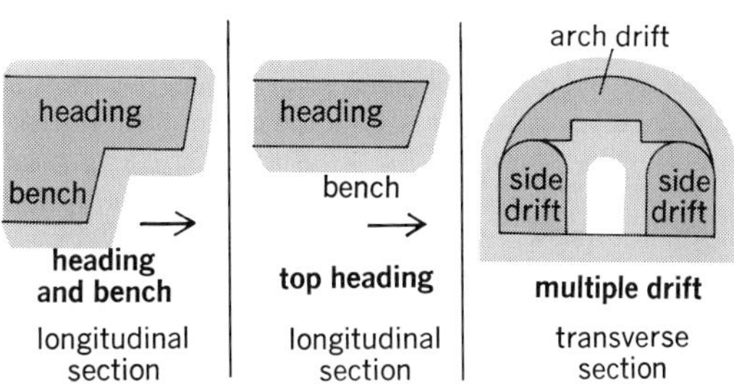

Fig. 2. Tunnel-driving methods.

Fig. 3. Steel rib and timber lagging supports. (*Bethlehem Steel Co.*)

and fastest, the tunnel is blasted out full size at each round of blasting. The required conditions for full-face operations are a rock type and tunnel section that permit self-support until after mucking has been completed. In the top-heading method the tunnel portion just below the arch, the heading, is driven full length first, after which the bottom portion, or bench, is removed separately. In the heading and bench method, the top portion is also driven first, but only a round ahead of the bench. This method is often preferred to the top-heading method, because there is less time for the roof loads to build up. The drift method is used where the rock is so poor that the face must be attacked in more than two steps. There are two common versions of the drift method: the side drift and the multiple drift. In the side-drift version, two small tunnels are driven at the side of the projected tunnel. Steel or timber posts or concrete abutments for the arch supports are then installed, after which the balance of the tunnel face is opened up. The multiple drift consists of excavating the arch drift after the side drifts, setting arch supports, and then removing the remaining core of rock.

Some tunnels in soft but sound rock, such as limestone or shale, can be excavated by means of special boring machines, which bore out a full cross section of the tunnel. This eliminates the standard drilling and blasting methods described.

Driving of tunnels through soft (nonrock) ground, as in vehicular tunnels under rivers, requires a completely different approach. The certainty of almost immediate arch collapse requires the installation of continuous circular-ring supports which abut each other for the full length of the tunnel (Fig. 3). The circular supports are installed under the cover of a tunnel shield. This is a steel cylinder slightly larger in diameter than the ring supports. The shield is moved ahead by jacking it against the previously placed ring supports. When a full jack stroke has been made, the jack piston is retracted and new rings are added. The material ahead of the shield is either pushed ahead or squeezed into the inside of the tunnel, where it is removed by special digging tools.

When the material being driven through is of fluid consistency, the inside of the tunnel must be kept under air pressure to balance the fluid pressure and prevent an uncontrolled flow of material into the tunnel. Under such conditions, a bulkhead with a door is placed just inside the shield edge to control the amount of fluid material squeezing into the tunnel. Tunnels driven under air pressure rarely exceed 110 ft in depth, because workmen cannot work under greater pressures. Men and materials must enter and leave the tunnel through air locks.

The high cost of constructing an underwater tunnel under air pressure has led to the use of the floated tunnel or trench construction method. Under this method, all work is done from the surface. A trench is first dredged under the river bed; then cylindrical tunnel sections, of either steel or reinforced concrete, are floated over the trench and sunk into place. The sections are joined together underwater by divers, covered over with fill, and then pumped out for completion of work inside the dry sections.

Tunnel linings. Concrete lining of a tunnel is required for several reasons. In water-supply tunnels, it provides better flow and therefore increases capacity. In vehicular tunnels, it is necessary for appearance and safety. A lining may also be required to help the arch ribs and roof bolts resist the vertical and lateral pressures that may develop as the rock adjusts itself to new load conditions caused by the driving of the tunnel. The surface of the concrete lining is formed by traveling steel forms 20 – 100 ft (6 – 30 m) in length. These forms may be either telescopic or nontelescopic.

In the telescopic system, which is standard for long tunnels, one set of forms is collapsed over a carriage which passes it under another set of forms in position. By providing enough sets of telescopic forms it is possible to place concrete continuously.

Nontelescopic forms must move as a unit. They are less expensive than telescopic forms, but concrete placing must be discontinued until the concrete gains sufficient strength to permit removal and reuse of the forms.

The concrete is placed between rock and form by compressed-air placers or piston-type pumps, which force the concrete through pipes to the top of the arch above the form. From there, the concrete flows down around the form. For long tunnels, the concrete is mixed near the form in the tunnel in a traveling mixer. Concrete materials are brought to the mixer by trucks or mine trains. After concreting is done, it is usually necessary to pump cement grout through small pipes left in the lining to seal any voids or gaps between rock and concrete.

[WILLIAM HERSHLEDER]

Turbine

A machine for generating rotary mechanical power from the energy in a stream of fluid. The energy, originally in the form of head or pressure energy, is converted to velocity energy by passing through a system of stationary and moving blades in the turbine. Changes in the magnitude and direction of the fluid velocity are made to cause tangential forces on the rotating blades, producing mechanical power via the turning rotor.

The fluids most commonly used in turbines are steam, hot air or combustion products, and water. Steam raised in fossil fuel-fired boilers or nuclear reactor systems is widely used in turbines for electrical power generation, ship propulsion, and mechanical drives. The combustion gas turbine has these applications in addition to important uses in aircraft propulsion. Water turbines are used for electrical power generation. Collectively,

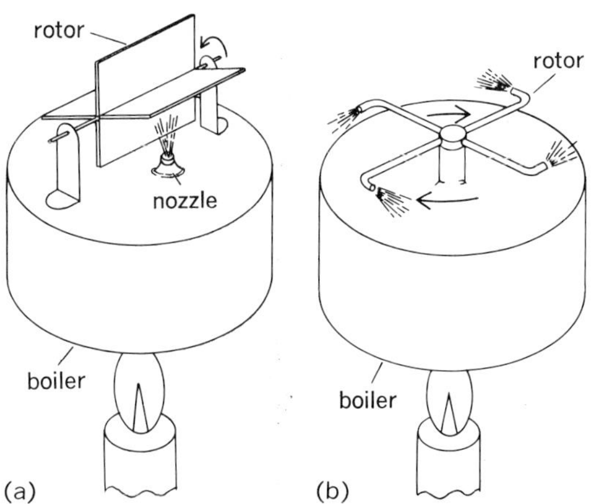

Turbine principles. (*a*) Impulse. (*b*) Reaction.

turbines drive over 95% of the electrical generating capacity in the world. *See* GAS TURBINE; HYDRAULIC TURBINE; STEAM TURBINE.

Turbines effect the conversion of fluid to mechanical energy through the principles of impulse, reaction, or a mixture of the two. Illustration *a* shows the impulse principle. High-pressure fluid at low velocity in the boiler is expanded through the stationary nozzle to low pressure and high velocity. The blades of the turning rotor reduce the velocity of the fluid jet at constant pressure, converting kinetic energy (velocity) to mechanical energy. *See* IMPULSE TURBINE; PELTON WHEEL.

The reaction principle is shown in illustration *b*. The nozzles are attached to the moving rotor. The acceleration of the fluid with respect to the nozzle causes a reaction force of opposite direction to be applied to the rotor. The combination of force and velocity in the rotor produces mechanical power. *See* REACTION TURBINE. [FREDERICK G. BAILY]

Turbodrill

A rotary tool used in drilling oil or gas wells in which the bit is rotated by a turbine motor inside the well. The principal difference between rotary and turbodrilling lies in the manner power is applied to the rotating bit or cutting tool. In the rotary method, the bit is attached to a drill pipe, which is rotated through power supplied on the surface. In the turbodrill method, power is generated at the bottom of the hole by means of a mud-operated turbine. *See* ROTARY TOOL DRILL.

The turbodrill (see illustration) consists of four basic components: the upper, or thrust, bearing; the turbine; the lower bearing; and the bit. Most turbodrills are about 30 ft (9 m) long, with shafts about 20 ft (6 m) long. The turbodrill is attached at its top to a drill collar, or heavy length of steel pipe, that makes up the bottom end of the drill pipe extending to the surface. Once the turbodrill passes below the well head, operations on the rig floor are the same as for rotary drilling. Rotation of the drill pipe is not necessary for turbodrilling, because rotation of the bit develops through the turbine on the lower end of the drill string. It is usual practice, however, to rotate the drill pipe above the turbine slowly, at 6–8 rpm, either by means of the rotary table on the derrick floor, or through torque of the turbine on the bottom. Rotation of the bit is much faster than in rotary drilling, and is usually between 500 rpm and 1000 rpm.

In operation, mud is pumped through the drill pipe, passing through the thrust bearing and into the turbine. In the turbine, stators attached to the body of the tool divert the mud flow onto rotors attached to the shaft. This causes the shaft, which is connected to the bit, to rotate. The mud passes through a hollow part of the shaft in the lower bearing and through the bit, as in rotary drilling, to remove cuttings, cool the bit, and perform the other functions of drilling fluid. Capacity of the mud pump, which is the power source, determines rotational speed.

Two basic types of turbodrills are used in the United States. One is a standard 100-stage unit (one rotor and one stator comprise a stage); the other is a tandem turbodrill, made up of two or three standard sections. Although turbodrills have been in wide use in the Soviet Union for several years, they are still relatively rare in the United

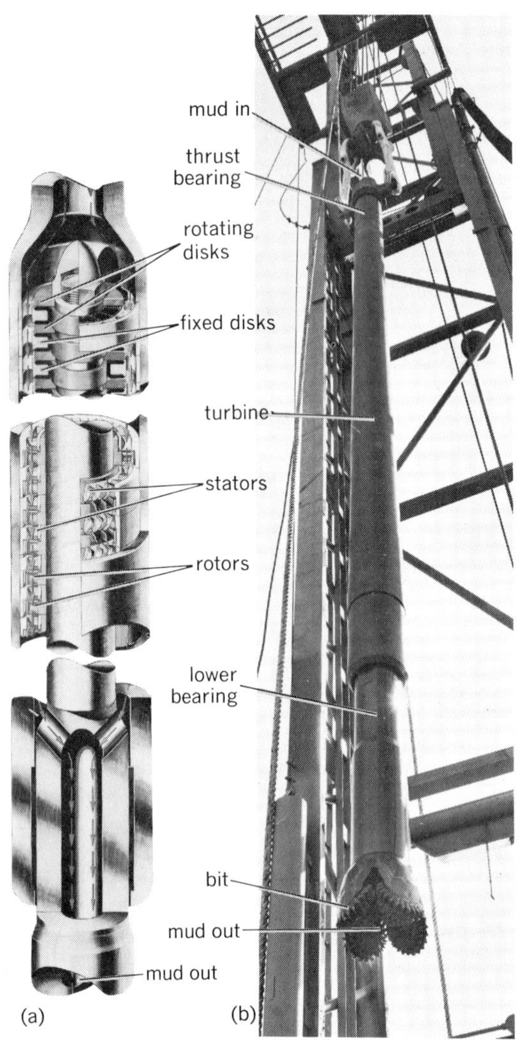

The components of a turbodrill. (*a*) Cutaway view of turbine, bearings, and bit. (*b*) Drill string suspended above hole. (*Dresser Industries*)

States, where the emphasis has been on the rotary tool method of drilling. Despite faster penetration with the turbodrill, and several other advantages, widespread use of the turbodrill in the United States has been limited principally because of the faster wear of the bits, necessitating time-consuming and costly round trips to remove the drill string from the hole and change the bit. *See* BORING AND DRILLING (MINERAL); OIL AND GAS WELL DRILLING.

[A. L. PONIKVAR]

Turnbuckle

A device for tightening a rod or wire rope. Its parts are a sleeve with a screwed connection at one end and a swivel at the other or, more commonly, a sleeve with screwed connections of opposite hands

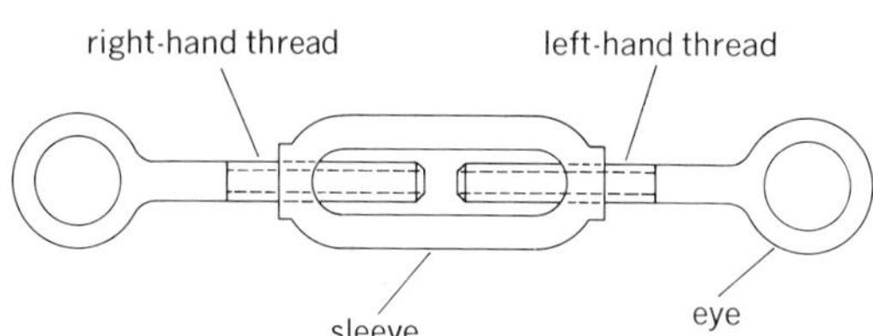

Turnbuckle with eyes.

(left and right) at each end so that by turning the sleeve, the connected parts will be drawn together, taking up slack and producing necessary tension (see illustration). Types of ends available are hook, eye, and clevis. The turnbuckle can be connected at any convenient place in the rod or rope, and several may be used in series if required. *See* SCREW THREADS.

[PAUL H. BLACK]

Ultrasonic machining

The removal of material by abrasive bombardment and crushing in which a flat-ended tool of soft alloy steel is made to vibrate at a frequency of about 20,000 Hz and an amplitude of 0.001–0.003 in. (25–75 μm) while a fine abrasive of silicon carbide, aluminum oxide, or boron carbide is carried by a liquid between tool and work. The tool does not touch the work.

A magnetostrictive transducer, mounted on a slide, is excited by an electronic generator at the desired ultrasonic frequency. The transducer converts the electric energy into mechanical oscillations transmitted to the tool. A tool of any desired shape and size is attached to the tool holder on the lower end of the transducer and drives the abrasive grit, suspended in the liquid, against the work, blasting away very fine particles. The penetration of the tool into the work is slow, only 0.001 in./min (25 μm/min). A wide range of operations can be done, such as drilling round or odd-shaped holes, coining, die sinking, and forming wire-drawing dies.

Impressions may be sunk economically in glass, ceramics, carbides, and hard brittle metals by this method. Also, holes 0.001–3.5 in. (0.025–90 mm) in diameter can be machined to depths of 1.5 in. (38 mm) or more. One cut on a surface or hole is often sufficient, but for greater accuracy and better finish, subsequent cuts with finer grains are used. A 280-grit boron carbide abrasive will produce a tolerance of less than 0.001 in. (25 μm) and a surface finish of 20–30 μin. (0.5–0.75 μm) root mean square. Finer grits, such as 800, cut more slowly but give closer tolerances and a smoother surface (8–12 μin. or 0.2–0.3 μm). Cutting speed is faster with brittle work materials, with large grit-size abrasives, with a small area between tool and work, and with high-power application. *See* MACHINING OPERATIONS.

Machines for commercial ultrasonic work are made in a number of model sizes, ranging in power from 200 to 2400 watts. Accessories are provided to accommodate a wide variety of machining operations. The ultrasonic method is also applied to surface and cylindrical grinding, with resulting lower temperatures and superior finish, but cost of the equipment is still too high to justify wide use. Single point turning tools have been vibrated normally and tangentially to the work with good results, but again the cost is not often justified.

[ORLAN W. BOSTON]

Bibliography: D. B. Dallas (ed.), The new look of ultrasonic machining, *Mfg. Eng. Manage.*, 64(1): 40–42, 1970; V. P. Severdenko et al. (eds.), *Ultrasonic Rolling and Drawing of Metals*, 1972.

Ultraviolet lamp

A mercury-vapor lamp designed to produce ultraviolet radiation. Also, some fluorescent lamps and mercury-vapor lamps that produce light are used for ultraviolet effects. *See* FLUORESCENT LAMP; MERCURY-VAPOR LAMP.

Near-ultraviolet lamps. Ultraviolet energy in the wavelength region from 320 to 400 nanometers is known as near ultraviolet, or black light. Fluorescent and mercury lamps can be filtered so that visible energy is absorbed and emission is primarily in the black-light spectrum. The ultraviolet energy emitted is used to excite fluorescent pigments in paints, dyes, or natural materials to produce dramatic effects in advertising, decoration, and the theater; in industrial inspection, fluorescent effects are often used to detect flaws in machined parts and other products, as well as invisible laundry marks.

Middle-ultraviolet lamps. Middle ultraviolet spans the wavelength band from 280 to 320 nm. Mercury-vapor lamps are sometimes designed with pressures that produce maximum radiation in this region, using special glass bulbs that freely transmit this energy.

One such lamp type is the sunlamp. The illustration shows the reflector sunlamp, with a self-contained filament ballast and starting mechanism.

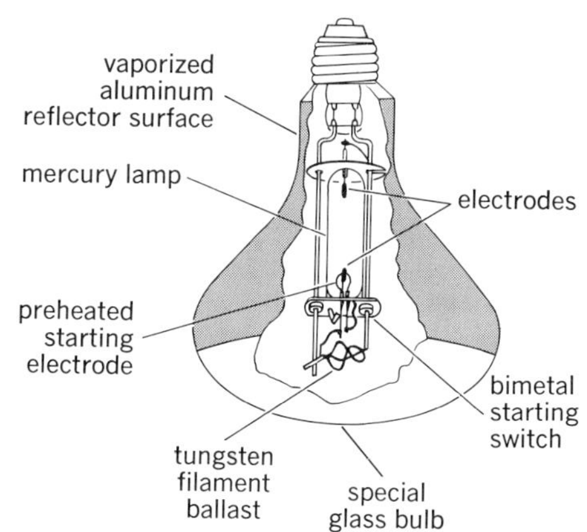

Ultraviolet lamp.

The reflector sunlamp combines the middle ultraviolet, which reddens the skin, with infrared energy and light from the filament to produce a suntanning effect with the sensations of warmth and brightness normally associated with sunshine.

Other lamps designed for middle-ultraviolet radiation are known as photochemical lamps. They are used for a variety of tasks, including mold destruction, inspection of sheet metal for pinholes, and black-and-white printing of engineering drawings.

Far-ultraviolet lamps. Some radiation in the 220–280-nm wavelength band has the capacity to destroy certain kinds of bacteria. Mercury lamps designed to produce energy in this region (the 253.7-nm mercury line) are electrically identical with fluorescent lamps; they differ from fluorescent lamps in the absence of a phosphor coating and in the use of glass tubes that transmit far ultraviolet. Germicidal lamps are sometimes used to reduce the airborne bacteria count and to kill certain organisms on or near the surface of perishable products in storage or on certain products in the pharmaceutical industry. [ALFRED MAKULEC]

Underground mining

An underground mine is a system of underground workings for the removal of ore from its place of occurrence to the surface, and involves the deployment of miners and services.

There are several basic physical elements in an underground mining system. The passageways (openings) in a mine are called drifts if they are parallel to the geological structure, and cross-cuts if they cut across it. They range in size about 60–200 ft^2 in cross section, depending on their functions. The workings on a level (horizontal plane) are joined with those on another level by passageways of similiar cross section, called raises if they are driven upward and winzes if driven downward.

The passageways give access to, and provide transportation routes from, the stopes, which are the excavations where the ore is mined. The stopes are between levels. There may be rooms on the level, such as pump rooms, service shops, and lunchrooms. This article discusses exploration of a mine site, methods of removing ore material, and design of underground openings for ore removal and mine facilities. For other aspects *see* COAL MINING; MINING EXCAVATION; MINING MACHINERY; MINING OPERATING FACILITIES.

EXPLORATION

A mine is designed and the mine openings specified after the exploration phase of a mine's history. Exploration in this context is not to be confused with prospecting. Exploration is the process of finding the characteristics of the mineralized rocks and the environmental rocks that make up the mine site. These attributes are absolute and unchanging, but they can only be predicted from sampling so there is always the risk that the predictions may be wrong. Some of the attributes of a mineral deposit and their limits of variation that are found by sampling and measurement are: shape, tabular or curvilinear; attitude, flat or vertical; dimensions, thick or thin, uniform or variable, long or short, or shallow or deep; physical character, hard or soft, strong or weak, laminar, jointed, or massive; mineralization, massive, globular, or disseminated, intense or sparse, or chemically stable or unstable: and surface and overlying formations, expendable or not expendable.

During the exploration there is a feedback from predictions of the revenue and expense that would result from operation. The end result of the exploration phase is a forecast of the grade (amount of valuable mineral per ton) and tonnage that can be mined at a specified rate. The ore grade acceptable could be different at another mining rate.

Parts of the risk involved are (1) a change in the mineralization or the environmental rocks, or both, as mining progresses, (2) drastic changes in the exchange value of the production relative to wages, equipment, and supplies, (3) the availability of new or better equipment, and (4) a change in the governmental attitude, such as in taxation. Any of these will affect the grade and the tonnage for the mine site. *See* PROSPECTING.

MINE OPERATION

A mine is designed, developed, and worked in blocks of levels and stopes. The size of the blocks may be determined by the amount of ore that has been sufficiently explored, by geological boundaries, or by the need for effective supervision. The design must meet ventilation requirements and the openings must be maintained as long as they are needed. When mining in a block is completed it may, if expedient, be cut off from the ventilating system and the workings may be allowed to collapse. This can be done only if the failure will not disrupt other operations, for example, by causing rock bursts. *See* ROCK BURST.

There are two basic plans of attack. The choice of plan depends on whether or not the surface, or the rocks overlying the ore, may be disturbed, and on stress redistribution problems.

Longwall. The principle of longwall mining is to advance in line all the stopes and pillars being mined in a block. No remnants are left either to support the back (overlying rocks) or to constitute stress concentrators. The line of attack may advance toward the shaft or other entrance or retreat from it. In the latter case, or if there are blocks beyond the current mining to be mined later, passageways must be maintained. Longwall mining is the method generally used to mine flat-lying deposits such as coal. Rooms with uniform dimensions separated by pillars with uniform dimensions are mined. The pillars support the backs. If the preservation of the backs or the surface is not a factor, the pillars may be systematically mined (robbed) in a longwall retreat or advance.

The result of mining with rooms and pillars is a cellular pattern. Most mining methods available to the mine designer involve the creation of a cellular pattern made up of stopes and pillars. Generally the ore from the stopes is won with less expense than is involved in recovering the pillars. Often the pillars are not mined because they are worth more as pillars than they would be as ore. They are stronger than any material that could economically be used to replace them, and they naturally fit better.

Fill. There are few situations in which it is possible to recover a worthwhile amount of pillar ore unless the adjacent stopes or rooms have been filled. No filling will support the overlying rocks as well as the ore or pillars can, but if there is some settlement onto the fill, it will be limited because the rock is dilated and occupies more volume than the solid rock does. The swell will finally support the back.

The fill may be any incombustible available material—waste rock, sand or gravel, or mill tailings. The tailings from the mill are treated to meet the mine specifications for settling and percolation rates by removing some of the fine sizes. They are then transferred from the mixing plant to the stopes as a slurry by a pipeline and distributed in the stopes through hoses. Fine-grained natural sand may be placed that way as well. The water must be taken out by decantation or by percolation, or both. Some operators add portland cement to form a weak concrete. A mixture of smelter slag and sulfide-bearing mill tailings has been used to form a weak rock by chemical action.

When fill is placed in a stope alongside a pillar that is to be recovered, a partition, usually of light timber, is installed between the fill and the pillar. Some operators use an enriched mixture of cement at the interface to form a concrete.

MINE DESIGN

Fundamentally, mining is materials handling, and a mine is designed accordingly. Three functions are involved: breaking the ore from the face, delivering it to the surface, and delivering supplies to where they are needed. Ancillary functions are getting miners and services, air, power, and water to where they are needed. Power may be electrical or compressed air.

Ore breaking and transporting together constitute 30–55% of the total cost of mining. If the ore has to be broken by explosives, the drilling and blasting cost is 10–20% of the total mine cost and the transportation portion is 20–35%.

Primary breaking cost varies inversely, and the transportation cost varies directly, with the size of the broken ore. There is an optimum size for the product of blasting. Each stage of the transportation phase has a limiting size that can be accommodated, so expense saved in the breaking phase may be exceeded by that of secondary breaking between stages in the transportation system.

A mine is designed around the method of primary breaking that is chosen, and the choice is governed by the forecasts from exploration. Stopes may be open or filled if the ore has to be broken by explosives. Caving may be used if conditions are favorable, and the ore will be broken by natural forces that make up the stress field in the ore and in the environmental rocks. Mining methods will be described under major headings, but since each ore body is unique and operators are ingenious, there are variations and hybrid methods.

Open stopes. There is a further qualification to this type of stope—with or without delayed filling. Using open stopes without delayed filling may be expedient, but if there are other extensive workings, the open stopes may redistribute the ground stresses in a manner which will interfere with subsequent work.

The mining method is chosen according to the thickness and inclination of the deposit. The breaking point between thick and thin in a tabu-

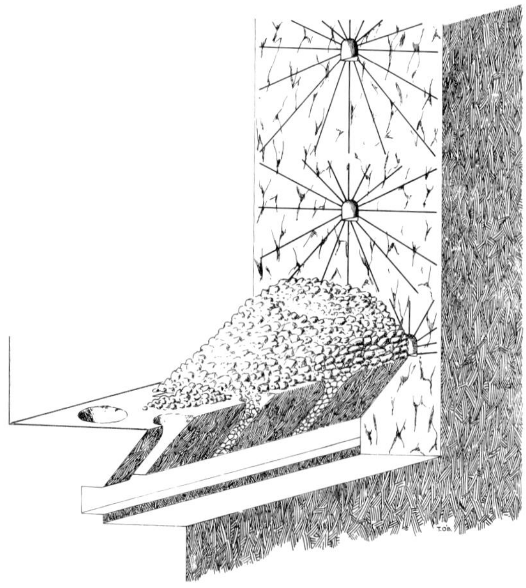

Fig. 1. Sublevel stoping.

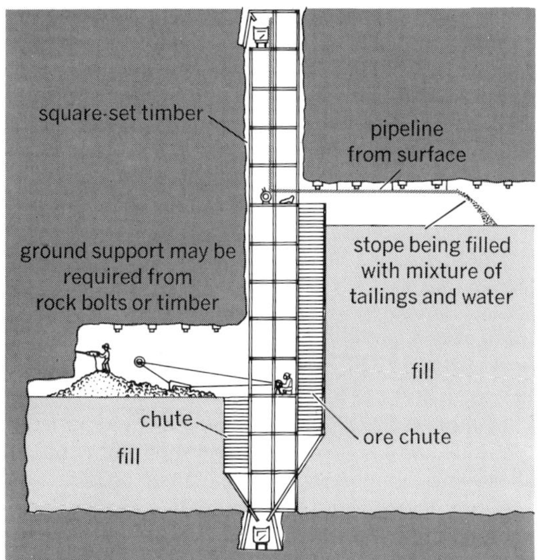

Fig. 2. Hydraulic-fill stope.

lar deposit is about 16 ft (4.9 m). That is about the maximum length of a stull (round timber) that can be handled conveniently in an open stope to give casual support where it is thought that loose rock might develop, or to provide a working platform for the miners. The breaking point between steep and flat is about 40° from horizontal, the limit at which rock will move by gravity on a rough surface.

The stope may be worked either overhand or underhand. If the ore will move by gravity to the drawpoints, the overhand attack is by advancing a breast (face) parallel to the level and so breaking out a slice. Because a platform to work off must be constructed with stulls and lagging or plank for each drilling site, the method is limited to thin deposits. The underhand attack is started from the top of the stope, usually from the access raise, and a block is broken out by drilling and blasting down holes. The bench must be cleaned off after each blast before drilling is resumed. This method has an advantage in gold mines, where coarse gold may lodge on the footwall and have to be swept out. Once cleaned, an area will not have to be cleaned again.

If the deposit is steep enough to deliver the broken ore to the drawpoints by gravity and thick enough to prevent the broken ore from arching over the opening, it probably should not be mined off platforms in an open stope.

Shrinkage stoping. An open stope requires successive platforms for the miners to work off. If the broken ore is drawn off just enough to give working room for the miners and they can work off the broken ore, the stope is called a shrinkage stope. The length of the stope is established by two raises, one at each end, which serve as manways and service entrances for air and water lines. By cribbing up with timber, they are maintained through the broken ore. The draw is from one-third to one-half of the break. The rest of the ore is retained in the stope until the stope is complete, and then the stope may be drawn empty through chutes which are installed during stope preparation.

The broken ore, which moves when drawn, has little ground-support capability. Shrinkage stoping may be used only up to the limit where the back is not self-supporting, or the walls will slab off and give unacceptable dilution. The limit may be extended by using casual timber support, or rock bolts for the back, and by rock-bolting potential slabs to the walls. Once a series of shrinkage stopes has been established and some of the stopes completed, the storage available provides flexibility in production beyond that of any other method.

If the ore body is too wide for the span of the back to be self-supporting or to be cheaply supported when mined parallel to the long dimension of the ore, it may be worked with transverse shrinkage stopes and intervening pillars. This involves delayed filling to permit mining the pillars.

Sublevel stoping. A deposit that is wide enough (about 40 ft or 12 m) and has walls and ore sufficiently strong to permit shrinkage stoping may be worked by sublevel stopes (Fig. 1).

A vertical face is maintained. At vertical inter-

Fig. 3. A cut-and-fill stope. (*a*) Ready for hydraulically placed sand fill, except for the burlap lining. (*b*) Placing the floor over the fill.

vals, spaced to accommodate the drilling method to be used for breaking, sublevels are driven in the long direction of the stope from the entry raise. Two procedures are available after the initial slot has been made, which is generally done by widening a raise to the width of the stope. Holes may be drilled radially from the sublevels to give an acceptable distribution of explosive behind the face or slice to be broken off, or a cut may be slashed out across the face from each sublevel and drilled with vertical holes, quarry fashion, to give a better explosive distribution. The choice depends in part on the equipment available and in part on the control wanted at the sides. When the stope is wide, or for better control of the sides (walls), the radial drilling may be done from each of two sublevels at the same level which have been driven on either side of the stope.

The stope may be longitudinal with respect to the long dimension of the ore body, or it may be transverse and separated from the adjacent stope by a pillar. It follows that it will be a delayed-fill stope if the pillar is to be recovered. Even delayed-fill stopes are not satisfactory if the pillar material is sufficiently valuable to require complete and clean recovery.

Filled stopes. When the wall rock or ore is not sufficiently strong to permit the use of one of the open-stope methods, or if clean pillar recovery is important, methods have been devised to mine with only a small area of unsupported rock or ore exposed.

Cut-and-fill stoping. The breaking phase of this method is not different from that which has been described for the overhand shrinkage stope, but the broken ore is removed after each breast (slice) is completed and a layer of fill is placed. The cycle is: breaking, removing the broken ore, picking up the floor, raising the fill level, and replacing the floor (Figs. 2 and 3). Before the new fill is placed, the manways and the ore chute are raised with cribbing. The cribbing is covered with burlap if hydraulic fill is used. The manway at one end of the stope is built with two compartments. One compartment is used as the ore chute, called a mill hole in filled stopes.

The back must be strong enough to be self-supporting over the span, but the method is sufficiently lower in cost than the alternative so that the back may be supported with casual timber or rock bolts, which are broken out with the new breast.

Square-set stopes. When the backs, and perhaps the walls, are not strong enough to permit cut-and-fill stoping, temporary support may be provided by carefully framed and placed timbers called sets. The sets are filled when no longer needed, at a rate that depends on the rate at which they take weight and might collapse (Fig. 4).

A set is made up of posts 8 or 9 ft (2.4–2.7 m) long, and caps and girts about 6 ft (1.8 m) long, cut to exact lengths and framed to give a good fit at the corners. The timber is usually about 8 in. (20 cm) square, though round timber may be used.

A set is installed in an opening just large enough to accommodate it, and then blocked against the surrounding unstable rock or ore. Little blasting is needed because of the characteristics of the rock that make square-setting the best choice. The problem is to hold back the broken rock until the set can be placed. This may be accomplished by extending boom timbers out over the caps.

Some of the ore may be moved manually. Generally, however, by retaining open sets in the fill and fitting them with inclined slides, provision is made for gathering the ore for scraping to the mill hole.

The sets may be installed either overhand or underhand, depending on the problem, or, if the choice is not critical, on the skill of the miners. The overhand technique is more common but in some camps the miners work better underhand.

Square-setting is the usual method for removing pillars if clean, complete extraction is needed. It is a flexible method and can be used to recover ore in offsets from the main deposit.

Caving. When the surface is expendable and other characteristics are favorable, one of the caving methods may be used.

Top slicing. In some respects this is like square-setting but it is less expensive after it is underway. It is used when the surface is expendable and the ore is too weak to stay in place over a useful span.

The mining block is developed by driving a two-compartment raise through the ore to serve as an access manway and a mill hole, and by driving a longitudinal drift from the raise, at the top of the ore, to the extremity of the ore or the end of the proposed stope.

The initial unit of mining is a timbered crosscut driven each way from the drift to the edge of the block to be mined. Subsequent units are crosscuts driven adjacent to each preceding crosscut to take out a slice of ore. As the face of the slice is retreated toward the mill hole, the timbers in abandoned workings (several sets back) are permitted to fail, or forced to fail by blasting. The routine is continued until the slice is completed to the raise. The

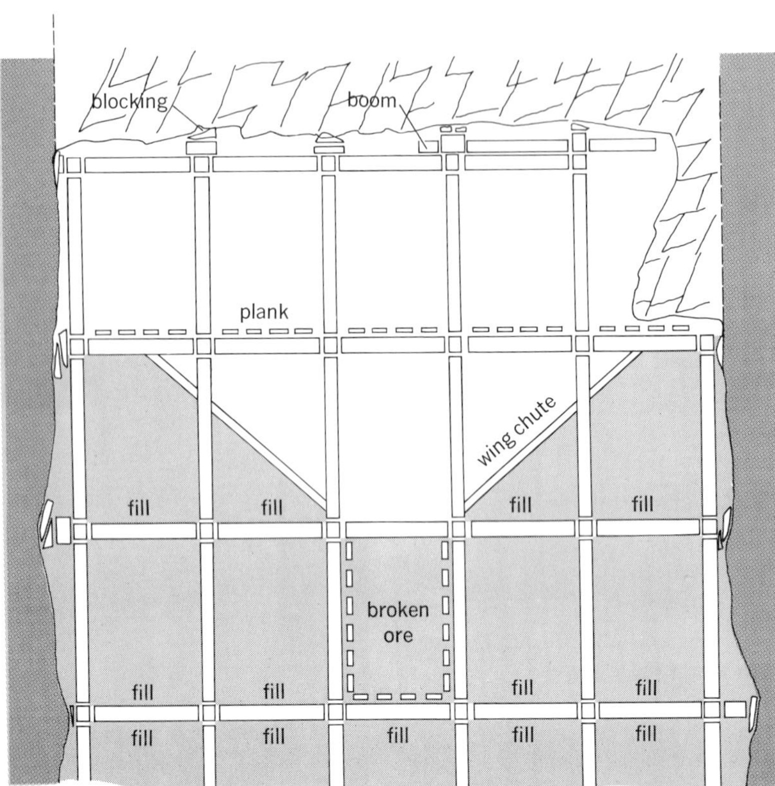

Fig. 4. Square-set stoping.

overlying formations collapse onto the broken timbers. As the routine is continued by taking successive slices, the broken timbers form a feltlike mat that has some tensile strength, and little timber support is needed in the crosscuts. Several slices are mined concurrently, step fashion. The overlying caving formations must follow the mat. No caverns can be left which could collapse and create an air blast.

Sublevel caving. If the ore is sufficiently strong, and after a timber mat has been developed, one or more slices may be omitted. The cantilever shelf formed when the next slice is taken will collapse under the load of caved material and its own weight. The broken ore is moved to the mill hole as in top slicing. Several slices are advanced simultaneously as in top slicing.

An adaption has been used in which the slices are taken out as sublevels in open stoping are taken, and the over lying formations are caved against the face. The sublevel slices are advanced in steps as in top slicing (Fig. 5). There is no mat and some ore is lost into the cave material. However, it is low-grade and the overall low cost of mining makes up for the loss.

Block-caving. If the ore texture (blockiness) and strength are suitable, and if the primitive stress field is favorable, an entire block 150–250 ft (45–76 m) on a side and several hundred feet high may be induced to cave after it is undercut. The broken ore is drawn off through bell-shaped drawpoints (Fig. 6).

The drawing cycle is critical. It must keep the undersurface of the block unstable and continuing to fail. The lateral dimensions of the block are controlled by weakening the perimeter with raises and lateral workings, or even short shrinkage stopes. No large cavities are permitted to develop. In the final phase, when caving has reached the overlying formation, care must be taken to avoid drawing it with the ore. There is no primary breaking expense but the cost of secondary breaking for transportation may be high.

Ground control. Mine openings must be kept open as long as they are needed. Mining engineers recognize that rock is not necessarily solid or inert. The study of the behavior of rocks when subjected to force is called rock mechanics. It is a comparatively new field, although knowledge of the phenomena under study has been utilized for years without formal analyses of what was going on.

The observations that rocks around a mine opening do not always behave in a manner that would be predictable by classical mechanics imply that there are other than gravitational forces involved, and that there is strong lateral component of strain energy. The source and reservoir of the strain energy have been less obscure since geologists have measured the rate of spreading of the North Atlantic Ocean floor and associated continental land masses (average 6 cm per year since Carboniferous times). The resultant force vector from the combination of gravitational force and tectonic force is referred to as primitive stress in this description of underground mining. The rock mass is in equilibrium until a mine opening is made; that is, it is in equilibrium for a relatively short time involved in the mine operation. The mine opening accepts no force and the force is diverted to around the opening.

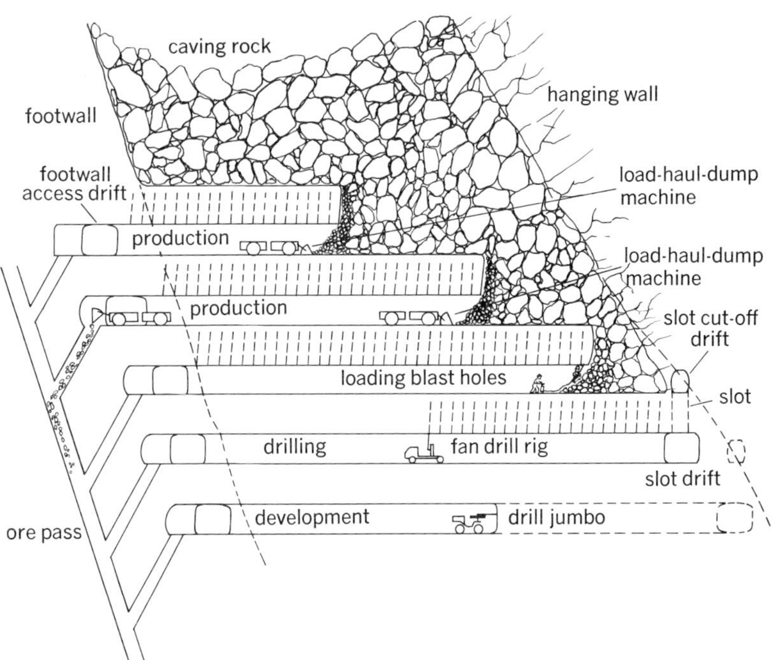

Fig. 5. An adaption of sublevel caving. Cutaway view shows progress of caving.

Ground support. The ideal support is a pillar of appropriate size, but the use of pillars is not always feasible.

Timber. Traditionally timber has been the usual support for the perimeters of mine openings. It is usually supplied as stulls or as lagging, depending on the slenderness ratio, diameter to length, and to some extent on the use. If it is slender, it is lagging. If a log (stull) is placed vertically, it is called a post. If it is placed nearly horizontally, it is usually called a stull, whether it is acting as a beam or as a column. Both posts and stulls are installed with lower ends in hitches in the rock and upper ends loosely fitted to the back or wall, depending on the location. The final fit is achieved by driving wooden wedges between the end of the timber and the rock. The hitch may be chiseled into the rock, but

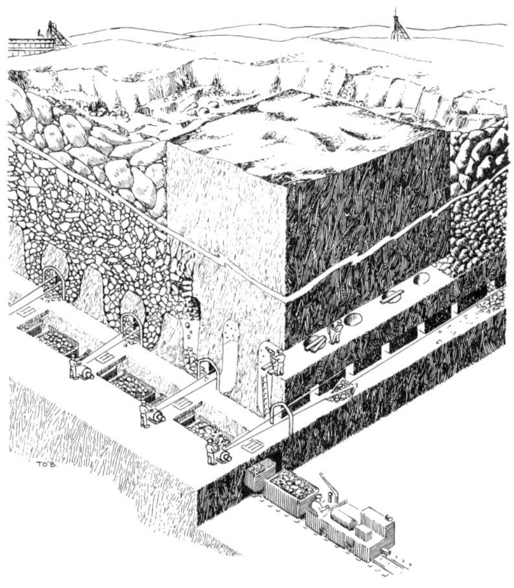

Fig. 6. Block-caving.

Fig. 7. The reinforcement of a mine opening with concrete. (*a*) A drift reinforced with gunite. (*b*) The same drift beyond the gunite.

in hard rock a natural recess is generally used.

When a lateral working requires timbered support, the stulls are usually framed to give a neat fit at the corners, and flatted on two sides to save space in the working. Sawn square timber is often used. The unit is two posts and a cap (stull), usually with a sill on the floor. Whether or not the sill is used depends on the expected loading. The posts and the cap are wedged tightly to the walls and the back at the corners. Lagging or plank is laid over the caps to provide overhead protection and placed behind the posts if a loose wall is expected. Raises and winzes are similarly protected unless they are to be used for hoisting and more precise timbering is needed. Steel is frequently used in the same manner as timber.

Concrete. Openings are frequently lined with concrete if permanence or added strength is needed, if the ventilation friction factor must be reduced, or if the operator does not trust timber because of the fire hazard. Generally the opening is made round or ovaloid so the concrete will be in compression. Forms and poured concrete are commonly used, generally with reinforcing bars. Circumferential steel reinforcement is not effective if the concrete is loaded in circumferential compression.

Concrete may be blown onto the rock surface with a cement gun (guniting). Sand and small-sized aggregate are mixed dry and blown through a hose to the face to be coated. Water is added as the mixture passes the nozzle. The low-moisture mixture hits the face and a portion of the aggregate falls out. A tight bond is formed at the concrete-rock interface. It is thought that the peining action of the aggregate helps to make the bond and to produce a dense concrete. In treacherous rock quite large rooms, such as underground hoist rooms 30 ft or more across, have been successfully secured in this way. The angle of impingement for the application is critical. The thickness of the coating is not more than a few inches (say 3 or 4 in., or 7.5 or 10 cm) over the depressions in the rock surface and thinner over the bumps (Fig. 7).

Rock bolts. The systematic use of rock bolts for rock reinforcement has increased rapidly. These are steel bolts about 3/4 in. (19 mm) in diameter and generally 3–5 ft (1.5–2.1 m) long, anchored at the bottoms of holes drilled at a right angle to the rock surface, and tensioned by a nut over a small plate at the rock surface. The anchorage is generally a split shell forced against the wall of the bore hole by a wedge as tension is applied at the bolt end. Some suppliers offer a method to anchor the bolt in an epoxy resin, and some others supply a bolt that is to be embedded in concrete or cement for its entire length.

There is no consensus as to the reason that rock bolts are effective, but it is agreed that they should be installed as soon as possible after an opening is made, and should be under high tension (Fig. 8). The mechanism offered most frequently for the effectiveness of bolts in a bedded formation is that a compound beam is built up by binding several laminar beds together to act as a single thick beam. For massive rocks that have no bedding it is commonly accepted that the bolts must extend

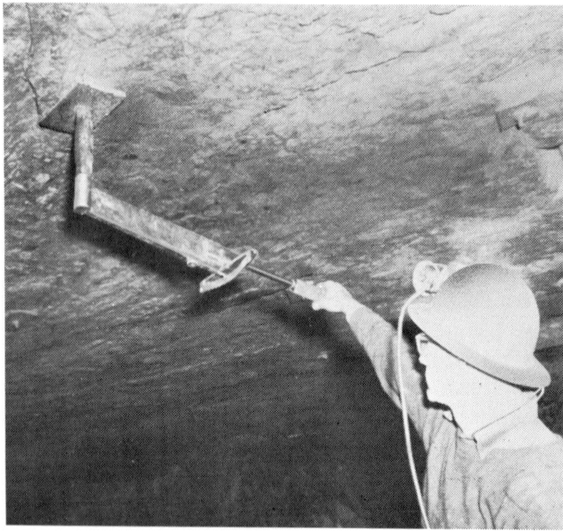

Fig. 8. Testing the tension on a rock bolt by using a simple instrument to measure torque.

into the compression arch that is postulated to be formed above the opening, and for that reason long bolts are often specified.

Actually, the abutments of the arch are restrained from moving outward, and the tendency is to move inward, especially if there is a high primitive stress lateral component. In narrow openings there is compression close to the skin of the opening and the function of the bolt in tension is to reduce the tendency for failure in oblique shear by preventing the thickening of the rock. At some width, as an opening is enlarged from narrow to wide, the compressive primitive stress is neutralized and the back goes into tension. Rock has little tensile strength because of discontinuities. If the rock is blocky, the blocks may be held together by bolts and form a flat, or nearly flat, voussoir arch.

Transportation. Gravity is used wherever it can be effective in the movement of ore toward the surface. Ore from open stopes that are steep enough for gravity flow is loaded through chutes directly into the level transportation units. When a stope is wider than about 25 ft (7.6 m), bell-shaped openings (drawpoints) are driven into the floor of the stope. If more than one row of them is needed, these drawpoints are driven on about 25-ft (7.6-m) centers on a regular pattern so one crosscut can serve the outlets of several drawpoints. When the ore is loaded into the haulage equipment, it is taken to an ore pass and moved by gravity either to a loading pocket at the shaft or to a crushing plant, and thence to the shaftpocket. An inclined ore pass will give considerable lateral movement. A mine will also have a system of waste passes.

A drawpoint may discharge through a chute into a haulage vehicle or into a short branch off the haulage line and be loaded into the main-line vehicle mechanically. If the stope is wide, the drawpoints may discharge onto the floor of a scraper drift at a higher elevation than the back of the haulage level. The ore is then delivered to the main-line vehicle by scraping (Fig. 6).

When a lode is too flat to permit the use of gravity, ore is moved to a central gathering point by a scraper, either in one stage or two. If the lode is flat enough to permit the use of wheeled or crawler-tracked vehicles, the broken ore may be loaded into a gathering vehicle and taken either to the ore pass or to the main-line transportation unit. An alternative is to use a load-haul-dump vehicle. Smaller versions of this type of machine are being introduced into large stopes. They are displacing the scraper, which in turn had displaced the small railcar and the wheelbarrow (Fig. 5).

Equipment is designed to do a specific job and its value in use beyond that job decreases rapidly. The primary gathering equipment is designed for a short haul.

Entry from surface. When the topography of the area has low relief, the entry will be by a shaft or a ramp. Sometimes both means are employed, the ramp being used for moving heavy, large equipment within the mine. If the relief is high, an adit (tunnel) may be used.

A shaft is usually located in the footwall far enough from the mine workings to avoid ground movement. It is designed for specific functions which determine the area (cross section) and shape, if the shape is not modified to accommodate ground stresses and sinking problems. It may be vertical or inclined, though the vertical shaft is the more common. Functionally either kind should be rectangular to accommodate the equipment used in it.

Many vertical shafts are circular or elliptical, but a rectangular framework is fitted in them to guide the shaft vehicles. There is an exception, not common in North America, when rope (steel-cable) guides are hung in the shaft to guide the vehicles. On the other hand, a round or elliptical shaft is better for ventilation because it may be smooth-lined and offers less air resistance.

A shaft is designed after its functions have been decided and the rock conditions have been forecast. It may be multipurpose and the cross section (plan) must include space for each of the functions, as well as a ladderway for an emergency exit. A shaft may be specialized, that is, designed exclusively for ore hoisting, for services, or for ventilation. A mine must have two shafts to provide alternate routes to the surface in case one shaft is out of commission in an emergency.

[A. V. CORLETT]

Bibliography: Engineering Mining Journal Magazine, *E/MJ Operating Handbook of Mineral Underground Mining*, vol. 3, 1979; R. S. Lewis and G. B. Clark, *Elements of Mining*, 3d ed., 1964; R. Peele (ed.), *Mining Engineers' Handbook*, vol. 1, 3d ed., 1941; Society of Mining Engineers, *SME Mining Engineering Handbook*, 1973; L. J. Thomas, *An Introduction to Mining: Exploration, Feasibility, Extraction, Rock Mechanics*, 1977.

Universal joint

A linkage that transmits rotation between two shafts whose axes are coplanar but not coinciding. The universal joint is used in almost every class of machinery: machine tools, instruments, control devices, and, most familiarly, automobiles.

Hooke's joint. A simple universal joint, known in English-speaking countries as Hooke's joint and in continental Europe as a Cardan joint, is shown in Fig. 1*a*. It consists of two yokes *A* and *B* (Fig. 1*b*) attached to their respective shafts and connected by means of spider *S*. Angle ϕ between the shafts may have any value up to approximately 35°, if angular velocity is moderate when angle ϕ is large. Although shaft *B* must make one revolution for each revolution of shaft *A*, the instantaneous angular displacement of shaft *B* is the same as that of shaft *A* only at the end of each 90° of shaft rotation. Thus, only at four positions during each revolution is angular velocity ω_B of shaft *B* the same as angular velocity ω_A of shaft *A*. Three curves for designated values of ϕ are plotted in Fig. 2. These curves show deviation of ω_B from a constant ω_A as shaft *A* is turned through 180°. *See* FOUR-BAR LINKAGE.

Double Hooke's joint. The variation in angular displacement and angular velocity between driving and driven shafts, which is objectionable in many mechanisms, can be eliminated by using two Hooke's joints, with an intermediate shaft (Fig. 1*c*). This arrangement is conventional for an automobile drive shaft. The axes of the driving and driven shafts need not intersect; however, it is necessary that the axes *y* and *y'* of the two yokes attached to intermediate shaft *b* lie respectively in planes containing the axes of adjoining shafts (*a*, *y*, and *b* in same plane; *b*, *y'*, and *c* in same plane), and that

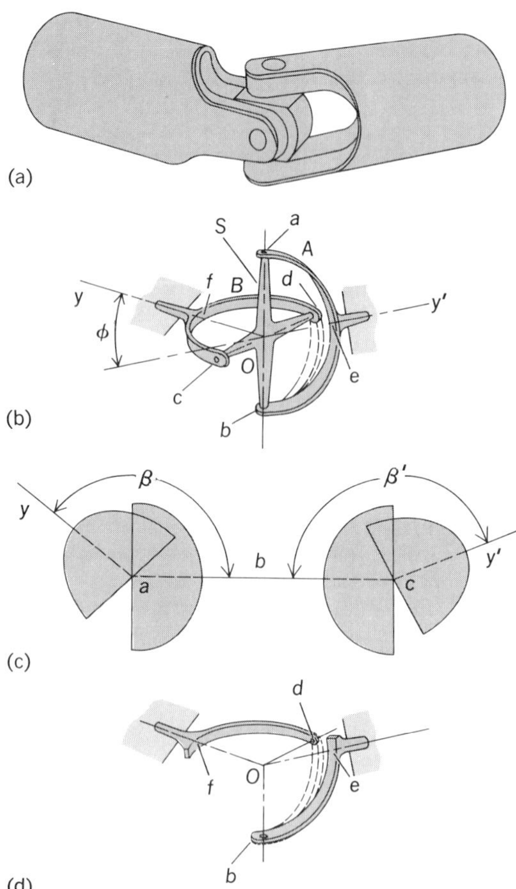

Fig. 1. Universal joints. (*a*) Simple. (*b*) Yoke and spider. (*c*) Double. (*d*) Four-bar conic linkage equivalent of yoke and spider. (*From C. W. Ham, E. J. Crane, and W. L. Rogers, Mechanics of Machinery, McGraw-Hill, 1958*)

angle β between the driving and intermediate shafts equal angle β' between the intermediate and driven shafts.

Bendix-Weiss joint. Two intersecting, thin, bent shafts (Fig. 3*a*) with plane Z through the point of contact bisecting the angle between the shafts and perpendicular to the plane containing the axes of the shafts will maintain the constant angular velocity.

One practical application of this principle, in which constant angular velocity is transmitted through a single universal joint, is the joint developed by Carl Weiss (or Rzeppa joint) and illustrated in Fig. 3*b*. Four large balls are transmitting elements, while a center ball acts as a spacer. The transmitting balls must lie in plane Z, as explained above. By means of milled grooves in the yoke attached to each shaft, the balls are maintained at all times in plane Z. This joint (Fig. 4) is used for a front-wheel automotive drive because it transmits unvarying angular displacement even when steering the vehicle requires a varying angle between driving and driven shafts.

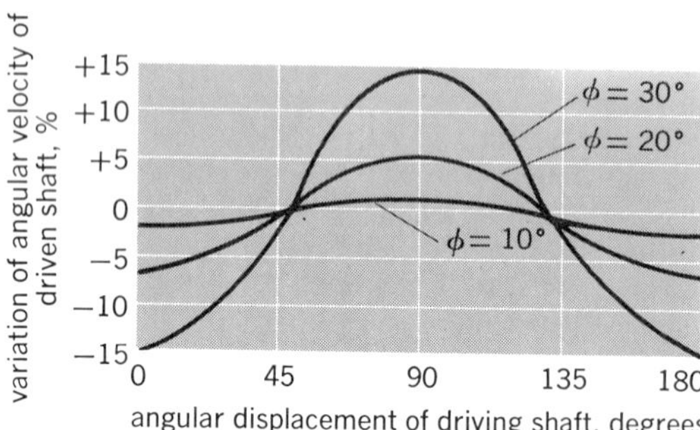

Fig. 2. Variation in the angular velocity of the driven shaft of a simple universal joint for the constant angular velocity of the driving shaft.

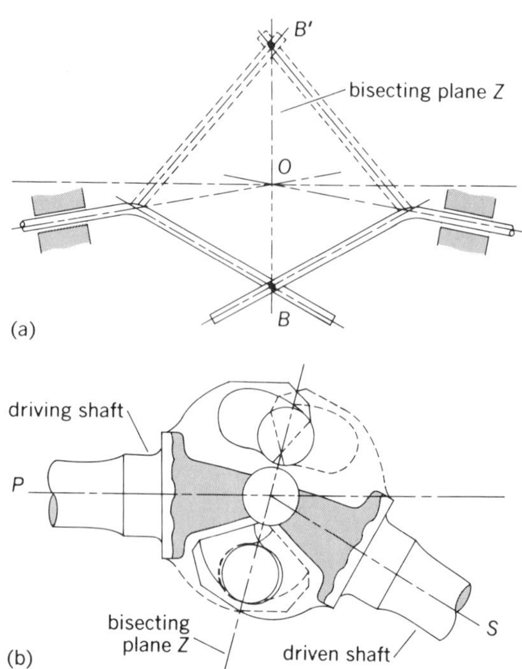

Fig. 3. Interacting shafts in sliding contact transmit constant angular velocity. (*a*) Basic configuration (*from R. T. Hinkle, Kinematics of Machines, Prentice-Hall, 1952*). (*b*) Cross section of ball-bearing adaptation.

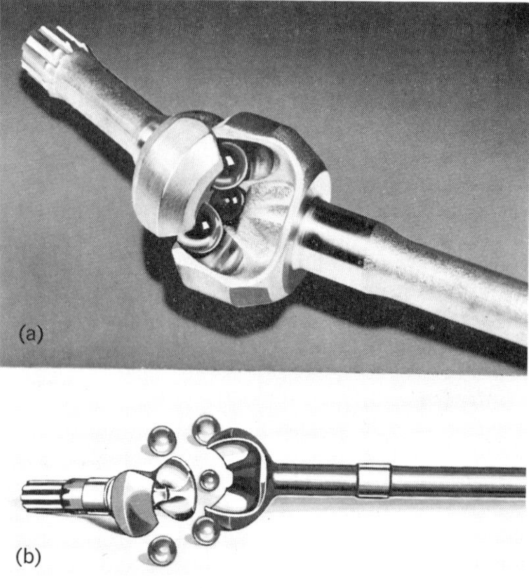

Fig. 4. Constant-velocity universal joint. (*a*) Partially separated cutaway. (*b*) Disassembled joint showing the arrangement; the right-hand member has been cut away. (*Bendix Aviation Corp.*)

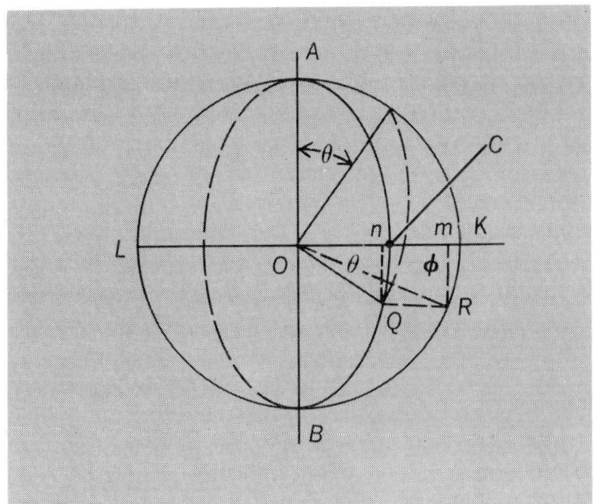

Fig. 5 Geometry of motion for spherical joint of Fig. 1*d*.

Spherical four-bar linkage. The universal joint is a spatial, spherical four-bar linkage. If only half of each fork is considered as *eb* of *A* and *fd* of *B* (Fig. 1*b*) and these are assumed to be connected by spherical link *db* equal to the fixed distance between the two adjacent points of spider *S*, a four-bar conic linkage is produced (Fig. 1*d*) in which the axes of all the turning pairs intersect in *O*. With this arrangement the fork could be omitted, and there would be a kinematic equivalent of the original mechanism.

Driver and follower make complete revolutions in the same time, but the velocity ratio is not constant throughout the revolution as can be shown by analysis. In Fig. 5, *COQ* is the projection of the real angle described by the follower. The real component of the motion *Qn* of the follower is in a direction parallel to *AB*, and line *AB* is the intersection of the planes of the driver's and the follower's path. The true angle ϕ described by the follower (while the driver describes the angle θ) can be found by revolving *OQ* about *AB* as an axis into the plane of circle *AKBL*. Then *OR* is the true length of *OQ*, and *ROK*, designated ϕ, equals the true angle that is projected as *COQ*, designated θ. With this notation $\tan \phi = Rm/Om$ and $\tan \theta = On/On$. But $Qn = Rm$; hence $\tan \theta / \tan \phi = Om/On = OK/OC = 1/\cos \beta$. [DOUGLAS P. ADAMS]

Bibliography: H. H. Mabie and F. W. Ocvirk, *Mechanisms and Dynamics of Machinery*, 3d ed., 1978.

Universal motor

A series motor built to operate on either alternating current (ac) or direct current (dc). It is normally designed for capacities less than 1 hp. It is usually

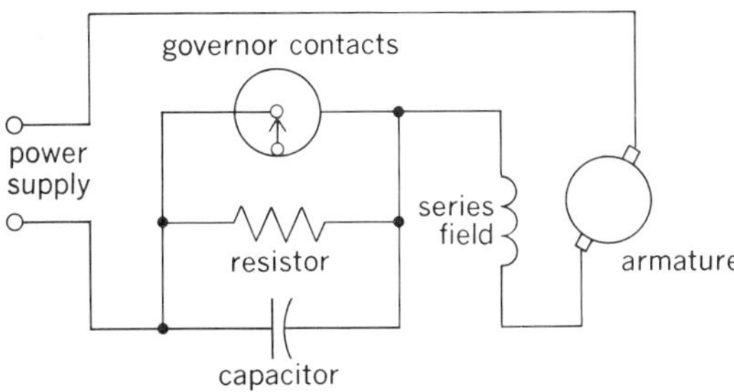

Fig. 1. Universal motor diagram.

operated at high speed, 3500 rpm loaded and 8000 to 10,000 rpm unloaded. For lower speeds, reduction gears are often employed, as in the case of electric hand drills or food mixers. As in all series motors, the rotor speed increases as the load decreases and the no-load speed is limited only by friction and windage. To obtain more constant speed with variations in load a contrifugal governor may be used to switch in or out a small resistor in series with the armature as in Fig. 1.

If an alternating current is applied to any dc series motor, the motor would still rotate. Since the current is reversed simultaneously in the armature and the field, the torque would pulsate but would not reverse direction. However, a universal motor designed to operate on ac should have certain modifications: laminated cores to avoid excessive eddy currents, fewer turns in the field coils than in a dc motor, and more poles and usually more commutator segments. *See* CORE LOSS; DIRECT-CURRENT MOTOR.

The series ac motor is an alternating-current commutator motor which has great flexibility of performance. It can be operated over a wide range of speeds and is readily controllable. The series ac commutator motor is in many respects similar to the dc series motor and the universal motor.

The ac series motor, like the dc series motor and the repulsion motor, consists fundamentally of these windings or their equivalent: (1) rotating armature winding, (2) stationary field winding, and (3) compensating winding (Fig. 2).

A major problem in larger (up to 1000 hp) ac series motors is in commutation. Because of the transformer action between the field and armature coils, voltage is produced in the armature coils which are short-circuited by the brushes as the commutator bars pass under them. The coils which are short-circuited act like a short-circuited secondary of a static transformer. The resulting

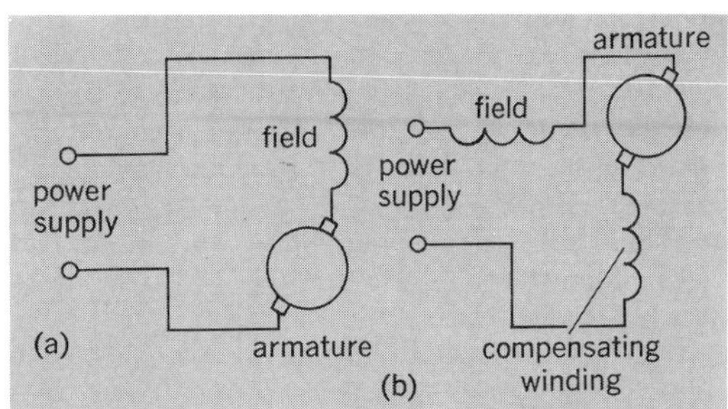

Fig. 2. Series motor diagram. (*a*) Without compensating winding. (*b*) With compensating winding.

large currents are interrupted as the bars pass the brushes, causing bad sparking. In addition, these induced currents reduce the magnetic flux of the field and reduce the torque of the motor. Interpoles shunted with noninductive resistance are required on ac series motors as in Fig. 3. *See* COMMUTATION.

The single-phase commutator motor usually has a large number of turns in the armature winding, more commutator segments, and a small number of turns in the field winding, as compared with the dc motor, which is designed for relatively strong

power supply
main field
armature
shunt resistance
compensating field winding
interpole winding

Fig. 3. Single-phase traction motor circuit diagram. (*General Electric Co.*)

field and weaker armature. The ac motor usually has more poles and operates at a lower voltage than its dc counterpart.

[IRVING L. KOSOW]

Bibliography: D. F. Fink and H. W. Beaty (eds.), *Standard Handbook for Electrical Engineers*, 11th ed., 1978; A. E. Fitzgerald, C. Kingsley, and A. Kusko, *Electric Machinery*, 3d ed., 1971; I. L. Kowsow, *Electric Machinery and Transformers*, 1972.

Unloader

A power device for removing bulk materials from railway freight cars or highway trucks. Especially in the case of railway cars, the car structure may aid the unloading, as in the hopper-bottom dumper; here the unloader may be a vibrator which improves flow of the bulk material into a storage bin or to a conveyor. Thus an unloader is the transitional device between interplant transportation means and intraplant materials-handling facilities. *See* TRANSPORTATION ENGINEERING; MATERIALS HANDLING.

UNLOADER

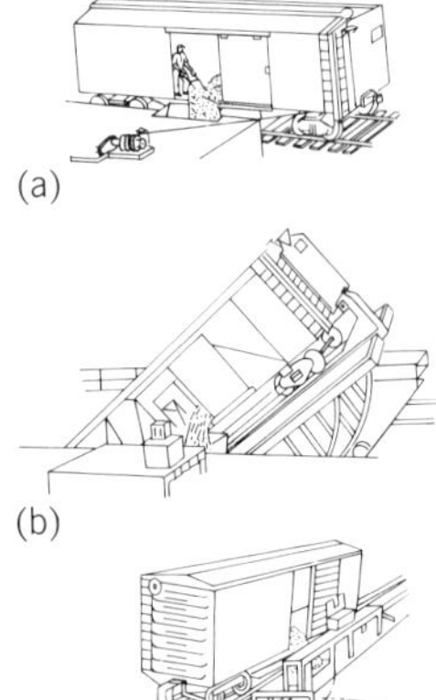

Unloader. (*a*) Power scoop. (*b*) Rocker. (c) Tilter. (*Link-Belt Co.*)

Basically, design considerations for an unloader are similar to those for a device for removing bulk from a storage bin or warehouse with low headroom and limited access. A special consideration is the possible need for an elevating platform to bring the car or truck floor to the level of the unloading dock and to maintain the alignment as material is removed and the vehicular springs expand. Unloading may then be accomplished by portable belt, drag-chain, flight, or similar conveyor (illustration *a*). *See* BULK-HANDLING MACHINES.

A fully mechanized unloader includes a section of track to which the car is clamped and then rocked or tilted (illustration *b* and *c*). In this way unloading rates of up to ten boxcars per hour are achieved. For less specialized unloading dock equipment such as power scoops may be used, or yard handling equipment such as boom unloaders may be adapted.

[FRANK H. ROCKETT]

Vacuum metallurgy

The making, shaping, and treating of metals, alloys, and intermetallic and refractory metal compounds in a gaseous environment where the composition and partial pressures of the various components of the gas phase are carefully controlled. In many instances, this environment is a vacuum ranging from subatmospheric to ultrahigh vacuum (less than 760 torr or 101 kPa to 10^{-12} torr or 10^{-10} Pa). In other cases, reactive gases are deliberately added to the environment to produce the desired reactions, such as in reactive evaporation and sputtering processes and chemical vapor deposition. The processes in vacuum metallurgy involve liquid/solid, vapor/solid, and vapor/liquid/solid transitions. In addition, they include testing of metals in controlled environments. The table lists the processes, pressure ranges, and gas environments.

The early applications of vacuum metallurgy were involved with the more conventional aspects, such as extractive metallurgy, melting, and heat treatment. Since the early 1970s, two important developments have occurred. First, other nonvacuum specialty melting processes, such as electroslag remelting (ESR) and plasma melting, have been developed. They have some unique aspects (for example, grain size refinement) which make them strong competitors with vacuum melting processes on the one hand and complementary melting processes on the other. For many applications, the final ingot may have been processed by a number of vacuum and nonvacuum melting processes. Second, in many high-technology applications, a single material, such as a metal or alloy, cannot meet the multiple requirements placed on it for a particular application. For example, the blades and vanes for the hot stage of a gas turbine engine have to be a composite material, with high-temperature nickel or cobalt base alloy for strength at elevated temperatures and a coating of an alloy such as Ni(or Co)-Cr-Al-Y to resist the severe corrosion environment. Thus a large and rapidly growing aspect of vacuum metallurgy consists of vapor deposition techniques.

Advantages. There are three basic reasons for vacuum processing of metals: elimination of contamination from the processing environment, reduction of the level of impurities in the product, and deposition with a minimum of impurities. Contamination from the processing environment includes the container for the metal and the gas phase surrounding the metal. For example, melting of titanium in a refractory crucible in air would severely contaminate the melt both from the material of the crucible and from the atmospheric oxygen, making it useless. Therefore titanium is processed in water-cooled copper crucibles (which do not react with the metal) and in vacuum. In the vacuum process, impurities, particularly oxygen, nitrogen, hydrogen, and carbon, are released from the molten metal and pumped away; and metals, alloys, and compounds are deposited with a minimum of entrained impurities. There are numerous and varied application areas for vacuum metallurgy; these are discussed below.

Extractive metallurgy. Included in extractive metallurgy are the beneficiation of ores, reduction of ores or compounds to crude metals, and subsequent refining of the high-purity metals. Examples are the reduction of dolomite with ferro-silicon, the refining of rare-earth metals by distillation, and the preparation of pure metals by halide dissociation on a heated wire.

Melting processes. The melting processes have several functions. Among them are the refining of crude metals, the consolidation of various compo-

Various metallurgical processes and their operating pressure regimes*

Processes	Approximate pressure regime, torr†	Gas environment‡
Extraction and refining of metals	Several atmospheres – 10^{-3} torr	V
Degassing of liquid steel	400 – 1	V
Vacuum induction melting	100 – 10^{-6}	V
Vacuum arc melting	1 – 10^{-3}	V, I
Electron-beam melting	10^{-3} – 10^{-8}	V
Levitation melting	760 – 10^{-5}	V, I
Melting on water-cooled hearths	760 – 10^{-5}	V, I
Zone refining	760 – 10^{-8}	V, I
Surface treatment (carburizing, nitriding)	10^{-1} – 10^{-3}	R
Heat treatment:		
High vacuum	1 – 10^{-6}	V
Ultrahigh vacuum	10^{-8} – 10^{-10}	V
Evaporation		
For purification and bulk deposits	10^{-2} – 10^{-7}	V
For thin-film deposition	10^{-1} – 10^{-10}	V, R
Sputtering to form thin-film or bulk deposits	10^{-2} – 10^{-5}	I, R
Chemical vapor deposition	760 – 10^{-2}	V, R
Electron-beam welding (except where beam is brought out to atmosphere)	$<10^{-4}$	V
Brazing	10^{-2} – 10^{-5}	V
Diffusion bonding	10^{-2} – 10^{-5}	V
Evaluation of physical and mechanical properties	Several atmospheres – 10^{-9} torr	V
Study of nature of surfaces and reactions occurring on clean surface	$<10^{-8}$	V
Mechanical working in vacuum	10^{-1} – 10^{-4}	V
Sintering of powders in vacuum	10^{-1} – 10^{-5}	V
Electron-beam machining	$<10^{-4}$	V

*From R. F. Bunshah, *Techniques of Metals Research*, vol. 1, copyright © 1968 by John Wiley and Sons, Inc.; used with permission.

¶1 torr = 133 Pa.

‡V = residual gases in a vacuum; I = inert gas added; R = reactive gas added.

nents into alloys, and the production of an ingot of desired shape.

The metal may be heated and melted by several heating methods which distinguish the various processes, for example, induction melting, consumable-electrode arc melting, and electron-beam melting. In addition, there is the levitation melting technique where no crucible is used, the metal being supported by an electromagnetic force.

Casting of shaped products. The molten metal is cast into intricate shapes by using a mold of refractory compounds usually made by the lost-wax process (see illustration). A good example of this process is the production of turbine buckets for aircraft jet engines by using a nickel-base high-temperature alloy. *See* METAL CASTING.

Vacuum degassing of molten steel. This is the largest single application of vacuum metallurgy. In this process, steel melted in conventional electric arc furnaces is treated in vacuum before being cast into ingots. There are several processes for vacuum degassing. During the process the gases in the molten metal are released and pumped away. *See* STEEL; STEEL MANUFACTURE.

Heat treatment. Heat treatment of refractory and reactive metals (titanium, columbium, tung-

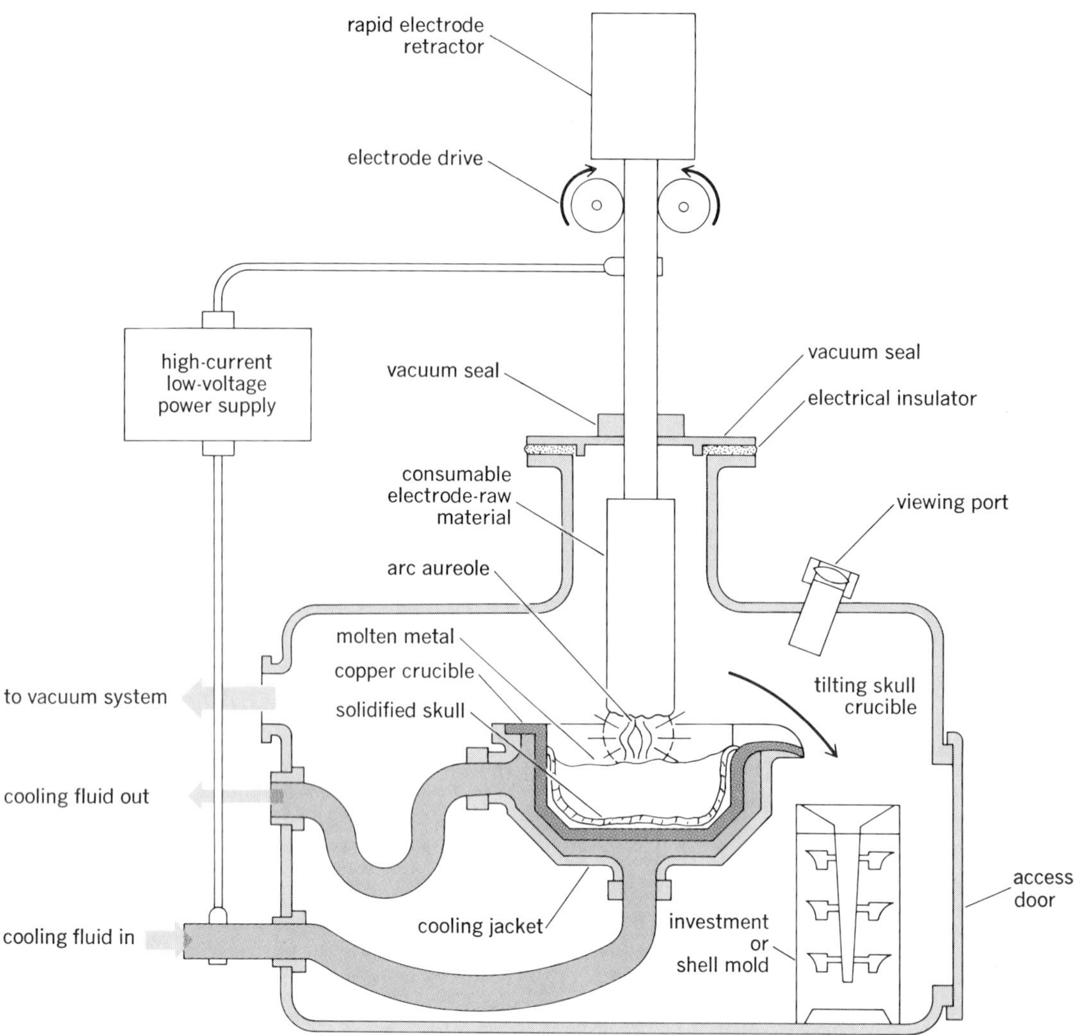

Vacuum metallurgy. Skull-crucible, consumable-electrode vacuum arc melting and casting furnace. (*From R. F. Bunshah, Techniques of Metals Research, vol. 1, copyright © 1968 by John Wiley and Sons, Inc.; used with permission*)

sten, and so forth) is carried out in vacuum to promote desired reactions in the solid metal while preventing recontamination from the atmosphere; an example is age-hardening of titanium alloys to get maximum strength.

Surface treatment. Modification of the surface properties of a material are carried out by diffusing various species such as carbon, nitrogen, boron, or aluminum from a solid, liquid, or gaseous environment. The effects of impurities in the gas phase will affect the surface properties, and hence the need for a controlled environment. In some instances ionization of the reactive gas permits the process to occur more rapidly and at lower temperatures, for example, ion nitriding and ion carburizing.

Joining. Joining processes include welding, brazing, and diffusion bonding. They are carried out in vacuum where the metals being joined would be contaminated by joining in air. A development of particular importance is electron-beam welding. *See* ARC WELDING; WELDING AND CUTTING OF METALS.

Vapor deposition. This type of process includes the physical vapor deposition—evaporation, ion plating, and sputtering—and the chemical vapor deposition. The products of these processes are coatings and freestanding shapes such as sheet, foil, and tubing of thickness ranging from 20 nm to 25 mm. These processes have widespread application in diverse applications in the metallurgical, chemical, engineering, and electronic industries. A large variety of products are made possible by using high-vacuum coating processes.

Decorative. Automotive trim (interior and exterior), toys, cosmetic packaging, pens and pencils, Christmas decorations, food and drink labels, costume jewelry, home hardware, eyeglass frames, packaging and wrapping materials, watch cases.

Optically functional. Laser optics (reflective and transmitting), architectural glazing, home mirrors, automotive rearview mirrors, eyeglass lenses, projector reflectors, camera lenses and filters, instrument optics, auto headlight reflectors, television camera optical elements, meter faces.

Electrically functional. Semiconductor devices, integrated circuits, capacitors, resistors, magnetic tape, disk memories, superconductors, electrostatic shielding, switch contacts, solar cells.

Mechanically functional. Aircraft engine parts, aircraft landing gear, solid film lubricants, tool bit hard coatings.

Chemically functional. Corrosion-resistant fasteners, gas turbine engine blades and vanes, battery strips, marine equipment.

Space processing. The space environment combines microgravity conditions and high vacuum. Results of experimental work in various space missions have demonstrated improvements in properties of metals, semiconductors, and glasses over similar material processed on Earth in a 1-*g* environment. The space environment also presents a more functional possibility of studying physical and chemical properties of materials. In a particular configuration entitled the molecular shield, an extreme vacuum environment (less than 10^{-13} torr) with a very high pumping throughput is available in space. Such an environment is not possible on Earth. When combined with the micro-*g* environment of space, the potential for producing entirely unique materials is large.

Other processes. Other metallurgical operations carried out in vacuum are the testing of metals, electron-beam machining, study of surfaces, powder metallurgy, and mechanical working. *See* METAL COATINGS; POWDER METALLURGY.

The products of vacuum processing include the entire gamut of materials, steels, reactive metals (titanium, zirconium, and so forth), refractory metals (tungsten, molybdenum, and others), their alloys, intermetallic compounds, and refractory compounds. Vacuum metallurgy will be a key part of the future space processing program. *See* PYROMETALLURGY; PYROMETALLURGY, NONFERROUS.

[ROINTAN F. BUNSHAH]

Bibliography: American Vacuum Society, *Transactions of the Vacuum Metallurgy Conference*, 1967, also other annual editions; J. A. Belk, *Vacuum Techniques in Metallurgy*, 1963; R. F. Bunshah, *Films and Coatings for High Technology Applications*, 1981; R. F. Bunshah, *Techniques of Metals Research*, vol. 1, pt. 2, 1968; R. F. Bunshah (ed.), *Vacuum Metallurgy*, 1958; R. F. Bunshah et al., Deposition technologies and their applications, in B. N. Chapman and J. C. Anderson (eds.), *Science and Technology of Surface Coatings*, 1974; *J. Vac. Sci. Technol.*, 1964 to date; *Proceedings of the International Conference on Metallurgical Coatings*, 1974 to date; O. Winkler and R. Bakish (eds.), *Vacuum Metallurgy*, 1971.

Vacuum pump

A device that reduces the pressure of a gas (usually air) in a container. When gas in a closed container is lowered from atmospheric pressure, the operation constitutes an increase in vacuum in this container. The unit used for vacuum is a millimeter (mm) column of mercury (Hg). Another term for the mm Hg unit is the torr. These terms are used interchangeably. The torr is named in honor of E. Torricelli, a pioneer in vacuum technology. The various vacuum regions are classified as follows:

Low vacuum, 760 (atmospheric pressure) to 25 mm Hg

Medium vacuum, 25 to 10^{-3} mm Hg

High vacuum, 10^{-3} to 10^{-6} mm Hg

Very high vacuum, 10^{-6} to 10^{-9} mm Hg

Ultrahigh vacuum, 10^{-10} mm Hg and beyond

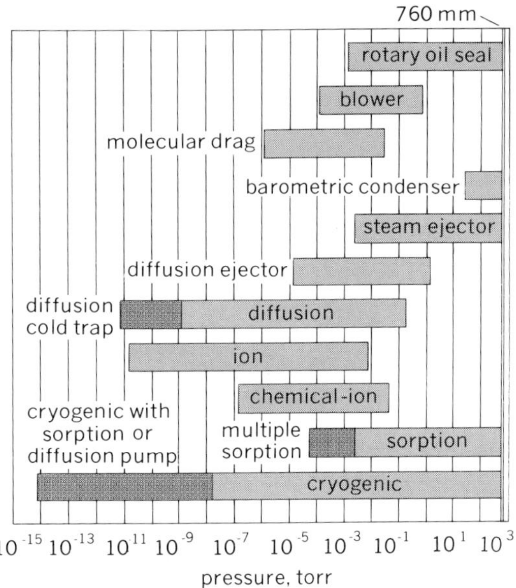

Fig. 1. Comparison of the degree of vacuum attained by important types of vacuum pumps.

A vacuum as high as 10^{-15} mm Hg or one-billionth of a billionth of an atmosphere has been reached. *See* PRESSURE.

Vacuum pumps are evaluated for the degree of vacuum they can attain and for how much gas they can pump in a unit of time. Figure 1 compares the degree of vacuum that the more important types of vacuum pumps can attain. The determination of the quantity of gas a vacuum pump handles is discussed later. At low vacuum, this quantity is usually expressed as pounds of air per hour. Any other gas is converted, via molecular weight relationship, to air equivalent for this purpose. For high vacuums, pounds of air per hour is expressed as torr liters (or mm Hg liters) per second. In practice, where high vacuum is required, two or more different types of pumps are used in series. For example, a rotary oil-seal pump or steam ejector can be used with a diffusion pump to have continuity in pumping from atmospheric pressure to the high or very high vacuum range.

Mechanical pumps. One of the early vacuum pumps was similar in design to the reciprocating steam engine. A power-driven piston working in a

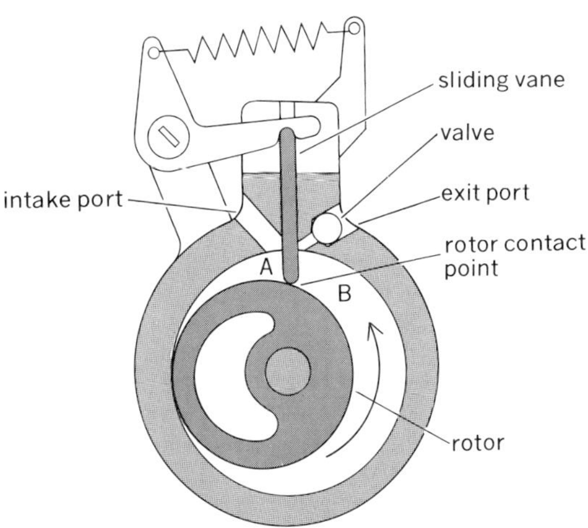

Fig. 2. Chief components of rotary oil-seal pump.

cylinder sucked the gas out of any attached container. A more recent development is the rotary oil-seal pump (Fig. 2). Gas is sucked into chamber A through the opening intake port by the rotor. A sliding vane partitions chamber A from chamber B. The compressed gas that has been moved from position A to position B is pushed out of the exit port through the valve, which prevents the gas from flowing back. The valve and the rotor contact point are oil-sealed. Since each revolution sweeps out a fixed volume, it is called a constant-displacement pump.

A rotary blower pump operates by the propelling action of one or more rapidly rotating lobelike vanes. It does not use oil as a sealing medium.

The molecular drag pump is one that operates at very high speeds, as much as 16,000 rpm. Pumping is accomplished by imparting a high momentum to the gas molecules by the impingement of the rapidly rotating body.

Ejector pumps. The earliest form of this type is the water aspirator (Fig. 3). When water is forced under pressure through the jet nozzle, it will force the gas in the inlet chamber to go through the diffuser, thus lowering the pressure in the inlet chamber. When high-pressure steam is used instead of water, it is called a steam ejector. Steam ejectors often are combined in multiple series with barometric condensers to reach lower pressures. A barometric condenser operates by injecting water into the upper end of a vertical 35-ft (10.7-m pipe whose exit is immersed in a catch basin called a hot well. The gas trapped in the inlet chamber is carried out with the water. Like the water aspirator, it can produce a vacuum (usually between 25 and 40 mm Hg) equal to the vapor pressure of the water. This type of pumping is widely used in the chemical and petroleum industries.

Diffusion pumps permit high pumping speed at very high vacuum. The vapor-jet principle of the ejector pumps has been modified by using oil or mercury vapor diffusing through a jet, then condensing the vapor on the cooled wall of the pump chamber. Figure 4 illustrates the diffusion pump operation. Figure 5 shows a small, 1-in. (2.5 cm)-diameter diffusion pump with pumping speeds as low as 1 liter/sec compared for size with one of the largest pumps, 48 in. (120 cm) in diameter having a pumping speed of 100,000 liters/sec. Sometimes to meet operating requirements one jet is built on the ejector principle. Then it is called a diffusion-ejector pump.

Specialty pumps. In recent years a number of new pumps have been developed which meet special pumping requirements. In all cases it is not the principles that are new, but the way they are used. The ion pump operates electronically. Electrons that are generated by a high voltage applied to an anode and a cathode are spiraled into a long orbit by a high-intensity magnetic field (Fig. 6). These electrons colliding with gas molecules ionize the molecules, imparting a positive charge to them. These are attracted to, and are collected on, the cathode. Thus a pumping action takes place. Ion pumps are used for evacuating nuclear accelerators, x-ray tubes, and other specialized equipment. In a chemical-ion pump, a metal (usually titanium) is evaporated by electrical power, and the metal vapor reacts with chemically active molecules (O_2 and N_2). The reaction products condense on the walls of the pump. The inert gases that remain are removed by devices similar to that of the ion pump. Sorption pumping is the removal of gases by adsorbing and absorbing them on a granular sorbent material such as a molecular sieve held in a metal container. When this sorbent-filled container is immersed in liquid nitrogen, the gas is "sorbed." This type of pump produces a quick,

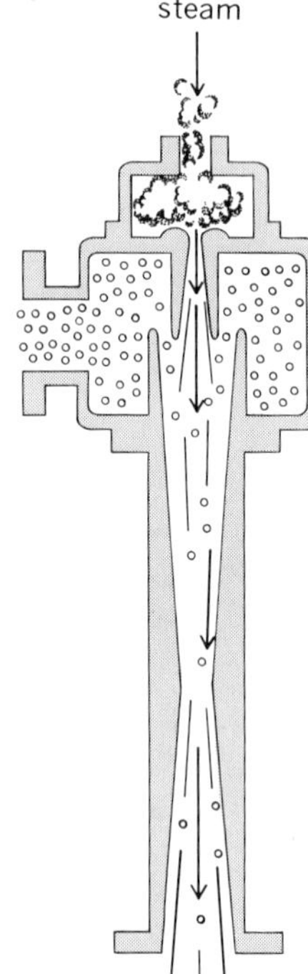

Fig. 3. Diagram of the water aspirator pump. Utilization of (*a*) water and (*b*) steam.

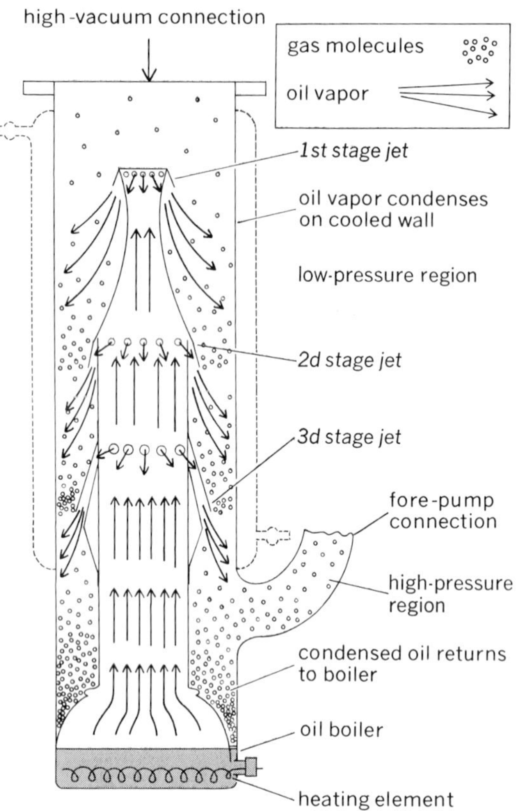

Fig. 4. Main operating features of diffusion pump.

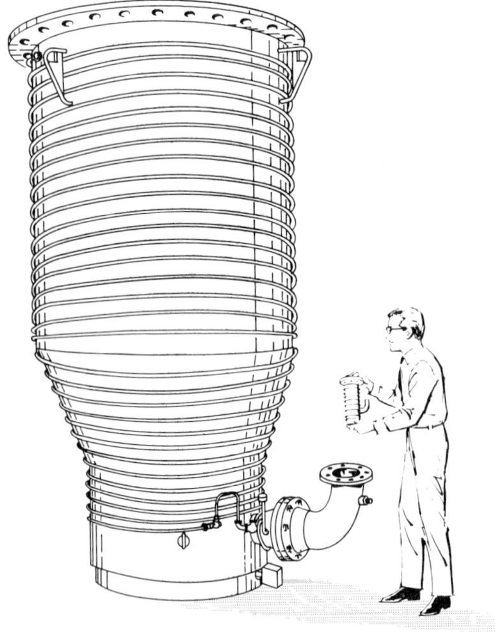

Fig. 5. Size comparison of two diffusion pumps.

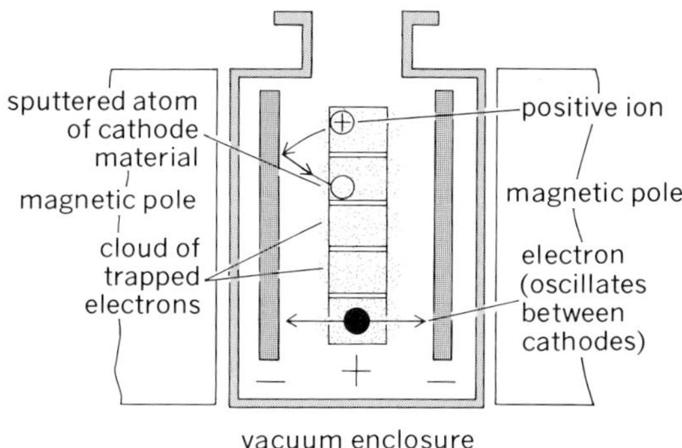

Fig. 6. The principal features of an ion pump.

clean vacuum. Cryogenic pumping is accomplished by condensing gases on surfaces that are at extremely low temperatures. This is usually attained by the use of liquid or gaseous hydrogen at around 20 K (−253°C) or liquid helium at 4 K. The equipment is very complex in design. However, it attains very high pumping speeds for low-density wind tunnels and space simulator chambers.

Vacuum pumping equations. There are two relations of practical use in vacuum technology. The first is $P_1V_1 = P_2V_2$ (at constant temperature). That is, when pressure P_1 is changed to a new pressure P_2, the volume of the gas at P_1 which is V_1 is changed to a new volume V_2.

The second is $Q = SP$. Here Q is the gas throughput (expressed as torr liters/second at a specified temperature), S is the speed of pumping (in this case liters/second), and P is the pressure (in torr) at the point where S is to be measured. The measurement of S requires a vacuum gage to obtain P and one of several methods for measuring Q. One way to determine S is to measure the amount of air at atmospheric pressure that is admitted into the vacuum chamber through a controlled, calibrated leak. This measurement gives Q, and S is then obtained from the second equation. A more precise method, especially useful for very-high and ultrahigh vacuum, is the measurement of a pressure drop across an orifice of known gas conductance. The gas conductance is a mathematically determined quantity obtained from the geometry of the test portion of the vacuum chamber.

[EDWARD S. BARNITZ]

Bibliography: R. R. La Pelle, *Practical Vacuum Systems*, 1972; A. Roth, *Vacuum Technology*, 1976; J. Yarwood (ed.), *Vacuum and Thin Film Technology*, 1978.

Value engineering

A thinking system (also called value management or value analysis) used to develop decision criteria when it is important to secure as much as possible of what is wanted from each unit of the resource used. The resource may be money, time, material, labor, space, energy, and so on. The system is unique in that it effectively uses both knowledge and creativity, and provides step-by-step techniques for maximizing the benefits from both. It promotes development of alternatives suitable for the future as well as the present. This is accomplished by identifying and studying each function that is wanted by the customer or user, then applying knowledge and creativity to achieve the desired function. Resources are converted into costs to achieve direct, meaningful comparisons. By using the methods of value engineering, 15 to 40% reduction in the required resources often results.

Application. Value engineering has applications in five broad areas: in design, purchase, and manufacture of products; in administrative groups, private or public, where the task is to achieve accomplishment through people; in all areas of social service work, such as hospitals, insurance services, or colleges; in architectural design and construction; and in development as well as research.

Approach—function study. The specific functions which the customer or user wants will be of two types: use function, which does something for the user; and esthetic function, which is selected because it pleases the user. Value engineering uses a planned approach for intensive and effective utilization of every applicable technique. It requires the development of sufficient skill in the application of enough techniques to bring into clear view a liberal number of value alternatives. Each function is systematically understood, identified, clarified, and named.

Value. Value is defined as the proper function for the least resources. Best value, then, is the attainment of the full function desired for the lowest cost achievable. The value of a function becomes a vital measure, being the lowest cost of securing the function. After arriving at this measure, high effort is made to achieve a function near its value.

The system is used to improve value in either or both of two situations: (1) The product or service as used or as planned may provide 100% of the functions the user wants, but lower costs may be needed. The system then holds those functions but achieves them at lower cost. (2) The product or service may have deficiencies, that is, it does not perform the desired functions or lacks quality, and so also lacks good value. The system aims at correcting those deficiencies, providing the functions wanted, while at the same time holding the use of resources (costs) at a minimum.

Evaluation of function by comparison. Once the functions have been identified, clarified, understood, and specified, the following question must be answered: What is the lowest cost which, under the present conditions, would provide the described function? The answer is developed by comparisons to the past. These values are established by other valid comparisons, such as: How might an important portion of the function be accomplished, and what would that cost? How would that function be accomplished in a different industry or a different country, under very different conditions, and what would it cost? These values have to be compared with larger, smaller, and similar items or services and their costs. If there is no comparison, there can be no evaluation.

Often this task of creatively evaluating the function of itself brings a good answer to the problem. For example, the Navy was building 1000 landing crafts. One function was to "contain 200 gallons of gasoline." A noncombat life of 8 years was desired. The best quotation for each landing craft was $520. The function was evaluated by comparisons. How else could 200 gal be contained and what would each cost? Four 50-gal drums would have cost $25. A standard 250-gal oil tank, often used with oil burners in homes, cost $30. The $30 figure is selected as a base. Additional costs were considered for some connections, piping, and perhaps coatings that would be needed, so the value of the function was selected to be $50. As a result of the evaluation the Navy elected to use four drums, separated into two groups of two, at a unit cost of $80. Thus, the cost for the job was $80,000 instead of $520,000.

Minimizing normal human negatives. The system, with its intense emphasis on functions, deep searches for knowledge, and constant effective creativity, built into step-by-step techniques, offsets many human traits which retard or prevent beneficial change. Some of the human negatives which act as retardants are: (1) Thoughts tend to follow habit patterns. (2) A decision to change, if proved wrong, may bring embarrassment. (3) Making a change may bring personal risk to the decision maker. (4) Decisions based upon sound general criteria often do not fit the specific instance. (5) The required good decision may be contrary to what is "normally" done. (6) Subjective coloring of attitudes of important people in the area makes good decisions difficult without sound objective data. (7) Decisions vital to profitability are often made by people not accountable for profits. (8) Obscure cause-and-effect relationships in some matters that create costs allow decisions to be made that will injure value. (9) While feelings are strictly personal, they may influence and often control decisions affecting value. (10) Most environments are hazardous to anything new, or to a change. Sound, objective alternatives developed by the value engineering problem solving system do much to overcome these human retardants.

Job plan. All minds must be "tuned" to work on exactly the same problem at exactly the same time. A job plan is thus developed in order to present the problems that will be faced and to establish the functions that should be accomplished. The job plan requires the following five steps: (1) gathering extensive information about the problem area; (2) analyzing the information for meaning and sense of direction; (3) doing the essential creative work; (4) judging the results; (5) creating a development planning program. The first four steps all require a different mental activity, and each has to be thoroughly completed before proceeding onto the next step.

Information step. In this step, all the facts and pertinent information (such as costs, quantities, and specifications) have to be obtained because only through complete understanding of the situation can valuable assessments be made. Assumptions have to be sorted out and reviewed to determine if they can be replaced or supported by facts. Long-standing assumptions have to be especially checked for validity.

Analysis step. This step involves the development of "function" thinking. From the information collected in the first step, functions are developed to answer such questions as: What are the problems involved, and which must be solved first? Are the solutions to these problems reasonable? What goals should be aimed for, and what steps should be taken in order to achieve them? Is any more information needed? Have any assumptions been overlooked, and are all assumptions already noted still valid? Have the best approaches to these problems been developed and, if so, what savings or benefits will result? Should better solutions be sought? What problems, if presented with better solutions, would produce even more beneficial results for the project?

Creativity step. After acquiring the relevant information and reaching an understanding of the problems through analysis, one must apply creative thinking, involving free use of the imagination. Two aspects of the mind are important here: the diverse knowledge bits that it has acquired over the years, and the ability to join these bits in various combinations constituting temporary "mental pictures" that may be unique. Thus a number of fresh alternative solutions to consider emerge in solving the overall problem or individual problems.

Judgment step. This step involves creatively studying the ideas presented in the previous step. No ideas should be thrown out. Rather, they should be developed and improved into better ones. Ideas involving monetary value should especially be studied closely and objectively in order to seek out their limitations and to try to lessen, overcome, or eliminate any negative aspects. It may be necessary to send certain ideas back to step one and to run them through the entire process again. Some ideas may have so many advantages that the lessening of their drawbacks may become the principal point of concern. And those ideas which appear to be completely thought through and seem capable of providing the greatest yield are sent on to step five.

Development planning step. In this step the best specialists and vendors are selected for consultation, and an investigation program is established to provide the most recent information on and the most current capabilities of any of the approaches that show potential.

VECP (value engineering change proposals). When this thinking system is used on government or military supply or construction situations, the same or better quality is usually secured for millions of dollars less in cost. To make it profitable for the manufacturer or contractor to hire value engineers and make this contribution to the government, a percentage of the saving, often 40 to 50%, for a time is paid on VECPs. Hundred of millions of tax dollars are being saved each year because of VECPs. In addition, resources are being conserved, and millions of dollars are being saved by owners, because of the work of the architectural firms and construction companies that have trained and qualified value engineers.

Training and qualification. Skilled consultants and some universities teach the techniques of value engineering. A professional society, the Society of American Value Engineers, with chapters in many cities, sets standards, gives examinations, and awards the citation of "Certified Value Specialist" (CVS) to qualified people.

World status. Extensive application of value engineering is growing in the United States, Japan, Germany, Sweden, France, Canada, and England. Important application is growing in Norway, Italy, Spain, Korea, Taiwan, South Africa, India, and other places. Professional societies exist in the United States, Japan, Scandinavia, France, and South Africa. *See* INDUSTRIAL COST CONTROL;

INDUSTRIAL ENGINEERING; METHODS ENGINEERING; OPERATIONS RESEARCH; PROCESS ENGINEERING; PRODUCTION ENGINEERING; PRODUCTION PLANNING.

[LAWRENCE D. MILES]

Bibliography: A. J. Dell'Isola, *Value Engineering in Construction*, 1975; C. Fallon, *Value Analysis*, 2d rev. ed.; W. L. Gage, *Value Analysis*, 1967; M. C. Macedo Jr., P. V. Dobrow, and J. J. O'Rourke, *Value Management for Construction*, 1978; L. D. Miles, *Cutting Costs by Analyzing Values*, 1952; L. D. Miles, *Techniques of Value Analysis and Engineering*, 2d ed., 1972; A. Mudge, *Value Engineering*, 1971; J. J. O'Brien, *Value Design and Construction*, 1976; *Value Engineering and Management Digest*, monthly.

Valve

A flow-control device. This article deals with valves for fluids, liquids, and gases. Valves are used to regulate the flow of fluids in piping systems and machinery. In machinery the flow phenomenon is frequently of a pulsating or intermittent character and the valve, with its associated gear, contributes a timing feature.

Pipe valves. The valves commonly used in piping systems are gate valves (Fig. 1), usually operated closed or wide open and seldom used for throttling; globe valves (Fig. 2), frequently fitted with a renewable disk and adaptable to throttling operations; check valves (Fig. 3), for automatically limiting flow in a piping system to a single direction; and plug cocks (Fig. 4), for operation in the open or closed position by turning the plug through 90° and with a shearing action to clear foreign matter from the seat.

Valves may have various structural features such as outside stem and yoke; packless construction; angle, as opposed to straightway flow; power instead of manual operation; and combined nonreturn and stop-valve arrangements. Valves are made in a wide assortment of materials, and a wide variety of trim, with brass or bronze for general service; cast iron for low steam pressures and temperatures (less than 250 psi or 1.7 MPa) and for hydraulic pressures below 800 psi (5.5 MPa); steel and alloy steels for the highest operating pressures and temperatures (such as 5000 psi or 34 MPa, 1200°F or 650°C steam); and selected metals for chemical and process applications. Most valves are manufactured and available as hardware and comply with the requirements of the ASTM, ANSI, and ASME as to material and dimensional standards. They are variously offered as flanged, screwed, welded, sweated, or compression-fitted for connection to pipe, machinery, and fittings.

Safety and relief valves are automatic protective devices for the relief of excess pressure. They are usually rigorously specified under the legal regulations of public authorities and insurance underwriters. They must open automatically when the pressure exceeds a predetermined value; they must allow the pressure to drop a predetermined amount before closing to avoid chattering, instability, and damage to the valve and the valve seat; they must have adjustment features for both the relieving and blowdown pressures; and they must be tamperproof after setting by responsible licensed operators. *See* SAFETY VALVE.

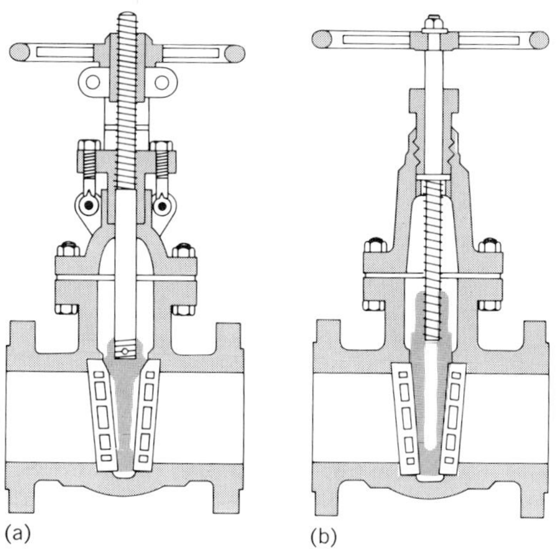

Fig. 1. Gate valves with disk gates shown in black. (*a*) Rising threaded stem shows when valve is open. (*b*) Nonrising stem valve requires less overhead.

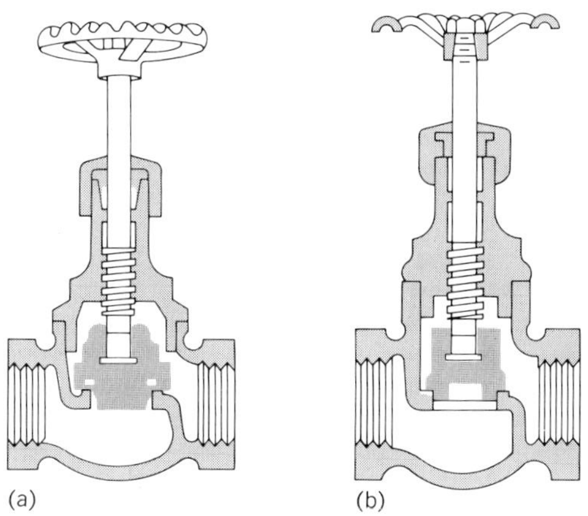

Fig. 2. Globe valves. (*a*) With gasket in disk. (*b*) With ground metal-faced disk.

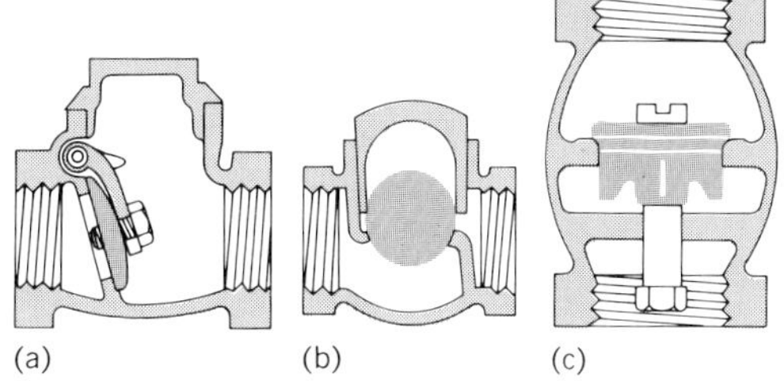

Fig. 3. Various types of straightway check valves. (*a*) Swing. (*b*) Ball. (*c*) Vertical.

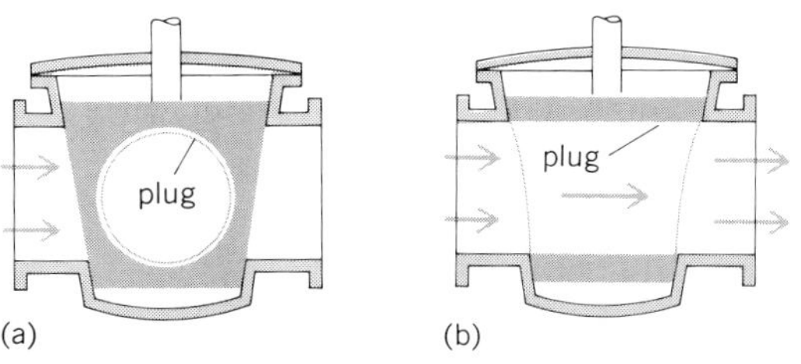

Fig. 4. Plug valve. (*a*) Closed. (*b*) Open.

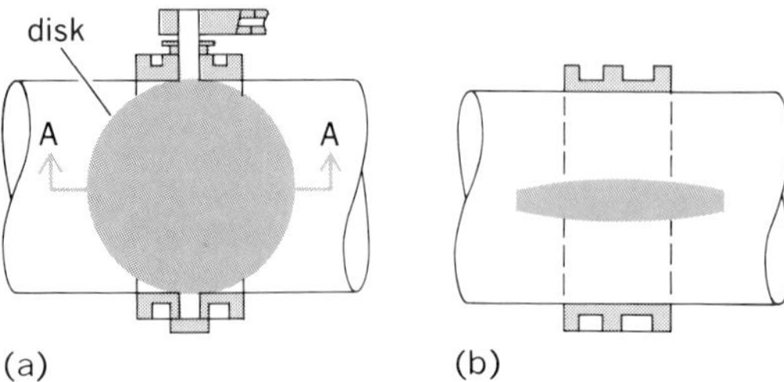

Fig. 5. Butterfly-type valve for penstock is typically 25 ft (7.6 m) in diameter. (*a*) Elevation. (*b*) Section A-A.

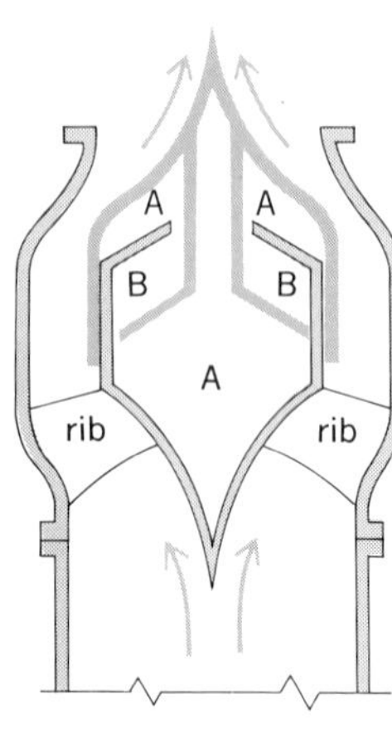

Fig. 6. Large needle valve is actuated hydraulically by pressure in chambers A to close and in annular chambers B to open.

Hydraulic turbine valves. For hydraulic turbines and hydroelectric systems, valves and gates control water flow for (1) regulation of power output at sustained efficiency and with minimum wastage of water, and (2) safety under the inertial flow conditions of large masses of water. Valve sizes are usually large (for example, 6 ft (1.8 m) in diameter) so that power operation is necessary. Carefully streamlined construction to minimize fluid dynamic losses must be accompanied by ample provision to withstand shock and damaging effects of hydraulic inertia. Gate, butterfly (Fig. 5), telescoping, and needle constructions (Fig. 6) are variously employed. Wicket or cylinder gates regulate the flow of water to a reaction turbine at the speed ring while a governor-operated needle valve regulates flow to a Pelton impulse unit. *See* HYDRAULIC TURBINE.

Steam-engine valves. To control the kinematics of the cycle, steam-engine valves range from simple D-slide and piston valves to multiported types. Slide valves control admission and release of steam to and from a double-acting cylinder by a single moving valve mechanism giving the necessary lap, lead, and angle of advance to accomplish the predetermined values of cutoff and compression. Multiported valves such as plug, Corliss, or poppet, valves provide four valves for a double-acting cylinder. Each valve serves a single purpose of admission or exhaust for the head or crank end. The uniflow construction uses a poppet valve for admission with a row of exhaust ports alternately covered and uncovered by the engine piston. Plug and slide valves are limited to low pressures and temperatures (200 psi or 1.4 MPa and 100°F or 55°C of superheat); poppet valves will operate under the maximum pressures and temperatures of steam-engine practice without warping. Many types of reversing gear have been perfected which use the same slide valve or piston valve for both forward and backward rotation of an engine, as in railroad and marine service. *See* STEAM ENGINE.

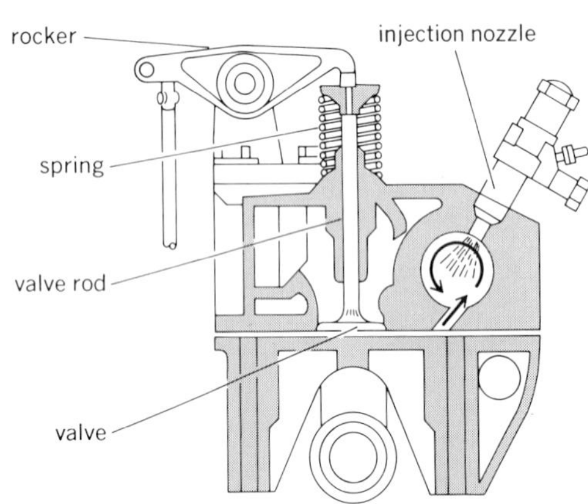

Fig. 7. Poppet valve for internal combustion engine. (*From T. Baumeister, ed., Standard Handbook for Mechanical Engineers, 7th ed., McGraw-Hill, 1967*)

Internal combustion engine valves. Poppet valves are used almost exclusively in internal combustion reciprocating engines because of the demands for tightness with high operating pressures and temperatures. The valves (Fig. 7) are generally 2 in. (5 cm) in diameter or smaller on high-speed automotive-type engines, are cam-operated and spring-loaded, with their lift a small fraction of an inch, and with a gas velocity of 200 – 300 ft/sec (60 – 90 m/sec). They are cooled by transfer of heat to the engine jacket mostly through the valve stem. In heavy-duty units, the stem may have a partial mercury or sodium filling which aids, by inertia, in conducting heat to the valve stem and guides. Exhaust valves are subject to the effects of extreme temperature and must accordingly be most carefully designed and constructed of alloy materials. Two-cycle engines utilize ports, alternately covered and uncovered by the main piston, for inlet or exhaust. *See* CAM MECHANISM; INTERNAL COMBUSTION ENGINE; VALVE TRAIN.

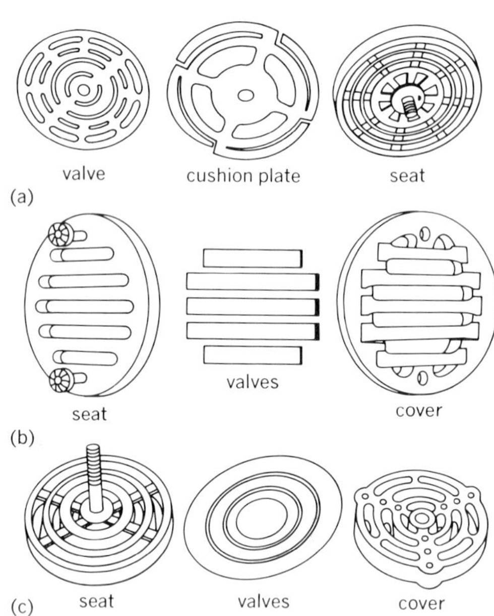

Fig. 8. Air and gas compressor valves. (*a*) Ingersoll-Rand ring plate. (*b*) Worthington feather valve. (*c*) Chicago pneumatic simplate valve. (*From T. Baumeister, ed., Standard Handbook for Mechanical Engineers, 7th ed., McGraw-Hill, 1967*)

Compressor valves. In compressors, valves are usually automatic, operating by pressure difference (<5 psi or 34 kPa) on the two sides of a movable, spring-loaded member and without any mechanical linkage to the moving parts of the compressor mechanism (Fig. 8). The practical objectives are timing, tightness, and life. Pressure drop should be a minimum. Moving parts must be light, but spring loadings should be heavy enough to avoid chatter and to give certainty to the valve action. Air speeds are in the order of 5000 ft/min

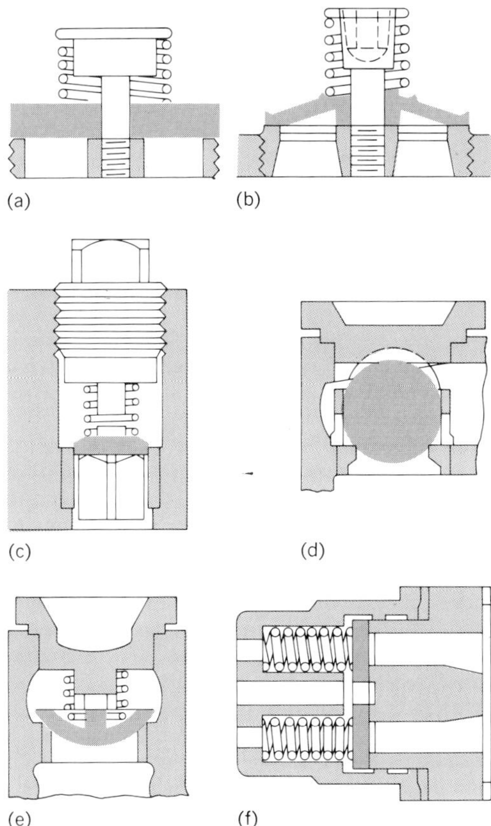

Fig. 9. Pump valves. (*a, b*) Disk valves are for moderate pressure. (*c*) Conical-faced wing valves are for high pressure (1000 psi± or 7 MPa±). (*d, e*) Ball and rounded valves are for viscous liquids. (*f*) Double-ported valves are for high-speed reciprocating pump.

(1500 m/min) through the valves. Valves for inlet and exhaust service are frequently made interchangeable in the interest of reduced investment. On high-pressure (>1000 psi or 7 MPa) service, automatic poppet valves are usually substituted for the more common plate valves to give greater strength and tightness under loading. *See* COMPRESSOR.

Pump valves. Like those for compressors, pump valves are usually of the automatic type operating by pressure difference. The service conditions, however, are very dissimilar because of the noncompressibility of liquids: the presence of entrained solids like grit, fibers, and sludge: the corrosive potential of chemicals like acids and alkalies; the high viscosity of many liquids; the vapor pressure at the pump-operating temperature; and inertia effects accompanying discontinuity of the liquid column. Fluid speeds are low (200–300 ft/min or 60–90 m/min with cold water and 100 ft/min or 30 m/min with viscous liquids). A multiplicity of small valves (4± in. or 10 cm± in diameter) arranged in docks with a positive suction is a common construction. Various constructions are used (Fig. 9). [THEODORE BAUMEISTER]

Bibliography: T. Baumeister (ed.), *Standard Handbook for Mechanical Engineers*, 8th ed., 1978.

Valve train

The valves and valve-operating mechanism by which an internal combustion engine takes air or a fuel-air mixture into the cylinders and discharges combustion products to the exhaust. *See* VALVE.

Mechanically, an internal combustion engine is a reciprocating pump, able to draw in a certain amount of air per minute. Since the fuel takes up little space but needs air with which to combine, the power output of an engine is limited by its air-pumping capacity. It is essential that the flow through the engine be restricted as little as possible. This is the first requirement for valves. The second is that they close off the cylinder firmly during the compression and power strokes. *See* INTERNAL COMBUSTION ENGINE.

Valve action. In most four-stroke engines the valves are of the inward-opening poppet type, with the valve head ground to fit a conical seat in the cylinder block or cylinder head (Fig. 1). The valve head is held concentric with its seat by a cylindrical stem running in a valve guide. The valve is held closed by a compressed helical spring. The spring must be strong enough to overcome the inertia of

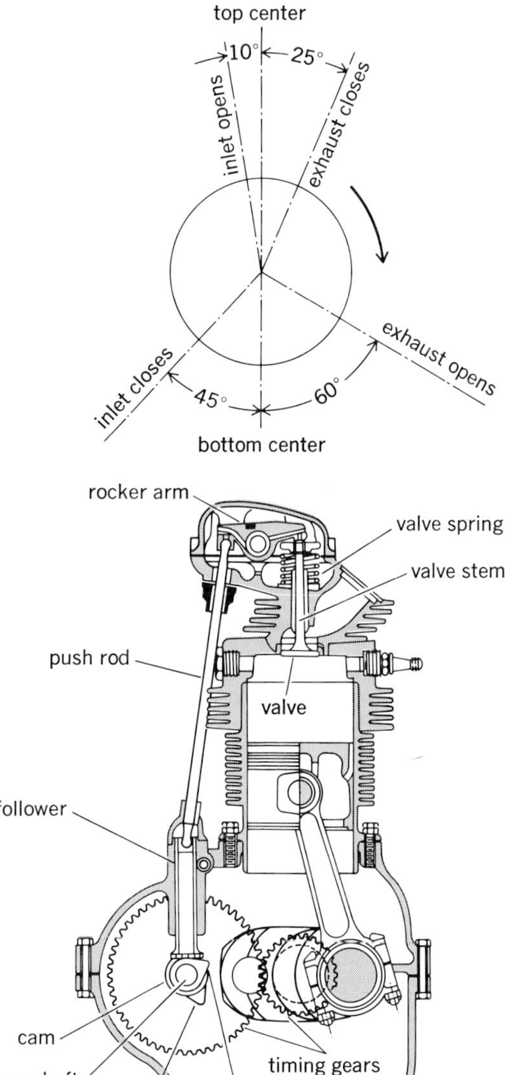

Fig. 1. Cross section of one-head cylinder and valve timing diagram of a four-cycle engine. (*From C. H. Chatfield, C. F. Taylor, and S. Ober, The Airplane and Its Engine, 5th ed., McGraw-Hill, 1949*)

the valve mechanism and to hold the valve follower on the cam at all times. The valve is opened wide by lifting it from its seat a distance equal to approximately 25% of the valve diameter. At this lift the cylindrical opening between valve and seat will be about as large as the cross section of the flow passage to the valve seat.

Valve action requires that the valve be streamlined and as large as possible to give maximum flow, yet be of low inertia so that it follows the prescribed motion at high engine speed. Valves are usually made of a stainless, nonscaling alloy which will keep its strength and shape at high temperature. The exhaust valve is sometimes made hollow and partially filled with metallic sodium. In operation the sodium melts and shakes back and forth, transferring heat from the valve head to the stem and valve guide. Valve seat inserts of tough heat-resistant alloys are used in cylinders of high performance engines or when the cylinder material alone is not strong enough for this purpose.

Cam action. Engine valves are usually opened by means of cams. Riding on each cam is a follower, or valve lifter, which may be a flat or slightly convex surface, or a roller (Fig. 2). The valve is opened by forces applied to the end of the valve stem through a mechanical linkage actuated by the cam follower. In overhead valve engines, the camshaft may be mounted on the cylinder head near the valves. In less expensive construction, the camshaft is placed in the crankcase. The operating linkage then consists of cam follower, push rod, and rocker arm. The push rod is a light rod or tube with ball ends, which carries the motion of the cam follower to the rocker arm. The rocker arm is a lever pivoted near its center so that, as the push rod raises one end, the other end depresses the valve stem, opening the valve.

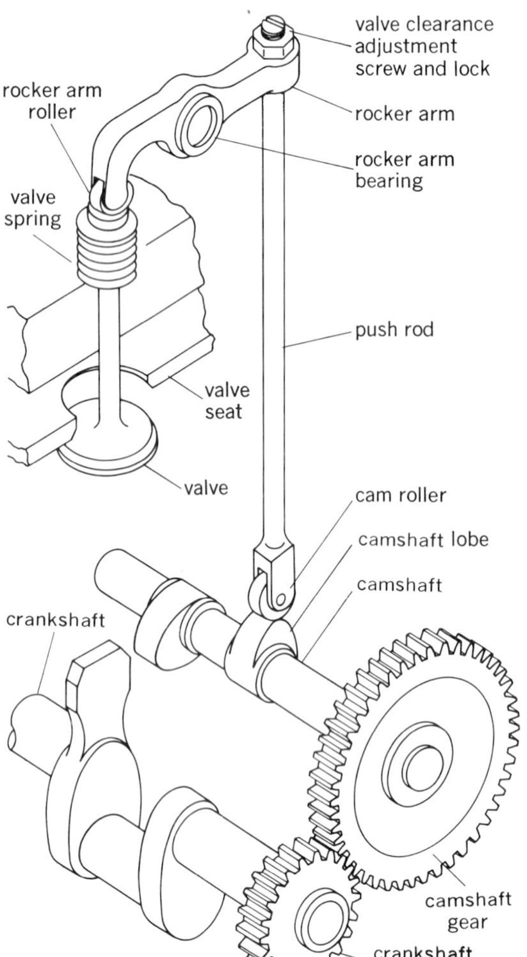

Fig. 2. Crankshaft gear meshes with the camshaft gear.

To ensure tight closing of the valve even when the valve stem lengthens from thermal expansion, the valve train is adjusted to provide some clearance when the follower is on the low part of the cam. The cam shape includes a ramp which reduces shock by starting the lift at about 2 ft (0.6 m) per second even though the clearance varies from time to time. To open the valve quickly, an acceleration of about 400 times gravity (for automotive-size engines) is used. Excessive acceleration deflects the valve linkage, giving false motion to the valve, causing it to close at high velocity before the closing ramp is reached. The cam surface between the opening and closing acceleration sections includes the point of maximum lift and maximum deceleration. If this part of the cam is sinusoidal and the valve spring is properly designed, the spring forces remain proportional to the forces required by the cam for the deceleration of the valve. The highest operating speed for a given lift can thus be obtained without the follower leaving the cam. Sometimes the valve is held at maximum lift for a time. This part of the cam is callled the dwell. *See* HYDRAULIC VALVE LIFTER.

Timing. At low piston speeds gas friction and inertia are small, and the valves should be opened or closed as the piston reaches the end of the appropriate stroke. At high piston speeds the inertia of the charge in the inlet passages causes the gas to continue to flow into the cylinder long after the piston has started up on the compression stroke. To trap the maximum amount of fresh charge in the cylinder, it is therefore necessary to delay the closure of the inlet valve up to about 60° after bottom center, depending upon the geometry of the inlet passages and the piston speed. Late inlet closing increases the torque or pulling power of the engine at high speeds, but reduces torque at low speeds, because of the loss of fresh charge back into the inlet system during compression. To reduce the work required to expel the products of combustion during the exhaust stroke, it is necessary, at high speeds, to open the exhaust valve considerably before bottom center so that most of the gas is blown out and the cylinder pressure has fallen to nearly atmospheric before the exhaust stroke begins. In supercharged engines the inlet valve often opens near top center before the exhaust valve closes. The time during which both valves are open is called the overlap period. Overlap permits scavenging of the combustion space with fresh charge for valve cooling and increased power, but results in poor idling characteristics because combustion products are drawn back through the engine by the low pressure of the inlet system when the throttle is nearly closed.

[AUGUSTUS R. ROGOWSKI]

Vapor condenser

A heat-transfer device that reduces a thermodynamic fluid from its vapor phase to its liquid phase. The vapor condenser extracts the latent heat of

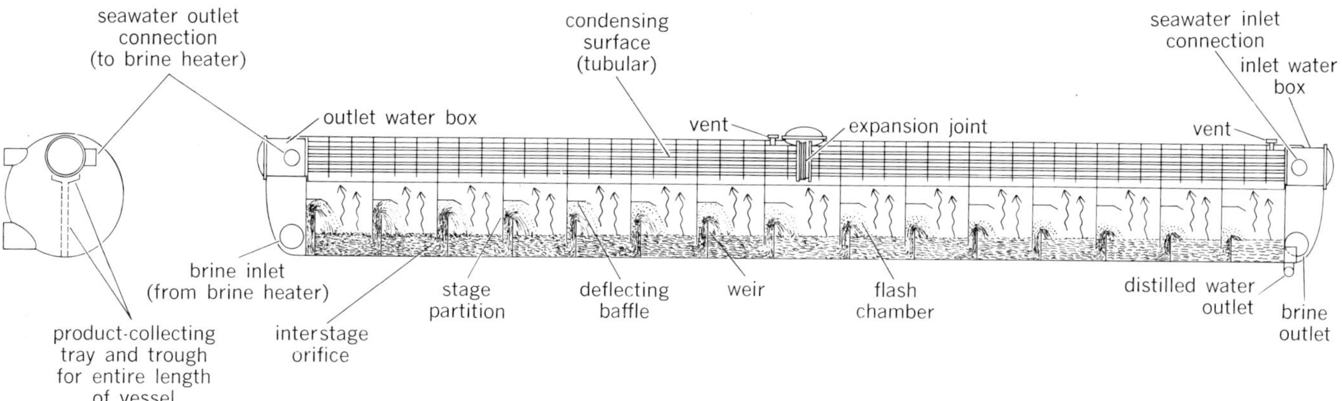

Fig. 1. Simplified views of multistage flash evaporator for production of potable water from sea water.

vaporization from the vapor, as a higher-temperature heat source, by absorption in a heat-receiving fluid of lower temperature. The vapor to be condensed may be wet, saturated, or superheated. The heat receiver is usually water but may be a fluid such as air, a process liquid, or a gas. When the condensing of vapor is primarily used to add heat to the heat-receiving fluid, the condensing device is called a heater and is not within the normal classification of a condenser.

Classification by use. Condensers may be divided into two major classes according to use: those used as part of a processing system, and those used for serving engines or turbines in a steam power plant cycle.

Process condensers. In a processing system condensers selectively recover liquid from a mixed vapor, recover pure liquid from the vapor of an impure liquid, recover noncondensables from a mixture of gas and vapor, or extract heat from heat-pump, refrigeration, and cryogenic cycles. The vapor condensed may be steam or the vapor from any nonaqueous liquid. Condensers used for any purpose except as vapor condensers in the power plant cycle, even though they provide the heat sink for a process cycle, are classified as process condensers.

One of the more important uses of process condensers is that of condensate recovery. Condensers are frequently used with fractionating and distillation columns in the production of hydrocarbon liquids and in processes for liquefying gases. Similarly they are used in the production of distilled water as distiller condensers in single- and multiple-effect evaporating systems. The multistage flash evaporator with integral vapor condensers has been established as one of the more economical means for the production of potable water from sea water. This apparatus contains a multiplicity of condensers in series, each condenser producing a successively lower pressure in which flashed vapor from sea-water brine is produced, condensed, and collected. Both distillate and the flashing brine streams reduce in temperature as they flow through the apparatus (Fig. 1).

Power cycle condensers. Used as part of the power plant cycle, condensers serve as the heat sink in the cycle and reduce the back pressure on the turbine or engine so that a maximum of heat energy becomes available as useful work. Steam (water) is the most common substance used for the power generation cycle. Other thermodynamic fluids can be used but seldom are. Another important function of a condenser in the steam power plant cycle is the recovery of condensate as the major source of pure boiler feedwater for the boilers. *See* FEEDWATER; THERMODYNAMIC CYCLE.

Classification of operation. Condensers may be further classified according to mode of operation as surface condensers or as contact condensers.

Surface condensers. In surface condensers the condensing vapor and the cooling fluid remain separated from each other by a dividing wall or walls, which form the heat-transfer surface. This heat-transfer surface is most often in the form of tubes but may also be plates or partitions of various geometries. Condensate, cooling fluid, and noncondensable gases are usually removed separately, although in some designs the condensate and noncondensable gases are withdrawn as a mixture. *See* SURFACE CONDENSER; VACUUM PUMP.

Contact condensers. In the contact condenser the vapor and cooling liquid come into direct contact with each other and are mixed in the condensing process. The condensed vapor and cooling liquid combine and are withdrawn from the apparatus together. The noncondensable gases are usually withdrawn separately, although in some contact condensers they are entrained in the mixed condensate and cooling liquid and removed through a common outlet. *See* CONTACT CONDENSER.

Removal of noncondensables. Condensers are required, almost without exception, to condense impure vapors, that is, vapors containing air or other noncondensable gases. Because most condensers operate at subatmospheric pressures, air leaking into the apparatus or system becomes a common cause for vapor contamination. Many process condensers are used to condense vapors whose noncondensable impurities are relatively independent of air that may leak into a vacuum system. Because the vapor supplied to the condenser is continuously reduced to liquid, the noncondensable gases in the gas-vapor mixture collect and concentrate. Their accumulation seriously affects heat transfer, and means must be provided to direct them to a suitable outlet. Most surface and contact condensers are arranged with a separate zone of heat-transfer surface within the condenser and located at the outlet end of the vapor flow path for efficient removal of the noncondensa-

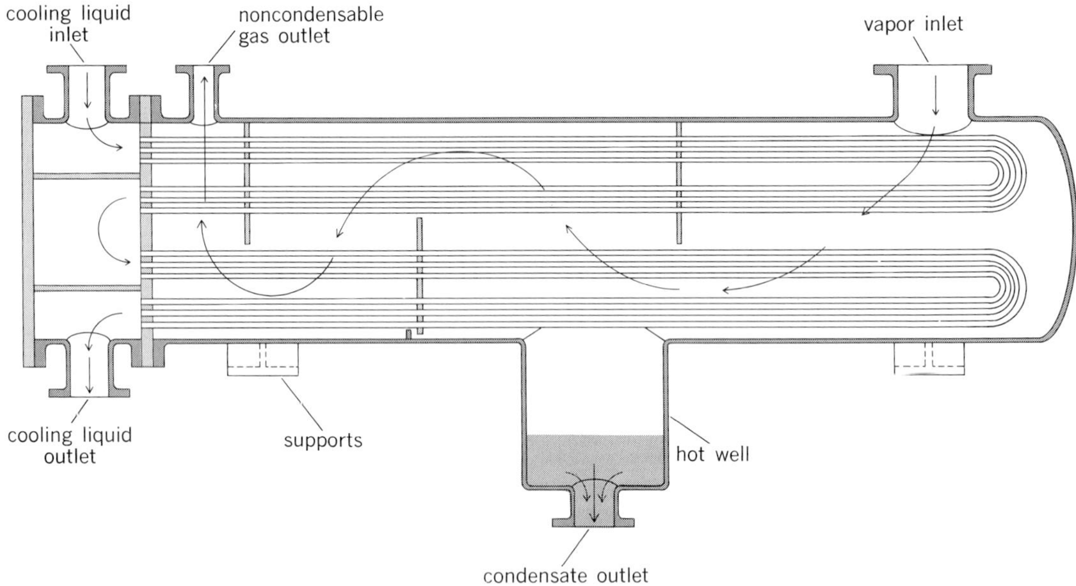

Fig. 2. Process condenser with one baffled-shell pass of condensing vapor and four tube passes of cooling liquid.

ble gases through dehumidification. A vapor flow path of the displacement type, free from zones of stagnation and short circuiting, is essential for achieving maximum condensing heat-transfer rates (Figs. 2 and 3).

Separate external vapor condensers arranged in series with the vapor flow path of the main condenser are used when the ratio of noncondensable gases to condensing vapor is high, or when the vapor content of the noncondensable gases must be reduced to low values.

Removal of noncondensables from condensers operating at subatmospheric pressure requires vacuum pumps. The noncondensables are removed from condensers operating above atmospheric pressure by venting to atmosphere or to aftercondensers to reduce further the moisture content of the removed gases before discharging them to atmosphere or to a recovery process.

Heat-receiving fluid. Water is the most commonly used liquid for absorbing heat from condensing vapors. Liquid hydrocarbons and other chemical compounds in liquid state are used primarily as heat receivers for condensers, used as a part of a petroleum-refining or chemical-manufacturing process. Because of natural evaporation, surface waters are normally at a lower temperature than ambient air when air temperatures exceed 32°F (0°C), and thus provide the lowest-temperature cooling medium readily available. High specific heat, ease in pumping, and rapid heat transfer characterize water as an excellent medium for use as a heat receiver in a condensing system. The tendency of water to corrode metal surfaces is not

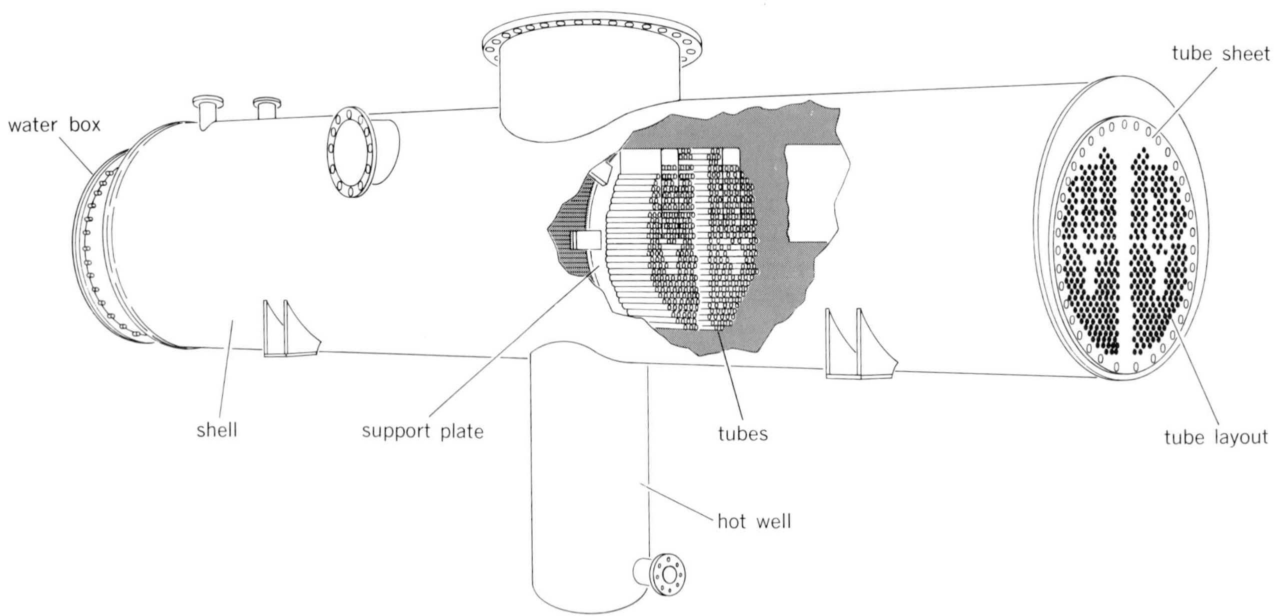

Fig. 3. Typical sections of small steam surface condenser.

usually a serious disadvantage. Thin-walled heat-transfer surfaces of the less costly corrosion-resistant alloys are generally satisfactory. Ferritic materials used for the water-containing parts of condensers can be made suitable for the corrosion environment by increasing thickness as a corrosion allowance or by the use of corrosion-resistant linings or by cathodic protection. *See* CORROSION.

Air is the gas most commonly used for absorbing heat from a condensing process. However, its low heat capacity, low density, and relatively low heat transfer preclude it from being an ideal fluid for this purpose. Its availability and the fact that air does not readily corrode or foul heat-transfer surfaces offer some advantages in its use. Shortages of water for cooling or for use as a heat sink and the danger of thermal pollution of surface water have markedly increased the use of air. Application of extended-surface tubes to process condensers, to process coolers, and to steam condensers serving turbines is reasonably economical and practical where water supplies are critical.

The cooling of heat-transfer surfaces by evaporation involves the use of both water and air. The heat-transfer surfaces are continuously wetted by water. Air, blown over the wetted surfaces, absorbs the water vapor as it is released from the evaporating cooling water, thereby removing the heat released in condensers.

Condensate cooling. In those process systems where recovery of condensate is of primary importance, it is also desirable to cool the condensate below the saturation temperature of its condensing vapor. Some condensate cooling results as condensate falls from tube to tube in the process of being collected and accumulated. If additional cooling is needed, condensate cooling sections in the condenser can be provided by flooding a selected amount of surface within the condenser with condensate. Flooding is done either by installing baffles in the condensate drain zone to flood the tubes in these sections or by effecting a similar result with a loop in the condensate drain piping. Baffles are most effective in process condensers of the horizontal type, and loops are more adaptable to process condensers of the vertical type.

Condensate reheating. Modern steam surface condensers for turbines are provided with means for condensate reheating (Fig. 4). Condensate falling from tube to tube becomes cooled below the temperature corresponding to condenser operating pressure; unless provision is made for reheating this condensate before it is removed from the condenser, a measurable amount of heat energy is lost from the cycle, and cycle efficiency is reduced. In addition, subcooled condensate absorbs air from the condensing vapor and, if allowed to remain in the boiler feedwater, may cause serious corrosion of the feed system and boilers.

The condensate is reheated by a portion of the incoming steam. Baffles direct this steam to the condenser hot well in such a manner that its velocity energy is converted to pressure (Fig. 5). As a result, the local static pressure and corresponding temperature become equal to, or greater than, the static pressure and corresponding temperature at the condenser stream inlet. Condensate falling from the tube bundle to the hot well is reheated, the degree of reheat depending on the height of fall and the flow rate per unit area. If vertical height is limited, effective reheating and the associated deaeration can be achieved by redirecting the condensate falling from the tubes over collecting baffles and a series of trays or plates. This higher zone of pressure and temperature is vented along with the noncondensables removed from the condensate into the condenser tube bank and finally to the air cooler joining the mainstream of noncondensables being cooled and expelled. Under favorable conditions the condensate may reach temperatures as much as 5°F (3°C) above the temperature corresponding to static pressure at the condenser steam inlet, and the dissolved oxygen content of the deaerated condensate may be consistently less than 0.01 cm³ per liter.

Condenser capacity. The quantity of cooling fluid required to condense a given quantity of vapor may be obtained by applying the general heat-balance equation. For a condensing vapor which may be superheated, saturated, or wet, with its condensate subcooled or reheated and with the noncondensable gases cooled, the relation is given by Eq. (1). (The standard engineering practice is to express quantities in lb/hr and enthalpies in Btu/lb.)

$$W_v(h_v - h_c) + W_n(h_{na} - h_{nb}) = W_f(h_{fb} - h_{fa}) \quad (1)$$

W_v = quantity of vapor
h_v = enthalpy of entering vapor
h_c = enthalpy of leaving condensate

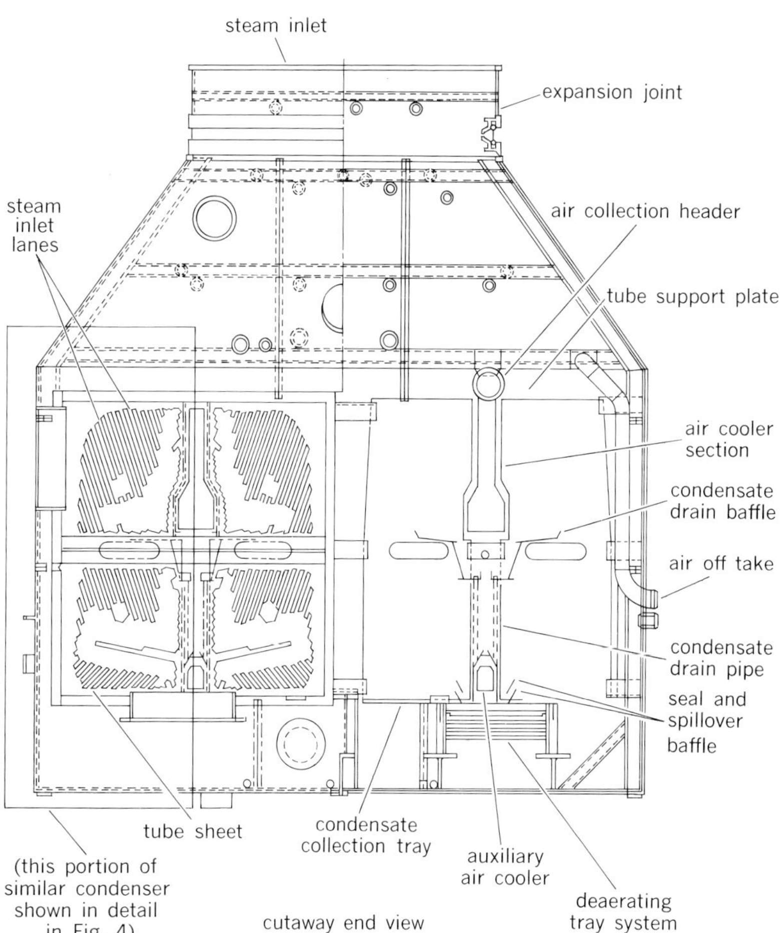

Fig. 4. Typical section of steam surface condenser. (*Worthington Corp.*)

W_n = quantity of noncondensable gases
h_{na} = enthalpy of entering noncondensables
h_{nb} = enthalpy of leaving noncondensables
W_f = quantity of cooling fluid
h_{fa} = enthalpy of entering cooling fluid
h_{fb} = enthalpy of leaving cooling fluid

The temperature of the leaving cooling fluid is usually selected to be 5–10°F (3–6°C) less than the condensing temperature of the saturated vapor. Ordinarily, low temperature differences are associated with contact condensers; larger temperature differences are characteristic of economic design for surface condensers. Small temperature differences require enormous amounts of condensing surface and are seldom a practical design criterion.

For both contact and surface condensers the conditions for determining the hourly heat transferred can be expressed by Eq. (2) for undirectional steady-state heat flow. Logarithmic mean temperature difference Δt_m is based on the assumption of constant specific heat and substantially constant condensing temperature with the cooling-fluid temperature increasing from inlet to outlet in its flow path. Overall heat-transfer coefficient U and surface A are generally applicable to surface condensers only.

$$Q = UA\,\Delta t_m \tag{2}$$

Q = hourly heat, Btu/hr
U = overall heat-transfer coefficient, Btu/(hr)(ft²)(°F)
A = area of heat-transfer surface, ft²
Δt_m = logarithmic mean-temperature difference, °F

For direct-contact condensers where surface is not readily determined, it is usual to consider heat transferred per unit volume of condensing space rather than per unit area. In this case Eq. (3) describes the heat flow, where K = volumetric heat-transfer coefficient, Btu/(hr)(ft³)(°F); V = volume of direct contact condensing space, ft³; and the other symbols are as previously defined.

$$Q = KV\,\Delta t_m \tag{3}$$

The magnitude of K depends on the physical properties of the condensing vapor and heat-receiving fluid, the type of direct-contact surface (spray, tray, or packing), and the operating temperature. It is ordinarily determined experimentally for each type of contact condenser.

The magnitude of U, for use in Eq. (2), can be obtained from the summation of Eq. (4) for the individual resistances in the heat-transfer system, resistance being expressed in (hr)(ft²)(°F)/Btu.

$$U = 1/(r_v + r_l + r_f + r_w) \tag{4}$$

$1/r$ = conductance across any boundary, Btu/(hr)(ft²)(°F)
r_v = condensing boundary resistance
r_l = cooling-fluid boundary resistance
r_f = fouling, dirt, or scale resistance
r_w = separating wall, resistance

Conductance of the condensing boundary $1/r_v$ may be determined analytically for pure vapors. For steam on the outside of horizontal tubes, $1/r_v$ may be 2000–2500 Btu/(hr)(ft²)(°F) [11,400–14,200 J/(s)(m²)(°C)] for well-designed multitube condensers. Conductance of the cooling-fluid boundary $1/r_l$, for fluids flowing inside of tubes, may be determined from the relation of the Nusselt number to the product of the Reynolds and Prandtl numbers with the use of empirical constants such as those suggested by W. H. McAdams. Excellent agreement between computed values and test values results from the use of these equations. For water inside of tubes $1/r_l$ may be in the order of 1000–1400 Btu/(hr)(ft²)(°F) [5700–8000 J/(s)(m²)(°C)] for the normal range of water velocities.

Values for conductance of the condensing boundary and the cooling-fluid boundary for some

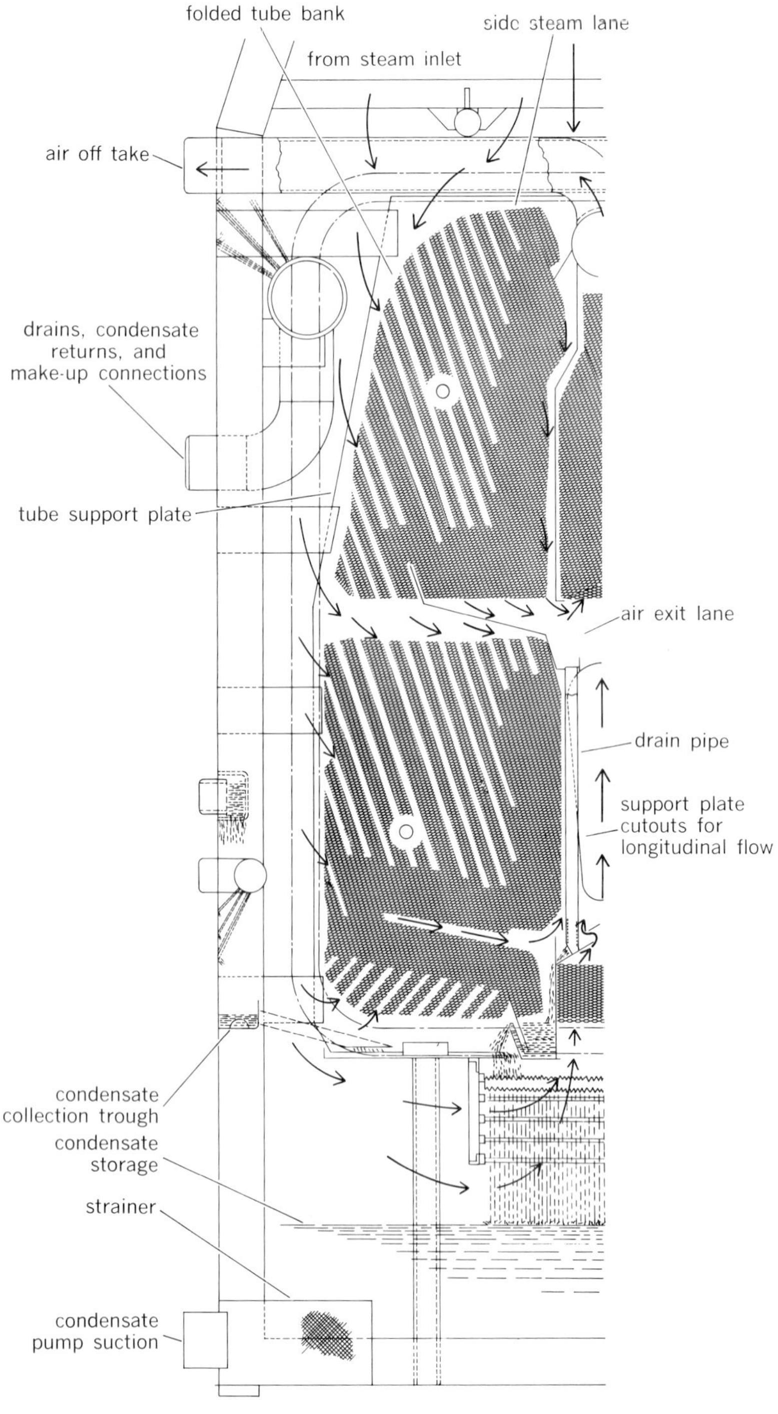

Fig. 5. Detail of portion of condenser similar to that of Fig. 4. (*Worthington Corp.*)

Conductance values for some thermodynamic fluids

Fluid	Condensing vapor conductance*	Cooling fluid conductance*
Isopropyl alcohol	400	360
Benzene	600	520
Water	2500	1400
Ammonia	3200	2300

*Btu/(hr)(ft²)(°F) = 5.8 J/(s)(m²)(°C).

common thermodynamic fluids are listed in the table for cooling-fluid velocities in the range of 6–7 ft per second (1.8–2.1 m/s), condensing temperatures in the range of 100°F (38°C), and condensing rates of approximately 10 lb/(ft²)(hr) [1.4×10^{-2} kg/(m²)(s)].

Conductance for fouling or dirt films $1/r_f$ depends on the characteristics of the cooling fluid and the condensing vapor in relation to the accumulation (of dirt and corrosion products on the separating wall or tube surfaces) in service. In addition, the oxide film on new clean tubes or other types of metal heat-transfer surfaces contributes to overall fouling resistance r_f. For water-cooled shell and tube steam condensers a conductance of 2000 Btu/(hr)(ft²)(°F) [11,400 J/(s)(m²)(°C)] is reasonable for copper-base alloy tubes after mechanical cleaning. Conductance of the separating wall $1/r_w$ may be computed from the thickness of the wall and the thermal conductivity of the material from which it is made. The overall heat-transfer coefficient varies with the physical properties, the flow rates of the fluids, and the geometry of the condenser and condensing surfaces. For steam condensers U may be in the order of 350–800 Btu/(hr)(ft²)(°F) [2000–4500 J/(s)(m²)(°C)]; with few exceptions, heat transfer is lower with condensing vapors other than steam.

Condenser components. The components of a representative contact condenser are shown in Fig. 6. The cooling-liquid distribution system shown consists of baffles and impingement surfaces to distribute the liquid uniformly and to allow it to cascade in counterflow relationship with the condensing vapor. In the illustrated condenser the cooling section for noncondensable gases coincides with the coolant distribution section where vapor is condensed and the noncondensable gases dehumidified and concentrated. Other constructions for distributing the coolant include rings or slats made of metal, plastic, or ceramic. Most process condensers of the contact type are of counterflow design. In jet condensers noncondensables are entrained in the cooling liquid and thereby removed. The tail pipe provides the barometric leg and discharges into a sealing well, thus eliminating the need for an extraction or tail pump. Low-level contact condensers, those without barometric legs, usually require pumps to extract the cooling water.

Components of typical surface condensers are shown in Figs. 2, 3, 4, 5, and 7. The condensing surface consists of tubes of 0.5–1.5-in. (13–38-mm) outside diameter made of copper base alloys, less frequently of aluminum, nickle alloys, chromium steel, chromium-nickel steels, and titanium. The cooling surface for noncondensables is ordinarily the same material as the main condensing surface, although special corrosion-resistant alloys may be used when the noncondensables are especially corrosive. Usual practice is to condense the vapors on the outside of tubes (Figs. 2, 3, 4, and 5), but when the vapors are especially corrosive they may be condensed on the inside of tubes (Fig. 7). Tubes are usually plain, but for condensing vapors from low-conductance fluids, finned tubes may be used. For this purpose low fins, approximately equal in height to basic tube wall thickness, are used on the condensing side spaced so that the condensate formed does not bridge the

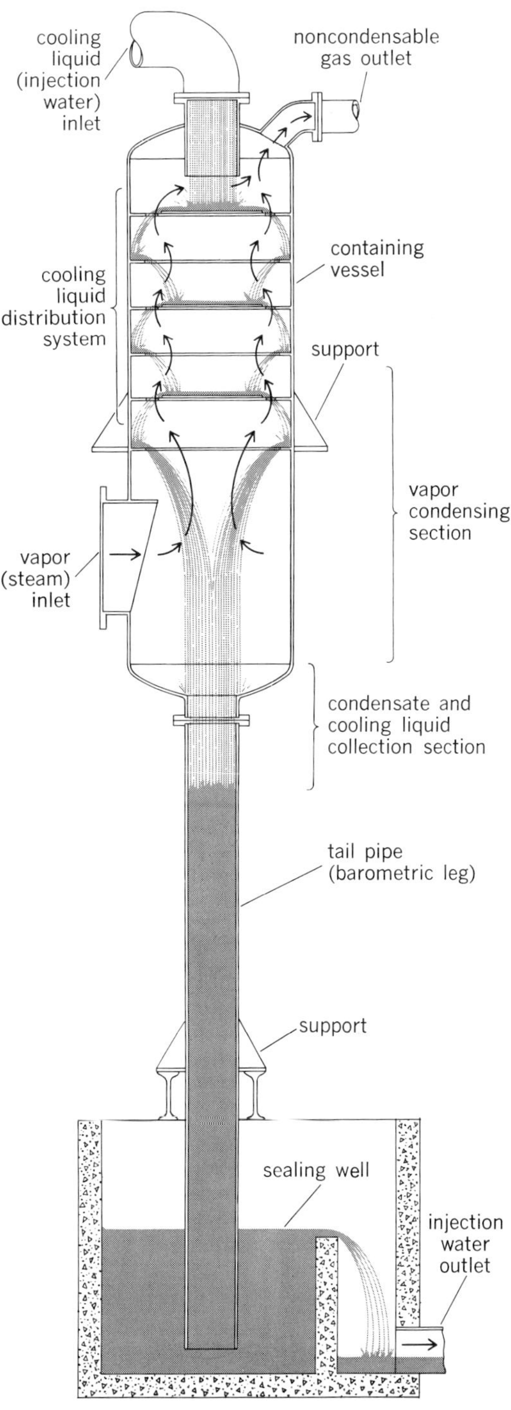

Fig. 6. Counterflow barometric-type contact condenser.

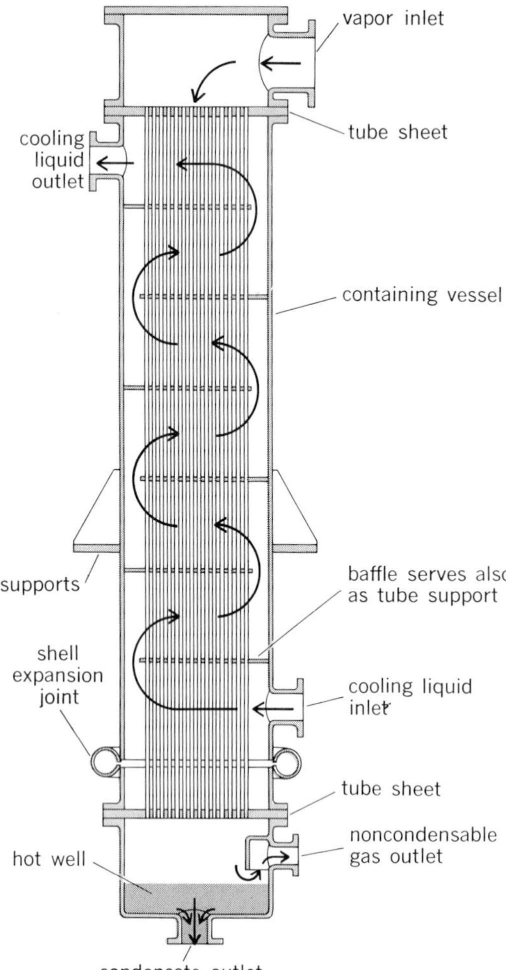

Fig. 7. Vertical-process surface condenser with condensing vapor inside the tubes and the cooling liquid outside the tubes, within the shell.

fins and act like insulation. Gas- or air-cooled condensers use high fins on the cooling side with the extended surface generally at least 10 times the equivalent plain tube area. Bimetal tubes are most frequently used in process condensers when the corrosion environment is severe and different on one side of the tube than on the other.

Tube vibration of magnitudes sufficient to cause tube failure is controlled by the spacing of tube support plates or baffles. Tube vibration seldom results from transmitted mechanical energy but rather from vapor velocity, causing tubes to deflect and vibrate at their natural frequency. With low-density vapors at high velocity, generally in the sonic range or greater, tube vibration is caused by the lift and drag effects of the vapor flowing around the tubes. Severe vibration is usually independent of the von Karman vortex street with low-density vapors, but may be associated with both the lift and drag effects of velocity and the von Karman vortex street at high-vapor densities even at relatively low velocities.

The tube sheets (into which tubes are welded, expanded, or packed) are generally of Muntz metal, naval brass, copper-silicon alloy, or other corrosion-resistant alloys. Carbon-steel tube sheets may be used with carbon or chromium-nickel steel tubes. In nuclear power plants steam condenser tube sheets may be double, that is, with an air gap between to detect leakage and to prevent cooling water from leaking into the steam space. The inner tube sheets are usually made of carbon steel and the outer sheets are made of copper alloy.

The cooling fluid system for surface condensers, when liquid is used, usually consists of chambers attached to the tube sheets and arranged with inlet and outlet connections for the circulation of the cooling medium. When this fluid is liquid, the chambers are designed to distribute the liquid over the face of the tube sheets and into the tubes with little or no cavitation.

Condensers may be designed with one or more liquid passes, dependent on the thermal design conditions, the quantity of cooling liquid available or desired, and the space conditions for installation. Steam surface condensers in large steam power plants are of single-pass design except where the cooling water supply is limited, in which case two-pass condensers are used. Large installations (those that require cooling towers, spray, or evaporation ponds as a means for controlling cooling-water temperature) ordinarily use two-pass condensers. Condensers with three or more passes seldom prove economical for serving engines or turbines. Process condensers are frequently designed with more than two cooling liquid passes and seldom with less.

Auxiliary equipment. Operation of condensers requires pumps (1) for injecting or circulating the cooling fluid, (2) for removing condensate or mixed condensate and injection water, and (3) for removing noncondensables.

Centrifugal pumps are usually used for injecting or circulating cooling water. Conventional volute pumps are used for circulating cooling water for small surface condensers and as injection water pumps for contact condensers. Mixed-flow volute pumps and axial-flow pumps of vertical design are best suited to large surface condensers.

Centrifugal pumps of the horizontal volute type, equipped with pressure-sealed stuffing boxes, are well suited for removing condensate. Vertical, multistage condensate pumps are more suited to large steam power plant installations, where they effect significant installation economics. Positive displacement pumps are used for withdrawing condensate from small condensers. Tail pumps of the conventional double-suction centrifugal type are used for pumping mixed condensate and injection water from low-level jet (contact) condensers. *See* PUMPING MACHINERY.

Pumps for removing noncondensable gases are classified as displacement and ejector types. The displacement machine is built either as a reciprocating vacuum pump, similar to a piston-type air compressor, or as a rotary machine, similar to a gear-pump or sliding-vane rotary compressor. The displacement-type vacuum pump is widely used on condensers of all sizes. It is not suitable for use at high vacuum with high noncondensable gas loads because it must be disproportionately large for such conditions. It is suitable for use with extremely large surface condensers serving turbines, because of the low noncondensable gas loads characteristic of these installations.

Steam jet ejectors are widely used as vacuum

pumps. Having no moving parts, they require little or no operating attention and are simple to install. They have excellent capacity characteristics at high vacuum and are especially suited to applications where the noncondensable gas load is high. When used with surface condensers serving turbines, steam jet ejectors are equipped with surface inter- and after-condensers. The exhaust steam is condensed and returned to the feed system, and the heat from the exhaust is also recovered in the feedwater. Thus they become highly efficient machines for removing noncondensable gases from condensers used in the generation of power. See CENTRIFUGAL PUMP; COOLING TOWER; DISPLACEMENT PUMP. [JOSEPH F. SEBALD]

Bibliography: *Feedwater Heater Workshop Proceedings*, Palo Alto, CA, sponsored by Electric Power Research Institute and Joseph Oat Corp., EPRI WS-78-133, July 1979; M. Jakob and G. A. Hawkins, *Elements of Heat Transfer*, 3d ed., 1957; W. H. McAdams, *Heat Transmission*, 3d ed., 1954; C. C. Peake, G. F. Gerstenkorn, and T. R. Arnold, Some reliability considerations for large surface condensers, *Proceedings of the American Power Conference*, Chicago, vol. 37, pp. 562–574, 1975; B. W. Pendrick, *The Surface Condenser*, 1935; *The Performance of Condensers in Nuclear and Fossil Power Plants*, pts. 1 and 2, Seminar (Columbus, OH) sponsored by Electric Power Research Institute, American Society of Mechanical Engineers, and Ohio State University, June 2–4, 1975; J. F. Sebald, Main and auxiliary condensers, *Marine Engineering*, chap. 13, Society of Naval Architects and Marine Engineers, 1971; J. F. Sebald and W. D. Nobles, Control of tube vibration in steam surface condensers, *Proceedings of the American Power Conference*, Chicago, vol. 24, pp. 630–643, 1962.

Vapor cycle

A thermodynamic cycle, operating as a heat engine or a heat pump, during which the working substance is in, or passes through, the vapor state. A vapor is a substance at or near its condensation point. It may be wet, dry, or slightly superheated. One hundred percent dryness is an exactly definable condition which is only transiently encountered in practice. Vapor behavior deviates so widely from the ideal gas laws that calculation requires the use of tables and graphs that give the experimentally determined properties of the fluid.

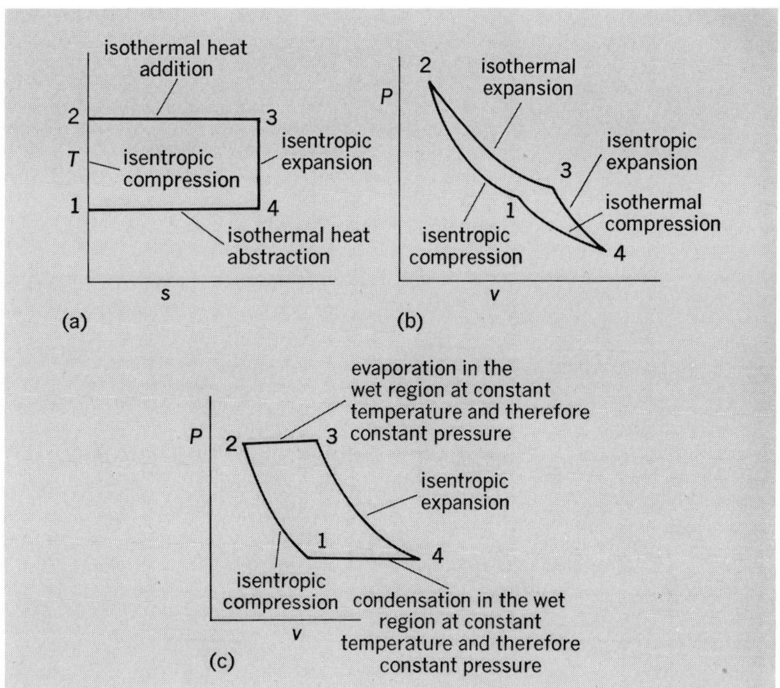

Fig. 3. Carnot cycle. (*a*) Temperature-entropy. (*b*) Pressure-volume for fixed gas. (*c*) Pressure-volume for vapor.

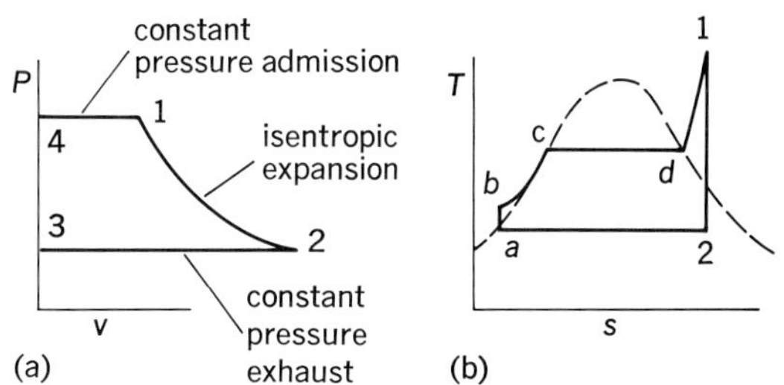

Fig. 4. Rankine cycle for heat-engine plant. (*a*) Pressure-volume diagram of prime mover. (*b*) Temperature-entropy diagram for the power plant.

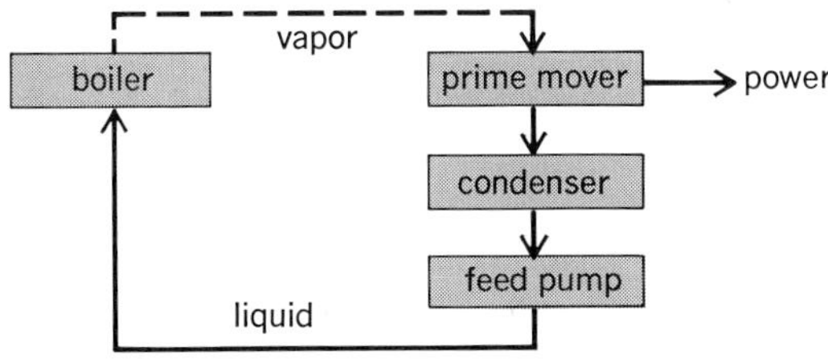

Fig. 1. Rudimentary steam power plant flow diagram.

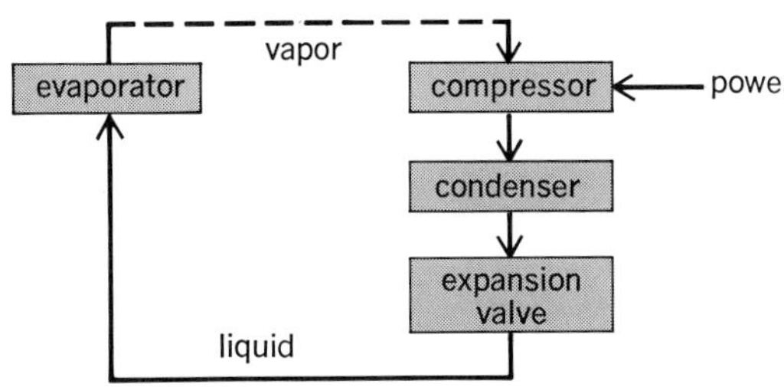

Fig. 2. Rudimentary vapor-compression refrigeration plant flow diagram.

Power and refrigeration plants. A steam power plant operates on a vapor cycle where steam is generated by boiling water at high pressure, expanding it in a prime mover, exhausting it to a condenser, where it is reduced to the liquid state at low pressure, and then returning the water by a pump to the boiler (Fig. 1).

In the customary vapor-compression refrigeration plant, the process is essentially reversed with the refrigerant evaporating at low temperature and pressure, being compressed to high pressure, condensed at elevated temperature, and returned as liquid refrigerant through an expansion valve to the evaporating coil (Fig. 2).

The Carnot cycle, between any two temperatures, gives the limit for the efficiency of the con-

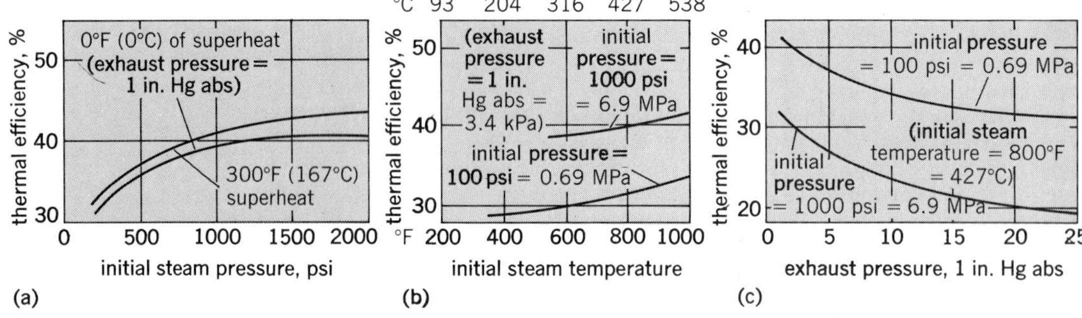

Fig. 5. Thermal efficiency of ideal Rankine steam cycle. (*a*) Steam pressure. 1 psi = 6.9 kPa. (*b*) Steam temperature. (*c*) Vacuum. 1 in. Hg abs = 3.4 kPa.

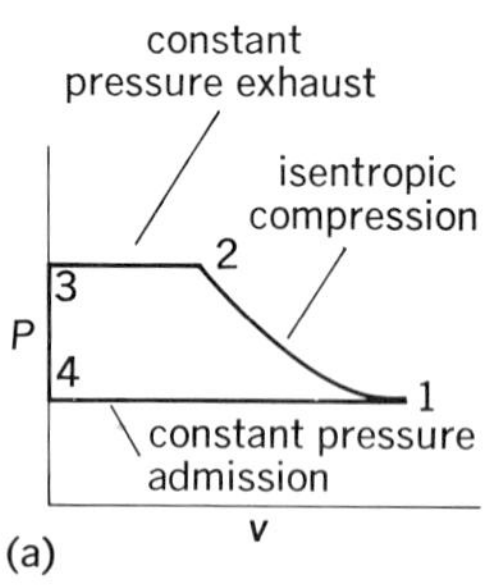

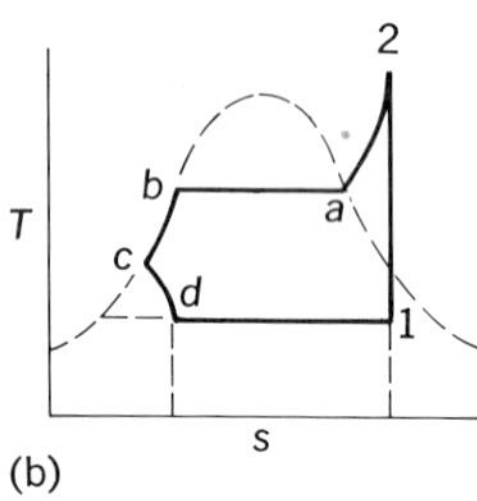

Fig. 6. Rankine cycle for heat-pump plant. (*a*) Pressure-volume diagram of compressor. (*b*) Temperature-entropy diagram for a vapor-compression refrigeration plant.

version of heat into work (Fig. 3). This efficiency is independent of the properties of the working fluid. Although the thermal efficiency is independent of the properties of the fluid, the mean effective pressure (and consequent physical dimensions of the engine) will be vitally influenced by choice of fluid (compare Fig. 3*b* with 3*c*). The Carnot cycle is not realistic for the evaluation of steam power plant performance because the cycle precludes the use of superheat and calls for the isentropic compression of vapor. It is useful, however, for specifying the limiting efficiency that a real cycle might approach. It is so used in judging performance of vapor-cycle heat engines and heat pumps. *See* CARNOT CYCLE.

The Rankine cycle is more realistic in describing the ideal performance of steam power plants and vapor-compression refrigeration systems.

Vapor steam plant. In the case of the steam power plant (Fig. 1), the Rankine cycle (Fig. 4) has two constant pressure phases joined by a reversible adiabatic (isentropic) phase 1–2. From the properties of the fluid, the work of the prime mover, ΔW_{PM}, is most conveniently evaluated as in Eq. (1), where h is the enthalpy, Btu/lb. The feed pump

$$\Delta W_{PM} = h_1 - h_2 \qquad (1)$$

uses some of this work, ΔW_{FP}, to return the water from the condenser to the boiler so that the net output of the cycle is $\Delta W_{PM} - \Delta W_{FP}$. This net output can be related to the heat that must be added to produce steam by consideration of the T-s diagram (Fig. 4*b*). The area under line *bcd*-1 is the heat supplied in the boiler (heat source); phase 1–2 is the isentropic expansion in the prime mover; the area under line 2-*a* is the heat rejected to the condenser (heat sink); phase *a*-*b* is the isentropic compression of the liquid in the feed pump. Thus, the thermal efficiency is given by Eq. (2).

$$\frac{\text{work done}}{\text{heat added}} = \frac{\Delta W_{PM} - \Delta W_{FP}}{h_1 - h_a - \Delta W_{FP}} \qquad (2)$$

There are many variables which influence the performance of the Rankine cycle. For steam the thermal efficiency is a function of pressure, temperature, and vacuum (Fig. 5). High pressure, high superheat, and high vacuum lead to high efficiency. *See* RANKINE CYCLE.

Vapor refrigeration plant. The Rankine cycle can be used to evaluate the performance of the vapor-compression system of refrigeration (Fig. 2). A counterclockwise path is followed on the P-v and T-s cycle diagrams (Fig. 6). The refrigerant enters the compressor as low-temperature, low-pressure vapor, (4–1); isentropic compression follows (1–2), and then high-pressure delivery (2–3). The work to drive the compressor is $h_2 - h_1$. The machine cooling coefficient of performance cp, which is essentially the reciprocal of thermal efficiency, is given by Eq. (3): for a warming machine, it is given by Eq. (4).

$$cp = \frac{\text{refrigeration}}{\text{work done}} = \frac{h_1 - h_d}{h_2 - h_1} \qquad (3)$$

$$cp = \frac{\text{heat delivered}}{\text{work done}} = \frac{h_2 - h_c}{h_2 - h_1} \qquad (4)$$

The difference $h_1 - h_d$ is heat removed in the

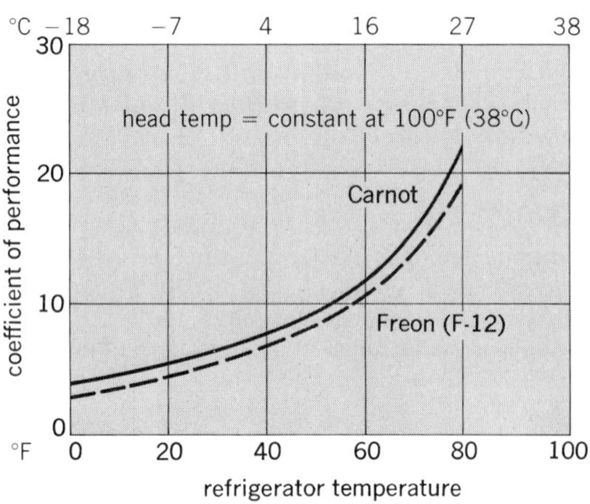

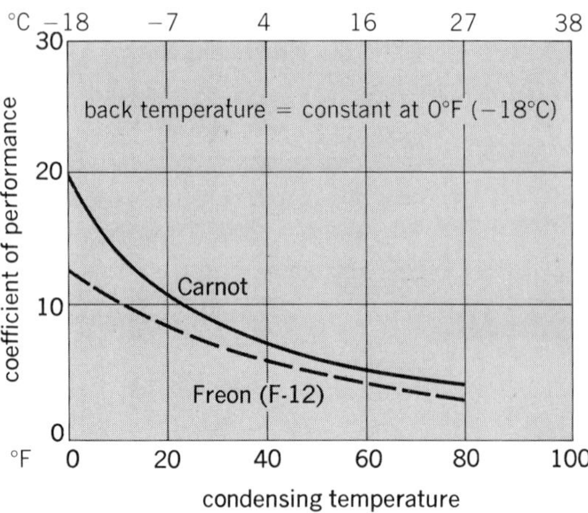

Fig. 7. Coefficient of performance of ideal Rankine and Carnot cycles as influenced by condensing and refrigerating temperatures.

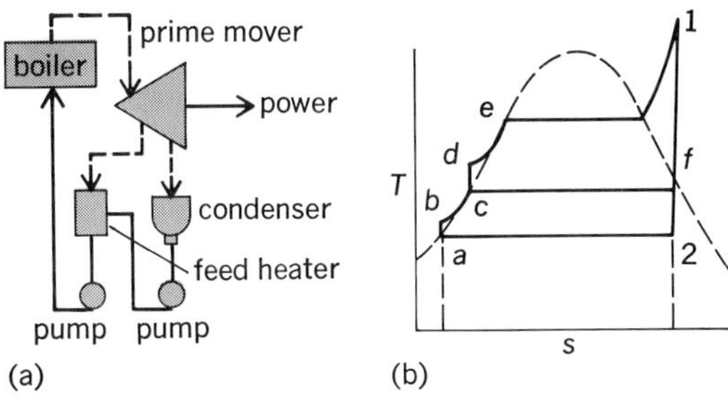

Fig. 8. Regenerative cycle. (a) Rudimentary steam power plant flow diagram with single-stage feed heating. (b) Temperature-entropy diagram.

refrigerating coils, the area under phase $d-1$ on the *T-s* diagram. The difference $h_2 - h_c$ is the heat delivered to the condensing coils, the area under phase (2*abc*). Because the flow is throttled through the expansion valve, the enthalpy is constant, $h_c = h_d$. Some ideal performance values of a vapor-compression refrigeration system are plotted in Fig. 7. For further details on heat-pump vapor cycles *see* HEAT PUMP.

The remainder of this article is concerned with the vapor cycle as it is applied to power-generation purposes.

Regenerative heat cycle. The regenerative cycle is a modification of the simple Rankine cycle. Feed water is heated by extracted steam (Fig. 8). As a result, less heat must be added in the boiler to evaporate a pound of steam and in turn, to deliver a kilowatt hour of work output from the associated steam engine.

A cycle with a single stage of regenerative heating can be viewed as two Rankine cycles superimposed on one another (Fig. 8*b*). In one the exhaust pressure and consequent temperature are substantially higher than in the other (point *f* versus point 2). The heat of condensation represented by the area under *c-f* can be used to raise the feed temperature from T_b to T_c. The area under phase *b-c* is smaller than under phase *c-f* so a fraction of a pound of steam is needed to raise 1 lb of water to the common temperature level T_c.

The principle of regeneration can be extended to

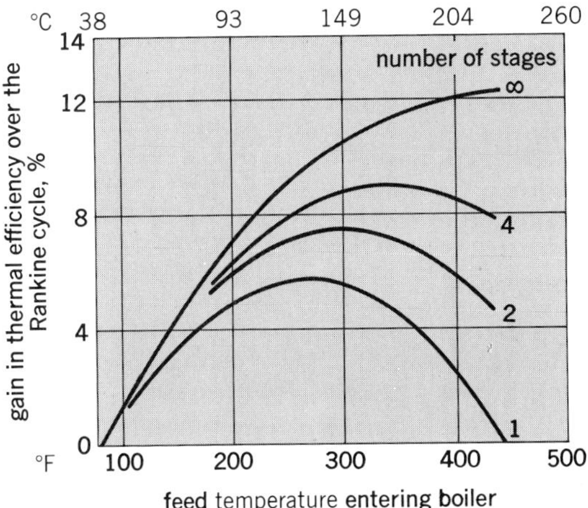

Fig. 9. Gain in thermal efficiency by use of regenerative instead of Rankine cycle; steam conditions, 400 psi (2.76 MPa) and 700°F (371°C); exhaust pressure, 1 in. Hg abs (3.4 kPa).

multiple-stage heating with different final feed temperatures and in the limit reaching the boiler saturation temperature T_e with an infinite number of heating stages. Some consequences of the process are reflected in the data of Fig. 9. The gain in thermal efficiency is the consequence of reducing the quantity of heat rejected to sink 2-*a* in Fig. 8. The weight of steam flow to the prime mover for the production of a kilowatt-hour is larger than with the simple Rankine cycle (Fig. 4). But the heat required to make a pound of steam is so much less that there is an overall thermodynamic gain per kilowatt-hour. The weight flow of steam, from the prime mover to the condenser, is less than with the simple Rankine cycle. It is this reduction in heat rejected to the thermodynamic sink that raises the overall thermal efficiency of the regenerative above the nonregenerative cycle. Modern steampower practice uses the steam turbine for up to ten stages of regenerative heating.

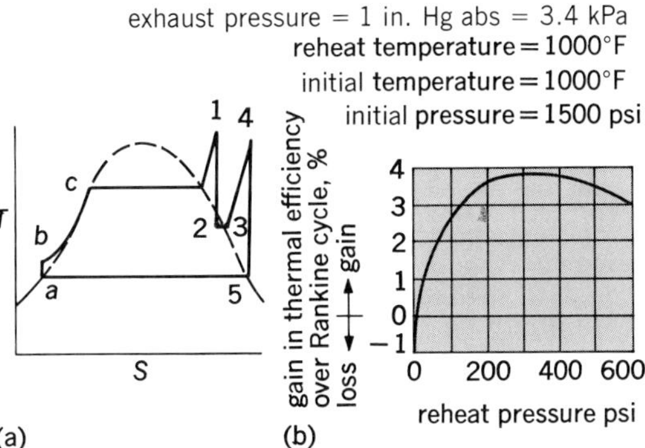

Fig. 10. Resuperheat or reheat cycle. (a) Temperature-entropy diagram. (b) Gain in thermal efficiency as function of reheat pressure; primary pressure, primary temperature, and reheat temperature constant at 1500 psi (10.3 MPa), 1000°F (538°C), and 1000°F (538°C), respectively. 1 psi = 6.9 kPa.

Reheat cycle. The resuperheat or reheat cycle is another improvement in vapor cycles favored in current central station practice. Steam expanding isentropically (Fig. 4*b*) grows wetter with consequent increased erosion of machinery parts and loss in mechanism efficiency: Superheat tends to correct these weaknesses but metallurgical limitations fix the maximum allowable steam temperature. Reheating the steam after a partial expansion (Fig. 10*a*) gives a practical correction. The reheating can be carried out at various pressure and temperature levels and in multiple stages. Thermal efficiency is improved over the simple Rankine cycle (Fig. 10*b*). Current practice uses a single reheat stage with temperatures approximately equal to primary steam temperature. Supercritical pressure plants favor two reheating stages.

Binary vapor cycle. Comparison of the Rankine (Fig. 4*b*), the reheat (Fig. 10*a*), and Carnot (Fig. 3*a*) cycles shows that there are considerable thermodynamic losses in the first two by failure to approach the rectangular *T-s* configuration of the Carnot cycle. The binary vapor cycle uses two fluids with totally different vapor pressures, such as mercury and water (Fig. 11). If a Rankine cycle using mercury is superimposed on that using steam, it is possible to operate the mercury con-

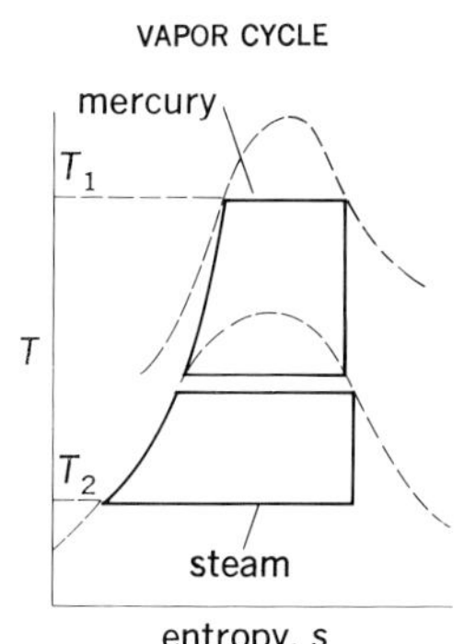

Fig. 11. Binary-vapor cycle using mercury and water, temperature-entropy diagram.

denser as the steam boiler, transferring the heat from the one fluid to the other. Because of the differences in the latent heats and specific heats of the two fluids, several pounds of mercury must be circulated to make 1 lb of steam. However, the combined *T-s* diagram (Fig. 11) approximates the rectangular specification with appreciable gain in thermal efficiency. For many years the binary vapor cycle was thought to have practical as well as theoretical advantages, but its economic difficulties have led to its eclipse by, and abandonment in favor of, the high-pressure, regenerative, reheat, steam cycles. *See* THERMODYNAMIC CYCLE.

[THEODORE BAUMEISTER]

Bibliography: T. Baumeister (ed.), *Standard Handbook for Mechanical Engineers*, 8th ed., 1978; V. M. Faires, *Thermodynamics*, 6th ed., 1978; W. C. Reynolds and H. C. Perkins, *Engineering Thermodynamics*, 2d ed., 1977; K. A. Rolle, *Introduction to Thermodynamics*, 2d ed., 1980.

Vapor lamp

A source of radiant energy excited by a supply of electricity which creates a current of ionized gas between electrodes in an enclosure that contains the arc while permitting transmission of the radiant energy. Gaseous-discharge lamps or vapor lamps are given various names relating to the element responsible for the majority of the radiation (mercury, sodium metal-halide, xenon), to the physical attribute of the lamp (short-arc, high-pressure), or, in the case of fluorescent lamps, to the way a phosphor on the bulb wall fluoresces as a result of the lamp's low-pressure mercury-vapor excitation.

Gaseous-discharge lamps are broadly and increasingly used throughout the world because the conversion of electric energy to radiant energy in a gaseous discharge provides radiation in narrow bands within the range of visible light in which the rods and cones of the eye are most sensitive. These light sources have high efficiency in conversion of electricity to light (see table). The popularity of high-pressure sodium lamps that emit yellow-white light for roadway lighting is due to high efficiency. This lamp's principal radiation is near the peak sensitivity of the eye, while radiation at other wavelengths is produced through a spectrum-line broadening due to high pressure.

Approximate efficiencies of popular lamp types

Lamp	Efficiency, lumens per watt
Mercury vapor	50
Fluorescent	70
Metal-halide	90
High-pressure sodium	110
Incandescent	20

The gaseous-discharge lamp requires a high voltage to start, but as the electric current increases, the resistance decreases, and some means must be available to limit the current and avoid lamp failure. Some gaseous-discharge lamps use a tungsten filament within the lamp envelope, which gets hot and limits the current. More commonly, a magnetic structure called a ballast (transformer) is used external to the lamp and limits the lamp current. *See* TRANSFORMER.

For discussions of common types of vapor lamps *see* FLUORESCENT LAMP; MERCURY-VAPOR LAMP; SODIUM-VAPOR LAMP.

[T. F. NEUBECKER]

Vapor lock

Interruption of fuel flow to an engine due to blockage of passages in the fuel system by fuel vapor.

To promote easy starting, all gasolines contain volatile constituents which under some conditions, such as high ambient temperature, tend to produce more vapor than the fuel pump and carburetor vents can handle. The very action of the fuel pump, in decreasing the pressure at its inlet, tends to vaporize the fuel. If the vapor forms faster than the pump can draw it from the fuel line, the flow of fuel to the carburetor is effectively stopped and the engine stalls. *See* CARBURETOR; INTERNAL COMBUSTION ENGINE.

The tendency to form vapor in a given fuel system has been correlated with the Reid vapor pressure and ASTM distillation curve of the gasoline. With a given fuel, vapor formation can be minimized by keeping all parts of the fuel system cool, eliminating sudden changes in cross section or direction of fuel lines, and using a fuel pump of adequate capacity at the lowest point in the fuel line. *See* FUEL SYSTEM.

[AUGUSTUS R. ROGOWSKI]

Velocity

The time rate of change of position of a body in a particular direction. Linear velocity is velocity along a straight line, and its magnitude is commonly measured in such units as meters per second (m/sec), feet per second (ft/sec), and miles per hour (mph). Since both a magnitude and a direction are implied in a measurement of velocity, velocity is a directed or vector quantity, and to specify a velocity completely, the direction must always

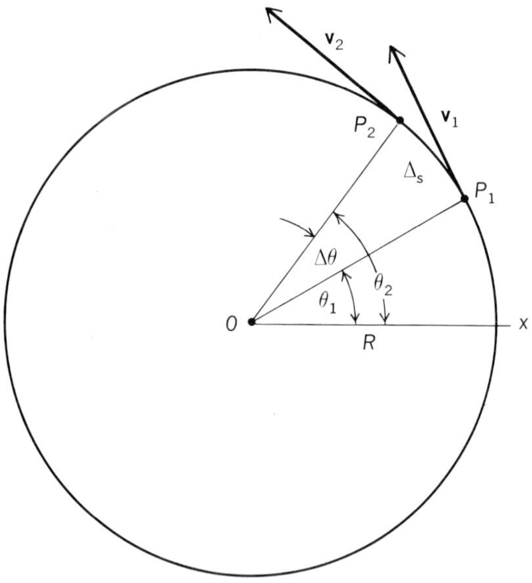

Fig. 1. Illustration of angular displacement, angular speed, and tangential velocity.

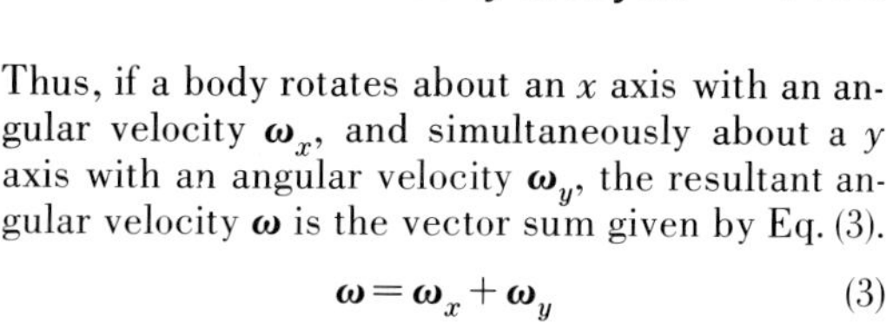

Fig. 2. The average velocity from P_1 to P_2 is $(s_f - s_0)/(t_f - t_0)$. The instantaneous velocity at point *P* is the limit of the ratio representing the average velocity as the interval approaches zero.

be given. The magnitude only is called the speed. *See* SPEED.

Linear velocity. A body need not move in a straight line path to possess linear velocity. The instantaneous velocity of any point of a body undergoing circular motion is a vector quantity, such as v_1 or v_2 in Fig. 1. When a body is constrained to move along a curved path (Fig. 2), it possesses at any point an instantaneous linear velocity in the direction of the tangent to the curve at that point. The average value of the linear velocity is defined as the ratio of the displacement to the elapsed time interval during which the displacement took place. The displacement of a body from an initial position s_0 to a final position s_f after time *t* is equal to $s_f - s_0$. The corresponding time interval is $t_f - t_0$. The magnitude of the average velocity is then given by Eq. (1), where Δs is displacement and Δt is the corresponding elapsed time.

$$\bar{v} = \frac{\text{displacement}}{\text{elapsed time}} = \frac{s_f - s_0}{t_f - t_0} = \frac{\Delta s}{\Delta t} \qquad (1)$$

The magnitude of the instantaneous velocity *v* of a body is the limiting value of the foregoing ratio as the interval approaches zero. In the notation of calculus, Eq. (2), ds/dt is the instantaneous time rate of change of displacement (Fig. 2).

$$v = \lim_{\Delta t \to \Delta} \frac{\Delta s}{\Delta t} = \frac{ds}{dt} \qquad (2)$$

The velocity of a body, like its position, can only be specified relative to a particular frame of reference. Consequently, all velocities are relative.

Angular velocity. The representation of angular velocity ω as a vector is shown in Fig. 3. The vector is taken along the axis of spin. Its length is proportional to the angular speed and its direction is that in which a right-hand screw would move. If a body rotates simultaneously about two or more rectangular axes, the resultant angular velocity is the vector sum of the individual angular velocities. Thus, if a body rotates about an *x* axis with an angular velocity $\boldsymbol{\omega}_x$, and simultaneously about a *y* axis with an angular velocity $\boldsymbol{\omega}_y$, the resultant angular velocity $\boldsymbol{\omega}$ is the vector sum given by Eq. (3).

$$\boldsymbol{\omega} = \boldsymbol{\omega}_x + \boldsymbol{\omega}_y \qquad (3)$$

It should be emphasized that whereas angular velocities are commutative in addition, that is, they may be added in any order, angular displacements are not commutative. *See* ROTATIONAL MOTION.

Angular displacement. Figure 1 represents a body rotating with circular motion about an axis through *O* perpendicular to the figure. Line OP_1 is the position of some radius in the body at a time t_1, with θ_1 being the angular displacement from a reference line. Line OP_2 is the position of the same radius at a later time t_2, with the angular displacement θ_2. Angular displacement may be measured in degrees, radians, or revolutions.

Angular speed. From Fig. 1, it is seen that the body has rotated through the angle $\Delta\theta = \theta_2 - \theta_1$ in the time $\Delta t = t_2 - t_1$. The average angular speed $\bar{\omega}$ is defined by $\bar{\omega} = \Delta\theta/\Delta t$, the instantaneous angular speed ω being $\omega = d\theta/dt$. Although it is customary in most scientific work to express angular speed in radians per second, it is common in engineering practice to use the units of revolutions per minute (rpm) or revolutions per second (rps).

Tangential velocity. When a particle rotates in a circular path of radius *R* through an angular distance $\Delta\theta$ in a time Δt, as in Fig. 1, it traverses a linear distance Δs.

VELOCITY

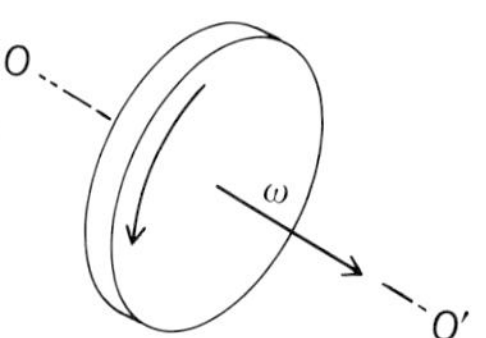

Fig. 3. Angular velocity shown as an axial vector.

The average linear speed $\bar{v}$ is given by Eq. (4),

$$\bar{v} = \frac{\Delta s}{\Delta t} = \frac{R\,\Delta\theta}{\Delta t} = \bar{\omega}R \qquad (4)$$

since $\Delta s = R\Delta\theta$. Similarly, the instantaneous speed *v* is given by $v = \omega R$. The direction of this instantaneous speed is tangential to the circular path at the point in question. Any vector **v** drawn in this direction represents the tangential velocity.

Combined velocities. A body may have combined linear and angular motions, as is the case when the wheel of a moving automobile rolls along the ground with an angular velocity about its axle which moves with a linear velocity parallel to the pavement. In this case, a point on the rim of the tire describes a curved path called a cycloid. If a circular body rolls on the surface of a sphere, a point on the periphery of the rotating body describes a curve called an epicycloid.

[ROGERS D. RUSK]

Bibliography: H. Goldstein, *Classical Mechanics*, 2d ed., 1980; C. Kittel, W. D. Knight, and M. A. Ruderman, *Mechanics*, vol. 1, 2d ed., 1973; R. Resnick and D. Halliday, *Physics*, pt. 1, 3d ed., 1977.

Velocity analysis

A technique for the determination of the velocities of the parts of a machine or mechanism. Both graphical and analytical analyses of plane mechanisms will be discussed in this article, but of the several methods of each type available, only one of each type will be described. The graphical method will be discussed first, since the visualization which is an inherent part of the graphical analysis generally gives a better physical feel for the problem than most purely analytical methods. Analyti-

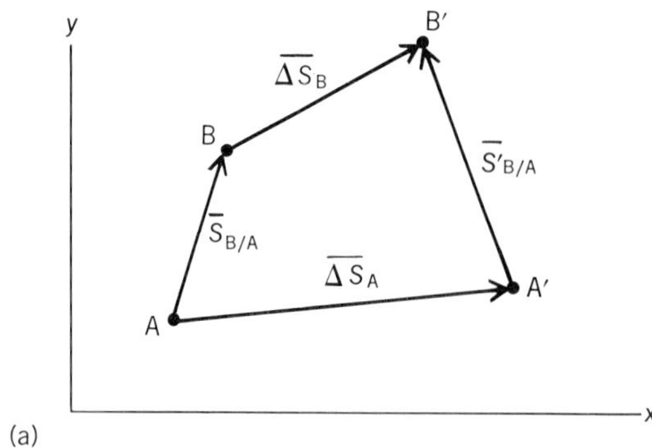

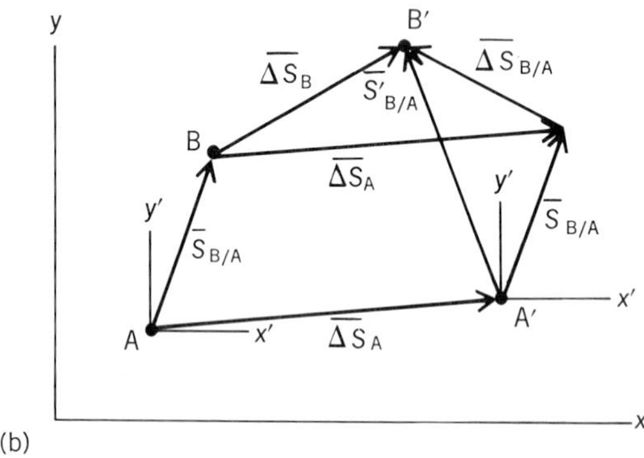

Fig. 1. Relative displacement. (*a*) Position vectors locating B with respect to A at beginning and end of a time increment. (*b*) Vectors for derivation of relative displacement equation.

cal methods, however, are necessary for computer analyses. *See* VELOCITY.

Purposes. In a high-speed machine it is important that the inertia forces be determined. This requires an acceleration analysis of the machine, and the first step in an acceleration analysis usually is a velocity analysis. In the analysis of some machines, the velocity of a particular point in the machine may itself be the important thing to be determined in the analysis—for example, the cutting speed and return speed of the cutting tool in a shaper, or the shuttle velocity in a textile machine. *See* ACCELERATION ANALYSIS.

Relative displacement. The method of relative velocities is widely used in velocity analysis, and the relative velocity equation is fundamental for this method of analysis. It is derived by first considering relative displacements. Referring to Fig. 1*a*, let A and B be any two points moving in the stationary x-y coordinate system shown. Point B is located with respect to point A by the position vector $\mathbf{S}_{B/A}$. Suppose that during a time interval Δt, point B has the vector displacement $\Delta\mathbf{S}_B$ and point A the displacement $\Delta\mathbf{S}_A$. At the end of the time interval Δt, B is located with respect to A by the vector $\mathbf{S}'_{B/A}$. The vector change in $\mathbf{S}_{B/A}$ during the time interval Δt is shown in Fig. 1*b* as $\Delta\mathbf{S}_{B/A}$. From Fig. 1*b* it is evident that Eq. (1) holds, which is the

$$\Delta\mathbf{S}_B = \Delta\mathbf{S}_A + \Delta\mathbf{S}_{B/A} \tag{1}$$

relative displacement equation. It is further noted from Fig. 1*b* that $\Delta\mathbf{S}_{B/A}$ is the displacement of B as measured in the nonrotating x'-y' coordinate system which is attached to and moves with A.

Relative velocity. If the time interval Δt approaches zero, the displacements shown in Fig. 1*b* become infinitesimal and the velocity vectors representing the velocities of the points are proportional to the infinitesimal displacement vectors as shown in Fig. 2. From Fig. 2 the relative velocity equation (2) can be written. Equation (2) may also

$$\mathbf{v}_B = \mathbf{v}_A + \mathbf{v}_{B/A} \tag{2}$$

be derived mathematically by dividing Eq. (1) through by Δt and taking the limit as Δt approaches zero, as shown by Eqs. (3) or (4). It must be clearly

$$\lim_{\Delta t \to 0} \frac{\Delta\mathbf{S}_B}{\Delta t} = \lim_{\Delta t \to 0} \frac{\Delta\mathbf{S}_A}{\Delta t} + \lim_{\Delta t \to 0} \frac{\Delta\mathbf{S}_{B/A}}{\Delta t} \tag{3}$$

$$\mathbf{v}_B = \mathbf{v}_A + \mathbf{v}_{B/A} \tag{4}$$

understood that $\mathbf{v}_{B/A}$ in Eq. (2) represents the velocity of B as measured in a nonrotating coordinate system attached to A. It is usually called the velocity of B relative to A, although there has been some objection to this terminology.

Two points on a rigid body. Let body 2 of Fig. 3 represent any mechanism link moving with general plane motion. Its angular velocity is ω (the vector representation of ω is a vector out of the paper). Equation (2) holds for any two points, and therefore holds for points A and B of the special case being considered. The relative velocity $\mathbf{v}_{B/A}$ is the velocity of B measured in the nonrotating x'-y' coordinate system attached to A. In this coordinate system B moves on a circular path with its center at

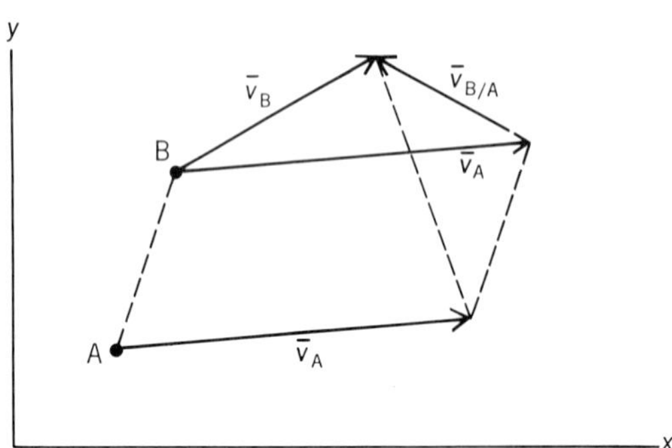

Fig. 2. Vectors for derivation of the relative velocity equation.

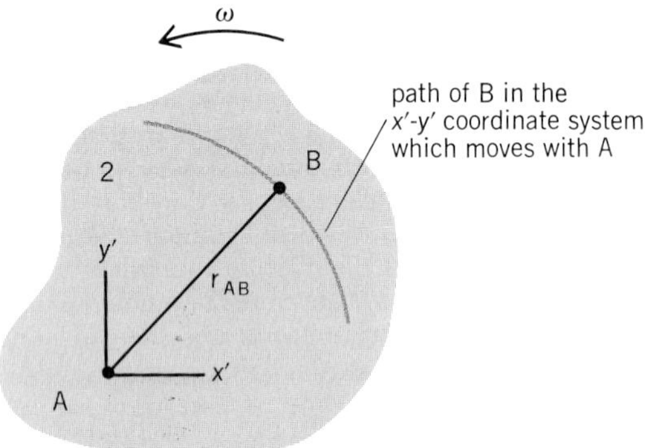

Fig. 3. Rigid body 2 moving with general plane motion.

A, and therefore its velocity as measured in this coordinate system has a magnitude equal to $r_{AB}\omega$ and a direction perpendicular to line AB.

Coincident points on two links. Consider the mechanism of Fig. 4*a* in which the pin at the end of link 2 is constrained to move in the slot of 3. Point P_2 is the center of the pin, and point P_3 is the point on link 3 coincident with P_2. There is no such actual physical point, but an extension built onto link 3 and overlapping the pin can be imagined. Equation (2) as applied to the coincident points P_2 and P_3 is Eq. (5). Assume the magnitude of $\mathbf{v}_{P_2}$ is

$$\mathbf{v}_{P_3} = \mathbf{v}_{P_2} + \mathbf{v}_{P_3/P_2} \qquad (5)$$

known. The directions of $\mathbf{v}_{P_2}$ and $\mathbf{v}_{P_3}$ are known. To solve a single-vector equation, there can be no more than two unknowns, so that the direction of $\mathbf{v}_{P_3/P_2}$ must also be known. There can be no relative motion of points P_2 and P_3 in the direction of line *n-n* normal to the slot, so that the only relative motion possible is along the slot. Therefore the direction of $\mathbf{v}_{P_3/P_2}$ is along the slot. Similarly for the direct-contact mechanism of Fig. 4*b*, the relative velocity $\mathbf{v}_{P_3/P_2}$ must be along the common tangent of the contacting surfaces, line *t-t'*.

Application example. The use of the relative velocity equation will be illustrated in the velocity analysis of a slider-crank mechanism. Slider-crank mechanisms are used in a wide variety of machines. Perhaps the most familiar example is the crank, connecting rod, and piston mechanism of the internal combustion engine. A skeleton drawing of such a mechanism is shown in Fig. 5*a*. The moving links are numbered 2, 3, and 4, with link 1 being the fixed frame. Key points are identified with letters. It is assumed that the angular velocity

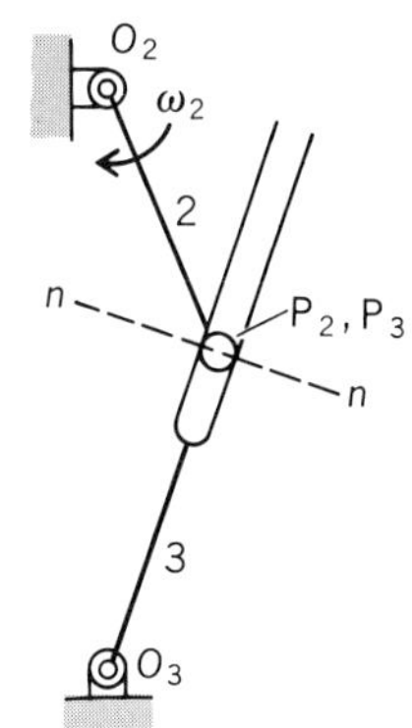

(a)

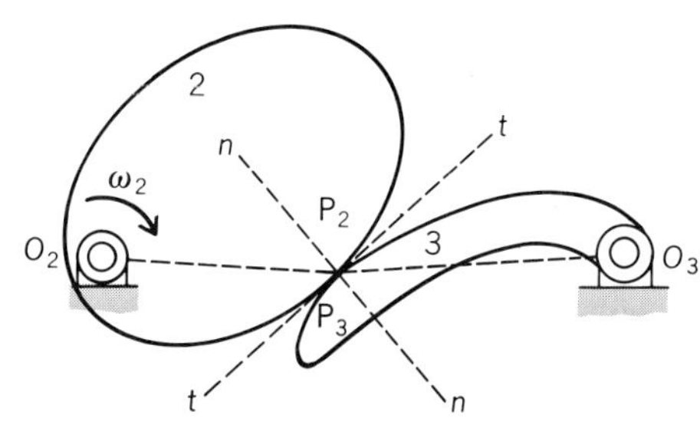

(b)

Fig. 4. Mechanisms *a* and *b* for which the relative velocity equation is applied to coincident points on different links.

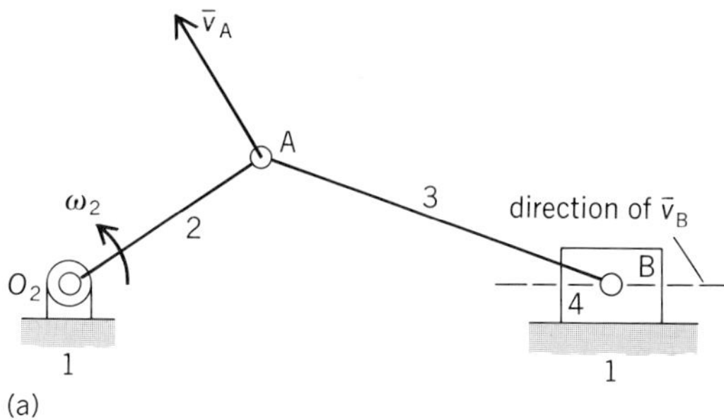

(a)

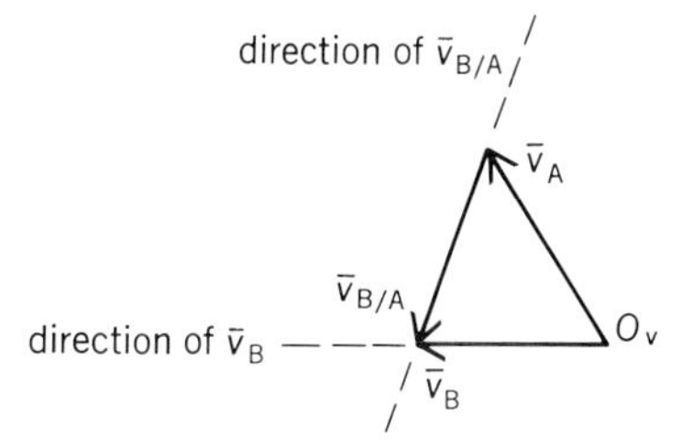

(b)

Fig. 5. Velocity analysis. (*a*) Slider-crank mechanism to be analyzed. (*b*) Velocity polygon.

of crank 2 is known. The magnitude of the velocity of point A is given by $v_A = r_{O_2A}\omega_2$. A vector representing this velocity is shown in Fig. 5*a* originating at point A, and also is laid off to scale from an origin O_v in Fig. 5*b*. The relative velocity equation, $\mathbf{v}_B = \mathbf{v}_A + \mathbf{v}_{B/A}$, is next applied to points A and B. Point B is constrained to move along a horizontal straight line so that the direction of $\mathbf{v}_B$ is known. Vector $\mathbf{v}_A$ is completely known; $\mathbf{v}_{B/A}$ is perpendicular to line AB, since A and B are two points on the same rigid body (link 3). A velocity polygon (triangle) can be drawn as in Fig. 5*b* to solve for the unknown quantities of the relative velocity equation, the magnitudes of $\mathbf{v}_B$ and $\mathbf{v}_{B/A}$. These magnitudes can be scaled directly from the polygon. Usually it would be necessary to make an analysis of this type for a number of positions of the mechanism in its motion cycle. Other mechanisms can be analyzed graphically for velocities in a similar manner. *See* SLIDER-CRANK MECHANISM.

Use of vector mathematics. In the graphical method of velocity analysis just discussed, the geometry of the mechanism is known in each phase of its motion cycle from the drawing of the mechanism in each position. In an analytical velocity analysis, the geometry is usually determined analytically; that is, a position analysis of the mechanism is performed by using trigonometry, vector mathematics, or some other analytical method as the first step in the velocity analysis.

Position analysis. One analytical method for the position analysis of mechanisms, developed by M. A. Chace, makes use of vector mathematics. It consists essentially of solving vector triangles containing two unknowns. For example, in Fig. 6 an offset slider-crank mechanism is shown with appropriate vectors placed on the mechanism. Assuming the position of crank 2 is known as the input to the mechanism, vectors **P** and **Q** are known and may be added together to obtain vector **C**. Vectors **C**, **r**, and **s** form a vector triangle containing two unknowns, the direction of vector **r** and

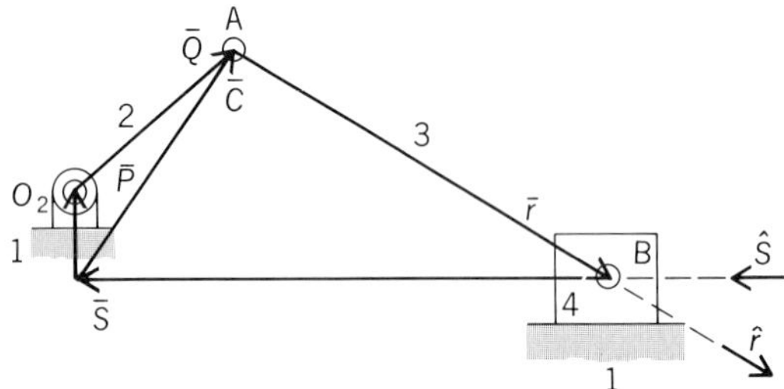

Fig. 6. Offset slider-crank mechanism showing vector triangle to be solved for position analysis of the mechanism.

the magnitude of vector **s**. That is, unit vector $\hat{\mathbf{r}}$ and s are the unknowns (carets are used in this article to designate unit vectors). The plane vector equation $\mathbf{C}+\mathbf{s}+\mathbf{r}=0$ (or $\mathbf{C}+s\hat{\mathbf{s}}+r\hat{\mathbf{r}}=0$) must be solved for the unknowns. Chace developed solutions for the plane vector equation for various cases using vector mathematics. Vector **C** is assumed always to be the known vector, and case 1 is the situation where two magnitudes, s and r, are unknown. Chace's equations for the vectors **s** and **r** for this case are Eqs. (6) and (7), where $\hat{\mathbf{k}}$ is the

$$\mathbf{s}=\frac{\mathbf{C}\cdot(\hat{\mathbf{r}}\times\hat{\mathbf{k}})}{\hat{\mathbf{r}}\cdot(\hat{\mathbf{s}}\times\hat{\mathbf{k}})}\hat{\mathbf{s}} \tag{6}$$

$$\mathbf{r}=\frac{\mathbf{C}\cdot(\hat{\mathbf{s}}\times\hat{\mathbf{k}})}{\hat{\mathbf{s}}\cdot(\hat{\mathbf{r}}\times\hat{\mathbf{k}})}\hat{\mathbf{r}} \tag{7}$$

unit vector in the z direction or out of the paper.

Case 2 is the situation illustrated in Fig. 6 where there is one unknown magnitude, s, and one unknown direction, $\hat{\mathbf{r}}$, in the vector triangle. Chace's equations for the vectors **r** and **s** for this case are Eqs. (8) and (9). The lower signs on the radicals

$$\mathbf{r}=-[\mathbf{C}\cdot(\hat{\mathbf{s}}\times\hat{\mathbf{k}})](\hat{\mathbf{s}}\times\hat{\mathbf{k}}) \pm\sqrt{r^2-[\mathbf{C}\cdot(\hat{\mathbf{s}}\times\hat{\mathbf{k}})]^2}\,\hat{\mathbf{s}} \tag{8}$$

$$\mathbf{s}=(-\mathbf{C}\cdot\hat{\mathbf{s}}\mp\sqrt{r^2-[\mathbf{C}\cdot(\hat{\mathbf{s}}\times\hat{\mathbf{k}})]^2})\hat{\mathbf{s}} \tag{9}$$

would be chosen for the analysis of this mechanism because **r** has a negative component in the $\hat{\mathbf{s}}$ direction for the complete mechanism cycle.

Case 3 is the situation illustrated in Fig. 7. Vector **P** connecting the fixed pivots is a known vector and can be expressed in the form $\mathbf{P}=x\hat{\imath}+y\hat{\jmath}$. The position of the input crank 2 is assumed to be known, so that vector **Q** is known and can be expressed in the same form as vector **P**. Vectors **P** and **Q** can be added together to get vector **C**. Vectors **C**, **s**, and **r** form a vector triangle in which the unknowns are unit vectors $\hat{\mathbf{s}}$ and $\hat{\mathbf{r}}$. Chace's solutions for vectors **r** and **s** for this case are Eqs. (10)

$$\mathbf{r}=\mp\sqrt{r^2-\left(\frac{r^2-s^2+C^2}{2C}\right)^2}(\hat{\mathbf{C}}\times\hat{\mathbf{k}}) -\left(\frac{r^2-s^2+C^2}{2C}\right)\hat{\mathbf{C}} \tag{10}$$

and (11). The upper signs on the radicals would be

$$\mathbf{s}=\pm\sqrt{r^2-\left(\frac{r^2-s^2+C^2}{2C}\right)^2}(\hat{\mathbf{C}}\times\hat{\mathbf{k}}) +\left(\frac{r^2-s^2+C^2}{2C}-C\right)\hat{\mathbf{C}} \tag{11}$$

selected for the analysis of the mechanism of Fig. 7, because **s** has a positive component in the $\hat{\mathbf{C}}\times\hat{\mathbf{k}}$ direction and **r** has a negative component in that direction. *See* FOUR-BAR LINKAGE.

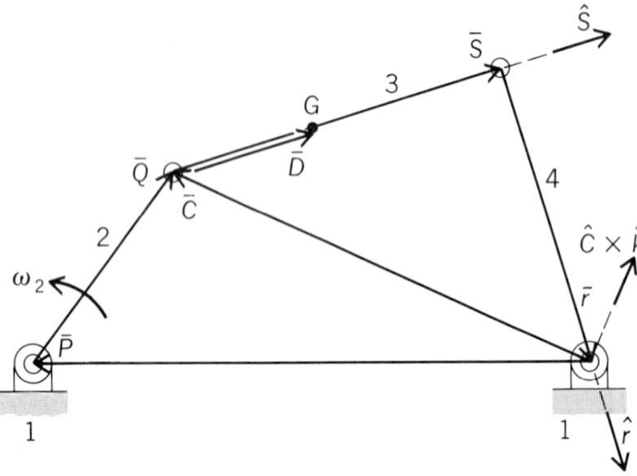

Fig. 7. Four-bar linkage showing vector triangle to be solved for the position analysis of the mechanism.

Time derivative of a vector. The time derivative of a vector originating from a fixed origin such as the one shown in Fig. 8 is the velocity of the point

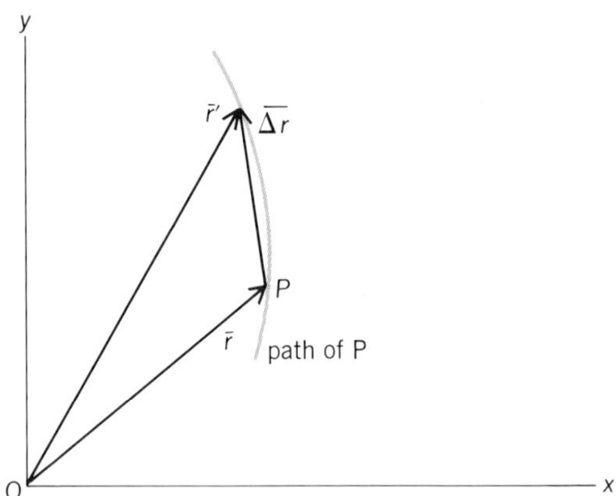

Fig. 8. Changing vector originating from a fixed origin.

at the tip of the vector; that is, Eq. (12) holds.

$$\mathbf{v}_P=\lim_{\Delta t\to 0}\frac{\Delta\mathbf{r}}{\Delta t}=\frac{d\mathbf{r}}{dt}=\dot{\mathbf{r}} \tag{12}$$

Time derivative of a fixed-length vector. If a vector **r** in a rotating body is of fixed length and originates on the axis of rotation of the body as shown in Fig. 9, its time derivative is $\boldsymbol{\omega}\times\mathbf{r}$, since the magnitude of $\boldsymbol{\omega}\times\mathbf{r}$ is $\omega r\sin\phi$, and it is evident from Fig. 9 that the velocity of point P (the time derivative of **r**) is $\omega r \sin\phi$ in the direction of $\boldsymbol{\omega}\times\mathbf{r}$. It may also be shown that the time derivative of any fixed-length vector **r** in a body is equal to $\boldsymbol{\omega}\times\mathbf{r}$ even though the vector **r** does not originate on the axis of rotation of the body. In this case, however, the derivative of the vector is not the velocity of the point at the tip of the vector, but the velocity of the point at the tip of the vector relative to the point at the origin or tail of the vector.

Four-bar mechanism. For the four-bar mechanism shown in Fig. 7, the vector loop equation (13)

$$\mathbf{P}+\mathbf{Q}+\mathbf{s}+\mathbf{r}=0 \tag{13}$$

may be written. Differentiating Eq. (13) with respect to time yields Eq. (14), since vectors **Q**, **s**,

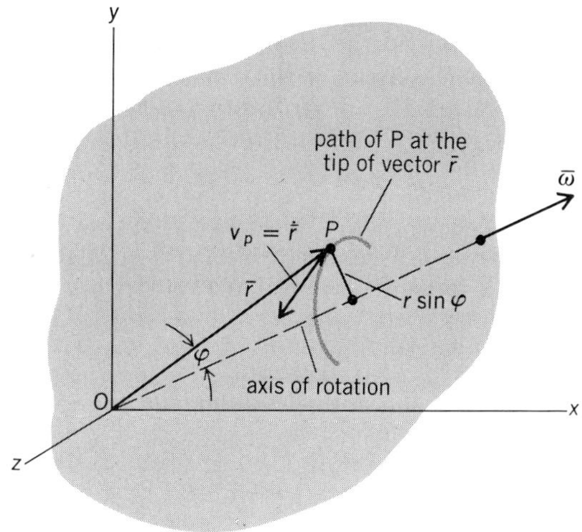

Fig. 9. Fixed-length vector in a body which originates on the axis of rotation of the body; $\mathbf{v}_p = \dot{\mathbf{r}} = \boldsymbol{\omega} \times \mathbf{r}$.

and **r** are all fixed-length vectors. In a plane motion

$$\boldsymbol{\omega}_2 \times \mathbf{Q} + \boldsymbol{\omega}_3 \times \mathbf{s} + \boldsymbol{\omega}_4 \times \mathbf{r} = 0 \quad (14)$$

mechanism the directions of $\boldsymbol{\omega}_3$ and $\boldsymbol{\omega}_4$ are known (direction $\hat{\mathbf{k}}$). The unknowns $\boldsymbol{\omega}_3$ and $\boldsymbol{\omega}_4$ may be determined by taking the dot product of each vector of Eq. (14) with $\hat{\mathbf{r}}$ and $\hat{\mathbf{s}}$, respectively, reducing Eq. (14) in each case to a scalar equation with a single unknown. The solutions thus determined for ω_3 and ω_4 are Eqs. (15) and (16). With these angular velocities known, the velocity of any point on the mechanism can be found. For example, the velocity of point G is given by Eq. (17).

$$\omega_3 = \frac{-\omega_2(\hat{\mathbf{k}} \times \mathbf{Q}) \cdot \hat{\mathbf{r}}}{s(\hat{\mathbf{k}} \times \hat{\mathbf{s}}) \cdot \hat{\mathbf{r}}} \quad (15)$$

$$\omega_4 = \frac{-\omega_2(\hat{\mathbf{k}} \times \mathbf{Q}) \cdot \hat{\mathbf{s}}}{r(\hat{\mathbf{k}} \times \hat{\mathbf{r}}) \cdot \hat{\mathbf{s}}} \quad (16)$$

$$\mathbf{v}_G = \boldsymbol{\omega}_2 \times \mathbf{Q} + \boldsymbol{\omega}_3 \times \mathbf{D} \quad (17)$$

Computer subprograms for the Chace solutions to the plane vector equation and for all the various vector operations have been written, thus making the position and velocity analyses (as well as the acceleration analysis) of mechanisms by computer using Chace's method quite convenient. Computer analyses are particularly useful when complex mechanisms must be analyzed in many positions of their cycle of motion. *See* STRAIGHT-LINE MECHANISM. [JAMES C. WOLFORD]

Bibliography: C. W. Ham, E. J. Crane, and W. L. Rogers, *Mechanics of Machinery*, 1958; B. Paul, *Kinematics and Dynamics of Planar Machinery*, 1979; J. Shigley and J. Uicker, *Theory of Machines and Mechanisms*, 1980.

Ventilation

The supplying of air motion in a space by circulation or by moving air through the space. Ventilation may be produced by any combination of natural or mechanical supply and exhaust. Such systems may include partial treatment such as heating, humidity control, filtering or purification, and, in some cases, evaporative cooling. More complete treatment of the air is generally called air conditioning. *See* AIR CONDITIONING.

Natural ventilation. Natural ventilation may be provided by wind force, convection, or a combination of the two. Although largely supplanted by mechanical ventilation and air conditioning, natural ventilation still is widely used in homes, schools, and commercial and industrial buildings. It is effective and economical in areas of prevailing winds and for high industrial buildings that have hot equipment which provides the motivating convective force.

Wind-force ventilation may be provided by direct force, such as wind blowing in one side of a building and out the other side or through a monitor. This is commonly known as cross ventilation. Because of the friction and velocity losses incurred through building openings, only about 25–60% of the wind velocity is available for ventilation, depending upon whether the wind direction is perpendicular or oblique to the building openings. The resulting negative pressures on the downwind side of the building (if open to the outside) and the sizes and locations of the exhaust openings all materially affect the flow of air.

Airflow around an object creates a negative pressure on the downwind side. This principle is used to advantage in the design of a large number of building ventilators and monitors to provide basic ventilation, to assist existing convective ventilation, or to prevent backdraft down into a building. Figure 1 illustrates the round, rotating head, and continuous roof-type ventilators. Because of the many complex forces involved, airflow capacities cannot be calculated but must be determined by testing.

The force for convection ventilation is created by the difference in weight of two air columns at different temperatures; the heavy, cool column attempts to displace the hot, light column. The pressure p exerted by a fluid column varies as the height h and density w of the fluid; that is, $p = hw$, expressed in consistent units. The basic equation for such flow is $V = \sqrt{2gh}$, again in consistent units for velocity V, height h, and gravity g. From these

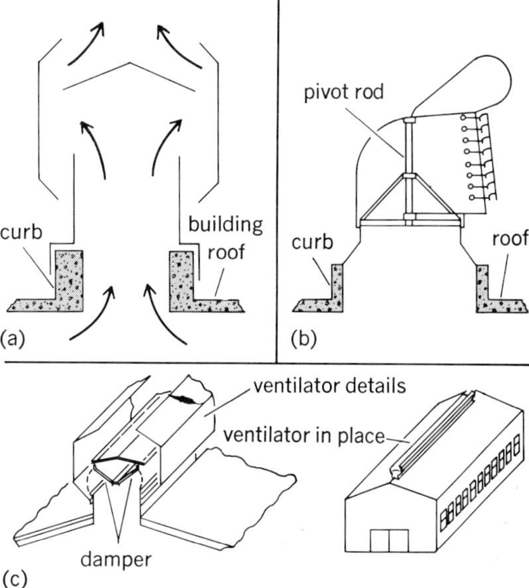

Fig. 1. Roof exhausts for natural ventilation. (*a*) Cross section of round ventilator and (*b*) of rotating-head ventilator. (*c*) Continuous roof ventilator.

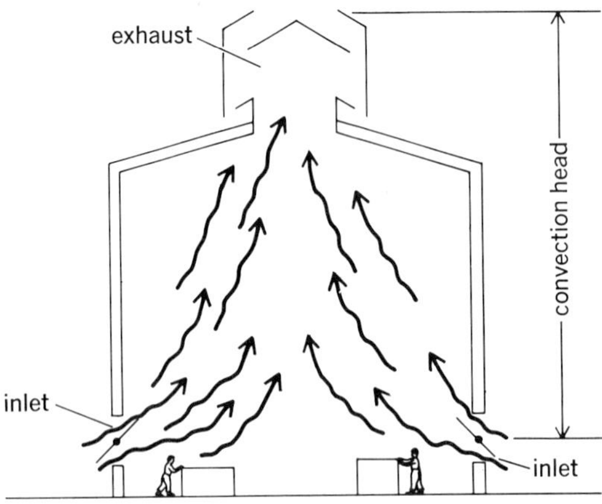

Fig. 2. Building cross section illustrating motivating convective head (height) for natural ventilation.

relations and because temperature has a direct relationship to density, the equation below may be used to estimate convective flow (Fig. 2).

$$Q = 9.4A\sqrt{h(t_i - t_o)}$$

Q = air passing through an opening by convective force, ft³/min
A = free (net) area of the opening, ft² (the inlet areas are assumed to equal the exhaust openings)
h = height from inlets to outlets, ft
t_i = average temperature of indoor air column in height h, °F
t_o = temperature of outdoor air, °F
9.4 = constant of proportionality, including value of 65% for effectiveness of openings

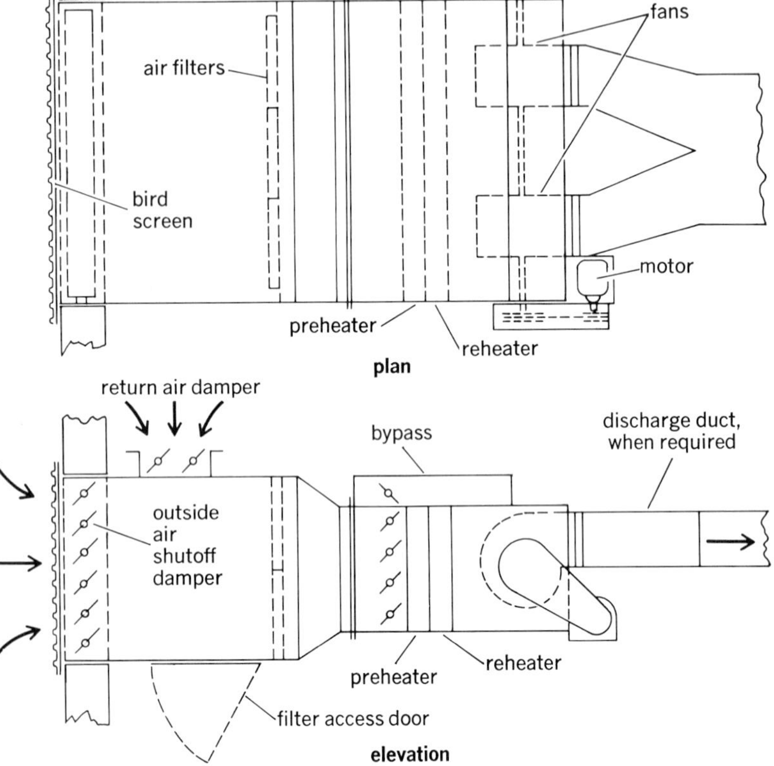

Fig. 3. Factory-assembled ventilating unit.

This equation indicates that convective velocities tend to be low. Consequently, high buildings having large exhaust (and inlet) openings and high internal heat loads are the most practical application for convection ventilation. The best flow would be obtained by a well-shaped hole in the roof, but, because weather protection must also be provided, it is customary to exhaust through roof monitor or baffle-type exhaust ventilators as in Figs. 1 and 2 for small volumes and Fig. 3 for large volumes.

Mechanical ventilation. Mechanical supply ventilation may be of the central type consisting of a central fan system with distributing ducts serving a large space or a number of spaces, or of the unitary type (Fig. 3) with little or no ductwork, serving a single space or a portion of a large space. Both types are employed for schools and for commercial and industrial applications.

Central system assemblies may be custom-built, factory-prefabricated for job site assembly, or factory-assembled. Ventilating units are factory-assembled with rare exceptions. The assemblies consist of a fan and usually include air filters and air heaters of the finned-tube type. The fans may be of the propeller, axial flow, or centrifugal type. Central systems require the last two fan types because of the higher static pressures usually encountered in the distributing ducts. *See* FAN.

The mechanically powered roof ventilator is a practical source of low-cost ventilation. For summer (nontempered) supply, such units consist of fans and weather hoods, or dampers (Fig. 4). For all-year ventilation, heaters or other equipment may be added.

Outside air connections are generally provided for all systems. Outside air is needed in controlled quantities to remove odors and to replace air exhausted from the various building spaces and equipment. The inlets are located to minimize the intake of fumes, dust, organic materials, and pollens. Because it is never possible to find completely clean air, ventilation air filters are usually provided in the system casings in all except a few industrial or summer relief applications. This prevents clogging and poor heat transfer for the air heaters and helps to reduce the pollen and dust in the occupied areas served by the systems.

Air distribution. Duct systems to distribute and disperse the air are required for all but the small unitary systems to avoid short-circuiting and to provide adequate ventilation to all parts of the space served. Such distribution permits desirable air movement in the space without undesirable draft and temperature stratification. Ventilation systems generally serve as conveyors for adding or removing heat and humidity. The distribution system must be adequate for this purpose also. The amount of air circulated is important because too small a volume requires uncomfortably high temperatures for heating or results in high building temperature for heat removal ventilation. Similar problems occur with humidity because of the limited amounts of moisture which can be conveyed without condensation problems. If a system is oversized, it becomes unnecessarily expensive. In addition, it may be difficult to discharge the air to the space within acceptable velocity limits.

Outlet grilles and diffusers of the rectangular,

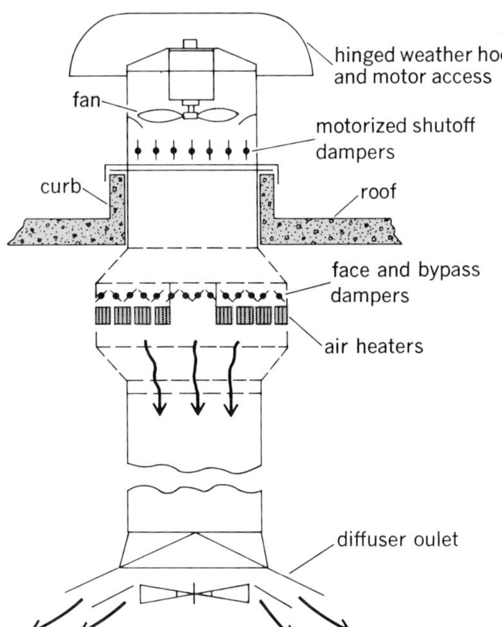

Fig. 4. Mechanically powered roof air supply unit. For summer ventilation, the portion shown in solid lines is used. Heater and dampers are for winter use. Diffuser outlet improves air distribution into ventilated space.

square, and round type are provided to further control and distribute the supply of air within the selected throw (blow length) for each outlet. Any air column acts as a pump and will entrain many times the primary volume of air. The outlets are designed and selected to obtain maximum entrainment (and mixing) because this greatly reduces air motion and temperature stratification, making possible greater comfort for the room occupants.

Exhaust ventilation systems. Exhaust ventilation is required to remove odors, fumes, dust, and heat from an enclosed occupied space. Such exhaust may be of the natural variety previously described or may be mechanical by means of roof or wall exhaust fans or mechanical exhaust systems. The mechanical systems may have minimal ductwork or none at all, or may be provided with extensive ductwork which is used to collect localized hot air, gases, fumes, or dust from process operations. Where it is possible to do so, the process operations are enclosed or hooded to provide maximum collection efficiency with the minimum requirement of exhaust air.

Because of the possibilities of recirculated or external air pollution, it is customary to remove dust and fumes where practical or where required by means of ordinary ventilation filters, more efficient washers or centrifugal collectors, or chemical scrubbers.

Where dust is conveyed in the exhaust ducts, the velocities must be adequate to lift and move the dust particles. Velocities of 3000–6000 ft/min (914–1828 m/min) are required for this purpose. Lower velocities are tenable for fume removal, but corrosion protection must be provided by selection of duct and equipment materials. The round duct is used in most dust and fume systems because of its lower friction and because of its better dust-handling characteristics.

Fans for ordinary ventilation exhaust or heat removal may be similar to supply fans. Fume and dust exhaust fans must be more rugged and are frequently of the radial-blade (paddle-wheel) type for this reason. Axial flow and conventional centrifugal fans are also used in these applications.

[JOHN H. CLARKE]

Vibration damping

The processes and techniques used for converting the mechanical vibrational energy of solids into heat energy. While vibration damping is helpful under conditions of resonance, it may be detrimental in many instances to a system at frequencies above the resonant point. This is due to the fact that the relative motion between the base of the vibration isolator and the mounted body tends to become smaller as the isolator becomes more efficient at the higher frequencies. With damping present, the force transmitted by the elastic element is unable to overcome the damping force; this leads to a resulting increase in transmissibility. *See* VIBRATION ISOLATION.

All metal springs which include structural members such as brackets and shelves have some damping. However, such damping is insufficient for vibration isolators and must be augmented by special damping devices.

Viscous damping. Several different types of damping devices have been developed and used successfully. Probably the most familiar is that used on automobiles, which, although known as a shock absorber, is in reality a damper, and functions as a limiter to the spring system of spring constant k. The system is shown in Fig. 1. A piston p is attached to the body m and is arranged to move vertically through the liquid in a cylinder c which is secured to the support s. As the piston moves, the force required to cause the liquid to flow from one side of the piston to the other is approximately proportional to the velocity of the piston in the cylinder. This type of damping is known as viscous damping. The damping force is controlled by the viscosity of the liquid and by the size of the orifice in the piston. There are several disadvantages to this type of damping; for example, it is unidirectional; it is affected by temperature changes; and because of the fact that the liquid is passed from one side of the piston to the other side through an opening, it is time conscious. The opening, whether it is an orifice or the clearance between piston and cylinder, can pass only so much liquid in a given length of time. If the body to which the piston is attached is caused to displace faster than the liquid can transfer, a bottoming effect occurs. This effect is experienced by riders in automobiles when a hole in the street is hit at too fast a rate; the springs may appear to bottom out, but actually it is the shock absorber or damper. *See* SHOCK ABSORBER.

VIBRATION DAMPING

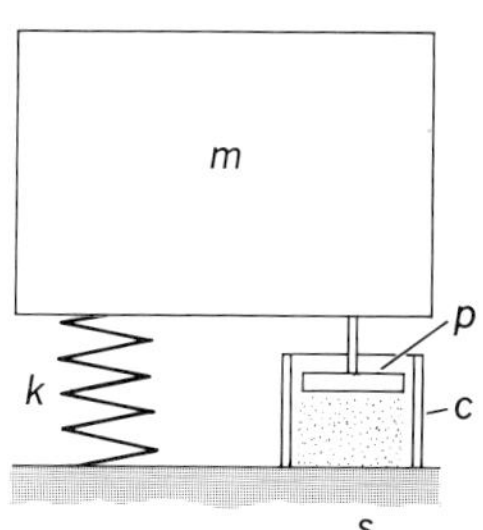

Fig. 1. Automobile shock absorber.

Some of the disadvantages of viscous damping may be overcome by using air instead of liquid as the damping medium. Air, being compressible, will add to the effective spring force with large displacements. If the air is housed within a flexible bellows, damping will be attainable horizontally as well as vertically. Such a system is illustrated in Fig. 2.

VIBRATION DAMPING

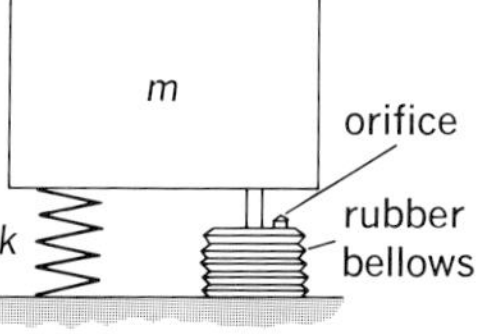

Fig. 2. System employing viscous damping with air.

This type of damping has proved very effective

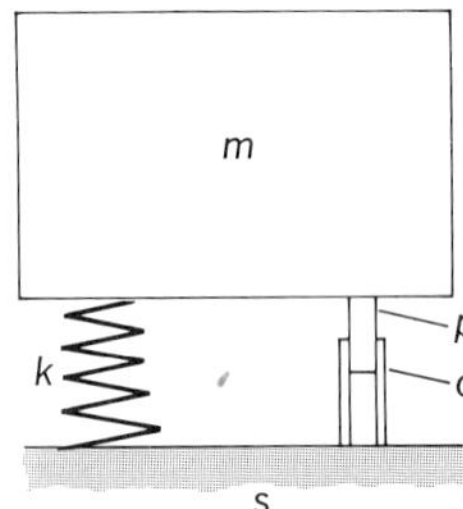

Fig. 3. System employing coulomb damping.

in vibration isolators. The primary disadvantage occurs under conditions of high and low temperature with the change in elasticity of the rubber bellows.

Friction damping. Damping forces may be generated by causing one dry member to slide on another. This is known as dry friction or coulomb damping. A damper of this type is shown in Fig. 3. A pin p inserted in cylinder c and attached to body m is arranged to slide between two vertical spring members which are attached to the support s. The force exerted by the damper in opposition to the motion of the body m is the product of the normal force and the coefficient of friction between the pin and vertical leaf springs. The damping force is usually constant; however, if the pin is tapered, a variable force may be obtained. Friction damping is used in several commercially available isolators because it provides a simple means to control the damping forces. If necessary, independent damping systems may be provided within the same unit for vertical and horizontal motions. Some frictional dampers are effective in both vertical and horizontal directions. One such system (Fig. 4) comprises a load-carrying concave-convex spring made of a metal screen consisting of two parts attached by an eyelet. A damping-coil spring encircles the two attached springs. Any motion vertically or horizontally will cause the load springs to deflect vertically or tilt horizontally. Any change in position of the load springs results in the damping spring's being forced out over the surface of the concave springs; this creates a frictional force. Since the surface contact of the damping spring increases with displacement, the damping force is approximately linear.

Inherent damping. There are many applications where vibration dampers of an external type such as those discussed cannot be used because of space limitations, economic considerations, or the fact that the system needs very little damping. These applications make use of the inherent damping, or internal hysteresis, of such materials as rubber, felt, and cork. Vibration isolation with inherent damping is most commonly used in applications with constant motor speeds, such as air compressors, generators, and grinders.

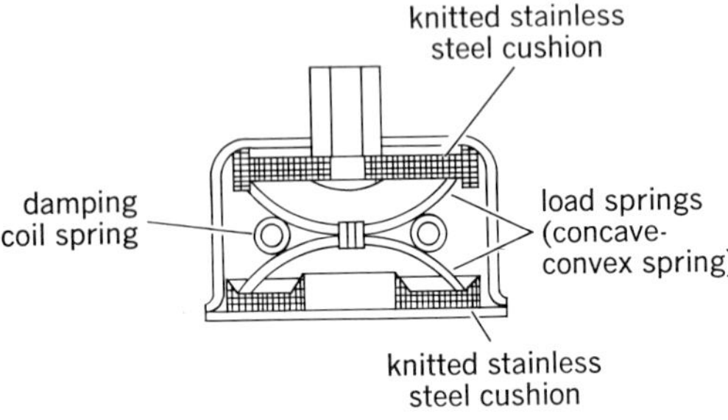

Fig. 4. Frictional damper.

Magnetic damping. This type of damping is attainable as a result of the electric current induced in a conductor moving through a magnetic field. The damping force can be made proportional to the velocity of the conductor moving through the field. Magnetic damping has not been used successfully in vibration isolators because its effectiveness is limited to a single direction.

Numerical values. Considerable emphasis is placed upon the numerical value for the damping force. Such a value is needed to predict the behavior of a damped system; however, difficulty is often encountered in such an analysis. The difficulty is that viscous damping, although susceptible to mathematical analysis and the establishment of a numerical value for the damping force, is seldom encountered in pure form in actual practice. The types of damping actually used are not well adapted to mathematical analysis. The damping value may be determined by the logarithmic decrement method, but care should be exercised when these values are used in equations of pure viscous damping. This is especially true of the two most common forms of damping—friction and inherent damping. With both these types the effect of damping will vary with the amplitude of vibration. Since the damping effect is different, the natural frequency of the system will change with amplitude of vibration. For double amplitudes of vibration, such as 0.060 in., the magnification at resonance may be one value and the resonance frequency may be so many cycles per second, while with a double amplitude of 0.020 in. the magnification may be different and the resonance period may be higher. This difference must be recognized when applying formulas derived from the analysis of pure viscous damping. *See* SPRING; VIBRATION MACHINE.

[K. W. JOHNSON]

Bibliography: R. E. Bishop et al., *Matrix Analysis of Vibration*, 1979; R. E. Bishop and D. C. Johnson, *Mechanics of Vibration*, 1979; C. E. Crede and C. M. Harris (eds.), *Shock and Vibration Control Handbook*, rev. ed., 1976; J. P. Den Hartog, *Mechanical Vibrations*, 4th ed., 1956; Institute for Power Systems, *Handbook of Noise and Vibration Control*, 4th ed., 1979; J. N. MacDuff and J. R. Currey, *Vibration Control*, 2d ed., 1981.

Vibration isolation

The isolation, in structures, of those vibrations or motions that are classified as mechanical vibration. Vibration isolation involves the control of the supporting structure, the placement and arrangement of isolators, and control of the internal construction of the equipment to be protected.

The simplest kind of mechanical vibration has the waveform of sinusoidal motion (Fig. 1). Vibrations in structures, although generally more complex in waveform, exist wherever movement takes place. Such movement may be caused, for example, by the engine in an automobile, by engines or wind buffeting in aircraft, or by a punch press in a building. Delicate electronic equipment and precision instruments must normally be isolat-

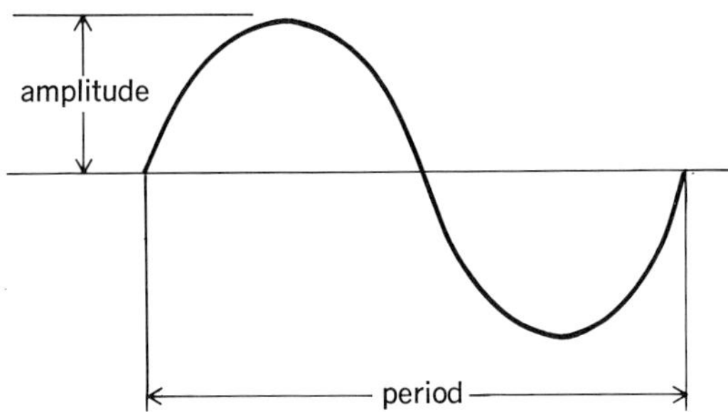

Fig. 1. Sinusoidal motion.

ed from these motions if accurate measurements are to be obtained. *See* MECHANICAL VIBRATION.

Vibration, in most cases, may be effectively isolated by placing a resilient medium, or vibration isolator, between the source of vibration and its surrounding area to reduce the magnitude of the force transmitted from a structure to its support or, alternatively, to reduce the magnitude of motion transmitted from a vibrating support to the structure. Isolating vibration at its source is commonly termed active or source isolation; isolating an instrument from its surroundings is known as passive isolation. In either case the vibration isolator is designed according to the same principles. Vibration isolators are available commercially, with published data covering their characteristics. These data include minimum and maximum rated load, natural frequency at rated load, transmissibility, damping characteristics, ultimate strength, sway space limits, and so on.

Natural frequency; resonance. The prime concern in the field of vibration isolation is the proper use of isolators under various load configurations with respect to their loading, the desired natural frequency, the position and location of the isolators, and the relationship to the structural response of the equipment to which they are attached. For the vibration isolator to be effective, the natural frequency should be approximately 0.4 times the frequency of the interfering source. The natural frequency is the frequency at which a freely vibrating mass system oscillates once it has been deflected. In the case where vibrations occur over a wide frequency range, such as in aircraft (5–2000+ Hz), the natural frequency of the isolator is established with respect to the cruising speed. This means that when the lower frequencies are traversed, such as during aircraft takeoff, the mounted equipment will pass through a condition known as resonance. Resonance is said to exist when the natural frequency of a spring (in this case an isolator) coincides with the frequency of the excitation forces. Resonance causes magnification of the input vibration and may be harmful to the equipment if not controlled within reasonable limits. To control the vibration magnification at resonance, the resilient element within the isolator must be damped. With suitable damping, the magnification factor may be held to 3 or less. *See* VIBRATION DAMPING.

The vibration isolator should be considered as only one part of the isolating system, the other parts being the supporting structure that lies below the isolator, and the internal structure of the equipment that is above the isolator. When isolators are selected for use where the period of resonance is critical, it should not be forgotten that the flexibility of the supporting structure and the flexibility of the isolators are in series, so that the resultant resonant frequency of the loaded system will therefore be inversely proportional to the square root of the sum of these two flexibilities. The additional flexibility of the structure will lower the natural frequency of the system and will also result in increased displacement during resonance, caused by the presence of the undamped structure. The flexibility of the supporting structure, including the structural linkages leading to such structures, should not exceed 25% of the flexibility of the isolator. To neglect the resiliency of the support structures in providing the desired natural frequency of the system, as so many textbooks do with the assumption of rigid support, is impractical.

There are several methods for determining the load at each isolator location; one is shown in Fig. 2 where the loads are as given in the following equations (W is the total weight).

$$A = W\left(\frac{b}{a+b}\right)\left(\frac{d}{c+d}\right) \qquad B = W\left(\frac{b}{a+b}\right)\left(\frac{c}{c+d}\right)$$

$$C = W\left(\frac{a}{a+b}\right)\left(\frac{c}{c+d}\right) \qquad D = W\left(\frac{a}{a+b}\right)\left(\frac{d}{c+d}\right)$$

The next step in vibration isolation is the positioning and arrangement of the isolators with regard to the geometry of the equipment.

Location of isolators. The vibration isolators may be positioned and arranged in many different ways, all variations of three basic types, each of which requires a definite amount of space: (1) isolators attached underneath equipment, known as an underneath mounting system; (2) isolators located in the plane of the center of gravity of the equipment, known as a center-of-gravity system; (3) mountings arranged four on each side in the plane of the radius of gyration, known as a double side-mounted system or radius-of-gyration system.

Textbooks generally treat problems relating to the mounting of equipment with resilient elements as masses in unlimited space. Under these undefined conditions any of the three systems listed may be used; however, when a space limitation is imposed, each system has a definite limitation and must be used accordingly.

Underneath mounting system. For an underneath mounting system, the most efficient for vibration isolation is one with a low natural frequency in both the vertical and horizontal axes. The most stable system is one with a high natural

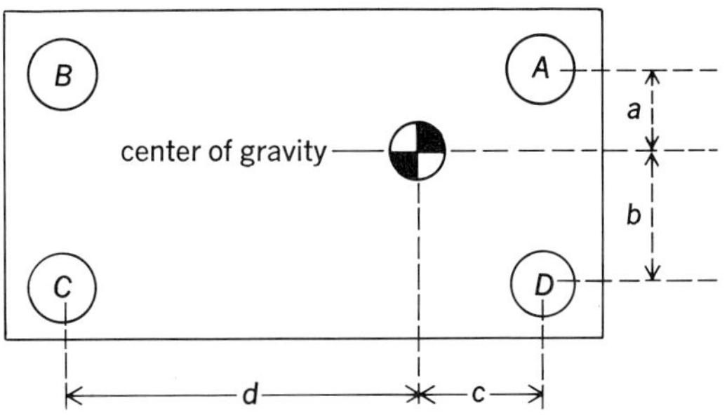

Fig. 2. Determination of isolator locations.

VIBRATION ISOLATION

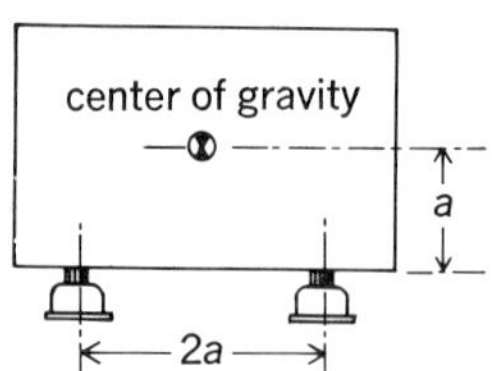

Fig. 3. Underneath mounting system.

frequency with respect to the disturbing frequency, but in this case the vibration characteristics will suffer. Stability may be achieved with low-frequency isolators by maintaining adequate spacing of the isolators with respect to the height of the center of gravity from the mounting plane. Tests show that the isolator spacing should not be less than twice the height of the center of gravity from the mounting plane. This condition is illustrated in Fig. 3. The use of an underneath mounting system for equipments that exceed the condition shown in Fig. 3 should be augmented by stabilizers to provide the needed stability.

Center-of-gravity system. Locating the isolators in the plane of the center of gravity has generally been considered the ideal system because of its ability to decouple the rotational modes of vibration. In actual use, however, such results are seldom achieved because of space limitations. The primary conditions are: (1) that the isolators be located in a plane passing through the center of gravity, (2) that the distance between isolators be twice the radius of gyration of the body, and (3) that the horizontal-to-vertical stiffness of the isolators be equal. The first and last of these conditions are easy to satisfy. The second condition is not always possible where a space limitation exists. The limit of the center-of-gravity system, especially in aircraft, is generally arrived at when the height-to-width ratio of the equipment exceeds 2.8.

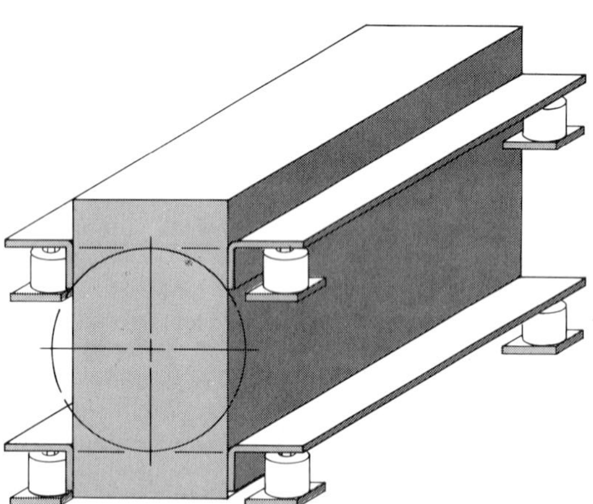

Fig. 4. Radius-of-gyration system.

Radius-of-gyration system. The third system that may be used when the limit of the center-of-gravity system has been reached is a double side-mounted system or radius-of-gyration system as illustrated in Fig. 4. Two sets of isolators are arranged on each side. For optimum results the isolators should be located in the plane of the radius of gyration. However, since it is often difficult to determine the exact radius, acceptable results will be obtained in most instances by assuming the body to be of uniform density. Satisfactory results have been obtained with bodies having height-to-width ratios up to 5. The limitation of this system will be reached when structural rigidity of the body is such that excessive bending occurs between the upper and lower isolator locations.

The next consideration in vibration isolation is the structural rigidity of the body to be isolated. This step is of importance in that the use of incorrect structures, particularly supporting brackets for component parts, can render the other two steps useless. Supporting brackets act as springs under a vibratory condition and become resonant at their natural frequency. Should resonance of the brackets coincide with the isolator resonance, damage may occur. The resonance point for the brackets and internal component should occur when the isolators are approaching their maximum efficiency so that the input vibration to the internal structure is at a low level. This level is normally reached at four times the natural frequency of the isolator.

The isolation of vibration may be accomplished in most instances by the use of commercial isolators. In some cases, such as in heavy industrial machinery where the supporting structure is concrete, less consideration need be given to the flexibility of the supporting structure. Also, where the body to be isolated is stationary, as in the case of compressors, the height-to-width ratio with respect to the isolator spacing is of little importance. *See* VIBRATION MACHINE. [K. W. JOHNSON]

Bibliography: *See* VIBRATION DAMPING.

Vibration machine

A device for subjecting a system to controlled and reproducible mechanical vibration. Vibration machines, commonly called shake tables, are widely used in vibration measurement and analysis. There are three types of vibration machines in general use. These are the mechanical direct-drive type, the mechanical reaction type, and the electrodynamic type. Other types, such as hydraulic excitation devices, resonant systems, piezoelectric vibration generators used for instrument calibration, and machines for testing packages, have only limited or specialized applications.

Direct-drive machine. The mechanical direct-drive vibration machine (sometimes referred to as the brute-force type) is a machine in which the vibration table is forced by a positive linkage to undergo a displacement. The linkage is driven by a direct attachment to eccentrics or crankshafts.

Fig. 1. Direct-drive vibration machine. (*L. A. B. Corp.*)

The degree of eccentricity or crank radius determines the amplitude of vibration of the table. The motion of the vibration table is approximately simple harmonic. The frequency is varied over a range of 5–60 Hz by a variable-drive unit. Automatic cycling is normally used to cycle the table to maximum frequency and return to the starting frequency. The period of cycling is usually 1–3 min.

The vibration table is usually designed, by the use of several cam linkages, to move in both vertical and horizontal directions. A machine of this type is pictured in Fig. 1.

Reaction-type machine. A vibration machine in which the forces exciting the vibrations are generated by rotating or reciprocating unbalanced masses is called a reaction-type machine. The system consists of a table which is suspended from a mounting frame attached to the floor. The unbalanced masses which are attached to a shaft are part of the floating table assembly. A separate variable-speed unit, which is attached to the mass-shaft assembly of the table by a flexible coupling, drives the system.

Fig. 2. Reaction-type vibration machine. (*L. A. B. Corp.*)

The amplitude of vibration is controlled by the degree of off-center setting of the rotating masses. Frequency is controlled by a rheostat in series with a dc motor. The frequency may be continuously varied over a range of 5–60 Hz. The primary advantage of this type of machine is the constant displacement provided by the rotating masses. Also, since the table is suspended, there is essentially no vibration at the floor attachment points. A reaction-type machine is pictured in Fig. 2.

Electrodynamic machine. The electrodynamic vibration machine derives its vibratory force from the action of a fixed magnetic field upon a coil of wire contained in it and excited by a suitable alternating current. The armature consists of those parts that are integral with the coil and the associated elements that move in the magnetic field.

Fig. 3. A complete vibration test system showing the electrodynamic shaker in the left foreground. In the rear, from left to right, are the power amplifier, control console, and instrumentation cabinets. (*Calidyne Co.*)

This type works on the same principle as that of an electrodynamic loudspeaker. The force output is proportional to the magnitude of the current. Automatic displacement or acceleration control may be had over a frequency range of 5–2000+ Hz.

The electrodynamic vibration machine provides motion of a sinusoidal waveform and a form commonly termed random vibration.

The primary advantage of this type of vibration machine is the wide frequency range it offers. Figure 3 pictures a machine complete with the control console. *See* MECHANICAL VIBRATION.

[K. W. JOHNSON]

Voltage regulation

The change in voltage magnitude that occurs when the load (at a specified power factor) is reduced from the rated or nominal value to zero, with no intentional manual readjustment of any voltage control, expressed in percent of nominal full-load voltage. Voltage regulation is a convenient measure of the sensitivity of a device to changes in loading. This concept is commonly used for transmission or distribution circuits as well as power supply equipment, including constant-voltage transformers, rectifiers, outdoor coupling capacitors, direct-current generators, induction frequency converters, synchronous generators, line regulation circuits, and thyristor converters. A common form of expressing voltage regulation in percent is shown in the equation below.

$$\text{Voltage regulation} = \frac{\text{No-load voltage} - \text{full-load voltage}}{\text{Full-load voltage}} \times 100$$

As an example of the use of this equation, one can compute the regulation of the supply voltage to an appliance, home, office, or an entire city. This gives a measure of the need for special voltage-regulating equipment, particularly if there are voltage-sensitive components in the load. *See* GENERATOR; TRANSFORMER.

[PAUL M. ANDERSON]

Bibliography: *IEEE Standard Dictionary of Electrical and Electronic Terms*, IEEE Std100-1977, 2d ed., 1977; J. R. Neuenswander, *Modern Power Systems*, 1971.

Volumetric efficiency

In describing an engine or gas compressor, the ratio of volume of working substance actually admitted, measured at specified temperature and pressure, to the full piston displacement volume. For a liquid-fuel engine, such as a diesel engine, volumetric efficiency is the ratio of volume of air drawn into a cylinder to the piston displacement. For a gas-fuel engine, such as a gasoline engine with carburetor, volumetric efficiency is based on the charge of fuel and air drawn into the cylinder.

Volumetric efficiency of naturally aspirated automobile and aircraft reciprocating engines may be 85–90% at rated speed. Supercharging increases volumetric efficiency, giving values over 100%. At low speeds and wide-open throttle, volumetric efficiency is high; at high speeds, resistance to airflow through the carburetor, the manifolds, and valve ports decreases volumetric efficiency. Air compressors, refrigerator compressors, and dry vacuum pumps are generally specified for a volumetric efficiency of 60–90%, dependent upon the compressor ratio, type of valve, valve gear, and machine speed. *See* INTERNAL COMBUSTION ENGINE. [NEIL MAC COULL]

Wage incentives

Wage payment plans designed to provide extra pay for extra effort over and above a "fair day's work." Base hourly, weekly, or monthly wage rates are paid to employees for normal effort, while incentives allow them to earn more through increased effort.

Incentives are used over a broad spectrum of the labor force. Sales commissions, executive bonuses, profit sharing, and suggestion systems are some of the more popular uses. However, the most widely used and the most effective incentives are in the manufacturing areas.

Plans. Some of the most used incentive devices are: the standard hour plan, the piecework plan, and factor plans.

Standard hour plan. In this plan, standards or goals (targets) are represented in a standard (hours) per unit of output. The number of units produced multiplied by the standard time value equals hours earned. Hours earned multiplied by the base rate equals pay for the period.

Example: A machine operator has a base rate of \$5.00/hour. This operator works all day on one job which has a standard of .0833 h/piece and produces 120 pieces. Earnings would be calculated as follows:

120 × .0833 = 10.0 h earned
10 × \$5.00 = \$50.00 earned for the day
8 × \$5.00 = \$40.00 base rate earnings
$\frac{\$50.00 - \$40.00}{\$40.00}$ = 25% earnings above the base rate (incentive)
\$50.00 ÷ 8 = \$6.25 average hourly earnings

Payment of at least the base rates is guaranteed to the employee. Performance of less than the standard output is paid at the base rate. The difference is a loss to the employer and is generally referred to as make-up pay.

The advantages of a standard hour plan are that it is easy to calculate and to understand, the standard need not be recalculated when base rates change (which they may frequently do), and the format lends itself to a number of cost-control reports which can be computer-generated from payroll information. One of the few major disadvantages of this plan is the need to train people to understand the decimal-hour system.

Piecework plan. These standards are usually expressed in dollars or cents per piece or per hundred pieces, such as 3c per piece or \$4.00 per 100 pieces. Historically there had been no base rate guarantees. However, the advent of minimum wage laws has required that at least the minimum wage be paid regardless of output. Very often, the guarantee will be at some point between the minimum wage and the earnings expectancy of the plan.

Example: An operator works all day (8 h) on one job with a piece rate of \$4.00/100 pieces and produces 1250 pieces.

12.5 hundred pieces × \$4 = \$50.00
Hours worked = 8
Average hourly earnings (50 ÷ 8) = \$6.25/h

The biggest advantage of this plan is ease of developing costs in situations where operators surpass the minimum wage. The biggest disadvantage is that the standards must be recalculated each time there is a change in the guaranteed base rate. For most companies this is time-consuming and expensive.

Factor plans. A number of wage incentives currently in popular usage might generally be lumped under this generic category because they are based on some factor or group of factors which are directly controllable by the employee. Some examples of these might be salespeople's commissions paid as a percentage of sales, executive bonuses paid as a percentage of base compensation in relation to net profits, or supervisory incentives which may be based on the factors controllable within the supervisor's department.

Group plans. In addition to the above, there are a number of group wage incentive plans which have often been successful catalysts toward improved performance. These include employee ownership or stock participation, profit sharing, value added, and various forms of participative management programs. These encourage employees to concern themselves with the overall general financial strength of the organization, and promise to share increased earnings with them. One popular participative plan is the Scanlon Plan, which organizes employees into formal subgroups to generate and consider improvement suggestions. The plan provides a formal means for implementing suggestions which are practical and cost-justified. Savings are shared on a monthly basis with employees, usually on a 60/40 basis, although this ratio may vary from company to company.

Effectiveness. Some basic principles for an effective incentive plan are that incentives should: pay for extra effort only; be individual wherever possible or for the smallest group that can be arranged; be based on factors which can be controlled by the employee; not be a substitute for management or management effort; and be high enough to provide the motivation to make the extra effort worthwhile. It is generally accepted that 25 to 35% earnings above base is an adequate range of opportunity.

Those considering the introduction of a wage incentive plan must recognize that the achievement of a successful plan requires time, effort, and expense in its development, installation, and administration; accurate and complete basic data to build it upon; and acceptance and support by both management and its participants. Also, the plan must be designed so as to pay only for overall beneficial results, that is, no payments should be made if the firm is not making a profit. Incentive plans are usually easier to start than to terminate.

Experience has shown that wage incentives are one of the most effective motivators to increased productivity that have as yet been devised. Productivity, which is the measure of effort expressed as 100% being normal, can often be doubled through the use of wage incentives. Typically, it has been found that organizations with no work measurement or incentives will operate at a level of 50–55% productivity. Work measurement, or the setting of goals and targets, can raise this to 75–80%, while incentives can bring this to 125–130%.

The illustration shows the effects of measurement and incentives on output. Translated into units per hour, with 100 defined as normal, this indicates that output per hour can often be raised from 55 pieces per hour to 135 pieces per hour, or a 145% increase. The effect on unit labor costs of such an increase in productivity is dramatic, although not of the same proportions.

Using the machine operator illustration in the example above, the following shows the unit labor cost for each of the three situations in the illustration:

(a) No measurement—Standard is 100 pieces per hour; the base rate is $5.00 per hour. Output is 55 pieces ÷ $5.00, or a cost per piece of $5.00 ÷ 55 = 9c each.

(b) Work measurement with no incentives—Output is 80 pieces, or a labor cost per unit of $5.00 ÷ 80 = 6.25c each.

(c) Wage incentives—Output is 135 pieces; earnings are $5.00 × 1.35 = $6.75. The unit labor cost is $6.75 ÷ 135 = 5c per piece.

Net reduction of unit labor cost in this instance is 4c, or a 44.4% reduction, from the 9c prior cost.

The widespread impression that the sole objective of wage incentives is to motivate people to work faster is not altogether true. Those who have made a career of productivity improvement have found that the biggest gains are achieved by: staying on the job for greater portions of the day; utilizing better human effort through improved methods; the introduction of more effective tools and equipment; and eliminating the causes of lost time or delays (normally humans work within a relatively narrow range of work pace of 80–120%). The other gains or losses are due to the outside factors mentioned above.

Of considerable importance is the fact that any incentive standards which are developed must give appropriate consideration to the needs of individuals for rest, relief, and personal requirements. Further consideration must also be given to safety and quality requirements. Many people believe that incentives promote unsafe practices and result in inferior product quality. There is no valid evidence to support this belief. In fact, where there has been a well-devised work-measurement program, it is invariably found that safety and quality are improved.

Incentive plans are people-oriented systems. As such, they are subject to abuses, such as cheating, far beyond those of other wage systems. Because of this it becomes even more important that the plans be well conceived, carefully monitored, and constantly maintained. Once an incentive plan gets out of hand, it is difficult to maintain its effectiveness. Nevertheless, the gains acquired from a well-operated incentive plan are so significant that the effort expended is very worthwhile. *See* PRODUCTIVITY. [WILLIAM J. OSHINSKI]

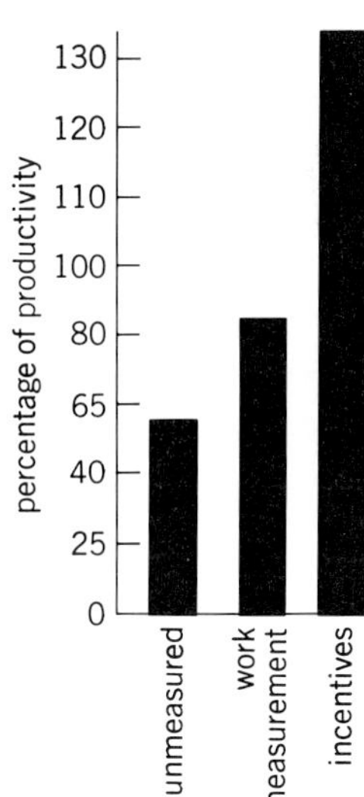

Effects of measurement and incentives on output.

Wall construction

An acceptable exterior wall for a building must be pleasing in appearance, structurally sound, and durable; it must provide insulation and fire resistance; and it must offer protection against condensation of water vapor. An interior wall must meet the same conditions but to a much lesser extent; it must also provide resistance to sound transmission.

In home construction, both the interior and exterior walls are formed around a basic core of wooden studs. The exterior exposed faces are covered with wood siding, wood or asbestos shingles, brick veneer, or other finish and insulated to retard heat flow. Solid and hollow concrete masonry and brick walls are used extensively for low buildings. Reinforced concrete walls, precast concrete walls, and precast concrete walls erected by the tilt-up method also offer a variety of surface finishes, low maintenance, strength, and durability.

The walls of many tall buildings consist principally of glass, the spandrels and pilasters being of steel or aluminum backed up by brick or tile masonry.

Plain or corrugated sheets of cement-asbestos and corrugated metal sheets are used for the walls of industrial buildings. Prefabricated panels consisting of two sheets of metal with a layer of insulation between provide a complete wall section which can be quickly attached to steel framing. A similar prefabricated unit using wood studs, foam insulation, and reinforced plastics in sheets is possible in home construction. Such prefabricated panels are light in weight and low in heat transmission.

The interior wall surface is finished in a number of different ways, depending on the use to which it will be put. Masonry surfaces require only paint, but like other types of wall construction, they may be plastered or finished with paneling, leather, or cloth. *See* BUILDINGS.

[CHARLES M. ANTONI]

Bibliography: Cities' plight spurs quest for new techniques, *Eng. News Rec.*, 180(4):42, 1968; *Civil Eng.*, vol. 36, no. 9, 1966; S. A. Erech, Cast in place concrete connects precast members, *Civil Eng.*, 36(3):64, 1966; F. R. Walker, *The Building Estimators Reference Book*, 16th ed., 1963; G. Winter et al., *Design of Concrete Structures*, 7th ed., 1964.

Warm-air heating system

In a general sense, a heating system which circulates warm air. Under this definition both a parlor stove and a steam blast coil circulate warm air. Strictly speaking, however, a warm-air system is one containing a direct-fired furnace surrounded by a bonnet through which air circulates to be heated (see illustration).

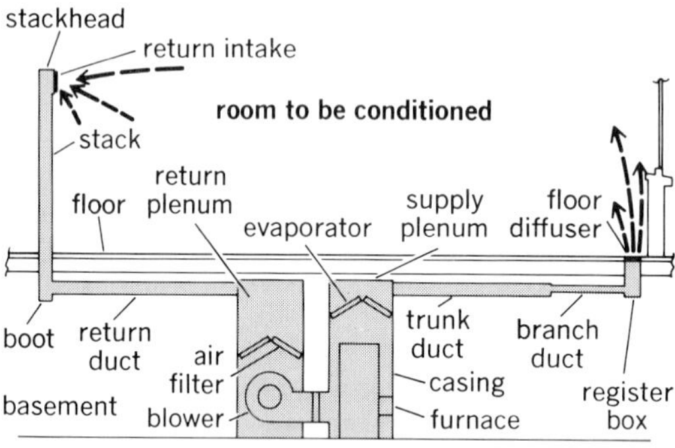

Air passage in a warm-air duct system. (*From S. Konzo, J. R. Carroll, and H. D. Bareither, Winter Air Conditioning, Industrial Press, 1958*)

When air circulation is obtained by natural gravity action, the system is referred to as a gravity warm-air system. If positive air circulation is provided by means of a centrifugal fan (referred to in the industry as a blower), the system is referred to as a forced-air heating system.

Direct-fired furnaces are available for burning of solid, liquid, or gaseous fuels, although in recent years oil and gas fuels have been most commonly used. Furnaces have also been designed which have air circulating over electrical resistance heaters. A completely equipped furnace-blower package consists of furnace, burner, bonnet, blower, filter, and accessories. The furnace shell is usually of welded steel. The burner supplies a positively metered rate of fuel and a proportionate amount of air for combustion. A casing, or jacket, encloses the furnace and provides a passage for the air to be circulated over the heated furnace shell. The casing is insulated and contains openings to which return-air and warm-air ducts can be attached. The blower circulates air against static pressures, usually less than 1.0-in. water gage. The air filter removes dust particles from the circulating air. The most common type of air filter is composed of 1- to 2-in. thick fibrous matting, although electrostatic precipitators are sometimes used. *See* OIL BURNER.

Accessories to assure effective operation include automatic electrical controls for operation of burner and blower and safety control devices for protection against (1) faulty ignition of burner and (2) excessive air temperatures.

Ratings of warm-air furnaces are established from tests made in laboratories under industry-specified conditions. The tests commonly include heat-input rate, bonnet capacity, and register delivery. Heat-input rate is the heat released inside the furnace by the combustion of fuel, in Btu/hr. Bonnet capacity refers to the heat transferred to the circulating air, in Btu/hr. Register delivery is the estimated heat available at the registers in the room after allowance for heat loss from the ducts has been made, in Btu/hr.

The recommended method for selection of a furnace is to estimate the total heat loss from the structure under design weather conditions, including the losses through the floor and from the basement, and to choose a furnace whose bonnet capacity rating is equal to, or greater than, the total design heat loss.

The complete forced-air heating system consists of the furnace-blower package unit; the return-air intake, or grille, together with return-air ducts leading from the grille to the return-air plenum chamber at the furnace; and the supply trunk duct and branch ducts leading to the registers located in the different spaces to be heated.

The forced-air system in recent years has no longer been confined to residential installations. The extreme flexibility of the system, as well as the diversity of furnace types, has resulted in widespread use of the forced-air furnace installations in the following types of installations, both domestic and commercial: residences with basement, crawl space, or with concrete floor slab; apartment buildings with individual furnaces for each apartment; churches with several furnaces for different zones of the building; commercial buildings with summer-winter arrangements; and industrial buildings with individual furnace-duct systems in each zone. *See* COMFORT HEATING. [SEICHI KONZO]

Bibliography: Air Conditioning Contractors of America, *Load Calculation for Residential Winter and Summer Air Conditioning*, Manual J, 1975; American Society of Heating, Refrigerating, and Air-Conditioning Engineers, *ASHRAE Handbook and Product Directory*: *Systems*, 1976; S. Konzo, J. R. Carroll, and H. D. Bareither, *Winter Air Conditioning*, 1958.

Washer

Lock washers. (*From W. J. Luzadder, Fundamentals of Engineering Drawing, 6th ed., Prentice-Hall, 1971*)

A flattened, ring-shaped device used to improve the tightness of a screw fastener. Three types of washer are in common use: plain, spring-lock, and antiturn (tooth-lock washers). Standard plain washers are used to protect a part from damage or to provide for a wider distribution of the load. Because a plain washer will not prevent a nut from turning, a locking-type washer should be used to prevent a bolt or nut from loosening under vibration (see illustration). For industrial applications, spring-lock washers are intended to compensate for possible loosening between assembled parts and to facilitate assembly and disassembly. Lock washers create a continuous pressure between the parts and the fastener. The antiturn-type washers may be externally serrated, internally serrated, or both. The bent teeth bite into the bearing surface to prevent the nut from turning and the fastening from loosening under vibration. To speed up assembly, a variety of permanent preassembled bolt-and-washer and nut-and-washer combinations are available. *See* SCREW FASTENER.

[WARREN J. LUZADDER]

Water pollution

Any change in natural waters which may impair their further use, caused by the introduction of organic or inorganic substances, or a change in temperature of the water. The growth of population and the concomitant expansion in industrial and agricultural activities have rapidly increased the importance of the field of water-pollution control. In the attack on environmental pollution, higher standards for water cleanliness are being adopted by state and Federal governments, as well as by interstate organizations.

Historical developments. Ancient humans joined into groups for protection. Later, they formed communities on watercourses or the seashore. The waterway provided a convenient means of transportation, and fresh waters provided a water supply. The watercourses then became receivers of wastewater along with contaminants. As industries developed, they added their discharges to those of the community. When the concentration of added substances became dangerous to humans or so degraded the water that it was unfit for further use, water-pollution control began. With development of wide areas, pollution of surface water supplies became more critical because wastewater of an upstream community became part of the water supply of the downstream community.

Serious epidemics of waterborne diseases such as cholera, dysentery, and typhoid fever were caused by underground seepage from privy vaults into town wells. Such direct bacterial infections through water systems can be traced back to the late 18th century, even though the germ or bacterium as the cause of disease was not proved for nearly another century. The well-documented epidemic of the Broad Street Pump in London during 1854 resulted from direct leakage from privies into the hand-pumped well which provided the neighborhood water supply. There were 616 deaths from cholera among the users of the well within 40 days.

Eventually, abandoning wells in such populated locations and providing piped water to buildings improved public health. Further, sewers for drainage of wastewater were constructed, but then infections between communities rather than between the residents of a single community became apparent. Modern public health protection is provided by highly refined and well-controlled plants both for the purification of the community water supply and treatment of the wastewater.

Relation to water supply. Water-pollution control is closely allied with the water supplies of communities and industries because both generally share the same water resources. There is great similarity in the pipe systems that bring water to each home or business property, and the systems of sewers or drains that subsequently collect the wastewater and conduct it to a treatment facility. Treatment should prepare the flow for return to the environment so that the receiving watercourse will be suitable for beneficial uses such as general recreation, and safe for subsequent use by downstream communities or industries.

The volume of wastewater, the used water that must be disposed of or treated, is a factor to be considered. Depending on the amount of water used for irrigation, the amount lost in pipe leakage, and the extent of water metering, the volume of wastewater may be 70–130% of the water drawn from the supply. In United States cities, wastewater quantities are usually 75–200 gal (284–757 liters) per capita daily. The higher figure applies to large cities with old systems, limited metering, and comparatively cheap water; the lower figure to smaller communities with little leakage and good metering. Probably the average in the United States for areas served by sewers is 125–150 gal (473–568 liters) of wastewater per person per day. Of course, industrial consumption in larger cities increases per capita quantities.

Related scientific disciplines. The field of water-pollution control encompasses a part of the broader field of sanitary or environmental engineering. It includes some aspects of chemistry, hydrology, biology, and bacteriology, in addition to public administration and management. These scientific disciplines evaluate problems and give the civil and sanitary engineer basic data for the designing of structures to solve the problems. The solutions usually require the collection of domestic and industrial wastewaters and treatment before discharge into receiving waters. *See* CIVIL ENGINEERING; SANITARY ENGINEERING.

Self-purification of natural waters. Any natural watercourse contains dissolved gases normally found in air in equilibrium with the atmosphere. In this way fish and other aquatic life obtain oxygen for their respiration. The amount of oxygen which the water holds at saturation depends on temperature and follows the law of decreased solubility of gases with a temperature increase. Because water temperature is high in the summer, oxygen dissolved in the water is then at a low point for the year.

Degradable or oxidizable substances in wastewaters deplete oxygen through the action of bacteria and related organisms which feed on organic waste materials, using available dissolved oxygen for their respiration. If this activity proceeds at a rate fast enough to depress seriously the oxygen level, the natural fauna of a stream is affected; if the oxygen is entirely used up, a condition of oxygen exhaustion occurs which suffocates aerobic organisms in the stream. Under such conditions the stream is said to be septic and is likely to become offensive to the sight and smell.

Domestic wastewaters. Domestic wastewaters result from the use of water in dwellings of all types, and include both water after use and the various waste materials added: body wastes, kitchen wastes, household cleaning agents, and laundry soaps and detergents. The solid content of such wastewater is numerically low and amounts to less than 1 kg per 1000 kg of domestic wastewater. Still, the character of these waste materials is such that they cause significant degradation of receiving waters, and they may be a major factor in spreading waterborne diseases, notably typhoid and dysentery.

Characteristics of domestic wastewater vary from one community to another and in the same community at different times. Physically, community wastewater usually has the grayish colloidal appearance of dishwater, with floating trash apparent. Chemically, it contains the numerous and complex nitrogen compounds in body wastes, as well as soaps and detergents and the chemicals normally present in the water supply. Biologically, bacteria and other microscopic life abound. Wastewaters from industrial activities may affect all of these characteristics materially.

Industrial wastewaters. In contrast to the general uniformity of substances found in domestic wastewaters, industrial wastewaters show increasing variation as the complexity of industrial

Table 1. General nature of industrial wastewaters

Industry	Processes or waste	Effect
Brewery and distillery	Malt and fermented liquors	Organic load
Chemical	General	Stable organics, phenols, inks
Dairy	Milk processing, bottling, butter and cheese making	Acid
Dyeing	Spent dye, sizings, bleach	Color, acid or alkaline
Food processing	Canning and freezing	Organic load
Laundry	Washing	Alkaline
Leather tanning	Leather cleaning and tanning	Organic load, acid and alkaline
Meat packing	Slaughter, preparation	Organic load
Paper	Pulp and paper manufacturing	Organic load, waste wood fibers
Steel	Pickling, plating, and so on	Acid
Textile manufacture	Wool scouring, dyeing	Organic load, alkaline

processes rises. Table 1 lists major industrial categories along with the undesirable characteristics of their wastewaters.

Because biological treatment processes are ordinarily employed in water-pollution control plants, large quantities of industrial wastewaters can interfere with the processes as well as the total load of a treatment plant. The organic matter present in many industrial effluents often equals or exceeds the amount from a community. Accommodations for such an increase in the load of a plant should be provided for in its design.

Discharge directly to watercourses. The industrial revolution in England and Germany and the subsequent similar development in the United States increased problems of water-pollution control enormously. The establishment of industries caused great migrations to the cities, the immediate result being a great increase in wastes from both population and industrial activity. For some years discharges were made directly to watercourses, the natural assimilative power of the receiving water being used to a level consistent with the required cleanliness of the watercourse. Early dilution ratios required for this method are shown in Table 2. Because of the more rapid absorption of oxygen from the air by a turbulent stream, it has a high rate of reaeration and a low dilution ratio; the converse is true of slow-flowing streams.

Development of treatment methods. With the passage of time, the waste loads imposed on streams exceeded the ability of the receiving water to assimilate them. The first attempts at wastewater treatment were made by artificially providing means for the purification of wastewaters as

Table 2. Dilution ratios for waterways

Type	Stream flow, $ft^3/(sec)$ (1000 population)*
Sluggish streams	7–10
Average streams	4–7
Swift turbulent streams	2–4

*1 $ft^3/sec = 28.3 \times 10^{-3}\ m^3$.

observed in nature. These forces included sedimentation and exposure to sunlight and atmospheric oxygen, either by agitated contact or by filling the interstices of large stone beds intermittently as a means of oxidation. However, practice soon outstripped theory because bacteriology was only then being born and there were many unknowns about the processes.

In later years testing stations were set up by municipalities and states for experimental work. Notable among these were the Chicago testing station and one established at Lawrence by the state of Massachusetts, a pioneer in the public health movement. From the results of these direct investigations, practices evolved which were gradually explained through the mechanisms of chemistry and biology in the 20th century.

Thermal pollution. An increasing amount of attention has been given to thermal pollution, the raising of the temperature of a waterway by heat discharged from the cooling system or effluent wastes of an industrial installation. This rise in temperature may sufficiently upset the ecological balance of the waterway to pose a threat to the native life-forms. This problem has been especially noted in the vicinity of nuclear power plants. Thermal pollution may be combated by allowing the wastewater to cool before it empties into the waterway. This is often done in large cooling towers.

Current status. Modern water-pollution engineers or chemists have a wealth of published information, both theoretical and practical, to assist them. While research necessarily will continue, they can draw on established practices for the solution to almost any problem. A challenging problem has been the handling of radioactive wastes. Reduction in volume, containment, and storage constitute the principal attack on this problem. Because of the fundamental characteristics of radioactive wastes, the development of other methods seems unlikely.

Public desire for complete water pollution control continues, but there is an increasing realization that solution to the problem is costly. While cities have had little concern about the initial construction cost because of the large Federal share, the expenditures for operation fall entirely on the local community. During the life of a project, operating costs may exceed the initial construction outlay, and with present rates of increase, they may become a major financial burden.

The control of 100% of the organic pollution reaching the watercourses is the goal of many people, but such an ideal cannot even be approached, since about one-third of the total is from nonpoint sources. Essentially, this third is from vegetable matter carried by surface drains and direct runoff. To achieve the public health protection which is the primary purpose of collection and treatment of wastewater, proper measures are essential, and should be the principal focus of a program of water pollution control. Interestingly, such was the original purpose of sewers and drains employed by ancient civilizations, as manifested by the Romans, who gave Venus the title of Goddess of the Sewers, in addition to her other titles associated with health and beauty. The pediment of her "lost" statue in the Roman Forum identifies her as *Venus Cloacinae.*

Table 3. Federal funds for wastewater treatment plant construction, 1973–1979, in $\$10^9$

Fiscal year	Authority	Appropriated	Obligated	Outlays
1973	5	2	1.531	0
1974	6	3	1.444	.159
1975	7	4	3.616	.874
1976*	0	9	4.814	2.563
1977†	1.48	1.48	6.664	2.710
1978	4.5	4.5	2.301	2.960
1979‡	5	4.2	.953	1.984

*Includes transition quarter, July-September 1976.
†Includes $480,000,000 under Public Works Employment Act.
‡Obligated and Outlays as of Apr. 30, 1979.

However, there are strong manifestations of improved quality in the waters throughout the United States. This is apparent not only in chemical and biological measurements, but in more readily observed effects such as better appearance, eliminated smells, and the return of fish life to watercourses which had become "biological deserts" because of the effects of pollution from municipal and industrial wastewaters. The overall results are a living tribute to the cooperative efforts of local citizens and local, state, and Federal agencies working together to improve the quality of the nation's waters. A measure of the activity in the field is indicated by the employment of nearly 90,000 in the local wastewater collection and treatment works of the United States.

Federal aid. Because of public demands and the actions of state legislatures and the Congress of the United States, there has been a surge of interest in, and a demand for, firm solutions to water-pollution problems. Although the Federal government granted aid for construction of municipal treatment plants as an employment relief measure in the 1930s, no comprehensive Federal legislation was enacted until 1948. This was supplemented by a major change in 1956, when the United States government again offered grants to municipalities to assist in the construction of water-pollution control facilities. These grants were further extended to small communities for the construction of both water and sewer systems.

Since 1965, Federal activity in water-pollution control has advanced from a minor activity in the Public Health Service, through the Water Pollution Control Administration in the Department of the Interior, to a major activity in the Environmental Protection Agency. In the 1972 act (P. L. 92-500) Congress authorized a massive attack on municipal pollution problems by a grant-in-aid program eclipsing any previous effort. Federal funds for 1973–1979 are given in Table 3.

State and Federal regulations are increasing constantly in severity. This tendency is expected to continue until the problem of water pollution is brought under complete control. Even then, water quality will be monitored to make certain that actual control is achieved on a day-to-day or even an hour-to-hour basis. *See* SEPTIC TANK; SEWAGE; SEWAGE COLLECTION SYSTEMS; SEWAGE DISPOSAL; SEWAGE SOLIDS; SEWAGE TREATMENT.

[RALPH E. FUHRMAN]

Bibliography: G. M. Fair et al., *Elements of Water Supply and Wastewater Disposal*, 1971; Federal Water Quality Administration, *Santa Barbara Oil Pollution, 1969: A Study of the Biological Effects of the Oil Spill Which Occurred at Santa Barbara, California, in 1969*, University of California, Santa Barbara, October 1970; R. E. McKinney, *Microbiology for Sanitary Engineers*, 1962; N. L. Nemerow, *Liquid Waste of Industry*, 1971; *Proceedings of the 1975 Conference on Prevention and Control of Oil Pollution*, San Francisco, Mar. 25–27, 1975, American Petroleum Institute, 1975; C. N. Sawyer and P. L. McCarty, *Chemistry for Sanitary Engineers*, 1962; J. E. Smith (ed.) *"Torrey Canyon" Pollution and Marine Life: A Report by the Plymouth Laboratory of the Marine Biological Association of the United Kingdom*, 1968; J. Snow, *Mode of Communication of Cholera*, 1855; Water Pollution Control Federation, *Careers in Water Pollution Control*.

Water supply engineering

A branch of civil engineering concerned with the development of sources of supply, transmission, distribution, and treatment of water. The term is used most frequently in regard to municipal waterworks, but applies also to water systems for industry, irrigation, and other purposes.

SOURCES OF WATER SUPPLY

Underground waters, rivers, lakes, and reservoirs, the primary sources of fresh water, are replenished by rainfall. Some of this water flows to the sea through surface and underground channels, some is taken up by vegetation, and some is lost by evaporation.

Groundwater. Water obtained from subsurface sources, such as sands and gravels and porous or fractured rocks, is called groundwater. Groundwater flows toward points of discharge in river valleys and, in some areas, along the seacoast. The flow takes place in water-bearing strata known as aquifers. The velocity may be a few feet to several miles per year, depending upon the permeability of the aquifer and the hydraulic gradient or slope. A steep gradient or slope indicates relatively high pressure, or head, forcing the water through the aquifer. When the gradient is flat, the pressure forcing the water is small. When the velocity is extremely low, the water is likely to be highly mineralized; if there is no movement, the water is rarely fit for use.

Permeability is a measure of the ease with which water flows through an aquifer. Coarse sands and gravels, and limestone with large solution passages, have high permeability. Fine sand, clay, silt, and dense rocks (unless badly fractured) have low permeability.

Water table. In an unconfined stratum the water table is the top or surface of the groundwater. It may be within a few inches of the ground surface or hundreds of feet below. Normally it follows the topography. Aquifers confined between impervious strata may carry water under pressure. If a well is sunk into such an aquifer and the pressure is sufficient, water may be forced to the surface, resulting in an artesian well. The water table elevation and artesian pressure may vary substantially with the seasons, depending upon the amount of rainfall recharging the aquifer and the amount of water taken from the aquifer. If pumpage exceeds recharge for an extended period, the aquifer is depleted and the water supply lost.

Salt-water intrusion. Normally the groundwater flow is toward the sea. This normal flow may be reversed, however, by overpumping and lowering of the water table or artesian pressure in an aquifer. Salt water flowing into the fresh-water aquifer being pumped is called salt-water intrusion.

Springs. Springs occur at the base of sloping ground or in depressions where the surface elevation is below the water table, or below the hydraulic gradient in an artesian aquifer from which the water can escape. Artesian springs are fed through cracks in the overburden or through other natural channels extending from the aquifer under pressure to the surface.

Wells. Wells are vertical openings, excavated or drilled, from the ground surface to a water-bearing stratum or aquifer. Pumping a well lowers the water level in it, which in turn forces water to flow from the aquifer. Thick, permeable aquifers may yield several million gallons daily with a drawdown (lowering) of only a few feet. Thin aquifers, or impermeable aquifers, may require several times as much drawdown for the same yield, and frequently yield only small supplies.

Dug wells, several feet in diameter, are frequently used to reach shallow aquifers, particularly for small domestic and farm supplies. They furnish small quantities of water, even if the soils penetrated are relatively impervious. Large-capacity dug wells or caisson wells, in coarse sand and gravel, are used frequently for municipal supplies. Drilled wells are sometimes several thousand feet deep.

The portion of a well above the aquifer is lined with concrete, stone, or steel casing, except where the well is through rock that stands without support. The portion of the well in the aquifer is built with open-joint masonry or screens to admit the water into the well. Metal screens, made of perforated sheets or of wire wound around supporting ribs, are used most frequently. The screens are galvanized iron, bronze, or stainless steel, depending upon the corrosiveness of the water and the expected life of the well. Plastic screens are sometimes used.

The distance between wells must be sufficient to avoid harmful interference when the wells are pumped. In general, economical well spacing varies directly with the quantity of water to be pumped, and inversely with the permeability and thickness of the aquifer. It may range from a few feet to a mile or more.

Infiltration galleries are shafts or passages extending horizontally through an aquifer to intercept the groundwater. They are equivalent to a row of closely spaced wells and are most successful in thin aquifers along the shore of rivers, at depths of less than 75 ft (23 m). The galleries are built in open cuts or by tunneling, usually with perforated or porous liners to screen out the aquifer material and to support the overburden.

Ranney wells consist of a center caisson with horizontal, perforated pipes extending radially into the aquifer. They are applicable to the development of thin aquifers at shallow depths.

Specially designed pumps, of small diameter to fit inside well casings, are used in all well installations, except in flowing artesian wells or where the water level in the well is high enough for direct suction lift by a pump on the surface (about 15 ft or 5 m maximum). Well pumps are set some distance below the water level, so that they are submerged even after the drawdown is established. Well-pump settings of 100 ft (30 m) are common, and they may exceed 300 ft (90 m) where the groundwater level is low. Multiple-stage centrifugal pumps are used most generally. They are driven by motors at the surface through vertical shafts, or by waterproof motors attached directly below the pumps. Wells are sometimes pumped by air lift, that is, by injecting compressed air through a pipe to the bottom of the well.

Surface water. Natural sources, such as rivers and lakes, and impounding reservoirs are sources of surface water. *See* DAM; RESERVOIR.

Water is withdrawn from rivers, lakes, and reservoirs through intakes. The simplest intakes are pipes extending from the shore into deep water, with or without a simple crib and screen over the outer end. Intakes for large municipal supplies may consist of large conduits or tunnels extending to elaborate cribs of wood or masonry containing screens, gates, and operating mechanisms. Intakes in reservoirs are frequently built as integral parts of the dam and may have multiple ports at several levels to permit selection of the best water. The location of intakes in rivers and lakes must take into consideration water quality, depth of water, likelihood of freezing, and possible interference with navigation. Reservoir intakes are usually designed for gravity flow through the dam or its abutments.

TRANSMISSION AND DISTRIBUTION

The water from the source must be transmitted to the community or area to be served and distributed to the individual customers.

Transmission mains. The major supply conduits, or feeders, from the source to the distribution system are called mains or aqueducts.

Canals. The oldest and simplest type of aqueducts, especially for transmitting large quantities of water, are canals. Canals are used where they can be built economically to follow the hydraulic gradient or slope of the flowing water. If the soil is suitable, the canals are excavated with sloping sides and are not lined. Otherwise, concrete or asphalt linings are used. Gravity canals are carried across streams or other low places by wooden or steel flumes, or under the streams by pressure pipes known as inverted siphons. *See* CANAL.

Tunnels. Used to transmit water through ridges or hills, tunnels may follow the hydraulic grade line and flow by gravity or may be built below the grade line to operate under considerable pressure. Rock tunnels may be lined to prevent the overburden from collapsing, to prevent leakage, or to reduce friction losses by providing a smooth interior. *See* TUNNEL.

Pipelines. Pipelines are a common type of transmission main, especially for moderate supplies not requiring large aqueducts or canals. Pipes are of cast iron, ductile iron, steel, reinforced concrete, cement-asbestos, or wood. Pipeline material is determined by cost, durability, ease of installation and maintenance, and resistance to corrosion. The pipeline must be large enough to deliver the required amount of water and strong enough to withstand the maximum gravity or pumping pressure. Pipelines are usually buried in the ground for protection and coolness. *See* PIPELINE.

Distribution system. Included in the distribution system are the network of smaller mains branching off from the transmission mains, the house services and meters, the fire hydrants, and the distribution storage reservoirs. The network is composed of transmission or feeder mains, usually 12 in. (30 cm) or more in diameter, and lateral mains along each street, or in some cities along alleys between the streets. The mains are installed in grids so that lateral mains can be fed from both ends where possible. Mains fed from one direction only are called dead ends; they are less reliable and do not furnish as much water for fire protection as do mains within the grid. Valves at intersections of mains permit a leaking or damaged section of pipe to be shut off with minimum interruption of water service to adjacent areas.

House services. The small pipes, usually of iron, copper, or plastic material, extending from the water main in the street to the customer's meter at the curb line or in the cellar are called house services. In many cities each service is metered, and the customer's bill is based on the water actually used.

Fire hydrants. Fire hydrants have a vertical barrel extending to the depth of the water main, a quick-opening valve with operating nut at the top, and connections threaded to receive fire hose. Hydrants must be reliable, and they must drain upon closing to prevent freezing.

Distribution reservoirs. These are used to supplement the source of supply and transmission system during peak demands, and to provide water during a temporary failure of the supply system. In small waterworks the reservoirs usually equal at least one day's water consumption; in larger systems the reservoirs are relatively smaller but adequate to meet fire-fighting demands. Ground storage reservoirs, elevated tanks, and standpipes are used for distribution reservoirs.

Ground storage reservoirs, if on high ground, can feed the distribution system by gravity, but otherwise it is necessary to pump water from the reservoir into the distribution system. Circular steel tanks and basins built of earth embankments, concrete, or rock masonry are used. Earth reservoirs are usually lined to prevent leakage and entrance of dirty water. The reservoirs should be covered to protect the water from dust, rubbish, bird droppings, and the growth of algae, but many older reservoirs without covers are in use.

Elevated storage reservoirs are tanks on towers, or high cylindrical standpipes resting on the ground. Storage reservoirs are built high enough so that the reservoir will maintain adequate pressure in the distribution system at all times.

Elevated tanks are usually of steel plate, mounted on steel towers. Wood is sometimes used for industrial and temporary installations. Standpipes are made of steel plate, strong enough to withstand the pressure of the column of water. The required capacity of a standpipe is greater than that of an elevated tank because only the upper portion of a standpipe is sufficiently elevated for normal use.

Distribution-system design. To assure the proper location and size of feeder mains and laterals to meet normal and peak water demands, a distribution system must be expertly designed. As the water flows from the source of supply or distribution reservoir across a city, the water pressure is lowered by the friction in the pipes. The pressures required for adequate service depend upon the height of buildings, need for fire protection, and other factors, but 40 psi (275 kPa) is the minimum for good service. Higher pressures for fire fighting are obtained by booster pumps on fire engines which take water from fire hydrants. In small towns adequate hydrant flows are the controlling factor in determining water-main size; in larger communities the peak demands for air conditioning and lawn sprinkling during the summer months control the size of main needed. The capacity of a distribution system is usually determined by opening fire hydrants and measuring simultaneously the discharge and the pressure drop in the system. The performance of the system when delivering more or less water than during the test can be computed from the pressure drops recorded.

An important factor in the economical operation of municipal water supplies is the quantity of water lost from distribution because of leaky joints, cracked water mains, and services abandoned but not properly shut off. Unaccounted-for water, including unavoidable slippage of customers' meters, may range from 10% in extremely well-managed systems to 30–40% in poor systems. The quantities flowing in feeder mains, the friction losses, and the amount of leakage are frequently measured by means of pitometer surveys. A pitometer is a portable meter that can be inserted in a water main under pressure to measure the velocity of flow, and thus the quantity of flow.

Pumping stations. Pumps are required wherever the source of supply is not high enough to provide gravity flow and adequate pressure in the distribution system. The pumps may be high or low head depending upon the topography and pressures required. Booster pumps are installed on pipelines to increase the pressure and discharge, and adjacent to ground storage tanks for pumping water into distribution systems. Pumping stations usually include two or more pumps, each of sufficient capacity to meet demands when one unit is down for repairs or maintenance. The station must also include piping and valves arranged so that a break can be isolated quickly without cutting the whole station out of service.

Centrifugal pumps have displaced steam-driven reciprocating pumps in modern practice, although many of the old units continue to give good service. The centrifugal pumps are driven by electric motors, steam turbines, or diesel engines, with gasoline engines frequently used for standby service. The centrifugal pumps that are used most commonly are designed so that the quantity of water delivered decreases as the pumping head or lift increases.

Both horizontal and vertical centrifugal pumps are available in a wide capacity range. In the horizontal type, the pump shaft is horizontal with the driving motor or engine at one end of the pump. Vertical pumps are driven by a vertical-shaft motor directly above the pump or are driven by a horizontal engine through a right-angle gear head. *See* PUMP.

Automatic control of pumping stations is provided to adjust pump operations to variations in water demand. The controls start and stop pumps of different capacity as required. In the event of mishap or failure of a unit, alarms are sounded. The controls are activated by the water level in a reservoir or tank, by the pressure in a water main, or by the rate of flow through a meter. Remote control of pumps is often used, with the signals transmitted over telephone wires. *See* WATER POLLUTION: WATER TREATMENT.

[RICHARD HAZEN]

Bibliography: M. A. Al-Layla et al., *Water Supply Engineering Design*, 1977; H. E. Babbitt, J. J. Doland, and J. H. Cleasby, *Water Supply Engineering*, 6th ed., 1962; C. V. Davis, *Handbook of Applied Hydraulics*, 2d ed., 1952; G. M. Fair et al., *Elements of Water Supply and Waste Water Disposal*, 2d ed., 1971; M. J. Hammer, *Water and Waste Water Technology*, 1975; E. W. Steel and T. McGhee, *Water Supply and Sewerage*, 5th ed., 1979.

Water treatment

Physical and chemical processes for making water suitable for human consumption and other purposes. Drinking water must be bacteriologically safe, free from toxic or harmful chemical or substances, and comparatively free of turbidity, color, and taste-producing substances. Excessive hardness and high concentration of dissolved solids are also undesirable, particularly for boiler feed and industrial purposes. The treatment processes of greatest importance are sedimentation, coagulation, filtration, disinfection, softening, and aeration.

Plain sedimentation. Silt, clay, and other fine material settle to the bottom if the water is allowed to stand or flow quietly at low velocities. Sedimentation occurs naturally in reservoirs and is accomplished in treatment plants by basins or settling tanks. The detention time in a settling basin may range from an hour to several days. The water may flow horizontally through the basin, with solids settling to the bottom, or may flow vertically upward at a low velocity so that the particles will settle through the rising water. Settling basins are most effective if shallow, and rarely exceed 10–20 ft (3–6 m) in depth. Plain sedimentation will not remove extremely fine or colloidal material within a reasonable time, and the process is used principally as a preliminary to other treatment methods.

Coagulation. Fine particles and colloidal material are combined into masses by coagulation. These masses, called floc, are large enough to settle in basins and to be caught on the surface of filters. Waters high in organic material and iron may coagulate naturally with gentle mixing. The term is usually applied to chemical coagulation, in which iron or aluminum salts are added to the water to form insoluble hydroxide floc. The floc is a feathery, highly absorbent substance to which color-producing colloids, bacteria, fine particles, and other substances become attached and are thus removed from the water.

The coagulant dose is a function of the physical and chemical character of the raw water, the adequacy of settling basins and filters, and the degree of purification required. Moderately turbid water coagulates more easily than perfectly clear water, but extremely turbid water requires more coagulant. Coagulation is more effective at higher temperatures. Lime, soda ash, or caustic soda may be required in addition to the coagulant to provide sufficient alkalinity for the formation of floc, and regulation of the pH (hydrogen-ion concentration) is usually desirable for best results. Powdered limestone, clay, bentonite, or silica are sometimes added as coagulant aids to strengthen and weight the floc, and a wide variety of polymers developed in recent years are used for the same purpose.

Filtration. Suspended solids, colloidal material, bacteria, and other organisms are filtered out by passing the water through a bed of sand or pulverized coal, or through a matrix of fibrous material supported on a perforated core. Filtration of turbid or highly colored water usually follows sedimentation or coagulation and sedimentation. Soluble materials such as salts and metals in ionic form are not removed by filtration.

Slow sand filters. Used first in England about 1850, slow sand filters consist of beds of sand 20–48 in. (51–122 cm) deep, through which the water is passed at fairly low rates—2,500,000 to 10,000,000 gal per acre (24,000–94,000 m^3/ha). The size of beds ranges from a fraction of an acre in small plants to several acres in large plants. An underdrain system of graded gravel and perforated pipes transmits the filtered water from the filters to the point of discharge. The sand is usually fine, ranging from 0.2 to 0.5 mm in diameter. The top of the filter clogs with use, and a thin layer of dirty sand is scraped from the filter periodically to maintain capacity.

Slow sand filters operate satisfactorily with reasonably clear waters but clog rapidly with turbid waters. The filters are covered in cold climates to prevent the formation of ice and to facilitate operation in the winter. In milder climates they are often open. Slow sand filters have a high bacteriological efficiency, but few have been built since the development of water disinfection, because of the large area required, the high construction cost, and the labor needed to clean the filters and to handle the filter sand. Slow sand filters are still used in many English and European cities, but have not been built in the United States since 1950, and few remain in operation.

Rapid sand filters. These operate at rates of 125,000,000 to 250,000,000 gal per acre (1,170,000–2,340,000 m^3/ha) per day; or 25 to 50 times the slow-sand-filter rates. The high rate of operation is made possible by the coagulation and sedimentation ahead of filtration to remove the heaviest part of the load, the use of fairly coarse sand, and facilities for backwashing the filter to keep the bed clean. The filter beds are small, generally ranging from 150 ft^2 (14 m^2) in small plants to 1500 ft^2

($140\ m^2$) in the largest filter plants. The filters consist of a layer of sand or, occasionally, crushed anthracite coal 18–24 in. (46–62 cm) deep, resting on graded layers of gravel above an underdrain system. The sand is coarse, 0.4–1.0 mm in diameter, depending upon the raw water quality and pretreatment, but the grain size must be fairly uniform to assure proper backwashing. The underdrain system serves both to collect the filtered water and to distribute the wash water under the filters when they are being washed. Several types of underdrains are used, including perforated pipes, perforated false bottoms of concrete, and tile and porous plates.

Filters are backwashed at rates 5–10 times the filtering rate. The wash water passes upward through the sand and out of the filters by way of wash-water gutters and drains. Washing agitates the sand bed and releases the dirt to flow out of the filter with the wash water. The quantity of water used for washing ranges from 1 to 10% of the total output, depending upon the turbidity of the water applied to the filters and the efficiency of the filter design. Combination air and water filter washes are popular in Europe, but are not often used in the United States.

Municipal and large-capacity filters for industry usually are built in concrete boxes or in open tanks of wood and steel. The flow through the sand may be caused by gravity, or the water may be forced through the sand under pressure by pumping. Pressure filters can be operated at higher rates than gravity filters, because of the greater head available to force the water through the sand. However, excessive pressure may increase the effluent turbidity, and bacteria may appear in the discharge water. For this reason, and because pressure filters are difficult to inspect and keep in good order, open gravity filters are favored for public water supplies.

Diatomaceous earth filters. Swimming-pool installations and small water supplies frequently use this type of filter. The filters consist of a medium or septum supporting a layer of diatomaceous earth through which the water is passed. A filter layer is built up by the addition of diatomaceous earth to the water. When the pressure loss becomes excessive, filters must be backwashed and a fresh layer of diatomaceous earth applied. Filter rates of $2\frac{1}{2}$–6 gal per minute per square foot ($1.7-4.1 \times 10^{-3}\ m^3/s\cdot m^2$) are attained.

Disinfection. There are several methods of treatment of water to kill living organisms, particularly pathogenic bacteria; the application of chlorine or chlorine compounds is the most common Less frequently used methods include the use of ultraviolet light, ozone, or silver ions. Boiling is the favorite household emergency measure.

Chlorination is simple and inexpensive and is practiced almost universally in public water supplies. It is often the sole treatment of clear, uncontaminated waters. In most plants it supplements coagulation and filtration. Chlorination is used also to protect against contamination of water in distribution mains and reservoirs after purification.

Chlorine gas is most economical and easiest to apply in large systems. For small works, calcium hypochlorite or sodium hypochlorite is frequently used. Regardless of which form is used, the dose varies with the water quality and degree of contamination. Clear, uncontaminated water can be disinfected with small doses, usually less than one part per million; contaminated water may require several times as much. The amount of chlorine taken up by organic matter and minerals in water is known as the chlorine demand. For proper disinfection the dose must exceed the demand so that free chlorine remains in the water.

Chlorination alone is not reliable for the treatment of contaminated or turbid water. A sudden increase in the chlorine demand may absorb the full dose and provide no residual chlorine for disinfection, and it cannot be assumed that the chlorine will penetrate particles of organic matter. Chlorine is applied before filtration, after filtration, and sometimes at both times.

Chlorine sometimes causes objectionable tastes or odors in water. This may be due to an excessive chlorine dose, but more frequently it is caused by a combination of chlorine and organic matter, such as algae, in the water. Some algae, relatively unobjectionable in the natural state, produce unbearable tastes after chlorination. In other cases, strong chlorine doses oxidize the organic matter completely and produce odor-free water. Excessive chlorine may be removed by dechlorination with sulfur dioxide. Also, ammonia is often added for taste control to reduce the concentration of free chlorine. Activated carbon is also effective in the reduction of natural and chlorine tastes, harmful organic compounds, heavy metals, and such.

Water softening. The "hardness" of water is due to the presence of calcium and magnesium salts. These salts make washing difficult, waste soap, and cause unpleasant scums and stains in households and laundries. They are especially harmful in boiler feedwater because of their tendency to form scales.

Municipal water softening is common where the natural water has a hardness in excess of 150 parts per million. Two methods are used: (1) The water is treated with lime and soda ash to precipitate the calcium and magnesium as carbonate and hydroxide, after which the water is filtered; (2) the water is passed through a porous cation exchanger which has the ability of substituting sodium ions in the exchange medium for calcium and magnesium in the water. The exchange medium may be a natural sand known as zeolite, or may be manufactured from organic resins. It must be recharged periodically by backwashing with brine.

For high-pressure steam boilers or some other industrial processes, almost complete deionization of water is needed, and treatment includes both cation and anion exchangers. Lime-soda plants are similar to water purification plants, with coagulation, settling, and filtration. Zeolite or cation-exchange plants are usually built of steel tanks with appurtenances for backwashing the media with salt brine. If the water is turbid, filtration ahead of zeolite softening may be required.

Aeration. Aeration is a process of exposing water to air by dividing the water into small drops, by forcing air through the water, or by a combination of both. The first method uses jets, fountains, waterfalls, and riffles; in the second, compressed air is admitted to the bottom of a tank through perforated pipes or porous plates; in the third, drops of water are met by a stream of air produced by a fan.

Aeration is used to add oxygen to water and to remove carbon dioxide, hydrogen sulfide, and taste-producing gases or vapors. Aeration is also used in iron-removal plants to oxidize the iron ahead of the sedimentation or filtration processes. *See* WATER POLLUTION; WATER SUPPLY ENGINEERING.

[RICHARD HAZEN]

Bibliography: R. W. Abbett, *American Civil Engineering Practice*, 3 vols., 1956; C. V. Davis (ed.), *Handbook of Applied Hydraulics*, 3d ed., 1969; G. M. Fair and J. C. Geyer, *Water Supply and Waste Water Disposal*, 1954; G. V. James, *Water Treatment*, 1965; L. G. Rich, *Unit Processes of Sanitary Engineering*, 1963.

Water-tube boiler

A steam boiler in which water circulates within tubes and heat is applied from outside the tubes. The outstanding feature of the water-tube boiler is the use of small tubes [usually 1–3-in. (2.5–7.5 cm) outside diameter] exposed to the products of combustion and connected to steam and water drums which are shielded from these high-temperature gases. Thus, possible failure of boiler parts exposed to direct heat transfer is restricted to the small-diameter tubes and, in the event of failure, the energy released is reduced and explosion hazards are minimized.

The water-tube construction facilitates greater boiler capacity by increasing the length and the number of tubes and by using higher pressure, since the relatively small-diameter tubes do not require an abnormal increase in thickness as the internal pressure is increased. In addition, water-tube boilers offer great versatility in arrangement, and this permits efficient use of the furnace, superheater, and other heat-recovery components.

There are many types of water-tube boilers but, in general, they can be grouped into two categories: the straight-tube and the bent-tube types. In essence, both types consist of banks of parallel tubes which are connected to, or by, headers or drums. The early water-tube boilers were fitted with refractory furnaces. However, most modern water-tube boilers utilize a water-cooled surface in the furnace, and this surface is an integral part of the boiler's circulatory system. Further, modern water-tube boilers generally incorporate the use of superheaters, economizers, or air heaters to utilize more efficiently the heat from the fuel and to provide steam at a high potential for useful work in an engine or turbine.

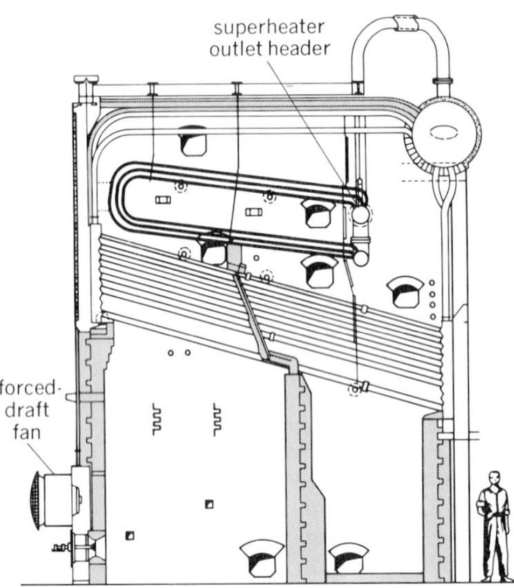

Fig. 1. Diagram of a straight-tube-type boiler.

Straight-tube boiler. The straight-tube boiler (Fig. 1), often called the header-type boiler, has the advantage of direct accessibility for internal inspection and cleaning through handholes, located opposite each tube end, in the headers. The steam-generating sections are joined to one or more steam-and-water drums located above and parallel or transverse to the boiler tube bank. The circulation of water in the downcomer headers and of the water-steam mixture in the boiler tubes and riser headers is the result of the differential density between the water in the downcomers and the steam and water mixture in the heated boiler tubes and riser portion of the circuit. In many designs the tube bank can be baffled to increase the rate of flow of the products of combustion and, thus, improve heat transfer and the resultant absorption efficiency.

The straight-tube boiler is not applicable to high-pressure designs because of header limitations, and capacity is restricted by space requirements. Thus, bent-tube instead of header-type boilers are used in most modern industrial installations.

Bent-tube boiler. In bent-tube boilers (Fig. 2), commonly referred to as drum-type boilers, the boiler tubes terminate in upper and lower steam and water drums which have few access openings. Although internal inspection is restricted, the tubes can be mechanically or chemically cleaned, and developments in water treatment and cleaning methods have overcome the early objections to the use of bent tubes. Circulation in drum-type boilers, as in header-type boilers, is due to the difference between the density of the water in the downcomer tubes and the water-steam mixtures in the riser tubes.

Because of the greater slope of tubes, or even the use of vertical tubes, in drum-type boilers, there is less possibility of steam pockets forming in the tubes and, thus, the rate of heat absorption can be increased appreciably. Consequently, drum-type boilers can be used for both low and high pressures and for capacities ranging from a few thousand to more than a million pounds of steam per hour. In addition, they can be designed to meet almost any space allotment and can readily accommodate heat-absorbing components such as superheaters, reheaters, economizers, and air heaters. An evolution of the bent-tube boiler is the so-called radiant-type boiler in which the boiler tubes (steam-generating surface) form the boundary of the furnace and the containment for the superheater, reheater, and economizer surface (Fig. 3). Such boilers can deliver several million pounds of steam per hour at high steam pressures and temperatures. Further, by eliminating the steam drum and using a once-through flow of water and steam,

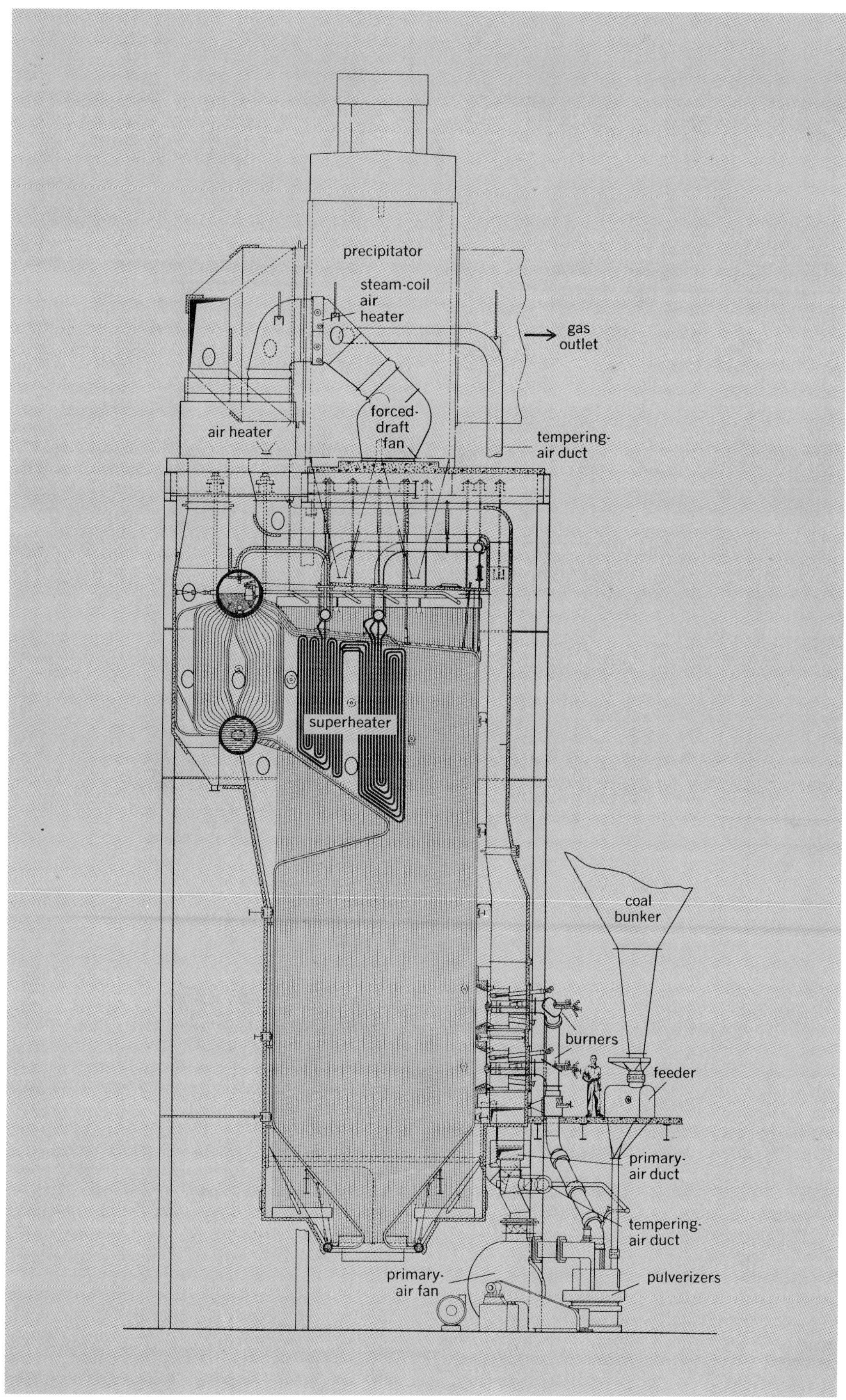

Fig. 2. Diagram of a bent-tube-type boiler.

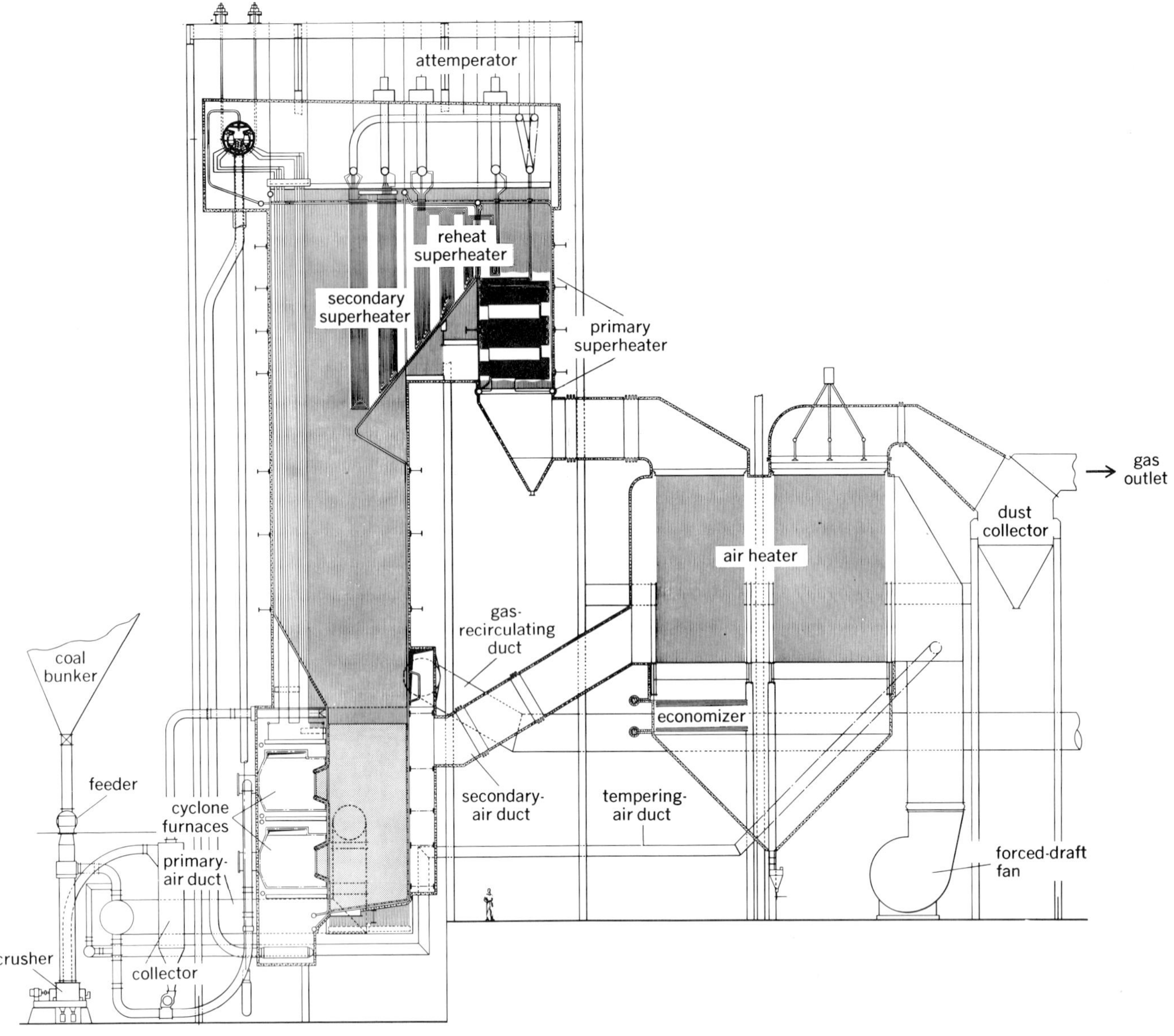

Fig. 3. Diagram of a radiant-type boiler.

they can operate at supercritical steam pressures. *See* BOILER; BOILER WATER; STEAM; STEAM-GENERATING FURNACE; STEAM-GENERATING UNIT.

[GEORGE W. KESSLER]

Bibliography: Babcock and Wilcox Co., *Steam: Its Generation and Use*, 39th ed., 1978; T. Baumeister (ed.), *Standard Handbook for Mechanical Engineers*, 8th ed., 1978; J. G. Singer (ed.), *Combustion/Fossil Power Systems*, 1981.

Waterpower

Power developed from movement of masses of water. Such movement is of two kinds: (1) the falling of streams through the force of gravity, and (2) the rising and falling of tides through lunar (and solar) gravitation.

While that part of solar energy expended to lift water vapor against Earth gravity is a minute fraction of the total, the absolute amount of energy that is theoretically recoverable from resulting streams is an enormous but unknown quantity. Of this, but a tiny portion is actually suitable for harnessing.

WATER RESOURCES

The capacity of world waterpower plants in use at the end of 1978 was about 408×10^6 kW, which produced in that year about 1558×10^9 kWh. This is about 21% of the total world electric power generated. Of this, the United States accounted for about one-fifth of the total. As of January 1980, the total conventional hydropower developed in the contiguous United States was 64,686 MW, provided by about 1400 installations. Almost one-half of this hydropower was installed in Washington, Oregon, and California. Under construction or planned were installations of some 24,000 MW, of which 90% was in the western United States.

In 1978 the waterpower plants of Europe totaled 138×10^6 kW; North America, 111×10^6; Asia, 62×10^6; Soviet Union, 41×10^6; leaving 56×10^6 for the remainder of the world. Thus the principal waterpower development of the future can be expected in Africa, Asia, South America, and island countries, whose potential has been little investigated.

The contribution of waterpower installations to the nation's electric power supply at the beginning of World War II was about 30%. While the output from hydroelectric plants has grown (to 274×10^9 kWh in 1978), their contribution to the total electric power has dropped to about 13%, because steam-electric plants have grown at a much more rapid rate.

As of March 1980, the Energy Information Agency of the Department of Energy estimated that by the end of 1993 the total developed capacity of waterpower installations could be 99,346 MW. However, that figure does not include a number of potential sites that could at some future time be considered. In theory, the Energy Information Agency estimates an eventual maximum potential of 187,000 MW, most of the undeveloped sites being in the Pacific Northwest and Alaska. Only a fraction of that will be developed for a variety of reasons. The most attractive sites have already been utilized. Hydro plants, with their initial high cost and generally long distances from major load centers, must compete with the large, efficient, fuel-fired stations, and the burgeoning, economical, large nuclear plants. Large dam sites usually must be justified not alone on the value of the power developed, but also on the benefits from flood control, irrigation, and recreation. Problems of migrating fish, conservation, and preservation of esthetic values are also factors. On the other hand, waterpower developments add greatly to power-system flexibility in meeting peak and emergency loads. Modern excavation and tunneling techniques are lowering construction costs. The economies of lowhead sites are improved by the new, efficient, axial-flow turbines of the tubular type. *See* ENERGY SOURCES.

Silting. The capacity of hydro plants cannot be counted on for perpetuity because of gradual filling of reservoirs with sediment. This effect is serious for irrigation, flood control, and navigation. Even when a lake behind a power dam becomes filled completely with silt, electric power can be generated on the run-of-the-river flow, although output would vary with stream flow.

The rate of silting varies widely with drainage basins. Because the Columbia River carries comparatively little silt, the reservoirs at Grand Coulee and Bonneville dams should have lives of many hundreds of years. The Colorado River, on the other hand, is muddy. In the first 13.7 years after Hoover Dam went into operation in 1935, 1,424,000 acre-feet (175,600 ha-m) of silt was dumped into Lake Mead. That is equivalent to a layer 1 ft deep over 2225 mi^2 (or 1 m deep over 1756 km^2). This inflow of silt has been diminished by about 22% by the construction of other dams upstream, for example, the Glen Canyon Dam. It is now expected that Lake Mead will have a useful life of more than 500 years.

Pumped storage. The most significant waterpower development of the 1960s was the rapid growth of pumped-storage hydroelectric systems. In these schemes, water is pumped from a stream or lake to a reservoir at a higher elevation. Pumping up to a storage reservoir is most commonly done by reversing the hydraulic turbine and generator. The generator becomes a motor driving the turbine as a pump. Power is drawn from the power system at night or on weekends when demand is low. It is not practical to shut down large, high-temperature steam stations or nuclear units for a few hours at night or even over a weekend. Because they must run anyway, the cost of pumping power is low, whereas the power generated from pumped storage at peak periods is valuable. Also, the pumped-storage system provides a means of supplying power quickly in an emergency situation, for example, during the failure of a large steam or nuclear unit. A pumped-storage system can be changed over from pumping to generation in 2 to 5 min.

Pumped-storage systems are not new. They date from 1928, but by 1972 only 13 pumped-storage plants of a capacity greater than 50 MW had been installed in the United States. These had a combined capacity of 3718.5 MW. By December 1979, pumped storage capacity had risen to 10.6 GW, with 10.6 GW additional planned or in construction.

Tidal power. A portion of the kinetic energy of the rotation of the Earth appears as ocean tides. The mean tide of all the oceans has been calculated as 2.1 ft (0.64 m), and the mean power as $54{,}000 \times 10^6$ hp (40 TW) or, on a yearly basis, the equivalent of 36×10^{12} kWh. Unfortunately, only a minute amount of this is likely to be harnessed for use. For tidal sites to be of sufficient engineering interest, the fall would have to be at least 15 ft (4.5 m). There are few such falls, and some of these are in remote areas. The only tide-power sites that have received serious attention are on the Severn River in England, the Rance River and Mont St. Michel in northern France, the San José and Deseado rivers of Argentina, the Petitcodiac and Memramcook estuaries in the Bay of Fundy, Canada, the Passamaquoddy River where Maine joins New Brunswick, Canada, and, lately, the Cambridge Gulf of Western Australia.

The Passamaquoddy site, with a potential of 1800 MW (peak), is the only important tidal-power prospect in the United States. However, as late as 1980 engineers did not consider its electrical output to be economically competitive with power produced by other means.

A second major handicap to tidal power is that, with a simple, single-basin installation, power is available only when there is a several-feet difference between levels in the sea and the basin. Thus, firm power is not available. Also, periods of generation occur in consonance with the tides—not necessarily when power is needed.

The only major tidal power plant in operation is the one near the mouth of the Rance River in Normandy, France. This plant operates on 40-ft (12-m) tides. It began operation in November, 1966. It consists of twenty-four 10-MW bulb-type turbine-generator units of novel design. The system embodies a reservoir into which sea water is pumped during off-peak hours. Turbines are then run as pumps, power being drawn from the French electrical grid. The plant produces 500×10^6 kWh annually, including a significant amount of firm power.

Tidal power is an appealing and dramatic technique, and some other large plants may be constructed. However, the total contribution of the tides to the world's energy supply will be miniscule.

[CHARLES A. SCARLOTT]

APPLICATION

The basic relation for power P in kilowatts from a hydrosite is $P=QH/11.1$, where Q is water flow in ft^3/sec, and H is head in feet. Actual power will be less as occasioned by inefficiencies such as (1) hydraulic losses in conduit and turbines; (2) mechanical losses in bearings; and (3) electrical losses in generators, station use, and transmission. Overall efficiency is always high, usually in excess of 80% to the station bus bars.

Choice of site. The competitive position of a hydro project must be judged by the cost and reliability of the output at the point of use or market. In most hydro developments, the bulk of the investment is in structures for the collection, control, regulation, and disposal of the water. Electrical transmission frequently adds a substantial financial burden because of remoteness of the hydrosite from the market. The incremental cost for waterwheels, generators, switches, yard, transformers, and water conduit is often a smaller fraction of the total investment than is the cost for the basic structures, real estate, and transmission facilities. Long life is characteristic of hydroelectric installations, and the annual carrying charges of 6–12% on the investment are a minimum for the power field. Operating and maintenance costs are lower than for other types of generating stations.

The fundamental elements of potential power, as given in the equation above, are runoff Q and head H. Despite the apparent basic simplicities of the relation, the technical and economic development of a hydrosite is a complex problem. No two sites are alike, so that the opportunity for standardization of structures and equipment is nearly nonexistent. The head would appear to be a simple surveying problem based largely on topography. However, geologic conditions, as revealed by core drillings, can eliminate an otherwise economically desirable site. Runoff is complicated, especially when records of flow are inadequate. Hydrology is basic to an understanding of water flow and its variations. Runoff must be related to precipitation and to the disposal of precipitation. It is vitally influenced by climatic conditions, seasonal changes, temperature and humidity of the atmosphere, meteorological phenomena, character of the watershed, infiltration, seepage, evaporation, percolation, and transpiration. Hydrographic data are essential to show the variations of runoff over a period of many years. Reservoirs, by providing storage, reduce the extremes of flow variation, which are often as high as 100 to 1 or occasionally 1000 to 1.

Economic factors. The economic factors affecting the capacity to be installed, which must be evaluated on any project, include load requirements, runoff, head, development cost, operating cost, value of output, alternative methods of generation, flood control, navigation, rights of other industries on the stream (such as fishing and lumbering), and national defense. Some of these factors are components of multipurpose developments with their attendant problems in the proper allocation of costs to the several purposes. The prevalence of government construction, ownership, and operation, with its subsidized financial formulas which are so different from those for investor-owned projects, further complicates economic evaluation. Many people and groups are parties of interest in the harnessing of hydrosites, and stringent government regulations prevail, including those of the U.S. Corps of Engineers, Federal Power Commission, Bureau of Reclamation, Geological Survey, and Securities and Exchange Commission.

Capacity. Prime capacity is that which is continuously available. Firm capacity is much larger and is dependent upon interconnection with other power plants and the extent to which load curves permit variable-capacity operation. The incremental cost for additional turbine-generator capacity is small, so that many alternatives for economic development of a site must be considered. The alternatives include a wide variety of base load, peak load, run-of-river, and pumped-storage plants. All are concerned with fitting installed capacity, runoff, and storage to the load curve of the power system and to give minimum cost over the life of the installation. In this evaluation it is essential clearly to distinguish capacity (kW) from energy (kWh) as they are not interchangeable. In any practical evaluation of water power in this electrical era it should be recognized that the most favorable economics will be found with an interconnected electric system where the different methods of generating power are complementary as well as competitive.

As noted above, there is an increasing tendency in many areas to allocate hydro capacity to peaking service and to foster pumped-water storage for the same objective. Pumped storage, to be practical, requires the use of two reservoirs for the storage of water—one at considerably higher elevation, say, 500 to 1000 ft (150 to 300 m). A reversible pump-turbine operates alternatively (1) to raise water from the lower to the upper reservoir during off-peak periods, and (2) to generate power during peak-load periods by letting the water flow in the opposite direction through the turbine. Proximity of favorable sites on an interconnected electrical transmission system reduces the investment burden. Under such circumstances the return of 2 kWh on-peak for 3 kWh pumping off-peak has been demonstrated to be an attractive method of economically utilizing interconnected fossil-fuel, nuclear-fuel, and hydro power plants. *See* ELECTRIC POWER GENERATION; HYDRAULIC TURBINE; POWER PLANT; NUCLEAR REACTOR.

[THEODORE BAUMEISTER]

Bibliography: Annual Statistical Report for 1980, *Elec. World*, 193(6):49–80, Mar. 15, 1980; T. Baumeister (ed.), *Standard Handbook for Mechanical Engineers*, 8th ed., 1978; Department of Agriculture, *Summary of Reservoir Sediment Deposition Surveys Made in the U.S. through 1960*, USDA Misc. Publ. no. 964, 1964; Department of the Interior, *Rate of Sediment Accumulation Drops at Lake Mead*, 1967; Edison Electric Institute, *Statistical Year Book of the Electric Utility Industry*, published annually; Federal Power Commission, *Development of Pumped Storage Facilities in the United States*, 1972, *Hydroelectric Power Resources of the United States*, Jan. 1, 1972, *The Role of Hydroelectric Developments in the Nation's Power Supply*, May 1974, *World Power Data*, 1966; D. G. Fink and H. W. Beaty, *Standard Handbook for Electrical Engineers*, 11th ed., 1978; French stem the tides, *Elec.*

World, 166:17, Nov. 7, 1966; Pumped storage power, at last, comes into its own, *Eng. News Rec.*, 182:22–25, Jan. 2, 1969; W. H. Hunt, Pumped storage: A major hydro power resource, *Civil Eng.*, 38:48–53, March, 1968; Rance tidal power station, *Smokeless Air*, no. 145, pp. 174–176, spring, 1968: W. A. Schoales, Prospects of tidal power in western Australia, *Elec. World*, 167:61–63, Feb. 20, 1967: Worldwide pumped storage projects, *Power Eng.*, p. 58, Oct., 1968.

Wear

The removal of material from a solid surface as a result of sliding action. It constitutes the main reason why the artifacts of society (automobiles, washing machines, tape recorders, cameras, clothing) become useless and have to be replaced. There are a few uses of the wear phenomenon, but in the great majority of cases wear is a nuisance, and a tremendous expenditure of human and material resources is required to overcome the effects.

For reasons which are hard to explain, wear did not emerge as a technical field until the 1950s. Before then, the feeling seems to have been that wear was an inevitable accompaniment of sliding, and that nothing could be done about it. Perhaps the key event which, rather indirectly, led to people's change in attitude was the building of nuclear reactors as part of the atomic bomb program in the 1940s. These reactors produced radioisotopes of the common engineering metals (iron, copper, chromium, and so on), and these allowed the use of extremely sensitive radiotracer techniques for measuring wear. For example, it became possible to measure the very small amount of wear which occurs when one metal is pressed with a force of a kilogram against another and then slides over it (see illustration).

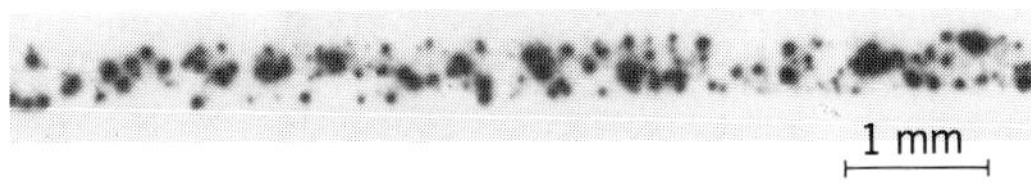

On sliding a radioactive copper rod once over an ordinary steel surface, tiny radioactive copper particles (black circles) are transferred to the steel. Total wear is 6×10^{-7} g.

Thus, for the first time it became possible to measure wear while it was occurring, rather than by before-and-after measurements. As result of this research, it was established by about 1950 that there were four principal forms of wear: adhesive, abrasive, corrosive, and surface fatigue; and it soon became possible to work out their mechanisms and to express the amount of wear in quantitative terms.

Adhesive wear. This is the only universal form of wear, and in many sliding systems it is also the most important. It arises from the fact that, during sliding, regions of adhesive bonding, called junctions, form between the sliding surfaces. If one of these junctions does not break along its original interface, then a chunk from one of the sliding surfaces will have been transferred to the other surface. In this way, an adhesive wear particle will have been formed. Initially adhering to the other surface, adhesive particles soon come off loose and can disappear from the sliding system. *See* FRICTION.

The volume of adhesive wear is governed by the equation $V = kLx/3p$, where V is the total volume of adhesive wear, k is a nondimensional constant called the wear coefficient, x is the total distance of sliding, and p is the indentation hardness of the surface expressed as a stress. Typical values of k are given in Table 1.

Table 1. Values of the wear coefficient for adhesive wear

Surface condition	Metal on metal	Nonmetal on metal, or nonmetal on nonmetal
Clean	10^{-2} to 10^{-4}	10^{-5}
Poor to fair lubricant	10^{-4} to 10^{-6}	3×10^{-6}
Good lubricant	10^{-6} to 10^{-8}	10^{-6}

A very great reduction in wear, by factors of up to a million, that can be produced in metallic sliding systems by using a good lubricant. Also, there is a great advantage in making unlubricated sliding systems nonmetallic, and well-lubricated systems metallic.

Adhesive wear causes two types of failure, one being a wear-out mode which occurs after long periods of sliding because too much material has been removed, and a seizure mode which occurs in systems which generate wear particles larger than the clearance, thus producing jamming. Since many material combinations give wear particles larger than 10 micrometers, it is dangerous to reduce the clearance of any sliding system below this value.

Abrasive wear. This is the wear produced by a hard, sharp surface sliding against a softer one and digging out a groove. The abrasive agent may be one of the surfaces (such as a file), or it may be a third component (such as sand particles in a bearing abrading material from each surface). Abrasive wear, like adhesive wear, obeys the equation given above; typical values of k are given in Table 2.

Table 2. Values of the wear coefficient for abrasive wear

Process	k value
Sharp file	2×10^{-1}
Sandpapering	5×10^{-2}
Loose abrasive grains	5×10^{-3}
Polishing	5×10^{-4}

It will be seen that abrasive wear coefficients are large compared to adhesive ones. Thus, the introduction of abrasive particles into a sliding system can greatly increase the wear rate; automobiles, for example, have air and oil filters to catch abrasive particles before they can produce damage.

Corrosive wear. This form of wear arises when a sliding surface is in a corrosive environment, and the sliding action continuously removes the protective corrosion product, thus exposing fresh surface to further corrosive attack. No satisfactory quantitative expression of corrosive wear yet exists, but when analyzing it in terms of the above equation, k values are obtained ranging all the way from less than 10^{-5} for surfaces in a gently corrosive environment, to above 10^{-2} for surfaces under severe corrosive attack.

Surface fatigue wear. This is the wear that occurs as result of the formation and growth of cracks. It is the main form of wear of rolling devices such as ball bearings, wheels on rails, and gears. During continued rolling, a crack forms at or just below the surface and gradually grows until a large particle is lifted right out of the surface.

Uses. Most manifestations of wear are highly objectionable, but the phenomenon does have a few uses. Thus, a number of systems for recording information (pencil and paper, chalk and blackboard) operate via a wear mechanism. A number of methods of preparing solid surfaces (filling, sandpapering, sandblasting) also make use of wear.

Wear is often used to study the operation of sliding systems. Thus, by looking at the wear of someone's shoes, it can be seen whether the wearer is walking properly. In analyzing accidents and failures, one of the main procedures is to look for signs of wear in the wrong place. Wear is specially useful in this case because it can be used in systems that are no longer operative.

[ERNEST RABINOWICZ]

Bibliography: S. Halling (ed.), *Principles of Tribology*, 1975; D. F. Moore, *Principles and Applications of Tribology*, 1975; J. J. O'Connor and J. Boyd (eds.), *Standard Handbook of Lubrication Engineering*, 1968; E. Rabinowicz, *Friction and Wear of Materials*, 1965; R. B. Waterhouse, *Fretting Corrosion*, 1972.

Wedge

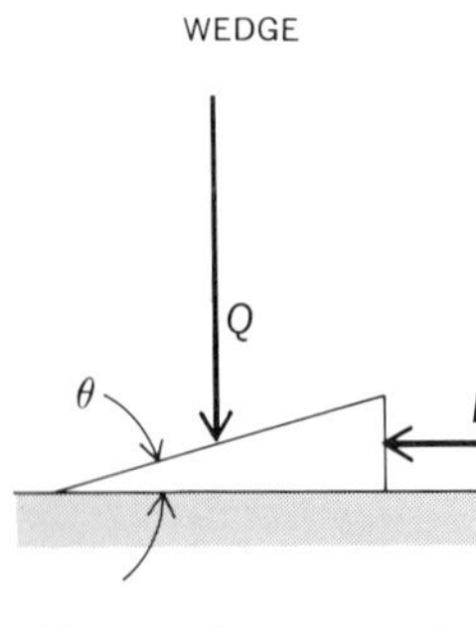

Forces acting on a wedge.

A piece of resistant material whose two major surface make an acute angle. It is closely related to the inclined plane and is used to multiply the applied force and to change the direction in which it acts (see illustration).

Force F is the smaller applied force and Q is the larger force to be exerted. In the absence of friction, forces must act normal to their surfaces; thus the actual force on the inclined surface is not Q but a larger force F_n. Summing up forces in the horizontal and vertical directions gives Eqs. (1).

$$F_n \sin\theta - F = 0 \qquad Q - F_n \cos\theta = 0 \tag{1}$$

Combining the expressions for F and Q and solving for F gives Eq. (2).

$$F = Q \tan\theta \tag{2}$$

If angle θ is small, the reaction of Q against F is exceeded by the friction between the face of the wedge and the adjacent body on which it rests. Thus the wedge tends to remain in position even when loaded by a large force Q.

Some applications of the wedge are in splitting wood, in raising the platform of low-lift platform trucks, in cone clutches, and in combination with friction to fasten parts together. *See* SIMPLE MACHINE.

[RICHARD M. PHELAN]

Welded joint

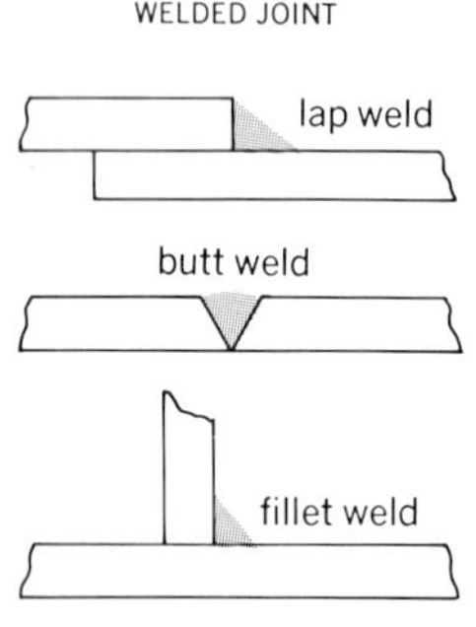

Fig. 1. Three types of welded joints.

The joining of two or more metallic components by introducing fused metal (welding rod) into a fillet between the components or by raising the temperature of their surfaces or edges to the fusion temperature and applying pressure (flash welding).

Types. In a lap weld, the edges of a plate are lapped one over the other and the edge of one is welded to the surface of the other (Fig. 1).

In a butt weld, the edge of one plate is brought in line with the edge of a second plate and the joint is filled with welding metal or the two edges are resistance-heated and pressed together to fuse.

For a fillet weld, the edge of one plate is brought against the surface of another not in the same plane and welding metal is fused in the corner between the two plates, thus forming a fillet. The joint can be welded on one or both sides.

A weldment is a cast steel, forged steel, or machined steel component that is assembled to plates or structural steel shapes or other weldments by welding to form a machine part.

A spot weld is an electrical-resistance lap weld wherein lapped surfaces are brought to high temperature by the resistance to a low-voltage, high-amperage current between a pair of water-cooled electrodes (Fig. 2). A spot weld is used primarily for thin sheet stock.

A seam weld is similar in production to a spot weld except that the electrodes are rollers or wheels rotating at constant speed; they produce a long narrow weld. The butted edges of pipe or tubing can be welded by a modified seam weld.

Strength. Because welded joints are usually exposed to a complex stress pattern as a result of the high temperature gradients present when the weld is made, it is customary to design joints by use of arbitrary and simplified equations and generous safety factors.

The force F of direct loading, and consequently the stress S, is applied directly along or across a weld. The stress-force equation is then simply $F = SA$, in which A is the area of the plane of failure (Fig. 3).

For eccentric loading, the force F causes longitudinal and transverse forces of varying magnitudes along the weld. The stress is found by assuming that rotation occurs around the centroid of the welded area and that the shear stress due to the torque about the centroid is vectorially added

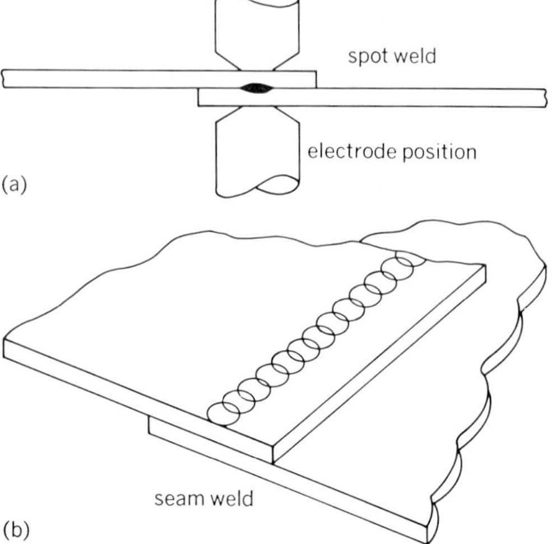

Fig. 2. Common electrical-resistance welds. (*a*) Spot weld, used mainly for thin sheet stock. (*b*) Seam weld, used to weld butted edges of pipe or tubing.

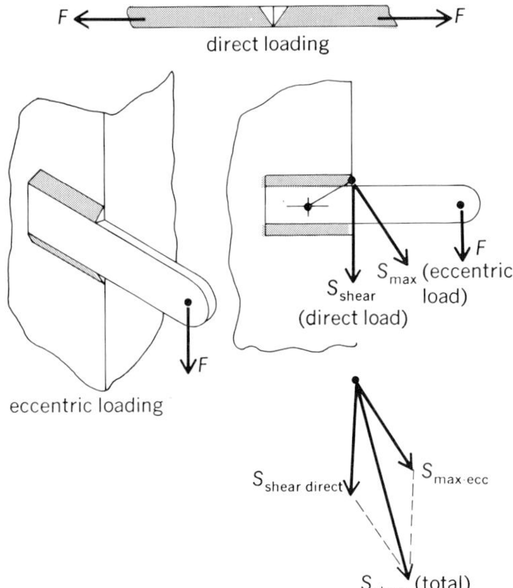

Fig. 3. Loading forces on a welded joint.

to the direct shear caused by the force. Several points in the welded area must be checked to establish the maximum vectorial summation of the direct and eccentric stress. Maximum shear stress S_{max} is $Fd\ r_{max}/J_{total}$, where J_{total} is total polar moment of inertia of weld failure sections about the centroid, d is the distance from force F to the centroid, and r_{max} is the radius from the centroid to the farthest part of the weld. *See* RIVETED JOINT; STRUCTURAL CONNECTIONS; WELDING AND CUTTING OF METALS. [L. SIGFRED LINDEROTH, JR.]

Bibliography: A. Althouse (ed.), *Modern Welding*, 1980; V. L. Doughtie and A. Vallance, *Design of Machine Members*, 4th ed., 1964; T. B. Jefferson (ed.), *The Welding Encyclopedia*, 1974.

Welding and cutting of metals

Processes based on heat to join and sever metals. Welding and cutting are grouped together because, in many manufacturing operations, severing precedes welding and involves the same production personnel. Welding is one of the joining processes, others being riveting, bolting, gluing, and adhesive bonding. *See* BOLTED JOINT; RIVETED JOINT; WELDED JOINT.

The American Welding Society's definition of welding is "a metal-joining process wherein coalescence is produced by heating to suitable temperatures with or without the application of pressure, and with or without the use of filler metal." Brazing is defined as "a group of welding processes wherein coalescence is produced by heating to suitable temperature and by using a filler metal, having a liquidus above 800°F (427°C) and below the solidus of the base metals. The filler metal is distributed between the closely fitted surfaces of the joint by capillary attraction." Soldering is similar in principle, except that the melting point of solder is below 800°F (427°C). The adhesion of solder depends not so much on alloying as on its keying into small irregularities in the surfaces to be joined. For comparison of metal joints *see* JOINT (STRUCTURES). *See also* BRAZING; SOLDERING.

Cutting is one of the severing and material-shaping processes, some others being sawing, drilling, and planning. Thermal cutting is defined as a group of cutting processes wherein the severing or removing of metals is effected by melting or by the chemical reaction of oxygen with the metal at elevated temperatures. Welding and cutting are widely used in building ships, machinery, boilers, structures, atomic reactors, aircraft, railroad cars, missiles, buses and trailers, and pressure vessels, as well as in constructing piping and storage tanks of steel, stainless steel, aluminum, nickel, copper, lead, titanium, tantalum, and their alloys. For many products, welding is the only joining process that achieves the desired economy and properties, particularly leak-tightness.

Nearly all industrial welding involves fusion. The edges or surfaces to be welded are brought to the molten state. The liquid metal bridges the gap between the parts. After the source of welding heat has been removed, the liquid solidifies, thus joining or welding the parts together. In the past there were three principal sources of heat for fusion welding: electric arc, electric resistance, and flame. However, new sources must be added to the above: light beams and electrons. Hence there are terms for five fusion welding processes: arc welding, resistance welding, gas welding, laser welding, and electron-beam welding. The flame in gas welding is provided by the combustion of a fuel gas.

Arc welding. The greatest volume of welding is done with arc welding processes.

Shielded metal-arc welding. This arc welding process is by far the most widely used of the various electric-arc welding processes. Like the other electric-arc welding processes, it employs the heat of the electric arc to bring the work to be welded and a consumable electrode to a molten state. The work is made part of an electric circuit known as the welding circuit. Arc welding with consumable electrodes is more widely practiced than welding with nonconsumable electrodes. A consumable electrode is melted continuously by the arc, one pole of which is the electrode, the other pole being the metal to be welded. Generally, the consumable electrode is a wire of the same chemical composition as the work. The arc melts the electrode and some of the base metal to form a pool of weld metal, which after freezing becomes the weld. A nonconsumable electrode is tungsten or carbon, both of which have high melting points and are not consumed, except slowly by vaporization (Fig. 1).

A covered electrode consists of a solid metal core; in some instances there are tubular-type

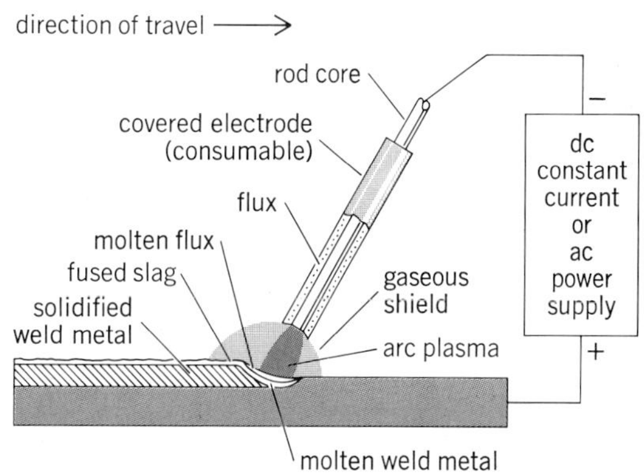

Fig. 1. Shielded metal-arc welding.

cores which are covered with material serving electrical and metallurgical purposes. Electrically, the covering insulates the metal core from accidental contact with adjacent material during welding and provides an arc free from interruptions. Metallurgically, the covering may provide gas- and slag-forming ingredients to protect the weld from the air, and it may supply deoxidizers or alloying elements to produce sound welds having specified chemical composition. In addition, the covered (or coated) electrode provides scavengers to refine the grain structure of the weld metal. As the arc consumes the electrode, the operator manually feeds it into the weld.

About 635,000,000 lb (288,000,000 kg) of covered electrodes are manufactured every year with diameters ranging from 1/16 to 1/4 in. (1.6 to 6.4 mm) in stainless steel, 5/64 to 5/16 in. (2.0 to 7.9 mm) in mild steel, 3/32 to 5/8 in. (2.4 to 15.9 mm) in tool steel, and comparable diameters for the nonferrous alloys. The weight of the covering may be 10–50% of the weight of the covered electrode. For welding mild steel, covered electrodes are made in a number of types, which have been classified by the American Welding Society. Class E-6010 has a covering consisting principally of cellulose. The electrode can be used readily to weld flat, vertical, and overhead surfaces. The prefix E designates an electrode. In the specification for mild-steel-covered arc welding electrodes, the first two digits stand for a minimum tensile strength of the deposited metal in the as-welded condition in 1000 psi (6.895 MPa). The third digit stands for the welding position or positions in which the electrode will make a satisfactory deposit, and the last digit classes the performance characteristics of the electrode and identifies the types of power sources on which it can be used. Class E-6012 electrodes have a thin covering high in rutile (titanium dioxide) and are adapted to welding across gaps up to 1/4 in. (6.4 mm). Class E-6016 electrodes are the low-hydrogen class. The calcium carbonate in the covering evolves carbon dioxide during welding. The covering is nearly free from hydrogen, and the weld, therefore, is free from defects due to hydrogen. Class E-6024 electrodes have about 40% iron powder in the covering, which makes possible high rates of depositing weld metal. The rate of depositing steel weld metal from steel electrodes in any arc welding process ranges from 0.03 to 0.06 lb/min/100 A [14 to 27 g/(min) (100 A)] of welding current.

Cored electrodes consist of a tube formed from strip and filled with slag-forming, arc-stabilizing, and alloying materials. For welding mild steel, the cored electrode is continuous. In automatic welding the arc is guided mechanically along the joint, whereas in semiautomatic welding the operator guides the arc manually by means of a torch. In some processes the arc is in air. In others a gas such as carbon dioxide protects the arc from the air.

Submerged-arc welding. This process is performed with a continuous electrode and a granular flux composed of silicates with or without deoxidizers and alloying elements. The flux is piled to a depth of 1/2 to 2 in. (12.7 to 50.8 mm) along the joint. The arc melts some of the flux and is submerged in the liquid slag so produced. Although submerged-arc welding often uses currents of about 40 nm, it can use as much as 400 nm, far above that usable in any other arc welding process. High currents enable the weld to penetrate more deeply below the surface of the work than in other arc welding processes. Welds of exceptionally good quality are produced in a wide range of metals greater than 1/16 in (1.6 mm) thick. Carbon, alloy, or stainless steels up to 1/2 in. (12.7 mm) thick are safely welded in one pass, while thicker materials are multipass-welded. Weld-metal deposition rates (pounds of metal deposited in 1 hr of operation), arc travel speeds, and weld completion rates are superior to any other process available today (Fig. 2).

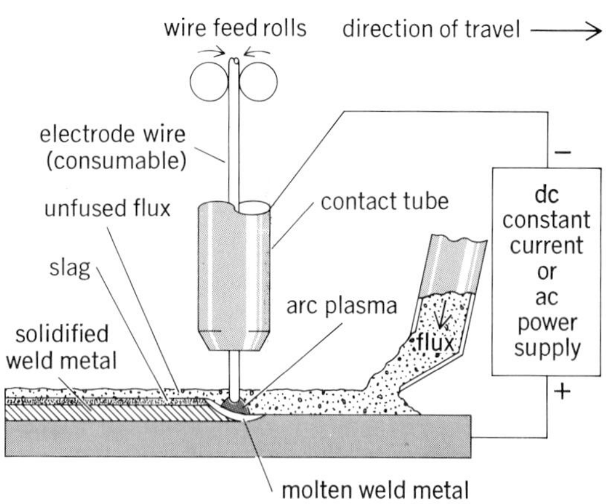

Fig. 2. Submerged-arc welding.

Gas metal-arc welding. Inert-gas-shielded metal-arc welding with consumable, continuous electrodes (called GMAW for gas metal-arc welding) requires no flux and produces welds without a slag cover. The arc is in an atmosphere of argon or helium supplied at 5–100 ft³/hr (0.14–2.8 m³/h) from the gas cup of the torch. Inert-gas shielding is particularly advantageous in welding reactive metals, such as titanium, which are susceptible to atmospheric contamination, and in welding metals, such as aluminum and stainless steel, which are susceptible to porosity (Fig. 3).

Carbon dioxide welding. Carbon dioxide welding with continuous electrodes is similar to GMAW, except that CO_2 is used as a low-cost shielding gas. The process is restricted to carbon and some low-alloy steels. To prevent porosity resulting from the formation of carbon monoxide by the reaction of the CO_2 with the carbon in the steel, electrodes for CO_2 welding contain deoxidizers such as manga-

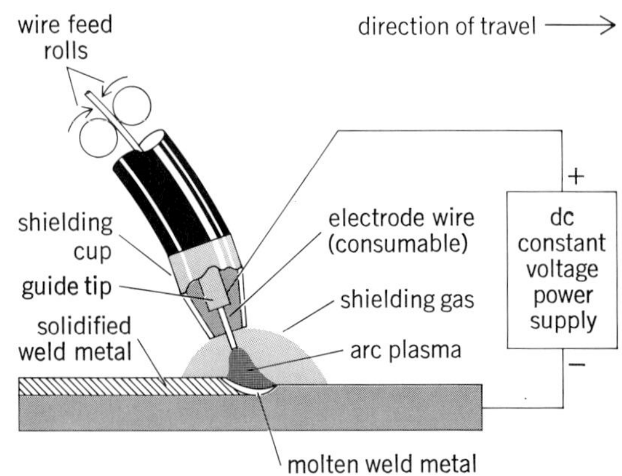

Fig. 3. Gas metal-arc welding.

nese, silicon, aluminum, titanium, and chromium.

Nonconsumable arc welding electrodes are not deposited as part of the weld metal. The electrode is one pole of the arc, usually the electron-emitting (and therefore cooler) negative pole. The other pole is the work, which is melted. If additional metal is required to fill the joint, a filler wire is fed into the arc. Only tungsten and carbon have sufficiently high melting points to provide requisite electron emission at the high currents used in arc welding. Carbon electrodes seldom are used for welding because they vaporize more rapidly than tungsten.

Gas tungsten-arc welding. Inert-gas shielding is essential with tungsten electrodes; consequently the term "gas tungsten-arc welding" (GTAW) is used. A common electrode diameter is 1/8 in. (3.2 mm) for 150 A. The process is adapted to welding thin material, 1/8 in. (3.2 mm) thick and less, and to the root, or first pass, because the operator can control penetration more readily than with most other arc welding processes. The process is especially adapted for welding light-gage work requiring the utmost in quality or finish, because of the precise heat control possible and the ability to weld with or without filler metal (Fig. 4). Direct current (electrode negative) is used in welding ferrous materials, copper, and nickel alloys. For aluminum and magnesium, alternating current is required because oxide film on the work is removed by reverse polarity (electrode positive). To prevent extinction of the arc as the electric potential passes through zero, a small high-frequency current (100,000 Hz) must be superimposed on the 60-Hz welding current.

Plasma-arc welding. This is a process which utilizes a plasma produced by the heat of a constricted electric-arc–gas mixture. Shielding gas may be an inert gas or a mixture of gases. Two arc modes are used to generate a plasma: transferred arc and nontransferred arc.

Plasma-arc welding resembles gas tungsten-arc welding in its use of an inert gas, but differs from it in the use of a constricting orifice. The transferred-arc mode is usually preferred. The advantages of plasma-arc welding have been observed primarily in material thicknesses greater than 3/32 in. (2.4 mm). In such thicknesses, a significant difference is observed in the weld puddle; it is known as the keyhole effect. Surface tension causes the molten metal to flow around the keyhole to form the weld. This keyhole can be observed during the welding operation and is an indication of complete and uniform weld penetration. A newer process, plasma-MIG, combines the features of plasma-arc and inert-gas metal-arc processes. The narrow arc is used for deep penetration welding of thick material or high-speed welding of thin plate.

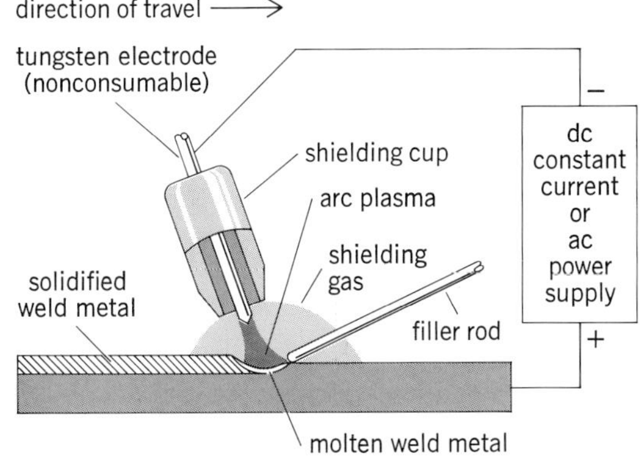

Fig. 4. Gas tungsten-arc welding.

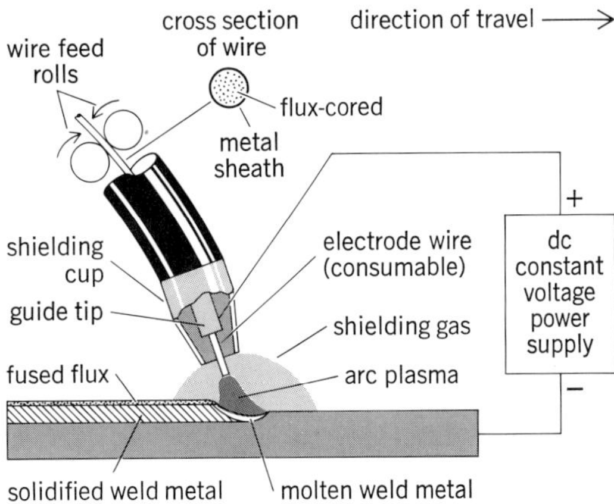

Fig. 5. Flux-cored wire welding.

Flux-cored arc welding. Another process, flux-cored arc welding, is somewhat like submerged-arc and shielded metal-arc welding except that the flux is encased in a metal sheath instead of being laid over the wire. Although wire is automatically fed from a coil, the equipment is portable and more versatile than submerged-arc. The weld metal is shielded by the melted flux and by a gaseous medium, either externally supplied or evolved from the flux. The two principal variations in the process employ a gas-shielded electrode that requires additional shielding in the form of CO_2 around the arc and weld puddle, or a self-shielded electrode that generates its own shielding (no external shielding gas is supplied; Fig. 5).

The classification system for flux-cored electrodes follows the general pattern used in other filler metal classifications. In a typical designation, E70T-1, the prefix E indicates an electrode as in other electrode classification systems; the number 70 refers to the minimum as-welded tensile strength in 1000-psi (6.895-MPa) units, the letter T indicates that the electrode is of tubular construction (is a flux-cored electrode), the suffix 1 places the electrode in a particular grouping based on the chemical composition of deposited weld metal, method of shielding, and adaptability of the electrode for single- or multiple-pass use.

Electroslag welding. This welding process is initiated much like the conventional submerged-arc process by starting an electric arc beneath a layer of granular welding flux. As soon as a sufficiently thick layer of hot molten slag is formed, all arc action stops and current passes from the electrode to the workpiece through the conductive slag. At this point, the process is truly electroslag welding.

Heat generated by the resistance to the current through the molten slag is sufficient to fuse the edges of the workpiece and the welding electrode. Since no arc exists, the welding action is quiet and spatter-free. The interior temperature of the bath is in the vicinity of 3500°F (1925°C). The liquid metal coming from the filler metal and the fused base metal collects in a pool beneath the slag bath and slowly solidifies to form the weld.

The wire electroslag process has been used for welding plates ranging in thickness from approximately 1/2 to 20 in. (1.25 to 50 cm). The welding of greater thicknesses is technically feasible, but requires custom installations. With each electrode, the process will deposit from 25 to 45 lb (11 to 20 kg) of filler metal per hour. The diameter of the electrode wire generally used is 1/8 in. (3.2 mm). Materials that have been joined by the electroslag welding process include mild and low-alloy carbon steels, stainless and high-alloy steels, and titanium.

Gas tungsten-arc spot welding. A combination of gas tungsten-arc and resistance spot welding has also been developed. Gas tungsten-arc spot welding is quite versatile and can be used to produce excellent-quality spot welds. Argon is used to shield the tungsten electrode and the face of the spot weld. On materials that require protection from air, the opposite side of the spot weld is shielded with either argon or helium. *See* ARC WELDING.

Resistance welding. Resistance welding processes are widely used in the manufacture of sheet metal assemblies, such as automobile bodies, aircraft, missiles, railroad cars, buses, and trailers. By definition, the required heat at the joints to be welded is generated by the resistance offered through the work parts to the relatively short-time flow of low-voltage, high-density electric current. Force is always applied before, during, and after applying current to assure a continuous electrical circuit and to forge the heated parts together.

In the most widely used processes—spot, seam, flash, and projection welding—the heat melts the surfaces to be welded. Every resistance weld involves a sequence of electrical energy and mechanical pressure. The sequence is provided by a control, which governs timing of both. For example, a control meters the 10 cycles of 60-Hz power at 10,000 A and also the electrode force at 500 lb (227 kg) required for spot welding two sheets of 0.04-in. (1-mm) steel. The welding operator merely presses the button that sets the control in operation (Fig. 6).

Spot welding. In spot welding, the sheets to be welded are held between electrodes of hard, high-conductivity copper alloy. They conduct heat away from the electrode-to-sheet contacts. The resistance heat at the sheet-to-sheet contact fuses the sheets together. Electrode pressure provides uniform sheet-to-sheet resistance and prevents expulsion of metal between the sheets. *See* SPOT WELDING.

Seam welding. Seam welding is a process for making overlapping spot welds by means of rotating electrode wheels. The joints can be made gas-tight and liquid-tight, as in tubing and stainless steel electric refrigerators.

Flash welding. In flash welding, the surfaces to be welded are held lightly in contact while a high current flows through the few contact points. Melting occurs instantaneously at these "bridges." Violent vaporization ensues, with expulsion of small particles of hot metal as visible flash. Flashing is continued until the surfaces are coated with a layer of molten metal. They are then squeezed together to form the weld. Flash welding is used to a considerable extent in the manufacture of automotive and aircraft products. The process is also used widely in the manufacture of household appli-

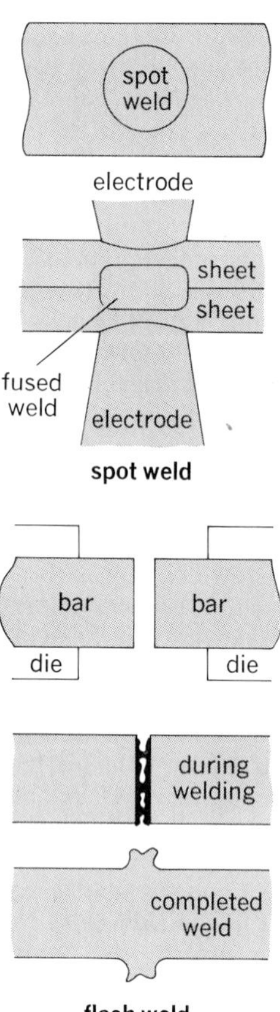

Fig. 6. Spot welding and flash welding.

ances, refrigerators, and farm implements. *See* FLASH WELDING.

Three other welding processes similar to flash welding should be mentioned: upset, upset butt and tube, and percussion welding. The greatest use of upset butt welding is in wire mills and in the welding of products made from wire. Equipment using the percussion welding process is found in several industries, mainly those in the electrical contact or component field.

Projection welding. In projection welding, a projection may be embossed on a sheet of metal. It may be produced on a solid piece of metal by machining, or it may be produced on an edge in a punch press in several ways. The height may be anywhere from a few thousandths of an inch to 1/8 in. or more, depending upon the job. The purpose of projections is to localize the current and pressure at predetermined points. In this modification of the spot welding process, the concentration of the welding current is determined by the preparation of the workpieces rather than by the size and shape of the electrodes. The major portion of the heat tends to develop in the part bearing the projections during the welding operation. Cross-wire welding, which is used extensively in making wire mesh and other forms of wire products, is a form of projection welding. *See* RESISTANCE WELDING.

Gas welding. A gas flame is a less concentrated and lower-temperature source of heat than an arc. For this reason, a welding torch is often used for brazing, for welding thin material, and for applications requiring low gradients of temperature to avoid cracking, as in welding cast iron. The oxyacetylene flame, using a mixture of equal volumes of oxygen and acetylene, C_2H_2, has a higher temperature than any other commercially available fuel gas combination. High flame temperature is essential for the rapid localized heating required in welding. Several sizes and types of torch tips are available to secure flames of different intensities and dispersions. The ratio of oxygen to acetylene governs the chemical action of the flame. If the ratio is higher than unity, for example 1.1:1, the flame is oxidizing and is useful for preventing hydrogen porosity in brass. With a ratio below unity, 0.95:1, an acetylene feather appears in the flame, carbon is picked up by molten steel, and mill scale on the surface of steel is reduced. A neutral flame with a ratio of 1:1 is used for steel.

Flux is not used in welding steel but is necessary for gas-welding nonferrous metals. The use of other fuel gases, such as hydrogen, is used in the welding of metals of low melting point, such as lead, aluminum, and magnesium.

Electron beam welding. Electron beam welding is a process wherein coalescence is produced by heat obtained from a concentrated beam composed primarily of high-velocity electrons impinging upon the surfaces to be joined. Electron beam welding equipment, which utilizes a heat source of electrons accelerated by an electric field to extremely high speeds and focused to a sharp beam by electrostatic or electromagnetic fields, is used for welding a wide range of metals in thicknesses ranging from foils to extremely thick sections. The welding equipment is manufactured for welding in high vacuum of thicknesses of 0.1 to 0.01 micrometer, in medium vacuum of 50 to 300 micrometers, and out of vacuum in an air or inert-gas atmosphere. The basic welding equipment is designed to operate in one of two voltage ranges. So-called low-voltage welding units operate at 15 to 60 kV, while high-voltage welders operate at 100 to 200 kV. A weld of very high purity may be obtained by this technique. Of all welding techniques, electron beam welding seems most suited for use in outer space, where it has been successfully demonstrated in United States and Soviet tests. *See* VACUUM METALLURGY.

[MEL M. SCHWARTZ]

Laser welding. In the continuous laser beam welding process, the laser beam is focused on the workpiece surface, and at the point of focus the metal vaporizes under the beam and a vapor column extending deep into the base metal is created. The vapor column is surrounded by a liquid pool in equilibrium with the vapor (Fig. 7). By traversing the path to be joined with the focused beam, autogenous deep-penetration fusion welds are created (Fig. 7). Depth-to-width ratios for these welds generally exceed four to one, and are often as great as ten to one.

Low-power lasers such as pulsed ruby or pulsed yttrium-aluminum-garnet (YAG) are suitable for welding thin-gage materials such as electronic components. These lasers can produce millisecond pulses of up to 100 kW for spot welding.

Deep-penetration autogenous welds are formed primarily with continuous multikilowatt gas laser systems, the most highly developed and widely used of which is the carbon dioxide laser, operating in the far-infrared region at a wavelength of 10,600 nm. These lasers, when focused, are capable of producing power densities adequate to initiate deep penetration, a phenomenon which was previously possible only with electron beams.

The laser welding process is ideally suited to automation, since the beam constitutes a clean, remote heat source which can be readily shaped and directed by using reflective optics. A vacuum environment is not required, and a choice of shielding gasses and gas shield configurations may be adapted to specific welding situations. Unlike the somewhat similar electron beam welding process, laser welding does not generate x-rays.

The laser welding process has been shown to be suitable for a variety of metals and alloys, including ferrous materials, aluminum alloys, titanium alloys, lead, copper and copper-nickel, and most superalloys and refractory metals. Laser welds have been fabricated in these materials, which are generally ductile and porosity-free, and have aver-

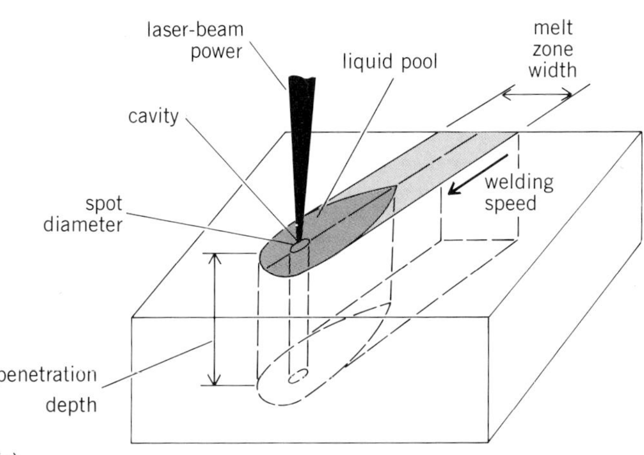

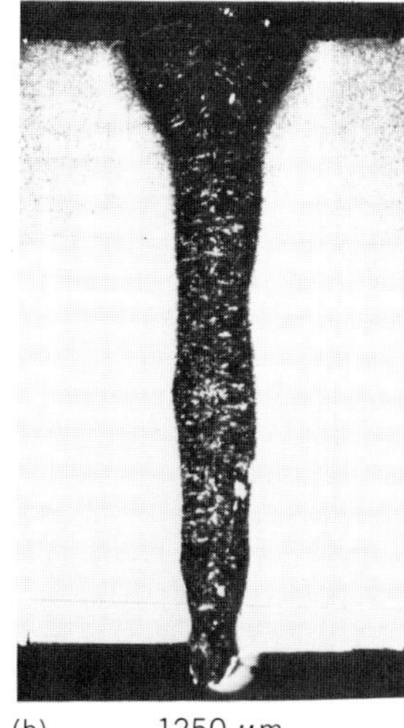

Fig. 7. Deep-penetration welding characteristics. (*a*) Schematic diagram of the deep-penetration laser welding process. The laser beam creates a vapor column (deep-penetration cavity) and deposits energy through the thickness of the workpiece. (*b*) Typical weld.

age tensile strengths equivalent to or exceeding those of the base metal. High weld strengths are the result of high cooling rates, which result from the low energy input per unit length of weld which is characteristic of the laser. These high cooling rates often are advantageous in that they produce desirable structures in the welds.

Laser beam welding typically requires good joint fit-up, and is a high-speed process ideally suited to automation. As a result of the generally high cost of laser welding equipment in excess of 1 kW power level, the process is most successfully applied to high-volume applications or to critical applications requiring unique weld characteristics. *See* LASER WELDING.

[EDWARD M. BREINAN]

Inertia welding. Inertia welding is one type of friction welding process which utilizes the frictional heat generated at a pair of rubbing surfaces to raise the temperature to such a degree that the two parts can be forged together and form a solid weld. The process is called inertia welding because the energy required for making the weld comes from a rotating flywheel system built into the head stock of the machine. Somewhat similar to an engine lathe, the inertia welding machine also has a tail stock where one side of the workpiece is held and can be pushed foward and retracted by a hydraulic ram.

During inertia welding, one part of the workpiece held in the spindle chuck of the head stock is accelerated rapidly to a preset spindle speed. The drive power is then cut off while the nonrotating workpiece on the tail stock is pushed against the rotating part under a large thrust force. The friction at the interface brings the spindle to full stop in a matter of seconds. In the meantime, the kinetic energy stored in the spindle-flywheel system is converted into frictional heat which causes the temperature at the weld to rise quickly so that a forge weld can be formed. Because of the high pressure, the original materials at the rubbing surfaces are first softened and then squeezed out of the interface to form a flash; thus, the process brings the nascent sublayers of the material close together and results in a metallurgical bond. The weld is formed at solid state without melting; it has a finer grain structure and narrower heat-affected zone as compared with any fusion welding process, for example, electric arc welding. This often results in higher strength at the weld than at the base material.

Inertia welding is commonly used to join dissimilar materials which cannot be achieved easily or economically by any other means. Material pairs of large varieties can be welded by this process, for example, steels to aluminum alloys or copper, copper to titanium or zirconium alloys, and many others. In addition to these advantages, inertia welding machines can be readily automated for mass production. The automotive and oil drill industries are the major users of the process. Typical applications include engine exhaust valves, shock absorber parts, universal joints, axle housings, hydraulic cylinder components, and oil drill pipes. The process is also applied to weld stainless steel to plain carbon steel for outboard motor shafts, and gas turbine parts in the aircraft industry. *See* INERTIA WELDING.

[K. K. WANG]

Other fusion welding processes. Among other fusion welding processes are high-frequency, thermit, stud, and plastic welding.

High-frequency welding may be divided into two categories: (1) high-frequency resistance welding, in which the high-frequency current is introduced into the work by direct electrical contact; and (2) high-frequency induction welding, in which the high-frequency current is induced in the work by an inductor coil, without electrical contact. The proximity effect concentrates the current in the edges to be joined, as in the manufacture of metal tubes.

In thermit welding, the joint is produced by heating with superheated liquid metal and slag resulting from a chemical reaction between a metal oxide and aluminum, with or without the application of pressure. Filler metal, when used, is obtained from the liquid metal.

There are two basic types of stud welding; arc stud welding and capacitor-discharge stud welding. Arc stud welding, still the more widely used of the two basic stud welding processes, is similar in many respects to manual shielded metal-arc welding. The heat necessary for end welding of studs is developed by passage of current through an arc from the stud (electrode) to the plate (work) to which the stud is to be welded. Capacitor-discharge stud welding derives its heat from an arc produced by a rapid discharge of stored electrical energy with pressure applied during or immediately following the electrical discharge.

There are several different ways of joining plastics to each other: hot-gas welding, friction welding, heated tool welding, and ultrasonic welding. Welding is a convenient method of joining most thermoplastic materials. The procedure is similar to that followed in metal welding when the oxyacetylene process is used. The strength of the weld, however, may vary according to the type of plastic material. The strongest joints are usually obtained from plastics with the highest degree of polymerization.

Nonfusion processes. In these processes, the surfaces to be welded coalesce in the solid state under the influence of pressure with or without heat. It is essential that the surfaces be free from films, such as oxides.

Electromagnetic joining is accomplished by induction-heating parts to about 200°F (93°C) below their melting point and then discharging a capacitor bank through the same induction coil. This induces up to 100,000 A in the parts, and the induced currents set up a pressure pulse of up to 50,000 psi (345 MPa) of attracting force for a few milliseconds. The process is reportedly capable of diffusion-welding up to 1100 parts per hour and is aimed at the automotive, appliance, ordnance, and nuclear fields.

Friction welding is a process in which all heat required for welding is the result of frictional heating caused by rotating the joint components against each other. As a mass-production tool, friction welding is practical because it requires low power and is readily automated.

In diffusion welding, intimately fitting faying surfaces of the parts to be joined are held in contact until bond occurs. Pressure and heat are usually used to speed bonding, and very little, if any, deformation occurs. Major applications have been concentrated in fabrication of aircraft and gas turbine parts.

The explosive welding process uses a detonating explosive to accelerate metal joint components to high-speed oblique collision. When collision conditions are within certain limits, jetting cleans the colliding surfaces of the metal components and results in welding. No external heating is needed, and it is essentially a cold process. Joints between dissimilar metals are possible with the process, and joint surfaces do not require extensive cleaning other than removal of heavy oxides and scale, and degreasing.

Cold welding is performed by pressure alone without heat, and soft-temper metals, such as aluminum and copper, are most easily cold-welded.

Ultrasonic welding also requires no heat and only light pressure. The workpieces are clamped together under moderately low static force, and ultrasonic energy is transmitted into the intended weld area. A sound metallurgical bond is produced without an arc or melting of the weld metal, and without the cast structure associated with melting. There is a minor thickness deformation. This joining process has demonstrated its versatility in applications involving bimetallic junctions and in producing a variety of joint configurations. It is in production use in the semiconductor microcircuit and electrical contact industries, in the manufacture of aluminum foil, and in the fabrication of aluminum products. It is uniquely useful in the packaging field for the encapsulation of materials such as explosives, pyrotechnics, and reactive chemicals, which require hermetic sealing for environmental protection and which cannot be processed by heat or electrical techniques. Many other ultrasonic welding applications are in the pilot-production stage as equipment is being engineered to manufacturing demands. The ultrasonic welding process is utilized in spot-type welding, ring welding, line welding, and continuous-seam welding.

Another process is the marriage of spot welding and adhesive bonding. In the process, two metals are adhesively bonded and then spot-welded. Joints are remarkably strong with outstanding fatigue characteristics. Procedures followed in weld bonding are very similar to those required in conventional resistance welding. Parts are cleaned, adhesive is applied to the joint, and the spot weld is made through the adhesive. Weld bonding has been applied to aluminum alloys, titanium alloys, and stainless steels. Most applications in aircraft use, as well as several commercial uses (buses and trailers) are in the prototype stages.

Welding metallurgy. The effects of welding on a product often govern its suitability for service. The weld metal, the zone close to the weld (called the heat-affected zone), and the thermal stresses and strains caused by welding may combine to yield a product different in strength, ductility, leak-tightness, and dimensional stability from that of the unwelded base metal. Weld metal frequently is compared with cast metal, with which it shares the characteristic of freezing in a columnar dendritic pattern in the direction opposite to the direction of withdrawal of heat. Unlike cast metal, weld metal usually has a free surface exposed to the air, and it is free from shrinkage cavities. Weld metal rapidly freezes compared with most castings and is comparatively fine-grained and free from chemical segregation. Porosity must be guarded against. It is a result of evolution of gas, such as nitrogen from steel, during freezing of the gas-saturated weld pool. Porosity also may result from a chemical reaction in the weld metal, such as the reaction between carbon and oxygen in liquid steel. Finally, porosity may occur as a result of inadequate inert-gas coverage during welding because of poor fixturing or poor welding conditions and procedures. Hot cracking of the weld metal during the last stages of solidification may be prevented either by fixturing to avoid strain during freezing or by adding suitable alloying elements to the weld, as in the addition of manganese to steel weld metal to combine with sulfur.

The heat-affected zone may be softer than the base metal, especially in alloys that have been cold-rolled or heat-treated for high strength. In some instances, for example the welding of annealed air-hardening steels, the heat-affected zone may be hard and brittle. Preheating and postwelding heat treatment may be required to avoid brittleness or restore strength. To achieve higher notch-impact value in steel weld metal than can be attained in a single pass, two or more passes must be deposited to take advantage of the grain-refining (hence toughening) effect of each pass on the heat-affected zone of the preceding pass. Welding heat causes distortion, which can be minimized by suitable choice of fixturing, joint design, and sequence of welding. The welded part also contains internal stresses, often up to yield strength locally in the vicinity of the weld. These may be reduced after welding by stress-relief heat treatment, which entails, in the case of steel, heating the welded structure to 1200°F (650°C) and cooling slowly. *See* HEAT TREATMENT (METALLURGY); METALLURGY.

Thermal cutting. Thermal cutting is used for a wide variety of work of the heavier type, such as structural work and shipbuilding, where the highest degree of dimensional accuracy is unimportant. Thermal cutting can be done manually or automatically, and in either case the torch can be brought to the work. The cut can be straight-line,

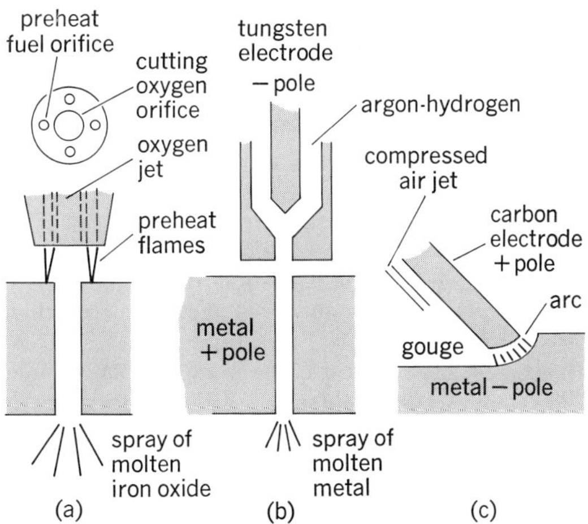

Fig. 8. Three thermal cutting processes. (*a*) Oxygen cutting. (*b*) Gas tungsten-arc cutting. (c) Air carbon-arc cutting.

curved, or beveled. The principal cutting process is oxygen cutting, which can be used for most mild steels. The cutting torch tip has a central orifice for oxygen and peripheral orifices for the preheat flame (Fig. 8). The flame preheats the steel to the ignition temperature, whereupon the oxygen converts the steel to iron oxide, which issues from the bottom of the cut as a molten spray. An oxygen cutting machine can cut a 1-in. (25.4-mm) steel plate at a speed of 18 in./min (46 cm/min), using an oxygen orifice of 0.0595-in. (1.51 mm) diameter and an oxygen pressure of 35 psi (240 kPa) and consuming oxygen at the rate of 150 ft^3/hr (4.25 m^3/hr). Oxygen lancing is similar to cutting except that an oxygen pipe replaces the torch for cutting steel up to 6 ft (1.8 m) thick. For highly alloyed and stainless steel cutting, it may be necessary to use a special process such as flux injection or powder cutting or some of the newer arc cutting processes described below. The selection of cutting process and type of operation (manual or mechanized) depends on the material that is being cut and the ultimate use of the product. In powder cutting, iron powder is injected into the oxygen in one process to increase the iron oxide content of the slag and thus lower its melting point.

There are seven cutting processes which utilize an arc. Some have been replaced by better and more efficient methods. These include oxygen arc cutting, air carbon-arc cutting, gas metal- and gas tungsten-arc cutting, and plasma arc cutting.

In oxygen arc cutting, metals are severed by a combination of melting, oxidizing, diluting, fluxing, and mechanical ejection. For oxidation-resistant metals, the cutting mechanism is more of a melting action. Oxygen arc cutting electrodes were developed primarily for use in underwater cutting and were later applied for cutting in air. In both applications, they can cut ferrous and nonferrous metals in practically any thickness and position.

Air carbon-arc cutting is essentially a progressive melting of material which is blown away by high-velocity jets of compressed air. Base-metal melting is much less than in oxygen cutting. The process is primarily physical rather than chemical; hence it can be used on most metals.

Gas metal-arc cutting was developed soon after the commercial introduction of the gas metal-arc welding process. Conventional gas metal-arc welding equipment can be used for cutting, although some modifications have been made in the equipment designed especially for cutting operations. Gas tungsten-arc cutting with a nonconstricted arc can be used to sever nonferrous metals and stainless steel in thicknesses up to 1/2 in. (12.7 mm), using standard gas tungsten-arc welding equipment. Metals cut include aluminum, magnesium, copper, silicon bronze, nickel, copper nickel, and various types of stainless steels. This method can be used for either manual or mechanized cutting. Good-quality cuts can be made with an argon-hydrogen gas mixture containing 65–80% argon. Nitrogen can also be used, but the quality of cut is not as good as that produced with an argon-hydrogen mixture.

Plasma arc cutting employs an extremely high-temperature, high-velocity constricted arc between the electrode in the plasma torch and the piece to be cut. Where inert gases are used, the cutting process depends upon thermal action alone. When cutting such materials as mild steel and cast iron, increased cutting speeds can be achieved by using oxygen-bearing cutting gases. The process can be used to cut any metal (see table). The plasma cutting process can be used in the general fabrication of all metals in many industries, such as shipyard, aircraft, chemical, nuclear, and pressure-vessel. It also finds application in the forging and casting industry. The process can be used in manual or mechanized operation and is especially adaptable to automation.

Typical conditions for plasma arc cutting of aluminum

Thickness, in.*	Speed ipm†	Orifice diameter, in.*	Current, A	Gas flow, cfh‡
1/4	300	1/8	300	80 A–40 H_2 or 130 N_2–30 H_2
1/2	200	1/8	300	65 A–35 H_2 or 140 N_2–60 H_2
1	90	5/32	400	65 A–35 H_2 or 140 N_2–60 H_2
2	20	5/32	400	65 A–35 H_2 or 140 N_2–60 H_2
3	15	3/16	450	130 A–70 H_2 or 140 N_2–60 H_2
4	12	3/16	450	130 A–70 H_2 or 140 N_2–60 H_2

*1 in. = 25.4 mm. †ipm = inches per minute. ‡cfh = cubic feet per hour; 1 ft^3 = 2.83 × 10^{-2} m^3.

The laser promises to be a valuable tool for drilling, cutting, and eventually milling of virtually any metal or ceramic or other inorganic solid. The mechanism by which a laser beam removes material from the surface being worked usually involves a combination of melting and evaporation, although with some materials, such as carbon and certain ceramics, the mechanism is purely one of evaporation. Any solid material can be cut with the laser beam. In addition to the pulsed ruby laser, continuous CO_2 and pulsed YAG lasers are of importance in drilling, cutting, and trimming. The potential for laser cutting of reactive metals has been greatly enhanced by a technique where oxygen is added to the continuous-laser cutting process. It was reported that on especially reactive material the cutting rate and penetration were greatly improved because of the exothermic reaction and creation of additional heat due to the oxidation of the reactive material.

[MEL M. SCHWARTZ]

Bibliography: American Society for Metals, *Metals Handbook*, 8th ed., vol. 6, 1971; American Welding Society, *Welding Handbook*, 7th ed., vol. 1, 1976, vol. 2, 1978; E. M. Breinan, C. M. Banas, and M. A. Greenfield, Laser welding: The present state of the art, *Proceeding of the Annual Assembly of the International Institute of Welding, Tel Aviv, Israel, 1975*, IIW Doc. no IV-181-75, 1975; M. M. Schwartz, *Metal Joining Manual*, 1979; M. M. Schwartz, *Modern Metal Joining Techniques*, 1969.

Wellpoint systems

A method of keeping an excavated area dry by intercepting the flow of groundwater with pipe wells located around the excavation area. Intercepting the flow before it reaches the excavated area also improves the stability of the edge of the excavation, permitting steeper bank slopes and often eliminating the need for supporting or shoring the banks. *See* CONSTRUCTION METHODS.

Wellpoint systems are most effective in coarse-grained soils, such as gravel or sand. They are not effective in fine soils, such as silts and clays, where the small size of the pores between grains restricts the flow of water.

The basic components of a wellpoint system are the wellpoint, the riser pipe, the header pipe or manifold, and the pump (Fig. 1).

The wellpoint consists of a perforated pipe about 4 ft (1.2 m) long and about 2 in. (5 cm) in diameter. It is equipped with a ball valve to regulate the flow of water, a screen to prevent the entry of sand during pumping, and a jetting tip. The steel riser pipe to which the wellpoint is attached is slightly smaller in diameter than the wellpoint. Its purpose is to bring the groundwater to the surface, where it is collected by the horizontal manifold pipe or header pipe, which is 6–10 in. (15–25 cm) in diameter. Wellpoints and riser pipes are spaced 3–6 ft (0.9–1.8 m) apart depending upon the characteristics of the ground. The wellpoint and riser pipe are usually driven into the ground to the required depth, generally 2–5 ft (0.6–1.5 m) below the excavation, by jetting, that is, by displacing the ground below it with water under pressure.

Where the groundwater stratum is overlaid with hard clay or heavy gravel, jetting may not be able to pierce the tough material and a special boring attachment is used to carry the wellpoint down.

The pumps are located above the water table and collect the water from the header pipes for discharge away from the excavation area. The pumps are centrifugal pumps, with vacuum pumps to increase the suction lift. The suction lift in each wellpoint rarely exceeds 18–22 ft (5.4–6.7 m). Consequently, where the depth of the wellpoint below ground level exceeds this depth, it is necessary to install two or more tiers of wellpoint systems, each with its own assembly of wellpoints, header pipes, and pumps. A three-tier wellpoint system is shown in Fig. 2. *See* PUMP; PUMPING MACHINERY. [WILLIAM HERSHLEDER]

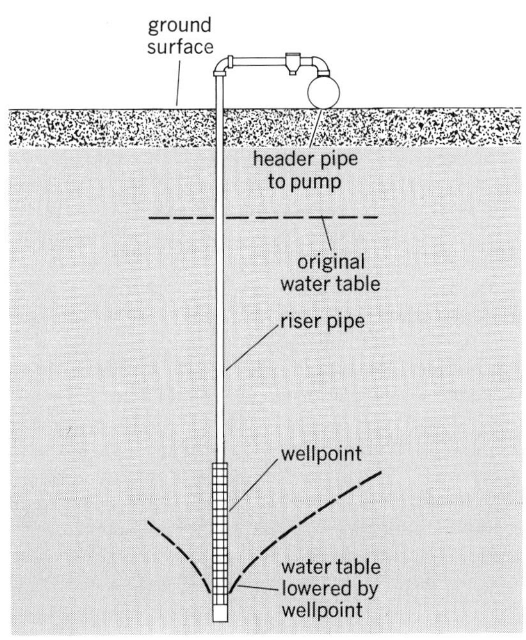

Fig. 1. Components of a wellpoint system.

Fig. 2. The layout of a three-tier wellpoint system. (*Griffin Wellpoint Co.*)

Wharf

A structure along a waterfront providing a berth for ships to load and discharge passengers and cargo. A marginal wharf, one parallel to shore, is generally known as a quay or bulkhead. When perpendicular or oblique to shore, a wharf is known as a pier. Construction is essentially that required for retaining earth fill or for embankment protection while at the same time providing a landing platform. Reinforced-concrete or masonry retaining walls are used as well as the familiar pile-supported platforms with sheet piling bulkheads of steel, timber, and concrete. Movement and settlement of marginal wharves and eventual failure are not uncommon. *See* COASTAL ENGINEERING; PIER. [EDWARD J. QUIRIN]

Wheel and axle

A wheel and its axle or, more generally, two wheels with different diameters or a wheel and drum, as in a windlass, rigidly connected together so that they rotate as a unit on a common axis. The principle of operation is the same as that of the lever in that, for static equilibrium, the summation of torques about the axis of rotation equals zero. Where flexible members, such as ropes, have been firmly attached to a wheel and drum (see illustration) and the machine is mounted on frictionless bearings, Eq. (1) applies.

$$F_1R_1 - F_2R_2 = 0 \qquad (1)$$

If F_1 is the input and F_2 the output, the mechanical advantage is given by Eq. (2).

$$\mathrm{MA} = \frac{F_2}{F_1} = \frac{R_1}{R_2} \qquad (2)$$

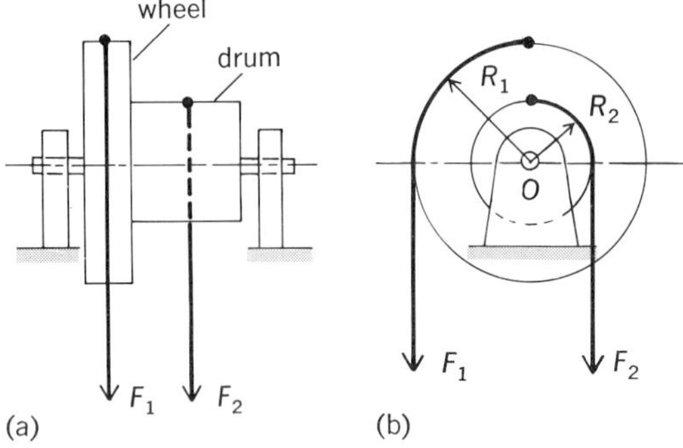

Wheel and axle. (*a*) Side view. (*b*) Front view.

If the bearings are not frictionless, a friction torque T_f will oppose the rotation. For the case with F_1 the input, the friction torque will act in the same direction as the load or output torque F_2R_2, and the equilibrium equation takes the form of Eq. (3), which may be solved for individual components as in Eq. (4).

$$F_1R_1 - F_2R_2 - T_f = 0 \qquad (3)$$

$$F_2 = \frac{F_1R_1 - T_f}{R_2} \qquad (4)$$

Because all members are rigidly connected, all torques act through the same angular displacement; the efficiency η is then given by Eq. (5).

$$\eta = \frac{F_2R_2}{F_1R_1} \times 100 = \frac{F_1R_1 - T_f}{F_1R_1} \times 100 = \left(1 - \frac{T_f}{F_1R_1}\right) 100 \qquad (5)$$

The main difference between the lever and the wheel and axle is that the wheel and axle permits the forces to operate through a much greater distance. In the illustration the wheel and drum could be allowed to rotate any number of revolutions if the ropes were wrapped the required number of times around each before they were attached.

The wheel and axle is seldom encountered in the exact form shown in the illustration, but the principle of the summation of torques about the axis of rotation equaling zero finds widespread applications in machines such as the belt drive, chain drive, friction drive, and gear drive. *See* FORCE; SIMPLE MACHINE. [RICHARD M. PHELAN]

Wheel base

The distance in the direction of travel from front to rear wheels of a vehicle, measured between centers of ground contact under each wheel. For a vehicle with two rear axles, the rear measuring point is on the ground midway between rear axles. Tread of a vehicle is the distance perpendicular to the direction of travel between front wheels, or between rear wheels, measured from centers of ground contact.

[FRANK H. ROCKETT]

Wind power

Kinetic energy in the Earth's atmosphere used to perform useful work. Total atmospheric wind power is of the order of 10^{14} kW. Annual kinetic energy is of the order of 10^{17} kWhr (1 kWhr = 3.60×10^6 J). Practical land-based wind generators could extract as much as 10^{14} kWhr of energy per year worldwide. Energy, and thus productivity of winds, varies markedly with geographic location. Annual energy available to a conversion machine at a site is very reproducible (±15% variability). Annual average power per square meter of vertical area at a 10-m height across and over water near the United States is estimated in Fig. 1. Southern Wyoming shows the greatest chance for productive wind power systems on land (>400 W/m^2), and the edge of the shelf off New England shows >800 W/m^2.

Wind power has lifted water for centuries, but there is new interest in wind-powered irrigation in the United States. Electricity was generated by wind power in 1880. Thousands of small wind generators and water pumps worked in the United States as late as 1940. Subsidized rural electrification, low fossil-fuel prices, and desire for more powerful farm machines caused near extinction of wind power systems in this country. In other nations, simple wind power systems are still the key to material sufficiency. The exponential energy appetite of the industrialized world consuming finite resources and the exponential growth in pollution associated with energy production has renewed interest in wind power.

Water lifting. Water spilled on agricultural soil by a random wind is still the best example of a storage subsystem buffering between the wind and the energy consumer. Most large-scale irrigation pumping in the United States today consumes natural gas, but wind power pumping could come back. Wind-powered pump-back could expand the capacity of hundreds of smaller hydroelectric installations.

Heating. At least 20% of United States energy is consumed in heating buildings. In colder climates, heating demand frequently matches high winds. Wind-generated electricity feeding thermal storage units, in some instances combined with solar thermal collectors feeding adjacent storage units, offers excellent potential for reducing fuel oil consumption. Large-scale wind-generator systems located off the Atlantic coast could take over a large fraction of building heating load in the largest urban areas.

Wind-generated electricity. Complete sets of hardware combining wind generators and electric storage batteries large enough to supply 500–1000 kWhr/month electricity on demand can be purchased. Where the generator can be placed at moderate height into a strong and persistent wind regime, delivered electricity is economically competitive. In milder wind regimes, electricity is still expensive compared with most utility prices. Larger centralized wind electricity systems sharing larger storage subsystems have a greater chance of being economic. When winds produce more electricity than the market demands, the energy must be stored, then recalled and used when demand exceeds windpower. Wind generators in large numbers located in productive winds and equipped with hydrogen generation, storage, and reconversion devices have been proposed. It has been estimated that winds available to the United States could generate as much as 2×10^{12} kWhr of firm power on demand, equivalent to the total 1975

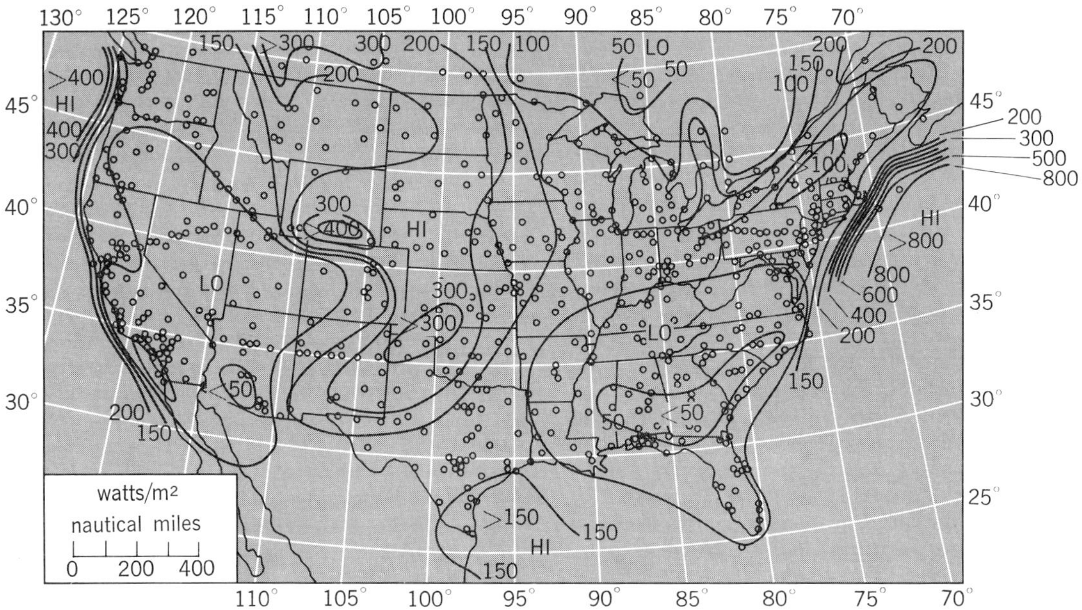

Fig. 1. Available wind power—annual average, in watts/m². 1 nautical mile = 1.852 km.

United States electricity consumption. *See* ELECTRIC POWER GENERATION.

Mechanics. A windmill is a rotating machine capable of interchanging (extracting) momentum with particles of air mass that flow through its swept area. Power available in the wind in a swept area varies with the size of that area, the density of the air, and the square of the velocity. Energy extractable from an oncoming wind stream over a period of time varies as the size of swept area, density, and cube of the velocity, as described in Eq. (1), where K.E. is the kinetic energy available,

$$\text{K.E.} = kDV^3 \tag{1}$$

k is the density of air, D is the sweeping blades' diameter, and V is the average wind speed. At sea level, K.E. = 0.000935 D^2V^3 lb-ft/s, corresponding to $1.7 \times 10^{-6}\ D^2V^3$ horsepower, for D in feet, and V in feet per second. Adolph Betz showed that no more than 59% of the energy in an oncoming stream tube of wind could be extracted without bypassing the momentum exchanger. The ability of any windmill to approach that 59% maximum extraction capability is thus an excellent indicator of its performance, and has been named the coefficient of performance, C_p. Thus, the realizable power P_R from a windmill is as shown in Eq. (2).

$$P_R = C_p \times 1.7 \times 10^{-6} D^2V^3 \text{ horsepower} \tag{2}$$

A C_p as high as 0.48 has been observed for a modern high-speed propeller-type wind machine, whereas the sheet-metal-bladed American Fan Mill for water pumping seldom achieves a C_p in excess of 0.30. Coefficient of performance is related to the aerodynamic features of a machine, particularly to the tip speed ratio, defined by Eq. (3).

$$\text{Tip speed ratio} = \frac{\text{tangential speed of blade tip}}{\text{speed of oncoming wind}} \tag{3}$$

Different types of wind machines work best at their own optimum tip speed ratio. The American Fan Mill wants a tip speed ratio near 1; the very-high-speed twisted and tapered two- or three-bladed propeller type wants a ratio between 7 and 12. Vertical-axis machines of the S-rotor (Savonius) type, cross-flow vertical-bladed machines, and the Darrieus twirling rope (troposkien) rotor operate best at a ratio between 1 and 2. It has been suggested that any configuration of horizontal-axis machine can be given characteristics that will permit it to have a high C_p, but economics and

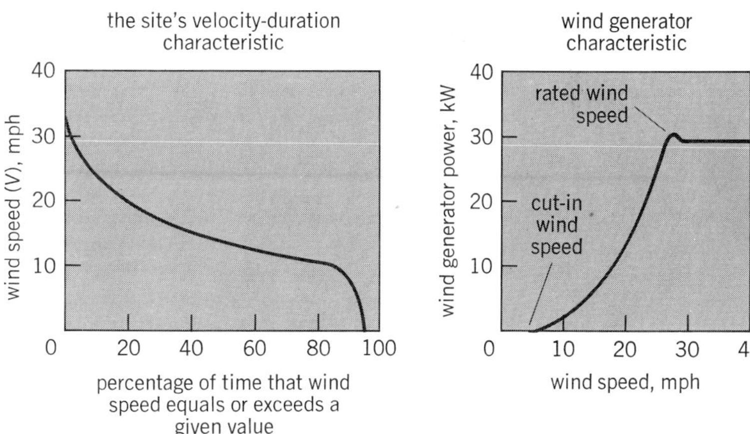

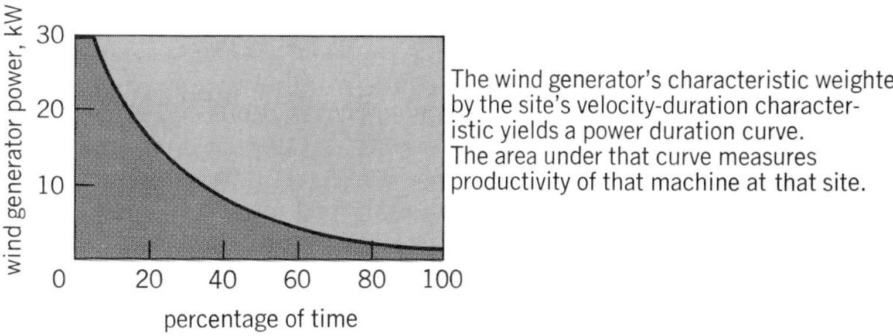

Fig. 2. Productivity of a wind generator is a function of wind regime, height, the machine's aero-mechanical-electrical characteristics, and the delivered product.

other practical considerations seem to favor the three-bladed twisted and tapered propeller type operating at a tip speed ratio of about 8. Advocates of the vertical-axis Darrieus type hope to prove that type most economic, however.

System and economic considerations. The utility and competitiveness of a wind power system depend upon the wind regime and height at which windmills are placed, size and characteristics of the machine, nature of delivered product, and productivity of that product. Figure 2 shows how a wind regime can be characterized by a velocity duration curve (for any specific time period) and the overall power out versus wind speed of the wind machine (in this case a generator of electricity). If the desired product is simply raw electrical energy, the product of delivered voltage times delivered amperage suffices. Figure 2 shows how a power duration curve measures that productivity. Wind systems delivering other products would be assessed differently, but with the same philosophy. Wind systems can thus be compared with one another and with other systems capable of delivering the same product. *See* ENERGY SOURCES.

[WILLIAM HERONEMUS]

Bibliography: A. Betz, *Windmills in the Light of Modern Research*, Nat. Advis. Comm. Aeronaut. Nat. Mem. no. 474 (available from NTIS), 1928; Department of Energy, *Federal Wind Energy Program: Program Summary*, 1979; F. R. Eldridge (ed.), *Proceedings of the 2d Workshop on Wind Energy Conversion Systems*, Washington, DC, 1975; H. Glauert, Windmills and fans, in W. F. Durand (ed.), *Aerodynamic Theory*, vol. 4, 1935; E. W. Golding, *Generation of Electricity by Windpower*, 2d ed., 1976; W. E. Heronemus, Pollution free energy from offshore winds, in *Preprints of the 8th Annual Conference and Exposition, Marine Technology Society*. Washington, DC, September, 1972; D. R. Inglis, *Wind Power and Other Energy Options*, 1978; P. C. Putnam, *Power from the Wind*, 1948.

Windings in electric machinery

Windings can be classified in two groups: armature windings and field windings. The armature winding is the main current-carrying winding in which the electromotive force (emf) or counter-emf of rotation is induced. The current in the armature winding is known as the armature current: The field winding produces the magnetic field in the machine. The current in the field winding is known as the field or exciting current. *See* ELECTRIC ROTATING MACHINERY; GENERATOR; MOTOR.

The location of the winding depends upon the type of machine. The armature windings of dc motors and generators are located on the rotor, since they must operate in conjunction with the commutator, and the field windings are mounted on stator field poles. *See* DIRECT-CURRENT GENERATOR; DIRECT-CURRENT MOTOR.

Alternating-current synchronous motors and generators are normally constructed with the armature winding on the stator and the field winding on the rotor. There is no clear distinction between the armature and field windings of ac induction motors or generators. One winding may carry the main current of the machine and also establish the magnetic field. It is customary to use the terms stator winding and rotor winding to identify induction motor windings. The word armature, when used with induction motors, applies to the winding connected to the power source (usually the stator). *See* ALTERNATING-CURRENT GENERATOR; ALTERNATING-CURRENT MOTOR; SYNCHRONOUS MOTOR.

Field windings. Field windings produce a magnetic pole fixed in space with respect to the magnetic structure on which they are mounted. If salient-pole construction is employed, the winding turns are concentrated around the pole core (Fig. 1). If cylindrical construction is employed, the field winding is constructed of elongated concentric loops embedded in slots cut in the surface of the field structure. In the case of dc motors and generators with compensating windings, both forms of construction may be found in one machine.

Concentrated field windings for salient-pole ac synchronous machines and for shunt fields of dc and series ac machines are form-wound of many turns of insulated copper wire. The completed coil is taped and impregnated to hold its shape and to fit around the pole core. Series field coils are normally concentrated. They are wound of larger wire, or of insulated copper strip, in order to carry the armature current without overheating.

Field coils for cylindrical construction are made of insulated rectangular copper or aluminum bars (Fig. 2). Individual coils are form-wound with several turns per coil. The dimensions of the coils are selected to provide a suitable space distribution of field flux. The coil is held in the slots by nonmagnetic wedges. The coil ends are held in place against mechanical forces by nonmagnetic bands.

Rotor field coils are connected to the external circuit through slip rings, on which carbon brushes rest to conduct the exciting current. *See* SLIP RINGS.

Armature windings. These windings carry ac current. They take many forms, depending upon the type and capacity of the machine and specific

Fig. 1. Salient-pole field winding on rotor of synchronous motor. (*Allis-Chalmers*)

design requirements. Armature windings have an active portion, which lies in slots in the magnetic circuit, and end turns and end connections, which are external to the airgap. The treatment of the end connection determines the appearance and operating characteristics of the winding. Usually the end turns are formed to make diamond-shaped coils for ease in inserting and bracing the windings. A coil may have a single turn, or it may have many turns. Each turn may have a single strand, or it may have several strands of copper electrically in parallel. A coil has two coil sides, which lie in slots approximately a pole pitch apart for dc machines, whereas most ac machines have short-pitch windings in which the coil pitch is less than 180 electrical degrees. The pitch of a coil is found by taking the quotient of the slots separating the coil sides and the slots in 180 electrical degrees. Windings may be single layer, one coil side per slot, or double layer, two coil sides per slot, in which case one side of each coil will rest in the bottom half of its slot and the other coil side at the top of its slot.

Direct-current armature windings. These may be lap, or parallel, windings, or they may be wave, or series, windings. Figure 3 shows a four-pole lap dc armature winding and a four-pole wave dc armature winding. The two positive brushes of the lap winding would be connected together and the two negative brushes would be connected together. The lap and wave windings have the following general properties:

1. Coil ends of lap windings are fastened to adjacent commutator bars. Coil ends of wave windings are fastened to commutator bars approximately two pole pitches apart.
2. The lap winding has as many parallel paths between the line terminals as there are poles, and the conductors of a path lie under adjacent poles. The wave winding has two parallel paths between the line terminals regardless of the number of poles, and conductors of either path lie under all poles.
3. The lap winding has as many sets of brushes as there are numbers of poles, whereas wave windings need only two brush sets. Usually, however, the machine carries as many brush sets as there are poles to provide greater current-carrying capacity for the commutator. Brush positions are such that the coils are short-circuited only when they are in the position of minimum induced voltage.

Lap windings are adapted to high-current machines because they may have more than two parallel paths, whereas the wave windings are adapted to small-capacity machines and high-voltage machines because of the series connection of the coils.

Alternating-current armature windings. These may be single or polyphase, full or fractional pitch, Y-connected or delta-connected. Practically all except small-capacity ac machines have three-phase windings, with coils distributed around the entire armature periphery for better utilization of space and material in the machine. Winding factors are used to evaluate the performance of a winding. The generated rms voltage per phase is $E = 4.44\ k_d k_p f N_{ph} \Phi$ where k_d is the distribution factor, k_p the pitch factor, f the frequency in hertz, N_{ph} the total turns in series per phase Φ the flux per pole in webers per square meter.

Fig. 2. Winding field on cylindrical rotor. (*National Electric Coil Co.*)

The distribution factor k_d of a winding gives the ratio of the voltage output of a distributed winding to the voltage output of a concentrated winding. This is shown by Eq. (1), where n is number of slots

$$k_d = \frac{\sin (n\gamma/2)}{n \sin (\gamma/2)} \tag{1}$$

over which one phase is distributed and γ is electrical angle between slots. Fractional-pitch windings are employed to reduce the magnitude of harmonics, to improve wave shape, to save copper, and to permit end connections between coils to be made more easily.

The pitch factor k_p of a winding gives the ratio of the voltage output of a fractional-pitch coil to the voltage output of a full-pitch coil. This is shown by Eq. (2), where ρ is electrical angle between coil sides.

$$k_p = \cos \frac{\pi - \rho}{2} \tag{2}$$

Since the angle γ and the angle ρ in the equations for k_d and k_p change with frequency, they may be selected to eliminate or minimize certain harmonic frequencies.

Figure 4 shows the active conductors of a distributed, three-phase, two-pole, fractional-pitch, armature winding wherein a, b, and c are conductors of three phases displaced 120 electrical degrees apart. Coil $a_1, -a_1$ spans 5/6 of a pole pitch, or 150 electrical degrees.

Stator and wound-rotor windings of polyphase induction motors conform to the above descrip-

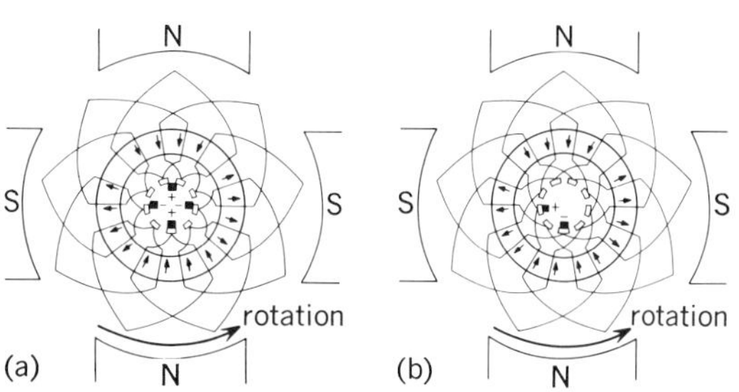

Fig. 3. Armature winding. (*a*) Simple four-pole lap winding. (*b*) Simple four-pole wave winding.

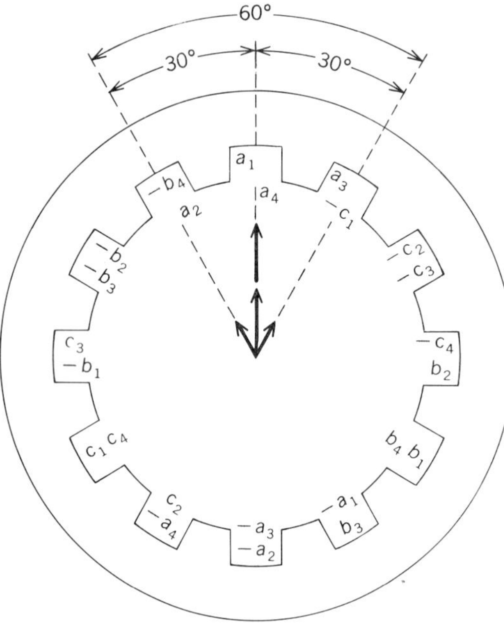

Fig. 4. Distributed, three-phase, two-pole, fractional-pitch armature winding. The voltage vector diagram is given. (*A. E. Fitzgerald and C. Kingsley, Electric Machinery, McGraw-Hill, 1952*)

tions. Single-phase induction motor stators may be wound in the same manner, except for the number of phases; or they may be formed of several concentric loops of wire. The windings are usually form-wound from one continuous length of wire, but are not taped. This facilitates the installation of the winding in small, partly closed slots. For additional information on these and squirrel-cage windings *see* INDUCTION MOTOR.

[ARTHUR R. ECKELS]

Bibliography: A. E. Fitzgerald, C. Kingsley, and A. Kusko, *Electric Machinery*, 3d ed., 1971; M. Liwschitz-Garik and C. Gentilini, *Winding Alternating Current Machines*, 1950, reprint 1975.

Wiring

A system of electric conductors, components, and apparatus for conveying electric power from source to the point of use. In general, electric wiring for light and power must convey energy safely and reliably with low power losses, and must deliver it to the point of use in adequate quantity at rated voltage. To accomplish this, many types of electric wiring systems and components are employed.

Electric wiring systems are designed to provide a practically constant voltage to the load within the capacity limits of the system. There are a few exceptions, notably series street-lighting circuits which operate at constant current.

In the United States the methods and materials used in the wiring of buildings are governed as to minimum requirements by the National Electrical Code, municipal ordinances, and, in a few instances, state laws. The National Electrical Code is a standard approved by the American National Standards Institute (ANSI). Most materials employed in wiring systems for light and power are tested and listed by Underwriters Laboratories, Inc. (UL). *See* ELECTRICAL CODES.

The building wiring system originates at a source of electric power, conventionally the distribution lines or network of an electric utility system. Power may also be supplied from a privately owned generating plant or, for emergency supply, a standby engine-generator or battery.

The connection from the supply to the building system through the metering devices, main disconnecting means, and main overcurrent protection is the service entrance. The conductors, cables or busways, are service conductors. The switch and fuse or circuit breaker serving as the disconnecting means and the main overcurrent protection are the service equipment. Up to six individual switches or circuit breakers may be used for the service entrance to a single building.

As a general rule only one set of service conductors to a building is permitted. Large industrial plants, commercial buildings, and institutions, however, are often served from more than one source. Separate service entrances are sometimes provided for emergency lighting, fire pumps, and similar loads.

Wiring systems. Wiring systems are generally three-phase to conform to the supply systems. Energy is transformed to the desired voltage levels by a bank of three single-phase transformers. The transformers may be connected in either a delta or Y connection. With a delta connection the ends of the transformer windings are connected together and line conductors connected to these points. A three-phase, three-wire system is thus formed, from which a single-phase line can be obtained from any two conductors. With the Y connection one end of each transformer winding is connected in a common connection, and line conductors are connected to the other ends of the transformer windings. This also forms a three-phase, three-wire system. A line wire is also often connected to the common point, forming a three-phase, four-wire system, from which a single-phase line may be obtained between the common wire and any other.

Service provided at the primary voltage of the utility distribution system, typically 13,800 or 4160 volts, is termed primary service. Service provided at secondary or utilization voltage, typically 120/208 or 277/480 volts, is called secondary service.

Primary service. Service at primary voltage levels is often provided for large industrial, commercial, and institutional buildings, where the higher voltage can be used to advantage for power distribution within the buildings.

Where primary service is provided, power is distributed at primary voltage from the main switchboard through feeders to load-center substations installed at appropriate locations throughout the building (Fig. 1). The load-center substation consists of a high-voltage disconnect switch, a set of transformers, and a low-voltage switchboard enclosed in a heavy sheet-metal housing.

In practice, several feeder arrangements are employed. These include (1) single primary, a single primary feeder serving several substations; (2) multi-primary, individual primary feeders to each substation; (3) loop primary, two primary feeders serving several substations, interconnected to form a ring or loop; and (4) primary selective, two primary feeders serving several substations, connectable to either feeder by a selector switch.

Wiring from the low-voltage-switchboard end of

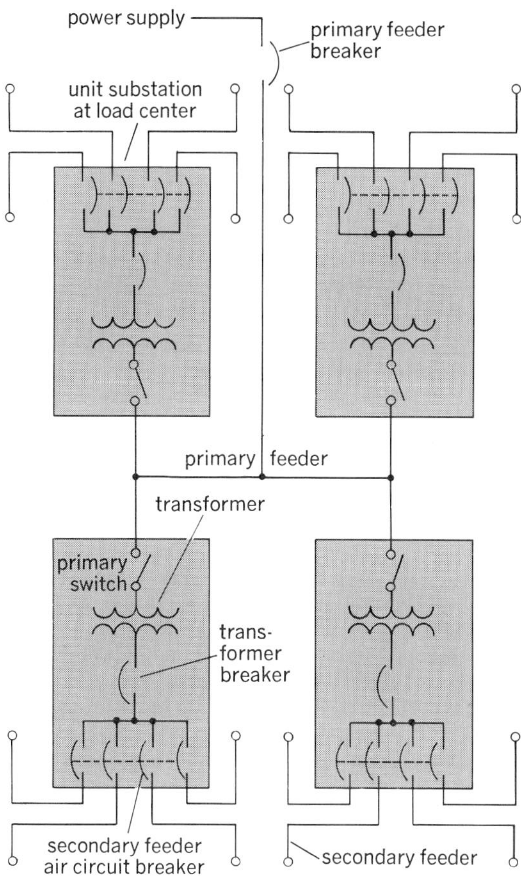

Fig. 1. Typical single primary distribution has a feeder from power supply to four load-center substations.

the substation to the load follows conventional utilization-voltage practice.

In large industrial plants, the secondary circuits of several substations are sometimes interconnected by feeder ties through switches or circuit breakers. The switches may be closed to form a network, provided that the transformers are suitable for parallel operation, or operated to transfer the load from one substation to another.

Enclosed bus-bar systems, called busways, are frequently used in the wiring system of industrial plants and large buildings. Busways are made and shipped in standard lengths with a wide variety of fittings. They are connected together on the job and installed as service-entrance conductors or feeders. One type of busway is designed to receive bus plugs at intervals along its length and thus functions both as a feeder and as a panelboard. The plugs which tap the feeder may contain fuses or circuit breakers, but the tap is often carried down to a more readily accessible overcurrent protective device.

Secondary service. This type of service supplies power to the building at utilization voltage. Most secondary services in the United States are 120/208 volts, three-phase, four-wire, or 120/240 volts, single-phase, three-wire serving both light and power. In some communities, separate light and power services are provided, typically 120/240 volts, single-phase for lighting and 440 volts, three-phase, delta for power.

Three-phase, four-wire services are almost universally Y with the neutral at ground potential. In some instances, however, a three-phase, four-wire service may be delta, with the neutral connected at the center tap of one phase. Such services provide typically 240 volts, three-phase, three-wire for power and 120/240 volts, single-phase, three-wire for lighting.

For relatively large buildings where the loads are predominantly fluorescent lighting and power (as for air conditioning), the service is often 277/480 volts, three-phase, four-wire, supplying 480 volts for power and 277 volts, phase-to-neutral, for the lighting fixtures.

Distribution switchboards. From the service entrance, power is carried in feeders to the main switchboard, then to distribution panelboards (Fig. 2). Smaller feeders extend from the distribution panelboards to light and power panelboards. Branch circuits then carry power to the outlets serving the various lighting fixtures, plug receptacles, motors, or other utilization devices.

The main distribution switchboard may also include the service equipment in its assembly. It consists of a group of switches and fuses or circuit breakers in a sheet-metal enclosure. It provides individual disconnecting and overcurrent protection for each feeder. In large buildings, additional distribution panelboards may be located at load centers.

Light and power panelboards provide individual disconnecting and overcurrent protection for the branch circuits. The circuit breakers of lighting panelboards are sometimes used as switches to operate the lighting circuits.

Plug-receptacle power at 120/240 volts, such as power for small appliances and business machines, in buildings provided with 277/480-volt supply is obtained from transformers. A feeder circuit from a 277/480-volt panelboard energizes the transformer primary. The secondary feeder serves a separate panelboard. The branch circuits serving the plug-receptacle outlets are conventional.

Wiring methods. Methods of wiring in common use for light and power circuits are as follows: (1) insulated wires and cables in raceways; (2) nonmetallic sheathed cables; (3) metallic armored cables; (4) busways; (5) copper-jacketed, mineral-insulated cables; (6) aluminum-sheathed cables; (7) nonmetallic sheathed and armored cables in

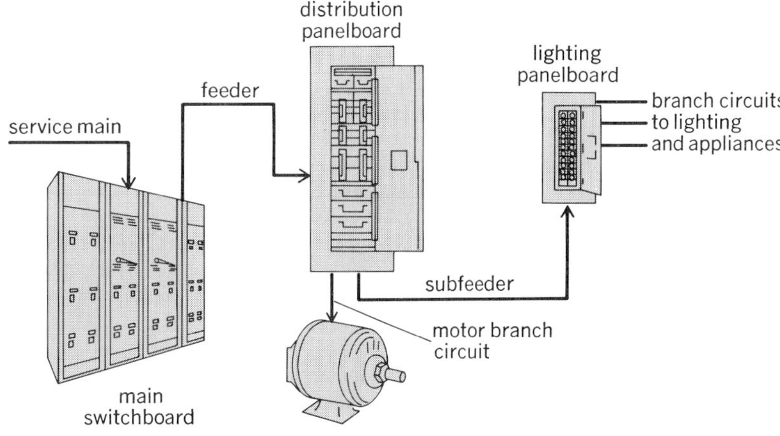

Fig. 2. Distribution wiring system has components to supply utilization voltage for several different types of load.

cable support systems; and (8) open insulated wiring on solid insulators (knob and tube).

Raceways in which insulated conductors may be installed are of several types: (1) rigid metal conduit; (2) electric metallic tubing (EMT); (3) flexible metal conduit; (4) liquid-tight flexible metal conduit; (5) surface metal raceway; (6) underfloor raceway; (7) cellular floor raceway; (8) rigid nonmetallic conduit; and (9) wireway.

The selection of the wiring method or methods is governed by a variety of considerations, which usually include code rules limiting the use of certain types of wiring materials; suitability for structural and environmental conditions; installation (exposed or concealed); accessibility for changes and alterations; and costs. Several methods may be employed together, for example, feeder busway risers in a multistory office building with the rest of the wiring in rigid conduit and underfloor raceways.

Circuit design. The design of a particular wiring system is developed by considering the various loads, establishing the branch-circuit and feeder requirements, and then determining the service-entrance requirements. Outlets for lighting fixtures, motors, portable appliances, and other utilization devices are indicated on the building plans and the load requirement of each outlet noted in watts or horsepower. Lighting fixtures and plug receptacles are then grouped on branch circuits and connections to the lighting panelboard indicated. Lighting and power panelboards are preferably located in the approximate center of the loads they serve; however, other considerations may require other locations. Panelboards in commercial and institutional buildings usually are located in corridors or electric closets. The size and number of panelboards are determined by the number of branch circuits to be served.

Lighting loads. Lighting branch circuits may be loaded to 80% of circuit capacity. However, there is a reasonable probability that the lighting equipment will be replaced at some future time by equipment of higher output and greater load. Therefore, in modern practice, lighting branch circuits are loaded only to about 50% capacity, typically not more than 1200 watts on a 20-ampere branch circuit.

Lighting branch circuits are usually rated at 20 A. Smaller 15-A branch circuits are used mostly in residences. Larger 30-, 40-, and 50-A branch circuits are limited to serving heavy-duty lampholders or appliances specially approved for connection to such circuits. The minimum conductor size for 20-A branch circuits is No. 12; however, on long runs larger conductors may be required to avoid excessive voltage drop. A common practice is to use No. 12 conductors between outlets and No. 10 conductors for the connection between the first outlet and the lighting panelboard.

Motor loads. These power loads are usually served by individual branch circuits, which must be rated at not less than 125% of the full-load rating of the motor.

Feeder and service-entrance design. The sum of the branch-circuit loads, including future load allowances, determines the feeder load. The National Electrical Code provides a table of factors for various occupancies giving the minimum unit load per square foot and demand factors that may be applied. Demand factors are applied in installations where the loads are diversified and not likely to occur at one time. The number of feeders and their loads determine the number and size of distribution panelboard circuit elements required. The sum of the feeder loads determines the size and capacity of the service-entrance conductors and equipment.

Conductor sizes. The size of wires and cables used in electrical wiring systems is expressed in terms of the American Wire Gage (AWG), known also as the Brown and Sharpe (B&S) gage. Size designations run from No. 14, the smallest size commonly used in wiring systems for light and power, to No. 4/0, the largest size in the gage. Sizes larger than No. 4/0 are designated by their cross-section areas expressed in circular mils. The largest size in practical usage is 2,000,000 circular mils (1013 mm^2). A circular mil is the area of a circle 0.001 in. (25.4 μm) in diameter.

Conductor capacity. The current-carrying capacity of wiring conductors is determined by the maximum insulation temperature that can be tolerated and the rate at which heat can be dissipated. All conductors offer some resistance to the flow of electric current. Consequently, heat is produced in the conductor by the flow of current through its resistance. The amount of heat is determined by the square of the current in amperes times the resistance in ohms (I^2R). Conductor heat is dispersed through the insulation and the surrounding raceway to the air. *See* ELECTRICAL INSULATION.

In practice, maximum current-carrying capacity of conductors is set forth in standard tables developed from laboratory tests and field experience. The National Electrical Code specifies the maximum current-carrying capacity of conductors. For any given size of conductor, the maximum capacity varies with the type of installation (exposed or in raceways) and the maximum safe temperature of the insulation. Approved values are reduced for high ambient temperatures and for more than three current-carrying conductors in a single raceway. *See* CONDUCTOR (ELECTRICITY).

Feeders supplying several motors must be rated at not less than 125% of the full-load current of the largest motor plus 100% of the full-load currents of the remaining motors.

Feeders serving continuous loads that are likely to operate for 3 hr or more (as office-building lighting) should not be loaded to more than 80% of rated capacity. The size of the feeder conductors will often be determined by the permissible voltage drop, which may require larger conductors than would be required by current-carrying capacity considerations alone.

Voltage drop. A drop in voltage along a conductor is a characteristic of all electric circuits. The voltage drop in a circuit causes the voltage at the load to be less than that applied to the circuit.

Wire or cable circuits are designed to carry a certain load but, whether the conductor be copper, aluminum, or other metal, the resistance characteristic impedes the flow of current.

Since conductor resistance is proportional to conductor length, longer circuits are especially susceptible to excessive voltage drops. The amount of voltage drop E_D is determined by the current I in amperes times the resistance R in ohms ($E_D = IR$).

Good practice in circuit design dictates the following percentage values for maximum voltage drop: 5% from service entrance to any motor load; 2% from service entrance to any panelboard; 2% from panelboard to any outlet on branch circuit; 4% in feeders and 1% in branch circuit to motor; and 2% total in conductors to electric heating equipment.

The feeder voltage drop is calculated by formulas derived from Ohm's law and the resistance of the conductor. For three-wire, three-phase circuits (neglecting inductance), the voltage drop E_D is given in the equation below, in which K is the resist-

$$E_D = \frac{1.732KIL}{CM}$$

ance of a circular mil-foot of wire (for copper, 10.8 ohms), I is the current in amperes, L is the circuit length (source to load) in feet, and CM is the cross-section area of the conductor in circular mils.

Copper loss. This characteristic of a circuit is related to voltage drop. It is a power loss, designated as the I^2R loss, and is often expressed in percentage as the ratio of the wattage loss to the wattage delivered to the circuit.

Circuit protection. In wiring systems of high capacity (typically 1200 A and above) supplied from utility networks of large capability, overcurrent protective devices of high interrupting capacity are required. Circuit breakers of special design or current-limiting fuses are employed in such installations. In some cases, current-limiting reactors or bus-bar arrangements presenting appreciable reactance under short-circuit conditions are inserted in the service conductors. *See* ELECTRIC POWER SYSTEMS; ELECTRIC PROTECTIVE DEVICES; TRANSMISSION LINES.

[WILLIAM T. STUART/J. F. MC PARTLAND]

Bibliography: D. L. Bellman, *Industrial Power Systems Handbook*, 1955; T. Croft, J. Watt, and W. I. Summers (eds.), *American Electricians Handbook*, 10th ed., 1981; J. Doyle, *Introduction to Electrical Wiring*, 1980; D. G. Fink and H. W. Beaty (eds.), *Standard Handbook for Electrical Engineers*, 11th ed., 1978; Institute of Electrical and Electronics Engineers, *The Recommended Practice for Electrical Power Systems in Commercial Buildings*, Publ. no. 241, 1974; J. F. McPartland, *The Handbook of Practical Electrical Design*, 1982; J. F. McPartland, *National Electrical Code Handbook*, 17th ed., 1981; *National Electrical Code*, NFPA 70-1981 (ANSI), 1981; J. E. Traister, *Handbook of Modern Electrical Wiring*, 1979.

Wiring diagram

A drawing illustrating electrical and mechanical relationships between parts on a component between which electrical wiring must be connected.

A wiring diagram is distinguished from an electrical schematic in that the arrangement of the schematic bears no necessary relationship to the mechanical arrangement of the electrical elements in the component. The wiring diagram provides an accurate picture of how the wiring on the components and between components should appear in order that the electrical wiring technician can install the wiring in the manner that will best contribute to the optimum performance of the device.

The degree of symbolism used in a wiring diagram depends on the extent of standardization in the particular field. For example, in telephone switchboard wiring, which consists of many standardized repetitive operations, extensive symbolism is used. On the other hand, when the exact physical location of wiring is important, as in radio-frequency devices where electromagnetic and electrostatic coupling between wires is appreciable, the diagram can be quite pictorial.

Wiring diagrams also include such information as type of wire, color coding, methods of wire termination, and methods of wire and cable clamping. *See* SCHEMATIC DRAWING. [ROBERT W. MANN]

Work

In physics, the term work refers to the transference of energy that occurs when a force is applied to a body that is moving in such a way that the force has a component in the direction of the body's motion. Thus work is done on a weight that is being lifted, or on a spring that is being stretched or compressed, or on a gas that is undergoing compression in a cylinder.

When the force acting on a moving body is constant in magnitude and direction, the amount of work done is defined as the product of just two factors: the component of the force in the direction of motion, and the distance moved by the point of application of the force. Thus the defining equation for work W is Eq. (1), where f and s are the

$$W = f \cos \phi \cdot s \qquad (1)$$

magnitudes of the force and displacement, respectively, and ϕ is the angle between these two vector quantities (Fig. 1). Because $f \cos \phi \cdot s = f \cdot s \cos \phi$, work may be defined alternatively as the product of the force and the component of the displacement in the direction of the force. In Fig. 2 the

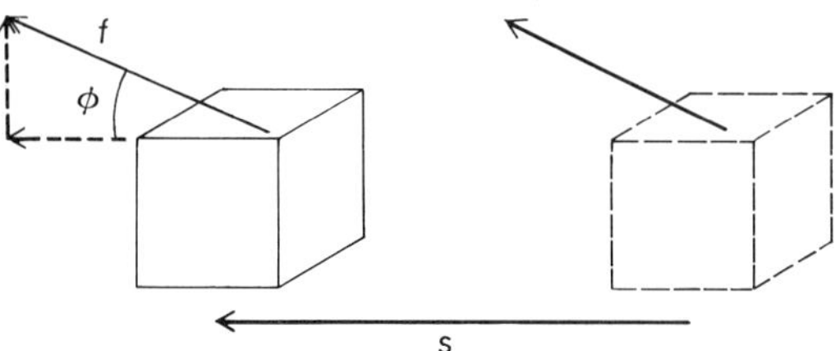

Fig. 1. Work of constant force f is $fs \cos \phi$.

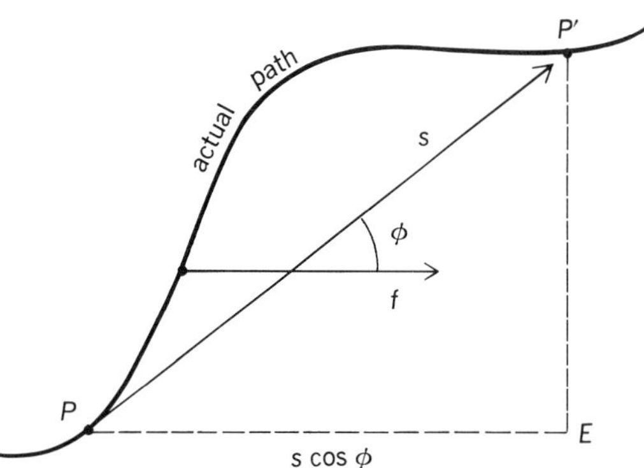

Fig. 2. The work done in traversing any path connecting points P and P' is $f \cdot \overline{PE}$, assuming the force f to be constant in magnitude and direction.

work of the constant force f when the application point moves along the curved path from P to P', and therefore undergoes the displacement $\overline{PP'}$, is $f \cdot \overline{PP'} \cos \phi$, or $f' \overline{PE}$.

Work is a scalar quantity. Consequently, to find the total work done on a moving body by several different forces, the work of each may be computed separately and the ordinary algebraic sum taken.

Examples and sign conventions. Suppose that a car slowly rolls forward a distance of 10 m along a straight driveway while a man pushes on it with a constant magnitude of 200 newtons of force (200 N) and let Eq. (1) be used to compute the work W done under each of the following circumstances: (1) If the man pushes straight forward, in the direction of the car's displacement, then $\phi = 0°$, $\cos \phi = 1$, and $W = 200\text{ N} \times 1 \times 10\text{ m} = 2000\text{ N} \cdot \text{m} = 2000$ joules; (2) if he pushes in a sideways direction making an angle ϕ of 60° with the displacement, then $\cos 60° = 0.50$ and $W = 1000$ joules; (3) if he pushes against the side of the car and therefore at right angles to the displacement, $\phi = 90°$, $\cos \phi = 0$, and $W = 0$; (4) if he pushes or pulls backward, in the direction opposite to the car's displacement, $\phi = 180°$, $\cos \phi = -1$, and $W = -2000$ joule.

Notice that the work done is positive in sign whenever the force or any component of it is in the same direction as the displacement; one then says that work is being done *by* the agent exerting the force (in the example, the man) and *on* the moving body (the car). The work is said to be negative whenever the direction of the force or force component is opposite to that of the displacement; then work is said to be done *on* the agent (the man) and *by* the moving body (the car). From the point of view of energy, an agent doing positive work is losing energy to the body on which the work is done, and one doing negative work is gaining energy from that body.

Units of work and energy. These consist of the product of any force unit and any distance unit. Units in common use are the foot-pound, the foot-poundal, the erg, and the joule. The product of any power unit and any time unit is also a unit of work or energy. Thus the horsepower-hour (hp-hr) is equivalent, in view of the definition of the horsepower, to 550 ft-lbf/sec × 3600 sec, or 1,980,000 ft-lbf, or (1,980,000)(0.3048 m)(4.45 N) = 2,684,520 joules. Similarly, the watt-hour is 1 joule/sec × 3600 sec, or 3600 joule; and the kilowatt-hour is 3,600,000 joule.

Work of a torque. When a body which is mounted on a fixed axis is acted upon by a constant torque of magnitude τ and turns through an angle θ (radians), the work done by the torque is $\tau\theta$.

Work principle. This principle, which is a generalization from experiments on many types of machines, asserts that, during any given time, the work of the forces applied to the machine is equal to the work of the forces resisting the motion of the machine, whether these resisting forces arise from gravity, friction, molecular interactions, or inertia. When the resisting force is gravity, the work of this force is mgh, where mg is the weight of the body and h is the vertical distance through which the body's center of gravity is raised. Note that if a body is moving in a horizontal direction, h is zero and no work is done by or against the gravitational force of the Earth. If a person holds an object or carries it across level ground, he does no net work against gravity; yet he becomes fatigued because his tensed muscles continually contract and relax in minute motions, and in walking he alternately raises and lowers the object and himself.

The resisting force may be due to molecular forces, as when a coiled elastic spring is being compressed or stretched. From Hooke's law, the average resisting force in the spring is $-\frac{1}{2}\,ks$, where k is the force constant of the spring and s is the displacement of the end of the spring from its normal position; hence the work of this elastic force is $-\frac{1}{2}\,ks^2$. *See* HOOKE'S LAW.

If a machine has any part of mass m that is undergoing an acceleration of magnitude a, the resisting force $-ma$ which the part offers because of its inertia involves work that must be taken into account; the same principle applies to the resisting torque $-I\alpha$ if any rotating part of moment of inertia I undergoes an angular acceleration α.

When the resisting force arises from friction between solid surfaces, the work of the frictional force is $-\mu f_n s$, where μ is the coefficient of friction for the pair of surfaces, f_n is the normal force pressing the two surfaces together, and s is the displacement of the one surface relative to the other during the time under consideration. The frictional force μf_n and the displacement s giving rise to it are always opposite in direction ($\phi = 180°$). *See* FRICTION.

The work done by any conservative force, such as a gravitational, elastic, or electrostatic force, during a displacement of a body from one point to another has the important property of being path-independent: Its value depends only on the initial and final positions of the body, not upon the path traversed between these two positions. On the other hand, the work done by any nonconservative force, such as friction due to air, depends on the path followed and not alone on the initial and final positions, for the direction of such a force varies with the path, being at every point of the path tangential to it. *See* FORCE.

Since work is a measure of energy transfer, it can be calculated from gains and losses of energy. It is useful, however, to define work in terms of forces and distances or torques and angles because these quantities are often easier to measure than energy changes, especially if energy changes are produced by nonconservative forces.

Work of a variable force. If the force varies in magnitude and direction along the path $\overline{PP'}$ of its point of application, one must first divide the whole path into parts of length Δs, each so short that the force component $f \cos \phi$ may be regarded as constant while the point of application traverses it (Fig. 3). Equation (1) can then be applied to each small part and the resulting increments of work added to find the total work done. Various devices are available for measuring the force component as a function of position along the path. Then a work diagram can be plotted (Fig. 4). The total work done between positions s_1 and s_2 is represented by the area under the resulting curve between s_1 and s_2 and can be computed by measuring this area, due allowance being made for the scale in which the diagram is drawn.

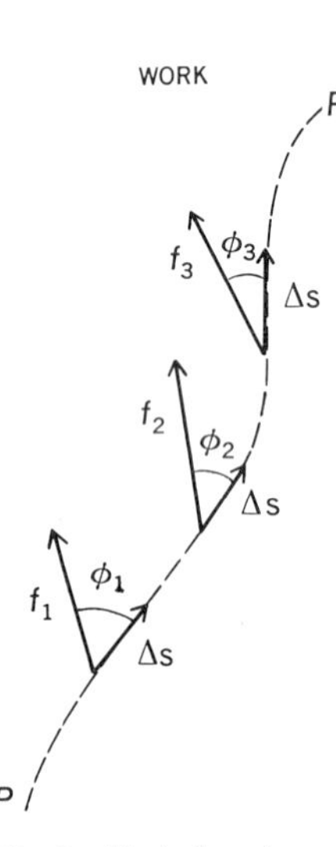

Fig. 3. Work done by a variable force.

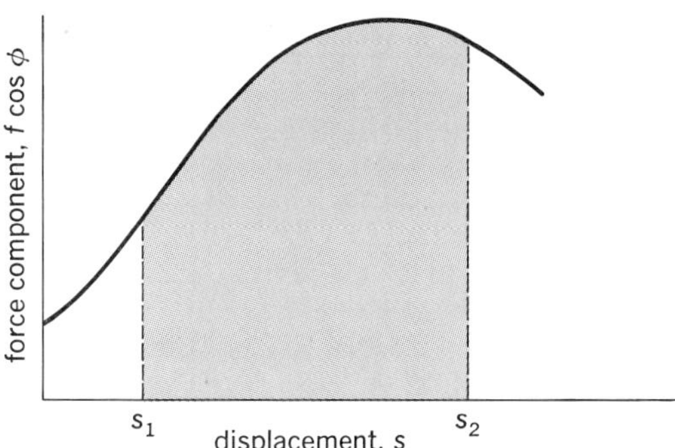

Fig. 4. A work diagram.

For an infinitely small displacement ds of the point of application of the force, the increment of work dW is given by Eq. (2), a differential expres-

$$dW = f \cos \phi \; ds \qquad (2)$$

sion that provides the most general definition of the concept of work. In the language of vector analysis, dW is the scalar product of the vector quantities **f** and $d\mathbf{s}$; Eq. (2) then takes the form $dW = \mathbf{f} \cdot d\mathbf{s}$. If the force is a known continuous function of the displacement, the total work done in a finite displacement from point P to point P' of the path is obtained by evaluating the line integral in Eq. (3).

$$W = \int_P^{P'} f \cos \phi \; ds = \int_P^{P'} \mathbf{f} \cdot d\mathbf{s} \qquad (3)$$

When a variable torque of magnitude τ acts on a body mounted on a fixed axis, the work done is given by $W = \int_{\theta_1}^{\theta_2} \tau \; d\theta$, where $\theta_2 - \theta_1$ is the total angular displacement expressed in radians. *See* ENERGY. [LEO NEDELSKY]

Bibliography: F. Bueche, *Understanding the World of Physics*, 1981; D. Halliday and R. Resnick, *Fundamentals of Physics*, 2d ed., 1981; E. M. Rogers, *Physics for the Inquiring Mind*, 1960.

Work measurement

The determination of a set of parameters associated with a task.

Rationale. There are four reasons, common to most organizations whether profit seeking or not, why time, effort, and money are spent to measure the amount of time a job takes. The fifth, pay by results, is used only by a minority of organizations.

Cost accounting. If the length of time it takes to do the job is not known, the charge for the job cannot be determined. Without knowing the time/job ratio, too much or too little may be charged. As a result, the product may be overpriced and, in the long run, sales will be lost, or if too little is charged, revenue will be lost.

Evaluation of alternatives. Without knowing the time (and thus the cost) of a job, decisions regarding whether to make or buy an item, whether to mechanize or not, whether to advertise or not, could be wrong.

Acceptable day's work. Suppose a worker makes 100 widgets a day. A standard of comparison is needed to determine whether this is superior, average, or poor performance. If the standard is 75 a day, the worker is to be praised; if the standard is 200 a day, the worker is to be questioned.

Scheduling. Managers need to know the time/job ratio to make reasonable decisions for production schedules, how many people to assign to the job, how much equipment is required, and so on.

Pay by results. In a small number of organizations, pay is based on the units produced. In most organizations, pay is by the hour, week, or month and, as long as output is "reasonable," pay continues. But if pay depends upon units produced, the organization needs to know how long the task takes so a reasonable pay rate/unit can be set. *See* WAGE INCENTIVES.

Given the decision to record the amount of time per job, it is important to emphasize that first the job should be properly designed. Recording time for a poorly designed job is a mistake. The job should be improved (that is, improve productivity) before doing work measurement. *See* PRODUCTIVITY.

There are three common ways to determine time per job: stopwatch time study (sequential observations), occurrence sampling (nonsequential observations), and standard data.

Stopwatch time study. Stopwatch time study has the great advantage of flexibility—it can be used for almost any existing job. Workers have become familiar with it. It is reasonable in cost and gives reasonable accuracy. However, it does require the worker to be rated. Once the initial cost of standard data system has been incurred, standard data may be the lowest-cost, most accurate, and most accepted technique.

There are eight steps to determine standard time:

1. Development of a good method to be timed.

2. Selection of an operator to time. In some cases, there is no choice as only one person does the job. If there is a choice, a typical or average worker should be used rather than the best. (The best is useful for productivity analysis.) An average worker is used because rating is more accurate for that worker than for an extreme worker and because other workers are more willing to accept a time standard determined on an average worker.

3. Preparation for timing. The job must be broken down into elements, indicating the termination points (see illustration). Elements permit reuse of the data, they give good internal consistency checks, they permit different ratings for different elements, and they improve methods descriptions.

4. Selection of a timing technique. There are four common alternatives: one watch with continuous hand movement; one watch with snapback; three watches with snapback; and an electronic watch with a hold circuit. The one watch with continuous hand movement formerly was the recommended procedure, but has been made obsolete by the improved technology and lower costs of the third and fourth methods. While the first method gives accurate results, it requires excessive clerical work. The last two methods give even better accuracy and substantially reduced clerical work, at a minor incremental equipment cost for the watch. The second method has low clerical costs, but has considerable potential for inaccuracy.

NO.	ELEMENTS	SPEED	FEED	UPPER LINE: SUBTRACTED TIME / LOWER LINE: READING 1	2	3	4	5	6	7	8	9	10	11	12	13	14	15	MIN. TIME	AV. TIME	Std. Dev.	OCC. PER CYCLE	EFFORT RATING	NORMAL TIME
1	Get pen; TP Ready to use			3.1																3.1		.25	100	.775
2	Get letter & envelope: TP Ready to check			2.6	2.7	3.5	3.0													2.95		1	90	2.655
3	Read letter: TP Read last letter: TP Read last word *			18.2 (100)	21.5 (125)	42.1 (225)	55.0 (205)	(A)				18.2/100 = .182/word 21.5/125 = .172 42.1/225 = .187								.180/word			100	.180/word
4	Sign and aside letter and envelope: TP RL paper			3.0	2.7	3.1	3.0																	
5	Return pen: TP RL pen						3.6													3.6		.25	110	.990
	* Number of words in bracket																							
																						4.42/.9 = 4.91		
																						.180/.9 = .2/word		

FOREIGN ELEMENTS:	TOOLS, JIGS, GAUGES, PATTERNS, ETC.:				
(A) Made 1 correction					
	OVERALL EFFORT RATING	BEGIN	END	ELAPSED	UNITS FINISHED	ACTUAL TIME PER PIECE

Elemental breakdown of a job—signing a letter. The end of the element is the termination point (TP). The times are seconds. It is assumed that four letters are considered for signature at a time, and a standard will assume no changes are needed in the letter. Elements 1, 2, 4, and 5 are constant elements. Element 3 is a variable element as the time varies with the number of words read. Allowances are assumed as 10%.

5. Determination of the number of observations. The number can be determined by using statistical techniques from standard tables used by the organization. For example, the organization may require 200 observations for any element time less than .002 hour, 175 for times from .002 to .003, and so on.

6. Elimination of foreign elements, irregular elements, and outliers from the data. Foreign elements are observations which are not allowed directly as part of the time standard—they may be included in some cases indirectly as allowances. Examples of foreign elements would be speaking to a supervisor, watching someone walk by, and lighting a cigarette. An irregular element (such as oiling of the machine) occurs at infrequent intervals. The data should be used, but the number of units between elements must be determined. An outlier is an abnormally high or low value; it can be eliminated only for specific known reasons or with standard statistical tests—never merely because it is high or low.

7. Performance rating. Completion of the first six steps gives the recorded time. The normal time is needed—the time that a typical experienced worker would take under ideal conditions and without a break: Recorded time × Rating = Normal time.

8. Allowances. Allowances generally fall into personal, fatigue, and delay allowances. Personal allowances would be for getting a drink of water; fatigue allowances are given for physically or mentally demanding work; and delay allowances are given for uncontrollable delays. Total allowances generally range from 5 to 20% of the standard time for the job or operation. Standard time is calculated from normal time divided by (1 − allowance percent).

Occurrence sampling standards. Occurrence sampling is also called work sampling or ratio-delay sampling. If time study is a "movie," then occurrence sampling is a "series of snapshots."

The primary advantage of this approach may be that occurrence sampling standards are obtained from data gathered over a relatively long time period, so the sample is likely to be representative of the universe. That is, the time from the study is likely to be representative of the long-run performance of the worker. Another advantage is that occurrence sampling can be used when "production" is not continuous. For example, when timing phone calls from customers, calls may occur only 5 to 10 times per day and at irregular intervals. Rating generally is not used as all work is assumed to be at a 100% pace. For work done at paces different from 100%, this will cause errors in the accuracy of the standard. Thus occurrence sampling standards probably should not be used for incentive pay purposes.

To set a standard using occurrence sampling, it

is necessary to do an occurrence sample plus record the units produced during the time the occurrence sample takes place. For example, consider the mechanic who does tune-ups. The individual should be observed 100 times over a 10-day period, with records of each period of idle time and of each truck or car tune-up. The output during the 10 days (say 5 trucks and 13 cars) is recorded as well as the scheduled work time (say 450 min/day).

Assume the study showed idle = 10%, truck tune-up = 36%, and car tune-up = 54%. Then working time was .9(4500) = 4050 min; a truck tune-up takes .36(4050)/5 = 292 min, and a car tune-up takes .54(4050)/13 = 168 min. The pace is assumed to be 100%, so normal time equals observed time. Giving 10% for allowances, a truck tune-up takes 292/.9 = 324 min, while a car tune-up takes 168/.9 = 187 min.

In making the occurrence sample there are two steps: getting a sample whose size gives the desired trade-off between cost of the study and risk of an inaccurate estimate, and obtaining a sample representative of the population.

The required number of observations in the sample can be determined from the formula:

$$A = z\sigma_p$$

where

$A = sp$ = absolute accuracy desired, decimal
s = relative accuracy desired, decimal
p = mean percent occurrence, decimal
z = number of standard deviations for confidence level desired (see Table 1)
n = number of observations
σ_p = standard deviation of a percent; $\sqrt{p(1-p)/n}$

Table 1. Confidence levels for z levels on occurrence sampling formula*

z (Number of standard deviations)	Corresponding confidence level, %
±1.0	68
±1.64	90
±1.96	95
±2.0	95.45
±3.0	99.73

*Use the table of the normal distribution (not shown here) for other values.

SOURCE: S. A. Konz, *Work Design*, published by Grid, 4666 Indianola Avenue, Columbus, OH 43214, 1979.

For example, the problem might be to determine whether or not to add an additional telephone line, as customers have been complaining the line is always busy. Management requests a study, indicating that they want a relative accuracy of ±10% and a confidence of 90%. The person doing the study must decide how many observations (samples) to take. First, from preliminary judgment, it is guessed that the lines are busy 60% of the time—that is, $p = .60$. Then $A = sp$ or .10 (.60) = .06. The number of standard deviations corresponding to a 90% confidence level is 1.64. The estimated value of p (.6) is then substituted into the formula for standard deviation of a percent. The resulting equation is $.06 = (1.64)\sqrt{(.6)(.4)/n}$. Solving for n, which is 179, and assuming 2 weeks would be needed to give a representative sample, 180 observations must be taken at the rate of 18/day for 10 days.

To make the sample representative, stratification, influence, and periodicity must be considered. Stratification means to divide the sample into strata (layers). Thus there may be 10 strata (5 days/week for 2 weeks), 20 strata (morning versus afternoon for each day of the sample), or some other division such as local calls versus long-distance. Influence means that it is not desirable that the behavior of the individual changes because of being observed. For this example, observations should be taken in such a way that neither customers nor employee change their use of the telephone because they are being sampled. Periodicity refers to existing patterns of behavior coincident with specific times of the day. The sample must not take too many or too few observations at these special times.

Assume that from the sample of 180 the lines were busy 90 times, so $p = 90/180 = .5$. Therefore, $A = 1.64\sqrt{(.5)(.5)/180}$ so $A = 6.1\%$. Thus the conclusion is that the phones are busy 50% (±6.1%) of the time with a confidence of 90%—if the situation does not change. The stratification information indicates that the percent busy was 40% on Monday and Wednesday, 50% on Tuesday, and 60% on Thursday and Friday.

To set a time standard for phone calls, all that is needed is to record the number of calls during the time of the study. It is already known that the line was busy 50% of the time of 480 min/day × 10 days × .50 = 2400 min. If there were 512 calls during the 2 weeks, then time/call is 2400/512 = 4.7 min/call.

From this information, several actions are possible, such as installing another telephone line, estimating the cost of dealing with customers over the telephone, and estimating how much extra time is needed for the telephone on Thursday and Friday (and how much is available on other days). However, a new phone line will change the situation, so the times from the sample may not be representative of the situation after the change.

Standard data standards. Reuse of previous times (standard data) is an alternative to measuring new times for an operation. Lower cost, consistency, and ahead-of-production are the three advantages. Looking up the time to walk 50 m from a table of standard times rather than doing a special study to measure the time to walk 50 m saves the time of the person setting the standard. However, this operating cost must be balanced by the capital cost of setting up the table of standard times for walking. Thus the capital cost must be justified by many uses—that is, standard data are economically justifiable only for standard repetitive elements. A second advantage is consistency. In setting the time to walk 50 m, a number of studies can be used, so any rating or measurement errors would tend to average out. In addition, since a table or formula is used, every analyst always gets the same answer; rating and judgment are minimized. A third advantage is that the time can be estimated prior to production. Timing requires an experienced operator, a work station with tools, and product. But times often are wanted ahead of pro-

Table 2. Three levels of detail for standard time systems

Microsystem (typical component time range from .01 to 1 s; MTM nomenclature)

Element	*Code*	*Time*
Reach	R10C	12.9 TMU*
Grasp	G4B	9.1
Move	M10B	12.2
Position	P1SE	5.6
Release	RL1	2.0

Elemental system (typical component time range from 1 to 1000 s)

Element	*Time*
Get equipment	1.5 min
Polish shoes	3.5
Put equipment away	2.0

Macro system (typical component times vary upward from 1000 s)

Element	*Time*
Load truck	2.5 h
Drive truck 200 km	4.0
Unload truck	3.4

*27.8 TMU = 1 s; 1 s = .036 TMU.
SOURCE: S. A. Konz, *Work Design*, published by Grid, 4666 Indianola Avenue, Columbus, OH 43214, 1979.

duction for determining which alternative work method to use and such.

There are three levels of detail: micro, elemental, and macro (Table 2). Micro-level systems have times of the smallest component ranging from about .01 to 1 s. Components usually come from a predetermined time system such as methods-time-measurement (MTM) or Work-Factor. Elemental level systems have the time of the smallest component, ranging from about 1 to 1000 s. Components come from time study or micro-level combinations. Macro-level systems have times ranging upward from about 1000 s. Components come from elemental-level combinations, from time studies, and from occurrence sampling.

For example, assume a standard time is needed for signing a business letter (see illustration). It may be decided to break the task into five elements: (1) get a pen, (2) get a letter and envelope, (3) read the letter, (4) sign and set aside the letter and envelope, and (5) return pen. For simplicity it may be assumed that all letters are examined in batches of four, and no corrections need to be made. Element 3, read the letter, is a variable element in that it varies with the letter length. The remaining elements are constant. Element 3 time might be $5.05N$, where N = the number of words in the letter. If the total time for elements 1, 2, 4, 5, and 6 was 150 time-measurement units (TMU), then total time for signing a letter is $150 + 5.05N$. Thus a 100-word letter takes $150 + 505 = 655$ TMU, and a 200-word letter takes $150 + 1010 = 1160$ TMU. Allowances should be added to both times, but rating is not necessary because it is built into the MTM time. *See* METHODS ENGINEERING; PERFORMANCE RATING.

[STEPHAN A. KONZ]

Bibliography: S. A. Konz, *Work Design*, 1979.

Work standardization

The establishment of uniformity of technical procedures, administrative procedures, working conditions, tools, equipment, workplace arrangements, operation and motion sequences, materials, quality requirements, and similar factors which affect the performance of work. It involves the concepts of design standardization applied to the performance of jobs or operations in industry or business. *See* DESIGN STANDARDS; WORK MEASUREMENT.

Work standardization is part of methods engineering and, where it is practiced, usually precedes the setting of time standards. There would be little point in establishing time standards until the method by which the work is to be done has been standardized. The objectives of work standardizations are lower costs, greater productivity, improved quality of workmanship, greater safety, and quicker and better development of skills among workers. Work standardization also tests the ingenuity of industrial engineers and of production or operations managers at all levels. Frequently, ideas leading to work standardization are derived from employee suggestions.

Because it concerns primarily the internal operations of an enterprise, work standardization is very much limited to individual firms or their divisions. Therefore, relatively little of such activity is found at the trade association or higher levels. The principal exception is the widespread standardization of certain tests of materials and products, many of which prescribe the procedures in some detail.

In some cases, work standardization has led to the establishment of new product lines to serve a market it created. One example is the burster form, in which information on an incoming order is recorded only once, with the various parts of the form serving every purpose from purchase of raw materials and production control to outgoing address label. This concept originated when forms were standardized so that anyone requiring them would be able to find the same information in the same location.

Work standardization often leads to simplification in that it presents opportunities for eliminating and consolidating unit operations of various kinds. Searching for a common approach also requires a careful definition of what needs to be done, which is also often a source of improvements.

One of the best known of the more formal techniques of work standardization is group technology. This is the careful description of a heterogeneous lot of machine or other piece parts with a view to discovering as many common features in materials and dimensions as can be identified. It is then possible to start a rather large lot of a basic part through the production process, doing the common operations on all of them. Any changes or additional operations required to produce the final different parts can then be made at a later stage. The economy is realized in being able to do the identical jobs at one time.

It may prove economical to do some of the common operations on all the parts even when they are not really required, rather than set the jobs up separately. Parts like shafts, cover plates, tooling components, or mounts for electronic circuitry are among the many applications of group technology. To implement this, a special form of "cellular" layout is required which parallels the branching characteristics of the production system. Opportunities also exist for this approach in the chemical processing industries in that basic mixes can be produced, with later additives or other unit operations providing the required product differentiation.

There has been considerable progress in computerized systems to facilitate group technology. The techniques are closely linked to computer-aided design and to formalized codes that permit the detailed description of many operations. This is a necessary prerequisite to later specification of the identities. In several of these systems, the total information is finally recorded on punched cards or other memory media, with some including a microfilm of the drawing of the part required.

Simpler and less formal forms of work standardization can occur just about everywhere, in both the private and public sector. Whenever a good idea is being put forth dealing with the more efficient performance of a detailed operation, it is worthwhile to study its possible application to a wider area. Conversely, if standard practice now exists in many parts of the business, it may be advantageous to apply it to activities hitherto left unstandardized.

However, the economics of work standardization must be carefully balanced against added costs such as new equipment or tooling, carrying greater inventories, and extra handling. The establishment of larger production batches does not necessarily produce economy of scale. Some forms of group technology, for instance, lose some of their advantages when a numerically controlled machine can quickly make changes in the parts it processes as it goes along.

There is also a possibility that work standardization may lead to repetitious and monotonous operations where previously there was a degree of flexibility and variety permitted; and, in such a case, become undesirable for a variety of operational and social reasons. *See* METHODS ENGINEERING; PERFORMANCE RATING. [JOHN E. ULLMANN]

Young's modulus

A constant designated E, the ratio of stress to corresponding strain when the material behaves elastically. Young's modulus is represented by the slope $E = \Delta S/\Delta\epsilon$ of the initial straight segment of the stress-strain diagram. More correctly, E is a measure of stiffness, having the same units as stress, pounds per square inch. When stress and strain are not directly proportional, E may be represented as the slope of the tangent or the slope of the secant connecting two points on the stress-strain curve. The modulus is then designated as tangent modulus or secant modulus at stated values of stress. The modulus of elasticity applying specifically to tension is called Young's modulus. Many materials have the same value in compression. *See* ELASTICITY; HOOKE'S LAW; STRESS AND STRAIN. [W. J. KREFELD/W. G. BOWMAN]

Zinc alloys

Combinations of zinc with one or more other metals. If zinc is the primary constituent of the alloy, it is a zinc-base alloy. Zinc also is commonly used in varying degrees as an alloying component with other base metals, such as copper, aluminum, and magnesium. A familiar example of the latter is the association of varying amounts of zinc (up to 45%) with copper to produce brass, the third largest use of zinc in the United States. *See* BRASS; COPPER ALLOYS.

Uses. Zinc-base alloys have two major uses: for casting and for wrought applications. Casting includes both die casting, which is the largest single market for zinc in the United States and second largest on a world basis, and gravity casting, which differs from die casting primarily in that no pressure is applied, except the force of gravity, in forcing the molten metal into the mold. *See* METAL CASTING.

Zinc alloys have historically been used for die casting primarily because of their relatively low melting temperatures. This means lower fabrication costs as well as longer tool life. Zinc-alloy die castings also have good strength and dimensional tolerance and exceptional castability and are easily plated and machined.

Alloys for casting. The zinc alloys used for die casting are well standardized and have been covered by American Society for Testing and Materials (ASTM) specifications in the United States since 1931. Equivalent specifications are in effect by similar organizations throughout the world, including the British Standards Institution (BSI), the Canadian Standards Association (CSA), and the International Standards Organization (ISO).

The presence of impurities above a certain specified maximum is deleterious in all types of alloys. Zinc die-casting alloys, like other alloys, are sensitive to small variations of certain impurities. The alloys, however, are not difficult to make, and most large die casters do their own alloying.

Lead, tin, and cadmium must be avoided as impurities in zinc, or as contaminants in the other alloying ingredients used. These impurities are high-density metals and have low melting points and, when present in amounts beyond the specification limits, render the alloys susceptible to intergranular corrosion, particularly on exposure to warm moist environments. To avoid these problems Special High-Grade Zinc, which is 99.99% pure, is used for die-casting applications. The great bulk of zinc die casting currently is carried out with two alloys, commonly identified as alloys 3 and 5, which have been in use for several decades. A third alloy has been added and is commonly designated as alloy 7. The entire group in the United States is often spoken of by its original trade name as Zamak alloys, and in England as the Mazak alloys. A fourth alloy that is especially suited for cold-chamber die-casting machines was introduced to the industry in the mid-1960s and is now being evaluated for many elevated temperature applications where its superior resistance to deformation over time is an important advantage. The composition and the mechanical and other properties of these alloys are given in the table, along with corresponding ASTM and Society of Automotive Engineers (SAE) designations.

Wrought zinc alloys. In the wrought zinc area, numerous compositions and alloys are used, depending on ultimate product requirements. Alloying metals can be used to improve various properties, such as stiffness, for special applications. The

Composition and properties of zinc alloy die castings

ASTM designation B86-64:	AG40A	AC41A	–	–
SAE designation:	903	925	903	–
General designation	3	5	ZAMAK 7	ILZRO 16
Composition % by weight				
Copper	.25 max	.75 – 1.25	.25 max	1.0 – 1.5
Aluminum	3.5 – 4.3	3.5 – 4.3	3.5 – 4.3	.01 – .04
Magnesium	.020 – .05	.03 – .08	.005 – .02	.01
Iron max	.100	.100	.075	–
Lead max	.005	.005	.0030	.004
Cadmium max	.004	.004	.0020	.003
Tin max	.003	.003	.0010	.002
Nickel	–	–	.005 – .020	–
Titanium	–	–	–	.15 – .25
Zinc (99.99+% purity)	Remainder	Remainder	Remainder	Remainder
Mechanical properties				
Charpy impact strength, ft-lb (N·m), 1/4 × 1/4 in. (6.35 × 6.35 mm) bar, as cast	43 (58)	48 (65)	40 (54)	–
Charpy impact strength, after 10 years of indoor aging	41 (56)	40 (54)	41 (56)	–
Charpy impact strength, after 20 years of indoor aging	39 (53)	20 (27)	–	–
Tensile strength psi (MPa), after 20 years of indoor aging	33,000 (228)	36,000 (248)	–	–
Elongation % in 2 in. (50.8 mm), after 20 years of indoor aging	20	12	–	–
Tensile strength psi (MPa), as cast	41,000 (283)	47,600 (328)	41,000 (283)	33,000 (228)
Tensile strength psi (MPa), after 10 years of indoor aging	35,000 (241)	39,300 (271)	35,000 (241)	–
Elongation % in 2 in. (50.8 mm), as cast	10	7	10	5 – 6
Elongation % in 2 in. (50.8 mm), after 10 years of indoor aging	16	13	16	–
Expansion (growth) inches per inch (cm per cm) after 10 years of indoor aging	.0001	.0001	.0001	–
Other properties (as cast)				
Brinell hardness	82	91	76	75 – 77
Compression strength, psi (MPa)	60,000 (414)	87,000 (600)	60,000 (414)	–
Electrical conductivity, mhos/cm³ at 20°C	157,000	153,000	157,000	119,000
Melting point, °C	386.6	386.1	386.6	418
Melting point, °F	727.9	727.0	727.9	785
Modulus of rupture, psi (MPa)	95,000 (655)	105,000 (724)	95,000 (655)	–
Shear strength, psi (MPa)	31,000 (214)	38,000 (262)	31,000 (214)	–
Solidification point, °C	380.6	380.4	380.6	416
Solidification point, °F	717.1	716.7	717.1	780
Solidification shrinkage, in./ft (cm/m)	.14 (1.2)	.14 (1.2)	.14 (1.2)	–
Specific gravity	6.6	6.7	6.6	7.1
Specific heat, cal/g/°C [J/(g·°C)]	0.10 (0.42)	0.10 (0.42)	0.10 (0.42)	–
Thermal conductivity, cal/sec/cm²/cm/°C [J/(s·cm·°C)] at 18°C	0.27 (1.13)	0.26 (1.09)	0.27 (1.13)	0.205 (0.86)
Transverse deflection, in.	0.27	0.16	0.27	–
Thermal expansion per °C	.0000274	.0000274	.0000274	.000027
Thermal expansion per °F	.0000152	.0000152	.0000152	.000015
Specific weight, lb/in.³ (g/cm³)	.24 (6.6)	.24 (6.6)	.24 (6.6)	.26 (7.2)

zinc-copper-titanium alloy has become the dominant wrought-zinc alloy for applications demanding superior performance. Composition of this alloy is 0.4–0.8% copper, 0.08–0.16% titanium, 0.3% max lead, 0.015% max iron, 0.01% max manganese, 0.01% max cadmium, 0.02% max chromium, and the balance Special High-Grade Zinc.

The conventional grades of slab or ingot zinc are rolled into sheets, strips, ribbons, foils, plates, or rods for a wide variety of uses. The zinc alloy also may be continuous-cast into rod or bar. Because of its properties, rolled zinc can be easily worked into various shapes and forms by common fabricating methods, including stamping, forming, and spinning. It can be polished and lacquered to retain its natural bright luster, or it can be plated or painted for other finishing effects. It can also be left to weather naturally, in which case it forms an attractive, nonstaining patina, making it especially suited for architectural applications.

The zinc-copper-titanium wrought-zinc alloy has all of these properties and on a comparative basis is stronger and more dent-resistant than some other metals of the same thickness. Tensile strength is 29,000 psi with the grain and 40,000 psi across the grain. Elongation over 2 in. is 26% with the grain and 14% across the grain. Rockwell hardness is 60–76. It is easily soldered, is nonmagnetic, and has an electrical conductivity equal to that of brass. It is produced in coils ranging 0.003–0.250 in. thick and $\frac{1}{4}$–19 in. wide, and in sheets ranging 0.020–0.125 in. thick, 20–60 in. wide, and to 120 in. long.

[A. L. PONIKVAR]

Bibliography: Comico Ltd., *Data and Facts: Zinc Alloy Die Castings;* N. Laugle, *Casting High-Strength Zinc in Permanent Molds*, 1968; C. H. Matthewson (ed.), *Zinc*, ACS Monogr. no. 142, 1959; D. O. Rausch et al. (eds.), *Lead–Zinc Update*, Society of Mining Engineers, 1977; Zinc Institute, *Zinc Die Casting—Molten Metal to Finished Part—Direct*.

Zone refining

One of a number of techniques used in the preparation of high-purity materials. The technique is capable of producing very low impurity levels, namely, parts per million or less in a wide range of materials, including metals, alloys, intermetallic

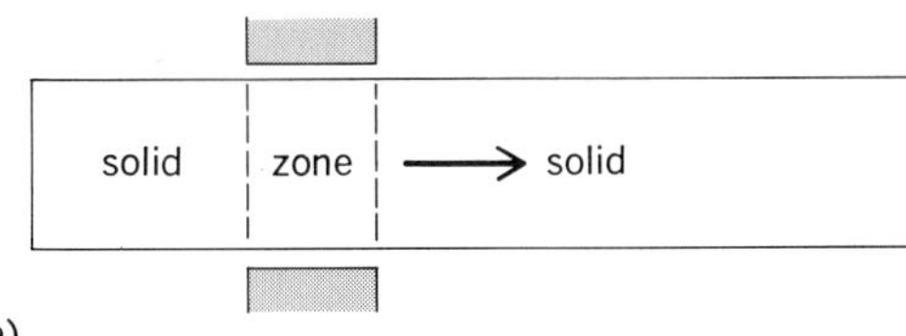

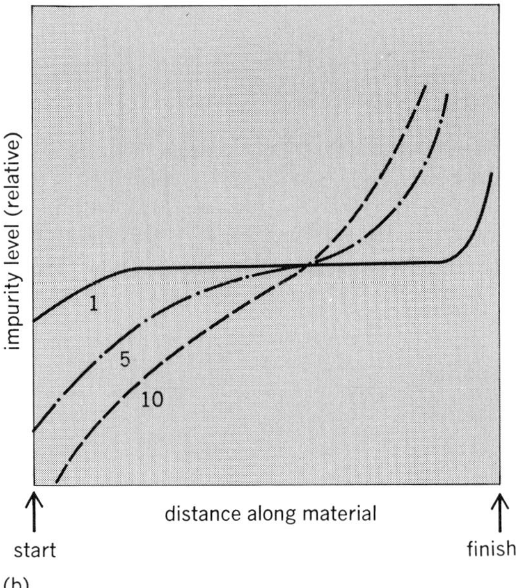

Fig. 1. Zone refining. (*a*) Passage of a molten zone along the material to be purified. (*b*) Effect of 1, 5, and 10 zone passes on the impurity distribution along the material.

compounds, semiconductors, and inorganic and organic chemical compounds. In principle, zone refining takes advantage of the fact that the solubility level of an impurity is different in the liquid and solid phases of the material being purified; it is therefore possible to segregate or redistribute an impurity within the material of interest. In practice, a narrow molten zone is moved slowly along the complete length of the specimen in order to bring about the impurity segregation.

Impurity atoms either raise or lower the melting point of the host material. There is also a difference in the concentration of the impurity in the liquid phase and in the solid phase when the liquid and solid exist together in equilibrium. In zone refining, advantage is taken of this difference, and the impurity atoms are gradually segregated to one end of the starting material. To do this, a molten zone is passed from one end of the impure material to the other, as in Fig. 1*a*, and the process is repeated several times. The end to which the impurities are segregated depends on whether the impurity raises or lowers the melting point of the pure material; a lowering of the melting point is more common, in which case impurities are moved in the direction of travel of the molten zone.

The extent or effectiveness of the impurity distribution is controlled by the distribution coefficient k, defined as the ratio of the impurity concentration in the solid to that in the liquid at a given temperature. The effect of multiple zone passes (in the same direction) on impurity content along the material is illustrated in Fig. 1*b*.

Zone leveling is a particular form of zone refining. Here zone passes are made in alternate directions so that a uniform (homogeneous) distribution of impurities (that is, alloying) is produced along the material. In this way desired levels of deliberately added impurities may be achieved in this form of zone refining.

In commercial application the object is to optimize the variables in order to produce the required amount of material having the necessary level of purity with a minimum of time and expense. Primary variables are zone length, interzone spacing in a multiple zone batch unit, number of zone passes, and travel rate of the zone along the material. The choice of zone length is governed by practical considerations and by the physical properties of the material; a narrow molten zone with a sharp demarcation between liquid and solid phases is desired. Small numbers of zone passes reduce the total time required but at the expense of the usable length of material. For large numbers of zone passes, the best segregation occurs for small zone lengths. Increasing the travel speed decreases the time per pass, but there is a concomitant decrease in the efficiency of segregation. Typical conditions for the zone refining of aluminum to a purity better than 99.9995% are a zone speed of 1 in./hr, 30 zone passes, and a zone length to material length ratio of 0.1.

The many designs of zone-refining apparatus center on factors such as the form of heating (or cooling), the form of container (material or shape), the travel mechanism, stirring of the liquid in the molten zone, and protective atmosphere. Common modes of heat input for material in a container are resistance heating and induction heating (3 Hz to 6 MHz in the 5- to 50-kW power output range). When the material to be purified melts below room tem-

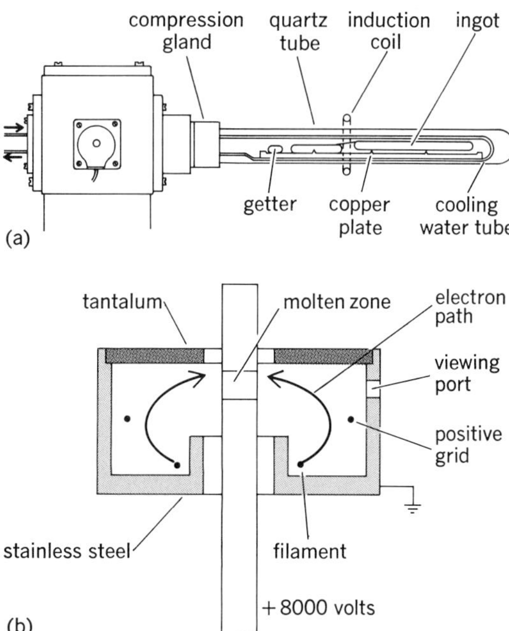

Fig. 2. Industrial techniques. (*a*) Zone refining on a water-cooled hearth. (*b*) Electron-beam zone-refining arrangement. (*From R. F. Bunshah, ed., Techniques of Metals Research, vol. 1, copyright © 1968 by John Wiley and Sons, Inc.; used with permission*)

perature, it is necessary to zone-refine in a refrigerated atmosphere or to traverse a solid zone along the bar or column of liquid material. Container materials in use are plastics, metals, glass, graphite, and oxides of aluminum, zirconium, magnesium, silicon, and beryllium. The choice becomes more restrictive as the temperature of the molten zone increases. The upper limit of melting point for metals and alloys to be zone refined in a container is ~1500°C, since excessive reaction occurs with the container above this temperature. Low-melting-point organic compounds are usually purified in glass containers or plastics. High-melting-point or reactive metals, compounds, and semiconductors may be zone-refined in a container if the latter is water-cooled, as in Fig. 2*a*. *See* RESISTANCE HEATING.

In floating zone refining, a molten zone is held in place by its own surface tension between two vertical colinear rods. Since no container is needed, contamination problems are avoided. The most common form of heating is electron bombardment, as in Fig. 2*b*, or induction heating. In this way refractory metals and compounds can be zone-refined. *See* INDUCTION HEATING.

The importance of the control and improvement of purity lies in the fact that, in most cases, the behavior of a material is a direct function of impurity content; the material behavior is manifested in mechanical, electrical, chemical, thermal, and optical properties. *See* METALLURGY.

[ALAN LAWLEY]

Bibliography: R. Bakish and O. Winkler, *Vacuum Metallurgy*, 1971; R. F. Bunshah (ed.), *Techniques of Metals Research*, vol. 1, 1968; W. G. Pfann and D. T. Hawkins, *Zone Melting*, 1966, reprint 1978; H. Schildknecht, *Zone Melting*, 1966.

Contributors

List of Contributors

A

Adams, Prof. Douglas P. *Department of Mechanical Engineering, Massachusetts Institute of Technology.* ACCELERATION ANALYSIS; CAM MECHANISM; FORCE ANALYSIS; PANTOGRAPH; UNIVERSAL JOINT; other articles.

Alger, Prof. Philip L. *Deceased; formerly, Consulting Professor of Electrical Engineering, Rensselaer Polytechnic Institute.* MAGNET WIRE (coauthored).

Allan, William. *Dean (retired), School of Engineering, City College of City University of New York.* HYDRAULICS.

Altamuro, Vincent M. *President, Management Research Consultants, Yonkers, New York.* CRITICAL PATH METHOD (CPM); OPERATOR TRAINING.

Anderson, Dr. Paul M. *Department of Electrical and Computer Engineering, Arizona State University.* SPEED REGULATION; VOLTAGE REGULATION.

Antoni, Dr. Charles M. *Department of Civil Engineering, Syracuse University.* GYPSUM PLANK; ROOF CONSTRUCTION; STORAGE TANK; STRUCTURAL MATERIALS; STRUCTURES; other articles.

Antonides, Lloyd E. *Development Engineer, Fenix and Scisson, Inc., Tulsa, OK.* MINING MACHINERY.

Apen, John R. *Bell Laboratories, Norcross, GA.* COAXIAL CABLE.

Applegate, Charles E. *Consulting Engineer, Weston, MA.* CIRCUIT TESTING (ELECTRICITY); ELECTRICAL RESISTANCE; INSULATION RESISTANCE TESTING; OHM'S LAW.

Archie, G. E. *Manager (retired), Petroleum Engineering Research, Shell Development Company, Houston, TX.* PETROLEUM PROSPECTING (in part).

Ash, Col. Simon H. *Consulting Civil and Mining Engineer, Santa Rosa, CA.* MINING OPERATING FACILITIES (in part).

Avallone, Prof. Eugene A. *Department of Mechanical Engineering, City University of New York.* ENGINEERING, SOCIAL IMPLICATIONS OF.

Averbach, Prof. Benjamin L. *Department of Materials Science and Engineering, Massachusetts Institute of Technology.* HEAT TREATMENT (METALLURGY).

B

Baer, Prof. Charles J. *Department of Mechanical Engineering, University of Kansas.* DRAFTING (in part); ENGINEERING DRAWING; PICTORIAL DRAWING.

Baily, Frederick G. *Large Steam Turbine-Generator Department, General Electric Company, Schenectady, NY.* STEAM TURBINE; TURBINE.

Balderes, Dr. T. *Engineering Specialist, Grumman Aerospace Corporation, Bethpage, NY.* FINITE ELEMENT METHOD.

Banfield, Dr. A. F. *Behre Dolbear and Company, Inc., New York, NY.* PROSPECTING.

Banner, Dr. Donald W. *John Marshall Law School, Chicago, IL.* PATENT.

Barker, Dr. Joseph W. *Chairman of the Board (retired), Research Corporation, New York, NY.* ENGINEERING.

Barnitz, Edward S. *Consultant in Vacuum Engineering, Rochester, NY.* VACUUM PUMP.

Barry, Brother B. Austin. *Civil Engineering Department, Manhattan College.* ALIDADE; LEVEL (SURVEYING); STADIA; SURVEYING; TRANSIT.

Baumeister, Prof. Theodore. *Deceased; formerly, Consulting Engineer; Stevens Professor of Mechanical Engineering, Emeritus, Columbia University; Editor in Chief, "Standard Handbook for Mechanical Engineers."* BOILER; CARNOT CYCLE; DIESEL ENGINE; ENERGY CONVERSION; STEAM-GENERATING UNIT; WATERPOWER; other articles.

Beatty, Ronald L. *Oak Ridge National Laboratory, Oak Ridge, TN.* NUCLEAR FUELS.

Beaty, H. Wayne. *"Electrical World," McGraw-Hill Publishing Company, New York, NY.* CONDUCTOR (ELECTRICITY).

Beggs, H. L. *Consultant, Fluid Processing Division, Selas Corporation of America, Bresher, PA.* FURNACE CONSTRUCTION (validated).

Beistline, Dr. Earl H. *Dean, College of Earth Sciences and Mineral Industry, University of Alaska.* PLACER MINING.

Bell, Morton A. *American Air Filter Company, Inc., New York, NY.* AIR FILTER.

Bell, Prof. Norman R. *Department of Electrical Engineering, North Carolina State University.* INDUCTION COIL.

Bennett, Dr. Orus L. *Supervisory Soil Scientist, Science and Education Administration, U.S. Department of Agriculture; Department of Plant Science, West Virginia University.* STRIP MINING.

Bennof, Anne O. *Manager, Educational and Informational Services, Association of American Railroads, Washington, DC.* RAILROAD ENGINEERING (in part).

Berg, Dr. Clyde. *Clyde Berg and Associates, Long Beach, CA.* SOLIDS PUMP.

Bernstein, Dr. I. M. *Department of Metallurgy and Materials Science, Carnegie-Mellon University.* EMBRITTLEMENT.

Bever, Prof. Michael B. *Department of Materials Science and Engineering, Massachusetts Institute of Technology.* METALLOGRAPHY (coauthored); METALLURGY; SURFACE HARDENING OF STEEL (coauthored).

Bewley, Dr. Loyal V. *Dean (retired), College of Engineering, Lehigh University.* HYSTERESIS MOTOR; RELUCTANCE MOTOR.

Biegel, Prof. John E. *Department of Industrial Engineering, Syracuse University.* PRODUCTION PLANNING.

Billington, Dr. Douglas S. *Senior Staff Advisor for Materials Science, Metals and Ceramics Division, Oak Ridge National Laboratory, Oak Ridge, TN.* RADIATION DAMAGE TO MATERIALS.

Black, Prof. Paul H. *Department of Mechanical Engineering, Ohio University.* ALLOWANCE; FORCE FIT; HARMONIC SPEED CHANGER; PACKING; PRESSURE SEAL; RIVET; other articles.

Blackburn, J. Lewis. *Consulting Engineer, Bothell, WA.* ELECTRIC PROTECTIVE DEVICES; RELAY.

Blair, Lewis N. *Anaconda Company, Butte, MT.* ENVIRONMENTAL ENGINEERING (coauthored).

Bobart, George F. *Manager, Induction Heating and Ultrasonic Cleaning Department, Westinghouse Electric Corporation, Sykesville, MD.* INDUCTION HEATING.

Bolster, R. H. *Deceased; formerly, Vice President, Clark Equipment Company, Buchanan, MI.* TRUCK AXLE.

Bolt, Prof. J. A. *Department of Mechanical Engineering, University of Michigan.* CARBURETOR; FUEL SYSTEM; GAS TURBINE—all validated.

Boshkov, Prof. Stefan H. *Henry Krumb School of Mines, Columbia University.* QUARRYING; ROCK BURST.

Boston, Dr. Orlan W. *Professor Emeritus, College of Engineering, University of Michigan.* ULTRASONIC MACHINING.

Bouton, George M. *Bell Telephone Laboratories, Murray Hill, NJ.* SOLDERING.

Bowman, Prof. Harry L. *Deceased; formerly, Dean, Drexel Institute of Technology, Philadelphia, PA.* TRUSS.

Bowman, Rush A. *Consulting Engineer, Memphis, TN.* COMPARATOR; GAGES; OPTICAL FLAT.

Bowman, Waldo G. *Editor (retired), "Engineering News Record."* HARDNESS SCALES; HOOKE'S LAW; LOADS, DYNAMIC; SAFETY FACTOR; other articles—all validated.

Bradley, Prof. William W. *Department of Civil Engineering and Institute of Colloid and Surface Science, Clarkson College of Technology.* METAL COATINGS.

Breinan, Dr. Edward M. *Materials Science Laboratory, United Technologies Research Center, East Hartford, CT.* LASER WELDING; WELDING AND CUTTING OF METALS (in part).

Breuer, Glenn D. *Manager, HVDC Transmission Engineering, General Electric Company, Schenectady, NY.* LIGHTNING AND SURGE PROTECTION; SURGE ARRESTER.

Brigham, Harry S. *Executive Vice President, Dixilyn Corporation, New Orleans, LA.* OIL AND GAS WELL COMPLETION.

Bright, Prof. James R. *(Retired) Graduated School of Business, University of Texas, Austin.* AUTOMATION.

Broadbent, Carl D. *Manager, Mining Engineering, Kennecott Copper Corporation, Salt Lake City, UT.* OPEN-PIT MINING.

Brown, Ronald D. *Stanley Consultants, Muscatine, IA.* RURAL ELECTRIFICATION (coauthored).

Bunshah, Dr. Rointan F. *Department of Materials, University of California, Los Angeles.* VACUUM METALLURGY.

Burnett, Peter G. *Petroleum Engineer, Chicago, IL.* OIL AND GAS STORAGE.

C

Cammack, R. L. *Manager, Plant Services Department, Westinghouse Corporation, Pittsburgh, PA.* JIG, FIXTURE, AND DIE DESIGN.

Campbell, Harold E. *Senior Engineer, Power Distribution Systems Engineering, General Electric Company, Schenectady, NY.* ELECTRIC-DISCHARGE MACHINING.

Carson, Robert W. *Formerly, Managing Editor, "Product Engineering," New York, NY.* CLADDING.

Carter, Archie N. *Baun, Carter & Associates, Inc., Minneapolis, MN.* HIGHWAY ENGINEERING; PAVEMENT; TRAFFIC ENGINEERING; TRANSPORTATION ENGINEERING.

Cheatham, Dr. J. B., Jr. *Professor of Mechanical Engineering and Chairman of the Department of Mechanical and Aerospace Engineering and Materials Science, Rice University.* OIL AND GAS WELL DRILLING.

Chellis, Robert D. *Deceased; formerly, Structural Engineer, Wellesley Hills, MA.* CAISSON FOUNDATION; RETAINING WALL.

Chen, Dr. Mo-Shing. *Department of Electrical Engineering, University of Texas, Arlington.* ELECTRIC HEATING.

Chenault, Roy L. *Chief Research Engineer (retired), Oilwell Division, U.S. Steel Corporation, Dallas, TX.* OIL AND GAS FIELD EXPLOITATION.

Chinitz, Dr. Wallace. *Department of Mechanical Engineering, Cooper Union, New York, NY.* ROTARY ENGINE.

Christie, Richard W. *Hardesty & Hanover, New York, NY.* BRIDGE (coauthored).

Clarke, John H. *Engineering Design Consultant, Air Conditioning and Power, Films-Packaging Division, Union Carbide Corporation, Chicago, IL.* VENTILATION.

Clements, John A. *Director, Corporate Product Evaluation, Gilette Company, Boston, MA.* CONTROL CHART; INSPECTION AND TESTING; QUALITY CONTROL.

Cleveland, Prof. L. F. *Department of Electrical Engineering, Northeastern University.* DIRECT-CURRENT MOTOR.

Conrad, Prof. Albert G. *Dean Emeritus and Professor of Electrical Engineering, College of Engineering, University of California, Santa Barbara.* ELECTRICAL ENGINEERING; INDUCTION MOTOR.

Cooper, Prof. W. Charles. *Department of Metallurgical Engineering, Queens University, Kingston, Canada.* HYDROMETALLURGY; PYROMETALLURGY, NONFERROUS.

Corlett, A. V. *Deceased; formerly, Mining Engineer, Kingston, Ontario, Canada.* MINING SAFETY; UNDERGROUND MINING.

Coulter, Earl E. *Manager, Mechanical Technology, Babcock and Wilcox Company, Barberton, OH.* STEAM SEPARATOR.

Crentz, William L. *Assistant Director, Energy, Bureau of Mines, U.S. Department of the Interior.* MINING OPERATING FACILITIES (coauthored).

Creson, William K. *Consulting Engineer, Ross Gear and Tool Company, Inc., Lafayette, IN.* ACKERMAN STEERING; AUTOMOTIVE STEERING.

Crittenden, Dr. Charles V. *Geographer, Economic Development Administration, U.S. Department of Commerce.* GAS FIELD AND GAS WELL.

Crossley, Dr. F. R. E. *Department of Mechanical Engineering, University of Massachusetts.* EFFICIENCY.

Crouse, William H. *Consulting Editor, Automotive Books, McGraw-Hill Book Company, New York, NY.* AUTOMOTIVE BRAKE (in part); AUTOMOTIVE SUSPENSION (in part); AUTOMOTIVE VOLTAGE REGULATOR; JOINT (STRUCTURES).

Cruickshank, Dr. Michael J. *Geological Survey, U.S. Department of the Interior, Reston, VA.* MARINE MINING.

Cumming, Donald C. *Headquarters Manufacturing Laboratory, Westinghouse Electric Corporation, Pittsburgh, PA.* ASSEMBLY METHODS.

D

Dallas, Daniel B. *Society of Manufacturing Engineering, Dearborn, MI.* MANUFACTURING ENGINEERING.

DeFacio, Prof. Brian. *Ames Laboratory, Iowa State University.* ENERGY; STATICS.

Dempsey, Dr. Jerry E. *President, Borg-Warner Corporation, Chicago, IL.* HEAT PUMP (in part).

Den Hartog, Prof. J. P. *(Retired) Department of Mechanical Engineering, Massachusetts Institute of Technology.* MECHANICAL VIBRATION (coauthored).

Derge, Prof. Gerhard. *Department of Metallurgy and Material Science, Carnegie-Mellon University.* FERROALLOY; STEEL MANUFACTURE (in part).

Deutscher, Prof. Guy. *Department of Physics, Tel Aviv University, Israel.* SUPERCONDUCTING DEVICES.

Digel, William. *Marketing Communications Department, E.I. du Pont de Nemours & Co., Wilmington, DE.* BLASTING.

Dodge, Dr. Donald D. *Principal Staff Engineer, Engineering and Research, Ford Motor Company, Dearborn, MI.* NONDESTRUCTIVE TESTING.

Dogramaci, Prof. Ali. *Department of Industrial and Management Engineering, Columbia University.* INVENTORY CONTROL.

Doscher, Dr. Todd M. *Department of Petroleum Engineering, University of Southern California.* PETROLEUM ENGINEERING; PETROLEUM ENHANCED RECOVERY; PETROLEUM RESERVES; PETROLEUM RESERVOIR ENGINEERING.

Duby, Prof. Paul F. *Henry Krumb School of Mines, Columbia University.* ELECTROMETALLURGY.

Duwez, Dr. Pol E. *W. M. Keck Laboratory of Engineering Materials, California Institute of Technology.* METALLIC GLASSES.

Dym, Prof. C. L. *Department of Civil Engineering, University of Massachusetts.* MECHANICAL VIBRATION (coauthored).

E

Eckels, Dr. Arthur R. *Department of Electrical Engineering, North Carolina State University.* ARMATURE; COMMUTATION; DYNAMO; ELECTRICAL DEGREE; REGENER-

ATIVE BRAKING; WINDINGS IN ELECTRIC MACHINERY; other articles.

Ellis, Dr. A. J. *Director, Chemistry Division, Department of Scientific and Industrial Research, Petone, New Zealand.* GEOTHERMAL POWER (in part).

Ely, F. G. *(Retired) Research and Development Division, Babcock and Wilcox Company.* RAW WATER.

Erdman, Dr. Arthur G. *Department of Mechanical Engineering, University of Minnesota.* BELT DRIVE (coauthored).

Everetts, Dr. John, Jr. *Professor of Architectural Engineering, Pennsylvania State University.* DEHUMIDIFIER.

F

Faw, Prof. Richard E. *Head, Department of Nuclear Engineering, Kansas State University.* RADIATION SHIELDING.

Felver, Prof. Richard I. *Designer, Carnegie Institute of Technology.* PRODUCT DESIGN; PRODUCTION ENGINEERING.

Fischer, Harold. *General Motors Corporation, Flint, MI.* DIFFERENTIAL.

Fischer, Henry W. *Hardesty & Hanover, New York, NY.* BRIDGE (coauthored).

Fisher, Prof. Russell A. *Department of Physics, Northwestern University.* KINETICS (CLASSICAL MECHANICS).

Fisk, Prof. Nelson S. *Department of Civil Engineering, Columbia University.* COUPLE; MOMENT OF INERTIA; TORQUE.

Fletcher, R. E. *International Manager, Spicer Transmission Division, Dana Corporation, Toledo, OH.* TRUCK TRANSMISSION.

Floe, Prof. Carl F. *Department of Metallurgy, Massachusetts Institute of Technology.* SURFACE HARDENING OF STEEL (coauthored).

Fraiman, Dr. Nelson. *Department of Industrial Engineering, Columbia University.* GANTT CHART.

Freeman, Prof. Ralph L. *(Retired) Department of Mechanical Engineering, Iowa State College.* EXTRUSION; SPINNING (METALS).

Fremed, Raymond F. *Burson-Marsteller Associates. New York, NY.* HEAT EXCHANGER.

Freudenstein, Dr. Ferdinand. *Department of Mechanical Engineering, Columbia University.* FOUR-BAR LINKAGE; MECHANISM.

Fuhrman, Dr. Ralph E. *Black and Veatch, Consulting Engineers, Washington, DC.* WATER POLLUTION.

Fuller, Prof. Dudley D. *Department of Mechanical Engineering, Columbia University.* ANTIFRICTION BEARING.

G

Gambs, Gerard C. *Vice President, Ford, Bacon, & Davis, Inc., New York, NY.* ENERGY SOURCES.

Gaylord, Prof. Charles N. *Chairman, Department of Civil Engineering, University of Virginia.* ARCH; FLOOR CONSTRUCTION.

Gibson, Dr. John E. *Dean of Engineering, Oakland University.* CLUTCH (coauthored).

Giese, Raymond. *Department of Mechanical Engineering, University of Minnesota.* BELT DRIVE (coauthored).

Gonser, Dr. Bruce W. *Battelle Memorial Institute, Columbus, OH.* TIN ALLOYS.

Goodheart, Prof. Clarence F. *Department of Electrical Engineering, Union College.* CIRCUIT (ELECTRICITY); OPEN CIRCUIT; PARALLEL CIRCUIT.

Gordon, Dr. William E. *Associate Professor of Physical Chemistry, Pennsylvania State University; and Consultant.* EXPLOSIVE FORMING

Gray, G. Ronald. *Director, Syncrude Canada, Ltd., Edmonton, Alberta, Canada.* OIL MINING (in part).

Green, W. E. *Clark Equipment Company, Buchanan, MI.* TRUCK AXLE.

Gulbransen, Dr. Earl A. *Department of Metallurgical and Materials Engineering, University of Pittsburgh.* CORROSION (in part).

H

Haber, Bernard. *Hardesty & Hanover, New York, NY.* BRIDGE (coauthored).

Haines, John E. *Deceased; formerly, Vice President, Minneapolis-Honeywell Regulator Company.* HUMIDISTAT.

Hale, Dr. Harry P. *Department of Mechanical Engineering, Wayne State University.* BRAKE.

Hardesty, Egbert R. *Hardesty & Hanover, New York, NY.* BRIDGE (coauthored).

Hardgrove, Ralph M. *Sales Engineer, Stock Equipment Company, Cleveland, OH.* CRUSHING AND PULVERIZING; GAS FURNACE; GRINDING MILL; PEBBLE MILL; ROLL MILL; TUMBLING MILL.

Harris, E. J. *Assistant Chief, Health and Safety Technical Support Center, Bureau of Mines, U.S. Department of the Interior, Pittsburgh, PA.* MINING OPERATING FACILITIES (in part).

Hartwig, Dr. William H. *Department of Electrical Engineering, University of Texas, Austin.* ELECTRIC ENERGY MEASUREMENT; ELECTRIC POWER MEASUREMENT.

Hassialis, M.D. *Krumb School of Mines, Columbia University.* MINING OPERATING FACILITIES (in part).

Hauf, Harold D. *Consulting Architect, Sun City, AZ.* ARCHITECTURAL ENGINEERING.

Hausner, Henry H. *Consulting Engineer, New York, NY.* CERMET.

Hayes, William C. *Editor in Chief, "Electrical World," McGraw-Hill Publications Company, New York, NY.* ELECTRIC POWER SYSTEMS.

Haynes, David O. *Deceased; formerly, Consulting Industrial Engineer, Tucson, AZ.* DERRICK.

Hazen, Richard. *Hazen and Sawyer, Consulting Engineers, New York, NY.* RESERVOIR; WATER SUPPLY ENGINEERING; WATER TREATMENT.

Hearmon, R. F. S. *Formerly, Timber Mechanics Section, Forest Products Research Laboratory, Princes Risborough, Bucks, England.* ELASTICITY; PLASTICITY.

Hedberg, Dr. Hollis D. *Department of Geology, Princeton University.* PETROLEUM PROSPECTING (in part).

Hedley, William J. *Assistant Vice President (retired), Norfolk and Western Railway, St. Louis, MO.* RAILROAD ENGINEERING (in part).

Herman, C. J. *Consulting Engineer, Advanced Processes and Products, Insulating Materials Department, General Electric Company, Schenectady, NY.* MAGNET WIRE (coauthored).

Heronemus, Prof. William. *Department of Engineering, University of Massachusetts.* WIND POWER.

Hershleder, William. *Consulting Construction Engineer, New York, NY.* CONSTRUCTION ENGINEERING; CONSTRUCTION METHODS; TUNNEL; WELLPOINT SYSTEMS.

Hicks, Dr. Philip E. *President, Hicks & Associates, Consulting Industrial Engineers, Orlando, FL.* PERFORMANCE RATING.

Hicks, Tyler G. *Publisher, Professional and Reference Book Division, McGraw-Hill Book Company, New York, NY.* COMPRESSOR.

Hill, James E. *Assistant Director (retired), Mining Research, U.S. Bureau of Mines.* MINING EXCAVATION.

Hingorani, Dr. Narain G. *Electric Power Research Institute, Palo Alto, CA.* ELECTRIC POWER SUBSTATION.

Holzman, Albert G. *Professor and Chairman, Department of Industrial Engineering, Systems Management Engineering and Operations Research, School of Engineering, University of Pittsburgh.* INDUSTRIAL ENGINEERING.

Howe, Prof. Carl E. *Deceased; formerly, Professor Emeritus of Physics, Oberlin College.* ACCELERATION; ROTATIONAL MOTION.

I

Ingram, William T. *Consulting Engineer, Whitestone, NY.* IMHOFF TANK; SANITARY ENGINEERING; SEPTIC TANK; SEWAGE TREATMENT; other articles.

Isbin, Prof. Herbert S. *Department of Chemical Engineering and Materials Science, University of Minnesota.* NUCLEAR POWER; NUCLEAR REACTOR.

J

Jackson, Dr. William D. *President, Energy Conservation, Inc., Washington, DC.* MAGNETOHYDRODYNAMIC POWER GENERATOR.

James, John W. *Vice President, Research, McDonnell and Miller, Inc., Chicago, IL.* STEAM HEATING.

James, Robert S. *McKeesport Campus, Pennsylvania State University.* MINING OPERATING FACILITIES (in part).

Jennings, Dr. Richard L. *Department of Civil Engineering, University of Virginia.* CIVIL ENGINEERING.

Johnson, K. W. *President, Vibra-Grip Corporation, Jamestown, OH.* DYNAMICAL ANALOGIES; SHOCK ISOLATION; VIBRATION DAMPING; VIBRATION ISOLATION; VIBRATION MACHINE.

Johnson, Peter K. *Metal Powder Industries Federation, American Powder Metallurgy Institute, Princeton, NJ.* POWDER METALLURGY.

Johnson, Wendell E. *Consulting Engineer, McLean, VA.* REVETMENT.

Jones, Dr. Edwin C., Jr. *Department of Electrical and Computer Engineering, Iowa State University.* COPPER LOSS.

Judd, Dr. William R. *School of Civil Engineering, Purdue University.* ENGINEERING GEOLOGY.

Just, Prof. Evan. *Department of Mining and Geology, Stanford University.* MINING.

K

Kalhammer, Dr. Fritz. *Electric Power Research Institute, Palo Alto, CA.* ENERGY STORAGE (coauthored).

Kalpakjian, Prof. Serope. *Mechanical and Aerospace Engineering Department, Illinois Institute of Technology.* DRAWING OF METAL; FORGING; METAL FORMING; METAL ROLLING; SHEET-METAL FORMING.

Kaplan, Robert A. *Vice President in Charge of Engineering, Automatic Burner Corporation, Chicago, IL.* OIL BURNER.

Karash, Joseph I. *Manufacturing Engineer (retired), Reliance Electric Company, Cleveland, OH.* TOOLING (in part).

Kaufmann, R. H. *Deceased; formerly, Consultant, Schenectady, NY.* GROUNDING.

Kayan, Prof. Carl F. *Deceased; formerly, Department of Mechanical Engineering, School of Engineering, Columbia University.* REFRIGERATION; REFRIGERATOR; TON OF REFRIGERATION.

Kellogg, Prof. Herbert H. *Stanley-Thompson Professor of Chemical Metallurgy, Krumb School of Mines, Columbia University.* PYROMETALLURGY.

Kerr, Prof. William. *Chairman, Department of Nuclear Engineering, University of Michigan.* NUCLEAR ENGINEERING.

Kessler, George W. *Vice President, Engineering and Technology, Power Generation Division, Babcock and Wilcox Company, Barberton, OH.* AIR HEATER; BOILER ECONOMIZER, CYCLONE FURNACE; FEEDWATER; STEAM TEMPERATURE CONTROL; other articles.

Khan, Dr. Fazlur R. *Partner, Skidmore, Owings & Merrill, Chicago, IL.* BUILDINGS.

Kilgore, Dr. Lee A. *Westinghouse Electric Corporation, East Pittsburgh, PA.* SYNCHRONOUS CAPACITOR.

Kimbark, Dr. Edward W. *Deceased; formerly, Head, Systems Analysis Group, Bonneville Power Administration, Portland, OR.* DIRECT-CURRENT TRANSMISSION; TRANSMISSION LINES (in part).

Kingery, Donald S. *Director, Health and Safety, Research and Testing Center, U.S. Bureau of Mines.* MINING OPERATING FACILITIES (in part).

Kinnard, Dr. Isaac F. *Deceased; formerly, Manager of Engineering, Instrument Department, General Electric Company, Lynn, MA.* INSTRUMENT TRANSFORMER.

Kinnier, Prof. Henry L. *School of Engineering, University of Virginia.* COMPOSITE BEAM.

Kirchmayer, Dr. L. K. *Manager, Advanced System Technology and Planning, Electric Utility Systems Engineering Department, General Electric Company, Schenectady, NY.* ELECTRIC POWER SYSTEMS ENGINEERING.

Koch, Leonard J. *Senior Engineer, Office of the Director, Argonne National Laboratory, Argonne, IL.* ELECTROMAGNETIC PUMPS.

Konz, Dr. Stephan A. *Department of Industrial Engineering, Kansas State University.* WORK MEASUREMENT.

Konzo, Prof. Seichi. *Department of Mechanical Engineering, University of Illinois.* AIR REGISTER; WARM-AIR HEATING SYSTEM.

Koral, Dr. Richard L. *Visiting Associate Professor, Pratt School of Architecture; Coordinator, Apartment House Institute, New York City Community College.* AIR CONDITIONING; DISTRICT HEATING; HUMIDISTAT; RADIATOR; other articles.

Kosow, Dr. Irving L. *Department of Electrical Technology, Staten Island Community College.* ALTERNATING-CURRENT MOTOR (in part); REPULSION MOTOR; UNIVERSAL MOTOR.

Kraehenbuehl, Prof. John O. *Professor Emeritus of Electrical Engineering, University of Illinois, Urbana.* ARC LAMP; INCANDESCENT LAMP.

Krauss, Prof. George. *Colorado School of Mines, Golden.* STEEL.

Krefeld, Prof. William J. *Deceased; formerly, Professor of Civil Engineering, Columbia University.* HARDNESS SCALES; HOOKE'S LAW; LOADS, DYNAMIC; SAFETY FACTOR; WORK STANDARDIZATION.

L

Larsen, Dr. Robert G. *Assistant to the Vice President, Shell Development Company, Emeryville, CA.* LUBRICANT.

Lawley, Dr. Alan. *College of Engineering, Drexel University.* METAL MATRIX COMPOSITE; ZONE REFINING.

Lawrence, Keith W. *Acting Chief, Marine Design Center, Department of the Army, Philadelphia District, U.S. Corps of Engineers.* DREDGE.

Lee, Dr. Thomas H. *Strategic Planning Operation, General Electric Company, Fairfield, CT.* BLOWOUT COIL; BUS-BAR; CIRCUIT BREAKER; ELECTRIC CONTACT; ELECTRIC SWITCH; FUSE (ELECTRICITY).

Lenel, Dr. F. V. *Professor Emeritus, Department of Materials Engineering, Rensselaer Polytechnic Institute.* SINTERING.

Leonard, Dr. Edward F. *Chemical Engineering Department, Columbia University.* PROCESS ENGINEERING (in part).

Leontis, Dr. Thomas E. *Magnesium Research Center, Battelle Columbus Laboratories, Columbus, OH.* MAGNESIUM ALLOYS.

Lesso, Dr. William G. *Department of Mechanical Engineering, University of Texas, Austin.* OPERATIONS RESEARCH.

Levin, Dr. Franklin K. *Esso Production Research Company, Houston, TX.* SEISMIC EXPLORATION FOR OIL AND GAS.

Limpel, Eugene J. *Consulting Electrical Engineer, Eagle River, WI.* FLASH WELDING; SPOT WELDING.

Lindberg, Prof. Roy A. *Department of Mechanical Engineering, University of Wisconsin.* MASS PRODUCTION; TOOLING (in part).

Linderoth, Prof. L. Sigfred, Jr. *Department of Mechanical Engineering, Duke University.* BOLTED JOINT; FLYWHEEL; SHOCK ABSORBER; SPRING (MACHINES); TORSION BAR; other articles.

Lutz, Dr. Raymond P. *School of Management and Administrative Science, University of Texas, Dallas.* INDUSTRIAL COST CONTROL.

Luzadder, Prof. Warren J. *Department of Engineering Graphics, Purdue University.* BOLT; NUT; SCREW THREADS; WASHER.

M

McClellan, Leslie N. *Consulting Engineer, Engineering Consultants, Inc., Denver, CO.* PIPELINE.

McClusky, V. G. *General Electric Company, Cleveland, OH.* SODIUM-VAPOR LAMP.

MacCoull, Neil *Deceased; formerly, Lecturer in Mechanical Engineering, Columbia University.* AUTOMOTIVE SUSPENSION (in part); COMPRESSION RATIO; ENGINE MANIFOLD; FUEL INJECTION; INTERNAL COMBUSTION ENGINE; VOLUMETRIC EFFICIENCY.

McCright, Dr. R. D. *Corrosion Center, Department of Metallurgical Engineering, Ohio State University.* CORROSION (in part).

McFarland, Forest R. *Engineering Consultant, Flint, MI.* AUTOMOTIVE TRANSMISSION; AXLE; GEAR LOADING; OVERDRIVE.

McGannon, Harold E. *Technical Editor, Research and Technology Division, United States Steel Corporation, Pittsburgh, PA.* IRON ALLOYS.

McPartland, Joseph F. *Editorial Director, "Electrical Construction and Maintenance," McGraw-Hill Publications Company, New York, NY.* BRANCH CIRCUIT (coauthored); ELECTRICAL CODES; WIRING.

Makulec, Alfred. *Product Planning and Application Specialist, Large Lamp Department, General Electric Company, Cleveland, OH.* FLUORESCENT LAMP; ULTRAVIOLET LAMP.

Mangion, Charles. *Technology Laboratory, Science and Technology Division, Systems Group, TRW, Inc., Redondo Beach, CA.* HYDRAULIC ACTUATOR.

Mann, Prof. Robert W. *Department of Mechanical Engineering, Massachusetts Institute of Technology.* ENGINEERING DESIGN; LAYOUT DRAWING; SCHEMATIC DRAWING; WIRING DIAGRAM.

Manning, Dr. Kenneth V. *Professor Emeritus, Pennsylvania State University.* INDUCTANCE.

Markus, John. *Consultant, Sunnyvale, CA.* ALTERNATOR; LAMP; QUADRATURE; SYNCHROSCOPE.

Mason, Willard F. *Consulting Engineer, Mason Engineering Company, Inc.* SLIP RINGS (in part).

Maynard, Dr. Harold B. *President, Maynard Research Council, Inc., Pittsburgh, PA.* TECHNOLOGY (validated).

Meijer, Dr. Roelof J. *Assistant Director of Research, Philips Research Laboratories, Eindhoven, Netherlands.* STIRLING ENGINE.

Meisel, Dr. Jerome. *Department of Electrical and Computer Engineering, Wayne State University.* ELECTROMAGNET; SOLENOID (ELECTRICITY).

Melman, Prof. Seymour. *Department of Industrial and Management Engineering, Columbia University.* PRODUCTIVITY.

Merritt, Frederick S. *Consulting Engineer, West Palm Beach, FL.* CONCRETE BEAM; CONCRETE COLUMN; CONCRETE SLAB; PRECAST CONCRETE; PRESTRESSED CONCRETE.

Miles, Lawrence D. *Engineering Consultant, Easton, MD.* VALUE ENGINEERING.

Miller, R. A. *Engineering Department, Babcock and Wilcox, Barberton, OH.* REHEATING.

Mississippi River Commission. RIVER ENGINEERING (coauthored).

Mock, Frank C. *Deceased; formerly, Bendix Products Division, Bendix Aviation Corporation, South Bend, IN.* CARBURETOR; FUEL SYSTEM.

Moder, Dr. Joseph J. *Chairman, Department of Management Science, University of Miami.* PERT.

Morgan, Dr. Karl Z. *Neely Professor, School of Nuclear Engineering, Georgia Institute of Technology.* RADIOACTIVE WASTE MANAGEMENT (in part).

Muffler, Dr. L. J. Patrick *Geologist, Branch of Field Geochemistry and Petrology, Geological Survey, U.S. Department of the Interior, Menlo Park, CA.* GEOTHERMAL POWER (in part).

Muther, Richard. *Executive Director, Richard Muther & Associates, Inc., Kansas City, MO.* MATERIALS HANDLING; PLANT FACILITIES (INDUSTRY); PRODUCTION METHODS.

N

Nedelsky, Prof. Leo. *Department of Physical Science, University of Chicago.* CONSERVATION OF ENERGY (validated); DENSITY; WORK.

Neubecker, T. F. *Manager, Engineering, High Intensity and Quartz Lamp Department, General Electric Company, Twinsburg, OH.* MERCURY-VAPOR LAMP; VAPOR LAMP.

Niebel, Benjamin W. *Professor Emeritus of Industrial Engineering, Pennsylvania State University; Industrial Engineering Consultant, State College, PA.* METHODS ENGINEERING; PROCESS ENGINEERING (in part).

Nippes, Prof. Ernest F. *Department of Materials Engineering, Rensselaer Polytechnic Institute.* RESISTANCE WELDING.

O

O'Connor, Dr. Donald J. *Department of Civil Engineering, Manhattan College.* SEWAGE DISPOSAL.

Old, Dr. Bruce S. *Senior Vice President, Arthur D. Little, Inc., Cambridge, MA.* PRESSURIZED BLAST FURNACE.

Oldenburger, Prof. Rufus. *Director, Automatic Control Center, Purdue University.* GOVERNOR; LINKAGE (MECHANISM).

Olishifski, Julian B. *Industrial Hygiene Consultant, Loss Control Department, Alliance of American Insurers, Chicago, IL.* INDUSTRIAL HEALTH AND SAFETY.

Olmsted, Leonard M. *Utility Consultant and Registered Professional Engineer, South Orange, NJ.* POWER PLANT (coauthored).

Oshinski, William J. *Executive Vice President, Patton Consultants Incorporated, Des Plaines, IL.* WAGE INCENTIVES.

Ott, Paul A., Jr. *Automotive Engineer (retired), Mt. Clemens, MI.* AUTOMOBILE; AUTOMOTIVE BRAKE; AUTOMOTIVE ENGINE; AUTOMOTIVE FRAME; other articles.

Otto, Dr. Norman C. *Ford Motor Research Laboratory, Dearborn, MI.* CATALYTIC CONVERTER.

P

Pake, Dr. George E. *Vice President, Xerox Corporation; General Manager, Xerox Palo Alto Research Center, Palo Alto, CA.* FORCE.

Parks, Prof. Roland D. *Department of Physics and Astronomy, University of Rochester.* MINING OPERATING FACILITIES (in part).

Paul, Prof. Burton *Department of Mechanical Engineering, University of Pennsylvania.* SHAFT BALANCING.

Peaslee, Dr. Robert L. *Vice President, Stainless Steel Division, Wall Colmonoy Corporation, Detroit, MI.* BRAZING.

Peirce, Dr. G. R. *Engineer, Champaign, IL.* INFRARED LAMP.

Perrin, Arthur M. *President, National Conveyors Company, Inc., Fairview, NJ.* BULK-HANDLING MACHINES; CONVEYOR; ELEVATING MACHINES; INDUSTRIAL TRUCKS; POWER SHOVEL; other articles.

Petershack, Dr. Victor D. *Rexnord, Inc., Milwaukee, WI.* CHAIN DRIVE.

Phelan, Prof. Richard M. *Department of Mechanical Systems and Design, Cornell University.* BLOCK AND TACKLE; GEAR DRIVE; HYDRAULIC PRESS; LEVER; MECHANICAL ADVANTAGE; WHEEL AND AXLE; other articles.

Phister, Montgomery, Jr. *Consultant, Santa Fe, NM.* DIGITAL COMPUTER.

Platt, Allison M. *Manager, Nuclear Waste Technology Department, Battelle Pacific Northwest Laboratories, Richland, WA.* RADIOACTIVE WASTE MANAGEMENT (in part).

Ponikvar, A. L. *Manager, Technical Publications, In-*

ternational Lead Zinc Research Organization, Inc., New York, NY. OIL MINING (in part); ROTARY TOOL DRILL; TURBODRILL; ZINC ALLOYS.

Pope, Michael. *Pope, Evans and Robbins, Consulting Engineers, New York, NY.* FLUIDIZED-BED COMBUSTION.

Price, Dr. Harvey S. *Vice President, Intercomp Resource Developing and Engineering, Inc., Houston, TX.* PETROLEUM RESERVOIR MODELS.

Priester, Gayle B. *Consulting Engineer, Baltimore, MD.* COMFORT HEATING.

Pritchard, Tram C. *Lockheed Missiles and Space Company, Sunnydale, CA.* DRAFTING (in part).

Pritchett, Prof. Wilson S. *Senior Project Engineer, Noller Control Systems, Inc., Richmond, CA.* SATURABLE REACTOR.

Pritsker, A. Alan B. *School of Industrial Engineering, Purdue University.* GERT.

Prouty, Dr. L. Fletcher *Senior Director, Corporate Communications, National Railroad Passenger Corporation, Washington, DC.* RAILROAD ENGINEERING (in part).

Pryke, John K. M. *Slocum and Fuller, New York, NY.* CENTRAL HEATING AND COOLING.

Puchstein, Albert F. *Deceased; formerly, Consulting Engineer, Columbus, OH.* ALTERNATING-CURRENT MOTOR (in part); MOTOR.

Pullen, Dr. Keats A., Jr. *Ballistic Research Laboratories, Aberdeen Proving Ground, MD.* TRANSFORMER.

Putz, Dr. Thomas J. *Deceased; formerly, Westinghouse Electric Corporation, Philadelphia, PA.* GAS TURBINE.

Q

Quirin, Edward J. *Consulting Engineer, Besier, Gibble & Quirin, Old Saybrook, CT.* BREAKWATER; COFFERDAM; PIER; SEA WALL; other articles.

R

Rabinowicz, Prof. Ernest. *Department of Mechanical Engineering, Massachusetts Institute of Technology.* FRICTION; WEAR.

Rabinowitz, Dr. Mario. *Senior Scientist, Electrical Systems Division, Electric Power Research Institute, Palo Alto, CA.* ELECTRICAL INSULATION.

Ramey, Dr. Robert L. *Department of Electrical Engineering, University of Virginia.* DIRECT-CURRENT CIRCUIT THEORY; SERIES CIRCUIT; SHORT CIRCUIT.

Rau, James L. *Design Engineer, Ross Gear Division, TRW, Inc., Lafayette, IN.* POWER STEERING.

Reed-Hill, Dr. Robert E. *Department of Metallurgical Science and Engineering, University of Florida.* ALLOY (coauthored).

Reilly, James D. *Vice President, Consolidation Coal Company, Pittsburgh, PA.* COAL MINING.

Reno, Horace T. *Bureau of Mines, U.S. Department of the Interior.* MINING OPERATING FACILITIES (coauthored).

Reynolds, Gardner M. *Dames and Moore, Los Angeles, CA.* FOUNDATIONS; PILE FOUNDATION.

Ricksecker, Ralph E. *Director of Metallurgy, Chase Brass and Copper Company, Cleveland, OH.* COPPER ALLOYS.

Robb, D. D. *D. D. Robb and Associates, Consulting Engineers, Electric Power Systems, Salina, KS.* DIRECT CURRENT.

Robbins, Nathaniel, Jr. *Director of Engineering, Residential Division, Honeywell, Inc., Minneapolis, MN.* THERMOSTAT.

Robertson, Prof. Burtis L. *Professor of Electrical Engineering (retired), University of California, Berkeley.* ELECTRICAL IMPEDANCE; POWER FACTOR; Q (ELECTRICITY); REACTANCE.

Robertson, Dr. Gordon I. *Control Automation, Inc., Princeton, NJ.* INDUSTRIAL ROBOTS.

Rockett, Frank H. *Engineering Consultant, Charlottesville, VA.* AIR BRAKE; BENDING MOMENT; BRASS; CANTILEVER; ENGINE; METAL; STEEL MANUFACTURE; other articles.

Roe, Kenneth A. *President, Burns and Roe, Oradell, NJ.* POWER PLANT (coauthored).

Rogers, Dr. Harry C. *Department of Metallurgical Engineering, Drexel University.* PLASTIC DEFORMATION OF METAL.

Rogowski, Prof. Augustus R. *Department of Mechanical Engineering, Massachusetts Institute of Technology.* COMBUSTION CHAMBER; DISTRIBUTOR; ENGINE COOLING; IGNITION SYSTEM; VALVE TRAIN; VAPOR LOCK; other articles.

Roller, Dr. Duane E. *Deceased; formerly, Harvey Mudd College.* CONSERVATION OF ENERGY.

Rosenberg, Leon T. *Senior Consultant, General Design, Generation, Installation, and Service, Allis-Chalmers Manufacturing Company, Milwaukee, WI.* ALTERNATING-CURRENT GENERATOR; CORE LOSS; ELECTRIC ROTATING MACHINERY; GENERATOR; HYDROELECTRIC GENERATOR.

Roth, Willard. *Engineer, Sunbeam Equipment Corporation, Meadville, PA.* RESISTANCE HEATING.

Rundman, Prof. Karl. *Department of Metallurgical Engineering, Michigan Technological University.* METAL CASTING.

Rusk, Dr. Rogers D. *Mount Holyoke College.* ACCELERATION (in part); DISPLACEMENT (MECHANICS); SPEED; VELOCITY.

Russell, Dr. Allen S. *Vice President, Alcoa Laboratories, Aluminum Company of America, Pittsburgh, PA.* ALUMINUM ALLOYS.

Ryan, Prof. James J. *Professor Emeritus of Mechanical Engineering, University of Minnesota.* BUSHING; MACHINE DESIGN; OLDHAM'S COUPLING; PRESSURE VESSEL; SHAFTING; other articles.

S

Scalzi, Dr. John B. *Program Director, National Science Foundation, Washington, DC.* BEAM; BRITTLENESS; LOADS, REPEATED; SHEAR; STRENGTH OF MATERIALS; STRUCTURAL ANALYSIS; other articles.

Scarlott, Charles A. *Manager of Publications Department (retired), Stanford Research Institute, Menlo Park, CA.* WATERPOWER (in part).

Schmidt, Dr. Paul W. *Department of Physics, University of Missouri.* MOMENTUM; POWER.

Schneider, Dr. Thomas R. *Electric Power Research Institute, Palo Alto, CA.* ENERGY STORAGE (coauthored).

Schoonmaker, G. R. *Vice President, Production-Exploration, Marathon Oil Company, Findlay, OH.* OIL AND GAS, OFFSHORE.

Schultz, Robert M. *General Manager, William Langer Jewel Bearing Plant, Bulova Watch Company, Inc., Rolla, ND.* JEWEL BEARING.

Schutt, H. C. *Deceased; formerly, Consulting Engineer.* FURNACE CONSTRUCTION.

Schwartz, Dr. Mel M. *Rohr Industries, Inc., Chula Vista, CA.* WELDING AND CUTTING OF METALS (in part).

Scott, Dave. *Industrial Engineering Manager, Modular Computer Systems, Inc., Fort Lauderdale, FL.* MEASURED DAYWORK.

Sebald, Joseph F. *Consulting Engineer; and President, Heat Power Products Corporation, Bloomfield, NJ.* CONTACT CONDENSER; COOLING TOWER; STEAM CONDENSER; STEAM JET EJECTOR; SURFACE CONDENSER; VAPOR CONDENSER.

Seifert, Dr. William W. *Professor of Electrical Engineering and Associate Director, Commodity Transportation and Economic Development Laboratory, Massachusetts Institute of Technology.* SLIP.

Shearer, Prof. J. Lowen. *Systems and Controls Laboratory, Pennsylvania State University.* MAGNETIC AMPLIFIER.

Sheridan, Prof. Thomas B. *Department of Mechanical Engineering, Massachusetts Institute of Technology.* HUMAN-FACTOR ENGINEERING; HUMAN-MACHINE SYSTEMS.

Sherwood, Robert S. *Deceased; formerly, Manager of Engineering, Steam Turbine Division, Worthington Corportation, Harrison, NY.* MACHINERY; MECHANICAL ENGINEERING; TECHNOLOGY.

Simon, Michael S. *Civil Engineer, New York, NY.* ENGINEERING AND ARCHITECTURAL CONTRACTS.

Sims, Dr. Chester T. *Gas Turbine, General Electric Company, Schenectady, NY.* HIGH-TEMPERATURE MATERIALS.

Skilling, Prof. Hugh H. *Department of Electrical Engineering, Stanford University.* ALTERNATING CURRENT; ALTERNATING-CURRENT CIRCUIT THEORY; RESONANCE (ALTERNATING-CURRENT CIRCUITS).

Skinner, E. N. *Assistant to the Director, Product Research and Development, International Nickel Company, Inc., New York NY.* NICKEL ALLOYS (coauthored).

Smith, Gaylord D. *Director, High Nickel Alloys and Precious Metals, Product Development and Research Division, International Nickel Company, New York, NY.* NICKEL ALLOYS (coauthored).

Smith, Dr. John A. *IBM Corporation, Lexington, KY.* STRAIGHT-LINE MECHANISM.

Smith, Philip H. *Consultant, Sherman, CT.* TRUCK.

Smith, Dr. Richard T. *Southwest Research Institute, San Antonio, TX.* SYNCHRONOUS MOTOR.

Snow, William W. *Consulting Engineer, William W. Snow Associates, Inc., Woodside, NY.* DRAFTING MACHINE; ELECTROPOLISHING.

Sorensen, Dr. Robert M. *Coastal Engineering Research Center, Department of the Army, Fort Belvoir, VA.* COASTAL ENGINEERING.

Spindler, John C. *Anaconda Company, Butte, MT.* ENVIRONMENTAL ENGINEERING (coauthored).

Spinrad, Dr. Bernard I. *Department of Nuclear Engineering, Oregon State University.* NUCLEONICS; REACTOR PHYSICS.

Stanley, C. Maxwell *Chairman of the Board, Stanley Consultants, Muscatine, IA.* RURAL ELECTRIFICATION (coauthored).

Starr, Dr. Eugene C. *Bonneville Power Administration, U.S. Department of the Interior, Portland, OR.* ELECTRIC POWER GENERATION.

Stefanovic, Dr. Victor R. *Electrical Engineering Department, University of Missouri.* REACTOR (ELECTRICITY).

Steindler, Dr. Martin J. *Chemical Engineering Division, Argonne National Laboratory, Argonne, IL.* NUCLEAR FUEL CYCLE; NUCLEAR FUELS REPROCESSING.

Steinert, Emil F. *(Retired) Arc Welding Division, Westinghouse Electric Corporation.* ARC WELDING.

Stephenson, Dr. R. J. *Department of Physics, Wooster College.* ACCELERATION; ROTATIONAL MOTION—both validated.

Stephenson, Prof. Richard M. *Department of Nuclear Engineering, University of Connecticut.* ISOTOPE (STABLE) SEPARATION.

Stewart, Dr. John W. *Department of Physics, University of Virginia.* IR DROP.

Stoloff, Dr. Norman S. *Department of Materials Engineering, Rensselaer Polytechnic Institute.* EUTECTICS; METAL, MECHANICAL PROPERTIES OF.

Stuart, William T. *Deceased; formerly, "Electrical Construction and Maintenance," McGraw-Hill Publications Company, New York, NY.* BRANCH CIRCUIT (coauthored); WIRING.

Sturges, Frank C. *Deceased; formerly, President, Pennsylvania Drilling Company, Pittsburgh, PA.* BORING AND DRILLING (MINERAL); CORE DRILLING.

Sullivan, John. *Lee Allan Associates, Sunnyvale, CA.* ELECTRIC UNINTERRUPTIBLE POWER SYSTEM.

Suryanarayana, Prof. N. V. *Department of Mechanical Engineering and Engineering Mechanics, College of Engineering, Michigan Technological University.* HEAT PUMP (in part).

Sutherland, J. R. *Power Transformer Department, General Electric Company, Pittsfield, MA.* AUTOTRANSFORMER; TRANSFORMER (in part).

T

Tang, Prof. K. Y. *Deceased; formerly, Department of Electrical Engineering, Ohio State University.* KIRCHHOFF'S LAWS OF ELECTRIC CIRCUITS.

Thompson, Jack R. *U.S. Army Corps of Engineers, Office of the Secretary of the Army, Washington, DC.* DAM.

Trump, Prof. John G. *Department of Electrical Engineering, Massachusetts Institute of Technology.* IMPULSE GENERATOR.

Tuttle, Alan H. *Vocational Division, State University of New York Agricultural and Technical Institute, Wellsville.* BORING; DRILLING MACHINE; GEAR CUTTING; LATHE; MACHINING OPERATIONS; SAWING; THREADING; other articles.

U

Ullmann, Dr. John E. *Department of Marketing—Management, Hofstra University.* DESIGN STANDARDS; WORK STANDARDIZATION (in part).

Underwood, Dr. Ervin E. *Alcoa Professor, Georgia Institute of Technology.* SUPERPLASTICITY.

Unger, Walter H. *Anaconda Company, Denver, CO.* ENVIRONMENTAL ENGINEERING (coauthored).

U.S. Army Corps of Engineers. CANAL; RIVER ENGINEERING—both coauthored.

U.S. Bureau of Reclamation. CANAL (coauthored).

V

Vander Sande, Prof. John B. *Department of Metallurgy and Materials Science, Massachusetts Institute of Technology.* METALLOGRAPHY (coauthored).

Vanick, James S. *Consultant, Foundry-Castings, Sea Girt, NJ.* CAST IRON.

Van Vlack, Prof. Lawrence H. *Department of Materials Engineering, University of Michigan.* ALLOY (coauthored).

Vokac, Charles W. *Marketing Administration, Whiting Corporation, Harvey, IL.* ARC HEATING.

W

Wang, Prof. K. K. *Sibley School of Mechanical and Aerospace Engineering, Cornell University.* INERTIA WELDING; WELDING AND CUTTING OF METALS (in part).

Weaver, Dr. Paul *(Retired) Texas A & M College.* OIL FIELD MODEL.

Weber, Erwin L. *Deceased; formerly, Trust Department, National Bank of Commerce, Seattle, WA.* HOT-WATER HEATING SYSTEM; PANEL HEATING AND COOLING; RADIANT HEATING; RADIATOR.

Weil, Robert T., Jr. *Deceased; formerly, Dean, School of Engineering, Manhattan College.* DIRECT-CURRENT GENERATOR; KIRCHHOFF'S LAWS OF ELECTRIC CIRCUITS (validated); MAGNETO.

Weil, Rolf. *Professor of Metallurgy, Stevens Institute of Technology.* ELECTROLESS PLATING; ELECTROPLATING OF METALS.

Weiss, Prof. Gerald. *Department of Electrical Engineering, Polytechnic Institute of New York.* DIFFERENTIAL TRANSFORMER.

Welch, Arthur A. *Engineering Consultant, Purcellville, VA.* BALLAST RESISTOR.

Wheeler, Prof. John A. *Department of Physics, Joseph Henry Laboratories, Princeton University.* CRITICAL MASS.

Williams, Prof. Dudley. *Department of Physics, Kansas State University.* LEAD ALLOYS; NEWTON'S LAWS OF MOTION; RIGID BODY.

Williams, Prof. Everard M. *Deceased; formerly, De-*

partment of Electrical Engineering, Carnegie-Mellon University. TRANSMISSION LINES (in part).

Winch, Prof. Ralph P. *Department of Physics, Williams College*. ELECTROMOTIVE FORCE (EMF).

Wirry, Henry J. *Product Manager, Hydraulic Drives, Twin Disc, Inc., Rockford, IL*. FLUID COUPLING; TORQUE CONVERTER.

Wisely, Prof. William H. *Department of Civil Engineering, University of Virginia*. CIVIL ENGINEERING (in part).

Wolford, Prof. James C. *Department of Mechanical Engineering, University of Nebraska*. VELOCITY ANALYSIS.

Woodall, James E. *Division Manufacturing Manager, Westinghouse Electric Corporation, Pittsburgh, PA*. PILOT PRODUCTION.

Wright, Elliott F. *Consulting Engineer (retired), Advanced Products Division, Studebaker-Worthington Corporation, Harrison, NJ*. CENTRIFUGAL PUMP; DISPLACEMENT PUMP; PUMP; PUMPING MACHINERY.

Wright, Prof. Roger N. *Department of Materials Engineering, Rensselaer Polytechnic Institute*. MACHINABILITY OF METALS.

Y–Z

Yankee, Prof. Herbert W. *Mechanical Engineering Department, Worcester Polytechnic Institute*. COMPUTER-AIDED DESIGN AND MANUFACTURING.

Yellott, John I. *Emeritus Professor, College of Architecture, Arizona State University*. SOLAR HEATING AND COOLING.

Young, Edward M. *Associate Editor, "Engineering News Record," McGraw-Hill Publications Company, New York, NY*. BULLDOZER; CONSTRUCTION EQUIPMENT; CRANE HOIST; EARTHMOVER; EXCAVATOR; GRAB BUCKET.

Zapffe, Dr. Carl A. *Consultant, Baltimore, MD*. STAINLESS STEEL.

Zifcak, John H. *The Foxboro Company, Foxboro, MA*. STRAIN GAGE.

Zimmerman, Dr. John R. *Department of Mechanical Engineering, Pennsylvania State University*. GEAR; GEAR TRAIN; PLANETARY GEAR TRAIN; PULLEY; ROLLING CONTACT.

Index

Index

Asterisks indicate page references to article titles

B

C

D

E

F

G

H

M

N

O

Q

R

S

T

U

V

W

Y-Z

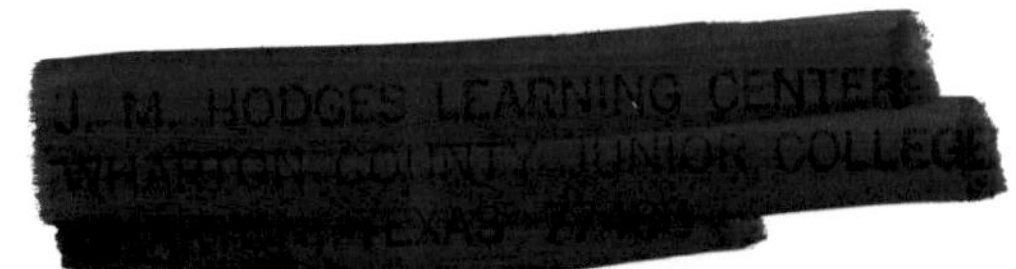